MW00720957

Biology

2nd edition

Biology

2nd edition

Australian focus on:

- **Biota**
- **Evolution**
- **Ecology**
- **Research**

Knox   Ladiges   Evans   Saint

McGraw Hill

Boston   Burr Ridge, IL   Dubuque, IA   Madison, WI   New York
San Francisco   St. Louis   Bangkok   Bogotá   Caracas   Kuala Lumpur
Lisbon   London   Madrid   Mexico City   Milan   Montreal   New Delhi
Santiago   Seoul   Singapore   Sydney   Taipei   Toronto

McGraw·Hill Australia

A Division of the **McGraw·Hill** *Companies*

Copyright © 2001 McGraw-Hill Book Company Australia Pty Limited
Additional owners of copyright are named in on-page credits.

Apart from any fair dealing for the purposes of study, research, criticism or review, as permitted under the *Copyright Act*, no part may be reproduced by any process without written permission. Enquiries should be made to the publisher, marked for the attention of the Publishing Manager, at the address below.

Every effort has been made to trace and acknowledge copyright material. Should any infringement have occurred accidentally the authors and publishers tender their apologies.

Copying for educational purposes

Under the copying provisions of the *Copyright Act*, copies of parts of this book may be made by an educational institution. An agreement exists between the Copyright Agency Limited (CAL) and the relevant educational authority (Department of Education, University, VET, etc.) to pay a licence fee for such copying. It is not necessary to keep records of copying except where the relevant educational authority has undertaken to do so by agreement with the Copyright Agency Limited. For further information on the CAL licence agreements with educational institutions, contact the Copyright Agency Limited, Level 19, 157 Liverpool Street, Sydney, NSW 2000. Where no such agreement exists, the copyright owner is entitled to claim payment in respect of any copies made.

National Library of Australia Cataloguing-in-Publication Data

Biology.

2nd ed.
Includes index.
ISBN 0 074 70833 3

1. Biology. I. Knox, R. Bruce.

574

Published in Australia by
McGraw-Hill Book Company Australia Pty Limited
4 Barcoo Street, Roseville, NSW 2069, Australia
Acquisitions Editor: Jae Chung
Production Manager: Jo Munnelly
Project Coordinator: Leanne Peters
Permissions Editors / Photo Researchers: Leanne Peters, Nathan Wilson, Susan Gentle
Marketing Manager: Debra James
Editor: Carolyn Pike
Designer: Di*kso*
Typesetter: Midland Typesetters
Illustrators: Lorenzo Lucia, Di Booth, Alan Laver, Stephen Francis
Cover photographer: David Paul
Proofreader: Tim Learner
Indexer: Michael Wyatt
Printed by Best Tri Colour Printing

Foreword

It is a great privilege to introduce this second edition of the award-winning Australian textbook *Biology*. It has been wonderful to see the excitement with which the first edition was received. "At last", teachers and students alike said, "a biology text is available that refers to our unique Australian flora and fauna to illustrate principles of biology". It is a major step forward in the development of biology in Australia that we have moved from being reliant on texts created for a Northern Hemisphere context and market to a truly Australian text. The references throughout this book to Australian biology set the framework for many young Australians to have a greater insight into the beauties and wonders in our continent that they observe when bushwalking, birdwatching, surfing and fishing and enjoying the many other outdoor activities that Australia offers. It highlights the richness of our biodiversity and how we may balance conservation of species and habitats with human use and activities in a sustainable way.

The text in this edition also gives a comprehensive coverage of the basis of modern molecular biology and genetics, which forms the basis of biotechnology. This is a fast-moving field. The fundamental knowledge is well established and many applications are now in everyday use. Biotechnology is undoubtedly one of the major technologies that will continue to transform our lives this century. It is a field in which Australia has excelled in the past and one in which we hope to build a significant industry in the future. This volume gives our biotechnologists of the future a wonderful grounding and a springboard for their future studies.

Adrienne E. Clarke, AO, FTSE, FAA
Laureate Professor, The University of Melbourne

E-Student

Welcome to the Online Learning Centre, packed with resources to aid your study from this text.

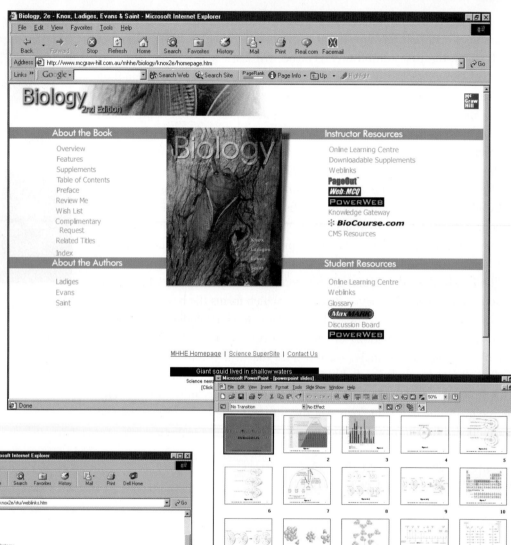

Weblinks

Over 30 weblinks to professional bodies and other biology-related organisations have been grouped. Use these to make sense of your study and enrich your learning.

PowerPoint Slides

PowerPoint figure sets will assist your studying techniques through visual learning.

Self-Assessment Quizzes

Do a quick overview of your knowledge of each chapter with the self-assessment quiz.

MaxMark

Unique to McGraw-Hill, MaxMark is a self-paced learning tool comprising forty-five interactive, multiple-choice questions for every chapter of the text. MaxMark is designed to help you 'maximise your marks' by allowing you to set time limits, randomise questions, and switch the extensive feedback on or off.

Access to MaxMark can be packaged with this text. Go to the Online Learning Centre or contact your bookstore for more information.

Feedback to explain each answer

There are 45 multiple-choice questions per chapter

More information from the text for further explanation

PowerWeb

PowerWeb is your online companion, providing journals articles on key current events in biology. PowerWeb includes up-to-the-hour biology news, weblinks, internet research facilities, and weekly updates. Access to PowerWeb may be packaged with this text or purchased independently. Go to the Online Learning Centre or contact your bookstore for more information.

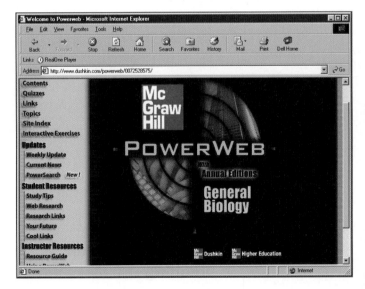

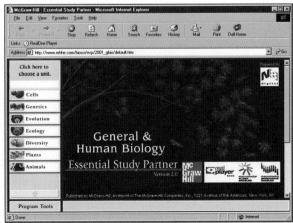

Essential Study Partner

The 'Essential Study Partner' is a general biology tool packed with additional learning materials, including informative text and images, animations, interactive learning activities, topic quizzes and module exams.

Study tools in the text

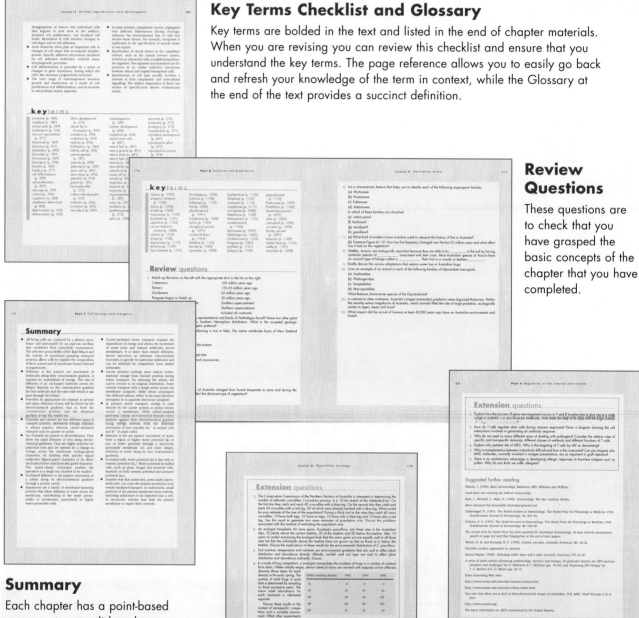

Key Terms Checklist and Glossary

Key terms are bolded in the text and listed in the end of chapter materials. When you are revising you can review this checklist and ensure that you understand the key terms. The page reference allows you to easily go back and refresh your knowledge of the term in context, while the Glossary at the end of the text provides a succinct definition.

Review Questions

These questions are to check that you have grasped the basic concepts of the chapter that you have completed.

Summary

Each chapter has a point-based summary to consolidate the information you have covered. The summary is also important when doing revision and exam preparation.

Suggested Further Reading

Suggestions for further reading are made at the end of each chapter so that you can expand your knowledge, find useful references for assignments, or explore a topic that interests you! A brief note about each reading helps to clarify your selection.

Extension Questions

These questions challenge you to apply the knowledge that you have acquired to answer more complex questions, and are great preparation for long answer and essay questions in tests and exams.

E-Instructor

A plethora of resources is available for instructors from the Online Learning Centre, including animations for Lectures, PowerPoint slides, and a test bank.

Instructor's Resource Manual

Create a customised unit outline from your teaching objectives. Challenge students with extension exercises, selected readings and internet sites. Provide feedback to students with answers to review questions. Utilise the testbank for class testing and examinations.

Artwork: available online or on CD

PowerPoint slides can be created using the artwork from the text, providing a visual tool to assist you in your teaching and maximise students understanding and learning potential.

Biocourse.com

Animations, Interactivities and PowerPoint overviews help bring science to life from a searchable database of more than 10,000 resources. Case studies and a range of online labs are readily available. Lecturer resources include teaching tips and basic information on pedagogy and assessment, reference searches and literature, as well as suggestions for classroom and lecture activities. Biocourse includes a generic testbank, which provides additional questions to the ones available with your textbook.

WebMCQ

WebMCQ provides an engine for online revision quizzes or tests with multiple choice, true / false, and essay questions, backed by powerful online tracking and reporting features. With WebMCQ, you can adapt each test to suit your individual class needs.

WebMCQ is exclusive to McGraw-Hill.

WebCT / Blackboard

The online resources to accompany this text are available in WebCT and Blackboard formats for online delivery. Contact your local sales representative for more information.

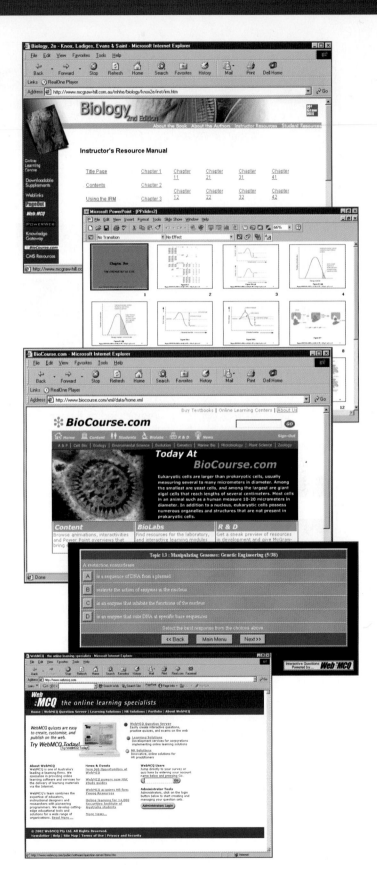

New to this edition

Introductory chapter on unifying themes in biology and the methods and practices of science.

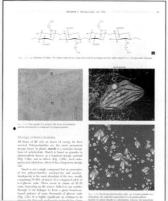

Significantly improved treatment of cell biology, including comprehensive updating and expansion of Chapters 1 (Molecules of Life) and 2 (Chemistry of Life).

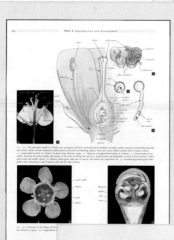

Part 3 (Reproduction and Development) has been restructured as three chapters focussing on animals, plants, and genes and development.

Significantly improved and expanded treatment of genetics, to place more emphasis on classical experiments as well as such recent developments as imprinting, telomere structure, genome analysis, and the use of DNA analysis in forensic science.

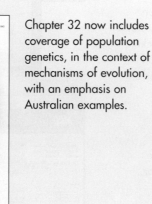

Chapter 32 now includes coverage of population genetics, in the context of mechanisms of evolution, with an emphasis on Australian examples.

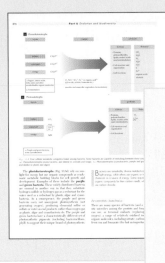

Chapter 33 has been expanded to provide greater coverage of the metabolic functions of bacteria.

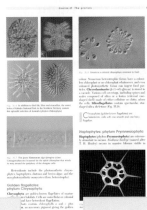

Chapter 35 contains up-to-date coverage of current research on the rapidly changing field of protists

Chapter 43 has been restructured with a stronger focus on marine examples.

Each chapter starts with a chapter contents list and finishes with a summary of the main points, list of key terms, revision and extension questions, and some suggested further reading, including websites.

Chapter 37 includes coverage of recent discoveries in Australia of a 'living fossil' conifer: the Wollemi pine.

Australian focus

Flora

The qualities of Australian flora are addressed in all relevant topics throughout the text.

Fauna

The unique nature of Australian fauna is covered across all relevant sections of the book.

Evolution

In addition to the chapter on Australian Biota, the discussion of evolution integrates uniquely Australian examples. (p.844)

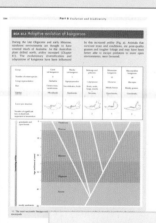

Human Impacts

Topics including salinity and erosion, fire management, introduced species, and the greenhouse effect are discussed in terms of the Australian experience.

Photographs

Topics are frequently illustrated with Australian images.

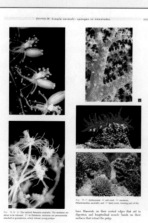

Australian Biota

An entire chapter is devoted to the coverage of Australian Biota and its unique features.

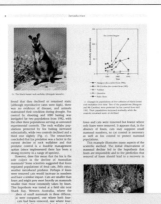

Scientific Method in Practice

The introduction to the text presents a case study to illustrate the practice of scientific method in an Australian case study.

Research

References to Australian research—such as the transgenic cotton trials, and noble prize winner J.C Eccles—highlight the contributions of Australian biologists.

Boxed Features

Boxed features present examples that reinforce concepts discussed in the text. These examples often focus on an Australian perspective of the topic, such as the discussion of bush medicines and a conservation strategy for Leadbeater's possum.

Living in Communities

This chapter includes discussion of ecological interactions that produce the patterns we see in some Australian communities, for example, adaptation to periodic fires.

Editors

Professor Pauline Ladiges

Pauline Ladiges is a Professor and Head of the School of Botany at The University of Melbourne. Her PhD on plant genecology is from the University of Melbourne and she is distinguished for her studies in taxonomy, biogeography and ecology of Australian flora. An expert on the genus *Eucalyptus*, Professor Ladiges was one of the first Australian biologists to adopt cladistic methodology and develop new methods for analysing biogeographic patterns.

In recognition of her scientific discoveries and leadership, she was awarded a Personal Chair at her University in 1992 and was elected Fellow of the Australian Academy of Science in 2002. In her 30 years of teaching, she has trained a strong team of postgraduate students, co-authored and edited several award-winning biology textbooks and served on a number of Boards and Advisory Committees for government and industry.

Professor Pauline Ladiges

Professor Barbara Evans

Barbara Evans is Dean of Graduate Studies at the University of Melbourne. Her PhD studies at the University of Melbourne described the effects of antihypertensive drugs on the function of the autonomic system in mammals. She went on to become an expert in comparative animal physiology, specialising in cardiovascular and respiratory regulation in animals such as crocodiles, ducks and the platypus, and worked in the UK and Canada. Her recent studies include the reproductive biology of Australian animals, including the platypus and echidna.

For many years a lecturer in first year biology, Barbara has contributed to the curriculum design of other tertiary and secondary level biology courses, and co-authored several highly successful biology texts for secondary schools in Victoria.

Among her other achievements, Barbara is a fully qualified secondary physical education teacher, and represented Australia in Olympic Gymnastics for a number of years.

Professor Barbara Evans

Professor Robert Saint

Robert Saint is Head of Molecular Genetics and Evolution at the Australian National University. He graduated from Adelaide University, where he studied gene structure in the earliest days of recombinant DNA technology and went on to study at Stanford University on the structure and expression of the Bithorax Complex of *Drosophila melanogaster*.

Professor Saint was one of the pioneers of research into cell cycle control in development. After being appointed Professor of Genetics at Adelaide University, he lobbied for an Australian Research Council Special Research Centre, becoming Director for the Molecular Genetics of Development, a position he currently holds.

His laboratory has discovered a new family of genes and demonstrated their role in development. His research interests continue to be in the molecular genetics of cell division and gene expression during *Drosophila* development, with additional recent interests in the genetics and development of coral.

Professor Robert Saint

Consultant reviewer

Professor David Patterson, Head, School of Biological Sciences, The University of Sydney

Contributing authors

Contributors to the second edition

Professor Russell Baudinette, University of Adelaide
Dr Michael Bennett, University of Queensland
Dr Mark Burgman, University of Melbourne
Dr Christina Cheers, University of Melbourne
Dr Andrew Drinnan, University of Melbourne
Dr Mark Elgar, University of Melbourne
Dr Bill Foley, Australian National University
Dr Peter Frappell, La Trobe University
Dr Don Gaff, Monash University
Professor Adrian Gibbs, Australian National University
Dr David Guest, University of Melbourne
Professor Adrienne Hardham, Australian National University
Professor David Hirst, University of Melbourne
Dr Jean Joss, Macquarie University
Dr Mick Keough, University of Melbourne
Dr Tim Littlejohn, The University of Sydney
Dr Bill Loneragan, University of Western Australia
Dr Geoffrey McFadden, University of Melbourne
Professor David Macmillan, University of Melbourne

Dr Steve Morton, Division of Wildlife and Ecology, CSIRO
Professor Gareth Nelson, University of Melbourne
Dr Edward Newbigin, University of Melbourne
Professor Roger Parish, La Trobe University
Professor Jeremy Pickett-Heaps, University of Melbourne
Professor Hugh Possingham, University of Queensland
Dr Roger Seymour, University of Adelaide
Dr Geoffrey Shaw, University of Melbourne
Professor Andrew Staehelin, University of Colorado
Dr Peter Stewart, Austalian National University
Professor Bruce Stone, La Trobe University
Dr Bernadette Vrhovski, New Children's Hospital
Dr Graeme Watson, University of Melbourne
Dr Rick Wetherbee, University of Melbourne
Dr Molly Whalen, Flinders University
Dr Paul Whitington, University of New England
Dr Jann Williams, Ecological Consultant, RMIT University
Dr Phil Withers, University of Western Australia
Dr Ian Woodrow, University of Melbourne

Contributors to the first edition

We would also like to acknowledge other contributing authors to the first edition, whose work provided the foundation of parts of this new book.

Professor Craig Atkins, Dr Marilyn Ball, Dr Prem Bhalla, Dr Mike Borowitzka, Dr Peter Crane, Dr Christa Critchley, Dr Rodney Devenish, Dr P. Finnegan, Dr Paul Fisher, Professor John Furness, Dr Michael Guppy, Dr Gustav Hallegraeff, Professor Barrie Jamieson, Dr Peter Kershaw, Dr Jill Landsberg, Dr Barry Lee, Dr Ian McDonald, Professor Jenny Marshall-Graves, Dr Don Metcalf, Dr Chris Moran, Dr Christina Morris, Dr David Morris, Professor Phillip Nagley, Dr Tom Neales, Dr Elbe Nielson, Mr Tim Offer, Professor Jim Reid, Dr Pat Rich, Dr John Ross, Dr Gordon Sanson, Dr Roland Scollay, Dr George Scott, Dr Lyn Selwood, Professor Andrew Smith, Dr David Smith, Dr Mark Tester, Dr Steve Tyerman, Dr Anthony Weiss, Dr Paul Willis.

Acknowledgments

Reviewers of the second edition

We would like to thank the following reviewers for their helpful comments and suggestions in the development of the second edition. We appreciate their time and effort in sharing their expertise with us.

Professor Russell Baudinette, University of Adelaide

Dr Paul Broady, University of Canterbury

Associate Professor Kenneth Brown, University of Technology, Sydney

Dr David Burritt, University of Otago

Dr Mike Calver, Murdoch University

Dr David Christophel, University of Adelaide

Associate Professor Bernie Dell, Murdoch University

Dr Ashley Garrill, University of Canterbury

Dr Paul Guy, University of Otago

Ms Julie Harris-Wetherbee, University of Melbourne

Dr Sue Jones, University of Tasmania

Associate Professor David Macey, Murdoch University

Dr Peter McGee, The University of Sydney

Ms Sheri McGrath, University of Melbourne

Dr Ben Oldroyd, The University of Sydney

Dr Robert Poulin, University of Otago

Ms Alex Pulkownik, University of Technology, Sydney

Dr Robert Reid, University of Adelaide

Ms Racheline Rogers, University of Adelaide

Dr Ken Sanderson, Flinders University

Dr Kathryn Schuller, Flinders University

Dr Clyde Smith, University of Auckland

Dr Trevor Stevenson, Royal Melbourne Institute of Technology University

Dr Fleur Tiver, University of South Australia

Ms Velta Vingelis, Adelaide University

Dr Graham Wallis, University of Otago

Mr Rob Wass, University of Otago

Dr Helen Wood, Charles Sturt University

Reviewers of the first edition

Dr Jim Akers, University of Southern Queensland

Dr Elizabeth Alexander, Curtin University

Dr Michael Augee, University of New South Wales

Br Robert Ballantyne, Charles Sturt University

Dr Bob Bennett, Murdoch University

Ian Bennett, Edith Cowan University (Mt Lawley Campus)

Dr Stuart Bradley, Murdoch University

Professor Don Bradshaw, University of Western Australia

Dr Neil Brink, Flinders University of South Australia

Dr Max Cake, Murdoch University

Dr Jim Campbell, Wollongong University

Dr Geraldine Chapman, The University of Sydney

Dr Andrew Collins, University of New South Wales

Dr Ron Crowden, University of Tasmania

Dr John Dearn, Canberra University

Dr Janet Gorst, University of Tasmania

Dr David Happold, Australia National University

Dr A. Chris Hayward, University of Queensland

Dr Chris Hill, James Cook University

Dr Sidney James, University of Western Australia

Dr Jacob John, Curtin University

Dr Michael Johnson, University of Western Australia

Dr Graham Kelly, Queensland University of Technology

Dr Kim Kohen, Macquarie University

Dr William A. Longeragan, University of Western Australia

Dr Hamish McCallum, University of Queensland

Dr Peter McGee, The University of Sydney

Dr David Morrison, The University of Technology, Sydney

Dr Ray Murdoch, Newcastle University

Dr Mary Peat, The University of Sydney

Dr Hugh Possingham, University of South Australia

Associate Professor Nallamilli Prakash, University of New England

Dr Rob Rippingale, Curtin University

Dr Liz Smith, Maquarie University

Associate Professor Ted Steele, Wollongong University

Dr Robert Vickery, University of New South Wales

Dr Helen Wood, Charles Sturt University

We would also like to thank the following Biology departments for assisting in reviewing the art program at various stages of development: Macquarie University, Newcastle University, The University of Sydney and Wollongong University.

The editors would like to acknowledge their friend and colleague the late Professor Bruce Knox for his role in the first edition. We thank Consultant Reviewer Professor David Patterson, School of Biological Sciences, The University of Sydney, and the many colleagues who gave constructive feedback to make this new edition what it is. We are very grateful to the superb efforts of all of our authors and to others who assisted us in the preparation of this book. They responded superbly to very short deadlines and maintained great enthusiasm for the idea of an 'Australian' text.

We are also most grateful to McGraw-Hill Book Company, Australia for their support in bringing this project to fruition. We particularly wish to thank Publisher Jae Chung for her support and encouragement, Production Editor Carolyn Pike for her patience and keen eye for detail, and Production Manager Jo Munnelly and her staff. Not least, we acknowledge the support of our families for their encouragement and patience during the many hours of writing and editing.

Brief Contents

Contents

Quizzing and further information on *CELL BIOLOGY AND
ENERGETICS* can be found on

MaxMARK   **POWERWEB**   OLC

**Part 2 Genetics and molecular
biology**   **201**

Quizzing and further information on *GENETICS AND MOLECULAR BIOLOGY* can be found on

Quizzing and further information on *REPRODUCTION AND DEVELOPMENT* can be found on

Quizzing and further information on *REGULATION OF THE
INTERNAL ENVIRONMENT* can be found on

Quizzing and further information on *RESPONSIVENESS AND
COORDINATION* can be found on

Part 6 Evolution and biodiversity 787

Quizzing and further information on *EVOLUTION AND BIODIVERSITY* can be found on

Quizzing and further information on *ECOLOGY* can
be found on

The nature of biology and science

Biology is the study of living organisms, from microscopic bacteria to kangaroos, gum trees and humans. Biology today is an enormous field and touches on our lives at various levels (Fig. I.1). Biochemists and molecular biologists are involved in discovering how organisms are structured and function at the chemical and cellular levels. Geneticists ask questions about how these features are coded for within the structure of DNA, how genes control the development of an organism, how cells divide and how inherited information is passed from one generation to the next—the continuity of life. At a different level of scale, biologists study whole organisms, how they function, how they interact with their environment and with one another to form ecosystems. Ecosystems also can be studied on a range of scales—from a small patch of heathland to the Great Barrier Reef of Queensland to the earth's biosphere where the carbon cycle, energy budgets and greenhouse effect need to be studied on a global scale.

Biology is presented in this book at each of these levels. Forty-five chapters, written by experts in their field, are grouped into seven parts based on these different levels. Different biology courses may emphasise one or more of these sections and you may not be required to study all chapters. Our book is designed to be a resource for you for whatever particular biology course you are studying. The fundamental concepts of modern biology are explained and we have generally presented these in a context that will be familiar to you. Australasian animals and plants, environments and ecosystems have many unique features. Their evolutionary history relates also to the southern continents more than to those of the Northern Hemisphere.

Biology is studied at various levels—molecules, genes, cells, whole organisms and ecosystems.

By observing and studying the many various kinds of organisms, even those that appear remarkably different from one another, biologists have discovered principles and processes that are fundamental to all living organisms. These unifying themes of biology, together with the process of how science works, are summarised in this opening chapter before we commence our studies in depth.

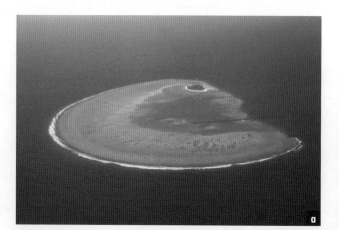

Fig. I.1 Biology is studied over a range of very different scales: at the level of **(a)** whole ecosystems, such as the Australian Great Barrier Reef, **(b)** individual organisms, such as this nautilus found in tropical waters, and **(c)** individual cells, such as this marine coccolithophorid, *Coccolithus pelagicus* (magnification × 6500)

Unifying themes in biology

Living organisms are made of cells

What do we mean by 'living'? Most of you have some notion of what it means to be 'living'; for an animal such as a human, it means being able to breathe, eat, move, respond to a touch or sound, grow and reproduce. But what about less familiar organisms such as bread mould or bacteria? The question becomes more difficult to answer when considering viruses (Chapter 34), which can be purified like chemicals and, in some cases, crystallised as apparently inert material.

Living organisms can be defined as cellular organisms, which excludes viruses. The **cell theory** is based on observations accumulated by scientists over a number of centuries and states that:

- all organisms are made of cells (Fig. I.2) and the products of cells
- all cells come from pre-existing cells
- the cell is the smallest organisational unit.

A single cell may perform many functions, but in multicellular organisms, cells differentiate in structure to perform specific functions—for exchange of materials, transport, photosynthesis, strength, protection, defence or communicating messages.

All cells have an outer cell membrane that encloses the fluid contents of the cell, the cytoplasm. The cytoplasm contains structures (known as organelles) that are characteristic of the type of cell. In the seventeenth century, the invention of the microscope allowed Robert Hooke (in 1665) to observe the dead cells (which appeared like tiny empty boxes) of a thin slice of bark that he had cut from a tree. Anton van Leeuwenhoek described many living cells such as sperm and blood cells. In the nineteenth century, largely due to the work of Matthias Schleiden (on plant tissues) and Theodor Schwann (on animal tissues), there was wide acceptance of the fundamental principle that all organisms are composed of cells.

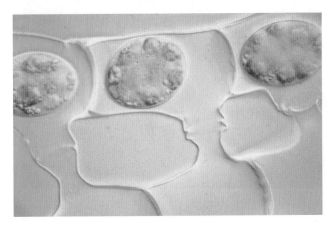

Fig. I.2 All organisms are made of cells, such as these from a plant

Towards the end of the nineteenth century, the second basic principle was formulated—that all cells come from pre-existing cells through cell division.

The electron microscope, invented in the twentieth century, opened up the development of new techniques for studying the fine structure of cells. A major discovery for modern biology is the recognition of three major cell types. On this basis, all of life is now classified into three 'super kingdoms': Bacteria, Archaea and Eukarya. Bacteria and Archaea are organisms called prokaryotes (cells without a nucleus and lacking other membrane-bound organelles) and the Eukarya are eukaryotes (cells with a nucleus and other membrane-bound organelles). Remarkably, eukaryotic cells evolved over billions of years through a process of symbiosis—one prokaryotic organism engulfing another, the engulfed cell evolving into an organelle (e.g. chloroplast or mitochondrion) within the host cell (Chapters 3 and 35). Multicellular organisms within the Eukarya include seaweeds, fungi, animals and plants.

Living organisms are made of cells, the smallest organisational unit. All cells come from pre-existing cells.

DNA is the basis of life

One characteristic of life is its remarkable complexity. Maintaining this complexity in the face of the thermodynamic drive towards disorder requires the input of energy and a high level of organisation, which, in turn, relies on stored information. This stored information is passed from cell to cell and generation to generation to control biological function and reproduction. This genetic information is normally stored in molecules of DNA that are carried in cells, transmitted during cell division and passed on to the next generation in the gametes. Most characteristics of organisms are dictated by a sequence of bases (cytosine, guanine, thymine, adenine) within the DNA molecule, and most differences between individuals and species result from differences in their DNA sequence.

The discovery of DNA as the store of genetic information and the elucidation of the genetic code by which the information is read has revolutionised biology. Entire sequences of the genomes of organisms are now being determined. These sequences can be modified to create altered or new gene function, tailoring the characteristics of organisms. These new technologies are being applied in a vast array of circumstances to produce pharmaceutical products, generate new agricultural varieties, develop genetic therapies for the treatment of disease and solve historical puzzles. An understanding of DNA and gene technology is essential for people to debate the issues surrounding the use of genetically modified organisms, for example, in the food industry or their release into the environment.

> Genetic information is normally stored in molecules of DNA in cells, transmitted during cell division and passed on to the next generation in gametes.

Life evolves

The enormous variety of life, the fossil record, organisms with apparently useless vestigial organs (e.g. rudimentary bones representing hind limbs in whales) and the geographic distribution patterns of organisms (e.g. *Banksia* plants and koalas in Australia but not in the Northern Hemisphere) posed enormous explanatory problems to early biologists who held a view of life (and earth) as unchangeable. These problems cannot only be explained by evolution but are predicted by it. Theodosius Dobzhansky summarised this view in the quote 'nothing in biology makes sense except in the light of evolution'. The increasing richness of the fossil record (Fig. I.3) and observations of the natural world have led to the idea of **evolution** becoming, in scientific terms, a 'fact' as firmly established as the idea of an atom or electron or the sun being at the centre of our solar system. There is still considerable debate about the *mechanisms* by which evolution occurs, with Darwin's hypothesis of evolution by natural selection competing with more recent theories that natural selection is complemented by a role for chance events that occur independently of natural selection. *How evolution occurs* is one of the most interesting and most lively areas of debate among biologists.

Evolutionary relationships are summarised in classifications

The discovery of DNA and techniques to sequence the order of bases has allowed biologists to compare organisms across the three super kingdoms. Together with more traditional observations of the structure of organisms (morphology), taxonomists are piecing together the entire evolutionary tree of life, discovering who is related to whom. Biologists have long recognised that all mammals are related based on the fact that they have hair and suckle their young, and that all vertebrates (fishes, amphibians, lizards, snakes, birds and mammals) are related because they have a backbone. But only recently have we discovered that the parasitic, unicellular organism that causes malaria is distantly related to plant-like photosynthetic organisms or that the so-called cinnamon 'fungus' that causes jarrah forest dieback disease is not a fungus at all but an oomycete, related to golden-brown algae (Chapter 35).

Discovery of how different organisms are related (phylogeny) is summarised in biological classifications. As you will read in more detail in Chapter 30, **classification** is a hierarchy of names (taxonomic ranks) that includes all organisms on earth (the world's biodiversity). Each type of organism is classified in a kingdom, phylum, class, order, family, genus and species. Scientific classification is not only an efficient, accurate way of communicating about organisms between scientists. When based on phylogenetic relationships, this kind of classification is *predictive* and useful for all areas of biology. For example, the discovery that the malaria parasite is related to photosynthetic organisms predicts that it is metabolically different from animal cells. This allows for the

Fig. I.3 The fossil record spanning billions of years is strong evidence for the theory of evolution. This fossil is a *Minmi paravertebra*, a small armoured dinosaur (ankylosaur). First discovered in the 1960s, this near-complete skeleton was found in north-west Queensland in 1990 and is dated at 110–100 million years old, from the early Cretaceous period

possibility of developing drugs that target the plant-like characteristic of the parasite and that at the same time do not harm the infected human host.

Evolution is an accepted theory of biology. Current knowledge of how organisms are related through evolutionary descent is summarised in biological classification—a hierarchical system of scientific names.

The practice of science

Scientific method

As multifarious as science has been and continues to be, a great deal about it can be explained by reference to just three elements: a desire to understand the world in which we live, the allocation of responsibility for one's contributions (both credit and blame) and the mutual checking of these contributions.

David Hull

Just as important as learning the facts and concepts of biology as we understand them today are the processes by which they were discovered. Without an understanding of **scientific method** we cannot go on to establish new knowledge or evaluate critically the work of those who claim to have made important new findings.

Making observations

I have had the strongest desire to understand or explain whatever I observed . . . [and] reflect or ponder . . . over any unexplained problem.

Charles Darwin

Science begins with careful **observation** and a desire to explain what is observed. Observation sounds deceptively simple but in reality it often requires planning, concentration and discipline to observe and record completely and accurately. Observations may be qualitative, meaning that they are recorded as descriptions. Quantitative observations employ numbers or units to answer questions of 'how much?' or 'how often?' and can be analysed statistically for a measurement of confidence in the results.

It is important to minimise value judgments in recording observations. This is much more difficult than it sounds because all language is replete with value connotations. It is important in the study of animal behaviour, for example, to avoid being anthropomorphic—attributing human characteristics to animals on the assumption that they feel the same way we do (happy, sad, etc.). It is also easy to confuse observations with interpretation, as in the sentence: 'The male

spider remained at the edge of the female's web *to avoid being eaten*'. The last phrase is an interpretation, not an observation, and should not be included with the observations.

Testing hypotheses by experiments

Experiments are conducted for the purpose of answering specific questions about nature. These questions ordinarily are stated as hypotheses, which are statements about how someone thinks that nature works. In other words, they contain implied predictions, and confirmation of those predictions is the most powerful means available to demonstrate their accuracy of our understanding of the world around us.

Nelson Hairston, Sr.

Having made observations, scientists suggest possible explanations. A possible explanation for an observed phenomenon is called an **hypothesis**. An hypothesis can be used to make certain predictions, which can then be tested by experiments. If the results of an experiment disagree with the hypothesis, the hypothesis is disproved or falsified (in other words, while an hypothesis can be disproved, it cannot be proved to be correct). If several hypotheses are proposed to explain a particular observation, then this process can be used to rule out individual hypotheses by disproving them until only one valid hypothesis remains that survives attempts to falsify it.

A number of different hypotheses can often be suggested to explain a particular situation. The simplest hypothesis is always preferred, a maxim known as 'Occam's razor' after the medieval philosopher who proposed it. Only if the simplest hypothesis is disproved is a more complex one tested.

Hypotheses that generate many predictions and survive tests to falsify them have high predictive power and may develop into a **theory**, such as 'the cell theory' mentioned earlier, and 'the theory of evolution'. Thus, in science, theories represent general principles about which we are very confident because they explain a lot and have survived various attempts to falsify them.

An hypothesis is a possible explanation of observations. Experiments are tests of the predictions arising from an hypothesis. If, after much testing, one hypothesis explains all the observations, it may be given the status of a theory or general principle.

Experimental design

Many different experimental designs are possible in biology and much of the intellectual challenge in research is in choosing the best one to test the hypothesis being studied. However, there are two fundamental principles of good **experimental design**. The first is that all experiments must have a **control** because many different factors or variables may influence the results. The second principle is that experiments should be replicated. There is always the possibility that the results from any single experiment were due to chance; if the same results are obtained by replicating the experiment, the probability that the results were due to chance are much less likely.

To decide on the impact of a particular variable, it is necessary to prepare at least two trials. In the first trial, one variable is manipulated to test the experimental hypothesis. In the second trial, the control, that variable is left unaltered. In all other respects the two trials are identical, so any difference that arises between the two trials must be caused by the single manipulated variable. For example, slow growth and the yellowing of leaves (called chlorosis) in the Australian plant *Eucalyptus obliqua* can be the result of iron deficiency in soil derived from coastal limestone sand. To test this, plants (replicates) were grown in iron-deficient soil and watered with a solution containing iron (Fe); the control was another set of plants grown under identical conditions and in the same soil type but watered with a solution lacking additional iron. The growth and leaf colour of the two sets of plants were compared—those treated with iron were green and significantly larger. The difference was clearly attributable to only one factor: the added iron. (Limestone soils are high in pH and Fe^{3+} ions have a low solubility under such conditions, becoming difficult for many plant roots to absorb.)

Sometimes the results from an experiment are extremely clear and leave no doubt about the correct interpretation. At other times, there may be uncertainty about whether or not an effect has been shown. In such cases, biologists apply statistical analysis to help them decide between the options. Statistical understanding is also helpful in designing valid experiments.

Experiments must include a control and be replicated since many factors or variables may influence the results.

Scientific publication

The publication of scientific results is a key process in all scholarship. As long as results remain unpublished there are in fact no results.

Nils Malmer

The experimental testing of the theory of spontaneous generation (Box I.1) spanned nearly two centuries and involved scientists from different countries. Pasteur could only develop the ideas and experiments of Redi

BOX I.1 Disproving spontaneous generation

The development of biological knowledge through successive experiments can be seen in the eventual disproving of the idea of spontaneous generation of life. Until the late seventeenth century, it was a widely held belief that living things could generate spontaneously from non-living matter. Thus, frogs arose from swamp mud, flies from carrion or manure and so on. In 1668, Francesco Redi tested the idea experimentally by exposing meat in open jars and in jars covered with gauze. Flies entered the open jars and the meat soon swarmed with maggots, while the meat in the gauze-covered jars remained free of maggots. Maggots clearly did not generate spontaneously from rotting meat. However, with the discovery of micro-organisms it was postulated that these tiny organisms could arise by spontaneous generation. In the late eighteenth century, Lazzaro Spallanzani tested this hypothesis by boiling flasks of nutrient broth to sterilise them. Some flasks were then cooled and sealed, while others were cooled and left open to the air. Microbial growth occurred only in the open flasks, suggesting that microbes did not generate spontaneously in the sterile broth. However, critics claimed that the boiling had spoiled the air in the flasks, making them unsuitable for generating microbes unless unspoiled air could enter.

In 1862, Louis Pasteur took up the challenge and showed Spallanzani to be right. Pasteur's experiment was similar to Spallanzani's, except that he used long, goosenecked flasks that remained open to the air. However, any microbes settled in the curve of the gooseneck and could not reach the broth. After boiling, the broth in these flasks grew no microbes. If the gooseneck on any flask was broken so that microbes could enter as well as air, microbial growth occurred. This experiment convinced most people that spontaneous generation did not take place under current conditions on the earth.

and Spallanzani because their findings were written down and communicated. Thus, **scientific publication** is a critical part of the scientific method.

In contemporary biology, findings are published in scientific journals. Scientists describe carefully their aims, experiments, results and conclusions and submit the written reports to a journal editor. The editor consults a range of experts in the field, seeking their opinions of the merit of the work. This process is called peer review and may result in the work being accepted for publication as it stands, being accepted after amendments are made, or rejected as faulty and unsuitable for publication. As well as considering the reports from peer review, editors often check work for compliance with agreed codes of practice such as animal ethics provisions. Unethical work that violates such codes is unlikely to be published. Once the work has been published, it is available widely and can influence the thinking and work of other scientists. Without publication, it might as well not exist.

BOX 1.2 An Australian case study

Seventeen of Australia's unique mammal species have become extinct since European settlement in 1788 and many more are currently listed as endangered. Most mammal extinctions have occurred in arid regions and are confined to species in the critical weight range of 35–5500 g (excluding bats). Given that the decline of many species began as the European red fox, *Vulpes vulpes* (Fig. a) was introduced into Australia and spread, it was blamed for the extinctions.

However, there was a lack of hard experimental evidence and a range of plausible alternative hypotheses. These included climatic change, habitat degradation caused by introduced herbivores such as mice and rabbits, disease, deleterious consequences of pastoralism, changes in fire-patterns when Aboriginal

(a) Extensive field experiments were necessary to establish the impact of introduced foxes (*Vulpes vulpes*) on Australian native mammals

Australians ceased traditional management, and human persecution of species seen as threats to stock or valued for hunting. One prediction of the fox predation hypothesis is that removal of foxes from a given area should lead to an increase in the population sizes of native mammals relative to control areas where fox numbers are undisturbed. The main difficulty in implementing such experiments was the lack of a method to remove foxes or greatly reduce their abundance over large areas.

Such a tool became available with the use of the poison sodium monofluoroacetate, commonly known by its commercial code of '1080'. 1080 was originally applied as a rodenticide during World War II to protect troops from diseases transmitted by rats and was later used extensively in the United States to control predators of stock.

Researchers from Western Australia's Agriculture Protection Board established that the poison sodium monofluoroacetate was present in some indigenous Australian plant genera, principally *Gastrolobium* spp. and *Oxylobium* spp. Therefore, many native vertebrates, which had a long evolutionary history of exposure to the toxin, had a high tolerance relative to introduced species, especially in south-western Australia where *Gastrolobium* spp. and *Oxylobium* spp. are abundant. Herbivores were exposed when feeding on the toxic plants, while carnivores were exposed by eating the tissues of prey that had recently eaten 1080. Native vertebrate populations from south-western Australia were found to have up to 100 times the tolerance to 1080 of those from eastern Australia, which had not been exposed to the 1080 plants. The fox, which is a very recent arrival in Australia in evolutionary history, is very susceptible to 1080. The researchers suggested that 1080 baits could be used as a target-specific poison against foxes to protect native animals in Western Australia.

Another team of scientists from Western Australia's Department of Conservation and Land Management applied this approach to test the hypothesis that the fox caused declines in native mammal populations. They studied fox impacts on the black-footed rock wallaby (*Petrogale lateralis*), which was very abundant in the early years of settlement but is now rare over much of south-western Australia (Fig. b). They monitored five remnant populations in the central wheat belt of Western Australia and

(b) The black-footed rock wallaby (*Petrogale lateralis*)

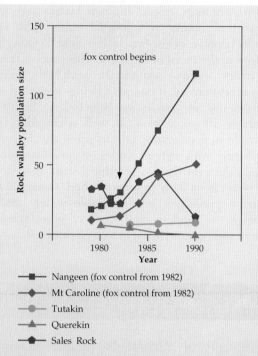

(c) Changes in populations of five colonies of black-footed rock wallabies over time. Two of the populations (Nangeen and Mt Caroline) were protected by fox control from mid 1982. Their populations increased markedly while the controls remained static or declined

found that they declined or remained static (although reproductive rates were high), there was no evidence of disease, and animals maintained their condition during drought. Fox control by shooting and 1080 baiting was instigated for two populations from 1982, with the other three populations serving as untreated experimental controls. The rock wallaby populations protected by fox baiting increased substantially, while two controls declined and a third rose slightly (Fig. c). The researchers concluded that fox predation was a factor in the current decline of rock wallabies and that predator control is a feasible management option (since implemented there has been a strong recovery in a range of species).

However, does this mean that the fox is the sole culprit in the decline of Australian mammals? Some scientists suggested that foxes regulated populations of feral cats, *Felis catus*, another introduced predator. Perhaps if foxes were removed cats would increase in numbers and have a similar impact. Cats are smaller than foxes and might prey most heavily on mammals smaller than those commonly taken by foxes. This hypothesis was tested at a field site near Shark Bay, Western Australia, where the numbers of small mammals in three different areas were compared, one where both foxes and cats had been removed, one where foxes but not cats had been removed, and one where neither predator had been removed. Small mammal numbers were highest where both

foxes and cats were removed but lowest where only foxes were removed. It appears that, in the absence of foxes, cats may suppress small mammal numbers, so cat control is necessary as well as fox control to protect mammal populations.

This example illustrates many aspects of the scientific method. The initial observations of mammal decline led to the hypothesis that foxes are responsible and to the prediction that removal of foxes should lead to a recovery in

(d) Researchers have established experimentally that feral cats suppress populations of small native mammals

the population sizes of medium-sized mammals. Experimental work supported the prediction and hypothesis. However, the solution to one problem often highlights a further question. In this case, it was the potential impact of cats and the issue was resolved through a further experiment. Critical to both experiments was the development of 1080 for target-specific control of introduced predators in south-western Australia. The interchange of information to achieve this was only possible because of communication between scientists through publication of research.

Ethics and social responsibility in science

Biological research is responsible for such developments as immunisation, antibiotics and a host of other discoveries of great benefit to humanity. Such research offers huge financial benefits to its backers. However, other research may threaten profitable activities by pointing out their adverse effects on people or the environment and powerful groups may oppose such research. For example, for several decades the tobacco industry contested the growing scientific evidence of the negative health impacts of smoking.

Attempts to suppress dissenting views in science show that science is not just a search for explanations of nature but also can be an exercise in social power in which those in powerful positions try to impose their views irrespective of scientific evidence. This is very different to the scientific disagreements that stimulate new ideas and test them with data. In confronting disagreements in science, it is often useful to follow codes of conduct (**ethics**). This requires disputants to outline clearly where they agree or disagree and to decide on what experiments or other data collection would resolve their disagreements.

While the great majority of scientists are honest and careful in their work, there are rare cases of scientific fraud. One of the most infamous is that of Sir Cyril Burt, who claimed on the basis of his studies of identical twins that intelligence had high heritability. He was knighted for his work and obtained the prestigious position of Professor of Psychology at University College, London. His findings were influential in restructuring the English educational system and in supporting arguments popular in the United States that social class was based on intelligence. After Burt's death, significant problems with his work were exposed by a detailed examination of his scientific method. Burt's scientific reports were so deficient in methodology that the findings were unreliable, while the consistent, close match of his results to his hypotheses left little doubt that his data were either invented or modified to fit the hypotheses. Given time, the scientific method is able to detect and expose such fraud and therefore contains a vital element of self-correction.

keyterms

cell theory (p. 3)	experimental control	hypothesis (p. 5)	scientific publication
classification (p. 4)	(p. 6)	observation (p. 5)	(p. 6)
ethics (p. 9)	experimental design	scientific method (p. 5)	theory (p. 5)
evolution (p. 4)	(p. 6)		

Review questions

1. Explain what is meant by an hypothesis and a theory.
2. Why is evolution considered a theory rather than an hypothesis?
3. State the cell theory.
4. What is meant by the term 'controlled experiment'? Why is it important to have a controlled experiment? Why are experiments replicated?
5. In disproving the idea of spontaneous generation, Louis Pasteur repeated experiments similar to those of Spallanzani's except that he used long, goosenecked flasks. What was the significance of this experimental design?
6. Cattle grazing is a contentious issue in the alpine regions of south-eastern Australia. Conservationists believe that cattle should be excluded from the region because of the impact they have on the environment, while graziers depend on the high country for grazing their stock. Design an experiment to test the hypothesis that cattle grazing decreases the species richness and abundance of grasses and herbs.
7. Many universities ban their researchers from accepting sponsorship money from tobacco companies. Why would this be so?

Suggested further reading

Chalmers, A. F. (1982). *What is This Thing Called Science?* Brisbane: University of Queensland Press.

A useful text for further discussion of what science is and how science proceeds.

Horwitz, P. and Calver, M. (1998). Credible science? Evaluating the Regional Forest Agreement Process in Western Australia. *Australian Journal of Environmental Management* 5; 213–25.

This article provides a case study for discussion and debate and is centred on the process of evaluating the conservation value of native forests versus their use for the forestry industry.

Part 1

Cell biology and energetics

CHAPTER

1

Molecules of life

Living organisms are composed almost entirely of 16–21 elements—16 are found in almost all organisms, while nine more are restricted to particular groups. With two exceptions, these are within the 30 lightest of the 93 natural elements (16 more have been produced artificially). More importantly, four elements, hydrogen (H), oxygen (O), nitrogen (N) and carbon (C), make up 99% of the living parts of organisms. Hydrogen is the most abundant element in the universe and hydrogen (15.40%) and oxygen (55.19%) constitute large fractions of the atoms in those parts of the earth accessible to living organisms, known as the **biosphere**. The biosphere comprises the hydrosphere (oceans and seas), the atmosphere and the surface of the lithosphere (the earth's crust) (Fig. 1.1).

On the other hand, carbon (0.44%) and nitrogen (0.16%), constitute only a small percentage of the accessible atoms in the biosphere. Two other 'biological' elements, phosphorus (P, 0.23%) and sulfur (S, 0.12%), must also be acquired from a small percentage of available atoms. It is clear, then, that the abundant elements in living organisms are not necessarily those most abundant in the biosphere. The relationship between abundance in the biosphere and abundance in living organisms is shown in Figure 1.2. The relative importance of these elements in living organisms is due to their particular fitness to impart critical properties upon which life forms depend.

What special distinction do the atoms of H, O, N, C, P and S have that led to their selection from the 93 naturally available atoms? To answer this we must first understand the nature of atoms and how they associate with one another to form molecules.

Elemental organisation of living organisms

Elements are made of atoms

An **element** is a substance made up of one type of atom only. An **atom** is the smallest part of an element that can exist and retain the properties of an element. Atoms themselves consist of three subatomic particles—a central nucleus consisting of positively charged **protons** and uncharged **neutrons**, and negatively charged **electrons**, which orbit the positively charged nucleus. In most atoms, the number of negatively charged electrons equals the number of positively charged protons, so that the atom is effectively neutral in electrical charge.

The number of protons in a nucleus is the **atomic number**, which is characteristic for each element. The **mass number** is the combined number of

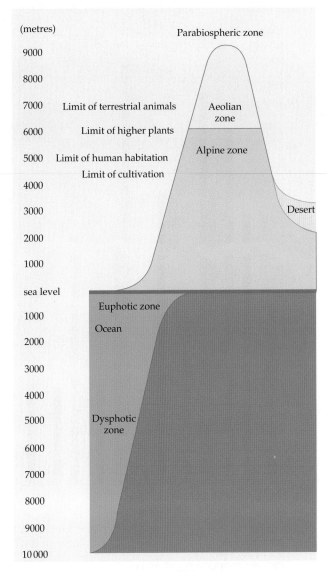

Fig. 1.1 Diagram showing the vertical extent of the biosphere, that is, that part of the earth where life exists. The biosphere is an irregular envelope surrounded by a zone where living, but not multiplying, forms (spores of bacteria and fungi) are wind carried. In the oceans, the zone of living forms where light may reach may be a few centimetres deep or extend to 100 m. But even at great depths (and pressures), organisms whose livelihood is independent of solar energy are present

protons and neutrons in each nucleus. For example, the common form of the element carbon has six protons and six neutrons, so the atomic number is 6 and the mass number is 12. This atomic form (isotope) of carbon is known as carbon-12. Other isotopes of carbon, which have the same chemical properties as carbon-12, have different numbers of neutrons and thus different mass numbers: carbon-13 has seven neutrons and carbon-14, rare in nature but very useful for tracing carbon atoms in biochemical reactions, has eight neutrons.

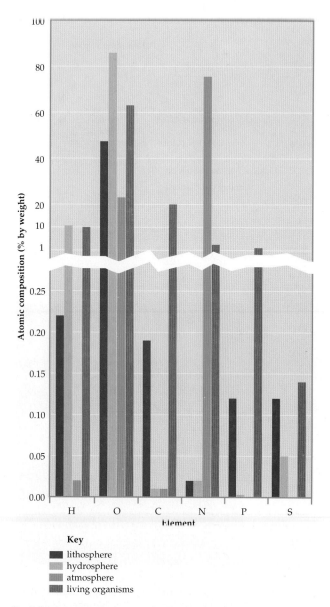

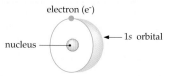

Fig. 1.3 Hydrogen atom. Model showing central nucleus of one proton and the orbital of the single electron

Fig. 1.2 Histogram comparing the relative abundance of some elements in the lithosphere (outer part of the earth's crust), hydrosphere, and atmosphere with that in living organisms

An element is made of one type of atom. Atoms consist of subatomic particles—a central nucleus of protons (positively charged) and neutrons (neutral charge), and orbiting electrons (negatively charged).

Energy levels and excitation of electrons

Electrons move around the nucleus in **orbitals**, which are zones of space in which electrons exist at any one moment. Thus, in an atom of hydrogen (Fig. 1.3), the single electron is in the 1*s* (spherical) orbital, the orbital next to the nucleus. This electron is at the lowest energy level. In atoms with more than two electrons,

the electrons exist in further orbitals at greater distances from the nucleus. A given orbital can contain no more than two electrons, however, there may be several orbitals within a given energy level or electron shell. For example, oxygen (Fig. 1.4b) has eight electrons orbiting the nucleus: two electrons in the orbit at the first energy level and six electrons at the second energy level (two in each of three orbitals at the same energy levels).

The innermost shell near the nucleus is filled before those that are further away. The first shell is complete when there are two electrons, the second shell is complete when there are eight electrons in four orbitals, the third has a maximum of 18 electrons in nine orbitals and the fourth has 32 electrons in 16 orbitals. The atomic structures of some biologically important elements are shown in Figure 1.4.

The further an electron level is from the nucleus, the greater is the energy of the electrons. A negatively charged electron can be moved from an inner to an outer orbital, that is, further from the positively charged nucleus, by providing it with more energy. If such a displaced electron moves back to an inner orbital, it will give up energy. Electrons move from one orbital to another—they do not occupy the space between orbitals. Thus, the energy required to move an electron from one orbital to another is discrete and is known as an energy quantum. Electrons can move to a higher energy level when a particular amount of energy is absorbed by an atom and to a lower energy level when the additional energy is released (Fig. 1.5). During photosynthesis by leaf cells in sunlight, electrons in pigment molecules absorb quanta of light energy, which shifts the electrons to higher energy levels. Electrons do not remain in this excited state for very long but fall back to their original orbital, with a release of energy that may be in the form of heat, or a flash of light (fluorescence), or may excite a neighbouring atom, or, in the case of photosynthesis, drive a chemical reaction in living cells.

Electrons orbit a nucleus in zones of space (orbitals). Electrons can move between orbitals of different energy levels, with the gain or loss of energy.

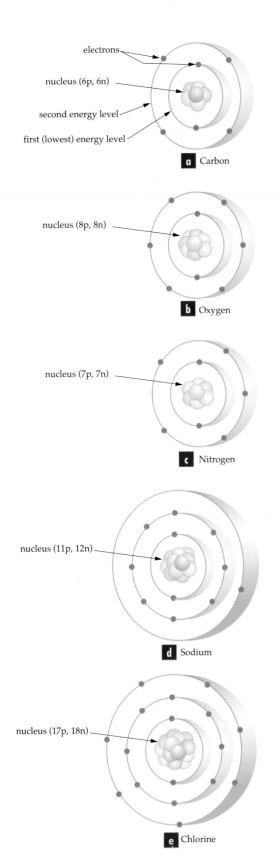

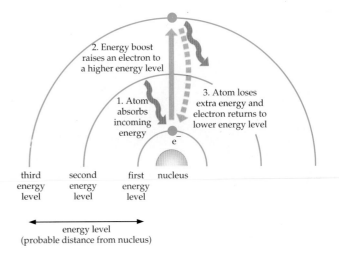

Fig. 1.5 An electron can move to a higher energy level when a particular amount of energy is absorbed by the atom or molecule (positions 1 to 2). An electron can move to a lower energy level when the additional energy is released from the atom (positions 2 to 3)

Chemical bonding of atoms makes molecules

The chemical properties of an element are determined primarily by the number and arrangement of the electrons in the outermost energy level (electron shell). When the outer shell of electrons of an atom contains fewer than eight electrons, the atom tends to lose, gain or share electrons to achieve an outer shell of eight electrons (or two in the case of hydrogen). Gaining electrons, in the form of sharing them with other atoms, is the means for making **chemical bonds** and so of making molecules.

Covalent bonds

The sharing of electrons creates a **covalent bond** (Fig. 1.6). A compound consisting of covalent bonds is called a covalent compound and the resulting stable association is called a **molecule**.

The special distinction of H, O, N and C is that they are the four smallest elements in the Periodic Table (Fig. 1.7) that achieve stable electronic configurations in their outermost shell of electrons by gaining, respectively, one, two, three and four electrons. The small size of the atoms of these elements makes for tight bonds and so the most stable molecules. The elements H, O and N form stable covalent molecules by sharing electrons with other atoms of the same element (Fig. 1.6).

Carbon has four electrons in its outer energy level (shell) available for covalent bonding. Thus, one carbon atom can share electrons with four hydrogen atoms to

Fig. 1.4 Structural representation of some biologically important atoms: **(a)** carbon, **(b)** oxygen, **(c)** nitrogen, **(d)** sodium and **(e)** chlorine. Each circle represents an energy level (shell). Dots on the circles represent electrons; p is a proton and n is a neutron

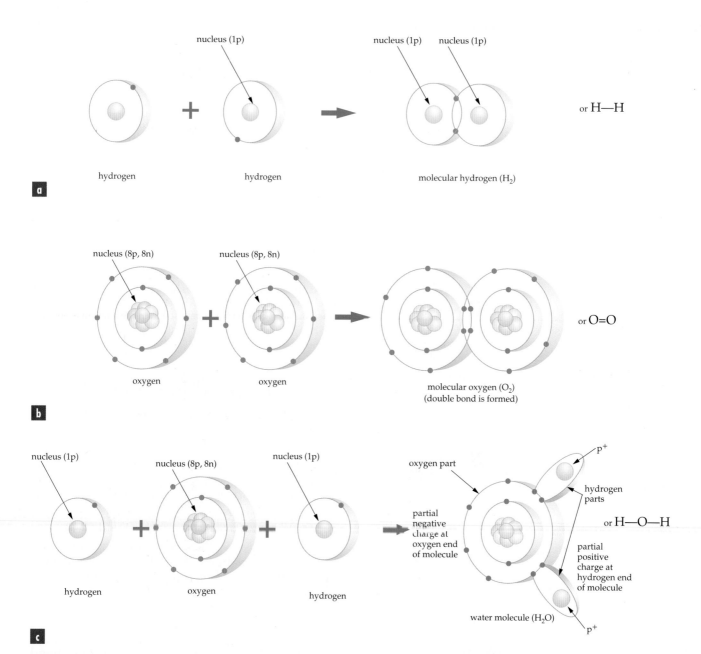

Fig. 1.6 Formation of covalent compounds. **(a)** Two hydrogen atoms achieve a stable electron shell by sharing electrons, thereby forming a molecule of hydrogen (dihydrogen)—the horizontal line in the formula H—H represents the paired electrons. **(b)** Two oxygen atoms share two pairs of electrons to give stable outer shells with eight electrons. Note that in dioxygen (O_2), a double bond is formed. Dinitrogen (N_2) is formed in a similar way. **(c)** When two hydrogen atoms share their electrons with an oxygen atom, a stable eight electron outer shell is formed, and the resultant molecule is water. In the water molecule, the electrons are not shared equally but tend to stay closer to the oxygen atom than the hydrogen nuclei. This results in the hydrogen atoms in water having a partial positive charge and the oxygen a partial negative charge. Although the water molecule is electrically neutral, it is a polar covalent compound

form the methane molecule, CH_4 (H—C—H with H above and below C), or with two oxygen atoms to give carbon dioxide, CO_2 (O=C=O). Nitrogen, with five electrons in its outer shell, can share electrons with three hydrogen atoms to form ammonia, NH_3 (H—N—H with H above).

The covalent water molecule is formed by an oxygen atom sharing the electrons in the outer shell with two hydrogen atoms (Fig. 1.6c).

Non-covalent bonds

A diverse group of forces, together known as non-covalent bonds, mediates the interaction of atoms and molecules with one another. They are collectively

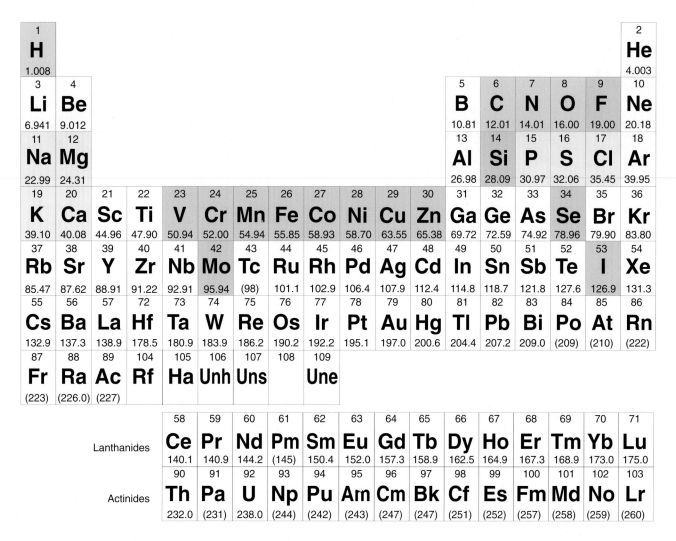

Fig. 1.7 Periodic Table of the elements showing the positions of the bioelements. The major bioelements are indicated in blue, important bioelements found in smaller amounts are shown in yellow and trace bioelements in green. The lanthanides (elements 58–71) and actinides (elements 90–103) play no role in biological systems

important in maintaining the shape of complex cellular molecules, in maintaining the stability of cell membranes and in recognition events between surfaces of molecules. Non-covalent bonds are relatively weak compared with covalent bonds, especially in the aqueous solution that characterises the cell's interior.

Ionic bonds

When atoms gain or lose electrons from their outer shell, they become charged particles called **ions**. Atoms with one, two or three electrons in their outer shell take up a stable electron configuration by losing electrons to other atoms. When atoms lose electrons they become positively charged as a result of excess protons in their nucleus. Positively charged ions are termed **cations**. Conversely, atoms with five, six or seven electrons in their outer shell tend to gain electrons from other atoms to give a stable outer electron shell and become

negatively charged particles known as **anions**. Ions are of utmost importance in biological systems and are essential for conduction of nerve impulses, muscular contraction and many other vital processes.

Substances called ionic compounds are composed of anions and cations bonded together through association of opposite charges. These bonds are known as **ionic bonds**. Such compounds are formed not by sharing electrons between atoms but by transfer of outermost electrons from one atom to another. The formation of ionic bonds is well illustrated in the interaction that occurs between sodium and chlorine atoms. A sodium atom has one electron in its outer shell, whereas the chlorine atom has seven (Fig. 1.8). Thus, by transfer of the single outer shell electron from the sodium atom to the chlorine atom, both atoms have complete outer shells. The loss of the electron from the sodium atom leaves it with a net positive charge. The chlorine atom with an extra electron has a net negative

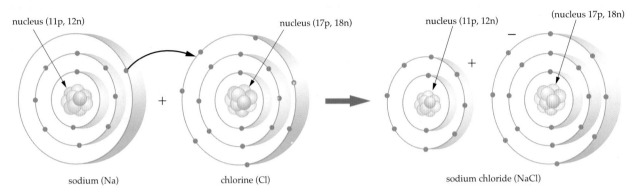

nucleus (11p, 12n)   nucleus (17p, 18n)   nucleus (11p, 12n)   (nucleus 17p, 18n)

sodium (Na)   chlorine (Cl)   sodium chloride (NaCl)

Fig. 1.8 Formation of an ionic bond by transfer of electrons between two atoms. Sodium donates the single electron in its outer shell to a chlorine atom with seven electrons in its outer shell. Both sodium and chlorine atoms now have complete (eight electron) outer shells but the two atoms are now oppositely charged and are attracted to one another, forming sodium chloride

charge. The ions attract one another because of their opposite charges. The resultant ionic compound is sodium chloride or common salt.

Ionic compounds, such as sodium chloride, tend to separate their respective anionic and cationic components when dissolved in water. The polar molecules cf water surround the charged ions and shield them from one another (Fig. 1.9). The ions are said to be hydrated.

van der Waals forces

van der Waals forces arise due to the non-uniform distribution of electric charge at any instant on an atom, although over a longer time the charge may appear to be uniformly distributed. The variation in charge distribution gives the atom a polarity so that one part will have, for example, a slightly negative charge. Neighbouring atoms may have an opposite polarity and the two charged atoms will attract one another. These van der Waals attractions are temporary and some 100 times weaker than covalent bonds. Nevertheless, they are of significance in interactions between complementary biological molecules, when thousands of these weak interactions achieve tight associations,

and in stabilising macromolecular shapes (e.g. proteins, p. 38).

Hydrogen bonds

As we have seen in a molecule of water (Fig. 1.6c), the hydrogen atoms are slightly more positively charged than the oxygen atom and the molecule is polar. The hydrogen on one molecule can be shared with an oxygen on another molecule to form networks of **hydrogen bonds** between water molecules (Fig. 1.10).

Hydrogen bonds tend to form between a hydrogen atom that is covalently bound to oxygen (as in water) or to nitrogen or other electronegative atoms.

Hydrogen bonds are weak and are readily formed and broken. They have a short lifetime of 10^{-4} seconds and, as one is broken, another is formed. The cumulative effect of large numbers of hydrogen bonds generates considerable strength, causing water molecules to cling together as a liquid at normal temperature and pressure. In many large biological molecules, for example, in cellulose microfibrils (p. 28), the α-helix and β-sheets of proteins (p. 39) and the DNA double helix (p. 46), hydrogen bonds are crucial for structural stability.

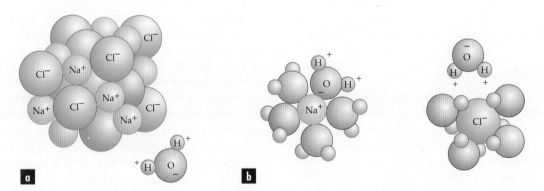

Fig. 1.9 Hydration of ionic compounds. **(a)** Crystals of sodium chloride are composed of regularly arranged sodium and chloride ions held together by ionic bonds. **(b)** When sodium chloride is added to water, the positively charged sodium ions attract the partially negatively charged oxygen end of the polar water molecules, separating them from the chloride ions. The negatively charged chloride ions attract the partially positively charged hydrogen end of the polar water molecules, separating them from the sodium ions

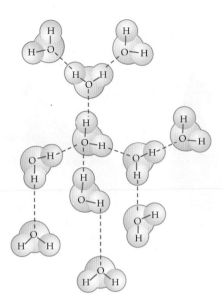

Fig. 1.10 Hydrogen bonding between water molecules. The water molecules interact with one another through their oppositely charged ends to form hydrogen bonds with a maximum of four neighbouring water molecules

Chemical bonds involve attraction between two atoms or molecules. Covalent bonds involve the sharing of electrons, while hydrogen bonds and ionic bonds involve attraction between opposite charges. van der Waals forces are temporary and 100 times weaker than covalent bonds.

Water

Life on earth exists only in an environment of water. Many organisms live in saline oceans, saline or freshwater pools, lakes and rivers (Fig. 1.11). Moreover, water is the most abundant molecule found in living organisms. Around 62% of our body weight is water, although bone has 20% water and brain cells are 85% water. Jellyfish are 96% water. The special physical and chemical properties of water provided a medium in which living organisms originated, survived and evolved.

Because of its special properties, especially its hydrogen bonding capacity, water is an ideal medium for living organisms, for example, as a carrier of nutrients, oxygen and waste products to and from cells. Hydrogen bonding between water molecules is responsible for many of the unusual and important physical properties of water, such as its solvent properties, high boiling point, high specific heat, high latent heats of fusion and vaporisation, cohesion and surface tension (see Box 1.1).

Water as a solvent

Water, being polar, is an excellent solvent for many substances. As we have seen, in aqueous solution, an

Fig. 1.11 Diversity of life in water. **(a)** Egg mass and newly hatched tadpoles of the spotted grass frog hanging below froth, with filamentous green algae in the freshwater pond and water reeds in the background. **(b)** Red snow algae at Paradise Bay, Antarctica. **(c)** This intertidal rock pool shows a diversity of invertebrate life, such as sea stars, anemones and sea urchins

BOX 1.1   Properties of water

The chemical processes of life evolved within a set of constraints imposed by the physicochemical properties of water. These properties include high specific heat, high heat of vaporisation, high heat of fusion, surface tension and cohesion, and density.

Specific heat

Water has a high specific heat, that is, the amount of heat required to raise the temperature of 1 g of a substance by 1°C. This is due to extensive hydrogen bonding between water molecules. The high specific heat means that water can absorb considerable amounts of heat with little change in temperature. Many organisms live in water and, since a considerable amount of heat is added to or lost from the aqueous environment, the surroundings remain relatively constant in temperature. Moreover, within a living organism, the heat generated by the vast array of chemical reactions occurring within cells, together with the heat absorbed from the external environment, could damage cells were it not for the 'heat-buffering' effect of water. The high water content of plant cells enables them to maintain a relatively constant internal temperature.

Heat of vaporisation

Water has a high heat of vaporisation, that is, the energy absorbed per gram as it changes from a liquid to a gas (water vapour). Water evaporating from a surface, such as the skin or the surface of a leaf, will draw heat from the surface, thereby cooling it. This property allows overheated organisms to lose heat to the environment (Fig. a).

Heat of fusion

Surprisingly, heat is liberated when liquid water is turned to ice. Thus, 1 g of water at 0°C loses eight times as much heat in turning to ice as it does in cooling from 1°C to 0°C. This property of water slows down the reduction in temperature of water within organisms during winter and allows the large masses of water on the earth's surface to resist the warming effect of heat and the cooling effect of low temperatures.

Cohesion and adhesion

Water molecules are strongly cohesive, that is, they tend to stick to one another due to hydrogen bonding. In addition, they tend to stick to other molecules that have charged surface groups to which the water molecules can hydrogen bond. These adhesive forces are responsible for the wetting

(a) Because of its high heat of vaporisation, evaporating water draws heat from surfaces. Kangaroos lick their forearms to lose heat in this way during hot weather

effect of water. Only surfaces composed of polar molecules are wetted; other surfaces, composed of non-polar molecules, such as waxy leaf surfaces, are water-repellent and stay dry (Fig. b).

At the interface between water and air, the cohesive forces between water molecules are much stronger than between water molecules and the molecules in air. The result is that the surface molecules tend to crowd together to produce a region at the air–water interface where there is a high degree of **surface tension**. Some insects, such as water striders, are able to use this surface tension to support themselves on water (Fig. c).

Cohesive and adhesive forces are together responsible for capillary action, which is the ability of water to rise in tubes of very narrow diameter. Capillarity also causes water to move through minute spaces between soil particles to plant roots and to rise through the stems of plants

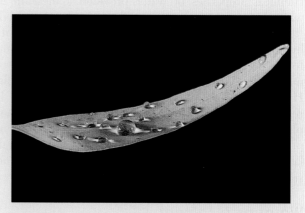

(b) Cohesion causes water to form droplets on the waxy surface of this eucalypt leaf

(c) Cohesion between water molecules at a surface allows these water striders to mate on the pond surface

Density

A critically important property for living organisms is that ice floats on liquid water (Fig. d). (In contrast, ammonia ice sinks in liquid ammonia.) Water, like most liquids on cooling, contracts, but is unique in that below 4°C it expands, so that at its freezing point (0°C) its density is lower than liquid water. The reason for this is that below 4°C water molecules are increasingly bonded to others by hydrogen bonds. At the freezing point, every hydrogen atom is engaged in both covalent and hydrogen bonding to adjacent oxygen atoms, holding all the water molecules rigidly in an open structure in which they are less densely packed than in liquid water (Fig. e).

If ice did not float, the water on the earth's surface would have long since frozen solid. What in fact happens in oceans, lakes and streams is

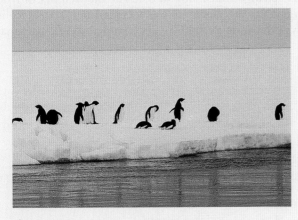

(d) Iceberg floating in the ocean

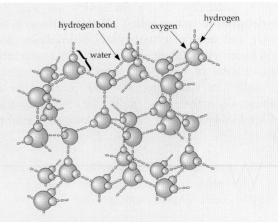

(e) When water freezes to form ice, each water molecule has formed hydrogen bonds with four other water molecules in a three-dimensional open lattice. The water molecules are further apart in ice than in liquid water

that, as the temperature drops, the colder water in the narrow temperature range, 0°C–4°C, where it is less dense, rises to the surface, where it freezes to form a layer of ice. This ice layer insulates the water beneath and prevents it from freezing, enabling aquatic organisms to survive.

Some aquatic organisms, for example, Antarctic ice fish, have chemicals (proteins) in their body fluids that effectively lower their freezing point, thus protecting them from the damaging effects of ice crystal formation. These chemicals function in the same way as does antifreeze, used in car radiators.

ion such as Na^+, K^+, Mg^{2+}, Ca^{2+} or Cl^- is surrounded by a loosely bound array of water molecules, forming a hydration sphere (Fig. 1.9). Hydration increases the effective size of an ion. The orientation of water molecules in the sphere will reflect the charge on a particular ion. Na^+ and Mg^{2+} attract the oxygen ends of water molecules, and Cl^- attracts the hydrogen ends.

Water also interacts with and dissolves polar molecules, such as ethanol, ammonia, simple carbohydrates and amino acids that carry groups that contain oxygen or nitrogen atoms (see p. 15), because they readily form hydrogen bonds with water molecules. Substances that dissolve readily in water are said to be **hydrophilic** (water loving).

Hydrophobic (water-hating) interactions

In contrast, non-polar molecules, such as hydrocarbons, oils and fats (see p. 31), in an aqueous environment cluster together and are insoluble. This is due to their exclusion by the strongly hydrogen-bonded, polar water molecules. These molecules are said to be **hydrophobic** (water hating). Both repulsion between polar and non-polar molecules (hydrophobic forces) and hydrogen bonding play crucial roles in the internal organisation of biologically important molecules, for example, in stabilising globular proteins in their biologically active shape (p. 38), and in the association of lipid molecules that make up biological membranes (p. 34).

W ater is an excellent solvent for polar molecules, which dissociate in solution to form ions (ionisation). Water repels non-polar molecules.

Acids, bases and buffers

Water molecules have a tendency to ionise, or dissociate, into **hydrogen ions** (H^+) and **hydroxyl ions** (OH^-). When a hydrogen atom dissociates from the water molecule, it leaves its electron behind to form the negatively charged hydroxyl ion and becomes, for an instant, a hydrogen ion (proton, H^+).

$$HOH = H^+ + OH^-$$

In pure water, the concentration of H^+ is 10^{-7} moles per litre and is exactly equal to the concentration of OH^-. (For simplicity, we refer to H^+, although in solution, most H^+ are actually **hydronium ions**, H_3O^+.) The number of such ions in pure water is very small. The concentrations of hydrogen ions and hydroxyl ions are equal and the solution is neutral, that is, neither acidic nor alkaline (basic).

Substances that dissociate, releasing hydrogen ions and an anion into solution, are **acids**. Thus, an acid is a proton donor.

$$Acid \rightarrow H^+ + conjugate\ base$$

e.g.
$$\underset{\displaystyle RCOH}{\overset{\displaystyle O \atop \displaystyle \|}{}} \rightarrow H^+ + \underset{\displaystyle RCO^-}{\overset{\displaystyle O \atop \displaystyle \|}{}}$$

Substances that accept hydrogen ions are **bases**.

Hydrochloric acid, HCl, which dissociates in water into a proton, H^+, and an anion, Cl^-, is a typical acid. The hydrogen ion (proton) from HCl then becomes attracted to a water molecule, forming a hydronium ion, H_3O^+.

$$HCl \rightarrow H^+ + Cl^-$$

$$H^+ + H_2O \rightarrow H_3O^+$$

By taking up a hydrogen ion, water has acted as a base. In cells, other powerful acceptors of hydrogen ions are also present. The hydronium ion acts as an acid when it donates its extra hydrogen to a more powerful acceptor. The most powerful of these acceptors of hydrogen ions is the hydroxide anion, OH^-.

S ubstances that release hydrogen ions into solution are acids, while those that accept hydrogen ions are bases.

pH

Cells are subject to changes in hydrogen ion concentrations resulting from the release or binding of hydrogen ions during continuing chemical reactions. Compared with pure water, an acidic solution has a high concentration of H^+, whereas a basic (alkaline) solution has a low concentration.

The concentration of H^+ in solution is represented on the pH scale, in which H^+ concentrations are expressed as the negative logarithm to the base 10 within a possible range of 0 to 14 (pH = $-\log[H^+]$). Because the pH scale is logarithmic, each unit represents a 10-fold change in H^+ concentration. The midpoint of the pH scale is 7, which is the pH value for pure water. Acid solutions have a pH lower than 7, and basic solutions have a pH greater than 7 (Fig. 1.12).

In a eukaryotic cell, the pH of the cytosol, the cytoplasm that occupies the space between the membrane-bounded organelles, is usually near neutral (pH 7.4). In an actively contracting muscle cell, the pH may change transiently to 6.4. The internal pH of lytic organelles in plant cells (vacuoles) and animal cells (lysosomes) is around 5. The internal pH of bacterial

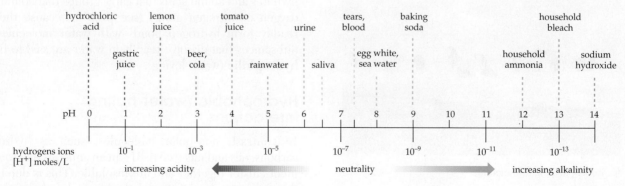

Fig. 1.12 The pH scale, showing the pH of some common substances

cells is also near neutrality, but for extreme acidophiles or alkaliphiles (see Chapter 33) the intracellular pH may vary by 1–1.5 units from neutrality. These bacteria are found in extremes of pH—1 for acidic to 11 for alkaline biotypes. The pH of the external environment of plants and animals may also be quite different from the internal pH of their cells. For example, for organisms living in a creek, the pH of the surrounding water may vary from 6.5 to 8.5. Sphagnum moss grows in acidic peat bogs of low pH (3) whereas the shrub wirilda, *Acacia retinodes*, grows in coastal limestone soils (pH 8.5) of southern Australia.

pH is the concentration of H^+ in solution. The pH scale is logarithmic and ranges from 0 to 14.

Buffers

Living cells have different ways of maintaining a relatively constant internal pH. One way in which pH is controlled involves **buffers**. These are substances that act as a reservoir for hydrogen ions, accepting them as pH falls and releasing them as pH rises. Consider a solution of carbonic acid, H_2CO_3, and bicarbonate ion, HCO_3^-, in water. Carbonic acid is formed, reversibly, in solution as follows:

$$CO_2 + H_2O \rightleftharpoons H_2CO_3 \rightleftharpoons H^+ + HCO_3^-$$

When excess H^+ ions are added in small amounts, bicarbonate combines with them to form carbonic acid. The carbonic acid is unstable and decomposes to CO_2 and H_2O and the pH of the solution is maintained.

If hydroxide ions are added, the buffer also acts to maintain the pH. The carbonic acid releases hydrogen ions, which react with hydroxyl ions to form water and bicarbonate ions.

$$OH^- + H_2CO_3 \rightleftharpoons H_2O + HCO_3^-$$

The carbonic acid–bicarbonate ion buffer system is the key buffer system in mammalian blood. Buffering blood against changes in pH is needed because many cellular activities are sensitive to changes in pH. If the blood becomes too acidic, coma and death may ensue; if it becomes too alkaline, it results in hyper-excitability of the nervous system and may lead to convulsions.

A buffer solution, such as a carbonic acid–bicarbonate ion solution, maintains relatively constant pH either by removing added hydrogen ions or by releasing them.

Biologically important molecules

So far the discussion of the molecules of living organisms has focused on the elementary particles, the atoms, and their behaviour in forming covalent and ionic compounds. The atoms of H, C, O, N, P and S form **organic molecules** by covalent bonding. Some 15 000 different unit organic molecules, building blocks of larger molecules, have been found in living organisms and more are being discovered. An estimated 13 million organic molecules have been synthesised in the laboratory. Remarkably, however, it turns out there are just 29 of the unit organic molecules needed to describe the basic molecular units found in all living organisms.

The important unit organic molecules include **glucose**, the major initial product of photosynthesis

---

BOX 1.2 Functional groups on biologically important molecules

Functional groups are combinations of atoms whose characteristic behaviour is maintained no matter to which molecule they are attached. The functional groups are shown in the table below.

Table Major functional groups

Alcohol	The —OH group is called an hydroxyl group.
Aldehyde, ketone	The $\overset{\mid}{\underset{\mid}{C}}$=O group is called a carbonyl group.
Carboxylic acid	The $-\overset{O}{\overset{\|}{C}}-OH$ group is called a carboxyl group. In water this group loses an H^+ ion to become $-\overset{O}{\overset{\|}{C}}-O^-$.
Esters	Esters are formed by combining an acid and an alcohol with the loss of water.
Sulfhydryl	The —SH is called a sulfhydryl group.
Amino	The $-NH_2$ is called an amino group.
Phosphates	Phosphoric acid, H_3PO_4, forms a stable phosphate ion. Sometimes abbreviated to P_i. Phosphate ions form esters with free hydroxyl groups. Phosphate ions can combine with carboxyl groups to give acid anhydrides or with other phosphate ions to give di- and triphosphate anhydrides.

and a major source of metabolic energy and hydrogen; the **triacylglycerides** (fats), which are the principal storage forms of energy in animals and some plant seeds; the **phospholipids** and **glycolipids**, which are important for their ability to form membranes; 20 **amino acids**, which are the units that form all proteins, including enzymes; and five **nitrogenous bases** (adenine, guanine, cytosine, uracil and thymine), which together with the monosaccharides **ribose**, and its simple derivative deoxyribose, and phosphoric acid, form all the nucleic acids (Table 1.1).

In the following section we will discuss how these unit organic molecules are assembled into macromolecules. These include the polysaccharide storage and structural molecules, the storage and structural lipids, enzymic and structural proteins, and the nucleic acids, which are concerned with the storage and transmission of information. With this background we can proceed to discuss cell structure and explore the complexities of energy metabolism.

Carbohydrates

Carbohydrates are the most abundant organic compounds in living organisms. They are a source of chemical energy for living organisms, form structural components and, in combination with other molecules, such as proteins and lipids, serve important recognition functions (Table 1.2). They are also one of the building units of nucleic acids.

Carbohydrates are composed of carbon, hydrogen and oxygen with the general formula $C_n(H_2O)_n$ (where n is the number of carbon atoms); hence the name 'hydrates' of carbon. They are polyhydroxy compounds that, in addition, contain a carbonyl group, involving a carbon–oxygen bond forming an aldehyde or ketone function (Box 1.2). The three main groups of carbohydrates are **monosaccharides**, **disaccharides** and **polysaccharides**.

Monosaccharides

The simplest carbohydrates found in biological systems are the monosaccharides, consisting of single molecules such as D-glucose, D-galactose and D-fructose. ('D' refers to one of the mirror image forms of glucose; the other is referred to as 'L'.) These six-carbon monosaccharides are known as hexoses. Glucose and galactose are aldohexoses and **fructose** is a ketohexose (Fig. 1.13). Galactose and fructose are **isomers** of glucose. They have the same number of the same kinds of atoms as glucose but they are arranged differently

Table 1.1 An alphabet of organisms. Living organisms are composed almost entirely of 16 elements. Nine other elements (in brackets) have a more restricted distribution in some groups

	Biomolecules	
Bioelements	Unit molecules	Macromolecules
Water: hydrogen (H), oxygen (O)	Monosaccharides (e.g. glucose)	Polysaccharides, glycoproteins and proteoglycans
Organic: hydrogen (H), carbon (C), nitrogen (N), oxygen (O), phosphorus (P), sulfur (S)	Triacylglycerols Phospholipids Glycolipids	
Ionic: sodium (Na), potassium (K), magnesium (Mg), calcium (Ca), chlorine (Cl)	Amino acids (20)	Proteins, glycoproteins and proteoglycans
	Nitrogenous bases (4) Ribose or deoxyribose Phosphate	Nucleic acids
Trace elements: manganese (Mn), iron (Fe), cobalt (Co), copper (Cu), zinc (Zn) [boron (B), aluminium (Al), vanadium (V), molybdenum (Mo), iodine (I), silicon (Si), selenium (Se), chromium (Cr), gallium (Ga), strontium (Sr), barium (Ba), arsenic (As), nickel (Ni), fluorine (F)][a]		

(a) Elements in square brackets are restricted to special groups of organisms; the others are generally distributed.

Table 1.2 Functions of carbohydrates in biological systems

Carbohydrate	Function
Monosaccharides	
Glucose	Transport form[a] of energy and carbon in animals and plants.
Fructose	Transport form of energy and carbon in foetal ungulates and plants.
Disaccharides	
Sucrose	Transport and storage form of energy in higher plants.
Lactose	Reserve form of carbohydrate in mammalian milk.
α,α′-Trehalose	Transport form of energy and carbon in insect haemolymph; storage form of energy and carbon in fungi.
Storage polysaccharides	Storage forms of energy and carbon.
Starch polymers (amylopectin and amylose)	Plant storage form.
Inulin	Plant storage form.
(1→3)-β-D-glucans	Storage form in protists.
Floridean starch	Storage form in red algae.
Glycogen	Storage form in animals, fungi and bacteria.
Structural polysaccharides	
Cellulose, mannans, arabinoxylans, xyloglucans, galactomannans, pectins	In cell walls of plants and some green algae.
Alginic acid, sulfated galactans (agarose, carrageenan)	In cell walls of algae.
Chitin	In cell walls of fungi and exoskeletons of insects and crustaceans.
(1→3)-β-D-glucans	In cell walls of fungi.

(a) 'Transport form' means a compound used to move material from one place to another in the body of an animal (as in blood or haemolymph) or a plant (as in phloem sap) or across biological membranes.

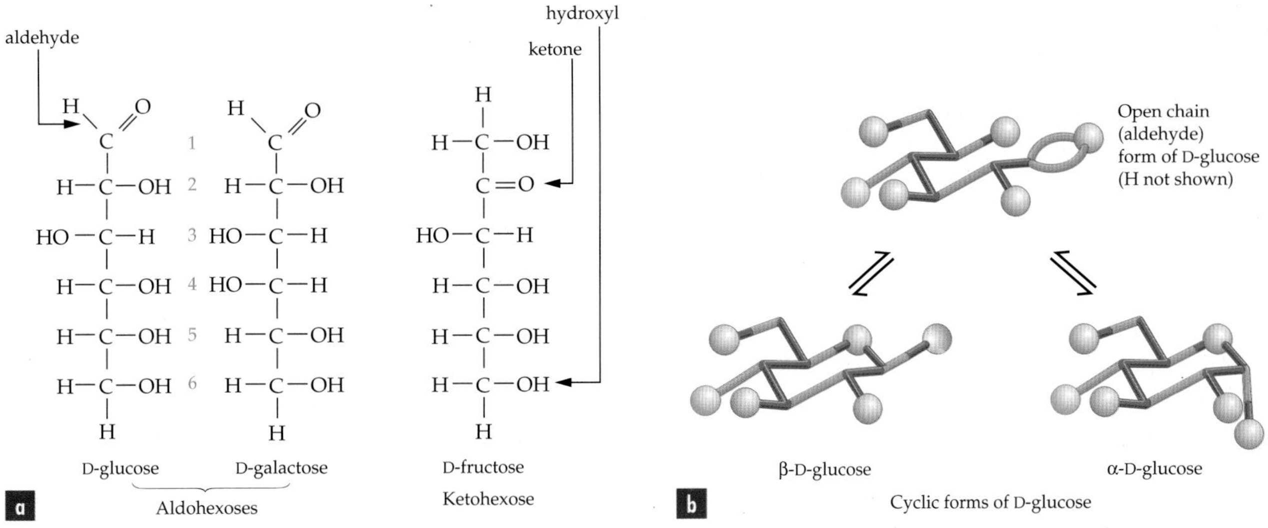

Fig. 1.13 (a) Structures of some naturally occurring monosaccharides. **(b)** Open chain and cyclic (ring) forms of D-glucose

(Fig. 1.13). Three-, four-, five- and seven-carbon (i.e. triose, tetrose, pentose and septanose) monosaccharides occur in living organisms.

Monosaccharides with five or more carbon atoms typically exist in ring (cyclic) structures, which in solution are in equilibrium with the open chain forms

(Fig. 1.13). There are two six-membered ring forms of D-glucose, termed α- and β-D-glucose, distinguished by the orientation of the hydroxyl at carbon 1 (Fig. 1.13). The monosaccharides are readily soluble in water due to their ability to form hydrogen bonds with water molecules through their oxygen atoms. They have varying degrees of sweetness. Fructose is the sweetest monosaccharide. Glucose is found in the blood of vertebrates, where it is a transport form of energy and carbon atoms. Fructose is found in the blood of ungulates and foetal ruminants, and as the sole monosaccharide in human and bull semen, as well as being abundant in fruits and plant nectars.

Disaccharides

Monosaccharides may link together in a variety of ways through glycosidic linkages to form various disaccharides. Thus, **lactose**, the disaccharide found in the milk of mammals, is formed by linking the hydroxyl at carbon 1 of β-D-galactose to the hydroxyl at carbon 4 of D-glucose, with the loss of water (Fig. 1.14). The disaccharide **sucrose** is formed by linking the hydroxyl at carbon 2 of β-D-fructose with the hydroxyl of carbon 1 of α-D-glucose, with the loss of water (Fig. 1.14). Sucrose is found in the sap of flowering plants as a transport form of the products of photosynthesis from leaves to other plant parts and as a storage molecule in the stems of sugar cane (Fig. 1.15). Sucrose is 0.75 times as sweet as fructose. The glucose disaccharide, α,α'-trehalose (Fig. 1.14) is found in insect haemolymph (blood) and in cells of fungi, including yeasts, as a transport or storage form of carbon and energy.

Polysaccharides

As the name implies, polysaccharides consist of chains of many, sometimes thousands, of monosaccharides. They have two major roles: as structural components in walls of plants, fungi and bacteria, and, in association with proteins, as structural components of connective tissue and exoskeletons of animals; and as storage reserves in all groups of organisms.

The properties of individual polysaccharides depend on the nature of the monosaccharides forming the chain, the linkages between them and the size and organisation of the chains. They may be linear, unbranched molecules or branched structures. Some polysaccharides are composed of a single monosaccharide type; others are more complex, with two or more monosaccharides in the molecule and often with different types of glycosidic linkages between the monosaccharide units. Examination of the chemical and physical properties of polysaccharides illustrates how structure and biological function are closely related.

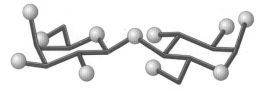

α-lactose (β-galactosyl (1→4) α-glucose)

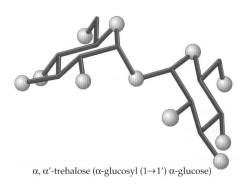

α, α'-trehalose (α-glucosyl (1→1') α-glucose)

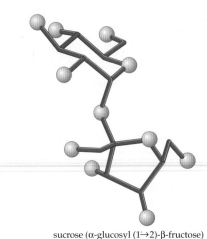

sucrose (α-glucosyl (1→2)-β-fructose)

Fig. 1.14 Structures of three biologically important disaccharides

Structural polysaccharides

Cellulose, the quintessential structural polysaccharide, is found in the walls of cells in plants, and brown, red and many green algae, but is absent from the cell walls of most fungi and, among animals, is found only in the tunic of sea squirts (ascidians, tunicates) and a few species of bacteria (Fig. 1.16). Cellulose is the most abundant polysaccharide in the biosphere and is produced at the prodigious rate of 1.7×10^{11} tonnes per year. Wood contains 50% cellulose and the walls of cotton seed hairs are 95% cellulose.

The molecules of cellulose are linear and composed solely of β-D-glucose monomers, all (1→4)-linked, that is, from the carbon 1 atom of one glucose molecule to the carbon 4 atom of an adjacent glucose molecule through the glycosidic oxygen bond (Fig. 1.17a).

Fig. 1.15 Sucrose is the product of photosynthesis in leaves and is a storage molecule in the stems of sugar cane

Individual chains of up to 15 000 glucose units have been recorded.

Due to the near planar shape of the β-D-glucose units (Fig. 1.17b), the chains have an overall ribbon-like form. Individual cellulose molecules associate through hydrogen bonds and van der Waals forces to form highly regular aggregates, termed microfibrils (Fig. 1.17c). Although the molecules of cellulose are composed of hydrophilic glucose units, the tight and regular hydrogen bonding between adjacent molecules makes extensive hydrogen bonding with water molecules impossible. Hence, cellulose is quite insoluble in water. Cellulose microfibrils have the highest tensile strength (a measure of the ability of a substance to withstand stretching) of any natural polymer and are, thus, ideal structural components of plant cell walls, which have to bear high tensile loads.

Other polysaccharides that accompany cellulose in cell walls of plants are also linear molecules but have side branches and may be charged. Their side branches are single monosaccharides or short chains of monosaccharides, which prevents their extensive lateral association compared with cellulose molecules. Despite these side branches, hydrogen bonding to the surfaces

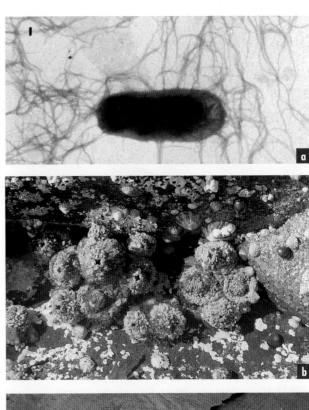

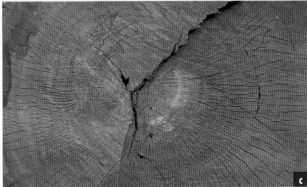

Fig. 1.16 Some cellulose producing organisms. **(a)** A cell of the bacterium, *Rhizobium leguminosarum*, producing thick bundles of cellulose microfibrils, which attach the bacterium to the root hairs of legumes. **(b)** The tunicate, *Pyura stolonifera*, has an outer covering, the tunic, that is reinforced by cellulose microfibrils. **(c)** The thick walls of cells constituting the woody tissues of trees, such as in karri, *Eucalyptus diversicolor*, from south-west Western Australia, contain up to 50% cellulose. **(d)** The cell walls of cotton seed hairs are more than 95% cellulose

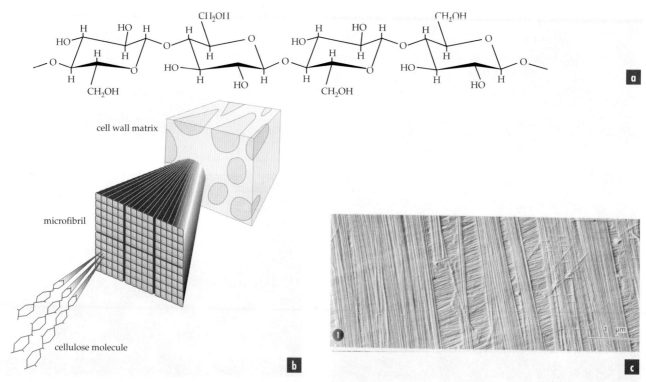

Fig. 1.17 (a) Structure of cellulose. The cellulose molecule is an unbranched chain of β-glucose units linked by (1→4) glycosidic bonds. Note its flat ribbon-like shape. **(b)** Cellulose molecules packed together into a microfibril. The adjacent molecules are associated into sheets by intermolecular hydrogen bonds. The sheets are stacked one on top of another and are associated by multiple van der Waals forces. In cell walls of plants, the microfibrils are embedded in a matrix of non-cellulosic polysaccharides and protein. **(c)** Cellulose microfibrils in the cell wall of the green alga, *Valonia ventricosa*, showing regular but different orientations of the microfibrils in two lamellae

of cellulose microfibrils over part of their length is possible and in cell walls these non-cellulosic polysaccharides form bridges between individual cellulose molecules (Fig. 6.14). Their backbone chains are built from (1→4)-linked β-D-glucose, β-D-mannose (an aldohexose), or β-D-xylose (an aldopentose) units.

Another group of polysaccharides in plant cell walls are the **pectins**. The backbone chains of pectins are formed from an acidic monosaccharide, galacturonic acid, often bearing quite complex side chains. Pectins are good gel-forming polysaccharides and, together with other non-cellulosic polysaccharides, form the matrix of plant cell walls (Fig. 6.14).

Chitin, a polysaccharide, which in many respects resembles cellulose, is a prominent structural component of walls of fungi and the exoskeleton of arthropods. Chitin chains are composed of (1→4)-linked β-*N*-acetylglucosamine units (Fig. 1.18b) and have the same ability as cellulose to associate laterally to form microfibrils.

Bacteria have cell walls composed of complex polysaccharides and individual bacterial cells are often surrounded by polysaccharide-rich coats or capsules (Fig. 1.19). Different species and strains produce structurally different capsular polysaccharides, which

are used to distinguish immunologically between bacterial types. The presence of the capsular polysaccharide is related to the pathogenicity of the bacterium.

Fig. 1.18 (a) The exoskeleton of a cicada is composed of chitin and proteins

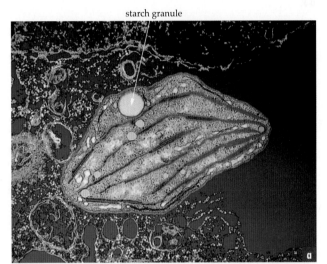

Fig. 1.18 (b) Structure of chitin. The chitin molecule is a long chain of β-*N*-acetylglucosamine units joined by (1→4) glycosidic linkages

starch granule

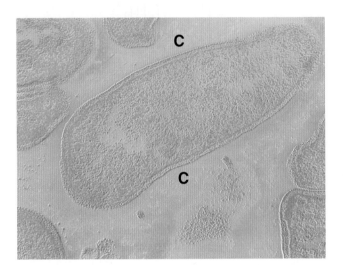

Fig. 1.19 The capsule (C) around cells of an *Acinetobacter* species of bacterium is composed of polysaccharides

Storage polysaccharides

All forms of life rely on stores of energy for their survival. Polysaccharides are the most prominent storage forms. In plants, **starch** is a common storage form of carbohydrate. Starch is found as granules in photosynthetic tissues, as a transient storage material (Fig. 1.20a), and in tubers (Fig. 1.20b), seed endosperm and cotyledons, where it has a long-term storage role.

Starch is not a single compound but an association of two polysaccharides, amylopectin and amylose. Amylopectin is the most abundant of the two, usually comprising 70–80% of starch. It is composed solely of α-D-glucose units. These occur in chains of 20–25 units, depending on the source, linked to one another through (1→6) linkages to form a giant branch-on-branch polymer of many thousands of glucose units (Fig. 1.21). It is highly significant, in relation to its storage role, that each amylopectin molecule has a large number of chain ends that are the site of release of monomer units during its mobilisation for use as an energy and carbon source.

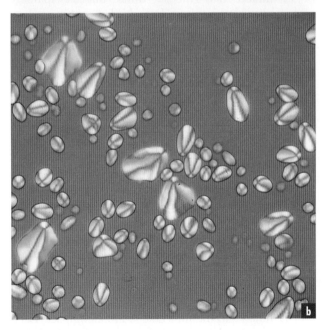

Fig. 1.20 Starch granules in plant cells. (a) A starch granule in a chloroplast, the organelle responsible for the photosynthetic fixation of carbon dioxide to carbohydrate. Glucose, the immediate product of photosynthesis, is temporarily stored as starch.
(b) Starch granules from the potato tuber in polarised light (magnification × 150). Starch provides a source of energy and carbon for the developing plant

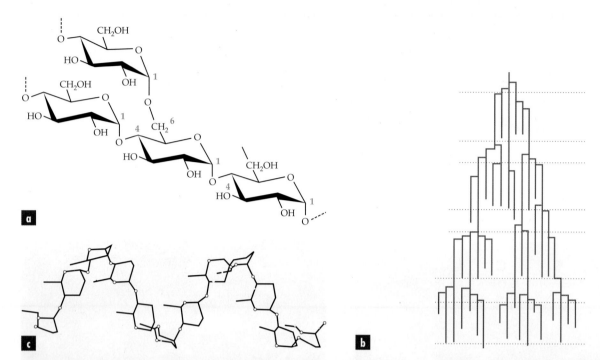

Fig. 1.21 Structure of the starch components, amylopectin and amylose. **(a)** Portion of an amylopectin molecule showing a chain of α-D-glucose units linked by (1→4) glycosidic linkages and a branch-point through a (1→6) glycosidic linkage. **(b)** Model of the amylopectin branch-on-branch structure. Vertical lines represent chains of (1→4)-linked α-D-glucose units; horizontal lines represent (1→6) glycosidic linkages. **(c)** Model of an amylose chain showing its helical structure. The hollow core of the helix is large enough to accommodate 'guest' molecules, such as iodine and fatty acids

The second starch component, amylose, comprises approximately 20–30% of starch: it is also a polymer of (1→4)-linked α-D-glucose units but is sparsely branched, with long inter-branch chains. An amylose molecule may have about 4000 glucose units. Amylose is relatively soluble in water and its chains are helical in shape due to the shape of the linkages between individual α-D-glucose units. Compare the (1→4)-β-linkage in cellulose (Fig. 1.17a) with the (1→4)-α-linkage in amylopectin (Fig. 1.21a). The intense blue colouration of starch in the presence of iodine is due to the inclusion of the iodine molecules in the amylose helix.

Two other related plant storage polysaccharides are the water-soluble fructose polymers: the inulins, found in dahlia tubers and Jerusalem artichokes, and the levans, found in temperate grasses.

Animals store carbohydrate in their tissues as **glycogen**. Like amylopectin, glycogen is composed solely of α-D-glucose residues and has a branch-on-branch structure similar to that of amylopectin (Fig. 1.22). However, the individual chains are shorter (10–12 units) and the molecule is more compact. Each glycogen molecule is attached to a core protein, called glycogenin. Glycogen occurs in all tissues but is particularly abundant in liver and muscle.

Some protistans and fungi have storage polysaccharides belonging to a family of (1→3)-β-D-glucans, variously named paramylon (*Euglena*), laminarin

(certain brown algae), mycolaminarin (water moulds), chrysolaminarin (yellow-green algae), and pachyman

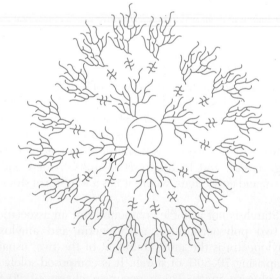

Fig. 1.22 Diagrammatic representation of branch-on-branch glycogen molecules attached to a glycogenin (protein) core. Each line represents a chain of (1→4)-linked α-D-glucose units. Branches are formed by joining chains of (1→4)-linked α-glucose units through their reducing ends to other chains through (1→6)-α-linkages. For clarity, parts of the molecule have been omitted and are represented by ⚡. The glycogen–glycogenin assembly may contain as many as 60 000 glucose units. Note the large number of outer, non-reducing chain ends, which are the sites for enzyme-catalysed chain extension or degradation

Fig. 1.23 (1→3)-β-D-glucan storage polymers. **(a)** *Euglena* species store the polysaccharide paramylon in granules. **(b)** *Xiphophora chondrophylla*, a brown alga, stores laminarin in its fronds. **(c)** The subterranean sclerotium of the fungus *Poria cocos* is largely composed of the polymer, pachyman, which is mobilised to form the fruiting body. **(d)** Ice algae, such as shown growing here in an Antarctic snow drift in Vestfold Hills, store the polysaccharide chrysolaminarin

(sclerotia of basidiomycete fungi) (Fig. 1.23). Other protistans have amylopectin-like storage polysaccharides, for example, floridean starch in red algae.

> Carbohydrates, including simple sugars and their polymers, are a source of energy for immediate use and storage, and have structural and recognition functions in living organisms.

Lipids

The term **lipid** is used to describe oily, greasy or waxy substances that can be extracted from biological materials using non-polar (hydrophobic) organic solvents, such as ether, benzene or hexane, but not by water. Lipids are composed principally of carbon, hydrogen and oxygen but differ from carbohydrates in having a much smaller proportion of oxygen. They may also contain other elements, particularly phosphorus and nitrogen.

Lipids perform many biologically important functions (Table 1.3). They are structural components of cell membranes, whereas others are important in biological fuel storage and transport, and in insulation. Lipids form protective and water-repellent coatings at the surface of many organisms, and yet others act as chemical messengers both within and between cells.

The most abundant lipids are the simple lipids, which are formed by joining an alcohol and a **fatty acid** (Fig. 1.24) with a long hydrocarbon chain through an ester linkage (Box 1.2).

> Lipids are water-insoluble, oily, greasy or waxy compounds composed principally of carbon, hydrogen and oxygen. They have many biological functions—as structural components of membranes, sources of energy, storage and transport, insulators and chemical messengers.

Simple lipids

Waxes are esters in which both the alcohol and fatty acid components have long hydrocarbon chains (Fig. 1.24).

Table 1.3 Functions of lipids in biological systems

Function	Type	Site or example
Storage	Triacylglycerols	Adipose tissue of mammals, endosperm of some seeds
Protective	Waxes	Surfaces of animals: hair, skin surfaces, above-ground surfaces of higher plants
Structural	Phospholipids and sterols	Components of all membranes
Informational (signalling molecules)	Steroid hormones in animals and plants	Sex hormones, moulting hormone in insects
	Glycolipids	Intercellular signalling, receptors for hormones
	Isoprenoids	Insect pheromones and fragrances of flowers
	Derivatives of unsaturated long chain fatty acids	Prostaglandins of animal tissues—involved in inflammatory action and uterine contraction
Pigments	Carotenoids	Plants
Photoreceptors	Carotenoids	Eyes of vertebrates and invertebrates, protistan photoreceptors, chloroplasts of photosynthetic organisms
Cofactors in biological reactions	Carotenoids	Vitamin A
	Steroids	Vitamin D, Vitamin E

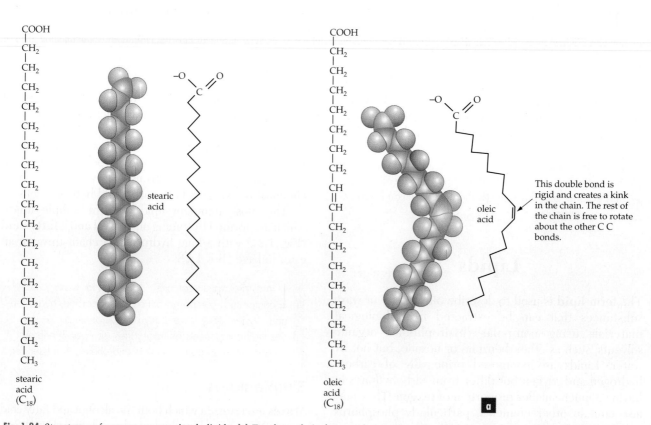

Fig. 1.24 Structures of some common simple lipids. **(a)** Two long-chain fatty acids: stearic acid and oleic acid (an unsaturated fatty acid)

oleyl alcohol

ester
linkage **b**

glycerol

C_{16} fatty acid: palmitic acid

C_{18} fatty acid: stearic acid

C_{18} unsaturated
fatty acid: oleic acid

c

Fig. 1.24 (b) A wax, an ester of a long-chain fatty alcohol and a long-chain fatty acid. **(c)** A triacylglycerol (fat), an ester of glycerol and three long-chain fatty acids

Fig. 1.25 Bees on a brood comb. The comb produced by bees to store honey is composed of wax secreted by abdominal glands of the insect

Waxes are totally insoluble in water due to the non-polar nature of their hydrocarbon components. Waxes are found as coatings on the surfaces of above-ground plant parts and are secreted from preen glands to coat the surfaces of feathers of aquatic birds, epidermal glands of mammals to coat fur and from subepidermal cells onto the exoskeletons of insects. In all these situations, waxes function as water-repelling and water-proofing agents (Fig. 1.25).

Triacylglycerols, or neutral fats, are simple lipids formed by esterification of the three-carbon alcohol, glycerol, which has three hydroxyl groups, with three fatty acids to form a triacylglycerol (Fig. 1.24c). The fatty acids may vary in chain length but usually have 16 or 18 carbon atoms and may contain unsaturated —C=C— bonds. The neutral fats are stored as droplets in special storage cells of adipose tissue (Fig. 1.26a) in mammals, where they act as energy reserves, and in spherosomes in cells of certain seeds (castor bean, canola, coconut, oil palm, flax) and fruits (avocado, olive) (Fig. 1.26b). The caloric density of neutral fats is 37.67 kilojoules per gram (kJ g^{-1}), compared with 16.74 kJ g^{-1} for both carbohydrates and proteins. This concentrated storage form is well-suited to the mobile lifestyle of animals and the need to package energy and carbon reserves in small seeds. Triacylglycerols are transported in mammalian blood in non-covalent associations with special transport proteins in the form of lipoprotein particles. The layer of adipose tissue beneath the skin of mammals has an insulating function, preventing heat loss. Adipose tissue around organs such as the kidney provides mechanical protection.

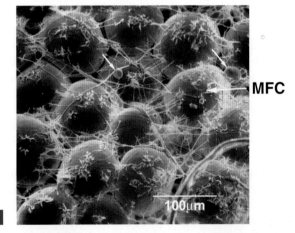

MFC

100μm

a

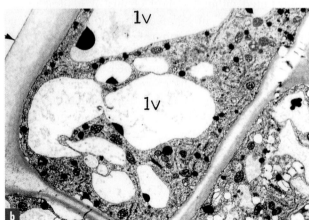

1v

1v

b

Fig. 1.26 Storage forms of triacylglycerols. **(a)** Animal fat storage cells, or adipocytes, which form a large part of adipose tissue. The cells are of two types: mature fat cells (MFC) and very small fat cells (arrow). **(b)** A part of the outer zone of the root cap of *Medicago sativa*, showing lipid vacuoles (1v), which store oil in the form of lipid droplets

Lipids in biomembranes: phospholipids and glycolipids

Two groups of complex lipids are found in **biomembranes**—the phospholipids and the glycolipids. The phospholipids are composed of long-chain fatty acids, esterified to two of the three hydroxyl groups of glycerol; the third is occupied by phosphate, which is, in turn, esterified to one of three polar (charged) nitrogenous compounds, serine, ethanolamine or choline. The phospholipid molecules (Fig. 1.27a) have a non-polar region, contributed by the hydrocarbon chains, which is *hydrophobic* and insoluble in water, and a *hydrophilic* region, contributed by the polar phosphate and nitrogen compounds, which is water-soluble. Such molecules are called **amphipathic**. When added to water, the molecules form stable structures. One of these structures takes the form of tiny droplets of the amphipathic lipid molecules suspended in the aqueous phase (Fig. 1.28a). The hydrophobic tails of the individual molecules interact in the body of the droplet to exclude water, whereas their hydrophilic (polar) heads are oriented to the droplet surface in contact with the aqueous phase. A second structure takes the form of a self-sealing, closed vesicle whose surface is a bilayer composed of amphipathic lipids (phospholipids and glycolipids) in which the hydrophilic heads are in contact with the external and internal aqueous phases. The interior of the bilayer is formed by association of the hydrophobic hydrocarbon chains (Fig. 1.28b). Phospholipid bilayers are the basic structural units of all biological membranes.

Another family of complex lipids, also found in cell membranes, are the glycolipids. In these amphipathic molecules, the hydrophobic region is again contributed by hydrocarbon chains but the hydrophilic region is contributed either by short chains of neutral monosaccharides or, in some cases, both neutral and negatively charged monosaccharides (Fig. 1.27b). In the membrane, the carbohydrate regions project away from the cell surface and function as recognition points for binding of signalling molecules, such as hormones, in cell–cell recognition, and also in the binding of bacteria to plant and animal cell surfaces in infection processes.

Polyisoprenoid lipids

Molecules of this diverse group of lipids are formed by polymerisation of isoprenoid building units (Fig. 1.29a). **Polyisoprenoid lipids** have roles in membrane structure, cell–cell signalling, light reception, reproduction and plant–animal signalling.

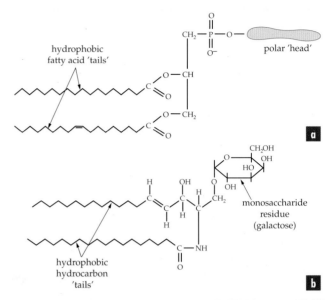

Fig. 1.27 General structures of **(a)** a phospholipid and **(b)** a glycolipid, showing their polar 'head' groups and the non-polar hydrocarbon 'tails'

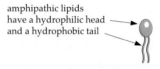

amphipathic lipids have a hydrophilic head and a hydrophobic tail

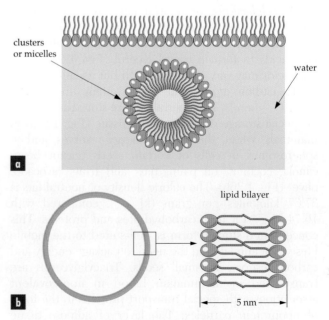

Fig. 1.28 Lipid aggregates. **(a)** In water, amphipathic lipids may form a surface film or small clusters, termed micelles. **(b)** Phospholipids and glycolipids can form self-sealing lipid bilayers, which are found in all cell membranes

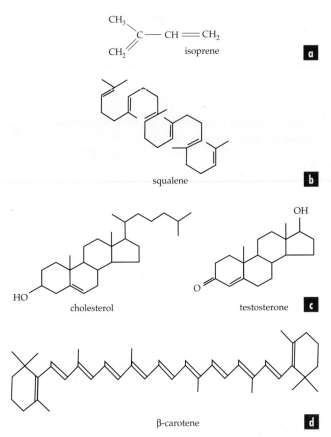

Fig. 1.29 Isoprenoid lipids. (**a**) The five-carbon isoprene building unit of polyisoprenoids. (**b**) Squalene, a linear polyisoprenoid, from which steroids are derived by chain folding and ring formation. (**c**) Two steroid molecules, cholesterol and testosterone, showing multiple ring structures. (**d**) β-carotene, a plant pigment, is a polyisoprenoid derivative

Among the biologically important isoprenoids are the steroid animal and plant hormones, carotenoid pigments and volatile scent compounds (known as essential oils), which are produced by flowers and leaves (lemon, jasmine, rose, geranium, violets) as insect-feeding attractants or, in other cases, as feeding deterrents (e.g. gossypol in cotton). Rubber, found in the latex of many plants, is an unbranched, polyiso-prenoid compound composed of 500–5000 isoprene units.

Steroids are isoprenoid derivatives and have a multiple ring structure formed by folding and ring formation of a long-chain isoprenoid precursor, squalene (Fig. 1.29b). The steroids are a family of molecules whose diverse biological properties and roles result from the different degrees of unsaturation of the steroid structure and the chemical groups at different positions on the rings.

Sterols are steroids that have an hydroxyl group attached to the ring and are components of animal, plant and fungal membranes. **Cholesterol** (Fig. 1.29c) is an amphipathic sterol that is abundant in animal membranes. It may represent 20% of the myelin sheath

around nerve cells. The multiple ring structure is flat and rigid and when inserted in membranes can alter their fluidity. Cholesterol is insoluble in water and is transported in blood, bound non-covalently to lipoproteins. Elevated levels of blood cholesterol are correlated with increased risk of cardiovascular disease. Cholesterol accumulates in atheromatous deposits (plaques) in arteries and contributes to the condition known as arteriosclerosis, where the diameter of the artery is narrowed as the fatty deposits build up. Cholesterol is also the precursor of the steroid hormones, testosterone (Fig. 1.29c), cortisol, ecdysone (the moulting hormone of insects), the brassino-steroids, which are plant growth regulators, and the D vitamins and bile salts.

Carotenoids are long-chain polyisoprenoids (Fig. 1.29d) involved with photoreception in the light-harvesting chloroplasts of photosynthetic organisms and in light-induced movement (phototropism, Chapter 24) in plants and many single-celled protists. They are also photoreceptors of the eyes of molluscs, insects and vertebrates. Vitamins A, E and K are carotenoid derivatives, as is ionine, which is responsible for the scent of violets. Carotenoids impart the characteristic colour to the tissues of carrots, sweet potatoes and other plants.

> The most abundant lipids are simple lipids formed by joining an alcohol and a fatty acid with a long hydrocarbon chain joined by an ester linkage. Lipids include storage waxes, found on the surfaces of many organisms, triacylglycerols found in adipose tissue and some seeds, phospholipids and glycolipids, found in membranes, and polyisoprenoid lipids, which have biological roles in membranes (cholesterol), cell signalling, light reception (carotenoid pigments) and reproduction (steroids).

Proteins

Proteins are the most functionally diverse of all biological molecules (see Table 1.4). Their foremost position among natural compounds is reflected in their name derived from the Greek *protos*, meaning first. They are constituted from amino acids joined in an unbranched chain. The chains may contain hundreds or even thousands of amino acids. Whatever the biological origin of the protein, the component amino acids are selected from a standard set of 20. All the amino acids found in proteins have the same general structure: a basic amino group ($-NH_2$), which may gain a proton and become $-NH_3^+$, and an acidic carboxyl group ($-\overset{\text{O}}{\overset{\|}{C}}-OH$), which may donate a proton and become $-\overset{\text{O}}{\overset{\|}{C}}-O^-$, and a variety of different side-chain groups,

Table 1.4 Classification of proteins into groups according to biological function, composition or solubility

Classification	Examples
Biological function	
Enzymes	Almost all enzymes are proteins.
Structural proteins	Proteins that form the fabric of the tissues, for example, collagen and elastin in connective tissue, keratin in hair and horn. Silk in insect cocoons and spiders' webs.
Contractile proteins	Muscle proteins: actin and myosin.
Hormones (signalling molecules)	Insulin, growth hormones.
Transport proteins	Haemoglobin and myoglobin (O_2 transport), serum albumin (fatty acids) etc.
Defence proteins	Immunoglobulins (antibodies), fibrinogen (blood clotting).
Storage proteins	Seed proteins (gluten, gliadins), serum albumin.
Composition	
Lipoproteins	Proteins containing lipids such as the lipoproteins of blood serum.
Glycoproteins and proteoglycans	Proteins with covalently attached oligosaccharide chains. Most proteins are glycoproteins, exceptions are insulin and serum albumin. Many connective tissue proteins have covalently attached polysaccharide chains. Glycogen is attached to a protein (glycogenin).
Nucleoproteins	Proteins non-covalently bound to DNA. Found in chromosomes and virus particles.
Haemproteins	Proteins containing an organic molecule complexed to an atom of iron (haem): haemoglobin, myoglobin and cytochromes.
Chlorophylls	Proteins containing an organic molecule complexed to magnesium.
Metalloproteins	Proteins containing one or more metal ions: cytochromes. Many enzymes and carbohydrate-binding proteins (lectins).
Solubility	
Globular proteins	Water-soluble proteins because the R-groups of the amino acids on their surfaces are polar. Cytoplasmic proteins, including many enzymes, are globular proteins. Serum transport proteins are globular proteins.
Membrane proteins	Membrane proteins have globular domains that are exposed on the inner or outer side (or both) of membranes but also have hydrophobic domains that span the phospholipid bilayer. Membrane proteins are generally not soluble in water but may be extracted by detergents to form micelles.
Fibrous proteins	These high molecular weight proteins (keratin, silk, collagen and elastin) are insoluble in water.

abbreviated as R (Fig. 1.30). The 20 different amino acids differ in their side-chain R-groups, which may be small, uncharged but polar, small uncharged and non-polar, large non-polar (hydrophobic), or positively or negatively charged (Fig. 1.31).

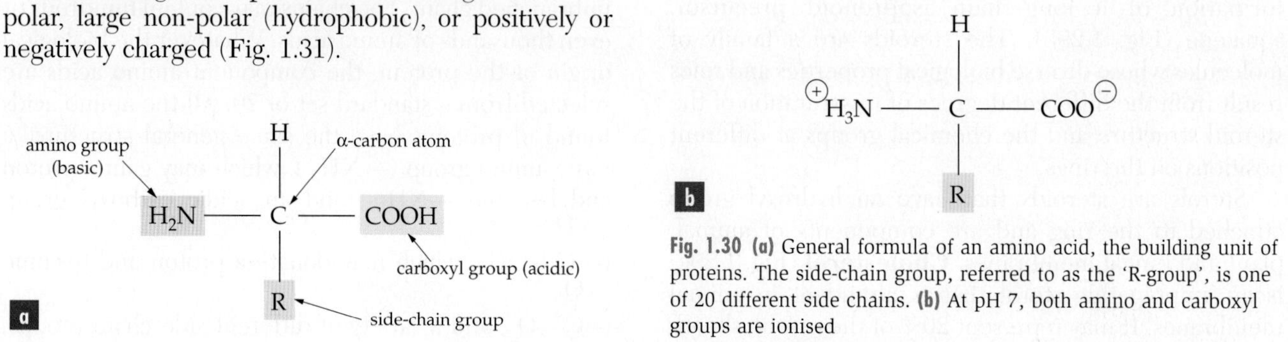

Fig. 1.30 (a) General formula of an amino acid, the building unit of proteins. The side-chain group, referred to as the 'R-group', is one of 20 different side chains. **(b)** At pH 7, both amino and carboxyl groups are ionised

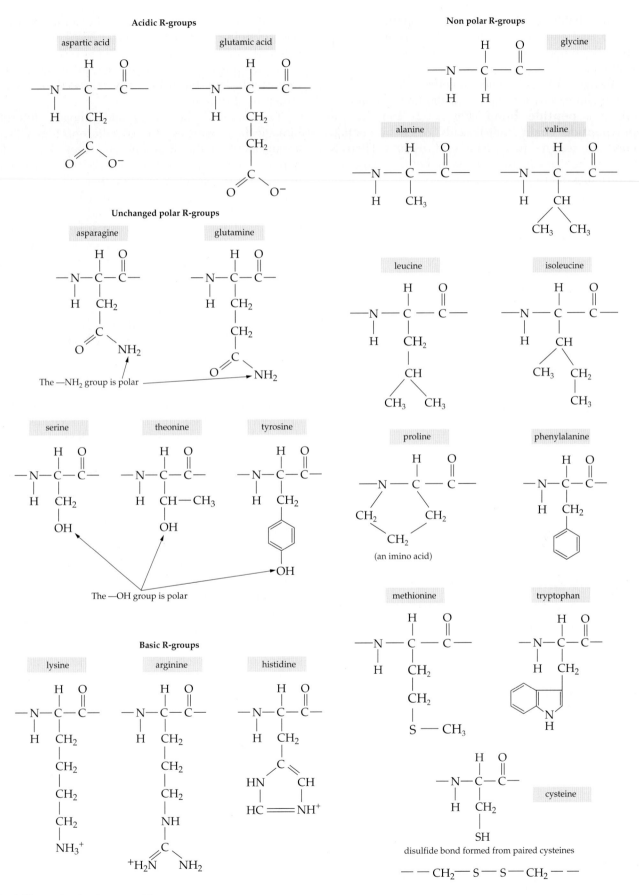

Fig. 1.31 The family of amino acids

The identity and unique chemical properties of each amino acid and the ways in which each amino acid affect the properties of a protein are determined by its R-group.

Amino acids are joined together through their amino and carboxyl groups, with the loss of water, to create a **peptide bond** (Fig. 1.32) and form a **polypeptide chain**. Amino acids joined by peptide bonds are referred to as amino acid residues. There is always a free amino group at one end of the chain, the N-terminus, and a free carboxyl group at the other end, the C-terminus.

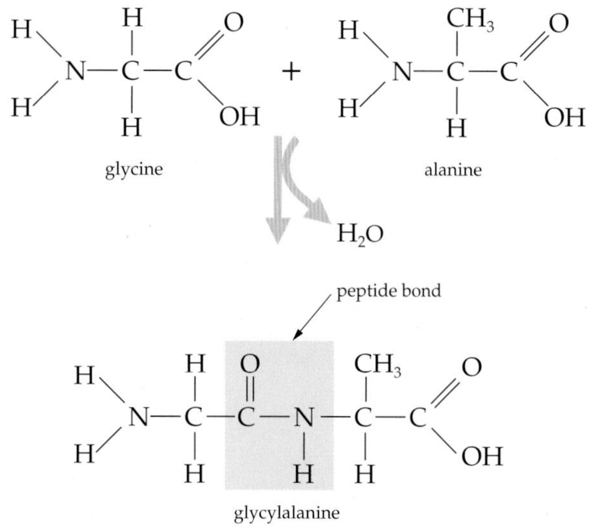

Fig. 1.32 Peptide bond formation. Amino acids are linked together in a protein by an amide linkage, called a peptide bond. Here, glycine and alanine are joined to give the dipeptide, glycylalanine

Structure and shape of proteins

Primary and secondary structure

The sequence of amino acids in the polypeptide chain of a specific protein is unique and all molecules of that protein will have the same sequence (Chapter 11). The sequence of the amino acids in a protein defines its **primary structure**.

Many proteins consist of a single polypeptide chain. In others, there is more than one chain. These chains may be joined covalently by cross-links through

disulfide bonds formed between the sulfur atoms of two cysteine residues (Fig. 1.33). Sometimes, disulfide bonds may link two cysteine residues within a single polypeptide chain. These linkages are important in stabilising the three-dimensional structure of proteins.

The order of the amino acids imparts different biological properties to specific proteins. The arrangement of the R-groups on either side of the peptide bonds along the polypeptide backbone determines the **secondary structure**. There are two common secondary structure motifs or patterns encountered in proteins: α-helix and β-sheets. Some stretches of the backbone do not have recognisable structures and are referred to as random coils. The secondary structures owe their existence to coiling of the backbone, as in the α-helix, or folding, as in β-sheets, to give structures that are stabilised by favourable hydrogen bonding between residues (Fig. 1.34). Special folding patterns, called β-turns, are encountered where the polypeptide chain changes direction. Two amino acids, glycine and proline (Fig. 1.31), have special functions in the formation of β-turns.

The common secondary folding patterns recur in different and often unrelated protein chains even though their primary structures are unique (Fig. 1.35).

Tertiary structure

The **tertiary structure** refers to the overall three-dimensional shape of a protein. Protein shapes may be either globular and compact, as in enzymes, antibodies, transport proteins and signalling (hormone) proteins, or extended rigid rods, as in **fibrous proteins** with mechanical or structural roles.

In **globular proteins**, the specific shape is created by the folding of the backbone so that the hydrophobic side chains are buried in the interior of the molecule out of contact with the aqueous environment. The surface of the protein will generally be rich in polar amino acids. The forces involved in stabilising the tertiary structure are non-covalent hydrogen bonds, ionic bonds and van der Waals forces (hydrophobic interactions) but may also involve covalent disulfide bond formation.

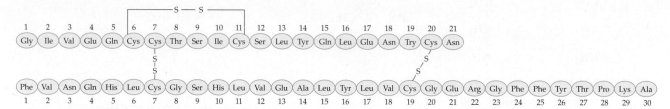

Fig. 1.33 Primary structure of the mammalian hormone insulin. The molecule of insulin has two polypeptide chains cross-linked by two inter-polypeptide disulfide bonds. There is also one intramolecular disulfide bond

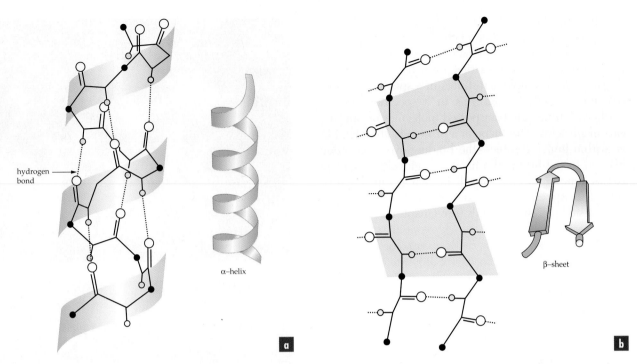

Fig. 1.34 Secondary structure motifs. **(a)** An α-helix held in place mainly by the hydrogen bonds between the carboxyl oxygen and the hydrogen on the amide nitrogen. For clarity, only the backbone atoms in the polypeptide chain are shown. **(b)** A β-sheet in which the polypeptide chain is stretched out into a zigzag structure. Neighbouring β-sheet chains are associated by hydrogen bonds between their peptide bonds. The individual strands in the β-sheets are connected by stretches of polypeptide chain incorporating glycine or proline amino acids that allow the chain to change direction

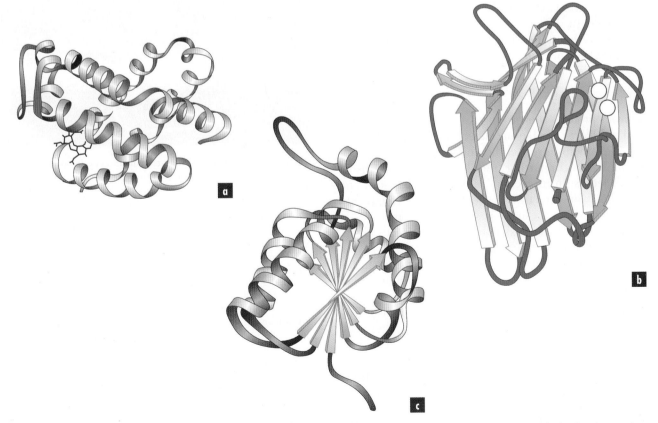

Fig. 1.35 Secondary structures of some proteins. **(a)** Myoglobin, an oxygen-carrying protein from muscle, showing the α-helical regions and the oxygen-binding haem group. **(b)** Concanavalin A, a carbohydrate-binding protein (a lectin) from jack bean seeds, showing the β-sheets (arrowed segments) stacked together. **(c)** Lactate dehydrogenase, an enzyme, showing a mixture of α-helical and β-sheet regions

The specific functions of globular proteins are determined by the folding of the surface into specific shapes, which are complementary to, and therefore recognised by, molecules that bind to their surfaces (ligands); for example, substrates in a specific enzymic reaction (Fig. 1.36), antigens in a specific antigen–antibody interaction, transported molecules on specific carrier proteins. The shape of a protein is not fixed but is, within limits, flexible—the binding event has been likened to the fitting of a hand in a glove! Changes in pH or temperature can result in changes in shape so that the specific functional sites are no longer recognisable by their ligands (see Chapter 2).

Proteins are polypeptide chains of amino acids linked together by peptide bonds. The sequence of amino acids defines the primary structure of a protein. The arrangement of R-groups on either side of the peptide bonds determines the secondary structure. The tertiary structure of a protein is the overall three-dimensional shape and is sensitive to the environment.

Quaternary structure

Several globular units can associate specifically to form a **quaternary structure**. The subunits may be of the same structure or may be composed of two different structures, as in haemoglobin (Fig. 1.37). Some enzymes, virus coats, cytoskeletal structures (p. 88) and ribosomes (p. 79) are aggregates of several different proteins with different functions. Their associations are self-directed through specific recognition sites on the polypeptide subcomponents.

Fibrous proteins

Fibrous proteins show a range of structures, from rigid and rod-like to flexible and sheet-like (Fig. 1.38). These diverse structures reflect their biological functions. For example, in imparting tensile strength to insect exoskeletons; in protection, as in hair, nails, horn and feathers; in imparting mechanical strength to connective tissue, such as bone and cartilage; in conferring elasticity to skin; or as in the structural and tension-bearing intracellular filament proteins of the cytoskeleton. The fibrous proteins are characteristically insoluble in water due to the high proportion of hydrophobic R-groups.

Keratins are a family of proteins found in hair, nail, horn, feathers and skin (Fig. 1.39). They are characterised by long sequences with an α-helical secondary structure. Keratins are composed of three α-helical molecules associated to form a protofibril. Eleven protofibrils are associated to form a microfibril, which is, in turn, associated with other microfibrils to constitute a macrofibril found in hair cells (Fig. 1.38a). The keratin molecule can be extended by applying tension at its ends. On removal of the tension, the keratin

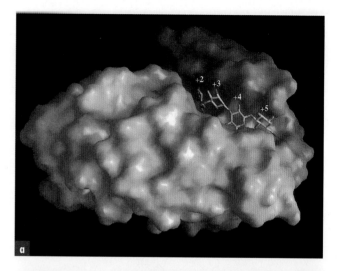

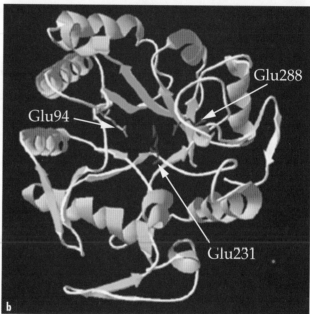

Fig. 1.36 Structure of a globular enzyme protein. **(a)** Molecular surface model of the enzyme (1→3)-β glucan hydrolase, from germinating barley. This globular protein presents a binding site for the substrate β-glucan (a polysaccharide) in a shallow groove on the enzyme's surface (shown in blue). The β-glucan substrate is shown as a wire model, with carbon atoms of the glucose residues of the polysaccharide in yellow, oxygen atoms in orange and the C(O)-6 atoms in red. **(b)** Model showing the secondary structure of the (1→3)-β glucan hydrolase polypeptide composed of α-helices (purple) and β-sheets (green) connected by coils (shown as grey rope). The polypeptide chain is folded into the globular tertiary structure shown in **(a)** with a shallow groove running across its surface, in which the acidic amino acids, glutamate 94, glutamate 231 and glutamate 288 are positioned to allow participation in the catalytic cleavage of glycosidic bonds in the polysaccharide substrate

resumes a compact or relaxed state. Some keratins have a high content of cysteine, which, when joined to cysteine on neighbouring chains, forms disulfide cross-links that reduce its extensibility. Tortoiseshell keratin contains 20% cysteine, producing a tough, hard material (Fig. 1.39e).

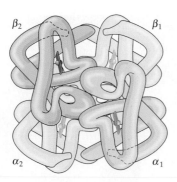

Fig. 1.37 Quaternary structure. Haemoglobin is composed of four separable globular subunits of two types—α and β—which are held together by polar interactions and hydrogen bonding between their opposing surfaces. The haem groups are indicated

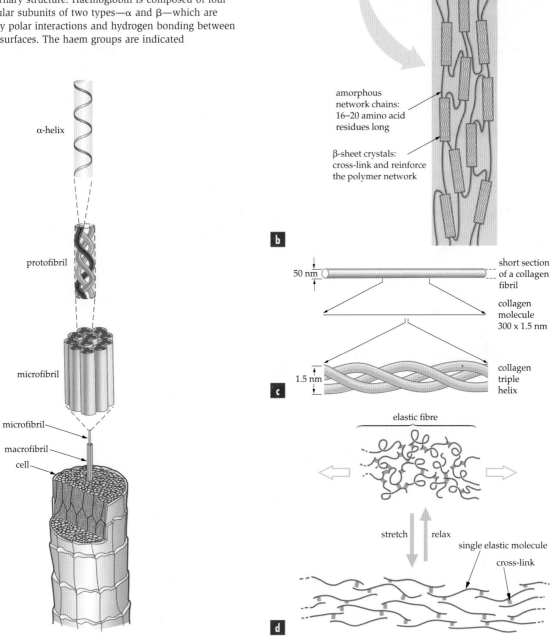

Fig. 1.38 Structures of fibrous proteins. **(a)** Diagrammatic structure of keratin, an α-helical protein, in a hair cell. **(b)** Silk fibre organisation, showing the structure of a nanofibril formed from β-pleated sheet crystals connected by protein chains with less ordered (amorphous) structures. **(c)** Collagen, a triple α-helix protein. In collagen, the rod-like, triple helices are cross-linked to each other to form rigid, high-tensile strength fibrils. **(d)** Elastin, an α-helical protein. The individual elastin molecules are able to unwind when the fibre is stretched and relax when the tension is relieved

Fig. 1.39 Keratins are found in a variety of forms: **(a)** in the furry coat of the brush-tail possum, *Trichosurus vulpecula vulpecula*; **(b)** in birds' feathers, such as this guineafowl feather (magnification × 40); **(c)** in spiders' webs; **(d)** in the horns of rhinoceros; and **(e)** in the shells of turtles, such as this green sea turtle, *Chelonia mydas*

Silks, or fibroins (Fig. 1.38b), in contrast to keratins, are extracellular proteins. They are secreted from special glands of moths and spiders, and are β-sheet proteins with high strength and flexibility but low extensibility. They are characterised by high contents of the three smallest amino acids, glycine, alanine and serine. There is no cysteine. The high tensile strength is due to the strongly hydrogen-bonded, β-sheet structure and the flexibility due to the ability of the β-sheets to slide across each other. This is possible as the β-sheets are only weakly associated by van der Waals forces. The super strength of spider's web silk is attributable to the architecture of the thread in which concentric layers of tiny coiled nanofibrils enclose the fibre's core. The angled nanofibrils absorb huge amounts of energy during stretching and it is only after they are straightened that they break.

Collagens comprise a third family of fibrous proteins. They are the major proteins in the connective tissue, bone, cartilage and skin of vertebrates. These proteins are found together with proteoglycans in the extracellular matrix that surrounds and connects cells and tissues. The collagen primary structure is rich in glycine, proline and lysine. After synthesis of the collagen backbone, some proline residues are hydroxylated to hydroxyproline and some lysines hydroxylated to hydroxylysine and substituted with a disaccharide. Cross-linking of collagen chains occurs through special lysine bridges. The collagen backbone is folded into an α-helix and three α-helices are entwined in a triple helix to give a rigid rod called tropocollagen (Fig. 1.38c). Tropocollagen can associate spontaneously to form collagen microfibrils. Members of the collagen family with distinctive structural features are found in specific connective tissue locations.

Elastin, another fibrous protein, is found in connective tissues of arteries, skin and ligaments, and in lungs, where stretching and relaxation is a necessary part of their function. Elastin has an α-helical structure with high glycine, proline and alanine content but is also rich in lysine. Special cross-links are formed through lysine residues on adjacent chains. There is an apparent lack of order in its three-dimensional structure (Fig. 1.38d). In the aqueous environment of the connective tissue, the cross-linked elastin can stretch in all directions. The cross-links ensure that relaxation occurs after tension is relieved.

Proteoglycans and glycoproteins

In animal connective tissues, for example, cartilage, tendon, skin and blood vessel walls, which form the supporting structures of the body, polysaccharide molecules are found covalently joined to a protein to form **proteoglycans**. The unbranched polysaccharide chains are composed of amino sugars and uronic acids and are invariably substituted with acidic sulfate groups. The proteoglycans are found in the extracellular matrix that surrounds all cells and tissues in vertebrates, where they are associated with the fibrous proteins, elastins and collagens.

Another connective tissue polymer is hyaluronic acid, a water-soluble, linear polysaccharide whose long, unbranched chains composed of a repeating disaccharide formed from an N-acetylglucosamine and charged glucuronic acid units. Hyaluronic acid has a very high water-holding capacity and forms clear gels, which are found in the aqueous humour of the eye. In synovial fluid, which lubricates joints, it has the unusual property of altering its viscosity in response to

forces applied to it. In connective tissue, a single hyaluronic acid molecule acts as an organising backbone for some 100 non-covalently bound proteoglycan molecules to form very large aggregates (molecular mass 10^8) equivalent in volume to a bacterial cell (Fig. 1.40).

Almost all proteins bear covalently linked saccharides, and are called **glycoproteins**. These may be simple disaccharides or large, complex oligosaccharides. The saccharides function as recognition signals and as modifiers of the shape and hydrophilicity of the proteins to which they are attached. Many membrane-bound proteins are glycosylated, the carbohydrate portion facing the cellular environment (Fig. 1.41).

Nucleic acids

Nucleic acids are found in all cells and are the molecules in which information on cell structure and function is stored and transferred from generation to generation. This genetic information is decoded (Chapter 11) to direct the formation of protein molecules, which execute the functions of cells through enzymic activities or provide the specific proteins that constitute the fabric of cells.

Nucleotides

The building unit of nucleic acids is the **nucleotide**, which has three parts: a nitrogenous base (so-called since it carries a proton at acidic and neutral pHs), a pentose monosaccharide (ribose or deoxyribose) and a phosphate group (Fig. 1.42a). The nitrogenous bases are of two types: a six-membered **pyrimidine** ring (uracil, U, cytosine, C, thymine, T) or a fused five- and six-membered **purine** ring (adenine, A, guanine, G) (Fig. 1.42b).

Besides being the building units of nucleic acids, these nucleotides, as their triphosphate esters (e.g. adenosine triphosphate, ATP), are the cellular currency of energy that drives chemical and physical processes in cells. Other nucleotide derivatives are the coenzyme carriers of hydrogen in cellular reactions, leading to the formation of ATP (Chapter 2), while others bear monosaccharide residues and are involved in polysaccharide synthesis, and yet others, such as cyclic adenosine monophosphate (cyclic AMP), are intracellular signalling molecules that provide for minute-by-minute modulation (regulation) of the rate of chemical reactions in the cell.

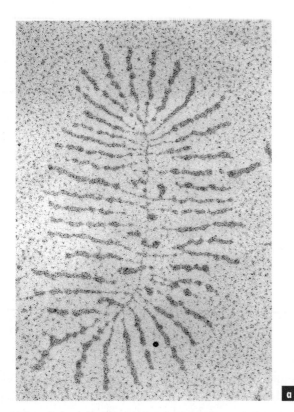
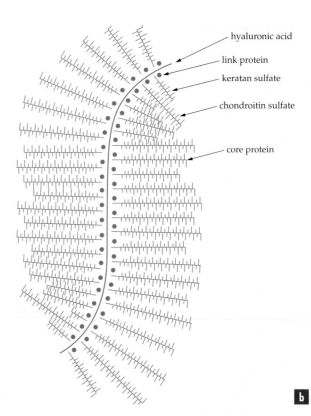

Fig. 1.40 Proteoglycan aggregate from cartilage, showing an array of proteoglycan chains linked non-covalently through one end of their core protein backbones via linker proteins to a stabilising hyaluronic acid molecule. **(a)** Electron micrograph of a proteoglycan aggregate. **(b)** Diagrammatic representation of an aggregate

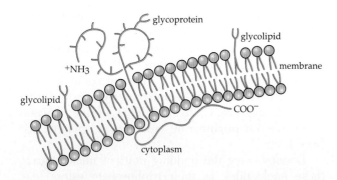

Fig. 1.41 Diagram showing the organisation of glycoproteins and glycolipids in a membrane bilayer. The carbohydrate portions of both types of molecules are located on the extracellular surfaces of plasma membranes

The building block of nucleic acids is the nucleotide, which has three parts, a nitrogenous base, a pentose monosaccharide (ribose or deoxyribose) and a phosphate group. Nitrogenous bases are pyrimidine rings (uracil, U, cytosine, C, thymine, T) or purine rings (adenine, A, guanine, G).

DNA and RNA

Nucleic acid molecules consist of long chains of nucleotides linked through phosphodiester linkages between the 3′ hydroxyl group on the monosaccharide and the 5′ phosphate group. In **deoxyribose nucleic acids (DNA)**, the monosaccharide is deoxyribose and the bases are A, T, G and C (Fig. 1.43). However, the **ribose nucleic acids (RNA)** have a ribose monosaccharide and the bases are A, U, G and C.

Each nucleic acid has a specific ordering of the bases, which represents the genetic information of the cell. The ability of the nucleic acid to receive or transfer this information is due to the ability of the bases on different nucleic acid chains to recognise each other by non-covalent interactions. These specific interactions are known as **base pairing**. Thus, C can pair only with G, and A only with T or U. In the cell, DNA exists as two long chains running in opposite directions, held together by complementary base pairs (Fig. 1.43b). The non-covalent interactions are through hydrogen bonds. There are two between A and T and three between C (or T) and G (Fig. 1.43a).

RNA is typically single-stranded. There are three main types of RNA molecules, each with different functions. Messenger RNA (mRNA) carries the information that specifies the amino acid sequence of a given polypeptide. Ribosomal RNA (rRNA) makes up a major part of ribosomes, the cytoplasmic structures on which polypeptides are assembled (Chapter 11). Transfer RNA (tRNA) carries specific amino acids to the ribosomes to add to growing polypeptides according to the base sequence in mRNA. Some RNA molecules are catalytic, that is, they function as enzymes (Chapter 2).

A more detailed description of the unique roles of DNA and the RNAs in carrying and interpreting hereditary information in living organisms is given in Chapter 11.

> DNA is a double-stranded, helical molecule composed of the bases C, T, A and G. It stores the genetic information of cells. RNA molecules are single-stranded, are composed of the bases U, C, A and G and are responsible for translation of the nucleotide sequence of DNA into the corresponding amino acid sequence in a given protein.

a

adenine (an amine)

phosphate group

ribose (a five-carbon sugar) or 2-deoxyribose

adenosine monophosphate (AMP)

b

thymine

adenine

cytosine

guanine

Purines

uracil

Pyrimidines

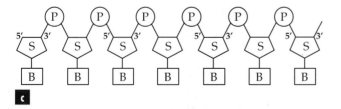

c

Fig. 1.42 (a) A nucleotide consists of three parts: a nitrogen-containing base (B), a monosaccharide (ribose or deoxyribose, S) and a phosphate group (P) covalently linked together.
(b) Structures of the nitrogenous bases. These are the pyrimidines (thymine, cytosine and uracil) and the purines (adenine and guanine). **(c)** The nucleotide units of nucleic acids are joined together by formation of phosphodiester linkages between their 5′ and 3′ carbon atoms to give a linear sequence

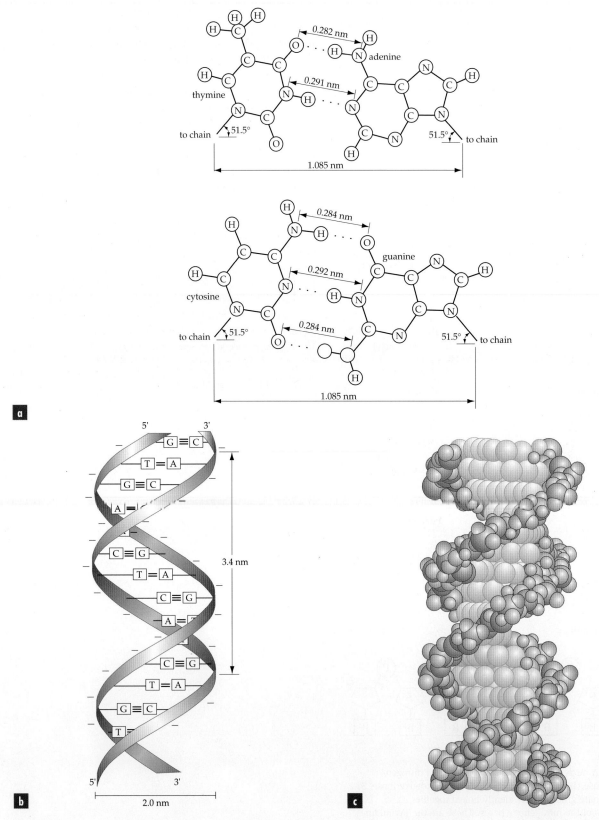

Fig. 1.43 Structure of deoxyribose nucleic acid (DNA) double helix. **(a)** Specific base pairing between thymine–adenine and cytosine–guanine, showing the three hydrogen bonds between the cytosine–guanine pair and the two hydrogen bonds between the thymine–adenine pair. The overall dimensions of the pairs are similar, allowing formation of a regular double helix. **(b)** Two DNA molecules with complementary base sequences forming a regular double helix by coiling. The structure is stabilised by hydrogen bonding between complementary bases.
(c) Model of a DNA double helix showing the stacks of paired bases forming the core of the structure and the charged phosphate residues occupying the surface of the duplex

Summary

- Living organisms are composed almost entirely of 16–21 elements. Four of these, hydrogen, oxygen, nitrogen and carbon, make up 99% of the living parts of organisms.

- An element is made of one type of atom. Atoms consist of subatomic particles: a central nucleus of protons (positvely charged) and neutrons (neutral charge), and orbiting electrons (negatively charged). Electrons orbit a nucleus in zones of space (orbitals). Electrons can move between orbitals of different energy levels, with the gain or loss of energy.

- Chemical bonds involve the attraction between two atoms or molecules. Covalent bonds involve the sharing of electrons, while hydrogen bonds and ionic bonds involve attraction between opposite charges. van der Waals forces are temporary and 100 times weaker than covalent bonds.

- The properties of water make it an ideal medium for cellular life. Water is an excellent solvent for polar molecules, some of which dissociate in solution to form ions (ionisation). Water repels non-polar molecules.

- Substances that release hydrogen ions into solution are acids, while those that accept hydrogen ions are bases. pH is the concentration of H^+ in solution. The pH scale is logarithmic and ranges from 0 to 14. A buffer solution, such as a carbonic acid–bicarbonate ion solution, maintains relatively constant pH, either by removing added hydrogen ions or by releasing them.

- Carbohydrates are the most abundant organic compounds in living organisms. They include simple sugars and their polymers, are a source of energy for immediate use or as a storage form, and have structural and recognition functions in living organisms.

- Lipids are water insoluble, oily, greasy or waxy compounds composed principally of carbon, hydrogen and oxygen. The most abundant lipids are simple lipids formed by joining an alcohol and a fatty acid with a long hydrocarbon chain by an ester linkage. They have many biological functions: as structural components of membranes (phospholipids, glycolipids, cholesterol), for energy storage (triacylglycerols, fats, oils) and transport, as insulators, for cell signalling, light reception (e.g. carotenoid pigments) and reproduction (e.g. steroids).

- Proteins are polypeptide chains of amino acids linked together by peptide bonds. The sequence of amino acids defines the primary structure of a protein. The arrangement of R-groups on either side of the peptide bonds determines the secondary structure. The tertiary structure of a protein is the overall three-dimensional shape and is sensitive to environmental changes, such as pH and temperature.

- Nucleic acids are found in all cellular organisms and in non-cellular infective agents, viruses, and are the molecules in which information is stored and passed from generation to generation. The building block of nucleic acids is the nucleotide, which has three parts: a nitrogenous base, a pentose monosaccharide (ribose or deoxyribose) and a phosphate group. Nitrogenous bases are pyrimidine rings (uracil, U, cytosine, C, thymine, T) or purine rings (adenine, A, guanine, G). DNA is a double-stranded, helical molecule composed of the bases C, T, A and G; it stores the genetic information of cells. RNA molecules are single-stranded, are composed of the bases U, C, A and G, and are responsible for translation of the nucleotide sequence of DNA into the corresponding amino acid sequence in a given protein.

keyterms

acid (p. 22)
amino acid (p. 24)
amphipathic (p. 34)
anion (p. 17)
atom (p. 13)
atomic number (p. 13)
base (p. 22)
base pairing (p. 44)
biomembrane (p. 34)
biosphere (p. 13)
buffer (p. 23)
carbohydrate (p. 24)
carotenoid (p. 35)
cation (p. 17)
cellulose (p. 26)
chemical bond (p. 15)
chitin (p. 28)
cholesterol (p. 35)
collagen (p. 43)
covalent bond (p. 15)
deoxyribose nucleic
   acid (DNA) (p. 44)
disaccharide (p. 24)

elastin (p. 43)
electron (p. 13)
element (p. 13)
fatty acid (p. 31)
fibrous protein (p. 38)
fructose (p. 24)
globular protein
   (p. 38)
glucose (p. 23)
glycogen (p. 30)
glycolipid (p. 24)
glycoprotein (p. 43)
hydrogen bond (p. 18)
hydrogen ion (p. 22)
hydronium ion (p. 22)
hydrophilic (p. 21)
hydrophobic (p. 22)
hydroxyl ion (p. 22)
ion (p. 17)
ionic bond (p. 17)
isomer (p. 24)
keratin (p. 40)
lactose (p. 26)

lipid (p. 31)
mass number (p. 13)
molecule (p. 15)
monosaccharide
   (p. 24)
neutron (p. 13)
nitrogenous base
   (p. 24)
nucleotide (p. 43)
orbital (p. 14)
organic molecule
   (p. 23)
pectin (p. 28)
peptide bond (p. 38)
phospholipid (p. 24)
polyisoprenoid lipid
   (p. 34)
polypeptide chain
   (p. 38)
polysaccharide (p. 24)
primary structure
   (p. 38)
protein (p. 35)

proteoglycan (p. 43)
proton (p. 13)
purine (p. 43)
pyrimidine (p. 43)
quaternary structure
   (p. 40)
ribose (p. 24)
ribose nucleic acid
   (RNA) (p. 44)
secondary structure
   (p. 38)
silk (p. 42)
starch (p. 29)
steroid (p. 35)
sucrose (p. 26)
surface tension (p. 20)
tertiary structure
   (p. 38)
triacylglycerol (p. 24)
van der Waals forces
   (p. 18)
wax (p. 31)

Review questions

1. What is an atom? What are the components of an atom?

2. Each element has an atomic number and an atomic weight. For example, oxygen has an atomic number of 8 and an atomic weight of 16. Explain the difference between these terms.

3. What are the characteristic atoms found in

   (a) galactose?

   (b) RNA?

   (c) keratin?

   (d) wax molecules?

4. Draw the atomic structures of several water molecules. Indicate on your diagram the forces holding the atoms together and the forces between the molecules in liquid water. Name the two types of bonding and describe the difference between them.

5. Explain how the special properties of water contribute to the functions of living organisms.

6. What is pH? How does a buffer work?

7. Construct a table summarising the properties of the four main types of macromolecules that are found in all organisms. List their basic subunits and chemical composition, and their biological functions.

8. What are the physical and chemical attributes of an amphipathic molecule? Draw the structures of three amphipathic molecules. Where in the cell would you find these molecules and what are their functions?

9. (a) What is the R-group of an amino acid?

   (b) Draw an example of one amino acid with a simple R-group and one with a complex R-group. Identify the functional groups on the two amino acids.

10. Why is aspartic acid considered an acidic amino acid and tryptophan a non-polar amino acid?

11. The tertiary structure of a globular protein relates to its:

    A amino acid sequence

    B peptide bonds

    C folding of the backbone of the molecule

    D ligands that bind to its surface

12. Distinguish between the structures and functions of fibrous proteins and globular proteins.

13. Nucleic acids are composed of nucleotides. The three components of a nucleotide are:

    A an amino acid, a pentose monosaccharide and a sulfate group

    B an amino acid, a pentose monosaccharide and a phosphate group

    C a nitrogenous base, a pentose monosaccharide and a sulfate group

    D a nitrogenous base, a pentose monosaccharide and a phosphate group

14. Compare the structures and functions of DNA and RNA.

Extension questions

1. Explain how the distribution of electrons in the orbitals of atoms can lead to the formation of a molecule and a compound.

2. Why don't the atoms of the elements helium (two electrons in the outer shell) and neon (two electrons in the inner shell and eight in the outer shell) form diatomic molecules with themselves or molecules with other elements?

3. Cellulose and glycogen have very different physical properties and biological functions, yet both are composed only of D-glucose. Explain these differences and show how their structures relate to their functions.

4. If the cytosolic pH of a muscle cell changes from 7.4 to 6.4 during contraction, what is the relative increase in H^+ ion concentration?

5. Plants store energy as carbohydrates, whereas animals and many seeds store energy as lipids. Suggest an explanation for the differences in energy storage forms.

Suggested further reading

Alberts, B., Bray, J., Lewis, J., et al. (1994). *Molecular Biology of the Cell.* 3rd edn. New York: Garland.

An excellent advanced text for students majoring in cell biology and biochemistry.

Garrett, R. H. and Grisham, C. M. (1999). *Biochemistry.* 2nd edn. Fort Worth: Saunders College Publishing.

Smith, C. A. and Wood, E. J. (1991). *Molecular and Cell Biochemistry. Biological Molecules.* London: Chapman and Hall.

Stryer, L. (1995). *Biochemistry.* 3rd edn. New York: Freeman & Co.

Zubay G. L., Parson, W. W., Vance, D. E. (1995). *Principles of Biochemistry.* Dubuque, IA: W. C. Brown.

Texts for the student majoring in biochemistry.

Harborne, J. B. (1988). *Introduction to Ecological Chemistry.* 3rd edn. London: Academic Press.

This book deals with the chemistry of plants, emphasising natural products and, for example, plant chemical defences; useful for taxonomy and ecology.

CHAPTER

2

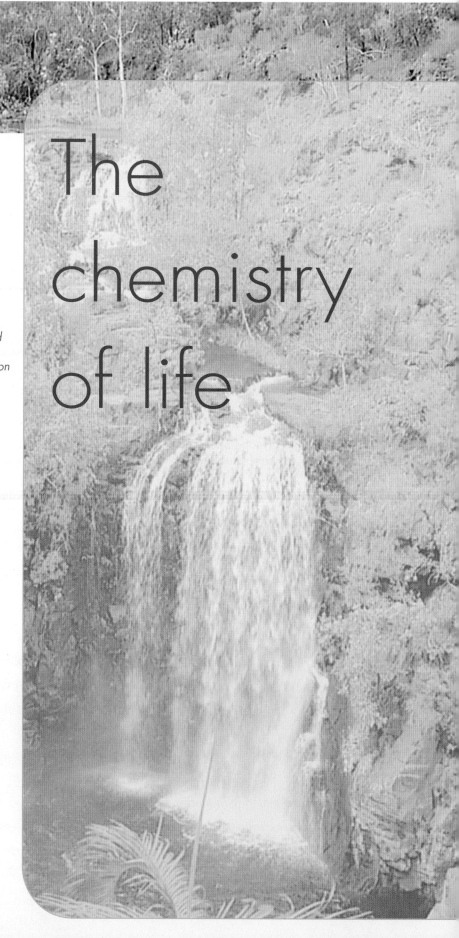

The chemistry of life

living organisms and their constituent cells have a high degree of organisation. To maintain this organisation against the tendency of any system to become disordered, energy must be continually expended. Energy is used for development, growth, feeding, moving, reproducing and responding. At the cellular level, energy is used to produce the biomolecules that form the fabric of the cell, to transport ions and molecules across membranes, to produce organised cell movement (cell division and movement of organelles), and to remove the end (waste) products of cellular chemical reactions and other toxic substances. The energy used arises from transformations of chemicals in cells, which either absorb (trap) energy from the surroundings or liberate energy to the surroundings. The chemical and energy transformations in cells are collectively known as **metabolism**. A striking feature of living organisms is that many metabolic transformations are achieved by basically the same sets of processes throughout all organisms.

In this chapter we will consider:

- the nature of energy and how it is involved in chemical reactions
- chemical reactions in cells
- ATP, the molecule that provides immediate energy in a form useable by cells
- **fuel molecules**, their structure and their potential chemical energy.

Fig. 2.1 The potential energy of water in this stream in Litchfield National Park, Northern Territory, is converted to kinetic energy as it tumbles to a lower level. This form of energy is harvested by hydroelectric schemes

Energy and its transformations

In simple terms, **energy** is the capacity to do work. Energy can exist in a number of forms, such as heat, electrical, mechanical, chemical and radiant energy. **Potential energy** is stored energy and is the energy usually involved in biological systems. Chemical energy is the potential energy that is stored in the bonds between atoms in molecules. **Kinetic energy** is energy that is expressed as movement, for example, in running water (Fig. 2.1). Heat or thermal energy is the kinetic energy of randomly moving molecules. Radiant energy is the energy of any form of electromagnetic radiation (light, ultraviolet radiation or X-ray and gamma-radiation). Much of the work performed within organisms involves the transformation of potential to kinetic energy and vice versa (Fig. 2.2). The study of energy transformations in living systems is referred to as bioenergetics. Transformations of energy, both in the physical and biological worlds, are governed by the laws of thermodynamics.

Laws of thermodynamics

The **first law of thermodynamics** states that *energy can be neither created nor destroyed*. According to this law, energy may be transformed from one form to another but the total energy of the universe remains constant. No energy transformation is 100% efficient. When energy changes from one form to another, some is always lost as heat (energy unavailable for work). In biological systems, the efficiency of energy conversions from stored to useable energy is never higher than 30%. For example, we all have experienced the rise in body temperature during exercise when the transformation of chemical energy to mechanical energy in muscle contraction dissipates a considerable amount of energy as heat to the surrounding tissues. As all forms of energy can ultimately be converted into heat, energy is usually measured in terms of its equivalent heat (in joules).

As we have seen, biological systems are highly organised, ordered structures that are formed and maintained as a result of a continual input of energy. In an organism, energy is continually lost as heat at every energy transformation and an input of energy is

Fig. 2.2 Examples of the transformation of potential energy to kinetic energy. **(a)** Australian wedge-tailed eagle swooping to catch its prey. **(b)** An emerging pea seedling converts potential chemical energy into the kinetic energy of movement as it pushes through soil towards the light

required to maintain the ordered state. Loss of energy during energy transformations results in increasing disorder of the universe, of which biological systems are part. **Entropy** is a measure of this disorder. This is the basis of the **second law of thermodynamics**, which states that *the entropy of the universe is increasing.*

E nergy is the capacity to do work and exists in two general forms: potential and kinetic energy. Energy transformations are described by the laws of thermodynamics.

Chemical equilibria

In a reversible chemical reaction, such as

$$A \rightleftharpoons B + C$$

the chemical potential that drives the reaction involves the intrinsic properties of the reactant and product molecules and their concentration; that is, the properties of A, B and C and how much there is of each. If more A is added, then more B and C are formed and vice versa. If B and C are removed as they are formed, then the reaction proceeds to the right and more products are formed.

The reaction is at **equilibrium** when there is no net change in the concentration of either reactants or products (Fig. 2.3). This is because the rates of the forward and reverse reactions are the same. Within the confines of the system in which they occur, chemical reactions at equilibrium are in a state of maximum disorder (possessing maximum entropy). Increased order can be achieved by expending energy to push a reaction away from its equilibrium. Energy is released from any reaction as it proceeds towards equilibrium, and energy is required to move a reaction away from equilibrium. Therefore, in a thermodynamic sense, a reaction that is not in equilibrium has potential energy, which can be released if the reaction proceeds spontaneously.

The equilibrium position of a chemical reaction is described by the thermodynamic **equilibrium constant** (K_{eq}):

$$K_{eq} = \frac{\text{concentration of product(s)}}{\text{concentration of reactant(s)}}$$

when the forward and reverse reaction rates are equal.

For the reaction

$$A \rightleftharpoons B + C$$

$$K_{eq} = \frac{[B][C]}{[A]}$$

If the reactants and products contain the same chemical energy per molecule, the K_{eq} is 1.0 (Fig. 2.3). For such a reaction to do work, that is, to be far enough away from equilibrium, there must be a high concentration of reactants or a low concentration of products. However, if the reactants or products contain very different amounts of chemical energy per molecule, the K_{eq} will be far higher or far lower than 1.0. The reaction will, thus, be out of equilibrium when the concentration of reactants and products are equal.

Cells have a limited solvent capacity and yet contain many compounds in solution (several thousand enzymes and chemical reactants) in relatively low concentrations. Therefore, given low

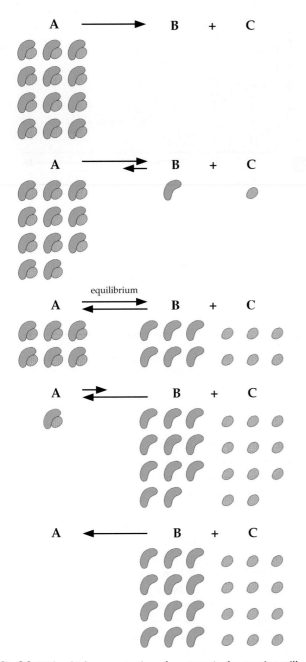

Fig. 2.3 With a high concentration of reactant A, the reaction will proceed strongly in a forward direction. With a high concentration of products B and C, the reaction will proceed strongly in the opposite direction. At equilibrium, the rates of the forward and reverse reactions are the same. (In this example, $K_{eq} = 1.0$)

and similar concentrations of reactants and products, reactions with a K_{eq} far higher or far lower than 1.0 are the most useful for doing work within the physical constraints of a cell.

> Reactions must be out of equilibrium in order to do work. The most useful reactions for providing energy are those with equilibrium constants far higher or far lower than 1.0.

Free energy and the equilibrium constant

We can now inquire about the amount of energy that can be obtained from a chemical reaction proceeding towards equilibrium.

The equilibrium constant describes quantitatively how far towards completion a reaction will proceed but gives no indication at all about the likelihood of the reaction occurring in the first place. The feasibility of a reaction occurring requires some estimate of change in the orderliness of the system in question. This is measured in terms of the degree of disorder of the system, entropy (S).

Reactions that are not at equilibrium have the capacity to do work and the extent of this work (or its value in energy terms) depends on the concentration differences between the reactants and products and the intrinsic energy content of each. The relation between the equilibrium constant for a reaction and the energy available to do work as the reaction proceeds is embodied in the concept of **free energy**.

Spontaneous chemical reactions and free energy

In cells of living organisms, thousands of chemical reactions are occurring at any one time. Some of these reactions are energetically unfavourable—they are the equivalent of pushing water up a hill—but these are the very reactions that are needed to build, for example, the highly ordered molecules, such as proteins and polysaccharides, that are essential to living systems. Before we can understand how such reactions are made possible, we must first consider what we mean by 'energetically unfavourable' reactions.

Application of the second law of thermodynamics indicates that chemical reactions will only occur spontaneously if they result in a net increase in the disorder (entropy) of their surroundings. In other words, the total entropy change is positive for every process. This disorder is created when the energy released is dissipated as heat. However, the entropy increase may not necessarily take place in the reacting system, for example, in a cell. Since

$$\Delta S_{universe} = \Delta S_{cell} + \Delta S_{surroundings}$$

the increase in total entropy may take place in any part of the universe. Thus, the surroundings of the cell may increase in entropy but within the cell the transformations of energy may lead to a state of decreased disorder or lower entropy.

Assessment of the feasibility that a reaction will occur could be measured in terms of the degree of disorder of the system. Although entropy always increases in a spontaneous process, its measurement is impractical. A more convenient measure for predicting the spontaneity of a process is free energy, G, (after the chemist Josiah Gibbs, the founder of the science of thermodynamics). Free energy represents the maximum amount of useful work obtainable from a reaction. The change in the value of G, designated ΔG, is the useable energy, or **chemical potential**, of a reaction. It turns out that the change in free energy can be defined as the algebraic difference between the change in heat content (ΔH), a measure of the internal energy of the reactants (determined by their number and kinds of chemical bonds), and the change in entropy (ΔS) at temperature T (measured in degrees Celsius above absolute zero) according to the equation:

$$\Delta G = \Delta H - T\Delta S$$

(*Note:* absolute zero is the thermodynamic temperature, $-273.15°C$, at which molecules no longer possess any kinetic (heat) energy and all molecular movement stops.)

Thus, the change in free energy of a reaction is a composite term incorporating the change in heat content (ΔH), determined largely by the making and breaking of chemical bonds, and the degree of disorder (ΔS), determined largely by the molecular organisation of the system. Its magnitude will depend on the total number of molecules reacting.

Free energy change can be related mathematically to the concentration of reactants and products. It can be shown that for the reaction

$$A \rightleftharpoons B + C$$

$$\Delta G = -RT \log_e K_{eq}$$

where R is the universal gas constant and

$$K_{eq} = \frac{[B][C]}{[A]}$$

Thus, the value of change in free energy (ΔG) may be regarded as a disguised equilibrium constant.

This equation is useful in predicting whether or not, and in which direction, a reaction will proceed. When ΔG is negative (i.e. < 0) and energy is released in a reaction, the reaction is **exergonic**; that is, the reaction is spontaneous or goes 'downhill' in energy terms (Fig. 2.4). When ΔG is positive (i.e. > 0) and energy is needed for a reaction to proceed, the reaction is **endergonic**; that is, the reaction cannot proceed without the input of extra free energy (Fig. 2.4). Endergonic reactions can be driven by being coupled with exergonic reactions so that one reaction supplies energy to the other. In biological systems, the coupling of such reactions results in chemical processes in which non-spontaneous, energy-requiring reactions are linked to spontaneous, energy-yielding reactions. This

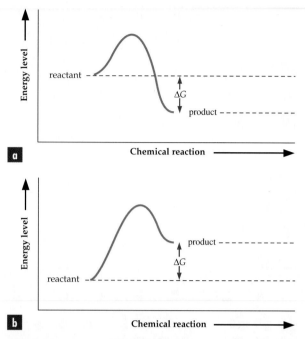

Fig. 2.4 (a) In an exergonic reaction, ΔG is negative; that is, energy is released. **(b)** In an endergonic reaction, ΔG is positive; that is, energy must be supplied in order for the reaction to proceed

is the key to the ability of organisms to perform energetically unfavourable (uphill) reactions.

The change in free energy (ΔG) is the energy that can be drawn upon to do work. It is the energy available above and beyond the energy changes due to the making and breaking of chemical bonds and the overall change in orderliness of the system. As long as ΔG is negative, the reaction will proceed spontaneously.

Since it is difficult to measure the concentrations of reactants and products in cells, the change in free energy under standard concentrations of reactants and products can be derived from the more easily accessible equilibrium constant (K_{eq}). Change in free energy can be calculated from the measured K_{eq} by the equation:

$$\Delta G^O = -RT \log_e K_{eq}$$

where ΔG^O is the standard free energy change (the gain or loss of free energy when 1 mole of reactant is converted to 1 mole of product under defined conditions in solution).

The amount of available energy from a reaction is its free energy. A reaction will proceed spontaneously if its products contain less free energy than the reactants (an exergonic reaction). In spontaneous reactions, ΔG is always negative. A reaction will not proceed without the addition of energy if its products have more energy than the reactants (an endergonic reaction).

Rates of chemical reactions, activation energy and catalysis

Many chemical reactions normally proceed far too slowly to be of use to organisms, especially at the temperatures at which most organisms function. The **rate** or velocity at which a chemical reaction proceeds towards equilibrium, or the kinetics of the reaction, is independent of the K_{eq} or ΔG^O and depends on the kinetic energy of the reacting molecules.

A population of molecules in solution may have a constant average energy at a given temperature but the individual molecules will have different energies (Fig. 2.5a). For a chemical reaction to occur between any two molecules, the pair must possess more than the minimal level of energy necessary to break existing bonds at the instant they collide. Those with less than this amount of energy will not react at all. The energy of the reactants required to initiate a reaction is termed the **activation energy** of that reaction.

The simplest way to increase the rate of a chemical reaction is to raise the temperature of the reactants. Heating increases the kinetic energy of molecules and, thus, increases the proportion of molecules in a

population whose energy exceeds the activation energy; thus, the rate of reaction increases.

Living systems do not rely on increasing their temperature to achieve useful reaction rates. Apart from anything else, their proteins would denature (change shape) at temperatures required to achieve useful reaction rates (see Chapter 1). The other option is to accelerate the reaction by **catalysis** (Fig. 2.5b).

Catalysis is the acceleration of the rate of a reaction (Fig. 2.6). A catalyst does this by reducing the activation energy of the reactants. Thus, a catalysed reaction can occur more readily at a lower temperature. There is no change in the equilibrium position, simply a change in the rate of approach to equilibrium. The reduction in activation energy by a catalyst accelerates both forward and reverse reactions by exactly the same amount without changing the final equilibrium position. At the end of a reaction, the catalyst itself remains unchanged and so can be used over and over again.

> The activation energy required to break existing bonds determines the rate of a reaction. Catalysis is the process of lowering the activation energy of the reactants so that a reaction occurs more readily.

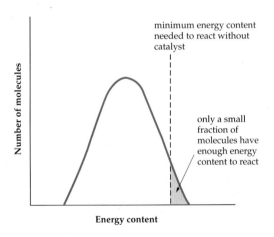

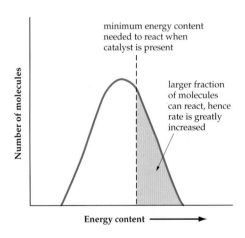

Fig. 2.5 (a) In a population of molecules in a solution, individual molecules have different energy levels and only those with high energy levels will react. **(b)** A catalyst increases the proportion of reacting molecules that have enough energy content to react

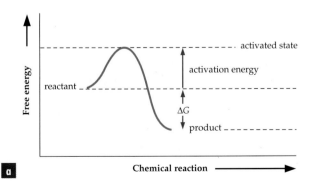

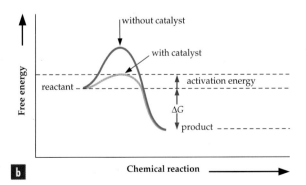

Fig. 2.6 Energy required for a chemical reaction. **(a)** For a reaction to occur, energy must be supplied to alter chemical bonds. This energy is the activation energy. **(b)** A catalyst speeds up a reaction by lowering the activation energy required

Enzymes are biological catalysts

Chemical reactions in living cells are accelerated by biological catalysts, known as **enzymes**, which, with few exceptions, are proteins (Chapter 1). An animal cell, for example, may contain up to 4000 different types of enzymes, each catalysing a different chemical reaction. Some enzymes are common to many types of cells, whereas others are found only in cells carrying out particular functions. In an enzyme reaction, the reactants are termed **substrates**. Most enzymes are specific in the way they function, acting selectively on only one substrate or a chemically related set of substrates. As catalysts, enzymes reduce the activation energy necessary for a reaction to occur at physiological temperatures (e.g. 37°C for most mammals).

Enzymes have extraordinary catalytic power. As an example, consider the reaction of carbon dioxide and water to form carbonic acid:

$$CO_2 + H_2O \rightleftharpoons H_2CO_3$$

In the absence of an enzyme, this reaction is very slow, perhaps producing only 200 molecules of carbonic acid per hour. In living cells, this reaction is catalysed by the enzyme carbonic anhydrase and 600 000 molecules of carbonic acid are formed every second; in other words, the reaction proceeds some 10 million times faster.

The first step in an enzyme-catalysed reaction is the reversible binding of the substrate or substrates onto the surface of the enzyme protein to form a transient enzyme–substrate complex. It is in this complex that the chemical reaction occurs. The products are then released from the enzyme surface.

The two steps in an enzyme–substrate reaction can be written in the form of the equations:

$$E \quad + \quad S \quad \rightleftharpoons \quad ES \qquad (1)$$
enzyme   substrate   enzyme–substrate
complex

$$ES \quad \rightarrow \quad E \quad + \quad P \qquad (2)$$
product

Two aspects of enzyme action must be explained: their specificity and their ability to accelerate the reaction.

To explain the specificity of an enzyme-catalysed reaction, we must consider the structure of an enzyme protein. As we saw in Chapter 1, enzymes are generally globular proteins. Their globular shape is a consequence of the folding of the polypeptide chain. In enzymes and other proteins that function by recognising and binding molecules, the polypeptide chains are folded in such a way as to form one or more pockets or grooves on the surface, creating a specialised region into which the substrate molecules can fit. This region of an enzyme is known as the **active site** (Fig. 2.7). An active site is formed from only a small proportion of an enzyme's total complement of amino acids. These amino acids may be adjacent to one another in the polypeptide chain or on different parts of the polypeptide chain that are brought together by folding to give the tertiary structure of the enzyme. The exposed R-groups of the amino acids (see Chapter 1) line the active site and their detailed arrangement in relation to one another determines the specificity of binding of the enzyme to its substrate(s) (Fig. 2.7).

Some of the R-groups lining the active site are concerned in the specific binding and orientation of the substrate molecules. These are termed **substrate-binding amino acids**. Their R-groups form charged, or uncharged, hydrophilic or hydrophobic surfaces in the active site, which bind precisely with particular parts of the substrate molecules. For example, if a substrate has a positive charge on a portion of its molecule that binds to the active site, then the corresponding part of the enzyme's active site will have a negative charge. In addition to such ionic interactions, non-covalent hydrogen bonds and van der Waals forces are also involved in specific binding and orientation of substrates in the active site.

Apart from the amino acids involved in substrate binding, other active-site amino acids are involved with the covalent bond making or breaking events. These are the **catalytic amino acids**, which participate in the chemical reaction being catalysed. In some enzymes the substrate may form a transient, covalent

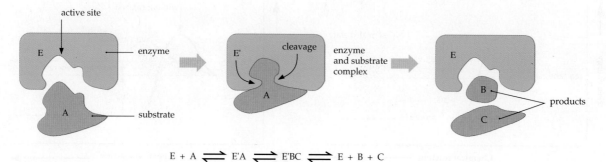

$$E + A \rightleftharpoons E'A \rightleftharpoons E'BC \rightleftharpoons E + B + C$$

Fig. 2.7 When a substrate molecule binds at the active site of an enzyme, it is brought into a transition state in which it is strained and distorted, so that existing bonds may be broken and the molecule dissociated into the products

intermediate with a catalytic amino acid in the active site.

For the enzyme to perform its catalytic function, it is imperative that the substrate undergoing the chemical change is positioned very precisely in relation to the catalytic amino acids, so that the rearrangement of atoms in the chemical reaction can occur. It is the role of the binding amino acids to direct the substrate to that position.

The specificity of an enzyme is attributed to the three-dimensional structure of its active site, which fits only one type of substrate molecule. The catalytic activity of the enzyme is made possible by the precise alignment of the substrate molecules in relation to the bond-breaking or bond-making amino acids in the active site.

Models of enzyme action

The specificity of enzymes was recognised a hundred years ago by the celebrated German chemist, Emil Fischer, who suggested that the specific relationship might result from an actual 'fit' between the substrate and the enzyme like a key in a lock (Fig. 2.8). Later, J. B. S. Haldane, the eminent British biologist, introduced the idea that the enzyme–substrate complex required a certain addition of energy of activation before reacting. He suggested that Fischer's lock and key hypothesis be amended to allow that the key does not fit perfectly in the lock but exercises a certain strain on it. The notion of substrate strain or distortion has been a part of enzyme description ever since.

The substrate molecules bound in the active site are brought into a state where electrons may flow, breaking existing bonds and creating new bonds. This intermediate state where the substrate molecules are poised ready for the reaction to occur is called the **transition state**. In this state the reactants are strained and distorted or have an unfavourable electronic structure.

Later, Linus Pauling presented the view that an enzyme is an essentially flexible molecular template that has evolved to be precisely complementary to its reactants in their activated transition state. This is

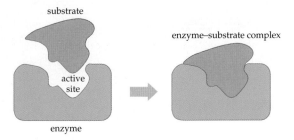

Fig. 2.8 In the lock and key model of enzyme action, the substrate molecule fits directly into the active site of the enzyme

BOX 2.1 How enzymes work

Enzymes can achieve their reaction rate enhancement effects through lowering the activation energy of the substrate molecules by several means. Individual enzymes may use more than one of these strategies.

Proximity effect

In uncatalysed reactions in solution involving more than one reactant, all the molecules must be brought together in the transition state. This is an unfavourable situation requiring the right number of molecules to collide simultaneously and in the correct orientation and explains why uncatalysed reactions proceed slowly.

In contrast, binding of substrate molecules at the active site in an enzyme–substrate complex ensures that substrate molecules are held close together in a precise and appropriate orientation for a reaction to occur. The binding of the substrates on the enzyme surface restricts their rotational and translational mobility and, the more rigidly they are held in place, with the appropriate stereospecificity, the faster they will react. The binding of the reactants to the enzyme surface results in increased order, that is, a lowering of the entropy of the system. In part, the rate enhancement is due to this entropy effect.

Flexibility of enzyme shape

The precise positioning of the reactants in the active site is fundamental to enzyme catalysis and we have seen that enzymes change their structure when a substrate is bound. In some cases this change, known as induced fit, is quite large. The substrate is forced to respond to directed electrostatic fields contributed by the functional groups on the enzyme surface that promote the formation of the transition state.

Enhancement of reactivity of susceptible bonds or groups

According to this model, substrate molecules bind with the active site of an enzyme by forming hydrogen and ionic bonds and, sometimes, covalent bonds. As a result, redistribution of electrons within the substrate molecules takes place, leading to weakening of susceptible bonds, which makes them more easily broken and thus more reactive.

described as **transition state stabilisation**. The catalytic power of enzymes results from the highly specific binding of their transition states. Pauling's concept is in line with the ideas of Koshland, who introduced the term **induced fit** (Fig. 2.9) to describe the change in shape of a protein when a ligand, in this case a substrate, was bound to it (see Chapter 1).

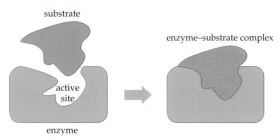

Fig. 2.9 In the induced-fit model of enzyme action, the substrate induces a shape change in the enzyme that positions the substrate so that a reaction will occur

Factors affecting enzyme activity

Enzymes can be extracted from cells or tissues and studied in an active form in the test tube. In this way, we can learn how enzymes catalyse their specific reactions and what effect the environment (pH, temperature, substrate concentration, etc.) has on the process. This study is known as enzymology.

The rate of an enzyme reaction is partly a function of the concentration of substrates, products and the enzyme itself. For example, in a solution of a given amount of enzyme, the higher the concentration of substrate molecules, the more frequently the substrate molecules are likely to encounter the active sites of enzyme molecules (Fig. 2.10). However, the reaction rate will eventually reach an upper limit when all active

sites of the enzyme molecules are occupied. The rate can only then be increased by raising the enzyme concentration (Fig. 2.10).

Apart from concentrations of enzyme, substrate and product molecules, environmental factors regulate enzyme activity. These include pH, temperature and cofactors.

pH

With change of pH, the shape of the active site of an enzyme may alter through changes in the interaction between negatively charged (acidic) and positively charged (basic) amino acid residues in the active site (p. 56), so that binding between the active site and substrate no longer occurs.

Most enzymes have an optimum pH in the range 6–8. However, pepsin, which initiates protein digestion in the acidic environment of the stomach, acts optimally at a very low pH (pH 2). Pepsin has an amino acid sequence that maintains the ionic and hydrogen bonds involved in folding even in the presence of strong acid.

Temperature

Hydrogen bonding and hydrophobic interactions help maintain the tertiary and quaternary structure of proteins, including enzymes. These bonds are easily disrupted by changes in temperature. Organisms usually have enzymes that function optimally at the temperature at which they live. In mammals, the optimum body temperature is 37°C. Hence, their enzymes do not function efficiently at temperatures lower than 37°C. Above 40°C, hydrogen and other weak bonds are unable to hold proteins together in their correct shape. Animals such as the spangled and sooty grunter fishes in northern Australia have proteins that can function at high temperatures and enable them to live in hot springs (Fig. 2.11). Also, there are many

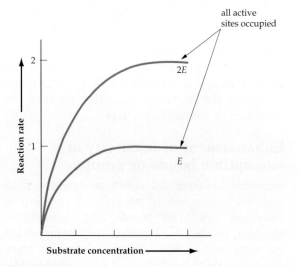

Fig. 2.10 The rate of an enzyme-catalysed reaction depends on the substrate concentration. At high substrate concentrations, the rate is limited by the amount of enzyme (*E*). Addition of more enzyme (2*E*) will increase the rate of the reaction

Fig. 2.11 Animals such as the sooty grunter fish, *Hephaestus fuliginosus,* in northern Australia have enzymes that can function at high temperatures and enable them to live in hot springs

thermophilic microorganisms with optimum temperatures for growth of more than 60°C and, for hyperthermophiles, more than 80°C. These organisms have adapted to the extreme environment by 'escaping' or 'compensating' the stress of the environment or by enhancing the stability of their enzyme proteins. It appears that enhanced thermostability is correlated with a consistent increase in the number of hydrogen bonds and ion pairs (see Chapter 1) and an increase in the polar surface area.

An enzyme that has lost its characteristic three-dimensional structure is said to be **denatured**. Enzymes that are partially denatured by heat have a slightly distorted structure and their polypeptide chains can regain their correct shape on cooling, but complete denaturation, as occurs to the protein albumin when an egg is boiled, is irreversible.

Cofactors and coenzymes

Many enzymes require an additional chemical component, a **cofactor**, to perform their catalytic function. Cofactors may be a metal ion or an organic molecule. When the cofactor is a non-protein, complex organic molecule, often with a vitamin as a building unit, it is known as a **coenzyme**.

Many enzymes have metal ions at their active site, which help to draw electrons away from substrate molecules, thereby helping to break bonds in the substrate molecule (Fig. 2.12). For example, the enzyme alcohol dehydrogenase, an oxidoreductase (Box 2.2), employs a zinc ion to help ionise a weaker molecule involved in the catalytic mechanism. In some cases, metal ions serve to hold the enzyme protein together.

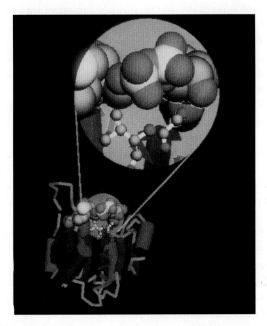

Fig. 2.12 Alcohol dehydrogenase is a zinc-requiring enzyme

Concentrations of enzyme, substrates and products, and pH and temperature are important in determining the rate of enzyme action. Metal ions and complex organic cofactors participate in many enzyme-catalysed reactions.

Chemical reactions in cells

In a living cell, chemical processes are usually the result of the action of enzymes. Enzyme activity is regulated so that appropriate amounts of products are made at the required rate. Underproduction would result in a slowing of cellular activity. Overproduction would result in wastage of both energy and raw materials.

In cells, short-term control of enzyme rate can be achieved in several ways. One way is by reversible, **covalent modification** of the enzyme polypeptide, for example, by phosphorylation at serine, threonine or tyrosine residues by specific **phosphorylation** enzymes. Depending on the system, this leads to an increase or decrease in the enzyme's activity. Specific dephosphorylating enzymes remove the phosphate residues and reverse the effect on enzyme activity.

Another mechanism for decreasing an enzyme's rate (or sometimes increasing it) is by the binding of organic molecules at specific sites on the enzyme surface other than the active site (Fig. 2.13). These molecules are often the end-product of the pathway in which the enzyme is part (end-product inhibition). This mechanism is known as **allosteric** (meaning different site) inhibition (or activation).

In both these rate-regulating mechanisms, the three-dimensional structure of the enzyme is modified, leading to changes in the shape of the active site with consequential changes in reaction rate. In this way, the flux of metabolites through a particular metabolic pathway can be altered. These mechanisms allow minute-by-minute control of the rate of enzyme action in a cell.

As we have seen (p. 58), in the test tube the rate of an enzyme-catalysed reaction depends, other things being equal, on the concentration of the enzyme. In cells, the concentration of an enzyme can be increased by synthesis of more enzyme protein, a process controlled by expression of the specific DNA coding for the enzyme polypeptide (see Chapter 11). Conversely, the amount of an enzyme can be decreased by specific breakdown of the enzyme protein. In eukaryotes, this is achieved by removal of the enzyme molecules to a special compartment of the cell, the lysosome, where they are degraded. In other cases, the enzyme protein is specifically marked for degradation by covalent association with a small polypeptide and then degraded. This type of control of enzyme concentration is mediated over a time scale measured in hours.

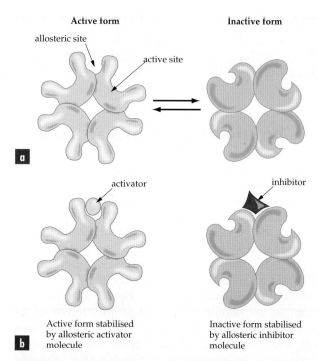

Fig. 2.13 (a) Allosteric enzymes are generally constructed from two or more subunits, each having its own active site. The enzyme can be in one of two states differing in shape, one active, one inactive. Remote from the active sites are allosteric sites that specifically bind regulators of the enzyme, which may be activators or inhibitors. **(b)** The opposing effects of an allosteric inhibitor and an activator on the shape of an enzyme with four subunits. Binding of the regulator to a single allosteric site affects the shape of all four subunits of the protein

Metabolism

All chemical processes in cells of living organisms involve the building up, maintenance or breaking down of tissue components. Together these reactions constitute cellular metabolism. Metabolic processes occur simultaneously, usually in different cellular compartments (organelles). As we have seen, the energy made available as some cellular components are broken down is used to build other cellular components.

The vast array of chemical reactions essential to life are catalysed by six different classes of enzymes (see Box 2.2). Two of these, the transferases and the ligases, are involved in the biosynthesis of cellular constituents. The transferase-catalysed reactions involve assembly of complex molecules from simple ones, such as polysaccharides from monosaccharides, triacylglycerols from fatty acids and glycerol, and polypeptide chains from amino acids. Ligases link together two molecules and these reactions are energetically unfavourable and require an input of energy (Fig. 2.14). This is provided by coupling the reaction with one that involves the breaking of the pyrophosphate bond in ATP (see p. 64).

The hydrolases, whose reactions involve water (see Box 2.2), are involved in the breakdown of complex

molecules to simple ones, such as di- and polysaccharides to monosaccharides, triacylglycerols to fatty acids and glycerol, and polypeptides to amino acids (Fig. 2.15). Functionally, these reactions are essential for the turnover of cellular constituents and are central to the conversion of polymeric nutrient molecules, such as starch, cellulose and proteins, to small soluble and diffusible molecules suitable for uptake into cells.

The lyases and isomerases are chiefly involved in steps in metabolic pathways that transform intermediate compounds into chemical forms that are substrates for oxidoreductases and ligases.

The oxidoreductases are involved in the process of trapping the potential energy of chemical compounds by coupling their reactions with the formation of ATP molecules. These, in turn, provide the driving force for the energetically unfavourable transferase and ligase reactions.

In Chapter 5, we discuss the major biochemical pathways associated with energy transformations. These are glycolysis and cellular respiration, which involve degradative reactions (breakdown of molecules or **catabolism**), and photosynthesis, which involves synthetic reactions (building of molecules or **anabolism**). ATP is produced in both glycolysis and cellular respiration. The latter process is dependent on the functioning of an **electron transport pathway**.

Electron transport pathways

An electron transport system is a group of membrane-bound enzymes and cofactors that operate sequentially in a highly organised manner. The basis of electron transport is that electrons are transferred stepwise from one molecule (a donor) to another (an acceptor). When an atom or molecule loses one or more electrons, it is said to be **oxidised**. Conversely, when it accepts one or more electrons, it is said to be **reduced**. Such electron transfer reactions are known as **oxidation–reduction reactions**.

When a fuel molecule, such as the monosaccharide, glucose, is degraded, molecular oxygen (O_2) combines with atoms of the fuel molecule to form oxides of carbon and hydrogen. For this reason, the fuel is said to have been oxidised. More accurately, as mentioned above, oxidation of a compound involves the removal of electrons, whereas in reduction, electrons are added (Fig. 2.16). Thus, when one substance is oxidised, another is reduced. The gain or loss of electrons may be accompanied by the gain or loss of protons (H^+) and the removal or addition of oxygen.

Breakdown of fuel molecules to release energy involves the oxidation of a C—H bond. Although C—C bonds have as much potential energy as C—H bonds, they cannot be oxidised directly. In degradative

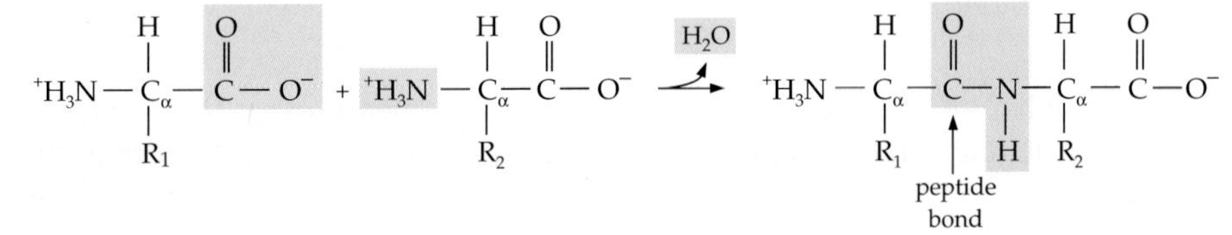

Fig. 2.14 In the ligase reaction for the formation of a peptide bond between two amino acids in a protein, a water molecule is liberated. This is an endergonic reaction and requires the input of energy. The energy is derived from the hydrolysis of ATP

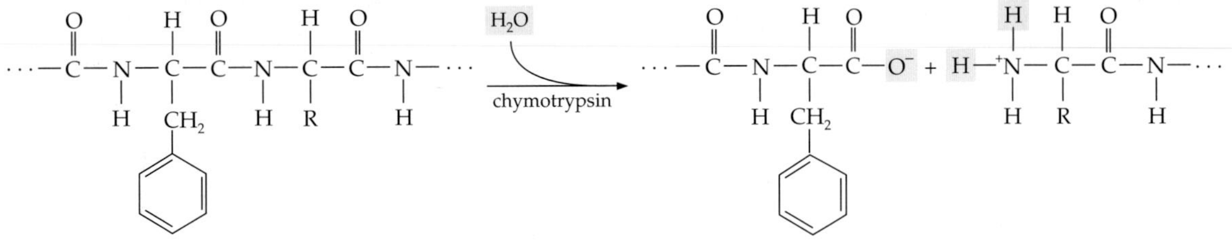

Fig. 2.15 In a hydrolysis reaction, a water molecule is used. For example, the enzyme chymotrypsin catalyses the hydrolysis of peptide bonds in proteins. This is an exergonic reaction

---

BOX 2.2 Six types of enzymes that catalyse chemical reactions

Six basic classes of enzymes can be distinguished according to the types of chemical reactions they catalyse.

Oxidoreductases

Oxidoreductases transfer electrons, usually in the form of hydride ions or hydrogen atoms.

$$AH_2 \quad + \quad B \quad \rightleftharpoons \quad A \quad + \quad BH_2$$

reduced substrate	oxidised substrate	oxidised product	reduced product

B may be an oxygen molecule, or a coenzyme such as NAD^+.

Transferases

Transferases move a chemical group, or a molecular unit, from a donor substrate to an acceptor substrate.

$$XA \quad + \quad Y \quad \rightleftharpoons \quad X \quad + \quad YA$$

donor   acceptor

The group transferred may be, for example, a methyl ($-CH_3$) group, a phosphate ($-PO_4^{3-}$) ion, a monosaccharide unit, and so on.

Hydrolases

Hydrolases cleave substrates, such as esters, disaccharides, polysaccharides, polypeptides and nucleic acids, to their molecular units. Water is the cosubstrate.

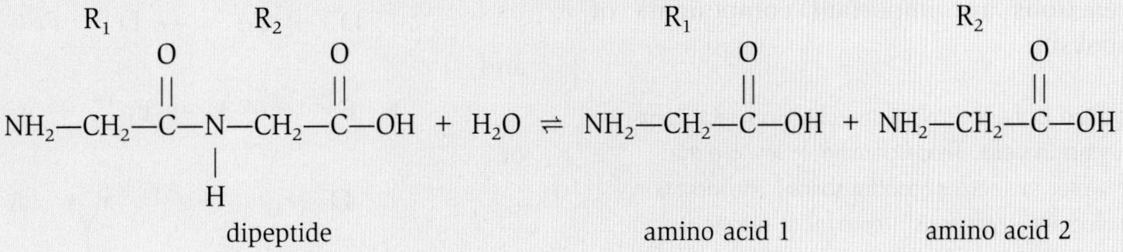

Lyases

Lyases act on substrates with C—C, C—O, C—N and other bonds by means other than hydrolysis or oxidation. Their activities include reactions that eliminate water, leaving double bonds, or conversely, add water to double bonds.

$$
\begin{array}{c}
\text{O}{=}\text{C}{-}\text{OH} \\
| \\
\text{CH} \\
|| \\
\text{CH} \\
| \\
\text{O}{=}\text{C}{-}\text{OH} \\
\text{fumarate}
\end{array}
+ \text{H}_2\text{O} \rightleftharpoons
\begin{array}{c}
\text{O}{=}\text{C}{-}\text{OH} \\
| \\
\text{CH}_2 \\
| \\
\text{CHOH} \\
| \\
\text{O}{=}\text{C}{-}\text{OH} \\
\text{malate} \\
\text{(hydration)}
\end{array}
$$

$$
\begin{array}{c}
\text{carbon dioxide} \\
\text{O}{=}\text{C}{-}\text{OH} \quad\quad \text{CO}_2 \\
| \quad\quad\quad\quad + \\
\text{C}{=}\text{O} \quad \rightleftharpoons \quad \text{H}{-}\text{C}{=}\text{O} \\
| \quad\quad\quad\quad\quad | \\
\text{CH}_3 \quad\quad\quad\quad \text{CH}_3 \\
\text{pyruvate} \quad \text{acetaldehyde} \\
\text{(decarboxylation)}
\end{array}
$$

Isomerases

Isomerases cause geometric or structural change (isomerisation) in a substrate molecule. For example:

glucose $\rightleftharpoons$ fructose   (see Fig. 1.13)   galactose $\rightleftharpoons$ glucose   (see Fig. 1.13)
(aldose)   (ketose)

Ligases

Ligases join together (ligate) two molecules to form C—C, C—S, C—O or C—N bonds coupled with the hydrolysis of the pyrophosphate bond in ATP or a similar triphosphate.

$$X + Y + ATP \rightleftharpoons XY + AMP + P{-}P,$$
(AMP—P—P)   pyrophosphate

where P = phosphate

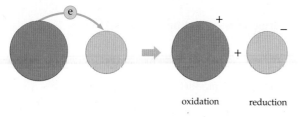

oxidation   reduction

Fig. 2.16 Oxidation is the loss of electrons and reduction is the gain of electrons

pathways, however, preparatory reactions convert C—C bonds to C—H bonds. The C—H bonds can then be oxidised.

Electrons removed from bonds are eventually shared with oxygen but this does not happen directly. Electrons are first transferred to dedicated electron-carrying organic compounds, which, in turn, transfer the electrons to other molecules and eventually to molecular oxygen. These types of oxidation and reduction reactions are important components of energy metabolism.

> Metabolism is the sum of all the chemical reactions occurring in cells. These commonly involve the removal of water molecules (condensation), the addition of water molecules (hydrolysis), removal of electrons (oxidation) and addition of electrons (reduction).

Oxidation and reduction reactions

The interconversion of ferrous (Fe^{2+}) and ferric (Fe^{3+}) ions is a simple example of an oxidation–reduction reaction, which involves the loss (or gain) of a single electron (e^-).

oxidation:

$$Fe^{2+} \rightarrow Fe^{3+} + e^-$$

and reduction:

$$Fe^{3+} + e^- \rightarrow Fe^{2+}$$

The electron removed in an oxidation–reduction reaction does not exist freely in solution but is passed directly from an electron donor (D^-) to ferric iron (Fe^{3+}) to give ferrous iron (Fe^{2+}), or from ferrous iron to an electron acceptor (A). More correctly, we should write:

$$D^- + Fe^{3+} \rightarrow D + Fe^{2+}$$

and

$$Fe^{2+} + A \rightarrow Fe^{3+} + A^-$$

or

$$
\begin{array}{ccc}
D^- & Fe^{3+} & A^- \\
& \diagdown\diagup \quad \diagup\diagdown & \\
D & Fe^{2+} & A
\end{array}
$$

Each donor has a tendency to give up electrons, while each acceptor has a tendency to take up electrons. The tendency to donate or accept electrons can be measured as the **electrical potential** of the redox couple, the **oxidation–reduction (or redox) potential**, which is expressed in volts or millivolts and indicated by the symbol E_0'. Any redox reaction is thermodynamically favourable (ΔG is negative) if electrons are transferred from a carrier with a more negative potential to one with a less negative (more positive) potential.

Some biological oxidation–reduction reactions together with their standard redox potentials are shown in Table 2.1. Cytochromes, flavin mono-nucleotide ($FMN/FMNH_2$) and nicotinamide adenine dinucleotide ($NAD^+/NADH$) participate in the oxidation reactions involved in cellular respiration (Chapter 5). Oxygen (with a high potential) will accept electrons from NADH (with a lower potential) (Table 2.1). The products of this exchange are water and NAD^+.

When a number of such reactions are linked together, they form a pathway, or chain, that transports electrons from one end to the other. In the following example, both electrons and protons are transferred together from donors to acceptors:

$$DH \quad \diagdown \diagup \quad A_1 \quad \diagdown\diagup \quad A_2H$$
$$D \quad \diagup\diagdown \quad A_1H \quad \diagup\diagdown \quad A_2$$

Or, using specific examples from Table 2.1, a possible electron transport chain, showing the E_0' voltage of the various components, is:

$$NADH + H^+ \quad FMN \qquad H_2O$$
$$\diagdown\diagup \qquad \diagdown\diagup$$
$$NAD^+ \qquad FMNH_2 \quad \tfrac{1}{2}O_2$$

E_0' (mV) –320   –120   +815

Two electrons and one proton are transferred along the chain from NADH to reduce oxygen and form water. In doing this, the two electrons move down the potential gradient from –320 mV to +815 mV ($\Delta E_0' = -1.135$ V).

The potential of such a reaction to do work is quantified by converting the electrical potential into the units of free energy, ΔG^O. The change in potential can be shown to be related to the change in free energy by the equation:

$$\Delta G^O = -nF\Delta E_0'$$

where n is the number of electrons transferred, F is a constant relating heat equivalent to electrical potential (the Faraday constant: 9.64 C mol^{-1}) and $\Delta E_0'$ is the change in standard redox potential.

In our example above, a $\Delta E_0'$ of 1.135 V corresponds to a ΔG^O of –218.8 kJ mol^{-1} ($-2 \times 96.4 \times 1.135$), which is a strongly exergonic reaction sequence. This energy would all be lost as heat were it not for the fact that biological systems are able to conserve the energy derived from the ordered and stepwise sequence of oxidation–reduction reactions in the electron transport pathway.

O xidation–reduction reactions involve the transfer of electrons from a donor molecule to an acceptor molecule and are measured in terms of their electrical (redox) potential.

Table 2.1 Standard reduction potentials for reactions (redox couples) of biological importance

Oxidation–reduction system	E_0' (mV) [a]
$2H^+ + 2e^- \rightleftharpoons H_2$	–420
$NAD^+ + 2H^+ \rightleftharpoons NADH + H^+$	–320
$FMN + 2H^+ \rightleftharpoons FMNH_2 + 2e^-$	–120
Coenzyme $Q_{ox} + 2e^- \rightleftharpoons$ Coenzyme Q_{red}	–170
Cytochrome b (Fe^{3+}) + $e^- \rightleftharpoons$ Cytochrome b (Fe^{2+})	+ 120
Cytochrome c (Fe^{3+}) + $e^- \rightleftharpoons$ Cytochrome c (Fe^{2+}) .	+ 220
Cytochrome a (Fe^{3+}) + $e^- \rightleftharpoons$ Cytochrome a (Fe^{2+})	+ 290
$\tfrac{1}{2}O_2 + 2H^+ \rightleftharpoons H_2O$	+ 815

(a) E_0' is the standard redox potential relative to that of the H_2 electrode at pH 7 (–420 mV).

Biological electron carriers

NAD^+ and FMN are examples of **biological electron carriers** in oxidation and reduction reactions. Biological electron carriers may act either as electron acceptors or donors. Some of these compounds only carry electrons and some carry both electrons and protons but it is the electron-carrying function that is important in terms of their redox potential.

Some electron-carrying compounds employ a metal atom, frequently Fe, as the electron transporter. The metal atom is bound into the structure of a large organic molecule, for example, haem (Fig. 2.17), which is associated with a protein. For example, cytochrome *c* is a haem-containing protein, normally located in membranes, which is involved in the ferrous/ferric oxidation–reduction reaction.

There are several different cytochromes, all of which have slightly different haem structures. The chemical structure of the haem in which the Fe atom is embedded alters the microenvironment around the Fe and, as a consequence, the redox potential of its reaction. Thus, different cytochromes, such as *a*, *c* and b_2, have different redox potentials (Table 2.1). These differences determine the unidirectional electron flow along a chain; for example, cytochrome b_2 is oxidised by cytochrome *c*, which in turn is oxidised by cytochrome *a* (Fig. 2.18). All of these cytochromes can oxidise *free* Fe^{2+}.

Many electron carriers accept and donate electrons without the participation of metal ions. These are usually much smaller molecules than cytochromes

although still quite complex. They can participate directly in reactions involving oxidation or reduction of C—C, C—H or C—N bonds. The most important of these is NAD^+ (Fig. 2.19). Although this large molecule has many potentially oxidisable or reducible covalent bonds, it is only the addition and removal of electrons and protons to and from the nicotinamide ring (Fig. 2.19) that function in electron transport.

> B iological electron carriers (e.g. NAD^+ and FMN) carry electrons between oxidation and reduction reactions.

ATP as an energy carrier

Reactions that oxidise reduced bonds in fuel molecules, releasing energy, are not coupled directly to reactions that require energy. The released energy is conserved in **adenosine triphosphate** (**ATP**). Hydrolysis of the terminal phosphate bond of ATP forms adenosine diphosphate (ADP) and inorganic phosphate (P_i) and release of this energy, which can be used to drive non-spontaneous (uphill) reactions, such as the attachment of a carbon dioxide molecule (CO_2) to a carbon compound (CH_3—R), as shown below. Hence, ATP is termed a 'high-energy' compound (Fig. 2.20).

$$\text{ATP} \qquad \text{ADP} + \text{P}_i + \text{H}^+$$

$$CO_2 + CH_3\text{—R} \qquad \overset{\text{O}}{\underset{\underset{\text{new C—C bond}}{\uparrow}}{\overset{\|}{^-\text{O—C—CH}_2\text{—R}}}}$$

ATP is well-suited to its role as a 'high-energy' intermediate or 'energy currency'. It has a number of biologically useful attributes.

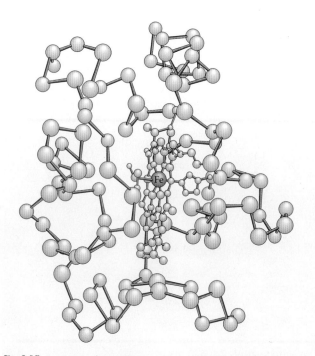

Fig. 2.17 Structure of cytochrome *c*. An iron-containing haem group is at the centre of the molecule

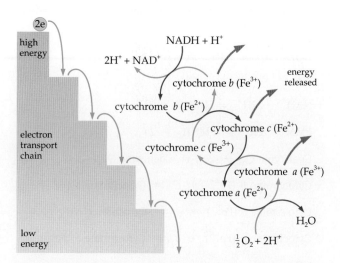

Fig. 2.18 An electron transport chain. Different cytochromes have different redox potentials and determine the unidirectional flow of electrons. As electrons move down the chain, energy that can be used for ATP synthesis is released

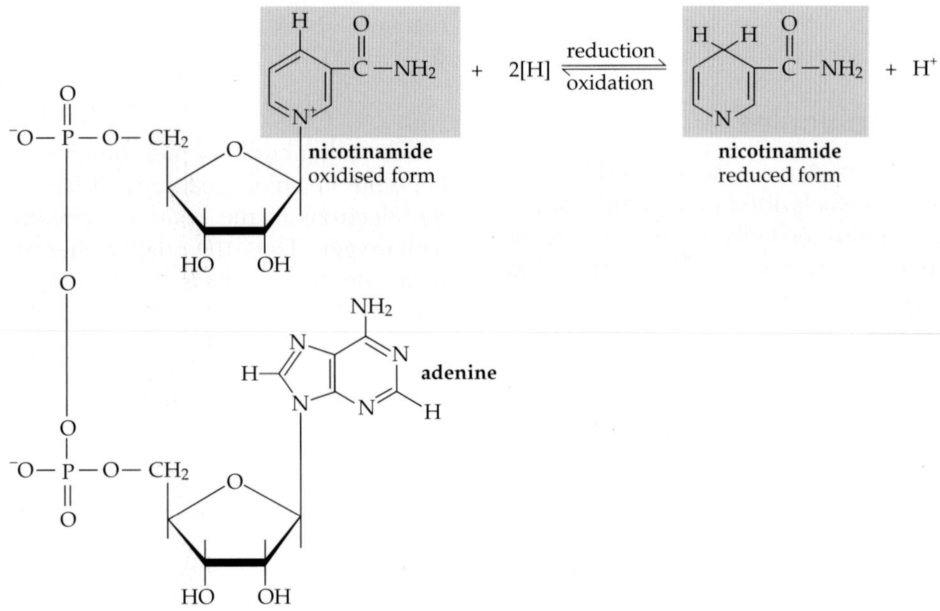

Fig. 2.19 The structure of the electron acceptor nicotinamide adenine dinucleotide (NAD$^+$), showing the oxidised and reduced forms that occur in cells. The nicotinamide portion of the molecule is an essential human dietary factor, the vitamin, niacin

1. The equilibrium constant of the ATP hydrolysis reaction is high. The ΔG^O for the hydrolysis of ATP to ADP is large and negative. Consequently, the reaction is out of equilibrium at the low concentrations of ATP, ADP and P$_i$ that occur in cells and can provide energy as it moves towards equilibrium.
2. ATP is formed at single steps in the pathways of glycolysis or cellular respiration.
3. ATP is a common intermediate between degradative and synthetic metabolic pathways. Many hundreds of reactions, including those in the synthesis of biological molecules, the active transport of molecules across cell membranes and in the generation of force and movement, are driven by ATP hydrolysis. Thus, the concentration of ATP is a regulator of metabolic pathways.

ATP is only one of several 'high-energy' phosphate compounds of biological importance. Another major phosphate compound, phosphocreatine, found in vertebrate striated muscle, has a higher energy of hydrolysis than ATP (Table 2.2). In the scale of ΔG^O for hydrolysis to produce inorganic phosphate, and therefore in energy transfer, ATP has an intermediate value. In this way, ATP can be formed from even higher energy compounds (e.g. phosphocreatine or 1,3-diphosphoglycerate) and can use this energy in the formation of phosphorylated compounds with a lower ΔG^O.

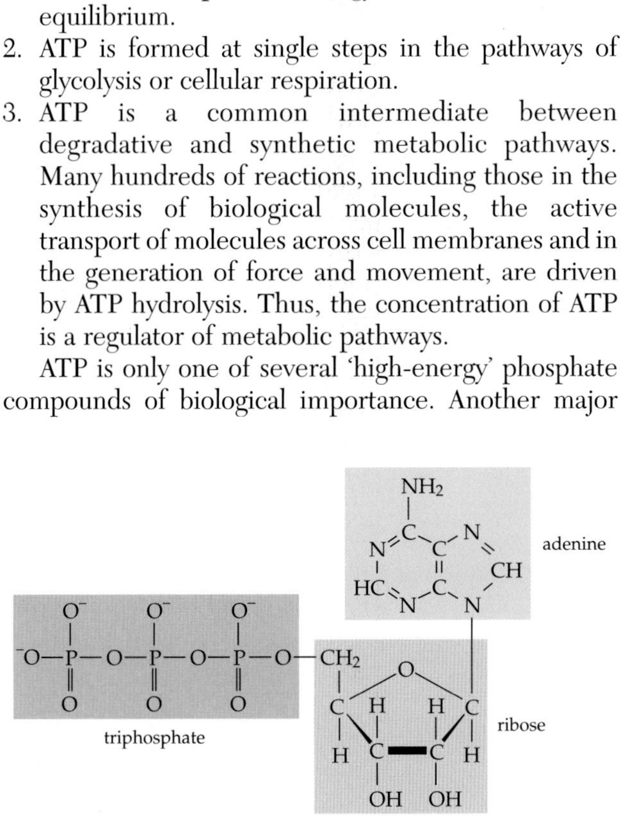

Table 2.2 Standard free energy of hydrolysis of phosphate compounds of biological importance[a]

Compound	ΔG^O (kJ mol^{-1})
Phosphoenolpyruvic acid	–53.56
1,3-diphosphoglyceric acid	–49.37
Phosphocreatine	–43.93
ATP	–29.29
Glucose 1-phosphate	–20.92
Fructose 6-phosphate	–15.90
3-phosphoglyceric acid	–12.97

(a) Adapted from A. L. Lehninger. (1965). *Bioenergetics. The Molecular Basis of Biological Energy Transformation.* New York: W. A. Benjamin Inc., p. 59.

Fig. 2.20 Structure of adenosine triphosphate (ATP)

ATP is a 'high-energy' compound. Energy is released when the terminal phosphate bond is hydrolysed.

Energy in fuel molecules

Carbohydrates, lipids and proteins can be used by cells as fuel molecules. These fuels differ in three main ways: firstly, in their C:O ratio; secondly, in the number of reduced bonds per carbon atom; and thirdly, in the presence or absence of nitrogen.

Although breakage of each bond in a molecule can yield energy, as far as a cell is concerned, energy can only be extracted from the C—C, C—H and C—N bonds of fuel molecules. These bonds are in a reduced state because their electrons are not being shared with oxygen. During metabolism of fuel molecules, electrons are removed from the reduced bonds to become part of a bond with oxygen, which has a much greater affinity for electrons than either carbon or hydrogen (Chapter 1). The reduced bonds contain vastly more extractable energy per molecule than do oxidised bonds. Oxidation reactions involving the reduced carbon bonds have high equilibrium constants and these reactions are out of equilibrium under all situations in the cell. Therefore, work can be done as a reaction moves towards equilibrium. The complete oxidation of glucose is a good example of such a reaction; its K_{eq} is approximately $10^{500\,000}$ and its $\Delta G^O = -2879$ kJ mol^{-1}.

$$C_6H_{12}O_6 + 6O_2 \rightarrow 6CO_2 + 6H_2O$$

The other carbon bond found in fuels is the C—O bond but in a biological sense it has no energy value as the electrons in the bond are already part of a bond with oxygen. Thus, the relative energy value of any fuel molecule to a cell is easily calculated by simply counting the number of C—C and C—H bonds.

The energy available to a cell depends on the type of fuel. For example, lipids have more energy per carbon atom than do carbohydrates because lipids have a higher proportion of energy-rich C—H bonds. If the energy in lipids is expressed in terms of unit weight, then lipids have an even greater advantage as fuel molecules. Lipid molecules with reduced bonds (C—H bonds), compared with carbohydrates with oxidised ones (C—O bonds), are less dense. Thus gram for gram, lipids represent a greater store of energy (see also Chapter 1).

Carbohydrates, lipids and proteins are fuel molecules. Extractable energy is contained in their C—C, C—H and C—N bonds. Lipids provide more energy per carbon atom than do carbohydrates.

Summary

- Energy is the capacity to do work and exists in two general forms: potential energy or stored energy and kinetic energy or the energy of movement. Energy transformations are governed by the laws of thermodynamics.

- The first law of thermodynamics states that energy can neither be created nor destroyed. However, energy can be transformed from one form to another. The second law of thermodynamics states that the entropy (disorder) of the universe is increasing.

- Chemical reactions that are not at equilibrium have the capacity to do work. The most useful reactions for providing energy are those with equilibrium constants far higher or far lower than 1.0.

- The amount of energy available in a molecule is known as free energy. A reaction will proceed spontaneously if its products contain less free energy than the reactants (an exergonic reaction). Reactions will not proceed without the addition of energy if their products have more energy than the reactants (an endergonic reaction).

- The rate of a reaction is determined by the activation energy required to break existing bonds. In catalysed reactions, the activation energy is lowered.

- Enzymes are the catalysts of living cells. The specificity of an enzyme is attributed to its active site, which fits only one type of substrate molecule. Temperature, pH and cofactors are important factors in determining the rate of enzyme-catalysed reactions.

- Enzymes may bring two substrates together in correct orientation by forming a temporary association with the substrates or by stressing particular chemical bonds of a substrate. Like all catalysts, enzymes can be used over and over again.

- Metabolism is the sum of all the chemical reactions occurring in cells. Six different types of enzymes are known: oxidoreductases, transferases, hydrolases, lyases, isomerases and ligases.

- Oxidation–reduction reactions involve the transfer of electrons from a donor molecule to an acceptor molecule and are measured in terms of their electrical (redox) potential. Biological electron carriers (e.g. NAD^+ and FMN) carry electrons between oxidation and reduction reactions.

- Cells use ATP, a 'high-energy' compound, to drive endergonic reactions. Useable energy is released when the terminal phosphate bond is hydrolysed.

- Carbohydrates, lipids and proteins are fuel molecules used in the synthesis of ATP. Extractable energy is available from their C—C, C—H and C—N bonds. Lipids provide more energy per carbon atom than do carbohydrates.

key terms

activation energy (p. 55)
active site (p. 56)
adenosine triphosphate (ATP) (p. 64)
allosteric (p. 59)
anabolism (p. 60)
biological electron carrier (p. 64)
catabolism (p. 60)
catalysis (p. 55)
catalytic amino acid (p. 56)
chemical potential (p. 54)

coenzyme (p. 59)
cofactor (p. 59)
covalent modification (p. 59)
denatured (p. 59)
electrical potential (p. 63)
electron transport pathway (p. 60)
endergonic (p. 54)
energy (p. 51)
entropy (p. 52)
enzyme (p. 56)
equilibrium (p. 52)
equilibrium constant (p. 52)

exergonic (p. 54)
first law of thermo-dynamics (p. 51)
free energy (p. 53)
fuel molecule (p. 51)
induced fit (p. 58)
kinetic energy (p. 51)
metabolism (p. 51)
oxidation–reduction (redox) potential (p. 63)
oxidation–reduction reaction (p. 60)
oxidised (p. 60)
phosphorylation (p. 59)

potential energy (p. 51)
rate (p. 55)
reduced (p. 60)
second law of thermodynamics (p. 52)
substrate-binding amino acid (p. 56)
substrate (p. 56)
transition state (p. 57)
transition state stabilisation (p. 58)

Review questions

1. State the first and second laws of thermodynamics.
2. (a) What does 'equilibrium' mean in a chemical reaction?
   (b) Why is it that the most useful reactions for providing energy are those with equilibrium constants far higher or lower than 1.0?
3. Define entropy. What is the ΔG^{O} of a reaction? On what factors does ΔG^{O} depend?
4. Define oxidation, reduction and hydrolysis. Give an example of each.
5. What are:
   (a) endergonic reactions?
   (b) exergonic reactions?
6. What is activation energy and why is it important for biological systems?
7. What is an enzyme? What is the basis for the specificity of enzyme action? Explain in terms of enzyme structure why boiling inactivates enzymes.
8. Match each term with an appropriate item from the list below (a–e).
   oxidation, reduction, anabolism, catabolism, cofactor, coenzyme, substrate, isomerase, hydrolase, lyase
   (a) Compound acted upon by an enzyme.
   (b) Type of enzyme that causes geometric or structural changes in a substrate molecule.
   (c) Metal ion required by many enzymes.
   (d) Conversion of ferrous (Fe^{2+}) ions to ferric ions (Fe^{3+}).
   (e) The breakdown of compounds in living cells.
9. What is an electron carrier? Name three electron carriers and indicate their reduced and oxidised forms.

Extension questions

1. How do living organisms maintain a high level of organisation and yet conform to the second law of thermodynamics? Why are enzymes needed in a cell?
2. Explain why ATP is well suited to its function as an energy intermediate in cells.
3. Organisms continually generate heat from their metabolic processes but are unable to use this form of energy to perform any useful work. Why is this so?
4. Why do triacylglycerols (fats) provide a more efficient store of energy than carbohydrates?
5. Calculate the net oxidation–reduction potential and the standard free energy change for each of the following reactions written from left to right. Indicate whether or not the reaction will tend to go spontaneously given the proper conditions. Use the equation $\Delta G^{O} = -nF\Delta E_0'$.
   (a) $NADH + H^+ + \frac{1}{2}O_2 \rightleftharpoons NAD^+ + H_2O$
   (b) coenzyme $Q_{ox} + 2H^+ + 2[\text{cytochrome } a \ (Fe^{2+})] \rightleftharpoons$ coenzyme $Q_{red} + 2[\text{cytochrome } a \ (Fe^{3+})]$

Suggested further reading

Alberts, B., Bray, J., Lewis, J., et al. (1994). *Molecular Biology of the Cell*. 3rd edn. New York: Garland.

An excellent advanced text for students majoring in cell biology and biochemistry.

Garrett, R. H. and Grisham, C. M. (1999). *Biochemistry*. 2nd edn. Fort Worth: Saunders College Publishing.

Smith, C. A. and Wood, E. J. (1991). *Energy in Biological Systems*. London: Chapman and Hall.

A clearly written text.

Stryer, L. (1995). *Biochemistry*. 3rd edn. New York: Freeman & Co.

Zubay G. L., Parson, W. W., Vance, D. E. (1995). *Principles of Biochemistry*. Dubuque, IA: W. C. Brown.

Texts for the student majoring in biochemistry.

Harborne, J. B. (1988). *Introduction to Ecological Chemistry*. 3rd edn. London: Academic Press.

This book deals with the chemistry of plants, emphasising natural products and, for example, plant chemical defences; useful for taxonomy and ecology.

CHAPTER

3

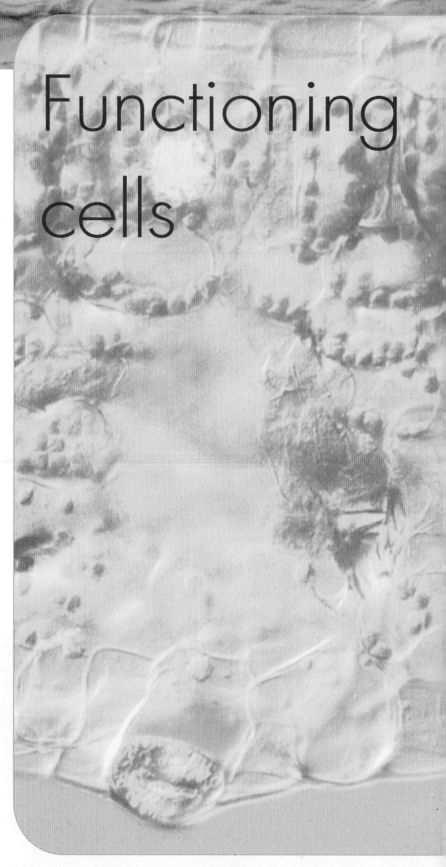

Functioning cells

All living organisms are made up of cells and the materials produced by cells (Fig. 3.1). Cells are small membrane-bound compartments in which most biological reactions occur. They contain a diverse range of ions and molecules, both in aqueous solution and organised into complex subcellular structures. These subcellular structures provide compartments in which particular chemical reactions can be contained. Localisation of enzymes and reactants within compartments increases the efficiency of reactions.

Some organisms consist of only one cell, while others are multicellular. Within the body of a multicellular organism, cells communicate and co-operate with one another so that their individual activities are integrated. The functions of an organism are the result of the activities of its cells.

During evolution, the internal complexity of cells increased greatly. The first cells to evolve were bacteria (see Chapter 30), which have a simple structure and lack internal compartments (Fig. 3.2). They are **prokaryotic cells** (from the Greek *pro*, before, and *karyon*, kernel or nucleus), in which the hereditary material, a double-helical strand of DNA, lies free within the cell. The more complex cells that arose from bacterial ancestors are **eukaryotic cells** (from the Greek *eu*, proper, and *karyon*, nucleus) (Fig. 3.3).

Eukaryotes include algae, fungi, plants and animals, and they range from single-celled organisms such as amoebae to the largest and most complex plants and animals.

Despite the enormous variety of outward form seen in eukaryotes, their cellular structure is fundamentally the same (Figs 3.4, 3.5). One of the main distinguishing features of eukaryotic cells is their possession of internal, membrane-bound components, **organelles**. The most important of these is the **nucleus**, the control centre of the cell and the compartment within which DNA is stored. Other organelles include mitochondria,

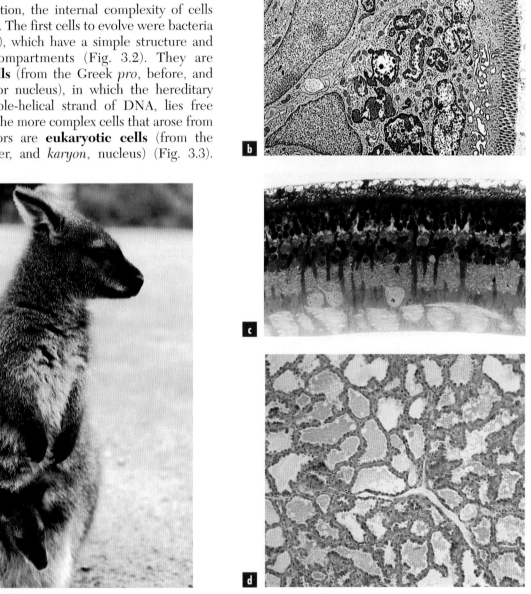

Fig. 3.1 All living organisms are made of cells. **(a)** Multicellular organisms, like this Bennett's wallaby, contain many millions of cells of many different types. The inserts show **(b)** intestinal epithelium (magnification × 4800), **(c)** retinal cells (magnification × 260) and **(d)** mammary gland tissue of the mother of *Macropus eugenii*

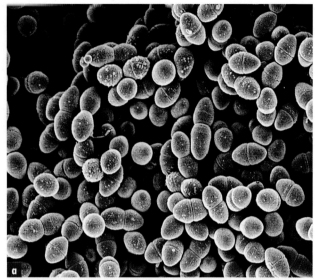

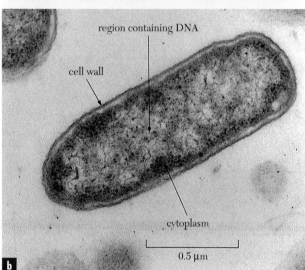

region containing DNA

cell wall

cytoplasm

0.5 µm

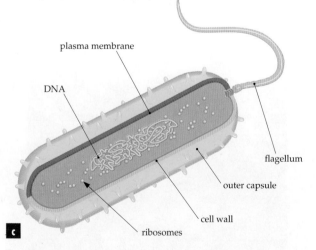

plasma membrane

DNA

flagellum

outer capsule

cell wall

ribosomes

Fig. 3.2 Prokaryotic cells. **(a)** Scanning electron micrograph of prokaryotic cells, *Streptococcus thermophilus*. **(b)** Transmission electron micrograph of a section through a prokaryotic cell, *Escherichia coli*, showing the cell wall, dense cytoplasm with numerous ribosomes and less dense regions containing DNA. **(c)** Diagram of a motile prokaryotic cell that has a single flagellum

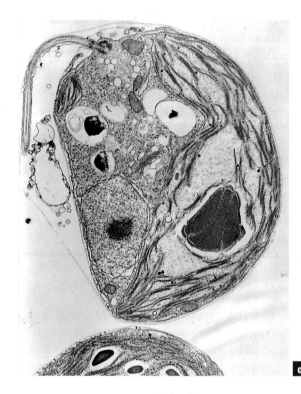

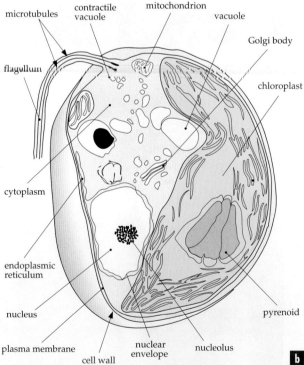

microtubules

contractile vacuole

mitochondrion

vacuole

Golgi body

flagellum

chloroplast

cytoplasm

endoplasmic reticulum

nucleus

pyrenoid

plasma membrane

nuclear envelope

cell wall

nucleolus

Fig. 3.3 Unicellular eukaryotic organism. **(a)** Transmission electron micrograph of the unicellular green alga, *Tetraspora* (magnification × 12 800). **(b)** Diagram showing structures in the *Tetraspora* cell. Note the presence of membrane-bound organelles and compartments, including a nucleus, compared to their absence in prokaryotic cells (Fig. 3.2)

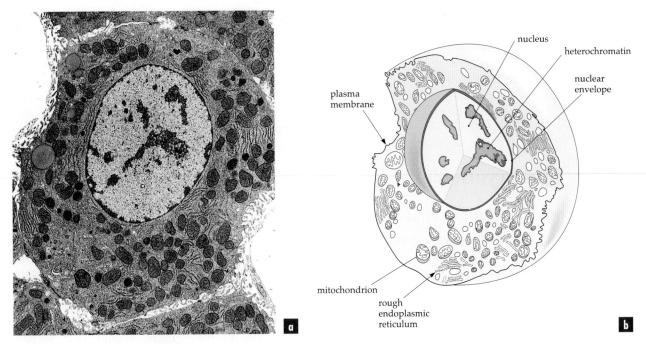

Fig. 3.4 (a) Transmission electron micrograph of a liver hepatocyte showing the typical features of an animal cell. **(b)** Model of an animal cell based on the section of the liver hepatocyte

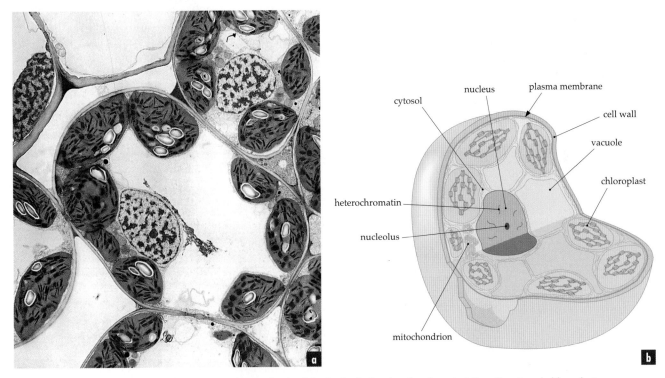

Fig. 3.5 A plant cell. **(a)** Transmission electron micrograph of a daisy leaf cell showing the characteristic cell wall and chloroplasts (magnification × 3000). **(b)** Model of a plant cell based on the section of the daisy leaf cell

plastids, vacuoles and endoplasmic reticulum. Eukaryotic cells also contain structures that are not surrounded by a membrane, such as ribosomes and microtubules. Each subcellular component carries out specific functions. The organelles and other components lie in the **cytosol**, an aqueous solution of molecules with a gel-like consistency. The cytosol and subcellular components, excluding the nucleus, constitute the **cytoplasm**. The cytoplasm and the nucleus together constitute the **protoplasm**.

In this chapter, we will first describe membranes and the variety of subcellular components found in eukaryotic cells. At the end of the chapter, we will compare the internal organisation of prokaryotic cells with eukaryotic cells.

> Living organisms are made up of cells and cell products. Eukaryotic cells contain intracellular, membrane-bound compartments, which separate different molecules and metabolic reactions.

Eukaryotic cells

Membranes: boundaries and barriers

All living cells are bounded by the **plasma membrane** (Figs 3.3–3.5), which separates a cell from its environment. Its most important function is to maintain the integrity of the cell by regulating the passage of molecules into and out of the cell (Chapter 4).

The basic structure of all membranes is a lipid bilayer composed mainly of phospholipids (Fig. 3.6). As discussed in Chapter 1, phospholipids are amphipathic molecules having a hydrophilic (water-loving) polar head and a hydrophobic (water-fearing) tail of fatty acids. In water, phospholipid molecules aggregate and align themselves so that the polar heads face outwards and interact with the polar water molecules. The

hydrophobic tails cluster together, away from contact with the water molecules. As a result, phospholipids tend to either bunch up into small spheres, micelles, or flatten out, forming a **lipid bilayer** (Fig. 3.6).

In addition to lipids, membranes contain two types of proteins (Fig. 3.7). Peripheral membrane proteins are loosely associated with the membrane surface by non-covalent interactions and can be removed with mild treatments such as washing in various salt solutions. Integral membrane proteins interact with the inner hydrophobic regions of the membrane, sometimes through covalent bonding to fatty acid chains or lipid molecules within the membrane. Some integral proteins extend right through the membrane and are known as transmembrane proteins. Integral membrane proteins can be extracted only by disrupting the membrane bilayer by washing with detergents.

The thickness of a membrane depends on its type. The plasma membrane is the thickest membrane in a cell (about 9 nm); membranes of the endoplasmic reticulum are the thinnest (about 6 nm). Membranes are usually, if not always, asymmetric, the properties on one side differing from those on the other. The lipid composition of each monolayer and the proteins exposed on each face are usually different. Carbohydrates, which may constitute 2% of the membrane mass, also contribute to membrane

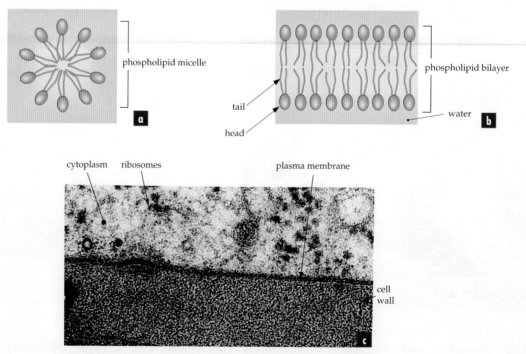

Fig. 3.6 Membrane lipids. **(a)** Phospholipids form micelles spontaneously. **(b)** In a membrane bilayer, the two monolayers of lipid molecules arrange themselves back to back in such a way that the polar heads face outwards to the water and the hydrophobic tails face each other in the centre of the bilayer. Phospholipids will behave this way in a test tube. **(c)** Electron micrograph of a root cell of *Eucalyptus sieberi* including a transverse section through the plasma membrane showing a 'tramline' pattern corresponding to polar and hydrophobic portions of the bilayer. The positions of hydrophilic–hydrophobic–hydrophilic regions across a membrane are seen as a dark–light–dark structure after staining for electron microscopy because the protein heads absorb more electrons than the lipid tails (magnification × 113 000)

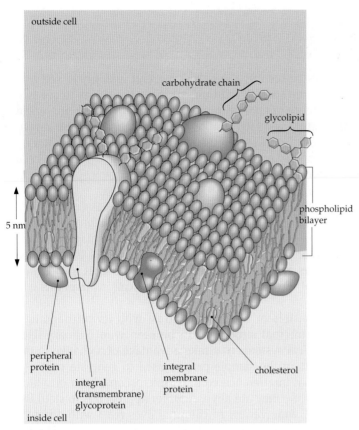

Fig. 3.7 Three-dimensional model of a membrane showing the phospholipid bilayer, cholesterol, glycolipids, integral proteins and peripheral proteins. Carbohydrate chains bind to some proteins and lipids, forming a glycocalyx or sugar coating on the outer surface of the membrane. In many membranes, about 50% of the membrane mass is made up of proteins, although this proportion can be as low as 25% (in myelin sheath of nerve cells) or as high as 75% (in internal membranes of mitochondria and chloroplasts)

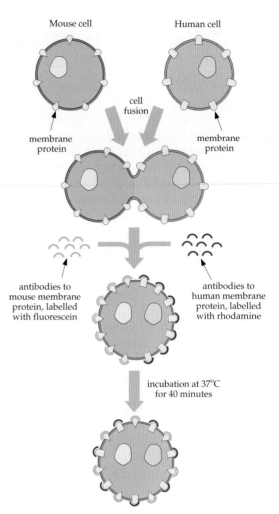

Fig. 3.8 An experiment in which human and mouse cells are fused in culture shows the mobility of cell-surface proteins. At first, the human and mouse marker proteins remain on their respective plasma membranes. After 40 minutes, the two sets of proteins intermingle. The proteins are detected by fluorescence microscopy

asymmetry (Fig. 3.7). Chains of sugar molecules may be attached to membrane lipids, forming glycolipids, or proteins, forming glycoproteins. The sugar chains occur on the non-cytosolic side of the membrane. For intracellular organelles, this means that the sugar chains face the inside (lumen) of the organelle. For the plasma membrane, carbohydrate chains on the outer surface of the cell form a **glycocalyx**, which is a cell coat.

Fluidity of membranes

Another important aspect of membranes is their *fluidity*: membranes are a **fluid mosaic**. The fluidity arises because lipid molecules can move laterally (Fig. 3.8); *mosaic* refers to the irregular arrangement of proteins throughout the lipid bilayer. Lipid membrane molecules can travel the length of a bacterial cell (2.5 μm in length) in just 1 or 2 seconds. In eukaryotic cells, cholesterol, a common membrane lipid, is a major determinant of fluidity. Some proteins move laterally within the membrane, while others are immobilised by

attachment to filaments or tubules either inside or outside the cell. Other proteins are immobilised in aggregates or because they are bound to proteins in adjacent membranes.

Lipid and protein molecules are also able to rotate, that is, flip about their axis perpendicular to the plane of the membrane. Movement of molecules from one monolayer to the other is a rare event. For a phospholipid molecule, movement from one monolayer to the other is so energetically unfavourable that it is unlikely to happen more than once a month! Cholesterol is one of the few molecules that can 'flip-flop' relatively easily.

Permeability of membranes

Membranes are *selectively permeable*. This means that some molecules can pass through the membrane while others cannot. Water, O_2 and CO_2 are able to diffuse freely across membranes but ions and other polar molecules, regardless of how small they are, are unable

BOX 3.1 Studying the structure of cells

Cells and subcellular components are usually too small to be seen by the naked eye. Our inability to see very small structures is a function of the limit of resolution of our eyes. Resolution, which is the ability to see two small particles as discrete objects, depends on the wavelength of light illuminating them and on the numerical aperture of the lens collecting the image. Numerical aperture is related to the angle subtended by the lens at the object. The human eye's limit of resolution is about 200 μm. Cells are usually only 10–20 μm in diameter and are thus well below our limit of resolution. We can, however, use microscopes to magnify images of cells and thus reveal details of their intracellular organisation. There are two main types of microscopes: light microscopes and electron microscopes (see Figs a–c).

Light microscopes use glass lenses to focus visible light onto an object and collect the light that passes through it. Using green light with a wavelength of about 550 nm, the limit of resolution of the light microscope is 250 nm. This is about 800 times better than can be achieved by eye. Thus, the upper limit for magnifications used in light microscopy should be about 800×. This can be obtained, for example, by using a 100× objective lens and an 8× ocular lens. At this magnification, particles at the limit of resolution of the light microscope will be magnified to a size close to the limit of resolution of the eye. Greater total magnifications than this will not reveal any further details of the object.

Another factor that determines whether an object is visible or not is its contrast with its surroundings. Many cells are quite transparent and no details of their internal structure can be seen with ordinary light microscopy. This problem can be solved in two ways. Optical systems that take advantage of a change in the phase of light waves passing through thick or dense parts of the sample can be used to generate contrast. Methods include phase contrast microscopy and differential interference contrast microscopy, both of which have the important advantage that cells can be observed while they are alive. Another approach is to stain the cell contents so that they contrast with the background. A wide variety of coloured dyes react with different types of organic molecules and these are used in light microscopy to visualise selected structures (see Fig. d).

In many cases where samples are too large to be observed intact, they can be embedded in a supporting resin or wax, which is cut into very thin slices (sections). Before doing this, tissues need to be treated in such

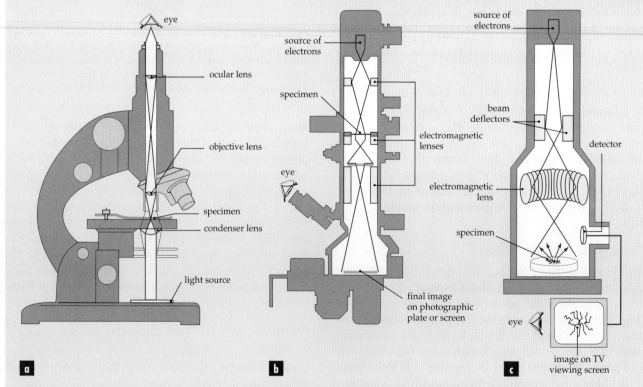

a   **b**   **c**

Comparison of different microscopes for viewing cells showing light and electron paths. **(a)** A simple light microscope (LM). **(b)** A transmission electron microscope (TEM). **(c)** A scanning electron microscope (SEM)

a way that the contents of their constituent cells are preserved in a form as close to that in the living cell as possible. The most common method is to kill cells in a chemical that cross-links proteins, such as formaldehyde or glutaraldehyde. Chemical fixatives do not act instantaneously and in some cells considerable disruption of cellular structure can occur during the fixation process. For this reason, it is important to observe samples that have been prepared by a variety of methods in order to identify fixation artefacts. An alternative to chemical fixation is rapid freezing of the tissue. Frozen samples may be observed either directly or after cutting frozen sections (cryosections). Alternatively, water in a sample can be removed while the tissue is frozen, the process of freeze-substitution. The dehydrated sample is embedded in resin, then sectioned and observed in the same way as chemically fixed samples.

Electron microscopes use magnets to focus a beam of electrons onto a sample. The wavelength of electrons in the beam is of the order of 0.004 nm, which is much lower than the wavelength of visible light. Electron microscopes thus have a theoretical resolution of 0.002 nm. In practice, the resolution normally attained is about 2 nm, which is still 100 times better than that of light microscopes.

In *transmission electron microscopes*, electrons pass through a sample and are projected onto a phosphorus screen. Thin sections of 70–100 nm are routinely used and are usually stained to enhance electron contrast. Electron-opaque dyes, which contain atoms such as lead, osmium or uranium, absorb electrons, leaving a shadow on the phosphorescent screen. The structures that react with the dyes are contrasted against surrounding electron-lucent areas (see Fig. e).

In *scanning electron microscopy*, the beam of electrons is scanned across the sample and electrons that bounce off the surface are collected and displayed on a video monitor. Scanning electron microscopes have a great depth of focus and reveal three-dimensional details of the surface (see Fig. f).

Selected cellular components can be stained by using specific antibodies. For light microscopy, the antibody can be coupled to a fluorescent dye that absorbs light of one wavelength and emits light at a longer wavelength. These specimens are viewed with a *fluorescence microscope*. Commonly used fluorescent dyes include fluorescein, which fluoresces a green colour, and rhodamine, which fluoresces a red colour. For electron microscopy, antibodies are usually coupled to electron-opaque *colloidal gold particles*. When viewed with an electron microscope, the gold particles appear as dark dots where the antibodies have bound to their targeted proteins.

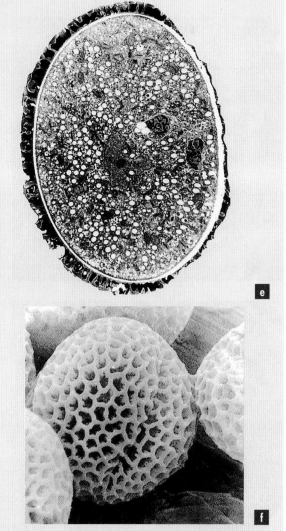

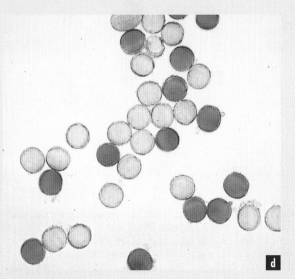

Examples of the same cell type viewed using different microscopes: pollen grains of oil seed rape (*Brassica*) viewed using (d) light microscopy, (e) transmission electron microscopy, (f) scanning electron microscopy

to diffuse across the hydrophobic bilayer. Many ions and polar molecules do, however, get into and out of cells; indeed, the life of a cell depends on their rapid passage. Their passage occurs through certain transmembrane proteins that form highly selective pores (Chapter 4).

Other membrane proteins are enzymes, which may be arranged in a linear sequence to drive a series of reactions in metabolic pathways, such as cellular respiration and photosynthesis (Chapter 5). Another class of membrane proteins functions as receptors for chemical signalling within and between cells (Chapter 7). Cell surface glycoproteins identify a cell as being of a particular type and are important in recognition processes (Chapter 23).

> Membranes are selectively permeable, fluid mosaic structures that are composed of a phospholipid bilayer containing proteins, glycoproteins and glycolipids.

The nucleus

The main feature of eukaryotic cells is the nucleus, which contains most of a cell's DNA (Figs 3.3–3.5). The nucleus is surrounded by a double membrane, the **nuclear envelope**, and usually contains one or several nucleoli, which are darkly staining regions that contain high concentrations of RNA and protein as well as DNA.

The two membranes of the nuclear envelope are separated by about 50 nm (Fig. 3.9). They are perforated at intervals by **nuclear pores**, which are channels that allow the movement of certain molecules between the cytoplasm and nucleoplasm. The outer membrane of the nuclear envelope is continuous with the endoplasmic reticulum, a system of membranes that branches throughout the cytoplasm. The inner membrane is continuous with the outer membrane at the nuclear pores. However, the inner and outer membranes have distinct chemical compositions.

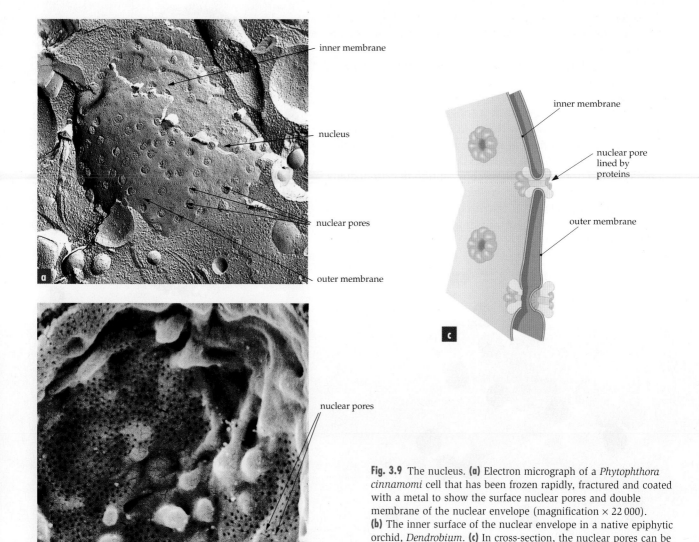

Fig. 3.9 The nucleus. **(a)** Electron micrograph of a *Phytophthora cinnamomi* cell that has been frozen rapidly, fractured and coated with a metal to show the surface nuclear pores and double membrane of the nuclear envelope (magnification × 22 000). **(b)** The inner surface of the nuclear envelope in a native epiphytic orchid, *Dendrobium*. **(c)** In cross-section, the nuclear pores can be seen to extend through the two membrane layers. Each nuclear pore is formed by a complex of many different proteins

The nuclear lamina is a scaffolding meshwork of lamins, a class of proteins found only on the inner surface of the nuclear envelope, giving it shape. Lamins form specific anchor sites that bind to DNA molecules and may prevent them becoming entangled.

Nuclear pores are not simple holes perforating the double membrane. Each nuclear pore is a precisely organised complex containing over one hundred different proteins. Some of these proteins form three octagonal rings, which delimit the size and shape of the pore. The diameter of the channel inside the rings is 65–75 nm. Molecules of up to 9 nm in diameter pass freely through the pores but, in general, the entry of larger molecules is prevented. However, some proteins possess a signal sequence of amino acids. When this signal interacts with a receptor on or near the pore, the pore enlarges, allowing molecules much larger than 9 nm to be actively transported into the nucleus. Similarly, large molecules inside the nucleus, such as messenger RNA (mRNA) molecules or ribosomal subunits (which may be 15 nm in diameter), can interact with the inner face of the pore to be actively exported to the cytoplasm.

> The nucleus contains DNA and is surrounded by a nuclear envelope, which is perforated by nuclear pores. Nuclear pores regulate the passage of proteins and RNA into and out of the nucleus.

Chromosomes

In the nucleus the long DNA molecules wind around clusters of **histone** molecules to form **nucleosomes**, resembling beads on a string (Fig. 3.10). This arrangement allows DNA to twist into a helix, forming a chromatin strand.

When a cell is not actively dividing, some chromatin strands aggregate, forming densely staining regions of **heterochromatin**, while others disperse, forming lightly staining regions of **euchromatin**. Euchromatic regions are sites of active gene transcription, where molecules of mRNA are manufactured. When a cell divides, the chromatin strands bunch up and thicken to form **chromosomes**, which stain densely and are large enough to be visible by light microscope (Fig. 3.10).

Nucleolus

The **nucleolus** is a suborganelle of the nucleus and is a darkly staining spherical region easily recognised under the light microscope (Figs 3.3, 3.5). It is the site of synthesis and maturation of ribosomal RNA (rRNA) and assembly of ribosomal subunits for export to the cytoplasm for protein synthesis. Cells typically make about 10 000 ribosomes per minute and, to do this, the nucleolus contains hundreds of copies of rRNA genes. The size of the nucleolus depends on the level of protein synthesis in the cell; it is small in quiescent cells

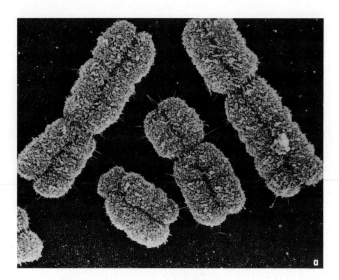

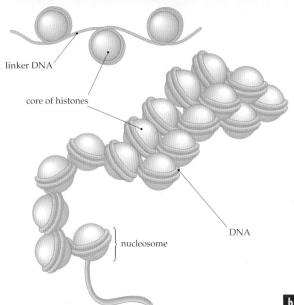

linker DNA

core of histones

DNA

nucleosome

Fig. 3.10 DNA and chromosomes of eukaryotic cells. **(a)** Scanning electron micrograph of human chromosomes. **(b)** Histones are associated with DNA molecules, forming nucleosomes like beads on a string. The DNA molecule winds tightly around histones. Histones are proteins rich in the basic amino acids, arginine and lysine. This allows them to interact with the acidic phosphate groups in DNA

and large in cells that are synthesising large amounts of proteins. During cell division, the nucleolus disappears as the chromosomes condense.

Ribosomes

The cytoplasm of eukaryotic cells contains several million **ribosomes**, small granular structures that are each 25–30 nm in diameter (Fig. 3.11). Ribosomes are composed of two subunits, each of which is assembled in the nucleolus from RNA and protein molecules. The subunits move through the nuclear pores into the cytosol, where they associate with an mRNA molecule,

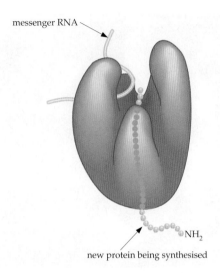

messenger RNA

NH₂

new protein being synthesised

Fig. 3.11 A ribosome. Ribosomes bind with mRNA and are the sites of protein synthesis

forming a functional ribosome that facilitates protein synthesis (Chapter 11).

As a ribosome moves along an mRNA molecule synthesising a polypeptide chain, sufficient room (about 80 nucleotides) becomes available for another pair of subunits to attach to the mRNA molecule, forming a second ribosome. As ribosomes continue to move along the mRNA strand, more ribosomes become bound, forming a polyribosome or **polysome**. A polysome may remain free in the cytosol or it may become attached to the surface of the endoplasmic reticulum. When cytosolic proteins are made, the polysome remains free in the cytosol.

Ribosome subunits are assembled in the nucleolus and move into the cytosol, where they associate with an mRNA molecule, forming a functional ribosome.

Endoplasmic reticulum

Endoplasmic reticulum (ER) is a network of membranous sacs (cisternae), which extends throughout the cytoplasm of eukaryotic cells (Fig. 3.12). It is continuous with the outer membrane of the nuclear envelope. Cisternae are usually flat and sheet-like but are often linked by tubular cisternae. The whole array forms an interconnected system composed of a continuous membrane enclosing a cisternal space. Although the cisternal space occupies only a small fraction of the cell volume, the surface-to-volume ratio of this compartment is so great that ER often constitutes over half of the total membrane in the cell. In plant cells, ER not only pervades the cell but also forms direct connections between adjacent cells as it passes through the centre of pore-like structures, plasmodesmata (Chapter 6), between cells.

Most ER in a cell is **rough endoplasmic reticulum**, which is ER with ribosomes bound to its surface giving it a 'rough' appearance (Fig. 3.12). These ribosomes are involved in the synthesis of proteins that will be incorporated into membranes, secreted at the cell surface or transported into vacuoles or lysosomes. Newly synthesised polypeptides pass into the lumen of the rough ER. Once inside the ER, protein folding is assisted by a chaperone protein known as binding protein (BiP). Disulfide bonds between cysteine residues exposed on the protein surface are formed through the action of protein disulfide isomerase and stabilise the three-dimensional structure of the protein. Disulfide bonds are unable to form in the cytosol because of the reducing environment there. In addition, carbohydrate chains may be added to those proteins whose sequence contains accessible asparagine residues. When a carbohydrate molecule attaches to the amino group on the side chain of asparagine, an N-linked glycoprotein forms (Fig. 3.12).

Smooth endoplasmic reticulum lacks attached ribosomes (Fig. 3.12). Although both rough and smooth ER can synthesise lipids, in cells that specialise in producing large amounts of particular lipids, such as steroid hormones produced in testicles, smooth ER proliferates and makes the majority of these compounds. Endoplasmic reticulum manufactures nearly all the phospholipids and cholesterol needed by the cell for membrane repair and synthesis. Fatty acid precursors occur in the cytosol and the ER enzymes involved in lipid synthesis have their active site facing the cytosol. Smooth ER also contains enzymes responsible for detoxifying lipid-soluble drugs and harmful metabolic products. Cytochrome P450 enzymes in the smooth ER, for example, convert water-insoluble compounds, which would be toxic if they accumulated in the cell, to water-soluble compounds, which can move out of the cell and are excreted in the urine.

Rough and smooth ER thus perform different functions even though their membranes are continuous. Different biosynthetic activities require different assemblages of membrane proteins. This means that the ER must be able to maintain regions that have different molecular compositions.

Endoplasmic reticulum is a membrane system, composed of cisternae, with a large surface area. It is involved in protein and lipid synthesis, and detoxification of harmful compounds.

The Golgi apparatus

The **Golgi apparatus** consists of stacks of four to 10 disc-shaped cisternae, which are 1–2 μm in diameter and positioned close together (Fig. 3.13). The cisternae may be flat or curved and are dilated at their margins.

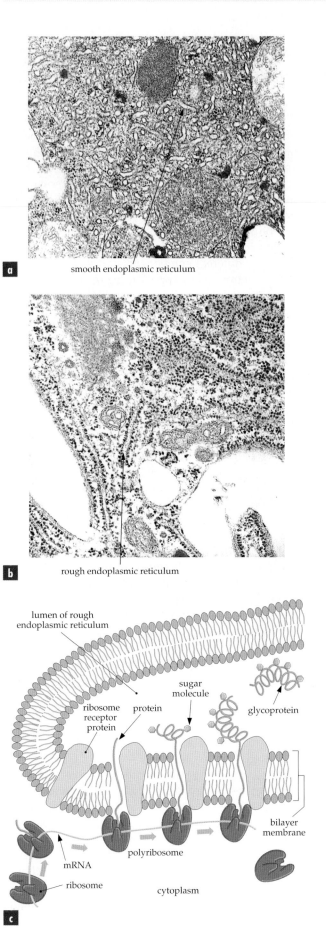

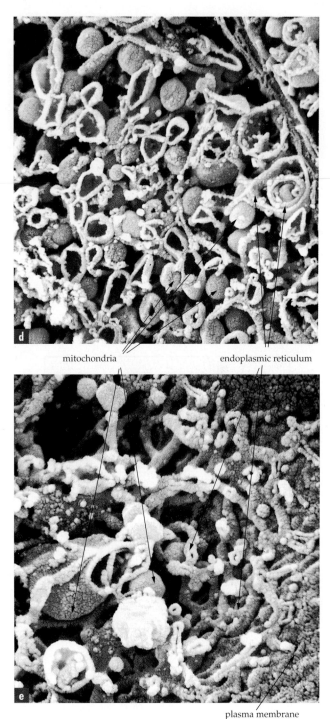

smooth endoplasmic reticulum

rough endoplasmic reticulum

lumen of rough endoplasmic reticulum

ribosome receptor protein

protein

sugar molecule

glycoprotein

bilayer membrane

polyribosome

mRNA

ribosome

cytoplasm

mitochondria

endoplasmic reticulum

plasma membrane

Fig. 3.12 Endoplasmic reticulum (ER). Transmission electron micrographs of **(a)** a hair cell of the alga *Bulbochaete* showing smooth ER (magnification × 41 000) and **(b)** a gland cell of *Primula kewensis* showing rough ER (with ribosomes; magnification × 25 000). **(c)** Proteins assembled on ribosomes of rough ER pass immediately into the lumen of the ER. Sugar chains may be added to the proteins within the ER to make glycoproteins. Scanning electron micrographs of developing pollen of the native epiphytic orchid, *Dendrobium*, show that **(d)** rough ER forms flattened sacs (cisternae), which occur in the cytoplasm among mitochondria, while **(e)** smooth ER is more tubular and is here seen in association with the plasma membrane

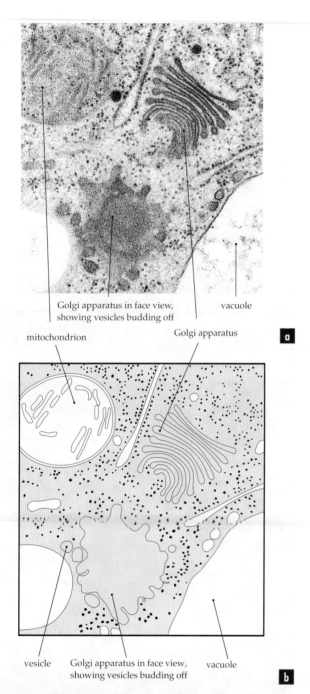

Golgi apparatus in face view,
showing vesicles budding off   vacuole

mitochondrion   Golgi apparatus   **a**

vesicle   Golgi apparatus in face view,   vacuole
showing vesicles budding off   **b**

Fig. 3.13 Golgi apparatus. **(a)** Electron micrograph of the Golgi apparatus in a root cell of *Eucalyptus sieberi*. The Golgi apparatus at the top of the picture is seen in profile and vesicles are budding off the ends of cisternae, while the Golgi apparatus at the bottom of the picture is seen in surface view and shows cisternae and vesicles at right angles to those in the upper structure (magnification × 43 000). **(b)** Diagram showing cell structures in the micrograph

Some algal cells have just one stack, while animal cells typically contain 10 stacks and plant cells may possess hundreds or even thousands of stacks. Each Golgi stack is usually surrounded by a cloud of small vesicles.

Each Golgi stack has a distinct polarity. One side of the stack faces a cisterna of ER. This is the *cis* face

BOX 3.2 Golgi structure

New research techniques drive scientific progress by providing the means for advancing the frontiers of human knowledge. In the field of cell biology, most of our current understanding of cellular architecture was produced by electron microscopists in the 1960s and 1970s, a period known as the golden years of electron microscopy. The rapid progress in ultrastructural research during that era can be traced to the introduction of better microscopes and better chemical fixatives for preserving the many different components of cells for viewing in the electron microscope. During the past decade, however, it has become evident that chemical fixatives do not always preserve membrane systems in a natural state, and that it is difficult to deduce high resolution three-dimensional information about the spatial relationship between different cellular structures based on two-dimensional electron microscopic images. This has stimulated researchers to look for new ways to preserve, visualise and display the three-dimensional architecture of cells.

The pictures of the Golgi apparatus of a kidney cell (see figures) demonstrate the type of three-dimensional information that can now be produced by electron microscopists using fixation by fast (millisecond) freezing to −100°C, substitution of the frozen water in the presence of a fixative, and tomography-based reconstruction of structures from tilt series of micrographs. The three-dimensional resolution of the Golgi model is about 6 nm, which represents a better than 10-fold improvement over what could be achieved using conventional methods.

What new information about the mammalian Golgi apparatus can be gleaned from this model? The reconstructed Golgi region contains two stacks of flattened cisternae that display a typical *cis*-to-*trans* polarity in their structural organisation (green through red cisternae). The laterally connected stacks are also surrounded by many free transport vesicles (light blue, green and golden), and the individual cisternae exhibit budding vesicle profiles (light blue and golden) as depicted in textbook models. Similarly, a *cis*-side ER (silver blue), which sends out newly produced secretory proteins in vesicles to the Golgi, is a well-known feature. In contrast, the extensive *trans*-side ER membrane system (silver blue) that is pressed against the two *trans*-most cisternae (brown and red) is a component of the Golgi apparatus whose function has yet to be elucidated.

An unexpected feature shown by this technique is the presence of only one type of budding vesicle type (golden buds) on the *trans*-most red cisterna, the *trans*-Golgi network (TGN) cisterna. According to current theories, the TGN is the centre where products destined for the cell surface and for the lysosomes or vacuoles are sorted and packaged. If this were true, then we should see two types of budding vesicles being formed along the edges of the red cisterna. This is not the case, which suggests that the sorting of the two types of products has to occur earlier, but exactly where this takes place is unknown.

The model also shows 394 colour-coded vesicles, as well as 76 blue- and 10 golden-coloured budding vesicle profiles. This new type of quantitative information is also important in that it allows researchers to test predictions about trafficking through the Golgi apparatus and to correlate these structures with biochemical measurements of function. Taken together, these and many other new findings highlight how

important it is to have precise architectural information about cellular compartments if we want to understand how the multitude of life-sustaining functional activities of cells are organised.

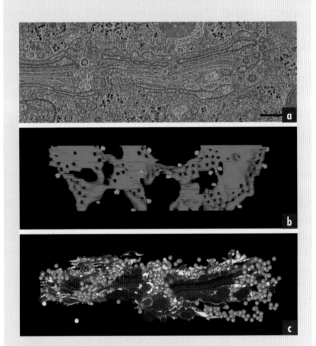

(forming or entry face) of the stack. The membrane of the ER that faces the *cis* face lacks ribosomes, and the cytosol between the ER cisterna and the *cis* cisterna of a Golgi stack is filled with small vesicles that are about 50 nm in diameter. The opposite side of the Golgi stack is the *trans* face (maturing or exit face). It is associated with a system of tubular membranes, the *trans*-Golgi network.

Proteins and glycoproteins enter each Golgi stack at the *cis* face. They are transferred from the ER in small transport vesicles that bud off the ER membrane and fuse with the membrane of the *cis* cisterna. As they progress through the stack, proteins and glycoproteins are modified in many ways. Sugar units of N-linked glycoproteins may be trimmed and other sugars added to the core of sugars that is left. O-linked glycoproteins are produced in the Golgi apparatus from proteins containing serine, threonine or hydroxyproline residues. Sugars are added to hydroxyl groups of side chains (i.e. O-linked) on these amino acids. The resulting molecules often have many sugar molecules added to them, thus forming proteoglycans, which form surface slimes and mucus. Polysaccharides are also formed in the Golgi apparatus. Polysaccharides and glycoproteins mature as they progress across the stack. Once mature products reach the *trans* face of the stack, they must be sorted and targeted to their correct destinations. Much of this sorting and packaging occurs in the *trans*-Golgi network.

The Golgi apparatus is involved in the processing and packaging of glycoproteins and polysaccharides.

Intracellular sorting and transport

The Golgi apparatus processes a large number of different products, which must be packaged and transported to different locations in the cell. Some molecules may be enzymes required for degradative processes within the cell. Some may be glycoproteins or polysaccharides for export across the plasma membrane.

Vesicle formation is the means by which these materials are sorted and distributed throughout the cell. Sorting of proteins within the cisternae or *trans*-Golgi network lumen is achieved by their association with cargo receptors on the inside face of the membrane. On the outside of the vesicle membrane, vesicle marker (v-SNARE) proteins function as

(a) Tomographic image of an NRK cell Golgi apparatus, in which the individual membrane compartments (cisternae) have been traced in different colours. (b) Reconstruction of the green cisternae seen in the top picture based on tracings from 387 serial tomographic images. Note the three-dimensional morphology of the cisternae and the presence of blue-coloured buds along the cisternal rims. These buds give rise to vesicles that recycle membrane-bound Golgi enzymes. (c) Complete model of the reconstructed Golgi stacks and of the surrounding 394 free transport vesicles

markers for different vesicle types. The formation of transport vesicles is aided by the binding of special scaffolding proteins to patches of membrane, which then round up and pinch off. The bound protein, clathrin, gives the vesicles a coated or spiny appearance. Once the coated vesicles are free in the cytoplasm, the coat of clathrin dissociates, exposing the v-SNARE marker proteins, which can then interact and fuse with target docking (t-SNARE) proteins on the target membrane (Fig. 3.14).

The sorting of molecules in the *trans*-Golgi network and the mechanisms of targeting vesicles in which molecules are packaged suggests that routing to the plasma membrane for export may not require any specific signalling mechanism, whereas routing to other destinations requires precise tagging. The best understood pathway is that to lysosomes.

Lysosomes

Lysosomes are membrane-bound organelles of animal cells involved in the breakdown of many types of molecules. They enable a cell to break down worn-out organelles, allowing molecules to be recycled to form new organelles. In this way, animal cells maintain their functions despite wear and tear. Lysosomes contain about 40 different hydrolytic enzymes, including nucleases, proteases, lipases and glycosidases. These enzymes are *N*-linked glycoproteins manufactured in

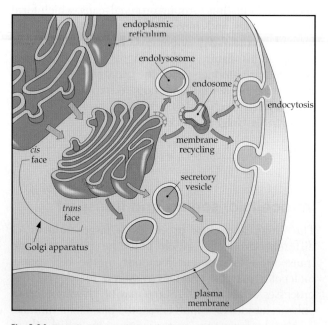

Fig. 3.14 Two-dimensional model showing the Golgi apparatus in an animal cell, and movement of vesicles (protein transport) from the endoplasmic reticulum, through the Golgi apparatus to the cell surface. The model shows unselected (constitutive) transport occurring by means of uncoated vesicles (blue arrows). The various kinds of signal-mediated transport are carried out by clathrin-coated vesicles (red arrows) budding from the *trans* face of the Golgi apparatus

the Golgi apparatus. The enzymes are tagged with mannose-6-phosphate residues attached to their oligosaccharide chains. This tagging, which happens in the *cis* cisternae of the Golgi apparatus, enables the glycoprotein enzymes to be sorted into specific vesicles in the *trans*-Golgi network (Fig. 3.14). Markers on the surface of these vesicles are recognised by receptors on the surface of other organelles, **endolysosomes**. The vesicles fuse with the endolysosome membrane and release their contents into the lumen of the endolysosome. This is accompanied by active uptake of hydrogen ions, which decreases the pH of the endolysosome, converting it to a mature lysosome. Only at a low pH do lysosome enzymes work at full capacity.

Exocytosis and endocytosis

Secretion (export) involves the movement of small transport vesicles in which material is packaged; the vesicles move to and fuse with the plasma membrane, releasing their contents to the outside of the cell. This process is **exocytosis**. Exocytosis of some components continues throughout the life of the cell and is referred to as **constitutive secretion**. Other molecules are stored in secretory vesicles and are exported only when a specific signal triggers the fusion of the vesicle with the plasma membrane, which is the process of **regulated secretion**. In plant cells, glycoproteins and proteins are also transported to vacuoles, which are large compartments surrounded by a single membrane. These molecules are stored in vacuoles in a highly concentrated form for later use as a food reserve.

During exocytosis, the surface area of the plasma membrane increases continually as newly arriving vesicle membranes are incorporated. To counter this increase in surface area, membrane is retrieved from the plasma membrane through the process of **endocytosis**, in which clathrin molecules associate with small areas of the plasma membrane, folding inwards so that a coated pit forms, which rounds up and pinches off, taking with it a portion of extracellular fluid (Fig. 3.14). In endocytosis, particular molecules that bind to specific receptors in the plasma membrane can also be transported. Once inside a cell, coated vesicles quickly lose their coat and fuse with an intermediate compartment, the **endosome**. Many molecules, including intact receptors, are retrieved from the endosome membrane and recycled. Others pass to endolysosomes and then to lysosomes, where they are degraded.

Small coated vesicles transport molecules and membranes between the ER, Golgi apparatus, cell surface, lysosomes, endosomes and vacuoles. Membrane is retrieved from the plasma membrane and specific molecules are taken into the cell by endocytosis.

Mitochondria

Mitochondria are very different from the organelles that we have looked at so far. They have their own DNA, make some of their own proteins and are able to grow and divide, although not without help from other components in the cell. Mitochondria are thought to have evolved from bacteria (prokaryotes) that were engulfed by ancestral eukaryotic cells (see Chapter 35).

Nearly all living eukaryotic cells contain mitochondria and it is within mitochondria that cellular respiration, which is the release of energy during the oxidation of sugars and fats, takes place. This process, also known as oxidative phosphorylation, is examined in detail in Chapter 5. The released energy is stored in molecules of ATP, which are used by cells to drive a vast range of chemical reactions. The need for a continuous supply of ATP by metabolically active cells, such as those of heart muscle, is reflected in the fact that these hard-working cells are richly endowed with well-developed mitochondria. Less-active cells have fewer and less elaborate mitochondria (Fig. 3.15).

Mitochondria are large enough to be seen using a light microscope and can be observed in living cells. Their size, shape and number varies widely, not only in different cell types, but also within a single cell. They also change during the cell cycle and in response to changing metabolic or environmental conditions. Mitochondria may be spherical or elongated and are able to fuse and fragment. Liver cells each contain 1000–2000 roughly ovoid mitochondria, whereas some small algal cells contain only one mitochondrion, which forms a branched reticulum throughout the cell. In some cells, mitochondria are anchored close to sites that require specially large amounts of ATP. In other cells, they are transported along cytoskeletal elements (p. 88) or as part of cytoplasmic flow.

Mitochondria are surrounded by a double membrane. The outer membrane is highly permeable, allowing ions and small molecules with a molecular weight of up to 10 kD to pass freely through abundant protein channels. The structure and chemical composition of the inner membrane contrasts sharply with that of the outer membrane. Whereas the outer membrane is smooth, the inner membrane is thrown into folds, **cristae**, which greatly increase the surface area of the membrane. The inner membrane is also highly impermeable, especially to ions. This property allows the generation of an electrochemical gradient due to unequal distribution of ions and uncharged molecules across the membrane. This gradient is a source of potential energy available for use by mitochondria (Chapters 2 and 5). Rapid import and export of molecules required for or produced by oxidative reactions is maintained by specific transport proteins within the inner membrane.

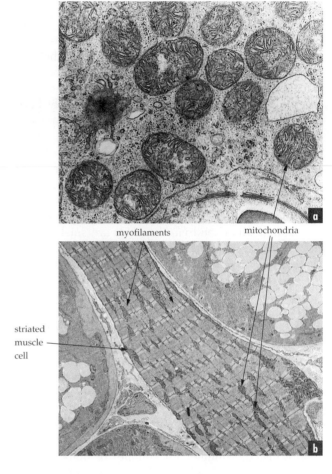

myofilaments   mitochondria

striated muscle cell

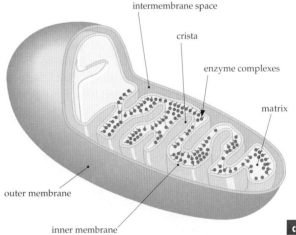

intermembrane space

crista

enzyme complexes

matrix

outer membrane

inner membrane

Fig. 3.15 Transmission electron micrographs showing mitochondria in **(a)** a plant cell, the broad bean, *Vicia faba* (magnification × 14 000) and **(b)** skeletal muscle fibre showing many mitochondria regularly arranged throughout the myofilaments. **(c)** Diagram of a mitochondrion showing double membrane, cristae and enzyme complexes (knob-like structures), which are responsible for ATP synthesis. The outer membrane is separated from the inner membrane by the intermembrane space

The inner surface of the inner mitochondrial membrane is lined with numerous knob-like structures. These are the enzyme complexes responsible for ATP synthesis (Fig. 3.15c). These enzymes use the electro-chemical gradient established across the inner membrane to generate ATP from ADP and inorganic phosphate. The number of enzyme complexes that can be accommodated within a mitochondrion depends on the surface area of the inner membrane. Thus, cells that synthesise large amounts of ATP not only have many mitochondria but also have numerous, extensively folded cristae within them.

The core of the mitochondrion is the matrix space. It contains mitochondrial ribosomes, one or more copies of mitochondrial DNA, and enzymes and structural proteins that are needed for oxidative reactions and DNA replication, transcription and translation. Mitochondrial ribosomes are smaller than those in the cytosol and have many properties similar to those of prokaryotic ribosomes. Mitochondrial DNA is a circular molecule, having features in common with prokaryotic DNA. Mitochondrial DNA codes for rRNA, transfer RNA (tRNA) and mRNA. Messenger RNA is translated on mitochondrial ribosomes in the matrix.

Although mitochondria possess their own DNA, their growth and function is under cellular control. During evolution, most of the ancestral bacterial genes became relocated to the cell nucleus. As a consequence, mitochondria manufacture few of their own proteins, importing 90% from the cytosol. While a cell is growing, the number (or volume) of mitochondria doubles in readiness for their partitioning to the daughter cells during cell division. In cells with only one mitochondrion, this divides before the cell itself divides. Mitochondria divide by simply pinching in two (fission) but they need cellular proteins to do so. Isolated mitochondria cannot divide but they can synthesise DNA, RNA and proteins for a brief period.

Mitochondria carry out cellular respiration and generate ATP for use throughout the cell. They are surrounded by two membranes, the inner one being highly impermeable and folded into cristae.

Plastids

Plant and algal cells contain **plastids**, which are a family of organelles, including chloroplasts, that, like mitochondria, have evolved from bacteria, which were engulfed by ancestral eukaryotic cells (Figs 3.16, 3.17). These bacteria were able to use energy from sunlight to incorporate carbon from CO_2 into organic molecules—the process of photosynthesis (Chapter 5). In this section, we look at the structure of a chloroplast and other plastids found in photosynthetic eukaryotes.

Plastids have several features in common with mitochondria. They are surrounded by a double membrane envelope and contain DNA, which is present as multiple copies of circular molecules, RNA and small ribosomes. The outer membrane is highly permeable to ions and small molecules. The inner membrane is more selective, containing carrier proteins responsible for transporting molecules into and out of a plastid. The matrix enclosed within the inner membrane of a plastid is the stroma. Within the stroma lies a third membrane system. This internal membrane system possesses structural and biochemical characteristics that distinguish plastids from mitochondria and also different types of plastids from one another.

Many of the proteins that are needed in plastids are encoded by nuclear genes and thus have to be imported from the cytosol into the stroma through contact sites at which the outer and inner membranes come close together, such as occurs in mitochondria. In addition to providing a plant or algal cell with energy and sugars,

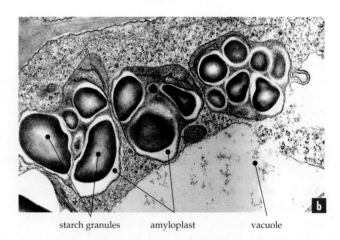

starch granules   amyloplast   vacuole

Fig. 3.16 Plastids are a family of organelles that occur in algae and plants, characterised by their own DNA and bounded by a double membrane. **(a)** Chloroplasts, the roughly spherical green structures, occur in mesophyll and guard cells but not in epidermal cells of a leaf of *Alocasia*. **(b)** Electron micrograph of amyloplasts in the soybean root cap cell. Starch reserves in amyloplasts are present as starch granules (magnification × 18 000)

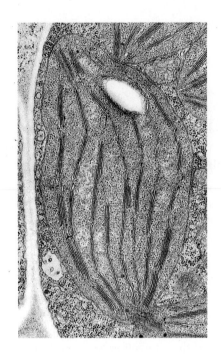

a

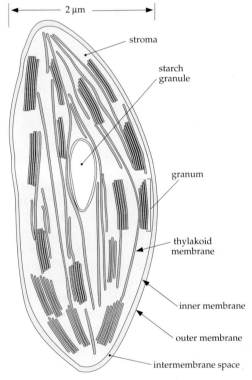

← 2 µm →

stroma

starch granule

granum

thylakoid membrane

inner membrane

outer membrane

intermembrane space

b

Fig. 3.17 (a) Transmission electron micrograph showing a chloroplast in a bean leaf cell (magnification × 18 200). **(b)** Diagram of a chloroplast

plastids also synthesise amino acids and most of a cell's fatty acids.

Proplastids are precursors of all types of plastids and occur in actively dividing cells. They are colourless, contain small starch granules and have only a rudimentary internal membrane system. There are usually 10–20 proplastids per cell. They divide by fission and their rate of division keeps pace with that of the cell. Differentiation of proplastids is under nuclear control and depends on cell type and environmental conditions.

Chloroplasts

Chloroplasts develop from proplastids in plant and algal cells grown in light (Fig. 3.17). Chloroplasts contain chlorophyll, a light-absorbing pigment used in photosynthesis. Cells containing chlorophyll generally appear green, although additional pigments in some groups of algae make them various shades of brown or red.

Chloroplasts have a highly developed internal membrane system. Some parts of the membrane form flattened disc-like sacs, **thylakoids**, which lie on top of each other forming stacks, **grana**. The thylakoids within a stack are joined to each other and to those in adjacent stacks by flattened tubular membranes. The whole system forms a continuous membrane network surrounding a single internal space within a thylakoid. This arrangement of internal membranes means that although chloroplasts have a relatively small volume, they contain a large area of thylakoid membrane. As chlorophyll molecules are assembled in arrays on the thylakoid membranes, the vast membrane area greatly increases a leaf's ability to capture light (Chapter 5).

Light energy trapped by chlorophyll molecules is used to establish an electrochemical gradient across the thylakoid membrane, between the thylakoid space and the stroma. As in mitochondria, this gradient is used to generate ATP. In chloroplasts, knob-like enzyme complexes involved in ATP synthesis project from the thylakoid membrane into the stroma, which is the region equivalent to the matrix space in mitochondria.

Other plastids

In underground storage organs of plants, such as roots and tubers, proplastids develop into **amyloplasts** (Fig. 3.16b). Amyloplasts lack coloured pigments, contain large reserves of starch and have very few, if any, membranes within the stroma. Proplastids may also develop into non-photosynthetic **chromoplasts**, which contain carotenoid pigments and are responsible for many of the yellow, red and orange hues of fruits, flowers and leaves.

Interconversions can occur between most plastid types. For example, amyloplasts can differentiate into **elaioplasts** (oleoplasts), which contain reserves of oil droplets in many seeds and pollen grains. In shoots grown in the dark, proplastids develop into **etioplasts**, which on exposure to light develop into normal chloroplasts.

Chloroplasts, the most important type of plastid, are responsible for the capture of energy from light during photosynthesis. Photosynthetic pigments and enzymes are assembled on internal thylakoid membranes.

Microbodies

Several organelles are involved in the breakdown of unwanted molecules. Lysosomes hydrolyse material originating both inside and outside a cell. Smooth ER detoxifies drugs and some metabolites. However, **microbodies** are the main organelle involved in the removal of compounds generated within cells.

Microbodies occur in all eukaryotic cells: a liver cell has about 1000. They are spherical to sausage-shaped with homogeneous, granular contents and sometimes crystalline inclusions. As with lysosomes, they are surrounded by a single membrane, contain a variety of degradative enzymes, and lack DNA, ribosomes and the internal membrane elaborations seen in chloroplasts and mitochondria. However, in contrast to the acidic interior of lysosomes, microbodies have a pH close to neutral and the types of enzymes within the two organelles also differ.

Lysosomes contain hydrolytic enzymes, whereas microbodies house oxidative enzymes that remove hydrogen from molecules to be degraded and couple it to oxygen. This reaction within microbodies generates hydrogen peroxide, H_2O_2, a very reactive molecule which, if not confined by a membrane, would cause considerable cell damage. Microbodies contain large amounts of an enzyme, catalase, which breaks down H_2O_2. Catalase constitutes about 40% of a microbody's protein and not only degrades excess H_2O_2 into water and oxygen but also harnesses the reactivity of H_2O_2, exploiting it to oxidise other molecules such as alcohols and phenols. Microbody membranes contain special transport channels for a large number of metabolites that are moved between microbodies and other organelles.

Microbodies can be either peroxisomes or glyoxysomes according to the enzymes they contain and the functions they perform. **Peroxisomes** oxidise amino acids and uric acid. In leaf cells, peroxisomes are involved in the oxidation of glycollate, a by-product of carbon fixation in the chloroplast (Chapter 5). **Glyoxysomes** occur in germinating seeds, whose energy reserves are stored in the form of lipids, and are involved in the conversion of fatty acids to sugars.

The exact complement of enzymes within a microbody depends on the cell type and cell activity. Growth of yeast cells on methanol induces an enlargement of peroxisomes and synthesis of enzymes involved in methanol metabolism. Transfer to a fatty acid source induces production of enzymes that break down fatty acids. As a young, growing plant uses up lipid stores in the seed, the types of enzymes in microbodies in the first leaf cells change and become more typical of leaf peroxisomes.

Peroxisomes and glyoxysomes are microbodies that contain different complements of enzymes involved in the oxidative degradation of molecules.

The cytoskeleton

So far we have looked at different organelles without particular reference to their location in the cell. Organelles are not scattered randomly throughout the cytoplasm; in general, there is considerable order in their arrangement. The nucleus may be suspended in the centre of the cell; Golgi stacks may cluster around the nucleus; and chloroplasts may be oriented with their grana facing towards a light source. Many organelles move within the cell in an orderly way.

Many aspects of cytoplasmic organisation and motility are a result of the interaction between organelles and three different elements of the **cytoskeleton**: microtubules (composed of tubulin), microfilaments (composed of actin) and intermediate filaments (several types, each composed of a distinct protein). Networks of these components interact to generate structure and order, to fix organelles in particular positions or to move individual organelles or the whole cytoplasm from one place to another. They also confer the ability to maintain and remodel cell shape.

Microfilaments

Microfilaments are fine fibres, 7–8 nm in diameter, composed of actin, a globular protein with a molecular weight of 42 kD (Fig. 3.18). Free actin molecules in solution are known as G-actin (globular actin). Each molecule of G-actin can interact with other actin molecules to form a chain or filament, F-actin. The two ends of a filament are different because each actin molecule in the filament is oriented in the same direction. Filaments thus have a distinct polarity (+ and −), which affects their biochemistry and function. For example, G-actin molecules add on to the positive end of the filament more quickly than they do to the negative end.

Microfilaments are highly dynamic structures. If there is a net addition of G-actin (polymerisation), microfilaments become longer; with a net loss of G-actin (depolymerisation), microfilaments become shorter. Cells control these two processes by regulating the number of actin molecules that are synthesised and hence the amount of G-actin available. In addition, filament length, stability and organisation are controlled by a wide range of *actin-binding proteins*.

Some actin-binding proteins allow net-like and bundled arrays of actin microfilaments to form,

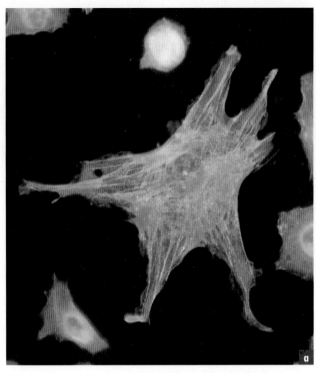

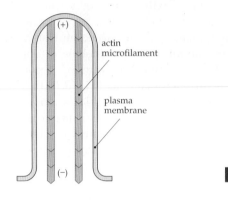

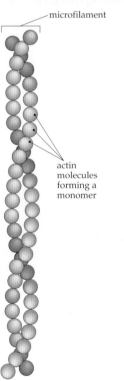

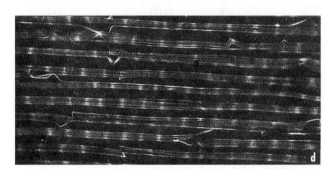

Fig. 3.18 Actin microfilaments of the cytoskeleton. **(a)** Fluorescence micrograph showing actin microfilaments in bundles (stress fibres) in cultured human fibroblasts, which form connective tissue. The green actin fibres provide a supporting network in the cell (magnification × 460). **(b)** A microfilament is composed of actin monomers. **(c)** Actin microfilaments are attached to the plasma membrane in intestinal microvilli, which are in constant movement. Attachment occurs at the (+) end of the microfilaments. **(d)** Fluorescence micrograph showing actin microfilaments formed into bundles in a large plant cell, *Chara corallina*. In giant cells, diffusion cannot ensure adequate mixing of metabolites. This is achieved by movement of the cytoplasm (cytoplasmic streaming) along actin cables, bundles of microfilaments all aligned with the same polarity. The streaming moves mitochondria, nuclei, internal membranes and cytosol, but chloroplasts are fixed in place beneath the plasma membrane by the microfilaments (magnification × 600)

while others break down existing arrays. The actin-binding protein, profilin, for example, binds to G-actin and makes the molecules unavailable for polymerisation. Another actin-binding protein, gelsolin, binds to and severs actin filaments, causing whole actin arrays to fragment. The association of other actin-binding proteins with F-actin helps to build up networks of microfilaments that hold small

organelles in place. For example, spectrin interacts with actin to form a network that is cross-linked to the plasma membrane, giving mechanical stability to the surface of many animal cells. Other actin-binding proteins link microfilaments into rigid bundles that support the fine projections of the plasma membrane (microvilli) in intestinal epithelial cells (Fig. 3.18c).

Interactions of actin microfilaments with myosin are the basis of many forms of cytoplasmic, organelle and cell movement. The sliding of actin and myosin filaments in muscle cells is the basis of muscle contraction (see Chapter 27). The association of myosin molecules with organelle membranes enables organelles to be moved along actin microfilaments. In large plant cells in which diffusion cannot ensure adequate mixing of metabolites, cytoplasm is moved at a rate of several micrometres per second along bundles of actin microfilaments (Fig. 3.18d). This cytoplasmic streaming may occur in thin strands of cytoplasm that traverse large vacuoles or it may occur throughout a thin layer of cytoplasm sandwiched between the plasma membrane and large central vacuoles.

> Microfilaments, which are part of the cytoskeleton, are composed of the protein actin. They contribute to the structural organisation of cells and facilitate intracellular movement and shape.

Microtubules

Microtubules are hollow cylinders, 25 nm in diameter. They are composed of **tubulin**, a globular protein that exists in two closely related forms, α-tubulin and β-tubulin. Both forms of tubulin have a molecular weight of 55 kD and in solution they associate into a dimer composed of one alpha and one beta subunit. Tubulin dimers bind to each other forming protofilaments, which are chains of alternating alpha and beta subunits (Fig. 3.19). Because the dimers all have the same orientation, microtubules are polar structures with the properties of one end being different from those of the other. For example, one end grows faster than the other.

Microtubules are repeatedly assembled and disassembled by a number of microtubule-associated proteins. Microtubule polymerisation usually begins at specific locations, *microtubule nucleating sites* (MTNSs). The slow-growing end of the microtubule is at the nucleating site. Dimer addition to the fast-growing end causes microtubules to grow out from the nucleating sites, forming radiating, bundled or parallel arrays. Individual microtubules in a cell may grow slowly for a while and then rapidly depolymerise; selected microtubules may become stabilised by special capping molecules.

Microtubules are more rigid than microfilaments and can provide greater mechanical support. Groups of microtubules support specialised appendages that project from a variety of cells. In nerve cells, for example, bundles of microtubules extend along the long thin axons. Within axons, numerous vesicles are transported rapidly along microtubules by the action of *microtubule-associated proteins* (MAPs), which are ATPases. Vesicles are moved towards the axon tip by

the action of the MAP, kinesin, and in the opposite direction by the MAP, dynein. MAPs also control the polymerisation of tubulin and stabilise microtubules by forming cross-bridges between adjacent microtubules.

Within a cell, the arrangement of microtubules depends not only on cell type but also on the stage of the cell cycle. In many animal cells that are not dividing, microtubules emerge from near the nucleus and radiate out through the cytoplasm towards the plasma membrane in an array that is constantly changing (Fig. 3.19). As cells begin to divide, the arrangement of their microtubule cytoskeleton alters dramatically. Microtubules form spindle fibres, which attach to chromosomes and draw them apart during cell division (Chapter 8).

> Microtubules are dynamic scaffolding elements in eukaryotic cells. Microtubule arrangement and stability are determined by microtubule-associated proteins, which initiate microtubule formation, cross-bridge microtubules into bundles and networks, and move organelles along microtubules.

Intermediate filaments

Intermediate filaments are 8–10 nm in diameter, making them intermediate in size between microfilaments and microtubules. The proteins of intermediate filaments vary from about 40 to 130 kD in molecular weight. They are fibrous rather than globular and more heterogeneous than actin or tubulin. Pairs of protein molecules (dimers) associate laterally in overlapping arrays to build up long filaments with a high tensile strength. Intermediate filaments are strong and resist stretching and, in general, do not undergo the rapid and repeated assembly and disassembly that occurs in microfilaments and microtubules. Arrays of intermediate filaments are stable and change little during cell growth.

Intermediate filaments provide mechanical support for the cell and the nucleus. For example, keratins are extremely stable and insoluble proteins in epithelial cells and form the basis of hair and nails (Fig. 3.20); desmin forms intermediate filaments in muscle cells; and neurofilaments occur in nerve cells.

Nuclear lamins are intermediate filaments that are cross-linked into a square lattice, the nuclear lamina. The nuclear lamina is disassembled and reassembled as the nuclear envelope breaks down and reforms during cell division.

> Intermediate filaments are strong, stable and resist stretching, and provide mechanical support.

Motility: cilia and flagella

In eukaryotes, cilia and flagella are fine, long cellular projections. They produce movement of a cell or of

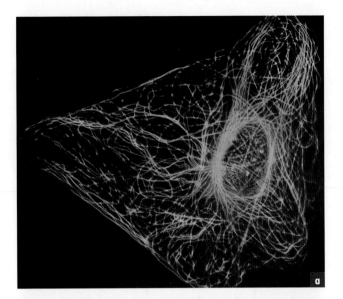

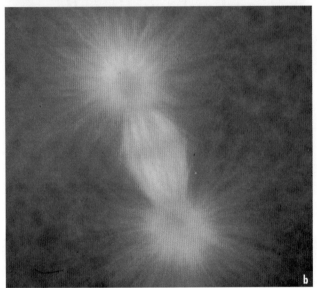

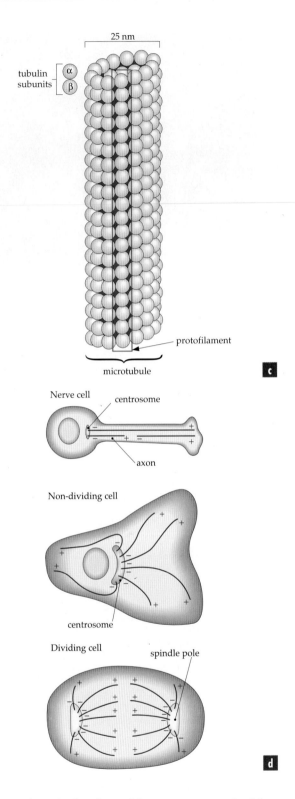

Fig. 3.19 Microtubules of the cytoskeleton. Microtubule arrangement during two stages of cell division in an animal cell: **(a)** in a cell before division (interphase) and **(b)** when the chromosomes are about to separate on the spindle, which comprises ropes of microtubules. **(c)** Structure of microtubules. Each microtubule, assembled from pairs of tubulin subunits (α- and β-tubulin), consists of 13 protofilaments aligned side by side. **(d)** Diagram showing microtubule polarity and different nucleating sites in animal cells. The negative ends of microtubules are usually embedded in the microtubule organising centres or nucleating sites and the positive ends are near the plasma membrane

liquid over a cell surface. Flagella are usually 20–100 μm in length and occur in small numbers (Fig. 3.21). Cilia have the same internal structure but are shorter (2–20 μm) and occur in much greater numbers, up to several thousands on the surface of some cells (Fig. 3.21).

Cilia and flagella are covered by plasma membrane and are supported by a precise array of microtubules, which forms the **axoneme** (Fig. 3.21). The axoneme consists of a cylinder formed by nine pairs or doublets of microtubules enclosing two central microtubules. Adjacent doublets are linked together by fine fibres and are connected to the inner sheath by radial spokes. On one side of each doublet are two short arms made up of the protein dynein.

Movement of cilia or flagella results from the sliding of adjacent microtubule doublets relative to each other.

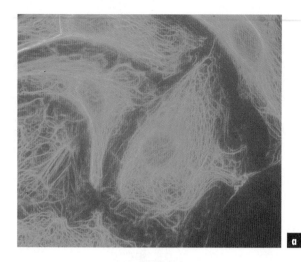

at the nucleation site and tubulin dimers add on to the distal end at the tip of the growing flagellum.

Basal bodies also occur as a pair of centrioles in animal cells. Centrioles lie at right angles to each other close to the nucleus. Before cell division, each centriole replicates, ensuring allocation of a pair of centrioles to each daughter cell (Chapter 8).

Cilia and flagella are composed of nine interconnected microtubule doublets. Cilia and flagella bending is a result of sliding of these adjacent doublets and is powered by dynein hydrolysis of ATP.

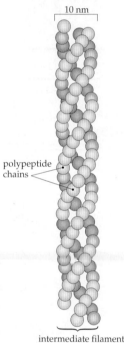

10 nm

polypeptide chains

intermediate filament

Fig. 3.20 Intermediate filaments of the cytoskeleton.
(a) Fluorescence micrograph showing intermediate filaments (vimentin) in kidney epithelial cells of the kangaroo rat. The filaments form three-dimensional networks involved in the movement of cell organelles. The dark spheres in the centre of the cells are the sites of the nucleus (magnification × 270).
(b) Structure of an intermediate filament

Dynein first attaches to the adjacent doublet, undergoes a conformational change and then dissociates. During this process, dynein hydrolyses ATP to provide energy. The conformational change in dynein causes sliding of adjacent doublets, resulting in a cilium or flagellum bending (Fig. 3.21).

Axoneme microtubules originate at a basal body, which is a highly structured microtubule nucleating site. The basal body is composed of a ring of nine triplet microtubules and acts as a template for the axoneme. The slow-growing end of the microtubules is attached

Comparison of prokaryotic and eukaryotic cells

The most ancient cells, the prokaryotes (bacteria) are smaller (about 1 μm in diameter) and have a simpler structure than eukaryotic cells. Bacteria have a semirigid cell wall that surrounds a plasma membrane. They have no membrane-bound organelles; that is, they lack a nucleus, endoplasmic reticulum, Golgi apparatus, vacuoles, mitochondria and chloroplasts. They also lack a cytoskeleton of microfilaments and microtubules.

Although some bacteria have flagella, which propel the cells, they have a completely different structure and mechanism of action (Fig. 3.22). Bacterial flagella are composed of flagellin (protein) fibrils that form a stiff coiled filament. The flagellum projects from a special structure in the bacterial membrane that serves as the motor for movement. Movement of the cell is generated by rotation of the curved filament, which thus acts like a propeller. To change the direction of movement, the direction of rotation is temporarily reversed.

The cytosol of bacteria contains a circular DNA molecule and a few hundred to a few thousand ribosomes, about 15 nm in diameter (compared with 25–30 nm in eukaryotes). Because DNA is not segregated into a nucleus, bacterial ribosomes can attach directly to mRNA molecules even while mRNA is being transcribed. In eukaryotes, mRNA molecules form in the nucleus and pass across the nuclear membrane into the cytoplasm before they interact with ribosomes during protein synthesis. In bacteria, the enzymes corresponding to those found in eukaryotic mitochondria are located in the plasma membrane, and thus a mitochondrion is similar to a whole bacterial cell. In cyanobacteria, internal membranes contain light-trapping pigments, and the eukaryotic chloroplast is similar to a cyanobacterial cell.

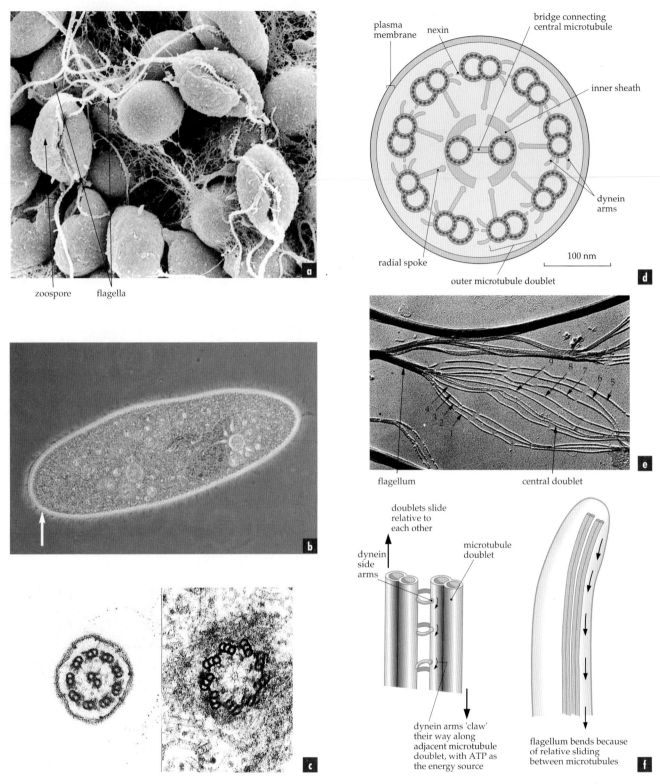

Fig. 3.21 Cilia and flagella. **(a)** Scanning electron micrograph of motile spores of the 'cinnamon fungus' *Phytophthora cinnamomi*, each with two flagella (magnification × 2400). **(b)** Living ciliated cell of *Paramecium multinucleatum*. The entire surface of the cell is covered by cilia (arrowed; magnification × 265). **(c)** Transmission electron micrographs of the single-celled green alga, *Chlamydomonas*, showing transverse section through a flagellum (left) and basal bodies (right). **(d)** Structure of axoneme of flagellum, composed of an outer circle of nine microtubule doublets (each with two dynein arms, a radial spoke, and nexin links), surrounding a central sheath containing a pair of microtubules. **(e)** An Australian first. Grigg and Hodges (1949) observed the nine doublets and central pair of the sperm flagellum three years before European scientists. This photograph shows a preparation of a fowl sperm flagellum. **(f)** Sliding of adjacent microtubule doublets causes bending of a cilium or flagellum, generating movement. This involves a type of 'climbing action' by the contractile dynein arms along the microtubules, using energy provided by the hydrolysis of ATP

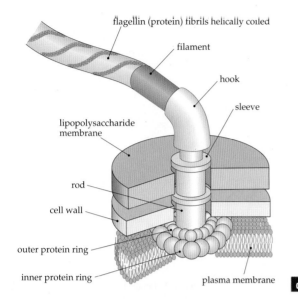

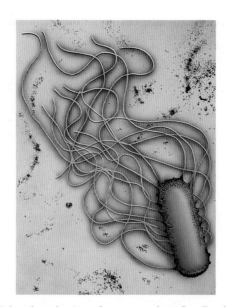

Fig. 3.22 **(a)** A flagellum in a bacterial cell has a different structure (made of flagellin) and mechanism of movement from flagella of eukaryotic cells. Movement of the bacterial cell is generated by rotation of the rigid spiral-shaped flagellum, which acts like a propeller. Changes in direction of movement are achieved by temporarily reversing the direction of rotation. Rotation is achieved by a motor-like structure embedded in the plasma membrane and wall of the bacterial cell. **(b)** Colour-enhanced scanning electron micrograph of *Salmonella* bacteria showing numerous flagellae

The structure and diversity of prokaryotes is considered in greater depth in Chapter 33, and theories of the origin of eukaryotes from prokaryote ancestors are discussed in Chapter 35.

Bacterial cells have a simple structure and lack membrane-bound organelles. Eukaryotic cells are more complex and often exist in organised assemblies that make up tissues.

Summary

- All living organisms are made up of cells and cell products. Cells are bounded by the plasma membrane. Eukaryotic cells contain intracellular membrane-bound compartments that segregate different molecules and metabolic reactions. Prokaryotic cells lack internal subdivisions.
- Membranes consist of two layers of phospholipid molecules and contain integral and peripheral proteins. Membranes form selectively permeable barriers, allowing only certain molecules to pass through them.
- In eukaryotic cells, the nucleus is a compartment surrounded by a double membrane and contains DNA in the form of chromosomes. Ribosomal subunits are assembled in the nucleolus and move into the cytosol through pores in the nuclear envelope.
- Proteins are synthesised on ribosomes associated with mRNA molecules. During protein synthesis, some ribosomes remain free in the cytoplasm, while others become associated with the surface of the endoplasmic reticulum.
- Endoplasmic reticulum is an extensive system of membranous cisternae in eukaryotic cells in which lipid synthesis, the initial processing of secretory and membrane proteins and glycoproteins, and detoxification of harmful compounds occur.
- The Golgi apparatus consists of stacks of cisternae in which polysaccharides are synthesised, and polysaccharides and glycoproteins are processed, sorted and packaged. Products leave the Golgi apparatus in coated vesicles. Products are secreted if the vesicles fuse with the plasma membrane or are stored within the cell if the vesicles are targeted to fuse with specific vacuoles. Membrane is retrieved and specific molecules are taken into cells by endocytosis.
- Mitochondria and plastids are surrounded by a double membrane. Although they contain small amounts of DNA, most of the proteins needed by these organelles are synthesised on cytosolic ribosomes. Mitochondria compartmentalise reactions that oxidise sugars and fats and synthesise ATP to be used as an energy source throughout the cell. Chloroplasts contain pigments that absorb energy from sunlight and enzymes that incorporate carbon from CO_2 into organic molecules.
- Microbodies contain oxidative enzymes involved in the digestion of unwanted molecules. Compartmentalisation of these enzymes and their reactive products protects other cell components from degradation.
- Three types of cytoskeletal elements—microfilaments, microtubules and intermediate filaments—form bundles and extensive networks throughout the cytoplasm of eukaryotic cells. They provide a structural framework that supports extended projections of the plasma membrane and they are responsible for transporting and positioning many subcellular components. Microtubules form the spindle, which separates chromosomes during cell division. Sliding of microtubule doublets within the axoneme of cilia and flagella results in bending and subsequent movement.

key terms

amyloplast (p. 87)
axoneme (p. 91)
chloroplast (p. 87)
chromoplast (p. 87)
chromosome (p. 79)
constitutive secretion
    (p. 84)
cristae (p. 85)
cytoplasm (p. 73)
cytoskeleton (p. 88)
cytosol (p. 73)
elaioplast (p. 87)
endocytosis (p. 84)
endolysosome (p. 84)
endoplasmic reticulum
    (p. 80)

endosome (p. 84)
etioplast (p. 87)
euchromatin (p. 79)
eukaryotic cell (p. 71)
exocytosis (p. 84)
fluid mosaic (p. 75)
glycocalyx (p. 75)
glyoxysome (p. 88)
Golgi apparatus
    (p. 80)
grana (p. 87)
heterochromatin
    (p. 79)
histone (p. 79)
intermediate filament
    (p. 90)

lipid bilayer (p. 74)
lysosome (p. 84)
microbody (p. 88)
microfilament (p. 88)
microtubule (p. 90)
mitochondrion (p. 85)
nuclear envelope
    (p. 78)
nuclear pore (p. 78)
nucleolus (p. 79)
nucleosome (p. 79)
nucleus (p. 71)
organelle (p. 71)
peroxisome (p. 88)
plasma membrane
    (p. 74)

plastid (p. 86)
polysome (p. 80)
prokaryotic cell
    (p. 71)
proplastid (p. 87)
protoplasm (p. 73)
regulated secretion
    (p. 84)
ribosome (p. 79)
rough endoplasmic
    reticulum (p. 80)
smooth endoplasmic
    reticulum (p. 80)
thylakoid (p. 87)
tubulin (p. 90)

Review questions

1. What is the chemical composition of a plasma membrane and why is it described as being a 'fluid mosaic'?

2. What are the distinguishing features of the nuclear membrane?

3. What is the functional significance of the large surface-area-to-volume ratio of the ER?

4. What are the basic differences between a prokaryotic and eukaryotic cell?

5. (a) What processes occur in the Golgi apparatus?

   (b) In which part of this organelle does each process occur?

6. Explain the importance of the limited permeability of the inner membranes of mitochondria and chloroplasts.

7. Contrast the different roles of lysosomes and microbodies within a cell.

8. (a) List the components of the cytoskeleton.

   (b) Describe the differences in their structure.

   (c) Explain the different functions they serve within the cell.

9. (a) What are MAPs and what do they do?

   (b) What are MTNSs and what do they do?

10. Contrast the structure and movement of prokaryotic and eukaryotic flagella.

11. Electron microscopes magnify the image of an object to a much higher level than light microscopes do. Is the higher magnification the most important difference between light and electron microscopes?

Extension questions

1. Describe the synthesis, transport and secretion of a typical glycoprotein from the process of gene transcription in the nucleus to that of exocytosis at the plasma membrane.

2. Discuss why the fluidity of membranes is important for eukaryotic cell function.

3. Discuss why the compartmentalisation of eukaryotic cells is so important.

4. Individual DNA molecules may be up to several centimetres in length, and many DNA molecules are packed into nuclei without becoming entangled or damaged. Describe the different levels of DNA organisation in the nucleus during the cell cycle and their role in DNA packaging, function and protection.

Suggested further reading

Alberts, B., Bray, J., Lewis, J., et al. (1998). *Essential Cell Biology*. New York: Garland Publishing, Inc.

This text gives a thorough account of our understanding of the molecular basis of cellular biology.

Gunning, B. E. S. and Steer, M. W. (1996). *Plant Cell Biology. Structure and Function*. Sudbury, MA: Jones and Bartlett Publishers.

This superbly illustrated text provides information on the structure and function of plant cells.

CHAPTER

4

Movement across membranes

All living cells exist in an aqueous medium. For a unicellular alga, this may be pond water; for a cell in a multicellular animal, the aqueous medium is the intercellular (extracellular) fluid. The contents of cells are physically separated from this aqueous environment by the plasma membrane (Figs 3.4, 3.5), which is composed of a phospholipid bilayer in which are embedded protein molecules, some of which span the bilayer. This membrane functions not as an impervious barrier but as a selective dynamic boundary to the cell. Without correct membrane functioning, cellular processes fail and the cells die.

Invariably, the composition of a living cell is different from that of its surrounding environment. The stability of the intracellular environment (homeostasis) must be maintained in the face of both physical and chemical fluctuations in the external environment. In addition, metabolic processes occurring within cells actively generate changes in the intracellular environment. Oxygen and fuel molecules are used and CO_2 is produced during cellular respiration, and consequently the pH is affected by the production of H^+ and HCO_3^-. Various solutes, both organic and inorganic, must enter cells so that they can be converted to other compounds. Waste products and compounds specially synthesised for export need to leave the cell.

Furthermore, in order for metabolism to proceed in a controlled and sustained manner, the concentrations of solutes within each cell must be kept within well-defined limits. Protein structure, and thus the activity of enzymes, for example, is affected by pH and by the concentration of ions such as Ca^{2+}, Na^+ and K^+. Therefore, these ions must be kept at relatively constant levels inside cells. Generally, the cytoplasm of all cells contains low levels of free Na^+, Cl^- and Ca^{2+} and high levels of K^+ at a pH of between 7 and 8 (Fig. 4.1). Ion gradients across the membrane are important sources of potential energy for many cellular activities, such as muscle contraction and conduction of nerve impulses. For each of these functions to occur, there must be selective movement of solutes across the plasma membrane.

Water is the most abundant component of cells and is the major determinant of cell volume. If too much water enters a cell, components dissolved in the cytosol may become too dilute for metabolism to occur efficiently. The reverse may also occur. Therefore, movement of water across the plasma membrane must also be controlled. Water moves into and out of cells passively along osmotic gradients (p. 108) and cells control osmotic gradients by regulating the internal quantity of low permeability solutes. In this way, they control the *net* movement of water across the plasma membrane.

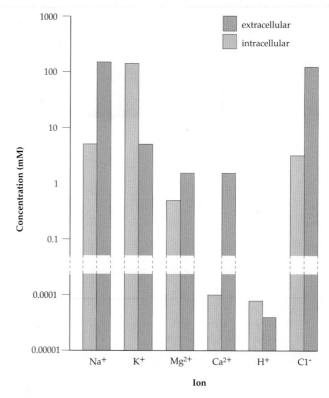

Fig. 4.1 Comparison of concentrations of free ions inside and outside a typical mammalian cell. Ions such as K^+ and Mg^+ have higher intracellular (cytosolic) concentrations than those externally, whereas the intracellular concentrations of Na^+ and Cl^- are lower than the extracellular concentrations

Within a cell, there are many different membrane-bound compartments (Figs. 3.4, 3.5), including mitochondria, endoplasmic reticulum and, especially in the case of plant cells, vacuoles. These compartments have different functions and, as a consequence, they contain different concentrations of solutes. Exchange of solutes between these compartments must therefore be controlled. For example, Ca^{2+} has a central role in controlling metabolism, and its movement between compartments is highly regulated.

Cells regulate the composition of the cytosol and of membrane-bound internal compartments within well-defined limits.

Cells are able to control the composition of the cytosol and of their internal compartments because of two features of biological membranes—**selective permeability** and the activities of **membrane-spanning transport proteins**. Selective permeability means that membranes function as a kind of molecular sieve. Because of the hydrophobic nature of the phospholipid bilayer, it is impermeable to most water-soluble molecules, in particular, ionised molecules. On the other hand, lipid-soluble substances pass freely through the bilayer. However, movement across membranes involves more than mere filtration; some water-soluble substances that pass from one side of a membrane to the other do so by direct passage through membrane-spanning proteins—channels or carriers. Lipid-insoluble particles may also cross the membrane by interacting with specific membrane-spanning proteins. Other substances are moved into or out of a cell by enclosing them in membrane-bound vesicles.

> Movement across membranes is controlled by the selective permeability of the phospholipid bilayer and the properties of membrane-spanning transport proteins.

Diffusion of solutes

As a result of their intrinsic kinetic energy, all molecules in a solution move about rapidly. At body temperature, water molecules travel at speeds of about 2500 km/h. Glucose, a larger molecule, travels more slowly at 850 km/h. Because there are so many molecules, all travelling at high speeds and in random directions, there are millions of collisions every second. This means that, despite the speed of individual molecules, the overall movement of molecules in any particular direction is quite slow.

If a drop of dye (the **solute**) is put in a glass of water (the **solvent**), the dye molecules will spread slowly through the water as a result of this random movement until the colour (the molecules) is evenly spread throughout. Thus, there has been a *net* movement of solute molecules from an area of high solute concentration (the drop of dye) to an area of low solute concentration (the rest of the water in the glass). This tendency for molecules to spread out evenly into the available space is called **diffusion** and leads to an increase in entropy (p. 53). The larger the difference in solute concentrations, that is, the **concentration gradient**, the more rapid the rate of diffusion. Note that increasing the temperature also increases the rate of diffusion because it increases the kinetic energy of all the particles in the solution. Note also that each molecule diffuses down its *own* concentration gradient, unaffected by the concentration gradients of other molecules.

Diffusion itself is a *passive* process; it takes place because there is a concentration difference and requires no further input of energy to occur. (The energy that drives diffusion comes from the energy that was used to set up the difference between the concentrations in the first place.)

> Diffusion is the passive net movement of molecules along their concentration gradient, from a region of high concentration to a region of low concentration. It requires no expenditure of energy.

Diffusion across the lipid bilayer

Diffusion can also occur across membranes. Because in a solution all molecules are moving about rapidly, if they can cross the membrane, there will also be movement of molecules back and forwards across the membrane. If the concentrations of a particular molecule are the same on either side of a membrane, there will always be about the same number moving across in either direction. That is, there will be *no net movement* from one side to the other and a dynamic equilibrium will exist. However, if the concentrations of a particular molecule are different on either side of the membrane, then more molecules will cross from the concentrated side to the dilute side than in the opposite direction. Thus, there will be a *net movement* of molecules from one side to the other. The *rate* at which diffusion across a membrane occurs is determined both by the concentration gradient and by the permeability of the membrane to a particular diffusing molecule (and to temperature as mentioned above).

Although diffusion across the lipid phase of membranes is not easily regulated, it is nevertheless the main means of exchange of certain solutes across membranes. For example, in all organisms, it is the way that O_2 and CO_2 are exchanged with the environment. In humans, diffusion is the means by which volatile anaesthetics are absorbed across the lungs and the way alcohol is rapidly absorbed into the bloodstream. Chemical pesticides, such as DDT, are highly soluble in lipids and are thus easily able to penetrate cells and accumulate in organisms.

> Any molecule that is able to pass through the lipid bilayer will cross passively by diffusion along its own concentration gradient.

Permeability coefficient

The **permeability coefficient** is a measure of the ease with which a molecule passes through the membrane. It depends upon both the properties of the membrane itself and of the diffusing molecule. For molecules that pass through the lipid bilayer, the most important factor affecting the permeability coefficient is the *lipid solubility* of the molecule. This determines how readily the molecule can dissolve in the lipid bilayer and, consequently, how rapidly it can diffuse across the membrane. The higher the lipid solubility, the greater the ability of a molecule to diffuse across membranes (Fig. 4.2). However, for some molecules other factors also influence permeability. In particular, very small polar molecules that are relatively insoluble in lipids, such as water and urea, appear to have a higher permeability coefficient than would be expected from their lipid solubility. Evidence suggests that these molecules simply pass through transient pores

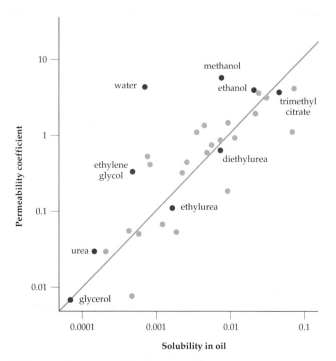

Fig. 4.2 The higher the lipid solubility of a molecule, the greater its ability to permeate a membrane by diffusion (permeability coefficient)

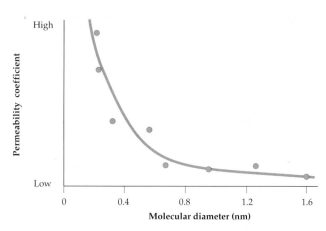

Fig. 4.3 This graph shows the relationship between molecular size and permeability coefficient for polar molecules. Small molecules are more permeable than large molecules. Molecules larger than 0.8 nm diameter do not readily cross the membrane

(discontinuities) in the lipid structure, or through membrane-spanning proteins that are embedded within the lipid bilayer (aquaporins, see p. 105). The increase in the permeability coefficient with decreasing molecular diameter, including the hydration sphere (p. 21), is shown in Figure 4.3.

The phospholipid bilayer of membranes is also highly permeable to a wide range of non-polar molecules, such as the dissolved gases O_2 and CO_2 (Fig. 4.4). The bilayer is not particularly permeable to larger molecules that have polar side groups, such as hydroxyl groups; examples of these molecules include

sugars, such as glucose and sucrose. As we have pointed out, the lipid bilayer itself is almost completely impermeable to charged compounds (ions) and effectively impermeable to macromolecules, such as proteins and nucleic acids.

> The rate of diffusion of a molecule across the lipid bilayer depends on its concentration gradient and the ease with which it can cross the bilayer (the permeability coefficient).

Diffusion through membrane-spanning proteins

Many molecules move across membranes through intrinsic proteins that span the lipid bilayer (Chapter 3) and facilitate the movement of particular molecules across the membrane. In contrast to diffusion across the lipid bilayer, rates of net movement of these

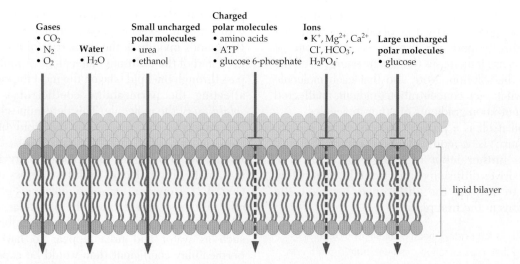

Fig. 4.4 Permeability of a lipid bilayer to different types of molecules. Small molecules, especially those that form few hydrogen bonds with water, diffuse more rapidly across a membrane than do larger and charged molecules

substances show little correlation with lipid solubility and do not increase linearly as the concentration of substrate increases (Fig. 4.5).

Channels and carriers

Channels and carriers (Fig. 4.6) are two different types of membrane-spanning transport proteins. **Channels** effectively form hydrophilic pores through the membrane and, when the channels are open, they allow very high rates of movement of certain solutes. Movement through channels is not directly coupled to an energy source, so movement through channels is *always* passive. Thus, an ion may diffuse across the membrane down its electrochemical gradient (see below) if an ion channel, specific for that particular ion, is present and 'open' in the membrane.

Molecules may also pass through the membrane if a specific **carrier** for that molecule is lodged in the membrane. Carrier-mediated transport involves the molecule interacting with the carrier protein. However, carriers tend to move solutes at a slower rate than channels and undergo a more radical conformational change during the passage of the solute. The energy for passage through a carrier may be provided by the concentration gradient (*facilitated diffusion*) or may require the expenditure of energy (*active transport*).

> Channels and carriers are two different types of transport proteins. Movement through channels is always passive, whereas carrier-mediated transport may be passive or active. Carriers undergo a more radical conformational change than channel proteins during the passage of a solute.

Ions diffuse along an electrochemical gradient

Many ions, unable to cross the lipid bilayer, are able to diffuse across the membrane provided an appropriate ion-selective channel is present and open. Similar considerations apply to the diffusion of charged solutes (ions) through these membrane-spanning proteins as apply to the diffusion of lipid-soluble solutes across the bilayer. However, because of their electrical charge, the energy driving diffusion of ions across a membrane depends on both the *concentration gradient* and the *electrical gradient* across the membrane.

Figure 4.7 illustrates the combined effect of concentration and **electrical gradients** on the movement of K^+ and Cl^- across a membrane. In this model, when electrochemical equilibrium is reached, a

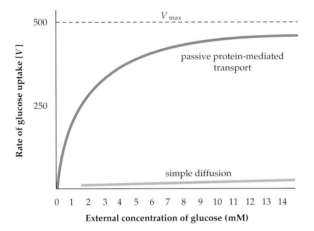

Fig. 4.5 The rate of transport of glucose (in micromolecules per millilitre of cells per hour) into red blood cells in relation to the external concentration of glucose illustrates the difference between diffusion and passive protein-mediated transport (facilitated diffusion). The rate of net diffusion into cells is low and is always proportional to substrate concentration. In passive-mediated transport, the rate of transport is much higher, reaching a maximum (V_{max}) as substrate concentration increases because the carrier proteins have become saturated

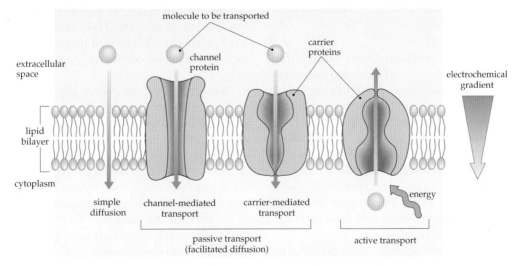

Fig. 4.6 Channels and carriers are two different types of transport proteins. Movement through channels is passive. Carriers undergo a more radical conformational change and may allow passive transport down an electrochemical gradient or be coupled to an energy source for active transport

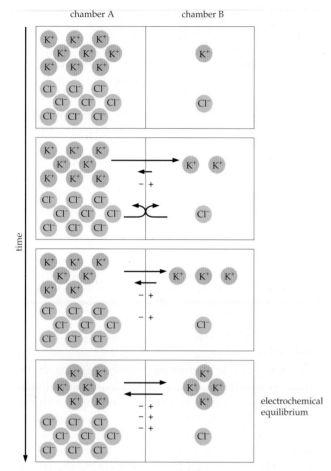

Fig. 4.7 This model illustrates the movement of K$^+$ in response to both concentration and electrical gradients across a membrane. Chamber A, containing high concentrations of K$^+$ and Cl$^-$, is separated from chamber B, containing low concentrations of these ions, by a membrane permeable to K$^+$ but not Cl$^-$. K$^+$ will diffuse from A to B along its concentration gradient. In the model, Cl$^-$ is unable to move, so chamber B becomes positively charged. This creates an electrical gradient favouring movement of K$^+$ in the opposite direction. Net movement ceases when the driving forces due to electrical and concentration differences are equal (electrochemical equilibrium)

difference in electrical potential exists across the membrane. In fact, a difference in electrical potential exists across the plasma membrane surrounding most cells, with the inside of the membrane being negatively charged with respect to the outside. This electrical potential is known as the **resting membrane potential** and it develops in a similar way to the model except that more ions are involved and they differ in their relative ability to cross the membrane (see Box 26.2). An additional factor involved in reaching equilibrium is the net negative charge on fixed proteins and nucleic acids inside cells.

For animal cells, the resting membrane potential is usually around –50 mV to –80 mV (or about one-twentieth of one volt), but in plants it is often much greater, at around –200 mV. A negative charge inside

the membrane relative to the outside means that positive ions are attracted to the inside of the cell, even though this may be against their concentration gradient.

The dual importance of concentration and charge on ions means that if the electrical component exceeds the concentration component, it is possible for ions to diffuse up a concentration gradient; that is, from a region of lower concentration to one of higher concentration. Thus, an ion diffuses passively through a protein channel along its **electrochemical gradient**. Clearly, if a particular ion-selective channel is absent, then that ion will not be able to pass through a membrane.

Ions diffuse along an electrochemical gradient. Provided an appropriate ion-selective channel is present and open, diffusion will be driven by both the concentration gradient and the electrical gradient across the membrane.

Diffusion of ions through channels

In 1991, the Nobel Prize for medicine was given to two German researchers, Erwin Neher and Bert Sakmann, for their pioneering research into channel proteins. They developed a technique enabling the formation of a very tight seal between a small glass pipette and a small patch of plasma membrane of cells (Fig. 4.8), thus enabling the measurement of any tiny electrical currents carried by ions flowing through channels in the membrane. They found that channel proteins behave like a switch in that they are either open or closed. When they are open, for a given set of conditions they allow reproducible movements of ions. They can flicker between open and closed states at very high speed. This is detected as a flickering of current passing through the membrane. Due to the extreme sensitivity of the patch-clamp technique developed by Neher and Sakmann, as well as the very high rates of ion movement permitted by ion channels, it is possible to observe a single channel protein turning on and off as it flickers from conducting to non-conducting mode (Fig. 4.9).

Ion-selective channels are complex proteins that lie in the membranes of cells. They can be thought of as a pipe that lies through the membrane. Charged particles diffuse through a membrane when it contains a membrane-spanning protein whose central region acts as a channel through which the particles can diffuse. As most channels are ion selective, the mouth of the pipe has a selectivity filter. Some channels are selective for K$^+$, others for Na$^+$ and yet others for Ca^{2+}. The selectivity, although profound, is not absolute. Channels that allow K$^+$ movement allow Na$^+$ movement but this occurs at as little as one ten-thousandth the rate that K$^+$ moves. Ca^{2+} channels allow Ca^{2+} movement far greater than the rate at which they allow Mg^{2+} movement. This

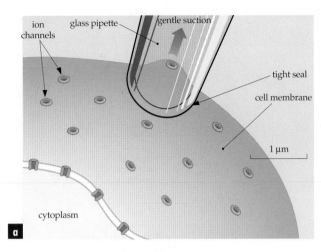

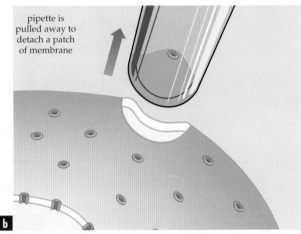

Fig. 4.8 In the Neher and Sakman technique, a tiny patch of membrane is sucked onto the tip of a micropipette, forming a tight seal. Electrical current can only pass across the tip through the ion channel. **(a)** The patch of membrane remains continuous with the plasma membrane. **(b)** The patch of membrane is detached, allowing modification of the solution on either side of the membrane for experimental purposes

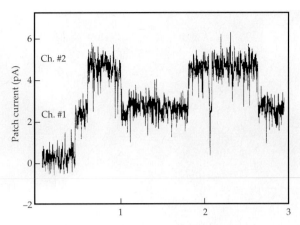

Fig. 4.9 Recordings of current through individual voltage-gated potassium channels in a patch of plasma membrane detached from a wheat root. The patch probably contained just two channels (Ch. #1 and Ch. #2). The membrane was abruptly depolarised from –50 mV to + 50 mV, causing the channels to open more frequently; each upward step in current flow represents the opening of a single channel; downward steps represent closure. When two channels open together in a patch, a current double the magnitude of the single channel is obtained (i.e. 5 picoamps instead of 2.5 picoamps). The minor fluctuations in the current arise from electrical noise

selectivity suggests an intimate association between the transported ion and the walls of the channel, and that any hydration sphere around the ion (p. 21) is shed during its passage. Ion movement through channels shows saturation kinetics.

By virtue of the remarkable speed by which channels transport, these proteins not only have a role in transport but also can be used by cells for the rapid transmission of messages. For example, the electrical impulses that move along nerves are caused by the rapid and selective movement of ions through channels (Chapter 26). Muscle contraction is initiated by an increase in intracellular Ca^{2+}, which results from the transport of Ca^{2+} through channels (Chapter 27).

Similarly, the movements of some plant parts are stimulated by the transport of Ca^{2+} through ion channels. An increase in Ca^{2+} levels in the cytoplasm opens other channels that allow the transport of Cl^- and K^+ out of the cell. The consequent decrease in the intracellular KCl concentration causes an increase in the osmotic (and water) potential inside the cell (p. 108). Water, therefore, leaves the cell by osmosis, thus reducing the **turgor** and allowing the collapse of particular cells and hence movement of tissue. This is the mechanism that underlies closing of the Venus fly trap and collapsing of the leaves of a *Mimosa* plant (Fig. 4.10). It is also the mechanism for the closure of pores in leaves (stomata) through which gases diffuse, in particular, through which water leaves the plant and CO_2 enters to be fixed by photosynthesis.

Many channels have 'open' and 'closed' states: movement, therefore, only occurs when the channels are open. Thus, the pipe across the membrane can be considered to have a trapdoor across it. Depending upon the nature of the channel, it will be opened or closed by a voltage change across the membrane, by an interaction between the channel and a specific chemical or by mechanical force. Thus, channels are conveniently divided into voltage-gated, ligand-gated and mechanically gated channels (Fig. 4.11).

Most **voltage-gated channels** are opened when the voltage across the membrane (the membrane potential) is moved in a positive direction (the inside becomes less negative), usually referred to as *depolarisation*. Changing the voltage difference across the membrane alters the three-dimensional configuration of voltage-gated channel proteins, causing the aqueous pore to open or close (Fig. 4.11a). Voltage-gated ion channels open very rapidly in response to a membrane potential change and ions flow rapidly through the channels. The speed of this

Fig. 4.10 The sensitive plant *Mimosa pudica*. Sudden movements of leaflets (opening and closing) are generated by the movement of ions through mechanically gated channels.

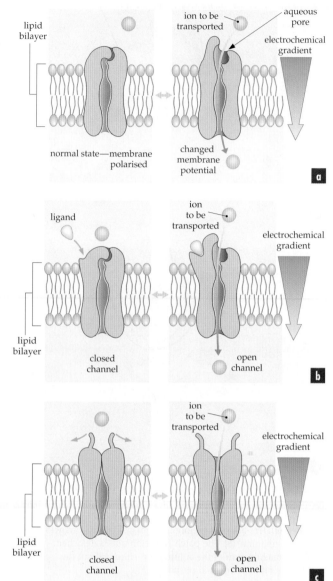

Fig. 4.11 Types of channels. **(a)** Voltage-gated channel. **(b)** Ligand-gated channel. **(c)** Mechanically gated channel

response is used by animals during signalling along nerve cells.

Ligand-gated channels have binding sites for specific signal molecules (ligands) (Fig. 4.11b). When the appropriate ligand binds to the binding site, the conformation of the channel protein alters and the channel opens. When the ligand unbinds, the channel closes. For example, ligand-gated channels are used when signalling between cells. Nerve cells release specific chemicals, neurotransmitters, which then bind to the binding site of ligand-gated channels on the adjacent cell, so activating the channels. Ions flow through the channels and the membrane potential of the recipient cell is changed.

In **mechanically gated channels**, mechanical force directly causes the channel to open. Every animal cell has mechanosensitive channels that are involved in the regulation of cell volume. Many also have specialised mechanosensitive channels linked to cytoskeletal elements within the cell or to the extra-cellular matrix protein for detecting particular forces (Fig. 4.11c); for example, the hair cells in the cochlea that respond to sound waves, and smooth muscle cells of

blood vessels responsible for autoregulation of vessel diameter. Some plant cells also have mechanosensitive channels, for example, the cells involved in the response of the sensitive plant *Mimosa* (Fig. 4.10) to touch.

Ion channels are present in all membranes. They allow the rapid movement of ions along electrochemical gradients. They are highly selective for particular ions. They include channels that are opened by a change in voltage across the membrane (voltage-gated channels), by binding of specific signalling molecules (ligand-gated channels) or by mechanical force (mechanically gated channels).

Facilitated diffusion through carriers

Many membrane-spanning carrier proteins allow passive movement down an electrochemical gradient—

facilitated diffusion. Facilitated diffusion is found where diffusion gradients are favourable but transport is required more rapidly than can occur by simple diffusion through the lipid bilayer. A wide variety of molecules, in particular sugars, amino acids and inorganic ions, move in this way. Carrier proteins that allow passive transport bind to a particular solute on one side of the membrane, change conformation and then release the solute on the opposite side of the membrane (Fig. 4.12). On releasing the solute, the carrier molecule returns to its original orientation within the membrane.

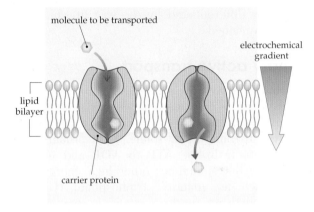

Fig. 4.12 Model of a carrier mediating passive transport as a result of a conformational change after binding to its substrate

Aquaporins

Water and urea diffuse through a family of membrane-spanning proteins termed **aquaporins**. These are responsible for the higher than expected levels of passive diffusion of water and urea across membranes. Tissues with high water permeability, such as the gall bladder and the proximal convoluted tubule of the kidney, have high levels of aquaporins in their plasma membranes. Other aquaporins are hormonally induced. For example, in the collecting tubule of the kidney, increased levels of adrenocorticotrophic hormone (ACTH) cause an increased deposition of these aquaporins in the membrane, thus transiently increasing its water permeability.

Aquaporins are a family of membrane-spanning proteins that contribute to the water permeability of membranes, particularly in highly water-permeable cells.

Active transport

Transport of many solutes is required at a faster rate than could occur by facilitated diffusion or must occur against the prevailing electrochemical gradient. (For uncharged solutes, this would be from an area of lower concentration to one of higher concentration.) The distinguishing feature of **active transport**, which occurs through carrier proteins, is that it requires the expenditure of energy. Energy is expended to move the required solute across cell membranes at a desirable rate or to move substances against a concentration or electrochemical gradient. In addition, it is worth remembering that the effect of passive ion movements across membranes is to reduce the driving electrochemical gradient across the membrane. Thus, active transport is required to regenerate the original ion gradients (p. 102). The importance of active transport is reflected by the fact that, in some cells, it can account for 30–50% of the total energy expenditure of the cell.

Active transport through membrane-spanning carrier proteins allows movement against the prevailing electrochemical gradient and requires the expenditure of energy.

To move solutes against an electrochemical gradient, carriers must be linked to a source of energy, such as the hydrolysis of ATP. Metabolic energy is used to drive the solute through the transport protein or 'pump'. Active transport of solutes can result in the accumulation of solutes inside cells; alternatively, solutes may be actively removed from cells. For example, in most plant cells, Cl^-, NO_3^- and $H_2PO_4^-$ are normally pumped in, whereas Na^+, Ca^{2+} and H^+ are pumped out. In mammalian cells, Na^+ and Ca^{2+} are expelled by cells and K^+ is pumped into cells.

Carrier proteins have many of the properties of enzymes They have binding sites to which the transported molecule must first bind. As the concentration of the transported molecule increases, more of the binding sites for the transported molecule become occupied and the rate of transport reaches a plateau, so showing the saturation kinetics similar to that of all enzyme-catalysed reactions. The reaction between solute and transport protein is usually quite specific. For example, the membranes of all mammalian cells have an ion transporter, which simultaneously moves K^+ into the cells, so setting up a transmembrane K^+ gradient, and moves Na^+ out of the cells. This ion transporter, often referred to as the Na^+–K^+ pump, is unable to transport carbohydrates across the cell. A different transporter, the glucose transporter, transports this carbohydrate. Both examples of transporter display further selectivity. The sodium pump transports Na^+ but cannot transport the cations Mg^{2+} or Ca^{2+}. The carbohydrate transporter in the plasma membrane of red blood cells favours the transport of glucose over similar sugars such as galactose and fructose.

Some carrier proteins transport only a single solute across the membrane (uniport), while others have two binding sites and cotransport two different solutes, either in the same direction (symport) or in opposite directions (antiport) (Fig. 4.13).

The general properties of carrier-mediated transport that distinguish it from simple diffusion are:

- transport is *faster*
- transport proteins become *saturated* as substrate concentration increases (Fig. 4.5)
- transport proteins are *specific* for particular substrates (or types of substrates)
- transport is inhibited by similar substrates that *compete* for the binding site.

Carrier-mediated transport is faster than would occur by diffusion, shows saturation as substrate concentration increases, is specific for particular substrates and can be inhibited by competition from similar substrates.

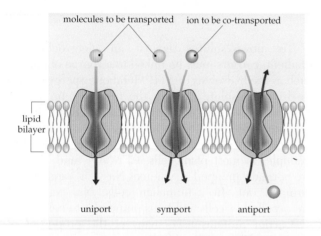

Fig. 4.13 Carriers may be uniporters, symporters or antiporters

Different cell types are characterised by different sets of transport proteins in their membranes. For example, red blood cells carry large amounts of CO_2 back to the lungs from active tissues. In blood, CO_2 is dissolved as HCO_3^-. In the plasma membrane of red blood cells, there is a protein that is responsible for rapidly transporting HCO_3^-, in exchange for Cl^- (to maintain electrical neutrality), into and out of these cells. This protein is more abundant in the plasma membrane of red blood cells than in any other membrane.

There are two main forms of active transport—primary active transport and solute-coupled exchange. These are differentiated by the way in which their energy is obtained.

Primary active transport

In **primary active transport**, the energy released by the hydrolysis of chemical bonds is used directly by the carrier to provide energy to undergo conformational change and transport its solute. Most primary active transporters hydrolyse ATP to ADP and inorganic phosphate, but sometimes other high energy molecules, such as guanosyl triphosphate (GTP) or pyrophosphate, are used.

Carriers that use ATP for energising transport are referred to as solute-translocating ATPases. There is a relatively small number of these. The best example is the *sodium–potassium pump* (Na^+–K^+ translocating ATPase), which is common in animal cells. This pump hydrolyses one ATP molecule in order to remove three Na^+ and acquire two K^+ (Fig. 4.14). The plasma membranes of plant and fungal cells are characterised

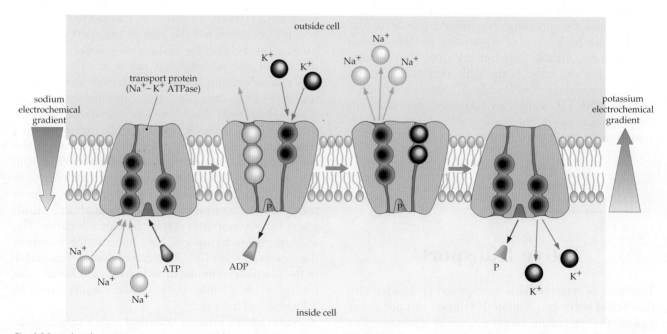

Fig. 4.14 Na^+–K^+ ATPase transports three Na^+ out of animal cells and two K^+ into the cell at a cost of the hydrolysis of one ATP molecule

by an ATPase that removes protons (H$^+$) from the cell—the *proton pump*. This pump removes protons produced internally (from the synthesis of organic acids, etc.) and protons that move into the cell. In doing so, the proton pump helps to regulate cellular pH, maintaining it just above neutral.

The ATPases described above belong to a group of closely related ATPases called P-type ATPases. They have similar amino acid sequences and it seems that all P-type ATPases evolved from a common ancestral ATPase. There are, however, two other types of ATPases with fundamentally different structures. V-type ATPases are found on some internal membranes of eukaryotic cells, such as on the tonoplast (the membrane surrounding the vacuole) of plant cells and on the membrane of coated vesicles. A V-type ATPase consists of many polypeptide chains, which together form a molecular mass of about 500 kD.

F-type ATPases are found in the plasma membrane of bacteria, the inner membrane of mitochondria and the thylakoids of chloroplasts. Instead of consuming ATP to pump solutes, these proteins use the passive movement of H$^+$ to drive the synthesis of ATP from ADP and inorganic phosphate. This reaction is the reverse of that catalysed by the other types of ATPase. The gradient of H$^+$ is initially established by a series of redox reactions (Chapters 2 and 5). The similarities between these reactions and between the structure of the ATPases of bacteria, chloroplasts and mitochondria forms part of the evidence to support the theory that mitochondria and chloroplasts originated by the non-destructive engulfment of smaller prokaryotes by larger prokaryotes (Chapter 35).

In primary active transport, energy is used directly by the carrier protein to pump solutes across a membrane.

Solute-coupled exchange

In some circumstances, an ion is transported against its electrochemical gradient with the energy being supplied by another ion gradient. In this way, different ions on either side of the membrane are exchanged—**solute-coupled exchange**—one ion moving against an electrochemical gradient and the other down an electrochemical gradient.

Both Na$^+$–K$^+$ and H$^+$ pumps 'push' positively charged ions out of eukaryotic cells. They therefore contribute to the difference in electrical potential found across most cell membranes, where the inside of cells is negative with respect to the outside. The pumps thus maintain an electrochemical potential difference for ions across the plasma membrane. The tendency for Na$^+$ and H$^+$ to move energetically 'downhill' and diffuse back into the cells from which they have been pumped is exploited in solute-coupled active transport.

In this form of transport, solutes are cotransported through transport proteins against an electrochemical gradient using energy derived from the 'downhill' movement of ions. Thus, the energy used initially to create the electrochemical gradient is used secondarily to power solute transport. (For this reason the term 'secondary active transport' is sometimes used.)

In animals, the driving ion for solute-coupled exchange is usually Na$^+$, and in plants H$^+$. An example in mammalian cells is the Na$^+$–Ca^{2+} exchanger. This protein transports Ca^{2+} from inside a cell to the outside. Thus, Ca^{2+} is moved against its concentration gradient, as $[Ca^{2+}]_o$ is greater than $[Ca^{2+}]_i$, and against the prevailing electrical gradient. The energy required is provided by the 'downhill' diffusion of Na$^+$ into the cell along its electrochemical gradient. The Na$^+$–Ca^{2+} exchanger binds three external Na$^+$ and one Ca^{2+}, and the energy provided by the Na$^+$ gradient 'flips' the exchanger protein, so expelling the Ca^{2+} and causing three Na$^+$ to enter. The entering Na$^+$ is quickly expelled by the Na$^+$–K$^+$ pump.

In fungi and bacteria, some organisms use H$^+$ and some use Na$^+$. As explained above, in eukaryotes these ions are pumped out of the cell through a primary pump, with the hydrolysis of ATP usually providing the necessary energy. In prokaryotes, the energy comes from a series of redox reactions. In all cells, a regulated number of these ions then moves energetically downhill back into the cell through specialised proteins, the solute-coupled transporters. This movement provides the necessary energy to transport other solutes against their electrochemical gradients (Fig. 4.15).

Solute-coupled exchange may involve a symport or an antiport. Important examples of symports include the Na$^+$–glucose pump in animal cells and the 2H$^+$–Cl$^-$ pump in plant cells. Both of these pumps are responsible for the accumulation of the solute (glucose or Cl$^-$), which moves in with the driving ion (Na$^+$ or H$^+$).

The use of energy gradients of ions across membranes is a universal process in organisms and occurs across most biological membranes. It was first described in bacteria in the early 1960s by Peter Mitchell. He proposed a role for H$^+$ gradients in both ATP production and transport of other solutes; for this, he was awarded the Nobel Prize for chemistry in 1978. The experimental evidence for his brilliant ideas was provided by others a decade or so later. Differences in energy levels of ions are used not only for the indirect harvesting of the energy of hydrolysis of ATP for the transporting of solutes but also for the *synthesis* of ATP from ADP and inorganic phosphate in mitochondria and chloroplasts, where the difference in the energy levels of H$^+$ built up by an electron transport chain is allowed to dissipate through ATP synthases (Chapter 5).

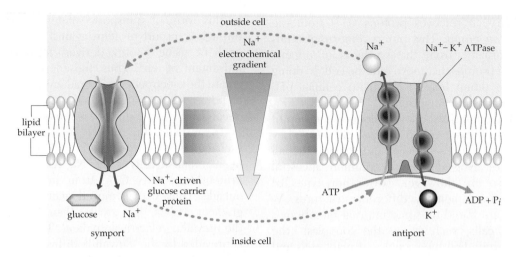

Fig. 4.15 In animal cells, the Na^+–K^+ ATPase (antiport) contributes to a strong electrochemical potential difference for Na^+. Downward movement of Na^+ through the solute-coupled transporter (symport) brings glucose with it into the cell

> With solute-coupled exchange, solutes are moved through carrier proteins against their energy gradient using energy derived from the 'downhill' movement of ions (usually Na^+ in animal cells and H^+ in plant cells).

Osmosis: passive movement of water

Water crosses cell membranes by a process termed **osmosis**. Osmosis is the passive *net* movement of water across a selectively permeable membrane, from an area where there are more free water molecules to an area where there are fewer free water molecules. Cell membranes are selectively permeable; they allow water molecules to cross at a considerably greater rate than the solutes dissolved in the water.

Water, like other molecules, will move from a region of higher energy to one of lower energy. In most biological systems, the two most important factors affecting the energy level of water are the *solutes dissolved* in the water and the *physical pressure* or tension exerted on the water (how much it is being squeezed or pulled). The resulting measure of the total energy level of water is **water potential**, ψ (psi), measured in units of pressure, megapascals (MPa). (The word 'potential' refers to potential energy, meaning the capacity to do work.)

Osmotic potential

The addition of solutes to pure water decreases its **osmotic potential** (ψ_π). This occurs for two reasons, both of which cause a reduction in the amount of freely diffusing water molecules in a particular volume (a reduction in the concentration of free water).

1. Solutes increase the volume of a solution without increasing the water content, so the concentration of water in the solution is reduced. In other words, solutes are diluting the water.
2. Solutes interact with water, reducing the ability of water to diffuse freely.

Note that ψ is standardised as 0 MPa for pure water, so solutions containing solutes (e.g. the cytosol) will normally have a negative osmotic potential (at atmospheric pressure). However, a few solutes may also interfere with the bonds between water molecules, increasing the ability of water molecules to diffuse freely. This occurs when urea is added to water, so that in this case the water potential is raised.

Pressure potential

When hydrostatic pressure is applied to water, its energy increases; when suction (or tension) is applied, its energy decreases. These two effects raise or lower the **pressure potential** (ψ_P) of water and therefore affect water potential (ψ).

Pressure potential, together with osmotic potential, gives the overall water potential,

$$\psi = \psi_\pi + \psi_P$$

Thus, using this terminology, osmosis is the net passive movement of water from a region of *higher water potential* to one of *lower water potential* through a selectively permeable membrane.

> Osmosis is the net passive movement of water from a region of higher water potential (ψ) to one of lower water potential through a selectively permeable membrane.

In contrast to plant cells, there is usually no pressure difference across the cell membrane of animal cells (otherwise they would shrink or explode). In this situation osmosis is simply the diffusion of water across the selectively permeable cell membrane along its *own* concentration gradient.

Osmosis and cells

Water will move into or out of cells depending on the difference in water potential across the plasma membrane. As water enters or leaves a cell by osmosis, this changes the internal solute concentration and thus changes the water potential.

If a cell has no cell wall, as in the case of animal cells, very little difference in water potential between the inside of cells and the surrounding solution can be tolerated because any slight difference will cause the cells to shrink or swell (Fig. 4.16). Thus, any differences in water potential across the plasma membrane in these cells will be due only to differences in osmotic potential, that is, the concentration of solutes.

In those cells constrained by a relatively rigid cell wall, the situation is different, primarily because the cell wall limits volume expansion. Thus, in plant, fungal and bacterial cells, *both* osmotic potential and pressure potential contribute to water potential. When walled cells are placed in a solution with a less negative water potential, water will enter. The cells can only expand by about 10% before pressure (turgor) starts to build up

inside them. At a particular pressure, the intracellular water potential will come to equal that of the external solution (Fig. 4.17). Turgor in cells is responsible for maintaining the shape of many plant parts. It also causes the crunch you hear when you bite into a piece of celery: the crunch is the explosion of pressurised cells as the walls are broken by your teeth. The action of some important antibiotics (e.g. penicillin, vancomycin, cephalosporins and bacitracin) is to weaken the wall structure of growing bacteria to the extent that they may burst under their own turgor.

The change in the shape of plants as they dry out and wilt is the result of a loss of turgor. Because the atmosphere has a very low water potential, water continually moves from cells to the air. If water is not replaced, the pressure potential (and hence turgor) reduces, allowing gravity to determine the shape.

If plant cells are placed into a solution that is more concentrated than their cellular contents, water will leave the cells and the pressure potential will move towards zero. Water may continue to leave until the volume of the cells has decreased to the point at which the internal and external osmotic potentials are equal, just as happens in the case of cells without walls. The shrinkage of the cell usually draws the cell membrane away from the cell wall, leaving a gap between the membrane and wall. This shrinkage is the process of

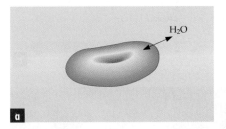 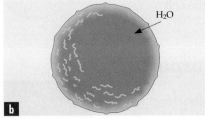 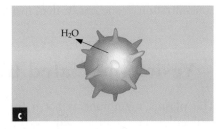

Fig. 4.16 Osmotic movement of water across the plasma membrane of a red blood cell placed in solutions of different osmotic potential. **(a)** The osmotic potentials inside and outside the cell are equal (iso-osmotic). There is no net movement of water so the solution is also isotonic. **(b)** The osmotic potential inside the cell is lower (a higher concentration of solutes) and so water enters the cell, causing it to swell and possibly burst (haemolysis). **(c)** The osmotic potential inside the cell is higher (a lower concentration of solutes) and water leaves the cell, causing it to shrink

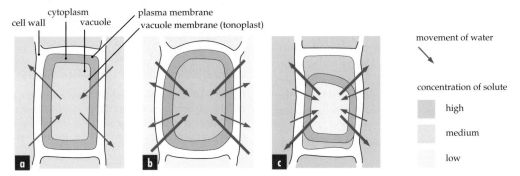

Fig. 4.17 Due to the presence of a cell wall, diffusion of water across the plasma membrane of a typical plant cell placed in different solutions is dependent on both osmotic potential (concentration of solutes) and pressure potential. **(a)** Water potential inside and outside the cell is equal. **(b)** The osmotic potential inside the cell is lower (a higher concentration of solutes) and so water enters the cell, increasing the hydrostatic pressure until the cell becomes more turgid. **(c)** The osmotic potential inside the cell is higher (a lower concentration of solutes) and so water leaves the cell, in this case causing the plasma membrane to pull away from the cell wall (plasmolysis)

plasmolysis (Fig. 4.17c). The point at which pressure potential reaches zero, without causing shrinkage, is the point of *incipient plasmolysis*. At this point, the osmotic potential of the external solution equals that of the cell when it was turgid and so provides a way to measure the osmotic potential of walled cells.

> Water moves into or out of cells depending on the difference in water potential across the plasma membrane. In animal cells, water potential is due only to osmotic potential. Water potential in walled cells, such as plant, fungal and bacterial cells, depends on both osmotic and pressure potentials.

Iso-osmotic and isotonic are terms that are often confused. Solutions are *iso-osmotic* if they contain the same overall concentrations of osmotically active particles (solutes)—thus their osmotic potential is the same. They are *isotonic* if, when placed on either side of a selectively permeable membrane, there is no net flow of water across the membrane. However, because membranes are not equally permeable to all osmotically active particles (some cross more freely than others), iso-osmotic solutions may not be isotonic. Thus, a cell will not swell or shrink if placed in an isotonic solution but it may with time if placed in an iso-osmotic solution. Hypertonic and hypotonic simply mean higher solute concentration and lower solute concentration, respectively, than inside the cell.

Vesicle-mediated transport

The plasma membrane is not a fixed structure; new molecules are synthesised and inserted into the membrane and existing components are retrieved and recycled or broken down. This dynamic process of insertion and retrieval is due to the pinching in of small portions of the membrane to form vesicles, the process of **endocytosis**, or the fusing to the plasma membrane of small vesicles constructed inside the cell, **exocytosis** (Figs 4.18, 4.19).

During endocytosis, a small area of plasma membrane enfolds or 'invaginates', enclosing substances that are outside the cell (Fig. 4.18). The invagination then pinches off to form a vesicle in which the substance is contained. The vesicle is then transported within the cell. This process brings about the inward movement of solids (phagocytosis) or of liquids (pinocytosis).

In some cases, endocytosis is stimulated by initial binding of the solute to a receptor molecule on the membrane. An example of receptor-mediated endocytosis is the uptake of low-density lipoproteins (LDL) into animal cells to be broken down to form cholesterol

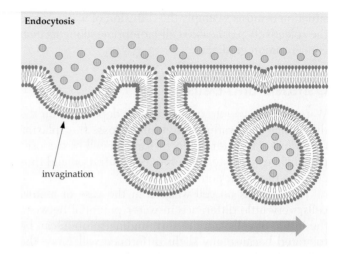

Fig. 4.18 Formation of a transport vesicle during endocytosis

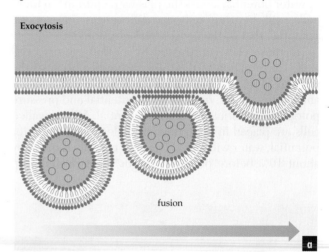

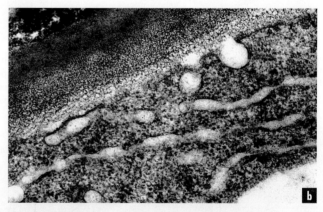

Fig. 4.19 (a) Fusion of a transport vesicle with the plasma membrane during exocytosis. **(b)** Transmission electron micrograph of a transport vesicle during secretion from a sugar-secreting cell in a gland on the foliage of the sunshine wattle, *Acacia terminalis*

(Fig. 4.20). Specific receptor proteins, present in the plasma membranes of animal cells, bind specifically to LDL particles. LDL-receptor complexes then cluster in depressions (pits) of the cell surface. The internal surface of the membrane at the pit is coated with the

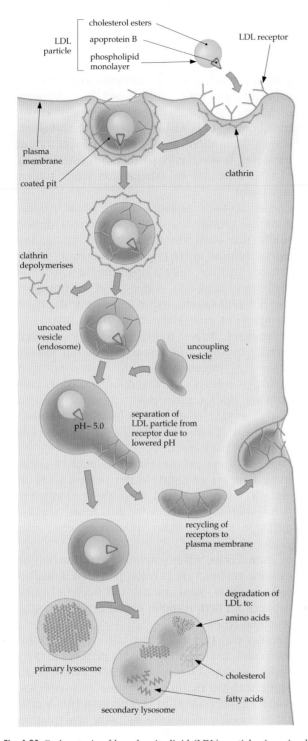

LDL particle
- cholesterol esters
- apoprotein B
- phospholipid monolayer

LDL receptor

plasma membrane

coated pit

clathrin

clathrin depolymerises

uncoated vesicle (endosome)

uncoupling vesicle

pH~ 5.0

separation of LDL particle from receptor due to lowered pH

recycling of receptors to plasma membrane

degradation of LDL to:

amino acids

primary lysosome

cholesterol

fatty acids

secondary lysosome

Fig. 4.20 Endocytosis of low-density lipid (LDL) particles in animal cells. LDL particles bind tightly and specifically to surface receptors and are then internalised as complexes in a clathrin-coated pit. Clathrin-coated vesicles are pinched off into the cytosol, where clathrin is depolymerised, forming an uncoated vesicle (endosome), which can fuse with special uncoupling vesicles. The pH of these vesicles is low (~5.0) and causes the dissociation of the receptor and the LDL particle. The receptor protein is recycled direct to the plasma membrane within its own vesicle, while the LDL particle enters the lysosome systems where the particles are degraded. Apoprotein B is broken down into amino acids, and cholesterol esters into fatty acids and cholesterol, which is incorporated into cellular membranes

protein clathrin (Chapter 3). Clathrin-coated vesicles containing LDL particles form by endocytosis.

In the process of exocytosis, intracellular vesicles fuse to the plasma membrane and the contents of the vesicle are deposited on the outside of the cell (Fig. 4.19). Many hormones and enzymes are released in this way. In the germinating seeds of many plants, the enzymes required to break down the storage compounds of the seed are secreted by exocytosis from a special layer of cells upon receiving a hormonal signal from the germinating embryo.

In endocytosis, small portions of the plasma membrane form vesicles, enclosing substances to be imported into a cell. In exocytosis, vesicles fuse with the plasma membrane to export their contents.

Summary

- All living cells are enclosed by a plasma membrane and surrounded by an aqueous medium that constitutes their immediate environment. The selective permeability of the lipid bilayer and the activity of membrane-spanning transport proteins allows cells to regulate the composition of their cytosol and of membrane-bound internal compartments.

- Diffusion is the passive net movement of molecules along their concentration gradient. It requires no expenditure of energy. The rate of diffusion of an uncharged molecule across the bilayer depends on the concentration gradient for that molecule and the ease with which it can pass through the bilayer.

- Provided an appropriate ion channel is present and open, diffusion of ions will be driven by the electrochemical gradient; that is, both the concentration gradient and the electrical gradient across the membrane.

- Channels and carriers are two different types of transport proteins. Movement through channels is always passive, whereas carrier-mediated transport may be passive or active.

- Ion channels are present in all membranes. They allow the rapid diffusion of ions along electrochemical gradients. They are highly selective for particular ions and are opened by a change in voltage across the membrane (voltage-gated channels), by binding with specific signal molecules (ligand-gated channels) or by direct mechanical force (mechanically gated channels). The 'patch-clamp' technique enables the operation of a single ion channel to be studied.

- Facilitated diffusion is the passive movement of a solute along its electrochemical gradient through a protein carrier.

- Aquaporins are a family of membrane-spanning proteins that allow diffusion of water across the membrane, contributing to the water permeability of membranes, particularly in highly water-permeable cells.

- Carrier-mediated active transport requires the expenditure of energy and allows the movement of some polar and ionised molecules across membranes. It is faster than simple diffusion, shows saturation as substrate concentration increases, is specific for particular substrates and can be inhibited by competition from similar substrates.

- Carrier proteins undergo more radical conformational change than channel proteins during solute transport. On releasing the solute, the carrier returns to its original orientation. Some carriers transport only a single solute across the membrane (uniport), while others cotransport two different solutes, either in the same direction (symport) or in opposite directions (antiport).

- In primary active transport, energy is used directly by the carrier protein to pump solutes across a membrane. With solute-coupled exchange, solutes are transported through carrier proteins against their electrochemical gradient using energy derived from the downhill movement of ions (usually Na^+ in animal cells and H^+ in plant cells).

- Osmosis is the net passive movement of water from a region of higher water potential (ψ) to one of lower potential through a selectively permeable membrane (or, put more simply, diffusion of water along its own concentration gradient).

- In animal cells, water potential (ψ) is due only to osmotic potential (ψ_π). Water potential in walled cells, such as plant, fungal and bacterial cells, depends on both osmotic potential and pressure potential (ψ_P).

- Another way that molecules, particularly macromolecules, can cross the plasma membrane is by vesicle-mediated transport. In endocytosis, small portions of the plasma membrane form vesicles, enclosing substances to be imported into a cell. In exocytosis, vesicles fuse with the plasma membrane to export their contents.

key terms

active transport (p. 105)	endocytosis (p. 110)	osmotic potential (p. 108)	selective permeability (p. 98)
aquaporin (p. 105)	exocytosis (p. 110)	permeability	solute (p. 99)
carrier (p. 101)	facilitated diffusion (p. 105)	coefficient (p. 99)	solute-coupled exchange (p. 107)
channel (p. 101)	ligand-gated channels (p. 104)	plasmolysis (p. 110)	solvent (p. 99)
concentration gradient (p. 99)	mechanically gated channels (p. 104)	pressure potential (p. 108)	turgor (p. 103)
diffusion (p. 99)	membrane-spanning transport protein (p. 98)	primary active transport (p. 106)	voltage-gated channels (p. 103)
electrical gradient (p. 101)	osmosis (p. 108)	resting membrane potential (p. 102)	water potential (p. 108)
electrochemical gradient (p. 102)			

Review questions

1. What is meant by selective permeability? Explain how the phospholipid bilayer and membrane-spanning proteins each contribute to movement across membranes.

2. Define concentration gradient and electrical gradient. How is each involved in the diffusion of solutes across membranes?

3. By what processes do each of the following substances enter a white blood cell:
   (a) O_2?
   (b) H_2O?
   (c) glucose?
   (d) a bacterium?

4. By means of diagrams, distinguish between the mechanisms of action of voltage-gated channels, ligand-gated channels and mechanically gated channels.

5. Outline the characteristic features of active transport through carrier proteins.

6. List three key features that distinguish primary active transport from solute-coupled exchangers.

7. What is water potential? In plant cells, what are its two components? Compare what happens when animal cells and plant cells are placed in distilled water.

Extension questions

1. Explain how passive protein-mediated transport enables glucose to pass through the membranes of red blood cells faster than by simple diffusion across the bilayer.

2. Describe the action of the Na^+–K^+ pump in animal cells. Explain how this is used indirectly to transport glucose into a cell. In your answer, distinguish between a symport and an antiport.

3. Using diagrams, describe the changes in water potential in a typical leaf cell occurring in a wilting plant growing in saline conditions compared with a similar plant in normal soil.

4. Discuss how the plasma membrane controls the internal stability and overall metabolism of cells.

Suggested further reading

Alberts, B., Bray, D., Lewis, J., et al. (1994). *Molecular Biology of the Cell*. 3rd edn. New York: Garland.

This text contains a very thorough and detailed account of the material covered in this chapter.

Hille, B. (1994). *Ionic Channels of Excitable Membranes*. Sunderland, MA: Sinauer Associates Inc.

This book provides a detailed introduction to channel behaviour, particularly the first four chapters.

CHAPTER

5

Harvesting energy

The earth formed about 4.5 billion years ago and, as it evolved (see Chapter 31), a crust solidified over the hot mantle and an atmosphere gradually developed. The primitive atmosphere contained water vapour, carbon dioxide, hydrogen, nitrogen, ammonia, methane, sulfur dioxide and hydrogen sulfide, much like volcanic gases do today. Because the atmosphere contained ample hydrogen and lacked oxygen, it was a reducing one, a condition that favoured the formation of molecules rich in C—C bonds. In 1953, S. L. Miller performed a classic experiment to show that an electrical discharge (such as lightning) in such an atmosphere could have led to the formation of complex carbon molecules. The waters of oceans or lakes, which formed from the steam of volcanoes, became a 'dilute soup' of carbon- and nitrogen-containing molecules— amino acids, carbohydrates and nucleic acids. Simple cell-like structures may have formed from aggregations of these molecules; for example, in water, phospholipid molecules can spontaneously form spheres (Chapter 3) and double-layered membranes. Although the exact way in which cells first formed is unknown, primitive self-replicating structures could have evolved from such aggregations of molecules.

The first organisms to evolve were prokaryotes. They were presumably **heterotrophs**, organisms unable to synthesise their own food. They would have used the complex carbon molecules in the 'soup' for raw materials and energy. However, there are no fossils of these heterotrophs and the oldest evidence of life, dating back 3.5 billion years, is of **autotrophs**, bacteria able to synthesise their own fuel molecules. The first of these (photosynthetic bacteria) had a non-oxygen requiring, or **anaerobic**, metabolism and had evolved the ability to trap the energy of sunlight and use it to synthesise carbohydrate directly from carbon dioxide and water; in the process, they produced oxygen as a by-product. The atmosphere thus gradually became oxygen-rich, destroying the reducing conditions of the earlier atmosphere and, hence, the very conditions that led to the formation of organic molecules and the evolution of life (Fig. 5.1).

Modern photosynthetic organisms include plants, algae and many bacteria (green, purple and cyano-bacteria). Modern heterotrophs include other forms of bacteria, protists, animals and fungi. Heterotrophs are ultimately dependent on the chemical energy produced by autotrophs. However, whether organisms produce organic molecules or feed on ready-made ones, all organisms use chemical energy in the form of ATP to drive metabolic reactions (Chapter 2). Therefore, this chapter deals first with how all organisms convert the chemical energy of fuel molecules to the useable energy of ATP. The second part of the chapter discusses how photosynthetic organisms trap the radiant energy of sunlight and store

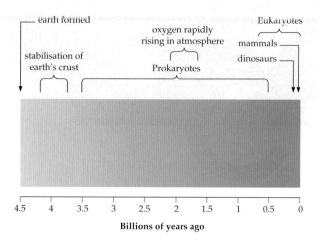

Fig. 5.1 Time scale showing the earth's history

it in the chemical bonds of carbohydrates such as sucrose and glucose.

The first organisms were heterotrophic and used molecules that formed spontaneously in the absence of oxygen as a source of raw materials and energy. Photosynthetic organisms evolved later, enriching the atmosphere with oxygen.

Harvesting chemical energy

Energy is released from fuel molecules along metabolic pathways that are initially different for carbohydrates and lipids. Carbohydrates, which are converted into glucose, are processed by glycolysis, while lipids are processed by β-oxidation. The products of both of these pathways can then act as the substrate for cellular respiration, the major process of energy extraction (Fig. 5.2). **Cellular respiration** involves oxidation of fuel molecules, that is, removal of electrons from the C—C and C—H bonds. Electrons extracted from these bonds are accepted by the coenzymes NAD$^+$ (Chapter 2) and FAD (flavin adenine dinucleotide) and passed down electron transport chains to the final electron acceptor, driving proton pumps that are coupled to the synthesis of ATP.

Glycolysis: the initial processing of glucose

Complex carbohydrates, such as starch (plants) and glycogen (animals), are broken down by catalysed hydrolysis into glucose subunits before entering energy-releasing pathways. In all cells, both prokaryotic and eukaryotic, glucose is processed initially in the cytosol by **glycolysis** (Fig. 5.3).

Glycolysis involves a three-stage rearrangement and splitting of the glucose molecule, energy production and recycling. In the first stage, two molecules of ATP

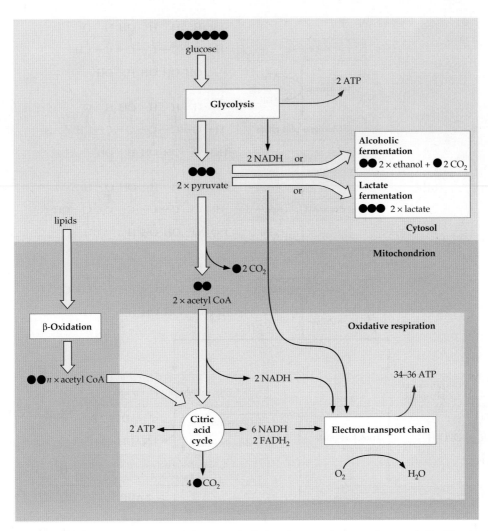

Fig. 5.2 Overview of the metabolic pathways for extracting energy from fuel molecules. Glucose is first metabolised by glycolysis, with a net production of two ATP molecules per glucose molecule. A 3-carbon molecule, pyruvate, from glycolysis can then enter different pathways depending on the organism and whether the environment is anaerobic or aerobic. In the absence of oxygen, pyruvate is converted to lactate or ethanol. In the presence of oxygen, the energy in pyruvate is converted to a further 34–36 ATP molecules. The energy extraction pathway is the citric acid cycle and an electron transport chain. Electrons are transferred to coenzymes NAD^+ and FAD and used to drive proton pumps on the inner membrane of mitochondria. Proton pumping is coupled to ATP synthesis.

In contrast to glucose, lipids are first metabolised by β-oxidation, with the product acetyl CoA being oxidised by the same pathway as pyruvate from glycolysis

are used to phosphorylate and change glucose in preparation for splitting it into two 3-carbon molecules (glyceraldehyde 3-phosphate). In the second stage, oxidation of glyceraldehyde 3-phosphate to **pyruvate** $(C_3H_3O_3^-)$ is coupled to ATP synthesis: four ATP molecules are produced giving a net energy profit of two ATP molecules. In addition, four electrons and two hydrogen atoms are transferred to two molecules of the electron acceptor NAD^+, producing two molecules of NADH.

The final stage depends on whether O_2 is present or not. If O_2 is absent (anaerobic conditions), pyruvate is reduced using the NADH produced in the second stage to form lactate or ethanol (Fig. 5.4). This recycles NAD^+ for reuse in the second stage. Both lactate and ethanol form during **fermentation** by microorganisms

(bacteria and yeasts). In animal tissues subjected to anaerobic conditions, lactate usually forms, whereas in plants, ethanol rather than lactate is produced. If O_2 is available (aerobic conditions), the 3-carbon pyruvate enters a mitochondrion and is oxidised to form a 2-carbon compound, **acetyl CoA** (Fig. 5.2). In this oxidation reaction, enzymes split off the COO^- end group of pyruvate and CO_2 is released (a decarboxylation reaction). A proton (H^+) and two electrons are accepted by NAD^+, which is reduced to NADH (used later to produce ATP). The 2-carbon molecule remaining is linked to coenzyme A, forming acetyl CoA, which is the substrate for the citric acid cycle.

Glycolysis was one of the earliest biochemical pathways to evolve, requiring no oxygen and occurring in the cytosol and not in any specific organelle. With

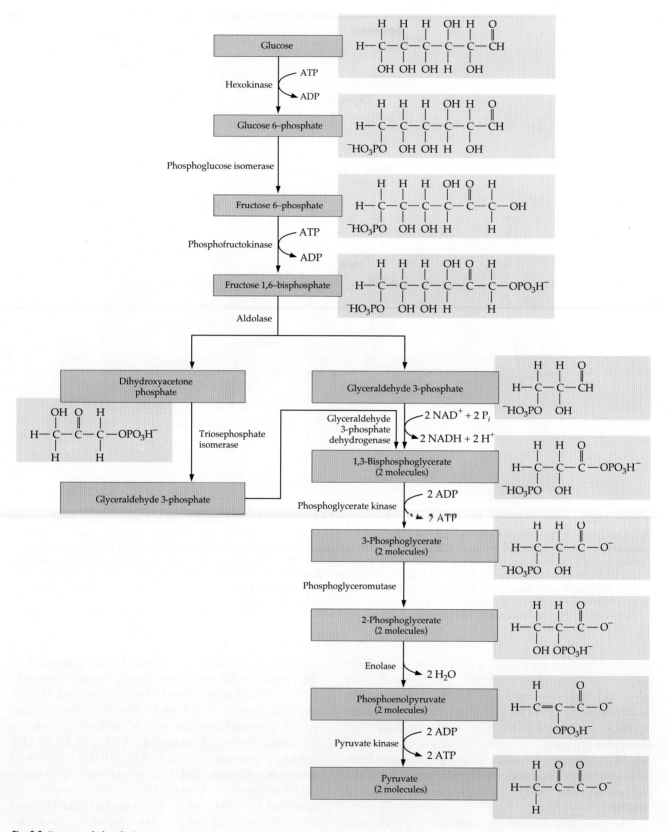

Fig. 5.3 Process of glycolysis

the production of only two molecules of ATP, it is not an efficient way to extract energy from glucose; two ATP molecules represent only about 2% of the chemical energy available. As the earth developed an oxygen-rich atmosphere, the evolution of pathways for **oxidative respiration** allowed the extraction of up to

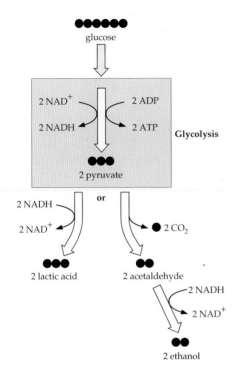

Fig. 5.4 Fermentation occurs under anaerobic conditions. In plants, ethanol is the end product, and in animals, lactate is the end product. Lactate is energy rich and, when oxygen becomes available, it may, with the use of ATP, be converted back to a substrate suitable for oxidative respiration

34 more molecules of ATP, giving a maximum of 36 molecules for every molecule of glucose. Nevertheless, glycolysis generates ATP very rapidly and remains the major source of useable energy in short bursts of activity, such as in a 100 m sprint by an athlete.

> In the cytosol, glycolysis produces two molecules of ATP. In the absence of O_2, the end product of glycolysis is ethanol or lactate. In the presence of O_2, pyruvate is converted to acetyl CoA, which is the substrate for the citric acid cycle.

β-Oxidation: the initial processing of lipids

In animals and some seeds, chemical energy is largely stored as lipids. Before entering the energy-releasing pathway, lipids are hydrolysed into free fatty acids and glycerol. The fatty acids, which have a long backbone of carbon atoms, are used as substrates for **β-oxidation**, and are broken down two carbon atoms at a time. In eukaryotes, β-oxidation occurs inside mitochondria and involves four reactions that oxidise the β-carbon and produce acetyl CoA (Figs 5.2, 5.5). The first three reactions, like glycolysis, are typical of those that make up the bulk of degradative pathways, oxidations and rearrangements. The two oxidations conserve some of

the energy available in the lipid substrate as reduced electron carriers. The last reaction is unique to β-oxidation and splits off acetyl CoA. The energy in the C—C bond that is broken is conserved as a C—H bond in acetyl CoA. In plants, β-oxidation also proceeds in other organelles (glyoxysomes) where the acetyl CoA is initially processed by the glyoxylate cycle and ultimately converted into sugars.

> β-oxidation degrades long-chain fatty acids by two carbon atoms at a time to form acetyl CoA, which enters the citric acid cycle.

The citric acid cycle: completing the oxidation of fuels

Acetyl CoA is the substrate for the **citric acid cycle** (also known as the **Krebs cycle** after the British biochemist, Sir Hans Krebs, who discovered the pathway in animal tissues). Acetyl CoA enters the citric acid cycle and combines with a 4-carbon molecule, oxaloacetate, releasing coenzyme A and forming a 6-carbon molecule, citrate (Fig. 5.6). Citrate is rearranged into the 6-carbon molecule isocitrate, which is the substrate for a series of oxidation reactions. Isocitrate is stripped of two electrons and one H^+, which are transferred to NAD^+ to form NADH. One molecule of CO_2 is also released. The resulting 5-carbon intermediate (α-ketoglutarate) is similarly stripped of electrons and H^+, forming another NADH and CO_2. The 4-carbon product, succinyl CoA, is converted in four reactions to oxaloacetate to complete the cycle. In these reactions, electrons and H^+ are transferred to form $FADH_2$ and NADH, and one molecule of ATP is produced.

> Oxidation of acetyl CoA yields CO_2, ATP molecules and energised electrons that are used to generate reduced electron carriers (three NADH and one $FADH_2$). In eukaryotes, the citric acid cycle occurs in mitochondria.

Recycling electron carriers

In the citric acid cycle, three of the four oxidation reactions require the electron carrier NAD^+ as a substrate. The quantity of NAD^+ in cells is very small when compared with the continuous flux of carbon through the cycle that is required to maintain metabolic rate. In the human brain, for instance, there is only enough NAD^+ for 3 minutes at resting metabolic rate. The secret is that the NAD^+ used to make NADH, and the FAD used to make $FADH_2$, cycle back after having been oxidised during the next stage, the electron transport system (Fig. 5.2). Recycling is characteristic of a number of biochemical pathways in cells.

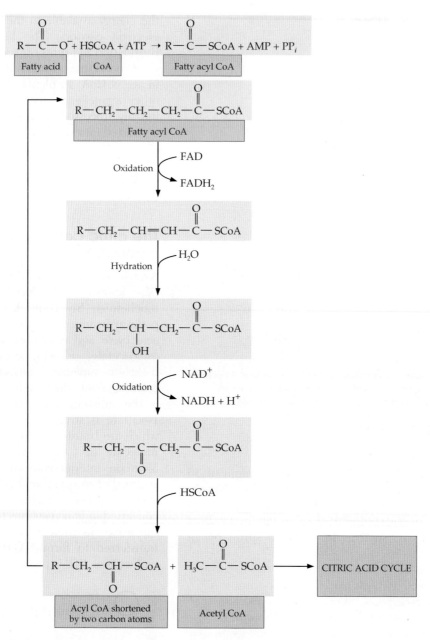

Fig. 5.5 β-Oxidation of fatty acids from lipids. In four reactions catalysed by enzymes, a fatty acid CoA molecule is converted to acetyl CoA and another fatty acyl CoA molecule with two fewer carbons. During oxidation, electrons are transferred to FAD and NAD$^+$ to produce FADH$_2$ and NADH. The pathway is repeated until fatty acids are completely converted to acetyl CoA, which enters the citric acid cycle

The electron transport system: generating ATP

The electrons released in glycolysis and the citric acid cycle are temporarily stored in the reduced electron carriers NADH and FADH$_2$. The energy now conserved in these molecules is then converted into ATP. The electron transport chain is responsible for 85% of the ATP that is ultimately produced from fuel molecules. The amount is variable because some of the energy released during the oxidation reactions is used for other purposes in mitochondria.

NADH and FADH$_2$ transfer electrons to special carrier proteins embedded in the plasma membrane of prokaryotic cells or the inner membrane of the mitochondrion of eukaryotic cells (Fig. 5.7). Here the energy of NADH and FADH$_2$ is harvested in an electron transport system to produce ATP. (The mechanism of electron transport and ATP production is described in Chapter 2.)

Electrons from NADH and FADH$_2$ are removed and passed through several protein complexes. Coupled to this transfer of electrons is the translocation of protons from the matrix to the outer

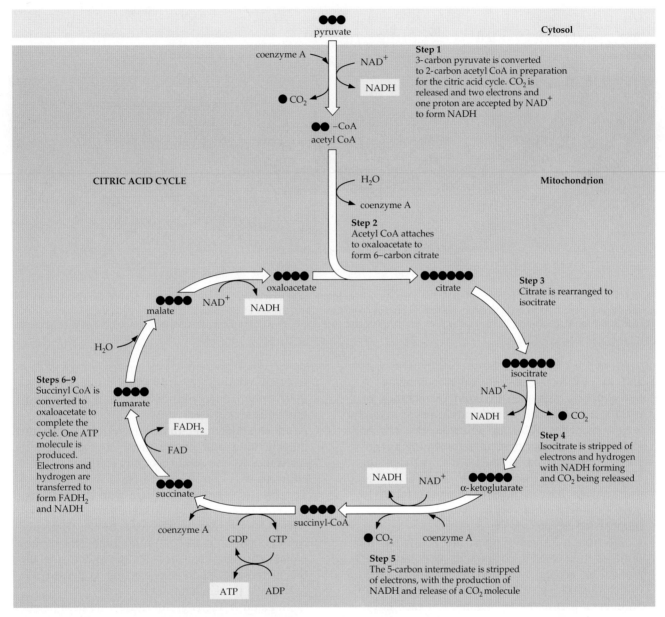

Fig. 5.6 The citric acid cycle. In eukaryotes, pyruvate enters the mitochondrion and is converted to acetyl CoA, which is progressively oxidised through a series of reactions with CO_2 released and ATP, NADH and $FADH_2$ produced

side of the inner membrane. The final protein complex (**cytochrome *c* oxidase**) uses four electrons and four protons to reduce one molecule of O_2 to two of H_2O. Oxygen is thus the final electron acceptor with water forming as the end product (Fig. 5.7). In the process, the oxidised forms of the coenzymes NAD^+ and FAD are regenerated.

The proton concentration gradient created by proton pumping provides the electrochemical force to drive ATP synthesis (Box 5.1). As protons move down the charge and concentration gradient, back across the mitochondrial membrane into the matrix, they pass through special protein channels (Fig. 5.7). The passage of protons drives the phosphorylation of ADP, resulting in the production of ATP.

In the electron transport chain, electrons donated by NADH and $FADH_2$ from the citric acid cycle drive proton pumps coupled to ATP synthesis.

BOX 5.1 Biological electron transport systems

Electron transport systems occur on and in membranes. In prokaryotes they occur on the plasma membrane or endoplasmic membranes,

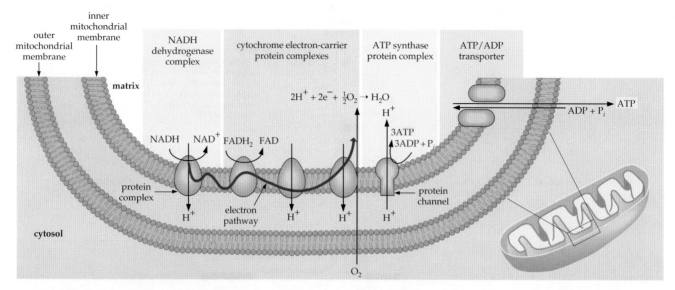

Fig. 5.7 The electron transport system and oxidative phosphorylation in the inner membrane of a mitochondrion. NADH and FADH$_2$ from glycolysis and the citric acid cycle transfer electrons to protein complexes. The transport of electrons is coupled to pumping protons to the outside of the inner membrane. Protons move back across the membrane and drive ATP synthesis. Oxygen acts as the final electron acceptor

and in eukaryotes on the inner membrane of mitochondria and the thylakoid membranes of chloroplasts (Fig. a). Mitochondria are relatively large organelles (Chapter 3), with the cristae (folds) of their inner membrane providing a large surface area for ATP synthesis. It is perhaps not surprising to find that heart and skeletal muscle cells, which have a high requirement for ATP, have mitochondria with particularly large numbers of cristae (see Fig. 3.15).

Electron transport systems are composed of proteins and smaller compounds that have the ability to accept and donate electrons (e.g. cytochrome *c*, Fig. 2.18). These components are an integral part of the membrane, and each has a specific affinity for electrons (redox potential, see Chapter 2). They interact with each other such that electrons move sequentially from high to low potential (Fig. b). This forms a thermodynamically favourable sequence, which achieves a 'flow' or transport of electrons down a gradient in potential, that is, from electronegative towards electropositive.

Electrons from the reduced electron carrier are passed to a component in the chain with a slightly higher affinity for electrons (less negative redox potential) than this carrier. They are then passed to the next component, which has a slightly higher affinity again, and so on until they are passed to the final acceptor, oxygen, which has the highest affinity of all. Water is the product. Each electron transfer can be considered as a reaction coming towards equilibrium because there is more energy in the reduced donor than in the reduced acceptor.

In the electron transport systems of both mitochondria and chloroplasts, H-group carrier molecules receive electrons from, and then donate electrons to, electron-carrying protein complexes. The sites of electron exchange with the proteins are on opposite sides of the membrane. Therefore, when an electron is received by an H-group carrier, a proton is picked up from that side of the membrane and deposited on the other side of the membrane where the electron is passed on to the next protein complex.

In Figure b, the H-group carriers are designated as B, D and F and the electron protein carriers as C, E and G.

With the transport of protons and electrons, the membrane becomes polarised with a higher concentration of H$^+$ (lower pH) on one side than the other. The two sides of the relatively impermeable membrane are in different compartments so that this pH difference (or proton motive force) cannot simply dissipate. In the case of the prokaryotic plasma membrane, the two compartments are the cytoplasm of the cell and the outside environment. In the case of mitochondria and chloroplasts, the inner membranes of the organelles form closed envelopes or vesicles.

As a result of the flow of electrons through the graded series of electron-carrier proteins in the membrane, a proton gradient is generated. In

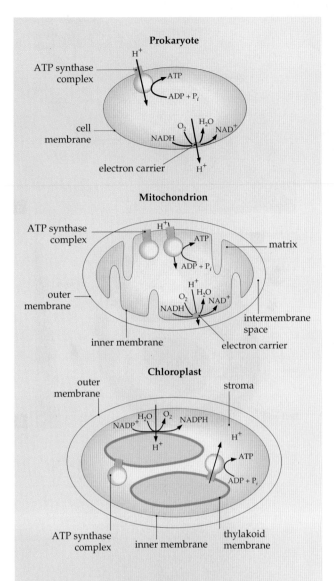

Prokaryote

Mitochondrion

Chloroplast

(a) Comparison of a prokaryote, a mitochondrion and a chloroplast showing the direction of proton (H^+) movement and the position of the ATP synthase complex where ATP is produced. The arrangement of electron carriers and the ATP synthase complex in the cell membrane of bacteria is very similar to the sequence found in the inner membrane of mitochondria

thermodynamic terms, the change in free energy as electrons traverse the series of carriers with decreasing redox potential is conserved as the proton gradient across the membrane.

Synthesis of ATP

Many transmembrane proteins are involved in the passage of protons. The protein is attached to an enzyme that can use proton translocation to synthesise ATP (Fig. 5.7). This enzyme, which has been found in a similar form in membranes of many kinds, is ATPase. As its name implies, it normally catalyses the reaction:

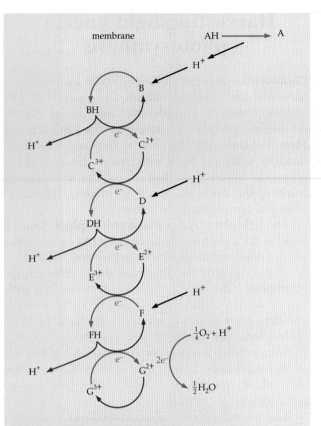

(b) Cytochrome electron transport chain. Electron carrier molecules in membranes interact with each other so that electrons move sequentially from high to low redox potential. This achieves a flow of electrons down a potential gradient

$$\text{ATPase}$$
$$\text{ATP} + H_2O \rightarrow \text{ADP} + HPO_4 + H^+$$

When ATP is hydrolysed, the reaction proceeds with a ΔG^O of -29.3 kJ mol^{-1} (see Chapter 2). However, because the enzyme in this case is physically attached to a proton channel, its catalytically active site has an almost unlimited supply of protons. The high concentration of protons effectively allows the reversal of the ATPase reaction to form ATP; the enzyme is thus functioning as an ATP synthase.

This 'chemiosmotic' coupling between electron transfer through localised carriers in membranes and the synthesis of ATP through utilisation of a proton gradient was originally proposed by Peter Mitchell in the 1950s and 1960s. It has proved to be a unifying concept in our understanding of energy conservation and transfer in living systems and, for this, Mitchell received the 1978 Nobel Prize for chemistry.

Harvesting light energy: photosynthesis

Photosynthesis is the process by which solar energy is harvested and used to convert CO_2 and H_2O into carbohydrates. A considerable amount of energy reaches the earth as radiation from the sun each day. Most is absorbed by the earth as heat and about one-third is reflected back into space. Photosynthetic organisms absorb less than 1% of the solar energy reaching the earth but use it to produce billions of tonnes of carbohydrate each year.

The light-absorbing pigment **chlorophyll** makes it possible for a photosynthetic organism to use light in a way no other organism can. In eukaryotes, chlorophyll is associated with the thylakoid membranes of the chloroplast (Chapter 3). Chloroplasts occur in a wide variety of sizes and shapes and in quite variable numbers per cell; mangroves typically have 20 to 30 chloroplasts per leaf cell whereas the green alga *Chlamydomonas* has only one (Fig. 5.8). A leaf the size of your hand can contain a total of three to five billion chloroplasts (roughly equivalent to the total human population on the earth!).

The overall process of photosynthesis involves a series of chemical reactions, which are summarised by the equation:

$$\overset{\text{visible light}}{6CO_2 + 12H_2O \rightarrow C_6H_{12}O_6 + 6O_2 + 6H_2O}$$

| carbon dioxide from atmosphere | water | sugar | oxygen from original water molecules | water |

In eukaryotes, there are three essential chemical processes in photosynthesis.

1. *Absorption of energy from sunlight by pigments— light-dependent reactions.* The harvesting of light energy for the light reactions of photosynthesis proceeds through two photosystems (**photosystem I and photosystem II**) operating concurrently. These photosystems consist of protein molecules in which pigments are held in very special arrangements. The photosystems are embedded in the thylakoid membranes of chloroplasts. Light energy is absorbed by the pigment molecules and transferred to a special part of the photosystems, the **reaction centre**. Here electrons are energised and removed to be used in a chain of reactions in which the energy is eventually stored in the molecules ATP and NADPH.
2. *Reactivation of reaction centres.* Reaction centres are replenished by replacing electrons removed

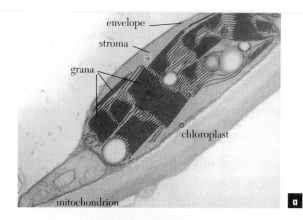

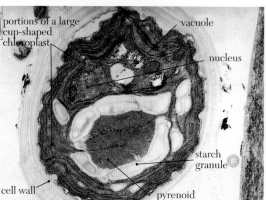

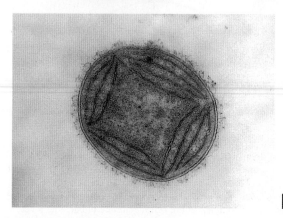

Fig. 5.8 Electron micrographs of cross-sections through **(a)** a chloroplast from a plant leaf, *Avicennia marina*, the white mangrove, **(b)** a unicellular alga, *Chlamydomonas reinhardii* and **(c)** a prokaryotic cell, *Prochloron*, a cyanobacterium

following light energy absorption. Water acts as a source of these electrons and oxygen is produced as a by-product.

3. *Carbon fixation to produce carbohydrates in 'dark reactions'.* These reactions are light-independent, using the energy (ATP and NADPH) captured in the light reactions to synthesise sucrose and starch. These reactions can function in darkness as well as in light provided that ATP and NADPH have been generated in the light reactions.

Light energy

Light is a form of electromagnetic radiation that has properties of both waves and particles. Its wave nature allows it to be described in terms of wavelength or frequency, just as can be done with sound waves. However, when it interacts with matter, light can be considered as discrete packages, **photons** (Gr. *phos* [gen. *photos*], light). Photons have different amounts of energy, the amount being inversely proportional to the wavelength of a photon. Blue photons, with a wavelength of 400 nm, have more energy than red photons, with a wavelength of about 700 nm.

Sunlight is a mixture of photons with a range of energy levels, some of which our eyes perceive and some of which are out of the visible range (Fig. 5.9). Visible light ranges from violet through the colours of the rainbow to red. Our ability to perceive colours relies on the fact that different substances vary in the degree to which they absorb or reflect different wavelengths of incident radiation. An object that we perceive as red absorbs in the green region of the spectrum and either reflects or transmits reddish hues. A black object absorbs all visible wavelengths of radiation and reflects none, whereas a white object reflects most wavelengths and absorbs few. The light used in photosynthesis falls within the visible spectrum (Fig. 5.9).

Below the lower limit of the visible spectrum (below about 380 nm) photons have progressively shorter wavelengths and greater energy, ranging from ultraviolet (UV) light, X-rays, gamma rays to cosmic rays. Beyond the red end of the visible spectrum (wavelengths greater than 700 nm) are infra-red radiation and long-wave radiowaves, which contain less energy than visible light. Infra-red radiation is emitted by warm bodies and is used by organisms such as some snakes to locate the small mammals upon which they prey.

Pigments: molecules that absorb light

Pigments are coloured molecules. We see them as coloured because they absorb photons with particular energy levels and reflect others. Chlorophyll, for example, is a rich green colour because it preferentially absorbs photons of blue and red light, and reflects photons in the green portion of the spectrum.

A pigment is best characterised by its pattern of absorption of photons, its **absorption spectrum**. The absorption spectrum of chlorophyll is similar to the wavelengths of light that activate photosynthesis (**action spectrum**) (Fig. 5.10), indicating that chlorophyll is the main pigment involved in photosynthesis. High values on the action spectrum indicate high rates of photosynthesis where photosynthetically active radiation is absorbed.

Chlorophyll absorbs light energy by a process of excitation, involving a central metal atom (magnesium) surrounded by alternating single and double bonds that form a **porphyrin ring** (Fig. 5.11). When photons are absorbed by the pigment molecule, electrons in the magnesium atom are excited and the energy is funnelled off through these bonds. Chlorophyll *a* is the principal photosynthetic pigment in prokaryotes and eukaryotes. Green algae and plants also contain

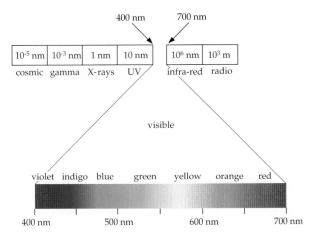

Fig. 5.9 The solar radiation spectrum is composed of extremely short (but high energy) wavelength cosmic, gamma, X-rays, ultraviolet (UV), visible, infra-red radiation and radio waves. Most of the sun's output ranges from UV to short wavelength infra-red. Photosynthetically active radiation is the visible spectrum and accounts for only a very narrow waveband, ranging in colour from violet to red (4×10^{-5} cm = 400 nm)

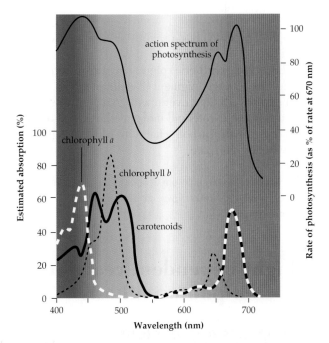

Fig. 5.10 Absorption spectra for chlorophyll *a* and *b* and carotenoids showing the particular wavelengths each pigment absorbs most. The sum of the absorption spectra corresponds to the action spectrum of photosynthesis

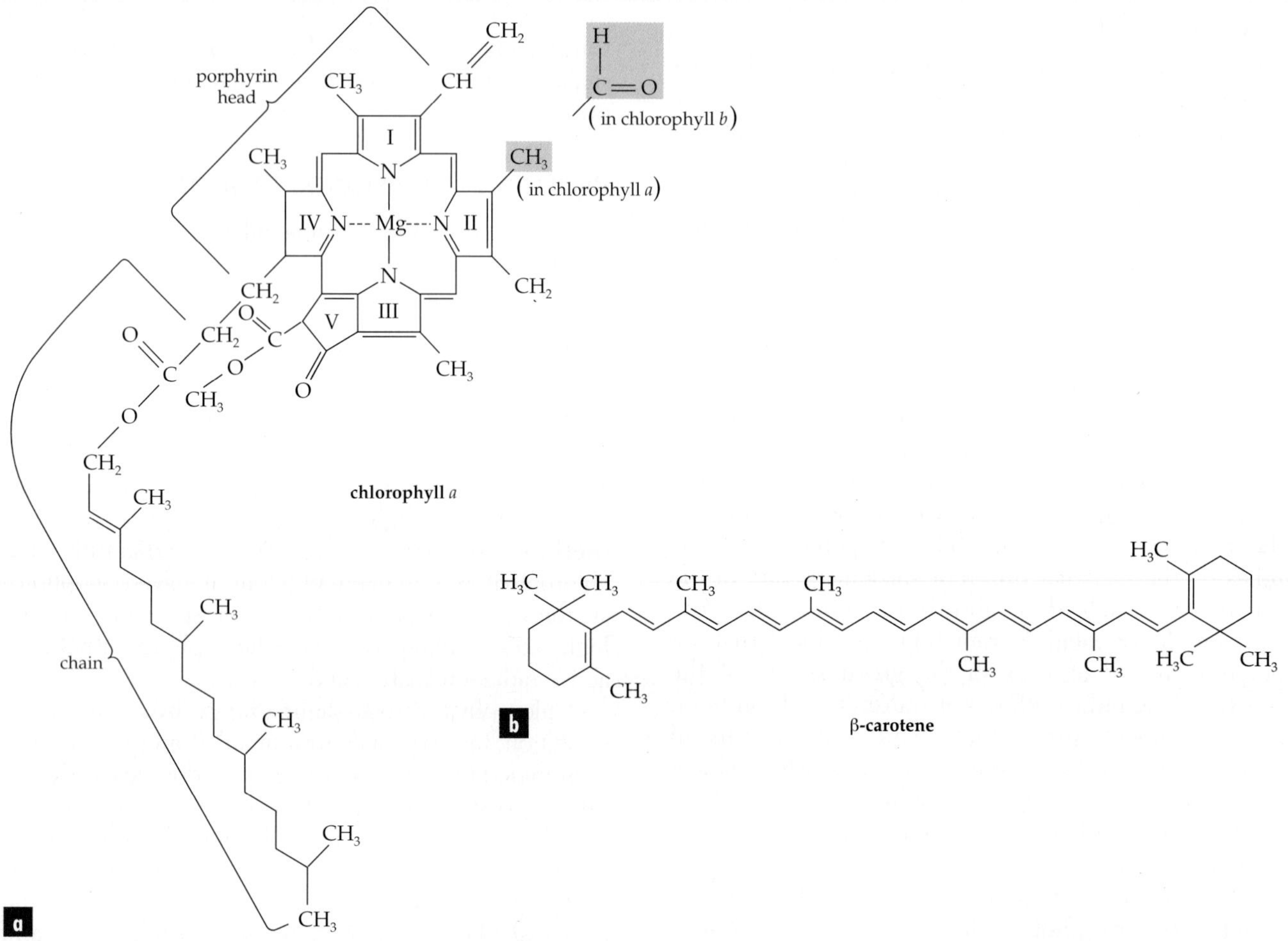

Fig. 5.11 Molecular structure of **(a)** chlorophyll *a* and **(b)** β-carotene. Chlorophyll consists of a porphyrin ring to which a phytol chain is attached. Chlorophyll *b* is different from chlorophyll *a* only by a substitution of the methyl group on ring II with an aldehyde group

chlorophyll *b*, which has a slightly different absorption spectrum.

Other photosynthetic pigments are carotenoids, such as the orange pigment β-carotene (Fig. 5.11), which can absorb light over a wide range of the visible spectrum. Each carotenoid molecule has a carbon ring and a long hydrocarbon chain in which single and double bonds alternate.

Photons of visible light are absorbed by chlorophyll and carotenoid pigments, providing energy for photosynthesis.

Light-dependent reactions of photosynthesis

Chloroplasts trap photons

In eukaryotes, the chlorophyll and carotenoid pigments that trap sunlight are located in chloroplasts (Fig. 5.12). You will remember from Chapter 3 that a chloroplast is bounded by a double membrane envelope. The membranes function as both border and barrier, defining the volume and shape of the organelle. They enclose the matrix (**stroma**) and control the movement of materials between the chloroplast and cytosol (see Chapter 4). Chloroplasts also have a third, innermost membrane system, the **thylakoid membrane** system, which is the site of conversion of light energy into electrical and then into chemical energy.

Thylakoid membranes form an elaborately folded, continuous membrane system (Fig. 5.13). Thylakoid membranes occur in two configurations: as stromal and as granal lamellae. Stromal lamellae are individual cisternae (sacs) traversing the chloroplast (Figs 5.8, 5.12) whereas granal membranes form stacks of from two to more than 50. Integrated into the thylakoid membrane are several types of protein complexes that together make up the photosynthetic electron transport system. The protein complexes include light-harvesting complexes, in which pigment molecules are bound to proteins, electron transport complexes and ATP-synthesising complexes. Associated with the light-

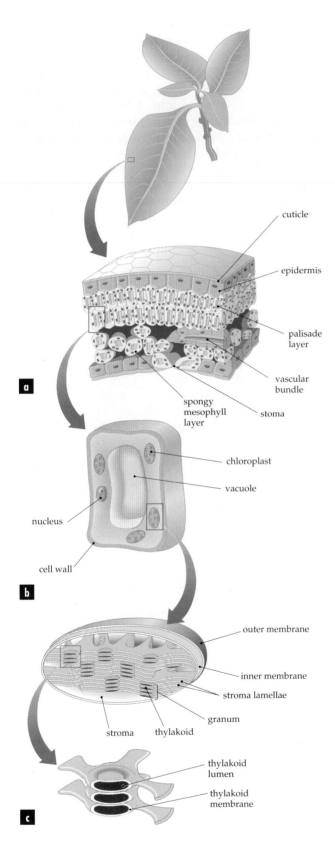

Fig. 5.12 Location of chloroplasts in a flowering plant. **(a)** In plants, chloroplasts occur in the palisade and spongy mesophyll cells of leaves. **(b)** Mesophyll cell. **(c)** Chloroplast showing the double outer membrane (envelope) and the grana and stroma lamellae embedded in the stroma

BOX 5.2 Chloroplasts in sun and shade

Australian researchers played a key role in the discovery of the structural and functional complexities of the thylakoid membrane. Dr Keith Boardman, former Chief Executive Officer of the CSIRO, and Dr Jan Anderson, from the CSIRO Division of Plant Industry in Canberra, were among the first in the 1960s and 1970s to describe the structural and functional differences between chloroplasts from leaves growing in the sun and in the shade. The figure shows the different morphological features of chloroplasts in the leaves of *Alocasia macrorrhiza* (cunjevoi) grown in a deeply shaded rainforest understorey or in an open, very sunny spot by a roadside. Note the

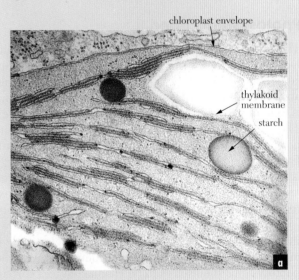

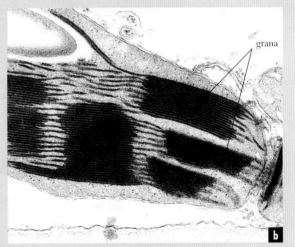

Chloroplasts in **(a)** sun and **(b)** shade leaves of cunjevoi, *Alocasia macrorrhiza*. In deep shade under the canopy of rainforest trees, chloroplasts have very large granal stacks

very large granal stacks in the shade chloroplast, which seem to pack the entire chloroplast.

Boardman and Anderson also provided the first evidence that all chlorophyll and other pigments involved in photosynthetic light capture and energy conversion are bound to proteins that are arranged in discrete supramolecular complexes, forming functional units in the thylakoid membrane.

harvesting complexes are two complexes known as photosystem I (PSI) and photosystem II (PSII).

The chloroplast envelope forms a selective barrier that regulates the transfer of molecules into and out of the chloroplast. Thylakoid membranes contain light-harvesting, electron transport and ATP-synthesising complexes.

Photosystems I and II

The protein-bound chlorophyll molecules of the light-harvesting complexes serve as light intercepting 'antennae' (Fig. 5.13). When a chlorophyll molecule absorbs a photon, it becomes excited (Chapter 1), that is, one of its electrons moves to a higher energy level. The excited chlorophyll molecule then interacts with its neighbour, transferring energy and exciting it in turn. Excited electrons are not transferred, only the energy is; the excited electrons return to their original energy level in the same molecules. This type of energy transfer ('excitation transfer') is extremely efficient.

The orientation and positioning of pigments allows the excitation energy of dozens of chlorophyll molecules in the light-harvesting complexes and the photosystems themselves to be channelled to a certain point within the associated photosystem complexes

(Fig. 5.14). Most of the absorbed energy is concentrated at a particular pair of chlorophyll molecules, the reaction centre. These chlorophyll molecules (called P680 in PSII and P700 in PSI) differ from others in that excitation causes them to expel an electron. An electron acceptor on the stromal side of the thylakoid membrane, but still within the photosystem, receives the electron, thereby producing a charge separation across the thylakoid membrane.

After absorption of a photon by the light-harvesting complexes associated with PSII or by the photosystem itself, and the loss of an electron from P680, an electron donor rapidly neutralises the positively charged reaction centre. This donor is then neutralised by

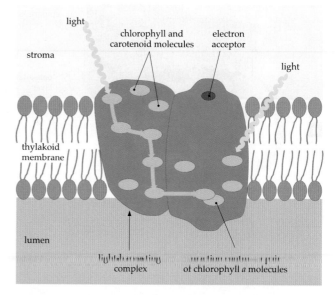

Fig. 5.14 A model of the light-harvesting complexes associated with both photosystems I and II. Pigments in the light-harvesting complex absorb photons, the energy of which is transferred to the reaction centre of a photosystem. Photons can also be absorbed by chlorophyll molecules within the photosystems themselves

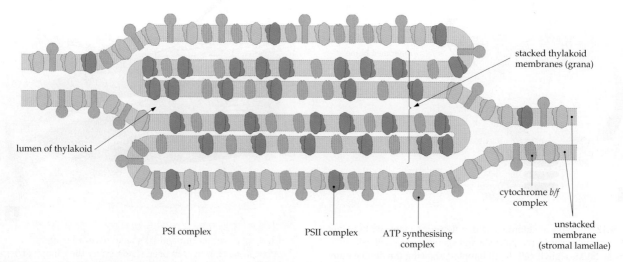

Fig. 5.13 A model showing the distribution of protein complexes (light-absorbing photosystem complexes I and II, cytochrome *b/f* complex and ATP-synthesising complex) on the thylakoid membrane system

removal of electrons from H_2O, producing O_2 and four protons for every four electrons displaced from the reaction centre. The protons accumulate in the thylakoid lumen (Fig. 5.15). The electron on the acceptor molecule of PSII on the stromal side of the membrane is then passed by an electron carrier to another protein complex, the **cytochrome b/f complex**, and then to the electron donor molecule of PSI. During the process, at least one more proton is pumped into the lumen of the thylakoids.

The light-harvesting complex associated with PSI or the photosystem itself absorbs an additional photon, the energy of which allows an electron from the reaction centre (P700) to move to an electron acceptor molecule on the stromal side of the membrane. Here electrons are removed from the PSI complex and given to a small acceptor protein, ferredoxin, which passes them on to $NADP^+$. Two electrons and one proton (H^+) reduce $NADP^+$ to NADPH (Fig. 5.15). Thus, electrons ultimately derived from H_2O are passed from PSII to PSI to produce NADPH for the light-independent reactions of photosynthesis. The two photosystems each energise electrons along the way. Some of this energy from the photosystems is used to pump protons to the thylakoid lumen, and the remainder is transferred to NADPH.

Photosystems I and II use solar energy to move electrons across thylakoid membranes. Electrons, ultimately derived from H_2O, are passed from PSII to PSI to produce NADPH for the light-independent reactions. Oxygen atoms from H_2O form O_2 as a by-product.

The protons that accumulate in the lumen of the thylakoids create an electrochemical proton (pH) gradient across the membrane. This provides potential energy that is utilised in the synthesis of ATP. Protons moving down the electrochemical gradient cross the membrane through ATP-synthesising protein complexes to the stroma. This is similar to the proton-driven ATP generation in mitochondria. (The process of using an electrochemical gradient to drive the synthesis of ATP is known as **chemiosmosis**.)

The electron flow from water through PSII and PSI to $NADP^+$ is one-way (Fig. 5.15) and is referred to as non-cyclic electron transport. The ATP synthesis coupled to this is **non-cyclic photophosphorylation**. Electrons excited by PSI, however, can take an alternative cyclic route. They can be transported by ferredoxin and the cytochrome b/f complex back to PSI. During this process protons are pumped across the membrane, providing energy for ATP synthesis, but the recycled electrons are not used for NADPH production. This process, **cyclic photophosphorylation** (Fig. 5.16), can vary the amount of ATP relative to NADPH produced by the electron transport chain. This flexibility is necessary to meet the varying demands of chloroplast metabolism for NADPH and ATP.

Protons, which accumulate in the lumen of the thylakoid sacs, move across the membrane to the stroma through proton channels, providing energy for the synthesis of ATP.

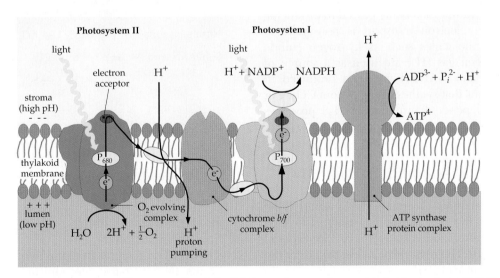

Fig. 5.15 Thylakoid membrane complexes. Photosystem II and the light-harvesting complex consist of at least a dozen individual proteins, all associated with each other to form a functional multiprotein complex. Bound to the proteins are pigments, such as chlorophylls a and b and carotenoids, and redox catalysts (such as quinones and manganese) for electron transfer. The cytochrome b/f complex consists of five proteins, which carry several prosthetic groups involved in electron transfers within the complex. Photosystem I consists of more than 10 proteins that bind pigments and electron transfer catalysts such as iron–sulfur centres and possible quinones. The ATP synthase (exclusively in stromal lamellae) consists of two parts: a proton channel and a catalytic site where ATP is synthesised

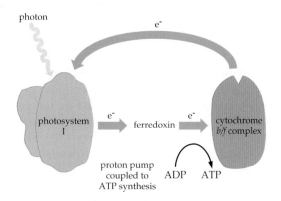

Fig. 5.16 ATP can be produced by cyclic photophosphorylation involving photosystem I. Excitation energy passes to ferredoxin and the cytochrome *b/f* complex, with the electron being returned to PSI

Prokaryotes and the evolution of photosystems I and II

In contrast to eukaryotes, photosynthetic prokaryotes lack chloroplasts. In primitive photosynthetic bacteria, such as green and purple bacteria, chlorophyll and the protein complexes associated with the light-dependent reactions of photosynthesis are located on the plasma membrane. In cyanobacteria, such as *Prochloron*, which is believed to be related to eukaryotic green algae, chlorophyll is located on thylakoid membranes within the cytoplasm (Fig. 5.8).

Green and purple bacteria lack PSII, do not use H_2O as a source of electrons and, therefore, do not produce oxygen. Electron flow is usually cyclic, with photosynthesis resulting in the production of energy in the form of ATP rather than being geared to producing NADPH. When electron transport is non-cyclic in these bacteria, molecules such as hydrogen sulfide (H_2S) or hydrogen gas (H_2) are used as a source of electrons. ATP and NADPH are then produced to drive the reactions that synthesise sugar from CO_2. For example, light-dependent oxidation of H_2S produces sulfur (S):

$$\text{light}$$
$$12H_2S + 6CO_2 \rightarrow C_6H_{12}O_6 + 12S + 6H_2O$$

With the evolution of cyanobacteria came PSII, which provided the mechanism for using H_2O as a source of electrons, and resulted in an atmosphere containing oxygen. The removal of an electron from the P680 reaction centre of PSII provides enough energy to split H_2O. The process is facilitated by the presence of a cluster of manganese ions bound to the reaction centre proteins. Removal of an electron from the P700 reaction centre of PSI only yields enough energy to split electrons from H_2S, not H_2O.

Green and purple bacteria lack PSII, do not use H_2O as a source of electrons and, therefore, do not produce oxygen. With the evolution of cyanobacteria came PSII, which provided the mechanism for using H_2O as a source of electrons, and provided the earth with an atmosphere of oxygen.

BOX 5.3   Chemosynthetic bacteria

Autotrophic bacteria have two basic sources of energy available to them for synthesising organic molecules. They can use sunlight in photosynthesis or they can use simple inorganic molecules in chemosynthesis. All autotrophic bacteria do, however, use carbon dioxide as their source of carbon and, as far as we know, all chemosynthetic bacteria use the Calvin–Benson cycle to fix carbon dioxide.

There are five groups of chemosynthetic bacteria, which use different inorganic substrates for chemosynthesis (see table). The enzyme-catalysed oxidation of these substrates by bacteria releases electrons that can be used by an electron transport chain to produce ATP. This electron transport is similar to mitochondrial oxidative electron transport in plants.

Chemosynthetic bacterial group	Half-reaction for substrate oxidation
Iron bacteria	$Fe^{2+} \rightarrow Fe^{3+} + e^-$
Ammonia bacteria	$NH_4^+ + 2H_2O \rightarrow NO_2^- + 8H^+ + 6e^-$
Nitrate bacteria	$NO_2^- + H_2O \rightarrow NO_3^- + 2H^+ + 2e^-$
Sulfur bacteria	$H_2S \rightarrow S + 2H^+ + 2e^-$
	$SO_3^- + H_2O \rightarrow SO_4^- + 2H^+ + 2e^-$
	$S + 3H_2O \rightarrow SO_3^- + 6H^+ + 5e^-$
Hydrogen bacteria	$H_2 \rightarrow 2H^+ + 2e^-$

The production of a pH gradient is coupled to the transport of electrons, and ATP is synthesised by a membrane-spanning enzyme that dissipates the proton gradient. Moreover, like plants, almost all chemosynthetic bacteria use oxygen as their final electron acceptor.

NADH production in chemosynthetic bacteria occurs by two basic mechanisms. In hydrogen bacteria, where the substrate has sufficient redox potential to reduce NAD^+ to NADH, it is thought that a hydrogenase enzyme catalyses NADH production from hydrogen gas (see figure). In

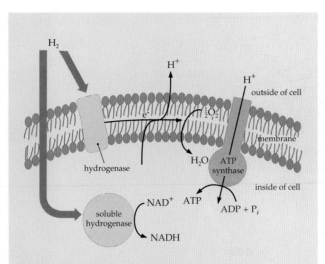

NADH and ATP production in a chemosynthetic bacterium, the hydrogen bacterium

other groups, however, it is doubtful that such direct NADH production can occur because the substrates do not have a high enough redox potential to reduce NADH. Bacteria in these groups employ an oxidative electron transport chain in reverse to produce NADH. This energetically unfavourable electron transport is driven in reverse by ATP hydrolysis and the consequent build-up of a proton gradient.

Chemosynthetic bacteria play a vital role in nutrient cycling (Chapter 44). Sulfur bacteria, for example, convert elemental sulfur into sulfate, which can be used by plants in protein synthesis. Ammonia and nitrate bacteria catalyse the process of nitrification. Ammonia, which is in equilibrium with ammonium ions, is converted into nitrate, the main form of nitrogen used by plants.

Light-independent reactions of photosynthesis

Carbon fixation and the Calvin–Benson cycle

The capture of atmospheric CO_2 and its incorporation into carbohydrates is **carbon fixation**. In eukaryotes it occurs in the stroma of chloroplasts. The stroma contains, among other things, multiple copies of the chloroplast genome, ribosomes and the enzymes needed for converting carbon dioxide to sugar, as well as those engaged in DNA duplication and the synthesis of RNA and protein. Not all of the proteins used in photosynthesis are, however, synthesised in the stroma (or encoded for in the chloroplast genome).

The first steps in CO_2 reduction is carbon fixation: the attachment of CO_2 to the 5-carbon sugar **ribulose bisphosphate** (RuBP). This carboxylation of RuBP is one of a series of reactions known as the Calvin–Benson cycle (Fig. 5.17), after its discoverers Melvin Calvin and Andrew Benson. A short-lived, 6-carbon intermediate is formed in a reaction catalysed by the enzyme **ribulose bisphosphate carboxylase–oxygenase** (**Rubisco**) (Fig. 5.18). Owing to it constituting up to 50% of the protein in chloroplasts, Rubisco is the most abundant protein on earth. The 6-carbon intermediate splits rapidly into two molecules of phosphoglyceric acid (PGA), a 3-carbon molecule (Fig. 5.17). From PGA, two enzymatic steps using NADPH and ATP take place. In the first step, PGA is phosphorylated using ATP, and in the second step the intermediate compound is reduced and dephosphorylated to form glyceraldehyde 3-phosphate (PGAL) in a reaction requiring NADPH.

PGAL, the primary carbohydrate product of the chloroplast, can follow three paths. In the first, up to two in every 12 PGAL molecules are exported from the chloroplast into the cytoplasm (Fig. 5.17). These PGAL molecules are combined and rearranged into fructose and glucose phosphates. These two intermediate compounds condense to form sucrose, the major transport form of carbohydrate in plants. Inorganic phosphate is imported into the chloroplast to replace that exported with PGAL. In the second path, up to two PGAL molecules are combined, rearranged and used in the synthesis of starch, which is stored in the chloroplast. In the final path, a series of reactions in the stroma, the remaining 10 PGAL molecules are used to form six RuBP molecules to complete the cycle. The regeneration of six RuBP molecules uses a further six ATP molecules.

Since the Calvin–Benson cycle is dependent upon an adequate supply of NADPH and ATP, its operation is closely linked with the electron transport chain. It is not surprising to find that the activity of certain key enzymes, such as sedoheptulose bisphosphatase, is light dependent. These enzymes function only when the electron transport chain is operating. The activity of the key enzyme Rubisco is itself regulated by a specific activating protein, activase.

The stroma is a matrix containing enzymes for carbon fixation as well as DNA, RNA, ribosomes and the machinery for protein synthesis. Carbon dioxide is fixed to RuBP in the first step of the Calvin–Benson cycle in a reaction catalysed by the enzyme Rubisco. Intermediate 3-carbon compounds are rearranged into sugar phosphates, some of which are used for carbohydrate synthesis and some for synthesis of new RuBP molecules.

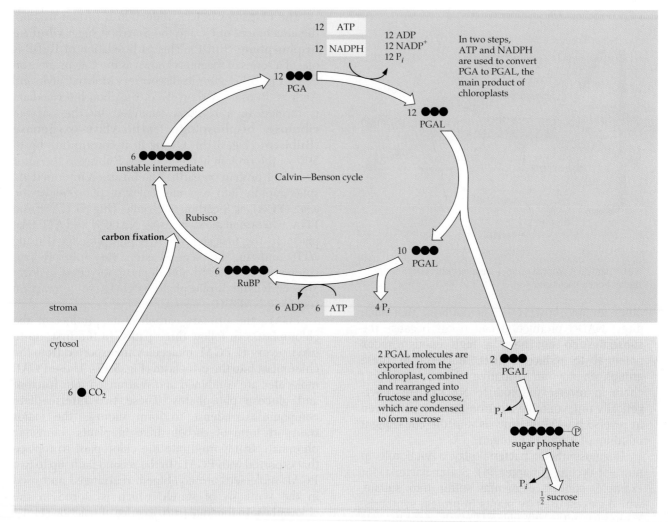

Fig. 5.17 The Calvin–Benson cycle in the stroma of chloroplasts. At each full turn of the cycle, three molecules of CO_2 enter; two turns are shown here. Six molecules of CO_2 combine with six molecules of the 5-carbon compound ribulose 1,5-bisphosphate (RuBP). The product, an unstable 6-carbon compound, splits to give the 3-carbon compound, phosphoglyceric acid (PGA). In a series of steps using the energy of ATP and NADPH, PGA is converted to glyceraldehyde 3-phosphate (PGAL). Two PGAL molecules are exported from the chloroplast and used to form sucrose, while 10 remain in the chloroplast to form six molecules of RuBP to continue the cycle. Instead of being used for sucrose production, the two PGAL molecules can be used for starch synthesis in the stroma

Photorespiration: competition between carbon dioxide and oxygen

The enzyme that catalyses carbon fixation, Rubisco, can also use oxygen as a substrate and oxygenate RuBP. This process, with the subsequent production of CO_2 from a product of oxygenation, is **photorespiration**. Under normal atmospheric conditions, O_2 is much more abundant than CO_2. Consequently, O_2 competes effectively for the binding site on the enzyme despite the enzyme having a higher affinity for CO_2 than for O_2. The process is called photorespiration because it occurs only in the light (it requires RuBP from the Calvin–Benson cycle) and, like respiration in mitochondria, it produces CO_2 and consumes O_2.

Photorespiration was not a problem billions of years ago when early photosynthetic organisms were evolving in an atmosphere lacking oxygen. However, in today's atmosphere, photorespiration results in CO_2 being released, undoing the binding of carbon that would otherwise result in carbohydrate synthesis. It also consumes ATP in the process. This means that photorespiration has the capacity to reduce significantly the efficiency of carbon fixation. In this way, many plants lose from 25% to 50% of the carbon that could be fixed during photosynthesis.

Photorespiration occurs because the enzyme Rubisco interacts with O_2 as well as CO_2. The process leads to less carbon being fixed during photosynthesis.

The C₄ pathway of photosynthesis

Plants have adapted to different environments and, thus, there are several variations in the photosynthetic

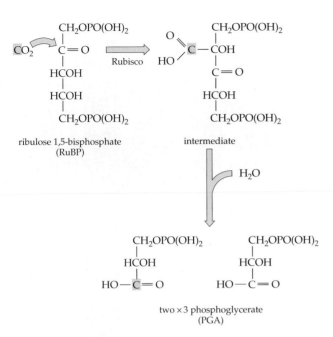

Fig. 5.18 Carbon fixation step showing the catalytic function of Rubisco. These reactions take place in the active site of Rubisco

pathway and the structure of chloroplasts. The process of carbon fixation that we have described so far is **C₃ photosynthesis** because the first stable product of carbon fixation is the 3-carbon compound, PGA (Fig. 5.17). However, many flowering plants, including tropical and subtropical grasses such as maize, sugarcane and millet, have a different carbon fixation pathway, **C₄ photosynthesis** (Fig. 5.19). While each C₄ plant group has its own characteristic enzymes, chloroplasts and leaf anatomy, in each case the first stable product of carbon fixation is a 4-carbon compound.

The C₄ photosynthetic process was first discovered independently in the 1960s by American and Russian workers who showed that a C₃ compound was not always the first stable product of carbon fixation in sugarcane and maize. Marshall (Hal) Hatch, an Australian, and Roger Slack, an Englishman, working at the time in the Colonial Sugar Refining Company's research laboratories in Brisbane, confirmed these findings and went on to resolve the basic mechanism of C₄ photosynthesis, identifying many of the enzymes involved in the pathway.

Plants with C₄ photosynthesis have a distinctive leaf anatomy ('Kranz anatomy') in which the vascular bundles are surrounded by a cylinder of bundle sheath cells and an outer layer of mesophyll cells (Fig. 5.20, see also Chapter 18). The bundle sheath and mesophyll cells contain chloroplasts that are different in structure and function.

Fig. 5.19 Sugarcane is an example of a plant that has C₄ photosynthesis

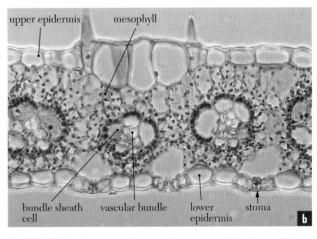

Fig. 5.20 (a) Distinctive leaf ('Krantz') anatomy of C₄ plants. The bundle sheath cells form tight wreaths around the vascular bundles. They have very thick cell walls, traversed by many plasmodesmata for metabolite exchange, which do not permit diffusion of CO_2, thus allowing high CO_2 concentrations to be achieved. The Calvin–Benson cycle occurs in the chloroplasts of these cells. Phosphoenolpyruvate carboxylation takes place in the mesophyll cells surrounding the bundle sheath cells.
(b) Micrograph of maize leaf showing bundle sheath cells

In C_4 plants, an additional carboxylation enzyme, phosphoenolpyruvate (PEP) carboxylase, operates in the cytoplasm of the leaf mesophyll cells. As its name implies, the enzyme catalyses the carboxylation of the C_3 compound, PEP. The product of the carboxylation reaction is a C_4 organic acid, oxaloacetate (Fig. 5.21), which is immediately converted into another C_4 compound. In sugarcane, this is malate, which is transported into bundle sheath cells where the Calvin–Benson cycle takes place. The mesophyll chloroplasts provide ATP and NADPH to make PEP and malate but they lack Rubisco; they have only the enzymes of the Calvin–Benson cycle that are needed to reduce PGA to PGAL.

Once in the chloroplasts of the bundle sheath cells (Fig. 5.22), malate (C_4) is decarboxylated to CO_2 and pyruvate (C_3). Carbon dioxide is then fixed into carbohydrates by Rubisco and the other Calvin–Benson cycle enzymes. The pyruvate is transported back into the mesophyll cells and there converted back into PEP to complete the cycle. The C_4 decarboxylation reaction generates NADPH, which is consumed in bundle sheath chloroplasts during PGA reduction. Bundle sheath chloroplasts often have only a few granal stacks and little PSII activity, and hence little capacity for light-dependent NADPH formation.

C_4 photosynthesis is a mechanism for concentrating CO_2 in bundle sheath cells, which are relatively impermeable to CO_2 and tend to hold it within them. The relatively high CO_2 concentration within these cells has the added advantage of inhibiting photo-respiration. Although the C_4 pathway uses more ATP than the C_3 pathway, the reduction of photorespiration offsets this cost, especially when light is abundant. In addition, because C_4 plants can concentrate CO_2, their

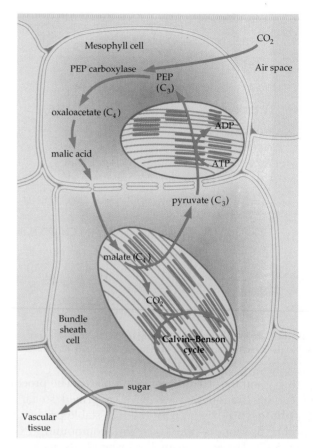

Fig. 5.22 Different cell types and chloroplasts involved in the metabolite transfers and gas exchange reactions in C_4 plant leaves

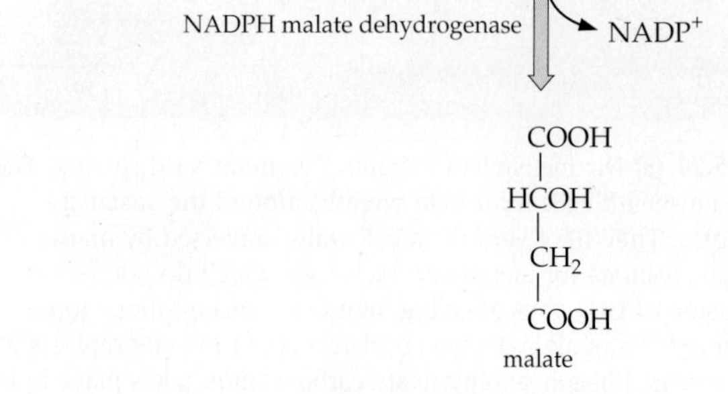

pores (stomata) are generally not as wide open as those of C_3 plants, and less water is consequently lost. This is particularly important under conditions of low humidity and high temperature, and when a lack of water limits growth. Many C_4 plants grow in hot climates and tropical crop plants such as sugarcane have a net rate of photosynthesis two to three times greater than temperate C_3 crop plants such as wheat.

In C_4 plants, carbon fixation occurs in mesophyll cells and CO_2 is then concentrated in bundle sheath cells, inhibiting photorespiration and increasing carboxylation efficiency.

Crassulacean acid metabolism

Crassulacean acid metabolism (**CAM**) is a variation on the C_4 pathway of photosynthesis and was first discovered in succulent plants of the flowering plant family Crassulaceae. It has evolved independently in other plants such as pineapple and 'Spanish moss', both members of the family Bromeliaceae (Fig. 5.23). Photosynthesis in CAM plants involves both the C_4 pathway and the Calvin–Benson cycle, but the reactions, which happen in the same cell, occur at different times.

Fig. 5.21 The start of the C_4 pathway of carbon fixation showing the carboxylation reaction and subsequent reduction reaction

Fig. 5.23 CAM plants occur in many families of flowering plants and include the **(a)** pineapple (*Ananas comosus*, family Bromeliaceae) and **(b)** the tropical epiphyte fern *Pyrrosia longifolia*. Measurements of net CO_2 exchange show carbon gain occurs via CO_2 dark fixation

CAM plants have the unusual ability to open their gas-exchange pores (stomata, located on the surface of leaves) and fix CO_2 into C_4 compounds at night rather than the day as in other plants. The C_4 compound produced in darkness is stored in the vacuole of the mesophyll cell and exported back into the cytoplasm of the same cell to be decarboxylated as soon as daylight appears. Carbon

dioxide released from the C_4 compound is then fixed in the chloroplasts in the normal manner via RuBP and the Calvin–Benson cycle (Fig. 5.24).

This peculiar biochemistry makes it possible for plants to survive under extremely hot and dry conditions. Closure of stomata during the day reduces water loss and their opening at night allows CO_2 uptake when evaporation of water is minimal. The accumulation and concentration of CO_2 during the night ensures efficient photosynthesis the next day.

CAM plants fix CO_2 at night and convert it to carbohydrate during the day. This allows them to close their gas-exchange pores during the day, minimising water loss.

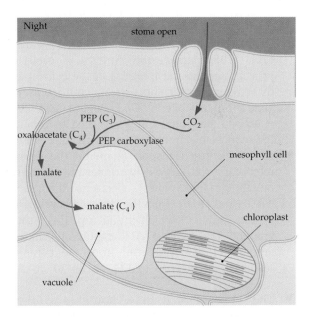

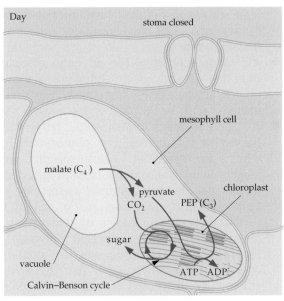

Fig. 5.24 CAM pathway in the cell. Gas exchange takes place at night

Summary

- The first organisms were heterotrophic and used molecules that formed spontaneously in the absence of oxygen as a source of raw materials and energy. Photosynthetic organisms evolved later, enriching the atmosphere with oxygen.

- All organisms extract energy from the oxidation of glucose. The first stage of energy extraction in the cytosol is glycolysis, which produces two molecules of ATP and two of pyruvate per molecule of glucose. In the presence of O_2, the end product is a 3-carbon molecule, pyruvate, which becomes a substrate for the citric acid cycle after conversion to acetyl CoA. In the absence of O_2, the end product of glycolysis is ethanol or lactate.

- Pyruvate from glycolysis is converted to acetyl CoA, which enters the citric acid cycle. In prokaryotes this cycle occurs in the cytosol and in eukaryotes it occurs in mitochondria. Oxidation of acetyl CoA yields ATP and energised electrons, which are used to generate the reduced electron carriers, NADH and $FADH_2$.

- In an electron transport chain on the plasma membrane (prokaryotes) or inner mitochondrial membrane (eukaryotes), electrons donated by NADH and $FADH_2$ from the citric acid cycle drive proton pumps, which build up the proton gradient that powers ATP synthesis. The net amount of ATP produced from the complete oxidation of glucose to CO_2 is up to 36 ATP molecules.

- The bulk of ATP in a non-photosynthetic eukaryotic cell is produced in mitochondria. The large surface area of the folded inner membrane of these organelles provides a large surface area for ATP synthesis.

- All energy-supporting life on earth originates from the sun. Only green land plants, algae and photosynthetic bacteria can make use of this energy directly by the process of photosynthesis.

- All photosynthetic organisms have light-harvesting complexes and photosystems in which pigments that absorb specific wavelengths of light are bound to proteins arranged in membranes. In photosynthetic eukaryotes, the protein-pigment complexes are located in the thylakoid membrane system of chloroplasts. A photon captured by a pigment energises an electron in the molecule and the excitation energy is transferred to a pair of specific chlorophyll molecules housed in the photosystem complex, the reaction centre.

- There are two types of reaction centre associated with the two photosystems (PSI and PSII) of higher plants. Following absorption of energy, an electron is then removed from the reaction centre, which drives electron transport from water (the electron donor) to $NADP^+$ (the electron acceptor). During electron transport, protons move across the thylakoid membrane, providing energy for the synthesis of ATP.

- Green and purple bacteria have only one photosystem, which is similar to PSI of higher plants. This photosystem donates an electron to an electron transport chain from where it is returned to P700. This cyclic flow of electrons results in the production of ATP (cyclic phosphorylation). When electron transport is non-cyclic in these bacteria, molecules such as hydrogen sulfide (H_2S) or hydrogen gas (H_2) are used as a source of electrons. NADPH and ATP are produced to drive the Calvin–Benson cycle reactions synthesising sugar from CO_2.

- With the evolution of cyanobacteria came a second photosystem (PSII), with the reaction centre P680. PSII provided a mechanism for stripping electrons from H_2O. The electrons are passed from PSII to PSI for the production of NADPH and ATP by non-cyclic photophosphorylation. The residual oxygen atoms from water form O_2 as a by-product.

- ATP and NADPH generated in the light-dependent reactions drive the Calvin–Benson cycle (located in the stroma of eukaryotic chloroplasts). Carbon dioxide is bound to RuBP in the first step of the Calvin–Benson cycle in a reaction catalysed by the enzyme Rubisco. Intermediate 3-carbon compounds are rearranged into sugar phosphates. It takes six turns of the Calvin–Benson cycle to produce one 6-carbon sugar phosphate because only one of every six glyceraldehyde 3-phosphate (PGAL) molecules is funnelled into carbohydrate synthesis. The remainder is used to form new RuBP molecules.

- Because the carbon-fixing enzyme Rubisco can use O_2 as well as CO_2 as a substrate, photorespiration is an inevitable consequence in an oxygen-rich atmosphere. Photorespiration leads to a loss of carbon fixed during photosynthesis and it consumes energy.

- Several variants of the carbon-fixation process have evolved in plants. C_4 photosynthesis and crassulacean acid metabolism (CAM) are characterised by an additional carboxylation reaction and carboxylation enzyme, phospho-enolpyruvate (PEP) carboxylase. Leaf anatomy and physiological properties allow spatial (C_4) or temporal (CAM) separation of the two carboxy-lation (C_3 and C_4) reactions. The advantage of C_4 photosynthesis and CAM is a CO_2-concentrating effect, leading to reduced photorespiration, higher carboxylation and better water use efficiency of photosynthesis. These features are important in arid environments, where combina-tions of high irradiation, high temperatures and limited water occur.

keyterms

absorption spectrum (p. 125)
acetyl CoA (p. 117)
action spectrum (p. 125)
anaerobic (p. 116)
autotroph (p. 116)
β-oxidation (p. 119)
C_3 photosynthesis (p. 133)
C_4 photosynthesis (p. 133)
carbon fixation (p. 131)
cellular respiration (p. 116)

chemiosmosis (p. 129)
chlorophyll (p. 124)
citric acid cycle (p. 119)
crassulacean acid metabolism (CAM) (p. 134)
cyclic photophosphoryl-ation (p. 129)
cytochrome *b/f* complex (p. 129)
cytochrome *c* oxidase (p. 121)

fermentation (p. 117)
glycolysis (p. 116)
heterotroph (p. 116)
Krebs cycle (p. 119)
non-cyclic photophosphoryl-ation (p. 129)
oxidative respiration (p. 118)
photon (p. 125)
photorespiration (p. 132)
photosynthesis (p. 124)
photosystems I and II (p. 124)

porphyrin ring (p. 125)
pyruvate (p. 117)
reaction centre (p. 124)
ribulose bisphosphate carboxylase–oxygenase (Rubisco) (p. 131)
ribulose bisphosphate (RuBP) (p. 131)
stroma (p. 126)
thylakoid membrane (p. 126)

Review questions

1. What is glycolysis and where does it occur in a cell? How much ATP is produced from glycolysis compared with oxidative respiration?

2. Explain how NADH and $FADH_2$ are used by the inner membrane of mitochondria to drive ATP synthesis. Of what significance are the cristae of mitochondria?

3. Summarise the different biochemical pathways by which energy is extracted from carbohydrates and lipids.

4. How do plants and animals differ in the way in which they metabolise lipids? (Choose the correct answer.)

   A They do not differ at all.

   B Plants metabolise lipids exclusively in glyoxysomes, whereas animals metabolise them in mitochondria.

   C Both plants and animals metabolise lipids in glyoxysomes.

   D Plants metabolise lipids in both mitochondria and glyoxysomes, whereas animals metabolise them exclusively in mitochondria.

5. What is the significance of fermentation reactions in cells? How does fermentation differ in a yeast cell and a muscle cell?

6. Which product of glycolysis is imported by mitochondria for ultimate use in the Krebs cycle? (Choose the correct answer.)

   A Acetyl CoA

   B Pyruvate

   C Lactate

   D Ethanol

7. Make a table showing the major differences between energy-harvesting pathways (glycolysis, oxidative respiration and photosynthesis) in prokaryotic and eukaryotic cells. Indicate where the different reactions occur.

8. Explain the relationship between the action spectrum of photosynthesis and the absorption spectra of chlorophylls and carotenoids.

9. How do pigment molecules in the light-harvesting complexes trap and transfer solar energy? What is the role of the reaction centres in the photosystems?

10. What organisms use water in the light-dependent reactions of photosynthesis? What is the role of water?

11. If ATP and NADPH were supplied externally to leaves, leaf cells or isolated intact chloroplasts, would you expect photosynthetic carbon fixation to continue in the dark? Explain your answer.

12. Why is the enzyme ribulose bisphosphate carboxylase–oxygenase so-named?

13. How does C_4 photosynthesis differ from CAM? What are the advantages of these two pathways for plants living in hot or dry environments?

Extension questions

1. Red and blue photons contain different amounts of energy. Do you think that they would sustain different rates of photosynthesis?

2. What organelles are involved in the process of photorespiration? Where is carbon dioxide released in this process?

3. What would be the significance for plants of increased carbon dioxide concentration and predicted increased temperature caused by the 'greenhouse effect'?

4. The oxidative electron transport system in plant mitochondria contains two terminal oxidases (one is cytochrome *c* oxidase). What is the function of the second 'alternative' oxidase?

5. What are the main forms of storage carbohydrate used to provide glucose for glycolysis? How do their structures and storage locations differ?

6. Carbon dioxide fixation in CAM plants proceeds in the dark via the enzyme PEP carboxylase. This enzyme uses PEP and carbon dioxide as substrates. How does the plant sustain a continuous supply of PEP?

7. Recent research shows that the ATP synthase enzyme of chloroplasts and mitochondria functions in some way similar to a rotational motor. What is the power source for this motor and how does it result in synthesis of ATP?

Suggested further reading

Atwell, B., Kriedemann, P., Turnbull, C. (1999). *Plants in Action*. Melbourne: Macmillan.

An excellent advanced text covering the material discussed in this chapter together with numerous Australian examples.

Taiz, L. and Zeiger, E. (1998). *Plant Physiology*. 2nd edn. Sunderland, MA: Sinauer.

An excellent advanced text covering the material discussed in this chapter and much more.

Walker, D. A. (1992). *Energy, Plants and Man*. Brighton, UK: Oxygraphics.

A clear and entertainingly written account of photosynthesis and energy.

CHAPTER

6

Cells and tissues

n multicellular organisms, such as animals and plants, cells are specialised for particular purposes and usually linked together into tissues, where they function in a co-ordinated way to meet the needs of the whole organism. In earlier chapters, we described the internal structure and functions of cells. This chapter is concerned with the associations between cells that are functioning together as a tissue. **Tissues** are groups of similar, differentiated cells and the associated extracellular matrix, which together carry out a particular function (Fig. 6.1): muscle tissue contracts, glands secrete and epidermis covers the surface of an animal or plant. Cells are organised into tissues in all animals except sponges (phylum Porifera, Chapter 38).

Plants are obviously different in appearance from animals. However, as we saw in Chapter 3, plant and animal cells have many similarities. The most obvious difference is that plant cells, like bacterial, algal and fungal cells, are encased in a cell wall external to the plasma membrane. It is the cell wall, more than any

other feature, that has influenced the evolution of plants. Cell walls affect how plant cells are organised, interact with each other and function. In fact, the fundamental differences between plants and animals in regard to mechanisms of nutrition, growth, reproduction and defence can be attributed largely to the presence of cell walls in plants.

Animal cells, by contrast, are surrounded by an extracellular matrix that forms a loose and flexible lattice. The matrix is quite different in nature to a cell wall and confers some of the unique properties of animal cells and their capacity to organise into different tissues. Unlike plant cells, the plasma membranes of adjacent cells may be in direct contact with each other. The surface adhesive properties of plasma membranes are important during developmental processes (Chapter 15). Chemical interactions between the surfaces of adjacent cells influence control of cellular functions, such as growth and division. Within some tissues, such as muscle, many cells are co-ordinated to function as a unit by direct chemical and electrical communication between them. In the absence of a cell wall, support through turgidity is not generally possible; larger animals have internal and external skeletal structures to provide support and protection for their bodies.

n multicellular organisms, cells are specialised for particular purposes and linked together into tissues.

Development of animal tissues

At an early stage of development, all cells of animal embryos have a similar appearance. As development continues, the structure and properties of different groups of cells diverge—cells **differentiate**. Under normal conditions, this divergence is irreversible. Even when isolated and grown in tissue culture for many generations, differentiated animal cells usually maintain their distinct characteristics (Fig. 6.2).

During development, cells with the same properties tend to adhere together to form tissues by a process of selective cohesion. In the same way that cells can aggregate to form tissues, so a number of tissues may combine to form a functional unit, an organ. In more complex animals, several organs may combine to form systems, which carry out entire processes, such as digestion or gas exchange.

Animal tissue are groups of similar, differentiated cells and their associated extracellular matrix, which together carry out a particular function.

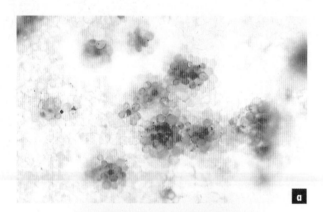

Fig. 6.1 The surface tissue of a rose petal is composed of two cell types: clusters of scent-releasing cells (stained red) and epidermal cells (unstained)

Differentiation of cells often involves considerable specialisation, that is, the development of certain

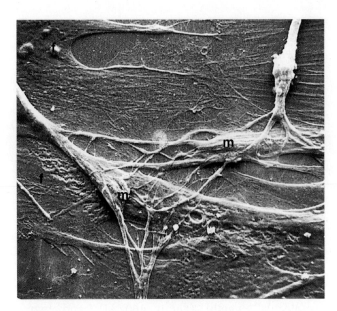

Fig. 6.2 When embryonic nerve and muscle cells are grown together in tissue culture, they develop many of the characteristics of normal differentiated nerve and muscle cells. Muscle cells (m) attach to each other and begin to contract and nerve cells develop long processes, which grow out and innervate the muscle cells

cellular functions at the expense of others. To take an extreme example, mature red blood cells in mammals have lost their nucleus and most of their cellular organelles; they are, essentially, a bag of haemoglobin with a limited life span. As a result of specialisation of function, some differentiated cells depend to a greater extent on the extracellular environment for the provision of necessary materials and appropriate physical conditions. The maintenance of this environment is, therefore, of vital importance to survival. In mammals, a diverse group of cells is involved in maintaining the extracellular environment. These include cells that secrete the extracellular matrix (fibroblasts), phagocytic cells that dispose of unwanted materials (macrophages), and white blood cells that help combat infection (leucocytes).

The extracellular environment is important in modulating the levels of activities of many types of cells. For example, circulating hormones can alter rates of cellular metabolic reactions, a change in diet may alter the rate of secretion of certain digestive enzymes, and increased stimulation by nerves can increase the size of muscle cells. In some tissues, the release of specific substances, such as growth factors, from adjacent tissues is important in the development and maintenance of fully functional, differentiated cells.

As a result of specialisation of function, some differentiated cells rely to a great extent on their extracellular environment.

The extracellular matrix

Within animals, the extracellular environment is a fluid containing the **extracellular matrix**, an extensive network of proteins and polysaccharides that fills the spaces between cells. The amount of matrix relative to the cells embedded in it and the type of proteins vary in different kinds of tissue. Tissues such as cartilage and bone have large amounts of matrix; muscle tissue has very little (Fig. 6.3).

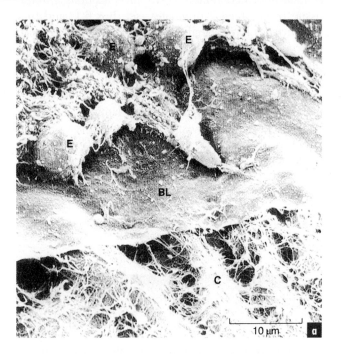

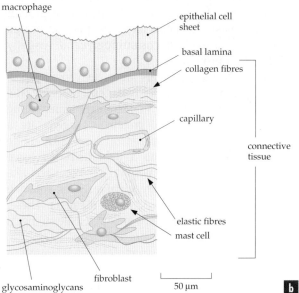

Fig. 6.3 (a) Scanning electron micrograph showing an epithelial basal lamina and underlying network of connective tissue microfilaments. **(b)** Extracellular matrix is abundant in connective tissue. The extracellular fluid gel contains scattered cells and an abundance of collagen fibres

There are two main forms of extracellular matrix—**interstitial matrix**, prominent in connective tissues, and **basal lamina**, which underlies epithelial cell layers. Both consist of strong protein fibres embedded in a highly hydrated polysaccharide gel. In both, large molecules are linked together by covalent and non-covalent bonds.

In the past, the extracellular matrix was regarded simply as a space filler, but recent research has changed this view dramatically. We now know that the extracellular matrix contains adhesive proteins unique to the tissue from which it originates (see Box 6.1) and that cell surfaces have specific receptors that bind to these matrix components. Strength and elasticity are provided by collagen and elastin (see Chapter 1). Hydration and porosity are generated by diverse proteoglycans (Chapter 1). Cohesion of cells to the matrix is mediated by fibronectin, laminin and other adhesive proteins. The extracellular matrix is a versatile structure that integrates cell activities and guides cell movement.

> The extracellular matrix contains proteins unique to each tissue, integrates cell activities and guides cell movement.

BOX 6.1 Components of the extracellular matrix

There are three main components in the extracellular matrix of animal cells: structural protein fibres, adhesive proteins and large polysaccharide molecules, called **glycosaminoglycans**, usually linked to a protein core.

The structural proteins are of two main types. The predominant one is **collagen**, probably the most abundant protein in mammals, constituting about 25% of the total protein. Collagens are a family of proteins composed of three polypeptide chains that entwine into a tight helix. In basal laminae, collagen associates into a strong sheet-like meshwork, whereas in interstitial matrices, collagen associates to form fibrils 30 nm in diameter. These, in turn, assemble into fibres of high-tensile strength, several micrometres in diameter.

The second type of structural protein found in some extracellular matrices is **elastin**. Elastin is an unusual protein because it remains in an unfolded, random coil configuration. The elastin forms a highly cross-linked network that can be stretched extensively but, when released, recoils to its original configuration.

The structural proteins, collagen and elastin, are embedded in a highly hydrated gel composed of complex polysaccharides, glycosaminoglycans. These are long, unbranched polysaccharides composed of repeating disaccharide units. The polysaccharides have a strong negative charge that attracts cations. This draws water molecules into the gel by osmosis, thus providing a cushioning against compressive forces in some connective tissues. Localised secretion of glycosaminoglycans can cause the extracellular matrix to swell due to gel hydration, and this assists cells in their migration through the matrix.

Adhesive proteins are the third major component of the extracellular matrix. The best characterised examples are **fibronectin**, which occurs in interstitial matrices, has a high molecular weight (about 460 kD) and two polypeptide chains, and **laminin**, which is found in basal lamina, and is a large glycoprotein nearly twice the size of fibronectin (about 850 kD).

Fibronectin and laminin molecules mediate attachment of cells to the matrix and, by so doing, affect cell polarity, growth, migration and differentiation. Both molecules contain multiple binding sites that interact with collagen, proteoglycans and specific receptors on the surfaces of cells (see also Box 15.2).

Maintenance of animal tissues

Tissues that compose various organ systems are maintained for the life of an animal in several ways. Some tissues, such as those that interact with the external environment, face continual wear and tear, and therefore require repair or replacement of cells. The skin and the lining of the intestine are two examples. Red blood cells also undergo considerable buffeting as they shuttle around the circulatory system. In general, the greater the wear and tear on a tissue, the greater the turnover rate of its cells (and therefore, incidentally, the greater susceptibility of these tissues to anti-cancer drugs that operate by inhibiting cell division).

Some cells, such as red blood cells, have become so highly specialised that they have lost their ability to undergo cell division; they therefore cannot replace themselves. However, the functions that they carry out are essential for continued survival of the animal. These tissues are replaced by division of small populations of relatively unspecialised cells known as **stem cells**.

> Tissues are maintained by repair and/or replacement throughout life.

Regenerating tissues

The individual cells of some tissues, for example, liver cells and endothelial cells lining blood vessels, are continually replaced by new cells as a result of simple cell division. The average life span of an endothelial cell is several months. As old cells become damaged or die, surviving differentiated endothelial cells are stimulated to divide, thereby maintaining the population of cells. In addition, when tissues surrounding a blood vessel are deprived of oxygen, they may release factors that stimulate endothelial cell division, leading to the formation of entirely new capillary networks.

In certain tissues, such as mammalian skin and blood, regeneration is not due to division of differentiated cells but to proliferation from a residual stem cell population. Stem cells are relatively unspecialised groups of cells that are able to divide repeatedly during the life of the animal. Although stem cells may not have a distinctive appearance, they have nevertheless undergone partial differentiation and their potential fate is determined (see Chapter 15). Stem cells give rise to cells that may either remain as stem cells or may continue development to become a fully differentiated cell. Some stem cells, such as skin stem cells, are **unipotent**: they produce only one type of differentiated cell. Blood-forming stem cells are **pluripotent**: they give rise to the whole range of blood cell types (Box 6.2).

Stem cell regeneration is found where there is a continuing need for replacement of cells but where the fully differentiated cell is itself either incapable of or unavailable for cell division. For example, red blood cells and keratinised skin cells have no nuclei, striated muscle cells are almost completely filled with actin and myosin fibres, mature sperm cells are released from the testes, and gut epithelial cells are lost from the tips of villi.

Differentiated skeletal muscle cells are unable to divide. The adult number of muscle fibres is laid down very early in development. Increase in the size of muscles, as seen in weight-lifters, occurs largely as a result of hypertrophy (increase in size) of individual cells. There is, however, a small population of muscle stem cells that can, when muscle tissue is damaged, give rise to new cells, which fuse to form mature multinucleate muscle fibres.

> Tissues may be regenerated by division of differentiated cells or by proliferation from a relatively unspecialised stem cell population.

'Permanent' cells

In some tissues there is no turnover of cells. The cells of these tissues are produced during development in

---

BOX 6.2 Blood cell formation

The formation of new mature blood cells (haemopoiesis) must occur continuously throughout adult life because they have a life span of only days or weeks. Most blood-forming tissue is located in scattered deposits within the marrow of various bones. There are eight major types of specialised blood cells (lineages) to be formed—red cells, granulocytes, monocyte–macrophages, eosinophils, megakaryocytes, mast cells, and T and B lymphocytes. These are all generated from a small common population of pluripotential stem cells, which form larger numbers of more mature haemopoietic cells in a series of steps (Fig. a).

Stem cells

Only one in every 10^5 haemopoietic cells is a stem cell. Stem cells are small to medium-sized mononuclear cells that are able to undergo cell division. One daughter cell becomes a stem cell but the other one is a committed progenitor cell that enters a pathway of irreversible differentiation to form different types of blood cells.

Committed progenitor cells

Committed progenitor cells constitute about 1% of haemopoietic cells and are medium to large undifferentiated cells (blast cells) that have the capacity to form very large numbers of maturing

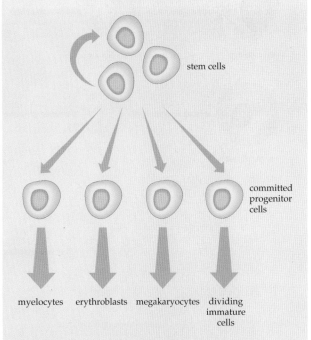

stem cells

committed progenitor cells

myelocytes   erythroblasts   megakaryocytes   dividing immature cells

(a) A schematic representation of the way stem cells generate committed progenitor cells, each of which then generates large numbers of maturing blood cells progeny in their particular lineage. Up to 10^5 cells are able to be produced by each progenitor cell. Progenitor cells cannot revert back to stem cells and cannot alter their differentiation commitment from one lineage to another. For example, erythroid progenitor cells (which lead to erythrocytes—red blood cells) cannot give rise to granulocytes (type of white blood cell). When progenitor cells divide, their progeny are soon able to be recognised from their morphology as immature blood cells of a particular type.

These immature cells are able to divide and, as they mature, the progeny develop characteristic features such as nuclear shape or granules in the cytoplasm. For example, committed granulocyte progenitor cells first form myeloblasts. These then generate promyelocytes, which in turn generate myelocytes (Fig. b).

Maturing blood cells

Myelocytes are the last cells in the granulocyte sequence able to undergo cell division. Their progeny are metamyelocytes, which are no longer capable of cell division (post-mitotic cells). These cells undergo extensive maturation changes to form fully mature granulocytes (neutrophils), which enter the peripheral blood, where they have a life span of, at most, a few days.

This pattern of continuous production of non-dividing mature cells with a limited life span also applies to the formation of red cells where maturation changes include extrusion of the nucleus from the cell before it is released to the blood. Platelets represent an even more extreme degree of maturation because they are merely membrane-bound portions of cytoplasm from mature megakaryocytes, which are pinched off and released to the circulation (Fig. b).

In some blood cell types, the apparently mature cells that enter the blood can be reactivated by appropriate stimuli and then undergo further cell divisions. This occurs with some T and B lymphocytes and, less frequently, can occur with some mast cells, eosinophils and macrophages.

Organisation of haemopoietic cells

In blood cell formation, cells at various stages of development in the different lineages are intermingled within the bone marrow. The properties of various ancestral cells and their progeny have only been able to be deduced by the development of cell sorting techniques, which use particular marker molecules expressed on

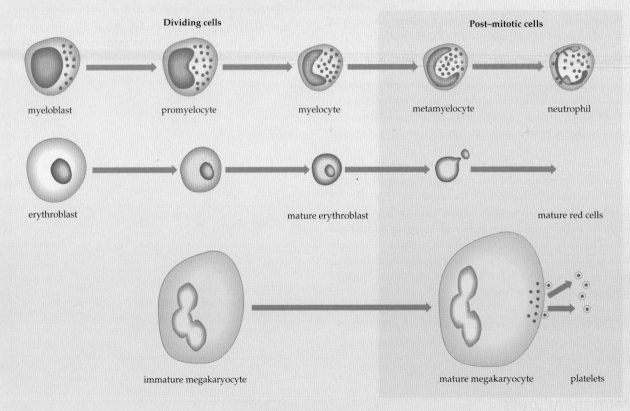

(b) Maturation occurs progressively in dividing haemopoietic cells and continues after the cells lose the capacity for further division. Only fully mature cells are released into the blood circulation

cell surfaces, combined with tissue culture of colonies of maturing blood cells. In this way, pure populations of each developmental stage can be obtained.

Cell division in haemopoietic cells follows stimulation by specific regulatory molecules. For each lineage, multiple regulators (glycoproteins active at pg/mL concentrations) interact to control development. So far, 20 of these have been discovered, the genes encoding them isolated and the regulators mass-produced by genetic engineering. It is suspected that possibly a total of 50 different regulators may exist to control the multiple events of cell division and maturation occurring in the eight lineages of haemopoietic cells. Co-ordination of cell production by haemopoietic tissues scattered throughout the body is achieved by the interactions of these haemopoietic regulators.

When populations of blood cells become depleted, for example, by loss of red blood cells during bleeding, or when additional cells such as granulocytes are needed to respond to an infection, haemopoietic regulators rapidly increase the rate of cell production by haemopoietic cells. In part, this accelerated blood cell production is achieved by shortening the cell cycle times of the dividing cells, but most of the amplification is achieved by increasing the numbers of progenitor cells derived from the stem cells. Where demand is extreme, additional haemopoietic tissue develops in the spleen and liver to allow production of the required number of cells.

This process of accelerated replacement or cell production is known as **hyperplasia**. The hyperplasia ceases when the numbers of cells have reached their required levels. For example, the need of tissues for a minimum level of oxygen requires a certain number of oxygen-carrying cells (red blood cells) to be present and, when that number is reached, blood oxygen levels return to normal and red cell formation returns to basal levels.

sufficient numbers to last the entire life of the animal. These cells are permanent and, if some die, they cannot be replaced. Examples include nerve cells, cells of the retina and lens (Fig. 6.4), oocytes and cardiac muscle cells. In the case of the nervous system, permanency of the developing complex network of nerves is understandably an advantage. Laying down this network during development is an ongoing interactive and sequential process. It is difficult to imagine how the precise relationships between individual nerve cells could be maintained, or how properties such as memory could be retained, if there were a significant turnover of nerve cells throughout the life of an animal. However, in the case of cardiac muscle cells, it is difficult to imagine any advantage in not having the ability to regenerate new cells.

Most permanent cells are able to repair themselves by turnover of cell components. With increased use, cardiac muscle cells become hypertrophied. Damaged nerve fibres may be able to regrow. Retinal cells, rods and cones continually replace the folded stack of photoreceptive membranes that form the sensory receptor that responds to light.

In some tissues there is no turnover of cells. Most permanent cells are able to repair themselves by turnover of cell components.

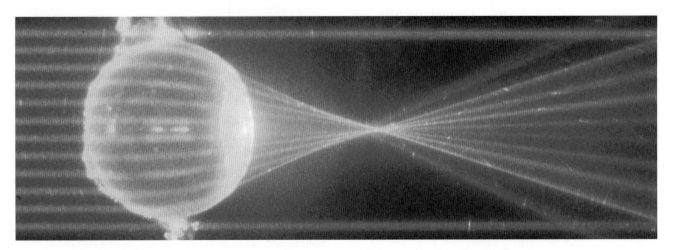

Fig. 6.4 Cells forming the lens of a vertebrate eye are highly specialised to allow unobstructed passage of light. Once the lens has formed, individual cells cannot be replaced. This lens is from a trout eye, with laser beams 0.5 mm apart shining through it

Intercellular connections

Cells in animal tissues are interdependent, bound together to some degree by a non-cellular matrix and in communication with each other for co-ordination and control. Chemical communication occurs by means of signal molecules that pass through the extracellular matrix. Light microscopy has shown that there are also physical connections between the plasma membranes of adjacent cells. The nature and variety of these connections were discovered with the advent of high-resolution electron microscopy.

Junctions between plasma membranes can be grouped according to their function. Examples of each of these types of connections can be seen in the layer of epithelial cells that lines the lumen of the intestine in mammals (Fig. 6.5).

- **Occluding junctions** are those where the plasma membranes of adjacent cells are fused tightly, preventing the passage of even small molecules through the extracellular space.
- **Anchoring junctions** provide sites of attachment for the mechanical support of tissues. They link adjacent cells, attach cells to the extracellular matrix, and form attachment sites for fibres of the cytoskeleton.
- **Communicating junctions** are those that are specialised for chemical and electrical communication between adjacent cells.

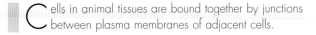

Cells in animal tissues are bound together by junctions between plasma membranes of adjacent cells.

Occluding junctions

These junctions, also known as tight junctions, are found linking epithelial cells, forming a barrier to the free movement of molecules between the cells of the epithelium. In the epithelial lining of the intestine, occluding junctions form a circumferential seal, preventing free movement of molecules between the lumen and interstitial space. Ultrastructural studies show that the plasma membranes are fused together by continuous bands of transmembrane junctional proteins (Fig. 6.6). The permeability of these junctions to small molecules varies in different epithelia. In the intestine, occluding junctions between epithelial cells are 10 000 times more leaky to ions than are those in the epithelium of the urinary bladder.

Interestingly, occluding junctions also serve to maintain the correct location of membrane protein channels involved in the transport of digested food from the gut lumen into the circulatory system. For example, active glucose channels are confined by occluding junctions to the luminal portion of the endothelial plasma membrane, whereas passive glucose

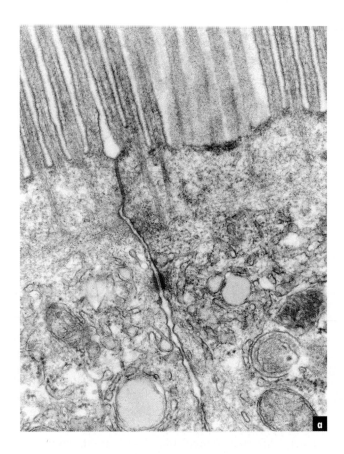

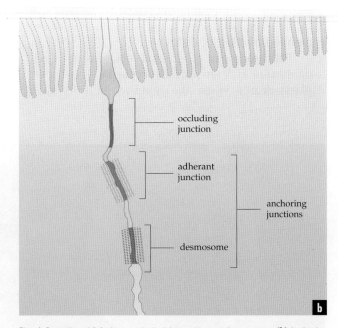

Fig. 6.5 Cells of **(a)** the epithelial layer lining the gut are **(b)** linked by occluding junctions, which form a bar to the movement of gut contents between the cells, and anchoring junctions that hold the cells together when physical strains are placed on the gut wall. Communicating junctions for ion exchange between cells are not obvious in this micrograph

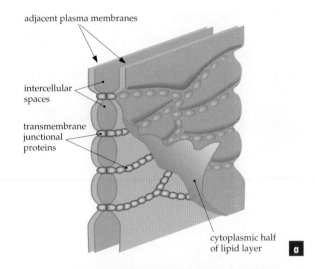

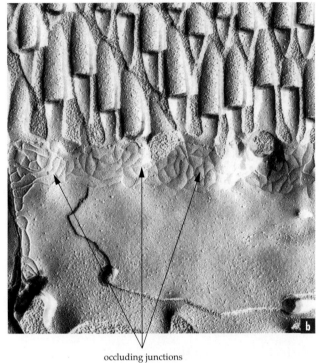

Fig. 6.6 (a) Frozen cells can be fractured in such a way that they split down the middle of the plasma membrane bilayer. This diagram illustrates how a fracture has exposed the occluding junctional proteins between two adjacent cells
(b) Occluding junctions of epithelial cells are shown in this freeze-fracture micrograph

carrier proteins are confined to the basolateral regions of the plasma membrane (Fig. 6.7).

> Occluding (tight) junctions form an impenetrable seal between cells and restrict the movement of membrane proteins.

Anchoring junctions

There are several types of anchoring junctions. Some have localised thickening due to the presence of a

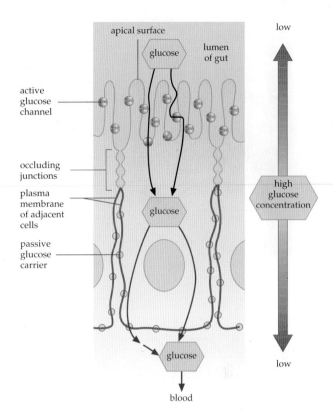

Fig. 6.7 The band of occluding junctions in the gut epithelium prevents active glucose transport channels moving away from the luminal surface. They actively transport glucose from the lumen against a concentration gradient, raising the intracellular glucose concentration. Glucose then passes from these cells into the blood via passive glucose carriers, which are, in this case, confined to the internal walls of the cells by the presence of occluding junctions

dense cytoplasmic layer (plaque) and a wider intercellular space that may show an intermediate line between the plasma membranes (Fig. 6.8). Intermediate filaments of the cytoskeleton (Chapter 3) attach to the plaque and pass back into the cytoplasm to connect with other structural elements within the cell. These junctions, also known as **desmosomes**, provide structural support to tissues by cross-linking between the cytoskeletal networks of adjacent cells. **Hemidesmosomes** anchor cells to the extracellular matrix, such as the basal lamina.

Another form of anchoring junction, **adherent junctions**, forms cross-links between contractile bundles of actin filaments (see Chapter 3) located in the cortical cytoplasm. The contractile nature of adherent junctions makes them important in the movements of cells and tissues that occur during embryonic development, for example, during the formation of the nervous system in vertebrates (see Chapter 15). Adherent junctions may be either focal or belt-like, for example, adhesion belts, which are bundles of actin filaments running parallel with and

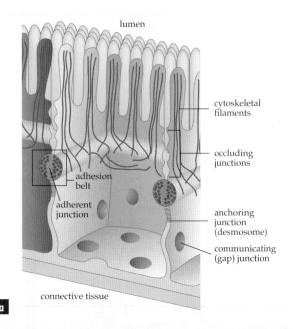

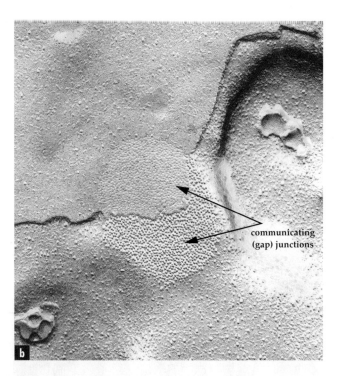

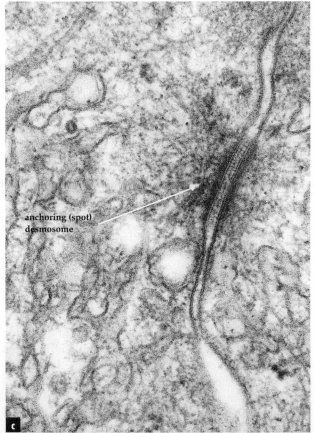

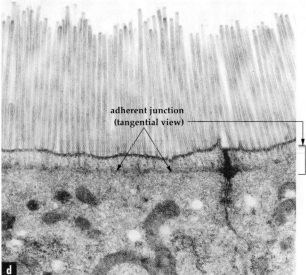

Fig. 6.8 (a) Several types of anchoring junctions (desmosomes, hemidesmosomes), communicating (gap) junctions and adherent junctions can be seen between gut epithelial cells and linking to cytoskeletal filaments. **(b)** Freeze fracture down the middle of the plasma membrane bilayer showing the concentration of protein subunits of a communicating (gap) junction. **(c)** Ornate structure of an anchoring (spot) desmosome. **(d)** An adherent junction seen in tangential view

connected to the plasma membrane. Adhesion belts of adjacent cells are connected by intercellular glycoproteins. Adherent junctions, like desmosomes, can also anchor to the extracellular matrix.

Anchoring junctions provide mechanical support and are important in developmental processes.

Communicating junctions

These junctions, also known as gap junctions, are specialised for electrical and chemical communication between cells. Because they allow ions to cross freely, they provide a pathway of low electrical resistance and permit rapid current spread from cell to cell. In the vertebrate heart, it is the presence of communicating junctions linking together all atrial cells that results in the co-ordinated contraction of both atria, expelling blood into the ventricles.

Communicating junctions are highly regular protein channels composed of six protein subunits that span the plasma membrane and link to a similar unit in the adjacent cell (Fig. 6.9). The diameter of the central aqueous channel appears to be about 1.5 nm, but, at least in some tissues, permeability can apparently be regulated.

Communicating (gap) junctions allow ions to pass freely between adjacent cells.

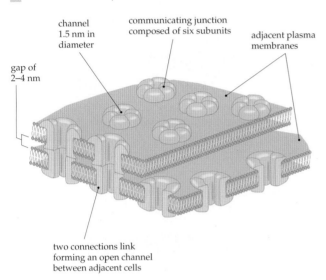

channel
1.5 nm in
diameter

communicating junction
composed of six subunits

adjacent plasma
membranes

gap of
2–4 nm

two connections link
forming an open channel
between adjacent cells

Fig. 6.9 Six membrane-bound protein subunits in each plasma membrane link across the intercellular space to form the channel of a communicating junction

Principal animal tissues

Most animals are constructed from four basic types of tissue. *Epithelial tissues* cover most surfaces and secrete various substances, *connective tissues* provide support and strength, *muscular tissues* enable movement and *nervous tissues* receive signals and conduct messages throughout the body.

Epithelia

Epithelia are a diverse group of tissues that, with rare exceptions, form continuous layers covering all surfaces, cavities and tubes, both external and internal. Epithelia therefore function as interfaces between adjacent biological compartments, and between the organism and its environment, where they provide protection and regulate exchange. In addition to forming barriers, epithelia give rise to the glandular tissue of endocrine and exocrine glands (Chapter 25) and parts of some sense organs. The epithelium of the gut, for example, is absorptive, secretes enzymes for digestion of food and secretes mucus for protection.

Epithelial cells are closely bound together by a variety of occluding and anchoring junctions and are supported by a basal lamina of variable thickness. The basal lamina is a thin, tough mat of connective tissue formed from a glycoprotein matrix, with a network of fibres that merge with those of the underlying connective tissue. Basal laminae are not penetrated by blood vessels, so epithelia are dependent on diffusion through the lamina from underlying blood vessels for exchange of nutrients and wastes with blood.

Epithelia are defined according to the number of layers and the shape of cells. They may be simple, forming a single layer, or stratified, composed of more than one layer; and, according to the shape of the cells, may be squamous, cuboidal or columnar (Fig. 6.10). Epithelia may also be categorised according to the presence of specialisations, such as cilia or deposition of keratin.

Epithelia function as interfaces between biological compartments and with the environment, and in secretion.

Connective tissue

In general, tissues that provide basic structural, metabolic and defensive support for other tissues of the body are **connective tissues**. These include bone and cartilage, which provide physical support; blood, which provides defence and transports important materials throughout an animal's body; adipose tissue, which is involved in storage and metabolism of fats; and extracellular fluids secreted by fibroblasts (Fig. 6.11). In vertebrates, the extracellular matrix is important in influencing the growth, migration and activity of cells with which it comes in contact, particularly during development (see Chapter 15).

In connective tissues, extracellular materials are usually more abundant than cells and characterise particular types of connective tissue. The extracellular matrix is produced by cells and generally contains a variety of polysaccharides and proteins in a meshwork of fibres (Box 6.1). The matrix can be calcified, as in bone and teeth, transparent, as in the cornea, or fluid, as in blood. Collagen forms the principal fibre type found in all connective tissues. Reticulin fibres predominate in the supporting reticular, connective tissues that enfold organs such as the kidney and liver. Elastin fibres are found in many connective tissues and they are prominent features of the elastic connective tissues of the bladder and dermis of the skin. Despite advertisements to the contrary, elastin fibres in the skin deteriorate with age and cannot be replaced by cosmetic creams.

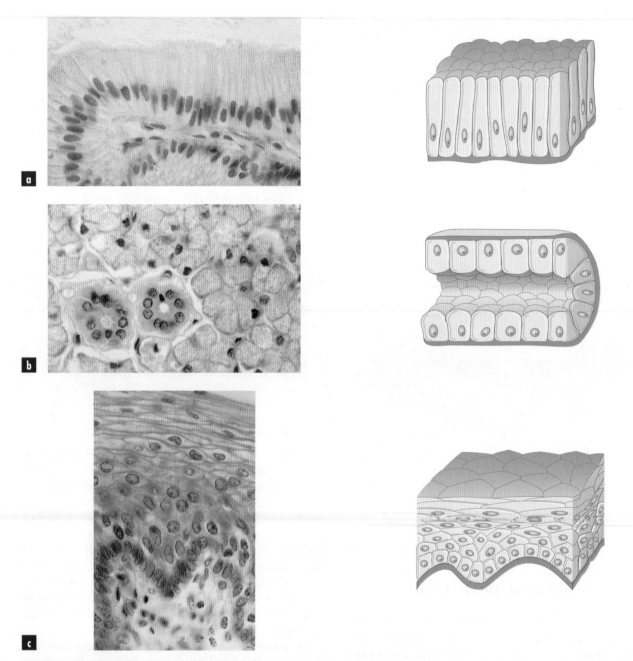

Fig. 6.10 Epithelial cells, as shown in these light micrographs, are usually described in terms of their histological organisation. **(a)** Simple columnar epithelium, composed of tall cells with an oval nucleus at the base, lines the human gall bladder. **(b)** Simple cuboidal epithelium forms the ducts of many glands, as shown in these two ducts in the human parotid gland. **(c)** Stratified epithelium, as in the human oesophagus, is a thick, many layered sheet of cells varying from columnar at the base, through to cuboidal and to squamous (flattened) on the outer surface. Stratified epithelium forms the skin of many vertebrates

Connective tissues are composed of cells, fibres and fluid matrix. Types of connective tissue differ primarily in the consistency of the matrix, which may be soft and fluid (adipose tissue), firm or hard (cartilage and bone) or liquid (blood and lymph).

Muscle

Movement, such as motility and change in cell shape, is an inherent property of all animal cells. Muscle is composed of specialised cells that contain highly organised fibrillar components (actin and myosin) and related biochemical components, which produce contraction or shortening of the cells. Muscles co-ordinate to move an animal in relation to its environment and move its internal organs to transport materials into, out of and around the body. There are two basic types of muscle cell organisation, striated and smooth, which reflect different degrees of organisation of the fibrillar network.

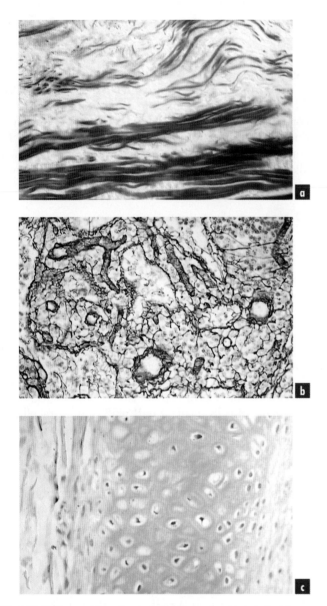

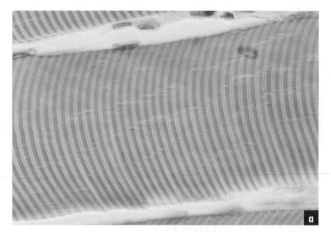

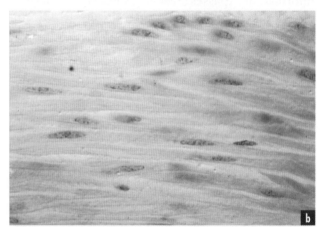

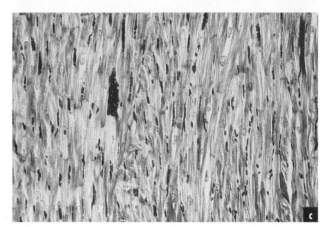

Fig. 6.11 Different types of connective tissue are characterised by differing types and proportions of fibres and matrix. Light micrographs of **(a)** elastic connective tissue from a ligament, **(b)** black-stained reticular fibres in a human lymph node, and **(c)** cartilage from the human ear

Fig. 6.12 Light micrographs of **(a)** human skeletal muscle showing the striations due to highly organised bands of fibrils forming the contractile mechanism, **(b)** smooth muscle, in which the contractile fibres are not regularly arranged and no striations are evident, and **(c)** cardiac muscle, a striated muscle composed of a branching network of individual cells linked through many communicating junctions

Striated muscle cells, found in skeletal and cardiac muscle, have a highly organised array of actin and myosin filaments (myofilaments), giving the appearance of cross-striations when viewed under the light microscope (Fig. 6.12; Chapter 27). **Skeletal muscle** consists of cylindrical cells (fibres) of variable length and diameter. These fibres are multinucleate, with the nuclei located peripherally, and are grouped into bundles within a connective tissue sheath. **Cardiac muscle** is composed of a branching network made up of individual cells linked through many communicating junctions.

The **smooth muscle** cells of internal (visceral) organs are spindle-shaped cells with a central nucleus.

They may be loosely scattered or aligned together to form dense bands or sheets. In the gut wall, individual smooth muscle cells are closely packed with many communicating junctions. In contrast to striated muscle, myofilaments in smooth muscle cells are less regularly arranged and there is no striated appearance

(Fig. 6.12). Although smooth muscle myofilaments are not visible under the light microscope, they can be discerned with the electron microscope.

> S triated muscle cells contain a highly organised array of actin and myosin filaments. Myofilaments in smooth muscle cells are less regularly arranged.

Nervous tissue

Nerve cells, **neurons**, are specialised to carry information rapidly and precisely from one part of an animal to another, often over long distances. Neurons form an interconnecting network through which information passes from sensory structures (receptors) through integrating circuits to effector structures such as muscles or glands (Chapter 26). The basic structure of neurons in all animals is similar. Information is received by relatively short cellular processes, **dendrites**, and passed into the nerve cell body (Fig. 6.13). If the nerve cell responds, a signal is passed down a single process, the **axon**. In some animals, such as whales and giraffes, the axon may be metres long.

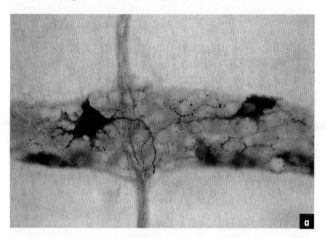

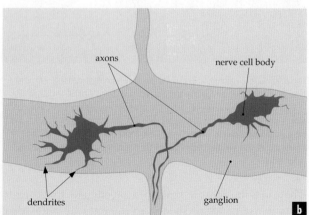

Fig. 6.13 Two nerve cells in a ganglion (cluster of nerve bodies) in the wall of the mammalian intestine. The short processes (dendrites) receive incoming information, and outgoing signals pass down the longer single process, the axon

Neurons are highly specialised cells that are unable to undergo cell division. They are only able to carry out their normal functions because of their association with a group of supporting cells called **glial cells**, which are derived from the same embryonic tissue as nerve cells. Glial cells are responsible for maintaining the composition of the extracellular environment, which is critical for the proper functioning of neurons. In vertebrates, glial cells also form myelin sheaths, which allow very rapid conduction of signals along the axons of some cells (Chapter 26).

> N erve cells are specialised to conduct signals rapidly and precisely throughout the body. They are unable to divide and are supported by glial cells.

Plant cells: the importance of the cell wall

Unlike animal cells, plant cells are encased in cell walls that form box-like compartments (Fig. 6.14). An important function of cell walls is in limiting the size and shape of cells. A plant **cell wall** is an extracellular matrix, and is thicker, stronger and more rigid than the extracellular matrix of animal cells. The characteristic component of plant cell walls is cellulose. The strength of the wall allows plant cells to exist surrounded by a hypotonic environment (with a lower concentration of solutes than the cytosol). Without a restraining wall, the inflow of water by osmosis (Chapter 4) would cause plant cells to burst.

Cell walls are also important in metabolic processes. They contain specific enzymes; act as a pathway for transport, absorption and secretion; and act as a defence barrier against pathogens. Recent discoveries indicate that some wall components are similar to plant hormones and thus can play a role in cell-to-cell communication.

> P lant cells characteristically have walls, which limit cell size and shape.

Cell wall structure

Molecules of cellulose are long, unbranched chains of $(1{\rightarrow}4)$-linked β-D-glucose molecules (Chapter 1). They are organised into **microfibrils** (Fig. 6.14). These microfibrils are embedded in and cross-linked to a porous, hydrated gel-like matrix of polysaccharides other than cellulose (*non-cellulosic polysaccharides*), pectin and proteins (Box 6.3). Non-cellulosic poly-saccharide molecules form bridges at regular intervals between adjacent cellulose microfibrils, spacing them about 20 nm apart. The pectins link with them to form

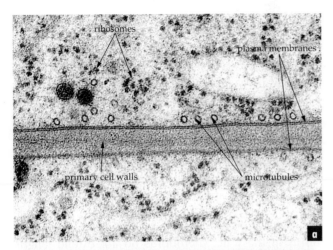

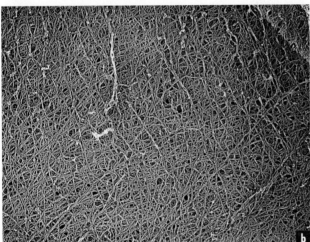

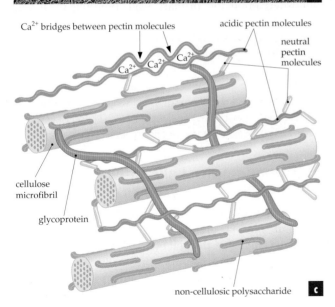

Ca²⁺ bridges between pectin molecules

acidic pectin molecules

neutral pectin molecules

Ca²⁺ Ca²⁺ Ca²⁺

cellulose microfibril

glycoprotein

non-cellulosic polysaccharide

Fig. 6.14 Plant cell walls. The cellulose forms microfibrils, which are interconnected by pectic polymers and proteins to complete the matrix. **(a)** Thin section of a primary cell wall of *Eucalyptus sieberi* prepared by rapid freezing in liquid nitrogen to retain the life-like appearance (magnification × 68 200). **(b)** Surface view of microfibrils of a cell wall prepared by fast freeze, deep etch, rotary shadowing. **(c)** Diagram showing the interaction between cellulose microfibrils, hemicelluloses and other cell wall molecules

BOX 6.3 Components of plant cell walls

The cell wall matrix in which cellulose microfibrils are embedded is composed of two main types of branched polysaccharides: pectins and non-cellulosic polysaccharides (Fig. 6.14).

Pectins have a backbone of galacturonic acid residues, to which other types of sugars are linked. Pectins have a strong negative charge that attracts cations (especially Ca^{2+}) and associated water molecules, making them highly hydrated molecules. Being divalent, calcium ions can form cross-bridges between two pectin molecules and this greatly strengthens the gel. Many pectins can be removed from cell walls by mild chemical treatments, but others remain tightly bound to cellulose microfibrils.

Non-cellulosic polysaccharides are removed from walls by harsher chemical treatment, such as with alkali. They have a backbone made of one type of sugar with short side chains of other sugars linked to it; for example, xyloglucans have a backbone of glucose units to which side branches of xylose are linked. The glucose backbone can form hydrogen bonds with cellulose microfibrils.

Sedimentation of cocoa powders in chocolate drinks is a common defect (right) that is usually alleviated by incorporating a stabilising agent, in this case, a plant cell wall carbohydrate (left). Stabilising agents increase viscosity of the milk and thereby slow sedimentation of particles

Many cell walls also contain small amounts of protein. Glycoproteins, such as extensin, contribute to the cross-linking of other components in the wall.

A variety of these cell wall components are of industrial and agricultural importance: for animal and human nutrition (mostly as dietary fibre), the building industry, paper and pulp industry. Pectins, common in the pith of citrus fruits, are used in the food industry; even adding lemon peel to home-made jam helps it set. Many processed foods (such as ice-cream) have plant or seaweed wall polysaccharides added to modify texture, taste and consistency (see figure). In Australia, plant biotechnologists are investigating the potential to develop industries based on exploiting certain cell wall polysaccharides. When grown in cell-suspension culture, plant cells secrete wall components directly into the surrounding medium. In this way, cell wall components can be harvested for commercial use.

a gel. This network forms a wall that has pores of about 5 nm, large enough to allow water and small molecules to diffuse freely but small enough to inhibit the movement of larger molecules such as proteins.

Cell walls of bacteria, fungi and some groups of algae differ from those of plant cells. In fungi, the main wall component is not cellulose but chitin, a polymer of (1→4)-linked β-*N*-acetylglucosamine. Bacterial walls are varied but all are composed of polysaccharides cross-linked by amino acids. In red algae (Chapter 35), cell walls are composed of microfibrils of cellulose (or another polysaccharide) and a mucilaginous matrix of components such as agar and carrageen.

> Plant cell walls comprise cellulose microfibrils embedded in and interconnected to a matrix containing non-cellulosic polysaccharides, pectins and proteins.

Development of plant cell walls

During the formation of cell walls, glucose precursor molecules are transported across the plasma membrane by specific carrier proteins (Chapter 4). Cellulose chains are assembled on the external face of the plasma membrane by large complexes of membrane-bound enzymes. As the chains are polymerised, they quickly become hydrogen-bonded together into microfibrils (Fig. 6.14).

New cell walls are laid down initially as partitions (cell plates) consisting of sheets of membranes formed after nuclear division in the parent cell (Fig. 6.15). The cell plate grows and fuses with the plasma membrane of

the parent cell. The two newly formed cells then add wall material on either side of the cell plate. As the common wall develops, the cell plate material is modified but persists as a thin layer, the **middle lamella**, between the walls of the adjacent cells. The middle lamella is rich in pectin molecules. Walls of growing cells are **primary cell walls**. Primary walls are relatively thin, typically about 100 nm in width, highly hydrated and extensible, allowing the cell to expand in size (Fig. 6.15).

As plant cells grow, the cell wall yields to the internal hydrostatic force of turgor pressure. The direction of expansion, and therefore growth, is determined by the arrangement of cellulose microfibrils in the wall. Microfibrils do not extend lengthwise, and any cell expansion occurs by their separation (Fig. 6.14). If, for example, cellulose microfibrils are deposited circumferentially around the side walls of a cell, lateral expansion will be inhibited and almost all growth will extend the cell lengthwise. The extent and direction of growth of cells is controlled so that connections between cells are not disrupted.

As plant cells differentiate and mature, primary cell walls may be added to, producing thicker, more rigid **secondary walls**. The main component of secondary walls is the polyphenolic compound **lignin**. Lignification makes walls very strong and rigid by cementing together and anchoring the cellulose microfibrils, thus greatly reducing pore size. Deposition of lignin can occur within the primary wall or new layers may be deposited upon the primary wall, building up a wall that may be several micrometres thick. Secondary walls may completely surround a cell or they may be formed only in localised regions.

> Cellulose is a major component of primary walls. Cellulose microfibrils give cell walls strength, and their orientation determines the direction of cell expansion. Lignin is characteristic of many secondary walls.

Plasmodesmata: links between plant cells

The rigid wall of plant cells limits contact between adjacent plasma membranes and the exchange of molecules between the cell cytoplasm and the environment and between neighbouring cells. Although cell walls allow small molecules and ions to pass between cells, plants also have special channels, **plasmodesmata** (Fig. 6.16), that directly link the plasma membranes and cytosols of adjacent cells. Plasmodesmata are similar to communicating junctions in animal cells.

Each plasmodesma is a fine cytoplasmic channel linking adjacent cells across a cell wall. The channel is bounded by plasma membrane and is cylindrical in

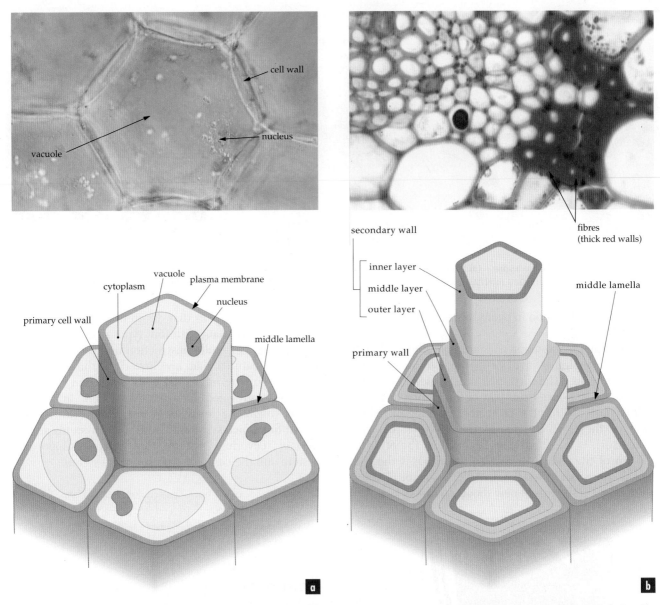

Fig. 6.15 Primary and secondary cell walls. **(a)** Transverse section of parenchyma cells from *Coleus* stem showing thin primary cellulosic cell walls and large vacuole-filled cell with nucleus and peripheral cytoplasm. **(b)** Secondary cell walls of fibres from a transverse section of *Hakea* leaf. The sequential lignin deposits (red stain) are added over the primary cell wall

shape. A narrower cylindrical structure, the desmotubule, runs through most plasmodesmata and is continuous with the endoplasmic reticulum of each cell.

Evidence that plasmodesmata are involved in transport comes from experiments using dyes that cannot cross the plasma membrane. When such a dye is injected into a plant cell, it passes readily to adjacent cells through plasmodesmata. Furthermore, since the plasma membranes of adjacent cells are continuous across plasmodesmata, they provide no barrier to ion movement and act as conduits for electrical signals that are transferred from one cell to another. In this way, plasmodesmata have a similar function to communicating junctions between animal cells.

Plant cell walls are penetrated by plasma-membrane-lined channels, plasmodesmata, which provide a means of direct communication and transport between cells.

Plant tissue systems

At the tip of a growing shoot or root is a specialised region of cells—the **apical meristem**. These small clusters of cells are continually dividing, each cell producing two daughter cells, one of which remains as part of the meristem and one that differentiates as part of the mature body of the plant. Meristematic cells

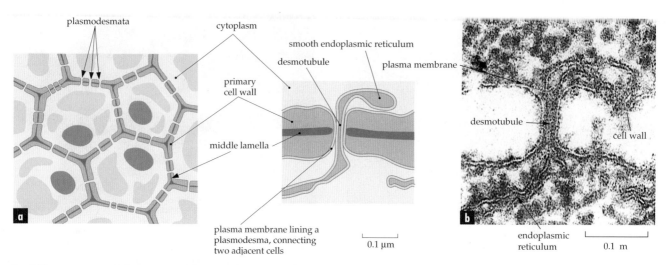

Fig. 6.16 Structure of plasmodesmata. **(a)** Plasmodesmata provide a direct connection between two cells. **(b)** This is achieved by means of a common plasma membrane and internal desmotubule; endoplasmic reticulum is commonly associated with the plasmodesmata

have a similar function in plants to stem cells in animals (p. 132).

Dividing meristematic cells are typically small, with a dense cytoplasm, small or no vacuole and a large, active nucleus. The first two apical meristems of a flowering plant develop as part of the embryo while it's still enclosed within the protective covering of a seed. One of these, at the tip of the epicotyl (or stem precursor), is the shoot apical meristem; the other, at the tip of the radicle (or root precursor), is the root apical meristem (Fig. 6.17). Additional apical meristems form at the tips of new shoots and roots as a plant grows and branches.

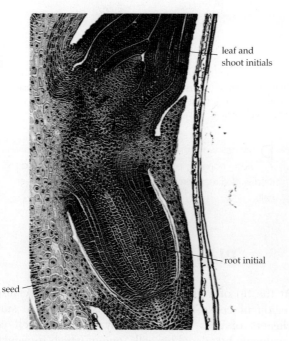

Fig. 6.17 Meristematic cells in an embryo of a cereal grain of wheat, showing the leaf initials around the epicotyl (upper) and the root initials of the radicle (lower)

The daughter cells that differentiate from an apical meristem form many different types of cells varying in size, shape, cell wall construction and organelle composition, features that are linked to the specialised functions of the individual cells. Different cell types become aggregated into tissues, which, in vascular plants (including ferns and seed plants), form three broad functional systems. These tissue systems are the *dermal* (covering) system, the *ground* (supporting) system and the *vascular* (transporting) system.

> Vascular plants have three major tissue systems: the dermal (covering), ground (supporting) and vascular (transport) systems. All these arise from apical meristem tissue.

Dermal tissue

The dermal system forms the outer covering of a plant and is the interface at which plants interact with the external environment. The major tissue type is the **epidermis** (Fig. 6.18), which consists of a layer (occasionally several layers) of closely packed cells that secrete a cover of water-resistant **cuticle**. The cuticle is composed of cutin, a water-insoluble polymer of fatty acids. A layer of wax platelets is deposited on its surface. The cuticle protects plants against evaporative water loss, physical damage and invasion by pathogens. Because the cuticle is also impermeable to air, the epidermis contains pores (stomata in vascular plants), which perforate the cuticle in leaves and stems and allow exchange of gases with the internal tissues. Stomata have two specialised epidermal cells, guard cells, which control opening and closure (Chapter 18).

Epidermal cells also develop as hairs or **trichomes**—unicellular or multicellular projections that help to trap moisture and reduce water loss or help to

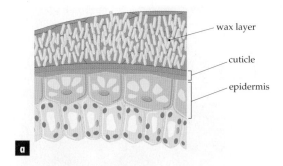

wax layer

cuticle

epidermis

deter insect herbivores (Fig. 6.18). Trichomes are sometimes associated with a gland producing toxic or unpalatable substances. Halophytes (plants tolerant of saline conditions) often have elaborate secretory glands to remove excessive salt from other internal tissues.

The major tissue of the dermal system is the epidermis, some cells of which are modified as guard cells for opening and closing stomata, hairs, trichomes and glands.

Ground tissue

The ground system consists of storage and structural tissues in which the vascular system is embedded. It is composed largely of three tissue types: parenchyma, collenchyma and sclerenchyma.

Parenchyma

Parenchyma tissue is composed of cells that are typically large, with a thin primary cell wall and well-defined pectin-rich middle lamella, a large vacuole that fills most of the cell and an active nucleus with dispersed chromatin. Parenchyma cells are commonly polyhedral (many-sided) and loosely packed together, separated by a network of intercellular spaces.

Different types of parenchyma have different functions. Photosynthetic parenchyma, often called **chlorenchyma**, contains chloroplasts and is found in leaves and the outer regions of photosynthetic stems. **Storage parenchyma** can contain either nutrient reserves in the form of starch granules or oil droplets (Fig. 6.19), or various metabolic products that accumulate in the vacuole. Aquatic plants have **aerenchyma**, a spongy parenchyma with a large network of air spaces, which allows aeration of tissues in an environment low in oxygen (Fig. 6.19).

A special type of parenchyma cell, a **transfer cell**, is characterised by ingrowths of the primary cell wall and associated plasma membrane (Fig. 6.19c). The increased surface area of the membrane allows rapid transfer of molecules to and from adjacent cells of the vascular system.

Parenchyma cells are living, vacuolated cells with thin primary cell walls. They form the major component of the ground tissue system.

Collenchyma

Collenchyma is a supporting tissue; it gives strength to plant parts where bending and flexibility are required and is commonly found in leaf stalks (petioles), leaf laminas and young stems. It is composed of strands of elongated cells (Fig. 6.20). Collenchyma cells have their cell walls thickened with additional quantities of cellulose. This extra thickening can be either in the corners of the cells, *angular collenchyma*, as in the

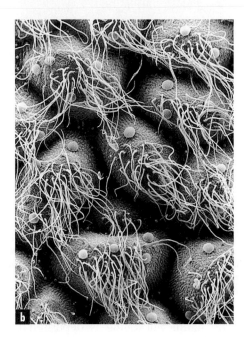

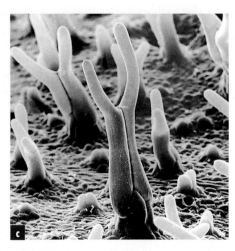

Fig. 6.18 Epidermis is the major tissue type of the dermal system that forms an outer covering. **(a)** The outermost wall layer is the cuticle, which lies over the primary cellulosic wall and is coated by wax platelets. **(b)** The herb sage (*Salvia*) has a leaf epidermis covered by long, filamentous hairs and globular oil glands, which give sage its characteristic odour (magnification × 42).
(c) Trichomes on juvenile leaves of *Eucalyptus wandoo* from Western Australia. A pair of long hairs are associated with an oil gland (magnification × 18) and probably play a role in deterring insect herbivores

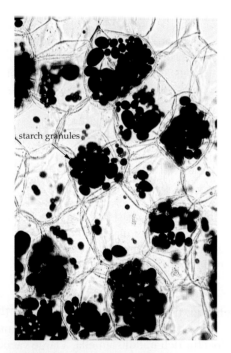

starch granules

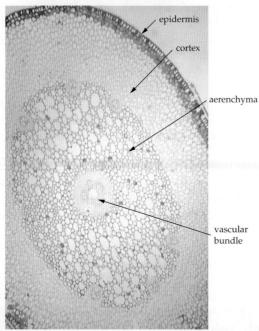

epidermis

cortex

aerenchyma

vascular bundle

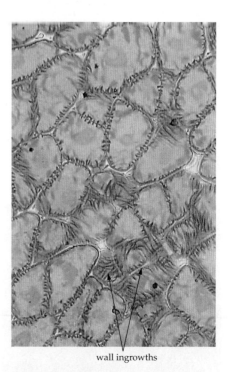

wall ingrowths

Fig. 6.19 Parenchyma cells form tissues with defined functions. **(a)** Parenchyma cells of potato tubers store starch granules (here stained blue-black with iodine). **(b)** Aerenchyma cells increase the circulation of gases in underground stems (rhizomes) of the seagrass *Amphibolis*. **(c)** Transfer cells with conspicuous wall ingrowths in an embryo of the seagrass *Amphibolis* increase the surface area of the plasma membrane for exchange of substances with parental tissue

stringy tissues of celery petioles, or along tangential walls, *lamellar collenchyma*, the latter more common in stems. Although thickened, the cell walls are flexible and can be stretched.

Collenchyma form strands of living cells with differentially thickened primary walls that are strong but flexible and have a support role.

Sclerenchyma

Sclerenchyma is also a supporting tissue but it imparts rigidity as well as strength. This rigidity is a property of the secondary cell walls, which are impregnated with lignin. The lignin thickening is deposited in layers, which can be distinguished as concentric rings or lamellae in the cell wall. Because the cell wall is so thick, in mature sclerenchyma cells the protoplasm usually degenerates to leave a gap or lumen in the centre of the cell.

Sclerenchyma includes two cell types—fibres and sclereids (Fig. 6.21). **Fibres** are elongated cells with tapering end walls that overlap adjacent cells. They are often arranged in a continuous layer in stems or in longitudinal bundles associated with vascular tissue. **Sclereids**, or stone cells, are branched or more or less even-shaped, and play a role of protection as much as support. They form the hard tissue of seed coats and give pears and some other fruits their gritty texture.

Sclerenchyma cells have rigid secondary walls with lignin deposited in concentric rings, forming elongated fibres or short sclereids.

Vascular tissue

The vascular system is composed of two tissues—xylem and phloem. Xylem transports water and dissolved minerals, such as nitrate and ammonia ions, from the point of uptake in roots to the point of loss in leaves

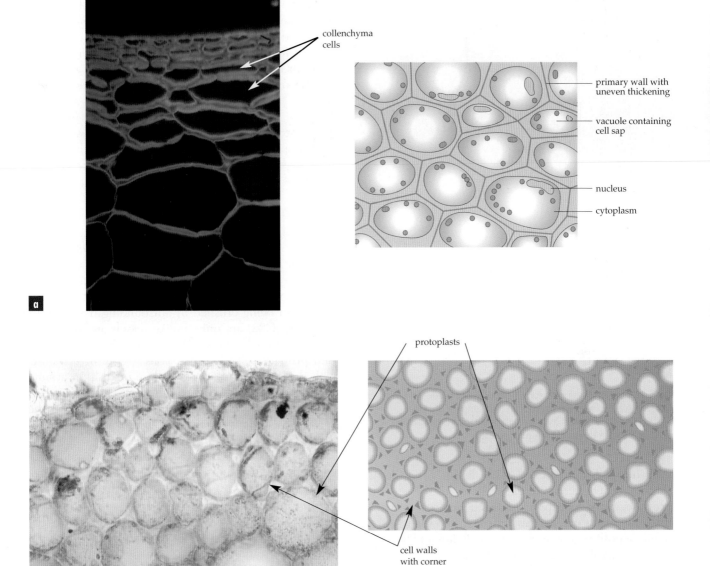

collenchyma cells

primary wall with uneven thickening

vacuole containing cell sap

nucleus

cytoplasm

protoplasts

cell walls with corner thickening

Fig. 6.20 Collenchyma cells form living support tissue in stems and fruits. **(a)** Lamellar collenchyma, with its wall thickening confined to the tangential walls, is responsible for the dermal system (skin) of fruit such as the grape. **(b)** Angular collenchyma, with its wall thickening uneven and confined to the corners, glistens silver in transverse sections of celery stem

(see Chapter 18). Phloem transports photosynthetic products, mostly sugars, from their place of production in leaves to their place of use or storage.

Xylem

Xylem is a complex tissue consisting of several cell types. Most distinctive are the water-conducting elements, vessels and tracheids (Fig. 6.22). In addition, sclerenchyma fibres, with a support role, and parenchyma cells, with a metabolic role, are usually prominent. Tracheids and vessels differ from most other cell types in that at maturity they are dead,

consisting only of hollow cell walls without any cell contents. In **vessels**, which are typical of flowering plants, the end walls are broken down to allow unimpeded flow of water from one cell to another. **Tracheids**, which are water-conducting cells of all vascular plants, have intact end walls and water moves from cell to cell through special pores in the cell wall. These pores often have an elaborate, overarching lip and are then referred to as **bordered pits** (Fig. 6.22).

Vessels and tracheids are produced by meristem divisions in the same way as other cells but differentiate early. They become elongated and develop a secondary

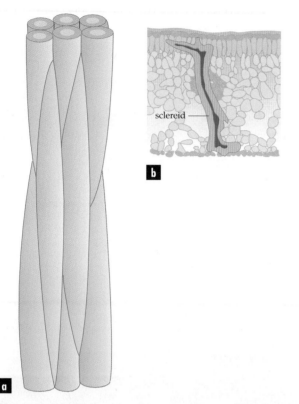

Fig. 6.21 Sclerenchyma cells form strengthening tissue. **(a)** Fibres are long tubular cells with secondary thickened walls (see Fig. 6.15b). **(b)** Sclereids form irregularly shaped cells usually scattered within tissue, shown here in a leaf

wall heavily impregnated with lignin. The lignin thickening acts as an internal skeleton for the cell and prevents it from collapsing under the internal pressure developed as water is drawn through the plant. After lignin deposition is complete, the cell dies and the wall assumes its role in water transport.

The pattern of lignin thickening in xylem vessels and tracheids varies depending on their position within the plant. At the tips of growing shoots and roots, the first formed primary xylem, **protoxylem**, has lignin deposited in a series of very closely spaced transverse rings (annular thickening). Protoxylem cells need some flexibility to accommodate elongation of the tissues around them. As the protoxylem is stretched, the rings of lignin separate but can still maintain their support function. Eventually they are stretched too far and they break and cease functioning. Water conduction is then taken over by slightly later maturing primary xylem, the **metaxylem**. Because metaxylem elements are formed in slightly more mature regions of stem and root, they do not require the same tolerance to stretching as protoxylem. Early metaxylem elements have lignin laid down in a spiral; they look very much like a spring and have a reasonable tolerance to stretching. Later

metaxylem elements have a reticulate (network) or mesh-like lignin ornamentation, and have a much more restricted ability to elongate. Xylem elements that mature in fully formed regions of the stem have no need for flexible walls and are provided with a uniform coating of lignin over the entire inner surface of the cell.

In the vascular system, xylem is a complex tissue composed of tracheids, vessels, parenchyma and sclerenchyma. When mature, tracheids and vessels consist only of cell walls without any cell contents.

Phloem

Like xylem, **phloem** is also a complex tissue comprising several cell types. The cells through which photosynthetic products are transported are **sieve cells**, and these are usually associated with parenchyma cells and bundles of sclerenchyma fibres that prevent the sieve cells from being crushed by surrounding tissue. Sieve cells (Fig. 6.22) are quite unlike vessels and tracheids, their functional counterparts in the xylem. They have thin primary cell walls that consist only of cellulose, with the exception of some callose deposition around pits that connect adjacent cells by their ends. These pits look like the holes in a sieve, hence the name of the cells. Sieve cells are elongated and form a continuous conducting pathway throughout a plant.

Sieve cells are living, retaining a plasma membrane and some cell contents. In flowering plants, each sieve cell is associated with a specialised parenchyma cell—a **companion cell**. Although sieve cells and companion cells are the products of a single mother cell that undergoes longitudinal division, their development and functions are quite different. As a sieve cell matures, the nucleus, vacuole, membrane system and most of the cell organelles degenerate to leave a dense, granular cytoplasm containing energy-producing mitochondria. It is through these cells that active (energy-requiring) transport occurs. Companion cells retain all their cytoplasmic contents and have a well-developed endoplasmic reticulum with ribosomes and a large, dense nucleus. The function of the companion cell is to control the activity of the sieve cell. At sites where products are being loaded into and out of sieve cells, companion cells have extensive ingrowths of the cell wall, which increase the surface area for cell-to-cell interchange.

Phloem tissue consists of sieve cells associated with parenchyma cells and bundles of sclerenchyma fibres. In flowering plants, sieve cells lack a nucleus but are associated with companion cells.

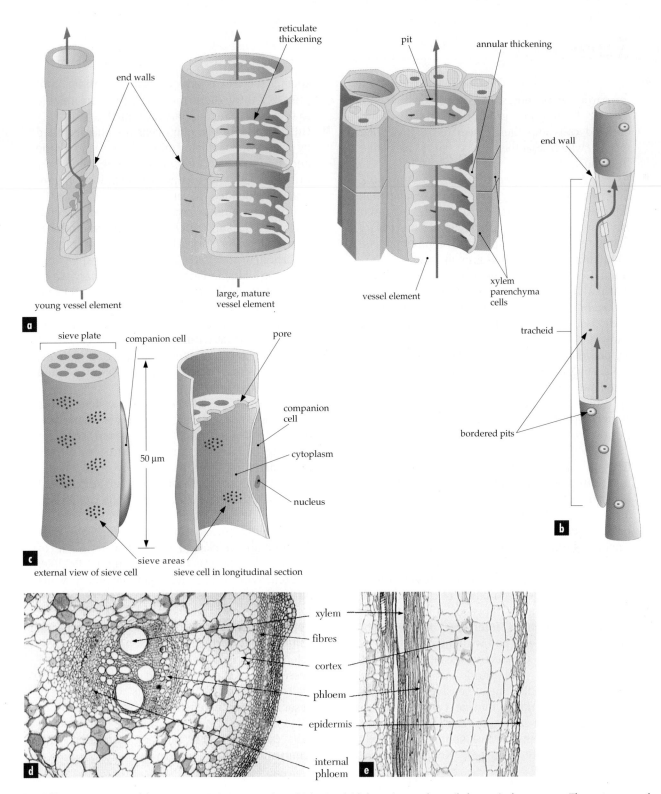

Fig. 6.22 Vascular tissue. **(a)** Xylem vessels have secondary thickening laid down in annular, coiled or reticular patterns. The mature vessels generally have open ends, so that a series of vessels forms a tube. The walls are perforated by small clusters of pits. **(b)** Tracheids differ from vessels in having rows of bordered pits, which provide the means of water transfer from tracheid to tracheid as they have closed pointed ends. **(c)** Phloem: sieve and companion cells. **(d)** Transverse section of a pumpkin stem, showing the site of the vascular strand central to the epidermis and cortex but external to the pith. **(e)** Longitudinal section of a pumpkin stem, showing the elongated phloem sieve cells (green stained) with conspicuous red-stained sieve plates, and annular and spiral patterns of xylem secondary thickening (red-stained walls)

Summary

- In multicellular organisms, cells are specialised for particular purposes and linked together into tissues. Tissues are groups of similar, differentiated cells and their associated extracellular matrix, which together carry out a particular function.

- As a result of specialisation of function, differentiated cells rely, to a greater extent than undifferentiated cells, on the extracellular environment for the provision of necessary materials, appropriate physical conditions and circulating hormones and growth factors.

- Animal cells are surrounded by an extracellular matrix that forms a loose and flexible lattice. The extracellular matrix contains proteins unique to each tissue and plays a vital role in regulating many aspects of cellular activity.

- Animal tissues are maintained by repair and/or replacement throughout life. Some tissues are regenerated by division of differentiated cells or by proliferation from a relatively unspecialised stem cell population. In other tissues, there is no turnover of cells; most 'permanent' cells repair themselves by turnover of cell components.

- Cells in animal tissues are bound together by junctions between plasma membranes of adjacent cells. Occluding (tight) junctions form an impenetrable seal between cells and restrict lateral movement of membrane proteins. Anchoring junctions provide mechanical support and are important in developmental processes. Communicating (gap) junctions provide direct communication between adjacent cells.

- There are four major types of animal tissues. Epithelia function as interfaces between biological compartments and with the environment, and in secretion. Connective tissues, composed of cells, fibres and fluid matrix, provide structural, metabolic and defensive support for other tissues. Striated muscle cells contain a highly organised array of actin and myosin filaments; myofilaments in smooth muscle cells are less regularly arranged. Nerve cells are specialised to conduct signals rapidly and precisely throughout the body; they are unable to divide and are supported by glial cells.

- Unlike animal cells, plant cells are characterised by a cell wall, which limits cell size and shape and is also important in metabolic processes. Primary cell walls comprise cellulose microfibrils embedded in and interconnected to a matrix containing non-cellulosic polysaccharides, pectins and proteins. Cellulose microfibrils give cell walls strength and their orientation determines the direction of cell expansion. Lignin is characteristic of many secondary walls.

- Plant cell walls are penetrated by plasma-membrane-lined channels, plasmodesmata, which provide a means of direct communication and transport between cells.

- Vascular plants have three major tissue systems: the dermal (covering), ground (supporting) and vascular (transport) systems.

- The major tissue of the dermal system is the epidermis, which lays down cuticle. Some epidermal cells are modified as guard cells for opening and closing stomata, hairs, trichomes and glands.

- The ground tissue system consists of storage and structural tissues. Parenchyma cells are living, vacuolated cells with thin primary walls. Collenchyma forms strands of living cells with differentially thickened primary walls made of cellulose and pectins that are strong but flexible. Sclerenchyma cells have rigid secondary walls with lignin deposited in concentric rings, forming elongated fibres or short sclereids.

- In the vascular system, xylem is a complex tissue composed of tracheids, vessels, parenchyma and sclerenchyma. When mature, tracheids and vessels lack cell contents, and consist only of cell walls. Phloem tissue consists of sieve cells associated with parenchyma cells and bundles of sclerenchyma fibres. In flowering plants, sieve cells lack a nucleus but are associated with companion cells.

keyterms

Review questions

1. What is a tissue? List the significant differences between plant and animal tissues.

2. What are stem cells? In what tissues are they found? How are stem cells involved in the maintenance of animal tissues?

3. Describe the types of connections that occur between animal cells. Use examples that illustrate the function of each type.

4. What are the principal features of:
   (a) epithelia?
   (b) connective tissue?
   (c) muscle?
   (d) nervous tissue?

5. What are the distinguishing features of parenchyma, collenchyma and sclerenchyma tissues in plants?

6. (a) What is the dermal tissue system of plants?
   (b) What is the role of the epidermis?
   (c) What are trichomes and what are their functions?

7. What types of cells compose xylem tissue? What are the structural and functional features of each?

8. Distinguish between phloem sieve cells and companion cells. Are these cells living or dead when mature?

Extension questions

1. In Chapter 3 we looked at the structure and function of individual cells. How do the functions of cells in tissues differ from the functions of an individual cell, such as a unicellular protist? If the function of a tissue was secretion, what types of organelles would dominate the cells in this tissue and why? Name another tissue and list the features of the individual cells that might help account for the overall function of that tissue.

2. Contrast the properties and functions of the extracellular matrix of animal tissues and plant cell walls, and use a drawing to illustrate the major components of each.

3. Describe and illustrate the structure of plasmodesmata and discuss how this structure relates to the controlled movement of materials between cells in either one or both directions. Although structurally very different, how are the communicating (gap) junctions of animal tissues similar to plasmodesmata in function, and how are they functionally different?

4. Why might the development of communicating junctions and plasmodesmata be required for the evolution of unicellular organisms into multicellular organisms?

Suggested further reading

Alberts, B., Bray, D., Lewis, J., et al. (1994). *Molecular Biology of the Cell*. 3rd edn. New York: Garland.

This text book is often used for more advanced courses in cell biology and includes an extensive account of tissues. Note Chapter 19 discusses cell junctions and the extracellular matrix of plants and animals in detail.

Esau, K. (1977). *Anatomy of Seed Plants*. 2nd edn. New York: John Wiley & Sons.

This classic text is an excellent introduction to plant cells and tissues.

Wolfe, S. L. (1993). *Molecular and Cellular Biology*. Belmont, CA: Wadsworth Publishing Co.

General text for university level cell biology courses.

CHAPTER

7

Responding to signals

All cells constantly gather information about their surroundings and use it to control their activities. In slime moulds, for example, chemical signals passing between cells co-ordinate the aggregation of free-living cells into a colony for feeding and reproduction (chemotaxis; Fig. 7.1). In the cellular slime mould *Dictyostelium discoideum*, starving amoebae produce and release cyclic adenosine monophosphate (cAMP), which stimulates neighbouring amoebae to do the same. Extracellular cAMP signals trigger cell aggregation. The amoebae change shape, become more motile and move towards the cAMP source. In mammals, information about heat levels, light and muscle length is constantly acquired and used to modify behaviour.

Receiving signals

Stimuli that act as signals for cells may be physical (e.g. light or heat) or chemical (e.g. food or hormones). These stimuli may come from the external environment or from other parts of the body of an organism. In multicellular organisms, certain cells are highly specialised to receive particular external stimuli. Other cells are specialised to send or receive internal signals to co-ordinate functions between different parts of their bodies. Most animals use a variety of internal chemical messengers in an elaborate communications network involving nerves and hormones to control the activities of body cells. Although much is known of the ways that animal cells and microorganisms receive signals, little is known of these mechanisms in plants.

To respond to a particular chemical or physical stimulus, cells must detect the **signal**. To do this, cells must contain the appropriate receptive protein (**receptor**), which undergoes some kind of change as a result of interaction with the incoming signal (Fig. 7.2). Examples of receptors are found in the specialised nerve cells that detect stimuli, such as: light—**photoreceptors**; heat—**thermoreceptors**; mechanical

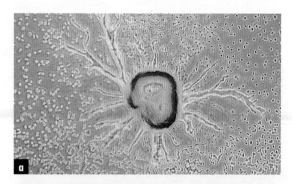

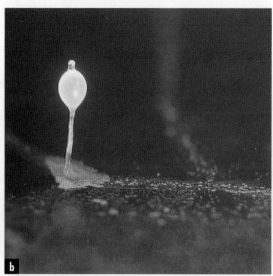

Fig. 7.1 (a) Individual cells of a slime mould stream toward one another and aggregate to form a motile 'slug'. **(b)** The 'slug' forms a stalked fruiting body for spore dispersal

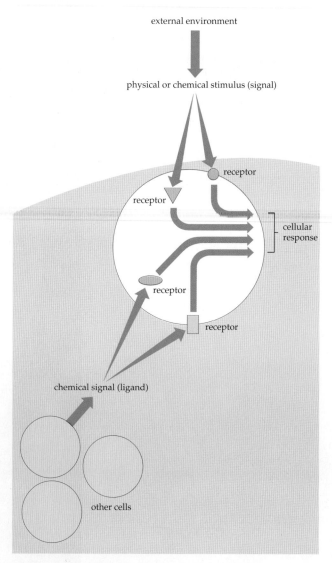

Fig. 7.2 A cell can respond to signals from the external environment and from other cells. Signals are received by specific receptors located on the surface of or inside the cell

pressure or stretch—**mechanoreceptors**; chemicals—**chemoreceptors**. Photoreceptors absorb light of a particular wavelength. Chemoreceptors bind with specific signal molecules, **ligands**, which have the appropriate molecular structure to interact with the receptor. In the case of light or chemical stimuli, the receptors are embedded in the plasma membrane. In the case of mechanoreceptors, the receptor may be part of an ion selective channel, such that when the receptor is distorted the channel moves from a closed to an open state.

Signalling pathways in cells may lead ultimately to the regulation of proteins involved in any of a variety of cellular activities, including cell division, differentiation and growth, metabolism, secretion and motility. Alternatively, they may simply generate an electrical signal, which is passed from the sensory receptor back to the central nervous system. The central nervous system, in turn, interprets the importance of the relayed signal. In this chapter, we will consider the different signals that cells receive and the mechanisms by which they process these signals to produce a response.

Cells regulate their activities in response to a range of chemical and physical stimuli. They have receptors, special proteins often embedded in the plasma membrane, which respond to particular signals.

Chemical stimuli

Cells receive chemical stimuli by direct interaction between the ligand and a specific receptor located in or on the surface of the responding cell. The chemical signal molecules may be lipid-soluble, water-soluble or bound to a surface, such as the plasma membrane of another cell or the extracellular matrix. Only lipid-soluble molecules are able to pass freely through the cell membrane to interact with receptors located inside the cell. Water-soluble and surface-bound signals (and a few lipid-soluble signals) interact with receptors located at the cell surface.

Lipid-soluble chemical signals

Steroid hormones and thyroid hormones are chemical signals for which the receptors are intracellular; they are lipid-soluble and pass freely into cells (Fig. 7.3a). Receptors for steroids are water-soluble proteins in the cytoplasm or nucleus of responsive cells. When binding occurs in the cytoplasm, a conformational change occurs, which then allows the ligand–receptor complex to pass into the nucleus through pores in the nuclear membrane. In the process of binding to a steroid molecule, the receptor changes its conformation in such a way that its affinity for specific nucleotide sequences in DNA increases. The ligand–receptor

complex binds to DNA, altering the expression of adjacent genes. The subsequent increase (or occasionally decrease) in the production of protein initiates a cellular response.

Water-soluble chemical signals

Most extracellular signal molecules are water-soluble and are thus unable to pass freely through the plasma membrane. Water-soluble signals are detected by receptors located on the cell surface (Fig. 7.3b). These receptors, like most other membrane proteins, are usually anchored in the cell membrane by one or more hydrophobic regions (Chapter 3). In most animals, water-soluble signals include **neurotransmitters**, which are released from nerve endings to act on other nerve cells, muscle cells or glands (Chapter 26). Water-soluble signals also include some local hormones, such as growth factors, which act on nearby responsive cells, some endocrine hormones, which are distributed through the bloodstream to cells throughout the body

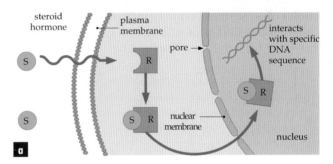

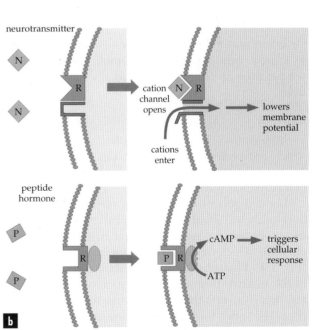

Fig. 7.3 (a) Steroid hormones pass through the plasma membrane, combine with a cytoplasmic receptor, and then pass into the nucleus to interact with DNA. **(b)** Peptide hormones and neurotransmitters cannot pass through the plasma membrane; they interact with a receptor on the surface

(Chapter 25), and metabolic products such as CO_2, which are released from metabolically active cells.

Unlike steroids, water-soluble hormones do not regulate protein production by DNA directly. They bind to surface receptors with high affinity. The receptors may be coupled directly to ion channels in the plasma membrane (Fig. 7.3b), linked to a specific type of protein, or may be associated with a specific enzyme, regulating catalytic activity. By these means, the signal is relayed into the cell, where it usually triggers one or more intracellular signals that produce the cellular response.

Surface-bound chemical signals

Some cells respond to molecules located either on the surface of other cells or bound to the extracellular matrix (Chapter 6). As the signal molecule binds to a receptor on the responding cell's surface, there is adhesion between the receptor-bearing cell and the ligand-bearing surface. For this reason, the receptors for such ligands are also known as **cell adhesion molecules**. Surface-bound ligands play important roles in development (Chapter 15) and in immune responses involving killer T cells (Chapter 23).

> Lipid-soluble chemical signals, such as steroid hormones, can enter cells freely and interact with intracellular receptors. Water-soluble and surface-bound ligands interact with receptors at the cell surface, which produces a response within the cell.

Physical stimuli

Cells can respond not only to chemical signals but also to a great variety of physical stimuli, including light, temperature, pressure, stretch and electric or magnetic fields.

Light

Light-sensitive responses involve photoreceptors, which are proteins containing a pigmented chemical group, a **chromophore**. A chromophore absorbs light of a particular wavelength only.

Light regulates many aspects of plant growth and development. Plants have several types of photoreceptors but the most important of these is **phytochrome**, which can exist in two interconvertible forms (Fig. 7.4). Phytochrome is synthesised and accumulates in the inactive form (P_r). Absorption of a quantum of red light converts the inactive form to an active form (P_{fr}). The active form can be reconverted rapidly to the inactive form by the absorption of a quantum of far red light or slowly and spontaneously in the dark. Most physiological responses are produced by the active form of phytochrome (P_{fr}) by a number of different mechanisms. For example, in fern spores, activated

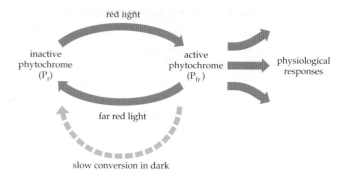

Fig. 7.4 In plants, most light-sensitive responses involve the active form of the pigmented molecule, phytochrome. The interconversion of the active and inactive forms depends on the absorption of red and far red light

phytochrome triggers a transient influx of Ca^{2+}, which increases internal levels of Ca^{2+} and stimulates germination. Activated phytochrome also controls the transcription of many light-activated genes (Chapter 24).

In animals, the photoreceptor whose structure and function is best understood is **rhodopsin**, which is found in rod cells in the retina of the vertebrate eye. Rhodopsin is associated with the chromophore, *retinal*, which undergoes a chemical rearrangement after absorbing light (Fig. 7.5). The new form of retinal then dissociates itself from rhodopsin, which alters its shape, triggering an interaction with a protein (the G-protein, transducin; p. 172) on the cytoplasmic side of the membrane. This leads to activation of phosphodiesterase enzymes. The increased phosphodiesterase activity increases the rate at which the cyclic nucleotide, cGMP, is destroyed. Consequently, cation selective channels in the plasma membrane, which had been held open by cGMP, close. This causes a reduction in depolarising current (making the membrane potential more positive), thus hyperpolarising the rod cell membrane (making the membrane potential more negative). This, in turn, reduces the release of neurotransmitter from the rod cell and so the signal is finally detected by the central nervous system. Retinal subsequently resumes its original form in a reaction that does not require light, after which it can reassociate with rhodopsin.

> Photoreceptors contain chromophores that absorb light of a particular wavelength and trigger a cellular response.

Mechanical stimuli

Many microbes and certain cells in animals and plants respond rapidly to physical contact or stretching by generating electrical signals (*generator potentials*). The generator potentials depolarise the membrane and trigger action potentials (Chapter 26), which conduct to the central nervous system. There are many different

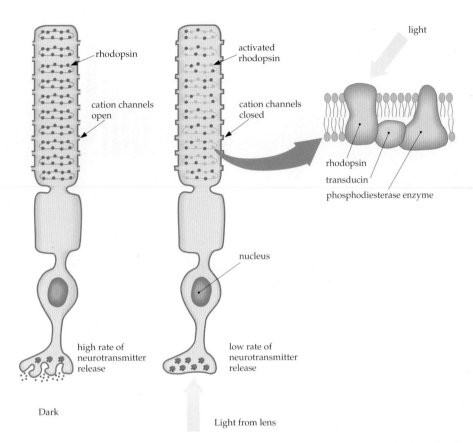

Fig. 7.5 Response to light of a rod photoreceptor cell in the retina of the human eye. Activated rhodopsin leads to the closure of cation channels, making the membrane potential more negative. This reduces the release of neurotransmitter from the rod cell

types of mechanosensory cells, each modified to respond best to a particular type of stimulus.

Specialised cells in the leaves of the sensitive plant, *Mimosa pudica* (Fig. 4.10), respond to touch by producing an electrical signal that triggers the opening of K⁺ channels in cells at the base of the leaf stalk. Potassium ions leave these cells, dragging water by osmosis, which leads to loss of turgor. This causes the leaves to fold and droop. The closure of the Venus fly trap (Fig. 7.6) when an insect touches sensory hairs involves a similar mechanism that causes the leaf to snap shut, trapping the insect within.

Animals possess a wide range of mechanoreceptors, ranging from simple, naked nerve endings that respond to force, as in skin pressure receptors, to the highly specialised non-neuronal hair cells of vertebrates (Figs 7.7, 7.8). Although the mechanisms involved in mechanosensitivity, particularly the ability of cells to differentiate between different stimuli, are not well understood, it seems that, in most, mechanosensitive channels are involved. Stretch sensitive channels respond to mechanical distortion by changing from a closed to an open state. As most channels appear to be cation selective, this effectively means there is an influx of Na⁺. As we have implied, the influx of a positively charged particle into a mammalian cell leads to

Fig. 7.6 The Venus fly trap responds to touch by producing an electrical signal that leads to loss of turgor and closing of the trap

membrane depolarisation. In humans, receptors sensitive to stretch are used to monitor everything from blood pressure to bladder distension, from balance to the position of limbs.

The sense of hearing is a specialised form of detection of a mechanical stimulus due to changes in air pressure. In mammals, pressure waves set up resonant vibrations in the eardrum, which are transmitted by a chain of bones in the middle ear to the fluid within the inner ear (Fig. 7.7). Cilia protruding from hair cells lining the cochlea of the inner ear are

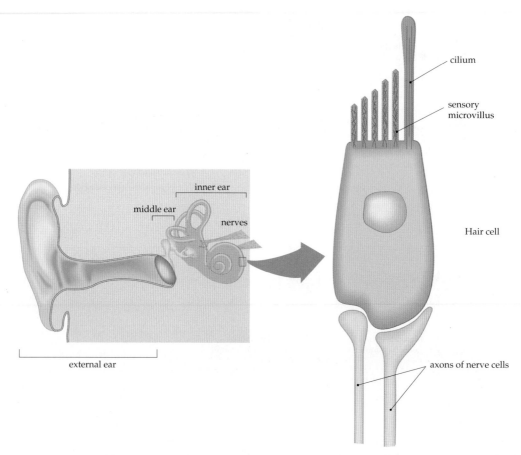

cilium

sensory microvillus

Hair cell

axons of nerve cells

Fig. 7.7 Sound waves are transmitted through the bones of the middle ear to the fluid-filled inner ear. Hair cells lining the cochlear are distorted by pressure waves

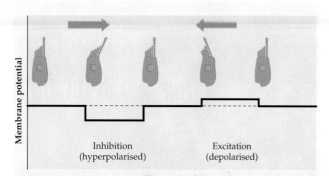

Inhibition (hyperpolarised)   Excitation (depolarised)

Membrane potential

Fig. 7.8 Bending towards the long cilium depolarises the hair cell, increasing the rate of action potentials in the sensory nerve (excitation). Bending in the other direction causes the opposite effect (inhibition)

distorted by the pressure waves and respond by increasing or decreasing the rate of nerve impulses passing along the auditory nerve. The cilia-bearing hair cells thus translate a mechanical stimulus into electrical signals that can be relayed by the auditory nerves to the brain.

Mechanoreceptors respond to physical contact or stretch, usually by opening ion channels in the plasma membrane.

Temperature

Temperature affects all biological activity in a general way because it determines the rates of chemical reactions and the activities of enzymes. It is therefore not surprising that many animals have specialised heat-sensing capabilities. Mammals, such as ourselves, have cold-sensitive nerve cells, which respond to decreasing temperature, and warm-sensitive nerve cells, which respond to increasing temperature. The response in both cases is an increased frequency of nerve impulses. Thermoreceptors may be temperature-sensitive channel proteins or active carriers (Na^+–K^+ ATPase, p. 96). In some organisms, a temperature-induced change in local fluidity of the cell membrane may regulate the activity of the membrane proteins.

Blood-sucking insects, ticks, leeches and some snakes, such as rattlesnakes and many Australian pythons, detect infra-red thermal radiation to locate their endothermic host or prey. The snakes do this by means of warm-sensitive thermoreceptors located in pits on their heads (Fig. 7.9). The high sensitivity of these receptors and their precise location on either side of the head provide directional information and the snakes are able to strike accurately at prey, even in total darkness.

Fig. 7.9 Thermoreceptors located in pits on either side of the lower jaw of the green Australian python, *Chondropython viridis*, allow the snake to strike accurately at prey, even in total darkness

Thermosensitive nerve cells respond to temperature by a change in frequency of nerve impulses.

Electric and magnetic fields

Sharks, skates and rays rely for navigation on electric fields generated by the magnetic field of the earth as they move through the water (Fig. 7.10). Other animals, such as homing pigeons and honey bees, are believed to use tiny particles of magnetic material (magnetite, Fe_3O_4) to navigate using the earth's magnetic field. Certain bacteria also contain magnetite crystals, which physically align the axis of the cell in a north–south direction, like a tiny compass needle. Such *magnetotactic* bacteria in the Northern Hemisphere move towards the north magnetic pole; in the Southern Hemisphere, they move southwards. From most points on the earth's surface, the shortest distance to either pole is through the earth; the bacteria, therefore, move downwards into the sediment, which, for them, is a good place to be.

Recent research has revealed that fishes are not the only animals that can detect electric fields: the Australian monotremes, the platypus and echidna, both have specialised nerve endings that can detect very weak electric fields. These electroreceptors, located in the bill of the platypus and snout of the echidna, assist these animals to catch their prey (Fig. 7.11). The freshwater crustaceans (yabbies), which form an important part of a platypus's diet, emit a small electric discharge when their tail muscles contract as they attempt to escape. The platypus, which dives with its eyes closed, uses these electrical signals to track and catch the yabby. Anatomical and physiological studies have revealed that the receptors are ion channels in neuronal membranes that are sensitive to changes in the electric field across the membrane.

Some animals are able to detect electric and magnetic fields by means of modified ion channels.

Signal processing

Once an extracellular signal has interacted with a receptor, it must be processed into information that produces the appropriate cellular response. Signal processing may be direct or may involve one or more intracellular molecular steps (such as G-proteins and second messengers), which pass the signal from the receptor to specific effector molecules that trigger the cell's response. Effector proteins are regulated by

Fig. 7.10 Specialised electroreceptors located in pits on the underside of the head of a shark are sufficiently sensitive to detect electric fields generated by the earth

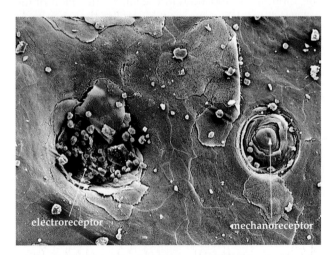

Fig. 7.11 Scanning electron micrograph of the skin of the bill of the platypus showing a mechanoreceptor (to the right) and an electroreceptor (to the left)

intracellular signalling, which frequently involves phosphorylation by specific kinases or binding of Ca^{2+}.

Direct receptor-mediated responses

The simplest signalling pathways are those in which the activated receptor acts directly to produce a cellular response. As we have stated, this occurs when a mechanoreceptor is distorted, triggering a local generator potential.

Steroid hormones also generate local responses. The hormone–receptor complex binds to a specific region of DNA and alters the rate of synthesis of a particular protein. For example, during egg production in hens, oestrogen stimulates the synthesis of the protein ovalbumin in the oviduct.

Steroid hormones can also produce delayed secondary responses in a target cell. These occur when the protein produced as a result of the primary ligand–receptor interaction binds to a different region of DNA to activate another gene. The same steroid can regulate different genes in different target cells. For example, the male steroid hormone, testosterone, is responsible for the development of the various male secondary sexual characteristics, such as deepening of the voice and beard growth. Testosterone always binds to the same receptor protein but produces different responses in different target cells.

. Other forms of direct receptor-mediated response involve regulation of dynamic changes of cytoskeletal proteins on the inner face of the plasma membrane by activation of cell adhesion receptors (Chapter 3) and channel-linked receptors. When activated, channel-linked receptors transiently open or close, regulating the movement of specific ions such as K^+ or Ca^{2+} across the plasma membrane. These are *ligand-gated channels* (p. 104). This alters the balances of ions across the membrane and therefore changes the membrane potential (p. 102). In excitable cells, the change in potential may be sufficient to affect other *voltage-gated channels* in plasma membranes (p. 103), through which ions such as Na^+ or Ca^{2+} pass, and cause hyper-polarisation (inhibition) or depolarisation (excitation). For example, acetylcholine is a neurotransmitter released by some nerves of vertebrates, which binds to several different receptor proteins. When acetylcholine released from one nerve cell binds to a nicotinic receptor on the membrane of the next nerve cell, a channel of 0.7 nm diameter opens in the post-synaptic membrane. This channel is a cation channel, which allows Na^+ influx and K^+ efflux. Because the Na^+ gradient is much larger than the K^+ gradient, the pre-dominant flow is an influx of Na^+, thus lowering the membrane potential, making it more positive (Fig. 7.3b). If the depolarisation is sufficient, it may initiate an action potential in the responding nerve cell.

The simplest mechanism of intracellular signal processing involves an activated receptor acting directly to produce a cellular response.

G-protein-linked receptors

Many signalling pathways require the action of intermediate proteins, most commonly G-proteins (guanosine triphosphate [GTP] binding regulatory proteins). The intracellular pathway leading to the aggregation of slime moulds involves a G-protein that changes cell shape and movement (p. 166).

G-proteins are coupled to receptors that have seven transmembrane domains. **G-protein-linked receptors** indirectly alter the activity of an ion channel or intracellular enzyme (Figs 7.12, 7.13). They usually do this by altering the concentration of small molecules, such as cAMP or Ca^{2+}, known as second messengers. For example, the photoreceptor rhodopsin is a G-protein-coupled receptor that brings about the closure of cation selective channels in rod cell membranes by activating the enzyme cGMP phospho-diesterase, thereby decreasing cGMP levels in the cell.

Second messengers

A common feature of G-protein-linked receptors is that their activity changes the concentration of one or more small intracellular signalling molecules (Fig. 7.12). These are known as **second messengers** since they are released in response to extracellular signals (the first messengers). We will consider several of these, including the cyclic nucleotides cAMP and cGMP, and Ca^{2+}. Second messengers are amplifying systems: the binding of a single first messenger molecule can activate many second messenger molecules.

Cyclic AMP (cAMP) was the first of the intracellular second messengers to be identified. Cyclic AMP is produced from ATP by the membrane-associated enzyme, adenylate cyclase, and degraded to AMP by cAMP phosphodiesterase (Fig. 7.12). Cyclic AMP concentration can be increased either by inhibition of phosphodiesterase or, more commonly, by activation of adenylate cyclase, the usual target of regulation by G-proteins. The increase in cAMP concentration leads to phosphorylation of a variety of target proteins by a specific protein kinase enzyme, a process that can regulate their activity.

Cyclic GMP (cGMP) is also a second messenger in some cells (Fig. 7.12). As with cAMP, the levels of cGMP may be regulated by external stimuli and many cells contain a protein kinase whose activity depends on cGMP.

Inositol trisphosphate, IP_3, is a second messenger formed when the enzyme phospholipase C is activated (Fig. 7.13). Phospholipase C hydrolyses the lipid precursor phosphatidylinositol 4,5-bisphosphate to

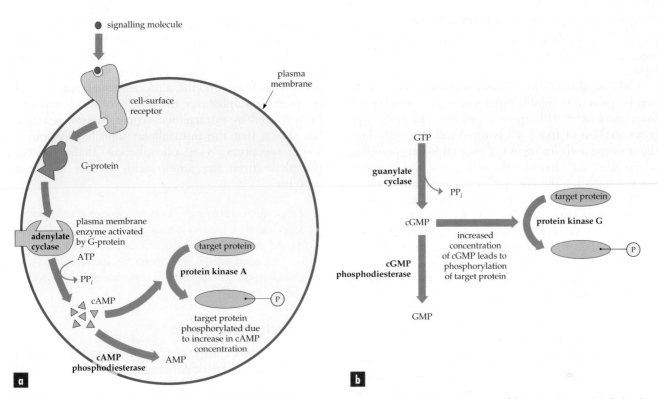

Fig. 7.12 Major pathways by which G-protein-linked cell surface receptors generate intracellular messengers. **(a)** A signalling molecule binds to a cell-surface receptor causing it to bind to a G-protein. The G-protein then activates a plasma membrane enzyme (adenylate cyclase), which produces the second messenger, cAMP. Increase in cAMP concentration leads to phosphorylation of a target protein by a specific enzyme, a protein kinase. cAMP is degraded to AMP by cAMP phosphodiesterase. **(b)** Cyclic GMP is also a second messenger, which can lead to phosphorylation of a target protein. However, the activity of the enzyme guanylate cyclase does not directly involve a G-protein as does adenylate cyclase

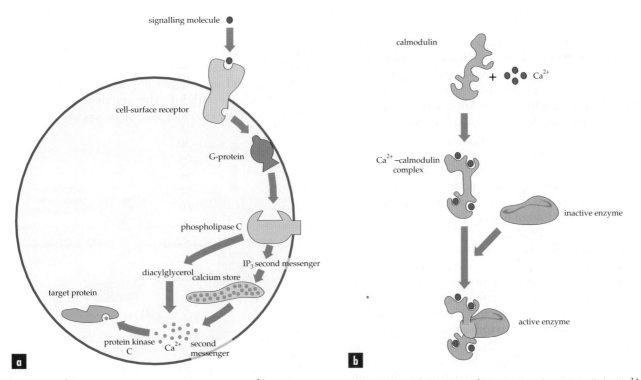

Fig. 7.13 Ca^{2+} acts in cells as a second messenger. Ca^{2+} regulates target proteins either **(a)** directly or **(b)** by means of an intracellular Ca^{2+} receptor, commonly called calmodulin. A calmodulin molecule has four Ca^{2+} binding sites. On binding with Ca^{2+}, calmodulin undergoes a conformational change, which allows it to interact with and activate other proteins (enzymes)

yield IP$_3$ and diacylglycerol (DAG). IP$_3$ has the important role of releasing Ca^{2+} from intracellular stores, so stimulating activity in many different cell types.

Calcium also acts in cells as a second messenger. It may be present in soluble form as free Ca^{2+} or in bound form associated with specific proteins. In cells, the concentration of free Ca^{2+} is regulated by controlling the opening and closing of Ca^{2+} selective ion channels, which allow Ca^{2+} to enter the cell. As we have pointed out in Chapter 4, the Na$^+$–Ca^{2+} exchanger expels Ca^{2+}. It may also be transported into internal stores, so removing it from the vicinity of its target protein. Alternatively, the entering Ca^{2+} can trigger the further release of Ca^{2+} from intracellular stores, so amplifying the effects of the Ca^{2+} selective channels.

In a cell, soluble Ca^{2+} can regulate the activity of enzymes directly. An example is protein kinase C, which is regulated by the Ca^{2+} along with the two second messengers it produces, IP$_3$ and DAG (see Fig. 7.13). Alternatively, Ca^{2+} can regulate enzyme activity by means of an intracellular Ca^{2+} receptor, most commonly calmodulin. The Ca^{2+}–calmodulin complex regulates protein activity either directly or through the activation of a specific Ca^{2+}–calmodulin-dependent protein kinase, which has the ability to phosphorylate a variety of target proteins.

Most of the calcium in a cell is either bound to calcium-binding proteins or stored inside the endoplasmic reticulum or mitochondria. The Ca^{2+} in these stores can be drawn upon in response to a stimulus and external Ca^{2+} may be drawn into the cell through transmembrane channels. The high binding and storage capacity of the cell for Ca^{2+} means that concentration increases can be localised to particular regions of the cell. Ca^{2+} does not diffuse freely to other parts of the cell because it is rapidly removed from the cytoplasm back into the intracellular Ca^{2+} stores.

Phosphorylating proteins

The mechanism of action of intracellular second messengers in many signalling pathways is to trigger protein phosphorylation. This is achieved by one of a large number of different protein kinases, each specific to a particular second messenger. The protein kinases operate by phosphorylating the amino acids serine and threonine in target proteins. The intracellular signalling regions of some receptors possess protein kinase activity, which enables phosphorylation of both the receptor itself and other proteins, adding phosphate groups to the amino acid tyrosine in the polypeptide chain of the protein. In many cases, phosphorylation causes conformational changes in the protein that are accompanied by changes in activity. In different situations, the active form of the target protein can be either the phosphorylated or dephosphorylated form.

If protein kinases were continuously active in cells, all target proteins in cells would be permanently phosphorylated and their activities would not change. Cells, therefore, need enzymes to remove the phosphates from proteins, a process usually carried out by specific phosphatases, the activity of which may also be regulated by extracellular signals. Recent research has shown that the intracellular signalling region of some receptors is a phosphatase that removes phosphate from the amino acid tyrosine in target proteins.

Intracellular processing of signals may involve activation of G-proteins, leading to changes in the concentrations of second messenger molecules such as cAMP, cGMP and Ca^{2+}. Second messengers change the activity of phosphorylating enzymes to increase or decrease protein phosphorylation.

Catalytic receptors

Some receptors consist of a single polypeptide having both an external ligand-binding region and an internal signalling region that exhibits tyrosine kinase enzyme activity (Fig. 7.14), a property that enables such receptors to attach phosphate groups to the amino acid tyrosine where it occurs in proteins. The catalytic activity of the receptor is triggered when the ligand binds to it. Catalytic receptors in mammals include the receptor for epidermal growth factor, which stimulates epidermal and many other cell types to divide, and the insulin receptor, which regulates glucose metabolism. Other kinds of catalytic receptor include those with guanylate cyclase activity (Fig. 7.12) and protein tyrosine phosphatase activity.

Some second messengers are activated directly by ligand binding without the mediation of any G-proteins. The enzyme guanylate cyclase, which

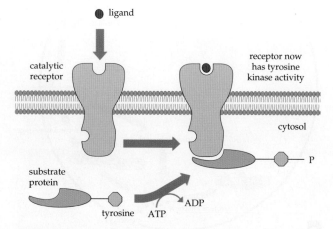

Fig. 7.14 Tyrosine kinase activity is triggered when the ligand binds to a catalytic receptor. The activated receptor phosphorylates the tyrosine side chain of a substrate protein, altering the activity of the protein

synthesises cGMP from GTP, is found in both a soluble and a membrane-associated form. The membrane-associated cyclase appears to be the signalling region of a receptor for some peptide hormones.

Intracellular processing of a signal can involve a change in the catalytic activity of the intracellular region of a receptor protein, by phosphorylation of specific amino acid residues, usually serine, tyrosine or threonine.

Producing a response

After the receipt of an input signal, the molecular events that are set in train are used by cells to generate a particular response. The result is a regulated change in cellular activities such as metabolism, growth, phagocytosis, secretion, cell division, movement or differentiation. Some dramatic changes occur as the result of a single stimulus, while other responses require stimulation of several receptors. Also, cells do not always produce the same response to a particular stimulus. A response may be affected by recent stimulation events or by other input signals for which the cell has appropriate receptors.

Signal amplification

How is it that the male *Bombyx mori* moth can produce a response to a single molecule of the sex pheromone, bombykol? (Sex pheromones are volatile chemical attractants emitted during courtship; Chapter 25.) The molecular events after activation of certain receptors can amplify (increase the strength of) the signal, thereby increasing the sensitivity of the system. Binding of a single molecule of bombykol to a receptor cell on the antennae of this moth causes a response that is amplified by intracellular molecular pathways. Photoreceptor cells in the vertebrate retina that respond to absorption of a single photon of light are another example. In both of these cases, activation of a single receptor leads to the closing or opening of many, perhaps thousands, of membrane ion channels.

How **signal amplification** occurs is well illustrated by rhodopsin (Fig. 7.15). Absorption of one photon of light can convert rhodopsin to its excited form; in this form rhodopsin can bind to the G-protein, transducin. After a molecular exchange, transducin dissociates from the receptor and, through a second messenger cGMP system, leads to closure of membrane cation channels. A single excited rhodopsin molecule can activate many transducin molecules before being converted back to its resting form. This causes a relatively large change in concentration of the second messenger cGMP and the closure of not one but many cation channels.

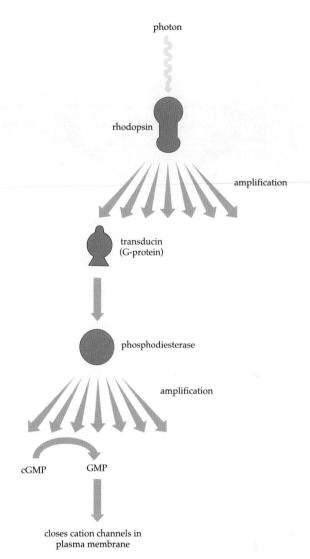

Fig. 7.15 Amplification of an extracellular signal. Absorption of a single photon of light by rhodopsin results in the activation of many transducin molecules, each of which activates the enzyme phosphodiesterase. Each molecule of phosphodiesterase catalyses the production of many GMP molecules, which then close cation channels in the plasma membrane

The sensitivity of some sensory pathways can be increased by signal amplification. For example, activation of a single receptor molecule can cause the opening or closing of many membrane ion channels.

Convergent pathways

Convergent pathways are those in which signals from separate receptors for different stimuli regulate the same intracellular event. This can allow the effects of two different extracellular signals to be integrated. An example of two convergent stimulatory pathways is seen in liver cells responding to the hormones adrenaline and glucagon (Fig. 7.16). These hormones bind to different receptors, each of which stimulates a G-protein to activate adenylate cyclase. The resulting

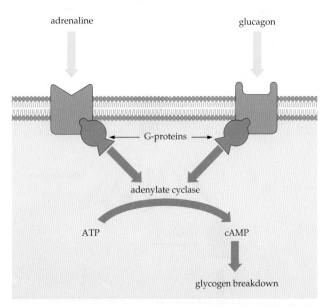

Fig. 7.16 Convergent pathways regulating glycogen metabolism

increase in cAMP levels stimulates glycogen break-down. At hormone concentrations that elicit less than maximal responses, the effect of both hormones is greater than it is for either hormone alone. However, if either hormone is present at a concentration high enough to evoke the maximal response (in terms of glycogen breakdown), presence of the second hormone has no additional effect.

In other cases where an inhibitory and a stimulatory signalling pathway converge, the two stimuli oppose each other and each can reduce the magnitude of the response to the other. When convergent stimulatory and inhibitory pathways are involved, the response can be different for different signal intensities. For example, many unicellular organisms move towards or away from light depending on light intensity. This can result from convergent pathways starting from different photoreceptors, one of which leads to movement towards light and the other to movement away from light.

Divergent pathways

Divergent signalling pathways allow a number of cellular responses to be regulated in parallel by the same stimulus. This involves intracellular signalling molecules, such as second messengers like cAMP, modulating the activities of a number of different cellular proteins. For example, neutrophils (a type of white blood cell) have receptors for specific protein breakdown products released by bacteria. Ligand-binding to these receptors regulates a variety of responses, including changes in cell adhesion, shape, motility, oxygen uptake and release of toxic oxygen metabolites (e.g. hydrogen peroxide) and release of proteases (enzymes that degrade proteins).

Convergent signalling pathways in cells involve integration of signals from separate receptors to regulate the same intracellular response pathways. Divergent pathways allow regulation of different cellular responses by the same stimulus.

Excitation and adaptation

For many extracellular signals, the responding cell is sensitive to relative differences in the level of the signal rather than to the absolute level. In bright light, more light is reflected from the *black* letters on this page than is reflected from the *white* background if the light is very dim. Yet, in both situations, the letters are perceived as black on a white background because the *relative* brightness of the page and the letters on it remains the same. If the book is taken from bright sunlight into a poorly lit room, the brightness of both the print and the paper decreases dramatically relative to what it was, and reading is difficult. It takes a few seconds or minutes for the eyes to adapt to the change in absolute stimulus intensity. This process of adaptation allows the photoreceptor cells in the retina to respond to the intensity of light from the page relative to the overall background intensity.

The process of photoreceptor adaptation has two aspects. Firstly, the cell responds similarly to relative differences in the signal (e.g. two-fold differences) over a wide range of absolute stimulus intensity. In the retina, each photoreceptor cell measures light intensity relative to some average of the light intensity over the whole retina. Relative to that average, the print is dark and the page is bright. Secondly, the cell responds initially to a change in signal intensity, but this response declines over time as the cell adapts to the new absolute stimulus levels. Until the retina adapts relative to that average, the perceived signal intensity is low everywhere when the book is taken from bright sunlight into dim room light. To behave this way, photoreceptor cells must compare, moment by moment, the signal that they are presently receiving with the background level.

The implication from this example is that, in photoreception, extracellular stimuli are usually transduced into two intracellular processes. One is **excitation**, a rapid measure of the intensity of the present stimulus. The second is **adaptation**, a slower estimate of the average background stimulus in the recent past. Both processes are activated by the same receptor so that, at some point, the signalling pathways for excitation and adaptation must diverge. The two pathways must also converge at some point because the present stimulus, measured by the excitation process, is compared with past stimuli, in the adaptation process.

Excitation pathways activate a cellular response while slower adaptation pathways inhibit excitation. Responding cells are thus sensitive to relative differences in stimulus intensity.

Summary

- The activities of cells are regulated by extracellular signals for which cells have specific receptors. Receptors that interact with light or chemical stimuli are proteins, usually embedded in plasma membranes.
- Extracellular chemical signals may be lipid-soluble, water-soluble or surface-bound. Lipid-soluble signals such as steroid hormones pass freely across the plasma membrane and bind to intracellular receptors. Water-soluble and surface-bound ligands interact with membrane protein receptors at the cell surface.
- Certain cells also respond to physical stimuli. Photoreceptors contain chromophores, which respond to a particular wavelength of light. Mechanoreceptors respond to pressure and stretch by altering ion flow across the cell membrane. Thermoreceptors respond to temperature change by altered nervous activity. Electric and magnetic fields are detected by modified ion channels.
- In eukaryotic cells, intracellular signal processing may involve: direct receptor-mediated responses; intermediate proteins, such as G-proteins, and second messengers, such as cAMP, cGMP and Ca^{2+}; protein kinases, which attach phosphate groups that may or may not be regulated by second messengers; and associated phosphatase enzymes, which remove phosphate groups.
- The molecular events that follow signal reception lead to a cellular response by a pathway that may involve: amplification of the signal to allow detection of very weak extracellular stimuli; convergence, allowing integration of inputs from several receptors; divergence, allowing parallel regulation of several responses; and adaptation pathways, which allow the present signal to be compared with recent signals, so that the cell can respond to a relative change in the stimulus.
- Responses that are regulated by signalling pathways include all of the major activities of the cell, including metabolism and growth, division, differentiation, secretion and motility.

key terms

adaptation (p. 176)
cell adhesion molecule (p. 168)
chemoreceptor (p. 167)
chromophore (p. 168)
excitation (p. 176)

G-protein-linked receptor (p. 172)
ligand (p. 167)
mechanoreceptor (p. 167)

neurotransmitter (p. 167)
photoreceptor (p. 166)
phytochrome (p. 168)
receptor (p. 166)
rhodopsin (p. 168)

second messenger (p. 172)
signal (p. 166)
signal amplification (p. 175)
thermoreceptor (p. 166)

Review questions

1. Define the terms signal and receptor in relation to cellular responsiveness. List the types of signals to which cells can respond.

2. How do photoreceptors respond to light?

3. What are the three main types of chemical signals to which cells respond? How does receptor interaction occur in each case?

4. What types of receptors are involved in hearing? Describe the way that sensory cells of the ear are involved in hearing.

5. What is a second messenger? Use an example to show how they are involved in intracellular signalling pathways.

6. What is meant by signal amplification? Give an example.

Extension questions

1. What is the function of:
   (a) phytochrome in plants?
   (b) rhodopsin in the retina of the eye?
2. Draw a simple diagram illustrating the differences between the signalling pathways of direct receptor-mediated receptors, G-protein-linked receptors and catalytic receptors.
3. Describe the role of Ca^{2+} in regulating the activity of cellular proteins.
4. Discuss the possible advantages of divergent signalling pathways in the response of a single-celled organism to an external stimulus.
5. Cells continuously receive many different stimuli from their environment. By what mechanisms do they integrate all this information to produce the most appropriate response?

Suggested further reading

Alberts, B., Bray, D., Lewis, J., et al. (1994). *Molecular Biology of the Cell.* 3rd edn. New York: Garland.

This text contains a very thorough and detailed account of the processes described in this chapter.

CHAPTER

8

Cell division

All cells are derived from other cells by cell division. Single-celled organisms, such as bacteria, protozoa and many algae, simply divide into two similar individuals that grow and divide again. The cells of multicellular organisms arise from a single cell as a result of **cell division**, followed by differentiation of these cells into the various tissues that make up the mature organism (Chapter 6). Cell division continues in most tissues to replace aged or damaged cells. The processes within the cell that bring about division are strictly controlled to ensure accurate transfer of DNA to daughter cells. The rate of division of cells in multicellular organisms must be regulated for the correct formation and subsequent maintenance of tissues. Also, the size of a mature organism, be it ant or elephant, is usually a function of cell number and cell volume. The mechanisms that control cell division in multicellular organisms are major areas of research in biology and medicine. Many aspects of abnormal development, disease, healing and ageing are related to changes in the abilities of cells to divide. When a cell divides, its cytoplasm and copies of its DNA are transmitted to each daughter cell. DNA molecules pass to daughter cells in an organised way so that each cell receives a full complement of genetic material. In eukaryotic cells, DNA molecules are organised into chromosomes, and **mitosis** is the process that ensures that each daughter cell receives one copy of every chromosome.

In most organisms, each cell is **diploid**, that is, it has two copies of each chromosome, one inherited from the egg and one from the sperm following sexual reproduction. The two copies of a chromosome are **homologous**, that is, similar but not identical. Each cell is described as $2n$, where n is the number of homologous pairs of chromosomes (the **haploid** number). A major advantage of diploidy is that every cell has two copies of each gene. Mutations are often deleterious, altering and perhaps destroying the function of gene products. When a cell has two copies of each gene, a mutation in one of these copies is not necessarily serious since the cell can survive by using the other copy of that gene.

During sexual reproduction in eukaryotes, two reproductive cells, usually sperm and egg, fuse to form the first cell (**zygote**) of a new generation. Reproductive cells are haploid and the products of **meiosis**, a special type of cell division that reduces the chromosome number to half that of the parent cell. After fusion of two haploid cells, the zygote will have the correct diploid number of chromosomes.

All cells arise from other cells by division. DNA molecules are accurately passed to daughter cells by the processes of mitosis and meiosis.

Cell division in prokaryotes

Cell division in prokaryotes such as bacteria is simple and rapid. Some bacteria in culture can divide every 10 minutes, providing that sufficient nutrients are available. Their single, circular molecule of DNA is attached to the plasma membrane at a specific point (Fig. 8.1). Between divisions, the double-stranded DNA molecule is replicated (Chapter 10) by a system of enzymes that produces an identical copy. When the growing cell reaches a certain size and after replication is complete, the new DNA molecule has a separate point of attachment to the plasma membrane. The cell then commences to divide into two. The two DNA molecules separate by growth of the plasma membrane and cell wall between their attachment points. The plasma membrane and wall grow inwards across the middle of the cell. This process of cleavage, **binary fission**, creates two equal-sized cells, each containing one copy of the genetic material and approximately half the cytoplasm. The two cells may separate or remain joined to form a growing filament of cells, depending on the type of bacterium.

In some bacteria, there are additional smaller, unattached circular molecules of DNA, **plasmids**, which also replicate so that several copies may be present in each cell. At fission, plasmids pass as part of the cytoplasm to the daughter cells.

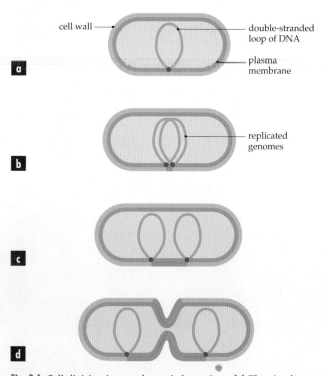

Fig. 8.1 Cell division in a prokaryotic bacterium. **(a)** The circular DNA molecule (chromosome) is attached to the cell membrane at a specific point. **(b)** The DNA replicates as the cell grows. **(c)** The two copies of DNA are separated by expansion of the membrane between the attachment points. **(d)** The membrane and wall furrow inwards to divide the cell in two

In a prokaryotic cell, the single circular molecule of DNA is attached to the plasma membrane. After replication, the attachment points of the two DNA copies are separated by growth of the plasma membrane and cell wall, and the cell divides by binary fission.

Cell division in eukaryotes

Eukaryotic cells are larger and have more internal structures than prokaryotes. Their cell division is correspondingly more complex, particularly since they contain more genetic material. They usually contain many molecules of DNA, each organised into a chromosome contained within the membrane-bound nucleus. Between cell divisions, chromosomes are invisible in the nucleus because this DNA appears as a fine three-dimensional network called **chromatin**. Just prior to cell division, the chromatin aggregates and then condenses into the complement of **chromosomes**. 'Condensation' of chromatin is an essential prerequisite for mitosis and very complex in detail. Each extremely long DNA molecule is looped and folded in a series of steps until it is transformed into the compact chromosome (see Fig. 10.4) visible through the microscope. Only then can it be organised and moved precisely about by the mitotic spindle, the structure that the cell assembles to bring about mitosis.

The appearance of chromosomes during cell division attracted the attention of the German microscopist, Walther Flemming, who first described their behaviour in 1882. Early microscopists studying mitosis soon established the link between the physical behaviour of chromosomes and the well-known passage of genetic information from cell to cell, and thus from organism to organism. Thus, the discovery of mitosis provided the cellular basis for understanding genetics. Before we discuss cell division itself, we will briefly consider what is known about its control.

The cell cycle

When cells are being formed by cell division, they pass through a regular series of events called the **cell cycle**. If these cells differentiate later, they may exit the cell cycle, either temporarily or permanently, in which event they will die, sooner or later. In actively growing eukaryotic cells, division itself occupies only a small part of the cell cycle (Fig. 8.2). For the rest of the time, cells are in **interphase**, during which DNA and most of the other molecules required by the cell are synthesised.

In interphase cells, the nucleus also contains one or more nucleoli, composed of RNA being assembled for export to the cytoplasm for protein synthesis. The first part of interphase, the **G₁ phase** (*first gap*), is often the

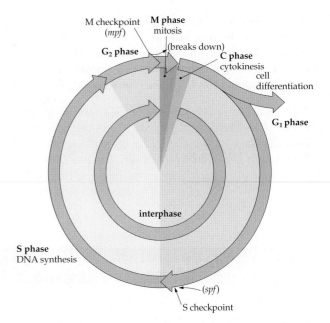

Fig. 8.2 The cell cycle. In actively growing cells, interphase alternates with cell division. After division, a cell enters the G₁ phase, when normal cellular activity, temporarily slowed down during division, resumes. If some of these cells differentiate, they exit the cell cycle during G₁ and stop dividing. The rest of these cells soon enter the S phase, during which DNA synthesis doubles the amount of DNA in the nucleus. The cell enters the G₂ phase before eventually dividing again (the M phase). Two important checkpoints are shown, which control entry into the M and S phases, respectively. Control is mediated through *mpf* and *spf* protein complexes. In cells with about a 24-hour cell cycle, mitosis lasts about 1 hour, G₁ phase about 9–12 hours, S phase 10 hours and G₂ phase about 2 hours

longest part of the cell cycle. In multicellular organisms, differentiated cells exit the cycle in this stage. Actively growing cells continuing through the cycle enter the **S phase** (*synthesis*), during which DNA is replicated. At the end of the S phase, the nucleus is appreciably larger and contains twice the amount of DNA. The cell then enters the **G₂ phase** (*second gap*), during which the main synthesis of other cellular molecules occurs.

Some time later, the cell prepares for division and enters the **M phase** (*mitosis*). On the basis of appearance alone, it is difficult to distinguish between cells that are in G₁, S or G₂ phases, but the beginning of mitosis is recognisable because of the condensation of chromosomes from chromatin within the nucleus. Chromosomes are the means by which the cell packages its long strands of DNA into compact manageable bodies. During condensation, DNA is repeatedly coiled and supercoiled around nucleosomes (histone proteins, p. 69). The end result is a set of chromosomes whose number and individual shape are specific for each organism; a few organisms (one kind of ant and some fungi) have only one pair of chromosomes, while some plants (such as ferns) have over a thousand. (In prokaryotes, the DNA molecule does not condense.) Mitosis is complete when two identical daughter nuclei have formed. After nuclear division, division of the whole cell occurs in what is sometimes called the **C phase** (*cytokinesis*).

> In the cell cycle of eukaryotes, the G_1, S and G_2 phases constitute interphase, during which DNA and other molecules are synthesised. Cells about to divide enter the M (mitosis) phase, during which long strands of DNA condense into visible chromosomes. Following nuclear division, the whole cell divides by cytokinesis (C phase).

Checkpoint proteins control the cell cycle

Working on a few select and simple cellular systems, biologists are slowly unravelling how the cell cycle and cell division are controlled. Genetic analysis of strains of cells whose cell cycles are blocked by single mutations has permitted the detection of a number of specific molecules that are critical to exerting control. Some of the most important are the *cdc* (from *c*ell *d*ivision *c*ycle) mutants, and the proteins involved are being characterised.

BOX 8.1   Cancer: cell division out of control

When something goes wrong with the control of cell division and differentiation, a mass of cells, a tumour, may form. Benign tumours, such as warts, remain localised and are not usually life-threatening. However, serious problems may arise with malignant tumours, which are highly invasive. In life-threatening forms of cancer, cells grow rapidly and uncontrollably at the expense of their neighbours, depriving them of nutrients and destroying the organisation and function of surrounding tissues. Cells of malignant tumours detach and spread throughout the organism, invading and disrupting other tissues (see figure).

Cancer cells can be identified by microscopy. They have a relatively large nucleus and prominent nucleoli, are often dividing and are not differentiated as are normal cells in surrounding tissue. Cancer cells often have abnormal numbers of chromosomes or chromosomal abnormalities. They also produce new proteins. For example, for malignant cells to penetrate tissues, they must secrete enzymes that digest the basal lamina that normally underlies epithelial surfaces.

The causes of cancer remain enigmatic but many things can stimulate cancerous behaviour, including constant injury or aggravation of tissue, mutagens and viruses. Cancer cells result from changes in the DNA of somatic (body) cells. Many cancers seem to start from a single aberrant cell, probably as a result of several successive changes within it. The changes transform a normally regulated cell to one that does not respond to signals that control the cell cycle, that is, the orderly progression from G_1, S, G_2 to the M phase.

When cells derived from normal animal tissues are grown in the laboratory in culture, they may proliferate and become independent of the cues that normally control them in tissues, and many such cell lines divide

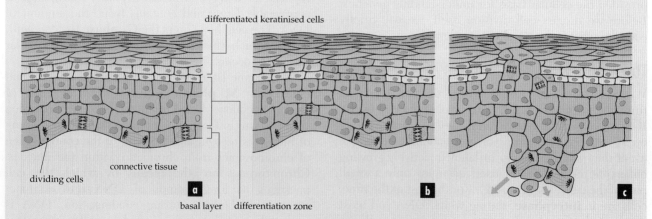

Tissue organisation in skin. **(a)** Normal cell division in the basal layer generates a steady supply of cells, which differentiate, become keratinised, die and thereby create the outer resistant layer of the skin. **(b)** During wound repair, there is temporary disruption in tissue organisation and stimulation of cell division to reform the normal tissue structure. Many of these cells re-enter the cell cycle. **(c)** In cancer, there is uncontrolled proliferation of cells, which do not differentiate properly and hence disrupt tissues. In dangerous cases, cancerous cells migrate away from where they are formed and invade other tissues (arrows) where they continue to grow repeatedly without ever reforming any tissue or performing the function they exhibited in the animal from which they were derived. These cells are therefore similar to cancer cells. Cells derived from cancerous tissues are often easier to grow in culture, apparently because they are already less dependent on the external growth signals that regulate normal cell growth.

The cell cycle is known to be dependent upon the appearance of **checkpoint proteins** at appropriate intervals. These keep all the sequential processes in order, ensuring that one phase of the cycle has been completed before the next is started. The two best known checkpoint proteins are those at the initiation of the S phase and M phase. In each case, several proteins are made, collectively called the **S phase promoting factor** (spf) and **mitosis promoting factor** (mpf). Only when these are assembled is the cell able to pass into the next phase. Both these factors contain two subunits, a type of protein called a **cyclin** and a protein **kinase**. Like many regulatory kinases involved in other systems, the protein kinase is physically altered when it is phosphorylated and dephosphorylated. *Mpf* phosphorylates other proteins when it starts mitosis as well as being controlled itself by phosphorylation. Thus, the checkpoint controls are complex in detail and still not fully understood.

Differentiating cells exit the cycle when in G_1. If they are to proceed through the cell cycle again, they must pass through the start checkpoint, which initiates the S phase. At this stage, the cell must have accumulated sufficient *spf,* which in turn requires a specific *cdc* protein called *cdk2*, which, along with certain other proteins, are dependent upon how big the cell has grown. As could be expected, the S phase cannot be initiated until the cell is becoming large enough to sustain cytokinesis. Later, after the G_2 phase, the mitosis checkpoint is activated to allow a cell with sufficient accumulation of *mpf* to initiate the events that will establish the mitotic spindle. *Mpf* is synthesised during interphase, steadily accumulating until it appears to be broken down rather quickly in the middle of mitosis.

The events within the mitotic cyle are apparently also controlled by checkpoints. The most obvious is the anaphase checkpoint, which delays the simultaneous splitting of the chromosomes and their anaphase separation until they have all become correctly attached to the mitotic spindle in metaphase. If the chromosomes are not all precisely attached, they will be unequally divided so that the newly formed cells will have an incorrect complement of chromosomes, with disastrous consequences.

The cell cycle is controlled by checkpoints. The two best known checkpoints are collections of proteins that initiate the S phase (S phase promoting factor, spf) and M phase (mitosis promoting factor, mpf).

Mitosis halves the chromosome number after replication

After the S phase of the cell cycle, when DNA synthesis occurs, each single chromosome has become replicated to produce two identical copies that are termed chromatids, that is, sister copies of the chromosome. The chromatids of a chromosome are held together at a constricted region, the **centromere** (Fig. 8.3a). Under the electron microscope, a centromere is seen to consist of two protein discs, **kinetochores**, into which are inserted **microtubules** (Fig. 8.3b, c).

The behaviour of chromosomes during mitosis is generated by and dependent upon the mitotic **spindle**. This is an elaborate cytoskeletal structure that provides the structural basis for:
- movement of chromosomes toward the equator of the cell, where they become aligned so that sister chromatids are attached to opposite poles
- the separation of all the chromatids so that each daughter cell receives a complete set of chromosomes.

The spindle contains numerous fibres visible under the light microscope after certain preparative techniques (Fig. 8.4). When examined with the electron microscope, these fibres are seen to be composed of microtubules. To construct a spindle, cells first mobilise proteins by breaking down most microtubules in the interphase cytoskeleton. These protein subunits, tubulin, are reassembled as the **spindle fibres**. After division, the spindle fibres are broken down in turn, and the tubulin recycled into the interphase cytoskeleton once more.

For mitosis, tubulin is mobilised and assembled into spindle fibres. The spindle is essential for correct organisation and separation of the chromosomes.

Mitosis, like the cell cycle, is conveniently subdivided into a number of phases that flow one into the next. These stages are shown in Figure 8.5.
- **Prophase**. Chromatin in the nucleus condenses to form visable chromosomes composed of paired chromatids.
- **Prometaphase**. The nuclear envelope enclosing the chromosomes breaks down, which allows the growing spindle to interact with and move the chromosomes.
- **Metaphase**. The spindle arranges all chromosomes precisely across the spindle so that sister chromatids will later move in opposite directions into the new daughter cells.

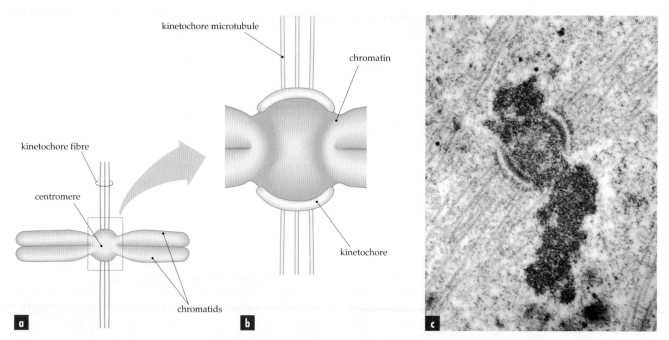

Fig. 8.3 The attachment region of the chromosomes to spindle fibres. **(a)** Kinetochore fibres attach to chromosomes, which, after DNA replication, consist of a pair of chromatids joined at the centromere. **(b)** Under the electron microscope, these kinetochore fibres are seen to consist of a number of microtubules, which terminate in the kinetochore itself. The chromatin in this region often shows no apparent differentiation into two chromatids. **(c)** Transmission electron micrograph of a chromosome in the green alga *Oedogonium*. This cell has particularly prominent kinetochores. Spindle microtubules are seen here inserted into the outer layers of the paired kinetochores (magnification × 24 000)

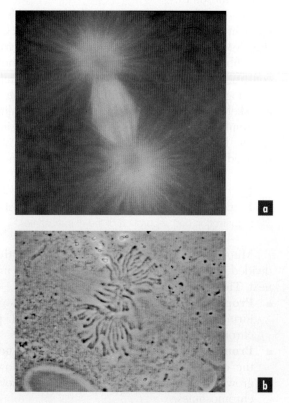

Fig. 8.4 Mitotic spindle in an animal cell at metaphase. **(a)** The yellow fluorescent stain has selectively bound to the spindle fibres, consisting of microtubules. The asters are also evident. **(b)** The same cell seen unstained in phase contrast. The chromosomes are now clearly visible, while the fibres can just be distinguished

- **Anaphase**. The two kinetochores of the centromere separate and sister chromatids move apart, forming two identical groups.
- **Telophase**. New nuclear envelopes form, separating the two groups of chromosomes.

Timing of these phases and duration of mitosis vary widely between different organisms. Some fungi and algae complete mitosis in a few minutes. More typically, mitosis lasts 30 to 60 minutes. In a typical mammalian cell, prophase slowly becomes obvious over about 20 minutes as the chromosomes thicken. Prophase finishes abruptly when the nuclear envelope disperses and chromosomes start moving about. Once the centromeres have moved to the middle of the spindle, the cell has reached metaphase. Cells stay in metaphase for some time (often about 20 minutes), after which they proceed through the anaphase checkpoint. Anaphase commences quite abruptly. Most of anaphase is completed in about 15 minutes. In many cells, telophase and the final stage of cell cleavage are slow and the two progeny cells become indistinguishable from their interphase neighbours.

Changes in the nucleus during mitosis have been studied for many years using tissues preserved, sectioned and stained for light microscopy. Less familiar to biologists is the behaviour of the living cell during these stages. It is difficult to study live dividing cells within tissues so one has first to find a source of flat, transparent cells (Fig. 8.6). Special optical

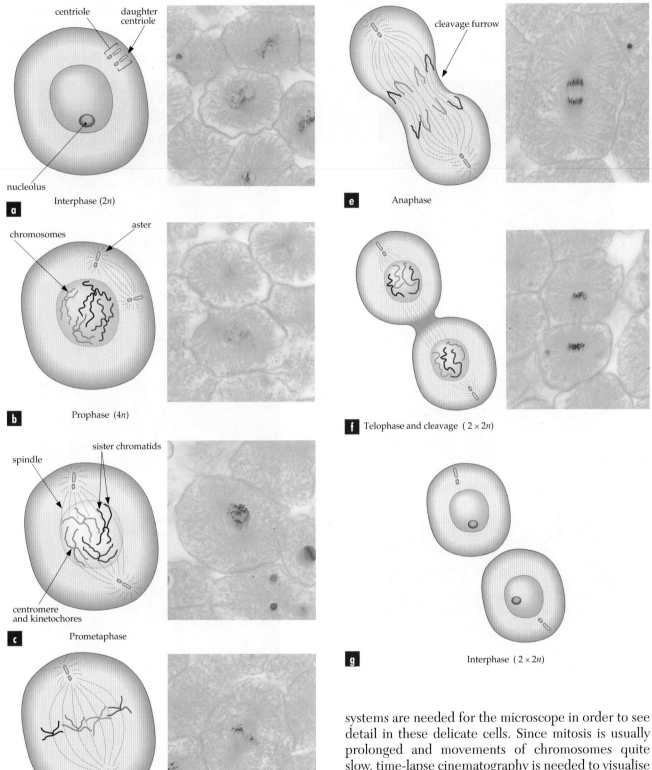

centriole daughter centriole

nucleolus

a Interphase (2*n*)

chromosomes aster

b Prophase (4*n*)

spindle sister chromatids

centromere and kinetochores

c Prometaphase

d Metaphase

cleavage furrow

e Anaphase

f Telophase and cleavage (2 × 2*n*)

g Interphase (2 × 2*n*)

Fig. 8.5 Stages of mitosis in animal (white fish) cells:
(a) interphase; **(b)** prophase; **(c)** prometaphase; **(d)** metaphase;
(e) anaphase; **(f)** telophase; and **(g)** cleavage and cell division
complete—forming two interphase cells

systems are needed for the microscope in order to see detail in these delicate cells. Since mitosis is usually prolonged and movements of chromosomes quite slow, time-lapse cinematography is needed to visualise clearly what is happening. The following description is based on living cells as well as fixed and stained material.

Mitosis in animal cells

The behaviour of chromosomes and spindle fibres in animal cells is summarised in Figure 8.7.

Prophase

During prophase, chromosomes slowly condense inside the nucleus. Earlier, during the G₁ phase, the **centrosome**, the microtubule organising centre of a typical interphase animal cell, contains a pair of **centrioles**. During S phase, the centrioles separate and two new centrioles arise alongside them. Thus, the centrosome has already replicated by the time the cell

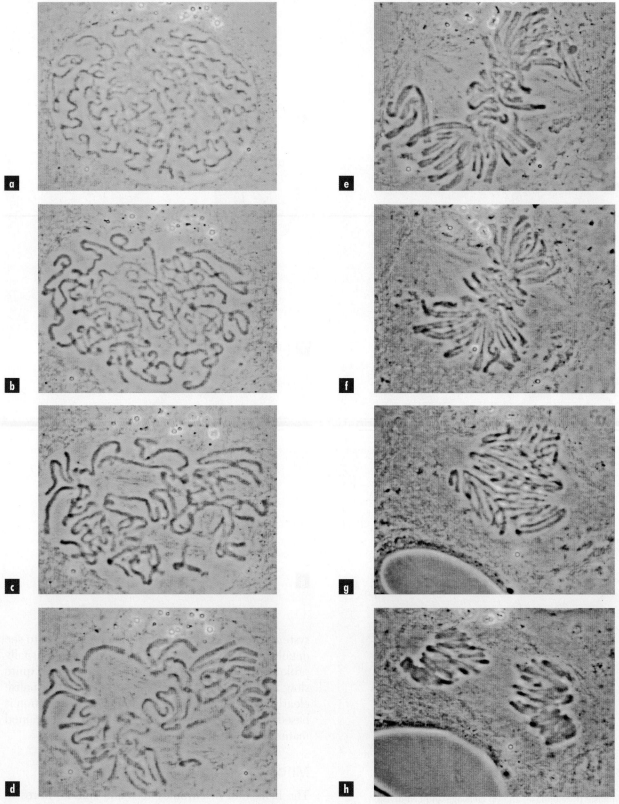

Fig. 8.6

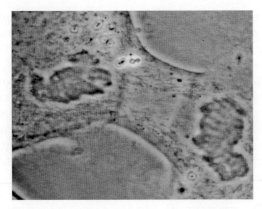
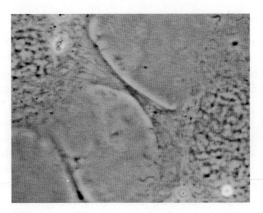

Fig. 8.6 Time-lapse sequence of living cultured newt lung cells showing a cell undergoing mitosis: **(a)** prophase; **(b)–(d)** prometaphase; **(e)** early metaphase; **(f)** metaphase; **(g)** anaphase; **(h)** late anaphase; **(i)** telophase; **(j)** cleavage

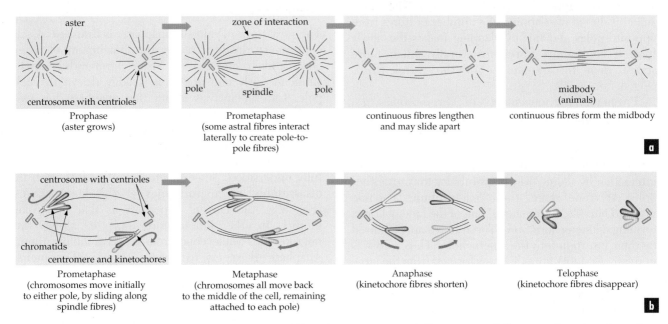

Fig. 8.7 This diagram shows the behaviour of the two sets of fibres that compose the spindle structure. All fibres grow from the poles and are initially identical. **(a)** Astral fibres, which interact laterally to create pole-to-pole fibres, hold the spindle poles together and define the axis of the spindle. **(b)** Fibres that interact with the kinetochores of the centromere become kinetochore fibres and are responsible for attaching and then moving the chromatids to the poles

proceeds into prophase. At the beginning of prophase, the centrosomes are activated to produce increasing numbers of fibres (bundles of microtubules). The centrosomes occupy the poles of the spindle for the rest of mitosis and their arrays of radial fibres are called **asters**. As the asters grow, the poles slowly move apart, apparently due to interaction between some of their fibres (Fig. 8.7a). At first, the spindle fibres are all identical. How they behave subsequently is dependent upon how they interact, either with each other (to form continuous fibres, running from pole to pole) or with kinetochores (to form **kinetochore fibres**, running from poles to kinetochores). In cells, these fibres are all intermingled.

It has recently been discovered that microtubules, particularly in asters, are far more dynamic than expected. Each microtubule grows steadily and then shrinks rapidly in what appears to be a remarkably irregular fashion. This phenomenon has been called '**dynamic instability**'. Both growth and shortening are due to tubulin subunits being added and removed from the end of the microtubule distant from the centrosome. The result of this activity during early mitosis is that the spindle fibres constantly probe among the chromosomes, ensuring that all kinetochores will eventually contact some microtubules of the aster.

During late prophase, RNA synthesis ceases and the prominent nucleoli shrink and disappear completely. They reappear at telophase.

During prophase, chromosomes slowly condense within the nucleus. Centrosomes move to opposite poles and form asters.

Prometaphase

During prophase, the asters form a cage-like array of microtubules over the surface of the nuclear envelope. As the size and activity of asters increase, the nuclear envelope immediately next to them becomes increasingly deformed inwards. Finally, the envelope ruptures and, in a few seconds, the whole membrane breaks down and the microtubules push into the nuclear region. This dramatic moment marks the beginning of prometaphase. As the growing fibres penetrate rapidly and deeply into the nucleus, many chromosomes react by randomly moving close to either centrosome (Fig. 8.7b). Asters also become firmly attached to each other by their fibres to form the biconical (twin cones) structure that gives the spindle its name. The interaction is due to overlapping of a number of fibres from opposite poles and this overlapping makes the newly formed continuous fibres structurally much more stable than the remainder of the astral fibres (Fig. 8.7a).

Prometaphase movements of chromosomes are difficult to follow. The function of the pair of kinetochores in the centromere of each chromosome is to interact with spindle fibres and thereby bring about chromosomal movement. Kinetochores commence this interaction by sliding over microtubules towards either pole. At some stage, these microtubules become inserted into (i.e. terminate in) one of the kinetochores, forming a kinetochore fibre. These fibres are stabilised by their insertion into the kinetochore. Thus, two sets of relatively stable fibres are now formed from the large number of microtubules from the aster—those attached to kinetochores, and those that overlap to form the fibres that run from pole to pole.

D uring prometaphase, the chromosomes begin to move as microtubules start attaching to kinetochores.

Metaphase

Having moved randomly to either pole, each chromosome oscillates gently around that pole. This movement is possibly a result of the population of rapidly shortening and elongating microtubules around them. Very soon each chromosome encounters some microtubules from the other pole and its second kinetochore attaches to these microtubules. Each chromosome then steadily moves to an equilibrium position midway between the poles. The central positioning of each chromosome on the spindle indicates that the chromatids (the two copies of each original chromosome) have become correctly attached to opposite poles. 'Correct' in this context means that when the chromatids split, they will move to opposite poles, an absolutely essential requirement for successful mitosis. As the chromosomes accumulate in the centre of the spindle, the cell reaches the metaphase configuration with kinetochores all in one plane. Meanwhile, the chromosomes continue to oscillate gently on either side of the central position.

Metaphase occupies a considerable proportion of mitosis. Just why this is so is not clear. Perhaps it is to give all chromosomes time to reach their correct orientation and position. By metaphase, the chromosomes have assembled in the centre of the spindle and the chromatids are attached by kinetochore fibres to opposite poles.

B y metaphase, all chromosomes have become attached to opposite poles and aligned across the spindle.

Anaphase

The next stage of mitosis, anaphase, is dramatic. Each centromere divides, freeing the attachment of sister chromatids to one another. Sister chromatids separate simultaneously and are then steadily drawn to each pole. Careful analysis shows that anaphase has two distinct components: anaphase A, which is movement of chromosomes to the pole; and anaphase B, when the poles move further apart, elongating the spindle. The motile mechanism that generates anaphase A is not fully understood. Since kinetochore fibres shorten by disassembly of microtubules at the kinetochores (Fig. 8.8), the current hypothesis is that molecular motors in the kinetochore pull it along the microtubules while concurrently disassembling them. Anaphase B seems to involve several phenomena. Overlapping spindle fibres elongate and/or slide apart.

D uring anaphase, the chromosomes simultaneously split and chromatids move rapidly to opposite poles.

Telophase

During telophase, chromosomes clump tightly near the poles of the elongated spindle as a new nuclear envelope appears around them. Then the whole dense mass of chromosomes starts swelling steadily as chromosomes decondense. Nucleoli reform and become conspicuous.

D uring telophase, new nuclei form from the separated groups of chromosomes.

Cytokinesis in animal cells

Once mitosis is complete, a cell divides by **cytokinesis**. In most cells, mitosis is so closely followed by cytokinesis that the two appear to be part of one event. In animal cells, for example, cytokinesis usually starts as cells enter anaphase. However, mitosis and cytokinesis are quite separate events, carefully synchronised in these cells but brought about by different cytoplasmic

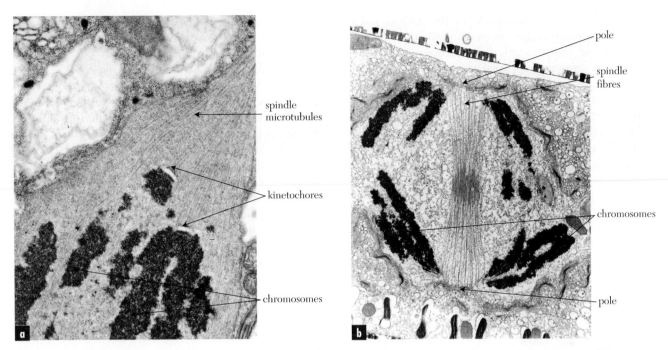

Fig. 8.8 Anaphase. **(a)** In the green alga *Oedogonium*, two single kinetochores, attached to their spindle microtubules, are seen while moving to the pole (magnification × 7700). **(b)** The spindle of the diatom *Nitzschia* is large and beautifully organised, in contrast with the spindle of most types of cells. Thus, it has been useful in analysing the structural aspects of how the spindle functions (magnification × 3900)

systems. In contrast, mitosis and cytokinesis are separated in some other types of cells, for example, during spore formation in algae and fungi; here, a number of cycles of mitosis can occur without cytokinesis, creating one large cell with many nuclei (Fig. 8.9). Cytokinesis in such cells involves the simultaneous cleavage of the whole cytoplasm into numerous tiny, mononucleate cells.

Animal cells do not have rigid cell walls and cytokinesis occurs simply by the cell pinching in half to form the two daughter cells. This constriction is brought about by a system of **actin** filaments forming a contractile ring (Chapter 6). For cell division, therefore, animal cells have to establish quite rapidly two separate cytoskeletal systems: the contractile ring and the spindle. Each system has to be put together in the

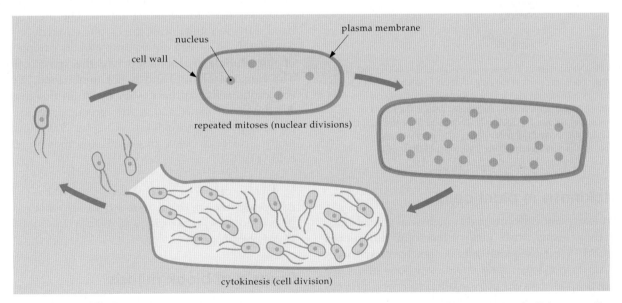

Fig. 8.9 A cell (e.g. certain algae and fungi) in which mitosis and cytokinesis are clearly separate events. A flagellated cell (e.g. a small swimming zoospore) undergoes rounds of mitosis, growing considerably. Accompanying cell growth generates a large cell (sometimes centimetres or more in length) containing hundreds or thousands of nuclei. A single round of cytokinesis cleaves the cytoplasm into uninucleate cells (shown here developing flagella), which are released from the parental wall for dispersal

right place so that cells divide in the correct plane between the two daughter nuclei, and the right time so that cleavage happens after chromosomes are separated. The remaining overlapping fibres that run between poles become aggregated during this constriction into a tight rod called the **midbody** (Fig. 8.10). The midbody is quite stable and often continues to hold the cell together for some time. The ingrowing cleavage furrow is lined with an even layer of actin filaments. Once these filaments start constricting the cell, they will continue to do so even if mitosis is stopped experimentally.

n preparation for cytokinesis, animal cells mobilise actin filaments into a contractile ring that cleaves the cell in two after mitosis.

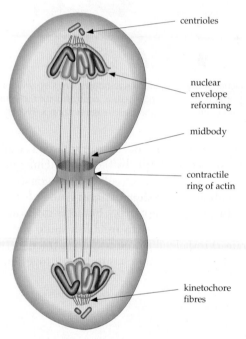

centrioles

nuclear envelope reforming

midbody

contractile ring of actin

kinetochore fibres

Fig. 8.10 Cytokinesis in animal cells. As anaphase is completed in an animal cell, the contractile ring of actin filaments constricts the cell and eventually divides it. Meanwhile, the continuous fibres remaining in the spindle are compressed into a single rod whose dense central region, the midbody, is derived from the region of overlap of the interacting polar fibres. The nuclear envelope reforms as the kinetochore fibres shorten and disappear

Cytokinesis in plant cells

Mitosis in plant cells proceeds much as in animal cells, but they display two major structural differences (Fig. 8.11). Firstly, plant cells do not have centrioles and so do not have astral spindles, whose fibres are focused upon discrete poles. (Interestingly, centrioles do appear in certain cells of lower land plants, such as mosses, ferns and cycads; these are the cells that will form flagellated sperm cells for sexual reproduction. In addition, many algal and fungal cells contain centrioles

and have astral spindles.) The poles of these spindles without centrioles are broad and thus the spindle is barrel-shaped.

Secondly, higher plant cells are enclosed in a rigid wall and do not undergo cleavage as animal cells do. Instead, as they go through anaphase, fibres remaining between the chromosomes thicken and proliferate. Soon, a densely fibrous **phragmoplast** forms (Fig. 8.12a). This steadily grows out laterally until it reaches the older walls of the cell. Inside the phragmoplast, tiny droplets appear and slide along the fibres, collecting halfway between the reforming telophase nuclei. These droplets coalesce into the **cell plate**, the new cross wall, which is flimsy at first but which soon thickens and separates the two new cells. Under the electron microscope, the cell plate is complex (Fig. 8.12b). It contains many tiny vesicles, probably derived from the Golgi apparatus, along with elements of the endoplasmic reticulum. Once the vesicles fuse together, fibrous wall material appears within them.

Plant cells also show another feature that has no equivalent in animal cells. Before prophase is clearly established, microtubules all group into the **pre-prophase band of microtubules** close to the wall. With remarkable accuracy, this band predicts exactly where the phragmoplast later will join the growing cell plate to the older walls. This predictive ability is particularly striking in highly asymmetric cell divisions involved in numerous differentiation events in plant tissues, such as those giving rise to root and epidermal hairs, and the guard cells of stomata. Thus, the preprophase band indicates that the plane of cell division has been decided upon well before the cell shows any sign of impending mitosis. The preprophase band starts forming during the G_2 phase and disappears completely as the cell enters prophase; its complement of microtubules is mobilised into the spindle fibres. The function of the band is not understood. Not all eukaryotic cells that have walls divide in this manner. All the land plants and a few green algae use the phragmoplast for cytokinesis (Chapter 37). However, many algal cells cleave in a similar fashion to animal cells (Fig. 8.13).

n plant cells, a preprophase band of microtubules predicts the position of the new cross wall, which is initiated during cytokinesis as the cell plate arising within the phragmoplast.

Drugs that block mitosis

Two types of drugs, **anti-microtubule** and **anti-actin drugs**, have proved of great use experimentally in investigating mitosis, cytokinesis and many other cellular activities. By using anti-mitotic drugs to break

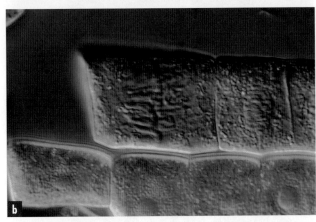

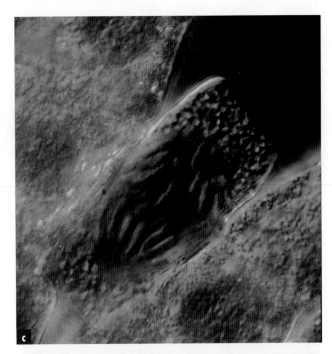

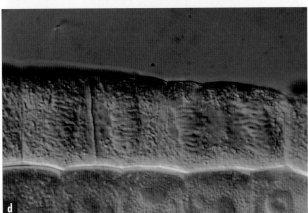

Fig. 8.11 Stages of mitosis in plant cells. These cells were fixed and then released from a wheat root tip by enzymatic digestion. They are unstained and photographed by differential interference optics: **(a)** two interphase cells; **(b)** a prophase and a metaphase cell; **(c)** anaphase; **(d)** two stages of telophase and cytokinesis (in one the cell plate is partly formed and in the other it is complete)

down the cytoskeleton, cell biologists can investigate the role of the cytoskeleton in the cell.

Anti-microtubule drugs cause disassembly of microtubules of the spindle. The best known of these, *colchicine*, is extracted from certain plants (autumn crocus, *Colchicum*). It is highly poisonous and functions by binding to tubulin, preventing its assembly and promoting disassembly. If colchicine or another anti-mitotic drug is applied before prophase, the cell undergoes chromosome condensation and normal chromosomes appear, while if it is applied during mitosis, the spindle rapidly dissolves and leaves chromosomes stranded. Later, each chromosome splits into two chromatids but, without a spindle, chromosomes cannot be separated correctly. In either case, the cell enters telophase and then interphase without division. The cell now has a double complement of chromosomes and DNA in its nucleus. If it were originally diploid, it is now tetraploid.

The second type of drug that interferes with cell division acts as an anti-actin agent; the best known are the **cytochalasins**, derived from certain fungi. These drugs specifically disrupt actin-based cytoskeletal systems and therefore stop cytokinesis. They also interfere with other actin-based phenomena, such as cytoplasmic streaming.

These two types of drugs show quite clearly that mitosis and cytokinesis are brought about by different cytoskeletal systems. Cytokinesis can proceed in the presence of colchicine or even, for example, when the spindle is mechanically removed from the cell with a micropipette. Conversely, events of mitosis proceed normally in the presence of cytochalasin.

Meiosis and the formation of haploid cells

Sexual reproduction involves the fusion of two reproductive cells (**gametes**) resulting in a doubling of

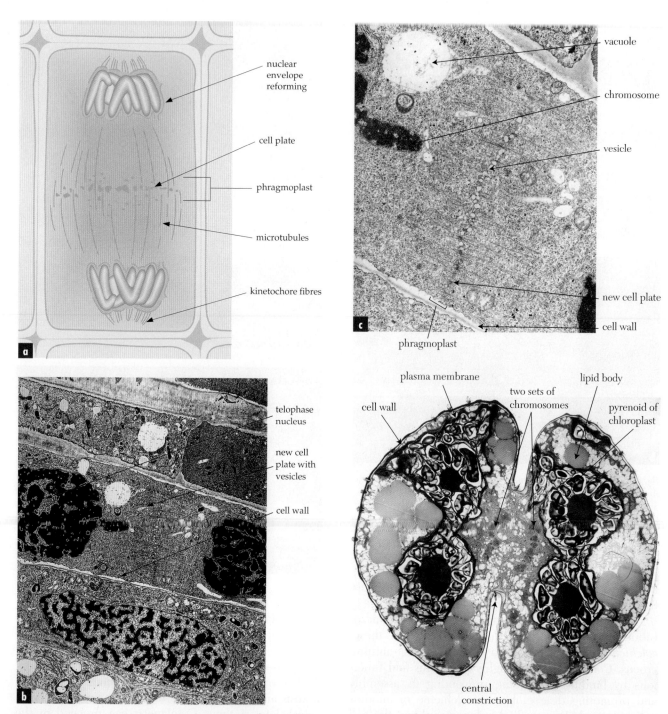

Fig. 8.12 (a) Cytokinesis in a typical higher plant cell. The overlapping fibres remaining between the chromosomes thicken and proliferate, forming the fibrous phragmoplast. Much cellular material (tiny vesicles, membranes etc.) accumulate along the midline of the phragmoplast and fuse to form the cell plate or new cross wall, which thickens and divides the cells. **(b)** A low (× 3400) and **(c)** a high (× 8400) magnification view of the phragmoplast in a wheat cell. Vesicles are collecting among the microtubules of the phragmoplast; these vesicles are guided into the correct position in the cell, whereupon they fuse together and form the cell plate

Fig. 8.13 Telophase in the green algae *Cosmarium*. The two sets of chromosomes are separated across the central constricted region of the cell, where cleavage is beginning. Many algae cleave in a similar fashion to animal cells. Only a few green algae (and all the land plants) use the phragmoplast for cytokinesis (magnification × 16 800)

the amount of genetic material in the zygote. At some stage in the life cycle before gamete formation, there has to be a corresponding halving in the amount of genetic material. This halving occurs by the special reduction division, meiosis. In meiosis, there is one round of DNA replication in a diploid cell ($2n$ to $4n$) as in the S phase of mitosis, followed by two separate cell divisions without any further DNA synthesis. The result is four haploid (n) cells (Fig. 8.14). Each haploid cell

Meiosis I

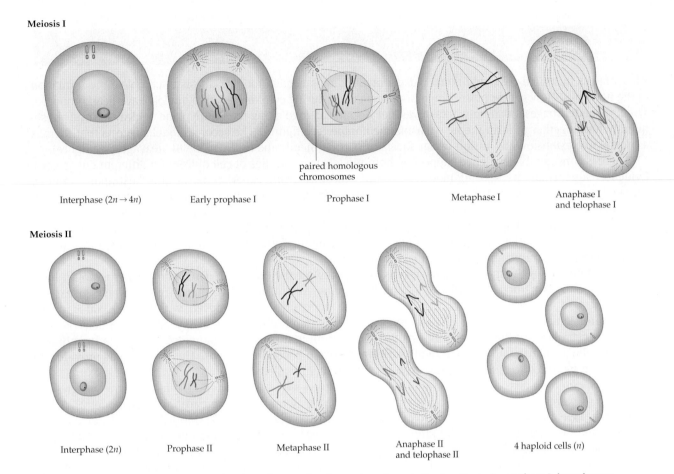

paired homologous chromosomes

Interphase ($2n \rightarrow 4n$)   Early prophase I   Prophase I   Metaphase I   Anaphase I and telophase I

Meiosis II

Interphase ($2n$)   Prophase II   Metaphase II   Anaphase II and telophase II   4 haploid cells (n)

Fig. 8.14 Meiosis in an animal cell. In meiosis I, interphase and early prophase I are similar to mitosis. In prophase I, homologous chromosomes, each consisting of two chromatids, pair and may exchange genetic material by crossing over at a chiasma (see Fig. 8.15). By metaphase I the chromosomes are aligned in the central plane of the spindle. During anaphase I, homologous chromosomes move to opposite poles. The two kinetochores of each centromere act as a unit, ensuring that the two chromatids of each chromosome remain attached and move to the same pole. Meiosis II is similar to a mitotic division except that there is no DNA replication. The end result of the two divisions of meiosis is four haploid cells

has one copy of each original homologous chromosome. The two divisions are called meiosis I and meiosis II.

Some organisms, for example, vertebrates and seed plants, are diploid for most of their life cycle. In animals, meiosis immediately precedes formation of male or female gametes. In seed plants, gamete formation occurs after two or three rounds of mitotic divisions of the haploid cell arising from meiosis. Other organisms, for example, mosses and many algae, are haploid for the larger part of their life cycle and diploid only briefly after sexual reproduction. In these, meiosis occurs well before the formation of gametes. Gametes are formed by mitosis during the haploid stage of the life cycle (Chapter 14).

In meiosis, there is one round of DNA replication in a diploid cell ($2n$ to $4n$) followed by two separate cell divisions resulting in four haploid cells (n).

Meiosis I

Prophase I

Prophase of meiosis I is prolonged and the events accompanying chromosome condensation are more complicated than in mitotic prophase. During mitosis, homologous chromosomes are completely independent of one another. However, during prophase of meiosis I, homologous chromosomes pair up precisely along their length (Fig. 8.15a). This pairing is called **synapsis**. As the chromatin condenses, the two sister chromatids, joined at the centromere, become apparent in each homologous chromosome, which often appear attached to the nuclear envelope. At this time, chromatids of homologous chromosomes may exchange portions of their genetic material by the process of crossing over (Fig. 8.15a). The DNA strands from a maternal and paternal chromatid are cut at the equivalent point and reconnected precisely. When crossing over occurs, new combinations of genetic information are formed.

BOX 8.2 Dividing a cell that has rigid walls

After mitosis in walled cells, additional plasma membrane is needed for cytokinesis to complete the formation of the new cells. How is new plasma membrane generated during cytokinesis? Two models have been proposed. The first model suggests that networks of vesicles become aligned along the future plane of cleavage and subsequently fuse together to form the new plasma membrane (see figure). The second model suggests that there is a progressive extension and fusion of flat sheets of plasma membrane in the plane of cleavage (see figure).

There is considerable evidence supporting both models from studies of cell division in different eukaryotes. Evidence supporting the second model has come recently from experiments at the Australian National University, where Dr Geoff Hyde and Dr Adrienne Hardham have used rapid freeze fixation techniques rather than traditional chemical fixation before transmission electron microscopy. Rapid freeze fixation is considered superior to chemical fixation in providing more life-like images of cells, their organelles and membrane systems. Chemical fixation can generate artefacts by altering the form of membrane components, for example, by breaking sheets of membranes into smaller droplets.

The production of the motile, infective cells (zoospores) of the oomycete, *Phytophthora cinnamomi*, have been studied. This pathogen attacks roots and kills many Australian native plants (e.g. dieback disease of jarrah forest in Western Australia, Chapter 44). The zoospores are produced within a special sac, the sporangium. Initially, as a result of numerous mitoses without cytokinesis, many zoospore nuclei lie free within the cytoplasm of the sporangium. As the sporangium matures, it is subdivided by plasma membranes and extracellular matrix into compartments surrounding each zoospore nucleus. This process has provided an ideal system in which to study the processes of cytokinesis in walled cells.

In *Phytophthora*, cell cleavage results not from aligned vesicles (model 1) but from the progressive extension of paired sheets of membrane, which interconnect along the future partition site (model 2). These sheets first appear as vesicles from the Golgi stacks and are initially clustered at one pole of each nucleus. The vesicles then become flattened membranous structures, before travelling along microtubules and coming to rest at the future partition site, where they fuse to form the new cell plate. This model of cytokinesis in *Phytophthora* may prove to be the mechanism of cell division in a variety of eukaryotes with walled cells.

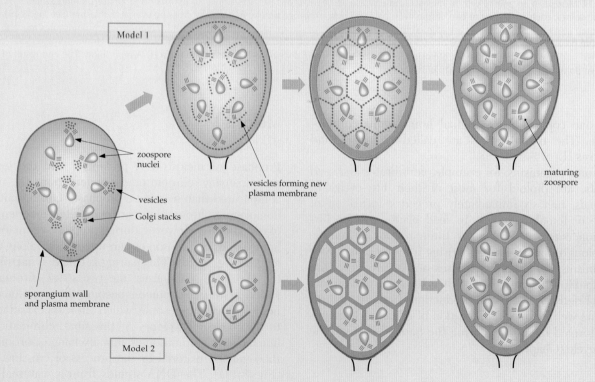

Two models of how new plasma membrane is generated during cytokinesis in the production of zoospores in the sporangium of the oomycete *Phytophthora cinnamomi*. In model 1, networks of vesicles align and fuse in the plane of cell cleavage. In model 2, flat sheets of membrane extend and fuse

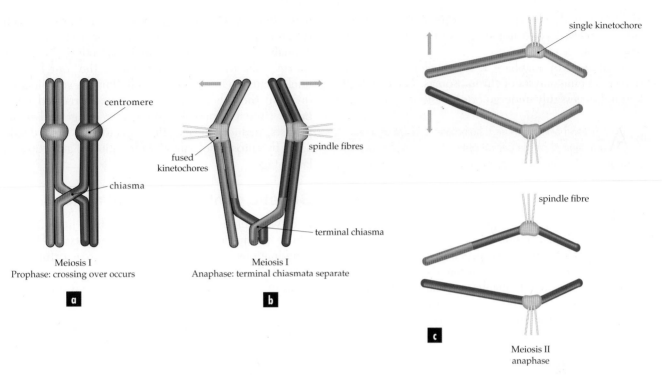

Fig. 8.15 Crossing over. **(a)** During synapsis in meiosis I, homologous chromosomes pair and then undergo crossing over at a chiasma. **(b)** Later, during anaphase I, sister kinetochores on each homologue act as single fused units, attaching to spindle fibres and moving to the poles together. The chiasma has moved along to the ends of the chromatids, becoming a terminal chiasma. **(c)** During meiosis II, these sister kinetochores now act individually and segregate as they would in normal mitosis

The point where crossing over occurs is a **chiasma** (pl. chiasmata). Under the electron microscope, chiasmata display a characteristic **synaptonemal complex**, consisting of a central core flanked by several layers between chromatids (Fig. 8.16). This is the molecular scaffold on which precisely controlled crossing over occurs. It consists of a long protein core, which the two homologous chromosomes align to, and has been likened to a zipper, with the DNA of each chromatid unwound and placed in register with the

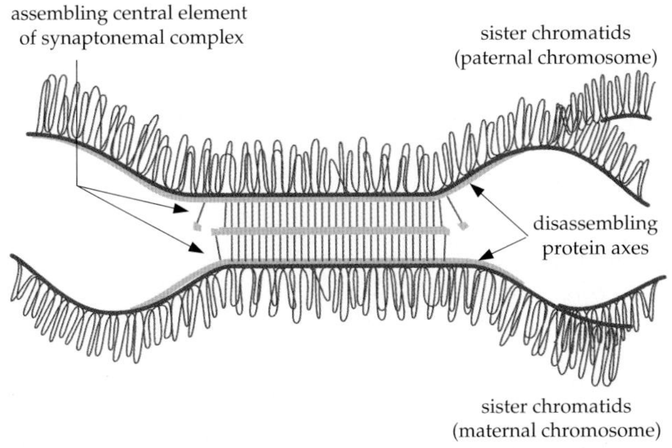

DNA of the homologue. Once crossing over at the DNA level is completed, this scaffold disappears and chromosomes separate from the nuclear envelope.

> During prophase I of meiosis, crossing over between chromatids of homologous chromosomes may occur, allowing the formation of new combinations of genetic information.

Metaphase I and anaphase I

As prophase I undergoes transition to metaphase I, the spindle forms as in mitosis (Fig. 8.14). However, again there are important functional differences between meiosis and mitosis at this stage. During mitosis, kinetochores of sister chromatids always end up connected to opposite poles to ensure accurate segregation of chromatids. In meiosis I, however, kinetochores in sister chromatids appear to be fused and act as a single unit. One homologous chromosome (with its two chromatids) attaches to one pole, while the other attaches to the opposite pole (Fig. 8.15b). The chiasmata continue to affect the behaviour of chromosomes, moving to the tips of the pairs of chromatids and becoming terminal chiasmata (Fig. 8.15b). By holding the chromosomes together, they assist in making sure that the two sets of paired kinetochores end up oriented towards opposite poles. During anaphase, the chromosomes eventually separate completely. The

Fig. 8.16 A synaptonemal complex is a ladder-like protein structure, which keeps homologous chromosomes together and closely aligned, and is involved in crossing over and thus recombination of genetic material

orientation and subsequent assortment of maternally and paternally derived homologues to the two poles is random (Fig. 8.15c).

At the end of anaphase I, two nuclei are formed, each containing one set of chromosomes. The cell itself does not necessarily undergo cytokinesis at this stage.

> At the end of metaphase I, homologous chromosomes are aligned on the central plane of the spindle. During anaphase I, homologous chromosomes move to opposite poles.

Meiosis II

Meiosis II is similar to mitosis except that there is no S phase of DNA synthesis preceding it. The two kinetochores of the sister chromatids, fused during meiosis I, separate at anaphase II and now act as normal single kinetochores, becoming attached to opposite poles of the meiosis II spindle. The end result, therefore, is four haploid sets of chromosomes. In many organisms, nuclear envelopes form around each set and the cytoplasm is divided to form four cells. In animals, these cells differentiate directly in males as sperm cells. In females, only one of the four haploid nuclei ends up in a large egg cell; during each meiotic division, the other nuclei form tiny short-lived polar bodies (see Chapter 15). In plants, fungi and many protists, including algae, the haploid cells may divide again by mitosis, delaying the formation of gametes to a later stage.

Genetic consequences of meiosis

The two meiotic divisions generate four haploid cells whose genetic complement consists of new combinations of parental genes. The reassortment of genes is a consequence of both crossing over and the random segregation of maternally and paternally derived chromosomes in each homologous pair. Meiosis, therefore, has important genetic consequences for sexually reproducing organisms.

Summary

- In prokaryotes, cell division is by binary fission. The single circular molecule of DNA in a bacterial cell is attached to the plasma membrane. Following DNA replication, the attachment points of the two DNA copies are separated by growth of the plasma membrane and cell wall, which invaginate and partition the cell into two daughter cells.

- Cell division in eukaryotes involves two separate processes: division of the nucleus (mitosis or meiosis) and division of the cell (cytokinesis).

- In the cell cycle of eukaryotes, the G_1, S and G_2 phases constitute interphase, during which DNA and other molecules are synthesised. Cells about to divide enter the M (mitosis) phase, during which long strands of DNA condense into chromosomes. After nuclear division, the whole cell divides by cytokinesis (C phase).

- The cell cycle and mitosis are subject to various controls, some of which involve specific checkpoint proteins.

- Chromosomes are essentially inert packages during mitosis. Their mitotically functional sites are kinetochores, which engage with spindle fibres to bring about alignment and subsequent separation of chromosomes into daughter cells.

- For mitosis, tubulin (the protein subunit of microtubules) is mobilised and assembled into spindle fibres. The spindle is formed from two sets of fibres that grow outwards from the poles. Some of these interact laterally with fibres from the other pole to create pole-to-pole fibres. These fibres define the axis of cell division. Others connect with kinetochores and are involved in moving chromosomes.

- During prophase, the nucleus is quiescent but chromosomes slowly condense. Centrioles move to opposite poles and form asters.

- Disappearance of the nuclear envelope marks the start of prometaphase. Microtubules become inserted into the kinetochores and the chromosomes begin to move, first to the poles and then steadily to the middle of the spindle. By metaphase, the chromosomes are all in the centre of the spindle. Each centromere is attached by kinetochore fibres leading to opposite poles. The centromere splits and chromatids are drawn towards each pole during anaphase. During telophase, chromosomes clump tightly near the poles as a new nuclear envelope appears around them.

- Cytokinesis, cell division, is a separate event from mitosis. It may immediately follow mitosis or be delayed. In animal cells, actin filaments are mobilised into a contractile ring that pinches the cell in two after mitosis. In plants, the cytoskeletal microtubules form a phragmoplast and a preprophase band predicts the position of the new cell plate that divides the parent cell.

- Meiosis differs from mitosis in that it brings about a halving of the number of chromosomes in a cell. It normally precedes the formation of gametes and sexual reproduction. Meiosis involves two separate meiotic divisions and differs from mitosis in three important aspects:
  □ during meiosis I, homologous chromosomes pair precisely and exchange genetic material by crossing over
  □ there is no DNA synthesis between meiosis I and II
  □ meiosis II is a reductive division, halving the number of chromatids in the progeny nuclei.

key terms

actin (p. 189)
anaphase (p. 184)
anti-actin drug
   (p. 190)
anti-microtubule drug
   (p. 190)
aster (p. 187)
binary fission (p. 180)
C phase (p. 182)
cell cycle (p. 181)
cell division (p. 180)
cell plate (p. 190)
centriole (p. 186)
centromere (p. 183)
centrosome (p. 186)
checkpoint protein
   (p. 183)

chiasma (p. 195)
chromatid (p. 183)
chromatin (p. 181)
chromosome (p. 181)
cyclin (p. 183)
cytochalasin (p. 191)
cytokinesis (p. 188)
diploid (p. 180)
dynamic instability
   (p. 187)
G_1 phase (p. 181)
G_2 phase (p. 181)
gamete (p. 191)
haploid (p. 180)
homologous
   chromosome
   (p. 180)

interphase (p. 181)
kinase (p. 183)
kinetochore (p. 183)
kinetochore fibre
   (p. 187)
M phase (p. 181)
meiosis (p. 180)
metaphase (p. 183)
microtubule (p. 183)
midbody (p. 190)
mitosis (p. 180)
mitosis promoting
   factor (*mpf*) (p. 183)
phragmoplast (p. 190)
plasmid (p. 180)

preprophase band of
   microtubules
   (p. 190)
prometaphase
   (p. 183)
prophase (p. 183)
S phase (p. 181)
S phase promoting
   factor (*spf*) (p. 183)
spindle (p. 183)
spindle fibre (p. 183)
synapsis (p. 193)
synaptonemal complex
   (p. 195)
telophase (p. 184)
zygote (p. 180)

Review questions

1. How does a bacterial cell divide?

2. Describe the following structures and their role in cell division: chromatid, centromere, kinetochore, centriole, aster, microtubule.

3. What are homologous chromosomes?

4. What is meant by haploid and diploid?

5. State when each of the following occur during the mitotic cell cycle of eukaryotes.

    (a) DNA replication

    (b) breakdown of the nuclear envelope

    (c) chromosome aggregation in the middle of the spindle

    (d) centromere division

6. Why is chromatin extended during interphase? What is the significance of chromatin condensation for mitosis?

7. Describe how sister chromatids separate from one another and move to opposite poles during anaphase.

8. Compare and contrast cytokinesis in animal and plant cells.

9. What major events occur during meiosis I? What is its end result? What is the significance of meiosis II and how does it differ from a mitotic division?

Extension questions

1. Redraw Figure 8.15 so that the sets of homologous chromosomes during meiosis I have two or three chiasmata instead of one. Draw the possible combinations that can occur in the haploid cells resulting from meiosis II.

2. What are the possible advantages of diploidy? Why aren't elephants and gum trees haploid?

3. Mitosis is very complicated yet it has to be faultless every time. How could the cell promote reliability in such a complex system?

4. Some drugs that are used to treat cancer interfere with microtubule formation. How does this help control cancer?

Suggested further reading

Alberts, B., Bray, D., Lewis, J., et al. (1994). *Molecular Biology of the Cell.* 3rd edn. New York: Garland.

Advanced text with a comprehensive treatment of cell division.

Hyams, J. S. and Brinkley, B. R. (1989). *Mitosis: Molecules and Mechanisms.* London: Academic Press.

General text for reviewing mitosis.

Lodish, H., Baltimore, D., Berk, A., Zipursky, S.L., Matsudaira, P., Darnell, J. (1995). *Molecular Cell Biology.* New York: W. H. Freeman & Co.

Advanced text with a comprehensive treatment of cell division.

McIntosh, J. R. and McDonald, K. L. (1989). The mitotic spindle. *Scientific American* 261(4): 48–56.

General review of mitosis.

Part 2

Genetics and molecular biology

CHAPTER 9

Inheritance

Genetics: the science of biological information

Living organisms are exquisitely complex. They cannot exist without information that controls their development and function. Genetics is the study of this information, its storage, the way it is replicated and how it is transmitted from cell to cell and from organism to organism. It is the study of the way that this information is read to influence the characteristics (traits) of organisms and the way that this information varies from individual to individual. Variation in biological information accounts for the diversity we see in living organisms and underpins evolutionary processes.

Throughout this section you will find analogies between the information that controls the development and function of biological organisms and the software that controls the functioning of computers. In a very real sense, genetics is the study of the 'biological software'. Unlike computers, in which the software is stored and read electronically, the biological software is stored and read chemically. Genetic processes, therefore, operate at a molecular level. However, their consequences can be seen throughout the scales of biological complexity, from cell structure and behaviour, through the characteristics of organisms and the properties of populations of organisms within ecosystems, through to the evolutionary history of life on earth. In this sense, genetics is the unifying discipline of biology. This section focuses primarily on the molecular and organism levels of genetics, while the role of genetics in populations and evolution is dealt with in later sections (Chapters 32, 42).

The foundation for understanding genetics

There is a great range of variation between living organisms but offspring tend to resemble their parents, that is, they appear to inherit certain features, or **traits**, from them. These features range from physical features (e.g. hair colour, eye colour) to behavioural characteristics (e.g. differences in the behaviour of dogs of different breeds). How are these traits inherited? We know that the traits themselves are not inherited (e.g. fertilised eggs do not have blonde hair or green eyes). Rather, the traits appear during the development of the organism. Characteristics can be remarkably similar, for example, in the case of identical twins. Each twin develops separately in a process that involves an extraordinarily complex series of cell divisions, cell death, cell shape changes, cell migration and tissue differentiation. Yet the outcome is that the twins are virtually indistinguishable. It is clear that there must be some form of information tightly regulating these

cellular processes. In the case of twins, the information is giving very similar instructions to the cells as they develop. Offspring resemble their parents because information that regulated development of the parents is passed to the offspring.

The first breakthrough in understanding the nature of biological information came not from a study of the stored information itself but from careful observations of the inheritance of characteristics. A monk, Gregor Mendel, studied the inheritance of a range of traits in different strains of peas grown in his garden at the Augustinian monastery in Brünn, Moravia (now Brno in the Czech Republic; Fig. 9.1).

Fig. 9.1 (a) Gregor Mendel and **(b)** the Augustinian monastery in Brünn (now Brno in the Czech Republic), where he established the principles of heredity using garden peas. The plot where he grew his peas lies in the angle of the building, underneath the second floor window of his room

Inheritance of a single gene

Mendel's experiments

Mendel performed a series of carefully controlled experiments that involved breeding from different varieties of peas. Pea plants are well suited to studies of inheritance because they can self-pollinate, are easy to cross artificially and produce large numbers of progeny. For artificial crossing, the young anthers of a flower are removed before their pollen is released. Pollen from a different plant can then be brushed on to the stigma to fertilise the eggs in the ovules (see Chapter 14 for a description of plant fertilisation).

Plant breeders had used self-pollination, or inbreeding, to generate varieties that were **pure-breeding** (also termed true breeding)—varieties in which all progeny had the same set of characteristics as their parents. Different pure-breeding varieties exhibited different characteristics (Fig. 9.2). Mendel succeeded in gaining an understanding of heredity where others had failed because he studied the inheritance of traits one at a time.

Monohybrid cross

The first question Mendel asked was if the traits of the parents differ, what do the progeny look like?. The consensus at the time was that characteristics in offspring were an average, or blend, of those of the parents. However, when a pure-breeding strain that produced yellow seeds and a pure-breeding strain that produced green seeds (Fig. 9.3) were cross-pollinated, the progeny seeds were always yellow. This occurred irrespective of which strain was used as the source of the pollen or the egg. This dramatic and unexpected result showed that characteristics of the offspring were not, in this case, a blend of those of the parents.

The progeny of a cross between two pure-breeding strains are referred to as the F_1 **progeny** (the first filial generation). In this example, the F_1 progeny all exhibited the yellow characteristic. Had the green characteristic disappeared or been masked? To examine this question, Mendel planted the yellow F_1 progeny seeds and allowed the resulting plants to self-pollinate, resulting in the F_2 **progeny** (the second filial generation). This time, as well as getting yellow seeds, some green seeds reappeared! Mendel counted 6022 yellow:2001 green seeds and noted that this ratio was very close to three-quarters yellow seeds to one-quarter green seeds. Importantly, it was clear that the 'green' characteristic had not been irretrievably lost from the F_1 progeny but had somehow been masked by the 'yellow' characteristic. Mendel reasoned that he was dealing with two factors that could determine the colour of the seeds.

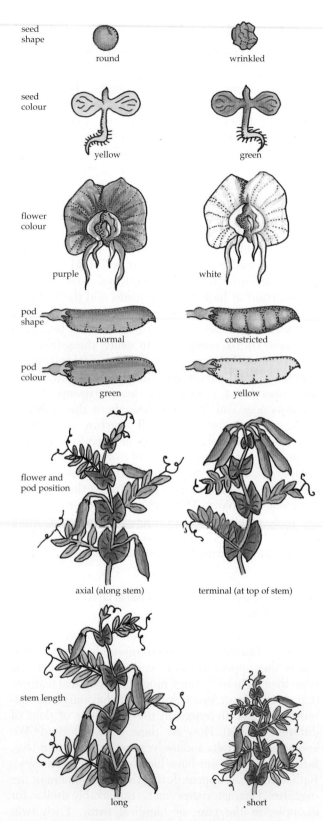

Fig. 9.2 Seven traits of garden peas studied by Mendel. Each trait occurs in two alternative forms, including seed colour (yellow or green), seed shape (round or wrinkled), flower colour (purple or white), pod shape (normal or constricted), pod colour (green or yellow), flower and pod position (along the stem or at the end) and stem length (long or short). In each case, the dominant form is listed on the left

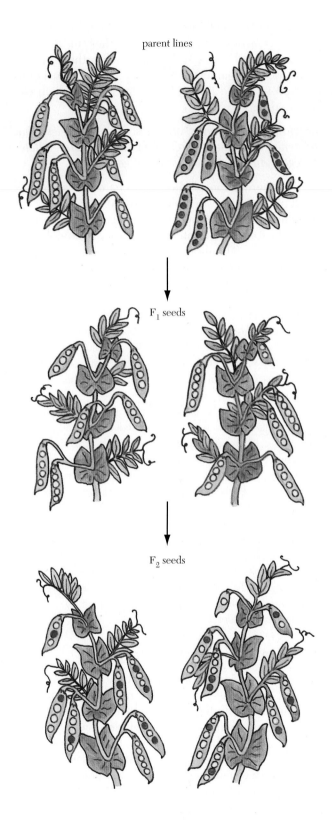

parent lines

F₁ seeds

F₂ seeds

Fig. 9.3 The results of Mendel's first type of experiment. He crossed two pure-breeding lines of peas, one with yellow seeds and one with green seeds. The progeny seeds resulting from such cross-fertilisations (F₁ seeds) were always yellow. These were grown into plants, which were allowed to self-fertilise to yield F₂ progeny seeds. On average, three-quarters of these seeds were yellow and one-quarter were green

Specifically, Mendel explained his results by proposing that:

- yellow colour in seeds is determined by a factor, which he termed Y, and green colour by a factor he termed y
- the yellow seed trait is **dominant** and the green seed trait is **recessive** (i.e. when the two factors are together in the one plant, the yellow trait masks the green trait and yellow seeds are produced)
- pea plants carry two of these factors, which may be the same (YY or yy) or different (Yy)
- pea plants pass only one of these two factors on to their pollen or egg cells (gametes) (i.e. the factors *segregate* [separate] from each other in the production of the sperm and egg cells)
- each gamete receives one or other of these factors with equal probability.

We now recognise that these factors are **genes**, which influence particular characteristics, in this case seed colour. Y and y are alternative forms of a gene that determines seed colour (Y, yellow; y, green). Different forms of a gene are termed **alleles**. The term **genotype** is used to describe the particular combination of alleles of an organism, in this example, YY, Yy or yy. The term **phenotype** is used to describe the set of characteristics, in this example, yellow or green seeds. Note that because certain phenotypes can be dominant (e.g. yellow) and others recessive (e.g. green), plants with different genotypes can have the same phenotype; in this example, YY and Yy plants both have yellow seeds. A cross that involves different alleles of a single gene is termed a **monohybrid cross**.

An individual carrying two copies of the same allele is said to be **homozygous** (e.g. YY) and will produce gametes carrying only one type of allele (in this case Y). Every individual receives one allele from each of its parents. In a cross between pure-breeding plants from the same strain (e.g. genotype YY, phenotype yellow), the progeny can only receive a Y allele from each parent and must therefore also be genotype YY and phenotype yellow, accounting for the pure-breeding nature of the strain.

Individuals carrying different alleles are said to be **heterozygous** (Yy). All of the F₁ progeny of the pure-breeding yellow plants (genotype YY) crossed to pure-breeding green plants (yy) must be heterozygous, as they must receive different alleles from each parent. Half of the gametes produced by these F₁ plants will carry one allele (Y) and half will carry the other (y). If these heterozygous F₁ plants are self-pollinated ($Yy \times Yy$), they will produce both yellow and green offspring in the ratio 3:1 (Fig. 9.4).

How does the 3:1 ratio arise? It is simply a consequence of two of the observations made above. The first is that F₂ progeny have an equal chance of receiving either the Y or y allele from each parent. The

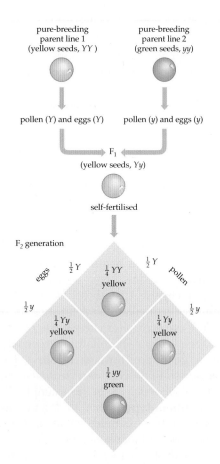

Fig. 9.4 Mendel's breeding program in which he followed the inheritance of seed colour in peas over two generations. This showed that the allelic forms of a gene controlling seed colour (*Y* and *y*) segregate from each other in equal numbers into the gametes formed by hybrid (F₁) plants. A matrix like this, which combines gametes into progeny, is termed a Punnett square, named after the British geneticist who first used it

outcome of this is that the F₂ progeny have the genotypic ratio of *YY:Yy:yy* of 1:2:1 (Fig. 9.4). The second observation is that in the heterozygote, the phenotype of one allele is dominant and the other recessive. In this example, it is the yellow phenotype that is dominant, that is, the *Yy* peas have a yellow colour. The phenotypic ratio is yellow (*YY* + *Yy*):green (*yy*), 3:1; that is, three-quarters of the seeds are yellow and one-quarter are green.

Mendel demonstrated that these principles applied equally to the inheritance of six other traits in pea plants (Fig. 9.2): seed shape, flower colour, pod shape, pod colour, flower and pod position, and stem length. He published the results of his experiments in 1866 but they were ignored until 1900 when other researchers performed similar experiments and realised the importance of Mendel's findings.

Mendel recognised that individual traits are determined by discrete factors (we now call these factors genes) rather than being a blend of influences from each parent, as breeders had assumed. Thus, unlike liquids, which blend together, the factors remain discrete entities, even when together in the one organism, and separate from each other in the production of gametes. Mendel's first conclusion is referred to as the *Principle of Segregation*, which can be stated as:

Individuals carry pairs of genes, termed alleles, that influence particular inherited traits. The alleles segregate during the formation of gametes.

This segregation of alleles into gametes is a direct result of the process of meiosis, described in Chapter 8. One allele is located on one homologous chromosome while the other is located on the other homologous chromosome. The homologous chromosomes segregate during meiosis, ensuring that the alleles are segregated into different gametes (Chapter 8).

Mendel showed that particular characteristics of an organism are influenced by discrete factors we call genes. Different genes influence different characteristics (e.g. the *Y* gene influences seed colour). Different forms of a gene (alleles) can have a different effect on the characteristic (e.g. yellow versus green seed colour). Individuals carry two alleles of each gene, which segregate from each other in the production of the gametes, each gamete carrying one allele. Each individual inherits one allele of each gene from each parent. The phenotype of one allele in a heterozygote can be dominant over the phenotype of the other, which is recessive. Patterns of inheritance can be explained by the segregation of the two alleles of each gene into gametes in equal proportion and the dominance of one phenotype over another. In a monohybrid cross involving alleles of a single gene, all F₁ progeny will have the same genotype and exhibit the dominant phenotype, while the F₂ progeny are produced in a phenotypic ratio of 3:1 and a genotypic ratio of 1:2:1.

Multiple effects of single genes

The gene for seed colour affects only one phenotypic trait but sometimes a single gene may affect more than one trait. In his studies of a gene that influenced flower colour, Mendel noted that the purple flower trait was dominant, while the white flower trait was recessive. However, Mendel also observed that plants with purple flowers always had reddish stems and grey seed coats and those with white flowers always had green stems and white seed coats. He proposed that the factor controlling flower colour also influenced the colour of the stem and seed coat. This phenomenon is termed **pleiotropy** (Fig. 9.5).

A pleiotropic gene affects different characteristics in the same organism.

Fig. 9.5 Pleiotropy can be seen occurring in this 'red damask' tea tree, with burgundy-coloured flowers, instead of white flowers, and dark reddish leaves

BOX 9.1 Key terms in genetics

The following summarises the terminology that is used in genetics.

1. Individuals in *pure-breeding* strains have the same phenotype and produce progeny with the same phenotype when bred with each other.
2. *Genes* are discrete hereditary factors that determine traits.
3. *Alleles* are different forms of a gene.
4. A *locus* is the position of a gene on a chromosome; a locus may be occupied by any one of the alleles of a gene.
5. Organisms in which the alleles are different, for example, *Yy*, are *heterozygous*, and those in which the alleles are the same, for example, *YY* or *yy*, are *homozygous*.
6. The *genotype* is the genetic make-up of an organism, that is, its specific allelic composition, for example, *Yy*.
7. The *phenotype* is the set of detectable properties or traits of an organism.
8. *Dominance* refers to the appearance of one of the homozygous phenotypes, such as yellow colour in seeds, in the heterozygous (*Yy*) organism. The phenotype of the homozygote that is absent in the heterozygote is termed *recessive*.
9. The *wild-type* is the phenotype found in most individuals in a population.
10. A *monohybrid cross* involves crossing organisms that are heterozygous at one locus, for example, *Yy* × *Yy*.
11. A *dihybrid cross* involves crossing individuals that are heterozygous at two different loci. This cross involves the segregation of alleles of two genes.
12. *Chromosomes* differ between males and females and are involved in sex determination. In organisms with an XX/XY sex-determining system, sex-linked genes occur on the X chromosome and are termed X-linked.

Codominance and blood groups

The study of blood groups illustrates two important points. The first is that more than two alleles of a gene can exist in populations, even though an individual can carry a maximum of two different alleles. The ABO blood group is an example of a gene that has multiple alleles in a population. In this case, three alleles, I^A, I^B and i, influence the blood types of individuals. Individuals carry, at most, two of these three different alleles. The second point is that the phenotype of one allele is not necessarily dominant over another in a heterozygote. In the earlier example of seed colour, the

BOX 9.2 Genetic notation

By convention, genes are represented by italicised symbols. To summarise genotypes, capital letters are used for alleles associated with dominant phenotypes (e.g. *Y*) and lower case letters for alleles associated with recessive phenotypes (e.g. *y*). The full genotype notation, *YYrr*, represents alleles of two separate genes. In a cross, such as *YYRR* × *yyrr*, the order of the two genes is consistent for the two parents. It is incorrect to present the cross in the order *YYRR* × *rryy*.

Conventions for genetic nomenclature are different for different organisms. For the vinegar fly, *Drosophila melanogaster*, for example, alleles associated with the **wild-type** phenotype are commonly shown as ' + ' and alleles are represented with a line separating them, while semi-colons separate genes on different chromosomes. A cross between a pure-breeding brown-eyed fly and a pure-breeding scarlet-eyed fly would therefore be represented as *bw/bw; + /+ × + /+ ;st/st*. The lower case *bw* and *st* indicate that the phenotypes associated with these alleles are recessive and the wild-type phenotypes are dominant.

phenotype of the yellow allele but not the green allele is expressed in the heterozygote. However, in the case of blood groups, two different alleles in a heterozygote can both be recognised as having a phenotypic effect. The phenotype, here, is the ability of the red blood cells to be clumped (agglutinated) by various sera.

In 1900, the same year in which Mendel's work was rediscovered, Karl Landsteiner began experimenting with human blood. He tested the effect of one person's serum (the non-cellular blood fluid that remains after clotting) on the red blood cells of another person. In some combinations, the red cells were clumped by the serum, but cells were never clumped by their own serum. Landsteiner found that individuals could be placed into one of four groups, A, B, AB and O, depending on the pattern of clumping of their red cells by other sera.

We now know that the four blood groups are due to the presence of genetically determined antigens on the surface of red blood cells (Chapter 23) and that the ABO gene is located on chromosome 9. Group A individuals have A antigen on the plasma membrane of their red cells and anti-B antibodies (Chapter 23) circulating in their serum (Table 9.1). Group B individuals have B antigen on their cells and anti-A antibodies. Group O individuals have neither A nor B antigen on their cells and both anti-A and anti-B antibodies circulating in their serum.

Antibodies in the serum will bind to their target antigen on the foreign blood cells, causing clumping. Thus, serum from a group A individual will clump cells from a group B individual but not cells from a group A individual and vice versa. AB individuals have both A

Table 9.1 Characteristics of the human ABO blood group system, illustrating the antigens and antibodies present in each blood type

Blood group	Genotype	Antigens on red blood cells	Antibodies in serum	Blood
A	$I^A I^A$, $I^A i$	A	anti-B	
B	$I^B I^B$, $I^B i$	B	anti-A	
AB	$I^A I^B$	A, B	—	
O	ii	—	anti-A, anti-B	

Example of antigen–antibody interaction (red cell clumping):

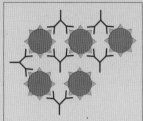

and B antigens on their red blood cells, but the serum of an AB individual has neither anti-A nor anti-B antibodies. Cells from an AB individual are clumped by serum from either an A or B individual, but AB serum does not clump other red blood cell types. Finally, group O individuals have neither antigen on their red cells, but have both antibodies in their serum. Their red cells are not clumped by any other serum, while group O serum will clump all other red cell types—A, B and AB.

As mentioned, ABO blood groups are determined by three alleles, I^A, I^B and i. The I^A allele is responsible for producing the A antigen, I^B the B antigen, and the i allele produces neither. Thus, individuals of genotype I^AI^A or I^Ai are type A phenotype, individuals of genotype I^BI^B or I^Bi are type B phenotype, individuals of genotype I^AI^B are type AB phenotype and individuals of genotype ii are type O phenotype, the recessive phenotype. Because the products of both the alleles I^A and I^B are equally recognisable in the heterozygote, they are said to be **codominant**.

It is important that blood group types are as closely matched as possible when giving blood transfusions. If they are not, then, for example, a large number of donor red cells bearing a foreign antigen may be introduced into serum containing antibody to that antigen. The incoming red cells will be clumped and the recipient may die. The matching of blood groups applies not only to blood transfusions. Organ transplantations require appropriate matching of antigen–antibody systems between donors and recipients (Chapter 23).

C odominance is the complete phenotypic expression of two alternative alleles in a heterozygote.

Inheritance of combinations of genes

Dihybrid cross

In addition to studying single traits, Mendel also studied the inheritance of pairs of different traits by carrying out crosses between pure-breeding lines differing in two unrelated traits, seed shape and seed colour. Plants with round yellow seeds were crossed with plants with wrinkled green seeds. As expected, the F₁ generation showed the dominant phenotypes of each allele pair, yellow and round seeds. However, on raising these F₁ plants and allowing them to self-pollinate (the **dihybrid cross**), two important observations were made. Firstly, four types of F₂ progeny seeds were observed (yellow wrinkled, green round, yellow round and green wrinkled, see Fig. 9.6). This shows that a dihybrid cross can produce new combinations of traits (yellow wrinkled and green

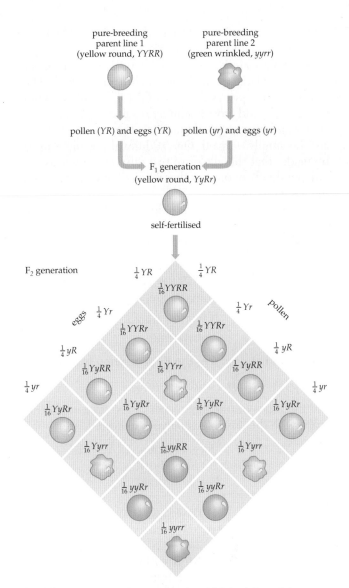

Fig. 9.6 Mendel's breeding program in which he followed the inheritance of both seed colour and seed shape in peas simultaneously. This showed that alleles of a gene controlling seed colour (Y and y) and alleles of a gene controlling seed shape (R and r) assort independently of each other into the gametes of hybrid (F₁) plants, giving one-quarter of each combination (YR, Yr, yR, yr). The resulting phenotypes have a 9:3:3:1 ratio

round) that were not present in the parents, suggesting that the alleles of the different genes were being shuffled with respect to each other. The second observation is that particular ratios of different phenotypes are consistently observed. For example, the numbers of each phenotype observed by Mendel were:

yellow round	315
yellow wrinkled	101
green round	108
green wrinkled	32
total	556

If the alleles of each gene are examined separately, both pairs of alleles exhibit close to the expected 3:1

ratio (yellow:green, 416:140; round:wrinkled, 423:133). The ratio for all four phenotypes is very close to a ratio of 9 yellow round:3 yellow wrinkled:3 green round:1 green wrinkled, but how can such a ratio arise? The answer is that, just as the ratio of 3:1 for F_2 phenotypes in a monohybrid cross is a direct result of segregation of alleles and phenotypic dominance, the ratio 9:3:3:1 can also be simply derived. One additional postulate must be made: that the alleles of two different genes assort independently of one another when they segregate into the gametes.

To understand this, it is helpful to summarise these crosses using the symbols Y for yellow, y for green, R for round and r for wrinkled seed:

Parents
$YYRR \times yyrr$
yellow round   green wrinkled

F_1 generation (all $YyRr$ genotype, yellow round phenotype)

The cross that produces the F_2 generation is therefore:

$YyRr \times YyRr$

and the ratios of the phenotypes of the progeny of this cross, with the possible genotypes indicated, are:

F_2 generation
$YYRR$ & $YYRr$ & $YyRR$ & $YyRr$
9 yellow round

$YYrr$ & $Yyrr$
3 yellow wrinkled

$yyRR$ & $yyRr$
3 green round

$yyrr$
1 green wrinkled

This ratio is expected if seed colour alleles and seed shape alleles are inherited independently of each other, the ratio 9:3:3:1 being effectively the product of two independent 3:1 segregations (Fig. 9.6). Mendel's results demonstrate that, during meiosis, the alleles for seed colour and seed shape that were inherited from one parent do not remain together but are free to assort independently of each other to form different allelic combinations. Mendel called his explanation of this pattern of inheritance the *Principle of Independent Assortment*:

Alleles of a gene controlling one trait assort into gametes independently of alleles of another gene controlling a different trait.

This shuffling of alleles of different genes with respect to one another is one reason for the enormous diversity of characteristics exhibited by individuals, even if they are brother or sister.

Backcrosses and testcrosses

Mendel then took a critical step in the validation of any scientific idea. He designed a specific cross, the outcome of which was predicted by his hypotheses. Instead of allowing the F_1 plants ($YyRr$) to self-fertilise, he crossed them back to the green wrinkled parent ($yyrr$). A cross such as this between F_1 progeny and either of their pure-breeding parents is termed a backcross. A cross to an organism that is homozygous for the alleles that express the recessive phenotype is termed a **testcross**. In this case, the green wrinkled parent in the testcross produces only one type of gamete—all yr. If the F_1 plant, when crossed with the parent with the recessive phenotype, produces the four types of gametes—YR, Yr, yR and yr—in equal proportions as predicted by Mendel's hypotheses, then the genotypes of the progeny should occur in the proportions ¼ $YyRr$, ¼ $Yyrr$, ¼ $yyRr$ and ¼ $yyrr$ and the phenotypes should be equal numbers of yellow round, yellow wrinkled, green round and green wrinkled peas. Mendel reported the following numbers of progeny for this cross:

yellow round	55
yellow wrinkled	49
green round	51
green wrinkled	53
total	208

This is close to the expected 1:1:1:1 ratio.

Mendel showed that this pattern, in which alleles of different genes are inherited independently of each other, was general and he was able to extend it to cases where inheritance of three traits was examined.

When two different genes (each with two alleles) are inherited independently, the traits appear in a 9:3:3:1 ratio in the F_2 generation or in a 1:1:1:1 ratio in a backcross between an F_1 individual and a pure-breeding parent of the recessive phenotype (a testcross). This shows that the alleles of one gene assort independently of the alleles of another gene during segregation of alleles into the gametes.

Mendelian inheritance in humans

The principles that Mendel formulated have been found to be generally applicable to plants and animals. A case of special interest to us is that of humans. Here the significance of Mendel's observations is realised every time a genetic disease is diagnosed. Mendel's principles provide the foundation for identifying genetic traits in humans, including genetic disease, and for counselling families in which genetic disease occurs.

Mendelian inheritance of disease alleles also permits the risk of transmission of disease alleles to be accurately determined.

Many human traits are known to be genetically determined. The majority of visible traits that exhibit Mendelian inheritance, such as cleft chin, hair on the middle segment of the fingers, polydactyly (additional fingers) and achondroplasia (dwarfism), are inherited as dominant phenotypes, while albinism is a recessive phenotype. Genetic disease phenotypes can also be dominant or recessive. Dominant diseases include Huntington disease, myotonic dystrophy and neurofibromatosis, while recessive diseases include cystic fibrosis, β-thalassaemia and mucopolysaccharidosis

(see Table 9.2). Dominant traits tend to be found in members of successive generations, while recessive traits, particularly recessive disease traits, are more likely to appear without a family history of the trait. Human geneticists draw family trees, or *pedigrees*, to represent the occurrence of a particular trait. Figure 9.7 gives examples of pedigrees, one showing a recessive characteristic (cystic fibrosis) and the other a dominant characteristic (Huntington disease).

Many human traits exhibit Mendelian patterns of inheritance. Both recessive and dominant traits have been identified, many of which were discovered because they correspond to genetic diseases.

Table 9.2 Examples of human traits and diseases determined by single autosomal or X-linked genes

Inheritance pattern	Frequency (1000 births)[a]	Abnormality
Autosomal dominant		
Huntington disease	0.1	Progressive nerve degeneration in brain from mid life
Achondroplasia	0.04	Dwarfism
Myotonic dystrophy	0.05	Progressive muscular weakness
Neurofibromatosis	0.25	Tumours of peripheral and other nerves
Polycystic kidney disease	1.0	Progressive kidney failure
Osteogenesis imperfecta	0.05	Brittle bones; blue sclera
Polydactyly	1.7	Additional fingers on the hand
Autosomal recessive		
Phenylketonuria	0.1	Mental retardation unless diet controlled
Cystic fibrosis	0.4	Viscous mucoid secretions, lung congestion
β-thalassaemia	0–10	Severe anaemia; frequency highest in Mediterranean and southern Asian people
Sickle cell anaemia	0–10	Haemolytic anaemia; highest in West African people
Albinism	0.1	No pigment; abnormal vision
Adrenal hyperplasia	0.1	Heterogeneous; abnormal genitalia
Mucopolysaccharidosis	0.05	Abnormal physical and mental development
X-linked[b]		
Red–green colour blindness	80	Red and green not distinguished as normal
Haemophilia A and B	0.2	Defect in blood clotting
Muscular dystrophy	0.3	Progressive muscular weakness
Fragile X syndrome	0.9	Mental retardation; some women carriers affected
Testicular feminisation	0.02	Sterile XY females; unresponsive to androgens

(a) Frequencies per 1000 births are approximate and, except for haemoglobin diseases, here relate to populations of northern European origin. In each case, X-linked diseases occur much more frequently in males. This is because in females a recessive disease phenotype is masked by the dominant wild-type phenotype in heterozygotes, whereas in males, which have only one copy of the X-linked gene, the recessive phenotype is expressed whenever the disease allele is present.

(b) Frequency is per 1000 male births.

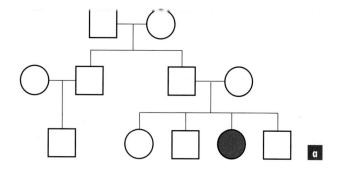

Fig. 9.7 Human pedigrees (family trees) showing patterns of inheritance of two genetic diseases: **(a)** cystic fibrosis, a recessive trait and **(b)** Huntington disease, a dominant trait. Males are represented as squares, females as circles. Filled squares or circles indicate individuals who have the disease

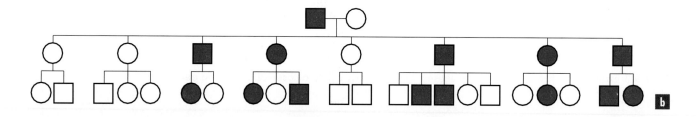

BOX 9.3 Cystic fibrosis, anaemia and heterozygote advantage

Some alleles show very different frequencies in different populations. One of the most striking and devastating examples is the recessive genetic disease, cystic fibrosis (Fig. 9.7a). Cystic fibrosis disease alleles exist at much higher levels in Caucasians than in any other racial group. Approximately one in 26 Caucasians are heterozygotes for (i.e. carriers of) a cystic fibrosis disease allele. Yet, until recent improvements in therapy, homozygotes for the disease allele usually died before they could reproduce. This should have resulted in a much lower incidence of the disease allele than we currently see. One likely explanation for the high allele frequency is that, at some time in the past, there was a survival and/or reproductive advantage to carriers of the disease allele.

The basis of such an advantage may have been discovered recently with the demonstration that there is a link between typhoid fever and the cystic fibrosis gene. Typhoid fever is caused by the bacterium *Salmonella typhi*. *S. typhi* has been found to require the protein encoded by the cystic fibrosis disease gene for entry into intestinal epithelial cells. Furthermore, it was shown that entry by *S. typhi* occurs at a lower level in cells that are heterozygous for the disease allele relative to cells that are homozygous wild-type. Individuals who are heterozygous for the cystic fibrosis disease allele are therefore likely to have had a greater chance of surviving typhoid fever than homozygous wild-type individuals. This is an example of heterozygote advantage, where the heterozygote is at a selective advantage over both homozygotes. The high level of cystic fibrosis disease alleles appears to be a now harmful relic of a time in the history of Caucasians when typhoid fever was rampant and heterozygotes for the cystic fibrosis disease allele had a greater chance of surviving infection.

A second example of heterozygote advantage in humans involves alleles for the serious blood diseases sickle cell anaemia and thalassaemia. The sickle cell allele results in a single amino acid change in the β chain of human haemoglobin, changing the tertiary structure of the protein. Individuals who are homozygous for this sickle cell allele suffer from severe life-threatening anaemia associated with distortion of their normally disc-shaped red cells into sickle-shaped forms. Heterozygous carriers are phenotypically normal under most conditions, although their red cells can be made to sickle under certain conditions in the laboratory. The potentially lethal allele occurs at high frequencies (up to 10%) in certain African populations. The distribution of the sickle cell allele closely matches the distribution of the often lethal form of malaria (see figure). Carriers of the sickle cell allele (heterozygotes) are less severely affected by malaria than are normal homozygotes. They have shorter bouts of fever, lower parasite counts in their blood and are hospitalised less often. It seems

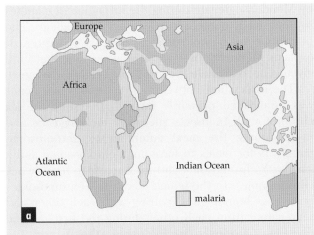

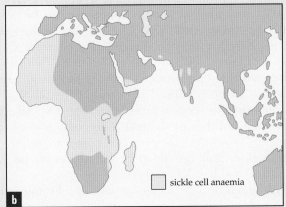

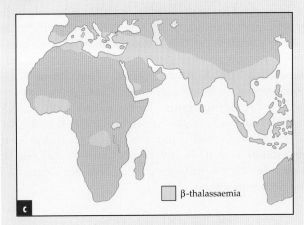

(a) Former and present distribution of falciparum malaria, showing closely paralleled distribution of alleles for the human blood diseases (b) sickle cell anaemia and (c) β-thalassaemia. Sickle cell anaemia and β-thalassaemia homozygotes are severely anaemic and may not survive to adulthood. These deleterious alleles are maintained in malarial environments because heterozygous carriers are more resistant to malaria than normal homozygotes and leave relatively more offspring

that red cells infected by the malarial parasite are rapidly removed from the blood in sickle cell heterozygotes. Thus, in malarial environments, sickle cell heterozygotes are at a selective advantage relative to both normal homozygotes (severely affected by malaria) and sickle cell homozygotes. The role of balanced polymorphisms in the genetics of populations and evolution is discussed further in Chapter 32.

High incidences of thalassaemias are found in populations distributed from the Mediterranean through the Middle East, to South Asia and South-East Asia and in parts of Africa, where malaria is (or was) endemic (see figures). Thalassaemia also affects haemoglobin. Homozygotes for some thalassaemia alleles are severely anaemic and may die unless they receive repeated blood transfusions. It seems likely that heterozygous carriers of thalassaemia alleles are more resistant to malaria than are normal homozygotes, thus keeping the frequency of these otherwise deleterious alleles high.

In a malaria-free environment, the sickle cell allele and thalassaemia alleles do not confer a heterozygote advantage and should ultimately be lost from the population. This is the case in emigrant populations, such as Afro-Americans and Australians of Mediterranean and Asian origin. In fact, loss of those deleterious alleles in Western populations is now accelerating through prenatal diagnosis. In couples who are both carriers, testing can determine if the fetus is an affected homozygote who, if born, would expect a short life of low quality. Parents may then choose to terminate the pregnancy.

The limits of Mendel's ideas

As well as providing two fundamental rules describing patterns of inheritance, Mendel's studies provided the basis for further investigation of heredity. As brilliant and important as Mendel's analysis was, there was still much to be discovered at the time that Mendel was elected Prelate of his Monastery and drawn from his scientific pursuits to new administrative duties. Since the rediscovery of Mendel's work, many discoveries have extended, refined and supplemented Mendel's observations. Genetics remains a frontier science, in that we continue to discover new and unexpected aspects of heredity. In the remainder of this chapter we will consider four extensions to Mendel's observations: sex linkage, autosomal linkage, X inactivation and imprinting.

Sex linkage and chromosomes

Sex-linked genes

None of the patterns of inheritance discussed above showed sex-specific effects. However, some genes are inherited differently in males and females. Such genes are termed *sex-linked genes*. The first sex-linked gene was discovered in the vinegar fly, *Drosophila melanogaster*. The eyes of wild-type *D. melanogaster* are a dark red colour but the appearance of a white-eyed male stimulated the developmental biologist, T. H. Morgan, to turn his attention to genetics. Morgan found that when a red-eyed female is crossed to a white-eyed male, only red-eyed F$_1$ progeny result, indicating that the phenotype of the wild-type allele (+) is dominant over that of the white allele (*w*), which is recessive. However, when male and female F$_1$ progeny were crossed, the F$_2$ progeny revealed a remarkable pattern of phenotypes. All of the females were red-eyed, but half of the males were white-eyed, the other half red-eyed. Clearly, the *w* allele did not behave according to Mendel's rules. In a testcross, the F$_1$ red-eyed females crossed to white-eyed males yielded progeny of which half were red-eyed and the other half white-eyed, irrespective of their sex. Remarkably, red-eyed males crossed to white-eyed females produce only red-eyed females and white-eyed males!

How could such a bizarre pattern of sex-related inheritance be explained? The answer came from the study of chromosomes.

Genes are on chromosomes

Chromosomes were introduced in Chapter 8 and are discussed in more detail in Chapter 10. Most eukaryotes have two types of chromosomes. **Autosomes** are the same in morphology (appearance) and number in males and females, while **sex chromosomes** differ in morphology, are present in different numbers in males and females and are involved in sex determination. With the exception of the gametes, cells of animals and plants are generally **diploid**, that is, they contain pairs of autosomes, one autosome of each chromosome type inherited from each parent. These pairs of autosomes are termed **homologous chromosomes**, or **homologues**. Humans have 22 pairs of homologues in their somatic cells. The connection between pairs of similar chromosomes (homologues) and pairs of alleles in an individual was first suggested by Walter Sutton who, in 1902, formulated the chromosome theory of inheritance. Sutton argued that genes are on chromosomes because, during meiosis (Chapter 8), homologues pair with each other and segregate to produce haploid gametes, just as the alleles described by Mendel segregate during the formation of gametes so that each gamete carries only one allele. The behaviour of homologues at meiosis (Chapter 8) explains Mendel's observations if we assume that for a given gene, one allele is carried on each homologue.

The association between chromosomes and genes was established primarily through the study of sex chromosomes. In insects, such as *D. melanogaster*, and in mammals, the most common sex chromosome pattern is for males to have an X chromosome and a smaller Y chromosome and for females to have two X chromosomes. In the female, the two X chromosomes behave as homologues, pairing with each other and segregating from each other during the first meiotic division. Curiously, in the male, the X and the Y chromosome, although appearing to be very different from each other, behave as homologues during the first meiotic division. They have a short region of homology that allows them to pair with each other and segregate from each other, the result being that the sperm produced by a male carry an X or Y chromosome with equal probability (Fig. 9.8). All eggs carry one or other of the two X chromosomes of the female. When these gametes come together at fertilisation, half the progeny will therefore be XX (females) and half XY (males).

Although the Y is an important chromosome, required in humans, for example, for determining maleness and for male fertility, it carries relatively little genetic information. For X-linked genes, there is usually no allele on the Y chomosome. Males, therefore, have the peculiar genetic property of having only one allele of X-linked genes. They cannot be homozygous or heterozygous for these genes but are said to be **hemizygous**. In organisms with an XX (female) and XY (male) sex-determining system, the alleles of sex-linked genes are passed on to the next generation as if there are two in females and one in males. In other words, sex-linked genes are located on the X chromosome.

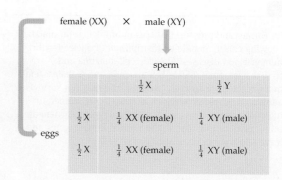

Fig. 9.8 Pattern of inheritance of sex chromosomes in humans. A male has an X and a Y chromosome, a female has two X chromosomes. Sperm have either the X or the Y. Eggs have an X but no Y. The result of fertilisation, then, is that on average half the progeny are male and half are female

The patterns of inheritance of the *Drosophila white* gene alleles follow the behaviour of X chromosomes. Each female offspring receives either of her mother's X-linked alleles with equal probability but always inherits her father's single X chromosome. Each male offspring also receives either of his mother's X-linked alleles with equal probability but always inherits his father's single Y chromosome. In the case where a white-eyed female (*w/w*) is crossed to a red-eyed male, for example, the female progeny are all heterozygotes (*+/w*) and are therefore red-eyed, while the males can only receive their X chromosome from their mother and are therefore all hemizygous for the recessive *w* allele (*w/Y*) and are white-eyed. The phenotype appears to have switched sexes between the generations!

Autosomes are chromosomes that are the same in males and females. The behaviour of pairs of autosomes (homologues) during meiosis explains the patterns of inheritance of alleles observed by Mendel. Sex chromosomes differ between males and females and are involved in sex determination. Behaviour of the sex chromosomes at meiosis explains the inheritance of sex-linked genes. Sex-linked genes on the X chromosome in organisms with an XX/XY sex-determining system are referred to as X-linked genes.

X-linked traits in humans

The most common X-linked trait in humans is red–green colour blindness (Fig. 9.9). Most of us can distinguish red and green colours but about 8% of Caucasian males have difficulty in doing so because they have an abnormality in the light-sensitive molecules of their retina. Colourblind males cannot transmit this allele to their sons but always transmit it to their daughters.

This inheritance pattern can be explained by proposing that the phenotype of the colourblind allele is recessive and that the gene for red–green colour blindness is on the X chromosome with no comparable gene on the Y chromosome (i.e. colour blindness is X-linked). Males, therefore, can be either *C* (normal) or *c* (colourblind) whereas females can be *CC*, *Cc* (both normal) or, rarely, *cc* (colourblind). Colourblind males will always pass on the Y chromosome (which does not carry the gene) to their sons. However, their daughters will always receive the *c* allele on their father's X chromosome. If the daughter receives an X chromosome bearing the *C* allele from her mother, she will be heterozygous and thus a carrier. If a mother is a heterozygote carrier (*Cc*), then her sons have a 50% chance of being colourblind. If the father is also colourblind (*c/Y*), half the daughters will also be affected. For a colourblind homozygote mother (*cc*) and normal father (*C/Y*), all of their sons will be

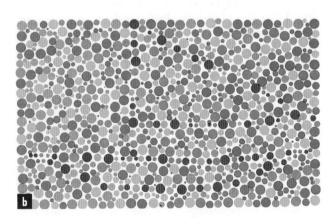

Fig. 9.9 (a) A pedigree showing inheritance of colour blindness. Colour blindness in humans is caused by a defect in pigments of the retina that are sensitive to red or green light. This leads to an inability to distinguish red and green colours in the normal way. The genes controlling red–green colour vision are X-linked, that is, they are on the X chromosome. An affected man always passes on the colourblind *c* allele to his daughters but never to his sons. His daughter may pass the *c* allele on to her children with a 50% chance. In the pedigree shown, the son is colourblind because he received his mother's X chromosome bearing the *c* allele rather than that bearing the wild-type *C* allele. **(b)** A test plate used for detecting colour blindness. Tests for colour blindness cannot be reliably conducted with this diagram. For accurate testing, the original plates should be used

colourblind and all of their daughters carriers. The inheritance of colour blindness from generation to generation can therefore be accounted for by the patterns of inheritance of X chromosomes.

Y-linkage and male sex determination

Traits influenced by Y-linked genes are passed from father to son and are never observed in females. Because the Y chromosome carries few genes, such a pattern of inheritance is very rarely seen. One Y-linked mammalian gene has particular significance, however. This is the recently identified sex-determining gene, SRY (sex-determining region Y), which switches development of the early fetus to the male pathway. If the Y chromosome (carrying this gene) is absent, the fetus becomes female. Once the particular sex-determining pathway has been activated, other genes that control male or female development take over.

Linkage on autosomes

Linkage and recombination

Organisms have many more genes than chromosomes. Humans, for example, are estimated to have approximately 40 000 genes, yet have only 22 different autosomes and the X and Y chromosomes. Many genes, therefore, reside on each chromosome. Homologous chromosomes carry the same set of genes, each carrying one allele of each gene. Homologues segregate at meiosis, as do the alleles that reside on the homologues, accounting for Mendel's Principle of Segregation. Alleles of different genes located on the same chromosome should be transmitted together, that is, they should be linked and should not follow Mendel's Principle of Independent Assortment. However, **linkage** between genes is never complete because of the crossing over between homologous chromosomes (associated with chiasmata, see Chapter 8). Crossing over leads to **recombination** between genes in a pair of homologous chromosomes. The level of recombination between different genes located on the same chromosome varies greatly. If the genes are very close together, alleles of the genes may recombine very rarely. Conversely, if separated far enough on the chromosome, the frequency of crossing over can mean that the alleles of the two genes appear to assort independently, as if they were on different chromosomes. When dealing with genes located on the same chromosome, it is useful to talk about their position, or **locus** (pl. loci), on the chromosome.

A direct test for independent assortment is to test cross a double heterozygote with the double recessive homozygote (*A/a;B/b* × *a/a;b/b*). From Mendel's

Principle of Independent Assortment, the four possible combinations of gametes arising from meiosis in the double heterozygote (*AB*, *Ab*, *aB* and *ab*) are expected to form in equal proportions. Thus, the four possible genotypes of the testcross progeny (*A/a;B/b*, *A/a;b/b*, *a/a;B/b* and *a/a;b/b*) are expected in the proportions 1:1:1:1 if the two genes assort independently. Deviations from this ratio indicate that independent assortment has not occurred. For example, in the Australian sheep blowfly, *Lucilia cuprina*, a pure-breeding wild-type strain has red eyes and straight bristles (genotype *W/W;CK/CK*), and a pure-breeding mutant strain has white eyes and crooked bristles (genotype *w/w;ck/ck*). When these two strains are crossed, the F_1 progeny (genotype *w/W;ck/CK*) have red eyes and straight bristles, indicating that the wild-type phenotypes are dominant. A typical set of results from a testcross between F_1 female flies and *w/w;ck/ck* males is shown in Table 9.3.

In this example, the expected four phenotypes are observed but not in the ratio 1:1:1:1 that we would expect from the Principle of Independent Assortment. Red eyes and straight bristles are more often inherited together, as are white eyes and crooked bristles. However, if we compare eye and bristle traits separately, then the ratio of red eyes (96 + 7) to white eyes (91 + 6) is 103:97, which is approximately 1:1. Similarly the ratio of straight bristles (96 + 6) to crooked bristles (91 + 7) is 102:98, also close to the expected 1:1 ratio.

The F_1 and testcross data are consistent with the hypothesis that there is a single gene for eye colour and a single gene for bristle type, with the wild-type phenotype being dominant in both cases. The Principle of Segregation applies to the alleles of each gene. However, the alleles of the eye colour gene are not assorting independently of the alleles of the bristle type gene, therefore defying Mendel's Principle of Independent Assortment. Specifically, the data show that the phenotypes of the parental strains are more likely to be inherited together. Thus, in this cross, red eyes and straight bristles are linked in inheritance, as

Table 9.3 Testcross offspring

Phenotype	Genotype	Number of individuals
Red eyes, straight bristles	*W/w; CK/ck*	96
White eyes, straight bristles	*w/w; CK/ck*	6
Red eyes, crooked bristles	*W/w; ck/ck*	7
White eyes, crooked bristles	*w/w; ck/ck*	91
Total		200

are white eyes and crooked bristles. Only in a few cases (13 out of 200 or 6.5%) do we see the non-parental combinations of white eyes and straight bristles or red eyes and crooked bristles being inherited together. The deviation from Mendel's second principle occurs because these eye colour and bristle shape genes are located on the same chromosome. Crossing over between the eye colour locus and the bristle type locus during meiosis in the F$_1$ individuals creates new combinations of alleles in the gametes and, therefore, in the progeny. The F$_2$ individuals carrying the non-parental phenotypes are referred to as **recombinants**.

Because linked genes are on the same chromosome, the genetic notation for them is different from that for genes on separate chromosomes. Linked genes are represented as:

$$\frac{AB}{ab}$$

often abbreviated to

$$\frac{AB}{ab}$$

where the alleles on one of the homologous chromosomes are represented together and separated by a line from the alleles on the other homologue. The crosses involving eye colour and bristle characteristics described above, which revealed linkage between the two genes, can now be represented more accurately in this way, as shown in Figure 9.10.

The frequency of recombination between two genes depends on the distance between the loci of the two genes. The closer the loci, the less often crossing over (recombination) occurs between them. If crossing over occurs very frequently, loci sufficiently far apart on the same chromosome may assort independently, that is, with a recombination frequency of 50%, as if they were on separate chromosomes.

> Linkage is the tendency of two or more genes on the same chromosome to be inherited together.

Chromosome mapping

Given that the probability of crossing over occurring between any two loci is related to the distance between the genes on the chromosome, the frequency of recombination can be used as a measure of this distance, allowing genes to be mapped on chromosomes in a linear order:

$$\frac{\text{Map}}{\text{distance}} = \frac{\text{Number of recombinant progeny}}{\text{Total number of progeny}} \times 100$$

A map unit is called a **centimorgan (cM)**, after T. H. Morgan, who discovered linkage. In our example,

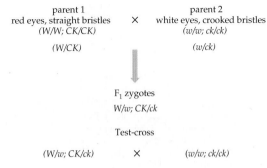

Fig. 9.10 A genetic diagram of the genotypes observed in testcross offspring involving the F$_1$ generation from an original cross between a pure-breeding wild-type (red eyes, straight bristles) strain of the Australian sheep blowfly, *Lucilia cuprina*, and a pure-breeding strain of white eyes and crooked bristles

the white eye and crooked bristle genes are 6.5 centimorgans apart.

Consider another gene, rusty brown colour (*ru*), which is linked to both white eye and crooked bristle. A testcross involving white eye and rusty colour shows a recombination frequency of 15%, indicating that white eye and rusty colour are 15 centimorgans apart. A further testcross involving rusty colour and crooked bristles shows a recombination frequency of 21.5%, indicating that the genes for rusty colour and crooked bristles are 21.5 centimorgans apart. Thus, the genetic map is:

$$ck \longleftarrow 6.5 \longrightarrow w \longleftarrow 15 \longrightarrow ru$$
(with 21.5 spanning the full distance above)

or

$$ru \longleftarrow 15 \longrightarrow w \longleftarrow 6.5 \longrightarrow ck$$

The order of these genes can be related to other linked genes to construct a map of the relative positions of all the known genes on a chromosome. Maps of some of the genes on a chromosome of *D. melanogaster* and a human chromosome are shown in Figure 9.11. The mapping of human genes is an important step in the isolation and characterisation of the genes (Chapter 12).

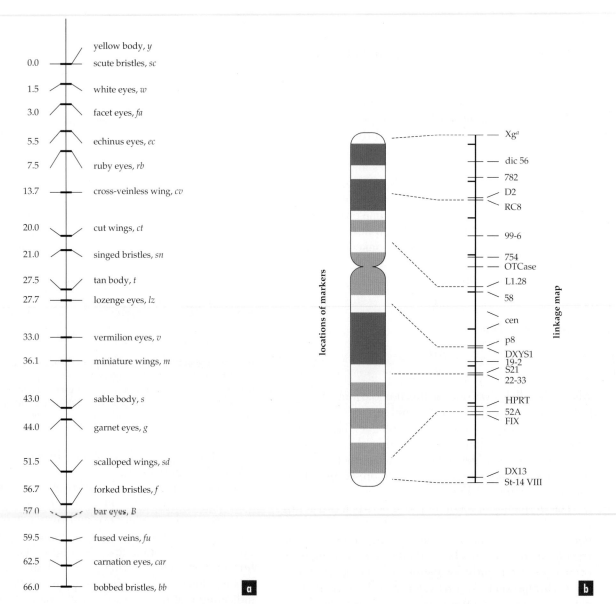

Fig. 9.11 **(a)** A linkage map of some of the genes on chromosome 1 (X chromosome) of *D. melanogaster*. The positions of genes are relative to the gene closest to the end of the chromosome, and are in the map units, centimorgans. **(b)** A linkage map of human chromosome 1 showing the distribution of genes and molecular markers (see Chapter 13)

More variations on Mendel's observations

Incomplete dominance

Absence of dominance of one phenotype over another occurs for many allele combinations. For example, in the snapdragon, *Antirrhinum*, flower colour is controlled by a single gene with two alleles. A cross between a homozygous red-flowered snapdragon and a homozygous white-flowered snapdragon will give pink F_1 heterozygotes (Fig. 9.12). When these pink F_1 plants are self-pollinated, they produce an F_2 generation containing a mixture of red-, pink- and white-flowered plants in the ratio 1:2:1.

Flower colour in snapdragons is referred to as partially or incompletely dominant because the phenotype of the heterozygote is intermediate between the phenotypes of the two homozygotes. **Incomplete dominance** may occur if the activity of one allele produces a certain amount of pigment, while the other allele (inactive) produces none. The homozygote for the active allele would produce two units of pigment, the heterozygote would produce one and the homozygote for the inactive allele would produce none.

A character is incompletely dominant when the phenotype of the heterozygote is intermediate between that of each homozygote.

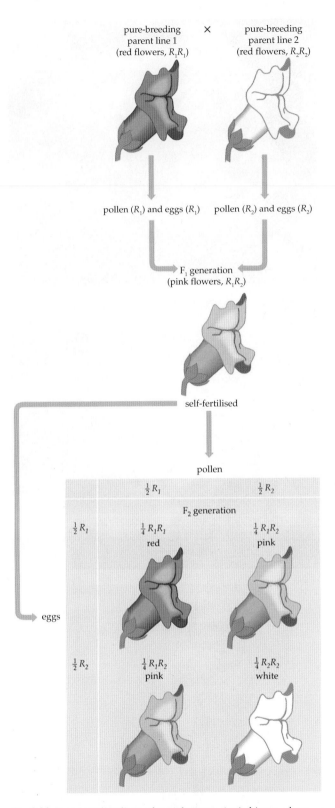

Fig. 9.12 Pure-breeding lines of snapdragons, *Antirrhinum*, show different flower colours—red, pink or white. When a red-flowered line is cross-pollinated with a white line (pollen parent), the F$_1$ generation is pink. When F$_1$ plants are self-pollinated, the phenotypes in the F$_2$ generation include all three colours. These data show that colours are controlled by two alleles, R_1 (red) and R_2 (white). Since heterozygous R_1R_2 plants have pink flowers, dominance is incomplete

Novel phenotypes

We have already seen that the inheritance of more than one pair of alleles can be followed in a series of crosses. For example, Mendel looked at the transmission of factors for both seed colour and seed shape in the same plants. The phenotypes could be examined separately because one phenotype did not affect the other phenotype. Occasionally, a combination of alleles can produce a phenotype not seen before, as seen for combinations of two separate genes controlling eye colour in *D. melanogaster* (Fig. 9.13). These are the genes *brown*, with two alleles, $+$ and *bw*, and *scarlet*, with two alleles, $+$ and *st*. As shown in Figure 9.13, the wild-type eye colour in *D. melanogaster* is reddish-brown due to the presence of two pigments, a brown pigment and a bright red pigment. Flies homozygous recessive for *brown* (*bw/bw;*$+$*/*$+$ or *bw/bw;*$+$*/st*) have brown-coloured eyes because they cannot make the red pigment, while flies homozygous recessive for scarlet ($+$*/*$+$*;st/st* or $+$*/bw;st/st*) have scarlet-coloured eyes because they cannot make the brown pigment. However, flies that are homozygous recessive for both brown and scarlet genes (*bw/bw;st/st*) have white eyes, as neither red nor brown pigment is made. (The X-linked *w* gene referred to earlier in this chapter is a

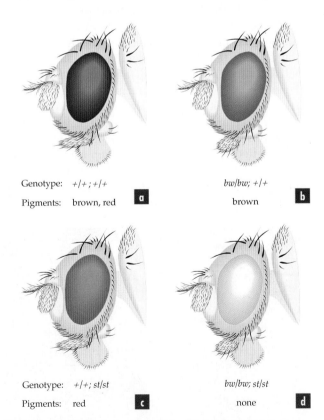

Fig. 9.13 Eye colour phenotypes of **(a)** wild-type, and two mutants **(b)** *brown* and **(c)** *scarlet* of *D. melanogaster*. **(d)** A different eye colour phenotype, *white*, occurs when the two mutants, both homozygous recessive, are bred together. This is an example of gene interaction

separate gene that prevents deposition of both red and brown pigments, independent of *brown* and *scarlet* function, resulting in a white eye colour.)

U nexpected phenotypes may arise as a result of interactions between the activities of different genes.

Phenotypes and environment

The effect of an allele can be masked or enhanced by variation in the environment. A well-known example of this is the coat colour mutation carried by Siamese cats (Fig. 9.14). The wild-type allele, *C*, results in fur being pigmented over all parts of an animal. In homozygotes for the mutant allele, *c*, pigmentation is restricted to fur at the extremities of the face, ears, legs and tail. The mutant allele can produce pigment only at lower body temperatures, which occur at these extremities. At the higher temperatures on the main body surface, the product of the mutant *c* allele is almost inactive. Thus, expression of the *c* allele is affected by an environmental factor, in this case, body temperature.

Another example is the hydrangea which, if grown in acidic soil has blue flowers, while a cutting from the same plant (same genotype) grown in alkaline soil has pink flowers (Fig. 9.15). The colour of the flamingo (Fig. 9.15c) is also influenced by its environment. These marshland birds are not born pink but turn this colour due to a red pigment that comes from the crabs and shellfish they eat. The pigment accumulates in the shafts of the birds' growing feathers. Although the pigment is not visible in the raw meat of crustacea, it is the same pigment that causes a prawn to turn pink when cooked. Flamingos in captivity are often fed shellfish rich in this pigment to make them more colourful.

The effect of a deleterious mutation may sometimes be overcome by changing the environment in which a mutant individual is raised. For example, the serious effects of the phenylketonuria mutation in humans

Fig. 9.14 A Siamese cat homozygous for a mutant pigment allele, *c*. This allele is more active at lower temperatures, such as occur at the extremities of the animal's body

Fig. 9.15 Flower colour in hydrangeas and coat colour in flamingos are influenced by the environment. Hydrangea flowers are **(a)** pink if the soil is alkaline and **(b)** blue if the soil is acidic. **(c)** Flamingos derive their pink colouration from a diet of small crustaceans

(Table 9.2) can be prevented by modifying the diet of affected individuals. The wild-type product of this gene is an enzyme that converts the amino acid phenylalanine into tyrosine, an amino acid used in the synthesis of proteins and the pigment melanin. In recessive homozygotes, phenylalanine (and its phenylketone breakdown products) accumulate. This affects normal brain development, leading to mental deficiency. As a pleiotropic effect of the mutant allele, affected individuals usually lack melanin pigment in their hair.

If the defect is detected early enough, an affected infant can be placed on a low phenylalanine, high tyrosine diet. Low phenylalanine intake reduces the amount of circulating toxic products, while the tyrosine makes up for the amino acid deficiency. A modified diet prevents brain damage occurring in early childhood and restores the individual to a phenotype not readily distinguishable from normal. This treatment is so successful that screening of all newborn infants is done routinely even though the frequency of affected individuals is only about 1 in 10 000 live births. A small sample of blood is tested for levels of phenylalanine. Babies with a higher than normal level are then tested further to determine if they are homozygous for a mutant form of this gene. If so, they are immediately placed on the modified diet.

The effect of some genes on the phenotype of an organism can be masked or enhanced by different environmental conditions.

Penetrance and expressivity

Genetically determined phenotypes can show variability or be difficult to recognise. Variation in phenotypic expression involves both expressivity and penetrance. **Expressivity** refers to the degree to which an allele is expressed phenotypically in an individual. For example, the dominant mutation *Antennapedia* causes the antennae of *D. melanogaster* to develop as legs (Chapter 16). The expressivity of the phenotype in heterozygotes can vary from fly to fly, some showing almost complete leg structures in place of antennae, others showing only a short thickening of the antennae. This phenotypic variation is largely due to genetic variation at other loci (referred to as the genetic background).

Penetrance refers to the proportion of individuals of a particular genotype that show a phenotypic effect. For example, in *D. melanogaster*, the phenotype of the *Curly* (*Cy*) allele is wings that curl upwards (Fig. 9.16). This phenotype is dominant in *Cy/+* heterozygotes. The degree of curvature is variable from individual to individual and can overlap with the wild-type phenotype. At a temperature of 25°C, most *Cy/+* flies will develop curly wings, but at 19°C more flies have the

Fig. 9.16 Phenotype of the mutant *Curly* (*Cy*) allele in *D. melanogaster*. The phenotype exhibits variation in expressivity (the degree of curling of wings varies) and incomplete penetrance (some flies in a population of *Cy/+* heterozygotes have an apparently normal wing shape)

wild-type phenotype. If 80% of *Cy/+* flies develop curly wings, then the gene is said to be showing 80% penetrance under these environmental conditions. Penetrance can be influenced by the age of onset of certain phenotypes. For example, Huntington disease, a dominant neurodegenerative disease, becomes apparent later in life and at different ages in different individuals. The penetrance would be considered low if only younger adults carrying the Huntington disease allele were examined but rises to quite high levels in older adults carrying the same disease allele.

Expressivity refers to the strength of the phenotype in an individual with a particular genotype. The penetrance refers to the proportion of individuals in a population with the genotype that exhibits the phenotype.

Epistasis

A different form of interaction between genes occurs if the phenotype controlled by one masks that of the other. This phenomenon is termed **epistasis**. For example, genes for eye colour in *D. melanogaster* cannot be expressed in a homozygous *eyeless* mutant fly as this fly has no eyes! In the plant *Arabidopsis thaliana*, one mutant with a recessive phenotype, *agamous* (without gametes), produces flowers that are remarkably transformed in structure. Petals form in place of stamens and carpels, resulting in an attractive double flower (but lacking male and female organs, Fig. 9.17). Another mutant, also with a recessive phenotype, *leafy*, lacks flowers almost completely (Fig. 9.17). F$_1$ plants heterozygous for both *leafy* and *agamous* have a normal phenotype. When F$_2$ progeny are raised from self-fertilised F$_1$ plants, the expected phenotypic ratios are 9 wild type:3 agamous:3 leafy:1 agamous leafy. However, it is not possible to

Fig. 9.17 Flowering stems of the model plant, *Arabidopsis thaliana*, a small, easily grown species of the mustard family: **(a)** wild type with normal flowers and a seed pod; **(b)** *agamous* mutant (recessive phenotype) with many extra petals in place of stamens; and **(c)** *leafy* mutant (recessive phenotype), in which normal flowers do not arise. It is not possible to observe the agamous phenotype in *leafy* plants as the leafy phenotype masks it. Therefore, *leafy* is epistatic to *agamous*

determine whether or not homozygous *leafy* plants have the *agamous* or wild-type phenotype, as they have no flowers. Thus, the *leafy* phenotype hides, or is epistatic to, the *agamous* phenotype. The modified ratio in the F_2 generation is therefore 9 wild type:3 agamous:4 leafy.

> Epistasis is an interaction between different phenotypes such that the phenotype of alleles of one gene interferes with or masks the phenotype of alleles of another gene.

Epigenetic regulation

One of the most intriguing recent developments in genetics has been the discovery that the activity of some genes can be modified and inherited through cell division or passed to offspring independent of the genotype. Such inherited regulation of gene activity is termed **epigenetic regulation**. Two examples are discussed here: X chromosome inactivation and imprinting.

X chromosome inactivation

Chromosome dosage is important for the functioning of organisms. Individuals that contain too many or too few chromosomes (e.g. as a result of errors in chromosome segregation during meiosis) are frequently inviable.

The existence of X and Y chromosomes, therefore, creates a problem. Female cells have twice the number of X chromosomes than male cells. This imbalance is overcome by mechanisms that reduce expression from the female X chromosomes or increase the level of expression from the male X chromosome. In the case of humans, expression is halved in females by a specific mechanism that causes one of the two female X chromosomes to be completely inactivated. In a random process that occurs early in embryogenesis, each cell inactivates one of the female X chromosomes. The X chromosome that is inactivated by this process remains inactive through all future cell divisions to produce a patch of cells that have the same X chromosome (either paternally or maternally derived) inactivated. The adult female is, therefore, a mosaic in which one or other of her X chromosomes is inactivated in each patch.

X inactivation can have a significant effect on the phenotype. A most striking example is that of the tortoiseshell cat (Fig. 9.18). Tortoiseshell cats have patches of red and black fur. The red and black fur patches are generated by alternative alleles of a gene that affects coat colour. This gene is located on the X chromosome, so is subject to X inactivation. Early in embryonic development, cells in females inactivate one of their X chromosomes, leaving either the red or the black allele active, but not both. The patch of cells that arises from each cell will produce either black or red fur. The result is the tortoiseshell cat with patches of red and black fur. The white patches that can be seen in these animals are the result of a separate gene that completely eliminates pigment production in certain parts of the coat. This gene is not X-linked and is not influenced by X inactivation, that is, both males and females will be found with white and black or white and red patches of fur, but only females can have the exquisite tortoiseshell pattern of black and red with white patches of fur.

Fig. 9.18 A tortoiseshell and white cat showing patches of white, red and black fur. The red and black patches are the result of an X-linked coat colour gene and X-inactivation

> Epigenetic regulation refers to modification of the activity of genes that is independent of the genotype and inherited through cell divisions. Inactivation of one of the two X chromosomes in the somatic cells of female mammals is an example of epigenetic regulation. X inactivation is required to ensure that X-linked genes are expressed in the same level in males and females.

Imprinting

Marsupial X inactivation differs from eutherian X inactivation in one remarkable respect. In eutherian (e.g. human, cat) females, one X chromosome is randomly inactivated in each somatic cell early in development. However, Australian geneticists have shown that marsupial females always inactivate the X chromosome that comes from their father (the paternal X chromosome), while the X chromosome from their mother (the maternal X chromosome) is always active. The conclusion is that either or both the paternal and maternal X chromosomes must be marked, or **imprinted**, before or during gamete formation, so that the two chromosomes can behave differently in the cells of the offspring.

X chromosome imprinting results in epigenetic regulation (inactivation) of an entire chromosome. Recently, a study of the inheritance patterns of two diseases, Angelman syndrome and Prader–Willi syndrome, have provided the surprising observation that imprinting can also affect specific genes on autosomes. The outcome of these observations is the realisation that, in some cases, an allele inherited from the mother can behave differently from an identical allele inherited from the father.

Angelman syndrome is characterised by seizures, sleep disorder, hyperactivity, severe mental retardation with lack of speech and a happy disposition with paroxysms of laughter. Prader–Willi syndrome is easily distinguishable from Angelman syndrome, being characterised by severe obesity, short stature, mental retardation and obsessive–compulsive disorder phenotypes. Both Angelman syndrome and Prader–Willi syndrome arise in individuals that are heterozygous for a deletion on the short arm of chromosome 15. However, different diseases occur depending on whether the deletion is inherited from the mother or the father. If the deleted chromosome is inherited from the father and a normal chromosome 15 is inherited from the mother, the individual will suffer Prader–Willi syndrome. However, if the same deletion is inherited from the mother and a normal chromosome 15 is inherited from the father, the individual will exhibit Angelman syndrome. Individuals with Prader–Willi syndrome and Angelman syndrome can, therefore, have identical genotypes but very different phenotypes because of the effect of maternal and paternal imprinting on different genes located on chromosome 15.

It is important to realise that imprinting is a normal process that results in the inactivation of genes in normal individuals. A normal chromosome 15 derived from the mother will have certain genes inactivated in normal offspring, while a normal chromosome 15 derived from the father will have different genes in the same region inactivated in normal offspring. This inactivation is not a problem in normal offspring, as imprinting of any one gene occurs only in one of the parents, so one active copy of the gene is derived from the other parent. It becomes a problem, however, if the gene is deleted in that parent. In deletion heterozygotes, the function of particular genes will be completely eliminated—by deletion of the gene on one homologue and by imprinting leading to inactivation of the gene on the other (Fig. 9.19). Because different genes are imprinted and inactivated on the maternally versus the paternally derived chromosomes, different genes are non-functional depending on whether the normal chromosome came from the father or mother (Fig. 9.19). The consequence is that a different phenotype (syndrome) arises depending on the maternal or paternal source of the deleted chromosome, despite the fact that the individuals with the different syndromes have identical genotypes.

In some cases, identical alleles behave differently depending on whether they are inherited from the mother or the father. Such genes are said to be imprinted. Some genes will be inactive if they were derived from the father, others inactive if derived from the mother. The result can be that individuals with the same genotype can exhibit markedly different phenotypes, as observed in the case of Angelman syndrome and Prader–Willi syndrome in humans.

Quantitative characters

Up to now we have examined the inheritance of genes by following their effects on characters that differ qualitatively between individuals. Qualitative characters can be used to group individuals into distinct phenotypic classes, for example, yellow or green seeds. Organisms also have quantitative characters, such as height and weight in humans, milk yield in cattle, and yield and growth rate in cereals and legumes. Quantitative characters exhibit continuous variation and phenotype is measured along a continuous scale.

Quantitative characters are influenced by the combined action of a number of genes, each of which produces only a small effect. Grain colour in wheat (Fig. 9.20), for example, is controlled by two different genes R_1 and R_2, each with two alleles R_1 and r_1, and R_2 and r_2. The R_1 and R_2 alleles produce red pigment; the r_1 and r_2 alleles do not. The effects of both the R_1 gene and the R_2 gene work additively in establishing seed colour, and a pure-breeding $R_1R_1R_2R_2$ line has dark red seeds because it has four doses of red pigment. A pure-breeding $r_1r_1r_2r_2$ line has white seeds due to a complete absence of pigment. When a pure-breeding $R_1R_1R_2R_2$ line is crossed with a pure-breeding $r_1r_1r_2r_2$ line, the F_1 progeny are $R_1r_1R_2r_2$ and have medium red grains, with two pigment doses, one from R_1 and one from R_2. The F_2 seeds from self-fertilised F_1 plants will display a range of five different colours, ranging from dark red to white depending on how many doses of pigment are present in each grain (Fig. 9.20).

The amount of pigment in a wheat grain may also depend to some extent on its environment; for example, more pigment may be produced if a plant is in

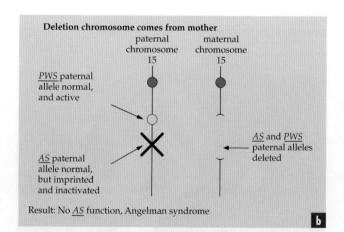

Fig. 9.19 Imprinting of genes on chromosome 15 explains the incidence of Angelman syndrome and Prader–Willi syndrome. Angelman syndrome occurs when neither copy of the *AS* gene is active, while Prader–Willi syndrome occurs when neither copy of the *PWS* gene is active. In the examples shown here, one allele of both *AS* and *PWS* is absent because of a deletion in the chromosome. The *AS* gene on the normal chromosome is imprinted and inactivated if it came from the father, while the *PWS* gene is imprinted and inactivated if it came from the mother. The identities of the *AS* and *PWS* genes have not been determined and there could be more than one gene in the deleted region involved in each syndrome

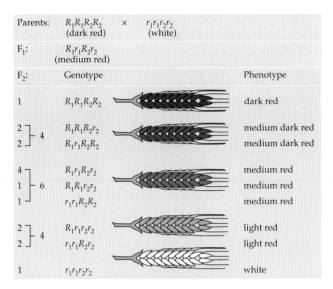

Parents:	$R_1R_1R_2R_2$ (dark red)	×	$r_1r_1r_2r_2$ (white)	
F_1:	$R_1r_1R_2r_2$ (medium red)			
F_2:	Genotype			Phenotype
1	$R_1R_1R_2R_2$			dark red
2 ⎤ 4 2 ⎦	$R_1R_1R_2r_2$ $R_1r_1R_2R_2$			medium dark red medium dark red
4 ⎤ 1 ⎥ 6 1 ⎦	$R_1r_1R_2r_2$ $R_1R_1r_2r_2$ $r_1r_1R_2R_2$			medium red medium red medium red
2 ⎤ 4 2 ⎦	$R_1r_1r_2r_2$ $r_1r_1R_2r_2$			light red light red
1	$r_1r_1r_2r_2$			white

Fig. 9.20 Genetic control of a quantitative character—the amount of red pigment in wheat seeds. Two genes are involved, each with two alleles (R_1 and r_1, and R_2 and r_2) which act without dominance. R_1R_1 produces twice as much red pigment as R_1r_1, while r_1r_1 produces none. The R_2 allele works similarly and additively with the R_1 allele. Thus, F_2 plants may have anything from none to four doses of pigment and range in phenotype from white to dark red

full sun than if it is in the shade. Thus, the amount of pigment in each of the five phenotypic classes of F_2 grains may vary. If environmental effects cause further variation in colour, the amount of pigment in each of the five classes may overlap, resulting in a continuous variation in the character seed colour.

Many quantitative characters are controlled by more than two genes that act additively. Quantitative characters that are influenced by multiple genes, each having a small additive effect, are termed **polygenic** characters. When plant or animal breeders want to generate a new breed or variety, they select individuals carrying the properties they desire and breed from them. If the selected individuals are better because of the genes they carry, the overall performance of the selected line will improve. In some cases, a breeder can fix the desired genes by inbreeding to make the new line homozygous (pure breeding). In the F_2 wheat grains in Figure 9.20, for example, a variety with dark red grains could be established by selecting and inbreeding only from the dark red individuals. In a more realistic case, a tomato breeder would select plants that have the desired characteristics of size, shape, growing period, response to fertilisers and resistance to fungus attack and inbreed from them. In this way, a new strain could be developed that has the desired polygenic characters.

> Quantitative characters show continuous variation and are controlled by both environment and multiple genes that each have a small additive effect on a character.

BOX 9.4 Quantitative trait loci

A gene that affects a polygenic character is termed a **quantitative trait locus** (QTL). In traditional breeding programs, the breeder selected for improvement in a particular trait by breeding from animals that exhibited the best characteristics. In so doing, the breeder was selecting for favourable alleles at quantitative trait loci. This approach to selection has been remarkably successful, generating domestic crops and animals that are vastly improved in the selected characteristics relative to the wild strains from which they were derived. However, the selection process can be made more efficient if the quantitative trait loci themselves can be identified and the desirable alleles followed directly in a breeding program. The discovery of DNA as the carrier of genetic information (Chapter 10) and the development of DNA analysis methods, which allow the molecular characterisation and manipulation of DNA sequences (Chapter 13), have led to greatly improved techniques for the mapping of quantitative trait loci and the analysis of the contribution that each locus makes to a particular quantitative trait.

Many human traits, such as asthma susceptibility, mood disorders (e.g. bipolar disorder), serum cholesterol levels, obesity, diabetes, bone mineral density, blood pressure, hypertension, epilepsies and drug sensitivities, are influenced by quantitative trait loci. It is not surprising, therefore, that the mapping and analysis of quantitative trait loci for traits such as these is an area of intense research and development. It is hoped that thorough genetic dissection of such traits and characterisation of the genes involved will lead to the development of drugs that modify these characteristics in a safer and more effective way because the drugs are tailored to the particular individual.

Understanding genetics as the study of biological information

Mendel identified discrete factors, which we now call genes, that influence particular characteristics (phenotypes). The characteristics themselves are not inherited but arise during the development of the organism. Genes provide information that determines or influences particular characteristics. Different alleles of a gene provide slightly different information that results in variation in those characteristics. This information is passed from generation to generation following the patterns of inheritance discussed in this chapter.

In summary, genes correspond to stored information that must be replicated, transmitted from cell to cell and generation to generation, and read during the development and functioning of organisms to generate the characteristics of the organism. Chapters 10 and 11 discuss the nature of these processes.

Summary

- The genotype is the genetic make-up of an organism and the phenotype is an organism's observable traits, which depend on both the genotype and the environment.
- Different inherited phenotypic traits are controlled by alternative forms of a gene (alleles). Individuals carry two alleles of each gene, which separate (segregate) into gametes. Half the gametes of an individual have one of the two alleles, half have the other. This is Mendel's Principle of Segregation.
- When gametes are formed, the segregation of alleles of one gene into gametes has no influence on the segregation of the alleles of another gene. This is Mendel's Principle of Independent Assortment.
- Phenotypes may be dominant, codominant or recessive. A dominant phenotype is expressed in a heterozygote. In codominance, there is full phenotypic expression of two different alleles in a heterozygote. A recessive phenotype is only evident in a homozygote.
- The expression of a gene can interact with the expression of other genes (epistasis) as well as the environment.
- Most genes are on autosomes, chromosomes that look the same in the two sexes. Some genes are on sex chromosomes. In organisms with an XX (female) and XY (male) sex-determining system, females have two alleles of an X-linked gene but males only have one copy. Characters determined by Y-linked genes are passed from father to son and never occur in females.
- Absence of independent assortment of pairs of alleles of different genes indicates linkage of genes on the same chromosome.
- Linked genes may be separated if crossing over occurs between them, resulting in recombinants.
- The frequency of recombination can be used to map genes on a chromosome.
- Traits that vary over a continuous scale may also be controlled by genes. Such quantitative (polygenic) traits result from the additive action of many genes (polygenes or quantitative trait loci), each with only a small effect.
- Phenotypes can be modified by a variety of factors, including the environment and interactions between genes.
- The activity of one X chromosome in female cells or of some genes on autosomes can be modified independently of their genotype in a process termed epigenetic regulation. In some cases, chromosomes or genes are imprinted so that a chromosome derived from the mother behaves differently from a chromosome derived from the father.

key terms

allele (p. 205)
autosome (p. 214)
centimorgan (cM) (p. 217)
codominant (p. 209)
dihybrid cross (p. 209)
diploid (p. 214)
dominant (p. 205)
epigenetic regulation (p. 222)
epistasis (p. 221)

expressivity (p. 221)
F_1 progeny (p. 204)
F_2 progeny (p. 204)
genes (p. 205)
genotype (p. 205)
hemizygous (p. 214)
heterozygous (p. 205)
homologous chromosomes (homologues) (p. 214)
homozygous (p. 205)

imprinting (p. 223)
incomplete dominance (p. 218)
linkage (p. 216)
locus (p. 216)
monohybrid cross (p. 205)
penetrance (p. 221)
phenotype (p. 205)
pleiotropy (p. 206)
polygenic (p. 225)

pure-breeding (p. 204)
quantitative trait locus (p. 225)
recessive (p. 205)
recombinant (p. 217)
recombination (p. 216)
sex chromosome (p. 214)
testcross (p. 210)
trait (p. 203)
wild-type (p. 207)

Review questions

1. When Siamese cats interbreed, the kittens have the Siamese pigment pattern. Kittens with this pattern can also arise from two parents each showing a non-Siamese pattern. Is the phenotype for Siamese pattern dominant or recessive? Explain your answer.

2. Polled cattle lack horns, a dominant phenotype arising as the result of the action of an autosomal gene *P*. Horned cattle are, thus, *pp* homozygotes. If a polled bull and a polled cow have a horned calf, what are the genotypes of the parents? If the same bull was mated with many horned cows, what proportions of polled and horned offspring are expected overall?

3. How are sex-linked genes inherited? Give an example.

4. If a gene occurs in three allelic forms, how many possible genotypes exist? If all alleles have codominant phenotypes, how many different phenotypes can there be? If two of the alleles have codominant phenotypes and the third a recessive phenotype, how many different phenotypes can there be?

5. A colourblind man of ABO blood group type O and a woman with normal vision and blood type AB have two children. The older child is a colourblind boy of blood type A. The younger child is a girl with normal colour vision also of blood type A. Draw a pedigree (family tree) showing the genotypes of all family members.

6. What sex should a tortoiseshell cat be?

7. What is a polygene? Do polygenes control qualitative or quantitative traits? Explain.

Extension questions

1. Design an experiment to test whether a new chemical increases the rate of mutation of the allele for the recessive phenotype green seed colour in peas to the allele for the dominant phenotype yellow seed colour. How would you measure the rate of occurrence of new mutations in the opposite direction?

2. Consider your characteristics and those of your parents. How do Mendel's two principles explain the inheritance of maternal and paternal characteristics?

3. Certain women exhibit a phenotype characterised by patches of skin that sweat normally and other patches that do not. Suggest a genetic explanation for this phenotype.

4. About one in 2500 Australians is born with cystic fibrosis, a recessive autosomally inherited genetic disease. Estimate the frequency of the cystic fibrosis allele. What is the proportion of carriers in the Australian population? What is the proportion of couples that are both carriers? What is the risk that such couples will have a child with cystic fibrosis?

Suggested further reading

Griffiths, A. J. F., Gelbart, W. M., Miller, J. H., Lewontin, R. C. (1999). *Modern Genetic Analysis*. New York: Freeman.

Hartwell, L. H., Hood, L., Goldberg, M. L. et al. (2000). *Genetics: From Genes to Genomes*. Boston: McGraw-Hill.

These are two excellent genetics textbooks.

Neel, J. V. (1994). *Physician to the Gene Pool*. New York: Wiley.

A fascinating insight into the birth and development of 20th century human genetics from one of its key contributors.

Wallace, B. (1992). *The Search for the Gene*. Ithaca: Cornell University Press.

A book aimed at the intelligent reader describing the history of inheritance and the gene from the earliest times through Mendel's discoveries to the age of recombinant DNA.

CHAPTER

10

Genes, chromosomes and DNA

The analysis of the patterns of inheritance described in Chapter 9 led to the concept of the gene, that is, an inherited unit that influences the characteristics of an organism. Genes are packets of information that control the development of particular characteristics. This information has several important properties.

1. The information must be held in a coded form because the characteristics themselves are not inherited but arise during development of the organism. The fertilised egg bears no resemblance to the mature organism, whose characteristics result from a genetically regulated program of cell division, cell death and cell differentiation during development.

2. The information must be replicated. We know this because organisms are able to produce large numbers of progeny, each of which requires this information.

3. The encoded information must be decoded, or 'read', to control cellular processes and produce the characteristics of the organism.

In this chapter we will examine the first two of these properties, the nature of the stored information and the way it is replicated.

Tracking the genetic material

Chromosome behaviour accounts for patterns of inheritance

The idea that chromosomes are directly involved in inheritance (the chromosome theory of heredity) was first formulated in 1902 by Walter Sutton, after observations of the pairing of chromosomes during meiosis (see Chapter 9). Sutton argued that genes are on chromosomes because he realised that homologue segregation during meiosis (Chapter 8) explained the patterns of inheritance of alleles observed by Mendel (Chapter 9). Later correlations between sex linkage and sex chromosomes, and between recombination (Chapter 9) and chiasmata (crossing over) between pairs of homologous chromosomes (Chapter 8), provided strong support for this theory.

Chromosome structure

Chromosomes were discovered as darkly staining bodies observed in eukaryotic cells during mitosis (Chapter 8). The structure of eukaryotic chromosomes varies greatly during a cell cycle. During interphase, individual chromosomes cannot be identified because they are decondensed, giving nuclei a diffuse unstructured appearance. During prophase, however, chromosomes condense to produce characteristic structures. Chromosomes can vary greatly in size but eukaryotic chromosomes are characteristically linear structures that

contain a centromere, which is responsible for attaching the chromosome to the mitotic spindle and moving them to the poles during mitosis (Chapter 8). The centromere can be located in the middle of the chromosome (metacentric chromosomes), at the end of the chromosome (telocentric chromosomes) or off centre (acrocentric chromosomes). The ends of the chromosomes are termed **telomeres** and have special properties, which are discussed on page 233. The compaction of mitotic chromosomes varies along the length of the chromosome so that, when stained, each chromosome has a characteristic banded appearance (Fig. 10.1). The combination of size, position of centromere and banding pattern in human chromosomes enables particular

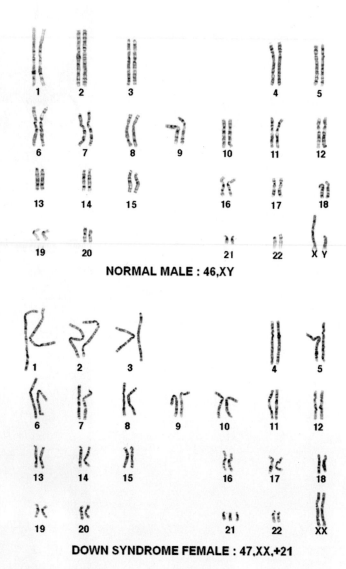

NORMAL MALE : 46,XY

DOWN SYNDROME FEMALE : 47,XX,+21

Fig. 10.1 Stained human chromosomes. The upper set shows the 46 chromosomes of a normal male consisting of 22 pairs of autosomes and the X and Y chromosomes. The lower set shows the 47 chromosomes in a Down syndrome individual. These individuals carry an additional chromosome 21

chromosomes to be uniquely identified. More recent methods using sophisticated DNA manipulation and cytogenetic methods to characterise chromosomes (Fig. 10.2) have enhanced the ability to identify chromosomes. Characterisation of chromosomes is a standard technique used in diagnosis. Some diseases exhibit dramatic changes in chromosome content in cells. For example, Down syndrome individuals carry an additional chromosome 21.

The realisation of the link between chromosomes and inheritance paved the way for the discovery of the molecular structure of genes and how genes function. In 1869, only three years after publication of Mendel's experiments, Friedrich Miescher isolated a substance, nuclein, from the nuclei of white blood cells present in the pus of wounded soldiers' bandages. Subsequent painstaking work by others led to the idea that nuclein was the material from which chromosomes were made. With this came the exciting realisation that if the precise composition of nuclein could be described, then it might be possible to explain inheritance at the molecular level. Analysis of nuclein revealed it to be primarily a mixture of DNA and proteins.

Chromosomes consist of protein and DNA and are assembled into nucleosomes

Eukaryotic chromosomes are made up of roughly twice as much protein as DNA. Together, they are referred to as chromatin. As we saw in Chapter 3, DNA is condensed around histones. Histones are small basic proteins (11 kD) with a high concentration of basic residues, such as lysine and arginine, which interact with the negatively charged sugar-phosphate backbone of DNA. Histones are abundant in the nucleus. They assemble in groups of eight to form a core upon which the DNA is bound. Double-stranded DNA (Chapter 1), of 146 base pairs (bp) in length, is coiled slightly less than twice around each histone core, forming a **nucleosome core particle** (Fig. 10.3). Nucleosome core particles are linked by sequences of DNA of variable length, up to 80 bp. The nucleosome core particle together with the linking DNA is referred to as a **nucleosome**.

During condensation, the nucleosomes themselves fold into regular, higher order structures, ultimately giving rise to the condensed mitotic chromosomes observed during mitosis (Fig. 10.4).

Genetic information is stored in DNA and not protein

For a long time DNA was regarded as too simple a molecule to convey the information of inheritance. DNA, composed of just four different chemical units termed bases (Chapter 1), was thought to act as some form of scaffold or support for proteins, which were believed to be the 'real' information carriers. Proteins were known to be made up of strings of 20 different amino acids. A string of only seven amino acids, for example, has over one billion possible sequence variants. As proteins can consist of strings of thousands of amino acids, they provide a potentially immense information storage capacity. However, in the biological

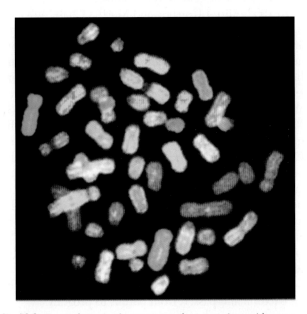

Fig. 10.2 Human mitotic chromosomes from a patient with acute promyelocytic leukemia (APL). A technique termed chromosome painting labels different chromosomes with different colours. This reveals rearrangements in chromosomes. Rearranged chromosomes are frequently observed in human cancers. Chromosomes exhibiting more than one colour have undergone some form of rearrangement

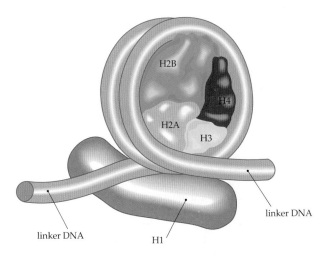

Fig. 10.3 Model of a nucleosome particle. DNA is wound around a histone octamer. This particle has been crystallised and its structure examined using X-ray diffraction

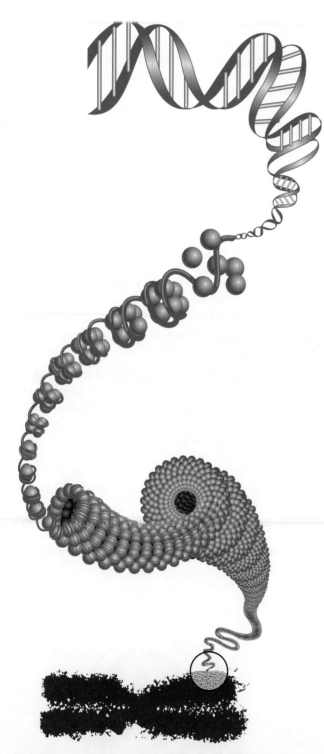

Fig. 10.4 A diagrammatic representation of a condensed chromosome in metaphase unravelled to reveal the way that DNA is wrapped within the chromosome. The diagram shows the highly condensed chromosome being unravelled to reveal higher order nucleosome packing, which unravels to individual nucleosomes and finally DNA

sciences, the most logical answer to a problem often proves to be wrong!

In 1944, Oswald Avery, Colin MacLeod and Maclyn McCarty performed an experiment that demonstrated

that the DNA component, and not protein, carried the genetic information. They built on an earlier observation about two different forms of the same *Streptococcus pneumoniae* bacterial strain, one of which formed capsules and was virulent (S, for smooth, strain) while the other was non-encapsulated and non-virulent (R, for rough, strain). Griffith had shown that mice injected with the non-virulent R strain mixed with heat-killed virulent S-strain material died (Fig. 10.5). The heat-killed S-strain material was able to 'transform' the R strain to the virulent form. Injection of either the R strain or the heat-killed bacteria alone was not sufficient to kill the mouse. The phenotype of the R strain bacteria was being altered by a component of the heat-killed bacteria. Furthermore, the phenotype was inherited by the progeny of the transformed bacteria, indicating that the genotype of the bacteria had been altered. The heat-killed extracts were providing a source of genetic information (a gene) that changed the phenotype of the R strain to that of an S strain.

Avery, MacLeod and McCarty used this observation to identify the chemical nature of this genetic information. They prepared protein-free, purified DNA from capsule-forming (S) bacteria after they had been killed by heat. Non-capsule forming R-strain *S. pneumoniae* were exposed to the DNA extracts and a proportion were found to be transformed into the capsule-forming virulent type. These changes were permanent and inherited by the daughter cells. This carefully designed experiment demonstrated that DNA was the carrier of genetic information.

In 1952, Alfred Hershey and Martha Chase provided further evidence by radioactively labelling a bacterial virus (bacteriophage). Bacteriophages produce more bacteriophages by using their own genes to reprogram the bacteria. DNA and protein in bacteriophages were selectively labelled using phosphorus-32 (^{32}P), which labels the phosphate backbone in DNA, and sulfur-35 (^{35}S), which selectively labels protein because of the presence of the sulfur-containing residue, cysteine (Chapter 1). When they infected bacteria with the bacteriophage and then disrupted the bacteriophage–bacteria mix using a blender, it was found that most of the ^{32}P incorporated into DNA remained with the bacteria, while most of the ^{35}S did not (Fig. 10.6). In addition, over 30% of the ^{32}P but less than 1% of the ^{35}S could be found in the bacteriophage progeny. This experiment demonstrated that the DNA component of the bacteriophage, and not the protein, is taken up by the bacteria and passed from parent to progeny.

The structure of DNA explains its capacity to store biological information

The double helix model of DNA, proposed by James Watson and Francis Crick in 1953, based substantially

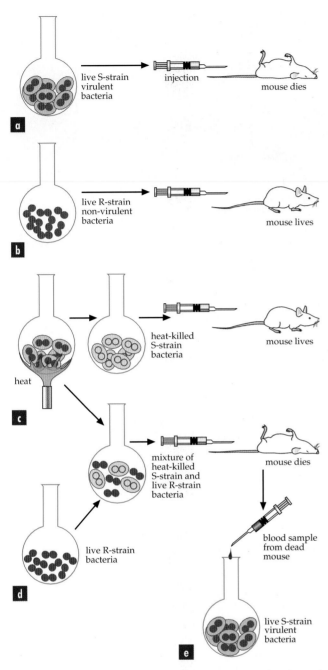

Fig. 10.5 Heat-killed material can 'transform' a non-virulent form of *Streptococcus pneumoniae* into a virulent form. Avery, MacLeod and McCarty purified the DNA component of the heat-killed material away from the protein component and showed that it maintained the ability to transform the non-virulent form, indicating that DNA held the genetic information

on X-ray crystallographic analysis by Rosalind Franklin, has been looked at in detail in Chapter 1 but will be summarised again here. DNA is composed of two strands, which are wound around each other to form a helix. Each strand is composed of a string of the building blocks, nucleotides. A nucleotide consists of a

deoxyribose sugar, a phosphate group and one of four nitrogen-containing bases: the purines, adenine (A) and guanine (G), and the pyrimidines, thymine (T) and cytosine (C) (Fig. 10.7; Chapter 1). Each strand consists of a sugar-phosphate backbone within which each deoxyribose sugar is connected via a phosphodiester linkage to the next sugar. The ends of each DNA strand are distinct, with a free phosphate at one end (the 5′ end) and a free hydroxyl at the other (the 3′ end). The two strands in the double helix molecule are arranged in an antiparallel fashion (i.e. the 5′ end of one strand is at the same end of the molecule as the 3′ end of the other strand). The two strands are held together by hydrogen bonds between the base pairs: A–T and G–C.

Despite consisting of just four different nucleotides, the potential number of different sequences of these four nucleotides in any stretch of DNA allows for an enormous amount of information to be stored. A length of DNA just three nucleotides long can be arranged in 4^3 different ways, giving rise to 64 different base pair sequences. Just 10 nucleotides can give 4^{10}, or over one million different combinations, and the human haploid genome consists of approximately three billion nucleotides! Given that an average gene is hundreds to thousands of nucleotides long, the potential number of different gene sequences is phenomenally large. The four nucleotides of DNA can provide all the information an organism needs for every aspect of its survival. Differences in the DNA sequence of different organisms generate the remarkable biological diversity that we observe in nature.

The integrity of DNA sequences relies on the fact that base pairing is not random but follows the strict order: T always pairs with A and G pairs with C. Two strands that can form a double-stranded helix (or **duplex**) are said to be **complementary**. The complementarity of base pairs in DNA suggested a possible mechanism for the replication of DNA. Because of the formation of base pairing between complementary nucleotides, one strand of DNA has the potential to specify the sequence of the other automatically. This arrangement allows for many copies of each DNA strand to be made without losing any information.

Eukaryotic chromosomes are made up of protein (principally histones) and DNA. The protein and DNA assemble into small structural units termed nucleosomes. Despite the simple structure, linear arrays of the four bases of DNA provide enough variation to store all the information needed for a biological organism to exist. The double-stranded nature of DNA allows it to be copied without any loss of the information stored as the nucleotide sequence.

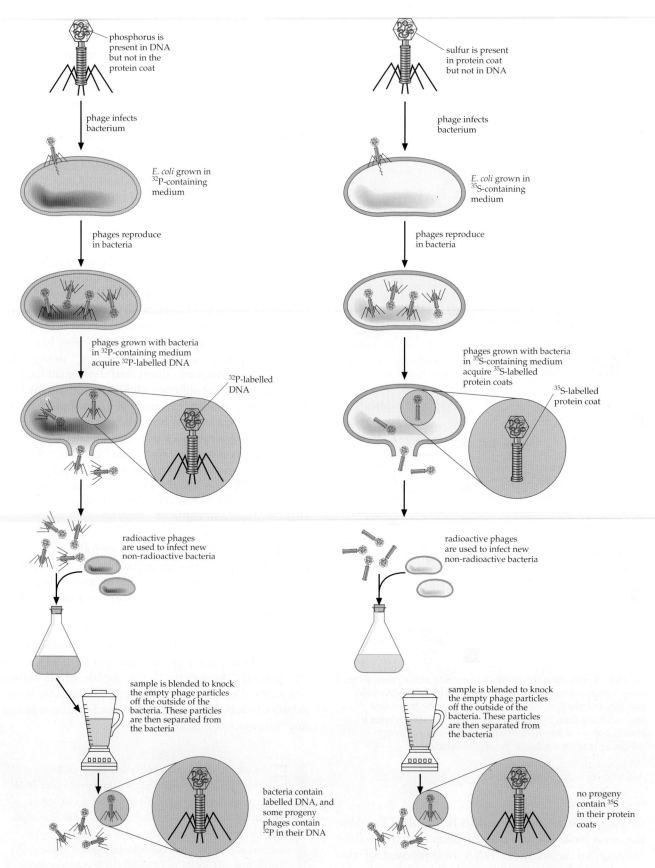

Fig. 10.6 Radioactive labelling of DNA with ^{32}P or protein with ^{35}S was used by Hershey and Chase to show that the DNA and not the protein of a bacteriophage entered the cell upon infection and could be found in progeny bacteriophage, supporting the idea that DNA was the genetic material

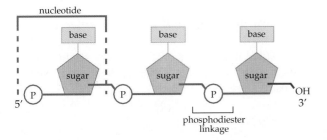

Fig. 10.7 Schematic outline of the molecular structure of DNA. One nucleotide consists of a sugar and phosphate (P) backbone with an attached base. A free phosphate is found at the 5′ end and a free hydroxyl (OH) at the 3′ end. A phosphodiester linkage connects adjacent sugar residues to one another

BOX 10.1 DNA, CD-ROMs and information storage

Ask any computer scientist when the first stored information program was developed and they will usually miss the mark by perhaps three billion years. DNA stores information that is used to program development and generate the characteristics of biological organisms. This information also programs the response of the organism to a variety of situations that might arise, such as the presence or absence of particular food sources (see Chapter 11).

The storage of biological information has some similarities to the storage of computer information. Both use strings of alternative states. For DNA, these are chemical alternatives along the rungs of the DNA double helix ladder. For stored computer information (e.g. on a CD-ROM), they are physical alternatives. Each step in DNA contains one of four possibilities: A, G, C or T. Each bit of information on a CD-ROM contains one of two possibilities, denoted as 0 or 1. The human haploid genome consists altogether of 3×10^9 sequential nucleotide pairs, any of which can be one of four different possibilities. The number of possible combinations of sequences in this length of DNA is, therefore, $4^{3\,000\,000\,000}$. By comparison, a standard CD-ROM holds 650 megabytes, each byte being eight bits of information (i.e. there are 5.2×10^9 bits of information), each having a 0 or 1 state. The number of possible combinations on a CD-ROM is therefore $2^{5\,200\,000\,000}$ or, to relate it to the DNA combinations given above, $4^{2\,600\,000\,000}$. You can see that either form of storage can hold vast amounts of information, yet one must marvel at the fact that the information in human DNA, which could be held in a little over one CD-ROM, can engineer the production of the extraordinarily complex and sophisticated organism that is a living human.

DNA synthesis: an overview

The DNA content in a cell is doubled by separating the two strands of the helix and using each strand as the **template** to synthesise a new complementary strand (see Box 10.2). This process is termed **semi-conservative replication**, that is, each of the original DNA strands remains unchanged but is paired with a newly synthesised strand (Fig. 10.8, see Box 10.2). Thus, one double helix is replicated to form two identical double helices for eventual distribution to the two progeny cells during mitosis or meiosis.

BOX 10.2 Discovering the nature of DNA replication

The double helix model of DNA immediately suggested that DNA replication could occur by the synthesis of new nucleotide strands opposite the old strands using the complementarity of base pairs to create a new DNA molecule exactly the same as the old one. However, replication could occur in one of several ways, illustrated in Figure a. In the first model, termed the conservative model, the original DNA molecule remains unaltered, having acted as a template for the two new strands which form the new DNA molecule. In the second model, termed the semi-conservative model, the original two strands are separated and new strands synthesised complementary to each. This would create two new DNA molecules, each containing one original template strand and one newly synthesised strand. In the third, termed the dispersive model, different parts of either strand are used to synthesise the new strand, so that following synthesis the original DNA and newly synthesised DNA are interspersed along each strand.

Matthew Meselson and Franklin Stahl devised an ingenious experiment that discriminated between these possibilities. They utilised the fact that DNA containing the heavy nitrogen isotope, ^{15}N, was more dense than DNA containing the normal ^{14}N. Bacteria were grown in ^{15}N for several generations so that all of the DNA strands were more dense than normal. The bacterial DNA was then allowed to replicate once or twice in normal ^{14}N (see Fig. b), so that there were one or two rounds of new strand synthesis. As Figure b shows, the density of DNA after one round of replication was intermediate between the heavy and normal DNA, indicating that the original heavy DNA was no longer present, but had been replicated in either a semi-conservative or dispersive way, ruling out the conservative replication hypothesis. Following the second round of replication in normal ^{14}N, DNA was found to have two densities, either intermediate

between the heavy and normal, or the same as normal DNA, that is, about half of the molecules contained none of the original heavy strands. This result confirmed that DNA replication is semi-conservative.

a

b

Growth in either ^{14}N or ^{15}N produces DNA of different densities, which can be separated by centrifugation in caesium chloride. The three models of DNA replication, conservative, semi-conservative and dispersive, all predict different outcomes for the density of 'heavy' ^{15}N DNA grown for two generations in ^{14}N. Only the semi-conservative replication model accounts for the appearance of a single band of intermediate density DNA in the first generation and bands of intermediate density and light (^{14}N) density in the second generation

BOX 10.3 Terms associated with DNA replication and repair

DNA ligase
Enzyme that joins the 3′ hydroxyl and 5′ phosphate ends of breaks in DNA strands by catalysing the reformation of a phosphodiester linkage.

DNA polymerase
Enzyme catalysing the addition of nucleotides to a growing strand of DNA in a 5′ to 3′ direction.

Endonuclease
Enzyme that cuts DNA internally by hydrolysing phosphodiester bonds.

Exonuclease
Enzyme that cuts DNA by removing bases sequentially from the ends of DNA strands by hydrolysing terminal phosphodiester bonds.

Leading strand
DNA strand that is synthesised continuously in a 5′ to 3′ direction.

Lagging strand
DNA strand that grows in an overall 3′ to 5′ direction, but is synthesised discontinuously in short fragments (5′ to 3′) that are later joined by DNA ligase.

Origin and terminus
Sequences of DNA at which replication is initiated and terminated, respectively.

Primer
Short sequence, usually of RNA, that pairs with a strand of DNA and provides a starting point (free 3′ OH end) for DNA polymerase to begin synthesis of the new DNA chain.

Replication fork
Point of separation of strands of duplex DNA where replication occurs.

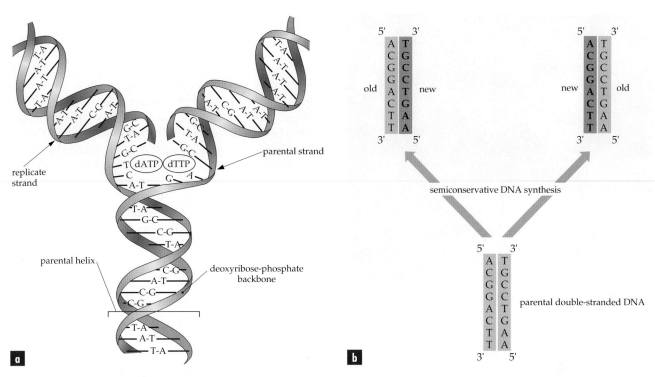

Fig. 10.8 (a) In semi-conservative replication, double-stranded DNA separates, each single strand acting as a template for the synthesis of a new strand. In this three-dimensional diagram, the lower portion is still in a duplex and is base paired. The upper portion is unwound and new DNA strands are associated with each parental strand by means of base pairing. **(b)** Sequence-based representation of replicating DNA, showing the 5′ and 3′ polarities of strands

As described in Chapter 1, DNA strands consist of covalently linked nucleotides. Each DNA chain contains a free 5′ phosphate group at one end and a free 3′ hydroxyl group at the other, giving each strand a polarity that is referred to, by convention, as 5′ to 3′. The strands of a double-stranded DNA molecule run antiparallel to each other. DNA synthesis occurs by the sequential addition of nucleotides from energy-charged building blocks. These building blocks are the **deoxyribonucleoside triphosphates** (dNTPs): dATP, dCTP, dGTP and dTTP. Synthesis of DNA is always in a 5′ to 3′ direction relative to the strand being synthesised. Deoxyribonucleotides are added to the 3′ end of the DNA strand via the formation of an ester linkage between the 5′ phosphate of the incoming dNTP and the 3′ hydroxyl (OH) of the end of the strand. This forms a phosphodiester bond between the adjacent sugars (deoxyribose), extends the sugar-phosphate backbone and leaves the 3′ OH group of the newly added deoxyribonucleotide free for the addition of the next deoxyribonucleotide. In this process, an enzyme splits off two of the three phosphates from the dNTP in a covalently linked form known as pyrophosphate. The release of the pyrophosphate provides energy to drive the reaction. Importantly, nucleotides are incorporated into the growing chain according to the rules of base pairing, so that the base of the new strand is complementary to the base in the template:

adenine opposite thymine and guanine opposite cytosine (Fig. 10.8a).

> Replication begins when the double-stranded DNA molecule unwinds, separates and undergoes semi-conservative duplication by the sequential addition of complementary nucleotides to the 3′ end of a DNA strand. Replication leads to the formation of two complete copies of the genome for eventual distribution to the two progeny cells.

DNA replication in prokaryotes

A bacterial chromosome, being simpler, has provided a model system that has helped our understanding of DNA replication in general. The circular chromosome of a prokaryote is about 1 mm in circumference when unwound. In the cell, it is compressed about a thousand-fold with the aid of special compression molecules, such as folding proteins and RNA, into the **nucleoid** (Fig. 10.9). Although highly condensed, this tiny genomic structure is in direct contact with the cytosol and is available for replication, transcription and repair (Chapter 33).

> The nucleoid contains the circular DNA molecule of prokaryotes. It is compressed with the aid of folding proteins and RNA and is located in the cytosol.

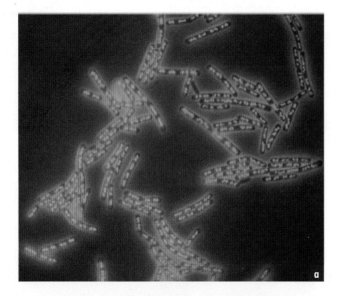

Fig. 10.9 (a) Fluorescence micrograph showing nucleoids containing DNA (blue) in prokaryotic cells (*Bacillus subtilis*). **(b)** Electron micrograph showing a partly unravelled nucleoid from the prokaryote *Bacillus licheniformis*

Initiation of replication

Bacterial chromosomes have a single **origin** of replication. Replication begins at the origin, where unwinding takes place simultaneously in both directions. The precise DNA sequences of origins from many bacteria are known and contain recognition sites for specific DNA-binding proteins, which are part of the initiation machinery. Initiation of replication at the origin is the stage where replication is controlled.

Replication is initiated by a cut (nick, i.e. a single break of a bond between adjacent nucleotides) in at least one strand. The cut is made by a specific enzyme, DNA gyrase. As the DNA unwinds, a **replication fork** is generated at each side of the origin by the initiation machinery. New strand assembly takes place at these replication forks as the double helix unwinds (Figs 10.10, 10.11). It is impractical within a cell for the entire length of the duplex DNA molecule of a chromosome to be unravelled at once, so the DNA is made single-stranded in small sections, while the rest of the chromosome remains double-stranded. These small sections of single-strandedness are where replication occurs, and involve the duplex 'peeling apart' or 'melting' into single strands immediately ahead of the replication fork. Replication forks move at an essentially constant rate.

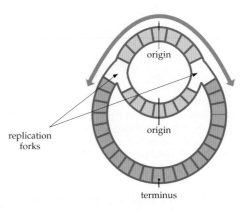

Fig. 10.10 DNA synthesis in circular chromosomes. Synthesis progresses by means of replication forks in both directions from the single origin around the circle until the growing forks meet at the terminus

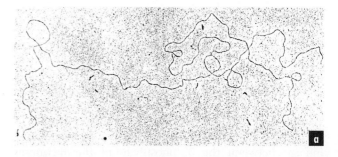

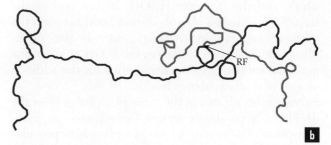

Fig. 10.11 (a) Electron micrograph showing a replication fork in a DNA preparation from the prokaryote *Bacillus subtilis*. **(b)** The parental DNA in **(a)** is shown as a blue line, the post-replicative DNA as red lines. RF indicates the replication fork

BOX 10.4 Building blocks for DNA synthesis

Deoxyribonucleoside triphosphates (dNTPs) are the 'building blocks', or precursors, for the biosynthesis of DNA, which is a polymer of deoxyribonucleotide units (i.e. deoxyribonucleoside monophosphates, usually abbreviated to nucleotides) containing four types of bases, two pyrimidine and two purine (Chapter 1). Because of the complementarity of base pairing, pyrimidines (dCTP and dTTP) and purines (dATP and dGTP) must be synthesised in the cell in equal amounts. The synthesis of purine and pyrimidine nucleotides is controlled by feedback mechanisms to ensure balanced levels of production. Both biosynthetic pathways utilise simple precursor molecules for the assembly of dNTPs.

Some drugs, such as mercaptopurine, are incorporated into DNA in place of the natural dNTPs and disrupt the genetic code by pairing with the 'wrong' nucleotides. Such changes result in mutations that can lead to cancer formation (Chapter 12).

Each replication fork contains a **leading strand** growing towards the fork and a **lagging strand** growing away from the fork (Fig. 10.12). The two forks generated at the origin proceed bidirectionally (in opposite directions) around the circular bacterial chromosome and eventually reach the **terminus**, where their movement is blocked (Fig. 10.10). Like the origin, the terminus of bacterial chromosomes has a specific sequence for the binding of terminator proteins.

By controlling replication at initiation, the cell controls replication with a minimum expenditure of energy. The initiation machinery assembles and allows the creation of the two replication forks only when it is appropriate to start a round of replication.

Replication begins at the origin. Unwinding of the double helix takes place simultaneously in both directions, forming two replication forks where new nucleotides are added until the terminus is reached.

Enzymes needed for replication

DNA synthesis at the replication fork involves the co-ordinated participation of many enzymes. Separation of the duplex is achieved by ATP-driven enzymes, DNA helicases, which peel apart the base-paired strands by breaking the hydrogen bonds between them (Fig. 10.12). Strands are kept apart and stabilised by single-strand binding proteins, which bind to single-stranded DNA, preventing the strands from rewinding and ensuring that templates are available for the synthesis of new complementary strands.

DNA normally occurs as a right-handed helix, with 10 base pairs per turn of the helix (Chapter 1). Double-stranded DNA is usually supercoiled, meaning that the double-stranded segments are further twisted around each other. The enzyme DNA gyrase makes a transient cut in one strand of the DNA, allowing the supercoils and twists to be unwound before it accurately rejoins the phosphodiester bond it initially cut (Fig. 10.12).

At the replication fork, dNTPs are added to the growing chain by **DNA polymerases**, enzymes with the ability to add nucleotides to the 3′ OH group of DNA and to remove incorrect bases with its own **exonuclease** activity (see Box 10.3). In *Escherichia coli*, DNA polymerase III is the major replication enzyme, attaching bases to DNA in a 5′ to 3′ orientation.

In a remarkably elegant process, DNA polymerase I, which has more than one type of enzyme activity, checks the added base and 'edits' any incorrect additions by means of its own 3′ to 5′ exonuclease activity. This editing removes any incorrect bases together with their attached sugar and phosphate, which diffuse away. DNA polymerase I then replaces the base with the correct one. DNA polymerase I differs from DNA polymerase III in functioning to remove primers and fill gaps accurately (Table 10.1).

Combined with the careful process of assembly, the editing process ensures that mistakes occur in less than about one in 10^9 cases, a remarkably low error rate. Thus, an entire bacterial chromosome of several millions of base pairs can be faithfully copied with only a small chance of insertion of an incorrect base.

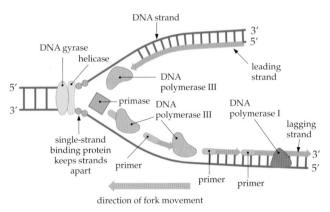

Fig. 10.12 The replication fork of *Escherichia coli*, including the various enzymes and accessory proteins that function in DNA synthesis. The enzymes DNA gyrase and DNA helicase carry out DNA unwinding, single-strand binding protein stabilises single-stranded DNA and primase synthesises the RNA primer. DNA ligase, required for ligation of the Okazaki fragments on the lagging strand, is not shown

Table 10.1 Properties of DNA polymerases of the bacterium *E. coli*

Property	DNA polymerase I	DNA polymerase III
Polymerisation 5′ to 3′	+	+
Exonuclease activity:		
3′ to 5′	+	+
5′ to 3′	+	–
Synthesis from:		
Intact DNA	–	–
Primed single strands	+	–
Primed single strands plus single-stranded binding proteins	+	+
In vitro chain elongation rate (nucleotides/min)	600	30 000
Molecules present per cell	400	10–20

Nucleotides are attached to a growing chain of DNA by DNA polymerases. These enzymes have different forms and can both add nucleotides and edit the DNA sequence.

DNA synthesis at the replication fork

DNA polymerase cannot start synthesis on a single-stranded template, rather, it can only add to a free 3′ OH group at the end of a strand. To initiate synthesis, then, there is a need for a **primer**, a starting tag. A primer is a short RNA sequence synthesised by a special RNA polymerase (see Chapter 11) termed **primase**. Unlike DNA polymerase, primase does not require a primer to synthesise RNA. The RNA primer created by the primase then allows DNA polymerase to commence synthesis from the 3′ end of the primer (Fig. 10.13). When it has finished its task, the primer is removed by an editing 5′ to 3′ exonuclease.

Each replication fork moves away from the origin of replication as it unwinds and 'melts' the DNA strands. As DNA synthesis occurs only in a 5′ to 3′ direction, DNA replication must occur in opposite directions on the two strands at each replication fork. One strand, termed the *leading strand*, grows towards the fork, while the other, termed the *lagging strand*, grows away from the fork. For the leading strand, in which the 5′ to 3′ synthesis occurs in the same direction as the movement of the replication fork, a single primer initiates synthesis at the replication origin and then DNA polymerase continues to add nucleotides sequentially, keeping pace with the replication fork.

The situation for the lagging strand is more complicated, as the 5′ to 3′ synthesis can only occur in the other direction, away from the fork. Yet, overall DNA synthesis on this strand, too, must be in the

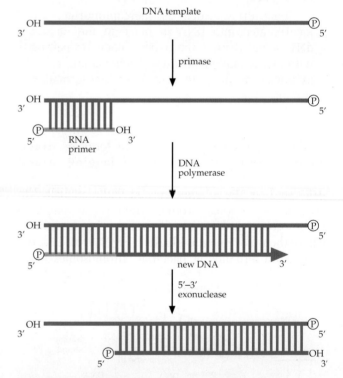

Fig. 10.13 Initiation of DNA synthesis. The enzyme, primase, synthesises a short complementary stretch of RNA. This RNA is the primer for synthesis of new DNA. The RNA part of the new chain is removed by a 5′ to 3′ exonuclease, leaving a gap, which is filled by the action of DNA polymerase I and DNA ligase. Synthesis of the lagging strand involves repeated priming and extension as the replication fork moves in the opposite direction

direction of the replication fork. This problem is solved by synthesis of the lagging strand as a series of short fragments, each of which initiates close to the replication fork and proceeds away from the fork until it meets the previous fragment. The result is that, in contrast to the leading strand, lagging strand synthesis is discontinuous, with primase laying down a series of primers as the replication fork progresses (Fig. 10.14). DNA polymerase III adds nucleotides to the 5′ end of the primers to lengthen them away from the fork until the next fragment is encountered, creating an approximately 1 kilobase (kb) piece known as an **Okazaki fragment**, named after its discoverer, Reiji Okazaki. When DNA polymerase III arrives at the 5′ end of a previously synthesised Okazaki fragment, it stops and departs. The process then restarts at the site of the next primer, which has been synthesised close to the replication fork. Synthesis on the lagging strand therefore occurs in short bursts.

The leading strand now consists of one unbroken chain, whereas the lagging strand is punctuated by regular breaks and attached short RNA primers. DNA polymerase I enters at this point and, by the action of its own 5′ to 3′ exonuclease, erases each primer. DNA polymerase I is an unusual enzyme because it then uses its polymerising ability to replace an erased primer with DNA. It is possible to do this because the 3′ OH end of the adjacent Okazaki fragment can be readily extended into the gap created by removal of the primer. DNA polymerase I also uses its other 3′ to 5′ exonuclease activity but only to check that it has attached the correct bases. The result is that DNA (produced by the editing enzyme itself and therefore with no errors in base sequence) has replaced the primer.

The last event in completing this portion of lagging strand synthesis is to join adjacent DNA fragments. This **ligation** (joining) is performed by **DNA ligase** and the result is a continuous sugar-phosphate backbone of DNA (Fig. 10.14e).

Before reaching the terminus, the partially replicated bacterial chromosome looks somewhat like the Greek letter theta, θ, because it is a circle that now contains two replication forks (Fig. 10.10). It is known as a **theta structure**. One of the convincing demonstrations of a theta structure was obtained with the bacterium *Bacillus subtilis* (Fig. 10.15). The terminus in this bacterium stops the clockwise replication fork before the arrival of the other fork. Termination is achieved by the binding of a small protein to a precise DNA sequence in this region, which acts to prevent movement of the replication fork, ensuring that termination and fusion of the two forks occurs in an ordered manner.

DNA polymerase requires an RNA primer (starting tag) made by an RNA polymerase (primase) before DNA synthesis commences. Two assembly mechanisms operate—continuous synthesis of the leading strand and discontinuous synthesis of the lagging strand.

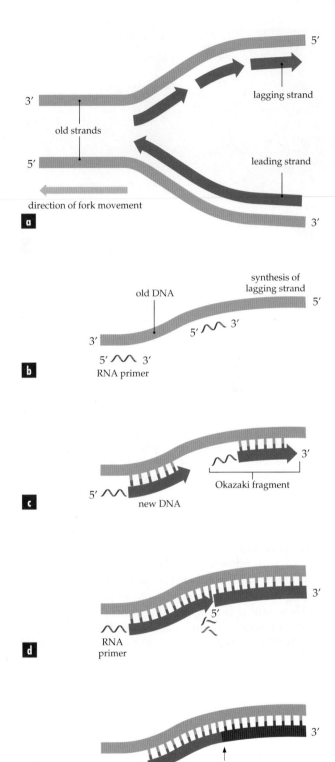

Fig. 10.14 (a) Organisation of the various elements in DNA synthesis in the leading and lagging strand at a replication fork. Catalysed by DNA polymerase III, synthesis of the leading strand depends on sequential addition of deoxyribonucleotides. The discontinuous lagging strand has several steps in its DNA synthesis: **(b)** Short stretches of RNA, copied from the DNA, act as a primer; **(c)** DNA polymerase III elongates RNA primers with newly synthesised DNA; **(d)** DNA polymerase I replaces the 5′ RNA primer with DNA; and **(e)** adjacent fragments are linked by DNA ligase

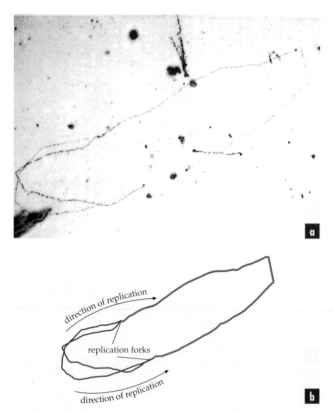

Fig. 10.15 **(a)** Replicating circular chromosome of *Bacillus subtilis*, showing a theta structure. **(b)** Interpretation of the DNA shown in **(a)**. Blue represents non-replicated DNA and red the newly replicated DNA

Replication in eukaryotes

In eukaryotes, chromosomes have many origins of replication (Fig. 10.16). Replication involves the formation of two replication forks at each origin. These replication forks are bidirectional, with leading and lagging strands, in the same way as in prokaryotes.

There are a number of differences in replication between prokaryotes and eukaryotes. Eukaryotes have smaller Okazaki fragments of 100 to 200 bases, about a tenth of the size of those in most prokaryotes. In human cells, replication of the leading and lagging strands is performed by different DNA polymerases (DNA polymerase α for the lagging strand, and DNA polymerase δ for the leading strand).

The size of the DNA molecule, and the fact that a replication fork in eukaryotes moves about 20 times more slowly than in prokaryotes, are major constraints in the replication of eukaryotic DNA. In the hypothetical case of a eukaryotic DNA molecule that is 30 times the size of the *E. coli* (bacterial) chromosome, replication would take more than two weeks if it occurred bidirectionally from a single origin. These difficulties are overcome by having multiple origins of replication along each chromosome. Each chromosome, thus, has many **replicons**, or unit regions of replication (Fig. 10.16). In contrast, a prokaryote chromosome has a single replicon (Fig. 10.10).

During the S phase (Chapter 8), each origin is used only once. As replication forks extend bidirectionally from an origin, they continue to move until they eventually join with replication forks from adjacent replicons, or until a replication fork approaches the end of the linear chromosome.

Replication of the genome of the vinegar fly, *Drosophila melanogaster* (165 Mb), takes about 7 minutes in early embryogenesis. There are thousands of origins, spaced about 10 kb apart (Fig. 10.17). In contrast, later in development, or in tissue culture where cells are grown *in vitro* in the laboratory, replication of the same genome takes hours rather than minutes because many fewer origins of replication are used. Replication of a mammalian genome *in vivo* typically takes about 8 hours.

Not all eukaryotic replicons are initiated at the same time. Groups of approximately 50 replicons appear to commence independently of other groups, so that a chromosome can be broadly divided into early, mid or late replicating regions. The time of replication appears

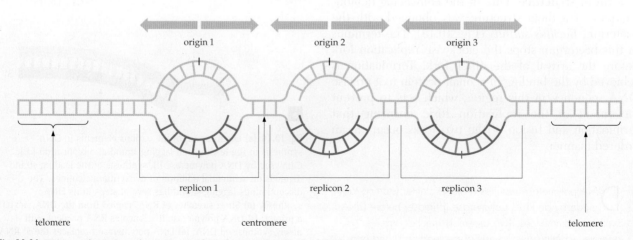

Fig. 10.16 DNA synthesis in a chromosome of a eukaryote. Synthesis of DNA involves many origins of replication. Each origin is a separate replicon and DNA synthesis is bidirectional from the origin

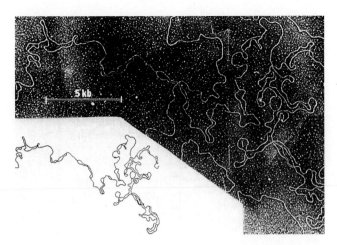

Fig. 10.17 Replication of *Drosophila melanogaster* chromosomal DNA. Many replication bubbles are present, due to multiple initiation sites

to be at least partly due to whether the region contains active genes or not. Active regions containing frequently transcribed genes (Chapter 11), such as those involved in the production of enzymes required for cell viability, tend to replicate early in the S phase. In contrast, inactive regions tend to replicate late in the S phase.

It is difficult to study replication processes when working with multicellular eukaryotes. For this reason, significant advances are being achieved by studying mutants of unicellular eukaryotes, such as the yeast *Saccharomyces cerevisiae*. Origin-binding proteins have been identified and this should lead to an understanding of the mechanism of initiation of replication. In mammals, clues are often obtained by observing events that go wrong rather than those that proceed smoothly (Box 10.5).

Eukaryotic chromosomes have multiple origins of replication, from which replication is bidirectional. This speeds up total replication in large chromosomes.

BOX 10.5 Replication and cancer therapy

In certain forms of cancer, chemotherapy using the drug methotrexate leads to the selection of a subpopulation of cells that are resistant to high concentrations of the drug. Methotrexate interferes with the action of an important enzyme involved in folate (vitamin B) metabolism.

Coenzymes derived from folic acid are essential for formation of adenine and thymine, which are required in DNA synthesis. More

rapidly growing cancer cells have a greater need for nucleotides and die when starved accordingly.

Cancer cells can acquire resistance to methotrexate as a result of a mutation leading to a change in the structure of the enzyme, or by selective amplification of the gene encoding the enzyme, thereby generating many copies of the gene and increasing the intracellular concentration of the enzyme (see figure). In the latter case, up to several hundred copies of the gene can be generated. This can happen because there is an origin of replication near the locus of this gene, which is repeatedly used in the absence of chromosomal replication elsewhere. It is the selective (and aberrant) use of this origin that causes the eventual accumulation of the enzyme and resulting resistance of these cells to methotrexate, rendering this form of chemotherapy useless.

Selective amplification of the chromosomal region containing the dihydrofolate reductase gene (bottom). The normal chromosome (top) lacks the expanded long arm

Telomeres during replication

The ends of eukaryotic chromosomes have sequences known as telomeres. Completion of replication at each end of linear eukaryotic chromosomes presents a special problem. Since DNA polymerases assemble DNA only in the 5′ to 3′ direction and need a primer, they cannot complete replication to the 5′ ends of strands because, after removal of the primer, there is nothing for the DNA polymerase to extend from. You can imagine that, after many rounds of cell division and hence replication, chromosomes would become progressively shorter because of the gap left after removal of the primer at the 5′ end of the leading strand (Fig. 10.18). DNA polymerase cannot fill the gap because there is no 3′ OH group present as a starting point.

This problem is overcome by the presence of the telomere, a terminal sequence consisting of repeated DNA sequences. For example, in the unicellular ciliate *Tetrahymena thermophila*, there are many repeats of TTGGGG, and in humans there are many repeats of TTAGGG, up to 10–15 kb long. These repeats are added to the ends of chromosomes by a special polymerase, termed **telomerase**. An RNA molecule within the telomerase complex provides the template to fill the gap and add more telomere sequences.

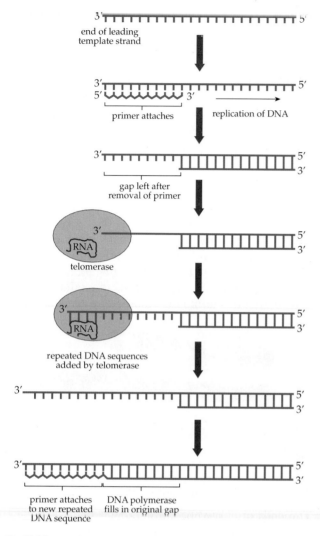

Fig. 10.18 Completion of replication at ends (telomeres) of eukaryotic chromosomes. A gap is left after removal of the primer at the 5′ end of the leading chain of DNA. Because of the absence of the 3′ OH group, DNA polymerase cannot fill the gap; instead, telomerases, using an RNA molecule as a template, add repeated sequences to fill the gap

Histones during replication

Origins of replication lack associated histones. This presumably makes it easier for initiation proteins to recognise these regions and thereby facilitate replication. One type of histone, H1, located outside of the nucleosome core particle, serves to attach the histone octamer firmly to the DNA (Fig. 10.3). After replication forks have been formed, an important part of eukaryotic replication is the unravelling of the chromatin by modifying histone H1 and disrupting interactions between nucleosomes. Histones bind more strongly to double-stranded than to single-stranded DNA. Because of this, the original histones associate with the leading strand of the replication fork, which remains substantially double-stranded during replication and so can associate with histones (Fig. 10.19).

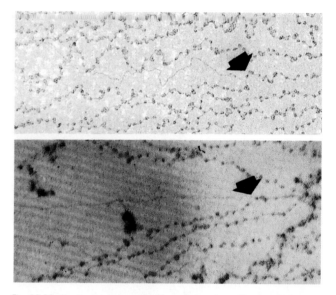

Fig. 10.19 Electron micrograph of a eukaryotic replication fork after cells were treated with cycloheximide to halt synthesis of new histones. Long strands are studded with beads (presumably old histones). One of the arms of the replication fork (arrow) has not been able to add any histones under these conditions, suggesting that old histones attach preferentially to one of the two arms

BOX 10.6 Telomerases and cellular ageing

Telomerases are required to overcome the inability of DNA polymerases to synthesise the very 5′ end of a chromosome. Curiously, although the germ cells of mammals contain telomerase, the somatic cells do not. The consequence is a progressive shortening of telomere sequences as the somatic cells proceed through cellular divisions. In cultured cells, the shortening eventually results in the loss of essential sequences that were adjacent to the telomeric repeats, causing the cells to die after a certain number of cell divisions. The shortening of telomeres from somatic cells during replication may contribute to the ageing process. It was recently shown that if telomerase expression can be induced in cultured mammalian cells, the cells no longer die as they normally would but continue to proliferate indefinitely. The telomerase has 'immortalised' these cells.

The recent cloning of the sheep, Dolly, from a somatic cell raised a lot of interest and controversy. As the somatic cell from which Dolly was generated should have had shortened telomeres, Dolly might be expected to have shorter telomeres than normal sheep at the same age. A recent study showed that, indeed, Dolly's telomeres were shorter than normal.

In contrast, the discontinuously synthesised lagging strand contains many single-stranded regions before it, too, becomes associated with nucleosomes.

The ends of chromosomes contain repetitive sequences called telomeres. Telomeres are maintained by the enzyme telomerase, which prevents the ends of chromosomes from shortening. During replication, original histones remain associated with the leading strand of the replication fork and new histones assemble on the lagging strand.

Summary

- Chromosome behaviour explains Mendel's laws of inheritance and other genetic phenomena such as linkage and sex linkage. Chromosomes, therefore, contain the genetic information.

- Chromosomes are made up of proteins and DNA. Proteins associated with chromosomal DNA include histones and non-histone proteins. Histones are small basic proteins that form the spool upon which DNA is wrapped to form chromatin. Chromatin appears as a string of duplex DNA coiled around histone 'beads', the nucleosomes.

- A number of early experiments demonstrated that DNA, not protein, is the carrier of genetic information. DNA contains information in the form of alternative nucleotides at each of the millions to billions of nucleotide pairs in the DNA of a particular genome. The order of nucleotide pairs provides all the information needed for a biological organism to develop and function.

- Prokaryotes have only a single chromosome condensed with the aid of folding proteins and RNA to form a nucleoid. There is a single origin sequence for replication.

- Eukaryotic chromosomes have multiple origin sequences.

- DNA replication begins when the double-stranded DNA molecule unwinds and undergoes semi-conservative duplication. Each parental strand of the pair remains unaltered, but serves as a template to enable a complementary strand to be synthesised. Replication begins at the origin, where unwinding of the double strands takes place simultaneously in both directions, forming two replication forks. At each replication fork, new nucleotides complementary to each strand are added until the terminus is reached. Nucleotides are attached to the growing chain by enzymes, DNA polymerases. These enzymes have different forms, one of which has many subunits so that the complex can both add bases and edit the DNA sequence.

- DNA polymerases require a primer (starting tag) made by a primase before DNA synthesis commences at the origin. Two assembly mechanisms operate—continuous synthesis on one strand and discontinuous on the other.

- Telomeres are repetitive sequences found at the ends of chromosomes. Telomerase extends the telomere sequences to maintain the ends of chromosomes, which would otherwise not be able to be replicated.

- Origins of replication lack associated histones. Original histones remain with the leading strand. Newly assembled histones are placed on the lagging strand.

key terms

complementary (p. 233)	exonuclease (p. 239)	Okazaki fragment (p. 241)	semi-conservative replication (p. 235)
deoxyribonucleoside triphosphate (p. 237)	lagging strand (p. 239)	origin (p. 238)	telomerase (p. 243)
DNA ligase (p. 241)	leading strand (p. 239)	primase (p. 240)	telomere (p. 230)
DNA polymerase (p. 239)	ligation (p. 241)	primer (p. 240)	template (p. 235)
duplex (p. 233)	nucleoid (p. 237)	replication fork (p. 238)	terminus (p. 239)
endonuclease (p. 236)	nucleosome (p. 231)	replicon (p. 242)	theta structure (p. 241)
	nucleosome core particle (p. 231)		

Review questions

1. Describe the experiments that showed that DNA and not protein was the carrier of genetic information.

2. How does the structure of DNA enable its role as information carrier?

3. (a) What are the names of the DNA sequences that start and end DNA replication?

   (b) Construct a flow diagram to show the four major steps in the complete DNA replication of a prokaryotic genome.

4. Using a diagram, explain what is meant by semi-conservative replication.

5. (a) What is a replication fork?

   (b) How does replication differ between prokaryotes and eukaryotes?

   (c) What is the advantage of the many sites of replication found in a eukaryotic chromosome?

6. Describe the various enzymes required for DNA synthesis in *E. coli*.

7. RNA synthesis is an important component of DNA replication. What role does it play?

8. Why is DNA ligase essential for DNA replication?

9. (a) Describe the differences and similarities between *E. coli* DNA polymerase I and III.

   (b) What different roles do they play during replication?

10. How do histones contribute to the construction of a eukaryotic chromosome and what happens to them during DNA replication?

11. (a) What are telomeres and why are they necessary?

    (b) How are they maintained?

12. Why is DNA gyrase needed during DNA replication?

13. Somatic cells of humans do not express telomerase, yet they continue to divide. What is the expected consequence of this?

Extension questions

1. How does the nature of the stored information used by biological organisms and computers differ?

2. In some differentiated cells of some organisms, a particular region of a chromosome can be present in a higher number of copies than the rest of the chromosome. Using your knowledge of DNA replication, suggest a mechanism that could result in amplification of a specific region of a chromosome.

3. Table 10.1 shows that DNA polymerase III can synthesise a DNA strand 50 times faster than DNA polymerase I. Using your knowledge of the roles and properties of these enzymes, suggest an explanation for this difference.

4. From your knowledge of telomeres and telomerases, would you expect the offspring of Dolly, the sheep cloned from a somatic cell, to have normal or shortened telomeres?

Suggested further reading

Kornberg, A. and Baker, T. (1991). *DNA Replication*. 2nd edn. San Francisco: W. H. Freeman.

This book provides an advanced description of the mechanisms of DNA replication. Arthur Kornberg won the Nobel Prize in 1959 for his work on DNA replication.

Watson, J. D. and Stent, G. S. (1981). *The Double Helix: A Personal Account of the Discovery of the Structure of the DNA*. New York: W. W. Norton.

An interesting description of the events that led to the discovery of DNA from the personal viewpoint of one of the discoverers.

CHAPTER 11

The genetic code and gene expression

Mendel's analysis of patterns of inheritance led to the concept of the gene, an inherited factor carrying genetic information that influences particular characteristics (Chapter 9). The study of chromosomes and their constituent molecules showed that DNA stores genetic information as a sequence of nucleotides (Chapter 10). The complementary nature of base pairing between the two strands of the DNA double helix permits this information to be faithfully copied during DNA replication (Chapter 10). The processes of mitosis and meiosis (Chapter 8) distribute the DNA copies to daughter cells and gametes.

DNA sequences are the information on which organisms are built. They are analogous to the stored digital information used by computers. Just as a CD-ROM is 'read' to control the functions of a computer, the genetic information, encrypted as long sequences of nucleotide pairs, must also be read to control the development and functioning of an organism.

The one gene–one polypeptide hypothesis

The one gene–one enzyme hypothesis

Neurospora crassa is a fungus capable of synthesising virtually all of its nutritional needs, including amino acids and vitamins, from inorganic materials. George Beadle and Edward Tatum generated mutant strains of *N. crassa*, which had lost the ability to make one or other of these compounds. The strains could not grow on the inorganic medium but could grow if a single amino acid or vitamin was added to the medium. Each strain, which carried a mutation in a single gene, required only one compound to be added to permit growth. Vitamins and amino acids are the end products of a series of enzymatic reactions. Mutations in different genes could lead to a requirement for the same compound, but in this case the different mutations were found to block different enzymatic steps in the biosynthesis of that compound. Many hundreds of nutritional mutations were identified, one for almost every enzymatically controlled reaction required for synthesis of amino acids and vitamins and other compounds. From this it was concluded that one gene encodes the information for one enzyme.

The one gene–one polypeptide hypothesis

Although enzymes are nearly always proteins, many proteins are not enzymes. A study of sickle cell anaemia and haemoglobin structure led to an expansion of the one gene–one enzyme hypothesis. Pedigree analysis showed that sickle cell anaemia results from a mutation in a single gene (Chapter 9). Electrophoresis (Chapter 13) of haemoglobin showed that diseased and normal individuals had different types of haemoglobin (heterozygotes having both types). Normal haemoglobin (Hb A) has a higher net negative charge than does sickle cell haemoglobin (Hb S). Further biochemical study showed that Hb S has a single amino acid change (valine for glutamic acid) in the β-globin chain. (Adult haemoglobin is composed of two α-globin and two β-globin chains; Chapter 20.) Thus, the Hb A and Hb S forms of β-globin correspond to the normal and sickle cell alleles, respectively, of a single gene that controls β-globin synthesis. The conclusion is that a single gene provides the genetic information for the formation of one polypeptide.

An obvious consequence of the one gene–one polypeptide hypothesis was that a specific relationship exists between the sequence of nucleotide pairs in a DNA molecule and the sequence of amino acids in proteins. The way the DNA sequence is related to the protein sequence cannot be simple for two reasons. Firstly, DNA exists as very long stretches of nucleotides, often hundreds of millions, whereas proteins are made up of much shorter stretches of amino acids, tens to thousands. Secondly, each nucleotide position in DNA can be one of four possibilities (A, C, G or T), while each residue in proteins can be one of 20 amino acids. This means that the relationship between the DNA sequence and the protein sequence must involve a form of encryption termed the *genetic code*. The nature and regulation of the process by which this code is read is termed *gene expression*. The reading of the genetic code is ultimately responsible for converting the genetic information encoded in DNA into the characteristics (phenotype) exhibited by an organism.

BOX 11.1 One gene, one polypeptide

Genes code for enzymes

A developmental study of eye colour in *Drosophila melanogaster* in the 1930s led to the conclusion that genes are responsible for the production of enzymes. Wild-type eye colour is a dark red as a result of the production of two different pigments, one coloured bright red and the other brown. Animals mutant for two different genes, *vermilion* and *cinnabar*, which have the same bright red-eyed phenotype (i.e. lacking the production of the brown pigment), were used in the larval

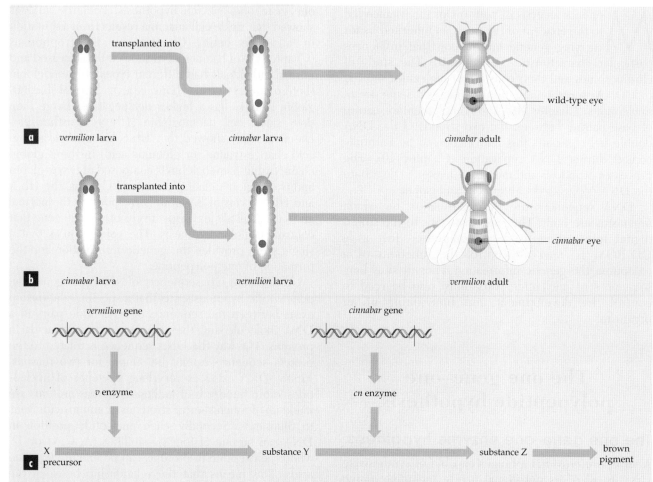

Transplantation experiments using eye discs of *Drosophila*. **(a)** When a *vermilion* mutant eye disc is transplanted into the embryonic abdomen of a *cinnabar* mutant larva, the disc synthesises wild-type pigment. **(b)** When a *cinnabar* mutant eye disc is transplanted into a *vermilion* mutant larva, no brown pigment is synthesised, giving a *cinnabar* phenotype. **(c)** This led the experimenters to deduce that in the biosynthesis of eye pigment, the *vermilion* gene product acts before the *cinnabar* gene product to catalyse sequential steps in the synthesis of brown pigment

transplantation experiments outlined in the figure.

When an eye disc from a *vermilion* mutant larva was transplanted to the abdomen of a *cinnabar* mutant larva, it could make the brown pigment and develop wild-type colour. However, when an eye disc from a *cinnabar* mutant larva was transplanted to the abdomen of a *vermilion* mutant larva, the transplanted *cinnabar* mutant eye could not synthesise the brown pigment and therefore developed the mutant cinnabar phenotype. The conclusions from these studies were that the biosynthesis of brown eye pigment involves at least two sequential steps which are controlled by different enzymes, and that the *cinnabar* and *vermilion* mutants each lack one of these enzymes.

The first step in the production of brown pigment involves an enzyme produced by the wild-type *vermilion* allele. It takes place in the general body tissues of a larva and produces a diffusible intermediate product. The second step, which can occur in the eye, involves the enzyme produced by the wild-type *cinnabar* allele. The second step uses the intermediate product produced in the first step to produce brown pigment. Production of brown pigment can occur when a *vermilion* mutant eye is transplanted into a *cinnabar* mutant larva because the intermediate product produced by the larval tissues (which are wild-type for the *vermilion* gene) can diffuse into the transplanted eye (which is wild-type for *cinnabar*) and undergo the second reaction to produce the brown pigment. However, when a *cinnabar* mutant eye is transplanted into a *vermilion* mutant larva, brown pigment cannot be made because the intermediate product is not made by the *vermilion* mutant larval tissue.

The central dogma

Gene expression is the process by which information encoded in genes is decoded into amino acid sequences during protein synthesis. The *central dogma* of gene expression is that DNA, which carries the genetic information, is transcribed into RNA, and RNA is translated to produce a polypeptide (Fig. 11.1). During gene expression, information must be transferred accurately from the DNA, through an RNA intermediate, to create a polypeptide of a specific amino acid sequence. The three-dimensional folding of polypeptide chains and their specific cellular functions depend on the sequence of their constituent amino acids (see Chapter 1).

Gene expression in cells is a two-stage process. Firstly, a DNA template directs the synthesis of an RNA molecule (the structure of RNA is described in Chapter 1). The nucleotide sequence of one strand (the template) of duplex DNA is copied to produce a complementary single-stranded RNA molecule. This process is termed **transcription** (the word transcription means to make a written copy). The RNA is a 'written copy' of the DNA, a string of ribonucleotides that differs in only minor ways from the deoxyribonucleotide sequence from which it was derived (see below).

The second stage in expression of a gene is the synthesis of a polypeptide with a specific sequence in a process termed **translation**. This is an enzymatic process that requires ribosomes binding to and moving along the RNA to 'read' the nucleotide sequence and convert it into a sequence of amino acids in a polypeptide chain. The term translation is used for this process because one 'language', the language of nucleotides in the RNA, is converted (translated) into a new 'language', the language of amino acids in a protein.

The basis of the one gene–one polypeptide hypothesis discussed above can now be understood. A gene is a specific DNA sequence that directs the synthesis of a polypeptide via an RNA intermediate. While all polypeptides, and hence the functional proteins they form, are coded for by genes, not all genes code for proteins. As we shall see below, some

genes code for RNA molecules (tRNA and rRNA), which are used as tools during the process of protein synthesis and are not themselves translated.

> Gene expression involves the accurate transcription of information from DNA to RNA and the translation of an RNA sequence into the amino acid sequence of a polypeptide.

Transcription

The first stage of gene expression, transcription, involves the synthesis of an RNA molecule whose sequence is complementary to the sequence of one strand of a segment of DNA. This process is similar to the synthesis of a new DNA strand during DNA replication (Chapter 10) in that copying involves separating the DNA strands and synthesising a new RNA strand in which complementary base pairing determines the base incorporated. In addition, growth of the synthesised RNA strand occurs as it does for DNA synthesis in a 5′ to 3′ direction (i.e. addition of ribonucleotides occurs at the free 3′ OH group forming a phosphodiester bond between the 3′ carbon of one ribose sugar and the 5′ carbon of the next).

However, there are also important differences. The structure of RNA differs from that of DNA in two ways. It is composed of ribonucleotides instead of deoxyribonucleotides (Chapter 1) and uracil (U) is used in place of thymine (T) as the pyrimidine base that pairs with adenine (A) (Table 11.1). A different enzyme, RNA polymerase, is used to synthesise RNA. RNA polymerase differs from DNA polymerase in that it does not need a primer to begin the synthesis of an RNA chain. Another difference is that during RNA synthesis, only one DNA strand is copied to produce a single-stranded RNA product.

The RNA sequence is complementary to the DNA sequence from which it was transcribed; for example, the sequence 3′-ATGCAA-5′ in a DNA template is transcribed into an RNA molecule with the sequence

Fig. 11.1 The central dogma of gene expression. This simple scheme summarises the read-out of genetic information in living cells. Some RNA molecules (ribosomal RNA and transfer RNA) are key components of the machinery of protein synthesis and do not contain information specifying a polypeptide

Table 11.1 Base complementarity in DNA and RNA

Base in the DNA template	Complementary base in RNA transcript
A	U
C	G
G	C
T	A

5′-UACGUU-3′. (Remember that complementary nucleic acid strands are antiparallel, so the orientation of the strands, either 5′ to 3′ or 3′ to 5′, is crucial in describing DNA and RNA synthesis and structure.)

Transcription involves transient separation of the DNA duplex strands and binding of an RNA polymerase, which then moves along the template DNA incorporating ribonucleotides and extending the growing RNA chain in a 5′ to 3′ direction. The nucleotide added by the RNA polymerase into the growing RNA chain must form the correct base pair with the corresponding nucleotide in the DNA template (Fig. 11.2). Genetic information stored as the nucleotide sequence of the DNA template is strictly maintained in the RNA transcript.

Initiation of transcription

As noted above, RNA polymerases are unlike DNA polymerases in that they do not require a primer to initiate synthesis. The RNA polymerase moves along the DNA double helix without making transcripts until it reaches and binds to a specific sequence called the *promoter*. The promoter itself is not transcribed but promotes initiation of transcription by the RNA polymerase from a position immediately 3′ to the promoter (Fig. 11.3). Initiation occurs by the base pairing of the first specified ribonucleoside triphosphate;

in our example (Fig. 11.4) this is uridine triphosphate, with its complementary base adenine on the DNA template. The next specified ribonucleoside triphosphate, guanosine triphosphate, then base pairs with its complementary base, cytosine, in DNA. RNA polymerase then catalyses formation of a phosphodiester bond between the two ribonucleotides and the new RNA molecule is initiated (Fig. 11.4).

Transcription continues with the sequential polymerisation of ribonucleotides complementary to the template DNA. The RNA product does not remain base paired to template DNA but dissociates as RNA polymerase moves along the template DNA (Fig. 11.2). This process continues until a specific termination signal on the DNA is reached. Like initiation, termination of transcription relies on a specific sequence of nucleotides in DNA. When a termination sequence is encountered, RNA polymerase ceases transcription and both the RNA polymerase and RNA product are released from the DNA template. The transcribed region of DNA, between promoter and termination sequences, defines a single **transcription unit** (Fig. 11.3) and its RNA product is termed the **primary transcript**. In eukaryotes, most primary transcripts undergo a series of maturation or processing reactions that produce a modified final RNA transcript.

Transcription generates three major forms of RNA: messenger RNA (mRNA), transfer RNA (tRNA) and

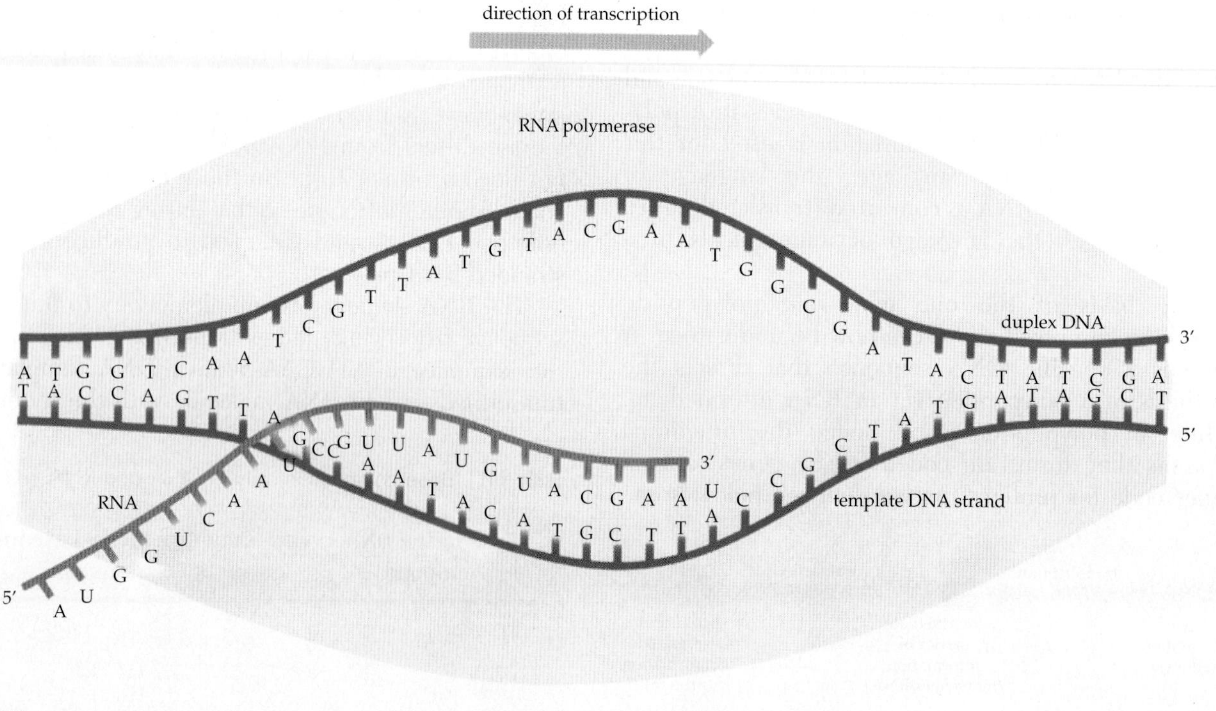

Fig. 11.2 Transcription of DNA by RNA polymerase. RNA is copied from only one strand (template) of a segment of DNA. Only a short region of the growing RNA chain remains attached to the DNA template by hydrogen-bonded base pairing. A molecule of UTP (U) is shown diagrammatically, ready to pair with the next available base (A) in the template strand and be added to the growing RNA chain. The continuously moving 'bubble' of unpaired DNA strands extends for about 17 nucleotides. The enzyme RNA polymerase, which catalyses RNA synthesis, may cover a longer segment of DNA, certainly more than 50 base pairs

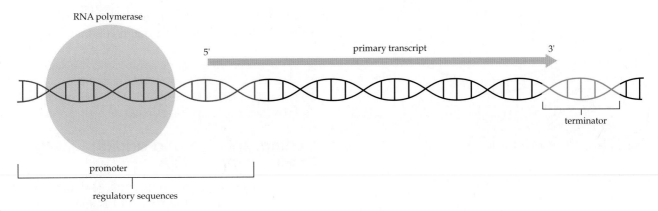

Fig. 11.3 The key features of a generalised bacterial transcription unit. The promoter and terminator represent the RNA polymerase binding site and stop signal in DNA, respectively, for the RNA polymerase that generates the primary transcript

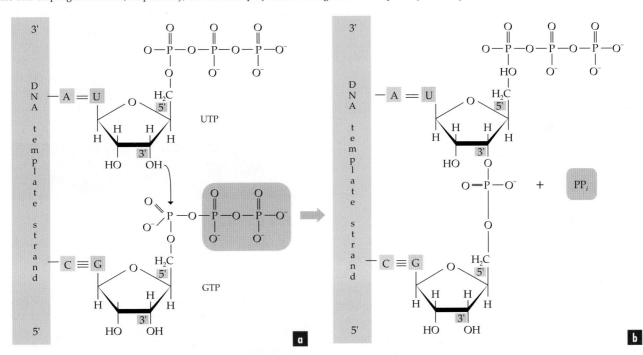

Fig. 11.4 Formation of the first phosphodiester bond to initiate transcription. In this example, the initiating nucleotide is UTP and is shown hydrogen bonded via its uracil base to the adenine base of its complementary nucleotide in the DNA template strand. The incoming guanine base of a ribonucleoside triphosphate (GTP) forms hydrogen bonds with the next base, cytosine, in the DNA template strand. A reaction occurs between the 3' OH group of the ribose sugar ring in the first ribonucleoside triphosphate (UTP) and the 5' phosphates of the second ribonucleoside triphosphate (GTP), leading to formation of a 3'–5' phosphodiester linkage between the two and release of pyrophosphate (PP_i)

ribosomal RNA (rRNA). mRNA molecules carry information, in the form of their nucleotide sequence, from DNA to the protein synthesis apparatus. The mRNA sequence determines the amino acid sequence of the protein synthesised during translation (see below). In contrast to mRNA, neither rRNA nor tRNA is translated. rRNA is an important component of ribosomes, while tRNA is also required for translation (see below).

> Transcription involves binding of RNA polymerase to the promoter sequence of DNA, initiation of synthesis, elongation of the RNA chain, termination and release of the RNA chain. Transcription generates three major forms of RNA: mRNA, rRNA and tRNA.

Prokaryotes: transcription in the cytoplasm

In bacteria, a single RNA polymerase transcribes all genes. Many bacterial transcription units encode more than one protein. They are said to be **polycistronic**. The protein-coding regions of polycistronic bacterial mRNAs are separated by non-coding sequences.

Since bacterial DNA is not separated from the cytosol by a nuclear membrane, ribosomes (the protein synthesis apparatus) can bind and initiate translation while transcription is still in progress (Fig. 11.5). In fact, during transcription, only the most recently incorporated 10 nucleotides remain base paired to the

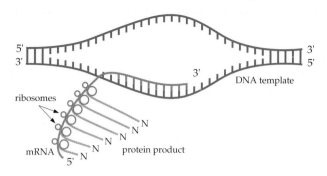

Fig. 11.5 Coincident transcription and translation in prokaryotes. Transcription is occurring as shown in Figure 11.2 and, at the same time, the newly synthesised mRNA has associated with ribosomes and translation of the protein product is underway

DNA template (Fig. 11.2). The rest of the transcript simply peels away and is available for translation.

Many bacterial mRNAs contain more than one polypeptide coding region. These mRNAs are said to be polycistronic. In prokaryotes, since bacterial DNA is not bound by a membrane, translation to produce a polypeptide can commence while mRNA is still being transcribed.

Eukaryotes: transcription in the nucleus

Unlike prokaryotic transcription, transcription of eukaryotic chromosomal DNA occurs within the membrane-bound nucleus, separated from the process of translation, which occurs in the cytosol. Also, the DNA of eukaryotic chromosomes is complexed with proteins (histones) to form nucleosomes, the basic unit of chromatin (Chapter 10). During mitosis (Chapter 8), the chromatin becomes highly condensed, giving the chromosomes their distinctive shape. During the remainder of the cell cycle, the chromatin is much less condensed, allowing RNA polymerase access to the DNA.

One striking feature of the eukaryotic nucleus is the appearance of a suborganelle, the nucleolus (Chapter 3). This organelle is associated with regions of chromosomes that have many tandemly repeated (i.e. head-to-tail) copies of genes, which encode rRNA. The nucleolus is the site of rRNA transcription.

In contrast to bacteria, which utilise a single RNA polymerase to transcribe all active genes, there are three types of RNA polymerases in the eukaryotic nucleus, each of which transcribes a specific set of genes. RNA polymerase I transcribes rRNA genes; RNA polymerase II transcribes mRNAs, the informational intermediates that code for proteins; and RNA polymerase III transcribes tRNAs, required for protein synthesis.

Although the sizes of particular mRNA transcripts vary widely, a typical eukaryotic mRNA is relatively small, of the order of 10 000 nucleotides, and encodes a single polypeptide product (i.e. it is monocistronic).

Transcription in eukaryotic cells occurs in the nucleus and involves three RNA polymerases responsible, respectively, for the synthesis of rRNA, mRNA and tRNA. Eukaryotic mRNAs are almost always monocistronic.

Cutting, splicing and adding to eukaryotic mRNA

Many eukaryotic primary transcripts undergo a series of modifications prior to export of the mature mRNA from the nucleus. Both the 5′ and 3′ ends of eukaryotic primary transcripts are modified in the nucleus before being exported to the cytoplasm. Firstly, a modified GTP residue is added to the 5′ nucleotide of the primary transcript (shown diagrammatically in Figure 11.6) to create a 'capped' RNA. Secondly, the 3′ end of the primary transcript is cleaved and A residues are added to the new 3′ end to create a poly A tail in a process termed **polyadenylation** (Fig. 11.6).

BOX 11.2 Terms associated with gene expression

Transcription

The process of synthesising a single-stranded RNA complementary to one of the strands of DNA.

Primary transcript

The initial transcript produced by RNA polymerase prior to RNA processing to produce the mature mRNA, rRNA or tRNA.

mRNA

The mature messenger RNA, containing one or more open reading frames, that is translated by ribosomes to produce polypeptides.

Intron

Sequences that are removed during processing of the primary transcript.

Exon

Sequences that are joined together during processing to form the mature mRNA.

Splicing

The processing of a primary transcript to remove introns.

5′ and 3′ untranslated sequences

Sequences in the mRNA, located either side of the open reading frame, that do not encode a polypeptide.

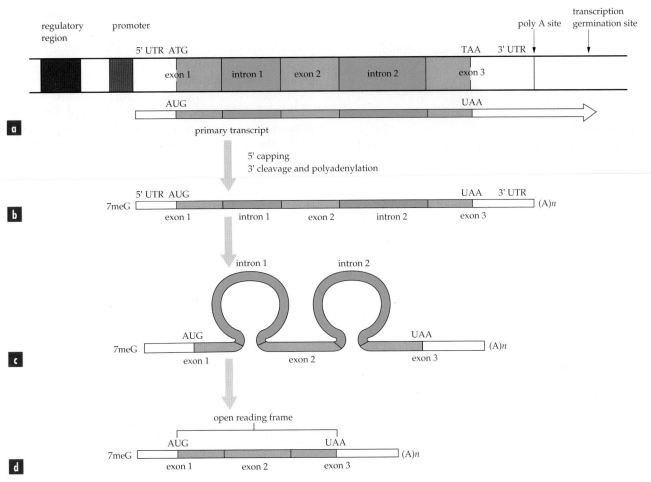

Fig. 11.6 Post-transcriptional processing of eukaryotic mRNA transcripts. **(a)** A double-stranded DNA region containing a generalised transcription unit. For convenience, a single regulatory region is indicated, located upstream of the promoter (RNA polymerase binding site). The coding region that specifies the mRNA includes three exons interrupted by two introns shown in green. The exons include sequences that encode the 5′ and 3′ untranslated regions (5′UTR, 3′UTR) as well as the open reading frame, which is interrupted by the introns and is indicated by the orange colour. The primary transcript is shown immediately below the DNA. **(b)** Following transcription, the cap structure (7-methyl guanine, 7-meG) is added to the 5′ end of the transcript. The 3′ end of the mature mRNA is created by cleavage of the primary transcript at the poly A site followed by addition of the poly A tail, (A)n, to the 3′ end of the transcript. **(c)** The final step in processing, namely, the splicing out of intron sequences and joining of the ends of each adjacent pair of exons to yield a mature mRNA. **(d)** The final mature mRNA, containing an open reading frame, is exported from the nucleus to the cytosol for translation on ribosomes

Translation

The synthesis of a polypeptide by a ribosome directed by the sequence of codons in an mRNA.

Open reading frame

A sequence of codons that begins with the AUG initiator codon, proceeds through a series of amino-acid-encoding codons and finishes with a termination codon.

Ribosome

The rRNA–protein complex that provides a scaffold for the assembly of mRNA, peptidyl tRNA and aminoacyl tRNA and catalyses peptide bond formation during protein synthesis.

tRNA

The adaptor RNA molecule that contains an anti-codon complementary to a codon in the mRNA.

Aminoacyl-tRNA

tRNA to which the appropriate amino acid has been covalently attached.

Operon

Sequences in bacterial DNA that encode a primary transcript and contain the *cis*-regulatory sequences required for regulated expression of that transcript.

Constitutive gene

A gene expressed constantly.

Inducible gene

A gene expressed only under certain conditions.

Promoter

The site to which RNA polymerase binds to initiate transcription.

Operator

A *cis*-acting sequence to which a transcriptional repressor binds.

Repressor

A protein that binds to an operator and prevents transcription.

Enhancer

A eukaryotic *cis*-acting regulatory sequence that controls expression of a gene, independent of its orientation or precise location with respect to the promoter for that gene.

Transcription factor

A protein that binds to enhancer sequences and regulates transcription.

A most surprising and remarkable modification involves **splicing** of the primary transcript. The primary transcripts of most eukaryotic genes are much longer than the mRNAs they produce because the genes include additional sequences that interrupt the sequences found in the mature mRNA. These additional sequences, which are removed from the primary transcript during splicing, are termed **introns**, while the sequences that are represented in the mature mRNA are termed **exons**. Introns are removed by cutting the primary transcript either side of the intron and joining the flanking exons together (Fig. 11.6). The intron RNA sequences are released and degraded.

The accurate and efficient removal of eukaryotic introns requires the activity of complexes of proteins and small RNA molecules termed snRNPs (small nuclear ribonucleoprotein particles). However, an alternative cutting and splicing mechanism has been found for rRNA primary transcripts in ciliates and slime moulds, and for primary transcripts of both rRNA and mRNA in yeast mitochondria. In these cases, the RNA transcripts are able to fold into specific conformations, which lead to self-splicing without catalysis

---

BOX 11.3 One gene, many proteins!

The original one gene–one enzyme hypothesis evolved into the one gene–one polypeptide proposal. The discovery of splicing, the analysis of gene structure and the characterisation of proteins have led to the realisation that more than one protein can be produced by one gene. In many cases, alternative splicing of the same primary transcript generates different mRNA forms and different protein products. Sometimes, the different spliced forms and protein products are produced in the same cells, but splicing can be regulated so that different products are produced in different cell types. A striking example is found in *Drosophila melanogaster*.

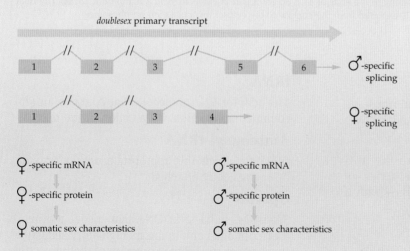

Differential splicing can produce different protein products from the same gene. The splicing of transcripts from the *Drosophila melanogaster* gene, *doublesex*, is different in females and males. The products determine female and male somatic sexual characteristics, respectively

Here, differences in the splicing of a key regulatory gene determine whether an individual develops into a male or female (see figure).

Proteins are also modified by a variety of post-translational mechanisms, including cleavage of the protein into smaller polypeptides or chemical modification of residues, for example, by the addition of phosphate groups, acetyl groups, lipids or sugar residues (glycosylation). The result is that the estimated 40 000–60 000 genes present in the human genome encode over a million chemically different proteins! In general, the different protein products of a single gene can be readily identified because most or all of their amino acid sequences are the same.

by protein enzymes. Such molecules are described as RNA enzymes or **ribozymes**.

Translation and the genetic code

The genetic code

How does the sequence of nucleotides in the mRNA specify the order of amino acids in a polypeptide? What mechanisms translate the four-letter nucleotide alphabet of mRNA (U, A, C and G) into the words (amino acids) of the protein language? Twenty amino acids are commonly found in proteins, so theoretically only 20 code words are required. A single nucleotide–single amino acid correspondence is not possible because there are not enough different nucleotides. If amino acids were determined by a sequence of two nucleotides, there would still only be enough information for a maximum of 16 amino acids. The minimum requirement, therefore, is a three-letter code for each amino acid, which provides 64 possible combinations—more than enough!

Codons

The process of synthesising a protein from an mRNA is termed translation. During translation, the ribosome moves along the mRNA and allows a triplet of nucleotides (a three-letter 'word'), termed a **codon**, to specify the addition of an amino acid to the growing polypeptide chain. This relationship between codons and amino acids is the genetic code.

There are 64 possible codon sequences but only 20 amino acids. In most cases, more than one codon specifies one amino acid (Fig. 11.7). For example, the sequences CCU, CCC, CCA and CCG all encode incorporation of a proline residue during protein synthesis. Thus, the genetic code is said to be *degenerate*. The degeneracy in the code is particularly evident in the third nucleotide of a codon, where changes often do not alter the amino acid that is specified. In Figure 11.7, about half the codes belong to this category; for example, a CU combination encodes leucine regardless of the nucleotide in the third position. The result of this degeneracy is that nearly all (61) of the 64 possible codons encode one or other of the 20 amino acids (Fig. 11.7). Codons that specify the same amino acid are termed **synonymous codons**. Only tryptophan and methionine are specified by a single codon each. Nevertheless, the code is unambiguous because each codon specifies only one amino acid.

Second nucleotide

First nucleotide	U	C	A	G	Third nucleotide
U	UUU Phe	UCU Ser	UAU Tyr	UGU Cys	U
U	UUC Phe	UCC Ser	UAC Tyr	UGC Cys	C
U	UUA Leu	UCA Ser	UAA stop	UGA stop	A
U	UUG Leu	UCG Ser	UAG stop	UGG Trp	G
C	CUU Leu	CCU Pro	CAU His	CGU Arg	U
C	CUC Leu	CCC Pro	CAC His	CGC Arg	C
C	CUA Leu	CCA Pro	CAA Gln	CGA Arg	A
C	CUG Leu	CCG Pro	CAG Gln	CGG Arg	G
A	AUU Ile	ACU Thr	AAU Asn	AGU Ser	U
A	AUC Ile	ACC Thr	AAC Asn	AGC Ser	C
A	AUA Ile	ACA Thr	AAA Lys	AGA Arg	A
A	AUG Met	ACG Thr	AAG Lys	AGG Arg	G
G	GUU Val	GCU Ala	GAU Asp	GGU Gly	U
G	GUC Val	GCC Ala	GAC Asp	GGC Gly	C
G	GUA Val	GCA Ala	GAA Glu	GGA Gly	A
G	GUG Val	GCG Ala	GAG Glu	GGG Gly	G

Fig. 11.7 The genetic code. Termination or 'stop' codons are shown in pink, and the usual 'start' codon, AUG, in blue. The codons GUG or UUG (purple) are occasionally used as start codons (see text). When any of these codons are used as start codons in bacteria, they direct the incorporation of N-formyl methionine as the initiating amino acid for the polypeptide. However, when the ribosome encounters an internal AUG, UUG and GUG residue, it incorporates methionine, leucine or valine, respectively, and not N-formyl methionine

Reading frames

Not all of the sequences of an mRNA molecule are read during translation. Normally, sequences at the 5′ and 3′ ends do not encode the protein. These sequences are termed the *5′ untranslated region* (*5′UTR*) and *3′ untranslated region* (*3′UTR*), respectively. In order to synthesise a protein, translation must start and stop at the appropriate places. The start and stop signals are themselves triplet codons. In most cases (see Box 11.4), an AUG codon specifying methionine (met or M) serves as the *start codon* (initiator codon). Translation ends at one of the three stop codons (UAA, UAG and UGA), which do not specify an amino acid (Fig. 11.7).

The region of the mRNA that is read during the synthesis of a protein is termed the **open reading frame** (**ORF**). The ORF consists of a start codon followed by a series of codons that specify the sequence of amino acids and concludes with a termination codon (normally UAA, UAG or UGA). Codons do not overlap, that is, the last nucleotide of one codon is adjacent to the first nucleotide of the next codon. Consider the 15-letter mRNA sequence, 5′-AUGAUCGCCACUUAA-3′. Triplets are read from the 5′ end (i.e. in this example from the AUG) to the 3′ end, so this mRNA codes for the polypeptide

BOX 11.4 Variations in codon usage

The genetic code that is common to most organisms is the end product of millions of years of evolution. Across the range of organisms that use the common genetic code, there are variations in the frequency with which alternative codons are used. Codon usage may even differ between different mRNAs in a single organism, depending on whether genes are expressed at a high level or not. For example, of six alternative codons for leucine in bakers' yeast, *Saccharomyces cerevisiae*, UUG is used most frequently in highly expressed genes, while the codons CUU, CUC and CUG are hardly used at all. In contrast, genes that are moderately expressed use all six codons at about the same frequencies.

Occasionally, the codon usage itself varies from that most commonly used. For example, a GUG codon, or even more rarely a UUG codon, is sometimes used as the initiator instead of AUG. When this occurs, GUG or UUG, like the initiator AUG, specifies N-formyl methionine, even though this same codon specifies valine when found at positions other than the site of initiation. This change in codon use does not, therefore, change the first amino acid of a newly translated protein, which remains methionine.

In the mitochondria of most eukaryotes (except plants), UGA codes for tryptophan, not stop (Fig. 11.7) and AUA codes for methionine, not isoleucine. Further examples have been found in *Mycoplasma* (a prokaryote), where UGA codes for tryptophan, and in eukaryotes such as the ciliated protists (e.g. *Paramecium*), where UAA and UAG both code for glutamine, not stop.

These deviations from the common genetic code tell us something of the evolution of eukaryotic cells and their organelles, providing evidence of the pathway along which evolution of a particular biological system has proceeded. One challenge that confronts biological scientists is to understand whether any underlying chemical relationships determined the specification of particular amino acids by certain codons at the time cellular life began on earth and the common code became established. Another challenge is to determine whether the variant codes predate the fixation of the common code or represent more recent changes away from the use of the common code.

methionine-isoleucine-alanine-threonine by the triplets AUG, AUC, GCC and ACU, while the UAA signals the end of polypeptide synthesis.

Because the codons do not overlap, nucleotides of an mRNA sequence have three possible and overlapping 'reading frames' that could specify the amino acids of a polypeptide (Fig. 11.8). Only one of these reading frames is used to generate the protein during translation. The reading frame used is set by the AUG initiation codon. Since DNA is double-stranded and mRNA could be made using either strand as a template, any DNA segment has six possible reading frames. In most cases, only one of the strands is used as a template to make the mRNA, and only one of the reading frames in the mRNA is used during translation. The result is that only one of the six potential reading frames in a DNA sequence actually encodes a protein. Rare exceptions to this general rule are found in some viruses, where different open reading frames that overlap are translated. The result is that more than one protein is produced from the same region of the DNA double helix.

Some chemical mutagens (Chapter 12), such as acridine dyes, cause mutations by the removal or addition of one or more nucleotides to a DNA sequence during DNA replication. The existence of this type of mutation provided early evidence for the triplet nature of the code. When a single nucleotide was added or removed, the reading frame for all subsequent triplets was changed (Fig. 11.9), destroying the function of the gene by changing the amino acid sequence of the encoded protein. When two additions or deletions occurred, the reading frame was different again and the gene product was non-functional. These mutations are termed **frameshift mutations**. However, if there was an addition plus a compensating deletion, or if the addition or deletion of three nucleotides occurred, the original correct reading frame was restored in the remainder of the gene and the activity of the gene was sometimes restored (Fig. 11.9).

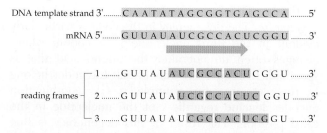

Fig. 11.8 The concept of reading frames. Bases are read sequentially in groups of three in the 5′ to 3′ direction (arrowed). In each strand of DNA, there are three possible phases of the codon units by which the base sequence could be read to code for a protein (the reading frame). The complementary strand (not shown here) can also specify a set of three possible reading frames. A given region of double-stranded DNA usually only contains one reading frame that codes for a particular protein

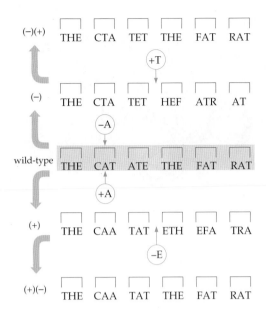

Fig. 11.9 Frameshift mutations. A series of three-letter words represent three-letter codons to illustrate the effect of a frameshift mutation. The addition (+) or deletion (−) of a single letter in the middle of a reading frame causes a frameshift mutation. All codons from that point on are altered. If there is both a deletion and an addition, the reading frame is restored in the sequences following the mutations, often restoring the function of the gene

Mutations that alter a codon so that it now specifies another amino acid are known as **missense mutations**. They lead to the production of an altered polypeptide. **Nonsense mutations** change the sequence of a codon to a stop codon (UAA, UAG or UGA). They lead to premature termination of translation.

The genetic code is almost universal, that is, the same basic codon dictionary is used to decipher mRNAs in most organisms. However, slight variations in the genetic code do occur (Box 11.4).

The common genetic code is made up of three-letter nucleotide code 'words' (codons). Each group of three nucleotides codes for an amino acid or serves as a start or stop signal in the production of a polypeptide. Codon sequences of mRNA specify the order of assembly of amino acids during protein synthesis. In eukaryotes, each mRNA codes for a single polypeptide (monocistronic), while prokaryotic mRNAs may code for several proteins (polycistronic).

Translation

Codons within an mRNA specify the exact order in which amino acids will be linked together. Translation is the process in which information in the nucleotide sequence of an mRNA molecule controls the ordered assembly of amino acids. In addition to the mRNA, two major components, ribosomes and RNA, are involved in translation.

Transfer RNA (tRNA)

To understand the role of tRNAs in the process of translation, it is important to realise that there appears to be no direct chemical affinity between codons and the amino acids they specify. The codon cannot itself bring a specific amino acid to the site of translation. There must, therefore, be an adaptor that mediates the relationship between a codon and the amino acid it specifies. tRNAs are small molecules, only 75 nucleotides in length, which act as the adaptors.

Each type of tRNA can be covalently attached to a single amino acid and has a three-ribonucleotide 'anticodon' sequence complementary to the codon that specifies the amino acid attached to the tRNA. tRNAs act as adaptor molecules in protein synthesis because they bring the correct amino acid to the codon during protein synthesis. Enzymatic mechanisms in the cell ensure that each type of tRNA becomes covalently linked to only one amino acid. The tRNAs are conventionally named for this amino acid; for example, $tRNA^{Trp}$ is the tRNA that is linked specifically to tryptophan. $tRNA^{Trp}$ will also carry the anticodon 5′-CCA-3′, which is the complement of 5′-UGG-3′, the codon for tryptophan. The anticodon, responsible for codon recognition, is found at the end of a loop in each tRNA molecule (Fig. 11.10). Complementary base pairing between the codon and anticodon ensures that the correct amino acid is brought to the site of protein synthesis.

tRNA molecules can be represented in two-dimensional drawings as clover-leaf structures because of the stem and loop structures in the molecule (Fig. 11.10a). Stem and loop structures arise from intramolecular base pairing between small segments of the tRNA nucleotide sequence. This pairing results in the formation of four double-stranded regions (stems) interspersed among single-stranded regions (loops). Loop 2, the anticodon loop (Fig. 11.10), contains the anticodon sequence. Other loops function in binding to the ribosome. The three-dimensional structure of tRNA shows that it is intricately folded (Fig. 11.10b).

The adenine residue at the 3′ end of a tRNA molecule becomes covalently attached to the appropriate amino acid by aminoacylation (Fig. 11.11), creating an aminoacyl-tRNA (referred to as a charged tRNA). For example, $tRNA^{Leu}$ is converted to leucyl-$tRNA^{Leu}$ (L-$tRNA^{Leu}$) by this process. Such reactions are catalysed by at least 20 different aminoacyl-tRNA synthetases, one required for the attachment of each amino acid to its specific tRNA.

There is a remarkable degree of specificity in this tRNA charging process. Each aminoacyl-tRNA synthetase enzyme puts one amino acid on to the tRNA molecule, whose anticodon subsequently pairs with the codon specifying the attached amino acid. Curiously,

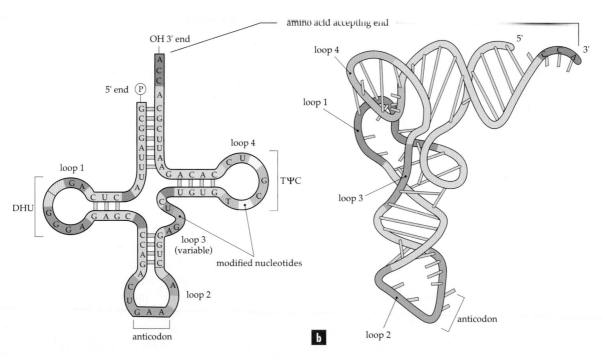

Fig. 11.10 Transfer RNA structure. **(a)** The base-paired regions in a typical tRNA are shown in two dimensions as a clover-leaf structure. Loop 2 contains the anticodon. The size of the variable loop 3 is different across a range of tRNAs. **(b)** An outline of the overall three-dimensional shape, as determined by X-ray diffraction

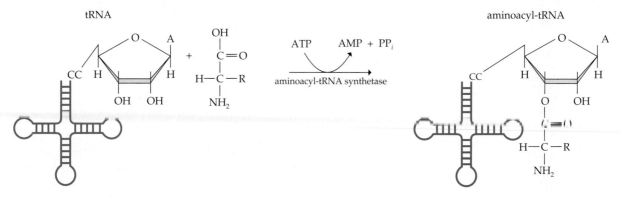

Fig. 11.11 Activation of tRNA by aminoacylation. The acylation of a tRNA molecule requires a specific aminoacyl-tRNA synthetase enzyme and energy from the hydrolysis of ATP. The end result is an aminoacyl-tRNA ready to participate in protein synthesis. The sequence CCA at the 3′ end of the tRNA is represented by the letters CC linked to the 5′ ribose sugar of the A residue. The ribose ring of the terminal A residue is drawn in full to show that its 3′ OH group becomes covalently linked to the carboxyl group of the amino acid. R indicates the side chain, which is different for each amino acid

the aminoacyl-tRNA synthetases derive their specificity through interactions with loop 1 of the tRNA and not the anticodon itself.

The genetic code has 61 amino-acid-specifying codons, but 61 different tRNAs are not required for translation. Some tRNAs recognise more than one codon because base pairs other than A–U and G–C can form between the third nucleotide of a codon and the first nucleotide of the corresponding anticodon. Although these non-standard base pairs have weaker bonds than standard base pairs, the bonds are strong enough to allow specific codon recognition. This ability of tRNAs to 'wobble' in their interaction at the third position of the codon decreases the number of tRNAs

required to decode an open reading frame. Eukaryotes have about 50 different tRNAs, while prokaryotes contain up to 40.

Each tRNA has a specific anticodon and is covalently linked to a specific amino acid, forming aminoacyl-tRNA. Each codon brings the correct amino acid to the site of polypeptide synthesis by base pairing with the anticodon of the tRNA, which is linked specifically to the amino acid specified by the codon.

Ribosomes

Ribosomes are made up of a specific set of RNA molecules and proteins (Chapter 3). The ribosome provides a scaffold for the mRNA and aminoacyl-tRNAs to assemble and provides a reaction chamber for the catalysis of polypeptide synthesis. Like tRNAs, each rRNA folds upon itself to form short, double-stranded regions. The resultant stem–loop structures in each rRNA give a characteristic three-dimensional structure, which is important for interactions with ribosomal proteins.

Mature ribosomes consist of two subunits, each containing different-sized rRNA molecules. The size of the subunits and their associated rRNAs were initially analysed by their sedimentation properties when subjected to centrifugal force. The unit of sedimentation is the Svedberg (S) and it is this unit that is used to describe the ribosomal subunits and their rRNA components. Each ribosome is composed of one large and one small subunit (Fig. 11.12). Prokaryotic ribosomes (50 S and 30 S subunits, which together make a 70 S ribosome) are smaller than those of eukaryotes (60 S and 40 S subunits, which together make an 80 S ribosome). (Note that, because the sedimentation property of a particle is a function of several parameters, including size and shape, the S value for an aggregate of particles is not simply the sum of the individual particles.) Each ribosome subunit contains a particular rRNA molecule or set of rRNAs. Ribosomes of prokaryotes contain 23 S, 16 S and 5 S rRNA molecules, while those of eukaryotes contain 28 S, 18 S, 5.8 S and 5 S rRNA molecules. When the large and small subunits combine, they undergo conformational changes, forming a groove in which the mRNA molecule sits (Fig. 11.13). Assembled ribosomes exist only when they are involved in the process of protein synthesis; otherwise they dissociate into their small and large subunits.

> Ribosomal RNA combines with a defined set of proteins to form ribosomes. In both prokaryotes and eukaryotes, each ribosome consists of a small and a large subunit.

Polypeptide synthesis

In the process of translation, amino acids are linked in a precise sequence to form a polypeptide according to the sequence of nucleotides in an mRNA molecule. Polypeptides are synthesised in three steps: initiation, elongation and termination (Figs 11.14, 11.15). The nucleotide sequence of the open reading frame in the mRNA is read from the 5′ end to the 3′ end as a triplet 'reading frame', starting with an AUG codon. The reading frame specifies the sequence of amino acids in the growing polypeptide from the amino (N) terminus to the carboxyl (C) terminus.

Initiation

Protein synthesis begins with the assembly of an initiation complex. Firstly, the small ribosome subunit,

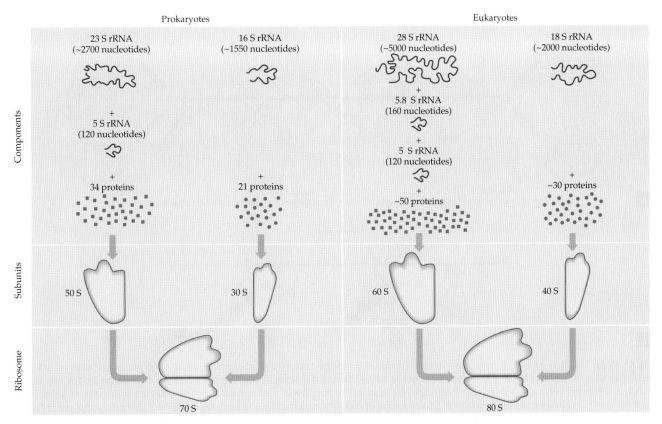

Fig. 11.12 Components of prokaryotic and eukaryotic ribosomes

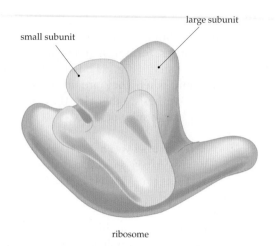

small subunit

large subunit

ribosome

Fig. 11.13 The ribosome of *Escherichia coli*. A three-dimensional model of the association of the small and large ribosomal subunits to form a functional ribosome. The subunits undergo a conformational change upon association with each other and with an mRNA molecule, forming a groove into which the mRNA molecule fits

the initiator aminoacyl-tRNA and a molecule of GTP assemble at the initiation codon of the mRNA. Assembly of these components requires a set of protein initiation factors and energy released by hydrolysis of the GTP. The first amino acid laid down during protein

synthesis is always N-formyl methionine, which is specified by the initiator AUG codon (but see Box 11.4). Once the anticodon of the initiating N-formyl methionine-charged tRNA has associated with the AUG initiation codon, the large ribosomal subunit binds and the initiation complex is complete and ready for the elongation reactions to begin.

The formation of an intact ribosome establishes two tRNA binding sites. One site, the A site, binds the incoming aminoacyl-tRNA, while the other site, the P site, accommodates the peptidyl-tRNA (the tRNA covalently linked to the growing polypeptide being synthesised). Upon association of the large ribosomal subunit with the pre-existing complex, the initiating methionyl-tRNA occupies the newly created P site, ready for the elongation reactions to begin.

> Protein synthesis begins when the mRNA to be translated associates with the initiating aminoacyl-tRNA and a ribosome.

Elongation

Elongation of a polypeptide is a repetitive process involving sequential addition of amino acids to the growing polypeptide chain (Fig. 11.15b). The elongation phase begins with synthesis of the first peptide bond

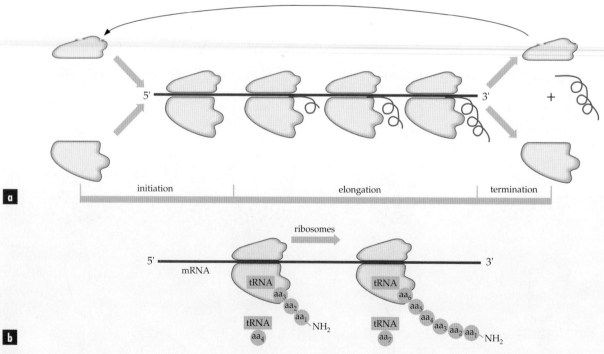

Fig. 11.14 (a) The three phases of protein synthesis. Initiation requires the formation of ribosomes from the two ribosomal subunits, which assemble with the initiator tRNA at the start codon of the mRNA. During the elongation phase, the ribosome decodes the mRNA sequence to produce a growing polypeptide chain. At termination, the synthesis of the protein is completed, the ribosome dissociates into its subunits and the newly synthesised protein is released. **(b)** Key events of protein synthesis. Ribosomes move in a 5′ to 3′ direction along the mRNA molecule. The growing protein is synthesised from its N-terminal amino acid to its C-terminal amino acid. Note that the growing polypeptide chain is always attached to a tRNA, which links it to the ribosome

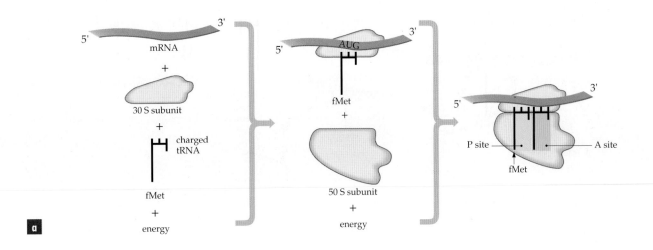

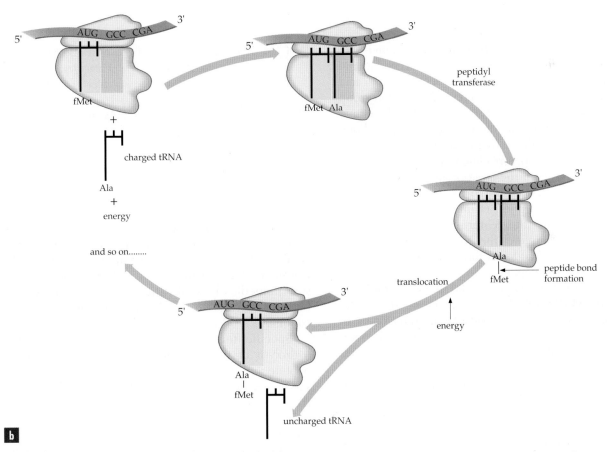

Fig. 11.15 Details of protein synthesis involving a bacterial ribosome (the process is essentially the same with eukaryotic ribosomes). **(a)** Initiation. Initiator tRNA, charged with N-formyl methionine (fMet), binds to a site on the small ribosomal subunit. The small ribosomal subunit binds to an mRNA molecule so that the AUG initiation codon pairs with the anticodon in the initiation tRNA. The large ribosomal subunit then joins to form a ribosome with two tRNA binding sites: A, which binds aminoacyl-tRNAs, and P, which binds peptidyl-tRNAs (tRNAs linked through their amino acid to the polypeptide being synthesised). Association of the large ribosomal subunit with the pre-existing complex causes fMet-tRNA^fMet to move into the P site, ready for elongation to begin. **(b)** Elongation. The second aminoacyl-tRNA specified by the next mRNA codon (in this example, GCC coding for alanine) binds to the A site. With both sites occupied, the enzyme peptidyl transferase transfers the fMet from the tRNA in the P site to the amino acid at the A site via formation of a peptide bond. The initiator tRNA then dissociates from the P site. The ribosome translocates relative to the mRNA to move the remaining tRNA (now with two amino acids attached) from the A site to the unoccupied P site, bringing the next codon to the now empty A site. The vacant A site now becomes occupied by the aminoacyl-tRNA specified by the codon at the A site and another cycle of peptide bond formation and translocation follows

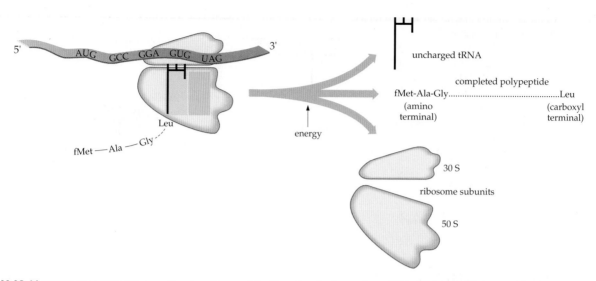

Fig. 11.15 (c) Termination. Elongation continues until one of the three termination codons (UAA, UAG or UGA) is positioned in the A site. Translation ceases and the aminoacyl bond between the polypeptide and the peptidyl-tRNA is hydrolysed, releasing the polypeptide from the ribosome

between the carboxyl group of the first amino acid (methionine) and the amino group of the second amino acid. Then the carboxyl group of the second amino acid in the chain forms a peptide bond with the amino group of a further amino acid, and so on, building the amino acid chain. This process of sequential peptide bond formation occurs as the ribosome moves from the 5′ end to the 3′ end of the open reading frame, matching each codon of mRNA with the appropriate anticodon of an aminoacyl-tRNA to place the correct amino acid at each position. The polypeptide chain grows from the N-terminus towards the C-terminus. The initiator methionine is, therefore, the N-terminal residue and the last amino acid added is the C-terminal residue of the completed polypeptide.

In order to form a peptide bond, there must initially be a peptidyl-tRNA in the P site on the ribosome. The appropriate incoming aminoacyl-tRNA then binds to the vacant A site on the ribosome, its anticodon matching the codon at this site. During the formation of the peptide bond, catalysed by the enzyme peptidyl transferase, the peptide attached to the tRNA in the P site is transferred onto the single amino acid attached to the tRNA in the A site. The new peptidyl-tRNA then moves to the P site on the ribosome (the *translocation reaction*; Fig. 11.15b), displacing the tRNA into the cytosol, where it can become recharged by its specific aminoacyl-tRNA synthetase. The translocation reaction, which moves the new peptidyl-tRNA to the P site, also causes the ribosome to move three bases along the mRNA, so that the new peptidyl-tRNA passes to the P site without the anticodon dissociating from the codon in the mRNA to which it is base paired. This brings the next codon in the mRNA to the vacant A site so that it can receive the next aminoacyl-tRNA.

Elongation of polypeptides during protein synthesis involves the sequential addition of amino acids, linked by peptide bonds, to a growing polypeptide chain. As the ribosome advances stepwise along the mRNA three nucleotides at a time, the mRNA codon sequence is translated into a precisely defined chain of polymerised amino acids.

Termination

The cycle of aminoacyl-tRNA delivery, peptide bond formation and translocation continues for each codon of the mRNA until one of the three stop codons (UAA, UAG or UGA) arrives at the A site on the ribosome. The positioning of a stop codon at the A site immediately stalls the process of translation because there is no tRNA that recognises such stop codons.

Specific termination factors then mediate the hydrolysis of the aminoacyl bond between the polypeptide and the final peptidyl-tRNA resident in the P site, releasing both the completed polypeptide and tRNA from the ribosome. Once the P site is vacant, the ribosome dissociates from the mRNA and disassembles into its two subunits, using the energy released by the hydrolysis of a final GTP molecule. This terminates polypeptide synthesis (Fig. 11.15c). The individual free ribosome subunits can be recycled for another round of protein synthesis.

Termination occurs when one of the three stop codons (UAA, UAG or UGA) reaches the A site of the ribosome. The newly synthesised polypeptide is released from the ribosome.

BOX 11.5 Putting a spanner in the ribosomal works

Many bacterial infections are treated by the use of antibiotics that inhibit protein synthesis and result in growth arrest or death of bacterial cells. These antibiotics selectively inhibit protein synthesis in bacterial cells rather than in surrounding eukaryotic cells. This specificity relies on structural and functional differences in the ribosomes of prokaryotic and eukaryotic cells. Certain steps of bacterial protein synthesis are inhibited by interactions between the antibiotic and specific protein components of the bacterial ribosome (see Table). These antibiotics do not react with the equivalent protein components of eukaryotic ribosomes and thus they do not affect protein synthesis in mammalian cells. Antibiotics have also proved to be useful tools for studying the mechanism of protein synthesis by allowing particular steps in the process to be identified and studied.

Table Selective inhibitors of bacterial protein synthesis

Antibiotic	Protein synthesis step blocked
Tetracycline	Binding of aminoacyl-tRNA molecules to the ribosome
Streptomycin	Pairing between aminoacyl and message codons
Chloramphenicol (Chloromycetin)	Peptidyl transferase reaction
Erythromycin	Translocation reaction

Protein targeting and processing

Genes are not fully expressed until their protein product is transported from its site of synthesis to its site of action (protein targeting) and, if necessary, converted into a functional protein (protein processing). Eukaryotic cells, which have many intracellular compartments, have highly specialised mechanisms that direct proteins to particular organelles or membrane systems of the cell (Fig. 11.16).

Polypeptides synthesised on 'free' ribosomes are released into the cytosol. From there, they may pass to the nucleus and organelles such as mitochondria and chloroplasts. Targeting of polypeptides to different organelles requires an 'address label', or **signal sequence** of residues within the polypeptide. Often the signal sequence is located close to the N-terminus and is deleted once the cellular site has been reached and the label is no longer needed. Sometimes the signal sequence is located within the coding region and is not removed (e.g. with nuclear import of proteins). The signal sequences are recognised and bound by specific receptors at the correct cellular site for the polypeptide.

Polypeptides synthesised on ribosomes bound to the endoplasmic reticulum (ER) follow a different pathway from polypeptides synthesised on free ribosomes (Fig. 11.16). Polypeptides entering the ER have an N-terminal signal sequence of 16 residues with a high content of hydrophobic amino acids (Chapter 1). This signal sequence is translated by a ribosome while the ribosome is still free in the cytosol. Specific receptors on the ER membrane recognise the signal sequence and insert it into the ER membrane, binding the ribosome to the ER. Translation of the remainder of the polypeptide proceeds, with the product being extruded across the membrane and into the ER lumen.

The N-terminal signal sequence embedded in the ER membrane may be cleaved from the newly synthesised polypeptide. This usually results in the protein being released into the lumen of the ER. These proteins usually proceed to the Golgi apparatus and then either to lysosomes, or to the plasma membrane or extracellular space via specialised membrane vesicles (Chapter 3). The N-terminal signal sequence is not trimmed from proteins destined to remain embedded in the membrane. Some of these membrane-bound proteins will become part of the plasma membrane of the cell.

> Polypeptides are targeted to specific cellular sites by signal sequences that bind receptors at the sites.

Regulation of gene expression: an introduction

Genetic information: beyond the genetic code

We use the term genetic code to refer to the rules by which a nucleotide sequence determines the sequence of amino acids in a polypeptide. However, there is much more to the information stored in a DNA sequence than just the encryption of amino acid sequences of polypeptides. There is also information that specifies when, where and under what conditions the proteins are synthesised. This information enables cells to respond in specific ways to particular environmental conditions. For example, in photosynthetic organisms, genes leading to the formation of an active

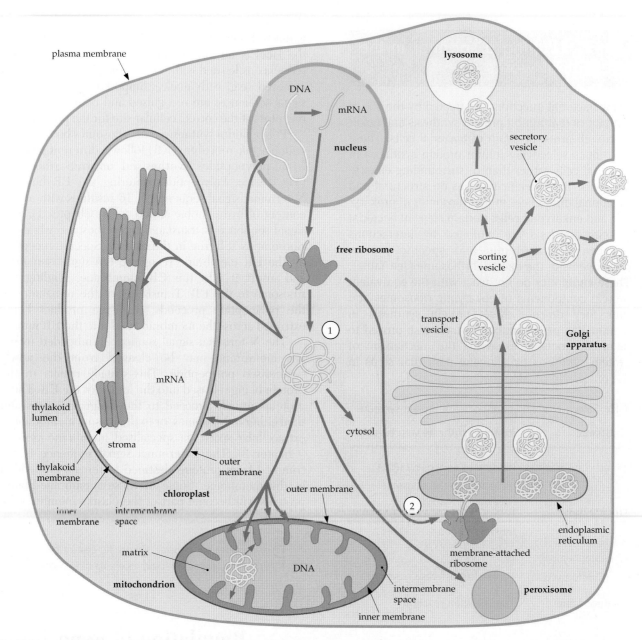

Fig. 11.16 Pathways of protein targeting in eukaryotic cells. Chromosomal genes in the nucleus encode mRNA molecules that move to the cytosol for translation by ribosomes. In one major targeting pathway (1), proteins synthesised on free ribosomes may remain in the cytosol or be directed to one of several destinations within membrane-enclosed organelles—the nucleus, peroxisomes, mitochondria or chloroplasts. A second major pathway (2) involves ribosomes that become bound to the endoplasmic reticulum (ER). Proteins pass into the lumen of the ER as they are synthesised. They may remain in the ER or move further to the Golgi apparatus, emerging within transport vesicles, then to sorting vesicles. Some proteins go to lysosomes, others to secretory vesicles for exocytosis. Some proteins may remain embedded in a membrane that ends up enclosing a particular compartment or becomes part of the plasma membrane itself

photosynthetic system are only expressed when the light intensity is sufficient for photosynthesis to occur.

Unicellular organisms regulate gene expression so that their metabolic and biosynthetic pathways change in response to changes in their environment. For example, bacteria such as *Escherichia coli*, growing in the presence of lactose, synthesise a set of enzymes that allow lactose to be utilised as an energy source. If lactose is replaced by another sugar, then a different set of metabolic genes are **induced** (expressed) and the genes required for lactose

metabolism are not synthesised (i.e. the genes are **repressed**). Enzymes involved in biosynthetic pathways for the production of molecules such as amino acids, nucleotide bases and vitamins are usually produced only when required. For example, if the amino acid tryptophan is available in the cellular environment, this source will be used and the genes required for the synthesis of tryptophan will be repressed.

Multicellular organisms have the additional capacity to regulate gene expression so that many genes are expressed in specific tissues. This important feature allows, for example, blood cells to synthesise globins, hair and skin cells to synthesise keratins and nerve cells to synthesise enzymes required to make neurotransmitters. The sequential activation of specific genes in different regions of the embryo drives the development of the embryo and the formation of the different tissue types (Chapter 16). Therefore, in addition to the genetic code that determines the sequence of amino acids in a polypeptide, a higher level of regulatory information also stored in the DNA sequence controls whether or not genes will be expressed in particular environmental and/or developmental situations.

Control points for gene expression

Regulation of gene expression may occur at a variety of stages during the processes of transcription, translation and formation of a functional protein.

- *Transcription.* When a protein is required, the gene encoding it may be transcribed at an increased rate. The resultant increase in mRNA levels leads to a corresponding increase in protein production. Regulation of transcription occurs extensively in cells of both unicellular and multicellular organisms.
- *Stability of mRNA transcripts.* Some mRNAs are more stable than others and stability can vary in different cellular circumstances, changing the levels of mRNAs present in the cells.
- *Translation.* Controls can operate on the rate at which particular mRNAs are translated.
- *Protein product.* Degradation of proteins may occur at vastly different rates under different conditions, thus regulating the abundance of particular proteins in a cell. Protein activity may also require posttranslational modification, which can also be regulated.

Gene expression is regulated and a cell uses only a subset of the available genes. Regulation of transcription, mRNA stability, translation and protein degradation or modification can all contribute to the level of gene expression.

Regulation of gene expression in prokaryotes

In prokaryotes, the most important regulatory point is transcription. Few prokaryotic mRNAs are subject to translational control.

Adjacent to the 5′ end of the transcription unit (the transcribed region), prokaryotic genes possess a sequence termed the promoter, which is the site of binding of RNA polymerase prior to the initiation of transcription (Fig. 11.17). The promoter region plays a critical role in transcriptional regulation, as RNA polymerase binding to, and progressing from, the promoter initiates transcription. In bacteria, the DNA sequences encompassing the transcription unit, which usually contains multiple open reading frames (see below), and adjacent sequences such as the promoter, required for transcriptional regulation, are termed an operon.

Genes that encode proteins that are required constantly for metabolism, cell growth and division are often expressed continuously, regardless of the chemical environment. They are said to be expressed at a **constitutive** level. Other genes are *inducible*, that is, they are only expressed in response to particular environmental conditions. The expression of inducible genes is modulated by the products of regulatory genes, which may either induce (**positive regulation**) or repress (**negative regulation**) the transcription rate of genes.

Regulation of the *lac* operon

One of the landmarks in our understanding of the regulation of gene expression came from studies of the lactose (*lac*) operon of *Escherichia coli* (Fig. 11.17) by Jacques Monod, François Jacob and their colleagues. The product of the *lac* operon is a polycistronic mRNA that encodes the genes *lacZ*, *lacY* and *lacA*. The *lacZ* gene encodes the enzyme β-galactosidase, required for the cleavage of lactose into its constituent sugars, glucose and galactose, both of which can be used by the cell as an energy source. The *lacY* gene encodes a transmembrane permease that increases the uptake of lactose into the cell. The role of the *lacA* gene in lactose metabolism is yet to be established.

The presence of lactose induces expression of *lacZ*, *lacY* and *lacA* by increasing the rate of transcription of the *lac* operon polycistronic mRNA (i.e. the *lac* operon is inducible). However, in the absence of lactose, transcription is repressed and the enzymes are not produced. This ensures the efficient use of lactose as an energy source, because the β-galactosidase and permease are produced only when lactose is available in the environment.

The *lac* operon also has a mechanism that prevents expression when lactose and a more efficient energy source such as glucose are available. In this situation, induction by lactose is blocked by a mechanism dependent on the presence of glucose. Here we will look at both of these processes.

Firstly, how does lactose induce expression of the *lac* operon? The key to solving this problem was the use of genetic analysis to examine the system of regulation. Mutations that affect the ability of the cell to

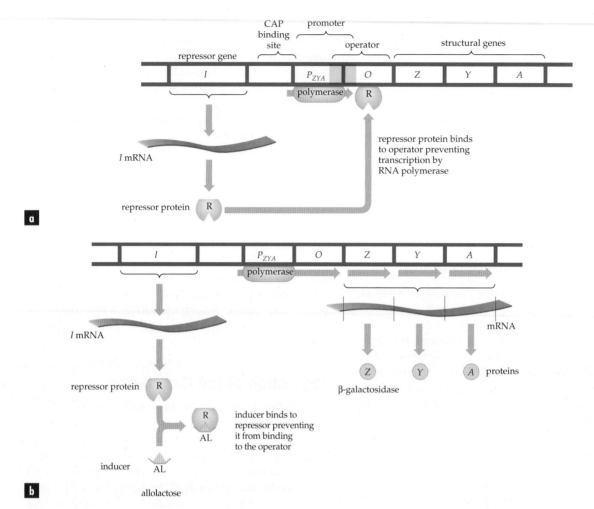

Fig. 11.17 Regulation of the lactose (*lac*) operon of *Escherichia coli*. **(a)** The repressor protein (R), expressed constitutively by the repressor gene (*I*), binds to the operator region (*O*). The operator is a short sequence positioned between the promoter (*P*) and the transcribed DNA. Binding of the repressor protein to the operator prevents RNA polymerase from binding to the promoter, thereby blocking transcription. **(b)** Allolactose acts as an inducer by binding to the repressor protein. With inducer bound, the repressor protein can no longer bind to the operator. This allows transcription to be initiated at the promoter. The synthesis of proteins needed for lactose metabolism, encoded by the *Z*, *Y* and *A* genes, then takes place

metabolise lactose were generated. For example, some mutations result in the constitutive expression of the *lac* operon (i.e. expression even in the absence of lactose) or complete absence of expression in the presence of lactose. Linkage analysis and the examination of phenotypes of different allele combinations (Chapter 9) were then used to locate the site of these mutations and study their interactions.

The *lac* operon and negative regulation: repression in the absence of lactose

The genetic analysis of *lac* expression identified two loci that are involved in regulation of *lac* operon transcription. Mutations at either locus cause constitutive expression of the *lac* operon proteins; that is, the mutations cause the *lac* operon to be transcribed and the *lac* operon proteins to be made at a steady rate

independent of the presence or absence of lactose. One of these loci, *lacI* (for inducibility), exerts its effect on *lac* operon expression regardless of its position in bacterial DNA. In genetic terms, this gene is said to act in *trans* (i.e. the gene does not need to be linked to the *lac* operon to exert its effect on *lac* operon expression). The ability to modify expression in *trans* implies that the *lacI* gene expresses a protein product that diffuses to the *lac* operon to *repress* transcription. This *trans*-acting product of the *lacI* gene is termed the **repressor protein**. The second regulatory locus, *lacO* (for operator), only exerts a regulatory effect on the adjacent *lac* operon. The *lacO^c* mutant allele, which leads to constitutive expression of the adjacent *lac* operon, does not affect a second unlinked *lac* operon introduced into the same cell. This locus is, therefore, said to act in *cis*. If *lacO* produced a protein product, we would expect that protein to be able to diffuse and act in *trans* to regulate any *lac* operon in the cell. The inference of the *cis*-regulatory action of the *lacO* locus is, therefore, that this locus does not produce a diffusible product. The *lacO* locus lies between the promoter, *P* (the RNA polymerase binding site), and the *lacZ*, *lacY* and *lacA* genes of the *lac* operon (Fig. 11.17).

The explanation for these observations is that the *lac* repressor is constitutively produced by the *lacI* gene and binds specifically to the *lacO* operator sequence (i.e. the *lac* repressor is a sequence-specific DNA binding protein). When bound, it prevents transcription of the *lac* operon. In the absence of lactose, this repressor binds to the *lacO* operator locus and represses transcription of the operon by physically blocking access of the RNA polymerase to the promoter. Thus, the repressor exerts *negative control* on transcription.

However, when lactose is available to the cell, it is metabolised to produce a molecule, allolactose, which induces transcription. This inducer binds to the *lac* repressor protein and induces a shape change that prevents the *lac* repressor from binding to the operator sequence. The interaction between the inducer and *lac* repressor, which causes a protein shape change so that it can no longer bind to the operator, is referred to as an **allosteric interaction**. The consequence of the presence of lactose, then, is that the operator, *lacO*, is not bound by the repressor–allolactose complex, permitting RNA polymerase to transcribe the *lacZ*, *lacY* and *lacA* genes, leading to the efficient metabolism of lactose (Fig. 11.17). If the level of lactose declines, free repressor will again become available to bind to the operator, preventing transcription. The consequence is that *lac* operon expression only occurs when lactose is present.

Accounting for the *lac* mutant phenotypes

This model of *lac* operon regulation explains the *lacI* and *lacO* mutant phenotypes. Most *lacI* and *lacO* mutant alleles result in constitutive expression of the *lac* operon (i.e. synthesis even in the absence of lactose). Most *lacI* mutations result in a non-functional repressor, which cannot bind to the operator. As a consequence, repression of the *lac* operon cannot occur. Conversely, most mutations in *lacO* change the operator sequence so that it can no longer bind the *lac* repressor. This also prevents *lac* repressor from blocking transcription. In either case, the result is constitutive *lac* operon expression.

Expression of *lac* operon genes is negatively regulated. The product of the *lacI* gene is a sequence-specific DNA binding protein, which binds to the operator and prevents transcription by RNA polymerase. In the presence of lactose, an inducer molecule forms a complex with the repressor and induces a shape change that prevents the repressor binding to the operator, allowing transcription of the *lac* operon.

Regulation of the *lac* operon: glucose repression and positive regulation

The *lacI* and *lacO* loci mediate repression of the *lac* operon when lactose is absent. However, the *lac* operon has a more sophisticated level of control. In the presence of the more efficient energy source, glucose, high levels of induction are not observed, even in the presence of lactose.

The *lacI* and *lacO* loci do not explain how the presence of glucose can override induction by lactose. The action of an additional factor, termed the catabolite activator protein (CAP), accounts for the effect of glucose on *lac* operon expression. Like the *lac* repressor, the CAP protein can form a complex with a small metabolite, in this case the molecule cyclic AMP (cAMP). Formation of this complex induces an allosteric interaction, but unlike the *lac* repressor and *lac* repressor–allolactose complex, the CAP–cAMP complex binds adjacent to the *lac* promoter, while CAP alone cannot (Fig. 11.18). Rather than block RNA polymerase progression, the bound CAP–cAMP complex enhances the binding of RNA polymerase, greatly increasing the rate of transcription and therefore expression of the *lac* operon. In the absence of cAMP, expression is very low irrespective of whether the *lac* repressor is bound to the operator. In the presence of cAMP and when lactose is also present (i.e. the *lac* repressor is not bound), the level of transcription is very high (Fig. 11.18).

How is it that cAMP can mediate regulation by glucose? The reason is that glucose availability lowers the concentration of cAMP in the cell. The CAP–cAMP complex does not form and the *lac* operon cannot be expressed at high levels, even in the presence of lactose. In the absence of glucose, cAMP levels rise and the CAP–cAMP complex acts to increase transcription, but only when lactose is present (i.e. when repressor is not bound). Induction of transcription by a protein complex bound to the locus is termed *positive regulation*.

The *lac* operon is expressed at high levels when lactose is present and glucose is absent. In the absence of glucose, a complex forms between the catabolite activator protein (CAP) and cAMP. This complex binds to the *lac* operon and acts to positively regulate *lac* operon transcription by increasing the rate at which RNA polymerase binds to the promoter.

Regulation of the *trp* operon

Regulation of the *trp* operon in *E. coli*, which expresses genes required for tryptophan biosynthesis, provides an important variation on the theme of transcriptional regulation illustrated using the *lac* operon. The *trp* operon consists of five genes required for tryptophan

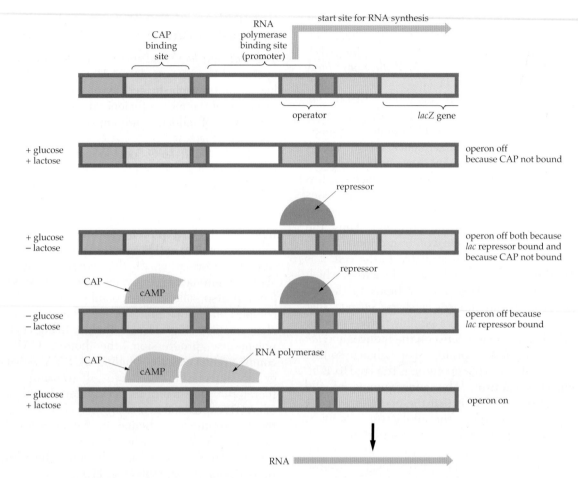

Fig. 11.18 Regulation of the *lac* operon by glucose. Absence of glucose raises the level of cAMP, which complexes with CAP and causes a conformational change that allows the CAP–cAMP complex to bind to the CAP binding site. The bound complex enhances the binding of RNA polymerase to the promoter and initiation of transcription (positive regulation). However, transcription can only occur if the repressor is not bound. The consequence of the action of the repressor and CAP regulatory proteins and their allosteric effectors is that the *lac* operon is transcribed only when lactose is present and glucose is absent

biosynthesis. If tryptophan is available in the environment, it is inefficient to continue *trp* operon expression. As a result, the *trp* operon is repressed in the presence of tryptophan and activated in its absence.

The mode of *trp* operon regulation contrasts with that of the *lac* operon in that it is the absence of tryptophan that leads to *trp* operon expression, whereas it is the presence of lactose that leads to *lac* operon expression. However the basic regulatory components are similar. The *trp* operon contains a promoter adjacent to the 5′ end of the transcription unit, a *cis*-acting operator and a repressor (Fig. 11.19). Tryptophan interacts allosterically with the *trp* repressor to cause a change in the conformation of the repressor. The tryptophan–*trp* repressor complex binds to the *trp* operator but the *trp* repressor alone does not. When bound (i.e. when tryptophan is present), the repressor complex prevents transcription (Fig. 11.19). When tryptophan levels are low, the repressor does not complex with tryptophan. As a result, it cannot bind to the *trp* operator and repress transcription. In low levels

of tryptophan, therefore, the *trp* operon is induced, resulting in the biosynthesis of tryptophan.

Repression of the *trp* operon in the presence of tryptophan is mediated by formation of a complex between the *trp* repressor and tryptophan. The binding of tryptophan causes a conformational change in the repressor that allows it to bind to the *trp* operator and block transcription. In the absence of tryptophan, the *trp* repressor cannot bind to the *trp* operator, resulting in transcription of the *trp* operon.

Regulation of gene expression in eukaryotes

Gene regulation plays an important role in the responses of eukaryotic cells to short- and long-term changes in their environment. In addition, the regulated expression of particular genes underlies the

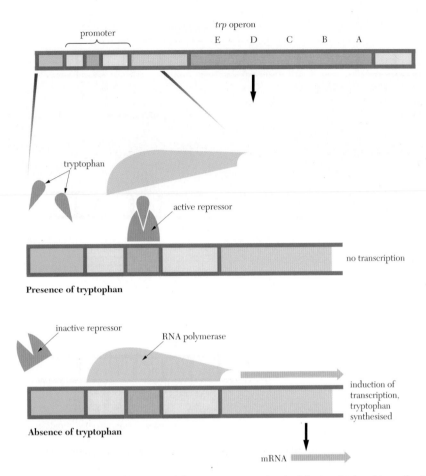

Fig. 11.19 Regulation of the *trp* operon of *E. coli*. The genes of the *trp* operon are required for tryptophan biosynthesis. When tryptophan is present, it binds to the *trp* repressor and causes a conformational change that allows the repressor to bind to DNA and block transcription. When tryptophan is absent, *trp* repressor cannot bind to the operator, allowing RNA polymerase to bind and transcribe the *trp* operon

development of embryos and the differentiation of cells in multicellular eukaryotes (Chapter 16). The expression of genes in the cells of multicellular organisms frequently occurs in response to intercellular signals. Some of these signals affect gene expression by binding to the cell surface and triggering signal transduction cascades (Chapter 16), while others must pass into the cytoplasm of the cell to elicit a response (Chapter 6).

Regulation of the initiation of transcription is a major control point for eukaryotic gene expression. The mechanisms of prokaryotic gene regulation provide important concepts that apply also to gene regulation in eukaryotes. For example, transcription of most eukaryotic protein-coding genes is started by an initiation complex consisting of RNA polymerase II and several other protein factors, which must bind efficiently to a promoter to activate transcription. In addition, there are many proteins that bind to specific DNA sequences to regulate transcription. These proteins are termed **transcription factors**, while the DNA sequences to which they bind are termed **enhancers** if they act to stimulate transcription, or **silencers** if they act to suppress transcription (Fig. 11.20). Tissue-specific gene

expression can usually be explained by the presence of different sets of transcription factors in different tissues.

However, there are also important differences between prokaryotic and eukaryotic mechanisms of transcription. The most important of these is that, unlike prokaryotic RNA polymerase, eukaryotic RNA polymerase II alone cannot initiate transcription. Positive regulation by transcription factors binding to enhancers is an essential part of gene expression in eukaryotes.

Active eukaryotic genes characteristically possess promoter elements to which complexes containing RNA polymerase bind. Transcriptional regulation requires activator and repressor proteins, termed transcription factors, which bind to regulatory elements termed enhancers and silencers.

Yeast *GAL* regulation: the modular nature of eukaryotic transcription factors

Molecular genetic analysis of the type used to elucidate *E. coli lac* and *trp* operon regulation has been applied

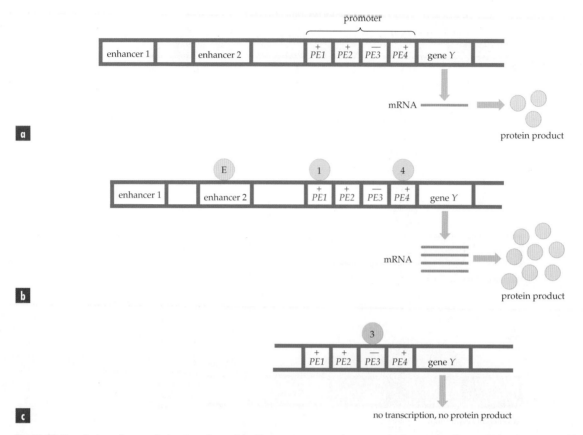

Fig. 11.20 Regulation of transcription in eukaryotic cells occurs at more than one regulatory site. Transcription is increased by transcription factors (E) that bind to enhancers, which may be up to 50 kb (50 000 base pairs) from the RNA start site and upstream or downstream of the promoter. The promoter may be comprised of several subsites, called promoter elements (PE1, PE2, etc.), at which different regulatory proteins bind. These proteins may act as activators (+) or repressors (−) of transcription. Three states are shown here. **(a)** A basal level of transcription occurs in the absence of any proteins binding at regulatory sites upstream of the gene. **(b)** Positive regulation occurs when proteins bind to an enhancer element and promoter elements (PE1 and PE4). **(c)** Negative regulation occurs when proteins are bound to a silencer element (PE3)

to the study of gene regulation in eukaryotes. Study of the budding yeast, *Saccharomyces cerevisiae* (known as brewer's or baker's yeast), has been at the forefront.

Regulation of the genes responsible for galactose metabolism provides an excellent example of transcriptional control in a eukaryote (Fig. 11.21). When yeast cells are grown in a medium containing galactose, the expression of five genes is induced. Rather than being grouped into a single cistron, as in prokaryotes, all five *GAL* genes are unlinked and are transcribed independently of one another. The expression of these genes is controlled by a positive regulator encoded by the *GAL4* gene and a negative regulator encoded by the *GAL80* gene, both of which are constitutively expressed. These regulatory genes were discovered because mutations in *GAL4* result in the co-ordinate loss of expression of the five *GAL* metabolism genes, while mutations in *GAL80* result in constitutive expression of these genes.

The product of the *GAL4* gene binds to sequences upstream of each of the five genes, inducing their

expression (Fig. 11.21). The *GAL4* binding sequence is referred to as an *upstream activator sequence* (UAS). The UAS upstream of each of the *GAL* genes provides a single DNA binding site for the *GAL4* activator protein and is sufficient to accelerate transcription of each of these genes. The protein encoded by the *GAL80* gene does not bind to DNA but represses transcription by binding directly to the *GAL4* protein and preventing it from acting as an inducer of transcription.

In the absence of galactose, therefore, the five galactose-inducible genes are inactive because the *GAL4* activator bound to the UAS sequence is complexed with the *GAL80* protein. When galactose is available to the cell, a metabolic product acts as an inducer by causing the *GAL80* protein to dissociate from the *GAL4*–*GAL80* complex. This allows the bound *GAL4* protein to activate transcription (Fig. 11.21). Protein–protein interactions such as those observed for *GAL4* and *GAL80* are important in the control of eukaryotic gene expression.

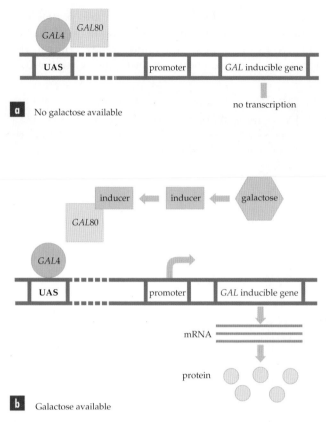

Fig. 11.21 Regulation of galactose (*GAL*)-inducible genes in yeast. A positive regulatory promoter element in yeast is called a UAS (upstream activator sequence). Two regulatory proteins, *GAL4* and *GAL80*, are synthesised constitutively. **(a)** In the absence of galactose, binding of the *GAL4*–*GAL80* complex to the UAS for the galactose-inducible gene prevents transcriptional activation. **(b)** In the presence of galactose, a metabolic product of galactose binds to *GAL80* and prevents it from binding to *GAL4*. *GAL4* is now free to activate transcription

> Regulation of galactose-metabolising enzymes is achieved in yeast by regulatory elements (UAS sequences) and regulatory proteins that bind to the UAS or complex with proteins bound at the UAS.

Enhancers and silencers: the modular nature of eukaryote gene expression

A characteristic feature of eukaryotic genes is that they have multiple regulatory elements, *enhancers* and *silencers*, that can independently regulate transcription of an adjacent transcription unit (Fig. 11.20). These regulatory elements act in a modular way, that is, they regulate the expression of a gene independent of any other enhancers that regulate the same gene. Enhancers and silencers differ from prokaryotic regulatory elements in that they can reside at varying distances from the promoter and do not need any particular orientation or proximity to the promoter to function. Enhancers and silencers can be found some distance from the 5′ or 3′ end of the transcription unit

or within introns. It is presumed that DNA folding brings the transcription factors, bound to the enhancers, to the site of the promoter, where initiation of transcription is stimulated. Different enhancers may drive expression of the same gene in different tissues. For example, in Figure 11.20, enhancer 1 may drive expression of the gene in one type of cell, because those cells have the transcription factors that bind to that enhancer. Enhancer 2 may drive expression of the same gene in a different cell type because of the presence of different transcription factors in that cell type. The modular nature of these regulatory sequences can be demonstrated using recombinant DNA (Chapter 13). When an enhancer from one gene is placed adjacent to a transcription unit of another gene, the enhancer exerts its regulatory effect on the foreign transcription unit.

Activation and repression: the importance of nucleosomes

The wrapping of DNA into nucleosomes (Chapter 10) hinders transcription by RNA polymerase II. A variety of mechanisms exist in eukaryotes to modify nucleosomes and facilitate transcription. One of the primary mechanisms involves the chemical modification of histones within the nucleosome, specifically by the addition or removal of negatively charged acetyl groups. DNA is highly negatively charged because of the phosphate backbone (Chapter 1). Histones, the proteins that form the nucleosome, are highly positively charged, enabling the formation of a nucleosome complex with tightly wrapped DNA (Chapter 10). The addition of acetyl groups to the histones makes them less positively charged, reducing the strength with which they bind to DNA. It is thought that this weaker interaction alleviates the repression exerted by tight binding of the DNA within nucleosomes. Enhancer-binding complexes that activate transcription often include enzymes that acetylate histones.

Repression is also an important component of gene expression in eukaryotes. The mechanisms by which genes are repressed are poorly understood. However, protein repression complexes that bind to silencers often contain enzymes that remove the acetyl groups from histones and permit the formation of more tightly bound nucleosomes, inhibiting transcription.

> Acetylation of histones within nucleosomes plays a critical role in eukaryotic gene expression. Acetylation lowers the overall charge of histones, decreasing the strength of interactions between the positively charged histones and negatively charged DNA, favouring transcription. Activation complexes frequently include enzymes that acetylate histones, while repression complexes frequently include enzymes that remove acetyl groups from histones.

Gene regulatory cascades in eukaryotes

Eukaryotic genes are monocistronic. Genes at different loci are co-ordinately regulated by the same transcription factors acting on equivalent enhancer and silencer sequences at each locus. If one of these genes itself encodes a transcription factor that regulates yet other genes, a cascade of gene expression occurs. A large number of genes can therefore be controlled by a few genes at each step in a cascade of regulation.

Regulatory cascades such as these are particularly important in the embryonic development of higher eukaryotes in which 'homeotic' genes control the formation of entire body parts or tissues (Chapter 16).

Regulatory elements can operate so that the products of one gene control the activity of other genes, some of which may themselves regulate yet other genes. Activation of one gene can therefore initiate a cascade of regulatory events.

Summary

- Gene expression involves the flow of information from DNA to an intermediate molecule, mRNA (the process termed transcription), which then specifies the amino acid sequence of a polypeptide and directs the assembly of that polypeptide at a ribosome (the process termed translation).

- The common genetic code is made up of three-letter code words (triplet codons). Each group of three nucleotides of DNA codes for an amino acid or signals termination of translation.

- Transcription begins when RNA polymerase binds to the initiation sequence in DNA. The region of transcribed DNA between initiation and termination signals defines a single transcription unit. The RNA product directly synthesised within a transcription unit is a primary transcript. The primary RNA transcript in eukaryotic cells is often much longer than the mature mRNA that is translated in the cytoplasm because of the removal of introns.

- In prokaryotes, mRNA is synthesised in the cytosol and translation may commence before transcription is complete. In eukaryotic cells, the mature mRNA moves from the nucleus to the cytoplasm, where translation occurs.

- Ribosomal RNA combines with a defined set of proteins to form ribosomes where polypeptide assembly takes place. Transfer RNA must be charged with the amino acid corresponding to its anticodon (aminoacyl-tRNA) before it can fulfil its function as an adaptor between a codon in the mRNA and an amino acid.

- Codons within mRNA specify the exact order in which amino acids will be linked together during protein synthesis. In eukaryotes, only one reading frame is utilised in each mRNA molecule (monocistronic). Bacterial mRNAs may contain two or more functional reading frames coding for different proteins (polycistronic).

- Elongation of the growing polypeptide involves the sequential addition of amino acids linked by peptide bonds as the ribosome moves along the mRNA. When a stop codon is reached, termination of translation and release of the polypeptide occur.

- Proteins are transported to their active sites by protein targeting mechanisms, which involve signal sequences within the proteins associating with receptors at the site to which the protein is targeted.

- Gene expression is regulated (i.e. cells use only a subset of the available genes at any one time or in any particular tissue). Regulation may occur at the point of transcription, in the stability of mRNA, during translation or even later in the lifetime of a protein.

- Control of *lac* operon transcription by the presence of lactose is an example of negative regulation of transcription. Transcription of the *lac* operon in the absence of glucose provides an example of positive regulation of transcription.

- Active eukaryotic genes possess a promoter at which the RNA polymerase complex assembles. Transcription requires activation by transcription factors, which bind to enhancer elements. Eukaryotic transcription may be controlled by the action of both positive and negative transcription factors.

key terms

allosteric interaction
   (p. 269)
codon (p. 257)
constitutive gene
   (p. 267)
enhancer (p. 271)
exon (p. 256)
frameshift mutation
   (p. 258)
induced gene (p. 266)
intron (p. 256)

missense mutation
   (p. 259)
negative regulation
   (p. 267)
nonsense mutation
   (p. 259)
open reading frame
   (p. 257)
polyadenylation
   (p. 254)
polycistronic (p. 253)

positive regulation
   (p. 267)
primary transcript
   (p. 252)
repressed gene
   (p. 266)
repressor protein
   (p. 268)
ribozyme (p. 257)
signal sequence
   (p. 265)

silencer (p. 271)
splicing (p. 256)
synonymous codon
   (p. 257)
transcription (p. 251)
transcription factor
   (p. 271)
transcription unit
   (p. 252)
translation (p. 251)

Review questions

1. Describe the key features of the common genetic code.
2. How are DNA and RNA sequences involved in the two stages of gene expression?
3. What is the function of a promoter?
4. How do the mRNA molecules synthesised in eukaryotes differ from those synthesised in prokaryotes?
5. Why is precise excision of intron sequences from eukaryotic transcripts important?
6. Describe the structure and function of tRNA molecules.
7. What are the roles of the P and A ribosomal sites in protein synthesis?
8. What is the relationship of a codon to an anticodon?
9. In the *lac* operon, when cells growing in a non-lactose-containing medium are switched to a medium containing only lactose, what is the effect on gene expression?
10. How do nucleosomes influence eukaryotic transcriptional regulation?
11. In eukaryotes, what sequence elements other than the promoter are important in regulating transcription?

Extension questions

1. The *lacI* gene encodes the *lac* repressor. Most mutations in this gene lead to the production of non-functional *lac* repressor and constitutive expression of the *lac* operon. Rare *lacI* mutations cause 'super-repression' of the *lac* operon, that is, they block the ability of lactose to induce transcription. How might a *lacI* mutation have such an effect?

2. Enhancer elements in eukaryotic genes regulate adjacent transcription units independent of their orientation or distance from the transcription unit. How might interactions between such enhancer elements and the promoter occur?

3. *Aspergillus nidulans* is a eukaryote that utilises a variety of compounds as energy sources. In the presence of glucose, genes that are required to metabolise less efficient sources of energy are repressed (similar to the situation for the *lac* operon of *E. coli*). A mutation in a single *A. nidulans* gene, *creA*, results in expression of these genes in the presence of glucose. What does this suggest about the mechanism of glucose-mediated repression in *A. nidulans*?

Suggested further reading

Cairns, J., Stent, G. S., Watson, J. D. (eds). (1992). *Phage and the Origins of Molecular Biology*. Cold Spring Harbor: Cold Spring Harbor Laboratory Press.

An interesting historical discussion of the beginnings of bacteriophage molecular genetics and the contribution these studies have made to our understanding of gene expression.

Lewin, B. (2000). *Genes VII*. Oxford: Oxford University Press.

A comprehensive and advanced textbook on gene structure and function.

Miller, J. H. (1996). *Discovering Molecular Genetics*. Cold Spring Harbor: Cold Spring Harbor Laboratory Press.

A modern look at the foundations of molecular genetics.

12

Genomes, mutation and cancer

Genomes

The sum of all the genetic information of an organism, stored as DNA sequence, is termed the **genome**. The genome encodes the instructions required for an organism to exist, develop and reproduce. The genome includes all genes carried by an organism and also contains DNA sequences that do not encode genes.

One of the most significant developments in the history of science is taking place now in the form of projects that are determining the complete sequence of genomes. Why is such a vast amount of money, time and effort being invested in these projects? The answer is that the genome controls or influences nearly all of the characteristics, or phenotype (Chapter 9), of an individual. The characteristics of different species are distinct because their genomes are different. Differences between individuals within a species are also, to a large extent, the result of genome differences. The similarity of monozygotic (identical) twins, even if reared apart, illustrates the influence of the genome on characteristics.

Many phenotypic differences between individuals, including height, weight, eye colour, susceptibility to particular diseases, sensitivity to certain drugs and differences in personality are all, to some extent, the result of differences between the genome sequences of different individuals. It is not only the nature of the inherited genome that is important. Genomes can change during the life of an organism. The ability of parasites of the genus *Trypanosoma* (the cause of sleeping sickness) to maintain debilitating long-term infections depends on regulated genome rearrangements, while our ability to fight infections also relies on the genome rearrangements that create antibody genes. Cancer arises because of changes that occur in the genome of somatic cells during our lifetimes. Understanding genomes and their relationships to phenotypes is therefore central to understanding organisms.

Genome evolution

The complexity of organisms ranges tremendously (consider the difference between a bacterium and a human) and is reflected in the structure and organisation of genomes. Related organisms have similarities in their genomes that underlie their overall biochemical, developmental and behavioural characteristics. The more related the organisms, the more similar are their genomes. For example, the genomes of humans and chimpanzees are 98% identical, while those of humans and the nematode, *Caenorhabditis elegans*, share only very small blocks of limited similarity. The discovery of genome sequence relationships has yielded, perhaps, the most significant recent evidence in support of the evolutionary origin of species, providing strong and independent support for the idea that related organisms share a common ancestor. The analysis of DNA sequences is providing a lot of additional information about the nature and course of evolution. Life has a history and, in a very real sense, that evolutionary history is written in our genomes.

Genome organisation: an overview

Genomes are typically organised into long stretches of linear or circular DNA contained in chromosomes (Chapter 11). Genomes can contain from one to a large number of different types of chromosomes. In the case of humans, there are 24 types of chromosomes, 22 autosomes and an X and a Y chromosome. In addition, the number of sets of chromosomes found in a cell can vary. Human somatic cells are diploid, that is, they have two copies of each autosome. Other species can have higher copy numbers. For instance, oilseed rape is tetraploid (four sets of chromosomes), while bread wheat is hexaploid (six copies of each chromosome type in every cell).

A eukaryotic cell contains more than one genome, as both chloroplast and mitochondrial organelles contain their own genomes separate from the nuclear genome. In addition, genomes can contain extra-chromosomal DNA. These are additional pieces of DNA, almost like miniature chromosomes, which may not be stably inherited. Prokaryotes, for example, can have plasmids, small circular double-stranded DNA molecules, which typically carry genes for special characteristics such as resistance to antibiotic drugs. Bacterial plasmids are useful in recombinant DNA technology (Chapter 13).

The era of genome analysis

The success of genome sequencing projects is changing the way biologists study living systems. In 1995, the first complete genome sequence of a free-living organism was determined, that of the bacterium *Haemophilus influenzae*. By the year 2000, complete DNA sequences of over 40 free-living organisms from the three major branches of life, the eubacteria, archaea and eukaryotes, had been determined. The challenges of the future lie in understanding how genomes encode information and influence phenotypes, and how genome differences account for variation between living organisms.

The genome is the term given to the sum of all the genetic material held by an organism or organelle. Differences between genomes are responsible for differences in characteristics between individuals of different species and for many of the differences between individuals of the same species. The more closely related the species, the more similar are their genome sequences.

The relative sizes of genomes

Genomes vary tremendously in size. Even viral genomes vary from a few thousand base pairs, such as the genome of the bacteriophage ØX174 (5375 nucleotides, the first ever complete genome sequence determined) to the 330 742 nucleotides of the chlorella virus, which infects *Paramecium bursaria*. Genomes are measured as the total number of base pairs of DNA in a single (haploid) set of different chromosomes of a particular organism. The most common units used to describe gene and genome sizes are base pairs (bp, or nucleotide pair), kilobase pairs (kb, or 1000 base pairs) and megabase pairs (Mb, 1000 kb or a million base pairs). Bacterial genomes vary in size from 500 kb to 10 Mb; for example, *Escherichia coli* has a genome of 4.7×10^6 bp or 4.7 Mb.

Humans have a haploid genome of about 3×10^9 bp (3000 Mb). Plant genomes are the largest known, ranging in size from a modest 100 Mb for *Arabidopsis thaliana* to 5300 Mb for barley. Examples of the genome sizes of representative organisms are given in Table 12.1.

Gene numbers in genomes

Genomes vary in the number of genes they contain. The bacterium *Mycoplasma genitalium* has the smallest known genome of any cell with only 483 genes, but it is not a free-living organism. The absolute minimum set of genes required to sustain a free-living organism is probably around 1000, which is the estimated gene complement in simple organisms such as bacteria of the genus *Rickettsia*, thought to be related to the ancestors of mitochondria. These organisms have genomes of around 1 Mb in size and thus a gene density (genes per kb) of approximately 1 gene/kb.

Although larger genomes tend to contain more genes, there is no strict relationship between genome size and gene number (Fig. 12.1). *E. coli* has about 4300 genes in a 4.6 Mb genome, while the yeast *Saccharomyces cerevisiae* has about 6200 genes in a 12 Mb genome. The *E. coli* gene density is, therefore, around twice that found in *S. cerevisiae*. Similarly, humans need nearly 1000 times as much DNA to encode about 10 times as many genes as *E. coli*. There is also no strict relationship between organism complexity and gene number. The free-living soil nematode *C. elegans* is a simple multicellular animal containing only 1031 somatic cells plus about 1000 germ cells. The genome of this organism encodes 19 100 genes in a 97 Mb genome. The larger and considerably more complex invertebrate, *Drosophila melanogaster*, complete with its two very different

Table 12.1 Genome sizes in selected organisms

Species	Number of chromosomes	Genome size (Mb)	Number of genes
Necturus maculosus (salamander)	12	81 300	ND
Zea mays (maize)	10	4500	ND
Homo sapiens (primate)	23	3000	40 000–60 000[a]
Oryza sativa (rice)	12	450	40 000[a]
Drosophila melanogaster (vinegar fly)	4	185	13 600
Arabidopsis thaliana (mustard)	5	100	20 000[a]
Caenorhabditis elegans (nematode worm)	6	97	19 100
Saccharomyces cerevisiae (yeast)	16	12	6200
Escherichia coli (Gram-negative bacteria)	1	4.6	4300
Bacillus subtilis (Gram-positive bacteria)	1	4.2	4100
Synechocystis (cyanobacterium)	1	3.6	3100
Haemophilus influenzae (Gram-negative bacteria)	1	1.8	1700
Mycoplasma genitalium	1	0.58	483

(*a*) Estimation.

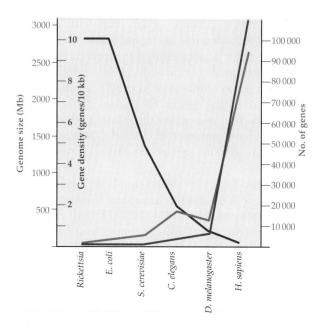

Fig. 12.1 The gene numbers, genome size and average gene densities (expressed here as genes/10 kb) in a range of organisms. There is a general correlation between complexity of the organism and size of the genome and an inverse correlation between genome size and gene density

larval and adult forms, has a genome that encodes 13 600 genes in a 185 Mb genome. The difference in gene density between organisms is due to the presence of introns (non-coding regions) within genes and large variations in the amount of DNA separating genes (intergenic regions). In most cases, it is not obvious why different organisms have different amounts of non-coding DNA.

Discovering genes

Humans have an estimated 40 000–60 000 genes in a 3000 Mb genome. One reason why there still exists a degree of uncertainty about the number of genes in a genome is that the tools used for identifying genes are still being developed. Traditional genetic analysis, using phenotypic variation and genetic mapping (Chapter 9), is slow and can only identify genes at a slow rate. Genomics analysis relies on computer-based approaches to predict the locations of genes from DNA sequences. While these approaches are fast and discover genes far more rapidly than traditional methods are able to, these computer methods are not yet always reliable. As genes encode mRNA transcripts, much effort is currently being placed on identifying genes by characterising their mRNA products. The approach is to make a DNA copy of the mRNA product (termed **complementary DNA** or **cDNA**) and determine its sequence using methods described in Chapter 13. Libraries of cDNAs from different tissues constitute the set of genes that are expressed in those tissues. Each different cDNA sequence is referred to as an **expressed sequence tag**

(**EST**). The generation of libraries of cDNA clones and the determination of their sequences to yield ESTs reveals the set of genes that are expressed in any particular tissue.

> The mRNA molecules that result from transcription can be converted into complementary DNA (cDNA) molecules and the nucleotide sequences determined. These sequences, termed expressed sequence tags (ESTs), correspond to the set of genes that are expressed in the cells from which the mRNA was derived.

Organelle genomes

One of the most exciting evolutionary stories is that of the acquisition of organelles by eukaryotes (Chapter 35). Eukaryotes obtained their organelles (mitochondria and chloroplasts) by the acquisition of bacteria around one billion years ago. It appears that an ancestral eukaryote 'swallowed' an entire bacterial cell in the formation of a symbiotic relationship, recruiting the energy production capacity of the bacterium. The bacterium, in turn, increasingly used the resources of the host cell for its needs, forming the mitochondria that we see today. Chloroplasts are thought to have arisen in a similar way from a cyanobacterial ancestor (Chapter 35). Both of these organelles have retained vestigial genomes, even though many protein components of these organelles are encoded by the nuclear genome and are synthesised in the cell cytosol before being imported into the organelle (Chapter 11). Because of the existence of more than one genome in a eukaryotic cell, the chromosomal DNA in the nucleus is termed the nuclear genome, while the organelle genomes are termed the mitochondrial and chloroplast genomes (Fig. 12.2).

The DNA sequences of over 100 organelle genomes have been completely determined. Mitochondrial genomes range in size from 13 kb for animals to 337 kb for the plant *Arabidopsis thaliana*. Mitochondria typically contain a single circular genome, although some mitochondrial genomes are linear (e.g. in the ciliate *Tetrahymena*). Mitochondrial genomes contain genes required for electron transport and coupled oxidative phosphorylation, translation (ribosomal proteins, rRNA and tRNA), and a small number of other functions. Chloroplast genomes range in size from 35 kb to 156 kb. Chloroplast DNA is highly conserved across the plant kingdom with respect to the number and type of genes (about 160) and their coding sequences (at least 70% sequence similarity). Chloroplast genomes contain rRNA genes, a functional set of tRNA genes using normal codons, some ribosomal proteins and proteins concerned with photosynthesis.

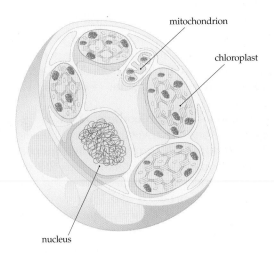

mitochondrion

chloroplast

nucleus

Fig. 12.2 The three eukaryote genomes—nuclear, mitochondrial and chloroplast—shown here diagrammatically in red, each encode their own transcription and translation machinery. Many mitochondrial and chloroplast proteins are encoded by nuclear genes and imported into the respective organelles. The sizes of these genomes vary enormously, nuclear genomes of eukaryotes being in the range of tens to many thousands of Mb in length, while mitochondrial and chloroplast genomes are 0.01–0.3 Mb in length

Chloroplasts and mitochondria contain genomes, typically as small, circular double-stranded DNA molecules. Organelle genomes are highly compact, with much of the DNA coding for rRNA, tRNA or for protein products. Organelles usually contain many copies of their genomes.

Types of DNA sequences

As previously discussed, genomes are composed of DNA molecules, that is, double-stranded strings of the four nucleotides termed A, C, G and T (Chapter 1). The DNA sequence of the genome is the store of information on which living organisms depend for their existence. The cellular machinery is able to identify and use the information in this DNA. This remarkable ability of a cell to recognise and express genes in specific circumstances is itself encoded in the DNA sequence of the genome (see Chapter 11). This information is not easily decoded by simply looking at it. If a human genome were to be printed out in 12 point font, it would cover 600 000 pages of paper. If this sequence were written as a single line, it would stretch from Sydney to Athens! Imagine staring at these three billion G, A, T and C residues and trying to work out where all the genes are (see Fig. 12.3) Making sense of this sequence and representing the data in an informative way is a major challenge to those determining and analysing genome sequences (see Box 12.1).

Coding and non-coding DNA

As you will see below, not all DNA sequences appear to carry information that is used by the organism.

```
   1 AAGCTTTAGT  TTTCCCTTGT  GCAAAATGGG  ACTAGTAATA  ATAGTGTCTA  TCTCGTACAA
  61 GTGTTATTAC  AAGATTATAA  GAGATGATGA  TCTTTAGAAC  AAGTCCTGGC  ACATAGTTCC
 121 CATAAATGTT  AGTTATTGTC  ATGATAGTAT  TATTAAAATA  TTTGTGTTGA  TTTCTTCCAA
 181 TAGACTCCAA  GCTCCATGGG  CTTTCATGTA  GGACCGTGTT  TGTCTTATTC  ATCATGAGCC
 241 TGGCACAGAG  CAGATGGTCT  GAACATATTT  AGGGGATGAA  GCTCTGGGCT  TCATTTTACA
 301 TATGAAGAAA  TGAAGTTCCA  GCAAGATTGA  ATGACTTGCC  CAAGGTCACA  CAGCAGAGGA
 361 GAGCCTGGGG  CTGGGCCCAG  GACTTCCCTT  TCTCTGTCCA  GTGTTCTAAC  ATATCCTGGA
 421 GGACAGATGC  AGAGAGCTAA  GACATTCTCT  TAGATTCTGC  TCTCAGACCC  AGCTTCTCAG
 481 GAATCTTCTG  GAACATGACA  TATGACCTGT  TCAGTTTCCT  CTCTGGGCTT  CAGGGTTCTT
 541 GCTGTTTATG  TTTTCATGGA  CTGAAGCCTC  CAGCTAGGGG  CTGAATTACG  AGAAGCAGAG
 601 TACAACCTGC  TGCACACCAG  GTGAAGCAGA  ACTTCTTGTT  CCCTTTAACT  GTCGCCGAGT
 661 TGAATCACTT  TGTTAAGCTC  ATGCGCTAGA  GTTACTTCCT  ATTTTAAACC  TTATGAATCT
 721 GGGTAGTGTG  GGGGGTTACA  AAATAAATCT  CCCCAGCCAA  GGTGACACCT  CAATTAGACC
 781 TCAAAGGCTT  GCAGTGAAGC  AGTTTGAAAA  TGGATGTTAA  GCACATGCTA  GTTCTCCTTT
 841 AGACTTGCAA  TAAAAAATAG  TGGAGTGATT  GCAAATGAAA  ATGGTTTTTC  AACTTCATTA
 901 AAAAAAAAAA  AAGCTCATCT  ATGTACAAAT  CCAGGAGACA  GGCATATCTT  TTTAAGTTGT
 961 TGCCATTTTA  TCCACATTGT  GCTCCGGCAT  TTACAGTACA  AAGAGATTTT  ACCAAGTTCA
1021 TAGCTTAGAG  CAACCTTGGG  CAGTCAGGAG  GTCTGAGGGA  GAGTCACATC  AGCTGTCGGG
1081 GTGGAATTGT  GGTCATATAT  AGCCTGTCTC  ACATTACAGT  GAGTCAGAGA  CTGACTGTCC
1141 CAAACAGTAT  TTTCTTCTAG  GATGGAAGCG  GTGTGAAGTT  ATCTTCAAGT  CTCTATCCAT
1201 CAGGACAGAT  GCAGGAATTA  TTTTAAAACG  TGTATGGTAA  GAATTTCAGC  CAATGTAGTT
1261 TAAAATAAGA  TGCCCAAGTC  ATTTTTGAGA  TGTAGTAAAA  AATATCGTGT  CTCGATTAAT
1321 AGCATATTTT  TAGCCTCACA  TAGTCAGTAA  TTGTTTTCAA  ATGTGCTTTG  TAATTGAATA
1381 TCCATATTTT  ATTTAGAAAT  CTAGTATGAG  CAGGGAGAAA  GAAACCATAG  AAATTGTTCC
1441 CATAACATCA  GATTACATTA  CCTATGTGTG  TGTATATATA  TATGCACACA  CATATATATA
1501 CATCATATAT  ATCATCAGAT  AAATTTATCA  TGTATGTGTT  TGTGTGTATA  TTCTTTATAT
1561 ATGCACAAAT  AATCTCTAAA  TAACTATCAC  TTCTTAACGT  TTTGAAGAAA  AGGCAGAGAC
1621 TTGGAGAGAT  TGATCTGTCA  AAGGAAATGA  ATTGTCTACT  TTTTGTTTGC  TTAACTGTTT
1681 TATTTCCTAC  TTCATTCCAA  GAGTGGTTTT  TAGTAGTTAC  GGAAATACAC  ACAATATGAT
1741 ATTTTTAGGT  AAAGTGGTA   ATTAAGGAAT  AATTAAGATT  AGGGAAAAAA  ATAAAGCCAC
1801 TCAGTTCTGT  AAGTAAAATA  CTTAACTTCA  AGGCCTAGGC  TGACAAATAC  AAATTATATA
1861 TATTCAAGGT  GTGCAACATG  ATGATTTAAT  ATTCGTATAC  TGTATTAGTC  CATTTTCACA
1921 CTGCTATAAA  GAATACCTGA  GACTGGGTAA  TTTATAAAGA  AAAGAGGTTT  AATTGACTTA
1981 CAGTTCTGCA  TGGCTGGGAG  GTCTCAGGAA  ACTTAAATC   ATGGCGGAAG  GCAAAAGGGA
2041 AGCAAGGCAC  GTCTTATATG  GCGGCAGGAG  AGAGAGAGAG  TGAGGGGGGA  ACTGTCAAAC
2101 ATAAACCATC  AGATCTCATG  AGAACTCACT  CACTATCATA  AGAACAGTAT  GGTGGGAAAC
2161 TCCCCCATGA  TCCAGTCACC  TCCCACCAGA  TCCCTCCTTC  CACACATGGG  GATTACAATT
2221 CAAGATGAAA  TTTGGGTGGG  GACACAGAGC  CAAGCCGTAT  CATACATACT  GTGTAATGAT
2281 TACTATGATC  AAAATAATGA  ACACATCTAT  CACCCTTAGT  TACGATTTGT  GTGTGTGTAT
2341 GTGTGTGGTG  AGGACACTTA  AAATCTGCTC  TCTTACCAAT  TTTTAAATAA  ACAATTCAGT
2401 ATTATTAACT  GTAGTTACCA  TACTTTACAT  TAGATCCCCA  GAACTTATTC  ATCTTATAAC
2461 TGAAAGTTTG  TACCCTTTGA  CCAACATAGA  AGTTTCTTAG  GAAAACCTGC  CAAAGTAAGA
2521 TGGAAATAGA  ATCTATTACA  TAAATTCACA  GTGTCCATAA  GAAAAAAGAC  AAAGGGCCAT
2581 TTGTTCAGGA  AATACTCAGC  GCTTCCCTAA  TACATTATGC  AATGTGAATC  ATGGCCTCAA
2641 AAATGTTCCT  ATAGTGAATA  GAAATATGGC  TCTGAGCTTC  CAGGAAGCCA  CAACAATAAA
2701 GGAAATACGA  TTATTAATGA  TTTATTGTGT  CCTCAAGATT  AAGTGAGTTT  CTTGGAAGCA
2761 CACCTTCCTG  GAGCTGAGAC  TTGATTTTTC  TTTCATAAAG  AGGACACTAT  AACATACTGA
2821 ACACCATTTT  ATTCCCAAAT  ATAGTAAGTT  TCTTAGAGCA  GTGTTTCCTT  CTAGAGGGAA
2881 AACATGAAAG  CCCAGAGTAG  AGGGTGGAGA  CCAGAGAAAG  TTCTTTTCCC  TAATTTTAAT
2941 TTAGACCACA  ATGACAAATC  CTAGCCGATTA  GGATTTCCCT  GTGGTTTAAA  GGCTATTCTT
```

a

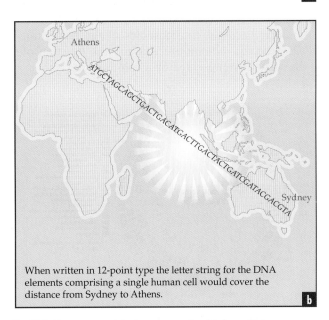

Athens

ATGCCTAGCAGCTGACTGACATGACTTGACTACTGATCGATACGACGTA

Sydney

When written in 12-point type the letter string for the DNA elements comprising a single human cell would cover the distance from Sydney to Athens.

b

Fig. 12.3 Genomes contain large amounts of information. **(a)** A small part of the sequence of chromosome 16 of the human genome. **(b)** A representation of the size of the human genome sequence

However, of those sequences that do carry information, most correspond to specific genes and the most common genes are protein-coding genes. The term **coding sequences** is given to DNA sequences that are ultimately translated into a functional polypeptide. In

BOX 12.1 Annotating complete genome maps

One of the great challenges involved in the analysis of genomes is the problem of dealing with enormous amounts of information. Although the human genome sequence is large (3000 Mb), it can be written to a small number of CD-ROMs. However, this is only the tip of the informational iceberg. To make sense of this information, all of the signals and sequences that constitute a gene, the mRNA products made from each gene, the tissues in which each gene is expressed and the biological roles of each gene need to be identified and described.

Currently, we understand the functions of only a proportion of the genes that have been revealed by DNA sequence analysis. These functions are often inferred from similarity to the sequence of a related gene for which a function has been determined. One way that is currently being used to represent genes and gene functions in completed genome sequences is shown in the figure. In this representation, the location of a gene is represented by a bar at the appropriate position on the DNA map (here shown for one of the two circular bacterial chromosomes of the bacterium *Deinococcus radiodurans* R1) and the function of the gene is represented by a colour code.

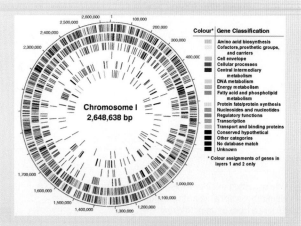

A schematic representation of the location and function of the genes of a circular chromosome found in the bacterium *Deinococcus radiodurans* R1. The genes are denoted by lines around the circular chromosome. The two outer rings show genes transcribed from the two DNA strands of the genome. The next rings, moving inwards, show repeated sequences, transposable elements, transporters and rRNA and tRNA genes, respectively. The colour code indicates the functions predicted for the genes. Note that a number of genes, shown in black, are yet to be associated with a function. *D. radiodurans* R1 was originally identified as a contaminant of irradiated canned meat. It is the most radiation-resistant organism known

eukaryotes, coding sequences are frequently interrupted by introns (Chapter 11), which are non-coding sequences that are interspersed in genomic coding DNA and which are removed from the mRNA product by splicing prior to translation (Figs 12.4, 12.5, see Chapter 11).

The amount of **non-coding DNA** in genomes varies enormously between different species but typically constitutes a surprisingly high proportion of the total genome sequence. As little as 2% of mammalian DNA codes for proteins. Part of the remaining 98% of the DNA encodes 5′ and 3′ untranslated regions (UTR), enhancer and silencer sequences, origins and termini of replication, centromeres and telomeres. However, even these sequences appear to account for only a small proportion of the non-coding DNA.

Two main explanations have been put forward to explain the remaining non-coding DNA sequences. The first is that the non-coding DNA plays a different type of role in the organism, for example, by acting as 'filler' DNA to produce nuclei with the right size and biophysical properties. The second proposes that non-coding sequences are of no value to the organism but nor do they do any harm. This hypothesis proposes that much of the non-coding sequences are 'junk' DNA. Genomes evolve remarkably rapidly and given the vast amount and variable nature of non-coding sequences in many genomes, it is hard to believe that all non-coding sequences provide a specific function to their host organisms. However, as we study genomes, functions are sometimes assigned to stretches of non-coding DNA, so the debate about the role of these sequences is far from over.

Repetitive DNA sequences

Eukaryotic cells contain small sequences of DNA that are repeated many times and have no known function. These are referred to as **repetitive DNA** (Fig. 12.6). As a result of these repetitive sequences (and other non-coding sequences), eukaryotic genes are spaced further apart compared with prokaryotic genes. An example of repetitive DNA in the human genome is the *Alu* family, which is composed of sequence elements of approximately 300 bp in length. These elements are present in about one million copies in the genome and represent about 6% of our total DNA.

In some cases, the repeats are arranged in tandem to form very long stretches of DNA made up of simple repeated sequences. Such DNA is termed **satellite DNA**. In some insects, such as *Drosophila virilis*, over a third of the genome consists of satellite DNA, with over 10 million copies of a few 7 bp sequences! Chromosomes stained with particular dyes typically show two different types of chromatin: a lighter staining and more decondensed form referred to as

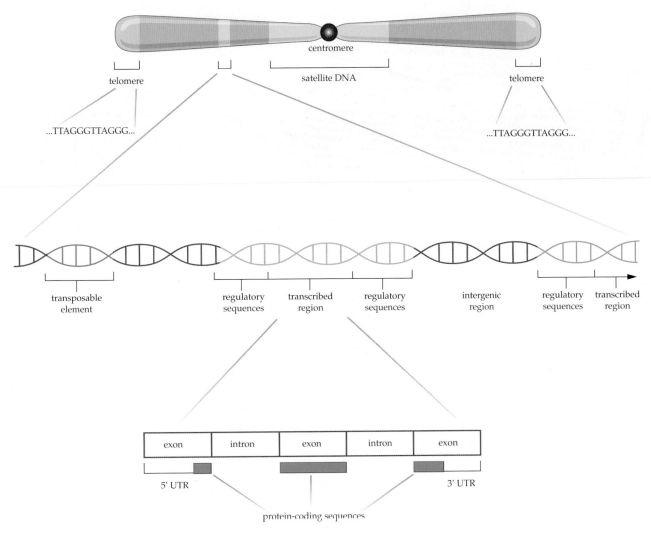

...TTAGGGTTAGGG...

...TTAGGGTTAGGG...

Fig. 12.4 A diagrammatic summary of some of the types of sequences found in a human chromosome. The diagrammatic representation of DNA is not to scale with respect to the sequence types indicated

euchromatin and a darker staining, more highly condensed form referred to as heterochromatin. Satellite DNA typically forms the bulk of the sequences found in heterochromatin, while protein-coding sequences are typically found in euchromatin. The function of satellite DNA is unknown but it may have structural and organisational roles.

Ribosomal DNA (rDNA)

DNA encoding ribosomal RNA genes (rDNA) constitutes a special type of repeated DNA sequence. rDNA encodes the RNA components of ribosomes and is located in the nucleolus (Chapter 3). In most eukaryotes, there are more than 100 tandem (i.e. head-to-tail) repeats of rDNA, each repeat encoding one primary transcript, which is processed to produce both the 18 S and 28 S rRNA molecules. In the oocytes of certain species of toads, there are more than a million copies of rDNA, which together make up about three-quarters of

the genome. Many of these help to produce the large number of ribosomes needed to synthesise the yolk component of egg cells. In contrast, in the *E. coli* chromosome, rDNA sequences are repeated seven times, constituting around 1% of the total genome.

Telomeres and centromeres

Telomere and centromere sequences are found only in eukaryotes. Telomeres are the DNA sequences found at the ends of chromosomes. They are important for the functional stability of linear chromosomes. Telomeric sequences are typically short repeated sequences, TTGGGG in *Tetrahymena thermophila* and TTAGGG in humans. There can be over 2000 of these repeats, generating a telomeric region of 10–15 kb in length.

A centromere is the site of attachment of a chromosome to the spindle. Centromeres are involved in the separation of sister chromatids during mitosis and

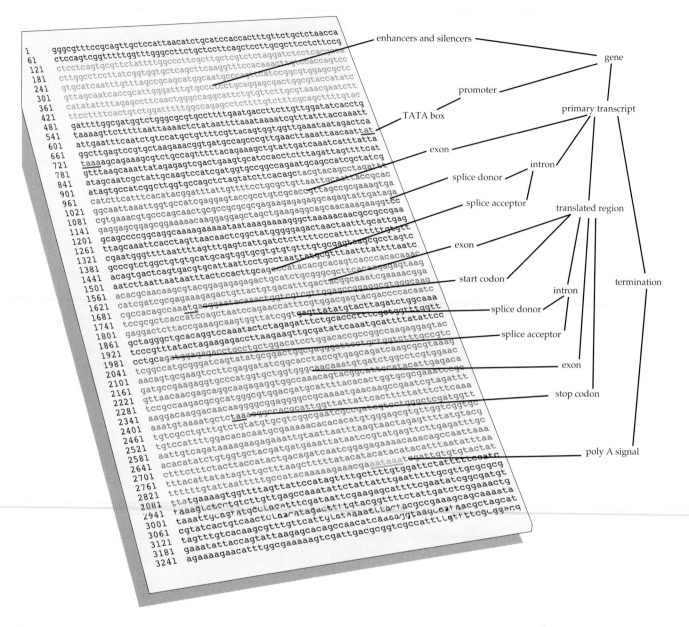

Fig. 12.5 Different parts of a gene, discussed in Chapter 11, can be defined as sequences in genomic DNA

homologues during meiosis (Chapter 8). Curiously, there appear to be no specific sequence requirements for a centromere in many organisms. In the mouse and many other eukaryotes, the centromere is flanked by satellite DNA, giving the centromere a typically hetero-chromatic appearance.

Prokaryotic counterparts to the centromere are termed partitioning sequences, which are attached by specific protein linkages to the growing plasma membrane in the region where the cell is to divide.

Transposable elements and genome plasticity

Transposable elements are DNA sequences that are able to replicate and insert into another location in the genome. These elements were discovered in maize by Barbara McClintock in the early 1940s and have become known as 'jumping genes'. Transposable elements are self-maintaining in the genome because they can replicate and transpose. They have become a valuable tool in genetic engineering (Chapter 13).

Transposable elements come in a variety of forms. In many cases, the elements encode enzymes required for their transposition to new sites in the DNA. These enzymes recognise sequences at the ends of the transposable element, cut the element from its current position and allow it to move to a new position. Other elements function very much like viruses, producing RNA copies of themselves, which are later copied back into DNA and inserted into a new site in the genome.

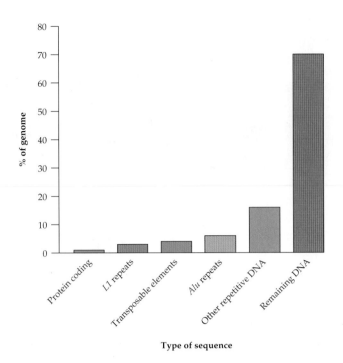

Fig. 12.6 Relative amounts of DNA sequences in the human genome. The proportion of protein-coding DNA is remarkably low, especially compared with the abundance of repetitive sequences

The discovery that particular DNA sequences have the capacity to transpose to new loci within a genome changed the view that genomes are inert. We now know that DNA can transpose in nature, moving from genome to genome and even crossing species boundaries. For example, a transposable element in *D. melanogaster*, termed the P element, appears to have 'jumped' from the genome of another fly species into the *D. melanogaster* genome within the last 100 years. The element then spread through natural populations of *D. melanogaster* across the world with remarkable speed.

The plasticity of genomes is not restricted to transposable elements. There are now many examples of programmed DNA rearrangements used by organisms for specific purposes. For example, some parasites rearrange their DNA to encode new protein sequences that allow them to evade the host immune defence mechanisms. In what appears to have been an evolutionary host-pathogen arms race, the mammalian immune system also rearranges its genes to produce the large repertoire of antibodies that fight infections (see Box 12.2).

Gene families

Most genes that encode proteins exist as single copies. However, some protein-coding genes have been duplicated one or more times and the sequences of the copies have diverged during the course of evolution. These sequence-related genes, such as the genes that

BOX 12.2 The immunological arms race and DNA plasticity

Finding food and mates is not sufficient to ensure survival and procreation. An individual must also survive the threat of becoming a source of food for others. While predators pose an external threat, legions of disease-causing organisms are waiting for the chance to infect individuals and go about their own procreation. Animals and plants have a variety of strategies to defend themselves against infection. Humans have a highly developed immune system, which patrols the body, recognises foreign organisms and destroys them (Chapter 23). The power of the immune system lies in its ability to produce a very large range of antibodies capable of recognising an enormous diversity of foreign molecules.

For about two decades after the structure of DNA was determined, it was thought that DNA molecules in cells were constant, that is, that the DNA sequence remained unchanged through replication and from generation to generation. The discovery of DNA elements that could transpose from site to site changed that idea. Another surprise was in store when immunoglobulin genes were identified and studied. The genomic DNA that encodes immunoglobulin proteins (Chapter 23) undergoes programmed rearrangements in differentiating immune system cells. Although the rearrangements are tightly controlled, they involve steps in which random sequences are generated within the protein-coding part of the gene. The sequence variation that is introduced results in the production of an enormous variety of immunoglobulins from genes that only assume their final sequence following DNA rearrangement.

Curiously, in battling with the immune system, some parasites have adopted the same tactic. In this case, however, the parasite uses rearrangement of its genome during the course of infection to create new coat proteins with new antigenicity, so that it evades the antibodies generated against earlier versions of the coat protein. The best characterised examples occur in the genus *Trypanosoma*, members of which are responsible for diseases such as sleeping sickness, a devastating disease of tropical Africa. Trypanosomes have a large number of genes (perhaps as many as 1000), that encode coat proteins termed variant surface glycoproteins (VSGs), but they only express one of these genes at a time. During infection, as the human immune system mounts its defence by producing antibodies directed against the first trypanosome VSG present during the infection, DNA sequence rearrangements in the trypanosome genome result in a switch in expression to other coat proteins (see figure). The organisms that have switched their coat protein are not recognised by the primary immune response and grow until a second immune response is mounted against the second VSG, by which time some trypanosomes will have switched to yet another type of coat protein. The result is that trypanosomes are able to mount a long-term debilitating infection that is not effectively dealt with by the human immune system.

There are now many more examples of DNA sequence rearrangements that operate during the normal life of organisms. In a simpler version of the immune evasion rearrangements used by trypanosomes, inversion of a DNA fragment in the genome of the bacterium *Salmonella typhimurium* causes switching in expression between two alternative genes. The yeast *S. cerevisiae* also has a sequence rearrangement mechanism that results in the expression of one of two possible alternative genes. The role of these yeast genes is not, however, to deal with the immune system of a host, but to permit haploid cells to switch between two mating types. In yeast, only cells of different mating types can fuse to form a diploid.

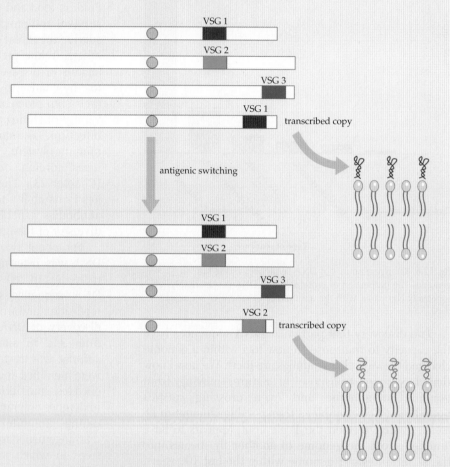

The genomes of trypanosomes have many silent and only one active variant surface glycoprotein (VSG) gene encoding the major surface protein. During growth, sequences from a silent VSG gene can replace those of the transcribed gene, so that a new type of VSG appears on the surface of the parasite, allowing it to evade the immune system

encode globin proteins, constitute a **gene family**. In some cases, the genes are found clustered together at a single locus, while in others, the genes are dispersed in the genome. The mammalian ß-globin genes constitute a clustered gene family in which an ancestral gene has duplicated (Fig. 12.7). The function of the duplicated genes has diverged so that particular β-globin genes function during embryonic, fetal or adult life, possessing biochemical properties that are suited to the different physiological conditions they encounter at the stages in which they are expressed. Duplication and specialisation of members of a gene family appear to have been an important part of genome change in the evolution of greater biological complexity.

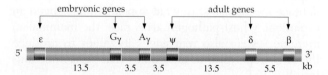

Fig. 12.7 The cluster of β-globin genes found on human chromosome 11. The genes are related in their sequence but serve different functions at different times. The epsilon (ε) and gamma (γ) genes are expressed in embryonic and fetal life, respectively, while the delta (δ) and beta (β) genes are expressed after birth. One of these globin genes (ψ) is an evolutionary remnant of an earlier duplication and, although it retains sequence similarity to all other β-globin genes, is not expressed. It is termed a pseudogene

Genome maps

The arrangement of genes along the length of a chromosome can be described by either genetic or physical maps. As we saw in Chapter 9, genetic maps are usually obtained from mating experiments. They provide a description of the relative positions of genes in terms of recombination distances. Although genetic loci may be mapped to a particular region of a chromosome, genetic maps do not necessarily identify the precise site of the locus. In contrast, a physical map describes the positions of genes on chromosomes in terms of nucleotide sequences.

Genetic maps describe the relative positions of genes in terms of recombination distances, while physical maps define distances between loci in terms of numbers of nucleotides.

Genes and genetic programs

Sets of genes respond to various situations

Conventional analysis of gene expression has shown that sets of genes are co-ordinately regulated. For example, the *GAL4* and *GAL80* genes of *S. cerevisiae* regulate a set of unlinked genes involved in galactose utilisation (Chapter 11). However, the availability of complete genome sequences has made it possible to study the response of the entire set of genes to particular circumstances. DNA sequences corresponding to each gene can be arrayed on a microscope slide and used to examine expression patterns, using the DNA hybridisation methods described in Chapter 13. From such analyses, it is clear that a change in cellular behaviour, or a cellular response to an external stimulus, involves a concerted change in the expression of discrete subsets of genes (Box 12.3).

Sets of genes regulate related biological processes in different organisms

The analysis of regulatory processes has revealed that related sets of genes carry out particular functions in very different organisms. For example, equivalent sets of genes are used to receive intercellular signals and transduce the signals within the cell during formation of photoreceptor cells in the vinegar fly, *D. melanogaster*, vulval cells in the nematode, *C. elegans*, and cancer cells in humans. It appears that the genes that encode the proteins involved in this pathway were present in the common ancestor of these organisms, more than 600 million years ago, the pathway being

BOX 12.3 DNA microarrays and measuring whole genome transcriptional responses

One of the great benefits of knowing the entire sequence of a genome is that the response of every gene in the genome to particular developmental events or environmental changes can be examined. The challenge is to design experimental approaches that can investigate expression of many thousands of genes at once when, traditionally, the expression of genes has been studied one at a time.

Recent technological developments have provided a solution to this problem. The approach involves spotting tiny dots of liquid, each dot containing the denatured (separated) strands of DNA corresponding to one gene, onto a microscope slide to create a 'microarray' of different DNA sequences. Thousands of dots corresponding to thousands of different genes can be arrayed onto a single slide.

To investigate patterns of expression, mRNA samples are copied into complementary DNA (cDNA) that is labelled with a fluorescent dye. When these sequences are incubated with the microarray slides, they base pair with DNA sequences that correspond to the gene from which they were derived. Dots that contain DNA encoding a gene that is not expressed in the tissue from which the mRNA was obtained will not base pair with any of the labelled cDNA and will therefore not fluoresce. Conversely, dots containing DNA-encoding genes that are expressed at high levels will fluoresce brightly.

To compare expression patterns of two tissues or of a tissue before and after a particular event, two mRNA samples are used, each labelled with a different fluorescent molecule, usually one that fluoresces green and another that fluoresces red. For example, mRNA can be extracted from *D. melanogaster* larval tissues before exposure to ecdysone, a hormone that induces pupation, and labelled with a green fluorescent dye. A second sample can be extracted after ecdysone treatment and labelled with a red fluorescent dye. When the mixture of these labelled mRNAs is incubated with the microdots and any non-bound mRNA washed off, genes that are expressed before exposure to ecdysone will fluoresce green but not red, while those expressed after exposure to ecdysone will fluoresce red but not green. Those that are expressed constitutively (i.e. constantly) will fluoresce both red and green, creating a yellow colour. The figure shows an example of a set of over 6000 DNA-containing microdots hybridised to two such samples.

In this example, the response of all genes in the genome to the moulting hormone, ecdysone, can be determined and genes that are induced or repressed identified. The microarray approach permits the comprehensive study of differences in gene expression, for example, between normal and tumour tissues, uninfected and infected tissues and growth-arrested and growth-stimulated tissues.

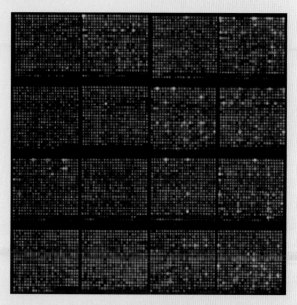

A microarray of *Drosophila melanogaster* genes. Each dot contains DNA corresponding to one gene. The 6400 dots shown here have been printed onto a microscope slide. The DNA in the dots has been incubated with different expressed sequences from before and after exposure to the hormone ecdysone. Dots that are green represent genes that are expressed before but not after the hormone treatment. Dots that are red correspond to genes that are expressed only after hormone treatment. The majority of dots are yellow, that is, they correspond to genes that are expressed both before and after treatment. In this way, all genes in the genome that respond to an ecdysone stimulus can be identified

adapted to different biological roles in the different organisms.

A second example of a conserved pathway is that involving the *hedgehog* and *patched* genes of *D. melanogaster*. These genes were discovered because of their roles in the formation of insect segments and limbs (Chapter 16). Sequence analysis revealed the presence of equivalent genes in a wide range of organisms. Recently, mutations in the human equivalent of the *patched* gene were shown to be responsible for the development of basal cell carcinoma, a form of skin cancer, in humans (Fig. 12.8).

Essential and non-essential genes

One of the most surprising discoveries to come out of the analysis of genomes has been the observation that a large proportion of the genes in a genome are not required for viability of the organism. For example, only 17.5% of the approximately 2000 *S. cerevisiae* genes that have so far been examined were found to be essential for viability. What are the other 82.5% of genes doing? It is important to remember that the survival of an organism and, ultimately, of the species, relies on the ability to cope with all of the various conditions found in natural environments. Some genes, therefore, deal with the response of the organism to environmental stresses. For example, the yeast *S. cerevisiae* contains a gene that is not necessary for normal growth and survival but becomes critically important in conditions where DNA damage has occurred. Cells are exposed to a wide variety of environmental stresses, including temperature variations and variations in the chemical nature of the environment, and may carry a large number of genes designed to cope with these environmental variations.

For higher organisms, other genes that are not required for viability include those that regulate behaviour. Both the nematode, *C. elegans*, and the vinegar fly, *D. melanogaster*, carry genes that influence feeding behaviours. These genes are not essential for viability of the organism in laboratory conditions but animals in the wild lacking these genes would not be able to compete effectively for food with normal animals. In this situation, the term 'essential' becomes a matter of definition. Such genes are not essential for survival of the organism in optimal conditions but are likely to confer better rates of survival and/or reproduction in the wild, either in terms of dealing with environmental variation or in terms of competing with other organisms for available resources.

Mutation, genetic variation and genome instability

As discussed above, the more closely related the species, the more similar are their genome sequences. Sequence variation occurs not only between species but also between individuals of the same species. It is the sequence variation within a species that creates much of the variation we see in individuals in populations. This variation is responsible for the allelic and phenotypic differences described in Chapter 9.

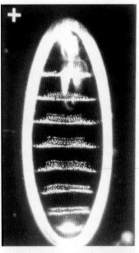

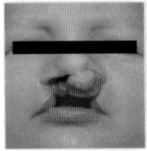
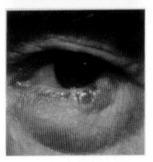

```
Drosophila:   24  DLYIRTSWVDAQVALDQIDKGKARGSRTAIYLRSVFQSHLETLGSSVQKHAGKVLFVAIL  83
                  |  |:  || ||||::| |||| | ::  || : :||: ||| | : ::||:||
Human:        48  DYLHRPSYCDAAFALEQISKGKATGRKAPLWLRAKFQRLLFLKGLGCYIQKNCGKFLVVGLL  107

Drosophila:   84  VLSTFCVGLKSAQIHSKVHQLWIQEGGRLEAELAYTQKTIGEDESATHQLLIQTTHDPNA  143
                  :  | |||||  : : :|::  |||:  ||  || ||::  ||||    ||:|||  : |
Human:       108  IFGAFAVGLKAANLETNVEELWVEVGGRVSRELNYTRQKIGEEAMFNPQLMIQTPKEEGA  167

Drosophila:  144  SVLHPQALLAHLEVLVKATAVKVHLYDTEWGLRDMCNMPSTPSFEGIYYIEQILRHLIPC  203
                  :||  :||||:| |||:|| | |:::   ||: :  |   :||::|:||| : ||:|| |
Human:       168  NVLTTEALLQHLDSALQASRVHVYMYNRQWKLEHLC-YKSGELITETGYMDQIIEYLYPC  226

Drosophila:  204  SIITPLDCFWEGSQLLGPESAVVIPGLNQRLLWTTLNPASVMQYMKQKMSEEKISFDFET  263
                  ||||||||||||::| :|    | || :| :: :| ||::  :  |:  :  :|
Human:       227  LIITPLDCFWEGAKL---QSGTAYLLGKPPLRWTNFDPLEFLEELK------KINYQVDS  277

Drosophila:  264  VEQYMKRAAIGSGYMEKPCLNPLNPNCPDTAPNKNSTQPPDVGAILSGGCYGYAAKHMHW  323
                  |: : :| :| |||::||| :||||::| ||| |||||||| |: :|:||||:: |:|||
Human:       278  WEEMLNKAEVGHGYMDRPCLNPADPDCPATAPNKNSTKPLDMALVLNGGCHGLSRKYMHW  337

Drosophila:  324  PEELIVGGRKRNRSGHLRKAQALQSVVQLMTEKEMYDQWQDNYKVHHLGWTQEKAAEVLN  383
                  |||||||  | :| | |||  |:| |:: ||:|::  |::|: :  : | :||| || |
Human:       338  QEELIVGGTVKNSTGKLVSAHALQTMFQLMTPKQMYEHFKGYEYVSHINWNEDKAAAILE  397

Drosophila:  384  AWQRNFSREVEQLLRKQSRIATNYDIYVFSSAALDDILAKFSHPSALSIVIGVAVTVLYA  443
                  ||||  :  | |::|  |     : |:: |||| || |  :  :::|||| | : :||
Human:       398  AWQRTYVEVVHQSVAQNS----TQKVLSFTTTTLDDILKSFSDVSVIRVASGYLLMLAYA  453

Drosophila:  444  FCTLLRWRDPVRGQSSVGVAGVLLMCFSTAAGLGLSALLGIVFNAASTQVVPFLALGLGV  503
                  |:||| |    :  :||:||||||| | |||| | |||| ||||| ||||:||||||| |
Human:       454  CLTMLRW-DCSKSQGAVGLAGVLLVALSVAAGLGLCSLIGISFNAATTQVLPFLALGVGV  512

Drosophila:  504  DHIFMLTAAYAES--NRR----EQTKLILKKVGPSILFSACSTAGSFFAAAFIPVPALKV  557
                  | :|:|  |::|: |:|     ::|  || |: :  |: ||   :|  :|| |||:||:|
Human:       513  DDVFLLAHAFSETGQNKRIPFEDRTGECLKRTGASVALTSISNVTAFFMAALIPIPALRA  572

Drosophila:  558  FCLQAAIVMCSNLAAALLVFPAMISLDLRRRTAGRADIFCC-CFPVWKEQPKVAPPVLPL  616
                  | ||||:|:  | |||: :|  |:||||  | || | |||| |  |   : |   :  ||
Human:       573  FSLQAAVVVVFNFAMVLLIFPAILSMDLYRREDRRLDIFCCFTSPCVSRVIQVEPQAYTD  632

Drosophila:  617  NNNNGRGARHPKSCNNN------------------------------------------  633
                  ::| | : |  :::
Human:       633  THDNTRYSPPPPYSSHSFAHETQITMQSTVQLRTEYDPHTHVYYTTAEPRSEISVQPVTV  692

Drosophila:  634  -RVPLPAQNP--------LLEQRADIPGSSHSL----ASFSLATFAFQHYTPFLMRSWVK  680
                  :  | |:|         || ||  | :||::|| :||   :|| :|: |:| |
Human:       693  TQDTLSCQSPESTSSTRDLLSQFSD--SSLHCLEPPCTKWTLSSFAEKHYAPFLLKPKAK  750

Drosophila:  681  FLTVMGFLAALISSLYASTRLQDGLDIIDLVPKDSNEHKFLDAQTRLFGFYSMYAVTQGN  740
                  :  || | |  | ||| :| :::||||| |:||::  |: |: |  | ||:|| |||
Human:       751  VVVIFLFLGLLGVSLYGTTRVRDGLDLTDIVPRETREYDFIAAQFKYFSFYNMYIVTQ-K  809

Drosophila:  741  FEYPTQQQLLRDYHDSFVRVPHVIKNDNGGLPDFWLLLFSEWLGNLQKIFDEEYRDGRLT  800
                  :||   |  ::  |: | | |  :|:  :| :| :| ||  :|  |:|  | :  ::
Human:       810  ADYPNIQHLLYDLHRSFSNVKYVMLEENKQLPKMWLHYFRDWLQGLQDAFDSDWETGKIM  869

Drosophila:  801  KECWFPNASSDAILAYKLIVQTGHVDNPVDKELVLTNRLVNSDGIINQRAFYNYLSAWAT  860
                  : | || :||: |  | :||  | :| | || :|:|||:|:||| ||::|||: :: |:
Human:       870  PNN-YKNGSDDGVLAYKLLVQTGSRDKPIDISQLTKQRLVDADGIINPSAFYIYLTAWVS  928

Drosophila:  861  NDVFAYGASQGKLYPEPRQYFHQPNEY----DLKIPKSLPLVYAQMPFYLHGLTDTSQIK  916
                  || || ||| :| :  |  :|  |  :|   |:|| |: ||| ||||||:|: |||
Human:       929  NDPVAYAASQANIRPHRPEWVHDKADYMPETRLRIPAAEPIEYAQFPFYLNGLRDTSDFV  988

Drosophila:  917  TLIGHIRDLSVKYEGFGLPNYPSGIPFIFWEQYMTLRSSLAMILACVLLAALVLSLLLL  976
                  || :| : | ||:   ||   :   ||||||| | |||::|: |||:||| | || ||
Human:       989  EAIEKVRTICSNYTSLGLSSYPNGYPFLFWEQYIGLRHWLLLFISVVLACTFLVCAVFLL  1048

Drosophila:  977  SVWAAVLVILSVLASLAQIFGAMTLLGIKLSAIPAVILILSVGMMLCFNVLISLGFMTSV  1036
                  : | | :::  :     :: || || |:||||||:| |||| ||| ||: | | ::|:|:
Human:      1049  NPWTAGIIVMVLALMTVELFGMMGLIGIKLSAVPVVILIASVGIGVEFTVHVALAFLTAI  1108

Drosophila:  1037  GNRQRRVQLSMQMSLGPLVHGMLTSGVAVFMLSTSPFEFVIRHFCWLLLVVLCVGACNSL  1096
                   |:: ||  | :|| |  : |||  :| ||| |: |:||:|: || |::: |:  :: |
Human:      1109  GDKNRRAVLALEHMFAPVLDGAVSTLLGVLMLAGSEFDFIVRYFFAVLAILTILGVLNGL  1168

Drosophila:  1097  LVFPILLSMVGPEAELVPLEHPDRISTPSPLP----VRSSKRSGKSYVVQGSRSSRGSCQ  1152
                   :: |:|||  ||  : |:  |  |:   ||||    ||:   ||: |  |:   | | |
Human:      1169  VLLPVLLSFFGPYPEVSPANGLNRLPTPSPEPPPSVVRFAMPPGHTHSGGSDSSDSEYSSQ  1228

Drosophila:  1153  KS---------HHHHHKDLNDPSLTTITEEPQSWKSSNSSIQMPNDWTY---QPREQRPA  1200
                   :  |        | | :  : ::  |::|  || :|::|:  | |   :  :  |:
Human:      1229  TTVSGLSEELRHYEAQQGAGGPAHQVIVEATENPVFAHSTVVHPESRHHPPSNPRQQPHL  1288

Drosophila:  1201  SYAAPPPAYHKAAAQQHHQHQGPPTTPPPP--FPTAYPPELQSIVVQPE  1247
                    : ||         | | |   ||  ||   :| | |   :  |
Human:      1289  DSGSLPPG---------RQGQQPRRDPPREGLWPPLYRPRRDAFEISTE  1328
```

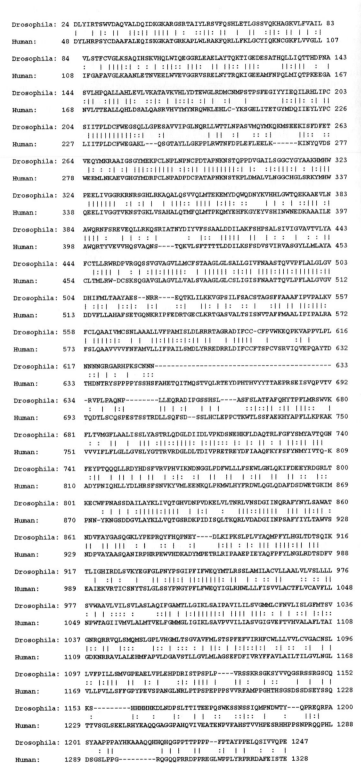

Fig. 12.8 Conservation of the sequence of the *patched* gene in the fly, *Drosophila melanogaster*, and humans. Identical amino acid residues are indicated by a vertical line, while similar amino acids are indicated by two dots. The top photograph shows the epidermal structures found in a wild-type (*+*) *Drosophila* embryo. The second photograph shows that mutation of the *patched* gene disrupts formation of the epidermis in the *Drosophila* embryo. The lower two photographs show that mutation of the *patched* homologue in humans also results in developmental defects in the epidermis

This variation is also the primary resource on which natural selection acts to drive evolution (Chapter 32).

To understand why it is critically important to consider genetic variation within populations, you only need to look at any group of people. The consequences of genetic variation are obvious. For humans, differences in the colour of hair, skin and eyes and differences in height and weight result, in large part, from genetic variation between individuals (i.e. from differences in the DNA sequence of their genomes). There are also less obvious differences. Some of the group may have a greater susceptibility to diseases such as asthma, some will have a higher blood pressure than others, some will have a higher level of cholesterol, some may react to particular drugs, some will be more adventurous than others or have a more reserved personality. Although the environment plays a role in these characteristics, differences are also caused by variation in the genome sequence inherited by each individual.

DNA sequence, gene expression and genetic variation

How does DNA sequence variation result in altered phenotypes? The genetic code and the nature of gene expression (Chapter 11) provide a molecular explanation of phenotypic variation that results from genome sequence differences. Here we will use the β-globin gene as an example. A gene can be defined as the set of chromosomal DNA sequences that encode the primary transcript and regulate its expression. Alterations to any of these sequences can cause a change in β-globin expression or function that alters the phenotype of the individual. In the extreme case, sequence changes that prevent production of a functional protein generate alleles that are lethal to the homozygous individual

(e.g. see the discussion of the thalassaemia alleles in Chapter 9). Other changes in the β-globin DNA sequence may have different outcomes in terms of gene expression or protein function and, consequently, phenotype.

> The genomes of all individuals, with the exception of identical twins, exhibit DNA sequence variation. Some of that variation is responsible, at least in part, for different characteristics of different individuals.

Mutation: the source of genetic variation

New genetic variation arises when alterations are made to the DNA sequence, a process termed **mutation**. Mutations can arise through errors in the incorporation of nucleotides during DNA replication or through chemical or irradiation-induced damage to DNA.

Mutation can occur by changes in a single or small number of base pairs of the DNA sequence. Such changes, termed **point mutations**, include **substitutions**, in which a single base in DNA is replaced by another, or **deletion** or **insertion** of one or more base pairs (Fig. 12.9). Any of these alterations can have a dramatic effect on the organism by changing the amino acid encoded by a triplet codon or by changing the reading frame (see Chapter 11). Deletions and insertions are frequently unfavourable because they tend to destroy gene function.

Mutations can also arise as large-scale rearrangements of chromosomes, for example, inverting large portions of the chromosome or translocating a region of one chromosome onto another. They may also arise by the insertion of transposable DNA. These types of mutation are considered further in Chapter 32.

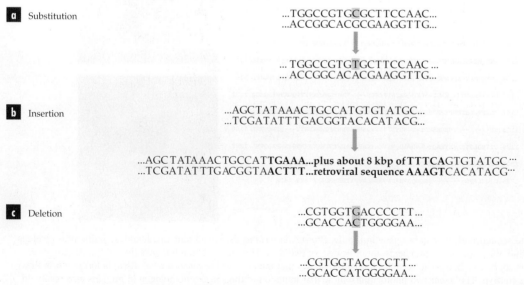

Fig. 12.9 DNA sequences can be mutated by substitution, insertion and deletion

Mutations are the source of new genetic variation. Point mutations are small changes in DNA sequence and include base substitutions, deletions and insertions. Chromosomal mutations involve large-scale rearrangements of chromosomes. The insertion of transposable elements is also a source of mutation.

DNA synthesis errors and spontaneous mutation

DNA replication is a remarkably faithful process but it is estimated that bacterial DNA synthesis involves an error rate of 10^{-8}–10^{-10} (i.e. about one incorrect nucleotide in about one billion nucleotides incorporated). Degradation of DNA, such as spontaneous deamination of cytosine to form uracil, can also occur. Probably the most frequent event is depurination due to spontaneous hydrolytic loss of guanine or adenine from the sugar–phosphate backbone of DNA. Any damage to DNA can affect its ability to function and lead to the transmission of altered and potentially defective genetic information to the next generation of cells or organisms.

Environmental mutagens

DNA is also susceptible to damage from environmental factors. DNA damage can result from exposure to radiation (e.g. X-rays, ultraviolet radiation and gamma rays) or chemical mutagens.

Somatic cells, germ cells and nuclear radiation

The immense human tragedy of the use of nuclear weapons during World War II has revealed a surprising difference in the ability of somatic cells and germ cells to deal with radiation-induced damage to DNA. Because of the relationship between DNA mutation and cancer (see below), people exposed to radiation in Hiroshima and Nagasaki developed cancer at an extremely high rate relative to members of the population who were not exposed, compounding the devastation inflicted by the explosions and radiation sickness. However, the appearance of mutations in the progeny of exposed people, a measure of the level of mutation in germ cells, has been found to be indistinguishable from that of the normal population. The reasons for this are unclear. Perhaps DNA repair mechanisms are far more efficient in germ cells or perhaps there are more highly developed mechanisms that can sense DNA damage and induce cell death if the damage is irreparable.

Chemical mutagens

Chemical mutagens are of three major types. Firstly, chemicals such as mustard gas react with DNA and alter or remove bases from the DNA backbone, leading to misincorporation of bases in the next cycle of replication. They also cause more drastic changes, including breakage of the DNA backbone. Chromosomal rearrangements result if the repair process rejoins the incorrect ends. Secondly, chemicals described as base analogues mimic some components of the structure of the four normal bases—A, C, G, T—but do not permit complementary pairing. Incorporation of these leads to mistakes during the next cycle of replication. An example is azidothymidine, better known as AZT, which is used as an antiviral agent in the treatment of the acquired immunodeficiency syndrome (AIDS). Thirdly, some chemicals that bind with DNA interfere with the normal process of DNA replication. An example is ethidium bromide, a fluorescent agent widely used for staining DNA, which inserts neatly into double-stranded DNA.

Mutagens can be used to increase the frequency of mutations and therefore increase genetic variability, for example, in plant breeding. The inherited effect of a new mutation does not usually show up until a later generation, if at all. This is because most mutations change a wild-type allele that produces an active product into an allele that does not produce an active product and is associated with a recessive phenotype. The new mutant allele is only revealed in homozygous individuals. For example, to create mutations in flower colour, plant breeders expose seeds to a mutagen, germinate them and self-pollinate the plants; mutant forms may appear in the second generation when individuals can be homozygous for a new mutant allele.

DNA repair mechanisms

Damage to a single strand of DNA can be repaired because genetic information is carried in both strands of the duplex DNA. Mechanisms that repair damaged DNA are primarily enzyme-based, and involve a process of recognition of damaged or altered DNA, followed by the removal and replacement of the damaged nucleotides.

Photoreactivation

When DNA is exposed to ultraviolet (UV) light, adjacent pyrimidines, especially thymines, undergo a chemical rearrangement leading to covalent bonding of these adjacent bases to form pyrimidine dimers (Fig. 12.10a). This type of structure must be repaired because it distorts the double helix and interferes with transcription and with DNA replication. Visible (blue) light induces a process of repair involving the photoreactivation enzyme, PRE (Fig. 12.10b). The enzyme associates with a dimer and, with the absorption of a

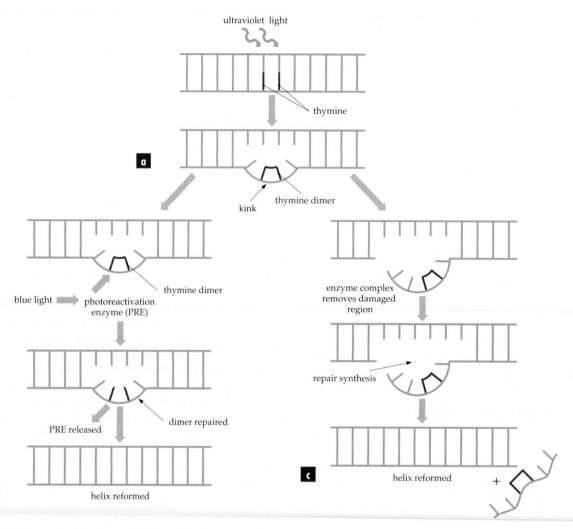

Fig. 12.10 DNA repair mechanisms. **(a)** Effects of a thymine dimer. Ultraviolet radiation can cause two adjacent thymines to link as dimers by formation of a covalent bond between them. The effect of a thymine dimer on the double helix is to make a kink in the duplex, which blocks DNA replication. **(b)** Photoreactivation repair mechanism of an ultraviolet-induced thymine dimer. **(c)** Excision repair mechanism of an ultraviolet-induced thymine dimer

photon of light, cleaves the bonds and reverses the effect of UV light.

Excision repair

A set of patrolling enzymes in the cell respond rapidly to DNA damage by excising the damaged bases along with some nucleotides on either side (Fig. 12.10c). Sites of mutation are recognised by an enzyme complex, which creates nicks (breaks in the phosphodiester linkages between adjacent nucleotides in the backbone of the DNA molecule) around the site of the damage. The damaged portion of DNA is excised by the enzyme complex. DNA polymerase I then fills in the gap by repair synthesis and the phosphodiester backbone is sealed by DNA ligase (Chapter 10). Unlike photoreactivation, excision repair does not require light.

DNA can be damaged by environmental factors. Repair mechanisms involve a number of enzymes that remove the damaged region and replace it with a new sequence.

Genomes, mutation and cancer

Cancer is a term used to describe a large number of diseases that share the common feature of cells growing in an unregulated fashion and invading and damaging other tissues. Cancer is a genetic disease in the sense that it results from a series of mutations that change the behaviour of the cell. Cancers can occur at a high frequency in some families. This results from the inheritance of alleles that predispose an individual towards cancer. However, the majority of cancers have no familial history and are referred to as sporadic cancers.

Cancer cells have modified growth and survival characteristics

The proliferation and survival of normal cells of the body is tightly regulated by intercellular signals produced by other cells (see also Chapter 16). Some of these signals act on the mechanisms that control the cell cycle (Chapter 8) to regulate the rate at which cells undergo mitosis. Other signals act as survival signals, preventing cells from undergoing **programmed cell death**, or **apoptosis**, a form of cellular suicide. Normal cells that do not receive the appropriate intercellular survival signals undergo apoptosis. Sometimes apoptosis is programmed as part of the normal developmental processes of cells (Chapter 16), but it should also occur if cells lose their normal controls and start to proliferate in an unregulated manner. For a cell to become tumorous, it must have mutations that both deregulate the cell cycle and prevent apoptosis from occurring.

Additional mutations are required to turn these proliferating cells into cancer. Cancerous cells are invasive, disrupting nearby tissues as they grow. They also frequently metastasise (i.e. move to new locations in the body). Both of these characteristics require major changes to the behaviour of the cells, changes which result from additional mutations. For a cell to migrate, for example, it must change its shape by modifying the normal cytoskeletal components such as actin filaments. It must also secrete enzymes that digest the extracellular matrix with which it is associated, or through which it must pass to reach the circulatory or lymphatic system.

Cancer formation is therefore a progressive process that involves a series of mutation events. Progressive changes to the cellular phenotype, leading to cancer, can be observed. Skin cancer (melanoma) may, for example, start with deregulation of cell proliferation and loss of the ability to undergo apoptosis, creating an abnormal, partially differentiated tissue appearance termed dysplasia. Later the cells may be found to be proliferating rapidly and to be showing no signs of differentiation. Finally, the cells become invasive and acquire the ability to metastasise.

Cancer is caused by cells which survive, proliferate and become invasive when normally they would not. This is a sequential process requiring a number of different mutations that cause loss of the normal cellular controls that regulate these characteristics.

Tumours are clonal in origin

All cells in a tumour taken from a female have the same inactive X chromosome (either the paternal or the maternal). This suggests that tumours do not arise from groups of cells. Furthermore, in any one patient all tumour cells carry identical genetic lesions, such as particular chromosomal rearrangements. These observations show that tumours arise from a single cell (i.e. they are clonal in origin). This means that it is intrinsic changes to a cell (i.e. changes occurring within the cell), rather than changes being induced by external signals, which ultimately result in tumour formation. What events drive a single cell to behave in such a destructive way?

Cancer is a genetic disease

It is now firmly established that cancer is a disease caused by genetic alterations to cells. This does not mean that cancer is necessarily an inherited disease. In fact, the majority of cancers are sporadic (i.e. there is no family history). In these cancers, genetic changes accumulate in somatic cells during the life of the individual, resulting in the development of cancer. As noted above, cancer formation requires a number of mutations to create the cancer cell phenotype of unregulated growth, survival and invasiveness. The study of genes that are mutated in cancer cells has revealed two different classes of genes, termed **dominant oncogenes** and **tumour suppressor genes**, based on the dominant or recessive nature of the mutant allele phenotypes.

Dominant oncogenes

Oncogenes are normal cellular genes (the normal cellular version being termed a **proto-oncogene**) that, when mutated, result in the development of tumours. As the name suggests, the cellular phenotype of this class of oncogenes is dominant, that is, only one mutated copy is required to cause the cancer phenotype. The dominant oncogene alleles cause overexpression, increased activity or aberrant activity of a normal gene, leading to tumour cell characteristics. In most cases, mutations in dominant oncogenes activate intercellular signalling or signal transduction pathways that mimic the normal growth signals or mimic normal pathways that block cell death.

For example, the *ras* oncogene encodes a component of a signal transduction pathway that is activated when intercellular signalling molecules such as growth factors bind to their receptors. The action of these intercellular signalling molecules is to stimulate rates of mitosis and hence increase the proliferation rate. Ras is normally activated when particular growth factors bind to their receptors (Fig. 12.11) but some mutations cause Ras to be constitutively active, that is, to behave as if the cell is constantly receiving the growth factor, even when it is not. The *constitutive activity* of Ras triggers continued proliferation of the

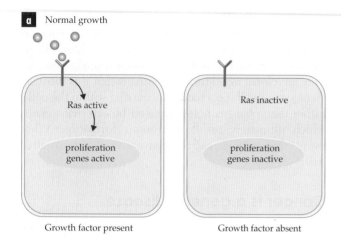

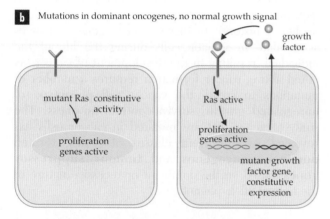

Fig. 12.11 Mutations in dominant oncogenes stimulate proliferation. **(a)** Proliferation is normally stimulated when intercellular growth factors bind to transmembrane receptors and activate an intracellular signal transduction mechanism, involving the Ras protein, which results in activation of genes required for cell cycle progression. When the growth factor is absent, the cell does not divide. **(b)** Two examples of mutations that generate a constitutive, cell-autonomous growth signal. In the first, a mutation in the gene encoding the Ras protein results in a constitutively active protein, which stimulates growth independent of any external growth signal. In the second, a mutation in a growth factor gene causes constitutive synthesis of the growth factor, which is received by the same cell, causing constitutive activation of the proliferation genes

cells even in the absence of the proliferation signal (Fig. 12.11). Similar effects are caused by mutations that cause abnormal expression of a growth factor, so that the cell is constantly supplying and responding to its own growth signal, or by mutations in the membrane-bound receptor that make it constitutively active. All of these mutations stimulate growth that is independent of the normal external signals.

The *Bcl2* gene is another example of a dominant oncogene. However, the role of *Bcl2* is not to stimulate proliferation but to block apoptosis. Mutations that cause *Bcl2* to be expressed where it would normally be silent can lead to the survival of cells that would normally die, for example, when tissue overgrowth

results in the loss of the normal survival signals from other cells.

Dominant oncogenes are normal cellular genes that are mutated during cancer formation so that they are expressed where they would normally be silent, expressed at higher than normal levels or produce a constitutively active protein product. Dominant oncogenes commonly play roles in stimulating proliferation or preventing apoptosis (programmed cell death) during normal development.

Tumour suppressor genes and cell cycle regulation

Tumour suppressor genes are normal cellular genes whose function is lost in the generation of tumours. The existence of such genes was inferred initially from cell fusion studies, which revealed that when a cancer cell is fused to a normal cell, the cancer cell phenotype of unregulated proliferation can be lost. This suggested that the tumour cell had lost the function of one or more genes that are present in the normal cell. This idea was supported by a brilliant insight into the nature of a familial cancer, retinoblastoma, by the geneticist Alfred Knudsen.

Familial cancer and Knudsen's two-hit hypothesis

Retinoblastoma is a genetic disease in which tumours of the retina and certain other tissues arise. Retinoblastoma is inherited as a dominant trait (see Chapter 9). However, only a very small number of cells in individuals who inherit the retinoblastoma susceptibility allele develop into tumours, while the vast majority remain normal. Non-familial (sporadic) cases of retinoblastoma also occur but, in these cases, single tumours form and there is no recurrence, unlike the familial form of the disease.

Knudsen offered an explanation based on his studies of the incidence of such tumours. He suggested that retinoblastoma susceptibility was caused by the inheritance of a non-functional allele of the *Retinoblastoma* (*Rb*) gene. During the life of the individual, tumours result when the second allele is also mutated, leaving the cell with no functional *Rb* gene (Fig. 12.12). Sporadic retinoblastoma tumours form when, much more rarely, both alleles are lost from a retinal cell in an individual who inherited two wild-type copies of the *Rb* gene (Fig. 12.12). The tumour phenotype is therefore a recessive phenotype at the cellular level, that is, a cell heterozygous for a wild type and mutant copy is normal, but a cell homozygous for the mutant allele is tumorous. The disease itself is inherited as a dominant phenotype because individuals who inherit one nonfunctional disease allele will almost certainly develop

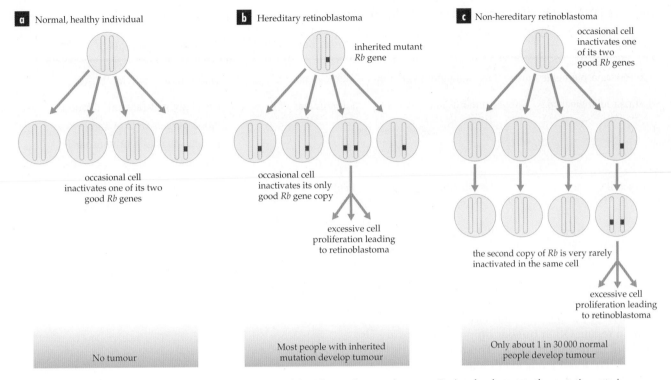

Fig. 12.12 (a) Normal individuals inherit a functional copy of the *Rb* gene from each parent. During development, the mutation rate is normally insufficient to cause loss of both alleles. **(b)** In familial retinoblastoma, the individual inherits one functional *Rb* allele and one mutant allele that produces no functional retinoblastoma. During development, loss of the normal allele results in the formation of a tumour. **(c)** In very rare cases, a sporadic retinoblastoma may arise when both normal alleles are lost during development. The majority of individuals who are heterozygous for the *Rb* mutant allele develop tumours, while sporadic retinoblastomas occur at a rate of approximately one in 30 000 individuals

tumours by loss of function of the other allele in some cells during development.

The application of recombinant DNA techniques combined with DNA sequence analysis (Chapter 13) led to the isolation of the *Rb* gene and confirmation that individuals who inherit retinoblastoma sensitivity have a mutation in one of their *Rb* alleles. Cells in tumours from these individuals, as well as tumours in sporadic cases of retinoblastoma, are mutant for both *Rb* alleles.

The *Rb* gene encodes an important component of cell cycle regulation

The *Rb* gene encodes a protein that acts to repress progression of a cell from G_1 phase into S phase. Loss of *Rb* gene function means that a key negative control mechanism of cell proliferation is lost. This constitutes one step in the creation of deregulated cell proliferation in cancer cells.

> Tumour suppressor genes are genes that are lost during cancer formation. Familial susceptibility to cancers is caused by inheritance of one non-functional allele of a tumour suppressor gene. Tumour susceptibility is inherited as a dominant phenotype but, at the cellular level, both copies of the tumour suppressor gene must be lost for a tumour to form. Sporadic tumours frequently have mutations in both alleles of one or more tumour suppressor genes.

Tumour suppressor genes and the integrity of the genome

The oncogenes discussed above are involved primarily in cell cycle control and the control of apoptosis. The study of the genetic basis of cancer has revealed an important additional class of tumour suppressor genes. These are genes responsible for maintaining the integrity of the genome. The genome is subject continually to mutation, from DNA replication errors and spontaneous mutation to changes induced by chemicals and irradiation (see above). The cell has several mechanisms for detecting and repairing DNA damage. These mechanisms dramatically reduce the mutation rate.

One of the hallmarks of cancer is the increased plasticity of cancer cell genomes. Cancer cells frequently show a wide range of chromosomal defects, from alterations in chromosome number to chromosomal rearrangements and deletions of regions of chromosomes. We now know that this plasticity is both a cause and consequence of cancer formation.

Tumour suppressor genes and DNA repair

One form of xeroderma pigmentosum (Table 12.2) is a rare skin disease in humans inherited as an autosomal recessive trait. The skin of an affected homozygote is

Table 12.2 DNA repair defects are associated with human diseases

Disease	Sensitivity	Cancer susceptibility	Symptoms and signs
Xeroderma pigmentosum	UV irradiation alkylation	Skin carcinoma	Skin photosensitivity
Ataxia telangiectasia	Gamma irradiation	Lymphoma	Unsteady gait (ataxia); dilation of blood vessels in skin and eyes (telangiectasia); chromosomal aberrations
Fanconi's anaemia	Cross-linking agents	Leukaemia	General decrease in numbers of blood cells; congenital anomalies
Bloom syndrome	UV irradiation	Leukaemia	Photosensitivity; defect in DNA ligase

very sensitive to sunlight and many sufferers die from skin cancer before the age of 30. The disease is caused by a defect in the endonuclease involved in DNA excision repair. Other tumour suppressor genes are also involved in repairing DNA. These include the gene responsible for Bloom syndrome, which encodes a DNA ligase, and genes mutated in the most common form of colon cancer, which encode DNA repair enzymes.

Tumour suppressor genes and checkpoint control

A cell not only needs to have specific DNA repair enzymes, it must also have a mechanism for arresting the cell cycle when DNA damage has occurred. Progression of the cell into S phase or mitosis with damaged DNA results in perpetuation of the mutations before they have had the chance to be repaired. For this reason, mechanisms exist that sense DNA damage and cause cell cycle arrest until the damage is repaired. These mechanisms are termed **checkpoint control** mechanisms, as they 'check' for DNA damage and induce cell cycle arrest if damage is present (Fig. 12.13). The three best characterised tumour suppressor genes involved in this process are all associated with familial forms of cancer. Ataxia telangiectasia (Table 12.2), which encodes a protein termed ATM, Li-Fraumeni syndrome, which encodes a protein termed p53, and a common form of familial breast cancer, which encodes the protein BRCA1, are all involved in the sensing of DNA damage and inducing cell cycle arrest. When both copies of any of these genes are lost, the cell loses the ability to arrest when DNA damage occurs, resulting in progression through the cell cycle with damaged DNA and a dramatic increase in mutation rate. Because of its function in maintaining genome integrity, the *p53* gene referred to above has been termed the 'guardian of the genome'.

In addition to DNA repair mechanisms, cells have mechanisms that induce cell cycle arrest if the DNA is damaged. These are termed checkpoint control mechanisms. Loss of any of these mechanisms by mutation of the genes involved increases the mutation rate by allowing cells to enter S phase or mitosis with damaged DNA.

An increased mutation rate appears to be essential for cancer formation

Inheritance of a mutant allele of the *p53* gene results in susceptibility to the tumours associated with Li-Fraumeni syndrome, which is multiple incidences, within a family, of sarcoma, breast cancer, primary brain tumour, leukaemia or adrenocortical carcinoma at young ages. However, the majority of cancers are sporadic and not familial. Investigation of a large number of sporadic cancers has revealed that both alleles of the *p53* gene are frequently mutated in cells of sporadic cancers. Mutations in other genes associated with checkpoint control or DNA damage repair mechanisms have also been found in cells from sporadic tumours. It now appears virtually certain that formation of a tumour requires at least one mutation in a checkpoint gene, indicating that a high mutation rate is essential for tumour formation.

Mutations that increase the mutation rate, either by inactivating checkpoint genes required for sensing DNA damage and inducing cell cycle arrest or genes encoding DNA repair enzymes, are an essential part of cancer formation.

The genome and cancer

We can now build a clear picture of the role of the genome in the formation of cancer. Cancer formation requires multiple genetic changes that disrupt the normal regulatory mechanisms of a cell that lead to survival and increased proliferation. Further changes

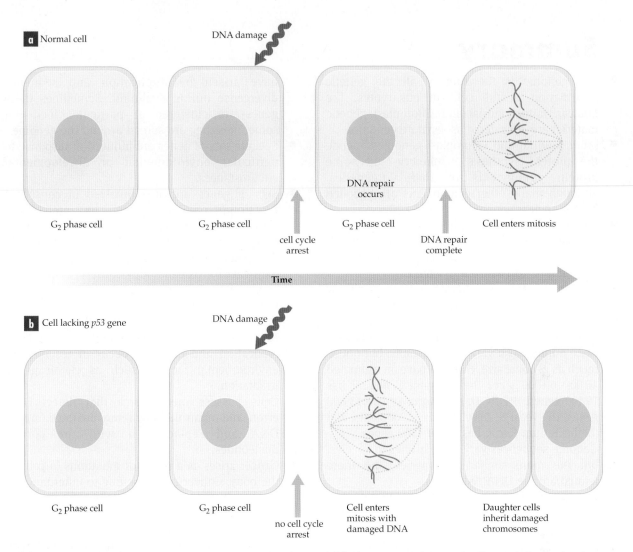

Fig. 12.13 Checkpoint genes cause cell cycle arrest when DNA is damaged. **(a)** The response of a normal cell to DNA damage involves a period of cell cycle arrest, allowing time for the repair mechanisms to work. The cell senses when the repair is complete and then proceeds into mitosis. **(b)** In a cell lacking the *p53* gene, the cell cycle arrest does not occur and the cell proceeds into mitosis, perpetuating the DNA damage into subsequent generations of cells

are required for invasiveness and metastasis. Normal mechanisms that sense and repair damage ensure that the mutation rate is too low to produce the multiple mutations required for cancer formation. However, a rare mutation that disrupts these mechanisms can dramatically increase the mutation rate, leading to mutations in additional tumour suppressor genes and proto-oncogenes, ultimately producing the cellular phenotype associated with cancer. Cancer is clonal because, at each stage of cancer development, the progeny of a single cell that undergoes a mutation that increases proliferation rate, survival rate or invasive characteristics will out-compete other cells in the generation of the tumour.

Summary

- The genome is the sum of all the genetic information in the cell. In eukaryotes, the nuclear, mitochondrial and chloroplast genetic material are considered as separate genomes.
- Mitochondrial and chloroplast genomes encode the protein synthesis machinery and some protein components of these organelles, although many other components are encoded by the nuclear genome.
- Some genomes include extrachromosomal elements, such as plasmids, which are small, usually circular DNA molecules, often the source of antibiotic resistance in bacteria.
- Genome sizes vary considerably. Cellular genomes range from a few thousand kilobase pairs (kb) for bacteria to many thousand megabase pairs (Mb) for mammals and plants.
- Gene number also varies considerably, but not as much as genome size. Humans have an estimated 40 000–60 000 genes, the nematode *C. elegans* has 19 100 genes, the fly *D. melanogaster* has 13 600 genes, the yeast *S. cerevisiae* has 6200 genes and the bacterium *E. coli* has about 4300 genes.
- Genome sequences vary from species to species, but the more similar the species, the more related are the DNA sequences.
- Genome sequences vary also from individual to individual within a species. Differences in these sequences are responsible for much of the phenotypic variation observed between individuals.
- The genome consists of DNA sequences that encode proteins (coding sequences) as well as sequences that encode introns, 5′ and 3′ untranslated sequences of mRNA, transcriptional regulatory sequences, centromeres and telomeres. The genome also contains DNA sequences of no apparent function, often termed 'junk' DNA.
- Genomes frequently contain repeated DNA sequences. These can be functional repeats, such as the ribosomal DNA repeats, or repeats with no known function, such as highly repeated satellite DNA. Telomeres are special structures at the end of eukaryotic chromosomes composed of simple repeated sequences.
- Genomes contain transposable elements, small segments of DNA that can move to new positions in the genome.
- Some genes within a genome share sequence similarity. Such genes form gene families that have arisen by duplication and sequence divergence during evolution. Sometimes these genes are clustered at particular loci, but sometimes they are spread around the genome.
- Specific sets of genes are induced in response to particular environmental or developmental circumstances.
- Sets of genes are also involved in conserved functional pathways, such as intercellular signalling and intracellular signal transduction. Equivalent sets of genes can be found in widely diverged species, indicating that basic molecular pathways have been highly conserved during the evolution of different species.
- Genome variation arises through the process of mutation (i.e. a change to the DNA sequence). Mutation occurs at a low frequency during DNA replication and can also occur in response to environmental factors, such as chemicals or irradiation.
- Cells possess mechanisms that detect DNA errors and repair them by removing the damaged DNA and replacing it with the correct nucleotides.
- Cancer arises as a result of mutations in cells of the body. Cancer can arise due to inheritance of a cancer susceptibility allele or where no previous family history exists (sporadic cancers).
- Cancer cells have lost the normal controls on their survival, proliferation and interaction with other cells and tissues. This results in uncontrolled growth, invasiveness and metastasis.
- A tumour arises from a single cell (i.e. it is clonal in origin) and arises from mutations in two classes of genes, dominant oncogenes and tumour suppressor genes.
- Dominant oncogenes are normal cellular genes (proto-oncogenes) that are mutated during cancer formation so that they are expressed where they would normally be silent, expressed at higher than normal levels or altered to produce a constitutively active protein product. Dominant oncogenes play roles in stimulating proliferation or preventing apoptosis (programmed cell death) during normal development.
- Tumour suppressor genes are genes that are lost during cancer formation. Familial susceptibility to cancers is caused by inheritance of a nonfunctional allele of a tumour suppressor gene.
- Tumour susceptibility is inherited as a dominant phenotype, but at the cellular level, both copies

of the tumour suppressor gene must be lost for a tumour to form. Sporadic tumours frequently have mutations in both alleles of one or more tumour suppressor genes.

- In addition to DNA repair mechanisms, cells have mechanisms that induce cell cycle arrest if the DNA is damaged. These are termed checkpoint control mechanisms. Loss of any of these mechanisms increases the mutation rate by allowing cells to enter S phase or mitosis with damaged DNA.

- Mutations that increase the mutation rate, either by inactivating genes required for DNA damage sensing and cell cycle arrest or by inactivating genes that encode DNA repair enzymes, are an essential part of cancer formation.

key terms

checkpoint control (p. 296)

coding sequence (p. 281)

complementary DNA (cDNA) (p. 280)

deletion (p. 290)

dominant oncogene (p. 293)

expressed sequence tag (EST) (p. 280)

gene family (p. 286)

genome (p. 278)

insertion (p. 290)

mutation (p. 290)

non-coding DNA (p. 282)

oncogene (p. 293)

point mutation (p. 290)

programmed cell death (apoptosis) (p. 293)

proto-oncogene (p. 293)

repetitive DNA (p. 282)

satellite DNA (p. 282)

substitution (p. 290)

transposable element (p. 284)

tumour suppressor gene (p. 293)

Review questions

1. A eukaryotic cell can have several genomes. What are they?

2. Is there a simple correlation between gene number and genome size for different genomes?

3. Not all DNA in the genome encodes amino acid sequences of proteins. What roles, if any, do other sequences play?

4. What is the nature of satellite DNA?

5. DNA molecules in living cells are no longer thought to be inert molecules in terms of possessing an unchanging DNA sequence. Why is this?

6. What is meant by the term 'gene family'?

7. What is the cause of variation in the DNA sequence?

8. How do mutations occur?

9. How are errors in DNA that arise during replication repaired?

10. How do cancer cells differ from normal cells?

11. How do dominant oncogenes and tumour suppressor genes differ?

12. What is the difference between a familial tumour and a sporadic tumour?

13. The retinoblastoma disease is inherited as a dominant genetic trait, yet both alleles of the gene must be mutated for retinoblastoma to occur. What is the explanation for this?

14. What is the role of the *p53* gene?

Extension questions

1. The genome sequences of humans and chimpanzees are approximately 98% identical. Do you expect the differences to be found primarily in coding sequences or non-coding sequences?
2. When a retinoblastoma tumour cell is fused with a normal cell, the fused cell behaves like the normal cell in terms of the ability to arrest during G_1 phase. Explain this observation using your knowledge of *Rb* gene function.

Suggested further reading

Keller, E. F. (1983). *A Feeling for the Organism: the Life and Work of Barbara McClintock*. San Francisco: W. H. Freeman.

A biography of Barbara McClintock, the geneticist who discovered transposable elements.

Lee, T. F. (1991). *The Human Genome Project: Cracking the Genetic Code of Life*. New York: Plenum Press.

Wills, C. (1991). *Exons, Introns and Talking Genes: the Science Behind the Human Genome Project*. Oxford: Oxford University Press.

Two books aimed at the intelligent general reader describing the events leading to the establishment of the program to sequence the entire human genome.

Singer, M. and Berg, P. (1997). *Exploring Genetic Mechanisms*. Mill Valley, CA: University Science Books.

A comprehensive textbook on the subject of genes and genomes.

CHAPTER 13

Manipulating genomes: genetic engineering

The last two decades have seen a revolution in biology, both in terms of advances in our understanding of fundamental biological processes and in our ability to modify living organisms for medical or commercial purposes. This revolution occurred primarily because of the development of genetic engineering techniques (see Box 13.1 for an overview), which have allowed biologists to isolate specific fragments of DNA that encode specific genes. As a result, the sequence of genes and of entire genomes can now be determined (Chapter 12), the positions of genes on chromosomes can be located and the function of the genes studied. These tools have also permitted *in vitro* modification of DNA sequences and the insertion of modified sequences into the genomes of organisms. The result has been a wide range of advances: new and more accurate means of diagnosing disease, new cures for human diseases, the efficient production of improved protein pharmaceuticals and the improvement of domestic animals and crop plants.

In this chapter, we will consider how genetic engineering makes it possible to study genes and manipulate gene function. Recombinant DNA technology, or genetic engineering, as it is commonly referred to, has generated much public anxiety about the safety of genetically modified organisms (see Box 13.2). However, there are already many instances in which the technology has provided important medical and agricultural advances and valuable research and diagnostic tools.

Recombinant DNA techniques enable the direct study of genes and manipulation of gene function.

BOX 13.1 Overview of recombinant DNA

The preceding chapters showed that DNA sequences contained within the genome are the store of information that controls formation of most of the characteristics of an organism. This information is held as the nucleotide sequence in DNA. Genomes typically contain many thousands of genes and a large amount of non-coding DNA. Viewed biochemically, all stretches of DNA look much the same, all being the same double-helical structure and all containing roughly similar amounts of the A, C, G and T nucleotides. It is impossible to purify and study specific genes by classical biochemical purification approaches.

Small segments of DNA that are easier to purify and study do exist, however, in the form of plasmids and the genomes of viruses. The essence of the initial recombinant DNA breakthrough was the ability to take small fragments of DNA from the very large plant and animal genomes and ligate them into plasmids and viral genomes to create recombinant DNA molecules. These molecules can be introduced into bacterial host cells, grown, purified in large amounts and studied.

Recombinant DNA technology makes use of a range of enzymes to create recombinant DNA molecules. The process involves the following series of steps. Firstly, eukaryotic DNA molecules are isolated and cleaved with a **restriction endonuclease** (also termed **restriction enzyme**) to produce segments of DNA (**donor DNA**). Self-replicating circular plasmid DNA (**vector DNA**), which contains a single cleavage site for a particular restriction endonuclease, is also prepared and cleaved to create a linear DNA molecule. A donor eukaryotic DNA fragment is joined to the vector plasmid DNA molecule using DNA ligase, forming a recombinant DNA molecule (see Fig. 13.3). The vector–donor recombinant DNA construct is taken up by a bacterial cell (host) in a process termed **transformation**. Once in the host cell, the recombinant DNA molecule replicates and the host cell proliferates, producing large amounts of the recombinant DNA molecule, which can be isolated, purified and characterised. In addition, the recombinant DNA in the cell can be constructed so that it is transcribed, the mRNA translated and the protein product purified and characterised.

Subsequent developments have vastly enhanced recombinant DNA capabilities. It is now possible to determine precisely the sequence of any DNA molecule (including entire genomes, see Chapter 12). It is also possible to modify a DNA molecule to any desired sequence and reintroduce that sequence back into living cells of an increasing number of organisms. These technologies have provided extraordinary advances in our understanding of fundamental aspects of biology, such as genes and gene function, population structures and evolutionary mechanisms. They have also provided the means to improve agricultural plant and animal characteristics, produce a variety of pharmaceutical products, discover the basis of genetic disease and develop new approaches to therapy.

BOX 13.2 Recombinant DNA: the public debate

Methods for cutting and joining DNA were first developed and used to make recombinant molecules in the 1970s. It was not long before scientists became concerned that the release of genetically modified organisms into the environment may have unforeseen consequences.

A group of scientists, including many who had led the development of these techniques, called for a moratorium on gene manipulation experiments until it could be shown that DNA technology posed no hazard to humans and other organisms. A lot of discussion ensued among scientists about the possible dangers of inadvertently creating disease organisms, adding a new organism to the natural environment or tinkering with natural genomes. Many scientists noted, with some concern, that the *Escherichia coli* host used in these experiments was originally isolated from the human gut. Would this mean that *E. coli* carrying recombinant plasmids with, say, a human cancer gene, could infect a careless research worker or pollute the water supply of a whole city? It did not take the media long to imagine frightening (and sometimes comic) scenarios of genetically engineered monsters taking over the world (see figure). The debate shifted from academic lecture theatres and scientific conferences to the front pages of newspapers.

The scientific debate resulted in a series of guidelines for the conduct of experiments involving recombinant DNA. At the time, these measures were hardly sufficient to allay public fears, but in retrospect, the guidelines were, if anything, overly cautious.

Relatively simple physical and biological containment rules apply to some kinds of recombinant DNA experiments, such as those involving DNA from organisms that could hybridise (cross) naturally. Cloning genes, such as β-globin or insulin, into plasmids presents no obvious hazard but must be performed under conditions that would prevent the escape of the recombinant organisms. This is done in laboratories designed for physical containment

and by selecting bacterial host cells that are weakened by mutation so that they cannot survive outside the laboratory. Recombinant DNA experiments involving genes that are known to make products dangerous to human health (e.g. bacterial toxins, human cancer genes) are subject to stringent regulation. Such experiments are permitted only if they can be shown to be necessary for our understanding of human disease or the design of drugs and vaccines.

In Australia, the Genetic Manipulation Advisory Committee oversees the conduct of all experiments using recombinant DNA technology and control is exercised at a local level in each institution where genetic engineering experiments are carried out. The regulation of the use of recombinant organisms is still evolving but trials have been permitted of genetically engineered bacteria that prevent frost damage to plants, potatoes engineered to destroy attacking viruses and cotton plants engineered to express a recombinant insecticide. Pharmaceuticals synthesised using genetic engineering are also now in widespread use.

'Crack out the liquid nitrogen, dumplings . . . we're on our way.' A cartoon representing contemporary public fears of genetic manipulation

Recombinant DNA technology

DNA-modifying enzymes

Recombinant DNA technology grew primarily out of two areas of study, the enzymology of DNA-modifying enzymes and bacterial genetics, particularly the molecular genetics of plasmids and bacterial viruses (bacteriophages). Enzymes that are capable of modifying DNA, such as DNA polymerases, DNA nucleases and DNA ligases, are involved in the normal processes of DNA replication and DNA repair (Chapters 10, 12). These enzymes can be purified and used to modify or join DNA molecules *in vitro*.

Restriction endonucleases

A key event in the development of genetic engineering technologies was the discovery of a set of DNA-modifying enzymes that cut DNA at precise sequences, producing DNA segments of particular sizes. These enzymes are termed restriction endonucleases (also restriction enzymes). The fragments that are generated when DNA is digested (i.e. cleaved) by a restriction endonuclease can be ligated together in new combinations using DNA ligase to create recombinant DNA. Although cutting and ligation occur naturally as a result of genetic mechanisms such as recombination, recombinant DNA technology can produce entirely novel combinations of DNA molecules or segments, for example, between eukaryotic DNA and bacterial plasmid DNA.

The normal role of restriction endonucleases is to cut, and therefore inactivate, foreign DNA molecules introduced into a bacterial cell. The names of restriction endonucleases are based on the bacterium from which they are isolated. For example, the restriction enzyme, *Eco*RI, was the first restriction endonuclease isolated from *E. coli* strain R. Restriction endonucleases cut double-stranded DNA at specific base sequences known as recognition sequences (Fig. 13.1). Many of these enzymes cut DNA precisely within the recognition sequence. Recognition sequences of these enzymes are palindromic, that is, the nucleotide sequence reads the same in the 5′ to 3′ direction along either strand. Some restriction endonucleases cut at four-base sequences, some at five- or six-base sequences (Fig. 13.1) and others at even longer recognition sequences.

Restriction enzyme mapping

All DNA molecules with the same sequence will be cut into the same size fragments because they all contain cleavage sequences in the same position along the molecule. For example, when DNA is cut by *Eco*RI, the molecule will be cut at every position that contains the sequence 5′-GAATTC-3′ (Fig. 13.1). A circular molecule with only one such sequence will be cut to produce a single linear fragment.

The lengths of fragments produced by the cleavage will depend on the location of the recognition sequences in the DNA molecule. If the same DNA is cut with a different restriction enzyme, for example, *Bam*HI, it will be cut at every position that contains 5′-GGATCC-3′. *Bam*HI fragments will be different to but overlap with the *Eco*RI fragments and thus cutting with both enzymes together will produce fragments whose lengths add up to those of the original *Eco*RI fragments or *Bam*HI fragments. This enables their relative positions in DNA to be mapped. DNA fragments can be separated from one another and their length determined by gel electrophoresis (Fig. 13.2). DNA has a negative charge so that, in an electric field, fragments move towards the anode according to their size, with small fragments moving faster than longer fragments.

Ligation of DNA molecules

Once separated, DNA fragments from different sources can be ligated together. Some restriction endonucleases, such as *Eco*RI, make a staggered cut in the DNA, leaving short 'sticky' single-stranded regions (because they can hydrogen bond with each other to form a duplex) at each end (Fig. 13.3). If one of these free ends encounters complementary bases on the free end of another molecule, the two may pair by forming hydrogen bonds. Although the pairing of such a short sequence is very weak, it is sufficient for a DNA ligase enzyme (see Chapter 10 for its role in DNA replication) to catalyse the formation of phosphodiester bonds to join the two fragments together (Fig. 13.3) in a process termed *ligation*.

The generation of recombinant DNA from different species is possible because the structure of DNA is the same in all organisms, so that the restriction endonucleases and ligases cannot distinguish the source of the DNA fragments. Any two fragments cut by the same restriction endonuclease can ligate together equally well regardless of whether they were derived from a bacterium or a human cell (Fig. 13.4).

> DNA molecules may be cut at specific sequences by restriction endonucleases and the resulting fragments separated using gel electrophoresis, allowing their purification. DNA fragments from different species can be ligated to form recombinant DNA molecules.

Genetic transformation using plasmid and bacteriophage vectors

In order to make many copies of a specific fragment of DNA, the fragment is ligated to a self-replicating vector

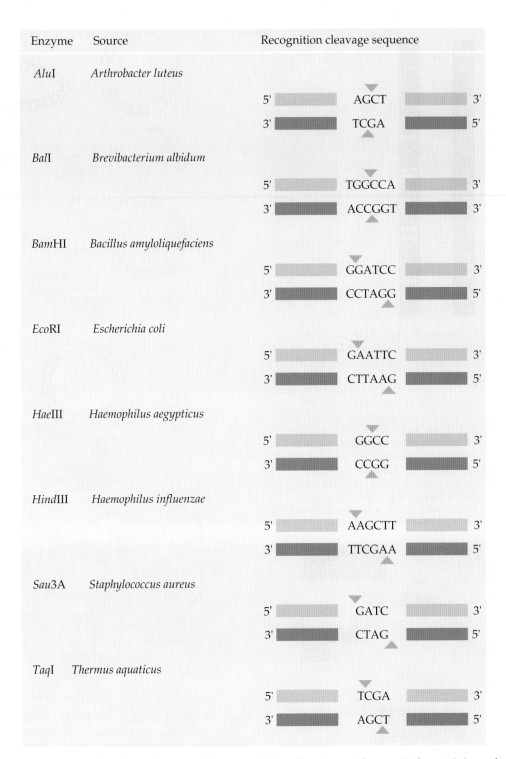

Fig. 13.1 Recognition sequences, cutting sites and sources of some restriction endonucleases. These particular restriction endonucleases recognise either a four-base or six-base sequence of DNA. The sites on each strand at which the phosphodiester bonds are broken by these enzymes are arrowed. Restriction enzymes can cut the two strands at the same site, producing a blunt-ended fragment (e.g. *Hae*III), or make staggered cuts to produce fragments with single-stranded sticky ends (e.g. *Bam*HI)

DNA sequence. Self-replicating DNA, such as plasmids, bacteriophages, cosmids and artificial chromosomes, are termed vectors, as they 'carry' the foreign DNA with them into cells. These vectors are self-replicating because they contain origins of replication that are used by the replication machinery of the host.

Plasmids are small double-stranded DNA molecules that exist in bacterial cells independent of the bacterial chromosome. In general terms, they can carry an insert of donor DNA of up to 10 kb. Plasmid vectors carry one or more antibiotic resistance genes, so that cells that carry the plasmid can be selected by growth in the

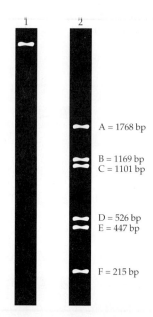

Fig. 13.2 Electrophoretic separation of fragments produced by cutting the circular SV40 virus genome with the restriction endonuclease, *Hind*III. Digested viral DNA in solution was loaded in a well at the top of the gel (origin), and subjected to an electric field (gel electrophoresis). The uncut virus genome of 5.2 kb forms a single band near the origin (lane 1). The restriction endonuclease cuts the circular genome at six sites, producing six fragments, which separate according to their size. DNA is visible as bright bands after staining with the fluorescent compound, ethidium bromide. Fragments of the same size migrate at the same position (band) in the gel. The distances of the bands from the origin, compared with the distances travelled by marker fragments of known size, are used to calculate the size of the fragment. Fragment sizes here are expressed as base pairs (bp) of DNA. Larger fragment sizes are usually expressed as kilobase pairs (kb)

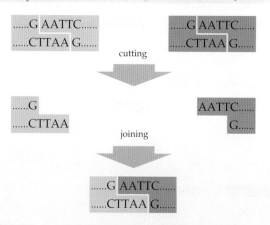

Fig. 13.3 Ligation of DNA fragments from different sources to make recombinant DNA molecules. Cleavage of DNA with the restriction endonuclease *Eco*RI produces a staggered cut, and leaves short, single-stranded ends. The single-stranded ends of different fragments may pair by hydrogen bonding and new phosphodiester bonds, created by DNA ligase, generate a recombinant molecule

presence of the antibiotic. Bacteriophages (bacterial viruses) such as λ (lambda) phage contain sequences in their small genomes that are not essential for infection and growth. These sequences can be removed and

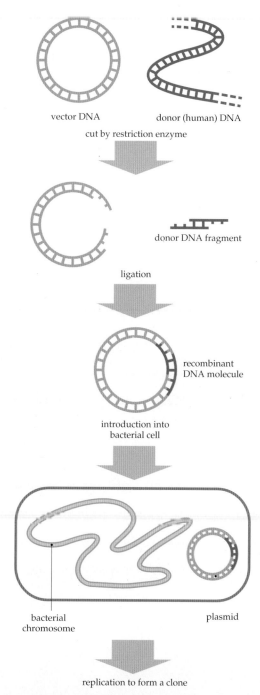

Fig. 13.4 Cloning a human gene. Donor (human) DNA is cut by a restriction endonuclease to leave sticky, single-stranded ends. The vector DNA (plasmid or virus) is cut once with the same restriction enzyme to open out the circle. Ligation of a restriction fragment of human DNA to the cleaved vector creates a recombinant DNA molecule. When this molecule is introduced into a bacterial cell by transformation, the molecule replicates within the cell and the cell proliferates to form a colony, each cell of which carries copies of the original human DNA fragment. Because all of the recombinant DNA molecules in this colony are derived from a single molecule transformed into a single cell, the recombinant DNA is said to be cloned

replaced with a fragment of donor DNA of up to 20 kb. Cosmids are hybrid vectors comprised of plasmid sequences that contain an antibiotic resistance gene and the terminal 'cos' sequences of λ phage, which are required for assembly of phage chromosomes into their protein coats. Cosmids allow the insertion of larger segments of donor DNA (up to 40 kb) and, when packaged into the coat protein and introduced into the host cell by the normal phage infection process, grow as a plasmid. Segments of yeast chromosomes (yeast artificial chromosomes, YACs) and bacterial chromosomes (bacterial artificial chromosomes, BACs) can be used as vectors for larger genomic fragments.

Replication of recombinant DNA occurs in living cells, so it is necessary to introduce the vector carrying the DNA into a host cell. Several microorganisms, especially bacteria and yeasts, have been employed as hosts for replication. Laboratory strains of *E. coli* that are genetically disabled and cannot survive outside the laboratory are commonly used. Different treatments can be used to make the cells **competent** to take up DNA from the external medium. As bacteria transformed with the recombinant plasmid grow and divide, the recombinant plasmids replicate. Growth on an antibiotic-containing medium selects for cells that carry plasmids. Each cell grown on a culture plate will grow to form a **colony** containing many millions of bacterial cells. The cells of each colony are a clone of genetically identical cells, each cell descended from the original transformant and containing identical copies of the original recombinant plasmid transformed into the cell (Fig. 13.4). Cells from this colony can be picked and grown to produce large amounts of identical copies of the recombinant plasmid.

If the cloning vector is a viral chromosome, such as the small chromosome of λ phage, donor DNA is ligated into the phage DNA and the recombinant phage DNA molecules are packaged into phage particles and allowed to infect bacterial cells. Once inside bacteria, the recombinant phage genomes are replicated by the bacterial replication machinery, producing many identical copies of the recombinant phage DNA. Ultimately, the host cell bursts (lyses), releasing many progeny phages, all carrying the inserted DNA. These infect neighbouring cells, which in turn lyse, releasing phages that infect nearby cells. Multiplication of the phages and death of the bacteria result in a **plaque**, a clear area visible against the host bacterial lawn growing on the culture plate. Each plaque contains a clone of phages, descended from a single original phage and containing identical copies of the original recombinant phage genome.

> Fragments of DNA can be ligated into self-replicating bacterial plasmid or phage vectors. In this way, recombinant bacteria or phages carrying recombinant DNA can be grown to produce large amounts of specific DNA fragments.

BOX 13.3 Some terms used in recombinant DNA technology

Autoradiography

Exposure to X-ray sensitive film to visualise radioactivity.

cDNA

Complementary DNA copies of mRNA transcripts.

cDNA clone

A single recombinant cDNA. The term clone is used because a single recombinant molecule in a host cell replicates and the cell proliferates to generate a colony, all cells of which carry multiple identical copies of the recombinant DNA molecule. As all of the product derives from a single initial recombinant DNA molecule, it is said to be a clone. Note, here, that the term clone has a very different meaning to the same term used to describe the process of using a single cell to generate a new organism.

cDNA library

A collection of cDNA clones transformed into host cells so that each cell contains only one of the many different cDNA types.

Gel electrophoresis

Migration of nucleic acids or proteins through a porous medium (gel) driven by a voltage drop across the gel. Different molecules migrate at different rates, depending on their characteristics. For DNA, larger fragments migrate more slowly in the gel.

Genomic clone

One particular recombinant genomic DNA fragment ligated into a vector and grown in host cells as described for a cDNA clone above.

Genomic DNA library

A complete set of DNA sequences of an organism cut into fragments, each fragment ligated to vector DNA and transformed into host cells.

Host cell

Typically a bacterial cell in which a recombinant vector can be replicated.

Hybridisation

The process of incubating one single-stranded DNA or RNA sample with another (one sample usually being labelled to enable detection) under conditions in which strands that have the complementary base sequence will base pair to form a double-stranded molecule.

Northern blot

A technique used to detect specific mRNA molecules that have been separated by gel electrophoresis, transferred by blotting onto a solid support (filter paper or membrane) and analysed by hybridisation with specific labelled DNA or RNA probes.

Polymerase chain reaction

Repeated amplification of a DNA segment by rounds of replication of the segment using specific primers.

Primer

Short single strands of DNA synthesised to be complementary to, and therefore hybridise to, the 3′ end of the DNA or RNA strand that is the template for new DNA synthesis in a polymerase chain reaction or DNA sequence analysis procedure.

Probe

Labelled DNA or RNA (either with radioactivity or with a chemical marker or enzyme) used to hybridise to other DNA samples (see Southern and northern blots) to detect the complementary sequences present.

Restriction endonuclease (restriction enzyme)

A nuclease enzyme that recognises a short sequence of nucleotides in a DNA molecule and cuts the DNA at this recognition sequence.

Reverse transcriptase

An enzyme that synthesises a complementary DNA copy of mRNA by adding deoxyribonucleotides to a primer on an mRNA template.

RFLP

Restriction fragment length polymorphism: a variation in fragment length generated by the presence or absence of restriction sites in different individuals.

Selectable marker

A gene that creates a selectable phenotype. Selectable markers permit selection for viability (e.g. antibiotic resistance genes) or selection by a visible phenotype (e.g. an enzyme that cleaves a substrate to produce a coloured product).

Sequence homology

The similarity of nucleotide sequences in different genes. It should be noted that the term homology should only be used when it has been clearly demonstrated that the similarity is the result of descent from a common ancestral gene. In practice, however, the term homology tends to be used whenever a clear similarity is detected between two gene sequences.

Southern blot

A technique used to detect specific DNA molecules that have been digested with a restriction endonuclease, separated by gel electrophoresis, transferred by blotting onto a solid support (filter paper or membrane) and analysed by hybridisation with specific labelled DNA or RNA probes.

Vector

A plasmid, bacteriophage, cosmid or chromosomal DNA molecule into which donor DNA can be ligated. It must have one or more unique restriction sites into which the donor DNA can be inserted, an origin of replication and a selectable marker.

Western blot

A technique used to detect specific protein molecules that have been separated by gel electrophoresis, transferred by blotting onto a solid support (filter paper or membrane) and analysed by incubation with antibodies that are specific for a particular protein.

Identifying and isolating individual genes

Selecting recombinant clones

Ligations between vector and donor DNA result in a mixture of products, some in which the donor DNA has inserted into the vector and some in which the ends of the vector have religated to reform the original plasmid. To identify and isolate any donor gene, it is necessary first to select the recombinant clones, that is, those in which the vector contains a donor DNA insert. A common way to achieve this is to use a restriction endonuclease site within a marker gene as the site of insertion. The most common marker gene used is the *lacZ* gene (Chapter 11), which encodes the enzyme β-galactosidase. When grown on an artificial substrate, the β-galactosidase, produced by cells with the intact *lacZ* gene, cleaves the substrate to create a blue precipitate. However, when a donor DNA fragment is ligated into the restriction endonuclease site of the vector, the *lacZ* gene is disrupted and β-galactosidase cannot be synthesised, resulting in white colonies (Fig. 13.5). Thus, blue/white colour selection provides an efficient system for identifying colonies carrying recombinant DNA molecules.

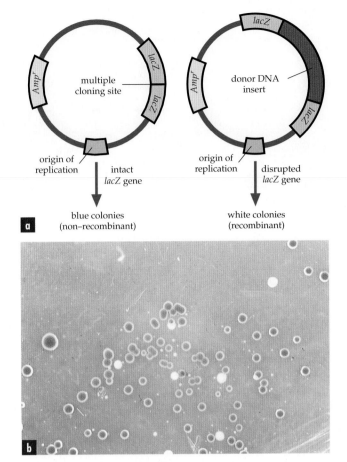

Fig. 13.5 (a) Structure of the vector pBluescript II, which is a plasmid DNA vector containing an origin of replication (for DNA replication in a host bacterial cell), the antibiotic resistance gene, *Amp^r*, and a *lacZ* gene containing a series of restriction endonuclease cleavage sites (the multiple cloning site). **(b)** Blue-coloured bacterial colonies (when grown on culture plates containing an artificial β-galactosidase substrate) are produced when the *lacZ* gene is intact, while white colonies are produced when the *lacZ* gene is interrupted (and inactivated) by insertion of donor DNA at the multiple cloning site

cDNA and genomic DNA libraries

A crucial concept in understanding the power of genetic engineering to isolate any gene from a complex genome is the concept of a *library* of cloned DNA fragments. The general idea is that a complex set of sequences, for example, those that constitute an entire genome of thousands of kilobase pairs (kb), is broken into fragments small enough to be cloned into vector DNA (either by restriction endonuclease cleavage or by physical shearing of the DNA). The fragments created are ligated to the vector DNA and transformed into host cells to generate millions of recombinant molecules, each containing one piece of the genome. Following ligation into the vector, the recombinant molecules are transformed into host cells and plated onto nutrient medium to produce millions of colonies, each of which contains one piece of the donor genome ligated to the vector. Each cell, which grows to

produce a colony (if the vector is a plasmid) or plaque (if the vector is a phage), takes up only one recombinant molecule. Different cells take up vectors containing different donor DNA inserts, so that each colony or plaque carries one segment of genomic DNA, while different colonies or plaques carry different segments. If the ligation and transformation generates millions of colonies or plaques, each carrying a random segment of DNA, then any sequence within a genome should be present as a recombinant DNA molecule in one or more of the clones of the library (Fig. 13.6).

The collection of genomic clones is termed a **genomic library**. For example, if DNA from the 3 000 000 kb human genome is sheared into 30 kb fragments, ligated to cosmid vector sequences and transformed to produce 1 000 000 colonies carrying recombinant DNA, any sequence in the human genome should, on average, be present in 10 of these colonies.

A second type of clone library is critically important in the analysis of eukaryotic genes. Eukaryotes have large genomes that include DNA with no coding function (Chapter 12) and eukaryotic transcripts are spliced in the process of making the mature mRNA (Chapter 11). To characterise genes it is necessary to make clone libraries that represent the mature mRNA sequences. RNA molecules cannot be ligated into DNA vectors, so it is necessary to make complementary DNA (cDNA) copies of mRNA transcripts using the enzyme *reverse transcriptase* (an RNA-dependent DNA polymerase, Fig. 13.7), which is made by a class of viruses termed retroviruses. The *in vivo* function of this enzyme is to make DNA copies of viral RNA genomes during replication (Chapter 34). Reverse transcriptase uses mRNA as a template to make DNA copies that are complementary in sequence to the mRNA (Box 13.4). The cDNA copies can now be ligated to vectors and transformed into host cells to produce a library of cDNA clones. Each clone will carry a cDNA complementary to one mRNA transcript. In this way, clones derived from all of the different mRNAs expressed in a typical eukaryotic cell can be represented in a cDNA library. Genomics terminology refers to the sequences of these cDNA clones as ESTs (expressed sequence tags, see Chapter 12).

Genomic libraries and cDNA libraries are collections of cloned DNA segments derived from genomic DNA or from mRNA. Each colony or plaque in a library contains one genomic segment or one cDNA sequence. Libraries can contain millions of different colonies or plaques so that every genomic sequence or mRNA transcript is represented in at least one colony or plaque within that library.

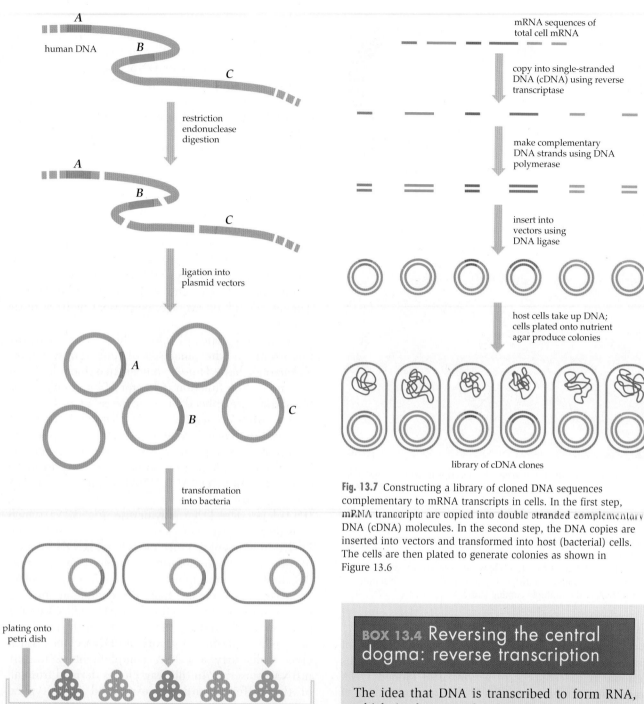

Fig. 13.6 Constructing a human genomic library. Human DNA is cut into fragments, containing genes *A*, *B*, *C* and so on. These fragments are each ligated to vector DNA (in this case a plasmid). The plasmids are transformed into bacteria, which are then plated on an agar culture plate. Each transformed bacterium grows separately on the plate and forms a bacterial colony. Each colony contains only one recombinant DNA molecule, but different colonies contain different DNA fragments cloned into the vector. The collection of thousands to millions of such colonies constitutes the clone library. Providing sufficient colonies are generated, any human genomic sequence should be represented in at least one colony

Fig. 13.7 Constructing a library of cloned DNA sequences complementary to mRNA transcripts in cells. In the first step, mRNA transcripts are copied into double-stranded complementary DNA (cDNA) molecules. In the second step, the DNA copies are inserted into vectors and transformed into host (bacterial) cells. The cells are then plated to generate colonies as shown in Figure 13.6

BOX 13.4 Reversing the central dogma: reverse transcription

The idea that DNA is transcribed to form RNA, which is then translated to produce a protein, became so firmly established that it was referred to as the 'central dogma' of gene expression. American scientists (and later Nobel laureates), Howard Temin and David Baltimore, worked separately on a class of tumour-causing viruses that infect vertebrates (see Chapter 34). They discovered that these viruses, whose genetic information is stored as RNA, convert this information 'back' into DNA inside the cell as part of their reproductive cycle. The conversion is carried out by a special enzyme capable of using

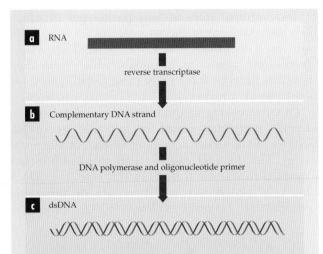

a RNA

reverse transcriptase

b Complementary DNA strand

DNA polymerase and oligonucleotide primer

c dsDNA

Generation of a double-stranded DNA copy from a template RNA molecule. The diagram shows only the major nucleic acid molecules involved. **(a)** Synthesis of a single-stranded DNA copy of template RNA using reverse transcriptase. This reaction requires a short priming DNA strand bound to the 3′ end of the template RNA. **(b)** Synthesis of the second strand of DNA. DNA polymerase utilises the single-stranded DNA as a template for polymerisation of the complementary DNA strand. This reaction also requires a priming oligonucleotide bound to the 3′ end of the template strand of single-stranded DNA. **(c)** The product, a double-stranded DNA molecule derived from the mRNA, is termed complementary DNA (cDNA)

an RNA template to make a DNA copy (see figure). The enzyme is an RNA-dependent DNA polymerase, termed **reverse transcriptase**, which was found to be associated with infectious RNA tumour virus particles. Since viruses harness the machinery of the cell to reproduce, they must use their RNA to reverse the central dogma by making a double-stranded DNA copy of their RNA genome inside the cell so that their genes can be expressed. As a consequence, RNA tumour viruses are now widely known as retroviruses. The DNA product becomes integrated into a host cell chromosome to become part of that host cell genome, essentially for the rest of the lifetime of that cell.

RNA tumour viruses not only provided the first glimpses of the process of reverse transcription in nature, but were also important in the development of our current knowledge of the molecular basis of cancer. Investigations of these viruses and the cellular genes whose function they affected identified the class of dominant oncogenes, whose products are able to change (transform) eukaryotic cells so that they grow in a way similar to tumour cells (see Chapter 12).

Identifying specific cloned sequences: hybridisation

Genomic clone libraries contain DNA fragments that are approximately gene size, making it possible to characterise specific genes in detail. However, in a library of human genomic fragments, for example, only one clone in 100 000 may carry the gene of interest. This is truly a needle in a haystack situation. How can the one colony or plaque in the 100 000 be identified? Here the ability of complementary sequences of DNA strands to base pair with each other, even if those strands come from different DNA molecules, comes to the rescue. The base pairing of complementary strands from different DNA molecules is termed **DNA hybridisation**. To identify sequences in cloned DNA, the recombinant DNA molecules in colonies or plaques are first *denatured* (i.e. the strands separated) and the strands immobilised on to filter paper (termed a membrane, Fig. 13.8). The membrane is then incubated with a specific DNA sequence, termed a **probe**, which has been labelled with radioactivity or with a chemical or enzyme tag. Only those colonies or plaques that contain DNA with sequences complementary to the probe will hybridise to the probe. After the non-hybridised probe is washed off, the colonies or plaques that contain the complementary sequences can be identified, in the case of a radioactive label, by autoradiography (i.e. exposure to X-ray sensitive film, Fig. 13.8).

The colony hybridisation method can therefore be summarised as involving the following steps.

1. Bacterial colonies or plaques, each containing a different recombinant plasmid (from the clone library) are grown on culture plates.
2. A sheet of nitrocellulose filter paper or nylon membrane is placed on to the colonies or plaques to produce a replica of the cells on the filter.
3. Cells on the filter are lysed and the DNA from the cells denatured and bound to the filter.
4. A radioactive DNA or RNA probe, which will hybridise to the gene of interest, is made.
5. The probe is allowed to hybridise to DNA on the filter. Unbound probe is washed off the filter.
6. Hybridisation to the DNA is detected by exposure to X-ray film.
7. Identified colonies are recovered from the original plate, grown and the recombinant DNA prepared.

One of the first probes used was mRNA. For example, two different mRNA species code for the two major components of adult haemoglobin, α-globin and β-globin. These mRNAs can be readily purified from precursor red blood cells. Radioactive mRNA was made from each and allowed to hybridise (so that complementary base pairing occurred) with the corresponding DNA in the colonies of the clone library. The

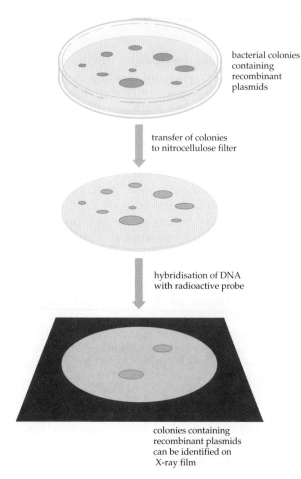

Fig. 13.8 Colony hybridisation method for screening DNA libraries for particular sequences. Bacterial colonies containing recombinant plasmids are grown on a culture plate. A replica of the colonies is prepared by transferring a portion of the colonies, *in situ*, onto a membrane. The DNA on the membrane is hybridised with a specific radioactive probe. Where a colony contains cloned DNA complementary to the probe, the probe is able to bind by complementary base pairing. Following washing of the membrane, radioactive spots corresponding to colonies containing the DNA of interest can be visualised by exposure to X-ray film

D. melanogaster from mutant phenotypes in which body parts develop in the wrong places (Chapter 16) were used to isolate genes with related sequences (homologous genes) in mammals, which were then shown to be critical for early mammalian embryonic development.

Other screening methods have been developed to identify particular genes in genomic or cDNA libraries. Differential screening is one of the most useful. This method can be used to identify genes expressed in specific tissues. In this case, the cDNA library is plated out and two sets of the same recombinant bacterial colonies (Fig. 13.9) are

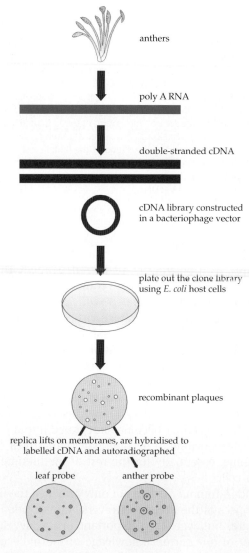

Fig. 13.9 Differential hybridisation as a means of screening DNA libraries for genes expressed in specific tissues. In this example, a cDNA library derived from anther mRNA is constructed and clones specific to the anther are selected by screening recombinant plaques, which have been transferred from an agar plate to membranes with a labelled anther-specific cDNA probe or a labelled leaf-specific cDNA probe. Those plaques detected only with the anther probe represent genes expressed specifically in the anther

hybrid complexes were detected by autoradiography (Fig. 13.8). In this way, both genomic and cDNA sequences corresponding to the α-globin and β-globin genes were isolated.

Genes that have an important general cellular function, and consequently whose DNA sequence is likely to have been conserved during evolution, can be cloned using the equivalent gene from another organism as a probe. For example, many genes important in development have been cloned from the relatively small genome of *Drosophila melanogaster*. Radioactive single-stranded DNA prepared from one of these cloned genes can be used to hybridise to similar sequences in human genomic or cDNA libraries in a process termed cross-hybridisation, so that the corresponding human gene can be isolated for study. For example, homeobox genes identified in

separately exposed to radioactively labelled cDNA from two different tissues. For example, screening of a clone library with cDNAs from anthers or leaves of oilseed rape plants identifies anther-specific genes, which can be recovered from the plate (Fig. 13.9). Oilseed rape plants are an important agricultural crop. Identifying and characterising anther-specific genes may be useful in controlling seed production.

Identifying specific cloned sequences using the protein product

When the protein product of the gene of interest is known, two methods can be used to isolate the gene encoding the protein. Firstly, part of the amino acid sequence of the protein can be determined by the technique of microsequencing. The nucleotide sequence encoding this peptide can be deduced from the genetic code table and then synthesised. The resultant oligonucleotide can be radioactively labelled and used to screen the library.

A second method relies on detection of the protein product of a gene expressed in bacterial colonies. Here the cDNA molecules are cloned into vectors that have been specially modified so that the insertion site is downstream of a promoter. These expression vectors drive transcription of the inserted DNA, which is then translated into protein. Antibodies specific for a protein are then used to detect the colonies expressing that protein.

Purifying recombinant DNA molecules

Cloned genes or DNA fragments can be purified from bacterial colonies. Total DNA is extracted from the colonies and vector DNA separated from the much larger bacterial genome. The inserted fragment can be cut out using the same restriction enzyme originally used to insert it, and purified away from vector DNA by gel electrophoresis. The result is an inexhaustible supply of a purified eukaryotic gene.

> A DNA library is a collection of clones of DNA from an organism. A gene of interest may be identified in a clone library by hybridisation (complementary base pairing) to a specific DNA or RNA probe, or by detection of the expression of a protein product. The DNA encoding the gene can then be isolated and used to study the structure and function of the gene.

Isolating genes by PCR amplification

Another useful technique, which more rapidly generates a specific DNA fragment without the need of living cells, is termed the polymerase chain reaction (PCR). This involves the selective and repeated amplification of a specific DNA sequence from a complex DNA mixture (Fig. 13.10). PCR makes use of the *in vitro* replication of DNA by purified DNA polymerase using synthetic

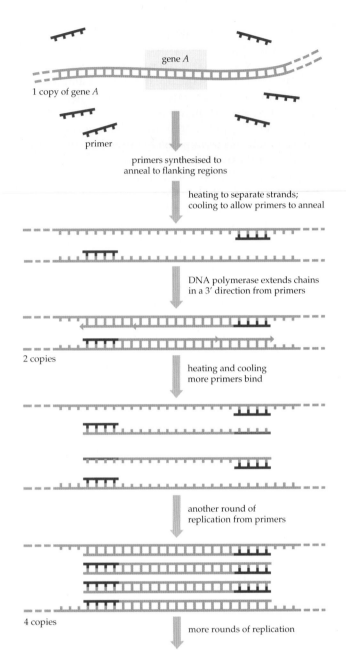

Fig. 13.10 Use of the polymerase chain reaction (PCR) to amplify a human gene. Short DNA primers complementary to sequences flanking the gene are chemically synthesised. Human template DNA is heated to denature (separate) the strands, then cooled, allowing the primers to anneal to the sequences on either side of the gene. DNA polymerase then extends the chain, replicating the gene from either direction. The cycle of heating and cooling is repeated, doubling the number of copies of the gene in each cycle. A heat-stable DNA polymerase, such as that from the bacterium *Thermus aquaticus*, is used because of its stability during the denaturation step

deoxynucleoside triphosphates. PCR allows the synthesis of a selected part of the genome between two regions whose sequences are known. The known sequences are used to design two synthetic DNA oligonucleotides, one complementary to the 3′ end of each strand of the DNA segment to be amplified (Fig. 13.10).

The synthetic oligonucleotides are added in excess to denatured genomic DNA at a temperature of approximately 50°C–55°C. The oligonucleotides hybridise with the correct sequence in the DNA and serve as primers for DNA synthesis *in vitro*. When deoxynucleotides and a high-temperature-tolerant DNA polymerase (*Taq* polymerase from *Thermus aquaticus*, a thermophilic bacterium isolated from hot springs) are added, synthesis is initiated from the primers (Fig. 13.10). At the end of the DNA synthesis reaction, the mixture is heated to 95°C, separating the newly formed DNA duplexes. When the temperature is lowered again, each separated strand acts as a template for a new cycle of DNA synthesis from the primers. At each cycle of synthesis and separation, the number of copies of the DNA segment is doubled. PCR is a remarkably specific process. In the course of only a few hours, a large amount of a specific DNA segment (the DNA between the two primers) can be synthesised from as little as a single DNA template molecule in the starting material.

PCR forms the basis of many diagnostic procedures. It has particular use in forensic science for identification based on traces of DNA in bloodstains, semen or even a single hair left at the scene of a crime (see below). PCR has even been used to amplify DNA sequences from traces of DNA of extinct organisms, and in human remains from prehistoric and Classical times.

> The polymerase chain reaction (PCR) enables synthesis of specific DNA segments from a complex mixture of DNA molecules in a test tube.

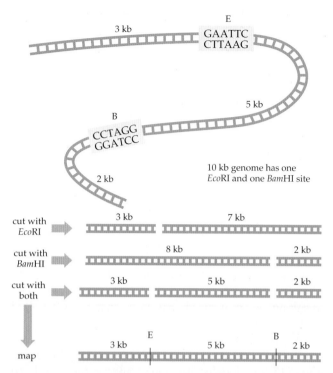

Fig. 13.11 Constructing a restriction map of the genome of a small virus. In this example, the linear viral DNA is 10 kb long and includes one *Eco*RI (E) and one *Bam*HI (B) site. When it is cut with *Eco*RI, fragments of 3 kb and 7 kb are seen on the gel. When the same genome is cut with *Bam*HI, fragments of 2 kb and 8 kb are released. When both are used together, the genome is cut into three fragments, of 2 kb, 3 kb and 5 kb. The 7 kb *Eco*RI fragment has disappeared, and has been replaced by 2 kb and 5 kb fragments. This must mean that the *Bam*HI site is within the 7 kb *Eco*RI fragment, and must be 2 kb from the end of the molecule

Characterising cloned DNA sequences

Recombinant DNA technology has made it possible to study the organisation and structure of eukaryotic genes, providing a new understanding of gene regulation. The following sections describe some of the methods used in the analysis of eukaryotic genes.

Restriction mapping

Restriction mapping is the term given to the mapping of restriction endonuclease cleavage sites within a DNA segment. Restriction maps can be derived from the nucleotide sequence of DNA where it is available, but in most cases, they must be deduced from the sizes of DNA fragments that are generated when the DNA is cut by restriction endonucleases. DNA fragment sizes are determined by gel electrophoresis. As described above, the length of a restriction fragment is inversely related to its migration through a gel in an electric field, so that DNA fragment sizes can be determined from their position in the gel. Figure 13.11 shows an example of restriction mapping of a small linear viral DNA

genome that has one *Eco*RI cleavage site and one *Bam*HI cleavage site. When the DNA is digested by *Eco*RI, the sizes of the two fragments generated allow the position of the *Eco*RI site to be mapped with respect to the two ends of the molecule. If the same DNA is digested with *Bam*HI, the position of the *Bam*HI site can be determined with reference to the ends of the molecule. If the DNA is digested with both *Eco*RI and *Bam*HI, the two restriction sites can be mapped with respect to one another.

Nucleotide sequence analysis

The sequence of bases (A, C, G, T) within any short stretch of purified DNA can be determined using *in vitro* DNA synthesis and gel electrophoresis. The key to sequence determination is the inclusion in the DNA synthesis reaction, together with normal deoxynucleoside triphosphates, of altered deoxynucleoside triphosphates that terminate synthesis when they are incorporated (Fig. 13.12). The altered deoxynucleotides used for this purpose lack the 3′ OH group required for the addition of the next deoxynucleotide

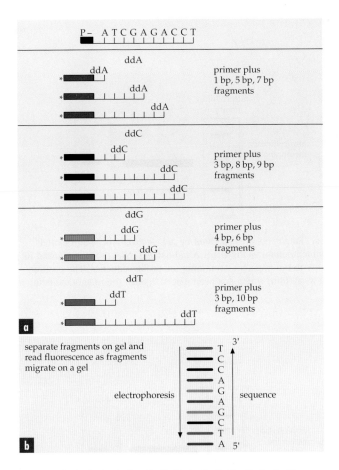

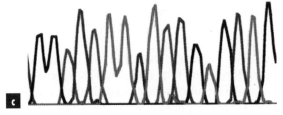

Fig. 13.12 Sequence determination of a short DNA fragment. **(a)** Synthesis begins with an oligonucleotide primer. The same primer is used in each of four reactions, but it is labelled with a different flourescent tag in each reaction (shown here as green, blue, black or red). A different dideoxynucleoside triphosphate is included in each reaction, together with the four deoxynucleoside triphosphates. For example, one mix contains the 'green'-labelled primer, ddATP and dATP, dCTP, dGTP and dTTP. Whenever the DNA polymerase needs to incoporate an A residue, it sometimes uses dATP and can continue on, or sometimes incorporates ddA and terminates. The reaction therefore has a mixture of fragments that all end in ddA. **(b)** The fragments generated in each of these four reactions are separated by electrophoresis on a gel and the migration of the fragments is detected by fluorescence imaging. The shortest fragments migrate fastest. The sequence can then be read directly by virtue of the fluorescent tag running at each position, starting at the smallest fragment. **(c)** The migration of the fragments through the gel is normally read automatically by a fluorescence reader and is represented as an electropherogram of the fluorescence

during DNA synthesis, that is, they are **dideoxynucleoside triphosphates (ddNTPs)**.

Sequence analysis involves four different synthesis reactions, each of which contains the four normal deoxynucleoside triphosphates and one of the four dideoxynucleoside triphosphates. This generates a family of fragments that extend varying lengths, but always end with the dideoxynucleoside triphosphate that was included in the reaction (Fig. 13.12). All reactions use the same primer, but the primer is tagged with a different fluorescent dye in each reaction so that the products can be distinguished after electrophoresis. In this way, families of fragments that end at any of the A residues are labelled with one colour, those that end in C are labelled with a second colour, those that end in G are labelled with a third colour and those that end in T are labelled with a fourth colour. These fragments are then separated by gel electrophoresis and the base sequence is simply read as a ladder of fluorescent dye bands (Fig. 13.12).

DNA, RNA and protein blotting

DNA molecules produced by recombinant methods can be analysed by hybridisation to detect specific genes or restriction fragments by the technique of DNA blotting, termed Southern blotting after its inventor, Edwin Southern. DNA is purified and cleaved by a restriction endonuclease and DNA fragments of different sizes separated by gel electrophoresis (Fig. 13.13). The fragments are denatured and then transferred from the gel onto a nitrocellulose membrane by 'blotting', that is, drawing liquid up through the gel on to the membrane. A cloned DNA probe is radioactively labelled, denatured and hybridised to the DNA on the filter. Unbound probe is washed away, and the bound probe detected using X-ray film (autoradiography). Of all the thousands of fragments into which genomic DNA is cut, the probe DNA will hybridise only to DNA fragments that have a sequence complementary to the probe (Fig. 13.13).

A related technique can be used to look for any mRNA transcripts produced from the cloned sequence (Fig. 13.14). RNA extracted from cells is separated by gel electrophoresis and the bands transferred by blotting to a nitrocellulose membrane. The RNA can then hybridise to a radioactively labelled probe via complementary base pairing and the bound probe detected by autoradiography. This method, known as northern or RNA blotting (in contrast to Southern blotting), reveals the presence and sizes of any mRNA transcripts that are derived from the cloned fragment. The range of tissues that express a particular gene can be determined using northern blot analysis of RNA extracted from a variety of different tissues (Fig. 13.15).

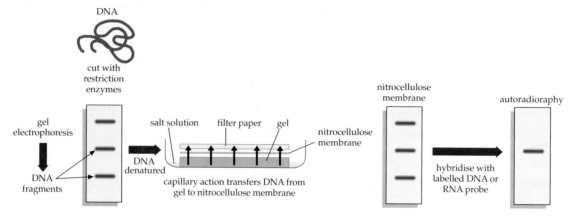

Fig. 13.13 Southern (DNA) blotting. DNA is cut by a restriction endonuclease. Fragments are separated by gel electrophoresis and blotted onto a nitrocellulose membrane after treatment with an alkaline solution, which denatures the DNA. A radioactive DNA probe is allowed to hybridise with single-stranded DNA trapped on the membrane and binds only to the fragment containing the probe sequence. The position of the radioactive fragment in the gel is detected as a darkened band on an X-ray film, which has been placed against the membrane before development (autoradiography)

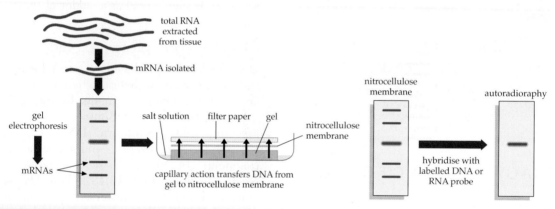

Fig. 13.14 Northern (RNA) blotting. mRNA is purified from tissues, separated according to size by gel electrophoresis and transferred to a nitrocellulose membrane. The membrane is incubated with a radioactive DNA or RNA probe, which detects the complementary sequences in the corresponding mRNA band as described in Figure 13.13. In this way, the molecular weight and quantity of the mRNA in a tissue can be determined

Restriction maps are constructed using the lengths of fragments generated by restriction endonuclease digestion to deduce the location of restriction cleavage sites. DNA sequences are deduced using DNA synthesis and chain termination at specific nucleotides to generate products of specific lengths that are separated by gel electrophoresis. DNA and RNA molecules separated by gel electrophoresis can be analysed by hybridisation to detect specific sequences.

Investigating genome structure in eukaryotes

Analyses of cloned genomic and cDNA fragments by gel electrophoresis, DNA–DNA or DNA–RNA hybridisation (complementary base pairing) and, ultimately, complete DNA sequence determination, are providing us with a clear picture of the structure and nature of genomes (Chapter 12). They have led, for example, to the discovery of introns, identification of transposable elements, isolation of human disease genes and the ability to manipulate DNA sequences.

The discovery of introns is a particularly interesting example of the application of recombinant DNA technology to the study of gene structure. β-globin mRNA was purified and its base sequence was shown to code for the known amino acid sequence of the β-globin protein. However, when a genomic clone of β-globin was obtained, it was possible to examine for the first time the gene itself rather than just its RNA transcript. Restriction mapping showed that two parts of the gene were present, sited more than a kilobase apart in the genomic clone (Fig. 13.16a). This was puzzling because the protein is only 146 amino acids long and should require an open reading frame of less than 500 base pairs. When a strand of the cloned genomic β-globin DNA was hybridised to the globin mRNA, three separate regions of the gene could form complementary base pairs with the mRNA (Fig. 13.16a). Two other regions were left as single-stranded loops. These regions

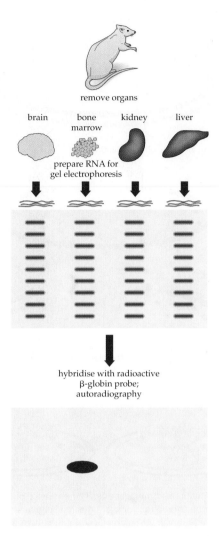

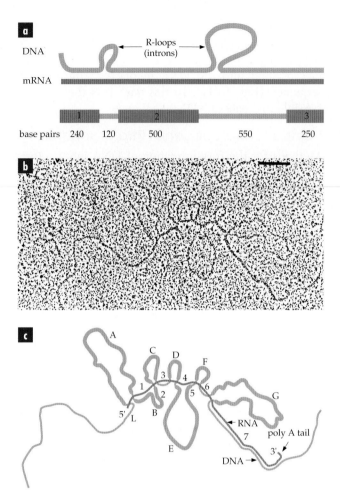

Fig. 13.15 Expression of the mouse β-globin gene in different tissues. RNA is prepared from samples of mouse brain, bone marrow, kidney and liver and electrophoresed to separate the mRNAs according to their size. The membrane is hybridised with a radioactive β-globin DNA probe, and the probe bound to RNA is detected by autoradiography. The probe hybridises only to RNA from bone marrow, showing that the β-globin gene is being transcribed only in this tissue

Fig. 13.16 (a) Structure of the β-globin gene. The sense strand of a cloned β-globin gene (blue) is hybridised to purified β-globin mRNA (red). Hybridisation occurs in three regions, separated by two loops of intervening sequence (termed R-loops). **(b)** Electron micrograph and **(c)** diagram of the ovalbumin gene, showing single-stranded DNA with R-loop structures obtained by hybridising the cloned ovalbumin gene to its mRNA. The gene is shown to contain seven large introns (A–G), which form R-loops, and eight exons (L, 1–7), which bind to complementary regions of the mRNA

had no sequences that were present in the globin mRNA! The conclusion is that the β-globin gene is an interrupted structure: a series of exons and introns (Chapter 11). Other genes, such as the ovalbumin gene, were also shown to have sequences that were not present in the mature mRNA (Fig. 13.16). DNA sequence analysis of the β-globin gene and of many other genes revealed the presence of introns in most eukaryotic genes.

Altering DNA sequences: site-directed mutagenesis

The availability of the complete nucleotide sequence of genes, combined with a knowledge of the genetic code and an understanding of the mechanisms for the regulation of gene expression, allows genes with new characteristics to be designed. There are many reasons why this is desirable. For example, a powerful way of studying protein function is to create proteins with altered amino acid sequences. It is extremely difficult to synthesise proteins chemically. However, it is straightforward to alter the sequence of the gene encoding the protein and to transform this sequence into a host cell, which then produces the altered protein.

Altered DNA fragments can be generated by restriction endonuclease cleavage and ligation of fragments that were not previously joined. However, this is a relatively crude method for manipulating DNA sequences. A more specific method for altering DNA sequences uses short synthetic oligonucleotides, which are largely complementary to the original sequence but which have specific alterations in the sequence. When such an oligonucleotide is used as a primer in a DNA synthesis reaction, the altered nucleotides become incorporated into the DNA in place of the original sequence (Fig. 13.17). In this way, it is theoretically possible to create any DNA sequence.

Reintroduction of genes into organisms

Methods for the transformation of DNA into bacteria and yeast cells were described briefly above. These methods have been extended to many organisms using a variety of approaches. The simplest approach is to introduce DNA into cells, allowing the DNA to become incorporated randomly into the host genome. Approaches include micro-injecting DNA into oocytes or cultured cells (Fig. 13.18), or coating microscopic ballistic pellets with DNA and shooting them into cells. In yeast and mouse cells, techniques exist for the precise replacement of existing sequences with related but modified sequences. In this way, the genomes of these organisms can be manipulated with complete precision. Transposable elements can also be used to introduce DNA into the genomes of organisms (Chapter 12). Insertion of a transposable element into the genome normally requires only transposase enzyme and the sequences found at the ends of the element. Foreign DNA inserted within the element will be carried into the genomic DNA when the element inserts. This approach has been used successfully for many insect and plant species.

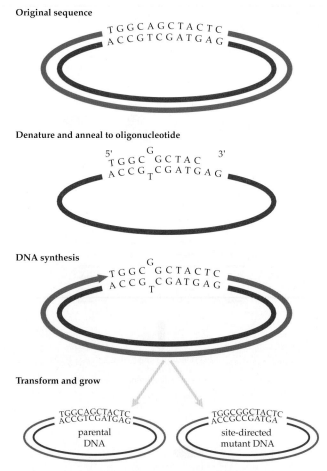

Fig. 13.17 Site-directed mutagenesis. The sequence of any segment of DNA can be altered to produce a desired sequence. In this example, the base pair A–T is replaced by the base pair G–C. A primer with the desired nucleotide sequence alteration is annealed to a denatured strand of vector DNA containing the recombinant DNA to be modified. The residue to be mutated, G, cannot base pair with the T, but the other complementary base pairs hold the oligonucleotide to the plasmid strand. The oligonucleotide is used to prime synthesis of the complementary strand and the resulting molecule, containing the G–T mismatch, is transformed into bacteria, where it is repaired and/or replicated. Some of the plasmid products contain the desired sequence change. Techniques exist, although not shown here, for preventing growth of the parental DNA so that the site-directed mutant DNA is produced with high efficiency

Applications of recombinant DNA technology

Analysing genetic variation

Identifying genetic variation

A number of methods have been developed to identify differences in DNA sequences between individuals. Such differences can be used as markers in recombination mapping experiments (Chapter 9) to isolate disease genes, or to study the genetic structure of natural populations. Early methods used Southern blots of restriction endonuclease digests to look for polymorphisms in restriction fragment lengths in different organisms (Fig. 13.19). If a base change has occurred and disrupted a restriction sequence in one allele, there will be one fewer cut in the DNA of that allele. The sizes of fragments will therefore be different between individuals, resulting in **restriction fragment length polymorphisms** (**RFLPs**) (Fig. 13.19).

Variation in the genomes of different individuals often occurs at the site of clusters of repeated sequences. Different alleles have fragments of different sizes because they contain **variable numbers of tandem repeats** (**VNTRs**) or variable numbers of very short (three base pair) repeated sequences, termed **microsatellites**. The different sizes can be detected by Southern blotting using the repeated sequence as a probe, or by PCR amplification using specific primers either side of the repeat region.

Identifying lesions in mutant alleles

Using DNA technology, we can identify base changes that are the cause of mutant alleles. This has been particularly important in the study of human genetic disease. Comparisons of the sequence of a gene may be

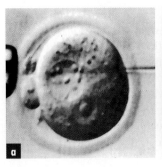

Fig. 13.18 Production of transgenic mice. **(a)** Cloned DNA is injected into a fertilised mouse egg through a fine glass micropipette. One or more copies of the gene may become inserted into one of the mouse chromosomes and may be expressed. **(b)** Insertion of the rat growth hormone gene into the fertilised mouse egg produced 'Supermouse' (left), shown with a normal litter mate

made between normal and affected family members in order to identify the mutation that results in the genetic disease.

The use of DNA technology has also been important in the study of the nature of oncogenes in tumour cells (Chapter 12). PCR amplification of sequences has made allele sequence analysis very efficient because DNA corresponding to a suspected disease allele can be readily amplified and its sequence determined. Analysis of gene sequences has also revealed that some base changes produce amino acid substitutions that do not affect protein function.

It is now routine to use PCR products to look for known mutations in disease alleles, even if the disease is caused by a single base difference between alleles. This can be done by constructing short primers that hybridise perfectly to the sequence of one allele, but not to that of another. Binding of the primers and amplification of the specific gene will take place only in a reaction where the primers match the DNA sequence perfectly.

Mutations that alter numbers of restriction sites are detected as restriction fragment length polymorphisms (RFLPs) by Southern blotting. Southern blotting or PCR amplification can be used to detect the highly variable numbers of tandem repeats present in VNTRs and microsatellites. Differences in the base sequences of normal and disease alleles can be identified by cloning and sequence analysis of the alleles.

DNA technology in forensic science

The ability to detect genetic variation between individuals at the level of the DNA sequence has been put to use in designing new techniques for solving crime. Whereas older techniques relied on the identification of blood group antigens or other proteins in body fluids using specific antibodies, the favoured material for identification is now DNA. For example, media attention is often given to cases in which a rapist has been positively identified by the DNA fingerprint left in semen found with his victim.

Two factors favour DNA analysis over older techniques. Firstly, sampling DNA is a lot more straightforward than sampling proteins. DNA is a rather robust molecule, whose structure can survive drying out. Thus, even old bloodstains or dried or decomposed tissue may contain useable DNA. In addition, PCR techniques require so little material for amplification—as little as a single molecule—that even trace amounts may be sufficient for analysis.

Secondly, there is much more variation at the DNA level than at the protein level. This is because a lot of eukaryotic DNA serves no coding function and variation in this sequence is much more frequent, presumably because it is not under stringent natural selection. A VNTR probe may detect up to 100 bands because these tandem repeats occur at many different locations in the genome. The number of repeats at each

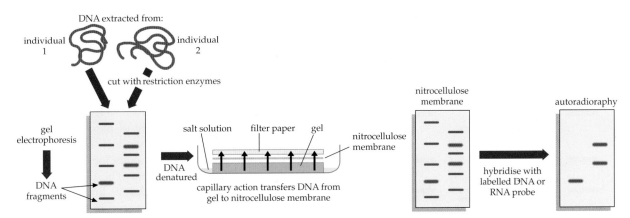

Fig. 13.19 Detection of restriction fragment length polymorphisms (RFLPs). DNA samples from two different individuals are extracted and digested with a restriction endonuclease and analysed by Southern blotting. The blot is incubated with a probe, for example, an oligonucleotide of known sequence or a cDNA clone, that hybridises to an RFLP, a restriction fragment whose size may vary between the two individuals

locus is variable (i.e. there are many alleles at each locus). The number of possible combinations of VNTR bands that any one individual can have is therefore in the billions. With the exception of monozygotic twins, the pattern of VNTR bands in an individual should not only be different to anyone else alive today, but should also be different to any human that has ever lived! Thus, banding patterns are essentially unique to an individual and serve as a molecular fingerprint of the individual. Traces of blood or a few hair follicles can therefore be used to identify the criminal (Fig. 13.20).

Several countries are introducing compulsory DNA testing for convicted criminals as a means of providing a database that can be used to solve future crimes. In Australia, the Crimtrack database is being introduced for this purpose. Criminology studies have shown that criminals who commit very serious crimes, such as serial rape, frequently have earlier convictions for less serious offences. The existence of the database would result in the identification of such rapists after the first attack.

The outcome of the use of such databases is very likely to be a major reduction in crimes because of the increase in the crime clear-up rate. In one region of England, where such a database was recently introduced, the crime rate has already been significantly reduced.

Paternity

DNA techniques are also more useful for establishing family relationships than the older method of blood group matching. For example, in a case of disputed paternity of a child of blood type O, only men with type AB could be excluded as the father. Even when a range of different blood group types are used, the chance of unrelated individuals having the same set of alleles is about 5%. However, each VNTR band of a child must have been inherited from either the mother or the father; thus, the father can be uniquely identified when his banding pattern is compared to that of the child and mother. The use of polymorphic DNA markers to investigate family relationships has also been extended to solving some interesting historical cases, such as those involving the Romanov family of the last Tsar of Russia and the nature of the relationship between Thomas Jefferson and his African-American servant (Box 13.5).

Mapping genes

Classically, recombination between genes was determined by the segregation of phenotypes in the offspring of a double heterozygous parent crossed to a double homozygous recessive parent (Chapter 9). This method served well for eukaryotes such as fungi, *D. melanogaster* and several plant species. However, linkage analysis is difficult in vertebrates, especially mammals, and almost impossible for humans because of the small family size and inability to control mating. The ability to detect DNA sequence polymorphisms has radically improved the assemblage of human linkage maps. RFLPs, microsatellites or any other DNA sequence polymorphisms can act as recombination markers along human chromosomes (Fig. 13.21). PCR typing can rapidly show whether recombination has occurred in the offspring of parents heterozygous for DNA marker alleles at two loci. Most importantly, DNA polymorphisms at particular loci can be tested for linkage to genetic disease loci. The discovery of DNA sequence polymorphisms closely linked to a genetic disease locus is an important step in the isolation and characterisation of the disease gene.

An entirely different approach to mapping genes uses the ability to fuse somatic cells from different

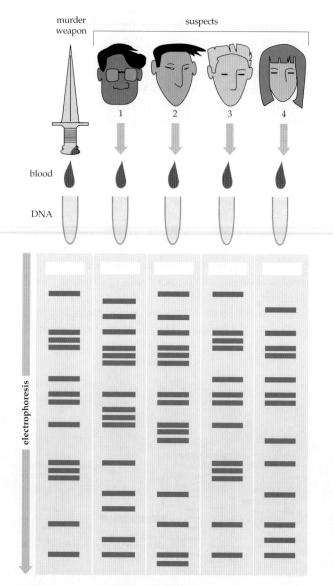

Fig. 13.20 Find the murderer! DNA fingerprints are obtained from four suspects and compared with the DNA fingerprint from blood from a murder weapon. One of the suspects gives a perfect match

BOX 13.5 Of tsars and presidents: solving historical puzzles with genetics

The ability to recognise individuals and family relationships from the pattern of DNA polymorphisms has had some fascinating applications. One case concerns the identification of the remains of Tsar Nicholas II of Russia. He and members of his family, together with their physician and three servants, were executed by firing squad on the night of 16 July 1917, during the Bolshevik uprising, ending almost 300 years of Romanov rule. Film-maker Geli Ryabov and geologist Alexander Avdonin used photographs and the report filed by one of the executioners, held in the Kremlin, to locate the likely grave near the town of Yekaterinburg. Almost 1000 bone fragments were assembled into five female and four male skeletons, a number that was consistent with the executioner's account.

Confirmation that the remains included those of the Tsar and his family came from DNA polymorphism analysis. On 15 September 1992, the skeletal remains were met at Heathrow Airport by a BBC television director, who hired a funeral hearse because he felt it was 'inappropriate to carry the Russian Imperial family in the boot of my Volvo'. Both nuclear and mitochondrial sequences were examined, revealing that five of the bodies were related and three were female siblings. The analysis of samples was extended to living relatives of the Romanovs and to the remains of Grand Duke Georgij Romanov, the 28-year-old brother of Tsar Nicholas II, who died of tuberculosis in 1899. These analyses confirmed that the remains were those of the Romanov family.

The remains of two of the Tsar's children have never been found. An American immigrant, Anne Anderson, claimed for many years to be Anastasia, the youngest daughter of Nicholas II. However, a series of DNA tests carried out after her death showed that she could not have been a member of the Romanov family and that her claim to the royal lineage of Russia was false.

A second interesting case concerns the relationship between Thomas Jefferson and his African-American slave, Sally Hemings. Jefferson is alleged to have fathered Eston Hemings, the last child of Sally Hemings, who was born shortly after Jefferson and Hemings returned from France.

Thomas Jefferson, American president 1801–9. DNA tests showed strong evidence that he was the father of Eston Hemings

Eston is said to have borne a striking resemblance to Thomas Jefferson and he was able to enter white society in Madison, Wisconsin, taking the name Eston Hemings Jefferson. To test whether Eston could have been Jefferson's son, sequences on the Y chromosome from male-line descendants of Field Jefferson, a paternal uncle of Thomas Jefferson (Field's line was used because Thomas had no known male-line descendants), were compared with Y chromosome sequences of male-line descendants of Eston Hemings Jefferson. Y chromosomes are passed from father to son (Chapter 9), so if Thomas Jefferson was the father of Eston Hemings, the Y chromosome sequences of Field Jefferson and of the male-line descendants of Eston Hemings should have shown the same sequence polymorphisms. The analysis indeed showed that Eston's Y chromosome and the Jefferson Y chromosome were derived from the same lineage. These results provide very strong evidence that Eston Hemings was indeed the son of Thomas Jefferson and his slave Sally Hemings.

species to form cell hybrids. Once fused, these hybrid cells tend to occasionally lose chromosomes (and the genes they carry) from one of the parental species. This makes it possible to correlate retention of particular genes with retention of particular chromosomes. This technique provides a rapid way to map genes to

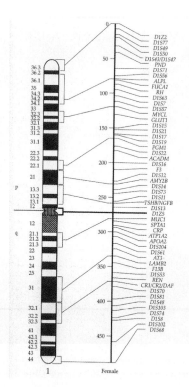

Fig. 13.21 A physical map of the human X chromosome. The marked locations indicate the positions of known DNA polymorphisms that can be used to map genes more accurately to specific locations on chromosomes

chromosomes. For example, human cells can be fused with mouse cells (Fig. 13.22). Mouse–human cell hybrids eliminate certain human chromosomes. Human and mouse β-globin genes can be distinguished by their specific Southern blot pattern when probed with a radioactive β-globin probe. Hybrids that retain a human chromosome 11 also retain the human β-globin gene. All hybrids that lack human chromosome 11 also lack the human β-globin gene. The β-globin gene must, therefore, be on chromosome 11 (Fig. 13.22).

More direct methods of physical gene mapping, which do not require allelic differences between parents nor a mating event, are now available. A probe labelled with a fluorescent dye is applied directly to a preparation of chromosomes in which the DNA has been denatured and can, therefore, hybridise to the probe. Figure 13.23 shows how this method can be used to determine the location of a gene. This method, termed FISH (fluorescence *in situ* hybridisation), gives a physical location for any cloned gene. If enough gene markers are used, genetic and molecular mapping methods can produce detailed maps.

The presence of DNA polymorphisms has greatly increased the resolution of linkage mapping in humans and other higher eukaryotes. Genes can also be assigned and mapped to chromosomes in mammals using cell hybrids and can also be mapped directly by *in situ* hybridisation.

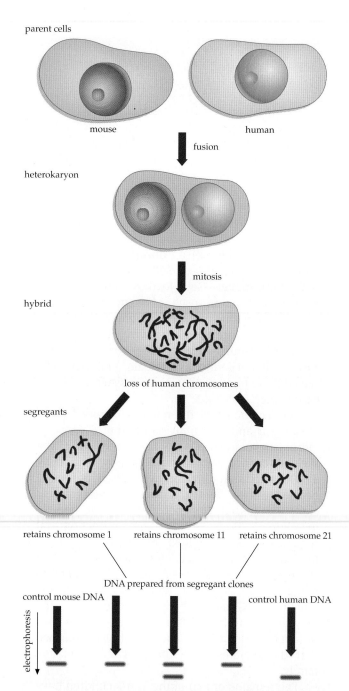

Fig. 13.22 Somatic cell genetic analysis. Mouse and human cells are fused to form a heterokaryon, a cell containing two different nuclei, one mouse (blue) and one human (red) nucleus. Following mitosis, the mouse and human chromosomes may combine in a single nucleus containing a full set of 40 mouse chromosomes (of which only eight are shown) and a full set of human chromosomes (six shown). This cell may proliferate to form a clone of hybrid cells. During growth, human chromosomes are randomly lost, forming segregant clones that contain a full set of mouse chromosomes, but only one or a few human chromosomes (e.g. lines retaining human chromosome 1, 11 or 21). DNA is prepared from each clone, and cut with a restriction enzyme. The presence or absence of the human β-globin gene is determined from the restriction fragment length on a Southern blot. It is found that whenever human chromosome 11 is present in the hybrid cells, the human β-globin is present. Thus, β-globin is located on chromosome 11

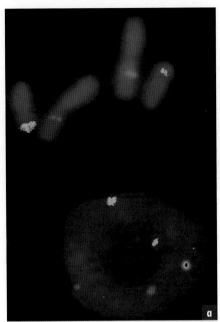

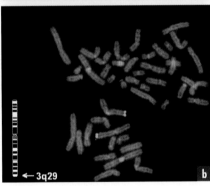

← 3q29

Fig. 13.23 Mapping genes to chromosomes by FISH (fluorescence *in situ* hybridisation). The condensed mitotic chromosomes from dividing cells are spread onto a microscope slide. The chromosomal DNA is denatured using alkali treatment. A DNA probe labelled with a fluorescent dye is incubated with the chromosomes. Only the locus, which contains sequences complementary to the probe, will hybridise to the probe. After unbound probe is washed off, the site of hybridisation can be observed using fluorescence microscopy. **(a)** Condensed mitotic chromosomes of the plant *Brachycome dichromosomatica*. The DNA in the chromosomes is stained blue and different loci on the different homologue pairs have hybridised to red and green fluorescing probes respectively. An interphase nucleus containing decondensed chromosomes also shows hybridisation to the two probes (lower half of picture). **(b)** Human metaphase chromosomes stained blue hybridised to a green-fluorescing probe specific for the terminal band of the long arm of chromosome 3

Cloning genes of unknown function

There are many inherited human diseases whose biochemical basis is unknown. Thus, the genes responsible for the disease cannot be cloned using the sequence of the affected protein products. Instead, the gene is first mapped by looking for linkage between the disease and a set of DNA markers whose chromosomal location is already known. When linkage is detected, the disease can then be assigned to a chromosome region. The position of the unknown gene is then further defined by measuring recombination between the disease locus and DNA markers across this region. Eventually, the linkage data refine the position of the disease locus to a DNA region that is small enough to look for the gene itself. Transcripts expressed from this region are identified and the DNA sequence of the transcripts in alleles from normal and diseased individuals determined. The disease gene is identified when alleles of the gene from diseased individuals contain mutations that would affect gene function. This approach was successful in isolating the gene responsible for cystic fibrosis (Box 13.6) and many other diseases.

BOX 13.6 **The cystic fibrosis gene**

The disease cystic fibrosis (CF) becomes apparent at a young age and affected children suffer persistent chest infections. The autosomal recessive pattern of its inheritance is well known; affected children have normal parents who are both carriers of the mutant *cf* allele. CF is the most common genetic disease in Caucasian populations, with the frequency of carriers being approximately 1 in 26. The pattern of abnormal mucus secretion and severe respiratory infections leading to early death is also well known. However, the primary biochemical cause of the disease was unknown until the advent of molecular cloning.

The first step in isolating the *CF* gene was to map it by genetic recombination by following the inheritance of the disease in families and correlating it with the inheritance of other genes and of polymorphic DNA markers. This turned out to be a most difficult task, with many false leads. In 1989, through a collaboration involving scientific groups in London and Toronto, the *CF* gene was shown to be linked to a gene (*MET*) known to lie on chromosome 7. The location of *MET* on chromosome 7 was not known precisely but it narrowed the search to DNA markers located on this chromosome.

The DNA of the *CF*-containing region was cloned by *chromosome walking*. In this technique, overlapping DNA clones are isolated to extend the cloned region (i.e. to 'walk') along the chromosome to an adjacent gene of interest. The polymorphic DNA marker that mapped closest to *CF* was used to screen a genomic library for clones that overlapped it (Fig. a). This clone, in turn, was used to screen for more overlapping clones. These clones were used to test for linkage with *CF*, so that the scientists could confirm that they were walking toward the *CF* gene, not away from it. In this way, the whole region between the flanking markers was cloned and the region most likely to contain the *CF* gene identified. By examining the genes within this region and determining the sequence of the genes in individuals suffering from CF, the *CF* gene was identified. CF disease alleles exhibit a variety of base changes compared with the sequence of the normal allele in unaffected children. In the most common disease allele, a small deletion of three base pairs has occurred in the open reading frame, so that a single phenylalanine residue is missing from the encoded protein. This small change to the CF protein is sufficient to cause the devastating disease.

Isolation of the *CF* gene permitted analysis of the *CF* gene product. The CF protein serves as a membrane-bound regulator of Cl⁻ ion exchange (i.e. it is a Cl⁻ ion channel, Chapter 4). This role explains the claggy mucus and abnormal salt balance of sweat in affected children.

DNA sequence analysis makes it possible to tell whether parents carry an abnormal allele (Fig. b). It also makes prenatal diagnosis possible, using DNA from embryonic cells of the placenta (chorionic villus sampling) or cells shed into the amniotic fluid (amniocentesis). With this knowledge, parents may then choose to terminate the pregnancy.

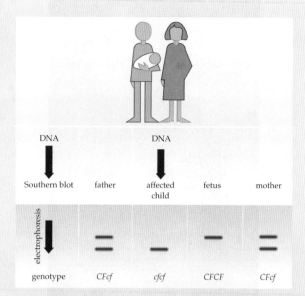

(b) Two normal parents have an affected infant. The mother is again pregnant and the parents are anxious to know whether their second child will be normal. Southern blot analysis shows that in the affected child homozygous for the disease allele, the *cf* gene is contained in a 7 kb fragment. The parents are heterozygous for this fragment, as well as for a 9 kb fragment that must contain the normal *CF* allele. DNA from the developing fetus can now be compared. There is good news for the couple, as their baby has inherited the normal allele from both parents and will therefore be normal

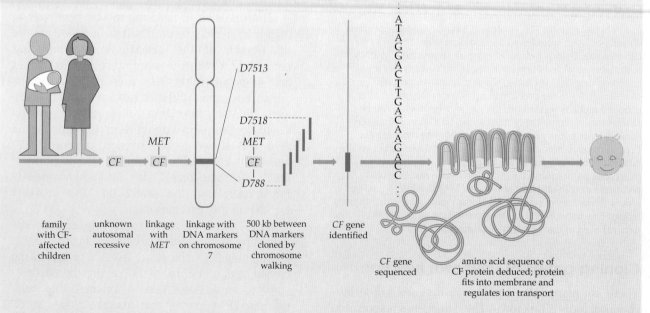

(a) Identifying the *CF* gene. From knowing little about the gene except its autosomal recessive mode of inheritance, it was possible to localise the *CF* gene to a region of chromosome 7 using genetic and molecular mapping techniques. The gene was mapped in relation to polymorphic DNA markers, its chromosomal location determined and its physical position progressively narrowed down. The region containing *CF* was cloned and the gene recognised and characterised. The DNA sequence of the mRNA product of the gene was used to deduce the amino acid sequence of the protein it makes, allowing predictions to be made about the function of the protein product

Does this discovery offer any hope to the individuals who suffer from cystic fibrosis? The possibility of using the cloned *CF* gene to cure affected children has been discussed for decades but, until recently, seemed to belong in the realm of science fiction. However, exciting results have recently been gained from experiments on rats and mice showing that an adenovirus vector containing a normal *CF* allele can infect lung cells, delivering the *CF* gene to the lungs. Infected cells can produce enough CF transport protein to correct the ion channel defect and alleviate the symptoms of the disease. These experiments raise hopes that humans with CF can be treated with gene therapy in this way.

Sometimes genetic abnormalities can be caused by major rearrangements of chromosomes. Missing pieces (deletions) and exchanges of large chromosome regions (translocations) can be observed in cytological preparations under the microscope and may provide a guide to the location of an unknown gene. For example, a study of the DNA of a number of patients with deletions or rearrangements of the Y chromosome led to the successful identification of the gene that controls male sexual development in humans (Box 13.7).

DNA technology, combined with new gene mapping methods and DNA polymorphisms, can be used to pinpoint a gene responsible for a disease, allowing the region to be cloned and the gene identified and studied.

BOX 13.7 Finding and isolating the human sex-determining gene

In humans, as in other mammals, sex is determined by the presence or absence of the Y chromosome (Chapter 9). It has long been supposed that this chromosome carries a critical gene that operates a switch in a five-week old human embryo, which turns undifferentiated gonads into testes. Once testes are formed, they produce hormones that influence other aspects of male development. The testis determining gene, *TDF*, has been inferred for a long time, but its sequence and how it functions as a genetic switch were unknown until recently.

Cloning and characterisation of *TDF* involved examination of a number of patients with abnormal sexual development. The position of *TDF* was narrowed down by gene mapping using individuals with chromosomal rearrangements. Individuals in one group are phenotypically male but carry two X chromosomes and part of the Y chromosome, containing *TDF*, which has been translocated onto the X chromosome. Individuals in the second group are phenotypically female, but have one X and one Y chromosome, the Y chromosome lacking *TDF*. By aligning the regions of the Y present in XX males and absent in XY females (deletion mapping, Fig. a), a region containing *TDF* was defined.

The first likely candidate gene identified in this region turned out not to correspond to *TDF*. Australian geneticists isolated a kangaroo homologue of this gene and found that it was not present on the kangaroo Y chromosome. This unexpected finding led to further mapping and cloning. Additional XX male humans were characterised and the location of *TDF* further restricted to the tip of the Y chromosome. This region was studied in great detail and a sequence, *SRY* (for sex region, Y gene), was found to be present in men and absent in women. This *SRY* gene is male-specific and located on the Y chromosome in all other mammals tested, including kangaroos.

SRY has all the hallmarks of a sex-determining gene. Its sequence is altered in at least some

segments of Y chromosome found in:

XX male

XY female

normal male

smallest region that must contain *TDF*

(a) Deletion mapping of the human Y chromosome. XX males who have only a small part of the Y chromosome all retain the region outlined in red. All XY females who have even large parts of the Y lack this region. The *TDF* gene can therefore be pinpointed to this region of the Y chromosome. Analysis of this 30 kb region led to the discovery of the *SRY* gene

individuals who have a Y chromosome and yet are female, presumably because they have a mutant *SRY* gene. The final proof was delivered when the cloned mouse *SRY* gene was introduced into the genome of a female mouse embryo. The animal that developed, nicknamed 'Randy', was phenotypically male and exhibited normal male mouse behaviours (Fig. b).

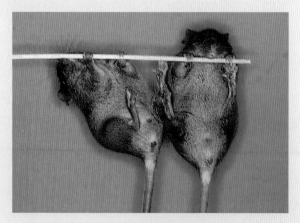

(b) An XX mouse embryo carrying the mouse *SRY* gene develops as a male. The transgenic 'Randy' (right) has normal testes and male genitalia indistinguishable from his normal male litter mate (left), but is sterile because he lacks other genes on the Y chromosome required for sperm production

Genetic engineering and industry

The techniques of genetic engineering make use of recombinant DNA technology for applications in the pharmaceutical, agricultural and veterinary industries. This technology has permitted genes to be modified and transgenic organisms (i.e. organisms that have foreign genes introduced into their genomes) to be produced.

Making pharmaceutical products

Many plasmid vectors are designed to express an inserted gene. Such vectors, termed expression vectors, carry efficient promoters such as that found in the *lac* operon (Chapter 11). These promoters can generate large amounts of mRNA, which is then translated to produce the protein product.

Genes encoding important human proteins can be cloned into expression vectors and used to produce proteins with amino acid sequences that are identical to the natural proteins. This overcomes the supply problem of therapeutic hormones and enzymes, which had been limited to the small amount that could be purified from donated blood or cadavers.

One of the best examples is that of insulin. Before the advent of genetic engineering technologies, insulin used by people with diabetes was extracted from pigs. Pig insulin functions in humans but the amino acid sequence is slightly different, creating problems arising from the reaction of the immune system to the foreign protein. Human insulin produced using recombinant DNA is identical to the human hormone because it is synthesised from the human gene. As a result, it is a more effective treatment for diabetes.

Another significant product is a genetically engineered blood clotting factor that is used to treat haemophilia. Here, the recombinant product is preferable to natural clotting factor purified from donated human blood because it is free of the risks of blood-borne diseases such as AIDS.

Modifying agricultural organisms

The history of agriculture is, to a large extent, a history of the genetic manipulation of plant and animal species. Initially this manipulation occurred by selecting, for breeding, the best plants growing in the fields or the best animals in the herd. The discovery of the rules of heredity (Chapter 9) permitted more effective breeding programs for the generation of improved strains. Breeding of this type led to the development of vastly improved rice strains, responsible for the so-called green revolution in agriculture. These strains had a large impact on the world in the latter part of the twentieth century. However, recombinant DNA technologies have created the capacity to alter the characteristics of domestic animals and plants in an even more targeted and specific way.

One significant but controversial development is the introduction into plants of genes that encode natural insecticides found in other organisms. Some of the cotton grown commercially carries such a gene. The hope is that it will be possible to reduce the use of chemical insecticides, which are more destructive to the environment because they leave residues and are relatively indiscriminate in their action against insects.

Gene therapy

Gene therapy in humans is in the earliest stages of development. The idea is to introduce a functional gene into patients who suffer from a particular genetic disease because they lack that gene. The technology for inserting foreign genes into cells and embryos is well advanced, but methods for incorporating genes into the particular cells that require the gene are yet to be developed. The ultimate goal, for example, is to target the β-globin gene specifically to red blood cell precursors in order to cure sickle cell anaemia, or target the Duchenne muscular dystrophy gene, *dystrophin*, to muscle cells, or target the cystic fibrosis gene to lung cells.

Recombinant DNA technology can be used to change the genetic make-up of organisms. Bacteria can be transformed to produce recombinant human proteins. Animals and plants may be genetically modified by the insertion of cloned genes.

Summary

- Recombinant DNA techniques are used to isolate genes or small segments of DNA from chromosomes.
- DNA molecules can be cut at specific sequences by restriction endonucleases. The fragments can be separated by gel electrophoresis and their sizes (in kilobases) determined. Fragments from different sources can be ligated together to form recombinant DNA molecules.
- A library of cloned DNA fragments of a eukaryotic genome can be generated by ligation of the fragments into self-replicating bacterial plasmids or viral vectors and transformation of the recombinant DNA molecules into host cells. Growth and isolation of a single bacterial or viral clone allows a single DNA fragment to be propagated and purified.
- Complementary DNA (cDNA) copies of mRNA transcripts can be cloned into plasmids or viral vectors, yielding a cDNA library. A cDNA clone corresponding to a gene of interest may be recognised within this library by the hybridisation of recombinant DNA in the bacterial colony or plaque to a specific DNA or RNA probe, or by expression of a particular protein product. The clone containing the gene of interest may then be isolated and the recombinant DNA purified.
- The polymerase chain reaction (PCR) technique offers a rapid means of preparing and isolating eukaryotic genes and DNA fragments from tiny amounts of DNA. PCR makes use of a heat-stable DNA polymerase to make copies of small, defined regions of the eukaryotic genome by amplifying DNA segments using repeated rounds of replication initiated from short, synthetic, primer DNA sequences.
- Any cloned fragment can be subjected to restriction mapping and DNA sequence analysis. Restriction maps are constructed using the lengths of fragments generated by restriction enzyme cleavage to deduce the position of restriction sites.

- Southern blotting enables the restriction fragment sizes of specific sequences within a complex genome to be determined. Northern blotting can be used to determine the size and tissue distribution of RNA transcripts derived from a particular gene.
- DNA sequences are deduced from the lengths of fragments that are synthesised from a specific primer in four separate reactions in which termination occurs at a different base in each reaction.
- The ability to purify, restriction map and sequence fragments of eukaryotic DNA has made it possible to examine the structure of genes in higher organisms.
- DNA technology permits us to look directly at the genome to identify genetic variation in terms of changes in base sequence. This can be done by PCR amplification or cloning and sequencing normal and mutant alleles. Changes that alter restriction sites can be detected as restriction fragment length polymorphisms (RFLPs) by Southern blotting. Repeated sequence probes may detect variable numbers of tandem repeats (VNTRs or microsatellites).
- The availability of a range of DNA polymorphisms has greatly increased the resolution of linkage mapping in humans and other higher eukaryotes. Somatic cell genetic techniques have allowed thousands of genes to be assigned to chromosomes in mammals. Genes can also be mapped directly by *in situ* hybridisation.
- DNA technology, combined with new gene mapping methods, can be used to pinpoint a gene responsible for a disease, allowing the region to be cloned and the gene to be identified.
- Recombinant DNA technology can be used to change the genetic make-up of organisms. Bacteria may be modified to produce any protein. Flowering plants and animals may also be genetically modified by the insertion of a cloned gene. Transgenesis may be used to improve domestic plant or animal species and treat human genetic disease.

keyterms

colony (p. 307)
competency (p. 307)
dideoxynucleoside
   triphosphate
   (dd NTP) (p. 315)
DNA hybridisation
   (p. 311)
donor DNA (p. 302)

genomic library
   (p. 309)
microsatellite (p. 318)
plaque (p. 307)
probe (p. 311)

restriction
   endonuclease
   (restriction enzyme)
   (p. 302)
restriction fragment
   length polymorphism
   (RFLP) (p. 318)

reverse transcriptase
   (p. 311)
transformation
   (p. 302)
variable numbers of
   tandem repeats
   (VNTRs) (p. 318)
vector DNA (p. 302)

Review questions

1. Describe the various types of vector used in constructing recombinant DNA molecules.

2. In making recombinant cDNA clones, what is the role of reverse transcriptase?

3. Several human genes produce protein products that have potential therapeutic uses. The cloned DNA can be inserted into an expression vector and the recombinant protein expressed in *E. coli*. However, human genes encode introns that cannot be excised in prokaryotes. How can this problem be overcome to produce human proteins in *E. coli*?

4. What is special about the function of *Taq* polymerase as a DNA polymerase in the polymerase chain reaction? How is DNA sequence amplification achieved?

5. What are the most useful sequence polymorphisms in forensic analysis? Why are they so useful?

6. Explain how polymorphic DNA markers and chromosome walking can be used to isolate eukaryotic genes.

7. What techniques are available for genetic transformation of animal or plant cells?

Extension questions

1. Limitations apply to the use of gene transformation in the application of gene therapy of inherited diseases. What are these?

2. What are the issues that are important to human society in terms of the ethics of the use of recombinant DNA technology?

Suggested further reading

Davis, B. D. (ed). (1991). *The Genetic Revolution: Scientific Prospects and Public Perceptions.* Baltimore: Johns Hopkins Press.

A collection of essays describing the science behind recombinant DNA and the social issues arising from it.

Judson, H. F. (1999). *The Eighth Day of Creation: Makers of the Revolution in Biology.* Cold Spring Harbor: Cold Spring Harbor Laboratory Press.

An excellent discussion of the history of molecular biology research leading to and including the genetic engineering revolution.

Watson, J. D., Gilman, M., Witkowski, J., Zoller, M. (1992). *Recombinant DNA.* New York: W. H. Freeman.

A very good description and explanation of the methods used in recombinant DNA technology.

Part 3

Reproduction and development

CHAPTER 14

Reproduction growth and development in plants

One of the most striking characteristics of living organisms is that they have the capacity to produce others of their own kind. For an individual, the process of reproduction ensures genetic continuity; that is, at least some of an organism's genes continue to exist in its offspring. More generally, reproduction of individuals ensures the continuation of the species. Organisms reproduce asexually, sexually or by both means.

In **sexual reproduction** in eukaryotes, a new individual forms from the fusion of haploid gametes that are typically from two different organisms. At some stage before gamete formation meiosis occurs, during which homologous chromosomes exchange genetic material to produce new combinations of parental genes (Chapter 8). By allowing DNA from one organism to be combined with that from another, sexual reproduction results in unique individuals and generates genetic variation within populations.

Some organisms reproduce asexually, with sexual reproduction occurring rarely or not at all. The fact that these organisms have survived for millions of years indicates that **asexual reproduction** (possibly punctuated by rare bouts of sexual reproduction) is a very successful strategy for survival. Asexual reproduction involves the production of new individuals by either mitotic cell division (e.g. unicellular algae) or binary fission (e.g. bacteria; Chapters 8). An advantage of asexual reproduction is that it allows offspring to form without the need to get two individuals of different sex into reproductive condition at the same place and the same time. A disadvantage is that because offspring are genetically identical to the parent and to each other, genetic diversity is limited. This constrains an organism's ability to adapt to new or changing environments.

> In asexual reproduction, new individuals arise from a single parent by mitotic cell division and are genetically identical to their parent. Sexual reproduction involves redistribution of parental genes into offspring and is the principal mechanism for generating genetic variation in eukaryotes.

Whether reproducing sexually or asexually, the life of a multicellular organism is also characterised by changes of shape as individuals reach their full size and form. These changes are complex and are controlled by both genetic and environmental influences. Broadly speaking, these changes can be viewed as either resulting from the appearance in the organism of new organs and tissues (**development**) or from an increase in the size of an organism (**growth**). Growth involves the production of new cells by repeated mitotic divisions. Development is the differentiation of these cells into the diverse cell types found in adult organisms, and the ordered assembly of the differentiated cells into the three-dimensional structures characteristic of mature organs. The formation of organs is genetically controlled by a number of developmental processes that are initiated in response to cues either produced by the organism itself (such as its age or a cell's location within the organism), or derived from some aspect of the external environment (such as day length or temperature). Developmental processes act by turning certain sets of genes in the nuclei of the different cell types either on or off in an ordered and hierarchical way (Chapter 16).

This chapter describes some of the important concepts of sexual reproduction and life cycles in plants and animals, and shows examples of different types of gametes. The life cycle of a typical flowering plant (angiosperm) is examined in detail. You can compare this to the life cycles of animals in the next chapter. This chapter also describes how many plants propagate themselves by vegetative means—without sex! A feature of plants that animals largely lack is the ability to produce new organs throughout their life. Indeed, an individual cell from a mature plant can be regrown into a complete new plant. This property, which is called totipotency, is the basis of plant tissue culture and genetic engineering technologies.

Sexual reproduction: some concepts

Sexual reproduction is characteristic of nearly all eukaryotes. It often involves a considerable cost to the parents in energy and building materials for the production of gametes and to ensure mature male and female gametes occur together. However, its most important advantage is the redistribution of parental genes into offspring. Advantageous genetic combinations may result from the random assortment of parental characters during meiosis and enhance the survival of individuals. Sexual reproduction that enhances genetic variation within populations has been exploited in selective breeding programs of domestic plants and animals.

Life cycles

A **life cycle** is the sequence of stages in the growth and development of organisms from zygote to reproduction, that is, from one generation to the next. Life cycles of sexually reproducing organisms show a pattern of alternation between diploid and haploid stages, with fertilisation alternating with meiosis (Fig. 14.1). However, the relative amount of time spent in diploid and haploid stages varies greatly between different types of organisms.

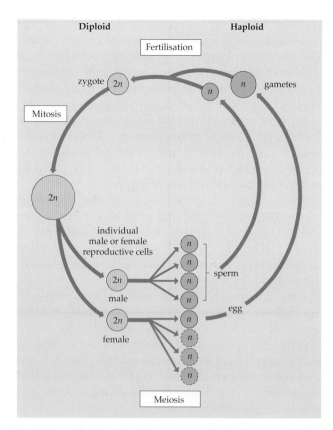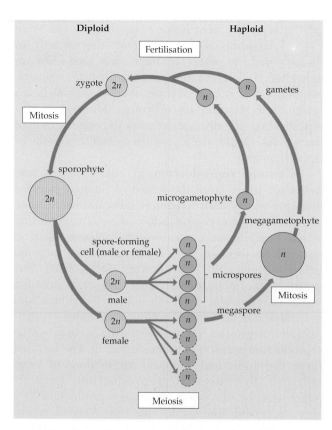

Fig. 14.1 In the life cycles of plants, there are from two to many mitotic divisions after meiosis before gamete production occurs. Meiosis produces female (often only one, with the other three products degenerating) and male haploid spores (four), each of which undergoes cell division by mitosis to produce the female and male gametophytes, which produce haploid gametes by mitotic divisions

In animals, fertilisation follows on directly from meiosis and therefore the haploid generation is simply the gamete stage (Fig. 15.4). Plants, in contrast, have from two to many mitotic divisions after meiosis before gamete production occurs (Fig. 14.1). In the life cycles of plants, meiosis produces female and male haploid spores, each of which undergoes cell division by mitosis to produce a multicellular haploid stage, the female or male **gametophyte**. These gametophytes produce haploid gametes by mitotic divisions. Two gametes fuse to form a diploid zygote, which then undergoes mitotic divisions to produce the diploid life stage, the **sporophyte**. In flowering plants, the sporophyte is the diploid vegetative structure that produces flowers. At some stage, diploid cells of the sporophyte undergo meiosis and produce haploid spores, so completing the cycle (see Chapter 37 for different kinds of plant life cycles).

> Life cycles of sexually reproducing organisms show a pattern of alternation between diploid and haploid stages.

Types of gametes

Gametes that fuse during sexual reproduction may be similar in structure, or distinctly different (Fig. 14.2).

Isogamy, in which the gametes are similar in appearance, is the simplest form of sexual reproduction in eukaryotes; this type of reproduction is found in certain algae and fungi. Although the gametes look the same, at the molecular level, they are of two mating types '+' and '−', and only gametes of opposite mating types can fuse.

In the unicellular alga *Chlamydomonas*, the two mating types—mt^+ and mt^-—are determined by recognition proteins located at the tips of the flagella, where initial fusion between opposite mating types occurs (Fig. 14.2a, b). A given cell is either mt^+ or mt^-. After meiosis, two of the four product cells are mt^+ and two are mt^-. Gamete formation occurs only when the alga grows under reduced nutritional conditions, suggesting that sexual reproduction is confined to less favourable environments. In good conditions, reproduction occurs asexually.

In the yeast *Hansenula*, sexual reproduction occurs when haploid cells of opposite mating types make contact and form aggregates of cells (Fig. 14.2c). Recognition involves sex-specific molecules on the surface of the cell walls, rather than contact between flagella as in *Chlamydomonas*.

Anisogamy involves the fusion of gametes that are different in appearance. In anisogamous systems, the

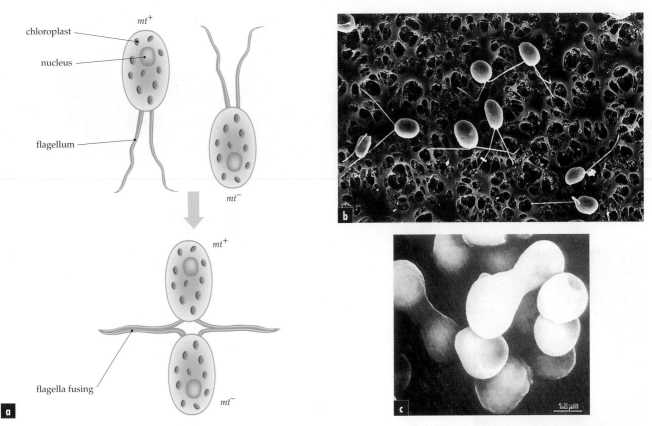

Fig. 14.2 Isogamy. **(a)** In the unicellular alga *Chlamydomonas*, the two mating types (mt^+ and mt^-) are determined by recognition proteins located at the tips of the pair of posterior flagella, where initial fusion between cells of opposite mating types occurs **(b)**. **(c)** In the yeast *Hansenula*, sexual reproduction occurs when cells of opposite mating types make contact and form aggregates of cells

smaller of the pair of gametes is considered male. Male gametes, **sperm**, are produced in large numbers and their function is associated with motility and dispersal. The large, non-motile female gametes, **eggs**, are produced in smaller numbers and contain food reserves. Most eukaryotes, including many algae, many fungi, and all plants and animals, are anisogamous.

> Isogamy involves the fusion of gametes that are morphologically similar but of opposite mating types ('+' /'−'). In anisogamy, male and female gametes differ in appearance.

The formation of gametes: gametogenesis

Gametogenesis in animals involves a special line of germ cells that is distinct very early in the development of the embryo and whose exclusive function is to undergo meiosis and form gametes. There is no evidence for the existence of such a germ line in flowering plants. On the contrary, cells of the developing flower of an adult plant form the reproductive organs and reproductive cells are determined by their central position within these organs. These cells undergo meiosis to form haploid spores, which then divide mitotically to produce the male and female gametophytes where sperm and egg cells differentiate. In order to facilitate comparison with the processes in animals (Chapter 15), gametogenesis is applied here in its widest sense to include the development of both the gametophyte and gametes.

The flower: the site of sexual reproduction

Flowers are the sites of sexual reproduction in flowering plants. A **flower** is a short shoot with transformed 'floral leaves', which grow in four whorls around the receptacle (tip of the flower stalk). The two outer whorls are the sepals, which are leaf-like, green and leathery but can be reduced to scales or may even be absent in some plants (Figs 14.3, 14.4), and the petals, which are usually brightly coloured and scented. Some petals have honey guides, lines that direct insects and other animal pollinators to nectary glands, which are sited on the floral axis (receptacle). Floral nectaries produce nectar (Fig. 14.3), a liquid secretion containing sugars, amino acids and other rewards for floral visitors (see also Chapter 37).

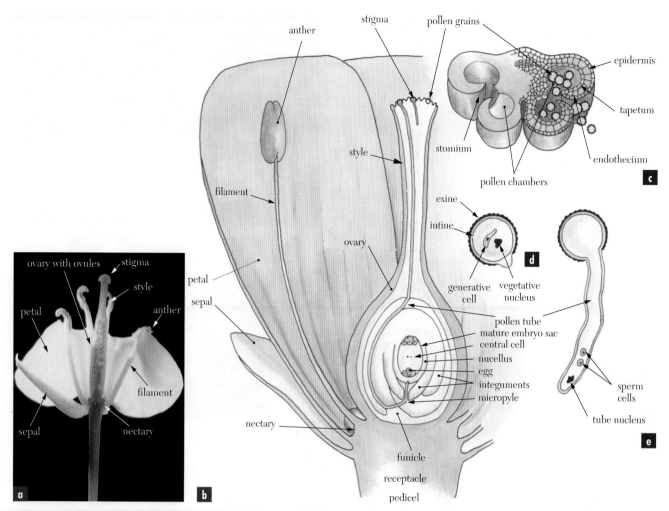

Fig. 14.3 The principal organs of a flower are arranged in whorls, consisting (from outside) of sepals, petals, stamens (comprising filament and anther, which contain numerous pollen grains) and pistil (comprising stigma, style and ovary, which contain rows of many ovules). **(a)** Longitudinal section of a flower of oilseed rape *Brassica napus*. **(b)** Diagram of longitudinal section of a flower. **(c)** Cross-section of an anther, showing the pollen grains and tissues of the wall, including the tapetum, endothecium and epidermis, as well as the stomium, which splits when the anther opens. **(d)** Mature pollen grain with pair of sperm cells within the vegetative cell. **(e)** Germinating pollen grain with pollen tube containing a pair of sperm cells and the tube nucleus

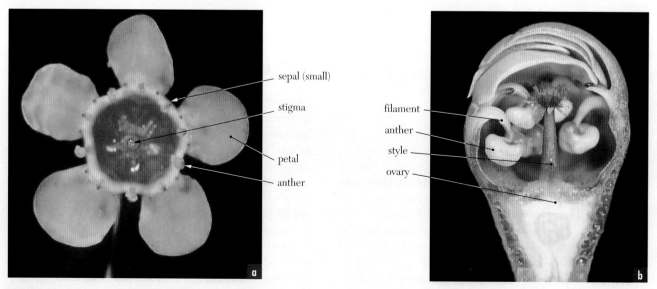

Fig. 14.4 Structure of the flower of Geraldton wax, *Chamaelaucium uncinatum* (family Myrtaceae). **(a)** Top view of the flower showing the four whorls of organs. **(b)** Longitudinal section of the flower

The two inner whorls comprise the reproductive organs, the whorl of stamens and another of pistils. **Stamens** are the male reproductive organs and consist of anthers borne on slender filaments. The **pistil** is the female reproductive organ and lies in the middle of the receptacle (Fig. 14.3). The whorls of sepals and petals surround and protect these reproductive organs.

In the life cycle of flowering plants, the male gametophytes are **pollen** grains, which form in the anthers. In most instances, pollen must be physically carried from the **anther** to the female pistil by air, water or animal vectors. Pollen lands on the **stigma** on top of the pistil and germinates to produce a pollen tube that grows down the **style** to the **ovary**. In this way, the male gametes are transferred to the female gametophyte, the **embryo sac**, which is contained within an **ovule**. After a short passage of time, fertilisation takes place and the ovule develops into a seed containing an embryo. The **seed** is the unit of dispersal.

The anther: the male reproductive organ

At flower opening, anthers are often exserted (extended above the petals) by elongation of the stamen filament (Fig. 14.5). Each anther contains two elongated pollen sacs joined together by a layer of connective tissue, which is an extension of the filament. The pollen sacs are usually divided into two chambers, which house the male gametophytes, pollen grains.

The anther wall typically contains three layers of cells when viewed in cross-section (Fig. 14.3c). The innermost layer is the **tapetum**, a layer of nurse cells that surrounds the pollen sac and supports the developing male gametophytes. Tapetal cells are secretory cells and are usually binucleate. They are metabolically very active and secrete enzymes, nutrients and wall materials into the pollen sac. The tapetum is prominent in early stages of anther development but typically degenerates during later stages.

The outer two layers are responsible for opening of the mature anther. The outer epidermis usually serves a protective function but can also be pigmented or specialised for scent production, as in the flowers of some Australian wattles. The **stomium** is a cluster of thin epidermal cells that allows the pollen sac to split (Fig. 14.3c). The endothecium is a tissue that aids splitting of the pollen sac. Its cells possess massive, bar-like secondary thickenings. When the anther is ready to split open, water is withdrawn from the endothecium into the vascular system of the filament. This results in the pollen sac splitting at the stomium and dispersal of the pollen.

Fig. 14.5 Flowers of the rye grass (*Lolium perenne*) exsert their anthers when flowers are open. The green glumes part and the anthers elongate on their slender filaments, so that the anthers hang from the flower. Pollen is released from slits at the base of anthers into air currents

Anthers are specialised for differentiation, nutrition, protection and dispersal of pollen. The tapetum has a nutritive function, while the endothecium and stomium are responsible for anther opening.

Development of microspores

Cells located centrally within each anther divide by mitosis to form diploid spore-forming cells, **microsporocytes** (Fig. 14.6). Most of these will undergo meiosis, giving rise to tetrads of haploid **microspores** (Fig. 14.6), unicellular structures that develop into pollen grains. Microspores undergo several changes. At the beginning of meiosis, most of the RNA in the microsporocyte cytoplasm is destroyed. New RNA required by the developing pollen grain is made after meiosis is over. Each tetrad of microspores is held together within a callose wall and is surrounded externally by the tapetum. Callose has different structural properties from cellulose (the main component of plant cell walls), although both are polymers of glucose (see Chapter 1). Callose prevents the passage of

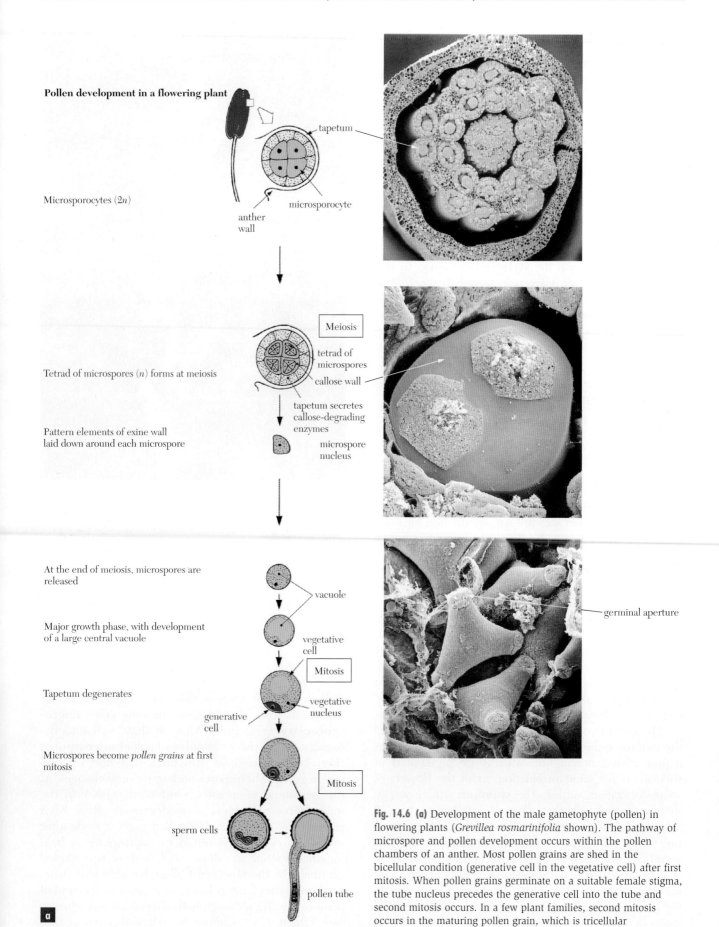

Pollen development in a flowering plant

tapetum

Microsporocytes (2*n*)

anther wall

microsporocyte

Meiosis

Tetrad of microspores (*n*) forms at meiosis

tetrad of microspores

callose wall

Pattern elements of exine wall laid down around each microspore

tapetum secretes callose-degrading enzymes

microspore nucleus

At the end of meiosis, microspores are released

vacuole

Major growth phase, with development of a large central vacuole

vegetative cell

germinal aperture

Mitosis

Tapetum degenerates

vegetative nucleus

generative cell

Microspores become *pollen grains* at first mitosis

Mitosis

sperm cells

pollen tube

Fig. 14.6 (a) Development of the male gametophyte (pollen) in flowering plants (*Grevillea rosmarinifolia* shown). The pathway of microspore and pollen development occurs within the pollen chambers of an anther. Most pollen grains are shed in the bicellular condition (generative cell in the vegetative cell) after first mitosis. When pollen grains germinate on a suitable female stigma, the tube nucleus precedes the generative cell into the tube and second mitosis occurs. In a few plant families, second mitosis occurs in the maturing pollen grain, which is tricellular

a

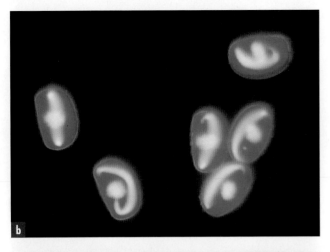

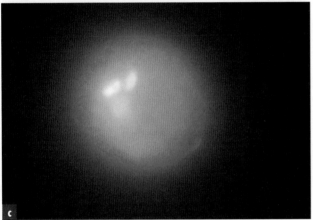

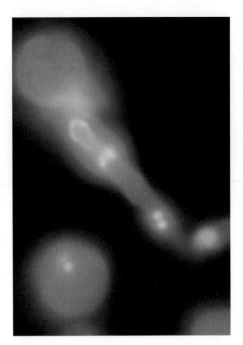

Fig. 14.6 (b) Bicellular *Tradescantia* pollen treated with a DNA fluorochrome so that the nuclei fluoresce blue, showing elongate generative nuclei and spherical vegetative nuclei. **(c)** Tricellular pollen of oilseed rape treated with the same fluorochrome, showing bright blue sperm nuclei and fainter vegetative nucleus. **(d)** The pollen tube of oilseed rape transports the pair of sperm cells and tube nucleus within the living cytoplasm at the tube tip. The pair of sperm nuclei and tube nucleus at the tip fluoresce blue, while yellow fluorescence shows the callose wall sealing off the tip from the rest of the grain

molecules larger than sugars into microspores, thus sealing off microspores from parental influences.

The end of meiosis is signalled by the dissolution of the callose wall, which is carried out by callase (1,3 β-glucanase), an enzyme that is secreted by the tapetal cells. This separates the microspores, which now lie free in the pollen sac. After release, the microspores become spherical and commence their major period of growth. Microspores grow exponentially and their volume increases about three times before levelling off. This major growth period is also marked by the appearance of a large vacuole within the microspore. The nucleus and cytoplasm become peripheral, which gives the microspores the characteristic appearance of a 'signet ring' (Fig. 14.6b). The microspore is unicellular until the end of the growth period (vacuolation), when typically the first mitosis occurs, forming the pollen grain.

Microspores are formed after meiosis. Each tetrad is enclosed in a wall of callose. At the end of meiosis, microspores are released from the tetrad and enter a major growth phase, developing a large central vacuole. The microspore becomes a pollen grain.

Formation of pollen and sperm cells

Pollen is formed when microspores undergo an asymmetric mitosis that cuts off a small **generative cell** from a larger **vegetative cell** (Fig. 14.6a). At first, the generative cell is attached to the perimeter wall, but this attachment is soon broken and the generative cell eventually moves to the centre of the vegetative cell, as a *cell within a cell*. Pollen maturation then proceeds with loss of the vacuole and the accumulation of storage reserves of lipid and carbohydrate. Mature pollen grains of most species are shed from the anther in this *bicellular* condition (Fig. 14.6b). When pollen lands on a stigma surface, a pollen tube forms. The nuclei and cytoplasm of the pollen grain enter the tube. The generative cell divides within the pollen tube tip to produce two sperm cells (Fig. 14.6a, d). The nucleus of the former vegetative cell has become the **tube nucleus**, which regulates pollen tube function. In other cases, pollen is shed in a *tricellular* condition and the mitotic division of the generative cell occurs in the maturing pollen grain within the anther (Fig. 14.6c).

In a pollen grain and tube, sperm cells are usually linked to each other and to the vegetative (tube) nucleus to form the **male germ unit**. The two sperm cells come to lie adjacent to each other, end to end, within a common sac, the inner pollen tube plasma membrane. Their close proximity ensures transfer of male DNA along the cytoskeleton of the pollen tube tip and may also lead to the correct physical placement of sperm cells for double fertilisation (p. 341).

The pollen grain wall

The *primexine* is a wall laid down external to the microspore plasma membrane but within the callose wall (Fig. 14.7a). The intricate pattern of the future pollen wall, the **exine**, is initiated within the primexine matrix. Within the outer exine wall, there is also a number of *germinal apertures*, which are slits or circular openings (pores) in the exine (Fig. 14.7a, b). The exine is made of **sporopollenin**, a polymer that is tough, plastic-like and extraordinarily resistant to degradation. There is no known enzyme in flowering plants that can degrade sporopollenin. Chemically, sporopollenin resembles lignin (Chapter 1) and is found only in exines and spore walls of certain algae and fungi. Pollen exines, preserved in fossil deposits, provide valuable clues to the history of plant life in past geological eras (Chapter 31).

The inner layer of the pollen grain wall is the **intine**, which is made up largely of pectic polysaccharides (Fig. 14.7c). This layer meets the surface of the grain only at the germinal apertures. When pollen germinates, the intine bulges out and the pollen tube grows out by tip extension from one of the germinal apertures. The pollen tube wall is comprised of an outer pectic layer and an inner callose layer.

Pollen grains vary widely in size (Fig. 14.8). Most pollen grains are spherical with an average size of about 35 μm. The smallest grains are about 3.5 μm in diameter (forget-me-not, *Myosotis*). Intricate and family-specific exine patterns are commonly seen. The largest pollen grains include those of pumpkin and the Turk's cap, *Malvaviscus* (the national flower of Hawaii), which are about 300 μm in diameter. An Australian seagrass, *Amphibolis*, holds the record for the longest pollen. *Amphibolis* pollen is more than 3000 μm in length and is filamentous, with hooks at the ends (Fig. 14.8c). Some pollen grains, such as those of *Acacia* and *Rhododendron*, are compound and are released from the anther as a tetrad in which the four microspores made by meiosis of a single microsporocyte remain united (Fig. 14.8d).

Most pollen is bicellular and contains a generative cell enclosed within a vegetative cell. The generative cell divides to form two sperm cells after germination within the growing pollen tube. In tricellular pollen, the generative cell divides within the pollen grain.

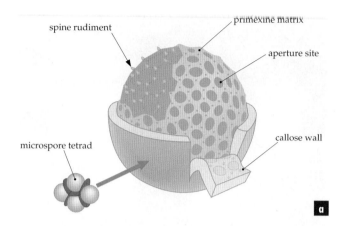

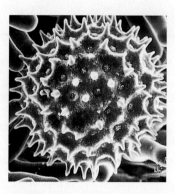

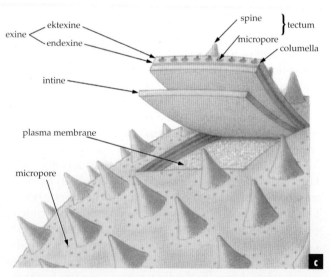

Fig. 14.7 The primexine establishes pollen wall pattern formation. **(a)** Tetrad of microspores of the morning glory, *Ipomoea*, dissected away to reveal the surface pattern of the developing microspore, showing the pattern of the primexine matrix, spine rudiments and sites of germinal apertures. **(b)** Scanning electron micrograph of mature pollen of *Ipomoea* showing the prominent spines encircling the many germinal apertures (pores). **(c)** Typical two-layered exine showing roof layer, tectum, supported by rod-like columellae over a foot layer, endexine, and the smooth inner intine layer

germinal apertures

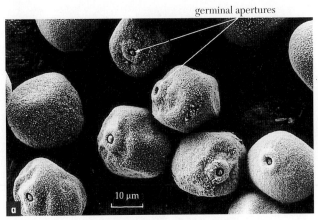

Profile of a rye-grass pollen grain

Size: 35 μm in diamater
Morphology: spherical with single germinal aperture
Mass: 22×10^{-9} g
Aerobiology: up to 800 grains. m^{-3} air; can remain in the atmosphere for up to three days
Cellular status: tricellular
Pollen tube growth rate: 240 μm per min (maize)

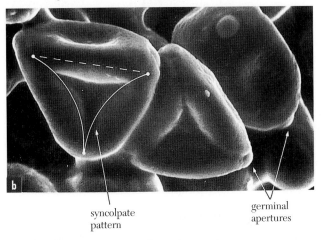

syncolpate pattern

germinal apertures

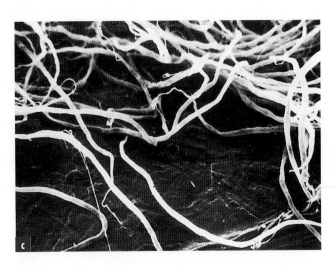

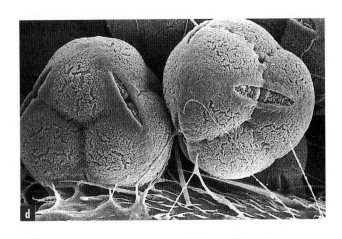

Fig. 14.8 Morphological features of different types of pollen. **(a)** Scanning electron micrograph of rye grass pollen grains showing spherical pollen with a single aperture (pore), some covered by a cap of exine. **(b)** Triporate pollen of *Eucalyptus* (family Myrtaceae) adheres together by pollen coat materials in the open anther. Three apertures occur in slits (colpi) at each corner of the triangular grains. *Eucalyptus* grains have a characteristic furrow pattern linking the colpi (syncolpate) (magnification × 1600). **(c)** Filamentous pollen of the Australian sea nymph, *Amphibolis* (family Cymodoceaceae), which is 3200 μm in length and lacks an exine, so that it can bend in seawater currents. **(d)** Structure of compound pollen grains, in which all the microspores in a meiotic tetrad remain joined at maturity. Pollen grains of *Rhododendron* (family Ericaceae) adhering by means of sporopollenin threads to hairs on the legs of a honeybee. These grains are held in tetrads and are arranged rhomboidally with one grain overlying three others. The slit-like germinal apertures form across adjacent grains

The pistil: the female reproductive organ

The stigma is at the tip of the pistil and its epidermal cells may be smooth or form elongate hairs, papillae, which serve to trap pollen and begin the screening process. A receptive stigma is one that is ready for pollination and the papillae, if present, are fully expanded. The stigma surface may be either dry or covered by a copious exudate (Fig. 14.9). Dry-type stigmas are usually covered with a membrane-like sticky layer, the **pellicle**. The pellicle has several functions in the interaction between pollen grains and stigma (see p. 349).

Beneath the stigma, the style forms a transmitting pathway or, in some cases, a central stylar canal, for pollen tubes to grow to the ovary. The ovary at the base of the pistil contains the ovules, each comprising a central nucellus (nurse tissue) within which the female gametophyte will develop (Fig. 14.10). During development, a pair of integuments usually form that envelop and protect the nucellus: these are known as the inner and outer integuments. The micropyle, a funnel-shaped entrance to the female gametophyte, develops at the tip of the ovule (Fig. 14.11). The micropyle is located at the side of the ovule, as in some types of eucalypts. This type of ovule is called a **hemitropous ovule** (Fig. 14.11c, d).

BOX 14.1 Grass pollen, hayfever and allergic asthma

Grass flowers open soon after sunrise and again in the afternoon. Their golden pollen is released into air currents and dispersed to other grass flowers. Unfortunately, their pollen grains also land on the eyes and nose, where they trigger the symptoms of hayfever. Special 'spore traps', which biologists use to sample the air, show that grass pollen is the major single component in the air in spring and early summer in cool temperate climates such as Melbourne (Fig. a). Studies of Melbourne schoolchildren who are sensitive to grass pollen show that asthma attacks occur coincidentally with the highest peaks of grass pollen.

Rye grass (*Lolium perenne*) pollen is the most abundant pollen of all grasses. It contains several different proteins that can trigger hayfever and allergic asthma in sensitive individuals (allergens, see Chapter 23). The function of pollen allergens is slowly being understood by isolating and cloning the genes that encode these proteins (Chapter 12). The genes can be sequenced and the triplets of nucleotides used to predict the amino acid sequence of each allergen.

Most grass pollen is too large to easily enter the bronchial passages. How, then, do grass pollen allergens cause asthma? One idea is based on the observation that, when placed in rainwater, grass pollen bursts by osmotic shock to release starch granules containing allergens (Fig. b). Rain could therefore cause the release of allergen-containing starch granules from grass pollen into the atmosphere. Indeed, during the grass pollen season, air samples in Melbourne have a staggering 54 000 granules in each cubic metre of air on days after rainfall, compared with only 1000 on sunny days. These starch granules are small enough (0.55 μm in diameter) to be respirable. They can enter the bronchi, where they may trigger allergic asthma. Wheezing is caused by constriction of the bronchi (Fig. c).

In fact, during November of most years, an epidemic of asthma sweeps south-eastern Australia within 24 hours of a major thunderstorm. In 1989, for example, there was a 10-fold increase in the number of asthmatics admitted to Melbourne hospitals after such a thunderstorm and laboratory studies of a sample of patients showed that all were sensitive to grass pollen. When lung function was tested with

aerosols containing isolated starch granules from rye grass pollen, these same patients showed bronchial constriction in every case.

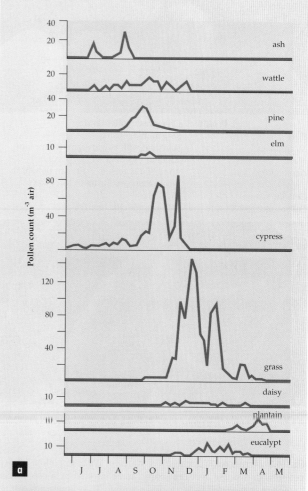

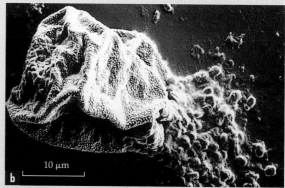

Association of atmospheric pollen content with hayfever and allergic asthma. **(a)** Pollen calendar for a cool temperate city (Melbourne), showing the seasonal progression of different types of pollen detected in spore traps. Many of the types are exotic to Australia, including ornamental trees and agricultural grasses. **(b)** Grass pollen grains burst by osmotic shock in rainwater, releasing many hundreds of starch granules

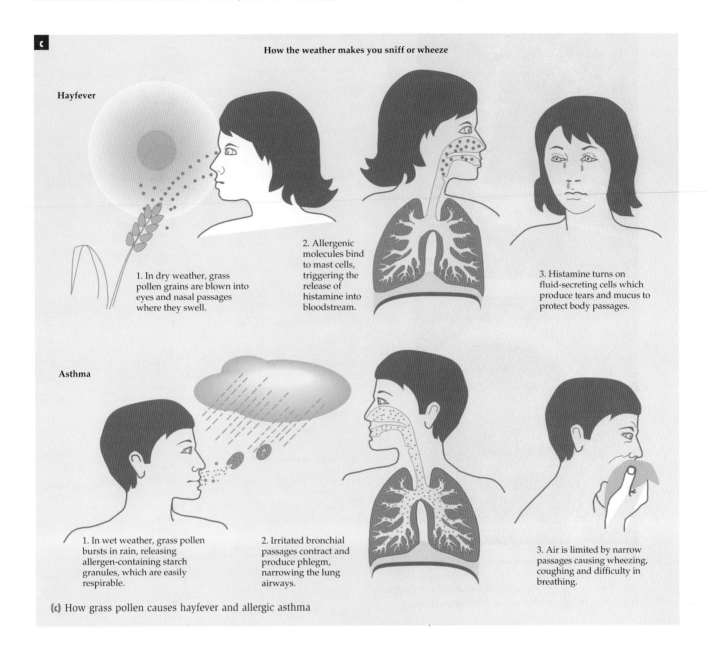

How the weather makes you sniff or wheeze

Hayfever

1. In dry weather, grass pollen grains are blown into eyes and nasal passages where they swell.

2. Allergenic molecules bind to mast cells, triggering the release of histamine into bloodstream.

3. Histamine turns on fluid-secreting cells which produce tears and mucus to protect body passages.

Asthma

1. In wet weather, grass pollen bursts in rain, releasing allergen-containing starch granules, which are easily respirable.

2. Irritated bronchial passages contract and produce phlegm, narrowing the lung airways.

3. Air is limited by narrow passages causing wheezing, coughing and difficulty in breathing.

(c) How grass pollen causes hayfever and allergic asthma

Commonly, the ovule bends back on itself so that the micropyle comes to lie adjacent to the funicle when mature, as in the **anatropous ovules** of passionfruit (Fig. 14.11a, b).

Within each ovule, a single diploid cell that lies centrally in the nucellus enlarges to form the diploid spore-forming cell, the **megasporocyte** (Fig. 14.10). The megasporocyte divides by meiosis, producing four haploid **megaspores**. These are located in a row that stretches towards the micropyle. Commonly, the three megaspores nearest the micropyle abort and the remaining megaspore is the first cell of the gametophyte generation (Fig. 14.10). In a few plants, such as the lily, all four megaspore nuclei participate in formation of the female gametophyte.

The female reproductive organ, the pistil, bears a stigma to receive and screen pollen grains, a style with transmitting tissue or central canal for growth of pollen tubes and a basal ovary containing ovules. The female gametophyte develops in the ovules.

Embryo sac formation and the egg cell

The surviving haploid megaspore undergoes three rounds of mitotic divisions to form a female gametophyte, the embryo sac, within each ovule (Fig. 14.10). Cell walls then develop, dividing the eight-nucleate embryo sac into a seven-celled female gametophyte. The egg cell, the female gamete, occurs centrally at the micropylar end (Fig. 14.10). Two

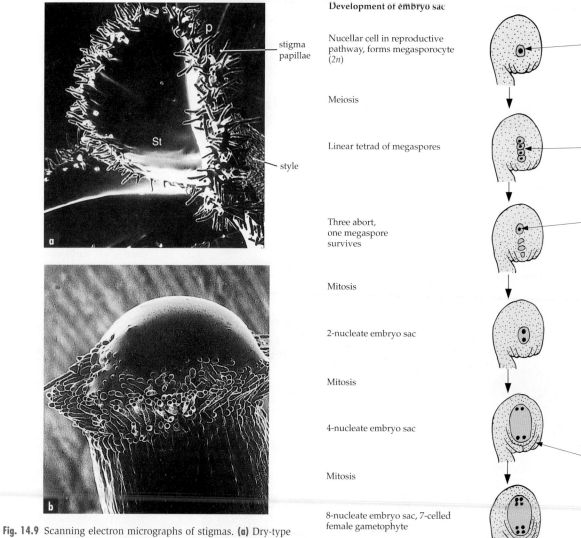

Fig. 14.9 Scanning electron micrographs of stigmas. **(a)** Dry-type stigma of *Gladiolus* showing the row of papillae along the edge of the style. **(b)** Wet-type stigmas of *Eucalyptus*, showing pollen grains germinating in the pool of exudate that covers the papillae

Development of embryo sac

Nucellar cell in reproductive pathway, forms megasporocyte (2n) — megasporocyte

Meiosis

Linear tetrad of megaspores — tetrad of megaspores

Three abort, one megaspore survives — megaspore

Mitosis

2-nucleate embryo sac

Mitosis

4-nucleate embryo sac — integuments

Mitosis

8-nucleate embryo sac, 7-celled female gametophyte

stigma
style
antipodal cell
ovary
central cell
egg
integuments
synergid
micropyle
funicle

Fig. 14.10 Pathway of development of the female gametophyte in flowering plants and the meiotic and mitotic divisions involved

synergids, which function to receive the pollen tube, lie one on either side of the egg cell, and three nutritive **antipodal cells** lie at the opposite end of the sac. A **central cell** is formed when one nucleus from the micropylar end and another from the opposite end associate as polar nuclei and later fuse to form a diploid nucleus. The central cell is the largest in the embryo sac, occupying its mid portion. This is the condition when the embryo sac is mature, awaiting fertilisation.

The embryo sac is formed from a diploid megasporocyte, which undergoes meiosis to produce four haploid megaspores. Typically, three abort and the remaining megaspore differentiates by three mitotic divisions, producing an eight-nucleate embryo sac.

A seven-celled female gametophyte is formed, containing an egg, two synergids, a central cell with two polar nuclei and three antipodal cells.

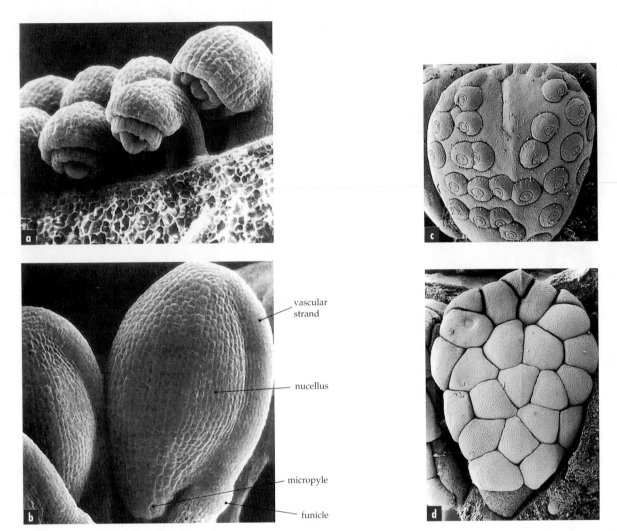

vascular
strand

nucellus

micropyle

funicle

Fig. 14.11 Scanning electron micrographs of **(a)** developing and **(b)** mature ovules of *Passiflora* (passionfruit), in which the micropyle is sited at the base of the ovule (anatropous), and **(c)** developing and **(d)** mature ovules of *Eucalyptus*, in which the micropyle is sited on the side of the ovule (hemitropous)

Reproductive strategies and fertilisation

The process of transferring pollen from an anther to a stigma is known as **pollination**. When pollen grains alight on a stigma, a series of interactions is set in train, which, if successful, may lead to **fertilisation** (Fig. 14.12). In the case of dry-type stigmas, such as grasses, pollen grains attach directly to the stigma pellicle. If chemical recognition occurs, fluid passes out from the stigma into the pollen grain. Consequently, the pollen grain becomes hydrated, swells, and finally germinates to produce a pollen tube that emerges through the germinal aperture. In the case of wet-type stigmas, such as eucalypts (Fig. 14.9b), the pollen grains become immersed in the stigma exudate that surrounds the papillae, where they swell and germinate. The stigma receives pollen from many different sources. Some pollen are from the plant itself or other

plants of the same species, while other pollen are from plants of different species. Plants have evolved several mechanisms that help them screen the range of pollen grains received by the stigma and let only certain types grow through the style to the ovaries. Since plants are essentially immobile, this is the only way plants can select their partners and optimise successful pollination. Selectivity operates in most species to ensure cross-fertilisation between different plants occurs and, in a few species, to ensure that only self-fertilisation occurs.

Reproductive mechanisms in flowering plants enable them to optimise pollination and allow the pistil to screen out undesirable pollen.

Cross-fertilisation

Cross-fertilisation involves the transfer of pollen from one plant to the female organs of a different plant of

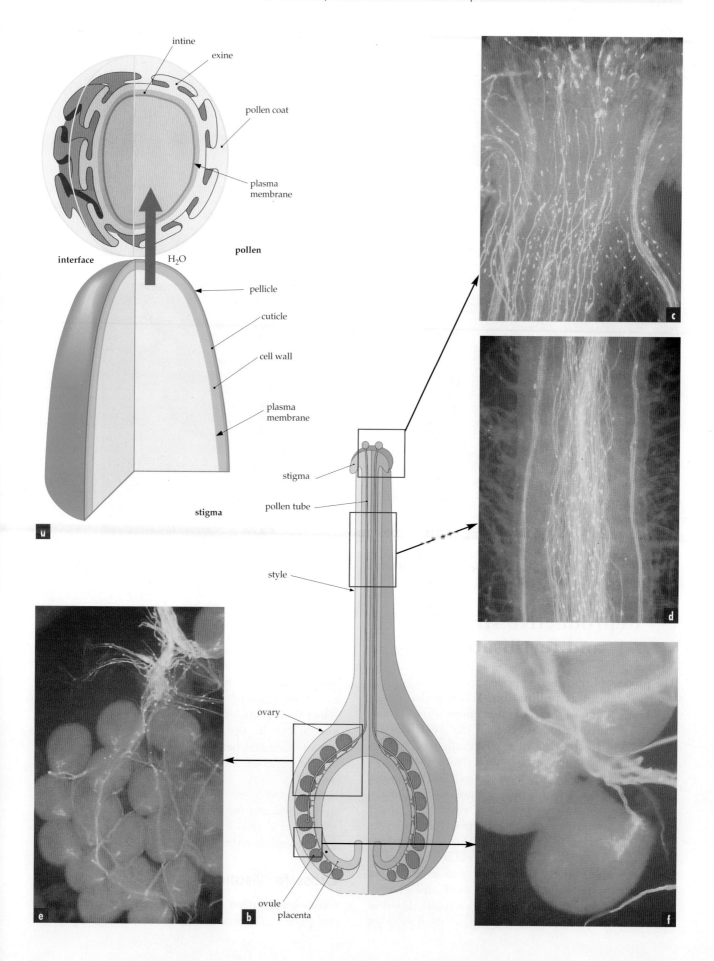

intine

exine

pollen coat

plasma
membrane

pollen

interface   H₂O

pellicle

cuticle

cell wall

plasma
membrane

stigma

a

stigma

pollen tube

style

ovary

ovule

placenta

b

c

d

e

f

Fig. 14.12 (opposite) Events of pollen–pistil interactions. **(a)** When the pollen grain arrives at the stigma surface (dry types), surface informational molecules of the pollen coat interact with the stigma surface. The pellicle initiates signals that permit *attachment* and release of fluid for pollen *hydration* in successful pollinations. **(b)** Pathway of pollen tube in pistil of wild tomato (wet type) and fluorescence micrographs showing pollen tubes in pistil of wild tomato after compatible pollination. **(c)** Pollen grains have germinated and tubes pass through the stigma (red zone). **(d)** Passage of tubes through part of the style. **(e)** Arrival of tubes in the ovary and passage of rope of tubes passing ovules (red spheres). **(f)** Entry of a pollen tube into each ovule via a micropyle

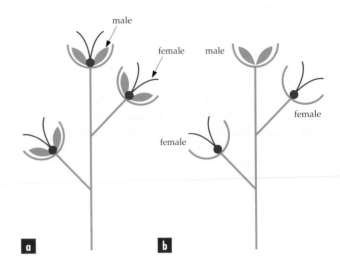

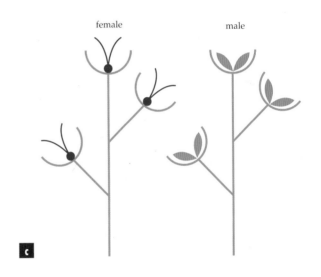

the same species. This maintains maximum genetic variability of offspring, which is an advantage in environments that are either unpredictable or changing. However, as shown above, the flowers of most species contain both male and female reproductive organs and are described as **bisexual**. Cross-fertilisation in some species with bisexual flowers is achieved by temporally separating the sexes by either protandry or protogyny (Fig. 14.13). Other species do not produce bisexual flowers and ensure cross-fertilisation by physically separating the sexes as in dioecious and monoecious species (Fig. 14.14).

In protandry and protogyny, flowers go through separate male and female stages. **Protandry** (male first) is seen in the bisexual flowers of eucalypts that shed pollen while the pistil is unreceptive. Upon flower opening, the stamen filaments extend and the anthers release their pollen. Some time later, the flower becomes female as the pistil matures and is able to be pollinated, usually by pollen from another plant that is in its male phase. In **protogyny** (female first), the reverse is true and the pistil becomes receptive to pollen before the anthers have opened. In wattle trees, for example, flower buds open with the pistil protruding, ready to receive pollen from another flower that is in its male phase. Pollen release is delayed until the following day.

Fig. 14.13 Protogyny occurs in the wattle tree, *Acacia retinodes* (wirilda), whose flowers are grouped in heads of 20 flowers. Flower buds open with the pistil protruding, ready to receive pollen (female phase). The male phase occurs the next day, when the flower heads are covered in anthers and resemble golden spheres. In this flowering shoot, female and male flowers are present at the same time

Fig. 14.14 (a) In hermaphroditic plants, the male (blue) and female (red) organs are housed within a single flower. **(b)** In monoecious plants, male and female organs are found in separate flowers on the same plant. **(c)** In dioecious plants, male and female flowers occur on separate plants. **(d)** Female plant of kikuyu grass, *Pennisetum clandestinum*, a common lawn grass. The female flowers are evident by their white stigmas (plumes). Male plants of this African grass have not been introduced into Australia

The word **dioecious** means 'two houses' and reflects the fact that in dioecious species, separate plants carry the male and female gametes. Examples of dioecious species are willows, *Cannabis* (from which marijuana and hemp are obtained), the Australian mountain pepper (*Tasmannia lanceolata*), and the grass, kikuyu (Fig. 14.14). **Monoecious** means 'one house' and refers to the presence of separate male and female flowers on the same plant. Familiar examples of monoecious species are maize and zucchini.

> Many adaptations have evolved to promote cross-fertilisation in flowering plants. Male and female gametes may be either physically separated as in monoecious and dioecious species, or temporally separated as in protogyny and protandry.

Self-incompatibility

In many flowering plants, self-fertilisation is not possible because of a **self-incompatibility** mechanism, a genetically controlled recognition system that stops eggs from being fertilised by pollen of the same plant. Pollen from a self-incompatible plant is, however, perfectly capable of fertilising eggs of most other plants of the same species. Self-incompatibility involves an exchange of information between the pollen grain and pistil. This information is often encoded by a single genetic locus, the S locus, which has numerous different alleles. Fertilisation will not occur if the same S allele is present in both pollen and pistil (Fig. 14.15).

An example of a self-incompatible plant is the cultivated sweet cherry, *Prunus avium*. Each sweet cherry cultivar has its own particular pair of S alleles and cultivars are exclusively propagated asexually by grafted cuttings.

We can illustrate the mechanism of self-incompatibility by considering the outcome of each combination of crosses between three different cultivars of cherry. In the most common type of self-incompatibility, pollen tube growth is slowed or arrested soon after the tube enters the pistil (Fig. 14.15). The growth of compatible pollen tubes, however, is not adversely affected. Figure 14.16 shows crosses between cultivars Bedford (genotype S_1S_2), Napoleon (S_3S_4) and Van (S_1S_3). When pollen from cultivar Bedford (haploid S_1 or S_2) lands on the stigma of Van (diploid S_1S_3), only that pollen containing the S_2 allele will grow through the pistil and fertilise an egg within an ovule. Because it matches one of the S alleles in Van, Bedford S_1 pollen is recognised as 'self' and its growth is markedly slowed.

Many flowering plants are self-incompatible, such as apple, almond and blueberry. Information obtained from reciprocal crosses of the kind described for cherry is important so that orchardists who plant self-incompatible plants can ensure that compatible pollination, and hence fruit set, occurs. Orchards are designed so that cross-pollinating varieties are interplanted. This is done by planting cross-pollinating varieties that flower at the same time together, and by placing hives of honeybees within the orchard so that

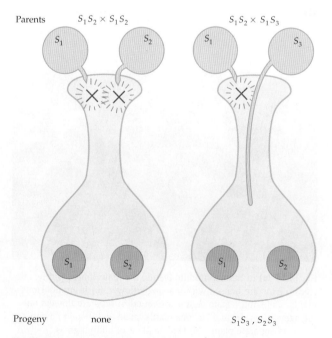

Fig. 14.15 Genetics of self-incompatibility. The fate of pollen (S_1 or S_2) on a self-pollinated pistil (S_1S_2, left) and the same kind of pistil cross-pollinated with S_1 or S_3 pollen (right)

	Female		
Male	Bedford S_1S_2	Van S_1S_3	Napoleon S_3S_4
Bedford S_1 or S_2	—	S_1S_2 S_2S_3	S_1S_3 S_2S_3 S_1S_4 S_2S_4
Van S_1 or S_3	S_1S_3 S_2S_3	—	S_1S_3 S_1S_4
Napoleon S_3 or S_4	S_1S_3 S_2S_3 S_1S_4 S_2S_4	S_1S_4 S_3S_4	—

Fig. 14.16 Offspring from self-incompatible matings. The checkerboard shows the reciprocal crosses between three different cultivars of sweet cherry. The female parent (top), the pistil, is diploid and carries two S alleles. The male parent (left), haploid pollen grains, has half the grains carrying one S allele and half the grains carrying the other S allele in each anther. Pollen tubes carrying an S allele in common with the stigma are arrested in the style and cannot produce progeny

pollen can be transferred between plants at flowering time. Self-incompatibility is not restricted to fruit trees and is also found in other flowering plant families, including wattles, *Acacia*, and grasses (family Poaceae).

S elf-incompatibility is found in many families of flowering plants. It is genetically controlled by the *S* locus and its effect is to prevent self-fertilisation.

Self-fertilisation

Self-fertilisation occurs when pollen lands on the stigma of the same flower or on a different flower of the same plant. Self-fertilisation has many advantages but it also has many disadvantages. For example, self-fertilisation ensures a high production of seeds in environments where pollen from other plants is not readily available. Moreover, there is little value in ensuring variation in the progeny if the environment is unchanging. However, one disadvantage of self-fertilisation is the loss of genetic variability and the appearance of recessive gene combinations that are either lethal or otherwise reduce the 'fitness' of inbred compared with cross-bred progeny. For example, a single generation of self-fertilisation in maize can reduce the yield of viable seeds by up to 30%, and in *Cacao*, the cocoa tree, embryos produced by self-pollination will often abort a few days after fertilisation.

In some plants, **cleistogamy** ensures self-fertilisation. Cleistogamy describes flowers that are still completely enclosed by petals when the anthers open and the stigma becomes receptive to pollen (Fig. 14.17). Examples of cleistogamous flowers include the violet, *Viola*, and *Impatiens*. Interestingly, some species of violet produce cross-fertilising flowers that open normally in spring and cleistogamous flowers in summer, when conditions for cross-pollination are less favourable. In a wallaby grass, *Danthonia spicata*, both closed cleistogamous and open cross-fertilising flowers form.

S elf-fertilisation enables genetic continuity in plants that are successful in their environment. Seed production is enhanced through cleistogamy in environments where cross-fertilisation is unreliable.

Screening of foreign pollen

The fate of foreign pollen from a different species varies enormously depending on the plant group. Species in some genera such as *Grevillea* hybridise readily, whereas others such as *Banksia* hybridise

Fig. 14.17 (a) Flowers of the violet open in spring. **(b)** In autumn, cleistogamous flowers of the violet develop in which the petals remain closed and self-fertilisation occurs within

rarely. In large genera such as *Eucalyptus*, there is variation within the genus with some closely related species crossing readily, whereas crosses between other distantly related species fail completely.

At each step of the pollen–pistil interaction, the pistil can screen out and arrest the development of incompatible pollen and pollen tubes. Some foreign pollen alighting on a stigma will fail to germinate and remains dehydrated on the stigma surface (Fig. 14.18). In other cases, pollen may germinate and produce a pollen tube but this is unable to penetrate the stigma surface and the process of fertilisation is arrested at this early point in the interaction.

Successful hybrid crosses have pollen tubes that penetrate the stigma and enter the ovules, whereas an unsuccessful combination results in abnormal pollen tubes with swollen tips that do not reach the ovary. The major barriers in the pistil are the upper style and the ovary. However, some crosses that show normal pollen tube growth and ovule penetration still fail to result in hybrid seed, indicating that barriers to hybridisation also exist in the ovule.

Compatible interactions

Pollination typically involves a large number of pollen grains. Pollen tubes produced by these grains compete to be first through the style to reach the ovary and fertilise an ovule. Pollen grains in compatible matings between the individuals of a species will hydrate and

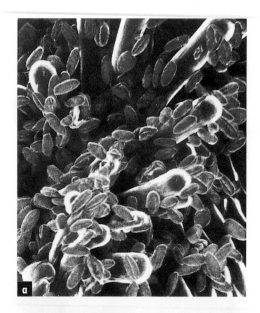

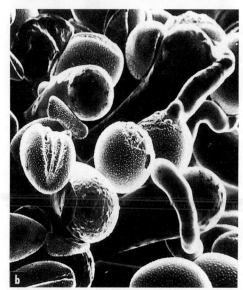

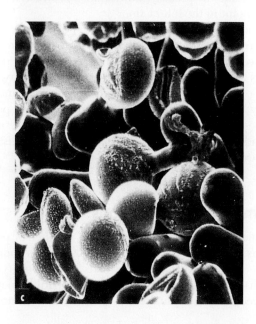

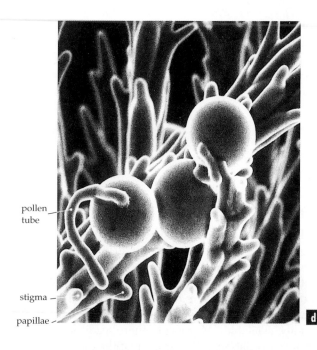

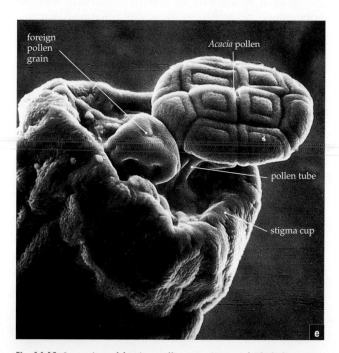

Fig. 14.18 Screening of foreign pollen on stigmas of *Gladiolus*. **(a)** Foreign pollination: lily pollen on *Gladiolus* stigmas (different families), showing that the lily pollen remains dehydrated. **(b)** Intergeneric pollination: *Crocosmia* pollen on *Gladiolus* stigmas (both members of family Iridaceae), showing that the *Crocosmia* pollen hydrates and germinates but the pollen tubes are unable to penetrate the stigma surface. **(c)** Compatible pollination: *Gladiolus* pollen on *Gladiolus* stigmas, showing the pollen germinating and the tubes penetrating stigma surface. **(d)** Grass pollen germinating and producing a pollen tube on feather-like stigma papillae. **(e)** Compound disk-like pollen of *Acacia* germinating in the stigma cup, also containing a foreign pollen grain (dehydrated). The *Acacia* pollen grains have germinated and pollen tubes can be seen emerging from under the disk and penetrating the stigma

germinate. The pollen tube penetrates the stigma surface and grows into the style, passing through the gelatinous walls of transmitting tissue or mucilage of the stylar canal.

In bicellular pollen systems, the time interval from pollination to fertilisation may be 16 to 48 hours, or even longer. This reflects the slow metabolism in this type of pollen. Pollen tubes of tricellular species, which have a high metabolic rate, can grow more than 20 times faster than pollen tubes of bicellular species. The time from pollination until fertilisation is only 20 minutes in grasses and 45 minutes in sunflowers.

The male reproductive cells and tube nucleus are transferred down the pistil within the growing pollen tube tip (Fig. 14.12c–f). As the tube grows through the style, the cytoplasm in the pollen tube tip is sealed off from the remainder of the pollen tube by a succession of callose wall plugs. During growth through the style, the tube nucleus usually precedes the sperm cells, lying just behind the growing tube tip.

Events of double fertilisation

As a pollen tube approaches an ovule, one of the synergids in the embryo sac degenerates (Fig. 14.19a, b). The first pollen tube to arrive penetrates the micropyle and enters the ovule. On reaching the embryo sac, the pollen tube passes into the degenerated synergid. The tube tip ruptures, releasing the pair of sperm cells and tube nucleus (Fig. 14.19c). In this position, there are no cell walls and male or female gametes are bounded only by their plasma membranes. One sperm fuses with the egg cell to form the diploid embryo, the first cell of the new sporophyte generation (Fig. 14.19d). The other sperm fuses with the central cell to form the **triple fusion nucleus**. This nucleus divides mitotically to produce a nutritive **endosperm**, which supports growth of the embryo and, in some cases, growth of the seedling as well. Thus, the process of double fertilisation in flowering plants involves true fertilisation (fusion of sperm and egg) plus an accessory fusion (between sperm and central cell).

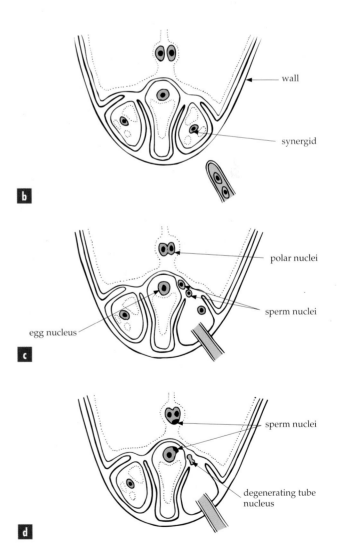

Fig. 14.19 Cellular processes in a typical double fertilisation event in flowering plants. **(a)** Embryo sac at fertilisation seen in longitudinal section. View of the micropylar end of the embryo sac, showing the sequential events of double fertilisation. **(b)** Pollen tube containing sperm cells and tube nucleus approaches the embryo sac (details of ovule omitted for clarity). **(c)** Pollen tube pushes through the wall of the embryo sac, enters the now degenerated synergid and releases sperm cells into the space between the plasma membranes of the egg and central cells. Cellular fusion follows. **(d)** Nuclear fusion follows when smaller sperm and egg nuclei fuse, and larger sperm and polar nuclei fuse. The tube nucleus degenerates

Double fertilisation appears to be a complex process and yet it has increased the rate and timing of reproduction in flowering plants compared with other vascular plants. For example, double fertilisation in grasses can take place in only 20 minutes compared with single fertilisation in conifers, which requires many months (Chapter 37).

> Fertilisation in flowering plants begins when pollen alights on the stigma. Pollen tubes emerge from the pollen grains and grow through the stigma and style to the ovary, where they penetrate the ovules. There, double fertilisation occurs. One of the pair of sperms in the pollen tube fuses with the egg to produce the diploid zygote, the other with the central cell to produce the triploid endosperm.

Reproduction without fertilisation: the unusual case of apomixis in plants

Apomixis is a form of asexual reproduction or parthenogenesis found in some flowering plants (Fig. 14.20). Developing diploid eggs do not undergo meiosis but instead begin to divide by mitosis to form an embryo. Apomictic embryos are genetically identical to the parent plant and pass through the same stages as a zygote. Because of the absence of meiosis, apomictic progeny do not show any new gene combinations and hence potentially adaptive variation. They provide a means for rapid multiplication of individual plants that are successful in their environment. In many dandelions and grasses, sexual reproduction is completely replaced by apomixis. In kangaroo grass, *Themeda triandra*, and in *Citrus*, both sexual reproduction and apomixis can occur side by side in the same flower. In the case of *Citrus*, cells near a sexually produced embryo are triggered to divide and produce diploid embryos. These somatic embryos compete with the single embryo produced by sexual reproduction and, being more numerous, usually form the seed.

> In apomixis, meiosis does not occur and embryos develop asexually. Apomixis allows for the rapid multiplication of plants.

Major events of embryonic development

A zygote undergoes rapid cell divisions to form an **embryo**, a young plant formed within a seed by the processes of embryogenesis. The first mitosis is typically in a transverse plane. The lower cell

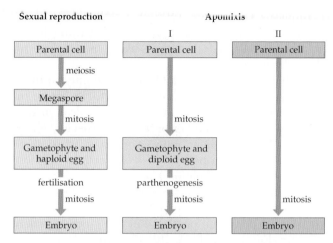

Fig. 14.20 Comparison of the developmental mechanism of sexual reproduction and apomixis in flowering plants. In sexual reproduction, the parental spore-producing cell is centrally located in the ovule and undergoes meiosis to produce a megaspore, which develops into the female gametophyte containing the female gamete, the egg. After fertilisation, the zygote develops by mitotic divisions into the diploid embryo. In the two types of apomixis shown here, a diploid cell of the ovule adjacent to the sexual pathway develops by mitosis as follows—type I: into a diploid female gametophyte containing a diploid egg, which develops by parthenogenesis into an embryo (dandelions and grasses); type II: directly into a diploid embryo (*Citrus*), which competes with the sexually derived embryo to produce a seed

subsequently forms a filament of cells, the **suspensor**, terminating in a basal cell (Fig. 14.21a). The suspensor provides nutrition to the developing embryo. In some plants, repeated mitoses of the upper apical cell gives rise to the embryo (Fig. 14.21b), while in others both upper and lower cells contribute to the embryo.

The early embryo of dicotyledonous (dicot) plants such as a bean is a globular mass of about 40 cells (Fig. 14.22). As the **cotyledons** (seed leaves) develop, the embryo becomes heart-shaped and then torpedo-shaped, at which time the axes of the embryo are established. The growing meristem of the shoot forms the **epicotyl** (the part of the axis above the cotyledons that forms leaves and lateral branches). The **hypocotyl** lies immediately below the cotyledons. The root meristem of an embryo grows in the opposite direction to the epicotyl, forming the **radicle**, the primary root of the seedling (Fig. 14.21c, d). This structure is at the suspensor end of the original embryo. Root and shoot meristems continue to function throughout the life of the plant, showing an unlimited potential for growth.

> The zygote divides to form the embryo and suspensor, which acts as a nutritive link. In dicotyledons, the embryo becomes heart-shaped as the two cotyledons develop. The shoot meristem forms the epicotyl and the root meristem forms the radicle. In monocotyledons, the single cotyledon is the scutellum.

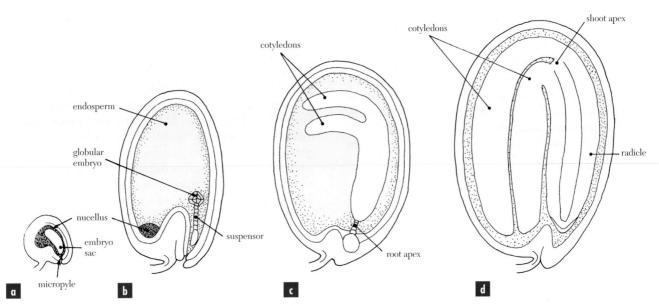

Fig. 14.21 Scheme showing embryogenesis in the dicotyledon *Arabidopsis*. **(a)** Ovule at the time of fertilisation, showing embryo sac within nucellus. **(b)** Globular embryo and suspensor in ovule, surrounded by endosperm, with nucellus now pushed to opposite end of embryo sac. **(c)** Torpedo-shaped embryo and suspensor in ovule. **(d)** Mature seed 0.5 mm in length containing a pair of cotyledons and radicle

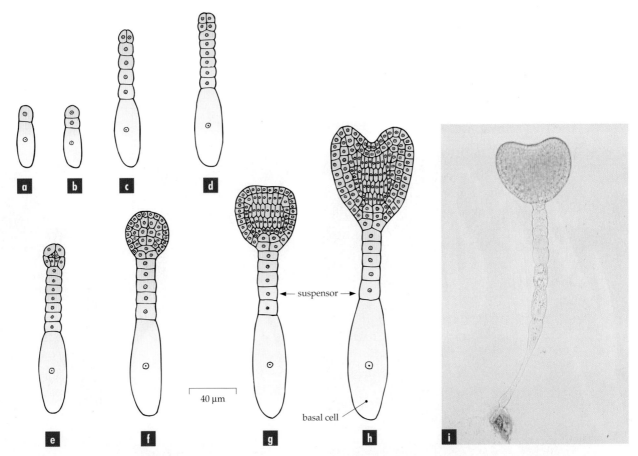

Fig. 14.22 Cellular structure of embryo and suspensor at different stages of development viewed in longitudinal sections. **(a)** Two-celled proembryo, after first (horizontal) division. **(b)** Three-celled proembryo. **(c)** A two-celled embryo proper is formed after a single vertical division in the apical cell. The suspensor elongates by horizontal divisions and enlargement of the basal cell. **(d)** A four-celled embryo is formed by horizontal divisions in the two-celled embryo. **(e)** An eight-celled embryo. Divisions parallel to the surface of the four-celled embryo form an epidermal layer. **(f, g)** Globular stages of embryogenesis. **(h)** Heart-shaped stage of embryogenesis. **(i)** Heart-shaped embryo and suspensor of *Brassica*

Endosperm formation

Mitotic divisions of the triploid endosperm nucleus begin about 24 hours after zygote formation and continue without wall formation to form liquid endosperm (as in the liquid from a green coconut). At about the time the proembryo is at the globular stage, the endosperm usually becomes cellular. It loses its liquid appearance as it forms cell walls around each nucleus and expands to form the major nutritive tissue of the seed, rich in storage reserves.

Seed and fruit development

Fertilisation initiates many changes. The petals fall off and the stigma and style wither. The fertilised ovules develop into seeds that vary in size from tiny spheres less than 0.1 mm in diameter, as in orchids, to 10 cm or larger, as in coconuts (Fig. 14.23). The dormant embryo, which originates from the fertilised egg and endosperm, is protected by a **seed coat** or **testa** that is derived from the integuments of the ovule and is often hard and composed of sclerenchyma (Chapter 6). The ovary wall forms a layer (the **pericarp**) around the seed or seeds to form a fruit. Fruits occur in a variety of shapes, textures and sizes (see Chapter 37).

In cereals and grasses (monocotyledons or monocots), the fruit contains a single fertilised ovule and develops into a grain. The integuments form the outer protective layers (bran) surrounding the embryo and endosperm. The single cotyledon, the **scutellum** (Fig. 14.24), is an interface organ between embryo and endosperm. In wheat or barley grains, the endosperm forms an outer layer, the **aleurone layer**, which functions in seed germination. In barley grains, the germinating embryo produces a signal hormone, gibberellin, which passes through the scutellum to the aleurone layer (Chapter 24). There the hormone triggers the production of enzymes that digest the starchy endosperm into soluble nutrients to sustain growth of the young seedling. In the embryo, a sheath of tissue, the **coleoptile**, protects the immature shoot meristems. The **coleorhiza** performs the same function for the immature root meristems.

In some plants, the triploid endosperm is consumed before seed maturity and development depends on food stored in the cotyledons (Fig. 14.21), which in legumes, including *Acacia*, become swollen and filled with storage reserves. The small seeds of some orchids have no endosperm because there is no sperm fusion with the central cell.

Fig. 14.23 (a) Several thousand seeds form in the fruit of the orchid *Phaius tancarvillae*. **(b)** Germinating coconut showing the embryonic shoot emerging through the hard, white cellular endosperm, brown shell (seed coat) and fibre

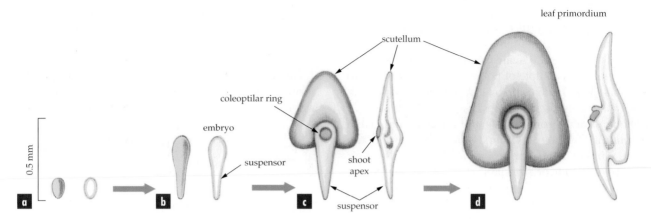

Fig. 14.24 Early stages of embryogenesis in the kernel of maize (monocotyledon) showing the origin of the single cotyledon, the scutellum. The diagrams show the embryo in face view (left) and in longitudinal section (right): **(a)** proembryo; **(b)** globular embryo and suspensor; **(c)** coleoptilar embryo showing origin of scutellum and coleoptilar ring around shoot apex; **(d)** later embryo showing development of leaf primordium (meristem) within the coleoptilar ring

> Seeds result from fertilised ovules and have an embryo from the fertilised egg and a seed coat derived from the integuments of the ovule. Endosperm is the nutritive tissue in many mature seeds; other seeds store food reserves in the cotyledons. The ovary develops into a fruit that encloses the seeds.

Reproductive success

Reproductive success is the ability of a plant to parent as many viable embryos as possible. Each plant allocates resources to its essential functions, which are vegetative growth, maintenance and reproduction. Maintenance costs include such activities as regulation of water movement, metabolism and avoidance of predation and pathogens. The cost of reproduction includes dispersal and selection of pollen, fertilisation and parental care. Reproductive effort represents the proportion of the total budget that is devoted to reproduction and hence into maturing embryos. Most organisms devote sufficient reproductive effort to maximise the chance of their offspring reaching maturity.

Female reproductive success depends on viable ovules. The female gametophyte is usually viable for only a short period. Under some weather conditions, ovules can abort before fertilisation occurs. For example, in cherry and peach, there are two ovules in each ovary (Fig. 14.25). The larger ovule is viable when the flower first opens, but soon aborts and the second ovule then becomes receptive. This extends the window of opportunity for fertilisation, especially during unfavourable weather, and is the reason why there are never two seeds in a cherry stone!

Some plants, such as the peach and avocado, have one or very few ovules in each ovary. Ovaries develop into large fruits that provide a substantial source of

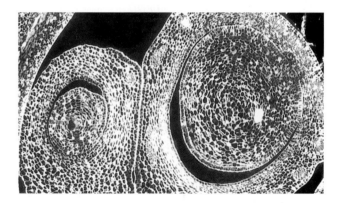

Fig. 14.25 Ovule viability. Ovary of a peach, showing a large viable receptive ovule (right), and a smaller ovule (left), which will become viable later when the ovule on the right has degenerated. This device ensures a longer period of ovule viability for fertilisation

nutrition for the new offspring. Other plants, such as orchids, have several thousand ovules in each ovary. Since orchid seeds are tiny and contain almost no nutrients, an orchid seedling must instead form an association with a particular fungus that provides a source of nutrients for growth.

Similarly, male reproductive success also depends on the quantity of gametes (pollen) produced and their viability. The number of pollen grains made by each anther is generally constant for a species but varies widely between species. For example, the exotic pasture grass, rye grass (*Lolium perenne*), produces 5400 pollen grains per anther. The anthers of Australian grasses, by contrast, produce less pollen: each anther of kangaroo grass (*Themeda triandra*), for example, releases around 2000 pollen grains, whereas a wallaby grass (*Danthonia caespitosa*) releases only 400 pollen grains per anther.

Pollen viability is defined as the ability of pollen to carry out successful fertilisation. However, this is difficult to estimate experimentally. Pollen quality, that is, whether the grains are living or not, is readily estimated by germinating a sample of pollen, which can take up to 24 hours (Fig. 14.26a), or by a rapid microscopic test, which takes only a few minutes (Fig. 14.26b).

Pollen grains from grasses and pumpkin are viable for only a few hours. Others from cabbage and the native climber *Pandorea jasminoides* slowly lose their viability over four or five days, after which only 8% of pollen are still alive. Up to 50% of pollen from *Banksia spinulosa* are still living after eight days but pollen from *Banksia menziesii* lives for only a day. Perhaps the longest-lived pollen comes from *Iris*, which can be stored for years at room temperature! *Eucalyptus* pollen is remarkably heat tolerant and remains viable at temperatures of 45°C or more, when most other types of pollen are killed at temperatures above 30°C. Most pollen types have to be stored frozen in liquid nitrogen at −176°C to maintain viability for prolonged periods for use in plant-breeding programs.

Reproductive success is the ability of an organism to parent as many viable embryos as possible.

Seed maturation and dormancy

As seeds mature, further development of the embryo is arrested. A lowering of metabolic rate marks the onset of dormancy in the embryo. Dormant seeds are dry and have a water content of about 10% (plant cells normally are more than 85% water). The seed coat is impervious to water and/or oxygen and is mechanically resistant to embryo enlargement. In some seeds, germination is also prevented for some time by chemical inhibitors. Dormancy ensures that seeds do not germinate while conditions are unsuitable for early plant development.

Seeds of many tropical plants do not show any dormancy as conditions are usually suitable for germination throughout most of the year because of the absence of marked seasonal variation.

Embryo development is arrested in the mature seed when it enters a period of dormancy, which is marked by a dehydrated state and low metabolic rate.

Triggering seed germination

The life span of seeds varies widely, from a few days (e.g. poplar seed) to a thousand years or more. In Japan, viable seeds of a water plant, *Nelumbo*, have been shown by radiocarbon dating of associated fragments of wood to be about 3000 years old.

Reinitiation of embryo development is triggered by environmental cues (see Chapter 24). Germination cannot take place until water and oxygen reach the embryo and this is initiated by fracture of the seed coat.

The radicle is the first organ to emerge through the seed coat; the epicotyl follows. Pea and cereal seeds show **hypogeal germination**, where the hypocotyl remains short, the cotyledons do not emerge from the seed and the epicotyl (young shoot) reaches the surface by its own growth (Fig. 14.27b). In contrast, bean and *Acacia* seeds show **epigeal germination**, in which the hypocotyl elongates into a U-shape, pushing cotyledons and epicotyl above ground (Fig. 14.27a). Consequently, when cotyledons are brought above ground in this way, they emerge from the encasing seed coat and function as leaves for a short time.

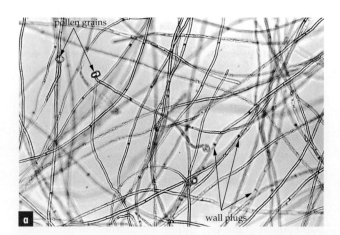
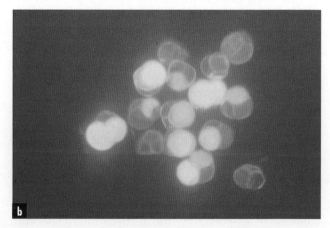

Fig. 14.26 Pollen viability. **(a)** Pollen germination test *in vitro*. Pollen grains of wild tobacco were cultured in a liquid medium and the extent of pollen tube growth was photographed after 4 hours. **(b)** Pollen quality can be tested using fluorescence microscopy, a procedure that takes only a few minutes to perform. This micrograph shows pollen of a hybrid *Rhododendron*, in which many of the clusters of four grains are sterile. In this procedure, pollen takes up a non-fluorescent compound, which is broken down by enzymes within the grain to form a fluorescent polar compound. This is retained within intact plasma membranes (green fluorescence), but diffuses out of dead grains, which have porous plasma membranes (no fluorescence, black)

Seed germination cannot take place until water and oxygen reach the embryo, usually by fracture of the seed coat. Cotyledons either remain in the seed with the radicle and epicotyl emerging (hypogeal germination) or the hypocotyl elongates and pushes the cotyledons above ground (epigeal germination).

Organogenesis

Growth and development of new organs occurs from the tips of shoots and roots. There is a specialised region of cells, the **apical meristem**, at the extreme tip of every shoot and root of a flowering plant. The function of these clusters of cells is to undergo rapid mitotic cycling. At each division, two daughter cells are produced, one of which remains part of the meristem and the other differentiates and contributes to the tissues that form part of the mature body of the plant. The first apical meristems are the root and shoot meristems of the embryo. Additional apical meristems form at the tips of new shoots and roots as the plant grows.

The fate of plant cells becomes determined during development. When development has proceeded beyond a certain point, cells are committed to a particular pathway; for example, a leaf primordium will only develop into a leaf (Fig. 14.28). Cells dividing and differentiating at the apical meristem are channelled into the 'leaf' pathway. A peg-like primordium elongates and the leaf lamina differentiates by increased cell division and elongation in this region.

The shoot apex

The shoot apical meristem produces new leaves. The apex of the shoot is a spherical dome of meristematic cells, which divide to produce leaf primordia on the flanks of the dome (Fig. 14.28a, b). The apex consists of two distinct regions of cells, the tunica and corpus. The **tunica** (*covering*) consists of one to three layers of cells that cover the apical dome. The **corpus** (*body*) is an area of cells below the tunica, occupying a central position in the meristem (Fig. 14.28c). Tunica and corpus divide in different ways to produce different tissues. Cells of the tunica divide in a regular manner to produce an outgrowth that becomes a leaf primordium. Cells of the corpus divide in an irregular manner to form the mature tissues of the stem. Divisions of both tunica and corpus lead to the production of a leafy shoot.

Axillary buds, which give rise to side branches, arise separately from a lateral meristem. Lateral meristems develop in the axil (angle between stem and leaf) of a leaf primordium (Fig. 14.28c).

In contrast to flowering plants, growth in ferns is derived from divisions of a single, large, **apical cell** situated at the tip of a stem or root. All tissues in a fern (Chapter 37) are derivatives of an apical cell.

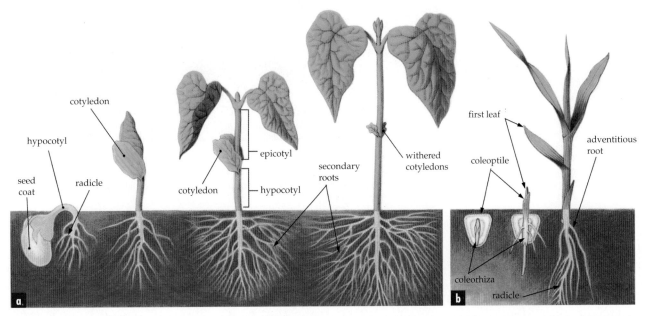

Fig. 14.27 Types of seed germination. **(a)** Epigeal germination, in which the elongating hypocotyl (U-shaped) pulls the cotyledons up into the air. **(b)** Hypogeal germination, in which the cotyledon remains underground, the epicotyl grows upwards into the air and the radicle grows downwards from the seed

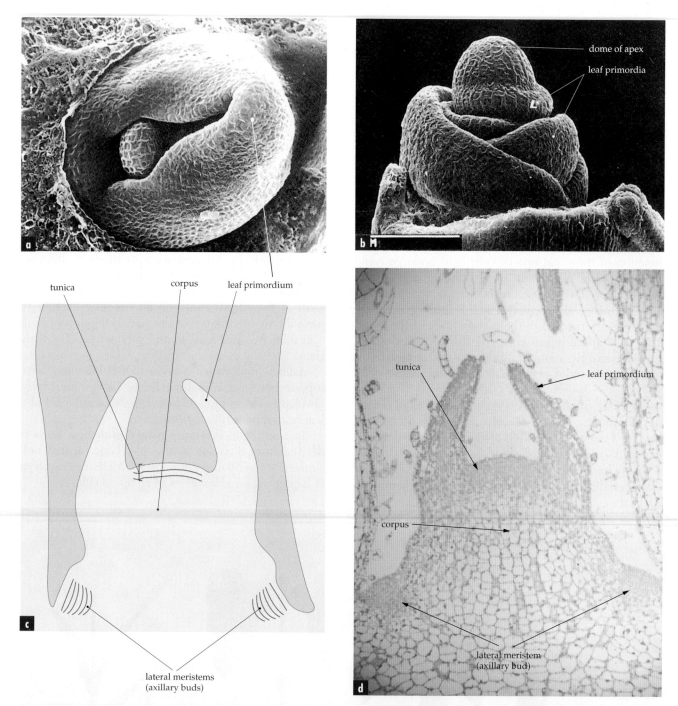

Fig. 14.28 Characteristic features of the shoot apical meristem of wheat viewed by scanning electron microscopy. **(a)** View of apex from above, with dome covered by expanding leaf primordium. **(b)** Side view showing apical dome with leaf primordia present on flanks of the apex. **(c)** Diagram showing tissues of shoot apex. **(d)** Lateral meristems in shoot apex of *Helianthus* as seen in longitudinal section

The root tip

Root apical meristems of flowering plants do not have a tunica–corpus structure. They contain a zone of rapidly dividing cells, which gives rise to all the mature tissues of the root. Roots do not produce lateral structures equivalent to the leaves on a stem. As we will see later (Chapter 17), root hairs arise by elongation of an

epidermal cell, while lateral roots arise deep within the tissues of more mature parts of the root, with no involvement of the root apex. Root tips produce a structure, the **root cap** (Fig. 14.29), that has no equivalent in the shoot. The root cap is a protective covering for the meristem as it grows through the soil. Root cap cells are produced from the meristem by cells in front of the apex.

Fig. 14.29 Scanning electron micrograph showing the structure of the root tip of wheat

Fig. 14.30 Spinifex grass, *Spinifex hirsutus*, has long stems (stolons) that grow along the surface of the soil, producing new plantlets (leaves and roots) at each node

The tissues of stems and roots produced by apical meristems are primary tissues. Later, in woody plants, secondary tissue, forming bark and wood, develops from specialised meristems once the primary tissues are mature (see Chapter 17).

Apical meristems of shoots and roots are clusters of rapidly dividing cells that give rise, directly or indirectly, to all cells of the plant.

Asexual reproduction in flowering plants

Vegetative reproduction

As any gardener knows, many plants can be propagated asexually (vegetatively) from small parts of a parent plant. For example, strawberry plants and spinifex grass (Fig. 14.30) have runners or **stolons**, long stems that grow horizontally along the surface of the soil. The stolon produces leaves and roots at regular intervals (at each node) and can be subdivided into a number of new plants. The weeping willow, *Salix babylonica*, reproduces asexually and disperses widely by means of broken branchlets that float along rivers and take root on riverbanks (Fig. 14.31). Known also as Napoleon's willow, the tree was introduced into eastern Australia from a single clone from the island of St Helena (where the Emperor Napoleon was exiled). In Australia, this willow reproduces only asexually because the introduced clone was female and there are no male trees.

Underground storage organs of plants, such as **corms** (swollen stems, e.g. gladioli), **bulbs** (swollen leaf bases, e.g. daffodils) and root **tubers** (e.g. potatoes and begonias; Fig. 14.32), also function in vegetative propagation. **Rhizomes** are underground stems that give rise to new shoots. They are characteristic of ginger, ferns such as bracken, many grasses and irises. 'Suckers' are new shoots that arise from roots. Trees and shrubs that sucker, such as reeds (*Phragmites australis*; Fig. 14.32b), wattles (*Acacia*) and blackberries (*Rubus*), can spread quickly into a vacant patch of habitat after disturbance.

Fig. 14.31 Weeping willow trees, *Salix babylonica*, reproduce asexually by means of broken branchlets that float along rivers, such as the Murray River, dispersing widely and taking root on riverbanks. These are also known as Napoleon's willow and reproduction has been asexual since only a single female clone was introduced

Fig. 14.32 (a) The potato develops new tubers from swollen regions of the stem. **(b)** The common reed, *Phragmites australis*, spreads rapidly by suckering through aquatic habitats

Budding is one of the more unusual forms of asexual reproduction seen in plants. Budding involves the development of a new individual as an outgrowth of the parent plant. For example, *Kalanchoe* produces buds along leaf margins, which can break off and begin life independently, eventually growing to adult size (Fig. 14.33).

> In asexual reproduction, new plants arise from a single parent by mitotic division and are genetically identical to their parent.

Totipotency

Unlike animal cells, plant cells have the ability to regenerate an entire plant from a single plant cell. This property is known as **totipotency** and was first demonstrated in 1958 by F. C. Steward, from Cornell University, using carrot cells growing in sterile culture. He took transverse sections of a carrot root and, under sterile culture conditions, cut out tiny pieces of secondary phloem tissue. These were then cultured in a shaking flask in a liquid medium containing nutrients and minerals, plus green coconut milk to stimulate growth. Free cells sloughed off from the pieces of

Fig. 14.33 Asexual reproduction by budding. **(a)** *Kalanchoe* generates tiny plantlets by budding along the margins of its leaves. These plantlets drop to the ground and take root. **(b)** The modified stems of the prickly pear, *Opuntia*, bear numerous buds, which detach easily and become new plants. This feature helped *Opuntia* colonise subtropical eastern Australia and make extensive areas of formerly productive rangeland unsuitable for grazing. The larvae of introduced moths now largely control *Opuntia* in Australia, one of the first successful examples of biological control

tissue and divided, forming embryo-like structures. These could be transplanted onto solid medium and grown in the light into new carrot plants.

Small pieces can be cut from tobacco stems and maintained by *in vitro* culture. Within a few weeks, massive growth of callus cells occurs from the site of wounding at the cut ends. These cells are part of the 'wound program' of cell division, which is initiated when an organ is injured. Callus cells can be subcultured and maintained indefinitely, providing a source of plant cells that are constantly undergoing mitosis. Folke Skoog, at the University of Wisconsin, designed an experiment to show that such callus cells are totipotent—able to differentiate along different pathways in different culture media. The ratio of certain plant hormones (small molecules that act as growth regulators) in the growth medium was found to be crucial (Fig. 14.34), as well as their absolute concentrations (see Chapter 24). No growth or differentiation took place when concentrations were low. However, when the hormone auxin was high, root formation occurred on the clumps of cells (*root program*). When the hormone cytokinin was high and auxin low, shoot formation was induced (*shoot program*). With moderate amounts of both hormones, growth of the callus clumps occurred without differentiation (*callus program*).

Thin cell layers of epidermis peeled from tobacco stem can be cultured and the cells induced directly to enter different programs of differentiation, for example, into vegetative buds, roots or callus. If the epidermal cells came from flowering shoots, then flowers can also be induced.

These experiments suggest that single plant cells, if in a suitable environment of nutrients and light, can undergo cell division and growth to generate an entire organism.

> Plant cells are totipotent and possess the potential to regenerate an entire organism.

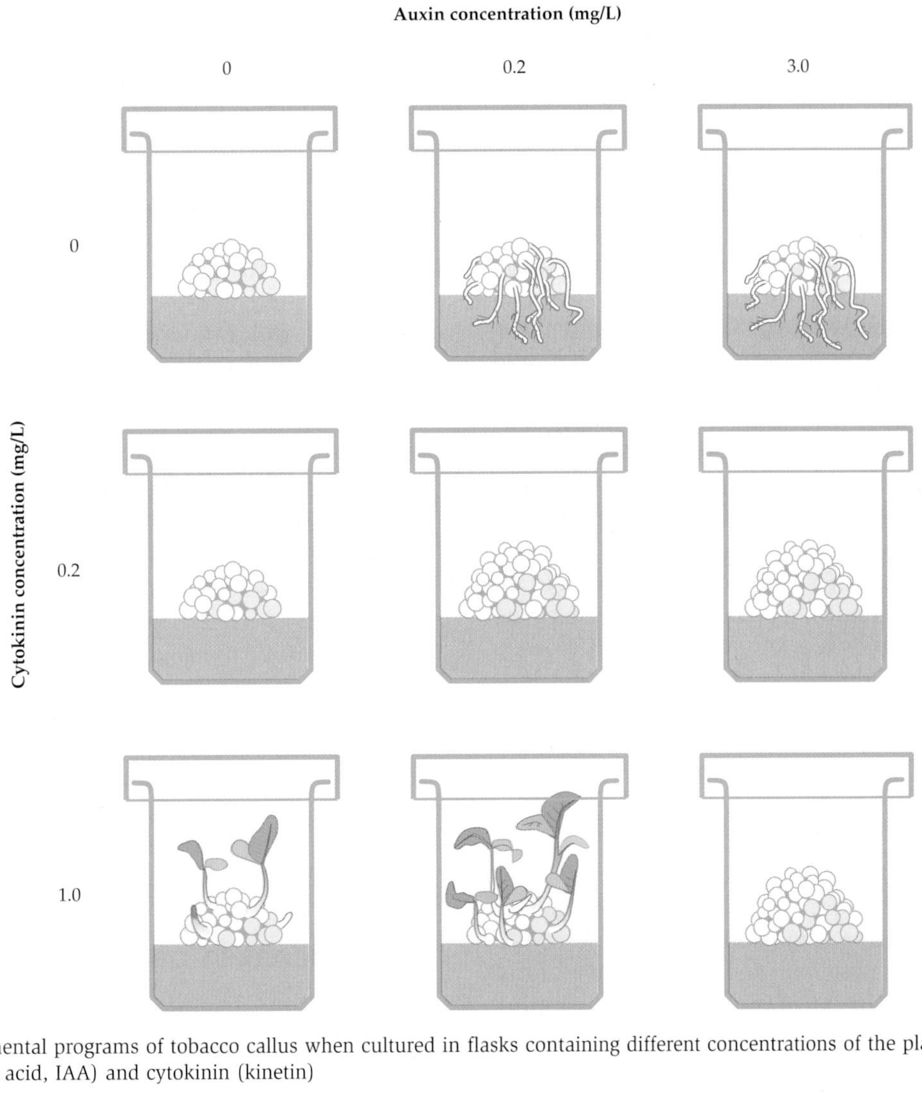

Fig. 14.34 Developmental programs of tobacco callus when cultured in flasks containing different concentrations of the plant hormones: auxin (indole acetic acid, IAA) and cytokinin (kinetin)

Tissue culture and plant biotechnology

Modern cell culture techniques have made use of microbiological culture methods, especially adapted for plant cells. Their use has resulted in dramatic advances in regenerating plants from individual cells, organs or tissues. Such techniques are an essential component of genetic engineering programs for gene transfer.

Currently, agriculturists and horticulturists clone plants with particularly desirable characteristics, such as the red-flowering form of *Eucalyptus ficifolia*, for commercial use by means of tissue culture. Tissue culture is a propagation technique in which one or a few cells are treated so that they give rise to a whole new plant. Instead of planting bulky potato tubers, in future, farmers will sow 'artificial seeds' made of clusters of potato cells grown in tissue culture and coated with a protective covering.

Plant tissue culture can begin with a fully organised structure, for example, apical and lateral meristems and young axillary shoots. 'Meristem culture' was established to make use of heat treatment to eliminate viruses from meristem regions. Increasing the temperature in meristem culture allows plant cells to divide faster than viruses can multiply, thus eliminating viruses from the tissue. This process has been widely used to produce strawberries and potatoes free of virus diseases.

During the past 20 years, callus cultures and suspension cultures have been widely used for mass propagation of plants. However, tissue culture of callus cells results in increased chromosome number and gene mutations. Plants regenerated from callus cultures often show structural and physiological abnormalities. In some cases, where crops were regenerated using these methods, there was a disastrous effect on the commercial viability of the crop. Mass propagated strawberries remained vegetative after field planting in the United States; after five years in the field in Malaysia and New Guinea, oil palms grown from tissue culture of roots produced sexually modified flowers with drastically reduced kernel yields. Methods involving induction of callus tissue are now avoided and have been replaced by techniques that give rise to somatic embryos directly.

Hybrid embryo rescue

Interspecific hybridisation provides the potential to generate new genetic combinations but, even with successful cross-fertilisation, the hybrids may form non-viable seeds. *In vitro* culture of developing embryos from fertilised ovules of hybrids is a useful method to rescue hybrid embryos. For example, hybrids between clovers (*Trifolium ambiguum* × *T. repens*) were given embryo rescue treatment in New Zealand with the aim of providing a source of new pasture legumes. When each of the parent clover species is self-fertilised, the embryos progress normally through the globular, heart-shaped and torpedo-shaped stages until mature embryo formation. However, in interspecific hybrid embryos, growth in the first two days after pollination is more rapid than in the parents, but then slows, with development within the seed being arrested at the heart-shaped stage, when the embryos have about 1000 cells. This developmental arrest is due to the action of deleterious genes, which appear to act by blocking the normal nutrient supply to the growing embryo. Several rescue techniques have been successfully used. Nurse cultures are initiated with hybrid embryos being cultured on a normal endosperm (Fig. 14.35a). Embryos can also be cultured directly on artificial media, in which the required nutrients are provided. These embryo rescue techniques successfully produced new hybrid clovers (Fig. 14.35b).

Cell line selection

Plants contain many secondary chemicals that are important in the cosmetic, food and pharmaceutical industries. Nearly all these useful molecules are produced in secondary metabolic pathways. Typically, such secondary compounds are produced in small amounts, for example, capsaicin, the hot substance in chilli peppers, or saffron, the yellow colouring from stigmas of wild crocuses. In a large population of cells, some cells will produce the desired metabolite at higher levels than others will. Cell line selection and cloning is used to grow the higher yielding lines.

Somatic embryos

Suspensions of carrot cells can be maintained indefinitely in a medium containing the plant growth substance 2,4-dichlorophenoxy acetic acid (2,4-D) (Chapter 24). If 2,4-D is removed, embryogenic development is induced. The embryos are *somatic* embryos: they are formed from cells of the body of the plant but they behave exactly like a *zygote*, faithfully replaying the developmental program leading to embryo-like structures. Somatic embryos can also be induced from epidermal cells of seedling hypocotyls when cultured in the presence of particular hormones, especially cytokinins (Fig. 14.36).

Applications in genetic engineering

Tissue culture techniques are a part of genetic engineering technologies. Firstly, the desired genes (e.g. disease-resistance genes) need to be isolated (see

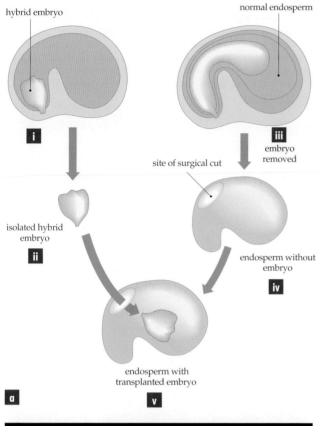

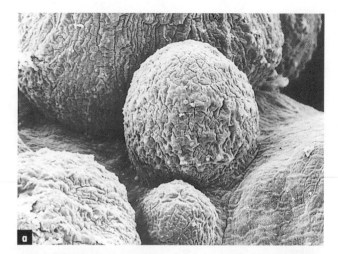

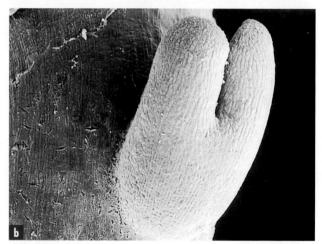

Fig. 14.36 Embryo culture: scanning electron micrographs of developing somatic embryos developing from the hypocotyl of clover *Trifolium*. **(a)** Globular embryos. **(b)** Heart-shaped embryos

Fig. 14.35 Hybrid embryo rescue. **(a)** Transplantation of a hybrid embryo into a de-embryonated normal endosperm for culture until maturity. **(b)** New clover hybrid raised by embryo rescue in New Zealand: the parents are *Trifolium ambiguum* (left) and *T. repens* (right), with the new hybrid (centre)

Chapter 12). Secondly, these genes need to be physically transferred into the host plant cell. The process by which the genetic make-up of a single cell is altered is *transformation* (see Chapter 13). Micropropagation techniques can then be used to regenerate a new plantlet with the altered phenotype from this cell.

Transformation is typically achieved using a vector to transfer the gene (Fig. 14.37). These are usually plasmids, small DNA molecules in the cytosol of bacteria that carry one or more genes and can replicate themselves in bacteria. The most common vector is the Ti plasmid of *Agrobacterium tumefaciens*, crown gall disease, which produces tumour-like growths in plants (but for transformation, the tumour-inducing genes are disabled). The critical feature is that the bacterium transfers part of the Ti plasmid to the plant nucleus (known as T-DNA), thus effecting transformation. The Ti plasmid is an efficient way of transferring one or two genes (~15 kb of DNA) into plant cells.

The use of *Agrobacterium* for transformation is mostly confined to certain dicotyledons, including tobacco and *Arabidopsis*. A vector is not always needed for transformation. Direct gene transfer can be achieved by using a biolistic DNA delivery system, which uses helium gas under pressure to fire DNA-coated tungsten microprojectiles into plant tissues.

Genetic engineering has some major advantages over traditional methods of cross-breeding. It is rapid and, as it involves the transfer of only a few exotic genes, the balance of the host plant genome is not

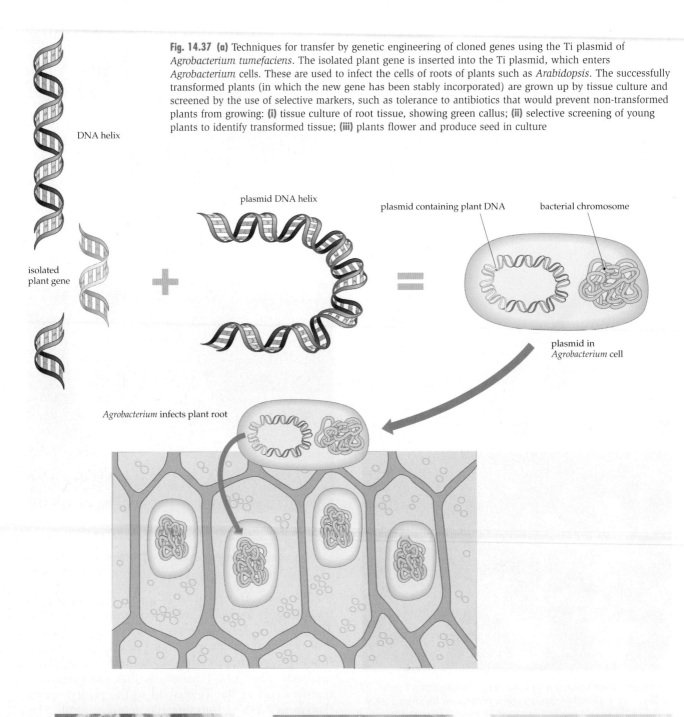

Fig. 14.37 (a) Techniques for transfer by genetic engineering of cloned genes using the Ti plasmid of *Agrobacterium tumefaciens*. The isolated plant gene is inserted into the Ti plasmid, which enters *Agrobacterium* cells. These are used to infect the cells of roots of plants such as *Arabidopsis*. The successfully transformed plants (in which the new gene has been stably incorporated) are grown up by tissue culture and screened by the use of selective markers, such as tolerance to antibiotics that would prevent non-transformed plants from growing: **(i)** tissue culture of root tissue, showing green callus; **(ii)** selective screening of young plants to identify transformed tissue; **(iii)** plants flower and produce seed in culture

DNA helix

plasmid DNA helix

plasmid containing plant DNA

bacterial chromosome

isolated
plant gene

plasmid in
Agrobacterium cell

Agrobacterium infects plant root

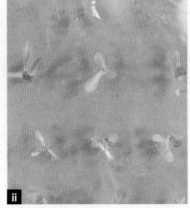

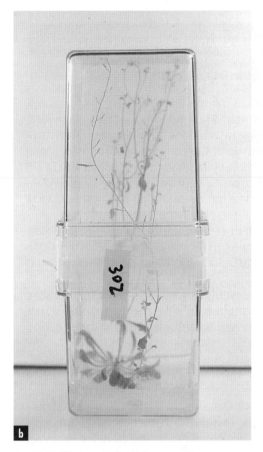

Fig. 14.37 (b) Growth of transformed *Arabidopsis* plant by tissue culture. **(c)** Testing of *Arabidopsis* plant for selectable marker. The plant gene being transferred is expressed only in anthers, as seen here in this flower using a marker enzyme that gives a blue colour after incubation in a suitable substrate. The presence of the blue colour in the anthers confirms that the gene has been successfully transferred

usually disturbed. This means that the special characteristics of the cultivar are conserved. There also appear to be few restrictions on the source of the gene, as even viral genes have been expressed in plants. An example is the genes encoding viral coat proteins in the *coat protein cross-protection* approach, a technique that provides protection against viral infection (see Chapter 13). Another example is the use of bacterial genes encoding an insecticidal protein, known as Bt toxin, from *Bacillus thuringiensis* (see Box 14.2).

One disadvantage is that there are still many plant genes whose functions are not well studied that presumably play important roles in regulating growth and development and in resisting diseases. The insertion of exotic DNA into plant chromosomes is apparently random, so position effects can occur when adjacent DNA sequences and genes modify the expression of the inserted gene. Genetic engineering is particularly well developed at present in crops such as potatoes and tomatoes, both members of the family Solanaceae, for the oilseed crop canola, a member of the Brassicaceae, for soybeans, a member of the Fabaceae, and for maize and rice, which are both members of the Poaceae.

Tissue and single cell culture techniques have resulted in the generation of plants from individual cells, organs or tissues. Such techniques are essential for gene transfer into host plant cells using a vector.

BOX 14.2 Ethics of genetic engineering: field trials of transgenic cotton in Australia

Cotton is one of Australia's main rural exports. In 1998, Australia's cotton exports were worth about $1400 million. But to protect the crop from insects, growers spend around $125 million a year on pesticides, a figure that is expected to rise as control becomes more difficult. Using insecticides on this scale also creates severe ecological problems. The pesticides are needed to control the larvae of the cotton budworm, which eats the leaves and flowers of cotton. The CSIRO Division of Plant Industry, in collaboration with Monsanto, a US agricultural company, has engineered cotton plants to express Bt toxin, a protein made by a bacterium commonly found in soil. The protein is a potent killer of budworm larvae but does not kill most other insects or spiders, and is harmless to humans and other animals.

Field trials of transgenic cotton plants expressing the Bt toxin began in 1992 in northern New South Wales. The potential biological and environmental risks associated with these trials were assessed by the Gene Manipulation Advisory Committee (GMAC), the regulatory body in Australia, and the trials were carried out according to guidelines laid down by GMAC.

One of the potential risks identified by GMAC was that the Bt gene would spread from the crop to native cotton species growing in the area. To overcome this, the transgenic cotton plants were grown within buffer zones of normal cotton varieties. The buffer zones were intended to prevent pollen from the transgenic cotton plants spreading outside the field trial. The chance this will occur is relatively low as cotton is largely self-fertilising. Experiments show that cross-fertilisation between adjacent rows of cotton plants is only 20%, and is less than 1% between plants separated by 7 m. Experiments also show that cultivated cotton and native species of cotton (*Gossypium*) do not readily cross-hybridise.

Commercial growing of genetically engineered cotton began in Australia in 1996. It was expected this technology would lead to a reduction in the use of chemical pesticides, and thus a reduction in the financial burden on growers and the chemical burden on the environment. This has proved to be the case. For example, the 1998–99 season was one of the worst for insect pests in many years. Despite this, farmers growing genetically engineered cotton used, on average, 38% less pesticide than farmers growing conventional cotton. Farmers in some areas reported using 50% less pesticide. A continuing concern, however, is that the budworm larvae will become resistant to Bt toxin. Researchers and farmers are working together to develop strategies to prevent this from happening. Farmers are asked to plant crops attractive to budworm larvae around their transgenic cotton fields. As the cotton budworms living in these 'refuges' will outnumber those that survive feeding on the transgenic cotton plants, the chances that large numbers of resistant insects will develop is thought to be low.

Summary

- In asexual reproduction, new individuals arise from a single parent by mitotic cell division and are genetically identical to their parent. Sexual reproduction involves redistribution of parental genes into offspring and is the principal mechanism for generating genetic variation in eukaryotes.
- Life cycles of sexually reproducing organisms show a pattern of alternation between diploid and haploid stages.
- Isogamy involves the fusion of gametes that are morphologically similar but of opposite mating types ('+'/'−'). In anisogamy, male and female gametes differ in appearance.
- In flowering plants, the sexual organs are within the flower. The male organs are the anthers, specialised for differentiation, nutrition, protection and dispersal of pollen, the male gametophyte. The tapetum of the anther has a nutritive function, while the endothecium and stomium are responsible for anther opening.
- Microspores are formed in tetrads within an anther at meiosis. Each tetrad is enclosed in a wall of callose. At the end of meiosis, microspores are released from the tetrad and enter a major growth phase, developing a large central vacuole. The microspore becomes a pollen grain after the first (asymmetric) mitosis.
- Most pollen is bicellular and contains a generative cell enclosed within a vegetative cell. The generative cell divides to form two sperm cells after pollen germination within the growing pollen tube in the style. In tricellular pollen, the generative cell divides within the pollen grain.
- The female reproductive organ, the pistil, bears a stigma to receive pollen, a style with transmitting tissue or central canal for growth of pollen tubes and a basal ovary containing ovules. An ovule contains the female gametophyte, the embryo sac.
- The embryo sac is formed from a diploid megasporocyte, which undergoes meiosis to produce four haploid megaspores. Typically, three abort and the remaining megaspore differentiates by three mitotic divisions, producing an eight-nucleate embryo sac. A seven-celled female gametophyte is formed, containing an egg, two synergids, central cell with two polar nuclei and three antipodal cells.
- Reproductive mechanisms in flowering plants enable them to optimise pollination and allow the pistil to screen out undesirable pollen. Many adaptations have evolved to promote cross-fertilisation in flowering plants. Male and female gametes may be either physically separated, as in monoecious and dioecious species, or temporally separated, as in protogyny and protandry. Self-incompatibility is found in many families of flowering plants. Its effect is to prevent self-fertilisation. Self-fertilisation enables genetic continuity in plants that are successful in their environment. Seed production is enhanced through cleistogamy in environments where cross-fertilisation is unreliable.
- Fertilisation in flowering plants begins when pollen alights on the stigma. Pollen tubes emerge from the pollen grain and grow through the stigma and style to the ovary, where they penetrate the ovules. There, double fertilisation occurs. The pair of sperm cells are usually linked together and to the tube nucleus as the male germ unit. One of the pair of sperm cells in the pollen tube fuses with the egg to produce the diploid zygote, the other with the central cell to produce the triploid endosperm. The zygote divides to form the embryo and a suspensor that acts as a nutritive link. In dicotyledons, the embryo develops into a globular, then heart-shaped and finally torpedo-shaped structure as the two cotyledons differentiate. In monocotyledons, the single cotyledon is the scutellum.
- Seeds result from fertilised ovules and have a seed coat derived from the integuments of the ovule and an embryo from the fertilised egg. Endosperm is the nutritive tissue in the mature seed; other seeds store food reserves in the cotyledons.
- In apomixis, meiosis does not occur and embryos develop asexually. This allows for the rapid multiplication of plants.
- Embryo development is arrested in the mature seed when it enters a period of dormancy, which is marked by a dehydrated state and low metabolic rate. Seed germination cannot take place until water and oxygen reach the embryo, usually by fracture of the seed coat. The radicle and epicotyl emerge (hypogeal germination) or the hypocotyl pushes the cotyledons above ground (epigeal germination).
- At the tips of every shoot and root are specialised apical meristems, clusters of rapidly dividing cells that give rise, directly or indirectly, to all cells of the plant.

- Plant cells are totipotent and possess the potential to regenerate an entire organism.
- In asexual reproduction, new plants arise from a single parent by mitotic division and are genetically identical to their parent. Vegetative propagation can be by corms, bulbs, tubers, stolons, rhizomes or budding.

- Tissue and single cell culture techniques have resulted in an ability to generate plants from individual cells, organs or tissues. Such techniques are essential for gene transfer into host plant cells using a vector.

key terms

aleurone layer (p. 354)	development (p. 333)	male germ unit (p. 339)	self-fertilisation (p. 349)
anatropous ovule (p. 343)	dioecious (p. 348)	megaspore (p. 343)	self-incompatibility (p. 348)
anisogamy (p. 334)	egg (p. 335)	megasporocyte (p. 343)	sexual reproduction (p. 333)
anther (p. 337)	embryo (p. 352)	microspore (p. 337)	sperm (p. 335)
antipodal cell (p. 344)	embryo sac (p. 337)	microsporocyte (p. 337)	sporophyte (p. 334)
apical cell (p. 357)	endosperm (p. 351)	monoecious (p. 348)	sporopollenin (p. 340)
apical meristem (p. 357)	epicotyl (p. 352)	ovary (p. 337)	stamen (p. 337)
apomixis (p. 352)	epigeal germination (p. 356)	ovule (p. 337)	stigma (p. 337)
asexual reproduction (p. 333)	exine (p. 340)	pellicle (p. 341)	stolon (p. 359)
bisexual (p. 347)	fertilisation (p. 345)	pericarp (p. 354)	stomium (p. 337)
budding (p. 360)	flower (p. 335)	pistil (p. 337)	style (p. 337)
bulb (p. 359)	gametogenesis (p. 335)	pollen (p. 337)	suspensor (p. 352)
central cell (p. 344)	gametophyte (p. 334)	pollination (p. 345)	synergid (p. 344)
cleistogamy (p. 349)	generative cell (p. 339)	protandry (p. 347)	tapetum (p. 337)
coleoptile (p. 354)	growth (p. 333)	protogyny (p. 347)	totipotency (p. 360)
coleorhiza (p. 354)	hemitropous ovule (p. 341)	radicle (p. 352)	triple fusion nucleus (p. 351)
corm (p. 359)	hypocotyl (p. 352)	rhizome (p. 359)	
corpus (p. 357)	hypogeal germination (p. 356)	root cap (p. 358)	tube nucleus (p. 339)
cotyledons (p. 352)	intine (p. 340)	scutellum (p. 354)	tuber (p. 359)
cross-fertilisation (p. 345)	isogamy (p. 334)	seed (p. 337)	tunica (p. 357)
	life cycle (p. 333)	seed coat (testa) (p. 354)	vegetative cell (p. 339)

Review questions

1. (a) What is the male gametophyte of a flowering plant?

   (b) Where is it located?

   (c) What is the function of the callose wall during microsporogenesis?

2. (a) What is the female gametophyte of a flowering plant?

   (b) Where is it located?

   (c) By means of diagrams, show how it forms by meiotic and mitotic divisions.

3. What are the special features of the location of the egg and central cells within the embryo sac? Draw a diagram to illustrate your conclusions.

4. (a) List the advantages and disadvantages of asexual reproduction and sexual reproduction.

   (b) What processes of sexual reproduction are avoided to produce seeds by apomixis?

5. Many bisexual plants prevent self-fertilisation. Describe the various ways by which they do this.

6. (a) What is self-incompatibility?
   (b) By means of a checkerboard, deduce the *S* alleles of the offspring when a self-incompatible plant, A (S_1S_2), is crossed reciprocally with another plant, B (S_1S_3).
7. By means of labelled diagrams, distinguish between epigeal and hypogeal forms of germination.
8. (a) In a dicotyledon embryo, what is:
       (i) an epicotyl?
       (ii) a hypocotyl?
       (iii) a radicle?
   (b) What is the scutellum of a monocotyledon embryo?
9. Plants grow from meristems.
   (a) What organs are derived from the tunica?
   (b) What is the main difference between apical meristems of ferns and flowering plants?
   (c) The shoot apical meristem of grasses is cone-shaped. What are the peg-like structures?

Extension questions

1. Describe the sequence of events that begins when a pollen grain lands on the surface of a compatible stigma and ends in double fertilisation.
2. Do flowering plants go through embryonic stages equivalent to those of animals? Compare and contrast embryo development in flowering plants with that in animals (Chapter 15).
3. Why is the cereal grain really a fruit?
4. In 1992, the CSIRO began field trials of genetically engineered cotton plants containing a bacterial gene expressing a pesticidal toxin that is expected to displace chemical sprays.
   (a) List five advantages that this new technology will have for cotton growers in Australia.
   (b) What are the ethical issues that need to be considered before plants containing exotic genes are released commercially?

Suggested further reading

Chrispeels, M. J. and Sadava, D. E. (1994). *Plants, Genes, and Agriculture*. Boston: Jones and Bartlett.

Contains extended discussions on plant development.

Conway, G. and Toenniessen, G. (1999). Feeding the world in the twenty-first century. *Nature* 402 (6761 supplement S): C55–C58.

A millennial essay discussing the challenges facing world agriculture and the role biotechnology may play.

Esau, K. (1977). *Anatomy of Seed Plants*. New York: John Wiley & Sons.

The book to read for insights into plant development—a classic.

http://genetech.csiro.au/'Gene technology in Australia'.

Further examples of gene technology in Australia, including genetic engineering of plants.

CHAPTER

15

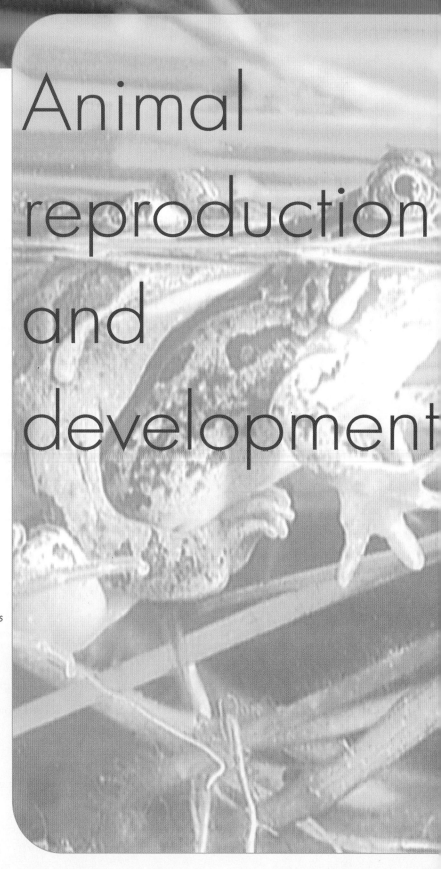

Animal reproduction and development

Animals reproduce asexually, sexually or by both means. In **asexual reproduction**, new individuals are clones, with one parent only, so they are genetically identical to that parent. In sexual reproduction, a new individual forms from the fusion of haploid gametes from two different organisms. Prior to gamete formation, meiosis occurs, during which homologous chromosomes exchange genetic material, producing new combinations of parental genes (Chapter 8). By allowing DNA from one organism to be combined with that from another, sexual reproduction results in unique individuals and generates genetic variation within populations.

Asexual reproduction

Most people are aware that plants can be propagated asexually, for example, by planting a cutting (Chapter 14). However, many animals also can reproduce asexually, with sexual reproduction occurring rarely or not at all. The fact that these organisms have survived for millions of years indicates that asexual reproduction (possibly punctuated by rare bouts of sexual reproduction) is a very successful strategy for survival.

Asexual reproduction involves the production of new individuals through mitotic cell divisions. An advantage of asexual reproduction is that it allows offspring to form without the need to get two individuals of different sex into reproductive condition at the same place and the same time. A disadvantage is that, because offspring are genetically identical to the parent and to each other, genetic diversity is limited. This limits the ability of organisms to adapt to new or changing environments.

> In asexual reproduction, new individuals arise from a single parent by mitotic cell division and are genetically identical to their parent.

In the life cycles of many parasites, some stages involve prolific asexual reproduction. These parasites include invertebrates such as tapeworms and flukes. The asexual stage of the life cycle dramatically increases the number of infective individuals, which enhances the chance of finding a new host. Asexual reproduction can occur by several different means, including regeneration, budding and parthenogenesis.

Regeneration

Regeneration involves the production and differentiation of new tissues and is normally responsible for replacement of damaged and missing parts of the body. It is highly developed in many invertebrate animals,

such as hydras, flatworms, annelids and echinoderms. In some of these animals, new individuals can arise from separated body parts. Starfish, for example, can regenerate a new individual from a single arm provided that part of the central disc is present (Fig. 15.1). In some species of aquatic annelids, the capacity to regenerate missing body parts has evolved into a means of reproduction by fragmentation. Periodically, these worms simply break into several parts, each of which then regenerates the missing parts to form a complete worm.

Reproduction by regeneration is little more than an extension of the normal processes of growth and tissue repair. It involves cellular replication by mitosis followed by differentiation of the tissues but no rearrangement of genetic material.

Budding

Budding involves the development of a new individual from outgrowths of the body wall of the parent. This is typical of *Hydra* (Fig. 15.2) and other polyps, such as corals, and some plants. Hydras produce buds that branch from the side of the body. They may break off to begin life independently, eventually growing to

Fig. 15.1 A single arm of the sea star *Linkia mulifora*, regenerating many new arms to form an entire new individual

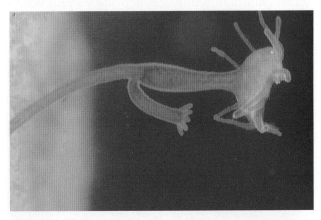

Fig. 15.2 Asexual reproduction by budding. In *Hydra*, a new individual develops from outgrowths of the body wall of the parent by producing a bud that branches from the side of the body. This is typical of *Hydra* and other polyps such as corals

normal adult size. Alternatively, a bud may remain attached and, together with other buds, contribute to the formation of a colony.

> Asexual reproduction can occur by means of regeneration or budding.

Parthenogenesis

In bees, wasps and ants, fertilised eggs develop into diploid females, while unfertilised eggs become males (Fig. 15.3). Unfertilised eggs become activated and develop by **parthenogenesis**, which is initiated by mitosis of the haploid egg nucleus that is not followed by cytokinesis. The two nuclei produced then fuse to form a diploid nucleus and the egg develops as if it had been fertilised.

In some species of aphids, during spring and summer, females produce eggs that develop parthenogenetically into new offspring. A female might have up to 10 parthenogenetic cycles in a season, producing three to 100 offspring per cycle. The offspring mature rapidly and within a few days may also be producing eggs. Thus, under favourable conditions, population numbers can rise dramatically. Most offspring produced in this way are female, but some are winged males, which can fly to new host plants. These males can mate with females to produce fertilised eggs, which are more able to survive harsh winter conditions than parthenogenetic eggs. Surviving fertilised eggs hatch in the next spring to provide a new, genetically varied population.

Parthenogenesis occurs in a few species of vertebrates, such as the common Australian gecko (see Chapter 32). The whiptail skink from eastern North America also has secondarily adopted a unisexual way of life. Only females are present in the population and eggs develop by parthenogenesis. Intriguingly, this species retains part of a previous sexual pattern of reproduction: ovulation only occurs after a stereotyped sequence of courtship behaviour

in which another female takes a male behavioural role.

> Parthenogenesis is a form of sexual reproduction in which egg cells develop into embryos without fertilisation.

Sexual reproduction

Sexual reproduction is characteristic of nearly all eukaryotes. It often involves a considerable cost to the parents in energy and building materials for the production of gametes and to ensure mature male and female gametes occur together. However, its most important advantage is the redistribution of parental genes into offspring. Advantageous genetic combinations may result from the random assortment of parental characters during meiosis (Chapter 9) and enhance the chances of survival of individuals. Sexual reproduction also generates variation within populations, which has been exploited in selective breeding of domesticated plants and animals.

> Sexual reproduction involves redistribution of parental genes into offspring and is the principal mechanism for generating genetic variation in eukaryotes.

A life cycle is the sequence of stages in the growth and development of organisms from zygote to reproduction, that is, from one generation to the next. Life cycles of sexually reproducing organisms show a pattern of alternation between diploid and haploid stages. Fertilisation, which combines two haploid genomes, alternates with meiosis, which divides the genome (Fig. 15.4). Combination of the haploid cells, the gametes, at fertilisation produces the diploid zygote. The new organism arises from the zygote by mitosis. Therefore, in animals, the haploid generation is simply the gamete stage.

In some simple eukaryotes the gametes are structurally similar (isogamy), but in animals they are distinctly different in structure (anisogamy) (Fig. 15.5). In anisogamous systems, the smaller of the pair of gametes is considered male.

Sex refers to the distinction between individuals of different mating types, usually referred to as male and female. In anisogamous organisms, individuals that produce only one form of gamete are often morphologically, physiologically and behaviourally distinct. Organisms producing only sperm are male and those producing only eggs are female. The distinctions in appearance and behaviour between male and female individuals of a species are known as secondary sexual characteristics.

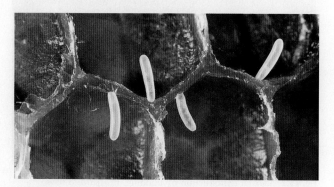

Fig. 15.3 In bees, fertilised eggs develop into female workers and unfertilised eggs into males. Bees breed in tree holes and, here, eggs in cells can be seen

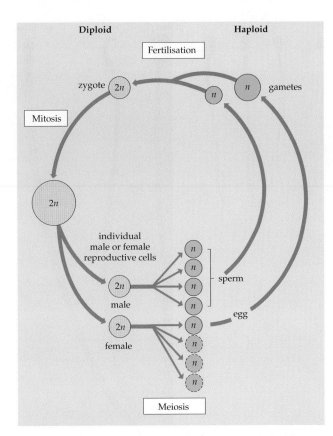

Diploid | Haploid

Fig. 15.4 Life cycles of sexually reproducing animals. After meiosis, reproductive cells produce either four haploid sperm or one egg (the others degenerate, shown by broken circles). Two gametes fuse to form a diploid zygote, which then undergoes mitotic divisions to produce the embryo, the beginning of the diploid life stage

> Life cycles of sexually reproducing organisms show a pattern of alternation between diploid and haploid stages. Sex is the distinction between male and female individuals of a species.

Hermaphroditism

In some animals, both female and male reproductive organs occur within the same individual. These species are **hermaphrodites** or monoecious ('one house'). This situation is common in a wide range of invertebrates including sponges, flatworms, annelids and molluscs. More commonly in the animal kingdom, species are dioecious ('two houses'), where sperm and eggs are produced by separate individuals.

> In dioecious organisms, male and female reproductive organs occur in separate individuals. In hermaphrodites, male and female reproductive organs are present in the same individual.

Preventing self-fertilisation

Although hermaphroditic organisms can produce both sperm and eggs, they usually do not fertilise themselves. Self-fertilised hermaphrodites lose the advantage of genetic variation in offspring that arises from cross-fertilisation. Indeed, in various species a variety of adaptations have evolved that prevent self-fertilisation.

In hermaphroditic animal species, where eggs and sperm are produced simultaneously, self-fertilisation may be prevented by anatomical separation of the gametes. In the earthworm, male and female reproductive openings are well separated. The worms have a complex mating procedure with another individual in which they fertilise each other's eggs (Chapter 39).

In many marine hermaphrodites that release both eggs and sperm simultaneously, the eggs become fertile some time after release, by which time the sperm are generally no longer capable of fertilisation. This mechanism ensures cross-fertilisation.

> Many adaptations have evolved to limit self-fertilisation in hermaphroditic organisms.

Changing sex

One reproductive strategy involves an individual alternating between male and female sex so that it does not produce both eggs and sperm at the same time. In some species, there is a single change of sex. An organism may start off as male, converting to female at some later stage, **protandry**, or as female, with a later conversion to male, **protogyny**.

In salmon, fish start life as males and over several years gradually get bigger until they exceed a threshold size, at which time their gonads transform into ovaries (Fig. 15.6). There is a clear advantage in this pattern of reproduction. Sperm are small, so even small males can produce sufficient sperm to fertilise vast numbers of eggs. However, because eggs are large and yolky, the larger a female is, the greater the number of eggs she can produce.

A coral reef fish, the blue wrasse, provides an example of protogyny. The large dominant male controls a harem of smaller, drab-coloured females. The male alternates his colour between green and blue. About an hour before spawning, the blue colour dominates and he commences mating (Fig. 15.7). This involves elaborate courtship behaviour followed by spawning with each female in turn. His colour then changes back to green. If the male is lost from a group, the largest female will undergo sex reversal and change colour to become the male with control of the harem. Protogyny in the blue wrasse maximises reproductive output, since all but one of the individuals is female and producing eggs, and the single male is able to fertilise eggs produced by all females in its harem.

> In protandry and protogyny, self-fertilisation is limited because individuals do not produce eggs and sperm at the same time.

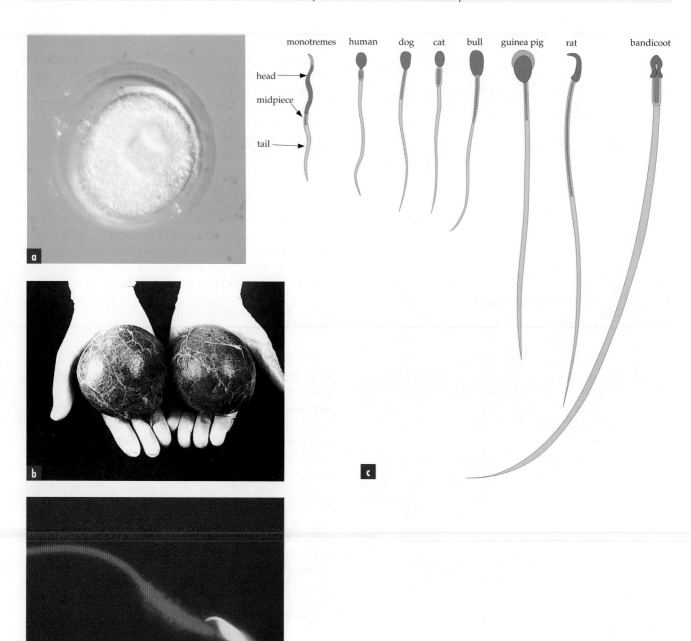

Fig. 15.5 Anisogamous gametes of different animal groups vary greatly in size and shape. **(a)** Egg of tammar wallaby. **(b)** These two eggs from the lobe-finned fish, *Latimeria chalumnae*, are over 9 cm in diameter. **(c)** Diversity of form of mammalian spermatozoa. **(d)** Sperm of tammar wallaby stained with acridine orange

Reproductive strategies

After fertilisation, the zygote undergoes a sequence of divisions and developmental changes that ultimately result in the formation of a new adult animal. This process, development, is dealt with in detail later in this chapter. The nature of development is related to the reproductive strategy of a particular species. Some animals show **indirect development**, with one or more intermediate larval forms before the adult form is attained. Others show **direct development**, in which the offspring hatch or are born in miniature adult form.

Indirect development

In many animals showing indirect development, fertilisation follows the random release of gametes into the environment. Since vast numbers of gametes are released, the amount of yolk that the female can invest in each individual egg is limited. Each egg has sufficient yolk to sustain only a brief period of embryonic development. The embryo then hatches into a larval form, which feeds and grows, accumulating food resources, before metamorphosing into the adult form.

Fig. 15.6 Salmon are protandrous. After reaching a threshold size, they change from male to female and begin producing thousands of yolky eggs

Fig. 15.7 Blue wrasse are protogynous. This large dominant male is blue in colour, signalling his readiness to fertilise the eggs spawned by all the females in his harem

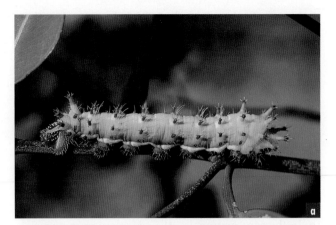

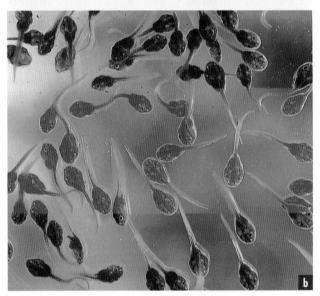

Fig. 15.8 Indirect development in animals involves larval stages, such as **(a)** an emperor gum caterpillar, *Antheraea eucalypti*, and **(b)** tadpoles

Indirect development is not restricted to aquatic animals with external fertilisation. For example, butterfly eggs hatch into caterpillars, which feed voraciously until sufficient reserves have been stored to support change into the adult form (Fig. 15.8a). The caterpillar secretes a silky cocoon about itself, within which it undergoes metamorphosis into a butterfly. In fact, all insect eggs develop into various larval stages, instars, which feed and grow before metamorphosis. Similarly, in frogs, eggs hatch into aquatic, herbivorous tadpoles (Fig. 15.8b). After a period of feeding and growth, a tadpole undergoes metamorphosis into a frog.

Direct development

Animals showing direct development are typically those where females produce smaller numbers of large yolky eggs containing sufficient food reserves for more elaborate development. Turtles and crocodiles lay clutches of eggs in sandbanks beside the sea or a river (Fig. 15.9). When an egg hatches, the offspring is essentially a small adult, able to crawl up from the buried nest to the surface to find its way to the water and to feed independently. In birds, hatchlings also have adult form but most are more dependent than turtles and

Fig. 15.9 Direct development in a yolky egg gives rise to tiny crocodiles, *Crocodylus porosus*

crocodiles. Most young birds depend on their parents for food and shelter, sometimes for an extended period.

Mammals also show direct development but their eggs contain relatively little yolk. Extended development is sustained by the transfer of nutrients directly from the mother to the embryo, either via the placenta or in milk during suckling.

> Some animals release large numbers of eggs with relatively little yolk, which typically undergo indirect development. Other animals have fewer eggs with large nutrient reserves or direct transfer of nutrients, which allow direct development.

Development in mammals

In eutherian mammals, the developing embryo is retained within the mother's body in a specialised region of the oviduct, the uterus, for the major part of its development, **gestation** (Box 15.1). Here the embryo receives the protection of the mother's body and an excellent supply of nutrients from both uterine secretions and maternal blood. A specialised organ, the **placenta**, facilitates this nutrient exchange.

At birth, young eutherians may be essentially independent. For example, guinea pigs are highly advanced at birth, with fully developed teeth and able to survive on solid foods (Fig. 15.10a). They can be weaned immediately after birth. Large herbivores, such as antelope or horses, have young that can stand and run soon after birth, though they are still dependent on milk for at least a while. At the other extreme, many shrews give birth to offspring at a very early stage of development, and even mice and rat pups are virtually helpless at birth (Fig. 15.10b). Most eutherians lie between these two extremes, with newborn offspring (neonates) dependent on parents for food and shelter for a short period relative to their overall life expectancy.

The other two groups of mammals, monotremes (platypus and echidna) and marsupials (Box 15.2), show

BOX 15.1 Human reproduction

As in other mammals, human reproductive systems are adapted for internal fertilisation and internal embryonic development. Sperm are produced by meiosis in seminiferous tubules in the testis (see Fig. a). Several hundred million sperm are produced daily and these pass into the epididymis, a convoluted tube lying beside the testis, where they complete their maturation and are stored. In humans, as in most other mammals, the testes must be kept several degrees below normal body temperature for normal sperm formation, so they are held outside the abdominal

cavity in the scrotum. A specialised set of muscles holds the testes closer to or further away from the body to regulate temperature. At ejaculation, muscular contractions cause sperm to pass through the vas deferens; then various secondary sex glands, including the prostate and Cowper's glands, add their secretions, producing a buffered, nutrient-rich seminal fluid. The seminal fluid then passes into the urethra and through the penis.

Seminal fluid passes into the female tract during copulation. The female reproductive system comprises paired ovaries, where, normally, a single egg is released each month. The ovulated egg is collected by the specialised ends of the oviduct, the Fallopian tube (see Fig. b). It is usually in the Fallopian tube that an egg meets and is fertilised by sperm. The zygote then passes into the uterus, a region of the oviduct specialised for nurturing embryo development. The young embryo implants in the wall of the uterus and forms the placenta, an organ specialised for nutrient exchange. The lower portion of the reproductive tract, the vagina, serves as a conduit for the sperm and as a birth canal for the developed young.

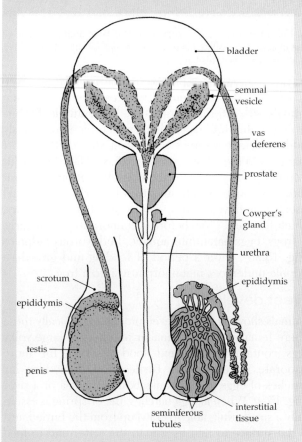

(a) The human male reproductive system

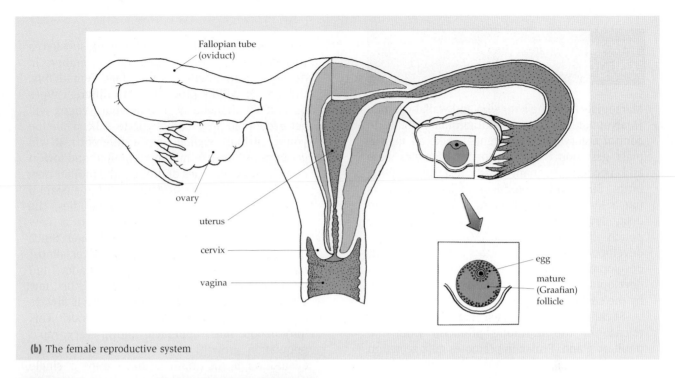

(b) The female reproductive system

Fig. 15.10 Different eutherian mammals are born at very different stages of development, for example, **(a)** independent newborn guinea pigs and **(b)** immature newborn mice

quite different patterns of development. Monotremes are oviparous (egg-laying) and marsupials are viviparous (bearing live young) but, in both cases,

embryos emerge at a very early stage of development. This short gestation followed by a prolonged lactation contrasts with most eutherians, which have longer internal development and shorter lactation.

Monotremes produce large, relatively yolky eggs. These are fertilised internally and undergo part of their embryonic development in the female reproductive tract before they are laid (Fig. 15.11). After about 10 days' incubation in the nest, during which time the embryo is nourished by yolk, the eggs hatch. The small, nearly helpless offspring emerge and remain dependent on milk from the mother for a substantial part of development.

In marsupials, very early development involves nutrient supply via a placenta but, like monotremes, the offspring are poorly developed at birth. Newborn honey possums weigh less than 10 mg and the offspring of the red kangaroo weighs only half a gram. These offspring undergo a major part of their development within the mother's pouch, totally dependent on milk for their nutrients.

Eutherian and marsupial mammals have yolk-poor eggs but direct development by means of nutrient transfer from mother to embryo during development. Monotremes are oviparous, marsupials and eutherians are viviparous.

The costs of sexual reproduction

Engaging in sexual reproduction involves changes in an organism's usual way of life; these changes are not without cost. The investment of parents in reproduction (**reproductive effort**) must be affordable in terms

BOX 15.2 Reproduction in the red kangaroo

Marsupials from the kangaroo and wallaby family (Macropodidae) have an extraordinary ability to control embryonic development. Like all marsupials, gestation is rather short and females mate and become pregnant again just after giving birth. Since milk production (lactation) lasts much longer than gestation, it is necessary to delay development of the new embryo until the pouch becomes vacant. Macropodids have developed the ability to completely halt development at a very early stage. This delay in development, embryonic diapause, is controlled by the suckling of the young in the pouch.

Suckling a young (see Fig. a) stimulates release of the hormone prolactin from the pituitary gland. Prolactin promotes milk secretion by the mammary gland and in kangaroos it also acts on the ovary to suppress production of the hormone progesterone. At weaning, the suckling stimulus wanes and prolactin levels decline, allowing increased progesterone production by the ovary. This increased progesterone stimulates the uterus to become more secretory, and the change in the uterine environment leads to reactivation of the diapausing embryo. Development of the embryo then leads to birth of a new young about a month later. Mating will follow soon after, producing a new diapausing embryo.

In red kangaroos (see Fig. b), reactivation of the diapausing embryo occurs before the young is fully weaned, so a female may have an older young out of the pouch but still being suckled, a newborn young in the pouch, and an embryo in diapause in the uterus. At this stage, remarkably, the mother will be producing two different types of milk simultaneously. The newborn young will get a low lipid, high carbohydrate milk, while the young 'at foot' feeding from a different teat will get large volumes of milk high in lipid and low in carbohydrate. In times of drought the mother may be unable to produce sufficient milk to sustain a growing pouch young. If the young dies, the diapausing embryo will reactivate so a new young will enter the pouch a month later. Since the newborn young is so small it involves almost no maternal investment during pregnancy, and needs only a tiny quantity of milk for the first few weeks in the pouch. This ensures there is always a young ready when the drought ends. This 'production line' approach allows very rapid population growth when conditions are good, even though the mothers have only a single young at a time.

(a) Newborn tammar wallaby, suckling in pouch

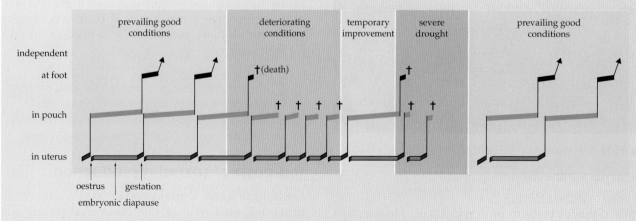

(b) Diagram illustrating patterns of reproduction in the red kangaroo under different environmental conditions

Fig. 15.11 Monotremes are the only oviparous mammals. This is the egg of an echidna, which is usually laid in pairs

Fig. 15.12 This male green and gold bellfrog, *Litoria aurea*, expends energy in calling to attract a mate. While calling, he risks attracting the attention of potential predators

of their own survival and future **reproductive success** (the cost) and must increase the chance of survival of their offspring (the benefit).

Reproductive costs vary greatly between species and can include devoting energy to the production of gametes, primary and secondary sexual organs, phero-mones, synchronising signals, and providing parental care to the offspring. With increasing parental invest-ment, the chances of individual gametes being fertilised and developing successfully might increase, but the increased cost reduces the number of offspring that can be supported. In regard to female costs, increased egg yolk, as seen in birds and reptiles, saps maternal reserves, but allows faster or further development of the embryo before hatching. The physiological costs of pregnancy in eutherians, including intrauterine development and suckling, are great, and place the mother at some risk. However, these mechanisms greatly increase the chance that an individual offspring will survive.

There are many other potential costs. Many animals reduce competition and predation by living a solitary lifestyle and using camouflage; but sexual reproduction may require individuals to find and recognise a mate. Thus, solitary animals may temporarily become gregar-ious and may adopt special breeding colours that transiently increase the chance of finding a mate at the expense of increased risk of predation. Frogs begin calling, which attracts both potential mates and predators (Fig. 15.12); the colourful breeding plumage of some birds attracts the attention of potential mates but conflicts with their usual protective camouflage.

Aggressive tendencies in predatory animals may need to be inhibited. It is pointless for a male and female to come together for reproduction if predatory instincts pre-empt the reproductive process and one individual kills the other.

The costs of reproduction are often very high. In the red deer, with the approach of the breeding season, or rut, adult males grow antlers that are sometimes massive. They are used in dominance-related fighting

and are shed after the rut. Males expend considerable energy and risk severe injury in fighting to secure a harem of females. Through the rut, dominant males are so involved with guarding the harem, warding off intruders and mating with the females, that they have little or no time to feed. As a consequence, their body condition falls dramatically. The toll is such that the chance that a dominant buck will survive the winter is reduced.

Male kangaroos also fight strenuously to secure their mating rights, using teeth, forefeet and hindfeet, and sometimes leaning back on their tail to deliver a two-footed attack (Fig. 15.13). Fights tend to be ritualised, with relatively few, highly damaging blows, and the skin is heavily thickened in the regions most prone to attack.

Males of the marsupial mouse *Antechinus* (Fig. 15.14) pay an extreme price for sexual repro-duction. Both males and females are sexually mature in their first year. Breeding occurs in mid winter with a mating period of two to three weeks. By the time the

Fig. 15.13 Male grey kangaroos fight to gain mating access to females. This behaviour involves considerable cost to the dominant male

Fig. 15.14 The marsupial mouse, *Antechinus*. The mottled appearance of this male indicates loss of condition as a result of stress towards the end of his first and only mating season

Fig. 15.15 Young froglets emerging from the mouth of a female gastric brooding frog after developing directly in their mother's stomach

females produce their litters, all the males are either dead or dying. The death of males occurs as a result of the following sequence of events. Levels of the male hormone testosterone increase dramatically in males as the photoperiod shortens with the onset of winter. High levels of testosterone cause increased territorial aggressiveness and fighting between males. The stress of fighting enlarges the adrenal glands, which produce large amounts of glucocorticoid stress hormones (Chapter 25). These high levels of stress hormones suppress the immune system, so wounds do not heal and the males quickly die from infection. If, however, males are kept isolated in captivity, without the stress of competition and mating, this sequence of events does not occur and they can live for several years.

Parental care

After fertilisation, further costs may be encountered. With eggs that develop externally, behaviour that protects offspring reduces a parent's ability to forage and increases the risk of its predation. Internal development drains the parent's body reserves and reduces mobility, so the risk of predation increases.

An extreme example of parental care is seen in the gastric brooding frog, *Rheobatrachus silus* (Fig. 15.15), which was discovered in forests north of Brisbane in 1974 and which may already have become extinct. It shows remarkable reproductive behaviour: eggs are shed and fertilised during **amplexus**, as in other amphibians but, after this, rather than leaving the eggs to develop alone and unprotected, the female swallows them. In the stomach, digestive secretions cease and the eggs settle into the stomach wall, where they are protected and apparently absorb nutrients from the parent. This gastric brooding appears to last six to seven weeks, during which time the female does not eat. When the young frogs are ready, they are regurgitated

through the mouth. Thus, these animals have external fertilisation but internal development.

In birds, fertilisation is internal but the fertilised egg undergoes the majority of its development externally within the egg. Parental care is often needed continuously. Most species brood their eggs, with parents taking turns so that each can go and feed during the incubation period. Other behaviours may also be used. The megapodid birds, such as Australia's brush turkey, build a mound from twigs, soil and leaf litter, into which they place their eggs (Fig. 15.16). Heat from decomposition of the leaf litter keeps the eggs warm but the male parent frequently tends the nest, adding or removing material to control the temperature of incubation.

> Sexual reproduction imposes significant costs in terms of increased energy use, decreased food intake, increased risks to survival, and in developing and maintaining specialised anatomical and physiological adaptations.

Fig. 15.16 A brush turkey maintains the temperature of its nest by adding or removing litter

Gametogenesis and fertilisation

Most animals reproduce sexually, although some species can reproduce either sexually or asexually, depending on environmental conditions. In a suitable environment, rotifers multiply asexually until crowding occurs or conditions deteriorate (Fig. 15.17). These circumstances stimulate a switch to sexual reproduction, generating a variety of starter clones, which then multiply asexually and compete for available resources. In corals (Box 15.3), asexual reproduction ensures growth locally, while sexual reproduction, which includes a free-swimming larval stage, provides a dispersal mechanism as well as a means of increasing the genetic diversity of offspring.

The primary sex organs in animals are **gonads**, which produce the gametes The male gonad is the **testis**. The female gonad is the **ovary**.

Animals usually have a number of secondary sex organs, including glands producing secretions that protect or nourish the gametes, and ducts that carry gametes from the gonads. These ducts may be highly specialised and serve storage, protective or nutritive functions. Primary and secondary sex organs together constitute an animal's reproductive system. A secondary but important role of gonads is the production of hormones that regulate development of gametes and secondary sex organs, produce sexual differences in appearance and behaviour, and act as pheromones or sex-attractants (Chapter 25).

> In animals, eggs and sperm are produced in ovaries and testes respectively. These gonads also produce hormones that regulate secondary sexual characteristics and behaviour.

The formation of gametes: gametogenesis

A long sequence of events, called **gametogenesis**, is involved in the production of the mature gametes. **Spermatogenesis** refers to the formation of male gametes and **oogenesis** to the formation of female gametes.

Migration of germ cells

From the earliest developmental stages of most animals, two broad types of cells are present: **somatic cells**, which will form the body of the animal; and **germ cells**, which alone have the ability to develop into gametes. In some cases it is possible to recognise germ cells very early in embryonic development as they separate from somatic cells (Fig 15.18). At this time they are called *primordial germ cells*. Usually, only relatively small numbers of these primordial germ cells are formed initially and, strangely, they first appear far removed from their final location, the gonads (ovary or testis). They make their way to that distant site reliably, moving through a complex terrain, in some cases by a process of active migration. How germ cells find their way is poorly understood.

Mitotic divisions

After the germ cells have migrated to the developing gonads, their numbers increase greatly by repeated rounds of mitotic cell divisions to generate the large number of cells that are needed for sexual reproduction. The timing of proliferation varies between animals and

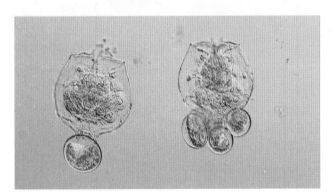

Fig. 15.17 Two female rotifers, *Brachionus angularis* Gosse, from a Murray River billabong. In this genus, most females reproduce parthenogenetically, as seen in the female on the left, which is carrying a single, mitotically produced (female, 2n) egg. Increasing population density stimulates the production of sexually producing females, which produce several smaller eggs by meiosis. One of these females carrying five smaller (male, n) eggs is shown on the right

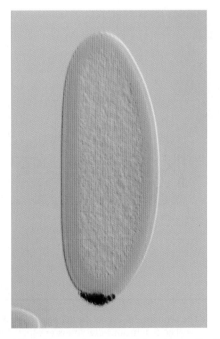

Fig. 15.18 A blastoderm stage embryo of the vinegar fly, *Drosophila melanogaster*, showing the germ cells (stained brown) in a discrete cluster at the posterior end of the embryo, separated from the somatic cells

between the sexes of any one species. For example, in mammals, mitotic divisions of germ cells begin during embryogenesis. In the female, they are completed before birth (by 20 weeks of embryonic development in humans): all of the egg cells that a female will ever possess are present in her ovaries before she is born. In

male mammals, a second round of mitotic divisions commences later in life after sexual maturation and continues for the remainder of the animal's reproductive life. During the phase of mitotic divisions, the female germ cells are **oogonia** (sing. oogonium) and male germ cells are **spermatogonia** (sing. spermatogonium) (Fig. 15.19).

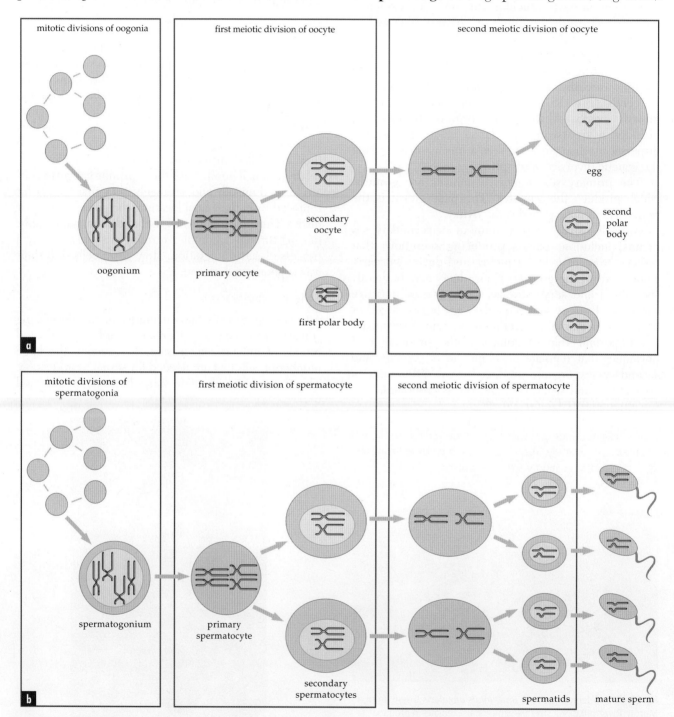

Fig. 15.19 (a) The sequence of events during oogenesis. A series of mitotic divisions generates a pool of diploid oogonia. These cells then undergo meiosis. The first meiotic division is unequal, generating one secondary oocyte and a small polar body, both of which are haploid. Each of these cells then undergoes the second meiotic division. The secondary oocyte divides unequally to produce another polar body and a haploid egg cell or ovum, and the first polar body divides equally to produce two additional polar bodies. **(b)** The sequence of events during spermatogenesis. Repeated mitotic divisions of the spermatogonia are followed by the two divisions of meiosis. The primary spermatocyte divides equally to produce four haploid spermatids. Differentiation of the spermatid into a mature sperm cell follows

BOX 15.3 Synchronised spawning

Reproduction of many marine organisms on coral reefs is triggered by environmental cues. Annual, lunar, tidal and daily rhythms modulated by temperature appear to regulate their biological clocks. Corals produce one of the biggest and most spectacular examples of this in their synchronised, annual mass spawning on the Great Barrier Reef, Queensland (see figures). In colonies of the large plate coral, *Acropora tenuis*, spawning occurs during spring and early summer, one night or so after a full moon. Spawning reaches its peak on the fourth, fifth and sixth nights during a period of neap tides and rising sea temperatures. Within one day, fertilised eggs develop, forming into swimming larvae, planulae. After a few days at the surface, the larvae descend to find a suitable site to form a new colony. Although millions of larvae are produced, almost all are eaten by predators. Of the few remaining, only a tiny fraction find a suitable substrate, settle and develop into adulthood.

The precise timing after a full moon ensures nights when the sea is characteristically calm. With neap tides, there is little tidal movement, so that less water will move away from the reef. Thus, there is less chance that the developing larvae will be carried away from suitable habitats.

Since multiple species of corals all spawn together, there must be precise mechanisms for species recognition between sperm and egg. Self-matings between male and female gametes from the same colony are uncommon and matings between male and female gametes from different species do not occur.

Flowering of seagrasses is also related to monthly and annual lunar cycles. The Australian sea nymph, *Amphibolis antarctica*, which is distributed on the southern and western seaboards of Australia, is dioecious and flowering occurs only once a year, in early summer. *Thalassia hemprichii* grows on the seaward side of wave-cut platforms of tropical reefs around the Indian Ocean and Northern Australia. It flowers regularly once each month in summer. Flowers emerge during the period of neap tides and pollen

Mass spawning of corals on the Great Barrier Reef. **(a)** Release of bundles, containing sperm and egg from a brain coral, *Platygyra* sp. **(b)** Detail of release of an egg bundle from the mouth of a coral polyp, *Goniastrea palauensis*. The bundle is 3–4 mm across, containing hundreds of eggs as it is launched into the sea. **(c)** On reaching the sea surface, the egg and sperm bundles of the large plate coral *Acropora tenuis* float and begin to break up, ready for fertilisation. **(d)** Two to three days later, the larva of *Acropora* is cylindrical, 2–3 mm in length. **(e)** After about three months, the larva has attached to the substrate and laid down its skeleton and grown into a tiny coral polyp that is capable of budding new polyps

is shed coincident with the following period of spring tides.

In certain tropical reef fishes, spawning can occur as frequently as once a day. Reproductive behaviour of the male moon wrasse, *Thalassoma lunare*, is centred around female spawning time, which occurs at high tide each day during summer. As the high tide starts to fall, surface currents sweep all the fertilised eggs away from the reef and its predators.

> During gametogenesis, primordial germ cells migrate to the gonads, where they undergo repeated mitotic divisions.

Meiotic division

The final divisions of oogenesis and spermatogenesis are *reduction divisions*. While undergoing meiosis, germ cells are known as **spermatocytes** and **oocytes**.

The timing of the meiotic division, like that of the mitotic divisions, differs between species and sexes. In female mammals, meiosis begins during embryogenesis but is commonly arrested at prophase I (Chapter 8). The primary oocytes remain in a dormant state in the ovary until after sexual maturation, when, in response to hormonal signals, they enter an active phase of growth and maturation (see below). Release of a mature egg from an ovary, **ovulation**, triggers the completion of meiosis I. In some species, the oocyte is arrested again in the second meiotic division and this arrest is released by the act of fertilisation.

The meiotic divisions of the oocytes are highly asymmetric (Fig. 15.19a). The first produces a large cell, the secondary oocyte, which contains the vast majority of the cytoplasm and a tiny cell, the first polar body. The secondary oocyte again divides unequally at the second meiotic division, generating a large mature egg cell and a diminutive second polar body. The polar bodies eventually degenerate and only one mature egg cell is produced from each primary oocyte.

In contrast to the stop–start style of meiotic division in female germ cells, male germ cells proceed through meiotic division smoothly and relatively quickly. The sequence of events can be seen in a cross-section of the seminiferous tubules of the mammalian testis (Fig. 15.20). Initially, spermatogonia lie in the outer regions of the tubule and, as they undergo their mitotic divisions, they move inwards towards its lumen. Further inward are meiotically dividing spermatocytes, and the innermost zone of the tubule is occupied by germ cells in the process of maturation. Unlike female germ cells, meiotic division of spermatocytes is symmetrical and four potential sperm cells are

produced from each (Fig. 15.19b). The whole sequence of events takes about eight weeks in humans.

Maturation of gametes

Both male and female germ cells differentiate into cells that are highly specialised to perform the functions of reproduction. Two of these functions, motility in order to meet another gamete and provision of stored materials to support the development of the embryo, are not easily combined in the one cell type.

As a spermatocyte matures, most of its cytoplasm is lost and it develops a long flagellum with a power supply in the form of a jacket of mitochondria (Fig. 15.21). The DNA condenses tightly into a compact nucleus located at the head of the sperm and a specialised secretory vesicle, the **acrosome**, forms between the nucleus and the cell membrane. The acrosome, which contains hydrolytic enzymes, plays an important role during fertilisation.

The egg contains the food required by the developing embryo and also other 'signal molecules' that are critically important in directing the early stages of development. As it matures, an egg grows enormously as nutrients such as yolk proteins and lipids, as well as organelles such as ribosomes, mitochondria and microtubules, are stockpiled in its cytoplasm for later use. Mature egg cells are generally at least 1000 times larger than a typical somatic cell. In some cases, these materials are produced within the egg cell itself, necessitating a substantial upgrading of the synthetic machinery of the oocyte. For example, there is a dramatic increase in the number of copies of ribosomal RNA genes in the oocytes of many amphibia, as evidenced by the huge number of nucleoli (several thousand) present in these cells. In other cases, materials produced elsewhere in the female's body are

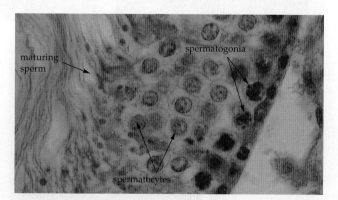

Fig. 15.20 Cross-section through a portion of the seminiferous tubule of the mouse, showing different stages in spermatogenesis. The dark-staining nuclei in the outermost region of the tubule (right-hand side) are spermatogonia undergoing mitosis. The lighter nuclei further to the left are spermatocytes in various stages of meiosis. The small, condensed nuclei of spermatids and the long tails of mature sperm can be seen adjacent to the lumen of the tubule (left-hand side)

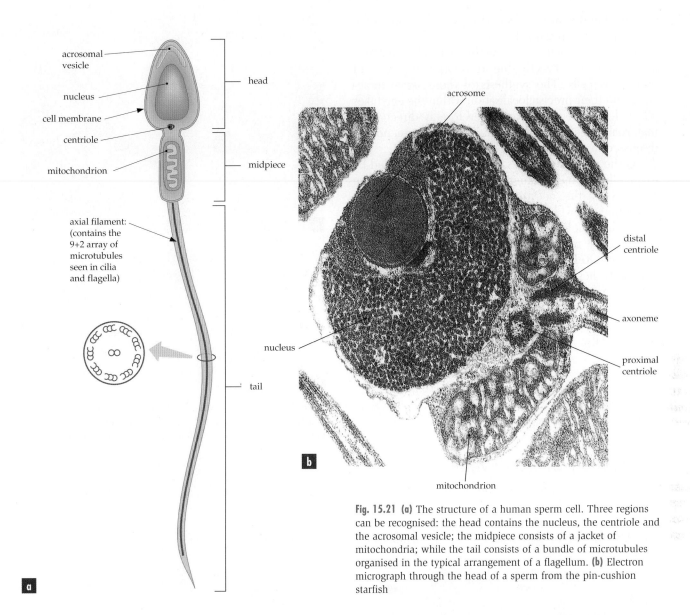

Fig. 15.21 (a) The structure of a human sperm cell. Three regions can be recognised: the head contains the nucleus, the centriole and the acrosomal vesicle; the midpiece consists of a jacket of mitochondria; while the tail consists of a bundle of microtubules organised in the typical arrangement of a flagellum. **(b)** Electron micrograph through the head of a sperm from the pin-cushion starfish

transported into the eggs. For example, virtually all of the yolk in the bird egg is first produced in the liver of the female and is then delivered to the oocytes in her ovaries via the circulatory system. In mammals, the oocyte is surrounded by layers of cells, called **follicle cells**, and one of the functions of these cells is to regulate the movement of substances such as yolk from the bloodstream to the oocyte (Fig. 15.22).

In many invertebrates, including some insects, annelids and molluscs, the oocyte is connected to a number of **nurse cells** by cytoplasmic bridges: the nurse cells and the oocyte are, in fact, sister cells, produced by repeated mitotic divisions of an oogonium in which cytokinesis (Chapter 8) is incomplete (Fig. 15.23). After these mitotic divisions are over, one of the cells develops into an oocyte and commences meiosis, while the others become nurse cells. RNA

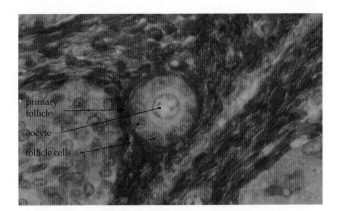

Fig. 15.22 A section through the ovary of a mouse. A primary follicle lies in the centre: it consists of a single oocyte, surrounded by a layer of smaller follicle cells. The follicle cells absorb materials from the circulatory system and transfer them to the oocyte

produced in the nurse cells is transported to the oocyte through the cytoplasmic bridges.

Mature egg cells also contain molecules that will later influence directly the developmental fate of embryonic cells. The synthesis of these signal molecules and their precise localisation within the egg takes place during oogenesis and therefore is directed by maternal genetic information. This phenomenon is discussed in more detail in Chapter 16.

As the egg cell matures, it becomes surrounded by extracellular membranes (envelopes), which are secreted by the oocyte and/or follicle cells. These membranes provide protection after the egg is released from the sheltered environment of the female reproductive tract. In most species, a thin yet tough layer, the **vitelline membrane** (or zona pellucida in mammals) lies closest to the plasma membrane of the egg. A variety of other envelopes may surround the vitelline membrane: the harsher the environment into which an egg is released, the more highly developed these are (Fig. 15.24).

After meiosis, germ cells mature into eggs and sperm. Oogenesis includes a phase of dramatic growth associated with storage of nutrients and signal molecules for use during embryogenesis.

Fertilisation: cellular events

Fusion of gametes, **fertilisation**, marks the beginning of the new organism. It involves two processes, *egg activation* and *nuclear fusion*. Firstly, the egg, which towards the end of oogenesis entered a metabolically

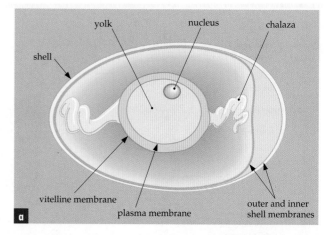
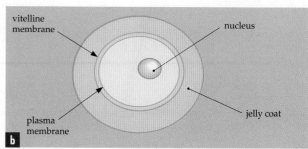

Fig. 15.24 Envelopes associated with the egg cell in **(a)** the chicken and **(b)** the frog. Chicken embryos, which develop in a terrestrial environment, have a more extensive and protective set of membranes than do frog embryos, which develop in water

dormant state, is activated by the fusion of egg and sperm plasma membranes to begin synthetic activity again. Secondly, fertilisation leads to the fusion of the sperm and egg nuclei (called pronuclei), thereby creating a diploid zygote.

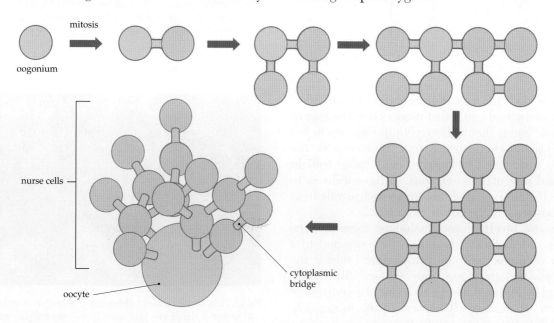

Fig. 15.23 In some insects, the oocyte and its associated 15 nurse cells are products of four rounds of mitotic divisions of a single oogonium. The cells remain connected to each other by cytoplasmic bridges, allowing materials to move from the nurse cells to the oocyte by direct cytoplasmic transfer

In the sea urchin, contact between sperm and the jelly coat that surrounds the egg triggers the acrosomal vesicle in the head of the sperm to release its contents (Fig. 15.25). These contents include hydrolytic enzymes, which soften the jelly coat and the vitelline membrane. A long process, formed at the head of the sperm cell as a result of the breakdown of the acrosome, bores through the jelly coat to the vitelline membrane (Fig. 15.25). Here, species-specific recognition occurs involving binding between molecules on the sperm plasma membrane and complementary molecules on the vitelline membrane. The acrosome process then breaks through the vitelline membrane and fusion of the egg and sperm cell membranes takes place (Fig. 15.25). The initial contact between egg and sperm cell membranes triggers a cascade of changes in the egg. These begin with a change in permeability of the egg plasma membrane to Na^+ and a rise in the intracellular Ca^{2+} concentration, and culminate in the activation of DNA and protein synthesis. The cell has commenced along the path of embryonic development.

Fusion of egg and sperm cell membranes allows the sperm pronucleus to pass into the egg cytoplasm and fuse with the egg pronucleus. Firstly, however, the sperm pronucleus must locate the egg pronucleus, a not inconsiderable task considering the size of the egg cell. In humans, a period of 20 hours elapses between entry of the sperm pronucleus into the egg and nuclear fusion. With pronuclear fusion, a diploid zygote is formed.

Curiously, in many species (with the notable exception of mammals), the contribution from the sperm, the set of *paternal* genes, plays little if any immediate role in embryonic development. This can be demonstrated in sea urchins and frogs by artificially activating an egg, such as by pricking its plasma membrane or changing the pH of the surrounding water. The activated egg undergoes development to the gastrula stage. However, development proceeds no further.

> Fertilisation involves fusion of the plasma membranes of sperm and egg, which activates the egg cell, and fusion of sperm and egg pronuclei, which restores the diploid number of chromosomes.

Methods of fertilisation

Complex physiological and behavioural mechanisms are often used to bring sperm and egg into close proximity to allow fertilisation to occur.

Many marine animals achieve fertilisation by simply shedding their gametes into the sea. Whether sperm and egg meet is a matter of chance, but two things occur to increase the likelihood of successful fertilisation. Firstly, the number of gametes released is often enormous. A single spawning oyster can release millions of eggs but the majority of the tiny larval offspring is unlikely to survive and develop into an adult. Secondly, within a population, gamete release tends to be synchronised. Environmental cues, such as water temperature, tides and day length, control the reproductive cycle, so that most individuals in a region reach reproductive condition at the same time. When one animal starts to spawn, pheromones released along with gametes will stimulate nearby individuals to spawn, which in turn release more pheromones, resulting in co-ordinated spawning over a wide area. During mass spawnings of coral on Australia's Great Barrier Reef, the number of gametes shed is so great that, for a time, the sea turns milky (Box 15.3). An

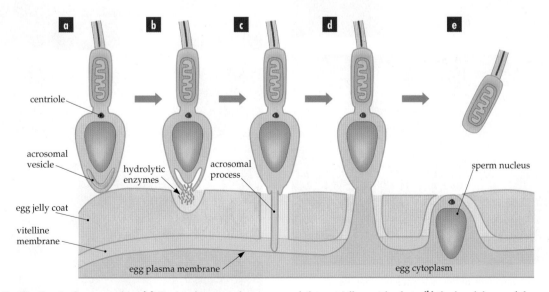

Fig. 15.25 Fertilisation in the sea urchin. **(a)** Contact between the sperm and the egg jelly coat leads to **(b)** the breakdown of the acrosomal vesicle. The hydrolytic enzymes thus released soften the jelly coat and **(c)** allow the acrosomal process to penetrate through the jelly coat and vitelline membrane. **(d)** Fusion of the plasma membranes of the sperm and egg activates the egg and **(e)** allows the sperm nucleus to move into the egg cytoplasm. The tail and midpiece of the sperm are left behind

additional benefit of mass spawning is that predators rapidly become sated by the sudden glut of planktonic food, and consequently a higher proportion of gametes and larvae escape predation.

Other species rely less on chance to get sperm and egg together. In most frogs, for example, males and females attract each other with special mating calls. Once together, the male climbs onto the back of the female, clasping the female behind the head. Males often develop special 'nuptial pads' on the side of their forelimbs to facilitate this grip (Fig. 15.26). The grip provides a behavioural cue for the female to lay her eggs and, as the eggs emerge, the male sheds sperm over them. This behaviour, amplexus, facilitates fertilisation by providing fresh sperm in close proximity to the eggs as they are laid.

These examples involve fertilisation of eggs in the external environment. External fertilisation requires a moist environment since unprotected sperm and eggs are prone to drying with consequent loss of viability. An alternative to this, internal fertilisation, is normally found in terrestrial animals but has also evolved in many aquatic species. Many variations are seen but usually sperm are placed directly into the female's reproductive ducts, often by a specialised organ, the **penis**. The penis is connected by a duct to the testes. There is usually a complex behavioural sequence, **copulation**, associated with penetration of the female reproductive duct to effect sperm transfer.

Copulation using a penis is not the only means of internal fertilisation. A wide variety of animals use packets of sperm, **spermatophores** (Fig. 15.27a). In some mites, the transfer of spermatophores is a chance affair. As the spermatophores mature, they are simply deposited by the male, and females pick them up whenever they happen to come across them. In many insects and birds, transfer follows a highly ritualised behavioural sequence (see Chapter 28).

More bizarre methods of fertilisation are seen in some leeches, which are dioecious. Copulating leeches become entangled and one injects a spermatophore directly through the body wall of the other, using the penis rather like a hypodermic needle. Sperm then migrate through the leech's body from the injection site to fertilise eggs in the ovary.

Fig. 15.26 (a) The forelimbs of the male common frog, *Rana temporaria*, showing dark nuptial pads used to clasp the female firmly during amplexus. **(b)** During amplexus, the male spotted grass frog, *Limnodynastes tasmaniensis* (on top), sheds sperm onto the freshly laid eggs

Fig. 15.27 (a) A spermatophore of the bush cricket. **(b)** Sedentary barnacles, *Balanus perforatus*, have an extra long penis capable of reaching considerable distances for fertilising neighbouring eggs

For species with a sedentary lifestyle, the ability to exchange gametes directly with others is limited. This limitation has been overcome in barnacles, for example, which have developed an extremely long penis and so can fertilise others some distance away (Fig. 15.27b).

> Fertilisation involves a variety of adaptations that ensure that mature gametes come into close proximity. External fertilisation usually involves the co-ordinated release of gametes into the environment; internal fertilisation follows transfer of sperm into the female reproductive tract.

Embryonic development

Multicellular organisms begin life as a single cell. Within a short period that may be days, weeks or, at most, months, this cell develops into a complex animal, such as an echidna, a mayfly or a wedge-tailed eagle. This transformation is particularly striking in animals because it is compressed into a relatively brief period. In most animals, body form is basically established during embryogenesis, the phase between fertilisation and hatching or birth.

Three main processes are involved in development:
1. **cell proliferation**—the production of millions of new cells by repeated mitotic divisions of the zygote
2. **cell differentiation**—specialisation of these cells into various types
3. **morphogenesis**—the ordered assembly of these cells to form complex organs.

Development is a progressive process, with complexity arising by degrees, building upon earlier events. For this reason, it is only possible to understand one event in the context of that which has gone before. Development proceeds under the instructions of genetic information. However, the connection between the genes and the end product of their activity—the parts of the body—involves interactions between molecules, organelles, individual cells, groups of cells, tissues and organs. To understand the control of development we need to examine these interactions.

> Animal development is a progressive process, involving cell proliferation, cell differentiation and morphogenesis.

Fertilisation sets in train the process of embryonic development, which appears very different in different animal species. Nonetheless, certain events or phases are found in all animal groups, namely **cleavage**, **gastrulation**, **organogenesis** and growth. The separation of embryogenesis into these phases is arbitrary as one stage leads imperceptibly into the next and they overlap in time in different parts of the embryo. However, the division is useful for outlining the developmental sequence.

Cleavage

Most egg cells are too large for efficient movement of molecules within the cell and between the cell and the environment. During cleavage, which is triggered by activation, the zygote undergoes a series of rapid cell divisions. Immediately after cytokinesis, cells enter the S phase and then move directly into the M phase; the G_1 and G_2 phases are eliminated (Chapter 8). As a result, the size of the embryonic cells, **blastomeres**, progressively decreases during cleavage, while the overall size of the embryo remains unchanged. The reduction in cell size is important as it reduces the cytoplasmic volume, restoring the surface area-to-volume ratio of blastomeres to a more normal somatic cell value.

In most cases, early cleavage divisions follow a rigid pattern and a regular time sequence. Each cleavage furrow is oriented in a predictable plane with respect to other divisions. As cleavage proceeds, this regularity in pattern and time is usually lost, although in some animals, divisions are rigidly stereotyped right through to hatching. In one such case, the nematode *Caenorhabditis elegans*, developmental biologists have been able to trace the origin and fate of every cell from zygote to adult worm to produce a complete cell lineage.

While all animal embryos undergo cleavage, the way in which it occurs varies greatly. One of the most important factors influencing the pattern of cleavage is the amount of yolk in the egg. Yolk displaces the spindle to an off-centre position and it also presents a physical barrier to the passage of the cleavage furrow, slowing down or stopping cytokinesis. In those animals with a small amount of yolk, such as the sea cucumber, yolk is evenly distributed within the egg and the mitotic spindle is centrally located. Early cleavage divisions, therefore, divide the egg into equal-sized blastomeres and subsequent divisions proceed at about the same rate in all parts of the embryo. As a result, the blastomeres in the **blastula**, as the embryo is known at the end of cleavage, are of approximately equal size (Fig. 15.28). The sea cucumber blastula shows a feature seen in many other animal groups—a fluid-filled central cavity, the **blastocoel**.

Amphibians have eggs with an intermediate amount of yolk, which is distributed in a graded fashion, ranging from a high concentration at one end of the egg, the **vegetal pole**, to a low concentration at the opposite end, the **animal pole** (Fig. 15.29). In amphibians, cytokinesis takes place throughout the egg but divisions proceed more slowly in the vegetal region than in the animal region due to the higher yolk concentration. As

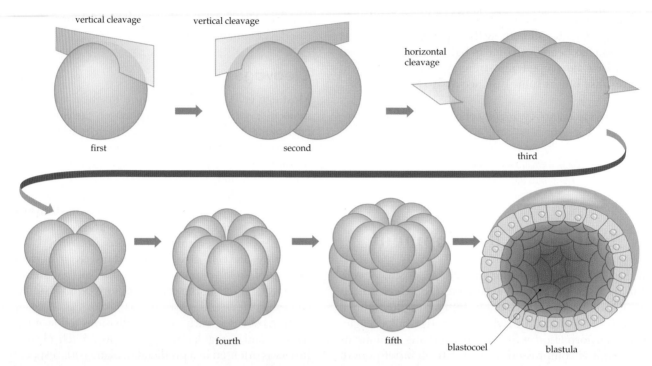

Fig. 15.28 The pattern of cleavage divisions in the sea cucumber embryo. The first and second divisions are vertical and at right angles to each other, the third is horizontal, the fourth vertical and the fifth horizontal. Precisely the same sequence of divisions takes place in every embryo. The end product of cleavage, the blastula, is a ball of equal-sized cells surrounding a central cavity, the blastocoel

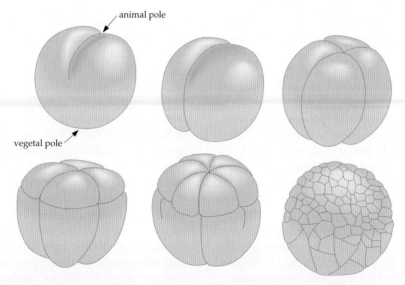

Fig. 15.29 Cleavage in the frog. The first two vertical cleavage divisions produce equal-sized blastomeres, but the third horizontal cleavage is displaced towards the animal pole because of the high yolk content in the vegetal hemisphere. As a result, the top tier of four blastomeres is smaller than the bottom tier. Later cleavage divisions in the yolk-filled vegetal region proceed more slowly than those in the animal hemisphere, resulting in a blastula in which the vegetal cells are larger than animal cells

a result, more divisions occur in the animal region and thus blastomeres are smaller in the animal region of the blastula than in the vegetal region.

Birds and insects have yolk-rich eggs. In birds, most of the egg cytoplasm and the spindle are displaced to a thin, yolk-free disc on top of a large mass of yolk (Fig. 15.30). Cytokinesis is confined to this superficial layer of cytoplasm and at the end of cleavage the

embryo takes the form of a cap of cells, a **blastodisc** or **blastoderm**, lying on top of an uncleaved yolk mass. In insects, cleavage produces a superficial layer of cells (blastoderm) enclosing a central yolk mass.

Development in eutherian mammals, for example, the mouse, shows a different pattern of cleavage. The eggs of most mammals are virtually yolk-free and almost all of the nutrients required by an embryo are provided

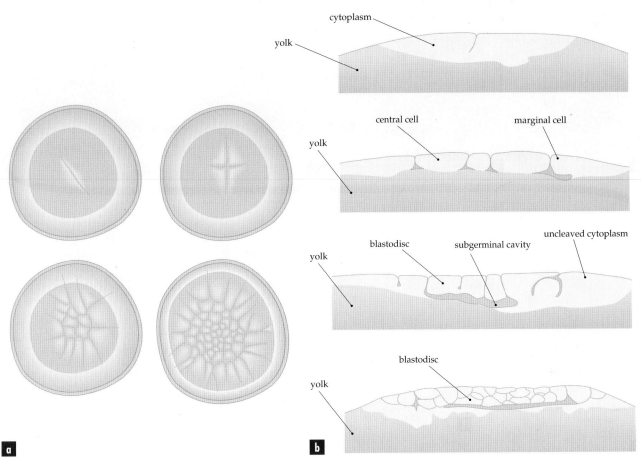

Fig. 15.30 Cleavage in the chicken showing **(a)** views from above the blastodisc and **(b)** views in longitudinal section. Cleavage furrows divide up the yolk-free patch of cytoplasm but do not penetrate the mass of yolk underneath. This results in a disc of cells, the blastodisc, separated from the uncleaved yolk mass beneath by the subgerminal cavity

by the placenta. Like in the sea cucumber, cleavage divisions are equal, but the orientation of early cleavage planes is quite different from that seen in other animals. The first cleavage is vertical, but at the second division, one blastomere divides vertically, while the other divides horizontally (Fig. 15.31). Furthermore, mammalian embryos cleave much more slowly than other animals and even the early divisions are often asynchronous. In mice, divisions are about 12–24 hours apart, whereas a sea urchin zygote develops into a tiny free-swimming larva in about 12 hours. The blastula of mammals is also somewhat different from that of other animals with yolk-poor eggs in that it possesses a cluster of cells, the inner cell mass, at one end of the blastula, which is called a **blastocyst** in mammals (Fig. 15.31). Only the cells of the inner cell mass give rise to embryonic tissues. The surrounding shell of trophoblast cells contributes to the formation of the chorion, which fuses with the uterine wall to form the placenta. Cleavage in marsupials proceeds in a different manner again (Box 15.4), a consequence of the increased amounts of yolk in the eggs of these animals.

As a result of cleavage, the contents of the egg cytoplasm are segregated into individual blastomeres. Because many components of the egg cytoplasm are not uniformly distributed throughout the egg, different blastomeres receive different types of molecules. This is visible in some species in which there are differently pigmented regions of the egg cytoplasm. As we shall see later, certain cytoplasmic materials influence the developmental fate of the blastomeres that receive them.

In many animals, the genes of the new individual do not participate in the events of cleavage. Cellular processes taking place during this stage, which are largely associated with mitosis, depend entirely upon maternal gene products stored in the egg during oogenesis. This idea is supported by two experimental observations. Firstly, fertilised sea urchin and frog eggs that have had their nucleus removed continue to cleave normally, showing that the presence of the nucleus is not necessary; and secondly, sea urchin embryos treated with the transcriptional inhibitor actinomycin-D cleave normally, showing that new synthesis of RNA is not needed.

Cleavage produces a multicellular embryo and restores a more usual nucleus-to-cytoplasm ratio in blastomeres. The pattern and timing of cleavage is affected by the amount of yolk in the egg.

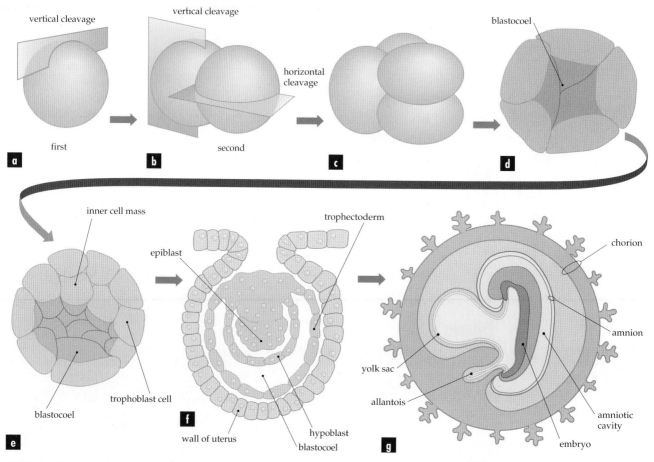

Fig. 15.31 Cleavage and tissue formation in the mouse. The first division **(a)** is vertical but the second **(b, c)** is vertical in one of the blastomeres and horizontal in the other. Furthermore, these divisions are not necessarily synchronous. Cleavage results in a single layer of trophoblast cells surrounding a blastocoel, with a cluster of cells, the inner cell mass, at one end. **(d, e)** The embryo subsequently sinks into the wall of the uterus and, as it does, the inner cell mass separates into various layers **(f)** Some of these give rise to the embryo proper, while others contribute, along with the trophectoderm cells, to the development of the allantois, amnion and chorion, a system of membranes that protects the embryo **(g)**

BOX 15.4 Cleavage in marsupials

Unlike eutherian mammals, marsupial eggs have three external envelopes: the zona pellucida, mucoid layer and shell (see figures). These envelopes enclose the marsupial embryo in its own microenvironment within which the embryo is constructed. The shell is porous to uterine secretions that nourish the embryo until it implants in the uterine wall at a relatively advanced stage of organogenesis.

Marsupial eggs are rich in yolk, which becomes localised to a particular region of the zygote cytoplasm during the activation process of fertilisation. This yolk-rich cytoplasm is separated from the remainder of the cytoplasm during the first two cleavage divisions (see figure). Marsupials show a variety of cleavage patterns related to the patterns of yolk localisation in the zygote and of yolk segregation during cleavage. In Australian marsupial mice and native cats, yolk is segregated into a single yolk mass at the first cleavage division.

The first three divisions are equal and vertical but the fourth is unequal and horizontal, producing two tiers of eight blastomeres. The tier of smaller blastomeres lies in the hemisphere containing the yolk mass. In marsupials, blastomeres do not clump together to form an inner cell mass as in many eutherian mammals. Instead, blastomeres line the zona pellucida to form the single-layered epithelium of the unilaminar blastocyst. A number of eutherian mammals such as primitive insectivores, pigs and goats also have a unilaminar blastocyst with no inner cell mass. Even though these mammals have no inner cell mass in a morphological sense, they do have a specific set of blastomeres, which go on to give rise to the tissues of the embryo proper. These cells, like the inner cell mass of mammals such as the mouse, are exposed to a different micro-environment to the rest of the embryonic cells, which may determine their particular developmental fate.

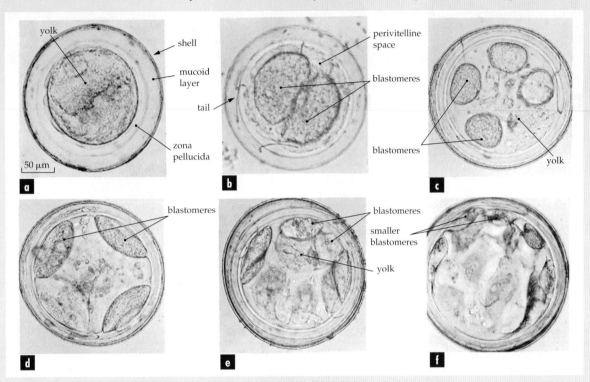

Stages in development of embryos of the brown *Antechinus*. **(a)** The zygote is surrounded by the zona pellucida, mucoid layer and shell. The yolk mass has become localised at one pole of the zygote. **(b)** Two-cell stage viewed from the yolk-poor pole. The two blastomeres have not yet moved apart and are concealing the yolk mass, which was pinched off from the blastomeres during the first division. The perivitelline space is opaque because it is filled with small yolk-like particles, which have been extruded by the blastomeres. The tail of the spermatozoan is trapped in the mucoid layer. **(c)** Oblique view of an arrested four-cell stage showing the four approximately equal-sized blastomeres lying apart from each other and slightly below the yolk mass. **(d)** Late four-cell stage when the four blastomeres are flattening on the zona pellucida before the third division. **(e)** An oblique view of the eight-cell stage at the end of the vertical third division. Six of the blastomeres have stretched longitudinally from the yolk mass towards the opposite pole. The remaining two blastomeres have completed the division and rounded up but have not yet flattened longitudinally. **(f)** Sixteen-cell stage viewed from the side. The yolk mass is out of focus at the top of the figure. The fourth division is an unequal one and divides the blastomeres into an upper tier of eight smaller blastomeres, which lie over the yolk mass, and a lower tier of eight larger blastomeres

Gastrulation

The blastula is structurally simple, consisting of a sphere or, in some cases, an irregular mass, of similar cells, often with a single internal cavity, the blastocoel. Contrast this form to the body plan of most mature animals, which have several tissue layers, many organs and two internal cavities. Also, there is a change in symmetry: a blastula is generally radially symmetrical, whereas most animals are bilaterally symmetrical.

During the next stage of embryogenesis, gastrulation, the embryo undergoes a wholescale rearrangement of its cells to form the **gastrula**, an embryonic stage that displays many of the basic features of the definitive body plan of the adult animal.

As with cleavage, gastrulation proceeds in very different ways in different animal groups. We shall compare gastrulation in an echinoderm, the sea urchin, and a vertebrate, the frog, *Xenopus laevis*.

Sea urchin gastrulation

The sea urchin blastula is spherical in shape and consists of a single layer of closely packed epithelial cells around a large central blastocoel. The first sign of gastrulation is a flattening of the vegetal pole region to form the **vegetal plate**, caused by a lengthening of epithelial cells in this region (Fig. 15.32). Then, a small group of cells, primary mesenchyme cells, detaches from the vegetal plate and moves to the interior of the embryo. These cells migrate along the inner surface of the blastula towards the animal pole and ultimately give rise to the skeleton of the embryo. Next there is a

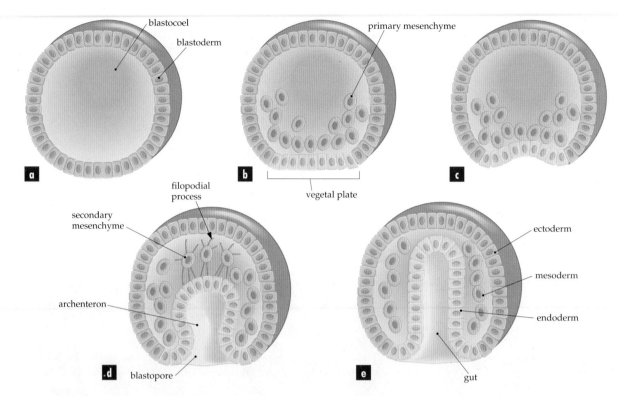

Fig. 15.32 Gastrulation in the sea urchin. **(a)** The blastula consists of a single-layered blastoderm surrounding a central blastocoel. **(b)** The vegetal plate flattens and, at the same time, primary mesenchyme cells detach from the vegetal plate and move into the blastocoel. **(c)** The vegetal plate begins to invaginate, creating a new cavity, the archenteron, which becomes the gut. **(d)** Secondary mesenchyme cells detach from the blastoderm and migrate into the blastocoel. Some send out long processes, filopodia, which adhere to the invaginating vegetal plate and to the inner wall of the blastoderm. **(e)** The invaginating vegetal plate contacts the blastoderm on the opposite side. The primary germ layers—ectoderm, endoderm and mesoderm—are now evident

change in shape of the epithelial cells of the vegetal plate from a columnar to pyramidal form. At the same time, the vegetal plate starts to buckle inwards (invaginates). A new cavity, the **archenteron**, forms as a result of this invagination.

As invagination continues, a new group of mesenchyme cells, secondary mesenchyme cells, moves from the vegetal plate region into the blastocoel. These cells send out long processes (filopodia), which contact the inner surface of the blastoderm in the animal hemisphere region. The vegetal plate continues to invaginate towards the animal pole, partly as a result of contraction of the filopodial processes and partly as a result of rearrangement of cells in the wall of the archenteron. Secondary mesenchyme eventually fills the blastocoel space.

The opening on the surface leading to the archenteron is the **blastopore** and forms the posterior end of the gut, the anus. The tip of the invaginating archenteron eventually touches the cells near the animal pole and the mouth of the sea urchin larva forms at this site. Gastrulation in the sea urchin larva therefore also marks a change from radial to bilateral symmetry with the appearance of the anteroposterior axis.

By the end of gastrulation three discrete tissue layers, the **germ layers**, can be identified:

1. **ectoderm**—the outer layer of cells, comprising that part of the blastoderm that did not migrate inwards or invaginate
2. **mesoderm**—the intermediate layer of cells, formed from mesenchyme cells and by an outpocketing of the archenteron (see below)
3. **endoderm**—the innermost layer, the wall of the archenteron.

The germ layers prefigure the layout of tissue layers in the body of the adult animal: ectoderm gives rise to the outer body covering and the nervous system, endoderm to the lining of the gut, and associated organs, and mesoderm to the tissues and organs that lie in between.

Another characteristic feature of the body plan of the sea urchin is the presence of a second body cavity, the **coelom** (a cavity that is entirely enclosed by mesoderm). In the sea urchin, the coelom develops from an outpocketing of the archenteron, which then expands, forming inner and outer mesoderm layers (Fig. 15.33). In essence, the body plan of a coelomate animal is a tube within a tube: an outer tube of ectoderm and mesoderm, forming the body wall, and an inner tube of mesoderm and endoderm, enclosing the gut cavity. The space between, the coelom, is completely lined by mesoderm.

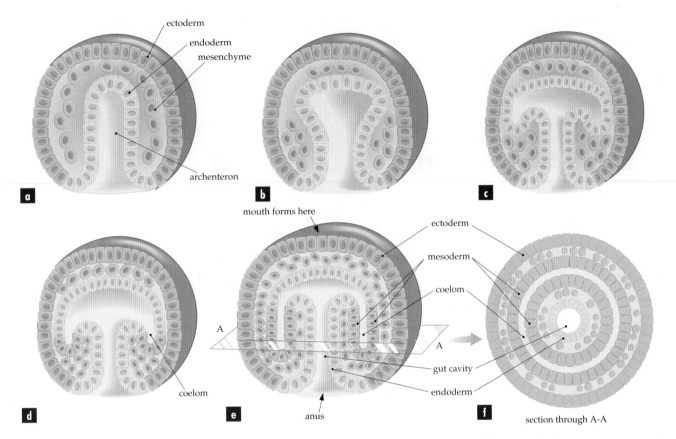

Fig. 15.33 (a)–(e) Coelom formation in the sea urchin. The coelom forms from an outpocket of the archenteron, which grows posteriorly and eventually separates from the archenteron. The inner and outer walls of this outpocket contribute to the mesoderm, which therefore lines the coelom on both sides. **(f)** A cross-section through the embryo at the level shown by plane A–A shows the definitive body plan of the sea urchin, with three germ layers and two body cavities

Xenopus gastrulation

Gastrulation in the frog *Xenopus* appears very different to that in the sea urchin. The small size of the frog blastocoel and the presence of a mass of large, yolk-filled cells in the vegetal hemisphere make gastrulation by simple invagination impossible. Instead, presumptive mesoderm (mesoderm-to-be) rolls into the interior of the embryo by a process known as **involution**, and presumptive endoderm is internalised by overgrowth of ectodermal cells, a process called **epiboly**.

In the frog embryo, the anteroposterior axis and the dorsoventral axis are evident at the beginning of gastrulation. The very first cells to undergo involution lie in the animal hemisphere just beneath the surface of the embryo along the future dorsal midline. These cells will give rise to the **notochord**, a longitudinal, rod-like structure characteristic of all chordate embryos (see Chapter 40). They also play a central role in the formation of the central nervous system and other dorsal structures (see below). They move deeper into the embryo by first migrating in a vegetal direction, then rolling under their neighbours and moving towards the animal pole (Fig. 15.34). A lip, the **dorsal lip of the blastopore**, forms at the site where the

presumptive notochord cells roll inside the embryo and this lip marks the boundary of the blastopore. Later, more lateral and then ventral cells roll inwards and spread out, eventually forming a continuous sheet of mesoderm under the outer layer of the embryo, and completing the formation of the blastopore. As in the sea urchin, the blastopore marks the site of the anus and the mouth breaks through at the opposite end of the archenteron; this pattern is characteristic of deuterostomes (echinoderms and chordates, Chapter 40).

While involution of presumptive notochord and mesoderm is taking place, yolk-rich cells in the vegetal region of the blastula are also being internalised. They come to lie in the interior of the embryo as a result of the lateral expansion of the superficial sheet of animal hemisphere cells. By the end of gastrulation, these animal hemisphere cells enclose the entire embryo, forming the ectoderm, and the internalised yolk-rich vegetal cells form the endoderm. The archenteron, lined by endodermal cells, forms as the presumptive endoderm moves inwards. A large mass of yolk-filled endodermal cells lies ventral to the archenteron and the mesoderm lies in between the endoderm and ectoderm. Soon a split appears within the mesoderm and enlarges to form the coelom.

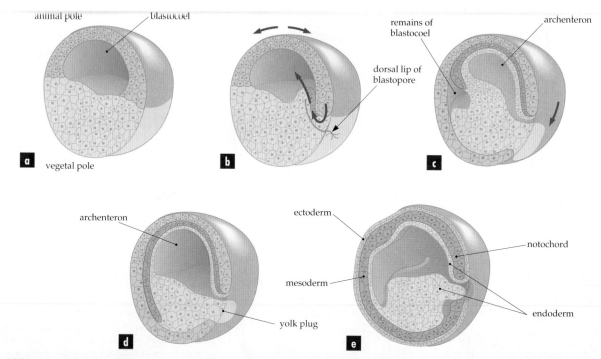

Fig. 15.34 Gastrulation in the frog *Xenopus*. (**a**) The blastula. (**b**) Involution of presumptive notochord cells around the dorsal lip of the blastopore. (**c**) Involution continues, the archenteron forms and the blastocoel is crowded out. Epiboly of the presumptive ectoderm takes place. (**d**) The endoderm has been internalised. (**e**) The completed gastrula showing the three germ layers

Gastrulation results in the formation of:
- the germ layers (ectoderm, mesoderm and endoderm)
- new body cavities (archenteron and coelom)
- in many cases, the basic axes of body symmetry.

Mechanisms of morphogenesis

The tissue and cell movements that generate the germ layers and internal cavities in the gastrulae of frogs and sea urchins are different: invagination and movement of single cells are important in the sea urchin whereas involution and epiboly play the dominant role in the frog. However, the end product of gastrulation in each case is the same: a three-layered, bilaterally symmetrical gastrula with an archenteron and coelom. Furthermore, the cellular processes that underlie the movements of gastrulation are similar in the two species, the most important being change in cell shape and change in cell adhesion. For example, the detachment of primary mesenchyme cells from the blastoderm of the sea urchin blastula is accompanied by a change in shape of these cells from a columnar to a tear-drop shape and by a disappearance of the cellular connections that link these cells with their neighbours (Fig. 15.35). Similarly, involution in the frog embryo is initiated by elongation and rearrangement of cells at the dorsal lip of the blastopore (Fig. 15.36), involving changes in cell shape and in cell adhesion. Similar changes underlie gastrulation movements in other animals and, as we

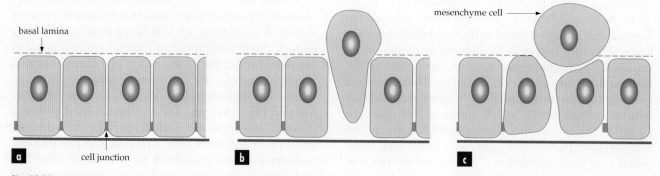

Fig. 15.35 Cellular changes underlying early gastrulation in the sea urchin. (**a**) Epithelial cells of the vegetal region of a late blastula. The cells are held together by intercellular junctions and an underlying basal lamina. (**b**) Cellular junctions and the basal lamina break down, freeing an epithelial cell, which changes from columnar to tear-drop shape. (**c**) The cell becomes a primary mesenchyme cell as it breaks away from the vegetal plate

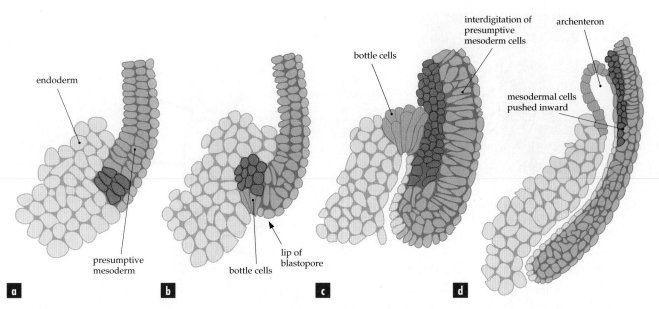

Fig. 15.36 Cellular changes underlying involution in *Xenopus*. **(a)** A cross-section through the blastopore region of the blastula. Presumptive mesodermal cells lie just beneath the surface. **(b)** Certain cells at the lip of the blastopore, 'bottle cells', take on a tear-drop shape. This shape change may initiate involution of the mesoderm. **(c)** The cells in the layers of presumptive mesoderm merge into a single layer, which becomes thinner and extends forward. **(d)** This extension may drive the continued involution of the mesoderm

shall see below, the complex morphogenetic changes occurring during organogenesis.

> Gastrulation is influenced by the structure of the blastula and varies widely between species; however, similar cellular processes are involved.

The cytoskeleton and cell shape

Of the three components of the cytoskeleton, microfilaments, microtubules and intermediate filaments, the most important in relation to cell shape changes during development are microfilaments. Microfilaments are composed of long actin rods associated with other proteins such as myosin. Myosin molecules can interact with actin filaments to produce forces for movement or shape change, as in the highly organised filaments of skeletal muscle (see Chapter 27). Bundles of actin filaments at the apical surface of epithelial cells, adhesion belts, are a feature of cell sheets that undergo folding. The invagination of the vegetal plate during gastrulation in the sea urchin may be generated by a wave of constriction of the band of microfilaments found at the apex of these epithelial cells (Fig. 15.37).

In many animal cells, a network of actin filaments found just beneath the plasma membrane of the cell has been implicated in cell shape change and particularly in the movement of whole cells. In migrating cells, bundles of microfilaments project into extensions of the plasma membrane called filopodia and lamellipodia (Fig. 15.38). These structures appear to act as feelers for exploring the environment and for moving

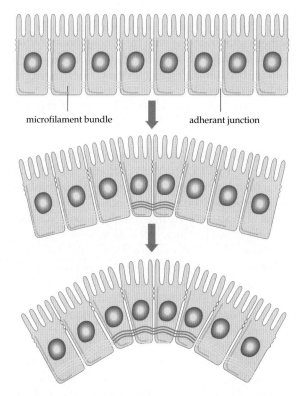

Fig. 15.37 When bundles of microfilaments in adhesion belts in certain epithelial cells contract, they cause a decrease in the diameter of the cells at that point. If these cells retain adhesive contacts with their neighbours (adherant junctions), the epithelial sheet as a whole will bend

the cell forward. We have already seen how filopodia are involved in the movements of mesenchyme cells in the sea urchin gastrula.

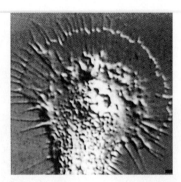

Fig. 15.38 The growing tip, or growth cone, of a nerve cell from the mollusc *Aplysia* in tissue culture; a thin veil, or lamellipodium, extends around the perimeter of the growth cone from which long, thin filopodia project

Cell–cell recognition and adhesion

In an embryo, migrating cells move along particular pathways and come to rest at precise locations. To do this they must be able to sense their environment and respond to it, either by adhering to surrounding cells or by sending out lamellipodia or filopodia in a directed manner. The importance of cell adhesion in morphogenesis is not restricted to cell migration. For example, cells are held together within an epithelial sheet partly by adhesive interactions with their neighbours. When a cell leaves a sheet, as do mesenchyme cells during sea urchin gastrulation, a change in this adhesive property must have occurred.

Biologists have now identified many of the molecules involved in such cell–cell recognition and adhesion. Cell adhesion molecules found to date are proteins projecting from the outer surface of cell membranes (Fig. 15.39). Many of them project through the cell membrane into the cytoplasm. The cytoplasmic region of the molecule often contains enzymatic sites that can activate cytoplasmic signalling molecules, such as protein kinases. Thus, cell–cell adhesion molecules can do more than stick cells together—they also enable cells to react to changes in their environment. Two broad classes, Ca^{2+}-dependent and Ca^{2+}-independent, have been identified. Many cell adhesion molecules belong to the immunoglobulin superfamily, a Ca^{2+}-independent type of cell adhesion molecule. The extracellular region of this family possesses a characteristic repeated motif, similar to that found in antibody molecules, which mediates binding to a molecule on another cell (Fig. 15.39). The cadherins are a family of Ca^{2+}-dependent cell adhesion molecules that includes the molecules N-cadherin, E-cadherin and P-cadherin. The cadherins are anchored to microfilaments by a group of proteins called catenins. This link provides a means by which signals outside the cell can alter the cytoskeleton and therefore the shape of a cell. The tissue distribution of cell–cell adhesion molecules within embryos and changes in their

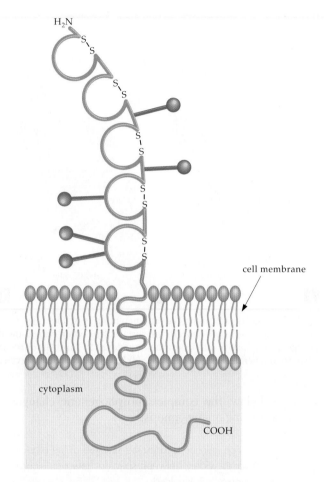

Fig. 15.39 A model of the immunoglobulin superfamily cell adhesion molecule, N-CAM. The extracellular region contains five loops, a structure similar to that seen in immunoglobulin molecules. Carbohydrate molecules, represented by balls on sticks, are associated with this extracellular region. The protein passes through the cell membrane and continues into a large cytoplasmic region

distribution are consistent with their involvement in morphogenesis (Box 15.5). Furthermore, knocking out the function of such molecules by genetic ablation has been shown to result in morphological changes in tissues and organs.

> Actin filaments often play an important role in changes of cell shape that accompany morphogenesis. Specific adhesive interactions, mediated by cell adhesion molecules, underlie many morphogenetic processes.

Organogenesis

While the basic body plan of an animal is laid down during gastrulation, the morphogenetic processes that mould tissues into complex three-dimensional organs, such as the heart, lungs and limbs, take place during the next phase of embryogenesis, organogenesis. Like gastrulation, this is a rapid and highly dynamic phase of development. It takes place in humans between three and six weeks after conception, during which time the

BOX 15.5 The neural crest

The development of the neural crest is a spectacular example of morphogenesis by cell rearrangement and cell migration, and it highlights the role of cell adhesion in these events.

Cell adhesion molecules and migration

Tests performed on cultured presumptive nerve tissue from chicken embryos show that cell–cell adhesion declines in the neural crest cell population at the time of onset of cell migration. Transmission electron microscopy shows that the cells lose specialised intercellular adherens junctions (Chapter 6) at this stage. Likewise, specific molecules belonging to known cell adhesion molecule families (p. 388) decrease on neural crest cells. One important group is the cadherin family of Ca^{2+}-dependent adhesion molecules, which are associated with adherens junctions. Using antibodies against various cadherins as markers, epidermal ectoderm cells can be shown to possess E-cadherin, whereas neural ectoderm cells possess N-cadherin. These enable cells with like cadherins to adhere. Neural crest cells initially show N-cadherin while they are part of the neural epithelium, but this is lost when they become mesenchyme (see Fig. a).

If the onset of migration of neural crest cells is due to a loss of cell–cell adhesion, artificially decreasing cell–cell adhesion before the onset of migration should provoke premature migration. Cadherins can be selectively inactivated by exploiting their extraordinary sensitivity to

protease enzymes in the absence of calcium. Applying this technique to cultured neural tissue from chicken embryos bears out the prediction that inactivation of cell adhesion molecules will stimulate the onset of cell migration. Premigratory neural crest cells commence migration immediately when briefly treated with protease.

Migration routes and signposts

Transplanting labelled neural crest mesenchyme cells to different regions in embryos shows that, although neural crest cells have highly developed migratory abilities, the pattern of their migration is controlled by the surrounding microenvironment. The early stage of neural crest cell migration occurs into cell-free spaces. Microenvironmental structures that could provide stable cell adhesion and a substrate for this migration include the extracellular matrix (ECM). Transmission electron microscopy has shown that neural crest cell extensions contact and adhere to the ECM, and antibody labelling has revealed that glycoproteins of the ECM include many adhesive molecules, such as fibronectin, laminin and collagens, which form fibrous meshworks. In addition, adhesion-inhibiting molecules, such as chondroitin sulfate proteoglycans, have also been found in the ECM. These are distributed broadly in and around the neural crest cell migration pathways, suggesting some role for the ECM in defining those pathways (see Fig. b).

ECM molecules can be purified and used as substrates for growth of neural crest cells in

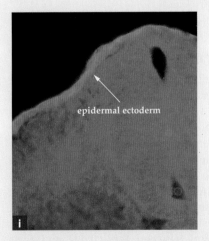
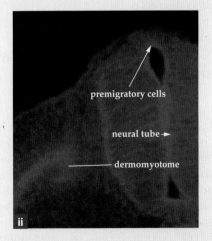
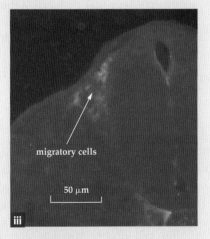

(a) Multiple immunolabelling of a single section of a chick embryo at the stage of early neural crest cell migration shows (i) E-cadherin (green) on the cells of the epidermal ectoderm and (ii) N-cadherin (red) on the cells of the neural tube and dermomyotome. (iii) In the neural crest population, migratory cells possess neither E- nor N-cadherin, while those cells still at the premigratory stage possess N-cadherin

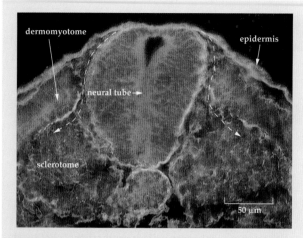

(b) Triple immunolabelling of a section of a chick embryo, with the major pathway of neural crest cell migration indicated by dotted lines. The preferred pathway between the dermomyotome and sclerotome shows a green colour, indicating a predominance of the adhesive ECM molecule fibronectin. The potential pathway between the dermomyotome and epidermis is not followed at this stage, and its blue colour indicates a predominance of an anti-adhesive ECM molecule. The red colour labels N-cadherin

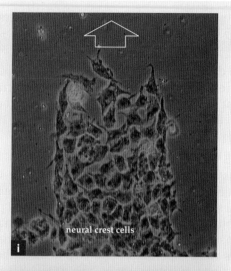

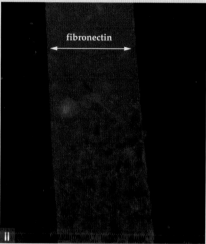

(c) (i) In tissue culture, neural crest cells, shown in phase contrast microscopy, migrating in the direction of the arrow, are confined to a narrow band. **(ii)** Immunofluorescent microscopy of the same field shows that cells are following a track of the adhesive ECM molecule, fibronectin

tissue culture. These experiments show, for example, that chicken neural crest cells adhere to fibronectin, laminin and collagens, and that migrating cells can accurately follow ECM molecule pathways (see Fig. c). Neural crest cells adhere to the ECM via cell-surface receptors, which link extracellular ECM molecules and the intracellular cytoskeleton.

Other cells are also potential substrates for the migration of neural crest cells. It is suspected that neural crest cells entering blocks of mesodermal cells known as somites may use the somite cells themselves as a migratory substrate since there is little ECM present. In addition, electron microscopy has revealed cell–cell adhesions between neural crest and somite cells, but the cell adhesion molecules involved in this case are not yet known.

Preventing cell adhesion inhibits migration

Cell adhesion can be disrupted using antibodies against the adhesive region of a ligand (e.g. an ECM molecule) that prevent the cell receptor from locking onto it. Alternatively, antibodies can be produced against the receptor molecule with similar effect. The receptors can also be blocked using small soluble peptides mimicking the ligand. Microinjection of such reagents known to inhibit cell–fibronectin adhesion in cell culture has been used successfully to curb neural crest cell migration in chicken embryos (see Fig. d).

This evidence confirms the importance of adhesive interactions in neural crest cell migration.

The end of migration

Neural crest cells eventually cease migrating. The best studied examples involve the formation of neural ganglia, where the cells form clumps within expanses of non-crest cells with which they were previously interspersed. Antibody labelling has shown the renewed appearance of cell adhesion molecules such as N-cadherin on cells in these ganglionic clusters and a reduction of fibronectin receptors. These observations are consistent with increased adhesion between neural crest cells and reduced adhesion between these cells and their surroundings, which could explain the reversion to cell aggregation and cessation of movement.

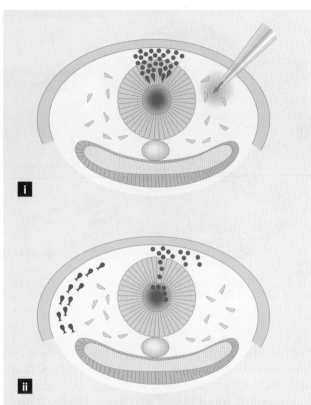

(d) Scheme showing that **(i)** an adhesion-perturbing antibody against the fibronectin receptor microinjected into the migration pathway of neural crest cells (black) on one side of the head of a chick embryo **(ii)** curtails later migration into this region. The uninjected side serves as a control. This provides strong evidence for the role of cell adhesions in cell migration

The overall conclusions drawn from these studies on neural crest systems are, firstly, that complex morphogenetic changes can be in part controlled by alterations in a relatively small number of cell adhesion molecules, ECM molecules and adhesion receptors. Secondly, these molecules act in concert and may have overlapping functions. Thirdly, since there is considerable similarity between neural crest cell behaviour and behaviour of other cells at different stages of development, regulation of cell adhesion is likely to be involved in morphogenesis in these other situations.

fetus is particularly susceptible to damage. Many agents known to cause birth defects, such as the drug thalidomide and the virus causing rubella (German measles), have their effects during this period.

Though the final shapes of organs are extraordinarily diverse, a limited number of morphogenetic mechanisms are involved in their formation. These mechanisms include thickening and folding of epithelial sheets; disaggregation of tissues into individual cells

and migration of those cells to new sites in the embryo; localised cell proliferation; and localised cell death. We shall examine the involvement of these morphogenetic mechanisms in the formation of two vertebrate organs, the nervous system and limbs.

Neurulation

The nervous system is the first major organ system to develop in a vertebrate embryo. **Neurulation**, the process by which the primordium of the central nervous system forms, so dominates the embryonic landscape that, during this stage, the embryo is called a **neurula**. The basic pattern of neurulation is similar in all vertebrates. In the chicken, the first stage of neurulation is seen as a thickening of the sheet of ectodermal cells in the midline, a region known as the **neural plate** (Fig. 15.40). The neural plate then buckles inwards along the midline forming a **neural groove**. Simultaneously, two **neural folds** rise upwards on either side of the neural groove and meet dorsally. The fusion of these folds forms a tube, the **neural tube**, which will become the central nervous system. These events occur first at the anterior end of the developing embryo and then continue towards the posterior end. The anterior part of the neural tube bulges out to form the brain, while the remainder becomes the spinal cord.

These movements appear to result from changes in the shape of the ectodermal cells of the neural plate. The thickening of the plate is caused by a change in epithelial cell shape from cuboidal to columnar by a process similar to that occurring during invagination of the vegetal plate in the sea urchin. The formation of the neural tube is an example of morphogenesis by thickening and folding of an epithelial sheet. Development of other organs, such as the pancreas, lungs, trachea and eye, involves similar mechanisms.

As the neural tube separates away from the overlying ectoderm, cells, originally from the crests of the neural folds, transform into mesenchyme and migrate away from the junction of the tube and the ectoderm. These are known as **neural crest cells** and they play an important role in the further development of vertebrates. They migrate *through* the tissues of the embryo as individual cells and lodge at certain sites, where they give rise to a variety of organs. For example, some do not migrate far and form groups of sensory nerve cells (ganglia). Others migrate further and form autonomic nerve cells, part of the adrenal glands, some sensory organs (such as hair cells, Chapter 7) and the sheaths around nerve axons. In addition, they are a major contributor to the soft and hard connective tissues of the skull, face, jaws, pharynx, gills and major arteries. During the evolution of the vertebrates, these structures enabled an increased uptake of oxygen and higher metabolic rates, faster swimming, better spatial

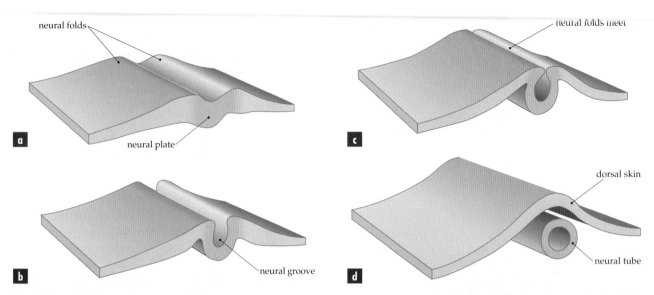

neural folds

neural plate

a

neural folds meet

c

neural groove

b

dorsal skin

neural tube

d

Fig. 15.40 Neurulation in the chicken. **(a)** Neural folds form on either side of the dorsal midline. **(b)** The folds rise upwards, creating a neural groove. **(c)** The folds continue to rise up and meet in the dorsal midline. **(d)** The neural tube separates from the overlying ectoderm, which forms the dorsal skin

orientation and detection of prey, and an increased ability to respond quickly and accurately to sensory information. The appearance of organs derived from neural crest cells largely allowed vertebrates to increase in size and adopt an active, predatory lifestyle in contrast to the slow-moving, filter-feeding invertebrates from which they evolved.

Limb formation

The formation of the vertebrate limb involves several different morphogenetic mechanisms. A limb first appears as a small bud on the flank of the embryo. It consists of a core of mesodermal cells covered by a layer of ectodermal cells. The ectoderm at the tip, the apical ectodermal ridge, is thicker than elsewhere and stimulates underlying mesodermal cells to divide. The limb bud elongates as a result of this localised proliferation of mesoderm (Fig. 15.41).

As the limb elongates, the divisions of mesodermal cells at the base slow and some of these cells aggregate to form cartilage plates that are the rudiments of the limb bones (Fig. 15.41). Simultaneously, other mesodermal cells condense around this cartilage, eventually to form muscles. The mesoderm that produces the muscle does not arise within the limb. Rather, it migrates into the limb bud as single cells that split off from blocks of mesoderm, **somites**, which surround the developing neural tube.

Initially, a limb bud is paddle-shaped, with no sign of the digits characteristic of a mature limb. In amphibians the rudiments of digits arise by a process of local proliferation of mesodermal cells. In birds and mammals, however, a different mechanism operates. Here, digits are sculpted from the paddle by a process of selective cell death: mesodermal cells in the zones between the digits start to degenerate and then die, thus separating the digits (Fig. 15.42). This cell death appears to be programmed, as mesodermal cells from the interdigit zones placed into tissue culture initially appear healthy but then die on schedule, at the same time as interdigit zone cells left in the limb bud.

The basic set of morphogenetic mechanisms, evident during nervous system and limb bud development, are used in different combinations and in different situations to produce the vast array of forms seen in the organs of animal embryos. These tissue changes are, in turn, generated by changes in a limited set of cellular properties, including some changes, such as in cell shape and cell adhesion, that were involved earlier in development, during gastrulation.

Morphogenetic changes during organogenesis are generated by thickening and folding of epithelial sheets; disaggregation of tissues into individual cells and single cell migration; localised cell proliferation; and localised cell death.

Cell differentiation

As organ rudiments take shape, another important developmental process, cell **differentiation**, is taking place within the cells composing them. As discussed in Chapter 16, differentiation involves changes in the pattern of gene expression of the cell. However, many changes in gene expression take place before a cell switches on the genes that reveal its final differentiated character. These may not result in any obvious change in the appearance of the cell, but they can restrict the cell's fate, a process called **determination**. This restriction in cell fate can be demonstrated by

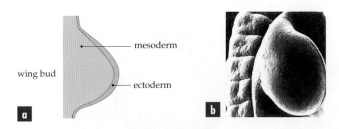

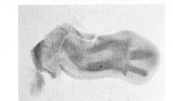

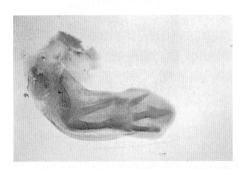

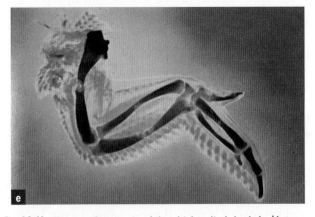

Fig. 15.41 The morphogenesis of the chicken limb bud. (a, b) A small bud of mesoderm, surrounded by a layer of ectoderm, develops on the flank of the embryo. (c)–(e) As the limb grows, mesodermal cells condense together to form the rudiments of bones

Fig. 15.42 A chick limb bud at the 'paddle-bud' stage. Cells within the dark blue regions die later in development, giving rise to the shapes of digits

The path to differentiation usually involves a series of steps in which the developmental potential of a cell is progressively determined. For example, while dorsal ectoderm becomes determined to form neural tissue during gastrulation, the precise type of neural tissue to be formed by individual cells is determined later.

Cell differentiation is preceded by a series of changes in gene expression, during which the cell's fate becomes progressively restricted.

Growth and maturation

The basic structure of an organism is laid down very early in development, for example, in the human embryo, when it is only about 12 mm long. In species that show direct development, the remainder of development largely involves gradual growth of component parts. Further cell proliferation, differentiation and expansion occurs, along with an increase in the extracellular matrix. Growth, like all developmental processes, is tightly regulated, but the mechanisms by which this occurs are not understood.

In indirect developing species, substantial developmental changes continue after the end of embryogenesis, as the larva is transformed into the adult during metamorphosis.

The later stage of embryogenesis involves growth and maturation as a result of cell proliferation and differentiation, and an increase in extracellular matrix materials.

transplanting cells to a different site in the developing embryo. If the transplanted cells develop in a similar way to their new neighbouring cells, the transplanted cells were undetermined at the time of transplantation. If, however, the transplanted cells continue to develop as they would have done at their original site, despite their new location, they were determined at the time of transplantation. The series of experiments shown in Figure 15.43 shows that presumptive neural ectoderm in amphibians becomes determined during the gastrulation stage.

Regulation of development

Reproduction almost always produces 'normal' offspring. For example, the frequency of abnormalities in newborn humans is very low. The reason is that developmental events occur sequentially and are very tightly controlled. Just how these processes are regulated has occupied the attention of biologists for well over a century and is today one of the most active research fields in biology.

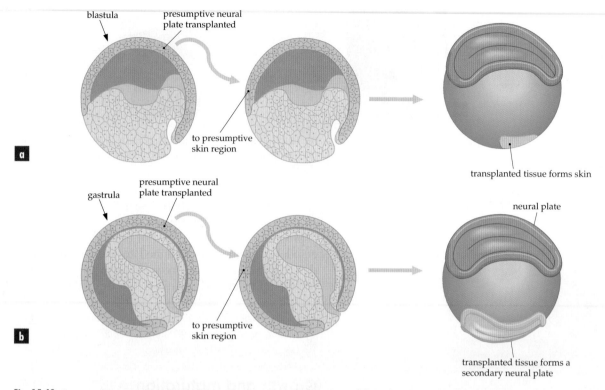

Fig. 15.43 An experiment to demonstrate determination in the frog embryo. **(a)** At the early gastrula stage, the presumptive neural plate tissue is undetermined and will give rise to skin if transplanted to the presumptive skin region of another gastrula. **(b)** If the same experiment is carried out at the late gastrula stage, the presumptive neural plate, which is now determined, gives rise to a secondary neural plate

Cells alter their activities during development in highly reproducible ways because they respond to specific signals emanating from either within the cell or the extracellular environment. We will consider some of what is known about the nature of these signals, their sources and the nature of the resulting cellular responses by examining two developmental systems.

Cytoplasmic specification in sea squirts

One of the earliest events during embryogenesis is the segregation of the cytoplasmic contents of the egg cell to cells of the blastula, the blastomeres. Since egg cytoplasm is not homogeneous, different blastomeres will inherit different cytoplasmic components, thus setting up differences between cells in different parts of the early embryo. Evidence that **cytoplasmic determinants** influence the developmental pathway taken by cells that receive them has been obtained in a number of embryonic systems. One of the most intensively studied systems concerns the determination of the muscle cell lineage in larval ascidians (sea squirts).

Early in the twentieth century, pioneering embryologists observed that clear differences in pigmentation could be seen in different regions of eggs of ascidians such as *Styela*. These cytoplasmic regions were reliably segregated to particular blastomeres, which in turn

gave rise to specific tissue types: a yellow-coloured cytoplasm, **myoplasm**, was separated out to the cells of the mesodermal lineage that produce muscle; a grey crescent portion went to the notochord and neural tube; the clear animal cytoplasm went to the ectoderm, which became the larval epidermis; and the grey vegetal region went to the endoderm, which gave rise to the gut (Fig. 15.44). It was suggested that factors associated with the differently coloured cytoplasmic regions might determine cell fate; that is, myoplasm might contain a 'muscle determination' factor. Indeed, if blastomeres containing yellow myoplasm are removed from an embryo early in cleavage and allowed to develop in isolation, they produce cells that differentiate into precisely the same cell type, muscle, that is produced by these blastomeres in the intact embryo (Fig. 15.45). Most other blastomeres show similar patterns of development, producing the same cell types in isolation that they produce in the whole embryo. The early sea squirt embryo seems to be a mosaic of parts that undergo independent determination; thus, this mode of development is known as **mosaic development**.

These findings are consistent with the idea that myoplasm directs cells that receive it to become muscle cells. Direct evidence in support of this hypothesis requires a different experiment, in which some myoplasm is relocated to blastomeres that normally have a

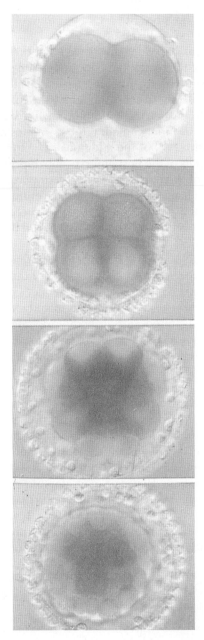

Fig. 15.44 The separation of different cytoplasmic regions in the egg of the ascidian *Styela* to different blastomeres during cleavage

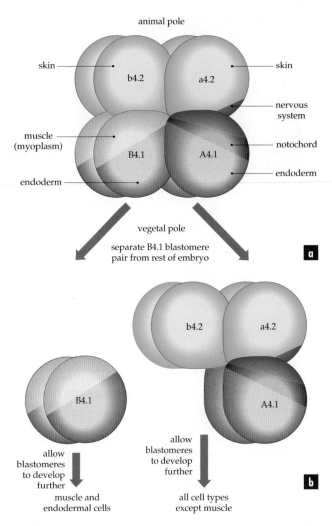

Fig. 15.45 (a) The normal fates of different regions of the eight-cell *Styela* embryo. **(b)** When separated from the rest of the *Styela* embryo, the B4.1 blastomeres give rise to muscle, mesenchyme and endodermal cells; this is the same range of cell types that is generated by these blastomeres in the whole embryo during normal development

different fate (such as skin), in another region of the embryo. If the hypothesis is correct, these cells should show characteristics of differentiated muscle cells.

This decisive experiment had to wait some 80 years after the original discovery of myoplasm and was carried out in 1982 by J. R. Whittaker, an American biologist. He pressed a glass needle against the vegetal blastomeres of the eight-cell stage *Styela* embryo, displacing some of the myoplasm, which normally is restricted to the vegetal B4.1 blastomere pair, into the animal b4.2 blastomeres. In addition to their normal ectodermal progeny, the animal blastomeres went on to generate muscle cells and expressed the muscle-specific biochemical marker, acetylcholinesterase

(Fig. 15.46). Hence, the presence of myoplasm imposed a new developmental fate on the animal cells that inherited it.

Many questions remain about the involvement of myoplasm in the specification of muscle cell lineage. The biochemical identity of myoplasm is unknown. We have little idea of how the myoplasm comes to lie in its characteristic position within the egg. Also, as we do not know what biochemical changes take place in the blastomeres when they become determined for a muscle fate, we can only guess at the role of the myoplasm in this process.

Finally, while these studies show that the specification of muscle fate involves cytoplasmic factors in blastomeres of the early embryo, they do not exclude a role for cell–cell interactions in this process. Such interactions could take place between cells that give rise to the B4.1 blastomere or between its progeny.

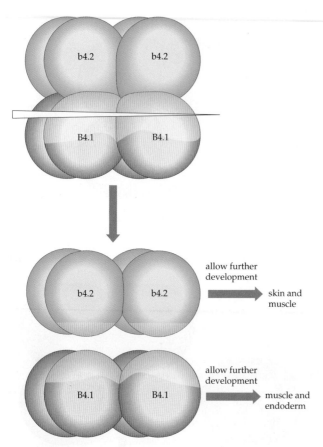

Fig. 15.46 When a *Styela* embryo is pushed with a glass needle during the third horizontal cleavage division, some of the myoplasm in the blastomere B4.1 pair moves into the animal hemisphere blastomeres b4.2. Some of the progeny of these b4.2 cells subsequently adopt a muscle fate

In some animals, cytoplasmic factors, segregated into different blastomeres during cleavage, influence the developmental fate of cells that receive those factors. For example, myoplasm is implicated in the specification of muscle tissue in sea squirts.

Primary induction and mesoderm induction in amphibia

Amphibians have long been favoured subjects for embryological research because of the ready availability, accessibility and large size of their eggs and embryos. Spemann and Mangold's 1924 study of the determination of the central nervous system in the amphibian embryo provides the classic example of cell fate being regulated by interactions between cells or tissues, the process of embryonic **induction**.

We saw previously that the first sign of gastrulation in the amphibian embryo is the involution of a group of cells at the position of the dorsal lip of the blastopore. These cells develop into the notochord, which lies underneath the presumptive nervous system. The appearance of the dorsal lip of the blastopore is also the first sign of a change from radial to bilateral symmetry:

the dorsal lip prescribes the position of the future dorsal midline of the embryo.

Spemann and Mangold showed that the cells of the dorsal lip of the blastopore have remarkable properties. If transplanted into the ventral ectoderm of another early gastrula, these cells can induce the formation of a second site of gastrulation and a second set of dorsal structures, including a neural tube, somites and dorsal gut structures. All of the additional structures develop from the cells of the host embryo, apparently under the influence of the dorsal lip of the blastopore, which has therefore changed the fate of the ventral ectoderm from belly skin to dorsal body structures. What eventually results is two embryos attached belly to belly (Fig. 15.47).

The conclusion drawn from these findings was that, in normal development, the dorsal lip of the blastopore and/or the presumptive notochord induces surrounding tissues (dorsal ectoderm, mesoderm and endoderm) to form an organised dorsal body side and hence it became known as the **organiser**. The action of the organiser represents an early example of embryonic induction, and hence it was called primary induction. As we shall see below, this is a misnomer because it is actually preceded by other inductive events.

Having recognised the powerful properties of the organiser, embryologists were eager to determine the identity of the chemical responsible for its action. Despite some 60 years of effort, there is still no answer to this question. However, considerable progress has been made in understanding the events that lead to the appearance of the organiser. In the toad *Xenopus laevis*, the organiser has been found to be the product of an earlier inductive interaction between animal and

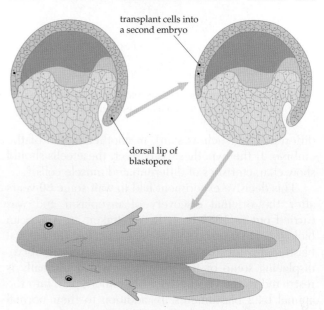

Fig. 15.47 When transplanted adjacent to the ventral ectoderm of another early gastrula, the dorsal lip of the blastopore of an amphibian gastrula induces the formation of a second dorsal axis

vegetal cells of the morula stage (32–48 cell) embryo. After removal from the embryo and culture in isolation, animal hemisphere cells produce only epidermis and vegetal hemisphere cells produce only gut tissue. However, if animal and vegetal hemisphere cells are cultured together, mesodermal tissues, including notochord and muscle, which are normally produced solely from equatorial cells, also appear (Fig. 15.48).

Experiments over the last 10 years have identified some strong candidates for the inducer for dorsal mesoderm. These molecules, activin and Vg1, belong to the TGF (transforming growth factor)-β family, and are small, secreted proteins. While they have been shown to have strong inductive effects when applied to isolated animal hemisphere blastomeres, definitive evidence that these are the signalling molecules actually used in the embryo is still lacking. TGF-β-like molecules have been implicated in a wide variety of processes involving cell–cell signalling, both during embryogenesis and later in the adult. They carry out this signalling function in animals as evolutionarily divergent as insects and mammals.

> Specification of dorsal tissues, such as the central nervous system, in the amphibian embryo involves an interaction with a neighbouring tissue, the organiser. The organiser is itself the product of an earlier inductive interaction between animal and vegetal hemisphere cells.

Cytoplasmic versus intercellular signalling

Two cases of developmental control that provide examples of regulatory signals arising from either within or outside the cell have been considered. Factors within the egg cytoplasm can play an important role in the early determination of the fate of blastomeres that inherit those factors. As the embryo develops and new anatomical relationships between cells arise, interactions between a cell and its environment (induction) become increasingly important. These two modes of cell fate specification should not be viewed as exclusive alternatives. The specification of cell fate is not a single event: rather it involves a whole series of complex molecular and cellular processes. When we trace the embryonic history of a given cell type from its progenitor to the fully differentiated state, we generally find that, irrespective of the organism involved, both modes of signalling—cytoplasmic and extracellular—play a role in the progressive restriction of the cell's fate.

There are, however, some evolutionary trends in the relative importance of cytoplasmic versus extracellular specification. Cytoplasmic specification is dominant in ascidians, nematodes, annelids and molluscs, which show mosaic development. Intercellular signalling is more important in mammals. Even at the four- and eight-cell stage, mammalian blastomeres are totipotent, retaining the ability to form an entire embryo. Mammalian embryos show **regulative development**: they can develop normally despite the removal of some cells at an early stage in development.

> Specification of cell type usually involves a mixture of both cytoplasmic and intercellular signalling. The relative importance of these two modes of specification shows evolutionary trends.

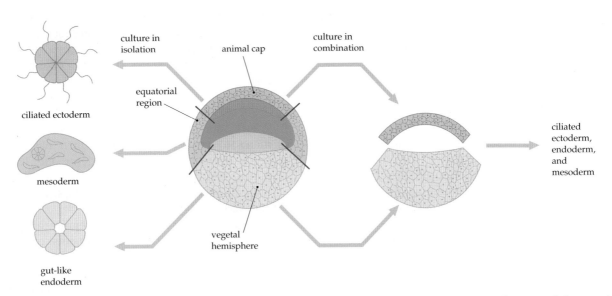

Fig. 15.48 Mesoderm induction in amphibia. When cultured in isolation, tissue from the animal cap, the equatorial region and the vegetal hemisphere of the blastula give rise to ciliated ectoderm, mesoderm and gut-like endoderm, respectively. However, if the equatorial region is removed and presumptive endoderm is cultured in combination with presumptive ectoderm, mesodermal tissue also develops

Summary

- In asexual reproduction, new individuals arise from a single parent by mitotic cell division and are genetically identical to that parent.
- Asexual reproduction can occur by means of vegetative growth (regeneration or budding).
- Parthenogenesis is a form of asexual reproduction in which egg cells develop into embryos without fertilisation.
- Sexual reproduction involves redistribution of parental genes into offspring and is the principal mechanism in eukaryotes for generating variation.
- Life cycles of sexually reproducing organisms alternate between diploid and haploid generations. In animals, the haploid stage is usually simply the gametes formed by meiosis.
- Sex is the distinction between male and female individuals of a dioecious species; that is, where male and female reproductive organs occur in separate individuals. In hermaphrodites, male and female reproductive organs are present in the same individual.
- Many mechanisms have evolved to limit self-fertilisation in hermaphroditic organisms. In protandry and protogyny, self-fertilisation is limited because individuals do not produce eggs and sperm at the same time.
- Sexual reproduction imposes significant costs in terms of increased energy use, decreased food intake, increased risks to survival, and in developing and maintaining specialised anatomical and physiological adaptations.
- In animals, eggs and sperm are produced in the ovaries and testes respectively. These gonads also produce hormones that regulate secondary sexual characteristics and behaviour.
- Gametes, the reproductive cells that fuse to form a zygote, are products of a sequence of events that includes migration of germ cells to the gonads; repeated mitotic divisions of germ cells to generate a stockpile of potential gametes; meiotic divisions to halve the chromosome number; and maturation to produce specialised egg and sperm cells. The timing of these events varies between species and between the sexes of any one species.
- Oogenesis includes a phase of dramatic cell growth, when various substances are stored in the maturing oocyte for use later in embryogenesis. These stored products may be produced within the oocyte itself or by surrounding maternal cells.
- Fertilisation involves penetration of the egg cell plasma membrane by the sperm, which activates egg cell metabolism and initiates embryogenesis, and fusion of the sperm and egg pronuclei, which restores the diploid number of chromosomes.
- Animals use a variety of adaptations to ensure that mature gametes come into close proximity. External fertilisation usually involves the co-ordinated release of gametes into the environment; internal fertilisation follows transfer of sperm into the female reproductive tract.
- Some species release large numbers of eggs with relatively little yolk and these typically undergo indirect development. Other species have fewer eggs with large nutrient reserves, or direct transfer of nutrients, and this allows direct development.
- Eutherian and marsupial mammals have yolk-poor eggs but direct development by means of nutrient transfer from mother to embryo during development. Monotremes are oviparous, marsupials and eutherians are viviparous.
- Reproductive success is the ability of an organism to parent as many viable embryos as possible. Reproductive effort represents the proportion of an organism's total energy and nutrient budget that is devoted to reproduction and hence into maturing embryos. Most organisms devote sufficient reproductive effort to maximise the chance of their offspring reaching maturity.
- Development of a mature animal from a single-celled zygote involves three main processes: cell proliferation, differentiation and morphogenesis.
- Cleavage is a rapid series of mitotic cell divisions that restores the nucleus-to-cytoplasmic ratio and forms the blastula. The pattern of cleavage is affected by the amount of yolk in the egg.
- Gastrulation involves rearrangement of blastomeres to form the gastrula, which has three germ layers (ectoderm, mesoderm and endoderm), new body cavities (archenteron and, if present, coelom) and, in many cases, the basic axes of body symmetry.
- Organogenesis involves the assembly of germ layer tissues to form the rudiments of body organs.
- Gastrulation and organogenesis involve a limited set of morphogenetic mechanisms, including: thickening and folding of epithelial sheets;

disaggregation of tissues into individual cells that migrate to new sites in the embryo; localised cell proliferation; and localised cell death. Movement of cells involves changes in cell shape and in cell adhesion.

- Actin filaments often play an important role in changes of cell shape that accompany morphogenesis. Specific adhesive interactions, mediated by cell adhesion molecules, underlie many morphogenetic processes.
- Cell differentiation is preceded by a series of changes in gene expression, during which the cell's fate becomes progressively restricted.
- The later stage of embryogenesis involves growth and maturation as a result of cell proliferation and differentiation, and an increase in extracellular matrix materials.

- In some animals, cytoplasmic factors, segregated into different blastomeres during cleavage, influence the developmental fate of cells that receive those factors. For example, myoplasm is implicated in the specification of muscle tissue in sea squirts.
- Specification of dorsal tissues in the amphibian embryo, such as the central nervous system, involves an interaction with a neighbouring tissue, the organiser. The organiser and mesoderm are the products of an earlier inductive interaction between animal and vegetal hemisphere cells.
- Specification of cell type usually involves a mixture of both cytoplasmic and intercellular signalling. The relative importance of these two modes of specification shows evolutionary trends.

keyterms

acrosome (p. 384)
amplexus (p. 380)
animal pole (p. 389)
archenteron (p. 394)
asexual reproduction (p. 371)
blastocoel (p. 389)
blastocyst (p. 391)
blastoderm (p. 390)
blastodisc (p. 390)
blastomere (p. 389)
blastopore (p. 394)
blastula (p. 389)
budding (p. 371)
cell differentiation (p. 389)
cell proliferation (p. 389)
cleavage (p. 389)
coelom (p. 394)
copulation (p. 388)
cytoplasmic determinant (p. 404)
determination (p. 402)
differentiation (p. 402)

direct development (p. 374)
dorsal lip of blastopore (p. 395)
ectoderm (p. 394)
endoderm (p. 394)
epiboly (p. 395)
fertilisation (p. 386)
follicle cell (p. 385)
gametogenesis (p. 381)
gastrula (p. 393)
gastrulation (p. 389)
germ cell (p. 381)
germ layer (p. 394)
gestation (p. 376)
gonad (p. 381)
hermaphrodite (p. 373)
indirect development (p. 374)
induction (p. 406)
involution (p. 395)
mesoderm (p. 394)

morphogenesis (p. 389)
mosaic development (p. 404)
myoplasm (p. 404)
neural crest cells (p. 401)
neural fold (p. 401)
neural groove (p. 401)
neural plate (p. 401)
neural tube (p. 401)
neurula (p. 401)
neurulation (p. 401)
notochord (p. 395)
nurse cell (p. 385)
oocyte (p. 384)
oogenesis (p. 381)
oogonia (p. 382)
organiser (p. 406)
organogenesis (p. 389)
ovary (p. 381)
ovulation (p. 384)
parthenogenesis (p. 372)
penis (p. 388)

placenta (p. 376)
protandry (p. 373)
protogyny (p. 373)
regeneration (p. 371)
regulative development (p. 407)
reproductive effort (p. 377)
reproductive success (p. 379)
sex (p. 372)
somatic cell (p. 381)
somites (p. 402)
spermatocyte (p. 384)
spermatogenesis (p. 381)
spermatogonia (p. 382)
spermatophore (p. 388)
testis (p. 381)
vegetal plate (p. 393)
vegetal pole (p. 389)
vitelline membrane (p. 386)

Review questions

1. List the advantages and disadvantages of asexual reproduction and sexual reproduction.

2. Why can parthenogenesis in the whiptail skink (p. 362) be considered a degenerate form of reproduction?

3. Explain why mutations that take place in muscle cells during the life of an organism are not passed down to the next generation.

4. What are the roles of follicle cells and nurse cells in oogenesis?

5. The mechanisms involved in bringing gametes together for fertilisation vary between species. Use examples to explain how these mechanisms are related to an organism's lifestyle.

6. Most hermaphroditic species prevent self-fertilisation. Describe the various ways by which they do this.

7. Explain why changing sex is an advantage for sexual reproduction in both salmon and the blue wrasse.

8. How is mass spawning of corals triggered? What advantages and disadvantages does it have for the organisms?

9. Explain why animals that produce large numbers of eggs often undergo indirect development, while those that produce fewer eggs often undergo direct development.

10. Outline some of the potential costs of sexual reproduction for animals.

11. (a) List four consequences of cleavage.

    (b) In what way do cleavage divisions differ from conventional mitotic divisions?

12. Explain why the blastomeres in the animal hemisphere of the frog blastula are smaller than those in the vegetal hemisphere.

13. 'Gastrulation usually establishes the basic body plan of an animal.' Illustrate this statement with reference to sea urchin and frog embryos.

14. List the four principal mechanisms of morphogenesis in animals and give an example of each.

15. (a) How are microfilaments involved in the folding of epithelial sheets?

    (b) How do changes in cell shape and cell adhesion underlie the morphogenetic movements of neural crest cells?

16. Explain the term 'induction' and give two examples of this phenomenon in amphibian development.

Extension questions

1. Both plants and animals have a life cycle in which there is an alternation of haploid and diploid cells. Compare and contrast animals and plants in this respect.

2. Under what sorts of circumstances is it better to adopt a reproductive strategy that produces many offspring, most of which may die without reproducing, compared to a reproductive strategy where fewer offspring are produced but more go on to reproduce?

3. Explain what is meant by mosaic development. Why does this phenomenon provide evidence for a role for cytoplasmic factors in cell determination?

4. Outline a direct experimental test of the hypothesis that a cytoplasmic factor such as the myoplasm is directly involved in determination of cell fate.

Suggested further reading

Alberts, B., Bray, D., Lewis, J., et al. (1995). *Molecular Biology of the Cell.* 3rd edn. New York: Garland.

An excellent and authoritative introduction to cell biology. It is up to date and presents the most recent findings in this field.

Austin, C. R. and Short, R. V. (eds). (1982–1987). *Reproduction in Mammals.* 2nd edn. Vols 1–5. Cambridge: Cambridge University Press.

A highly accessible, internationally renowned series of booklets providing an introduction to all aspects of reproduction in mammals.

Gilbert, S. F. (1997). *Developmental Biology.* 5th edn. Sunderland, MA: Sinauer Press.

An excellent, in-depth treatment of most aspects of modern developmental biology.

Johnson, M. and Everitt, B. (2000). *Essential Reproduction.* 5th edn. Oxford: Blackwell.

The new edition of this well-recognised and readily available text provides an up-to-date account of the various aspects of reproduction.

Wolpert, L., Beddington, R., Brockes, J., et al. (1998). *Principles of Development.* Oxford: Oxford University Press.

A highly accessible, well-illustrated text that conveys much of the excitement on modern developmental biology.

CHAPTER

16

Genes and development

G enes regulate development

Genes regulate development

The instructions for making an organism are contained within the nucleus of the fertilised egg. How are these instructions read out? In the last decade great advances have been made in answering this longstanding question.

The process of development involves the programmed spatial and temporal regulation of gene expression, culminating in the formation of each differentiated cell type characterised by expression of a unique set of genes. For example, red blood cells uniquely express haemoglobin, white blood cells express immunoglobulins and skin cells express keratins in order to carry out their respective functions. The events that result in the formation of these tissue types in a precise pattern in the developing embryo are orchestrated not by some mysterious external conductor but by information in the organism's own genome.

Regulated gene expression is essential for embryonic development. However, the sequence of regulatory interactions is not a simple linear one; it involves multiple parallel pathways and complex networks that generate expression of particular genes in certain positions of the embryo at certain times. The regulatory interactions become progressively more complex with time as divergent patterns of activity are set up in different regions of the embryo and more and more tissue types are formed. For example, the formation of heart muscle is the result of a sequence of events in which cells are directed along particular developmental pathways by the expression of specific genes. The first developmental decision made is a choice between ectoderm, mesoderm or endoderm fates (Chapter 15). Some cells become committed to mesoderm lineages because they express mesoderm-specific genes. Later they become further subdivided into, for example, fat cells or muscle cells. Those cells that express genes that induce a muscle cell fate are then further subdivided into skeletal muscle or visceral (gut) muscle or cardiac (heart) muscle by the expression of yet other specific genes.

Gene regulation is therefore central to this process of tissue formation and can occur because many genes act during development to regulate the expression of other genes. Many of the genes involved in the developmental regulation of gene activity have been identified and their mode of action elucidated. Some examples will be given later in this chapter. However, our understanding of the overall scheme of genetic interactions during development is still limited.

> Development involves a program of differential gene expression that is genetically regulated to occur in a spatial and temporal specific manner and that results in the formation of different tissue types in specific locations.

Differentiation does not involve the loss of genetic information

Development and differentiation involve the expression of different genes in different cell types. Theoretically, one way to accomplish this might be to eliminate genes from cells in which those genes are not expressed. If this mechanism existed, the genomes of differentiated cells would not be the same as the genome of the zygote from which they are derived, that is, they would not be genomically equivalent.

Genomic equivalence of cells has been tested in a number of ways. Biochemical analysis of DNA extracted from different cells of an organism shows that DNA type and content are the same in most somatic cells of that organism at all stages of development.

Other experiments have shown that genes are not irreversibly altered during development and that differentiated cells contain a complete set of potentially functional genes. Differentiated plant cells are totipotent, that is, they are capable of giving rise to a complete organism under appropriate conditions. For example in the carrot, *Daucus carota*, differentiated cells in root tissue can be induced to develop into a mature plant. When phloem cells are isolated from roots and grown in tissue culture, they divide and form groups of cells with 'embryo-like' properties (Fig. 16.1). In the presence of certain hormones, some of these groups of cells will give rise to differentiated plantlets with stems, leaves and roots.

Similar experiments have been attempted with animal cells but it has not been possible to induce a fully differentiated somatic cell to behave like a zygote. However, it has been possible to show that, at the level of the nucleus, the process of differentiation is reversible. In two species of frog, the grass frog, *Rana pipiens*, and the African clawed toad, *Xenopus laevis*, nuclei were transferred from a differentiated cell into an egg of the same species with its nucleus removed, either surgically or by irradiation (Fig. 16.2). Some of the eggs with transplanted nuclei continued to develop through a number of stages and, occasionally, normal tadpoles were produced.

We can conclude from these experiments that at least some nuclei of differentiated cells are totipotent and that the genome has not changed irreversibly as a consequence of differentiation. Genes that were inactive in one situation could be expressed in a different cellular environment and nuclei from some differentiated animal cells can support full embryonic development. These results indicate that there is no whole-scale loss of genes during differentiation. Genomic equivalence means that regulated gene

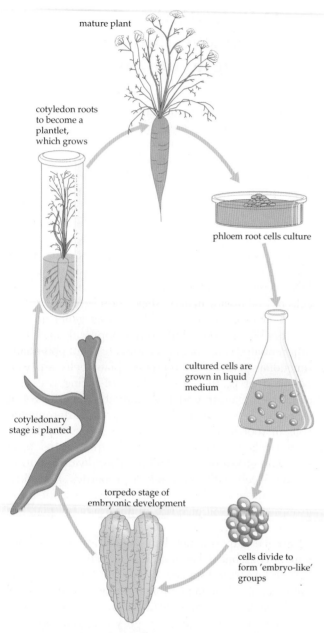

Fig. 16.1 Development of a carrot plant from differentiated root cells. Phloem cells from the root are isolated and cultured in liquid medium. As the cells divide they become differentiated and form 'embryo-like' groups (somatic embryos). Transfer of the somatic embryos to appropriate solid media stimulates them to form plantlets, which can then be grown into mature plants

Labels in figure:
mature plant
cotyledon roots to become a plantlet, which grows
phloem root cells culture
cultured cells are grown in liquid medium
cotyledonary stage is planted
torpedo stage of embryonic development
cells divide to form 'embryo-like' groups

expression and not gene elimination accounts for the expression of different genes in different tissues.

The original nuclear transplantation experiments were carried out in amphibians and, until recently, it was unclear whether nuclei from differentiated mammalian cells could be reprogrammed to support development from the egg to adult. However, early in 1997, a British group reported the successful cloning of a sheep, christened Dolly, by transplanting the nucleus from a differentiated adult mammary gland cell into an enucleated oocyte (Fig. 16.3). This finding raises the possibility that cloning may one day be possible in humans. Society must now confront the ethical dilemmas associated with this prospect.

> Development and differentiation are not generally associated with loss of genetic material. Nuclei from differentiated cells can be reprogrammed to support full embryonic development.

Gene expression can be regulated at a number of different levels

It is clear that differentiation is achieved not by the selective loss of genes but by the selective expression of genes in different cell types. How is this brought about?

As discussed in Chapter 11, regulation of gene expression can operate at a number of different levels, including transcription, stability of mRNA transcripts and translation. Examples of all of these levels of control can be found in development but we will consider only transcriptional and translational control here.

Transcriptional control

Many genes are known whose protein products regulate transcription during development. The products of these genes are known as **transcription factors**. A very important family of transcription factors is the family of **homeodomain proteins**. The homeodomain, a sequence of 60 amino acids, was first recognised in proteins involved in specifying segment identity in the vinegar fly, *Drosophila melanogaster* (see *Hox* genes, p. 426). It has subsequently been found in transcription factors that play many other regulatory roles in development. The mechanism of transcriptional regulation by homeodomain proteins appears to be the same in all cases: the homeodomain binds to a particular site in the DNA in the regulatory region of a specific gene, activating or repressing transcription of that gene (Fig. 16.4).

Another family of transcription factors is the group of **basic helix-loop-helix (bHLH) proteins**. One of these, the MyoD protein, activates muscle-specific genes in vertebrate myoblasts (muscle progenitor cells), causing them to adopt a muscle cell fate. Insertion of a *MyoD* gene that has been modified to be constitutively (permanently) active converts pigment cells, nerve cells or fat cells into muscle-like cells. (Fig. 16.5).

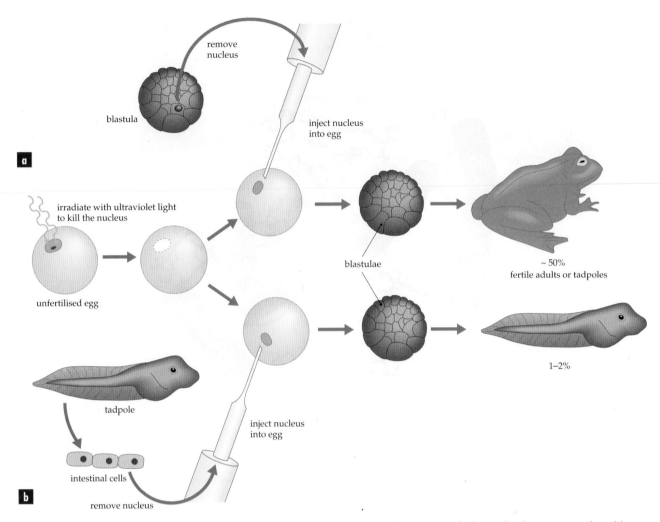

Fig. 16.2 The nucleus of a differentiated amphibian cell can support development from egg to tadpole. Fertilised eggs are enucleated by killing the nucleus with ultraviolet light. They are then injected with differentiated nuclei. The success of these in supporting development depends on the developmental stage of the transplanted nuclei. **(a)** If the nucleus is taken from the early developmental stage, the blastula, then there is a high probability that the transplanted nucleus will direct development to produce fertile adults. **(b)** If, on the other hand, the nucleus is taken from a differentiated tissue, such as tadpole intestinal cells, most transplants are unsuccessful, but in a few cases the transplanted nucleus can direct development to the tadpole stage, indicating that the necessary genes are present and active

Fig. 16.3 The sheep 'Dolly', the first mammal to be cloned by transplantation of nuclei from adult, differentiated cells into an enucleated egg cell

Translational control

In Chapter 15, you learnt that mRNA molecules can be stored in the cytoplasm of the developing oocyte as it matures in the ovary. These gene products can influence the development of the embryo after the mature oocyte is fertilised. In some cases, these mRNAs specify the fate of cells that inherit them. In such cases, the phenotype of the embryo is determined by mRNAs transcribed from the genes of its mother and not from its own genes. This phenomenon is therefore called a maternal effect. Genes that are transcribed during oogenesis but are translated and active during embryonic development are called **maternal-effect genes** (see p. 422).

Since maternal mRNA molecules are transcribed long before they are needed, there must be a mechanism to prevent their premature translation. One

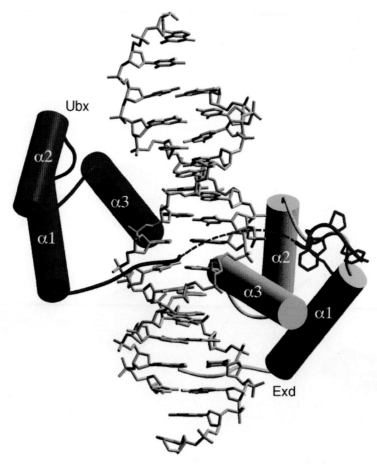

Fig. 16.4 Representation of the three-dimensional structures of the homeodomains from the Ultrabithorax and Extradenticle proteins bound to DNA, showing how one of the helical regions fits into a major groove on the double-helical DNA molecule

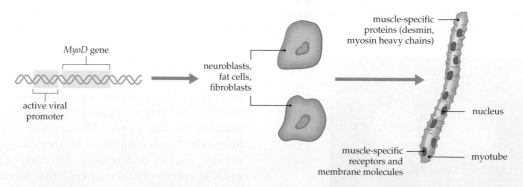

Fig. 16.5 Transfection of non-muscle cells (e.g. fat cells, neuroblasts, fibroblasts) with a continuously expressed *MyoD* gene causes these cells to differentiate as muscle cells

means of repressing their translation is to remove the long chain of adenine nucleotides, the poly A tail, which is normally found at the 3′ untranslated end of active mRNAs (Chapter 11). Clipping of the poly A tail takes place during oogenesis as the mRNA moves into the cytoplasm after transcription. The addition of adenine nucleotides to the end of the mRNA later, during embryogenesis, allows it to be translated.

Gene activity during development can be regulated at a number of different levels, including transcription and translation. Maternal-effect genes are transcribed during oogenesis and translated after fertilisation to regulate development.

Gene expression and pattern formation during development

To understand how an organism develops, we must account for the fact that genes are expressed in a definite *temporal order* and in precise *spatial patterns* during development. Cells in one region of the embryo express different genes and therefore engage in different activities from cells in another region (Fig. 16.6). These patterns of gene activity change continuously during development and are responsible both for morphogenesis and for cell differentiation.

The process by which embryonic cells form ordered spatial arrangements of different tissues is termed **pattern formation**. One of the crowning achievements of developmental biology has been to explain how the distribution of gene products within the early *Drosophila* embryo can regulate gene expression in a way that accounts for the genesis of a pattern as fundamental as the head–tail body axis (see p. 423).

Precise regulation of gene activity in space and time allows correct cell types to arise in the correct position in the embryo. This process is termed pattern formation.

Fig. 16.6 Differing expression patterns of the two proteins, Odd and Sox, in the mouse embryo, revealed by immunohistochemical labelling

Mutation is a powerful means of identifying developmental genes

A necessary first step in understanding how genes regulate development is to identify the genes that act during development. One of the most powerful strategies for achieving this is to induce mutations in organisms by exposing them to chemical mutagens or X-rays and then to screen for mutant individuals that show abnormal development. The assumption is that if a gene plays an important role in development, an embryo that lacks the function of that gene will not be able to develop normally.

Developmental geneticists using this approach have concentrated on two invertebrate animals, the vinegar fly *D. melanogaster* and the nematode, *Caenorhabditis elegans*, a plant, *Arabidopsis thaliana*, and, more recently, a vertebrate, the freshwater fish *Danio rerio*. All have the desirable characteristics of a short life cycle, ease of breeding in large numbers and a well-defined genetic map. Hundreds of mutants showing developmental defects have been isolated in these organisms. You can view lists of some of the *Drosophila* mutants at 'The interactive fly' website http://sdb.bio.purdue.edu/fly/aimain/1aahome.htm.

Some of these mutants are defective in morphogenesis or the final stages of cell differentiation. For example, the cells of flies homozygous for the *Drosophila* mutant *singed*, which has been known since 1922, develop gnarled, kinky bristles instead of the normal long curved bristles. It was recently discovered that the *singed* gene encodes a cytoskeletal protein termed Fascin, which bundles actin filaments into large, tightly packed hexagonal arrays that support cellular structures such as microvilli and filopodia (Chapter 15). Another example of a mutation that affects terminal differentiation is the appropriately named *shotgun* mutant, which results in holes in the cuticle. The *shotgun* gene encodes a *Drosophila* homologue of the vertebrate cadherin family of cell adhesion molecules (see Chapter 15).

However, two other classes of mutants are of particular interest because they reveal higher levels of developmental control.

Mutations that produce changes in cell fate

Some mutations cause cells to change their fate and develop into the wrong cell type. Two examples from *Drosophila* are the *Notch* mutant, in which epidermal cells adopt a neural fate (i.e. become neurons when they would normally form the epidermis) and the *cut* mutant, in which external sensory organs are transformed into internal, sensory organs (Fig. 16.7). The genes affected in this class of mutants appear to act in 'switch' mechanisms, directing cells into one developmental pathway, such as an epidermal **cell fate**, rather than another, such as a neural cell fate. The vertebrate *MyoD* gene discussed above is another example of such a master switch gene.

Mutations that result in changes in the pattern of body structures

Another particularly fascinating mutant phenotype involves a change in fate, not of a single cell or cell type, but of entire organs or body structures. For example, pair-rule mutants in *Drosophila* lack every alternate body segment (see p. 425). In homeotic mutants in insects and vertebrates, there is a switch in segmental identity (e.g. cervical vertebrae develop in place of thoracic vertebrae, see p. 428). In the *tolloid* mutant in the zebrafish, there is a disruption in dorsal–ventral axis formation, so that structures such as the ventral fin fail to form (Fig. 16.8). The fascinating property of these mutations is that they do not affect the ability of cells to adopt particular fates, as the normal range of cell fates can be observed. Rather, it is the pattern of the cell types that is disrupted. These genes must therefore be

Fig. 16.8 Adult zebrafish mutant for the *tolloid* gene. The mutation results in the loss of ventral tail tissues, including the embryonic ventral tail fin, which gives rise to the adult anal and tail fins

acting as spatial organisers, determining where and how particular structures will form.

Both cell fate and patterning genes regulate the activity of numerous other 'downstream' genes. They sit high up in a hierarchy of developmental control, in some cases several steps removed from their ultimate targets, the genes involved in final morphogenesis and cell differentiation. The segmentation genes in *Drosophila* (see below) provide a good example of a regulatory hierarchy.

> Screens for developmental mutants in a few select model organisms have uncovered both high level regulatory genes and downstream genes involved in morphogenesis and cell differentiation.

Developmental regulatory genes usually play multiple roles

An important general principle that has emerged from studies of mutants is that individual genes usually have diverse roles, being involved in the development of several different organs. For example, the *Notch* gene in *Drosophila* is involved in the development of the eye, the central and peripheral nervous systems, the body epidermis, the ovary and the wing. Geneticists have long recognised, from analysis of mutant phenotypes, that many genes affect different, seemingly unrelated characteristics and have called this phenomenon pleiotropy (see Chapter 9).

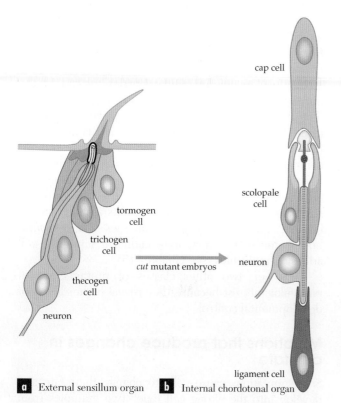

a External sensillum organ   **b** Internal chordotonal organ

Fig. 16.7 Mutations in the *cut* gene result in a change to the developmental fates of the four cells of the **(a)** external sensillum organ, which detects movement of the external bristle, so that they develop into **(b)** the four cells of the internal chordotonal organs, which sense internal stresses and movements

Mutations in developmental regulatory genes are generally pleiotropic.

Finding developmental genes in genetically intractable species

Not all organisms lend themselves to the mutation strategy for finding genes. For example, most vertebrates have long generation times and are difficult to breed in the large numbers required for a mutagenesis experiment. However, once a gene has been found in one species, such as *D. melanogaster* or *C. elegans*, it is possible to search for its equivalent in other, genetically intractable species using molecular biological techniques. The approach taken is to screen a genomic or cDNA library of the new species with a DNA probe made from the gene in the original species (see Chapter 13). Alternatively, the gene in the new species can be cloned by PCR using primers designed from the sequence of the original gene (see Chapter 13). Genome sequence analysis (Chapter 12) is the most comprehensive way of identifying such homologues.

This strategy of **homology cloning** works because most of the genes that play important regulatory roles in development are strongly conserved in evolution. For example, *Hox* genes have been found in a wide variety of animal phyla and in all cases appear to be involved in specifying the regional identity of body parts along the head–tail axis (see p. 426). Indeed, *Hox* genes were first identified in vertebrates using homology cloning.

Homology cloning or genome sequence analysis provides a means for identifying a developmental regulatory gene in a new species, starting with a known gene in a different species.

Developmental gene regulation often involves cell signalling pathways

Development involves a sequence of gene expression in which proteins encoded by genes regulate the expression of other genes. However, these regulatory mechanisms can be indirect. For example, expression of a gene in one cell may result in induction of gene expression in a neighbouring cell. To do this, there must be production of an extracellular signalling molecule, a ligand, from one cell, which binds to a receptor protein on the other cell. If the receptor is an integral membrane protein, binding of the ligand must in turn activate a **signal transduction cascade**, which results in induction of transcription of the second gene (see Chapter 7).

It has become clear that such signalling cascades tend to occur in specific pathways, often beginning with binding of a ligand to a cell-surface receptor and ending with activation or repression of a gene in that cell (Fig. 16.9). There appears to be a limited set of such cellular signalling pathways. Many of them are widely conserved in animal evolution and, like other developmental genes, can be used in a variety of different developmental contexts. For example, the receptor tyrosine kinase (RTK)–Ras pathway is involved in developmental processes as diverse as the development of both the eye and the tracheal system of *Drosophila*, the vulva of *C. elegans* and cartilage cells in mammals.

The cellular signalling pathways that are activated by binding of a ligand to a plasma membrane receptor mediate interactions between the cell and its environment. The ligand may be a diffusible factor released by other cells or a molecule bound to the extracellular matrix or on the surface of neighbouring cells. In any case, to explain how gene activity is regulated in the target cell, we need to consider events occurring outside, as well as inside, that cell (see Boxes 16.1, 16.2). In Chapter 15 we saw that, in most cases, both cytoplasmic and extracellular signalling play a role in the specification of the fate of any given cell type.

Gene activation often occurs in response to intercellular signalling.

The molecular genetics of segmentation in *Drosophila*

The molecular genetics of segmentation in *Drosophila* represents one of the best documented examples of the role of developmental gene regulation in pattern formation. Both larval and adult *Drosophila* have a segmented body plan, comprising a head, three thoracic segments and eight abdominal segments (Fig. 16.10). We now understand in large part how gene activity sets up the anterior–posterior axis and the body segments of this organism.

Early embryonic development in *Drosophila*

Early development in *Drosophila* takes place within the egg. The zygotic nucleus divides and 13 rapid mitotic divisions takes place without cytokinesis to form a

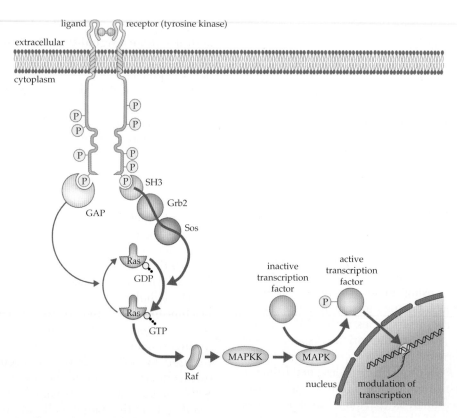

Fig. 16.9 The RTK–Ras signalling pathway. When the ligand binds to the receptor, the receptor is activated. This in turn induces phosphorylation of the intracellular portion of the receptor and sequential activation of a series of proteins, resulting ultimately in altered gene expression in the nucleus. Grb2, Sos, Ras, Raf, MAPKK and MAPK are all intermediates that play a role in transducing the extracellular signal to the nucleus

---

BOX 16.1 *Sonic hedgehog* and the vertebrate neural tube: cell fate specification and pattern formation by an extracellular signal

The vertebrate neural tube shows a characteristic dorsal–ventral pattern of cell differentiation: specialised glial cells, floor plate cells, develop in the ventral-most region at the midline, motor neurons in the ventral third, relay neurons in the middle third, smaller interneurons in the dorsal third, while the most dorsal region produces neural crest cells that later migrate to give rise to the majority of neurons in the peripheral nervous system (Fig. a). Each cell type is characterised by the expression of different sets of genes.

The development of this pattern is dependent on signals from the notochord, which lies directly beneath the neural tube. Early removal of the notochord results in the ventral cell types failing to develop (Fig. b), while implantation of an extra notochord to the lateral or dorsal side of the neural tube causes additional floor plate cells and motor neurons to develop in these regions (Fig. c).

This is a classic example of the phenomenon of embryonic induction (see Chapter 15), where a signal arising from one tissue, in this case the notochord, induces particular developmental fates in a second tissue, in this case the neural tube.

The molecule responsible for the induction of both floor plate cells and motor neurons is a protein called Sonic hedgehog (SHH). The *Sonic hedgehog* (*Shh*) gene was first identified in vertebrates by homology cloning from its *Drosophila* counterpart *hedgehog* (*hh*), which is involved in cell–cell signalling during segmentation and limb development in the fly. *Shh* is normally expressed in the notochord at a time when the neural tube is undergoing patterning. When misexpressed in the dorsal part of the neural tube, it leads to abnormal floor plate and motor neuron differentiation in these regions.

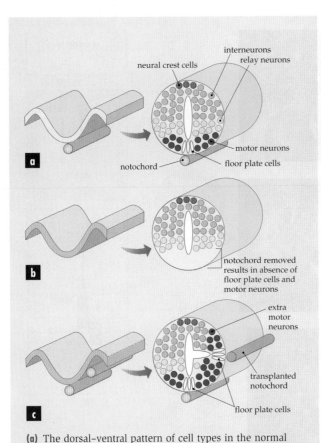

(a) The dorsal–ventral pattern of cell types in the normal vertebrate neural tube. Floor plate cells develop at the most ventral position, then in successively more dorsal locations: motor neurons, relay neurons, smaller interneurons and neural crest cells. **(b)** Removal of the notochord leads to the absence of both floor plate and motor neurons.
(c) Transplantation of an additional notochord to the side of the neural tube leads to the appearance of an additional set of floor plate cells and motor neurons at that site

SHH is involved in many other inductive processes during vertebrate development, including establishment of asymmetry in the left–right body axis, limb formation, patterning of mesoderm in the somites, the formation of the pancreas, patterning of scales, feathers or hairs and tooth development. It is an excellent example of a pleiotropic gene.

SHH is an extracellular signalling molecule (see Chapter 15). Binding of SHH to its receptor triggers an intracellular signalling pathway, which leads to the transcriptional activation of a specific set of downstream genes. This pathway, like the SHH protein itself, appears to be conserved between insects and vertebrates.

BOX 16.2 *numb/Notch* signalling in the central nervous system of *Drosophila*: cell fate specification by both intracellular and extracellular factors

The *Drosophila numb* gene provides a well-documented example of cell fate determination by a cytoplasmic protein. Within the developing central nervous system, a cell termed the MP2 cell divides to give rise to two neurons, the dMP2 and vMP2 neurons (the d and v terms derived from the dorsal and ventral positions of the respective neurons). These two neurons have different axon morphologies: the dMP2 axon projects posteriorly, while the vMP2 axon runs anteriorly (Fig. a).

Numb protein is produced within the MP2 cell and, prior to division, Numb protein is localised at one pole of the MP2 cell, such that it is segregated exclusively to the dMP2 daughter cell. In a *numb* mutant embryo, the dMP2 neuron is transformed into a vMP2 neuron. Conversely, inducing abnormal expression of Numb protein in the vMP2 neuron causes it to take on a dMP2 fate (Fig. a). These results suggest that the Numb protein acts as a cytoplasmic determinant, specifying dMP2 cell fate.

Yet even in this system involving specification by an intrinsic cytoplasmic determinant, factors external to the MP2 progeny cells have been shown to be important in determining their fate. The genes involved are *Delta* and *Notch*.

Notch is a cell-surface receptor expressed on the surface of all central neurons, including dMP2 and vMP2. Delta is the ligand for the Notch receptor and is expressed, not on dMP2 or vMP2, but on mesodermal cells that lie in contact with these neurons. Both *Notch* and *Delta* mutants show an opposite phenotype to the *numb* mutant, namely a transformation of the vMP2 neuron to a dMP2 fate. This shows that signalling between the MP2 progeny and neighbouring mesodermal cells is also involved in determining their fate. The phenotype of double *Notch/numb* mutants (transformation of vMP2 into dMP2 fate) suggests that Numb protein may normally antagonise Notch signalling (Fig. b). Specification of d/vMP2 fate by *numb/Notch* is, therefore, an example of an intrinsic factor interacting with an extrinsic signalling pathway.

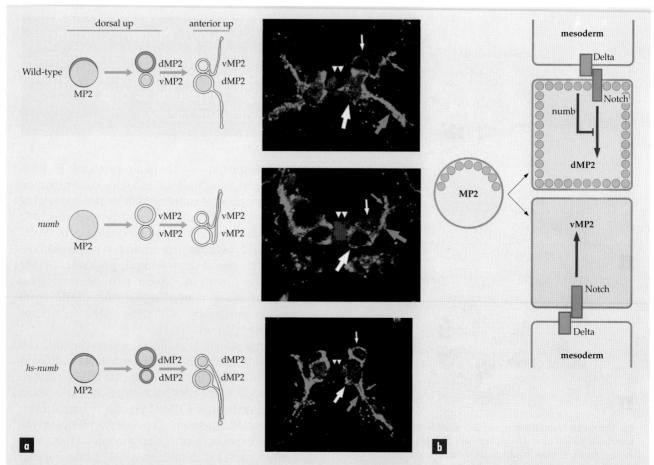

(a) Division of the MP cell in the central nervous system of the *Drosophila* embryo produces two progeny—the vMP2 neuron, whose axon runs anteriorly, and the dMP2 neuron, which runs posteriorly. In a *numb* mutant, the dMP2 neuron takes on the vMP2 fate, sending its axon anteriorly, while in an embryo in which the wild-type *numb* gene is expressed in all cells, vMP2 is converted to a dMP2 fate. **(b)** A model for the interaction between the Numb cytoplasmic determinant and the Delta/Notch signalling pathway in d/vMP2 neurons in *Drosophila*. Delta/Notch signalling is blocked by Numb, causing this cell to adopt the dMP2 fate, rather than the vMP2 fate

syncytium (i.e. a cell containing multiple nuclei, Fig. 16.11). The first 10 of these divisions take less than 10 minutes each and produce a cluster of nuclei within the egg. Most of the nuclei then migrate to the periphery of the egg where after another three divisions, cell membranes form around each of the nuclei. During this phase the germ line is formed from about 10 cells, the **pole cells**, which are segregated at the posterior end of the embryo.

The earliest stages of development are controlled by maternal-effect genes. Mutations in maternal-effect genes are readily identified in genetic experiments because homozygous mutant individuals, from heterozygous parents, can develop but the homozygous females lay eggs that cannot develop. If we carry out the following crosses (m^- is a recessive maternal-effect mutant and m^+ is the wild-type or normal allele), then:

$$\frac{m^-}{m^+} \times \frac{m^-}{m^+} \rightarrow \frac{m^-}{m^-} \text{ (normal development)}$$

but female $\dfrac{m^-}{m^-} \times$ any male $\rightarrow$ embryos with abnormal development

The homozygous $\dfrac{m^-}{m^-}$ females produce eggs that are not capable of normal development because they cannot express the gene and deposit the mRNA product into their eggs to be translated following fertilisation. Thus, as described earlier in the chapter, the wild-type gene product from the mother is necessary for normal development of the zygote.

A maternal-effect gene is acting when the developmental phenotype of the embryo is determined by the genotype of the mother and not the embryo itself.

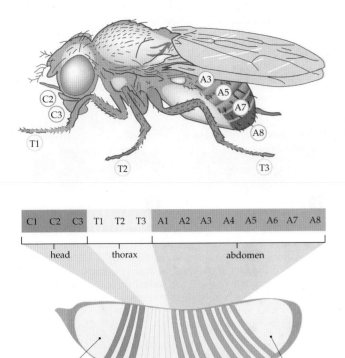

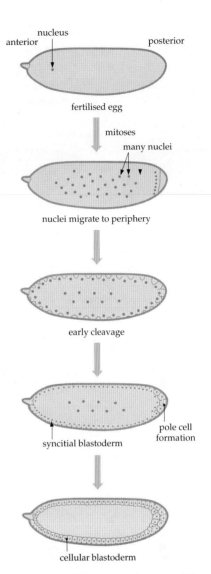

Fig. 16.10 Relationship between larval and adult segmentation in *Drosophila*. The segmented body plan is laid out in the embryo. Segments C1–C3 produce the head, while segments T1–T3 produce the thorax. Each thoracic segment carries a pair of legs. The wings develop on the second thoracic segment (T2) and the halteres (flight balancers) on the third thoracic segment (T3). Segments A1–A8 give rise to the abdomen

Specifying the anterior–posterior axis

In *Drosophila*, analysis of maternal-effect mutants shows that maternal-effect genes produce substances, stored in the egg cytoplasm, that establish the polarity of the *Drosophila* egg before fertilisation takes place. They determine which parts of the egg are dorsal or ventral and which are anterior or posterior, and thus determine the basic axes of the embryo.

If a small amount of the cytoplasm is removed from the anterior region of certain insect eggs, the embryo develops with its head and thorax missing. This observation led to the idea that some factor, an *anterior determinant*, which specifies both the head and thoracic segments, is released from the anterior pole of the egg. A search for mutants whose phenotype resembled embryos produced by removal of the anterior pole cytoplasm led to the discovery of the *bicoid* mutant (Fig. 16.12a). The *bicoid* mutant phenotype can be reversed (or 'rescued') by injecting anterior pole cytoplasm from a normal wild-type embryo into a *bicoid* egg, suggesting that the wild-type product of the *bicoid* gene may be the anterior determinant (Fig. 16.12b).

Fig. 16.11 Early developmental stages in the *Drosophila* embryo

The *bicoid* mRNA accumulates at the anterior pole of the oocyte during oogenesis. After fertilisation, *bicoid* mRNA remains confined to the anterior pole of the egg. The Bicoid protein, formed by translation of this mRNA, diffuses away from the anterior pole and a concentration gradient is established. The peak of this gradient is at the anterior pole and the low point about two-thirds of the way to the posterior pole (Fig. 16.13). During this phase, the *Drosophila* embryo is a syncytium. This means that the Bicoid protein has free access to nuclei in the regions in which it occurs and can interact with genes in them.

Injection of pure *bicoid* mRNA into a mutant *bicoid* egg causes a full set of head and thoracic structures to form at the site of injection (Fig. 16.14). This experiment shows conclusively that the concentration gradient of Bicoid protein dictates the spatial pattern of body structures along the anterior–posterior axis: nuclei exposed to different concentrations of Bicoid protein

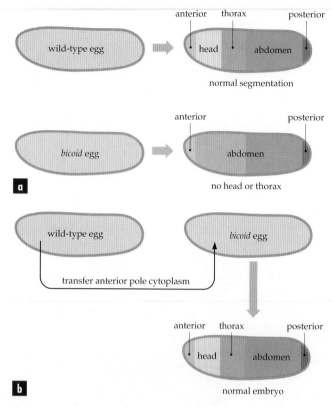

Fig. 16.12 Effects of the *bicoid* mutation (a maternal-effect gene) in *Drosophila*. **(a)** Structures produced in the anterior–posterior axis in wild-type embryos and in embryos derived from *bicoid* mutant females. In the mutant, no head or thorax segments develop. **(b)** 'Rescue' of the *bicoid* mutant by injection of anterior pole cytoplasm from wild-type eggs

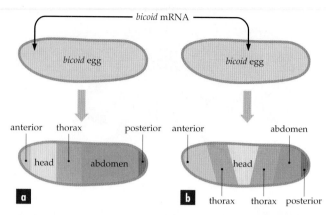

Fig. 16.14 The effects of injection of wild-type *bicoid* mRNA into eggs derived from *bicoid* mutant females. Injection of *bicoid* mRNA causes anterior structures to form, with the site of injection defining the location of the head. **(a)** When the mRNA is injected into the anterior end of the egg, a normal set of segments forms; **(b)** whereas when it is injected into the middle of the egg, head structures form at the middle with a duplicated set of thoracic structures on either side

express different sets of genes. Developmental regulatory molecules that act in this concentration-dependent fashion are called **morphogens**.

Over 20 genes and their products are involved in the formation of the anterior body segments, so the Bicoid protein does not itself cause these anterior segments to form but regulates other genes lower in the hierarchy of interactions. The Bicoid protein possesses a homeodomain and is a transcription factor, so it directly regulates the expression of the genes that act next in the hierarchy.

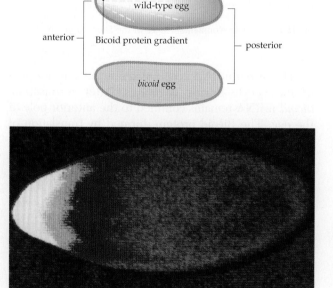

Fig. 16.13 The Bicoid protein in *Drosophila*. The distribution of the Bicoid protein in the wild-type egg peaks at the anterior pole and forms a gradient to about two-thirds of the distance to the posterior pole. This gradient is absent in eggs derived from *bicoid* mutant females

> The *bicoid* gene is an example of a maternal-effect gene, which is transcribed during oogenesis but not translated until after fertilisation. Bicoid protein acts as a morphogen: the concentration gradient of Bicoid protein specifies the anterior–posterior body axis of the embryo.

Zygotic segmentation genes

The next set of genes to act in segment formation are transcribed from the zygotic genome and are therefore called **zygotic genes**. These begin to be transcribed just before cellularisation of the embryo. The zygotic segmentation genes respond to the maternal morphogens and act to determine the position and pattern of the segments in the embryo. Segmentation genes act more or less sequentially to produce increasing levels of organisation in the embryo. These genes fall into one of three classes (Fig. 16.15).

Gap genes

The first segmentation genes induced by the maternal morphogens such as Bicoid protein are the **gap genes**, which act in regions along the embryo to interpret the

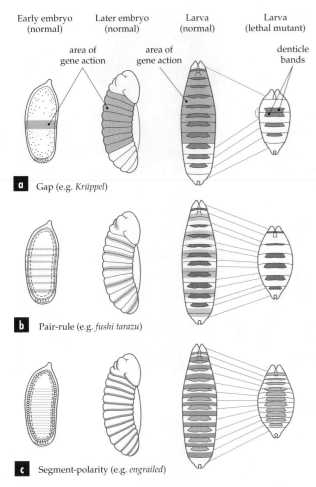

Early embryo (normal) | Later embryo (normal) | Larva (normal) | Larva (lethal mutant)

area of gene action

area of gene action

denticle bands

a Gap (e.g. *Krüppel*)

b Pair-rule (e.g. *fushi tarazu*)

c Segment-polarity (e.g. *engrailed*)

Fig. 16.15 Expression regions of examples of the three different classes of segmentation genes (gap, pair-rule and segment-polarity) and morphological defects seen in mutants of each of those genes

anterior–posterior information and establish a spatial organisation that will lead to segmentation. Mutations in gap genes result in an embryo that lacks one or a group of segments, so that gaps appear in the normal structural pattern of the embryo. Expression of the gap genes is directly regulated by the maternal-effect genes. For example, Bicoid protein activates transcription of the gap gene *hunchback* by binding to specific DNA sequences in its promoter region (Fig. 16.16a). *hunchback* is expressed in the anterior half of the embryo because this is where high levels of Bicoid protein are found (Fig. 16.16b).

Pair-rule and segment-polarity genes

The other two classes of segmentation genes act in the formation of all segments. For example, the **pair-rule genes** pattern the embryo into discrete segments; mutations in pair-rule genes cause deletion of every second segment. The mutation *even-skipped* affects even-numbered segments, while *odd-skipped* affects only odd-numbered segments. Finally, there are **segment-polarity genes**, which react to signals to determine the pattern of development within each segment of the embryo. Mutations in segment-polarity genes often lead to a normal number of segments, but part of each segment may be deleted and replaced by a symmetrical mirror-image duplication of the portion that is retained.

A complex set of regulatory interactions operates between each of these classes of genes. There is a hierarchy of interactions: maternal-effect genes can directly regulate the activity of gap, pair-rule and segment-polarity genes but not vice versa, while gap genes regulate the activity of pair-rule and segment-polarity genes. However, cross-regulation occurs within a level of the hierarchy (e.g. gap genes

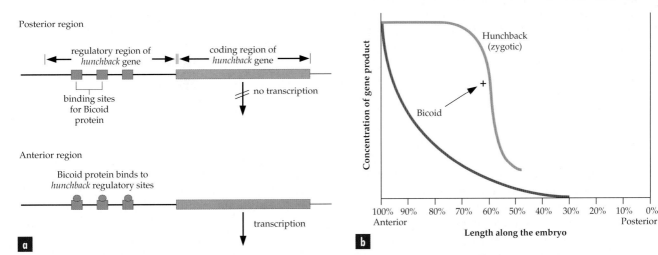

Posterior region

regulatory region of *hunchback* gene

coding region of *hunchback* gene

no transcription

binding sites for Bicoid protein

Anterior region

Bicoid protein binds to *hunchback* regulatory sites

transcription

a

b

Hunchback (zygotic)

Bicoid

+

Concentration of gene product

100% 90% 80% 70% 60% 50% 40% 30% 20% 10% 0%
Anterior — Posterior

Length along the embryo

Fig. 16.16 (a) Binding of Bicoid protein to specific sites within the regulatory region for the *hunchback* gene leads to activation of *hunchback* transcription. **(b)** Graph showing the levels of Bicoid and Hunchback protein expression along the length of the embryo (100%, anterior end; 0%, posterior end). There is a threshold concentration for activation of *hunchback* by Bicoid. *hunchback* is expressed only in the anterior 40% of the embryo where Bicoid protein levels are relatively high

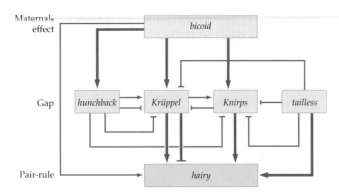

Fig. 16.17 Interactions between maternal-effect and segmentation genes. This is not an exhaustive list of all maternal-effect and segmentation genes, nor are all known interactions depicted. Activation is indicated by →, inhibition by ⊣

regulate the activity of other gap genes, Fig. 16.17). The result of the gene activity that follows fertilisation is the establishment of segmentally repeated patterns of gene expression along the anterior–posterior axis, which in turn establish the pattern of body segments.

> **Z**ygotic segmentation genes extend the developmental program beyond that established by the maternal-effect genes to create the segmental structure of the embryo.

Hox genes

The net result of the expression of the maternal-effect, gap, pair-rule and segment-polarity genes is the subdivision of the body of the *Drosophila* embryo into a series of segments. However, this is not the end of the story. *Drosophila*, like all other arthropods, shows the phenomenon of **tagmatisation**: the organisation of segments into groups with differing structure and function (Fig. 16.10, Chapter 39). For example, the segments of the thorax bear appendages (legs and wings) that are absent on abdominal segments.

Segments acquire their unique identities through the action of the **homeotic genes**. Mutations in homeotic genes transform one body part into another, producing some very striking changes in morphology. For example, *Antennapedia* mutants have legs on the head where the antennae would normally be (Fig. 16.18). Another example is the mutant *bithorax*, in which the halteres (flight balancers) are converted into a second pair of wings on an extra thoracic segment (Fig. 16.18c).

The homeotic genes in *Drosophila* are arranged in two separate clusters on a single chromosome. One cluster, the Antennapedia complex, is composed of five genes, while the other, the Bithorax complex, is composed of three genes (Fig. 16.19a). The order of these genes on the chromosome mirrors the order

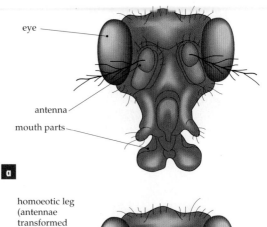

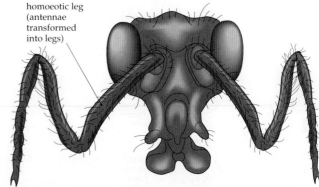

Fig. 16.18 Examples of homeotic mutants that transform one body part to another. **(a)** The head of a wild-type *Drosophila*. **(b)** *Drosophila* in which the antennae are transformed into legs (*Antennapedia* mutant). **(c)** An adult *Drosophila* with four wings produced by mutations in the Bithorax complex. The mutants convert the third thoracic segment into the second thoracic segment, and the halteres normally present on the third thoracic segment become converted into a second pair of wings

along the anterior–posterior body axis in which they are expressed (Fig. 16.19b). The identity of individual segments is generally specified by the combined activity of several homeotic genes, rather than a single gene. For example, the most posterior four abdominal segments are specified by the expression of three genes—*Ubx*, *abd-A* and *Abd-B*.

All of the proteins encoded by the homeotic genes contain the characteristic homeodomain sequence of 60 amino acids. The homeodomain binds to DNA sequences in the promoter region of target genes and

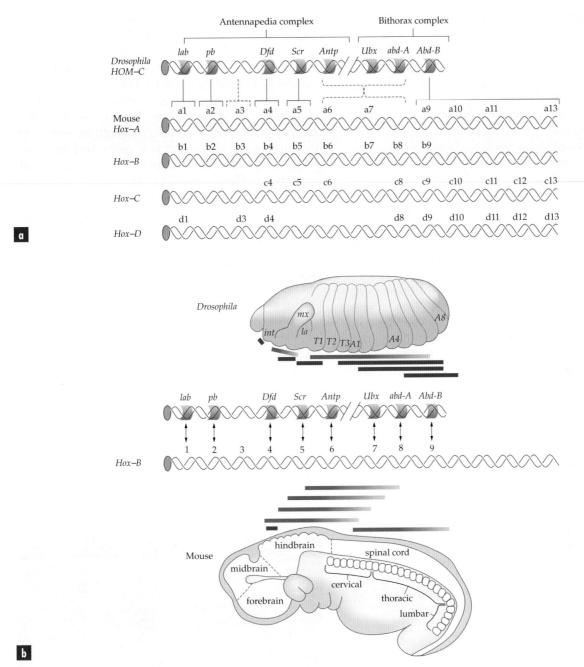

Fig. 16.19 **(a)** Conservation in the order on the chromosome of *Drosophila* homeotic genes and the equivalent *Hox* genes in the mouse. There are four duplicated sets of *Hox* genes in the mouse (*Hox-A*, *Hox-B*, *Hox-C* and *Hox-D*), each containing between nine and 11 genes, eight of which have *Drosophila* homologues. **(b)** A comparison of the pattern of transcription of the *Drosophila* homeotic genes and the mouse *Hox-B* genes at mid-embryonic stages. In both the vinegar fly and the mouse, the order of these genes on the chromosome corresponds to their order of expression in the anterior–posterior axis of the body

activates or represses transcription of those genes. The activity of the homeotic genes themselves is regulated by the segmentation genes.

Hox genes and the evolution of body patterns

Soon after the first homeotic genes were cloned in *Drosophila*, a search was made for homologous genes in other animals. Homologues were found in a wide variety of animals, including vertebrates, where they are called *Hox* genes. There are vertebrate equivalents of each of the *Drosophila* homeotic genes (numbered 1, 2, 4, 5, 6, 7, 8 and 9), judged from similarity in nucleotide sequence. However, vertebrates possess some *Hox* genes that are not present in *Drosophila* (numbered 3, 10, 11, 12 and 13). Furthermore, whereas *Drosophila* has only one set of homeotic genes, vertebrates contain four sets—*Hox-A*, *Hox-B*, *Hox-C* and *Hox-D*—each set containing between nine and 11 members of the *Hox* gene family (Fig. 16.19a). The order of vertebrate *Hox* genes along the chromosome is the same as in *Drosophila* and, as in the fly, accords with their order of expression along the anterior–posterior axis of the embryo (Fig. 16.19b). There is, therefore, strong conservation in the structure, chromosomal arrangement and expression pattern of *Hox* genes between insects and vertebrates.

Has *Hox* gene function been similarly conserved in evolution? While vertebrates are not externally segmented in the same way as insects, many body parts show patterning along the anterior–posterior axis. For example, the hindbrain neural tube becomes divided into segmental units, called **rhombomeres**. There is a chain of cranial ganglia in the head and a series of branchial arches of varying structure (see Chapter 40), formed in large part from neural crest cells that migrate out from the hindbrain neural tube (see Chapter 15). The morphology of vertebrae varies along the anterior–posterior axis: the division of the spine into cervical, thoracic and lumbar regions recalls the tagmatisation seen in arthropods.

We now have good evidence that *Hox* genes are involved in each of these examples of patterning along the vertebrate head–tail axis. For example, in a mouse lacking *Hoxa-5* function, the seventh cervical vertebra is converted into a thoracic vertebra. Mice lacking *Hoxa-3* function show abnormal differentiation of the thymus, thyroid and parathyroid glands, together with defective neck cartilage morphology; all of these structures develop from neural crest cells that populate the fourth and sixth branchial arches.

Hox genes have now been isolated from all of the major phyla of animals. In each case it appears that they are responsible for specifying regional identity along the anterior–posterior axis.

> Homeotic genes encode transcription factors that specify regional identity along the anterior–posterior axis in animals as diverse as insects and mammals by being differentially expressed in specific groups of segments.

Flower development in *Arabidopsis*

While much of our understanding of how genes act to regulate development has come from studies in animals, studies of the genetic regulation of development in plants are proceeding apace. For example, it has been found that homeotic genes play a regulatory role in plants, as in animals. During flower formation,

these genes control the identity of the whorls of floral organs as they develop at the floral meristem. The model plant, *Arabidopsis*, has been used to study genes that regulate flower development because of its small genome size, rapid generation time from embryo to seed set and ease of use in generating mutants.

Mutants are produced by treating seeds with a chemical mutagen, such as ethylmethane sulfonate. Seedlings of this M_1 (mutation 1) generation are grown to flowering, self-pollinated and their seeds grown (M_2 generation). Floral mutations are visible in the M_2 generation in a small proportion of plants. Most of the mutants obtained are recessive and only produce a visible phenotype in the M_2 generation when in the double-recessive condition.

The war of the whorls

In *Arabidopsis* flowers, there are four whorls of organs. Outermost are the four sepals (whorl 1, Fig. 16.20), followed by the four petals (whorl 2) internal to and alternating with the growing sepals, then a whorl of six stamens (whorl 3) and finally a central pistil formed from two carpels (whorl 4). These are precisely sited in relation to each other because they are formed sequentially during development: sepals initiated first and carpels last.

Studies of floral mutants have identified a class of whorl identity genes (Fig. 16.21). These are homeotic genes, that is, regulatory genes whose actions determine the fate of each whorl. This can be seen in the floral mutants. For example, the *agamous* mutant has

Fig. 16.20 Flowers of a wild-type *Arabidopsis thaliana* plant, showing whorls of sepals, petals, stamens and carpels

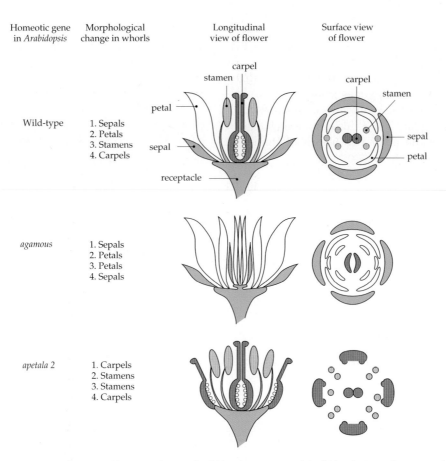

Homeotic gene in *Arabidopsis*	Morphological change in whorls	Longitudinal view of flower	Surface view of flower

Fig. 16.21 Flower developmental patterns in wild-type and two whorl identity mutants of *Arabidopsis*. In each mutant, the two outer and inner whorls have been switched as indicated. The *agamous* flowers are shown in Figure 9.18b

flowers that do not develop stamens and carpels. Instead, they have sepals (whorl 1), petals (whorl 2), petals (whorl 3) and sepals (whorl 4), forming 'double' but sterile flowers (Fig. 9.13). Another mutant, *apetala 2*, lacks sepal and petal whorls, which are replaced by carpels and stamens, that is, flowers have carpels (whorl 1), stamens (whorl 2), stamens (whorl 3) and carpels (whorl 4). These examples show that mutations always affect not just one whorl but a pair of adjacent whorls.

A genetic model has been proposed to explain these observations in *Arabidopsis*. The four whorls of a wild-type flower are specified by three sets of homeotic genes, which are active in three overlapping regions of a floral meristem (Fig. 16.22). These regions, A, B and C, each express a specific set of organ identity genes. The set of genes active in each region specifies two adjacent whorls. If one region is missing as a result of mutation, the organs specified will differ from wild type. A specific combination of genes specifies each whorl because the expression regions are overlapping. Sepals are specified by expression of region A organ identity genes acting alone; petals arise if there is overlapping expression of region A and B genes; stamens from overlapping expression of region B and C genes; and carpels only from region C gene expression.

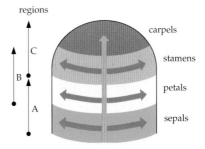

Fig. 16.22 Whorl identity genes regulate the sequential development of floral organs in flowers of *Arabidopsis*. These genes are active in three overlapping regions of the floral meristem (arrows on left). As the floral meristem develops, sepals are initiated when the region A genes are expressed alone, petals are initiated when region A and B genes are both expressed, stamens when regions B and C genes are both expressed, and carpels when region C genes are expressed alone. Regions A and C are mutually antagonistic, and a mutation in, for example, a region A gene means that region C genes would be expressed throughout the floral organ, changing the fate of two whorls

In order to explain the phenotypes of many mutants that have been observed, we need to assume, in addition, that the activities of regions A and C are mutually antagonistic. This means that when a mutant affects a gene that is expressed in region A (i.e. gene set A is missing), genes from region C are expressed in all four whorls. *apetala 2* is therefore a mutation of a homeotic gene expressed in region A, and this mutation changes the pattern of development of the outer two whorls of sepals and petals. When the mutant affects a gene that is expressed in the C region, genes from region A are expressed in all four whorls. *agamous* is a mutation of a homeotic gene in region C, and this mutation changes the development of the inner two whorls of stamens and carpels.

Interactions between these homeotic genes controlling different regions of the flower have been termed the 'war of the whorls'.

Genes that regulate flower development have been identified by analysis of homeotic mutations affecting flower structure. Mutations affect the identity of floral organs in adjacent whorls, switching them to alternative types of organs.

Summary

- Development involves a program of differential gene expression that is genetically regulated to occur in a spatial and temporal specific manner and which results in the formation of different tissue types in specific locations.
- Development and differentiation are not generally associated with loss of genetic material. Nuclei from differentiated cells can be reprogrammed to support full embryonic development.
- Gene activity during development can be regulated at a number of different levels, including transcription and translation.
- Gene activity during development is precisely regulated in space and time.
- Screens for developmental mutants in model organisms have uncovered both high level regulatory genes and downstream genes involved in morphogenesis and cell differentiation.
- Mutations in developmental regulatory genes are generally pleiotropic.
- Homology cloning provides a means for isolating a gene in a new species, starting with a known gene in a different species.
- Gene activation is often preceded by a series of protein–protein interactions in a cellular signalling pathway. These earlier events can involve the reception of molecular signals from other cells.

- Products of maternal-effect genes are synthesised in cells of the mother, located in egg cytoplasm and are essential for normal development of the zygote following fertilisation.
- *bicoid* is an example of a maternal-effect gene, which is transcribed during oogenesis but not translated until after fertilisation. Bicoid protein acts as a morphogen: the concentration gradient of Bicoid protein specifies the anterior–posterior body axis of the embryo.
- Zygotic (segmentation) genes extend the developmental program beyond that established by the maternal-effect genes to create segmentally repeated patterns of gene expression.
- Homeotic genes encode transcription factors that specify regional identity along the anterior–posterior axis in animals as diverse as insects and mammals by being differentially expressed in specific groups of segments.
- Genes that regulate flower development have been identified by analysis of homeotic mutations affecting flower structure. Mutations affect the identity of floral organs in adjacent whorls, switching them to alternative types of organs.

key terms

basic helix-loop-helix (bHLH) protein (p. 414)
cell fate (p. 418)
gap gene (p. 424)
genomic equivalence (p. 413)

homeodomain protein (p. 414)
homeotic gene (p. 426)
homology cloning (p. 419)
maternal-effect gene (p. 415)

morphogen (p. 424)
pair-rule gene (p. 425)
pattern formation (p. 417)
pole cell (p. 422)
rhombomere (p. 428)
segment-polarity gene (p. 425)

signal transduction cascade (p. 419)
tagmatisation (p. 426)
transcription factor (p. 414)
zygotic gene (p. 424)

Review questions

1. How could one test by experiment whether differentiation of a cell type involves the loss of genetic information in that cell?

2. Would you expect differentiated muscle cells to contain the gene for haemoglobin? Justify your answer.

3. Differentiated muscle cells do not contain haemoglobin mRNA. Which mechanism for regulation of haemoglobin gene activity operates in muscle cells?

4. What are transcription factors and how do they operate?

5. What is the homeodomain? Give an example of a protein that contains a homeodomain.

6. What are maternal-effect genes? Give a molecular explanation for the action of maternal-effect genes.

7. Describe one mechanism for translational regulation of gene activity.

8. Why would screening for mutations be an inappropriate method for identifying new genes involved in the development of the elephant? Suggest an alternative method for searching for developmental genes in this species.

9. Give an example of a developmental gene that controls cell fate. Draw a diagram to show where such a gene operates in a hierarchy of genetic interactions.

10. The genetic control of the pattern of a morphological structure can be independent of the control of the differentiation of the cells in that structure. What line of evidence has led developmental geneticists to that conclusion? Give an example of a patterning gene.

11. Give an example of a developmental gene that is involved in cell–cell signalling. What sort of protein does this gene encode and how does it act at a molecular level?

12. Give an example of a gene that acts intrinsically (i.e. within the cell in which it is expressed) to determine cell fate.

13. Explain how the *bicoid* gene specifies the anterior–posterior axis in the *Drosophila* embryo.

14. Draw a diagram to show regulatory interactions between genes involved in segmentation in *Drosophila*.

15. Draw diagrams to show the whorls of organs in flowers of two different homeotic mutations in *Arabidopsis*. Explain the simple model of the role of homeotic genes in regulating floral development.

Extension questions

1. Mutation of regulatory genes that are expressed very early in embryonic development are more likely to result in major defects in embryonic development than mutations in regulatory genes that are expressed late in development. Suggest why.

2. The nucleus from a differentiated cell is transplanted to an enucleated, fertilised egg. The egg starts to cleave but development ceases shortly after gastrulation. Can one justifiably conclude from this result that differentiation in this cell type involves the loss of genetic information? If not, why not?

3. If you were screening for a novel gene involved in segmentation it would be advisable to perform the screen at an embryonic rather than the adult stage. Explain why.

4. 'The formation of each organ in the body is under the control of a set of genes that operate exclusively in that organ. For example, there is a set of genes which is exclusively responsible for heart formation, a separate set of genes for limb formation and so on.' How do we know that this view of the genetic control of development is false?

5. 'Extrinsic versus intrinsic specification of cell fate are mutually exclusive alternatives.' Give an example to show why this statement is false.

6. Compare and contrast *Hox* gene structure and function in *Drosophila* and vertebrates.

Suggested further reading

Gilbert, S. F. (1997). *Developmental Biology*. 5th edn. Sunderland, MA: Sinauer Press.

An excellent advanced developmental biology text.

Lawrence, P. A. (1992). *The Making of a Fly*. London: Blackwell Scientific.

An advanced discussion of the patterning mechanisms that operate during embryogenesis in D. melanogaster. *Very good descriptions of segment formation mechanisms.*

Wolpert, L. (1991). *The Triumph of the Embryo*. Oxford: Oxford University Press.

A book, aimed at the general reader, that describes the way embryos develop.

Wolpert, L., Beddington, R., Brockes, J., et al. (1998). *Principles of Development*. Oxford: Oxford University Press.

Another excellent developmental biology text.

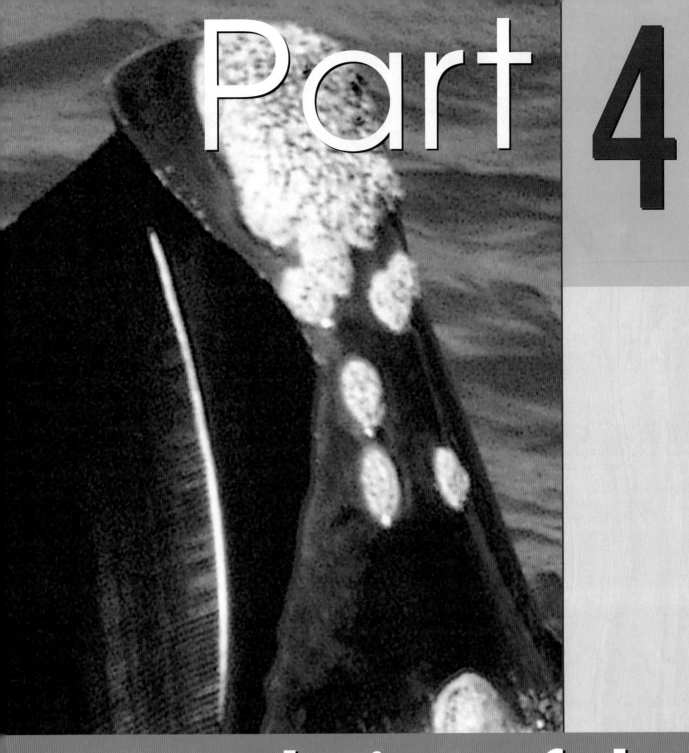

Part 4

4

Regulation of the internal environment

CHAPTER
17

Structure
of plants

Many structural features of plants are related to the requirements of living on land. These include preventing desiccation, acquiring nutrients and water from the soil, and providing physical support in an aerial environment. Plants have an aerial shoot system for photosynthesis and an underground root system for absorption and anchorage. Vascular plants, such as ferns, conifers and flowering plants, have a transport system linking the shoot and root.

While there are basic similarities in structure common to all vascular plants, there are also differences between taxonomic groups. For example, eucalypts (*Eucalyptus*) are easily recognisable by their characteristic branching and foliage patterns (Fig. 17.1): the crown of the tree is open, with branchlets in clusters and leaves that hang vertically. In contrast, a palm tree has a single trunk topped with large umbrella-like leaves (Fig. 17.2). There are also differences between individual plants of the same taxonomic group adapted to different environments.

In this chapter, we will consider the structure of vegetative organs of flowering plants, that is, stems, roots and leaves. The tissues and cells that make up

Fig. 17.2 An Australian rainforest palm has a non-branched trunk topped with large compound leaves

these organs have been described in Chapter 6. In Chapter 18, we describe the physiological functions of plants.

Stem structure

Aerial shoots consist of **stems** with attached leaves. The stem provides support, contains vascular tissues for water movement and sugar transport and, in some plants, is an important storage organ. The shoot grows from an apical meristem, which produces new cells and leaf primordia (Chapter 14). Leaves are attached at **nodes**, while the portions of stem between successive nodes are **internodes** (Fig. 17.3). Anatomically, the stem consists of a core of supporting and transport tissues surrounded by a mass of ground tissue, the whole structure protected from the external environment by a layer of epidermal tissue.

Primary growth

Vascular bundles are the conspicuous structural features in young stems of dicotyledon flowering plants

Fig. 17.1 The eucalypt, *Eucalyptus papuana*, has an open crown with branchlets in clusters

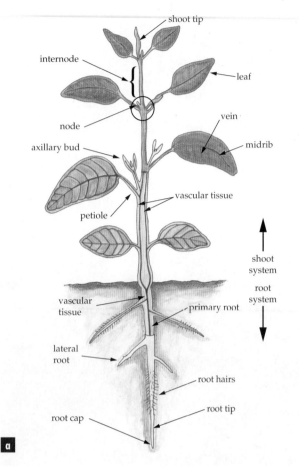

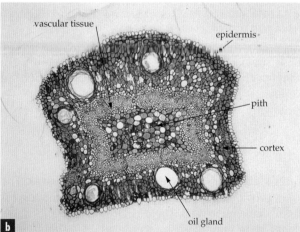

Fig. 17.3 The structure of a flowering plant. **(a)** Arrangement of the organs of a eucalypt, with primary stem and root (portion cut away to show vascular bundles). **(b)** Cross-section of a stem internode of *Eucalyptus preissiana*, showing major tissue types, including the ring of vascular tissue

double rings occur in some plants (e.g. pumpkin, *Cucurbita*). The vascular ring divides the stem into outer **cortex** and inner **pith**. The cortex is ground tissue, mainly parenchyma, usually containing chloroplasts. The pith also contains parenchyma, often with prominent intercellular spaces, and usually disappears during stem growth.

Vascular bundles consist of xylem and phloem tissues (Chapter 6). The first-formed and earliest maturing xylem, *protoxylem*, develops as thin strands towards the centre of the stem. To the outside of protoxylem, larger, thicker walled xylem elements, *metaxylem*, develop. This pattern of primary xylem development, in which new xylem is added to the

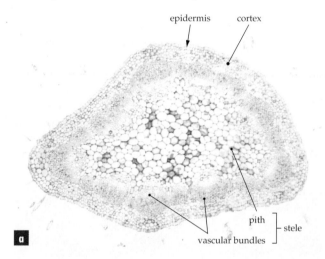

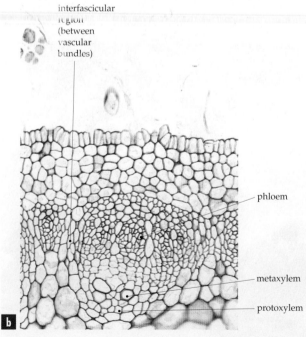

Fig. 17.4 Primary stem structure in the dicotyledon *Acacia*. Cross-sections showing **(a)** ring of vascular bundles dividing the stem into cortex and pith and **(b)** vascular bundle showing protoxylem and metaxylem elements forming an endarch xylem. Phloem is to the outside of the xylem

(Chapter 37), such as eucalypts (Fig. 17.1), wattles (*Acacia*) and beans (e.g. *Vicia*). Together with parenchyma, the vascular bundles form a central cylinder, the **stele** (Fig. 17.4a). Bundles diverge regularly from the stele and pass through the outer tissues into the leaves. Vascular bundles of the stem of a dicotyledon are usually arranged in a single ring, although

outside of the protoxylem, is known as **endarch xylem**. Primary phloem forms outside the xylem on the same radius (Fig. 17.4b).

In contrast to dicotyledons, vascular bundles in the stems of monocotyledons, such as grasses, lilies and palms, are scattered throughout the ground tissue so that there is no division into cortex and pith (Fig. 17.5). Each vascular bundle comprises xylem and phloem surrounded by sclerenchyma fibres.

> Stems of dicotyledons have a prominent ring of vascular bundles containing endarch xylem. Stems of monocotyledons have vascular bundles scattered in ground tissue.

Secondary growth

The mature stem of most herbaceous (non-woody) plants consists only of primary tissues as we have just described. However, in woody dicotyledons (shrubs and trees), there is another, quite different type of growth and cell production, secondary growth, which involves **cambium**, a secondary meristem. Unlike an apical meristem, which produces new cells behind it and continually adds length to a shoot (axial growth), cambium produces sheets of new cells *laterally*, thus adding girth (radial growth). Furthermore, whereas cells in primary tissues are irregularly organised (e.g. Fig. 17.5), cells formed by secondary meristems, at least initially, show regular radial alignment (e.g.

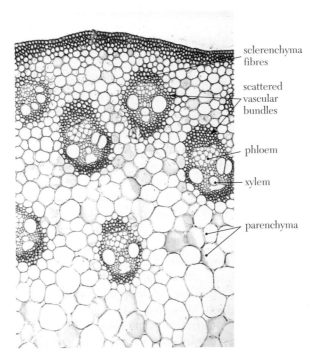

sclerenchyma fibres

scattered vascular bundles

phloem

xylem

parenchyma

Fig. 17.5 Stem structure of a monocotyledon. Cross-section of a stem of maize, *Zea mays*, showing scattered vascular bundles, with outer bundles surrounded by fibres. In contrast to dicotyledons, there is no division into cortex and pith

Fig. 17.6b,d). There are two types of cambium: **vascular cambium**, which produces wood, and **cork cambium**, which produces bark. Together these meristems produce the majority of tissue in the stems of dicotyledonous shrubs and trees.

In 'woody' tissues of monocotyledons, such as coconut palm, there is no vascular cambium and no secondary growth; strengthening of the stem results from many individual vascular bundles and their increased density towards the outer region (Fig. 17.5). A few monocotyledons, for example the New Zealand cabbage tree or *Ti Kauka* (*Cordyline australis*) and Australian grass trees (*Xanthorrhoea*), show secondary growth from a cambium. However, unlike dicotyledons, this cambium produces complete vascular bundles and parenchyma to the inside of the stem.

Vascular cambium

In most trees and shrubs, vascular cambium is a continuous, multilayered cylinder of living meristem between the cortex and pith, and is responsible for the production of the secondary vascular tissues (Fig. 17.6). Like primary vascular tissue, secondary vascular tissues consist of a number of cell types, most prominently tracheids and vessels of **secondary xylem**, and sieve cells of **secondary phloem**. Cell divisions of vascular cambium are generally periclinal, that is, the plane of division is parallel to the stem surface. When a cambial cell divides, daughter cells form to the inside and the outside of the cambium.

The vascular cambium contains two types of meristematic cells: fusiform and ray initials (Fig. 17.6c). **Fusiform initials** are greatly elongated cells aligned longitudinally in the stem, producing xylem vessels, tracheids and sclerenchyma fibres to the inside, and phloem sieve cells and sclerenchyma fibres to the outside. **Ray initials** produce parenchyma cells that are aligned radially in the stem and aggregated into clusters, forming wood rays. Ray parenchyma cells are the only living cells in the mature secondary xylem and maintain links with the vascular cambium. Parenchyma cells of the xylem can remain functional, storing starch and other inclusions, for several years after they are produced.

Secondary growth is most rapid in spring, and the xylem vessels produced are large and thin walled. As the growing season nears its end, environmental factors, such as reduced water supply, shorter day lengths and lower temperatures, slow growth so that smaller and thicker walled xylem is produced. Eventually growth ceases until the following spring. Consequently, there is an abrupt contrast between the last-formed xylem of one season and the new season's growth, which is visible in wood as **annual rings** (Fig. 17.6a). The age of many trees can be estimated by counting these annual rings, although allowances have

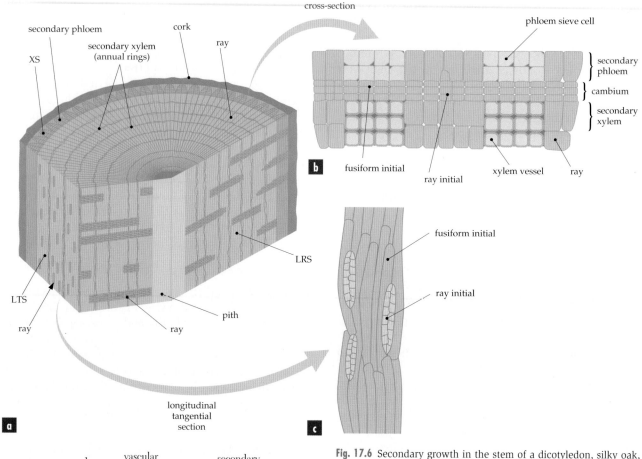

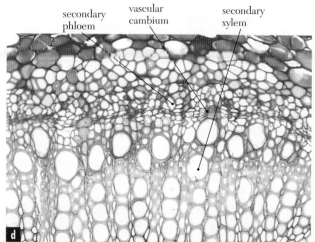

Fig. 17.6 Secondary growth in the stem of a dicotyledon, silky oak, *Grevillea robusta*. **(a)** Block of wood showing organisation of secondary growth in three planes: cross-section (XS), longitudinal tangential section (LTS) and longitudinal radial section (LRS). **(b)** Cambium and secondary xylem and phloem formation shown in cross-section. Fusiform initials of the cambium give rise to xylem vessels and phloem sieve cells. Ray initials give rise to ray cells, parenchyma cells that elongate radially. **(c)** Organisation of the fusiform and ray initials seen in longitudinal tangential section. **(d)** Cross-section of secondary xylem and phloem in the stem of a eucalypt

to be made for drought years when growth may be interrupted and resumed later in the same season. Also, climatic changes during a tree's life can be deduced from the appearance of growth rings. However, where growth seasons are erratic, such as in the Australian mallee, it is difficult to age eucalypts, for example, by counting growth rings.

When the secondary xylem matures, xylem vessels become infiltrated with organic compounds, such as oils, gums, resins, tannins, aromatic materials and pigments. At this time, the parenchyma cells of the rays

die, and the central region of xylem loses its water-conducting capacity and becomes **heartwood**. The outer sheath of xylem, **sapwood**, continues to function. The strength of heartwood compared with sapwood makes it useful in the building industry.

Cork cambium

Periderm is a protective tissue that replaces the epidermis in older stems and forms corky tissue (Fig. 17.7a). Periderm develops from the cork cambium (**phellogen**), which consists of a series of small, temporary meristems. Individual meristems first form as disc-shaped sheets of cells just underneath the epidermis in the outer stem cortex. Like other meristematic cells, they are continually dividing. Toward the outside of a stem, they produce **cork**, or phellem (Fig. 17.7b), and to the inside, small amounts of thin-walled parenchyma

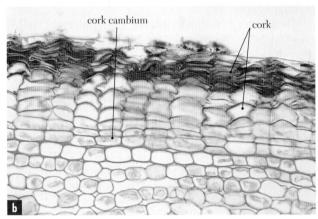

cork cambium

cork

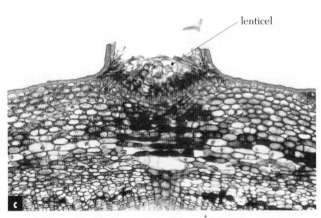

lenticel

Fig. 17.7 The periderm or outer protective layer of plants. **(a)** Bark is the protective covering on the trunk of this eucalypt, *Eucalyptus peltata*. **(b)** Cross-section of bark showing the periderm (cork) and cork cambium. **(c)** A lenticel in the bark of a tree forms an aeration pore

(phelloderm). Cork consists of closely packed dead cells whose cell walls are heavily impregnated with suberin, a very resistant and impermeable material. Suberin protects the delicate inner tissues of stems and roots. **Bark** is a general term for all tissue outside the vascular cambium and includes secondary phloem, cork cambium and cork. The bark of cork oak, *Quercus suber*, is impermeable to gases and liquids, and hence one of its uses is as stoppers for wine bottles.

Each individual cork meristem is short-lived. New discs of meristematic cells continually form deeper in the cortex. As these new cambia produce cork, older tissues to the outside are shed from the plant, for example, as ribbons of bark in the ribbon gum, *Eucalyptus viminalis*. Eventually all of the original primary tissues of the cortex are discarded, and all subsequent cork cambia arise in the no-longer functional outer layers of the secondary phloem. The reason that large quantities of secondary phloem never accumulate in the stem is because older phloem is continually shed by cork cambial action.

Like leaves (p. 452), young stems have special sites for gas exchange, stomata, in the epidermis. Because stomata of the stem are shed with the cortex during the earliest activity of cork cambia, new sites of gas exchange between the atmosphere and inner living tissues of a woody stem are needed. These are **lenticels**, tiny regions of the periderm consisting of loosely arranged cells with an extensive network of intercellular spaces (Fig. 17.7c). Lenticels often appear as raised areas or dots on stems and the skins of fruits; in old stems, they occur in furrows in the bark. However, unlike stomata, lenticels have no opening and closing mechanism.

Cambia are secondary meristems, sheets of cells forming secondary tissues. The vascular cambium produces wood and cork cambium produces periderm (corky tissue of bark).

Special functions of stems

Underground stems

Not all stems are aerial; some grow horizontally underground as **rhizomes**, from which aerial branches or leaves arise. Rhizomes often function in vegetative propagation, for example, in grasses (Chapter 14). Underground stems also function as food storage organs. A potato **tuber** is actually a swollen underground stem, complete with scale leaves and buds, produced in a single growing season. The edible parts of ginger (a monocotyledon; Fig. 17.8) and Jerusalem artichokes (a dicotyledon related to sunflower) are long-lived perennial rhizomes, with many short, thick nodes bearing papery, scale-like leaves. The **corms** of gladioli are also swollen underground stems, bearing buds and membranous leaves on the upper surface and adventitious roots below (see p. 443 and Fig. 17.13).

Stems of desert plants

In certain desert or semi-desert plants, where foliage would be a liability in terms of excessive heat absorption or water loss, leaves are tiny or absent and their role in photosynthesis is assumed by the stem. In she-oaks, *Allocasuarina*, leaves are reduced to small, scale-like

Fig. 17.8 Ginger rhizome, showing the short thick nodes bearing brown, papery, scale-like leaves

structures in a whorl at the nodes of stems or **cladodes** (Fig. 17.9), which are the photosynthetic organs. The cladodes have stomata located in grooves that extend the length of the internodes. Some desert plants, such as many cacti, are entirely leafless. Swollen, succulent stems of plants in the family Euphorbiaceae have special photosynthetic parenchyma (chlorenchyma)

together with water-storing parenchyma in the cortex. The epidermis of these plants is often multilayered and covered by a thick cuticle.

Stems of marsh and aquatic plants

Plants inhabiting saline environments, which, like deserts, induce water stress in plants, have developed a similar growth form. Leaves of the saltmarsh plant, *Sarcocornia*, are reduced to scales (Fig. 17.10) and the fleshy stem cortex consists of large palisade cells, which store water as well as carry out photosynthesis. Marsh soils are also waterlogged. Oxygen levels are low and plant stems often contain extensive aerenchyma, consisting of parenchyma cells with large, gas-filled intercellular spaces (Fig. 17.11).

In submerged aquatic plants, the epidermal wall is covered by a thin cuticle, which allows gas exchange directly with the surrounding water. Epidermal cells are rich in chloroplasts and carry out photosynthesis in addition to leaves. In seagrasses, the vascular system is greatly reduced and the xylem is not lignified.

> Stems show specific adaptations to the environment in which the plant lives. Some underground stems are storage organs, stems of desert plants can be photosynthetic, while stems of aquatic plants are adapted for living submerged in water.

Fig. 17.9 Stems of the she-oak, *Allocasuarina*, are cladodes and function as photosynthetic organs. Leaves are reduced to scales at each node

Fig. 17.10 Stems of the saltmarsh plant, *Sarcocornia* (samphire), are thick and fleshy for water storage

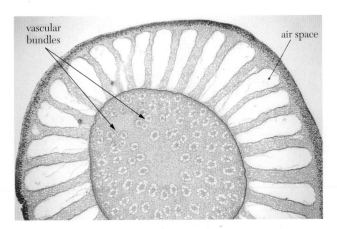

Fig. 17.11 Stems of the marsh plant rush, *Juncus*, have large air spaces in the cortex, formed by cell breakdown. The central area contains scattered vascular bundles typical of monocotyledonous plants

vascular bundles

air space

Root structure

Roots are the underground organs of vascular plants and grow from apical meristems at their tips (Fig. 17.12). They grow downwards by tip growth in response to gravity. Unlike stems, roots lack nodes and internodes. They do, however, have lateral branches, which originate from internal tissues. Roots and their surfaces have several important functions, including nutrient and water uptake from the soil, anchorage and support, synthesis of plant hormones and storage of nutritional reserves (Chapter 18). Roots of some plants are also modified for special functions, including aerial roots for oxygen uptake in salt marshes and swamps, clasping roots in climbing plants, prop roots for support and contractile roots to pull the plant firmly into its substrate.

The first root of a plant develops from the radicle of the embryo (Chapter 14). As this root enlarges, it produces lateral roots, which in turn divide to form a branched root system. This type of root system, centred around a main taproot, is the common type in dicotyledons (Fig. 17.13a, b). In some plants, roots can also arise from deep within the tissues of stems. These are **adventitious roots**, which are common at the nodes of grasses and other monocotyledons such as palms. In these plants, the taproot degenerates early and the functional root system is formed entirely from adventitious roots (Fig. 17.13c). In the Western Australian grass tree, *Kingia australis*, which can be up to 8 m tall, adventitious roots arise just below the stem apex and grow downwards, proliferating between the persistent leaf bases. This peculiar form of root arrangement explains why this species is especially difficult to transplant; replacement roots cannot develop fast enough to avoid desiccation of the plant.

A mature plant develops a large number of roots with enormous total length and surface area. A single, four-month old rye plant (*Secale cereale*) may possess 14 million roots with a total length of 630 km (about the distance from Melbourne to Adelaide). A single cubic centimetre of soil from under a Kentucky bluegrass plant (*Poa pratensis*) may contain a total root surface area of 400 cm^2.

> The main root produced from the radicle is the taproot; lateral roots develop to form a highly branched root system. Adventitious roots arise from stems.

Developmental processes

Lateral root formation
Meristem intiation

Specialisation zone
Cell differentiation

Elongation zone
Cell expansion

Meristematic zone
Cell division
Cell polarity establishment
Cell determination

Environmental stimuli

Light

Nutrients

Temperature

Aeration

Water

Obstacles

Microorganisms

Gravity

Adjacent roots

Fig. 17.12 The structure and functions of a root are affected by both developmental and environmental stimuli. Here we see the root of a radish, growing downwards by tip growth. Unlike stems, it lacks nodes and appendages, and develops root hairs and later lateral root branches. The internal (developmental) and external (environmental) factors that control root structure and development are shown

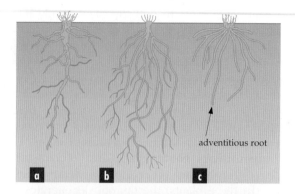

Fig. 17.13 Types of root systems. (a) and (b) show primary roots with lateral roots and (c) shows adventitious roots, which are characteristic of grasses such as (d) annual meadow grass, *Poa annua*, in which many lateral roots form at the base of the stem

Primary growth

A root is covered by an epidermis on the outside, with a well-developed cortex and an inner vascular cylinder (Fig. 17.14). A unique feature of roots is the root cap at the tip, which protects the apical meristem (Chapter 14).

Root cap

The **root cap** (Fig. 17.15) comprises large parenchyma cells, which secrete a polysaccharide, mucigel, containing a mucilaginous matrix and sloughed-off living cells. Mucigel acts as a lubricant, enabling the root to grow between soil particles without damage. It also binds soil particles together and may function as a culture medium for bacteria and other microorganisms associated with the root.

Root cap cells frequently contain masses of starch-containing amyloplasts. These are typically sited at the lower side of each cell, where they have accumulated under the force of gravity (Fig. 17.15). It has been

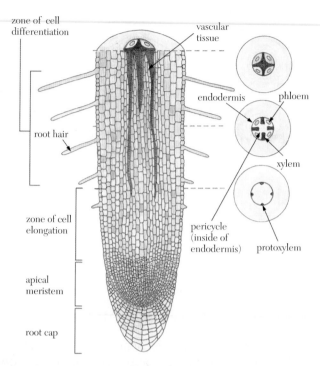

Fig. 17.14 Organisation of roots. Tissues in a root of a dicotyledon, showing the root cap, apical meristem, zone of cell elongation and zone of cell differentiation. Cross-sections of the root are shown for three regions

suggested that these starch granules act as statoliths, that is, cellular gravity sensors (Chapter 24).

Epidermis and root hairs

The extensive root surface area, required for uptake of water and nutrients from soil, is enhanced by a zone of numerous fine root hairs, finger-like extensions of the epidermal cells just behind the growing tip (Figs 17.12, 17.14). Root hairs usually possess a thin cuticle, which offers little resistance to the uptake of water and minerals. The soil immediately surrounding this absorptive portion of the root is the **rhizosphere**. In

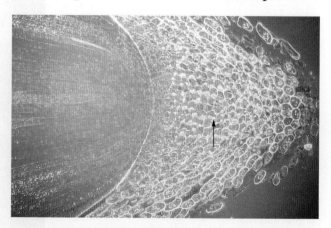

Fig. 17.15 Root cap of maize viewed in longitudinal section. Starch granules lie to one side of each cell (arrow), suggesting that they may act as the cellular sensors of gravity. Can you guess in which direction this root was growing?

this region, interactions take place between the plant and its soil environment, and with free-living and symbiotic microorganisms. Behind this root hair zone, lateral roots may develop.

Vascular cylinder

Most roots, including those of dicotyledons, have vascular tissue at the centre; there is no pith or vascular bundles as in stems. Roots have **exarch xylem**, in which xylem forms from the outside towards the inside of the root. A stele is formed with a star-shaped central core of metaxylem with protoxylem points (protostele, Fig. 17.14). Areas of primary phloem form separately and alternate with the protoxylem on different radii; in stems they are on the same radii.

In monocotyledon roots, such as wheat or rice, metaxylem does not completely fill the centre of the root. The xylem is **polyarch**, divided into many ridge-like projections (arches) of protoxylem and metaxylem vessels surrounding a large pith (Fig. 17.16). Phloem alternates with the protoxylem points.

Pericycle

The vascular cylinder is surrounded by the **pericycle** (Fig. 17.17), a layer of one or more cells thick from which lateral roots arise. Lateral roots grow from the pericycle through the cortex, physically forcing and digesting their way to the root surface and the soil (Fig. 17.17a).

Endodermis

Immediately outside the pericycle is the **endodermis**, a specialised, one-cell-thick layer that plays a critical role in water uptake (Fig. 17.17b). The four radial walls of each endodermal cell are impregnated with a strip of suberin, which makes the cell wall impermeable and water-resistant (Fig. 17.18). This suberin layer in the endodermis is known as the **Casparian strip**. Most water movement from the soil through the epidermis

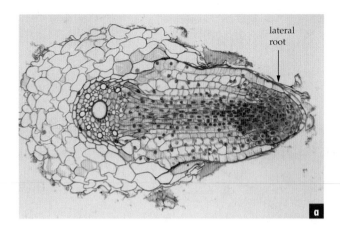

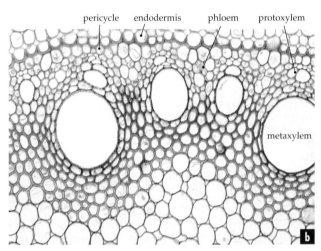

Fig. 17.17 (a) Thick section of a root showing the development of a lateral root (stained red). Cell divisions within the pericycle lead to the development of a new root meristem, which grows and pushes through the cortex and epidermis. **(b)** Endodermis in the root of a monocotyledon, maize, *Zea mays*. Semi-thin section of the root, showing the thick suberised radial and inner tangential wall (stained red), overlying the pericycle, the outer layer of vascular tissue. Phloem alternates with the xylem, which shows both protoxylem and metaxylem elements

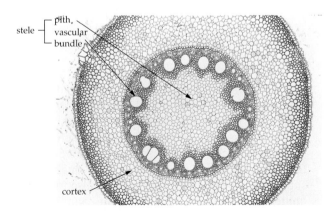

Fig. 17.16 Root of a monocotyledon, *Zea*, viewed in cross-section, showing the central vascular bundles (polyarch) and pith

and cortex of the root takes the path of least resistance through cell walls and intercellular spaces, the **apoplastic pathway**. Because this route does not involve crossing plasma membranes, the plant does not exert any physiological control over water movement. However, once water reaches the endodermis, it is prevented from continuing along the cell wall pathway by the Casparian strip. Instead, it must pass across the plasma membrane and into the cytoplasm of the endodermal cells, the **symplastic pathway**.

Because of the presence of the Casparian strip, ions must be actively transported through living cells of the stele before they can enter the non-living cells of the xylem (Fig. 17.18). Inward ion movement causes osmotic changes and the water potential (Chapter 4) in the stele decreases. Water enters, increasing turgor pressure in the stele. The endodermis is important

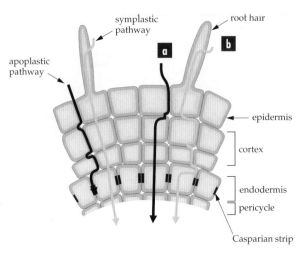

Fig. 17.18 The endodermis and paths of water and mineral movement in the root epidermis and outer cortex. There are several possible routes: through cell walls (apoplastic) with a switch at the Casparian strip; through cytoplasm (symplastic) after entry through root hair. **(a)** The effect of a switch from an apoplastic pathway to symplastic pathway. **(b)** The effect of a switch from a symplastic pathway to an apoplastic pathway

because it enables a plant to control the intake of water and ions. The Casparian strip prevents backflow of water from the vascular tissue into the root cortex.

Along most of the root, the endodermis also prevents the loss of sucrose from the phloem along the apoplastic pathway and sugar unloading occurs through the symplastic pathway. However, near the root tip, where the endodermis has yet to develop, there is no barrier and sugars are unloaded along the apoplastic pathway.

As a root ages, the endodermis becomes progressively modified to maintain turgor pressure in the stele. Suberin is laid down, covering all the primary wall except for sites of plasmodesmata. Additional thickening with cellulose is accompanied by lignification, which further toughens and seals the endodermis except for the plasmodesmata. The outer tangential wall of the endodermis is not usually suberised. In addition, some endodermal cells opposite the xylem are passage cells, remaining thin-walled and allowing some movement of materials. Despite suberisation, endodermal cells remain living cells.

Since the endodermis is internal to the cortex, its position leaves the cortex exposed to ions diffusing from the soil along the apoplastic pathway. In some roots (e.g. *Citrus*), an **exodermis** is formed at the junction of the epidermis and cortex. This layer appears to have many of the properties of an endodermis, but the Casparian strip is less well developed.

In most roots, vascular tissue forms a solid central core, surrounded by pericycle and endodermis. Endodermis is a specialised cell layer that controls water uptake.

Secondary growth

Secondary growth of roots, like stems, is characteristic of dicotyledonous flowering plants and other woody plants, such as pine trees. It results from the formation of secondary xylem and phloem from a vascular cambium in the stele. Periderm forms at the cortex as bark. Because of the presence of bark, roots with secondary growth function primarily in transport and not water absorption; water uptake from the soil only occurs near the tips of primary roots where there are root hairs. As a result of secondary growth, adjacent roots that come in contact with one another sometimes become united. Such root grafts commonly occur between adjacent trees and are one way that infectious diseases are transmitted from tree to tree in forests.

BOX 17.1 How palms grow into trees

There are few plants as distinctive or striking as palm trees, which are a common feature of tropical Australian rainforests. Their majestic unbranched trunks and dense crown of large, frond-like leaves makes them a favoured landscape tree in public gardens, beachfront reserves and stately avenues. Palms belong to the family Arecaceae. Together with several other distant relatives, such as *Pandanus* (screw palm), yucca and bananas, they are referred to as tree-forming (arborescent) monocotyledons. The fact that they can grow into trees is quite remarkable because, like other monocotyledons, they do not produce a vascular cambium and so cannot increase their stem diameter by secondary growth. All of the cells and tissues in a palm tree trunk, including the xylem and phloem, are produced by the apical meristem of the shoot and must remain functional for the life of the tree. This is in stark contrast to woody dicotyledons, in which the primary tissues soon degenerate and the structural and metabolic functions of the stem are assumed by secondary tissues.

When a palm seedling emerges from a seed, the apical meristem is relatively small and enlarges gradually as the stem grows. Mature palm trees have massive apical meristems, up to 30 cm across. This results in a trunk of increasing diameter, which is narrowest at its base. Such a trunk is inherently unstable and is unable to support a large tree. The narrow base also forms a bottleneck, severely restricting the amount of water that can be conducted into the stem. The seedlings of most species of palm, such as coconuts and dates, grow downward into the soil for some distance before then growing toward the surface. The widening of the apex is achieved during this period of underground growth, ensuring that the narrow base of the trunk is firmly embedded in the soil (Fig. a). A few palms and arborescent monocotyledons, such as *Pandanus*, overcome the problems of instability and water transport by a different but equally unique

solution. Adventitious roots arise from the stem and grow down to the ground (Fig. b). These prop roots are thick rigid organs that stabilise the base of the tree, but their name belies their other major function. They effectively bypass the narrow base of the stem by delivering water from the soil directly to the thicker portions of the trunk.

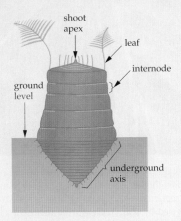

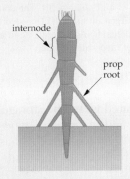

(a) Most palms have an underground stem that progressively widens with short internodes; growth in height is achieved by development of longer internodes formed after the diameter of the adult stem is attained

(b) Screw palm, *Pandanus*, with prop roots. Growth in height begins early and the widening stem is supported by prop roots

Special functions of roots

Roots of aquatic plants

Plants grow in many different environments, and their roots show specific modifications, especially those associated with aquatic habitats. For example, the mangrove *Rhizophora* has upright aerial roots, **pneumatophores** (Fig. 17.19a), which are exposed at low tide and function in gas exchange for aerobic respiration. They are negatively gravitropic and grow upwards toward the light. Pneumatophores have a cortex containing aerenchyma through which, at low tide, oxygen diffuses from the air down to roots buried

in the anaerobic mud. Similarly in rice, cortical root cells break down to form large air spaces (Fig. 17.19b), which are required for gas exchange when the soil is flooded.

Aquatic plants, especially those that are totally submerged, suffer little water loss by transpiration, so they have a substantially lower requirement than terrestrial plants for water uptake. As a consequence, their root systems are generally simple in structure. Seagrass roots, in particular, have very weakly developed vascular tissues, lacking lignified xylem. Like mangroves, they also have cortical aerenchyma with intercellular spaces for gas exchange because the concentration of oxygen in water is relatively low.

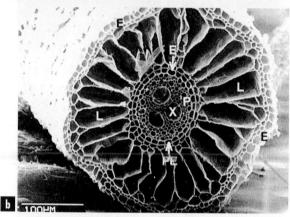

Fig. 17.19 Special functions of roots. **(a)** Aerial prop roots and pneumatophores, the latter functioning in gas exchange, of the mangrove *Rhizophora stylosa*. **(b)** Appearance and structure of the root of rice viewed by scanning electron microscopy. The cross-section of the root shows epidermis, cortex, endodermis (E), pericycle (PE), xylem (X) and phloem (P). The cortical cells have broken down to form larger air spaces (L) needed for gas exchange in waterlogged paddy fields. **(c)** Aerial roots of an epiphytic orchid covered by a thick outer layer, the velamen

Other types of aerial root are found in epiphytic orchids, which grow on other plants for support (Chapter 43). The epidermis, known as the *velamen*, covers all but the absorptive tip of the orchid root and is thick and multilayered, preventing water loss (Fig. 17.19c).

Roots and salinity

Root structure plays a role in determining salt tolerance in river red gums, *Eucalyptus camaldulensis*. River red gum roots have an exodermis (Fig. 17.20), a suberised layer beneath the epidermis that develops as roots mature. In trees susceptible to salinity, there is always a zone of non-suberised hypodermal cells near the root apex. In salt-tolerant trees, suberised exodermis develops earlier and near the root tip, minimising the zone of non-suberised cells. The presence of the exodermis is correlated with the root's ability to exclude chloride ions, and this barrier thus acts to protect the plant from saline conditions.

Storage roots

Adaptations associated with food storage often involve a swollen taproot and an unusual type of secondary growth. Carrot, *Daucus carota*, has a vascular cambium that produces a moderate amount of

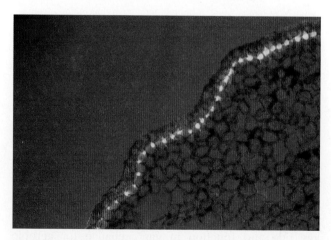

Fig. 17.20 Root tip of river red gum, *Eucalyptus camaldulensis*, in longitudinal section, showing the presence of exodermis, a suberised endodermal-like layer surrounding the cortex. This layer develops close to the root tip, enabling some plants to be salt tolerant since salt cannot pass across the exodermis

secondary xylem interspersed with a large amount of storage parenchyma. Roots of the sweet potato, *Ipomoea batatas*, develop additional vascular cambia that mainly produce parenchyma to form the swollen tuber, an organ of vegetative reproduction that forms roots below and stems above. The fleshy root of beetroot is also a product of multiple, concentric vascular cambia. Each new cambium forms a ring of parenchyma (dark bands) alternating with vascular tissue (light bands) to form the fleshy texture and familiar pattern of beetroot slices.

> Roots of some plants are modified for growth in aquatic or saline conditions. Storage roots involve unusual secondary growth.

Root adaptations and nutrient supply

A wide range of plants, especially legumes (family Fabaceae) and wattles (family Mimosaceae), have **root nodules** (Fig. 17.21a). Root nodules contain nitrogen-fixing bacteria, which convert inert gaseous nitrogen from the atmosphere to organic nitrogenous compounds that are then available to the host plant (Chapter 44).

Soil bacteria, for example, *Rhizobium* in legume roots or *Frankia* in non-legume roots such as the she-oak, *Allocasuarina*, enter roots via root hairs. They then form an infection thread that passes through several cell layers and into the root cortex. The infection stimulates rapid cell division of the root, resulting in the formation of a root swelling or nodule, which joins with the conducting system of the host plant. Infected cells contain numerous *bacteroids*, which are modified bacterial cells formed within the root nodule cells. The nodules have a distinct pink colour (Fig. 17.21b) due to the presence of the transport protein haemoglobin (Chapter 20), which is usually produced in the host cell to maintain a high flux of oxygen.

Coralloid roots are found in the roots of she-oaks, alder (*Alnus*) and cycads (e.g. *Macrozamia*). These result from root hair infection by the bacterium *Frankia* and cyanobacteria, producing a coral-like growth of roots, which grow upwards with a characteristic dichotomous branching structure. The cells of apical meristems are uninfected, while growing cortical cells are invaded by thin filaments of the micro-organisms. The cortical cells develop vesicles in which nitrogen fixation takes place. The infected outer cortex contains periderm and tannin-filled cells, while the inner cortex and vascular system are uninfected.

Root clusters (**proteoid roots**) are groups of hairy rootlets that form dense mats at the soil surface in *Banksia* and *Dryandra* (family Proteaceae) and in some legumes (*Viminaria* and *Daviesia*, family Fabaceae). They result from massive proliferation of rows of lateral rootlets, up to 1000 per centimetre, along the parent root (Fig. 17.22). Masses of proteoid roots have a white glistening appearance and feature extensive development of vascular tissue xylem inside the root. Their function is to enhance nutrient uptake in nutrient-poor soils.

Mycorrhizae are examples of symbiotic associations between roots and fungi (Chapter 36). Such symbiosis aids the root in the absorption of nutrients, particularly phosphorus, from the soil, and mycorrhizae are often essential for successful plant growth. They are a feature of most terrestrial plants, particularly

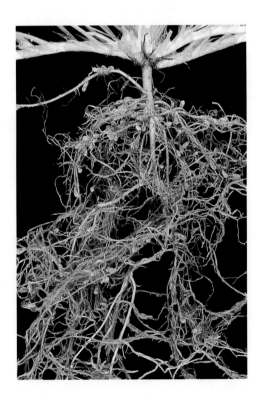

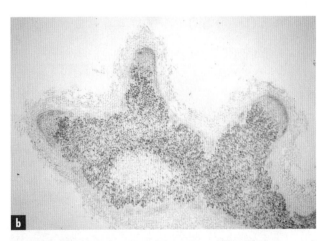

Fig. 17.21 Structure of nitrogen-fixing root nodules. **(a)** The roots of white clover show the presence of nodules, in which symbiotic, nitrogen-fixing bacteria live. **(b)** In cross-section, the infected cells (central in the nodule) are a distinct pink colour because haemoglobin is usually produced in the host cell

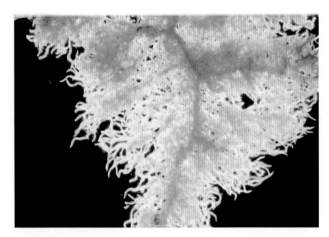

Fig. 17.22 Root clusters (proteoid roots) of *Banksia* result from a proliferation of lateral rootlets, enhancing nutrient uptake

epiphytic orchids and heath plants (Chapter 41). There are two types of mycorrhizae. The fungal symbiont of *ectomycorrhizae* produces a thread-like mass or mycelium (Chapter 36) on the root surface, which is partly in the rhizosphere and partly in the root (Fig. 17.23a). When the fungal mycelium penetrates the cortex, the fungus grows between the cells forming the Hartig net (see Fig. 36.23c). The fungus is typically restricted to the cortex and does not penetrate the endodermis. Infected roots are short and usually dichotomously branched. In contrast, in *endomycorrhizae*, common in orchid roots, the plasma membrane of a host plant cell surrounds the fungal hyphae, which form *intracellular* coils of hyphae (Fig. 17.23b).

> R oots show several modifications associated with increasing the supply of nutrients: root nodules and coralloid roots fix nitrogen, while root clusters (proteoid roots) and mycorrhizae improve phosphorus uptake.

Leaf structure

Leaves are the prominent photosynthetic organs of vascular plants. They have an enormous diversity of shapes and sizes depending on the species and the environment in which a plant is growing. Most leaves consist of a narrow stalk (*petiole*) and a thin, flat, expanded blade (*lamina*) and are attached to stems at nodes (Fig. 17.3). The structure and orientation of leaves maximises the interception of sunlight while minimising water loss, and allows carbon dioxide to be extracted from the atmosphere. Leaves also have secondary roles such as the storage of toxic compounds, which deter herbivores (Chapter 19). Some leaves are modified to perform particular functions, such as protection in the form of bud scales and spines, carbohydrate storage in fleshy bulbs and support as tendrils.

Leaf arrangement and life span

Leaves are arranged on the stem in an ordered and regular pattern. In most plants this pattern is a spiral, but leaves in pairs, whorls or some other configuration are also common. The pattern of leaf arrangement, **phyllotaxy**, is largely determined when leaf primordia are formed at the shoot apex, and ensures minimum shading of one leaf by another and maximum exposure to sunlight.

Individual leaves have a limited life span, generally about one growing season, but in pines the leaves ('needles') commonly persist for two to four years. Evergreen plants, which bear leaves all year round, such as most Australian trees, continuously shed their oldest leaves and replace them with new ones, thus maintaining a full complement of leaves on the tree at all times. In northern Australia, some deciduous trees shed their leaves at the onset of the dry season

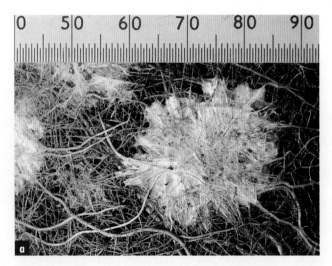

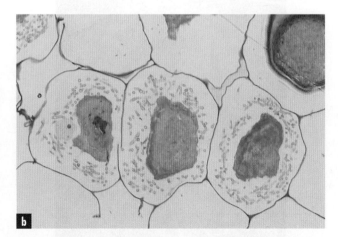

Fig. 17.23 Mycorrhizae—a fungal–root association. **(a)** Ectomycorrhizae produce a fungal sheath on the surface of pine roots, partly in the rhizosphere and in the root, which grows between the cells of the cortex but does not penetrate the endodermis. **(b)** Endomycorrhizae occur in orchid roots, in which the coils of hyphae are intracellular, forming pelotons (blue-staining regions) in the outer tissues

(Fig. 17.24). This is usually at the end of the growing season, so that a complete set of foliage is renewed the following spring. In the forests of Europe and North America, which are dominated by deciduous oaks, birches, beeches and maples, trees shed their leaves in autumn before the onset of the cold winter.

Leaves of herbaceous plants typically wither on the plant and decay, but leaves of woody plants are actively shed. Shedding occurs as a result of changes in a special **abscission zone**. This is located at the base of the leaf petiole and comprises several layers of small, cube-like cells, which secrete enzymes that digest cell walls and cause leaf fall (abscission). This process is regulated by hormones, such as ethylene, and by several environmental factors, including drought, low temperature, low light intensity and short day lengths (Chapter 24).

Leaves are photosynthetic organs and have a limited life span: evergreen leaves are shed continuously while deciduous trees shed their leaves before the onset of a cold or dry period.

Leaf shape and orientation

Leaf size ranges from several metres in palms to tiny scales of a few millimetres. Leaves with a single lamina are **simple leaves**, while those divided into many leaflets, each with their own stalk, are **compound leaves** (Fig. 17.25). Clover leaves are an example of compound leaves that have three leaflets arising from the same point (*trifoliolate*). Compound leaves of silver wattle,

Fig. 17.24 Young trees of *Brachychiton megaphylla* shed their large leaves at the onset of the dry season in Kakadu National Park, Northern Territory

Fig. 17.25 Organisation of compound leaves of a legume, showing the main rachis (R) with a swollen pulvinus (P) at the base, which controls the angle of the leaf to the stem, and the many leaflets (L) attached to each secondary rachis forming the bipinnate leaf

Acacia dealbata, have leaflets borne in rows and are *bipinnate* (Chapter 41). In bipinnate leaves of legumes, the base is slightly swollen (the **pulvinus**). In some species, the pulvinus acts as a motility organ that can change the position of the leaflet. In *Albizia*, the pulvinus operates so that each leaflet is held open horizontally during the day, and closed in a vertical position at night as a kind of sleep movement. In most Australian wattles, compound leaves are replaced by phyllodes as the plant grows. A **phyllode** develops by expansion of the petiole and rachis (midrib of a compound leaf) to form a photosynthetic lamina (Chapter 41).

In most plants, leaves are displayed horizontally and are **dorsiventral**, that is, with a definite upper (dorsal) and lower (ventral) surface. Some plants, however, have leaves that are held vertically. In these leaves, both surfaces are the same, and the leaf is **isobilateral** or *bifacial*. Leaves of adult eucalypts are typically pendant and hang vertically from a branch, while those of the geebung, *Persoonia*, stand erect. Both are adaptations for equal illumination on both sides of the blade while at the same time maintaining an orientation that reduces heat absorption.

Juvenile and adult leaf structure

Many plants produce a number of different leaf types during growth and maturation from a seedling to an adult. Juvenile and adult leaves can be markedly different in size, shape, colour, arrangement and even anatomy. In *Eucalyptus calcicola* from Western Australia, the juvenile leaves are shiny, bright green and lack a petiole (Fig. 17.26a). They are arranged in opposite pairs on the stem and are held horizontally for increased light interception. In contrast, the adult plant has the typical isobilateral, petiolate leaf of eucalypts (Fig. 17.26b). The juvenile leaves of the blue gum, *Eucalyptus globulus*, are waxy and blue-grey and strikingly different from those of the adult. Some eucalypts, such as spinning gum, *Eucalyptus perreniana*, reach full size, flower and set fruit, but their foliage never undergoes a transition to adult leaves. This results in a mature tree bearing only juvenile leaves. Such foliage is attractive and used in the horticultural export trade.

Simple leaves have a single lamina while compound leaves have many leaflets. Juvenile leaves can differ from adult foliage in colour and form.

Fig. 17.26 Juvenile and adult foliage of *Eucalyptus calcicola*. **(a)** Shiny green juvenile leaves are heart-shaped and held horizontally on the stem, without a petiole. **(b)** Adult leaves are waxy, with a petiole and are pendant

Leaf structure and organisation

A leaf blade typically comprises upper and lower epidermis, enclosing a mesophyll containing chloroplasts. Vascular bundles enter the petiole from the stem and branch out into the lamina to form a network of **veins**, the conducting tissue of the lamina. The leaves of dicotyledons usually have a distinct central vein (midrib, Fig. 17.3), prominent lateral veins and a network of smaller veins that form a web-like reticulate pattern. Leaves of most monocotyledons have veins running parallel to each other along the axis of the blade.

Epidermal cells

The leaf epidermis is typically a single layer of cells that are rectangular in profile (Fig. 17.27). The epidermal cells are covered with a thick cuticle and particles of wax, and contain specialised pores, **stomata**, which allow uptake of carbon dioxide from the atmosphere (Chapter 5). Each stomatal apparatus consists of two kidney-shaped **guard cells** that are joined at the ends of their abutting walls. These walls are impregnated with suberin, and when the guard cells fill with water and become turgid, the differential thickening of their walls causes a gap to open between them, the stomatal pore, exposing internal cells of the leaf to the atmosphere. The stomatal opening is very small, typically 7–40 μm long, and 3–12 μm wide when fully open. Guard cells contain chloroplasts, but other epidermal cells are devoid of photosynthetic pigments.

The number of stomata on leaf surfaces varies. On the lower surface, numbers of 15–1000 per mm^2 are typical, the lower number referring to plants with large stomata. For example, cabbage, a leafy vegetable, has about 600 per mm^2. In Australian plants, the number of stomata varies from about 28 per mm^2 in geebung, *Persoonia*, to 100–340 per mm^2 in eucalypts.

As well as stomata, there may be one or more types of specialised cells present in the epidermis, such as leaf hairs (Fig. 17.28), secretory gland cells and crystal-containing cells. Grass leaves possess several unique cell types. Dumbbell-shaped cells with silicate crystals in their walls form in rows on the epidermis, giving many grasses a knife-edge quality and abrasive texture, which deters grazing animals. Bubble-shaped bulliform cells are long and thin-walled, and contain a large vacuole (Fig. 17.29). Change in turgor of these cells appears to cause the lateral rolling and unrolling of a grass leaf in response to environmental conditions.

Mesophyll

Mesophyll is the ground tissue of leaves. It consists of chloroplast-containing parenchyma cells, which are the sites of photosynthesis. In dorsiventral leaves, the shape of mesophyll cells is different in the upper and lower

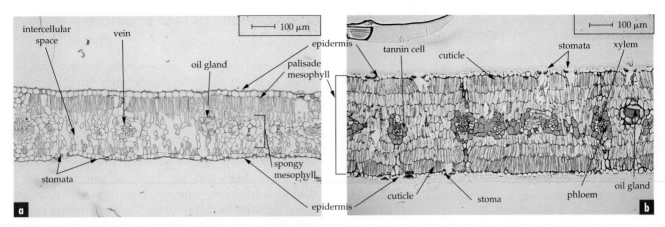

Fig. 17.27 Leaf anatomy of *Eucalyptus globulus*, viewed in cross-section. **(a)** Dorsiventral juvenile leaf. **(b)** Isobilateral adult leaf

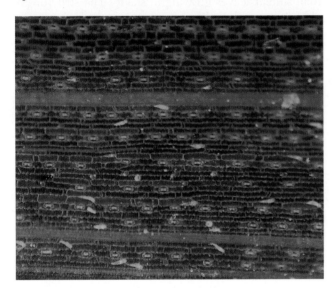

Fig. 17.28 Fluorescent micrograph of the surface of a leaf of kikuyu grass, a monocotyledon, showing parallel veins, stomata (closed) and small hairs. Patterns of epidermal cells can also be seen

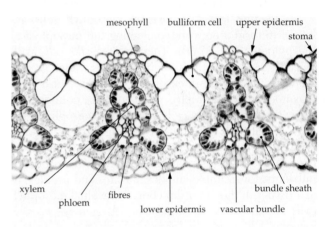

Fig. 17.29 Structure of the leaf of love grass, *Eragrostis brownii*, a monocotyledon, viewed in cross-section. The vascular bundle (xylem above phloem) is surrounded by a prominent bundle sheath, comprising large cells containing chloroplasts, indicating a C_4 pathway of photosynthesis. There are large bulliform cells on the upper surface responsible for rolling and unrolling the leaf

halves of the leaf (Fig. 17.27a). Toward the upper surface, mesophyll cells are elongated, with their long axis perpendicular to the leaf surface and chloroplasts arranged just inside the vertical walls for maximum light penetration; this is the **palisade mesophyll**. These mesophyll cells have the highest numbers of chloroplasts and intercept most of the sunlight. Toward the lower surface, mesophyll cells are more or less regular in shape (isodiametric) and loosely packed, forming the **spongy mesophyll**. The extensive intercellular space in spongy mesophyll allows free diffusion of carbon dioxide from the stomata, which are located predominantly in the lower epidermis.

Isobilateral adult leaves of eucalypts have palisade mesophyll developed at both surfaces, with a small amount of spongy mesophyll restricted to the centre of the leaf (Fig. 17.27b). Eucalypt leaves are also equally thickly cutinised on both sides (Fig. 17.27b). The cuticle arches over the stomata so that they are sunken beneath the epidermis, which reduces water loss. The

veins are surrounded by tannin cells, which may serve as an ultraviolet filter.

Leaves of many plants, especially the myrtle family Myrtaceae, which includes eucalypts and clove, and the mint family Labiatae, have oil glands. In *Eucalyptus*, oil glands are lined with cells that secrete oils, which accumulate in a central cavity (Fig. 17.27). Each gland can fill a space as large as the entire width of the upper or lower palisade. Oils are a mixture of toxic compounds that make the leaves less palatable to herbivorous insects, but they are also the feature that make herbs and spices valuable culinary plants. Some aromatic flavourings are extracted from glandular hairs, for example, menthol and peppermint oil from the mint *Mentha piperita*.

Vascular system of leaves

Veins (vascular bundles) of the leaves of flowering plants form an interconnected branching system. Leaf vascular bundles contain xylem located toward the upper surface and phloem toward the lower surface (Figs 17.27, 17.29, 17.30). The minor veins in dicotyledons and the veins of monocotyledon leaves are surrounded by a **bundle sheath**, a layer of tightly packed parenchyma cells (Fig. 17.29). These cells are usually elongated in shape (in the direction of the vein axis) and may contain chloroplasts. In most grasses, the bundle sheath cells link with fibres that connect with the epidermis to provide increased rigidity (Fig. 17.29). In some tropical grasses, which have the C_4 pathway of photosynthesis (Chapter 5), the bundle sheath cells are enlarged and contain many chloroplasts (Fig. 17.29). In some plants, especially mountain and alpine peppers, *Tasmannia* (family Winteraceae), the sheath comprises sclerenchyma, which makes the leaf more rigid.

In legume leaves, paraveinal mesophyll cells are linked to form a network connecting the mesophyll to the phloem (Fig. 17.30). They are *transfer cells* with conspicuous ingrowths of their cell walls, which greatly increase the surface area of the cell. They also have plasmodesmata, which allow direct connections between neighbouring cells. Paraveinal mesophyll cells form a pathway for transfer of sugars from photosynthetically active tissue to the vascular system in one direction, and distribution of xylem-transported nutrients, for example, amides, nitrates and ureides, to leaf tissues in the other direction. Paraveinal mesophyll cells are the sites for storage of nitrogen reserves as proteins. Clover leaves, for example, are rich in protein and are important in the diet of grazing animals.

The leaf lamina consists of palisade and spongy mesophyll, both of which contain chloroplasts. The leaf epidermis has specialised cells, including stomata for gas exchange.

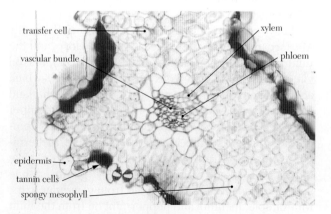

Fig. 17.30 Cross-section of a phyllode of a wattle, *Acacia cardiophylla*. Spongy mesophyll is the photosynthetic tissue, and many of the cells are transfer cells with wall ingrowths (red-stained patches). The transfer cells form part of the paraveinal mesophyll cell system that connects the phloem with the photosynthetic tissue. The layer of cells beneath the epidermis are tannin cells (stained dark purple)

Modifications of leaf structure

Environmental effects

Leaf size and shape are usually similar within a species, which indicates that the major underlying developmental processes are under genetic control. However, a number of aspects of leaf morphology and anatomy are directly influenced by the environment. For example, differences in leaf size and structure are evident in parts of the same plant exposed to different illumination levels (high and low light, or sun and shade). Shaded leaves are generally larger in size than are leaves exposed to full sunlight. Leaves growing in full sunlight have substantially more palisade mesophyll than similar leaves growing in shade. The structure of chloroplasts also differs in sun and shade leaves (Box 5.2).

The thickness of cuticle deposited on the epidermis is strongly influenced by the environment in which the plant grows. Plants grown outdoors can have a leaf cuticle 10 times thicker than that of similar plants grown in a glasshouse. The cuticle is needed to prevent water loss, protect the photosynthetic surface from the sun's rays and prevent the entry of pathogens.

Leaf adaptations

Plants growing in particular environments show special adaptations of their leaves. For example, plants in arid and semi-arid environments often have small, scale-like leaves (Fig. 17.9). Unusual pore-like structures, **hydathodes**, occur on leaves of certain rainforest plants, along their margins or at the tips. These permit water to be extruded by guttation, in which water is forced out of the hydathodes by root pressure (hydrostatic pressure in the xylem). Guttation typically occurs early in the morning when root pressure is high, for example, in the calla lily. It helps maintain water movement and therefore ion transport through a plant that is growing under conditions of high relative humidity when transpiration is low. A considerable number of tropical plants also have leaves with 'drip tips' (Fig. 17.31a), an adaptation that allows runoff of rainwater in wet tropical rainforests.

Succulents such as pigface, *Disphyma*, have swollen leaves that contain water-storage tissue—large parenchyma cells filled with a central vacuole containing hydrophilic colloids. Succulence is an adaptation for drought tolerance in dry habitats.

Some climbing plants have leaves modified as tendrils, which twine around the substratum for support. For example, the climbing rattan palms of Australian and South-East Asian tropical rainforests (Fig. 17.31b) have whip-like extensions of the midrib covered with spines that enable the plants to climb to the upper canopy of the forest (and inflict pain on the unsuspecting visitor who walks by!). In the grape vine, *Vitis*,

Fig. 17.31 Adaptations of leaves. **(a)** Drip tips on the leaves of the giant stinging nettle tree ensure that excess rainwater runs off the leaves. **(b)** Rattan palms have hooked tendrils, which enable them to climb over rainforest vegetation. **(c)** Tendrils on compound leaves of the garden pea, *Pisum*. **(d)** Needle-like leaf of Geraldton wax, *Chamelaucium uncinatum*, viewed in cross-section, showing the epidermis covered by a thick cuticle with stomata, and photosynthetic mesophyll

the entire leaf is a tendril, while in the garden pea, *Pisum*, only the terminal leaflet is modified (Fig. 17.31c).

Many Australian flowering plants, particularly those that live in seasonally dry or arid habitats, have stiff, hard leaves (**sclerophyll**). For example, the needle-like leaves of *Hakea* and Geraldton wax, *Chamelaucium uncinatum*, have an epidermis that is heavily cutinised with sunken stomata (Fig. 17.31d). Palisade mesophyll is arranged in concentric layers around the leaf and contains characteristic sclereids that are dumbbell-shaped, while the central spongy mesophyll contains isolated vascular bundles and areas of sclerenchyma. These features combine to confer rigidity and drought resistance on the leaf. Some plants in dry environments also have leaves covered by wax plates, giving them a dull, glaucous appearance.

Surface hairs or trichomes can have several functions and are important in certain environments. Their physical presence reduces water loss by creating a barrier to evaporation, as in the case of 'woolly' leaves that are covered in dense hairs. They may also provide an unusual source of nutrition, as in carnivorous plants, where the hairs secrete a mucilage to trap insects and surface glands secrete enzymes that digest the captured prey. They may secrete chemicals to deter herbivores. For example, the substances secreted by glandular hairs in potatoes, tomatoes and sunflowers give protection from attack by aphids, caterpillars and other herbivores. Tomato leaves have six types of trichomes; one of these (type IV) can secrete more than 22 different small molecules (sesquiterpene lactones). These molecules have the remarkable property that they can be detected by the nervous system of insects through special hairs on the insect's surface, causing the insect to leave. These chemicals thus act as natural insect repellents.

Plants growing in particular environments show special leaf adaptations. In dry environments, leaves may be needle-like, have stomata sunken in pits and be covered with a thick cuticle and special wax plates. Hairs or trichomes on leaves may secrete small molecules as insect repellents.

Summary

- Plant structure is a compromise between competing demands for mechanical support, food production, transport of water and nutrients, and reproduction.
- Plants have an aerial shoot system, consisting of stem and leaves, for the function of photosynthesis, and a root system for the functions of absorption and support.
- Primary plant structure differs between dicotyledons and monocotyledons. Stems of dicotyledons are characterised by a prominent ring of vascular bundles containing endarch xylem, a cortex and pith. Stems of monocotyledons have vascular bundles scattered in ground tissue.
- The structure of woody dicotyledons (trees and shrubs) results from secondary meristem (cambia) activity. The vascular cambium produces wood and cork cambium produces the corky tissue of bark. Cambial activity is influenced by climate, leading to the formation of annual rings in wood.
- Some monocotyledons (palms) also show a tree-like habit but this results from primary growth.
- Stems of many plants have special functions related to the environment in which they live. Some stems are storage organs and others may contain enlarged parenchyma cells specialised for photosynthetic and water-storage functions.
- The main root produced from the radicle is the taproot from which lateral roots develop to form a highly branched root system. In most roots, including dicotyledons, the vascular tissue containing exarch xylem forms a solid central core within the cortex, bounded by the endodermis and pericycle. In monocotyledons such as grasses, the roots usually arise adventitiously; the vascular tissue is in a ring surrounding the central pith.
- Like stems, root structures may also have special functions related to their environment. The roots of some plants are modified to grow in aquatic or saline conditions. Storage roots involve unusual secondary growth. Some plants have roots specialised for improving nutrient supply. Root nodules, coralloid roots and mycorrhizal roots involve symbiotic associations with microorganisms.
- Leaves normally have a limited life span: evergreen leaves are continuously being shed while deciduous trees shed their leaves in autumn before the onset of cold conditions. Some tropical eucalypts shed their leaves before the dry season.
- Simple leaves have a single lamina while compound leaves have a number of leaflets. The leaf lamina consists of palisade and spongy mesophyll, both of which contain chloroplasts arranged to optimise light capture. The leaf epidermis has specialised cells, including stomata for gas exchange. Plants produce different types of leaves as they develop from seedling to adult.
- Plants growing in particular environments show special leaf adaptations. In dry environments, leaves may be needle-like, have stomata sunken in pits and be covered with a thick cuticle and special wax plates. Hairs or trichomes on leaves help reduce water loss and may secrete small molecules as insect repellents.

keyterms

abscission zone (p. 451)	corm (p. 441)	pericycle (p. 445)	secondary phloem (p. 439)
adventitious roots (p. 443)	cortex (p. 438)	periderm (p. 440)	secondary xylem (p. 439)
annual rings (p. 439)	dorsiventral (p. 451)	phellogen (p. 440)	simple leaves (p. 451)
apoplastic pathway (p. 445)	endarch xylem (p. 439)	phyllode (p. 451)	spongy mesophyll (p. 453)
bark (p. 441)	endodermis (p. 445)	phyllotaxy (p. 450)	stele (p. 438)
bundle sheath (p. 454)	exarch xylem (p. 445)	pith (p. 438)	stem (p. 437)
cambium (p. 439)	exodermis (p. 446)	pneumatophore (p. 447)	stomata (p. 452)
Casparian strip (p. 445)	fusiform initials (p. 439)	polyarch xylem (p. 445)	symplastic pathway (p. 445)
cladodes (p. 442)	guard cells (p. 452)	proteoid root (p. 449)	tuber (p. 441)
compound leaves (p. 451)	heartwood (p. 440)	pulvinus (p. 451)	vascular bundle (p. 437)
coralloid roots (p. 449)	hydathodes (p. 454)	ray initial (p. 439)	vascular cambium (p. 439)
cork (p. 440)	internode (p. 437)	rhizome (p. 441)	vein (p. 452)
cork cambium (p. 439)	isobilateral (p. 451)	rhizosphere (p. 444)	
	lenticel (p. 441)	root cap (p. 444)	
	mycorrhizae (p. 449)	root nodule (p. 449)	
	node (p. 437)	sapwood (p. 440)	
	palisade mesophyll (p. 453)	sclerophyll (p. 455)	

Review questions

1. By means of diagrams, show two different types of xylem organisation in stems.
2. How do cladodes differ from phyllodes?
3. Describe the function of cork cambium. What is periderm and what is bark?
4. What is the function of a root cap? Why is it covered in mucilaginous slime?
5. What is the function of the pericycle in a root?
6. How does the endodermis limit flow of water through a root?
7. (a) What do root hairs and leaf hairs have in common?
   (b) What are the functions of each type of hair?
8. How can you distinguish the upper side from the lower side in a dorsiventral leaf?
9. Describe the structure of an isobilateral leaf. In what environment would you expect to find such a leaf? Give an example.

Extension questions

1. Explain why annual rings are likely to be more evident in trees growing in temperate rather than in tropical climates?

2. (a) What is a stem growth ring?

   (b) How can the study of the growth rings of a fossil tree from Antarctica help scientists determine past climate?

3. Suggest environmental conditions that might advantage plants with underground stems.

4. Would you describe stomata and lenticels as homologous or analogous structures (see also Chapter 30)?

5. (a) What is the main difference in leaf venation between a dicotyledon and a monocotyledon?

   (b) In what ways does the *Eucalyptus* adult leaf differ from that of a grass?

   (c) What are the possible functional aspects of these differences?

6. (a) What are the possible advantages of different developmental leaf stages in plants?

   (b) How would you expect the features shown in Figure 17.26 to relate to the different environments you might expect a seedling and a mature tree to encounter?

Suggested further reading

Cambridge, M. and Kuo, J. (1982). Morphology, anatomy and histochemistry of the Australian seagrasses of the genus *Posidonia* König (Posidoniaceae). III. *Posidonia sinuosa. Aquatic Botany* 14: 1–14.

The ocean waters present a very different environment to that on land. Nonetheless, they have their own important angiosperm flora. This paper provides an introduction to some of the special features of marine angiosperms.

Fahn, A. (1982). *Plant Anatomy*. 3rd edn. Oxford: Pergamon Press.

A comprehensive text on plant anatomy in the 'classical' (structural) style. Differs from other similar texts in its use of some Australian and Mediterranean-type examples.

Groom, P. K., Lamont, B. B., Markey, A. S. (1997). Influence of leaf type and plant age on leaf structure and sclerophylly in *Hakea* (Proteaceae). *Australian Journal of Botany* 45: 827–38.

This paper describes morphological and anatomical features contributing to the sclerophyllous character of leaves in the genus Hakea (family Proteaceae). Use the listed references for further insights into the anatomy of proteaceous plants.

Latz, P. (1995). *Bushfires and Bushtucker: Aboriginal Plant Use in Central Australia*. Alice Springs: IRD Press.

While not an anatomical text, this book provides interesting material on morphological features of central Australian plants and how the various plant organs have been used by Aboriginal Australians for food, medicine and construction.

Turner, I. M. (1994). Sclerophylly: primarily protective? *Functional Ecology* 8: 669–75.

This paper reviews the concept of sclerophylly and evidence related to three hypotheses to explain its functional significance: water conservation; nutrient conservation; and prevention of damage.

Wilkins, M. (1988). *Plant Watching: How Plants Live, Feel and Work*. London: Macmillan.

Describes anatomy in a functional context. The excellent colour photographs and diagrams aid in interpretation.

CHAPTER 18

Plant nutrition, transport and adaptation to stress

n addition to light, carbon dioxide and water, which are required for photosynthesis (Chapter 5), plants use other raw materials to synthesise the organic compounds they need for growth. These nutrients are inorganic elements, 14 of which are essential, especially nitrogen (N), phosphorus (P), potassium (K) and magnesium (Mg). Compared with animals (Chapter 19), the nutrient requirements of plants are relatively simple.

A unicellular green alga, such as *Chlamydomonas* (Chapter 35), has little need for a transport system: nutrients, such as nitrate, phosphate, calcium and carbon dioxide, are dissolved in the water that surrounds the cell. The distances over which these nutrients move are sufficiently small that simple diffusion is adequate. In contrast, multicellular plants obtain nutrients from their environment by root uptake generally from soil. These nutrients are required throughout the plant and, in plants that are more than a few centimetres tall, they are distributed by a transport system (Chapter 17). In a vascular land plant, such as a tall eucalypt tree, mineral nutrients absorbed from the soil solution are transported in the xylem from the roots up the stem to more than 30 m above ground. Similarly, the products of photosynthesis are transported in the phloem from the leaves to all parts of the plant.

Land plants operate under a further constraint, one that is at least as significant as the need to transport inorganic nutrients and the products of photosynthesis. Sometimes referred to as the 'dilemma' of land plants, this constraint arises from the conflicting demands of having to obtain carbon dioxide from surrounding air while at the same time ensuring that cells remain hydrated. When stomata of leaves are open to allow carbon dioxide to enter for photosynthesis, water vapour escapes from the leaf through the same open stomata. In other words, growth, which requires a supply of atmospheric carbon, cannot occur without some water loss. To manufacture 1 g of dry weight, a plant uses more than 200 g of water! This dilemma limits the types of plants that can grow in particular environments, especially arid habitats where the supply of fresh water is limited. The ways by which plants get around this dilemma are diverse and are discussed in this chapter.

> n addition to light, carbon dioxide and water, plants require inorganic nutrients, which they obtain from their environment and transport to all parts of the plant. Because photosynthesis requires uptake of carbon dioxide through stomatal pores in leaves, growth in plants is inevitably accompanied by evaporative water loss through the stomatal pores in the opposite direction.

Nutrition of plants

Although plants are composed largely of water and organic compounds containing the elements carbon, hydrogen and oxygen, plants also synthesise compounds, such as amino acids and vitamins, thus requiring nitrogen, calcium, zinc and other elements (Table 18.1). All plants appear to have very similar nutrient requirements, although the amounts required vary between individual plants and between species. In general, **macronutrients** are required in large amounts and **micronutrients**, **trace elements**, in small amounts.

Essential minerals

The 14 **mineral elements** identified as essential for plant growth were discovered from the early 1800s to the 1940s. Experiments designed to identify an essential element involve growing plants either in sand or water culture (**hydroponics**) and not in soil. Controlled amounts of highly purified nutrients from chemical sources are then added. Complete nutrient solutions include the elements shown in Table 18.1. Various ions can be omitted and the resulting plant deficiency symptoms observed (Fig. 18.1). Plants

Table 18.1 Essential elements required by plants [a]

Element	Form absorbed by plants	Relative concentration in dry tissues
Macronutrients		
Nitrogen (N)	NO_3^-, NH_4^+	1000
Potassium (K)	K	250
Calcium (Ca)	Ca^{2+}	125
Magnesium (Mg)	Mg^{2+}	80
Phosphorus (P)	$H_2PO_4^-$, HPO_4^{2-}	60
Sulfur (S)	SO_4^{2-}	30
Micronutrients		
Chlorine (Cl)	Cl^-	3.0
Boron (B)	H_3BO_3	2.0
Iron (Fe)	Fe^{2+}, Fe^{3+}	2.0
Manganese (Mn)	Mn^{2+}	1.0
Zinc (Zn)	Zn^{2+}	0.30
Copper (Cu)	Cu^+, Cu^{2+}	0.10
Molybdenum (Mo)	MoO_4^{2-}	0.001
Nickel (Ni)	Ni^{2+}	0.001

(*a*) Some plants also require sodium (Na) and silicon (Si).

Fig. 18.1 Boron is an essential trace element for plants but not animals. These flax plants have been growing in a nutrient solution for three weeks without boron (left) or with the addition of 0.1 ppm boric acid, HBO_2 (right). Glass-distilled water cannot be used in these experiments because it is contaminated with boron from borosilicate glassware used in the construction of stills

growing on soil deficient in iron, for example, typically become yellow (**chlorotic**) and may die (Fig. 18.2).

Mineral nutrients in the form of ions serve a *general* function in plants in that they affect the ionic balance of cells and the regulation of water balance. Minerals also have *specific* functions; for example, Mg^{2+} is a component of chlorophyll, K^+ affects the conformation of a number of proteins, and Ca^{2+} is important for

Fig. 18.2 Seedlings of *Eucalyptus viminalis* grown on alkaline calcareous soil are severely chlorotic (yellow) due to the low availability of iron at a high soil pH

BOX 18.1 Australia: a trace element desert?

Scientific research identifying the trace element requirements of plants and animals has been of great benefit to Australia's agricultural industry. Before the 1930s, large tracts of bushland were unsuitable for agriculture. Even when the major deficiency of phosphorus in soil was alleviated by the application of superphosphate, crops simply would not grow and grazing animals suffered nutritional deficiencies. Australian soils are geologically old and many nutrients important to plants have been leached far below the soil surface.

Pioneering research in South Australia and Western Australia showed that the soils of many areas were deficient in the trace elements Cu, Zn, Mo and Mn. It was demonstrated in the early 1940s that application of as little as 5 g of Molybdenum per hectare greatly improved the growth of clovers in many Victorian pastures. These legumes supplied the necessary nitrogen for pasture grasses, which then supported a greatly increased number of grazing animals.

In other areas, animals did not thrive without additions of trace elements to the soil because of deficiencies in copper and cobalt (Co) (see figure). Once the soils of these areas had been treated with very small amounts of these trace elements, together with superphosphate fertiliser, it was possible to establish farms.

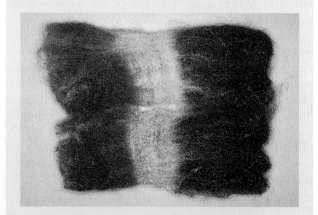

A sheep with black fleece fed on pasture growing on a copper-deficient soil. Copper deficiency in the sheep is indicated by the band of pale colour in the fleece

maintenance of the physical properties of membranes and as a component of primary cell walls. A key role of trace elements is their catalytic function as part of enzymes that are essential in metabolism. The trace elements Mn, Mo, Cu, Zn and Fe are involved in redox reactions (Chapter 2); iron, which can exist in the forms Fe^{2+} and Fe^{3+}, is a constituent of cytochromes, which are involved in the electron transport reactions of photosynthesis and respiration (Chapter 5).

> Fourteen mineral elements are essential for the growth of plants: N, P, K, Mg, Ca and S are required in large amounts and Fe, Cl, Mn, Cu, Zn, B, Mo and Ni are required in trace amounts.

Where do essential nutrients come from?

Plants obtain essential minerals from the soil (Fig. 18.3a), which is formed from the weathering of underlying rocks (the lithosphere). However, the supply of nutrients to plants does not come solely from this source. Consider the supply of phosphorus necessary for the growth of a forest tree (Fig. 18.3b). Within a forest ecosystem, pools of phosphorus exist within living trees and understorey shrubs, in dead plant litter on the ground and in the soil. The total phosphorus in this ecosystem moves between these pools as bacteria decompose litter and plants absorb the released nutrients. This is an example of a nutrient cycle (Chapter 44). The supply of phosphorus available for tree growth is more dependent upon the amounts of phosphorus in the various pools and the rates of movement between them, than on the very slow release of phosphorus to the soil from the underlying rocks.

Another exception is nitrogen, which is absent from the lithosphere. Besides being the most important of the essential elements (Table 18.1), nitrogen is the major gas of the atmosphere. Nitrogen (N_2) is 'fixed' to form ammonium (NH_4^+) and eventually nitrate (NO_3^-) by certain bacteria (Chapter 44), which is then absorbed by plant roots. These bacteria may live free in the soil or in association with plants, such as the *Rhizobium* bacteria in the root nodules of legumes, or actinomycete nitrogen-fixing organisms in the roots of casuarinas (Box 18.2).

> Plants obtain essential nutrients primarily from soil. Nutrients cycle from plants, to decomposers, to soil and back into plants. Nitrogen in the atmosphere becomes available to plants as nitrate and ammonium by nitrogen-fixing bacteria.

Pathways and mechanisms of transport

Non-gaseous substances that are transported into and within plants are water and solutes such as ions, amino acids and sucrose. There are a number of pathways:

- between solutions in the external environment and plant roots

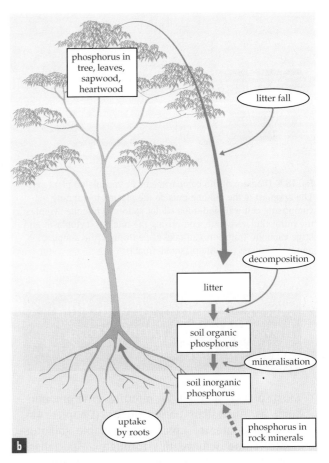

Fig. 18.3 (a) Terra rossa soil (red earth) derived from limestone at Buchan, eastern Victoria, is relatively fertile. (b) Plants not only obtain nutrients from the lithosphere, but depend on nutrient cycling. The cyclic movement of phosphorus in a forest involves annual leaf fall, decomposition of litter by soil organisms and return of nutrients to the soil

- between cells either along the apoplastic pathway or the symplastic pathway via plasmodesmata (Fig. 17.17)
- between compartments within a cell, for example, between vacuole and cytoplasm, which are separated by a membrane, the **tonoplast** (Fig. 18.4)
- long-distance transport in xylem and phloem.

Transport mechanisms include both passive and active processes (Chapter 4). Passive processes in plants include diffusion, mass flow and osmosis, and active processes include uptake or movement of ions and sugars with the expenditure of energy. Mass-flow transport occurs when solutes are carried in a flow of solution driven by gradients of hydrostatic pressure, as in a hose pipe. Such gradients of pressure occur in xylem and phloem, and are responsible for the movement of water and sugars.

> In plants, cell-to-cell movement of water and solutes occurs via apoplastic and/or symplastic pathways. Long-distance transport occurs in xylem and phloem. Transport mechanisms include passive processes, such as diffusion, mass flow and osmosis, and active processes, which require energy from cell metabolism.

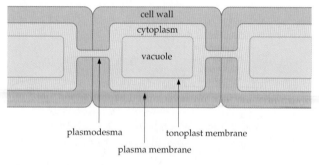

Fig. 18.4 Diagram of the compartments in vacuolated plant cells. The apoplast is the volume outside the plasma membrane, comprising cell wall and intercellular spaces, while the symplast is the intracellular volume, comprising the nucleus, cytoplasm and large vacuole. Transport can take place through the apoplast and/or symplast (via the plasmodesmata)

BOX 18.2 Doing a lot with little: how Australian plants cope with low nutrients

Most plants adapted to infertile environments, such as Australian native plants (Chapter 41), have high rates of uptake of available soil nutrients. Some achieve this through a specialised root structure, for example, by having many fine roots and/or root hairs, or by having a high ratio

of roots to shoots. Some may even be able to exploit immobilised forms of key nutrients such as phosphorus. Red bloodwood, *Corymbia gummifera*, for example, grows on coastal sands that are low in nutrients and may be able to utilise insoluble iron and aluminium phosphates. There is evidence that exudates from the plant's roots may convert insoluble forms of phosphate to soluble forms, which the roots can absorb.

Other more common ways of enhancing nutrient absorption usually involve some form of interaction between roots and soil microorganisms. The most widespread microbial symbioses are those between roots and **mycorrhizal fungi**, which are extremely common in woody plants. Many different fungi are involved and both external (ectomycorrhizal) and internal (endomycorrhizal) associations occur (Chapter 17).

So important are mycorrhizal associations that it is common practice to inoculate many exotic agricultural and forestry crops with fungal spores at the time of planting to ensure optimum rates of growth. Development of mycorrhizae involves a considerable increase in the surface area of roots and associated fungal hyphae, thus greatly increasing the absorption surface for the uptake of minerals. It is also probable that mycorrhizae play an important role in nutrient cycling because the fungal hyphae readily penetrate litter and compete with other decomposers for phosphorus and other nutrients.

Another form of specialised microbial association, **proteoid roots** (Chapter 17), is more restricted in its occurrence. Proteoid roots are highly absorbent and their growth is stimulated by particular microorganisms in the root zone. They are particularly well developed in the family Proteaceae (e.g. many species of *Banksia*, *Grevillea* and *Hakea*). This family is commonly found in Australia and South Africa, particularly

Heath plants are adapted to soils of low fertility, such as this sand sheet, near the Gippsland Lakes, Victoria

in heaths growing on sandy or rocky soils of very low nutrient status (see figure); the phosphorus contents of some of these heath soils are the lowest in the world.

For most plants inhabiting low-nutrient environments, the ways of coping with restricted soil reserves are two-fold. As well as maximising rates of nutrient uptake, these plants also are very conservative in their internal usage of nutrients. This may be achieved in part by matching rates of growth to rates of uptake. Very low growth rates and an evergreen perennial habit are common characteristics of plants growing in nutrient-deficient environments. The relatively small, hard (sclerophyllous) and long-lived leaves that are typical of many Australian plants are generally capable of producing much more plant material (biomass) for a given input of phosphorus or other key nutrients than are the larger, softer and shorter-lived leaves of deciduous trees more common in Northern Hemisphere forests. Economies of use are also achieved by internal recycling (e.g. withdrawal of nutrients from leaves before they are shed, a strategy common to many eucalypts) and internal storage (e.g. in stems, trunks, lignotubers and roots) to allow retranslocation of nutrients when needed by actively growing plant parts.

BOX 18.3 Organic versus artificial fertilisers

The present high yields of agricultural crops depend to a great extent on 'artificial' fertilisers. These are chemicals manufactured by the fertiliser industry and used to augment nutrients, particularly nitrogen, phosphorus and potassium, which may be below optimal levels in the soil for plant growth. Artificial fertilisers include urea, ammonium sulfate, superphosphate and potassium chloride. In contrast, growers who produce 'organically grown' food (see figure) do not use artificial fertilisers but use organic manures and compost, a farming method used for centuries.

An ammonium ion (NH_4^+), however, is chemically the same whether it is derived from animal manure or from ammonium sulfate. Similarly, a phosphate ion derived from a rotting bone does not differ chemically from one derived from superphosphate. Plant roots cannot distinguish between these different sources of ions in the soil solution. It is doubtful, therefore, that the source of the inorganic essential nutrients affects the quality of plant produce.

If there is a difference in quality, for example, between hydroponically grown and organically grown tomatoes or lettuces, it is not related to the origin of the nutrient ions in the soil solution, but may relate to other factors, such as water supply or care of the crop. Although there is much evidence that the structure and water-holding capacity of many soils is improved by the use of organic manures and compost, the case against the use of artificial fertilisers as inferior suppliers of essential nutrients is unsubstantiated.

Organically grown tomatoes

Water transport

The movement of water molecules from soil into roots, or from leaf mesophyll cells through stomata to the atmosphere, takes place down gradients of **water potential** (ψ, psi), that is, down a gradient of decreasing free energy (Chapter 4). The free energy of water molecules in the soil is normally greater than in the plant, which, in turn, is greater than the free energy of water molecules in the air (Fig. 18.5). Hence, water moves from soil into roots and up a plant.

Values of ψ in soil, plant or air are expressed relative to the energy status of pure water at the same temperature and pressure. For pure water, ψ is set at 0, just as the temperature of a water–ice mixture on the Celsius scale is set at 0°C. The units in which ψ is measured are those of pressure (MPa, megapascals),

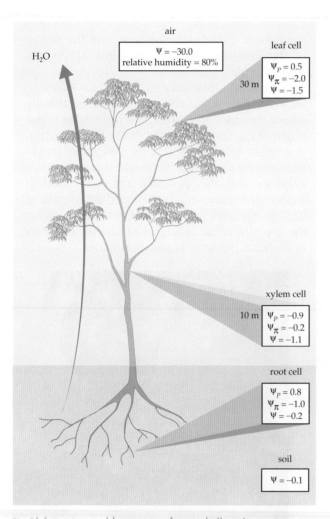

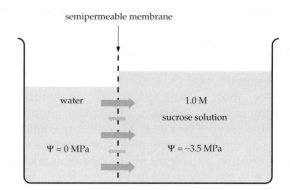

Fig. 18.6 The net flow of water through a semipermeable membrane and down a gradient of water potential (ψ), from pure water to a 1 M sucrose solution

Fig. 18.5 Gradients of free energy of water, indicated as water potential (ψ), in the pathway of water flow from the soil, through the plant to the atmosphere. Values of ψ become progressively more negative from the soil to the atmosphere. In the vascular system, the xylem sap is under tension. Also, the xylem sap is more dilute than that in cell vacuoles and hence osmotic potential (ψ_π) is higher (less negative) in the xylem compartment compared with that in root and leaf cells. The units of ψ, ψ_π, and ψ_P are in MPa

which is also equivalent to energy per unit volume. (One megapascal equals approximately 10 bars on other scales of pressure.)

Factors that affect water potential

In soil and within a plant, ψ is less than 0, that is, it has a negative value. This is because the free energy of water molecules is decreased below that of pure water by the presence of solutes as well as solids that adsorb water molecules. For example, a 1 M sucrose solution has a water potential, ψ_solution, of −3.5 MPa (Fig. 18.6). The free energy of the water molecules in this case is reduced below 0 by the presence of the sucrose molecules.

Water potential is affected by hydrostatic pressure. It is decreased when a fluid is under tension (negative pressure), such as occurs in xylem, whereas it is

increased by the application of positive pressure, such as occurs in a turgid plant cell. A plant cell wall has elastic properties and it can swell and contract as the water content and volume of the cell change. In this way, the elastic wall of a turgid cell exerts a positive hydrostatic or **turgor pressure** on the cell contents.

Thus, the water potential of a plant cell (ψ_cell) is the sum of the positive effect of cell turgor pressure (**pressure potential**, ψ_P) and the negative osmotic effect (**osmotic potential**, ψ_π) of solutes (Fig. 18.7):

$$\psi_{cell} = \psi_P + \psi_\pi$$

If the ψ_π of a solution in a plant cell is −1.2 MPa and ψ_P is 0.5 MPa, then ψ_cell is −0.7 MPa. Where a gradient of water potential exists across a semipermeable membrane, such as that bounding the protoplasm of a root cortex cell or a leaf mesophyll cell, a net osmotic flow of water molecules will occur across the membrane (from higher to lower water potential). Where there is no semipermeable membrane, as in open-ended xylem vessels or in phloem sieve tubes with open sieve plates, fluid movement is by mass flow along a gradient of hydrostatic pressure.

> In plants, water potential (ψ) is usually negative but may increase to 0 in water saturated conditions. Water potential is the sum of the osmotic potential (ψ_π) and pressure potential (ψ_P).

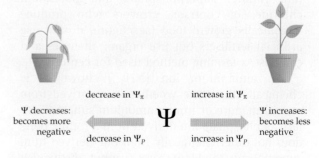

Fig. 18.7 The water potential of a plant cell (ψ) is affected by changes in solute concentrations (affecting ψ_π) and hydrostatic pressure (affecting ψ_P)

Water uptake by plant cells

How do these properties of water movement relate to the ability of a plant to take up water from soil and to maintain a high degree of hydration? We have shown that the water potential of a cell determines the capacity of that cell to take up or lose water from its surroundings. When plant cells exchange water with their environment, they shrink or swell to a limit imposed by the cell wall. These changes are accompanied by changes in pressure potential (ψ_P) and solute concentration, and therefore osmotic potential (ψ_π). If cell water content and volume increase, cell walls distend more and cell turgor, ψ_P, increases, the solutes are diluted, ψ_π decreases in magnitude (i.e. is less negative) and ψ_{cell} becomes less negative (Fig. 18.8). When a cell is fully turgid, $\psi_P = \psi_\pi$ and $\psi_{cell} = 0$. The fully turgid cell has no capacity to absorb water, even from distilled water.

Consider, however, what happens to a leaf cell on a hot day. Water vapour lost through stomata exceeds water taken up by roots. Cells lose water and their volume decreases. As water content and cell volume decrease, ψ_P decreases, ψ_π increases in magnitude (i.e. is more negative) and ψ_{cell} becomes more negative. At the point of incipient plasmolysis, there is no wall pressure ($\psi_P = 0$), and $\psi_{cell} = \psi_\pi$. Thus, the capacity of the leaf cell to absorb water from surrounding cells, such as the xylem, is greatly increased and water moves into the cell.

Change in turgor is a mechanism by which plants adjust their water-retaining and water-absorbing capacities in drought conditions. When water availability increases, as at night when the rate of evaporation from leaves is low, cells have the capacity to take up water and regain turgidity (Fig. 18.9a). Changes in turgidity have many important consequences: for example, when turgor has been lost, growth by cell expansion cannot occur and, in leaves, stomatal closure takes place, restricting photosynthesis. With a loss of turgor, thus pressure potential, tissues also lose mechanical strength or rigidity and leaves will wilt and may suffer permanent damage (Fig. 18.9b). Most plants die if their water potential drops below about -6 MPa.

> When a cell loses water and decreases in volume, pressure potential and osmotic potential both decrease, with the result that water potential also decreases (i.e. cells are more able to absorb water).

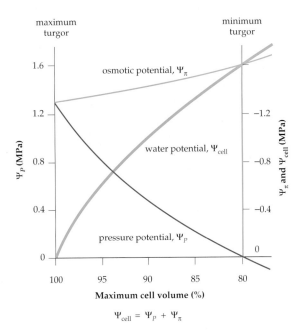

Fig. 18.8 This graph shows how the water potential of a cell becomes more negative with increased loss of water. Changes in cell water potential (ψ_{cell}), osmotic potential (ψ_π) and pressure potential (ψ_P) are compared with cell volume changes. The cell is at maximum turgor and volume when the leaf is well-supplied with water (left). However, on the right side of the graph, the cell is at zero turgor ($\psi_P = 0$) and at incipient plasmolysis because the leaf is suffering from drought stress. The values of ψ_{cell} and ψ_π are negative, whereas values of ψ_P are positive

Fig. 18.9 Wilting of leaves of a cucumber plant. (a) Leaves are wilted at noon on a hot day when their cells have lost turgor. (b) Leaves have regained turgor at 8 p.m. the same day

Osmotic adjustment in plant cells

Loss of turgor and wilting can occur very rapidly, in a matter of minutes, resulting in equally rapid decreases in cell water potential. When plants are subjected to *persistent* soil water deficits over days or weeks (i.e. drought stress), they may respond in a different way. In this case, the amount of solutes in cell vacuoles increases, which has the effect of decreasing osmotic potential and reducing cell water potential without adversely affecting cell turgor and growth. This response to drought is *osmotic adjustment*. Osmotic adjustment allows growth and photosynthesis to continue in drier conditions. Solutes that contribute to increased osmotic potential of the vacuole include ions, sugars and organic solutes, such as the amino acid proline. Osmotic adjustment also occurs in plants growing in wet but saline soils. This is achieved by accumulation in cell vacuoles of Na^+ and Cl^- from the soil solution. Salt-tolerant plants (**halophytes**, Fig. 18.10) may accumulate sufficient inorganic ions in their cells to generate an osmotic potential as low as -2.5 to -3.0 MPa.

Fig. 18.10 *Disphyma australe* is a halophyte (salt-tolerant plant). Plants growing in wet saline soils are able to osmotically adjust their cells to retain and absorb water

The capacity of a cell to retain and to absorb water is increased by osmotic adjustment. Plants respond to longer term soil water deficits or salinity by accumulating solutes in the vacuole, which allows turgor and growth to be maintained in dry or saline conditions.

Transpiration

Transpiration is the loss of water by evaporation from leaves, and is a process that uses energy from incoming solar radiation to vaporise water (Fig. 18.11). Water vapour passes through both stomatal pores and, to a lesser extent, the leaf cuticle. In a well-watered growing plant, transpired water is replaced by water drawn up from the roots through the xylem. On a sunny day a single leaf from a herb, such as a sunflower, may lose an

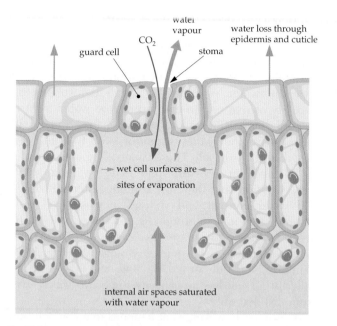

Fig. 18.11 Pathway of the movement of carbon dioxide and water vapour into and out of a leaf

amount equivalent to its entire water content in just 3 hours. In a mountain ash tree, *Eucalyptus regnans*, growing in a moist environment, the amount of water transpired on a summer day can be as much as 300 L. On a hot day, a hectare of wheat can lose up to 100 tonnes of water each day, which is equivalent to 10 mm of rainfall. Transpiration is a physical process and its rate is highly dependent upon climatic conditions.

The overall gradient of water potential from soil to air is maintained during the day by the negative water potential of the surrounding air (Fig. 18.5). Even at a relative humidity of 90% and temperature of 20°C, air has a water potential of approximately -14 MPa, which is much more negative than the water potential of plant cells. Moving along this gradient, water enters the root system close to the tip region of young roots, where root hairs are in close contact with the soil solution. Water and solutes move radially from the root epidermis, across the cortex, to the endodermis and pericycle, and then into the xylem elements. Water moves through the apoplast by mass flow but, at the endodermis, the apoplastic pathway is obstructed by the Casparian strip (Chapter 17). At this point, the whole flux of water and solutes must travel through the symplast of cells via plasmodesmata to enter the stele of the root (Fig. 17.18).

Transpiration in plants is the loss of water vapour by evaporation from leaf surfaces.

Water movement in xylem

In flowering plants, xylem cells include vessels and tracheids. A vessel consists of elements joined end to end to form a tube that may be quite short or may extend up to 15 m long, as in one species of oak tree. The diameter of large vessels is about 50 μm. Water and solutes, but not gas bubbles or solid particles, pass through pits in the walls of both vessels and tracheids. Both vessels and tracheids contain a continuous column of xylem sap, extending from the root to the veins of a leaf.

Xylem sap is a dilute solution of inorganic ions and some organic nitrogen compounds (amino acids and amides, Fig. 18.12). The column of sap in the xylem of a transpiring plant is under negative hydrostatic pressure. When the stem of a transpiring plant is cut, the fine xylem vessels are so narrow that the surface tension forces in the cut vessels are strong enough to prevent the sap from leaking out. There are, however, exceptions. Aboriginal Australians living in the Australian desert discovered that, when cut, mallee roots will provide a source of water (Fig. 18.13). Cut mallee roots release sap because surface tension forces

are small in the cut root xylem vessels, which are of large diameter and offer little resistance to flow. Water flows out when cut roots are up-ended.

Root pressure

Under certain conditions, such as very moist soil and high atmospheric humidity, when the transpiration rate is zero or very low, root xylem sap may be under positive pressure. Pressure produced by roots, **root pressure**, is responsible for the exudation of sap from the cut stems of some vines, walnut trees and North American sugar maple in early spring. It does not account for the ascent of xylem sap in tall plants. Root pressure is probably a result of an osmotic flow of water generated in roots due to an accumulation of solutes in the root xylem following uptake of mineral nutrients.

> Positive pressure occurs in the roots of some plants under certain conditions but is insufficient to support rapid transpiration.

How does water reach the top of tall trees?

So far, we have established that the driving force for mass flow of sap in xylem is a gradient of water potential, and that this gradient, which is increasingly negative towards the leaf and air, 'pulls' the xylem fluid upwards. This idea assumes that there is a continuous water column in narrow xylem elements, and that this column exists under tension without rupturing.

Atmospheric pressure is only sufficient to hold a column of water up to a height of 10 m. However, water columns much higher than 10 m, as in a tree 100 m high, can be maintained in very narrow tubes such as xylem vessels. This is because molecular forces will hold water molecules together and to the sides of a tube (capillary attraction), resulting in a column of water under tension. Removal of water by evaporation in leaves at the top of the column results in a flow of water from the roots through a plant. This explanation for how water rises to the top of a tall tree is called the **cohesion theory**.

The cohesion theory hypothesis received support from the experiments of John Milburn, a plant physiologist, who used acoustic devices to 'listen' to stems of water-stressed plants. When the tension in a column of water becomes very high, as in a plant suffering from drought stress, the column sometimes breaks—**cavitation**. The spaces formed in the column, momentarily a vacuum, fill with water vapour, forming an air bubble, or embolism (Fig. 18.14). Milburn noted that high tension in the xylem produced audible 'clicks', which became more frequent as the degree of water stress was increased. The clicks were the sound made by the breaking of the column of xylem sap under tension, resulting in an embolism. The spread of an

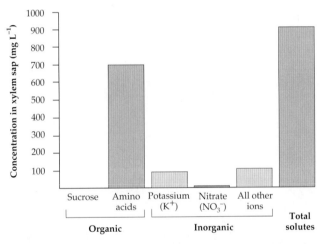

Fig. 18.12 Chemical composition of xylem sap from white lupin, *Lupinus albus*

Fig. 18.13 Xylem sap is tapped from roots of mallee gums by Aboriginal Australians at the Yalata Reserve, South Australia

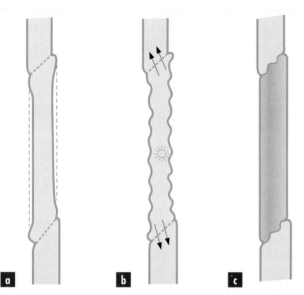

Fig. 18.14 Cavitation in a xylem vessel. **(a)** Under tension, xylem sap strains the wall of the vessel, pulling it inwards.
(b) A cavitation bubble forms and the xylem walls vibrate, resulting in a 'click'. Water moves out of the vessel element, which **(c)** becomes filled with air or water vapour

embolism is restricted by bordered pits, which act as valves in the vessel walls. Plants recover from these embolisms at night, when the tension in the xylem decreases, possibly assisted by root pressure.

> A continuous column of water ascending in the xylem of trees is dependent on strong adhesion of water molecules to the walls and intermolecular cohesive forces that maintain the flow.

Water movement from leaves

Leaf mesophyll cells obtain water from leaf veins and are surrounded by air spaces, which may occupy up to 50% of the volume of a leaf (Fig. 18.11). Because these air spaces are in contact with wet mesophyll cells, they are almost saturated with water vapour (relative humidity 100%). The movement of water vapour from leaves meets two points of resistance. The first is the cuticle, which offers the greatest resistance, and stomata, through which 90% of the flux of water vapour passes. The resistance due to stomata depends upon the mean aperture of all the stomatal pores on the leaf surface; when the stomata are closed, resistance is high. The second point of resistance is the leaf boundary layer, that is, the layer of still air just outside the leaf surface (Fig. 18.15). Water vapour tends to accumulate around a leaf, increasing humidity relative to the atmosphere. The thickness of the boundary layer depends on the shape of the leaf and the presence of leaf hairs, which trap water vapour. Wind increases transpiration rate by decreasing the thickness of this boundary layer.

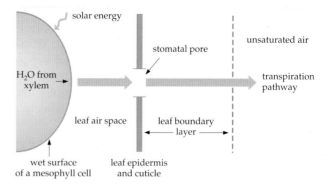

Fig. 18.15 Transpiration. Water evaporating from a mesophyll cell meets two points of resistance to movement: the cuticle and stomatal pores, and the leaf boundary layer, where there is a gradient of water vapour concentration to the outside air. These two resistances reduce potential transpiration rates

Stomata: turgor-operated valves

Although a leaf of *Eucalyptus globulus* may have more than 30 000 stomata per square centimetre, the proportion of the leaf surface area occupied by stomatal pores is only about 1% (Fig. 18.16). Despite this, the amount of transpiration from a leaf with open stomata is nearly the same as evaporation from an open wet

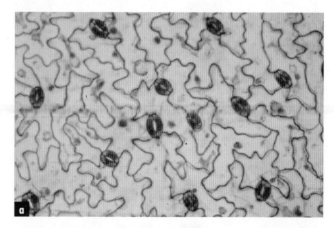

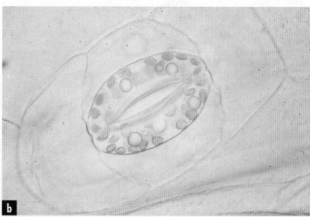

Fig. 18.16 Stomata in the leaf epidermis of a garden pea.
(a) Stomata are stained with neutral red, a dye that accumulates in cell vacuoles. **(b)** A living, half-closed stoma showing chloroplasts in the guard cells

surface; gas movement through small pores is remarkably efficient. By controlling the size of stomatal pores, a plant can regulate the exchange of carbon dioxide, oxygen and water vapour. Regulation of stomatal pores allows a trade-off between carbon dioxide uptake required for photosynthesis and potentially damaging water loss.

The size of a stomatal pore is dependent upon the turgor of the two guard cells, which may be likened to two sausage-shaped balloons with less extensible walls on the inner surfaces. When the guard cells become turgid, they expand outwardly, opening the pore, which can have a diameter of up to 10 μm. When they lose turgor, water moves out of the guard cells to the surrounding subsidiary and epidermal cells. The guard cells deflate, closing the pore. Guard cells are, therefore, *turgor-regulated valves.*

Control of stomatal opening and closing

What are the causes of changes in guard cell turgor that are associated with stomatal opening and closing? Increase in guard cell turgor (pressure potential) is caused by uptake of water by osmosis from surrounding cells. This occurs when the water potential of a guard cell is more negative than that of adjacent cells. Stomatal opening, therefore, involves a decrease in guard cell water potential, which is brought about by importing inorganic solutes into the guard cell (Fig. 18.17) and by release of organic solutes from the hydrolysis of the water-insoluble starch in guard cells.

Stimuli that induce stomatal opening activate membrane-bound H^+–ATPase antiports (Chapter 4). These antiports actively transport H^+ out of the guard cells (Fig. 18.18), generating an H^+ gradient that is coupled to the import of K^+ and Cl^-. Guard cells of open stomata contain about 20 times more K^+ ions than do those of closed ones (Fig. 18.18). There are also increases in concentration of the anion, malate, which together with Cl^- balances the cation, K^+. Malate is generated within the guard cell by the activity of the enzyme, PEP carboxylase (Chapter 5). An outward flux of these ions results in guard cell osmotic potential becoming less negative, so water moves out, turgor is lost and stomata close.

> Stomatal opening and closing is the result of movement of solutes, particularly K^+ and Cl^-, into and out of guard cells. This results in changes in osmotic potential, water potential and thus cell turgor.

Diurnal stomatal rhythms

When transpiration and stomatal behaviour are measured in plants growing in well-watered conditions, a diurnal stomatal rhythm is observed: stomata are open by day and closed by night. CAM plants

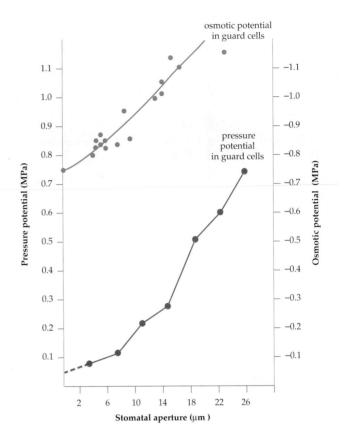

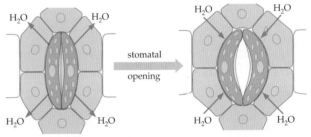

Fig. 18.17 This graph shows the results from an experiment with *Tradescantia* leaves. As stomata opened, measurements were made of the size of the stomatal aperture, and the osmotic and pressure potentials in guard cells. Stomata opened as osmotic potential became more negative, causing guard cells to take up water (increasing turgor or pressure potential)

(Chapter 5) do the reverse, opening their stomata by night, closing them early in the day and, in moist conditions, reopening them in the afternoon. This stomatal behaviour is related to the ability of CAM plants, such as pineapple and succulent plants, to assimilate carbon dioxide at night as well as by day (Chapter 5).

The environmental stimuli that affect stomatal opening and closing are: light, intercellular carbon dioxide concentration, air humidity, and plant and soil water deficits.

- *Light intensity*. Stomata open rapidly when a plant that has been in the dark is provided with light (Fig. 18.19). Blue light is most effective.

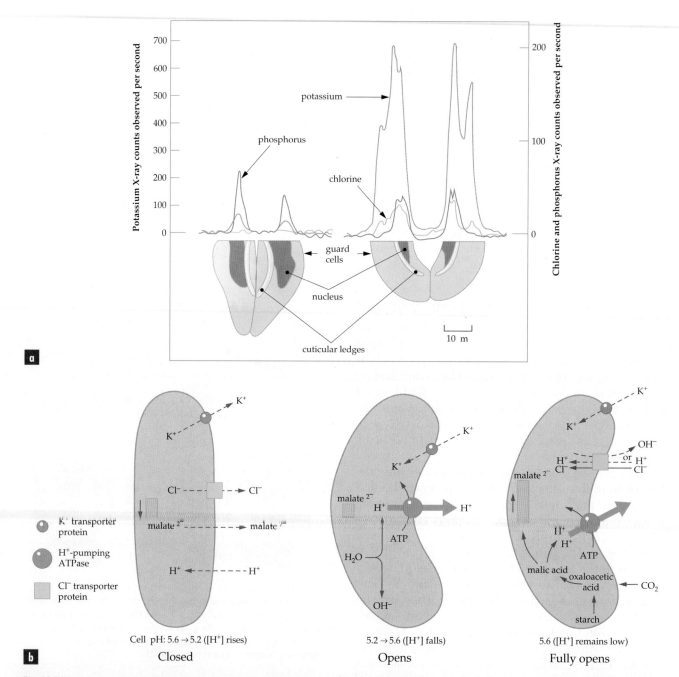

Fig. 18.18 Changes in amounts of various ions in closed and open stomata. **(a)** Profiles across stomata of the elements potassium, chlorine and phosphorus were obtained by scanning electron microscopy using an electron probe, which detects each element by emission of X-rays. **(b)** Ion movements into and out of guard cells during stomatal closure and opening depend on proton-pumping of ATPase, which provides the proton gradients that are coupled to other secondary active transport mechanisms for K^+ and Cl^-

- *Carbon dioxide concentration.* Stomata are sensitive to CO_2 concentration in the surrounding air. This CO_2 signal appears to be sensed by the plasma membranes of the guard cells. Decreases in CO_2 concentration cause stomata to open and increases cause stomata to close. In C_3 plants that are photosynthesising, the CO_2 concentration inside a leaf may be 100 ppm less than that in air, which is sufficient to cause stomata to open during the day. At night, when there is no photosynthesis, leaf

respiration causes an increase in internal CO_2 concentration sufficient to decrease stomatal aperture. In CAM plants, assimilation of CO_2 at night causes the CO_2 concentration to decrease and stomata to open at night (Fig. 18.20).

- *Air humidity.* If air flowing over a leaf changes from high to low humidity, the transpiration rate of a plant will increase, guard cells will lose turgor and stomatal aperture will decrease. By reducing stomatal aperture, the leaf can restrict water loss.

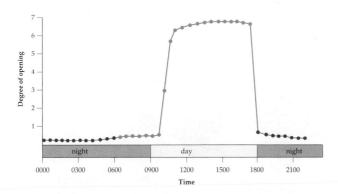

Fig. 18.19 Stomatal opening and closing in response to changes from dark to light and light to dark in a cabbage leaf. The response is rapid, stomata closing completely in about 30 minutes when the lights are turned off and opening in about 60 minutes when the lights are switched on

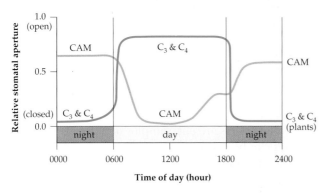

Fig. 18.20 Stomatal opening and closing in C_3 and C_4 plants compared with CAM plants. The graph shows the results of experiments under standard growth conditions with a day length of 12 hours

- *Soil and plant water deficits.* When water deficits are imposed slowly, such as during the onset of drought conditions, plants respond to gradual decreases in cell turgor by producing the hormone abscisic acid (ABA, Chapter 24). In the early stages of drought, plants generate an ABA signal in their roots. The hormone travels in the xylem stream to the leaves, where it induces stomatal closure. The ABA content of leaves may increase 40 times as leaf turgor decreases. This has the effect of conserving soil water, which is important for survival during a drought.

S tomatal opening and closing is affected by environmental stimuli, including light intensity, carbon dioxide concentration, air humidity, and soil and leaf water deficits. During drought, plant roots generate a hormonal signal (ABA), which induces leaf stomatal closure.

Translocation of assimilates

The raw materials for the growth of all parts of a plant come from the transport, **translocation**, of assimilates,

the end products of photosynthesis. The main sources of assimilates are leaves. Other non-photosynthetic tissues act as sinks, that is, importers of assimilates. A storage organ, however, which imports and exports carbohydrates, can act as both a source and a sink. In a wheat plant, it has been shown that carbohydrates in the grain, the sink, are mainly derived from the photosynthetic activity of the two upper leaves, the source. Thus, sinks tend to be supplied by the nearest source. The growth of an apple, for example, depends greatly on assimilates from the nearest leaves.

Pathway of assimilate transport

How are assimilates transported from a source to a sink? Evidence of assimilate transport in phloem comes from observing ring-barked trees. In the widespread clearing of forests in Australia in the nineteenth and early twentieth centuries, settlers who wished to establish farms in place of forests often ring-barked trees (Fig. 18.21). Ring-barking involves cutting through the bark to the wood around the base of a tree trunk. This causes the tree to die gradually, often over one to two years. Experiments show that leaves do not wilt after ring-barking, indicating that transport in xylem is unimpaired. Above the ring, cell division and growth occur and sucrose levels are higher than below the ring. This indicates that assimilates move downwards in a tissue of the bark. Thus, ring-barked trees die because all the tissues below the cut portion, including the roots, are starved of assimilates. These

Fig. 18.21 Ring-barking of trees kills the phloem and hence the tree. This forest in north-east Victoria was cleared by ring-barking to create grazing land

observations implicate phloem as the assimilate-transporting tissue of bark.

Evidence also comes from experiments using radioactive carbon dioxide. When leaves photosynthesise in the presence of carbon-14-labelled carbon dioxide, which is radioactive and emits β-radiation, the pathways by which assimilates move in the plant can be traced. Figure 18.22 shows the distribution of carbon-14-labelled assimilates in a single source leaf. The assimilates moved out of the source leaf through the stem. The sinks to which assimilates moved are the roots, the youngest expanding leaves and the shoot tip. No movement occurred into fully expanded leaves. A cross-section of a leaf petiole cut 2 hours after feeding the leaf with radioactive carbon dioxide showed that the carbon-14 label was confined to the phloem, within sieve cells and companion cells.

> Assimilates are transported in phloem from their source in leaves to sinks in other parts of the plant.

What substances are transported in phloem?

Phloem sap can be collected from cut stems or from aphids, which are sucking insects that use their mouthparts (stylets) to penetrate phloem sieve tubes for sugar (Fig. 18.23). Because there is a positive

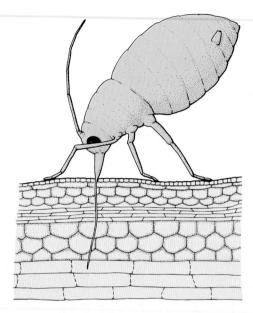

Fig. 18.23 Phloem-feeding aphids are used to tap the contents of sieve tubes

hydrostatic pressure in sieve tubes (up to 1 MPa), the sugar solution flows into the insect's stylet. If the stylets of feeding aphids are cut, samples of phloem sap exude out and can be collected for analysis.

Phloem sap is a concentrated solution of solutes (Fig. 18.24), predominantly sucrose at a concentration

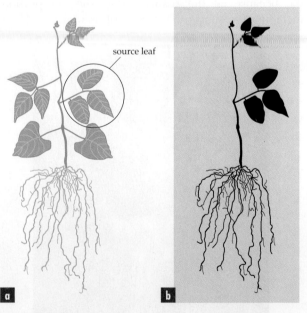

Fig. 18.22 Transport of carbon-14-labelled assimilates from a single source leaf of French bean, *Phaseolus vulgaris*. **(a)** The source leaf has been fed radioactive carbon dioxide for 4 hours. In this experiment, autoradiography was used for locating carbon-14 in the plant: radioactive carbon dioxide produces an image on unexposed photographic film. **(b)** An autoradiograph of the plant shows the distribution of the carbon-14 label from the source leaf to sinks. Three expanded leaves show no carbon-14 label and thus are not sinks

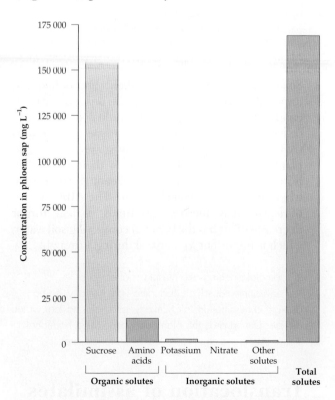

Fig. 18.24 Concentration of solutes in the phloem of white lupin (*Lupinus albus*). The total concentration of solutes in the phloem is 186 times that in the xylem. Sucrose is the main phloem solute and is virtually absent from the xylem

of about 0.5 M. However, the trisaccharide raffinose, rather than sucrose, occurs in white ash, and mannitol occurs in apple and apricot. Phloem sap is a sweet liquid, which, when fermented, is the basis of many alcoholic beverages. For example, the phloem sap of the century plant, *Agave americana*, is collected and fermented in Mexico to make pulque and tequila.

Sucrose in the sieve tubes is not always directly derived from current assimilation in leaves but may also come from the breakdown of stored carbohydrates. For example, about 20% of the carbohydrates in the grain of wheat are derived from the remobilisation of stores in the upper parts of the flowering stem. In growing tubers of Jerusalem artichoke, *Helianthus tuberosus*, carbohydrate comes from redistribution of storage reserves in the stem and not directly from the current products of photosynthesis.

In most plants, phloem sap is under positive pressure and contains a high concentration of sucrose.

Rate of phloem transport

The rate of phloem transport can be measured by tracing the speed of a pulse of radioactively labelled assimilates as it passes down a petiole and stem, away from a leaf. Transport in phloem (unlike xylem) involves living cells and the rate of transport is sensitive to temperature. The rate, for example, slows when a leaf petiole is cooled to 0°C. When the petiole is killed by heat treatment, movement of assimilates down the petiole stops (Fig. 18.25).

The concentration of sucrose and the hydrostatic pressure in sieve tubes is greater near a source than at a sink and a gradient exists between them. CSIRO scientists found that in wheat, for example, the rate of translocation varies from 39 cm per hour down the leaf sheath to 87 cm per hour in the stem. These rates of

transport greatly exceed (by a factor of about 10^4) those predicted by a simple diffusion model. Diffusion, therefore, cannot be the mechanism by which sucrose is transported in the phloem.

The rate of transport in phloem ranges from approximately 40 cm to 100 cm per hour.

Mass-flow mechanism of transport

Sucrose transport from a source, such as mesophyll cells of a leaf, to a sink, such as growing cells at a root tip, involves three processes:

- phloem loading, that is, active transport from the mesophyll to the sieve tubes in the veins of a leaf
- long-distance transport of sucrose in the stem and root phloem
- phloem unloading, that is, passive transport from the sieve tubes to the cells at the root tip.

The pathway of the loading of sucrose, from a photosynthesising mesophyll cell to sieve tubes, is shown in Figure 18.26. Sugar moves from the photosynthesising cell into a chain of paraveinal mesophyll cells, transfer cells (Chapter 17), to the phloem of a vascular bundle. Transfer cells have many projections and folds on their inner walls, which increase the surface area for transport.

The concentration of sucrose in a photosynthesising cell is approximately 50 mM. In contrast, concentrations in the sieve element at the site of loading may be 10 to 20 times higher. Phloem loading, therefore, involves transport against a concentration gradient and is an active process.

The loading of sucrose is accompanied by a pH increase. This indicates that proton transport is associated with active sucrose uptake. The site of this sucrose pump is thought to be the plasma membrane of the sieve element. The driving force for the active movement is a membrane-bound ATPase. The discharge of the proton gradient is coupled to a protein

Fig. 18.25 Translocation can be blocked by controlled heat-killing of a small section of phloem in the vascular tissue of a petiole. This does not prevent transport in the xylem. This can be done by filling with hot wax a device placed around the petiole, as shown

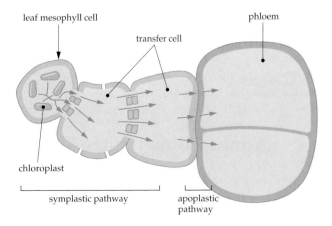

Fig. 18.26 Pathway of active sucrose transport leading to phloem loading

(sucrose translocator) that transports both sucrose and H^+ into the sieve element. Such a process may also be coupled to the transfer of K^+ inwards, accounting for the high concentration of both sucrose and K^+ in the phloem sap. Phloem unloading takes place passively down a concentration gradient of sucrose, and transfer cells are often present at unloading sites.

Movement of phloem sap in sieve tubes between sites of loading and unloading is by **mass-flow transport** (Fig. 18.27), that is, driven by a gradient of hydrostatic pressure. As a consequence of the increased concentration of sugar at the loading site, water potential decreases and water enters the sieve tube by osmosis. This increases hydrostatic pressure in the source end of the transport column so that it is greater than at the sink and movement occurs in the direction of source to sink. This model accounts for the transport of solutes in phloem of different vascular bundles occurring in different directions.

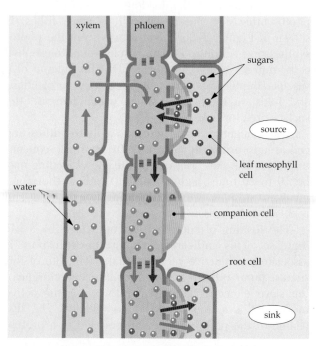

Fig. 18.27 Mass-flow transport of phloem sap between source and sink. As sucrose is actively loaded into phloem, water also enters by osmosis, creating a hydrostatic pressure that moves the sap

S ucrose is transported in sieve tubes by a pressure-driven, mass-flow mechanism.

Adaptations to stress

The evolution of the vascular transport sytems of xylem and phloem has allowed plants to penetrate high into the aerial environment while they still exploit water and inorganic nutrient resources of the soil. Under normal conditions, the water carried in the xylem replaces water lost by transpiration from the leaves. The impact of transport systems on the evolution of plant form can be seen in the striking difference between the small, ground-encrusting non-vascular liverwort plants and the majestic 100 m trees of *Eucalyptus regnans*, the wood of whose trunk is composed of tightly packed xylem. However, drying of the soil or excessively high transpiration (e.g. under hot dry winds) may lead to water loss exceeding water transport into the leaves and the foliage suffers a shortage of water. The aerial environment also exposes the plant to other environmental stresses, particularly rapid and extreme temperature changes.

Water stress

Water stress or drought stress are the terms used to describe the condition of impaired function in plants due to lack of water. Terrestrial plant biomass, productivity and distribution often correlate with supply of water (Chapter 44). The effect on plant growth is shown in Figures 18.28 and 18.29 for some arid zone plants in South Australia. For short-lived plants, rainfall in the previous 12 months correlates with growth. For bladder saltbush, *Atriplex vesicaria*, which is slow-growing and has deep roots, growth is better correlated with rainfall over the previous two years. The close link between water supply and plant growth arises because of the inevitable water vapour loss associated with uptake of CO_2 for photosynthesis.

Water stress generally occurs due to drought but can also occur if some other factor prevents the efficient absorption of water by roots. Low soil temperature and low O_2 concentration can induce water stress by increasing the resistance to water flow through roots, presumably by affecting membranes or the connections between cells (plasmodesmata, Chapter 6) or by blockages arising in the xylem.

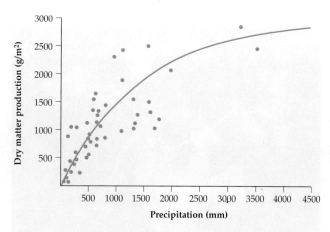

Fig. 18.28 Net primary productivity as a function of mean annual rainfall. The curve flattens at high rainfall because more water exits the sources available to plants by run-off and deep drainage. At low rainfall (less than 500 mm) each additional amount of rainfall results in a proportional increase in production

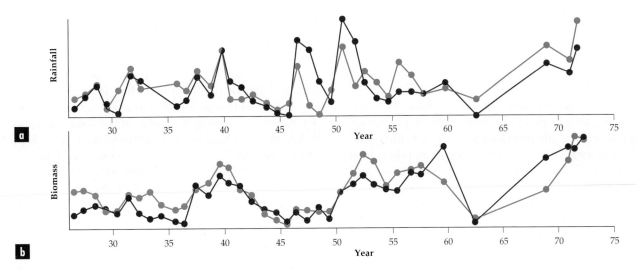

Fig. 18.29 Changes in relative biomass (red) and rainfall (blue) for arid zone vegetation at Koonamore in South Australia. **(a)** The biomass of herbs, annuals and dwarf shrubs is correlated with rainfall in the previous 12 months. **(b)** The biomass of bladder saltbush, *Atriplex vesicaria*, is better correlated with rainfall in the previous 48 months

The movement of water from the soil through the plant to the atmosphere is governed by the water potential gradients in the soil–plant–atmosphere continuum. If a plant is to extract water from the soil, it must generate a lower water potential than that of the soil. However, measurements of water potential alone will not indicate how a plant is performing since two plants may have identical water potentials yet one can be obviously wilted while the other can be turgid and still growing. Many plants, such as the white mangrove, *Avicennia marina*, can lower their water potential to below −3.0 MPa and still carry out photosynthesis while more mesophytic plants wilt and

cease photosynthesising at around −1.5 MPa. In mangroves, turgor pressure of leaf cells is above zero because cells are able to generate a high internal osmotic pressure. It is the maintenance of positive turgor that is correlated with maintained photosynthesis, transpiration and growth. The ability of a plant to maintain turgor pressure as its water potential decreases means that photosynthesis (and transpiration) can continue, although probably at a reduced rate, while water can continue to be extracted from drying soil. Plants show differences in their ability to function as their water potential decreases, even when growing in the same environment (Fig. 18.30).

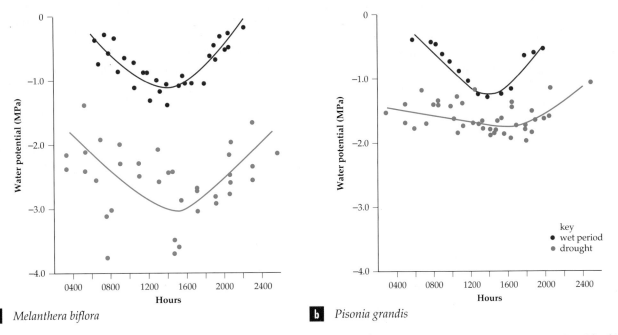

Fig. 18.30 Daily variation in water potentials for **(a)** *Melanthera biflora* (daisy) and **(b)** *Pisonia grandis* (a tree) growing on One Tree Island in droughts and wet periods. The daisy's water potential falls below −3 MPa during droughts

The extent to which a plant can continue to function during water stress depends on its ability to maintain a positive turgor pressure at low water potentials.

Ways that plants cope with drought

There are numerous ways in which plants live and survive in drought-prone environments. A plant may escape drought while in a dormant part of its life cycle, such as a seed. For example, when water becomes available in deserts, annual plants quickly grow and flower to produce more seeds. Sturt's desert pea grows quickly after rain, reaching seed set as quickly as six weeks from germination and producing seed that is capable of withstanding dry periods and high temperatures.

In an arid zone, perennial plants are *drought-tolerant*. Dehydration is the main difficulty for these plants. Most drought-tolerant plants avoid dehydration, but a few, for example, some native grasses such as *Tripogon loliiformis* and desert mosses, can withstand total dehydration.

Avoiding dehydration

Plants avoid dehydration by either increasing water absorption, reducing water loss by transpiration, or both. A deep and extensive root system will be able to extract water from a large soil volume. If taproots can reach the watertable, then transpiration and photosynthesis can proceed normally. River red gums, *Eucalyptus camaldulensis*, growing along seemingly dry creek beds exploit underground water via deep taproots.

The proportion of root compared with shoot (root:shoot ratio) is often high for arid zone plants and plants subjected to water stress. A high root:shoot ratio means that the absorptive surface relative to the evaporative surface is large. Many plants respond to water stress by a change in growth pattern or by reducing the size of the shoot as a result of leaf shedding, as in blue bush, *Maireana sedifolia*. The size of the root system may be increased at the expense of the shoot by redirecting sugars to roots.

The soil under the canopy of vegetation is modified by the accumulation of leaf litter and by the shape of plants. As a result, rainwater penetrates the soil more effectively and evaporation from the soil is reduced. Due to their shape, leaves and branches may intercept water during rainfall and channel it down the trunk to the roots, as in mulga, *Acacia aneura*, with its upward-pointing foliage.

The fleshy water-filled tissue of cacti and other succulent plants provides water storage. The cells composing this succulent tissue have large vacuoles and, by virtue of their flexible cell walls, are able to give up water to photosynthetic cells that are more dependent on maintaining a constant volume as water becomes scarce. In *Opuntia ficus-indica*, up to 82% of the water in water-storage tissue can be lost without irreversible tissue damage.

Control of water loss is closely related to control of leaf temperature. The shape of leaves enables them to capture light for photosynthesis but, in so doing, they also collect long-wavelength infra-red radiation. Most of the absorbed energy is reradiated but some must also be lost by other means if leaf temperature is to be kept within favourable limits. Transpiration from leaves is particularly important in dissipating such heat. If transpiration is reduced as stomata close during drought, leaf temperature increases and plants can suffer heat damage.

Leaf characteristics that reduce the absorption of radiation aid water conservation because less reliance needs to be placed on evaporative cooling. Leaf hairiness and increased reflectance (glaucousness), for example, due to surface waxes, reduce temperature and transpiration rate. Some saltbushes, *Atriplex*, change leaf reflectance during the course of leaf development such that higher reflectance coincides with summer. Varieties of wheat that have more reflective leaves have been selected and planted in hot environments, resulting in a reduction in absorbed radiation. However, reducing the amount of radiation absorbed may be at the cost of reduced photosynthesis.

Leaf orientation is also important. Leaves that hang vertically, like those of eucalypts and some salt-tolerant mangroves, avoid a high radiation load maximum in the middle of the day. Other plants are able to move their leaves to track the sun (heliotropic leaves). Siratro, *Macroptilium atropurpureum*, a pasture legume grown in Queensland, tracks the sun in different ways depending on whether the plant is water-stressed or not. Unstressed leaves face the sun through the day; water-stressed leaves orientate parallel to the sun's rays, thereby reducing the sun's heating effect and consequently transpiration from leaves (Fig. 18.31).

Heat loss is greater for small leaves or highly dissected leaves than it is for large leaves because small leaves are more efficient at losing heat to moving air. This is a physical effect related to the high perimeter–surface area relationship, and to the average thickness of the boundary layer of air around a leaf and how it varies with the velocity of air movement along the leaf (Fig. 18.32). It is not surprising that many arid zone plants have small leaves.

Plants able to cope with water stress may do so by entering a more tolerant stage of their life cycle (seeds) to escape dry conditions, or by either tolerating or avoiding dehydration.

Water-use efficiency

Minimising water loss under arid conditions is only half the story. Plants that can minimise water loss yet

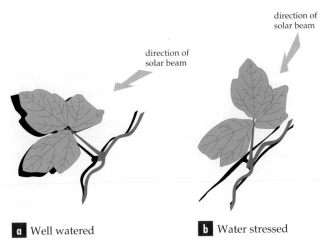

a Well watered   **b** Water stressed

Fig. 18.31 Changes in leaf orientation with respect to the direction of the sun's rays in response to water stress in Siratro, *Macroptilium atropurpureum*, a pasture legume used in Queensland: **(a)** leaf from a well-watered plant; **(b)** leaf from a water-stressed plant

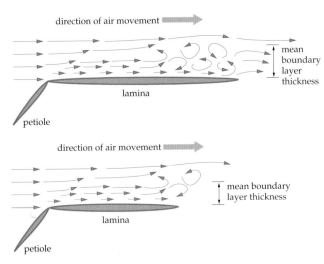

Fig. 18.32 Longitudinal section along leaves of different length showing air movement on the surface of the leaf and the average boundary layer thickness. The boundary layer is thinnest at the leading edge of the leaf and gets thicker towards the centre. At the trailing edge, more turbulent flow occurs. The thickness of the boundary layer is greatly exaggerated and is normally less than a few millimetres. The longer the length of the leaf in the direction of air movement or the lower the velocity of air movement, the larger is the average boundary layer thickness. Heat loss from the leaf is rapid when the boundary layer is thin or the temperature difference between moving air and the leaf is large

simultaneously maximise the gain of CO_2 should perform well in dry conditions. We can assess this performance by measuring the efficiency of water use, that is, the carbon gained (e.g. dry weight) divided by the amount of water transpired in gaining the carbon:

$$\text{water-use efficiency} = \frac{\text{carbon gained}}{\text{water lost}}$$

Water-use efficiency varies widely among different species and within a species. Within-species

variation can be due to different phenotypes resulting from different conditions during growth. Also, measurement of water-use efficiency for an individual crop plant grown in isolation may be very different from that measured for a whole crop on a farm where competition between plants occurs and the micro-climate is changed by the behaviour of the crop. A measure of water-use efficiency can be used to select for varieties of agricultural plants that may be more productive in Australia's arid climate.

> In arid climates, perennial plants have high water-use efficiency, conserving water and maximising the amount of CO_2 gained.

Responsive stomata

The largest drop in water potential occurs at stomata, between the inside of the leaf and the atmosphere. Therefore, to a large extent, the pattern of stomatal opening and closing over time controls the water-use efficiency of a plant. For example, if a leaf is water-stressed, it is better for stomata to close a little, thereby saving water but not greatly reducing the rate of photosynthesis as a result of lowered CO_2 uptake. Optimisation of carbon gain versus water loss accounts for the midday closure of stomata often seen in plants in summer (Fig. 18.33).

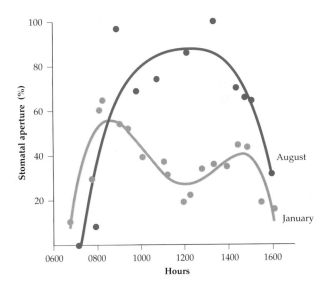

Fig. 18.33 Variation in stomatal aperture during the day for the same black box trees, *Eucalyptus largiflorens*, during January and August on the Chowilla floodplain near Renmark in South Australia. The stomatal aperture is relative to the maximum observed in the data times 100%. During summer (January) the stomata close in the middle of the day, giving a double peak in the pattern. Despite the near vertical leaf orientation in these trees, the high temperature and low humidity in the middle of the day in summer results in a large evaporative demand. It is thought that the stomata close in response to low external humidity, thus preventing too much evaporation. Even if the stomata were to open to the values of winter, it is unlikely that any extra gain in photosynthesis (CO_2 uptake) would occur but there would be a large increase in transpiration

Stomatal control optimises water use and carbon gain on a daily basis for a particular set of conditions. To optimise the use of soil water over a longer term, a plant may have to limit water loss, perhaps at the expense of carbon gain, so that it does not run out of water before it flowers and reproduces or before the next seasonal cycle of rainfall. There is good evidence that roots sense soil dryness and send a signal to the leaves, which affects stomatal aperture; for example, plants often close stomata as the soil dries before there is any detectable change in the water potential of the plant. The signal may be the hormone ABA (Chapter 24) or a compound that determines its synthesis, which is released by roots into the transpiration stream under these conditions. ABA in minute quantities closes stomata and is also produced by mesophyll cells in wilted leaves.

Various controls are exerted on stomata in order to optimise the use of water. These involve an ability to sense water vapour flux and CO_2 concentrations, and the hormone ABA, which is released from roots and mesophyll cells in wilting leaves.

Photosynthesis and water stress

Evolutionary modifications to the basic type of photosynthesis (C_3, C_4 or CAM; Chapter 5) have enabled some plants to improve their water-use efficiencies considerably, particularly at high temperatures and high light intensities typical of arid environments. This allows photosynthesis to continue to some degree when stomata are almost closed; high stomatal resistance restricts water loss greatly and also inward diffusion of carbon dioxide in amounts that are sufficient to maintain the low concentrations (inside the leaf) that support photosynthesis in C_4 plants but are too low for photosynthesis in C_3 plants.

The CAM mode of photosynthesis results in an even more spectacular water-use efficiency, but not high rates of photosynthesis (Chapter 5), because uptake of CO_2 occurs at night when evaporation is lowest. Some CAM plants have the requirement of low night-time temperatures for stomata to open, ensuring low evaporation rates. Some plants only use the CAM mode at certain times. These facultative CAM plants use C_3 photosynthesis when water is plentiful and, when water becomes limiting, CAM is invoked. The ice plant, *Mesembryanthemum crystallinum*, is such a facultative CAM plant. The switch to CAM is probably triggered by low turgor pressure in the leaf cells, a trigger that is also important in other plants as an emergency mechanism to close stomata under extreme water stress.

C_4 and CAM modifications of photosynthesis allow plants to tolerate drought and high temperature by increasing water-use efficiency.

Salinity and mineral stress

High salt concentrations in the root zone of a plant lower soil water potential and may lead to water stress (p. 476). Sea water has a water potential of about -2.5 MPa and, for water to be drawn from this solution, plants need to generate an even lower (more negative) water potential. Some ground waters near the Murray River have salinity levels higher than sea water. The ground water has risen in some regions due to locking on the Murray; as a result, black box trees, *Eucalyptus largiflorens*, are suffering (Fig. 18.34). Water stress may occur, in this case, because the volume of soil from which fresh water can be obtained has been reduced by the rising saline watertable. Roots submerged in the ground water may suffer both osmotic stress (due to the low water potential of the saline water) and salt toxicity.

Salt (NaCl) can be toxic to most plants. Enzymes are inhibited by Na^+ at concentrations above about 0.1 M. This is so for all types of plants, halophytes included. Excess Na^+ taken into a cell must be excluded from the cytoplasm and it is generally sequestered in the vacuole. The osmotic potential of the cytoplasm is balanced to that of the vacuole by the accumulation in the cytoplasm of solutes more compatible to metabolism, such as the amino acid, proline (p. 466). Proline is accumulated by a variety of plants and not just under salt stress. Compartmentalisation within the cell is, however, just one of the ways that may lead to resistance to salinity stress in plants.

Salt that enters the transpiration stream will be carried to leaves, where it will become concentrated as water evaporates to the atmosphere. Even if only a very

Fig. 18.34 Salt-affected black box trees, *Eucalyptus largiflorens*, on the Chowilla floodplain near Renmark in South Australia. Two apparent varieties of box tree are observed based on leaf colour. The green variety (tree on the left) appears to be more tolerant of the conditions that have led to the demise of the grey variety on the right. This area has been subject to a rise in height of the saline ground water to within a few metres of the surface

small fraction of the salt in the soil solution enters xylem, salt may build up to toxic levels in leaves, eventually killing them. Some halophytes, for example, white mangrove, *Avicennia marina*, have special salt glands in their leaves that excrete salt that is not excluded by roots. In contrast, most plant species are **glycophytes**; that is, they cannot tolerate saline soils or salt accumulation in their leaves.

Since transpiration delivers salt to the shoot, efficient control of transpiration is one way to reduce the salt burden on leaves. For example, although mangroves are C₃ plants, they have very high water-use efficiencies, which would reduce the salt load to the shoot. Low transpiration also means that there is less evaporative cooling of leaves. Some mangroves have small leaves vertically inclined that would reduce their radiation load and potential transpiration.

> High salt concentrations in soil can lead to water stress and Na⁺ toxicity. Compartmentalisation of ions within cells and within the plant, the ability to exclude salt at roots and leaves, and the balance of transpiration and leaf growth with ion uptake all contribute to salt tolerance.

Another form of salt stress occurs from salt-laden air. Salt crystals can collect on leaves after evaporation of small water droplets. Some coastal plants have a mesh of water-repellent cuticular wax over stomata, which prevents small water droplets from entering the leaf. Tiny amounts of detergent from sewage outfalls decrease the surface tension of droplets of sea spray, allowing the saline fluid to pass through the mesh of

stomatal wax platelets and enter the leaf. Norfolk Island pine has suffered salt damage as a result of this wind-borne detergent pollution (Fig. 18.35).

Other types of ions besides Na⁺ and Cl⁻ that may be in excess in the soil can be toxic to plants. Soil acidity (excess hydrogen ions, H⁺), either naturally occurring or induced by the use of ammonia- and amide-containing fertilisers, can release aluminium (Al) in a water-soluble form that is toxic to plants. Aluminium inhibits root growth and reduces uptake of minerals such as calcium. Problems associated with aluminium toxicity in acid soils currently cost farmers in New South Wales and Victoria an estimated $100 million a year from lost crop and pasture. Adding lime to acid soils can eliminate aluminium toxicity in topsoil, but liming subsoil is not feasible economically or technically. Development by plant breeding or genetic manipulation of aluminium-tolerant geno-types, for example in wheat, may be an alternative means of improving yields on acidic, aluminium-containing soils, and reducing costs for lime application.

Deficiencies of particular nutrients also produce stress symptoms in plants, either directly or indirectly (p. 462). A deficiency of nitrogen affects photosynthesis as nearly 80% of the nitrogen in plants is used in the photosynthetic apparatus, particularly the enzyme ribulose bisphosphate carboxylase, which fixes CO₂ in all plants (Chapter 5). Mineral deficiencies may also lead to plants being more susceptible to disease, as is thought to be the case for a fungal disease in wheat where there is a strong correlation between zinc deficiency and presence of the disease.

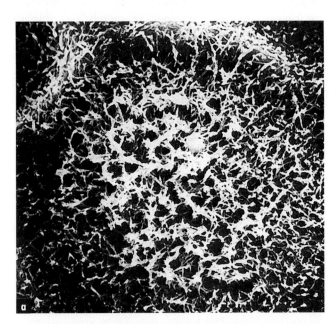
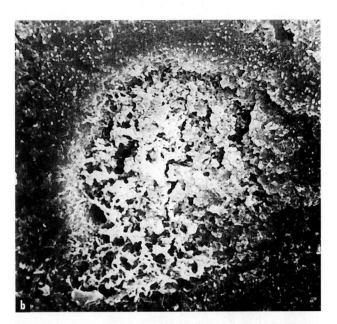

Fig. 18.35 The effect of detergent on the waxy fibrils that protect the stomata from the ingress of salt spray into leaves of the Norfolk Island pine, *Araucaria heterophylla*. **(a)** The normal condition. **(b)** Treatment with sea water spray containing detergent (30 ppm) causes the condition where the waxy fibrils are coalesced to form plates. The salt can traverse the stomata after such treatment

on toxicity can result from an increase in concentration of ions to toxic levels at low soil pH or from the effect of pollutants that facilitate entry of toxic ions into a plant. Deficiencies of particular nutrients also cause stress in plants.

Lack of oxygen around roots

Roots require O_2 for cellular respiration (Chapter 5), which generates ATP used by roots to actively accumulate particular ions, such as NO_3^-, or to excrete unwanted solutes, such as Na^+. When soil becomes flooded, air is displaced and, because of the slow diffusion of O_2 in water, the root environment quickly becomes anaerobic. ATP production in roots then occurs only by glycolysis, which yields much less ATP than does aerobic respiration and which leads to a rapid inefficient use of carbohydrate reserves. Thus, under waterlogged conditions, root growth, transpiration and translocation may decrease or even cease completely.

Toxic substances are also produced under anaerobic soil conditions due to microorganisms using alternative electron acceptors for their respiration. These end products include hydrogen sulfide, methane and Fe^{2+} and Mn^{2+}. An immediate effect of anaerobiosis is an increase in the resistance to water flow across the root (an effect on the cell membranes), which can lead to water stress in shoots if the plant continues to transpire. There are also effects on membrane permeability and ion selectivity in uptake.

Many plants, such as mangroves, seagrasses and rice, are naturally adapted to low O_2 contents in soils (Chapter 17). Adaptations include anatomical features, such as air canals in the roots (Fig. 18.36) or the production of lateral roots on the soil surface, which allow improved O_2 transport to root tissues; biochemical features, such as an increased ability to

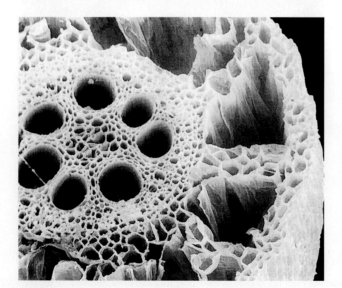

Fig. 18.36 Air canals produced in *Zea mays* as a result of several days' exposure of the roots to a solution of low oxygen (1%)

sustain anaerobic fermentation in root and an increased resistance to toxic substances produced in anaerobic soil; and hormonal signals from roots to shoots, which slow growth, nutrient uptake and water use (Chapter 24). These adaptations allow roots in low O_2 conditions to sustain the aboveground parts.

Lack of O_2 in waterlogged soil decreases cellular respiration by root cells. This impairs root growth, water uptake and nutrient uptake. Also, toxic substances are produced by anaerobic respiration of soil microbes.

Temperature stress

Different parts of the leaf canopy and root environment may fluctuate in temperature so that each part of a plant is subjected to a different temperature regimen. Temperature differences between different parts of a plant can be as much as 30°C. Temperatures fluctuate most widely near the soil surface, which affects seedling establishment. As mentioned earlier, water stress, and leaf orientation and reflectance properties also influence leaf temperature.

Growth rate of plants shows a characteristic bell-shaped curve in response to temperature (Chapter 42). Temperature affects biochemical reactions in two quite different ways. It affects the rates of reactions due to its effect on the kinetic energy of reacting molecules, and it affects the tertiary structure of enzymes and membranes (Chapter 2). At low temperatures, membrane fluidity can abruptly change, resulting in a 'frozen membrane' and disruption of the bilayer structure. Many important enzymes in photosynthesis and respiration are embedded in membranes and require a fluid membrane for proper function. Damage to membranes due to chilling or freezing is often indicated by leakage of solutes and water from cells into intercellular spaces giving the tissue a water-soaked appearance.

Some plants survive freezing temperatures as low as −50°C by preventing intracellular ice formation. Ice formation usually occurs only in extracellular fluid because solute concentrations and lack of ice nucleation sites in cells tend to prevent intracellular freezing. Extracellular ice formation causes cells to dehydrate since water can easily be drawn out of the cell across the plasma membrane and into the forming ice crystal with its very low water potential. The problem then becomes one of tolerating dehydration and, when thawing occurs, the ability of a cell to be able to re-expand without damage to the cell membrane.

Damage to plants at extremes of temperature is due to alteration in the tertiary structure of proteins, to changes in membrane fluidity or to disruption of biochemical processes. Dehydration can occur as a result of high or freezing temperatures.

Summary

- In addition to light, carbon dioxide and water, plants require inorganic nutrients (minerals) for growth, which they obtain from their environment and transport to all parts of the plant. Because photosynthesis requires uptake of carbon dioxide through open stomata, growth in plants is inevitably accompanied by evaporative water loss.

- Fourteen mineral elements are essential for the growth of plants: N, P, K, Mg, Ca and S are required in large amounts and Fe, Cl, Mn, Cu, Zn, B, Mo and Ni are required in trace amounts. These minerals are obtained primarily from soil, and cycle from plants to decomposers and back to the soil. The ultimate source of nitrogen, however, is the atmosphere, which becomes available to plants as nitrate and ammonium due to the action of nitrogen-fixing bacteria.

- In plants, cell-to-cell movement of water and dissolved nutrients occurs via apoplastic and/or symplastic pathways. Long-distance transport of water and solutes occurs in xylem and sugars are transported in phloem. Transport mechanisms include passive processes, such as diffusion, mass flow and osmosis, or active processes, which require energy.

- The movement of water in plants is down gradients of water potential. Water potential (ψ) is negative and in plant cells is the sum of osmotic potential (ψ_π) and pressure potential (ψ_P):

$$\psi_{cell} = \psi_P + \psi_\pi$$

- When a plant cell loses water and decreases in volume, pressure potential and osmotic potential decrease, with the result that water potential decreases.

- The capacity of a cell to retain and to absorb water is increased by osmotic adjustment. Plants respond to longer term soil water deficits by accumulating solutes in the vacuole, which allows turgor and growth to be maintained in dry or saline conditions.

- Plants lose water by evaporation from leaf surfaces, the process of transpiration, which draws on solar energy to vaporise water. In a transpiring plant, xylem sap is under negative hydrostatic pressure or tension. In some plants when transpiration is low, a positive pressure can occur in roots (root pressure).

- Assimilates are transported from their source in the leaf via phloem to sinks in other parts of a plant. Mature (source) leaves do not import assimilates from another source.

- At a source, sugar is actively loaded into phloem against a concentration gradient. Water also enters the phloem by osmosis, creating a positive hydrostatic pressure. This pressure drives the mass movement of phloem sap in sieve tubes to sites of unloading. At a sink, sugar is unloaded by passive diffusion.

- Responses of individual plants to environmental stress may involve anatomical, physiological and behavioural changes in phenotype, within limits defined by their genotype.

- Different environmental conditions, such as low temperature, low O_2 levels, drought and salinity, may lead to the same stress in an organism, for example, dehydration.

- The extent to which plants can continue to grow during water stress depends on their ability to maintain a positive turgor pressure at low water potentials. Plants able to cope with water stress do so by entering a more tolerant stage of their life cycle (seeds) to escape dry conditions, or by either tolerating or avoiding dehydration. In arid climates, perennial plants must conserve water and maximise the amount of CO_2 gained.

- Stomata are regulated to optimise water use. This involves the ability to sense water vapour flux and CO_2 concentrations, and the hormone ABA, which is released from roots and mesophyll cells in wilting leaves.

- C_4 and CAM modifications of photosynthesis allow plants to tolerate drought and high temperature by increasing water-use efficiency.

- Lack of O_2 in waterlogged soil decreases cellular respiration by root cells. This impairs root growth, water uptake and nutrient uptake. Also, toxic substances are produced by anaerobic respiration of soil microbes.

- High salt concentrations in soil can lead to water stress and ion toxicity. Compartmentalisation of ions within cells and within the plant, and the ability to exclude salt at the roots all contribute to salt tolerance.

- Ion toxicity can result from an increase in concentration of ions to toxic levels at low soil pH, or from the effect of pollutants that facilitate entry of toxic ions into a plant.

- Damage to plants at extremes of temperature is due to alteration in the tertiary structure of proteins, to changes in membrane fluidity or to disruption of biochemical processes. Dehydration can occur as a result of high or freezing temperatures.

keyterms

cavitation (p. 469)
chlorotic (p. 462)
cohesion theory
(p. 469)
glycophyte (p. 481)
halophyte (p. 468)
hydroponics (p. 461)
macronutrient (p. 461)

mass-flow transport
(p. 476)
micronutrient (p. 461)
mineral element
(p. 461)
mycorrhizal fungi
(p. 464)

osmotic potential
(p. 466)
pressure potential
(p. 466)
proteoid root (p. 464)
root pressure (p. 469)
tonoplast (p. 464)
trace element (p. 461)

translocation (p. 473)
transpiration (p. 468)
turgor pressure
(p. 466)
water potential
(p. 465)
water-use efficiency
(p. 479)

Review questions

1. What essential inorganic nutrients does a plant require for growth? Which of these nutrients are required in trace amounts? What are some of the general functions of inorganic nutrients?

2. Describe, with examples, two ways in which some Australian plants are able to survive on low-nutrient soils.

3. Explain in terms of changes in cell water potential (ψ_{cell}) how guard cells function as turgor-regulated valves.

4. Explain why salt makes soil water less available to a plant and has similar effects to a drought.

5. What is the composition of phloem sap and how is it different from that of xylem sap?

6. Explain the mechanisms by which assimilates are translocated in phloem from a source to a sink.

7. (a) What is transpiration?

   (b) Why is it a good practice to remove some leaves when transplanting a plant?

Extension questions

1. (a) Trace the two possible pathways of a water molecule moving from the soil into the stele of a root.

   (b) By what processes do roots take up ions?

2. (a) What happens to the transpiration rate and stomata of a plant when it is exposed to a hot dry wind?

   (b) What is the role of the hormone ABA during prolonged periods of water stress?

3. Explain how soil water can be transported to the top of a tall tree. What special structural features of xylem enable this to occur?

4. Explain how (a) radioactive tracers and (b) aphids have been used to study the transport and distribution of carbohydrates in plants.

5. Why is efficient water use in a terrestrial plant beneficial to salt tolerance?

6. Why is it unlikely that a single gene is responsible for either drought or salt tolerance in plants?

7. A leaf is detached from a plant and kept in its natural orientation. What could you conclude regarding its temperature regulation if the leaf's temperature remained constant and only slightly above ambient temperature?

8. In Figure 18.30, by approximately how much would the osmotic potential of the cell sap of the daisy have to increase in order for turgor pressure to be maintained constant during drought?

Suggested further reading

Atwell, B., Kriedemann, P., Turnbull, C. (eds). (1999). *Plants in Action*. Melbourne: Macmillan Educational Australia.

An in-depth modern account of how plants function that emphasises examples in Australia. A good source of further detail on all subject areas of Chapter 18.

Pate, J. S. and McComb, A. J. (eds). (1981). *The Biology of Australian Plants*. Perth: University of Western Australia Press.

This work contains interesting chapters on mycorrhizae and other adaptations of Australian plants for collection and storage of nutrients.

Cloudsley-Thompson, J. L. and Chadwick, M. J. (1964). *Life in Deserts*. London: Foulis.

A well-crafted overview of the wide range of adaptations that have evolved in desert plants and which allow them to cope with shortage of water.

CHAPTER 19

Heterotrophic nutrition in animals

Animals are *heterotrophs*—they are unable to manufacture organic compounds from inorganic molecules and must obtain them by eating other organisms. During the evolution of animals, many different kinds of systems for capturing, processing and digesting food have arisen. The aim of each is to obtain and break down large molecules into molecules small enough for absorption. Some animals also depend on symbiotic relationships with other organisms for certain nutrients. For example, some corals and the giant clams found throughout the Great Barrier Reef contain algae, zooxanthellae, in their tissues (Fig. 19.1a). Zooxanthellae are photosynthetic and synthesise carbohydrates, some of which are released and utilised by the host's cells. Other animals, such as the koala (Fig. 19.1b), have bacteria in their gut, which produce enzymes that digest cellulose. Still other animals, such as intestinal tapeworms and marine pognophoran worms (Fig. 19.1c), have no gut at all, and rely totally on other organisms for their nutrition.

The activities associated with obtaining and processing food do not occur in isolation from other behaviour. Feeding in some animals is limited by environmental influences, such as day length, tidal level or temperature. For example, animals may confine food gathering and processing activities to times when the risk of being caught by a predator is lowest, or when a food source is most abundant (Chapter 28). The study of nutrition and digestion in a broad context is known as nutritional ecology.

What nutrients do animals need?

Besides organic compounds, such as carbohydrates and lipids, which are needed as a source of chemical energy, animals require particular chemical compounds—amino acids, fatty acids, vitamins, minerals and water. These nutritional requirements may vary diurnally, seasonally and over an animal's lifetime. The nutrient content of a food and the ease with which the animal can access these nutrients are both important. Some nutrients cannot be stored in an animal's body, so these must be eaten regularly and there is no value in eating them in excess amounts. Also, because different metabolic pathways are interconnected, an abundant nutrient may be poorly utilised if another nutrient is in short supply. For most animals, the major sources of energy are **carbohydrates** (sugars, starches and fibre) and **lipids** (fats and oils). These substrates can be digested and the products oxidised to yield energy for the synthesis of ATP (Chapter 5). Compounds that normally serve other needs in an animal also may be oxidised for energy. For example,

Fig. 19.1 Animals have many different ways of obtaining nutrition. **(a)** This giant reef clam filters microscopic organisms from the water, but it also has autotrophs (algae) incorporated into the blue mantle tissue, which provide additional nutrients. **(b)** The koala eats *Eucalyptus* leaves and relies on gut microorganisms to break down cellulose. **(c)** These pognophoran worms at a deep-sea thermal vent have no gut and obtain all necessary nutrients from chemotrophic bacteria inside their body

amino acids required for protein synthesis also yield energy upon oxidation.

Essential **amino acids** are those that cannot be manufactured by an organism and so must come from a dietary source. Many animals obtain some essential amino acids from microorganisms that live symbiotically in the host gut or tissues. Different amino acids are essential for different animals. The actual amino acids required depend on an animal's metabolism and body form. For example, sheep have a greater need than do many other animals for the sulfur-containing amino acid methionine because it is an important component of wool.

Whereas particular requirements vary, almost all animals, except some insects, require certain **fatty acids** in their diet—these are essential fatty acids. Fatty acids form part of the phospholipids of cell membranes and contribute to the structure of other biologically active compounds, such as hormones.

Vitamins (Table 19.1) are usually described as organic compounds that are required by animals in small amounts for normal growth and maintenance. Vitamin C (ascorbic acid) is an example of a vitamin that can be synthesised by most animals, but humans, other primates and fruit bats, for example, lack one of the enzymes necessary for synthesis. Vitamins may be water soluble (B group, C) or fat soluble (A, D, E, K). Water-soluble vitamins are transported as free compounds in the blood. They cannot be stored and need to be replenished continually from a dietary source by those animals that cannot synthesise them. In contrast, fat soluble vitamins are transported as complexes with lipids or proteins and can be stored in varying amounts. Consequently, deficiencies of fat-soluble vitamins are unusual, at least in adults. In excess, some vitamins can be toxic; for example, several starving polar explorers were poisoned by excess vitamin A after eating the livers of seals and sled dogs!

Mineral elements (Table 19.2) are essential nutrients for all animals. Macroelements are those that are needed in relatively large amounts: sodium, chlorine, potassium, calcium and phosphorus. Sodium is the major cation and chlorine the major anion in body fluids; calcium is a major component of bone; phosphorus is a central element in ATP and in genetic material. Microelements are required in much smaller amounts and are often components of particular enzymes or transport systems such as iron in haemoglobin. Other microelements include cobalt, copper, molybdenum, zinc and iodine.

Although not strictly a nutrient, water is required by all animals and is obtained from a variety of sources. Animals produce metabolic water in their bodies during the complete oxidation of carbohydrates and fats (Chapter 5). Many animals drink, especially vertebrates, and water is naturally present in most foods.

Even apparently dry seeds contain 10% water, depending on the prevailing humidity.

> Animals need organic compounds for chemical energy, amino acids, fatty acids, vitamins, minerals and water.

Mineral nutrition of animals: a comparison with plants

Although the mineral requirements of animals have some similarities with plants (Chapter 18), there are important differences. Animals are unable to use inorganic nitrogen to build proteins and they need more sodium and chlorine than do plants. Plant tissues are rich in potassium and, normally, not in sodium. Animals do not have a boron requirement, whereas plants do, while many vertebrates (e.g. sheep and cattle) need sources of iodine, selenium and cobalt in their diet. Plants may contain sufficient concentrations of particular minerals for their own growth but too little of the mineral to meet the needs of the animals that eat them.

Nutrient composition of foods

Plant tissues

In general, the nutrient composition of plant tissues is more variable than animal tissues (Fig. 19.2). For example, the nutritional quality of north Australian tropical grasses declines markedly between the beginning and end of the northern dry season (Fig. 19.3). Most plant tissues are poor sources of protein, which has led to suggestions that protein availability is the major limiting factor for **herbivores**. However, optimal nutrition depends not on a single nutrient group, but on the right mix of nutrients. For example, proteins will be poorly utilised by herbivores if their diet does not contain adequate amounts of carbohydrates.

Although plant tissues eaten by animals are predominantly carbohydrate, the type of carbohydrate varies. Monosaccharides, disaccharides and starch are digested relatively easily but some of the more complex polysaccharides, such as cellulose and pectins of plant cell walls, are not. These components of plant fibre are important in digestion because of their ability to form bulk. Fibre takes up water and swells, and many plant cells retain their shape and size, even after their contents have been digested. In humans, one effect of dietary fibre is to stimulate movement of food by gut muscle. Diets of highly processed foods are low in fibre and may cause constipation, which can lead to a number of serious bowel diseases. On the other hand, for some herbivorous animals that eat high-fibre diets, the bulky fibre filling the gut can limit the absorption of nutrients.

Table 19.1 Vitamins: common sources, actions and effects of deficiency

Vitamin	Common sources	Actions	Effects of deficiency
Fat-soluble vitamins			
Vitamin A (retinol)	Yellow and green vegetables, yellow fruits, (fish) liver, egg	Used in synthesis of retinal pigments, regulation of bone cell activity, maintenance of epithelial tissue	Scaly skin, (night) blindness, growth failure
Vitamin D (calciferol)	Made in human skin (when exposed to sunlight), dairy products, egg	Increases calcium absorption, aids in bone growth and mineralisation	Bone deformities in children (rickets), bone softening in adults.
Vitamin E (tocopherol)	Vegetable oils, whole grains, seeds	Antioxidant, helps prevent damage to cell membranes	Lysis of red blood cells
Vitamin K (phylloquinone)	Formed by intestinal bacteria; green vegetables, cabbage, tea	Necessary for blood clotting	Severe bleeding (haemorrhage) due to defective blood clotting
Water-soluble vitamins			
Vitamin B_1 (thiamin)	Meat, whole grains, green leafy vegetables, legumes	Coenzyme in many enzyme systems	Beri-beri (heart, nervous system and digestive tract disorders)
Vitamin B_2 (riboflavin)	Dairy products, green leafy vegetables, meat, fish	Component of coenzymes essential in cellular respiration	Skin inflammation and lesion, mental depression
Niacin	Meat, nuts, green leafy vegetables, fish	Component of coenzymes essential in cellular respiration	Pellagra (skin, gut and nervous system disorders)
Vitamin B_6 (pyridoxine)	Legumes, meat	Coenzyme in many reactions in amino acid metabolism	Anaemia, digestive tract disturbances, irritability, skin damage
Pantothenic acid	Most foods	Component of coenzyme A (important in carbohydrate and fat metabolism)	Nausea, fatigue, headaches
Folic acid (folacin)	Formed by intestinal bacteria; green vegetables, whole grains, yeast, meat	Coenzyme in nucleic acid synthesis and amino acid metabolism	Anaemia, gastrointestinal disturbances
Biotin	Formed by intestinal bacteria; meat, legumes, chocolate	Coenzyme in carbon dioxide fixation	Scaly skin, mental depression
Vitamin B_{12} (pyridoxine)	Meat, fish, dairy products	Coenzyme in nucleic acid metabolism	Anaemia, impaired nerve function
Vitamin C (ascorbic acid)	Fruits and vegetables (especially citrus fruits, berries, tomatoes, broccoli, cabbage)	Used for synthesis of intercellular substances, aids in resistance to infection, used in carbohydrate metabolism	Scurvy (degeneration of skin, teeth, blood vessels), slow wound healing, impaired immunity

Table 19.2 Minerals: common sources, actions and effects of deficiency

Mineral	Common sources	Actions	Effects of deficiency
Calcium (Ca)	Dairy products, green vegetables, legumes	Components of bone and teeth, blood clotting, needed for muscle and nerve function	Retarded growth, osteoporosis
Chlorine (Cl)	Table salt	Formation of HCl in stomach, important in acid–base and osmotic balance, nerve function	Muscle cramps, reduced appetite
Copper (Cu)	Seafood, meat, beans, nuts, dark chocolate	Component of many enzymes (for synthesis of melanin, haemoglobin, transport chain components; iron metabolism)	Anaemia, bone and blood disorders
Fluorine (F)	Water, tea	Maintenance of bone and teeth	Tooth decay
Iodine (I)	Seafood, iodised salt, dairy products	Component of thyroid hormones	Enlarged thyroid (goitre)
Iron (Fe)	Meat, liver, nuts, egg, whole grains, green leafy vegetables	Component of haemoglobin, myoglobin, cytochromes and electron carriers in energy metabolism	Anaemia, impaired immune system
Magnesium (Mg)	Whole grains, dairy products, nuts, legumes	Needed for muscle and nerve function, component of coenzymes	Disturbances in nerve and muscle function
Phosphorus (P)	Meat, dairy products, whole grains, legumes, nuts	Component of bone, ATP, DNA and RNA; role in acid–base balance	Muscular weakness, loss of minerals from bone
Potassium (K)	Fruit, vegetables, meat, whole grains	Neural function, role in acid–base balance and water balance	Muscular weakness, heart failure, paralysis
Sodium (Na)	Table salt	Important in acid–base balance and water balance, neural function	Muscle cramps, reduced appetite
Sulfur (S)	As part of many proteins in diet	Component of many body proteins	Protein deficiency
Zinc (Zn)	Seafood, meat, legumes, whole grains	Component of many (mainly digestive) enzymes	Growth failure, impaired immune function, reproductive failure, scaly skin inflammation

The lipids of plants are composed mostly of unsaturated fatty acids and so are liquids at room temperature (e.g. olive oil and sunflower oil) but there are some exceptions where saturated fatty acids predominate (e.g. coconut oil is solid at room temperature). Although animals may eat unsaturated fatty acids in plant food, these are often converted to saturated fats in the digestive tract.

The mineral content of plant tissue may vary greatly between different geographic regions. In general, animals do not show specific appetites for mineral elements, apart from sodium. A hunger for salt (sodium chloride) has a major effect on the behaviour of many animals (including humans). For example, on the higher slopes of the Snowy Mountains region of Australia, much of the sodium has been leached out of

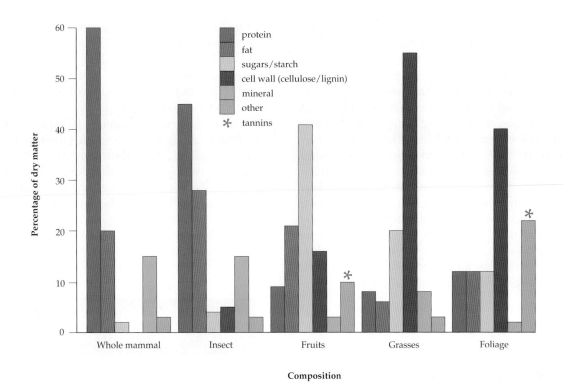

Fig. 19.2 The composition of some foods eaten by animals

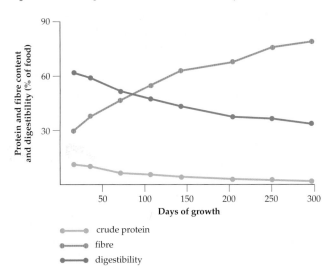

Fig. 19.3 The composition of many plant foods changes seasonally. Spear grass, *Heteropogon contortus*, in tropical Australia, declines in protein content and increases in fibre content as it ages. The combination of these means that the nutritive value (measured as percentage digestibility) also declines throughout the year

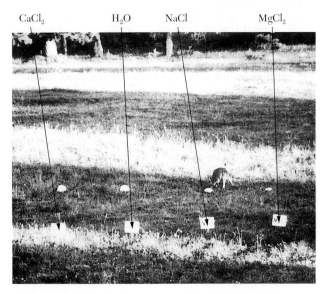

Fig. 19.4 Wild mammals have a specific appetite for sodium in the alpine regions of Australia, where sodium is sparse. The blocks impregnated with sodium chloride were eaten while those soaked in water or calcium or magnesium chloride were ignored

the soil, and animals, such as wombats, eastern grey kangaroos and feral rabbits, are chronically exposed to sodium shortage. These animals show a range of mechanisms for retaining sodium in their bodies, and will avidly chew on any material that contains sodium (Fig. 19.4).

The nature and availability of plant matter in aquatic and terrestrial environments are different. In aquatic environments, most autotrophs are algae,

which produce cellulose and storage carbohydrates but, unlike most land plants, have no secondary thickening. In aquatic environments, plants are physically supported by water and even vascular plants lack secondary cell walls. This means that aquatic plants and algae are softer than land plants and easier for an animal to digest. Seagrasses, for example, are much easier to crush than terrestrial grasses. Dugongs, major consumers of seagrasses in the Great Barrier Reef Region, have horny pads for crushing their plant food, whereas horses have hard enamel teeth with well-developed sharp ridges for chewing grasses.

Terrestrial vascular plants have complex cell walls and often lignified woody tissues (Chapter 17), which are difficult to digest. Rupturing plant cell walls makes the entire cell contents available. This requires hard jaws or teeth that can grind against each other. Terrestrial herbivores, such as molluscs, arthropods and vertebrates, all have mechanisms incorporating hard tissues that enable the animal to bite or chew pieces of plants. The digestive actions of fungi tend to make plant detritus more accessible. An added bonus for **detritivores** (animals that feed on detritus, dead organic matter) is that any fungi and bacteria feeding on the decaying wood will also be consumed.

> Because of a lack of secondary thickening, tissues from aquatic plants are generally easier to digest than are those from terrestrial plants. Plant matter is high in carbohydrates, low in protein, contains mostly unsaturated fats and has variable mineral content.

Animal tissues

Most animal tissues contain large amounts of protein, which is relatively easy to digest, but little carbohydrate. Given that carbohydrates, glucose in particular, are important sources of energy for most animals, an absence of dietary carbohydrates could pose serious problems. The brain, for example, uses glucose as its main energy source and a constant supply is critical. In the absence of dietary sources of glucose, animals produce glucose from other substrates, such as fats and proteins, through a process known as **gluconeogenesis**. The types of fats found in animals depend largely on the food eaten but tend to be dominated by saturated fatty acids with the exception of fish, which contain mostly unsaturated fatty acids and are therefore liquids at room temperature. The mineral composition of animal matter is essentially constant.

> Animal tissues are rich in protein, poor in carbohydrates, contain mostly saturated fats (except fish) and have a relatively constant mineral composition. They are relatively easy to digest.

How much food is required?

The nutrient requirements of a particular animal depend on its **metabolic rate**, age and reproductive state. Metabolic rate varies with level of activity, body mass and prevailing environmental conditions. Activity is probably the most important factor affecting metabolic rate: increased activity increases metabolic rate in both ectothermic and endothermic organisms. In addition, many animals show pronounced daily or seasonal changes in activity, metabolic rate and body weight. Box 19.1 discusses some of the recent findings about exercise and weight gain in humans.

For endothermic animals (Chapter 29), basal metabolic rate is the rate of metabolism of a quiet inactive animal in a thermoneutral environment. Metabolic rate increases when ambient temperature rises or falls below an endothermic animal's thermoneutral range. For ectothermic animals, standard metabolic rate is that of a quiet inactive animal and is temperature-dependent; increasing ambient temperature increases metabolic rate.

BOX 19.1 The new biology of body weight regulation

Between the ages of 20 and 40 a person will consume between 10 and 20 tonnes of food, yet the vast majority will maintain a relatively constant body mass. Body mass represents the balance between energy intake and energy expenditure. If there were even a 1% discrepancy between your energy intake and energy expenditure, then at age 40 a person would weigh about 100 kg more than they did at age 20—clearly this doesn't happen! This suggests that our bodies have precise, long-term mechanisms to detect and regulate energy balance. That is not to say that we can't gain weight in the short-term—many of us do—but on average over the longer term we should and do maintain a relatively constant body mass.

Studies in the 1950s argued that body fat was kept relatively constant over the longer term by a feedback system involving body fat and appetite. The idea was that fat cells secreted some substance that was transported in the blood and that somehow suppressed appetite. This idea became known as the lipostatic (constant fat) theory of body weight regulation. This is an example of a negative feedback. These early ideas were based on studies of rats that became obese as a result of single gene mutations—in particular a mutation in a gene called *Ob*. However, it was not until 1996 that the protein encoded by this gene was isolated and characterised. The protein has been called leptin (*leptos* means thin) and when it is administered to obese mice with a mutation in the *Ob* gene, they rapidly lose up to

30% of their body mass. In humans, leptin is secreted by the fat cells and its concentration in the blood is directly proportional to body fat. Women have significantly more leptin than men, but this relates to the fact that women have a higher body fat content than men. Leptin appears to affect appetite by a complex interaction with the appetite stimulating neurochemical called neuropeptide Y (NPY). The more leptin is produced, the less NPY is released. Thus, as body fat declines, the amount of leptin in the blood declines and so the more active NPY can be in stimulating appetite. This partially explains why many diets don't work—a short-term weight loss is countered by an increased appetite and so in the longer term the weight lost is regained. In humans, mutations in the *Ob* gene are extremely rare and obesity appears to be more related to poor communication between leptin and the appetite centres.

But food intake is only one part of the body weight equation and body weight will only remain constant if energy intake from food is balanced by energy expenditure. Energy expenditure is made up of our basal metabolism, the energy required for digestion and the energy we use in activity. Some of us seem to gain weight relatively easily in the short term whereas others seem to be very resistant to short-term weight gain even without increasing voluntary exercise. Recent studies have shown that much of the difference between 'easy gainers' and 'hard gainers', in the short term at least, is due to all the subconscious minor actions that we do—best described as how 'fidgety' we are. Physiologists distinguish between conscious, voluntary exercise (e.g. jogging, swimming and cycling) and subconscious activities. These non-exercise activities (or more precisely non-exercise activity thermogenesis, NEAT) comprise such things as posture, fidgeting and spontaneous muscle contractions. In controlled studies, volunteers were fed excess energy and exercised similarly but differences among them in the degree of 'fidgetiness' or NEAT explained 77% of the differences in fat gain. As humans overeat, those individuals who can dissipate the excess energy through subconscious activity will not be likely to gain fat whereas those with lesser degrees of non-exercise activity will be likely to have greater fat gain and be predisposed to develop obesity.

Body weight regulation is a complex topic with a number of different regulatory systems operating at different time scales. Nonetheless, these recent findings suggest that we are just starting to unravel the elaborate interactions between our stores of energy and the neurochemical basis of appetite and energy expenditure.

Metabolic rate and body mass

When metabolic rate is plotted against body mass for a wide range of mammals, it can be seen that daily energy requirements increase with increasing size. However, plotting metabolic rate per kilogram body mass shows an opposite trend: the energy requirements of an animal increase exponentially as body mass falls (Fig. 19.5). Thus, an elephant requires more energy per day than a mouse; but a mouse requires more energy per unit of body mass. To put it another way, 5 tonnes of mice require a lot more energy in a day than a 5 tonne elephant! **Mass-specific food intake** (food intake per unit body mass) also increases exponentially as body mass declines. This pattern holds true for all vertebrate and invertebrate groups that have been examined (Fig. 19.5).

The relationship between body mass and energy requirements may be related to the types of diets that animals select. For example, consider the Australian macropod marsupials—kangaroos, wallabies and rat kangaroos. More than 60 species, ranging in size from 0.6 kg to 90 kg, have a variety of diets (Fig. 19.6). The diets of small animals, such as the musky rat kangaroo, are composed of somewhat scarce but nutritious items, such as insects, or contain easily digestible material, such as starch-rich tubers. They consume this food at a relatively rapid rate but they do not need too much and can afford to spend time searching for these high-quality items. In contrast, large animals such as the red

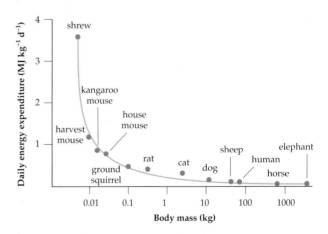

Fig. 19.5 Although large animals use more energy each day than small animals, the energy expenditure of animals per unit mass of tissue (plotted here) increases exponentially as animals get smaller. This is the mouse–elephant curve. Similar relationships exist for other animals, such as insects and marsupials

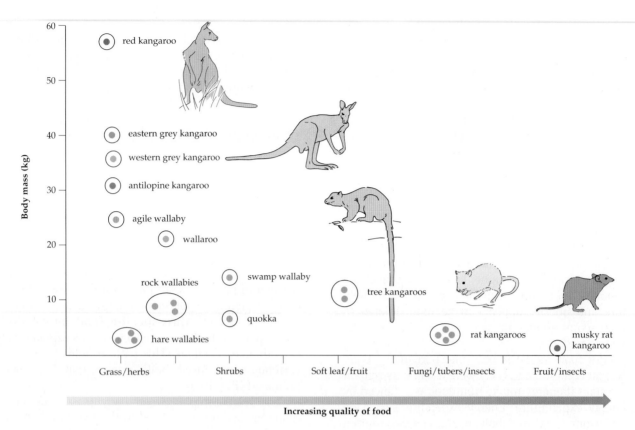

Fig. 19.6 The relationship between the nutritional quality of preferred foods and the body mass of some kangaroos, wallabies and rat kangaroos. Generally, smaller bodied species eat better quality foods, although this relationship is not perfect

kangaroo, which has adapted to grassland habitats, eat poor-quality bulky foods, mainly grasses. They need a large total amount of energy every day and cannot afford to spend time searching for rare, high-quality foods. They take what is easily obtainable and eat a lot of it. Not all kangaroos fit this pattern. Some have diets that, in theory, would appear to be so poor that they should suit only larger animals. The challenge for nutritional ecologists is to understand how, for example, small animals such as hare wallabies survive on a poor diet.

> Small animals need more energy per kilogram of body mass than do large animals. Small animals usually eat better quality foods.

The digestive process

Selecting or catching food is only the first step in obtaining necessary nutrients. Large organic molecules, such as fats, proteins and carbohydrates, must be broken down into molecules small enough to be transported across the gut epithelium. To do this within a reasonable period of time requires the action of digestive enzymes, often preceded by the physical breakdown of food, which increases the surface area available for enzymatic attack.

Physical digestion

Physical digestion, or the physical breakdown of food into small particles by grinding or chewing, occurs in different regions of the gut in different animals. It may take place upon ingestion, after storage or during chemical digestion. For example, the action of teeth and jaws in vertebrates and the rasping action of the radula in snails (Chapter 39) break up food items before they enter the gut. In some birds, a highly muscular gizzard, containing swallowed stones or shells, is used to grind food before it enters the stomach. Ruminants (p. 509) repeatedly regurgitate and rechew plants to break down cell walls.

> For many animals, the physical breakdown of food into small particles is necessary for efficient chemical digestion. This is particularly true for animals that eat terrestrial plants.

Enzymatic digestion

Although there are many different feeding mechanisms, the process of **enzymatic digestion** is basically similar in all animal groups. It involves the breakdown of complex molecules by hydrolytic enzymes, usually secreted into the gut lumen. Table 19.3 shows the location and function of a range of **digestive enzymes**.

Table 19.3 Important enzymes and their products and sites of action in the human digestive tract

Site	Protein digestion	Carbohydrate digestion	Nucleic acid digestion	Fat digestion
Oral cavity, pharynx, oesophagus		Polysaccharides (starch, glycogen) ↓ *Salivary amylase* Smaller polysaccharides, maltose		
Stomach	Proteins ↓ *Pepsin* Small polypeptides			
Lumen of small intestine	Polypeptides ↓ *Trypsin, chymotrypsin* Smaller polypeptides ↓ *Aminopeptidase, Carboxypeptidase* Amino acids	Polysaccharides ↓ *Pancreatic amylases* Maltose and other disaccharides	DNA, RNA ↓ *Nucleases* Nucleotides	Fat globules ↓ *Bile salts* Fat droplets (emulsified) ↓ *Lipase* Glycerol, fatty acids, glycerides
Epithelium of small intestine (brush border)	Small peptides ↓ *Dipeptidases* Amino acids	↓ *Disaccharidases* Monosaccharides	↓ *Nucleotidases* Nucleosides ↓ *Nucleosidases* Nitrogenous bases, sugars, phosphates	

Not surprisingly, there is a good correspondence between the types of food an animal eats and the types of digestive enzymes it produces. For example, the introduced passerine bird, the starling, does not eat fruits rich in sucrose and lacks the enzyme sucrase. People who are vegetarian tend to have higher salivary amylase activity (which splits starches to sugars) than those who eat a lot of meat. An interesting human example involves galactosidase, which splits the main milk sugar lactose. Although this enzyme is present in most babies, it is not secreted in some adults and so they are unable to digest milk adequately. People of European origin tend to continue to consume and digest lactose, whereas many of Asian and African origin are intolerant to lactose in milk.

Digestive enzymes are catalysts and their rate of reaction is affected by temperature and the pH of the medium in which they operate (Chapter 2). For example, human salivary amylase works best in an environment with a pH of 6.5, which means that it acts well in the mouth. However, pancreatic amylase, which enters the gut in the duodenum, has a higher optimum pH and acts on starches that have already been exposed to the acid conditions of the stomach.

In contrast to many other enzymes, digestive enzymes tend to have a lower level of specificity (Chapter 2). They tend to be specific for types of molecules, such as proteins or carbohydrates, rather than for specific bonds. Some enzymes may act on particular bonds within a molecule, such as terminal peptide bonds of a protein. Complete enzymatic breakdown of food usually involves the sequential secretion of different digestive enzymes along the length of the gut. In vertebrates, nervous and hormonal control mechanisms are responsible for ensuring that this occurs.

Control of digestive secretion in humans

The composition of digestive enzymes largely reflects the composition of the normal diet. Digestive enzymes tend to have a lower level of specificity than do other enzymes.

Saliva secreted into the buccal cavity lubricates food for its passage through the gut, contains enzymes, such as amylase (Table 19.3), that are involved in initial chemical digestion, and may contain toxins for use in defence or attack. Secretion of saliva is generally associated with ingestion of food. The Russian physiologist Ivan Pavlov conditioned dogs to salivate by ringing a bell (Chapter 28), showing that salivation was under nervous control. Enzyme secretion in the stomach and duodenum are under hormonal control (Chapter 25). Food reaching the stomach stimulates the secretion of the hormone gastrin from mucosal cells in the stomach (Fig. 19.7). Gastrin circulates in the blood and stimulates the release of hydrochloric acid and pepsinogen from the stomach mucosa. Pepsinogen is a **zymogen**, an inactive precursor of a protease, in this case, pepsin. Pepsinogen is activated to pepsin by either acid or existing pepsin. Cells secrete zymogens rather than active proteases to prevent damage to the cells themselves.

As the acidic stomach contents move into the small intestine, they stimulate intestinal cells to release another hormone, secretin. Secretin stimulates the pancreas to release hydrogen carbonate ions, which neutralise the acidity of the partially digested food. It also stimulates the release of bile from the gall bladder. Bile emulsifies fat, increasing the surface area of fat droplets. This is necessary because fat-digesting enzymes (lipases), like all digestive enzymes, are water-soluble and work in an aqueous environment. Without bile, lipases would have little access to fat. Another hormone, cholecystokinin, released from the duodenum, stimulates the pancreas to release other zymogens, including trypsinogen, which is activated to the protease trypsin. A third hormone, enterogastrone, inhibits the continual release of stomach acid and enzymes by a feedback system so that these are present only when required.

Enzymatic digestion is the co-ordinated, sequential breakdown of large molecules by enzymes secreted along the length of the gut. In most animals, secretion of these enzymes involves both hormonal and nervous control.

The time that digestive enzymes have to act depends on the speed at which food moves though the digestive tract. Some materials, such as cellulose, are resistant to hydrolysis and need a long time for digestion, whereas others, such as sucrose and starch, are readily hydrolysed. Increasing the length of the gut and expanding regions for storage allow longer periods for enzymatic breakdown. However, if food remains in the gut for long periods and yields fewer nutrients than

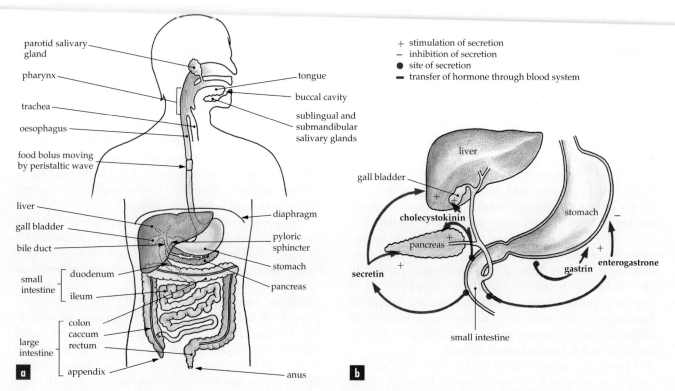

Fig. 19.7 The digestive system in humans: **(a)** major features of the gastrointestinal tract; **(b)** hormonal control of human digestive secretions

BOX 19.2 How does the stomach protect itself against digestion?

Biologists have long wondered why enzymes in the stomach can digest meat and other foods and yet, generally, do not digest the stomach lining itself. Painful gastric ulcers are evidence that breakdown of protection can occur and medication for peptic ulcers is one of the major sources of income for the world pharmaceutical industry. Many of these medicines are intended to counter the natural acidity of the stomach, but new approaches are continually sought. The discovery of Australian frogs that incubate their young in the stomach (the gastric brooding frog, Fig. 15.15) prompted speculation that the mechanisms used by these frogs to turn off stomach secretions could be useful in the treatment of ulcers. However, prolonged suppression of acid production in the stomach has undesirable side-effects in humans, such as weight gain and the formation of cancer-promoting agents in the gut.

During investigation of the mechanism that naturally protects the cells lining the stomach, Australian scientists have found a surface-active phospholipid (SAPL) in the stomach, chemically similar to some corrosion inhibitors that protect metal against acid. Further experiments have shown that these SAPL molecules make the stomach wall hydrophobic but that this property is lost when exposed to aggressive ulcer-inducing agents, such as bile salts and alcohol. Some bacteria, for example, *Helicobacter pylori*, are also major agents inducing gastric ulcers. All these agents tend to remove the SAPL by dissolving it. It has now been shown that these SAPL molecules protect the stomach itself against the digestive processes.

Anti-ulcer drugs currently in use are very effective in curing ulcers in the short term but the rate of relapse is 800% over two years if medication is discontinued. Lamellar structures similar to the SAPL observed in the stomach have been found in bananas, eggs and unpasteurised milk. The use of these simple foods might prevent ulcers at a fraction of the cost of the medication currently prescribed.

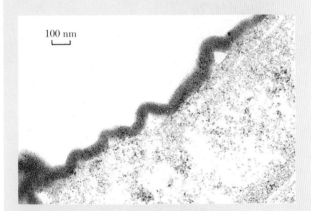

100 nm

The gastric mucosal barrier. A transmission electron micrograph from the epithelial surface of an oxynic duct in the stomach of the rat. In this region, there is no mucous lining, and the pH is about 1. This is the most corrosive environment in the stomach. The layered structures protecting the stomach epithelium are clearly visible. Some invasive microorganisms use similar surfactant coatings to protect themselves from digestion

new food would provide, there is no nutritional advantage. The right balance between digesting food for long periods or making room for new food depends largely on the metabolic rate of an animal. Small animals, with high **mass-specific metabolic rates**, tend to feed on foods that are digested relatively easily and that pass through the gut relatively rapidly, whereas in larger animals with lower mass-specific metabolic rates, the opposite tends to occur (Fig. 19.8).

> The speed at which food moves through the digestive tract governs the time that digestive enzymes have to act upon it.

Once foods have been digested into small molecules, such as amino acids and monosaccharides, they are absorbed across the epithelium of the gut wall. This may occur by simple diffusion, but more often it occurs against a concentration gradient by active transport (Chapter 4). Separate protein channels transport particular types of compounds. There is a close correlation between the natural diet of an animal and the numbers and types of **transport channels** in the gut. For example, fruit- and nectar-eating birds have a greater capacity to take up glucose than they do amino acids, whereas carnivorous birds have a greater capacity for amino acid uptake relative to their capacity for glucose uptake.

The folding and finger-like projections (villi) of the gut mucosa, and microvilli on epithelial cells, which can be seen in the small intestine (Fig. 19.9), increase the surface area of gut epithelium. This increases the number of transport channels and enables more rapid absorption of nutrients. The relative surface area of the small intestine is related to an animal's energy requirements. For example, free-living mammals require about 20 times as much energy per day as do lizards of similar size. Comparison of the intestines

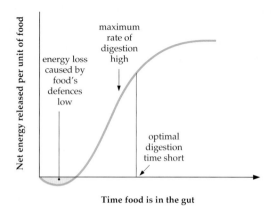

a High-quality food

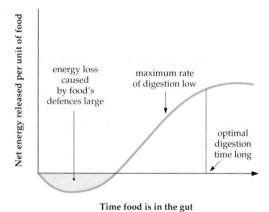

b Low-quality food

Fig. 19.8 An idealised model showing predicted patterns of digestion for **(a)** high- and **(b)** low quality foods. The amount of energy released from a food will initially be negative because the animal needs to expend energy in chewing. Initially, digestion is rapid because of release of material from the cell contents but, thereafter, it becomes progressively harder to digest the structural material. The maximum rate of digestion is lower on the lower quality food but the time that the food is kept in the gut is greater. If food is filling the gut, then the animal cannot continue to eat more. Animals must trade-off the time taken to further digest food already in the gut against the possibility of finding and eating new food

from similar-sized mammals and lizards shows that, although the overall dimensions of the small intestines are similar, the amount of folding of the lining of a mammal's small intestine is much greater, increasing its surface area by up to 50 times. This allows mammals to take up nutrients much faster than lizards and so fuel their greater metabolic rate.

The size and **absorptive capacity** of the gut can change in response to change in diet or when energy requirements change, such as during lactation or exposure to cold. For example, during winter times and during lactation, most mammals seem to grow more intestine; some parts of the gut can become 40% larger. This allows animals to eat more food while extracting the same proportion of nutrients from it.

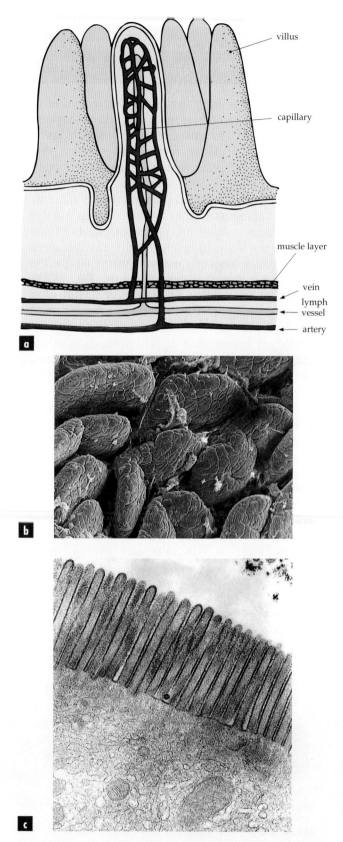

Fig. 19.9 The surface area of the small intestine is covered by folds and projections called villi. **(a)** A diagrammatic view of several villi and their vascular supply. **(b)** Scanning electron micrograph of villi in the rat intestine. **(c)** Transmission electron micrograph showing microvilli on the epithelial surface of the villi

Nutrients are actively absorbed from the gut lumen by special transport proteins. The type and number of these transport channels and the absorptive surface area of the intestine reflect an animal's dietary requirements.

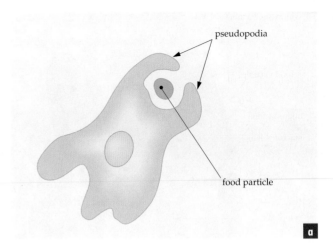

Evolution and diversity of digestive systems

The evolution of multicellularity allowed division of labour between specialised cells and led to the evolution of organs involved in food procurement, storage, digestion and absorption. Different patterns of feeding also evolved as organ systems increased in complexity.

Intracellular and extracellular digestion

Unicellular organisms, sponges and cells lining the gut of some multicellular animals ingest food by **phagocytosis**, a form of endocytosis where a food particle is engulfed in a membrane-bound food vacuole. Food ingested in this way must be considerably smaller than the cells themselves. Digestion is intracellular, although the digestive process is separated from the cytoplasm by a membrane (Figs 19.10, 19.11). Enzymes produced by the endoplasmic reticulum and the Golgi apparatus are stored in membrane-bound lysosomes (Chapter 3). A food vacuole and a lysosome fuse to form a digestive vacuole in which complex molecules are broken down into molecules small enough to cross the membrane into the cytoplasm (Fig. 19.10). Undigested residue is then expelled from the cell by exocytosis (Chapter 4).

Most animals use extracellular digestion, which involves the secretion of digestive enzymes onto potential food either externally, as in many spiders and starfishes (p. 501), or into the lumen of a gut, as in vertebrates. Chemical breakdown occurs and the resultant simple molecules are absorbed across the epithelial surface and into the animal. For this reason, food in the lumen of the gut is, in a digestive context, still considered to be outside the animal.

Simple digestive cavities

Corals, anemones and jellyfishes (cnidarians), and free-living flatworms (turbellarians), have a simple, **sac-like gut** into which enzymes are secreted and in which extracellular digestion occurs. In cnidarians, enzymes are diluted by considerable amounts of water also taken into the cavity. Tiny food particles may also be engulfed directly by endocytosis and digested intracellularly. Waste products are released back into the cavity and then out through the mouth. Digestive systems of this

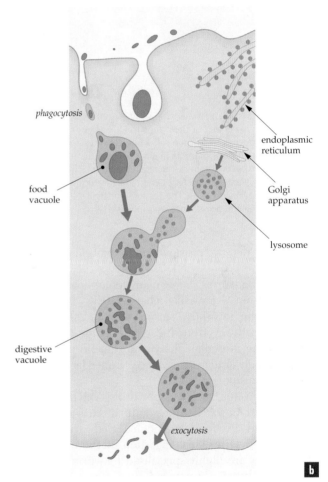

Fig. 19.10 (a) Amoeba engulfing a food particle with pseudopodia. **(b)** Intracellular digestion of a food particle in a protozoan. Food material that the cell takes in by phagocytosis is enclosed in a food vacuole, which fuses with a lysosome containing digestive enzymes. Digestion takes place within the composite structure thus formed (digestive vacuole), and the products of digestion are absorbed across the vacuolar membrane. The vacuole eventually fuses with the cell membrane and then ruptures, expelling digestive wastes to the outside

kind are relatively inefficient because both food and waste products pass through the same opening and, in

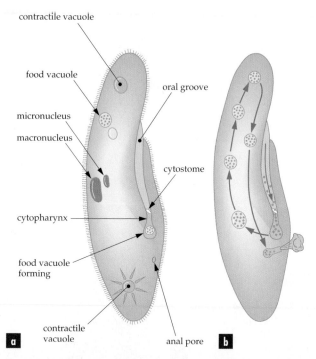

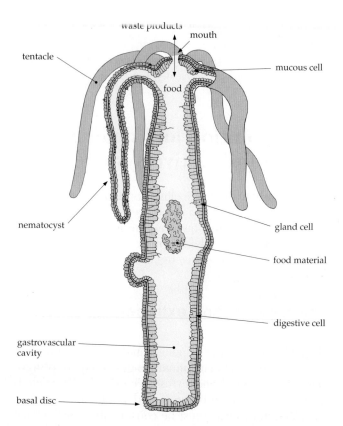

Fig. 19.11 (a) Major structures of a paramecium. **(b)** A food vacuole formed at the lower end of the cytopharynx separates and moves towards the anterior end of the cell while enzymes are secreted into it. Digestion takes place as in amoebae and the products of digestion are absorbed into the general cytoplasm. The vacuole then moves towards the posterior end, attaches to the anal pore and expels digestive wastes. The vacuole undergoes several changes in size and appearance as it moves

Fig. 19.12 Gastrovascular cavity of *Hydra*. The cavity contains food material. The gland cells secrete digestive enzymes and the digestive cells absorb the products of digestion. The nematocysts are specialised cells that help capture prey

cnidarians, digestion is slow due to dilution of enzymes (Fig. 19.12).

Some flatworms have many complex blind sacs opening from a muscular pharynx (Fig. 19.13). This increases the absorptive surface area and shortens the diffusion pathway from the gut to other cells of the animal. The muscular pharynx is capable of ingesting small animals or tearing off pieces from a larger, usually dead, organism.

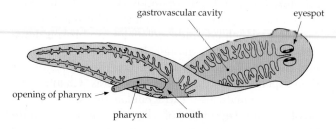

Fig. 19.13 A planarian, showing the highly branched gastrovascular cavity and muscular pharynx extruding through the mouth

Gut parasites, such as tapeworms, which live in the digestive tract of another animal, are continually surrounded by food molecules that are the end products of extracellular digestion by the host animal. Tapeworms are highly specialised flatworms that have lost their digestive tract and simply absorb required nutrients directly through their body wall.

Two openings: one-way movement of food

The next major advance in the evolution of digestive systems is seen in simple worms such as nemerteans and nematodes (Chapter 38). These worms have a **one-way digestive tract**, with a mouth at one end and an anus at the other. A one-way gut allows the independent elimination of wastes and regional specialisation

involving sequential secretion of different enzymes. The gut is usually only one cell thick, with no muscular lining. Food is forced along the tract by external body pressure or pressure of following food. One group, the rotifers, has jaws that are able to grind food, although it is not clear how effective they are.

> A one-way gut with a mouth and anus allows regional specialisation of the gut for more efficient digestion and waste elimination.

Muscular gut wall: coelomates

The evolution of a **coelom** (Chapter 15) in more complex groups, such as annelids (Fig. 19.14),

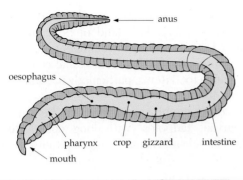

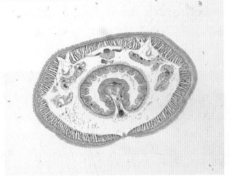

Fig. 19.14 (a) Digestive system of an earthworm showing the one-way system where food is ingested by the mouth and waste products are eliminated by the anus. **(b)** Cross-section of an earthworm in the intestinal region. The typhlosole, which projects into the cavity of the intestine, greatly increases the surface area available for absorption of food

arthropods, molluscs, echinoderms and chordates, led to a marked increase in length of the gut and development of a muscular gut wall, with its own blood supply. Waves of sequential contraction and relaxation, peristalsis, move food through the gut independently of whole body movements. Different rates of food movement are possible depending on the size of food particles and how difficult they are to digest. There is a capacity to store ingested food for later digestion, allowing feeding to be discontinuous, leading to an increased capacity to utilise localised food sources. Water absorption from the gut lumen is particularly important for terrestrial animals.

This basic pattern of gut structure and function is found in all complex animals. With increased specialisation of the gut, animals are able to eat and digest a wider range of foods, including larger or hard to digest items.

| Increased gut specialisation allowed animals to eat and digest a wider range of foods.

Chitinous mouthparts: arthropods

In the arthropods, the evolution of a hard chitinous exoskeleton leading to the development of **paired mouthparts** that can work against each other has allowed a diverse array of feeding types to evolve. The arthropods are usually divided into the chelicerates and the mandibulates. The chelicerates, of which the spiders are common examples, tend to pierce and inject digestive enzymes into their prey and then suck out the predigested food; their gut is short. The aquatic mandibulate crustaceans are generally filter feeders, predators or scavengers and their mouthparts are adapted for ingestion and crushing of the food.

The predominantly terrestrial insects show the greatest diversity of feeding types; they have crushing jaws or piercing mouthparts (Chapter 39). The prey of carnivorous insects includes soft-bodied worms and larvae from the soil, other arthropods, which have a resistant exoskeleton, and vertebrates with an endo-skeleton. Most herbivorous insects feed on sap or plant cell contents. The size of an insect, such as an aphid, compared with a plant cell allows it to pierce the phloem tubes or other cells with specialised mouth-parts (Fig. 18.23). The mouthparts of an aphid for sucking plant sap are in many ways similar to those used by a female mosquito for sucking vertebrate blood (Fig. 19.15). Plant cell contents are readily digestible and so an aphid's gut is short and uncomplicated as it is in carnivorous insects.

Insects that feed on plant tissue and ingest cell wall material have cutting or grinding mandibles, often with specially hardened cutting edges. At moulting, the cuticle, including the mandibles, is lost and then replaced. Consequently, a new set of mandibles is produced at each moult. This is particularly important to those insects that eat abrasive foods, such as grass, because mechanical processing of plant tissue wears the leading edges of the mandibles. Many of these insects do not digest cellulose but extract the cell contents and then eliminate the residue. These animals tend to have relatively short guts combined with

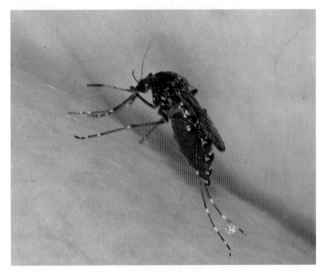

Fig. 19.15 Female mosquito using elongated mouthparts to suck blood

complex crushing and grinding mandibles and often a muscular grinding gizzard. Other insects hold bacteria and colonies of protists in their caecae where fermentation of the cell wall takes place; their gut is more complicated and gut retention time is longer. The host insect utilises the by-products of this fermentation process for its own metabolism. We discuss the digestion of plant cell walls in more detail below.

> Insects with an exoskeleton that can be modified into complex mouthparts have developed an enormous array of dietary specialisations.

Jaws and teeth: vertebrates

In vertebrates, jaws supporting hard enamel-covered teeth provide an efficient mechanical processing system that allows improved access to large, hard or tough foods. The earliest vertebrates to evolve were the fishes (Chapter 40). Today, cartilaginous and bony fishes (teleosts) include specialist filter feeders, detrital bottom feeders and **carnivores**. Herbivores, which consume algae and seagrasses, are only found among the bony fishes. The contrasting jaw and gut structures associated with the two dietary extremes, carnivory and herbivory, are well illustrated when the carnivorous flathead is compared with the herbivorous luderick. The flathead has sharp grasping needle-like teeth, a large but simple stomach and a very short almost straight intestine. The luderick has flat crushing teeth and a long coiled gut.

The skull structure of higher bony fishes is complex with a series of hinged bones supporting the teeth,

which can be moved relative to other parts of the skull (Figs 19.16, 19.17). This kind of bone movement is known as **kinesis** and its development has produced a diverse range of specialist feeders that can manipulate often strange food items, particularly in coral reef fishes (Chapter 40).

All adult amphibians appear to be carnivorous. Frog larvae, tadpoles, with long coiled guts feed primarily on algae. At metamorphosis into the carnivorous adult, the gut shortens markedly (Fig. 19.18), illustrating a general pattern seen in many other animals—herbivores have longer intestines than carnivores.

Fig. 19.17 Movable or extendible mouthparts are an important feeding adaptation in many fishes. This slingjaw wrasse, *Epibulus insidiator*, can rapidly extend the whole jaw and catch unsuspecting prey

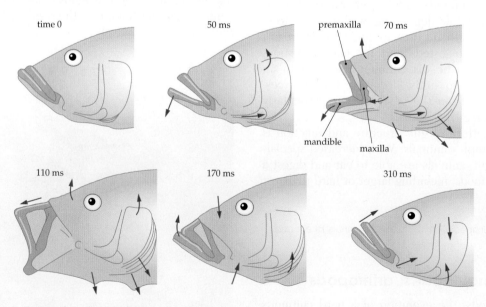

Fig. 19.16 Feeding in *Serranachromis*, a cichlid fish. The mouth at rest is closed and conforms to the streamlined shape of the body. The mouth can be rapidly opened to a very large gape and the jaws protruded to grasp prey. The mouth is then rapidly closed. Only the premaxilla and mandible support teeth. The maxilla acts as a lever to raise the premaxilla. Herbivorous fish or fish that prey on slow moving animals do not have such wide gapes but have powerful jaw-closing muscles

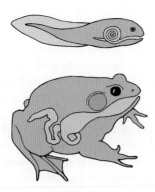

Fig. 19.18 Gastrointestinal tract of an adult frog and tadpole. The much-coiled intestine of the herbivorous tadpole is far longer relative to the size of the animal than is the intestine of the adult frog

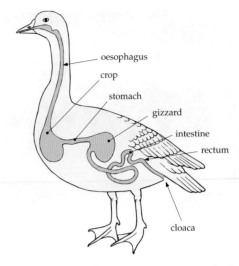

Fig. 19.19 Digestive system of a bird. The crop stores food and the chamber for mechanical break-up, the gizzard, is located posterior to the acid-secreting stomach, which means that the food is pretreated before it enters the gizzard. Note the similarity to the earthworm

Ectothermic reptiles have lower energy requirements than endothermic birds and mammals. This would seem to allow herbivorous reptiles the time required to ferment plant cell walls. However, this advantage may be offset to some extent by their lower body temperature, at least at night, and by poor mechanical preparation of food since reptiles have relatively uncomplicated teeth and poor chewing mechanisms.

Aquatic reptiles are mainly carnivorous, although a few, such as marine iguanas and green turtles, consume algae and seagrass. Some carnivores swallow the prey whole without dismemberment. Snakes disarticulate the lower jaw to accommodate the broadest part of the prey. However, feeding in this way means that a large amount of non-nutritive material may also be ingested.

Birds are constrained in what they eat by their mode of locomotion. Flight requires a lot of energy and there is a high cost in having heavy organ systems. Many birds consume high energy foods such as insects, nectar or fruit. A relatively large gut filled with a heavy mass of fermenting vegetation would prohibit flight, and fermentation does not provide energy rapidly enough. The few herbivorous birds tend to be larger, ground-loving forms or may, like the emu and the other ratites, have lost the power of flight altogether. Modern birds have lost their teeth, presumably a weight-saving adaptation, but the cost is a loss of mechanical processing capacity. For predators such as owls, which swallow prey whole, feathers and bones ingested are regurgitated after the easily digestible material has been processed. In some grain-eating species, such as chickens, a muscular part of the stomach known as the gizzard has developed a mechanical processing role (Fig. 19.19).

> Birds generally eat high-energy, easily digestible foods to fuel their high metabolic rates and to keep weight low.

Mammals

Mammals maintain high constant body temperatures and have complicated very hard teeth held in jaws that can move independently in precisely controlled ways. Their teeth have become an integral part of the digestive system, and both teeth and gut have become specialised for particular diets. The parallels between tooth structure and digestive function in mammals of diverse phylogenetic origins, but similar diets, are striking. However, if one component of an animal's diet needs specialised teeth or gut, the animal tends to show these adaptations, even though the dietary component may be only a minor part of its diet. It is partly for this reason that one can be misled when attempting to deduce diet from functional anatomy. The best example of this is the giant panda, which has a carnivore's gut but a herbivore's diet.

Mammalian teeth have become specialised into four main types with different functions. Incisors grasp and hold food, and tend to remain similar and unspecialised in all mammals (Fig. 19.20). Canines are long conical teeth, which may be used for stabbing and gripping prey, and which are usually well developed in carnivores and lost in herbivores. Post-canine teeth, the cheek teeth, are involved with mechanical processing and the lower teeth work against the upper teeth in either cutting or grinding actions. The first set of cheek teeth, premolars, together with incisors and canines, are deciduous teeth and are replaced in the adult animal by permanent, usually larger teeth. The posterior teeth, molars, are not replaced, only one set being produced in the life of most mammals. For animals with an abrasive diet, tooth wear may reduce effective feeding (Fig. 19.21).

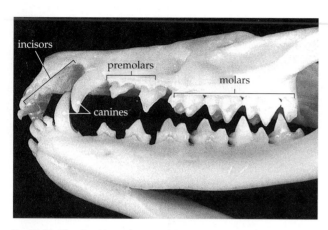

Fig. 19.20 The dentition of an insectivorous mammal, the marsupial native cat. The incisors, canines, premolars and molars are indicated. The molars have two functions: to puncture and then crush insects with a rigid exoskeleton

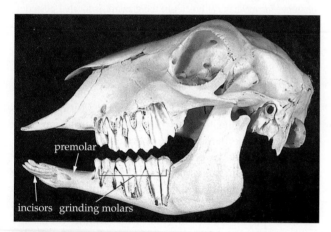

Fig. 19.21 The dentition of a mammalian herbivore, the sheep. The grinding molars occupy most of the jaw. The teeth have high crowns, which are held in the body of the jaw and have continually growing roots. This means that for a good part of the animal's life the enamel crown is continually being replaced as it is worn away by abrasive foods

Small mammals require and are generally restricted to high-energy foods that are usually quick to digest. Their main source of high-energy food is energy-rich plant products, such as nectar, pollen, fruit and flowers, and arthropods, which have an exoskeleton of tough chitin. The smallest mammalian herbivores utilise plant tissue with little cell wall material and with energy-rich, starch-filled cell contents. They have efficient teeth with relatively large surface areas, powered by large jaw muscles. The gut is relatively simple and short, usually only several times the body length. In general, larger mammalian herbivores incorporate more structural material in their diet than smaller mammals.

Small insectivores (carnivores for which insects form a major part of the diet) use teeth to penetrate the tough cuticle of their prey and expose the nutrient-rich haemolymph. The molar teeth of virtually all small insectivorous mammals are well suited to this, initially puncturing and crushing the exoskeleton followed by fine shearing of the inner tissues. Large carnivores cannot survive on insects unless they are concentrated in large numbers, as are ants and termites, because it takes too long to catch them in sufficient quantities. Consequently, large terrestrial mammalian carnivores are limited to preying upon other vertebrates that have an endoskeleton. Among carnivores such as dogs, cats and marsupial lions, scissor-like carnassial teeth have evolved independently several times (Fig. 19.22). These teeth are adapted for shearing chunks of meat off the bone. Some carnivores, such as the canids and especially the hyenas, have specialised molars for crushing bone. Cats, which tend to be specialist flesh-eaters, have lost nearly all teeth except the canines and carnassials (Fig. 19.22). The toothed whales (although only possessing simple teeth) and most seals are predators, mostly on large single prey items such as fish or other mammals. The gut of carnivores remains relatively unspecialised with a simple stomach, short intestine and often a reduced or missing caecum.

S mall mammals tend to consume easily digested foods, such as nectar, fruit and insects. Small insectivorous mammals have simple guts and complex teeth for puncturing and crushing insect exoskeletons. Carnivorous animals have short digestive systems, while herbivores have long, often complex digestive systems.

Most herbivores utilise only the cell contents, thereby avoiding less digestible cell walls that require slow enzymatic breakdown or fermentation. Fermentation of plant cell walls is enhanced by high temperature, large body size (accommodating a large gut and contents) and efficient initial mechanical processing of food. Large mammals are thus well adapted to process structural plant tissue. Herbivorous mammals typically have broad crushing teeth adapted to an abrasive fibrous plant diet.

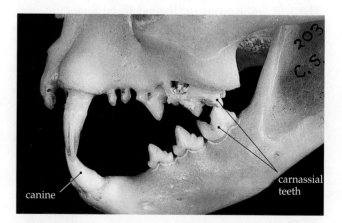

Fig. 19.22 The carnassial dentition of a cat, which largely consumes other vertebrates with an endoskeleton. The carnassial tooth is specialised for shearing flesh only

Herbivores such as the koala (p. 510), which rely on the cell contents being exposed and digested in the small intestine before the food reaches the bacteria in the hindgut, also need very efficient teeth. Herbivores that use bacteria in the stomach as a means of fermenting plant cell walls rely on their teeth to damage cell walls for bacterial attack rather than to expose the cell contents for digestion. If the cell wall is degraded, then the contents will spill out and become available in any case.

The advantage of size can be seen in the common ringtail possum and koala, which both consume eucalypt leaves. The possum is small and selects a diet lower in fibre than does the larger koala. Both animals eat the same leaves but the possum, with its smaller mouth, can bite out parts of the leaf between major veins. It thus avoids lignified fibre, and the mesophyll and epidermal cells that it ingests are relatively easy to crush between its teeth. The koala, with its larger mouth size, cannot avoid the lignin and must deal with it in its diet. It has teeth with more pronounced cutting edges to process the resistant fibre bundles.

Omnivorous animals can utilise a variety of different food types. We might presume that **omnivores** are not as efficient at every task as specialists but are compensated by having a greater variety and abundance of resources available to them. The problem with this argument is that it does not take into account processing time within the animal and the limitations of gut fill. If an animal cannot process food fast enough, and hence efficiently enough, to satisfy its requirements, an abundance of food may not be an adequate recompense. It is interesting that omnivores tend not to be found among the smallest mammals, where rate of energy acquisition is most important. Those that do occur are likely to eat high-energy, easily digestible plant and animal products, such as nectar and insects (these animals tend to be considered as insectivores).

Filter feeding: eating tiny particles

This method of feeding is represented in most animal phyla. It limits an animal to tiny particles suspended in water and it is no surprise that **filter feeding** is predominantly found in organisms that feed in aquatic systems (Fig. 19.23). Vertebrates that filter small particles of food from large volumes of water include mammals such as the baleen whales (e.g. the humpback whale of the Australian east coast) and birds such as flamingos.

Filter feeding presents a number of difficulties, particularly for intertidal animals. For example, intertidal animals can feed only when they are submerged. Perhaps not surprisingly, individuals from the high intertidal zone consume more prey and absorb more nutrients in a given time period than do those from the

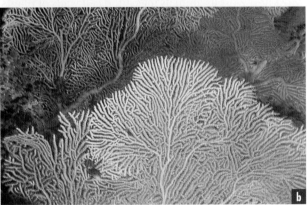

Fig. 19.23 Sifting the seas—many animals feed by filtering miroscopic plant and animal life from large volumes of water. **(a)** Baleen whales feed using large sieve-like plates. **(b)** Corals such as this gorgonian coral intercept the water current in a variety of ways that allow the individual polyps to capture suspended food

low intertidal zone. Generally, filter-feeding animals ingest whatever material happens to pass by; the only discrimination possible is on the basis of the size of a particle. The wide range of prey ingested (e.g. bacteria and plankton, and also inorganic particles) means that the animals need a wide range of digestive enzymes to be able to use what is captured.

Filter feeding is one of the most common feeding mechanisms among sessile or mobile aquatic invertebrates. Most invertebrates lack the structures to reduce the size of large prey. In both freshwater and oceans, detritus particles, and living and dead phytoplankton (such as diatoms), are abundant. The mechanism for trapping small particles varies from sheets of mucus (e.g. tunicates or 'sea squirts') to modified tube feet (sea cucumbers). One of the most widespread is the use of cilia, which may set up local currents to trap microscopic food particles in a mucous stream to pass this towards the mouth. This type of pattern is commonly found in bivalves such as oysters and mussels.

Generally, shallow oceanic waters contain more particulate matter than the open ocean but filter feeders must cope with a wide variation in food supply. Studies of larvae of the crown-of-thorns starfish have shown just how important food supplies can be for filter-feeding animals. The natural levels of particles in Great Barrier Reef waters are generally too low or marginal for the development of the crown-of-thorns' larvae. However, at certain times there are sufficient phytoplankton to ensure a moderate survival of larvae and this may contribute to subsequent outbreaks of the starfish on the Great Barrier Reef.

Dealing with plant cell walls

Most herbivorous animals ingest the cell walls of plants. Cell walls are difficult to digest and impede access to the soluble sugars and proteins of the cell contents. In response, many herbivorous species have highly modified digestive tracts that house symbiotic populations of microorganisms that help digest the cell walls and provide the host animal with access to this rich energy source. Much is known about these processes because of the importance of herbivorous species to agriculture and because many high profile Australian animals such as koalas and kangaroos are herbivorous.

Some herbivores solve the problem posed by plant cell walls by either spitting them out or else passing it rapidly through the gut, utilising only the easily digested parts of the plant. For example, in northern Australia at the beginning of the rainy season, flying foxes eat leaves of leguminous trees. They chew the leaves, swallow the juices (soluble proteins, sugars and starches) and then spit out the fibrous portion. Giant pandas swallow entire bamboo leaves but they have a short uncomplicated gut through which fibre passes rapidly. Consequently, pandas extract only the proteins and carbohydrates from cell contents and must therefore eat large amounts of food.

> Plants provide a diet of readily digestible cell contents surrounded by a tough cell wall.

Digesting cellulose

The cell walls of plants are a rich source of energy for animals that can utilise cellulose. Cellulose is the most abundant carbohydrate on earth, but very few animals can produce the necessary enzymes to hydrolyse cellulose molecules. To release the energy of the chemical bonds of cellulose, animals need a group of enzymes known broadly as **cellulases**. It used to be thought that no animals produced their own cellulases, but we now know that several Australian insects,

including some termites and the wood cockroach, and some land snails can produce cellulase. Nonetheless, it is likely that all animals (certainly all vertebrates) that digest cellulose rely primarily on cellulases produced by **symbiotic microorganisms** (Box 19.3). Since larger animals do not need to acquire energy as rapidly as smaller animals, they can afford the time necessary for symbiotic microorganisms to slowly digest the cellulose.

BOX 19.3 Microbes in the gut

Most animals harbour a range of different microorganisms in the gut, including protists, bacteria and even fungi. For example, in the eastern grey kangaroo, there are about 5×10^5 protists per millilitre of digestive fluid and about 3×10^9 bacteria. Protists, although in lower numbers, form up to 40% of the microbial biomass of the gut. The smaller bacteria consist of a diverse group ranging from those that digest cell wall carbohydrates (cellulose and pectin) to those that specialise on soluble sugars. Others require substrates (e.g. formic acid) that are formed entirely by other bacteria.

Many of these microorganisms in the gut attach themselves to particles of plant matter and then secrete extracellular enzymes to digest away parts of the plant. The microbes generally attach to broken tissues and so initial mechanical disruption of food by the host's teeth, or by other means, is very important.

Some microbes in the gut can also detoxify plant toxins. For example, cattle in northern Australia could not eat the otherwise-nutritious legume *Leucaena leucocephala* because it contains a toxic amino acid, mimosine. Australian scientists observed that, in Hawaii, goats were able to eat this plant without problems and suggested that the Hawaiian animals had a specific bacterium that could break down the toxin. This proved correct and, after some digestive fluid from Hawaiian goats was brought back to Australia, all animals treated were able to eat *Leucaena* safely. The bacterium was spread naturally and *Leucaena* is now an important part of our northern grazing industries.

There are two broad patterns of microbial fermentation. In foregut fermentation, the microbial population is situated anterior to the true stomach (the region where acid and pepsin are secreted). Examples include all but one of the kangaroos, sheep, cattle and their allies (deer, antelope, goats), some primates and tree sloths from South America. The other pattern, hindgut

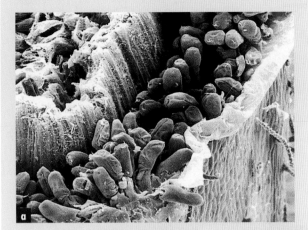

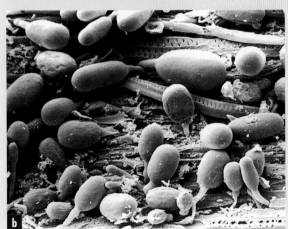

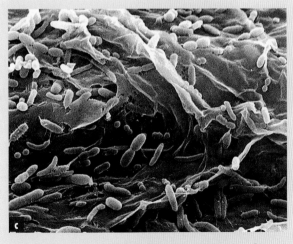

Microbial fermentative digestion in action. Scanning electron micrographs showing **(a)** protists on the cut end of a lucerne stem (densely packed between the epidermis and the vascular bundle) after 15 minutes in the rumen of a sheep; **(b)** sporangia of anaerobic fungi found on fragments of lucerne in the rumen of sheep; **(c)** bacterial populations on pieces of *Eucalyptus* leaf from the caecum of a brushtail possum; **(d)** bacteria attached to plant tissues by means of extracellular material (from caecum of brushtail possum)

fermentation, is more common and the microbes are located after the true stomach. Animals that show this pattern include termites, koalas, ringtail possums, wombats, horses and related forms, many rodents, and many herbivorous fishes. Generally, foregut fermentation is restricted to relatively large animals, while hindgut fermentation spans a wide range of body size.

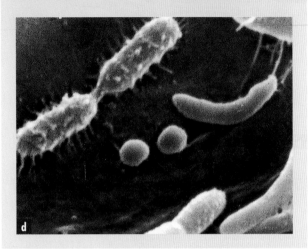

Digestion of cellulose by **microbial fermentation** occurs in nearly all vertebrate herbivores and a wide range of invertebrate herbivores. The microbes are usually housed in an expanded region of the gut, which protects them from being continually lost with the passage of food, and provides an anaerobic environment at relatively constant temperature and pH. The microbes secrete cellulases, which hydrolyse cellulose into its constituent sugars. These sugars are then rapidly used by the microbes, which release short-chain fatty acids (acetic acid, propionic acid and butyric acid) as wastes. It is these latter molecules that are then absorbed by the host animal and used as a source of chemical energy.

Cell walls are a rich source of energy if a herbivore has the enzymes to digest them. Cellulose is usually digested by microbial symbionts living in the gut. Gut microbes hydrolyse cellulose to glucose, which they use. They release short-chain fatty acids, which can be absorbed by the host and used for energy.

Foregut fermentation

In the eastern grey kangaroo, *Macropus giganteus*, food is chewed and mixed with copious amounts of saliva. The first part of the stomach is a large sac containing billions of bacteria and protists, and partially digested food particles (Fig. 19.24). The microbes attack the food and the soluble part (sugars and starches) is quickly digested and absorbed by them. Digesting the

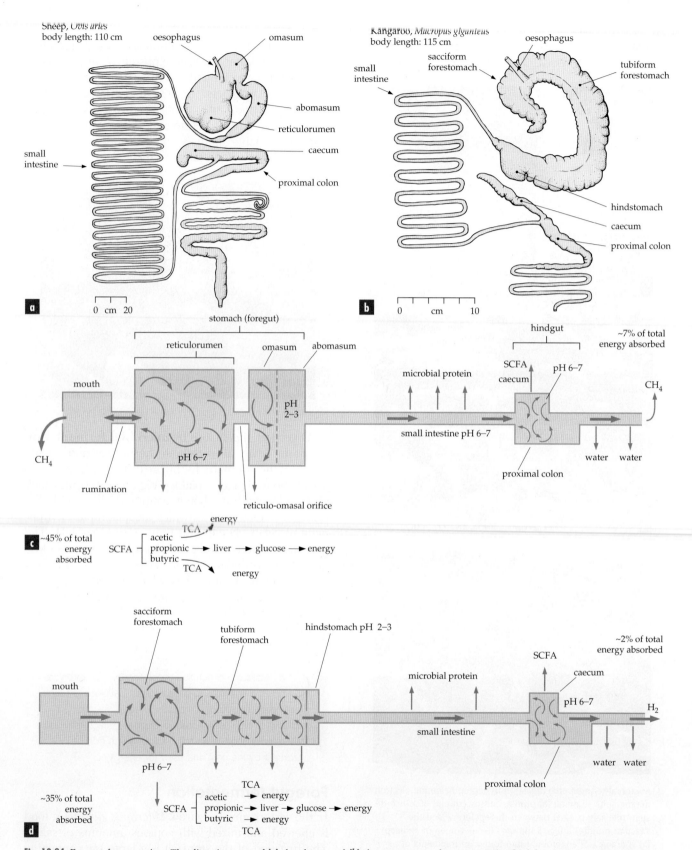

Fig. 19.24 Foregut fermentation. The digestive tract of **(a)** the sheep and **(b)** the eastern grey kangaroo. Microbial fermentation occurs in the forestomach of both species. Note the difference between the structure of the stomachs. The kangaroo stomach is dominated by a tubiform region, whereas the sheep stomach is dominated by the sacciform reticulorumen. In both species, there is a second, smaller area of microbial fermentation in the caecum. Schematic representation of the digestive processes occurring in the digestive tract of **(c)** the sheep and **(d)** the eastern grey kangaroo. Small arrows indicate mixing of ingesta; bold arrows indicate one-way flow

fibre takes longer. The microbes attach to the fibre (Box 19.3) and break the chemical bonds between the glucose molecules that make up cellulose. They then further metabolise the glucose for their own use. So far, all the nutrients in the grass have been captured by the microbes and there is little return for the kangaroo.

However, the microbes in the kangaroo's stomach excrete fatty acids that the kangaroo absorbs and uses as an energy source (Fig. 19.25). Bicarbonate ions in saliva help to buffer these fatty acids so that the pH of the microbial broth stays relatively constant. The end result is that while cellulose is not utilised directly by the host animal, the microbes convert it into usable molecules.

This seems to be an efficient process with benefits for both kangaroo and microbes, but it is not without some cost to the kangaroo. As we said, the kangaroo cannot directly digest the cellulose but it could directly digest the soluble sugars of the grass. However, the microbes have first access to these and the kangaroo must rely on using the microbes' organic wastes. Energy-transforming processes are never 100% efficient (Chapter 2); the kangaroo gets about 75% of the original energy value of the soluble sugars.

> Foregut fermentation allows for extensive degradation of cellulose, but soluble sugars and starches are first used by microbes.

Similarly, proteins are available to microbes before becoming available to the host. Proteins in grass are quickly fermented. The nitrogen is absorbed by the microbes as ammonium and used for microbial growth. However, the microbes themselves are an excellent source of essential amino acids for the host. As food passes through the digestive tract, microbes are washed out of the saccular part of the stomach into the final, acid part of the stomach and are a high-quality source of complete protein that is digested in the small intestine.

> In foregut fermenters, dietary protein is used by microbes and then the host digests the microbes, which are a rich source of essential amino acids.

Ruminant foregut fermenters

Sheep and cattle, which form the basis of our extensive livestock industries, and related wild herbivores, are **foregut fermenters** that **ruminate**; that is, they regurgitate the contents of the first part of the stomach and rechew it, further reducing the size of food particles. Ruminants have a stomach composed of four parts: rumen, reticulum, omasum and the true stomach, abomasum (Fig. 19.24). Microbial fermentation takes place largely in the rumen and reticulum. Food can only pass from the reticulum to the omasum after it has been reduced to a certain size by rumination. In sheep, for example, only particles less than 1.05 mm are likely to pass into the omasum through the reticulo-omasal orifice. This means that a ruminant animal keeps food in its fermentation chambers for prolonged periods, giving the microbes plenty of time to digest the cellulose. However, this may become a disadvantage if the quality of the pasture becomes very poor. As grasses mature, the amount of soluble sugars and proteins declines and the proportion of indigestible fibre increases markedly (Fig. 19.3). This means that an animal must spend more and more time rechewing ingested food before it is small enough to pass into the omasum. Eventually, processing

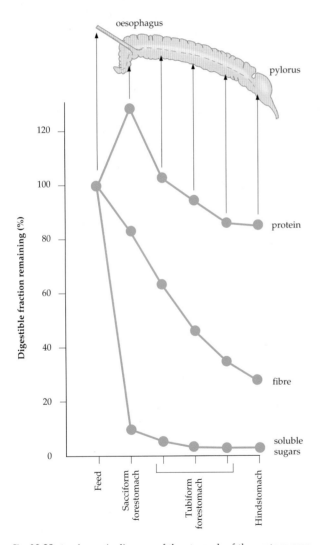

Fig. 19.25 A schematic diagram of the stomach of the eastern grey kangaroo showing the disappearance of certain components of the diet as food passes through the stomach. Note how the soluble sugars are fermented and disappear immediately. The fibrous part of the diet is fermented and digested more slowly throughout the stomach. Protein content actually increases because the microorganisms that attach to the food are themselves rich in protein

indigestible material can take so long that an animal can starve, gaining no further nutrients from ruminating but, because its stomach is still full, unable to ingest new food with available soluble nutrients.

All animals that eat fibrous diets are limited to some extent by the gut-filling effect of dietary fibre, and the rumen system can be a digestive bottleneck under certain conditions. Kangaroos, however, do not have to rechew the food to pass it out of the stomach because they do not have an equivalent to the reticulo-omasal filter found in the sheep. As pastures decline in quality, kangaroos simply eat more and digest less (Fig. 19.26).

Hindgut fermentation

The nature of microbial fermentation in the koala, a **hindgut fermenter**, is similar to that of foregut fermenters such as the eastern grey kangaroo, except that fermentation occurs in the hindgut (Fig. 19.27). Chewed eucalypt leaves are mixed with saliva but the saliva has a very different composition from that found in foregut fermenters.

Soluble sugars and proteins, directly accessible to the koala by chewing, are hydrolysed by the koala's own digestive enzymes and glucose and amino acids are absorbed in the small intestine. The microbial populations in koalas and kangaroos act on different substrates. In koalas, microbes utilise mainly cell walls using similar digestive reactions to those in the kangaroo. Cell walls are broken down, with glucose being absorbed by the microbes; the fatty acids they produce are absorbed and utilised by the koala. The major difference between the two systems is that microbes washed out of the koala hindgut are not able to be digested as a high-quality protein source but are lost in the faeces.

Common ringtail possums make use of the rich source of microbial protein in their faeces. During the day, when they are normally resting in nests or tree hollows, they stop eliminating normal faeces and start to pass out unchanged caecal contents (Fig. 19.28).

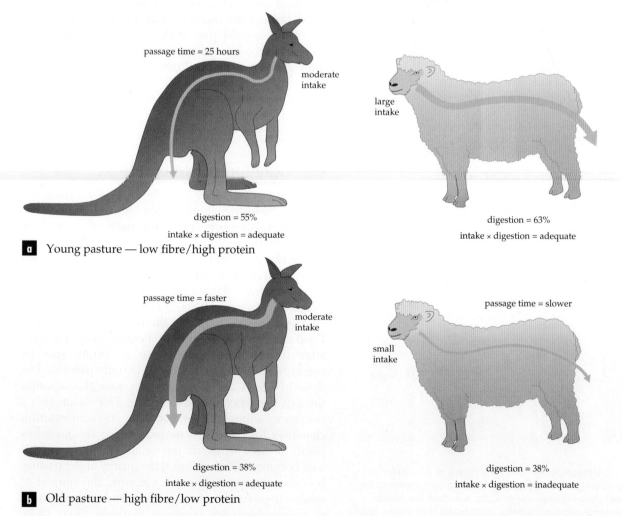

passage time = 25 hours

moderate intake

digestion = 55%

intake × digestion = adequate

a Young pasture — low fibre/high protein

large intake

digestion = 63%

intake × digestion = adequate

passage time = faster

moderate intake

digestion = 38%

intake × digestion = adequate

b Old pasture — high fibre/low protein

passage time = slower

small intake

digestion = 38%

intake × digestion = inadequate

Fig. 19.26 When pasture quality is good, eastern grey kangaroos eat and digest less than sheep. When pasture quality is poor, kangaroos eat at a slightly lower level but have a quicker passage of food through the gut. In contrast, sheep cannot increase their food intake because they must continually rechew the tough food before they can pass it from the gut. In effect, sheep have a rate-limiting step (the reticulo-omasal orifice) whereas kangaroos do not (see Figure 19.25). Kangaroos can usually maintain themselves longer on poor quality pastures

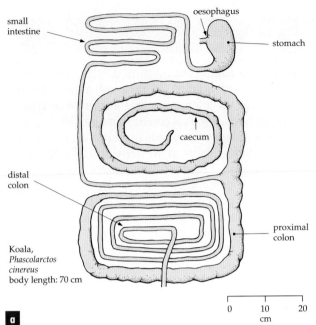

Koala,
*Phascolarctos
cinereus*
body length: 70 cm

0   10   20
cm

a

This material is very rich in microbes and the animal eats the pellets as they are expelled. The microbes are then digested and the nutrients absorbed in the stomach and small intestine. This behaviour is known as **coprophagy** or caecotrophy, and is possible because of a complicated system of particle sorting that occurs in the possum hindgut. This behaviour is restricted to relatively small animals such as rabbits and many small rodents. Without doubt, the ability to use this extra source of protein is one reason rabbits were able to colonise vast areas of forests and grasslands in Australia.

Hindgut fermentation is similar to foregut fermentation, except that most animals cannot utilise high-quality microbial protein. Some small mammals do so by eating special faecal pellets.

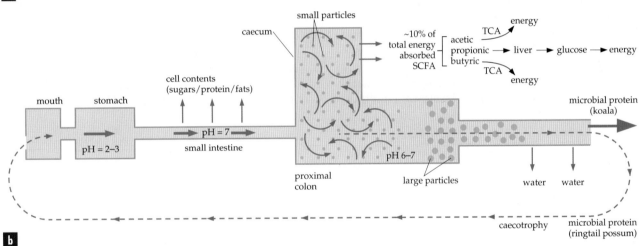

b

Fig. 19.27 Hindgut fermentation. **(a)** The digestive tract of the koala. Note the simple stomach but greatly expanded caecum and proximal colon where microbial fermentation occurs. **(b)** Schematic diagram of the digestive processes in the koala gut. Small arrows indicate mixing of ingesta; bold arrows indicate one-way flow. The dotted line indicates the reingestion of special faecal pellets, which does not occur in the koala but only in the smaller common ringtail possum

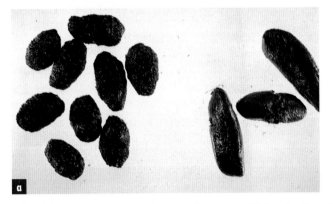

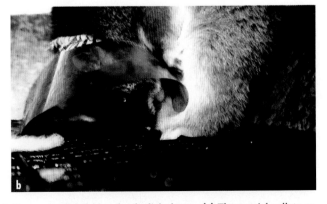

Fig. 19.28 The common ringtail possum ingests special faecal pellets (called caecotrophs) during the daylight hours. **(a)** The special pellets on the right contain higher levels of protein and B-group vitamins than do the normal faecal pellets on the left. **(b)** The pellets are eaten directly from the anus. The possum here is wearing a plastic collar to try to prevent the behaviour so that samples of special pellets could be collected for analysis

BOX 19.4 The poisoned platter

Along with useful nutrients, herbivores also often ingest potentially harmful or toxic compounds found in plants. These chemicals are known as secondary plant compounds because they were thought to have no role in the primary metabolism of plants. We now believe that many of them have a deterrent effect on herbivores and that they are important in disease resistance.

Tannins and phenolics are one of the most important groups. Tannins can precipitate proteins. When 'tanning' an animal hide, plant tannins react with the skin's proteins to form leather. When an animal eats a tannin-rich plant, released plant proteins may be precipitated and so their availability to the animal is reduced. Alternatively, tannins can bind with digestive enzymes and so inhibit the digestion of other food constituents. For example, common brushtail possums are less able to digest *Eucalyptus* leaves containing active tannins than leaves in which the tannins have been deactivated. By developing varieties of grain that contain higher than normal levels of tannins, plant breeders have turned the deterrent properties of tannins to their advantage. Such 'bird-resistant grains' are less attractive to crop-raiding birds, such as corellas and galahs.

Other plant compounds exert more direct toxic effects. Some plants in Western Australia contain a compound called sodium monofluoroacetate, which, as compound 1080, is used around the world as a poison for vertebrate pests. The native mammal fauna of Western Australia have been exposed to sodium monofluoroacetate for millions of years and have developed a remarkable resistance to its effects. Toxicity testing showed that the lethal dose for a brushtail possum in Western Australia was about 100 times that of the same species in eastern Australia, where plants do not contain 1080. This provides land-management authorities in Western Australia with a wider safety margin when controlling feral animals such as foxes, cats and pigs. Recent efforts to re-establish small marsupials in areas from which they have disappeared depend on controlling these introduced animals.

Summary

- Heterotrophic animals are unable to manufacture their own organic compounds and must obtain them from other organisms. All animals need energy, nitrogen-containing compounds, fats, vitamins, minerals and water.

- Plant tissues are rich sources of carbohydrates but generally poor sources of protein. Animal tissues are the reverse.

- Large animals eat more food than do small animals but small animals actually need more food per unit body mass. This means that small animals often feed on better quality diets than do large animals.

- Digestion is a stepwise process that usually starts with the physical breakdown of food, providing a large surface area for the action of digestive enzymes that break down complex molecules. The type of digestive enzymes present usually reflects the composition of the normal diet.

- Enzymatic digestion normally involves the sequential secretion of different digestive enzymes along the length of the gut. These processes are controlled by both nervous reflexes and hormonal actions.

- Digestive systems became more complex as animals evolved into large multicellular forms.

- The simplest digestive systems involve engulfing of food particles into a membrane-bound vacuole and intracellular digestion.

- Extracellular digestion, in contrast, involves the secretion of digestive enzymes onto food. In the simplest forms, this occurs in simple sac-like guts but the development of guts with two openings allows for foods and wastes to be separated and for sequential digestion along the gut.

- Animals with coeloms developed longer guts with muscular gut walls and their own blood supply. These larger guts meant the food could be ingested and stored for later digestion. Increasing gut specialisation meant that animals could feed on a wider range of foods.

- Since many foods are hard, physical processing is necessary before digestion can occur. The chitinous mouthparts of insects and the jaws of vertebrates have allowed an enormous range of dietary specialisations.

- Similar design patterns are found in a wide range of animals. Carnivores tend to have short and relatively simple digestive systems whereas herbivores have long, large and complex guts.

- Filter feeding is practised by a wide range of animals and clearly demonstrates the close match between the nature of the food and the design of feeding and digestive systems.

- Not all foods can be digested by an animal's own enzymes. Plant cell walls are a potentially rich source of energy but most animals need enzymes from symbiotic microbial organisms to hydrolyse the cellulose and hemicellulose. The host animal benefits from the waste products (mostly short-chain fatty acids) excreted from the microorganisms.

- In some animals (e.g. kangaroos and sheep), the microbes are located before the true acid-secreting stomach. This is foregut fermentation. In hindgut fermenters (e.g. possums and horses), the microbes are located posterior to the region of acid secretion.

- Foregut fermenters benefit by being able to digest some of the actual microbes, which provides them with a rich source of essential amino acids. However, the microbes digest any sugars and proteins in the diet and although some of these nutrients will eventually be made available to the host animal, the transformations are inefficient.

- Hindgut fermenters benefit by being able to use the sugars and proteins in the diet directly but are unable to digest microbial protein, except by eating faeces, coprophagy. This is a common behaviour in small herbivorous mammals.

keyterms

absorptive capacity (p. 498)	fatty acid (p. 488)	mass-specific metabolic rate (p. 497)	phagocytosis (p. 499)
amino acid (p. 488)	filter feeding (p. 505)		physical digestion (p. 494)
carbohydrate (p. 487)	foregut fermenter (p. 509)	metabolic rate (p. 492)	ruminate (p. 509)
carnivore (p. 502)	gluconeogenesis (p. 492)	microbial fermentation (p. 507)	sac-like gut (p. 499)
cellulase (p. 506)	herbivore (p. 488)	mineral element (p. 488)	symbiotic microorganism (p. 506)
coelom (p. 500)	hindgut fermenter (p. 510)	omnivore (p. 505)	transport channel (p. 497)
coprophagy (p. 511)	kinesis (p. 502)	one-way digestive tract (p. 500)	vitamin (p. 488)
detritivore (p. 492)	lipid (p. 487)	paired mouthparts (p. 501)	zymogen (p. 496)
digestive enzyme (p. 494)	mass-specific food intake (p. 493)		
enzymatic digestion (p. 494)			

Review questions

1. Explain why 'total carbohydrate' is unhelpful when it appears in a statement of food composition.
2. Explain the terms 'essential amino acids' and 'essential fatty acids'.
3. Why do small animals tend to eat better quality foods than do large animals?
4. What are zymogens and why are they important?
5. Describe some of the advantages and disadvantages of microbial fermentation in the foregut and the hindgut of herbivorous mammals.
6. Outline the dietary differences faced by aquatic herbivores and terrestrial herbivores.
7. What are the advantages of a one-way digestive tract?
8. Why are there few birds that rely on fermentation of plant cell walls as their major energy source?
9. Grasses are rich in fibre and abrasive particles. How have the teeth of grazing mammals adapted to these independent factors?

Extension questions

1. If a person lacked the enzyme amylase in his saliva, what effect would this have on the digestion of starch?
2. Many people with stomach ulcers are prescribed drugs that inhibit the production of acid in the stomach. What effects would inhibition of acid production have on digestion of a hamburger?
3. Relate differences in the structure of the gut of sheep and kangaroos to the likely responses of each to poor-quality pastures.
4. Grasses are rich in fibre and abrasive particles. How have the teeth of grazing mammals adapted to these independent factors?
5. What problems do blood-sucking and sap-sucking insects face? How do they overcome the response of their prey?
6. How might you expect the digestive system to respond to (a) prolonged periods without food and (b) increased energy demands such as exposure to cold conditions or during lactation?

Suggested further reading

BOOKS

Hume, I. D. *Marsupial Nutrition*. Cambridge: Cambridge University Press.

This book describes the diversity of food acquisition strategies found among marsupials and emphasises how the digestive system matches the food eaten and the animals' nutrient requirements.

Stevens, C. E. and Hume, I. D. (1995), *Comparative Physiology of the Vertebrate Digestive System*. Cambridge: Cambridge University Press.

This more advanced book surveys the digestive system in a wide range of mammals and provides a thorough review integrating structure and function.

Van Soest, P. J. (1994), *Nutritional Ecology of the Ruminant*, 2nd edn. Ithaca, NY: Cornell University Press.

This text is designed for agriculture and animal science students seeking a more detailed understanding of the nutrition and digestive physiology of ruminant livestock. But it also includes an excellent discussion of the nutritional quality of forages applicable to other animals such as horses.

WEB LINKS

There is an enormous amount of information on human nutrition on the web. Much of it is subjective, based on incorrect understanding of human physiology and biochemistry. Another large part is promoted by particular interest groups and may be biased. Some advertised diets and weight loss programmes are dangerous. In others, the recipes may promote weight loss but the rationale behind them is faulty.

The following list of sites are in the main, reliable sources of information about nutrition. I have listed mainly Australian sites that link to Australian Tables of Food Composition.

'Healthy Eating'

The premier Australian Nutrition Site is 'Healthy Eating'—this is the home page of a multi-government programme based at Monash University in Melbourne.

It contains a great monthly newsletter and archives of current nutrition research and trends. Very easy to browse—facilities for having the monthly update emailed to you. Links to the most up-to-date Australian Food Composition Tables.

http://www.healthyeating.org/

The Healthy Eating website also contains two online nutrition textbooks written by Mark Wahlqvist and David Briggs.

'Food Questions and Answers'—common misconceptions about food with easy to read explanations and descriptions. Very accessible and informative.

http://www.healthyeating.org/books/fqa-book/

'Food Facts'—an excellent online Australian textbook focussing on nutrient composition of Australian foods, energy balance and effects of food processing.

http://www.healthyeating.org/books/foodfacts/index.html

'Nutriquest from Cornell University'

The Department of Nutritional Sciences at Cornell University in the USA have an excellent Q&A section called Nutriquest. Unfortunately it is not being updated but the archive contains a large amount of reliable and credible nutritional information.

http://www.nutrition.cornell.edu/nutriquest/nqhome.html

'Pathophysiology of the Digestive Tract'

The best web site on digestive physiology with excellent images, animations and self-assessment tests. Easy to get the basics but to explore a topic in depth.

http://arbl.cymbs.colostate.edu/hbooks/pathphys/digestion/

CHAPTER

20

Gas
exchange
in animals

Gas exchange supports cellular respiration by supplying a fuel, oxygen (O_2), and removing a by-product, carbon dioxide (CO_2; Chapter 5). Gas exchange involves the movement of these gases between the environment and mitochondria, the sites of cellular respiration. In unicellular organisms, O_2 reaches mitochondria simply by diffusing through the plasma membrane and cytosol. In multicellular organisms, diffusion is usually inadequate and a variety of mechanisms that ensure an adequate supply of O_2 and removal of CO_2 have evolved. The nature of these mechanisms reflects the size of an animal, its level of activity and whether or not the external respiratory medium is water or air. Animals inhabit a diverse array of habitats. Usually, terrestrial animals obtain their O_2 from the atmosphere, while aquatic animals extract O_2 from water. Some animals, such as frogs, breathe water as larvae and air as adults.

Gas exchange involves the supply of oxygen for cellular respiration and the removal of carbon dioxide.

Air and water as respiratory media

As respiratory media, air and water have very different properties. Compared with air, water has a greater density, is more viscous and presents a greater resistance to diffusion of both O_2 and CO_2. In a respiratory context, the most important property is how much O_2 or CO_2 can be contained in a given volume of air or water. The concentration of O_2 in a litre of air is 20.9% by volume. For water, the concentration (shown by square brackets, [ ]) depends on the solubility of O_2 (amount physically and chemically bound) and the **partial pressure** of O_2 (the proportion of the total pressure of a gas mixture provided by O_2) as shown in the equation:

$$[O_2] = \beta o_2 \cdot Po_2$$

where β is a constant related to solubility (**capacitance** of water for O_2), and P is the partial pressure of O_2 in water. A similar relationship holds for CO_2.

Figure 20.1 shows the concentrations of O_2 and CO_2 plotted as a function of their partial pressures in distilled water and air. The slope of the line of concentration versus partial pressure corresponds to the capacitance of the fluid for the gas. Two important properties are illustrated in this figure: the capacitance of air for O_2 and CO_2 is identical, and the capacitance of water for O_2 is far lower than for CO_2. Thus, water, in equilibrium with air, contains some 20 times less O_2 than does the air. The precise amount of O_2 contained

in water also depends on factors such as temperature, salinity and distance from the air–water interface. An increase in either temperature or salinity decreases the capacitance of water for gases and hence decreases gas concentrations. On the other hand, because of the higher capacitance of water for CO_2, CO_2 diffuses through water some 20 times faster than does O_2. These differences in the capacity of water for O_2 and CO_2 have significant consequences for gas exchange.

The capacitance of air for O_2 and CO_2 are the same, whereas the capacitance of water for O_2 is 20 times less than for CO_2. Increased temperature and salinity reduce the capacitance of water for dissolved gases.

There are several other factors that influence O_2 availability in aquatic habitats. In the sea, despite the low content of O_2, there is usually enough O_2 at the surface to meet the aerobic demands of animal life. The same is true for deeper regions of the ocean, where lower temperatures raise the capacitance for O_2 and replenishment is assisted by ocean currents. The O_2 content of fresh water can also be maintained provided that there is sufficient water circulation to achieve effective aeration. In situations where water circulation is reduced or lost, as in swamps, ponds or tidal pools, O_2 deprivation may occur. However, this may be offset by the production of O_2 by photosynthetic organisms. The balance between removal of O_2 by cellular respiration and addition of O_2 by photosynthesis can vary daily and seasonally.

Except for caves and burrows (see Chapter 29), where air circulation is reduced, the composition of air is relatively stable. The amount of O_2 available in air does, however, depend on altitude (see Chapter 29). With increase in altitude, total atmospheric pressure, and thus the partial pressure of O_2, decreases. Although

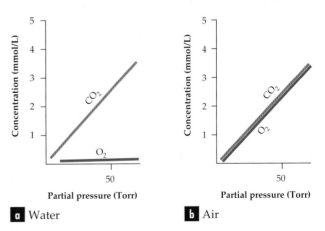

Fig. 20.1 Concentrations of O_2 and CO_2 versus their partial pressures in (a) water and (b) air. The slope of the line is related to the capacitance for a gas. The capacitance of CO_2 in water (solubility) is far greater than that for O_2, while the capacitance for both gases is the same in air

the *proportion* of O_2 remains the same, O_2 concentration decreases with increased altitude.

When compared with aquatic dwellers, air breathers live in environments that tend to be more stable in their composition of gases and contain more O_2. However, unlike water breathers, air breathers are faced with the risk of water loss and desiccation (Chapter 22).

> The composition of air is relatively stable and O_2 content is high compared with water; however, air breathers often face problems of desiccation.

Exchanging gases

Mitochondria, the sites of utilisation of O_2, are isolated within the cytosol by mitochondrial membranes. In multicellular organisms, cells are separated from interstitial fluid by plasma membranes and often from the external environment by an outer covering of skin, scales or exoskeleton. In order to get from the atmosphere into body fluids, O_2 molecules must first dissolve in water, either the water surrounding an aquatic organism or, in the case of air breathers, the film of water that covers gas-exchange surfaces (respiratory surfaces are by nature moist). Once in solution, O_2 diffuses across successive membrane barriers provided there is a favourable partial pressure gradient. Net diffusion at any of the barriers will cease if the partial pressure difference becomes zero. Cellular respiration reduces intra-mitochondrial Po_2 towards zero, so there is clearly an overall gradient of Po_2 from the atmosphere to the inside of mitochondria.

The rate of diffusion of a gas across a membrane not only depends upon the partial pressure gradient but also on properties of the membrane. Fick's law shows that the amount of gas transferred per unit of time across a membrane (**conductance**, dM/dt) is directly proportional to its *permeability* (chemical and physical properties, ϕ), *surface area* (A), and the *difference in partial pressure* of the gas on either side of the membrane (Δp), and inversely proportional to its thickness (*diffusion distance, d*).

$$\frac{dM}{dt} \propto \frac{\phi \cdot A \cdot \Delta p}{d}$$

In diseases in which fluid accumulates in the alveoli, such as emphysema and pneumonia, the diffusion distance between air and blood becomes so great that the rate of O_2 exchange is inadequate. This causes breathing to become laboured and the patient is unable to tackle even moderate exercise without great distress. The nature of the diffusion barrier is also important in gas exchange. For example, vertebrate embryos are often encased in protective structures (shells) that pose potential problems for gas exchange (Box 20.1).

> Diffusion across a membrane is directly proportional to the permeability, surface area and partial pressure gradient, and inversely proportional to the thickness of the membrane.

BOX 20.1 Gas exchange across the surface of eggs

Fish and amphibian eggs are often surrounded by gelatinous capsules, while those of most reptiles, all birds and monotremes are surrounded by shells (Fig. a). The rate at which gas diffuses through an egg covering, its conductance, is described by Fick's law.

The capsules of fish and amphibian eggs are composed mainly of water, across which dissolved gases diffuse slowly. Under these circumstances, it is impossible to achieve high gas conductances; thus, these embryos are small (always less than 1 g) with low metabolic rates and gas-exchange requirements. Bird eggs, on the other hand, are surrounded by a shell pierced by thousands of tiny air-filled pores through which gas diffusion occurs easily. Such shells permit

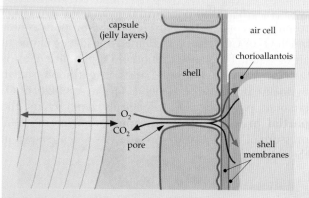

(a) Pathways of oxygen and carbon dioxide diffusion through the jelly capsule in an amphibian egg (left) and an avian eggshell (right). Respiratory gases diffuse slowly through the aqueous layers of the capsule and support relatively low metabolic rates in amphibian embryos. Inside the capsule, gas exchange occurs across the embryo's skin and gills. In contrast, high rates of metabolism in bird embryos are supported by rapid diffusion through air-filled pores in the shell and shell membranes. Oxygen is taken up and carbon dioxide is given off by blood flowing through the chorioallantois, the respiratory membrane that grows from the embryo. The inner and outer shell membranes separate at the blunt end to form an air cell where the embryo begins to breathe shortly before hatching

high rates of gas exchange, so embryos can be larger, up to several kilograms in flightless birds (and dinosaurs). Embryos of this size depend upon internal transport systems (Chapter 21) to achieve adequate rates of O_2 delivery and CO_2 removal to cells. In bird eggs ranging in size from the hummingbird to the emu, conductance and metabolic rate increase in parallel. The result is that levels of O_2 and CO_2 inside the shell are about the same in all birds.

Conductance is also adapted to match conditions prevailing in unusual incubation environments. The Australian mound-building birds (mallee fowl and brush turkey) lay their eggs underground where humidity and CO_2 levels are high, and O_2 availability is low (Fig. b). In compensation for this rather stuffy subterranean atmosphere, the shells of these birds are unusually thin, with very high conductances. The incubation mound protects the fragile shells from breaking and gas tensions inside the shell are similar to those in chicks of other species.

Given the rigid construction of a bird's egg, shell conductance of most bird eggs remains fairly constant throughout incubation. However, conductance of frog egg capsules changes during development. Recent studies of the eggs of the Australian terrestrial toadlet *Pseudophryne bibronii*, which lives and lays its eggs under leaf litter, have shown that the embryo gradually absorbs water from the environment into the space beneath the capsule. This makes the capsule swell, decreasing its thickness and

therefore greatly increasing its conductance (Fig. c). The fact that conductance increases with increasing metabolic rate means that the oxygen level near the embryo is kept high and constant throughout development.

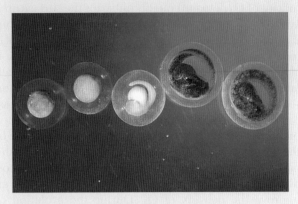

(c) Eggs of the Australian toadlet, *Pseudophryne bibronii*, at 3, 6, 15, 37 and 54 days of incubation, showing the development of the embryos and the increase in fluid volume within the jelly capsule

Surface area and volume

It has been estimated that, for a spherical animal with a metabolic rate appropriate for a single-celled organism surrounded by air-equilibrated water, the maximum diameter that can be supported by diffusion alone is about 2 mm. There are several reasons for this limitation, the most important of which is the animal's surface area-to-volume ratio. This is a simple consequence of geometric relationships—as an object increases in size, its surface area-to-volume ratio decreases.

This relationship results because volume (V) is related to the cube of the length (l), whereas surface area (A) is related to the square of the length. As a result, surface area is proportional to volume to the two-thirds power. Thus, if one species is twice the length of another in each direction, then it is likely to have eight times the volume but only four times the surface area. This can be seen graphically by taking a cube and doubling its length in each dimension (see below). The resulting cube will contain eight cubes of

(b) An unhatched egg and a hatchling chick of the Australian brush turkey found inside an incubation mound. The mound is constructed from decaying leaves and twigs from the forest floor. The eggs lie about 60 cm below the surface and chicks require about 2 days to dig their way out

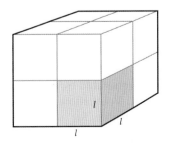

$V \propto l^3$ i.e. $l \propto V^{\frac{1}{3}}$

$A \propto l^2$

$\therefore A \propto \left(V^{\frac{1}{3}}\right)^2 \propto V^{\frac{2}{3}}$

the original size but only have four times the external surface area.

This means that the bigger an animal is, the less surface area it has available for exchange of respiratory gases (or heat for that matter). As an animal grows, the mass of metabolically active cells (the prime determinant of O_2 demand) rapidly outstrips the surface area available to supply O_2. In animals thicker than about 2 mm, the surface area is insufficient and diffusion is too slow for adequate O_2 to be supplied by diffusion alone. In general, if an active animal is to grow much larger than a few millimetres in size, it must rely on some other method of enhancing the diffusion process.

We know from Fick's law that diffusion across a membrane can be improved by altering one or other of the variables in the equation. Thus, an increase in the partial pressure difference, a decrease in the diffusion distance or an increase in the surface area will all improve the delivery of O_2. During the evolution of animals, a variety of mechanisms incorporating changes in one or more of these factors has arisen.

As an object increases in size, its surface area-to-volume ratio decreases. Animals larger than a few millimetres in size have evolved ways of enhancing gaseous exchange with the environment.

Ventilation and convection

A small organism such as an amoeba is limited in obtaining sufficient O_2 by the presence of a layer of stagnant water around it, the boundary layer. As O_2 is taken up from this layer, the Po_2 adjacent to the amoeba decreases, thus decreasing the partial pressure gradient for diffusion across the membrane. Replenishment occurs as O_2 diffuses into the **boundary layer** from surrounding water. However, far more effective replenishment occurs if the boundary layer is stirred or flushed away by currents in the pond water.

Rather than relying upon natural currents to renew the external medium at the gas-exchange surface, many organisms generate their own currents by means of cilia or flagella, or make use of various paddle-like appendages to promote water flow. In animals, this is commonly associated with restriction of gas-exchange surfaces to a reduced region of the body surface. This frees other regions to become specialised for locomotory, protective or sensory functions.

The bulk flow of a fluid (air or water) is known as **convection**. When convection of the external medium at the site of gas exchange is generated by the animal, it is referred to as **ventilation**. Ventilation occurs in all larger animals and is responsible for renewing the external medium, thereby maintaining a favourable

partial pressure gradient and maximising the rate of diffusion at the site of gas exchange.

The value of convection in maintaining a favourable partial pressure gradient applies equally well to the internal side of the gradient—body fluids. Many animals possess some means of circulating body fluids (Chapter 21), although an efficient internal transport system is first seen in the segmented worms, annelids (Chapter 39). In terms of gas exchange, internal convection, **perfusion**, is responsible for transporting O_2 from the site of gas exchange to cells, thereby maintaining the Po_2 gradient across the respiratory surface. For CO_2 a similar situation occurs in reverse: internal convection moves CO_2 from cells to the site of gas exchange, thereby maintaining favourable gradients for the loss of CO_2 at both sites.

Thus, in terms of gas exchange, the evolution of larger animals was dependent upon the development of two systems (Fig. 20.2):

1. a gas exchange system, often involving ventilation of the respiratory medium
2. an internal transport system.

Both systems depend upon convection and diffusion to transfer gases. Diffusion is responsible for conveying both O_2 and CO_2 between the two systems and from the transport system to mitochondria. Circulatory systems are considered further in Chapter 21.

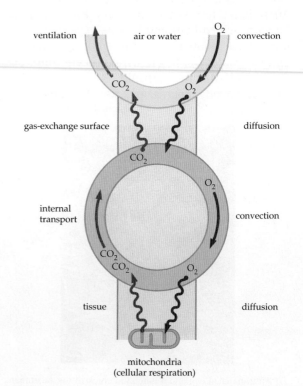

Fig. 20.2 The pathways for O_2 and CO_2 in an animal. O_2 is transferred from the environment to the mitochondria whereas CO_2, a waste product, is transferred in the opposite direction. Whether transfer is by convection or diffusion is shown on the right of the diagram

Ventilation renews the external medium at the gas-exchange surface, thereby maintaining favourable partial pressure gradients for O_2 and CO_2. Internal convection maintains partial pressure gradients for O_2 and CO_2 across the gas-exchange surface, and transports O_2 and CO_2 to and from cells.

Gas-exchange organs

Water breathers

Ventilation of the gas-exchange surface is particularly important in water because of its low O_2 content and diffusibility of O_2. Furthermore, the thickness of the boundary layer adjacent to the exchange surface decreases with increasing flow of water past the gas-exchange surface (Fig. 20.3). Maintenance of a good convective flow of water past the gas-exchange surface both supplies fresh oxygenated water and reduces the boundary layer.

The simplest multicellular animals are the sponges (phylum Porifera). While they have no special organs for gas exchange, their body is permeated with pores, canals and chambers through which a water current is established to bring in food and O_2, and to remove waste products, including CO_2 (Fig. 20.4). Specialised cells lining the inside walls, choanocytes, establish the water current by the beating of their flagella. The beating of the flagella has no particular synchrony so the flow of water is regulated by regulating the size of the osculum and closing the incurrent pores.

Cutaneous exchange

Aquatic oligochaetes depend on ·the exchange of gases across the general body surface—**cutaneous exchange**. Replacement of the surrounding water is

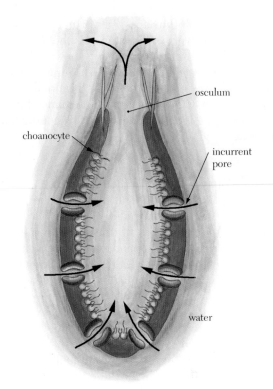

Fig. 20.4 The flow of water through a sponge, generated by the beating of the choanocytes, is regulated by opening and closing the osculum and incurrent pores

either actively maintained or relies upon natural convection. A good example of active ventilation is found in *Tubifex*, a freshwater oligochaete. This worm lives in burrows but achieves ventilation by the rhythmic waving of its well-vascularised posterior end. The frequency of waving increases as the amount of O_2 in the surrounding water decreases. Annelids (e.g. worms) also possess a circulatory system for convection of blood carrying O_2 and CO_2 between tissues and skin.

As larger animals evolved, protective outer surfaces developed, which had a reduced ability to exchange gases. This, together with a decreasing surface area-to-volume ratio, led to the evolution of ventilatory organs specialised for gas exchange. Nevertheless, many fishes and amphibians rely on cutaneous exchange, although often only as a supplement to the main form of gas exchange.

Cutaneous exchange is limited in larger organisms by protective outer surfaces and a low surface area-to-volume ratio.

Simple gas-exchange structures

In echinoderms, a variety of structures are involved in gas exchange. Sea stars (asteroids) have many tiny outgrowths of the body wall (papulae), which aid gas exchange by increasing the surface area available, decreasing diffusion distance and by the internal movement of coelomic fluid (Fig. 20.5a). In sea

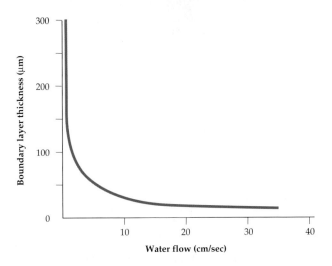

Fig. 20.3 The relationship between the boundary layer formed and the speed of water past a surface

cucumbers (holothurians), a primitive pumping mechanism delivers sea water via the cloaca to a specialised gas-exchange surface, known as a rectal respiratory tree (Fig. 20.5b, c). The branches of the respiratory tree end in thin-walled ampullae, across which O_2 and CO_2 diffuse into coelomic fluid. The respiratory tree is actively ventilated by contractions of the cloaca.

Invertebrate gills

Gills are outgrowths of the body surface. Perhaps the best example of a simple gill occurs in the primitive crustacean family Anaspididae, all of which are restricted to Tasmania. In this family, the gill is essentially a flattened appendage (pleopod), which is perfused by haemolymph, the fluid circulated in animals with an open circulatory system (Chapter 21), circulating through channels around the edge and across the centre (Fig. 20.6a). The gill is exposed and ventilation is created by the rhythmic beating of swimming appendages.

In other aquatic crustaceans and molluscs, the gills hang freely in the water. In these animals, increasing either the number or complexity of the gills, or both,

has increased the total respiratory surface area. Most species, however, possess gills enclosed within a gill chamber. While the gill chamber protects the delicate gills from physical damage, active ventilation of the gill is a necessity if a good supply of oxygen is to be achieved.

> Ventilation is necessary for gills housed in a gill chamber.

In the majority of molluscs, except for cephalopods and land-dwelling gastropods (pulmonates), ventilation of the gill chamber (mantle cavity) is created by cilia covering the surface of each gill and the mantle itself. For crustaceans with gills housed within chambers (branchial chambers), ventilation of the gills is maintained by rhythmic beating of a specialised pair of appendages, the scaphognathites (Fig. 20.6). In both

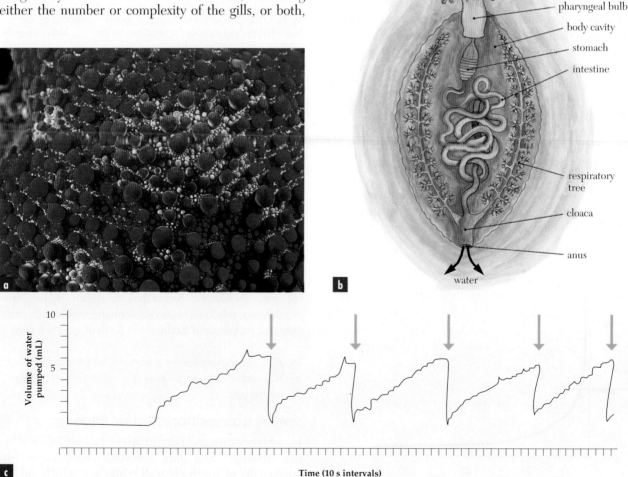

Fig. 20.5 (a) Close-up view of the centre surface of a sea star, showing papulae (skin gills). **(b)** The respiratory tree of a holothurian. A number of small contractions of the cloaca, represented by each small upward movement on **(c)** the chart, results in water being drawn into the respiratory tree. Water is then expelled in a single expiratory contraction, marked by the arrows

cases, the flow of water is, on the whole, unidirectional in that it has separate routes of entry and exit from the gill chamber.

The relationship between water flow (ventilation) and haemolymph flow (perfusion) for the crayfish and crab is shown in Figure 20.6b, c. An enormous advantage for gas exchange exists for the crab, in which ventilation draws water inwards over the gill filament in the opposite direction to that of haemolymph perfusing within the filament. Such an arrangement, **counter-current flow**, greatly increases the efficiency of gas exchange when compared with water flowing in the same direction as haemolymph—**co-current flow**.

The benefit of countercurrent flow arises because haemolymph entering the gill filament has a low P_{O_2} and first encounters water leaving the gill, which has already given up some of its O_2 (Fig. 20.7a). Nevertheless, the partial pressure gradient results in O_2 diffusing

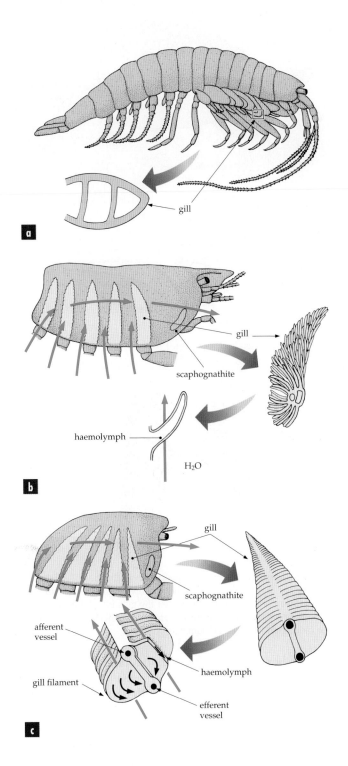

Fig. 20.6 Representative gills of three types of crustacea: **(a)** Anaspididae; **(b)** crayfish; and **(c)** crab. Ventilation in Anaspididae is achieved by the beating of the swimming appendages, whereas in the two decapod crustaceans, in which the gills are housed within brachial chambers, ventilation (blue arrows) is unidirectional and achieved by the rhythmic beating of the scaphognathites. The relationship between ventilation and perfusion is shown for the two decapod crustaceans. In the crab, the flow of water is in the opposite direction to the flow of haemolymph, permitting countercurrent gas exchange

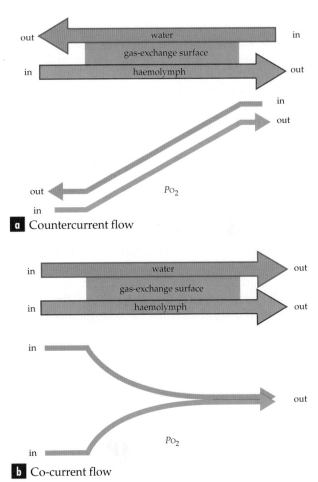

a Countercurrent flow

b Co-current flow

Fig. 20.7 Functional models for **(a)** countercurrent flow and **(b)** co-current flow. The advantage of countercurrent flow, in which water and haemolymph flow in opposite directions through the gas exchanger, is that the partial pressure of oxygen (P_{O_2}) of the haemolymph leaving the gas exchanger is in near equilibrium with the P_{O_2} of the water entering. With co-current flow, water and haemolymph reach equilibrium and leave the gas exchanger at an intermediate P_{O_2}

from water into haemolymph. As the haemolymph continues through the filament, it meets fresher water, with progressively higher O_2 levels; therefore, the partial pressure gradient between water and haemolymph is maintained and O_2 continues to diffuse into the haemolymph. The result is that the haemolymph leaves the gill with almost the same partial pressure of O_2 as the surrounding fresh water. By contrast, in co-current flow, with water flowing in the same direction as the haemolymph, equilibration of O_2 would eventually occur at a partial pressure somewhere between those of the incoming haemolymph and water (Fig. 20.7b).

> The directions in which water and blood (or haemolymph) flow relative to one another has a profound effect on the efficiency of gas exchange, with the greatest efficiency being achieved by a countercurrent arrangement.

Vertebrate gills

Gill structure in fishes is similar in principle to that in the crab except that the surface area is greatly increased and the gills are better vascularised. Most fishes have four gill arches on either side of the mouth, each arch bearing a large number of gill filaments (Fig. 20.8). The surface area of each filament is further enlarged by the addition of vertically stacked lamellae on both the dorsal and ventral surfaces. Each lamella is perfused with blood from a series of branchial arches derived from the ventral aorta.

Most fish actively ventilate their gills by means of a double-action pump involving sequential contractions and expansions of the buccal and opercular cavities (Fig. 20.9). The gills present a resistance to water flow between the two chambers, effectively separating the two, and the mouth and pair of opercula operate as valves. During inspiration, the mouth opens, the opercula remain closed and the floor of the mouth lowers, increasing the volume of both the buccal and opercular cavities. Water is drawn through the mouth into the buccal cavity and through the gill curtain into the opercular chamber. On expiration, the mouth closes, the opercula open, and the floor of the mouth is raised. This movement forces water from the buccal cavity into the opercular cavity and out through the open opercula. By co-ordinating the movement of the mouth and opercula, the fish maintains a hydrostatic pressure difference across the gills, which drives water across them. This is an open-flow or unidirectional ventilatory system. Blood and water flows at the level of the secondary lamellae (the gas-exchange structures) are countercurrent, with the expected advantage for exchange efficiency (Fig. 20.7).

> Fishes have a unidirectional ventilatory system and countercurrent flow in the secondary lamellae.

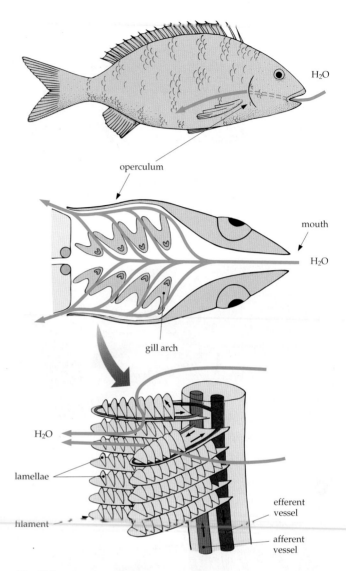

Fig. 20.8 Direction of water and blood flow in the gills of a fish. The section of a fish head from above shows the gill arches and the pathway of water. Water flow past the gills is unidirectional, permitting a countercurrent flow to exist between the water and the blood perfusing the gill lamellae

Another advantage of a unidirectional water flow pattern is that some fishes can effectively ventilate their gills simply by opening their mouth and opercula while swimming. This **ram ventilation** is observed in many fast-swimming fishes, such as sharks and tuna. In fact, tuna are unable to pump ventilate their gills and as a result must swim continuously. Ventilation in ram ventilators is powered solely by the powerful locomotory musculature of the body and does not require additional energy expenditure by buccal and opercular muscle systems. Ram ventilation has the advantage in that it removes the braking effects due to drag created by the cyclic opening of the mouth and opercula. Fishes that ram ventilate can accelerate to faster swimming speeds without increasing the energetic cost of swimming.

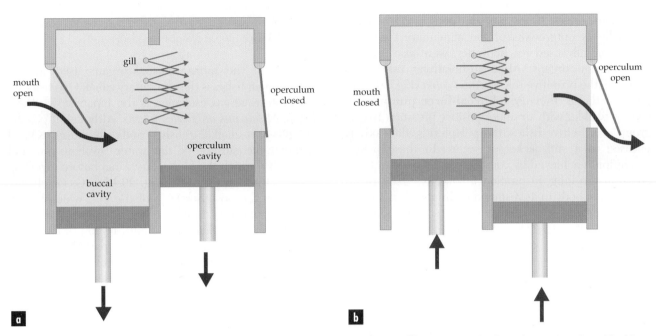

Fig. 20.9 A model of how teleost fishes ventilate their gills showing **(a)** inspiration and **(b)** expiration. The flow of water is indicated by blue arrows and the movement of the floor of the mouth is shown by red arrows

The transition from water to air breathing

Gills are thin-walled vascularised outgrowths of the body surface and are not particularly well suited to function in air. Their physical weight, together with the powerful effect of surface tension, would cause them to collapse from lack of support and stick together in air, reducing the surface area available for gas exchange. To prevent gill collapse, many intertidal and terrestrial crabs strengthen their gills with a chitinous covering over part of the gill surface and increase the spacing between the gill filaments. These changes reduce the effective surface area available for gas exchange, but this is easily compensated for by the greater availability of O_2 in air.

In terms of gas exchange, terrestrial crabs are adapted to cope with the aerial environment. The advantage of O_2-rich air is offset to some extent by the difficulties associated with water loss from the gills. Because of their large surface area and the short diffusion distances, aerial gas-exchange surfaces are particularly vulnerable to water loss. In terrestrial crabs, gills are housed within a chamber to minimise water loss and to provide physical protection for the gills. In general, internally housed gas-exchange surfaces were a prerequisite for terrestrial existence.

Some terrestrial crabs, such as robber crabs, *Birgus* (Fig. 20.10), have very reduced gills that play only a minor role in gas exchange. In *Birgus*, gill reduction is associated with an increase in the size and vascularisation of the gill chamber and some blood vessels in the walls of the gill chamber are extended into tufted outgrowths, increasing the surface available for gas

Fig. 20.10 A robber crab, *Birgus latro*

exchange. Even though O_2 and CO_2 diffuse more readily in air, the chamber must be ventilated. Movement of the scaphognathites usually achieves this.

In air breathers, gas-exchange surfaces are usually housed internally for physical protection and to minimise water loss.

In vertebrates, the evolutionary transition from aquatic to aerial gas exchange probably began in fishes and occurred independently in a number of lineages. Air breathing offered an advantage for fishes in habitats with a shortage of O_2 in water. Survival probably depended upon a fish's ability to gulp a bubble of air and draw upon this as a source of O_2.

In some teleost fishes, **gas bladders**, which evolved originally as outgrowths of the alimentary tract and were generally used for buoyancy, were also used as gas-exchange organs. Most air-breathing fishes, for example, the primitive teleost *Amia calva* (Fig. 20.11a), gulp air; that is, they use a **buccal-force pump** to fill the gas bladder with air under positive pressure. In one group of primitive fishes, the polypterids, the body is encased in a stiff jacket of scales. In these fishes, expiration is achieved by constricting the body to force air from the lung; inspiration follows as elastic recoil restores the body to its resting dimensions. This creates a negative pressure within the lungs and air is drawn into them (aspirated) (Fig. 20.11b).

> Most air-breathing fishes use a buccal-force pump to force air into the gas bladder.

One of the more unusual organs used for air breathing in fishes is the alimentary canal. One fish that uses intestinal gas exchange is the Japanese weather-loach, *Misgurnus anguillicaudatus*. Although it possesses gills, this small (2 g) fish periodically shoves its head through the surface of the water, swallows air and passes it through the alimentary canal before expelling the gas through the vent (Fig. 20.12a). Gas exchange occurs in the posterior 60% of the gut, which has an epithelial layer so thin that the diffusion distance is considerably less than that of gills. Its use of the intestine for gas exchange depends on the partial pressure of O_2 in the surrounding water. As water Po_2 decreases, frequency of air gulping increases, increasing gas exchange across the intestine and supplementing gill gas exchange (Fig. 20.12b).

Air breathing: lungs

The onset of aerial breathing in fishes paved the way for early tetrapods and the beginning of the vertebrate invasion of the land (Chapter 40). The air-breathing organs of vertebrates, **lungs**, developed as outgrowths of the gut.

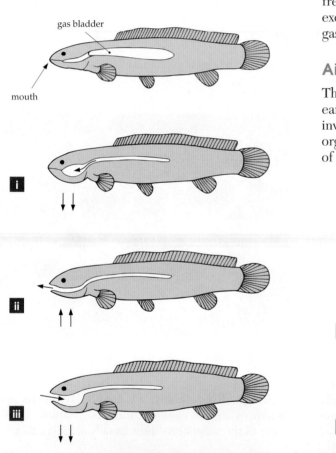

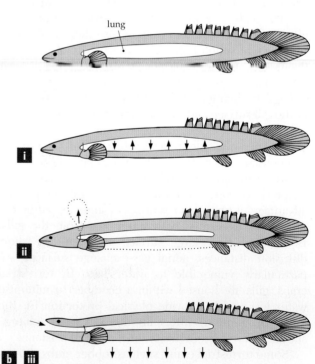

Fig. 20.11 Pattern of air flow in two types of air-breathing fish. **(a)** *Amia* uses a buccal-force pump. **(i)** With the mouth closed, air is drawn from the gas bladder into the buccal cavity by lowering of the mouth floor. **(ii)** With the mouth open, buccal cavity elevation forces air out. **(iii)** The floor of the mouth is again lowered, drawing air into the buccal cavity. **(iv)** The mouth closes and buccal cavity elevation compresses the air in the buccal cavity, which is then forced under positive pressure into the gas bladder. **(b)** *Polypterus* breathes by aspiration. **(i)** The body is squeezed, resulting in air within the lungs being compressed. **(ii)** The air is forced out of the lungs. **(iii)** With the mouth open, the passive recoil of the body creates a negative pressure within the lung that results in air being sucked into the lungs

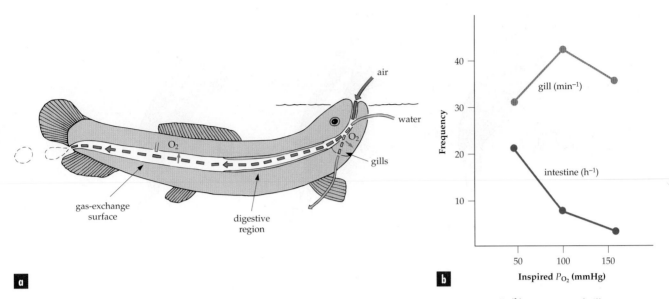

Fig. 20.12 (a) Intestinal air breathing in the Japanese weatherloach supplements gas exchange across the gill. **(b)** Frequency of gill or intestinal ventilation as a function of ambient P_{O_2}

Amphibians

Most amphibians can and do breathe through their skin; characteristically, the skin is the major site for CO_2 loss and the lungs are for O_2 absorption. The simple sac-like form of the amphibian lung appears to have departed little from that of its ancestors. The gas-exchange surface is usually well vascularised with a network of capillaries, the degree of vascularisation varying with the overall importance of the lung in gas exchange.

Amphibians, like most air-breathing fishes, ventilate their lungs by a buccal-force pump. Several breathing patterns have been described in frogs and toads (anurans). In general, they depend upon co-ordinated action of the nostrils and glottis in relation to changes in volume of the buccal cavity, which functions like a bellows to inflate the lungs. The totally aquatic South African clawed toad, *Xenopus laevis*, ventilates by first allowing its lungs and buccal cavity to empty passively through the open glottis and nostrils. It then closes the glottis, lowers the floor of the mouth and draws fresh air into the buccal cavity. Finally, it closes the nostrils, opens the glottis and forces air under positive pressure into the lungs (Fig. 20.13a).

A somewhat different pattern is observed in the frogs *Rana* and *Bufo*. These frogs ventilate the buccal cavity continually by pumping air in and out through the nostrils with the glottis closed. Intermittently, ventilation of the lung commences with a large buccal inspiration drawing fresh air into the buccal cavity. The glottis then opens and air moves passively from the lungs and mixes with buccal gas, with a little escaping through the nostrils as they close. The gas within the buccal cavity is then forced back into the lungs as the buccal cavity is compressed. The nostrils then open and

the glottis closes, and ventilation of the buccal cavity recommences (Fig. 20.13b).

Pumping air into and out of internally housed lungs results in ventilation that is periodic and tidal. With **tidal ventilation**, it is inevitable that some air that has been in contact with the gas-exchange surface will be rebreathed, in contrast to open-flow ventilation, where only fresh medium is passed across the gas-exchange surface. In terms of O_2 exchange, rebreathing results in mixing fresh air with air that has already lost some O_2; thus, lung P_{O_2} is reduced. However, this presents no problem given the far greater quantity of O_2 in air than in water, and can be compensated for by reducing the diffusion distance and increasing the surface area available for gas exchange. More importantly, as a result of tidal ventilation and rebreathing, CO_2 accumulates in the lungs, which is significant in relation to the control of ventilation (see later).

> Amphibians use the buccal cavity to force air into the lungs under positive pressure. Tidal ventilation inevitably involves rebreathing air, which lowers the P_{O_2} and raises the P_{CO_2} in the lungs.

Reptiles

Unlike amphibians, where ventilation is achieved using a force pump, reptiles use an **aspirating pump**, that is, a pump that draws air into the lungs under negative pressure. Aspiration is achieved in much the same way as in polypterid fishes by actively moving the body wall out, resulting in a negative pressure that draws air into the lung (inspiration) and passively or actively expelling air from the lung (expiration). Unlike in polypterid fishes, where the body stiffness required for aspiration is supplied by thick scales, the ribs supply

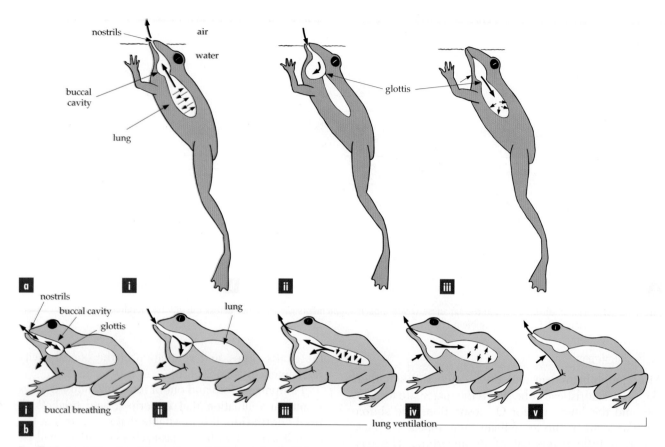

Fig. 20.13 Amphibian breathing patterns. **(a)** *Xenopus laevis*. **(i)** Lungs are emptied out through the open glottis and nostrils. **(ii)** With the glottis closed, the floor of the buccal cavity is lowered and air is drawn into the buccal cavity through the nostrils. **(iii)** The nostrils close and the glottis opens; then the floor of the buccal cavity is raised, forcing air under positive pressure into lungs. **(b)** *Bufo*. **(i)** Buccal breathing. With the nostrils open, repeated lowering and raising of the floor of the buccal cavity moves air in and out of the buccal cavity continually. **(ii)**–**(v)** Lung ventilation. **(ii)** With glottis closed, a large buccal inspiration draws air into the buccal cavity through the open nostrils. **(iii)** The glottis then opens and air moves from the lungs into the mouth, with a little escaping as the nostrils close. **(iv)** With the nostrils closed, the buccal cavity floor is elevated and air is forced under positive pressure into the lungs. **(v)** The nostrils then open, the glottis closes and buccal ventilation recommences

body stiffness in reptiles. As in amphibians, ventilation is tidal.

Reptilian lung structure is relatively simple, although it differs between groups in regard to the number of chambers and distribution of the gas-exchange surface. For example, in a simple sac-like lung, as found in the Australian central netted dragon, *Ctenopherous*, most of the gas-exchange surface is located in the anterior section of the lung (Fig. 20.14). In some species of reptiles, for example, goannas, the gas-exchange surfaces are raised and further subdivided so as to increase the surface area. This results in an increase in the overall thickness of the lung wall.

The physical consequences of such differences in lung structure among reptiles affect the mechanics of lung ventilation. Airflow resistance is inversely related to the radius raised to the fourth power, so small changes in the radius of subdivisions have a large influence on resistance. In addition, an increase in the overall thickness of the lung wall makes the lung less **compliant**, so a greater pressure will need to be developed to achieve a given change in volume. Increased airway resistance and/or decreased compliance contribute to the metabolic cost of ventilation in that more energy is required to ventilate the lungs.

> Aspiration involves the expansion of a relatively stiff chamber, creating a negative pressure that draws air into the lungs. The mechanics of lung ventilation are governed by airway resistance and compliance.

Mammals

All mammals, except some of the very small newborn marsupials (Box 20.2), rely on their lungs for gas exchange. In mammalian lungs, a single tube, the trachea, branches repeatedly until it terminates in an array of cup-shaped chambers, the alveoli (Fig. 20.15). Gas exchange occurs in **alveoli** and also in **alveolar ducts**, which, being thin-walled and better ventilated than the alveoli, are at least as important as alveoli in gas exchange. Alveoli have extremely small diameters, for example 40 μm in mice and 250 μm in humans. The

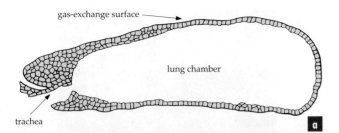

Fig. 20.14 The lung of the Australian central netted dragon, *Ctenophorous nuchalis*, **(a)** showing the distribution of the gas-exchange surface. **(b)** Scanning electron micrograph showing the gas-exchange surface stretching from the very thin external wall of the lung to an internal network of supporting muscular trabeculae (T)

BOX 20.2 Breathing through the skin in mammals

The exchange of oxygen and carbon dioxide through the skin is generally insignificant in most mammals because they have high metabolic rates and diffusion through the skin is poor. Nevertheless, small amounts of CO_2, up to about 12%, have been recorded as being lost across the membranous wings of bats and have been attributed to the large surface area and thin nature of the wings. More recently, in one of the smallest newborns of any mammal, the Julia Creek dunnart (*Sminthopsis douglasi*), gas exchange across the skin was found to be the predominant form of O_2 and CO_2 transfer in the first few days of pouch life. The newborn of this species, a small dasyurid marsupial found in the Downs country of Queensland, weighs only approximately 17 mg and is born after a gestation lasting 13 days. At birth, the skeleton is entirely cartilaginous and the internal organs are visible through the translucent skin. The lungs appear as spherical air sacs (Fig. a).

By placing a tiny mask over the mouth and nose of the newborn joey, researchers at La Trobe

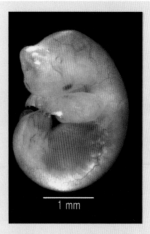

(a) Joey of a Julia Creek dunnart at 1 day old. The lungs are visible as 'air bubbles' in the thorax. Despite the presence of ribs and a diaphragm, no respiratory movements can be detected

University were able to measure gaseous exchange through the skin and lungs separately. During the first week of pouch life, the skin's contribution to gas exchange exceeded that through the lungs (Fig. b). In fact, in animals 2–3 days old, breathing was not apparent and spontaneous body movements did not appreciably expand the air sacs. This suggests that in the early days of pouch life, lung convection is inefficient. By sustaining oxygen demands though the skin, this very small newborn can be born before the respiratory apparatus is fully functional. This is only possible because the newborn has a very low oxygen demand (partly due to being born in a warm pouch and not needing to generate its own heat, see Chapter 29) and a very high surface area-to-volume ratio.

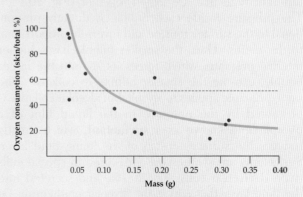

(b) Ratio between skin and total (skin plus lung) gas exchange, measured as the rate of oxygen consumption, in the pouch young of the Julia Creek dunnart maintained at pouch temperature (36°C). For the first week after birth, skin gas exchange exceeds 50% of the total oxygen consumption (broken line)

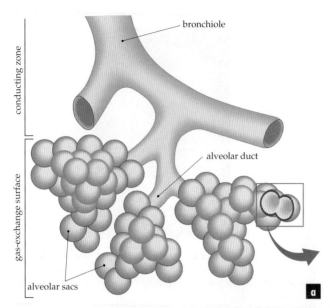

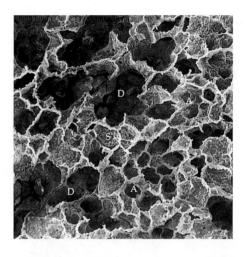

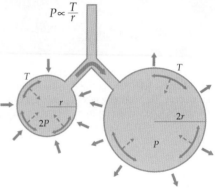

$$P \propto \frac{T}{r}$$

Fig. 20.15 (a) The terminal portion of the mammalian lung showing the cup-shaped alveoli. **(b)** A scanning electron micrograph of the lung of a red kangaroo, showing individual alveoli (A) opening onto an alveolar duct (D). Alveoli are separated by a thin septum (S). **(c)** If two bubbles of different radii (r) are joined together, the surface tension (T) in the smaller bubble will create a higher internal pressure (P), resulting in air moving from the smaller bubble to the larger one and leading to the inevitable collapse of the smaller bubble

walls of alveoli comprise only capillaries and a small amount of tissue. The mammalian lung is little more than a dense capillary network encased by a thin epithelial layer and exposed on all sides. The diffusion distance between air and blood is very small; in humans it is 0.62 μm but it is as small as 0.25 μm in the smallest mammal known, the Etruscan shrew.

Alveoli are lined with a thin film of liquid and the resultant **surface tension** produces a force that tends to cause some to collapse. Consider two bubbles of different radii that are joined together (Fig. 20.15c). The pressure inside each bubble is directly proportional to the surface tension and inversely proportional to the radius. Thus, the smaller bubble has a higher internal pressure, which forces its air into the larger bubble, causing the smaller bubble to collapse. The same principle applies to alveoli, but alveolar collapse is prevented by the presence in the liquid film of a phospholipid known as a **surfactant**, which greatly reduces the surface tension of the lining fluid layer. Surfactants are synthesised and stored in special cells (type-II alveolar cells; Fig. 20.16a) and secreted into the fluid layer that covers the alveolar epithelium. Of more importance is the fact that surfactants reduce the surface tension to a greater degree as the area of the surface layer (the radius of the alveolus) decreases (Fig. 20.16b). Not only does this prevent alveoli from collapsing but permits continuous differences in their dimensions, a requirement for ventilation. Premature

infants whose lungs fail to secrete surfactants are unable to inflate their lungs due to the high surface tension: the resultant 'respiratory distress syndrome' can be treated by application of synthetic surfactant

> Surfactants lower the surface tension at the air–water interface, preventing the collapse of the alveoli during expiration.

In humans, 23 successive branchings occur between the trachea and alveolar ducts. The volume of these conducting airways represents a dead space, in which air remains after expiration. At the end of inspiration, air in the dead space does not contribute to gas exchange. This volume of dead space is by no means insignificant. In a dog, it is about 150 mL, whereas in a whale it may be up to 300 L. On average, it accounts for about one-third of each breath. If dead space is artificially increased, such as when a diver breathes through a snorkel, the build-up of stale air can be so great that a diver may lose consciousness. Modern snorkels have been designed to have the smallest possible internal volume in order to minimise this risk.

In compensation for the low compliance of a highly divided lung, mammals have a more efficient breathing mechanism than simple movement of the body wall, as is found in many reptiles. The lungs of mammals are housed within the thorax, which is separated from the

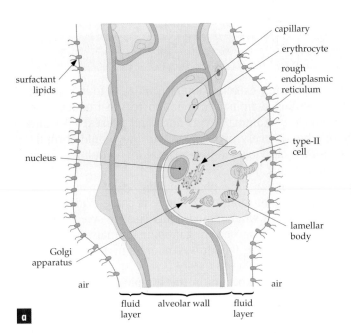

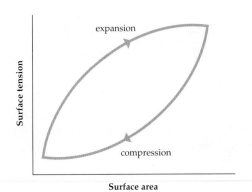

Fig. 20.16 (a) Cross-section of the alveolar wall showing capillaries and a type-II cell. The synthesis, storage and secretion pathway for surfactants in a type-II cell is shown. On secretion, the surfactant lipids form a layer on the surface of the fluid lining the alveolar wall. **(b)** Relationship between surface tension and surface area. Surfactants prevent alveolar collapse by altering the surface tension of the surface film. As the surface area of the alveolus decreases, the surface tension in the film layer decreases

abdomen by the **diaphragm** (a muscular and tendonous sheet). Expansion of the rib cage and contraction of the diaphragm create a negative pressure within the thorax, which draws air in through the airways to fill the lungs.

Except in large mammals, such as elephants and possibly horses, the lung is not physically attached to the thoracic wall but is separated by a thin fluid-filled space, the **pleural cavity** (Fig. 20.17). Negative pressure within the pleural cavity keeps the lung surface closely applied to the thoracic wall and prevents the lung from collapsing. If the thoracic wall is punctured, allowing air to enter the pleural space, a state known as pneumothorax, the lungs will collapse. Inspiration is an active muscular process in mammals; expiration generally results from passive elastic recoil of the thoracic wall. Some mammals synchronise ventilation with locomotion (see Box 20.3).

Mammals have stiff lungs with a low compliance, which are ventilated with the aid of a diaphragm. Inspiration is achieved with a negative pressure.

Birds

In the bird lung, ventilation is driven by muscles acting on highly compliant chambers, air sacs, which have no gas-exchange surfaces and function solely as bellows (Fig. 20.18a). The high compliance enables rib movements alone to produce very efficient air movement in bird lungs. Additionally, because the gas-exchange regions are separated from the air sacs, the gas-exchange region is immobile. Because the volume of the gas-exchange region does not vary, its surface area-to-volume ratio can be greatly increased and air–blood diffusion distances can be extremely short.

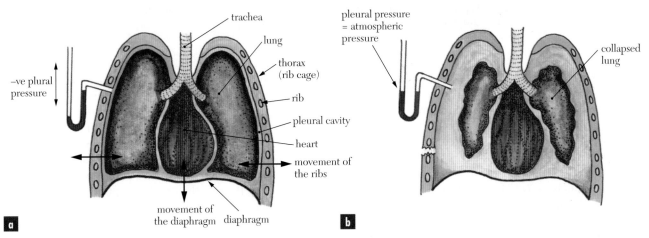

Fig. 20.17 The thoracic cavity of a mammal. Ventilation is achieved by movement of the ribs and diaphragm. **(a)** The lungs are held in close contact with the chest wall and prevented from collapse by the negative pressure within the pleural cavity. This negative pressure is the result of the ribs' tendency to recoil out and the lungs' tendency to recoil in (small arrows). **(b)** Puncture of the thoracic wall results in this negative pressure being lost and the inevitable collapse of the lungs

BOX 20.3 Breathing and running

In some animals, the mechanical basis of lung ventilation during exercise differs from that found during resting. Studies of the frequency of limb movement in running mammals in relation to the frequency of breathing have shown that, at least in some gaits, six types of mammals (e.g. dogs and horses) take one breath per stride. Trained athletes were found to take one breath every two strides (but they could abruptly change breathing frequency). This phenomenon is known as 'locomotory–respiratory coupling'. Such coupling has been also described in fishes, and flying and running birds.

Locomotory–respiratory coupling is an example of entrainment, the phase-locking of two oscillating systems. Little is known of the mechanisms controlling entrainment in animals. There is some evidence for the role of information from mechanoreceptors in joints, but in cases where limb frequencies are synchronised with breathing in a 1:1 ratio, mechanical explanations have been proposed. In galloping animals, flexion of the lumbar region tends to push the abdominal viscera and diaphragm forwards, compressing the lungs and causing exhalation; extending the back achieves the opposite.

One of the clearest examples of locomotory–respiratory coupling occurs in the hopping wallabies (see figure). This was first observed when researchers from Flinders Medical Centre and the University of British Columbia observed that as tammar wallabies, *Macropus eugenii*, hopped on a chilly morning, a jet of vapour would leave the nostrils with each hop. Their proposition was confirmed when, on a motor-driven treadmill, hopping wallabies showed a phase-locked 1:1 ratio between respiratory and locomotory frequencies at all hopping speeds. The data were explained by a piston mechanism, with inspiration beginning just as the animal leaves the ground. The animal's lungs are fullest just before it reaches the highest point in the hop. Kangaroos and wallabies have a central tendon in the diaphragm that may aid the piston displacement. In addition, the viscera are rather loosely slung, which would help them move within the abdominal cavity.

Locomotory–respiratory coupling. **(a)** At the beginning of a hop, lungs are compressed by the weight of the viscera. **(b)** Leaping forward, viscera attached to the diaphragm are thrust backward by inertia, helping to draw air into the lungs. **(c)** At landing, viscera move forward, pushing air out of the lungs

The gas-exchange region in birds is composed of a parallel series of tubes, the **parabronchi**. The lumen of each parabronchus gives rise to narrow conducting tubes, which in turn give rise to numerous **air capillaries** (Fig. 20.18b). The diameter of these air capillaries may be as small as 3 μm (much smaller than the alveolus of a mammal). Such a small diameter is possible only because the air capillary is rigid and prevented from collapsing. The air capillaries have an intricate network of blood capillaries separated from the air capillaries by an exceedingly small diffusion distance (e.g. 0.12 μm in the pigeon).

In birds, airflow through the parabronchi of the gas-exchange region is unidirectional. During inspiration, most air passes along the bronchus and into the posterior air sacs; the remainder enters the posterior side of the lung (Fig. 20.19a). Once in the lung, air passes through the parabronchi and into the anterior air sacs. Air is prevented from entering the junction between the bronchus and the anterior side of the lung during inspiration by pressure differences that create zones of no flow. At the onset of expiration, the pressure differences change (Fig. 20.19b). Air can now pass between the anterior side of the lung and the bronchus, but is prevented from passing between the junction of the bronchus and the posterior side of the lung. As a result, air from the posterior air sacs is directed into the posterior side of the lung through the parabronchi,

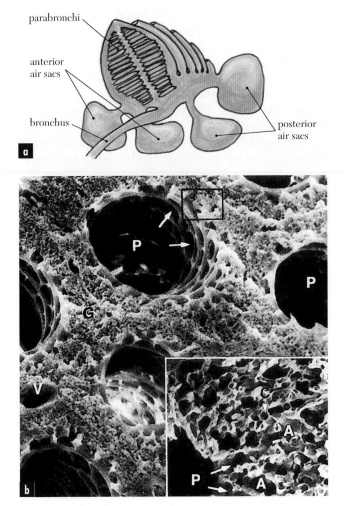

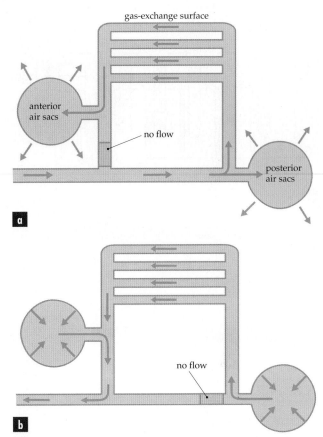

Fig. 20.18 The bird lung and its fine structure. (a) The gas-exchange region (parabronchi) and the air sacs. (b) The lumen of a parabronchus showing the air capillaries. The space between the air capillaries is filled with a network of blood capillaries

Fig. 20.19 The flow of air through the bird lung. Zones where airflow is prevented because of pressure differences are shown in yellow. The patterns of (a) inspiration and (b) expiration are explained in the text

where it joins air being squeezed from the anterior air sacs. This air, which has passed through the gas-exchange region of the lung, is directed through the anterior junction of the bronchus and on out of the system. Thus, airflow through the gas-exchange region is unidirectional and occurs during both inspiration and expiration.

The unidirectional airflow through parabronchi and the fact that blood flow in the blood capillaries is towards the lumen of the parabronchus suggest the possibility of countercurrent flow between air and blood. However, closer examination shows that this is not the case (Fig. 20.20). Movement of air in air capillaries only occurs by diffusion. Starting at the entrance to a parabronchus, the partial pressure of O_2 is essentially the same as outside air. Due to the uptake of O_2 as air progresses through the parabronchus, the Po_2 falls in subsequent air capillaries and the amount of O_2 extracted by the blood progressively declines. Blood leaving the parabronchus is a mixture drawn

from all capillaries and usually has a Po_2 greater than that of the air leaving the parabronchus. This pattern of blood and airflow has been termed **cross-current exchange**.

Birds have highly compliant chambers (air sacs) used to ventilate stiff gas-exchange surfaces. Airflow through the bird lung is unidirectional and occurs during both inspiration and expiration. Cross-current exchange is more efficient than co-current exchange but less efficient than countercurrent exchange.

Air breathing: tracheae

In insects, which are successful terrestrial arthropods, air passes internally to cells through tubular ingrowths of the body surface, **tracheae**. Tracheae usually arise at openings, **spiracles**, of which there are normally 10 pairs on the thorax and abdomen (Fig. 20.21a). Tracheae branch repeatedly and terminate in blind-ending tubules, **tracheoles**, which contain liquid (Fig. 20.21b). In some insects with exceptionally long legs, having additional spiracles on the legs circumvents possible diffusion problems over such distances.

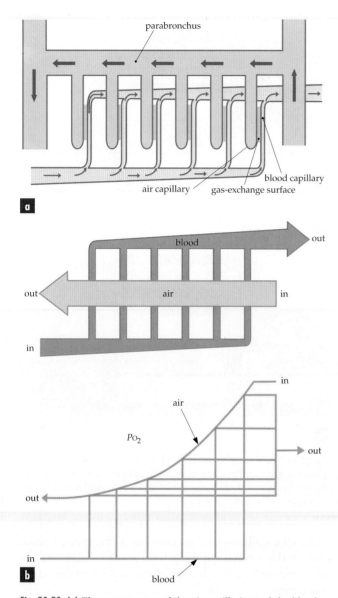

a

b

Fig. 20.20 (a) The arrangement of the air capillaries and the blood capillaries in the bird lung. **(b)** A functional model for cross-current flow. In cross-current flow, capillary blood flow is perpendicular to the main flow of air. In most cases, this permits the blood leaving the gas exchanger to have a higher partial pressure of O_2 (Po_2) than the air leaving the organ

of tracheoles are filled with liquid through which exchange occurs between tracheoles and cells. With such small diameters, one might expect a tendency for the tracheoles to collapse due to surface tension, as occurs with alveoli. This does not occur because tracheole walls are strengthened with chitin.

Osmotic forces (Chapter 4) are used to reduce diffusion distance within tracheole liquid during exercise. During increased muscle activity, metabolites are released into the interstitial fluid, raising the tissue osmotic pressure. Water is drawn osmotically from the tracheoles to the tissue. Given the greater diffusion of gases in air, such a mechanism has obvious advantages. The removal of fluid from the tracheole during increased activity permits gaseous air to travel further down the tracheole, reducing the diffusion barrier (Fig. 20.21). At the end of activity, the metabolites are dispersed and water moves back into the tracheoles.

> In insects, air passes directly to tissues through a network of tracheae and is generally renewed by simple diffusion.

Gases may also move through the tracheal system by bulk flow. Numerous large insects actively ventilate by movements of the abdomen. Expiration is brought about by muscular activity, whereas in many species inspiration relies upon elastic recoil of the cuticular exoskeleton. Many flying insects make use of flight movements to ventilate tracheae supplying the flight muscles through open spiracles.

Ventilation in insects is usually tidal, although some beetles (order Coleoptera) are equipped with two pairs of large-diameter tracheae that run the length of the thorax. From these, numerous secondary tracheae branch and invade the flight muscles. During flight, air is ducted through the large tracheae as a result of ram ventilation and the flow is therefore unidirectional.

Not all insects have spiracles. In some aquatic insects, the tracheal system has no external connection. Whether their tracheal systems are sealed or open, aquatic insects show a number of modifications to the tracheal system, which permits gas exchange in water (Box 20.4).

Most spiders and scorpions have tracheae, but also possess an additional site for gas exchange, the **book lung**. This structure consists of numerous horizontal air spaces, connected to the outside air through a spiracle, which alternate with haemolymph spaces, giving the appearance of pages in a book and hence the term book lung (Fig. 20.22). Such an arrangement brings haemolymph into very close contact with air and provides a large surface area for exchange. Arachnids have an open circulatory system. Nevertheless, almost all haemolymph returning to the heart from the abdomen must pass through the book lung where it is oxygenated. From the heart, oxygenated haemolymph is

Spiracles can be opened or closed, which assists in minimising water loss. Tracheae taper to a diameter of 0.51 μm and deliver O_2 to within a few micrometres of the cell, often indenting the cellular membrane and, in some cases, passing near to mitochondria. Tracheal volume varies greatly between species. In locusts, the tracheal system may occupy almost half of the insect's body volume. Tracheal systems of varying complexity are also found in most spiders and scorpions (arachnids) and in terrestrial isopods (crustaceans).

Diffusion is usually adequate for the renewal of gas along tracheae, especially for small insects or the inactive stages of larger species. However, the blind tips

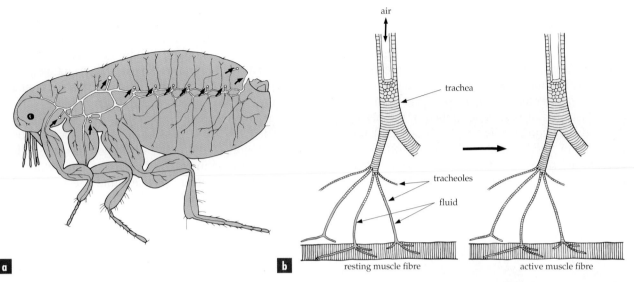

Fig. 20.21 Insect tracheal system. **(a)** The tracheal stem of the flea showing the openings of the spiracles (arrows). **(b)** The terminal portions of the tracheoles are filled with fluid during rest. During activity, osmotic forces draw fluid from the tracheoles and permit air to advance further down the tracheoles

BOX 20.4 Gas exchange in aquatic insects

Aquatic insects have, in terms of evolution, returned to water from the land and their tracheal systems are modified to permit air-filled tracheae to operate in water. There are essentially two patterns of gas exchange: the air-filled tracheal system may be in open contact with some form of gas source or it may be completely internal.

Modification of a spiracle into a tube that pierces the surface of the water permits some aquatic insects, such as mosquito larvae (Fig. a), to form a direct link with air above the water surface. When the larva is suspended at the surface, air moves by diffusion through the tube and into the tracheal system. Interestingly, the larvae of one genus of mosquito, *Mansonia*, replenish their air supply by piercing aquatic plants with the tip of their spiracle tube and drawing air from aerenchyma tissue (Fig. b).

When they visit the surface, diving beetles and bugs trap bubbles of air and their tracheae open into these bubbles. As O_2 is absorbed by the insect from the bubble, an O_2 diffusion gradient is established and more O_2 diffuses in from the surrounding water. The high solubility of CO_2 in water means that CO_2 is easily lost to the surrounding water and the PCO_2 inside the bubble is negligible. Therefore, we might expect that an insect with a bubble would never need to surface.

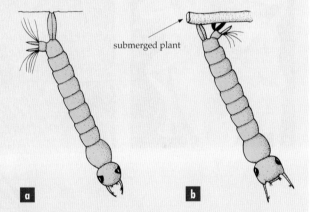

(a) A mosquito larva obtaining air at the water surface. **(b)** Another mosquito, *Mansonia*, obtains O_2 by piercing air-filled aquatic plant tissue

However, as a result of a decrease in the partial pressure of O_2 within the bubble, the partial pressure of nitrogen (N_2) increases, favouring loss of N_2 from the bubble to the water. Also, hydrostatic pressure increases with depth. The deeper a beetle dives, the higher the gas pressure inside the gas bubble. This increases the loss of N_2 and O_2 from the bubble. As a result, the bubble inevitably collapses and the insect must surface to replenish its air supply.

If the bubble were prevented from collapsing, the animal would not need to surface as O_2 would continue to diffuse from the water into the bubble (N_2 being maintained in equilibrium with the water). Some diving bugs and beetles prevent bubble collapse by covering part of their surface with very fine submicroscopic hairs that trap air between and beneath them. Physical factors, such as surface tension, prevent hydrostatic compression of the bubble. The air is in contact with a spiracular system that has fine hairs to prevent water entering. Such a structure is known as a plastron and is well-developed in the beetle *Simsonia tasmanica* (Fig. c). In these bugs, the majority of the ventral and dorsal surface is covered in many submicroscopic hairs, up to 2.5 million/mm^2. Because the bubble is prevented from decreasing in size, bugs with a plastron can descend to depths of several metres.

In some aquatic insect larvae, the gas-filled tracheal system is sealed, having no openings to external air supplies. Abdominal appendages, tracheal gills, form the external gas-exchange site, which relies on diffusion between gas contained in the tracheal system and the surrounding water (Fig. d). Ventilation over the exposed gills is created by water passing over the body or by undulation of the body. In dragonflies, the gills are housed in an enlarged region of the rectum, a 'branchial chamber' (Fig. e), and are ventilated by muscular pumping actions.

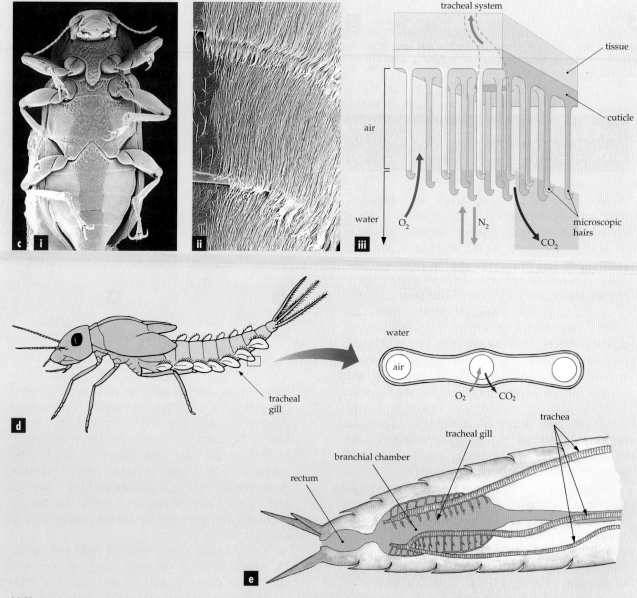

(c) (i) General view and (ii) fine view of the ventral surface of an aquatic beetle, *Simsonia tasmanica*, showing the fine hairs of the plastron. (iii) Diagram illustrating the function of a plastron. (d) Tracheal gills in a mayfly larva. (e) Dragonfly larva showing rectal branchial chamber

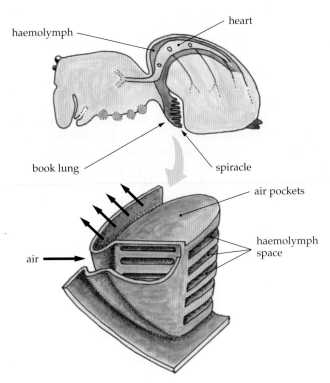

Fig. 20.22 The book lung of the spider. The book lung consists of numerous air spaces, which alternate with haemolymph spaces (see inset). The air spaces are connected to the outside by a spiracle. Oxygen diffuses from the air spaces to the haemolymph, returning to the heart. Once in the heart, this oxygenated haemolymph is distributed to all parts of the body

conducted to all parts of the body. Book lungs rely upon diffusion for the replenishment of oxygen and removal of carbon dioxide in the lung chamber. Changing the diameter of the spiracle controls the exchange of gases in the book lung, thereby affecting diffusion.

Gas transport

Transport of oxygen

Because of its low solubility, the content of O_2 in water and physiological fluids is low. Most animals therefore improve the transport of O_2 within the body by the addition of specialised O_2-carrying molecules, **respiratory pigments**, such as haemoglobin and haemocyanin. These pigments (coloured molecules) substantially increase the maximum amount of O_2 that can be carried by a fluid—**oxygen-carrying capacity**—and hence increase the transport of O_2.

Respiratory pigments are found in the blood of animals with closed circulatory systems (e.g. nematodes, annelids, cephalopods and vertebrates), the haemolymph of animals with open circulatory systems (e.g. arthropods and molluscs) and, sometimes, in coelomic fluid (e.g. some annelids and echinoderms).

The pigment may be either directly dissolved in the fluid or contained within specialised cells (corpuscles, erythrocytes or coelomocytes) suspended in the fluid. Pigment-containing cells are found in some annelids, molluscs, echinoderms and all vertebrates except the ice fish.

In mammals, the respiratory pigment is haemoglobin and is located entirely within specialised cells, **erythrocytes**, which do not contain a nucleus or mitochondria. Packaging haemoglobin into specialised cells avoids problems that would be caused by the presence of this osmotically active material in plasma, and provides the pigment with a stable and regulated environment with the appropriate ions and enzymes.

R espiratory pigments increase the O_2-carrying capacity of a fluid.

Oxygen-carrying capacity

When a pigment is present, the total O_2 content of blood is the sum of that transported in solution and that bound to the pigment (Fig. 20.23). Besides having the ability to bind with O_2, a respiratory pigment must be able to release O_2 under certain conditions. The ability of a pigment to load and unload O_2 (reversible binding) is partly dependent on the local P_{O_2}.

The relationship between P_{O_2} and total O_2 content is known as the **oxygen equilibrium curve**, sometimes called the oxygen dissociation curve. Examination of such a curve for haemoglobin reveals that the amount of O_2 bound to a pigment is not a simple relationship (Fig. 20.23). The S-shaped (sigmoid) curve demonstrates that, in a region of low P_{O_2}, haemoglobin contains very little O_2 and that this rapidly increases

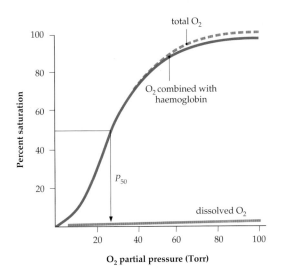

Fig. 20.23 The amount of O_2 in the blood is dependent upon the partial pressure of oxygen. The amount of O_2 binding to a respiratory pigment, in this case haemoglobin, usually exceeds that in solution to an extent where it is the major source of O_2

with increasing P_{O_2} until at high P_{O_2} the pigment is fully saturated and the curve reaches a plateau.

A respiratory pigment must be able to combine reversibly with O_2. The amount of O_2 bound to a respiratory pigment increases with increases in the partial pressure of O_2.

Not all O_2 equilibrium curves are sigmoidal because the shape for a given pigment depends on the number of sites to which O_2 can bind (Fig. 20.24). Where there is a single binding site, as in myoglobin (a respiratory pigment found in muscle), the curve is hyperbolic (curve A). A sigmoidal curve (curves B and C) results from the fact that binding of O_2 at one site increases the affinity for O_2 at the remaining sites (co-operative binding). The advantage in a pigment having a steep dissociation curve is that such a pigment will change from loading to unloading with minimal change in P_{O_2}. This tends to maximise diffusion gradients for O_2 exchange at gas-exchange organs and tissues.

Co-operative binding, which results in a sigmoidal oxygen equilibrium curve, permits the loading and unloading of O_2 over relatively small changes in P_{O_2}.

Oxygen affinity

Different pigments have different affinities for O_2, expressed in terms of the P_{O_2} at which the pigment is half saturated, P_{50}. A pigment has low affinity for O_2 if it requires a high P_{50} to become saturated and vice versa. The functional significance of pigments with

different O_2 affinities can be illustrated by considering mammals during development. In eutherian mammals, the developing fetus must take up O_2 from the blood of its mother. Since maternal tissues also use O_2, the partial pressure of O_2 in maternal blood falls and therefore fetal O_2 uptake must occur at lower partial pressures of O_2 than those for the mother taking up O_2 from air. This is possible because the oxygen equilibrium curve for fetal haemoglobin lies to the left of the maternal curve, that is, fetal haemoglobin has a low P_{50}. For a given partial pressure of O_2, fetal haemoglobin saturation is higher than for maternal haemoglobin (Fig. 20.25).

The affinity of haemoglobin for O_2 is partly dependent upon the prevailing P_{O_2}. Oxygen affinity can also be altered by the presence of various molecules, such as organic phosphates, by pH and by temperature, which effectively shift the O_2 P_{50} horizontally. In most cases, O_2 affinity decreases as pH decreases. This is the **Bohr effect** and it assists in both loading O_2 in the lungs or gills (where pH is slightly higher) and unloading it in the tissues (where accumulated CO_2 tends to make the region slightly more acidic). Oxygen affinity of mammalian haemoglobin is reduced by CO_2 binding reversibly with the pigment (at a different site on the molecule to the O_2 binding site)—the **Root effect**. An increase in temperature also reduces O_2 affinity by directly affecting the reaction between haemoglobin and O_2, and to a small extent by affecting pH.

The functional significance of the effects of pH, CO_2 and temperature on O_2 affinity becomes clear if we consider an actively metabolising tissue. Such a tissue produces heat and CO_2, and may also affect pH

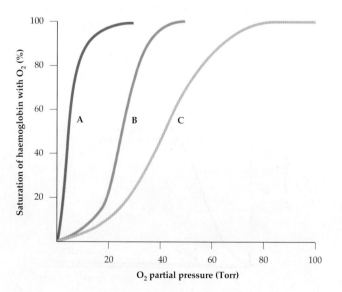

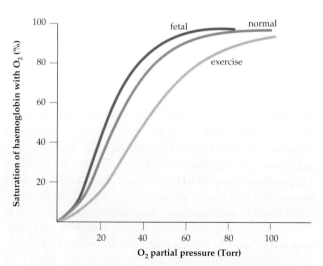

Fig. 20.24 The shape of the oxygen equilibrium curve for respiratory pigments reflects the number of binding sites for oxygen. A single binding site results in a hyperbolic curve (curve A), while more than one binding site results in a sigmoidal curve (curves B and C); the more binding sites, the steeper the curve. The steeper the curve the lower the P_{O_2} for saturation

Fig. 20.25 The oxygen equilibrium curve for haemoglobin. The curve for fetal haemoglobin is shifted to the left of the maternal (normal) curve, showing that fetal haemoglobin has a greater affinity for oxygen than does maternal haemoglobin. The rightward shift of the normal curve caused by exercise results in a lowering of the oxygen affinity of haemoglobin, favouring unloading of oxygen

by the production of lactic acid (Chapter 5). As blood passes through an active tissue, these products enter the blood and lower the O_2 affinity of haemoglobin, which results in more O_2 being unloaded (Fig. 20.25).

> The oxygen affinity of a respiratory pigment decreases with a decrease in pH and with an increase in CO_2 or temperature.

Transport of carbon dioxide

Carbon dioxide, a by-product of cellular respiration, must be removed. The main pathways for CO_2 transport in mammals are given in Figure 20.26. After its formation in the cell, CO_2 diffuses rapidly into blood. In air-breathing vertebrates, some CO_2 dissolves in plasma but the majority diffuses from plasma into erythrocytes. Regardless of whether CO_2 is in plasma or erythrocytes, most of it is hydrated to form carbonic acid (H_2CO_3), which in turn is dissociated into hydrogen carbonate (HCO_3^-) and hydrogen ions (H^+). The hydration step proceeds very slowly and, despite the fact that dissociation of H_2CO_3 is a rapid reaction, few HCO_3^- ions are formed in plasma. In erythrocytes, the enzyme carbonic anhydrase greatly speeds up the hydration reaction. As a result, a large HCO_3^- concentration gradient exists between an erythrocyte and plasma, and HCO_3^- diffuses from erythrocytes to plasma. To maintain electrical balance within an erythrocyte, Cl^- moves in the opposite direction. At the lungs, opposite gradients exist and HCO_3^- diffuses out

of erythrocytes and is exchanged for Cl^-, which moves in. The exchange of Cl^- between plasma and erythrocytes, known as the chloride shift, permits substantial exchange of HCO_3^- ions between the two.

> In mammals, most CO_2 is transported in the form of hydrogen carbonate ions. The chloride shift allows large amounts of hydrogen carbonate to be exchanged between erythrocytes and plasma.

In mammals, some of the CO_2 entering an erythrocyte binds directly with haemoglobin, forming carbaminohaemoglobin. The more deoxygenated the haemoglobin, the better is its ability to take up CO_2, a shift known as the Haldane effect. Up to 10% of CO_2 can be transported this way. A side product of this reaction is the production of H^+. To prevent changes in blood pH, these H^+, together with those produced with HCO_3^-, must be removed. Haemoglobin can take up or give off H^+ at particular sites; that is, haemoglobin can act as a buffer. In addition, the more deoxygenated the haemoglobin, the better it binds with H^+. Buffering of H^+ in plasma is also achieved by some proteins in the plasma. Thus, CO_2 diffusing into capillaries of tissues is transported either as HCO_3^-, carbaminohaemoglobin or dissolved in solution. Furthermore, almost all the H^+ produced from the HCO_3^- and carbaminohaemoglobin reactions is buffered by haemoglobin. In the lungs, the above reactions are reversed as CO_2 moves from the blood to the alveoli (Fig. 20.26). As a result of the buffering capacity of haemoglobin, the pH difference between arterial and venous blood is generally very small.

In invertebrates, the mechanisms involved in CO_2 transport appear similar to those in vertebrates. The enzyme carbonic anhydrase is found in the gills of polychaetes, gastropods, cephalopods and some crustaceans, where it assists in CO_2 elimination. However, in aquatic organisms, CO_2 transport is more complex because the ions involved, HCO_3^- and H^+, can be exchanged directly across gills using active membrane transport.

Temperature also affects CO_2 transport. An increase in temperature reduces the amount of H^+ that is buffered by haemoglobin. As a result, there is an accumulation of H^+ as free ions, which slows down the CO_2 hydration process, producing less HCO_3^-. The effects of temperature on H^+ buffering are particularly important in poikilothermic animals whose body temperatures vary with environmental fluctuations.

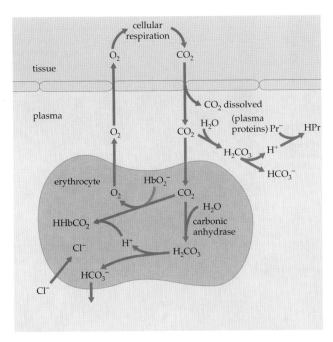

Fig. 20.26 The main pathways describing CO_2 transport in mammals. The diagram shows the movement of CO_2 into blood from tissues. The reverse occurs as CO_2-laden blood passes through the lungs

Control of ventilation

Ventilation with respect to oxygen consumption (convection requirement) is shown in Figure 20.27 for a

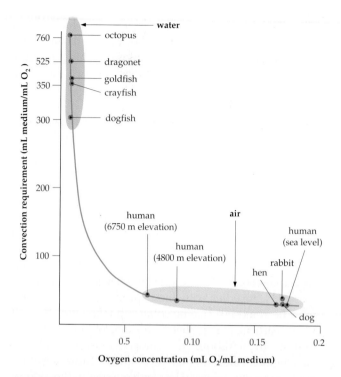

Fig. 20.27 Convection requirements against concentration of oxygen for a number of air and water breathers

For a given oxygen consumption, water breathers need to ventilate around 10–30 times more than do air breathers because of the low concentration of oxygen in water. Ventilation in water breathers is sensitive to changes in oxygen. Ventilation in air breathers is particularly sensitive to changes in carbon dioxide or pH.

In order to use O_2 or CO_2/pH levels to regulate ventilation, an animal must have some means of detecting these. Chemoreceptive tissue responsible for the predominant effects of CO_2 and pH on ventilation in mammals is located in the medulla region of the brain. These central receptors respond exclusively to changes in the pH of the cerebral interstitial fluid that surrounds them. The effects of CO_2 on these receptors is apparently mediated by its effect on pH. Since CO_2 can cross the blood–brain barrier but H^+ cannot, changes in pH of cerebral interstitial fluid reflect changes in the partial pressure of CO_2. The response of central receptors overrides input from other chemoreceptors, such as those in the carotid bodies.

In mammals, peripheral chemoreceptors sensitive to O_2 and CO_2/pH levels are found in the **carotid bodies**, tiny organs located at the bifurcation of the common carotid arteries into internal and external branches. The carotid body has an extensive capillary network and contains a specialised tissue composed of **glomus cells**, and efferent and afferent nerve endings (Fig. 20.28). Glomus cells contain a variety of neurotransmitter substances, which, when released in response to changes in partial pressures of O_2 or CO_2,

number of water and air breathers at different oxygen concentrations. It can be seen that the convection requirement (i.e. ventilation for a given oxygen consumption) in water breathers is some 10–30 times higher than that of air breathers. Also shown is the effect of altered oxygen concentration on the convection requirement for a person at sea level and at two elevated altitudes. As altitude increases, the barometric pressure decreases and the partial pressure, hence concentration, of oxygen also declines. Ventilation must therefore increase to maintain a given level of oxygen consumption.

Given the high ventilation rates of water breathers and the high solubility of CO_2 in water, CO_2 is easily removed as a waste product from water breathers. As a consequence, the difference in partial pressure of CO_2 between inspired and expired water is very small and, because CO_2 diffuses easily, internal partial pressures for CO_2 are not very different from external partial pressures. The main problem for water breathers is to obtain sufficient oxygen and it is not surprising that they regulate ventilation in response to changes in O_2.

Air breathers face a different problem. Tidal volume can be reduced because of the higher availability of O_2 in air compared with water. As a consequence, such animals tend to have higher internal levels of CO_2. The main difficulty for air breathers is to control the level of CO_2 to protect against pH imbalances. Ventilation in air breathers, although sensitive to changes in O_2, is found to be particularly sensitive to CO_2 or pH (which is directly affected by CO_2 levels).

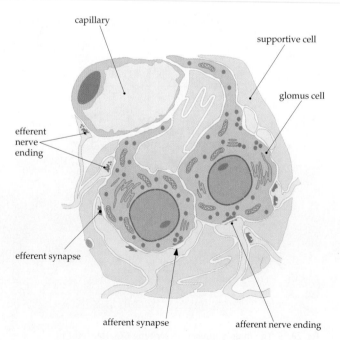

Fig. 20.28 In the carotid body, chemoreceptive cells known as glomus cells lie in close proximity to both blood capillaries and nerve endings

or change in pH, regulate the rate of firing of the chemoreceptor nerve endings. An increase in P_{CO_2} or a decrease in pH or P_{O_2} result in an increased rate of firing of the chemoreceptors. This is transmitted to the central nervous system and brings about an increase in ventilation. Carotid bodies are primarily concerned with transmitting information about O_2 levels, but also respond to changes in CO_2 and pH.

Central chemoreceptors respond to changes in pH (and thus P_{CO_2}). Peripheral chemoreceptors respond primarily to change in P_{O_2}.

In fishes, oxygen-sensitive chemoreceptors are located in the gills. An external receptor responds to changes in ambient oxygen levels whereas the internal receptors monitor blood oxygen content and flow. Both receptors can initiate compensatory changes in breathing and/or heart rate.

There are other receptors important in the control of ventilation, such as mechanoreceptors in fish gills, which monitor the water flow and are responsible for altering gill geometry and adjusting breathing pattern; lung stretch receptors in air-breathing vertebrates, which provide feedback about changes in lung volume, pressure or wall tension; and mechanical receptors, which are involved in defence reflexes that protect the respiratory tract.

Output from all receptors is fed via neural pathways to a central pattern generator that adjusts the breathing pattern to maintain homeostasis or to integrate breathing movements with other activities, such as feeding, talking or locomotion.

Summary

- Gas exchange in animals involves the supply of O_2 for cellular respiration and the removal of CO_2, a product of cellular respiration.
- Water contains 20 times less O_2 than air, which has important consequences for gas exchange.
- Gas exchange involves gaseous diffusion across membranes. It is dependent upon the concentration gradient, and the permeability, thickness and surface area of the membrane. Convection of external medium (ventilation) and internal medium (perfusion) maintains a favourable concentration gradient.
- Gills are outgrowths of the body surface that, when housed in gill chambers for protection, require ventilation.
- Air breathers have gas-exchange surfaces housed internally to minimise water loss. In most cases, this results in tidal breathing. Amphibians ventilate lungs using a positive pressure buccal pump. Reptiles, birds and mammals generate a negative pressure within the lungs by expanding the thorax, and in mammals lowering the diaphragm, to draw air into the lungs.
- Lungs have a tendency to collapse because of the surface tension at the air–liquid interface. This tendency can be overcome by surfactant.
- Airflow through the gas-exchange region of a bird lung is unidirectional and is cross current to flow in the blood capillaries.

- Insects conduct air directly to the tissues by means of tubular tracheae, the tips of which are fluid-filled. Air in tracheae is generally replaced by diffusion. During exercise, fluid is osmotically withdrawn from tracheole tips, decreasing the diffusion barrier.
- Blood transport of O_2 generally involves a respiratory pigment, which reversibly binds with O_2 according to the partial pressure of O_2. The affinity of haemoglobin for O_2 decreases with a decrease in pH and an increase in CO_2, temperature or organic phosphates.
- In air-breathing vertebrates, most CO_2 is hydrated to form carbonic acid, which dissociates and is transported as hydrogen carbonate ions. The enzyme carbonic anhydrase, located in erythrocytes, speeds up the hydration step. Haemoglobin acts as a buffer to counteract the increase in H^+ that occurs with the dissociation of carbonic acid into bicarbonate ions.
- In general, ventilation in water breathers is more sensitive to Po_2 whereas ventilation in air breathers is particularly sensitive to Pco_2. On the whole, changes in Po_2 are detected by peripheral chemoreceptors, whereas central chemoreceptors respond to changes in pH (as a consequence of changes in Pco_2).

keyterms

air capillaries (p. 532)
alveolar duct (p. 528)
alveoli (p. 528)
aspirating pump (p. 527)
Bohr effect (p. 538)
book lung (p. 534)
boundary layer (p. 520)
buccal-force pump (p. 526)
capacitance (p. 517)
carotid bodies (p. 540)

co-current flow (p. 523)
compliance (p. 528)
conductance (p. 518)
convection (p. 520)
countercurrent flow (p. 523)
cross-current exchange (p. 533)
cutaneous exchange (p. 521)
diaphragm (p. 531)
erythrocytes (p. 537)
gas bladders (p. 526)

gills (p. 522)
glomus cells (p. 540)
lungs (p. 526)
oxygen-carrying capacity (p. 537)
oxygen equilibrium curve (p. 537)
parabronchi (p. 532)
partial pressure (p. 517)
perfusion (p. 520)
pleural cavity (p. 531)
ram ventilation (p. 524)

respiratory pigments (p. 537)
Root effect (p. 538)
spiracles (p. 533)
surface tension (p. 530)
surfactant (p. 530)
tidal ventilation (p. 527)
tracheae (p. 533)
tracheole (p. 533)
ventilation (p. 520)

Review questions

1. What are the two systems that are responsible for the supply and removal of O_2 and CO_2 in most animals? What two processes are fundamental to the operation of these two systems?
2. Describe and explain with examples the difference between aspiration and buccal-force pump ventilation.
3. Contrast the mechanisms and efficiency of ventilation in a bird and a mammal.
4. Why are surfactants important in lungs?
5. Explain the difference between co-current, countercurrent and cross-current exchange.
6. Why are respiratory pigments important? What factors affect the transport of O_2?
7. (a) Explain the reason for the differences in the shape of oxygen equilibrium curves.
   (b) What advantages are offered by a respiratory pigment with a steeper curve and a lower P_{50}?
8. (a) Outline the transport of CO_2 in mammals.
   (b) How does the transport of CO_2 affect pH regulation in a mammal?

Extension questions

1. In air breathers, gas-exchange surfaces provide an avenue for water loss. Reduction of water loss was an important factor in the evolution of animals able to inhabit terrestrial environments. How is this reflected in the gas-exchange organs of terrestrial animals?
2. When compared with air, water is more viscous and dense, and is lower in O_2 content. As a result, what design constraints have been imposed on gas exchange in water breathers?
3. Surface tension forces at an air–water interface tend to collapse structures with a small radius. Discuss ways in which different groups of animals have overcome this problem while at the same time maintaining viable gas-exchange surfaces.
4. Water and air breathers differ in their convection requirements. What relevance is this to the control of ventilation?

Suggested further reading

Randall, D., Burggren, W., French, K. (1997). *Animal Physiology: Mechanisms and Adaptations*. New York: W. H. Freeman.

A standard text for the student studying animal function. This book goes a long way in explaining the basic mechanisms of animal physiology and the adaptations of animals and deals directly with the major concepts. The book is easy to read and well illustrated.

Schmidt-Nielsen K. (1997). *Animal Physiology: Adaptation and Environment*. 5th edn. Cambridge: Cambridge University Press.

Excellent reading—written with extreme clarity. Students in zoology will learn a lot from this book on animal function and at the same time will enjoy reading it.

West, J. B. (1999). *Respiratory Physiology—the Essentials*. 6th edn. Baltimore: Lippincott, Williams & Wilkins.

An advanced text for the student in physiology with a particular interest in respiratory physiology. This book is very readable, clear and didactic.

Circulation

Cells must exchange materials with their surroundings to stay alive. As we have seen in previous chapters, oxygen (O_2) must be supplied to mitochondria within cells, and carbon dioxide (CO_2) must be eliminated; nutrients and waste products of cellular metabolism must similarly move into and out of cells. These substances are transported between the exchange surfaces and the cells of the body. Within cells, movement occurs by diffusion and by a form of intracellular circulation, cytoplasmic streaming. Diffusion also occurs from cell to cell, but it is a slow process that can transport substances effectively only over relatively short distances, usually less than a millimetre or so. Thus, the only animals that rely exclusively on diffusion are usually small or thin, for example, flatworms. Larger animals require circulatory systems that speed up the internal transport of materials.

Circulation involves *convection*, the bulk movement of fluid and any substances it contains. It brings dissolved substances close enough to cells or exchange surfaces so that diffusion can be effective, and it enhances diffusion by reducing the thickness of the *boundary layer*, which is a stagnant layer of fluid near the exchange surface through which diffusion occurs (Chapter 20).

Circulatory systems have three general functions.
1. They provide mass transport of substances and blood cells throughout the body.
2. They transport heat between different parts of the body, for example, to or from the external surface.
3. They allow transmission of force, which is used, for example, for locomotion by many worms and molluscs.

A cardiovascular system that provides sufficient convective transport of O_2 almost invariably is more than adequate in providing for the other functions. Therefore, levels of respiratory gases in the blood are important factors in the regulation of cardiovascular function (Chapter 20).

Circulatory systems provide internal convective transport of substances, cells and heat, and transmission of force.

Types of circulatory systems

We normally think of a circulatory system circulating blood, but there are many animal systems that circulate other body fluids. Extracellular fluid compartments include coelomic spaces, vessels (blood and lymphatic) and interstitial spaces between cells. The arrangements of these fluid compartments, and the patterns of fluid movement through them, are extremely diverse, especially among invertebrates.

Circulatory systems are often described as being *open* or *closed*, referring to whether or not the circulated fluid is always contained within a system of vessels. Of course, to carry out their exchange functions, circulatory systems are never totally 'closed'. Open circulatory systems, found in many invertebrates such as crustaceans (Fig. 21.1a) and some molluscs, are those in which a heart or hearts pump fluid through vessels that finally open into interstitial spaces. The fluid then percolates among the cells and makes its way rather slowly back to the heart, often by way of gills that oxygenate the fluid on its return journey. Thus, in open systems, the circulated fluid is indistinguishable from interstitial fluid.

Closed circulatory systems occur in a few invertebrate groups, for example, annelid worms (Fig. 21.1b) and cephalopod molluscs, and in vertebrates. In these animals, blood pumped from the heart (or hearts) passes around the body within a series of branching **arteries** leading to thin-walled **capillaries** and then into **veins**, which return blood to the heart. In closed systems, therefore, the circulated fluid, **blood**, is anatomically separated from interstitial fluid throughout its circuit.

Differences in the organisation of circulatory systems generally reflect differences in the level of activity of animals. It is interesting to consider the costs and benefits of having a closed circulatory system. Open circulatory systems do not circulate fluid very

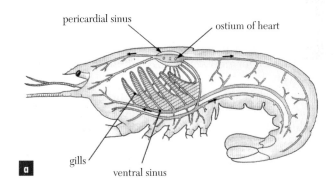

a

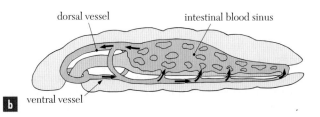

b

Fig. 21.1 (a) Open circulation in a lobster. Blood enters the heart from the pericardial sinus through ostia and is pumped through the body in large vessels. It leaves the vessels, percolates through the tissues and enters the ventral sinus, leading back to the heart by way of the gills. **(b)** Closed circulation in a primitive annelid worm. Blood tends to move anteriorly in the dorsal vessel and posteriorly in the ventral vessel. It surrounds the gut and exchanges material in a large intestinal sinus before returning to the dorsal vessel

rapidly, so only low rates of metabolism can be supported. More active animals with higher metabolic rates are better served by closed systems in which blood flows rapidly throughout the body within a system of smooth vessels. However, in a closed circulatory system more energy is required because the heart must pump more strongly to overcome the increased resistance to flow as blood moves through small vessels.

I n open systems, the circulated fluid is indistinguishable from interstitial fluid. In closed systems, the circulated fluid, blood, is enclosed in a system of vessels and may be circulated more rapidly.

Vertebrate circulatory systems

From a comparative point of view, the general arrangement of major arteries in different vertebrate groups is believed to reflect changes that occurred during their evolution. Similarly, the changes that occur during development of the circulatory system in embryonic vertebrates suggest a history of evolutionary change (Box 21.1). Within vertebrates, there is a trend towards reduction in the number of major arteries. In early vertebrates, the circulatory system is thought to have included a series of gill vessels (arches) supplied with blood directly from the heart (Fig. 21.2). This primitive arrangement is only slightly different in modern fishes, with the loss of some of the anterior gill arches. With the evolution of lungs in terrestrial vertebrates, gills were no longer necessary. Most gill arches were progressively lost; those that remained became involved with other functions. Some became the major systemic arteries, such as the aorta, and others supply blood to the lungs (Fig. 21.2).

D uring the evolution of vertebrates, there has been a trend towards reduction in the number of major arteries.

Dramatic changes also occurred in the pattern of blood flow through the heart. In fishes, the heart is a series of four chambers (**sinus venosus**, **atrium**, **ventricle**, and **conus arteriosus**). The chambers are separated by valves that prevent reverse flow; the chambers fill and empty in sequence, moving the blood forward (Fig. 21.3). The sequence of contraction begins in the sinus venosus, but most of the propulsive force originates in the highly muscular ventricle. Circulation in fishes is a single circuit, with the heart, gills and body tissues supplied with blood in that order (Fig. 21.4). Thus, blood pressures in the gas-exchange organs (gills) are greater than in the rest of the body. The reverse is

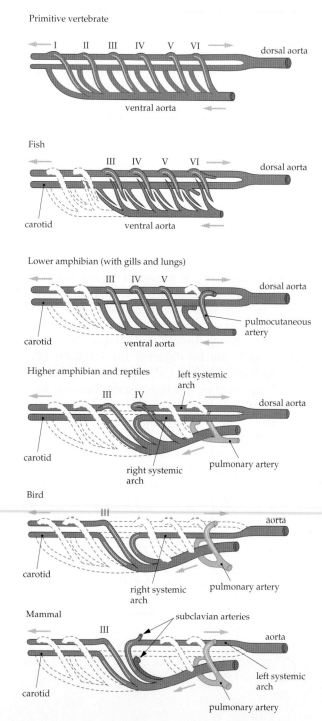

Fig. 21.2 Arrangement of the major arteries in vertebrate animals. All of the vessels are thought to have evolved from six pairs of aortic arches in primitive vertebrates. Originally these arches perfused the gills, but beginning with the first air-breathing amphibians, arch VI became the pulmonary artery. Notice that the aorta in mammals is derived from the left aortic arch IV, while in birds it is derived from the right aortic arch IV, indicative of the independent origins of birds and mammals from reptilian ancestors that had both arches

true in other vertebrates that have lungs. These have a low-pressure **pulmonary circuit** to the lungs and a higher pressure **systemic circuit** to the rest of the body.

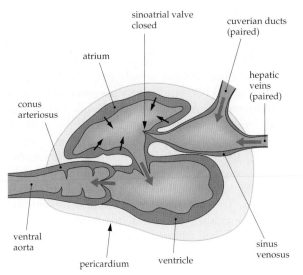

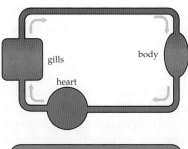

a Fish

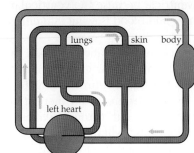

b Anuran amphibian

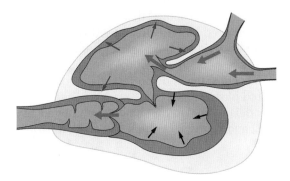

b Ventricular contraction

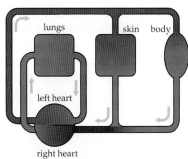

c Non-crocodilian reptile

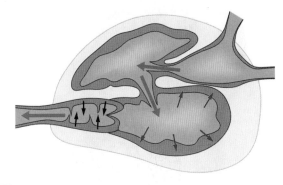

c Conal contraction

Fig. 21.3 The sequence of contraction in a fish's heart propels blood from the atrium to the ventricle **(a)**, to the conus arteriosus **(b)**, and to the ventral aorta **(c)**. The pericardial compartment around the heart chambers is rigid, so contraction of one of the chambers tends to expand the others

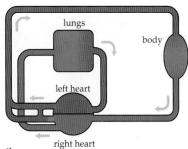

d Crocodilian reptile

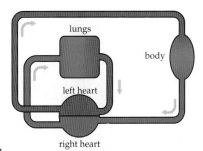

e Bird, mammal

Fig. 21.4 Patterns of circulation in vertebrate animals. **(a)** Fish have a single circuit with complete separation of deoxygenated (blue) and oxygenated (red) blood. **(b–d)** Amphibians and reptiles have a partly divided double circuit. **(e)** Birds and mammals have a completely divided double circuit

Although living lungfishes are not ancestral to amphibians, their hearts have characteristics that offer insight into what might have been the condition in early amphibious fishes. In the heart of the Australian lungfish, *Neoceratodus*, for example, the atrium is almost completely divided by a septum and the ventricle and conus arteriosus are also partly divided by septa (Fig. 21.5). A large fibrous plug attached to the septa acts as a valve between the atrium and ventricle. Although incomplete, the dividing septa tend to separate blood from different origins. Surprisingly little mixing of oxygenated and deoxygenated blood occurs in the ventricle because the ventricular chamber is like a sponge, and blood flowing into either side tends not to mix.

Functional separation of blood through the heart of a lungfish makes gas exchange more efficient. Poorly oxygenated blood returns from the body to the right side of the atrium. From there it passes through the ventricle and out through channels leading to the posterior gills, where it is oxygenated. Meanwhile, oxygenated blood from the lung enters the left side of the atrium, passes through the ventricle and out through another channel to the anterior gill arches. The anterior arches are simple tubes without gill capillaries, so oxygen-rich blood from the lung can be distributed to the body without the danger of loss of oxygen to the surrounding water. Lungfishes can use their lungs or gills for gas exchange and the pattern of blood flow through the heart can change, depending on which gas exchanger is in use.

The hearts of modern amphibians have two separate atria, the right atrium, which collects deoxygenated blood from the body (and some oxygenated blood from the skin), and the left atrium, which collects oxygenated blood from the lungs (Fig. 21.6). Both atria drain through one valved opening into a single ventricle. Although these animals have no trace of a ventricular septum, mixing in the ventricle is minimised by folds in its internal spongy wall and by a spiral fold in the conus arteriosus. Thus, deoxygenated blood is moved preferentially towards the pulmocutaneous arteries bound for the lung and skin, where it can obtain oxygen, and most oxygenated blood is moved to the body (Fig. 21.4).

As in amphibian hearts, most reptilian hearts have two atria and a single ventricle. However, the ventricle is partially divided into three chambers by two ridges. Recent studies have shown that mixing of oxygenated and deoxygenated blood is minimised because, during contraction, the ridges close against the walls, isolating the two types of blood into separate chambers (Fig. 21.7). Thus, the body preferentially receives well-oxygenated blood (Fig. 21.4).

Crocodilians are unique among reptiles in having both atria and ventricles completely divided. However,

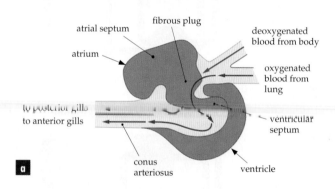

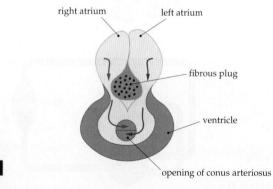

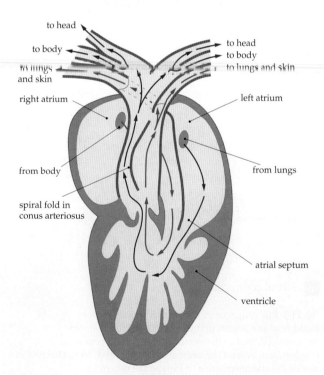

Fig. 21.5 (a) Longitudinal and **(b)** transverse sections through the heart of the Australian lungfish, *Neoceratodus forsteri*. The blood from the body and lung divides on either side of the atrial septum and flows past the fibrous plug that acts as a valve between the atrium and ventricle. A partial ventricular septum hangs down from the ventricle roof. Oxygenated and deoxygenated bloods are largely separated as they enter the divided conus arteriosus

Fig. 21.6 The three-chambered heart of a frog. Oxygenated and deoxygenated bloods are separated in two atria, but there is potentially some mixing of blood in the ventricle. However, mixing is minimised by separation of blood in the spongy walls of the ventricle and also by streaming of blood in the conus arteriosus. Deoxygenated blood is sent preferentially to the lungs and skin

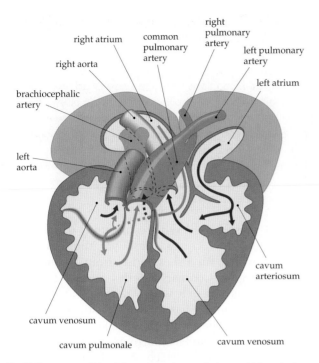

Fig. 21.7 Pattern of blood flow through a turtle heart. Although the ventricle consists of three incompletely divided chambers, oxygenated and deoxygenated blood remain somewhat separate because of laminar flow and a valving effect during the cardiac cycle

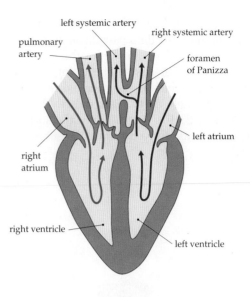

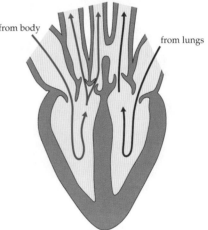

Fig. 21.8 The crocodilian heart. **(a)** Normally there is no mixing of oxygenated and deoxygenated blood because the valve at the beginning of the left systemic artery remains closed, and oxygenated blood flows through the foramen of Panizza. **(b)** During long dives, however, contraction of the pulmonary artery causes right ventricular pressure to rise and open the valve. Thus, some of the blood from the body bypasses the lung and returns to the body through the left systemic arch

the left aortic arch originates from the right ventricle along with the pulmonary artery, and the right aortic arch arises from the left ventricle (Fig. 21.8). This odd arrangement potentially recirculates deoxygenated blood to the body through the left aortic arch. However, when a crocodile is breathing, pressure is lower in pulmonary vessels and therefore in the right ventricle, so the valve between the right ventricle and the left aortic arch remains closed. Blood under high pressure from the left ventricle enters the left aortic arch through the foramen of Panizza, thus ensuring separation of the two bloods (Figs 21.4, 21.8). Recent work on the Australian salt-water crocodile shows that when the animal submerges, resistance increases in the valve leading to the pulmonary artery, so right ventricular pressure increases, thereby opening the valve into the left aortic arch. Thus, blood is able to bypass the lung and pass back to the body (Fig. 21.8).

The conclusions of these and other recent studies are that mixing of oxygenated and deoxygenated blood through the heart in reptiles and probably amphibians is regulated and that it is advantageous under certain circumstances. Fishes have relatively low metabolic rates, but they breathe water that is very low in O_2 content, so they need to breathe fairly continuously. Amphibians and reptiles occupy an interesting position. They can breathe air, which is rich in O_2, but they are ectothermic and have relatively low metabolic rates. This means that they do not need to breathe

continuously to provide sufficient O_2. When they are not breathing, such as during long periods under water, there is no advantage in using energy to pump all of the blood through the pulmonary circulation. But when they do breathe, there is an advantage in passing deoxygenated blood through the lungs at an increased rate to pick up O_2. Hence, the ability to shunt blood either towards or away from the pulmonary circuit is an advantage for amphibians and reptiles.

The same argument does not hold for birds and mammals. Although they also breathe air, birds and mammals are endothermic and usually ventilate their lungs continuously to provide sufficient O_2 for their needs. There is no mixing of oxygenated and deoxygenated blood in their four-chambered hearts

(Fig. 21.9) and all blood that passes through the systemic circuit must then pass through the pulmonary circuit (Fig. 21.4). The sinus venosus has been reduced to a small node of tissue in the right atrium, the sinoatrial node, which is the pacemaker region of the heart, and a single aorta leads directly from the left ventricle (Fig. 21.10). Interestingly, the aortae of birds and mammals are derived from different aortic arches of their reptilian ancestors, birds retaining the fourth right arch and mammals retaining the fourth left arch (Fig. 21.3).

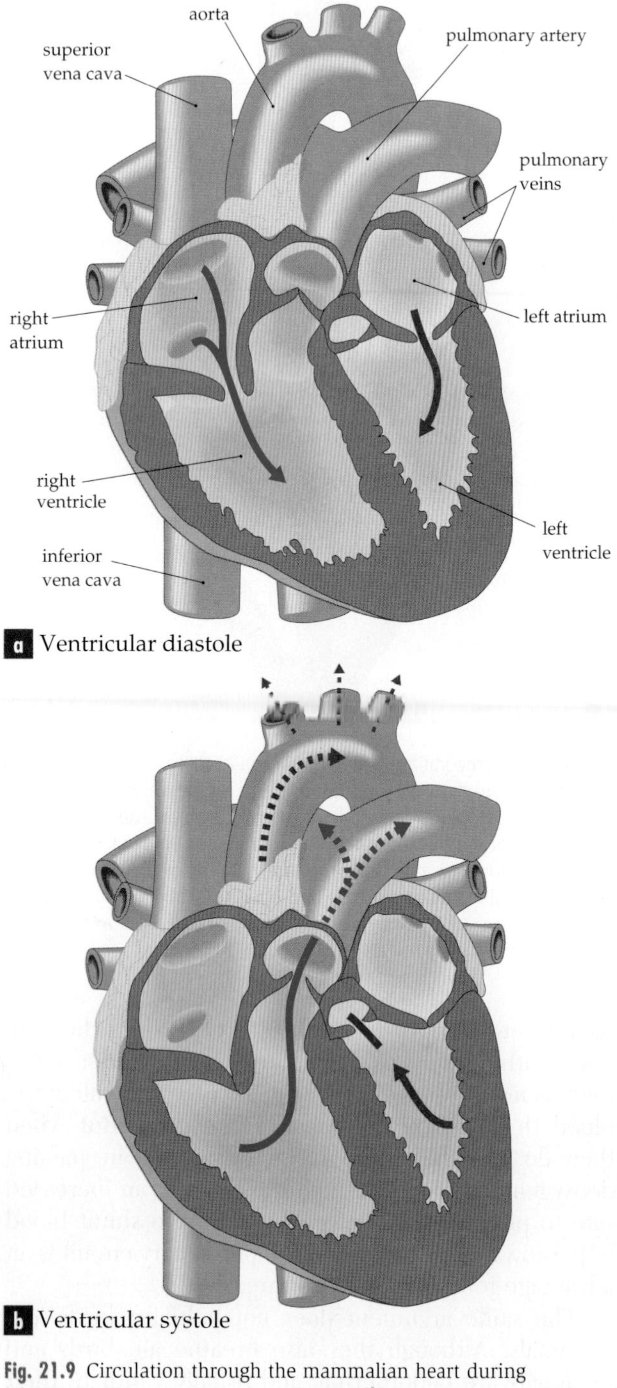

In fishes, blood flows through a single circuit, from heart to gills to body. In amphibians and most reptiles, atria are divided, flow is regulated to some degree through an undivided ventricle and blood may be shunted towards or away from the lungs. In birds and mammals, there is complete separation of oxygenated and deoxygenated blood, and all blood passes through the pulmonary circuit and then the systemic circuit.

Heart muscle itself requires a supply of nutrients and oxygen. It receives these via the **coronary arteries**, which, in fishes, arise from the arteries leaving the gills. Coronary arteries are not found in amphibians and are only weakly developed in reptiles, but their heart muscle has a spongy construction, which permits blood to permeate the tissue. In contrast, the heart walls of birds and mammals are so compact that a coronary blood supply, which arises from the base of the aorta, is essential.

The mammalian heart

The mammalian heart consists of four separate chambers that alternately contract, **systole** (pronounced sis-toh-lee) and relax, **diastole** (pronounced dye-as-toh-lee; Fig. 21.10). Blood from the body enters the right atrium, moves through the **tricuspid valve** into the right ventricle, and then through the pulmonary **semilunar valve** to the lung. At the same time, blood returning from the lung enters the left atrium, goes through the **mitral valve** into the left ventricle, and then out through the aortic semilunar valve into the **aorta**. The tricuspid and mitral valves are strengthened by fibrous attachments, **chordae tendinae**, to the ventricular wall, but the semilunar valves are not. The closing of the two sets of valves causes the familiar heart sounds, 'lubb–dupp'. 'Lubb' is due to simultaneous closure of the tricuspid and mitral valves when the ventricles begin to contract. The semilunar valves open silently and blood is forced into the two major arteries. At the end of ventricular systole, pressure in the arteries causes the semilunar valves to abruptly close, and the 'dupp' sound is heard.

In humans, atrial contraction does not contribute greatly to filling of the ventricles. One reason is that there are no valves to prevent blood from passing back into the veins as there are in fishes. Another is that the ventricles largely fill themselves by creating a suction during diastole, when the relaxing ventricular walls passively expand. Indeed, in the human heart, if disease

a Ventricular diastole

(labels: aorta, superior vena cava, pulmonary artery, pulmonary veins, right atrium, left atrium, right ventricle, left ventricle, inferior vena cava)

b Ventricular systole

Fig. 21.9 Circulation through the mammalian heart during **(a)** ventricular relaxation (diastole) and **(b)** contraction (systole). There is complete separation of deoxygenated blood in the right heart from oxygenated blood in the left heart. The heart is essentially the same in birds and mammals

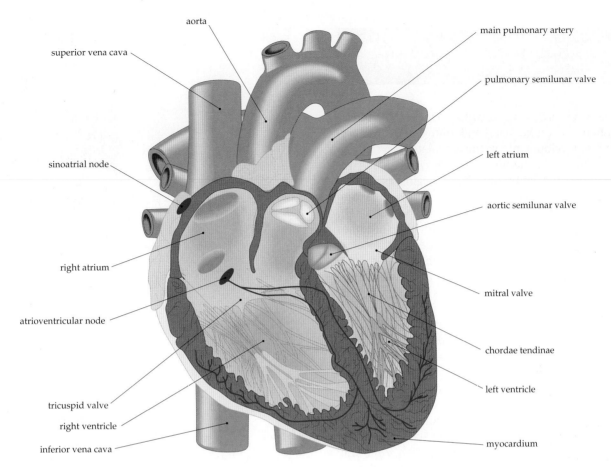

Fig. 21.10 The human heart

causes the atria to stop contracting entirely, the ventricles continue to fill almost normally and there is little change in circulation. It is perhaps more surprising that an adequate circulation can be maintained even if the right ventricle also ceases to function.

The heart wall is composed of three layers of cardiac muscle, collectively called the **myocardium**. Both ventricles pump the same amount of blood but the right ventricle wall is thinner than the left because pulmonary pressure is much less than systemic pressure (Fig. 21.10). The bundles of muscle fibres run spirally over the ventricles and, when they contract, blood is squeezed from the bottom of the heart (the apex) up towards the major arteries.

Blood is pumped by sequential contractions of the atria and ventricles. Valves between the atria, ventricles and major arteries prevent back flow and cause the familiar heart sounds when they close.

Electrical activity

The heartbeat of many invertebrate hearts is **neurogenic**, that is, the heart contracts only in response to stimulation by a nerve. In contrast, the heartbeat of vertebrates is **myogenic**, that is, it originates in the heart muscle itself. If the heart of a vertebrate is removed from the body, it continues to beat; even if the heart is cut apart, the individual cells spontaneously contract, but at different rates. For the whole heart to function correctly, the contractions of the atria and ventricles must be co-ordinated. Both atria must contract together and then, after a short delay, the ventricles contract.

The mammalian heart cycle is initiated in the right atrium at the **sinoatrial node**, or 'pacemaker' (Fig. 21.10). This structure is derived from the sinus venosus, which initiated the sequence of contraction in early vertebrate hearts. The sinoatrial node is a small group of non-contractile cardiac muscle cells that nevertheless show spontaneous rhythmic electrical activity at a rate faster than those of all other cells of the heart. The pacemaker cells initiate an action potential, which is a change in the electrical polarity of their walls that, like a ripple in a pond, sweeps through all of the cells of the atria (Fig. 21.11). This wave of depolarisation moves at about 30 cm/sec, causing both atria to contract and push blood into the ventricles. The wave is prevented from entering the ventricles by a non-conducting layer, except at the **atrioventricular node**,

BOX 21.1 Embryonic circulation in mammals and birds

There are obvious similarities between the phylo-genetic (evolutionary) and ontogenetic (embry-onic) development of the circulatory system in vertebrates (Chapter 40). In mammals, for example, the embryonic heart begins as a four-chambered tube. During growth and differentiation, the tube bends upon itself, forming an S-shape, and gradually becomes divided into pulmonary (right side) and systemic (left side) circuits by the appearance of intra-atrial and intraventricular septa (Fig. a). However, the separation of embryonic circuits is not complete. At this stage, the lungs are not functional and all exchanges occur across the placenta (Fig. b). The fluid-filled lungs offer a high resistance to blood flow through pulmonary vessels and therefore most blood ejected from the right ventricle is forced through the ductus arteriosus, which connects the pulmonary artery directly to the aorta. Meanwhile, blood returning to the heart from the body and placenta into the right atrium moves across to the left atrium through a one-way valve, the foramen ovale. These two pathways enable blood to bypass the lungs and supply the most oxygenated blood to the body.

At birth, however, dramatic changes occur. When a newborn mammal inflates its lungs with its first gasps, pulmonary vascular resistance drops and more blood flows to the lungs. As this blood returns from the lungs, pressure in the left atrium increases, closing the foramen ovale. The ductus arteriosus gradually constricts and closes (as do the umbilical vessels to and from the placenta). Failure of the ductus arteriosus or the foramen ovale to close results in mixing of blood between the pulmonary and systemic circuits, and circulation of inadequately oxygenated blood explains the 'blue' appearance of human babies born with these problems.

Mammals have a single extraembryonic cir-culation to the placenta but, in embryonic birds, there are two circulatory loops—one goes to the yolk sac and the other to the chorioallantois (Fig. c). Yolk sac circulation is mainly involved with the uptake of nutrients from the yolk, while the chorioallantoic circulation largely involves gas exchange and control of fluid volumes in the egg compartments. Near hatching, circulatory changes occur in a way similar to those in

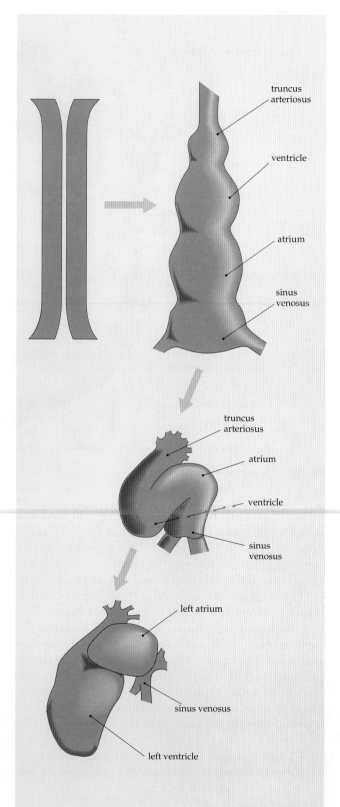

(a) The development of the human heart. In the early embryo, blood forms within two parallel tubes. These fuse into the four-chambered linear heart, which begins pumping blood. Later the heart flexes upon itself, and divides into two halves again, resulting in two, two-chambered hearts in parallel

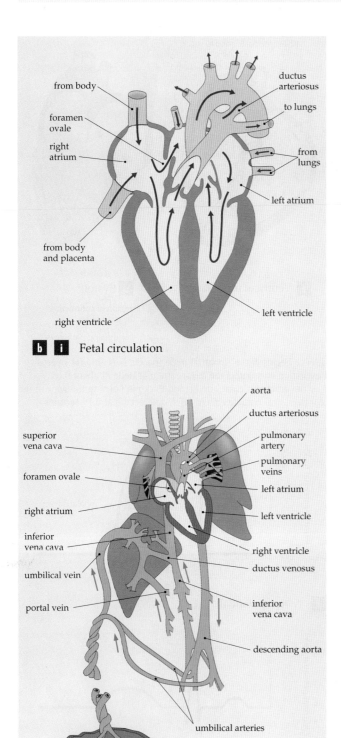

b i Fetal circulation

ii Placenta

(b) (i) In human fetal circulation, most of the blood flowing through the heart effectively bypasses the lungs by shunting through the ductus arteriosus and foramen ovale.

(ii) Circulation to the placenta occurs through two umbilical arteries branching from the femoral arteries, and it returns via the umbilical vein, which feeds into the liver and ductus venosus

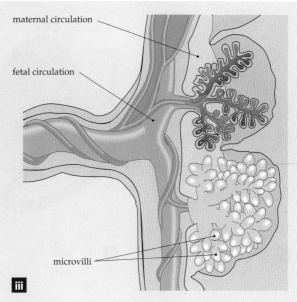

iii

(iii) At the placenta, fetal blood enters small microvilli, where it comes into equilibrium with maternal blood

mammals, only much more slowly. The yolk sac is gradually absorbed into the body cavity of the chick over several days. During the last day of incubation, the bird begins breathing gas that has collected in the air cell inside the shell. As it gradually aerates its lungs, blood flow slowly shifts away from the chorioallantois towards the pulmonary vessels. At hatching, therefore, the lungs are functional and chorioallantoic circulation has all but stopped.

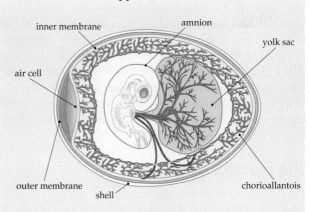

(c) Circulation in the embryonic chick. The chorioallantois is a well-vascularised embryonic membrane that functions in gas exchange and solute regulation while the yolk sac blood vessels supply nutrients to the embryo

a group of specialised muscle cells located at the junction between the right atrium and ventricle. At this node, the speed of conduction of the action potential is slowed to about 5 cm/sec due to the particular structure of the muscle cells (they have very small diameters and

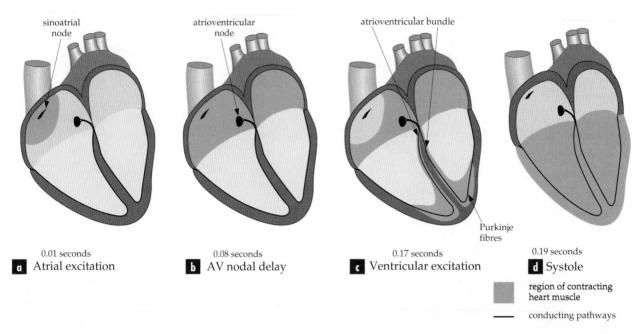

0.01 seconds	0.08 seconds	0.17 seconds	0.19 seconds
a Atrial excitation	**b** AV nodal delay	**c** Ventricular excitation	**d** Systole

region of contracting heart muscle

conducting pathways

Fig. 21.11 The pattern of electrical excitation in the human heart. **(a)** The sequence begins at the sinoatrial node (pacemaker) and **(b)** sweeps over the atria in about 0.08 seconds, causing atrial contraction. The wave of excitation encounters the atrioventricular node at about 0.03 seconds but it is delayed for 0.13 seconds before **(c)** being rapidly conducted by the atrioventricular bundle and Purkinje fibres to the apex of the ventricles. Ventricular excitation begins at 0.17 seconds, well after the atrial contraction is completed and **(d)** systole occurs at 0.19 seconds

few electrical connections). The slow passage through this node allows time for the atrial contraction to finish before ventricular contraction begins.

After leaving the atrioventricular node, the action potential is conducted very quickly (1500 cm/sec) through the **atrioventricular bundle** (bundle of His), which runs down the interventricular septum towards the apex of the heart. In this case, the conducting muscle fibres have large diameters and many electrical connections, resulting in rapid conduction velocities. On its way, the atrioventricular bundle first divides into two main branches, which lead to both ventricles, and then into many **Purkinje fibres**, which invade ventricular muscle. This anatomical arrangement ensures a synchronous ventricular contraction, which begins at the apex of the heart.

The electrocardiogram

Cardiac muscle fibres are similar to skeletal muscle fibres in being striated and covered with an electrically active membrane, the sarcolemma (Chapter 27). When stimulated, the membrane produces an action potential during which its polarity reverses briefly (depolarisation) and then returns to normal (repolarisation). In skeletal muscle, an action potential is complete in a fraction of a millisecond, but the cardiac action potential lasts for about 250 milliseconds, time enough for the heart to complete its contraction and eject blood (Fig. 21.12).

The electrical activity that sweeps over the heart during contraction is strong because it involves a large

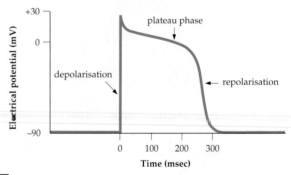

a Action potential

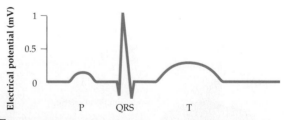

b Electrocardiogram

Fig. 21.12 (a) An action potential recorded in a single ventricular muscle fibre. The cell membrane is initially polarised at about −90 millivolts. When stimulated, it quickly reverses polarity, remains depolarised for 250 milliseconds and then quickly repolarises. **(b)** These electrical changes can be detected by electrodes placed in contact with the skin, yielding the typical electrocardiogram

number of cells operating in unison. Some electrical current generated by the heart is passively conducted throughout the salty fluids of the body. This 'leakage' can be detected by electrodes placed on the body surface, resulting in an **electrocardiogram** (ECG or EKG) (Fig. 21.12). The ECG is usually divided into three parts: the P wave, QRS complex and T wave. The P wave represents atrial depolarisation, the QRS complex results from ventricular depolarisation, and the T wave comes from ventricular repolarisation. Atrial repolarisation is obscured by the QRS complex. The ECG is of enormous diagnostic value for cardiac physicians—abnormalities in the timing of this sequence can reveal damage to the conducting pathway, which results in poor co-ordination of cardiac events and reduced pumping ability.

> The heartbeat is highly co-ordinated. Contraction is initiated in the pacemaker region, conducted through atria and, via the atrioventricular node, into the ventricles. Rapid conduction through the atrioventricular bundle triggers a co-ordinated contraction of the ventricles.

Conducting vessels

Blood circulates through the body in arteries, arterioles, capillaries, venules and veins. In addition, there is a system of lymphatic vessels, which returns tissue fluid to the circulation.

Arteries carry blood away from the heart, either to the body in the systemic arteries or to the lungs in the pulmonary arteries. Artery walls are relatively thick, with four layers of tissue, including an inner endothelial layer surrounded by a thin connective tissue layer, a thick layer of smooth muscle and elastic connective tissue, and a thinner fibrous outer layer (Fig. 21.13). Elastic membranes may separate these layers. The abundance of elastin helps smooth out pulsations and maintains a high blood pressure. During systole, arteries are passively expanded by the higher systolic blood pressure. Thus, some of the energy expended by the heart during systole is stored in the stretched arterial walls rather than in greatly increasing arterial blood pressure. The outer layer of fibrous collagen tissue limits the maximum diameter of the artery and prevents overexpansion. During diastole, the stored energy is transferred back to the blood and maintains blood pressure. The result is that pressure in the arterial system remains relatively high and fluctuations are strongly damped (Fig. 21.14). If the walls become less elastic with age or disease, blood pressure fluctuations can become extreme. German organ makers used the same principle (a 'Windkessel' chamber) to reduce pressure oscillations from the bellows; thus the arteries are sometimes called **windkessel vessels**.

Veins collect blood from capillaries and return it to the heart. They have a thin layer of smooth muscle

between the outer connective tissue and inner endothelium (Fig. 21.13). This layer is thicker in larger veins, particularly in veins of the legs, which may

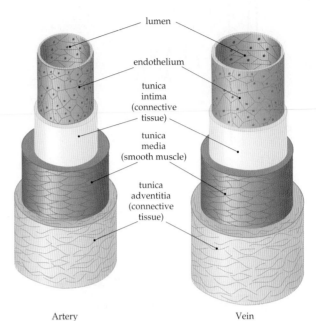

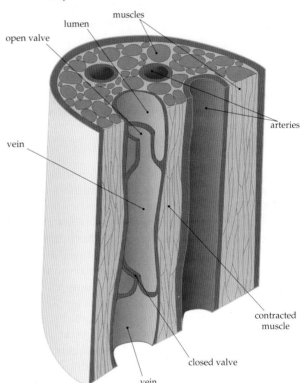

Fig. 21.13 Arteries and veins consist of four layers of cells. Outside of a single layer of endothelial cells is the tunica intima (collagenous and elastic fibres), the tunica media (elastic and muscle fibres), and tunica adventitia (tough collagenous fibres). The walls of arteries are thicker and have more elastic and muscle fibres than those of veins at any location in the body because the blood in arteries is under higher pressure. Veins often contain valves that permit one-way flow only and create a pumping action when external muscles compress the segment between adjacent valves

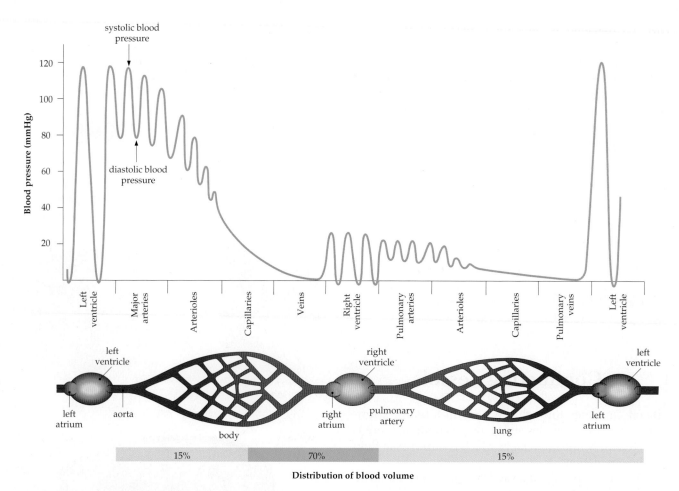

Distribution of blood volume

Fig. 21.14 Blood pressure and volume in the systemic and pulmonary circulation. Large pressure changes during systole and diastole in the ventricle are dampened by the elastic arteries. The major drop in blood pressure occurs in the arterioles and capillaries. There is not much change in pressure along the veins because they are numerous and offer little resistance to blood flow. Although the capillaries normally contain only 5% of the total blood volume, they represent over 90% of the vascular surface area

become as muscular as arteries. Interestingly, when veins are transplanted to replace diseased coronary arteries, they grow to resemble arteries in a few weeks.

Venous blood pressure is low and is not sufficient on its own to raise blood back to the heart against gravity. Fortunately, return of venous blood is assisted by external compression of veins due to contraction of the adjacent muscles, and intravenous valves that prevent reverse flow (Fig. 21.13). Veins are distensible and, if venous pumping is prevented, for example, in soldiers standing completely motionless at attention, blood may pool in the legs and consequently reduce blood flow to the head. This effect is intensified in hot weather because blood flow to the skin is high, and it is not unusual for a soldier to faint under these conditions. Most of the blood volume (about 75%) lies in the veins, which are referred to as **capacitance vessels**. Some of this volume can be shifted to other parts of the circulation when required, for example, to fill the vascular beds of active tissues, or to maintain arterial blood pressure following excessive bleeding. Contraction of veins increases the blood supply to the

heart, thereby increasing cardiac output and arterial pressure.

Arteries carry blood away from the heart. Their elastic muscular walls store some pressure energy and reduce fluctuations in arterial blood pressure. Veins return blood to the heart. At any time, most blood is held in the veins.

The microcirculation

Vessels between the smallest arteries and venules constitute the **microcirculation** (Fig. 21.15). These include **arterioles**, which are completely surrounded by smooth muscle, **metarterioles**, which are surrounded by discontinuous smooth muscle, capillaries, which have no muscle except for **precapillary sphincters**, and non-muscular **venules**, which are a single layer of endothelial cells surrounded by a layer of collagen.

Arterioles are very small arteries (about 30 μm diameter) that lead to capillaries. Arterioles impose the major resistance to blood flow through the circulation

BOX 21.2 Circulation and thermoregulation

Circulating blood can transport heat between different parts of the body and is instrumental in regulating body temperature (Chapter 29). In amphibians and reptiles that bask in the sun, heat from warmed skin is transferred to the body core, much as a solar water heater provides hot water to a house. Basking birds and mammals can also obtain solar heat this way, but they usually use their circulatory systems to control heat loss.

When a mammal is overheated, blood flow is increased to the skin and other evaporative surfaces where excess heat can be lost; when metabolic heat needs to be retained, circulation to the skin and appendages is greatly reduced. Many native people living in cold environments show cardiovascular adaptations to cold weather. Aboriginal Australians undergo pronounced vasoconstriction of their skin vessels and sleep comfortably on very cold nights. The hands of Alaskan Inuit (Eskimos) are often exposed to freezing temperatures without the pain and distress felt by Europeans. Periodic bouts of vasodilation, during which a pulse of warm blood flows through their extremities, protects them from frostbite.

Circulation to the limbs in some birds and mammals is adapted to conserve heat by countercurrent exchange, where heat is transferred between two bloodstreams flowing in opposite directions. As arteries enter a limb, they run parallel with veins returning blood to the body. Warm arterial blood loses most of its heat to cooler venous blood, which carries heat back into the body rather than out into the limb. The best examples of this mechanism come from aquatic birds and mammals with fins or flippers. Blood flows to the fins of marine mammals in central arteries that are completely surrounded by veins (Fig. a). When in cold water, arterial heat moves to the veins and is retained. However, when an animal is in warm water, metabolic heat may need to be lost. At such times, the central veins close and superficial veins open, eliminating the countercurrent exchange and allowing venous blood to lose heat to the water.

In other animals, instead of having single arteries surrounded by veins, the arteries divide into a series of small parallel vessels, which interdigitate with branching veins before entering the limb. This arrangement increases the surface area available for countercurrent exchanges so that transfer is virtually complete. In cross-section, the vessel walls of this structure resemble a net (Fig. c), so it is usually called a **rete** or *rete mirabile*, which means 'wonderful net'. Retes also occur in the cranial circulation of some mammals and birds, where they maintain

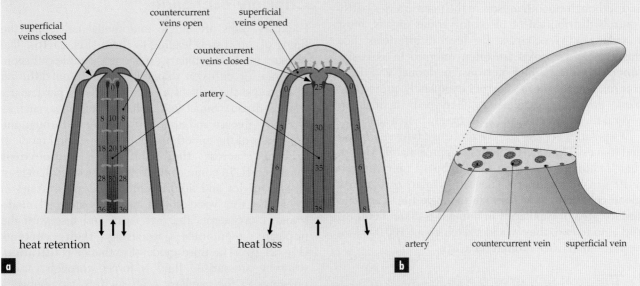

(a) Countercurrent heat exchange in the fin of a cetacean. When heat needs to be retained, warm blood from the body core loses its heat to the venous blood returning to the body in countercurrent veins that surround each artery. When heat needs to be lost, the blood returns in the superficial veins, where the heat is transferred to the sea water. The fin circulation acts as a 'thermal window', which can be opened or closed to heat. **(b)** Cross-section through a fin, showing the location of vessels described in **(a)**

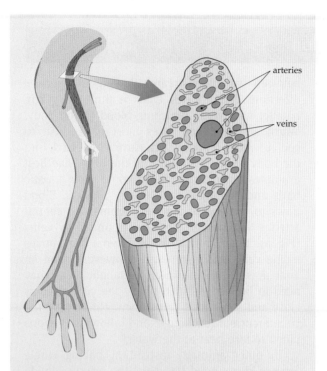

(c) Cross-section through the rete in the limb of a small mammal (the loris). The arteries and veins interdigitate and facilitate countercurrent heat exchange in the vessels at the base of the limb

brain temperature if the animal is overheated. Certain 'warm-blooded', fast-swimming fishes, such as tuna and some sharks, also use retes to isolate their warm active muscles from the cold blood returning from their gills.

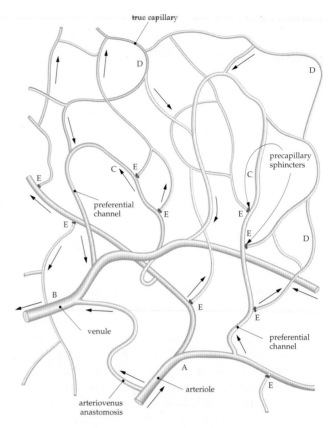

Fig. 21.15 The microcirculation. Blood enters the tissue in fine, muscle bound, arterioles (A) and is collected in venules (B). At rest, most of the blood travels through preferential channels (C), but it is allowed into the true capillaries (D) when the precapillary sphincters (E) open in response to increased cellular activity. In some tissues, blood can bypass the exchange area through an arteriovenous anastomosis.

and, although smooth muscle in the main arteries partly controls distribution of blood to various organs, most vascular control occurs at the level of arterioles. For this reason, arterioles are often referred to as **resistance vessels**. Under nervous, hormonal and local control, arterioles can constrict completely, preventing blood flow to a particular capillary bed, or dilate to several times their normal size, greatly increasing flow.

Capillaries are the sites of exchange between blood and tissues. They are numerous (approximately 10^{10} of them in humans) and quite small (about 1 mm long and 8–10 μm in diameter). They are so narrow that red blood cells must pass through in single file. The total surface area of capillaries in humans is about 6000 m^2 and, end to end, they would stretch about 100 000 km, or more than twice around the earth.

> The microcirculation consists of the small vessels lying between arteries and veins. Muscular arterioles and capillary sphincters control blood flow to vascular beds. Capillaries are the sites of exchange between blood and interstitial fluid.

Capillary exchange

Capillary walls are single endothelial cells, like the lining of the vessels leading to them (Fig. 21.16). The primary mode of capillary exchange is simple diffusion. The rate of diffusion depends inversely on diffusion distance, so the cells of capillary walls are particularly thin (1–3 μm). Small lipid-soluble molecules, such as O_2, CO_2, glucose and urea, diffuse directly through the lipid walls of the endothelial cells. The second mode of exchange across the capillary is **pinocytosis** (Chapter 4). In some capillaries, the slow exchange of large particles and lipid-insoluble materials occurs in numerous tiny vesicles that form on one cell membrane, move across the cytoplasm and fuse with the opposite membrane, releasing their contents (Fig. 21.16). The third mode of exchange is **filtration**, whereby pressurised fluid is forced through various openings in the capillary wall, such as gaps between adjacent cells, and specialised regions known as fenestrae (Fig. 21.16). Carried with the fluid are substances small enough to pass through the openings. The relative area involved is quite small, less than a

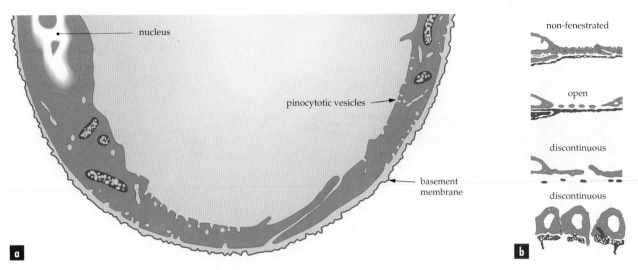

Fig. 21.16 Capillary in cross-section. **(a)** Diffusion occurs through the thin endothelial walls of the capillary. Larger molecules can move across by pinocytosis in vesicles, and water can be filtered through gaps in the lining. **(b)** The occurrence of fenestrae (windows) and discontinuities in the wall is related to the requirement to exchange large molecules

thousandth of the surface area of the capillary walls. However, this varies from tissue to tissue. Particularly permeable capillaries are found in tissues where filtration of fluid and rapid transport of large molecules is important, for example, in the kidney glomerulus, intestinal villi, spleen, bone marrow, liver and endocrine glands. On the other hand, non-fenestrated capillaries are found in muscle, skin, fat and nervous tissue, where little filtration occurs and small molecules are exchanged mainly by diffusion or active transport. The maximum size of particles that can move through fenestrations is about 4 nm (1 nm = 10^{-9} m). Thus, red blood cells and large plasma proteins, principally albumin, are not filtered, but smaller molecules, for example, amino acids and vitamins, are. The filtrate enters interstitial fluid spaces where exchange with cells takes place. Tissue fluid balance is maintained because most of the filtered fluid is reabsorbed into capillaries by osmosis and the remainder is removed by the lymphatic system.

Understanding of the balance between filtration and reabsorption is attributed to Ernest Starling, who proposed that the *net* fluid movement between the capillary and interstitial fluid is determined by the balance between two forces—**hydrostatic** (fluid) **pressures** and **colloid osmotic** ('oncotic') **pressures** across the capillary wall (Fig. 21.17). This is now called the '**Starling principle**'. On the one hand, the hydrostatic (blood) pressure in the capillary tends to filter water from the capillary. On the other hand, the colloid osmotic pressure of large proteins ('colloids') in the blood tends to absorb water from the interstitial fluid into the capillary. If the hydrostatic pressure gradient is greater than the osmotic gradient, then a net filtration occurs; if the osmotic gradient is stronger, then a net reabsorption of fluid occurs. As blood flows

along a capillary, its hydrostatic pressure decreases from about 30 mmHg at the arteriolar end to about 15 mmHg at the venular end, but the colloid osmotic pressure gradient and the hydrostatic pressure of the interstitial fluid remain fairly constant. Thus, filtration predominates at the arterial end of the capillary because blood pressure is highest there; at the venular end, absorption prevails because colloid osmotic pressure is highest. Filtration and reabsorption do not necessarily occur equally in the same capillary. Some capillaries may largely filter while others largely absorb, the balance depending on the pressure gradients in each one. Overall, however, in any tissue, most filtered fluid is reabsorbed.

Two important features of circulation put the Starling principle into perspective. Firstly, it has been estimated that, of the 8000–9000 L of blood pumped from the human heart each day, only 20 L (0.25%) are filtered from capillaries into tissues. With regard to exchange of respiratory gases, nutrients and wastes, therefore, filtration is of negligible importance. Secondly, because of the imbalance between net outward and net inward pressures, not all filtered fluid is reabsorbed back into capillaries. About 2–4 L (10–20% of filtered fluid) is collected from tissues by the lymphatic system and returned to the blood via the great veins near the heart. Failure of the lymphatics to return excess filtrate to the blood results in an increase in interstitial fluid and swelling of tissue, **oedema**. For example, in tropical regions, filarial nematode worms can block lymphatic vessels in humans and cause enormous swelling of the affected parts, a condition known as elephantiasis (Chapter 38).

Oedema is particularly dangerous in the lung because increased interstitial fluid may leak into the alveoli and greatly increase the diffusion barrier

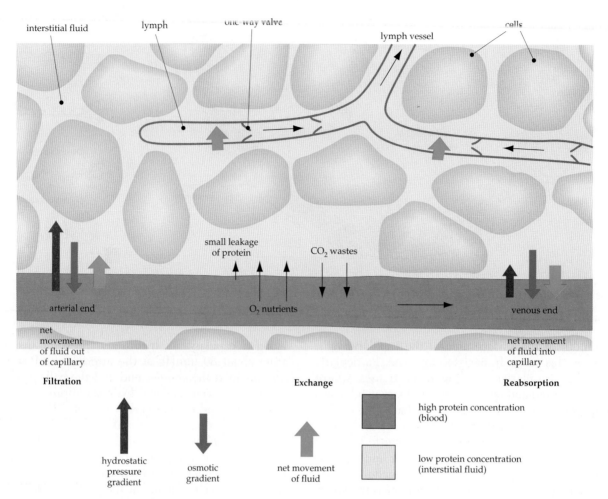

Fig. 21.17 Filtration and reabsorption in the capillary. Balance of pressures across capillary walls and resultant movement of fluid between blood plasma and interstitial fluid. Net filtration occurs due to the high blood pressure at the arterial end of the capillary. As blood pressure decreases along the capillary, the colloid osmotic pressure eventually exceeds hydrostatic pressure and osmotic reabsorption occurs. Dissolved substances diffuse along their concentration gradients, and are also carried with filtered and reabsorbed water. Excess interstitial fluid is returned to the blood vascular system via the lymphatic channels

between air and blood (Chapter 20). In normal circumstances, however, excess filtration in the lungs is unlikely because pulmonary blood pressure is much lower than systemic blood pressure. Thus, in the lung, hydrostatic pressure is normally less than colloid osmotic pressure and any water that gets into the lung is immediately absorbed into the blood. However, in the case of heart failure involving the left ventricle, a backup of blood in pulmonary vessels causes a rise in pulmonary blood pressure and fluid may filter into the lungs.

Exchange at the capillary occurs mainly by diffusion, but also by pinocytosis and filtration. Filtration results from a balance between colloid osmotic and hydrostatic pressures of interstitial fluid and blood.

The lymphatics

Mingled among cells in interstitial spaces are primary **lymphatic capillaries** (Fig. 21.17). Like blood capillaries, they are simple endothelial channels but, in contrast, they are blind-ended. Interstitial fluid enters lymphatic capillaries through gaps between endothelial cells, some of which appear to have valve-like flaps that prevent reverse flow back into the tissues. **Lymph** is moved as a result of contraction of smooth muscle in walls of larger lymph vessels and by indirect external pressure on lymph vessels due to contraction of adjacent skeletal muscles. In some fishes and amphibians, lymph flow is also assisted directly by muscular 'lymph hearts'. As in veins, one-way valves cause lymph to flow in one direction only. In humans, lymph passes through many lymph nodes (Chapter 23), finally entering the venous system from either the right lymphatic duct or the thoracic duct. Some tissues lack lymphatic vessels (e.g. the central nervous system, bone marrow and lung), but they are abundant in other tissues (e.g. gut, skin and upper airways).

The lymphatic system has a particularly important role in tissue fluid balance. A small leakage of protein occurs from capillaries into the interstitial fluid. If allowed to accumulate in interstitial fluid, its colloid osmotic pressure would rise, reducing the net absorption pressure and causing oedema. In terms of tissue fluid balance, the main role of the lymphatic system is to return these escaped plasma proteins, along with excess interstitial fluid, back to the blood.

The lymphatic system also transports other materials. For example, it is the major route of lipid transport from the small intestine to the blood and it carries certain products of the liver (Chapter 19). Another important function of the lymphatic system is its role in defence, which is considered in Chapter 23.

> Fluid that is not reabsorbed back into capillaries, and proteins that leak from capillaries, are returned to the circulation by way of the lymphatic system.

Regulation of blood flow and pressure

The vertebrate circulatory system is essentially a pump that propels blood around a circuit of vascular tubing. Whether it is a single circuit, as in fishes, or a completely double circuit, as in the systemic and pulmonary circuits of birds and mammals, the heart works against friction in moving blood through the vessels. By analogy with the flow of current in an electrical circuit, we can describe blood flow by Ohm's law, which states that current is equal to the ratio of voltage over resistance. In our case, rate of blood flow (V_b) is equal to the pressure difference between the arteries and veins ($P_a - P_v$) over the resistance of the vessels (R; Fig. 21.18).

In mammals and birds, **cardiac output** (V_b) is the amount of blood pumped by the left ventricle into the aorta every minute (mL/min). In contrast, in vertebrates with undivided ventricles, cardiac output refers to the entire output of the heart. Blood pressure (P) is traditionally measured in units of millimetres of mercury (mmHg). V_b and ($P_a - P_v$) are dependent on the peripheral resistance (R, mmHg·min/mL), which is regulated by controlling the diameter of arterioles. In the homeostatic condition, any changes in overall arteriole diameter are matched by reciprocal changes in cardiac output so that blood pressure is regulated at a more or less constant level.

A closer look at the 'plumbing' of the circulatory system (Fig. 21.19), and we see that it consists of many organs usually arranged in parallel rather than in series. The liver is one exception in that part of its circulation comes from the gut. Parallel blood supply ensures that organs are supplied with oxygenated blood delivered at high pressure by the systemic arteries. It also permits changes in blood flow to individual organs to be

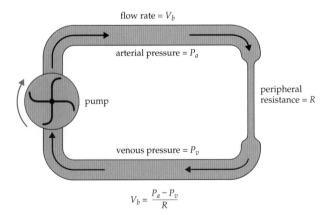

$$V_b = \frac{P_a - P_v}{R}$$

Fig. 21.18 Blood flow in the circulatory system depends on the difference in blood pressure between the main arteries and veins, and also on the resistance to blood in the smaller vessels (peripheral vessels)

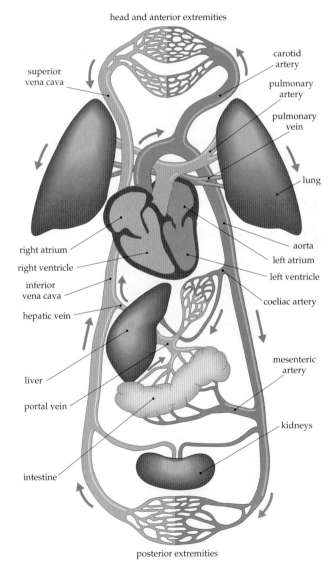

Fig. 21.19 Diagram of the double circulation of mammals. The vascular beds are arranged in parallel, except for the intestine and liver, which are connected in series by the portal vein

BOX 21.3 Bernoulli, gravity and blood flow in dinosaur necks

It takes energy to move fluid through tubing. Much of the work that the heart performs is used to generate pressure energy to overcome friction as blood is moved around the circuit. Work is also needed to get the blood moving, that is, to impart kinetic energy, and work is required to raise blood to organs above the heart, increasing its gravitational potential energy. Fluid moves according to differences in total fluid energy, which is the sum of these three components of energy (Fig. a). This relationship was described in 1738 by Daniel Bernoulli, the famous Swiss mathematician, physicist and physiologist.

In most vertebrates, pressure energy is the important component. Blood velocity is generally so low that kinetic energy is small, and the vertical distance of the head above the heart is relatively short, so potential energy changes are minor. However, the gravitational potential energy component can become important in particularly tall animals, such as the giraffe (Fig. b). In a fully grown 4 m giraffe, the head may be about 2 m above the heart. The column of blood in the neck produces a pressure of about 170 mmHg at the heart, simply as a result of gravity. Therefore, the giraffe's heart has to produce a mean pressure equal to 170 mmHg to

support the blood column, plus about 80 mmHg to overcome frictional losses in the small blood vessels. Direct measurements confirm that the mean blood pressure in the giraffe's aorta is about 250 mmHg. In comparison to many mammals in which mean aortic pressure is about 100 mmHg, the giraffe certainly has a high blood pressure. To produce such a pressure, the walls of the giraffe's left ventricle are extremely thick.

Giraffes, at about 4 m, are the tallest living animals, but some of the long-necked dinosaurs (the sauropods, often called brontosaurs) were much taller. Modern reconstructions show giants such as *Brachiosaurus* or *Barosaurus* towering 12 m above the ground and feeding from tall trees. Similar calculations to those for giraffes provide an estimate of required blood pressure of 500–750 mmHg at the heart. Such high pressures seem impossible because the heart muscle would have to be so thick that it would be mechanically inefficient and simply too large to fit inside the ribs of the animal. On the other hand, if sauropods did not lift their heads high, they would not have needed high blood pressures or large hearts. Their long necks would have been useful to browse on low vegetation without having to move the bulky body around. Alternatively, they could have been

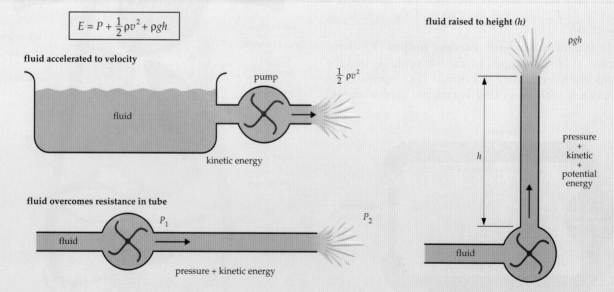

(a) The Bernoulli equation. The energy that a pump must produce to move fluid is the sum of three components: the pressure energy (*P*), the kinetic energy of fluid motion ($\frac{1}{2}\rho v^2$), and the gravitational potential energy (ρgh). In the equation, ρ (rho) is the density of the fluid, *v* is the velocity, *P* is the hydrostatic pressure, *g* is the gravitational acceleration constant, and *h* is the height of the fluid column. In blood circulatory systems, the kinetic energy component is small because the velocity is low. Sufficient fluid pressure is important, however, to overcome the resistance to flow in the tubing. The heart must work against the pressure developed in the vertical blood column above it

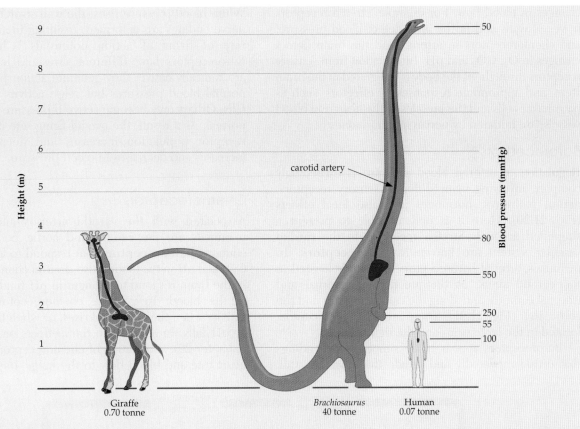

(b) Blood pressures measured at the levels of the heart and head in a human and an adult giraffe, compared with that calculated in a sauropod dinosaur

aquatic, floating in lakes and using the neck to reach deep aquatic vegetation. In water, gravity is not a problem for the cardiovascular system because gravitational pressures in blood vessels are essentially matched by pressures in water surrounding the animal. In any case, these huge animals doubtless came on land, perhaps to lay eggs, but they almost certainly did not forage in tall trees as often depicted.

somewhat independent of flow through the whole system. If organs were arranged in series, all would be obliged to receive the same blood flow, and those first in line would receive the freshest blood.

B lood flow through vessels is governed by blood pressure and resistance to flow in the vessels. Blood supply to most organs is arranged in parallel, rather than in series, providing fresh blood to most organs.

Cardiovascular regulation in vertebrates

Despite the parallel arrangement of blood supply to most organs, any system of fluid reticulation supplied by a common pump tends to suffer from the fact that a change in flow to one part of the system affects flow to the other parts. For example, someone turning on a water tap in the kitchen can affect water flow to the shower. This is less likely if there is good water pressure

in the mains supplying the house, and would be even less likely if mains water pressure were regulated to stay the same in the face of changing demands. Circulatory systems in vertebrates have both these features: they generate relatively high pressures and have negative feedback mechanisms that regulate arterial blood pressure to compensate for redistribution of flow. As a result, in ordinary circumstances, an increase in blood flow to one organ need not decrease flow to the rest of the body.

A negative feedback system consists of three parts:
1. *Sensors* (receptors), which are able to detect change or disturbance and pass this information on.
2. An *integrating centre*, which evaluates the information and responds to it.
3. *Effectors*, which produce an appropriate compensatory response that reduces the change.

The body has no receptors capable of measuring blood flow. However, there are several types of sensors that provide information on variables directly related to changes in blood flow. For example, stretch receptors in vessel walls and in the atria detect blood pressure, and chemoreceptors in arteries and the brain detect changes in O_2, CO_2 and pH. Information from sensory receptors throughout the body is integrated in the brain stem, and appropriate responses by effectors, such as pacemaker cells and the muscles of the heart and blood vessels, are initiated by nerves and hormones.

Baroreceptors

Regulation of arterial blood pressure is exceedingly complex, involving many different sensory receptors, neural pathways, hormones and also local effects (Fig. 21.20). Here it is only possible to present a simplified account. Stretch receptors in the cardiovascular system are known as **baroreceptors**. In mammals, arterial baroreceptors are mainly located in the carotid sinus, at the junction of internal and external carotids, and along the carotid arteries and the aortic arch (Fig. 21.21). There are also baroreceptors located in the large veins and the right atrium.

Baroreceptors appear to be simple nerve endings that divide repeatedly and invade the muscular wall.

When blood pressure rises, the wall stretches and the nerve endings are deformed, resulting in an increased rate of firing of action potentials (Chapter 26). Baroreceptors have different threshold levels. Some are tonically active: they generate action potentials at normal blood pressures but cease activity if pressure falls. Others only become active if pressure rises above normal. As a result, the overall firing rate of the baroreceptor population increases and decreases with increases and decreases in blood pressure.

Chemoreceptors

Associated with the carotid arteries and aorta in mammals are the carotid and aortic bodies, which contain **chemoreceptors** that respond to levels of O_2, CO_2 and pH in the blood. Also, central chemoreceptors in the brain respond to changes in pH (and thus CO_2) of the blood. In general, chemoreceptors increase activity when the CO_2 level rises, or when the O_2 level or pH falls. In addition to their effects on ventilation (Chapter 20), stimulation of chemoreceptors increases heart rate and blood flow to the lungs, improving gas

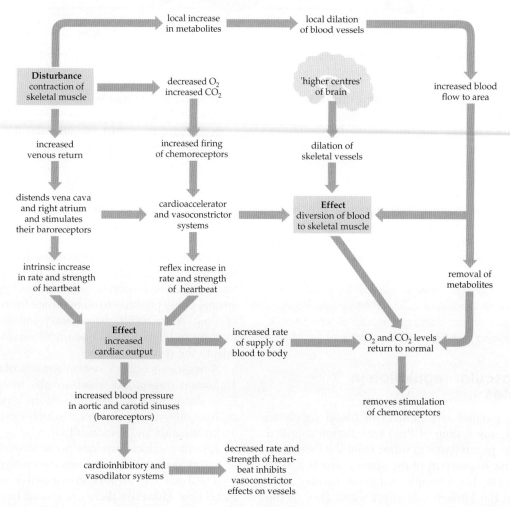

Fig. 21.20 Some of the pathways involved in the homeostatic control of blood flow and pressure during exercise (the disturbance) in a mammal

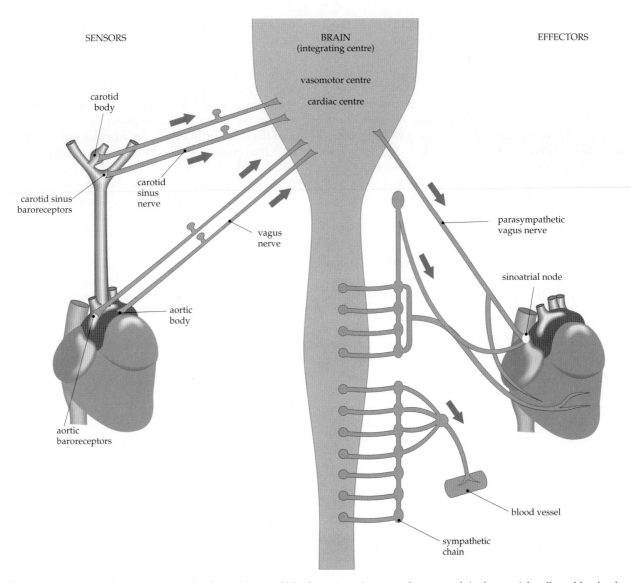

Fig. 21.21 Major neuronal feedback systems for the regulation of blood pressure. Receptors detect stretch in the arterial walls and levels of blood gases and pH. Sensory nerves relay this information to the brain, where it is processed in the vasomotor and cardiac centres. Effector nerves can change the power and rate of the heartbeat and also alter peripheral resistance by affecting smooth muscles in the arterioles

exchange. They also influence systemic blood flow and pressure.

Integration

Information from baroreceptors is sent to the **vasomotor centre** and the **cardiac centre** in the brain stem (medulla and pons). The vasomotor centre controls the state of contraction of the smooth muscle of arterioles, and the cardiac centre controls the rate and force of beating of the heart. Nervous impulses from chemoreceptors pass to the brain stem, where they primarily influence the respiratory centres controlling breathing, but they also have an impact on the cardiovascular centres.

Responses

In vertebrates, both cardiac output and regional blood flow are regulated by the autonomic nervous system, including parasympathetic and sympathetic nerves (Chapter 26), by circulating hormones and by local (intrinsic) factors. In general, if arterial blood pressure drops, output from the vasomotor and cardiac centres causes an increase in activity of sympathetic nerves and a decrease in activity of parasympathetic nerves. This causes most peripheral arteries and arterioles to contract further, increasing overall peripheral resistance. Increased sympathetic input and decreased parasympathetic (vagal) input to the heart cause the pacemaker to produce action potentials more quickly, increase the speed of conduction through the atrio-ventricular node and increase the force of ventricular contraction. Increased sympathetic nerve activity to venous blood vessels causes them to contract and force more venous blood towards the heart. Increased venous return stretches baroreceptors in the right atrium, causing a reflex increase in heart rate. Increased venous return also stretches cardiac muscle fibres directly, causing them to contract more forcefully, raising the volume of blood pumped. All of these responses raise blood pressure. If arterial blood pressure rises above normal, opposite responses occur to again return blood pressure towards normal. These nervous responses, particularly those involving parasympathetic nerves, are extremely fast, and compensatory adjustments can occur within the time of a single heartbeat. This is important for blood pressure regulation during postural changes; if they were slower, we would faint every time we stood up.

Several hormones also have significant cardiovascular effects. For example, you have probably felt your heart start pounding in stressful situations. This is caused by *adrenaline*, a hormone produced and released from the adrenal glands, particularly under conditions of excitement or stress. Like *noradrenaline*, which is released from sympathetic nerves, adrenaline speeds the heart and causes arterioles to constrict. Other small peptides, such as *vasopressin*, produced by the posterior pituitary, and *angiotensin*, produced in the blood through the influence of renin from the kidney, cause vasoconstriction. Still other substances, for example, *kinins*, *histamines*, and *prostaglandins*, have distinct effects on the heart and blood vessels.

> Negative feedback control involves baroreceptors and chemoreceptors, integration in both the vasomotor and cardiac centres of the brain, and sympathetic and parasympathetic outputs that control the timing and strength of the heartbeat and the state of contraction of muscles in vessel walls. Regulation is also affected by blood-borne hormones.

Regulation of blood flow to tissues

The rate of blood flow to particular capillary beds is controlled by intrinsic mechanisms (local factors within the tissue itself) and extrinsic mechanisms (influences from outside the tissue), including nerves and hormones. In different tissues, different mechanisms may predominate. For example, coronary blood flow is almost entirely under intrinsic control, whereas control of flow to skeletal muscle or skin has a strong neural component.

Intrinsic control: autoregulation

Blood flow through capillaries is largely under the control of precapillary sphincters (Fig. 21.15), which are responsive to local conditions in tissues. There is an inherent level of contraction (**tone**) in the smooth muscle of precapillary sphincters and the level of this tone tends to increase if a vessel is stretched by an increase in pressure. Precapillary sphincters are also

responsive to increased levels of metabolites, such as lactic acid and ATP breakdown products. When metabolic activity of a tissue increases, levels of CO_2, lactic acid, ADP and AMP increase, and any one of these may cause opening of the precapillary sphincters, increasing blood flow to the area. Vasoconstrictor and vasodilator substances produced in the endothelium also have powerful effects on vascular smooth muscle.

Local control of blood flow, **autoregulation**, is also responsible for the distribution of blood within a capillary bed. When a tissue is at rest, it receives a minimal rate of blood flow to sustain resting metabolic rate. Blood perfusing the tissue does not pass through all capillaries because, at any time, about 80% of them are closed by their precapillary sphincters. Blood normally flows in larger capillaries, **preferential channels**, also known as thoroughfare channels (Fig. 21.15), which remain open continually and allow blood to pass through at a velocity appropriate for exchanging gases and substrates with the tissues. Preferential channels are relatively far apart, but exchange with them is sufficient to meet the needs of resting tissue. During activity, however, precapillary sphincters open up and allow blood to flow through true capillaries. This not only allows a greater blood flow through the tissue but it ensures that blood passes more closely to active cells, thus decreasing the diffusion distance and increasing the rate of exchange.

The distribution of blood flow through the lungs appears to be dependent on local factors but in the opposite direction to that in systemic tissues. Although the lungs of mammals receive all blood flowing from the right ventricle, blood tends to flow preferentially through areas of the lungs containing the freshest air. Here, under the influence of high levels of O_2, pulmonary arterioles dilate rather than constrict. In lower vertebrates with incompletely divided ventricles and intraventricular shunting (Chapter 40), the fraction of blood sent to the lungs can be changed according to the level of O_2 in the lung and the requirements of the animal.

Extrinsic control

Extrinsic regulation of vessel diameter is brought about by nerves and hormones. Arterial vessels are innervated by sympathetic nerves (Chapter 26), which are always active, releasing noradrenaline (and possibly other neurotransmitters), usually producing a vasoconstrictor tone. However, in the same vascular bed, the response can change direction depending on the level of activity in the nerves. In skeletal muscle, for example, vasodilation occurs at low levels of sympathetic input when noradrenaline stimulates a set of receptors (beta-receptors) on arterial smooth muscle. However, at higher levels of sympathetic nerve activity, noradrenaline begins to stimulate a different set of receptors (alpha-receptors), which cause vasoconstriction. In some species, there are also cholinergic vasodilator nerves innervating vessels supplying skeletal muscle.

Blood flow to specific organs can also be modified by neural output from higher (conscious) centres of the brain. Familiar examples include blushing of the skin with emotion, skin blanching with shock, and erection of genital organs during sexual arousal.

Hormonal regulation of vascular smooth muscle involves circulating hormones such as adrenaline, vasopressin, angiotensin, renin, kinins, histamines and prostaglandins.

> Blood flow within a tissue is controlled by arterioles and precapillary sphincters, which are responsive to both intrinsic mechanisms (stretch, metabolites and other locally released vasoactive substances) and extrinsic mechanisms (effects of nerves and hormones).

Blood

Blood has numerous functions, but its most significant roles are the transport of respiratory gases (Chapter 20), nutrients, waste products and hormones. It is also central to the defence mechanisms of the immune system (Chapter 23), and regulation of fluid balance in tissues (Chapter 22) and body temperature

(Chapter 29). The formation of mammalian blood is described in Chapter 6. Here we will examine some of its general properties.

In vertebrates, whole blood consists of two fractions: **plasma**, a clear, slightly yellowish fluid (containing a significant amount of protein, chiefly albumins and globulins) and suspended cells or fragments of cells (Fig. 21.22), produced in the bone marrow. When whole blood is spun in a centrifuge, the cells pack at the bottom of the tube and their volume, as a proportion of the total volume, is the **haematocrit**. In healthy humans, the haematocrit is around 40–45%. Red blood cells (**erythrocytes**), which transport O_2, are the most numerous. White blood cells (**leucocytes**) are less common (about 1 per 600 erythrocytes), but their roles in protecting the body against disease and removing cellular debris are essential and are considered in detail in Chapter 23. Finally, there are **thrombocytes**, which, in lower vertebrates, are small nucleated cells but, in mammals, are cell fragments called **platelets**. Platelets are disc-shaped, membrane-bound cell fragments, about half the diameter of an erythrocyte. They do not contain a nucleus or DNA. Huge cells, **megakaryocytes**, in the bone marrow, pinch off thousands of platelets during their life. During their 1-day life after release into the circulation, platelets may take part in wound healing and clot formation (see below).

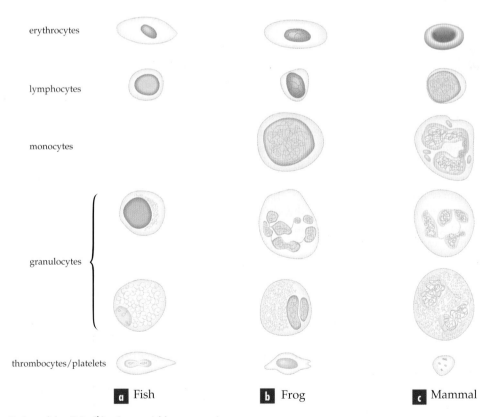

	erythrocytes
	lymphocytes
	monocytes
	granulocytes
	thrombocytes/platelets

a Fish **b** Frog **c** Mammal

Fig. 21.22 Blood cells from **(a)** a fish, **(b)** a frog and **(c)** a mammal

Vertebrate blood contains cells suspended in a fluid plasma. In mammals, the haematocrit is about 45% of the total blood volume.

Erythrocytes

In humans, erythrocytes are round discs, about 7–8 μm in diameter, with thick edges, giving them a biconcave appearance. Although only 2.1 μm in maximum thickness, if all the red blood cells in your body (about 30 million million) were stacked like plates, the stack would measure 63 000 km, or about one and a half times the circumference of the earth. Their combined surface area is about 3840 m^2 or about half the area of a soccer field. Mammalian erythrocytes are the smallest among vertebrates, lack nuclei in the mature form and have low metabolic activity. On the other hand, erythrocytes of other vertebrates are larger (up to 78 × 46 μm in a salamander), nucleated and have higher metabolic activity. The significance of these differences is unclear. However, smaller cells have a larger relative surface area for gaseous exchange, and a low metabolism reduces the amount of oxygen consumed by the cells themselves on their way to the tissues.

Erythrocytes owe their colour to haemoglobin, which increases the O_2-carrying capacity of the blood (Chapter 20). Among vertebrates, the amount of haemoglobin and the number of erythrocytes tend to reflect the metabolic rate of the animal. Certain Antarctic fishes, which have very low metabolic rates in the frigid water, have no haemoglobin at all. The highest haematocrits known, about 60%, are found in some diving mammals, which store large amounts of O_2 in their blood (Chapter 29).

In many invertebrate animals, respiratory pigments circulate freely in the blood. In some invertebrates and all vertebrates, pigment molecules are enclosed in cells. Packaging the pigment in cells has mechanical and physiological advantages. For example, it facilitates blood flow by decreasing the viscosity of blood in small arterioles where resistance to flow is highest. Some capillaries of mammals are smaller in diameter than erythrocytes and red cells deform as they pass through the capillary, pressing against the capillary wall. This reduces the distance over which gas diffusion must occur and sweeps stagnant plasma away from the wall. Finally, O_2 binding by haemoglobin is influenced by various substances in the blood (e.g. organic phosphates such as 2,3-diphosphoglycerate and the enzyme, carbonic anhydrase; Chapter 20). The concentrations of these substances inside the cell can be regulated at higher levels than would be possible if they were all freely dissolved in the blood.

In mammals, erythrocytes are formed in bone marrow (Chapter 6). They are incapable of dividing or repairing themselves and they simply wear out. In humans, they circulate for about 120 days. Before they actually break down, old erythrocytes are phagocytised in the spleen, liver and bone marrow. Proteins are broken down into amino acids and recycled, and the iron from haemoglobin is retained for synthesis into new haemoglobin. Part of the haemoglobin molecule is converted by the liver into bilirubin, which leaves in the liver bile.

In humans, about two and a half million erythrocytes are formed and destroyed every second. However, the number of erythrocytes in the blood remains remarkably constant, reflecting a highly efficient regulatory system. The hormone **erythropoietin** is continually produced by the kidney (an organ that usually receives a high and constant blood flow in order to carry out its excretory role) and released into the blood. When erythropoietin reaches bone marrow, it stimulates red cell production (erythropoiesis, Chapter 6). The amount of erythropoietin released depends on the oxygenation of the blood. If blood has a low O_2 content, erythropoietin production increases. A low blood O_2 concentration can result from anaemia (low levels of haemoglobin), respiratory or cardiovascular disease, or living at high altitude. When a person donates a quantity of blood, he or she has less haemoglobin and oxygen levels decline. This causes erythropoietin production, which stimulates erythropoiesis until haemoglobin and therefore O_2 levels have returned to normal. On the other hand, excessive oxygenation of the blood reduces the rate of erythropoietin, and hence erythrocyte, production. For example, if someone with haemoglobin-rich blood living at high altitude descends to sea level, the kidney produces less erythropoietin and the rate of erythrocyte production drops below normal until the appropriate number of circulating erythrocytes is attained.

Red blood cell production is regulated by erythropoietin, which is produced by the kidneys in response to decreases in O_2 levels of the blood.

Blood clotting

When blood is exposed to air, even in a test tube, it forms a clot, **thrombus**, within a few minutes. The yellowish fluid outside the clot is **serum**, which is plasma minus some of the protein constituents that help form the clot. The reactions leading to clot formation are extremely complex. They involve a cascade of perhaps 12 sequential enzymatic reactions, with fail-safe thresholds at each stage to prevent accidental triggering of clot formation. Each reaction converts an inactive form of an enzyme into an active form, which is capable of catalysing the next reaction of the cascade and producing hundreds of active products. Thus, the number of products is multiplied at each step. The final reaction involves the enzyme **thrombin**, which converts soluble plasma protein, **fibrinogen**, into fine strands of insoluble **fibrin**. These fibres form a meshwork that traps erythrocytes and platelets to form a clot (Fig. 21.23). The enzyme sequence of the clotting cascade has been studied in humans with variations of the genetic disease haemophilia, which prevents the blood from clotting normally.

In damaged tissue, factors other than clotting are also involved in preventing blood loss. Firstly, traumatised vessels usually contract, dramatically reducing blood flow through the vessel and therefore blood loss. When a vessel is broken, the endothelial layer is breached, exposing collagen fibres in the deeper layers. Platelets immediately attach to the collagen and each other. Thrombin, which forms as outlined above, has three effects on platelets. It causes them to:

1. become sticky and attach to fibrinogen
2. become fragile and release ADP, which enhances linking with fibrinogen
3. form long pseudopod-like processes that attach to adjacent platelets.

Thrombin converts fibrinogen to fibrin, which further strengthens the clot. The gap in the vessel quickly fills with a mat of fibrin and aggregated platelets. Thrombin apparently stimulates platelets to contract, squeezing serum from the clot and drawing the broken edges of the vessel together. Platelets also release a substance that makes the surrounding blood vessels contract, further reducing the blood flow to the area.

Tiny breaks in vessel walls occur all the time as a result of normal activity. These are rapidly sealed by platelets, which plug holes, usually without preventing blood flow through the vessel. Platelets are so important in preventing blood loss that, when their production is inhibited, as in some leukemias, numerous tiny bruises appear over the body.

Blood clotting involves a series of enzyme-controlled reactions leading to production of the insoluble protein, fibrin. Platelets attach to the exposed edges of a wound, to each other and to fibrin, and help close damaged vessels.

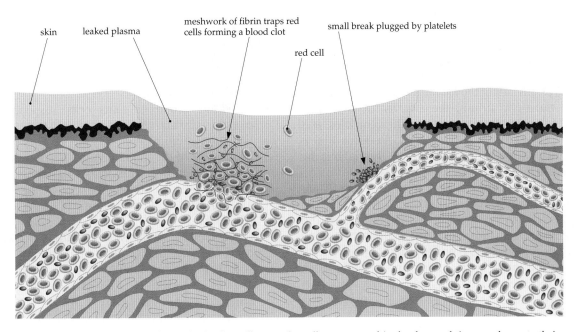

Fig. 21.23 Clot formation in a broken blood vessel. Platelets adhere to the collagen exposed in the damaged tissue and secrete their contents. Other platelets adhere to them and help close the wound. Thrombin is also produced and it converts fibrinogen to strands of fibrin, which entangle red blood cells, forming a clot

Summary

- Circulatory systems are internal convective systems that transport materials between the environment and cells, and from one part of the body to another.
- In open systems, the circulated fluid is indistinguishable from interstitial fluid. In closed systems, the circulated fluid, blood, is enclosed in a system of vessels and is distinct from interstitial fluid. Closed systems favour more rapid flow rates and fluid with higher O_2-carrying capacity.
- Fishes have circulatory systems with a single circuit—heart, gills and body—with a low-pressure systemic blood supply. In lungfishes, amphibians and reptiles, mixing of oxygenated and deoxygenated blood in the ventricle is regulated by septa, spongy heart muscle and/or valves. Complete separation of oxygenated and deoxygenated blood and high pressure systemic flow is found in birds and mammals.
- In vertebrates, cardiac contraction is myogenic, that is, initiated in the striated heart muscle itself. Excitation is conducted through the heart by a pathway of specialised cardiac muscle fibres, including the sinoatrial node (pacemaker) in the right atrium, the atrioventricular node, which delays conduction into the ventricles, and the atrioventricular bundle and Purkinje fibres, which conduct the signal rapidly throughout the ventricles.
- Arteries carry blood away from the heart. Their elastic, muscular walls store some of the pressure energy from the pumping heart and reduce fluctuations in arterial blood pressure. Veins collect the blood from the tissues and return it to the heart. At any time, most blood is held in the veins. The microcirculation consists of the small vessels lying between arteries and veins.
- Capillaries are the sites of exchange between blood and extracellular fluid. Exchange occurs by diffusion, pinocytosis and filtration. Net movement of fluid across the capillary wall depends on the balance between colloid osmotic and hydrostatic pressures of interstitial fluid and blood. Fluid that is not reabsorbed into capillaries, and proteins that leak from capillaries, are returned to the circulation by way of the lymphatic system.
- Regulation of both cardiac output and regional blood flow is the result of both intrinsic and extrinsic mechanisms. Negative feedback control involves baroreceptors and chemoreceptors, integration in the vasomotor and cardiac centres of the brain, and sympathetic and parasympathetic outputs that control the timing and strength of the heartbeat and the state of contraction of muscles in vessel walls. Regulation is also affected by blood-borne hormones.
- Blood flow within a tissue is controlled by circular muscle fibres in arterioles and precapillary sphincters. These smooth muscles are responsive to both intrinsic (stretch, metabolites and other locally released vasoactive substances) and extrinsic mechanisms (effects of nerves and hormones).
- Vertebrate blood contains cells suspended in a fluid plasma. In mammals, haemoglobin-containing erythrocytes (red blood cells) occupy about 45% of the blood volume. The production of erythrocytes, erythropoiesis, is regulated by erythropoietin, which is produced by the kidney in response to decreases in O_2 levels of the blood.
- Blood clotting involves a cascade of enzyme-controlled reactions leading to the production of strands of the insoluble protein fibrin. Platelets attach to the exposed edges of a wound, to each other and to fibrin. Under the influence of thrombin, platelets undergo a series of changes, and assist in closing the wound and reducing blood flow in adjacent vessels.

keyterms

aorta (p. 550)
arteriole (p. 556)
artery (p. 545)
atrioventricular bundle
    (p. 554)
atrioventricular node
    (p. 551)
atrium (p. 546)
autoregulation
    (p. 566)
baroreceptor (p. 564)
blood (p. 545)
capacitance vessel
    (p. 556)
capillary (p. 545)
cardiac centre
    (p. 565)
cardiac output
    (p. 561)
chemoreceptor
    (p. 564)
chordae tendinae
    (p. 550)
colloid osmotic
    pressure (p. 559)

conus arteriosus
    (p. 546)
coronary artery
    (p. 550)
diastole (p. 550)
electrocardiogram
    (p. 555)
erythrocyte (p. 567)
erythropoietin (p. 568)
fibrin (p. 569)
fibrinogen (p. 569)
filtration (p. 558)
haematocrit (p. 567)
hydrostatic pressure
    (p. 559)
leucocyte (p. 567)
lymph (p. 560)
lymphatic capillary
    (p. 560)
megakaryocyte
    (p. 567)
metarteriole (p. 556)
microcirculation
    (p. 556)
mitral valve (p. 550)

myocardium (p. 551)
myogenic (p. 551)
neurogenic (p. 551)
oedema (p. 559)
pacemaker (p. 551)
pinocytosis (p. 558)
plasma (p. 567)
platelet (p. 567)
precapillary sphincter
    (p. 556)
preferential channel
    (p. 566)
pulmonary circuit
    (p. 546)
Purkinje fibres
    (p. 554)
resistance vessel
    (p. 558)
rete (p. 557)
semilunar valve
    (p. 550)
serum (p. 568)
sinoatrial node
    (p. 551)
sinus venosus (p. 546)

Starling principle
    (p. 559)
systemic circuit
    (p. 546)
systole (p. 550)
thrombin (p. 569)
thrombocyte (p. 567)
thrombus (p. 568)
tone (p. 566)
tricuspid valve
    (p. 550)
vasomotor centre
    (p. 565)
vein (p. 545)
ventricle (p. 546)
venule (p. 556)
windkessel vessel
    (p. 555)

Review questions

1. Why are circulatory systems an inevitable consequence of the evolution of large animals?

2. Distinguish between open and closed circulatory systems.

3. Mixing of blood in the hearts of some air-breathing fish, amphibians and reptiles is possible because the ventricle is not completely divided. Nevertheless, how it is possible for them to prevent complete mixing in the heart?

4. Describe the sequence of electrical and mechanical events that occur during the cardiac cycle of a mammal.

5. Explain how heart sounds and an electrocardiogram can be useful in diagnosis of heart disease.

6. Compare the functions of arteries, veins and capillaries.

7. What are the mechanisms of exchange of materials at the capillary level?

8. How are blood pressure and flow regulated in mammals (a) in the circulatory system as a whole and (b) at the level of individual tissues?

9. What is the haematocrit and how is it regulated?

10. Describe the processes that prevent blood loss from a wound.

Extension questions

1. What circulatory changes occur (a) around the birth of a mammal or (b) during the hatching of a bird?

2. Describe the circulatory mechanism that prevents loss of heat from the feet of a penguin.

3. Why is the arterial blood pressure of a giraffe higher than that of most mammals?

4. If an animal is wounded and loses blood, its blood pressure drops until some of the lost blood volume is regained from the interstitial fluid. Describe the mechanism that underlies this shift in fluid balance.

5. What are the functions and properties of the windkessel vessels, the capacitance vessels, the resistance vessels and the exchange vessels?

6. How is blood flow controlled during a change from rest to exercise?

Suggested further reading

Berne, R. M. and Levy, M. N. (1993). *Cardiovascular Physiology*. 7th edn. St Louis: Mosby.

This is a standard text for human cardiovascular physiology used principally by medical students.

Schmidt-Nielsen, K. (1997). *Animal Physiology: Adaptation and Environment*. 5th edn. Cambridge: Cambridge University Press.

This is a text on all aspects of comparative physiology but includes circulation in relation to environment and gas exchange.

Vogel, S. (1993). *Vital Circuits: On Pumps, Pipes and the Workings of Circulatory Systems*. New York: Oxford University Press.

This is an interesting treatment of the biophysics of circulation.

CHAPTER

22

Water, solutes and excretion

For most animals it is essential for the normal functioning of cellular processes that the water and solute composition of body fluids is maintained relatively constant, despite the constant fluxes of water and solutes into and out of the animal. **Homeostasis** is the maintenance of relatively constant internal conditions with respect to one or more variables, such as salt or water levels, body temperature or levels of oxygen in the blood. Controlling the intake, or ingestion, of substances from the environment and regulating the loss, or excretion, of solutes and water, result in homeostasis of the water and solute composition of body fluids.

Excretion refers to the removal of substances that once formed part of the body. Excretory systems generally produce a specialised fluid, urine, which is expelled to the outside of the animal. There are also other avenues for excretion. Carbon dioxide is excreted across respiratory surfaces (skin, gills or lungs; Chapter 20). Salts and water are excreted in sweat and other skin secretions. **Elimination** is the removal of unabsorbed food in faeces (Fig. 22.1), and should not be confused with excretion. Unabsorbed food has never been part of the metabolic pool of the body. However, the gut can have a role in excretion. For example, bile pigments produced by the liver as metabolic breakdown products of haemoglobin are secreted into the gut for excretion. In some animals, such as insects, reptiles and birds, urine is mixed with faeces and is highly modified by the hindgut before excretion.

The excretory systems of animals contribute to the regulation of the internal environment in three main ways:

1. control of normal body water content
2. maintenance of normal solute composition
3. excretion of metabolic waste products and unwanted substances that may have been inadvertently absorbed, such as dyes and toxins.

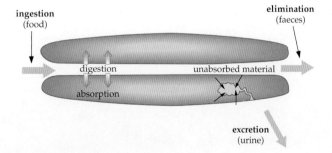

Fig. 22.1 Excretion is the discharge of water and solutes from the body fluids; these materials have been absorbed into the body pool of water and metabolites. Elimination is the discharge from the digestive tract of unabsorbed materials that have never been part of the body pool of metabolites. Note that there are other avenues for excretion (e.g. CO_2 by the lungs, sweat by the skin) and that some materials are secreted into the gut for excretion (e.g. bile pigments)

Excretion is the removal of substances that have been assimilated into the body; it regulates body fluid solute concentrations and water content. Elimination voids unabsorbed foods from the digestive tract.

Water and solutes

Water is essential for life: about 70–80% of most animals is water. One role of water is as a solvent to dissolve solutes and provide their mobility for chemical reactions to occur readily. Only a few animals, and often just their early developmental stages, can survive the loss of all body water; they survive desiccation in a state of suspended animation called **anhydrobiosis**.

Life evolved in sea water and the solute composition of the body fluids of animals reflects their marine origin. Various ions and organic molecules are invariably present in the body fluids of animals and are an essential part of the internal cellular environment in which all biological processes occur. These solutes include ions such as sodium, potassium, chloride, calcium, magnesium, sulfate and phosphate, and organic molecules such as glucose, amino acids and proteins, and nitrogenous wastes. The concentrations of particular solutes in solution are usually expressed as molar concentrations, whereas the total concentration of all dissolved solutes is expressed as the osmolar concentration (Box 22.1).

BOX 22.1 Measuring solute concentrations

A solute is an inorganic ion or organic molecule that is dissolved in a solvent, which is water for biological systems. A solution is the resultant mixture of solute and solvent.

The concentration of solutes in solution is usually expressed as molar concentration, the number of moles of that solute per litre of solution (**molarity**, M). A mole of solute is 6.02×10^{23} molecules, or the number of grams of solute equal to its molecular weight. For low concentrations, either millimolar (mM) or micromolar (μM) units are more convenient than molar concentration. The molal concentration of a solution (**molality**) is the number of moles of solute per kilogram of solvent. In very dilute solutions, molarity and molality are practically equal, but in more concentrated solutions, especially in solutions with a very high protein content, the difference becomes important.

The total concentration of all osmotically active particles is the osmotic concentration. It is measured as osmoles per litre (**osmolarity**, Osm) or osmolals per kilogram of water (**osmolality**). An osmole is a mole of solute particles, regardless of the actual nature of the solutes. For solutes that dissociate when dissolved in water, the osmotic concentration is not the same as the molar concentration. For example, sodium chloride (NaCl) dissociates in water into two osmotically active particles, the ions Na^+ and Cl^-, so the molar concentration is 1 M NaCl when 1 mole of sodium chloride is dissolved to 1 L of solution but the osmotic concentration is 2 Osm.

Exchange with the environment

There is considerable exchange of water and solutes, including waste products, between an animal's body fluids and its environment. This occurs as a result of food and fluid intake, and respiratory, urinary and faecal losses (Fig. 22.2).

For aquatic animals, water and solutes are gained by drinking and eating, and lost in urine and faeces. Water is also produced by cellular metabolism, but this is generally an insignificant source of water. There may also be active transport of salts into or out of the body fluids, for example, across gills. In addition, solutes are lost or gained by passive diffusion across the body surface, depending on the concentrations of the body fluids and external medium, and water is gained or lost by osmosis.

For terrestrial animals, water and ions are gained by drinking and eating, and lost in urine and faeces, but there is no exchange of solutes across the body surface. However, the loss of water across the skin and respiratory surface by *evaporation* is a new problem for terrestrial animals. The skin of many terrestrial animals is relatively impermeable to limit the evaporation of water, but some is inevitably lost from the skin and respiratory surfaces. There may also be loss of water and solutes by skin secretions, such as sweat in mammals. Although evaporation is an inevitable avenue of water loss for most animals, a few can actually absorb water vapour from the air using highly specialised oral or rectal structures (see below).

Solutes may be passively exchanged by diffusion down a concentration gradient, and many solutes are exchanged more rapidly or against a concentration

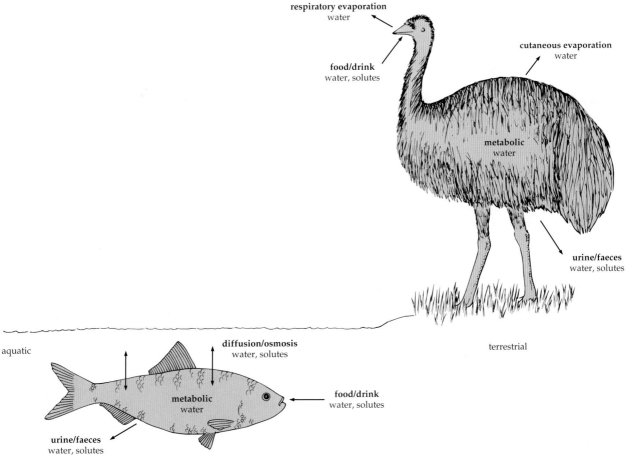

Fig. 22.2 The primary avenues for water and solute exchange by aquatic and terrestrial animals

gradient by active transport. Water molecules readily move through plasma membranes and across epithelia by osmosis, the passive diffusion of water along an osmotic concentration gradient from low to high osmotic concentration (or from high water potential to low water potential, ψ, Chapter 4). There is no mechanism for active transport of water molecules, but water can move across epithelia against a considerable osmotic gradient as a passive osmotic consequence of active solute transport. For example, the blowfly *Calliphora* has rectal pads that can withdraw water from the rectal contents against a considerable osmotic gradient (see Fig. 22.16).

E xchange of water and solutes between an animal and its environment occurs across many surfaces, including excretory organs, gut, respiratory surfaces and skin, by passive or active processes.

Extracellular and intracellular environments

In multicellular animals, body water and solutes are located in two different 'compartments'; the *intracellular compartment* consists of fluid inside cells, and the *extracellular compartment* consists of fluid between cells. For animals with a closed circulatory system (Chapter 21), extracellular fluid is further compartmentalised into fluid inside the circulatory system (plasma) and fluid between the cells and outside the circulatory system (interstitial fluid).

In animal cells, the osmotic concentration of extracellular fluid is the same as intracellular fluid. This must be so because the flexible plasma membrane of a cell is permeable to water and is unable to sustain a hydrostatic pressure difference across the membrane to balance an osmotic concentration difference (unlike plant cells, which have a rigid cell wall). Therefore,

water moves across the plasma membrane until intracellular fluid has the same osmotic concentration as extracellular fluid (is **iso-osmotic**) and has the same osmotic water potential (ψ_π). However, the solute composition of the intracellular fluid is typically very different from the extracellular environment (see also Chapter 4). Intracellular fluid always has higher K^+ and lower Na^+ and Cl^- concentrations than extracellular fluid (Table 22.1). These ionic imbalances provide the basis for the membrane voltage potential and action potentials (Chapter 26).

Life evolved in the oceans, and the ionic and osmotic composition of the first animals was compatible with sea water as the external medium. The solute composition of modern marine protists, many invertebrates, and one primitive group of jawless vertebrates, the hagfishes, reflects this marine origin; their osmotic concentration is the same as sea water, and their extracellular fluid has similar Na^+, K^+ and Cl^- concentrations (Table 22.1). All other vertebrates, including lampreys, which are the other primitive group of jawless vertebrates, have much lower Na^+ and Cl^- concentrations. The low ion concentrations of all vertebrates (except hagfishes) suggests a freshwater origin for vertebrates.

W ater and solutes are distributed in intracellular and extracellular compartments, which have the same osmotic concentration but different solute concentrations.

Patterns of ionic and osmotic balance

There are two basic patterns of how body fluid osmotic concentration is related to the osmotic concentration of the environment. Some animals simply allow the

Table 22.1 Comparison of the osmotic and ionic concentrations of sea water, a marine protist (*Uronema*), a marine horseshoe crab (*Limulus*), a marine hagfish (*Myxine*) and a marine bony fish (*Muraena*)

	Sea water	Protist (intracellular)	Horseshoe crab Intracellular	Horseshoe crab Extracellular	Hagfish Intracellular	Hagfish Extracellular	Bony fish Intracellular	Bony fish Extracellular
Osmotic concentration (mOsm)	1033	950	≈1050	≈1050	≈1100	≈1100	≈350	≈350
Na^+ (mM)	480	84	126	445	32	487	25	212
K^+ (mM)	10	134	100	12	142	8	165	2
Cl^- (mM)	560	16	159	514	41	510	24	188

osmotic concentration of their extracellular fluids to equal that of their surroundings; these animals **osmoconform** (Fig. 22.3). In contrast, many animals regulate the osmotic concentration of their extracellular fluids to be higher or lower than the surrounding water; these animals **osmoregulate**. All freshwater animals osmoregulate because it is impossible for them to have body fluids as dilute as fresh water.

There are also two basic patterns of how the body fluid ionic concentrations are related to the ionic concentrations of the environment. Many marine animals have ion concentrations of their extracellular fluid similar to sea water; they **ionoconform**. Other marine animals use ion pumps to maintain their extracellular ion concentrations substantially below sea water; they **ionoregulate**. Freshwater animals must ionoregulate and maintain their extracellular ion concentrations substantially higher than fresh water.

We might expect those animals that osmoconform to also ionoconform, and those that osmoregulate to also ionoregulate. This is generally true for most marine osmoconformers, but even these animals actually ionoregulate to some extent as the concentration of most ions differs slightly from sea water, and the concentration of a few ions can differ substantially from sea water.

Thus, there are three basic patterns of osmotic and ionic balance for animals:
1. some osmoconform and ionoconform
2. some osmoconform but ionoregulate
3. some osmoregulate and ionoregulate.

The pattern for any group of animals largely reflects its phylogenetic and environmental history. To osmoconform and ionoconform is the primitive pattern, with ionoregulation a derived pattern probably being a freshwater or brackish water adaptation. Freshwater or brackish water animals that return to sea water can either continue to ionoregulate and osmoregulate or can revert to osmoconforming and either ionoconform or ionoregulate but osmoconform using non-ionic solutes. If vertebrates evolved in fresh water, then all three patterns are evident for groups that returned to sea water: osmoregulate and ionoregulate (bony fishes), osmoconform but ionoregulate (cartilaginous fishes), and osmoconform and ionoconform (hagfishes).

> Some animals osmoconform and ionoconform, some osmoconform and ionoregulate, and some osmoregulate and ionoregulate. The pattern for any group of animals reflects its phylogenetic and environmental history.

Marine animals

Most marine invertebrates osmoconform and ionoconform. They evolved in sea water and have remained in sea water, and retain the primitive and expedient pattern of osmoconforming and ionoconforming. Only one primitive group of marine vertebrates, the hagfishes, osmoconforms and ionoconforms. This might reflect a marine origin for vertebrates or an unusual (for vertebrates) return to the primitive pattern of osmoconforming and ionoconforming when hagfishes reinvaded the sea. Because marine osmoconformers are in osmotic balance with their environment, they have essentially no water flux across their skin or gills but do have a low rate of water intake (by eating and drinking) and loss (by excretion). Because these marine ionoconformers are in ionic balance with their environment, they do not expend energy in maintaining substantial ionic gradients, although these animals are capable of limited ionoregulation. For example, some regulate a low concentration of heavy ions (such as sulfate) for buoyancy control.

Marine cartilaginous fishes (sharks, rays and chimaeras) osmoconform but ionoregulate. They are strong ionoregulators of Na$^+$, K$^+$ and Cl$^-$ and maintain a similar ionic composition of body fluids to other vertebrates, perhaps reflecting a freshwater ancestry of vertebrates. However, they avoid the need to drink by osmoconforming to sea water. They are able to osmoconform but ionoregulate by accumulating high levels of urea (Box 22.2). The marine coelacanth (a lobe-finned fish) and a semi-marine amphibian (the crab-eating frog) also osmoconform but ionoregulate, using urea as a balancing solute.

Many marine animals osmoregulate and ionoregulate. A few marine invertebrates regulate their

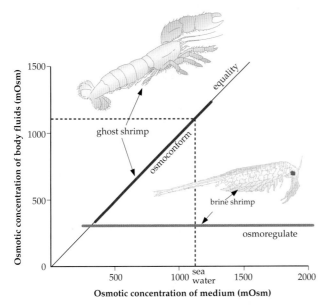

Fig. 22.3 Patterns of change in the body fluid osmotic concentration and the osmotic concentration of the medium for an osmoconformer, the ghost shrimp, *Callianassa*, and an osmoregulator, the brine shrimp, *Artemia*

BOX 22.2 Urea in marine osmoconformers/ionoregulators

Marine chondrichthyean fishes (elasmobranchs—sharks, skates and rays—and the rat fishes) osmoconform; they have almost the same osmotic concentration as sea water (1100 mOsm). Their body fluid osmotic concentration is actually a little higher than the osmotic concentration of the sea water and so some water diffuses into these fishes by osmosis.

The extracellular Na^+ and Cl^- concentrations of these chondrichthyean fishes are much lower than sea water despite the osmotic concentration being the same. The total ion concentration is about 600 mOsm. This means that there is an 'osmotic gap' of about 500 mOsm between the total ion concentration and the osmotic concentration. These fishes synthesise and actively retain urea (which is normally a nitrogenous waste product of terrestrial animals) and also trimethylamine oxide (TMAO) as balancing osmolytes to fill the 'osmotic gap'. Consequently, these chondrichthyean fishes are said to **ureo-osmoconform** because urea is a major osmolyte

that allows them to osmoconform to sea water. The high urea concentrations of chondrichthyean fishes would be toxic to many other vertebrates but their metabolic physiology is adapted to function normally with high urea levels.

The ureo-osmoconforming strategy of chondrichthyeans is also used by two other marine vertebrates. The coelacanth, *Latimeria chalumnae*, a relict lobe-finned crossopterygian fish related to the evolutionary line leading to tetrapod vertebrates (Chapter 40), is also a ureo-osmoconformer. The crab-eating frog, *Rana cancrivora*, is an unusual amphibian because it can survive in sea water. Adult crab-eating frogs are ureo-osmoconformers. Their tadpoles also survive in sea water but are not ureo-osmoconformers; rather, they are ionoregulators and osmoregulators, like marine bony fishes. The crab-eating frog is not, however, completely marine; its eggs require fresh water to develop and its tadpoles need fresh water to metamorphose successfully.

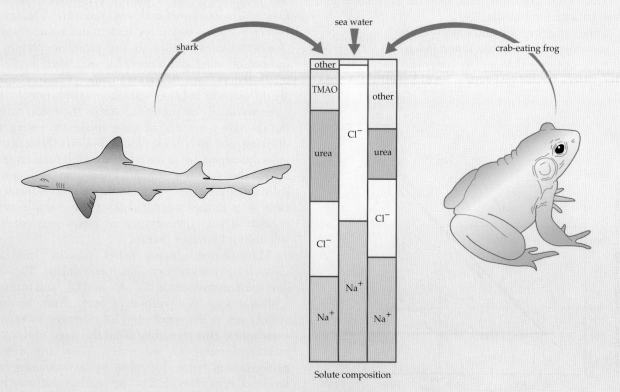

Comparison of the ion, urea and osmotic concentrations of two ureo-osmoconforming marine vertebrates, a shark and the crab-eating frog

osmotic concentration slightly lower than sea water. Marine lampreys, bony fishes, reptiles, birds and

mammals regulate their body fluids at about one-third of sea water (Fig. 22.4). These marine osmoregulators

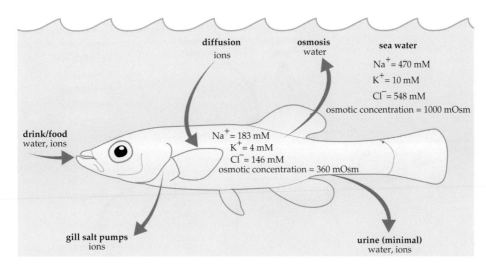

Fig. 22.4 A marine bony fish is an ionoregulator and osmoregulator at a body fluid osmotic concentration of about 360 mOsm. Consequently, there is a constant influx of ions by diffusion and a loss of water by osmosis. Water is replaced by drinking (and from food), and active gill pumps remove excess salts. There is some loss of water and ions by the production of small amounts of urine

lose water by osmosis from their body fluids to sea water across the skin and gills (except for air-breathing marine vertebrates). Water loss can only be replenished by drinking sea water, which markedly increases salt gain. Ion gain by drinking and by diffusion across the skin (and gills, if present) is countered by active ion excretion across the gills, via the kidneys, or by salt glands.

> Most marine invertebrates and one vertebrate group, the hagfishes, osmoconform and ionoconform. Cartilaginous fishes, the coelacanth and the crab-eating frog ionoregulate but osmoconform using urea. Marine lampreys, bony fishes, reptiles, birds and mammals ionoregulate and osmoregulate.

Hypersaline animals

While many animals are able to live in sea water, only a select few can survive in **hypersaline** water, such as natural salt lakes formed by salt leaching from the soil or water evaporation, and the evaporation ponds of salt works. Some bony fishes can live in salt lakes, estuaries and pools, with a concentration up to two or three times sea water. The brine shrimp, *Artemia*, can survive in hypersaline water over five times as concentrated as sea water.

These animals that live in hypersaline water must ionoregulate and osmoregulate. Excessive diffusion of ions into their extracellular body fluids or osmotic loss of body water would disrupt normal cellular functions. Salts that are ingested or diffuse through their body wall are actively excreted by salt pumps located on gills in fishes, and on the appendages of brine shrimps (see Fig. 22.9). The body water lost by osmosis can only be replenished by drinking the salty water, even though this markedly increases the salt load that must then be excreted.

> Animals living in hypersaline water must ionoregulate and osmoregulate.

Freshwater animals

Animals that live in fresh water gain water by osmosis and lose salts by diffusion to their environment. They must ionoregulate and osmoregulate because their cells require some ions and other solutes for normal function. Nevertheless, many freshwater animals have remarkably dilute body fluids compared with marine animals. For example, the intracellular osmotic concentration of many freshwater protists is only about 50 mOsm and the body fluids of some freshwater invertebrates are also very dilute (sponges, 55 mOsm; molluscs, 40–100 mOsm; worms, 100–200 mOsm).

Freshwater bony fishes have a similar extracellular composition to marine bony fishes, with a total osmotic concentration of about 300 mOsm (Fig. 22.5). They gain water by osmosis and drinking, and surplus water is excreted as copious and dilute urine. Salts lost by diffusion across the skin and gills and in urine are replenished by ions absorbed by the gut and by active uptake of ions across the gills. The few freshwater cartilaginous fishes ionoregulate and osmoregulate, just like freshwater bony fishes. Unlike marine cartilaginous fishes, they do not accumulate urea in their body fluids but excrete ammonia instead.

> Freshwater animals must ionoregulate and osmoregulate, but many have remarkably dilute body fluids.

Terrestrial animals

Terrestrial animals invariably ionoregulate and osmoregulate; their body fluid osmotic concentration is generally 300–600 mOsm.

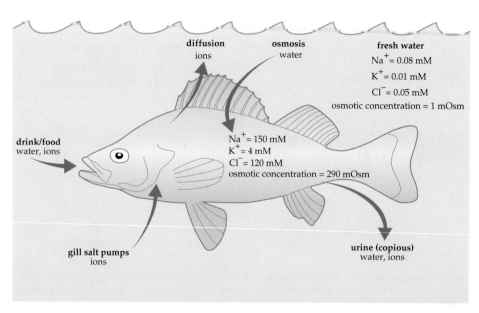

Fig. 22.5 Freshwater bony fishes have higher body fluid ion and osmotic concentrations than does their freshwater medium. Water is gained by osmosis, by drinking and in food, and ions are lost by diffusion and in urine. Producing a copious amount of dilute urine excretes water. Ions are actively absorbed by gill salt pumps and are obtained in food

Water is lost by terrestrial animals in their urine and faeces, and by evaporation from their body and lung surfaces (Fig. 22.2). To maintain balance, this water loss must be replenished through drinking, by preformed water contained in food and by metabolic water synthesised by cellular metabolism (Chapter 5).

For many terrestrial animals, drinking is the most important avenue of water gain. One lizard, the Australian thorny devil, *Moloch horridus*, has evolved a remarkable mechanism to facilitate drinking; its skin acts as a 'blotter' to absorb rain and dew and then directs the water to its mouth (Box 22.3). However, if drinking water is unavailable, water balance must be maintained by preformed and metabolic water. For example, the Australian hopping mouse, *Notomys alexis*, a desert seed-eating rodent, does not have to drink although it will if free water is available. Under dry conditions, the hopping mouse can obtain sufficient water from the seeds it eats and from metabolic water production to cover water losses (Table 22.2).

Water loss must be minimised if water is not freely available. Consequently, terrestrial animals often have physiological mechanisms and behaviours that minimise evaporative, urinary and faecal water loss. Many desert-adapted animals (like *Notomys*) are able to reduce their evaporative, urinary and faecal water losses to such an extent that they can survive by relying on metabolic water production as their main water source. On the other hand, animals that consume nectar or plant sap must excrete the excess water without compromising their solute balance, and produce copious amounts of very dilute urine.

The ion balance of terrestrial animals is relatively straightforward because there is no significant exchange

Table 22.2 Routes of water gain and loss for the Australian hopping mouse, *Notomys alexis*, which can survive without drinking

Route	Amount of water (mL per day)
Water gain	
Drinking	0.0
Food	0.7
Metabolic	0.9
Total gain	1.6
Water loss	
Evaporation	1.21
Urine	0.33
Faeces	0.06
Total loss	1.6
Net gain/loss	0

across the body surface by diffusion or active transport. Ions are only gained in food and drink, and only lost in urine and faeces.

Terrestrial animals ionoregulate and osmoregulate. Drinking, preformed water and metabolic water are avenues of water gain, and evaporation, urine and faeces are avenues of water loss.

A few terrestrial arthropods are able to supplement their normal avenues for water gain (food, drink and metabolism) by absorbing water vapour directly from

BOX 22.3 Water uptake by the thorny devil

Desert animals must take maximum advantage of the infrequent rainfall. Many desert reptiles lick rainwater or fog condensate from their skin, but a remarkable Australian lizard, the thorny devil, *Moloch horridus*, has an unusual skin texture that facilitates drinking of dew and rain.

The skin of the thorny devil, like that of all lizards, is covered with scales, but many of the thorny devil's scales are small and surrounded by a narrow channel about 10 μm wide. These inter-scalar channels attract water by capillary action and the skin absorbs water like a piece of blotting paper. However, water is not absorbed into the body directly through the skin. Rather, the interconnected network of channels directs this water towards the mouth, where it is drunk. Rain that falls on the skin can be drunk, as can dew. Water can also be absorbed onto the skin from wet soil after rain, and conveyed to the mouth.

(a) The thorny devil

(b) The thorny devil has numerous small scales, surrounded by narrow channels, which absorb water by capillarity

unsaturated air. For example, some isopods can absorb water vapour from the air if the relative humidity is greater than 93%, and the desert silverfish *Thermobia* can absorb water from air as dry as 45% relative humidity. The site of water vapour absorption is always the mouth or the anus and there are a variety of mechanisms. Some ticks extrude a droplet of KCl-rich saliva, which is hygroscopic and absorbs water vapour from the air. Mealworm larvae of *Tenebrio molitor* absorb water vapour through the rectum; active pumping of K^+ into the blind ends of the Malpighian tubules creates a steep concentration gradient that osmotically removes water from the rectal contents and forms very dry faeces as well as absorbing water vapour (Fig. 22.6). Other mechanisms for water vapour absorption are less well understood.

> A few terrestrial arthropods can absorb water vapour from unsaturated air.

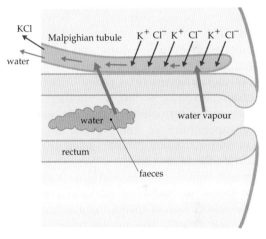

Fig. 22.6 The hindgut of the mealworm larva, *Tenebrio molitor*, fuses with the blind ends of the Malpighian tubules to form the complex cryptonephridial system. This system reabsorbs water from the faeces in the hindgut, and can absorb water vapour from air inside the hindgut at relative humidities greater than 88%

Nitrogenous wastes

Cellular metabolism produces a variety of waste products that must be excreted. Water and carbon dioxide are the only waste products from the complete aerobic metabolism of carbohydrates and lipids. However, metabolism of protein requires excretion of the nitrogen (N) component of its constituent amino acids (and to a much lesser extent the excretion of sulfur). The high rate of protein turnover of cells, and the high N content of protein, means that there is a significant requirement for excretion of N as a nitrogenous waste.

Ammonia (NH_3) is the nitrogenous waste initially formed by metabolism of protein through deamination of amino acids to keto acids, but it can be converted to a number of other nitrogenous wastes such as urea, uric acid and guanine (Fig. 22.7). The primary nitrogenous waste can be ammonia (this pattern is called **ammonotely**), urea (**ureotely**), or a purine (**purinotely**), which is usually uric acid (**uricotely**) but can be guanine (**guanotely**).

A mmonia, urea, uric acid and guanine are the primary nitrogenous waste products of animals.

Ammonia is very soluble in water but is also highly toxic. Although many aquatic animals excrete ammonia as their nitrogenous waste, terrestrial animals generally convert the ammonia to a less toxic nitrogenous waste. Invertebrates such as earthworms and pulmonate snails, and vertebrates such as amphibians and mammals, convert ammonia to urea (CON_2H_4). A biochemical cycle of reactions, called the urea cycle, condenses ammonia with carbon dioxide to form urea in a process that requires energy. Urea is less toxic than ammonia but is still very soluble. Some terrestrial animals, such as insects, reptiles and birds, convert ammonia to an even less toxic and highly insoluble nitrogen waste product, uric acid ($C_5H_4O_3N_4$), by a series of metabolic reactions called uricogenesis. This conversion involves a larger number of specific enzymes and intermediate reactions, and an even greater expenditure of energy than urea synthesis. Spiders and scorpions excrete guanine ($C_5H_5ON_5$), which is similar to uric acid.

The metabolism of nucleic acid bases is another source of nitrogenous wastes. Ammonia is directly produced from the metabolism of the pyrimidine bases (thymine, cytosine and uracil). The purine bases (adenine and guanine) are degraded by a complex series of reactions

and excreted as guanine (spiders and scorpions), uric acid (terrestrial snails, reptiles, birds, humans, apes and Dalmatian dogs), urea (many mammals), or ammonia (many aquatic animals). The amount of nitrogenous waste formed from nucleic acid metabolism is small relative to that from protein metabolism, and is not necessarily in the same form. For example, humans excrete a little uric acid as their nitrogenous waste from purine metabolism but urea from protein metabolism is the predominant nitrogenous waste.

There are relative advantages and disadvantages associated with ammonia, urea, uric acid and guanine as a nitrogenous waste. Ammonia contains only one nitrogen molecule per atom, so every nitrogen atom is equivalent to an osmotic particle that requires water for excretion, and it is also very toxic. However, it is extremely soluble and no energy is expended in its synthesis. Urea is less toxic than ammonia and the excretion of nitrogen as urea requires less water than ammonia because each molecule of urea contains two nitrogen atoms. However, the synthesis of urea from ammonia requires the expenditure of energy, so there is a metabolic cost to urea excretion. Uric acid contains four nitrogen atoms per molecule, so its excretion conserves even more water; it is also highly insoluble and non-toxic but its synthesis requires even more energy expenditure than urea synthesis. Guanine, another purine, is also nearly insoluble and, like uric acid, can be excreted with little water loss.

Phylogenetic and environmental patterns

Not surprisingly, the pattern of nitrogenous waste excretion seen in any particular animal group generally

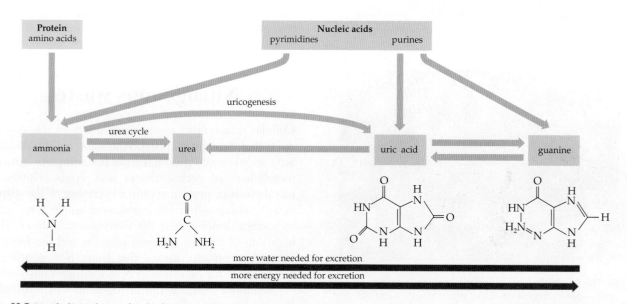

Fig. 22.7 Metabolic pathways for the formation of various nitrogenous waste products from protein and nucleic acid metabolism (purple arrows) and the synthesis of complex nitrogenous wastes from ammonia (yellow arrows)

reflects the availability of water as well as its phylogeny. A survey of nitrogenous wastes in animals (Fig. 22.8) shows a general environmental pattern with aquatic animals excreting ammonia and terrestrial animals excreting urea or purines. For terrestrial animals there are phylogenetic patterns: annelids, amphibians and mammals excrete urea whereas molluscs, arthropods, reptiles and birds excrete uric acid or guanine. However, there are some notable exceptions and flexible patterns.

Most aquatic animals excrete ammonia across their body or gill surface. Although it is very toxic, ammonia is also very soluble in water and diffuses into the external water so readily that it does not accumulate in the body to toxic levels. Freshwater cartilaginous fishes excrete ammonia but marine forms and the lobe-finned coelacanth excrete urea because it is a major body fluid solute for osmoregulation (Box 22.2 and Chapter 40). Aquatic amphibians excrete ammonia, as do some aquatic reptiles (tortoises, turtles, crocodilians).

Terrestrial animals generally do not excrete ammonia. Because it is so soluble, ammonia would have to accumulate to toxic levels in the body fluids for it to be excreted across their body surface as a gas and too much water would be required for its excretion in urine. Nevertheless, ammonia is the primary nitrogenous waste for terrestrial isopods and a few other crustaceans, molluscs and insects, which can excrete gaseous ammonia across their body surface or via faeces.

Terrestrial animals generally convert ammonia to a less toxic form, either highly soluble urea or insoluble purines (uric acid, guanine). Terrestrial worms and some snails excrete urea. Lungfishes produce urea rather than ammonia when they become dormant (aestivate) out of water. Most terrestrial amphibians excrete urea and a few semi-aquatic tortoises excrete urea. Mammals excrete urea. Most terrestrial insects and snails, reptiles and birds, and some crabs, excrete uric acid whereas spiders and scorpions excrete guanine. In purinotelic animals, urine is often emptied into the hindgut so that excess water can be reabsorbed and pasty urine is then eliminated (mixed with the faeces). Two remarkable genera of frogs, *Chiromantis* and *Phyllomedusa*, have independently evolved uric acid excretion.

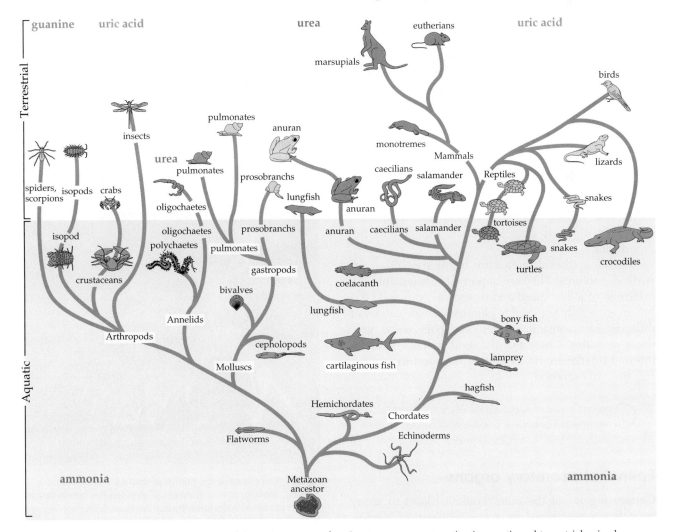

Fig. 22.8 Schematic summary of the evolution of the major patterns for nitrogenous waste excretion in aquatic and terrestrial animals

Although there are environmental and phylogenetic patterns to nitrogenous waste excretion, some groups of animals are remarkably flexible in changing their primary nitrogenous waste. For example, aquatic tortoises can be primarily ammonotelic whereas semi-aquatic tortoises are ureotelic and terrestrial tortoises are uricotelic. Even individual animals can change their pattern of nitrogenous waste. Lungfishes are ammonotelic in water and ureotelic when aestivating in air; amphibians are ammonotelic when aquatic and ureotelic when terrestrial.

> Aquatic animals generally excrete ammonia, but cartilaginous fishes and the coelacanth excrete urea. Terrestrial animals generally excrete urea or purines (uric acid, guanine) but a few excrete gaseous ammonia.

Excretion

The role of excretion is to maintain the normal solute composition of body fluids, to regulate the body water content and to remove nitrogenous wastes from the body fluids.

Some freshwater protists, such as the amoeba and paramecium, and most cells of simple freshwater sponges, excrete water and solutes by the action of an intracellular organelle, the **contractile vacuole**. In more complex animals, specialised excretory organs are present. There is considerable diversity in the structure of excretory organs but they can be classified into two general types:
1. epithelial solute pumps located at the surface of the animal, which transport specific solutes across the body surface
2. internal tubular excretory organs that form then modify a liquid urine, which is expelled to the exterior.

Surface epithelial pumps are highly specialised for particular solutes. They are important in regulating the exchange of a few specific ions, such as Na^+ and Cl^-, but are generally not used to eliminate a wide variety of different ions, organic solutes, metabolic waste products or water. Consequently, most animals have an internal tubular excretory system that performs these functions of excretion.

> Contractile vacuoles excrete excess intracellular fluid to regulate cell volume in unicells. Multicellular animals have epithelial and/or tubular excretory organs.

Epithelial excretory organs

Certain regions of the outer epithelial layer of many animals are specialised for active transport of solutes, either into or out of the animal. For example, brine shrimp living in hypersaline water gain salts by diffusion and drinking. They have epithelial salt glands (Fig. 22.9) that actively excrete Cl^-, and Na^+ moves passively along with the Cl^- to maintain electrical neutrality.

Freshwater and marine bony fishes also maintain salt balance using salt pumps located on their gills. These pumps actively excrete Cl^- (and Na^+ passively exits) in marine species, and actively absorb Cl^- (and Na^+ passively enters) in freshwater species. Amphibian skin actively transports Na^+ into the body fluids (and Cl^- passively enters).

If the movement of ions results in a change in the osmotic gradient across the epithelium, then water will also tend to move by osmosis. Thus, through active ion transport and regulation of permeability of an epithelium to water, specific ions such as Na^+ and Cl^-, and water, can be transported across an epithelium.

> Surface epithelial salt pumps excrete or absorb specific ions for ionoregulation. Water may also be passively transported.

Tubular excretory organs

Tubular excretory organs have a number of functional roles, including urine formation, reabsorption, secretion, and sometimes osmodilution or osmoconcentration (Fig. 22.10). All excretory tubules form a fluid at their inner end, usually by filtration of coelomic fluid or blood. As this fluid passes along the excretory tubule, its composition may be substantially modified by reabsorption of solutes and water, or by solute secretion. Active reabsorption of useful solutes and water

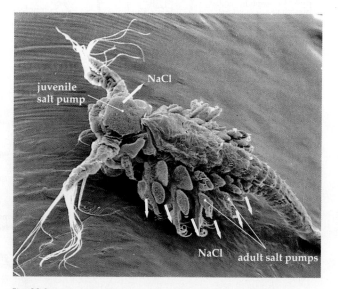

juvenile salt pump

NaCl

NaCl adult salt pumps

Fig. 22.9 The epithelial salt pumps of immature brine shrimp, *Artemia*, are located in a juvenile salt gland in the neck whereas the salt pumps of adult brine shrimp are located on the appendages. Intermediate stages of brine shrimp, as shown here, have ion pumps in both locations

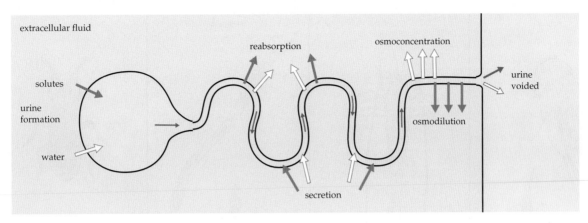

Fig. 22.10 Tubular excretory organs form urine, subsequently modify it by reabsorption and secretion, and then void it to the exterior

is important because the initial fluid has essentially the same solute composition as blood plasma, and contains many ions, organic solutes and vitamins. Active secretion of specific waste products into the urine occurs. In some animals, to excrete or conserve water the tubules also modify the osmoconcentration of urine by the active addition of solutes and/or osmotic withdrawal of water.

> Tubular excretory organs form, modify and expel urine to excrete a wide variety of substances.

There are three basic types of tubular excretory organs in animals—nephridia, coelomoducts and Malpighian tubules. The nephrons of the vertebrate kidney are derived from coelomoducts. **Nephridia** develop as ingrowths from the body towards the body cavity. Protonephridia are blind-ended whereas metanephridia have a ciliated opening, the nephridiostome (Fig. 22.11). The cells at the blind end of protonephridia have numerous cilia, which draw fluid into the tubule via perforations in its wall. These cells are commonly called 'flame cells' because the continual beating motions of their cilia resemble the flickering of a flame. The terminal cells also form the perforations that allow coelomic fluid to be drawn into the tubule lumen by the negative hydrostatic pressure that the cilia produce. Protonephridia are present in many invertebrates, for example, rotifers. Metanephridia are found in more complex invertebrates. A metanephridium has a terminal ciliated opening, the nephridiostome, which is continuous with the coelomic space, and coelomic fluid is drawn into the metanephridial lumen by ciliary beating. The coelomic fluid is formed elsewhere in the coelom, generally by filtration of fluid out of blood vessels.

Coelomoducts are present in many invertebrates and also primitive vertebrates such as hagfishes. Coelomoducts resemble metanephridia in general structure but unlike nephridia they develop outwards, from the coelomic lining towards the external body surface. Coelomic fluid is drawn into a ciliated funnel-like opening, the coelomostome, then passes along the tubule and is eliminated from the coelomopore. Coelomic fluid is formed elsewhere in the coelom by filtration of fluid from blood vessels. In advanced vertebrates, the roles of the fluid-forming blood vessels and the coelomoducts are combined into a single structure, consisting of a vascular glomerulus and the nephron tubule, which is the functional unit of the kidney.

Many arthropods, such as insects and some spiders, have **Malpighian tubules**. These are linear or branched blind-ended tubules that open into the digestive tract at the junction of the midgut and hindgut. Fluid and solutes are drawn into the blind ends by the active transport of K^+ from the extracellular fluid into the tubule; other solutes and water passively follow K^+ into the tubule. The 'urine' formed by the Malpighian tubules is emptied into the hindgut, where it is subsequently modified by the rectal epithelium, then excreted with the faeces.

> There are three types of tubular excretory organs— nephridia, coelomoducts and Malpighian tubules.

Invertebrates

Excretory organs

There are many kinds of excretory organs in invertebrates. Protists and primitive animals such as freshwater sponges rely on ion pumps for ionoregulation and the contractile vacuole for water regulation by excretion of excess intracellular fluid. These cells face a continual osmotic uptake of water that must be excreted from the cell if it is to maintain a constant volume. A series of small tubules, the spongiome, collects intracellular fluid and delivers it to a spherical vesicle that contracts and expels the fluid from the cell through a pore. The rate of contractile vacuole activity is inversely proportional to the osmotic concentration of the external medium so that water is excreted faster in dilute media.

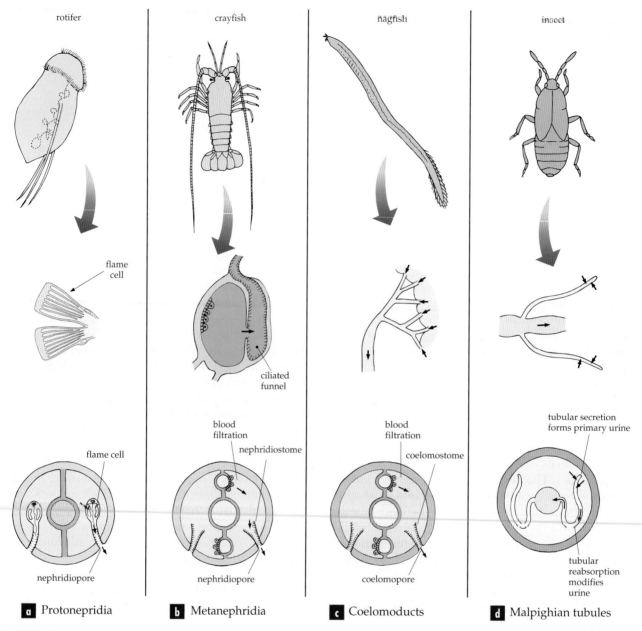

Fig. 22.11 Different forms of tubular excretory organs. **(a)** Protonephridia have a terminal flame cell, which transports fluid from the coelomic cavity into the tubule by ciliary action. **(b)** Metanephridia have a ciliated funnel that transports coelomic fluid into the tubule. **(c)** Coelomoducts are similar to metanephridia in structure and function. **(d)** Malpighian tubules of arthropods form a filtrate by the active secretion of K^+ into the tubule; the tubular fluid is emptied into the hindgut

Multicellular invertebrates, such as flatworms, nemerteans, pseudocoelomate worms and some annelids, have protonephridia. These collect body fluid to form a primary filtrate that is modified to form urine. For example, the freshwater rotifer *Asplanchna* excretes urine (40 mOsm) that is more dilute than its body fluids (80 mOsm) but more concentrated than lake water (18 mOsm). The larvae of many higher invertebrates also have protonephridia.

Many of the more complex invertebrates have metanephridia or coelomoducts rather than protonephridia, particularly as adults. Annelid worms have metanephridia that filter coelomic fluid from the more anterior body segment (Fig. 22.12). The osmotic concentration of the urine declines markedly as it passes through the tubule of freshwater annelids because of solute reabsorption (mainly active Na^+ and passive Cl^- reabsorption). The **antennal glands** (or green glands) of crustaceans are paired metanephridial excretory organs. A coelomic end sac forms urine by filtration of fluid from blood vessels and solutes are reabsorbed during passage along a nephridial canal to form dilute urine; a bladder stores urine and reabsorbs solutes to further dilute the urine before it is voided

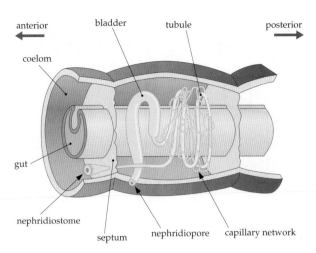

Fig. 22.12 The excretory system of the earthworm, *Lumbricus*, consists of paired, segmentally arranged metanephridia (coloured green), which collect coelomic fluid from the more anterior body segment, modify the urine by active solute reabsorption, then void it to the exterior via the nephridiopore

through an excretory pore (Fig. 22.13). Marine crustaceans form urine that is iso-osmotic with blood and similar in ion concentration; the nephridial canal may be short or absent.

Many arthropods have both metanephridial excretory tubules (called coxal glands) and Malpighian tubules (Fig. 22.14). Primitive spiders have two pairs of coxal glands, each consisting of a thin-walled spherical sac that filters haemolymph by suction, an excretory tubule that modifies the urine and an excretory pore on the coxal part of the leg. More advanced spiders rely

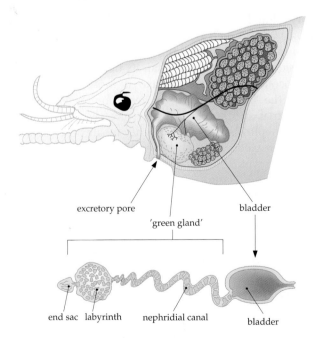

Fig. 22.13 The crustacean antennal gland is a metanephridium. The coelomosac forms urine by filtration, which is then diluted as it passes along the tubule by active solute reabsorption

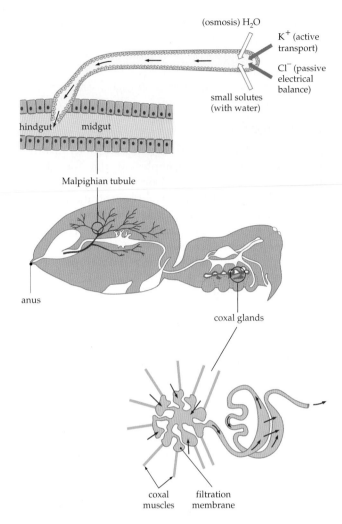

Fig. 22.14 Primitive spiders have two pairs of metanephridial coxal glands and branching Malpighian tubules. The muscles of the coxal glands stretch the spherical sac and suck fluid and small solutes into the sac; the fluid is modified as it passes along the excretory tubule. Ions are pumped into the blind end of the Malpighian tubules and this osmotically draws in water and other small solutes; the fluid is modified as it passes along the tubule and then empties into the hindgut for further modification

primarily on two branched Malpighian tubules for excretion.

Insects have a variable number of Malpighian tubules. Urine is formed in the blind ends of the tubules and is modified as it passes along the tubules by the reabsorption of ions and the precipitation of uric acid. It is emptied into the hindgut for further modification by reabsorption of solutes and water to form a very dry urate paste. The rectal reabsorption of water is coupled in some insects with the blind ends of the Malpighian tubules (Fig. 22.6) where the high solute concentration osmotically removes water from the rectal contents, thus forming very dry faeces. This also allows absorption of water vapour from air in the rectum down to a relative humidity of about 88% in *Tenebrio*.

Protists and freshwater sponges have ion pumps for ionoregulation and contractile vacuoles for water regulation. Primitive multicellular animals use protonephridia for excretion. More complex invertebrates have metanephridia, coelomoducts or Malpighian tubules for excretion.

Tubular function

In reviewing the excretory organs of various invertebrates, we have seen examples of the four basic functions of tubular excretory organs—filtration, reabsorption, secretion and osmoconcentration.

Filtration of blood or coelomic fluid to form urine can be accomplished in a variety of ways. The flame cells of protonephridia create a negative pressure in the internal blind end, which draws fluid through slit-shaped structures in the protonephridial wall; the filtration force is hydrostatic pressure and a thin plasma membrane that covers the slits is the filter. The coxal glands of spiders work in a similar way but on a larger scale. Contraction of the coxal muscles expands the saccular region, creating a negative internal pressure that draws fluid into the saccule; the filtration force is hydrostatic pressure and the saccular membrane is the filter. Fluid 'filtration' in metanephridia and coelomoducts is not really filtration at all; the beating action of cilia at the internal nephridiostome/coelomostome simply draws coelomic fluid, with all of its solute constituents, into the tubule. The blind end of Malpighian tubules creates an osmotic gradient by pumping K^+ into the lumen of the tubule (Cl^- follows passively). The osmotic gradient draws water into the lumen through the plasma membranes and also between cells, and the influx of water 'drags' in solutes that are small enough to pass through the cell membranes or intercellular spaces. The filtration force is an osmotic pressure, and the tubular plasma membranes and intercellular spaces are the filter that allows passage of small solutes (e.g. glucose, amino acids, nitrogenous wastes) but excludes large solutes (e.g. most proteins) and blood cells from passing into the urine.

An important role for all tubular excretory organs is to reabsorb nutrients and other useful solutes (e.g. glucose, amino acids, protein) because the filtration mechanism is selective for molecular size and not identity of specific solutes. Reabsorption of water or solutes almost invariably occurs as fluid moves along the tubular excretory organ, but the balance of water and ion reabsorption varies for marine and freshwater animals. Freshwater invertebrates, such as crayfish, reabsorb large amounts of major body fluid ions, such as Cl^- (Fig. 22.15). Marine invertebrates often do not reabsorb water, but some ions are reabsorbed (e.g. K^+ and Ca^{2+}).

Secretion of solutes by tubular excretory organs is not so important in general terms as reabsorption but

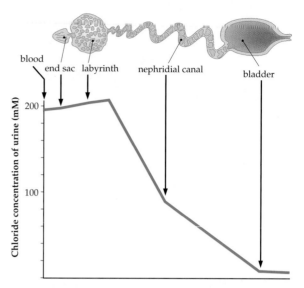

Fig. 22.15 The urine of freshwater crustaceans such as *Potamobius* is diluted considerably by active Cl^- reabsorption as it passes along the tubule of the antennal gland; the osmotic concentration also declines

can be very important for specific solutes. For example, excretory organs of many invertebrates (and also vertebrates) secrete complex organic molecules such as para-aminohippuric acid and the dye phenol red, and presumably many organic waste products or toxins are similarly secreted into the urine. Ions such as Mg^{2+}, SO_4^{2-} and H^+ are often secreted into the urine.

Modification by osmodilution of urine is also an important tubular function in many, especially freshwater, animals. To counter the osmotic influx of water, freshwater animals produce copious urine; the urine is very dilute because of the extensive reabsorption of the major solutes such as Cl^- and Na^+, but not water. For example, the urine of the earthworm *Lumbricus* is considerably more dilute than its blood. The ability to osmoconcentrate urine (make it more concentrated than the blood) is rare in invertebrates except the terrestrial arthropods. We have already seen how the cryptonephridial system of the mealworm larva of *Tenebrio molitor* can absorb water vapour (Fig. 22.6), and the cryptonephridial system also osmoconcentrates urine in the rectum. The epithelium of the rectal pads of some flies actively reabsorbs ions from the rectal lumen and water is passively withdrawn by osmosis. Ions are then actively removed from the reabsorbed fluid, leaving it relatively dilute and the rectal fluid relatively concentrated (Fig. 22.16).

Vertebrates

The primary excretory organ of vertebrates is the **kidney** but various other organs can be involved with solute and water regulation, including the skin, gills, gut and salt glands.

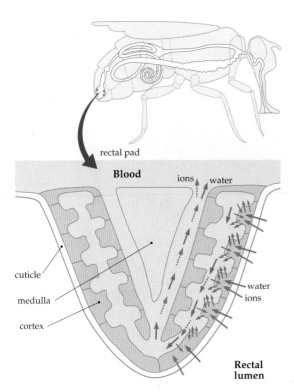

Fig. 22.16 The rectal pads of some insects are able to absorb a hypo-osmotic fluid from the rectal contents by a complex arrangement of active ion pumps. Ions are actively absorbed from the rectal lumen and water passively follows by osmosis; ions are subsequently removed from the absorbate by active transport

Aquatic vertebrates exchange ions and water across the gills and excrete them via the kidney. Freshwater bony fishes and amphibians are in constant danger of becoming overhydrated by the osmotic influx of water from the dilute medium, and hypoionic by diffusional loss of ions. Their kidneys excrete the excess water as copious dilute urine and salt pumps on the gills (fishes) or skin (fishes and amphibians) absorb salts from the dilute medium. The kidneys of freshwater fishes are relatively large, reflecting their need to have a high urine flow. In contrast, marine bony fishes gain ions by diffusion; these ions are excreted by gill or skin salt pumps. They lose water by osmosis and as urine, although the urinary water loss is low because the kidneys are small and they often lack glomeruli (see below). Marine cartilaginous fishes and the coelacanth excrete ions via their kidneys and a special rectal gland located in the posterior gut.

Terrestrial vertebrates rely primarily on their kidneys for ionoregulation and osmoregulation, although some reptiles and birds have salt glands for NaCl or KCl excretion.

The kidney is the principal excretory organ of vertebrates.

Kidneys

Vertebrate kidneys are paired organs that range in shape from ovoid and compact structures to elongated and more diffuse organs, especially in fishes and amphibians (Fig. 22.17). Each kidney consists of thousands, or even millions, of individual **nephrons**. The nephrons of primitive vertebrates, such as hagfishes, are coelomoducts but the nephrons of more advanced vertebrates are structurally modified so that the urine formed by the nephron is not collected from the coelomic space but is formed by filtration of blood from a tuft of capillaries, the **glomerulus**. Some of the nephrons of amphibians have both ciliated coelomostomes and glomeruli (Fig. 22.17).

In glomerular nephrons, the proximal end of the nephron is dilated and invaginated to form a hollow, double-walled cup. The outer wall of cells forms the renal or Bowman's capsule (Fig. 22.18). The inner wall of cells is highly modified as podocytes, which cover the glomerular capillaries but leave many slit-like spaces.

The nephron tubule, which extends from the renal capsule, consists of an initial **proximal convoluted tubule** and a subsequent **distal convoluted tubule**. Nephrons of lower vertebrates have a ciliated neck segment near the glomerulus and an intermediate ciliated segment between the proximal and distal convoluted tubules; the amphibian nephron well illustrates this general structure (Fig. 22.17). In mammal and some bird nephrons, the intermediate segment forms a hairpin loop, the **loop of Henle**, resulting in descending and ascending segments lying parallel and close together. This arrangement greatly improves the kidney's ability to reabsorb water from urine and is extremely important for terrestrial mammals living in arid environments. The distal convoluted tubule runs into the **collecting duct**, which connects to the **ureter** that drains urine from the kidney.

The vertebrate nephron has a glomerulus, renal capsule, proximal and distal convoluted tubules, and a collecting duct. Mammalian and avian nephrons also have a loop of Henle.

The vertebrate nephron is the functional unit of the kidney. Like the tubular excretory organ of invertebrates, it has four main functions: formation of fluid for excretion by filtration, and subsequent modification of the fluid before it is excreted by reabsorption, secretion and osmodilution/concentration.

Filtration

Hydrostatic pressure forces fluid from glomerular capillaries into the renal capsule. Glomerular capillary cells are perforated by numerous tiny holes (fenestrae) that allow ready loss of fluid and solutes from the

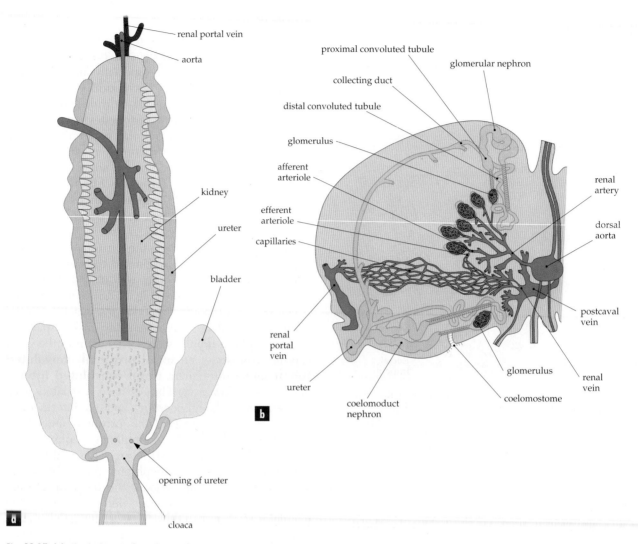

Fig. 22.17 (a) The kidneys of a salamander, *Necturus maculosus* (a urodele amphibian), are paired structures with an arterial blood supply (dorsal aorta and renal arteries), a venous portal blood supply (renal portal veins) and venous drainage (renal veins and postcaval vein). **(b)** The kidneys contain two types of glomerular nephrons, those with a coelomostome and those typical of tetrapod nephrons that lack a coelomostome

capillaries. The podocytes of the renal capsule support the glomerular capillaries but their numerous branches (pedicels) interdigitate and form filtration slits (Fig. 22.18). The basal lamina of the capillary is continuous across the capillary fenestrae and filtration slits, and blood plasma is filtered across it to form the glomerular filtrate. The larger blood constituents, for example, blood cells (8–1000 μm diameter) and proteins (molecular weight greater than 60 000 daltons), are retained in the capillaries. Otherwise, glomerular filtrate has essentially the same composition as blood.

In glomerular capillaries, blood pressure is relatively high, near arterial blood pressure, because the efferent arteriole is narrower than the afferent arteriole. This high pressure forces fluid through the glomerular filtration membrane. However, a complex balance of forces determines the overall filtration pressure that forms glomerular filtrate (Fig. 22.19). Although the main force causing filtration is the high blood pressure within the glomerular capillaries, there is also a smaller hydrostatic pressure in the Bowman's capsule. This is a result of filtered fluid entering the capsule, and is required to force fluid along the nephron tubule. In addition, the difference in protein concentration between the glomerular capillary blood and the glomerular filtrate also affects filtration. The high protein concentration of glomerular blood effectively creates a 'suction pressure' (the blood colloid osmotic pressure), which tends to draw water into the capillaries by osmosis. The glomerular filtrate also has a colloid osmotic pressure, tending to draw water from blood into the Bowman's capsule, but the colloid osmotic pressure of the glomerular filtrate is negligible because of its low protein concentration (the filtration membrane is essentially impermeable to proteins). The

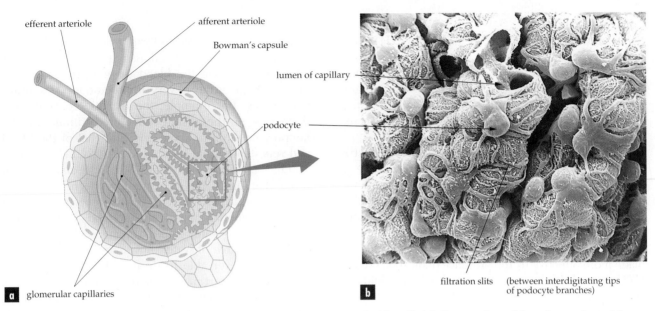

efferent arteriole

afferent arteriole

Bowman's capsule

lumen of capillary

podocyte

a glomerular capillaries

b

filtration slits   (between interdigitating tips
of podocyte branches)

Fig. 22.18 (a) The glomerulus is a spherical tuft of capillaries surrounded by a double-walled, hollow cup formed from the terminus of the nephron tubule. The Bowman's capsule is formed from the outer wall of capsule cells. **(b)** The inner wall of capsule cells forms highly modified podocytes, which support the glomerular capillaries and form filtration slits, as shown in this SEM of the glomerulus of a rat kidney

overall filtration pressure due to the balance of these forces is:

$$\begin{pmatrix} \text{glomerular} \\ \text{hydrostatic} & + & \begin{matrix}\text{filtrate}\\\text{colloid}\\\text{osmotic}\end{matrix} \\ \text{pressure} & & \text{pressure} \end{pmatrix} - \begin{pmatrix} \text{capsule} \\ \text{hydrostatic} & + & \begin{matrix}\text{blood}\\\text{colloid}\\\text{osmotic}\end{matrix} \\ \text{pressure} & & \text{pressure} \end{pmatrix}$$

In humans, for example, the overall balance of forces is a filtration pressure of about 10 mmHg (Fig. 22.19).

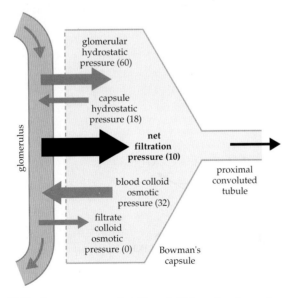

Fig. 22.19 The net pressure that filters fluid from the glomerular capillaries into the Bowman's capsule is the sum of the hydrostatic and colloid osmotic pressures acting across the filtration membrane. The approximate pressures (in mmHg) for the human nephron are shown

F ormation of the glomerular filtrate is due to the balance of hydrostatic and colloid osmotic forces.

Reabsorption

The rate of glomerular filtration is much higher than the rate of urine excretion. For example, in a resting human, about 1200 mL of blood (about one-quarter of the cardiac output) enters the kidneys per minute; about 60% of this (720 mL min^{-1}) is plasma flow. About one-fifth of the renal plasma flow is filtered; the glomerular filtration rate is about 125 mL min^{-1} (or 180 L per day). Obviously most of this glomerular filtrate must be reabsorbed! Urine flow rate is actually less than 1% of the glomerular filtration rate, about 1 mL min^{-1}, because over 99% of the glomerular filtrate, including most of its solutes, is reabsorbed. Each day, the human kidney reabsorbs about 178 L of water, 1200 g of salt and 250 g of glucose. The kidney has a high metabolic rate relative to its weight, reflecting its active absorption and secretion of solutes.

Reabsorption allows the recovery of important ions such as Na$^+$ and Cl$^-$, and nutrients such as glucose, amino acids and vitamins. The chemical composition of body fluids is precisely regulated by the control of solute reabsorption from the glomerular filtrate. Much of the reabsorption occurs in the proximal convoluted tubule, and water and solutes are returned into the blood by diffusion and osmosis into the peritubular capillary bed that surrounds the nephron.

M ost of the glomerular filtrate, particularly water and important body fluid solutes, such as amino acids and glucose, is reabsorbed by the nephron tubules.

Secretion

Some substances, including H^+, K^+ and NH_4^+, are secreted by active transport from the peritubular capillary blood into the renal tubules. In addition, a number of specific organic molecules, such as para-aminohippuric acid, the dye phenol red, and drugs such as penicillin, are actively secreted. Secretion occurs mainly in the distal convoluted tubule.

Secretion of H^+ is an important mechanism for the regulation of blood and urine pH. Carbon dioxide in the epithelial cells of the distal convoluted tubule combines with water in the presence of carbonic anhydrase to form HCO_3^- and H^+, which is actively transported into the renal tubule in exchange for Na^+ (Fig. 22.20). The excretion of H^+ in this manner is made more effective by the presence of buffers, such as HPO_4^{2-} and $H_2PO_4^-$ in the renal tubular fluid. In addition, the H^+ combines with NH_3 to form NH_4^+, thus 'trapping' the H^+ in the urine. Buffering and trapping H^+ in the urine can excrete excess metabolic H^+ in the urine at only moderately low pH.

Marine bony fishes typically have small kidneys and their nephrons are often modified to minimise urine flow and conserve water. The glomeruli are often small to restrict filtration and the nephrons of some marine fishes have no glomeruli at all—they are aglomerular! Urine is formed by these aglomerular kidneys only through active secretion of solutes, followed by the passive osmotic inflow of water and other solutes into the tubules (like insect Malpighian tubules).

> Ions such as H^+, K^+ and NH_4^+ and many organic solutes are actively secreted by the nephron.

Osmodilution and osmoconcentration

All vertebrates are able to form urine that is the same concentration as the blood (iso-osmotic) or a lower concentration than the blood (hypo-osmotic). Excretion of dilute urine is very important for freshwater vertebrates. For example, the osmotic concentration decreases in the distal convoluted tubule of the lamprey nephron to about one fifth that of the blood and very dilute urine is excreted (Fig. 22.21).

Only mammals and some birds are able to form urine that has a higher osmotic concentration than body fluids (hyperosmotic). This is important because terrestrial animals must rely on food, drinking and metabolism for water gain. Drinking is often the most important source of water, but many desert mammals are almost never able to drink water. It is therefore essential for desert mammals to minimise their urinary water losses, while at the same time maintaining excretion of their waste solutes. Many can excrete very concentrated urine, as high as 5000–9000 mOsm.

The mammalian kidney is structurally and functionally divided into two regions, the outer cortex and the inner medulla (Fig. 22.22). The cortex contains major blood vessels, the glomeruli and the proximal and distal tubules. The medulla contains the loops of Henle and the collecting ducts. The collecting ducts open into the renal pelvis, which drains into the ureter. Urine passes along the ureter to the bladder, where it is stored until voided.

There are two types of mammalian nephrons. Cortical nephrons are located near the outer cortex of the kidney. They are important for filtration, reabsorption and secretion, but have a short loop of Henle and do not contribute to the osmotic concentrating mechanism. Juxtamedullary nephrons are located nearer the medulla

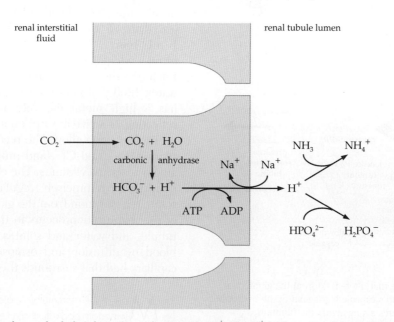

Fig. 22.20 Secretion of H^+ by the renal tubule is by active exchange with Na^+. The H^+ is then trapped in the lumen by buffering with HPO_4^{2-} and NH_3

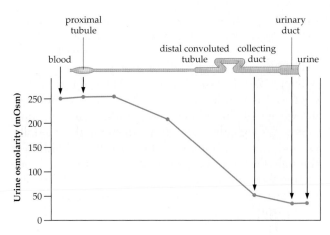

Fig. 22.21 The distal convoluted tubule of the nephron of a freshwater fish, a lamprey, can markedly reduce the osmotic concentration of the tubular fluid, hence very dilute urine is excreted

and have a long loop of Henle. These long loops of Henle descend into the renal medulla, make a sharp hairpin bend at the tip of the medulla and ascend back into the cortex. It is the juxtamedullary nephrons that establish the osmotic concentrating mechanism.

> All vertebrates can produce urine that is the same concentration as blood or more dilute. Mammals and some birds can osmotically concentrate their urine using juxtamedullary nephrons, which have long loops of Henle.

Since active removal of water from urine is not possible, urine osmoconcentration must utilise active transport of solutes and passive osmotic removal of water (Chapter 4). The loop of Henle establishes an osmotic concentration gradient in the renal medulla by the countercurrent exchange of solutes and water between the thin descending and thick ascending limbs. There are marked structural and functional differences between the descending and ascending limbs. The ascending portion of the loop of Henle has a relatively thick epithelium and actively transports Cl^- out of the nephron. Na^+ passively follows the Cl^- to maintain electrical neutrality but water does not osmotically follow because the tubule wall is impermeable to water. The effect of these Cl^- pumps is therefore to increase the solute concentration outside the ascending limb of the loop of Henle and to dilute the fluid remaining inside the ascending limb as it passes along the tubule. In contrast, the thin descending limb lacks Cl^- pumps and is permeable to water. Because it runs in close contact with the ascending limb but in the opposite (counter) direction, there is countercurrent exchange between the descending and ascending limbs. As fluid passes through the descending limb, it encounters regions of greater osmotic concentration because of the solutes pumped out of the ascending limb. Consequently, water is drawn by osmosis out of the descending limb and some solutes may enter.

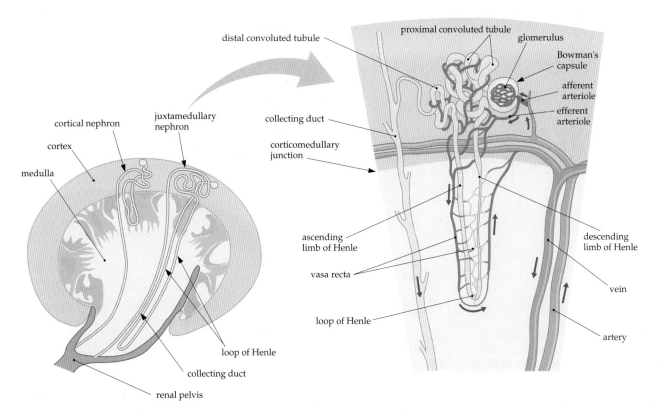

Fig. 22.22 The kidney of a rodent illustrates the general structure of the mammalian nephron and the arrangement of cortical and juxtamedullary nephrons in the renal cortex and medulla

The net effect of Cl⁻ pumping and countercurrent exchange in the loop of Henle is to establish an osmotic gradient in the renal medulla. This gradient increases from normal body fluid concentration at the cortex–medulla boundary to a high osmotic concentration at the renal medulla tip, where the loops of Henle make their hairpin bends (Fig. 22.23). Tubular fluid is only about 200 mOsm when it leaves the loop of Henle and would remain dilute as it passed along the collecting duct if there were no solute or water exchange. The urine can be made even more dilute by active reabsorption of ions from the collecting duct fluid. Although the loops of Henle do not themselves osmoconcentrate the urine (in fact, renal tubule fluid leaving the loop of Henle is actually more dilute than body fluids), the osmotic gradient that they establish provides the mechanism for urine osmoconcentration in the collecting duct.

A countercurrent arrangement of the blood supply to the medulla (the vasa recta) avoids dissipation of the renal medullary osmotic gradient by blood flow.

> The role of the loop of Henle is to establish an osmotic concentration gradient in the renal medulla.

Osmoconcentration of urine can occur as fluid flows through the collecting ducts, which pass through the osmotic concentration gradient of the renal medulla to their openings into the renal pelvis and then the ureter. The collecting duct wall is essentially impermeable to

the passive movement of ions, so there is no passive ion exchange. However, the water permeability of the collecting duct is controlled by antidiuretic hormone (ADH), which is secreted by the posterior pituitary (Chapter 25). In the presence of ADH, the collecting duct wall is permeable to water, which is osmotically drawn out of the tubular fluid as the collecting duct passes through the osmotic gradient of the medulla. Osmotic concentration increases as tubular fluid flows through the collecting duct and can become much higher than the body fluid concentration. ADH also increases the permeability of the collecting duct to urea, which passively diffuses into the medullary interstitium and further contributes to the renal medullary osmotic gradient.

> Osmotic concentration of urine occurs in the collecting ducts as they traverse the renal medullary osmotic gradient in the presence of antidiuretic hormone, which increases their permeability to water.

The extent to which urine can be osmotically concentrated depends on the length of the loop of Henle, the activity of the Cl⁻ pumps of the thick ascending limb, the magnitude of the effect of ADH on the water permeability of the collecting ducts, and the rate of fluid flow through nephrons. The longer the loop of Henle and the more active the Cl⁻ pumps, the higher is the osmotic gradient that can be established at the hairpin bend. The higher the ADH concentration, the greater is the water permeability and the higher the osmotic concentration of the urine. The maximal urine osmotic concentration is 500–1000 mOsm for some mammals and birds, but can be as high as 2000 mOsm for some birds and much higher for desert-adapted mammals. For example, the kidney of the Australian hopping mouse, *Notomys alexis*, has relatively long loops of Henle (Fig. 22.24) and can excrete urine as concentrated as 9000 mOsm.

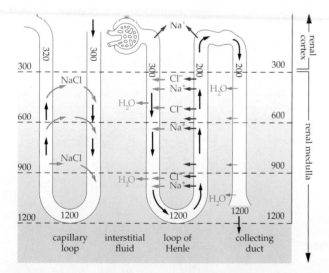

Fig. 22.23 The ascending limb of the loop of Henle actively transports Cl⁻ out of the tubule. The descending limb is permeable to water and solutes. Countercurrent exchange of water and solutes between the ascending and descending limbs establishes an osmotic concentration gradient in the renal medulla. This osmotic gradient is used to osmoconcentrate urine as it passes through the collecting duct. Blood in the capillaries to this part of the nephron also is osmotically concentrated by passive equilibration of the blood with the medullary osmotic gradient. Active transport shown by red arrows, passive transport by blue arrows

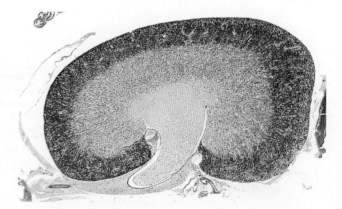

Fig. 22.24 The kidney of the Australian hopping mouse, *Notomys alexis*, has an extensive medulla with a very elongate papilla. The relatively long loops of Henle are able to establish a very extreme osmotic gradient and urine can be concentrated to 8000–9000 mOsm

Salt glands

Salt glands excrete a solution of either NaCl or KCl, which is considerably more concentrated than their body fluids. Cartilaginous fishes and the coelacanth have a rectal salt gland and some crocodiles, turtles, snakes, lizards and birds have lingual or orbital salt glands. The orbital salt gland of the herring gull, for example, can secrete a solution of about 1600 mOsm NaCl. This is more concentrated than sea water (1100 mOsm; Fig. 22.25), so the gull can drink sea water to replenish water lost by evaporation and urine, and its urine can be fairly dilute because the kidney does not have to excrete the drinking salt load.

The salt gland of some fishes, reptiles and birds can excrete a concentrated NaCl or KCl solution.

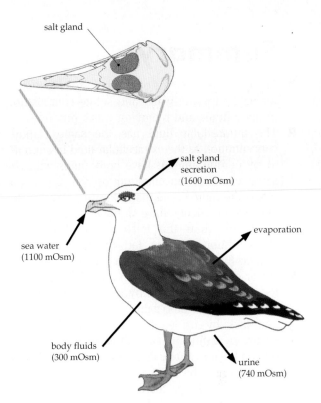

Fig. 22.25 The orbital salt gland of the herring gull, *Larus*, excretes a hyperosmotic NaCl solution (up to 1600 mOsm), which is more concentrated than body fluids (300 mOsm) and sea water (1100 mOsm). This enables the gull to drink sea water to replenish the water lost in urine and faeces and by evaporation

Summary

- The excretory systems of animals are responsible for regulating the water and solute composition of body fluids and removing waste products.
- The intracellular fluid has the same osmotic concentration as the extracellular fluid in animals. Many marine animals have body fluids with the same osmotic concentration as sea water but others are more dilute than sea water. Freshwater animals invariably have a higher osmotic concentration than the fresh water. The solute concentrations of the intracellular and extracellular fluids invariably are markedly different.
- Water exchange between an aquatic animal and its environment occurs by osmosis through the body and gut walls, gain by drinking and food, and loss by urine and faeces. Ion exchange occurs by diffusion and active transport across the body and gut walls, gain by drinking and food, and loss by urine and faeces. For terrestrial animals, drinking, food, urine and faeces are avenues for water and ion exchange; water is also lost by evaporation.
- Most marine invertebrates but only one group of vertebrates, hagfishes, have similar extracellular ion and osmotic concentrations to sea water. All other marine vertebrates are ionoregulators; many also are osmoregulators but some (cartilaginous fishes, coelacanth) are osmoconformers. Freshwater animals invariably are strong ionoregulators and osmoregulators. Terrestrial animals are also ionoregulators and osmoregulators.
- Metabolism of protein and nucleic acids produces nitrogenous waste solutes. Ammonia is the nitrogen waste for most aquatic animals. Urea, uric acid or guanine is the nitrogenous waste commonly excreted by terrestrial animals.
- Many freshwater protists and cells of freshwater sponges have contractile vacuoles for excretion of excess intracellular water. More complex animals generally have two types of excretory organs in animals: surface epithelia with ion pumps, and tubular excretory organs. Surface epithelial organs excrete or absorb specific ions (e.g. Na^+ and Cl^-), and may exchange water. Tubular excretory organs, such as nephridia, coelomoducts and Malpighian tubules, regulate a wide range of solutes and water. These tubular excretory organs form urine by filtration, and subsequently modify its composition by reabsorption of solutes and water, and secretion.
- The major excretory organ of vertebrates is the kidneys, but skin, gills and salt glands are also involved in water and solute balance. The functional unit of the kidney of higher vertebrates consists of a glomerulus capillary tuft and a nephron tubule. The balance of hydrostatic and colloid osmotic pressures between the glomerulus and Bowman's capsule forms a glomerular filtrate, which is modified as it passes through the nephron by reabsorption of ions, other solutes and water, and by secretion of specific wastes.
- Nephrons of mammals and some birds can also osmotically concentrate urine. Their juxtamedullary nephrons have a long loop of Henle, which establishes an osmotic gradient in the medulla of the kidney. This osmotic gradient allows the osmotic concentration of urine as it flows through the collecting ducts. The water permeability of the collecting ducts and therefore the concentration of urine is controlled by antidiuretic hormone.
- Some fishes, reptiles and birds have salt glands to excrete a highly concentrated and pure salt solution.

keyterms

ammonotely (p. 582)	elimination (p. 574)	Malpighian tubules (p. 585)	proximal convoluted tubule (p. 589)
anhydrobiosis (p. 574)	excretion (p. 574)	molality (p. 574)	purinotely (p. 582)
antennal gland (p. 586)	glomerulus (p. 589)	molarity (p. 574)	salt gland (p. 595)
coelomoduct (p. 585)	guanotely (p. 582)	nephridia (p. 585)	ureo-osmoconform (p. 578)
collecting duct (p. 589)	homeostasis (p. 574)	nephron (p. 589)	ureotely (p. 582)
contractile vacuole (p. 584)	hypersaline (p. 579)	osmoconform (p. 577)	ureter (p. 589)
distal convoluted tubule (p. 589)	ionoconform (p. 577)	osmolality (p. 575)	uricotely (p. 582)
	ionoregulate (p. 577)	osmolarity (p. 575)	
	iso-osmotic (p. 576)	osmoregulate (p. 577)	
	kidney (p. 588)		
	loop of Henle (p. 589)		

Review questions

1. What is the difference between excretion and elimination?

2. What are the principal differences in the solute and osmotic concentrations of the intracellular and extracellular fluids of animals? Why?

3. Define osmoconform, osmoregulate, ionoconform and ionoregulate. Generally speaking, which animals osmoconform and ionoconform, which osmoconform and ionoregulate, and which osmoregulate and ionoregulate?

4. How do brine shrimp, which live in very hypersaline solutions, solve their ionic and osmotic problems?

5. Compare the advantages and disadvantages of ammonia, urea and uric acid as the nitrogenous waste product for aquatic and terrestrial animals.

6. How is urine formed by protonephridia, metanephridia, Malpighian tubules and nephrons? How does the composition of the initial urine differ from that of the body fluids?

7. What processes are used to modify the composition of urine after it is formed by excretory tubules?

8. Compare and contrast the structure and function of the excretory organs of earthworms, crustaceans and insects.

9. Describe the structure of a mammalian nephron. What is the primary physiological function of each part of the nephron?

10. Explain how the loop of Henle establishes an osmotic concentration gradient in the renal medulla of the mammalian kidney. How is this osmotic gradient used to osmotically concentrate urine?

11. Desert mammals such as the Australian hopping mouse, *Notomys alexis*, do not have to drink. What are their avenues for water gain and loss, and what are the physiological adaptations of these animals for maintaining water homeostasis?

Extension questions

1. There is no evidence of an active transport mechanism for water in animals. Explain how water is apparently absorbed against an osmotic gradient in the rectal pads of *Calliphora* and by mammalian nephrons to form a concentrated excretory fluid.

2. It has been suggested that vertebrate animals evolved in fresh water and that marine species are derived from freshwater species that invaded the oceans. If this is true, in how many different ways have marine vertebrates adapted their patterns of ionic and osmotic balance to the marine environment?

3. Identify which invertebrates and vertebrates are able to form urine that is more concentrated than their blood. Compare and contrast the mechanisms for forming concentrated urine.

4. It would be advantageous for terrestrial animals to be able to produce urine that is more concentrated than their blood, but few are able to. Why don't terrestrial animals simply concentrate their urine by pumping ions into the urine?

5. Some herbivorous lizards have salt glands. What might their role be?

Suggested further reading

Griffith, R. W. (1987). Fresh water or marine origin of the vertebrates? *Comparative Biochemistry and Physiology* 87A: 523–31.

This research article clearly examines the interesting question of whether vertebrates evolved in fresh water or sea water from a physiological perspective.

Krogh, A. (1965). *Osmotic Regulation in Aquatic Animals*. New York: Dover Publications.

This classic book by one of the pioneers of comparative physiology, August Krogh, examines the general osmoregulatory and ionoregulatory physiology of aquatic animals.

Maloiy, G. M. O. (1979). *Comparative Physiology of Osmoregulation in Animals*. New York: Academic Press.

This comprehensive review provides a modern description of osmoregulation and ionoregulation in aquatic and terrestrial animals, with various chapters written by experts in their field.

Potts, W. T. W. and Parry, G. (1964). *Osmotic and Ionic Regulation in Animals*. Oxford: Pergamon Press.

This classic book clearly and succinctly describes the concepts and mechanisms for osmoregulation and ionoregulation in a wide variety of animals.

Schmidt-Nielsen, K. (1997). *Animal Physiology*. Cambridge: Cambridge University Press.

This popular textbook on comparative physiology of animals provides a clear overview of osmoregulation and ionoregulation in animals as part of its general treatment of comparative animal physiology.

Smith, H. W. (1959). *From Fish to Philosopher*. New York: Doubleday.

This interesting and enjoyable book provides a broad (but now somewhat dated) overview by the eminent comparative renal physiologist, Homer Smith, of the role of the kidney in the adaptation of vertebrate animals to diverse aquatic and terrestrial habitats, and speculates on how iono- and osmoregulation of the internal environment allowed the development from primitive fishes of more complex organisms, culminating in the consciousness and thought of philosophers.

Withers, P. C. (1992). *Comparative Animal Physiology*. Philadelphia: Saunders College Publishing.

This popular textbook of comparative physiology provides a detailed and quantitative survey of the physical principles, mechanisms and patterns of osmoregulation and ionoregulation in aquatic and terrestrial animals, as well as other aspects of comparative physiology.

CHAPTER 23

The immune system

All organisms must be able to defend themselves against damage stemming from their environments. Damage to an organism can occur at three levels:

1. *physical damage*, such as occurs through predation or accidental injury
2. *damage resulting from extreme environmental conditions*, such as excessive or limiting levels of water, oxygen or salts, or extremes of pH or temperature
3. *damage from parasites, microorganisms and macromolecules*—it is at this level that the immune system is of key importance in defence.

The role of the immune system is to provide protection against viruses, bacteria, fungi and other small parasites and their toxins that threaten the integrity of the organism. As life has evolved, so too have immune systems, and the increasing complexity of organisms is reflected in the possession of more complex and refined immune systems. In this chapter we shall deal mainly with the mammalian immune system, primarily because it is among the most complex and advanced but also because it is the best understood. Immunity in non-mammalian animals and in plants will be discussed at the end of the chapter.

There are four important concepts around which much of this chapter revolves.

1. Substances that an immune system reacts against, for example, molecules on the surface of a bacterium, are known as **antigens**.
2. The cells in the body that carry out many of the functions of the immune system are **lymphocytes**. There are two main types: B lymphocytes and T lymphocytes.
3. B lymphocytes make and release **antibodies**, which are important defence molecules that react with and help destroy antigens.
4. T lymphocytes regulate immune responses and can directly attack and kill foreign organisms or infected cells.

BOX 23.1 Key terms in immunology

Antibody

The protein molecule produced by B lymphocytes and plasma cells in response to antigen and which reacts specifically with that antigen. Each different antibody is the result of a random genetic rearrangement event occurring during development of the B lymphocyte.

Antigen

Any molecule (usually organic) that can be recognised by (i.e. bind to) one of the specific molecules (antibodies or T-cell receptors) of the immune system. These include foreign (non-self) and self antigens. Antigens are not necessarily immunogenic (i.e. do not necessarily induce an immune response).

Auto-

From the same or a genetically identical individual (e.g. in autoimmune, autoantigen, autograft).

Epitope

The actual portion of an antigenic molecule that is recognised by a receptor. A large protein antigen may have numerous different epitopes, each of which could induce a unique response.

Haemopoietic stem cells

The ancestral cells of all blood cells, including lymphocytes and phagocytic cells.

Immunogen

An antigen that stimulates immune responses—either antibody production or the generation of effector T cells.

Immunoglobulin (Ig)

Another term for antibody.

Lymphocyte (lymphoid cell)

Mononuclear cells that are the predominant cells in immune organs such as lymph nodes, spleen, thymus, Peyer's patches and tonsils. There are two types of lymphocyte: B lymphocyte (B cell)—Produced mainly in the bone marrow and differentiates into antibody-producing plasma cell; T lymphocyte (T cell)—Matures in the thymus, recognises antigen by means of the T-cell receptor and functions independently to kill microorganisms. It is also crucial in controlling B-cell responses.

Lymphokine

Glycoprotein messenger molecule produced mainly by T lymphocytes and which controls the behaviour of other cells.

Lymphoid tissue

Tissue rich in lymphocytes. Divided into primary lymphoid tissue, such as thymus and bone marrow, where lymphocytes mature, and secondary lymphoid tissue, where mature lymphocytes mediate an immune response.

Plasma cell

Mature antibody-producing cell.

Phagocytic cell or phagocyte

Cell that engulfs particles such as bacteria. Examples are macrophages (large mononuclear cells) and polymorphonuclear granulocytes (smaller cells with a multilobed nucleus, abbreviated to polymorphs).

T-cell receptor (TCR)

The glycoprotein molecule on the surface of T cells that recognises an antigen. Like antibodies, each of these molecules is the result of a genetic rearrangement event.

Non-specific defence mechanisms

Organisms also have defences that are not part of the specific immune system but which provide non-specific protection against foreign invaders (Fig. 23.1). At the simplest level these include 'external' defences, such as enzymes in secretions that can damage or destroy microorganisms (e.g. lysozyme in saliva and tears), low pH in the stomach, sticky mucus on many surfaces, and

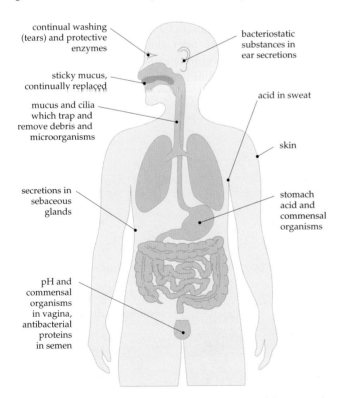

continual washing (tears) and protective enzymes

sticky mucus, continually replaced

mucus and cilia which trap and remove debris and microorganisms

secretions in sebaceous glands

pH and commensal organisms in vagina, antibacterial proteins in semen

bacteriostatic substances in ear secretions

acid in sweat

skin

stomach acid and commensal organisms

Fig. 23.1 Simple non-specific defence systems. Some of the external physical and chemical defence mechanisms that prevent foreign organisms from gaining access to the body. Remember that the inside of a mammalian body—warm, wet and rich in nutrients—is an ideal environment for the growth of many microorganisms, so the best first line of defence is not to let them in

the presence in the gut and genital tract of commensal organisms that are able to prevent the growth of pathogenic organisms. The last example explains some of the gastrointestinal side-effects of antibiotics that kill harmless or beneficial gut flora, while allowing more dangerous microorganisms to grow.

Internal non-specific defences include molecules that are activated or increase in number during the course of an infection and protect against a range of organisms. These substances include **interferons**, which, among other effects, induce resistance to viral infection in a variety of cells, molecules of the **complement system** (see Box 23.4), which enhance the removal of bacteria by phagocytic cells, and the phagocytic cells themselves. These non-specific mechanisms are a vital first line of defence. They are always present at low levels and are rapidly activated early in an infection or after tissue damage. They reduce or slow down the development of infections during the several days that it takes for the immune system to develop a specific response.

Non-specific defences play a vital role in slowing or even preventing the early progress of infection until the onset of the specific immune response. Certain non-specific mechanisms also act in conjunction with specific responses to finally remove pathogens.

Specific immune responses

Specific immune responses can be distinguished from non-specific defences by their specificity. Immunisation with a polio vaccine results in immunity to polio but not to other diseases. A key part of specific immunity is that it involves **immunological memory**. When a person is immunised against polio, a **primary response** occurs. The immune system 'remembers' this and, if the person comes into contact with the polio virus again, a **secondary response** occurs, which is usually bigger and faster than the primary response and sufficient to prevent infection the second time. However, if the same individual is immunised later against a different organism (e.g. rubella virus, which causes measles), this results in a new primary response that is unaffected by the previous response to polio. Immune responses are not always absolutely specific to a particular organism; protection may be conferred against closely related organisms, as in the famous example where infection with the cowpox virus also provides immunity against the smallpox virus (Chapter 34). In addition, there may be accidental cross-reactions, as occurs between red blood cell types and carbohydrates on gut bacteria (see Box 23.3).

Why is the immune system so complex?

Consider the following situation. During the lifetime of an individual, the immune system has the difficult task of protecting against a huge and unpredictable variety of foreign molecules or organisms. No organism can possibly predict which foreign invaders it may encounter or which new ones may evolve during its lifetime. The task is even more difficult because a complex animal is itself composed of a very large number of cells and molecules that must not be attacked by its own immune system. Inappropriate immune responses against these 'self' molecules (**autoimmune** responses) can have very damaging consequences. An immune system therefore needs a large array of very specific defences, which operate at the level of molecular recognition.

How can an immune system meet these requirements? One way would be to have millions of genes coding for millions of defensive molecules that do not react with self. But, even in mammalian genomes, with an estimated 100 000 genes for an entire organism, the millions of additional genes required would take an impossible amount of genetic space. Evolution has resulted in a novel solution to the problem. A relatively small number of inherited genetic elements are randomly shuffled, or rearranged during the differentiation of the specific immune cells in such a way that a vast number of different defence molecules is generated. This solution is very efficient in that it uses little genetic space, but it has the disadvantage that the range of defence molecules generated is unpredictable and will include many that may react with self components. As we shall see, many of the complexities of the immune system result from the subsequent need to screen the unpredictable range of defence molecules in order to remove any that are potentially self-destructive.

> The immune system must have the potential to respond to an extraordinarily large and unpredictable array of foreign molecules, although only a relatively small number will be encountered in the lifetime of any individual. At the same time, it must ignore 'self' molecules.

Cells of the immune system

Cells that take part in immune responses can be separated into three categories (Fig. 23.2):
1. *specifically acting cells*—**T** and **B lymphocytes**
2. *non-specifically acting cells*—**phagocytic cells** (**macrophages** and **granulocytes**)
3. several groups of *partly specific cells*—for example, **natural killer (NK) cells**, which can kill tumour cells or virus-infected cells but not normal cells.

Most cells found in blood, bone marrow and lymphoid tissues are derived from multipotent **haemopoietic stem cells** that reside mainly in the bone marrow (see Box 6.2). Further development of T cells occurs after migration of stem cells to the **thymus**; B-cell development occurs in various tissues in different species but mostly in the **bone marrow** in humans.

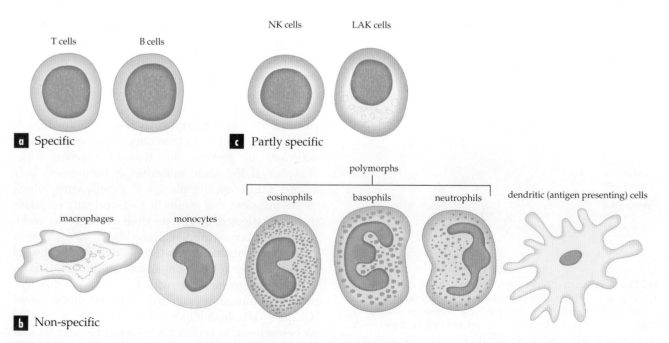

Fig. 23.2 Cells of the immune system: **(a)** specific; **(b)** non-specific; and **(c)** partly specific

Lymphocytes and specificity

Lymphocytes are small, round mononuclear cells (Fig. 23.3), which are generated in large numbers each day. They make up the majority of white blood cells and cells in the organs of the immune system. T and B lymphocytes, although superficially similar in appearance, can be distinguished on the basis of their development and functions. Individual lymphocytes are different from their neighbours because each carries a distinct antigen receptor that is the result of gene rearrangement (Box 23.2). Thus, only one or a very small number of cells has a particular arrangement and hence particular surface receptor, of which it expresses many copies.

In principle, each receptor recognises a different antigen, although among a large number of randomly generated receptors it is likely that several different receptors could interact with the same antigen. A normal human has approximately 10^{12} lymphocytes bearing at least several hundred million different specificities referred to as its 'immune repertoire'.

During an immune response, any cell that encounters its specific antigen is stimulated to proliferate, generating a clone of daughter cells, each with the same receptor specificity, each capable of responding to the same antigen. This clonal expansion boosts the response to the stimulating antigen by increasing the number of cells reacting to it. The key features of this **clonal selection** are outlined in Figures 23.4 and 23.5.

BOX 23.2 Generation of diversity and gene rearrangement

The complex rearrangement events that generate diversity among antibody (Ig) and TCR molecules occur during development of B and T lymphocytes in bone marrow and spleen respectively. Each Ig and TCR molecule is made up of two types of protein chains, each bearing a variable portion that contributes to the antigen-binding site (see Figs 23.17, 23.18). The potential diversity of a population of whole molecules is thus the product of the potential diversity of each chain. Part of this diversity comes from the variety of variable (V) genes that can be used in each chain, but much more comes from the way the genetic elements are joined. Between the V and constant (C) genes are small elements known

as diversity (D) and joining (J) elements. Immense variation can occur as different combinations of V, D and J segments are combined.

The approximate contribution of each of these sources to total variation for Ig and TCR molecules is shown in the table. As you can see, there may be 10^{11} possible antibody molecules and 10^{15} possible TCR molecules, with each mature lymphocyte expressing a single specificity. Since mammals contain 10^{8}–10^{12} lymphocytes, it is obvious that there are many more possibilities than will ever occur in any one individual. The figure (p. 604) shows a scheme of these events for a typical rearrangement process.

Table Genetic sources of diversity in Ig and TCR molecules

	Ig		TCR	
	Light chain	Heavy chain	α chain	β chain
V genes[a]	250	250–1000	100	25
D segments	0	10	0	2
J segments	4	4	50	12
N-region addition[b]	none	V–D, D–J	V–J	V–D, D–J
Possible combinations[c]	~ 10^{11}		~ 10^{15}	

(a) Number of gene segments in DNA of genes for the two chains of immunoglobulin and the TCR.

(b) Random insertion of nucleotides at the point of joining between segments.

(c) Estimated number of different antibody molecules that could be generated by different combinations of V, D and J segments, along with N-region effects.

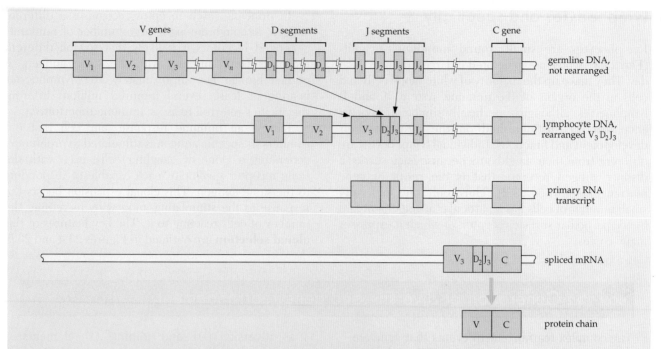

Rearrangement events in a hypothetical Ig or TCR chain. The gene segments used in this case were V_3, D_2 and J_3. During the rearrangement of the DNA, the section between V_3 and D_2 (V_4–n and D_1) was excised, as was the section between D_2 and J_3 (D_3–n and $J_{1,2}$). The primary RNA transcript includes everything between the rearranged V gene and the C gene, but this RNA is later spliced to bring VDJ beside the C gene. Note that some chains only have V, J and C segments (Ig light chain and TCR α chain), while others have V, D, J and C (Ig heavy chain and TCR β chain). Some chains also have several C genes to select from, but these do not affect diversity of the antigen-binding site

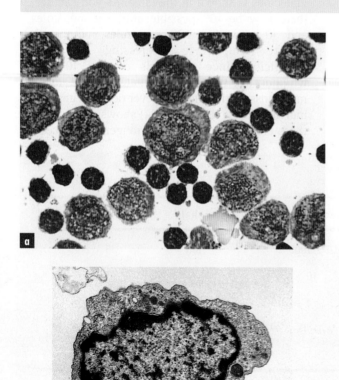

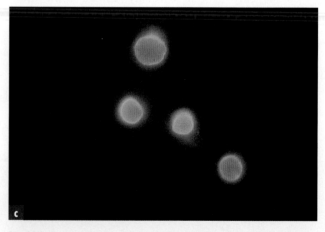

Fig. 23.3 The lymphocyte. **(a)** A smear of a population of lymphocytes taken from the efferent lymph leaving a stimulated lymph node. The small, round cells with little cytoplasm are small lymphocytes, the large ones are activated, dividing lymphocytes responding to antigen. There are both T and B cells among the small and large cells but they cannot be distinguished by visual criteria alone. **(b)** An electron micrograph of a small lymphocyte, with little cytoplasm and few outstanding features or organelles. Again this could be a T or B cell. **(c)** A group of lymphocytes labelled on the surface with an antibody to a molecule called Thy 1, which is only expressed on T cells. The antibody is marked with a green fluorescent tag, so only T cells show up as green. B cells in the picture are not visible as they are not labelled

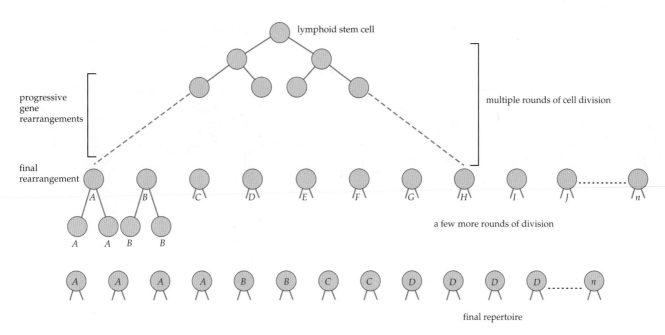

Fig. 23.4 Generating the preimmune repertoire. The timing of gene rearrangement and cell proliferation gives rise to a large number of small clones of cells, each clone expressing a different antigen receptor (Ig for a B cell, TCR for a T cell). This process is antigen independent and occurs in the primary lymphoid tissues. The final repertoire is the range of clones available before any response to antigen has occurred. *A, B, C...n* are receptors of different specificities

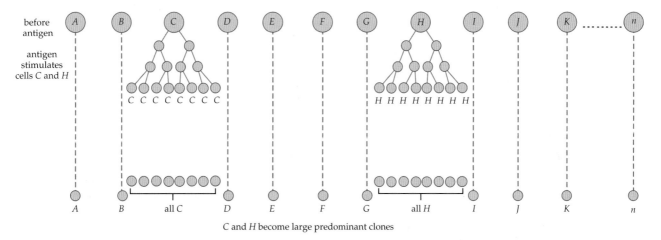

Fig. 23.5 Clonal expansion and the immune repertoire. Some of the cells from within the preimmune repertoire will eventually encounter the antigen that their surface receptor recognises or can interact with. These cells will be stimulated to proliferate and carry out effector functions. The proliferation will result in an increase in the number of these cells, so they will make up a larger proportion of all the lymphocytes. If the same antigen is encountered again, there will already be large numbers of cells present, and only activation will be required. All the time required for proliferation will be saved, allowing much faster responses on secondary stimulation

Each lymphocyte carries a different surface receptor. During an immune response, those few cells that interact with an antigen proliferate, increasing the number of cells reacting to that antigen. This process is known as clonal selection.

Cellular and humoral immunity

There are two kinds of specific response to foreign antigens—cellular and humoral (Fig. 23.6). Both can be triggered during an infection, but one or other tends to predominate and be most effective in any particular infection. The cellular immune response is most effective against viral infections and other intracellular parasites, whereas the humoral response is most effective against the extracellular phases of bacterial and viral infections.

Cellular immunity involves active destruction of foreign organisms or of the body's own virus-infected cells by T cells. T cells also regulate most reactions in the immune system, including antibody formation, and self–not-self discrimination.

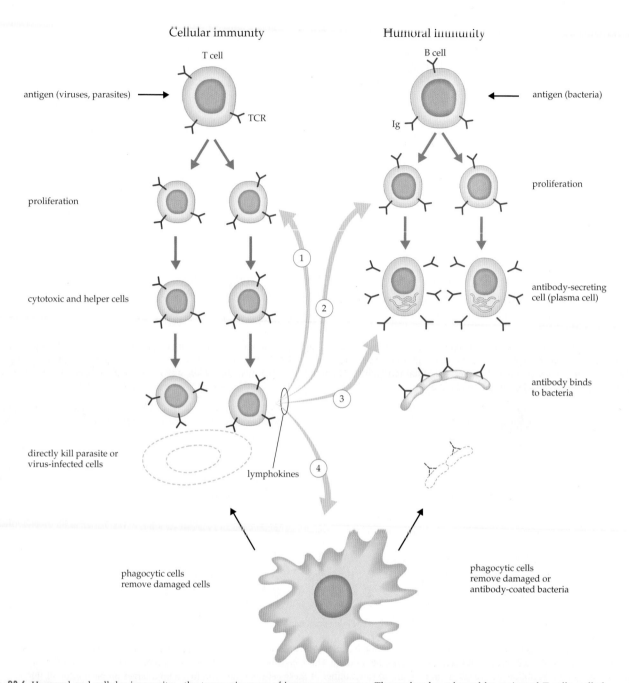

Fig. 23.6 Humoral and cellular immunity—the two main arms of immune responses. The molecules released by activated T cells, called lymphokines, are important in helping both T and B cells to proliferate (1 and 2), and B cells to make antibody (3). Lymphokines also activate macrophages (4), increasing the speed of removal of debris

Humoral immunity is mediated by soluble antibody molecules secreted by B cells in the serum or other body fluids. Once formed, antibodies act independently of B cells, and pure cell-free serum retains its humoral activity. Unlike cellular immunity, antibodies are reasonably stable outside the body. Antibodies are present in breast milk and are important in protecting a baby from infection. Antibodies can also be effective across species and are used for providing 'passive' immunity against diseases such as hepatitis.

There are two distinct kinds of immune response that are usually effective against different kinds of microorganisms: a cellular response, mediated by T cells, and a humoral response (in serum), mediated by antibodies produced by B cells.

T lymphocytes (T cells) and the thymus

T cells mature in the thymus (Fig. 23.7). If the thymus is absent, as happens in certain genetic abnormalities, or removed early in foetal development, no T cells are

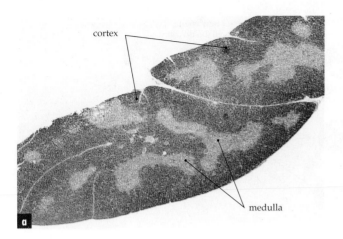

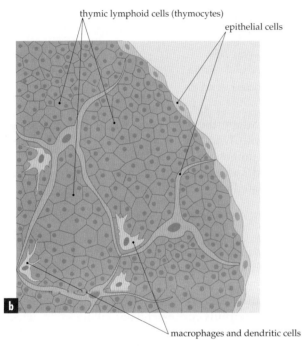

Fig. 23.7 The structure of the thymus. **(a)** A section of a mouse thymus. The darker outer zone of densely packed cells is the cortex where most of the lymphocyte proliferation and differentiation occurs. The light-stained central areas are medulla, which contains mainly mature lymphocytes waiting to migrate out to the lymphoid organs, such as spleen and lymph nodes. **(b)** Detailed structure of the thymus, with a meshwork of fixed epithelial cells supporting scattered bone-marrow derived macrophages and dendritic cells, and all the spaces packed full of thymic lymphocytes, often called thymocytes

made and profound *immunodeficiency* results. Haemopoietic stem cells enter the thymus and proliferate rapidly. During this time the genes coding for the T-cell receptor (TCR) for antigen are being rearranged (Box 23.2) and, eventually, TCR proteins are expressed on the cell surface. At this point, all cells must be screened so that only useful cells, that is, those cells that do not recognise self antigens, continue development. Only about 1% of cells survive screening. Those that do survive go on to mature, finally leaving the thymus and

populating the lymphoid organs; the remaining 99% die within the thymus.

T cells that leave the thymus fall into two functionally distinct subclasses: **helper cells** (T_H cells) and **cytotoxic cells** (T_C cells). T_H cells are regulatory cells that produce and secrete **lymphokines**, molecules that control the development and function of other T and B cells, as well as accessory cells such as macrophages. T_C cells are effector cells that, when stimulated by antigen and lymphokines produced by T_H cells, directly **lyse** or kill target cells recognised by the T_C cells on the basis of their particular antigen.

T cells proliferate in the thymus in very large numbers. Only a few useful, non-self-reactive T cells survive the screening processes and are exported to the lymphoid organs as mature T cells. The exported T cells are of two types: helper cells (T_H cells), which mainly regulate other immune cells, and cytotoxic cells (T_C cells), which kill foreign organisms or infected body cells.

Lymphokines

Many of the functions of T cells, particularly T_H cells, are mediated by hormone-like molecules released by these cells after stimulation by an antigen. These molecules, usually glycoproteins, are variously called lymphokines, cytokines or interleukins. Some lymphokines act only in the local environment, so only other cells near the T cell are affected. Others may travel in the bloodstream, stimulating cells at distant sites or exerting generalised effects that cause fever or 'malaise', nausea or changes in blood pressure.

The effects of lymphokines are varied. Some help B cells make antibodies, others help T cells develop. Still others stimulate the production or activation of macrophages and granulocytes, red cells or platelets. Yet others increase the permeability of blood vessels, causing swelling that may accompany inflammation. Overall, lymphokines are key players in almost every immune response, relaying signals between T cells and other cells, and communicating between the immune system and the endocrine and nervous systems.

B lymphocytes (B cells) and their development

B cell development parallels that of T cells but it occurs at different sites. In humans, B cells develop mainly in the bone marrow. There is extensive proliferation as the cells progressively differentiate and, during this time, the genes that encode antibodies (also known as **immunoglobulins** or Ig) are rearranged (Box 23.2).

Mature B cells manufacture antibodies that are expressed on their surface and function as receptors (Fig. 23.8). New B cells die within a few days if they do not interact with their specific antigen, which is the fate

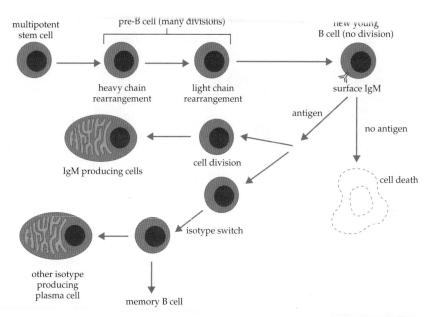

Fig. 23.8 B-cell development, showing some of the steps in development from a multipotent stem cell

of the majority of B cells produced. When a membrane-bound antibody receptor binds to its specific antigen in the presence of stimulated T_H cells, the receptor triggers further differentiation of the B cell. Some B cells will become long-lived **memory cells**, ready to respond to future encounters with the same antigen even some years later; others differentiate further into **plasma cells**, non-dividing antibody factories that produce and release large quantities of antibody molecules over their short life span of only a few days.

> In humans, B cells are made in the bone marrow. They have antibodies on their surface that, on binding with their specific antigen in the presence of T_H cells, trigger further differentiation into memory cells or plasma cells.

Phagocytic cells

About a hundred years ago, Elie Metchnikoff made the first observation of white blood cells (leucocytes) engulfing microorganisms and he called the process *phagocytosis* or cellular eating. There are two main types of phagocytic cells—mononuclear phagocytes (macrophages and monocytes), and polymorphonuclear granulocytes, usually known as polymorphs or granulocytes (Fig. 23.2).

Macrophages and monocytes

These cells play an important role in body maintenance. They engulf and digest dead and damaged cells, old red cells, debris, antibody-coated microorganisms and damaged fatty particles or molecules. They participate in wound healing, and in both acute and chronic inflammation. Almost all of these activities

increase when the cell becomes activated through contact with lymphokines released by T cells undergoing an immune response. **Monocytes** are found in the blood, whereas macrophages are a more fully differentiated phagocyte found in the tissues.

Polymorphonuclear granulocytes

There are three types of polymorphonuclear granulocytes, so-called because of their multilobed nucleus and their many cytoplasmic granules. The most common of these encountered under normal conditions are **neutrophils**, which are predominantly phagocytic cells. **Eosinophils** are essentially killing cells, which are especially important in the removal of larger parasitic animals, such as flukes and worms. However, the destructive activities of eosinophils can also cause tissue damage at sites of allergic reactions, so they are potentially dangerous. Circulating **basophils** and stationary **mast cells** are mainly involved in acute inflammation and allergy.

Dendritic cells

There are many cells in the immune system, including macrophages and B cells, that can present antigens to T and B cells. Most effective of all are **dendritic cells**, a poorly understood but ubiquitous cell type, which are able to break down foreign molecules and present them to lymphocytes. Dendritic cells, so-called because they have many long thin processes (dendrites) that spread out between surrounding cells (Fig. 23.9), are located in all lymphatic tissues, including the thymus, where they play an important role in the screening process. They are also found in blood, in skin (Langerhans cells) where they are key players in many inflammatory

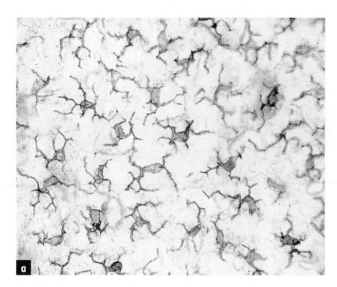 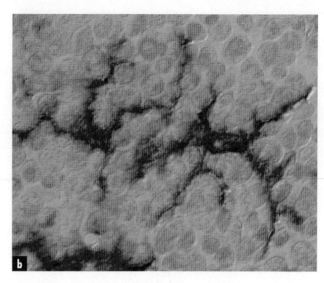

Fig. 23.9 Dendritic cells play a sentinel role in peripheral tissues, in particular at mucosal surfaces in contact with the external environment, where their prime function is capture of incoming antigens, which are subsequently transported to central lymphoid organs for presentation to the immune system. **(a)** Epidermal Langerhans cells stained for ATPase. **(b)** Airway intraepithelial dendritic cells stained for MHC

reactions, and at mucosal surfaces in contact with the outside environment, such as the gastrointestinal and respiratory tracts.

NK cells

NK cells or natural killer cells resemble lymphocytes in appearance but lack the surface receptors (TCR or antibody) that allow those cells to recognise specific antigens. They kill (lyse) virally infected cells or tumour cells, which they distinguish from healthy self cells by changes in the *expression of carbohydrates* on the cell surface.

> There are several cell types that assist lymphocyte functions. Phagocytic cells (macrophages and granulocytes) remove damaged or foreign cells and can kill parasites. Macrophages, B cells and dendritic cells can present foreign molecules to lymphocytes.

Secondary lymphoid tissues and the immune system

To understand how the immune system functions in maintaining the health of an individual, it is necessary to appreciate the structure and behaviour of the immune system in the whole body (Fig. 23.10).

The immune system is made up of **primary lymphoid organs** (thymus and bone marrow), which are largely producers of new lymphocytes, and **secondary lymphoid organs** (lymph nodes, spleen, tonsils, etc.), where immune responses take place. These are linked together by lymphatic vessels and

blood vessels. Since a wound, infection or dying cell might be anywhere in the body, the immune system needs to survey the entire organism continually. Surveillance is achieved by keeping the cells of the immune system on the move around the body. Lymphocytes are the body's great travellers and the highways for their travelling are the network of lymphatic vessels.

The lymphatic network

The network of lymphatic vessels has two important roles:
1. to clear proteins and fluids from tissues (see Chapter 21)
2. to transport cells.

The finest **lymphatic vessels**, which drain from the tissues, join up to form larger and larger vessels, eventually flowing into a small number of major lymphatic vessels, which return fluid back to the blood (Fig. 23.11). Lymphatic vessels pick up lymphocytes and macrophages that are wandering through tissues, and collect microorganisms or debris from inflamed or infected sites.

Lymph nodes, located at numerous points along the lymphatic vessels, act as filters and are also highly organised lymphoid tissues capable of producing a vigorous immune response. Although an infection results in local **inflammation**, or influx of immune cells, all the cells and debris and antigens from this site will eventually be carried by the lymphatic vessels to the local lymph node, where the strongest immune response, well organised and distant from the inflammation, occurs. Hence, an infected hand results in enlarged and tender glands (lymph nodes, Fig. 23.12)

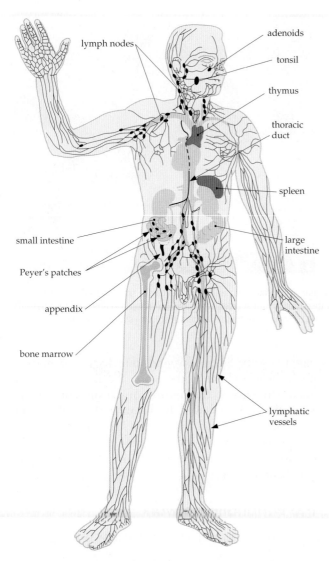

Fig. 23.10 The immune system of the human. This comprises primary lymphoid organs, which produce new lymphocytes (thymus and bone marrow), secondary lymphoid organs where organised immune responses occur (lymph nodes, spleen, adenoids, tonsils, appendix, Peyer's patches), the network of blood and lymphatic vessels that link all the tissues and, of course, the lymphocytes themselves, which migrate around the body and reside in the lymphoid tissues

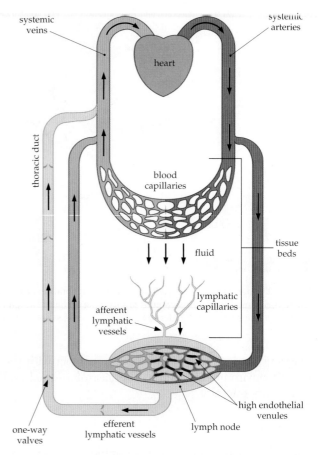

Fig. 23.11 Lymphocyte recirculation. Blood circulation is shown in red and purple, lymphatic circulation in yellow. The arrows show the direction of flow of fluid and cells. Blood circulation is driven by the heart. Afferent lymphatic vessels collect fluid and cells through osmotic and fluid pressure, while lymph moves through the efferent vessels by squeezing past one-way valves as the muscles move and contract. Hence, prolonged inactivity in a standing or sitting position can lead to swollen hands or feet as lymph accumulates and there is insufficient muscle movement to massage the lymph up the long lymphatic vessels of the arms and legs

in the armpit, the site of the main lymphatic filters for the lymphatics in the hand and arm. Other lymphoid tissues, such as the tonsils, Peyer's patches and adenoids, function in a similar way.

Even in the absence of infection there is a major traffic of lymphocytes from blood through specialised blood vessels directly into the lymph nodes (Figs 23.13, 23.14). After leaving the bloodstream, these cells migrate through the nodal tissue back into the lymphatic vessels and to the blood again. The majority of T cells and to a lesser extent B cells are constantly recirculating—going round and round this circuit, about one trip per day, and passing through many lymph nodes each week.

Antibodies formed in an activated lymph node will be carried to the blood through lymphatic vessels and dispersed around the body. Similarly, immune-activated cells and memory cells will eventually leave the node and impart immunity to all parts of the body. Some of these will leave the blood again at the site of inflammation, carrying effector cells right to the cause of the trouble.

Blood is filtered in a similar way by the spleen (Fig. 23.15). Should an infection get into the bloodstream, the spleen will filter out the foreign organisms and co-ordinate an organised immune response involving T and B cells and antigens.

Lymphocytes continually migrate around the body, circulating from blood to lymphoid tissue and back to blood through lymphatic vessels. This allows lymphocytes to monitor a wide range of sites around the body. Secondary lymphoid organs act as filters and sites where immune responses can be organised.

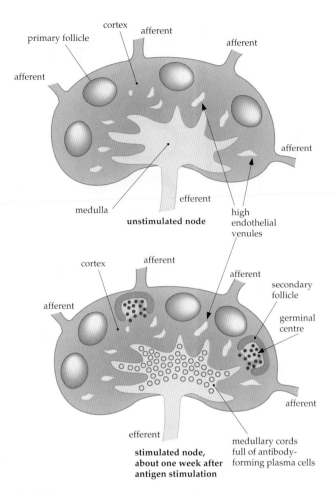

Fig. 23.12 Lymph node structure. B-cell areas (follicles and medulla) are indicated in red, while the T-cell areas (cortex) are blue

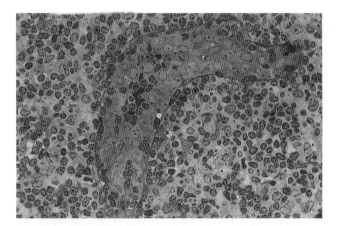

Fig. 23.13 A histological section of a high endothelial venule in the cortex of a lymph node. Note the lymphocytes migrating between the high endothelial cells

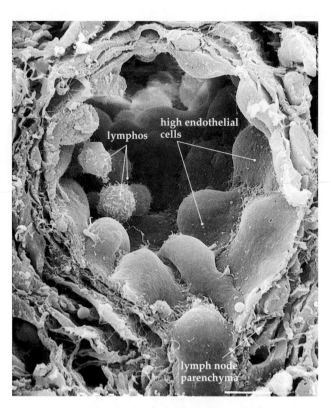

Fig. 23.14 Scanning electron micrograph of the inside of a blood vessel in a lymph node. Small round lymphocytes (lymphos) adhere to the dome-shaped high endothelial cells that make up the walls of this high endothelial venule, a specialised blood vessel, which allows the emigration of lymphocytes from the blood into lymphoid tissues. The lymphocytes adhere to the endothelial cells, then migrate between them, into the body (the parenchyma) of the node, and then on out into the efferent lymph to continue their recirculation back to the blood

Molecules of the immune system

The important molecules of the immune system are *antigens*, molecules against which the immune system will react, *antibodies*, produced by the immune system to help get rid of foreign antigens, **T-cell receptors**, molecules on the surface of T cells that interact with antigen, and **major histocompatibility complex (MHC) molecules**, which help to present antigens to the immune system.

Antigens

An antigen is a molecule that can react with and bind to the variable region of an antibody or T-cell receptor molecule. If this binding stimulates a lymphocyte to make an immune response, then the antigen is an **immunogen** and is said to be **immunogenic**. B cells can respond to a wide variety of antigens, including proteins and carbohydrates. T cells respond predominantly to protein antigens.

Proteins are the best understood and most diverse antigens, and the smallest molecules that can function as antigens are peptides of about 10 amino acids. Molecules any smaller than this are invisible to antibodies or TCRs. A typical large protein molecule of hundreds or thousands of amino acids will contain

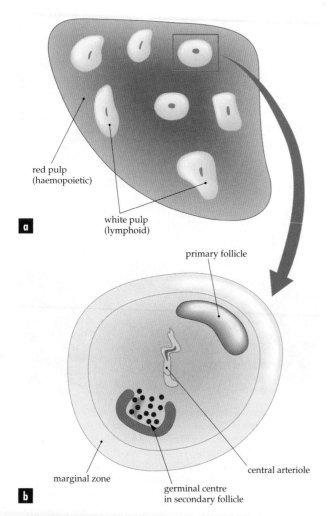

Fig. 23.15 Spleen structure. **(a)** The spleen is made up of red pulp, coloured red because of the high erythrocyte content, and islands of white pulp, coloured white because there are few erythrocytes, mainly lymphocytes. **(b)** White pulp enlarged. The T-cell area is shown in blue, the B-cell areas in red. The white pulp always has a blood vessel running through the centre and a marginal zone, rich in antigen-presenting cells, around the outside. As in lymph nodes, primary follicles mature into secondary follicles and germinal centres upon antigen stimulation. In the spleen, plasma cells are often found scattered through the red pulp

many 'minimal' peptides, or **epitopes**, that are potentially antigenic (Fig. 23.16).

Protein antigens

Large immunogenic protein antigens are broken down by antigen-presenting cells, such as macrophages, into small peptides. These peptides are the units displayed or presented by cell-surface MHC molecules (see below) to T cells. T cells are stimulated only by MHC-bound peptides and they recognise peptides derived from protein antigens that may not have been located on the surface of the intact molecule. In contrast, antibodies and the B cells that produce them can be stimulated directly by whole antigen molecules. Thus, antibodies are much more

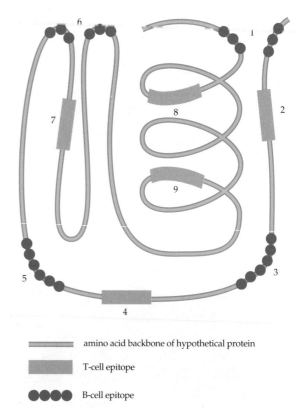

amino acid backbone of hypothetical protein

T-cell epitope

B-cell epitope

Fig. 23.16 Hypothetical protein molecule showing T- and B-cell epitopes (the actual small piece of an antigen that the antibody or TCR binds to), which can be quite distinct. B-cell epitopes, those bound by antibody molecules, are usually on the surface of a molecule and may be dependent on the structure of the molecule (e.g. epitopes 1 and 6), or on a string of amino acids, often one that stands out on the surface (epitopes 3 and 5). T-cell epitopes are entirely dependent on the short sequence of amino acids since they are always presented after the amino acid chain is broken into small bits. They can therefore be internal to the whole molecule (epitopes 7, 8 and 9), and cannot be dependent on the structure of the whole molecule

likely to recognise epitopes from the surface of the whole antigen molecule.

Carbohydrate antigens

The second important group of antigens are carbohydrate antigens. As for proteins, only large complex polysaccharides function as antigens. In general, carbohydrates are more likely to stimulate B cells to make antibodies than they are to stimulate T cells. Some important bacterial and viral antigens are polysaccharides. Another well-known group of carbohydrate antigens is the human **ABO blood group antigens**. Most humans (except those of blood type AB) have antibodies against one of these antigens in their serum (see Box 23.3).

Both proteins and carbohydrates can be antigens. T-cell receptors and B-cell antibodies 'see' antigens quite differently. Antibodies react with an antigenic molecule as a whole, whereas TCRs only recognise small peptides that are presented by MHC molecules on the surface of antigen-presenting cells.

BOX 23.3 Blood group antigens

Blood group antigens are predominantly poly-saccharides and are found on the surface of red blood cells. The most important and best known blood group antigens are the human ABO and **Rhesus antigens**. Individuals who have group A blood have A antigen on their red cells. Those with group B blood have B antigen on their red cells. Those with group AB have both and those with O blood have neither antigen on their surface.

Curiously, everyone with blood group A carries antibodies directed against blood group B, and vice versa. Group O individuals carry antibodies against both antigens, while group AB individuals make neither (see table). This apparently spontaneous existence of antibodies against group A or group B antigens probably results from the common occurrence of similar antigens in bacteria. So, for example, antibodies are produced when a person is exposed, usually early in life, to bacteria with these antigens in their structure, but they would not develop against any antigens similar to the person's own antigens, which induce tolerance.

When a recipient of a blood transfusion carries antibodies to the transfused blood, the incoming blood cells are destroyed. The products of destruction lead to kidney and liver damage, fever and clotting defects. The recipient's own cells are not harmed by donor antibody because it is diluted in the larger volume of recipient blood. Thus, a group O person, who carries antibodies to A and B, can nevertheless donate blood to anyone since they have no AB antigens on their red blood cells to act as a target.

Table Human ABO blood groups

Blood group	Red cell antigen	Antibody present in serum	Transfusion status
O	O	Anti-A and Anti-B	Universal donor
A	A	Anti-B	Selective
B	B	Anti-A	Selective
AB	AB	None	Universal recipient

Antibodies (immunoglobulins)

The specificity of the immune response resides in the B and T lymphocytes. Each has its own distinct but related molecule that mediates this specificity. For B cells this molecule is an antibody and for T cells it is the T-cell receptor.

Antibodies are either membrane-bound or secreted free in solution. For any particular B cell, the antibody it secretes has exactly the same antigen-binding region as the membrane-bound antibody. Antibodies are glycoproteins, all of which have a similar underlying structure. Each molecule is made of two identical *heavy* chains and two identical *light* chains (Fig. 23.17). Each antibody molecule therefore has *two* identical antigen-binding regions at the ends of its two arms.

Both heavy and light chains can be divided into distinct regions: a *variable* region, the portion coded for by the rearranged part of the DNA (Box 23.2), which contains the antigen-binding site, and a *constant* region which, as the name suggests, is relatively constant from antibody to antibody. The variable region is the part that binds to the specific antigen (e.g. on the surface of a bacterium), but it is the constant region that carries out most of the other functions. For instance, antibody can aid phagocytosis because receptors on phagocytic cells bind to the constant region, providing a link between the phagocyte and the bacterium that has bound to the antibody. The segment of antibody that triggers the complement cascade is also found in the constant regions of IgG and IgM. Differences in the heavy chain constant regions are used to divide immunoglobulins into various *classes* (or isotypes), known as IgM, IgD, IgG, IgA and IgE, which are found in different sites and have different functions.

Antibody molecules function as antigen receptors on B cells and as effector molecules in all body fluids. They have a variable region, which recognises and binds to antigen, and a constant region, which is responsible for most other functions.

The T-cell receptor (TCR)

Unlike antibodies, which may be membrane-bound or free in body fluids, TCR molecules are always membrane-bound. Like antibodies, TCRs are composed of variable and constant regions, with the antigen-binding site a product of the joining of the two chains (Fig. 23.18). Antibodies and TCRs have many features in common, especially the genetic rearrangement events that generate their diverse specificities (Box 23.2). In contrast to antibodies, which have two antigen-binding sites, each TCR has only one antigen-binding site. The differences between TCRs and antibodies are shown in Table 23.1.

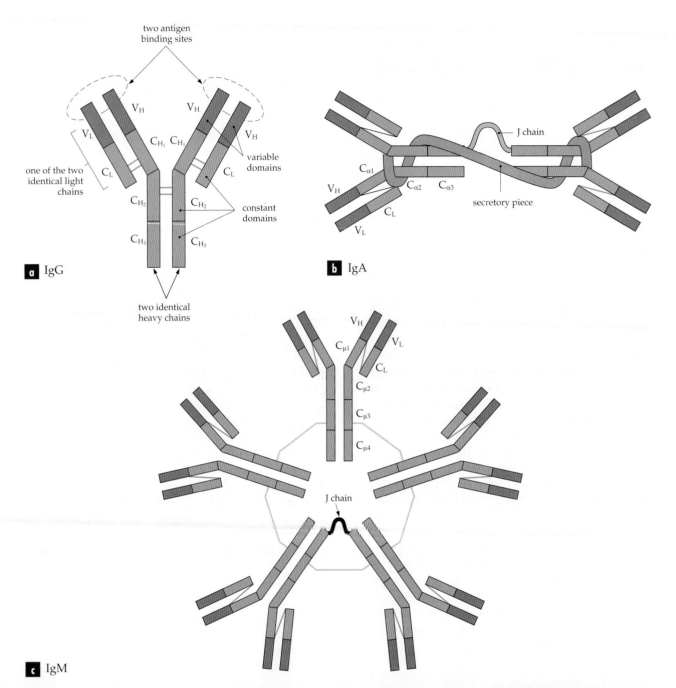

Fig. 23.17 Different classes (isotypes) of antibody molecules: The variable domains (V) are in red, the constant domains (C) in blue. V domains bind antigen, and differ for each specific antibody. The C domains differ for different classes of antibody, allowing each to have specialised functions, although they have nothing to do with specificity for antigen. **(a)** Schematic diagram of an IgG molecule, the commonest antibody class in blood. **(b)** Schematic diagram of the dimeric IgA molecule. The secretory piece functions to transport the molecule across mucous membranes. The J chain holds the two monomers together. **(c)** Schematic diagram of IgM. The pentameric molecule is again held together by a J chain. Because IgM has 10 binding sites, it strongly cross-links antigens, allowing better clearance

The TCR functions primarily as a tool for inter-action between T cells and other cells at their surfaces. The TCR is one of a large group of molecules involved in this interaction or in the responses that follow (see Figs 23.18, 23.20), but the specificity of the interaction is determined by the variable region of the TCR itself.

The TCR does not interact with free antigen. In fact, it only recognises antigen in the form of peptides bound to the surface of molecules of the MHC (see below). Once the TCR recognises a foreign peptide/MHC complex, a complicated, rather poorly understood series of events happens, resulting in a response within the T cell. The response involves activation, proliferation and differentiation of the T cells, for example, to become killer cells or to produce lymphokines.

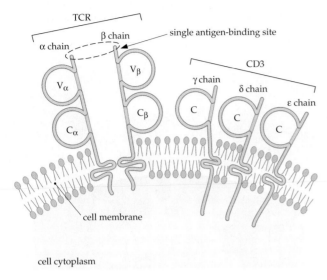

Fig. 23.18 Schematic diagram of the T-cell receptor and the associated chains of the CD3 molecule, which is essential for the appearance of TCR chains in the membrane, and for the transduction of signals when TCRs bind to antigens. Notice that the domains of the CD3 chains outside the membrane are similar to the C domains of the TCR molecule or of immunoglobulin

TCRs are always bound to the T-cell surface. The TCR does not react with free antigen. Each TCR has only one antigen-binding site, unlike antibodies, which have at least two.

The major histocompatibility complex (MHC)

MHC molecules are the only molecules that can present (show) antigen to T cells. They are expressed on all cells of the body but particularly on those cells known as *professional antigen-presenting cells*, such as dendritic cells, macrophages and B cells. Since T cells regulate most reactions in the immune system,

including antibody formation and self–not-self discrimination, MHC molecules play a key role in every aspect of immunity. B cells, as we have seen, can see antigens directly without needing MHC molecules to present them, but since B cells usually need help from T cells, the production of antibodies is also dependent on MHC molecules. Rejection of tissue grafts is another immune reaction in which the role of T cells and MHC molecules is central.

Antigens detected by T cells are small peptides having only eight to 20 amino acids derived from whole proteins. These peptides are only recognised as antigens by T cells when they are displayed on the end of the MHC molecules (Fig. 23.19). The peptide, together with the two neighbouring portions of the MHC molecule, is the structure recognised by the TCR (Fig. 23.20). The peptide is 'picked up' as the MHC molecule is synthesised inside the cells. Peptides may be derived from other molecules being synthesised by the cell (e.g. viral antigens), or from molecules external to the cell that have been phagocytosed and broken down.

The genes coding for MHC molecules are part of a *multigene complex*, which is a region in the genome containing many kinds of related genes. In the case of the MHC, some of the genes are associated with tissue graft rejection, and most of these occur in many allelic forms, with up to 100 alleles present in a given population. As the name implies, this is the most important set of genes in the determination of graft rejection. If two individuals have different alleles for *any* of the many MHC genes, then tissue grafts between them will be rejected as foreign material.

MHC restriction

A key concept of modern immunology is the concept of MHC restriction. Interactions of T-cell receptors with

Table 23.1 Differences between antibodies and TCR molecules

Feature	Antibody (immunoglobulin)	T-cell receptor
Number of chains	2 heavy, 2 light	1α, 1β (or 1γ, 1δ)
Antigen-binding sites	2	1
Number of subtypes	10 (IgM, IgG, etc.)	2 (αβ, γδ)
Localisation	Cell surface and soluble	Cell surface
Polymerisation	Up to 5 molecules (IgM)	No
'Sees' soluble antigen	Yes	No
Variable part	Yes	Yes
Constant part	1–4 per chain	1 per chain

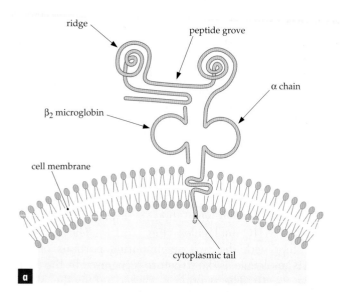

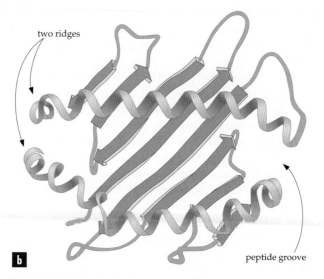

Fig. 23.19 MHC molecule. **(a)** Structure of the MHC class I molecule, attached in the cell membrane. It has two chains, a large α chain and a smaller β microglobulin chain. **(b)** View of the peptide groove and ridges from above, giving a more accurate picture of how the amino acid chain (i.e. the ribbon) coils and crisscrosses to make the platform and ridge structure

antigen are **MHC restricted**, which simply means that the antigen is recognised in association with an MHC molecule and not alone. This is not surprising since we know that the antigen is presented by MHC molecules, but the key point is that the MHC itself provides part of the entity that is recognised. Thus, if an MHC-restricted T cell recognises peptide X presented by an MHC molecule of a particular type, for example, MHC^a, then this T cell will only recognise (MHC^a + peptide X), not (MHC^a + peptide Y) or (MHC^b + peptide X), or MHC^a alone or peptide X alone. The TCR sees the complex of MHC and peptide, both components contributing to the specificity and both being required to trigger the particular T cell (see Table 23.2).

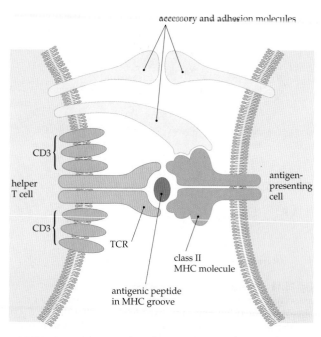

Fig. 23.20 Schematic representation of the TCR-mediated interaction between a T cell and an antigen-presenting cell. The interaction between the two cells is complex, involving many surface molecules (accessory and adhesion molecules) on each cell. However, the interaction is initiated only when the variable portion of the TCR recognises an antigen/MHC structure on the antigen-presenting cell

MHC molecules present to T cells a sample of either internal synthesised antigens or external scavenged antigens. The MHC molecule is part of the structure that the T cell recognises and interacts with.

Special features of the immune system

Tolerance in B and T cells

It is important that the immune system does not respond inappropriately to certain antigens, particularly self antigens. If the immune system fails to respond to a potentially immunogenic antigen, a state of **immunological tolerance** is said to exist. Tolerance occurs when lymphocytes come in contact with self antigens at a particular early stage of development—in the thymus for T cells, in the bone marrow for B cells. As a result, mature lymphocytes do not normally initiate damaging responses against self antigens. Normally, mature lymphocytes cannot be made tolerant otherwise tolerance could develop to a foreign potentially dangerous organism.

Autoimmunity

The processes that give rise to self-tolerance result in the removal, inactivation or suppression of most cells

Table 23.2 MHC restriction and the specificity of the TCR

Antigen recognised by TCR	Antigen being presented	T-cell response
MHC^a + peptide X	MHC^a + peptide X	Yes
MHC^a + peptide X	MHC^b + peptide X	No
MHC^a + peptide X	MHC^a + peptide Y	No
MHC^a + peptide X	MHC^a alone	No
MHC^a + peptide X	peptide X alone	No
MHC^a + peptide Y	MHC^a + peptide Y	Yes
MHC^a + peptide Y	MHC^a + peptide X	No

with the potential to react against self antigens. This process is not perfect and, in some individuals, immune responses develop against the individual's own antigens, giving rise to **autoimmune diseases**. These diseases are usually chronic and typically strike people between 15 and 35 years of age. They are second only to cancer in their cost to the health-care system.

Autoimmune diseases affect many tissues, including lungs, kidney, stomach, skin, joints, adrenal glands, thyroid or pancreas (Fig. 23.21). Damage to the pancreas results in a reduction in the capacity of the tissue to make insulin, resulting in diabetes. Auto-immune diseases are usually treated with anti-inflammatory drugs (especially in the case of arthritis) or, in severe cases, with immunosuppressive drugs to dampen the autoimmune response. This, of course, carries the risk of infection because other immune responses are also depressed.

> B oth T and B cells can be made tolerant if they contact self antigens at critical early stages of their development. Normally, mature lymphocytes cannot be made tolerant. When this process is faulty, autoimmune disease may develop.

Nature of immune responses

Humoral responses

Imagine bacteria entering the bloodstream through a cut or scratch and beginning to proliferate. Macrophages at the tissue site or in the spleen break down some of the bacteria and recycle some bacterial peptides back to the cell surface on MHC molecules. T$_H$ cells in the spleen or local lymph node that recognise these peptides respond to stimulation by

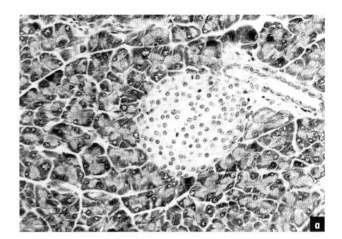
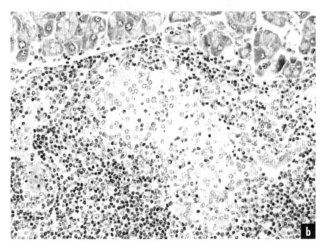

Fig. 23.21 Autoimmune diabetes. **(a)** A section of a normal pancreatic islet, which is the part of the pancreas that makes insulin, embedded in the exocrine pancreas. **(b)** Severe invasion of the area of the islet with lymphocytes (small round darkly stained cells), and a breakdown in the structure of the islet. Eventually this islet and others like it will be too badly damaged to make enough insulin to regulate blood sugar levels, and diabetes will result. Notice that there is essentially no invasion of the exocrine tissue. The invading T cells are quite specific for the islets

dividing and releasing lymphokines. At the same time, B cells come into contact with bacteria and those that have a surface antibody that recognises a bacterial surface antigen are stimulated. If the B cells also receive a signal from lymphokines made by T_H cells, they will proliferate and begin to release antibodies. These antibodies will bind to the bacteria and, in the presence of complement (Box 23.4), kill them and/or make it easier for macrophages and polymorphs to dispose of them. This will provide further stimulus for T_H cells, further lymphokines and more B cells to respond. As increasing numbers of cells circulate through the spleen, more cells reactive to the antigens on this particular bacterium can become stimulated and more B cells are available to produce antibody. Soon some B cells will become plasma cells, pumping out huge amounts of antibody against the bacteria. Eventually, the increasing levels of antibody will swamp the remaining bacteria, resulting in their elimination by macrophages and polymorphs.

Once the bacteria are gone, there is no more antigen to stimulate the T or B cells, so they eventually stop proliferating and making lymphokine or antibody and the system goes back to a resting state, the threat of bacterial invasion averted. However, the proliferation results in many more T cells able to recognise the particular bacterium and many more B cells that can make antibody to it scattered throughout the lymphoid tissues as memory cells. This is the immunised state. Should the same bacterium come along again, a second immune response (secondary or memory response) will be much quicker and more effective.

> When both B cells and helper T cells are stimulated by a foreign antigen, T_H cells will help the B cells make antibody, which will help kill and remove the pathogen. Once the antigen has been cleared, there is no further stimulus to the T or B cells and the response subsides.

Cellular responses

If the initial infection is a virus that proliferates within body cells, some viral antigens would appear as peptides in MHC molecules on the surface of infected cells. Antigens of this kind are more likely to stimulate T_C cells, which attack and kill any cell with viral antigen on the surface. Some free viral antigens might also be taken up by macrophages, and peptides from them presented by MHC molecules to T_H cells, stimulating them to respond and further helping the T cells to react. The stimulated T_H cells will also release a range of lymphokines that will attract and stimulate other cells and may cause local inflammation or fever. Some

---

BOX 23.4 The complement system

Complement is the collective term for a series of about 20 serum proteins that exist in the fluids of the body in an inactive state. If the first component in this series is turned into the active form, a complex cascade of reactions occurs, in which each component, once activated, activates the next. The first component can be directly activated by substances on the outer surface of certain bacteria or parasites, by some serum proteins bound to a variety of microorganisms, and by antibodies that are bound to their antigens (antigen–antibody complexes). Complement can thus be activated directly by pathogens or more specifically by antibody. Several of the components along the cascade, when activated, have important effects on the immune system.

The activation of the complement cascade has three main consequences:

1. attraction to the site of infection of a range of cell types involved in inflammation and defence, including macrophages, granulocytes and mast cells
2. facilitation of the engulfment by phagocytic cells of microbes or cells with complement bound to them
3. lysis of cells and damage to bacteria with complement bound to their surface (by hole-punching).

In addition to these effects, complement plays a role in the induction of local inflammation, increasing vascular permeability at the sites of infection, inducing contraction of smooth muscle (blood vessel walls) and attracting granulocytes to sites where complement is bound.

B cells might also respond to the viral antigens and, in concert with the T_H cells, make antibodies against some viral components. Since the whole virus is inside the cell, there is little opportunity for B cells to be directly stimulated by it, or for antibodies to reach it, since antibodies cannot normally get inside cells. However, antibodies may prevent the spread of virus from cell to cell.

Again, proliferation of T cells means that more cells will be available to protect the individual from a subsequent infection and, if antibodies have been made, they might also protect against initial infection while the virus is outside the cells.

Cellular responses are also largely responsible for rejection of foreign tissues (Box 23.5). Here the foreign

BOX 23.5 Tissue transplantation

Skin grafts, widely used for cosmetic surgery, usually involve transfer of skin from one part of the body to another (**autografts**). The tissue is 'self' and so there is no immune response against it; it is not rejected. Had the skin come from another non-identical individual (anyone but an identical twin), it would have been rejected within a week or two due to an immune response mounted against the MHC molecules it carried. However, with improvements in surgical techniques, tissue matching (MHC and blood-group typing) and immunosuppressive drugs, which dampen the rejection process, a number of tissues can now be transplanted between individuals with a good chance of success.

In Australia, there are 500–600 kidney transplants per year (mostly cadaver donations), with a one-year success rate of 80%. There are about 100 heart transplants and 50 liver transplants, both with about the same 80% one-year success rate. Another tissue frequently transplanted is the cornea, in an operation that is now fairly routine and with a success rate (when the recipient eye is not inflamed or severely damaged) well over 95%.

A major difficulty in transplantation is the risk of infection when immunosuppressive drugs are used. These suppress the response to pathogens as well as to the transplant, so immunosuppressed patients are often kept isolated to reduce the risk of infection. Much current research is aimed at finding ways to suppress immune responses to particular antigens (i.e. on the transplant) without causing generalised immunosuppression.

MHC antigens of the grafted tissue, be it heart, skin or kidney, stimulate both T_H and T_C cells, giving rise to cytotoxic cells, which can destroy the graft, or to lymphokines, which induce macrophages and granulocytes to cause inflammation, also destructive to the graft.

When antigens, such as viral antigens, are synthesised by cells and presented on MHC molecules, protective responses involve direct killing of infected cells by T_C cells, activation of phagocytic cells and inflammation due to lymphokines produced by stimulated T_H cells.

Regulation of immune responses

As we have seen, immune responses can take several courses, depending on the nature of the stimulus, but in all cases there is a lot of co-operation between cells. Responses can include an antibody response, a T_H response and/or a T_C cell response. T_H cells can make different lymphokines in different circumstances. The balance of these various responses will vary widely, although as you might expect, a response completely restricted to one component would be unusual. Much research is underway to understand the subtle processes that alter the balance of these responses, since every disease is different in terms of the kinds of immune response that best provide protection or recovery. For example, it is no use making a vaccine that stimulates a strong antibody response if T_C cells are needed.

The regulation of immune responses is not always perfect and in some conditions responses may overshoot or undershoot, causing damage to self tissue, or not clearing the pathogen effectively. Secondary damage to surrounding tissues caused as a result of an immune response is immunopathological (Box 23.6), as distinct from damage caused by the pathogen that caused the response. In some cases, the immune response does more damage than the disease during its attempts to remove a relatively benign foreign organism.

BOX 23.6 Immunopathology and allergy

Immunopathology and allergy are both situations where an immune response to a foreign antigen becomes excessive and results in damage to surrounding tissues, or causes life-threatening physiological responses.

Allergic responses occur when initial contact with an antigen (e.g. pollens or house dust) results in production of IgE antibody, which binds to the surface of mast cells. Later contact with the same antigen results in the antigen binding to the IgE on the mast cells, stimulating mast cells to release substances that cause acute inflammation. Since mast cells are common in the respiratory tract, it is here that allergic reactions often occur, resulting in constriction of the airways and, in extreme cases, suffocation. An allergen that gets into the bloodstream, such as bee venom, also has dramatic effects in allergic individuals, and the factors released by mast cells and basophils cause constriction of the airways and loss of blood pressure through dilation of blood vessels. Both these potentially life-threatening symptoms are reversed by adrenaline, which is the usual treatment for acute allergic attack.

Other immunopathologies result when immune responses become disregulated or antigens persist in tissues, resulting in prolonged immune responses, inflammation and damage. For example, contact with certain chemicals results in the binding of the chemical to molecules in the skin. Immune responses to the bound chemical result in local inflammation as T cells release lymphokines, which stimulate macrophages and other inflammatory cells. The antigen is not easily cleared because it is bound to self tissues, so the immune response is prolonged and difficult to contain and considerable tissue damage may ensue.

Infection by parasites may also cause immunopathology. Macrophages or other cells can become persistently infected with intracellular parasites and constantly activated by surrounding T cells. Hyperactive macrophages multiply and enlarge to cause granulomas, which are densely packed large swellings in the skin or internal organs. This response may ultimately be protective if the parasite is cleared, but persistence of the parasite can result in chronic granulomas that may become life-threatening.

Antibodies too can cause damage, not only by binding to foreign antigens expressed on tissues but also because antibody attached to its antigen (an immune complex) can form precipitates that cause damage to blood vessels and kidneys. This most often occurs with persistent antigens, such as antigens from uncleared infections or self antigens in autoimmune disease. Precipitated complexes can activate complement, causing local inflammation at sites of accumulation.

Multiplicity of immune responses

Pathogenic organisms may contain many antigenic molecules with tens or even hundreds of potential T- and B-cell epitopes. Usually only some of these will actually stimulate a response during a normal immune reaction. Some may stimulate B cells and others may stimulate T cells. In the overall immune response to any microorganism, there may be dozens of different antibodies being produced in varying levels, as well as T-cell responses to many different epitopes. Which antigens are responded to and the nature of the responses to them can be of vital importance in determining the outcome of an infection. A typical immune response will have many different components, some of which may be protective, while others may be not

useful or even harmful, stimulating allergy or other immunopathology.

> There are many possible arms to the immune responses and many potential antigens in most foreign organisms. Different antigens from the same organism might stimulate different aspects of the immune response.

Stress and immune responses

These days there is considerable talk of stress and the effect this can have on health. An example is the 'overtraining syndrome' where elite athletes find themselves subject to repeated infections. How does stress or, more broadly, mental state affect the immune system? There is good evidence that acute stress, such as a sudden accident or injury, can have a positive effect on an immune response that follows it. However, when either mental or physical stress is prolonged there can be marked reductions in the ability to develop immune responses, partly through prolonged exposure to stress hormones (Chapter 25).

There is a growing understanding of the interaction of the immune response with our hormonal and nervous systems. The nervous system directly or indirectly influences the release of many of our hormones, including corticosteroids and adrenaline, which are released during stress. Under conditions of stress, the adrenal glands are stimulated to increase their secretions, including the corticosteroids. Lymphocytes carry *receptor molecules* on their membranes, which lock onto and react to these hormones. Corticosteroids are major mediators of the suppressed immune response that accompanies prolonged stress. Indeed, cortisone is used medically when immunosuppression is needed, as in the prevention of graft rejection and treatment of autoimmune disease.

The immune system can also affect our nervous state. Fever and exhaustion that accompany infection

---

BOX 23.7 Immunodeficiency and AIDS

The immune system is critical for the protection of an individual against infection. In the absence of immune protection, fatal infection almost invariably follows. In a clinical setting, the immune system may be inhibited by immunosuppressive drugs after transplantation or by radiation or drugs used in the treatment of cancer. In these cases, great care and the use of antibiotics will usually see the patient through until the immune system recovers. There are, however, a number of diseases in which babies are born with defective immune systems. Di George syndrome is a T-cell deficiency resulting from the absence of a thymus. Severe combined immunodeficiency (SCID) results when genetic abnormalities result in the failure of both T and B cells to develop. In the past, children with these diseases died soon after birth due to severe viral, bacterial or fungal infections. Even the best modern medicine cannot save these children indefinitely without the help of an immune system. In recent years, some of these diseases have been successfully treated by transplanting normal MHC-matched bone marrow into the children, replacing the abnormal marrow and allowing the immune system to develop from the stem cells in the normal marrow transplant.

There are also a number of diseases that result in damage to the immune system. The best known example is AIDS (acquired immunodeficiency syndrome), a disease caused by the human immunodeficiency virus (HIV). This virus is normally difficult to transmit since its target cells are mainly lymphocytes, which are not accessible from outside the body. Direct contact of body fluids through blood transfusion or as a result of sexual intercourse is necessary. Once inside the body, the virus binds to a molecule called CD4, which is predominantly expressed on T_H cells. After binding to the CD4 on the surface of the T cell, the virus, like all retroviruses (Chapter 34), enters the cell and integrates into the DNA, essentially becoming a part of the cell. It then uses the cell's own replication mechanisms to multiply, eventually damaging or killing the T cell. Over a long period of time, the virus slowly spreads through the immune system, severely reducing the number of T cells, making the immune system less and less effective. With time, secondary infections with other pathogens become more and more serious, eventually causing the death of the patient. In the absence of these secondary infections, HIV would not itself be lethal.

The difficulty in treating patients with AIDS comes from the fact that the virus is so completely integrated into the host cell genome and becomes part of the cell itself. Furthermore, the very cells that would help initiate an immune response to the HIV virus are themselves being destroyed. However, greater understanding of how the virus proliferates in the cells is helping us to develop drugs that target the virus without damaging the cells of the host. Unfortunately, a vaccine against AIDS still presents difficulties because the virus can readily change its antigenic coat to trick the immune response.

are triggered by some of the lymphokines released during the immune response. Excess production of lymphokines also causes the loss of weight that accompanies prolonged infection or cancer. There is currently speculation that these factors may account for the symptoms of chronic fatigue syndrome.

Acute stress may stimulate immune responses but chronic stress can be inhibitory. Inhibition can be caused by stress hormones, whose release is governed by the central nervous system.

Immunity to viruses, bacteria and fungi

Viruses are diverse and cause a wide range of diseases in many organs (Chapter 34). They are dependent on growing within cells. Each type of virus grows only in certain types of cell. Influenza virus, for example, only grows in respiratory epithelium, while polio virus grows in the nervous system. Infections may be acute, chronic, recurrent, latent (present but not easily detected) or subclinical (detectable virus but no obvious disease). The role of the immune system in many of these conditions is unclear, although it plays a vital role in acute infections, such as influenza. Early in an infection, non-specific defence mechanisms can restrain the disease. Interferon, which interferes with the replication of virus within cells, and NK cells, which kill virally infected cells, are of particular importance. If general viraemia occurs, antibodies can prevent invasion of host cells, but once the virus is inside its target cell, T_C cells are probably essential to kill the infected cells. Thus, the importance of antibody or T cells varies with the particular disease and the stage of infection. Antibodies may be useful in preventing spread or reinfection by attacking the virus while it is outside the cells, but T_C cells are probably required to clear an existing infection (Fig. 23.22). Some viral diseases, such as polio, measles and mumps, are easily controlled by vaccination (see Box 23.8). Others, such as herpes or AIDS (Box 23.7), are not yet easily controlled.

Within a host, most bacteria are extracellular and so are easier to deal with than are viruses. The main sources of protection are intact skin and mucosal surfaces. When these are breached, rapid proliferation of bacteria may initially outrun the immune system. Rapid but relatively non-specific defences, such as phagocytosis and killing by macrophages (Fig. 23.23), polymorphs and complement (see Box 23.4), are crucial at this stage. Many pathogenic bacteria have evolved means of evading phagocytosis by acquiring capsules on their surface, which repel the membrane of the phagocyte. In this case they need to be coated with antibody or complement to be more efficiently phagocytosed by macrophages, which finally kill the bacteria.

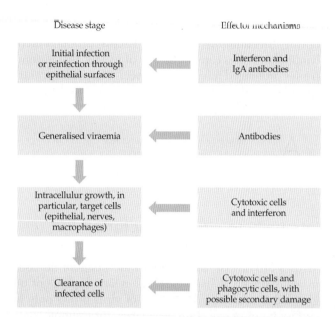

Fig. 23.22 Immune mechanisms that protect against viral infections as a viral disease progresses from initial infection to recovery

---

BOX 23.8 Vaccines

Vaccination against infection made a tremendous contribution to public health over the twentieth century. One disease, smallpox, has been completely eradicated from the world, while the World Health Organization is aiming to eradicate polio by the year 2005. Other diseases whose incidences have been dramatically curbed by vaccination include diphtheria, measles, chickenpox, meningitis and whooping cough.

Vaccines have traditionally been made using one of three approaches:

1. Purified protein toxoids (chemically detoxified toxin) for use against bacterial infections where the main damage is mediated by a toxin (e.g. diphtheria, tetanus). These act by inducing antibodies that neutralise the toxin.
2. Killed whole organisms, again good inducers of antibodies. These antibodies may be directed towards preventing the initial invasion of cells by a virus or enhancing phagocytosis of a bacterium, or activating the complement pathway. Examples include the Salk polio vaccine and whooping cough vaccine.
3. Live attenuated vaccines. These organisms have been rendered less virulent (attenuated) usually by years of culture in the test tube. These vaccines are particularly good at

inducing cellular immunity, although they can also induce antibody. They include the Sabin oral vaccine against polio, measles, mumps and rubella vaccines and BCG against tuberculosis.

Table Comparison of the merits of live and killed vaccines

Live	Killed
Must be attenuated by prolonged culture	Produced from fully virulent microorganisms
Given as single dose	Given in multiple doses
Smaller number of microorganisms needed	Large number of microorganisms needed
Necessary for induction of cellular immunity	Do not induce cellular immunity
Can induce response at site of infection	Generally induce serum antibodies
Tend to be less stable	Tend to be more stable
Possibility of spread to unvaccinated individuals	Spread is not possible

Much research is now focused on the production of new vaccines based on gene cloning, with the following aims:

*   Cloning the genes for antigens that would protect against a number of infections into a single vaccine.
*   Production of better-defined attenuated organisms by disabling a particular virulence gene.
*   Production of purified protein vaccines to replace the less well-defined killed whole organisms.

Some bacteria, like those that cause tuberculosis or leprosy, have evolved the means of surviving and living within phagocytic macrophages. These diseases require production of lymphokines by T_H cells to boost the killing mechanisms of the macrophages.

Fungi mostly cause superficial infections and do not usually enter tissues or blood of healthy people. Common fungal diseases include athlete's foot, ringworm and candida, a common disease of female genital mucosa. Immunity to these diseases is not well understood and, being largely external, the organisms are out of reach of T cells and antibodies. Some resistance, probably T-cell mediated, does develop and

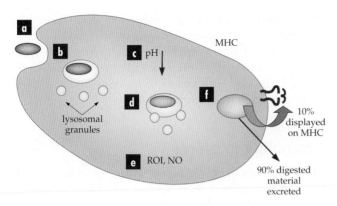

Fig. 23.23 Phagocytosis and killing of bacteria. **(a)** Bacterial cell contacts surface of phagocyte, stimulating formation of pseudopodia. **(b)** Pseudopodia embrace the bacterium, which is enclosed in a phagocytic vacuole. **(c)** pH in the phagocyte drops, creating suitable conditions for the action of enzymes released when **(d)** lysosomes fuse with the phagocytic vacuole. **(e)** Bacteria-killing metabolites of oxygen (reactive **o**xygen **i**ntermediates, ROI), such as hydrogen peroxide (H_2O_2), and nitric oxide (NO) are produced. **(f)** The bacteria are digested and 90% of the product is excreted. An important fraction of the bacterial peptides is passed to the surface of the cell on the MHC for presentation to T cells. Immunity to 'difficult' bacteria is helped by antibody and complement, which enhance phagocytosis of capsulated bacteria, and lymphokines from T_H cells, which activate the killing mechanisms against the bacteria of tuberculosis and leprosy

is manifested through local release of lymphokines, which activate macrophages and polymorphs to clear infectious material.

Immunity to protozoa and parasitic worms

Protozoa and parasitic worms are more difficult for the immune system to deal with because they are larger and more complex than are viruses or bacteria; some intestinal worms can be metres long. Also, their life cycles are often complex, with many rapidly changing stages, and a variety of mechanisms that subvert or avoid the immune system have evolved. This difficulty is apparent from the high levels of chronic infection in some human populations, particularly in the tropical world. Diseases such as malaria and sleeping sickness (protist parasites, Chapter 35) and schistosomiasis and hookworm (round-worm parasites, Chapter 38) each affect over 100 million people worldwide. Not surprisingly, resistance to these parasites involves almost every aspect of the specific and non-specific elements of the immune system.

T cells are important in controlling parasites, particularly because the lymphokines they release stimulate macrophages, polymorphs and mast cells into states where they can kill parasites. Antibodies can also play a vital role, attracting macrophages or complement to cause lysis. In many cases, a combination of all these mechanisms is necessary to rid the body of a parasitic organism.

Immunity to tumours

There is a widely held belief that cancers are tissues that have escaped the control of the immune system; that is, the surveillance function of lymphocytes has broken down. This, however, is not the case: tumours are made up of our own tissue, admittedly growing in an unregulated manner, and are generally recognised as 'self' by the immune system and ignored. The major fatal human cancers of the lung, gut, breast and urogenital tract proliferate unchecked because their cells are not immunogenic, not because the immune system has failed.

Nevertheless, there may be some instances where a tumour could express antigens that are 'not self' and hence be potential targets for immune attack. These include viral antigens, since some tumours are thought to be virus-induced, or antigens normally expressed only in the fetus, which are re-expressed in tumours of adults. These could possibly be the targets of a vaccine or immune therapy since they are unique to the tumour in the adult, and if an immune response to them could be induced, it would not damage other tissues in the individual.

> Different immune responses are effective in protecting against different diseases. It is important to understand which aspects are protective in order to design effective vaccines.

Evolution of immune responses

Immunity in animals

Clearly, every organism must have mechanisms for keeping itself distinct from its environment or from other organisms. These mechanisms become more complex as organisms become more complex. Invertebrates do not have organised lymphoid tissues or lymphocytes or any truly specific immune responses. However, all multicellular invertebrates have some kind of roving phagocytic cell that is able to engulf and destroy damaged or foreign material (Fig. 23.24). Some invertebrates can reject **allografts** (tissues from other individuals of the same species) and some have molecules able to bind to foreign material and enhance phagocytosis. In molluscs, annelids and arthropods, these molecules may be induced by damage or infection. In no invertebrate studied to date is there evidence of any kind of memory or secondary response.

Each of the above reactions requires that the defensive molecules or cells are able to distinguish self from non-self, even though specificity is lacking in the sense of being able to distinguish between different

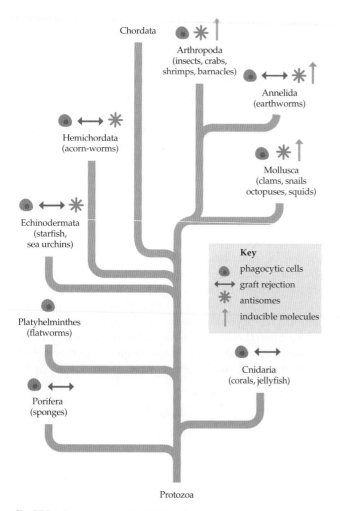

Chordata

Arthropoda
(insects, crabs,
shrimps, barnacles)

Annelida
(earthworms)

Hemichordata
(acorn-worms)

Mollusca
(clams, snails
octopuses, squids)

Echinodermata
(starfish,
sea urchins)

Key
phagocytic cells
graft rejection
antisomes
inducible molecules

Platyhelminthes
(flatworms)

Cnidaria
(corals, jellyfish)

Porifera
(sponges)

Protozoa

Fig 23.24 Schematic diagram showing the increasing complexity of 'immune' responses as organisms become more complex. Antisomes are molecules that are able, in a general way, to attach to foreign (non-self) material, making destruction or clearance easier. They function like antibodies but are not highly specific and may or may not be inducible (i.e. produced when needed in the event of an infection)

more primitive fishes, most vertebrates have something resembling the mammalian system, with antibodies or similar molecules, as well as spleen, thymus and T and B cells. Nonetheless, there is a gradual increase in the complexity and refinement of immune systems as we move through the vertebrates.

Immunity in plants

Like all other living organisms, plants must have mechanisms that protect them against invasion and, indeed, a number of complex processes have evolved in plants to provide such protection. Plants differ from most animals in that they have no circulatory system or migrating cells, so immune mechanisms must be local; each cell must fend for itself.

The first line of protection is, of course, the cell wall but, as in animals, physical barriers are never sufficient. Plant cells have both 'humoral' and 'cellular' mechanisms for protecting themselves from pathogens that penetrate the cell. Both these mechanisms are inducible. Humoral factors include antibiotics and a variety of enzymes that can destroy pathogens. Cellular mechanisms essentially involve 'self-destruction', as each infected or damaged cell breaks itself down, leaving neighbouring healthy cells intact.

Both these processes require that a plant cell be able to distinguish self from non-self, that is, to recognise a pathogen as foreign. At least in some cases, plants have individual genes in the genome that code directly for molecules able to recognise a particular pathogen, and indeed have a number of genes for different pathogens. This is quite distinct from vertebrates, which, as we have seen, use a combination of gene segments to generate a large random repertoire of defence molecules. In that case, there is no pre-existing gene for any particular pathogen, although there may be a rearranged gene in one or a number of clones of B or T cells. Obviously, plants cannot have genes for all pathogens since they would have the same problem as animals in using up too much genetic space and in being unable to deal with new pathogens. We must assume that they have evolved matching genes for important common pathogens but must have some additional, as yet unknown, mechanisms for distinguishing self from non-self more generally.

non-self molecules. It is not clear how this recognition of non-self occurs, although it is likely that substances released from damaged self cells play some role. Removal of foreign invaders in this case may involve removal by engulfment of any damaged self tissue.

Essentially, all birds and mammals (including marsupials and monotremes) have all the characteristics of the immune system that we have described in this chapter (Fig. 23.25). With the exception of the

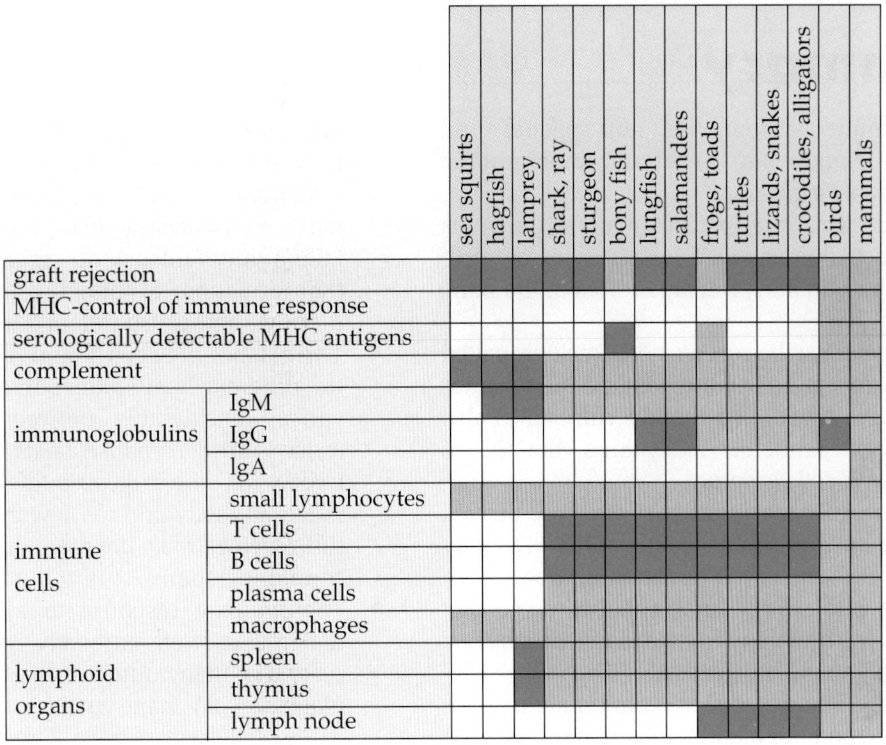

	sea squirts	hagfish	lamprey	shark, ray	sturgeon	bony fish	lungfish	salamanders	frogs, toads	turtles	lizards, snakes	crocodiles, alligators	birds	mammals
graft rejection														
MHC-control of immune response														
serologically detectable MHC antigens														
complement														
immunoglobulins — IgM														
immunoglobulins — IgG														
immunoglobulins — IgA														
immune cells — small lymphocytes														
immune cells — T cells														
immune cells — B cells														
immune cells — plasma cells														
immune cells — macrophages														
lymphoid organs — spleen														
lymphoid organs — thymus														
lymphoid organs — lymph node														

Fig. 23.25 Changes in immune responses with increasing complexity in the Chordata. Pink boxes indicate the presence of immune functions similar to those of mammals. Green boxes indicate the presence of atypical or only partly developed traits. White boxes show cases where the particular trait is absent, or there is no information available

Summary

- The immune system has non-specific and specific components. Both are important in defence against disease.
- Cells that carry out specific immune functions are called lymphocytes. T lymphocytes are produced in the thymus, B lymphocytes in the bone marrow.
- The antigen-specific receptor on T cells is called the T-cell receptor. The antigen-specific receptor on B cells is called an antibody. Antibodies also function independently in solution, but this is not true of T-cell receptors.
- Antigen-specific receptors are extremely diverse in their antigen-binding site. This diversity comes from the random shuffling (rearrangement) of a relatively small number of genetic elements, rather than from a large number of pre-existing complete genes.
- Cells of the immune system continually migrate around the body in blood and lymphatic vessels, passing regularly through organised lymphoid tissues, such as spleen and lymph nodes. Immune responses occur largely in these organised tissues, but can also occur at sites of inflammation.
- Antigens for T cells are mostly proteins, and these must be broken down and presented to the T cell in the form of short peptides attached to an MHC molecule. Antigens for B cells can be proteins or carbohydrates, and the whole or part of the antigen can interact with an antibody.
- The primary role of B cells is to make antibodies in large quantity, to kill directly or assist in the killing of foreign infectious agents by macrophages and granulocytes.
- T cells have diverse functions. They can: kill antigen-bearing target cells directly, but they also release a wide range of soluble mediators, called lymphokines, that help B cells proliferate and make antibodies; activate the phagocytosis and killing mechanisms of macrophages; enhance the formation of blood cells from the bone marrow; and cause inflammation and fever.
- The balance between T- and B-cell responses and the different lymphokines that can be produced varies considerably between infections. The exact form a response takes can have a major impact on the outcome of an infection. Too much antibody and too few cytotoxic T cells (or the reverse) may do more harm than good in certain infections.
- The immune response to a particular foreign material is often extremely complex, usually to several or many different antigens on the foreign material, and made up of interactions between many cell types and a large array of soluble mediators.
- The inability to respond to certain antigens is called tolerance. The establishment of tolerance to self antigens is an important aspect of the development of lymphocytes. A breakdown in self-tolerance can result in destructive immune responses against self tissues, resulting in autoimmune disease.
- The evolution of the immune system parallels the evolution of more complex life in general. All organisms need some form of defence against infection; however, specific immune responses, organised tissues and specialised cells and molecules exist only in vertebrates. The immune system seems to be at its most complex in birds and mammals.

keyterms

ABO blood group
   antigen (p. 612)
allografts (p. 623)
antibody (p. 600)
antigen (p. 600)
autograft (p. 619)
autoimmune disease
   (p. 617)
autoimmunity (p. 602)
basophil (p. 608)
B lymphocyte (p. 602)
bone marrow (p. 602)
cellular immunity
   (p. 605)
clonal selection
   (p. 603)
complement system
   (p. 601)
cytotoxic (T_C) cell
   (p. 607)

dendritic cell (p. 608)
eosinophil (p. 608)
epitope (p. 612)
granulocyte (p. 602)
haemopoietic stem cell
   (p. 602)
helper (T_H) cell
   (p. 607)
humoral immunity
   (p. 606)
immunogen (p. 611)
immunogenic (p. 611)
immunoglobulin
   (p. 607)
immunological
   memory (p. 601)
immunological
   tolerance (p. 616)
immunopathology
   (p. 619)

inflammation (p. 609)
interferon (p. 601)
lymphatic vessel
   (p. 609)
lymphocyte (p. 600)
lymphokine (p. 607)
lysis (p. 607)
macrophage (p. 602)
major
   histocompatibility
   complex (MHC)
   molecule (p. 611)
mast cell (p. 608)
memory cell (p. 608)
MHC restriction
   (p. 616)
monocyte (p. 608)
natural killer (NK) cell
   (p. 602)
neutrophil (p. 608)

phagocytic cell
   (phagocyte) (p. 602)
plasma cell (p. 608)
primary lymphoid
   organ (p. 609)
primary response
   (p. 601)
Rhesus antigen
   (p. 613)
secondary lymphoid
   organ (p. 609)
secondary response
   (p. 601)
T-cell receptor (TCR)
   (p. 611)
thymus (p. 602)
T lymphocyte (p. 602)

Review questions

1. What are the differences between specific and non-specific components of immune responses? Give examples of each.

2. How does the specific immune response protect against (a) viral and (b) bacterial infections? How can we use this knowledge to build better vaccines?

3. What is immune tolerance? What happens if it fails?

4. What are the essential differences between the antigens recognised by T cells and B cells? Why?

5. Explain how immunological memory works. Why, in rare cases, does immunological memory to one organism provide cross-protection against another?

6. Why and how do lymphocytes circulate around the body? Draw a map of the routes they take.

Extension questions

1. Explain how the process of gene rearrangement occurs in T and B lymphocytes to give rise to a wide range of different TCR and antibody molecules. How does this help us to cope with the huge number of antigens in our environment?

2. How do T cells regulate other cells during immune responses? Draw a diagram showing the cell interactions involved in generating an antibody response.

3. Why do we need so many different ways of dealing with pathogens? Consider the relative roles of specific and non-specific immunity, different classes of antibody and different functions of T cells.

4. Explain why patients die of AIDS. Why is the targeting of T cells by HIV so devastating?

5. Why is transplantation between individuals difficult and how is this overcome? Can you imagine why MHC molecules, normally involved in antigen presentation, are so important in graft rejection?

6. There is no evolutionary advantage in developing allergic responses to harmless antigens such as pollen. Why do you think we suffer allergies?

Suggested further reading

Sharon, J. (1998). *Basic Immunology.* Baltimore, MD: Williams and Wilkins.

Good basic text covering the field of immunology.

Roitt, I., Brostoff, J., Male, D. (1998). *Immunology.* 5th edn. London: Mosby.

More advanced but beautifully illustrated general text.

Zinkernagel, R. (1997). The Nobel lectures in Immunology. The Nobel Prize for Physiology or Medicine 1996. *Scandinavian Journal of Immunology.* 46: 423–36.

Doherty, P. C. (1997). The Nobel lectures in Immunology. The Nobel Prize for Physiology or Medicine 1996. *Scandinavian Journal of Immunology.* 46: 528–40.

The second time the Nobel Prize has been awarded for Australian Immunology. At least read the presentation speech on page 422 and their biographies at the end of their papers.

Weiner, D. B. and Kennedy, R. C. (1999). Genetic vaccines. *Scientific American* 281: 34–41.

Describes modern approaches to vaccines.

Special Report: (1998). Defeating AIDS: what will it take? *Scientific American* 279: 61–87.

A series of seven articles discussing epidemiology, vaccines and therapy. Of particular interest are 'HIV vaccines: prospects and challenges' by D. Baltimore & C. Heilman (pp. 78–83) and 'Improving HIV therapy' by J. G. Bartlett & R. D. Moore (pp. 64–7).

Some interesting Web sites:

http://www.umass.edu/microbio/rasmol/rasclass.htm

http://www.umass.edu/microbio/chime/index.html

Two sites that allow you to look at three-dimensional images of antibodies, TCR, MHC. Need Netscape 3.01 or later.

http://www.unaids.org/

The latest information on AIDS maintained by the United Nations.

Part 5

Responsiveness and co-ordination

CHAPTER
24

Plant hormones and growth responses

Plants compensate for their relative immobility by modifying their growth in response to stimuli from the environment in which they live. For example, shoots grow and bend towards a light source, roots grow down into soil and flowers are produced at times of the year suitable for seed production. Thus, environmental factors, such as light, day length, temperature and gravity, can exert dramatic effects on plant growth and development. Nevertheless, some developmental changes are largely programmed genetically and hence occur regardless of environmental factors.

Responses of plants to both internal and external influences may involve changes at the molecular level through altered gene expression; the cellular level, where cells may either undergo division or growth and differentiation; and the whole plant level, where the growth pattern of a root or shoot is altered or new organs are produced. Many plant responses that are triggered by environmental stimuli are brought about by the action of hormones. **Plant hormones** are molecules that affect plant growth and development at very low levels, sometimes at concentrations below one part per billion. This may involve action at some distance from the site of synthesis but frequently plant hormones are produced in the same tissue in which they exert a response. However, whether growth responses are regulated primarily by changes in the *levels* of plant hormones or by changes in the *sensitivity* of tissues to these substances is still a matter of debate among plant physiologists. Both mechanisms probably occur.

Our understanding of the action of plant hormones has been aided by the use of modern methods of analysis, such as gas chromatography, mass spectrometry and immunological assays, and by the use of single-gene mutants that markedly alter growth and development compared with wild-type plants. If exogenous (external) application of a particular hormone to such a mutant induces development of the wild-type phenotype then, by implication, that hormone is probably involved in the normal process of growth and development (Fig. 24.1).

In this chapter, we will discuss five main groups of plant hormones—auxins, gibberellins, cytokinins, abscisic acid and ethylene—the physiological processes that they influence, and the growth responses that are triggered by environmental stimuli.

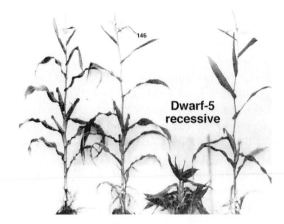

Fig. 24.1 Effect of the plant hormone, gibberellic acid (GA), on normal and dwarf corn. Left to right: normal plant (control); normal plus GA; dwarf plant (control); dwarf plus GA

Auxin

Phototropism

Auxin was the first plant hormone to be discovered. In 1881, Charles Darwin and his son Francis performed experiments on coleoptiles, the sheaths enclosing young leaves in germinating grass seedlings (Fig. 24.2). They exposed coleoptiles to light from a unidirectional source and observed that they bend towards the light. This phenomenon is termed **phototropism**. A tropic response is one in which growth is directed by an environmental factor and usually involves bending of the organ involved. By covering various parts of the coleoptile with light-impermeable metal foil, the Darwins discovered that light is detected by the coleoptile tip, but that bending occurs below the tip in the

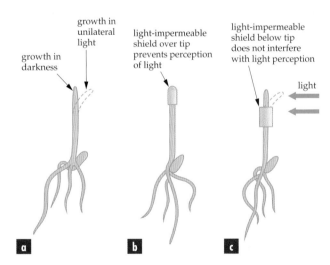

Fig. 24.2 (a) Growth of a coleoptile in darkness and unilateral light. **(b)** The tip of a coleoptile is the site of perception of unilateral light. If the tip is covered with a metal foil, there is no response. **(c)** Covering the zone below the tip does not prevent bending towards the light source

Plant growth and development are controlled by many environmental and genetic factors. Many responses are brought about by the action of hormones—molecules active at very low concentrations.

zone of **cell elongation**. They proposed that a messenger is transmitted in a downward direction from the tip of the coleoptile.

In the 1920s, Friedrich Went, a plant physiologist from The Netherlands, showed that a chemical messenger diffuses from coleoptile tips. When small agar blocks containing messenger from coleoptile tips were transferred to the cut ends of similar coleoptiles, they caused the coleoptiles to bend (Fig. 24.3). Went later proposed that the messenger substance is a growth-promoting hormone, which he named **auxin**, that becomes asymmetrically distributed in the bending region. He suggested that auxin is at a higher concentration in the shaded side, promoting cell elongation, which results in a coleoptile bending towards the light.

Since these experiments were carried out, the existence of auxin and its tendency to move downwards from the tip of a coleoptile have been confirmed, and Went's original experiment with agar blocks has been repeated with similar results. However, doubts now exist about the exact role of auxin in coleoptile bending. Determinations of auxin distribution using modern analytical techniques have shown that approximately equal levels of auxin occur on the illuminated and shaded sides of coleoptiles and stems. Furthermore, it has been reported that the shaded side contains lower levels of growth-inhibiting substance(s). This, of course, would exert an effect similar to an accumulation of growth-promoting auxin on the shaded side. The original method of determining the growth-promoting potential of illuminated and shaded tissue, by means of bioassays of coleoptiles, would not distinguish between a *low* level of auxin and an *accumulation* of growth-inhibiting substances. Thus, bending of coleoptiles may be regulated by auxin combined with growth inhibitors.

Auxins are growth-promoting hormones; they induce bending of coleoptiles towards light, an effect that may also involve growth inhibitors.

Indole-3-acetic acid

The main auxin occurring naturally in plants is **indole-3-acetic acid** (IAA), which is a small molecule consisting of an indole group with an acetic acid side chain (Fig. 24.4). When applied to intact plants, IAA does not elicit large responses but is very effective in promoting growth of portions of stem or coleoptile segments. Concentrations of IAA as low as 0.2 parts per million (approximately 10^{-7} M) will produce a response (Fig. 24.5). IAA exerts this effect by promoting elongation of cells rather than by increasing the number of cells present in the segment.

Within a plant, IAA may be synthesised primarily from the amino acid tryptophan. The level of active IAA in tissue can be influenced by the formation or hydrolysis of conjugates, which are composite molecules comprising IAA bonded to another molecule, such as a sugar. The level of IAA in a particular organ

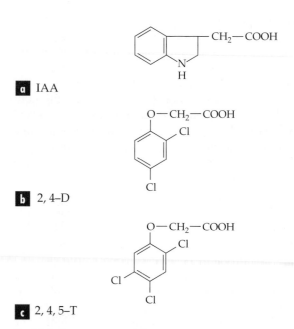

a IAA

b 2, 4–D

c 2, 4, 5–T

Fig. 24.4 The structure of **(a)** indole-3-acetic acid (IAA) and the synthetic auxins **(b)** 2,4-dichlorophenoxyacetic acid (2,4-D) and **(c)** 2,4,5-trichlorophenoxyacetic acid (2,4,5-T). These synthetic auxins are used as herbicides because at high levels they disrupt normal growth

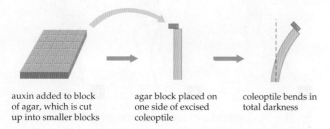

auxin added to block of agar, which is cut up into smaller blocks

agar block placed on one side of excised coleoptile

coleoptile bends in total darkness

Fig. 24.3 Went's experiment, showing that the application of auxin to one side of a coleoptile with the tip removed will show a growth response in darkness

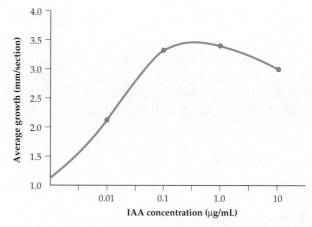

Fig. 24.5 The effect of IAA on elongation of segments of pea epicotyl

can also increase as a result of transport from another part of the plant. IAA shows **polar transport**, that is, the hormone is transported from the top to the bottom of stem segments much more readily than in the reverse direction.

IAA moves at a velocity of between 5 and 20 mm per hour in shoots and coleoptiles. Polar transport is believed to reflect the action of influx and efflux carrier proteins for IAA in the plasma membrane of transporting cells. The efflux carrier is asymmetrically localised to the basal side of the cells. Recently, candidate influx and efflux carriers have been identified. The AUX1 protein in *Arabidopsis*, for example, is a membrane protein similar to plant amino acid permeases and, like them, auxin uptake may involve a proton cotransport mechanism (Chapter 4).

In roots, two polar IAA transport streams are present. IAA moves from the shoot to the root tip in cells adjacent to or within the stele, while IAA moves from the root tip to the top of the root via epidermal and cortical cells.

Several other naturally occurring auxins have been discovered, including 4-chloro-indoleacetic acid, which has been identified in extracts from certain legumes. There are also many synthetic auxins, not found in nature, that are used in agriculture as **herbicides** (Fig. 24.4) or as root-promoting substances in plant propagation (Fig. 24.6). For example, both 2,4-dichlorophenoxyacetic acid (2,4-D) and 2,4,5-trichlorophenoxyacetic acid (2,4,5-T) are well-known herbicides that were used during the Vietnam war. 2,4,5-T was a component of Agent Orange, used as a defoliant in jungle warfare. Unfortunately, dioxins, which are among the most toxic substances known, were present as trace contaminants in herbicides containing 2,4,5-T and the use of these substances is now banned.

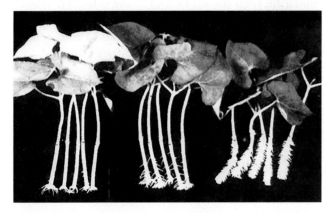

Fig. 24.6 The effect of an auxin on root initiation. Leaves of bean cuttings were either untreated (i.e. a control, left), or treated with 5 mg/L (middle) or 50 mg/L (right) of the synthetic auxin naphthalene acetic acid (NAA). Increasing concentrations of NAA induce the formation of roots on the stems

Auxin and cell elongation

One mechanism proposed for the effect of auxin on cell elongation is the **acid growth hypothesis**. The hypothesis is that auxin stimulates release of H^+ from within a cell (Table 24.1), which results in a loosening of bonds in the cell wall, making it more flexible. As a result of turgor pressure, cells elongate. This hypothesis explains two experimental observations: excised stem segments floating on a solution of low pH show enhanced elongation; and the pH of a solution in which auxin-treated stem segments are floating becomes lower.

Auxin may also affect plant development by regulating gene expression. For example, in strawberry fruits, high levels of a specific mRNA are associated with a block in fruit development. Auxin within the plant represses the level of this specific mRNA, allowing fruit to develop normally.

> The acid growth hypothesis for the action of auxin on cell elongation suggests that the release of H^+ from cells into the walls causes cell-wall loosening. This leads to cell expansion under turgor pressure. Auxin may also regulate gene expression.

Table 24.1 Biochemical and molecular responses to auxin

Growth response	Response time (min)	Tissue
Increases in:		
Respiration rate	5	Maize coleoptile
mRNA sequences	5	Soybean
	10–20	Garden pea
H^+ extrusion	7–8	Maize coleoptile
Membrane potential	12	Oat coleoptile
β-glucan synthase activity	10–15	Garden pea
K^+ uptake	30	Oat coleoptile
Protein synthesis	30	Tobacco protoplast
Decreases in:		
ATP:ADP ratio	5	Oat coleoptile
Cytosolic pH	5	Maize coleoptile

Auxin and apical dominance

Many plants consist of a single dominant shoot because the apical bud inhibits the growth of axillary buds further down the stem. On removal of the apical bud of such a shoot, axillary buds will grow out to form branches. However, if IAA is applied to the cut surface of a stem after removal of the apical bud, lateral bud growth remains suppressed. IAA mimics the effect of the apical bud and, since shoot tips are important sites of IAA production, IAA is therefore implicated in the maintenance of **apical dominance**. This is supported by the increased apical dominance shown in certain plants genetically engineered to overproduce IAA. For example, when a gene for IAA production was transferred from *Agrobacterium tumefaciens* (a bacterium widely used in genetic transformation) into petunias, IAA levels were increased and the degree of branching decreased. But how can IAA, a growth promoter, act as a growth inhibitor? It may do so by diffusing downwards from the apical bud and influencing cells in the nodes adjacent to the lateral buds to produce a second factor, which suppresses the growth of the lateral buds.

> I AA can substitute for the apical bud in the maintenance of apical dominance.

Gravitropism

In **gravitropism**, roots bend towards the earth's gravitational field while stems and coleoptiles grow away from it. Gravitropic bending of a root or shoot placed in a horizontal position results from differential growth on upper and lower sides. In roots, the root cap detects the stimulus of gravity, whereas actual bending occurs in the elongation zone behind the root cap. In shoots, however, there does not seem to be such a clear demarcation: the capacity to detect the direction of gravity appears much more uniformly spread along a coleoptile or stem.

Detection of gravity appears to involve the sedimentation of plastids (statoliths), usually **amyloplasts**, which contain starch granules (Chapter 17). Evidence for the role of amyloplasts comes, for example, from observing plants grown in spacecraft where they are exposed to only a weak gravitational field. These plants have roots that are orientated randomly, as are the amyloplasts within their root cap cells. Also, some starch-deficient mutant plants show impaired gravitropism. In a normal root or rhizoid growing vertically, amyloplasts settle against transverse cell walls (Fig. 24.7). When a root is moved to a horizontal position, amyloplasts settle on the side of cells that were originally vertical. The movement of amyloplasts may result in a gradient of growth hormones within cells, causing growth and bending of the root in the direction of the force of gravity.

A role for auxin in mediating gravitropism was proposed in the 1930s by the Russian plant physiologist Cholodny. Support has come from recent work on soybean hypocotyls, which show an asymmetrical distribution of auxin-regulated mRNAs correlated with

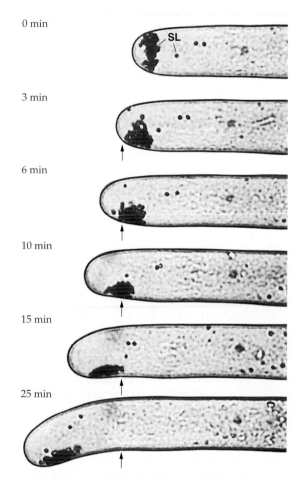

Fig. 24.7 Time-lapse photographs of a rhizoid of *Chara*, a green alga related to land plants (Chapter 37). The rhizoid, originally vertical, was displaced to a horizontal position. The photographs show sedimentation of statoliths (SL) followed by bending downwards of the rhizoid after 25 minutes. The arrow indicates the same point on the cell wall in each photograph

the gravitropic response. However, there is no apparent auxin gradient in the direction of the force of gravity and the auxin level in coleoptiles placed horizontally is similar on both the upper and lower sides. Therefore, rather than alter the *level* of auxin, gravity may somehow influence the *sensitivity* of tissue to endogenous auxin and thus affect the production of mRNA.

There is evidence that Ca^{2+} is also an important messenger in gravitropic responses. It has been suggested that, in horizontally orientated roots, a downward movement of Ca^{2+} towards the lower side is responsible for the subsequent bending of the root.

> G ravitropism results in bending of stems and roots in response to gravity. In roots, the response seems to involve auxin and Ca^{2+} gradients.

Auxin binding proteins

The primary response to auxin is probably mediated by a receptor (Chapter 7). This receptor has not been unequivocally identified. However, an **auxin binding protein** has been isolated, which may fulfil such a role. Auxin binding protein (ABP) was originally isolated from corn and homologous proteins were subsequently found to be widespread among plants. These ABPs are hydrophilic (soluble) proteins and mainly located in the endoplasmic reticulum, although small amounts are extracellular. After having bound auxin, the extracellular ABP is postulated to then recognise and bind to a transmembrane protein ('docking protein'). The docking protein then activates an unknown signalling pathway in the cell, leading to the appropriate response. Evidence for this hypothesis comes from work with antibodies that specifically interact with the auxin binding site on ABP. These antibodies are auxin agonists. When bound by them, ABP responds as though it had bound auxin. When such antibodies are added to cells, they mimic the effects of added auxin. For example, an increase in membrane potential occurs (hyperpolarisation) and stomata will open. Such experiments show that extracellular ABP can induce these responses since the antibody molecules are unable to access the intracellular ABP.

Gibberellins

Gibberellins are another group of plant hormones that influence growth, including stem elongation, seed germination, mobilisation of starch reserves in seeds and leaf expansion. They occur in low concentrations in vegetative tissues (about 10 parts per billion) but in higher concentrations in developing seeds (more than 1 part per million).

More than 80 different gibberellins have been isolated from a range of plants and fungi. All are small molecules (diterpenes) of similar basic molecular structure, with 19 or 20 carbon atoms (Fig. 24.8). They were first recognised in the 1930s as a product of a pathogenic fungus, *Gibberella fujikuroi*, which infects rice plants, making them tall and spindly. The molecular structure of the active compound, **gibberellic acid (GA)**, was identified later in 1956. The major endogenous active gibberellin that causes stem elongation in many plants is GA (Fig. 24.8).

Gibberellins and stem elongation

When Mendel's dwarf (*le*) pea plants (Chapter 9) are treated with 1 mg of GA, normal *stem elongation* and growth are restored (Fig. 24.9). Dwarfness is due to the *le* mutation, which blocks the last step in the biosynthetic pathway leading to GA synthesis. Stems and leaves of tall (wild-type) peas contain about 15 ng of GA_1 per gram of fresh weight, while shoots from dwarf

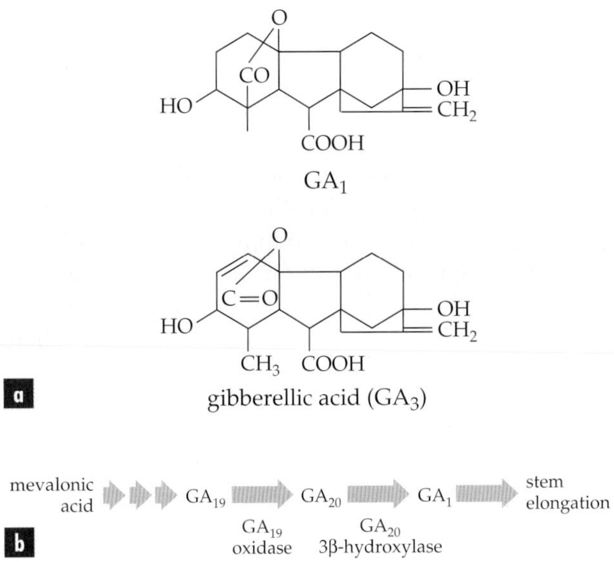

Fig. 24.8 (a) Structure of two active gibberellins, GA_1 and GA_3. **(b)** The major biosynthetic pathway in the vegetative tissue of many higher plants leading to the formation of active gibberellin, GA_1, which is responsible for shoot elongation

Fig. 24.9 The response of dwarf peas to applied GA_1. The control plant is on the right. The plant second from the right has been treated with 0.01 mg GA_1, with increasing GA_1 doses in two-fold steps to 24 mg of GA_1 (plant on left)

le peas contain only about 1 ng per gram. However, dwarf *le* peas do contain high concentrations of GA_{20} (about 70 ng per gram), the precursor of GA_1. Since tall plants have only a third of this, the *le* mutation seems to result in an inability to convert GA_{20} to GA_1 (Fig. 24.8). This was confirmed experimentally when GA_{20} was labelled with a radioactive isotope and fed to dwarf and tall peas. Only tall peas produced significant quantities of labelled GA_1, indicating that the *le* mutation blocks the step converting GA_{20} to GA_1. The response of dwarf peas to gibberellin involves both cell division and elongation. The response is dependent on the level of GA_1, the biologically active form of the hormone, and not on the total level of gibberellins.

Gibberellins are also involved in **bolting**, the rapid shoot elongation of many rosette plants before flowering. This process is normally induced by either the onset of longer day lengths or the completion of a period of exposure to low temperatures (p. 647). For example, in spinach, transfer from short-day lengths to long-day lengths results in the accumulation of GA_1 and its immediate precursor GA_{20} due to increased activity of the enzyme, GA_{19} oxidase (Fig. 24.8).

Evidence for the role of gibberellins in breaking **seed dormancy** and stimulating **germination** has come from studying *Arabidopsis* mutants (Fig. 9.18). Seeds of mutants deficient in gibberellins do not germinate, even when provided with light, water and prechilling. The only treatment that results in germination is application of exogenous active GA. Release from dormancy in wild-type *Arabidopsis* appears to involve an increase in the sensitivity of seeds to endogenous GA. It has also been suggested that light, often required for germination, may increase the biosynthesis of endogenous gibberellins.

Gibberellins and mobilisation of seed storage products

In cereal crops, the **endosperm** of grain contains carbohydrate and protein reserves upon which the developing embryo depends for energy and nutrition until it commences photosynthesis. The reserves in the endosperm must therefore be mobilised and transported to the embryo (Fig. 24.10). The breakdown of endosperm starch and proteins into smaller, more easily transported molecules, such as sugars and amino acids, results from the action of a range of hydrolytic and proteolytic enzymes, including α-amylase, which breaks down starch molecules into maltose (a disaccharide consisting of two glucose molecules).

In barley, enzymes are produced specifically by cells in the outermost layer of endosperm, the **aleurone layer**. The secretion of these enzymes is under the control of the embryo, since if the embryo is removed from the seed before soaking the seed for germination, no endosperm hydrolysis takes place (Fig. 24.10). GA_1 is the messenger that moves from the embryo to the aleurone layer, inducing the transcription of various genes, increasing the level of mRNAs coding for α-amylase production.

GA_1 induces the production of a regulatory protein (called GAMYB) that binds to an element in the promoter (Fig. 11.3) of the α-amylase gene and activates its transcription.

Thus, a plant hormone (GA_1) produced in one tissue (the embryo) may regulate gene expression in another

BOX 24.1 Synthetic growth retardants

Synthetic growth retardants, such as paclobutrazol, reduce vegetative plant growth and have important applications in agriculture and horticulture. Paclobutrazol, which is a substituted pyrimidine, acts by blocking the plant's synthesis of gibberellin. Treatment of plants with such growth retardants can result in reduced lodging (falling over) in small grain crops; reduced need for pruning, and increased flowering and fruiting in orchard trees; reduced growth in ornamental trees (e.g. for use near power lines); and more compact shoot growth in herbaceous potted plants. Less frequent mowing of turf grasses also may be a potential application for these compounds.

Fenarimol is another substituted pyrimidine, related to paclobutrazol, but with more activity against fungi than plants. In fungi, it blocks the biosynthesis of ergosterol, an essential requirement for growth, and is thus used as a fungicide to control plant diseases. At high doses, however, it can affect the growth of plants as well. In apple trees, overuse of fenarimol causes reduced growth of leaf petioles, stunted shoot growth, flower and fruit abortion, all symptoms of reduced gibberellin biosynthesis.

tissue (in this case, the aleurone layer). Another plant hormone, abscisic acid (p. 638), also regulates the transcription of these genes, but has the opposite effect to GA_1, resulting in reduced mRNA levels. An understanding of this system has allowed gibberellins to be used commercially in the brewing industry, which depends on the breakdown of endosperm reserves for the malting process.

Gibberellins are a group of growth-promoting hormones that stimulate both cell division and elongation, thus influencing stem growth and bolting. In germinating cereal seeds, GA_1 produced in the embryo moves to the aleurone layer, where it activates genes that control the synthesis of the enzyme that converts starch to sugar.

Cytokinins

Cytokinins are plant hormones identified by their ability to stimulate cell division in plant tissue cultures.

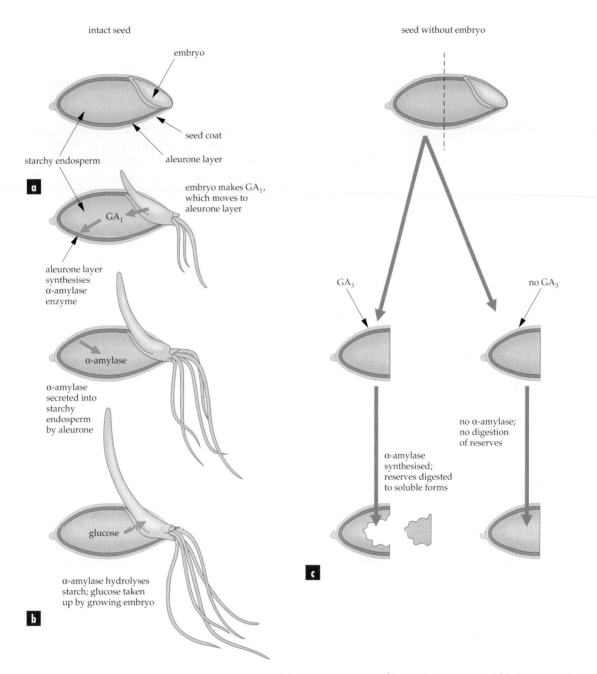

intact seed

embryo

seed coat

aleurone layer

starchy endosperm

a

embryo makes GA$_1$, which moves to aleurone layer

GA$_1$

aleurone layer synthesises α-amylase enzyme

α-amylase

α-amylase secreted into starchy endosperm by aleurone

glucose

α-amylase hydrolyses starch; glucose taken up by growing embryo

b

seed without embryo

GA$_3$

no GA$_3$

no α-amylase; no digestion of reserves

α-amylase synthesised; reserves digested to soluble forms

c

Fig. 24.10 Longitudinal sections through barley grains indicating the **(a)** major tissue types, **(b)** growth responses and **(c)** the action of gibberellin and the embryo on α-amylase production. When the embryo is removed from the seed, no hydrolysis takes place

Most cytokinins are derivatives of the purine adenine (Fig. 24.11). Roots and developing fruits are major sites of cytokinin synthesis in flowering plants. **Zeatin**, isolated from corn seed, is the most active, known, naturally occurring cytokinin.

Cytokinins may regulate senescence (yellowing) of leaves on an intact plant. In experiments where leaves are removed from certain plants and floated on water containing cytokinin, leaves take longer to senesce than do leaves in control treatments not provided with hormone. In other words, the longevity of cut leaves is enhanced by cytokinin.

Cytokinins may also be involved in the regulation of the growth of axillary buds since bud growth is promoted by direct application of cytokinin (Fig. 24.12). Exogenous cytokinin and auxin are therefore antagonistic in their effects on axillary bud growth.

Cytokinins are used commercially to induce organogenesis in tissue cultures. Small pieces of plant material, for example, from leaf or stem tissue where cell division has ceased, are excised and placed on a medium containing sugar, vitamins, various salts, and various concentrations of auxin and cytokinin. Under these conditions, the piece of plant material may

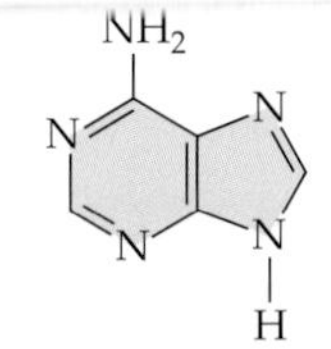

a Adenine (6-amine purine)

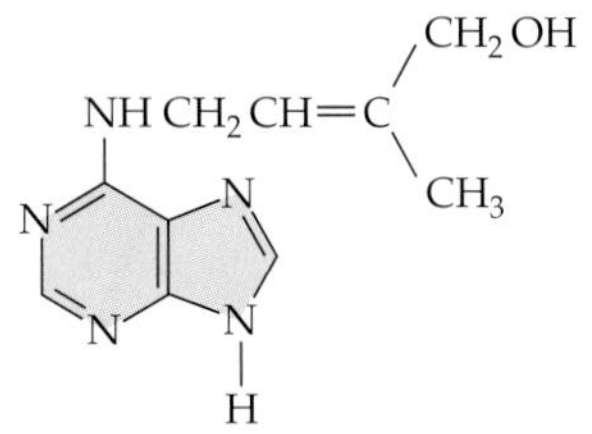

b Zeatin

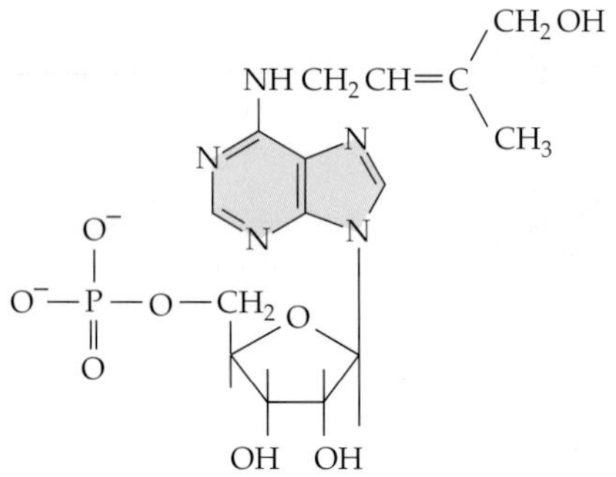

c Zeatin riboside

Fig. 24.11 Structure of **(a)** the purine adenine and the cytokinins, **(b)** zeatin and **(c)** zeatin ribotide

Fig. 24.12 The effect of a cytokinin on axillary bud growth. The bud of the plant on the right was treated with 300 parts per million of the synthetic cytokinin, kinetin, three days previously. The plant on the left is a control

give rise to a mass of callus (meristematic) cells (Chapter 14). From this callus, shoots and roots may be regenerated. Within certain concentration ranges, a high cytokinin-to-auxin ratio results in the production of new shoots, whereas a high ratio of auxin to cytokinin is conducive to root formation (Fig. 24.13). Cytokinins

are also widely used in micropropagation, where axillary or apical buds of shoots are used as the starting material for the production of many new cloned shoots, each of which may give rise to a new plant. Native Australian plants that have been propagated in this way include river red gum, *Eucalyptus camaldulensis*, kangaroo paws, *Anigozanthos*, and Christmas bells, *Blandfordia grandiflora*.

In *Agrobacterium tumefaciens*, a gene has been identified that controls an important step in cytokinin production. This gene has been transferred to tobacco plants and the transgenic plants have increased levels of certain cytokinins and increased lateral bud outgrowth. The reason scientists transfer genes from an organism such as *A. tumefaciens* is largely because there are no known single-gene mutants in flowering plants that specifically cause large changes in the levels of cytokinins. However, this is not the case in mosses. One moss, *Physcomitella patens*, has a cytokinin-producing mutant with much higher levels of certain cytokinins than the wild-type.

> Cytokinins are plant hormones that stimulate cell division in tissue cultures and may be involved in the control of growth of lateral branches and leaf senescence.

Abscisic acid

In addition to growth stimulators, such as auxins, gibberellins and cytokinins, plants also have growth inhibitors. These assist them to tolerate or avoid periods of adverse conditions, such as drought, salinity and low temperatures. Responses to stressful environments (Chapter 18) may involve modification at a morphological level (e.g. leaf drop and the formation of dormant buds in deciduous trees), changes at a physiological level (e.g. closure of stomata) or changes at a biochemical level (e.g. increase in frost resistance). The search for regulatory compounds involved with plant responses to adverse conditions occurred during the 1950s and 1960s and resulted in the isolation of the growth inhibitory compound **abscisic acid**.

Abscisic acid contains 15 carbon atoms (Fig. 24.14) and is synthesised from mevalonic acid via violaxanthin, a carotenoid intermediate. This pathway was indicated by work with maize mutants with blockages at various steps in the pathway for carotenoid synthesis. These mutants are almost white and possess markedly reduced levels of ABA.

ABA and drought resistance

Three mutations of tomato are known that cause plants to be ABA deficient so that even the mildest water

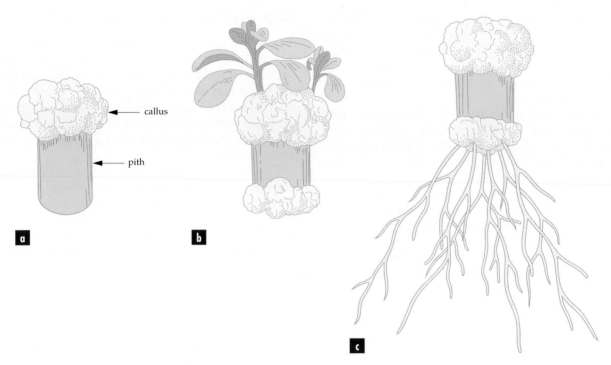

Fig. 24.13 The effects of various levels of a synthetic cytokinin (kinetin) and of auxin (IAA) on the growth of shoots or roots from tobacco callus cultured on nutrient agar: **(a)** kinetin 0.2 mg/L, auxin 0.18 mg/L; **(b)** kinetin 1 mg/L, auxin 0.03 mg/L; and **(c)** no kinetin, auxin 0.18 mg/L

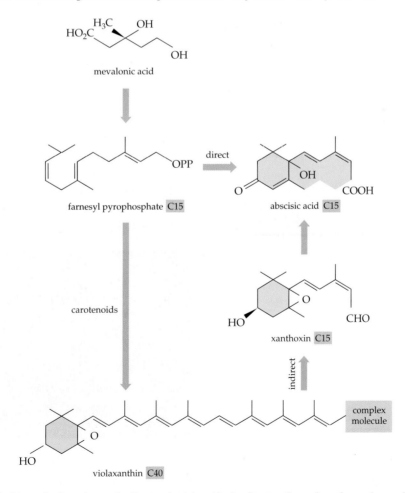

Fig. 24.14 The two possible biosynthetic pathways leading to abscisic acid: the direct pathway from farnesyl pyrophosphate and the indirect pathway, which occurs via the breakdown of carotenoid pigments

stress results in a tendency for wilting to occur. There is a close correspondence between the severity of the wilting phenotype and ABA concentration (Table 24.2). Addition of ABA can cause the mutants to revert to wild type, confirming that this hormone controls the tendency to wilt.

Due to ABA deficiency, the *stomatal aperture* remains relatively large in mutants and thus these plants have much higher rates of transpiration. In wild-type plants, leaf ABA levels rise dramatically after water stress, resulting in rapid closure of stomata. In stomatal guard cells, this increase in ABA level may be 20-fold. An even quicker response (in terms of stomatal closure) may be mediated by the redistribution of ABA already present within leaves. These results suggest that modification of drought tolerance in breeding programs may be possible by selecting for modified ABA levels or sensitivity. This could be of benefit to arid countries such as Australia.

Osmotic solute efflux from guard cells, causing a decline in cell turgor, is responsible for stomatal closure (Chapter 18). Notably, K^+ and anions are lost from the vacuole and cytoplasm across the plasma membrane. In response to ABA, the cytoplasm is alkalinised (0.2–0.4 pH units within 5–15 minutes) and this is both necessary and sufficient to account for the activation of K^+ efflux channels. At the same time, an increase in *cytoplasmic pH* inactivates K^+ influx channels. Hence, both types of K^+ channels in the plasma membrane respond to changes in proton concentration in the cytoplasm. Protons act as a second messenger regulated by ABA.

Table 24.2 Abscisic acid levels regulate wilting in tomatoes, as shown by specific mutants[a]

Genotype	Phenotype	ABA (ng g^{-1} fresh weight)
Flc/Flc; Not/Not; Sit/Sit	Normal	110
Flc/Flc; not/not; Sit/Sit	Mildly wilty	52
flc/flc; Not/Not; Sit/Sit	Wilty	23
Flc/Flc; Not/Not; sit/sit	Wilty	9
flc/flc; Not/Not; sit/sit	Wilty	10
flc/flc; not/not; Sit/Sit	Extremely wilty	4
Flc/Flc; not/not; sit/sit	Extremely wilty	3

(a) Mutants: *flacca* (*flc*), *notabilis* (*not*) and *sitiens* (*sit*). Plants were grown in a controlled environment at 95% relative humidity for six weeks.

ABA and frost tolerance

Frost tolerance is correlated with elevated ABA levels. For example, ABA levels increase transiently during exposure to cold temperatures in potatoes, while an ABA-deficient mutant of *Arabidopsis* will not become cold-tolerant. Cycloheximide, an inhibitor of protein synthesis, can prevent increase in frost resistance, suggesting that the synthesis of new proteins is involved in the response.

There is now clear evidence that ABA mediates responses to *environmental stress* by regulating gene expression. Certain genes are up-regulated by ABA while others are down-regulated. Understanding the genetic and biochemical regulation of plant responses to environmental stress is an important part of research programs aimed at maximising crop production under periodically unfavourable conditions.

Seed dormancy

Once a seed matures and development of the embryo ceases, dehydration follows (Chapter 14). The seed may then either be **dormant** or quiescent. A dormant seed will not germinate immediately when rehydrated. A quiescent seed will germinate if rehydrated because of its higher metabolic state. For many plants, seed dormancy is advantageous since at the time of seed release conditions are often inappropriate for growth and survival of young seedlings. This is particularly so in climates where seed dispersal in late summer or autumn is followed by winters of sufficient severity to kill young, recently germinated seedlings.

Dormancy is often broken by a period of exposure to low temperature, often less than 5°C, over a period of weeks or months in the presence of moisture (Fig. 24.15). A period of cold, which breaks dormancy and causes germination, is known as **stratification** or prechilling. By the time the prechilling requirement

Fig. 24.15 The seeds of high-altitude plants, such as this snow daisy *Celmisia aseriifolia*, growing in the Australian alps, require the cold of winter to break dormancy, ensuring that seed germination occurs under more favourable conditions in spring or summer

has been met, winter will have passed and then germination may proceed in the more favourable conditions of spring. The optimum temperature for the breaking of dormancy (1–5°C) is quite different from the optimum temperature for subsequent germination and seedling growth (usually 15–25°C).

Just as it may be disadvantageous for seeds to germinate during winter, problems for emergent seedlings may also occur if seeds germinate while they are buried too deeply in soil, covered by litter or beneath an extremely dense canopy. In these cases, a seedling may exhaust its food reserves before its leaves are able to begin harvesting energy from sunlight. Many species have a mechanism to prevent such premature germination: their seeds are **photodormant**, that is, they will not germinate unless they are exposed to appropriate levels of red light. The effect of light on germination appears to be under the control of phytochrome. In addition to the effects of low temperature and light, dormancy in some seeds is reduced by the effect of water, which may leach out inhibitors. In desert plants, seeds often do not germinate after a light shower, but only after steady rain that washes out sufficient inhibitor. This ensures that seedlings only emerge when enough rain has fallen to provide adequate moisture for growth. Seed dormancy is also reduced with the passage of time, possibly as inhibitors break down.

Comparative studies of mutant and wild-type plants have been important in discovering the role of ABA in the control of seed dormancy. For example, dormancy of freshly harvested seeds of wild-type *Arabidopsis* can be broken by exposure to 2°C for several days followed by exposure to light at 24°C in the presence of water. Such treatment results in the germination of 40–70% of seeds. Without the cold treatment, fresh seeds will not germinate. In contrast, seeds of the mutant *aba* form, which is deficient in abscisic acid, do not show dormancy. Mutants insensitive to abscisic acid, which appear incapable of responding to their own (endogenous) abscisic acid, also have reduced seed dormancy. However, even in wild-type plants of *Arabidopsis*, dormancy diminishes with time in dry storage, so that after a period of several weeks up to 75% of seeds are capable of germinating in the presence of light and water without a cold treatment. It thus appears that *development* of dormancy requires a certain level of abscisic acid but that the hormone is not responsible for *maintenance* of dormancy over time.

Abscisic acid is a growth-inhibitory hormone that affects plant responses to stress, for example, drought and frost, through modified gene expression. The development of seed dormancy requires the presence of abscisic acid and dormancy is often broken by a period of low temperature (stratification) or by red light.

Ethylene

Alone among plant hormones, **ethylene** (C_2H_4) is a gas of low molecular mass. Its effects on plant growth have been known since the nineteenth century, when it was observed that gas leaks from underground pipes caused **leaf abscission** in shade trees growing along streets. However, it was not until the 1960s that ethylene was shown to be produced naturally by higher plants and also to influence plant development at hormonal levels (as low as 0.01 mL/L). The gaseous nature of ethylene means that it readily diffuses through intercellular spaces from the site of production and that it may diffuse out of a plant completely. This allows a rapid equilibrium to be established between production and loss, with consequently rapid changes in the plant's response.

Applied ethylene influences a wide range of developmental processes from seed germination to shoot elongation, flowering, fruit ripening, abscission and senescence. However, plants seem to be able to grow and reproduce without ethylene since mutants insensitive to ethylene and plants grown in the presence of specific inhibitors of ethylene synthesis grow normally. Ethylene appears to be involved primarily in plant responses to certain environmental stresses and wounding.

Signal transduction

Several transmembrane proteins have been identified that bind to ethylene at the cell surface and function as signal transducers (Fig. 24.16). These ethylene receptors all resemble two-component protein kinases in bacteria and consist of a 'sensor' with histidine kinase activity and a 'response regulator'. Each protein functions as a dimer. The histidine kinase is autophosphorylated on a conserved histidine. When ethylene binds to the N-terminal sequence, which is in the extracellular region of the dimer, a conserved histidine in the histidine kinase domain is phosphorylated. The phosphatase group is then transferred to an aspartate in the response regulator. The receptor then activates a protein kinase (Chapter 7) called CTR1. Via several unknown steps, CTR1 activates a DNA-binding protein (EIN3), which binds to ethylene-response elements in the promoters of genes regulated by ethylene. These genes are then transcribed (expressed).

Fruit ripening

As fruits ripen under natural conditions they undergo an integrated set of changes, including a decline in organic acids, an increase in sugar content, softening and often a colour change (Fig. 24.17). These all involve active metabolic processes and, in many fruits,

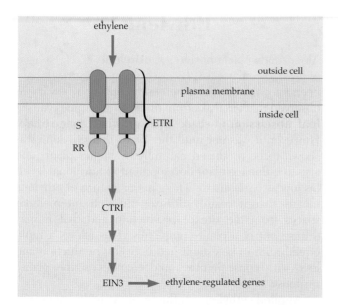

Fig. 24.16 The signal transduction chain for the ethylene response in *Arabidopsis thaliana*. The *ETHYLENE RESISTANCE 1* gene encodes an ethylene receptor (ETR1). ETR1 is a two-component-like protein kinase and acts as a dimer. It is composed of a 'sensor' (S) with histidine kinase activity and a 'response regulator' (RR) containing an aspartate that is phosphorylated by the histidine kinase. Binding of ethylene leads to autophosphorylation of a conserved histidine in the histidine kinase (S) domain and the phosphate is then transferred to the aspartate of the response regulator domain. The latter directly or indirectly activates the CTR1 protein kinase (encoded by the *CONSTITUTIVE ETHYLENE RESPONSE 1* gene). After a series of unknown steps, transcription factors such as EIN3 (encoded by the *ETHYLENE INSENSITIVE 3* gene) are activated and regulate the expression of a variety of genes

Fig. 24.17 Fruit ripening in banana. Green unripe and yellow ripe fruits of twin bananas. Ethylene is used commercially to ripen bananas that have been transported in the unripe state

coincide with a period of increased respiration, the respiratory **climacteric**. During the climacteric there is also a dramatic increase in ethylene production. In some fruits (e.g. melon), this natural increase in ethylene precedes the rise in respiration but in others it coincides with or follows it, suggesting that ethylene levels may not be the endogenous controlling factor in fruit ripening. Rather, changes in tissue sensitivity to ethylene or a fall in the level of a ripening inhibitor may be involved.

Applied ethylene can initiate the climacteric in certain fruits and is used commercially to ripen tomatoes, avocados, melons, kiwi fruit and bananas. This allows fruit to be harvested green and ripened 'on demand'. It also explains the age-old saying that 'one bad apple can spoil the barrel', since the bad apple produces ethylene, which triggers ripening in the other fruit. The commercially available compound, ethephon, which breaks down to release ethylene once inside a plant, is sprayed on tomatoes to ensure uniform ripening. In contrast, inhibitors of ethylene action, such as CO_2, are used to retard ripening of stored fruit.

Shoot growth and flowering

Applied ethylene also influences shoot growth. In dark-grown seedlings, which are etiolated, application of ethylene can result in reduced elongation of the stem, bending of the stem (i.e. the loss of the normal gravitropic response) and swelling of the epicotyl or hypocotyl. The combination of these responses is referred to as the triple response and is frequently observed when many seedlings are grown together in a confined, unventilated space or a laboratory with gas outlets.

One of the best examined and most startling results of ethylene application is the rapid elongation of the shoots of some aquatic plants such as deep-water rice. These plants elongate rapidly when submerged (Fig. 24.10). Elongation is caused by an increase in the level of ethylene, which builds up because diffusion of the gas out of the plant is retarded in a submerged shoot. The endogenous role of ethylene can be demonstrated by the application of ethylene antagonists, such as silver ions (Ag^+), which inhibit the elongation response.

Ethylene can induce flowering in plants of the pineapple family and this is an important commercial application. Ethylene also promotes flower senescence in plants such as petunias, carnations (Fig. 24.19) and peas. Alternatively, flower senescence can be inhibited by treatment with the ethylene antagonist, Ag^+. This is of practical benefit to the horticulture trade, where it is used to prolong the shelf-life of certain cut flowers.

Interactions of auxin with ethylene

High auxin levels can induce ethylene production. Consequently, responses to high doses of auxin can produce ethylene responses such as stem swelling and reduced stem elongation. Such effects are perhaps the clearest indication of how plant development and responses depend on interactions between hormones and should be taken into account when examining both ethylene and auxin responses.

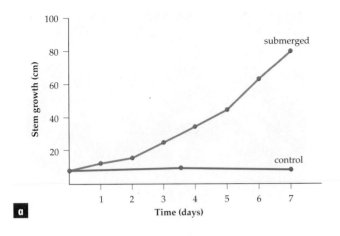

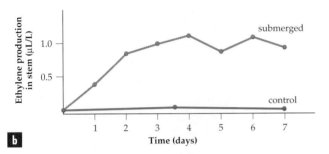

Fig. 24.18 Effect of submergence in water on **(a)** stem growth and **(b)** ethylene production within the stem of deep-water rice plants. Control plants were not submerged. **(c)** Deep-water rice (right) can elongate at rates up to 30 cm per day during flooding and can grow in water up to 4 m deep; plants here were growing in 1.5 m of water. Irrigated rice (left) usually grows at 0.1–0.2 m water depths

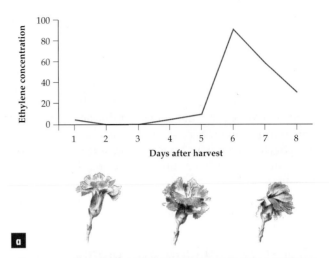

Fig. 24.19 Senescence in cut flowers of carnation. **(a)** Petals senesce when levels of ethylene produced in the petals rise about six days after flowers are cut. **(b)** Improved post-harvest life of the flowers is achieved through genetic engineering. A Melbourne plant biotechnology company, Florigene Pty Ltd, has 'silenced' expression of a gene (*EFE*) associated with ethylene production in carnation petals, so that the flowers remain fresh for many days longer. Here, flowers of a normal carnation (left) are compared with those of a genetically engineered carnation (right) on day 6

Ethylene is a gaseous plant hormone that influences a wide range of developmental processes, from seed germination to shoot elongation, flowering, fruit ripening, abscission and senescence.

Photoperiodism and control of flowering

The cycle of night and day is an ever-present environmental stimulus that regulates the growth and development of certain plants. Response to the length of light and dark periods in a 24-hour cycle, **photoperiodism**, allows plants to reproduce synchronously and in the appropriate season. For example, in temperate climates, fruits and dormant seeds are produced before the onset of unfavourable winter conditions.

Unequivocal evidence of the importance of photo-period (day length) in flowering was obtained in 1920, when it was discovered that certain varieties of soybeans always flowered on the same day at a particular latitude regardless of when they were planted. Similarly, it was shown that, in the glasshouse, a new tobacco variety, Maryland Mammoth, did not flower during summer but only during winter. Several environmental factors were investigated, including temperature, nutrition and soil moisture, to see what controls flowering. It was found that both soybean and tobacco would flower provided that the photoperiod was reduced by a few hours relative to summer conditions.

Short-day and long-day plants

There are two major types of responses to photoperiod: plants may be **short-day plants** or **long-day plants**. Short-day plants flower when the photoperiod becomes less than a particular day length, called the *critical day length*, which is usually between 12 and 14 hours (Fig. 24.20). Thus, short-day plants are typically autumn-flowering plants, such as the chrysanthemum, and are frequently of tropical or subtropical origin. By comparison, long-day plants flower when the photo-period exceeds a critical day length and typically include many spring and early summer flowering plants of temperate origin (Box 24.2). Plants that do not show a photoperiod response for flower initiation are **day-neutral plants**.

It is actually the length of the *dark period*, rather than the length of the light period, that determines when flowering occurs. Soybean, for example, is a short-day plant that will not flower unless it receives greater than 10 hours of darkness (Fig. 24.21). (A period of high-intensity light is also required by short-day plants in order to meet their photosynthetic requirements.) In long-day plants, the reverse occurs. The dark period must be shorter than some critical length. In both long- and short-day plants the interruption of the dark period by a short period of light (as little as 1 minute in some particularly sensitive short-day plants, but usually an hour or more in long-day plants) negates the effect of the dark period (Fig. 24.22). Consequently, long-day plants are really short-night plants and short-day plants are long-night plants.

Environmentally induced flowering responses are frequently under simple genetic control; for example, in garden peas, a single mutation (*Sn* to *sn*) converts a long-day type into a day-neutral type. In some species that have a wide geographic range, different varieties (ecotypes) have evolved that are suited to local environ-mental conditions. Kangaroo grass, *Themeda triandra* (Fig. 24.23a), includes short-day, long-day and day-neutral ecotypes. Short-day ecotypes occur in tropical regions in New Guinea and the Northern Territory

Fig. 24.20 (a) Effect of photoperiod on flowering in the short-day plant, *Pharbitis nil*, Japanese morning glory. The plant on the left (with flower) was grown under conditions of short days while the vegetative plant on the right was grown under conditions of long days. **(b)** Effect of photoperiod on flowering in the long-day plant, *Brassica campestris*, rapeseed. The plant grown under conditions of short days (bottom) has remained vegetative while exposure to long days (top) has caused flowering, accompanied by the pronounced stem extension known as bolting

from latitudes 6°S–15°S (Fig. 24.23b). Long-day ecotypes extend from southern New Guinea down the

BOX 24.2 Examples of plants with special requirements for flowering

Short-day plants

Chenopodium rubrum
Chrysanthemum morifolium
Xanthium strumarium
Coffee arabica (coffee)
Glycine max (soybeans)
Cannabis sativa (hemp)
Oryza sativa (rice)[a]
Gossypium hirsutum (cotton)
Zea mays (corn)[a]
Fragaria (strawberry)

Day-neutral plants

Nicotiana tabacum (tobacco)
Pisum sativum (peas)
Oryza sativa (rice)[a]
Zea mays (corn)[a]
Lycopersicon esculentum (tomatoes)
Vicia faba (broad beans)
Solanum tuberosum (potatoes)

Long-day plants

Avena sativa (oats)
Arabidopsis thaliana
Lolium temulentum (rye grass)
Hordeum vulgare (barley)
Pisum sativum (peas)
Triticum aestivum (wheat)
Spinacia oleracea (spinach)
Brassica rapa (turnip)
Vicia faba (broad beans)

Vernalisation

Chrysanthemum morifolium
Pisum sativum (peas)
Triticum aestivum (winter wheat)
Arabidopsis thaliana
Hordeum vulgare (winter barley)
Spinacia oleracea (spinach)

(*a*) Particular varieties.

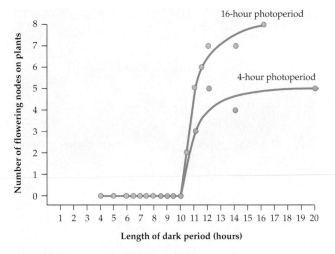

Fig. 24.21 The effect of different periods of darkness on the flowering of the short-day plant, soybean, *Glycine max*, exposed to either 4-hour or 16-hour light periods. Plants need more than 10

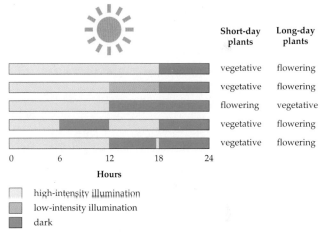

Fig. 24.22 The flowering behaviour of short-day and long-day plants subjected to various photoperiods. A critical photoperiod of 13 hours is assumed for both response types

Some plants are very sensitive to photoperiod, allowing very precise timing of the onset of flowering. For example, a variety of rye grass, *Lolium temulentum*, and the Chicago strain of cocklebur, *Xanthium strumarium*, flower in response to a single cycle of the appropriate photoperiod. Some plants can even detect changes in photoperiod as small as 15 minutes.

> Plants in which flowering is affected by photoperiod are either long-day or short-day plants. The length of the dark period, rather than the length of the light period, is the controlling factor. Plants not affected by day length are day-neutral.

Detection of photoperiod

The photoperiod response is detected by leaves. If leaves, but not the shoot apex, are exposed to the appropriate photoperiod, flowering will occur. When

east coast of Australia to temperate regions in Tasmania (43°S). In the Australian alps, where harsh winters occur, kangaroo grass requires vernalisation (p. 647) and long days before flowering. Ecotypes from dry inland sites, where rainfall is sporadic, are day-neutral, which may be advantageous if they are to flower and fruit as soon as adequate water is available, regardless of season.

Fig. 24.23 (a) The Australian native kangaroo grass, *Themeda triandra*, includes several ecotypes, which differ in their environmental requirements for flowering. **(b)** Distribution of the photoperiod ecotypes: day-neutral (DNP); long-day plants (LDP); short-day plants (SDP); and plants responding to vernalisation (V)

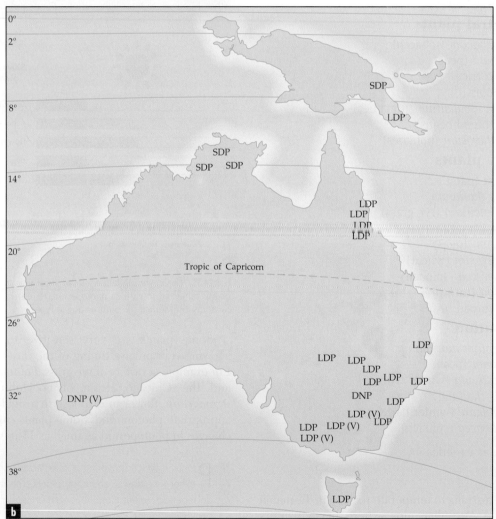

the shoot apex, but not the leaves, is exposed to the inductive photoperiod, no flowering occurs. This simple result implies that a signal must pass from the leaves, where induction of flowering occurs, to the apex, where flower initiation occurs. Grafting studies also indicate that a transmissible signal is involved. In some cases the signal appears to be a flower-promoting signal, often referred to as the hypothetical hormone **florigen** but, in other instances, flower-inhibiting substances appear to be involved. Plant physiologists have been able to transfer the promoting signal between long-day and short-day plants, but its chemical identity has not been determined, despite nearly 50 years of intensive research. Discovering the hormone(s) responsible for flowering remains one of the major challenges of plant science research, since artificial control of flowering would be of major benefit to agriculture and horticulture.

How do photoperiodic plants measure the length of the dark period? Light is detected by the pigment phytochrome, which exists in two forms (Chapter 7). By exposure to red light the P_r form is converted to the P_{fr} form. The P_{fr} form, which is the biologically active form, is converted back to P_r by absorption of far red light:

$$P_r \xrightleftharpoons[\text{far red light}]{\text{red light}} P_{fr} \text{ (active)}$$

During sunlight much of the phytochrome in a plant exists as P_{fr}. In darkness, as in the night, P_{fr} levels decline. Thus, the relative levels of the two forms of the pigment give a plant a way of detecting the light–dark transition. However, the simple decline of P_{fr} levels in darkness is not the mechanism by which plants measure the *length* of the photoperiod.

One suggested timing mechanism is an *hour-glass mechanism*, where the amount of some substance (other than P_{fr}) provides a measure of the length of the dark period between two light periods. A second idea is that *endogenous circadian rhythms* act as the timing mechanism. It is suggested that rhythms of approximately 24 hours occur within plants, even in continuous light or continuous dark. These rhythms, or internal biological clocks, may be controlled by light-on or light-off signals, and may enable a plant to measure the length of the dark period. Endogenous rhythms have been demonstrated for many other plant responses in addition to flowering, and include leaf movements, flower opening and closing, enzyme activities, cell division and growth rate. Examples of phytochrome-controlled responses in addition to flowering are described in Box 24.3.

Phytochrome pigment in leaves enables plants to detect light and darkness. This pigment interacts with an internal clock mechanism to measure the length of the dark period.

Vernalisation and flowering

A second major environmental cue that induces flowering in many species is a period of low temperature.

Induction of flowering by low temperature is termed **vernalisation**. Like long-day plants, plants with a vernalisation response frequently flower in spring. Some of these plants grow as compact rosettes and their stems only elongate, that is, bolt (p. 636), when the plant is about to flower (Fig. 24.20b). Some plants require both a particular photoperiod and vernalisation to stimulate flowering (Box 24.2). In others, a particular photoperiod or vernalisation is not essential but hastens flowering.

Temperatures that are effective in vernalisation range from –1°C to 9°C and are usually required for at least four weeks. The site of detection of low temperature is usually the shoot apex, but may be in the leaves in some plants. Where leaves are sensitive to a period of low temperature, a graft-transmissible signal can be demonstrated as for photoperiodic effects. Vernalisation can be reversed (devernalisation) by relatively short periods of high temperatures (two to four days at 25°C–40°C) provided that these conditions immediately follow the cool period.

The flowering of certain plants is promoted by exposure to a period of low temperature: vernalisation.

Monocarpic senescence

In annual plants, once flowering and fruiting has taken place, the death of the entire plant normally follows. This is an example of monocarpic senescence. This phenomenon can present a dramatic sight when whole fields of wheat or maize die *en masse* within a few days. Biennial plants grow vegetatively for one year, over winter, and then flower once before dying.

The most unusual plants that undergo monocarpic senescence are long-lived species, such as bamboo (Fig. 24.24), which may thrive in the vegetative state for in excess of 30 years before flowering and fruiting once, then senescing. All the bamboo plants in one area may flower and senesce in the one year, but how this is controlled is an intriguing question. Development of seeds within the fruit appears to be crucial for the onset of monocarpic senescence since it is retarded by removal of fruit. Young seeds may simply outcompete vegetative parental tissue for essential but limited nutrients, consequently leading to the death of the parent. An alternative theory is that seeds may produce a hormonal signal that results in death of the vegetative plant. Whatever the mechanism, monocarpic senescence is the last step in growth and development and is a genetically controlled and programmed event.

BOX 24.3 Phytochrome-controlled responses

The table lists some of the many responses that are phytochrome-controlled and which exhibit the characteristic red/far red reversibility. The nature of the last irradiation received by the plant determines the response. The responses include short-term changes with response times of less than 5 seconds, changes in gene expression and enzyme levels with response times of a few hours, and long-term responses, such as flower initiation, where the change may not be evident at the morphological level until many weeks later.

Two responses illustrate the benefits that a plant may gain from sensing the light environment. When a seedling is transferred from dark to white light, its internodes become shorter, the rate of leaf expansion increases, the apical hook uncurls, leaves expand and chloroplast and chlorophyll development takes place. All of these changes are advantageous to the plant, since it has to develop the capacity to utilise light as an energy source. The dark-grown plant is etiolated and has an apical hook, elongated internodes, reduced leaf expansion and lacks chlorophyll. These characteristics are equally advantageous to a seedling growing up through the soil to reach the light (Fig. a).

The second example deals with established plants growing in competition with other plants. Light beneath a canopy is not as suitable for photosynthesis (and hence growth), as direct sunlight. The canopy filters out most of the visible wavelengths, especially in the blue and red regions, leaving a green-tinged light that also contains a high portion of far red light. The $P_{fr}:P_r$ ratio falls from approximately 0.67 in direct sunlight to about 0.45 under a canopy (Fig. b). This change can result in modified growth. For example, plants adapted to growth in potentially shaded environments respond by producing longer internodes, smaller and thinner leaves and altered chlorophyll levels. These changes are of adaptive value under conditions of shade. In addition, changes in P_{fr} levels due to shading may allow plants to detect the proximity of other plants. Thus, phytochrome pigment not only allows plants to sense qualitative changes in the light environment (such as from light to darkness) but also to perceive quantitative changes in light quality.

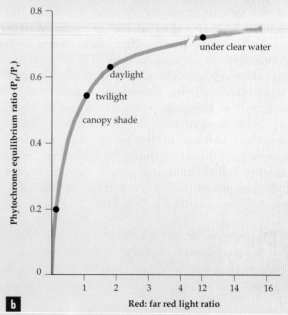

(a) The effect of darkness on pea plants 15 days old. The plant on the left was grown in complete darkness while the plant on the right was exposed to natural light. **(b)** The relationship between the ratio of red to far red light and the proportion of phytochrome in the P_{fr} form. Values are given for conditions of shade, twilight, natural daylight and under water

Table Examples of phytochrome-controlled processes

Chloroplast movement in algae

Protonema growth in bryophytes

Spore germination

Electrical potential changes

Membrane permeability

Anthocyanin synthesis

Chlorophyll synthesis

Hypocotyl hook formation

Internode elongation

Flower initiation

Seed germination

Fig. 24.24 Bamboo plant in flower

Summary

- The development of a plant from seed to maturity involves seed germination, vegetative growth, flowering, fruit development, senescence and seed dormancy. All these phases are strongly influenced by environmental and endogenous factors that may act by changing the level of, or tissue sensitivity to, a hormone.

- Experiments in which a hormone is applied to an intact plant, or to an excised part of a plant, have provided a basis for determining the endogenous roles of plant hormones. However, single-gene mutations that block the biosynthesis of a hormone have proved crucial in determining how plant growth and development is controlled by plant hormones.

- The best known plant hormones are auxins, gibberellins, cytokinins, abscisic acid and ethylene. They are all effective at very low concentrations.

- Auxin (IAA) promotes growth of excised stem or coleoptile segments. The primary action of auxin is on cell elongation and may involve the release of H^+ from cells into walls, softening the walls and allowing cells to expand (acid growth hypothesis). Auxin also regulates gene expression.

- Over 80 gibberellins have been identified. The active members of this group are growth-promoting hormones that stimulate both cell division and cell elongation. They influence stem growth, cause stem bolting and promote germination. In germinating cereal seeds, GA_1 produced in the embryo moves to the aleurone layer, where it activates genes that control the synthesis of the enzyme that converts starch to sugar.

- Cytokinins are plant hormones that stimulate cell division in tissue cultures and appear to be involved in the control of growth of lateral branches and leaf senescence.

- Abscisic acid is a growth-inhibitory hormone that affects plant responses to stress and modifies gene expression. It is involved in the control of stomata and transpiration, frost resistance and seed dormancy. Seed dormancy is often broken by a period of low temperature (stratification) or by red light.

- Ethylene is a gaseous hormone that influences a wide range of developmental processes, from seed germination to shoot elongation, flowering, fruit ripening, abscission and senescence. In fruit ripening, ethylene production is associated with a period of increased respiration, the respiratory climacteric.

- In many plants, flowering is controlled by photoperiod, but the length of the dark period, rather than the length of the light period, is the controlling factor. The pigment phytochrome enables leaves to sense light and darkness. The measurement of the length of the photoperiod appears to depend on the interaction of phytochrome with an internal clock mechanism. Flowering may also be promoted by exposure to a period of low temperature—vernalisation.

key terms

abscisic acid (p. 638)
acid growth hypothesis (p. 633)
aleurone layer (p. 636)
amyloplasts (p. 634)
apical dominance (p. 634)
auxin (p. 632)
auxin binding protein (ABP) (p. 635)
bolting (p. 636)
cell elongation (p. 632)

climacteric (p. 642)
cytokinins (p. 636)
day-neutral plant (p. 644)
dormant (p. 640)
endosperm (p. 636)
ethylene (p. 641)
florigen (p. 646)
germination (p. 636)
gibberellic acid (GA) (p. 635)
gibberellins (p. 635)
gravitropism (p. 634)
herbicides (p. 633)

indole-3-acetic acid (IAA) (p. 632)
leaf abscission (p. 641)
long-day plant (p. 644)
photodormant (p. 641)
photoperiodism (p. 643)
phototropism (p. 631)
plant hormones (p. 631)
polar transport (p. 633)

seed dormancy (p. 636)
short-day plant (p. 644)
stratification (p. 640)
vernalisation (p. 647)
zeatin (p. 637)

Review questions

1. Define the terms 'hormone' and 'tropism'. List the major types of plant hormones.
2. (a) What is the best known, naturally occurring auxin in plants?

   (b) Explain the acid growth hypothesis for the action of auxin on cell elongation.
3. Explain how roots probably detect gravity.
4. Synthetic hormones are used in agriculture. What, for example, are 2,4-D and 2,4,5-T and what have been their uses?
5. A plant is described as a long-day plant with an absolute requirement for vernalisation. Explain in simple terms what this means.

Extension questions

1. The mobilisation of food reserves in cereal crops such as barley is an excellent example of hormonal control in plants. Explain whether this is a valid statement.
2. Of what value to a plant is the capacity to synthesise a growth-inhibitory substance? Give an example of the role of abscisic acid as such a growth inhibitor.
3. What advantages are conferred on a plant by the gaseous nature of the hormone ethylene? Give an example of the role of ethylene in the response of certain plants to flooding.
4. In order to show that a certain developmental trait is controlled by a given hormonal substance, it is insufficient to simply demonstrate that application of that substance affects the characteristic in question. What additional evidence is necessary?
5. How have the techniques of genetic engineering become valuable to studies of plant hormones? What are the benefits of using single-gene mutations to study the control of plant growth and development?
6. In an herbaceous species of plant, a mutant has arisen that branches more than does the wild type. Cytokinins are thought to promote such branching. Possibly the shoot of the mutant is more sensitive than the wild type to root-produced cytokinins. Alternatively, the mutant may produce more cytokinins in its roots, which move into the shoot, than does the wild type. Explain how a grafting experiment may differentiate between these two possibilities.

Suggested further reading

Davies, P. J. (ed.). (1995). *Plant Hormones: Physiology, Biochemistry and Molecular Biology.* 2nd edn. Boston: Kluwer Academic.

An advanced book with 35 chapters on all aspects of plant hormones written by specialists in the fields.

Hopkins, W. G. (1999). *Introduction to Plant Physiology.* 2nd edn. New York: John Wiley & Sons.

A basic text in plant physiology with a good introduction to plant hormones.

Raven, P. H., Evert, R. F., Eichorn, S. E. (1999). *Biology of Plants.* 6th edn. New York: Worth Publishers.

A good general introduction to plant science with an up-to-date section on plant hormones.

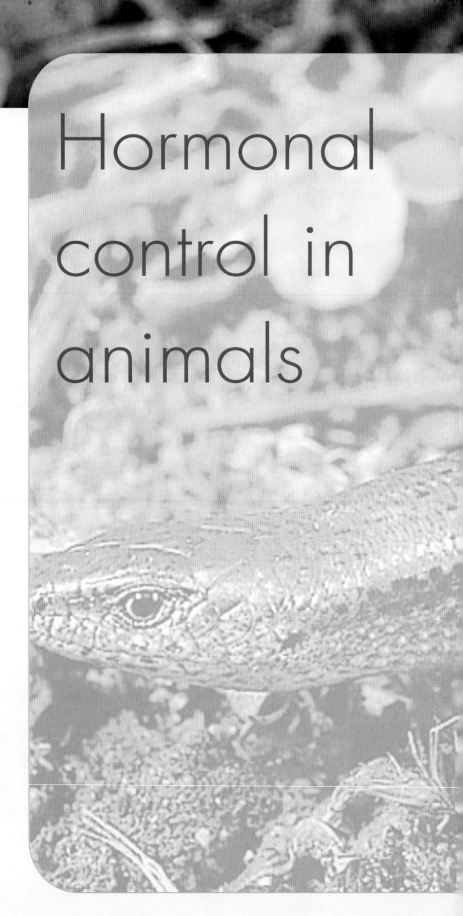

CHAPTER

25

Hormonal control in animals

Hormonal control systems, with widely differing degrees of complexity, are found in representatives of a range of organisms, including protists, fungi, plants and all animals. With the increasing complexity of hormonal control systems that has occurred during evolution, the chemical nature of hormones appears to have been strongly conserved. Hormones with the same, or closely related, chemical structures have been found in all animals so far studied. However, the biological roles of these hormones vary widely between different animal groups.

As we saw in Chapter 24, **hormones** are chemical messengers secreted by cells of an organism in response to specific stimuli. Their presence in the watery 'internal environment' modifies the activity of cells in various organs as a result of interaction with specific receptors, so providing an appropriate co-ordinated response to the stimulus.

Animal hormones

The word 'hormone' (meaning 'to stir up') was introduced by the physiologist E. H. Starling in 1905 to categorise a substance released by cells in the duodenum during digestion. This substance, which he named secretin, is released into the circulating blood of mammals in response to the presence of acidic stomach contents entering the duodenum after a meal (Fig. 25.1). The presence of secretin in blood stimulates another digestive organ, the pancreas, to secrete an alkaline juice into the duodenum. This neutralises the acid, providing a suitable environment for further digestion of food entering from the stomach. This study introduced the concept of non-neural control of biological functions.

The existence of animal hormones was first recognised in mammals and the organs that produced them were called 'glands of internal secretion' or **endocrine glands** to distinguish them from glands of external secretion, **exocrine glands**, such as salivary glands, sweat glands and digestive glands. Initially, the term 'hormone' was restricted to the regulatory products of discrete endocrine glands that are released into circulating blood and that exert their action at a distance from the site of their secretion. However, it is now clear that hormones are secreted by a wide variety of tissues, not necessarily organised into endocrine glands, and that they may reach their site of action by simple diffusion.

Historically, hormones have been considered to differ from another class of chemical messengers found

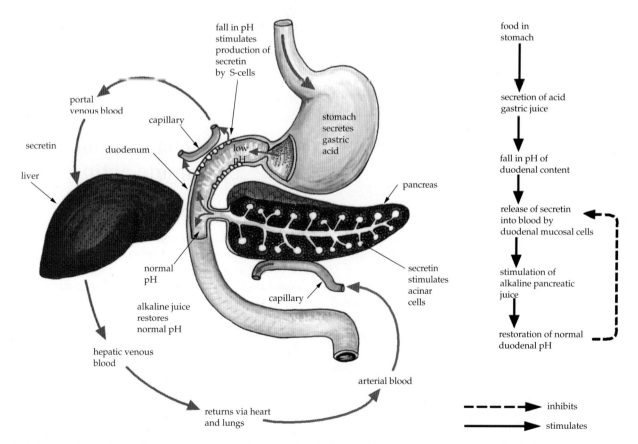

Fig. 25.1 Diagram illustrating a simple form of hormonal control of a physiological function, using, as an example, the first hormone to be recognised—secretin

in animals, *neurotransmitters*, which are released at the synapse between the terminals of a nerve cell axon and its effector cell (Chapter 26). While hormones and neurotransmitters both interact with specific receptors to produce their effects, neurotransmitters usually act only locally at the site of release. However, given our increasing knowledge of the secretion of hormones by nerve cells, **neurosecretion**, the distinction between hormones and neurotransmitters is no longer clear.

A nimal hormones are secreted in response to specific stimuli. They diffuse or are transported through the extracellular environment to their site of action, where they modify the activity of particular cells to provide an appropriate co-ordinated response to the stimulus.

Sites of action

One way to group hormones is in terms of the distance over which they travel to exert their effect (Fig. 25.2).

Effects on distant cells: endocrine hormones

Endocrine hormones are usually secreted into circulating blood, so that a hormone can reach its target organ rapidly and at an adequate concentration. In some invertebrates, endocrine hormones reach their target organs via circulating haemolymph or by diffusion through extracellular fluid.

Effects on nearby cells: paracrine hormones

Paracrine hormones usually act over very short distances and travel to their site of action by diffusion through extracellular fluid. However, they may also diffuse into blood and exert an effect on more distant organs.

Effects on the same cell: autocrine hormones

Some cells can be stimulated to secrete **autocrine hormones**, which interact with receptors on their own surface to produce a response, usually cell division.

Effects on another animal: pheromones

To the above three groups can be added an additional form of chemical communication. **Pheromones** are released into the *external environment* and provide chemical communication *between* individuals through senses such as smell and taste. They are usually highly volatile compounds that are released by exocrine glands and are detected at extremely small concentrations by chemical receptors on the surface of the recipient, for example, the nasal epithelium of mammals or the antennae of insects. There are also less volatile and more persistent pheromones that may be excreted in

urine, faeces or saliva, as well as from specialised surface glands. These pheromones are deposited on objects as territorial or mating signals, and to induce reproductive activity. These effects serve to synchronise reproductive activity and maximise reproductive success, particularly in dispersed and solitary species.

E ndocrine, paracrine and autocrine hormones and pheromones are grouped according to the distance over which they travel to exert their effect.

General functions

Hormones do not appear to initiate any unique cellular activities. Rather, they *modify* the rates of existing activities, usually through the induction or repression of enzymes within cells. In some cases, they act at the nucleus of a cell to influence the activity or expression of genes; in other cases, they influence the permeability of cells to solutes or the activity of cytoplasmic enzymes.

Through their control of a wide variety of physiological processes, the major actions of hormones are to influence reproduction, growth and development, metabolism, osmoregulation and mineral exchange. A most important consequence of these actions is the ability hormones confer on an organism to respond appropriately to changes in its environment, particularly in the long term.

H ormones have regulatory biological functions that enable organisms to respond appropriately to changes in their environment.

Mechanisms of hormone action

As with plant hormones, animal hormones exert their effects at very low concentrations, of the order of 10^{-12}–10^{-9} M. Although the term hormone implies excitation, the effects of hormones on cells may, in fact, be either stimulatory or inhibitory. Cells or organs that respond specifically to a particular hormone are known as the **target cells** or *organs* for that hormone. Specificity is achieved by the properties of specialised molecules, *receptors*, located either at the surface of the cell membrane or inside the cell, depending on the chemical nature of the hormone (Chapter 7). These receptors reversibly bind the hormone with high degrees of specificity and affinity.

There are two general chemical classes of hormones—water-soluble and lipid-soluble hormones. Hormones that are readily soluble in water are either derivatives of amino acids (catecholamines, peptides and

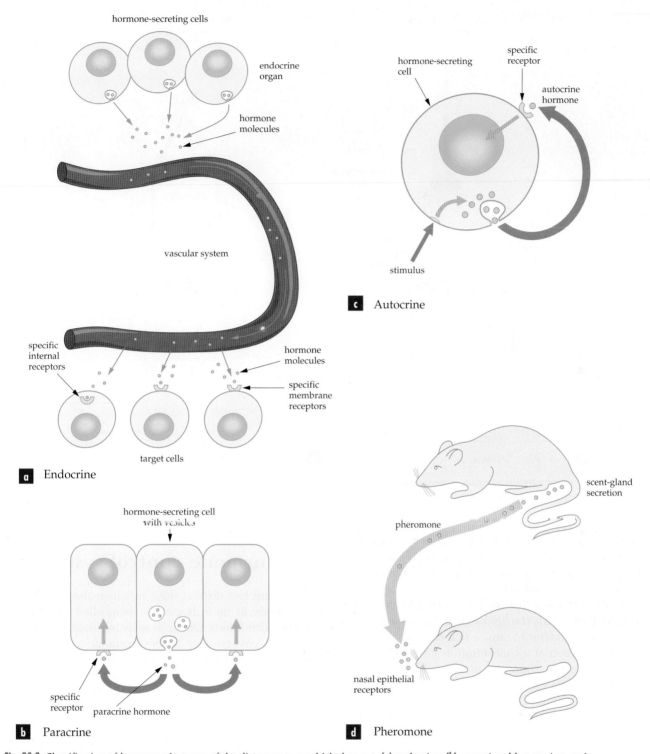

Fig. 25.2 Classification of hormones in terms of the distances over which they act: **(a)** endocrine; **(b)** paracrine; **(c)** autocrine; and **(d)** pheromone

proteins) or derivatives of fatty acids (eicosinoids). Hormones that are soluble in lipid solvents are either steroids (adrenocortical and gonadal steroids in vertebrates, ecdysones and juvenile hormones in invertebrates) or iodinated derivatives of the amino acid tyrosine (thyroid hormones of vertebrates). Solubility in water or lipid determines the way in which a hormone acts on

cells. Water-soluble hormones usually interact at the cell surface with receptors that span the cell membrane and frequently interact with a G-linked protein to produce a 'second messenger' (e.g. cAMP) within the cell; lipid-soluble hormones usually interact with intracellular receptors (Fig. 25.3 and see Figs 7.3, 7.12, 7.13). Some protein hormones, such as insulin, once bound to their

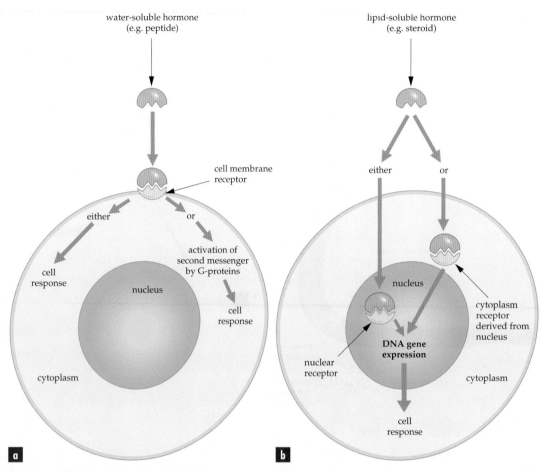

Fig. 25.3 Different mechanisms of action of the two major chemical classes of hormones: **(a)** water-soluble and **(b)** lipid-soluble hormone (see also Chapter 7). Water-soluble hormones interact with receptors on the external cell surface, whereas lipid-soluble hormones pass through the cell membrane and interact with nuclear-derived intracellular receptors.

membrane receptors, can be internalised by endocytosis of the portion of the membrane containing the hormone–receptor complex. The significance of this internalisation is not yet fully understood.

Many hormones that are transported in circulatory systems, particularly the lipid-soluble hormones, need a water-soluble carrier to ensure that they are carried to the site of action at a concentration sufficient to exert an effect. Such carriers are proteins. They reversibly bind the hormone, taking it up where concentrations are highest (at the site of release) and giving it up as concentrations decrease (at the target sites). They may be non-specific (usually albumins, with a high binding capacity) or highly specific to a particular class of hormone (usually globulins, with a low binding capacity). Some water-soluble hormones, for example, the posterior pituitary hormones of mammals, are secreted together with a specific binding protein, which probably protects it from breakdown while it is being transported in blood.

Hormones interact with receptors on target cells in very low concentrations with high specificity and affinity.

Hormone control systems

There are two distinct ways in which the secretion of hormones in animals can be controlled by environmental conditions. One is to relay information from the external environment to the hormone-secreting cells via the nervous system, either by direct innervation or by *neurosecretory cells*, specialised nerve cells that secrete hormones. The other is for the hormone-secreting cells themselves to respond directly to changes in the local environment. Both mechanisms can be integrated to provide a complex and precise hormonal control of biological function.

From an evolutionary point of view, neural control of hormonal secretion seems to have been the first to appear. Neurosecretory cells are the only hormone-secreting cells to be found in the simplest invertebrates, where their hormones influence growth and maturation to adulthood. As animals evolved further, this basic form of hormonal control was retained, with the addition of non-neural endocrine organs and with increasing complexity.

Neurosecretion

Neurosecretory cells derived from the embryonic source of nerve cells are an important source of hormones. They have been identified by selective staining reactions in most species of animals so far investigated. Neurosecretory cells may be scattered diffusely through all parts of the nervous system or they may be aggregated into discrete, highly vascularised **neurohaemal organs** (Fig. 25.4).

Neurosecretory cells look like nerve cells and are innervated by the axons of other nerve cells via contacts (synapses) with their dendrites (Chapter 26). However, the cell bodies of neurosecretory cells are often larger than in most neurons and their cytoplasm contains large secretory granules. The axon of a neurosecretory cell is usually densely packed with secretory granules of the order of 1000–2000 nm in diameter. These granules are membrane-bound vesicles containing hormone (or its precursor), which is usually a protein. Characteristically, neurosecretory axons do not form synapses with other nerve cells. Instead, they release their hormone by exocytosis (Chapter 4) into the local environment, usually in the vicinity of a blood capillary (Fig. 25.4). Neurosecretions then influence effector organs either *directly* or *indirectly* through the stimulation of other hormone-secreting cells (Fig. 25.5).

> Neurosecretion involves the release of hormones by exocytosis from nervous tissue into the local environment or a blood vessel.

Non-neural hormone secretion

Non-neural hormone-secreting cells are derived mostly from tissues in the digestive and excretory systems. These cells are usually organised into discrete endocrine glands, characterised by a rich blood supply, which respond either indirectly, through the nervous system, or directly to changes in the external or internal environment. Changes in the external environment detected by the nervous system may be relayed to an endocrine gland either by direct innervation of the

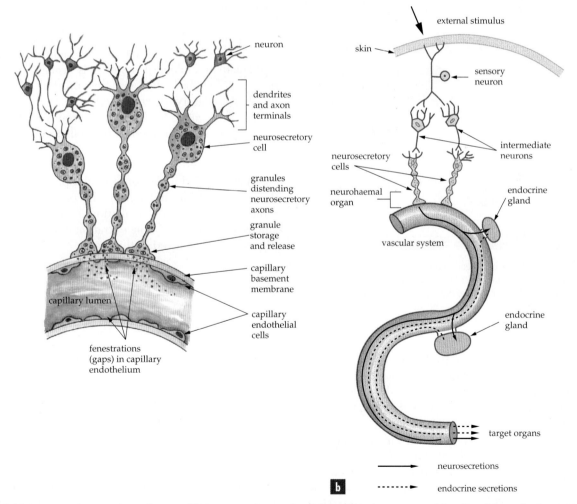

Fig. 25.4 (a) Structure of a neurohaemal organ. **(b)** Neurosecretions and endocrine secretions may act either directly on a final target organ, or they may stimulate or inhibit secretions from another endocrine gland, which in turn may also act directly on a target organ, or on the secretion from yet another endocrine gland, whose secretion affects the final target organ

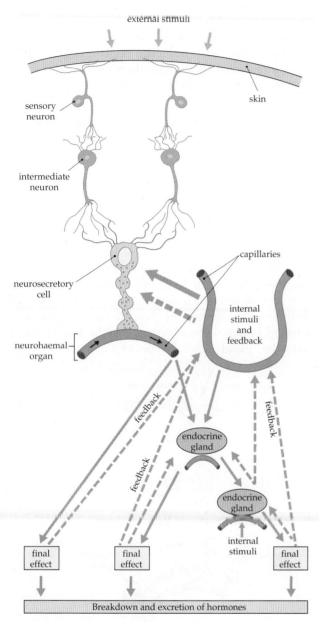

Fig. 25.5 General scheme of a neuroendocrine control system. Continuous arrows indicate direct hormone actions, while dotted arrows indicate feedback actions. Both actions may be stimulatory or inhibitory, but feedback actions are mostly inhibitory

pH of the duodenal contents. If these become acid, secretin is produced and stimulates secretion of alkali by the pancreas. The pH is restored to normal and secretin production ceases.

Non-neural secretion of hormones usually occurs from tissues of the digestive and excretory systems, organised into a discrete endocrine gland with a rich blood supply.

Feedback control

Neurosecretory and endocrine cells that can respond to change in the external environment also respond to changes in their internal environment. They may respond to the presence of specific stimulatory hormones in the blood as well as to changes in the internal environment resulting from the actions of their hormones. For example, many hormone-secreting cells are sensitive to the concentration of their own hormone in the blood or local environment and adjust their hormone output accordingly. They may also respond to change in concentration of hormone(s) secreted by their target organs. Such interactions result in precise hormonal control systems, sensitive to changes in both the external and internal environment (Fig. 25.5). Feedback in biological systems is usually inhibitory (negative feedback) but in some instances it can be briefly stimulatory (positive feedback).

Simple negative feedback is inherently unstable and the controlled variable tends to oscillate widely unless some extra form of control is present. In many cases, the action of one hormone is opposed by the action of another, secreted in response to an opposite change in the same variable and having the opposite effect, so reducing the size of oscillations. Control of blood glucose or blood Ca^{2+} concentration in vertebrates (see Fig. 25.21) is achieved in this way.

Hormone release is often under negative feedback control.

Hormones in invertebrates

Research on insects has led the way in studies of invertebrate hormones and in some cases (e.g. mode of action of steroid hormones) has even provided models on which our understanding of vertebrate systems subsequently developed. We know far more about insect hormones than about those of any other invertebrate groups—for example, a literature search on 'hormone and insect' turned up over 900 references, while 'hormone' with various other invertebrate groups seldom turned up as many as 10.

endocrine gland or by a multistage 'cascade' of stimulating (or inhibiting) hormones (Fig. 25.5).

Some non-neural endocrine glands themselves monitor the local concentration of the variable they control, responding directly to a change in its concentration by secreting hormones that activate mechanisms to restore the variable to its previous concentration. As this is achieved, secretion of the hormone declines. Such endocrine glands are usually not influenced by other hormones. This is an example of **negative feedback control**; secretin is the classic example of such control (Fig. 25.1). The regulated variable is the

nsect hormones are much better known than those of other invertebrates. There are probably two reasons for this dominant position of insect studies: insects are economically and medically important, and also, insects are easy to keep.

Work on invertebrate hormones commenced around the 1930s with physiological experiments: where it was suspected that a function might be hormonally controlled, the suspected source of the hormone was surgically removed and the effects on physiological function were observed. If the physiological function was abolished, extracts of the supposed hormone-producing tissue were applied or injected. Restoration of function was good evidence that a hormone was indeed produced by the tissue and controlled the function. At that time, it was not possible to identify the active molecules chemically.

Often, the hormone-producing tissue proved to be part of the nervous system. Techniques were developed that enabled scientists to recognise specialised, usually large, nervous tissue cells that were termed 'neurosecretory cells' because of their role in hormone production. Some very elegant physiological experiments confirmed their role in metabolic functions. Cells with the morphological characteristics of neurosecretory cells have been identified in all phyla with differentiated nerve tissue, although we still seldom know the precise nature of the hormones or their functions.

In higher invertebrates, separate endocrine glands are found, as well as neurosecretory hormonal systems. Since insects are best known, this account will concentrate on them.

The main neuroendocrine and endocrine organs of insects consist of groups of neurosecretory cells in the brain and other nerve ganglia, plus the *corpora cardiaca*, *corpora allata* and *prothoracic glands* (Fig. 25.6). The corpora cardiaca are paired structures just behind the brain and act both as neurohaemal organs, where axons from brain neurosecretory cells release their products to the blood, and as endocrine organs synthesising their own hormones. The corpora allata are endocrine glands just behind the corpora cardiaca, synthesising **juvenile hormones** (epoxides of methyl farnesoate derivatives, Fig. 25.7b). Neurosecretory axons proceed from the brain, through the corpora cardiaca, to the corpora allata, and other neurosecretory axons reach it from the *suboesophageal ganglion*. The prothoracic glands, which lie freely in the thorax or head, synthesise the steroid hormone **ecdysone** (Fig. 25.7a) (we now know that many other tissues also synthesise ecdysone).

Hormonal control of development

The process of moulting the exoskeleton and secreting a new one is a characteristic of insects and other arthropods. Insects also undergo metamorphosis, where the body of the juvenile stages changes more or less markedly at the moult to the adult form, at the very least completing the development of its reproductive structures and usually completing the development of wings. In the higher insects, such as flies, moths, beetles and wasps, the change is much greater, as the larval stage bears very little resemblance to the adult. In these cases, rather than just completing the growth of adult structures, almost the whole larval body is destroyed during the pupal stage and new adult structures develop from cell clusters set aside to increase in number without differentiation during the larval stage. These amazing transformations are controlled by the hormones we have mentioned above in response to environmental and internal factors. The nervous system contributes a range of hormones to the process.

The first step in an insect moult is the synthesis of a brain hormone (known as **prothoracicotropic hormone**, or PTTH) in certain neuroendocrine cells in the front of the brain. This hormone is a protein of around 66 kDa (in *Drosophila*). Its release depends on internal factors such as size, or in some insects on abdominal stretching following a large meal. Even if the correct size has been reached, the PTTH is released only within a short period of time in the circadian cycle—usually just around the expected time of darkness falling. This phenomenon is termed **gating**. PTTH travels along the neuroendocrine cell axons to the corpora cardiaca, where it is released into the blood. It stimulates the prothoracic gland to synthesise and release ecdysone, which is hydroxylated at the target tissues to the active form, 20-hydroxy-ecdysone. Before the moult there is a small 'commitment peak' of ecdysone: at this time a decision is made on the type of moult that will follow. The level of juvenile hormone in the blood determines this. Then follows a large peak of 20-hydroxy-ecdysone. This binds to nuclear receptors (protein in nature) and the bound complex permits transcription of genes required for the moulting process. The processes include the breakdown of non-sclerotised parts of the old cuticle, and determination of the pattern and secretion of new cuticle. Ecdysone also controls the programmed death of larval tissues.

The secretion of juvenile hormone is considered to be controlled by peptides originating in neurosecretory cells: **allatostatins** inhibit juvenile hormone production, while **allatotropins** stimulate it. These hormones are among the very numerous small peptide hormones, of five to 20 amino acids, now known from insects. Their amino acid sequence has been determined for several insect species. It appears that they have functions other than their role in corpus allatum control since both are present in axons associated with the gut. Juvenile hormone itself is a multifunction hormone,

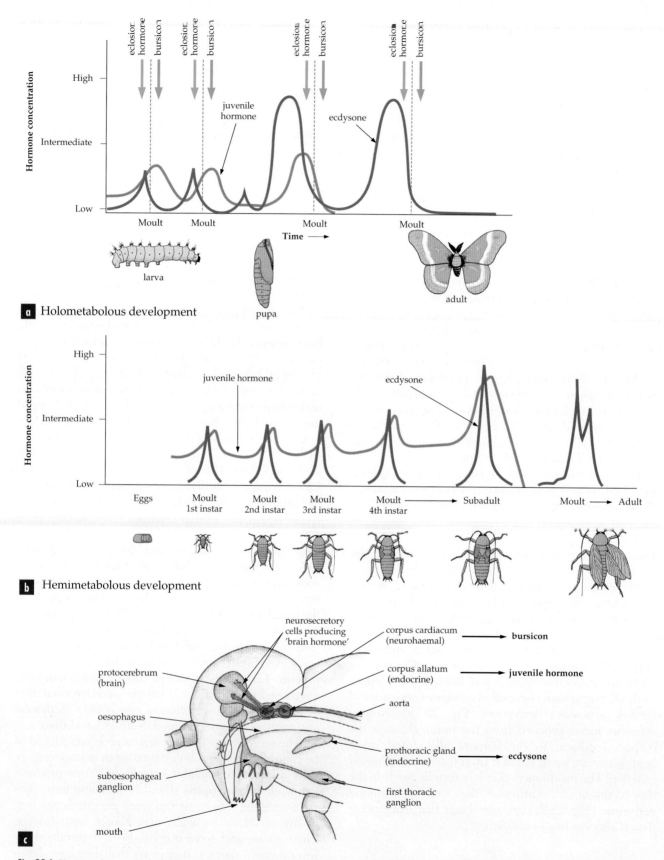

Fig. 25.6 Hormonal control of moulting. Juvenile hormone is a long-chain fatty acid that prevents adult development. Ecdysone is a steroid that acts on nuclear chromosomes to initiate gene transcription. Eclosion hormone and bursicon are produced for a brief period before and after each ecdysis (moult). **(a)** Holometabolous insects: those that include metamorphosis in their development. **(b)** Hemimetabolous insects: those with direct development. **(c)** Sagittal section through the head of an insect to show the origin of hormones involved in moulting

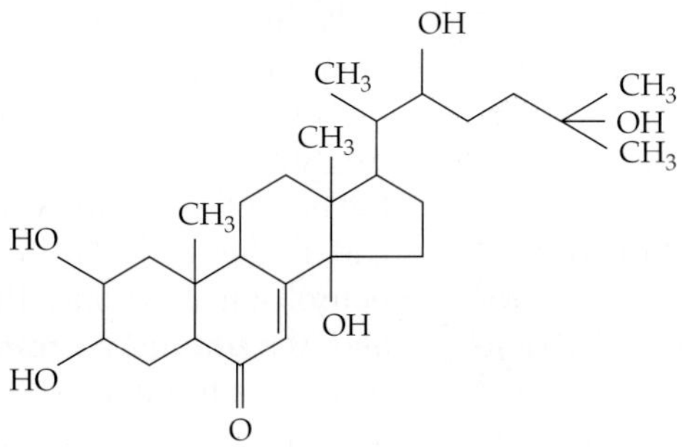

a Ecdysone

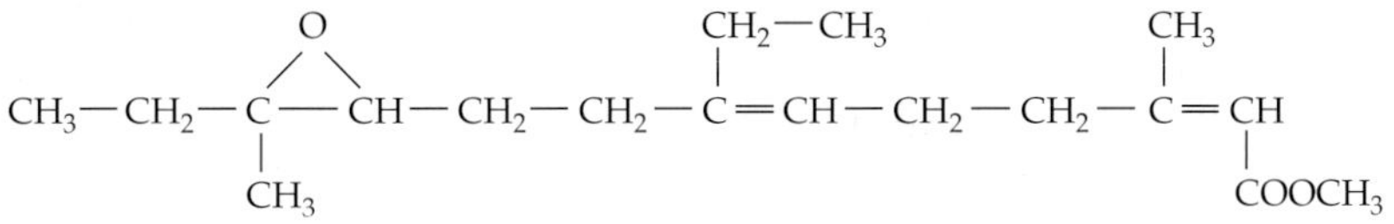

b Juvenile hormone

Fig. 25.7 Chemical structures of two non-neurosecretory hormones in arthropods: (a) ecdysone (a steroid) and (b) a juvenile hormone (an epoxide of methyl farnesoate), a derivative of fatty acids

with roles in diapause control, reproduction, cell death, control of mitotic activity and caste determination.

In more recent years, several important peptide hormones have been implicated in the moulting sequence. These include **ecdysis-triggering hormone** (ETH), which stimulates excitability of neurons producing the **eclosion hormone** (EH), and unlocks the sequence of behaviours that precede moulting. EH in its turn increases excitability of **crustacean cardiactive peptide** (CCAP) neurons. As its name suggests, this peptide was first identified from crustaceans, but in insects facilitates the commencement of the moulting process. This is an example of a *neuroendocrine cascade*.

Bursicon is the last of the hormones well known for its part in moulting. Produced by lateral neurosecretory cells of the abdominal ganglia, it is a 30 kDa protein that brings about the hardening of the new cuticle by cross-linking protein chains.

> In insects, moulting is an important process during development. Moulting is controlled by a suite of integrated hormones, which include juvenile hormone, ecdysone, ecdysis-triggering hormone, eclosion hormone, crustacean cardiactive peptide and bursicon.

Other invertebrate hormones

Our knowledge of insect hormones goes well beyond those hormones controlling moulting and metamorphosis. Just one other example is the family of *adipokinetic hormones*, produced by the corpora cardiaca: these peptides mobilise stored fuel reserves such as fat and glycogen at the onset of flight.

Recent technical advances have allowed us to identify and localise specific hormone activity in many neurosecretory systems throughout the invertebrates by reacting sections or whole mounts with antibodies to particular peptides. For example, we now know that invertebrates down to cnidarians contain products that react with antibodies to vertebrate gonadotropins and gonadotropin-releasing hormones. In some cases, rather similar functions have also been demonstrated. An example within the invertebrates is the occurrence of cells reacting to allatostatin antibodies, again found throughout the invertebrates from cnidarians up to, but not including, the urochordate *Ciona*. **Molecular cloning** has enabled us to sequence genes responsible for coding the large protein precursors of many of the small peptide hormones. The individual active peptides are produced by cleavage of the large proteins and they are finally modified (e.g. by C-terminal amidation) so that they will be resistant to peptidases. Another research area is in the cloning of genes for hormone receptor molecules. Frequently, proteins with the structure of receptor molecules are identified but the nature of the hormone that would react with them is unknown. These are called *orphan receptors* and their existence is one of many reminders that, despite enormous progress over recent years, we still have only scratched the surface of understanding invertebrate hormone systems. Particularly important will be a linkage of physiological with molecular techniques, so that the functions of new hormones can be investigated in living tissues.

Hormones in vertebrates

In vertebrates, there is a greater diversity of non-neural hormone-secreting organs, derived mostly from the embryonic digestive tract. Most of these are controlled by neurosecretions, but some act independently, responding directly to changes in blood and tissue fluid composition.

Neurohaemal organs

The major neurosecretory region is found at the base of the brain in a mass of nervous tissue, the **hypothalamus**, surrounding the third brain ventricle (Fig. 25.8). The hypothalamus is an important neural control centre, receiving information from most parts of the nervous system and providing central control of many functions by the autonomic nervous system.

The **pituitary gland**, also known as the **hypophysis**, comprises an outgrowth of neural tissue (the pituitary stalk) from the base of the hypothalamus to which a non-neural endocrine gland, derived from an outgrowth of the digestive tract, is closely attached (Fig. 25.9). The two structures together form a distinct appendage to the base of the brain, which is embedded in a cavity in the base of the skull. The neural element contains two major neurohaemal organs, the **posterior pituitary gland** and the **median eminence**. Collectively, these two organs are termed the *neurohypophysis*. The non-neural anterior part is the **anterior pituitary gland** or **adenohypophysis** (Fig. 25.8).

Two other important neurohaemal structures in vertebrates are the **pineal gland** and the **chromaffin organs**.

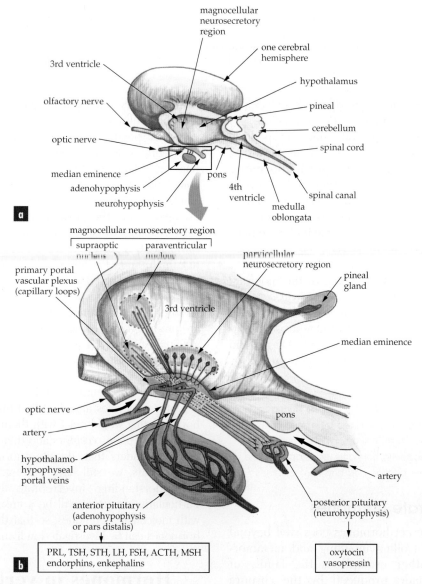

Fig. 25.8 (a) Diagram showing the location of the major neurohaemal organs in the lateral walls and floor of the third brain ventricle (the hypothalamus) of a mammal. The neurosecretory regions are highly vascular. **(b)** In the region of the median eminence, an array of capillary loops, derived from nearby arteries, coalesce into the hypothalamo-hypophyseal portal veins, which supply the hormone-secreting cells of the anterior pituitary gland. Neurosecretions released at the median eminence reach the anterior pituitary gland at high concentration and stimulate or inhibit the secretion of anterior pituitary hormones

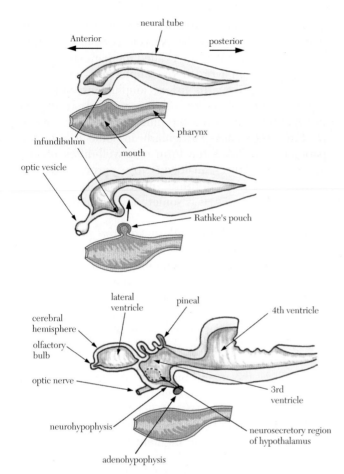

Fig. 25.9 Diagram to illustrate the embryonic development of the neural tube in vertebrates and the location of the central neurosecretory structures. An outgrowth of the roof of the mouth buds off to form the adenohypophysis (anterior pituitary gland) and the neurohypophysis forms from a downgrowth of the developing neural tube

> In vertebrates, the pituitary gland has distinct neurohaemal and non-neural endocrine regions. Neurosecretory tissue is found in the hypothalamus, neurohypophysis (posterior pituitary gland and median eminence), pineal gland and chromaffin organs.

Posterior pituitary gland

In most vertebrates, the posterior pituitary gland is a distinct appendage connected to the brain by a *neural stalk* or *infundibular process*. The neurosecretory cells that send axons to the posterior pituitary gland (*magnocellular neurons*) are located bilaterally in two nuclei in the walls of the third ventricle (the supraoptic and paraventricular nuclei). The axons, which contain dense neurosecretory granules, pass into the posterior pituitary and terminate in the vicinity of highly permeable blood capillaries.

In mammals, the posterior pituitary hormones are the nine-amino-acid peptides (nonapeptides) **vasopressin (antidiuretic hormone)** and **oxytocin**, which have the same ring (formed by a sulfur bridge between cysteines at positions 1 and 6 to form a cystine) with a side chain configuration but differ from each other in the amino acid composition of the side chain (Fig. 25.10). Marsupials have an amino acid substitution in the second position of the ring in their vasopressin, which is called *phenypressin*. Vasopressin influences blood pressure in high concentrations but is more important as an antidiuretic, and oxytocin affects reproductive functions (lactation and parturition). In all the vertebrates so far studied, the nonapeptides are similar with only minor differences in the amino acid composition of the side chain, such as vasotocin, the ancestral hormone and principal posterior pituitary hormone of non-mammalian vertebrates.

> The posterior pituitary secretes peptides that affect water balance and reproduction.

Median eminence

In all vertebrates except teleost fishes, axons of a set of neurosecretory cells (*parvicellular neurons*) in the inferior and lateral walls of the third ventricle terminate in the vicinity of a dense group of capillary loops in a neurohaemal organ, the median eminence. This is a small, highly vascularised prominence on the floor of

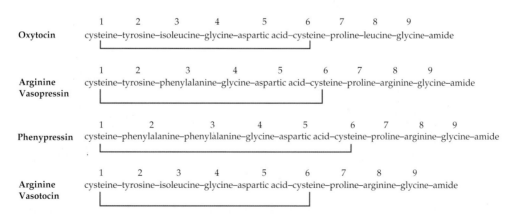

Fig. 25.10 Chemical structures of some common neurohypophyseal hormones: oxytocin; vasopressin; phenypressin and vasotocin. Note the minor changes in amino acid composition that confer differences in biological activity

the third ventricle, just anterior to the posterior pituitary gland. Capillaries emerging from the median eminence coalesce to form veins that constitute the **hypothalamo-hypophyseal portal system**. These veins transport the neurosecretions released at the median eminence directly, at high concentration, to an adjacent endocrine gland, the anterior pituitary gland. There, the veins divide into a second network of capillaries, which perfuse the secretory cells of the anterior pituitary (Fig. 25.8).

The neurosecretions from the median eminence act specifically to stimulate or inhibit the secretion of particular anterior pituitary hormones, providing a chemical link between the neural and non-neural elements of the hormonal control system. This arrangement of a neurosecretory system with a direct vascular link to a non-neural endocrine gland is analogous to the corpus cardiacum–corpus allatum relationship in insects.

Teleost fishes lack a median eminence. The axons of the neurosecretory cells pass directly to the anterior pituitary, where they release their secretions in the vicinity of specifically responsive cells. In this group of animals, the neural control of the anterior pituitary is thus of a *paracrine* rather than an *endocrine* nature.

Pineal gland

The pineal gland is an outgrowth in the midline of the roof of the third brain ventricle, between the cerebral hemispheres and the cerebellum (Fig. 25.11). It varies greatly in size and complexity between vertebrates. In many vertebrates, associated tissue becomes differentiated into a separate organ called the parapineal organ (in fishes), frontal organ (in amphibians) or parietal organ (in reptiles). In fishes, amphibians and some reptiles (but not snakes) the parapineal, frontal or parietal organ and even the pineal gland itself contain photoreceptor cells. Particularly in some lizards, the parietal organ takes the form of a well-differentiated eye with structures analogous to a cornea and lens, which serves a definite photosensory function (Fig. 25.11). Hence, it is sometimes referred to as the 'third eye'.

In mammals, birds and snakes, the photosensory cells have become modified into **pinealocytes**, which have a secretory function. The major secretory product is **melatonin**, which is derived from the neurotransmitter, serotonin. The first function described for melatonin was its ability to aggregate the pigment within melanophores in the skin of amphibians, so causing blanching of the skin. While this function has since been found for many other vertebrates (Fig. 25.12), the main function of melatonin is one of time-keeping. Melatonin secretion is inhibited by light and stimulated in the dark. Hence its release is capable of signalling onset of dark to all other tissues and organs in the body, such as the pigment cells in the skin. In addition to diurnal time-keeping, melatonin can also signal seasonal changes in day length. This is one mechanism by which seasonal reproduction is controlled, particularly in some mammals (Chapter 15).

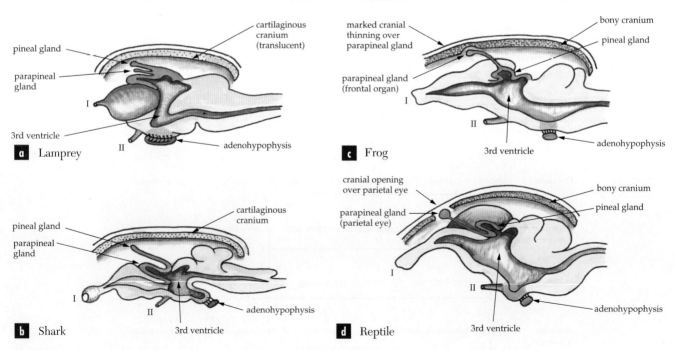

Fig. 25.11 Arrangement of the pineal–parapineal complex in four different species of vertebrates (I: olfactory nerve; II: optic nerve): **(a)** lamprey; **(b)** shark; **(c)** frog; and **(d)** reptile. Note how, in some reptiles, the parapineal becomes a functioning eye. The neurosecretion of the pineal gland is melatonin

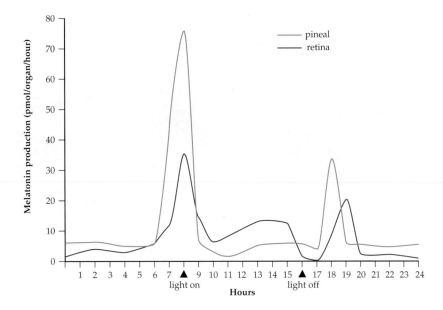

a

b

Fig. 25.12 Synthesis of melatonin is associated with 'dawn' and 'dusk' changes in environmental lighting in the scincid lizard. **(a)** Graph illustrating this association for melatonin synthesised by the pineal gland and the retina of the eye. Only pineal synthesis anticipates lights off, suggesting that the pineal is a time-keeper in this lizard. **(b)** The scincid lizard, *Lampropholas guichenoti*

Melatonin secretion by pinealocytes in mammals and birds appears to be under direct nervous control and the pinealocytes secrete their product directly into blood capillaries without the intervening extension of an axon. In mammals and birds, the innervation appears to be exclusively from the sympathetic division of the autonomic nervous system, which, in turn, receives information about environmental lighting from the lateral eyes, relayed through the supra-chiasmatic nucleus and superior cervical ganglion of the sympathetic nervous system. In other vertebrates, the photoreceptors in the pineal complex respond directly to environmental darkening by secreting melatonin.

The pineal gland secretes melatonin in response to absence of light, which influences diurnal and seasonal rhythms of pigmentation and reproductive activity in a wide variety of vertebrates. It is under the influence of light through photoreceptors or the nervous system.

Chromaffin organs

Chromaffin organs are groups of cells containing small cytoplasmic granules that take up chromate stains, hence the name. These cells aggregate around capillary vascular sinuses in the vicinity of the kidneys (Fig. 25.13) and are innervated by preganglionic fibres of the autonomic nervous system (Chapter 26). They are considered to be post-ganglionic sympathetic neurons that have been modified to become neuro-secretory cells.

In mammals, chromaffin tissue forms the **adrenal medulla**, which is surrounded by a cortex of non-neural steroidogenic cells to form the adrenal gland. Adrenal chromaffin tissue secretes the catecholamines adrenaline, noradrenaline and dopamine, all derived from the amino acid tyrosine. These are, in fact, classified as neurotransmitters because they are also released by the terminals of sympathetic axons and are involved in neurotransmission in the central nervous system. However, when released into the bloodstream from organs such as the adrenal medulla, they behave as hormones, exerting their effects on the metabolism or contractility of cells at a distance from the site of secretion. The major effects of catecholamines as hormones are to mobilise energy reserves in the form of glucose from stored carbohydrate in the liver and fatty acids from fat stores, as well as to influence cardiac contractility and blood pressure.

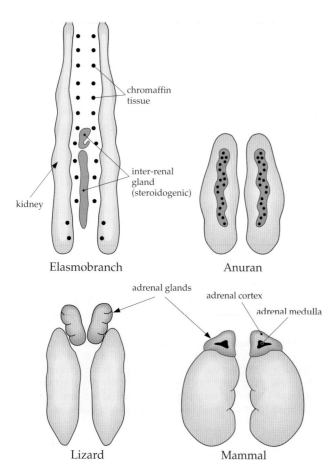

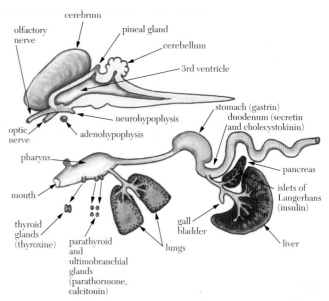

Fig. 25.14 Origins of some non-neural endocrine organs from the developing digestive tract of vertebrates

Fig. 25.13 Distribution of chromaffin and steroidogenic tissues in different vertebrates. In mammals, the two types of tissue have combined to form a discrete endocrine gland, the adrenal gland, with a cortex and medulla of quite different origins and subserving distinct functions

Non-neural endocrine glands

The non-neural endocrine glands of vertebrates are derived mainly from the digestive and excretory tubes (Fig. 25.14). They may be grouped according to their ability to respond either to changes in the external environment through neurosecretory hormones—anterior pituitary, thyroid gland, adrenal glands, and gonads—or in direct response to changes in the internal environment—parathyroid and ultimobranchial glands, islets of Langerhans and gastrointestinal mucosa.

Anterior pituitary gland

The anterior pituitary gland secretes three double-chain glycoprotein hormones (TSH, FSH, LH), two single-chain protein hormones (GH, PRL) and two peptide hormones (ACTH, MSH) derived from the same large precursor. The glycoprotein hormones and one of the peptide hormones, ACTH, stimulate the activity of other endocrine glands. The remaining three hormones influence metabolism, growth, reproduction and water balance (Table 25.1). The secretion of most, possibly all, anterior pituitary hormones is subject to

hormonal feedback control, by which the presence in the blood of the anterior pituitary hormone itself or the target organ hormone inhibits the secretion of either the releasing hormone at the hypothalamic level, or the anterior pituitary hormone at the site of its secretion.

> The anterior pituitary secretes protein-based hormones that either control the secretion of other endocrine glands or act directly on cells of the body. Secretion is under the influence of stimulatory or inhibitory neurosecretions from the hypothalamus, which perfuse the anterior pituitary via the portal vessels of the median eminence.

Thyroid glands

The thyroid glands are paired glands (Fig. 25.15), located in the pharynx or neck (when present). Each thyroid gland is composed of follicles, which have large lumina and simple cuboidal cells when inactive and much smaller lumina and columnar cells when active. The production of the thyroid hormones **thyroxine (T_4)** and **triiodothyronine (T_3)** takes place in the cells of the follicle, after which they are transferred to the lumen of the follicle, where they are stored still attached to thyroglobulin. Thyroid hormone synthesis is highly dependent on an adequate intake of iodine in the diet and is under the control of TSH from the anterior pituitary, which itself is subject to neuro-secretory control from a hypothalamic-releasing hormone (TRH in mammals and birds) and negative feedback control by the thyroid hormones. When stimulated, thyroid hormones pass from the lumen of the follicle back through the cells surrounding the follicle, where they are released from the thyroglobulin

Table 25.1 The anterior pituitary hormones, their hypothalamic control and their target organs

Hypothalamus	Anterior pituitary gland	Target organ
Thyrotrophin-releasing hormone (TRH)	Thyrotrophic hormone (TSH) (in mammals and birds)	Thyroid gland
Gonadotrophin-releasing hormone (GnRH)	Follicle stimulating hormone (FSH) Luteinising hormone (LH)	Ovary and testis
Corticotrophin-releasing hormone (CRH)	Adrenocorticotrophic hormone (ACTH)	Adrenal cortex
Growth hormone-releasing hormone (GHRH), somatostatin (GH-inhibiting hormone; GHRIH)	Growth hormone (GH)	Liver and many other cells in the body
Dopamine (PRL-inhibiting hormone)	Prolactin (PRL)	Mammary glands, gonads, secretory epithelia, etc.
?	Melanophore stimulating hormone (MSH)	Melanophores

to pass out of the cells and into the surrounding capillaries (Fig. 25.15c).

The two thyroid hormones are lipid-soluble and penetrate cell membranes, T_4 more easily than T_3. The thyroid predominantly produces T_4, unless iodine is limited in the diet, which travels in the blood bound to carrier proteins. In the target cells of the body, enzymes convert T_4 to T_3. The T_3 then binds to the receptor in the nucleus. The hormone–receptor complex binds to a specific region of the DNA called the hormone-response element. This initiates the production of proteins that are essential for normal growth and maturation of immature individuals and stimulation of the metabolic rate of cells. One of the most dramatic examples of thyroid hormone action is its role in the metamorphosis of an anuran tadpole to an adult frog, which is shown in Figure 25.16. The high levels of circulating thyroid hormones required to complete metamorphosis are dependent on stimulation of the pituitary TSH by a hypothalamic-releasing hormone (possibly CRH), which in turn depends on the development of the vascular link between the hypothalamus and the pituitary.

Several disease states in humans are directly related to malfunction of the thyroid and its hormones. Signs of excessive thyroid hormone activity, a condition known as **thyrotoxicosis**, include elevated metabolic rate, rapid heart rate and dilated pupils. Lack of iodine in the diet leads to a decrease in thyroid hormone secretion because there is no iodine to incorporate into thyroglobulin. This, in turn, causes increased TSH secretion because of decreased feedback inhibition. The increased TSH, unable to increase synthesis of thyroid

hormones in the absence of iodine, causes an enlargement of the thyroid gland, **goitre**. Lack of thyroid hormones can impair growth and development, particularly of the nervous system, in young vertebrates (**cretinism** in humans) and slow metabolism and nervous activity in adults. In adult humans, lack of thyroid hormones can also cause abnormal accumulation of water in subcutaneous tissues, **myxoedema**.

Thyroid hormones stimulate growth and development in immature vertebrates and metabolic rate in mature vertebrates.

Adrenal cortices

In mammals and birds, the **adrenal cortex** is the outer layer of the adrenal gland (located at the anterior, cephalic, end of the kidneys). Less well-defined groups of similar cells are present in the kidneys of other vertebrates (Fig. 25.13). These cells secrete two major classes of steroid hormones (corticosteroids): **glucocorticoids**, such as cortisol and corticosterone, which influence carbohydrate and protein metabolism; and **mineralocorticoids**, such as aldosterone, which influence Na^+ and K^+ balance (Fig. 25.17). The chemical nature of the principal corticosteroids is very similar among vertebrates, with only minor variations between groups.

Like thyroid hormones, corticosteroids diffuse readily into cells, where they bind to a cytoplasmic receptor derived from the nucleus. The receptor–steroid complex moves into the nucleus, where it binds to its hormone-response element on DNA, inducing the production of cytoplasmic enzymes. The steroid is

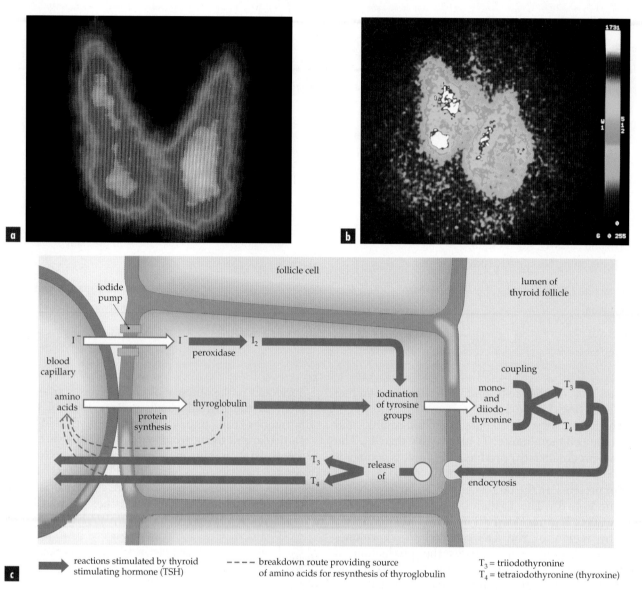

Fig. 25.15 Coloured computer-enhanced gamma scan of **(a)** a healthy human thyroid gland and **(b)** an enlarged thyroid gland which is suffering from overproliferation of normal cells (hyperplasia). **(c)** The mechanism of thyroid hormone production by follicle cells. Circulating iodide is 'trapped' by an iodide pump in the cell membrane. Within the follicle cell it is converted to free iodine, which combines with the tyrosine residues on thyroglobulin synthesised by the follicle cells. The iodinated thyroglobulin is stored extracellularly in the lumen of the thyroid follicle. Stimulation of thyroid hormone release results in endocytosis of iodinated thyroglobulin back into the follicle cell, where the coupled iodinated tyrosine residues are released from the thyroglobulin as thyroid hormones, which are then secreted from the follicle cells to be picked up by the capillaries surrounding each thyroid follicle

then removed and the receptor migrates back into the cytoplasm.

Glucocorticoid secretion is increased by adreno-corticotrophic hormone (ACTH) in response to some metabolic disturbances, and to stressful disturbances of the environment that require mobilisation of energy reserves. Mineralocorticoid secretion is not so influenced by ACTH. It is primarily controlled by plasma K^+ levels and **angiotensin II**, a peptide hormone derived from the action of a kidney enzyme, renin (released in response to reduced vascular pressure and low sodium), on a blood protein, angiotensinogen, produced in the liver.

In most vertebrates, adrenocortical secretions, particularly mineralocorticoids, are essential for survival. Death usually follows removal of the adrenal glands and is a consequence of the loss of body Na^+, with dehydration and circulatory failure. Many vertebrates, such as rats, red kangaroos and possibly North American opossums, can survive complete removal of both adrenal glands provided they have access to sufficient salt to replace that lost through the kidneys as a result of lack of mineralocorticoids. Echidnas appear to be unaffected by removal of both adrenal glands provided the external environment is stable. However, when exposed to a cold environment, an adrenalectomised echidna is unable to

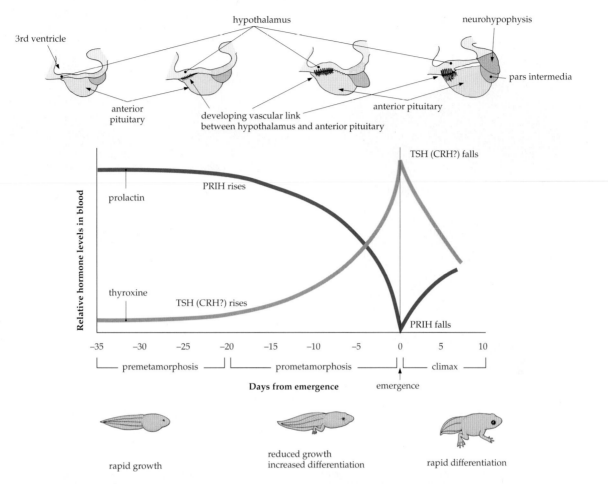

Fig. 25.16 The role of hormones in the growth and development of the tadpole of a frog. Prolactin secretion by the anterior pituitary declines due to secretion of prolactin-inhibitory hormone (dopamine) from the hypothalamus. At the same time, secretion of a TSH-releasing hormone (possibly CRH in amphibians) rises and stimulates TSH secretion from the anterior pituitary, which in turn stimulates thyroxine secretion by the thyroid gland

increase thermogenesis and maintain body temperature. The ability to thermoregulate is restored by injection of cortisol. This is a good example of the *permissive* nature of some hormone actions; that is, a lack of the hormone does not appear to influence body functions in the normal state but its presence is essential for regulatory responses to a disturbance of the environment.

The glucocorticoid hormones, as their name implies, influence sugar metabolism. They are important in the induction of enzymes that synthesise glucose from non-carbohydrate sources, such as amino acids, a process known as **gluconeogenesis**. They also cause the breakdown of cell proteins to their constituent amino acids. The combination of these two actions results in the production of glucose, as a source of energy, at the expense of tissue protein. Two important tissues susceptible to this catabolic action are muscle and lymphoid tissues, both of which reduce in mass if high glucocorticoid levels are sustained in the blood.

Among the lymphoid cells affected by glucocorticoids are the thymus-derived lymphocytes (T cells,

Chapter 23), so that prolonged increase in glucocorticoid secretion, as a result of stress, can impair the ability of an animal to withstand infections. Other actions of the glucocorticoids include inhibition of release of the tissue factors that cause the inflammatory response to injury. Because of this, glucocorticoids can delay wound healing. All the adverse effects of glucocorticoids can occur as a consequence of prolonged exposure to a stressful environment.

Glucocorticoids also enhance the actions of catecholamines on cardiac contractility and blood pressure, a property useful in the management of shock in severely injured humans.

> The adrenal cortex secretes two major classes of steroid hormones (corticosteroids)—glucocorticoids, which influence carbohydrate and protein metabolism, and mineralocorticoids, which influence Na$^+$ and K$^+$ balance.

Gonads

In addition to their basic function of germ-cell production, ovaries and testes, located near the genital tracts,

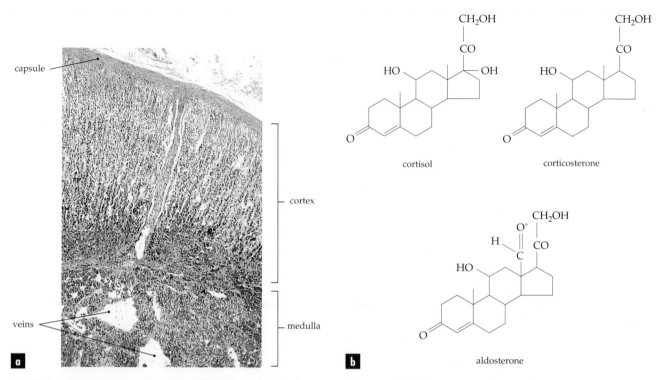

Fig. 25.17 (a) Section through the adrenal cortex and medulla of a mammal showing the different zones of the cortex. Blood flows from the outer cortex through the medulla to the central adrenal vein. **(b)** The chemical structures of the major adrenal steroids found in vertebrates. The major glucocorticoid is either cortisol (fishes and mammals other than rodents) or corticosterone (reptiles, birds and rodents). The major mineralocorticoid in all tetrapods (non-fish vertebrates) is aldosterone. Cortisol appears to have both roles in many fish

BOX 25.1 Endocrinology of reproduction

Reproduction in vertebrates involves considerable synchronisation of events, all of which are predominantly controlled by hormones. Most vertebrates are seasonal breeders. Environmental cues such as day length and temperature are monitored by the central nervous system to trigger a release of gonadotrophin-releasing hormone into the vessels of the hypothalamo-hypophyseal portal system in the median eminence. This hormone, in turn, stimulates the *gonadotrophic cells* in the anterior pituitary to release gonadotrophins (FSH and LH) into the general circulation. The gonadotrophins are responsible for the maturation, or recrudescence, of the gonads (testes in males and ovaries in females). These events are synchronised in males and females of the same species so that ripe gametes (sperm and ova) are produced at the same time.

The same hypothalamic–pituitary–gonadal axis again comes into play to process the visual/auditory/tactile/chemical stimuli that will ensure that males and females will come into close proximity at the time of gamete maturity and to ensure that mating will occur. These signals, which frequently involve secondary sexual characters (e.g. deer antlers, bird plumage, fish colouration) are predominantly controlled by the gonadal steroid hormones, androgens (e.g. testosterone) in males and oestrogens (e.g. oestradiol-17β in females). Reproductive behaviour, courtship and mating are also primarily controlled by these same gonadal steroids. Mating is the final event in reproduction for many vertebrates, especially fish and amphibians that produce very large numbers of eggs (see Chapter 15). For those that incubate a smaller number of eggs, either externally or internally, another pituitary hormone, prolactin, and an array of ovarian and/or placental hormones, which include progesterone, come into play to maintain the appropriate environment for the egg to develop and complete the process of reproduction.

contain hormone-secreting cells. In females, the ovarian follicular cells secrete steroidal **oestrogens**. In males, interstitial cells of the testes secrete steroidal **androgens**, the most common being testosterone. These steroidal hormones diffuse into the cells of their target organs where, like thyroxin, they are usually converted to a more potent form, which binds to a specific nuclear receptor, which then binds to its hormone-response element on DNA. In the target brain cells of most vertebrates, testosterone is converted to an oestrogen (oestradiol) and then combines with an oestrogen receptor in the nucleus. In many other target tissues, testosterone is converted to dihydrotestosterone, which is a more potent activator of the androgen receptor.

Oestrogens and androgens influence growth, development of male and female sexual characteristics and behaviour, development of gametes, and ovulation (Fig. 25.18). Their effects on metabolic exchange are **anabolic**, that is, they favour increase of tissue mass. They are under the control of the two pituitary gonadotrophic hormones. Follicle-stimulating hormone (FSH) stimulates ovarian follicles to proliferate and secrete oestrogens. Luteinising hormone (LH) acts in concert with FSH to cause ovulation. LH then stimulates the cells of the ovulated follicle to proliferate and form a corpus luteum, which secretes a further steroid hormone, progesterone. Progesterone prepares the reproductive tract either to secrete an appropriate coating for the eggs in oviparous species or to receive an embryo and maintain its smooth muscle quiescent through pregnancy in viviparous animals (Fig. 25.18).

In males, FSH stimulates spermatogenesis and LH stimulates the secretion of the steroidal androgens. Mammalian gonads also produce a peptide hormone inhibin, which inhibits secretion of FSH and not LH. In this way, differential stimulation of the two gonadotrophins can be attained through the single hypothalamic-releasing hormone (GnRH). Inhibin may prove useful for male contraception.

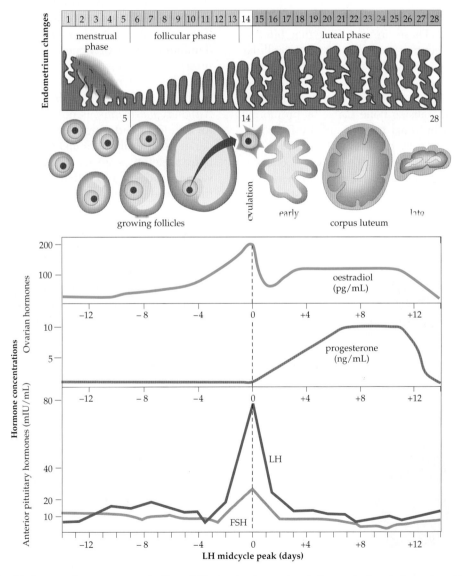

Fig. 25.18 The relationship between hormone concentrations, follicular development and changes in the uterine endometrium during the menstrual cycle in a human female

The ovaries and testes secrete oestrogens and androgens in females and males respectively. These steroidal hormones influence growth, development of male and female sexual characteristics and behaviour, development of gametes, and ovulation.

Parathyroid and ultimobranchial glands

Parathyroid and ultimobranchial glands are located either associated with the thyroid or just posterior to it. Like the thyroid, they are outgrowths of the *branchial clefts* of the embryonic pharynx (Fig. 25.14) and form discrete clusters of cells usually close to, or incorporated in, the thyroid glands (Fig. 25.19). In mammals, ultimobranchial cells are incorporated into the thyroid glands and are called *thyroidal C-cells*. Both glands are concerned with the regulation of extracellular Ca^{2+} concentration and their functions are best understood in birds and mammals.

The parathyroid glands secrete **parathyroid hormone** (PTH) and the ultimobranchial glands secrete **calcitonin**. These are peptides that bind to cell-membrane receptors and activate intracellular messengers to stimulate existing enzyme activity. Their three main sites of action are bone, the intestine and the kidneys. PTH increases Ca^{2+} concentration in blood plasma by mobilising bone Ca^{2+}, by promoting Ca^{2+} reabsorption by renal tubules and by indirectly increasing Ca^{2+} uptake by the gut. Calcitonin has the opposite effects. It depresses plasma Ca^{2+} concentration by promoting its uptake by bone and increasing its excretion by the kidney. PTH also stimulates synthesis of a dihydroxylated form of vitamin D, which enhances Ca^{2+} absorption from the gut and to a lesser extent Ca^{2+} mobilisation from bone.

These glands respond directly to changes in plasma Ca^{2+} concentration. Increase in Ca^{2+} concentration in blood plasma increases calcitonin secretion and decreases PTH secretion, so that Ca^{2+} is removed from the blood. Decrease in plasma Ca^{2+} concentration has the opposite effect, so that Ca^{2+} is added to the blood.

The opposing actions of calcitonin and PTH produce a very precise control of the extracellular Ca^{2+} concentration, which is necessary because of the important role Ca^{2+} plays in neuromuscular transmission (Chapter 27). In humans, a small fall in extracellular Ca^{2+} concentration causes a marked increase in neuromuscular irritability, a condition known as *tetany*, which can lead to convulsions and death. The occasional development of tetany after removal of a goitrous thyroid gland in humans led to the discovery of the parathyroid glands and their role in Ca^{2+} balance. In most vertebrates, the bony skeleton is a major reservoir of exchangeable Ca^{2+} and a major site of action of Ca^{2+}-regulating hormones. For example, excessive PTH secretion leads to demineralisation of bones, which become weak and prone to fracture.

Ultimobranchial glands are well defined in both cartilaginous and bony fishes. The Mexican cavefish, *Astyanax mexicanus*, is normally exposed to dim light for a short time during the day. If it is kept in complete darkness, the ultimobranchial bodies enlarge and the bony skeleton becomes decalcified. It was this observation, made in 1954, that led to the later realisation that the ultimobranchial glands were concerned with regulation of Ca^{2+} balance. It also shows that, in this species of fish at least, secretion by the ultimobranchial gland can be influenced by a change in the external environment—light.

Parathyroid glands appear to have evolved first in amphibians, where together with the ultimobranchial glands, they appear to influence Ca^{2+} exchange in much the same way as described for mammals. In birds and some reptiles, the parathyroid and ultimobranchial glands have an additional function related to the production of large eggs with calcified shells. The large amount of Ca^{2+} required, especially in birds, appears to be derived from bone under the influence of PTH. Oestrogens are also important in the regulation of Ca^{2+}. They appear to inhibit the ability of PTH to deplete bone Ca^{2+}. **Osteoporosis** is a fairly common consequence of a reduction in gonadal oestrogen production after menopause in human females, especially if Ca^{2+} intake is low.

Parathyroid and ultimobranchial glands secrete parathyroid hormone and calcitonin respectively. These hormones regulate extracellular calcium concentration through opposing actions on uptake, mobilisation and excretion of calcium.

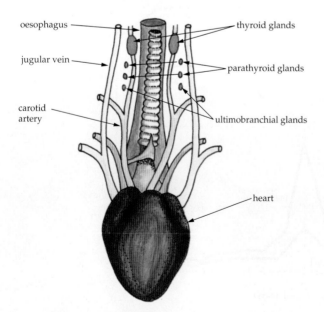

oesophagus
thyroid glands
jugular vein
parathyroid glands
carotid artery
ultimobranchial glands
heart

Fig. 25.19 The location of the thyroid, parathyroid and ultimobranchial glands in the neck of a bird

Islets of Langerhans

These cells are embedded within the outgrowth of the digestive tube that forms the exocrine **pancreas** (Fig. 25.14). The exocrine (zymogen) cells release digestive juices into ducts, which open directly into the duodenum, whereas the islet cells have only an endocrine function, releasing their secretions into the circulating blood (Fig. 25.20). There are three major cell types in these islets—alpha, beta and gamma—which secrete respectively the peptides **glucagon**, **insulin** and **somatostatin**. Insulin is the most important regulator of carbohydrate metabolism in vertebrates. It is a protein that occurs, with minor chemical modifications, in most animals. It has even been found in unicellular organisms, including bacteria, although its functions, if any, in such organisms are unknown.

Insulin is secreted in response to a rise in blood glucose concentration, for example, after ingestion of carbohydrate. It acts by binding to a transmembrane receptor on its target cells, inducing a cascade of events

that lead to an increase in the permeability of the cell membrane to glucose and amino acids, an increase in cytoplasmic glucose transporters, and the induction of metabolic enzymes. This results in increased glucose utilisation and increased storage of glucose as glycogen in liver and muscle. These actions cause a fall in blood glucose concentration and increased uptake of amino acids with increased protein synthesis. Insulin also acts on fat cells to promote the conversion of glucose to glycerol and pyruvate, resulting in increased fat production (lipogenesis).

The human disease **insulin-dependent diabetes mellitus** (type I) is characterised by a high blood glucose concentration and excretion of glucose in the urine, and is due to destruction of the pancreatic islet β-cells, possibly by an autoimmune reaction after a viral infection. The disease can be controlled by administration of insulin. A more common form of diabetes is **non-insulin-dependent** or *maturity onset* **diabetes** (type II). Type II diabetes is caused by a relative lack of insulin and is a problem of secretion rather than synthesis. It is usually not diagnosed until the individual is middle-aged and is commonly associated with obesity. It can be controlled by diet.

The effects of insulin are **hypoglycaemic** (blood glucose-lowering) and *anabolic* (favouring synthesis of tissue). They are opposed by one of the other islet cell hormones, glucagon, which is secreted when blood glucose levels fall. Glucagon binds to a transmembrane receptor leading to a decrease in glucose oxidation by cells and an increase in breakdown of glycogen to glucose in liver. It also inhibits lipogenesis and stimulates production of gluconeogenic enzymes, which synthesise glucose from amino acids derived from structural proteins. These effects are **hyperglycaemic** (blood glucose-raising) and **catabolic** (favouring tissue breakdown).

While there are many other hormones that participate in the control of blood glucose and metabolism, such as glucocorticoids from the adrenal cortex, adrenaline from the adrenal medulla (chromaffin cells) and somatotrophin from the pituitary, the blood glucose level is normally accurately controlled by the interaction between insulin and glucagon effects (Fig. 25.21). The exact role of somatostatin is less well understood but it is capable of inhibiting both alpha and beta cells in the pancreas and appears to do so in a paracrine fashion.

Islet cells respond directly to changes in extracellular glucose concentration. Insulin secretion increases with rising extracellular glucose concentration, causing increased uptake of glucose by cells, which counters the rise. Glucagon secretion increases when extracellular glucose falls, causing mobilisation of glucose stores in the liver, so countering the fall. In this way, insulin and glucagon control extracellular glucose

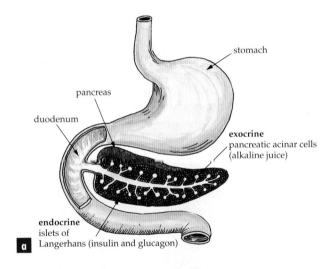

stomach

pancreas

duodenum

exocrine
pancreatic acinar cells
(alkaline juice)

endocrine
islets of
Langerhans (insulin and glucagon)

a

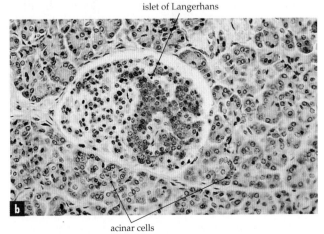

islet of Langerhans

acinar cells

b

Fig. 25.20 (a) Islets of Langerhans are located throughout the exocrine pancreas. **(b)** An islet, in which the insulin-containing β-cells have been stained brown, surrounded by exocrine acini

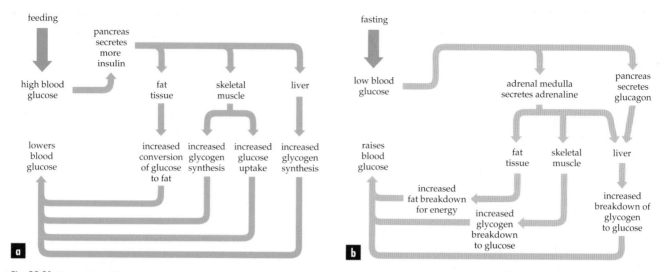

Fig. 25.21 Summary of hormonal control of blood glucose concentration in mammals in the **(a)** feeding and **(b)** fasting states

levels in much the same way as calcitonin and PTH control extracellular Ca^{2+} concentration.

> Insulin and glucagon are secreted by the beta and alpha cells respectively of the islets of Langerhans in the pancreas. Insulin increases tissue uptake and storage of glucose, and its secretion is stimulated by a rising blood glucose concentration. Glucagon has the opposite effects. Between them they control normal blood glucose concentration.

Gastrointestinal mucosa

The cells that secrete gastrointestinal hormones are scattered in loose aggregations throughout the gut mucosa. They are bipolar, the side on the mucosal surface detecting a change in the chemical environment provided by the contents of the digestive tract, while the internal side secretes the hormone into local blood vessels in response to an appropriate change in gut content (Fig. 25.22).

These hormones control the secretion of digestive juices. There are three major hormones in this group, **secretin**, **gastrin** and **pancreozymin/cholecystokinin** (CCK). They are all released into the *portal vein* to be carried in the blood through the liver to their target organs (specific cells of the digestive tract, bile duct or pancreas), where they optimise the composition of the digestive juices to ensure adequate digestion of ingested food, and control the motility of gut smooth muscle and blood flow to the gut (Fig. 25.22).

Polypeptide growth factors

This is a group of polypeptide hormones, secreted by a wide variety of cells, that act as **mitogens**; that is, they stimulate mitosis of cells in the environment in which they are released. Some examples of these are *epidermal growth factor*, *nerve growth factor* and

transforming growth factor. They bind with high affinity to the extracellular domain of specific protein receptors on plasma membranes and activate intracellular messengers to stimulate mitosis. They therefore have important functions in the regulation of growth and differentiation. It is now thought that viruses can induce cancer by producing analogues of these locally acting growth factors. A considerable research effort is being made to understand the functions of these hormones with a view to controlling cancers.

BOX 25.2 Autocrine and paracrine hormones

These hormones constitute a diverse group of biologically active agents released by a wide variety of tissues in response to local stimuli, which act on cells in the vicinity of their release.

Eicosanoids are the most widely studied members of this group of hormones and most of the knowledge applies to eutherian mammals. They are all derived from the unsaturated fatty acid *arachidonic acid* and all have 20 carbon atoms in their molecule. There are four subgroups—prostaglandins, prostacyclins, thromboxanes and leukotrienes.

All mammalian cells, except erythrocytes, contain eicosanoids. They act at extremely low concentration in the same environment in which they are synthesised. They also appear in the blood, where they can be measured, but their functions as blood-borne hormones are uncertain. In mammals, eicosanoids are involved in a wide

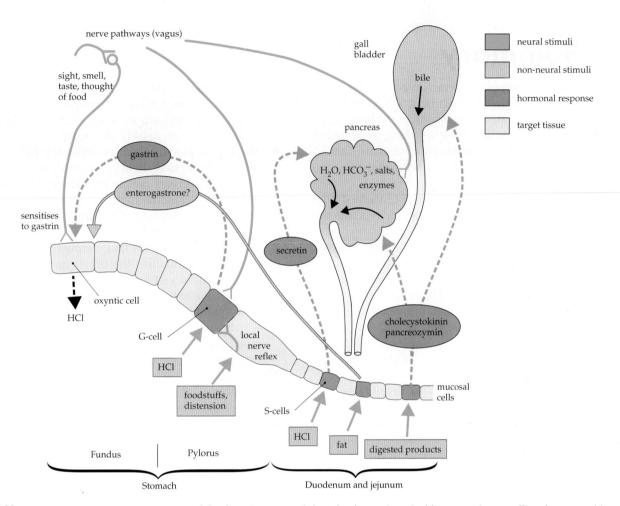

Fig. 25.22 The major hormone-secreting regions of the digestive tract and the role of gastrointestinal hormones in controlling the composition of the digestive juices

variety of functions, all mediated by cellular cAMP. These include:
- induction of local increase in blood flow and increased capillary permeability in response to injury of cells—the inflammatory response
- stimulation of nerve fibres subserving the sensation of pain
- induction of fever by affecting the thermoregulating region of the brain
- participation in blood pressure control by acting on vascular smooth muscle
- participation in the induction of blood clotting at a site of injury

- control of several reproductive functions, including induction of labour, lysis of the corpus luteum and some aspects of reproductive behaviour
- participation in regulation of the sleep–wakefulness cycle.

Although little is known about their evolutionary history, it is likely that these compounds are widely distributed in vertebrates and are important in reproduction and in mediating local responses to injury.

Summary

- Different hormones interact to achieve control of growth, maturation, reproduction, metabolism and the inorganic composition of the body fluids.
- Neurosecretion is the earliest form of hormonal control to have evolved in animals, and is probably the major form of hormonal control in invertebrates.
- Neurosecretory hormones are concerned with regulation of growth, maturation, reproduction and metabolism.
- Non-neural hormone-secreting organs controlled by the nervous system evolved later and add to the effectiveness and range of actions of the neurosecretory control of body functions.
- Hormonal control systems in invertebrates, with the exception of insects, are not well understood. In insects, there are complex cascades of hormonal interactions controlling life history events such as moulting.
- Hormonal control systems in vertebrates have become increasingly complex and regulate most physiological functions. However, the basic ground plan of neurosecretory response to environmental change, first evident in lower invertebrates, is retained.
- The increasing complexity of vertebrate physiological control systems is the result of the evolution of additional hormonal mechanisms responding to and controlling variables in the internal environment.
- In the vertebrate pituitary, neurosecretory tissue is found in the posterior pituitary (neurohypophysis) and non-neural endocrine tissue in the anterior pituitary (adenohypophysis). The posterior pituitary secretes peptides affecting water balance and reproduction. The anterior pituitary secretes protein hormones that either control the secretion of other endocrine glands or act directly on cells of the body.
- The pineal gland, which is under the influence of environmental light perceived by photoreceptors, secretes melatonin, which influences diurnal and seasonal rhythms, particularly those involved in pigmentation and reproductive activity in many vertebrates.
- Thyroid hormones stimulate growth and development in immature vertebrates and metabolic rate in mature vertebrates.
- Parathyroid and ultimobranchial glands secrete parathyroid hormone and calcitonin respectively, which, through opposing actions on uptake, mobilisation and excretion of calcium, regulate extracellular Ca^{2+} concentration.
- The adrenal cortex secretes corticosteroids—glucocorticoids, which influence carbohydrate and protein metabolism and mineralocorticoids, which influence Na^+ and K^+ balance.
- The testes secrete androgen in males and the ovaries secrete oestrogen and progesterone in females. These steroidal hormones influence growth and development of male and female sexual characteristics and behaviour, development of gametes, ovulation, pregnancy and many other aspects of reproduction in vertebrates.
- Insulin and glucagon are secreted by the beta and alpha cells, respectively, of the islets of Langerhans in the pancreas. Insulin increases tissue uptake and storage of glucose and glucagon has the opposite effects. Between them, they control blood glucose concentration.

key terms

adrenal cortex
(p. 667)
adrenal medulla
(p. 665)
allatostatin (p. 659)
allatotropin (p. 659)
anabolic (p. 671)
androgen (p. 670)
angiotensin II (p. 668)
anterior pituitary
gland
(adenohypophysis)
(p. 662)
autocrine hormone
(p. 654)
bursicon (p. 661)
calcitonin (p. 672)
catabolic (p. 673)
chromaffin organ
(p. 662)
cretinism (p. 667)
crustacean cardiactive
peptide (p. 661)
ecdysis-triggering
hormone (p. 661)
ecdysone (p. 659)
eclosion hormone
(p. 661)

eicosanoids (p. 674)
endocrine gland
(p. 653)
exocrine gland
(p. 653)
gastrin (p. 674)
gating (p. 659)
glucagon (p. 673)
glucocorticoid
(p. 667)
gluconeogenesis
(p. 669)
goitre (p. 667)
hormone (p. 653)
hyperglycaemic
(p. 673)
hypoglycaemic
(p. 673)
hypothalamo-
hypophyseal portal
system (p. 664)
hypothalamus (p. 662)
insulin (p. 673)
insulin-dependent
diabetes mellitus
(p. 673)
juvenile hormone
(p. 659)

median eminence
(p. 662)
melatonin (p. 664)
mineralocorticoid
(p. 667)
mitogen (p. 674)
molecular cloning
(p. 661)
myxoedema (p. 667)
negative feedback
control (p. 658)
neurohaemal organ
(p. 657)
neurosecretion
(p. 654)
non-insulin-dependent
diabetes mellitus
(p. 673)
oestrogen (p. 670)
osteoporosis (p. 672)
oxytocin (p. 663)
pancreas (p. 673)
pancreozymin/
cholecystokinin
(p. 674)
paracrine hormone
(p. 654)

parathyroid hormone
(p. 672)
pheromone (p. 654)
pineal gland (p. 662)
pinealocyte (p. 664)
pituitary gland
(hypophysis)
(p. 662)
posterior pituitary
gland
(neurohypophysis)
(p. 662)
prothoracicotropic
hormone (p. 659)
secretin (p. 674)
somatostatin (p. 673)
target cell (p. 654)
thyrotoxicosis (p. 667)
thyroxine (T_4) (p. 666)
triiodothyronine (T_3)
(p. 666)
vasopressin
(antidiuretic
hormone) (p. 663)

Review questions

1. Describe two basic differences in the modes of action of water-soluble hormones and lipid-soluble hormones.

2. With reference to hormone actions, what is meant by the terms autocrine, paracrine and endocrine?

3. What is meant by neuroendocrine control? Describe (a) a neurosecretory cell and (b) a neurohaemal organ.

4. In vertebrates, explain the developmental origins of the pituitary.

5. What are the general properties of target organs and physiological functions of (a) posterior pituitary hormones and (b) anterior pituitary hormones?

6. (a) What is the significance of dietary iodine to thyroid function?

   (b) What physiological processes do thyroid hormones influence?

7. (a) What is the functional difference between a mineralocorticoid and a glucocorticoid?

   (b) What is the effect of ACTH on the adrenal gland?

8. (a) Where are the ultimobranchial and parathyroid glands?

   (b) Draw a flow diagram to show how the hormones they secrete interact to control blood Ca^{2+} levels.

9. What are the endocrine cell types in the islets of Langerhans and what are their functions?

10. In considering control of moulting in insects, give an example of a neuroendocrine cascade.
11. What are the functions of (a) ecdysone and (b) bursicon in insect development?
12. Draw a diagram to illustrate the hormonal control of ovulation in a mammal.

Extension questions

1. In considering the action of a water-soluble hormone, explain what is meant by a 'cascade' from the hormone–receptor complex to the final response of the cell.
2. What are the major functions of progesterone in (a) mammals and (b) non-mammalian vertebrates?
3. What does it mean when a hormone's role is described as 'permissive'? Which vertebrate hormones are frequently described as permissive?
4. What is an orphan receptor? How have they been discovered?

Suggested further reading

Brook, C. G. D. and Marshall, N. J. (1996). *Essential Endocrinology*. Oxford: Blackwell Science.

General endocrine text with a primary human focus. It provides a balance between basic science and clinical practice. It also includes chapters describing basic endocrine techniques.

Norman, A. W. and Litwack, G. (1998). *Hormones*. 2nd edn. San Diego, CA: Academic Press.

This text provides a comprehensive treatment of hormones viewed from the context of modern theories of their action in the framework of current understanding of the molecular, subcellular and cellular architecture as well as classic organ physiology.

Norris, D. O. (1996). *Vertebrate Endocrinology*. 3rd edn. San Diego, CA: Academic Press.

This is a text that covers all classes of vertebrates. It provides a basic background in cellular and organismic endocrinology and includes interfaces of classical endocrinology with neurobiology, immunology, cellular biology and molecular biology.

Chapman, R. F. (1998). *The Insects, Structure and Function*. 4th edn. Cambridge: Cambridge University Press.

This text contains chapters on the current understanding of insect endocrinology.

Bentley, P. J. (1998). *Comparative Vertebrate Endocrinology*. 3rd edn. Cambridge: Cambridge University Press.

This book deals with the nature, actions and roles of hormones among vertebrate animals, with special emphasis on the evolution and origins of hormones and their receptors, on the role of hormones in the physiological co-ordination of vertebrates, and on dealing with the endocrine process in the context of the organism's physiology, ecology and evolution.

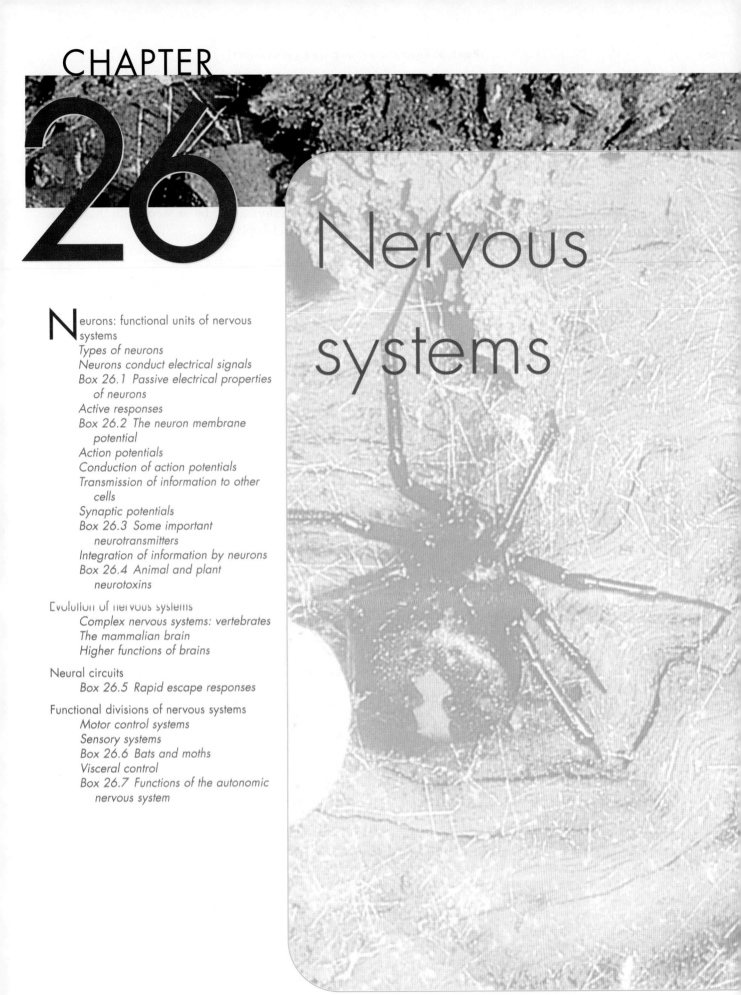

CHAPTER

26

Nervous systems

Co-ordination of function is achieved in all animals by the passage of information within cells and between cells, tissues and organs. This is accomplished in two main ways—by means of hormones, as explained in Chapter 25, and by the nervous system. The nervous system has specific roles in integrating information about the state of an organism and in providing rapid and precise signalling between parts of the body. In both simple and complex organisms, the nervous system is responsible for stereotyped activities, such as avoidance behaviours and movements of internal organs. In mammals, it provides remarkable computational power, which is expressed in such phenomena as visual memory and fine movement control. The human nervous system has evolved further to allow speech and abstract thought.

Communication in nervous systems depends on the functions of **neurons**, cells specialised for conducting information. Neurons, together with supporting **glial cells**, form the nervous system (Table 26.1).

Table 26.1 Features of two cell types found in nervous systems

Neurons	Glial cells
Are electrically excitable	Provide mechanical support for neurons
Receive inputs from the environment (sensory neurons) or from other neurons (interneurons and motor neurons)	Provide electrical insulation for neurons
	Maintain the extracellular environment of neurons
Provide output to other neurons or to effectors (mostly muscle and gland)	Guide neuronal development and repair

Neurons: functional units of nervous systems

The function of a neuron is to receive information, process that information and transmit the processed information to other cells. Most neurons have three major structural regions (Fig. 26.1a): **dendrites**, processes that are generally short and receive inputs; the **soma**, the cell body containing the nucleus; and an **axon**, a long process that carries the output. Neurons are often quite long cells; for example, neurons whose axons reach your feet have their cell bodies located in the spinal cord at about the level of your kidneys, and

axons in the arms of giant squids run for more than 10 m. Some neurons in simple animals, such as jellyfish (phylum Cnidaria), are unspecialised in structure and pass information in any direction (Fig. 26.1b). In more complex animals, neurons of many shapes have evolved, with regions specialised for receiving, processing or passing on information (Fig. 26.1b). Transduction of information by sensory cells has been considered in Chapter 7.

Neurons receive, process and transmit information. Glial cells provide both physical and nutritional support, and insulation.

Types of neurons

In simple animals, a single neuron can carry out several functions. It can detect incoming stimuli, transfer information between other pairs of neurons and activate muscles. This arrangement is not very specific for processing and routing signals, and is not conducive to integration of responses. The neurons of most animals are more specialised; they can be grouped into three functional classes: sensory neurons, interneurons and motor neurons. **Sensory neurons** are associated with specialised detectors, known as sensory receptors (Chapter 7). Receptors that detect signals from the external environment, such as sensory neurons for touch or smell, are known as **exteroreceptors**. Those that monitor internal states, such as blood pressure or core temperature, are **enteroreceptors**. Sensory neurons pass their information on to other specialised neurons.

Interneurons receive information from one group of neurons and pass it on to another. They have major roles in the integration of information, that is, sorting, comparing, combining and discarding information from various sources, and in providing appropriate outputs.

Motor neurons provide the final output of the nervous system to the muscles and glands that it controls. The cells or tissues innervated by motor neurons are called *effectors*.

Pathways carrying information *towards* the central nervous system are called *afferent pathways*, whereas those carrying it *away* are called *efferent pathways* (the same terms are used for other structures, such as blood vessels: afferent means passing towards a point or area, efferent means passing away from it). Thus, sensory neurons are often referred to as afferent neurons and motor neurons as efferent neurons.

Sensory neurons detect signals in the internal and external environment. Interneurons receive, process and transmit information from one group of neurons to another. Motor neurons influence the activities of muscles and glands.

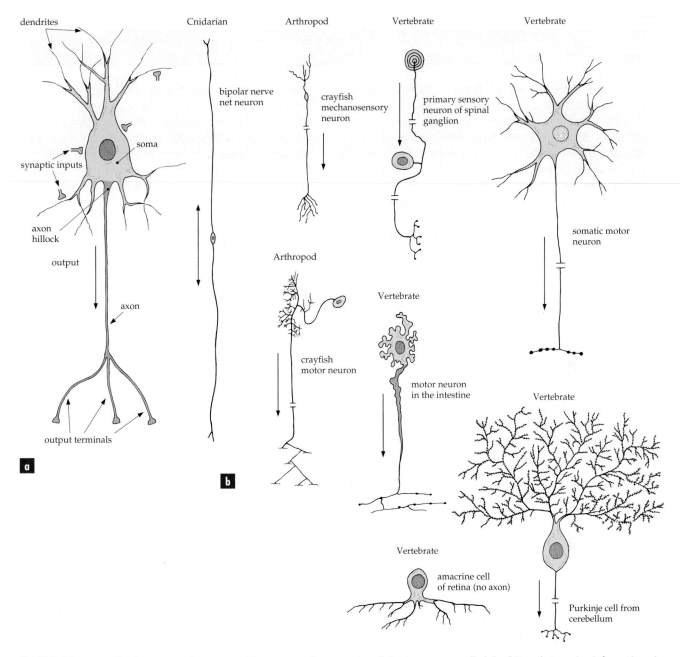

Fig. 26.1 (a) A generalised representative neuron. Most neurons have a series of short processes, called dendrites, that receive information via synaptic inputs from other neurons. Excitation of the cell body (soma) causes an electrical event, the action potential, to travel along the axon towards the output terminals of the neuron, which release chemical transmitters that influence other neurons or effectors such as muscles or glands. **(b)** A comparison of the morphology of a number of different neurons from a variety of animals and locations within nervous systems. Arrows show the direction of information flow

Neurons conduct electrical signals

The manner in which information passes along a neuron depends on the passage of electrical currents across its plasma membrane. A current will travel when there is a difference in voltage between two points, and a conducting path along which a current can flow. In a neuron, electrical communication depends on the voltage difference across the plasma membrane, with the inside of a neuron being negative with respect to the outside when the neuron is at rest (Chapter 4). Local changes in membrane voltage may be conducted away passively, that is, their amplitudes decrease with distance and time (Box 26.1).

> nformation in the form of electrical signals is conducted along the membranes of neurons.

The voltage difference between the inside and outside of a neuron depends on transmembrane proteins in the neuronal membrane that function as ion

BOX 26.1 Passive electrical properties of neurons

The plasma membrane of nerve cells, like all plasma membranes, is a protein and lipid bilayer that incorporates integral membrane proteins (Chapter 3). Membranes have electrical capacitance and resistance, which are depicted in the electrical equivalent circuit in Figure a. Electrical charge and the insulating lipid together create the capacitance component. The charge is present partly as ions in aqueous solutions on either side of the membrane and partly as polar groups on the membrane surface. Ion selective channels composed of integral proteins (Chapter 4) allow ions to traverse the membrane but offer some opposition to their passage; this is the membrane resistance represented in Figure a. There is also resistance to flow of current through the cytoplasm in nerve cell processes.

If the voltage across the membrane is changed at a point, the current generated will leak both across the membrane at that point and lengthways along the process before leaking across the membrane at other points. The greater the resistance of the membrane, per unit length (R_M, ohm/cm) relative to the cytoplasmic resistance, per unit length (R_L ohm/cm), the greater the tendency of the current to follow the length of the process. The measure of the conductance along the process in relation to leakage across the membrane is the length constant, more generally called the **space constant**, which is represented by λ, the distance for an applied voltage to fall to $\frac{1}{e}$ (about 37%) of its original value. The space constant, λ, is equal to $\sqrt{R_M/R_L}$. The capacitance per unit length of the membrane C_M, coulomb/volt cm) delays the effect of an applied current on the voltage across a nerve fibre membrane because the current must first displace capacitative charge before it generates the full voltage change. The rate of discharge of the capacitance is measured as the **time constant**, τ, which is the time to change the difference between the initial and final state by a factor of $1 - \frac{1}{e}$ (about 63%); τ equals $R_M \cdot C_M$. The electrical properties of the membrane determine how quickly the peak of an electrical event is conducted passively in time and space. This speed is λ/τ.

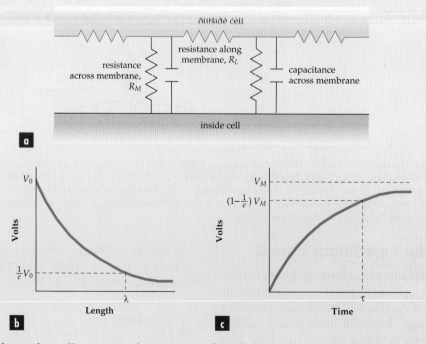

(a) Electrical model of a membrane (R_M represents the resistance to flow of current across the membrane and R_L the resistance to flow of current through the cytoplasm). (b) Length–voltage relationship for an axon: charge leaks through the membrane (V_0 is voltage at zero length; λ is length constant). (c) Time–voltage relationship: an imposed current takes time to change the voltage because it must displace capacitative charge (V_M is maximum voltage attained; τ is time constant)

BOX 26.2 The neuron membrane potential

A voltage difference exists across the plasma membrane because it is differentially permeable to ions and separates solutions of different ionic compositions (Chapter 4). Each type of ion tends to move down its concentration gradient, but the passages of different ions are hampered to different extents. It is this tendency of ions to move across the membrane that creates the voltage difference.

Firstly, consider K^+. The plasma membrane is permeable to K^+ and the concentration of K^+ inside the cell is greater than that outside ($[K]_i > [K]_o$), which means that there will be a net diffusional movement of K^+ outwards. Outward K^+ movement makes the cell interior more negative; the membrane is thus equivalent to a battery, with its negative pole inside and its positive pole outside the cell.

In solution, ions behave like an ideal gas; that is, the work that the gas or ions in solution can do by moving down a concentration gradient depends on the concentration difference. From the gas laws, the diffusion energy available for movement from concentration [K] to concentration zero is

$$RT \log_e [K] \text{ joules/mole}$$

where R is the gas constant (in J mol^{-1} degree^{-1}). For a gas, the equation is usually expressed in terms of pressure, which is proportional to the concentration of gas molecules; the concentration of solute molecules is proportional to diffusional pressure.

The difference in diffusion energy across the membrane (driving K^+ out) is the difference in the diffusional energy of the inside and outside solutions:

$$RT \log_e [K]_i - RT \log_e [K]_o$$

This difference in diffusional energy must equal the electrical energy that the battery, created as a result of the concentration difference, can deliver. Battery energy (the work that a battery would need to do to cause movement of a single ion) is equal to the strength of the battery (V; voltage) multiplied by the charge on the ion (qZ; elementary charge × valency), that is,

$$VqZ \text{ volts coulomb}$$

We need to put this in molar terms. To do this we use Faraday's constant, F (the number of coulombs in a mole), giving

$$VFZ \text{ volts coulomb mole}^{-1}$$

Because the two energies—the battery energy and the energy for diffusion—must be equal,

$$VFZ = RT \log_e[K]_i - RT \log_e [K]_o$$
$$= RT (\log_e [K]_i - \log_e [K]_o)$$
$$= RT \log_e \frac{[K]_i}{[K]_o}$$

That is, the voltage across the membrane is

$$V = \frac{RT}{FZ} \log_e \frac{[K]_i}{[K]_o}$$

This is known as the Nernst equation and V is the equilibrium potential for K^+, E_K; that is, the potential that would be generated by the distribution of this ion if the membrane were not permeable to any other ions. Because the ability of K^+ to diffuse across the membrane determines E_K, it is also called the diffusion potential for K^+. In thinking about this equation, remember that the logarithm of a number greater than 1 is positive, the logarithm of 1 is zero and the logarithm of a number less than 1 is negative. A positive value of V means the outside cell is positive and the inside cell is negative.

Now imagine a membrane permeable to Na^+, K^+ and Cl^-. Each will provide a driving potential due to differences in its concentrations across the membrane and charge. The effect of the driving potential (equilibrium potential) of any ion on the resting membrane potential will depend on its ability to cross the membrane—only ions that can cross contribute to the diffusion potential.

In fact, the membrane of the cell resists the movements of all ions to some extent. The effect of this resistance is to reduce the effective diffusional energy supplied by a given concentration of ions. This is equivalent, in effect, to lowering the concentration. To take into account membrane resistance, we should replace [K] by $[K]_i$/resistance. In practice, we use the inverse of resistance to flow of ions, a term known as conductance (g_K). Thus, the Nernst equation can be rewritten as

$$E_K = \frac{RT}{F} \log_e \frac{g_K [K]_i}{g_K [K]_o}$$

Of course, the g_K terms cancel out, but considering that there are other ions being 'pushed' by their concentration differences through different conductances, adding in the terms for these ions we get

$$E_M = \frac{RT}{F} \log_e \frac{g_K [K]_i + g_{Na} [Na]_i + g_{Cl} [Cl]_o}{g_K [K]_o + g_{Na} [Na]_o + g_{Cl} [Cl]_i}$$

This is known as the Goldman equation. Note the inversion of the Cl^- concentrations because the Cl^- charge, and hence the electrical equivalence of its diffusional drive, is opposite to Na^+ and K^+. Simply looking at the equation shows that the greater the value of g_K relative to other conductances, the closer the Goldman equation comes to the Nernst equation for K^+, in other words, the closer the membrane potential (E_M) approaches the K^+ equilibrium potential (E_K).

Test yourself on the questions below. Cover the right column, answer the questions on the left and then compare your responses with those provided.

Questions	Answers
What happens to E_M if $[K]_o$ increases?	E_M becomes more positive (i.e. the membrane depolarises).
What happens if g_{Na} increases?	E_M becomes more positive, it moves towards E_{Na}.
What happens to E_M if all the conductances increase, but the ratios between them stay the same?	E_M does not change.
In this last case, what happens to overall conductance?	It increases.
What would be the relative effects on E_M of the diffusion potential for a fourth ion at rest after all the other conductances (but not its conductance) are increased?	The effect of the fourth ion would be diminished.

pumps, using energy from ATP (Chapter 4). These pumps transport sodium (Na^+) to the outside and potassium (K^+) to the inside of the cell, creating and maintaining concentration gradients for these ions. In the resting state, the neuronal membrane is more permeable (leaky) to K^+ than to Na^+; K^+ moves down its concentration gradient from inside to outside, leaving the inside of the neuron electrically negative compared with the outside (Box 26.2). The polarisation of a membrane can be measured using a microelectrode inserted through the neuronal membrane (Fig. 26.2). Such measurements show that the electrical polarisation of most neuronal membranes is about 80 mV; that is, the cell has a **resting membrane potential** of -70 to -80 mV (see also Chapter 4).

> The surface membrane of a neuron is electrically polarised, with the inside negative compared with the outside.

Active responses

When the membrane potential inside the neuron becomes less negative (depolarised) a local feedback can amplify the potential change, giving rise to an event called an **active response**. Active responses eventually die away with distance from their points of initiation. An active response involves charged particles (ions) moving through channels in the membrane. Two cations, Na^+ and K^+, are important for active responses.

Sufficient depolarisation to generate an active response can be caused physiologically by excitation from another neuron or a sensory stimulus, or by a current passed through a microelectrode (Fig. 26.2). The potential at which depolarisation is sufficient to trigger an active response is the **threshold potential** for that membrane. At this potential, conformational changes occur in certain transmembrane proteins, increasing their permeability to external Na^+ (and occasionally Ca^{2+}). These proteins are voltage-dependent channels because the voltage across the membrane determines whether they are open or closed (Chapter 7). When they open, Na^+ diffuses down its concentration gradient, making the inside of the neuron even less negative. This change in membrane potential causes additional voltage-dependent Na^+ channels to open. The active response terminates when the membrane depolarises sufficiently for a different set of voltage-dependent channels to open, in this case, K^+ channels, allowing K^+ to leave the cell. This moves the potential back towards its resting level.

The distance that an active response spreads along a neuron process and the time it lasts depend on both passive and dynamic characteristics of the membrane. In some neurons, voltage change spreads rapidly, with little decrease in amplitude over large areas of membrane; in others, it affects only the membrane immediately around the stimulus site. Different parts of the membrane of the same neuron can have different characteristics. Local active responses are important in transmission of information over short distances.

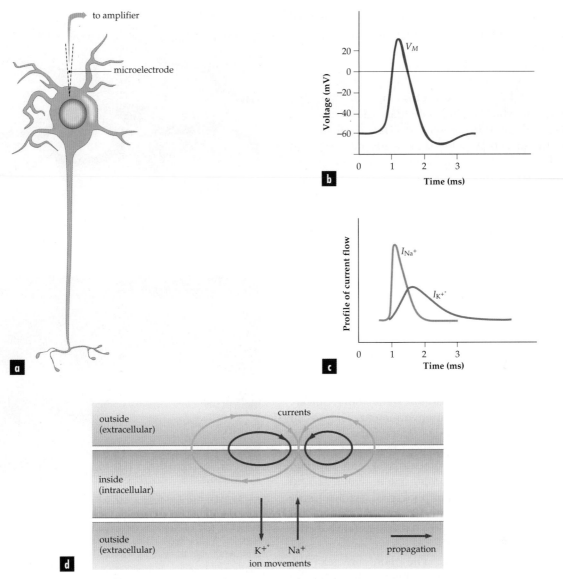

Fig. 26.2 The neuronal action potential. **(a)** Action potentials are recorded by microelectrodes inserted into neurons. **(b)** The action potential consists of a brief depolarising potential, with a rapid onset and a slightly slower decline, followed by a slight undershoot. **(c)** The voltage changes are the consequence of brief changes in permeability to Na$^+$ and K$^+$, producing an inward current (I_{Na^+}) and an outward K$^+$ current (I_{K^+}). **(d)** The movement of Na$^+$ and K$^+$ during the action potential causes local current circuits. The electrical disturbance propagates along the axon. After the action potential has passed, the usual ionic gradients are restored by the Na$^+$–K$^+$ pump

The active response of a neuronal membrane is dependent on the dynamic properties of the voltage-dependent channels, and time and space constants of the membrane.

Action potentials

Active responses are not suitable for conducting information along processes of neurons that are from several millimetres to metres long. Long-distance communication involves the conduction of an electrical event that does not die out with distance. This event is the **action potential** or spike. Action potentials travel the full lengths of neuron processes without loss of amplitude because they regenerate themselves at successive points. The regenerative nature of action potentials allows information to be carried over long distances without distortion or loss.

In neuronal membranes that generate action potentials, the number of voltage-dependent Na$^+$ channels opening at the threshold potential increases rapidly and in a non-linear fashion. This creates a positive feedback loop in which more channels opening induce yet more to open so that the potential inside the membrane becomes rapidly positive. The positive feedback loop is broken when high levels of depolarisation cause the Na$^+$ channels to close and the voltage-dependent K$^+$ channels to open. All this happens within a few milliseconds, giving rise to a rapid inward Na^+ current followed by a short-lasting outward K^+ current (Fig. 26.2).

For a brief period after each action potential, the membrane becomes less excitable. This period of reduced excitability, the **refractory period**, limits the frequency at which action potentials can occur in a neuron to about 100 per second. The depolarisation caused by an action potential spreads into adjacent areas of membrane in the same way as local active responses. When this occurs in an axon, the depolarisation is effectively channelled along the cable, opening voltage-dependent Na^+ channels in the adjacent region of the neuronal membrane where the action potential is regenerated. Sequential activation regenerates the action potential at contiguous points along the axon, that is, it is propagated or *conducted*.

> Some membranes are capable of producing regenerative active responses called action potentials.

Conduction of action potentials

The speed of **conduction** of action potentials along axons determines, to a large extent, how quickly animals can respond to their environment. There is an advantage in rapid conduction in those parts of nervous systems involved in fast responses that affect survival, for example, escape from predators or prey capture. Axon conduction speeds vary from about 0.5 to 120 m s^{-1} and depend on morphological characteristics of the axon that determine its effectiveness as an electrical cable, particularly its diameter and the insulating properties of the plasma membrane and closely adjacent glial cells. The more efficient the insulation and the greater the diameter, the faster the conduction.

Based on these factors, two ways of increasing conduction velocity in neurons have evolved. Vertebrate axons that conduct at speeds greater than 5–10 m s^{-1} are invariably wrapped in an insulating layer called **myelin** (Fig. 26.3a). Myelin is formed by glial cells (Schwann cells in the peripheral nervous system or processes of oligodendrocytes in the central nervous system). A region of the glial cell wraps around the axon, excluding the cytoplasm, and unique myelin proteins are laid down. **Nodes of Ranvier** are small regions of bare axon, 1–1.5 mm long, where one myelin-forming cell abuts another. Action potentials skip from node to node in a process known as **saltatory** (meaning leaping) **conduction**.

In many invertebrates, rapid conduction is achieved by means of large axon diameters. Speed of conduction is proportional to the axonal diameter, increasing as the diameter increases. In the squid, axons more than 1 mm in diameter supply mantle muscles, which

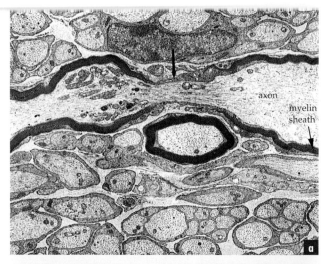

Fig. 26.3 (a) Fast conducting axons in vertebrates are wrapped in many layers of glial cell plasma membrane, the myelin sheath, as shown in this electron micrograph. A gap in the myelin (a node of Ranvier) can be seen at the arrow. **(b)** Much of the original work on the function of neurons was carried out on giant axons from squid. These axons are responsible for producing the very rapid jet propulsion shown by these animals under certain circumstances

contract during the jet propulsion used for escape and prey capture (Fig. 26.3b).

> Speed of conduction of an action potential along an axon is increased by increase in diameter and/or insulation of the axon.

Transmission of information to other cells

Neurons transmit information directly to other neurons, or to muscles, at small areas of close contact, **synapses**. Transmission can increase activity in the post-synaptic cell (*excite*) or reduce it (*inhibit*). At **chemical synapses**, which are most common, there are narrow separations about 20–50 nm wide between presynaptic axon terminals and the surfaces of the post-synaptic cells they influence. Signals are transmitted across these gaps via the release of chemical signalling substances, **neurotransmitters** (Fig. 26.4). At **electrical synapses**, the membranes of the pre- and post-synaptic cells are apposed and communicating junctions are present, allowing electrical signals to pass across directly (Chapter 6). Some electrical synapses pass signals in both directions, others only one way.

Within the presynaptic nerve ending at a chemical synapse are small membrane-bound structures, **synaptic vesicles**, which contain chemical transmitter molecules. When an action potential arrives at the presynaptic nerve ending, it causes the opening of voltage-dependent Ca^{2+} channels. Ca^{2+} then rushes down its concentration gradient into the nerve ending. This triggers a series of biochemical events that culminate in the synaptic vesicle membrane coalescing with the presynaptic plasma membrane and releasing the transmitter into the synaptic cleft by exocytosis (Chapter 4). Embedded in the post-synaptic membrane are specialised receptors to which the transmitter chemical binds. This binding triggers a post-synaptic response. For example, activation of receptors could cause muscle to contract, glands to secrete, or neurons to be excited or inhibited. It is important to note that the same transmitter chemical can cause excitation of one cell and inhibition of another depending on the receptor and the post-receptor mechanism on which it acts (Chapter 7).

One recently discovered chemical transmitter does not act in the way just described. This is nitric oxide, which is not stored but is rapidly synthesised when Ca^{2+} enters the presynaptic nerve ending. The nitric oxide then rapidly diffuses through the presynaptic membrane, across the synaptic cleft and into the post-synaptic cell, where it stimulates responsive enzymes.

> Signals, which can excite or inhibit the post-synaptic cell, are transmitted by a neuron either by release of a neurotransmitter or by direct electrical communication.

Synaptic potentials

By changing the permeability of the post-synaptic membrane to one or more ions, a neurotransmitter can change the post-synaptic membrane potential. When the membrane potential becomes more negative, the neuron is **hyperpolarised**; that is, the voltage difference between the inside and the outside (membrane polarisation) is increased. In this circumstance, activation of the neuron is *inhibited*. Conversely, when the membrane potential of a neuron is made more positive, the neuron is **depolarised**; that is, the membrane potential is brought closer to the threshold level for action potential initiation. The neuron is thus *excited*. Muscle cells are similarly inhibited or excited by membrane potential changes.

During chemical neurotransmission, a series of events links the arrival of the neurotransmitter to the generation of a post-synaptic potential (Fig. 26.5). Firstly, the transmitter selectively binds to a receptor in the post-synaptic membrane. This binding is brief, usually lasting less than a millisecond, and causes a conformational change (alteration in shape or electrical charge distribution, usually both) in the receptor protein. This alteration of the receptor protein can elicit an electrical change in a number of ways (Chapter 7). The receptor protein may be an ion channel, with the conformational change leading to altered ionic permeability, or the receptor may be linked to an ion channel whose conformation is

presynaptic axon

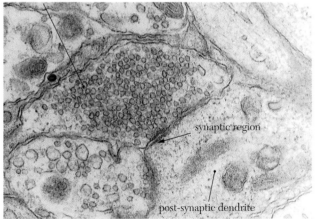

Fig. 26.4 A chemical synapse as it is seen in the electron microscope. The major features are the presynaptic accumulation of vesicles and the pre- and post-synaptic membrane thickenings

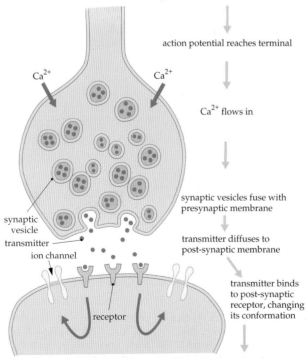

Fig. 26.5 Transmitter release and action at a chemical synapse

changed by the altered receptor. Alternatively, receptor changes may trigger intracellular messenger systems, such as the adenylate cyclase system or the inositol triphosphate system, which, via protein phosphorylation, change the permeability of ion channels. The transmitter nitric oxide does not act on a membrane receptor but passes through the membrane to act on an intracellular enzyme, usually guanylate cyclase.

The electrical event caused by transmission at an excitatory synapse between two neurons is a depolarisation of the post-synaptic neuron and is known as an excitatory post-synaptic potential (***epsp***; Fig. 26.6). The response at an inhibitory synapse is almost always a hyperpolarisation, an inhibitory post-synaptic potential (***ipsp***; Fig. 26.6).

The dominant ionic permeability change underlying an *epsp* is usually an increase in Na^+ conductance (movement of Na^+ across the membrane), but a decrease in K^+ conductance can also cause an excitatory potential change (Box 26.2). An *ipsp* can result from a transient increase in K^+ permeability or an increase in Cl^- conductance. Although the equilibrium potential for Cl^- (E_{Cl}) is close to the resting membrane potential, a significant increase in Cl^- permeability increases the influence of E_{Cl} on the membrane potential (Box 26.2). As a result, the concomitant opening of channels for other ions has less effect than usual and inhibition can occur with little change in membrane potential. One of the most important inhibitory transmitters of the vertebrate central nervous system, gamma-aminobutyric acid (GABA), acts via the opening of Cl^- channels.

Many different molecules function as neurotransmitters (Box 26.3). Each transmitter can produce either an excitatory or inhibitory response according to the nature of the post-receptor mechanism (Chapter 7). In general, as with hormones (Chapter 25), evolution has been highly conservative and all groups of animals utilise a selection from a common group of neurotransmitters.

N eurotransmitters bind to receptors in the post-synaptic membrane, causing post-synaptic potentials that may be excitatory or inhibitory.

Integration of information by neurons

There are two major levels of **integration** in the nervous system: integration that occurs at the level of a single neuron and integration that involves a number of neurons. A deceptively simple act, such as catching a ball in flight, involves integration of information involving thousands of neurons. Even at the level of a single neuron, integration of information is complex. The role that any one synaptic input plays in generating an output by the post-synaptic cell depends on the *activity* of that synapse, the *location* of the synapse on the post-synaptic neuron, and the *timing* of input activity in relation to the activity at other synapses on the same neuron.

The location of synapses on the post-synaptic neuron may be axodendritic, axosomatic or axoaxonal. The space constant of the post-synaptic membrane, together with the active response characteristics of the

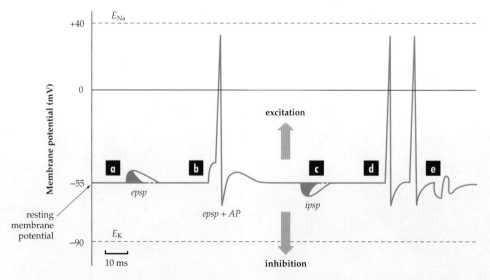

Fig. 26.6 Depiction of the types of electrical events that occur in neurons. Presynaptic inputs cause excitatory post-synaptic potentials (*epsp*'s) and inhibitory post-synaptic potentials (*ipsp*'s) in the post-synaptic neuron. **(a)** An *epsp* that does not reach threshold to evoke an action potential. The time course of the current underlying the *epsp* is indicated by the shading. **(b)** An *epsp* that reaches threshold to generate an action potential (AP). **(c)** An *ipsp* with its current indicated by shading. **(d)** Action potentials. **(e)** An *ipsp* prevents an action potential that invades the region of the *ipsp* from reaching threshold; only a small depolarisation is now caused by the action potential. The equilibrium potentials for Na^+ (E_{Na}) and K^+ (E_K), the resting membrane potential and zero potential are indicated. The values of these potentials are representative; they differ slightly between neurons

BOX 26.3 Some important neurotransmitters

Acetylcholine

Excitatory transmitter to skeletal muscle in vertebrates, inhibitory transmitter to vertebrate heart. Excitatory transmitter at many neuro-neuronal synapses.

Adrenaline

Excitatory transmitter to the heart, blood vessels and various viscera in vertebrates. In mammals, the same role is served by the closely related compound, noradrenaline.

Glutamate

An amino acid that is a commonly used excitatory transmitter in the brain. In arthropods, glutamate is an excitatory transmitter to muscle.

Gamma-aminobutyric acid (GABA)

An amino acid that is an inhibitory transmitter in the vertebrate brain and at some arthropod neuro-muscular junctions.

Serotonin

A transmitter to heart muscle and in the ganglia of some invertebrates. A transmitter in the vertebrate brain.

Peptides

Peptides, which are chains of three to over 30 amino acids, are involved in transmission in many invertebrate and vertebrate species. Over 50 neurotransmitter peptides have been identified. Closely related or identical peptides are found in the neurons of taxonomically widely separated species (e.g. *Hydra* and humans).

Nitric oxide

A free radical that is a transmitter to some visceral muscles of vertebrates and may be involved in memory formation.

The first part of the axon, the **axon hillock** or proximal process (Fig. 26.1a), is a key region because it is the site at which the active response characteristics of the membrane permit the generation of an action potential for transmission into the axon. Inhibitory inputs at the axon hillock can block the propagation of an action potential and therefore cancel out the effects of even massive excitatory inputs onto the dendrites or cell body. In neurons where the cell body is situated on the end of a process, away from the dendrites and axons, as occurs in vertebrate sensory neurons and in many invertebrate motor neurons (Fig. 26.1b), the membrane of the cell body plays little or no part in the signal processing function of the neuron.

In some cases, an axon terminal forms an inhibitory synapse on the terminal of a second axon, near its synapse. Activation of the inhibitory synapse can reduce transmitter output from the terminal of the second axon and so decrease or even block transmission across that synapse. This process is known as presynaptic inhibition.

The combination of synaptic inputs, together with the characteristics of the neuron membrane, determine the final levels of depolarisation in different parts of a neuron. This, in turn, determines its output in terms of rate of action potential firing and amount of neurotransmitter release.

Integration occurs both at the level of a single neuron and in assemblies of neurons. The influence of each synapse on signal output depends on its level of activity, its location on the post-synaptic cell and the electrical characteristics of the membrane.

Evolution of nervous systems

The basic properties of neurons from all animals appear to be the same. The evolution of more flexible and complex behaviours, however, has been accompanied by an increasing complexity of the nervous system.

Almost all cells have some sort of potential difference between the inside and the outside. The development and use of this potential difference for signalling and for control of effectors appears to have occurred early in evolution. Intracellular recordings from single-celled protists, such as *Paramecium*, reveal a resting potential similar to that of cells in multicellular animals. Moreover, potential changes are part of cell-signalling in protists. Depolarisation of the plasma membrane in *Paramecium*, for example, causes a reversal in the direction of the ciliary beat used for locomotion.

surrounding membrane, determine how far the effect of synaptic potential changes will travel. In general, synapses on distal dendrites have relatively smaller individual influences on the membrane potential of the cell body than do more proximal synapses. The summation of effects of numerous excitatory and inhibitory axodendritic and axosomatic synapses contribute to the net membrane potential of the cell body and hence to the firing pattern of the neuron.

BOX 26.4 Animal and plant neurotoxins

Nervous systems play an important role in the movement, co-ordination and survival of most animals. It is therefore not surprising that chemicals that cause nervous systems to malfunction have evolved as offensive or defensive weapons in both plants and animals. For example, two Australian animals, the red-back spider, *Latrodectus mactans hasselti* (Fig. a), and the tropical taipan, *Oxyuranus scutellatus* (Fig. b), produce neurotoxins that cause an excessive outpouring of transmitters from neurons. As a consequence, bites from these animals can cause paralysis in both prey and predators. Many plants produce alkaloids and other chemicals that poison or deter animal predators. The Australian corkwood, *Duboisia* (Fig. c), contains in its bark a substance called hyoscine, which prevents certain receptors for the neurotransmitter acetylcholine being activated. An almost identical toxin is found in the unrelated European plant, deadly nightshade, *Atropa belladonna*. It has the name belladonna (beautiful lady) because, when placed in the eye, the toxin causes dilation of the pupil by interfering with the acetylcholine receptors and preventing activation of the muscles that constrict the pupil in the eye.

Another interesting neurotoxin is strychnine, which comes from the seeds of an Indian tree, *Strychnos nuxvomica*. It was shown by the Australian neuroscientist and Nobel Prize winner, J. C. Eccles, that strychnine acts by blocking inhibitory synaptic potentials in the central nervous system, causing powerful convulsions. Strychnine blocks receptors for the neurotransmitter glycine.

The golden orb-weaving spider, *Argiope*, one of the spiders responsible for the beautiful, intricate orb webs we find in our gardens, produces a

(a) The red-back spider, *Latrodactus mactans*

(b) The tropical taipan, *Oxyuranus scutellatus*

(c) The corkwood tree, *Duboisia hopwoodii*

toxin that rapidly paralyses insects by blocking the excitatory receptors on their muscles. The toxin only works, however, on channels already opened by the excitatory transmitter, glutamic acid. This means that the more an insect struggles to escape, the more severe the paralysis becomes. The insect, still alive but paralysed, is wrapped in silk and hung up for storage. The poison in its system ensures that it will reparalyse itself whenever it starts to struggle. Thus, the spider has a secure source of fresh food.

In primitive metazoans, such as cnidarians, a number of levels of electrical communication are found. For example, some muscle cells are in direct electrical communication with each other via communicating junctions (Chapter 6). Although the process is slow and rather imprecise, because excitation tends to move outwards in all directions from the source like ripples in a pond, it does play a role in the co-ordination and behaviour of the whole organism. Cnidarians and platyhelminthes also have nerve nets. These are networks of neurons connected by synapses to form diffuse networks (Fig. 26.7). Signals can pass through the network in many directions but tend to conduct radially and decrease gradually with distance. These simple methods of conduction have the obvious disadvantages of being slow and diffuse. It is therefore not surprising to find that, although a number of groups of animals have retained nerve nets for local responses, they have also evolved faster, more directional systems to complement them. Large jellyfish, for example, have defined neuronal pathways consisting of larger neurons dedicated to co-ordinating the swimming bell; these neurons are not part of the nerve net.

Other trends in neural properties associated with more complex behaviour are apparent. As we have seen, integration of electrical information occurs ultimately at the membrane level. The time and space constant characteristics of neuronal membranes place physical limits on the distances over which this integration can occur. It is therefore effective for cells that exchange and integrate information to be close together (Fig. 26.7). Flatworms (phylum Platyhelminthes) have simple aggregations of nerve cells along the through-pathways down each side of the animal. Transverse pathways link these with the equivalent groupings on the other side, producing a ladder-like nervous system. Similarly, even more highly organised groupings of neurons, known as ganglia, occur in all other metazoan phyla. In many invertebrates, local information can be processed within ganglia as simple voltage changes without the need for action potentials (this information is said to be amplitude- or analogue-coded), but is converted into an action potential code for transfer over greater distances, to other ganglia or to effectors, such as muscles (this information is said to be frequency- or digitally-coded).

A number of organisms, such as flatworms and annelids, move through the environment with a preferred end forward. This allows specialisation of locomotory structures for more efficient movement. In these animals, the advantages of localising food sensing and gathering structures at the anterior end are immediately apparent. The anterior concentrations of neuronal cell bodies that accompany this trend, **cerebral ganglia**, are the forerunners of brains. The simplest ones receive information from chemical and movement sensors, and often primitive light detectors, and may direct co-ordination of feeding structures and body movement. The progressive aggregation of nerve cells in groups at the anterior end of an animal is **encephalisation** (Fig. 26.7).

Over the course of evolution of arthropods, the ganglia on either side of the animal have moved closer together and eventually fused. In metamerically segmented animals (Chapter 39), a segmental ganglion is responsible for co-ordinating the movements within each segment and for passing essential information to ganglia of adjacent segments and, where necessary, to the cerebral ganglia.

The nervous system of echinoderms is interesting because these animals are bilaterally symmetrical as larvae but metamorphose into radially symmetrical adults. Because of the radial symmetry, the nervous system shows little tendency to be concentrated in one place. In starfish, the nervous system has equally sized ganglia at the base of each arm joined by a nerve-fibre ring around the mouth.

The animal kingdom can be divided on the basis of the early development of its members into two large groups or 'superphyla'—the *protostomes* and the *deuterostomes* (Chapter 38). The layout of the nervous system in the two groups is fundamentally different. Protostomes, exemplified by arthropods, have a dorsal brain connected by a pair of connectives passing either side of the oesophagus to a suboesophageal ganglion and a ventral nerve cord with segmental ganglia. Deuterostomes, including chordates, have a dorsal brain and a dorsal segmentally ganglionated nerve cord. The parallel evolution of encephalisation and coalescence of local ganglia in the two groups suggests that the evolutionary advantages were considerable.

Although the basic properties of neurons are the same in all species, increasingly complex behaviour is accompanied by increases in nervous system complexity, including the formation of ganglia linked by nerve cords, and by encephalisation.

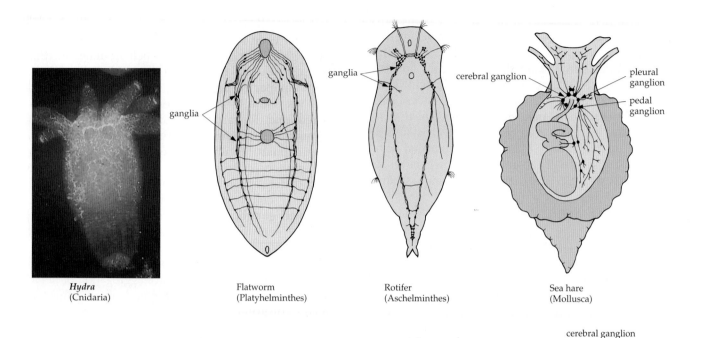

Hydra
(Cnidaria)

Flatworm
(Platyhelminthes)

Rotifer
(Aschelminthes)

Sea hare
(Mollusca)

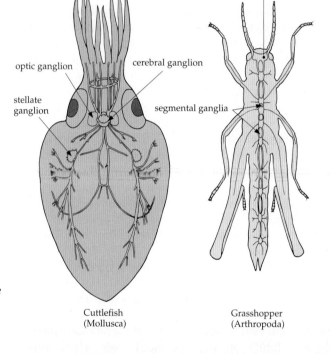

Cuttlefish
(Mollusca)

Grasshopper
(Arthropoda)

Fig. 26.7 Invertebrate nervous systems. These examples show the increasing complexity of nervous systems in more complicated animals and the formation of anterior aggregations of nerve cells (encephalisation). In coelenterates, illustrated here by *Hydra*, isolated neurons form a network without any significant aggregation of nerve cell bodies. In addition to a nerve net, platyhelminthes also have nerve cells gathered into longitudinal trunks with ladder-like cross connections, and concentrations at the front of the animal. In the rotifer, groupings of nerve cells in primitive ganglia can be seen. In the drawings of higher invertebrates, molluscs and arthropods, individual nerve cells are no longer represented. In these animals, tens of thousands of nerve cells are gathered in ganglia, in the head, viscera and foot of the sea hare, and in the head of the cuttlefish and grasshopper

Complex nervous systems: vertebrates

From simple to complex animals, the total number of neurons increases dramatically; there are about 300 in a rotifer, about 500 000 in an arthropod, 10^8 in some cephalopod molluscs and 10^{10} in mammals. This vast increase in numbers is associated with increased behavioural complexity and adaptability. Most neurons in vertebrates have their cell bodies located in the brain and spinal cord, which together constitute the **central nervous system**. As vertebrates evolved, the *brain* became larger in relation to the *spinal cord*.

In the most primitive chordates, such as the lancelet amphioxus (Chapter 40), the pattern of nervous system

organisation typical of vertebrates can be seen. In such species, the central nervous system develops by the infolding of the dorsal ectoderm and differentiation of the cells that surround the canal formed by this infolding (Chapter 16). Further differentiation occurs, particularly at the head end, resulting in the typical hollow dorsal central nervous system of vertebrates. The major regions of the vertebrate brain are present in primitive eel-like fish called cyclostomes, where the brain is divided into forebrain, midbrain and hindbrain, and is continuous with the spinal cord.

In addition to the central nervous system, vertebrates have numerous peripheral ganglia. Many of these ganglia are involved in control of the viscera, such as the heart and the digestive system. Signals are relayed between the central nervous system and the organs of the body via the **peripheral nervous system**.

> Vertebrates have a central nervous system, which contains the majority of their neurons, and a peripheral nervous system comprising numerous ganglia and connecting nerves.

The mammalian brain

The dominant features of mammalian brains are the external layers—the cerebral and cerebellar cortices (sing. cortex; Fig. 26.8). The *cerebral cortex* is part of the forebrain. The degree of folding of the cerebral cortex tends to be greater in larger animals. In humans, it is very highly developed and its growth outstrips the rest of the brain so much that it folds to maintain close apposition to the underlying brain stem. In mammals such as the rabbit (Fig. 26.9), rat or guinea pig, the cerebral cortex is smooth and may not completely obscure the upper brain stem from a lateral view. These species also tend to have prominent olfactory bulbs in relation to the rest of the brain.

The cerebral cortex is concerned with the control of movement and with what are known as higher nervous functions. In humans, these include learned behaviours, memory and recall, pattern recognition, emotional

responses, abstract thought and language. Many attempts have been made to localise these higher functions in the cerebral cortices; these have been only partly successful.

The major regions of the human cortex, and the functions these regions subserve, are shown in Figure 26.10. In some cases, lateralisation of cortical function occurs, meaning that one side of the brain predominates in controlling that function. Motor commands from the cortex are lateralised and crossed; that is, the cortex on the right side commands muscles on the left side and vice versa. A most remarkable lateralisation is that of human speech. If part of the left temporal lobe is injured, then speech can be entirely lost in most people, whereas any deficit caused by a comparable injury to the right side is much less obvious. Patients with speech loss due to left lobe injury can understand written or spoken speech perfectly well and can make correct word associations, as demonstrated by their ability to give written responses to questions.

The *cerebellar cortex* (or cerebellum) is a part of the hindbrain. It plays a crucial role in the co-ordination of balance and movement. The cerebellum is one of the brain regions most obviously susceptible to alcohol, whose earliest effects are thus on balance and eye–hand co-ordination.

Underlying the cerebral and cerebellar cortices is the *brain stem*, which is derived from structures that arose earlier in vertebrate evolution. It is composed of the remainder of the forebrain, the small midbrain and the hindbrain. The major regions of the brain stem are the *thalamus*, which is concerned with relaying and processing information related to locomotor activity between the cortex and lower centres; the *hypothalamus*, which is concerned with integrating visceral activities, such as appetite and body temperature control; the *pons*, which includes auditory processing circuits; and the *medulla*, which includes centres for control of specific viscera, such as the cardiovascular and digestive systems. Twelve pairs of *cranial nerves* emerge from the brain stem; through these, information is received from or passed to more peripheral sites.

The brain stem is continuous with the spinal cord, which lies within the vertebral column and has a very characteristic structure when viewed in cross-section (Fig. 26.11). Except for the head and some structures in the neck, the majority of motor commands pass out from the central nervous system via the spinal cord. Moreover, a lot of sensory information, including all sensations from the limbs and from the surface of the trunk, enters via the cord. Everything that occurs in the spinal cord is able to be controlled or modified by higher brain centres. This includes spinal reflexes, such as the tendon jerk reflex, which occurs when the tendon below the knee is struck.

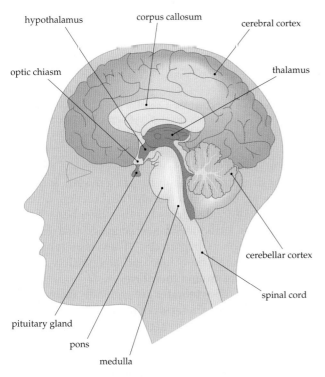

hypothalamus

corpus callosum

cerebral cortex

optic chiasm

thalamus

cerebellar cortex

spinal cord

pituitary gland

pons

medulla

Fig. 26.8 Major structural divisions of the mammalian brain, in this case, the human. The brain is drawn with its left side cut away, so the deeper structures are not hidden by the cerebral cortex. Its appearance with the cortex intact is shown in Figure 26.9

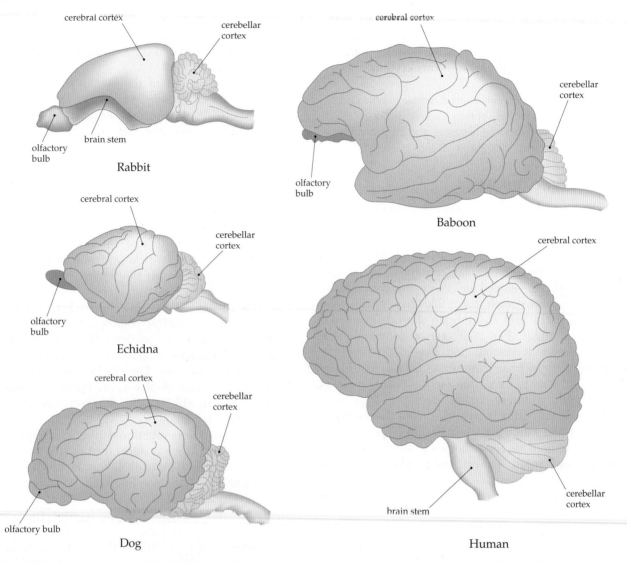

Fig. 26.9 Differences in the structure of mammalian brains seen in lateral view. In the simplest of mammalian brains, represented here by the rabbit, the cerebral cortex is smooth; it only partly obscures the upper brain stem and does not cover the cerebellar cortex. The olfactory bulb is prominent. As brains become more complex, convolutions of the cerebral cortex progressively develop, the upper brain stem and cerebellar cortex become enveloped by the cerebral cortex and the olfactory lobes become relatively less prominent. In humans, the angle of the lower brain stem changes to accommodate upright posture

The main regions of the mammalian brain are the cerebral and cerebellar cortices, the thalamus and hypothalamus, and the pons and medulla.

Higher functions of brains

All species that show behaviour involving complex computations, such as those required for learning and language, have large arrays of highly ordered neurons characterised by repeating patterns. The best-known examples outside the vertebrates are those found in insects (Fig. 26.12) and octopuses. The brains of such animals have fewer cells available for such computations than do those of mammals and birds, but they can nevertheless generate some surprisingly sophisticated behaviour. Octopuses, for example, can learn to

distinguish abstract symbols to gain a reward or avoid a punishment. Bees communicate the direction, distance and quality of food sources to other members of their hive using an elaborate dance language that encodes the information (Chapter 28). They also remember landmarks around their hive and are possibly able to map the locality around their hive.

The brains of vertebrates are generally many orders of magnitude larger in terms of neuronal numbers, and the complexity of their activity reflects this. Their higher level functions can be divided into two groups: those of thought (cognition) and those of emotion (affect). In line with this, psychiatric conditions are divided into affective and cognitive disorders. The cerebral cortices have major roles in all higher nervous functions. These functions can be localised to specific

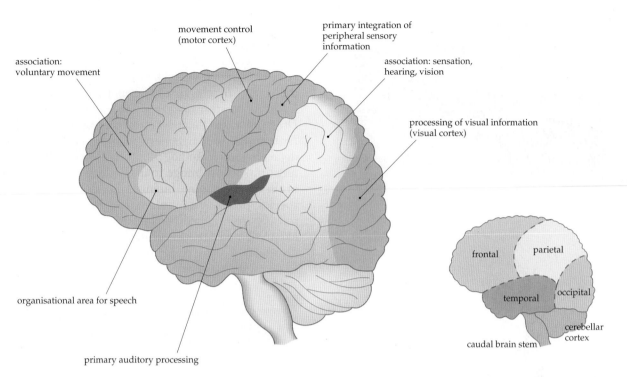

Fig. 26.10 Lateral view of the human brain, indicating the positions of some of the major functional regions of the cerebral cortex and the four cortical lobes

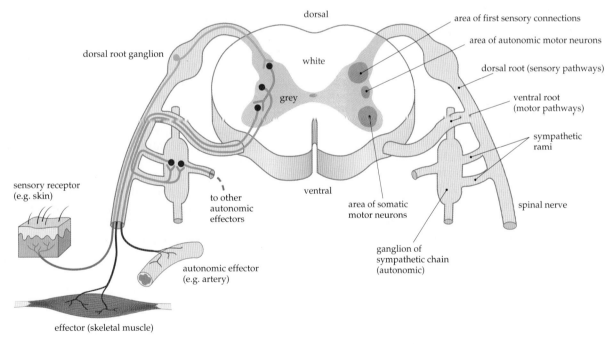

Fig. 26.11 Representation of a cross-section through the mammalian spinal cord. The spinal cord has a central grey region, consisting of nerve cell bodies, their dendrites, initial parts of axons and synaptic inputs, and a white region, consisting of axons that pass up and down the cord. At each vertebral level, bundles of axons, the dorsal and ventral roots, connect with the spinal cord. These join to form spinal (segmental) nerves, small branches from which join the sympathetic chains at thoracic and upper lumbar levels

cortical areas, or groups of areas, and can be altered by drugs that modify synaptic transmission. It is a tenet of modern neurobiology that what we refer to as 'mind' is the outcome of the functioning and interaction of neural circuits of the cortex. Examples of cognitive functions include the comprehension and formulation of speech, learning and memory, anticipation of future consequences of existing conditions, and abstract thought. The formulation of speech is localised in an area of the left frontal lobe, known as Broca's area,

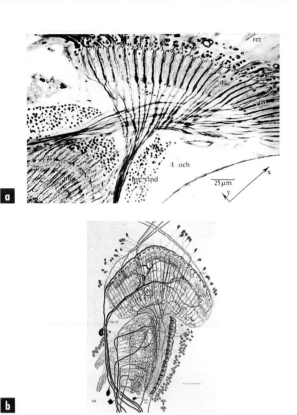

Fig. 26.12 Highly ordered patterns of neuronal organisation are seen in animals capable of complex behaviours, such as learning and language. **(a)** Photograph and **(b)** composite drawing showing the complex organisation of nerve cells in the visual system of the insect brain

whereas the site of speech comprehension is in the temporal lobe. Certain emotions are also centred in the temporal lobe, and patients with temporal lobe epilepsy may have disorders of affect. The older parts of the brain, in evolutionary terms, for example, the thalamus and the pons, are also important in storing and implementing emotional responses. In the immensely complex cortex, with its multiple inter-connections between areas, regions associated with higher functions often have parallel and serial inter-connections. This has made it particularly difficult to define precisely which areas are associated with particular neural processes.

Higher nervous system functions include thought (cognition) and emotion (affect).

Neural circuits

In order to understand how nervous systems produce co-ordinated behaviour, it is necessary to unravel individual neural circuits. This is usually done either by tracing a circuit backwards from the motor side or by following the path forward from the sensory side.

Extensive studies of this kind have been carried out in invertebrates, where the smaller number of neurons makes access and interpretation easier. The nervous systems of molluscs, such as the sea hare, *Aplysia*, and the snail, *Helix*, and arthropods, such as locusts, *Locusta*, crayfish, *Procambarus*, and the vinegar fly, *Drosophila*, are the most often investigated (Fig. 26.7). *Drosophila* is particularly important because our detailed understanding of its genetics and develop-ment provides a bridge to understanding these aspects of nervous system biology. In many invertebrate species, precisely equivalent neurons can be identified reliably from animal to animal and the physiological, chemical and structural attributes of these neurons investigated (Box 26.5). Identifiable neurons provide

BOX 26.5 Rapid escape responses

Invertebrates and vertebrates have evolved rapidly responding nerve networks for avoiding predators. If you have ever tried to catch a crayfish (yabby) in a net you will know that they usually try to escape by flicking their tails and propelling themselves away rapidly. Neurobiologists have discovered that this escape response is controlled by two sets of giant neurons with axons running the entire length of the nerve cord (Fig. a). They are termed lateral or medial giant axons according to their position in the nerve cord. The medial giant axons are activated by abrupt visual or touch stimuli to the animal's head. They conduct action potentials rapidly (15–20 m s^{-1}) backwards along the body and directly stimulate neurons activating a selected group of muscles in the legs and tail. The muscles in the legs cause them to flex in alongside the trunk, reducing the drag on the body as it moves through the water, and the tail muscles produce a strong abdominal flexion that propels the animal backwards (Fig. a). The lateral giant fibres function similarly but with distinct and interesting differences. They are activated by mechanical stimuli from behind, particularly abrupt waterborne vibrations, such as occur when something lunges at the crayfish from behind. They conduct action potentials forwards to a different set of muscles that cause the animal to somersault away from the stimulus (Fig. a). In this case, the leg muscles are not activated; their position does not affect the efficiency of the somersault.

Vertebrates also have large neurons with functions dedicated to escape. Some fishes have a pair of giant cells, called Mauthner neurons, after the biologist who first described them (Fig. b). These cells are activated by a mechanical stimulus to the side of the fish; which one of the pair is stimulated is determined by the site of the stimulus. Water movements with an intensity equivalent to those associated with the predatory strike of another fish are sufficient to trigger the cells. When a Mauthner cell fires, it initiates a series of complicated movements involving the trunk, fin, eye, jaw and opercula muscles. Because the axon of each Mauthner neuron crosses over to the other side of the body, the initial body contraction is on the side away from the stimulus so that the fish's body rotates between 30° and 100° about its centre of mass in 15 milliseconds. This rapid turn away from the stimulus positions the fish for a powerful acceleration out of the area (Fig. b).

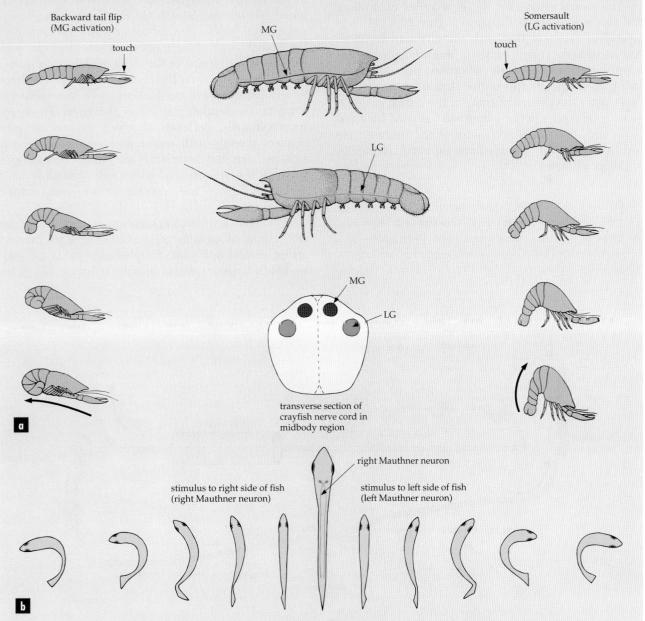

(a) There are two sets of giant neurons in the crayfish nerve cord: a medial pair and a lateral pair called the medial giant (MG) axons and lateral giant (LG) axons respectively. The MG axons are activated by a stimulus to the anterior end of the animal and cause abdominal contractions that cause the animal to move backwards. The LG axons are activated by a sharp stimulus to the tail and activate muscles that cause a somersault. The silhouettes of the animals are shown at 10 ms intervals. (b) Giant-sized Mauthner neurons in many varieties of fishes are responsible for an abrupt body turn away from a lateral stimulus. Body profiles here are shown at approximately 5 ms intervals

the opportunity to examine general principles of circuit operation, which aids in establishing guidelines for the vastly more difficult task of analysing vertebrate circuits.

Vertebrate neural circuits often consist of thousands or millions of neurons. Even simple reflex circuits usually involve far more neurons than could be included in a diagrammatic representation. In order to explain such circuits, stylised diagrams are drawn in which large numbers of neurons are represented by single morphologically simplified neurons.

The neural circuit for the light reflex controlling the mammalian iris is shown in Figure 26.13. This is typical of the way that such circuits are depicted. Approximately one billion photoreceptors are represented by a single photoreceptor and the integration of photoreceptor information, which determines the pattern of firing of the retinal ganglion cells, is ignored. On the output side, several hundred pre- and postganglionic autonomic neurons are involved to provide a graded and eucentric change in pupillary diameter.

> In many invertebrates, specific neurons can be identified reliably and the physiological, chemical and structural attributes of these neurons investigated. Vertebrate neural circuits often consist of thousands or millions of neurons.

Functional divisions of nervous systems

Motor control systems

Motor or somatic control systems are concerned with the control of posture and movement and direct all movements under voluntary control. Movement can be initiated by the central nervous system or by reflexes via peripheral sensory inputs. The system is more or less hierarchically arranged. At the lowest end of the hierarchy are simple reflexes that pass through segmental ganglia of invertebrates or restricted regions of the spinal cord of vertebrates. These include such reflexes as the closing of the claws in crabs in response to stroking of hairs on the chelae, and tendon reflexes in vertebrates that cause the leg to 'jerk' in response to a tap to the tendon just below the knee. These are **monosynaptic reflexes** in which sensory neurons connect directly with motor neurons. Monosynaptic reflexes are not common and most connections between the sensory and motor side, including most spinal reflexes, are disynaptic or polysynaptic (Fig. 26.11).

The next level of organisation is responsible for integration of movements involving sets of muscles acting around one joint. Simple hinge joints are controlled by opposing sets of muscles (Chapter 27), *flexor*

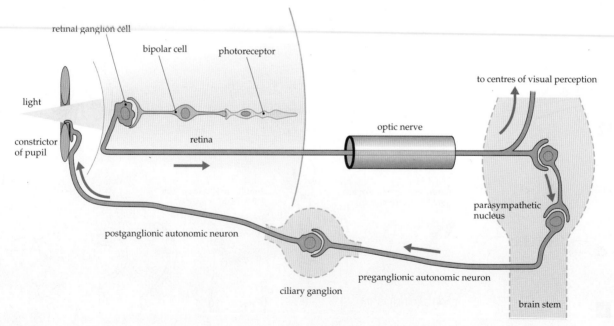

Fig. 26.13 Representation of a neuronal circuit: the light (pupilloconstrictor) circuit. Neuronal circuits are usually very much simplified when represented in diagrams. In reality, wherever there are neuroneuronal or neuroeffector junctions (six in this circuit), there is integration of information. Wherever a single neuron is drawn, it represents many parallel pathways. The light reflex is an example of a simple negative feedback circuit; if the amount of light falling on the retina is increased, the pupil constricts, and if the light level drops, the pupil will dilate. Neurotoxins that block transmission from the final pupilloconstrictor neurons cause the pupil to enlarge. Neurons are usually represented schematically, as in this diagram, as a round cell body without depicting dendrites, a line for the axon and a Y-shaped structure at the synapse

muscles that bend the joints and *extensor* muscles that straighten them. Most arthropod joints are of this type so that the muscles are capable of producing only one sort of movement and the circuitry is relatively simple. Vertebrate joints are usually more complicated because the muscles can produce different actions depending on which other muscles are active at the same time and the integration required is necessarily more complicated. An example is the shoulder joint, which can extend in several directions and rotate.

Limb movements involve co-ordination of several or many joints. Integration for this is more complicated again. It is still carried out in local ganglia or spinal segments but more of them interact and there is direction from the cerebral ganglion or brain. When you lift an object by bending the elbow, for example, integrated muscle action is needed. In this case, both flexor and extensor muscles of the wrist are contracted so that the wrist is held firm, the elbow flexors are contracted and the elbow extensor muscles are inhibited from contracting. More and more local centres are activated and co-ordinated as other limbs become involved. In both insects and cats, for example, reflexes producing strong extension in one leg can produce flexion in an adjacent one. If you step on a nail, there will be a reflex withdrawal of the injured foot and, at the same time, a reflex extension of your other leg, bracing it to take all your weight.

In both vertebrates and invertebrates, parts of rhythmically co-ordinated patterns, such as those involved in breathing, walking, flying and swimming, can be produced by neural circuits in the absence of input from the cerebral cortex or cerebral ganglion, although the decision about whether to turn them on or off may come from these higher centres. Flight is one of the most complicated motor behaviours. The basic flight rhythm is produced by rhythmic generators but their output is constantly modified by inputs from other centres, such as those co-ordinating wind- and ground-speed detection, gravity, and balance in the yaw, pitch and roll planes (Chapter 27).

Co-ordination involving consciously driven patterns of movement, such as those used to hit the keys on the piano with correct order, force and interval in response to the music sheet in front of your eyes, require the higest levels of and most widely distributed neuronal interactions. This is because inputs from higher brain centres and sensors in different areas all over the body have immediate and constantly changing involvement. For production of the optimal output, adjustments are being made continuously at a number of levels.

Motor control systems are hierarchically organised. They are concerned with posture and movement, and direct all movements under voluntary control.

Sensory systems

The environments of the earth contain energy of many different kinds. Most of it is of little relevance to animals (e.g. radiowaves). Animals have sensory receptors that enable them to respond selectively to those aspects of their environment that impact on their ability to survive and reproduce. The types of sensory receptors present, and the sensitivity, discrimination and range of these receptors, differ substantially between animals. For example, a wombat has a lower visual ability to distinguish small objects (visual acuity) than a kookaburra; dogs use chemical scents extensively for communication and have far greater olfactory discrimination than humans; some moths have chemoreceptors that can detect one molecule of pheromone; some bats emit and can hear ultrasonic sounds beyond the range of most animals, except the moths they hunt with ultrasound (Box 26.6). Many animals respond to parts of the energy spectrum undetectable to humans. For example, a number of animals, including snakes, can detect and use infra-red radiation, bees see ultraviolet light, some fishes detect weak electric currents, and sensitivity to magnetic fields occurs in a wide range of animals, including birds and insects (Chapter 7).

The key elements of sensory reception are receptors specialised for detecting particular forms (modalities) of environmental energy: parts of the electromagnetic spectrum (light, heat), mechanical displacement (movement, sound) and chemicals (odours, tastes). Sensory receptors distributed widely over the surface or within the body are the basis for the *general senses* (e.g. touch, pain and joint position). Aggregations of receptors into specialised organs provide for the *special senses* (e.g. sight and hearing). Those receptors that detect internal states, such as blood (or fluid) pressure and fluid chemistry (e.g. oxygen and carbon dioxide tension) are known as *visceral receptors* or enteroreceptors.

The special senses are characterised by receptor organs in which the neural receptor cells are associated with each other and with accessory cells of various types that assist in the collection and primary processing of a specific stimulus energy (Fig. 26.14). In the case of light, photoreceptor cells that contain light-sensitive pigments (Chapter 7) may be associated with lenses and pigment cells controlling the light pathway. In the case of chemoreception, cells sensitive to different chemicals may be grouped and the groups supported inside structures that maximise the probability of contact. For sound, the vibration-sensitive neurons may be attached to larger vibrating structures that select, filter and sometimes amplify the frequencies that are important to the animal.

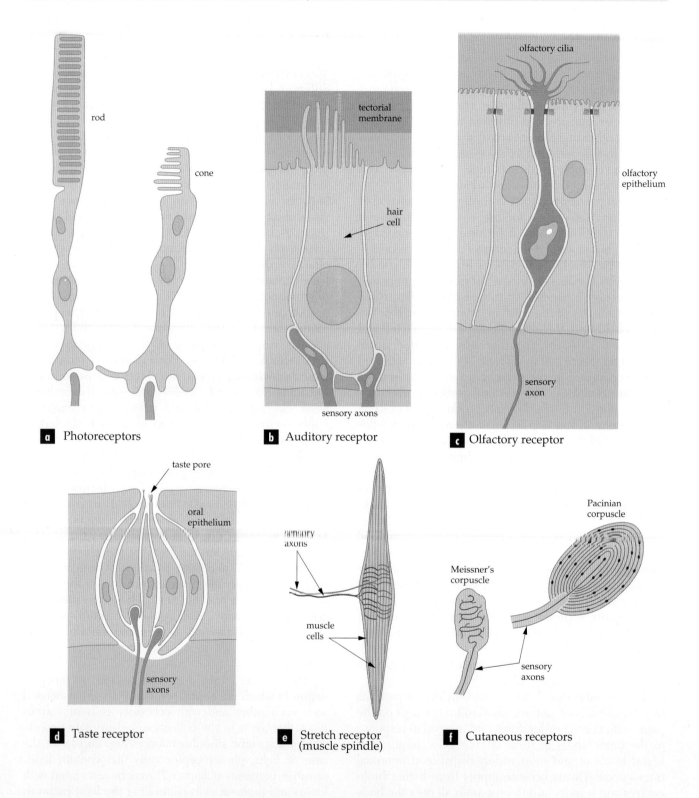

a Photoreceptors

b Auditory receptor

c Olfactory receptor

d Taste receptor

e Stretch receptor (muscle spindle)

f Cutaneous receptors

Fig. 26.14 Examples of sensory receptors. **(a)** Photoreceptors are elongated cells with disc-like structures at one end. Within these discs are light-sensitive pigments. The interaction of light with the pigment causes an electrical change, causing the release of a chemical transmitter to activate neurons in the retina (see also Chapter 7). **(b)** The receptors for sound are hair cells where cilia are embedded in the tectorial membrane, which sound waves move relative to the hair cell. **(c)** Olfactory neurons have chemosensitive cilia that penetrate the mucus of the nasal passages. **(d)** Taste buds lie in the oral epithelium, principally of the tongue. They contain chemosensitive cells that form excitatory synapses with taste sensory axons. **(e)** Endings of axons wrap muscle fibres to form the muscle spindle. **(f)** The ends of axons of the two cutaneous axons that are illustrated here are encapsulated. Mechanical distortion of the stretch receptor or the cutaneous corpuscles causes their axons to be excited

BOX 26.6 Bats and moths

In 1793 an Italian biologist, Lazzaro Spallanzani, showed that owls cannot avoid objects when flying in complete darkness whereas some bats can. We now know that these bats emit ultrasonic sound pulses (frequencies that may vary from 10 kHz to 100 kHz in different bat species) while flying, and detect the extremely faint echoes that return from objects in the immediate vicinity. Bats that can do this have ears specialised for detecting the direction of the returning echoes (Fig. a) and auditory pathways in the nervous system specialised for selecting the frequencies necessary for echo analysis. The bat determines the distance to an object by measuring the time taken for the echo to return; echoes from distant objects take longer to arrive than do those from close ones. The echo detection system is sufficiently accurate that the bats can use it to hunt flying insects that make up their diet, and much of the investigative work has been done on bats trained to catch mealworms tossed into the air.

Observations on some of the insects hunted by insectivorous bats show that they have developed specialisations for avoiding bat predation. Some moths in the family Noctuidae (Fig. b) have ears on either side of the thorax that are tuned to the bat emission frequency. When noctuid moths detect the sound of a hunting bat, they exhibit characteristic avoidance behaviour. The details have been analysed by playing bat calls over loudspeakers set up in the open and observing the behaviour of free-flying moths to the onset of the artificial bat calls (Fig. c). The experimenter can control the amplitude of the signal and also decide when to activate it. When the signal is faint, equivalent to a distant bat, the moth turns so that the source of the sound is directly behind it and flies away. If the signal is loud, representing a bat close by, the moth flutters erratically to the ground, or folds its wings and drops vertically (Fig. c).

(b) The noctuid moth, *Heliothis armigera*, which forms part of the bat's diet

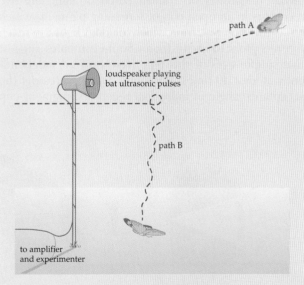

(c) Experiment showing that moths detect bat ultrasound and respond to it by altering their behaviour. Bat calls were played over a speaker set up at night in a clearing. A spotlight at the edge of the clearing allowed the experimenter to see moths approaching the speaker. A low-intensity sound, equivalent to a distant bat, caused moths to fly out of the area (path A). A high-intensity sound, equivalent to a bat close by, caused moths to take rapid evasive action (path B)

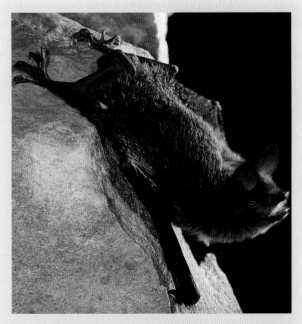

(a) The mouse-eared bat, *Myotis adversus*, has specialised ears for detecting faint echoes from ultrasonic pulses emitted when the bat is hunting

A series of experiments of this type have shown that the point at which a moth's behaviour switches from steady flight out of the area to rapid evasive action is roughly equivalent to the point at which a bat can detect the presence of the moth. This suggests that at least some aspects of moth hearing evolved specifically in response to the selective evolutionary pressure imposed by hunting bats.

> Sensory receptors monitor conditions in an animal's internal and external environment and provide information that enhances the animal's ability to survive and reproduce.

Vision

There is evidence that light-sensitive pigments appeared very early in animal evolution because the photosensitive pigments, called opsins, are common to all light-detecting animals. Light is detected when it interacts with visual pigments that respond by changing their conformation; the conformational change is used to trigger a cascade of events (Chapter 7) that eventually results in an electrical signal and the particular reaction of that species to light.

The most primitive type of eye, an eyespot or pigment-cup eye, is found in almost all the major animal phyla. It consists of a small patch of light-sensitive receptors in an open cup of screening pigment, which limits the direction from which incident light can reach the receptors to excite them. Some estimates suggest that simple eyes of this type have evolved independently at least 40 times. Without a device to restrict the acceptance angle of individual receptors, pigment-cup eyes do not resolve images well enough for pattern recognition. The development of an optical system that focuses the light arose much later in evolution and in fewer groups of animals. About 10 different ways of forming an image have evolved and these fall into one of two categories: *simple eyes*, single-chambered eyes constructed on the same principle as a camera, and *compound eyes*, in which a number of lens systems form images that are then combined physically or by neural integration to provide pattern information (Fig. 26.15). The most elementary simple eye is found in the cephalopod mollusc *Nautilus*, which has a pinhole instead of a lens and no cornea. All other simple eyes, including our own, use a lens, a cornea or both to form an image on the photoreceptors of the retina. The commonest and probably most primitive form of compound eye is the *apposition eye*. Each lens unit of the eye, called an ommatidium, forms an image onto a photopigmented area. Sections of the image are taken through layers of neurons that reconstruct one image for the whole visual field. Thus, the final image is constructed by the nervous system from 'pixels' contributed by a number of ommatidia.

Interesting visual specialisations are encountered. For example, a number of animals, including cats and some deep-sea fishes, have surfaces at the back of the eye that reflect light back into the photoreceptor layer after it has passed through it once. This increases the light sensitivity of the eye but at the expense of resolution. Eagles and jumping spiders, both active hunting species, have good visual acuity for sighting prey. Jumping spiders have long tubular principle eyes whose operation has been compared to a telephoto lens. Bees and some birds can detect ultraviolet light, invisible to most animals, and ultraviolet lines are an important component of the colour patterns on the petals of

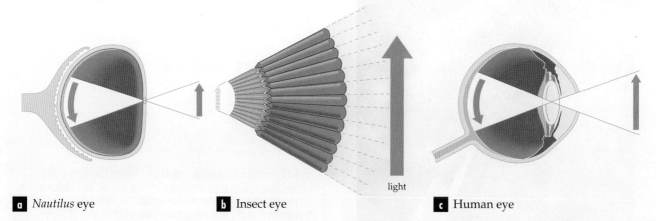

| **a** *Nautilus* eye | **b** Insect eye | **c** Human eye |

Fig. 26.15 Comparison of mechanisms of visual detection. **(a)** The mollusc, *Nautilus*, has an eye cup, open to the sea through a small hole. The eye acts like a pinhole camera. It thus has the ability to form an image, but suffers from a poor light-gathering ability. **(b)** The compound eye of the insect has thousands of individual detectors, like small eyes. Each of these provides an input to the nervous system. These eyes sacrifice resolution: the resolution can be no better than the area of each unit. **(c)** The vertebrate eye, in this case a human eye, has a focusing lens and a fine-grained retina, giving greater light sensitivity than *Nautilus* and better resolution than the insects. All these groups use the same chemical, *cis*-retinal, to detect photons of light

bee-pollinated flowers (Chapter 37). A number of animals, including birds and arthropods, can detect the polarisation of the light coming from the sky, which can be used as a navigational clue.

Hearing

Hearing is a specialised form of *mechanoreception*. Sound travels through air, water and solids as pressure waves (small compressions and rarefactions). Animals detect these with mechanoreceptors specialised for detecting and magnifying minute vibrations. Naked hairs, such as those on the cerci of insects such as cockroaches, can detect air currents. Very sensitive ones can detect airborne sound vibrations but only at low frequencies. Hearing organs that detect a range of frequencies have mechanical devices that amplify the movements generated by those frequencies useful to the animal to the point where mechanoreceptors can detect them. The commonest mechanical amplifier for airborne sounds is a membrane or tympanum. The tympanal organs of insects are found on the legs, body or wings, while those of vertebrates are found in the ears.

In most terrestrial vertebrates, vibrations of the membrane are further amplified by bony levers and transmitted to a fluid-filled canal (Fig. 26.16). Within the canal are hair cells that have evolved from those that detect water movements around fish and which are sensitive to waterborne vibrations (Chapter 7).

Chemoreception

Chemoreception is the detection of specific chemicals in the environment. It occurs throughout the animal kingdom and is thought to provide the basis for the evolution of inter- and intraspecific communication between simple organisms. Aquatic animals commonly have chemoreceptors over the body surface, with concentrations in some places for specific purposes, such as food or mate detection. Some chemoreceptors are extremely sensitive; catfish can detect some amino acids at concentrations less than one part in a million. Terrestrial animals have specialised organs for olfaction (smell), detecting airborne chemicals, and taste, detecting those derived from food. In addition to receptors for common amino acids and sugars, which would be useful to most animals searching for food, many animals have evolved receptors for specific chemicals emitted by prey or predators or pheromone chemicals released by others of their species signalling states such as sexual readiness. Receptors in the antenna of the male silk moth, for example, can detect single molecules of the appropriate sexual pheromone released by the female.

Mechanoreception

A range of receptors inform animals about their mechanical interactions with their environment and provide information to the brain about such things as

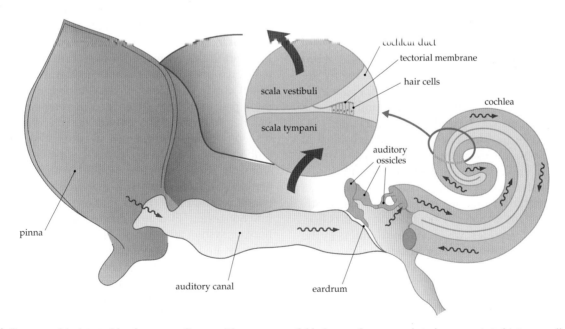

Fig. 26.16 How sound is detected by the mammalian ear. The energy available in soundwaves needs to be concentrated into a smaller area so that it is sufficient to excite the hair cells. This is done by an external ear (pinna) that funnels soundwaves into the auditory canal at the end of which they cause the eardrum to vibrate. Via a series of small bones (the auditory ossicles: the hammer, the anvil and the stirrup), which act as levers, the vibration is conducted to a fluid-filled canal, the cochlea, in which sit rows of hair cells (inset). The intrinsic frequency of vibration of the membrane on which the hair cells sit varies along the cochlea. This allows different tones to be distinguished. The inset is a diagram of the cochlea in cross-section. Soundwaves run up the scala tympani and down the scala vestibuli, as indicated by the arrows. In so doing, they vibrate the membranes within the cochlea and distort the hair cells. This distortion causes a chemical to be released from the hair cell, which excites the adjacent sensory nerve ending

joint position and tension in the walls of viscera, such as the lungs or stomach. Cutaneous mechanoreceptors include stress detectors in animals with chitinous exoskeletons, detectors of pressure in animals with elastic integuments and detectors that monitor the deflection of hairs. Some mechanoreceptors have specialised transducer structures, such as sensory corpuscles or scolopales at the end of the sensory axon (Fig. 26.14f), whereas in others, the **transduction** of movement to an electrical signal occurs in the endings of the axonal membrane.

Pain

All animals with nervous systems avoid encounters with noxious external stimuli. However, it is not clear how complicated a nervous system has to be before an animal experiences something akin to pain felt by humans. Vertebrates that vocalise in response to stimuli that humans interpret as painful certainly behave as if their experience is similar. At present we have no way of answering questions of the general type 'Do worms feel pain?'.

In humans, painful sensations generally come from the body surfaces, which are supplied by three types of pain receptors. Mechanical pain receptors respond to strong stimuli, most effectively from sharp objects. Heat pain receptors respond when the skin is heated to 45°C or more. Polymodal pain receptors respond to noxious mechanical, heat or chemical stimuli. For all pain sensations, the nerve endings are believed to be activated by chemicals released from the damaged or irritated tissue. The signal is carried to the spinal cord or brain stem by the axons of sensory neurons. A series of relay neurons then carry the information to the cerebral cortex, which must be functional for pain to be perceived. Psychological and behavioural influences on pain are pronounced. Individuals show marked differences in pain threshold and the same individual in different circumstances can experience pain differently. A striking example is the suppression of pain during vigorous physical activity or in a crisis situation.

Drugs related to opium suppress pain, apparently by mimicking the actions of peptides, known collectively as opioid peptides, which are transmitters in the central nervous system. The best known of these transmitters are enkephalin and endorphin.

Visceral control

Organs whose functions are not consciously controlled are the visceral organs, including the heart, stomach, spleen and pancreas, blood vessels, specialised glands, such as sweat glands, and smooth muscles of the eye. Complex animals, both vertebrate and invertebrate, have a specialised neural network for this purpose, which is termed the visceral or autonomic nervous system (Fig. 26.17). It integrates closely with the endocrine system (Chapter 25) to co-ordinate the involuntary or regulatory physiological functions of the body and to stabilise the internal environment. The major functions influenced by the **autonomic nervous systems** are listed in Box 26.7.

> The autonomic nervous system innervates the visceral organs of the body; their functions are not consciously controlled.

The vertebrate autonomic nervous system

The vertebrate autonomic nervous system consists of a central portion within the brain stem and spinal cord, and a peripheral part comprising ganglia and connecting nerves. Autonomic axons that leave the central nervous system, with the exception of those innervating the adrenal gland, make synaptic connections with nerve cell bodies that are grouped in peripheral ganglia. Autonomic neurons whose axons project to peripheral ganglia from cell bodies in the central nervous system are known as *preganglionic neurons* and the neurons with which they connect are *postganglionic neurons*. Autonomic ganglia are small centres for integration of neuronal information. Their output consists of patterns of action potentials in axons that supply smooth muscle, cardiac muscle, glands or other autonomic neurons.

The peripheral nerves and ganglia are classified into three subsystems, known as the **sympathetic**, **parasympathetic** and **enteric** divisions of the autonomic nervous system (Fig. 26.17). The sympathetic division is that part of the autonomic nervous system whose motor pathways emerge from thoracic and lumbar parts of the spinal cord. The parasympathetic division consists of those pathways arising from the brain stem and from the sacral spinal cord. The enteric nervous system is the division of the autonomic nervous system that is embedded in the walls of digestive organs, notably the stomach, small intestine, colon, pancreas and gall bladder. The enteric nervous system is unique in that it contains complete reflex circuits and thus it can function when all neural connections with the central nervous system are severed. The enteric nervous system is innervated by the sympathetic and parasympathetic divisions of the autonomic nervous system.

Many autonomic pathways are tonically active, which means that some groups of autonomic neurons fire action potentials continually and therefore control the level of activity of the organ they innervate. This is rather like having the accelerator of a car partly on; by the use of a single controlling device, a greater or lesser activity can be obtained. Sympathetic nerves control the diameter of arteries in this way. If you are sitting at rest at a comfortable temperature, the arteries to your skin will be slightly constricted due to sympathetic

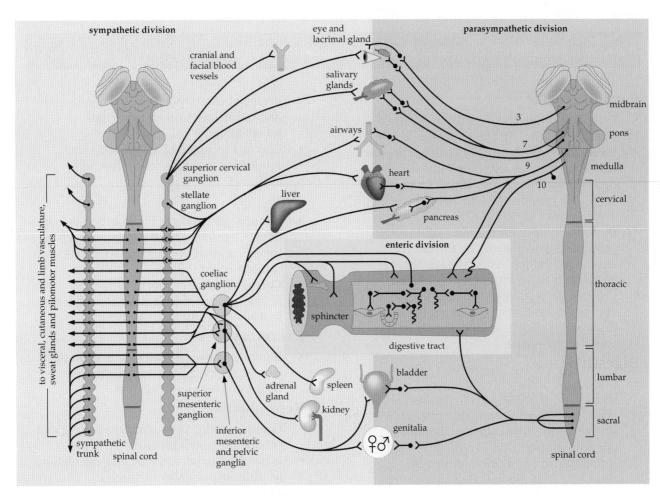

Fig. 26.17 Diagrammatic representation of the autonomic nervous system. This system has three divisions through which organs of the thoracic, abdominal and pelvic cavities, glands of the head and skin, peripheral blood vessels and ocular smooth muscles are controlled. For clarity of presentation, the brain stem and spinal cord are represented twice, on the left to show the sympathetic outflows and on the right to show parasympathetic pathways. The enteric division of the autonomic nervous system is embedded in the wall of the digestive tract. Enteric reflexes are modulated, and sometimes overridden, by inputs from the sympathetic or parasympathetic divisions. The cranial nerves that include autonomic motor pathways are numbered (3, oculomotor; 7, facial; 9, glossopharyngeal; and 10, vagus). Visceral sensory neurons connecting with the digestive tract and within the enteric division of the system are marked by asterisks. There are many other visceral sensory neurons that are not included in this diagram

vasoconstrictor neurons being active. If the room becomes warmer, these vasoconstrictor neurons will decrease activity, the arteries will dilate and the increased flow of warm blood will lose heat to the surrounding environment. This heat loss can be enhanced by evaporative cooling due to increased sweat production, also caused by changed autonomic activity. Conversely, if you become cold, sympathetic neurons to skin arteries become more active, constricting the vessels and reducing blood flow to the skin. Vascular changes of other organs are similarly controlled.

Other examples of tonic activity are the vagal (parasympathetic) neurons to the heart, which in humans maintain the heart rate below its intrinsic or 'free-running' level (Chapter 21), and the parasympathetic neurons that keep the pupil partly constricted at normal daylight levels (Fig. 26.13).

Together with the endocrine system, the autonomic nervous system regulates the activities of internal organs to produce a stable internal environment.

BOX 26.7 Functions of the autonomic nervous system

Cardiovascular

- Controls the rate and strength of the heart beat of the heart. The heart muscle of many invertebrates, such as crabs, requires specific excitatory nervous stimuli to initiate each contraction. A small ganglion of around nine to 15 cells controls the rate and order of contraction of the muscle fibres of the crab heart. In vertebrates, the heart beat is generated spontaneously by the cardiac muscle but the frequency and strength of beat is influenced by autonomic nerves.
- Controls the relative distribution of blood flow to different organs by changing the diameters of their arteries of supply at key points in the system.

Digestive

- Controls mixing and propulsion of food through the gut. Some invertebrates, such as crabs, have complex stomachs that grind and filter food. The complex co-ordination of the dozens of small muscles required for this is done by an aggregation of some 30 neurons, called the stomatogastric ganglion. The strength and frequency of the mixing and propulsive movements of the muscle layers in the digestive tract of vertebrates is controlled by autonomic nerves.
- Controls digestive secretions into various parts of the digestive tract and transport of electrolytes, hence controlling water balance.
- Controls secretion from the salivary glands.

Respiratory

- Controls variation in diameter of major airways of lungs in vertebrates, and the degree of contraction or relaxation in gills and respiratory trees in fishes and invertebrates.
- Controls secretion of mucus and other lubricating materials over respiratory surfaces.

Eye

- Controls mechanisms for adjusting the light intensity reaching photoreceptors, such as the diameter of the pupil in vertebrate and octopus eyes.
- Controls lubricatory secretions.
- Is involved in focusing mechanisms, such as movement or distortion of lenses.

Excretory

- Controls emptying contractions of waste storage reservoirs, such as bladders, and filtration pressures in kidneys.

Reproductive

- Controls contractions of organs storing and conducting eggs, sperm and embryos.
- Is involved in intromission and ejaculation in males of a number of species.

Metabolic

- Controls the formation and release of hormones affecting the overall metabolism of the organism.

Temperature

- Controls systems used to cool or warm animals, such as cutaneous blood flow and sweating in vertebrates and spiracular opening in insects.

Summary

- The plasma membranes of almost all cells show the basic physical properties and ionic selectivity that are the basis for evolution of electrical signalling within a specialised type of cells, neurons.
- Nervous systems are made up of networks of neurons and are capable of providing precise and rapid co-ordination of cellular function, movement and behaviour.
- Sensory neurons convert other forms of energy to electrical signals, allowing the nervous system to monitor changes in the external and internal environments. Interneurons pass information from neuron to neuron and aid in the integration of responses. Motor neurons influence the excitability of tissues, such as muscles or glands.
- Conduction along the processes of neurons is electrical; electrical signals can be conducted

passively or actively over short distances but must be actively conducted over longer distances. Regenerating active responses that are conducted over long distances are called action potentials.
- Transmission from neurons to other cells is usually chemical but, in some cases, is electrical.
- The same range of chemical transmitters is found across the animal kingdom although different phyla and classes have characteristic groupings of these.
- The more complex and adaptable the behaviour of an organism, the more extensive and complex is its nervous system.
- The motor outputs of the vertebrate nervous system occur by way of two subsystems: the somatic system, which is under voluntary control, and the autonomic system, which automatically adjusts the activity of internal organs.

key terms

action potential (p. 685)
active response (p. 684)
autonomic nervous system (p. 704)
axon (p. 680)
axon hillock (p. 689)
central nervous system (p. 692)
cerebral ganglia (p. 691)
chemical synapse (p. 686)
conduction (p. 686)
dendrite (p. 680)
depolarised (p. 687)
electrical synapse (p. 687)

encephalisation (p. 691)
enteric nervous system (p. 704)
enteroreceptor (p. 680)
epsp (p. 688)
exteroreceptor (p. 680)
glial cell (p. 680)
hyperpolarised (p. 687)
integration (p. 688)
interneuron (p. 680)
ipsp (p. 688)
monosynaptic reflex (p. 698)
motor neuron (p. 680)
myelin (p. 686)

neuron (p. 680)
neurotransmitter (p. 687)
node of Ranvier (p. 686)
parasympathetic nervous system (p. 704)
peripheral nervous system (p. 693)
refractory period (p. 686)
resting membrane potential (p. 684)
saltatory conduction (p. 686)
sensory neuron (p. 680)
soma (p. 680)

space constant (p. 682)
sympathetic nervous system (p. 704)
synapse (p. 686)
synaptic vesicle (p. 687)
threshold potential (p. 684)
time constant (p. 682)
transduction (p. 704)

Review questions

1. Describe the basis of the resting membrane potential.
2. How is an action potential transmitted along an axon?
3. What are the differences between an electrical synapse and a chemical synapse?
4. How do synaptic transmitters excite or inhibit post-synaptic cells?
5. What are the main changes we observe during evolution of more complex nervous systems and why are they thought to have occurred?
6. What are the two divisions of the autonomic nervous system? List some of the ways in which some body functions are affected by each.
7. What do we mean by the higher functions of the brain? In what way are animals that evolve higher functions advantaged?

Extension questions

1. Why are are the time constant and space constant of a neuron so important for its integrative properties?
2. Motor control can occur at different levels in the nervous system depending on the complexity of the behaviour produced. Compare what is involved in a reflex, such as withdrawal from a hot surface, with the co-ordination necessary to catch a ball.
3. Give examples of some of the ways in which animal and plant toxins have evolved to act on the nervous systems of prey or predators.

Suggested further reading

Delcomyn, F. (1998). *Foundations of Neurobiology*. New York: Freeman.

A readable introductory text to nervous system function that uses examples from a wide range of vertebrate and invertebrate animals.

Simmons, P. and Young, D. (1999). *Nerve Cells and Insect Behaviour*. 2nd edn. Cambridge: Cambridge University Press.

A good introduction to the way in which the relationship between the nervous system and behaviour is analysed and understood.

Animal movement

Locomotion as a key to animal life

Locomotion refers to movement of an animal from one place to another. All animals move. It is a fundamental requirement for life, from protists and sponges through to mammals and birds. The reasons to move are manyfold but can be broadly summarised by a need to find food, escape from predators, find a mate, disperse or find a suitable place to live.

The majority of animal species alive in the world today are able to move throughout their lives, although a number of phyla contain species that are sessile as adults. These marine or freshwater species (e.g. corals, many bivalve molluscs, tunicates) have motile planktonic stages in early life to facilitate the dispersal of the species. They subsequently settle onto suitable substrates, where they attach and remain for the rest of their lives.

The modes of locomotion displayed by animals are extremely diverse, with modern species found burrowing into, moving along the bottom of and swimming at all depths in lakes, rivers and seas. Many can cross the interfaces between water and land, land and air, and air and water. They may move with or without the aid of specialised limbs and have occupied every niche on earth—all due to the fundamental ability to move.

Locomotion requires force to be exerted on the surrounding environment and the resultant movement is always in accordance with Newton's three Laws of Motion. The production of forces upon which locomotion depends can be broadly divided into two categories—muscular and non-muscular.

Muscular movement involves elongated cells that are specialised for contraction, and is the principal method of force generation in higher animals. Once contracted, a muscle cell cannot extend itself. It must be extended by an external force before being able to contract again. Non-muscular locomotion is limited to small animals or the motile larval stages of larger organisms. In the majority of these, locomotion is achieved using flagella or cilia, which are motile organelles (Chapter 3).

The mechanisms of force production in both types of locomotion are examined later in this chapter. Firstly, we will look at some of the different forms of locomotion used by animals in water, on land and in the air.

A ll forms of locomotion require force to be exerted on the surrounding environment. Locomotion can be broadly divided into two categories—muscular and non-muscular.

Aquatic locomotion

Animals use many methods to exert a force on the surrounding water, accelerating the water in one direction and the body in the opposite direction. Some mechanisms are effective only if an animal is small, others operate best for large animals. We shall examine aquatic locomotion, not in a phylogenetic framework, but instead by grouping animals according to their method of swimming. This approach is used because similar ways of moving in water have evolved in animals from widely separated phyla. But firstly we need to consider the property of buoyancy in aquatic animals.

A quatic locomotion involves an animal exerting a force on the surrounding water. This accelerates water one way and the body in the opposite direction.

Buoyancy

Buoyancy is the tendency for objects to float. An animal that rises when at rest is positively buoyant, whereas one that sinks is negatively buoyant. If there is neither a tendency to rise nor sink, then an animal is neutrally buoyant. Neutral buoyancy occurs where the downward-directed force, due to gravity acting on the mass of the body, is opposed by an equal and opposite force produced by upthrust from the surrounding water. Many aquatic organisms are, or are very nearly, neutrally buoyant. Neutral buoyancy allows animals to hover in mid-water without having to swim constantly. In addition, a neutrally buoyant animal requires less **power** (work per unit time) to swim at a particular speed than if it were negatively buoyant. Neutral buoyancy allows energy savings in both these circumstances.

Positive buoyancy reaches an extreme in *Physalia*, the Portuguese man-of-war or 'bluebottle' of temperate and tropical seas. Each 'bluebottle' is actually a colony of many individuals, one of which is highly modified and forms the conspicuous gas-filled sac that acts as a float.

Negative buoyancy occurs when the overall density of an animal is greater than the density of the surrounding water. All benthic organisms, those that live on the bottom of oceans and lakes, are negatively buoyant. If the animal ceases to swim, it sinks.

Nautilus, the only surviving genus of the subclass Nautiloidea (Fig. 27.1), is found in coastal waters of the south-west Pacific. This animal has a spiral shell, divided into numerous chambers. All of these chambers are gas-filled, with the exception of the largest chamber at the mouth of the shell, which is occupied by its body. The gas-filled chambers are much less dense than the surrounding water, compensating for the greater density of the shell and body tissues. As a result, the animal is neutrally buoyant. Many teleost fishes use a gas-filled sac, the **swim bladder**, to control buoyancy. Gas is secreted

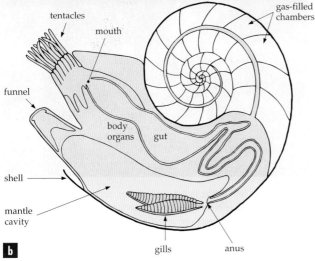

Fig. 27.1 *Nautilus*, the only surviving shelled cephalopod. This animal achieves neutral buoyancy by secreting gas into the spaces within its shell. Swimming is achieved by expelling water from the funnel, with the direction of swimming dependent on which way the funnel is pointing during expulsion. When feeding, it swims forwards, tentacles extended to capture prey items, but it swims backwards at other times

tendency to sink by using fins, particularly their pectoral fins, as **hydrofoils**. These act like the wings of aircraft and produce an upward-directed force, known as **lift**. The caudal (tail) fin also produces some lift and provides the motive force. Thus, a swimming shark has the potential to cruise in mid-water without sinking, whereas a stationary shark will sink.

Highly mobile aquatic animals are usually neutrally buoyant. This permits individuals to occupy any level in the water column with minimum energy expenditure. Neutral buoyancy is achieved by incorporation of low-density substances into the body.

Non-muscular locomotion

Cilia and flagella

Cilia and **flagella** are long thin extensions of cytoplasm able to undergo vigorous bending movements (Chapter 3). Typical flagellate locomotion occurs when a wave of bending travels from the tip of a long flagellum to its base, or from the base to the tip, forcing water in the opposite direction. Animals that have shorter (and more numerous) cilia tend to be larger than flagellates and can move faster. In cilia, the locomotory forces are produced by a powerstroke, where a rigid cilium is moved by bending at its base alone. The cilium then becomes limp and is drawn forward in preparation for the next powerstroke (Fig. 27.2). The

into the swim bladder from the bloodstream, and can be removed depending on the species of fish, either via the venous system or by a direct physical connection to the pharynx. Adjusting the swim bladder volume allows fishes to vary their buoyancy. Even a moderately small swim bladder makes a relatively large difference to the overall density of a fish, as it increases body volume without appreciably increasing body weight.

Certain sharks approach the condition of neutral buoyancy by having very large livers that contain lipids that are less dense than sea water. These lipids compensate for other more dense body tissues, which results in the overall body density being close to that of sea water. Many sharks are slightly more dense than sea water and therefore have a tendency to sink. Even so, they are found throughout the water column, from the surface waters to the sea bed. They counteract the

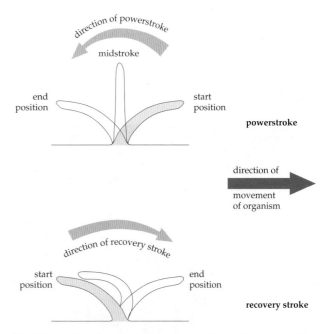

Fig. 27.2 Beating of a cilium. The powerstroke provides the motile force and is followed by a recovery stroke. Ciliated animals use numerous beating cilia to move through fluids. Larger animals often use cilia to move fluids within their bodies or across their surface. Our respiratory tract has numerous cilia, which move fluids and debris away from the lungs

co-ordinated beating of numerous cilia can be seen as a visible wave propagated across a field of cilia.

Non-muscular locomotion involving cilia and flagella is only practical for animals of small size. Many of the single-celled protists use these structures to produce propulsive forces, and the majority of invertebrate phyla (with the apparent exception of the Arthropoda and some of the more obscure phyla that compose the lesser protostomes) contain species that have a ciliated larval stage.

Muscular locomotion

Jet propulsion

The cephalopod molluscs—octopuses, squids, cuttlefish and *Nautilus*—have the ability to move by **jet propulsion**. This involves a forceful contraction of circular muscles in the wall of the mantle cavity, expelling water rapidly through the funnel and accelerating the animal in the opposite direction (Fig. 27.3). These animals move forwards or backwards depending on which way they point the funnel, but they generally travel backwards when moving fast.

Cephalopods are not alone in using jet propulsion. Many jellyfish, such as the Indo-Pacific sea wasp or box jellyfish, *Chironex fleckeri*, found in the warm coastal regions of Australia, produce thrust by contracting circular subumbrellar muscles (Fig. 27.4), which results in the ejection of a pulse of water from the bell. They are not high-performance swimmers though, primarily because of the small amount of muscle involved in the contraction process.

Most bivalve molluscs are sedentary. However, scallops and file shells can hop and swim by clapping their shells together. They may do this to escape from predators, particularly starfish or sea stars, or may simply clap their way to a new site on the sea bed.

Fig. 27.3 Cuttlefish use undulations of their fins for steering and propulsion at low speeds, but resort to jet propulsion for fast locomotion. Squids use similar modes of swimming. One group, the 'flying squids', Onycoteuthidae, can shoot out of the water and glide for considerable distances

Fig. 27.4 The lethal box jellyfish, *Chironex fleckeri*. Although jet-propelled, these jellyfish do not approach the speed and manoeuvrability of the more heavily muscled cephalopods

Muscles attached to the internal surfaces of the shells provide the power by pulling the shells shut, ejecting water rapidly. An elastic hinge ligament, which is compressed when the valves are pulled shut, acts as the antagonist (Fig. 27.5). Thus, when the muscles relax, the energy contained in the compressed hinge causes the valves to spring apart.

Underwater flying

Penguins and turtles have wings or flippers with cross-sections like those of the wings of flying birds. By flapping the limbs up and down they generate lift (Box 27.1), which drives them forward through the water. Some fishes, like tunnys and tunas, swim by oscillating their stiff, wing-like caudal fins from side to side to generate lift forces. However, most fishes swim by propagating waves along their body (see Fig. 27.8).

Rowing

In **rowing**, limbs push against the surrounding water. In the backwardly directed powerstroke, the extended limbs present a large surface area to the water, rather like the blade of an oar. The forward recovery stroke presents a smaller profile to the water so that, over the whole limb cycle, the net effect is to move water

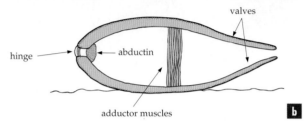

Fig. 27.5 Scallops have a block of a rubber-like protein called abductin in the hinge region of the shell valves. Adductor muscles close the valves, squeezing the abductin. Muscle relaxation allows the compressed abductin to 'spring' the valves open

backwards, thus propelling the animal forwards. Water beetles achieve this by having hairs that fan out during the powerstroke, increasing the size of the effective 'oar'. The hairs lie flat during the recovery stroke, reducing the profile presented to the water. As less power (work per unit time) is required to accelerate large masses of water to a low speed than to accelerate small masses to a high speed, rowing animals, like water beetles, tend to have relatively large 'oar blades'.

Plesiosaurs, aquatic reptiles of the dinosaur era, may have used a form of rowing that involved their four large, paddle-shaped limbs. Alternatively, the paddle may have had a cross-section much like that of a bird's wing and plesiosaurs may have 'flown' underwater. In each case, the limb would have to be rotated between the powerstroke and the recovery stroke to alter its effect on the surrounding water. Another hypothesis is that they used a combination of the two techniques, resulting in paddle movements similar to those of sea lions (Fig. 27.6). The simple fact that the question is under debate illustrates the plausibility of rowing as a locomotor style that can be used by animals differing in size by orders of magnitude.

Swimming using body undulations

The majority of fishes use trunk musculature and the caudal fin to propel themselves through the water using an **undulatory swimming** action. Locomotion powered by fins other than the caudal fin is found in many groups of fish. The weedy seadragon, *Phyllopteryx*

taeniolatus, of south-eastern Australian waters, manoeuvres by propagating a wave along its dorsal fin; trigger fish, *Balistes*, use both the dorsal and anal fins; skates and rays use highly modified pectoral fins (Fig. 27.7), the latter reaching its extreme manifestation in the huge manta rays.

Aquatic mammals, such as whales, dolphins, porpoises and sirenians (manatees and dugongs), use large tail flukes, and others, such as seals and sea lions, use modified hind limbs in a similar way to tail flukes. A major difference between fishes and aquatic mammals is that fishes employ a side-to-side sweep of the tail while mammals use vertical oscillations. Eels and snakes have slender elongated bodies and swim by lateral wave-like movements, which travel backwards along the body, the amplitude of the wave increasing as it travels towards the tail. This is known as anguilliform locomotion (Fig. 27.8).

> Methods of aquatic locomotion include ciliar and flagellar motion, jet propulsion, rowing, swimming by body or fin undulation, and underwater flying.

Aerial locomotion

The majority of animals that fly use muscle-powered wings and are capable of sustained flight. Other flying animals either have different sources of power, or use unpowered flight, such as gliding.

> Aerial locomotion can be divided into muscle-powered flight and unpowered or gliding flight.

Unpowered (gliding) flight

Gliding flight is more economical in energetic terms than flapping flight. Muscular involvement is limited to the maintenance of a particular posture. The wings, or their equivalent, are outstretched and held motionless relative to the body, apart from constant small adjustments in response to changing air conditions or flight requirements. Some animals are only capable of gliding for a few seconds, others for hours; some have maximum gliding speeds in the order of metres per second, others in the order of centimetres per second. In all cases, in order to stay aloft, an animal has to produce enough upward-directed force, lift, to counteract the effect of gravity (Box 27.1). This lift depends predominantly upon the speed of flight, the size and geometry of the wing, or **aerofoil**, and how the wing is tilted relative to the direction of travel, the **angle of attack**.

> An aerofoil is used to generate lift in all types of flight. Lift generally opposes the effect of gravity. The size, shape and orientation of the aerofoil determines flight performance.

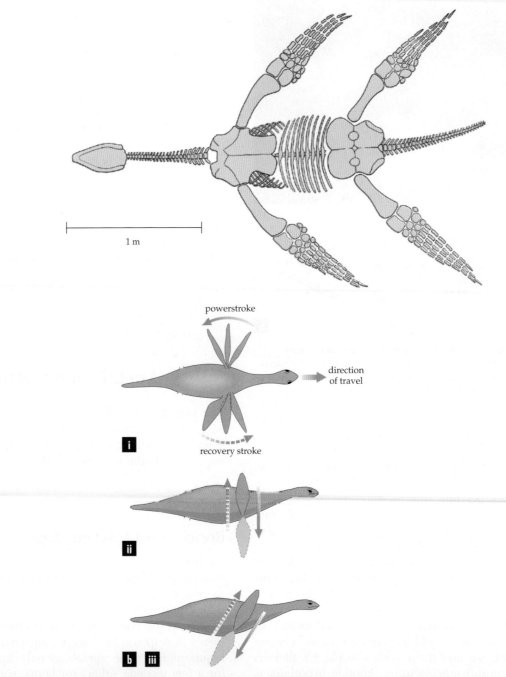

Fig. 27.6 Plesiosaurs. Did they row themselves through the water using their large paired fins as paddles, or were they able to 'fly' underwater using their limbs as hydrofoils? **(a)** Skeleton of a long-necked plesiosaur shown from below. **(b)** Three possible swimming techniques showing how flippers could have been moved relative to the body if the animal moved by rowing **(i)** or by either of two forms of underwater flight **(ii** and **iii)**

Many of the larger sea birds and birds of prey (raptors) with large wingspans can stay aloft by gliding for long periods of time. The wandering albatross, *Diomedea exulans*, has a wingspan of about 3.2 m and the wedge-tailed eagle, *Aquila audax*, has a wingspan of 2.5 m. Predictions made using flight equations suggest that the narrow wings of the albatross, compared with those of the eagle, would result in a higher wing loading

and thus a greater gliding speed (Box 27.1). Gliding performance for a variety of animals is shown in Figure 27.9.

Most, if not all, gliding animals can alter the size, shape and angle of their wing or flight membrane to modify gliding performance. Gliding speeds are generally reduced by postural adjustments immediately before landing to reduce the chance of injury, whereas

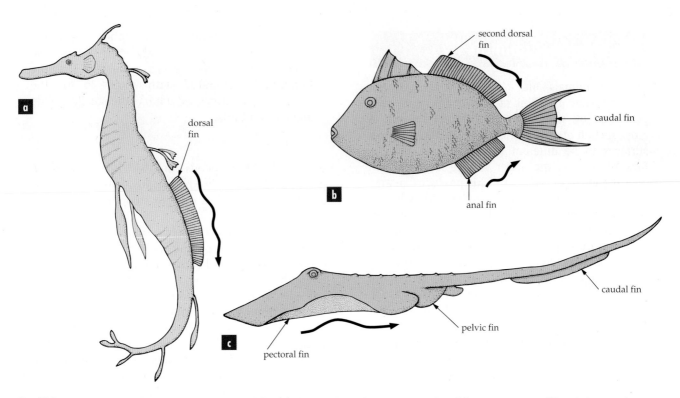

Fig. 27.7 Swimming using fins other than the caudal fin. **(a)** The weedy seadragon uses its dorsal fin to manoeuvre. **(b)** Undulations of the second dorsal and anal fins power the low-speed swimming of trigger fish, while burst swimming uses the body musculature and caudal fin. **(c)** Waves of motion along the highly enlarged pectoral fins of rays provides the thrust for swimming. The caudal fin and tail may act as a rudder or may be modified as a defensive weapon, as in the stingrays. Pelvic fins of rays do not play an important role in locomotion

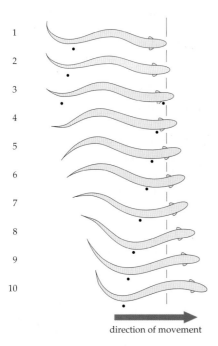

Fig. 27.8 Anguilliform locomotion is often used by long thin animals when swimming. The head does not deviate to either side by very much, but by the time the wave of contraction reaches the tail it has a very large amplitude. The wave is produced by co-ordinated contraction of muscle blocks on each side of the body. The dots indicate wave crests and can be seen to move backwards along the body. This sequence is taken from 10 photographs taken at 50 ms intervals

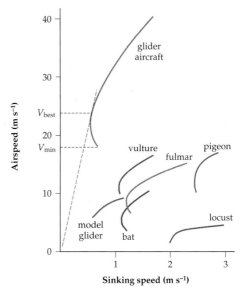

Fig. 27.9 Gliding performance of some animals and glider aircraft. Sinking speed is the rate at which height is lost during flight and is related to the airspeed. Stalling will occur below the minimum gliding speed (V_{min}). A line drawn as a tangent to the curve from the origin of the graph gives the 'best glide ratio', that is, the speed (V_{best}) at which maximum range can be achieved for any given starting height

BOX 27.1 How animals fly

In flying animals, lift, a force that acts at right angles to the direction of movement, is usually generated by wings moving through the air. A cross-section of a wing shows the upper and lower surfaces to be different (asymmetrical) and the flow of air over the upper surface is faster than over the lower one (Fig. a). This produces an area of lower air pressure above the wing and higher air pressure beneath. In addition, as an object moves through a fluid, such as air or water, it is subject to retarding **drag** forces due to friction. Drag always acts backwards along the direction of motion.

The angle of attack (Fig. b) affects the amount of lift and drag a wing creates. Lift increases as the angle of attack rises from 0° to about 20°, but then declines. Further increases in this angle cause the aerofoil, or wing, to stall as the airflow over the wing becomes turbulent. Drag increases as the angle of attack increases. Narrow wings with a large span produce the greatest amount of lift for the smallest amount of drag. Hence the design of glider aircraft.

Simple analysis of gliding flight has revealed a number of important general principles. For example, gliding speed is proportional to the square root of wing loading (body weight divided by wing area), which means that animals with a high wing loading are better suited for faster flight than are those with low ones. Another important parameter is the **aspect ratio** (tip-to-tip wing length divided by average width). Animals with large aspect ratios are able to glide at shallow angles relative to the ground. This is another reason why glider aircraft have extremely long thin wings. They can fly at glide angles as small as 1.3° to the horizontal, compared with about 3° for the wandering albatross, which has the smallest glide angle of all birds.

increased gliding speeds can assist prey capture. The peregrine falcon, *Falco peregrinus*, folds its wings back and then enters a fast dive, or stoop, to catch birds in mid air.

Soaring is an extremely energy-efficient form of gliding. In normal gliding, height is slowly lost and eventually the animal comes to ground or has to regain height by active flapping. In soaring, an animal uses air-currents to remain airborne, even gaining height in certain circumstances without flapping its wings. One of the most obvious forms of soaring uses **thermals**, which are upward movements of air produced by irregular heating of the ground by the sun. A wedge-tailed eagle, for example, could enter an upward-moving mass of air and gain altitude. Within a thermal, an eagle glides in its normal fashion, losing height relative to the surrounding air, but if the air column is rising fast enough, then it will gain height relative to the ground.

Kestrels and a number of the smaller kites are adept at slope soaring (in addition to their ability to hover using powered flight). They are often seen facing into the wind, remaining stationary relative to sloping ground, and moving their wings only fractionally. These birds are soaring in wind that is angled upwards relative to the horizontal. If this angle is equal to the normal gliding angle of the bird, then the bird can remain stationary relative to the ground as long as the wind speed is equal to the normal gliding speed (Fig. 27.10). Many sea birds are good slope soarers, using the updrafts that occur along the coastline, especially at cliffs.

A variety of animals other than birds, including a number of Australasian marsupial species, are known to glide. These marsupials move between trees by gliding, with outstretched limbs extending the gliding membrane into a large 'wing' (Fig. 27.11). Flight performance is not particularly good, as their aspect ratios are small and the relatively large body produces considerable drag which slows them down. They use steep glide angles of up to 27°, which, although poor compared with most birds, are sufficient for the requirements of gliding between trees in open forest.

There are also a number of fishes, amphibians and reptiles capable of unpowered flight. Flying fish belong to the family Exocoetidae, and one species in particular, *Exocoetus volitans*, has greatly enlarged pectoral fins

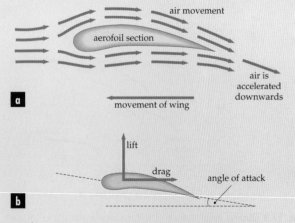

(**a**) Schematic diagram of the airflow past a wing that is producing lift. (**b**) Lift acts at right angles to the direction of motion, whereas drag acts directly backwards along the direction of motion. The lift and drag depend on the angle of attack of the wing

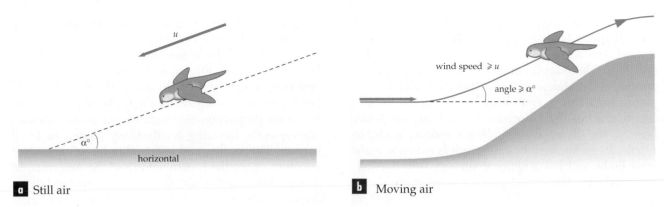

a Still air   **b** Moving air

Fig. 27.10 Hovering kites and the kestrel are often observed hanging motionless in the air while looking for prey. **(a)** If a bird can glide downwards through still air at a speed u and glide angle $\alpha°$, then **(b)** it will be able to remain stationary relative to the ground if it faces into the wind, moving at speed u, as long as the wind is directed upwards by at least angle $\alpha°$

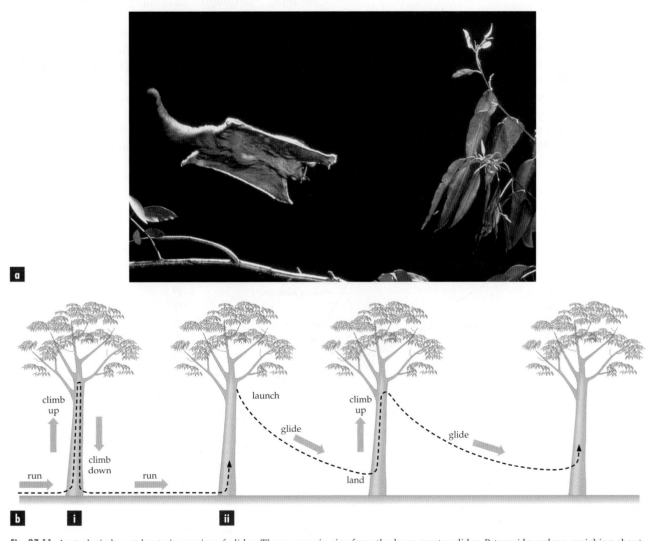

Fig. 27.11 Australasia has at least six species of glider. These range in size from the large greater glider, *Petauroides volans*, weighing about 1200 g, to the diminutive feathertail glider, *Acrobates pygmaeus*, of about 15 g. **(a)** This is a sugar glider, *Petaurus breviceps*, about to land. **(b)** To move between trees, it is energetically more expensive to **(i)** climb down, run to the next and climb up, than it is to **(ii)** glide between trees. Gliders use the second type of locomotion for tree-to-tree movement when foraging. Powered flight in the vertebrates may have evolved via gliding in tree-living animals. This arboreal or 'trees-down' theory was proposed by Darwin in 1859. An alternative theory is that active flight evolved in ground-living animals via a run–jump–glide–fly progression. This is the cursorial or 'ground-up' theory

that it uses as wings. These fish swim until a high enough speed is attained, at which point they break through the water surface, extend their pectoral fins and glide. There are a number of 'flying frogs', *Rhacophorus*, in the Malaysian region that glide, albeit rather poorly, from tree to tree. They have enlarged webbed feet that act as flight surfaces. South-East Asia is also host to the flying dragon, *Draco volans*, a gliding reptile. Its gliding membrane extends between fore and hind limbs and has additional support from laterally extended ribs. The extinct pterosaurs are probably the best-known group of gliding reptiles, the largest of which was *Quetzalcoatlus northropi*, which had a wingspan of about 12 m.

Powered flight

In flapping flight, strong muscular contraction is needed to power wing movements. Active flight is much more complex than gliding flight. In most birds, it is the downstroke that produces most of the useful lift and thrust, which support body weight and propel the animal forwards respectively. The upstroke normally generates little or no lift (and may even produce negative lift), but it does return the wings to a position from which the next downstroke can be made. Wings tend to be fully extended during the downstroke to

generate maximum lift and are brought closer to the body during the upstroke to reduce drag forces.

Various small birds (and bats) are able to produce large amounts of lift on both the downstrokes and upstrokes, enabling them to hover. Hovering flight is, however, energetically demanding. The power requirements are proportionately higher for large animals and, consequently, hovering is effectively limited to small animals. Dragonflies and many other insects are adept hoverers, as are the hummingbirds of the Americas. In general, birds up to about the size of pigeons and doves are able to hover for short periods of time, but larger birds cannot generate sufficient power. Although active flight is extremely energy-demanding, requiring a high output of power, it is an efficient way of moving a unit mass over a unit distance. For a given body mass, active flight is a far cheaper way to travel than walking or running, although swimming is cheaper still (Fig. 27.12).

Birds (Table 27.1) and bats provide good examples of how wing characteristics are linked with ecology. Normal flight behaviour, foraging behaviour and habitat preference can be shown to be strongly influenced by the size and shape of bat wings. Species with low wing loading are very agile and are able to change direction rapidly, whereas those with high wing loading are not. More agile species often occupy 'cluttered' environments and are capable of hawking for insects in

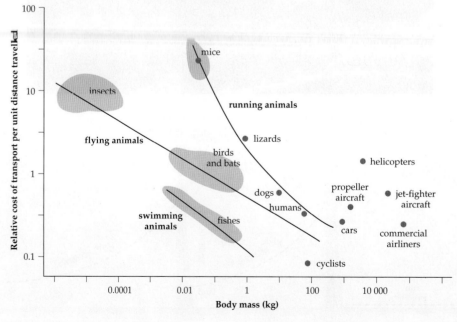

Fig. 27.12 Metabolic costs of transport. Measurements of oxygen consumption can be used to reveal how much metabolic energy is used in moving an animal of a given size a given distance. The graph shows that, for animals of the same mass, swimming is the most efficient mode of transport (note logarithmic axes). The cost of swimming is lower than flying, which is in turn lower than legged locomotion. (These curves generally refer to animals that are specialised for the particular form of locomotion represented here.) Swimming is efficient because the body is supported by water. In terrestrial locomotion, energy is consumed in order to counteract the force of gravity and, in addition, energy is required to accelerate and decelerate legs during a stride cycle. Notice that all the curves slope in the same direction. Large animals use less energy per kilogram body mass than do small animals when moving the same distance. Bicycle travel is very efficient, primarily because no work is done against gravity. Energy is primarily used in overcoming air resistance

Table 27.1 Relationships between flight mode, wing characteristics and flight performance of various birds

| Flight and foraging modes | Example | Observed wing characters | | | Flight performance predicted for the observed wing characters based on aerodynamic theory |
		Aspect ratio	Wing loading	Wingspan length	
Cruising and hawking in open spaces	Swifts, swallows, hawks, falcons	High	High	High	Relatively slow and enduring flight. Relatively high manoeuvrability
Cruising flight in search of prey in open spaces	Albatrosses, harriers, gulls, pelicans	High	Low	High	Slow enduring flight, soaring ability
Soaring or cruising flight in search of prey, often perching	Sea-eagle, wedge-tailed eagle	Low	Low	High	Slow flight, soaring ability
Perching and hawking within vegetation	Flycatchers	Low	Low	Low	Slow, highly manoeuvrable, but rather expensive flight
Foraging in open water	Ducks, geese, swans	High	High	Low	Rapid enduring flight adaptations to migration and commuting, low manoeuvrability. Usually need taxiing run for take-off

the complex three-dimensional space. Less agile bats tend to occupy more open habitats where good turning ability is less crucial.

> Flight is energetically expensive but is the cheapest mode of transport per unit distance for non-aquatic animals.

Terrestrial locomotion

Movements performed by applying force against the ground include burrowing, crawling, walking, running and jumping.

Locomotion without legs

Amoeboid movement

Single-celled organisms, such as *Amoeba*, and motile cells within higher animals, such as some vertebrate white blood cells, use this slow form of locomotion, which involves cytoplasmic streaming. Progression is made by extending finger-like projections (**pseudopodia**) forwards and attaching them to a surface. The fluid endoplasm then streams forwards ahead of this

adhesion point through a tube of more solid cytoplasm, the outer ectoplasm. At the leading edge it is transformed to the stiffer gel-like ectoplasm. At the rear, the cell 'shrinks' as the ectoplasm is converted back to endoplasm, which then streams forwards (Fig. 27.13).

Peristalsis

Worms move on or through soil by a form of crawling that uses **peristaltic** contractions. Their bodies are

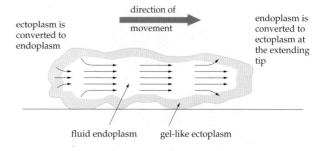

Fig. 27.13 Microfilaments in the cytoplasm are thought to power amoeboid locomotion. This diagram of a vertical section through a crawling *Amoeba* demonstrates how these types of cell move by relocating the material that forms their 'tail end' to the 'front' of the cell. Cells may be free-living, as the amoebas, or may be motile cells within an organism, such as phagocytic white blood cells in our own bodies

divided up into segments, each of which has two sets of opposing muscles (circular and longitudinal) in the body wall. As each segment is fluid-filled, its volume is constant. Contraction of longitudinal muscles shortens a segment and increases its diameter, and stiff hairs (setae) are extended to anchor with the substrate. When the circular muscle contracts, diameter reduces, the setae are retracted and the segment elongates. Progression is made by successive waves of contraction of each muscle layer, extending the body segment and anchoring it in turn (Fig. 27.14). The direction of these co-ordinated contraction waves can be reversed, enabling worms to crawl backwards.

Pedal waves

Molluscs, such as snails, slugs and chitons, seemingly glide across the ground without obvious movement of

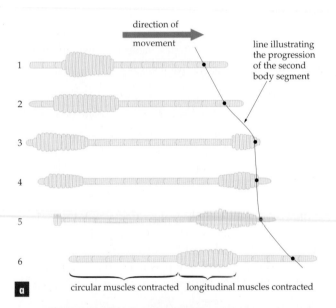

Fig. 27.14 (a) Worms generally burrow through the soil, moving by alternately contracting circular and longitudinal body wall muscles. The six positions shown represent body positions after short intervals of time. **(b)** The giant earthworm, *Megascolides* (class Oligochaeta), of Australia reaches lengths of about 3 m

their broad flat foot. Some species of snail, particularly small snails and those that live on soft substrates, move by ciliary propulsion, but most use waves of muscular contraction of the foot to propel themselves along. These pedal waves can easily be seen by observing snails crawling up the inside of an aquarium. The mucus that coats the ventral surface of the foot has peculiar physical properties. Beneath non-moving parts of the foot, the mucus appears to behave like a solid and effectively sticks the foot to the ground. This enables other portions of the foot to pull against these 'anchored' regions. The mucus beneath the actively extending regions of the foot has markedly different properties. It behaves like a viscous fluid, enabling these moving portions of the foot to slide over the ground with ease.

> Crawling includes amoeboid movement, peristaltic and pedal wave forms of locomotion.

Locomotion with legs

Stability

Legs raise an animal above the ground, greatly reducing friction during locomotion. When animals are stationary, legs provide support and static stability. Stability increases as the height of the centre of mass of the body above the ground decreases, as feet become relatively larger and as the area enclosed by those feet that are in contact with the ground increases. Humans are bipedal, walking upright on their hind limbs. When we stand still, it is essential to maintain our centre of gravity directly above the area bounded by the outlines of our feet, or we topple over. We accomplish this without conscious effort by feedback mechanisms involving sight, balance, muscles and special receptors in the joints and tissues of the body (Chapter 26). Animals with four or more feet in contact with the ground are more stable as it is harder to 'accidentally' cause the centre of gravity to move outside of the support base. In a similar way, squat animals with outstretched limbs are more stable than are tall animals with limbs held beneath the body. This latter point is of particular importance to animals of low mass, such as insects and spiders, where there is a considerable risk of being blown over by air currents. Similar factors affect aquatic legged animals, such as crabs, which are prone to being overturned by water currents. In both instances, the limbs tend to project away from the body enough to ensure stability without compromising the animal's ability to move.

In legged animals, movement is accomplished by lifting individual legs, moving them forwards and placing them back on the ground. Animals with four or more legs move their legs in a particular order to maximise stability and to minimise the chances of one leg interfering with another. Quadrupeds move their feet in the order shown in Box 27.2 when walking slowly. Insects tend to use an alternating tripod **gait**, where three legs are moved while the remaining three legs form a stable tripod (Fig. 27.15). These examples illustrate stable gaits, where the centre of mass remains within the triangle of support at all times. Stability becomes less of a problem as the number of legs increases, but for multi-legged animals the potential for 'tripping over' their own legs becomes greater. Precise co-ordination of leg movement prevents interlimb interference, as demonstrated by the wave of leg movement seen in centipedes and millipedes.

> Support and stability are important for legged animals. To ensure stability when walking, limbs are moved in a precise sequence. This varies according to the animal and the number of legs possessed.

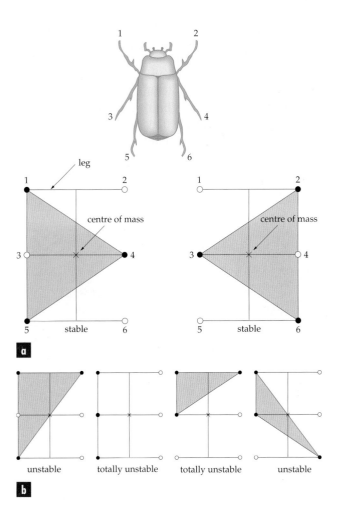

a

b

Fig. 27.15 (a) Insects use a stable tripod gait, where three of their six feet are in contact with the ground at any one time. Feet on the ground are represented by solid circles. The centre of mass stays well within the blue triangle formed by the supporting legs. **(b)** The animal would topple over if it used these combinations of foot–ground contact as the centre of mass falls outside or on the outer border of the triangle of support

Walking and running

Animals often change gait as they increase speed. Horses and many other quadrupedal mammals walk, trot and then gallop as they increase speed. Kangaroos use a curious pentapedal form of locomotion when moving slowly, with the tail acting as a 'fifth limb', but use a bipedal hopping gait at higher speeds. The reasons why animals change between different gaits at certain speeds appears to be related to the energetic cost of locomotion and simple physical constraints that would otherwise limit speed.

We will examine the first reason further, using the examples of mammals moving on motorised treadmills. Studies on a wide range of placental mammals, from mice to elephants, have shown that the rate of consumption of oxygen (mL O_2 kg^{-1} s^{-1}) increases linearly with speed—the faster you run the more oxygen (= energy) you use per unit time. Interestingly, in studies using kangaroos and wallabies, the results are quite different. While slow-moving kangaroos using non-hopping gaits show a similar relationship between the rate of oxygen consumption and speed, when they change to a bipedal hopping gait this is accompanied by a levelling-off of the cost of locomotion (Fig. 27.16a). This 'uncoupling' of the metabolic cost of locomotion from the speed of travel appears to be unique among mammals. It suggests that kangaroos are more efficient at moving at moderate or high speeds than placental mammals of similar size. This is believed to be due to an increase in the importance of energy storage and return in tendons of the hind limbs.

An elegant study, using horses trained to change their gait on command, allows us to examine the energetic cost of using different gaits in an individual animal. Analysis of the metabolic energy cost of transport shows that the cost of moving unit distance (mL O_2 kg^{-1} m^{-1}) varies with speed within each gait (Fig. 27.16b). However, there is a particular speed within the walk, trot and gallop at which the energetic cost is minimal. When the horses were allowed to move freely, they chose speeds within each gait that corresponded to the lowest energetic cost of travelling unit distance. Thus, it appears that changing gaits can allow animals to move at different speed while minimising the metabolic energy required to move unit distance.

Jumping

Animals may use **jumping** as the normal form of locomotion (a kangaroo hop is, in effect, a jump) or, more often, as a mechanism of escape. In most instances, a forceful and rapid extension of limbs, powered by contraction of skeletal muscle, propels the individual at high speed away from its original position. This may be sufficient to escape from a predator.

BOX 27.2 Animal gaits

A gait can be defined as a distinctive form of locomotion. Gaits can thus be defined for aquatic, aerial and terrestrial locomotion, although gaits are most commonly considered in association with the latter.

As an example, consider the changes in gait that occur as a horse gradually increases speed. The first and slowest gait is the walk, which is followed by the trot, the canter and finally the fastest gait, the gallop. Galloping is not just a faster version of walking but a distinctly different form of locomotion. It is often easiest to describe a gait diagrammatically or by using numerical values of various parameters, such as the relative phase of a foot (defined as the time at which it is set down expressed as a fraction of the total stride time). The relative phases used by walking, trotting, cantering and galloping horses are shown in the figure. When walking and galloping, each foot hits the ground at a different time; when trotting, the diagonally opposite feet hit the ground simultaneously; and in the canter only one diagonal pair of feet hit simultaneously. This is why the sounds of

trotting, cantering and galloping horses have quite different rhythms.

Camels and long-legged dogs often use a modified trot, the pace, where the limbs on one side of the body are moved in phase with each other. This reduces the likelihood of the fore and hind limbs interfering with one another. Small mammals (possums, gliders, rabbits) tend to use a bound or half-bound instead of a gallop (figure).

A gait may change because of a simple inability to move faster in the one currently used. Thus, it is analogous to changing gear in a car. Experimental evidence suggests that gaits change before the top speed is reached in any one gait. This may be linked to the observation that horses choose speeds of locomotion within a gait that are the most energetically economical. It has also been noticed that the forces on limb bones rise with speed within a gait, but fall at gait transitions. It may be that changing gaits is a method of ensuring that the skeleton is not exposed to potentially damaging forces during normal locomotion.

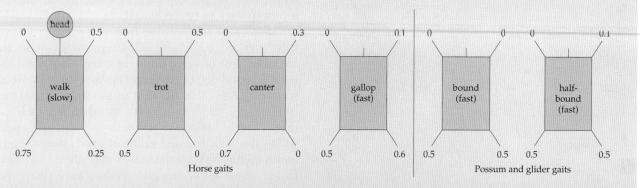

Diagram showing the relative phases of each foot for different gaits in horses and possums and gliders. The front left limb starts the cycle (or stride) and always has a relative phase of zero

Fleas are able to leap about 0.1 m vertically, which may seem poor when compared with humans who can raise the centre of mass (located at about the level of the navel) about 1 m, but it represents a leap of about 100 times body height compared with our half body height jump. In general terms, the actual height of a jump depends on the speed and angle of take-off, and the interrelationship between the animal's mass and its air resistance. Small animals tend to have large surface

areas relative to their volume (or mass) and the effects of drag are relatively larger than in larger and heavier animals. Therefore, for any given speed of take-off, the small animal will slow down faster than the large one and will not jump as high.

Click-beetles, *Elateridae*, also jump but not using their legs. They throw themselves into the air by rapidly flexing their body at a joint between two thoracic segments. This movement involves a catapult mechanism

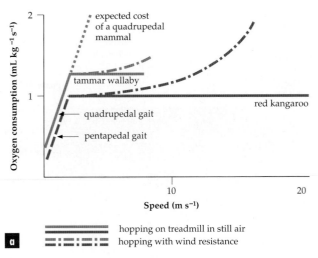
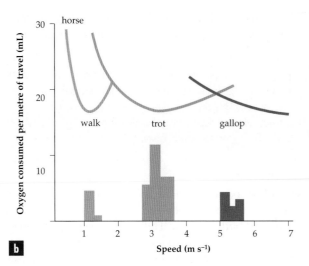

Fig. 27.16 Energetic cost of treadmill locomotion in kangaroos and horses. **(a)** Results from experiments with wallabies and kangaroos suggest that their oxygen consumption is independent of the speed of locomotion when they are hopping. How this occurs is still a mystery but is thought to involve their efficient utilisation of energy stored in their hind limb tendons—analogous to bouncing along like a rubber ball. Note that the expected rate of oxygen consumption for a placental mammal of the same size as the tammar wallaby increases linearly with speed. The tammar wallaby, unlike the red kangaroo, uses a form of quadrupedal bounding gait when moving slowly. **(b)** The oxygen cost to move unit distance against speed for a horse. The histograms show the speeds that were used when the horse moved freely

whereby muscles contract over a relatively long period of time, straining part of the thorax. A catch mechanism releases all of the stored strain energy 'instantaneously' to produce a relatively large leap. Catapult-type mechanisms, where energy is stored in stretched or compressed tissues (of the limbs), also power the jumps of grasshoppers, locusts and fleas. In the Orthoptera (grasshoppers and crickets), the energy is stored in the cuticle of the limb and the tendon (apodeme) of the leg extensor muscle, whereas fleas utilise the properties of a protein rubber called resilin. The limbs directly involved with jumping are often relatively long in animals that leap. This allows feet to remain in contact with the ground for a relatively long period of time, during which limb extension accelerates the body upwards.

> Jumping animals use fast powerful movements, often in the form of a quick extension of the supporting limbs, in order to throw themselves into the air. In many cases, energy is stored in stretched or compressed structures and then released 'instantaneously' to produce large forces.

Structure and function of musculoskeletal systems

Skeletons

In neutrally buoyant aquatic animals, the downward force of gravity is counteracted by an upthrust due to displaced water. In many cases as a result, skeletal structures have become adapted for locomotion rather than for support against gravity. If these animals are removed from their aquatic environment and consequently lose this support, they may collapse under their own weight or will be incapable of movement.

Terrestrial animals need to support themselves against the constant downward-directed gravitational acceleration without the assistance of a supporting fluid medium. Skeletons provide support against gravity in addition to their involvement in locomotion. There are three main types of skeletons—hydrostatic skeletons, rigid internal skeletons and rigid external skeletons. Each type, as we will see, has its advantages and disadvantages.

> The three main skeletal forms are hydrostatic skeletons, rigid external skeletons and rigid internal skeletons.

Hydrostatic skeletons

Fluid-filled body compartments form the basis of the **hydrostatic skeleton** in various invertebrates (cnidarians, flatworms, nemerteans, nematodes and annelids). Since water is non-compressible, a shortening of a cylindrical compartment results in an increase in its diameter. As we saw in annelids (p. 709), circular and longitudinal muscles of each segment act on a hydroskeleton to produce peristaltic locomotion.

Locomotion may involve the propagation of undulatory waves of contraction along the body, producing eel-like swimming if in a fluid medium, or a crawling progression if on a solid substrate (see below). Animals that have hydrostatic skeletons are generally small. This is due, in part, to the lack of protection afforded by the skeleton and because it provides only limited mechanical support, especially in the terrestrial environment.

> Hydrostatic skeletons are fluid-filled compartments, often under pressure. Movement is mediated by muscle contraction. This skeletal form offers little protection, provides limited support against the effects of gravity and is basically limited to 'small' animals.

Exoskeletons

Exoskeletons are found only in invertebrates. They are formed when hard skeletal material is deposited on the external body surface. The chemical composition of the skeleton varies considerably in different invertebrates. For instance, the shells of gastropod molluscs are composed of calcium carbonate and an organic material in the ratio of about 2:1, whereas in arthropods, including insects, crustaceans and spiders, the principle component in the exoskeleton, or cuticle, is the polysaccharide **chitin**.

As animals with exoskeletons grow, they need to either continuously enlarge the skeleton to accommodate the growing soft tissues (e.g. limpets and snails) or periodically replace it with a new larger one (e.g. moulting in crabs and spiders). Exoskeletons may be one piece (a snail shell), two piece (bivalve molluscs, such as clams) or multipiece, as in the case of the exoskeletons of the arthropods. A one-piece skeleton does not play a direct role in locomotion but provides good protection for soft body parts. Two-piece skeletons are used also primarily for protection as they are usually capable of completely enclosing the soft parts, but they may also be involved in locomotion, as seen in the jet-propelled scallops. In arthropods, the entire body is covered by a chitinous exoskeleton and movement is possible because of joints between each rigid section of the body. The arthropod skeleton is secreted by the epidermis and consists of an outer epicuticle, which acts as a waterproofing membrane, an intermediate exocuticle, a tough layer of tanned chitin and protein, and an inner endocuticle of untanned chitin and protein (Fig. 27.17). The stiff exocuticle layer is absent at the joints, allowing flexibility.

Exoskeletons provide excellent protection and they can act as almost impermeable barriers to water. This is important in reducing the risk of desiccation in land animals and is likely to have been an important factor in the highly successful colonisation of land by insects. Disadvantages of exoskeletons include the weight of the skeleton, the limitation it imposes on body size and, importantly, the vulnerability of an animal immediately after moulting when the new exoskeleton is soft. Very heavily armoured invertebrates are aquatic, where their effective weight is less due to support from the surrounding water.

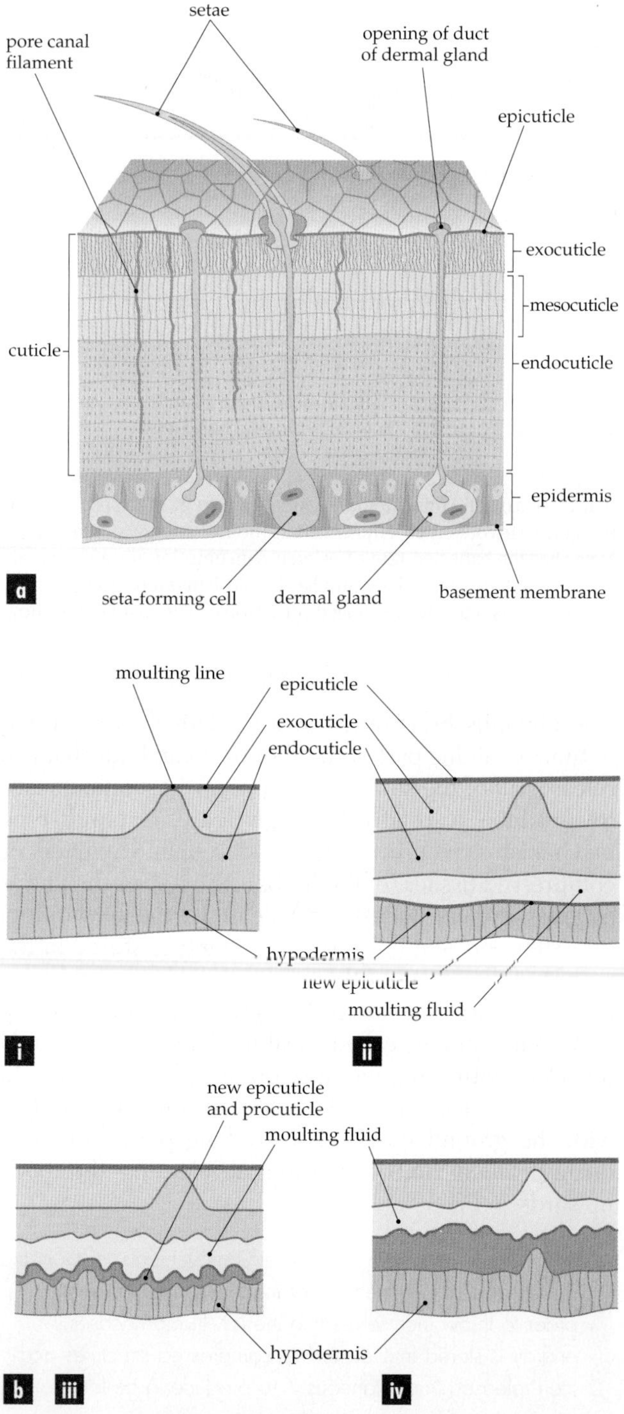

Fig. 27.17 (a) Diagrammatic section through the integument of an arthropod. **(b)** Representation of the moulting sequence in an arthropod. **(i)** Normal appearance of the exoskeleton in intermoult. **(ii)** The epidermal layer (hypodermis) separates from the exoskeleton and secretes a new epicuticle and moulting fluid. **(iii)** The untanned endocuticle is eroded by the moulting fluid and a new procuticle is formed. **(iv)** The animal is encased in the old and new skeleton immediately before moulting. The old skeleton breaks along predetermined lines and the animal emerges with its new pliable exoskeleton. Uptake of water or air enlarges the animal before hardening of the cuticle occurs

E xoskeletons provide good protection but are often heavy and/or limit animal size. Some forms are replaced periodically (via moulting), whereas others are enlarged as the individual grows.

Endoskeletons

Vertebrate animals have an **endoskeleton**. It is constructed primarily of **bone**, except in chondrichthyan fishes (sharks, rays and chimeras) where it is cartilaginous. The functions of the skeleton are numerous: it is involved in movement, mechanical support of the body, protection of parts of the body, mineral storage and regulation (Chapter 25), and in the production of white and red blood cells (haemopoiesis, Chapter 6).

The major constituents of bone are collagen and a complex form of calcium phosphate (hydroxyapatite). Collagen, a structural protein, is strong in tension and gives bones a degree of flexibility, whereas the inorganic component, hydroxyapatite, is strong in compression and gives bone much of its strength.

New bone is usually produced by deposition on a pre-existing surface or in a non-calcified substrate (cartilage) during normal growth and development of the vertebrate skeleton. The material that forms the bone shaft is compact and dense in comparison with the 'honeycomb' configuration found in the expanded ends of bones (Fig. 27.18), with these two morphologies being related to the functional requirements of bony structures. Woven-fibred bone differs from most forms of new bone in that it does not require a surface or non-calcified model for deposition. It is found in regions of skeletal repair and is the type of bone involved in the callus formation after bone fracture.

Bone is a living tissue and, as such, is capable of adaptive change. Large changes occur during growth, for example, the long bones of the forearm of a human adult are much larger than are those of an infant, both in terms of bone lengths and diameters. Even when full adult stature is achieved, bones are still capable of change. New bone layers can be deposited or old bone removed at the surfaces of bones, and the internal bony structure can be altered by a process known as secondary remodelling. Thus, the skeleton can adapt to new situations where the pattern and/or magnitude of forces acting on bones is altered. For example, professional players of racquet sports often have thicker and heavier bones in their racquet arm than in the other arm. This bone hypertrophy is in direct response to the large repetitive forces the racquet arm is exposed to. When a player retires from the game, within a moderate period of time the skeletal asymmetry declines as a result of remodelling and net bone resorption in the bones of the racquet arm. When an injured arm is immobilised in a plaster cast, net resorption and remodelling occurs in response to the abnormally low levels of bone

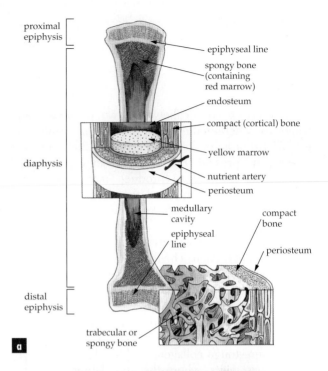

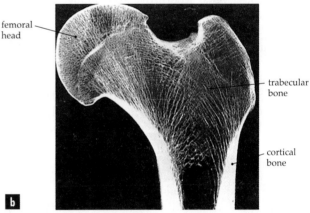

Fig. 27.18 Mammalian long bones have an approximately tubular shaft (diaphysis) composed of compact (cortical) bone and containing bone marrow. At each end of the diaphysis there are expansions of the bone, the epiphyses. Within each epiphysis there is a complex 'honeycomb' of supporting struts, the trabeculae. The orientation of this trabecular bone is thought to be intimately related to the forces that the bone is subjected to in normal life. Increases in bone length are achieved by growth at the epiphysial plates, which remain unossified until full skeletal maturity is reached. Changes to the thickness of the walls of bones occur by the deposition or removal of bone at their external (periosteal) or internal (endosteal) surfaces. **(a)** Diagram of the structure of a typical mammalian bone (human tibia). **(b)** Section through the head of a human femur

loading. Similarly, large reductions in the strength and mineral content of astronauts' bones after prolonged periods of weightlessness have been recorded. These are partially due to changes in the mechanical environment, although in these cases the causes of bone loss are complex and the hormonal status of individuals also plays an important role.

Internal skeletons provide good support, allow growth to be continuous and are remodelled according to use throughout life.

Joints

If motion is to occur between two or more rigid structures, whether they form part of an exoskeleton (Fig. 27.19) or an endoskeleton (Fig. 27.20), there has to be a **joint** present (Box 27.3). Muscles and skeletal components interact to produce movement, with muscles producing force and the skeleton acting as levers, articulating with one another at joints (Box 27.3).

In humans, joints are classified according to their structure and function. The three functional groups are immovable, such as the joints between the bones of the skull (sutures); slightly movable, such as the discs between the vertebrae of the spinal column; and freely movable joints, like most of the joints in the arms and legs. Structurally, joints can be grouped into fibrous joints, where tough collagenous fibres link two bones; cartilaginous joints, where the articulating bones are bound together by cartilage; and lastly, synovial joints.

In synovial joints the articulating bone surfaces, which are covered with cartilage, are located in an articular cavity containing synovial fluid (Fig. 27.20a). This fluid acts as an extremely effective lubricant, resulting in almost frictionless movement. The geometry of the articulating surfaces effectively determines the types of movements available at any particular joint. In addition, the shape of the surfaces, the position of ligaments around the joint and the muscles and tendons that cross the joint, are all involved in determining how much movement is possible at that articulation. Ligaments are, with a few exceptions, strong bands of a collagenous material attached to bone on either side of joints. They keep the joint surfaces pressed against each other and are positioned so as to allow specific movements of adjacent bones. The strength and positioning of ligaments act to stabilise joints, and damage to ligamentous structures often results in a large loss of stability. If you badly twist your ankle, for example, the stretched or torn ligaments will take a long time to repair. Until the ligaments are fully healed, further ankle injury is likely because of compromised joint stability.

Skeletons provide the means to transfer forces generated by muscles to the environment. In the case of rigid skeletons, the bones or body segments are hinged at the joints and act as levers.

Size and the skeleton

Many valuable insights into 'how and why' animals have evolved into the forms we see today can be achieved by the application of mechanical engineering principles to biology—the discipline of biomechanics. The skeleton provides us with numerous examples of the inter-relationships between skeletal structure and function. One such example is the interaction between animal size and skeletal structure.

The bones of small mammals tend to appear rather delicate, whereas large mammals have relatively robust skeletons. Imagine a small macropod marsupial, such as a potoroo, magnified so it becomes the same size as a red kangaroo. The bones of the hind limb will be about the same length in both animals, but the diameter of these bones will be different. The femora (thigh bones), for example, would have a smaller diameter in an 'enlarged potoroo' than in the red kangaroo. The reason is simple. For animals of the same shape, a limb bone's cross-sectional area will increase as the square of the linear dimensions and its mass (derived from the volume) will increase by the cube of the linear

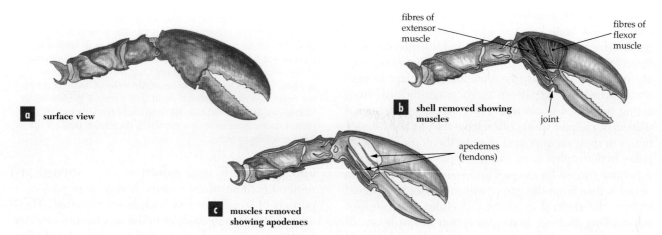

a surface view

b shell removed showing muscles

fibres of extensor muscle

fibres of flexor muscle

joint

apedemes (tendons)

c muscles removed showing apodemes

Fig. 27.19 In a crab claw, the muscle fibres originate over the inner surfaces of the exoskeleton and insert onto tendon-like apodemes, which increase the surface area available for insertion. The apodemes direct vectors of the contractile force onto the site of action where the apodeme attaches to the skeleton

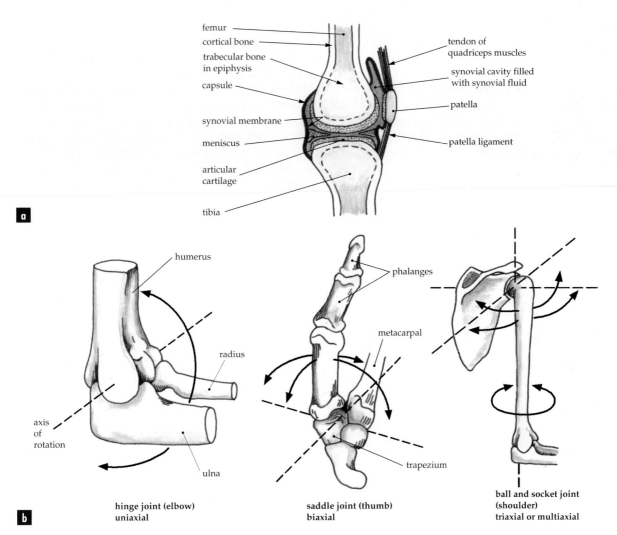

Fig. 27.20 (a) Synovial joint structure, illustrated by the human knee joint. The expanded ends of bones support a layer of smooth articular cartilage and the whole joint is bounded by a fibrous joint capsule, the inside of which is lined with a synovial membrane that secretes synovial fluid. This serves to lubricate the joint surfaces, resulting in an almost frictionless articulation. Ligaments (not shown) hold the joint surfaces together and passively 'control and limit' joint movement. The menisci help to transfer forces across the knee joint. **(b)** Joint movement is often described using easily understood terms. Joints in the forelimb illustrate this

BOX 27.3 Bones as levers

A lever is a rigid structure that rotates about a fulcrum when a force is applied. In a joint, muscles produce the force, bones act as levers and the joint itself is the fulcrum (Fig. a).

The precise arrangement and proportions of these components are important in determining how the system operates. A hypothetical change in the site of insertion of a muscle alters the lever system of the human forearm (Fig. b). With the arrangement found in *Homo sapiens*, the insertion of the tendon of biceps brachii onto the radius is

close to the elbow joint. This results in the muscle having to generate large amounts of tension to support even quite modest loads held in the hand. This may be viewed as being a disadvantage. However, the advantage of this arrangement is seen when the speed and range of movement is considered. Small length changes of the biceps brachii muscle result in large displacements of the forearm and these movements can occur quickly. A muscle inserting more distally on the radius would significantly change the properties of the lever system. The muscular tension required to support loads carried in the hand will be considerably less than in the above case. Thus, larger loads could be carried, or the same loads carried with less effort. This arrangement of bones, joints and muscle would have the disadvantages, in comparison with the true situation, of having a smaller range of motion and a reduced speed of forearm flexion.

The lengths of limb bones are generally related to function. Echidnas and moles, for example, have short limb bones, which, together with the associated musculature, can perform highly efficient and powerful digging movements. By analogy, this system of levers is like that shown in Figure b(ii). Kangaroos, on the other hand, use a lever system more like that shown in Figure b(i) and are adapted for moving at speed by virtue of their elongated hind limbs. Long legs allow for longer strides to be taken, which can result in greater speeds being attained.

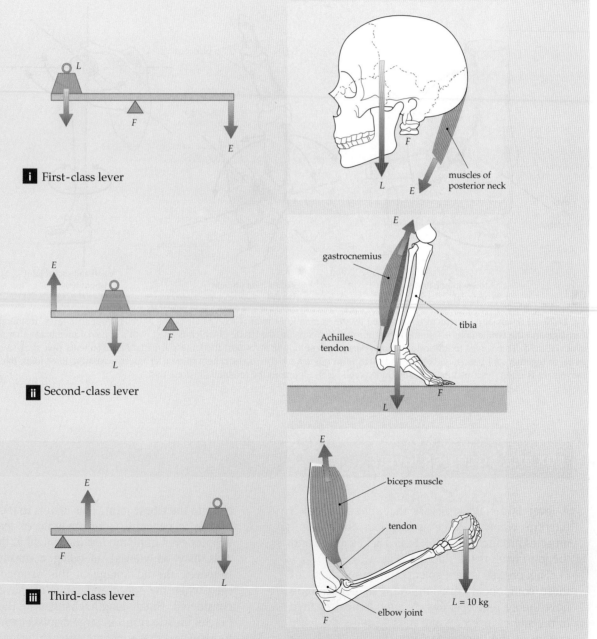

i First-class lever

ii Second-class lever

iii Third-class lever

muscles of posterior neck

gastrocnemius

Achilles tendon

tibia

biceps muscle

tendon

elbow joint

$L = 10$ kg

(a) Bones as levers. There are three classes of levers, which are defined by the relative positioning of a pivot (fulcrum, F), a muscular force (effort, E) and a resistance or passive load (L). **(i)** A first-class lever has the fulcrum between the effort and the load. **(ii)** A second-class lever has the load between the effort and the fulcrum. **(iii)** A third-class lever has the effort between the fulcrum and load

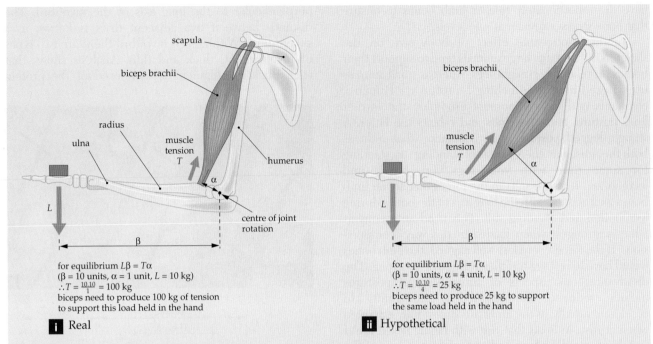

for equilibrium $L\beta = T\alpha$
(β = 10 units, α = 1 unit, L = 10 kg)
$\therefore T = \frac{10.10}{1} = 100$ kg
biceps need to produce 100 kg of tension
to support this load held in the hand

i Real

for equilibrium $L\beta = T\alpha$
(β = 10 units, α = 4 unit, L = 10 kg)
$\therefore T = \frac{10.10}{4} = 25$ kg
biceps need to produce 25 kg to support
the same load held in the hand

ii Hypothetical

(b) The biceps brachii produces flexion at the elbow joint. **(i)** The muscle inserts, via a tendon, onto the radius of the forearm at a point close to the elbow. This arrangement is well adapted for rapid movement as a small shortening of the muscle will flex the forearm through a large arc. **(ii)** If the muscle were attached to a more distal point, the degree of flexion would be much less than in **(i)** for the same shortening of the biceps brachii. The latter arrangement is better adapted for forceful movements of the forearm, demonstrated by balancing moments around the elbow. Tension in the biceps, T, acting at distance α from the elbow, acts in a clockwise direction and has to equal the load, L, acting at a distance β from the elbow in an anticlockwise direction if equilibrium is to be maintained. It is obvious that in the situation illustrated in **(i)**, the muscle is in a poor position to support large loads held in the hand

dimensions (Fig. 27.21). The **stress** (force per unit area) applied to the limb bones can be seen to have increased in the 'enlarged potoroo' and would be undesirable. This is because of a reduction in the difference between a load that the bones could withstand and one that would break them. As a consequence, large animals

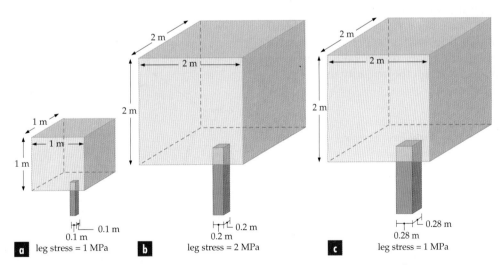

Fig. 27.21 (a) A hypothetical 'one-legged animal'. Its body tissues have a uniform density of 1000 kg m^{-3}. It has a volume of 1 m^3 and thus has a body mass of 1000 kg. The leg has a cross-sectional area of 0.01 m^2. The stress (force per unit area) acting in the leg is (1000 kg g/0.01 m^2) $\approx 1.10^6$ newtons per square metre (or 1 MPa), where g is gravitational acceleration (9.81 m s^{-1}). **(b)** If the 'animal' doubles each of its linear dimensions, it would increase its mass to 8000 kg and its leg cross-sectional area would now be 0.04 m^2, resulting in a stress of 2 MPa. This doubling of stress in the leg increases the likelihood that it may break. Consequently, in real animals, leg bones of large animals are proportionally thicker than are those of small animals in order to keep bone stresses low. **(c)** The leg size required to keep the stress at 1 MPa

have disproportionately thicker limb bones to ensure that bone stresses remain at safe levels.

Another noticeable effect of body size on the skeleton is that many large animals tend to support their bodies on relatively straight legs, whereas small animals use flexed-limb (crouching) postures. Humans, elephants and large dinosaurs exemplify the former limb posture, and mice, rats and rabbits the latter. A straight-legged posture is an adaptation that reduces both the muscular energy costs of standing (try standing for some time with your legs very bent!) and locomotion. It also results in the forces that act on limb bones well aligned with the long axis of the bones during locomotion and standing still. This is advantageous as bone is strongest when loaded in this way (compression). If the forces were less well aligned, the limb bones would have to be bigger if they were to be as strong. The disadvantageous aspects of straight-legged postures include reduced manoeuvrability and ability to accelerate rapidly, as the limbs are almost fully extended when at rest. Animals that rest with highly flexed limbs are able to produce long and powerful extensions of the limbs to produce rapid body acceleration. This is why human sprinters start a race from a crouched position on racing blocks—to improve their acceleration.

Muscle

Muscle tissue has a unique property in that it can contract. It is the force produced by shortening that powers locomotion. In vertebrates, there are three types of muscle tissue—skeletal, cardiac and smooth muscle—each of which is adapted to perform particular tasks within the body. They are all capable of converting chemical energy into mechanical energy in order to produce force, but it is skeletal muscle that is directly involved in locomotion and maintenance of posture. Some invertebrates use smooth muscle to power locomotion, although the majority use forms of striated (skeletal) muscle.

A|| vertebrates and many invertebrates use striated (skeletal) muscle to power their locomotor movements.

Muscle structure

Skeletal muscle cells, or muscle fibres, are roughly cylindrical, contain many peripherally located nuclei and have alternating light and dark bands crossing them, called striations.

The intracellular structure of a skeletal muscle cell can be examined using transmission electron microscopy. Photomicrographs taken at high magnification show a regular arrangement of filaments contained within intracellular structures called myofibrils (Fig. 27.22). The myofibrils are tubular and are divided into compartments by discs (Z-discs), which lie

perpendicular to the long axis of the myofibril. The distance between two adjacent discs is known as a **sarcomere**. Within each myofibril there are two types of myofilament, thick and thin. Analysis shows that each thick myofilament is composed of the protein

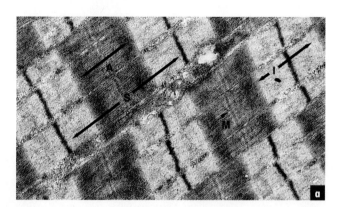

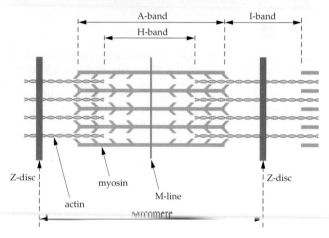

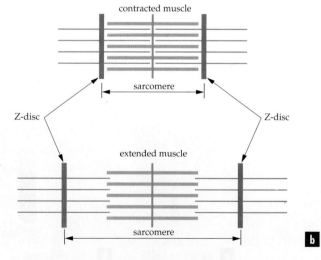

Fig. 27.22 (a) Transmission electron micrograph of mammalian skeletal muscle. At this magnification it is possible to see the regions that correspond to the actin and myosin filaments. This muscle is in about the middle of its contractile range.
(b) Representation of skeletal muscle at the level of the myofibrilar filaments. The actin (A) and myosin (M) filaments can slide relative to each other. Note the changes in the width of bands between contracted and extended muscle

myosin and the thin myofilaments are predominantly the protein **actin**. A myosin filament consists of a bundle of hundreds of myosin molecules, each of which is composed of two identical proteins. Each protein is shaped like a golf club, with the 'shafts' lying parallel to the myosin filament with the double heads projecting laterally. The actin filament is composed of two strands of fibrous actin, tropomyosin and troponin molecules (Fig. 27.22). Each sarcomere contains two sets of actin filaments extending from the Z-discs, and one set of myosin filaments in the centre of the sarcomere. These thread-like structures are arranged parallel to one another and, in regions where the two populations of filaments overlap, actin filaments surround each myosin filament in a hexagonal arrangement.

The next level of organisation consists of the myofibrils grouped together in bundles to form the muscle fibres (cells). The cell membrane of a muscle fibre is the sarcolemma, and a specialised endoplasmic reticulum in muscle, known as the sarcoplasmic reticulum, forms a network of membranous channels throughout the sarcoplasm (the muscle fibre's cytoplasm) in which the myofibrils are embedded. In addition, there is a specialised membrane system known as the transverse (T) tubule system, formed by the invagination of the sarcolemma, which runs to the core of the fibre (Fig. 27.23). Although the T-tubules and the sarcoplasmic reticulum are in contact where the two sets of tubules cross, there is no direct connection between them; therefore, there is no exchange of fluid between the two types of tubules.

Connective tissue binds the components of the muscle together, supports the capillary and nerve networks of the muscle and acts to transfer the force from the muscle fibres to the skeleton. The connective tissue sheath that surrounds individual fibres is the endomysium. Groups of muscle fibres (also termed fascicles) are surrounded by the perimysium, and the muscle belly as a whole is bounded by the epimysium (Fig. 27.23).

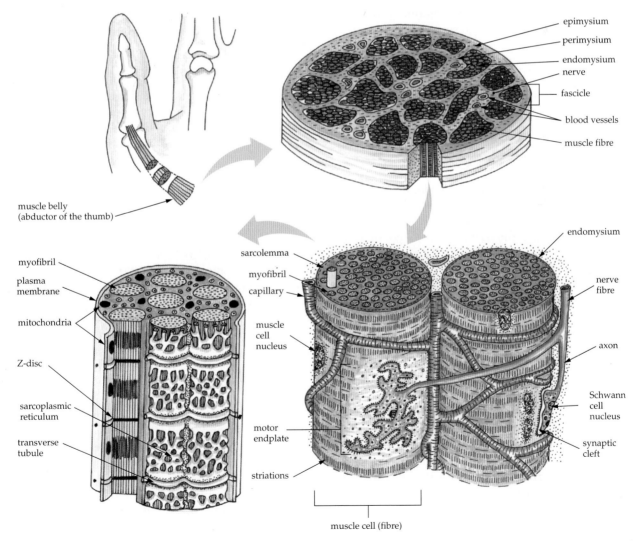

Fig. 27.23 Schematic diagram to show the hierarchical structure of muscle from the whole belly to individual microfibrils, and the structural relationship between the myofibrils and the surrounding sarcoplasmic reticulum and transverse (T) tubule system

Muscles involved in movement of the skeleton insert into the skeleton either directly or indirectly via tendons (apodemes in arthropods). It is common to find a tendon at either end of a muscle belly, attaching it to bones. The attachment of a muscle that moves less during contraction is called the origin, whereas the one that moves most is called the insertion. In muscles associated with moving the limbs, the origin is usually the proximal attachment point and the insertion is the distal attachment point.

Cardiac muscle is the tissue of which the heart is predominantly made. The muscle cells join in an irregular manner to form a branching network. This form of striated muscle is very resistant to fatigue and, unlike skeletal muscle, has a mechanism by which electrical potentials can pass from cell to cell. It is innervated by the autonomic nervous system (Fig. 26.17).

Smooth muscle (also called visceral muscle) is found in numerous sites throughout the vertebrate body: in the walls of blood vessels, bronchioles of the lungs, the bladder and associated ureters, uterine tubes, the ductus deferens and the digestive tract. Individual smooth muscle cells have a single nucleus and are elongated, and usually pointed at each end. This type of muscle produces relatively slow but forceful contractions and can produce force over a wide range of lengths. Innervation is by the autonomic nervous system (Fig. 26.17).

How skeletal muscle works

The mechanism of contraction is explained by the **sliding filament model**. In this model it is proposed that the actin filaments, which slide relative to the myosin filaments, move towards the middle of the sarcomere and pull the Z-discs, to which they are attached, towards one another. This results in a shortening of the sarcomere and hence the muscle contracts. Note that individual filaments do not change length or contract but simply slide relative to one another. The whole process requires an input of energy, which is derived from the breakdown of ATP.

The basis for the sliding filaments is the interaction between the heads of myosin molecules and specific sites on adjacent actin filaments, for which myosin heads have a high affinity (Fig. 27.24). When a muscle is in its resting state, these active sites on the actin filament are obscured by tropomyosin, a regulatory protein associated with the actin filament. In this condition, the myosin heads are unable to form attachments or **cross-bridges** with the actin. Muscle contraction can only occur when these binding sites are exposed, and this requires the presence of calcium Ca^{2+} ions, which are normally stored within modified endoplasmic reticulum, the sarcoplasmic reticulum. Upon neural stimulation great enough to produce muscle contraction, an action potential spreads across the muscle cell membrane and along the T-tubule (transverse tubule) system formed by invagination of the plasma membrane. The T-tubules conduct the action potential deep into the fibre, causing a wave of depolarisation to cross the membrane of the sarcoplasmic reticulum. This increases membrane permeability to Ca^{2+}, thus releasing these ions into the sarcoplasm. The Ca^{2+} binds to troponin and causes a conformational change in the tropomyosin–troponin complex, which uncovers the binding sites, allowing contraction to occur.

Myosin heads now bind to actin filaments, forming cross-bridges. A subsequent change in shape of the myosin heads results in the actin filament being pulled towards the centre of the sarcomere. (It is at this stage of the cycle that mechanical work occurs, that is, the muscle shortens and/or generates force.) The binding of ATP to a myosin head causes it to dissociate from the actin filament and to return it to its original 'primed' condition before reattachment to another binding site further along the actin filament (Fig. 27.24). This cycle, which lasts for tenths or hundredths of a second, can then be repeated, resulting in further muscle shortening. The energy needed to fuel one complete cycle is ultimately derived from the hydrolysis of one ATP molecule. Muscle relaxation also requires an input of ATP, but in this case it is used in the active transport of Ca^{2+} back into the sarcoplasmic reticulum. A reduction in the Ca^{2+} concentration within the sarcoplasm allows the tropomyosin–troponin complex to re-establish its position, blocking the actin binding sites once again and preventing the establishment of new cross-bridges.

After death, intracellular levels of ATP decline. As a result, Ca^{2+} leaks into the muscle cells and causes the uncovering of active sites. Cross-bridges form and remain until the tissues start to decay, there being insufficient ATP available to dissociate the myosin heads from the active sites. The development of muscle rigidity after death is termed rigor mortis.

The force a muscle fibre can produce is related to the length of its sarcomeres. If the sarcomeres are in a highly shortened or a highly extended state, then the muscle fibre cannot generate much force (Fig. 27.25). In addition, the rate at which muscles can contract is limited by the rate at which cross-bridges can attach, pull, release and reattach. The faster the rate of shortening of a muscle, the smaller the force it can produce. At the fastest rate of shortening, a muscle can produce no force. On the other hand, when a muscle is successfully stimulated but is held at a constant length (isometric contraction), it produces a large force. In life, muscles usually operate between these two extremes, with muscles shortening while producing reasonable forces.

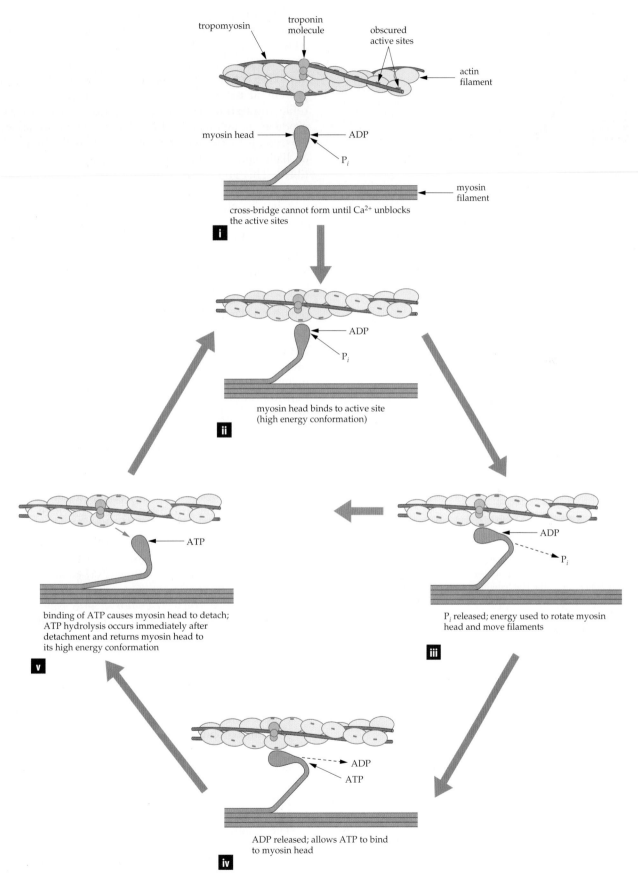

Fig. 27.24 Schematic representation of the mechanism of muscle contraction

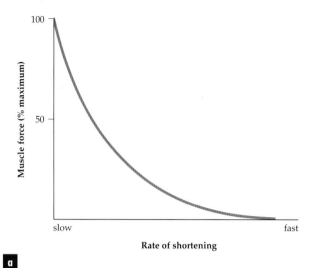

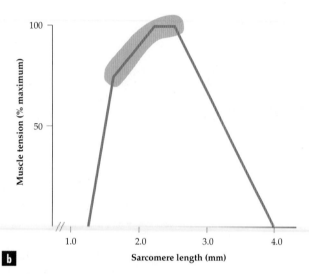

Fig. 27.25 (a) Relationship between the speed of muscle contraction and the force it can produce. Muscles that contract quickly cannot generate much force. **(b)** The length–tension curve for skeletal muscle. Most muscles are thought to operate in the shaded region of the curve

> Skeletal muscle derives its energy ultimately from the splitting of ATP. The relative motion between actin and myosin myofilaments within muscle cells produces shortening of individual cells, resulting in muscle contraction.

Muscle energetics

The direct source of energy for muscular contraction is ATP and is produced by aerobic or anaerobic respiration (Chapter 5). The amount of ATP stored in a resting muscle cell is low but other energy stores are present within the cell. Glucose is converted into glycogen and this is stored in the cell. When muscle activity occurs, glycogen can be broken down to yield glucose. Creatine phosphate is an important energy storage compound in muscle and is formed by the

transfer of a high-energy phosphate from ATP to creatine. It can be utilised for the rapid conversion of ADP to ATP in conditions of sudden demand.

Using the example of human runners, short periods of intense activity (sprinting) exhaust the supply of ATP present in resting muscle almost immediately. Anaerobic respiration and creatine phosphate breakdown produce enough ATP for further contraction (for about 20 seconds). One outcome of sprinting is the build-up of high concentrations of lactic acid within muscle fibres. This and the exhaustion of creatine phosphate stores effectively limit the time over which sprinting can be maintained. Aerobic respiration is used to produce ATP in long-distance running, where the muscular effort is sustained at submaximal levels for long periods of time. Under these conditions, fatty acids form a more important energy source than does glucose, although glucose is also metabolised.

Skeletal muscle fibre types

Muscle fibres, even within the belly of a single muscle, are not all alike. Histological and histochemical observations have identified three distinct fibre types in mammals (although some animals appear to have a fourth fibre type (type IIX), which is intermediate between type IIA and type IIB fibres).

1. *Slow oxidative (type 1) fibres*, also called slow-twitch fibres. These contain numerous mitochondria and appear red due to their high vascularity and large myoglobin content. They have a great capacity for generating ATP by oxidative (aerobic) activity and are difficult to fatigue because the high blood flow to the fibres delivers oxygen and nutrients at a rate sufficient to keep pace with the breakdown of ATP. They have a relatively slow contraction velocity due to the slow rate of splitting of ATP and produce relatively little tension.

2. *Fast oxidative (type IIA) fibres*, also called fast-twitch, fatigue-resistant fibres. These are red fibres with a fast contraction speed resulting from an ability to split ATP at a rapid rate. They have a moderate capacity for both oxidative and glycolytic (anaerobic) activity.

3. *Fast glycolytic (type IIB) fibres*, also called fast-twitch, fatigable fibres. These are large fibres with a small myoglobin content and few capillaries, giving them a white appearance. They are adapted to generate ATP by the anaerobic process of glycolysis, using local glycogen stores. These stores are rapidly exhausted and there is a rapid build-up of lactic acid, causing these fibres to fatigue rapidly.

The relative proportions of these fibre types in a muscle belly can often be related to the normal tasks of that muscle. Type I fibres are ideal for maintaining posture and are found in relatively high proportions in the soleus (calf) muscle, that helps us maintain our balance when standing and in neck muscles that hold the head up. In contrast, type IIB fibres are found in high proportions in arm and shoulder musculature, which are often used for rapid and powerful movements. Type IIA fibres are found in large proportions in the leg muscles of sprinters, with the specialist sprinter, the cheetah (*Acinonyx jubatus*), having an extremely high proportion of this fibre type in its limb musculature. It has a fibre composition, glycolytic enzyme capacity and mitochondrial content at the extreme of values measured in other runners, such as greyhounds, race horses and elite human sprinters.

Fibre types are probably determined genetically and an analysis of elite athletes shows that the percentage of fibre types differs from the normal population according to their chosen sport. Long-distance runners and cross-country skiers have a large proportion of type I fibres for endurance, and the performance of endurance-trained cyclists has been shown to be related to the proportion of type I fibres in their muscles. Sprinters have more type II fibres for speed and power. Similar observations can be made between different species of animal according to the lifestyles and behaviours commonly used.

Summary

- Many different methods of moving have evolved in animals in order to feed, avoid various dangers, find a suitable place to live and find a mate for reproduction.
- Aquatic locomotion is achieved by an animal exerting a force on the surrounding water. This accelerates water one way and the body in the opposite direction. Many highly mobile aquatic animals are neutrally buoyant, which permits them to occupy any part of the water column from the surface to the bottom. They achieve neutral buoyancy by incorporating low-density substances in the body, either in discrete structures (e.g. the fish swim bladder) or within the general body tissues.
- Methods of aquatic locomotion include ciliar and flagellar motion, jet propulsion, underwater flying, rowing, and swimming by body undulation or fin undulation.
- Aerial locomotion can be divided into muscle-powered flight and different forms of unpowered, or gliding, flight. All types of flight use an aerofoil, such as wings, skinfolds or webbed feet, to generate lift, an upward force that opposes the effect of gravity. The size, shape and orientation of the aerofoil determines flight performance and can be used to predict how an animal will fly.
- Flight is energetically expensive per unit time but works out to be a cheap mode of transport per unit distance.
- Methods of terrestrial locomotion by a very diverse group of animals include walking, running, jumping, burrowing and crawling. Crawling includes amoeboid movement, peristaltic and pedal wave forms of locomotion.
- Support and stability are important for legged animals whose bodies are raised up from the ground. To ensure continuous stability when walking, it is necessary for limbs to be moved in a particular sequence depending on the animal and the number of legs it possesses.
- Jumping animals use fast powerful movements, often in the form of a quick extension of supporting limbs, to throw the body into the air. In many cases, energy is stored in stretched or compressed structures and then released 'instantaneously' to produce large forces.
- Hydrostatic skeletons consist of fluid-filled compartments, often under pressure. Movement is mediated by muscle contraction. This skeletal form is basically limited to 'small' animals. It offers little protection and provides limited support against the effects of gravity.
- Exoskeletons provide good protection but are often heavy and/or limit animal size. Some forms are replaced periodically (by moulting), whereas others are enlarged as the individual grows.
- Internal skeletons allow growth to be continuous, provide good support and are important in blood cell production and the maintenance of blood Ca^{2+} levels.
- Skeletons provide the means to transfer forces generated by muscles to the environment. In the case of rigid skeletons, the bones or body segments act as levers, being hinged at the joints.
- All vertebrates and many invertebrates use striated (skeletal) muscle to power their locomotor movements.
- Skeletal muscle derives its energy ultimately from the splitting of ATP. The relative motion between actin and myosin myofilaments within muscle cells produces shortening of individual cells, resulting in muscle contraction.

keyterms

actin (p. 731)
aerofoil (p. 713)
angle of attack (p. 713)
aspect ratio (p. 716)
bone (p. 725)
buoyancy (p. 710)
chitin (p. 724)
cilia (p. 711)
cross-bridge (p. 732)

drag (p. 716)
endoskeleton (p. 725)
exoskeleton (p. 724)
flagella (p. 711)
gait (p. 721)
gliding (p. 713)
hydrofoil (p. 711)
hydrostatic skeleton (p. 723)
jet propulsion (p. 711)

joint (p. 726)
jumping (p. 721)
lift (p. 711)
myosin (p. 731)
peristalsis (p. 719)
power (p. 710)
pseudopodia (p. 719)
rowing (p. 711)
sarcomere (p. 730)

sliding filament model (p. 732)
stress (p. 729)
swim bladder (p. 710)
thermals (p. 716)
undulatory swimming (p. 713)

Review questions

1. For rapid movement of a part of the skeleton, is it better to have a muscle inserting close to or far away from the joint over which it passes and has its effect?

2. What are the basic components of skeletal muscle and how do these interact during a muscle shortening event?

3. What is the energy source for muscle contraction? What happens when muscle operates anaerobically?

4. Why is it that most large animals tend to stand and move using more upright limb postures compared to small animals?

5. What is the general response of living bone to increased loading? Why is this a desirable response?

6. Why do many jumping animals have relatively long limbs?

7. Why is a bipedal stance intrinsically less stable than a quadrupedal one?

8. What is the 'aspect ratio' of a bird and what influence does it have on gliding performance? Why is the ability to hover limited to small animals?

9. Which groups of animals use jet propulsion to locomote?

10. Define buoyancy. Explain how it might be achieved and the possible benefits gained by neutrally buoyant animals.

Extension questions

1. How is elastic strain energy used in (a) jumping by a grasshopper and (b) scallops swimming? How might the use of stored strain energy in elastic tissues effectively enhance the jumping performance compared with that achieved by muscle contraction alone?

2. By referring to Table 27.1 and from general observation of common birds in your neighbourhood, can you identify wing and tail features that are important for (a) gliding flight, (b) manoevrability (c) fast, dashing flight, and (d) catching insects?

3. What are the advantages of being neutrally buoyant in water? What happens to the gas in the swim bladder of a teleost fish if it swims rapidly from the surface (where there is 100 kPa [1 bar] of pressure) to a depth of 10 m (where the pressure is 200 kPa [2 bar])? How do fishes with swim bladders cope with such pressure changes? (An understanding of Boyle's Law and swim bladder anatomy and physiology is needed.)

4. Skeletal muscle. What are the advantages and disadvantages of muscle fibres that use aerobic and anaerobic respiration? Consider the properties of different fibre types and try and work out which ones would be used in different parts of flying and swimming vertebrates.

5. In reference to terrestrial locomotion of placental mammals, why is the curve describing the rate of oxygen consumption (mL O_2 kg^{-1} s^{-1}) different from that describing the cost of transport (mL O_2 kg^{-1} m^{-1})?

6. In considering the structure and function of vertebrate joints, why aren't all joints of the 'ball-and-socket' type, which allows for a high degree of mobility? Why do we have synovial joints that have vastly different ranges of movement available to them?

Suggested further reading

Alexander, R. McN. (1983). *Animal Mechanics*. 2nd edn. Oxford: Blackwell Scientific.

A citation classic, introducing the fundamentals of animal structure and function from a biomechanical perspective. Eminently readable.

Biewener, A. A. (1992). *Biomechanics, Structures and Systems: A Practical Approach*. Oxford: Oxford University Press.

A book that provides good advice and information about 'hands-on experimental approaches' to working in the field of animal mechanics.

Norberg, U. M. (1990). *Vertebrate Flight*. Berlin: Springer-Verlag.

This is a comprehensive account about the mechanics, physiology, morphology, ecology and evolution of flight. It is most suitable for more advanced students.

Schmidt-Nielsen, K. (1984). *Scaling: Why is Animal Size so Important?* Cambridge: Cambridge University Press.

A great and readable account of the effects of size on biological structures and organisms.

Vogel, S. (1994). *Life in Moving Fluids: The Physical Biology of Flow*. 2nd edn. Princeton, NJ: Princeton University Press.

An excellent book on the biology of life in water from a fluid-flow perspective.

CHAPTER

28

Animal behaviour

During spring, male satin bowerbirds, *Ptilonorhynchus violaceus*, build bowers consisting of two parallel rows of vertical twigs. The area immediately around the bower is decorated with flowers, feathers, snail shells and sometimes items such as bottle tops and toothbrushes (Fig. 28.1). Each male displays to attract females to his bower, and mating usually takes place at or near the bower.

Understanding why a male bowerbird takes so much trouble to attract the attention of a female is one of many questions that is addressed in **ethology**, the study of animal behaviour. Ethology is not simply documenting the activities of animals; it aims to explain *why* animals behave in the ways they do. Niko Tinbergen, one of the founders of modern ethology, believed that ethology should try to explain the causation, development, adaptive value and evolution of any behaviour under investigation. In response to the question 'Why do male bowerbirds construct and decorate bowers?', for example, we might suggest answers in terms of:

- *Causation*. The hours of daylight increase during springtime, and this might trigger changes in hormone levels that stimulate the male to build bowers.
- *Development*. Males have learned the behaviour from their parents or neighbours.
- *Adaptive value*. Males build bowers in order to attract females for breeding; males without bowers are unable to attract and mate with females.
- *Evolutionary history*. Complex bower-building behaviour may have evolved from more simple nest-building behaviour in the ancestors of bowerbirds.

These explanations fall into two main categories. **Proximate explanations** are those that consider the causation of a particular behaviour at any particular time and **ultimate explanations** are those that are concerned with the evolution and function of a particular behaviour. A proximate explanation for bower-building might emphasise the learning abilities of bowerbirds or their responses to objects of different colours, while ultimate explanations consider the effects of bowers and decorations on the reproductive success of males. Proximate and ultimate explanations are not alternative explanations but rather answers to different kinds of questions. This chapter mostly addresses the evolution of behaviour and therefore emphasises ultimate explanations. Nevertheless, it is often necessary to consider causal or proximate explanations in order to understand the mechanisms underlying ultimate explanations. The 'internal' mechanism by which behaviour is controlled, such as neural pathways and the influence of hormones, are the subject of Chapters 25–27.

Genetics and the evolution of behaviour

An animal's behaviour is an integral part of its phenotype. It has a genetic basis and consequently is under the influence of **natural selection** (Chapter 32) and sexual selection (p. 741). For example, a marsupial mouse, *Antechinus*, that forages in exposed areas during the day may be more likely to be captured by a predator than if it were to forage under the cover of bushes at night. Similarly, an *Antechinus* that forages for food that yields less energy than the animal expends obtaining it will not survive. It is therefore possible to attribute survival value to an animal's behaviour in the same way that it is attributed to an animal's anatomical or physiological characteristics. Frequently, an animal's behaviour reflects a balance between several alternative behaviours; an *Antechinus* that spends excessive time hiding from predators rather than foraging may be as much at risk of death, through starvation, as is one that ignores the risk of predation. Understanding how animals balance these conflicting demands is a common theme in contemporary studies of animal behaviour.

Natural selection can act only on genetic differences between individuals within a population. Consequently, an animal's behaviour, like other aspects of its phenotype, can evolve only if: there is, or was in the past, variation in the behaviour of individuals in the population; the differences in behaviour are heritable or genetic in origin; and some behaviours result in greater survival or reproductive success than others. The genetic basis of animal behaviour can be illustrated by examining genetic markers, by artificial selection experiments and by comparing populations with genetic differences.

Fig. 28.1 A male satin bowerbird, *Ptilonorhynchus violaceus*, beside a bower, which he constructs to attract females

Genetic markers

Male and female vinegar flies, *Drosophila melanogaster*, usually copulate for about 20 minutes. However, the duration of copulation can be altered by subjecting the flies to agents that produce genetic mutations. In one mutant, the male does not disengage from the female after the usual time, while in another mutant the male disengages from the female after only 10 minutes and thus fails to fertilise any eggs. These differences in mating behaviour are the result of changes in the sensory receptors and muscles of the flies, which can be attributed directly to genetic mutations.

Selection experiments

It is possible to select artificially for particular behaviours in the same way that plant or animal breeders select for physical attributes of flowers, crops or domesticated animals. For example, some males of the field cricket, *Gryllus integer*, call to attract females, while others remain silent and attempt to intercept females that are attracted to the calling males. It is possible to produce populations of males that call very frequently by allowing only frequently calling males to breed. Similarly, a population of almost silent males will result if frequently calling males are prevented from breeding.

Populations with genetic differences

Individuals from populations that are geographically separated may differ in their morphology and behaviour, often as a result of different selection pressures imposed by different ecological conditions. For example, the garter snake, *Thamnophis elegans*, is found in separate populations on the coast and inland in the south-western United States. Inland snakes are mostly aquatic and feed on frogs and small fish, while coastal snakes are terrestrial and feed mainly on slugs. Food-choice experiments with newborn snakes reared in the laboratory showed that individuals derived from coastal populations had a stronger preference for slugs than did snakes from inland populations. Furthermore, offspring obtained by matings between individuals from coastal and inland populations had intermediate preferences for slugs. These experiments indicate that the food preference of snakes from the two populations has a genetic basis and is associated with local ecological conditions.

> All forms of animal behaviour have a genetic basis, which means that, like anatomical and physiological characteristics, behaviour is under the influence of natural selection.

Learning and the development of behaviour

The undignified tricks that some animals are obliged to perform in circuses and zoos are examples of learning. But learning is an integral, and frequently important, feature of the behaviour of animals under natural conditions. Any change in an individual's behaviour that is due to its experience can be referred to as **learning**. One advantage of learning is that it provides a greater potential for changing behaviour according to changes in an individual's circumstances. For example, a worker honeybee, *Apis mellifera*, can learn the location of her hive and the location and quality of several flower crops during her brief three-week life span. This means that she can fly directly to the most productive crops rather than randomly 'discover' them every time she leaves the hive.

The behaviour of an organism capable of learning is unlikely to be constant throughout its life but will change over time. Thus, it is important to distinguish learning from the **development of behaviour**. For example, barnacles (Cirripedia) have a wide range of behaviours that are specific to their age. A barnacle changes from a planktonic, swimming, grazing nauplius to a swimming and then crawling cyprid, which seeks out a settlement site, attaches itself to the substrate and metamorphoses into a sessile, filter-feeding adult, which behaves in a different way again. Newly hatched chicks of the domestic hen peck more accurately at grains on the ground as they grow older. Although this may be partly due to experience, the improvement will also occur irrespective of how frequently the chicks are allowed to peck at food. In other words, older chicks are more accurate as a result of their maturity, regardless of how much practice they have had. The reduction in play behaviour with increasing development, common in many animals, including humans, is another example of a change in behaviour with age.

Imprinting is a feature of the early development of some kinds of behaviour. When a young gosling hatches, it will recognise the first moving object as something to follow for the next few weeks. Usually, this will be its mother, and consequently the gosling will *imprint* on its mother. However, a chick will imprint on a human if the human is the first moving object the chick sees (Fig. 28.2). Imprinting usually occurs only during a *sensitive period*, the exact duration of which varies between species.

While certain behaviours simply develop with age, others clearly include an element of learning. For example, the songs of many birds are learned from their parents or other nearby birds. A young bird's first attempt at song usually occurs some months after hatching. This song, called a subsong, resembles the

Fig. 28.2 These young geese have imprinted on the famous ethologist Konrad Lorenz and behave as if he were their mother

There are several ways in which animals learn behaviour and these can be roughly placed into two categories: *associative* and *non-associative* learning. The difference relates to whether the animal learns the behaviour in association with some *stimulus* or property of the environment. The classic example of associative learning is Pavlov's conditioning experiments with dogs. Pavlov had noticed that dogs salivate when presented with meat or a meat-smelling substance. In one experiment, Pavlov rang a bell every time he presented a dog with meat. The dog eventually learned to associate the sound of the bell with the meat, and Pavlov was able to make the dog salivate simply by ringing the bell. Learning by trial and error is another form of associative learning. For example, white-winged choughs, *Corcorax melanorhamphos*, obtain their food by digging into leaf litter and soil. This method of foraging is both time- and labour-intensive, and difficult for the young bird to master; juvenile choughs can take up to eight months before they learn the technique and can forage independently. *Habituation* is a form of non-associative learning. For example, fish may fail to show an escape response to a shadow passing overhead if this stimulus is repeated every few minutes.

adult song in length, pitch and tonal quality, but is variable, imprecise and lacks specific features that are typical of the adult song. For example, a young white-crowned sparrow, *Zonotrichia leucophrys*, raised in isolation, will be able to sing the subsong but will fail to develop the normal adult song. In contrast, young birds that are raised with their parents or other birds will learn to sing the adult song (Fig. 28.3).

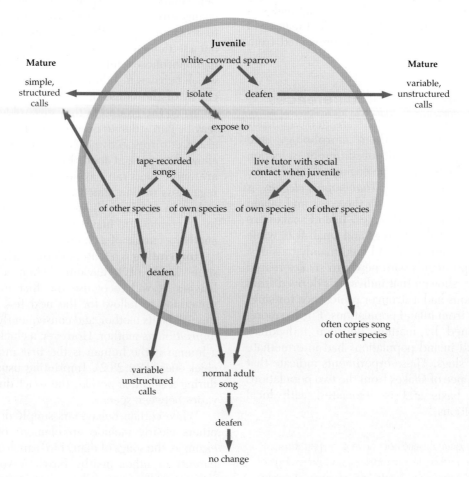

Fig. 28.3 This diagram illustrates the results of numerous experiments that have examined the factors responsible for the development of song in the white-crowned sparrow, *Zonotrichia leucophrys*

Clearly, some behaviour is learned while other behaviour is not. However, it is by no means the case that complex behaviour is always learned and simple behaviour is not. For example, the pallid cuckoo, *Cuculus pallidus*, is a nest parasite; it lays its eggs in the nests of other species, especially the superb fairy wren, *Malurus cyaneus*. The cuckoo chick usually hatches before the wren chicks and shortly afterwards ejects the wren eggs from the nest, an activity that requires considerable dexterity. The behaviour of the cuckoo chick could not be learned from either its cuckoo parents or its foster parents. Similarly, young ogre-faced net-casting spiders, *Dinopis*, not only construct a web between their legs, but also recognise prey and snare them with the net. This behaviour cannot have been learned from parents or other adult spiders (Fig. 28.4) because they never come in contact with each other.

For many years there was controversy among ethologists over whether most animal behaviour is innate (determined by the genotype) or learned (determined by environmental experiences). However, these views represent two extremes of a continuum; most animal behaviour is the result of an interaction between genetic instructions and learning from the environment.

Learning refers to changes in an individual's behaviour that are due to experience, and it provides a great potential for changing behaviour under different circumstances. In contrast, the development of behaviour involves changes that occur as a result of age, not experience.

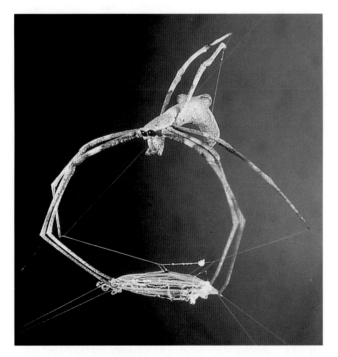

Fig. 28.4 The ogre-faced net-casting spider, *Dinopis subrufa*, builds a silk web between its legs and then casts this web onto a passing insect. The insect is unable to escape from this sticky 'net'

Understanding complex behaviour

There are three methods of obtaining data that are used to understand the evolutionary basis of animal behaviour: these are intraspecific comparisons, manipulative experiments and interspecific comparisons.

The most common way in which animal behaviour is studied is by comparison of the behaviour of individuals within a species. Although this approach has provided numerous insights, interpreting the observed patterns can be difficult. For example, observations of wallabies may reveal that they have a different foraging behaviour in grassland and woodland habitats. If the two habitats vary in both the kinds of available food and the risk of predation, it would not be possible to decide which of the two factors is responsible for the differences in the feeding behaviour of the wallabies. Indeed, it may be another, unknown, factor that is responsible.

Experimental manipulations provide a less ambiguous indication of the selective pressures that may have led to a particular behaviour but they are more difficult to undertake. A simple but effective experiment was performed by Niko Tinbergen, who wanted to know why black-headed gulls, *Larus ridibundus*, removed the empty eggshell from the vicinity of their nests shortly after a chick had hatched. The eggs and hatchlings of these gulls are well camouflaged among the grass and twigs around the nest, which is on the ground. However, the inside of the shell is white and highly conspicuous. Tinbergen thought that the parents' eggshell-removal behaviour might reduce the risk of predation on eggs and chicks. He painted hens' eggs to resemble gulls' eggs and placed them throughout the colony. By placing broken gulls' eggs beside some of these painted hens' eggs, Tinbergen was able to show that the camouflaged hens' eggs were more likely to be discovered and eaten by predators if they had broken eggs near them. The parents' removal of eggshells takes only a few minutes each year, but it is crucial for the survival of chicks because it ensures that both chicks and nests remain inconspicuous.

Most ideas about the evolutionary significance of particular behaviours are derived from comparisons *between* species. The generality of an explanation of why a certain behaviour is found in one species may be derived from an understanding of why it is not found in another species. For example, a comparative study could examine whether the ability to sing complex songs is associated with territorial behaviour (p. 748) by comparing the songs of territorial species with non-territorial species.

omparison of the behaviour of individuals within a species, and between species, and experimental manipulations are three ways in which the behaviour of animals can be studied.

Obtaining food

Animals obtain their food in extraordinarily diverse ways (Fig. 28.5). Some species, such as the cheetah, *Acinonyx jubatus*, and jumping spiders (Salticidae) are *'free ranging'* or *active foragers*, searching for and then pursuing their prey. Other species, such as the death adder, *Acanthophis antarcticus*, or praying mantids (Mantodea) are *sit-and-wait predators*, remaining relatively immobile and attacking only those prey items that venture within striking distance. These predatory strategies represent two extremes and the foraging behaviour of many species may include elements of both. Active foragers also include animals that move considerable distances in search of pollen, seeds and other plant matter.

Despite the diversity of ways in which food is obtained, the foraging behaviour of animals has an important common feature. In general, animals do not forage in a haphazard way but rather individuals make *foraging choices* that maximise their growth and survival. These choices (or decisions) might include the proportions of time spent resting and foraging, whether to forage in one place rather than another, the size and kind of food to eat or the time spent foraging in one location. It is convenient to think of a foraging animal as a 'decision maker', although this does not necessarily imply that this is a conscious decision. Rather, natural selection will favour those 'decisions' that represent the best balance of costs and benefits associated with each choice: some food items may be time-consuming to obtain but rich in nutrients, and some locations may have a rich supply of food but may entail greater risk of predation. Viewed in this way, it is possible to evaluate these costs and benefits mathematically and predict which decisions might be expected under different circumstances. This approach is the basis of **foraging theory** (Box 28.1).

Honeybees, *Apis mellifera*, foraging on nectar must, at some stage, return to their nest and empty their crops. Interestingly, bees frequently return to the nest with less than the maximum load of nectar that they can carry in their crops. In fact, a bee carrying a large load expends more energy flying than does one carrying a small load. Thus, while the gross quantity of nectar increases as the bee continues to forage, the net return on a foraging trip may decrease as a result of the additional energy expended in carrying the load. Consequently, the behaviour of bees is expected to reflect a

Fig. 28.5 Examples of the ways in which animals obtain food: **(a)** a praying mantid feeding on a grasshopper; **(b)** a cheetah chasing a springbok; and **(c)** a sea slug, *Cyerce nigricans*, grazing on turtle weed

BOX 28.1 Foraging theory: a simple prey-choice model

Foraging animals are often confronted by many kinds of prey and an important decision is whether to specialise on only one kind of prey (at the exclusion of others) or to take every type of food as it becomes available. The decision will depend upon the costs and benefits of each type of prey. For example, some prey may be easy to find and process but provide little nutrition. Foraging on this prey may exclude the opportunity of obtaining other prey types that provide greater returns. The costs and benefits of prey types, and their influence on the foraging decisions of animals, can be evaluated mathematically.

Imagine a foraging animal that can feed on two types of prey that contain E_1 and E_2 kilojoules of energy, and where handling times (e.g. the amount of time taken to husk a seed) are h_1 and h_2 seconds. The profitability (or benefits) of these prey therefore provide E_1/h_1 and E_2/h_2 energy returns for the time taken to process each. The animal searches for a total of T_s seconds and encounters these two prey types at rates of m_1 and m_2 prey per second. If the foraging animal eats both types of prey, it will obtain the following energy in T_s seconds:

$$E = T_s (m_1 E_1 + m_2 E_2)$$

The total time to obtain this food will be the searching time plus the handling time:

$$T = T_s + T_s (m_1 E_1 + m_2 E_2)$$

The forager's rate of food intake is simply the total energy divided by the total time:

$$\frac{E}{T} = \frac{(m_1 E_1 + m_2 E_2)}{(1 + m_1 h_1 + m_2 h_2)}$$

Note that the foraging time, T_s, has cancelled out.

If the forager is to specialise on one prey type, then the energy gain (E/T) from foraging only on this prey must be greater than if it were to forage on both types. In other words, if it is to forage only on prey type 1, then E/T from eating prey type 1 must be greater than E/T from eating both prey items:

$$\frac{(m_1 E_1)}{(1 + m_1 h_1)} > \frac{(m_1 E_1 + m_2 E_2)}{(1 + m_1 h_1 + m_2 h_2)}$$

This can be simply rearranged to give:

$$\frac{1}{m_1} < \frac{(h_2 - h_1) E_1}{E_2}$$

Thus, a forager should specialise on prey type 1 if the time taken to encounter it is less than the difference in handling times multiplied by the ratio of kilojoules obtained for each. Notice that if prey type 2 takes less time to handle than prey type 1, an animal should never specialise on prey type 1 irrespective of the encounter rate or energy content of this prey.

balance between the benefits of obtaining additional amounts of food against the energetic costs of carrying it. Experiments that involved the manipulation of the distances bees had to travel to artificial flowers showed that this was the case: bees returned to the hive with smaller loads when they were forced to fly greater distances to the artificial flowers.

This kind of analysis has been remarkably successful in predicting the foraging behaviour of animals, both in the laboratory and in natural field populations. However, while providing insight, the approach has some limitations. In particular, earlier studies made the common assumption that animals forage in a way that maximises their rate of food acquisition. This means that other influences, such as the risk of predation, were ignored, which may not be a realistic assumption for many foraging animals. For example, the risk of predation may vary between food sites, such that a site with high-quality food may also have a higher risk of predation than, for example, a site with low-quality food. Thus, the foraging animal may have to balance the relative costs and benefits of different sites according to the quality of food and risk of predation at each site. North American grey squirrels, *Sciurus carolinensis*, are sometimes faced with this problem and field experiments show that they apparently balance these costs and benefits. If food items (biscuits in these experiments) were placed near trees in which the squirrels could seek refuge from predators, they usually took each biscuit back to the trees and ate them there. But when the biscuits were placed further from the trees, the squirrels brought only large biscuits back to the trees and ate the smaller biscuits in the open field. The energetic costs of travelling to and from the trees in order to feed on food that has low energetic returns outweighs the costs associated with increased risk of predation. A similar pattern of behaviour can be observed in brush-tailed possums, *Trichosurus vulpecula*, foraging in urban parks. These decisions may be further complicated by an animal's energetic requirements. A small animal with low food reserves may have to forage in an area of higher predation risk if alternative, less risky, sites provide insufficient food.

A nimals forage for food in ways that maximise the benefits of particular kinds of food against the costs of obtaining these food items.

Avoiding being eaten

Natural selection has favoured numerous anatomical, physiological and behavioural characteristics that are **defence mechanisms** against predators and parasites. Essentially, these adaptations increase the costs to a predator of detecting, identifying, attacking, pursuing, handling and consuming prey. An obvious anti-predator behaviour is the ability to flee quickly from predators, thus increasing a predator's pursuit time and energy expenditure. Greater escape speed may mean that the cost of capturing the prey exceeds the energetic return and the predator should pursue another victim. However, escape from predators is not the only way that animals avoid predation. Many insects produce a noxious chemical when attacked, which deters the predator, and lizards may lose part of their tail, which may distract the predator while the lizard escapes.

Some species are well camouflaged or **cryptic** and blend into the background substrate; this deception works only if the animal remains motionless or moves in an appropriate way. Leaf-insects (Phylliinae), for example, look like leaves, and are even more convincing when they move slightly, as if blown by the wind. Being cryptic increases the time taken for the predator to locate the prey, and consequently it may search for other victims.

Many animals resemble other unrelated species. These resemblances may be visual, chemical, behavioural or acoustic and are usually referred to as **mimicry**. Mimicry can be defensive or aggressive. *Defensive mimicry* functions to reduce the probability of predation; the palatable (or edible) mimic resembles an unpalatable model. For example, very few predators feed on ants because they are distasteful, and many species of arthropods mimic ants, thereby reducing the probability of predation (Fig. 28.6). In contrast, *aggressive mimicry* has evolved as a mechanism by which predators capture their prey. Adult bolas spiders (Araneidae) prey exclusively on male moths. The spider swings a droplet of adhesive (the bolas) that is attached to the end of a silk thread (Fig. 28.7). The bolas attracts the male moths because it contains a chemical substance that mimics the mate-attracting pheromone of a female moth. Interestingly, the spider does not swing the bolas continuously, but only when a moth flies nearby. Apparently, the swinging behaviour of the spider is elicited in response to the vibrations generated by the actions of the moth's wings.

The relatively immobile and nutritious larvae of butterflies are particularly vulnerable to predators and

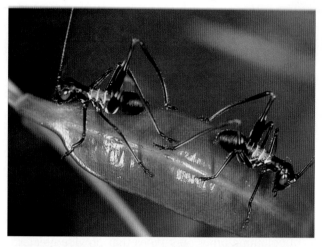

Fig. 28.6 Many different kinds of animals resemble ants in appearance and behaviour, such as these juvenile long-horn grasshoppers. These ant-mimics may be less likely to be victims of predators because few animals are predators of ants

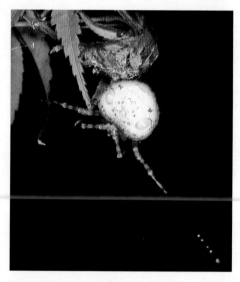

Fig. 28.7 The bolas spider, *Dichrostichus*, although belonging to the orb-web-building family Araneidae, does not build a web but instead captures male moths on the end of the sticky bolas. The moths are attracted to the bolas because it contains a chemical that mimics the sex-attracting pheromone produced by female moths

parasites. Consequently, a number of defence mechanisms have evolved in this taxon: larvae may be cryptic, hiding during the day and only feeding at night, or they may be highly distasteful by incorporating toxins from the plants they feed on. One intriguing solution has evolved in the lycaenid, or blue butterflies. The larvae of many species of this large family are tended by ants, which swarm over each larva, touching it with their antennae. Numerous experiments with the Australian imperial blue butterfly, *Jalmenus evagoras* (Fig. 28.8), which is tended by the small meat ant, *Iridomyrmex*, have demonstrated that large numbers of larvae were killed by predators and parasitoids if the ants were artificially removed. The ants tend the larva, rather

Fig. 28.8 The larvae of the imperial blue butterfly, *Jalmenus evagoras*, an Australian lycaenid whose larvae and pupae are tended by various species of *Iridomyrmex* meat ants. The ants provide protection against predators and parasitoids, and in return obtain food in the form of carbohydrates and amino acids

than attack and eat it, because the larva produces carbohydrates and amino acids harvested by the ants. The relationship is *mutualistic* (Chapter 43); the lycaenid gains protection from the ants, which gain a profitable source of food. The larvae may also produce chemicals that mimic part of the ants' communication system, perhaps eliciting tending rather than aggressive behaviour in the ants. Indeed, workers of the Australian green tree ant, *Oecophylla smaragdina*, carry young larvae of the lycaenid, *Liphyra brassolis*, into their nest. Once in the nest, the lycaenid starts feeding on the eggs and larvae of the host ants. This parasitic relationship is most likely maintained because the young butterfly larvae chemically fool the ants.

Living in groups

Many animals reduce the risk of predation by living in a group. There are several ways in which group living can reduce the risk of predation. The most obvious is that if a predator makes a single attack on a group of prey, and all individuals within the group are equally likely to be caught, then the probability that any individual will be captured is simply the inverse of the group size. This probability is 0.5 for a group of two prey individuals, 0.1 for a group of 10 individuals and 0.01 for a group of 100 individuals. The dilution effect of living in a group is clearly reduced if groups are more conspicuous than solitary individuals, or if predators make multiple attacks on individuals within the group. Furthermore, group members may not gain equal protection if the predator selects particular kinds of individuals, such as juveniles or unhealthy individuals.

An individual at the centre of a group can be protected by others if the predator takes prey from the edge rather than the middle of a group. This may

explain why many group-living animals, such as birds and fish, commonly form tight flocks or schools when a predator approaches. Such behaviour may also produce a *confusion effect* because it is more difficult for the predator to focus on any one individual in a tight group of rapidly moving individuals. In some species, members of a group may actively defend themselves from predators by *mobbing*. Goannas, *Varanus*, are often mobbed or harassed by sulphur-crested cockatoos, *Cacatua galerita*, if they get too close to the birds' nests and the goanna usually leaves without destroying the eggs.

Group living may also provide an advantage in detecting predators. The success of many predators depends upon surprise attack, but the chance of a successful attack is reduced if the predator is detected early. An individual could avoid predation by remaining constantly alert, but this is impossible if other activities, such as pecking for food, are incompatible with looking out for predators. Prey individuals must therefore partition their time between looking out for predators and feeding. Early detection of the predator, and hence escape, depends upon how much of the time the potential prey is alert, or vigilant, for predators. By living in a group, each individual can take advantage of the vigilance behaviour of the others. Studies of many species of birds and mammals have shown that a predator is more quickly detected by individuals in groups than by those foraging alone.

If there are so many benefits to living in a group, why do many animals live alone? The answer is that there are costs associated with living in a group. One of the most important is that group living increases the opportunity for competition over resources, such as food. Another is that parasites and pathogens are more likely to be transmitted when individuals live in groups than alone.

> Swift locomotion, camouflage, mimicry and interactions with other species are examples of defence mechanisms against enemies, such as predators or parasites. Another common way of reducing the risk of predation is by living in groups.

Competition and territorial behaviour

When two people on opposite sides of a pond throw bread to ducks, the ducks are likely to distribute themselves between the two sources of food according to the rates at which the bread is thrown into the pond. If one person throws in twice as much food as the other, then that person is likely to attract twice as many ducks as the other. This can be called an **ideal free**

distribution of ducks; free because the ducks are free to choose between resource sites, and ideal because each duck goes to the place that provides the highest returns. For the ducks, the site with the higher rate of food delivery clearly provides more food, but if all the ducks foraged there, each of them would obtain less food because of competition. Clearly, it would pay some ducks to move to the other site where there is a lower delivery rate but less competition. Eventually, in an ideal free distribution, the ducks distribute themselves in the pond in proportion to the food delivery rate.

In this kind of competition, access to the resource depends only upon the number of other individuals competing for it. Later arrivals to the resource are not excluded, and no one individual attempts to monopolise the entire resource.

In some species, it is very common for one individual to exclude actively others from the resource. This may be achieved by establishing a *territory* or area that contains the resource, and defending it against other individuals (Fig. 28.9).

Territories often have clearly defined borders; birds, for example, may sing along the borders of their territory and chase off any intruder that crosses that border. The kinds of resources defended within a territory vary; for example, male frogs and butterflies may defend areas in which females lay eggs, birds defend nesting sites, and some ants defend areas that contain rich food resources.

Territorial behaviour

The reason why some species are **territorial** and others are not may be found in terms of the costs and benefits of defending a territory. Territory-holders do not have to share their resource, but instead must endure the costs of excluding others, including the time

Fig. 28.9 Magpies, *Gymnorhina tibicen*, are vigorous defenders of their territories

and energy required to chase off intruders. Both costs and benefits are likely to increase with larger territories, but not necessarily at the same rate. Thus, there may be a territory size that maximises the net benefit (Fig. 28.10). Larger territories may increase benefits, but at some point the additional benefits will be outweighed by higher costs.

Territories are not always defended by one individual. In many species, a male and female may jointly defend a territory, and in some species, such as the Australian grey-crowned babbler, *Pomatostomus temporalis*, family groups defend territories. Joint territoriality is frequently associated with breeding and not necessarily related to the economic defence of resources for each individual.

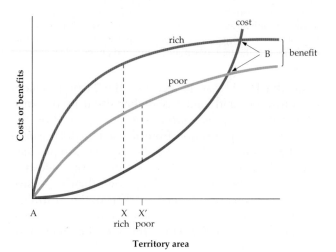

Fig. 28.10 The costs and benefits of defending a territory increase with territory size but at different rates. The benefits of defending territories of different sizes will depend upon the quality of resources within the area; smaller territories in quality habitats may provide as much benefit as larger territories in poor-quality habitats. There is likely to be a territory size X and X' at which the net benefit (benefits minus costs) is maximised. The resource is economically defendable between A and B

A nimals may defend areas containing food, nesting sites or other resources, thereby forming a territory. These territories are only defended if the benefits of exclusive use of the resource exceed the costs, in time and energy, of excluding others.

Animal contests

Competition between individuals of the same species for particular resources, such as food, nesting sites and sexual partners, frequently involves direct conflicts or **contests** between individuals. Some contests, such as fights between male red kangaroos, *Macropus rufus*, are obvious (Fig. 28.11), and in some species may result in serious injury. For example, between 5% and 10% of

Fig. 28.11 A contest between two male red kangaroos, *Megaleia rufa*. The males grab each other with their front legs and attempt to kick each other with their hind legs while using their tails for balance. The claws on the hind legs are capable of causing a serious injury to the opponent

musk ox bulls, *Ovibos moschatus*, die each year as a result of injuries sustained through fights. However, not all contests between individuals are as dramatic or obvious. Sea anemones, while appearing to be sedentary, often engage in fights using specialised tentacles, called acrorhagii, as weapons against each other.

Animal contests that result in physical injury are not common and most disputes are resolved before there is serious injury to either contestant. Examination of the costs and benefits of different kinds of contests reveals why this is the case. Winning a contest will provide benefits in terms of the resources obtained but may also incur costs in terms of time or injury. For example, if the costs of injury to a contestant that defends its resource are greater than the value of that resource, the contestant should relinquish the resource before the fight escalates. The relative costs and benefits of the different behaviours can be examined using ideas developed in a mathematical technique called **game theory** (Box 28.2), which is also used commonly in economics. Game theory models show why contestants usually spend a lot of time displaying their relative

strength, and contests involving physical injury are relatively rare.

In reality, the outcome of contests between individuals depends upon a variety of factors, including the fighting ability of the contestants, the value of the resource to each contestant, and which party currently owns the resource. Clearly, it would be to a contestant's advantage to be able to assess the fighting ability of another individual before engaging in combat, and there are numerous examples of contestants advertising their fighting ability by singing, calling, roaring and so on, before the fight escalates. The display may escalate into more costly behaviour, in terms of energy expenditure or risk of physical injury, if the contestants are more evenly matched and the earlier displays did not reveal their relative strengths. Further escalation may occur if the value of the resource exceeds the cost of fighting, but either contestant may leave if the risk of injury exceeds the value of the resource.

Competition between individuals for scarce resources can sometimes result in direct conflict or physical contest. These contests can result in physical injury, but most disputes are resolved before this occurs.

Courtship and mating behaviour

A crucial requirement for all sexually reproducing animals is finding and mating with a partner of the same species. Males and females usually come together as a result of one sex advertising its location to the other. This is achieved using a variety of chemical, auditory and/or visual cues: female moths produce a sex pheromone that attracts males; male frogs attract females by calling; and males of many species of fireflies attract females by producing flashes of light. An important feature of these sex-attracting mechanisms is that they are species-specific. Thus, species of frogs have calls that are distinct from the calls of other species that are found nearby, and the chemical composition of moth sex pheromones differs between species. These differences ensure that the advertising cue attracts individuals of the same species, thereby reducing the possibility of either sex wasting time or energy attempting to fertilise eggs with the sperm of another species. This is particularly important for the female, who may lose her entire reproductive output, whereas the male can at least attempt to mate with another female.

Courtship

In many animals, **courtship behaviour** takes place after the two sexes have located each other. Courtship

BOX 28.2 Hawks and doves: fighting behaviour and evolutionarily stable strategies

Assume that two contestants can engage in a contest in one of two ways. A 'hawk' strategy means an individual will fight until its opponent either dies or is too seriously injured to continue, and a 'dove' strategy means an individual only displays and never engages in serious fights. When a 'hawk' encounters a 'dove', the 'hawk' always wins because the 'dove' always retreats without fighting. When a 'hawk' encounters a 'hawk', they fight and we assume that each wins on half the occasions. Similarly, a 'dove' encountering another 'dove' always displays and wins on half the occasions. Although imaginary, these two 'strategies' represent the two possible extremes of fighting behaviour.

We can investigate whether each of these strategies can be maintained in the population by assigning scores to the outcomes (winning or losing) and the costs (of injury and time wasted in a display) of fights between individuals. As an illustration, we can assign the following imaginary numbers that reflect some measure of fitness as a result of the conflict. Let the value of the resource equal 40. Therefore, the winner scores +40 and the loser 0. The cost of injury equals −60; and the cost of displaying equals −10. We will further assume that hawks and doves reproduce their own kind, and the number of offspring produced is in proportion to their overall scores. A higher score at the end of the contest is the equivalent of producing more offspring. The exact values do not matter; the game could be analysed using algebra. The pay-offs to the attacker for each strategy played against the other strategies can be shown as a matrix:

	Opponent	
Attacker	Hawk	Dove
Hawk	0.5(40) + 0.5(−60) = −10	+40
Dove	0	0.5(40−10) + 0.5(−10) = +10

It is now possible to examine what would happen in the population if individuals played different strategies. If all individuals played 'dove', then the average pay-off for each individual would be +10. However, a mutant 'hawk' individual introduced into this population would be at a great advantage, and the 'hawk' strategy would rapidly spread. But once the population comprises mostly hawks, doves would be at an advantage because the average pay-off for the 'hawk' strategy is −10, which is less than 0, the average pay-off when a dove encounters a hawk.

Clearly, it is best to be a hawk in a population comprising mostly doves and a dove in a population consisting mostly of hawks. In each situation, selection will favour the rarer strategy. Nevertheless, a stable mixture of hawks and doves can occur, when the average pay-off for hawks equals that for doves. The stable mixture can be calculated quite simply. Let h be the proportion of hawks. The proportion of doves will then be $(1-h)$. The average pay-off is the pay-off for each type of fight multiplied by the probability of meeting that type of opponent:

Average pay-off to hawks, $H = -10h + 40(1-h)$

Average pay-off to doves, $D = 0h + 10(1-h)$

The stable mixture of hawks and doves occurs when $H = D$, that is,

$$-10h + 40(1-h) = 0h + 10(1-h)$$

which is true when $h = 0.75$. Thus, the population is stable when three-quarters of the population are hawks and one-quarter are doves. This mixture could arise in two ways. The population could comprise individuals that played pure strategies of either hawk or dove, with three-quarters of the population being hawks and one-quarter doves. Alternatively, the population could comprise individuals that played mixed strategies of hawks and doves, with every individual playing hawk three-quarters of the time and dove one-quarter of the time. Either scenario would result in an **evolutionarily stable strategy**, which is a strategy that cannot be beaten by any other strategy in the game. The important point about this very simple model is that the behaviour of one individual can depend critically upon the behaviour of other individuals within the population.

behaviour varies dramatically between species, ranging from little more than the male moving directly towards the female, to an elaborate display of vocalisations and body movements. Courtship behaviour may provide additional confirmation that a male and female are of the same species if there are particular aspects of the behaviour that are species-specific, such as the leg-waving behaviour of many jumping spiders. The female's response to a male's courtship behaviour can also indicate to the male whether the female is ready to mate.

The courtship displays of many animals are far more complex and elaborate than might be expected if their sole function were to indicate species identification and readiness to mate. For example, the courtship display of the raggiana bird of paradise, *Paradisea raggiana*, involves an extraordinary combination of colour, movement and vocalisations (Fig. 28.12). The male Prince Rupert's blue bird of paradise, *Paradisea rudolphi*, sings from a perch and then slowly rotates backwards. Once he is hanging upside down, he shakes himself with repeated movements, extending his plumage and making a soft monotonous song.

The extraordinary courtship displays of many animals have most likely evolved as a result of **sexual selection** (Box 28.3). Sexual selection favours those characteristics in males that improve their chances of

Fig. 28.12 A male raggiana bird of paradise, *Paradisea raggiana*. The female's plumage is quite dull in comparison. The colourful plumage and extravagant courtship behaviour of the male has evolved through sexual selection by female choice

BOX 28.3 Sexual selection and secondary sexual characteristics

The theory of sexual selection was first proposed by Charles Darwin, who was interested in explaining the extravagant and often brightly coloured adornments that are characteristic of the males of many species, such as the peacock, *Pavo cristatus* (Fig. a). The problem for Darwin was that these adornments seemed unlikely to have evolved through natural selection (Chapter 32) as a result of improved survivorship for the male. In order to understand why these adornments may have evolved (and why it is mostly males that possess them) it is necessary to appreciate the fundamental difference between the sexes: males produce sperm and females produce eggs. A single egg is more expensive to produce than a single sperm, and a male has the potential to fertilise far more eggs than a female can produce. This means that while female reproductive success depends primarily on her ability to produce eggs, male reproductive success depends on the number of eggs he can fertilise, which will increase with the number of mates he can secure. Females are a limiting resource for male reproductive success because males are generally capable of mating more females than are available.

The difference between the sexes means that males may compete with each other for access to female partners. The rewards for male success in this competition, in terms of reproductive success, are high and hence selection for male ability to secure fertilisations is strong. Selection of characteristics that increase mating success is

(a) The stunning appearance of the male peacock, *Pavo cristatus*, is thought to have evolved as a result of positive feedback between female choice and elaborate plumage in males referred to as sexual selection. Sexual selection can occur in two ways. The first is *intrasexual selection*, which favours the ability of one sex (usually, but not always, the male) to compete directly with other members of the same sex. The other way is *intersexual selection*, in which some characteristics in one sex are favoured by the other sex. Darwin's original idea was concerned with those characteristics responsible for mating success but it is now recognised that sexual selection can occur after copulation has commenced, particularly if females mate with more than one male and their eggs can be fertilised by any of these males.

Intrasexual selection favours any trait that increases a male's competitive ability and it is often referred to as male–male competition. The most obvious traits are weapons for fighting, such as horns on stag beetles and many ungulates (Fig. b). These traits allow the male to monopolise the female and prevent other males from gaining access to her. Intersexual selection occurs when one sex shows a preference for certain individuals of the other sex. The sex that provides the greatest material investment in offspring is the one that shows the preference. In most species, the female provides the greater investment and hence

(b) A red deer stag, *Cervus elaphus*, and a male stag beetle: both have armaments that are used in contests with other males over access to females

intersexual selection is often referred to as *female choice*. Females may select males on the basis of their ability to provide resources, such as food and oviposition sites, and on the basis of their genetic qualities (in addition to their ability to provide resources).

The elaborate adornments and courtship displays of males of many species (Fig. a) are thought to have evolved because they are an accurate indication or *honest signal* of male quality. For example, the differences among male barn swallows, *Hirundo rustica*, in their resistance to parasites and pathogens has a genetic basis, and males with fewer parasites tend to have longer tail feathers than those with higher parasite loads. Thus, theory would predict that female barn swallows should prefer males with longer tail feathers, and this was confirmed by experiments that manipulated the length of male tail feathers. In barn swallows, tail feather length is an honest signal of male quality because only males with few parasites have long tails. Female choice could create a positive feedback system because females that prefer to mate with males with elaborate adornments will produce sons with relatively elaborate adornments and daughters with a relatively greater preference for these adornments. The opposing effect of natural selection on male survival may explain why the extravagant tail feathers of some birds are not even more extravagant.

Sexual selection does not necessarily cease after the male has mated if the female mates with other males and their ejaculates compete for fertilisation of her ova. This competition between ejaculates is called **sperm competition** and an extraordinary diversity of male traits has evolved to ensure that it is prevented or avoided. Males of the Australian big greasy butterfly, *Cressida cressida*, deposit a substance over the genital opening of the female while mating. The substance hardens to form a permanent structure, known as a sphragis, that prevents other males mating with her, and thereby ensures that only his sperm fertilises her eggs. Female choice may also continue after mating for a multiple-mating female because she could ensure that the sperm of a favoured male fertilises most of her eggs. Females of the orb-web spider *Argiope keyserlingi* can terminate copulation by capturing and eating the male. Experiments in which females mated with two males showed that the females delay cannibalising relatively smaller males, and these smaller males fertilise most of her eggs.

fertilising the eggs of the female, through the process of either **male–male competition** or **female choice**. The former process does not involve courtship, but the latter refers to female discrimination between males on the basis of the male's secondary sexual characteristics, which may include courtship displays. Birds seem to have an especially spectacular range of courtship displays. For example, the male Australian musk duck, *Biziura lobata*, courts a female by leaning back and spraying her with water using his legs, while male superb fairy wrens, *Malurus cyaneus*, present potential female mates with a flower petal. Male superb lyrebirds, *Menura novaehollandiae* attract females by making calls that mimic those of other species of birds or other noises, including those made by humans.

In some species, courting males provide females with a *nuptial gift*. These gifts may be a prey item or a 'package' containing material synthesised by the male. Males of several species of birds provide the female with a prey item (Fig. 28.13) and this may reflect the ability of the male to provide for the offspring later. Male lygaeid bugs improve their reproductive success by providing females with a nuptial gift that contains nutrients that the females use to produce more or larger eggs (Fig. 28.14). Courting males of most species of empiid midges can only mate with the female if they present her with an insect gift that is wrapped in silk. However, in one species, the male cheats by presenting the female with an empty silk cocoon. Although there is no nutritional value to this gift, the male must still present it if he is to mate with the female.

Courtship may also provide a mechanism for reducing the risk of *sexual cannibalism*, although only when the female attempts to consume the male before mating takes place. Sexual cannibalism that takes place after copulation has commenced, which occurs in nudibranchs, praying mantids and certain spiders (most notably the red-back spider, *Latrodectus*) may be a way

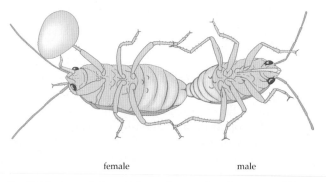

female   male

Fig. 28.14 The male lygaeid bug collects and predigests an edible seed as a nuptial gift for his partner. After giving his gift to the female, she permits him to copulate while she consumes the gift

in which the male or female manipulates the paternity of her offspring (see Box 28.3). It is often stated that sexual cannibalism in praying mantids is necessary for the male to copulate effectively but, in fact, they can mate successfully without losing their lives. Sexual cannibalism in these species most likely occurs when the female has been deprived of food and devouring the male may increase her fecundity.

Males of numerous species often associate with the female but without displaying courtship behaviours. This **mate-guarding behaviour** may prevent rival males from mating with the female or fertilising her eggs. Mate guarding can occur before or after mating has taken place. Pre-mate-guarding of virgin females is more likely if the sperm from the first male to mate with the female fertilises most of the eggs, whereas males are more likely to guard females after mating if the sperm from the last male to mate fertilises most of the eggs. For example, in damselflies, the male guards the female by grasping her thorax with the tip of his abdomen in the so-called tandem or wheel position. This prevents other males from subsequently mating with the female and thus protects his sperm. Males of the linyphiid spider, *Linyphia*, destroy the female's web before mating takes place. The silk of the female's web contains sex pheromones that attract other males, so, by destroying her web, the male reduces the possibility that other males find and mate with her.

Males often exhibit elaborate and complicated courtship behaviours after a female has been located. These behaviours can enable each sex to determine whether they are of the right species and provide the female with a basis from which to choose between males.

Parental care

In many animals, one or both parents may provide some form of **parental care** for their offspring, such as feeding them, or protecting them from predators.

Fig. 28.13 Male azure kingfishers, *Alcedo azurea*, offer prey items to the female during courtship

Whether the female alone, the male alone, or both parents care for the offspring depends on the species. For example, in seahorses, it is the male that provides parental care by carrying the fertilised eggs in a special brood pouch (Fig. 28.15). The amount of care a parent might provide depends upon a variety of factors but frequently reflects a balance between the costs (in terms of lost opportunities to reproduce) and the benefits (in terms of increased survival of the offspring). Clearly, infant mammals would have little chance of surviving if their mothers failed to provide sufficient milk, while the female of some thomisid and eresid spiders has little to lose by allowing her offspring to devour her since she will not survive the winter. For males, the time spent caring for offspring may impinge on the time spent seeking copulations with females other than their social partner (*extra-pair copulations*). Male fairy martins, *Hirundo ariel*, adjust the time they spend incubating their social partner's eggs according to the opportunity to mate with other females in the colony. When there is a high proportion of fertile females, the male spends less time in his nest and more time seeking matings with other females.

The conflict between providing care and seeking additional matings is at least partially reflected in the relationship between which parent provides parental care and the **mating system** of the species. Mating systems are defined according to the number of partners each sex may have during its lifetime or during the mating season. There are four main categories of mating system: *monogamy*, where the male and female form a relatively long-lasting pair bond; *polygyny*, where a male mates with several females but a female mates with only one male; *polyandry*, the reverse of polygyny; and *promiscuity*, where both males and females may mate with several individuals. These definitions more properly describe social relationships; in most monogamous birds, the male may help his partner raise their chicks but will also have obtained extra-pair matings with other females.

The comparative relationship between the provider of parental care and the mating system is shown in Table 28.1. Of course, there are many exceptions to this pattern but, in essence, the sex that does not provide care deserts its partner and mates with other individuals. In mammals, females are predisposed to providing care: the offspring have a prolonged gestation and, after birth, the mother must provide the infant with milk. There is little parental care that the male can provide other than protecting the female and her young and, in general, mammals are polygynous.

In most species of birds, the number of offspring that can be reared successfully depends upon the amount of food that is provided by the parents. Since two parents can provide more food than one, both parents could increase their reproductive success by staying together. If one parent deserted, its reproductive output may not only be reduced but it would also have to spend additional time finding a new partner. Thus, in birds, monogamy and care provided by both parents is the general rule. These monogamous relationships can last a lifetime in some species, particularly many seabirds (Fig. 28.16).

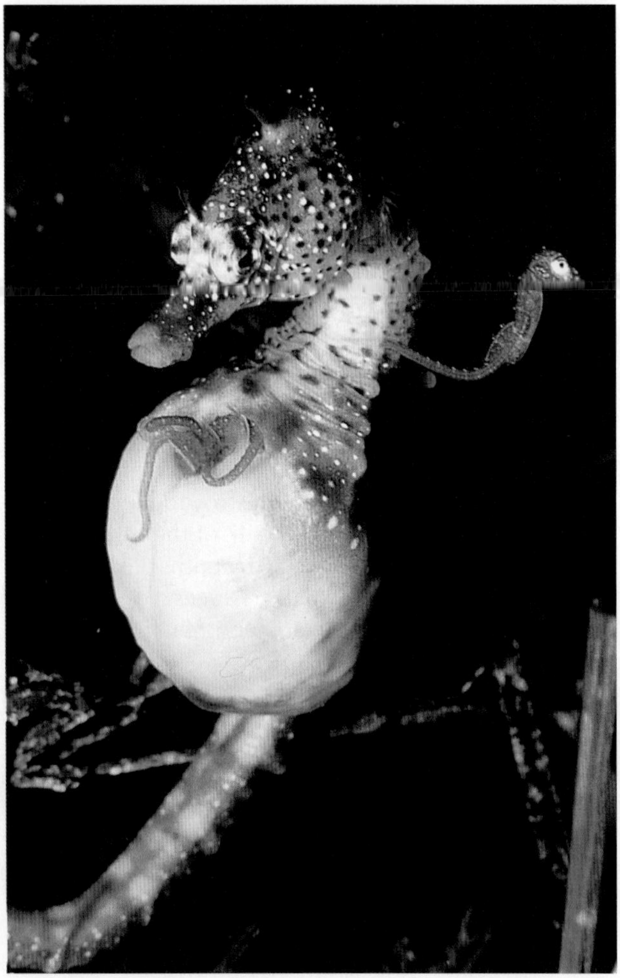

Fig. 28.15 Male seahorses, *Hippocampus*, brood care for their developing offspring by maintaining and feeding them in a special pouch. Here the offspring are being born

Table 28.1 Broad associations between vertebrate mating systems and parental care

Parental care	Mating system	Taxonomic example
Male and female	Monogamous	Many birds
Only females	Polygynous	Most mammals
Only males	Polyandrous/ promiscuous	Some fishes

Fig. 28.16 Male and female wandering albatrosses may court for several years before mating and attempting to produce their first clutch of eggs. They will then remain mates for the rest of their lives

Guarding and fanning eggs and brood are the primary forms of parental care shown by fish and can be performed by either parent alone. The question of which sex provides care seems to depend partly upon whether the species has internal or external fertilisation. If fertilisation is internal, the female cares for the offspring, while the male cares for the eggs if fertilisation is external. This pattern may arise because external fertilisation gives males a high degree of certainty about the paternity of the eggs. If fertilisation is internal, the female may have mated with other males and thus the male may care for eggs that he has not fertilised. In contrast, the paternity of eggs fertilised externally will depend upon the number of males in close proximity and the male can adjust his level of care accordingly.

Parental care increases the survival prospects of offspring and can be provided by either or both parents. The amount of care provided may depend on a balance between the survivorship of the offspring and future reproductive opportunities for the parents.

Social organisation and co-operative behaviour

In most species, individuals live a generally solitary life, only briefly pairing up with a member of the opposite sex for mating and reproduction. In others, individuals may form loose aggregations as a result of their attraction to a particular resource, such as a water hole or rich supply of food, rather than to each other. Vinegar flies, *Drosophila*, which congregate around rotting fruit, are an example of this kind of aggregation. The size and membership of these aggregations usually varies over time and although individuals within these groups may frequently interact with each other, these interactions are likely to be competitive in nature.

Aggregations, flocks or schools of individuals can become established because individuals benefit more by being in a group than by being alone. One advantage of living in a group is that individuals gain protection against predators (p. 747). Another is that individuals in a group may gain protection from the cold (Fig. 28.17). Group living may also provide benefits in terms of obtaining food. Individuals can take advantage of the food-finding abilities of others or, by foraging together, may increase the chance of capturing prey. The latter is a common feature of the foraging behaviour of social carnivores, such as lions, hyenas and hunting dogs; some individuals may drive prey towards others that are concealed by vegetation, or may share in the task of chasing the prey until it is exhausted.

More complex *social behaviour* is found among individuals that form essentially permanent groups. Individuals in such groups usually *co-operate* in finding, hunting and processing food, defending themselves against predators and competitors, and in rearing young. Social behaviour occurs in a diversity of animals, including spiders, crabs, insects, birds and mammals. Individuals of the Australian social thomisid spider, *Diaea*, co-operate in building nests by weaving together *Eucalyptus* leaves with their silk. Nests may contain over 50 individuals, which shelter in the nest by day and co-operate in capturing insect prey that walk on or near the nest at night. Individuals live together in social groups because their co-operative activities result in individuals obtaining more prey and because they are less likely to be victims of predators than if they lived alone.

The composition of *social groups* varies between species, resulting in considerable diversity in the ways in which individuals interact with each other. Interactions between individuals within a *Diaea* nest, for example, do not seem to vary, perhaps because the interests of each individual are similar. In contrast, the

Fig. 28.17 Emperor penguins huddle in groups to reduce the chill factors of Antarctic winds

interactions within groups of primates, such as chimpanzees or baboons, are considerably more variable, leading to a more complex social organisation.

The relationships between individuals within a group primarily reflect the conflicts and congruences of interests that arise within these groups. Social groups usually comprise one or a few breeding individuals, infants, juveniles, and in many cases a number of other adults that do not raise their own offspring (see below). There are exceptions: some groups do not contain non-breeding adults and, in some species, males may, during part of their lives, form bachelor groups. The number of breeding individuals varies, ranging from one male and one female, through one male and several females, to a few males and numerous females. The principal reason for the conflicts of interests within a group is the difference between individuals in the opportunity for breeding.

Dominance relationships between individuals in a group arise when one individual gains access to a contested resource at the expense of another. Domestic chickens, for example, have a linear dominance hierarchy in which one individual is the most dominant in the group and is capable of displacing all others. Below her is a second-ranking bird who can dominate all but the top-ranking bird, and so on until the most subordinate bird, which is displaced by all other birds. Chickens are unusual in having this kind of dominance hierarchy. Primates, for example, may have more complex hierarchies, such that individual A may dominate individual B and B dominate C, but A may not necessarily dominate C. An important feature of dominance hierarchies is that, while there may be considerable antagonistic or aggressive behaviour between individuals during the establishment of the hierarchy, this behaviour rarely continues.

Although individuals within a social group co-operate in various ways, some adults in the group may not themselves reproduce but instead help raise the young of the breeding individuals. This behaviour is called co-operative breeding and is found in insects, birds and mammals.

> ndividuals of many species form permanent groups and exhibit a variety of social behaviours, including co-operative hunting, defence and rearing of offspring. Social groups also generate competition over resources and opportunities for breeding.

Co-operative breeding in birds and mammals

Species of birds and mammals in which adult individuals regularly help rear the young of other individuals are called **co-operative breeders**. Helpers may provide the young with food or protection from

predators. In some species, such as the Florida scrub jay, *Aphelocoma coerulescens*, or the silver-backed jackal, *Canis mesolmelas*, in Tanzania, helpers are young individuals that do not disperse from their parents' territory but instead help their parents raise the next brood of offspring. Although this is probably the most common system, other species have more complex social organisations. In some species, such as the Australian bell miner, *Manorina melanophrys*, groups are larger and comprise several 'nuclear' families that live together in a single large territory. Other co-operative breeders, such as the white-fronted bee-eater, *Merops bullockoides*, are highly gregarious and live in large colonies. Smaller extended family groups, including helpers, exist within the colony but, unlike miners, helpers attach themselves to a single nest. The social organisation is more complex in species in which the breeding individuals are not monogamous but males and females mate with several partners. This forms the basis of the social organisation of some primates, such as saddle-backed tamarins, *Sanguinus fuscicollis*.

Co-operative breeding occurs in about 3% of all the species of birds in the world but, curiously, over 22% of the Australian passerine (perching) birds breed co-operatively (Fig. 28.18). Why is co-operative breeding so common in the Australian bird fauna compared with the rest of the world? The answer most likely lies

Fig. 28.18 The co-operatively breeding superb fairy wren, *Malurus cyaneus*. This and several other co-operatively breeding Australian birds, including magpies, butcherbirds and some honeyeaters, are commonly seen in urban areas. This male has just fed the chicks in the nest

in the biogeographical history of Australian birds (Chapter 41) rather than any peculiar feature of the Australian environment that might favour co-operative breeding. Recent genetic evidence indicates that the Australian passerines have radiated from two groups: an old endemic radiation called the Corvi and a more recent group of immigrants from Asia (Chapter 41). The Australian cooperative breeders all belong to the Corvi, and co-operative breeding is not found in any of the comparatively recent invaders. Thus, the high frequency of co-operatively breeding birds in Australia is not the result of repeated evolution among several taxa but rather of speciation from a common, co-operatively breeding ancestor.

The most extraordinary co-operatively breeding vertebrate is the naked mole-rat, *Heterocephalus glaber*, of Africa, whose social system was first reported in 1981. Naked mole-rats live underground in colonies of up to 300 individuals. The colony comprises one breeding female, who mates with several males. These males provide most of the direct care of the young. Other individuals in the colony, including the non-breeding females, defend and maintain the underground tunnel system. Younger individuals bring food to the breeding nests and generally keep the tunnels clean, while older individuals excavate new tunnels and defend the colony from predators, such as snakes. Why this lifestyle has evolved in the naked mole-rat is still a matter of contention but this species is the closest mammalian analogue of a form of social organisation that is found among two groups of insects: the Hymenoptera (bees, ants and wasps) and termites.

Insect societies

The social systems found among insects are among the most diverse and spectacular. The most simple social system involves groups of offspring that are cared for by their mother for extended periods, as in the female intertidal beetle, *Bledius spectabilis*, which remains in her mud-burrow with her young in order to prevent them from suffocating. At the other end of the continuum are the complex eusocial societies of many species of bees, wasps, ants and all termites. **Eusocial societies** are characterised by overlapping generations of individuals and reproductive division of labour, where some individuals reproduce and others do not. Such societies can be vast. A large ant or termite colony can contain as many as 10^5–10^7 individuals and occupy a complex home that is 10^9 times as large as any of the individuals within it (Fig. 28.19). Typically, these colonies are based on the reproductive output of a single *queen*. Her daughters, called *workers*, undertake duties such as maintenance of the colony, obtaining food and caring for the queen's brood. Not all eusocial species have a single queen; many species have several queens

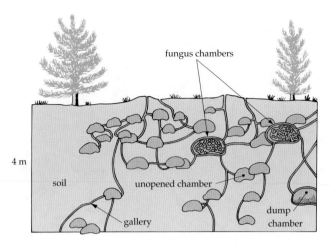

Fig. 28.19 The Texas leaf-cutter ant, *Atta texana*, constructs very large nests. One in the southern United States, excavated by a bulldozer, was over 4 metres deep and contained 110 chambers. The ants take leaf fragments below ground as a substrate for culturing fungus. The ants eat parts of the fungus and remove wastes to dump chambers

in one colony. Colonies of some species of Australian *Iridomyrmex* ants can be very large, comprising several widely separated nests connected by trails. The queens can be found either in one central nest or distributed throughout the smaller satellite nests. Among ants, bees and wasps, workers are always female and males make no contribution to nest maintenance.

Typically, a colony is founded by a singly mated female, whose eggs hatch into workers that both forage and build the nest. The founding queen continues to produce workers and the colony increases in size. When the number of individuals in the colony reaches a certain size, which varies from species to species, the queen starts producing a few females that will eventually leave the nest as 'reproductives', or female *alates*. Males are also produced at this time and the line is continued when these male and female alates leave the nest to mate and found new colonies.

In some species, individuals within the colony have different morphologies and perform different tasks. These different morphs are called *castes*, and the range and number of castes within a species can be quite spectacular (Fig. 28.20). Soldiers are usually larger than others in the nest and their large mandibles are effective in defending the nest from enemies. In some species, such as honey-pot ants (e.g. *Camponotus*), some individuals are essentially food containers hanging from the roof of the nest with their abdomens distended with food. These ants live in arid zones where food availability is unpredictable. Foraging ants return to the nest and regurgitate their food to the 'honey-pots', which then effectively store it in their bodies. During times of food shortage, the members of the colony can obtain food from the 'honey-pots' through regurgitation (Fig. 28.21).

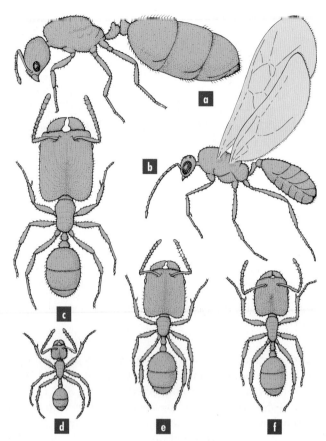

Fig. 28.20 The different castes of the myrmicine ant, *Pheidole tepicana*. The proportions of the body (and especially the head) change dramatically with size from the queen **(a)**, male **(b)**, and four kinds of workers **(c–f)**

Fig. 28.21 Honey-pot ants, *Camponotus*

The organisation of colonies of eusocial insects, including the care of young, construction and maintenance of the nest, collection of food and provisioning of colony members, is mediated by a complex communication system. When enemies threaten the nest, some individuals release *alarm pheromones* and soldiers rush to the location of the threat, while other workers will move the queen to the safest place in the nest. The location of food is indicated by chemical trails that workers leave on the substrate. Not all species rely entirely on chemical pheromones;

honeybees communicate to nest mates the precise direction, distance and nature of food sources using a complex system of dances (Fig. 28.22).

Superficially, the nests of these eusocial insects may resemble a superorganism, in which the individuals are analogous to the cells of a single organism. However, while individuals within a nest are engaged in extraordinarily co-operative relationships, the opportunity for behaving in more selfish ways, such as cannibalising eggs, occurs frequently in many eusocial species.

Evolution of co-operation

Individuals that co-operate to help others raise young are apparently forfeiting their own reproductive success while enhancing the reproductive success of others. The evolution of this apparently **altruistic** helping behaviour is puzzling because natural selection is not expected to favour behaviour that reduces an individual's reproductive success. The puzzle may be resolved for co-operatively breeding vertebrates by asking first why individuals delay breeding and remain in their natal territory, and secondly why they help raise the offspring of other individuals. Individuals often delay breeding because there are insufficient breeding territories available and these individuals must wait for a vacancy before leaving their natal territory. The non-dispersing individuals may subsequently help in order to gain experience in raising offspring, which benefits them later in life. Alternatively, the parents may eject them from the territory if they do not help.

Explanations for the evolution of eusociality in insects are more complex because, with few exceptions, the workers remain workers and never become queens. Unlike co-operatively breeding birds or mammals, workers do not delay reproduction, they simply do not reproduce. This is an extreme form of altruism, and we still do not understand fully the evolution of eusocial behaviour in insects. Three kinds of explanation are currently favoured and essentially they take the altruism out of altruistic behaviour. The first suggests that there is a mutual benefit of helping to both the parents and workers. This idea is equivalent to the explanation of co-operative breeding in vertebrates and only applies to those species in which workers are capable of producing eggs. The second explanation is that worker progeny are manipulated into helping by their parents. The mechanism by which parents are able to prevent workers from dispersing and reproducing but instead to remain in their natal nest as helpers is not fully understood. The important point is that the workers' behaviour is not necessarily co-operative but rather the result of selection favouring parental self-interest.

The third explanation emphasises genetic considerations and is derived from the concepts of inclusive

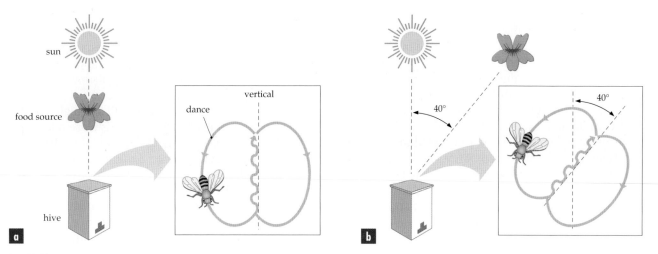

Fig. 28.22 The dance of the honeybee, *Apis mellifera*, is a complex behaviour that conveys a variety of information about the source, location and quality of different food sources

fitness and kin selection (see Box 28.4). These concepts reveal that co-operative behaviour between individuals that are close genetic relatives is likely to evolve because they share a higher proportion of genes, including those favouring this behaviour, than do individuals that are not closely related. Hence, an allele that favours co-operative behaviour can spread in the population. Eusociality has evolved independently several times in Hymenoptera and this may be due to their peculiar haplodiploid genetic system, in which females are diploid but males are unfertilised and therefore haploid. If the queen only mates with one male, then her daughters are more closely related to each other than they are to the queen. Thus, selection for helping raise sisters, at the expense of one's own offspring, may be more easily favoured.

These different kinds of explanations are not mutually exclusive and the importance of each will depend upon the biology of each species. Nevertheless, the emerging consensus is that this extreme form of co-operative behaviour is not such an evolutionary puzzle.

At best, the non-reproductive workers (or helpers at the nest) gain at least some level of reproductive success, while at worst their behaviour is the result of parental manipulation. The outcome is that among these apparently co-operative individuals there may be considerable conflicts of interest. One conflict arises over the sex ratio of the reproductives, because the preferred ratio of sons to daughters may not be the same for the queen and workers. Additionally, if the queen mates with more than one male, workers prefer rearing sons produced by the queen rather than by other workers. In these circumstances, the workers may actively prevent other workers from laying eggs, a behaviour termed *worker policing*.

The evolution of co-operative breeding is intriguing because natural selection is not expected to favour behaviour that reduces an individual's reproductive success. Several plausible explanations for this behaviour have been proposed.

BOX 28.4 Inclusive fitness, kin selection and co-operation

Inclusive fitness is a concept of evolutionary theory and particularly the evolution of co-operative behaviour. The logic behind this concept is that selection acts at the level of the gene rather than the individual; only genes are passed on from generation to generation. Individuals can be thought of simply as vehicles for genes to be represented in the next generation. For example, genes for parental care are likely to spread in the population because genes that ensure their

'bodies' are more effective at caring for their young will perpetuate themselves in the bodies of the cared-for young. However, while the only way that genes can be passed from generation to generation is by parents to offspring, care of one's own offspring is not the only behaviour that can ensure that genes for parental care will spread in the population. Genes for helping a relative raise offspring can also spread in the population. This is because close relatives share a higher proportion of genes than two distantly related individuals, and consequently have a higher chance of sharing the gene for this behaviour. Genes for helping sisters or brothers thus help the same genes carried by these relatives and can perpetuate themselves through the children of these relatives. Selection that takes account of relatives as well as direct descendants is termed **kin selection**.

Helping a relative reproduce will only be favoured if the benefit of helping is greater than the cost (i.e. reduction in reproductive success) to the helper. Helping a sister raise offspring is unlikely to evolve if the help does not result in an increase in the sister's reproductive output and it also reduces the reproductive output of the helper. Clearly, the degree of relatedness between individuals is important. Strategies that cause an individual to forfeit reproduction to help another unrelated individual raise its offspring are unlikely to be maintained in the population because the gene for this behaviour is unlikely to be passed on to the next generation.

W. D. Hamilton argued that selection tends to maximise the rate at which alleles are transmitted to the next generation. Individuals can pass on copies of their alleles in two ways: directly by producing their own offspring and indirectly by increasing the production and/or survival of relatives such as brothers and sisters. Fitness realised by direct reproduction is called *direct fitness*, whereas that realised through collateral relatives is called *indirect fitness*. These two fitness components, when added together, are referred to as *inclusive fitness*. Natural selection tends to maximise the direct fitness of individuals,

while kin selection tends to maximise their inclusive fitness. Investing in offspring leaves fewer resources for investing in other relatives (and vice versa) because individuals have limited time or resources.

Generally, indirect fitness will be high when the altruistic aid is directed towards close relatives and when these relatives are able to benefit greatly from this aid (e.g. a young sibling being protected from a predator). Altruism can also be favoured if the cost to the altruist is low (e.g. if defending a sibling entails little risk of being killed or injured). Balancing the costs and benefits of altruism can be summarised by Hamilton's Rule, which states that a donor is expected to help a recipient when $rb - c > 0$, where r is a measure of the relatedness, or genetic similarity, between donor and recipient; b is the fitness benefit of the aid to the recipient; and c is the fitness cost of the behaviour to the donor.

Molecular genetic techniques provide good estimates of the genetic relationships between different individuals. However, measuring the costs and benefits of co-operative behaviour is more difficult. For example, it is possible to measure the benefits of helping by comparing the number of offspring produced by parents with and without helpers, but it is not always possible to establish the number of offspring a helper might have produced had it not helped. Nevertheless, the ideas of kin selection and inclusive fitness have been extremely important in helping understand the evolution of co-operative behaviour and the social organisations that derive from this behaviour.

Summary

- Ethology is the study of animal behaviour and its aims are to explain animal behaviour in terms of causation, development, adaptive value and evolutionary history.
- Behaviour, being an integral part of an animal's phenotype, has a genetic basis and is consequently subject to the influence of natural selection.
- An animal's behaviour may change during its lifetime; some changes develop with age while others occur as a result of learning.
- Animals obtain food in diverse ways and for many species they must make choices about where, what and when to eat particular food items. These choices often reflect a trade-off between the benefits of particular foods and the costs of obtaining them.
- Numerous mechanisms have evolved in response to reducing the risk of predation, including camouflage, mimicry, living in groups, or even forming mutualistic associations with other species.
- Animals often compete for resources but individuals can ensure exclusive use of these resources by defending territories from intruders. The size of a territory and its owner's tenure will depend on the balance between the benefits of the resource and the costs of its defence.

- Conflicts over access for resources can result in physical contests; however, they rarely result in serious injury to either contestant if the benefits from obtaining the resource are exceeded by the cost of injury.
- Males and females locate each other for mating using a variety of auditory, chemical and visual cues. Courtship behaviour subsequently takes place and may function to allow species' identification and the female to choose her mating partner.
- Parental care increases the survival prospects of offspring and can be provided by either or both parents. The amount of care provided may depend on a balance between the survivorship of the offspring and future reproductive opportunities for the parents.
- Complex social behaviour is found in species that form essentially permanent groups, and individuals in these groups may co-operate in obtaining food, defence against predators, constructing places in which to live and caring for offspring.
- Individuals of some species forfeit breeding in order to help raise the offspring of other individuals. The evolution of this behaviour is still not fully understood.

keyterms

altruism (p. 758)
animal contests (p. 748)
co-operative breeding (p. 756)
courtship behaviour (p. 749)
cryptic (p. 746)
defence mechanisms (p. 746)
development of behaviour (p. 741)

dominance relationship (p. 756)
ethology (p. 740)
eusociality (p. 757)
evolutionarily stable strategies (p. 750)
female choice (p. 753)
foraging theory (p. 744)
game theory (p. 749)
ideal free distribution (p. 747)
imprinting (p. 741)

inclusive fitness (p. 759)
kin selection (p. 760)
learning (p. 741)
male–male competition (p. 753)
mate-guarding behaviour (p. 753)
mating system (p. 754)
mimicry (p. 746)
natural selection (p. 740)

parental care (p. 753)
proximate explanation (p. 740)
sexual selection (p. 751)
sperm competition (p. 752)
territorial behaviour (p. 748)
ultimate explanation (p. 740)

Review questions

1. Define the four suggested answers to the question 'Why do male bowerbirds construct and decorate bowers?' (p. 740) as either ultimate or proximate explanations. Explain your reasoning.
2. How can behaviour that has been learned be distinguished from that which has developed?
3. What kinds of 'decisions' are made by foraging animals? Illustrate your answer with examples.
4. What are the advantages of foraging in a group? Are all individuals in the group similarly advantaged?
5. Explain why animal contests are usually resolved before serious injury occurs.
6. Explain the difference between intrasexual and intersexual selection.
7. Why do helpers of co-operatively breeding birds remain with the parents rather than attempt to breed on their own?

Extension questions

1. Do you think females that mate more than once might exert choice over which sperm fertilises their eggs?
2. What evidence is required to demonstrate that a particular behaviour has a genetic basis?
3. Compare the advantages of experimental approaches to understanding animal behaviour and purely observational studies.
4. Imagine supermarket shoppers queuing at two check-outs. What features of the shoppers and check-out attendants might the shoppers use if they were to queue at each check-out according to the ideal free distribution?
5. Why is it usually the case that the male courts the female rather than the other way around?
6. Why is the naked mole rat often referred to as a vertebrate analogue of termites?

Suggested further reading

Alcock, J. (1993). *Animal Behaviour: An Evolutionary Approach.* 5th edn. Sunderland, MA: Sinauer.

This classic text covers a wide range of topics in behaviour, often taking an evolutionary perspective, and is written for second and third year students.

Drickamer, L. C., Vessey, S. H., Meikle, D. (1996). *Animal Behaviour.* 4th edn. Boston: William C. Brown Publishers.

This text offers a broader approach to animal behaviour, covering a wide range of topics, and is written for first year students.

Krebs, J. R. and Davies, N. B. (1993). *An Introduction to Behavioural Ecology.* 3rd edn. Oxford: Blackwell Scientific Publications.

This highly readable textbook, which takes an evolutionary approach to understanding complex animal behaviour, is written for second and third year students.

CHAPTER 29

Animals responding to environmental stress

Organisms are well suited to their usual environments as a result of evolutionary change by natural selection; that is, adaptation (Chapter 32). They have heritable anatomical and physiological characteristics that allow them to grow under particular environmental conditions. Some of these inherited characteristics also allow individual organisms to respond to environmental change in a way that increases their own chances of survival.

Under optimal conditions, an organism should attain its full growth and reproductive potential as determined by its genotype. However, such optimal conditions may occur rarely in natural environments. Suboptimal conditions can occur on a daily basis, seasonally or at random. In a biological sense, stress refers to disturbances in the normal functioning of an organism that are induced either by environmental factors, such as other organisms, or chemical or physical factors, or by internal factors, such as disease.

As a result of natural selection, organisms have heritable anatomical and physiological characteristics that allow them to grow under particular environmental conditions. They are adapted to their environments.

Phenotypic plasticity

Changes occurring in an individual organism during its development, or in response to an environmental factor, have been termed 'phenotypic plasticity', 'acclimatisation' or 'acclimation'. **Acclimatisation** refers to the response of an organism to changes in several aspects of its natural environment, such as occur seasonally or geographically, whereas **acclimation** refers to the response of an organism to change in a single environmental factor, usually within a laboratory. These changes in phenotype occur without any change in the genotype of the individual and take place within the limits set by the genotype. However, the capacity for such plasticity is genetically based and therefore subject to selection.

Interaction between the environment and the genotype of an organism determines the expression of some adaptive features of the phenotype. Phenotypic plasticity is best developed in plants, presumably because of their sedentary growth. For plants, acclimation is often called 'hardening'; for example, frost-hardening, where a period at low temperature, above freezing, can increase the tolerance of a plant to subsequent freezing temperatures.

There are many examples in animals of this type of response to environmental change. Young rats raised in large hypobaric (low pressure) chambers have been found, as adults, to have larger total alveolar surface areas available for gas exchange than do control animals raised at sea level barometric pressure. This change, termed 'alveolar proliferation', is also a response to low oxygen (O_2) levels at high altitude and can occur in all mammals if exposure to low O_2 levels occurs when an animal is young.

Adult mammals exposed to high altitude for even short periods of time show increases in the number of red blood cells, and therefore the amount of haemoglobin available to transport O_2, as a response to the stress of low O_2 levels. Another example is seen in the wide range of animals that have the capacity to switch reversibly between enzymic systems with different temperature optima as the temperature of their environment changes seasonally. We will look at an example of this later in this chapter.

Responses of individual animals and plants to environmental stress may involve a change in phenotype, known as acclimatisation, acclimation or hardening.

For an animal, behavioural responses to environmental stress may also be important. For example, many animals dig shelters, such as elaborate burrows with a means of ventilation, which allows the animal to reduce the fluctuations in surface temperature and humidity to which it would otherwise be exposed. A striking example of a behavioural response to environmental stress is migration. Many large mammals of East Africa migrate to follow patterns of seasonal rainfall, and many whale species migrate seasonally before breeding. The muttonbird or short-tailed shearwater breeds in south-eastern Australia but migrates to the north Pacific each year.

Many insects undertake equally impressive migrations. The bogong moth, *Agrotis infusa*, occurs on the western slopes of the Great Dividing Range in New South Wales. In spring, most of the annual foods on which it feeds are no longer available and the moths migrate for hundreds of kilometres to congregate in caves over the summer (Fig. 29.1). In autumn, the moths move back to the breeding grounds to lay eggs.

Many organisms show a response to a particular seasonal stress. What stimulus could initiate such a response? A common characteristic of animals and plants is the ability to sense day length (photoperiod) and even the rate at which day length changes (Chapter 7). Biologists term photoperiod a proximate factor, one that triggers an organism to initiate adaptive responses. In tropical areas, where day length is virtually constant throughout the year, or in cases where an environmental stress is unpredictable, the stress itself may provide the appropriate cues.

Fig. 29.1 Bogong moths, *Agrotis infusa*, sheltering between rocks in summer in Canberra

The responses of an organism to changes in its environment can involve anatomical, physiological and behavioural changes within limits defined by its genotype.

In this chapter, we look at selected examples of responses of animals to environmental stress. These include animals living in Australian arid regions and at high altitude. One of the most common environmental stresses is an extreme of temperature, so we also discuss how body temperature and metabolic rate vary with environmental (ambient) temperatures. In particular, we look at some of the differences in temperature physiology between reptiles and mammals.

Fig. 29.2 An eastern grey kangaroo, *Macropus giganteus*, sheltering during the heat of the day

Animals in the Australian arid zone

The Australian arid zone includes about 70% of the continent and is characterised by high temperatures, long droughts and flooding rains (Chapter 41). Many desert animals, particularly small mammals, reptiles and invertebrates, avoid the extremes of desert temperatures by seeking shelter in burrows or cracks in the soil during the day and limiting their foraging to the cool evenings. Larger mammals, such as kangaroos, seek shelter in caves or shady areas to avoid the direct heat of the sun (Fig. 29.2). Thus, most animals in the desert avoid the direct impact of high temperatures.

The unpredictability of rainfall and therefore food availability is another critical stress for most Australian desert animals. In fact, it is this unpredictability that sets Australian deserts apart from other arid areas of the world. Few desert species of animals in arid Australia regularly require drinking water. CSIRO scientists have estimated that only about 4% of mammal species and 10% of bird species in arid Australia require drinking water. Of the 210 reptiles

and thousands of insects that inhabit the arid zone, probably none require liquid water to drink.

Most desert animals avoid the direct impact of high temperatures and few animal species in arid Australia regularly require drinking water.

Termites in arid Australia

One group of animals that has adapted to living in the arid zone is termites. Because there are few resident herbivores in arid Australia, much of the plant production is immediately available to the abundant and diverse termite populations. These insects have symbiotic flagellates in their hindgut that have the capacity to hydrolyse this source of cellulose (Chapter 19).

Fluctuations in temperature inside termite nests are reduced by the insulative properties of the wood-pulp material of the walls (Fig. 29.3). 'Magnetic' termites, *Amiterms meridionalis*, which live in the Northern Territory, build their nest or mound with a north–south orientation, which exposes the maximum area to the sun when the air temperature is low but reduces the area exposed to the hot sun in the middle of the day.

Fig. 29.3 The termite nest is a good example of complex social behaviour resulting in a controlled microclimate that isolates the animals from an otherwise harsh climate

The air passages in these structures also aid air circulation. The relative humidity inside the nest is always around 95% and is maintained by water transported by the animals from the watertable and stored in the matrix of the nest walls.

Nests built by termites provide a controlled microclimate that isolates the animals from an otherwise harsh climate.

Animals living at low temperatures

The upper altitudinal limit to forested areas, the 'timber line', reflects a marked change in habitat. Here the vegetation changes from trees to low shrubs and herbs, lichens and mosses. The annual mean temperature decreases, on average, 6.5°C for every 1000 m in altitude, and the reduction in dust and haze particles causes an increase in the amount of radiation reaching the ground. High altitudes are characterised by distinct seasonal changes in temperature but even in summer temperatures can change rapidly and marked local climatic differences can occur. In alpine environments, we might expect to find animals with a dependence on

behavioural temperature regulation in summer and a tolerance of freezing in winter. Similar properties are found in animals of polar regions.

The ultimate cause of death in animals exposed to low temperatures is complex and not fully understood. Ice formation within cells causes the cytoplasm to become more concentrated and the pH may change. Ice may also form in the extracellular space, resulting in the osmotic withdrawal of water from the cell. The resultant dehydration of proteins is likely to be the chief cause of freezing damage.

Animals living at low temperatures are likely to be tolerant of freezing.

Do animals freeze?

The likelihood of survival for an over-wintering animal in alpine or polar regions ultimately depends on its ability to prevent intracellular freezing. However, many invertebrates, for example, some insects and intertidal molluscs, are able to tolerate freezing of at least a portion of their extracellular fluids. A number of these animals have nucleating agents that promote a slow rate of freezing and thus reduce the possibility of intracellular ice formation. A few species of frogs and reptiles are also able to tolerate freezing of extracellular fluid, especially in the limbs and in subcutaneous tissues. Some animals prevent freezing by **supercooling**, a process that involves a reduction in the temperature of a fluid below its usual point of freezing. Ice will not form unless a nucleus (seed) for its formation is present. In animals employing this mechanism, the digestive system is emptied of food material that could act as sites of nucleation. In some cases, antifreeze compounds synthesised within the body inhibit ice formation. As a result of these mechanisms in insects and nematodes, whole body temperatures as low as −50°C have been reached experimentally. Much lesser degrees of supercooling (to −10°C) are found in benthic Arctic fishes, reptiles and one mammal, the Arctic ground squirrel.

A major disadvantage with supercooling is that contact with ice 'seeds' supercooled fluids, thus freezing the whole animal. Antarctic fishes avoid this problem by producing antifreeze proteins or glycoproteins in the blood that lower the freezing point as well as preventing ice formation. These molecules are rich in polar groups and interact with ice crystals, stopping them from growing. This is analogous to using antifreeze compounds in cooling systems of cars in the snow.

Few animals tolerate freezing of tissues. Survival at very low temperatures usually involves supercooling or biological compounds that lower the freezing point and prevent ice formation.

Insects at low temperatures

One of the most striking features among high-altitude insects is their small size when compared with related species living at lower altitudes. This reduction in size enables these animals to take advantage of sheltered microclimates, which would not be accessible to larger forms. Selection pressures associated with high winds may have also resulted in evolution of small size, especially in butterflies that forage in vegetation boundary layers. In the Himalayas above 6000 m, 60% of insects have reduced wings or are wingless (apterous), and species of flightless grasshoppers live in the Australian Alps. Insects from cold climates also tend to be darker in colour than those from lowlands. This colouration, 'thermal melanism', leads to increased body temperature due to an increased absorption of radiation. Various patterns of thermoregulatory basking behaviour have been described in butterflies, for example, the genus *Pieris*, which orient the dorsal thorax towards the sun and use their white wings to reflect radiation onto the body. Apart from such behaviours, alpine insects seek warm microclimates, including parabolic corollas of some alpine plants (Fig. 29.4).

Insects from cold climates tend to be small, darker in colour and use various forms of basking behaviour.

The Q_{10} concept

Most biochemical processes are temperature sensitive: as temperatures increase their rates increase. A simple explanation is that the kinetic energy of molecules increases and so the probability of reactions occurring increases. A measure that has been used to describe this sensitivity is the Q_{10}, which is defined as the increase in the rate of a physiological process or reaction for a 10°C rise in temperature. It is calculated by dividing the rate at a certain temperature (RT) by that at a temperature 10°C lower ($RT - 10$).

$$Q_{10} = \frac{RT}{RT - 10}$$

The values of Q_{10} range from 1, for processes that are temperature insensitive, to most biological examples, which lie between 2 and 3 (Fig. 29.5). Q_{10} is a useful concept because it can be used to quantify the effect of temperature on chemical reactions as well as more complex processes such as breathing and levels of metabolism in whole animals.

Temperature regulation and metabolism in animals

In the past, the rather simple terms 'cold-blooded' and 'warm-blooded' were used to describe how animals controlled their body temperature. Mammals and birds fitted the latter category as they spent the day at a constant temperature; however, the definition failed in animals that underwent periods of hibernation or **torpor**. Reptiles, amphibians and invertebrates were assigned the term 'cold-blooded', however, most reptiles, when active during the day, have body temperatures approaching mammalian levels, as do several groups of insects during flight. Biologists now prefer to use the term **ectotherms** for animals that largely depend on external sources of heat and **endotherms**, for animals that rely on metabolic or internal sources of heat. Endotherms include mammals, birds and some insects, leaving fishes, reptiles, amphibians and invertebrates to be primarily ectothermic in their regulation of body temperature.

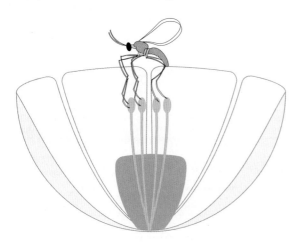

Fig. 29.4 The geometry of the flower of the Arctic plant *Dryas*, showing the parabolic inside cup that focuses the sun's rays

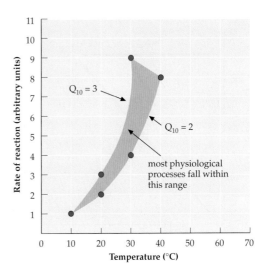

Fig. 29.5 The relationship between the rate of a physiological or biochemical process and temperature. The shaded area indicates the range of most reactions

The terms ectotherm and endotherm are functional descriptions and it is interesting to see how groups assigned to these categories differ in laboratory experiments. If we place a lizard and a mammal (mouse) of the same body mass in a chamber, we can set up an experiment in which we can simultaneously measure body temperature and the rate of oxygen consumption. In a further addition to the experiment, we can change the temperature of the chamber from 10°C to 35°C. The results obtained from the mouse differ from those of the lizard (Fig. 29.6). The metabolic rate of the lizard increases with temperature in much the same way as the Q_{10} relationship in Figure 29.5. The mouse has the ability to increase its rate of metabolism, and therefore heat production, at low temperatures but at around 30°C it can maintain a constant level of metabolism. The range of environmental temperatures over which metabolic rate does not change is termed the 'thermal neutral zone'. It tends to be broader in larger animals compared with small ones and in animals that live in cold habitats rather than tropical ones. At higher temperatures above 30°C, the rate of metabolism increases again. In Figure 29.6 we can see that there are also differences in the way in which body temperature varies with ambient temperature. The body temperature of the mouse remains at 37°C whereas that of the lizard stays the same as the chamber.

The conclusions from this simple laboratory experiment confirm that lizards have lower rates of metabolism than mammals and that they cannot regulate their body temperature. This finding differs from field experiments conducted on lizards, in which it was found that body temperatures were reasonably constant during the day. These experiments were done by surgically implanting a radio telemetry unit inside the lizard's body and monitoring its temperature on a radio receiver as it engaged in its normal behaviour. By basking, orienting its body at various angles to the sun, seeking out microclimates such as tree branches or burrows with different temperatures, or even by changing skin colour, lizards can achieve nearly the same control over body temperature as mammals (Fig. 29.7). The difference is that they do it primarily by behavioural adjustments, whereas birds and mammals have the ability to change metabolic rates.

The term ectothermy describes the process where the body temperature of an animal is derived from heat gained from the environment. Animals such as reptiles adopt this means of temperature regulation through adjustments in their behaviour. Endotherms derive heat from internal or metabolic sources. Endotherms include birds and mammals.

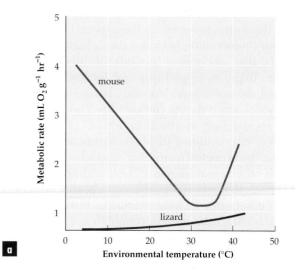

a

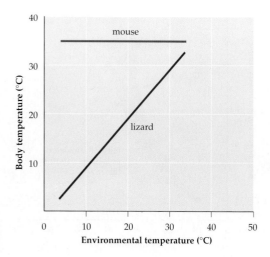

b

Fig. 29.6 The effects of ambient temperature on **(a)** metabolic rate and **(b)** body temperature for a mouse and a lizard of the same body mass

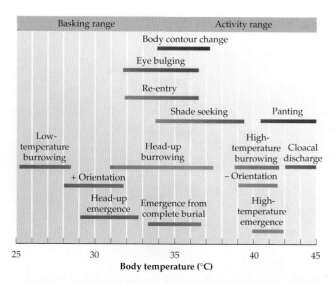

Fig. 29.7 Patterns of behavioural thermoregulation in a lizard

BOX 29.1 Temperature regulation in heat-producing flowers

Birds and mammals are well known to regulate their body temperatures at a fairly stable level despite changes in the temperature of their surroundings. They do this by increasing their metabolic rate, and hence their rate of heat production, as their environment gets colder. Often this involves shivering, but some mammals produce the needed heat in special heater cells known as brown adipose tissue. At high environmental temperatures, the animal reduces its metabolic heat production and may initiate evaporative cooling by sweating or panting to prevent the body temperature rising. These heating and cooling mechanisms are the physiological basis for thermoregulation. Temperature regulation also occurs in some flying insects, for example, many moths, bees and beetles. Unless these insects raise their body temperatures to a certain level (usually 30°C or more), they are unable to fly.

Thermoregulation is usually thought to occur only in animals, so it is startling to discover that a few flowers warm up and regulate their temperature too. One example occurs in northern Australia and throughout Asia—the sacred lotus (*Nelumbo nucifera*) (see figures a and b). The flower of this aquatic plant is about the size of two cupped hands. It starts heating as a bud and reaches about 32°C when the petals just begin to open. It remains at about this temperature for two to four days despite changes in environmental temperature between 7°C and 45°C (Fig. c). Like a mammal during cold nights, it increases its metabolic heat production, measured as the rate of oxygen consumption (Fig. c). During hot days, evaporative cooling takes over and the flower can be several degrees cooler than its surroundings. The precision of thermoregulation in this flower is astounding given the fact that it lacks most of the physiological mechanisms necessary for thermoregulation in animals (temperature sensors, nervous integration in the brain and activities such as shivering, sweating or panting).

An intriguing unanswered question is 'why' a flower would warm up and thermoregulate. Although it has been proposed that warming increases the scent of the flower and makes it more attractive to insects, this does not explain thermoregulation. Another possibility is that flying insects (mainly beetles and bees) attracted

The flower of the sacred lotus, *Nelumbo nucifera*, **(a)** whole and **(b)** in section

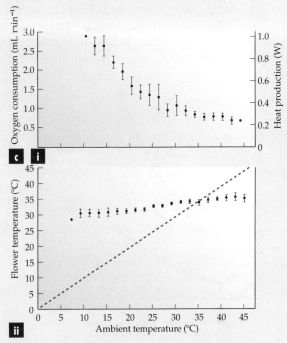

(c) The relationship between the **(i)** rate of oxygen consumption (or heat production) and **(ii)** flower temperature with ambient temperature in the flower of the sacred lotus

to the flower are rewarded with a warm, stable environment in a large chamber enclosed by the petals. After the insects have brought in pollen from another flower and accomplished cross-pollination, some remain in the chamber over-night. During their stay, the warm and constant temperature of the chamber facilitates feeding, mating and preparing for their departure flight. When the petals open widely, the flower cools, and the departing insects are provided with pollen to take to another warm flower. The advantage to the plant is that it ensures cross-pollination, and the advantage to the insect is that the flower keeps it warm and active without having to produce the heat on its own. Thus, the flower may thermo-regulate as a direct energy reward for the insect.

Thermal acclimation

A particularly interesting feature of the relationship between metabolic rate and ambient temperature in ectotherms is that it can change with season. From what we now know about ectotherms such as fish, we might expect them to be more sluggish in cold water during winter than in the warmer creeks of summer. This is not usually the case. If you were to bring a fish that you caught from a pond at 25°C in summer into the laboratory and measured how much O_2 it consumed as you varied the temperature, you might find data similar to that in Figure 29.8. It has the typical pattern of a process with a Q_{10} of 2. If you repeated the

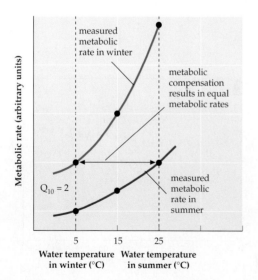

Fig. 29.8 Metabolic compensation for seasonal differences in water temperature in a fish acclimatises the animal to its changing environment. Successful acclimatisation results in the animal having an equivalent metabolic rate in summer and winter when measured at the respective environmental temperature

measurements in winter when the water was 5°C, you might expect that the fish would show a similar pattern. In many examples where this experiment has been done on fish and frogs it has been found that the curve differs between summer and winter. This means that the animal's metabolic rate in winter is always higher than that measured at the same experimental temperature in the fish caught during summer. The surprising thing is that metabolic rates are very similar at 5°C in the winter fish as they are at 25°C in the summer fish, explaining why fish in cold winter streams are not necessarily easier to catch by hand than those in summer. What we are seeing is a process of *thermal acclimation* that involves seasonal metabolic compensation.

Biologists now have an understanding of the mechanisms that underlie this compensation. Many animals have different sets of enzymes that differ in their temperature requirements for optimal function. In the example we have just discussed, it is a reasonable hypothesis that the fish had two sets of enzymes, one for summer and one for winter. The result of this adaptation is that it can be active throughout the year even though its immediate environment undergoes seasonal changes.

> Temperature acclimation is common in ectotherms and results from the induction of enzymes with different temperature optima.

Metabolic depression in animals

A reduction in metabolic rate, or oxygen consumption, in response to environmental extremes such as tem-perature, desiccation, low oxygen levels and starvation, is seen in all major animal phyla. The phenomenon was probably first recorded by the developer of the microscope, van Leeuwenhoek, who found that dried animals called rotifers resumed normal behaviour when they were rehydrated. This form of 'suspended animation' has been described for many species of invertebrates, particularly early egg or larval develop-ment stages, and has been termed *cryptobiosis*.

Perhaps the best examples are nematodes, which have been known to survive low temperatures and desiccating conditions for several years. In the case of a parasitic nematode from wheat galls, *Anguine triatic*, the metabolic rate can drop to 0.0006 of the resting level. In some cases, such as desert snails and frogs, the animals prepare for seasonal hot and dry conditions by becoming inactive or burrowing. Several desert-adapted Australian frog species form a cocoon in which they burrow for months in the bottom of dried water holes. Some earthworms also form a dry, mucous cocoon and aestivate in a spherical shape. Terrestrial snails form an epiphragm over the shell opening to reduce evaporative water loss and there is a report in the scientific literature of a species surviving over 23 years of dormancy. In snails, the reduction in metabolic rate is to about 20% of the levels at rest. Among reptiles, dormancy in response to environmental stress is also well known. Turtles and crocodiles, including the Australian freshwater crocodile *Crocodylus porosus*, respond to starvation or dry conditions by reducing their metabolism by around 30%. These examples are of 'intrinsic' **metabolic depression** (Table 29.1). This means that it is an active process, which, because of its wide occurrence in animals, may have some common biochemical features, such as

changes in cellular pH and a reduction in rates of protein synthesis. Scientists are now only beginning to understand some of these processes.

Many mammals and birds have the ability to hibernate, aestivate or enter shorter periods of torpor in which the metabolic level drops to around 1% of the normal active level. This is discussed below. This ability, common in animals such as small marsupials, ground squirrels and hummingbirds, has the advantage of saving energy on a daily or seasonal basis. It appears as if this form of metabolic depression may have a different basis from that discussed above. Mammals and birds, being endothermic, have a high basal metabolic rate and regulate their body temperatures between 35°C and 42°C. During torpor this level of body

Table 29.1 Summary of metabolic depression, expressed as the ratio of depressed (*D*) to resting (*R*) metabolic rate for animals undergoing stress

	Metabolic depression (*D/R*)	State
Nematode, *Anguina tritici*	0.0006	Anhydrobiotic larva
Leech, *Nephelopsis obscura*	0.0014	Long-term anaerobic
Crustacean, *Artemia* spp.	0.015	Cyst
Cnidarians	0.047	Anaerobic
Nematode, *Aphelenchus avenae*	0.059	Anhydrobiotic adult
Cephalopod, *Nautilus pompilius*	0.13	Hypoxic
Turtle, *Chrysemys scripta*	0.15	Anaerobic
Snails	0.18	Aestivating
Fishes	0.20	Aestivating
Annelid worms	0.21	Anaerobic
Crustacean, *Holthusiana transversus*	0.28	Aestivating
Goldfish, *Carassius auratus*	0.29	Anaerobic
Amphibians	0.29	Aestivating
Sponge, *Ephydatia mülleri*	0.34	Gemmule
Reptiles	0.41	Aestivating
Nematode, *Eustrongylides ignotus*	0.46	Starved larva
Insects	0.48	Dormant larvae
Insect, *Ips acuminatus*	0.53	Over-wintering adult
Slug, *Limax flavus*	0.70	Starved
Reptile, *Crocodylus porosus*	0.71	Starved

temperature is reset to a lower level, usually between 10°C and 20°C. The drop in body temperature is linked to a drop in metabolism through the Q_{10} effect (see Fig. 29.5) and in most cases there is little or no contribution from any intrinsic metabolic depression as seen in many other animals.

> Metabolic depression occurs in response to environmental stress or, in the case of some animals, in anticipation of stress that may have a seasonal basis. In invertebrates and ectothermic vertebrates, the depression has been termed 'intrinsic' and is an active metabolic process. In mammals and birds, the ability to reduce metabolic demands by this means appears limited and entry into torpor appears to be primarily a Q_{10} effect.

Hypothermia and the concept of torpor

Torpor is used to describe natural **hypothermia** involving a lowering of metabolic rate and a lessened responsiveness to external stimuli. Vertebrate ectotherms—fishes, amphibians and reptiles—may enter a period of torpor in which they are inactive or dormant. Hibernation is a specific term used to refer to long-term torpor in cold weather. Its use in describing the lethargic state of bears in winter, in which body temperature decreases by only a few degrees, and the seasonal dormancy of reptiles and invertebrates, is misleading.

The major problems encountered during torpor are the dangers of freezing and desiccation, a lack of energy supply and, for animals that are under water, a lack of oxygen. Like over-wintering insects, animals in torpor avoid freezing by the use of **cryoprotectants**. Cryoprotectants are organic molecules, such as sugars, that protect membranes and enzymes against cold-induced damage. Others tolerate ice formation in extracellular fluids, including blood. Energy demands during this time are met by lipids, economically stored as fat bodies in muscle or liver; a major involvement of anaerobic pathways has been suggested. Novel anaerobic end products, such as ethanol in fishes, may also be used in addition to the more usual products of anaerobic metabolism, such as lactic acid.

Tolerance levels during hibernation or dormancy are impressive. Freshwater tortoises are dormant for up to six months of the year, some of them in water. Among fishes, the crucian carp, *Carassius carassius*, one of the few fishes to occur in the ponds of northern Europe, is tolerant of anoxia for periods of months. Occasionally, however, there are problems for hibernating aquatic animals, and in Northern Hemisphere temperate lakes the phenomenon of 'winterkill' occurs. Frogs and fishes may die during a winterkill, resulting in thousands of dead animals in the spring thaw. Lakes

in which this occurs are usually shallow and rich in organic matter. If snow on the frozen surface reduces light penetration so that photosynthesis ceases, O_2 levels are reduced and microorganisms in sediments produce ammonia and hydrogen sulfide in sufficient quantities to be toxic.

> Torpor is hypothermia involving a lowering of metabolic rate and reduced responsiveness to external stimuli. Long-term torpor in cold weather (hibernation) is associated with the dangers of freezing, dehydration and lack of energy supply.

Torpor in mammals and birds

Some small mammals, including rodents, marsupials and bats, and birds, such as hummingbirds and swifts, respond to seasonal changes in climate and food availability by reducing their rate of metabolism and body temperature and entering a period of torpor. In these animals, torpor is not an abandonment of body temperature regulation but a controlled state that may involve an initial period of fattening, the preparation of a nest and several preliminary test drops in body temperature. In many species, temperature alarms are set off if the body temperature goes below a certain point and arousal begins (Fig. 29.9). Bouts of torpor are often interrupted by days of arousal before the animal again drops its body temperature. Figure 29.10 shows measurements of metabolic rate in a small marsupial during entry into torpor and a subsequent period of

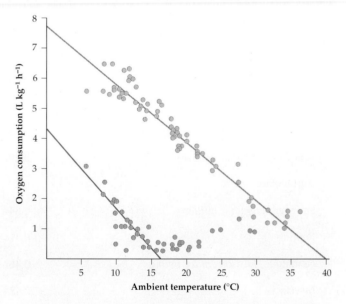

Fig. 29.9 The rate of O_2 consumption in a small marsupial, the fat-tailed dunnart, *Sminthopsis crassicaudata*, at different ambient temperatures. The pink circles show the responses of animals that were not in torpor, the blue circles show data from torpid animals. At ambient temperatures below 15°C, the animals remain in torpor but their oxygen consumption shows an increase to prevent body temperature dropping to dangerously low levels

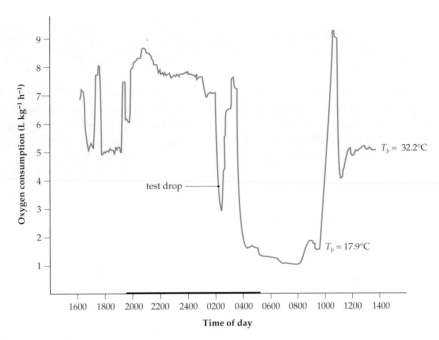

Fig. 29.10 The rate of O_2 consumption in a small marsupial, the fat-tailed dunnart, *Sminthopsis crassicaudata*, measured over a 21-hour period at a constant ambient temperature of 12°C. The dark bar represents the period of darkness; the body temperature is shown as T_b. The trace shows a 'test drop' in metabolism before the final decrease in metabolism during the 6-hour period the animal was in torpor

arousal. This pattern is also typical of rodents that show daily torpor.

Dramatic circulatory changes occur when mammals allow their body temperatures to drop as they enter hibernation. The lowering of body temperature is associated with a lowered metabolic rate and heart rate. The decrease in cardiac output is associated with strong peripheral vasoconstriction, and blood flow is maintained to the brain, heart and lungs, rather similar to the diving response (Box 29.2). However, during arousal, blood is also sent to the brown fat and adrenal glands. Adrenaline released from the adrenal glands causes the brown fat to produce a lot of heat, which is carried by the blood to the heart, increasing cardiac output and therefore blood flow to brown fat, and shivering of skeletal muscle. The cycle continues until the animal returns to normal body temperature.

In mammals and birds, torpor is not an abandonment of body temperature regulation but a controlled state that may involve an initial period of fattening, the preparation of a nest and several preliminary 'test drops' in body temperature.

Environmental oxygen stress

The proportion of O_2 in air or well-mixed surface waters at any altitude remains relatively constant (approximately 21%), although for the same partial pressure of O_2 (P_{O_2}) the concentration of O_2 in water is substantially less than in air (see Chapter 20). However, in both air and water the P_{O_2} decreases with increasing altitude as a result of a fall in atmospheric pressure (Fig. 29.11). P_{O_2} may also decrease in situations where poor mixing occurs and the atmospheric O_2 level is depleted by O_2 consumption. A decrease in P_{O_2} from normal O_2 levels is known as **hypoxia**.

Poorly ventilated environments are often associated with life underground, for example, in burrows or in caves. In these habitats, exchange of air or water with the outside is reduced and the rate of O_2 consumption by the inhabitants may exceed the rate at which renewal of the medium takes place. In terrestrial burrows, most air exchange takes place through the walls of the burrow. Plugging of burrow openings will therefore have little effect on the composition of gases in the burrow and many animals, such as the platypus, show this behaviour. However, if soil should become waterlogged, a considerable barrier to gas exchange develops, as the rate of diffusion of a gas in water is thousands of times slower than in air. In addition, metabolic activity of microorganisms within the soil can decrease the effective replacement of burrow gases.

Partial pressures of O_2 as low as 45 mmHg (equivalent to 6% O_2, compared to 21% in atmospheric air) have been recorded in the burrows of pocket gophers, whereas in some species of marine fish and invertebrates that burrow, values close to zero have been recorded. In isolated water bodies, such as ponds or lakes, O_2 consumption of the inhabitants can reach such high levels that replenishment of gas at the surface

BOX 29.2 Diving

For mammals and birds, so dependent on high levels of O_2, diving is a considerable challenge. However, early studies showed that many could tolerate submersion for longer periods than would be expected on the basis of their O_2 stores and metabolic rates. More recent studies, centred on the diving behaviour of animals in their natural habitat, have shown that most voluntary dives are of a much shorter duration than the maximum of which the animal is capable. For example, the sperm whale is capable of remaining submerged for periods of 1 hour but sonar reports show that most dives are less than 10 minutes. Similarly, tufted ducks can dive for periods of 1 minute but the preferred diving time is 20 seconds.

During most dives in the field, well-adapted divers take enough O_2 down with them to provide for the needs of their tissues. But there is always the danger of being trapped under water for an unexpectedly long period, for example, while escaping a predator, dealing with food, or looking for a breathing hole in ice. Under these conditions, there may be insufficient O_2 for the whole body to remain aerobic. In birds and mammals, the heart and brain are particularly sensitive to lack of O_2 and permanent damage can occur in just a few minutes. So it is not surprising to find that these

animals have a complex mechanism, the diving response, that can be invoked to reserve precious O_2 stores for these organs. This powerful cardio-vascular response maintains circulation to the heart and brain (and lungs, of course) but greatly reduces circulation to all other organs of the body (Fig. a). The eminent physiologist, Per Scholander, called the mechanism 'the master switch of life'.

Although most vertebrates, including humans, show elements of a diving response, it is best developed in dedicated divers, such as marine mammals and diving birds. The primary reaction is an intense slowing of the heart, bradycardia, due to increased activity in the cardiac branch of the vagus nerve. In seals, heart rate can drop to less than a tenth of the normal rate, beating only five or six times a minute. At the same time, a powerful vasoconstriction of the major peripheral arteries of the body greatly slows or actually stops blood flow to the muscles, skin and visceral organs. The decrease in blood flow is called ischaemia and results in almost complete lack of O_2 (anoxia) in the tissues involved. Anaerobic metabolism in ischaemic tissues results in an accumulation of lactic acid during the dive; but, because there is no blood flow to these organs,

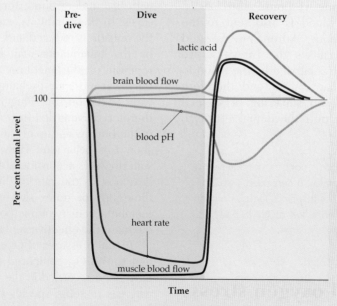

(a) The diving response. During the dive, heart rate decreases but blood flow to the lungs and brain is maintained at the expense of flow to the skeletal muscles and gut. Lactic acid is produced by anaerobic metabolism in the ischaemic muscles and it is flushed out into the circulation, causing a decrease in blood pH after the dive

very little lactic acid appears in the blood. Meanwhile, the O_2 stores of the blood and lungs are available to the heart and brain. Central arterial blood pressure is maintained and blood flow to the brain and eyes may actually increase, despite the intense bradycardia.

After the dive is over and the animal breathes again, normal heart rate and circulation are re-established. Blood flowing to the previously ischaemic tissues flushes out the built-up lactic acid into the general circulation and the blood pH drops. It requires many minutes of recovery for the acid to be removed and the situation to return to normal. For this reason, it is obvious that the diving response is an unusual event; quickly repeated dives involving the diving response are not possible. In fact, recordings of heart rate and blood parameters in Antarctic Weddell seals show no diving response during repeated routine dives at sea but the response can appear strongly if, for example, the animal is prevented from returning to its breathing hole in the ice.

The diving response is a pattern of reflexes under both conscious and unconscious control. It seems to be triggered when there is some doubt concerning the length of a dive. Animals forcibly submerged always show it, but when they are allowed to dive naturally they show it only if they are unfamiliar with the surroundings.

In the experiment shown in Figure b, a freely diving duck decreases its heart rate only when it becomes aware that it cannot surface at the usual place. Seals that have been trained to put their noses under water on command show the response because they do not know how long the trainer wants them to stay under. However, if they are trained to dive for particular periods of time, they show a level of bradycardia from the outset of the dive that is appropriate for the length of the dive. It is interesting that even fishes show a 'diving response' when lifted from water (gills cannot take up sufficient O_2 from air). Apparently the response is a general life-saving one, shown by vertebrates when removed from their normal respiratory environment.

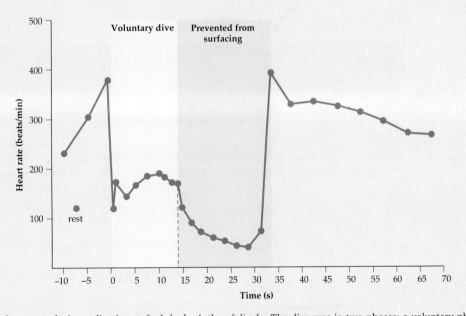

(b) The average heart rate during a dive in a tufted duck, *Aythya fuligula*. The dive was in two phases: a voluntary phase, and a second phase in which the duck apparently became aware that it was temporarily being prevented from surfacing and decreased its heart rate to a lower level

Is exceeded and a situation arises where Po_2 levels fall below atmospheric levels.

Swamps, tidal pools and some coastal estuarine environments are often poorly mixed and large diurnal oscillations in respiratory gases occur. While the usual problem encountered by animals is obtaining enough O_2, it is possible to have too much. Photosynthesis by macrophytes and algae during daylight hours can generate large amounts of O_2 such that the Po_2 in the water of ponds and lakes can rise higher than that of the atmospheric air at the surface. A good example where

the Po_2 levels fluctuate and depend on the amount of O_2 consumed or produced in the water can be found in tidal pools (Fig. 29.12). Besides photosynthesis, high Po_2 can also occur as a result of the effects of increased hydrostatic pressure on a confined body of gas, such as in a lung of a diving mammal or bird or plastron of a diving insect (see Box 20.2). High partial pressures of O_2 are often associated with O_2 toxicity, which can result in overstimulation of the central nervous system and convulsions.

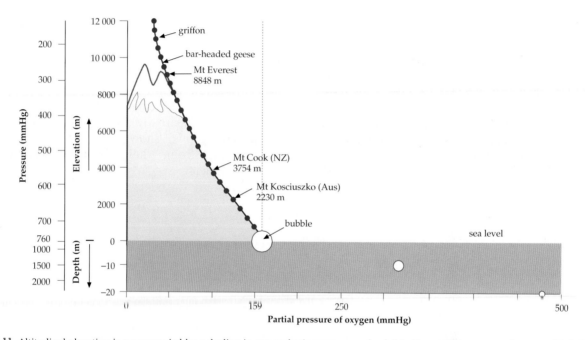

Fig. 29.11 Altitudinal elevation is accompanied by a decline in atmospheric pressure and a fall in the partial pressure of oxygen (Po_2). Well-mixed water has the same Po_2 as the atmospheric air above it. At sea level, the Po_2 of the air and well-mixed water is 159 mmHg. Because hydrostatic pressure in water increases 760 mmHg for every 10 m of depth, the volume of a gas bubble will decrease as it is subjected to increasing pressure

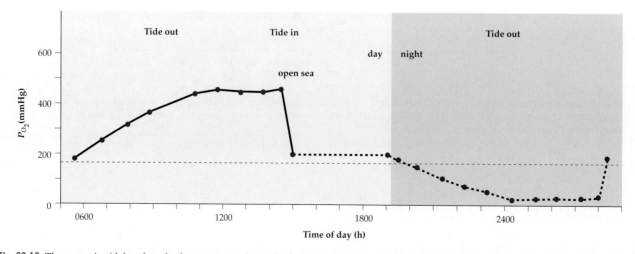

Fig. 29.12 The water in tidal rockpools shows great variation in dissolved O_2 levels when it is isolated from the sea at low tide. These values were recorded in a rockpool off the coast of France. Normal values of O_2 in the open sea are given by the dotted lines

Adaptations to low levels of oxygen

The problem with exposure to low P_{O_2} is that diffusion of O_2 across membranes is restricted and O_2 availability to tissues is reduced. Adaptations to living in a hypoxic environment are a combination of physiological and behavioural characteristics that compensate for this problem. Burrowing species usually have low rates of O_2 consumption compared with animals of similar body mass. This would confer an advantage in that O_2 consumption would be reduced in a situation where the hypoxia in the burrow is a result of a poorly ventilated environment. Burrowing fish, on the other hand, move in and out of their burrows pushing water ahead of them, and this piston-like convection is known to increase burrow O_2 levels. A similar shunting behaviour is observed in the wombat, a large burrowing mammal, but whether this improves burrow ventilation is yet to be determined. One of the best examples of burrow ventilation occurs in the North American prairie dog. These animals construct burrows with two openings and, because the openings are at different heights, wind blowing across the ground induces a convective air current within the burrow, known as viscous entrainment (Fig. 29.13).

Animals that in general are normally exposed to hypoxia possess a haemoglobin with a greater affinity for O_2 and one that is therefore fully saturated at lower O_2 partial pressures (Chapter 20). This is true of animals that burrow, live at high altitude or dive (Fig. 29.14) but is also found in animals, including humans, who sojourn to high altitudes. Not only do these groups of animals possess a haemoglobin with a greater affinity for O_2 but they tend to have higher red cell counts, hence higher haemoglobin concentrations.

The increase in the concentration of haemoglobin can be substantial: in humans native to 5000 m it may be 50% more than in sea level inhabitants. Long-term exposure to hypoxia can also be associated with an increase in the number of capillaries in the tissues.

Normally hypoxia is a stimulus to breathing in most animals, whether water- or air-breathing, but in burrowing forms, at least in mammals such as the echidna, this response seems to be reduced. A blunted response to environmental hypoxia is also expected in animals, and is known for humans, native to high altitude.

Responses to high altitude in humans

Investigations into altitude exposure probably date back to the Chinese, with accounts of the Great Headache Mountain around 30 BC, and continue to the present, when all the peaks above 8000 m in altitude have been climbed without supplementary O_2 (Table 29.2). In between were descriptions of mountain sickness by the Spanish Conquistadors and Jesuit priests in South America, and the dramatic flights of early balloonists. Heroic experiments were conducted in the late 1800s by the Italian physiologist, Mosso, who placed his laboratory assistant in a chamber and exposed him for 33 minutes to an O_2 level equivalent to an altitude of 6500 m. Low fertility among people living at high altitude is common, and the Spanish arriving in South America and Chinese immigrants to Tibet had high rates of *in utero* deaths of infants due to reduced O_2 transfer across the placenta.

As anyone who has ever ventured to high altitude knows, sorroche, or acute mountain sickness, invariably follows too quick an ascent. Mild sorroche is accompanied by headache, nausea, dizziness,

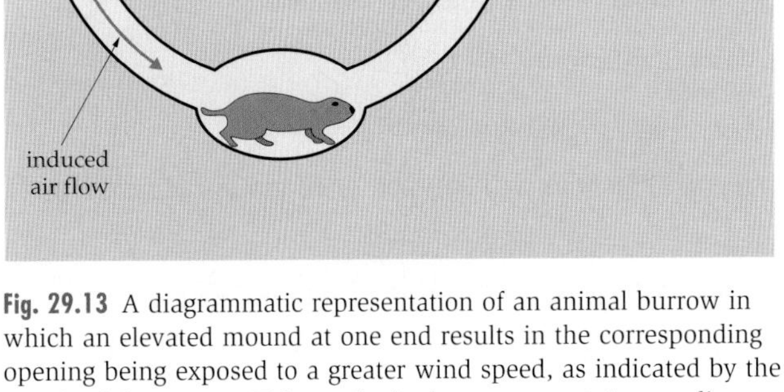

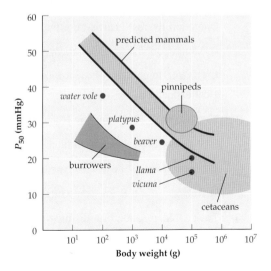

Fig. 29.13 A diagrammatic representation of an animal burrow in which an elevated mound at one end results in the corresponding opening being exposed to a greater wind speed, as indicated by the length of the arrows. This results in the pressure at A exceeding that at B and air moves from A to B, thus ventilating the burrow

Fig. 29.14 The affinity of haemoglobin for O_2, defined as the partial pressure at which haemoglobin is half saturated (P_{50}), varies with body weight and lifestyle. Note that animals that burrow or live at altitude have lower P_{50} values than equivalent-sized counterparts

Table 29.2 Some milestones in the study of high-altitude physiology

Year	Event
c. 30 BC	Reference to the Great Headache Mountain and Little Headache Mountain in the *Tseen Han Shoo* (classical Chinese history)
1590	Publication of the first edition (Spanish) of *Naturall and morall historie of the East and West Indies* by Joseph de Acosta, with an account of mountain sickness
1644	First description of mercury barometer by Torricelli
1648	Demonstration of fall in barometric pressure at high altitude in an experiment devised by Pascal
1783	Montgolfier brothers introduce balloon ascents
1786	First ascent of Mont Blanc by Balmat and Paccard
1878	Publication of *La Pression Barométrique* by Paul Bert
1890	Viault describes increase in red blood cells at high altitude
1894	Angelo Mosso completes the high-altitude station, Capanna Regina Margherita, on a summit of Monte Rosa at 4560 m
1909	The Duke of Abruzzi reaches 7500 m in the Karakoram without supplementary O_2
1910	Zuntz organises an international high-altitude expedition to Tenerife; members included C. G. Douglas and Joseph Barcroft
1911	Anglo-American Pikes Peak expedition 4300 m
1913	T. H. Ravenhill publishes *Some experiences of mountain sickness in the Andes*, describing in detail the symptoms of *puna*
1920	Barcroft et al. publish the results of the experiment carried out in a glass chamber in which Barcroft lived in a hypoxic atmosphere for six days
1921	A. M. Kellas publishes *Sur les possibilités de faire l'ascension du Mount Everest* (Congrés de l'Alpinisme, Monaco, 1920)
1924	E. F. Norton ascends to 8500 m on Mt Everest without supplementary O_2
1946	Operation Everest I carried out by C. S. Houston and R. L. Riley
1948	C. Monge, publishes *Acclimatization in the Andes*, describing the permanent residents of the Peruvian Andes
1952	L. G. C. E. Pugh and colleagues carry out experiments on Cho Oyu near Mt Everest in preparation for the 1953 expedition
1953	First ascent of Mt Everest by Hillary and Tensing (with supplementary O_2)
1978	First ascent of Mt Everest without supplementary O_2 by Reinhold Messner and Peter Habeler
1981	American Medical Research Expedition to Mt Everest, Scientific Leader J. B. West
1987	First ascent of Mt Everest in winter (December) by Sherpa Ang Rita without supplementary O_2
1990–98	The characterisation of a distinct high-altitude phenotype in Quechuas and Sherpas—two groups of people in which these adaptations probably arose independently
1997	The publication of *Into Thin Air: a Personal Account of the Mount Everest Disaster* by John Krakauer. This is a frightening account of the deadliest season in the history of Everest and removes any idea that climbing Mt Everest is now 'easy'

palpitations, insomnia and loss of appetite and is attributable to internal hypoxia (hypoxaemia) and alkalosis, the result of hyperventilating. **Hyperventilation** is one of the most obvious and useful responses to high altitude as the increase in breathing results in an increase in the P_{O_2} in the lung. However, it is associated with a concomitant fall in the partial pressure of CO_2 in the lung, which, until bicarbonate levels are adjusted by renal excretion, results in the pH of the blood becoming alkalotic. In severe cases of sorroche, pulmonary or cerebral oedema (swelling) results from a redistribution of body fluid and in many cases proves fatal. Despite the problems associated with high altitude, about 15 million people live at altitudes greater than 3000 m and are therefore permanently exposed to atmospheric oxygen levels equivalent to 14% (P_{O_2} 110 mmHg). Many populations live and work at even greater altitudes. Miners in the Andes and the Himalayas can be found at altitudes in excess of 5000 m. In Peru, 70 000 people live permanently in the mining city of Cerro de Pasco at an elevation of 4300 m.

As altitude increases, the atmospheric pressure, and therefore the partial pressure of O_2, decreases by about one-half every 5500 m. While the proportion of O_2 in air remains constant at 20.95%, the total amount available for exchange is less. The problem with exposure to high altitude is that the O_2 partial pressure gradient is reduced between different compartments of the total O_2 delivery system (Fig. 29.15). Partial pressure differences determine the rate of diffusion of a gas across membranes (Chapter 20); hence, exposure to high altitude ultimately reduces the O_2 availability to tissues.

The responses of humans to altitude hypoxia range from changes that occur within a couple of hours to adaptations, occurring over many generations, that have a genetic basis. Some individual responses, such as an acute increase in heart and respiration rates, and an increased production of red blood cells, are later reversed as other changes occur. Slower-developing changes include an increase in the number of capillaries in various tissues and a decrease in the ventilatory response to low O_2 levels (Fig. 29.16).

Initial responses of individual humans to altitude hypoxia include an increased heart rate and ventilation rate, and an increase in the number of red blood cells. Later changes include an increased number of capillaries supplying tissues and a decreased ventilatory response to low O_2 levels.

Other animals at high altitude

While high altitude is usually considered in terms of terrestrial animals, it should not be forgotten that some birds and insects are found at altitudes higher than Mt

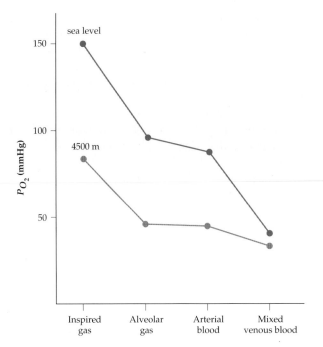

Fig. 29.15 Scheme of O_2 levels from inspired air to mixed venous blood taken from one group of people resident at sea level and a second group resident at an altitude of about 4500 m in the Andes

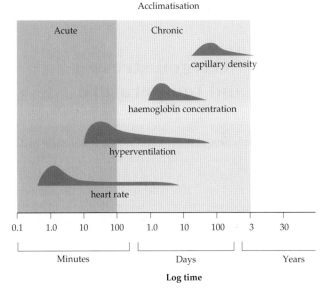

Fig. 29.16 The time courses of a number of acclimatisation and adaptive changes that occur in people during exposure to high altitude. The shape of the curve denotes the rate of change of each response

Everest. Bar-headed geese routinely fly over the Himalayas at altitudes in excess of 9000 m, but the record is probably a griffon (type of vulture) that was reported by an airline pilot at 11 000 m, which presumably had been carried on air thermals to such an altitude. The ability to reach such high altitudes may relate to the structure of the respiratory system in birds (Chapter 20).

High altitude poses a particular challenge for the embryos of birds and mammals. In mammals, an increase in ventilation of the mother helps to maintain arterial Po_2, while modifications in the maternal circulation and placental morphology increase both the rate of delivery and the surface area for gas exchange with the fetus. High-altitude infants are also born with a higher haemoglobin concentration than their sea-level counterparts. Eggs of birds laid at high altitude tend to be smaller than those at sea level and their total pore area for a given shell surface area is decreased (see Box 20.1). Nevertheless, the overall diffusive conductance of high altitude and sea-level eggs are similar because the coefficient of gas diffusion increases at high altitude. While this property would favour O_2 delivery, the accompanying increase in water vapour diffusion presents a danger of desiccation. Despite similar diffusive conductances, the Po_2 at the chorio-allantoic blood is low in the high-altitude embryo, the result being that O_2 consumption is depressed and hatchability decreased. In the domestic fowl, hatchability is severely restricted at elevations greater than 4000 m. As such, the fowls seen in Cerro de Pasco (elevation 4300 m) are transported to the city after hatching at sea level.

Oxygen stress and temperature

A large number of animals lower their body temperature when acutely exposed to low levels of O_2 (hypoxia). This is achieved behaviourally in ectotherms, whereas low levels of O_2 can abolish thermogenesis in endotherms and may also lead to behavioural changes. Many animals, regardless of whether they are ecto-thermic or endothermic, given the choice, will choose a temperature that they prefer to live in, that is, they exhibit some form of behavioural thermoregulation. Almost all animals studied show a preference for a lower environmental temperature when placed in a thermal gradient and subjected to hypoxia. This behavioural adjustment is observed from *Paramecium* through to mammals. A good example can be seen with a population of microscopic crustaceans, *Daphnia*, placed in a thermal gradient. In well-oxygenated water, the population prefers water with a temperature of about 23°C, however, on exposure to hypoxia the population will shift to a lower temperature (Fig. 29.17).

Acute (short-term exposure to) hypoxia is known to depress thermogenesis in all its forms in endotherms, and this is particularly apparent in small or newborn animals. Accompanying the depression is a lowering of

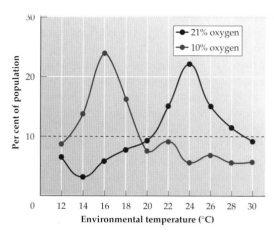

Fig. 29.17 The distribution of *Daphnia carinata* in a thermal gradient when exposed to aerated water (21% O_2) or water made hypoxic (10% O_2). Exposure to hypoxia results in a decrease in the thermal preference of the population

body temperature, the exact extent as to the drop in body temperature will depend on the environmental temperature. Despite a depression in thermogenesis, thermoregulatory ability is not abandoned altogether and it now appears that the thermal set point deciding body temperature has been reset to a lower temperature (Fig. 29.18). Larger animals (>50 kg) are not nearly as affected. Firstly, as a result of their large size, they possess a smaller surface area with respect to their volume and therefore have lower thermal conductances. Secondly, they possess a lower mass specific metabolic rate and as such a considerably smaller proportion of O_2 consumption is used for maintaining body temperature.

What possible benefits are to be gained by an animal lowering its body temperature when challenged with hypoxia? Principally, the greatest advantage is the associated drop in metabolic rate that is brought about by the direct effect of decreasing temperature (Q_{10} effect). For every 1°C reduction in body temperature when Q_{10} is 2.5, there is a corresponding 11% decrease in metabolic rate, which means the animal's need for oxygen is reduced. In addition, a reduction in temperature results in an increase in the affinity of haemoglobin for oxygen. Finally, the energetically expensive responses to hypoxia, for example, increased ventilation and cardiac output, may be avoided. A good example of the benefit offered by the hypothermia that accompanies hypoxia is an increase in the survival rates of neonatal mammals. Congenitally asphyxiated (extreme situation of hypoxia and high CO_2) human babies clinically treated by hypothermia have a mortality rate of approximately 11% compared with 48% when hypothermia is not used.

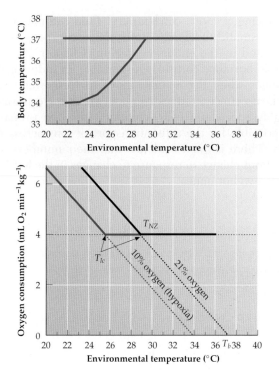

Fig. 29.18 Schematic representation of changes in O_2 consumption as a function of environmental temperature in the adult rat. Above the lower critical temperature (T_{lc}), O_2 consumption remains constant as temperature increases (T_{NZ}). Below T_{lc}, oxygen consumption increases linearly with decreasing ambient temperature. The straight line describing thermogenesis extrapolates to the x-axis to intersect at the temperature of the body (T_b). A resetting of the thermal set point in hypoxia reduces T_{lc} and lowers T_b. Parallel slopes of the two lines indicate that conductance has not altered

It should now be apparent that parameters such as O_2 consumption and body temperature that were previously thought to be tightly protected can be made to alter when environmental conditions are unfavourable. When oxygen supply is reduced, two responses are possible. Compensatory mechanisms that attempt to increase O_2 delivery, such as those displayed by animals who sojourn to high altitude (e.g. hyperventilation, increased haemoglobin concentration), enable the animal to function in its new environment. On the other hand, a reduction in oxygen consumption and an accompanying drop in body temperature when Po_2 falls offers an alternative strategy that at least in the short term ensures survival of the animal.

Responses to osmotic stress in aquatic animals

Marine invertebrates have blood osmotic concentrations close to that of sea water (1000 mOsmol/L) and live in the open sea where they encounter no osmotic stress. Many are **stenohaline**, that is, they can tolerate

little change in external salinity. However, invertebrates in intertidal zones and estuaries are exposed to periodic fluctuations in salinity. They show a range of behavioural and physiological adaptations that allow them to tolerate a wide range of salinities, that is, they are **euryhaline** (Chapter 22).

Behavioural adaptations of euryhaline invertebrates involve either burrowing in the substrate where ambient salinities are more stable, or seeking refuge in a closed shell. Burrowing strategies are adopted by annelid worms and crabs, and by this behaviour they also avoid the effects of currents. Hard-shelled species, such as mussels, close their valves as sea water concentrations fall and maintain salinities close to that of sea water within their mantle cavity. Mussels have sodium and magnesium sensors on the tentacles associated with the incurrent siphon that control this response.

> Stenohaline animals tolerate little change in external salinity; euryhaline animals tolerate a wide range of salinities.

Osmotic conformers and intracellular osmotic regulation

In some aquatic animals, experiments have shown that exposure to a range of osmotic concentrations results in corresponding changes in body fluids. These animals have been termed **osmotic conformers** (Chapter 22). If extracellular osmolarity decreases we would expect cells to take up water and consequently swell. With an increase in extracellular osmotic levels, water would move out of cells and they would decrease in volume. In the short term, such changes do occur in osmoconformers to some extent, however, they are followed by a slow recovery of cell volume to the initial level. This recovery phase is due to changes in the levels of low molecular weight organic molecules, such as nonessential amino acids, polyhydric alcohols and mono- and disaccharides. The control of this intracellular solute pool is responsible for reducing any osmotic gradient between the intracellular and extracellular fluid compartments, and appears to be the primary adaptation permitting osmotic conformers to inhabit areas of varying salinity. Recently it has been shown that a similar process occurs in mammalian cells, particularly in the regulation of brain cell volume. Although changes in plasma osmolarity of vertebrates do not show the degree of change seen in invertebrates, it is interesting that vertebrates may have retained this system from simpler animals.

> In some animals the composition of an intracellular solute pool may be varied to reduce the osmotic gradient across the cell membrane.

Migratory fishes

Most teleost fishes have a limited ability to move between the sea and fresh water. In fresh water, osmosis brings water into the body, which is then removed by the production of large volumes of dilute urine; solutes are lost by diffusion and are replaced by active transport across the gills and from food (Chapter 22). In sea water, fish are hypo-osmotic and the opposite water and solute gradients occur; water loss is replaced by drinking and production of more concentrated urine and salts are actively excreted across the gills.

Some fishes, including lampreys, salmon and eel, migrate between the sea and freshwater streams as part of their normal breeding cycle. To compensate for this osmotic stress, the direction of salt transport across the gills can be reversed and the kidneys can change from producing copious amounts of dilute urine to producing a more concentrated fluid (although still hypo-osmotic to blood). Both of these processes are regulated by the endocrine system (Chapter 25). The steroid hormone cortisol is responsible for inducing the changes when the animals move to fresh water and prolactin concentrations increase when in fresh water. Highly specialised cells, chloride cells, are primarily responsible for the active transport of ions across the gills. There is good evidence that their numbers and functions change in response to changes in the levels of these two hormones.

Fishes that migrate between sea and fresh water can change the direction of salt transport across the gills and adjust the concentration of their urine. Both changes are under endocrine control.

Summary

- Responses of individual animals to environmental stress may involve anatomical, physiological and behavioural changes in phenotype within limits defined by their genotype.
- In animals, physiological adjustments to the environment are linked with behaviour that allows an animal to select microhabitats within its tolerance limits.
- Different environmental conditions, such as low temperature, low O_2 levels, drought and salinity, may lead to the same stress in an organism, for example, dehydration.
- Few organisms tolerate freezing of tissues. Survival of animals at very low temperatures usually involves supercooling or biological compounds that lower the freezing point and prevent ice formation.
- Endotherms are animals such as birds and mammals in which body temperature is derived from metabolic heat. Ectotherms are animals that derive their body temperature from heat gained from the environment.
- Torpor is hypothermia involving a controlled lowering of metabolic rate and reduced responsiveness to external stimuli. Long-term torpor in cold weather (hibernation) is associated with the dangers of freezing, dehydration and lack of energy supply.

- Environmental stress often results in a reduction in the metabolic rate of animals. This appears to have a different basis than that involved in torpor.
- Most desert animals avoid the direct impact of high temperatures and few animal species in arid Australia regularly require drinking water.
- As the altitude increases, the partial pressure of O_2 decreases by about one-half every 5500 m. Animals that are adapted to altitude generally possess a haemoglobin with a high affinity for O_2.
- Initial responses of humans to altitude hypoxia include an increased heart rate and ventilation rate, and an increase in the number of red blood cells. Later changes include an increased number of capillaries supplying tissues and decreased ventilatory response to low O_2 levels.
- Many animals when acutely exposed to low levels of O_2 lower their body temperatures. This can be a behavioural adjustment or a drop in metabolic rate.
- Stenohaline animals tolerate little change in external salinity; euryhaline animals tolerate a wide range of salinities. Fishes that migrate between the sea and fresh water can change the direction of salt transport across the gills and adjust the concentration of their urine. Both changes are under endocrine control.

keyterms

acclimation (p. 764)
acclimatisation
  (p. 764)
cryoprotectants
  (p. 772)
ectotherm (p. 767)

endotherm (p. 767)
euryhaline (p. 781)
hyperventilation
  (p. 779)
hypothermia (p. 772)
hypoxia (p. 773)

metabolic depression
  (p. 771)
osmotic conformers
  (p. 781)
Q_{10} (p. 767)
stenohaline (p. 781)

supercooling (p. 766)
torpor (p. 767)

Review questions

1. Differentiate between the terms 'acclimation' and 'acclimatisation'.

2. What are biological antifreezes and how do they work?

3. If you took a saw, cut a termite mound in half and rotated the severed end through 90°, describe how the microhabitat of the insects would change.

4. Describe the circulatory changes that occur when a diving mammal remains under water for an unexpectedly long period and explain why they are advantageous.

5. Explain why some athletes move to high altitude for some months to train before returning to sea level to compete. What are the short- and long-term changes that occur when they move to high altitude and when they return to sea level?

6. Lake Titicaca in the Andes of South America is at an elevation of 3812 m with a maximum water depth of 281 m. The temperature of the surface water averages 10°C, with less than a 4°C annual range. A frog, *Telmatobius*, lives in the lake. What would you expect to be its major adaptations to this habitat?

Extension questions

1. An Indian python, wrapped around eggs, can function as an endotherm. Discuss the mechanisms it may use to generate and retain heat.

2. Low rates of fertility are recorded from people who live at high altitude. Suggest reasons for this phenomenon.

3. An annelid worm is collected from mud rich in hydrogen sulfide. Suggest the properties of its haemoglobin.

4. Suggest the advantages of an animal selecting a low body temperature when exposed to hypoxic conditions.

Suggested further reading

Guppy, M. and Withers, P. (1999). Metabolic depression in animals: physiological perspectives and biochemical generalities. *Biological Review* 71: 1–40.

This is a fascinating review article, which is sometimes difficult but will introduce you to the idea that the depression of metabolic rate is present in most animal phyla in response to environmental stress.

Heinrich, B. (1993). *The Hot Blooded Insects.* Cambridge, MA: Harvard University Press.

A good example of curiosity-based research described by one of the best writers in science.

Louw, G. (1993). *Physiological Animal Ecology.* London: Longman.

A book written to bridge the gap between the teaching of ecology and physiology. It is well organised around the themes of temperature, water, nutrition and reproduction.

Randall, D., Burggren, W., French, K. (1997). *Animal Physiology: Mechanisms and Adaptations.* New York: W. H. Freeman.

An excellent source of material that blends the basic principles of animal physiology with a discussion of the adaptations that allow them to live in a range of habitats. Particularly good on cellular and membrane material.

Schmidt-Nielsen, K. (1997). *Animal Physiology: Adaptation and Environment.* 5th edn. Cambridge: Cambridge University Press.

This was one of the first books to place animal function in relation to environment. A wonderful and easy to read account of animals and how they function in their environment by an author who was one of the pioneers in the field.

Wilmer, P., Stone, G., Johnson, I. (2000). *Environmental Physiology of Animals.* Oxford: Blackwell Scientific.

This book attempts to integrate comparative and environmental physiology in an evolutionary perspective. A good treatment on the nature and mechanism of adaptation.

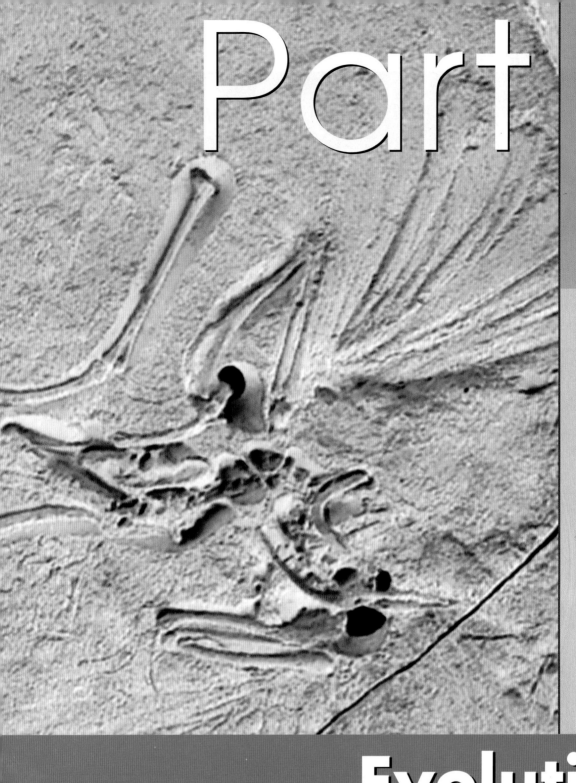

Part 6

6

Evolution and biodiversity

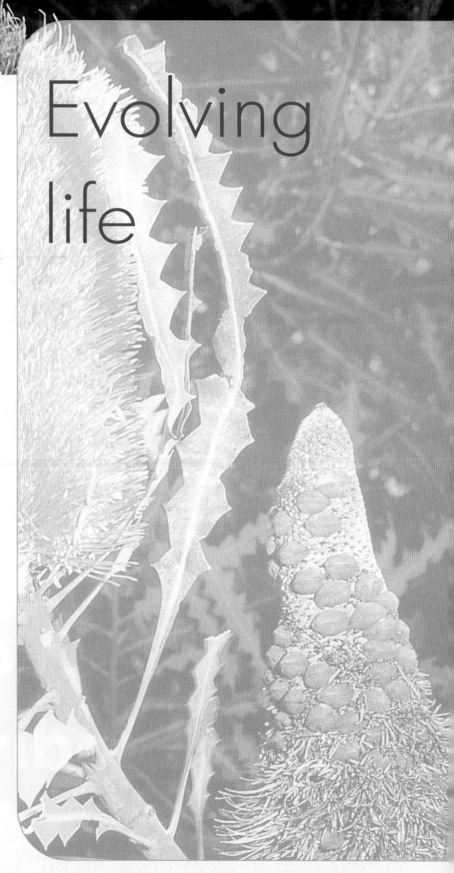

CHAPTER

30

Evolving life

During evolution, new kinds of organisms arise and others disappear through competition, climatic changes or natural catastrophes. Today, despite the fact that we realise that biological diversity is one of our most precious resources, species are being lost at a rate that is probably 1000 to 10 000 times the rate before human intervention. We do not know the exact rate of extinction of species, but in Australia since European settlement in 1788, 75% of rainforests, 99% of south-eastern temperate, lowland grasslands and 60% of south-eastern coastal vegetation have been cleared.

The dramatic loss of species caused by human activity worldwide is known as the *biodiversity crisis* (see also Chapter 45). Each species represents an immense amount of genetic information, from 1000 to 400 000 genes, and each species is represented by many individuals, from less than a few thousand to many millions. When a species is lost it can never reappear. Extinction is forever. Every species interacts with others; some, such as corals, are keystone species upon which whole communities depend. We do not know what will be the ecological impact of a continuing loss of species. Nor have we explored the possible uses of many species facing extinction, let alone even discovered their existence.

Of the 6–10 million species conservatively estimated for the world, only about 1.5 million have names and have been documented. Information about biology, relationship (classification) and distribution is available for only this small proportion of species. Among the species that have been named, the best known are birds and mammals, but all the vertebrates together add up to only 0.4% of species. Insects make up 65% and fungi 8% of the world's species, yet we have described and named only a small fraction of these and other groups, such as the soil-inhabiting nematodes, protozoa, microscopic algae and bacteria. Although not classified as cellular organisms, there are also many forms of viruses that have yet to be described. The least-known biotas are also among the most species-rich, including those in the tropics and subtropics.

At an international level, biologists are working together to discover and record the world's **biodiversity**, so that it can be synthesised into a classification system that will reflect our knowledge of the **phylogeny** of organisms, that is, their history and evolutionary relationships.

BOX 30.1 Australia's biodiversity

Australia has a very rich biota—somewhere between 7% and 8% of the world's total species diversity. We do not know exactly how many species occur in Australia, but there are well over 800 000. Some groups are relatively well known: there are nearly 25 000 species of vascular plants, 70% of which are known and named, and there are about 5400 vertebrates, 90% of which are known. However, the vertebrates represent less than 1% of Australia's biota and there are many other groups of organisms that are much more abundant and not well known (see table), particularly in soil and in the marine environment. It is estimated that Australia has nearly 250 000 species of fungi, of which only 10% have been named, at least 225 000 species of insects and related groups, and tens of thousands of molluscs (only 30% of invertebrates having been named). The insect order Coleoptera (beetles) alone constitutes over 50% of Australia's named fauna (see figure).

Often the only knowledge we have of Australian insects, other invertebrates and lower plants is that associated with the specimens held in biological research collections, such as our state museums, herbaria or the National Insect Collection in Canberra. At the National Insect Collection, for example, CSIRO biologists have surveyed the insects of rainforests that occur in the McIlwraith Range on Cape York Peninsula. This study is the first major attempt to sample insects in that region and, although far from complete, a very rich fauna has already been demonstrated. In a three-week survey during the beginning of the dry season, more than 2000 species of moths were found. Several species previously recorded only from New Guinea were found and several species of large moths were seen for the first time.

A very large proportion of all species occurring in Australia, perhaps over 80%, are **endemic** (having evolved there and occurring nowhere else). For example, there are 2830 endemic plant species in the eucalypt forests, woodlands and heathlands of south-western Western Australia.

The moth family Oecophoridae, known as mallee moths, has approximately 6000 species in Australia. Many of these species occur in mallee and in dry sclerophyll eucalypt forest where their larvae, which feed on dead eucalypt leaves, play an important role in the breakdown of sclerophyllous leaf litter to humus. In comparison, only approximately 100 species of Oecophoridae occur in the whole of Europe and in America north of Mexico. Australian oecophorids probably diversified with the eucalypts. Another group of insects, the ancient moth family Lophocoronidae with six species, is endemic to the mallee and dry sclerophyll forests of southern Australia. This moth family, which probably had a wider distribution in the past (in the Cretaceous), is now known to be the closest relative of all the higher moths and butterflies, nearly 250 000 species worldwide. The question of the value of species is a difficult one but it seems that the conservation of this unique, endemic family Lophocoronidae must have some priority.

Table **Estimated species richness and endemism in Australia**

800 000 species, being about 7–8% of the world's biota

25 000 species of vascular plants, 85% endemic

850 species of birds, 45% endemic

276 species of mammals, 84% endemic

146 species of marsupials, 52% of the world's total with 90% endemic

174 species of amphibians, 93% endemic

700 species of reptiles, 89% endemic

3600 species of fishes, most endemic

225 000 species of insects, many endemic

tens of thousands of species of molluscs, most endemic

250 000 species of fungi, including 5000 'mushrooms'

A collection of beetles showing diversity of form

Discovering phylogeny

Evolutionary biology aims to discover the history of life, to answer the question 'How have living organisms come to be the way they are?'. One major task is to discover the evolutionary tree of life (the pattern of evolution) and another is to determine the mechanisms of evolution that lead to change (the process of evolution). Before we look at how evolution happens (Chapter 32), we first need to understand methods that biologists use to discover the evolutionary relationships of organisms, called phylogeny.

How to discover a phylogenetic tree

The phylogeny of a group of organisms can be described by the pattern of branching depicting their evolutionary relationships. For example, the branching pattern in Figure 30.1 shows the relationship of the platypus, koala and dingo to one another. It indicates that the koala and dingo are more closely related than either is to the platypus. This kind of branching diagram is called a **phylogenetic tree** or **cladogram** (Gr. *clados* meaning branch).

How do we discover the order of branches in a phylogenetic tree? Comparison of the characteristics of these three species shows that they share a number of general features. All three have hair, internal fertilisation and suckle their young with milk from mammary glands. Both the koala and dingo have separate anal and urogenital openings as adults, mammary glands with teats and produce eggs that lack a shell. The platypus has only one opening, lacks teats and lays an egg with a thin papery shell. These characteristics of the platypus are more general, being shared with a larger group of organisms, including snakes, lizards and birds. A more general character is considered ancestral or **primitive** (called **plesiomorphic**), with an inheritance that is more remote. Characters that are less general, such as those shared by the koala and dingo but not the platypus, are derived or **advanced** (called **apomorphic**), and are considered to have evolved more recently.

Thus, each branch point in a phylogenetic tree or cladogram can be defined by the advanced characters that are shared by the various organisms on that branch. (*Shared* advanced characters are termed *synapomorphies*). Of course, each species also has its own unique characters (*autapomorphies*) because each lineage, each branch in the tree, has undergone evolutionary change after divergence. An adult platypus lacks teeth, has a duck-like bill and webbed feet used for swimming; the koala is a highly specialised mammal with a gut adapted to digesting eucalypt leaves; the dingo is a carnivore specialised for hunting prey.

The phylogenetic tree (the pattern of relationship) can also be the basis for hypothesising how these organisms evolved (the process of evolution). One interpretation of our example in Figure 30.1 would be that the platypus, koala and dingo all descended from a common ancestral species, which split into two lineages, with one leading to the modern platypus and the other to the common ancestor of the koala and dingo, which later split.

A phylogenetic tree, or cladogram, is based on discovering shared, advanced characters (synapomorphies) for each branch point. These characters indicate that all the species at a particular branch point on the tree shared a common ancestor.

Comparing morphology

As the example in Figure 30.1 illustrates, hypotheses about evolutionary relationships may be based on **comparative morphology**, the comparison of body form, including the study of embryology.

Comparative morphology as a scientific discipline has a long history and biologists of the nineteenth century showed it to be a key to reconstructing phylogeny (see Box 30.2). For example, if the early stages of development in vertebrates are compared, gill slits are observed to form in all embryos, even though they do not persist in adults other than fishes. The striking similarity of the embryos of fishes, frogs,

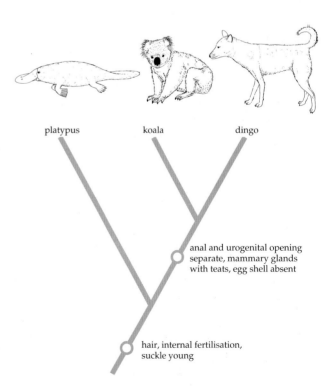

Fig. 30.1 A phylogenetic tree showing the relationships of the platypus, koala and dingo

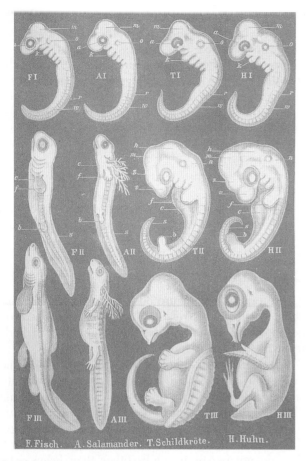

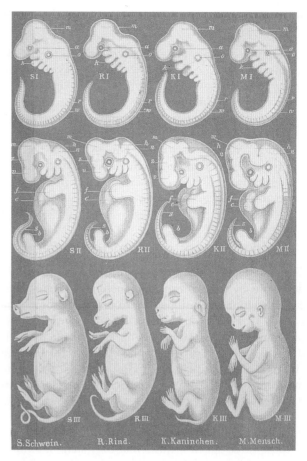

Fig. 30.2 There is a striking similarity between the embryos of vertebrates, from fishes to mammals. These drawings are from Ernst Haeckel's 1896 publication *The Evolution of Man*

lizards, birds and mammals (Fig. 30.2) is one reason why we hypothesise that these organisms had a single line of descent and why we classify them as vertebrates. The similarity of their early embryonic stages indicates that all vertebrates share a fundamental step in their developmental programs. Modification of the later stages of development accounts for the great diversity of form within the vertebrates; for example, birds develop feathers and humans have a brain larger than that of other primates.

Vestigial organs, that is, rudimentary organs with no apparent function, are also clues to evolutionary relationships when comparing organisms (Fig. 30.3). For example, humans have a vestigial tail (coccyx) similar to the prehensile tail of monkeys. An adult baleen whale lacks teeth but develops them as an embryo. The best explanation of vestigial organs is that they were functional and present in ancestral species and, thus, are indicators of phylogeny.

Fossils can also be compared with living organisms and included in phylogenetic trees, although often only the hard parts of organisms (such as bone and wood) are preserved and available for comparative study (see also Chapter 32). Importantly, the fossil record supplies evidence of both the forms of life in the past and their

geological ages. However, there is one note of caution in interpreting fossil data. There is no way of knowing whether a fossil is a *direct ancestor* of a more recent species or represents a related line of descent (lineage) that simply became extinct.

Fig. 30.3 All Australian pythons, family Boidae, have vestigial legs, known as cloacal spurs. A spur is visible on this black-headed python, *Aspidites melanocephalus*, as it lays an egg, which is stretching the skin. The spur appears as a spike

BOX 30.2 Unity of plan

The concept of homology

That all of life is based on the same fundamental principle was an important idea of the French zoologist Etienne Geoffroy Saint-Hilaire (1772–1844). He tried to see the 'unity of plan' first through comparison of adult and then of embryonic structures of animals. He recognised corresponding parts of different animals with reference to their 'connections', namely by observing the position of the part relative to all other parts of the animal. He believed, for example, that the middle ear bones of mammals—malleus, incus and stapes—correspond to the opercular (cheek) bones of fishes. Today we use the term 'homologous', after the usage established by the English palaeontologist Richard Owen (1804–92), for parts with the same connections, which is seen today to mean the same evolutionary origin. Many of Geoffroy's comparisons were improved by later workers who found, for example, that the jaw and not the opercular bones of fishes—articular, quadrate and hyomandibular—are the correct homologues of the mammalian ear bones.

Geoffroy's students believed that homologies existed between adult parts of animals as diverse as molluscs, insects and fishes, a point of view opposed by Geoffroy's famous contemporary

Georges Cuvier (1769–1832). Today, no biologist imagines that the chitinous exoskeleton of insects and the bony skeleton of vertebrates are homologous. Geoffroy's 'unity of plan' led him eventually to believe that all life evolved through descent with modification, as first suggested in a scientific context by another of Geoffroy's famous contemporaries, Jean-Baptiste de Lamarck (1744–1829). Biologists of today see a 'unity of plan' (or 'unity of composition') at the molecular level because of the universal occurrence of DNA. They use Geoffroy's principle of connections to recognise homologous positions in base sequences of nucleic acids, and in amino acid sequences in proteins.

Homologous features are evidence of relationships between divergent forms

Comparative morphology focuses on the basic similarity of organisms as evidence of how they diverged during evolution and departed from a common ancestral form (**divergent evolution**). For example, the flightless emu and a sparrow are different in many ways but both have feathers, indicating that they are related as 'birds'. Similarly, the comparative study of bones reveals that the flipper of a whale has the same basic structure as the front leg of a frog or crocodile and the wing of a bird or bat (Fig. 30.4). Structures that have the same basic plan but not necessarily the same function are **homologous**. Homologous structures indicate common inheritance and they allow us to recognise that frogs, crocodiles, birds, bats and whales are related as tetrapods (four-limbed vertebrates).

Analogous features are evidence of convergence

By **convergent evolution**, organisms from different, *unrelated* (or distantly related) lineages come to resemble one another superficially. Structures that have a similar function as a result of convergence are called **analogous**. For example, plants of the families Cactaceae in the New World and Euphorbiaceae in the Old World look remarkably similar with their succulent, leafless stems (Fig. 30.5). Both groups of plants are adapted to living in arid environments and their succulence is an analogous feature that serves as a store of water. The Australian echidna, with its protective covering of spines, resembles the European hedgehog; the Australian lemuroid ringtail possum earned its common name because it is so monkey-like,

Etienne Geoffroy Saint-Hilaire (1772–1844)

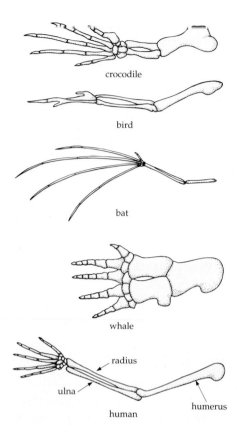

Fig. 30.4 The forelimb of a crocodile, bird, bat and whale have the same basic structure and are homologous

leaping from tree to tree, porpoises and seals have become streamlined for swimming, as have fishes; social behaviour has evolved a number of times among bees and ants.

Closely related organisms also independently evolve similar features, termed **parallel evolution**. For example, within the plant genus *Banksia*, the ability to resprout and regenerate vegetatively appears to have evolved a number of times and is not a feature of all *Banksia* species.

Convergent and parallel evolution result in features that are so similar that we could be misled in our search for phylogeny if other features did not indicate that the organisms are unrelated. Thus, the different flowers of cacti and euphorbs indicate that these two plant groups are not closely related; porpoises and seals have the characteristics of other mammals and are clearly not fishes. In discovering phylogeny, only homologous features are used, not analogous features.

Fig. 30.5 Convergence. Succulence occurs in different organs and different plant groups that inhabit arid environments throughout the world. Plants with leafless, succulent stems include **(a)** cacti (family Cactaceae) and **(b)** euphorbs (Euphorbiaceae). **(c)** Plants with succulent leaves include *Aloe*, a member of the monocot family Liliaceae

BOX 30.3 Riddle of the pinnipeds

Pinnipeds (seals, walrus and sea lions) have posed a persistent problem for evolutionary biologists. As early as the nineteenth century, some biologists considered them to be two separate groups among the dog-like carnivores descended from different terrestrial ancestors, the sea lions and walrus being descended from bear-like ancestors and seals from otter-like ancestors. The argument for two groups was based on comparison of bones, principally cranial features. Features shared by pinnipeds were interpreted as convergent, relating to adaptations to an aquatic lifestyle. Other biologists considered that pinnipeds are a single group, descended from a common marine ancestor.

New and compelling evidence favours the hypothesis that the pinnipeds are a single group and comes from molecular biology, biochemistry, genetics and parasitology. The eye lens protein, α-lens crystallin A, shows two amino acid substitutions (at positions 51 and 52), shared by sea lions and seals, that are unique among vertebrates. Seals, walruses and sea lions also share another amino acid substitution that occurs in a few other animals (whale and chicken) outside the carnivores. Pinnipeds also share a unique bile acid (phocaecholic acid) found in no other animal yet sampled.

DNA–DNA hybridisation data also support the hypothesis of one group. DNA–DNA hybridisation is a technique where single strands of DNA from two organisms are combined to form hybrid double helices. The melting point of this hybrid DNA is directly related to how closely the nucleotide sequences match and gives an overall measure of the genetic similarity of a pair of organisms; the lower the melting point the greater the difference between the two strands of DNA. Seals, walrus and sea lions are more similar to each other than they are to other dog-like carnivores. The karyotypes of the pinnipeds are similar; diploid numbers are 32, 34 or 36, and chromosome banding patterns appear to be homologous.

Finally, seals, walrus and sea lions share closely related ectoparasites. All three kinds of pinnipeds are hosts to a particular group of parasitic lice. These different data together with comparative anatomy support the phylogeny shown on page 796.

(a) A sea lion characteristically has external ears

(b) The Weddell seal from Antarctic waters keeps its hind limbs pointed posteriorly

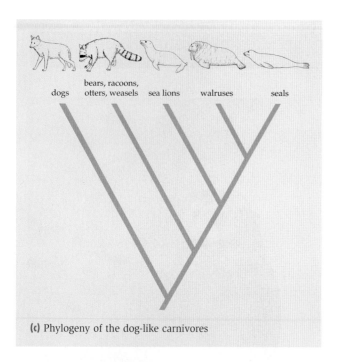

(c) Phylogeny of the dog-like carnivores

Hypotheses about evolutionary relationships (phylogeny) are traditionally based on comparative morphology and the discovery of homologous structures. Homologous structures are indications of common inheritance. Analogous structures, which have similar functions, result from convergence and are misleading in the search for phylogeny.

Comparing molecules

Homology exists also at the level of molecules and comparative study of DNA or its protein products from different species now contributes importantly to our knowledge of phylogeny. With the development of techniques for **sequencing** amino acids in proteins and nucleotides in DNA molecules, it is now possible to compare organisms at the most basic level. Furthermore, unlike morphology, molecules allow us to compare organisms that have no apparent morphological homologies, for example, a eucalypt tree and a kangaroo! Because all organisms, ranging from bacteria to fungi, algae, flowering plants, insects, birds and mammals, have nucleic acids, DNA sequencing promises to provide data for discovering the tree of all life.

Amino acid sequences

With the understanding of the role of DNA in protein synthesis (Chapter 11), it became clear that the sequence of amino acids in a polypeptide chain is a molecular record of the genetic information and evolutionary history of an organism. In the 1960s and 1970s, it was shown that humans and chimpanzees have identical amino acid sequences in the respiratory enzyme cytochrome *c* and in α- and β-haemoglobins, while they have one amino acid difference in myoglobin. However, not all point mutations in DNA cause amino acid substitutions, for the genetic code is degenerate. The same amino acid may be coded by different base triplets. For example, valine is specified by CAA, CAG, CAT or CAC, and any change in the third position of each of these codons is silent. For this reason, and because it is now technically easier and cheaper to sequence nucleic acids than proteins, DNA sequences today are the preferred type of data.

Nucleotide sequences

Nucleotide sequencing has been carried out on mitochondrial DNA, chloroplast DNA and nuclear DNA of various organisms (Table 30.1). The amount of information is potentially enormous because every nucleotide base pair in a DNA sequence is a separate character. The following is a simple example to show how the sequence of bases in a piece of DNA can be used to discover the relationships of organisms to one another.

Table 30.2 shows three nucleotide positions (at positions 1156, 2311 and 3257) in the sequence of the 28 S ribosomal gene of three vertebrates—mouse, cockatoo and frog—and an insect.

There are three possible phylogenetic trees for the three vertebrates (Fig. 30.6). Each tree seems to be supported by one of these three base substitutions. For example, mouse and frog share T at position 1156, suggesting that they are related (tree 1). On the other hand, cockatoo and frog share G at position 2311 (tree 2) and mouse and cockatoo share G at position 3257 (tree 3).

If we include the sequence for the insect *Drosophila*, which we assume is an earlier branch on the phylogenetic tree below the vertebrates, we can determine which base positions are phylogenetically informative within the vertebrates. We assume that a base that is in both the *Drosophila* sequence and any of the vertebrate sequences (e.g. A at position 3257) must have been inherited from a more remote common ancestor. A different base (G) at the same position in the vertebrate sequences must have evolved within that group. Thus, in this case, G is a derived character (G substituted for A). G at position 3257 for mouse and

Table 30.1 Part of the DNA sequence for a γ-haemoglobin gene, which is homologous for the five mammal species shown. The sequences are aligned by matching the parts that are identical. The bases that differ from humans are in bold

Human	TGACAAGAACA—GTTAGAG—TGTCCGAGGACCAACAGA...
Orang-utan	T**C**AC**G**AGAACA—GTTAGAG—TGTCCGAGGACCAACAGA...
Rhesus monkey	TGAC**G**AGAACA**A**GTTAGAG—TGTCCGAGGACCAACAGA...
Lemur	T**A**AC**G**ATAACA**GG**ATAGAG—T**A**TC**T**GAGGACCAACAGA...
Rabbit	TG**GTG**ATAACAAGAC**A**GAG **A** TATCCGAGGACCA**G**CAGA...

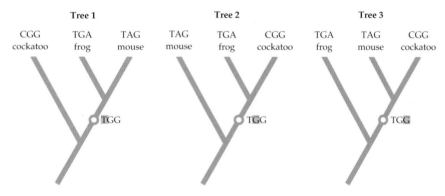

Fig. 30.6 Three possible trees for the frog, mouse and cockatoo

Table 30.2 A comparison of three nucleotide positions in the 28 S ribosomal gene of three vertebrates and a more distantly related organism (called an outgroup) (see Fig. 30.6). The ancestral condition is the base shown for the outgroup, the insect, and the advanced condition is shown in bold

	Base position		
	1156	**2311**	**3257**
Mouse, *Mus*	T	**A**	G
Cockatoo, *Cacatua*	**C**	G	G
Frog, *Xenopus*	T	G	A
Insect, *Drosophila*	T	G	A

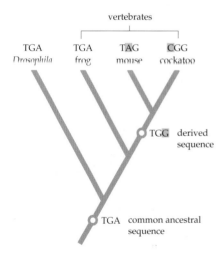

Fig. 30.7 The tree that best fits the DNA data when we include the outgroup, *Drosophila*. The frog has the ancestral sequence (same as *Drosophila*). The mouse and cockatoo share a base substitution (G substituted for A) at the third position, indicating their relatedness. The mouse also has one unique substitution (A for G at the second position), as does the cockatoo (C for T) at the first position

cockatoo, therefore, tells us that tree 3 best fits the data (Fig. 30.7). The bases at the other two positions are not informative of a vertebrate relationship because only cockatoo has the derived character (C) at 1156, and only mouse (A) at 2311.

This example uses only three base positions, but in reality a biologist would compare hundreds or thousands of positions (e.g. by sequencing a whole gene) to be more confident of finding the correct phylogenetic tree.

Sequencing amino acids in proteins and nucleotides in DNA is contributing to our knowledge of phylogeny.

Mitochondrial DNA

The early sequencing work was done on hominid mitochondrial DNA (mtDNA). The entire mitochondrial DNA, about 15 000 bases, has now been sequenced. The sequence data show, with high confidence, that African apes (gorilla and chimpanzees) and humans share a common ancestor, and that chimpanzees are our closest living relative. The orang-utan diverged earlier (Fig. 30.8). (These results are also supported by the sequencing of the β-haemoglobin gene.)

One problem with mtDNA is that **transitions**, which are changes from one purine to another and one pyrimidine to another (Chapter 1) accumulate so rapidly that within about 20 million years all readily substituted positions have been changed and further changes result in back mutations. **Transversions**, changes between purines and pyrimidines, occur less frequently (by a factor of eight in hominids). Thus, sequence evolution of mtDNA can generally give meaningful results only for organisms that have diverged from one another in a relatively short time (about 20 million years). However, certain aspects of mtDNA are highly conserved (including secondary structures of tRNA and rRNA sequences, gene order and some coding regions) and can be used for higher level phylogenetic studies.

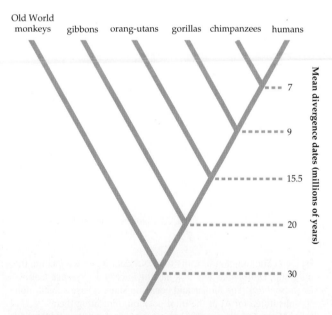

Fig. 30.8 Phylogeny of the great apes and humans. Nucleotide sequencing of mitochondrial DNA shows that African apes (gorilla and chimpanzee) and humans are closely related and that orang-utan is a lineage that diverged earlier. Sequencing of the β-haemoglobin gene supports these findings that chimpanzees are the closest living relatives of humans, *Homo sapiens*. In contrast, two morphological characters, hind limb morphology associated with knuckle walking and enamel structure of teeth, have been used as evidence to link gorillas and chimpanzees as the closest relatives

Chloroplast DNA

The genome of chloroplasts (cpDNA) also provides sequence information for photosynthetic organisms. Sequencing of the chloroplast *rbc*L gene is allowing botanists to discover the relationships among the seed plants, particularly the major groups within the flowering plants (angiosperm phylogeny). More rapidly evolving, non-coding regions of cpDNA are being used to study more closely related species and the origin, for example, of crop plants from wild species. The entire chloroplast genome has been sequenced for some species.

Nuclear-encoded RNA and molecular clocks

Many studies, such as in our example of the frog, mouse and cockatoo, have been made on nuclear DNA that encodes for ribosomal RNA (rRNA). There are three genes (in vertebrates they are 18 S, 5.8 S and 28 S rRNA genes) that form an array (unit) that is repeated many times along a chromosome. The three rRNA genes have evolved at a very slow rate and contain the most highly conserved sequences of DNA that occur in living organisms, and portions of these genes have been used to reconstruct phylogeny back to the origin of life nearly four billion years ago.

The arrays show regions of greater variation as well: the genes are separated by transcribed spacer regions (Fig. 30.9), which evolve at a rate faster than that of the genes. Adjacent arrays are separated also by non-transcribed spacers (silent DNA), which evolve even more rapidly than the transcribed spacers.

As you might expect, nucleotide changes (mutations) that adversely affect the function of a critical gene are eliminated by natural selection; on the other hand, changes that are *neutral* (do not affect function) accumulate. This means that there is a variety of molecular clocks, some ticking slowly and some ticking quickly. The more variable spacers can be used to reconstruct the phylogeny of organisms back to 1 million years ago, whereas rRNA genes can be used to reconstruct phylogenies back to four billion years ago. Thus, the various regions of ribosomal DNA (genes, transcribed spacers and non-transcribed spacers) can be used to construct phylogeny over the entire history of life (Fig. 30.10).

DNA sequences evolve at different rates. Regions of DNA where base substitutions accumulate rapidly are useful for studying the phylogeny of closely related organisms. Highly conserved regions, such as nuclear genes that encode for ribosomal RNA, can be used to study organisms dating back to the origin of life, four billion years ago.

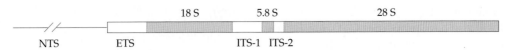

Fig. 30.9 The ribosomal DNA repeat unit of vertebrates consists of three genes (18 S, 5.8 S and 28 S) separated by internal spacer regions (ITS) and an external transcribed spacer (ETS). Each unit is separated by a non-transcribed spacer (NTS)

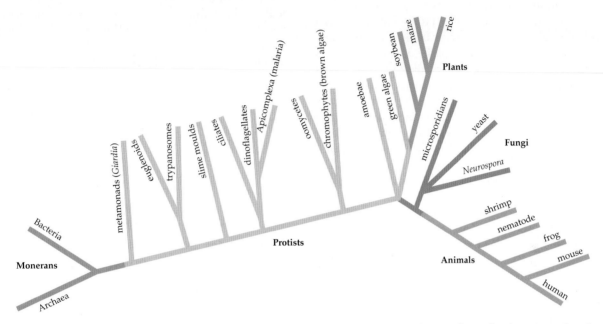

Fig. 30.10 A tree of major groups of life based on homologous parts of ribosomal RNA genes in the cytoplasm of prokaryotes and nucleus of eukaryotes. The base, or root of the tree, is not specified. The two groups of bacteria are either treated together as a single kingdom (Monera) or as two separate 'super kingdoms' or domains—the Bacteria and Archaea. In the latter case, the other kingdoms are grouped together as a third 'super kingdom' or domain, the Eukarya. The tree is interesting because it places oomycetes (water moulds), previously classified as fungi, near photosynthetic organisms (chromophytes; see Chapter 35). Ciliates, traditionally considered animal-like, are placed in a more remote position, near the photosynthetic dinoflagellates, and fungi are related to animals

Classification

In discovering phylogeny, we recognise groups of organisms that have characters in common (homologies). On page 791, we grouped the platypus, koala and dingo together because they have hair, internal fertilisation and suckle their young. In fact, we can make the general statement that all mammals have these features. We can also make less general statements about subgroups within the mammals: koalas, wombats, kangaroos and possums all raise their young in a pouch and are classified as marsupials; dingoes together with shrews, bats, primates, rodents, whales, cows and horses are 'so-called placental' mammals. Thus, groups and subgroups of organisms can be defined, placed in a **classification** and given scientific names as **taxa** (sing. taxon). What groups should be recognised and named as taxa are the primary concerns of classification.

Classification is an old notion dating back to the early Greeks around the time of the fourth century BC.

Centuries later the Swiss botanist de Candolle (1813) coined the term taxonomy for plant classification. Today, the word **taxonomy** is used to refer to the methods and principles of classification and rules that govern the naming of all organisms.

Purposes of classification

By referring to a name we can indicate a group of organisms without having to list all the properties known to occur in each of the organisms belonging to the group. Thus, given the vast diversity of organisms on earth (see p. 789), classification obviously performs an essential function in information storage and retrieval. Considerable information is subsumed into single words such as mammal or monocotyledon.

Classification, however, does more than store information; it serves also to predict information that we do not yet have. For example, if a set of organisms shares characteristics A, B, C, D and E that no other organisms have, and we find another organism about which all we know is that it has characteristics A and B, we can predict that it will also have properties C, D and E. We can make such predictions because our classifications reflect the order that exists in nature—we have based them on our discovery of phylogenies.

Being able to predict information is useful. For example, the legume Moreton Bay chestnut or black bean, *Castanospermum australe* (family Fabaceae), is a tree of Australian rainforest river banks. It is the only species in the genus and is easily identified by its distinctive orange-red pea flowers, dark pinnate leaves that smell of cucumber and large seeds rich in starch (Fig. 30.11). The seeds, although highly poisonous, were gathered by Aboriginal Australians for food; the toxins were removed by a process of grating, soaking and baking. One of the toxins is an alkaloid, castanospermine (Fig. 30.11), which has been found to control cancer cells in rats and inhibit production of the AIDS virus glycoprotein, an essential element in HIV infectivity. In a search for other sources of this medically important alkaloid, researchers turned to taxonomy to reveal related plants, in particular the genus *Alexa*, which is native to wetlands of tropical South America, including the Brazilian Amazon Basin. *Alexa* and *Castanospermum* are known to have very similar and distinct pollen within the family Fabaceae, and their close relationship was confirmed by the finding of the alkaloid castanospermine in *Alexa*. This compound has not yet been found in any other plant genus.

Naming the hierarchy of life

In the biological system of classification, a group is recognised as a taxon and is given a Latinised name; for example, Plantae (multicellular land plants), Magnoliophyta (flowering plants) and Liliopsida (monocotyledons: flowering plants with a single seed leaf in the embryo). Since the time of the Swedish naturalist Carl Linnaeus (1707–78), the biological system has been conceived as a hierarchy with specified levels or ranks. Thus, the taxon Plantae has the rank of kingdom, Magnoliophyta is a phylum and Liliopsida a class.

The rank order of taxa commonly used today is:

 kingdom
  phylum
   class
    order
     family
      genus
       species

Thus, **species** are grouped into **genera**, genera into **families**, families into **orders**, orders into **classes**, classes into **phyla** (or divisions) and phyla into **kingdoms**. Intermediate ranks can be designated by 'sub', such as subfamily or subclass. The platypus, together with two species of echidnas, are formally classified in subclass Prototheria (monotremes) within class Mammalia, while the koala and dingo and other related mammals are in subclass Theria (marsupials and 'placentals') within class Mammalia (Table 30.3).

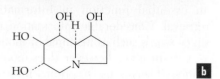

Fig. 30.11 (a) Moreton Bay chestnut, *Castanospermum australe*, contains **(b)** the alkaloid castanospermine, which has been found to inhibit production of the AIDS virus glycoprotein

BOX 30.4 **Bush medicines**

Bush medicines have always been important to Aboriginal Australians and European colonists were not slow to try them. Native currants, *Leptomeria acida*, gave some relief from scurvy, oil in the leaves of peppermint trees, *Eucalyptus*, aided digestion, and chewing the leaves of native pepper, *Piper novaehollandiae*, treated sore gums.

Bush medicines became important during World War II when normal supplies of drugs diminished. An alkaloid extracted from corkwood, *Duboisia myoporoides*, a tree known to Aboriginal Australians for its narcotic effects, was used to treat soldiers who suffered shell-shock and air sickness. After the war, 4000 species of Australian plants were surveyed in a major search for alkaloids that might prove useful to medicine. Although 500 alkaloids were discovered, many in rainforest plants and many previously unknown to science, the survey did not directly lead to the commercial development of new drugs. However, in recent years there has been a renewed search for phytochemicals in rainforest species for contraceptives and drugs for the treatment of cancer and AIDS.

Family Solanaceae

Plant steroidal saponines, which closely resemble human hormones, can be converted into synthetic hormones that have been used in the manufacture of the contraceptive pill. Eastern European manufacturers first used Mexican species of yams, *Dioscorea*, but with the decline in this resource due to habitat loss they turned, with success, to Australian plants in the family Solanaceae as a source of precursors for synthetic hormones. The kangaroo apple, *Solanum laciniatum* (Fig. a), which occurs naturally in Australia and New Zealand, contains the alkaloid salasodine, very similar to disogenin in the yams and easily

converted to a steroid. Extensive plantations of kangaroo apple are now grown industrially in Eastern Europe.

The family Solanaceae includes food plants such as potatoes, tomatoes, eggplant and peppers, as well as tobacco and hallucinogenic plants, such as deadly nightshade and *Datura*. Plants in this family contain a vast range of phytochemicals and are a 'good bet' in the search for medicinal drugs. Harvesting corkwood, also a member of the Solanaceae, has been a commercial success. Each year Australia exports large amounts of dried powdered leaves, representing half the world's market of the alkaloid hyoscine, used to treat motion sickness, stomach disorders and side-effects of cancer treatment.

In addition to the common corkwood, *Duboisia myoporoides*, which occurs on the margins of rainforests in Australia, New Guinea and New Caledonia, there is a rarer species, *D. leichhardtii*, restricted to vine thickets and scrub in southern Queensland. Both species produce hyoscine but the greatest yields are produced by hybrids of the two species (Fig. b), which are now cultivated in Queensland.

The potential use of other native plant species as 'bush medicines' is one argument for the conservation of Australia's remnant patches of rainforest.

(a) The potato family Solanaceae includes the kangaroo apples (i) *Solanum laciniatum* and (ii) *S. aviculare*, which are similar shrubs with dark green leaves, purple flowers and orange or scarlet berries. They are sources of steroids for the contraceptive pill

(b) The corkwood hybrid, *Duboisia leichhardtii* × *myoporoides*, is grown commercially in southern Queensland for hyoscine, an alkaloid used to treat motion sickness and side-effects of cancer treatment

Table 30.3 Biological classifications

Rank	Platypus	Koala	Dingo	Blue gum	Kangaroo paw
Kingdom	Animalia	Animalia	Animalia	Plantae	Plantae
Phylum	Chordata	Chordata	Chordata	Magnoliophyta	Magnoliophyta
Class	Mammalia	Mammalia	Mammalia	Magnoliopsida (dicots)	Liliopsida (monocots)
Subclass	Prototheria	Theria	Theria	Rosidae	Liliidae
Order	Monotremata	Marsupialia	Carnivora	Myrtales	Liliales
Family	Ornithorhynchidae	Phascolarctidae	Canidae	Myrtaceae	Haemodoraceae
Genus	*Ornithorhynchus*	*Phascolarctos*	*Canis*	*Eucalyptus*	*Anigozanthos*
Species	*O. anatinus*	*P. cinereus*	*C. lupus* subsp. *dingo*	*E. globulus*	*A. manglesii*

The hierarchical classification of the platypus, koala and dingo expresses the branching pattern of the phylogeny of these three organisms that we constructed on page 791. Class Mammalia is the branch within the vertebrates that includes all three species (Fig. 30.12). Subclasses Prototheria and Theria are more recent branches within the mammals.

Taxa recognised in this way are termed **monophyletic**, meaning an entire branch on a phylogenetic tree. However, many traditional classifications include taxa that have been found not to be monophyletic. For example, in the 1860s, Ernst Haeckel placed orang-utans, gorillas and chimpanzees in one family, now called Pongidae, separate from the family for humans,

now called Hominidae. Modern biologists agree that gorillas and chimpanzees are related more closely to humans than to orang-utans, as indicated by the branching pattern of the phylogeny in Figure 30.8. A classification based on phylogeny would group gorillas, chimpanzees and humans (one branch of the tree) in one monophyletic taxon (Fig. 30.13). A non-monophyletic group such as Pongidae is called a **paraphyletic group**. A paraphyletic group excludes some of the descendants of a common ancestor and, in a phylogenetic sense, has little meaning.

A **polyphyletic group** is one where unrelated organisms have been grouped together based on superficial, non-homologous resemblance due to convergent evolution (see p. 793). As we discover homologous characters, there are some taxa that we now know are polyphyletic, and these are being eliminated from classification schemes. For many of the unicellular eukaryotes, including protozoa and algae, relationships and monophyletic groups are only just being discovered using DNA sequencing and study of the ultra-structure of cells using electron microscopy.

> In discovering phylogeny, we identify monophyletic groups of organisms. A classification results when we name these groups as taxa, at the rank of kingdom, phylum, class, order, family, genus and species. Essential functions of classification are information storage and retrieval, and prediction.

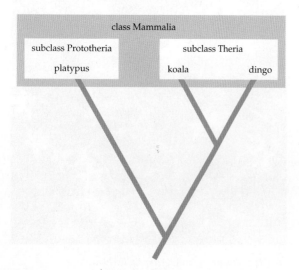

Fig. 30.12 A classification of the platypus, koala and dingo based on the branching pattern of their phylogeny. Taxa recognised in this way are termed monophyletic, that is, a taxon including all lineages that can be traced back to a common branch point

The binomial system

Naming taxa is governed by international sets of rules called **codes of nomenclature**, one each for bacteria

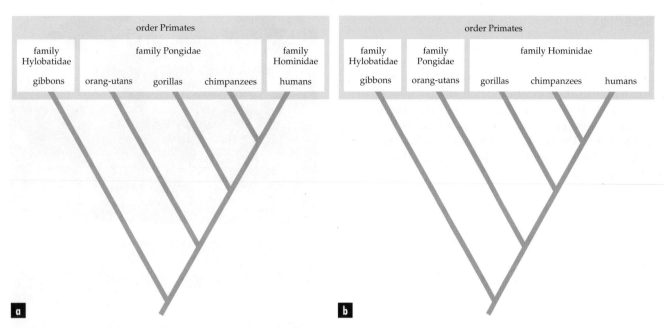

Fig. 30.13 (a) According to traditional classifications, great apes are grouped in the family Pongidae, separate from humans, family Hominidae, but this is not consistent with our knowledge of phylogeny (see Fig. 30.8) and recognises a paraphyletic group. **(b)** An alternative classification would recognise the monophyletic taxon that includes gorilla, chimpanzees and humans (which share a common branch point), at the rank of family or even genus as originally proposed by Linnaeus

(and viruses), animals, plants (and fungi) and cultivated plants. Having a number of sets of rules does lead to some confusion. There is no consistency between zoology and botany in the word ending of different taxonomic ranks; for example, botanical family names end in 'aceae', while zoological family names end in 'idae' (see Table 30.3). However, there are many similarities in the codes.

The ranks or categories of genus and species have a special significance because they form the basis of the **binomial system**, perfected by Linnaeus. In this system, the name of each kind of organism consists of two parts. *Homo sapiens* (humans), *Pan troglodytes* (the more common species of chimpanzee), *Gorilla gorilla* (gorilla), *Acacia melanoxylon* (blackwood) and *Trypanosoma cruzi* (a parasitic flagellate infecting humans) are each an example of a binomial. Binomials, once created, do not necessarily remain constant. Although *Homo sapiens* is the only living species in the genus *Homo*, in the future chimpanzees and gorillas could conceivably join humans in the genus *Homo*—as *Homo troglodytes* and *Homo gorilla*! It is interesting to note that, in 1758, Linnaeus placed chimpanzees in *Homo*.

In a binomial, the genus name is always written first, as in *Eucalyptus viminalis* or *Eucalyptus globulus*. The second word (the specific epithet), *viminalis*, is meaningless alone because different species in different genera may have the same specific epithet; for example, *Eucalyptus viminalis* is the manna gum tree

and *Callistemon viminalis* is a bottlebrush. Names given to taxa may be in honour of people. The genus *Banksia* was named after Sir Joseph Banks, a wealthy naturalist who voyaged to Australia with Captain Cook on the HMS *Endeavour* in 1770; the tropical plant genus *Aristolochia* was named after the celebrated philosopher Aristotle. Names may also be descriptive. *Banksia serrata* (saw-toothed banksia) has distinctive leaves with a serrated edge; *viminalis* refers to the willow-like habit of both the eucalypt and callistemon; *punctata* means spotted, as in *Sillaginodes punctata*, the southern Australian King George whiting, a fish easily distinguished by its spotted colour pattern.

Binomials are useful for unambiguous communication about organisms. Although common names exist for many taxa, they mean different things to different people. Thus, the Australian magpie, *Gymnorhina tibicen*, is not the same as the American magpie, *Pica pica*, and the prickly pear, the cactus *Opuntia stricta*, is very different from pears that are cultivated for their edible fruit, species of the genus *Pyrus*. The Australian salmon, *Arripis trutta*, is not a salmon. The marsupial frog, *Assa darlingtoni*, of Australian rainforests is obviously not a marsupial, but an extraordinary frog in which the males form pouches on their flanks in which the young tadpoles are carried (Fig. 41.29).

> In the binomial system, the name of each kind of organism consists of two parts based on the ranks of genus and species.

Species

As a taxonomic category, the species is the lowest rank in the Linnaean hierarchy to which all organisms must be classified. In practice, samples of organisms that a biologist can distinguish and tell others how to distinguish or diagnose are considered to represent different species. The specimens of three *Banksia* species illustrated in Figure 30.14 differ in the shape and size of leaves, flowers and fruits, characteristics that a plant taxonomist uses for identification. Some species, however, are morphologically very similar but differ in other ways; the fruitflies *Drosophila pseudoobscura* and *D. persimilis* were first recognised as different species by their chromosomes and ecological behaviour, and only later were subtle differences in the male genitalia observed.

Within many species, individuals are very different in form, such as males and females, and young and adults. Male birds are often more brightly coloured than females, and caterpillars are strikingly different from adult butterflies. The sample of organisms that a biologist believes represents a species includes all parts of the life cycle, because males and females together, and eggs, larvae, pupae and adults together, involve reproduction—self-perpetuation.

In an evolutionary context, the question 'What is a species?' has been discussed extensively by biologists. Numerous different species' concepts have been proposed. One traditional concept, dating from the botanist John Ray (1627–1705), is of *biological species*, which the ornithologist Ernst Mayr described in the twentieth century as 'groups of actually or potentially interbreeding natural populations, which are reproductively isolated from other such groups'. This species definition is based on the idea that, if organisms of two species freely exchange genes, the difference between the species will break down. The definition, however, does not work well in some animal groups and many plant groups, such as ferns, eucalypts and oaks, that form fertile hybrids. Neither does it seem appropriate for organisms that reproduce asexually, such as bacteria and many unicellular eukaryotes. For evolutionary biologists, the discussion about species' definitions is concerned with discovering the processes of evolution—the origin of species and higher taxa—which we shall return to in Chapter 32.

As a taxon, the term species refers to the lowest rank in classifications. In an evolutionary context, the species concept has been discussed extensively and there is no single definition that is agreed upon. The biological species concept emphasises reproductive isolation and the general lack or reduction of interbreeding between sexually reproductive species.

Fig. 30.14 These specimens are all members of the Australian genus *Banksia*, which is classified in the Southern Hemisphere family Proteaceae, together with *Protea, Macadamia, Persoonia* and *Grevillea*. **(a)** *B. oreophila*, **(b)** *B. coccinea* and **(c)** *B. ashbyi*. The genus is characterised by its flowers and fruits, which are woody follicles. Although all three specimens share characteristics typical of the genus, there are clear differences in the shape and size of their leaves, flowers and fruits by which botanists recognise them as different species

Kingdoms of life

Linnaeus classified living organisms into two kingdoms—plants and animals—and Georges Cuvier classified all animals into just four phyla. However, with the development last century of microscopy and biochemistry, our knowledge of the kinds of organisms on earth has increased remarkably. Cell biology has revealed fundamental differences between prokaryotes and eukaryotes. Bacterial classification is changing with the discovery of differences in ribosomal DNA between major groups. Observations of ultrastructure, cell wall biochemistry and photosynthetic pigments of algae show them to be an assemblage of organisms with diverse relationships. With the recognition that the oomycetes (water moulds, white rusts and downy mildews) have cellulose cell walls and flagellated zoospores, it appears that not all organisms traditionally called fungi are fungi!

How the major taxa are related to one another is still being discovered. Slime moulds have amoeboid, animal-like stages but their reproductive structures are like fungi—to what are they related? This uncertainty of relationships is reflected in continuing changes to classification, including the recognition of **super kingdoms**, also called **domains**, and the expansion of the number of kingdoms (five or more) and phyla (nearly 100).

Many biology books follow a traditional system that recognises five kingdoms of cellular organisms (Fig. 30.15): Monera (prokaryotes), and Protista, Fungi, Plantae and Animalia (all eukaryotes). As you saw in Chapter 3, prokaryotes, **Monera**, are microorganisms that lack a nuclear membrane, mitochondria and chloroplasts. Two major evolutionary lineages are now recognised among the prokaryotes, and they are classified by some taxonomists as the 'super kingdoms' **Bacteria** (also termed the Eubacteria or true bacteria) and the **Archaea** (ancient bacteria, also termed the Archaeobacteria; see Chapter 33). The other four eukaryotic kingdoms are grouped together as the 'super kingdom' **Eukarya**. **Protista** are mostly unicellular, including photosynthetic and non-photosynthetic forms. **Fungi**, **Plantae** and **Animalia** are multicellular organisms with different modes of nutrition: fungi absorb organic molecules from their surroundings, plants are producers of organic molecules and animals are consumers that ingest other organisms for food. Because viruses are not cellular organisms, they are not included in these kingdoms (see Chapter 34).

The five kingdom classification is not completely satisfactory. There are particular problems, for example, with respect to the Protista, which are an assemblage of a number of different lineages and are not a monophyletic kingdom (see Fig. 30.10). Some of the green algae, for example, are closely related to land plants, yet this relationship is not apparent in the five kingdom classification, which groups green algae with other protists. We will explore such problems, and the biology and evolutionary history of living organisms, in more detail in Chapters 31 to 41. You should consult the Appendix, which gives one classification scheme of the various groups of living organisms.

> Five kingdoms of cellular organisms are usually recognised: Monera (prokaryotes), Protista, Fungi, Plantae and Animalia, although classification schemes are changing. With the recognition of two major evolutionary groups among the prokaryotes, they are often classified as two 'super kingdoms' or domains—Bacteria and Archaea. The other four eukaryotic kingdoms are grouped together as the 'super kingdom', or domain, Eukarya.

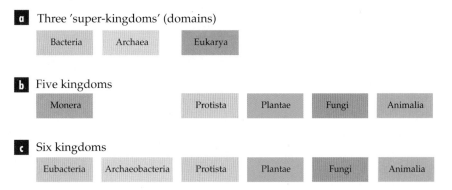

Fig. 30.15 Three different schemes for the classification of the major groups of cellular life. In other systems, the Protista are further split into a number of groups

Summary

- Hypotheses about evolutionary relationships (phylogeny) are traditionally based on comparative morphology and discovery of homologous structures.

- Homologous structures are indicators of common inheritance and allow us to discover phylogeny. A phylogenetic tree (cladogram) is based on discovering shared, advanced, homologous characters (synapomorphies) for each branch point. These characters indicate that all the organisms at a particular branch point on the tree shared a common ancestor. Analogous structures, which have similar functions but result from convergence, are misleading in the search for phylogeny.

- Sequencing of amino acids in proteins and nucleotides in DNA (in the genomes of mitochondria, chloroplasts and the nucleus) contributes to our knowledge of phylogeny. Different DNA sequences evolve at different rates. Regions of DNA where base substitutions accumulate rapidly are useful for studying the phylogeny of closely related organisms. Highly conserved regions, such as nuclear genes that encode for ribosomal RNA, can be used to study organisms dating back to the origin of life, four billion years ago.

- In discovering phylogeny, we recognise groups of organisms, such as mammals or flowering plants, whose members share many characters. Naming these groups (taxa) gives a classification. The science that deals with the methods and principles of classification is called taxonomy. Essential functions of classifications are communication, information storage and retrieval, and prediction.

- Classifications, like the evolution of organisms, are hierarchical and include the ranks of kingdom, phylum, class, order, family, genus and species. Taxa that include all the lineages that can be traced back to a common branch point are termed monophyletic, but not all taxa in traditional classifications are monophyletic.

- Naming of taxa is governed by international codes of nomenclature. The ranks of genus and species form the basis of the binomial system. As a taxon, the term species refers to the lowest rank in the Linnaean hierarchy. In practice, if a biologist can distinguish between samples of organisms, including all stages in the life cycle, they are considered to represent different species. In an evolutionary context, the species' concept has been discussed extensively and there is no single definition that is agreed upon. The biological species' concept emphasises reproductive isolation and the lack of interbreeding between sexually reproducing species.

- There are 6–10 million species of cellular organisms classified in five kingdoms and nearly 100 phyla. The five kingdoms usually recognised are: Monera (prokaryotes), Protista, Fungi, Plantae and Animalia (all eukaryotes), although classification schemes are changing. With the recognition of two major evolutionary groups among the prokaryotes, they are often classified as two 'super kingdoms' or domains—Bacteria and Archaea. The other four eukaryotic kingdoms are grouped together as the 'super kingdom', or domain, Eukarya.

key terms

advanced (p. 791)
analogous (p. 793)
Animalia (p. 805)
apomorphic (p. 791)
Archaea (p. 805)
Bacteria (p. 805)
binomial system
  (p. 803)
biodiversity (p. 789)
cladogram (p. 791)
class (p. 800)
classification (p. 799)
codes of nomenclature
  (p. 802)
comparative
  morphology (p. 791)

convergent evolution
  (p. 793)
divergent evolution
  (p. 793)
DNA–DNA
  hybridisation
  (p. 795)
domain (p. 805)
endemic (species)
  (p. 789)
Eukarya (p. 805)
family (p. 800)
Fungi (p. 805)
genus (p. 800)
homologous (p. 793)
kingdom (p. 800)

Monera (p. 805)
monophyletic group
  (p. 802)
order (p. 800)
parallel evolution
  (p. 794)
paraphyletic group
  (p. 802)
phylogenetic tree
  (p. 791)
phylogeny (p. 789)
phylum (p. 800)
Plantae (p. 805)
plesiomorphic (p. 791)
polyphyletic group
  (p. 802)

primitive (p. 791)
Protista (p. 805)
sequencing
  (nucleotides and
  amino acids)
  (p. 796)
species (p. 800)
super kingdom
  (p. 805)
taxon (p. 799)
taxonomy (p. 799)
transitions (p. 798)
transversions (p. 798)
vestigial organ
  (p. 792)

Review questions

1. What kinds of evidence support the theory that life evolves?

2. (a) What proportion of the world's biodiversity is found in Australia?

   (b) In terms of discovering, describing and naming species, about which groups of organisms do we know the least?

3. Approximately what proportion of Australia's

   (a) insects

   (b) marsupials

   (c) vascular plants

   are considered to be endemic? (Make sure that you know what the terms biodiversity and endemic mean.)

4. (a) Distinguish between the terms 'homologous' and 'analogous', giving examples.

   (b) What sort of features of similarity are the most useful for discovering evolutionary relationships?

   (c) Give one example of homology at the level of morphology and one at the level of DNA.

5. For the group of three mammals—platypus, koala and dingo (p. 781)—state whether the following features are plesiomorphic or apomorphic characters:

   (a) laying eggs with a shell

   (b) mammary glands with teats

   (c) webbed feet.

6. Flowers of many grevilleas (*Grevillea*) and waratahs (*Telopea*) are red, which attracts birds as pollinators. Is this colouration an example of convergent or divergent evolution? Why?

7. Why is it useful to have a classification of organisms? Why is it most useful to have a classification that is based on phylogeny?

8. (a) What is meant by the terms 'taxon' and 'taxonomy'?

   (b) List in order from highest to lowest the ranks commonly used in biological classification.

9. If eucalypts and tea trees belong to the same family, then they must also belong to the same:

(a) phylum?

(b) class?

(c) order?

(d) genus?

(e) species?

(More than one of these may be correct.)

10. Name the five kingdoms and list the major characteristics of each.

Extension questions

1. From your knowledge of the cellular characteristics of a bacterium, cyanobacterium (blue-green alga) and a green plant, construct a branching diagram to illustrate the possible phylogeny of these organisms, as was done for the platypus, koala and dingo on page 781.

2. Given your knowledge of DNA, the genetic code and proteins, explain why some parts of DNA accumulate mutations and evolve rapidly, while other parts are more conserved and evolve slowly. How can this variation in the rate of base substitutions be used to discover phylogeny across a range of organisms?

3. On what basis do traditional classifications place humans in a family (Hominidae) separate from the great apes (Pongidae)? Why do some taxonomists want to change the classification, and what do they mean when they recommend recognising only 'monophyletic taxa'?

4. Some biologists believe that species (e.g. the dog *Canis familiaris*) are real but that higher taxa (e.g. Mammalia) are unreal, being defined arbitrarily by taxonomists as convenient units of classification. Other biologists disagree and view all taxa (species, genera, families, etc.) as real in the sense that they all have evolved. What could be the arguments for each of these opposing views?

5. (a) How is the kingdom Protista problematic as a taxon?

(b) Given the phylogenetic tree in Figure 30.10, and by only naming monophyletic groups, how many kingdoms would you recognise? What groups of organisms would they include? (Try making a case for recognising three kingdoms instead of five.)

Suggested further reading

Burgman, M. A. and Lindenmayer, D. B. (1998). *Conservation Biology.* Sydney: Surrey Beatty and Sons.

Introductory chapters provide useful information on Australia's biota and levels of endemism.

Panchen, A. L. (1992). *Classification, Evolution and the Nature of Biology.* Cambridge: Cambridge University Press.

Advanced text.

Patterson, C. (1999). *Evolution.* 2nd edn. London: The Natural History Museum.

A very readable, up-to-date text that covers both patterns of evolution and processes.

Search on the Internet for the 'Tree of Life Project' (http://phylogeny.arizona.edu/tree/phylogeny.html)

Knowledge of the tree of life is accumulating each year, and this website regularly updates phylogenies.

CHAPTER 31

Evolving earth

How is the history of the earth relevant to the evolution of life? We saw in Chapter 5 that the environment of the early earth was conducive to the evolution of the first prokaryotes, cells that used ready-made organic molecules as a source of raw materials and energy. After the evolution of photosynthetic prokaryotes, which produce oxygen as a by-product of photosynthesis, the earth's atmosphere became oxygen-rich, destroying the conditions that first allowed life to evolve. Through time the physical and chemical environment of the earth has influenced the evolution of life: most importantly, the movements of continents continue to alter the distribution and area of exposed land and sea and, therefore, the environment of organisms.

In this chapter, we follow the fossil record of marine and terrestrial organisms, from the first multicellular animals and plants to the appearance of humans in relation to the earth's history. The characteristics of modern, extant groups of organisms, particularly Australian flora and fauna, are described in more detail in Chapters 33–41.

Geologic time scale

Several centuries ago, before scientists were able to measure the age of rocks, geologists and palaeontologists devised a *relative* time scale for the evolution of the earth and life. This relative scale was based on the observation that layers of sedimentary rocks, which contain characteristic fossils, are deposited one on top of another. It was assumed that the deepest rock strata and their fossils are the oldest, and that the uppermost strata and their fossils are the youngest. Thus, a geologic feature was established as older than, equal to, or younger than another feature.

Sediments of similar composition, laid down during the same geologic event, are sometimes found in different geographic regions, providing a time reference point. The best example of a sediment deposited simultaneously over large areas is a thin bed of volcanic ash blown over a large area from a single volcanic eruption. Sometimes the same stratigraphic sequence of fossils is found in different areas. Within northwestern Europe, marine Jurassic strata are divided into numerous age zones on the basis of an orderly sequence of forms of ammonoids, an extinct group of molluscs (see Chapter 39). The same sequence of ammonoids occurs wherever these animals lived in that general region.

With the discovery of radioactivity at the end of the nineteenth century, absolute measurements of geologic age became possible (Box 31.1). The modern geologic time scale (Fig. 31.1) incorporates both actual age measurements (expressed in thousands or millions of years), as well as sequences of relative ages. One problem with measuring the age of ancient rocks is that an error of only 1% for rocks that are 500 million years old represents an error of 5 million years.

The geologic time scale is divided into eras and eras are subdivided into periods. Eras, from oldest to youngest, are the **Precambrian**, **Palaeozoic**, **Mesozoic**

BOX 31.1 How old is that rock?

Radiometric dating

Quantitative determination of geologic age is made possible by means of radioactivity (**radiometric dating**). Atoms of radioactive elements emit radiation as they spontaneously decay. Each radioactive isotope has its own constant rate of decay. This constant rate of decay allows definition of a geologic clock, although different methods give ages with different degrees of uncertainty.

Because the decay of individual radioactive atoms is random, the time required for all radioactive atoms of a given sample to decay is indefinite. For this reason, scientists calculate the time required for half of the atoms to decay, the **half-life**. Of the radioactive elements that occur naturally on earth, some are continuously being produced by neutron bombardment and have a relatively short half-life; others were inherited when the earth was formed and have a long half-life. Potassium-40, for example, has a half-life of 1.25 billion years, decaying to argon-40. By measuring the proportion of potassium-40 and argon-40 in a rock that has been undisturbed over geologic time, we can estimate when the rock formed. If the rock contains the radioactive isotope and decay product in the ratio of 1:1, then it is 1.25 billion years old. If there is proportionately less argon-40, then the rock is younger; if more, then the rock is older.

Potassium-40, rubidium-87 (half-life of 50 billion years; decays to strontium-87), thorium-232 (half-life of 14 billion years; decays to lead-208) and uranium-235 (half-life of 4.5 billion years; decays to lead-207) all allow ancient rocks to be dated. The oldest rocks dated so far are 3.8 billion years old from west Greenland.

Carbon-14 has a relatively short half-life of 5730 years and decays to nitrogen-14. Carbon-14 is produced in the atmosphere above an altitude of about 10 000 m. Some carbon-14 combines with oxygen to form carbon dioxide, which is fixed by photosynthetic organisms and moves through food chains. Because of rapid decay, the ratio of carbon-14 to nitrogen-14 is only useful for dating fossil material less than 40 thousand years old.

Magnetic reversals

The earth's magnetic field occasionally reverses, being either in *normal* (present day) or *reversed polarity* (magnetic north becomes south and south becomes north) for a period of thousands or millions of years. The cause of **magnetic reversals** is not known, although it is known that the time required for a reversal to take place is about 5000 years. While volcanic and sedimentary rocks are forming, they become magnetised—their constituent magnetic minerals becoming aligned with the earth's magnetic field, either normal or reversed. A record of past reversals is contained in the magnetism of the rock sequence formed throughout geologic time.

Rock magnetism may be measured directly from samples in the laboratory, or indirectly through measurement of the strength of the earth's magnetic field over extensive rock beds such as the sea floor. Variations in strength of the magnetic field, either more or less than an average value, are termed *anomalies*. Positive anomalies occur when the magnetism of rock beds is normal, locally enhancing the strength of the earth's magnetic field. Negative anomalies occur when magnetism is reversed, reducing the strength as measured today. Coherent patterns of magnetic anomalies are clearest in areas of the ocean. Symmetric patterns occur in areas to either side of a spreading ridge.

During the last 160 million years there have been two long periods when reversals were frequent, separated by a 40 million year magnetically quiet period (see figure). Where a sequence of four to six reversals is determined, the unique pattern of its reversals correlates with corresponding patterns elsewhere and identifies the time when the rock sequences were formed.

Reversals of the earth's magnetic field through the Cenozoic (C) and late Mesozoic (M). Normally magnetised intervals extend to the right of the polarity graph; reversely magnetised intervals extend to the left. The time between two successive reversals is a polarity interval, and one or more intervals, depending on length, is called a *chron*. Numerous chrons have been named over the last 160 million years of geologic history: for example, C1N for the Pleistocene, and C34N spanning more than 40 million years during the Cretaceous

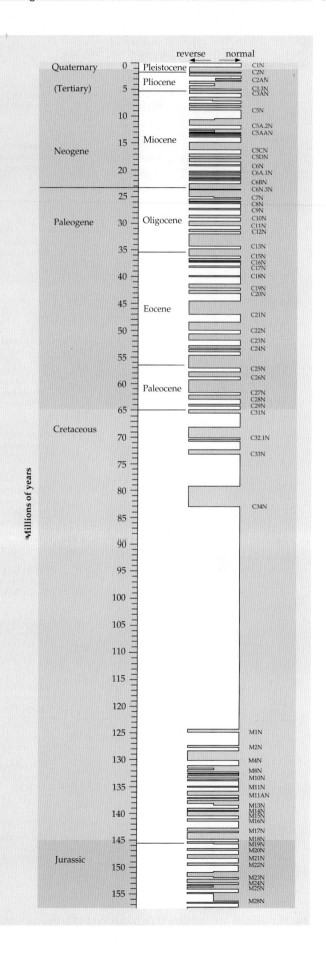

Eras of time	Periods of time	Epochs of time (Cenozoic era only)	Age (millions of years)	Major biological events
Cenozoic	Quaternary	Recent (Holocene)		Humans expand in range Major ice ages and extinction of large animals in Northern Hemisphere
		Pleistocene		
	Neogene	Pliocene	1.6	
		Miocene	23	Extensive radiation of flowering plants and mammals Dominance of gastropods
	(Tertiary)	Oligocene		
		Eocene		
	Paleogene	Paleocene	65	
Mesozoic	Cretaceous			First flowering plants Extinction of ammonites, marine and aerial reptiles
			146	
	Jurassic			Cycads, conifers, ginkgoes, dinosaurs dominant First birds, flying reptiles, marine reptiles
			208	
	Triassic			First dinosaurs and mammals Dominance of mammal-like reptiles Dominance of ammonites
			251	
Palaeozoic	Permian			Extinction of trilobites and many invertebrates Reptiles more abundant as amphibians decline
			290	
	Carboniferous			Coal swamp forests Amphibians on land First reptiles Algal-sponge reefs Echinoderms and bryozoans dominant
			362	
	Devonian			Oldest land vertebrates Radiation of land plants and fishes Corals, brachiopods and echinoderms
			408	
	Silurian			Oldest life on land: plants, scorpions First jawed fishes
			439	
	Ordovician			Diverse marine communities: brachiopods, bryozoans, corals, graptolites, nautiloids First jawless fishes
			510	
	Cambrian			Metazoan animals with skeletons Dominance of trilobites
			545	
Precambrian				First metazoan animals Origin of eukaryotes Origin of prokaryotes
			4560	

Fig. 31.1 Modern geologic time scale. Today, geologists use the terms Neogene and Paleogene instead of Tertiary

and **Cenozoic**. Boundaries between eras were initially recognised as abrupt changes in the fossil record, marking important biological events. The Precambrian extends from the origin of the earth through the formation of the earth's crust, oceans and atmosphere, the origin of life as prokaryotic cells, and the evolution of the first eukaryotes and multicellular organisms. The beginning of the Palaeozoic is marked by the appearance of a diversity of animals (the Cambrian explosion); its end is marked by the great Permian extinction. The Cretaceous extinction ended the Mesozoic.

The modern geologic time scale is based on relative ages of sequences of sedimentary rocks and fossils and quantitative measurements based on radiometric dating.

Movements of the earth

By the time the oldest rocks were formed, portions of the outer earth began to move. The modern geologic theory of **plate tectonics** recognises that the earth's crust and part of the upper mantle (together the lithosphere) are divided into a number of plates (Fig. 31.2), and that these plates move relative to one another.

Plate tectonics

Lava upwells from part of the earth's mantle at **oceanic ridges**. New crust of sea-floor basaltic rocks moves apart on either side of a ridge and, as the sea floor spreads, continents, which are made of lighter (sialic) rocks, are carried away from the ridge. For example, the Mid-Atlantic Ridge is centrally located in the Atlantic Ocean and is the site of sea-floor spreading that separated the continents that now lie on either side of this ocean basin (Fig. 31.3).

If new crust forms and flows laterally, it must disappear somewhere else. **Deep-sea trenches** (Fig. 31.3) are the sites where sea floor descends back into the mantle in a process called **subduction**. Because continents are of lower density than the crust, they do not readily descend into the mantle and tend to remain on the surface. Most trenches of the world today encircle the Pacific Ocean, where the large Pacific plate is moving north-west relative to surrounding plates. Trenches mark sites of earthquakes and volcanoes, and explain the 'Ring of Fire' around the rim of the Pacific. Subduction explains why no modern ocean (Pacific, Indian and Atlantic) has any part of its floor older than about 200 million years (Jurassic); older parts have been destroyed through subduction at plate margins.

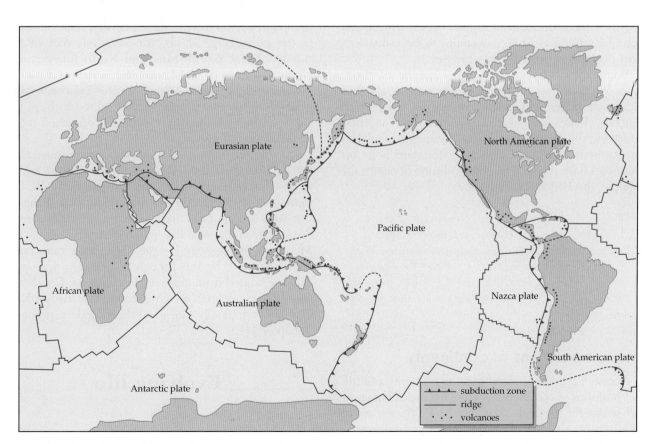

Fig. 31.2 The distribution of plates over the earth's surface. Ocean ridges, where plates are moving apart, and subduction zones, where one plate is sliding under another, are shown. Volcanoes around the rim of the Pacific form the 'Ring of Fire'

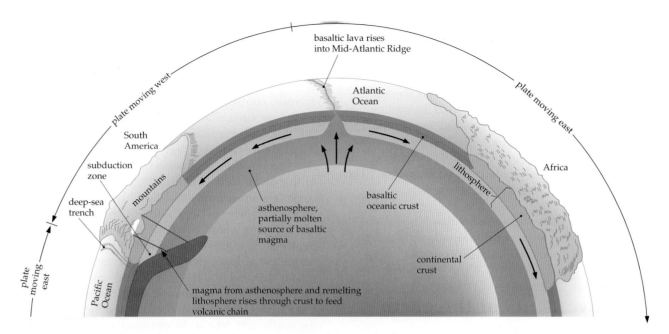

Fig. 31.3 Cross-section of the lithosphere and asthenosphere (partially molten mantle) in the Atlantic Ocean, showing the separation of South America (moving west) and Africa (moving east). South America meets the eastern part of the Nazca plate along a deep-sea trench

Immobile points at the surface of the earth's mantle where a column of hot upwelling asthenosphere rises are called **hot spots**. As a plate moves over a hot spot, a chain of volcanic islands may form, such as the Hawaiian Islands. A chain of successive volcanoes is a record of the movement of a plate over a fixed point and allows rates of plate movement to be estimated. Most plates move about 5 cm per year!

Boundaries between two plates can be of three types. We have seen that they can be moving apart at mid-ocean ridges and moving under one another at trenches. Plates can also scrape past one another, deforming the earth's crust, a classic example being the San Andreas Fault in California. Mountain belts on continents have resulted from the collision of plates, for example, the Himalayas and Andes.

T he earth's lithosphere consists of a number of plates, which develop at oceanic ridges, slide laterally and descend into the mantle again at deep-sea trenches. Plate movements deform the earth's crust, causing mountains to form and oceans to open and close. Continents, made of lighter rocks, ride on plates and shift position over geologic time.

Ancient positions of continents

Geologists have learnt about plate movements and the past positions of continents by studying rocks, hot spots and magnetic reversals (see Box 31.1). As we move back in time, the task of working out how land masses have shifted becomes harder because subduction and erosion destroy evidence of the past. Precambrian rocks, which represent the first 90% of the earth's history, form less than 20% of the rocks now exposed on the earth's surface.

In the early Palaeozoic (late Cambrian), the positions of the continents were different from today, with most land masses lying in the equatorial zone and none at the poles (Fig. 31.4a). Australia was part of the supercontinent **Gondwana**, and North America and Greenland were part of Laurentia. Europe was part of a separate land mass, Baltica. As the Palaeozoic progressed, the northern land masses united to form the supercontinent **Laurasia**. Gondwana remained intact and drifted southwards.

At the beginning of the Mesozoic, more than 200 million years ago, Laurasia and Gondwana joined to form a single supercontinent, **Pangaea** (Fig. 31.4b), uniting all northern and southern continents. Pangaea was such a large land mass, with inland areas a great distance from the sea, that much of it was arid. As the Mesozoic progressed, Pangaea separated into many fragments and, by the Jurassic, Gondwana was once again separated from northern land masses. Gondwana started to break up in the Cretaceous, and the southern continents gradually moved to their modern positions (Chapter 41).

Evolving life

Fossils

Just as rocks tell us about the history of the earth, fossils in rocks tell us something about past forms of life on

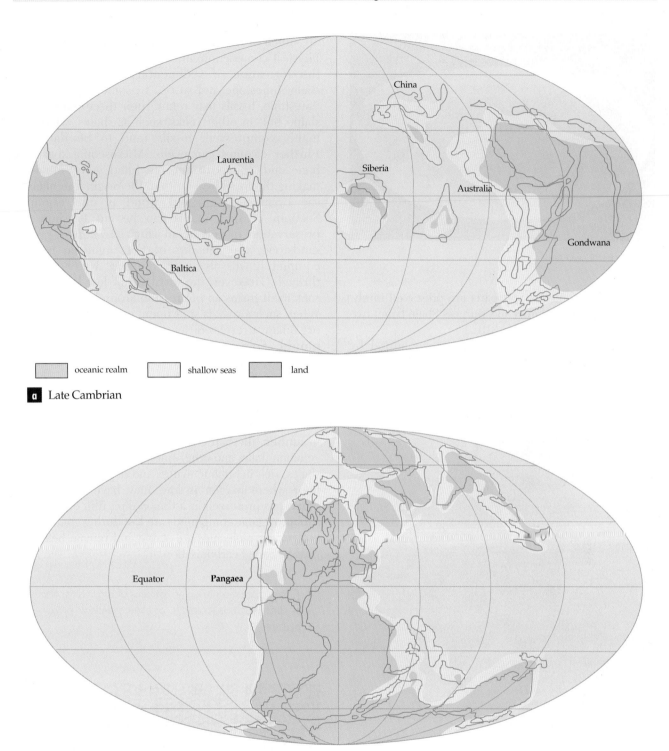

oceanic realm · shallow seas · land

a Late Cambrian

oceanic realm · shallow seas · land

b Late Permian

Fig. 31.4 Position of the world's land masses in the Palaeozoic. **(a)** Early in the Palaeozoic era, in the late Cambrian, no continents were at the poles and many were inundated by shallow seas. **(b)** Towards the end of the era, in the late Permian, the continents united to form one single supercontinent, Pangaea

earth. How do fossils form and where do we find them? **Fossils** are the preserved remains of organisms (Fig. 31.5) or traces of them, such as dinosaur footprints, trilobite tracks or the tubes of soft-bodied worms—**trace fossils**—or even organic compounds produced by them—**chemical fossils**. Hard parts of organisms are most often fossilised, including teeth, bone and external shells of animals, or leaves and

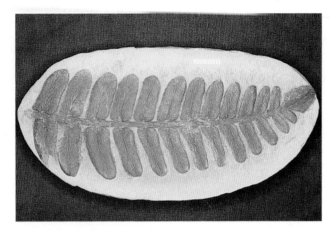

Fig. 31.5 Fossil seed fern

woody parts of plants. Soft parts are preserved rarely, with the Ediacaran fauna and Burgess Shale being two special exceptions (pp. 818, 819).

To be preserved, organisms must be buried by sediments of sand, silt or clay (Fig. 31.6) because burial prevents decay by bacteria. As sediments build up, much of the water they contain is squeezed out. Physical compaction and chemical changes result in the formation of sedimentary rocks—sandstone, shale, limestone and so on. In some rocks, such as limestone, fossils may retain their three-dimensional shape but, in other rocks, such as shales, which are highly compressed, fossils tend to be flattened. Further changes can occur. Microscopic spaces in bone and wood can become filled in by minerals, making the fossil heavier than the original remains of the organism. Petrified wood is one example. Sections of such fossils can be cut to show the preservation of cell detail (Fig. 31.7). In a porous sandstone, the remains of the organism can be completely dissolved by ground water passing through. However, the fossil is not lost because the rock itself forms an impression around the organism. When such a rock is split open, each piece shows the organism preserved as an impression or **fossil mould**. Sometimes the mould is filled by other materials, such as silica (e.g. opal) or phosphate, forming a **fossil cast**. Silicified fossil casts can be exposed from limestone by application of acid, revealing intricate external details.

Fossils can be formed in other ways. Insects trapped in **amber**, the hardened resin of conifers, show wing venation, hairs and other fine details (Fig. 31.8). Black shales, rich in organic matter and deposited on the sea floor where oxygen is low, may have soft parts of organisms preserved as a thin carbon film on the rock. Mobile organic compounds are lost during compaction, but carbon molecules are stable and remain behind, a process called carbonisation.

Fossils are the preserved remains of organisms, traces of them, such as footprints, or even chemicals produced by them.

Fig. 31.6 The remains of a baby diprotodon that has fallen into a muddy pool will be preserved as fossil bones

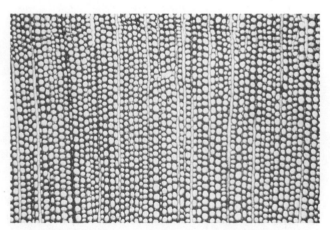

Fig. 31.7 Fossil wood of a 'progymnosperm', *Callixylon trifilieri* from the late Devonian

Fig. 31.8 Fossil insect trapped in amber

Precambrian life

The Precambrian era comprises two eons: the Archean (4500–2500 million years ago) and the Proterozoic (2500–545 million years). Before 1950, only a few fossils were known from Precambrian rocks. The mystery of early life (Fig. 31.9) unravelled as geologists and palaeontologists learned just what kinds of rocks to search in for such fossils and how to recognise these fossils. Most Precambrian fossils have since been found in black **chert**, formed from gels of silica that precipitated on the surface of ancient sea floors. The gels trapped primitive organisms and hardened them into rock. The oldest organisms known from fossils are from Archean cherts of the Warrawoona Group in the Pilbara district of Western Australia, dated at 3.3–3.5 billion years. These fossils resemble modern bacteria and cyanobacteria.

Other Precambrian fossils are **stromatolites**, which are concentrically layered rocks, the layers being formed by sediment trapped by successive growth of

thin mats of cyanobacteria, one on top of the other (see Box 31.2). Even in the absence of fossil remains of organisms, these rocks provide evidence of life. Stromatolites found in the Pilbara of Western Australia are also dated at 3.3–3.5 billion years.

> ### BOX 31.2 Stromatolites—evidence of prokaryotic life
>
> In the calm waters of Shark Bay, Western Australia, living cyanobacteria build stromatolites as they have done for 3.5 billion years. Stromatolites are hard, dome-shaped rocks formed by a continuing process of sediment trapping, mineral precipitation and hardening. The cyanobacteria in Shark Bay form mats on the surface of the stromatolites, trapping sediment (Fig. a). Motile, photosynthetic and responsive to light, they move through the fine sediment that accumulates around them and continue to live on top of the growing stromatolites. The cyanobacteria cause precipitation of the mineral calcite, producing

(a) Stromatolites at Shark Bay, Western Australia

Fig. 31.9 Artist's view of the earth 3.5 billion years ago

sediment layers that harden. Stromatolites grow slowly, less than 1 mm per year, and individual domes reach a diameter of about 200 cm and a height of 50 cm.

Various cyanobacteria build stromatolites (Fig. b). Coccoidal species build stromatolites with a coarse internal structure, and filamentous species in deeper water build stromatolites with a smooth structure. Micro-algae (eukaryotes) and other small organisms live with them, forming an intertidal community. Stromatolites generally form in shallow, saline water, such as the Persian Gulf, the Bahamas and Shark Bay, but they also form in hot springs, such as at Yellowstone National Park in the United States, and in freshwater lagoons rich in calcium carbonate, which provides the minerals for their structural framework.

The fossil record indicates that between 2.5 billion and 545 million years before the present (BP), stromatolites were abundant and diverse in shape, being associated with a great variety of prokaryotes. Subsequent decline in importance of stromatolites in the fossil record coincides with the radiation of marine multicellular organisms.

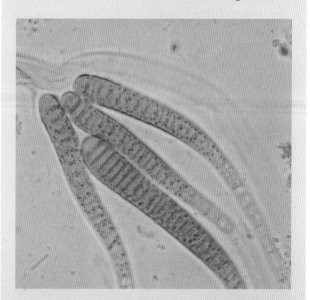

(b) *Dichothrix*, a cyanobacterium that forms benthic microbial communities that produce stromatolites in Lake Clifton, Western Australia

The oldest fossil organisms are from Precambrian rocks of Western Australia dated at 3.3–3.5 billion years. These fossils resemble modern bacteria and cyanobacteria. Stromatolites (rocks) are also an indication of life 3.3–3.5 billion years ago.

Precambrian organisms were prokaryotes and some at least were photosynthetic, as evidenced by the presence in black cherts of pristane and phytane, stable degradation products of chlorophyll. Banded iron-stones, rich in iron oxides and ranging in age from 1.8 to 2.3 billion years, give an indication of when the atmosphere became oxygen-rich through photosynthesis. Increase in oxygen in the atmosphere must have led to the extinction of many anaerobic forms of bacteria, although some lineages have persisted for 3.5 billion years to the present. These include modern forms of bacteria (Chapter 33) that live in anaerobic environments.

Younger Precambrian rocks reveal a variety of small prokaryotes, some spheroidal cells and some filamentous strands of cells. Prokaryotes are the only organisms recorded for the next two billion years of the fossil record. The oldest eukaryotes, which have chloroplasts and resemble green algae, are in 1.4 billion year old rocks from North America, China and India.

The discovery of eukaryotes is significant because it marks the evolution of organisms with a more complex cell structure (organised nucleus, chromosomes, chloroplasts, mitochondria), meiosis and sexual reproduction. How some of these characteristics evolved is unknown, but molecular data indicate that parts of the eukaryotic cell evolved by endosymbiosis (Chapters 3 and 35).

Ediacaran fauna—evidence of multicellular organisms

All phyla of shelly invertebrate animals, with the possible exception of the Bryozoa, are present as fossils in the Cambrian, the beginning of the Palaeozoic era, about 545 million years BP. The 'sudden' appearance of all the basic body plans of metazoans (animals) led palaeontologists to search for older fossils in Precambrian rocks.

Fossil traces on sea floors of tracks and burrows, and impressions of soft-bodied animals (Fig. 31.10), possibly marine worms, jellyfishes and anemones, have

Fig. 31.10 A fossil animal, *Spriggina flindersi*, of the Precambrian Ediacaran fauna of South Australia

been dated at 640–680 million years BP, although recent work suggests dates of 570–590 million years BP. Some are unlike any living animals and may represent extinct lineages. Collectively, these fossils are named the **Ediacaran fauna** after the beds in which they were found in the Ediacara Hills in South Australia, but the fauna is widespread, being found on all continents. The fauna flourished for a relatively short time and disappeared before the appearance of the phyla of shelly invertebrates in the Cambrian. Ediacaran fauna are the best evidence that multicellular animals, with specialised tissues and organs allowing larger body size, had evolved in the Precambrian.

In contrast to animals, the fossil record of multicellular photosynthetic organisms—metaphytes—is poor and little is recorded for Precambrian rocks. The oldest metaphyte fossil is a red alga from Canada dated at 1260–950 million years BP.

The earliest eukaryotes, resembling green algae, are fossils from rocks dated at 1.4 billion years old. The Ediacaran fauna is evidence that multicellular animals had evolved in the Precambrian, perhaps by 640–680 million years BP.

Palaeozoic life: ancient life

Palaeozoic means 'ancient life'. The Palaeozoic (545–251 million years BP) includes six periods: the Cambrian, Ordovician, Silurian, Devonian, Carboniferous and Permian (see Fig. 31.1).

Marine life

Ancient marine communities depended on phytoplankton—microscopic, photosynthetic organisms that float near the water surface. Phytoplankton fossils of cyanobacteria and green algae have been found from the Precambrian. In the Lower and Middle Palaeozoic (from the Cambrian to the Devonian) phytoplankton fossils are mainly **acritarchs**, tiny spherical cells with walls (algal cysts), some with small spines. Not until the beginning of the Mesozoic (in the Triassic) do we find fossil phytoplankton resembling modern taxa, such as diatoms and coccoliths (Chapter 35).

Marine microscopic zooplankton are not well represented as fossils because they lack hard components. One group of microscopic zooplankton, radiolarians, however, has an intricate skeleton of silica, and is recorded as far back as the Cambrian. Modern radiolarians occur today in offshore waters. Larger zooplankton included **graptolites** (see Fig. 31.14), extinct colonial animals with skeletons of chitin.

Trilobites are the most common marine multicellular animals of the early Cambrian fossils (Fig. 31.11). These are an extinct group of arthropods (Chapter 38), which had hard skeletal parts, large eyes,

Fig. 31.11 A trilobite, an extinct group of arthropods known from the early Cambrian

long antennae, and appendages for swimming, walking and feeding. Other fossils include brachiopods, with two shells, and rare specimens of echinoderms, living examples of which are starfishes.

During much of the Ordovician, shallow seas were widespread on the continents. Marine communities at the beginning of this period were dominated by sessile, filter feeders, especially brachiopods, bryozoans and stalked echinoderms (Chapters 39 and 40). Echinoderms were once a diverse group, including more than 20 classes, only five of which are alive today. The stalked **crinoids** were one of the most successful groups (Fig. 31.12).

Two main groups of marine predators in the Ordovician were corals and cephalopods. The earliest

BOX 31.3 The Burgess Shale

One of the most famous fossil sites in the world is the **Burgess Shale** in the Rocky Mountains of Canada. A crowd of marine animals was preserved during the Middle Cambrian in a deposit of black, organic-rich mud that lacked oxygen, preventing decay. Trilobites and brachiopods, with their hard shells, are well preserved, but more important is the great diversity of soft-bodied animals in the deposit (Fig. a). Unusual organisms include *Hallucigenia*, a segmented animal apparently with spikes for legs, an indefinite head, and projections from the back of each segment (Fig. b). Some animals were clearly predators, having a mouth with teeth.

The Burgess fossils suggest that the fossil record is generally incomplete with respect to soft-bodied organisms and that Cambrian marine communities were probably quite diverse.

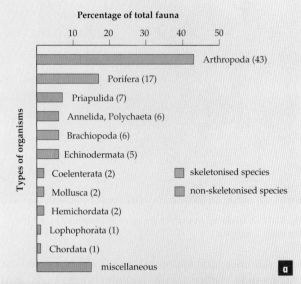

(a) Bar graph showing the major animal groups represented in the Burgess Shale. **(b)** *Hallucigenia*, a segmented animal from the Cambrian Burgess Shale (Rocky Mountains, North America). The fossil animal was interpreted as having 14 stilt-like legs and seven tentacles, placing it in an extinct group unlike any modern phylum. However, scientists studying similar fossils from China interpret *Hallucigenia* the other way up, making it an animal with seven legs and many spines, and classifying it in the arthropod phylum

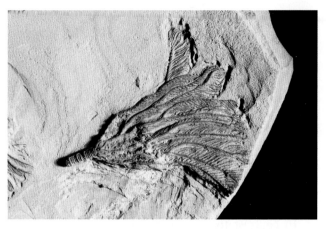

Fig. 31.12 A stalked crinoid from the Ordovician

Fig. 31.13 A Jurassic ammonoid found near Geraldton, Western Australia

larger armoured jawed fishes were active predators (see Box 31.4). The Devonian, often called the 'Age of Fishes', was a time of rapid evolution of fishes, including sharks and bony fishes, groups which include nearly all modern fishes. Among the bony fishes were **fleshy-finned fishes** (rhipidistians, coelacanths and lungfishes), which are related to the first four-legged vertebrates that colonised the land.

The Palaeozoic era (545–251 million years BP) means 'ancient life' and includes six periods: the Cambrian, Ordovician, Silurian, Devonian, Carboniferous and Permian. Marine communities of the Palaeozoic show a sequence of organisms over time, including the first fishes in the Ordovician.

Terrestrial life

The earliest fossil record in terrestrial rocks is of Silurian age. These fossils are invertebrates—scorpion-like arthropods. The earliest known land vertebrates are amphibians, found in Upper Devonian rocks of Greenland and Canada. By the Devonian, fishes lived not only in marine environments but in freshwater ponds, streams (see Box 31.4) and deltas. As climate changed to alternating wet and dry periods and some

cephalopods (molluscs, see Chapter 39) were **nautiloids**, related to the living *Nautilus*. Most nautiloids had a long, tapering, cone-shaped shell, which contained chambers. Another group of cephalopods, the **ammonoids**, became dominant later by the Devonian (Fig. 31.13).

Ordovician jawless fishes were the first vertebrate animals. The oldest fossils are small, only a few centimetres in length; the mouth was a simple hole at the front end of the gut and they were probably filter or detritus feeders. By the late Silurian and Devonian,

BOX 31.4 Australia in the Palaeozoic

When the Palaeozoic fauna of Australia were alive, Australia was further north, near the equator, at times closer to China; it later moved south towards the South Pole. The climate was warm and seas covered much of the eastern half of the continent, slicing it in half during the Ordovician (Fig. a). In early Ordovician times, about 480 million years ago, this inland shallow sea was inhabited by a species-rich marine fauna. One early vertebrate, *Arandaspis prionotolepis* (Fig. b), a member of the armoured, jawless fishes (ostracoderms), was a sluggish, bottom-dwelling animal.

There are many other fossil Palaeozoic fishes described from Australia. Ray-finned fishes and lungfishes, for example, have been superbly preserved in Upper Devonian deposits from a reef complex that once fringed the Kimberleys in north-western Australia (the Gogo Formation). Such ancient seas eventually retreated and the climate cooled until late in the Palaeozoic when glaciation occurred.

(b) Armoured, jawless fish, *Arandaspis prionotolepis*, lived on the bottom of the early Ordovician sea

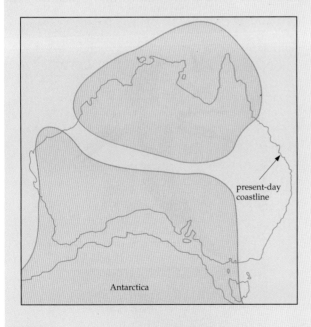

    present-day
    coastline

    Antarctica

☐ land in the Ordovician

☐ sea in the Ordovician

(a) Much of Australia was covered by shallow seas in the Ordovician period

aquatic habitats dried up in times of droughts, some fleshy-finned fishes were able to move on land. The skull and skeleton of Devonian fleshy-finned fishes are almost identical to the earliest fossil amphibians except for the modification of fins into stubby limbs for locomotion on land.

The oldest land plant fossils are of late Silurian age, found mostly in Australia. The oldest terrestrial flora includes *Baragwanathia* (Fig. 31.14); younger fossils include Psilophyta and *Rhynia*. All of these fossils are small, vascular plants, having the main stem with a central core of xylem and phloem tissue, allowing transport of water and nutrients (Chapter 37). They lacked leaves and roots, and were anchored in marshy, wet bogs by rhizomes. Small non-vascular plants, liverworts and mosses may be older than vascular plants but there are no fossils of them in the Middle Palaeozoic. Abundant vegetation provided an energy source for terrestrial fungi. Hyphae and spores of ascomycetes (Chapter 36) are recorded for the Silurian and Lower

Fig. 31.14 Early Devonian terrestrial plant fossil, *Baragwanathia longifolia*, from Victoria, Australia, preserved with graptolites (planktonic marine organisms)

Devonian and resting spores in the Rhynie Chert in Aberdeenshire are suggestive of mycorrhizal associations between fungi and Devonian land plants.

By the end of the Devonian, land plants became well established. During the Carboniferous, forests of large tree-sized plants (lycopods and sphenopods, Chapter 37) with woody stems evolved, providing habitats for a variety of terrestrial animals. Carboniferous forests (Fig. 31.15) of coastal swampy areas in the Northern Hemisphere were later buried by marine sediments deposited by seas that moved in, forming extensive coal beds that are mined today.

The Permian marked the close of the Palaeozoic. It was a time when the continents coalesced as Pangaea, changing the distribution of seas and land, and affecting climate. It was a period of dwindling rainfall in some regions and extremes of temperature. Evolution of seeds by the end of the Devonian was of great significance to the flora, allowing land plants to survive harsher conditions. Reptiles, which first appeared in the Carboniferous, became more abundant in the Permian, and amphibians, more dependent on water, declined in dominance. At the end of the Permian there was a progressive loss of shallow seas around the edges of the continents, which appears to have contributed to **mass extinction** of animal groups dominant for millions of years, including trilobites and some

groups of brachiopods, bryozoans, ammonoids and crinoids.

Scorpion-like arthropods of the Silurian are the earliest fossil record of life on land. The earliest known land vertebrates are amphibians, found in Upper Devonian rocks. By the end of the Palaeozoic, every major group of land plant, except for flowering plants, is found as fossils.

Mesozoic life: age of dinosaurs

The Mesozoic era ('middle life', 251–65 million years BP) includes the Triassic, Jurassic and Cretaceous periods, and is often described as the 'Age of Reptiles'.

During much of the Triassic the dominant vertebrates were the so-called mammal-like reptiles, a diverse and abundant group of bulky, large-headed animals with a sprawling posture. They did not look much like mammals but it is argued that one group of them gave rise to mammals in the late Triassic. Living alongside these reptiles were forms that had limbs that were under the body, allowing more support of body weight and more mobility. Some were bipedal, with their forelimbs free for other functions, such as grasping prey or flying. Included among them were flying pterosaurs (Fig. 31.16), now extinct, crocodiles and dinosaurs. The major groups that evolved are shown in Figure 31.17.

Dinosaurs, which included some of the largest animals ever to have lived on earth, reigned over the

Fig. 31.16 A marine turtle and a flying reptile (pterosaur), inhabitants of the great inland sea that covered much of Australia in early Cretaceous times. Flying reptiles are extremely rare in the Australian fossil record

Fig. 31.15 Reconstruction of a Carboniferous forest

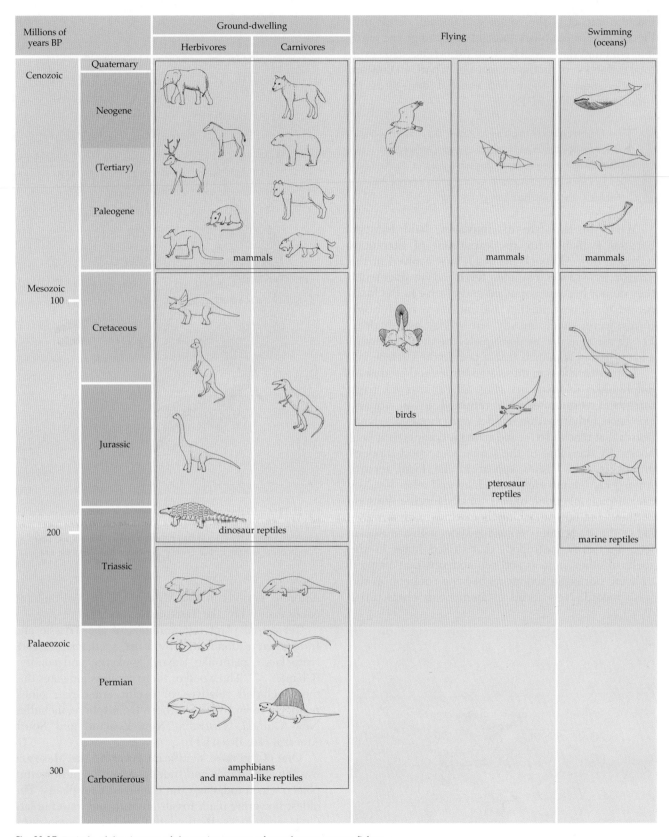

Fig. 31.17 Periods of dominance of the major groups of vertebrates, except fishes

land throughout the Jurassic and Cretaceous. They became extinct rather abruptly at the end of the Cretaceous. The reasons for this and other mass extinctions are uncertain. Some scientists point to geologic evidence of a large comet or asteroid that struck the earth 65 million years ago. They consider that the impact of the comet sent enough debris into the atmosphere to block sunlight and plunge the earth into cold and dark, which led to extinctions. However, total extinction of dinosaurs may be, in part, an illusion, because some of them had evolved into birds by Jurassic times (Fig. 31.18).

> The Mesozoic era ('middle life', 251–65 million years BP) includes the Triassic, Jurassic and Cretaceous periods, and is often referred to as the 'Age of Reptiles'. Dinosaurs reigned but became extinct at the end of the Cretaceous, marking the close of the era.

Whatever the fate of dinosaurs, land habitats became available for the expansion of mammals (Chapter 40). Although mammals first evolved in the Triassic, as evidenced by a few teeth and jaw fragments, they do not become abundant and diverse in the fossil record for almost another 100 million years. Jurassic forms were small, mostly shrew-like insectivores. Fossils recognisable as marsupials (the most common) and eutherian mammals first appear in the Cretaceous.

In marine communities, large turtles and predatory, dolphin-like ichthyosaurs (Fig. 31.19) evolved from terrestrial vertebrates that returned to an aquatic existence. Other marine reptiles were mosasaurs, large lizards that reached 8 m, and giant plesiosaurs with long necks. By the close of the Cretaceous, most modern groups of bony fishes appear in the fossil record. Among marine invertebrates, clams and other bivalves (inconspicuous components of Palaeozoic faunas) became common, attaching themselves to, or burrowing in, the sea bottom, or moving about with a muscular foot. Predatory gastropods that drilled holes in the shells of other animals and echinoderms that adapted to living buried in sediments became common in the fossil record. New forms of ammonoids characterise

Fig. 31.19 An ichthyosaur from the Australian Cretaceous period in shallow seas. An agile swimmer, this species, *Platypterygius australis*, grew to 6–7 m; it has caught a squid-like animal that later became extinct in the Eocene period of the Cenozoic era

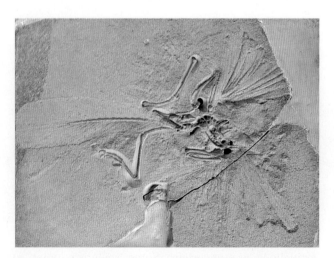

Fig. 31.18 The Jurassic bird, *Archeopteryx*, from southern Germany. A nearly complete fossil shows the impression of wing and tail feathers; the skeleton has teeth (unlike modern birds)

different periods in the Mesozoic; and reef-building corals evolved by the Triassic.

Terrestrial plant communities of the Triassic and Jurassic were dominated by ferns, seed ferns, cycads (with large, palm-like leaves), ginkgoes and conifers (Chapter 37). The wood of these conifers indicates that they included relatives of extant araucarian pines, restricted today to the Southern Hemisphere, including New Guinea, Australia, New Zealand and South America (see Box 41.1).

One of the most significant events of the Mesozoic was the appearance and sudden expansion of flowering plants, which today are the dominant land plants. The oldest flowering plant fossils are rare and found in late Cretaceous rocks. By the late Cretaceous, angiosperm fossils are more abundant throughout the world. Herbaceous forms rapidly filled the understorey of plant communities, where ferns and mosses dominated for most of the Mesozoic. Flowering plants evolved a highly successful mechanism of fertilisation

(Chapters 14 and 37) that involved insects rather than wind for the transport of pollen. Such features explain the success of flowering plants, which continued into the Cenozoic.

D uring the Mesozoic, land habitats became available for the expansion of mammals, which are first recorded as fossils in the Triassic. Also significant in the Cretaceous were the appearance and sudden expansion of flowering plants.

Cenozoic life: the beginning of modern life

The Cenozoic era ('modern life', 65 million years BP to the present) includes the Tertiary (now officially classified by geologists as two periods—the Paleogene and Neogene) and Quaternary periods, bringing us to today. Throughout the Cenozoic we can recognise among the fossils more and more modern families and genera of flowering plants and biogeographic patterns with groups characteristic of either the Northern or Southern Hemisphere.

Mammals became abundant during the Cenozoic. The mammalian groups that flourished during the Paleocene, Eocene and Oligocene epochs of the Tertiary are now largely extinct but, by the Miocene and Pliocene epochs, modern groups of mammals had evolved, most of which are present today. Early primates (prosimians) were common in the early Cenozoic but decreased markedly in importance when true monkeys and apes evolved. Miocene and Pliocene deposits in Africa, Europe and Asia have yielded many fossil teeth and jaw fragments of apes, and fossil footprints are evidence of hominids at least four million years ago in Africa. *Homo* expanded in range rapidly in the last two million years, experiencing several ice ages during the Pleistocene. The history of modern humans is a story of increased use of tools, language, culture and development of agriculture and domestication of animals (Chapter 40).

T he Cenozoic era ('modern life', 65 million years BP to the present) includes the Tertiary and Quaternary periods. Mammals and flowering plants became abundant, with more and more modern groups becoming recognisable throughout the era. Hominid fossils date back at least four million years.

Biogeographic regions of the modern earth

The result of evolution, both of earth and life together, is the world we see today, with its great variety of

organisms living in their great variety of habitats, in sum, the earth's present biodiversity. Since the eighteenth century, biogeographers (those who study the distribution and abundance of organisms) have attempted to characterise regions, called **biogeographic regions**, inhabited by unique forms of native life to understand the evolutionary history of life and earth (Box 31.5).

The main continental regions recognised today by botanists and zoologists are very similar (Table 31.1; Fig. 31.20). The names of the regions are based partly on the division into old (palaeo-) and new (neo-) worlds, with Australia treated as if it were a world apart.

The **Boreal** region includes the northern fir forests, which are very similar in Europe, Asia and North America. Also known as the Holarctic region, it is sometimes divided into Nearctic (North America) and Palaearctic (Eurasian) regions. **Palaeotropical** plants occur in Africa (Ethiopian region) and in India and South-East Asia (Oriental region), in association with elephants and rhinoceroses among other animals. The **Neotropical** region includes all of South America and lower Central America to about the 200 km wide Isthmus of Tehuantepec in Mexico. Finally, the **Australian region** includes both the mainland as well as islands on the continental shelf, such as Tasmania in the south and New Guinea and its own neighbouring islands in the north.

Marine organisms, too, are diverse in geographic distribution. Marine biogeographers commonly begin with a division between shallow water (shelf) and open ocean (pelagic) realms. Life of the **continental shelf** is most diverse in the tropics, particularly in coral reef communities (Fig. 31.20). Bounded by 20°C isotherms for the coldest months of the year, tropical marine regions include the Indo-West Pacific, East Pacific, West Atlantic and East Atlantic. The Indo-West Pacific

Table 31.1 Geographic regions recognised for plants and animals

Botanical regions	Zoological regions
Boreal	Holarctic
	Palaearctic
	Nearctic
Palaeotropical	Ethiopian
	Oriental
Neotropical	Neotropical
Australian	Australian

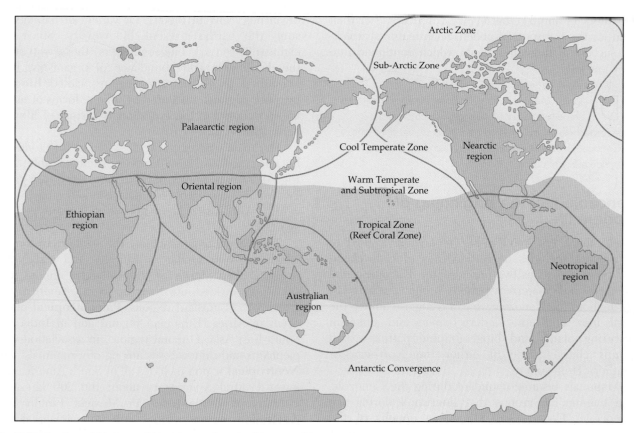

Fig. 31.20 Continental biogeographic regions and marine biogeographic zones

is the largest region, embracing the entire tropical Indian Ocean and the West and Central Pacific, including archipelagoes such as Hawaii and the Marquesas. Northern and Southern Temperate regions are commonly inhabited by organisms of related taxa, such as the commercially important pilchards (*Sardina*, *Sardinops*). Of six species of pilchards in the world, one lives in waters around southern Australia and New Zealand, and the other related species occur off the coast of southern Africa, Peru, California, Japan and northern Europe (Fig. 31.21).

Arctic and Antarctic seas have their own unique life forms but these, too, sometimes have their nearest relatives at the opposite poles. The phylum Priapulida, for example, consists of only eight species and six genera of cucumber-shaped animals, penis worms, that live buried in bottom sand or mud. The genus *Priapulus* has a **bipolar distribution**, one species living in the Arctic and another in the Antarctic (Fig. 31.22). The phylum is believed to be an ancient group.

The **pelagic realm** is home to planktonic organisms typically confined to surface water (to 1000 m). The pelagic realm is governed by ocean currents, themselves a product of prevailing winds and ultimately

of the earth's rotation. Major currents divide the marine environment into water masses, which tend to have their own physical properties of temperature and salinity. Water masses are bounded by major currents and their interactions. Currents separate at a divergence and come together at a convergence. A major feature of the Southern Ocean is the Antarctic Convergence (Fig. 31.20), which encircles the globe, where very cold Antarctic water sinks below warmer water to the north.

In the sea, as on continents, few species are cosmopolitan, that is, worldwide. Most species occupy only a relatively small area where they are native. Similar distributions of different organisms suggest common underlying causes, which in part relate to previous evolutionary history and in part to existing environmental conditions.

The world is divided into a number of continental and marine biogeographic regions, which are characterised by endemic organisms. Similar distributions of taxa are explained by present-day environmental factors as well as previous evolutionary history.

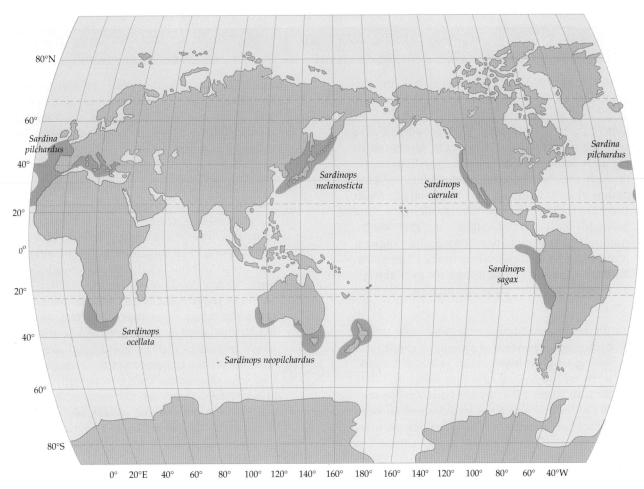

Fig. 31.21 Northern and Southern Temperate regions are commonly inhabited by related organisms as illustrated by the distribution of six species of pilchards (*Sardina* and *Sardinops*)

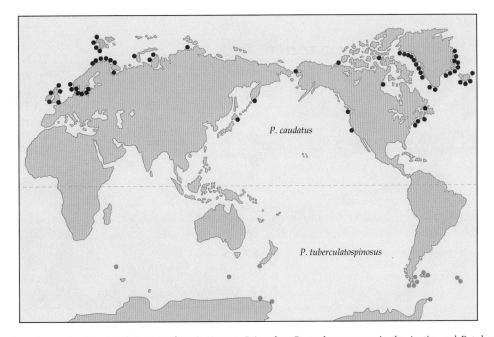

Fig. 31.22 Distribution of two closely related species of penis worms, *Priapulus*: *P. caudatus* occurs in the Arctic, and *P. tuberculatospinosus* in the Antarctic. This type of geographic distribution is termed 'bipolar'

BOX 31.5 Wallace's Line

Alfred Russel Wallace (1823–1913) was the co-discoverer of the principle of evolution by means of natural selection (Chapter 32) with Charles Robert Darwin (1809–82). Wallace worked as a self-employed collector of museum specimens in the East Indies (1854–62) and travelled widely through the islands, often under primitive conditions of great hardship. He found that the Asian and Australian faunas divide along a line between Bali and Lombok in the south, continuing north between Borneo and Sulawesi (Celebes), and continuing east of the Philippines. **Wallace's Line** has been recognised ever since.

In his book *The Malay Archipelago*, Wallace wrote:

> The great contrast between the two divisions of the Archipelago is nowhere so abruptly exhibited as on passing from the island of Bali to that of Lombok, where the two regions are in close proximity. In Bali we have barbets, fruit thrushes and woodpeckers; on passing over to Lombok these are seen no more, but we have abundance of cockatoos, honeysuckers and brush-turkeys, which are equally unknown in Bali, or any island further west. The strait is here only fifteen miles wide, so that we may pass in two hours from one great division of the earth to another, differing as essentially in their animal life as Europe does from America.

Wallace noted that:

> This division of the Archipelago into two regions, characterised by a striking diversity in their natural productions, does not in any way correspond to the main physical or climatal divisions of the surface. The great volcanic chain runs through both parts...Borneo closely resembles New Guinea, not only in its vast size and its freedom from volcanoes, but in the variety of geological structure, its uniformity of climate, and the general aspect of the forest vegetation that clothes its surface. The Moluccas are the counterpart of the Philippines in their volcanic structure, their extreme fertility, their luxuriant forests, and their frequent earthquakes; and Bali with the east end of Java has a climate almost as arid as that of Timor. Yet between these corresponding groups of islands constructed as it

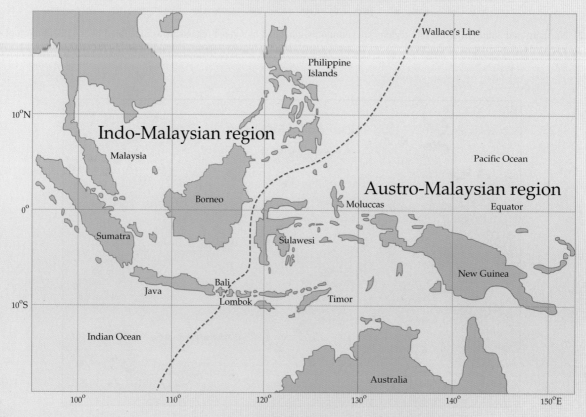

Wallace's Line, showing the Indo-Malaysian region to the west and the Austro-Malaysian region to the east

were after the same pattern, there exists the greatest possible contrasts when we compare their animal production. Nowhere does the ancient doctrine—that differences or similarities in the various forms of life that inhabit different countries are due to corresponding physical differences or similarities in the countries themselves—meet with so direct and palpable a contradiction.

Modern interpretation of Wallace's Line is based on the discovery that Australia, New Guinea and associated islands moved north during the last 60 million years. Wallace's Line approximates, but not exactly, the collision zone between Asian and Australian crustal plates.

Summary

- The evolution of life has been influenced by the history of the earth: by the composition of the earth's early atmosphere and the movement of the earth's crust. The first organisms evolved in the absence of oxygen, in oceans and lakes, where they used free molecules in the environment as a source of raw materials and energy. Photosynthetic organisms evolved later, enriching the atmosphere with oxygen.

- The modern geologic time scale is based on relative ages of sequences of sedimentary rocks and fossils, and absolute measurements based on radiometric dating. The main divisions of the geologic time scale are the Precambrian, Palaeozoic, Mesozoic and Cenozoic eras.

- Over geologic time, continents have shifted position. According to the theory of plate tectonics, the earth's lithosphere is divided into a number of plates, which arise at oceanic ridges, slide laterally and descend into the mantle again at deep-sea trenches. Continents, which are made of lighter rocks, ride on plates but do not descend into the mantle. For hundreds of millions of years Australia was part of the supercontinent Gondwana, lying near the equator in the early Palaeozoic and to the south by the end of the era, when all of the continents coalesced to form one single supercontinent Pangaea. Gondwana later separated from the northern land masses and began to break up in the Cretaceous, with southern continents drifting to their current positions.

- Fossils are the preserved remains of organisms, traces of them, such as footprints, or even chemicals produced by them.

- Stromatolites are fossils formed by the activities of cyanobacteria and indicate that life existed at least 3.5 billion years ago. The oldest organisms, resembling modern bacteria and cyanobacteria, are preserved as fossils in Precambrian rocks dated at 3.3–3.5 billion years. The earliest eukaryotes, resembling green algae, are fossils from rocks dated at 1.4 billion years. Ediacaran fauna are evidence that multicellular animals, with specialised tissues and organs, allowing larger body size, had evolved in the Precambrian, perhaps by 640–680 million years ago.

- Palaeozoic means 'ancient life'. The Palaeozoic era (545–251 million years BP) includes six periods: the Cambrian, Ordovician, Silurian, Devonian, Carboniferous and Permian. Marine communities show a sequence of organisms over time, including the evolution of the first fishes in the Ordovician. The earliest fossil record of life on land is of Silurian age. These fossils are invertebrates: scorpion-like arthropods. The earliest known land vertebrates are amphibians, found in Upper Devonian rocks. By the end of this era, every major group of land plant, except for flowering plants, is found as fossils.

- The Mesozoic era ('middle life', 251–65 million years BP) includes the Triassic, Jurassic and Cretaceous periods, and is often referred to as the 'Age of Reptiles'. Dinosaurs reigned but became extinct during the Cretaceous, marking the close of the era. Land habitats became available for the expansion of mammals, which are first recorded as fossils in the Triassic. Also significant in the Cretaceous was the appearance and sudden expansion of flowering plants.

- The Cenozoic era ('modern life', 65 million years BP to the present) includes the Tertiary (Paleogene and Neogene) and Quaternary periods. Mammals and flowering plants became abundant, with more and more modern groups becoming recognisable throughout the Cenozoic. Hominid fossils date back at least four million years.

- The modern world is divided into a number of continental and marine biogeographic regions, characterised by endemic organisms. Australia is recognised as a unique region. Similar distributions of taxa are explained by present-day environmental factors as well as previous evolutionary history.

keyterms

acritarch (p. 819)	continental shelf (p. 825)	half-life (p. 810)	Pangaea (p. 814)
amber (p. 816)	crinoid (p. 819)	hot spot (p. 814)	pelagic realm (p. 826)
ammonoid (p. 820)	deep-sea trench (p. 813)	Laurasia (p. 814)	plate tectonics (p. 813)
Australian region (p. 825)	dinosaurs (p. 822)	magnetic reversal (p. 811)	Precambrian (p. 810)
biogeographic region (p. 825)	Ediacaran fauna (p. 819)	mass extinction (p. 822)	radiometric dating (p. 810)
bipolar distribution (p. 826)	fleshy-finned fish (p. 820)	Mesozoic (p. 810)	stromatolite (p. 817)
Boreal region (p. 825)	fossil (p. 815)	nautiloid (p. 820)	subduction (p. 813)
Burgess Shale (p. 819)	fossil cast (p. 816)	Neotropical region (p. 825)	trace fossil (p. 815)
Cenozoic (p. 813)	fossil mould (p. 816)	oceanic ridge (p. 813)	trilobite (p. 819)
chemical fossil (p. 815)	Gondwana (p. 814)	Palaeotropical region (p. 825)	Wallace's Line (p. 828)
chert (p. 817)	graptolite (p. 819)	Palaeozoic (p. 810)	

Review questions

1. What sort of data are used as a basis for the modern geologic time scale?

2. Potassium-40 has a half-life of 1.25 billion years and decays to argon-40. If no elements have been leached in or out, and the ratio of potassium-40 to argon-40 in a rock is determined to be 1:1, how old is that rock: greater than 1.25 billion years, 1.25 billion years, or 0.625 billion years? (See Box 31.1.)

3. Geologic features that help us to recognise past movements of continents include magnetic reversals, deep-sea trenches, oceanic ridges and hot spots. Briefly explain each of these terms.

4. Briefly explain how the earth's crustal plates move and what happens at their boundaries.

5. What was the single supercontinent with which all continents were associated that formed by the Upper Permian period: Laurasia, Gondwana or Pangaea?

6. (a) What is a fossil?

   (b) How do fossils form? Give examples.

7. Name the four major geologic eras. What significant biological events characterise each of them?

8. Match each of the following fossils to its *approximate* age from a list of ages shown below (not all ages shown will match): oldest prokaryotes; first multicellular animals; first land animals; first land plants; dominance of flowering plants; dominance of mammals; first hominids. (Check your answers against Figs 31.1 and 31.17.)

   - 4 billion years
   - 2 billion years
   - 1 billion years
   - 600 million years
   - 400 million years
   - 200 million years
   - 100 million years
   - 10 million years
   - 4 million years
   - 400 000 years

9. Briefly describe the Ediacaran fauna, where it is found and of what significance it is.

Extension questions

1. At Shark Bay, in Western Australia, living prokaryotes build stromatolites. What is a stromatolite and what does its structure indicate about the growth of its associated microorganisms? Explain the historical significance of stromatolites.

2. What are the major continental regions and marine zones of the modern world that are recognised by biogeographers? How can you explain the existence of such regions and zones?

3. What is 'Wallace's Line' and what is its significance?

4. Examine Figure 31.4(b) when the continents were joined together during the Permian. If this happened again, what impact might this have on Australia—its climate, its terrestrial biota and its marine biota associated with the continental shelf?

Suggested further reading

Cox, B. C. and Moore, P. D. (1993). *Biogeography—An Ecological and Evolutionary Approach.* 5th edn. London: Blackwell Scientific.

General text that offers explanation of the various distribution patterns of life.

Long, J. (1995). *The Rise of Fishes.* Sydney: University of New South Wales Press.

Fossil history and evolution of fishes described in detail, with introductory sections on earth history, plate tectonics, dating rocks, stromatolites and a geologic time chart, and an account of the evolution of lungs and the first land vertebrates.

Rich, P. V., van Tets, G. F., Knight, F. (1985). *Kadimakara—Extinct Vertebrates of Australia.* Melbourne: Pioneer Design Studio.

This book tells the stories of how vertebrate fossils have been discovered in Australia and gives an illustrated account of the various groups of vertebrates that have lived in the past, including Australia's extinct megafauna.

White, M. (1986). *The Greening of Gondwana.* Sydney: Reed Books.

A book that details and illustrates 400 million years of history of plants in Australia.

White, M. (1988). *Australia's Fossil Plants—A Handbook on Prehistoric Environments and Vegetation throughout Geological Time.* Sydney: Reed Books.

A short account summarising the more detailed information in The Greening of Gondwana.

CHAPTER 32

Mechanisms of evolution

n simple terms, **evolution** can be defined as 'change in the genetic composition of a population over generations'. A **population** may be defined as a group of interbreeding individuals (or in the case of asexually reproducing organisms, a population may be recognised on the basis of shared similarity between individuals). This definition of evolution focuses on two aspects—how does genetic variation arise in populations and what processes cause the genetic composition of populations to change?

Evolutionary changes may result from selective processes (natural selection) or from non-selective (neutral) processes. Evolutionary changes are also thought to be responsible for divergence among populations and ultimately for differences among taxa we see today. They are reflected in the patterns of relationships among taxa, patterns of geographic distribution, homologous structures, the unity of life at the molecular level and the fossil record, as we have seen in Chapters 30 and 31.

The most widely accepted modern theory of evolution is often described as neo-Darwinian because it is based both on the idea of **natural selection** proposed by Charles Darwin and Alfred Wallace and new knowledge of genetics discovered after Darwin (Fig. 32.1). Darwin published *The Origin of Species by*

Fig. 32.1 Charles Darwin (1809–82). After being trained for medicine and the clergy, Darwin sailed as naturalist on the voyage of HMS *Beagle* (1831–36). He spent many years at his home at Down House in Kent, England, formulating his theory of evolution by natural selection. His work may never have been published if Alfred Russel Wallace (see Chapter 31) had not sent him a manuscript in which he had independently arrived at the same concept of natural selection. They jointly published a paper in 1858, and in 1859 Darwin published his book *On the Origin of Species by Means of Natural Selection*

Means of Natural Selection in 1859 and in it he presented abundant evidence to support his theory that extant species had descended, with modification, from ancestral species. He also described how organisms could become adapted to different environments through the process of natural selection. Darwin's observations and theories relating to evolution by natural selection were constrained by the lack of a valid theory of inheritance at the time. Our understanding of the genetic mechanism of inheritance has vastly expanded since Darwin's time, beginning with the work of Gregor Mendel, published in 1866, on the transmission of simple traits in peas (Chapter 9). Subsequent developments in population and quantitative genetics, together with the recognition of DNA as the genetic material and the elucidation of how genetic information is encoded, have greatly enhanced our understanding of evolutionary processes.

Before reading this chapter on evolutionary mechanisms, you may find it helpful to review the genetics section in Part 2.

Selection and evolution

Natural selection and adaptation

Charles Darwin was particularly concerned with **selective** or **adaptive evolution**, which refers to the processes by which organisms that are particularly well suited (well adapted) to their environment are favoured relative to individuals that are less well suited. As a result of natural selection, organisms acquire inherited traits (called **adaptations**) that benefit them by enabling them to survive and reproduce better than others in a particular environment. One of Darwin's important contributions was to propose a mechanism by which this process could occur.

Darwin's theory of natural selection was based on the following premises:
- Individuals within populations vary, and much of this variability is inherited.
- Organisms are capable of producing many more offspring than can survive because environmental resources, such as food or space, are limiting.
- Those individuals of a population that have characteristics particularly suited to the environment in which they occur are more likely to survive and reproduce successfully than are those with characteristics that are less suitable.
- If the characters contributing to the differing success of individuals are heritable, then the genetic composition of the population changes with time. Environments vary over space and time, and populations differentiate and diverge as local

populations become adapted to their own particular environments.

In summary, the survival and reproduction of organisms depends on their suitability to their particular environment. Individuals that are better suited to their environment will be naturally selected and will leave more descendants than others. Thus, natural selection results in a change in the genetic composition of the population over time—evolution. Darwin's theory concludes that divergence of the descendants of a common ancestor can lead to the formation of new species and hence to the tremendous diversity of living organisms in the world today.

Artificial selection

In formulating his ideas on evolution and natural selection, Darwin not only made observations on natural populations and fossils but also on domesticated plants and animals. Darwin kept and bred pigeons and gathered information from stock breeders. His thinking about the potential importance of natural selection was influenced by the success of **artificial selection**, in which humans decide on which organisms will breed in order to select for characteristics that are considered useful or otherwise desirable.

An example of an Australian plant species with horticultural potential is that of Sturt's desert pea (*Swainsona formosa*), which occurs in the Australian arid zone. Plants are very striking when in flower (Fig. 32.2). Variation in floral characteristics, for example, colour, number and arrangement of flowers, and vegetative characters, such as growth form, which varies from prostrate to upright, is found within and between natural populations of Sturt's desert pea. The potential that this variation provides for horticulturalists to breed plants for different uses has been recognised and some selective breeding for container plants has been undertaken.

Lamarckism: inheritance of acquired characters

Jean-Baptiste Lamarck (1744–1829) was a French naturalist who published the first modern theory of evolution. His theory, known as **Lamarckism**, was that traits acquired during the lifetime of an organism are inherited by subsequent generations. Acquired traits, such as increased muscular strength due to exercise, were believed to be transmissible to offspring. Experiments performed at the end of the nineteenth and early twentieth centuries, including one in which mice in 20 successive generations had their tails cut off to see whether the offspring were affected, did not support the Lamarckian theory.

Fig. 32.2 Variation in flower colour and arrangement in Sturt's desert pea

Our current understanding of molecular genetics also suggests that inheritance of acquired traits is impossible because this would require a reversal of the flow of genetic information: from protein to RNA to DNA. The central dogma of molecular genetics is that the flow of genetic information is from DNA to RNA to protein. Exceptions to this general rule are provided by some RNA viruses, which do not have DNA and thus the first step in this chain. More important exceptions are retroviruses (Chapter 34), which have added an extra step, reverse-transcribing their RNA genome into DNA before it is inserted into the genome of the host cell, where it behaves like host DNA. Retroviruses can reverse the transmission of information from RNA to DNA and sometimes they accidentally do this to host RNA sequences as well.

Recent attempts to revive Lamarck's theory postulate a role for retroviruses, suggesting that they permit movement along part of the path backwards from an organ to the genetic material. However, there is no example and no obvious mechanism of transmission of adaptive information backwards from protein to RNA or to the encoding DNA. Modifications of proteins by environmental factors during the lifetime of an organism are not known to be incorporated into the genetic blueprint for the protein.

> While the inheritance of traits gained during the lifetime of an organism was once proposed as a mechanism of evolutionary change—Lamarckism—it is now known that the mechanism of storage and expression of genetic information precludes the inheritance of acquired characters.

How does genetic variation arise?

Both selective (adaptive) and neutral (random) processes in evolution depend on the existence of heritable genetic variation within populations. Thus, it is important to understand how genetic variation arises. Ultimately, the appearance of novelties or new traits in populations originates from mutations.

Mutations are the source of new variation

Genetic information is stored in DNA sequences. A **mutation** is a change in the sequence or arrangement of the DNA of an organism. The effect of a mutation depends on what it is and where it is. As the raw

material for evolution, only mutations that are heritable—that occur in the germ line, in the cells that produce eggs and sperm—can be passed on to the next generation. Mutations in non-germ-line cells (somatic cells) only affect the individuals in which they occur and are not passed on to future generations. Some mutations, although altering the genotype, may have no detectable effect on the phenotype. For example, a change in, or deletion of, one tandem repeat in non-coding, highly repeated DNA, may have no impact. On the other hand, even a single nucleotide change in a coding or regulatory region can have a dramatic effect on the survival or reproduction of an organism.

There are two main types of mutation. Small changes in DNA sequence are **point mutations**. Point mutations include base substitutions, insertions and deletions (see Chapter 12). Changes causing breakage and rejoining of chromosomes are **chromosome mutations**.

Chromosome mutations

Changes can occur in the structure, size and number of chromosomes. Chromosomal rearrangements include *inversions* of segments within a chromosome, where a segment remains in the same position but is turned around (Fig. 32.3b). Pericentric inversions include the centromere and paracentric inversions do not. *Translocations* occur when segments are exchanged between non-homologous chromosomes. The number of chromosomes can decrease by *fusion*, when chromosomes join end to end, or increase by *fission* after chromosomes break. (Changes in chromosome number involving polyploidy are discussed below.) Chromosome mutations also include duplications of regions and deletions.

Inversions and translocations are common types of chromosome mutations. Duplication of a segment of

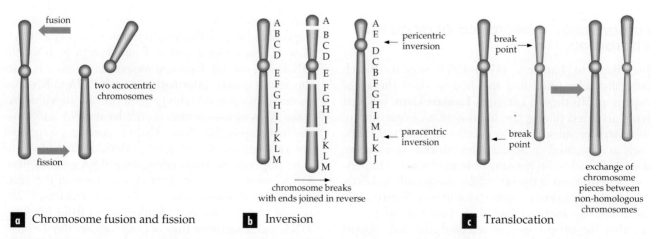

a Chromosome fusion and fission   **b** Inversion   **c** Translocation

Fig. 32.3 Major types of chromosomal mutations, causing variation in number, shape and size of chromosomes

DNA may be an important way that new evolutionary possibilities are realised rapidly. The variety of globins, including myoglobin and haemoglobins in vertebrates, is thought to have evolved after duplication of the ancestral globin gene (Fig. 32.4). Different copies of the gene accumulated different mutations and evolved different functions. Chromosome fusions and fissions also seem to have played a part in evolution. For example, related species often have different chromosome numbers (humans 46, chimpanzee and gorilla, 48).

Chromosomal analyses have revealed the existence of sibling species, which show no morphological divergence but considerable chromosomal divergence. Chromosomal differences may be barriers to interbreeding and thus hinder or prevent genetic exchange between diverging populations.

Chromosome mutations involve changes in large segments of DNA or whole chromosomes during cell division. They include duplications and deletions, inversion and translocation of segments, and the fusion and fission of whole chromosomes.

Transposable elements

Mutations can be the result of radiation and chemical agents (see Chapter 12) or the movement of sequences of DNA known as **transposable elements**. Transposable elements are mobile and capable of moving to new locations, frequently taking other genes with them. They were first discovered in maize by Barbara McClintock and occur both in prokaryotes and eukaryotes (Chapter 10). Although the probability of

movement is very small for an individual element, the large number of elements means that substantial amounts of transposition occur over time.

Mutations due to transposition occur in several ways. Firstly, if a transposable element inserts itself into or near a coding or regulatory part of a gene, the sequence of the gene is disrupted, resulting in modification or total abolition of its expression. The precise consequences of an insertion are difficult to predict but evidence from numerous organisms attests to their importance. In *Drosophila melanogaster*, the majority of spontaneous visible mutations affecting traits such as eye colour and wing shape are due to the insertion of a range of different transposable elements. One of these, the P element, is particularly interesting because it has invaded populations of *D. melanogaster* throughout the world in the past 40 or 50 years. Even the wrinkled seed trait of peas studied by Gregor Mendel (Chapter 9) is now known to be the result of a mutation caused by insertion of a transposable element into the gene coding for a starch branching enzyme.

The second way in which transposable elements cause mutations is by excision or movement away from their original position. Most transposable elements are capable of precise excision from the DNA of the host. Generally, precise excision will lead to reversion, as the sequence and function of the gene will be restored approximately to the original. However, on many occasions they excise imprecisely, taking flanking DNA sequences with them or sometimes even taking captive sequences to new locations in the genome. Deletion mutations, like those induced by radiation, can result.

Transposition can also cause chromosomal inversions and translocations. Transposable elements must be able to nick DNA to insert or excise. When multiple nicks are made in the DNA of the same or different chromosomes, the incorrect ends may rejoin.

Mutations can result from the movement of transposable elements.

Recombination and gene flow

While mutations are ultimately important in the formation of new alleles and new genotypes, their frequency of occurrence is relatively low. A significant amount of variation within sexually reproducing populations results from *recombination* of existing alleles. As a result of random segregation of homologous chromosomes and recombination due to crossing-over in meiosis (see Chapter 8), a tremendous variety of gametes with different combinations of parental genes can be generated. These processes can thus generate much genetic variation in the offspring of such populations.

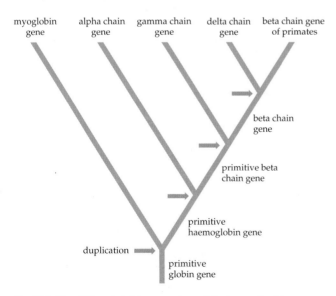

Fig. 32.4 The different globin genes in vertebrates evolved by gene duplication and divergence

Genetic variation can also result from gene flow between populations. **Gene flow** is genetic exchange resulting, for example, from the movement of animals from one population to another. In plants, pollen or seed can be blown by wind, moved in water currents or carried by animals between populations in different localities and environments.

Genes in populations

If a new allele arises through mutation or is introduced by gene flow, what will be its fate in a population? Will it disappear over time or increase in frequency between generations? What will be the effects of selective evolution or neutral evolution?

Allele frequencies

Genes may exist in one or more allelic forms within a population. When more than one allelic form of a gene is present, the population is said to be genetically **polymorphic** at that locus. Different genes show different levels of polymorphism. The most common allele at a polymorphic locus may be referred to as the **wild-type allele**.

We can estimate the frequency of different alleles of a gene in a population by examining the proportion of individuals that have particular phenotypes and therefore particular genotypes. For example, the human MN blood group is controlled by two codominant alleles, L^M and L^N. The heterozygote is blood type MN and the homozygotes are blood types M or N.

Consider a population of Aboriginal people from Elcho Island in Australia's Northern Territory (Table 32.1). To calculate the frequency of the L^M allele, remember that each member of the population with M type blood (28 individuals) is carrying two L^M alleles and each member with MN type blood (129) is carrying one. Therefore, the number of L^M alleles in the population is $2 \times 28 + 129 = 185$. The L^M allele frequency is the proportion of this allele relative to the total number of alleles in the population. For a population of 352 individuals there are a total of 704 alleles (2×352). The L allele frequency is therefore $185 \div 704 = 0.26$. The L^N allele frequency is 0.74 (Table 32.1) since allele frequencies for a gene in a population must add to 1. Allele frequencies are usually abbreviated p and q, with $p + q = 1$.

> G enetic polymorphism is the presence in a population of more than one allelic form of a gene.

The Hardy–Weinberg principle

Before discussing how evolution results in changes in allele frequencies within populations, we will first consider the behaviour of **allele** and **genotype frequencies** in populations that are not undergoing evolution. In 1908, G. H. Hardy and W. Weinberg independently formulated a model for the behaviour of allele frequencies in a population that was not affected by neutral or selective evolution. Their model predicts the genotype frequencies expected in the offspring generation from the allele frequencies in the parental generation. Frequencies of genotypes in a population can be calculated by dividing the number of individuals of each type by the total population size (Table 32.1). We can also calculate genotype frequencies expected in the next generation by using allele frequencies calculated for the present generation. Imagine that the next generation is a result of the random combination of

Table 32.1 The frequency of MN blood group alleles in a population of Aboriginal Australians from Elcho Island

| | Blood group | | | |
	M	MN	N	Total
Genotype	$L^M L^M$	$L^M L^N$	$L^N L^N$	
Number of individuals	28	129	195	352
Number of L^M alleles	56	129	0	185
Number of L^N alleles	0	129	390	519
Total number of alleles	56	258	390	704

Frequency of L^M allele = 185/704 = 0.26 = p.
Frequency of L^N allele = 519/704 = 0.74 = q.

gametes. Figure 32.5 shows the possible combinations of genotypes in a Punnett square. With a frequency p of gametes of one type and q of the other, the three genotypes arise in the next generation in the proportions p^2, $2pq$ and q^2. The proportions add to 1 and can be deduced from expansion of the binomial expression $(p + q)^2$. These genotype frequencies are called **Hardy–Weinberg frequencies** after G. H. Hardy and W. Weinberg.

Fig. 32.5 Calculating genotype frequencies in the next generation. Two alleles A and a are present in a population at frequencies p and q respectively. Progeny occur in the next generation in Hardy–Weinberg genotype frequencies: p^2 for AA, $2pq$ for Aa and q^2 for aa

For our MN blood group example, 26% of the gametes in the population will carry the L^M allele ($p = 0.26$) and 74% will carry the L^N allele ($q = 0.74$). Thus, in the next generation, 0.07 ($p^2 = 0.26^2$) will be blood type M, 0.39 ($2pq = 2 \times 0.26 \times 0.74$) will be type MN, and 0.54 ($q^2 = 0.74^2$) will be type N. These Hardy–Weinberg genotype frequencies expected for the next generation are close to those of the current population.

The assumptions of the Hardy–Weinberg model are that:

- the interbreeding population is infinitely large (so that it is large enough not to be influenced by chance effects)
- alleles mutate at the same rate
- genotype frequencies are not affected by migration in and out of the population
- individuals mate at random
- the population is not experiencing selection (i.e. all individuals have the same Darwinian fitness).

Given the above assumptions, the genotype frequencies are expected to remain constant from generation to generation. If one or more of these assumptions is not met, allele frequencies in a population may change and the population may evolve.

Population size

The first assumption is that the population is extremely large. If a population is very small, or if it goes through periodic 'bottlenecks' of low numbers, alleles in the population may drift to high or low frequencies purely by chance. Random change in allele frequencies in small populations is termed **genetic drift** (p. 843). This

leads to rapid fluctuations in allele frequencies, with the possibility that the frequency of a particular allele will be reduced to 0 or become fixed at 1, the population becoming homozygous.

Mutation rate

The second assumption underlying Hardy–Weinberg genotype frequencies is that different alleles mutate at the same rate. Mutation ultimately provides genetic diversity, the raw material of evolution. However, mutation rates are usually very low and, even if different alleles mutated at different rates, this by itself is unlikely to lead to rapid change in populations.

Migration

The third assumption is that individuals with particular alleles do not migrate at different rates. This may not be the case. Australia's human population has undergone great changes in genetic composition over time as a result of migration. Among European immigrants, the I^B allele of the ABO blood group alleles occurs at a frequency of about 0.1 (10%), while in recent migrants from Asia it is even higher (0.2 or 20%). Since Australian Aboriginal populations, with the exception of a few individuals from the Cape York area, completely lack the I^B allele, the overall frequency of the I^B allele among Australians has increased over the past 200 years. The types and relative numbers of alleles of other genes are also changing.

Random mating

The fourth assumption is that random mating occurs. Non-random mating may arise from positive assortative mating (in which individuals tend to mate with other individuals that have similar phenotypic traits, e.g. size or colour) or from inbreeding (in which individuals tend to mate with related individuals). Both of these types of non-random mating will distort Hardy–Weinberg genotype frequencies, so that there will be fewer heterozygotes and more homozygotes overall.

Darwinian fitness

Finally, constant Hardy–Weinberg frequencies will occur only if there are no differences in the Darwinian fitness of individual genotypes. **Fitness** is measured as the relative contribution of offspring to subsequent generations due to differential survival and/or reproduction. If one genotype leaves relatively more fertile offspring per head than others, the genotype frequencies in the next generation will not be in Hardy–Weinberg proportions.

For example, consider the case in which a recessive phenotype (homozygote) leaves no offspring. Relatively few of the recessive alleles will be transmitted to the

BOX 32.1 Application of the Hardy–Weinberg principle—the frequency of carriers of genetic disease

Calculations of Hardy–Weinberg genotype frequencies can be useful in predicting frequencies of alleles that are difficult to distinguish in the heterozygote because of dominance. For example, in the human genetic disease phenylketonuria, affected children are born in Australia at the rate of about 1 in 10 000 live births (a frequency of 0.0001). If the wild-type allele occurs in the population at the frequency p and the mutant allele at the frequency q, then we expect affected phenylketonuria zygotes to arise at the frequency q^2 under Hardy–Weinberg assumptions: that is, q^2 is 0.0001 and q is 0.01. Heterozygotes are likely to be present at the frequency $2pq$, or $2(0.01 \times 0.99) = 0.02$. In the Australian population, about 2% (1 in 50) of the population are likely to be carriers of a mutant phenylketonuria allele. There is a substantial reservoir of mutant phenylketonuria alleles in the Australian population, mostly carried silently in heterozygotes even though they come together to form homozygous affected individuals in only about 1 in 10 000 children.

next generation and their frequency will fall (all other factors being equal). In this way, deleterious alleles remain at low frequencies, with their loss usually being balanced by their creation through new mutation events. This is likely to be the case with severe human genetic diseases in which the affected individual does not reproduce, for example, the metabolic disorder phenylketonuria before its cause was recognised and treated successfully by diet control.

A ccording to the Hardy–Weinberg principle, allele frequencies remain constant from one generation to the next and genotypes occur in populations in predictable frequencies, given several assumptions: large population size, random mating and no differences in allele mutation rates, phenotype migration rates or Darwinian fitness. Evolutionary change may result if these assumptions do not apply.

Evolution in action

The Hardy–Weinberg principle is important because it demonstrates that genetic variation and heritability

cannot alone cause evolution. As we have seen, if all conditions of the Hardy–Weinberg principle are met, allele frequencies in a population should not change and evolution will not occur. Of course, these conditions are never completely met in any natural population. By determining which conditions are not being met, we can get a better understanding of how evolution occurs and which processes are operating—*genetic drift* (in the case of small population size and random changes), *gene flow* (where there is significant immigration or emigration of individuals in and out of populations), or *selection* (where there are differences in fitness).

Thus, the utility of the Hardy–Weinberg principle is that it provides a baseline for comparison with real populations. Since the Hardy–Weinberg principle assumes the absence of selective or neutral evolution, if observed genotype frequencies differ significantly from those predicted by the Hardy–Weinberg model, these differences may be due to selective or neutral evolution.

Adaptive (selective) evolution

As we have seen above, those individuals in a population that leave a greater relative proportion of offspring to the next generation are said to have a higher fitness. If differences in fitness are heritable, then as a result of differential survival and reproduction of phenotypes, a change in genotype frequencies will occur in a population. This is the process of natural selection. In this manner, genotype frequencies and allele frequencies can change through time and selective evolution can occur.

With selection operating, the fate of a mutation depends on its effect on fitness (Fig. 32.6). A mutation can cause change in the fitness of a phenotype, either increasing or decreasing its fitness. A mutation with a favourable effect on fitness will increase in frequency within a population (Fig. 32.6), whereas one with an unfavourable effect will decrease and eventually may be eliminated from the population.

Natural selection, the peppered moth and industrial melanism

If environments change, selection may lead to changes in the frequency of different phenotypes. A well-known example involves that of evolutionary change in the peppered moth, *Biston betularia*, in industrial Britain (Fig. 32.7). In pre-industrial times, the most common form of this moth had light speckled patterning on the wings. A form of the moth with uniformly dark wings was also seen occasionally. This dark phenotype resulted from the production of a dark pigment (melanin) by a dominant mutant gene. As industrialsation proceeded

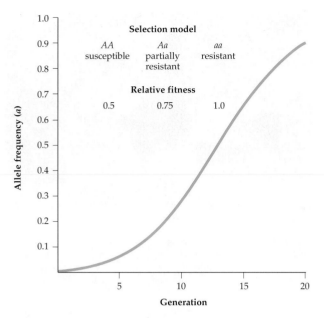

Fig. 32.6 Change in allele frequency in a large population due to selection. The plot shows the increase in allele frequency for an initially rare allele (1%), such as an insecticide resistance allele, which becomes favourable because of the introduction of a new chemical. The selection model is shown in the diagram. The relative fitness indicates the proportion of offspring transmitted to the next generation by the different genotypes

Fig. 32.7 Peppered moths, *Biston betularia*, from industrial Britain. The speckled form and the dark melanic form are shown together on dark, pollution-affected bark

and air pollution increased, the frequency of the dark form near Manchester increased from very low levels to 98% in the 45 years from 1850 to 1895. As levels of air pollution have fallen in recent years, so has the frequency of the dark (melanic) form. Thus, an ecological change has apparently led to an evolutionary change in the moth population.

These changes in the frequency of melanic forms of the peppered moth have been interpreted as resulting

BOX 32.2 Insecticide resistance in *Lucilia cuprina*

Almost all adaptively significant new mutations have adverse effects. However, if there is at least one good attribute associated with a mutation, other genes that modify or mitigate the adverse effects may be selected. A good example is provided by insecticide resistance in the Australian sheep blowfly, *Lucilia cuprina*.

The sheep blowfly is a major problem for the sheep industry in Australia because female flies lay eggs in wounds and moist fleece. Larvae burrow into the skin of sheep, causing distress, loss of production and even death if untreated. Chemicals, such as dieldrin and organophosphates, have been used extensively to treat and prevent fly strike. However, chance mutations conferring resistance to insecticides have arisen so that the effectiveness of these chemical control agents is greatly reduced.

Many newly arisen insecticide resistance genes are disadvantageous in the absence of insecticide and decrease to low frequency if the insecticide is withdrawn from use. Thus, rotation of use of different types of insecticides hinders the development of resistance. However, persistent use of an insecticide may cause the selection of modifier genes that reduce the disadvantageous effects of the resistance genes. A modifier gene, which removes the deleterious effects of the organophosphate resistance gene, has been recognised and characterised in Australian *L. cuprina*. This modifier gene enables the resistance gene to persist at high frequency after withdrawal of the insecticide, so that the insecticide can never be used effectively again.

Sheep blowfly *Lucilia cuprina*

from natural selection mediated by predation by birds. The moth rests during the day on the trunks and branches of trees. Before the Industrial Revolution, the light speckled patterns on the wings of most moths made them well camouflaged against the light-coloured background of lichens on the tree branches and would have helped to hide them from birds. At this time, the dark phenotype would have been more conspicuous to predators, leading to a relatively low contribution by melanic moths to the next generation (low Darwinian fitness). However, as atmospheric pollution from factories increased, the peppered moths became more visible on the soot-covered branches and the dark moths became less conspicuous to birds and increased in frequency. The dark phenotypes of the moths were now at a selective advantage and their Darwinian fitness was consequently higher than that of the lighter speckled form.

More recent research has revealed additional complexity that indicates that selection caused by predators does not completely explain the frequency of these forms in natural populations and that additional factors may also contribute to the relative fitnesses and frequencies of melanic and peppered forms. For example, there may be differences in survival of larval stages (caterpillars), and black moths may be more efficient at absorbing heat than the lighter coloured forms in an atmosphere clouded by smoke.

Modification or loss of structures

The example in Box 32.3 illustrates adaptive evolution and divergence of kangaroos. Increased aridity over the last 25 million years in Australia led to an expansion of open forests, woodlands and grasslands and a contraction of wetter forests (Chapter 41). At the beginning of this period, kangaroos were small and probably omnivorous or browsers on relatively soft foliage. They had rather unspecialised dentitions, foot structures and diets. Today, the musky rat-kangaroo, *Hypsiprymnodon moschatus*, which retains many ancestral features, is found only in rainforests of northern Queensland. Larger browsing and grazing forms with specialised changes in dentition and foot structure evolved from ancestral types as the physical environment and vegetation changed. Expansion of grassy forests and grasslands eventually favoured the grazing lineage.

Adaptive evolution leads to organisms possessing characteristics that suit them to their particular environments. This may lead to the modification and elaboration of structures or to their reduction or loss (Fig. 32.8). As an example of the latter case, in some deep-sea fishes where location of mates is difficult, the male is reduced to little more than a 'parasitic gonad' permanently attached to the female. In parasitic

Fig. 32.8 In addition to increased complexity, adaptive evolution can lead to the loss of structures. **(a)** *Rhizanthella gardneri* is an orchid that lives and flowers underground and lacks leaves. **(b)** Pygopodids, the legless lizards of Australia, live in cracks and crevices and some species even burrow through the soil. In these circumstances, an elongated body with reduced limbs is advantageous. Forelimbs are completely absent and the hindlimbs are reduced to scale-like vestiges

barnacles, the crustacean body plan is lost in the adults. The body is reduced to a branching structure resembling a fungus that penetrates the tissues of crab and lobster hosts so that nutrients can be extracted directly without the need for a complex alimentary system. Many plants, too, have lost complex structures during evolution. Underground orchids of the genus *Rhizanthella* from Western and eastern Australia lack leaves and photosynthesis (Fig. 32.8). These remarkable orchids even flower underground. They obtain all nutrients from mycorrhizal fungi (Chapter 36), which are also associated with the roots of trees and shrubs.

Adaptive evolution is evolution that occurs as a result of natural selection. Natural selection results in organisms becoming better suited (better adapted) than others to their particular environment and may lead to divergence among populations that occupy different environments.

Neutral evolution and genetic drift

Not all change in populations can be explained by natural selection. Genes that have a neutral effect on survival and reproduction will change in frequency at random. If a neutral change occurs in one member of a population, the number of future descendants carrying it will be a matter of chance. **Neutral evolution** refers to evolutionary change that occurs as a result of such random, rather than selective, processes.

In the case of *genetic drift*, allele frequencies in populations change as a result of chance sampling events. For example, the gametes that actually form zygotes usually represent a small sample from a much larger pool of available gametes and the allele frequencies in this sample may not be representative of the larger pool. Small random samples tend to be less representative of the overall gene pool of a population than large samples. Thus, the effects of genetic drift are greater in small populations than in large populations (Fig. 32.9).

To help understand why this is so, it may be useful to consider the following hypothetical situation. If you were to draw two marbles randomly from a large bag containing equal numbers of green and blue marbles, it would not be too surprising if both marbles were blue or both were green (this is analogous to fixation of one allele and loss of another). However, if you were to draw 40 marbles at random from the bag, you would be very surprised if all 40 were blue or all 40 were green.

In small populations, a neutral change may increase in frequency quite rapidly and even become universal, thus leading to divergence among populations. In large populations, there is less tendency for this to occur.

The remaining world population of cheetahs is homozygous for many genes, suggesting that small population size and genetic drift may have led to homozygosity. Genetic drift also seems to account for high levels of homozygosity in small populations of the Australian bush rat, *Rattus fuscipes greyi*, on islands off Eyre Peninsula in South Australia (Fig. 32.10). Animals

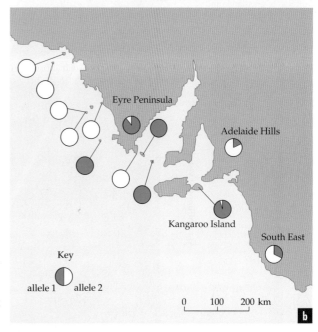

Fig. 32.10 (a) Bush rat, *Rattus fuscipes greyi*, from South Australia. **(b)** Thirteen populations of bush rats were screened for alleles of a gene controlling an enzyme present in heart and kidney cells. Allele frequencies are shown as pie diagrams. Three mainland populations and one large island population contained both alleles (i.e. they were polymorphic). Nine small island populations contained only one or other of the alleles, probably because of small population size and genetic drift

Fig. 32.9 Computer simulation of random drift in allele frequency in small populations. Five populations, each consisting of 10 breeding individuals, start out with an initial allele frequency of 50%. In the absence of selection, allele frequencies wander randomly and can increase to 100% or decrease to 0%. Differences between populations due to random drift constitute neutral evolution

BOX 32.3 Adaptive evolution of kangaroos

During the late Oligocene and early Miocene, rainforest environments are thought to have covered much of Australia. As the Australian plate drifted north, aridity increased (Chapter 41). The evolutionary diversification and adaptations of kangaroos have been influenced by this increased aridity (Fig. a). Animals that survived more arid conditions, ate poor-quality grasses and tougher foliage and may have been better able to escape predators in more open environments, were favoured.

Group	Giant rat-kangaroos	Musky rat-kangaroo	Bettongs and potoroos	Sthenurine kangaroos	Macropodine kangaroos
Number of extant species	0	1	9	10	49
Group representative	*Ekaltadeta*	*Hypsiprymnodon*	*Caloprymnus*	*Sthenurus*	*Macropus*
Diet	Omnivorous, carnivorous	Invertebrates, fruits	Roots, seeds, fungi, insects	Mainly leaves	Mainly grasses
Habitat	?Woodlands	Rainforests	Savanna, grasslands, forests	Open forests, savanna	Grasslands, savanna forests
Lower jaw structure					
Number of significant toes on hind foot, important in locomotion	?	3	2	1	2

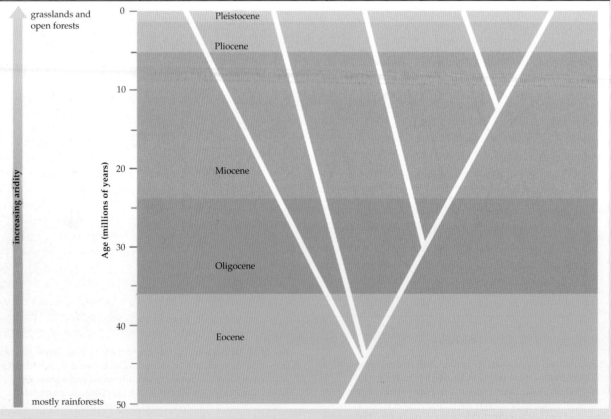

(a) The most successful lineages today are those kangaroos adapted to increasing aridity in Australia, in particular, the grazing macropods

(b) *Hypsiprymnodon moschatus*, the musky rat-kangaroo

Diet and dentition

The living kangaroo most similar to the ancestor of all kangaroos is the musky rat-kangaroo, *Hypsiprymnodon moschatus* (Fig. b). *Hypsiprymnodon* lives in rainforests and eats a variety of foods, including small invertebrates and fruits. It has specialised, blade-like premolars for slicing open the exoskeletons of insects, but the molars have simple, rounded surfaces for crushing soft food items into more easily digested pieces. The musky rat-kangaroo's diet is relatively nutritious and food is abundant on the rainforest floor. It meets its nutritional needs easily and there is no need to grind food to extract all the nutritional value. Its stomach is a simple sac, not divided into several cavities and probably not capable of cellulose digestion. Grasses are relatively poor-quality foods and kangaroos that feed on them must be able to extract a high proportion of the nutrients available. Species of *Macropus*, such as the red kangaroo (Fig. c), have high-crested molar teeth that efficiently shear and grind food into a paste, allowing nutrients to be extracted more efficiently in the gut, which is also adapted to less digestible plant material. Grazers (grass-eaters) have less need for a blade-like premolar and this tooth is reduced in size.

Locomotion and habitat

Kangaroos have a hopping (saltatory) form of locomotion, and can achieve high speeds as a result of modification of the foot structure, by reduction on the hind feet of the number of toes and increase in the length of the fourth toe, which functions as a long lever. In *Macropus* and other true kangaroos (family Macropodidae), the first toe has been lost and the slender second and third toes are joined together and used for grooming. *Hypsiprymnodon* does not hop

(c) *Macropus rufus*, the red kangaroo

bipedally and has a less specialised foot structure, differing from all other kangaroos in retaining the first toe, which is nail-less and movable, and which can be spread away from the other toes (as in possums). In grasslands and more open habitats, animals may be more conspicuous to predators than on the floor of a rainforest and speed is probably more important for escaping predators. As with the large hoofed, grazing animals of other parts of the world, large kangaroos attain speeds of more than 50 km per hour over short distances.

The evolutionary history of kangaroos is an active area of research about which much remains to be learned and the phylogenetic tree shown here represents a simplified view of a current hypothesis regarding the relationships of kangaroos. The fossil record for kangaroos at present extends back to about 25 million years ago. The divergence dates shown are estimates only (e.g.

taking into account fossils where available and molecular data).

The number of living species of grazing kangaroos and the smaller bettongs and potoroos reflects their current success following contraction of rainforests and expansion of drier open forests and grasslands. Although bettongs and potoroos do not eat grass, they have other adaptations to more open environments. The extinct browsing (leaf-cutting) sthenurine kangaroos (with possibly one living species) were once common but declined, possibly due to reduced availability of low-level browse. Giant rat-kangaroos were never very common and may have been omnivorous to at least partly carnivorous. These kangaroos had large, blade-like premolars, which may have functioned for slicing up flesh.

have been stranded on these islands by rising sea levels in the past and isolated for many generations.

Genetic drift has apparently been important in a number of human populations that are small and have been isolated by geographic, cultural or religious boundaries. For example, on Pingelap Atoll in the Pacific Ocean, 5% of the population have a form of colour-blindness (achromatopsia, where people only see one colour), yet this condition is very rare in other populations. The population of Pingelap Atoll was greatly reduced in 1775 following a typhoon and famine. By chance, one of the survivors was heterozygous for the achromatopsia gene, and he was a chief who had many children.

Populations with low levels of genetic variation may be at risk if their environment changes and if genotypes fixed by chance are less well adapted to the new conditions. In biological conservation, it is important to assess the level of genetic variation in natural populations to determine whether variation is sufficient to allow them to respond to future environmental changes. Conserving genetic variation is particularly important for populations of many Australian mammals, such as the numbat, that now exist only in isolated fragments of formerly wider distributions.

Between the extremes of genetic variants with no effect on fitness and those with a large effect on fitness, there is a complex interaction between fitness and population size. In small populations, genetic drift can swamp the influence of fitness, and a mutation with only a small favourable effect on fitness may be lost by chance. In large populations, the effects of differences in selection are more predictable and even selectively neutral variants have less tendency to drift in gene frequency.

Neutral mutations have no effect on fitness and chance alone determines the fate of such mutations within a population. Genetic drift is changes in allele frequencies between generations that result from random processes. Random genetic drift is particularly important in small populations, and alleles can increase in frequency to fixation or decrease in frequency and eventually become lost from the population.

Selective and neutral evolution at the molecular level

A lot of new data on variation at the molecular level has become available in the last few decades as a result of the development of new technologies. The role of selective and neutral (random) processes in evolution at the molecular level is currently a matter of active research and debate. DNA base sequences are analogous to architectural plans. When we carefully compare two sets of building plans, we recognise differences that affect the building specifications; these are analogous to differences in the DNA of organisms specifying adaptive changes. If the plans differ only in the colour of the ink used for the drawings, then the specifications are unaffected. These differences in colour are analogous to neutral evolutionary changes, which have no functional impact. For example, an amino acid may be coded by a number of different triplets (Chapter 11) and a change from TAT to TAG does not affect the amino acid specified. The change from T to G in the third position is an apparently neutral change. Furthermore, a protein nearly always occurs in a number of variant forms within a population as a result of single amino acid substitutions (e.g. there are many variants of human haemoglobin). Many of these variants are deleterious, but some are selectively advantageous under certain conditions and many others appear to be neutral.

The most convincing evidence for selection at the level of DNA is sequence conservation. Comparison of DNA sequences of the same gene in different species shows that non-coding and putatively non-functional regions (introns, Chapter 10) rapidly lose sequence similarity, whereas similarity is conserved in the coding (exon) regions. For example, comparison of the DNA sequence of the gene for scarlet eye colour in *Drosophila melanogaster*, the cosmopolitan vinegar fly, *D. buzzatii*, a species introduced into Australia on prickly pear, and *Lucilia cuprina*, the sheep blowfly of Australia, reveals that the length of the introns is not conserved and the sequence similarity is never greater than about 30% among these species. Thus, changes of intron sequences appear to occur largely under the influence of neutral evolution.

In contrast, the exons are highly conserved. There are severe functional constraints on the amino acid substitutions, which are consistent with function of the protein. In the nucleotide sequence of the scarlet eye colour gene, *D. buzzatii* has 77% similarity with *D. melanogaster*, and 70% similarity with the more distantly related *L. cuprina*. At the amino acid level, the similarity is 81% in each case, further indication of the functional constraints on exon evolution. At this level, selection removes maladaptive mutations and is described as **stabilising selection**. However, a small proportion of sequence changes are favourable and permit adaptive evolution in response to changing environmental circumstances.

Speciation

Like Lamarck and Darwin, modern evolutionary biologists are interested in the formation of new species—**speciation**. Speciation is a process that has occurred over hundreds of millions of years and is occurring now. Evidence from the fossil record, from comparative studies of morphology and development, DNA and proteins, from cytogenetics and from biogeography is used to test ideas and theories about speciation. Comparative and experimental studies are also used to test hypotheses about the evolution of reproductive isolation.

Among modern evolutionary biologists, there is much interest, research and debate regarding how speciation occurs and the relative importance of different mechanisms of speciation. Debate among biologists about speciation mechanisms relates, in part, to debate about species concepts.

What is a species?

The species is a taxonomic category in biological classification (Chapter 30). Different taxonomic species can generally be recognised by character differences; for example, *Banksia serrata* has leaves with a serrated (saw-toothed) margin compared with *B. marginata*, which has an entire (smooth) margin. As we saw in Chapter 30, there is not a single species concept that is generally accepted, but rather various species concepts have been proposed that differ in their emphasis on the criteria that are considered important for defining species.

The *taxonomic* or *morphological species* concept emphasises the phenotypic distinctiveness of species, and morphology in particular has been very important in the recognition of taxonomic species. In practice, most species are recognised on the basis of morphological characteristics. While species defined in this way are generally considered to reflect the operation of evolutionary processes, the following species concepts define species in terms of evolutionary processes.

Biological species, following Ernst Mayr's definition (Chapter 30), are 'groups of actually or potentially interbreeding natural populations, which are reproductively isolated from other such groups' (Box 32.4). Thus, the biological species concept emphasises both the distinctiveness of the **gene pools** (the total of all the genes and alleles in a species) of different species and the importance of interbreeding within species. Thus, the biological species concept does not apply to organisms that reproduce asexually or to extinct organisms. In a practical sense, it is often difficult to determine the capacity of populations to interbreed under natural conditions.

Some species may be more or less reproductively isolated but still occasionally exchange genes and form **hybrids**. Exchange of genes can occur between closely related and very similar species or between more distantly related and dissimilar species. For example, *Eucalyptus obliqua* (which is classified in the 'ash' group of eucalypts) hybridises with *E. baxteri* (a member of the 'stringybark' group) where the habitats of these two species overlap (Fig. 32.11). On morphological evidence, the two species are not each other's closest relative and are recognisably different. Their ability to interbreed is not used as a criterion for 'lumping' them together as a single species. Cases such as this have led some biologists to criticise the biological species concept, pointing out that species that are recognised as distinct taxonomic species and that maintain their evolutionary distinctiveness may not be completely reproductively isolated. This has raised the question of just how strongly reproductively isolated populations must be in order to be considered distinct species.

Species have been defined as groups sharing a common, specific mate recognition system (the *recognition species* concept). This concept emphasises mate recognition within species rather than reproductive isolation. Neither the recognition nor the biological species concept is relevant to organisms that normally reproduce asexually.

An *evolutionary species* has been defined as a lineage of populations bound by a common ancestry and an ability to maintain its integrity with respect to other evolutionary species. The evolutionary species concept is not concerned with mechanisms of speciation or how integrity is maintained. While this is a broad concept, which applies to both sexually and asexually reproducing organisms, it has been criticised for being difficult to apply; for example, how is integrity actually determined?

BOX 32.4 Reproductive isolation

The biological species concept, as we have seen, emphasises the importance of reproductive isolation in defining the limits of species. How does reproductive isolation between species come about? What role does reproductive isolation play in maintaining the evolutionary distinctiveness of related species that co-occur in the same area?

Barriers that inhibit or prevent gene flow between species and that are dependent on differences in the species involved are variously referred to as **reproductive isolating mechanisms** or barriers. Thus, simple geographical isolation is not considered to be a type of isolating mechanism. Different types of reproductive barriers are recognised and these are generally broadly classified as prezygotic or postzygotic. Prezygotic mechanisms, as the name implies, operate before the stage of zygote formation, whereas postzygotic mechanisms operate after zygote formation. Species may be isolated by a combination of barriers and individual barriers may vary in the extent to which they restrict gene flow.

Ecological isolation

Species may occur in the same area but not hybridise because they are ecologically isolated and occupy different habitats. Differences in habitat preferences can serve to restrict gene flow between species by reducing opportunities for contact and interbreeding. For example, two plant species may occur in the same geographical area but gene flow between them may be restricted because they grow on different soil types. Their differing soil preferences can serve to restrict gene flow between them by reducing spatial contact and thus opportunities for interspecific gene exchange.

Temporal isolation

Isolation may occur as a result of differences between species in the timing of reproduction. For example, related, co-occurring plant species may have partially or completely different flowering times, thus reducing the opportunity for pollen flow between them. Examples of differences in flowering time between related plant species can be found in a variety of well-known Australian plant genera including *Eucalyptus*, *Leptospermum* (tea trees), and *Banksia*. Temporal separation of flowering of two *Leptospermum* species has been observed at sites where they co-occur in Victorian coastal heathlands. The differences in flowering time would serve to inhibit pollen transfer between these species, which have flowers that are similar in structure and which attract similar pollinators.

Ethological isolation

Ethological isolation results from differences between species in behaviours that affect courtship and mating. A diverse array of courtship behaviours has evolved that enable organisms to recognise and respond to suitable mates. Among insects and frogs, mating songs and calls are widely used in courtship. These signals differ between species and may play an important role in species recognition and hence reproductive isolation.

Biologists have studied the role of mating calls in the reproduction of two frog species, *Litoria ewingi* and *L. verreauxi* in south-eastern Australia. The ranges of these frogs overlap in some areas. The calls of frogs of both species were found to be more similar in areas in which only one species occurred than in areas where they co-occurred. The differentiation of the mating calls in the area of overlap may allow females to better distinguish the calls of males of their own species and hence acts as a reproductive barrier (Fig. a).

The term ethological isolation is sometimes used broadly to include impediments to gene exchange between animal-pollinated plants that arise from the behaviour of pollinators. For example, two flowering plant species may differ in floral features, such as colour and odour, and attract different pollinators, thus reducing opportunities for pollen transfer between them. Even if two plant species attract the same pollinator(s), individual pollinators may restrict their visits to members of only one of the two species, possibly because this results in more efficient foraging.

The plant genus *Anigozanthos* (kangaroo paws and catspaws) occurs in south-western Western Australia and birds are important pollinators of the colourful flowers. More than one *Anigozanthos* species may co-exist in a particular site. For example, at Gingin, two species, *A. humilis* (Fig. b) and *A. manglesii* (Fig. c), grow together. (A small number of hybrids are also present.) One of the avian pollinators, red wattlebirds, in one year were observed to visit flowers of both species but visited the flowers of *A. humilis* earlier in the season (during its flowering peak) and later concentrated almost entirely on *A. manglesii* (during its flowering peak). Thus, while there

L. ewingi

L. verreauxi

(a) Oscillogram recording of the mating calls of *Litoria ewingi* and *L. verreauxi* from the region of species overlap

was some overlap between the two species in flowering time, birds tended to visit only one species during each of these periods. The two species differ in flowering stem height and red wattlebirds fed differently on the two species. To obtain nectar from flowers of *A. humilis*, which has shorter flowering stems, birds were able to move about on the ground, whereas to obtain nectar from flowers of *A. manglesii*, which has considerably taller flowering stems, birds perched on the stems. Red wattlebirds presumably increased their foraging efficiency by foraging preferentially on one species or the other. This behaviour would reduce opportunities for interspecific pollen flow.

Mechanical isolation

Barriers to gamete or pollen transfer may arise due to differences in reproductive structures between species. For example, differences in flower size and structure can serve to reduce or prevent pollen transfer between plant species that share the same pollinator(s) (e.g. due to differing placement of pollen on the pollinator). In the

case of *A. humilis* and *A. manglesii*, the latter species has larger flowers and its pollen tends to be placed on the crown and neck of red wattlebirds, whereas pollen of *A. humilis* tends to be placed on the bill and nearby feathers.

Gametic isolation

Gametic isolation may result from a lack of attraction between gametes of different species. In some animals, mortality of male gametes in the reproductive tract of females of a different species may occur prior to fertilisation. In flowering plants, pollen–pistil incompatibility between species may occur. The pollen of one species may not germinate on the stigma of flowers of a different species. Alternatively, the pollen tubes of foreign pollen may display abnormal growth in the style. Abnormal pollen tube development has been observed in some interspecific crosses between various eucalypt species.

Postzygotic isolation

Barriers we have examined up to this point are prezygotic barriers. Postzygotic barriers are manifested in a number of ways. Hybrids may be

(b) *Anigozanthos humilis.* (c) *A. manglesii*

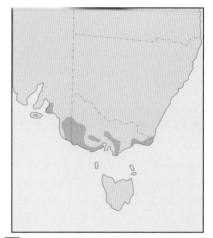

i *Eucalyptus baxteri*

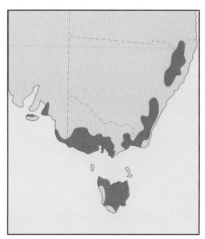

a   ii *Eucalyptus obliqua*

inviable or show reduced viability (e.g. following fertilisation an embryo may fail to develop normally). Hybrids may reach reproductive age but exhibit partial or complete sterility. Hybrid sterility may result from abnormal development of reproductive structures (e.g. anthers may abort) and from production of inviable gametes or gametophytes (e.g. a hybrid plant may produce inviable pollen). Another possibility is that of hybrid breakdown, where hybrids produce offspring that show reduced fitness (e.g. reduced viability or fertility).

Evolution of reproductive isolation

One way in which reproductive barriers are commonly thought to evolve is as an incidental outcome of the divergence and the independent evolution of populations that are geographically isolated. It has also been argued that when partially differentiated populations come into contact, selection may directly favour the evolution of prezygotic reproductive isolation to reduce the production of inferior hybrids. The importance of this mechanism in nature is a matter of considerable debate.

Fig. 32.11 Hybrids occur in some mixed populations of *Eucalyptus obliqua* and *E. baxteri* in coastal Victoria where the species' **(a)** geographic ranges and flowering time overlap. **(b)** Hybrids (centre) have fruits and leaves intermediate between the two parental species, *E. baxteri* (left) and *E. obliqua* (right)

The *cohesion species* concept emphasises mechanisms involved in maintaining the phenotypic cohesion of species. The importance of a wide variety of factors in maintaining the cohesion of species, including gene flow, selection and developmental and historical constraints, is recognised. As in the case of the evolutionary species concept, some biologists have argued that this concept can be difficult to apply.

Thus, the biological, recognition, evolutionary and cohesion concepts recognise species as distinct evolutionary entities but differ with regard to the criteria that are emphasised in defining species. Some biologists think that in view of the diversity of living organisms, it may be unrealistic to expect that a single species concept will be universally applicable to all organisms. However, despite differences in species definitions, models of speciation emphasise restriction of gene flow between populations so that divergence may occur.

> There is not a single species concept that is universally accepted but rather a number of different concepts that differ in their emphasis.

Modes of speciation

We can now reconsider the question posed by Darwin's book *On The Origin of Species by Means of Natural Selection*—how do species originate? Despite the major contributions of Darwin and later researchers, this is still an area of very active research in evolutionary biology. Modern evolutionary biologists place great importance on the question of how reproductive barriers arise in the formation of new species. They also emphasise the different geographical contexts—allopatric, parapatric, and sympatric—in which speciation can take place.

Allopatric speciation

In nature, species are made up of local populations. Characteristics of individuals, such as their dispersal ability, and the spatial relationships of populations within species will influence the extent of gene flow between different populations. The amount of gene flow between populations within a species is one factor that influences the degree to which differentiation between populations may occur. Opportunities for gene exchange between members of populations that are widely separated may be limited or non-existent. Furthermore, populations of a widespread species may occur over a range of different environments. If different populations are exposed to different environmental conditions, they may develop different adaptations and diverge over time, giving rise to variation within species on a geographical scale.

The most widely accepted mode of speciation is that of **allopatric speciation** (Fig. 32.12), whereby populations of an ancestral species become geographically

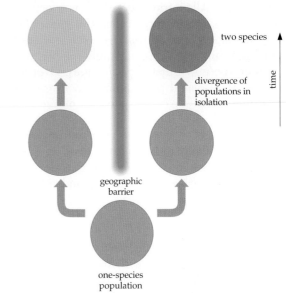

a Allopatric speciation

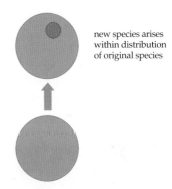

b Sympatric speciation

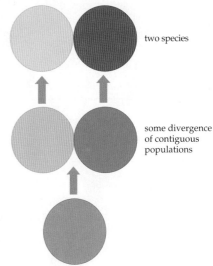

c Parapatric speciation

Fig. 32.12 Models of speciation

isolated and diverge to such an extent that they become reproductively isolated and ultimately speciate. Both natural selection and genetic drift may play a role in the divergence of separated (allopatric) populations. Allopatric speciation may occur if the distribution of a widespread ancestral species becomes split by some *geographic barrier*, such as a desert or a mountain range, which completely inhibits migration between the now isolated populations. Over a period of time, the isolated populations may diverge, so that eventually two distinct species emerge that are incapable of freely interbreeding if they subsequently come in contact if conditions change. Another way in which geographic isolation may arise is if isolated populations become established outside the main range of a species by colonising individuals. Such isolated populations may be exposed to different environmental conditions and selective pressures. If peripheral populations are small, they may also be subject to divergence via genetic drift.

Allopatric speciation is thought to have occurred among rock wallabies of the genus *Petrogale* (Fig. 32.13a). The habitat of these animals consists of rocky outcrops, which are frequently separated from other areas of suitable habitats by large distances (Fig. 32.13b). Wallabies are small animals with a tight social structure, and there may commonly be restricted dispersal between such disjunct populations. No less than 20 chromosomally and genetically distinct forms, 11 of which are classified as species, are recognisable

within Australia. Many of the lineages are reproductively isolated, although some species hybridise where their ranges currently overlap.

Wallaroos and euros are much larger relatives of the rock wallabies and also are adapted to similar habitats—rocky outcrops. However, these animals disperse widely and are less likely to form isolated populations. Only a few subspecies of *Macropus robustus*, including the wallaroo *M. robustus robustus* and the euro *M. robustus erubescens*, are recognised, and there is considerable intergradation between them.

For species with good powers of dispersal, major geographic barriers may be more necessary for isolation and divergence to occur. The eastern grey kangaroo, *M. giganteus*, is found on the east coast of Australia, whereas the related western grey, *M. fuliginosus*, is found in southern and south-western regions. Their distributions overlap in western Victoria and New South Wales but it seems likely that they evolved in isolation in the eastern and western parts of the continent respectively and subsequently expanded their range. Numerous other organisms have eastern and western counterparts, including bandicoots of the genus *Perameles*, frogs of the genus *Litoria*, whipbirds and plants such as banksias.

> Geographic isolation provides an effective mechanism for preventing gene exchange between populations. Divergence of geographically isolated populations may ultimately lead to speciation (allopatric speciation).

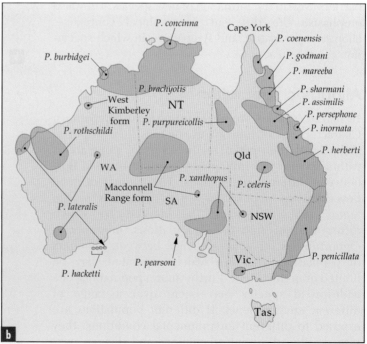

Fig. 32.13 (a) A typical rock wallaby of the genus *Petrogale* on a rock outcrop. **(b)** Distribution of rock wallabies in Australia, which suggests an allopatric model of speciation

Sympatric and parapatric speciation

Some biologists have argued that geographic isolation is necessary for speciation because all but a small amount of migration or exchange of genes between populations (gene flow) would prevent divergence. Other modes of speciation have been proposed that do not require full geographical isolation. In the case of one such alternative, **sympatric speciation**, it is proposed that populations can diverge and speciate without any geographic separation (Fig. 32.12b). One mechanism of sympatric speciation involves the formation of polyploids, which have more than two sets of chromosomes. **Polyploidy** is common in plants and has been important in plant evolution, but is much less common in animals.

One type of polyploidy that may result in speciation is **autopolyploidy**. In autopolyploids the chromosome sets originate from the same species. For example, in a population of diploid ($2n$) individuals, some diploid gametes may be produced as a result of abnormal meiosis (Chapter 8). Subsequently, two diploid gametes may unite to form an offspring that is an autotetraploid ($4n$). Autotetraploids may experience problems during meiosis, and hence reduced fertility, as four chromosomes may associate during meiosis rather than the normal pair in a diploid. This problem may be circumvented if, for example, the tetraploids can reproduce asexually.

In another type of polyploidy, **allopolyploidy**, the sets of chromosomes in the polyploid originate from different species. For example, where two related diploid species, with chromosome sets AA and BB respectively, coexist, AB hybrids may be produced. These hybrids may be sterile due to problems of pairing of A chromosomes with B chromosomes. In such cases fertility would be restored if chromosome doubling occurs and an allotetraploid ($AABB$) is produced. In the allotetraploid, A chromosomes will pair only with A chromosomes and B with B, and regular meiosis can occur so that normal sexual reproduction is possible. The allotetraploids ($AABB$) produced in this way usually cannot successfully reproduce with either their AA or BB parents because when they cross with them they will produce triploids that typically are sterile or have low fertility. Thus, the allotetraploids will be reproductively isolated from both parental diploid species. In this way, new allopolyploid species may be formed.

Polyploidy is common in the grass family, where about 70% of species are believed to be polyploid. Polyploidy has occurred in kangaroo grass, *Themeda triandra*, in which diploid ($2n = 20$) and polyploid ($4n = 40$) forms are found (Fig. 32.14). The most famous examples of allopolyploidy are found in wheat, another member of the grass family. Bread wheat, *Triticum aestivum*, is hexaploid, with 42 chromosomes

Fig. 32.14 Kangaroo grass, *Themeda triandra*, in which there are diploid and polyploid forms

derived from three diploid ancestors (Fig. 32.15). Two diploid ancestors, each with a diploid number of 14, initially hybridised to form a tetraploid species, which hybridised with another diploid to generate the hexaploid.

Another way in which sympatric speciation might occur is if populations of an ancestral species specialise on different resources, and perhaps as a result they

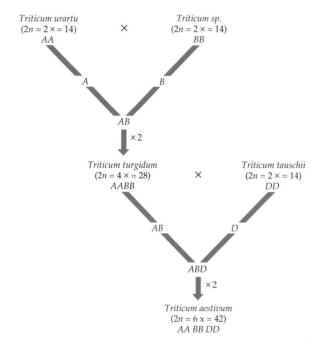

Fig. 32.15 Origin of the genomes of hexaploid wheat, *Triticum aestivum*, by hybridisation and chromosome doubling. The *A* genome is derived from *T. urartu*, the *B* genome from *T. speltoides* or a related species and the *D* genome from *T. tauschii*

also may begin to breed in different local habitats or at different times so that gene flow is restricted. Thus, divergence may occur without the need for geographic isolation. Evidence that has been considered suggestive of how sympatric speciation might occur is provided by host races of the North American fruit fly, *Rhagoletis pomonella*. Host races are confined to different host plant species, such as apples and hawthorn. These host races are believed to have arisen in the same geographical area. Divergence is thought to have been possible because populations became established on a new host (apple), which differs somewhat in the timing of fruit production. Populations of flies adjusted to the timing and life cycle of their host diverged and are now recognised as separate host races. It has been proposed that such a process could ultimately lead to speciation. It has been suggested that this type of sympatric speciation may have played a role in the evolution of some groups of host-specific, plant-feeding insects. However, the relative evolutionary importance of this type of sympatric speciation remains a matter of debate. For proposed cases of this type of sympatric speciation, it is generally difficult to exclude the possibility that the population divergence and speciation could have taken place during a period of geographic separation that was not observed and that the two species have subsequently come into contact. In other words, in such cases it is still possible to propose a plausible allopatric model of speciation.

Parapatric speciation is a mode of speciation in which divergence occurs among populations that have contiguous distributions and hence that are incompletely geographically separated (Fig. 32.12c). As a result of divergent selection, gene flow between the populations becomes progressively reduced and eventually the differentiated populations become reproductively isolated. In practice, it is difficult to determine whether speciation has occurred allopatrically with subsequent (secondary) contact or parapatrically without a period of complete geographical separation, and the evolutionary importance of parapatric speciation is a matter of debate.

> S ympatric and parapatric modes of speciation assume that speciation takes place without complete geographical isolation. Mechanisms of sympatric speciation proposed include sympatric speciation via polyploidy or by populations specialising on different resources and diverging.

Speciation and chromosomal rearrangements

There are many cases where related species differ by chromosomal rearrangements but are otherwise very

similar genetically. Several examples are found among the rock wallabies, *Petrogale*. The low level of genetic divergence indicates that the species may have been isolated for a short time. Furthermore, the chromosomal rearrangements that distinguish the species are often 'incompatible' and cannot persist in the same population for a long period of time. An individual with both forms of the chromosomes will be less fit than an individual with only one of the forms. An Australian geneticist, the late M. J. D. White, proposed that chromosomal rearrangements have played an important role in establishing reproductive isolation between some species. A brief period of isolation of a small population may have permitted a new chromosomal mutation to become fixed, which has contributed to reproductive isolation when contact was re-established with the original population.

Hybridisation and asexual reproduction

In some animal and plant groups, hybrids may be formed that persist and spread by asexual reproduction. In some animal species, *parthenogenesis* (a type of asexual reproduction in which eggs give rise to embryos without fertilisation) bypasses meiosis and problems of chromosome pairing in hybrids (Chapter 15). Gecko lizards belonging to the *Heteronotia binoei* complex are widespread in Australia. This complex is considered to consist of several unnamed, chromosomally distinct diploid species and a triploid ($3n$) parthenogenetic form, which occur in arid western and central Australia. The triploid all-female form appears to have arisen as a result of repeated hybridisation events between two chromosomally distinct sexual forms, forming a diploid hybrid occasionally producing diploid eggs (Fig. 32.16). The hybrids then mated with one of the parental species to produce a triploid hybrid form that is self-perpetuating by parthenogenesis; triploid females produce triploid female offspring. The Australian morabine grasshopper, *Warramaba virgo*, arose in a similar fashion from two sexually reproducing ancestral species but the parthenogenetic species in this case is diploid.

Although parthenogenetic forms may be highly adapted and very successful under a particular set of environmental circumstances, their long-term evolutionary potential may be limited, perhaps because asexual forms do not produce a diverse array of offspring genotypes. Consequently they may be more susceptible to attack by parasites or other biological enemies. In *H. binoei*, the parthenogenetic forms carry a greater load of ectoparasitic mites and are more often infected than the sexual forms.

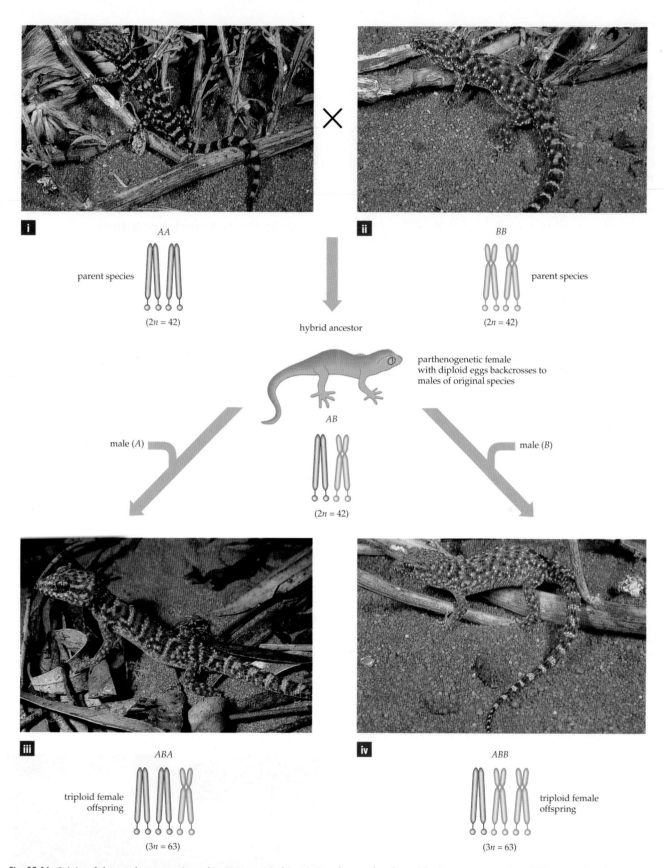

Fig. 32.16 Origin of the parthenogenetic gecko, *Heteronotia binoei*. Two forms of gecko, with chromosomes *AA* and *BB*, crossed to form a hybrid, *AB*. At least some female hybrids produced eggs with diploid sets of chromosomes. When these females back-crossed with males of the original two forms, their diploid eggs, *AB*, were fertilised with haploid sperm (either *A* or *B*), resulting in triploid offspring (*ABA* and *ABB*), which were all female

Rates of evolution

There are differing views regarding the rates and uniformity with which evolutionary change occurs, and controversy about rates of evolutionary change dates back to the time of Darwin. With respect to speciation, one view, **gradualism**, proposes that evolutionary change and speciation occur slowly and gradually over time. Another view, **punctuated equilibrium**, proposes that species undergo rapid evolution when they are formed and subsequently remain relatively constant over long periods of time, with change limited by the constraints of the developmental system, stabilising selection or the availability of genetic variation. Perhaps in response to a changing environment or as a result of change in small, isolated populations, sudden and dramatic change and rapid formation of new species occurs. In other words, the usual situation of static equilibrium (stasis) is punctuated by a burst of change.

The fossil record provides examples of taxa that have remained morphologically constant over very long periods of time, only to be apparently suddenly replaced by divergent forms. However, because the fossil record is inevitably incomplete and patchy, an alternative interpretation is that the bursts of rapid speciation are more apparent than real. Further, there are examples in the fossil record where gradual morphological change can be observed in series of fossils over a period of geologic time. These viewpoints represent two ends of a spectrum, and another view is simply that rates of evolutionary change and speciation vary among organisms and through time.

There are also differing views regarding rates of evolution of characters. One view proposes that evolutionary change in most characters tends to occur slowly and gradually over time. Yet complex or highly specialised features of organisms may initially seem difficult to imagine as evolving by gradual steps, giving rise to questions about the possible use of features such as half-developed eyes or lungs. Darwin himself was troubled initially about how to explain the evolution of complex eyes but realised that even small intermediate changes can be advantageous. There are many forms of eyes, ranging in complexity from simple eyespots, which detect the presence of light, to complex eyes, which can focus and perceive differences in colour. Also, a feature may have one function in the early stages of its evolution and another function in later stages. The swim bladder, which is found in many modern fishes, probably functioned as a lung in the ancestral fish. Another view of evolutionary change, saltationism, is that features may arise and evolve suddenly through mutations that have a drastic effect and that such evolutionary changes are important in the origin of new taxa. However, many biologists argue that such drastic changes are usually likely to be disadvantageous and therefore of relatively minor evolutionary significance.

> Gradualism is the idea that evolutionary change occurs slowly by the accumulation of small changes over a long period of time. One view of speciation is that it occurs slowly and gradually. An alternative view, that of punctuated equilibrium, is that speciation occurs rapidly followed by long periods of time in which there is relatively little change (stasis).

Development and evolution

We have considered above how complex features may evolve gradually and how features may change function as they evolve. Mutations that change the *timing* or *rate of developmental processes* in relation to each other can lead to the appearance of quite markedly different forms. Developmental mutations may generate organisms with novel characteristics that are subject to natural selection. Such mutations may be quite minor at the DNA level, involving little nucleotide change, but may have dramatic consequences at the level of the organism. An example of such a developmental change can be seen in axolotls. Axolotls reproduce while retaining the aquatic, gilled larval stage, unlike most other salamanders, which metamorphose and reproduce as terrestrial amphibians with lungs (Fig. 32.17). Although axolotls retain the potential to develop into the terrestrial form, since metamorphosis can be induced experimentally by the hormone thyroxine and terrestrial forms sometimes occur in nature, natural selection appears to have favoured the aquatic form.

The retention of juvenile features in reproducing adults is termed **neoteny**. Some biologists have suggested that neoteny has played an important role in human evolution. Although on genetic criteria humans are very closely related to chimpanzees and gorillas, we are quite distinct in external appearance. Many of our distinguishing features may have arisen as a result of retention as adults of juvenile characteristics. These include a high ratio of brain weight to body weight, little hair, thin skull bones, small jaws and late eruption of teeth.

Fig. 32.17 Axolotls are neotenous, being reproductive as juvenile (larval) forms

Summary

- Evolution is the process of genetic change in populations and taxa. Two models of evolution are recognised: selective and neutral evolution.

- Selective evolution is evolutionary change that results from variation among organisms in their relative ability to survive and reproduce. Those individuals most suited to a particular environment leave more descendants and are 'naturally selected'. Populations exposed to different environmental conditions may diverge and become adapted to their particular environments. This divergence may ultimately result in the formation of new species (speciation).

- Neutral evolution refers to evolutionary changes that occur as a result of chance processes. Changes in allele frequencies due to chance are called genetic drift. Genetic drift is particularly important in small populations.

- Mutations provide the variation that underlies both selective and neutral evolution. Point mutations are small changes in DNA sequences. Chromosome mutations involve structural changes in segments of DNA or whole chromosomes during cell division. They include duplications and deletions, inversion and translocation of segments, and the fusion and fission of whole chromosomes.

- Genes often occur in more than one allelic form in a population (polymorphism). The frequencies of each genotype can be predicted from allele frequencies, based on the Hardy–Weinberg principle, if the following assumptions are met: the population must be large; individuals must mate at random; and genotypic frequencies must not be subject to the effects of genetic drift, differential mutation, migration or natural selection.

- Evolutionary change in a population can be viewed as resulting from changes in allele frequency. Such changes can result from genetic drift, differential migration of genotypes, and differences in Darwinian fitness between genotypes.

- Various species concepts have been proposed which differ in their emphasis. For example, the biological species concept emphasises the importance of reproductive isolation in maintaining species integrity.

- Divergence of populations and speciation is affected by levels of gene flow and interbreeding, with restriction of gene flow between populations favouring divergence. Geographic isolation provides an effective mechanism for preventing gene exchange. Divergence of geographically isolated populations may ultimately lead to speciation (allopatric speciation). In sympatric speciation, populations diverge without geographic separation (e.g. via polyploidy or by populations specialising on different resources). In parapatric speciation, divergence between contiguous populations occurs.

- Gradualism is the idea that evolutionary change occurs slowly by the accumulation of small changes over a long period of time. One view of speciation is that it occurs slowly and gradually, while an alternative view (punctuated equilibrium) is that speciation involves periods of sudden changes between which taxa remain constant for long periods of time.

key terms

adaptation (p. 834)	evolution (p. 834)	neoteny (p. 856)	selective evolution
adaptive evolution	fitness (p. 839)	neutral evolution	(p. 834)
(p. 834)	gene flow (p. 838)	(p. 843)	speciation (p. 847)
allele frequency	gene pool (p. 847)	parapatric speciation	stabilising selection
(p. 838)	genetic drift (p. 839)	(p. 854)	(p. 847)
allopatric speciation	genotype frequency	point mutation	sympatric speciation
(p. 851)	(p. 838)	(p. 836)	(p. 853)
allopolyploidy	gradualism (p. 856)	polymorphism (p. 838)	transposable element
(p. 853)	Hardy–Weinberg	polyploidy (p. 853)	(p. 837)
artificial selection	frequencies (p. 839)	population (p. 834)	wild-type allele
(p. 835)	hybrid (p. 847)	punctuated equilibrium	(p. 838)
autopolyploidy	Lamarckism (p. 835)	(p. 856)	
(p. 853)	mutation (p. 836)	reproductive isolating	
chromosome mutation	natural selection	mechanism (p. 848)	
(p. 836)	(p. 834)		

Review questions

1. Darwin based his case for evolution by natural selection on a number of observations. One of these is that organisms have the potential to produce many more organisms than they actually do. Explain why this is central to his argument.

2. Explain what is meant by the terms biological fitness and adaptation. Give an example of each.

3. Mutations are the 'raw material' for natural selection to operate on. Explain why the fate of a mutation will depend on:

   (a) chance

   (b) whether the mutant gene is dominant, recessive or intermediate

   (c) whether its effects are harmful, beneficial or neutral.

4. Are changes in allele frequency always due to the action of selection? Briefly explain your answer.

5. Which of the following is not an assumption of the Hardy–Weinberg principle?

   A large population size

   B no migration

   C presence of natural selection

   D random mating

6. Of the five assumptions that underlie Hardy–Weinberg genotype frequencies, which assumptions break down in populations showing:

   (a) genetic drift?

   (b) assortative mating?

   (c) reduced reproductive success of one genotype?

7. Which of the following species concepts places particular emphasis on reproductive isolation between species?

   A taxonomic species concept

   B evolutionary species concept

   C cohesion species concept

   D biological species concept

8. Summarise and contrast allopatric and sympatric modes of speciation, giving an example of each.
9. Explain how new species may arise via autopolyploidy and allopolyploidy.

Extension questions

1. How has the application of modern molecular genetics enhanced our ability to study evolutionary change?

2. Soils surrounding copper mines have a high level of copper, which is generally toxic to plants. Some grass species have been able to establish on copper mine sites by natural spread of seed from surrounding areas. It has been shown that grass plants established on a copper mine site (population A) have a much higher tolerance to copper than their relatives on a non-toxic adjacent soil (population B). The copper-tolerant plants (A) also tend to be slower growing than B. Explain these observations in terms of genetic variation within populations and selection. If there is gene flow between populations A and B, via pollen and seed dispersal, why does the population differentiation not break down?

3. Discuss the possible role of developmental mutations in the origin of novel forms.

4. 'Knowledge of the mechanisms of evolution is useful in applied biology.' Discuss this statement in reference to pesticide resistance and methods of genetic control of pest species, such as insects that damage crops.

Suggested further reading

Darwin, C. (1859). *On The Origin of Species by Means of Natural Selection.*

Six editions were produced during Darwin's life and many afterwards. Read any edition in your library; one by Penguin is readily available.

Dawkins, R. (1986). *The Blind Watchmaker.* London: Penguin Books.

Dawkins, R. (1996). *Climbing Mount Improbable.* London: Viking.

Lucidly written and engaging discussions of natural selection and the evolution of complex adaptations.

Desmond, A. and Moore, J. (1992). *Darwin.* London: Penguin Books.

Highly readable account of Darwin's life, which places Darwin's works in the context of his society and times.

Jones, S. (1999). *Almost Like a Whale.* London: Doubleday.

Stimulating and entertaining rewrite of On The Origin of Species *in the light of modern knowledge of evolutionary biology.*

Patterson, C. (1999). *Evolution.* 2nd edn. London: Natural History Museum.

Succinct, readable, modern account of genetic theory, molecular evolution, origin of species, evolution and humanity, evolution as science.

Futuyma, D. (1998). *Evolutionary Biology.* 3rd edn. Sunderland, MA: Sinauer.

Ridley, M. (1996). *Evolution.* 2nd edn. Cambridge, MA: Blackwell Science.

Skelton, P. ed. (1993). *Evolution. A Biological and Palaeontological Approach.* Wokingham, UK: Addison-Wesley.

These three advanced texts all provide clear and comprehensive treatments of current concepts in evolutionary biology.

CHAPTER
33

Bacteria

Around three hundred years ago, the Dutch scientist Antonie van Leeuwenhoek first peered through his new invention, the microscope. Looking at a drop of pond water, he saw what he called 'little animalicules' vibrating and swimming within. Many of these microscopic organisms, we now understand, were bacteria. For a long time bacteria were classified as microscopic animals or plants and it was not until the 1960s that they were discovered to be a quite different form of life. We now know that **bacteria** are the smallest forms of *cellular* life on earth (Fig. 33.1). (As the figure shows, and as will be explained in Chapter 34, viruses are generally smaller than bacteria but are not cellular organisms. The life plan of viruses is quite different from that of cells.)

Bacteria generally exist as single cells, with simple shapes (rods, spheres, spirals, filaments; Fig. 33.2). Individually they can be seen only with a microscope because they are only 0.4–5 μm long. Their presence is often detected and their biochemical properties measured as a result of extensive 'colonial' growth, involving many millions or billions of cells grown together in culture (i.e. in liquid media or on agar gels containing the necessary nutrients for a particular bacterium).

Bacteria may have beneficial, harmful or indifferent effects on other organisms, including humans. Certain species of bacteria are the agents of many important industrial and agricultural processes, while others are the cause of many medically and agriculturally important diseases. Their greatest significance to life on earth is in the enormous variety of natural biochemical transformations that they carry out, transformations of material and energy that go on in soil and water, and in and on animals and plants, without which other life on earth would grind to a halt (see, for example, Box 44.4). Some examples of the more obvious beneficial and harmful influences of bacteria on humans are given in Table 33.1 but these hardly begin to describe the immensity and diversity of the impact that bacteria have as ubiquitous members of the biosphere.

Bacteria are prokaryotes

The way scientists perceive and classify living things has changed remarkably in the past few decades. The classification of cellular organisms (microscopic and macroscopic) into **prokaryotes** and **eukaryotes**, and the further division of prokaryotes into two major evolutionary subgroups, has greatly clarified our understanding of the deepest roots of cellular evolution, extending back 3.5 billion years, about a billion years after the formation of planet earth itself.

The application of modern techniques such as electron microscopy showed that the internal structure of prokaryotic cells differs from that of eukaryotic plant or animal cells (Chapter 3). Prokaryotic cells are not only smaller than eukaryotic cells, but they are simpler in structure (Fig. 33.2); most prokaryotes have a semirigid cell wall, located just outside their plasma membrane; they generally have no membrane-bound organelles— no nucleus (hence their name: 'pro' meaning before, 'karyote' meaning 'nucleus'), endoplasmic reticulum, Golgi apparatus, mitochondria or chloroplasts. Nor do they have a cytoskeleton of microfilaments and microtubules. Biochemists and molecular biologists have confirmed these structural differences, uncovering important metabolic and genetic differences between prokaryotes and eukaryotes and between major groups of prokaryotes.

Despite the important structural and metabolic differences between these major evolutionary cell types (as well as viruses), the differences are much less significant than the similarities—all life on earth is based on similar biochemical patterns of metabolism and synthesis, and the expression and replication of genetic information. This similarity is referred to as the *biochemical unity of life*.

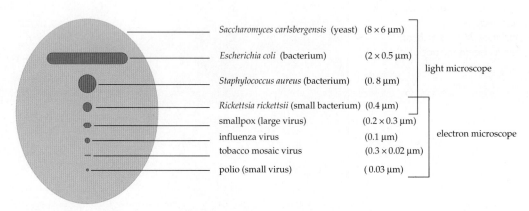

Saccharomyces carlsbergensis (yeast)   (8 × 6 μm)

Escherichia coli (bacterium)   (2 × 0.5 μm)

light microscope

Staphylococcus aureus (bacterium)   (0.8 μm)

Rickettsia rickettsii (small bacterium) (0.4 μm)
smallpox (large virus)   (0.2 × 0.3 μm)
influenza virus   (0.1 μm)
tobacco mosaic virus   (0.3 × 0.02 μm)
polio (small virus)   (0.03 μm)

electron microscope

Fig. 33.1 Relative sizes of some microbes and effective range of microscopes. (1 μm = 10^{-6} m)

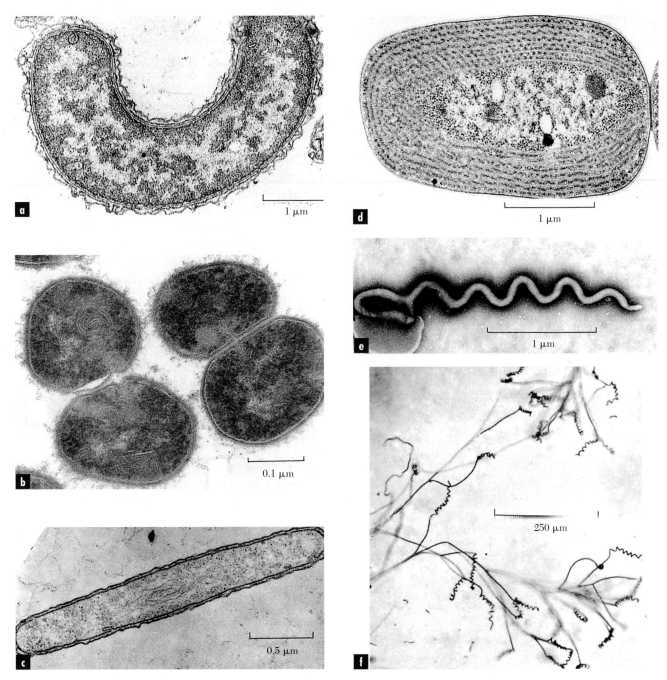

Fig. 33.2 Examples of bacterial cells and cell shapes: **(a)** a circular or helical rod, *Microcyclus major*; **(b)** a dividing coccus or sphere, *Gemella haemolysans*; **(c)** a long rod, *Nitrospina gracilis*; **(d)** one cell of *Plectonema* sp., which is a cyanobacterium containing stacked photosynthetic membranes; **(e)** a spiral bacterium, *Spiroplasma* sp., and **(f)** a filamentous type, *Streptomyces* sp. **(a)–(d)** Electron micrographs of cut sections of cells lightly stained with an electron dense substance such as lead; **(e)** is an electron micrograph of an uncut and negatively stained cell (stained with a uranium salt to stain the immediate surroundings more intensely than the cell itself); and **(f)** is a light microscope image of an unstained live sample

Bacteria are the smallest forms of cellular life on earth. They are prokaryotes, which differ in important ways structurally, biochemically and metabolically from eukaryotes. These differences, however, are not as great as the common features of life's blueprint in the three major cell groups that evolved, suggesting that cellular life on earth had a common origin but then evolved along different pathways.

Bacteria were the first cellular life on earth

The formation of cellular structures that contained living cytoplasm was an early event in the evolution of life, probably because the cell unit provided the means to keep everything in one place and at a concentration

Table 33.1 Activities of bacteria that benefit or harm humans

Beneficial activities	Harmful activities
Food production and technology	
Cheese and yoghurt	Spoilage of fresh foods and beverages
Meat preservation	Food poisoning
Vinegar	
Wine and beer	
Vitamins, amino acids	
Medicine	
Synthesis of the majority of antibiotics (some are also produced by fungi)	Causation of: cholera, typhoid, dysentery, typhus, meningitis, gonorrhoea, syphilis, pneumonia, toxic shock, tetanus, botulism, tuberculosis, legionnaires' disease, wound infections and many other important infectious diseases
Agriculture	
Biological insect control	Many animal diseases (mastitis, foot rot, fly strike,
Nitrogen fixation in soil	botulism, lumpy jaw, abscesses, wound infections)
Humus formation in soil (breakdown of dead plant and animal tissues)	Some plant diseases (wilt, damping off, galls)
	Frost damage to plants (by ice nucleation)
Recycling and transformation of carbon, sulfur, phosphorus and nitrogen compounds in soil and water	Nitrogen and sulfur loss from soil (N_2, NH_3, H_2S production from non-volatile compounds of these elements)
Industry	
Sewage treatment and nutrient recycling from wastes	Steel and concrete pipe corrosion (acid-producing bacteria)
Ethanol production from carbohydrates	
Biogas (methane plus H_2) from organic wastes	Contamination of fuels and oils

high enough for the biochemical reactions of life to work efficiently.

By 3.5 billion years ago, parts of the earth's surface had cooled enough to permit biochemical reactions to take place in liquid water, that is, below about 100°C, although temperatures were generally hotter and seas saltier than today. The earliest microscopic cellular life resembled forms similar to some modern bacteria. These early lines of prokaryotes continued to evolve for almost three billion years before multicellular macroorganisms appeared—geologically speaking as 'recently' as 0.6 billion years ago (see Chapter 31). It is interesting that, among existing bacteria, we still see many species that thrive under the hot and salty conditions that probably characterised early living conditions on earth, conditions that most other modern bacteria and eukaryotes cannot tolerate (p. 873).

Early photosynthetic bacteria

Two other features of the early physical and chemical environment on earth were important in the origin and early evolution of life. Firstly, the primeval atmosphere was anaerobic (without molecular oxygen), though much atomic oxygen was present as oxides and carbonates in rocks and as carbon dioxide in the atmosphere. The general conditions in the atmosphere were likely to have been highly reducing, perhaps with an abundance of the reduced forms of hydrogen (H_2), nitrogen (NH_3) and sulfur (H_2S). Second, the earth was bathed in a solar radiation far more intense than that which exists today. Earth's atmosphere then contained little ozone, which today forms a layer high in the atmosphere, absorbing a large fraction of incoming ultraviolet (UV) radiation. This protects life on earth by limiting the lethal and mutational damage of UV radiation on living cells.

Since there was at that time a limited supply of chemical energy in the form of organic carbon compounds ('fixed' carbon), and little oxygen or other oxidants, the evolution of photosynthesis to trap light energy was an important part of the early history of life. Early photosynthesis was anoxygenic (not generating oxygen); in fact, oxygen would have been very toxic to the early **photosynthetic bacteria**. Bacteria that carry out **anoxygenic photosynthesis** also exist today and are thought to retain features of those older bacteria. Furthermore, because the amount of UV radiation arriving at the earth's surface was high, photo-synthesising bacteria either must have been equipped in some way to prevent or overcome damage from UV radiation, or they evolved ecological strategies that allowed them to avoid direct sunlight. For example, they may have lived in deep aquatic habitats, which filter out damaging UV radiation without affecting the visible radiation needed for photosynthesis, or shaded habitats, out of the direct rays of the sun, with light scattering removing most of the UV radiation but not the visible component.

The earth's surface has cooled considerably since those primeval days. Molecular oxygen has also appeared in abundance in the atmosphere as a consequence of the development of **oxygenic photo-synthesis**, a process that probably first appeared in bacteria (as the forerunners of the modern group known as cyanobacteria) about 2.5 billion years ago. This type of photosynthesis, in which oxygen is a product of the overall reaction to fix carbon dioxide as sugar (Chapter 5), is also characteristic of modern eukaryotic algae and plants because the chloroplasts in which this photosynthesis occurs are themselves the evolutionary remnants of symbiotic cyanobacteria. (This is the endosymbiosis theory: that certain specialised bacteria formed symbiotic relationships with primitive eukaryotes, ultimately to become important cellular organelles; Chapters 3 and 35). The appearance of abundant molecular oxygen in the atmosphere would have permitted the evolution of cellular metabolic pathways for the oxidative meta-bolism of organic carbon and reduced inorganic substrates (e.g. H_2, H_2S, NH_3, Fe^{2+}) of the sort that are so common in most living organisms today.

Two major evolutionary lineages of bacteria—super kingdoms Bacteria and Archaea

Except for massive colonial aggregations, as seen, for example, in stromatolites (Chapter 31), micro-organisms rarely leave any fossil record because they have no hard structures that will calcify, as do bones, and because their size makes them difficult to distinguish among sedimentary rock grains and particles. In the absence of such morphological evidence, DNA and RNA molecules of modern types of bacteria have been used to discover evolutionary relatedness and thus to devise an evolutionary 'tree'. In Chapter 30 we saw that by using the sequences of bases in DNA or RNA, an evolutionary tree can be constructed, which summarises how different groups (taxa) are related to one another and thus how far back in time the evolutionary paths of modern taxa diverged from each other.

The comparison of nucleic acid sequences by molecular methods has brought about a revolution in the way we now perceive the evolution of cellular life on this planet. Figure 33.3 summarises the view that is now generally held about the evolutionary relationships between prokaryotes (bacteria) and eukaryotes. This form of evolutionary tree is 'unrooted' and is like a real tree viewed from above. We can see the twigs and branches but we can infer little about the root of the tree, that is, the nature of the ancestral taxon that gave rise to all cellular forms of living organisms now known to us (and in which the mostly common biochemical mechanisms of all living things on earth had their beginning).

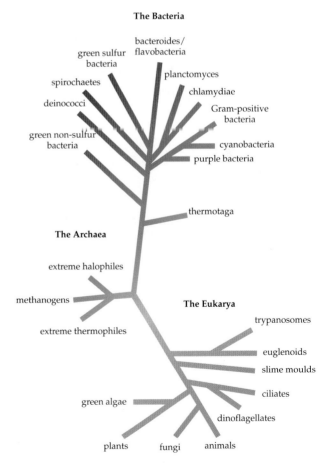

Fig. 33.3 The broad evolutionary relationships of prokaryotes and eukaryotes, showing two major groups of bacteria—Bacteria and Archaea. The root of the tree is unknown. The analysis is based on ribosomal RNA base sequences

The phylogenetic tree shows three major branches of cellular life. The prokaryotes divide into two major lineages, the **Bacteria** (previously called the Eubacteria or 'true' bacteria) and the **Archaea** (meaning 'very old life forms' and previously called the Archaeobacteria); the other lineage includes all the eukaryotes, collectively called the Eukarya. These three groups are sometimes referred to as super kingdoms (or domains), but the more conventional, five kingdom system of classification combines the Bacteria and the Archaea together as the kingdom Monera, and splits the Eukarya into the four kingdoms Protista, Plantae, Fungi and Animalia (see Chapter 30).

Interestingly, the Archaea appear to be more closely related in evolutionary terms to the Eukarya than to the Bacteria. Comparison of DNA sequences also confirms the endosymbiotic origin of mitochondria and chloroplasts of modern eukaryotes.

Many other biochemical and metabolic features of organisms support the recognition of the three super kingdoms (Table 33.2). These biochemical differences have important applications. Pathogenic species of the Bacteria, which cause diseases of humans and important agricultural plants and animals, can be chemically targeted without harming the cells of their eukaryotic hosts. The search for such target-specific agents (or 'magic bullets' as they were once known) began almost a hundred years ago and continues actively today. It led to the identification of many naturally occurring inhibitors of bacteria, part of a class of substances called **antibiotics** (see Table 33.2). Antibacterial antibiotics work in practice through inhibiting a range of processes in members of the Bacteria at concentrations that do not affect similar or other processes in eukaryotes. Perhaps the best known example of this is the use of the penicillins and cephalosporins, targeted at the synthesis of cell wall peptidoglycan, which occurs only in the Bacteria. The synthesis of DNA, RNA and protein in the Bacteria can likewise be differentially inhibited by such antibiotics as nalidixic acid, rifampicin, and streptomycin or tetracycline, respectively. Thus the antibacterial antibiotics are not toxic to eukaryotes, such as ourselves, but will inhibit or kill bacteria within or on us.

Table 33.2 A comparison of structural and biochemical features of typical members of the three super kingdoms Bacteria, Archaea and Eukarya

Characteristic	Bacteria	Archaea	Eukarya
Membrane-bound nucleus and organelles	Absent	Absent	Present
Plasmids	Present	Present	Rare
Introns in genes	Absent	Absent	Present
RNA polymerases	Single	Multiple	Multiple
Ribosome size	70 S	70 S	80 S
Protein synthesis–initiation tRNA	Formylmethionine	Methionine	Methionine
Protein synthesis sensitive to:			
• diphtheria toxin	No	Yes	Yes
• streptomycin	Yes	No	No
• cycloheximide	No	No	Yes
Peptidoglycan as the major cell wall polymer	Yes	No	No
Membrane lipids	Esters	Ethers	Esters
Methanogenesis	No	Yes (some species)	No
Nitrification	Yes (some species)	No	No
Nitrogen fixation	Yes (some species)	Yes (some species)	No
Photosynthesis	Yes (some species)	No	Yes (some species)
Chemoautotrophy on H_2, S, Fe^{2+}	Yes (some species)	Yes (some species)	No

Phylogenetic classifications based on comparison of DNA sequences recognise three major groups of cellular life, sometimes considered super kingdoms. They are two groups of bacteria (prokaryotes)—the Archaea and the Bacteria—and the Eukarya (all eukaroytes).

Classifying and identifying bacteria

The problems for the microbiologist of identifying and classifying bacteria are of a different dimension to those confronted by botanists or zoologists studying plants and animals. This is because bacteria offer few morphological criteria, apart from cell size and shape, cell motility and *staining reactions* of cells. While these limited morphological criteria are useful, most of the tests traditionally used in the identification and classification of bacterial species are biochemical, physiological or immunological (Box 33.1), features that are subject to a considerable degree of genetic variation. The propensity of bacteria to mix genes, to vary genes, and to gain and lose genes (see pp. 881–884) must be taken into account and classification schemes must not be so rigid as to exclude this sort of natural variability.

From extensive testing, bacteriologists have constructed systematic identification procedures that are convenient and practical and therefore useful for laboratories where it is necessary, for example, to identify a pathogen quickly in order to deal with the disease that it causes. For such utilitarian purposes, an evolutionary classification is not important. What is important is to identify the organism quickly so that the nature of the disease can be assessed and treatment begun. This also applies to identifying bacteria that inhabit a particular ecosystem and pose a health risk, such as a city beach used for recreation, a country stream used to supply town drinking water, or the cooling system of a large air-conditioning system in a city building.

Molecular methods, based on nucleic acid probe hybridisation (Chapter 13), the polymerase chain reaction (Chapter 13) or specific antibody reactions (Chapter 23), are being used increasingly for more rapid and specific identification of bacteria. These methods detect unique genes (as DNA or RNA sequences), or specific external proteins or polysaccharides on bacterial cells, often against a high background of other contaminating strains or species.

Bacteria are a diverse group and phenotypic classifications are used for practical purposes of identification.

The most commonly used taxonomic catalogue of bacteria, Bergey's classification, is outlined in Box 33.2. The sections listed are based on classic phenotypic and ecological characteristics and are not claimed to show evolutionary relationships. The phylogeny shown in Figure 33.3, by contrast, is based on genetic criteria (nucleic acid sequences) and thus shows evolutionary relationships clearly and precisely. However, the

---

BOX 33.1 Identifying bacteria found on or in humans

1. Isolation

Purify unknown bacteria (from a throat swab or contaminated wound, for example) by plating onto agar-gel growth medium (Fig. a). Visible colonies appear after incubation at 37°C for one or more days; each colony contains millions of cells that have grown from one original cell. Examine and record the size, shape, texture and colour of colonies (Fig. b).

2. Staining and microscopy

Gram stain with crystal violet and iodine, wash with acetone–alcohol and counterstain with carbol fuchsin (purple—**Gram-positive bacteria**; pink—**Gram-negative bacteria**; Fig. c). Examine cell shape (coccus, rod, spiral, other), cell grouping (chains, tetrads, pairs, clusters) and special structures (spores, capsules). Perform motility testing of live cells. Greater detail of cell size, shape and structure can be obtained using electron microscopy.

3. Physiological testing

Test for: growth on different carbon sources (e.g. sugars), nitrogen sources (e.g. protein, amino acids, ammonium or nitrate), sulfur compounds; requirement for oxygen; tolerance of salts or high osmolality; optimum temperature for growth.

4. Biochemical and genetic testing

Test for specific enzymes, metabolic pathways, resistance to antibiotics and other inhibitors. Isolate DNA and test for proportion of guanine and cytosine (GC%). Hybridise DNA with DNA from known organisms to estimate genetic similarity. Carry out polymerase chain reaction to identify specific genes or DNA base sequences characteristic of a particular species or strain. Carry out immunological reactions to identify specific external proteins or polysaccharides on bacterial cells.

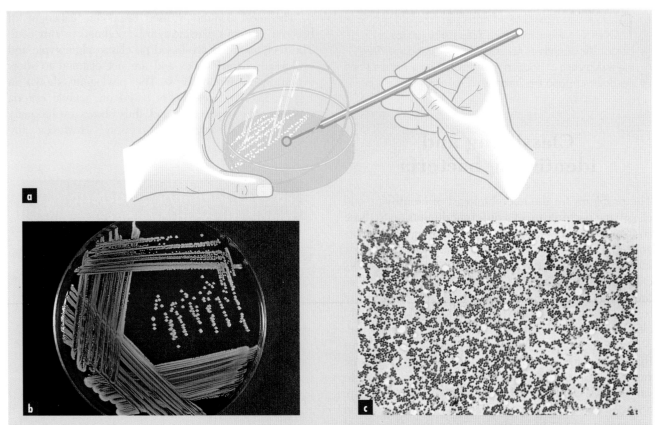

(a) Streaking bacteria to obtain single colonies. Material containing bacterial cells is collected with a sterile wire loop, then streaked onto the agar medium in the petri dish so that individual cells are laid along the surface of the agar. Incubation at growth temperature results in many divisions of these individual cells to give rise to a colony (a million cells or more, all derived from the one individual cell originally at that point on the agar). Each colony is thus a clone (a collection of identical individuals) of the original cell and can be used as pure starting material for the subsequent testing of that bacterium. (b) Individual bacterial colonies of *Staphylococcus* sp. (and confluent or unseparated colonies) generated by dilution streaking. This streaking technique provides clonally pure material for subsequent identification testing. The purity of a culture is also checked by this means. Pure cultures of a bacterium will give rise to colonies of the same size, shape, texture, colour and contour, while mixed or impure cultures give rise to mixtures of colonial types. (c) Stages in the identification of *Staphylococcus aureus*. Bacterial cells in a stained sample of sputum from a pneumonia case. Bacterial cells from these colonies yield Gram-positive (purple) cocci, clustered in bunches, typical of staphylococci. If necessary, scanning electron microscopy can be used to show cell shape in more detail. Further biochemical testing can be done (mannitol fermentation, growth in high concentrations of salt, serum coagulation, catalase activity) to confirm that the isolated bacterium is indeed *Staphylococcus aureus*

evolutionary tree is incomplete and is therefore less detailed than Bergey's traditional classification. It will be interesting to see whether, as more molecular genetic data become available, the two schemes converge and agree on the boundaries of major groups and genera of bacteria.

Bacteria are a diverse group of organisms despite their limited variation in size, shape and other microscopic characteristics. Molecular identification methods are proving to be particularly advantageous where (as in clinical medicine and community health) rapid and precise identification of potentially dangerous bacteria is necessary.

Super kingdom Bacteria

In Box 33.2, all but sections 25 and 33 (the Archaea plus other genera of uncertain type) are members of the super kingdom Bacteria (Fig. 33.3). The Bacteria are an enormously diverse group that share many environments with, or live on and in humans and other animals and plants. These are habitats of moderate temperature with water freely available, low in salt or other solutes, and where sunlight or organic compounds are plentiful. Oxygen is not so important since many of the Bacteria have powerful fermentation capabilities, or can replace molecular oxygen as a metabolic oxidant with oxidised anions such as nitrate and sulfate. Some are also capable of substituting, as a source of energy, reduced inorganic carbon compounds, such as sulfide or ferrous ions, or gaseous hydrogen in place of organic carbon. In fact, the Bacteria contains almost every variety and combination of biochemical energy extraction and carbon fixation that is thought to be feasible on the basis of the molecular composition of the biosphere (those few types not represented in the Bacteria are found instead among the Archaea). These matters are taken up in greater detail below. Only a few plastics and organo-chlorine synthetics used as insecticides and herbicides have proven resistant to the metabolic capabilities of bacteria.

However, this group also contains some of the most fastidious and sensitive cellular organisms known. These are often pathogens that have become highly adapted to special environments within animal or plant hosts. The bacteria causing sexually transmitted diseases in humans (gonorrhoea, syphilis, chlamydial infections) are often as difficult to grow in the laboratory as they are easy to catch in bed! Rickettsias, which cause typhus, cannot be grown away from host cells, and *Legionella pneumophila*, which causes the respiratory infection known as legionnaires' disease, has a fastidious requirement for ferrous ions and organic sulfur compounds when grown alone, yet grows easily in mixed culture with certain free-living protozoans (unicellular eukaryotes) that occur in the soil or in water reservoirs and tanks.

Cyanobacteria—photosynthetic bacteria containing chlorophyll

Cyanobacteria are an important subgroup of the Bacteria. They resemble algae and plants in that they contain chlorophyll *a* and generate molecular oxygen during photosynthesis (Chapter 5). Most other photosynthetic bacteria do not do this. Modern cyanobacteria have evolved from the early photosynthetic bacteria that established the high concentrations of oxygen in the atmosphere. As we saw in Chapter 31, stromatolites, which formed from the activities of large aggregations of cyanobacteria or more primitive filamentous phototropic bacteria, are the oldest fossil evidence of life on earth.

Cyanobacteria occur as individual, often spheroidal, cells or as filamentous aggregates of many individual cells joined end to end. In addition to chlorophyll *a*, cyanobacteria contain water-soluble pigments called **phycobilins**; blue phycobilins present in most species give them a blue-green appearance. A slimy material may be present on the outside of the cell walls, forming a sheath.

Cyanobacteria are often more highly differentiated than other bacteria in having specialised cells such as akinetes and heterocysts (Fig. 33.4). An **akinete** is a **spore** that develops from a cell that becomes enlarged and filled with food reserves. The spore can remain dormant and then germinate to produce a new filament. A **heterocyst** is relatively colourless, has a thick, transparent cell wall, may be involved in asexual

reproduction and (together with other cells) is a site of nitrogen fixation. The ability of some cyanobacteria to fix nitrogen (see Chapter 44) is used in rice cultivation, where the growth of such cyanobacteria is encouraged in rice paddies.

Cyanobacteria also often form dense mats of growth in shallow marine or estuarine environments (Fig. 33.5), or they produce scum-like **algal blooms**, which are extensive surface growths of cells in still lakes or slowly flowing rivers (Box 33.3).

Endospore-forming bacteria

The Bacteria also contain several genera (of which *Bacillus* and *Clostridium* are most familiar) that produce **endospores**, the most resistant form of life known (Fig. 33.6). Endospores are so-called because they form inside the mother cell (Chapter 16) rather than by budding from it, as occurs in spore formation in fungi. They form under conditions of nutrient

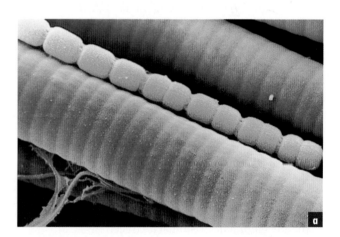

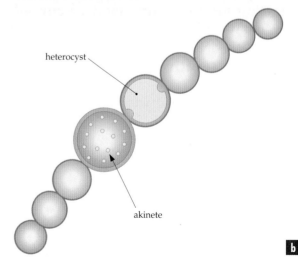

Fig. 33.4 (a) Two types of cyanobacteria. The large cylindrical form is *Microcoleus* and the smaller bead-like filament of cells is *Anabaena*, common in lakes and rivers. **(b)** Filament of *Anabaena* showing cell differentiation, with one akinete and one heterocyst

BOX 33.2 The Bergey classification of bacteria

The **Bergey classification** of bacteria is a practical reference system used by bacterial taxonomists which provides a relatively straightforward means of identifying bacteria. The nomenclature used is a standard binomial Linnaean scheme (Chapter 30). A **bacterial species** is defined as a group of bacteria with many common phenotypic characteristics, occupying similar habitats, whose DNA shows no major compositional differences (expressed as the percentage of the bases guanine plus cytosine), and whose 16 S ribosomal RNA sequences are 97% or more similar. A genus is defined as a group of similar species. Beyond this, genera with similar physiological and ecological features are placed together in 33 'sections'. A brief description of these 33 sections, together with significant examples, is given below.

1. Spirochaetes

Spiral shaped cells; e.g. *Treponema*: cause of syphilis, yaws.

spiral cell

2. Aerobic motile helical Gram-negative bacteria

Helicobacter: gastritis, peptic ulcers; implicated in the development of stomach cancer.

3. Non-motile Gram-negative curved bacteria

A little-known bacterial group.

4. Gram-negative aerobic rods and cocci

Pseudomonas: soil and aquatic organisms, burns infections.
Rhizobium: legume nodulation and nitrogen fixation.

5. Facultatively anaerobic Gram-negative rods

Many enteric (gut) bacteria, including *Escherichia*, *Salmonella* (food poisoning, typhoid), *Shigella* (dysentery), and *Vibrio* (cholera).

6. Anaerobic Gram-negative rods

Bacteroides and *Fusobacterium*, major members of the human gut and faecal microbiota.

7. Sulfate- or sulfur-reducing bacteria

Soil and aquatic bacteria, which reduce sulfate to sulfur and hydrogen sulfide in rotting organic matter.

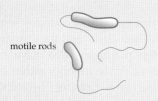

motile rods

8. Anaerobic Gram-negative cocci

Abundant as part of the normal and harmless microbiota of the human gut and mouth.

9. Rickettsias and chlamydias

Bacteria that live only as parasites inside animal cells.
Rickettsia: typhus, and scrub and spotted fevers.
Chlamydia: trachoma, venereal infections.

10. Mycoplasmas

Bacteria without a cell wall.
Mycoplasma: respiratory tract infections.

11. Endosymbionts

Bacteria that are symbiotically associated with and grow only inside the cells of host animals and fungi.

12. Gram-positive cocci

Staphylococcus and *Streptococcus* cause many serious infections of humans and animals; also include industrially important species used in milk and meat processing.

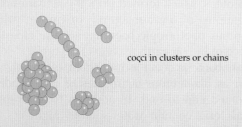

cocci in clusters or chains

13. Endospore-forming Gram-positive rods and cocci

This group forms bacterial endospores, which are the most resistant and enduring forms of life known—they resist heat, chemicals, drying and radiation (see p. 869). *Bacillus* and *Clostridium* are abundant in soil, dust and air, and cause food poisoning, tetanus, anthrax and gangrene.

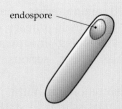

endospore

14. Non-sporing Gram-positive rods

Lactobacillus: milk processing (yoghurt, buttermilk) and silage production.

15. Irregular non-sporing Gram-positive rods

Corynebacterium: diphtheria in humans and plant diseases.
Eubacterium: a dominant type in the human gut.
Actinomyces: filamentous, branching cells.
Cellulomonas: degrades cellulose (a rare but important activity in nature).

16. Mycobacteria

Mycobacterium: tuberculosis and leprosy.

17. Nocardioforms

Partly filamentous cell growth common.
Nocardia: includes producers of important antibiotics and hydrocarbon-utilising species.

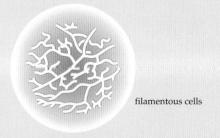

filamentous cells

18. Anoxygenic phototrophic bacteria

Photosynthesising bacteria that do not produce oxygen as an end product; found in sunny but anaerobic environments (e.g. undisturbed ponds, water-saturated soils).

19. Oxygenic photosynthesising bacteria

Photosynthesising bacteria that (like eukaryotic algae and plants) produce oxygen as an end product; include cyanobacteria (e.g. *Anabaena*, *Spirulina*), some of which also fix atmospheric nitrogen; important in nitrogen and energy cycling in soils and aquatic habitats.

20. Aerobic chemotrophic bacteria

Oxidise reduced inorganic compounds (e.g. H_2, NH_3, Fe^{2+}, H_2S) in place of organic substrates (e.g. sugars) for energy; important in the recycling and transformation of mineral nutrients in soil and water.

21. Budding and/or appendaged bacteria

Morphologically unusual bacteria, usually found in soil and water.

22. Sheathed bacteria

Appear filamentous but are short, rod-shaped cells packed within a common sheath or tube synthesised by the cells, and on which iron or manganese oxides may be deposited.

23. Non-fruiting, gliding bacteria

Bacteria whose cells move by gliding (creeping motion across solid surfaces, the mechanism of which is not well understood).

24. Gliding, fruiting bacteria

Bacteria (e.g. *Myxococcus*) that move by gliding and cells mass together to form extensive, structured aggregates (fruiting bodies) that contain resistant spore-like cells.

25. The Archaea (previously known as the Archaeobacteria)

Bacteria that are phylogenetically remote from the Bacteria (i.e. all other sections described here); includes methane producers, and types that generally prefer high temperatures or acidic or salty environments.

26. Nocardioform actinomycetes

27. Actinomycetes with multilocular sporangia

28. Actinoplanetes

29. Streptomyces and related genera

30. Maduromycetes

31. Thermomonospora and related genera

32. Thermoactinomycetes

Sections 26–32 are mainly filamentous bacteria with some overlap with section 17; many produce 'spores', but these are not the highly resistant endospores produced by bacteria from section 13; *Streptomyces* includes many important antibiotic-producing species.

33. Other genera

A small group with no known affinity with bacteria in the other sections listed.

Fig. 33.5 Cyanobacterial mats formed in the intertidal zone of a mudflat in Spencer's Gulf, South Australia. In these mats, layers of cyanobacteria and other bacteria (which utilise sulfur compounds from the underlying mud) build up as flat strata rather than as the mounds that form stromatolites (see Chapter 31)

BOX 33.3 Cyanobacteria and 'blue-green algal' blooms

In the summer of 1991–2, extensive cyano-bacterial blooms hundreds of kilometres long were reported on the Darling River. Towns and farms along the river had to be supplied with water from uncontaminated sources by tanker or had to install purification systems at considerable cost.

These blooms are usually referred to by the media as 'blue-green algae' or 'toxic algae', a con-sequence of early misidentification of these organ-isms. Species of cyanobacteria, such as *Microcystis aeruginosa* and *Anabaena flosaquae*, which form freshwater blooms of this sort, also produce **toxins** that are capable of causing illness and even death to animals and humans that drink water containing the bacteria. These toxins are small

peptides (5–10 amino acids chemically linked together) and have their primary effects on liver metabolism; swimming in toxin-contaminated water may also cause serious skin and eye reactions.

High concentrations of nutrients (particularly phosphate and potassium) in stationary or slow-moving water bodies (so that cells remain in the surface layers, capturing maximum light) promote blooms of this sort. The disposal into lakes, dams and slow-flowing streams of sewage and other wastes, or surface runoff from over-fertilised pastures and crops, sets the scene for serious cyanobacterial blooms.

Cyanobacterial bloom in Lake Burley Griffin, Canberra. Blooms of *Microcystis* often form in inland lakes, dams and rivers in Australia, and can be a problem for towns and farms, which rely on them for drinking or recreational water. Lake Burley Griffin catches street-water runoff from Canberra, rich in organic material and nutrients from animal droppings and from gardens and lawns. On calm sunny days, conditions are then ripe for development of a cyanobacterial (or 'algal') bloom in the lake. Natural agitation of the lake surface by wind or rain is usually sufficient to disturb the conditions needed for the organism's growth, so these blooms usually do not persist for long periods. In dry, calm conditions, a bloom may last for weeks

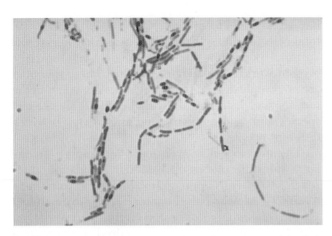

Fig. 33.6 Spore stain of *Bacillus* sp. The oval spores are red (carbol fuschin) and the vegetative rod is blue (methylen blue). The shape and position of the spore in the rod may be used in identifying the species. In this species, the spore is oval and subterminal (magnification × 1000)

depletion or as a result of other environmental signals forewarning difficult times ahead for the bacterium. Like the spores formed by fungi and cysts made by protozoans, endospores appear to have evolved to provide a resistant, metabolically quiescent cell type that increases the probability of survival under conditions of excessive cold or heat or desiccation. Bacterial endospores are remarkably resistant to a wide range of physical assault (high temperature, high flux of many types of radiation), and to many chemicals (including detergents and disinfectants), which would rapidly destroy all other living cells, including eukaryotic spores and cysts.

Bacterial endospores are the benchmark against which microbiologists measure sterilisation methods (i.e. procedures to kill living cells completely). Because species of *Bacillus* (*B. cereus*, *B. anthracis*) and *Clostridium* (*C. botulinum*, *C. welchii*) are important food poisoning and pathogenic agents, their complete elimination during processes such as food canning and surgical sterilising is critical to the safety of these procedures. If the elimination of endospores of these organisms can be guaranteed, then the elimination of all other pathogens and contaminants is also ensured.

> The super kingdom Bacteria contains most of the common bacteria that affect humans directly (through industry, agriculture, and medicine) and our cultivated plants and animals. This major group also includes cyanobacteria, a large and important group of oxygenic photosynthetic bacteria, and endospore-forming bacteria, whose spores are the most resistant life forms known.

Super kingdom Archaea

The super kingdom Archaea (Fig. 33.3; Bergey's section 25, Box 33.2) is a more specialised group than the super kingdom Bacteria and therefore Archaea are more restricted in the environments they inhabit. They might be thought of as the primitive forebears of modern bacteria but they are not, in fact, primitive—they have been evolving for as long and at similar rates as the Bacteria and the Eukarya. However, because they have largely continued to occupy specialised habitats (e.g. anaerobic, hot and salty places) they have retained their 'primitive' features and have not developed the biochemical diversity of the Bacteria or the intracellular compartmentation and multicellular style of the eukaryotes.

The Archaea include **halophiles** and **acidophiles** (salt- and acid-loving types) and **thermophiles** (preferring high temperatures). These tend to be found in extreme environments not generally colonised by the Bacteria or the Eukarya, for example, hot thermal springs, where the water may be salty or acidic (Fig. 33.7), or near volcanic vents on the deep-sea floor. Although temperatures in the latter environment may

Fig. 33.7 (a) This thermal valley near Rotorua, New Zealand, is a habitat for heat-tolerant cyanobacteria, seen here as zones of green. (The white areas are silica dioxide and the brown areas are iron oxides and hydroxides.) **(b)** Close up of mats of cyanobacteria

exceed 250°C, the water is present as super-heated liquid rather than as steam because of the high pressures that prevail at the depths concerned. Indeed, it is likely that bacteria have evolved to grow in the hottest environments on earth, provided that appropriate nutrients and water in liquid form are available. Thus, on the earth's surface, some thermophilic bacteria will grow at temperatures up to a few degrees above that of boiling water. The archaean genus *Pyrodictium*, found in an undersea volcanic region near Italy, can even grow at 110°C!

Another important and unique group within the Archaea are the **methanogenic bacteria**, which produce methane (an important greenhouse gas) as a metabolic end product. The methanogens do not occupy extreme environments but are strict anaerobes with a unique metabolism (see p. 876).

> The super kingdom Archaea includes many species that live in extreme conditions, such as salty, acidic or hot environments. This group also contains the methanogenic bacteria.

The remarkable abundance and metabolic diversity of bacteria

Bacterial populations are often very large and dense

The lives of bacteria are more clearly understood when they are contrasted with the macroscopic eukaryotes. Apart from the diversity in metabolism and ecology referred to elsewhere in this chapter, two other aspects show remarkable contrast: the size of typical bacterial populations, a feature often shared with other microorganisms; and the ease with which genetic information varies and mixes within large, heterogeneous bacterial populations. An important outcome of this latter aspect of bacterial life is that new genes and gene combinations appear and spread rapidly in response to environmental change and challenge to bacteria. As populations, they are remarkably adaptive, a fact that humans have learnt at great practical cost or benefit in medicine, agriculture and technology.

The density and dimensions of bacterial populations in favourable habitats is astonishing. Our own personal resident populations provide good examples. On the skin of the average clean-living person there are about 100 000 (10^5) bacterial cells (of many species) per square centimetre. The total skin surface of the average person is about 2 m^2 or 2×10^4 cm^2. Thus, the skin microbial population of a single

human amounts to some $2 \times 10^4 \times 10^5$, or 2×10^9 total cells, meaning that each of us carries an external bacterial population approximating to one-third the present world population of humans! These microbes are not evenly distributed, being more abundant in the moister and hairier realms of the body. The mix of species and total numbers tend also to vary between different people.

The human gut is even more astounding. In the large intestine or bowel, bacterial populations may reach 10^{11} cells per gram of faecal contents. Assuming a lower bowel content of approximately 1 kg, the total number of bacteria in this part of the gut alone is about 10^{14} cells (one hundred trillion). On a weight basis, this amounts to about one-third of the mass of faecal material. Higher up the gut the numbers of bacteria are fewer, and at the top of the digestive tract, that is, the stomach, there are relatively few bacteria (a few million) or any other microorganisms present, largely because of the acidity of the stomach (pH 1). In the gut generally there is an enormous diversity of bacterial species, many of which have yet to be described adequately. Like the skin, the composition of the gut microbial population is characteristic of each individual human, and in turn varies according to diet, age, general health, transient infections and so on.

And so it goes for the many environments in which bacteria flourish—soil, lakes and streams, hide and gut of animals, plant surfaces, rotting materials, spoiled foods, and foods in which controlled bacterial growth is encouraged (cheese, yoghurt, salami, sauerkraut, silage, to name a few).

The large natural populations of bacteria on the skin and in the gut of animals is probably important in limiting opportunities for incoming pathogens to become established. This is a consequence of many factors—physical crowding, competition for nutrients, and production of natural inhibitors such as viruses and antibiotics. A thriving natural population of bacteria in the gut and on the skin of probably all animals is an important aspect of normal healthiness. It is why, when antibiotics are taken by humans, they sometimes experience what are called opportunistic infections, caused by other microorganisms (including fungi) previously held in check by the normal natural population now depleted by the action of the antibiotic being used.

> Bacterial populations may grow to great numbers and there is great diversity of species. These natural populations in and on humans are often important in keeping potentially pathogenic bacteria and other infectious microbes at bay and thus are part of a person's normal health.

The metabolic diversity of bacteria

The impact of bacteria on their environment, and thus on the earth's biosphere, is remarkable. The bacteria conjure up no dramatic images compared with tropical rainforests soaking up sunlight and removing carbon dioxide from the atmosphere in mind-boggling quantities, nor of big cats on the veldt tackling buffalo or zebra twice or more their weight for the week's meal. But in ocean, soil and gut, on rocks, leaves and skin, hidden because of their small size, bacteria chemically transform, reduce and remake, finding energy and cell-building materials in every chemical substance made by other living organisms, and much that is not living but which is present in the original rocks or soil that make up our earth.

To understand this **metabolic diversity**, this ability to transform and reconstitute, it is useful to compare what bacteria do with what macroorganisms, such as plants and animals, do to harvest energy and nutrients needed for growth and reproduction.

Classifying nutritional types

Classification of organisms into nutritional types is based on the organisms' source of both carbon and energy.

All organisms require the chemical element carbon in some form as a nutrient for growth. Like animals and fungi, those bacteria that use organic (carbon-based) compounds as their major carbon source are called heterotrophs ('hetero-' meaning from others). These molecules are found in the food of animals, mostly as complex polymers—polysaccharide or protein or lipid typically. In contrast, like plants, those bacteria that use carbon dioxide as their major source or sole source of carbon are called autotrophs.

A **chemotroph** is an organism that gets its energy chemically, ultimately by oxidation–reduction reactions, implying two chemical components—electron donor (reductant) and electron acceptor (oxidant). For the cheetah on the veldt, for the unicellular protozoan in a pond, for fungi growing in a field or forest, and for us, the reductant is a reduced organic molecule, such as a sugar, an amino acid or a fatty acid. The oxidant or electron acceptor, for multicellular animals like ourselves and the cheetah, is oxygen. Thus, sugars and amino acids and fatty acids are said to be 'burnt' in animal metabolism in the same overall way (chemically speaking) that wood (mostly the polysaccharide cellulose) is burnt in a fire. The energy released in animal cell metabolism is captured mainly as ATP and NADH, as explained in Chapter 5. All animals thus are **chemoheterotrophs** because they use organic substances for both carbon and energy.

Those organisms that are able to use radiant energy (light) are **phototrophs**. Thus, plants are **photoautotrophs**—'photo' because they get their primary energy from sunlight and 'auto' because they get their carbon from carbon dioxide. During photosynthesis, carbon dioxide is reduced ('fixed') to form organic carbon compounds, phosphoglycerate and glucose being the primary products (Chapter 5). The reductant ultimately used by plants to fix carbon dioxide, and thus to make organic molecules, is water.

The world of macroorganisms consists of just these two metabolic types—chemoheterotrophs (animals and fungi) and photoautotrophs (plants and algae). Bacteria also include chemoheterotrophs and photoautotrophs but the metabolic options available are much wider than these two types (Fig. 33.8).

> There are four general nutritional categories of bacteria— chemoheterotrophs and photoautotrophs (similar to eukaryotes), and chemoautotrophs and photoheterotrophs (not seen among eukaryotes).

Four nutritional types of bacteria— chemoheterotrophs, photoautotrophs, chemoautotrophs and photoheterotrophs

Among the chemoheterotrophs are the common bacteria that we find on and in ourselves and other animals—enteric bacteria, skin bacteria, the pathogens that cause diseases—or those that we cultivate for commercial purposes—the bacteria used in cheese and vinegar and antibiotic production, for example.

Among the bacterial photoautotrophs are the cyanobacteria that are more or less identical to plants and algae in their photosynthetic capabilities. However, other photoautotrophic bacteria do not use water as the reductant to reduce carbon dioxide and therefore do not produce oxygen as an end product of photosynthesis, as plants, algae and cyanobacteria do. Nevertheless, the principle of using light for energy and carbon dioxide for carbon, plus a reductant, still applies to these bacterial photoautotrophs.

Certain bacterial species combine these energy- and carbon-capture strategies in other ways. The **chemoautotrophs** (Fig. 33.8c) use reduced inorganic substrates as sources of energy and reduce carbon dioxide to organic carbon (using water or hydrogen gas as a reductant). These bacteria are important in many transformations of inorganic compounds in soil and aquatic habitats, for example, in the conversion of ammonia to nitrate, sulfide to sulfate, ferrous to ferric iron, and hydrogen gas to water. There are no known chemoautotrophs that oxidise organic (reduced carbon) compounds while at the same time fixing carbon to form organic carbon.

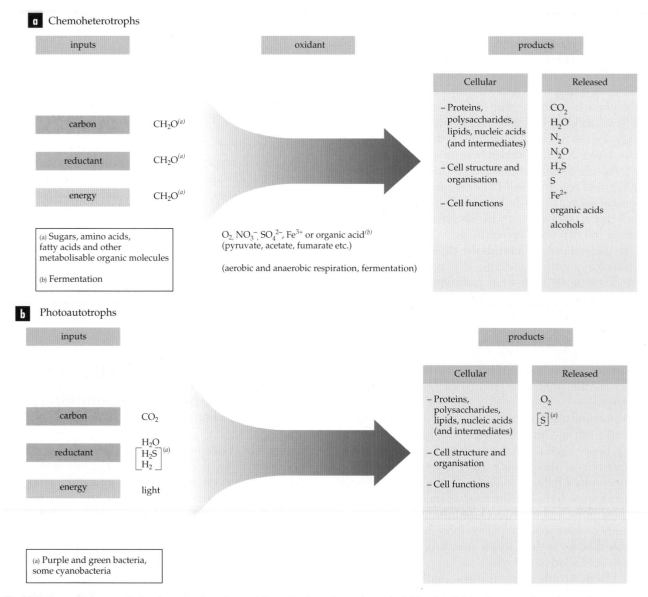

a Chemoheterotrophs

inputs		oxidant	products	

inputs

carbon	$CH_2O^{(a)}$
reductant	$CH_2O^{(a)}$
energy	$CH_2O^{(a)}$

(a) Sugars, amino acids, fatty acids and other metabolisable organic molecules

(b) Fermentation

oxidant

O_2, NO_3^-, SO_4^{2-}, Fe^{3+} or organic acid$^{(b)}$ (pyruvate, acetate, fumarate etc.)

(aerobic and anaerobic respiration, fermentation)

products

Cellular	Released
– Proteins, polysaccharides, lipids, nucleic acids (and intermediates)	CO_2 H_2O N_2 N_2O H_2S
– Cell structure and organisation	S Fe^{2+}
– Cell functions	organic acids alcohols

b Photoautotrophs

inputs	products

inputs

carbon	CO_2
reductant	$\begin{bmatrix} H_2O \\ H_2S \\ H_2 \end{bmatrix}^{(a)}$
energy	light

(a) Purple and green bacteria, some cyanobacteria

products

Cellular	Released
– Proteins, polysaccharides, lipids, nucleic acids (and intermediates)	O_2 $[S]^{(a)}$
– Cell structure and organisation	
– Cell functions	

Fig. 33.8 Four cellular metabolic categories found among bacteria. Some bacteria are capable of switching between these categories. **(a)** Chemoheterotrophs (many bacteria, and similar to animals and fungi). **(b)** Photoautotrophs (cyanobacteria, purple and green bacteria, and similar to plants and algae)

The **photoheterotrophs** (Fig. 33.8d) rely on sunlight for energy but use organic compounds as ready-made metabolic building blocks for cell growth and development. Examples of these include the **purple and green bacteria**. These widely distributed bacteria are unusual in another way in that they substitute hydrogen sulfide or hydrogen gas as a reductant for the water used as a reductant by plants, algae and cyanobacteria. As a consequence, the purple and green bacteria carry out anoxygenic photosynthesis (not generating oxygen), producing elemental sulfur or sulfate or water as end products, rather than oxygen gas as plants, algae and cyanobacteria do. The purple and green bacteria have a characteristically different set of photosynthetic pigments (including bacteriochlorophyll) to support their unique brand of photosynthesis.

Bacteria are remarkably diverse metabolically. Some use light energy, while others use organic or inorganic chemicals as a source of energy. Some require complex organic compounds for their carbon needs, while others use carbon dioxide.

Anaerobic bacteria

There are many species of bacteria (and a few eukaryotic microbes among the protists and fungi, too) that can use, as terminal oxidants (replacing molecular oxygen), a range of relatively oxidised inorganic and organic molecules, including nitrate, carbonate, sulfate, ferric ion and fumarate (the last an important metabolic

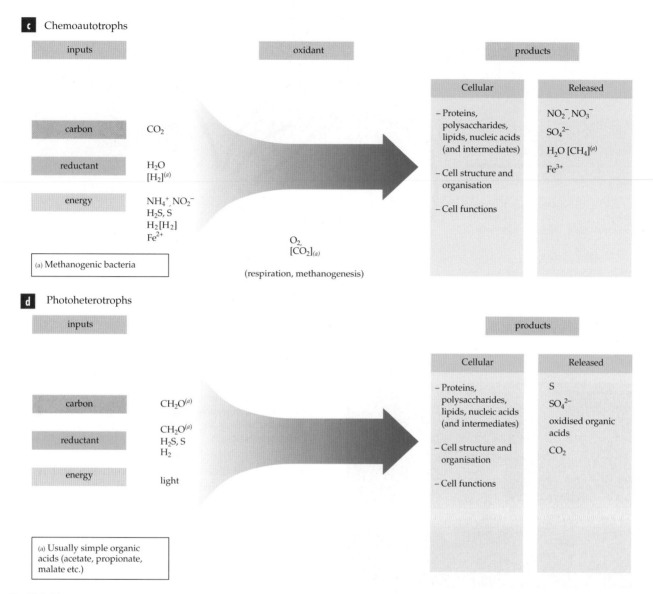

c Chemoautotrophs

Fig. 33.8 (c) Chemoautotrophs (nitrifying bacteria, hydrogen bacteria, sulfur bacteria, iron-oxidising bacteria and methanogenic bacteria; see also Chapter 5). These bacteria are also sometimes called lithoautotrophs (meaning living on rocks) because their inputs (carbon, reductant and energy) are all inorganic. **(d)** Photoheterotrophs (purple and green bacteria, growing anaerobically). This group of bacteria is highly metabolically diverse and flexible. Some will also grow as photoheterotrophs anaerobically and as chemoheterotrophs or chemoautotrophs aerobically or in the dark (non-photosynthetic conditions)!

intermediate of normal respiratory metabolism in cells, see Chapter 5). These bacteria are said to carry out **anaerobic respiration** because of the overall similarity of their oxidative metabolism to oxygen-dependent (aerobic) respiration:

Anaerobic respiration

$$CH_2O \text{ (organic carbon)} + NO_3^- \rightarrow CO_2 + N_2$$
$$\text{(reaction not balanced)}$$

(or SO_4^{2-} or HCO_3^- (or S or CH_4 or
or Fe^{3+} or fumarate) Fe^{2+} or succinate)

Aerobic respiration

$$CH_2O + O_2 \rightarrow CO_2 + H_2O$$

Combining these reactions with those carried out by the chemoautotrophs outlined earlier explains how inorganic nutrients involving elemental nitrogen, sulfur and iron are cycled and recycled by bacteria in soil and in aquatic habitats, such as oceans and lakes. The nitrogen cycle, discussed in its global context in Chapter 44, is one such cycle where the key transformative roles of bacteria are particularly clear. We now look more closely at this important complex of bacterial activities.

The nitrogen cycle

In this cycle, bacteria play major roles both as competitors and as symbionts in the nitrogen relations of plants (Fig. 33.9). An understanding of how this cycle

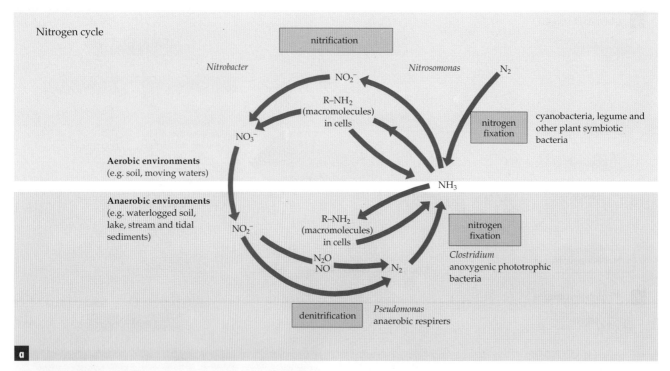

Nitrogen cycle

Fig. 33.9 (a) Role of bacteria in the nitrogen cycle (see also Chapter 44). **(b)** In legumes, such as clover and wattles (*Acacia*), root nodules, such as are shown here on *Acacia silvestris*, contain nitrogen-fixing bacteria

functions in nature has been crucial in the development and practice of modern agriculture, an understanding in which Australian scientists have played an important role.

The first sector of the nitrogen cycle involves **nitrogen-fixing bacteria**. These may be part of plant–bacteria symbioses of a range of types (the bacteria concerned include *Rhizobium*, *Frankia*, cyanobacteria), or may be due to free-living (non-symbiotic), photosynthetic and non-photosynthetic, aerobic and anaerobic bacteria.

Only bacteria are capable of fixing molecular nitrogen; the activity is widespread among them. All have a **nitrogenase** enzyme complex, which catalyses the highly energy-consuming reduction of molecular nitrogen to ammonium ion:

$$N_2 + 8H^+ + 6\,e^- \rightarrow 2NH_4^+$$
(reaction consumes 18–24 moles of ATP)

The nitrogenase reaction is multistep and is very sensitive to molecular oxygen or other strong oxidants. It thus takes place in a highly reducing or anaerobic environment within the cell.

The ammonium ion is used in amination reactions with glutamate and 2-oxoglutarate to form the amino acids glutamine and glutamate inside the bacterial cell. In this way, the nitrogen becomes 'fixed' to organic molecules and available for further nitrogen metabolism, in its many ramifications, within the bacterial cell. Nitrogen-fixing bacteria thus produce ammonium ions or amino acids, which are then transported into the host plant (in the case of symbiotic bacteria) or are released into the local environment (in the case of free-living bacteria) for utilisation by plants (after oxidation to nitrate by nitrifying bacteria, see below) or other microbes.

In the second part of the nitrogen cycle are the **denitrifying bacteria**, mainly chemoheterotrophs, which carry out anaerobic respiration. They use nitrite and nitrate as terminal electron acceptors in place of oxygen, producing gaseous nitrous oxide and molecular nitrogen, which escape to the atmosphere and are thus lost to the biosphere. These bacteria thus deplete their environment of available nitrogen, and in certain conditions (anaerobic waterlogged soils, for example) may limit plant and other microbial growth by virtue of this activity.

Joining these two groups, and making up the third part of the nitrogen cycle, are the **nitrifying bacteria**, which are chemoautotrophs utilising the ammonium ion as an energy substrate and reductant. Ammonium is oxidised to nitrite (by *Nitrosomonas*) and nitrate (by *Nitrobacter*). Because nitrate is the preferred form of inorganic nitrogen for plant nutrition, the nitrifying bacteria are key organisms in transforming the fixed nitrogen (primarily NH_4^+) from the nitrogen fixers, or organic nitrogen mineralised to NH_4^+ from dead and decomposing organisms, into a form available to plants.

Cycles of bacterial nutrient transformation can also be described for sulfur, iron, phosphorus and a number of micronutrients vital to life. Again, an understanding of these processes helps greatly in understanding how farming practice might be modified to optimise plant growth. For example, acidification of soil (Chapter 45), which results when H_2SO_4 is produced in the chemoautotrophic oxidation of S or H_2S, can affect plant growth directly or cause the insolubilisation of other nutrients such as iron, or bring about the release of potentially toxic levels of aluminium in soil.

> Some bacteria have the unique ability to fix molecular nitrogen, and some use electron acceptors other than oxygen (anaerobic respiration, shared with some protists and fungi). Through these activities bacteria transform and recycle inorganic and organic forms of carbon, nitrogen, sulfur, phosphorus and metals in a way that is vital to complex ecosystems, including farming.

Bacterial fermentations

Fermentative metabolism is physiologically evident to humans as the lactic acid build-up in our muscles and blood when we exert ourselves physically at such intensity that oxygen cannot be supplied sufficiently quickly to our muscles (Chapter 27). Under these conditions of near anaerobiosis in muscle cells, pyruvate generated within cells is utilised as the oxidant, in place of oxygen from the blood, to support glycolytic production of ATP (Chapter 5). The consequent reduction of pyruvate yields lactic acid, which accumulates in muscles and the blood, ultimately retarding muscle function.

Many bacteria also produce lactic acid when grown under anaerobic conditions. The best known of these are species of *Lactobacillus* and *Lactococcus*, which are used in yoghurt and cheese production from milk (Fig. 33.10). The acid produced, and the action of certain enzymes secreted by these bacteria into the milk, results in the precipitation of casein, the major protein of milk, and thus the formation of a 'curd'.

A variant of lactic acid **fermentation** is the ethanol fermentation carried out by some bacteria and by brewers' yeast (in the case of wine and beer making; see Chapter 36). In this case, pyruvate is first

Fig. 33.10 *Lactobacillus casei* Shirota is used in the production of Yakult, a yoghurt-based drink

decarboxylated to form acetate, which is then reduced to form ethanol.

Apart from these commercially important fermentations, there are many other types of fermentation known (and probably as many awaiting description) among bacteria. In principle, these fermentations are biochemically similar—in the absence of an external inorganic oxidant such as oxygen (or sometimes because oxidative enzymes are absent from the bacterium), an internally generated organic oxidant acts as terminal electron acceptor to produce the corresponding reduced molecule. Fermentations may thus be defined in terms of their oxidation/reduction pairs. Some examples are:

- acetate/ethanol
- lactate/propionate
- butyrate/butanol
- fumarate/succinate
- pyruvate/lactate
- hydroxybutyrate/butyrate
- acetoacetate (acetone)/isopropanol
- CO_2/formate.

In many bacteria, combinations of pathways are used, yielding complex mixtures of organic acid and alcohol products. Many of these compounds are important as solvents and reactants in the chemical industries. Half a century ago, the bacteria producing these organic acids and alcohols were used industrially for these products. These days, chemical syntheses of short-chain acids and alcohols from petroleum feedstocks and refinery by-products have replaced the microbial products. However, these fermentation end products remain hugely important in ecosystems, in agriculture and in food production. Mixtures of these alcohols and acids (and their intermediate aldehydes) also define the epicurean qualities, the aroma and taste of cheese, yoghurt and salami. Some significant examples of bacterial fermentations are given in Table 33.3.

> Bacteria carry out a wide range of biochemical fermentations (anaerobic energy metabolism), which are important in agriculture and in food and alcohol production.

Methane producing bacteria: methanogens

The biospheric production of methane by bacteria is an important aspect of global warming (Chapter 45). Methane levels in the atmosphere have approximately doubled in the past century, a consequence of both non-biological production (e.g. release from oil and gas wells, deforestation by burning) and biological production associated particularly with the expansion and intensification of agriculture (e.g. ruminant animals, rice paddies).

A major group of methanogenic bacteria (**methanogens**) utilise hydrogen gas and carbon dioxide to generate energy and make sugars. They are chemoautotrophs, oxidising hydrogen for energy and using hydrogen to reduce carbon dioxide to form organic compounds for growth and development. Methane is a major end product of their metabolism, which can be summarised in the following reaction:

$$4H_2 + CO_2 \rightarrow CH_4 + 2H_2O$$
(some methanogens use acetate or formate in place of CO_2)

The biochemistry of this apparently simple reaction is complex and involves enzymes and chemical cofactors not found in any other organisms. The methanogenic bacteria are abundantly represented throughout the biosphere, even though they are strict anaerobes. There are many anaerobic environments rich in organic material that support the fermentative production of H_2 and CO_2 by other bacteria and microorganisms. These environments will invariably contain methanogens: in the rumen, in the large intestine of mammals, in peat bogs, marshes and waterlogged soils, in the muddy bottoms of deep lakes and reservoirs and in urban garbage tips. Methanogens live

Table 33.3 Bacterial fermentations of industrial and agricultural significance

Type	Overall reaction	Bacterium
Alcohol (fuel)	Hexoses $\rightarrow$ ethanol + CO_2	*Zymomonas* sp.
Lactic acid (yoghurt)	Hexoses $\rightarrow$ lactic acid	*Lactococcus* (previously named *Streptococcus*) spp. *Lactobacillus* spp.
Propionic acid (cheese)	Lactic acid $\rightarrow$ propionate + acetate + CO_2	*Propionibacterium* spp. *Clostridium propionicum*
Butyric acid (silage)[a]	Hexoses $\rightarrow$ butyric acid + acetate + CO_2 + H_2	*Clostridium butyricum*
Butanol (industrial solvents)	Hexoses $\rightarrow$ butanol + acetate + acetone + ethanol + CO_2 + H_2	*Clostridium acetobutylicum*
Caproic acid (silage)[a]	Ethanol + acetate + CO_2 $\rightarrow$ caproic acid + butyric acid + H_2	*Clostridium kluyveri* *Clostridium aceticum*
Methane (biogas)	$H_2 + CO_2$ (or acetic or formic acid) $\rightarrow$ $CH_4 + H_2O$	*Methanothrix* sp. *Methanosarcina* sp. *Methanobacterium* sp.

(a) Silage is a form of 'pickled grass' made by farmers for stock feed. Typically, freshly cut grass and legume is placed in pits or stacks and covered. Under anaerobiosis, elevated temperatures (up to 60°C) develop. Thermophilic anaerobic bacteria carry out the fermentations shown, using sugars released from plant polysaccharides (mostly cellulose). The high temperatures and low pH that develop in the grass–legume mix kill or inhibit growth of spoilage microorganisms, thus preserving ('pickling') the grass for later use as stock feed. Cattle find silage a very attractive food for much the same reasons that pickled vegetables (such as sauerkraut and artichokes) are tasty for humans.

in close association with other bacteria and eukaryotes, utilising the end products of the fermentation of organic material by these other organisms. The methanogens are a prime example of how bacteria have evolved to extract energy and nutrients from what seem to be the most 'wasted' of biological end products.

> The methanogens are a unique group of bacteria, producing methane (a greenhouse gas) in large quantity in anaerobic environments using the metabolic 'waste' products of a range of other microorganisms.

Genetic systems of bacteria

In eukaryotes, meiosis promotes genetic recombination and thus generates an important part of the genetic variation on which evolution depends. However, meiosis does not occur in bacteria (Chapter 8). Bacteria engage in a simple mitotic or vegetative division of cells, in which the progeny cells contain chromosomes that are identical to those in the original single parental cell. How, then, is genetic variation generated in bacteria? How do bacteria respond genetically, speedily and appropriately to change of environment? If they did not vary genetically, and evolve through natural selection, then bacteria would not have persisted on this ever-changing planet for as long as they have since the dawn of biological time. A special aspect of this question is how **resistance** to chemical agents, such as antibiotics and **antiseptics**, arises among bacteria, and how this resistance may spread so speedily through hospitals and surgeries.

Firstly, genetic variation due to mutation (Chapter 32) is much more important in bacteria than in eukaryotes, if only for the reason that bacteria are haploid and therefore mutations in genes will usually be expressed in the cell in which they occur; a mutated allele is not obscured by the presence of an unmutated allele as it may be in diploid or polyploid organisms. But mutation alone is not sufficient.

The second option for bacteria has been provided by the evolution of other mechanisms of genetic variation that differ from eukaryotic meiosis and sexual reproduction but have the same outcome—the introduction into a cell of new and sometimes very different DNA, its integration or recombination into the genome of the recipient cell, its genetic expression as a change of phenotype, and its vertical and horizontal transmission to progeny and to other members of the same population (including sometimes to cells of different species). The phenotypic variation thus generated provides the population with variant individuals able to take advantage of new or changing environments.

Three major gene transfer mechanisms operate in bacteria: transformation, conjugation and transduction.

Transformation: gene transfer by free molecules of DNA

Many bacteria are able to take into their cells molecules of DNA that may occur free in their environment, released from cells that have died and disintegrated. Once inside the recipient bacterium, such DNA might be broken down by restriction enzymes if these recognise the DNA as 'foreign' to the cell. Alternatively, if the DNA is similar to that which is already present in the bacterium, then the incoming DNA may either recombine with the chromosomal DNA, or with plasmid DNA (**plasmids** are small, circular molecules of DNA separate from the bacterial chromosome; Chapter 10). Or, if the incoming DNA is able to control its own replication, it may establish itself as a plasmid without recombining with existing DNA in the cell. Whichever of these occurs, that is, recombination or establishment as a plasmid, new genetic information is introduced into the cell and therefore variation of the cell phenotype may result. Selection and multiplication of this new genetic variant will follow if the new phenotype is better adapted to conditions prevailing in that particular environment. **Transformation** is thus the process of taking up free DNA and recombining or establishing this DNA stably in the recipient cell (Fig. 33.11a).

Bacterial transformation is probably not a common process in nature because, in the environments in which bacteria abound, free DNA is rare. Free DNA is highly susceptible to breakdown by enzymes (DNAases) that many microorganisms secrete in their never-ending search for nutrients. However, transformation appears to be an important process for some pathogenic bacteria, such as members of the genus *Neisseria*, which includes the cause of gonorrhoea (*N. gonorrhoeae*) and meningitis (*N. meningitidis*), and it was first demonstrated for a strain of *Streptococcus pneumoniae* causing pneumonia in mice. Transformation is also used extensively in genetic engineering laboratories because it is a very convenient practical method for inserting synthetically constructed recombinant DNA molecules into a range of bacteria (Chapter 13).

Conjugation: gene transfer by plasmids

The second means of incorporating new DNA into bacterial cells is **conjugation**, the process whereby DNA is transferred directly from one bacterial cell to another (Chapter 15). Conjugation requires physical contact between cells and the presence of special plasmids in the donor bacterial cell (Fig. 33.11b).

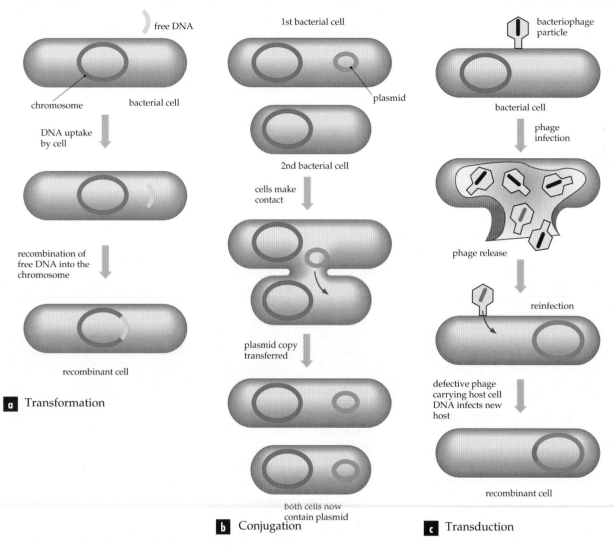

Fig. 33.11 Methods of gene transfer. **(a)** Transformation involves the uptake of free DNA by living cells. The incoming DNA recombines into the cell's chromosome or becomes a plasmid within the cell. Whichever path is taken, new genetic information is brought into the cell so that it becomes a genetic variant of the original cell (a recombinant). The new genetic information may be important in the adaptation of the recombinant to new environments in which the original cell type could not exist. **(b)** In conjugation, pairs or groups of bacterial cells make contact and a plasmid copy is passed from one to another. Once a plasmid is established in its new host, it then excludes the transfer of another copy of the same plasmid or a similar plasmid. However, a copy of this newly received plasmid can be transferred in turn to cells without plasmids with which contact is made. Some plasmids can also integrate themselves into the bacterial chromosome and they then make parts of the surrounding chromosome behave as a plasmid (i.e. the surrounding genes become transmissible). Alternatively, genes on the surrounding chromosome may become part of a new recombinant plasmid when it reverses out of the chromosome and becomes a plasmid again. Transfer of the recombinant plasmid will result in transfer of genes from one bacterium to another, so that a recombinant cell is generated; this results in genetic variation in the population. **(c)** In transduction, DNA from one bacterial cell is moved to another as a result of infection of the first by a virus (bacteriophage), and the subsequent infection of the second cell by virus particles released after infection of the first. During the infection cycle of the first cell, a small number of newly forming virus particles will pick up host cell DNA in error. These particles then transmit these genes to the second cell and because the virus particles contain host cell genes in place of virus genes, they are defective in that they cannot complete the virus life cycle and thus do not kill the second cell. This second cell is again a recombinant bacterium, having received genes from another cell. Once again, genetic variation occurs in the bacterial population due to this gene transfer system

Plasmids either transfer themselves between bacterial cells, or they act as *vectors* to transfer other DNA picked up from the chromosome or from other plasmids in the donor cell. By a process not fully understood, these plasmids transfer a copy of themselves (together with any hitchhiking DNA they may be carrying) from the donor cell across the contact junction into the recipient cell. There, as in transformation, the incoming DNA may recombine or integrate into the chromosome or into a resident plasmid, or it may establish itself as a new independent plasmid in the recipient cell. Whichever of these alternatives is used, new DNA is introduced into the recipient bacterial cell and the prospect of phenotypic variation and selection is possible once again. The spread of plasmids (and the genes they carry) in a population of bacteria can be quite rapid; in this sense plasmids are sometimes referred to as 'infectious'.

Transduction: gene transfer by bacteriophages

The third general method of DNA transfer between bacteria is **transduction** (Fig. 33.11c). Bacteria, though the smallest cellular organisms known, are themselves parasitised by smaller living organisms—viruses.

Viruses that multiply in bacterial cells are called **bacteriophages** (meaning 'bacteria-eating'), commonly abbreviated to phages (Chapter 34). Some phages are temperate and their presence in a bacterial cell does not necessarily result in cell death. Many temperate phages that have DNA as their genetic material (some phages use RNA, Chapter 34) integrate their DNA entirely into the host cell's chromosome and exist in this state indefinitely. However, with appropriate external stimuli (usually a physical or chemical change that could be threatening to the life of the host cell) many temperate phages become virulent to their host cell. They then multiply rapidly within the cell and kill and lyse it (make the cell burst), with the resultant release of a hundred or more phage particles or *virions* (see Chapter 34). These virions, in turn, infect any susceptible new bacterial cell with which they come into contact and another cycle of infection (temperate or virulent) is begun.

During the formation of new virions within the host bacterium in the virulent phase of infection, the assembling virions may, in error and usually quite rarely, pick up pieces of host cell DNA rather than their own virus DNA. The virion carrying host DNA is then released during lysis along with the normal virions. Infection of a new bacterial host by a virion carrying host cell DNA rather than viral DNA will thus result in delivery of the original host's DNA into the newly infected bacterial cell. Recombination or integration of the DNA into the host chromosome takes place and a change of phenotype again may result. The analogy with plasmid vectors of DNA in conjugation is a close one, except that in most cases host cell death is required for release of the transducing phage virion, whereas the transfer of a copy of a conjugative plasmid does not kill the donor cell.

Gene transfer and genetic variation in bacteria by each of the three processes described above usually involves small numbers of genes, largely because the dimensions of the DNA molecules transferred are small compared with the size of the bacterial chromosome (10–1000 times larger). In this respect, these processes differ fundamentally from sexual reproduction in eukaryotes, where whole chromosomes and sets of chromosomes recombine and reassort to generate entirely new chromosomes and sets of chromosomes.

> Bacteria achieve an important part of their natural genetic variation, not by sexual reproduction as in eukaryotes, but by other genetic mechanisms:
> - transformation (taking up free DNA from the environment)
> - conjugation (transfer of DNA from cell to cell by plasmids)
> - transduction (DNA transfer by viruses—bacteriophages).

The importance of plasmids and phages to bacterial populations

Transfer of genes by both conjugation and transduction are thought to be widespread in natural populations of bacteria because plasmids and temperate phages abound in these populations. The circulation and mixing of genes by these vectors can be observed in natural environments and can be demonstrated readily in the laboratory. A wide range of genes of bacterial origin are found as part of phage and plasmid DNA. Some of these genes are particularly obvious in medicine because they often cause the host bacterium to become resistant to antibiotics and antiseptics, or they provide additional genes (e.g. for toxins), which make the host bacterium more virulent. Examples of important genes that are found on mobile genetic elements in bacterial populations are shown in Table 33.4.

Plasmids and phages are often referred to as **accessory genetic elements** since they are not vital to the host bacterium's normal capacity to exist and sometimes they can be lost from the host cell relatively easily. Their presence in a population of bacteria, even in a tiny minority of cells, provides a small pool of variant cells able to expand rapidly as the host bacteria move into new habitats, for example, if the variants are antibiotic resistant throughout the body or wounds of a person receiving antibiotic therapy.

The host range of plasmids and phages is quite variable. Many will exploit different bacterial genera as hosts, while others enter only members of the same species or even subspecies of bacterium. There is much yet to be learnt about host range and compatibility between plasmids and phages and their host bacteria. Nevertheless, the abundance of these mobile accessory genetic elements and the wide overall host range they have mean that the circulation of genes and groups of genes in the bacterial world occurs relatively easily.

The bacterial world, although lacking the processes of meiotic recombination and gamete fusion, which ensure extensive phenotypic variation in eukaryotes, obtains its variation through unique processes of genetic transfer and genetic 'promiscuity', which have

Table 33.4 New properties given to bacteria as a result of carriage of plasmids or phages

Bacterium	Plasmid or phage	Property
Many species that cause human and animal disease	Plasmids or phages	Antibiotic and disinfectant resistance; toxin production; attachment to animal cells; production of tissue-damaging enzymes
Pseudomonas spp.	Plasmid	Degrade and metabolise complex synthetic chemicals (e.g. DDT), which would otherwise accumulate in the environment
Rhizobium spp.	Plasmid	Nodulate roots of legumes and fix nitrogen
Agrobacterium spp.	Plasmid	Plant galls
Streptomyces spp.	Plasmid	Antibiotic synthesis
Bacillus thuringiensis	Plasmid	Insecticidal toxin synthesis
Many species	Phage	Resistance of the host bacterium to attack by other virulent phages

the same end result. The DNA transferred in these processes is usually considerably less than a complete genome, by contrast with eukaryotic genetic recombination, in which complete genome complements are usually involved.

All modern bacteria are the result of genetic variation, adaptation and evolution in response to changes that have occurred as the earth and its living organisms have evolved. Exploitation of highly competitive niches (e.g. the gut or soil) or chemically and physically extreme niches (e.g. salty or acidic thermal springs) is easier for prokaryotes than for eukaryotes, probably because of the simpler cellular structures and greater genetic flexibility and thus adaptability of prokaryotes.

The abundance of plasmids and phages carried by bacteria in natural populations permits the ready transfer and reassortment of genes among different bacteria. Such transfer of genetic information is important in the genetic variation and evolution of bacteria and in part explains their adaptability to the most extreme and challenging of environments for life on earth (and perhaps even elsewhere).

Summary

- Bacteria (prokaryotes) are microscopic, unicellular organisms, with aspects of cell structure, biochemistry and genetic systems that distinguish them from eukaryotic cells.

- There are two major evolutionary groups of bacteria: the Bacteria and the Archaea. Classification into species within these two groups can be based on classic bacteriological criteria including cell size and shape, staining reactions, motility, spore formation, physiological preferences and biochemical (metabolic) reactions. More recent evolutionary classification methods are based on base sequence analysis of cellular DNA and RNA, and on amino acid sequences of proteins produced by different species.

- Bacteria are ubiquitous, occupying not only all the habitats that eukaryotes do but also extreme environments. Bacteria live in environments that are hotter, saltier and more acid than is tolerated by other life forms on earth. Some species produce endospores, the most enduring and resistant form of life known.

- Some bacteria are photosynthetic and an important group, the cyanobacteria, produce oxygen during photosynthesis in the same way as plants and eukaryotic algae. The cyanobacteria are thought to have been responsible for the earth's atmosphere becoming oxygen-rich more than two billion years ago. Some modern cyanobacteria cause toxic blooms in aquatic habitats.

- Bacteria are remarkably diverse metabolically. They have many biochemical capabilities not seen among animals, plants or fungi. These activities are important in the fixation and cycling of nitrogen and other inorganic nutrients in the biosphere, in the production of organic acids and alcohols important in agriculture and food production, and in methane production, an important greenhouse gas. Bacteria are also important as 'biochemical machines' in the production and transformation of high value chemicals such as vitamins, antibiotics, hormones and vaccines.

- Bacteria are major agents of disease in animals and plants. Their transmission by, or excessive growth in, water and food, may make the water or food hazardous to other organisms, such as humans. On the other hand, the natural abundance of bacteria in the gut and on the skin and mucosa of humans is an important element in protecting the human body from invasion by pathogenic bacteria and other abnormal microorganisms.

- Bacteria are thus agents both of harm and of benefit for humans. Moreover, without them the biosphere would lose much of its biological and chemical diversity. As a result, processes of decay and renewal would slow or cease, and life as humans know it would change for the worse in many ways.

- Bacteria achieve genetic variation not by sexual reproduction as in eukaryotes but by transformation (taking up free DNA from the environment), conjugation (transfer of DNA from cell to cell by plasmids) and transduction (DNA transfer by bacteriophages [bacterial viruses]). The freedom with which these gene transfer and exchange processes occur across species boundaries of the bacteria accounts, in part, for their remarkable evolutionary adaptability. This genetic fluidity and flux also accounts for the difficulties in defining species within the bacterial world and thus for continuing uncertainties in bacterial taxonomy.

key terms

accessory genetic
  elements (p. 883)
acidophile (p. 873)
akinete (p. 869)
algal bloom (p. 869)
anaerobic respiration
  (p. 877)
anoxygenic
  photosynthesis
  (p. 865)
antibiotic (p. 866)
antiseptic (p. 881)
Archaea (p. 866)
bacteria (p. 862)
Bacteria (p. 866)
bacterial species
  (p. 870)
bacteriophage
  (p. 883)

Bergey classification
  (p. 870)
chemoautotroph
  (p. 875)
chemoheterotroph
  (p. 875)
chemotroph (p. 875)
conjugation (p. 881)
cyanobacteria
  (p. 869)
denitrifying bacteria
  (p. 878)
endospore (p. 869)
eukaryote (p. 862)
fermentation (p. 879)
Gram stain (p. 867)
Gram-negative
  bacteria (p. 867)
Gram-positive bacteria
  (p. 867)

green bacteria
  (p. 876)
halophile (p. 873)
heterocyst (p. 869)
metabolic diversity
  (p. 875)
methanogen (p. 880)
methanogenic bacteria
  (p. 874)
nitrifying bacteria
  (p. 879)
nitrogen-fixing
  bacteria (p. 878)
nitrogenase (p. 878)
oxygenic
  photosynthesis
  (p. 865)
photoautotroph
  (p. 875)

photoheterotroph
  (p. 876)
photosynthetic bacteria
  (p. 865)
phototroph (p. 875)
phycobilin (p. 869)
plasmid (p. 881)
prokaryote (p. 862)
purple bacteria
  (p. 876)
resistance (p. 881)
spore (p. 869)
thermophile (p. 873)
toxin (p. 872)
transduction (p. 883)
transformation
  (p. 881)

Review questions

1. Which of the following structures appear in both eukaryotic and prokaryotic cells, and which appear in only one of these: nucleus, chromosome, mitochondrion, ribosome, chloroplast, cell wall, plasma membrane?

2. How long ago is it generally accepted by scientists that the following first appeared on earth: oxygenic photosynthesis, macroscopic fossils, prokaryotic cells?

3. Early evolutionary connections between eukaryotes and prokaryotes are most effectively deduced from:

   A the age of sediments in which their fossils are found

   B microscopic structural differences

   C sequence relationships of molecules containing genetic information

   D extent of UV damage to fossils found in sediments

4. Traditional methods of identification and classification of bacteria (such as Bergey's scheme) are based on:

   A phenotypic (expressed) characteristics

   B cell shape and size

   C sexual compatibility between possible species

   D gene sequence data

5. Name the two major super kingdoms of bacteria. Which of these appears most closely related evolutionarily to eukaryotes? To which of the super kingdoms do the cyanobacteria belong?

6. Which of the following characteristics are common among the cyanobacteria: anoxygenic photosynthesis, CO_2 fixation, nitrogen fixation, sugar fermentation, cell specialisation, recent (less than 0.6 billion years) appearance in evolution?

7. 'Some bacteria can grow in environments where the temperature is slightly above that of boiling water.' True or false? What appears to be an important physical determinant in the maximum temperature at which life occurs on earth?

8. Which is more accurate about the bacterial population normally present on the human skin?

   A the total number is about 10^5

   B the number is greater than is found in the gut

   C many different species are present

   D the number and composition is constant for different persons

9. In the term 'chemoheterotroph', what do 'chemo' and 'hetero' refer to?

10. What is the overall or summary reaction for the fixation of atmospheric nitrogen by bacteria? What is the name of the enzyme that catalyses this multistep reaction?

11. Name two commercially important processes dependent on bacterial metabolic fermentations.

12. In what sort of natural environments are methanogens likely to be found growing?

13. Name the processes of DNA transfer among bacteria in which:

    (a) cell-to-cell contact is required

    (b) viruses are required

    (c) free molecules of DNA are involved.

Extension questions

1. References to bacteria in the media often depict them as 'enemies' of humans. Respond on behalf of the bacterial world illustrating how, through their natural activities or their technological exploitation, many of them are really 'friends' of humans.

2. Outline the internal structural features that distinguish the prokaryotic cell from the eukaryotic cell.

3. Outline two important ways in which bacterial identification and classification differ from that for plants and animals

4. Complete the following table.

Nutritional type	Carbon source	Energy source	Examples (bacteria and eukaryotes)
Chemoautotroph			
Chemoheterotroph			
Photoautotroph			
Photoheterotroph			

5. What types of specialised bacteria might have been the evolutionary progenitors of (a) the mitochondria and (b) the chloroplasts of modern eukaryotic cells?

6. In January 1992 there were a number of reports in the media of 'blue-green algal blooms'. What organisms cause these blooms and what are the general features of the blooms and the implicated organisms? Why are these blooms significant to humans and why do they tend to occur seasonally?

7. If a bacterium such as *Staphylococcus aureus* (which has a doubling or generation time of 30 minutes in many moist, carbohydrate-rich foods) produces sufficient toxin to cause food poisoning when it is present at one thousand cells per gram of food, how long, to the nearest hour, would it take a single cell introduced into food to become a hazard to consumers of the food? How long if one million cells per gram were required?

8. How do the mechanisms of genetic variation in bacteria differ from those in eukaryotes?

9. In a hospital outbreak of bacterial infection that does not respond as expected to an antibiotic, it is noticed that resistance to the antibiotic is developing in bacterial species different from the one in which it was first identified. Describe in genetic terms how this interspecies spread might happen.

10. Mobile genetic elements such as plasmids and bacteriophages are important in transmitting genes between cells in bacterial populations. Give examples of the sorts of biochemical changes that may spread in this way and describe how these changes could provide evolutionary advantage to particular bacteria.

11. Describe the consequences for crop plants of loss of microbial:

    (a) nitrification activity

    (b) denitrification activity

    (c) nitrogen fixation activity in soils on an 'organic farm' where addition of chemical fertilisers is not practised.

Suggested further reading

Madigan, M. T., Martinko, J. M., Parker, J. (1997). *Brock: Biology of Microorganisms.* 8th edn. New Jersey: Prentice-Hall.

This text has a strong environmental emphasis. It also includes excellent summary descriptions of current ideas on the evolution of cellular life on earth, the impact of bacterial life on the biosphere and the importance of associations among bacteria and other microorganisms in nature. An ideal starting point from which to explore the diversity of bacterial biochemistry and metabolism.

Microbiology Australia (published five times a year by the Australian Society for Microbiology)

Deals with important and topical bacteriological, virological and other microbiological issues, often in the form of short, readable reviews. For example, the March 1999 issue covers the development of bacterial resistance as a consequence of the clinical overuse of antibiotics and the feeding of antibiotics to farm animals to improve growth rates. The May 1999 issue is dedicated to general microbiological problems related to the quality and safety of public water resources in Australia.

Pelczar, M. J. Jr, Chan, E. C. S., Kreig, N. R. (1993). *Microbiology, Concepts and Applications.* New York: McGraw-Hill.

This book provides a detailed coverage of a range of topics: bacterial nutrition, metabolism, diversity, genetics, microbial diseases, ecology and industrial applications.

CHAPTER

34

Viruses

Viruses are subcellular genetic parasites that reproduce only in the cells of susceptible hosts. Viruses have been isolated from animals, plants, fungi, protists and bacteria, indeed from almost all cellular organisms that have been properly tested. There are even some viruses, called satellite viruses, that only reproduce in cells already infected with another virus. Viruses probably have origins as ancient as life itself and it is likely that they originated more than once, that is, they are *polyphyletic* in origin. Perhaps the best known viruses are those that infect human beings and cause the common cold, influenza, AIDS and various childhood diseases such as mumps, measles and chickenpox; but others cause diseases as diverse as foot-and-mouth disease of cattle and **myxomatosis** of rabbits (Fig. 34.1a), lysis of bacteria, La France disease of mushrooms and various yellowing and mosaic diseases of plants (Fig. 34.1b).

The discovery of viruses

There are many historical records of viruses: there are scars, probably of **smallpox** (Box 34.1), on the faces of Egyptian mummies 3500 years old; a person with a withered leg, typical of poliomyelitis, is part of a bas relief on an Eygptian tomb 3000 years old; and flowers with viral symptoms are common in Dutch paintings from the sixteenth century.

It was not until the end of the nineteenth century that the unusual nature of viruses was realised. At that time many diseases of humans, domesticated animals and plants had been shown to be caused by bacteria or fungi, most of which could be grown in nutrient media, and all of which had cells that could be seen in light microscopes. However, there remained several diseases that seemed to be caused by invisible pathogens. No characteristic particles could be seen in material from diseased individuals and no pathogens could be cultured from them. One of these diseases was **tobacco mosaic** (Fig. 34.2), which could be transmitted from plant to plant by pricking sap from a diseased plant into the stem of a healthy one.

Experiments by several plant pathologists in the 1890s found that the tobacco mosaic pathogen had rather unusual properties. It diffused from sap into a block of gelatin jelly, could be precipitated with ethanol but remained infective and, most importantly, could pass through a porcelain filter with pores small enough to retain all known bacteria. These and other tests suggested that the pathogen was very small, and W. M. Beijerinck, a Dutch pathologist, concluded that tobacco mosaic was caused by an unusual pathogen, which he described as a *contagium vivum fluidum*, namely, a contagious living fluid.

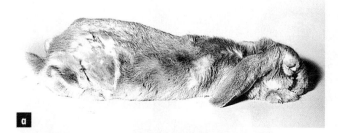

Fig. 34.1 (a) European rabbits suffering from myxomatosis caused by myxoma poxvirus. The natural hosts of this virus are the cottontail rabbits of the Americas, in which it causes localised warts, often on the ears or neck, where its mosquito or flea vectors feed. When European rabbits were taken to South America in the 1890s for laboratory use, some became infected with the virus and developed the very serious disease myxomatosis. The potential for using myxoma virus to control rabbit plagues was realised but it was not until the 1950s that the idea was tested in Australia. The first releases in Victoria were over-effective as the virus killed all the rabbits infected and it did not spread. The following wet year there were large populations of mosquitoes and less virulent variants of the virus spread rapidly along the Murray–Darling river system and killed most of the rabbit population. In the mallee of Victoria, where there are epidemics every year, rabbits that resist myxoma have appeared, but so too have strains of the virus that are more virulent. In 1996, the CSIRO released the rabbit calicivirus (RCD) for further viral-control of the European rabbit in Australia. **(b)** White clover leaf infected with white clover mosaic potexvirus. This virus is common wherever white clover grows

The filtration test was also used at about the same time by Loeffler and Frosch in Germany to test the saliva of cattle suffering from foot-and-mouth disease. They showed that healthy cattle became infected when saliva from infected animals was sprayed into their nostrils, even if it had been filtered. Soon, many other atypical pathogens of animals, plants and even bacteria were found to behave similarly, and became known as 'ultra-filterable viruses' and then just 'viruses' from the Latin word, *virus*, meaning slimy liquid or poison, although those that infect bacteria (Chapter 33) have retained the name **bacteriophage**, meaning bacterium eater, or just phage.

The nature of viruses, however, remained a mystery until the 1930s, when they were studied by various new

BOX 34.1 Smallpox

Variola (or smallpox) virus, with its ability to kill up to a third of the people it infects, has been a scourge of humanity for many centuries. Human communities free from variola virus, like those in the Americas before Europeans invaded in the sixteenth century, were particularly susceptible. When variola virus was inadvertently introduced into Mexico in 1520 by the Spanish invaders led by Cortés, millions of the indigenous people of the Americas were killed. Even in Roman times it was known that those who recovered from smallpox rarely became infected again, and so children were deliberately infected with mild strains of the virus, a risky technique called variolation. Safety was greatly improved when Edward Jenner, a British physician, reported in 1798 that deliberate infection with vaccinia (or cowpox) virus gave stable immunity to variola virus. This technique became known as **vaccination**. It was used in an **immunisation** campaign organised by the World Health Organization and led to the worldwide eradication of the virus in the human population by 1979. This unique operation was successful because:

- no animal reservoir of the virus was available
- a stable freeze-dried vaccine was available that could be administered simply by scratching and gave lifelong immunity
- the virus spread obviously but relatively slowly, often only to members of a family, so that 'contacts' of each infected person could be vaccinated and protected from infection, thus breaking the infection cycle.

Recently, ways have been found to transfer selected genes into the genome of vaccinia virus, so that they are expressed along with the genes of the virus. Vaccinia virus clones carrying the genes encoding the immunising antigens of a range of other microbes are being tested as multivalent vaccines for controlling diseases of humans and other animals, including rinderpest of cattle and even rabies in wild foxes.

Fig. 34.2 Capsicum leaf infected with a tobamovirus, probably pepper mild mottle virus

biochemical techniques developed in the branch of science now called molecular biology. In fact, most of the important advances of molecular biology have involved work with viruses.

In the 1930s, serological tests showed that the sap of tobacco plants with mosaic disease contained a novel antigen. The serum of rabbits immunised with the sap of infected plants reacted specifically with the sap of infected plants but not with sap from healthy plants. It was also shown that the sap of infected plants, like some shampoos, showed streaming birefringence in that, when sap that had been clarified by heating and centrifuging was stirred, the plane of polarisation of light passing through it was changed. This indicated that the sap contained particles that were either disc- or rod-shaped. No particles could be seen in this sap using a light microscope, indicating that the particles were less than about 400 nm in diameter, as this is the limit of resolution of a light microscope. However, minute particles had been seen in fluids from chickens infected with fowlpox, indicating that the particles of that virus might be larger than those of tobacco mosaic virus (TMV).

In the late 1930s, the electron microscope was invented and, with its greatly increased resolution, the sap of tobacco plants with mosaic disease was always found to contain characteristic rod-shaped particles, the structure of which was soon determined using the new technique of X-ray diffractometry. In the 1950s and 1960s, the characteristic particles, or **virions**, of a great range of viruses were purified and their beauty and complexity revealed (Fig. 34.3). Virions were shown to contain nucleic acids, proteins and, in some, lipids. It was experiments with T2 phage and turnip yellow mosaic virus that first showed that the nucleic acids, not the proteins, contained the viral genetic information.

Viruses: subcellular parasites

Features distinguishing viruses from cellular organisms

The fact that virions were like chemicals and could be purified, and some even crystallised, led to much speculation about whether viruses were living or dead. Biochemical work showed that virions are metabolically inert and consist of the genome of the virus in a protective and infective package ready to be transported to another susceptible cell; they are not equivalent to the cells of cellular organisms. They can be crystallised merely because they are of uniform shape and size.

When a virion enters the cell of a suitable host, it disassembles partly or wholly, the genome it contains

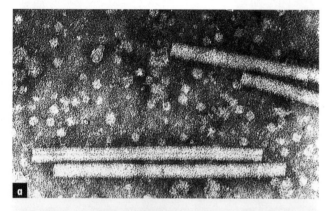
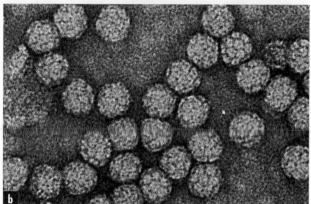
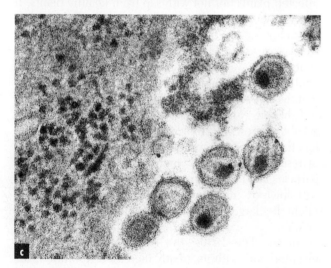
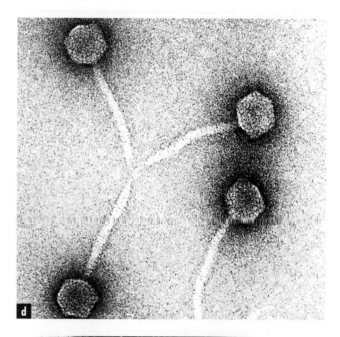
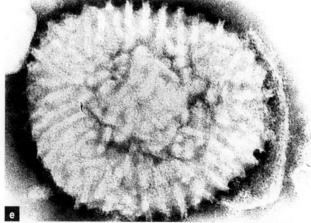

Fig. 34.3 Virions of: **(a)** tobacco mosaic tobamovirus (300 nm long); **(b)** turnip yellow mosaic tymovirus (28 nm in diameter); **(c)** human immunodeficiency lentivirus (HIV) (about 100 nm in diameter); **(d)** lambda phage (head diameter about 60 nm, tail about 150 nm long); **(e)** vaccinia poxvirus (about 300 nm in diameter)

usurps the host's own genes and diverts the metabolism of the host into reproducing the virus. Thus, viruses differ from cellular organisms in having two clearly defined phases in their biochemical life cycle. Firstly, there are the characteristic but metabolically inert virions, which are the transmission phase of the virus. These alternate with the reproductive phase, in which the virus consists of metabolically active viral genes. The viral genes use the metabolic systems of the host to replicate themselves and, using host ribosomes, produce a range of viral proteins; the progeny genomes assemble with virion proteins to form virions. By contrast, cellular organisms in all stages of their life cycles consist of cells that are bounded by cellular membranes and contain complete and largely independent metabolic systems that include mitochondria and ribosomes (Chapter 3).

Viruses are subcellular genetic parasites that infect all types of cellular organisms and many cause diseases.

Virions: the transmission phase

The simplest virions are regular geometric structures built from many copies of a virus-encoded protein, usually called the virion, or coat, protein and enclosing and protecting the viral genome.

The virions, for example of TMV (Fig. 34.3a), are rod-shaped and consist of a tube made of about 2100 helically arranged copies of a single type of coat protein (Fig. 34.4a). The protein subunits are grooved where

they contact one another, so that a molecular tube is formed, and wound into this is the genome of the virus, which is a single molecule of ribonucleic acid about 6400 nucleotides long; three nucleotides are tucked into and between each protein subunit. Many other viruses, including turnip yellow mosaic tymovirus (Fig. 34.3b), have isometric virions with the genome centrally folded within a protein shell made from 180 protein subunits arranged as an icosahedron. An **icosahedron** has a surface with 20 triangular facets (Fig. 34.4b). These two basic structures (rod and icosahedron) recur, in various forms, in the virions of most viruses. In viruses of animals, virions often have an outer lipid envelope (Fig. 34.4c). This contains viral proteins and is acquired as the virion buds from the surface of the infected cell.

The largest virions, those of poxviruses (Fig. 34.3e), are brick-shaped and, at about 300 nm in size, are just visible in a light microscope. They contain over 100 different protein species and are as genetically complex as the simplest bacteria.

Bacteriophages

The most obviously complex virions are those of the T2 and T4 phages. Each virion has a large rounded 'head', which contains the genome, and a complex tail with which it attaches specifically to a bacterial host cell by terminal fibres (Fig. 34.5). The outer layer of the tail then shortens and the core of the tail penetrates the host wall, like a hypodermic syringe, so that the viral genome can enter the bacterial (host) cell. Other

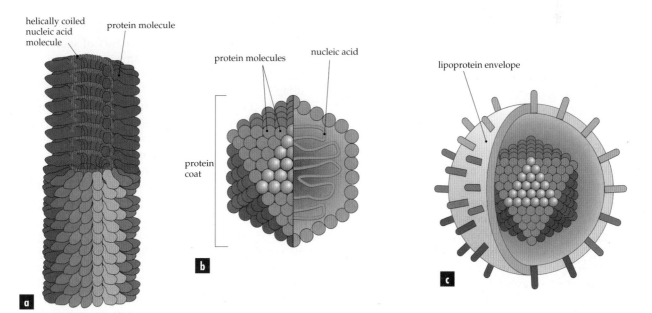

Fig. 34.4 Virions have two basic structures. **(a)** Rod-shaped virions, such as those of tobamoviruses, are tubes constructed of helically arranged protein subunits protecting a helically coiled nucleic acid molecule. **(b)** Isometric virions, such as those of the picornaviruses or turnip yellow mosaic tymovirus, have a nucleic acid core with a protective surface protein coat constructed of many protein subunits arranged as an icosahedron (20 triangular faces). **(c)** Virions of many viruses, especially those of animals, have a rod-shaped or isometric core enclosed in a lipoprotein envelope

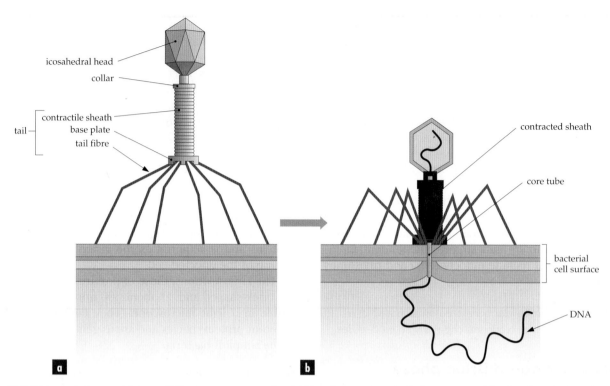

Fig. 34.5 Virions of phages with large DNA genomes are complex. **(a)** They have an icosahedrally constructed head and a tubular tail with a contractile sheath and terminal fibres for attachment to the host. **(b)** The sheath contracts during infection, enabling the genome to pass through the tail into the bacterium

phages, such as lambda phage (Fig. 34.3d), have particles of similar size but their tails are not contractile. Many of the simplest virions, such as those of TMV, can be disassembled into their constituent parts and reassembled in vitro; however, the larger virions, such as those of the T2 phage, cannot as their assembly involves several irreversible steps.

> As virions, viruses are metabolically inert, contain the genome in a protective package and are infective. When in a metabolically active phase, they consist of viral nucleic acids and proteins that replicate in a host cell.

Viral genomes

The genome of a virus is the minimum set of genes required to cause infection. In most viruses, the genome is packaged in a single infectious virion, but some viruses with divided genomes package the parts separately. Viral genomes are much more varied than those of cellular organisms; although smaller, they differ much more in size and composition. The genomes of some viruses are RNA, while others are DNA. Some, of both types, are single-stranded molecules, others are double-stranded. Many of the smallest viruses, such as TMV, have genomes that are single molecules of about 6000 nucleotides of single-stranded RNA (ssRNA). The largest genomes, such as those of the poxviruses of vertebrates, are single

BOX 34.2 **Genomes and bacteriophages**

Bacteriophages are diverse. Many have been involved in pioneering experiments of molecular biology and are used in many of the standard techniques of biotechnology. T2 and lambda phages have large dsDNA genomes: that of lambda phage is of 48 500 base pairs and that of T4 is four times larger. The phages with DNA genomes, together with plasmids, are an integral part of the gene pool of bacteria; genes are known to move between the different types of genome. Phages of this sort have been isolated from the Archaea as well as the Bacteria (Chapter 33). Some, such as T2 phage, infect, replicate in and lyse host cells. Therefore, they require a sustained succession of susceptible host cells to survive, whereas others, such as lambda phage, sometimes also integrate their own genome into that of their bacterial host and remain latent for several cell cycles but later lyse their host and become virulent. The virions of the tailed phages are constructed from subassembled components encoded and controlled by large numbers of genes (see figure). There are also other types of phage. M13 phage infects certain bacteria by attaching to their sex pili. It has a circular ssDNA genome of only 6407 nucleotides and variants of it are used by molecular biologists for cloning and sequencing genes. There are also other phages, such as MS2 and Qβ phages, specific to bacteria with sex pili, which have isometric virions and ssRNA genomes of only about 3500 nucleotides.

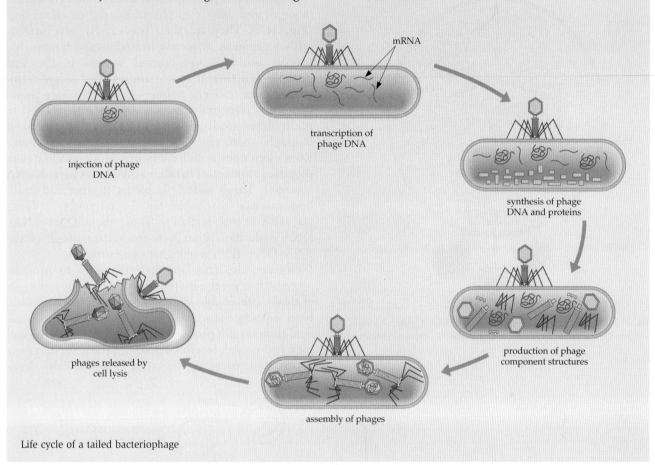

Life cycle of a tailed bacteriophage

molecules of 160 000–190 000 base pairs of double-stranded DNA (dsDNA), and those of the poxviruses of insects may be even larger. Some of the smaller ssDNA genomes are circular and some of the larger dsDNAs have the ends of the complementary strands, at each end of the genome, covalently linked so that they too are, in essence, circular. Many of the ssRNA and all the dsRNA genomes are divided into several parts.

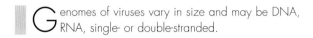

Genomes of viruses vary in size and may be DNA, RNA, single- or double-stranded.

Replication

The replication strategies of viruses are as varied as their genomes. They reproduce asexually (Fig. 34.6) but in almost all groups of viruses there is clear evidence of genetic recombination, or of reassortment of the parts of divided genomes.

Large double-stranded genomes replicate using biochemical pathways closely similar to those used by the genomes of the cellular organisms they infect. Viruses with single-stranded genomes replicate after first producing a nucleic acid strand complementary to the genome. For example, the ssRNA genomes of TMV and many other viruses, including foot-and-mouth disease virus and poliovirus, are transcribed into complementary strands by replicase complexes, which include viral and host proteins. These complementary strands then are transcribed repeatedly to produce progeny genomes. Some genomes, called plus-stranded genomes, are translated directly, although some of their genes may be translated from subgenomic mRNAs, and these genomes can infect cells even after being chemically deproteinised. By contrast, the genomes of other viruses, such as influenza and rabies, are negative strands. They are not infectious after being chemically extracted from virions because, before their genes can be translated, they must be transcribed into a

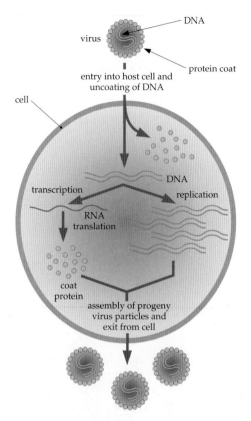

Fig. 34.6 Life cycle of a simple virus with a dsDNA genome and a single virion protein. In reality, viral replication is more complex

complementary plus strand. This is done by a viral replicase that is carried in infectious virions.

The most bizarre replication strategy is that of the **retroviruses**, which include human immunodeficiency lentivirus (HIV). Retroviruses have ssRNA genomes which, upon infection, are transcribed into dsDNA and incorporated into the chromosomes of their host (Fig. 34.7). They are then transcribed into progeny ssRNA genomes. It is easy to understand how such a process could disrupt control systems in the host genome and thus it is no surprise that many retroviruses cause cancers. Sequencing studies have found that some retroviral enzymes are distantly related to transposons (small sequences of DNA that can insert themselves into chromosomes at numerous locations). Even more unexpected, retrovirus genomes show clear sequence similarities to other viruses that have dsDNA genomes. These include hepatitis B virus and cauliflower mosaic virus, which have been found to replicate via ssRNA intermediates. Thus, their DNA–RNA–DNA replication strategy is the mirror image of the RNA–DNA–RNA replication of retroviruses.

Viruses also translate their genomes to produce proteins in a great variety of ways. Many, including all of those with double-stranded genomes, are translated from mRNAs transcribed from the genome, but some viral proteins are produced as large **polyproteins** and then hydrolysed, by virus-encoded proteases, into their constituent parts. Some viral genomes also have

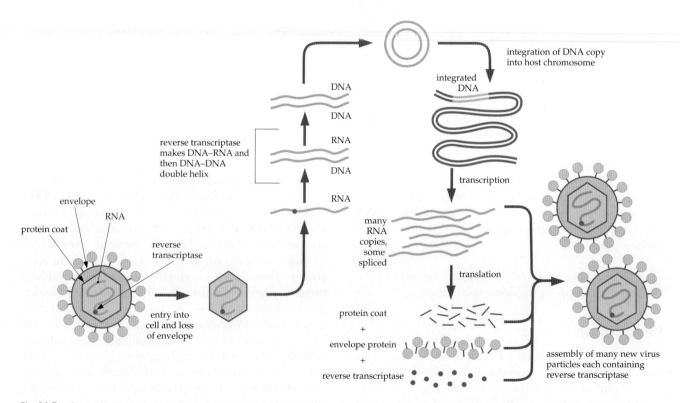

Fig. 34.7 Life cycle of a retrovirus. Reverse transcriptase from the infecting virion transcribes the RNA genome into dsDNA, which is then integrated into the host genome before being transcribed and translated to produce the constituents for progeny virions

overlapping genes, which are read from a single stretch of genome but using different reading frames (i.e. they start translating from nucleotides that are not separated by multiples of three nucleotides). Others, both dsDNA and ssRNA genomes, are ambisense in that different parts of one genome, or genome segment, are translated in opposite directions.

Viruses reproduce asexually but there is evidence of genetic recombination. They use a variety of mechanisms for genome replication and protein production.

Ecology of viruses

In addition to their diverse biochemical life cycles, viruses have ecological life cycles that are similarly diverse. Many animal viruses rely on the active lifestyle of their hosts for dispersal. They spread directly: coughing and sneezing produce infective droplets containing influenza and common cold viruses; faecal contamination of food and water spreads poliovirus and many others; **Epstein-Barr virus**, the cause of glandular fever, is spread during kissing; the transfer of body fluids during sexual intercourse spreads HIV; and some viruses are spread congenitally, or in milk, from mother to offspring. Few plant viruses (Box 34.3) are transmitted directly by contact, but some of these, such as TMV, are important in crop and glasshouse crops. More plant viruses are transmitted by pollen and by seed, but most, together with a significant proportion of the

picornaviruses. Interestingly, some potyvirus and picornavirus proteins have similar sequences. Potyviruses cause leaf mosaics, and some cause attractive colour changes in flowers that may be prized; in this way, for example, tulip flower-breaking potyvirus is useful in the horticultural industry (see Fig. a).

Luteoviruses cause many important plant diseases, especially of temperate cereals. They have isometric virions and an ssRNA genome. Like the potyviruses, they are transmitted by aphids but neither replicate in those vectors. However, whereas potyviruses are spread by probing aphids flitting from plant to plant in search of a suitable host and are rapidly lost by aphids, luteoviruses only infect phloem cells and are thus only spread by aphids after long feeding periods and may survive in aphids for their lifetime. This difference in vector relations affects the pattern of spread of these viruses: luteoviruses can be transmitted over long distances but may be controlled by insecticides, whereas potyviruses only spread over short distances and are not controlled by many persistent insecticides.

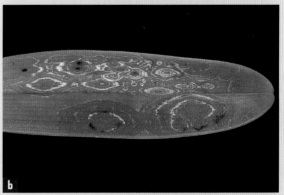

(a) Tulip flower showing symptoms of tulip flower-breaking potyvirus. (b) *Cymbidium* mosaic potexvirus symptoms in a leaf of the native orchid *Dendrobium speciosum*

BOX 34.3 Plant viruses

Tobacco mosaic tobamovirus (TMV), as described earlier (Figs 34.2, 34.3a), has been the subject of many pioneering viral studies. It and other tobamoviruses are transmitted when plants touch, and some, such as tomato mosaic tobamovirus (ToMV), are transmitted on seeds. Nowadays, TMV and ToMV are controlled in crops by using varieties into which natural resistance genes from wild relatives have been transferred. Viruses continue to be important crop pathogens despite success with controlling some. Most problems come from **potyviruses**, named after the type species, potato virus Y, and the **luteoviruses**, which cause yellowing symptoms. Potyviruses are the commonest plant viruses and over 300 have been described. They have filamentous virions that contain the ssRNA genome, which is translated as a single polyprotein, like that of the viruses of mammals and birds, are transmitted by **vector** organisms, which are usually mobile pests or parasites of the viral host.

Vectors of viruses of vertebrates, such as mammals or birds, include mosquitoes, ticks and midges, whereas those of plants are mostly aphids, beetles, leaf-hoppers, mites, thrips, whitefly and other arthropods above ground, and nematodes and fungi below. Some vectors act merely as mobile needles but others are infected by the virus they are carrying, and hence the distinction between host and vector is blurred. Some viruses of sap-feeding aphids and leaf-hoppers spread between individuals through their plant host without replicating in it; thus, the plant is the vector!

Symptoms of virus infection

Virus infections cause a great variety of symptoms. These are not merely a sign of illness but are a specific feature of the ecology of each virus and are important in its dispersal. For example, the coughing and sneezing caused by the common cold and influenza viruses are essential for their spread. Mosquito-borne viruses often cause fevers in their vertebrate hosts; the resultant increased body temperature and production of carbon dioxide both attract mosquitoes and ensure continued spread. Similarly, viruses of plants that are transmitted by aerial vectors, such as aphids, whitefly and leaf-hoppers, often cause bright yellowing symptoms which, it is known, attract these flying vectors.

All viruses have a defined and specific ecological life cycle. Many animal viruses rely on the active lifestyle of their hosts for dispersal and spread directly through the environment, such as droplets produced by coughing. Most plant viruses are spread by vectors, which are usually migratory pests or parasites such as aphids.

Classification and relationships of viruses

Viruses, like other organisms, fall into more or less well-defined species. These then fall into larger groups or genera, and some genera form even larger groupings (Box 34.4). However, there is no universal phylogenetic tree joining all viruses, indicating that they are probably polyphyletic in origin.

The virus isolates that form a single species are closely related in all their features so they usually cause similar symptoms in the same host species and are transmitted in the same way. Their genomes may differ in nucleotide sequence but by no more than a few percentage points, and so the proteins they encode are closely related in serological tests.

So what is it, in the absence of sexual reproduction, that holds a viral species together? The answer came first from experiments with an isolate of Qβ bacteriophage, which had been passaged many times without change in its genomic sequence. The Qβ genome is ssRNA, which has a much larger replication error frequency than DNA. It was found that although new variants of Qβ phage were constantly arising in the apparently stable population, one genome sequence, the 'master copy', was favoured during competition between variants. Thus, stabilising selection (Chapter 32), sometimes called purifying selection, operating on variants of a genome inherited from a common ancestor, maintains well-defined viral species.

The several viral species that form each genus or group usually share all the major features of their biochemical and ecological life cycles, such as the structure and composition of their virions, their replication strategy and mode of spread. They differ most often in host specificities; for example, tobacco and tomato are the preferred natural hosts of the closely related tobacco and tomato mosaic viruses, and herpes simplex virus I and II are similarly close but infect the lips and genital regions respectively.

Viruses have been isolated from almost all cellular organisms that have been carefully examined and, from

---

BOX 34.4 Viruses infecting humans

Viruses that infect humans and other animals are classified into a number of genera and families, the names of which may describe the group's characteristics. For example, picornaviruses have 'pico' (small) virions and have an 'rna' genome, whereas many poxviruses, including smallpox and fowlpox, cause pox symptoms (see figures). In humans, viruses may cause death (AIDS), permanent disability (polio) or be relatively harmless (warts) depending on the type of virus, the cells it infects and its ecology and hence survival strategy. The following are a few examples.

'Flu virus

Influenza orthomyxovirus causes a disease that has been known for several centuries and gets its name from the Italian word meaning 'influence', as the Italians thought the disease was caused by an evil influence of the stars, and from the Greek word, *muxa*, meaning mucus. This virus infects the lungs and respiratory tract of human beings. It is spread by droplets produced during coughing and causes epidemics in winter when people congregate indoors. The number of people infected and the severity of the disease vary from year to year; in the 1918 world epidemic, millions died. Influenza virions (Fig. c) have a lipid outer membrane containing two types of surface protein: haemagglutinin (HA) and neuraminidase (NA). Infected individuals develop antibodies to HA and NA and, as a result, resist infection. However, the HA and NA genes evolve by nucleotide changes, and hence the proteins vary antigenically from year to year, so the virus is able to escape neutralisation by antibodies resulting from past infections and people may be reinfected.

Two species of influenza virus are common in the human population, the most important of these being influenza A virus. This virus is also common worldwide in wild water birds, especially ducks and gulls, in which it usually causes a symptomless gut infection. Influenza A is very variable and isolates from different hosts, or collected on different occasions, often have antigenically different HAs and NAs. The genome of influenza A is in eight parts and these may reassort in mixed infections to produce strains with novel combinations of HA and NA. Such reassortants are able to infect hosts that had recovered from previous infections with other strains. Reassortants of human and other influenza As often appear in South-East Asia, where farmers live close to their stock, especially ducks and pigs. This is the likely source of the viruses that produced the great epidemics of Hong Kong and Asian influenza A in 1957 and 1968.

AIDS

Human immunodeficiency lentivirus (HIV) is the cause of the acquired immunodeficiency syndrome (**AIDS**) and perhaps the best known member of the retrovirus family (Fig. e). These viruses are called retroviruses because, during replication, they transcribe their ssRNA genome into dsDNA. HIV is a species of the genus *Lentivirus*, which includes many viruses that infect monkeys and other mammals chronically. HIV is spread during sexual intercourse, by blood products and from mother to fetus. It is not surprising that there has been little immediate success in designing a vaccine to control HIV as its major long-term effect is to destroy parts of the immune system. Since the early 1980s it has spread widely among humans, probably from monkeys in Africa.

Murray Valley encephalitis

Murray Valley encephalitis flavivirus (MVEV) is an arbovirus (arthropod-borne virus). There are several unrelated genera of arboviruses, all with an ecological life cycle that involves alternately infecting and replicating in blood-feeding arthropods, usually mosquitoes or ticks, and vertebrates, usually birds or mammals. MVEV is a flavivirus, named after the type species, yellow fever virus; the Latin for yellow is *flavus*. It is a virus of river systems and the tropical north of Australia, and normally infects birds and mosquitoes without causing obvious symptoms. When human beings are infected, most develop a fever but some develop a severe encephalitis that may be lethal. **Ross River alphavirus** is also an arbovirus but from another viral genus. It is widespread throughout Australia and infects mosquitoes and mammals, probably rodents and flying foxes. In humans it causes fever and an arthritis-like swelling of the joints.

Chickenpox and shingles

Varicella zoster herpesvirus (VZV) causes the diseases chickenpox and shingles, being named after the medical names of the two diseases it causes. Like other **herpesviruses** (Fig. d), VZV has complex isometric virions with a loose envelope and a dsDNA genome of 125 000 nucleotide pairs. Chickenpox typically occurs in children and shingles in adults. When children recover from chickenpox, the virus remains in one or more of the dorsal root ganglia of their spinal cords. In later life, stress activates one of these latent infections and the virus spreads through the nerves, causing a painful rash of the skin area supplied with nerves from that particular ganglion. The virus then spreads from the adults with shingles to susceptible children. Thus, a long period of latency in a ganglion, protected from the immune system of the host, enables the ecological life cycles of VZV and its host to be matched in time despite them having biochemical life cycles measured in minutes and years respectively.

Herpesviruses and cancer

Among the herpesviruses that affect humans are the **herpes simplex viruses**. Herpes simplex virus type 1 (HSV 1) is the cause of cold sores but it has been implicated in lip cancers. Herpes simplex virus type 2 (HSV 2) causes the majority of genital herpes infections and has been linked to cervical cancer in women. Epstein-Barr virus (EBV), another type of herpesvirus, causes lymphocytes to proliferate in the blood. It is linked to Burkitt's lymphoma (a cancer of the lymphoid system) and Hodgkin's disease (another type of lymphoma).

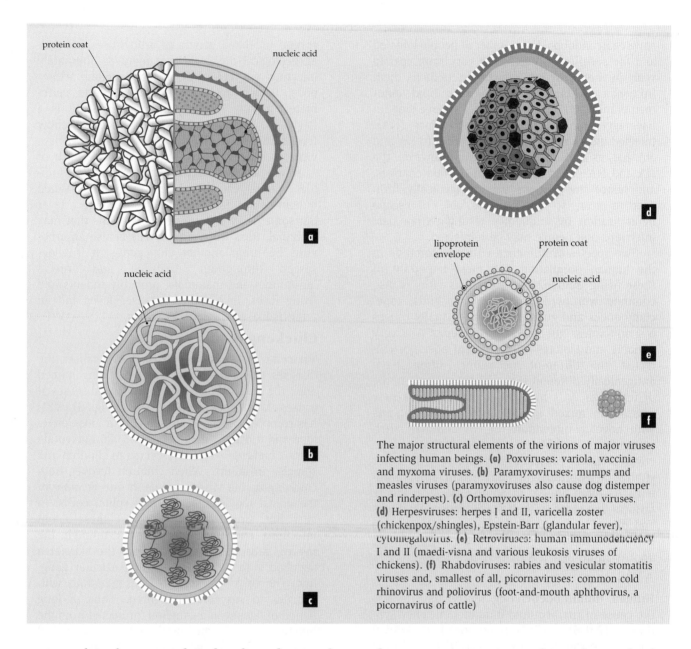

protein coat

nucleic acid

nucleic acid

lipoprotein envelope

protein coat

nucleic acid

The major structural elements of the virions of major viruses infecting human beings. **(a)** Poxviruses: variola, vaccinia and myxoma viruses. **(b)** Paramyxoviruses: mumps and measles viruses (paramyxoviruses also cause dog distemper and rinderpest). **(c)** Orthomyxoviruses: influenza viruses. **(d)** Herpesviruses: herpes I and II, varicella zoster (chickenpox/shingles), Epstein-Barr (glandular fever), cytomegalovirus. **(e)** Retroviruses: human immunodeficiency I and II (maedi-visna and various leukosis viruses of chickens). **(f)** Rhabdoviruses: rabies and vesicular stomatitis viruses and, smallest of all, picornaviruses: common cold rhinovirus and poliovirus (foot-and-mouth aphthovirus, a picornavirus of cattle)

some, such as humans and *Escherichia coli*, several dozen have been isolated. Most viruses seem to be confined to a single host species in nature and thus there are probably more species of viruses than of cellular organisms. However, at present only about 4000 of the 10 000 named viruses are well-studied enough to be recognised by the International Committee on Taxonomy of Viruses, and these fall into about 250 genera, some of which are grouped into about 60 families and three orders.

Recent work on the nucleotide sequences of viral genomes and the amino acid sequences of viral proteins has started to reveal the complex interrelationships of viral genera and their origins. It has been found that many viral genes, in particular the genes that encode the essential 'housekeeping' proteins, such as viral enzymes involved in genome replication, viral proteases

and virion proteins, are members of gene families (Table 34.1). Some of these gene families span viruses of plants, animals and bacteria. Some are encoded in DNA genomes and others in RNA genomes, and some are clearly related to host genes. Thus, the gene families are clearly very ancient. By contrast, some genes are only found in particular virus groups or individual viruses, and it is likely that many of these have arisen recently. These genes include the overlapping regulatory genes of some retroviruses.

The relationships indicated by one gene frequently do not coincide with those indicated by others. This indicates that viral genera have probably arisen by genetic recombination between genes from diverse sources and that genetic recombination has had a much greater influence on the evolution of viruses, at all taxonomic levels, than during the evolution of cellular

Table 34.1 Amino acid sequences of part of the replicase proteins of various viruses. The triplet GDD (gly-asp-asp) sequence, aligned here, is shared by almost all viral replicases[a]

Tobacco mosaic tobamovirus	QRKSGDVTTFIGNTVIIAACLASMLPMEKIIKGAFC**GDD**SLLYFPKGCEFPDVQHSA
Cucumber mosaic cucumovirus	QRRTGDAFTYFGNTIVTMAEFAWCYDTDQFDRLLFS**GDD**SLAFSKLPPVGDPSKFTT
Sindbis alphavirus	KSGMFLTLFVNTVLNVVIASRVLEERLKTSRCAAFI**GDD**NIIHGVVSDKEMAERCAT
Poliovirus	SGTSIFNSMINNLIIRTLLLKTYKGIDLDHLKMIAY**GDD**VIASYPHEVDASLLAQSG
Potato Y potyvirus	TVVDNSLMVVLAMHYALIKECVEFEEIDSTCVFFVN**GDD**LLIAVNPEKESILDRMSQ
Qβ levivirus (phage)	SMGNGYTFELESLIFASLARSVCEILDLDSSEVTVY**GDD**IILPSCAVPALREVFKYV

(*a*) Note that the sequences are grouped by their similarities. The top three are the most similar but include two plant viruses and one animal virus, the second group is of one animal and one plant virus, and the third is a phage. It is uncertain whether the -GDD- motif of all six viruses indicates that their polymerase genes have a common ancestor or merely that they have the same function, but those of the top two, and probably the third, are so alike that they are probably phylogenetically related.

organisms, where recombination mostly only occurs between the individuals of single species. This is not surprising given the fact that one host cell may have the genomes of several different viruses mingling within it, whereas the genomes of cellular organisms are almost always separated from one another by cell membranes.

Although viruses are probably polyphyletic in origin, species closely related to one another are classified into genera and higher taxa. Stabilising selection operating on variants of a genome inherited from a common ancestor maintains well-defined viral species. Genetic recombination is as dominant in the evolution of viruses as in the evolution of bacteria, whereas in eukaryotes, recombination rarely occurs except among individuals of a single species.

Other virus-like agents

Satellite viruses

In addition to the conventional types of viruses discussed so far, there are **satellite viruses**, which are only able to replicate in cells also infected with a specific helper virus. The first one described was the satellite virus (STNV) of tobacco necrosis virus (TNV). STNV has a genome of only 1200 nucleotides that encodes just one protein, the virion protein. It relies for all its other functions on TNV and the host plant. Simpler still are the **satellite nucleic acids**. These rely on a helper virus for all their proteins and are transmitted within the virions of the helper virus.

Most satellites depress the replication of their helper viruses. Many of the RNA satellites of helper viruses with ssRNA genomes are around 350 nucleotides in length and are ssRNA circles. They reproduce by transcribing the complementary genomic strand into a long ssRNA consisting of many copies of the genome. This is then cut into unit satellite genomes by a folded part of each genome. This fold is called a *ribozyme* (Chapter 11), and it attaches by base pairing to several nucleotides on either side of those where it hydrolyses; thus, ribozymes are very specific and there is much interest in designing ribozymes for specific biotechnological uses. For example, DNA encoding a ribozyme to cut the genome of a virus might protect the host of that virus against infection.

Satellite viruses are only able to replicate in cells also infected with a specific helper virus.

Viroids and prions: virus-like infectious agents

Finally, there are some unusual infectious agents that are structurally different from viruses but similar or smaller in size. **Viroids** are virus-like but lack a protein coat. They resemble satellite RNAs in structure but are able to replicate alone in a suitable host. They cause important plant diseases, especially of tropical trees, and are possibly spread by insects.

Most peculiar of all pathogens are **prions** (proteinaceous infectious particles). They have some similarity to viruses in that they are small and infectious but they also sometimes arise spontaneously. They are the cause of the human disease **kuru** and **Jakob-Creutzfeldt disease** (CJD), as well as *mad cow disease* (**bovine spongiform encephalopathy**, BSE) and scrapie of sheep, and similar agents have been found in yeast. These diseases may spread between species; 'early onset CJD' may be caused by BSE.

The diseases are caused by the accumulation of the prion protein, as amyloid plaques, in the nervous system, which then degenerates slowly. Prion proteins appear to lack nucleic acid and consist only of protein, and are encoded in the healthy host's genome. There is still no total agreement on the cause of these diseases, however, the majority view is that the healthy cellular prion proteins become altered in some specific structural way when the host is infected with already altered prion protein. Thus, the alteration, although only structural, is 'infectious' and causes their deadly accumulation.

Infectious agents that are structurally different from viruses but similar or smaller in size are viroids and prions. Viroids are virus-like but lack a protein coat. Prions are proteinaceous infectious particles.

BOX 34.5 Virus control

Viruses are best controlled by attacking a vulnerable step, or steps, of their biology, namely their ecological or biochemical life cycles. Unfortunately, the full details of the biology of most viruses is not known but the biology of closely related viruses is often similar and so useful extrapolations can be made for designing control strategies.

Viruses are often most easily controlled by stopping their spread to healthy hosts. Virus sources, such as infected plants, should be destroyed if possible. The spread of contagious viruses can be controlled directly. For example, smokers should wash their hands before gardening as tobacco products sometimes contain infective tobacco mosaic virus and this gets onto hands and may infect tomato or tobacco plants when they are handled and hair cells are wounded. Similarly, washing hands slows the spread of the common cold virus because this virus usually spreads along the 'nasal mucosa–sneezes–hands–door knobs–hands–nasal mucosa' pathway. Barriers like condoms protect against the spread of HIV during sexual intercourse and face masks protect against 'flu. Mosquito-borne viruses are most easily controlled by controlling the vector population, not only in the wild by draining swamps, but also around houses by disposing of empty tins and old tyres that collect rain and by putting fish in water tubs. Similarly, towns provide a protective garden environment for weeds that harbour aphids and viruses over winter. Vector-borne viruses can also be controlled by avoidance, for example, arboviruses, like Murray Valley encephalitis flavivirus and Ross River alphavirus, are spread by mosquitoes that mostly feed around sunset, so adequate clothing, fly screens and insect repellents are important at that time. Handling fruit bats is dangerous at all times because they carry rabies and paramyxoviruses, which can be transmitted when they bite or scratch.

A viral infection is difficult to stop by biochemical means without harming the host because viruses use the biochemical systems of their host for replication. By contrast parasitic bacteria or fungi are biochemically distinct from animal or plant hosts and so can be controlled by antibiotics or antifungal compounds that are specific to the parasite. However, replication of some viruses can be usefully slowed by treatment with purine or pyrimidine analogues, such as inhibitor AZT (for HIV) and acyclovir (for cold sore; herpesvirus).

Viral hosts differ in their resistance to infection; some strains of plants and animals are naturally more resistant to infection than others and so can be selected for use. Vertebrates have a complex immune system (see Chapter 23) for controlling infections, and hence they can be immunised by injecting non-infectious (e.g. formaldehyde treated) or live but attenuated virions or proteins obtained from them. Immunisation with vaccinia virions (i.e. vaccination) was used to eradicate smallpox (see Box 34.1) and has also been widely and successfully used with a range of other viruses, including yellow fever flavivirus, poliovirus, rabies and rubella (German measles), but has not yet been successful against HIV (see Box 34.4). Plants do not have an immune system like that of vertebrates, however, it has been long known that a plant infected with a virus usually resists infection by the same virus or its relatives; a phenomenon called 'cross-protection' or 'pathogen-derived resistance'. This resistance also occurs when individual viral genes or parts of them are transferred to a plant and shows promise as a way of producing virus-resistant crops, although there is the danger that such crops may also be a source of genes for recombination with other viruses.

Summary

- Viruses are subcellular genetic parasites and have been found infecting all sorts of cellular organisms. Some cause important diseases of humans and domesticated animals, plants, bacteria and fungi.
- Viruses are probably polyphyletic in origin, not so much a family but more a way of life that is probably as ancient as life itself.
- Viruses have a characteristic two-phase life cycle. One phase consists of particles or virions, which are metabolically inert and contain the genome in a protective and infective package. The alternate phase consists of metabolically active viral nucleic acids and proteins replicating in the cell of a susceptible host.
- Viruses have genomes of diverse size and composition, some being single-stranded, others double-stranded, some are DNA, others RNA. Genomes range from 1000 nucleotides to over 300 000 base pairs in size.

- Viruses spread from one host individual to another in various ways. Many animal viruses rely on the active life of their hosts for dispersal. Other plant and animal viruses are transported by vectors, particularly insects. Some viruses of both animals and plants replicate in both the vector and the main host. Viruses are often best controlled by ecological intervention.
- Other virus-like agents include the satellite viruses, which are only able to replicate in cells also infected with a specific helper virus. More unusual infectious agents that are structurally different from viruses but similar or smaller in size are viroids and prions.
- Viroids are virus-like but lack a protein coat. Prions are proteinaceous infectious particles, which cause diseases in animals by the accumulation of prion protein in the nervous system.

key terms

AIDS (p. 899)
bacteriophage (p. 890)
bovine spongiform encephalopathy (p. 901)
Epstein-Barr virus (p. 897)
herpes simplex viruses (p. 899)
herpesviruses (p. 899)
human immunodeficiency lentivirus (HIV) (p. 899)

icosahedron (p. 893)
immunisation (p. 891)
influenza orthomyxovirus (p. 898)
Jakob-Creutzfeldt disease (p. 901)
kuru (p. 901)
luteovirus (p. 897)
Murray Valley encephalitis flavivirus (p. 899)
myxomatosis (p. 890)
polyprotein (p. 896)
potyvirus (p. 897)

prion (p. 901)
retrovirus (p. 896)
Ross River alphavirus (p. 899)
satellite nucleic acid (p. 901)
satellite virus (p. 901)
smallpox (p. 890)
tobacco mosaic disease (p. 890)
vaccination (p. 891)
varicella zoster herpesvirus (p. 899)
vector (p. 898)
viroid (p. 901)

virion (p. 892)
virus (p. 890)

Review questions

1. (a) What features distinguish viruses from cellular organisms?
   (b) What features do viruses and cellular organisms share?
   (c) Make a case for and against viruses being considered living organisms.
2. The filtration test (p. 880) was used in the early studies that led to the discovery of viruses. What was this test and what did it show?

3. Viruses are thought to be polyphyletic in origin. What does this mean?
4. What is a bacteriophage? Describe the structure of a phage virion.
5. Distinguish between a virus, a satellite virus, a viroid and a prion.
6. (a) Name a disease caused by a prion.
   (b) Describe the characteristic features of diseases caused by prions.
7. What do the acronyms HIV and AIDS mean? What is the name of the virus genus to which HIV belongs?
8. Describe two viruses that are beneficial to humans.
9. What do herpes simplex viruses cause in humans?
10. What might be the importance, to a virion, of having a protein coat?
11. How do plant viruses spread from one host individual to another? How are viruses of plants controlled?
12. What measures can be used to control viruses of mammals? Why are antibiotics ineffective for controlling virus infections?

Extension questions

1. Describe how retroviruses replicate. Given this form of replication, suggest an explanation for why many retroviruses cause cancer.
2. Why was it possible to eradicate variola (smallpox) virus from the human population and why is it unlikely that influenza virus and HIV will also be eradicated?
3. Search for information about emerging viruses on the World Wide Web; try any search engine with the word 'outbreak'. Then choose one of those viruses and from knowledge you can glean of its biology or that of its relatives:
   * design strategies for controlling the damage that virus might cause
   * rank those strategies in order of their likely success (give reasons)
   * discuss those crucial aspects of the biology of the virus that are required for planning its control but seem to be unknown at present
   * discuss the sorts of experiments that might be done to acquire the missing data.

Suggested further reading

Granoff, A. and Webster, R. G. (1999). *Encyclopedia of Virology*. 2nd edn. 3 vols. New York: Academic Press.

A resource for more detailed information.

'All the Virology on the www' at http://www.virology.net/

This website provides the latest information on viruses.

Pelczar, M. J. Jr, Chan, E. C. S., Kreig, N. R. (1993). *Microbiology, Concepts and Applications*. New York: McGraw-Hill.

This book provides a detailed coverage of viruses, including classification, structure, replication, methods of cultivation and diseases.

CHAPTER
35

The protists

Kingdom Protista includes a weird and wonderful potpourri of eukaryotic organisms that few people ever see. Most protists are unicellular and live in aquatic habitats. There are at least 100 000 species and new ones are being discovered continually. Photosynthetic protists are a major source of primary productivity in lakes, rivers and oceans, producing at least 30% of the planet's oxygen. Herbivorous protists are the link in food chains between algal primary producers and larger animal consumers, such as fishes and invertebrates. Parasitic protists are responsible for serious illnesses in humans, such as malaria, sleeping sickness and certain types of dysentery. Protists also parasitise animals and plants, causing agricultural losses.

Protists are diverse. Comparing two protistan phyla is analogous to comparing elephants with moths or eels with tomatoes. In the past, phyla were grouped together based on their form of nutrition—whether they were autotrophic (able to produce food by photosynthesis) or heterotrophic (consumers of organic substances or other organisms). Photosynthetic protists were known as algae, protists that ate smaller organisms were known as protozoa (simple animals), and some protists that absorbed small food molecules from the environment were considered to be fungi. It is now obvious that this system is far too simplistic. Numerous photosynthetic protists, for example, swim about like animals and even capture smaller cells and eat them. These organisms are both animal-like and plant-like and cannot be classified on the basis of nutrition. A more natural classification based on morphological, biochemical and molecular features is now emerging. Some natural groups include organisms with various modes of nutrition. Alveolates, for example, have photosynthetic, parasitic and predatory members, but all are close relatives based on ultrastructure and DNA sequence data.

From the evolutionary tree in Figure 30.10, you can see that protists are not a monophyletic group. The kingdom is essentially a default taxon, containing all the eukaryotic organisms that are not plants, animals or fungi. Kingdom Protista is thus artificial and many members are now known to be more closely related to other kingdoms than to each other. Green algae, for example, are the closest relatives of land plants (Chapter 37) and microsporidia are close relatives of fungi. So why do we put them all together in one chapter as though they were one evolutionary lineage? The answer is partly historical and partly practical. There are still groups of unicellular eukaryotes of unknown evolutionary relationships, some not even named. For convenience, these organisms are temporarily collected together under the banner of protists—a vernacular term that serves to describe them as well as any.

Protists may be photosynthetic, parasitic, predatory or absorb small food molecules from the environment. Relationships within protists are still unclear but they are a diverse range of eukaryotic cell types.

Origin of eukaryotic cells

The oldest fossils of eukaryotic organisms do not appear until about 1.5 billion years ago. Since fossils of prokaryotes are older, it is generally thought that eukaryotes evolved from prokaryotic organisms.

As we have seen in earlier chapters, prokaryotic and eukaryotic cells share many cellular processes but the internal layout of their cells is different. Prokaryotic cells are essentially one single compartment, whereas eukaryotic cells contain several membrane-bound subcompartments. So how did these subcompartments originate? The answers turn out to be quite a surprise.

Origin of the nucleus

The eukaryotic nucleus differs from the prokaryotic nucleoid in numerous respects. Two major distinctions are the nuclear envelope and multiple linear chromosomes of eukaryotes. Prokaryotes lack a nuclear envelope and usually have a single circular chromosome. Transformation from a circular chromosome to linear chromosomes might have arisen from a break in the circle and duplication of the linearised chromosome to give multiple copies.

The origin of the nuclear envelope can be explained by the accumulation of vesicles resulting from cell membrane invaginations around the prokaryotic nucleoid. If the vesicles flatten around the nucleoid, as shown in Figure 35.1, they then form a rudimentary double envelope complete with gaps or nuclear pores. Such accumulations of membrane vesicles around the nucleoid are known to occur in certain cyanobacteria (Chapter 33).

Origin of the endomembrane system

The endomembrane system forms a conduit from the nuclear envelope to various subcellular compartments and also to the exterior of the cell via the plasma membrane. It probably evolved as a means of sorting and transporting proteins and glycoproteins in large eukaryotic cells. The endoplasmic reticulum probably developed from protrusions of the nuclear envelope, to which it still remains attached (Fig. 35.1). Interestingly, the plasma membrane of prokaryotes bears ribosomes for secretion of proteins and internalisation of a ribosome-bearing membrane would create a rudimentary rough endoplasmic reticulum that could secrete proteins into its lumen. These protrusions could then

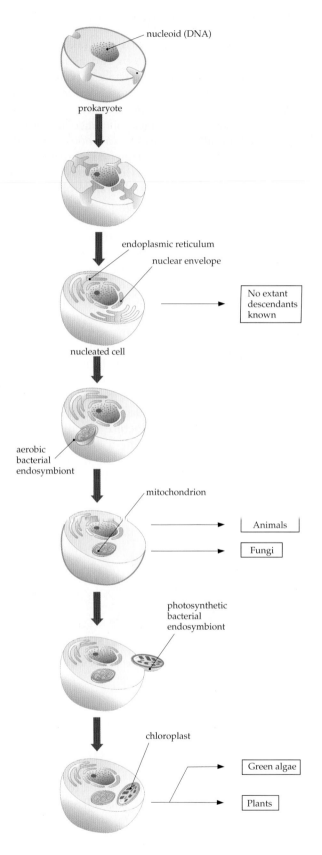

nucleoid (DNA)

prokaryote

endoplasmic reticulum

nuclear envelope

No extant
descendants
known

nucleated cell

aerobic
bacterial
endosymbiont

mitochondrion

Animals

Fungi

photosynthetic
bacterial
endosymbiont

chloroplast

Green algae

Plants

Fig. 35.1 The evolution of eukaryotic cells. The origin of the
nucleus and endomembrane system. Mitochondria originated from
an aerobic bacterial endosymbiont. Chloroplasts originated from a
photosynthetic bacterial endosymbiont

have become elaborated into the Golgi apparatus and
other components of the endomembrane network.

> The nuclear membrane and endomembrane system
> probably evolved from invaginations of the bacterial
> cell membrane that enveloped the nucleoid.

Origin of mitochondria and chloroplasts: endosymbiotic theory

Mitochondria and chloroplasts arose by an extraordinary process known as **endosymbiosis**, which
refers to an organism living inside another ('endo',
inside, 'symbiosis', living together). Chloroplasts and
mitochondria have long been recognised as having a
degree of autonomy within the cell. They divide before
the rest of the cell by fission, like bacteria. This led
nineteenth century microscopists to remark that
chloroplasts were reminiscent of cyanobacterial cells
living inside plant cells. The organelles also have
membranes separating them from the main cell
compartment. These ideas did not achieve much
acceptance, though, until the 1960s when it was
discovered that chloroplasts and mitochondria contain
DNA. With the revelation that the DNA in chloroplasts
and mitochondria are circular chromosomes (Chapter
10) and that the organelle genes were typically
prokaryotic, the endosymbiotic theory of the origin of
these organelles gained almost universal acceptance.

In fact, the more we look at chloroplasts and
mitochondria, the more convincing is the argument.
Chloroplasts and mitochondria have 70 S ribosomes
that contain rRNAs with nucleotide sequences similar
to bacteria. Like bacterial ribosomes, ribosomes of
chloroplasts and mitochondria are sensitive to the antibiotic chloramphenicol but insensitive to cycloheximide, which stops translation in eukaryotic cytoplasmic
ribosomes. Phylogenetic trees based on nucleotide
sequences of rRNAs actually group mitochondria and
chloroplasts with bacteria, not with eukaryotes.
Chloroplasts relate to cyanobacteria and mitochondria
to purple bacteria.

The circular chromosomes of chloroplasts and
mitochondria are considerably smaller than those of
their bacterial counterparts. They are so small that their
DNA can only encode a fraction of the proteins needed
in the organelle. The remaining proteins are encoded
by nuclear genes. Messenger RNAs (mRNAs) from
these nuclear genes are translated on 80 S ribosomes in
the cytoplasm and the proteins are translocated into the
chloroplast or mitochondrion. This was initially rather
puzzling but it is now believed that many of the
endosymbiont's genes moved from the organelle's
chromosome into the nucleus of the host. Exactly why
this should have occurred remains unknown but it
certainly serves to 'hobble' the endosymbiont by

making it absolutely dependent on the host for its survival.

One common feature of chloroplasts and mitochondria is the presence of a double membrane. The two membranes most probably derive from the two membranes that surround Gram-negative bacteria. The host plasma membrane that surrounded the endosymbiont during engulfment has apparently been lost.

The evolutionary tree in Figure 35.2 shows some of the major lines of descent. An endosymbiotic origin of eukaryotic organelles means that the tree actually has two grafts joining the prokaryotic line of descent to the eukaryotic line in at least two places: one for the mitochondrion of all eukaryotes and a second for the chloroplast of plants.

> Chloroplasts and mitochondria are derived from endosymbiotic bacteria that have become organelles in eukaryotic cells.

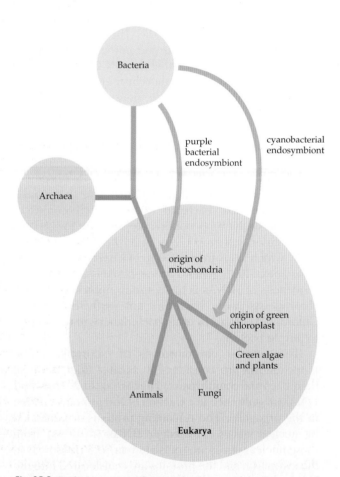

Fig. 35.2 Evolutionary tree showing the descent of the Bacteria and Archaea, animals, fungi and plants. Grafts joining lines of descent are formed by eukaryotic cells engulfing Bacteria (see Fig. 35.1), once for the origin of mitochondria, a second time for the origin of chloroplasts. Animal and fungal cells are chimaeras (derived from cells of two different organisms) of two evolutionary lineages and plant cells are chimaeras of three lineages

Origin of cilia or flagella

Cilia or flagella occur in most eukaryotic organisms. Although they are referred to by two names (cilia in animals and animal-like cells; flagella in plants, sperm, algae and flagellates), the two organelles are homologous, derived from a common ancestral structure (Fig. 35.3). Bacterial and eukaryotic flagella, however, are fundamentally different in both chemical composition and structure (Chapter 33) and appear not to be homologous. They are a case of convergent evolution: two similar solutions to the one problem—how to get around in a liquid medium.

So where did eukaryotic cilia and flagella come from? This is presently one of the most contentious questions in evolutionary cell biology. One school of biologists suggests that cilia or flagella arose as extensions of the cytoskeleton. A second school suggests that flagella or cilia are derived by endosymbiosis, one organism living inside another, in this case a spirochaete bacterium living within a eukaryotic cell. Some controversial experimental work suggests that cilia and flagella contain DNA. If this work can be substantiated, it would support the notion that cilia and flagella were originally organisms in their own right. A major component of cilia and flagella is tubulin protein. Recent studies of protein structure demonstrate that tubulin probably evolved from a bacterial protein known as FtsZ, which has a key role in bacterial cell division.

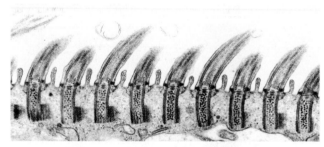

Fig. 35.3 Opalinid flagella. With the invention of the electron microscope it was discovered that cilia and flagella are essentially identical and differ only in length

Simple eukaryotes

To understand our own origins we need to know what the first eukaryotic cell was like. This cell, which existed more than one billion years ago, would presumably have been rather simple, fairly small, and might have lacked most of the structures currently recognised as hallmarks of eukaryotism. Does such a cell still exist today? Probably not. Although a number of protists that fit the above description have been regarded as potentially primitive examples of eukaryotes, it has recently emerged that these models, which were often known as the Archaezoa (oldest animals), have in fact undergone reversion from a complex state to a more simple cell organisation. For example, several protists such as microsporidia, diplomonads and trichomonads that lack mitochondria and were proposed to have diverged from the eukaryotic lineage prior to mitochondrial acquisition by endosymbiosis have now been revealed to have a cryptic mitochondrion. These discoveries have caused biologists to revise their models of early eukaryotes and the phylogenetic relationships of various protistan groups. The earliest lineages of eukaryotes are probably extinct. It is unlikely that small, ephemeral organisms will have left much trace in the fossil record.

P rimitive eukaryotes may no longer exist. Simple protists lacking major eukaryotic organelles are now recognised as having lost organelles and reverted to a simpler structure.

Photosynthetic protists

It appears likely that a single endosymbiosis produced the many different coloured chloroplasts observed in the protists. From this you might expect that all chloroplast-containing protists are closely related (descendants of the original host cell that acquired an endosymbiont) but the story is not that simple. The original chloroplast has apparently been faithfully handed down through hundreds of millions of years of evolution to the modern green algae and their descendants, the plants. However, other protist groups are now recognised to have stolen this chloroplast from the primarily photosynthetic lineage. They did this by simply engulfing algal cells and retaining them within their cells, much like the endosymbiosis of a cyanobacterium but this time with a eukaryotic endosymbiont. This means that heterotrophic eukaryotes can convert to autotrophy by taking the photosynthetic organelle from a distant relative. From this you can see that it is not valid to unite all chloroplast-containing protists into one group, traditionally labelled algae, because they do not share a common ancestor, only an acquired organelle.

The phyla discussed here have chloroplasts and are primarily photosynthetic. In some organisms the chloroplasts lack photosynthetic pigment and are called **leucoplasts**. Even though species with leucoplasts obtain their nutrition by means other than photosynthesis, they still retain bleached chloroplasts, suggesting that the organelle provides something to the cell in addition to food.

Flagellates with a photosynthetic endosymbiont: phylum Glaucacystophyta

Glaucacystophyta are living examples of an intermediate stage in the evolution of a chloroplast from a photosynthetic prokaryotic endosymbiont. Chloroplasts of glaucacystophytes are **cyanelles**, which have a **peptidoglycan** wall as do bacteria. The presence of the wall is evidence that the cyanelle was once a bacterium before it took up residence in the host cell. Cyanelles contain chlorophyll *a* and phycobilin pigments identical to cyanobacteria. Like a cyanobacterium, a cyanelle has a circular chromosome but it is no longer fully autonomous, having lost genes to the nucleus during the endosymbiotic relationship. Some genes for producing peptidoglycan have been found on the cyanelle chromosome, which is otherwise the same as a chloroplast chromosome. Cyanelles are thus partially dependent on the host cell and cannot survive independently. Host cells are typically flagellates with two laterally inserted smooth flagella.

G laucacystophytes are photosynthetic flagellates with chloroplasts, cyanelles, that have a peptidoglycan wall, as do bacteria.

Red algae: phylum Rhodophyta

Red algae (phylum **Rhodophyta**) are common seaweeds on rocky seashores around the world. There are some 4000 species, many of which are endemic to Australia (Chapter 41). Red seaweeds are of commercial importance in the production of **agar** for microbiology and molecular biology, and as food in the Orient, North America and Ireland. Sushi is prepared with dried *Porphyra*, Japanese *nori*. About 60 000 hectares of *nori* are grown by mariculture around the Japanese coast. **Carrageenan** from red algae is used also as a stabilising agent in confectionery, ice cream, cosmetics and pet foods. The red seaweed industry is worth about $1 billion per annum worldwide.

Most red algae are multicellular, adjacent cells often being attached by **pit plugs** (Fig. 35.4), and a few are unicellular. They form a plant with branches and blades plus extensions attaching the plant to the substrate. Red algae have complex life histories with alternating stages that are often markedly different in morphology. Some red algae are calcified and are known as coralline red algae because they were mistakenly thought to be coral animals.

Chloroplasts contain chlorophyll *a* and phycobilin pigments—phycocyanin and phycoerythrin (the latter producing the typical red colouration). Red algae absorb long wavelength blue and green light that penetrates deepest into the ocean, allowing them to

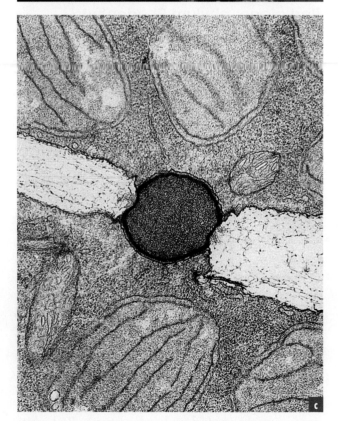

Fig. 35.4 Red algae range from **(a)** fine feathery structures to **(b)** crusty calcified plants resembling corals. **(c)** Adjacent cells are often attached by pit plugs

photosynthesise at depths of 250 m below the surface. The product of photosynthesis is stored in the cytoplasm as α-(1→4)-glucan.

Red algae lack flagella and basal bodies. Because their sperm cannot swim, they rely on the randomness of ocean currents to bring sperm to the female part of the **thallus** containing the egg. When a sperm does contact an egg to form a zygote, the alga capitalises on the event by distributing copies of the diploid nucleus to other female parts of the thallus. Thus, from a single fertilisation event, multiple spores can be produced for the next generation.

The lack of flagella and basal bodies was originally interpreted as a primitive character suggesting that red algae are ancient. Molecular analysis has failed to confirm this view, instead showing that red algae are advanced organisms that have lost the ability to produce flagella.

Red algae (phylum Rhodophyta) are familiar seaweeds. Most are multicellular and macroscopic and they lack flagella. They contain chlorophyll *a* and phycobilin pigments.

Green algae: phylum Chlorophyta

Green algae are a large group (about 16 000 species) including unicellular, colonial and multicellular forms. Their chloroplasts are grass green and contain the same pigments as land plant chloroplasts—chlorophylls *a* and *b*, β-carotene and other carotenoid derivatives. Like land plants, the product of photosynthesis of green algae is stored as starch (an α-(1→4)-glucan) within the chloroplast and the cell walls are primarily cellulose (β-(1→4)-glucan). These and other similarities leave us in no doubt that green algae are closely related to land plants (Chapter 37).

Green algae are common in most habitats and fix an estimated 1 billion tonnes of carbon from the atmosphere per annum. They are used as food (*Spirogyra* as vitamin supplement tablets) and are being tested in biotechnological applications (Box 35.1).

Green algae were classified traditionally on the basis of their form—unicellular, colonial, filamentous, **coenocytic** (technically unicellular but multinucleate and greatly enlarged to form a macroscopic thallus) and multicellular three-dimensional forms. Closer investigation with the electron microscope shows these categories to be somewhat artificial, with several cases of convergent evolution. Studies of mitosis, for example, have shown that two species from the filamentous genus *Klebsormidium* actually belong in different classes. Although superficially similar, the two species of *Klebsormidium* have different types of mitosis (the **phragmoplast** and **phycoplast** types described in Chapter 37) and fundamentally different

BOX 35.1 Green algae and biotechnology

Not only do green algae grow in a wide range of habitats, such as fresh water, oceans, salt lakes and snow, but they also show a great diversity in their chemistry. It is this chemical diversity, combined with the ability of some species to grow in extreme environments, that makes green algae attractive to biotechnologists.

Early studies in the 1970s focused on using algae, such as *Chlorella, Scenedesmus* and *Oocystis,* as sources of protein. Unfortunately, the algal protein proved to be expensive and was not always palatable. Since the early 1980s, the focus of algal biotechnology has shifted to the commercial production of high value chemicals, such as carotenoids, lipids, fatty acids and pharmaceuticals.

An important alga is *Dunaliella salina.* When grown at high salinity (about 10 times the concentration of sea water) and with high light intensity, *D. salina* accumulates large amounts of an orange-red carotenoid, β,β-carotene. This pigment compound is used to colour products, such as margarine, noodles and soft drinks, and as a vitamin supplement because it is readily converted to vitamin A. There is also evidence that β,β-carotene may help prevent lung cancer. Pure β,β-carotene is worth more than $600 per kilogram. Production of β,β-carotene from *D. salina* means growing and harvesting vast quantities of algae in 'farms'. The world's largest algal farms are at Hutt Lagoon in Western Australia (see figure) and Whyalla in South Australia.

With its wide flat spaces and intense sunshine, Australia is the perfect place for algal farms producing food, fuel and pharmaceuticals. These ponds of *Dunaliella salina* at Hutt Lagoon, Western Australia, range in colour from green to brick red depending on how much of the valuable β,β-carotene cells have accumulated

Another alga under study is the freshwater chlorophyte *Haematococcus pluvialis,* which is the best natural source of the carotenoid astaxanthin. Astaxanthin is used in aquaculture as a fish food additive to ensure trout and salmon flesh is the natural pink colour. Fish food currently contains synthetic carotenoids and astaxanthin is a desirable natural alternative.

Green algae may also be a future source of alternative fuels. *Botryococcus braunii* produces long-chain hydrocarbons similar to crude oils, and these can be cracked in a refinery to produce petrol and other useful fractions. *Tetraselmis* species accumulate fats and oils and, once extracted, the lipids can be used as a diesel fuel substitute.

The scope for algal use in producing pharmaceuticals, antibiotics, fuels and foods, and as an adjunct in waste treatment is enormous. Manipulation of strains by genetic engineering will contribute to the production of useful natural substances.

---

motile cells. A filamentous thallus therefore seems to have evolved more than once in the green algae. Current classification recognises five classes: Prasinophyceae, Chlorophyceae, Ulvophyceae, Charophyceae and Conjugatophyceae.

Primitive green algae: class Prasinophyceae

Prasinophytes are the most primitive green algae. Most are unicellular and have no cell wall. Instead they are covered with layers of delicate **scales** that have elaborate shapes and ornamentation. Fossils of prasinophyte cysts found in Tasmania show that prasinophytes are at least 600 million years old.

Chlamydomonas and relatives: class Chlorophyceae

The flagellate *Chlamydomonas* and green algae that produce motile cells similar to *Chlamydomonas* are included in this class. Several levels of organisation, including unicellular, filamentous, colonial and parenchymatous, occur among the Chlorophyceae. A unique form of crystalline glycoprotein cell wall has evolved in this class.

Common forms include *Dunaliella*, a unicellular flagellate (Box 35.1), and *Volvox*, a spherical, colonial form composed of *Chlamydomonas*-like cells. *Chlamydomonas* is a model organism for cell biology. It is readily grown in the laboratory, reproduces sexually, and produces a range of mutants able to be mapped by classic and molecular genetic techniques.

Sea lettuce and *miso* soup: class Ulvophyceae

The sea lettuce *Ulva* (Fig. 35.5a) is the best known member of this class, which includes most of the green seaweeds occurring along coastlines around the world. Most Ulvophyceae are multicellular or coenocytic, up to 3 m in size, although typically much smaller. Motile cells are produced during sexual and asexual reproduction. Common forms include the sea cactus, *Caulerpa* (Fig. 35.5b), and *Codium*. *Ulva* is used as a garnish for Japanese *miso* soup. The Ulvophyceae diverged from the other green algae early on.

Stoneworts: class Charophyceae

Charophytes are essentially restricted to freshwater habitats. They are delicate and typically small (2–30 cm in length) with some, the **stoneworts**, encrusted with $CaCO_3$ (calcite). Gametes are asymmetrical and mitosis involves a phragmoplast—characteristics that identify charophytes as the closest relatives of the land plants (Chapter 37).

Desmids and relatives: class Conjugatophyceae

These green algae have no flagellated cells, not even the gametes. They reproduce by conjugation, a process involving formation of an interconnecting tube between two strains of the one species through which an amoeboid gamete crawls to fuse with its opposite gamete. Most conjugatophytes live in fresh water and are either unicellular (Fig. 35.6) or filamentous (Fig. 35.7). The unicells are **desmids**, which have beautiful symmetrical shapes comprising two semicells connected by an isthmus of cytoplasm passing through a central constriction. There are more than 10 000 species of desmids (Fig. 35.7).

> Green algae (phylum Chlorophyta) are unicellular, colonial or multicellular, and one group is the closest relative of land plants. Chloroplasts contain chlorophyll *a* and *b*, the product of photosynthesis is stored as starch, and cell walls are cellulosic.

Heterokont protists

Heterokonts are characterised by flagellar architecture. They typically have one smooth flagellum directed posteriorly and one hairy flagellum directed anteriorly. The hairy flagellum has numerous thin, tubular appendages that alter the direction of thrust produced by the flagellar beat. The beat of the hairy flagellum thus drags the cell through the water. If the cell happens to be fixed in place, the flagellar beat draws the water down and over the cell.

Fig. 35.5 Green algae. **(a)** The sea lettuce *Ulva lactuca* is used as a food but **(b)** its relative *Caulerpa* can be poisonous. Both are common members of the class Ulvophyceae found on rocky shores around the south-eastern coast of Australia

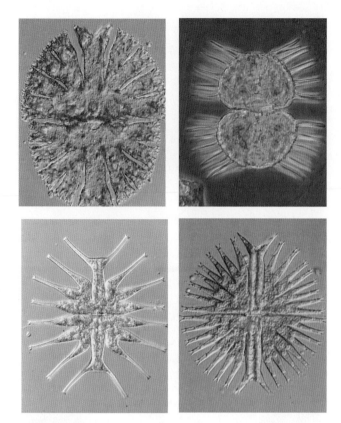

Fig. 35.6 In addition to bird life, lilies and crocodiles, the water holes of Kakadu National Park in the Northern Territory contain this splendid selection of desmids (phylum Chlorophyta)

Fig. 35.7 The green filamentous alga *Spirogyra* (class Conjugatophyceae) is named for the spiral chloroplast that winds its way around the periphery of the elongate cells

Heterokonts include the photosynthetic chrysophytes, haptophytes, diatoms and brown algae, and the non-photosynthetic oomycetes (thus, heterotrophs).

Golden flagellates: phylum Chrysophyta

Chrysophyta are golden-brown flagellates of marine and freshwater habitats. Cells are unicellular or colonial (Fig. 35.8) and have heterokont flagellation.

Chloroplasts contain chlorophylls *a* and *c* plus **fucoxanthin**, an accessory pigment giving the golden

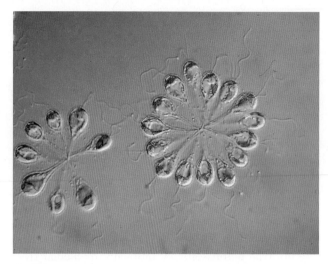

Fig. 35.8 *Synura* is a colonial chrysophyte common in fresh water

colour. Numerous heterotrophic forms have a colourless chloroplast or no chloroplast whatsoever, and even coloured photosynthetic forms can ingest food particles. **Chrysolaminarin** (β-(1→3)-glucan) is stored in a vacuole. Various cell coverings, including spines and scales composed of silica or a lorica (external vase-shaped shell) made of either cellulose or chitin, adorn the cells. **Silicoflagellates** contain spectacular, star-shaped silica skeletons (Fig. 35.9).

> Chrysophytes (golden-brown flagellates) are heterokonts, cells with one smooth and one hairy flagellum.

Haptophytes: phylum Prymnesiophyta

Haptophytes (phylum **Prymnesiophyta**) are extremely abundant in oceans. *Emiliana huxleyi* (named after T. H. Huxley) occurs in massive blooms visible in

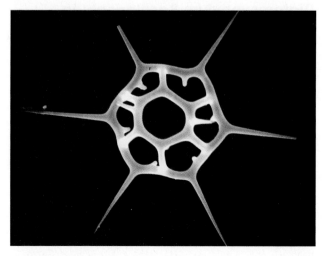

Fig. 35.9 Chrysophytes have various cell coverings. The beautiful star-shaped skeleton of a silicoflagellate is made from silica

satellite photographs. A global correlation between these satellite photographs and surface truthing (water samples taken from oceanographic vessels at the same time) indicates that *E. huxleyi* may have the largest biomass of any single species on earth. Enormous chalk deposits, such as the white cliffs of Dover, were formed from haptophytes and other protist skeletons. Several haptophytes are toxic to fish and shellfish, and blooms of these algae can result in total decimation of marine life over great areas.

Haptophytes are thought to be close relatives of the chrysophytes because they have similar chloroplasts and mitochondria. Their flagella are, however, quite different. The two flagella of haptophytes are both smooth and lack hairs, which means that haptophytes are not true heterokonts. Additionally, molecular trees do not show any close relationship between haptophytes and true heterokonts, so the similarity between their chloroplasts may be coincidental.

The name haptophytes refers to the curious **haptonema**, a thread-like (filiform) extension situated between the two flagella. The haptonema can move, either bending or coiling, and can capture prey, drawing them down to a 'mouth' on the posterior of the cell for ingestion.

A major group of haptophytes is the **coccolithophorids** (Fig. 35.10), which are covered with intricately sculptured calcite plates, **coccoliths**. Coccoliths form by crystallisation of $CaCO_3$ within the cell and are extruded onto the cell surface in overlapping arrays. The function of these elaborate investments is unknown.

> Haptophytes are unicellular and have chloroplasts similar to chrysophytes. Although classified as heterokonts, they have two identical, smooth flagella. Between the flagella is a haptonema for capturing prey.

Diatoms: phylum Bacillariophyta

Diatoms (phylum **Bacillariophyta**) are unicellular, golden-brown algae with siliceous walls (Fig. 35.11). They are ubiquitous in aquatic environments and are important producers. Chloroplasts and storage products of diatoms are the same as their close relatives, the chrysophytes.

Diatoms have a unique **silica** cell wall. Each cell has two silica dishes, **valves** or **frustules**, interconnected by silica hoops, girdle bands. The valves are highly ornamented with pores and spines, creating some remarkable patterns (Fig. 35.12). The valves and bands are perhaps derived from silica scales of an ancestor resembling modern-day chrysophytes. The silica valves form some of the best preserved fossils of any protists but, in older deposits, they have been converted to formless chert, destroying early diatom fossils. Massive

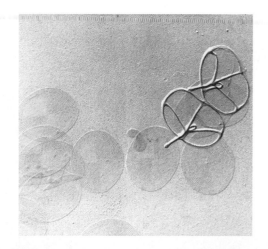

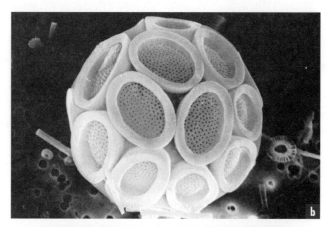

Fig. 35.10 Haptophytes. **(a)** The filmy scales of *Chrysocromulina* are only visible by high-resolution electron microscopy. The calcium carbonate armour plating of coccolithophorids can vary in shape from **(b)** flat discs, as in *Pontosphaera,* to **(c)** the elaborate trumpet-shaped structures of *Discosphaera tubifera*

deposits of diatom valves (diatomaceous earth) exist in recent strata and are mined for use as a very fine, high-grade filtration material or as an abrasive in toothpaste and metal polishes.

Diatoms are classified into two groups—**centrics**, radially symmetrical, and **pennates**, bilaterally symmetrical (Fig. 35.11). Many pennate forms have a

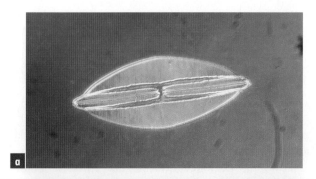

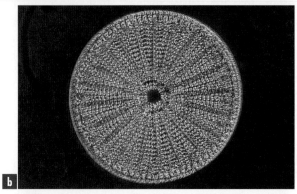

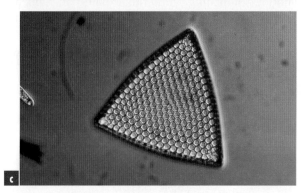

Fig. 35.11 Diatoms are typically either **(a)** pennate, such as *Navicula lyra*, or radially symmetrical, such as **(b)** *Arachnoidiscus* and **(c)** *Triceratium*. The silica valves have an opalescent appearance in the light microscope

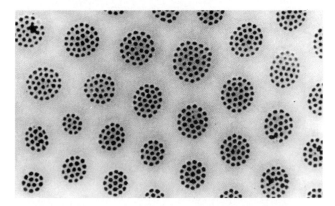

Fig. 35.12 Seen in detail under the scanning electron microscope, the markings observed on diatom valves by light microscopy are revealed to be small, regularly shaped pores in the silica. The pores allow transfer of materials through the cell's otherwise impervious, glass-like case

longitudinal slit, a **raphe**, in the valve, which enables them to move by crawling along the substrate. Wall-less, motile sperm with a single flagellum are released during sexual reproduction by some species. Although these sperm only have one flagellum, they are, nevertheless, heterokont because one flagellum simply fails to develop.

> Diatoms (phylum Bacillariophyta) are unicellular golden-brown algae with a unique silica wall that forms two valves.

Brown algae: phylum Phaeophyta

There are about 900 species of **brown algae** (phylum **Phaeophyta**), nearly all of which are marine and multicellular. They include the giant **kelps**, such as *Macrocystis pyrifera*, growing off the coast of California, which are as long as a blue whale and as tall as the biggest mountain ash trees in south-west Tasmania. Kelps form underwater forests that are home to a variety of temperate ocean marine life (see Box 41.2). They are also a source of alginic acid, a gelling agent used in foods, adhesives, paint and explosives. The large thallus of kelp is differentiated into a **holdfast**, which attaches to the substrate, a **stipe** and *blades* (Fig. 35.13). This organisation parallels that of terrestrial plants and kelps were once regarded as 'underwater trees' that were the marine ancestors of land plants.

While kelps are large and highly visible, many other brown algae are small inconspicuous tufts or simple filaments barely visible to the naked eye. Even some of the larger kelps have a microscopic filamentous life form as one of their alternating generations.

Brown algae have chloroplasts with the same pigments as chrysophytes. The storage product **laminarin**, a β-(1→3)-glucan, is similar to chryso-laminarin. The heterokont motile cells released as

Fig. 35.13 Phaeophytes (brown algae). **(a)** Bull kelp, *Durvillea potatorum*, occurs on southern Australian rocky shores subject to high surge action. The disc-shaped holdfast adheres tenaciously to rocks, preventing the thallus from being ripped away by waves

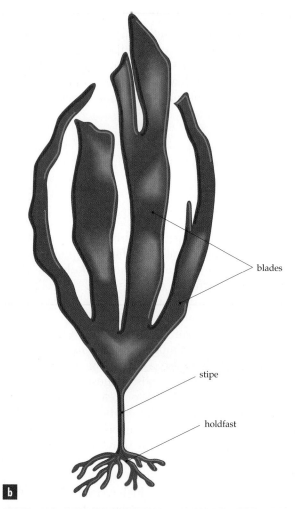

blades

stipe

holdfast

b

Fig. 35.13 (b) Diagram of the thallus of a kelp. The stipe contains a vascular system that translocates material down from the photosynthetic blades to the holdfast, which may be many metres below the surface

gametes or zoospores closely resemble chrysophyte flagellates and it is probable that multicellular brown algae evolved from unicellular chrysophytes.

> **B**rown algae (phylum Phaeophyta) include the largest protists with a differentiated, multicellular thallus. Pigments and the storage product, laminarin, are similar to chrysophytes.

Water moulds and downy mildews: phylum Oomycota

Water moulds and downy mildews, oomycetes (phylum **Oomycota**), have a superficial resemblance to fungi (Chapter 36) since they produce a network of filaments (hyphae) that permeate their food substrate. The hyphae are coenocytic, having no septa (cross-walls). Oomycetes are different from fungi, however, in that cell walls are cellulosic rather than chitinous. Oomycetes are so-named for their distinctive oogonium, the

BOX 35.2 Neptune's necklace, *Hormosira banksii*

If you poke around in the tide pools on the eastern coast of Australia, you will almost certainly find short strings of drab, olive-coloured beads splayed over the rocks. These beads are the brown alga Neptune's necklace, *Hormosira banksii*. Like other intertidal life forms, *H. banksii* must withstand exposure to the air twice daily, and the leathery, fluid-filled beads, termed **receptacles**, are resistant to desiccation. Supported by sea water on the flood tide, the floppy strings of beads fan up and out to sway back and forth in the surging waves.

H. banksii is dioecious. Reproductive structures are found within small warty growths, **conceptacles**, that stud the surface of the receptacles. Within the conceptacles on the male plant are two types of hairs: long, unbranched paraphyses and shorter, branching antheridial hairs on which sperm-producing antheridia develop. Each

Hormosira banksii

antheridium undergoes meiosis and several subsequent rounds of mitosis to produce 64 sperm cells. Motile sperm are biflagellate heterokonts (having one smooth and one hairy flagellum) and bear an orange eyespot. At low tide, an orange ooze of antheridia exudes from the conceptacles on the male thallus. Sperm are released on the flood tide.

Eggs are produced by oogonia on a female thallus. Like antheridia, oogonia develop in conceptacles, which also contain **paraphyses**. Four eggs (**ova**) are released from each oogonium. Ova have no flagella and drift motionless on the incoming tide. Sperm are attracted to a secretion produced by the ovum and cluster around the ovum until one successfully fertilises it. The zygote settles and, if it finds a suitable location, immediately develops into a new, diploid, male or female thallus. The gametes are the only haploid stage of the life cycle.

BOX 35.3 Dieback disease

In the 1920s there were a number of reports of mysterious deaths of jarrah trees, *Eucalyptus marginata*, in Western Australian forests (Fig. a). Tree deaths appeared to follow bush tracks and logging sites and were at first attributed to soil disturbance. When sand and gravel from these cleared areas was transported to other regions, trees at these sites also died.

It was not until the late 1960s that the cause of the forest dieback was identified as the oomycete, *Phytophthora cinnamomi*. This pathogen attacks the roots of susceptible plants, causing problems in water uptake and translocation. Infected trees show symptoms of water stress, with leaf yellowing and dieback of upper branches. Spread of the disease occurs underground by movement of flagellated zoospores, which are able to swim through moist soil. Zoospores seek a host rootlet, attach themselves and produce hyphae that invade the plant's root system (Fig. b). This mechanism of disease transfer explains how transport of contaminated soil or flushing of floodwater spreads the disease.

P. cinnamomi is thought to originate from cinnamon trees in Sumatra and was probably introduced to Australia by European colonists. Many endemic plants have no apparent resistance to dieback and some highly susceptible *Banksia*

species are threatened with extinction. The massive scale of the problem prevents the use of fungicide and outbreaks of the disease must usually run their course before natural antagonistic soil microbes bring the epidemic under control.

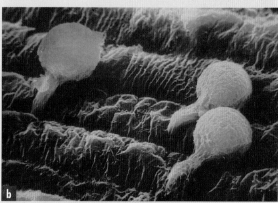

(a) Dieback of jarrah trees in Western Australia caused by the oomycete *Phytophthora cinnamomi*. **(b)** Cysts of *P. megasperma* germinating on a plant rootlet. *P. cinnamomi* and *P. megasperma* zoospores swim through soil water and encyst when they contact a plant root. The cyst then germinates to produce hyphae, which penetrate the root and invade the vascular system of the host, eventually causing dieback

female reproductive structure containing oogonia. Male gametes are produced in nearby antheridia and non-motile 'sperm' are brought to the oogonium through a fertilisation tube. Fusion of gametes (syngamy) produces a diploid oospore within which meiosis usually occurs to produce zoospores with heterokont flagella. These zoospores are remarkably similar to chrysophytes (golden algae) and comparison of gene sequences from oomycetes and chrysophytes indicates that they are related.

Oomycetes are of considerable commercial and environmental importance. *Phytophthora infestans*, which causes late blight of potatoes, destroyed potato crops in the 1840s in Ireland. Potatoes, which were introduced from South America, had become the staple food of workers in Europe. The average Irish farm worker ate 5 kg of potatoes—boiled, mashed, roasted or fried—every day. However, due to cool, damp summer weather, the potatoes became infected with *P. infestans* and all rotted. During the resultant famine, one million people perished, prompting many Irish to seek a new life in the United States and Australia. Also in the nineteenth century, another oomycete, *Plasmopara viticola*, attacked French grapevines and almost obliterated the French wine industry in a single season.

> Oomycetes have coenocytic hyphae with cellulosic walls. The gametes are non-motile and sperm are brought to the female reproductive organ, an oogonium, through a fertilisation tube. Oomycetes are related to chrysophytes.

Euglenoids and kinetoplasts

This group includes flagellated unicells that are photosynthetic or heterotrophic, some being parasitic,

others free-living. They are currently classified in different phyla but are related. They all have an anterior depression, **gullet**, from which the flagella emerge. Some of the heterotrophic forms ingest food particles through this anterior gullet.

Euglenoid flagellates: phylum Euglenophyta

Euglenoids (phylum **Euglenophyta**) (Fig. 35.14) are flagellates of both marine and freshwater habitats. There are about 800 species, a third of which are photosynthetic. The other species lack chloroplasts and are heterotrophic. Even some of the chloroplast-containing forms are facultative heterotrophs and when kept in darkness their chloroplasts shrivel and they revert to heterotrophy, engulfing prey through the gullet.

The euglenoid chloroplast is similar to that of green algae and land plants in that it contains chlorophylls *a* and *b* and β-carotene. However, the euglenoid chloroplast is bounded by three membranes and the organisation of genes on the chloroplast chromosome is unique. Unlike green algae, euglenoids do not store any starch in the chloroplast. It is not clear how photosynthetic euglenoids came by their chloroplasts. Some

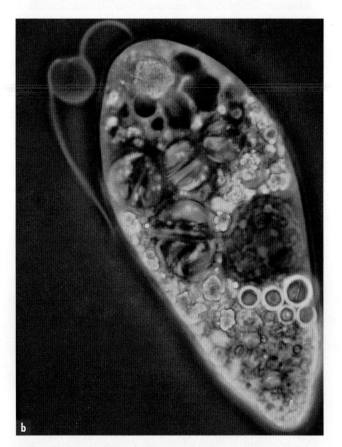

Fig. 35.14 (a) Diagram of a euglenoid. These flagellates can possess chloroplasts but are often heterotrophic. The helical proteinaceous strip forms a spiral pellicle, giving the cell its shape. Several paramylon granules are distributed throughout the cell. **(b)** This specimen of *Euglena* is photosynthetic but can also live heterotrophically

biologists think euglenoids stole their chloroplasts from green algae but others think that they acquired them at an early stage of evolution. Products of photosynthesis are stored as **paramylon**, a β-(1→3)-glucan, which forms solid granules in the cytoplasm. At the anterior end of the cell is a red stigma involved in the detection of light. Euglenoids usually swim with one long flagellum (a second short flagellum does not usually emerge from the gullet) and many species perform a sinuous gyration or crawling motion known as **metaboly**. Euglenoids are technically naked, having no cell wall or ornamentation outside the plasma membrane, but many species have an elaborate proteinaceous **pellicle**, comprising overlocking helical strips that interslide as the cell moves. Reproduction is principally by asexual division.

Flagellate parasites: kinetoplasts

Kinetoplasts are flagellate parasites (phylum Zoomastigina, class Kinetoplastida) known as **trypanosomes** and **leishmanias**, which are disease-causing organisms of major medical and veterinary significance. Species of *Phytomonas* infect plants and are a major problem in coconut palms, oil palms, coffee trees and various fruit crops in Latin America.

Kinetoplasts are unicellular biflagellates with an apical depression into which the flagella are inserted. The name kinetoplast refers to the large mass of DNA present in the single mitochondrion at the base of the flagella. A kinetoplast is composed of thousands of catenated DNA mini-circles (linked together like a chain), often forming an elongated rod-shaped structure in the mitochondrion (Fig. 35.15). The kinetoplast DNA also contains normal circular mitochondrial chromosomes.

Trypanosomes cause African sleeping sickness and nagana. These parasitic flagellates are free-swimming in the blood of humans and other vertebrates. Infection is usually transmitted by the blood-sucking tsetse fly. Occasional cross-infection occurs through bites from vampire bats. The trypanosome that causes Chagas' disease in South and Central America (Fig. 35.16) invades the heart and other muscles. About 10–12 million people are infected. Again, the disease is transmitted by blood-sucking insects. Charles Darwin may have contracted Chagas' disease while deliberately observing ticks suck his blood in South America.

Leishmaniasis is an infection of macrophage cells (white blood cells that normally ingest foreign particles in the bloodstream) caused by the kinetoplast parasite *Leishmania*. Disease transmission is by sandflies and the parasite occurs in South and Central America, Africa, the Middle East, the Mediterranean and Asia. Relatively benign forms cause cutaneous lesions (Fig. 35.17), but visceral leishmaniasis attacks

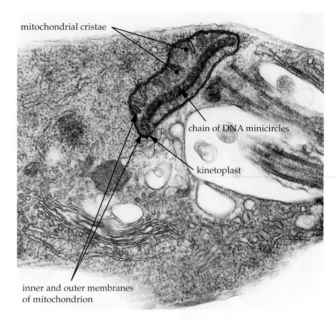

Fig. 35.15 Kinetoplasts are parasitic flagellates. The name kinetoplast refers to a specialised mitochondrion containing thousands of DNA mini-circles, which are visible here as a tangled mass of threads forming an elongate body. The kinetoplast lies at the base of the flagellum

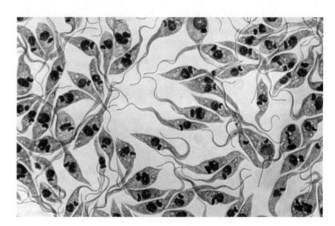

Fig. 35.16 *Trypanosoma cruzi*, a kinetoplast from Central and South America, causes Chagas' disease

macrophages of the liver, spleen and bone marrow, often resulting in fatal anaemia.

A remarkable feature of trypanosomes, such as *Leishmania*, is their ability to survive in the host's bloodstream and avoid elimination by the immune system. Trypanosomes do this by constantly changing the molecules on their surface. Thus, no sooner does the host mount an immune response (Chapter 23) to the invader, than the trypanosomes slip into another 'jacket' that the immune system cannot yet 'see'. The parasite has up to 1000 different versions of surface molecules that it produces by sequentially rearranging the genes that code for surface glycoproteins. In this way, the

Fig. 35.17 In Costa Rica, the small scars produced by cutaneous leishmaniasis, caused by the kinetoplast parasite *Leishmania*, are known as the 'seal of the forest'. More severe forms of leishmaniasis often result in death

parasite can stay one step ahead of the host's immune system. Morphologically, kinetoplasts are rather similar to euglenoids and studies of rRNA confirm that these groups are related. Euglenoids differ in that they are free-living, can have chloroplasts and never have kinetoplast DNA.

> Euglenoids and kinetoplasts are related flagellated cells. Euglenoids are free-living, some of which have chloroplasts and some of which engulf prey through an anterior gullet. Kinetoplasts are parasitic with a unique mitochondrion.

Amoebae lacking mitochondria: pelobiontids

Pelobiontids (presently classified as phylum Zoomastigina, order Pelobiontida) are amoebae that lack both mitochondria and an endomembrane system but they do have rudimentary flagella. The DNA is surrounded by a nuclear envelope but it is not known whether the nucleus divides mitotically with a microtubular spindle, or whether it simply pinches in two, such as a prokaryotic cell. The best known amitochondriate amoeba is *Pelomyxa palustris*, a giant, multinucleate, free-living, herbivorous cell. *P. palustris* may compensate for its lack of mitochondria by harbouring numerous endosymbiotic bacteria that apparently perform oxidative phosphorylation for the host. In any case, it seems to live only at the bottom of ponds where oxygen is fairly scarce.

Parasitic flagellates that contaminate water supplies: diplomonads

Diplomonads (phylum Zoomastigina, class Diplomonadida) are unicellular, heterotrophic flagellates. The name refers to the presence of two nuclei, each of which is associated with a pair of flagella. Diplomonads inhabit the gut of various animals, where they can attach and feed by the sucker-like, ventral disc. They lack obvious mitochondria and are restricted to an anaerobic environment.

Giardia, an intestinal parasite causing severe dysentery, is the best known diplomonad (Fig. 35.18). It is one of the first protists on record, accurately described by van Leeuwenhoek in 1681 from his own diarrhoeic stools. *Giardia* caused a major health scare in Australia in 1998 when it was discovered in Sydney drinking water reservoirs.

Alveolates

Members of the alveolates all have distinctive vesicles, **cortical alveoli**, just beneath the plasma membrane. The alveoli are flat sacs of endoplasmic reticulum. In some species, the cortical alveoli are involved in the formation of the cell's covering, such as plates and scales. Although they are a diverse group, including photosynthetic, parasitic and predatory organisms, DNA sequence data confirm that they are monophyletic.

Dinoflagellates: phylum Dinophyta

Dinoflagellates are an extremely diverse phylum of unicells. About half the species are photosynthetic and major primary producers in tropical seas (Fig. 35.19). Some cause red tides, which may be toxic (Box 35.4). Their name refers to the characteristic spinning motion

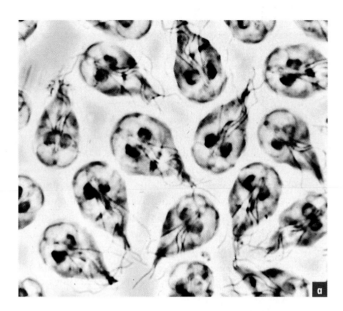

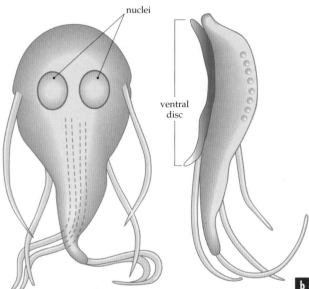

nuclei

ventral
disc

Fig. 35.18 (a) *Giardia* is a simple eukaryote (a diplomonad) that parasitises humans and other animals. **(b)** Cells have two nuclei (*n*), each of which is associated with a set of flagella. On the ventral side of the cell is a disc through which the cell attaches to the host's gut lining. Infection is spread by cysts excreted in faeces, either animal or human. The cysts, which remain viable in water for several months, can infect the gut of animals drinking from the contaminated water source. *Giardia* is not restricted to polluted waters and can occur in metropolitan water supplies or even in wilderness streams. The most effective means of purification is to boil the water; cysts are resistant to iodine and chlorine

of the cells as they swim through the water. By protist standards, dinoflagellates are quite vigorous swimmers and can swim at speeds of 1 m per hour. Cells have one posteriorly directed flagellum that steers the cell, plus a unique **transverse flagellum** positioned in a **girdle** encircling the cell (Fig. 35.19c). This transverse flagellum is corkscrew-shaped and its beat causes the cell to spin and aids forward movement.

Chloroplasts of photosynthetic dinoflagellates contain chlorophylls *a* and *c*, plus a xanthophyll, **peridinin**. Three membranes, rather than the usual two, surround the chloroplast. Starch is stored in the cytoplasm. Some dinoflagellates are naked, some have scales, and some are armoured with cellulosic plates. Dinoflagellates such as *Noctiluca* (night light) are bioluminescent and congregate in the surf, creating phosphorescence. The luminescence is perhaps a mechanism to startle would-be predators. Dinoflagellates known as **zooxanthellae** are endosymbionts in the tissues of corals, sea anemones and molluscs, supplying the host animal with nutrition in return for protection and a supply of nitrogen from the animal's excretory products.

Many dinoflagellate species lacking chloroplasts are predatory, capturing other cells. Several predatory species have feeding tentacles that pierce prey and suck out the contents. An extraordinary feature of certain dinoflagellates is their 'eye'. The eye-like structure has a refractile lens that changes shape, seeming to focus images onto a light-sensitive retinoid. Dinoflagellates may be able to 'see' their prey.

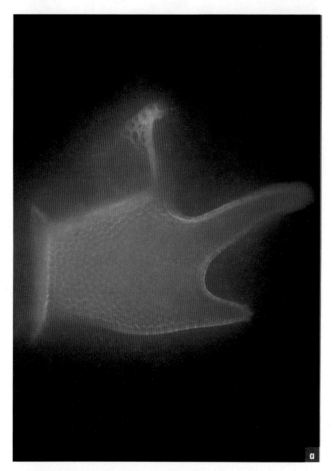

Fig. 35.19 Dinoflagellates. **(a)** The cellulose armour plating of dinoflagellates glows an eerie blue when stained with a fluorescent dye and viewed with an ultraviolet microscope

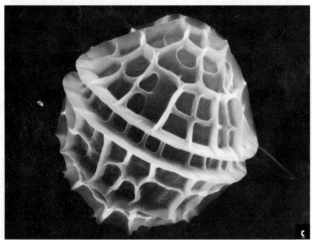

Fig. 35.19 (b) This dinoflagellate from the Coral Sea has wing like extensions of its plates that are believed to act like sails and catch water currents, moving the cell through the ocean. **(c)** The distinctive girdle formed by a constriction in the mid-region of dinoflagellates is where the spiral transverse flagellum is normally located. In this cell, prepared for scanning electron microscopy, the delicate flagellum is lost

Several characteristics distinguish dinoflagellates as a protist phylum. Dinoflagellate DNA appears to be permanently condensed (like a prophase nucleus) and is complexed with proteins that are different from typical eukaryotic histones. Originally thought to be a primitive feature described as mesokaryotic (intermediate between prokaryotic and eukaryotic), it is now thought that dinoflagellates lost their histones secondarily. Molecular studies of rRNA sequences suggest that dinoflagellates are closely related to ciliates and Apicomplexa (see below).

Dinoflagellates are alveolates with two flagella, one of which encircles the cell. Many are photosynthetic, containing chlorophylls *a* and *c*, and some are predatory.

BOX 35.4 Toxic dinoflagellates

Red tides occur when the concentration of dinoflagellates in sea water becomes so high that they discolour the surface of the sea (see fig.). The explosive burst of growth results in millions of cells per litre and is induced by a particular set of environmental conditions, such as high temperatures, excess nutrients, and a stratified, stable water column. Most red tides, such as those caused by the bioluminescent dinoflagellate *Noctiluca scintillans*, appear to be harmless events. However, under exceptional conditions, blooms of dinoflagellates can cause severe problems. Sometimes the algae become so densely concentrated that they generate anoxic conditions, suffocating fish and invertebrates in sheltered bays. Other dinoflagellates such as *Gymnodinium mikimotoi* cause serious damage to fish in intensive aquaculture systems, either by the production of mucus, which causes mechanical damage to fish gills, or by the production of haemolytic substances that destroy red blood cells in gill tissues.

About 30 species of dinoflagellates produce potent toxins that move through food chains via fish or shellfish to humans. Dinoflagellate toxins are so potent that a pinhead-size quantity (about 500 µg), an amount easily accumulated in just one 100 g serving of shellfish, could be fatal to humans. The toxins involved rarely affect the nervous systems of fish or shellfish but they evoke a variety of gastrointestinal and neurological symptoms in humans. The resulting illnesses are known as paralytic shellfish poisoning (PSP), diarrhoetic shellfish poisoning (DSP) and ciguatera fish food poisoning.

Algal blooms seem to be increasing in frequency and geographic spread. On a global scale, close to 2000 cases of human poisoning through dinoflagellate toxins occur each year. While toxic plankton blooms appear regularly on a seasonal basis in temperate waters of Europe, North America and Japan, until recently they were unknown in Australian waters. Tasmania was the first Australian state to suffer problems with toxic dinoflagellates contaminating the shellfish industry. In 1986, dense blooms of the chain-forming species *Gymnodinium catenatum*, a species causing PSP, resulted in the temporary closure of 15 Tasmanian shellfish farms. In 1988, the dinoflagellate *Alexandrium catenella*, which causes PSP, caused limited toxicity in wild mussels from Port Phillip Bay but fortunately no commercial shellfish farms were affected. Since 1986, blooms of *A. minutum* have occurred annually in the Port River area near metropolitan Adelaide. Ciguatera poisoning caused by the coral reef dinoflagellate *Gambierdiscus toxicus* poses an increasing danger in the Great Barrier Reef region.

Four possible explanations for this apparent increase in red tide problems have been suggested: increased scientific awareness of toxic species; increased utilisation of coastal waters for aquaculture; stimulation of plankton blooms by coastal eutrophication; and accidental introduction of toxic dinoflagellates to new areas. The recent discovery of resistant resting cysts in ships' ballast water suggests that international shipping could be distributing toxic dinoflagellates around the world. Introduction of toxic species may also be associated with the movement of shellfish stocks from one area to another.

Red tides occur when explosive plankton growth produces so many algal cells that they discolour the water. This bloom of the harmless dinoflagellate *Noctiluca scintillans* occurred in Lake Macquarie, New South Wales. Blooms of toxic algae have recently caused alarm in Port Phillip Bay and the Gippsland Lakes, Victoria

Small but deadly: phylum Apicomplexa

There are at least 5000 species of **Apicomplexa**, most of which are endoparasites of animals. Previously classified as Sporozoa, Apicomplexa are now named for their **apical complex**, a structure involved in the penetration of host cells. The apical complex is a conical arrangement of microtubules and secretory structures. The parasite attaches to the host at the apical complex and then forces its way into the host.

Apicomplexa include **gregarines**, **coccidia** and **haematozoa**. Gregarines, which only parasitise invertebrates, are probably the most primitive apicomplexans. Coccidia, which cause coccidiosis and toxoplasmosis in humans, can infect both invertebrates

and vertebrates. Some coccidia alternate between a vertebrate host and an invertebrate host. Coccidia leave one host as spores in the faeces and remain in the open environment until they can infect the second host. Humans can contract toxoplasmosis by ingesting spores of *Toxoplasma* present on the fur of cats carrying the infection. Toxoplasmosis is a common infection but is usually only a problem during pregnancy (when it can affect the developing fetus) or for immunocompromised individuals (such as HIV sufferers) who cannot combat infection and often die. Haematozoa, which are probably the most derived Apicomplexa, invade blood cells of vertebrates, where they feed on haemoglobin. The most notorious haematozoan is *Plasmodium*, the causal agent of malaria (Box 35.5). Like coccidia, haematozoa also alternate between vertebrate and invertebrate hosts and have efficient ways of effecting cross-transfer between host species.

BOX 35.5 Malaria

An estimated 500 million people suffer from **malaria** and each year the disease kills about two to three million people, mostly infants in Africa and South America. Malaria is caused by apicomplexan protists belonging to the genus *Plasmodium*. The *Plasmodium* life cycle involves two different hosts: a vertebrate and a blood-sucking insect. The parasite is transferred from one host to the other when insects suck blood from vertebrates. Humans become infected with *Plasmodium* by the mosquito *Anopheles*.

As **sporozoites** (a stage that has the apical complex), *Plasmodium* cells pass into the human bloodstream from the salivary glands of the mosquito. The sporozoites quickly move to the liver and undergo asexual reproduction to produce numerous **merozoites**, which invade red blood cells. Merozoites transform into **trophozoites**, which eat the contents of the red blood cells, then divide synchronously, and every 48 or 72 hours (depending on the type of malaria) induce the lysis of red blood cells, causing the release of toxins and hence the cycles of fever and chills characteristic of malaria. After several rounds of replication in red blood cells the parasite converts to the next cell type, **gametocytes**. Gametocytes are ingested by mosquitoes sucking blood from a malaria sufferer. They pass into the mosquito's gut, where they develop into sperm and eggs, which fuse to form a zygote. The zygote then undergoes meiosis to produce the oocysts in the

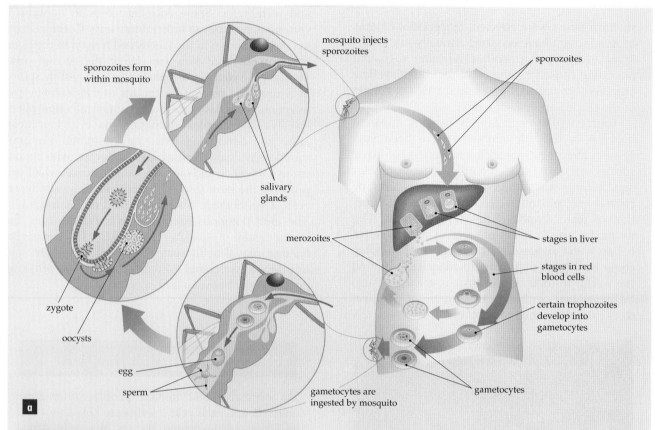

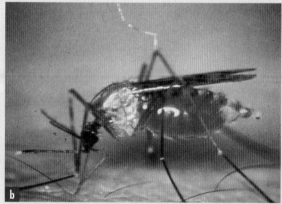

(a) Life cycle of *Plasmodium*, the apicomplexan parasite that causes malaria. **(b)** Blood-sucking insects, like this *Anopheles* mosquito, are both host and vector for parasitic haematozoans

mosquito's gut and eventually new haploid sporozoites, which move into the mosquito's salivary glands to complete the cycle.

The transfer from one host to the other can be risky and it is here that malaria parasites excel. When a female mosquito takes her meal of blood from a vertebrate, she infects the vertebrate with parasites. The parasites multiply in the vertebrate and are then available to back-infect the next generation of mosquitoes. By exploiting the relationship between blood-sucking insects and vertebrates, haematozoa such as malaria ensure their own reproduction and distribution. They also reduce the defence strategies available to hosts since only a part of the life cycle occurs in each host.

The protistology world was recently stunned by the discovery that apicomplexan parasites have a chloroplast similar to those of plants and algae. The chloroplast of these parasites is small and lacks chlorophyll but it contains a circular DNA genome similar to that of all other chloroplasts. This discovery tells us that these organisms were once photo-synthetic but converted to a parasitic lifestyle. Why they kept the chloroplast remains a mystery but parasitologists are hopeful that the chloroplast might

be the Achilles heel they have long searched for. Many processes in chloroplasts can be blocked with drugs and, because humans lack a chloroplast, these types of drugs will probably not have any side-effects on the patient.

Apicomplexa are endoparasites of animals. They often parasitise two hosts and cause diseases such as malaria. An apical complex is used to penetrate host cells.

Ciliates: phylum Ciliophora

Ciliates are unicellular organisms, ranging from 10 µm to 3 mm in length, with numerous cilia on the surface (Fig. 35.20). Cilia are often arranged in clusters, cirri, that beat synchronously and work as paddles or feet for the cell. Sometimes cilia are arranged evenly over the surface and beat in rhythmic waves to propel cells. Cilia are all embedded in a cortex of protein fibres that cross-link the network. Elements of the cortex are contractile and, like miniature muscles, they can modify shape in certain species.

There are about 7500 species of ciliates and most are predatory, bacteria being the favoured prey. Prey are driven into an invagination, the **buccal cavity**, through which they are ingested. Undigested material is excreted through the **cytoproct** (Fig. 35.21).

Ciliates have two types of nuclei: a **micronucleus** and a **macronucleus**. Cells usually have one of each type but some cells have several macronuclei. The micronucleus is diploid, contains normal chromosomes with all the ciliate's genes and divides mitotically. The macronucleus develops from the micronucleus and contains multiple copies of genes on relatively short pieces of DNA, some of which are circular. Macronuclei divide amitotically, simply by pinching into two approximately equal halves. Because the macronucleus contains multiple copies of the ciliate's genes (sometimes several thousand copies), it is not essential that the genetic material be accurately apportioned, as occurs in mitosis. Messenger RNAs are transcribed from the multiple gene copies in the macronucleus while the master genes in the micronucleus are not transcribed. The macronucleus seems to be the working copy of the DNA blueprint, while the micronucleus

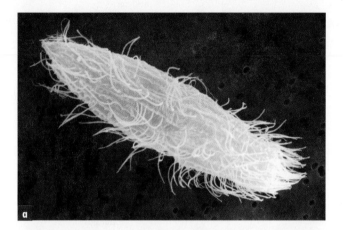
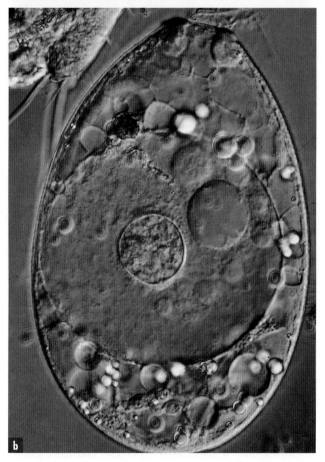

Fig. 35.20 Ciliates. **(a)** Scanning electron micrograph of a ciliate from the Pacific Ocean. The cilia are distributed uniformly across the surface. **(b)** A ciliate is a unicell that is capable of a variety of activities. It moves about, senses the environment, responds to stimuli and captures prey. Unlike multicellular animals, which accomplish these types of activities by use of complex organs, a ciliate must manage all these functions as an individual cell

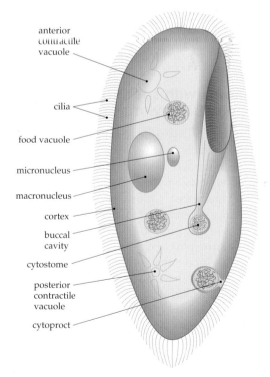

Fig. 35.21 Diagram of a ciliate. Food particles are wafted into the buccal cavity by the surface cilia and ingested through the cytostome into food vacuoles for digestion. Undigested material is ejected via the cytoproct

is retained as a master copy to be passed on during the sexual process. This is analogous to making several working copies of a computer program for daily use, while keeping the original program safely stored away.

Ciliates reproduce asexually by binary fission along the short axis of the cell. Sexual reproduction is by conjugation. Two cells of opposite mating type become attached and over a period of several hours they exchange haploid versions of their micronuclei produced by meiosis. The haploid micronuclei fuse to regenerate diploid micronuclei. After conjugation, the macronuclei degenerate and new macronuclei are produced from the recombinant micronuclei. Conjugation in ciliates is essentially a form of sex in which haploid nuclei from two individuals are brought together without ever needing to form gamete cells.

> Ciliates are predatory unicells with numerous cilia and two types of nuclei. They reproduce sexually by conjugation.

Protistan pirates

Recent work has demonstrated that at least two groups of protists have stolen the ability to photosynthesise from chloroplast-bearing cells. By cannibalising parts from photosynthetic prey, **secondary endosymbiosis**, they acquired chloroplasts and adopted an autotrophic way of life. Heterokonts, haptophytes, euglenoids, dinoflagellates and apicomplexans probably also acquired their plastids in this way too but the evidence is less clear cut.

Flagellates with second-hand chloroplasts: phylum Cryptophyta

Cryptomonads have a small anterior invagination (the 'crypt') into which their two flagella are inserted. They are unicellular and usually reproduce asexually. All genera, except *Goniomonas*, which is heterotrophic, possess a chloroplast. *Chilomonas* is also heterotrophic but contains a leucoplast. Chloroplasts have chlorophylls *a* and *c*2 plus a phycobilin pigment, either phycocyanin or phycoerythrin. The product of photosynthesis is stored outside the chloroplast as starch.

Cryptomonads have a second small nucleus associated with the chloroplast. They apparently obtained their chloroplasts by endosymbiosis, like other photosynthesising protists. However, the endosymbiont was not a prokaryote but a photosynthetic eukaryote. Cryptomonads appear to have acquired the capacity to photosynthesise second-hand by cannibalising a eukaryote (probably a red alga) that had already formed a permanent association with a prokaryote (Fig. 35.22). The much reduced second nucleus associated with the

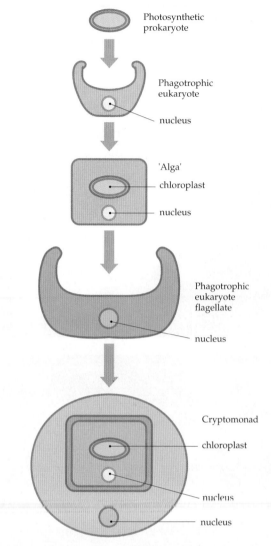

Fig. 35.22 Cryptomonads are animal-like flagellates that acquired the capacity to photosynthesise by enslaving an algal cell, which now exists as a permanent endosymbiont within the cryptomonad host. Double endosymbiosis explains the origin of cryptomonad chloroplasts. The primary endosymbiosis involved a eukaryote engulfing a prokaryote to create a eukaryotic alga. The secondary endosymbiosis resulted in the alga being engulfed by another eukaryotic phagotroph. The second nucleus in the cryptomonad cell is apparently the remnants of the algal endosymbiont's nucleus

cryptomonad chloroplast is the remnant of the eukaryotic endosymbiont's nucleus.

Amoebae with second-hand chloroplasts: phylum Chlorarachnida

Phylum **Chlorarachnida** is a small phylum that includes only a handful of species. The best known genus is *Chlorarachnion* (Fig. 35.23), which exists principally as an amoeboid **plasmodium**, individual cells being linked by a network of cytoplasmic strands, **reticulopodia**. The plasmodial network captures small prey, which are ingested. When starved, the plasmodium separates and forms individual walled cysts

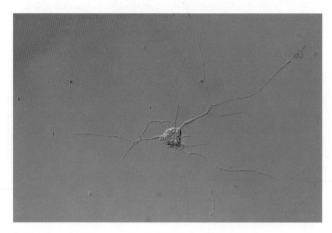

Fig. 35.23 With its amoeboid movement, *Chlorarachnion reptans* (phylum Chlorarachnida) is more like an animal than a plant but the cells have grass-green chloroplasts with which they are able to photosynthesise

that release uniflagellate swarmers. These swarmers regenerate a plasmodium.

Chlorarachnion is also photosynthetic and each amoeba has several grass-green chloroplasts containing chlorophylls *a* and *b*, like green algae. *Chlorarachnion* chloroplasts are different to green algal chloroplasts, however, in that they are bounded by four membranes rather than two and no starch is stored within the chloroplast. Associated with the *Chlorarachnion* chloroplast is a nucleus-like structure, suggesting that *Chlorarachnion* acquired its chloroplast by engulfing a photosynthetic eukaryote, presumably a green alga.

Not all eukaryotes with chloroplasts are close relatives. Photosynthetic cryptomonads and *Chlorarachnion* have captured and enslaved an algal prey cell, which they now use to photosynthesise.

Slime moulds

Slime moulds are amoeboid protists that produce fruiting bodies, **sorocarps**, as part of their life history. They were often classified with fungi because they absorb nutrients directly from the environment but this is the only similarity to fungi. The term slime mould refers to the habit of the most conspicuous part of the life cycle, which is a small slimy mass.

Cellular slime moulds

You could perhaps mistake a cellular slime mould for a minute slug if you found one creeping across the forest floor. The 'slug' or **pseudoplasmodium** is a mass of amoebae that have aggregated to form a single perambulatory colony. The amoebae, which are normally free-living individuals that prey on bacteria, congregate when their food supply runs short and move off collectively as a 'slug'. Having found a suitable location, the slug differentiates into a fruiting body that produces numerous spores (Fig. 35.24). Spores are released and eventually produce amoebae, completing the life cycle.

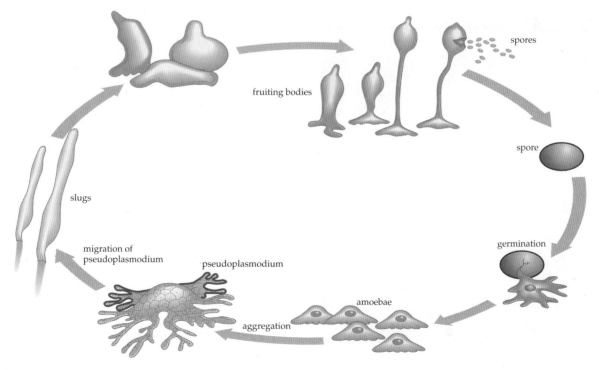

Fig. 35.24 Life cycle of the cellular slime mould *Dictyostelium discoideum*. Amoebae aggregate to form a pseudoplasmodium. They move off together as a slug, eventually forming a fruiting body in which spores are produced

Cellular slime moulds (phylum Acrasea, phylum Dictyostelida and the informal group protostelids) inhabit damp places in forests and gardens where they are usually found on rotting plant material or animal dung. The amoebae are often referred to as **myxamoebae** (slime amoebae) to distinguish them from normal amoebae, but the two may be closely related. Most cellular slime moulds do not have flagella.

Acellular slime moulds: phylum Myxomycota

Whereas the pseudoplasmodium of cellular slime moulds consists of numerous individual cells aggregated together, the plasmodium of a myxomycete (phylum **Myxomycota**) is one large multinucleate cell. The plasmodium resembles a slimy scum, sometimes vivid yellow or orange in colour (Fig. 35.25), and is the major feeding stage, absorbing organic matter and ingesting bacteria and other microorganisms. Should the plasmodium encounter a nutrient-poor region or other adverse environmental conditions, it differentiates into a fruiting body or sporangium (Fig. 35.26),

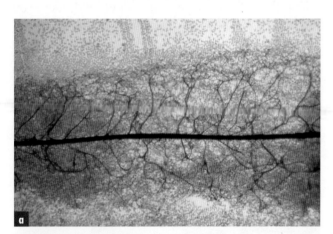

Fig. 35.25 Acellular slime moulds. **(a)** Plasmodium of the acellular slime mould *Stemonitis fusca*. **(b)** The brilliant yellow plasmodium of *Physarum polycephalum* coats plant leaves

with cells dividing by meiosis to produce haploid spores. Spores germinate to produce haploid myxamoebae, which are the gamete stage. In the presence of sufficient water, they convert to biflagellate forms. Two amoebae (or two biflagellates) fuse to form a zygote. The diploid nucleus of the zygote divides mitotically but no cell membranes separate the daughter nuclei, resulting in a multinucleate plasmodium.

> S lime moulds are amoeboid protists, which aggregate to form colonies, either cellular or acellular, with fruiting bodies that produce spores.

Sponge-like protists

'Collar' flagellates: choanoflagellates

Choanoflagellates are free-living, usually unicellular heterotrophs found in marine, brackish-water and freshwater environments. The cell has a single flagellum, which is surrounded by a ring of tentacles. If the choanoflagellate is sessile (attached to a substrate by a stalk), the flagellar beat draws water through the tentacular ring, where any small bacterial cells or detritus particles are captured and ingested. Some choanoflagellates swim freely using the flagellum to push them through the water. Cells are small (less than 10 μm) but they are surrounded by a basket-shaped enclosure, the lorica. The choanoflagellate lorica is composed of several silica strips cemented together and surrounded by a membranous web (Fig. 35.27). Reproduction is asexual and the parent cell releases a smaller juvenile cell. In some forms the juvenile cell inherits the silica strips from the parent lorica and uses them to commence construction of its own lorica.

Collar cells (choanocytes) of sponges (Chapter 38) bear a striking resemblance to choanoflagellates and sequence data show that they are close relatives, so this group of protists shares a common ancestor with us and may be representative of the earliest lines of animal evolution.

> C hoanoflagellates are marine protists that eat bacteria and detrital particles. They resemble sponge collar cells and choanoflagellates and sponges are close relatives.

Protists of unknown affinity

A number of protist groups are not known to be allied with any other eukaryotes. They may represent a

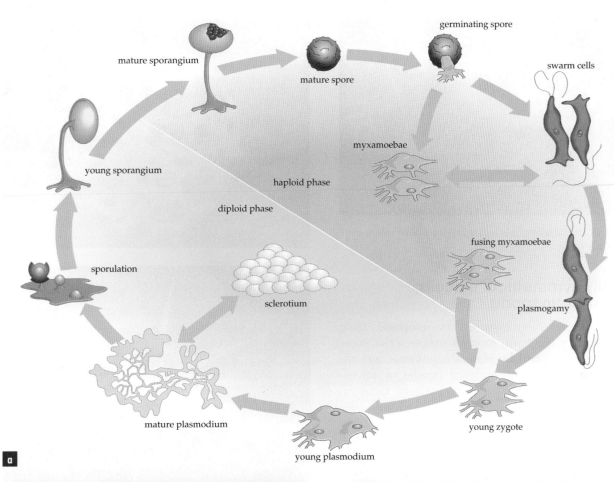

Fig. 35.26 Life cycle of an acellular slime mould (myxomycete). **(a)** A plasmodium is multinucleate and, in response to lack of food, forms a fruiting body (sporangium). Cells divide by meiosis to form haploid spores, which germinate as amoebae. They convert to biflagellate cells, which fuse to form a diploid zygote. **(b)** The sporangia of *Stemonitis fusca* take the form of tufts of brown threads on a log of wood. **(c)** Fruiting bodies of *Arcyria* are brilliant orange and of **(d)** *Trichia* look like little cups

number of early, divergent eukaryotic lineages that have been evolving independently for hundreds of millions of years. They have a variety of different cell forms and modes of nutrition.

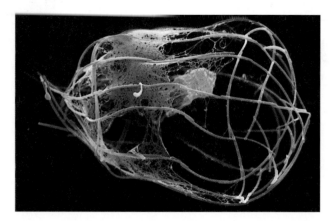

Fig. 35.27 Choanoflagellate cells (collar flagellates) live within a basket of silica strips

Radially symmetrical unicells: phylum Actinopoda

Actinopoda are characterised by **axopods**, long slender radial projections. Axopods contain a thin layer of cytoplasm bounded by plasma membrane and are reinforced with a highly ordered bundle of microtubules. Axopod microtubules collectively form an *axoneme*, which should not be confused with the microtubules of flagella and cilia given the same name. Axopod microtubules do not interslide to create bending.

The main function of axopods is prey capture. Food particles stick to their surface and are transported to the cell for ingestion. In one group (*Sticholonche*), axopods are modified to function as oars and 'row' the cell through the water. The axoneme microtubules of these oar-like axopods are attached to the nucleus by ball-and-socket articulations and the axopod is moved by co-ordinated contraction/relaxation of non-actin fibres that interconnect the axopods.

The cells of actinopods are highly variable in organisation and are often partitioned into inner and outer zones. The outer zone can harbour zooxanthellae. Some actinopods are amoeboid and others produce flagellate swarmers. Skeletons can be composed of organic material, accreted sand particles and diatom valves, celestite (strontium sulfate) or silica with traces of magnesium, copper and calcium, depending on the class of actinopod. Skeletons form fossils and huge deposits of 'radiolarian ooze', a sludge found on the ocean floor. Like diatom valves, actinopod skeletons are metamorphosed to chert with time and no extremely old fossils are known. The best known actinopods are radiolarians (Fig. 35.28), which are called sun animalcules because they resemble a minuscule sun with radiating rays.

> Actinopods have radial skeletons and projections known as axopods with which they capture food.

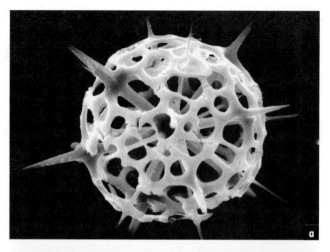

Fig. 35.28 Actinopods include radiolarians, such as **(a)** *Dictyacantha* and **(b)** *Trizona*, which produce spectacular siliceous skeletons that accumulate on the sea floor forming a radiolarian ooze

Amoebae: subphylum Sarcodina

Sarcodines are **amoebae** that are able to transiently produce extensions of the cell surface, *pseudopodia* ('false feet'), involved in locomotion or feeding. One of the first amoebae to be named was *Amoeba proteus* (Fig. 35.29) after the sea god Proteus of Greek mythology, who could change his shape at will (Gr. *amoeba* meaning change). Many sarcodines are naked but some produce internal or external skeletons. Most species are unicellular and uninucleate. Sarcodines are ubiquitous in aquatic habitats, where they prey on bacteria and other protists. The parasitic form *Entamoeba* lives in the human alimentary canal, where it causes amoebic dysentery. Another form, *Acanthamoeba*, was recently shown to infect the human brain, causing encephalitis, and the eye, causing keratitis. Amoebae in soil can play host to pathogenic bacteria such as *Legionella*.

There are two main groups of sarcodines: **rhizopods** and **foraminiferans**, which have different pseudopodia. Foraminiferan pseudopodia are reticulated (one

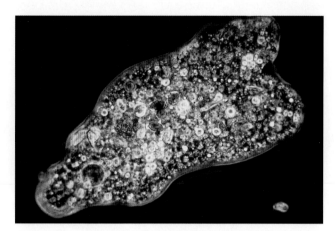

Fig. 35.29 A characteristic trait of sarcodines (amoebae) is their ability to transiently alter cell shape to produce pseudopodia. *Amoeba proteus* has several pseudopodia (false feet) projecting from the cell in different directions

pseudopod connects to others) and contain small granules.

Foraminiferans are mostly marine heterotrophs that produce calcareous (CaCO$_3$) shells. Sometimes the shell (also known as a *test*) can incorporate sand grains. Tests can be quite elaborate and have multiple chambers (Gr. *forum* meaning opening) for flotation. The chambers are often arranged in a spiral. Some foraminiferans are particularly large (up to 12 cm in diameter in the case of *Nummulites*) and contain many symbiotic algae. The shells are mini-greenhouses with algal endosymbionts housed in thin-windowed chambers around the surface to collect light. The spiny foraminiferan *Globigerina* (Fig. 35.30) acts as 'shepherd' to a 'flock' of dinoflagellate symbionts. At night, algae are harboured safely inside the foraminiferan's shell but each morning they venture out along the spines into the sunlight to photosynthesise.

About 40 000 species have been described, of which 90% are extinct and known only from fossil shells up to 600 million years old. Foraminiferans were once so numerous that deposition of their skeletons produced large chalk deposits. When building the great pyramids, Egyptian engineers noticed that the limestone blocks contained numerous nummulites (Fig. 35.31), fossil remnants of the large foraminiferan *Nummulites gizehensis*. Foraminiferan fossils are indicators of geological strata and are used extensively by the petroleum industry to characterise sediments in the search for fossil fuels. Foraminiferans occur in great abundance both in plankton and on the sea bed down to 10 000 m deep.

Sarcodines (amoebae) can transiently alter their shape. Most are heterotrophs but some have algal endosymbionts. Foraminiferans are marine amoebae with shells, often housing photosynthetic symbionts.

Living as commensals: opalinids and proteromonads

Opalinids (phylum Zoomastigina, class Opalinata) are large, unicellular, multinucleate and have numerous flagella. They are shaped like cigars or the fin of a surfboard (scalene triangle), and their surface is covered by rows of short flagella (Fig. 35.32). Between the flagellar rows are surface folds supported by ribbons of microtubules. First found in frog faeces by van Leeuwenhoek, opalinids live mostly in the rectum of ectothermic vertebrates. The relationship between host and opalinid is *commensal*, meaning that the opalinid uses the host as a place to live without apparently causing harm (Chapter 43). Opalinids feed by pinocytosis.

Proteromonads (class Proteromonadida) have either two or four flagella and are uninucleate spindle-shaped cells. They occur as commensals in the rectum

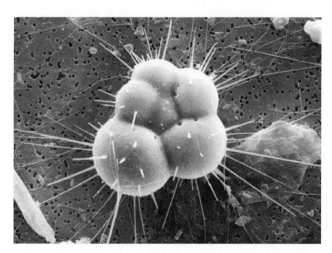

Fig. 35.30 The foraminiferan *Globigerina* has a calcareous shell through which pseudopods extend

Fig. 35.31 Nummulites are coin-shaped foraminiferans. These examples with holes in the middle are from Great Keppel Island beach and measure approximately 1 cm in diameter but much larger forms are known

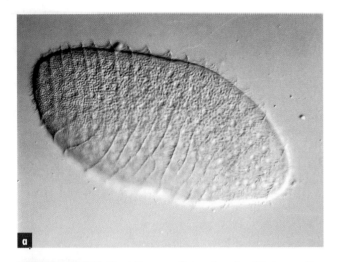

Fig. 35.32 Opalinids are covered with **(a)** rows of flagella, between which are **(b)** numerous flaps of plasma membrane. Opalinids superficially resemble ciliates but do not have a cortex interconnecting the flagella

of amphibians, reptiles and rodents. The surface of the cell has microtubule-supported folds like opalinids. On the cell surface are hairs similar to the flagellar hairs of heterokonts and opalinids.

Symbionts *par excellence*: parabasalids

Parabasalids (phylum Zoomastigina, class Parabasalia) are uninucleate flagellates involved in commensal or parasitic relationships with animals. They typically have a parabasal body, a large Golgi-type complex beside the basal body. An **axostyle**, a stiff rod-like bunch of microtubules, runs the length of the cell. *Trichomonas vaginalis* is a parabasalid that infects the human genital tract. A relatively benign sexually transmitted disease, *Trichomonas* is estimated to infect 3.5% of the world's population. Two types of parabasalids (*Triconympha* and *Mixotricha*) are symbionts in termite guts, where they are responsible for the digestion of wood. *Trichonympha* has several thousand flagella. *Mixotricha* has only four eukaryotic flagella but also has thousands of spirochaete bacteria attached to its surface that create propulsion (Fig. 35.33).

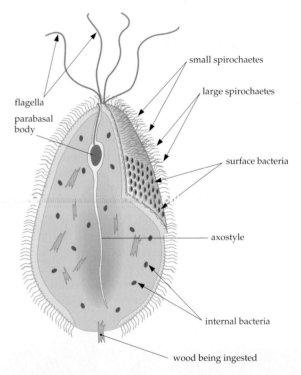

small spirochaetes

large spirochaetes

flagella

parabasal body

surface bacteria

axostyle

internal bacteria

wood being ingested

Fig. 35.33 The parabasalid *Mixotricha paradoxa* is a symbiont *par excellence*. The cell is actually a co-operative, involving as many as 500 000 individual organisms. The host cell is a quadriflagellate eukaryote. On the surface there are two forms of spirochaete bacteria that propel the cell. The spirochaetes attach to the cell surface via anchor bacteria embedded in the host cell membrane. Numerous internal bacteria within the host cell aid metabolism. The entire collection is an endosymbiont within the gut of Australian termites and is responsible for the digestion of wood

Summary

- Protists are eukaryotes not belonging to the plant, animal or fungal kingdoms. The three main kingdoms of eukaryotes probably arose from protist ancestors.
- Protists can be unicellular, colonial or multi-cellular.
- Protists take many cellular forms, including flagellates, amoebae, cysts, plasmodia and multi-cellular filaments and parenchymatous tissue.
- The majority of protists are aquatic and have flagella or cilia. Protists have a wide variety of ways of gaining nutrition, including photosynthesis, parasitism, predation and absorption.
- The first eukaryotic organisms were probably similar to modern-day protists. The nuclear membrane and endomembrane system probably evolved from invaginations of the bacterial cell membrane that enveloped the nucleoid. Chloroplasts and mitochondria are almost certainly derived from endosymbiotic bacteria that have become organelles in eukaryotic cells.
- Photosynthetic protists are diverse and are not all related. Glaucacystophytes are photosynthetic flagellates with apparently primitive chloroplasts (cyanelles) that have a peptidoglycan wall like bacteria.
- Red algae (phylum Rhodophyta) are familiar seaweeds. Most are multicellular and macroscopic and they lack flagella. They contain chlorophyll a and phycobilin pigments.
- Green algae (phylum Chlorophyta) include unicellular, colonial and multicellular forms, one group of which is the closest relative of land plants. Chloroplasts contain chlorophyll a and b, the product of photosynthesis is stored as starch and the cell walls are cellulosic.
- Heterokonts are characterised by flagella architecture. They have one smooth flagellum directed posteriorly and one hairy flagellum directed anteriorly. They include a number of phyla: Chrysophyta (golden flagellates), Haptophyta (such as coccoliths), Bacillariophyta (diatoms), Phaeophyta (brown algae, the largest protistans) and Oomycota (water moulds and downy mildews). Most are photosynthetic, with chlorophylls a and c, but oomycetes absorb food through filamentous hyphae.
- Euglenoids and kinetoplasts are related, flagellated cells. Euglenoids are free-living, some of which have chloroplasts and some of which engulf prey through an anterior gullet. Kinetoplasts are parasitic with unique mitochondria.
- Alveolates are unicells with distinctive vesicles, cortical alveoli, beneath the cell membrane. They include dinoflagellates, apicomplexans and ciliates. Dinoflagellates have two flagella, one of which encircles the cell. Many are photosynthetic, containing chlorophyll a and c, and some are predatory. Apicomplexans are endoparasites of animals, causing diseases such as malaria. An apical complex is used to penetrate host cells. Ciliates are predatory unicells characterised by two types of nuclei.
- Photosynthetic cryptomonads (biflagellates) and chlorarachniophytes (amoebae) are not related to other photosynthetic protists. They stole chloroplasts from eukaryotic prey cells that they ingested.
- Of the non-photosynthetic protists, slime moulds are amoeboid and aggregate to form colonies (cellular or acellular) with fruiting bodies that produce spores.
- Choanoflagellates are marine protists that eat bacteria and detrital particles. Their resemblance to sponge collar cells suggests that choanoflagellates and sponges are close relatives.
- Protists of unknown affinity include actinopods, sarcodines, opalinids and parabasalids. Actinopods are cells with radial skeletons and projections (axopods) with which they capture food. Sarcodines are amoebae, most of which are heterotrophs but some, such as foraminiferans, have algal endosymbionts. Opalinids and parabasalids are commensals and parasites.

keyterms

Actinopoda (p. 930)	cryptomonad (p. 926)	leucoplast (p. 909)	raphe (p. 915)
agar (p. 909)	cyanelle (p. 909)	macronucleus (p. 925)	receptacle (p. 916)
amoeba (p. 930)	cytoproct (p. 925)	malaria (p. 923)	red alga (p. 909)
antheridium (p. 917)	desmid (p. 912)	merozoite (p. 923)	red tide (p. 922)
apical complex	diatom (p. 914)	metaboly (p. 919)	reticulopodium
(p. 923)	dinoflagellate (p. 920)	micronucleus (p. 925)	(p. 926)
Apicomplexa (p. 923)	diplomonad (p. 920)	myxamoeba (p. 928)	rhizopod (p. 930)
axopod (p. 930)	endosymbiosis	Myxomycota (p. 928)	Rhodophyta (p. 909)
axostyle (p. 932)	(p. 907)	Oomycota (p. 916)	Sarcodina (p. 930)
Bacillariophyta	euglenoid (p. 918)	opalinid (p. 931)	scale (p. 911)
(p. 914)	Euglenophyta (p. 918)	ovum (p. 917)	secondary
brown alga (p. 915)	foraminiferan (p. 930)	parabasalid (p. 932)	endosymbiosis
buccal cavity (p. 925)	frustule (p. 914)	paramylon (p. 919)	(p. 926)
carrageenan (p. 909)	fucoxanthin (p. 913)	paraphysis (p. 917)	silica (p. 914)
centric (p. 914)	gametocyte (p. 923)	pellicle (p. 919)	silicoflagellate
Chlorarachnida	girdle (p. 921)	pelobiontid (p. 920)	(p. 913)
(p. 926)	Glaucacystophyta	pennate (p. 914)	slime mould (p. 927)
choanoflagellate	(p. 909)	peptidoglycan	sorocarp (p. 927)
(p. 928)	green alga (p. 910)	(p. 909)	sporozoite (p. 923)
chrysolaminarin	gregarines (p. 923)	peridinin (p. 921)	stipe (p. 915)
(p. 913)	gullet (p. 918)	Phaeophyta (p. 915)	stonewort (p. 912)
Chrysophyta (p. 913)	haematozoa (p. 923)	phragmoplast (p. 910)	thallus (p. 910)
ciliate (p. 925)	haptonema (p. 914)	phycoplast (p. 910)	transverse flagellum
coccidia (p. 923)	haptophyte (p. 913)	pit plug (p. 909)	(p. 921)
coccolith (p. 914)	heterokont (p. 912)	plasmodium (p. 926)	trophozoite (p. 923)
coccolithophorid	holdfast (p. 915)	proteromonad	trypanosome (p. 919)
(p. 914)	kelp (p. 915)	(p. 931)	valve (p. 914)
coenocytic (p. 910)	kinetoplast (p. 919)	Prymnesiophyta	zooxanthella (p. 921)
conceptacle (p. 916)	laminarin (p. 915)	(p. 913)	
cortical alveoli	leishmanias (p. 919)	pseudoplasmodium	
(p. 920)	leishmaniasis (p. 919)	(p. 927)	

Review questions

1. What major technological advances have allowed biologists to discover great diversity and previously unrecognised evolutionary relationships in the protists? Why is the kingdom Protista considered to be an artificial taxon?

2. Make a table comparing the main distinguishing features of red, green and brown algae. What useful substances are obtained from red, green and brown algae?

3. (a) Why are brown algae called heterokonts and what are their closest relatives?

    (b) What organisms are green algae most closely related to?

4. What features would enable you to identify a protist as (a) a euglenoid or (b) a dinoflagellate? Describe how each of these protists moves and how they gain their nutrition.

5. What are red tides? Why have they become a problem in Australia?

6. What is unusual about the genetic material in ciliates? How do ciliates reproduce sexually?

7. What type of protist is a trypanosome? How do these parasites avoid being eliminated by the immune system when they are in the bloodstream of their host?

8. Which of the following is correct? All members of the green algae have the pigment combination:

   A chlorophylls *a* and *b*

   B chlorophylls *a* and *c*

   C chlorophyll *a* and phycobilins

   D chlorophylls *a*, *b* and *c*

9. Flagellates with one smooth flagellum directed posteriorly and one hairy flagellum directed anteriorly are known as:

   A trypanosomes

   B dinoflagellates

   C heterokonts

   D euglenoids

10. The storage product in euglenoids is:

    A paramylon

    B starch

    C chrysolaminarin

    D glycogen

11. The flattened sacs of membrane known as alveoli beneath the plasma membrane are characteristic of which protists?

    A ciliates

    B apicomplexans

    C dinoflagellates

    D all of the above

12. A pseudoplasmodium is:

    A an aggregration of individual cells

    B a large multinucleate cell

    C a false form of malaria

    D a reproductive structure of red algae

13. Fucoxanthin is an accessory pigment found in:

    A haptophytes

    B red algae

    C green algae

    D protostelids

Extension questions

1. Which protist causes jarrah dieback? What features of this protist explain the extensive spread of the disease through Australian forests? How would you go about controlling the spread of this disease in a logging area or National Park?

2. In this chapter you have seen that 'flagellate' and 'amoeboid' cell types occur in photosynthetic, fungal-like and animal-like categories of protists. This suggests that these cell types are either primitive or that they evolved more than once. Discuss the role of endosymbiosis in the origin of chloroplasts and mitochondria and in the diversification of protists.

3. Describe the life cycle of the protist that causes malaria and suggest why parasitising two different hosts may have been a selective advantage.

Suggested further reading

Canter-Lund, H. and Lund, J. W. G. (1995). *Freshwater Algae*. Bristol: Biopress.

Excellent pictures of commonly found algae.

Entwisle, T. J., Sonneman, J. A., Lewis, S. H. (1997). *Freshwater Algae in Australia*. Sydney: Sainty & Associates.

A guide to collecting and studying freshwater algae in Australia, including easy to use keys and descriptions and information on freshwater management.

Margulis, L., Corliss, J. O., Melkonian, M., Chapman, D. J. (1989). *Handbook of Protoctista*. Boston: Jones & Bartlett.

This book is the authority on protists (called Protoctista in this book) and provides details of all major groups.

O'Kelly, C. J. and Littlejohn, T. (1999). Protist Image Database: http://megasun.bch.umontreal.ca/protists/protists.html

These are on-line images of many protists and the website links to other resources.

Patterson, D. J. (1996). *Free-living Freshwater Protozoa*. New York: John Wiley & Sons.

Excellent pictures of commonly found protozoa.

Triemer, R. E. and Farmer, M. A. (1999). The euglenid project. http://lifesci.rutgers.edu/~triemer/index.htm

A comprehensive website for euglenoids with exquisite movies of active cells.

CHAPTER

36

Fungi

Have you ever returned from holidays to find that your fruit bowl or refrigerator has turned into some kind of living science experiment? Have you wondered about those bright-coloured mushrooms that grow on rotting logs in the forest? Have you eaten bread and wondered why it is full of air bubbles? How do bubbles get into beer and champagne? What produces alcohol in beer and wine? Was Joan of Arc's religious experience intensified by eating bread made from flour containing hallucinogenic drugs? The answers to all of these questions involve members of the kingdom Fungi.

What is a fungus?

Fungi are eukaryotic heterotrophs that digest their food externally. Because of external digestion, fungi are secretory organisms—they secrete enzymes onto their substrate for digestion as well as a variety of other substances. These secretions may increase their competitive ability in environments usually teeming with other organisms. Many of these secretions have immense importance to humans as antibiotics and toxins.

The body of a fungus, the **mycelium**, generally grows as filamentous **hyphae** (sing. hypha), which are microscopic tubes of cytoplasm bounded by tough, waterproof cell walls (Fig. 36.1). Hyphae extend apically, branch and, theoretically, have an unlimited life. Fungi reproduce by spores (Fig. 36.2). During their life cycles, fungi may undergo several changes in ploidy and morphology. Their vegetative cells usually contain more than one haploid nucleus, and the diploid state is less common and in most species very brief. During mitosis the nuclear envelope remains intact.

Fungi occupy all habitats in the biosphere, including land, sea, water and air. They tolerate a wide range of environmental extremes, which means that they can be found in any environment where life is possible. They may be **saprophytes**, which play a crucial role in litter decomposition and nutrient recycling, or **parasites**, including many important plant and animal pathogens and mutualists.

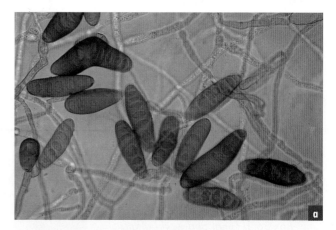

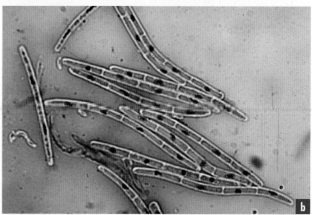

Fig. 36.2 Fungal spores come in many shapes: **(a)** *Bipolaris sorokiniana* (phylum Deuteromycota) conidia; **(b)** *Septoria tritici* (phylum Deuteromycota) conidia stained to reveal nuclei; **(c)** *Fusarium graminearum* (phylum Deuteromycota) conidia

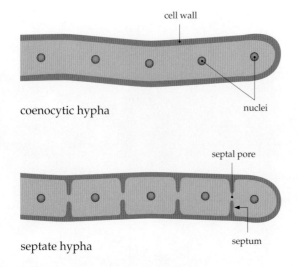

cell wall

coenocytic hypha

nuclei

septal pore

septate hypha

septum

Fig. 36.1 Coenocytic and septate fungal hyphae

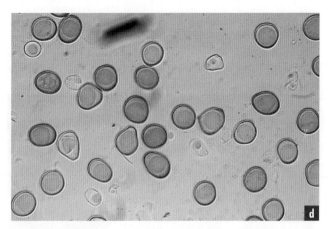

Fig. 36.2 (d) *Tilletia foetida* (phylum Basidiomycota) urediniospores

Evolution of fungi

There are at least 69 000 named species of fungi, although this probably represents only 5% of the real number, estimated at about 1.5 million species. Fungi are classified into six groups: four natural groups recognised by their different sexual life cycles, the phyla **Chytridiomycota**, **Zygomycota**, **Ascomycota** and **Basidiomycota**; one group of anamorphs that has no sexual stage, the phylum Deuteromycota; and one group, the lichens, which only exist in a mutualistic symbiosis, the phylum Mycophycota.

The most widely accepted view is that the kingdom Fungi represents a monophyletic lineage, related to animals with which they shared a choanoflagellate-like ancestor (Fig. 30.10 and Chapter 35). Chytrids, previously classified in the kingdom Protista because they release motile cells, are now thought to be an early

diverging lineage of the Fungi (Fig. 36.3). The few chytrids studied in detail are very similar to fungi in their molecular, biochemical and cell structure. Zygomycetes diverged later, developing spindle pole bodies and losing centrioles and the flagellum. The ascomycetes and basidiomycetes evolved regular septa in their mycelium and a dikaryotic (cells with two nuclei, p. 943) phase. These dikaryotic fungi often produce macroscopic, partially differentiated reproductive structures and are the most complex members of the kingdom. Morphological similarities with other organisms, such as the red algae (phylum Rhodophyta) and oomycetes (phylum Oomycota) (Chapter 35), demonstrate convergent evolution (Chapter 30) towards the fungal way of life rather than any evolutionary affinities.

> The kingdom Fungi is a discrete group of eukaryotic heterotrophs that digest their food externally and reproduce by forming spores.

Fungal growth

Vegetative growth

Fungal cells are bound by a tubular cell wall that encloses the cell membrane, nucleus, organelles and cytoplasm (Fig. 36.4). They contain all the usual

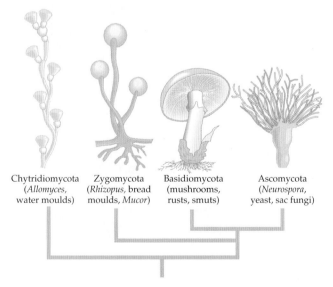

| Chytridiomycota (*Allomyces*, water moulds) | Zygomycota (*Rhizopus*, bread moulds, *Mucor*) | Basidiomycota (mushrooms, rusts, smuts) | Ascomycota (*Neurospora*, yeast, sac fungi) |

Fig. 36.3 Cladogram (see Chapter 30) of fungal phyla showing their evolutionary relationships

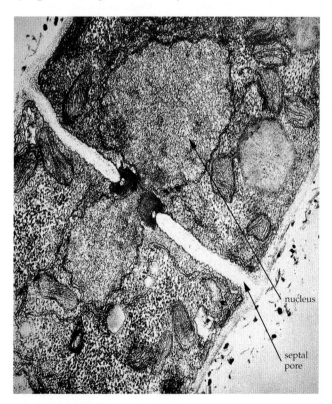

Fig. 36.4 Electron micrograph showing some ultrastructural features of *Neurospora crassa* (phylum Ascomycota). A nucleus is migrating through the simple pore in the septum

eukaryotic cell machinery except chloroplasts. Except in chytrids, yeasts and some specialised structures, fungal hyphae grow as strands one cell thick.

Fungi grow and explore their substrates by extension from the hyphal tip. As a hypha absorbs nutrients and water, it swells and expands at the tip, where the cell wall is incomplete and elastic. Branches also develop at the growing tip of hyphae. As a consequence of this type of tip growth, fungal colonies grow uniformly outwards by radial extension.

A mass of branched and apparently tangled hyphae is a mycelium. Macrofungi, such as lichens, mushrooms, puffballs and bracket fungi, produce a partially differentiated pseudoparenchyma. Yeasts and some chytrids grow as single cells rather than as hyphae. Yeasts grow by budding.

Fungal cell walls are produced from prefabricated subunits that are synthesised in the Golgi apparatus. These subunits are enclosed in membrane-bound vesicles, which bud off from the Golgi apparatus and migrate to the hyphal tip. The vesicles and their contents fuse with the existing membrane and wall at the tip and the cell elongates. The walls are made of layers of chitin (a polymer built from *N*-acetyl-D-glucosamine) microfibrils embedded in a matrix of non-glucose polysaccharides (mannans, galactans and glucomannans). Although similar in general structure to plant cell walls, the presence of chitin rather than cellulose microfibrils makes fungal cell walls rigid and waterproof, which has important physiological implications. Chitin is also used by arthropods (such as insects and crustaceans) to form their tough external skeletons and has a number of industrial uses (e.g. speaker cones) because of its strength and lightness.

Septa (sing. septum) are cross-walls that form after mitosis. Septa grow centripetally inwards from the cell wall, leaving a pore at the centre. These pores allow for nuclear migration and cytoplasmic continuity (Fig. 36.4). In this sense, fungi are not truly multicellular. Complete septa form only to separate spores, sexual reproductive structures or old hyphae from the actively growing mycelium.

> A mycelium grows as filamentous hyphae, which extend at the tip. Fungal cell walls are composed of chitin microfibrils embedded in a matrix of polysaccharides.

Hyphal aggregates

Nearly all fungi produce microscopic spores (Fig. 36.2), sometimes on very beautiful and elaborate fruiting bodies, such as mushrooms and toadstools (Fig. 36.5), or on a mat of hyphae called a stroma. These fruiting bodies are composed of pseudo-parenchyma tissue, a tight mesh of hyphal strands constructed like fibreglass.

Some soil-inhabiting fungi produce hard resting bodies, composed of tightly compacted mycelium, which are able to survive dormant in soil for several years. These bodies are **sclerotia** (sing. sclerotium) if

Fig. 36.5 Macroscopic fungal fruiting bodies, all from the phylum Basidiomycota. **(a)** *Amanita muscaria*, the fly agaric, an ectomycorrhizal fungus on many forest trees. **(b)** *Mycena pullata*, a saprophytic species growing in temperate forests. **(c)** *Coprinus disseminatus* growing on the forest floor

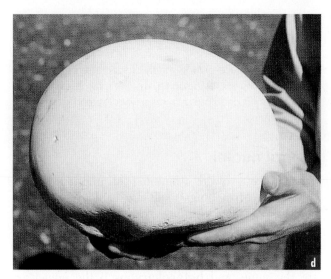

Fig. 36.5 (d) A giant puff-ball, *Calvatia gigantea*, from open eucalypt woodland

they are composed only of fungal tissue, or **pseudo-sclerotia** if they contain host tissue as well. Sclerotia germinate to form a mycelium or sexual fruiting bodies that produce many small spores (Fig. 36.6).

Armillaria luteobubalina (phylum Basidiomycota) is a pathogen that attacks the roots of forest trees. It is known as the bootlace fungus because of the thick, dark **rhizomorphs** it produces. Rhizomorphs are made of many hyphae growing together like rope with a tough, melanised outer sheath that makes them resistant to drying. They grow as much as a few centimetres—much faster than individual hyphae. Their rapid growth allows the fungus to spread from tree to tree and their strength allows them to penetrate the bark of roots (Fig. 36.7).

Fig. 36.6 The dark resting structure, called a sclerotium, of *Sclerotinia sclerotiorum* (phylum Ascomycota), which is embedded in a fine web of mycelium, has germinated to produce sexual fruiting bodies called apothecia. These apothecia bear the asci and ascospores

Fig. 36.7 The bootlace fungus, *Armillaria luteobubalina* (phylum Basidiomycota), is a serious root-rotting pathogen of forest trees. **(a)** It grows between trees as coarse, dark hyphal aggregates called rhizomorphs, which are the bootlace-like threads growing on the surface of this pine trunk. **(b)** The sexual fruiting bodies, or basidiocarps, bear millions of basidiospores

Specialised hyphae

Plant parasitic fungi often form **appressoria**, which are swellings at the tip of hyphae that adhere to the surface of the host by secreting firstly glues, then cuticle and plant cell wall degrading enzymes. Osmotic pressure inside the appressorium (as much as 10 MPa) forces a stiletto-like infection peg to emerge from the undersurface, puncturing the enzymically weakened plant cell wall. Hyphae may then grow through or between the plant cells, digesting and absorbing nutrients as they go. Some parasites penetrate plant cell walls to form delicate, wall-less **haustoria** that invaginate, but do not rupture, the host plasma membrane (Fig. 36.8). The haustorial membrane is rich in ATPase (Chapter 4), an enzyme that facilitates active uptake of nutrients across the plant membrane–haustorial membrane interface. This ability to feed from the living host cell without killing it is vital for obligate parasites.

Some saprophytic fungi form **rhizoids**, which are short, thin, root-like hyphal branches that anchor hyphae to their substrate and absorb nutrients. Fine

individual stalks that bear spores, **conidiophores**, are also examples of specialised hyphae (Fig. 36.9).

Fungi produce a wide variety of specialised structures, including spores, sclerotia, rhizomorphs, rhizoids, appressoria and haustoria, for survival, spread or parasitism.

Fungal nuclei

Fungi have evolved a number of unusual nuclear conditions. Most fungal cells are haploid and multi-nucleate. Coenocytic fungi, such as the chytrids and zygomycetes, produce multinucleate vegetative cells through successive mitotic divisions. Multinucleate vegetative cells are called **homokaryons** if the nuclei are the same, and **heterokaryons** if the nuclei are different (Figs 36.10, 36.11). Heterokaryons are most

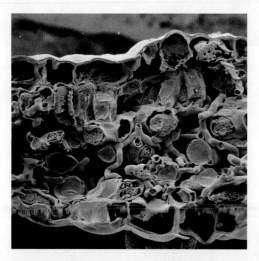

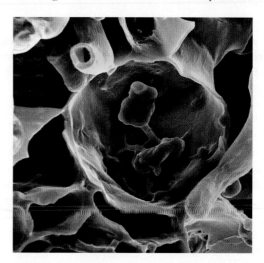

Fig. 36.8 (a) Hyphae of the rust fungus, *Hemileia vastatrix* (phylum Basidiomycota) growing through a coffee leaf, viewed with the scanning electron microscope. **(b)** Like many parasitic fungi, this species uses fine membrane-bound structures, haustoria (seen here as club-like intrusions in the mesophyll cells), to penetrate the cells of its host and absorb nutrients without causing severe disruption

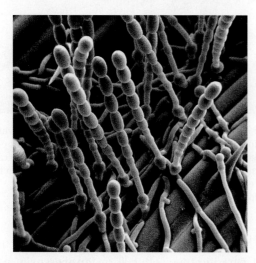

Fig. 36.9 Electron micrograph of conidiophores of the cereal powdery mildew fungus, *Blúmeria graminis* (phylum Ascomycota) bearing conidia

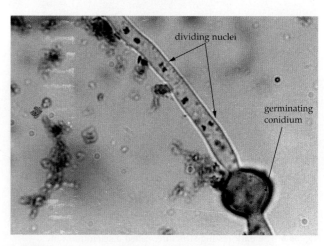

Fig. 36.10 Germinating conidia of *Botrytis cinerea* (phylum Deuteromycota). The dividing nuclei, in different stages of mitosis, are stained with aceto-orcein. The cell has several nuclei—it is heterokaryotic

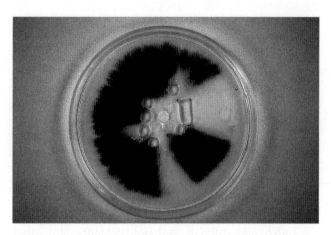

Fig. 36.11 A colony of *Verticillium dahliae* (phylum Deuteromycota) growing in artificial culture. The different coloured sectors are due to hyphal branches that developed from different nuclei in the same heterokaryotic mother cell

common in ascomycete hyphae, where they form by one of two mechanisms. In the first, hyphae from two compatible individuals may meet and fuse, or anastomose. The nucleus and cytoplasm then migrates from one individual into another. Interestingly, the genes that control anastomosis are different from the genes that control sexual compatibility. In the second mechanism, subsequent mitotic divisions propagate a mutation in one of the nuclei in a homokaryotic cell.

Another unusual nuclear condition is common in fungi. In most eukaryotes, cytoplasmic fusion (*plasmogamy*) is immediately followed by nuclear fusion (*karyogamy*). In fungi, karyogamy is delayed, sometimes indefinitely, forming a cell containing a pair of sexually compatible nuclei, a **dikaryon**.

Fungal mitosis

Mitosis (Chapter 8) is difficult to study in fungi because of their small nuclei and indistinct chromosomes, making the sequence of events difficult to follow under the light microscope. For these reasons, details of the mitotic processes in fungi are not yet completely understood.

In chytrids, zygomycetes and ascomycetes, the nuclear envelope remains intact and the nucleus remains visible during mitosis. The nucleolus also remains visible throughout most of mitosis. In basidiomycetes, the nucleus sometimes disappears when viewed under the light microscope, suggesting that the nuclear envelope may be disrupted at some stage of mitosis.

In fungi, chromosomes do not line up in a metaphase plate, but rather seem to separate asynchronously along the mitotic spindle, giving a distinctive 'double track' appearance (Fig. 36.10). Dense spindle pole bodies, attached to the nuclear membrane at the spindle poles, are connected to the kinetophores of each chromosome by microtubules. These spindle pole bodies replace the centrioles found in chytrids and other eukaryotes.

Fungal nutrition

Fungi are the primary decomposers of organic matter; decomposition not only releases minerals to the various nutrient cycles (Chapter 44) but also saves us from being buried under a pile of dead bodies. Without decomposing fungi and bacteria, the entire land surface of the earth would soon be covered by metres of dead plants and animals.

Nearly all fungi, whether saprophytes or parasites, can use glucose as a carbon source. Many species are also able to use more complex carbon molecules, requiring the induction of special enzymes. Because fungi can use different carbon sources, there are often complex successions of different fungal species on natural substrates, such as dung. The first fungi to appear are usually fast growing zygomycetes that reproduce rapidly. Slower growing species, usually ascomycetes or their anamorphs, which use cellulose as well as simple sugars, reproduce next. Finally, the slowest growing species, usually basidiomycetes, which use simple sugars, cellulose and lignin, are able to grow. Once established, these slower growing species inhibit the growth of competitors by secreting antibiotics.

Basidiomycetes and some ascomycetes are practically the only organisms able to digest lignin, which means they are very important in the decomposition of plant material. Evidence of wood-rotting fungi are found in fossils dating from the early Devonian period (Chapter 31). *Phanerochaete* (phylum Basidiomycota), the white rot fungus of wood, produces lignin-degrading enzymes when exposed to nutritional stress, such as would occur in decomposing wood after all the easily obtained carbon and nitrogen sources had been exhausted. Even then, lignin is only used if traces of another carbon source, such as cellulose, are available. It may be that lignin is only degraded to allow the fungus access to cellulose.

Amorphotheca resinae (phylum Ascomycota) can utilise hydrocarbons such as kerosene jet fuel as a carbon source and was once responsible for grounding the Indonesian Air Force when it contaminated the fuel tanks of jet aircraft. Related species are now exploited to help clear oil spills and to detoxify chemical pollutants such as chlorinated hydrocarbons. Other species attack computer floppy discs or even glass lenses (Box 36.1).

Trichophyton (phylum Deuteromycota) produces keratinases, enzymes that degrade the tough protein keratin in hair and skin. Most species are saprophytes

BOX 36.1 Fungi and world affairs

Biodeterioration and World War II

Fungi play an unwanted role in biodeterioration, often visible as mould on damp surfaces or materials, or less obviously as timber rots. Moulds, especially in the warm, humid tropics, attack nearly any organic, and some inorganic, materials. Tropical heat and humidity, encountered by soldiers during the Pacific war, provided ideal conditions for the rapid growth of many fungi. Dust particles on lenses of optical equipment, such as bombsights, field glasses and cameras, provide a perfect substrate for fungi. The mycelium not only blurred the image but the organic acids they produced also etched and destroyed the lenses. Working at The University of Melbourne, Professor J. S. Turner and his team developed effective antifungal coatings for polished glass lenses, which were eventually adopted by all Allied units.

Deadly Derek and a fungus combine to defeat Australia

Microdochium nivale (phylum Deuteromycota) attacks grasses during very cold weather. In Northern Hemisphere cereal crops and turf on sporting fields, which can be covered by snow over winter, this fungus causes snow mould, a serious disease that becomes obvious after the spring thaw. This fungus is notorious for its role in deciding the fourth cricket test at Headingley, Leeds, in England in 1972. The wicket had been covered for several cold, wet days before the match and, when the covers were removed, snow mould had killed the turf, leaving a soft, sticky, substandard surface. The bowling of Derek Underwood, the talented English left-arm spin bowler, was almost unplayable. He took 10 wickets and Australia lost by nine wickets.

that live off dead tissue or tissue remains, but some species cause skin, nail and hair diseases. House dust mites depend on these fungi to predigest skin flakes and dandruff, compounding the problem facing asthmatics who are allergic to both moulds and dust mites.

Apart from carbon, nitrogen is the most common limiting nutrient for fungi. Fungi that decompose leaf litter grow fastest when the carbon:nitrogen ratio is between 10:1 and 20:1. In dry leaf litter this ratio is about 30:1, so nitrogen is probably a limiting factor for fungal growth. A handful of nitrogenous fertiliser or manure will speed up decomposition in compost heaps. Fungi also require potassium, phosphorus, magnesium, sulfur, calcium and trace amounts of iron, copper, manganese, zinc and molybdenum, and generally grow fastest on slightly acidic substrates. They are able to synthesise their own vitamins, a feature that has made fungi useful in biotechnology.

Parasitic fungi are often fastidious and many have never been cultured outside their hosts. Those that have often require unusual amino acids or vitamins and even then do not sporulate properly outside their host. This specialisation reflects the degree of dependence resulting from coevolution of host and parasite.

Fungi, like animals and unlike plants, store their energy reserves as glycogen rather than as starch. They also store the unusual sugars trehalose and mannitol, glycerol, fats and oils.

> Nearly all fungi use glucose as a carbon source but specially adapted species, with novel enzymes, digest other carbon molecules. Most fungi are able to use inorganic sources of other nutrients and can synthesise their own amino acids and vitamins. They store energy reserves as glycogen, trehalose, mannitol, glycerol, fats and oils.

Environmental conditions for fungal growth

Fungi tolerate a wide range of environmental extremes and can be found wherever there are other organisms — in air, soil, freshwater and marine environments over the entire globe.

Water

Fungi are able to grow at extremely low water potentials compared with plants, animals or bacteria. **Xerotolerance** is achieved through the interconversion of storage compounds, releasing water:

$$\text{mannitol} \rightleftharpoons \text{glycerol} + \text{water}$$

or through the use of water released from the hydrolysis of cellulose by dry-rot fungi:

$$\text{cellulose} \rightarrow \text{glucose} + \text{water}$$

Temperature

Most species of fungi are **mesophiles** that grow best between 10°C and 30°C, where most life is found. **Thermophiles**, which grow best between 30°C and 50°C, are found in hot mineral springs, soil and compost heaps, while some are animal pathogens. They are able to adjust the lipid composition of their membranes to maintain constant membrane fluidity at high temperatures. Some fungi are able to tolerate temperatures down to –5°C, although they grow best at about 5°C. These **psychrophiles** are found in refrigerators, cold rooms or in other cold environments (Box 36.1).

Temperature may also be important in breaking the dormancy of spores. For example, teliospores of the rust fungus, *Puccinia graminis* (phylum Basidiomycota), require a cold shock to germinate, while the ascospores of many ascomycetes only germinate after a heat shock.

Light

Fungi do not require light, as do photosynthetic organisms, and most fungi thrive when kept in the dark. However, ultraviolet (near UV, UV-A or 'black' light; wavelengths between 315 and 400 nm) and visible light may be necessary to induce the formation, release and germination of spores. Some fungi have light sensors that help them to release their spores into the air, where they can be picked up and scattered by wind currents.

Gravity

Mushrooms and all macrofungi have gravity-sensing mechanisms, which enable them to orient their fruiting bodies vertically, so that falling spores are picked up and dispersed by air turbulence (Fig. 36.5).

Topography

Some fungal leaf pathogens invade their hosts through stomata (Fig. 36.12), in many cases utilising the surface topography of the leaf to guide them. Stomata are also located by a chemical attraction to unidentified photosynthetic by-products released through the stomata.

> Fungi tolerate a wide range of environmental extremes. They can be found wherever other organisms are in air, soil, freshwater and marine environments.

Fig. 36.12 Hyphae from germinating fungal spores often follow the topography of the surface of leaves in their search for stomata

Fungal reproduction

Spores

Spores are protectively packaged genomes with the two primary functions of survival and dispersal. Spores of the wheat stem rust fungus, *Puccinia graminis* (phylum Basidiomycota), released from rusted wheat leaves in Africa, are blown across the Indian Ocean to infect wheat crops in Australia. This journey involves a climb of about 10 km in thermal updraughts to the jet stream and a journey of about 13 000 km. These spores have melanised cell walls, which protect the genome from harmful UV radiation, are dehydrated to resist desiccation and freezing, and are produced in millions on each rust-infected plant. The probability that a spore would travel so far, survive the extreme conditions faced during its journey and find a susceptible wheat plant on which to land and germinate—all matters of chance—is infinitesimal, yet it happens. Every cubic metre of air over a city like Melbourne, for example, contains several thousands of spores, and even higher numbers are present in agricultural areas.

The form of a spore reflects its particular function. Spores may look the same as other cells, or they may be single or multicelled, have thick or thin walls, be pigmented, or they may be 'wet' (enveloped in a mucilage) or dry (Fig. 36.2). Because their morphology and method of formation is so much more distinctive than hyphae, spores and spore-forming structures are widely used characteristics in fungal classification and identification.

Physiologically, a switch from vegetative to reproductive growth induces sporulation. This switch may be triggered by nutrient depletion or the accumulation of staling products as the fungus exhausts its substrate, or an environmental stimulus. Hormones are known to mediate the transformation in at least some species. For example, the common mould, *Phycomyces blakesleeanus* (phylum Zygomycota), has two sexually compatible strains (Fig. 36.13). Trisporic acid, released by compatible hyphae, stimulates the differentiation of gametangia (where gametes form). In ascomycetes, a family of hormones induces the change from vegetative growth to reproductive growth. However, very little is known about hormones in other fungi. The potential application of fungal hormones as 'designer' fungicides, which could be used to suppress sporulation selectively in pathogenic species, has not escaped the attention of mycologists.

Many fungi produce spores both asexually and sexually during their life cycles. Only one type of sexual spore is produced by any one species but some fungi produce up to four different types of asexual spores or **conidiospores** (**conidia**). Sexual spores commonly function as survival spores, **memnospores**, which

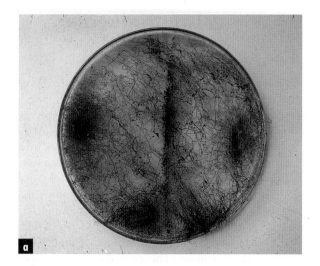

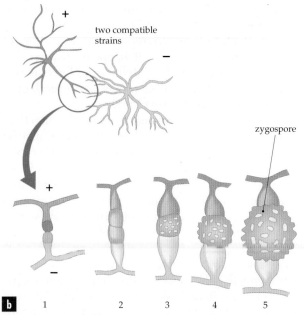

Fig. 36.13 Sexual reproduction in *Phycomyces blakesleeanus*, a heterothallic zygomycete. **(a)** Two compatible strains, '+' and '−', grow together to form a distinct line of zygospores where they meet. **(b)** Zygospores in different stages of development (1–5), from the initial contact bridge (1) to the spiny, mature zygospore (5). Compare these stages to the sequence shown in Figure 36.16

assist the fungus in persisting over unfavourable seasons. **Xenospores** are asexual spores that are produced in great numbers and assist in disseminating the fungus rapidly during periods of favourable environmental conditions.

Life without sex

Fungi produce asexual spores more commonly than sexual spores. Many fungi do not produce sexual spores, and a few, the **Mycelia Sterilia**, produce no spores at all. Sexual reproduction provides genetic diversity through recombination during meiosis. Without this source of diversity, most eukaryotes can

only reproduce clones of themselves and the species is less able to respond to changes in the environment (Chapter 15). However, those fungi that have dispensed with sexual reproduction generate sufficient genetic diversity by the release of incredibly large numbers of asexual spores. Mutations and rare mitotic recombinations result in enough variant propagules that their chances of survival are real. Heterokaryons are also able to 'carry' mutations that are only expressed under unusual selection pressures when they may be more 'fit' than the wild-type nucleus. A gene for fungicide resistance, for example, may reduce the competitiveness of the organism if expressed under normal conditions but can be hidden and 'kept in reserve' until the fungus is exposed to a fungicide.

> Fungi reproduce by spores, produced asexually and sexually. Spores are protectively packaged genomes with the two primary functions of survival and dissemination. Genetic diversity in asexual fungi is generated by random mutations and rare mitotic recombination, is harboured in heterokaryons and is propagated by the vast numbers of asexual spores produced.

Fungal phyla

Of the six groups of Fungi, the Chytridiomycota, Zygomycota, Ascomycota and Basidiomycota (Table 36.1) are 'natural' or monophyletic taxa (Chapter 30). They are characterised by their type of sexual reproduction. The Deuteromycota is an informal grouping of anamorphic fungi, organisms that do not reproduce sexually but have obvious affinities with one of the four natural phyla. The final group is the lichenised fungi, Mycophycota, which are 'dual organisms', where a fungus and an alga or cyanobacterium live together closely in a mutualistic association. The fungal partner is usually an ascomycete, rarely a basidiomycete, and is unable to live outside the relationship: it is an *obligate symbiont*.

Chytrids: phylum Chytridiomycota

Chytrids are unicellular or form short chains of cells and some produce a limited coenocytic mycelium buried in their substrate. Until recently, chytrids were included in the kingdom Protista because they form motile zoospores. However, like the Fungi, they produce lysine by the aminoadipic acid pathway, have flat mitochondrial cristae (in aerobic species) and contain chitin microfibrils in their cell walls. Recent molecular sequence data and ultrastructural evidence indicates beyond doubt that they are true Fungi.

Table 36.1 Four natural fungal phyla

Characteristic	Phylum			
	Chytridiomycota	Zygomycota	Ascomycota	Basidiomycota
Sexual spore	resting sporangium	zygospore ($2n$)	ascospore (n)	basidiospore (n)
Asexual spores	zoospores	sporangiospores (common)	conidia (common)	conidia (uncommon)
Septa	absent	absent	frequent; simple pores	frequent; complex pores and clamp connections
Vegetative hyphae	anucleate rhizoids or multinucleate	multinucleate ($n + n... + n$)	heterokaryon ($n_a + n_b...n_j$)	dikaryon ($n_a + n_b$)
Size	unicells or chains	microfungi	micro- and macrofungi	micro- and macrofungi

Chytrid zoospores have a single, smooth, posterior flagellum. A second basal body, without a flagellum, is located near the functional basal body. Although hyphae are coenocytic, a septum forms to delineate the sporangium. In species where sexual reproduction has been described, male and female gametangia produce haploid uniflagellate gametes that fuse to form the diploid zygote (Fig. 36.14). This zygote undergoes meiosis as it germinates to release haploid zoospores, often after a long dormancy.

Most of the thousand or so species of chytrids are saprophytes but some are pathogens of phytoplankton, plants, animals and fungi. In agriculture, *Synchytrium endobioticum* causes potato wart disease, while *Olpidium brassicae* infects the roots of many plants without significant damage, apart from the ability of its zoospores to act as vectors for several plant viruses. *Neocallimastix* is a chytrid that lives anaerobically in the gut of vertebrate herbivores, including ruminants and marsupials, where it digests plant fibre. Diseases caused by chytrids have recently been implicated in the decline of frog species in tropical Australia and Central America.

Chytrids are unicellular fungi with limited mycelial growth and motile cells. They include saprophytes, plant and animal pathogens and symbionts of herbivorous vertebrates (digesting cellulose in the gut).

Moulds, mycorrhizae and insect pathogens: phylum Zygomycota

The phylum Zygomycota probably evolved earlier than the Ascomycota and Basidiomycota. Zygomycetes are distinguished from other fungal phyla by: coenocytic hyphae, which have no septa, except to seal off reproductive structures; sexual zygospores; and asexual spores that are formed inside a **sporangium**, or spore-forming cell.

There is great diversity within the 600 named species. Zygomycetes include fast-growing saprophytic moulds that are the first visible colonisers of decomposing organic matter (Fig. 36.15), insect pathogens, some of which are potential biological control agents (Chapter 42), and the fungal partners of arbuscular mycorrhizae (p. 955).

Zygomycetes may be **homothallic** (sexually self-compatible) or **heterothallic** (self-incompatible). When compatible strains meet each other, they produce short hyphal branches that develop single celled gametangia, which fuse at their tips (Figs 36.13, 36.16). Nuclear exchange takes place, followed soon after by nuclear fusion. A thick wall develops around the zygote, resulting in the formation of a dormant **zygospore**. Upon germination, meiotic reduction division forms haploid multinucleate mycelium or a sporangium.

The cytoplasm of specialised hyphal branches that develop as sporangia cleaves in a series of mitotic divisions to produce haploid **sporangiospores** contained within the sporangial wall. The sporangium is separated from the vegetative mycelium by a simple septum and is often swollen at its tip. Eventually, the sporangial wall ruptures to release thousands of spores. Often these spores are surrounded by mucilage, in which case they depend on insects or water for dissemination. A few species, such as the bread mould *Rhizopus nigricans* (Fig. 36.15), produce dry spores that are dispersed by wind.

The hyphae of zygomycetes are coenocytic tubes of cytoplasm with many haploid nuclei. Zygomycetes produce both sexual spores (zygospores) and asexual spores (sporangiospores). Unlike other fungi, hyphal fusions are rare.

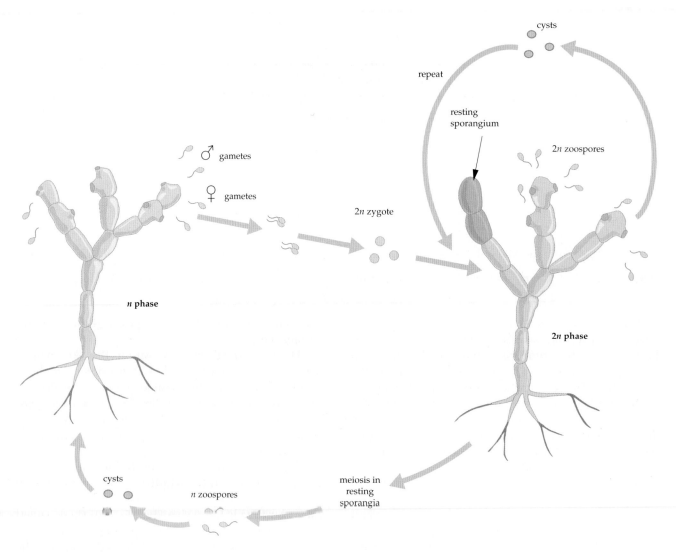

cysts

repeat

resting
sporangium

2*n* zoospores

♂ gametes

♀ gametes

2*n* zygote

n phase

2*n* phase

cysts

n zoospores

meiosis in
resting
sporangia

Fig. 36.14 Life cycle of a chytrid

Fig. 36.15 *Rhizopus nigricans* (phylum Zygomycota), a common spoilage fungus, growing on a tomato

Yeasts to truffles: phylum Ascomycota

The Ascomycota is the largest group of fungi (about 30 000 described species). They are the most important cellulose degraders in ecosystems and are also important plant and animal pathogens. We use yeast in food and beverage preparation, and, if we are lucky, might even savour the exquisite flavours of morels and truffles.

Ascomycetes have septate hyphae, produce sexual **ascospores** inside a sac-like cell, the **ascus** (pl. asci), and produce abundant asexual conidia. Ascospores create genetic diversity and often function as memnospores, while conidia function as xenospores.

In many ascomycetes, the female gametangium, the **ascogonium**, forms a short bridge that fuses with the male gametangium, the antheridium (Fig. 36.17).

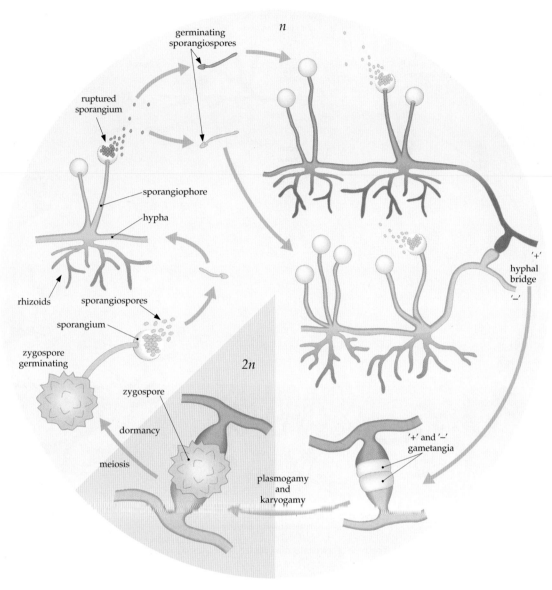

Fig. 36.16 Life cycle of a typical Zygomycota

Haploid nuclei migrate into the ascogonium but do not fuse immediately. Dikaryotic ascogenous hyphae, containing one nucleus from each parent, grow from the ascogonium and curl to form **croziers**. Mitosis occurs and asci form as short branches from the crozier. Thus, each young ascus contains two sexually compatible haploid nuclei. Successive crozier formation ensures that many asci develop from each fertilised ascogonium. Karyogamy within the ascus produces a zygote but the diploid nucleus divides immediately by meiosis to form four haploid daughter nuclei. Mitosis and cytoplasmic cleavage usually produces eight ascospores within each ascus (Fig. 36.18). In this way, ascomycetes are able to maximise the amount of genetic variation possible from each fertilisation, as each fertilised ascogonium zygote produces many asci, each containing meiotic offspring. This gives ascomycetes the ability to adapt to altered environments.

Vegetative hyphae around the ascogonium may form a fruiting body, the **ascocarp**. Classification of ascomycetes is based on the structure of the ascus and the presence and type of ascocarp.

The tip of the ascus wall often ruptures violently, releasing ascospores explosively, sometimes ejecting them 30 cm into the air. They can then be picked up by wind currents and scattered widely. Since fungi are non-motile, their ecological success depends on the production of vast numbers of spores, which, even when widely dispersed, have a very small chance of landing on a suitable substrate.

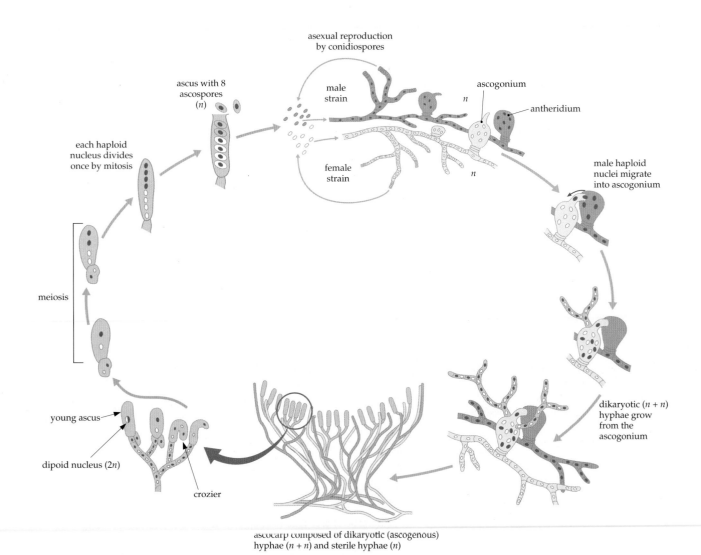

Fig. 36.17 Life cycle of a typical Ascomycota. The type of ascocarp represented here is an apothecium, similar to the ones in Figure 36.6

The sexual spore of ascomycetes is an ascospore, formed inside an ascus. Asexual conidiospores are borne directly on hyphae. The vegetative mycelium is haploid, multinucleate and heterokaryotic, and has septa with simple pores. Hyphal fusions are common and some species form macroscopic structures.

Smuts to mushrooms: phylum Basidiomycota

Basidiomycota include large mushrooms, puffballs and bracket fungi, as well as rust and smut pathogens of plants (Box 36.2). There are about 25 000 described species. The life cycle of basidiomycetes (Fig. 36.19) differs from that of ascomycetes because, while nuclear fusion is delayed in both groups, dikaryotic cells in the ascomycetes are only found in the ascogenous hyphae. In basidiomycetes, the normal vegetative mycelium is dikaryotic. Karyogamy, and the formation of diploid nuclei, occurs in a reproductive cell, the **basidium**, and

is immediately followed by meiosis to form four **basidiospores**. The basidium may be borne on a large **basidiocarp**, such as a mushroom or bracket. Basidiomycetes are classified by the structure of the basidium and the presence and type of basidiocarp produced. Basidiospores are produced in great profusion; for example, every gill in a mushroom cap is covered with millions of basidia.

Microscopically, two features distinguish the basidiomycetes from ascomycetes. **Clamp connections** are formed in mitotically dividing cells to preserve the dikaryon. As a result, the daughter cell always gets one of each of the haploid nuclei present in the parent cell (Fig. 36.20a). Many basidiomycetes also have complex **dolipore** septa, rather than the simple pores of ascomycetes (Fig. 36.20b).

Asexual reproduction is less common than in the Ascomycota but when it occurs it can be part of a very complex life cycle. *Puccinia graminis tritici*, the stem rust pathogen of wheat, produces five morphologically

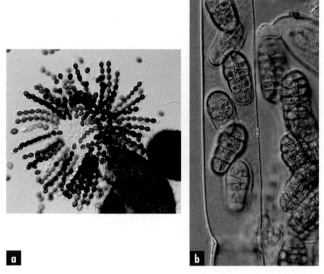

Fig. 36.18 Mature asci, each containing eight ascospores, of **(a)** *Sordaria macrospora* (phylum Ascomycota) and **(b)** *Pleospora herbarum* (phylum Ascomycota)

different spore types on two plant hosts, wheat and barberry, between which it alternates. One spore type will infect only barberry, another only wheat. The others germinate to form mycelium upon which another type of spore is formed (Fig. 36.21). Think for a moment about the precise control of gene expression that is required for this organism to complete its life cycle. The fungus never completes its sexual life cycle in Australia because barberry, although introduced by European settlers as an ornamental plant, never became widespread, and because our winters are not cold enough to break the dormancy of the spore that produces the basidium. Nevertheless, stem rust remains a dangerous pathogen because the disease cycle in Australia involves only one spore type, the **urediniospore**, which continually infects wheat growing in winter and spring crops, and as volunteer plants over summer and autumn.

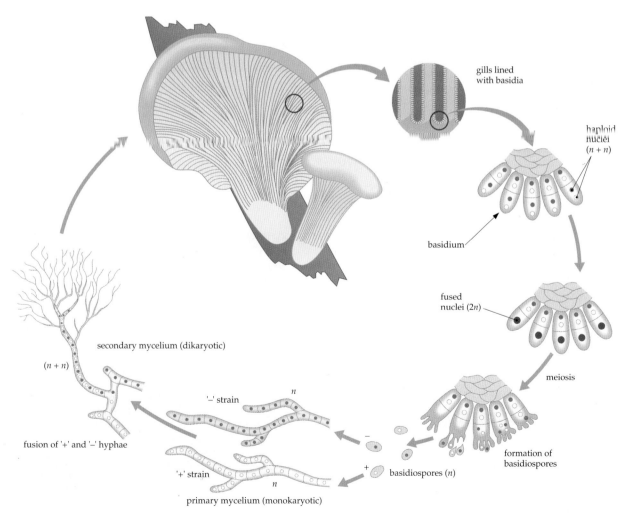

Fig. 36.19 Life cycle of a typical Basidiomycota

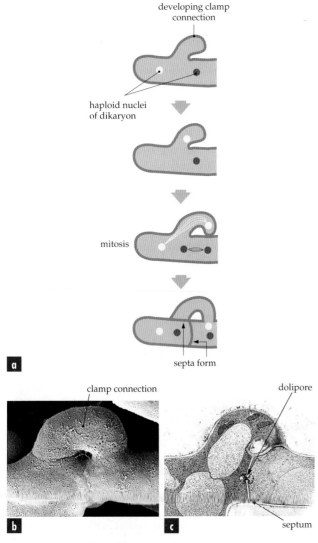

developing clamp connection

haploid nuclei of dikaryon

mitosis

septa form

a

clamp connection

dolipore

b

c

septum

Fig. 36.20 Two microscopic features used by basidiomycetes to maintain the dikaryon during vegetative growth: clamp connections ensure the distribution of one of each of the parental nuclei to each daughter cell; the complex dolipore septum restricts nuclear migration. **(a)** Mitosis and clamp connection formation. **(b)** SEM of a mature clamp connection. **(c)** Section through a clamp connection showing new septa and dolipore

Basidiomycetes produce basidiospores from a basidium, often borne on a mushroom or toadstool (a basidiocarp), and conidiospores. The vegetative mycelium is dikaryotic and divided by complex septa. Hyphal fusions are common. Some basidiomycetes have complex life cycles with more than one type of asexual spore and host.

BOX 36.2 Rust fungi and why the English drink tea

Rust diseases get their name because the pathogens, all basidiomycetes, produce masses of red

spores on their hosts. There are about 7000 species of rusts and they are all plant pathogens, infecting leaves and stems. Hyphae grow between host cells but push membrane-bound haustoria into cells. These haustoria become intimately associated with the host plasma membrane, forming a surface across which nutrients are transferred to the fungus. Nutrients are diverted away from seeds, fruits and other growing parts of the plant, causing the leaves to eventually wither and drop off. The fungus sporulates heavily producing red, brown or black pustules.

Rusts have played an interesting role in human affairs throughout history. Before the nineteenth century, the most common drink in polite English society was coffee. Coffee is produced from the berries of two species of *Coffea* trees, *C. arabica*, which is native to the area around Ethiopia, and *C. canephora*, a native of west and central Africa. English coffee came mostly from extensive plantations in what was then the colony of Ceylon. In 1869, coffee rust, caused by the basidiomycete *Hemileia vastatrix*, reached Ceylon from Africa, probably on infected propagating material. Airborne spores of the pathogen spread the disease throughout the island, causing severe defoliation and eventually the death of coffee bushes. By the 1880s, all the former coffee plantations were re-planted with tea, which is not affected by coffee rust. As no other British colonies produced coffee, tea became the standard English beverage.

Coffee rust caused by *Hemileia vastatrix* (Basidiomycota)

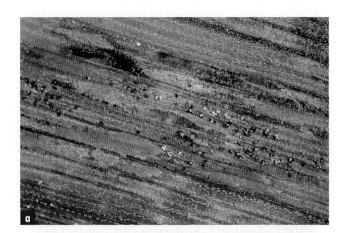

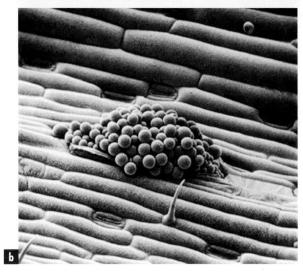

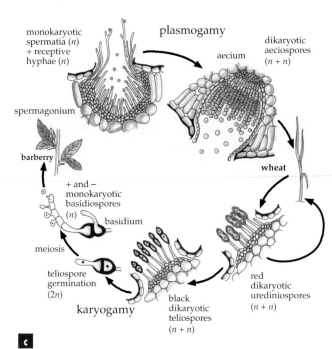

Fig. 36.21 (a) Stem rust of wheat, *Puccinia graminis tritici*, forms red pustules on wheat leaves and stems and causes the grains to shrivel. These red pustules are the asexual fruiting bodies that release millions of urediniospores that can infect healthy plants. **(b)** Rust spores. **(c)** The full life cycle involves two hosts and several spore types. In Australia, the disease cycle involves continuous infection of wheat by only one spore type, the urediniospore. Stem rust remains the most serious threat to wheat production in warm countries like Australia but a concerted program of wheat breeding to produce rust-resistant varieties has controlled it. While some varieties remain resistant for many years, rapid changes in the genetic composition of the pathogen population due to the selection of random mutants and migration of exotic genotypes over vast distances has 'broken down' the resistance of most varieties within a few years

> Classification of deuteromycetes is based on conidiospore formation because these organisms do not produce sexual spores. Otherwise, deuteromycetes are like ascomycetes and basidiomycetes.

The anamorphs

Anamorphic fungi, phylum **Deuteromycota**, sometimes called **fungi imperfecti**, include about 25 000 named species, many of which are economically important. They are a polyphyletic assemblage containing fungi with diverse evolutionary backgrounds. Anamorphic fungi have either completely lost the ability to reproduce sexually or have a teleomorph lurking somewhere, unknown to taxonomists. Even though they may have obvious morphological similarities to members of one of the natural fungal phyla, usually the ascomycetes, they cannot be formally classified until a sexual state is described. Informal classification of anamorphs is based on the type of asexual spores. However, members of the subgroup Mycelia Sterilia produce no spores at all, making the taxonomist's job very difficult.

Because some fungi named as Deuteromycota were later discovered to produce sexual spores, some species have two names (Box 36.3). Some Mycelia Sterilia, such as the common soil fungus, *Rhizoctonia*, have been found to have teleomorphs that form basidia (*Tulasnella* or *Thanetephorus*) or asci (*Trichophaea*). This has always caused confusion among mycologists.

BOX 36.3 Some fungi have two names

The rarity of sexual reproduction in some species of fungi has caused taxonomic problems. Many fungi have two names—one for the asexual, or **anamorphic**, form, and another for the sexual, or **teleomorphic**, form. The asexual state, which is usually more common in nature, was often described before the sexual state was discovered. The asexual state is often called the imperfect state because mycologists considered that an organism that does not reproduce sexually was clearly defective. Correspondingly, the sexual state is called the perfect state. In many instances, it was some time before the two were recognised as different forms of the same organism. After all, if all you can see are microscopic spores, two fungi that have different spores would reasonably be expected to be different species unless you saw an individual producing both types at once.

One of many examples is *Talaromyces flavus* (phylum Ascomycota), the teleomorph of *Penicillium dangeardii* (phylum Deuteromycota). The entire fungus, consisting of both anamorphic and teleomorphic states as described in the literature, is the **holomorph** and, under article 59 of the Botanical Code, takes the name of the teleomorph. In practice, unless the sexual spores are observed, it is common to use the name of the anamorph.

Fungal mutualisms

As heterotrophs, fungi depend for their existence on other organisms, with which they have developed intimate, often interdependent, relationships. The level of interdependence ranges from the dependence of plants on relatively unspecialised saprophytic decomposers involved in nutrient recycling to the exquisitely balanced mutualisms formed by lichenised fungi with their photosynthetic partners, or by mycorrhizal, endophytic and obligately parasitic fungi with plants. Further, macrofungi are valuable sources of nutrition for invertebrates and vertebrates, while microfungi assist in the digestion of complex food sources.

Lichens

The **lichens**, sometimes grouped as the phylum **Mycophycota**, are not so much a taxonomically distinct group of fungi as an ecological grouping of inseparable dual organisms. There are 18 000 species of lichenised fungi, all but about 20 of which are ascomycetes, the rest being basidiomycetes. The fungal partner provides most of the substance of the lichen; the photosynthetic partner, either a chlorophyte or a cyanobacterium, is embedded in the mass of mycelium. The photosynthetic partner captures light energy to produce sugars, while the fungus absorbs minerals and water. Lichens containing cyanobacteria, such as *Nostoc*, are also able to fix atmospheric nitrogen. The fungus controls the metabolism of its partner using biochemical signals, making it produce foods that only the fungus can use. The photosynthetic partner is sometimes able to live separately from the fungus but

the fungus is always dependent on the sugars provided by its photosynthetic companion.

Different morphological forms of lichens include: **crustose** ('crusty') lichens, which adhere very closely to their substrate; **foliose** ('leafy') lichens, which have more loosely attached leaf-like structures; and **fruticose** ('bushy') lichens (Fig. 36.22).

Lichen reproduction may simply be a matter of fragmentation, or they may produce **soredia**, made of algal cells embedded in fungal hyphae. Each partner reproduces independently, the fungus usually producing ascospores sexually and the photosynthetic partner usually reproducing asexually. The two partners come together again under favourable conditions to form a new lichen.

Because of their structure, lichens are among the hardiest of organisms, and are, for example, found on rocks or as part of the 'biotic crust' on soils in Australian deserts, in the Snowy Mountains and the New Zealand Alps, and to 86°30'S in Antarctica, as well as in the rainforests of eastern Australia and South-East Asia. They are also extremely slow growers, but because of their efficiency in absorbing minerals, they are important contributors to nutrient cycles. This efficiency also makes them sensitive to pollution. Lichens almost disappeared from forests around European cities after the Industrial Revolution, although as a result of improved pollution control they are now beginning to return. Reindeer in Northern Europe accumulated very high doses of radioactivity after the explosion at the Chernobyl nuclear reactor in the Ukraine in 1986 because their winter diet consists exclusively of lichens, which absorbed very high levels of radioactive minerals.

Lichens are mutualistic partnerships between an ascomycete (or some basidiomycetes) and a chlorophyte or cyanobacterium. They are classified on the basis of the fungal partner and their appearance, which may be crustose, foliose or fruticose. Lichens reproduce by simple fragmentation or as spores—soredia—an algal cell surrounded by hyphae.

Mycorrhizae

Some fungi have evolved the ability to use both saprophytic and parasitic modes of nutrition by forming a 'nutrient bridge' between the soil and plant roots. Mycorrhizal fungi infect about 90% of land plant species.

A mycorrhizal fungus absorbs and transfers to the plant mineral nutrients such as phosphorus, zinc and copper. Fungi are more efficient absorbers of mineral nutrients, enabling mycorrhizal plants to grow on soils that would otherwise be deficient in these nutrients (Chapter 18). One estimate is that mycorrhizal fungi

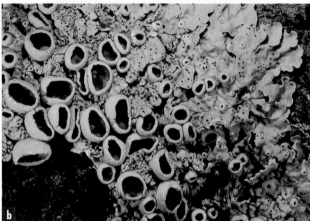

Fig. 36.22 Morphological forms of lichens: **(a)** crustose; **(b)** foliose; **(c)** fruticose

increase the efficiency of phosphorus uptake in wheat plants growing on phosphate-deficient Australian soils by a factor of 10 000. In return for these nutrients, the plant usually supplies the fungus with organic carbon. Plants infected with mycorrhizal fungi are also much more tolerant to environmental extremes, such as high temperature, drought, disease and even pollution. Mycorrhizal associations are probably almost ubiquitous because of the tremendous advantages to both partners.

Mycorrhizae are well represented in the fossil record and there is evidence that the successful colonisation of land by plants was assisted by the ability of mycorrhizae to extract nutrients from the almost totally inorganic soils of the time (see Chapter 31).

There are several types of mycorrhizae, including *arbuscular mycorrhizae, orchid mycorrhizae, epacrid mycorrhizae* and *ectomycorrhizae*. The most wide-spread are the arbuscular mycorrhizae, involving over 150 species of the phylum Zygomycota, which infect the cortical cells of the roots of at least 200 000 species or about 85% of land plants. The association is characterised by a network of fine, branched hyphae within cortical cells of the plant root. These **arbuscules**, with their large surface area, function as the organ across which nutrient exchange with the plant cell occurs. A sparse network of very fine hyphae extends from the root into the soil, but otherwise root morphology is not radically changed. Up to 3 m of hyphae may grow out of each centimetre of root, exploring the soil beyond the nutrient-depleted zone around the root. Hyphal swellings or spores, vesicles, may also be formed inside and between cortical cells and outside the root. These function as storage organs.

Epacrid mycorrhizae enable sclerophyllous heaths to grow on acid, nutrient-depleted soils in Australia (Chapters 18, 41). Mycorrhizae are essential for the germination and growth of many orchids, some of which have no chlorophyll and depend totally on the ability of their fungal partner to parasitise a second plant host that ultimately supplies the organic carbon. This could explain why some orchids occur in association with certain trees.

Ectomycorrhizae, or sheathing mycorrhizae, usually involve basidiomycetes and sometimes ascomycetes or zygomycetes. The fungi form much more selective symbioses, mainly with temperate forest trees, such as eucalypts and pines, but also with many Australian herbs. They are recognisable as many of the mushrooms and toadstools found growing in forests. The morphology of ectomycorrhizal roots is very different from that of uninfected roots. The fungus forms a sheath or **mantle** of mycelium over the root. The fungus forms a **Hartig net** of mycelium, which grows between the epidermal cells, facilitating nutrient transfer (Fig. 36.23). Around 15% of Australian plants form both arbuscular and ectomycorrhizal associations.

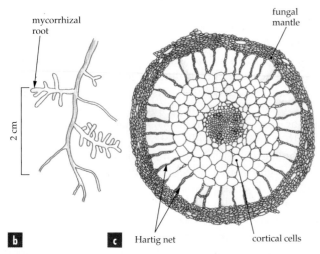

Fig. 36.23 Mycorrhizal roots. **(a)** A network of ectomycorrhizal hyphae under the leaf litter on a forest floor. **(b)** An ectomycorrhizal root. **(c)** Transverse section of a mycorrhizal root showing the Hartig net

Endophytes

Endophytic fungi parasitise living plant tissues, usually stems, leaves or fruits, without obvious detriment to the host. In a sense they occupy a niche in aerial parts of plants similar to that occupied by mycorrhizal fungi in plant roots. Plants growing in undisturbed environments harbour complex and balanced communities of fungi without showing disease symptoms, including up to 150 species of fungal endophytes, mycorrhizae and epiphytes on a single plant. If these communities are disturbed, as when plants are grown outside their natural environments or when fungicides are applied, the absence of balanced mutualistic relationships can lead to disastrous outbreaks of pests and disease.

Many grasses are systemically colonised by fungi, including species of *Neotyphodium*, that produce metabolically active secondary metabolites, including compounds that affect herbivory and mycotoxins (Box 36.4). Endophytes influence the competitive interactions of infected plants by defending them against herbivorous animals and insects, but endophytes may

also make infected pasture grasses poisonous to stock and humans. Some endophytes may become pathogenic when the host is stressed or when colonised fruit is ripening. Once again, while fruit rots are a nuisance for humans, they play an important role in natural ecosystems by exposing the mature seed and supplying nutrients from the rotted flesh for the seedling.

Fungi and insects

A number of insects have evolved mutualisms with fungi by exploiting their ability to digest cellulose, keratin or other difficult substrates. Fruit flies, leaf-cutting ants, termites and wood wasps eat fungi that saprophytically decompose plant cell walls, in some cases actively 'farming' the fungi. Ambrosia and bark beetles transport fungi in specialised pouches, *mycangia*, and feed off a combination of tree tissue and fungus. The elm bark beetle (*Scolytus*) transmits spores of the deadly Dutch elm pathogens (*Ophiostoma ulmi* and *O. novo-ulmi*) from tree to tree. Female beetles lay their eggs in the bark of newly killed trees. The hatched larvae pick up fungal spores from the gallery walls and the young adult beetles transmit the pathogen to new trees in the following spring. Dutch elm disease effectively eradicated elms in their natural habitats in the Northern Hemisphere early in the twentieth century and the world's healthiest elm trees now survive in the elegant avenues of Southern Hemisphere cities that remain free of the pathogen. Forebodingly, however, Dutch elm disease is present in parts of New Zealand, and the elm bark beetle was discovered in Australia, near Melbourne, in the 1980s.

> Fungi are involved in ecologically important mutualisms, including mycorrhizal and endophytic associations with plants, and mutually beneficial relationships with vertebrates and invertebrates.

Fungi and humans

Eating and drinking fungi

Like other vertebrates, humans enjoy eating mushrooms, the fruiting bodies of a number of species of basidiomycetes and ascomycetes (Fig. 36.24). Their nutritive value is limited because of their high water content but they are a useful source of easily digestible proteins, vitamins and minerals.

Some examples of commonly eaten fungi include the cultivated mushroom, *Agaricus bisporus* (phylum Basidiomycota); padi straw mushroom, *Volvariella volvacea* (phylum Basidiomycota); oyster mushroom, *Pleurotis ostreatus* (phylum Basidiomycota); shiitake mushroom, *Lentinus edodes* (phylum Basidiomycota);

Fig. 36.24 A number of fungi are eaten directly as food or used in the manufacture of food products

truffles, *Tuber* spp (phylum Ascomycota), and morels, *Morchella* spp. (phylum Ascomycota).

Fungi can be eaten directly, like mushrooms, or they can be used to increase the nutritive value or palatability of foods. Soybeans, although high in protein, are almost totally indigestible for humans. However, when they are fermented by *Aspergillus oryzae* (phylum Ascomycota) to make soy sauce and miso, or by *Rhizopus oligosporus* (phylum Zygomycota) to make bean curd or tempeh, soybeans become quite palatable. Properly cooked, they can be delicious.

The basic process of cheese-making involves bacteria and fungi are used during ripening to impart the delicate, or not so delicate, flavours to fancy cheeses. Camembert and brie are ripened by two aerobic fungi, *Endomyces geotrichum* (phylum Ascomycota), which removes the tart lactic acid from the curd, and *Penicillium camembertii* (phylum Deuteromycota), which partially hydrolyses the proteins, creating a smooth, creamy texture. Because both ripening fungi are aerobes, these cheeses are made flat to allow air and fungal enzymes to penetrate and ripen the cheese internally. Blue cheeses, such as roquefort, stilton and gorgonzola, are large, crumbly cheeses with blue veins of spores of *Penicillium roquefortii* (phylum

Deuteromycota), a species that tolerates low oxygen conditions. Raw cheese is inoculated with stainless steel spikes covered in fungal spores. Air enters the holes that remain, allowing fungal growth and the development of conidia that produce the characteristic blue veins. The flavour is due to the action of fungal lipase enzymes acting on fatty acids in the cheese.

Certain types of fungi, the yeasts, *Saccharomyces cerevisiae* (phylum Ascomycota), ferment sugars to alcohol and carbon dioxide. In bread-making, the carbon dioxide is trapped and causes the dough to rise, while the alcohol evaporates during baking. In traditional beer brewing, the carbon dioxide produced is contained under pressure in a sealed bottle, making bubbles, and the alcohol is produced by anaerobic fermentation (Chapter 5). Yeast cells can be seen as sediment in home-brewed and some commercial beers. In most beers, the yeast is filtered out before bottling and the residual yeast is used in making Vegemite. Every brewery has its own closely guarded, prized culture of yeast, often the result of years of trial and error.

In wine-making, the carbon dioxide is allowed to escape, except during the secondary fermentation of sparkling wine. Traditionally, natural yeasts living on the grapes were used but today more highly controlled fermentations using added yeasts and cool temperatures are responsible for a more consistent wine. Bacterial contamination of wines causes them to undergo another fermentation, which converts the ethanol to acetic acid, making vinegar.

Poisonous fungi

Some fungi are not so good to eat. Many toadstools contain poisonous alkaloids that affect the human nervous system, sometimes with fatal results. Throughout history, toadstools have been associated with witches and magic because of their hallucinogenic properties. For example, *Amanita muscaria* (phylum Basidiomycota; Fig. 36.5a), the fly agaric, has been used by shamans and witchdoctors in religious ceremonies in many cultures. In small doses it is a powerful hallucinogen, which the shamans used to assist them in their spiritual work and healing. In slightly higher doses it can be fatal. It is called fly agaric because in Europe it was mashed and mixed with milk and left to stand in dishes. The milky mash attracts flies, which are killed by the poison. Other hallucinogenic mushrooms include the common genus *Psiloscybe* (phylum Basidiomycota), the 'blue meanies' and gold tops prized by hippies (Fig. 36.25). These are rarely fatal to humans but may trigger psychotic reactions and should therefore be treated with caution.

Claviceps purpurea (phylum Ascomycota) infects the flowers of wheat, rye and other grasses. The fungus

Fig. 36.25 The hallucinogenic mushroom, *Psilocybe subaeruginosa* (phylum Basidiomycota)

replaces the seed with its own sclerotia or resting bodies known as ergots. Ergots contain alkaloids including lysergic acid, precursor to the hallucinogen, LSD (lysergic acid diethylamide). Other alkaloids from the ergot fungus cause veins and arteries to constrict and are used in medicine to assist in childbirth and to treat migraines caused by hypotension. Because they are the same size as cereal grains, ergots may contaminate flour made from infected crops. Epidemics of ergot poisoning are frequent throughout history and have been linked to unusual phenomena (such as Joan of Arc's vision, St Anthony's fire and outbreaks of witchcraft). Epidemics still occur, for example, in France in the 1970s, but livestock are more common victims than humans. In a recent case in New South Wales, over 300 cattle and sheep were killed after eating ergot-infested rye-grass.

Allergens and fungal diseases

Allergens are agents that provoke an over-reaction of our immune systems, causing respiratory diseases such as hayfever and asthma (see Chapter 14). Fungal spores may trigger allergic responses in sensitised people or in people exposed to unusually large doses of spores. These may be farmers handling bales of mouldy hay, people who sleep rough on grass contaminated with fungal spores or people living in damp, mouldy houses. The saprophytic species *Alternaria* and *Cladosporium* (phylum Deuteromycota) grow on fallen leaves, dead grass or on indoor surfaces where there is moisture or humidity, and release millions of spores into the air, particularly in late summer and autumn. When inhaled, these spores trigger severe allergic reactions in sensitive individuals.

A number of important human and animal diseases are caused by pathogenic fungi. Most are cosmetic diseases of the skin, such as ringworm and tinea, that are caused by keratin-digesting fungi such as *Trichophyton* and *Microsporum* spp. (phylum Deuteromycota) (Fig. 36.26). These are not usually life-threatening

---

BOX 36.4 Fungicides or mycotoxins in your food?

One poisonous fungus is *Aspergillus flavus* (Deuteromycota). It is a common mould on stale peanuts and produces aflatoxins. Aflatoxins are converted in animal livers to extremely potent carcinogens. As a result, the permitted levels of these toxins in peanuts are measured in parts per billion. In the early 1960s, grain-fed turkeys in Britain were fed mouldy peanut meal from South America. Most died and there was a critical shortage of Christmas turkeys that year.

Many other fungi produce poisons (**mycotoxins**) that affect both humans and animals. New Zealand sheep are regularly affected by outbreaks of diarrhoea, liver damage, facial eczema and photophobia caused by *Pithomyces chartarum* (Deuteromycota), a saprophyte that grows on dead leaves of pasture grasses. The fungus produces a mycotoxin, sporodesmin, that causes liver damage. Residues of chlorophyll, normally excreted by the liver, accumulate, causing eczema and photophobia.

Most of the fruit and vegetables we eat are treated with one or more fungicides to suppress the growth of fungi before and after harvest. The use of agricultural chemicals is coming under increasing scrutiny because of their potential toxicity to the environment and to humans. In Australia, agricultural chemicals must pass very strict safety tests for toxic and chronic effects on human health before they can be used. Although most of us would prefer to eat pesticide-free food, the judicious use of pesticides may actually make food safer. Many fungi that grow on our food produce potent toxins and carcinogens. There is no doubt that there are also many fungal secretions of unknown toxicity in much of our food—heads they win, tails you lose.

---

because the body's immune system mounts an effective response against fungi and fungi do not generally tolerate low-oxygen conditions. A few species are able to cause systemic infections of bone tissue and some anaerobic fungi cause fatal lung diseases. Normally mild fungal pathogens, such as *Candida albicans* (phylum Deuteromycota), the cause of thrush, may become lethal if the immune system is suppressed, for example, by HIV, the cause of AIDS, or by immunosuppressants given to organ transplant patients. The

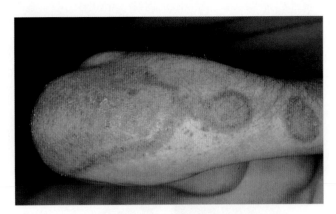

Fig. 36.26 A skin infection caused by ringworm, *Trichophyton rubrum* (phylum Deuteromycota)

immunosuppressants used in medicine are themselves fungal products, including **cyclosporin**, produced by *Tolypocladium inflatum* (phylum Deuteromycota).

Agriculture

Fungi are the most important causes of plant diseases, which account directly for losses of about 10% to the world's food production, plus an unknown loss due to storage decay. Fungal epiphytotics (plant disease epidemics) in agriculture result from the way we grow our crops as extensive monocultures of uniform genotypes. Once a virulent pathogen establishes on a monoculture of susceptible hosts, it is very difficult to stop. Monocultures are unusual in nature, where plants grow in mixed, genetically diverse communities (see Chapter 41), which usually prevent the build-up and rapid spread of a virulent pathogen. Selection pressure is for mild virulence, so that host and pathogen populations are in equilibrium and sudden changes are buffered. Occasionally, epiphytotics develop in natural plant communities but even these can usually be traced to human intervention.

Humans have also played an important part in the intercontinental spread of plant disease by inadvertently introducing new encounters between crops and pathogenic fungi, many of which have been disastrous (Box 36.2).

Many fungi, mostly ascomycetes that cause leaf spots, attack eucalypts. Although they sometimes cause a lot of damage, they are rarely destructive. *Armillaria luteobubalina* (phylum Basidiomycota) causes dieback of primary branches, crown thinning, wilting and epicormic shoot growth of eucalypts, melaleucas, grevilleas and other forest trees (Fig. 36.7). Death may follow quickly. A white mat of fungal mycelium grows under the bark of infected roots, causing a white rot of the sapwood. Sometimes thick, dark rhizomorphs may also be seen under the bark, on roots, stumps and dead trees. In autumn and early winter, dense clusters of yellow-brown mushrooms, between 4 cm and 15 cm in

diameter, appear at the base of infected trees. Clouds of basidiospores are released and carried by the wind to infect other trees. This type of dieback is scattered throughout temperate, subtropical and tropical Australian forests, woodlands, orchards, gardens and parks, and is often most severe when the topsoil has been compacted by roads or carparks.

Biological control

Fungi are also used in the biological control of weeds and pests. Blackberries, skeleton weed, Paterson's curse and heliotrope are major weeds introduced to Australia from Mediterranean Europe. The CSIRO runs a program based in Montpellier, France, that searches for pathogens of these weeds in their native habitats and screens them for their selectivity and virulence. All four weeds are now being attacked by specific rust fungi selected from this program. Although these fungi do not kill their host plants, they reduce their vigour and thus aggressiveness as weeds.

Some soil fungi trap nematodes using sticky branches or rings of hyphae that act like lassoes. Once trapped, the unfortunate nematode is invaded by hyphae and then digested by enzymes secreted by the killer fungus (Fig. 36.27). Stimulating the activity of these predatory fungi could reduce the severity of plant diseases caused by nematodes.

Fungi and biotechnology

Fungi are model eukaryotes for biotechnological research and application. Because they are eukaryotes, they have the potential to be 'engineered' to produce large amounts of useful biochemicals. Many fungi are easily cultured on a range of substrates and, because they naturally secrete enzymes and secondary metabolites, including antibiotics and toxins, they have the ability to convert low-value wastes into useful

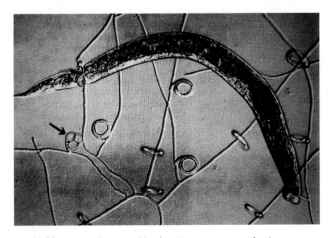

Fig. 36.27 One method used by fungi to trap nematodes is lasso-like loops

BOX 36.5 Coconuts, rhinoceros beetles and the green muscardine fungus

The coconut tree is so important in Asia and the Pacific that it is known as 'the tree of life'. Rhinoceros beetles, *Oryctes rhinoceros*, are one of the most important pests of coconut trees because they cut the leaves and eat the growing bud and inflorescences, sometimes killing a tree. It is thought that they also transmit propagules of the oomycete that causes bud rot, *Phytophthora palmivora*. Insecticide sprays are expensive and environmentally dangerous and not very effective because the beetles are often hidden in the foliage.

However, one feature of their life cycle makes them vulnerable to biological control using the green muscardine fungus, *Metarhizium* (phylum Deuteromycota). Rhinoceros beetles congregate in dead tree trunks or piles of sawdust to breed. One or two infected larvae or spores of the green muscardine fungus released into a breeding log are sufficient to start an epidemic that decimates the population.

(b) Rhinoceros beetle larva attacked by the green muscardine fungus (right). Scale bar = mm

(a) Coconut palm damaged by rhinoceros beetle

compounds, such as vitamins, hormones and antibiotics. The use of fungi for this purpose is 'Industrial mycology' and, in fact, the use of fungal cultures in large industrial vats in this way has pioneered the development of biotechnology. Some examples have already been given—the production of Vegemite, penicillin and cyclosporin, for example. Organic food acids such as oxalic and citric acids are produced by strains of *Aspergillus niger* (phylum Deuteromycota). Penicillin is produced by strains of *Penicillium chrysogenum* (phylum Deuteromycota).

Selection and mutation, and more recently direct gene transfer, improve the efficiency and potency of fungal strains used in industrial mycology. Once a eukaryotic gene has been identified and cloned, it is a relatively simple matter to incorporate it into the haploid genome of a fungus and have the gene expressed in large quantities. For example, the human insulin gene has been spliced into the genome of baker's yeast, *Saccharomyces cerevisiae* (phylum Ascomycota), and insulin for use by diabetics is now produced from bulk cultures of this fungus.

Fungi are directly important to humans as foods, as causes of food spoilage, plant disease and famine, as producers of mycotoxins and medicines, and as causes of human diseases and allergies.

Summary

- Fungi are filamentous eukaryotes that, apart from the chytrids, form non-motile spores by meiosis (sexual spores) or mitosis (asexual spores). In chytrids the spores may be motile. A fungal mycelium grows as filaments called hyphae, the cell walls of which are composed of microfibrils of chitin embedded in a matrix of complex carbohydrates. Fungi probably evolved from a conjugating protist-like ancestor similar to chytrids.

- Fungi are food absorbers secreting enzymes that digest their food externally. Fungi are either saprophytes (feeding on dead organisms) or parasites (feeding on living host cells). Some parasites are pathogens (they cause a disease on their host), while others are partners in mutualistic associations, including lichens, mycorrhizae and endophytes.

- Nuclei in vegetative mycelium are haploid. Sexual reproduction involves the conjugation of hyphae (plasmogamy), although karyogamy (nuclear fusion) is usually delayed with the life cycle of many fungi including a prolonged dikaryotic stage. When it finally occurs, karyogamy is usually followed immediately by meiosis.

- Within the kingdom Fungi there are six phyla. Four—Chytridiomycota, Zygomycota, Ascomycota and Basidiomycota—are natural groups that differ in their hyphal structure, sexual and asexual reproduction and size.

- Chytridiomycotes (chytrids) are typically unicellular fungi attached to their substrate through a limited coenocytic mycelium. Resistant sporangia release haploid, uniflagellate, asexual zoospores. Male and female gametangia release uniflagellate gametes that copulate to produce a zygote.

- Zygomycetes have coenocytic hyphae and produce diploid zygospores (sexual) and haploid sporangiospores (asexual).

- Ascomycete hyphae have septa with simple pores and anastomosis (vegetative hyphal fusion) is common, forming heterokaryotic hyphae. They produce asexual conidiospores. Haploid ascospores are formed following karyogamy and meiosis within an ascus.

- Basidiomycete hyphae have complex septal pores that function to maintain a dikaryotic vegetative mycelium. Karyogamy and meiosis within a basidium produce externally borne, haploid basidiopores. Asexual conidiopores are less common than in the ascomycetes.

- Both ascomycetes and basidiomycetes may produce macroscopic sexual fruiting bodies, commonly known as mushrooms and toadstools.

- Within the kingdom there is an evolutionary trend towards reliance on asexual reproduction. Anamorphic fungi (phylum Deuteromycota) are exclusively asexual but are otherwise similar to ascomycetes and basidiomycetes. Genetic diversity is generated through random mutations that are preserved in heterokaryotic hyphae and become apparent when vast numbers of conidia are exposed to natural (and artificial) selection. Some fungi also undergo a parasexual cycle, involving a sequence of rare mitotic 'mistakes'.

- Fungi, as biodegraders, mutualists, pathogens and contaminants form interdependent relationships with other organisms.

- Lichens (phylum Mycophycota) are mutualistic partnerships between an ascomycete and a chlorophyte, xanthophyte or cyanobacterium. They are classified according to their appearance—crustose, foliose or fruticose.

- Nearly all land plants harbour fungi in their roots (mycorrhizae) and shoots (endophytes). Insects and vertebrates also form intimate relationships with fungi. These fungi play important roles in the nutrition and ecology of their hosts.

- Fungi are important to humans as foods, decomposers, plant and animal pathogens, in biological control and in biotechnology.

key terms

Review questions

1. What features distinguish fungi from other organisms?

2. What are the similarities and key differences between fungal-like protists, red algae (phylum Rhodophyta) and fungi?

3. What evidence suggests that fungi evolved from a protistan ancestor not unlike the chytrids?

4. What evidence suggests a close affinity between the ascomycetes and the basidiomycetes?

5. Describe some of the unusual features of nuclear behaviour in fungi.

6. What features allow fungi to survive in all environments where life is possible? Why is this important to us?

7. What is a 'fungal succession'? Give an example.

8. How do imperfect fungi generate variability and diversity without sexual reproduction?

9. What features do fungal symbioses have in common? Why are fungi particularly well suited to their roles in these partnerships?

Extension questions

1. You are an environmental consultant engaged to advise on the redevelopment of an old petro-chemical factory site. You find that the soil has high levels of contamination with organochlorines but you have been able to isolate a range of fungi that appear in laboratory tests to be able to break down the organochlorines. List the criteria you would use to select the best isolates and what you could do to maximise their biodegrading activity.

2. In what ways have plant diseases affected the way we live? Think of both historical and present examples.

3. What are the benefits and risks of using fungi in biological control programs?

Suggested further reading

Alexopoulos, C. J., Mims, C. W., Blackwell, M. (1996). *Introductory Mycology.* 4th edn. New York: John Wiley & Sons.

The definitive mycological text with thorough descriptions of each of the fungal phyla as well as the fungal protists.

Deacon, J. W. (1997). *Modern Mycology.* 3rd edn. Oxford: Blackwell Science.

A comprehensive introductory text covering the physiology and ecology of fungi. It gives a good coverage of fungal cell biology and molecular genetics.

Jennings, D. H. and Lysek, G. (1996). *Fungal Biology: Understanding the Fungal Lifestyle.* Oxford: Bios Scientific Publishers.

A very readable text that concentrates on the physiology, ecology and environmental biology of fungi.

Kendrick, B. (1992). *The Fifth Kingdom.* 2nd edn. Newburyport, Mass: Focus Texts.

An entertaining and comprehensive treatment of the fungi from a systematic, functional and human perspective. Full of wonderful mycological anecdotes, cooking hints and recipes.

Young, A. M. (1994). *Common Australian Fungi—a Naturalists Guide.* Sydney: UNSW Press.

One of a number of useful field guides with descriptions, information, drawings and colour photographs of the common species of Australian macrofungi.

CHAPTER

37

Plants

This chapter summarises the characteristics of the major groups of land plants, emphasising their evolutionary history and characteristics. Before reading this chapter, you should first be familiar with the material in Chapters 14 and 17 which detail plant reproduction, architecture and structure. Remember also from Chapter 31 that over geologic time, the earth and organisms have changed together and continue to do so. The movement of continents affects climate and terrestrial environments and has played a part in the rise and fall of plant groups through time.

Evolution of plants

Plants, kingdom Plantae, are those photosynthetic organisms that have adapted to life on land (Fig. 37.1), and include liverworts, mosses, ferns, conifers and flowering plants among the major modern groups (Table 37.1). They are multicellular and have cells specialised to form tissues and organs.

Plants first colonised the land about 410 million years ago, towards the end of the Silurian period (see Chapter 31). At this time, the land surface was mostly bare and the earliest plants, which were small and inconspicuous, were restricted to the margins of wetlands and waterways.

The first characteristics important in the transition of plants from water to land were a cuticular covering of the plant body and a protective layer around the sex organs, both of which reduced loss of water by evaporation and prevented desiccation. The evolution of an internal transport system of xylem and phloem, specialised organs, such as leaves and roots, and woody tissue allowed vascular plants to attain greater body size than smaller non-vascular plants, which rely on simple diffusion for transport of water and nutrients. Transport of male gametes, in pollen, to the female reproductive organ by wind or animals rather than by water finally freed plants from their dependence on a moist environment.

Although the earliest vascular plants were relatively small and probably confined to swampy areas, the Devonian was a time of rapid diversification of plants. By the end of the period, large tree-sized plants formed complex forests. These Palaeozoic forests (360 million years BP) would have looked very different from the forests we know today. Many of the early plant groups are extinct or represented by only a few living taxa today.

Green algae and the origin of land plants

Green algae (phylum Chlorophyta, Chapter 35), which grow in marine habitats, fresh water or on soil, include the closest relatives of land plants. Green algae and land plants have similar pigments (chlorophylls *a* and *b*), chloroplast structure and cell wall chemistry (cellulose) and both have starch as a storage material. However, there has been considerable disagreement over which particular green algal group is most closely related to land plants. Electron microscopy has resolved this debate.

From the study of algal cell divisions in a diversity of green algae, it was found that their dividing nucleus is enclosed within an intact nuclear envelope and, during cytokinesis, their spindle collapses and is replaced by a system of fibres (microtubules) that are all oriented in the plane of cell division. The latter system is called the phycoplast. In comparison, the mitotic spindle of land plants is 'open' (i.e. the nuclear envelope disappears) and the new cross wall is formed in the phragmoplast, the structure containing the remnants of spindle fibres oriented at right angles to the new cross wall (Chapter 8; Fig. 37.2).

The significance of these differences soon became obvious. A few green algae, including *Coleochaete* and the order Charales (commonly called charophytes), were found to have a phragmoplast and open spindle. *Coleochaete* (Fig. 37.3) is an odd alga, with additional characteristics that are typical of land plants. For example, it has a covering of protective cells around its delicate reproductive structures. *Chara* (Fig. 37.4) is a common example of the Charales. It grows on the bottom of shallow, freshwater lakes and dams. Some species become encrusted with lime so that when the plant dies a calcareous 'skeleton' remains intact, hence the common name stonewort.

These algae, as well as liverworts, mosses, ferns, cycads and even the tree *Ginkgo* (see p. 975), all have similar asymmetric, motile flagellated cells. In contrast, all those green algae with a phycoplast produce symmetric motile cells during vegetative or sexual reproduction. Charophytes and land plants both have the enzyme glycolate oxidase, equivalent to, but different from, the enzyme glycolate dehydrogenase

Fig. 37.1 Australian heathlands are rich in plant species

Table 37.1 Major groups of land plants living today[a]

Phylum	First appearance in fossil record	Evolution of important characteristics for living on land
Non-vascular plants **Hepatophyta**: liverworts **Anthocerotophyta**: hornworts **Bryophyta**: mosses	Late Devonian (362 million years BP) first unequivocal liverwort fossil; late Silurian fossils of unidentified thalloid plants	Cuticular covering; resistant spores; small body size; stomata on sporophyte of hornworts and some mosses
Vascular plants	Late Silurian (408 million years BP)	Sporophyte dominant, vascular tissue
Psilophyta: *Psilotum* and *Tmesipteris*	No Palaeozoic fossils recognisable as these extant forms	Stems and underground rhizomes with simple vascular cylinder of phloem and xylem; lignin; stomata
Lycophyta: clubmosses and quillworts	Late Silurian fossils are non-woody forms (e.g. *Baragwanathia*); early Carboniferous (362–290 million years BP) tree forms	Roots, stems and leaves (small—microphylls); heterospory in some; reproductive structures in strobili (cones)
Sphenophyta: horsetails (*Equisetum*)	Late Devonian; many fossils; only one living genus	Roots, stems, small leaves
Filicophyta: ferns	Early–middle Devonian with increased diversity by late Carboniferous	Leaves large megaphylls
Seed plants	Late Devonian fossil ovules	Ovules and seeds with dormancy; pollen and pollination
Cycadophyta: cycads	Early Permian (290–251 million years BP)	Seeds borne on scales in cones; modern forms insect pollinated; some secondary growth
Ginkgophyta: *Ginkgo*	Late Carboniferous to early Permian	Secondary growth
Coniferophyta: conifers	Late Carboniferous primitive forms; Triassic (251–208 million years BP)—modern conifers	Secondary growth; pollen tube deposits non-flagellate sperm
Gnetophyta	Triassic to early Cretaceous (130 million years BP)	Vessels with circular perforations; flower-like cones; modern forms insect pollinated
Magnoliophyta: angiosperms or flowering plants	Early Cretaceous (130 million years BP) monosulcate pollen (single germination aperture) suggestive of magnoliid dicots or monocots	Vessels with ladder-like perforations; flowers; double fertilisation; seeds in fruits; seeds with nutritive endosperm; diversity of animal pollinators

(a) See Chapter 31 for geologic time scale.

found in other green algae. More recently, molecular biology has shown that charophytes and land plants share similarities in their 5 S rRNA (see Chapter 30). It is now apparent that of the several lines of evolution in the green algae, *Coleochaete* and Charales are two of the closest relatives of the land plants (Fig. 37.5). The different groups of green algae must have diverged a long time ago, well before the colonisation of the land.

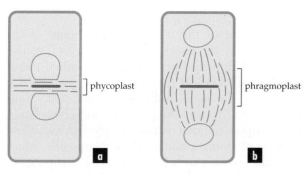

Fig. 37.2 (a) Cell division of many green algae involves a system of microtubules called a phycoplast. **(b)** Those green algae thought to be the closest relatives of the land plants have a different cell division, involving a phragmoplast

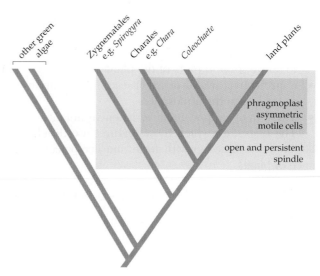

Fig. 37.5 Green algae include a number of evolutionary lines, with *Coleochaete* and Charales being the closest relatives of the land plants. Of the other major lines of evolution in the green algae shown here, the large order Zygnematales (which includes the familiar *Spirogyra*) have persistent spindles like those of *Coleochaete* and *Chara*. In *Spirogyra*, the persistent spindle even develops into a rudimentary phragmoplast. The Zygnematales have completely lost the flagellum, which is unusual for organisms living in water. They undergo sexual reproduction by conjugation, a process analogous to that seen in terrestrial fungi (see Chapter 36). Perhaps these algae adapted to terrestrial existence to some extent and then moved back into water—the algal equivalent of whales, penguins, seals and seagrasses!

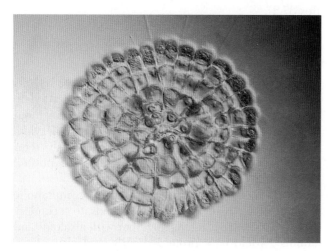

Fig. 37.3 *Coleochaete* is a small green alga that grows on other aquatic plants. It is a disc generally one cell thick

Plants (kingdom Plantae) are multicellular photosynthetic organisms adapted to living on land. Their closest relatives are green algae. The oldest plant fossils date back 410 million years.

Living on land

The increase in complexity and specialisation of vegetative and reproductive features associated with colonisation of the land surface is the key to understanding the remarkable diversity of plant life on earth. The differences between living in water and living on land are enormous, particularly with respect to two critical aspects of plant metabolic function—water balance and gas exchange. Many of the evolutionary modifications of plants relate to these factors.

Algae are surrounded by water. Passive diffusion across cell walls keeps them in equilibrium with the surrounding environment; there is no need for special water-absorbing or water-conducting structures and desiccation is not a problem except when seaweeds are exposed on rock platforms at low tide. In contrast, land plants inhabit an environment where water is limiting and usually confined to the soil in which they grow. They require specialised organs and complex tissues for several functions:

Fig. 37.4 *Chara* is a common member of the Charales, found in fresh water

- roots for extracting water and dissolved nutrients from soil
- vascular tissue for transporting water and nutrients to above-ground parts of the plant
- a protective, water-resistant coating to minimise water loss to the atmosphere
- a system of support in an aerial environment.

The degree to which these features are developed largely determines growth form and ecological tolerance of plants.

Specialisation of vegetative features

Cuticle: a waterproof barrier

Plant cuticle, composed of insoluble lipid polymers and waxes, is deposited over the entire above-ground surface of plants and was essential for land colonisation. It is completely impermeable and acts as a waterproof barrier to prevent diffusion of water from turgid plant cells to the drier atmosphere. Cuticle also filters a substantial component of UV radiation from sunlight.

Sporopollenin: protecting spores

Sporopollenin is a polymer, tougher even than lignin but with similar properties, composed chiefly of carotenoids. Some green algal cells and spores have a sheath of, or are impregnated with, sporopollenin. This protects them from attack by grazing organisms and microorganisms. It also makes spores of land plants (Fig. 37.6) tough and flexible and resistant to biodegradation; for example, it is found in the wall of pollen grains of seed plants and is the reason that pollen grains preserve well as fossils.

Stomata: controlling gas exchange and water loss

In aquatic environments, CO_2 is in solution (as HCO_3) and diffuses across cell walls for algal photosynthesis. In terrestrial environments, CO_2 is a gaseous component of air. Because of the water-resistant nature of the cuticle covering the outer plant surfaces, CO_2 cannot be absorbed directly from the air. Instead, it enters through small pores in the plant surface, where it is absorbed by the moist walls of internal cells. Some liverworts (p. 970) have simple pores for gas exchange that also expose internal parts of the plant to water loss. Evolution of stomata bordered by guard cells, which control the opening and closure of pores (Chapter 18), must have allowed plants to expand their range and inhabit drier environments. Stomata are present in modern hornworts, mosses and vascular plants, as well as simple, ancient fossil plants such as *Rhynia* (see Box 37.2).

Vascular supply and lignin: transport and support

The vascular supply of ferns, fern allies and seed plants consists of phloem for the transport of sugar and xylem for the transport of water and mineral ions throughout the plant (Chapters 17, 18). The cell walls of xylem tracheids, vessels and fibres are conspicuously reinforced by a rigid layer of lignin that is deposited between the cellulose microfibrils.

Biochemical conversion of primary metabolic products, such as sugars and amino acids, to secondary compounds, such as lignin (together with alkaloids and tannins), would have protected early land plants from UV radiation and pathogenic fungi. The ability to form lignin proved to be of great importance because it gave strength to upright terrestrial plants. Lignin prevents xylem cells from collapsing under the hydrostatic pressure that is developed around them. Greater structural support, together with a conducting system, allowed plant size to increase. The largest land plants, trees, produce large amounts of wood through secondary growth.

Stems, roots and leaves: division of labour

Early vascular land plants (e.g. *Rhynia*, Box 37.2) were small, rootless and leafless. They had simple upright, photosynthetic stems and simple underground stems, **rhizomes**, with extensions of epidermal cells, **rhizoids**, for anchorage to the substrate and possibly absorption of water and dissolved nutrients. True roots with a vascular system and aerial leaves for photosynthesis developed later (Fig. 37.7).

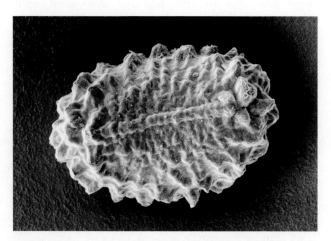

Fig. 37.6 Spores of land plants, for example, *Polypodium formosanum* (grub fern or caterpillar fern), are resistant to desiccation and biodegradation

At various stages in their evolution, land plants developed cuticles, resistant spores, vascular tissue, lignin, stomata, stems, roots and leaves, which overcame the problems of water balance, gas exchange and structural support.

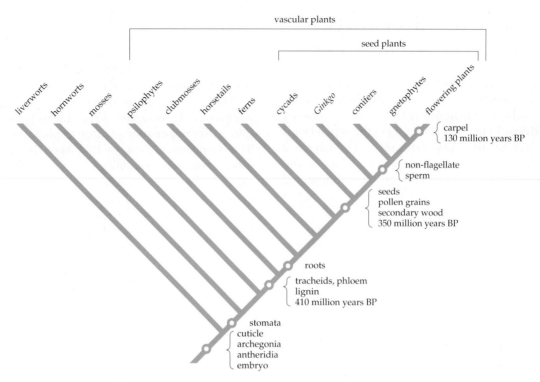

Fig. 37.7 Phylogeny of living groups of land plants, showing the evolution of important features and the approximate age of the oldest known fossils for some groups

Alternating generations and sexual reproduction

Remember from Chapter 14 that there are two **alternating generations** in the life cycle of plants. The **gametophyte** generation, which is *haploid*, produces male and female gametes by mitosis. Male and female gametes fuse and the zygote grows into the *diploid* **sporophyte** generation. The diploid sporophyte produces spores by meiosis, which are dispersed and germinate into new haploid gametophytes.

The successful transition from water to land was associated with the sperm-producing and egg-producing cells being protected by a layer of sterile jacket cells, forming the sex organs **antheridia** (male) and **archegonia** (female). Unlike their green-algal relatives, land plants also retain the zygote on the gametophyte, where the developing embryo is protected and nourished.

In land plants, the haploid and diploid generations are markedly heteromorphic, that is, gametophyte and sporophyte differ in size and shape. There is a distinct evolutionary trend regarding the contribution of each generation to the life cycle. In bryophytes, the most primitive plants, the gametophyte is the dominant, photosynthetic generation. The sporophyte is simple in structure, short-lived and totally dependent on the gametophyte for nutrition. In seed plants, the opposite is true. The conspicuous plant is the sporophyte generation. The gametophyte is very reduced, commonly just a few cells. Ferns have a dominant sporophyte but a gametophyte that is not as reduced as those of seed plants.

Evolution of pollen and seeds

In the life cycle of some ferns, fern allies and seed plants (see Table 37.1), two types of spore are produced by the sporophyte: **megaspores** in **megasporangia** and **microspores** in **microsporangia**. Megaspores germinate into female gametophytes, which bear egg cells in archegonia; microspores develop into male gametophytes, which produce sperm in antheridia. Having two sorts of spores is referred to as **heterospory** (having one type of spore is **homospory**). Heterospory gave rise to separate male and female gametophytes. This allowed the evolution of pollen, enabling wind and animal dispersal of male gametes, and the ovule, which houses an egg cell and, after fertilisation, develops into a seed.

In land plants, haploid and diploid phases are markedly different. Haploid gametophytes produce sex organs, antheridia and archegonia, while diploid sporophytes produce spores by meiosis. Heterospory (development of two types of spores) allowed the evolution of pollen and ovules.

Bryophytes: liverworts, hornworts and mosses

Non-vascular plants, commonly called bryophytes, include **liverworts** (phylum **Hepatophyta**), **hornworts** (phylum **Anthocerophyta**) and **mosses** (phylum **Bryophyta**). The similarity of these plants lies in their lack of true vascular tissue. They have a photosynthetic, free-living gametophyte, which is the dominant generation compared with their simple and largely 'parasitic' sporophyte. These features suggest bryophytes are early lineages of land plants (Fig. 37.7). However, fossils of non-vascular land plants, clearly recognisable as bryophytes (Fig. 37.8), are not as old as are those of vascular plants. Bryophytes first appear in the late Devonian, about 360 million years ago. Because non-vascular plants are not easily preserved, they are rare and older fossils may yet be found.

Gametophytes: the haploid generation

Gametophytes are the conspicuous generation of the bryophyte life cycle (Fig. 37.9) but rarely grow larger than several centimetres in height. They are usually attached to the substrate by elongated cells or filaments of cells, rhizoids. Male gametes (antherozoids or spermatozoids) are released from sac-like antheridia (Fig. 37.10a) and swim using their flagella to the archegonia (Fig. 37.10b), entering through the neck and fertilising the non-motile egg cell in situ. Sexual reproduction involving motile sperm requires the presence of free water (rain, dew, etc.) and, together with the absence of vascular tissue, explains why the larger bryophytes are mainly confined to moist environments.

Liverworts have either thalloid or leafy gametophytes (Fig. 37.11). The common garden and pot-plant weeds *Marchantia* (Fig. 37.11a) and *Lunularia* both have a relatively large and conspicuous thallus (plant body) that is anchored to the soil by a mat of rhizoids.

Fig. 37.8 Early Cretaceous fossil (120 million years old) from south-west Victoria, clearly recognisable as a thalloid liverwort

They are easily recognised by the circular or crescent-shaped cups on their surface that contain small, asexual, vegetative propagules, **gemmae**. In favourable conditions, they produce sexual reproductive organs on peculiar umbrella-shaped structures.

Marchantia and its relatives are among the most complex of the liverworts. The thallus is differentiated into a thin upper spongy layer of photosynthetic cells. These enclose air spaces that open by pores to the external environment, allowing CO_2 uptake. Below the photosynthetic tissue is a thick lower layer of more closely packed storage cells (Fig. 37.11b). However, the majority of liverworts are smaller, with a delicate leafy habit. Although they are divided into stem and simple leaves, they are far less complex with respect to tissue differentiation and sex organ production than *Marchantia*.

The half dozen genera of hornworts resemble thalloid liverworts (Fig. 37.12). Mosses, on the other hand, have leafy gametophytes but the leaves are extremely diverse in relation to their function and the environment.

Leaf cells of peat moss, *Sphagnum*, are of two types: a network of small, photosynthetic cells entwine large, perforated cells that are devoid of contents and act in a water-holding capacity, much like a sponge (Fig. 37.13). This water-holding capacity, which is an important ecological feature of peat bogs in the highlands of Australia, also makes *Sphagnum* a favoured component of horticultural potting mix. The upper leaf surface of *Polytrichum* and its relatives is ornamented with parallel, longitudinal files of cells several cells high, which increase the photosynthetic area of the leaf (Fig. 37.14).

Although mosses lack specialised strengthening and supporting tissue and are generally small in stature, some species are quite stiff and robust. The largest moss, *Dawsonia superba* (Fig. 37.15a) grows up to about 40 cm in height in moist eucalypt forests of south-eastern Australia. In *Dawsonia* and most other mosses, antheridia and archegonia are borne in separate clusters at the tips of gametophytes (Fig. 37.15b). Antheridia form in leafy rosettes and are interspersed with sterile hairs, **paraphyses**, that presumably function in trapping moisture and preventing desiccation of the sex organs (see Fig. 37.10).

Sporophytes: the diploid generation

Bryophytes retain the egg cell within the archegonium on the gametophyte (Fig. 37.9). Only the sperm are released to rely on free water (rain, dew, etc.) to swim to the archegonia and down the neck canal to the egg cell, attracted by chemical substances produced during the breakdown of the archegonial neck canal cells. After fertilisation, the sporophyte is retained and

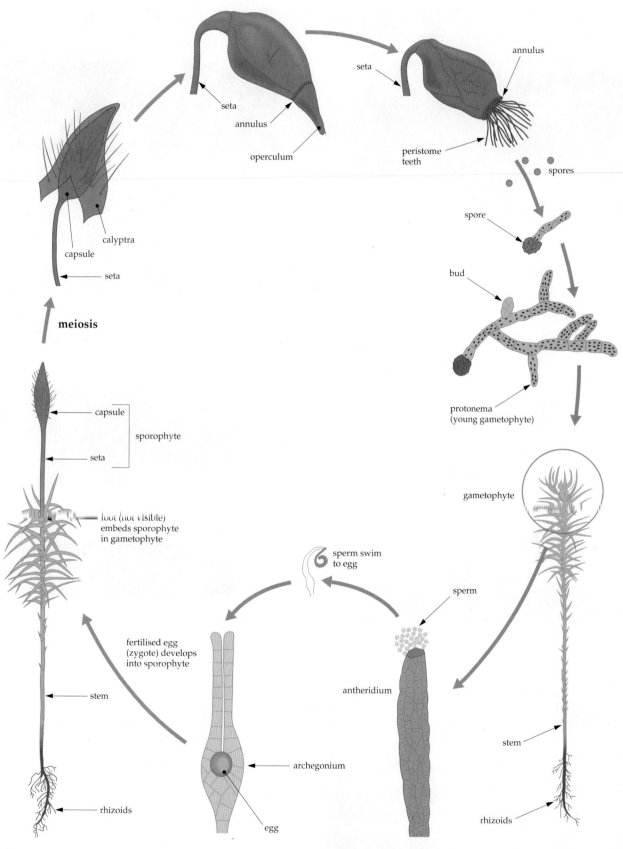

Fig. 37.9 Life cycle of a moss (phylum Bryophyta). Sperm and egg cells are produced in antheridia and archegonia respectively on the gametophyte (haploid) plant. Sperm swim to the neck of an archegonium, where one fertilises the single egg cell. The zygote divides and develops into the sporophyte (diploid) plant, which consists of a seta and capsule, and is parasitic on the gametophyte. Haploid spores form by meiosis, are released from the capsule and germinate into filamentous protonemas, from which leafy gametophytes grow

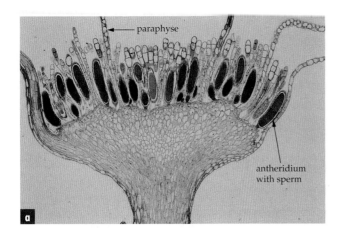

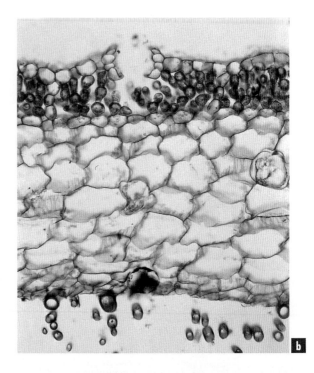

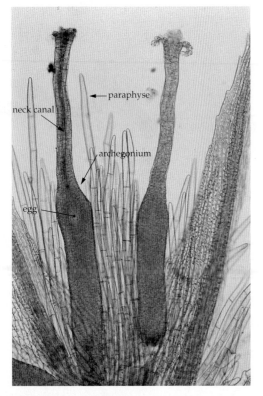

Fig. 37.10 (a) Moss male antheridia, which produce sperm and **(b)** archegonia, each of which produces an egg cell

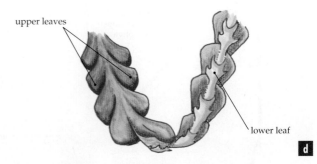

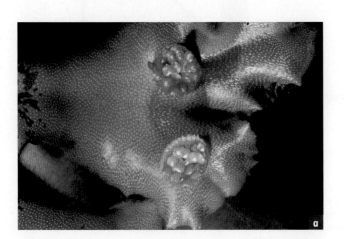

Fig. 37.11 (a) *Marchantia*, a thalloid liverwort, has circular-shaped cups on its surface that contain asexual propagules, gemmae. **(b)** A cross-section of *Marchantia* shows photosynthetic cells near the upper surface, which also has simple pores for gas exchange. **(c)** A leafy liverwort. Leaves are simple, consisting of a single layer of undifferentiated cells, with no midrib; the leaves are generally arranged in two rows **(d)** often with a third row of reduced leaves on the undersurface

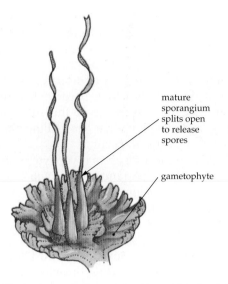

mature sporangium splits open to release spores

gametophyte

Fig. 37.12 The hornwort *Anthoceros*. Cavities within the thallus are filled with mucilage and house cyanobacteria (*Nostoc*), which fix atmospheric nitrogen

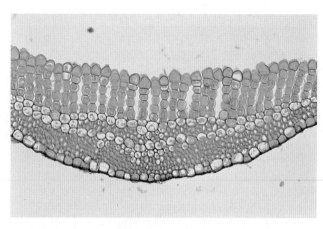

Fig. 37.14 Transverse section of a leaf of the moss *Polytrichum*, showing the upper leaf surface with longitudinal files of cells that are thought to increase the photosynthetic area and trap water

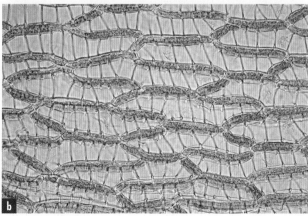

Fig. 37.13 (a) Peat moss, *Sphagnum*, has leaves with: **(b)** two types of cells—small and photosynthetic or large and lacking cytoplasm. These large cells have an enormous water-holding capacity, which is why peat bogs in the highlands of south-eastern Australia are an important store of water

Fig. 37.15 (a) The largest moss, *Dawsonia superba*, grows in moist mountain ash (*Eucalyptus regnans*) forest. **(b)** Antheridia (male), shown here, and archegonia (female) are usually clustered at the tips of gametophytes in leafy rosettes (moss heads)

BOX 37.1 Bryophytes in Australian environments

Although bryophytes occur in virtually every habitat where vascular plants are found, except the sea, their anatomical and physiological properties make them ecologically very different.

Although they lack lignified vascular tissue, many have a central strand of elongated cells that are capable of conducting water. However, mostly they are *ectohydric*, conducting water up the outside of the plant by capillarity among the small spaces between leaves and stem, like a wick. Fine hair-like rhizoids, which anchor them to the ground and often assist this capillary conduction by coating parts of the stem like cottonwool, are generally not responsible for absorbing and conducting nutrients from the substrate. That function is performed by the whole plant, which soaks up, like a sponge, the mineral nutrients in rainwater or dissolved from dust blown onto the plant surface.

Apart from rainforest species, the majority of Australian bryophytes are *poikilohydric*, that is, they are resurrection plants capable of withstanding desiccation to dryness, sometimes for months on end, but rehydrate whenever rain returns. They assume full metabolism within minutes, or at most a few hours, only to dry up and suspend activity again when dry conditions resume. This opportunistic physiology is the secret of their success in intermittently dry habitats, such as rock surfaces, tree trunks and even city buildings, where 'rain tracks' are generally colonised by bryophytes (see figure). In the Mallee and similar areas of Australia, this characteristic permits mosses and liverworts, together with lichens, cyanobacteria and the roots of dead flowering plants, to form tenacious soil crusts that hold the soil in place even in times of drought. The importance of this role has long been underestimated and it was the disruption of the soil crust by clearing and ploughing that allowed the notorious Mallee dust storms to occur.

One further feature gives bryophytes unexpected ecological importance. Although they are surface plants seldom penetrating soil to more than a few centimetres depth, in bogs and forests they are capable of building up a substantial layer of living and dead tissue above ground, forming a mat through which rainwater has to filter before reaching plant roots. These mats have cation-exchange properties, preferentially retaining some ions while releasing others, and thus modifying the nutrition of the entire ecosystem.

Perhaps the most remarkable of all bryophyte properties is their totipotency, that is, their ability to regenerate new plants from casual fragments even as small as one or two cells (Chapter 14). There are many bryophytes for which sexual reproduction and spore formation are unknown and may never occur; these bryophytes, and even most others, rely on vegetative reproduction from detached fragments blown or washed away to new habitats where they can establish. This extraordinary power of asexual reproduction, which can be shown easily by air-drying a handful of mosses, grinding them to a powder and sowing on moist earth or peat, probably explains the prevalence of mosses and liverworts throughout the world today. Furthermore, in the case of prostrate species, especially thallose liverworts but also many other mosses and liverworts, new plant material is produced at the tip as the product of the apical cell and its divisions. As tissue ages and dies at one end of the plant, new growth extends at the other, so that the plant slowly creeps forwards over the ground, subdividing as it goes.

Moss on exposed rock, Mt Baw Baw, Victoria

develops on the gametophyte. The sporophyte remains dependent upon the gametophyte for nutritional requirements.

The sporophyte of bryophytes is simple in structure, consisting of a single sporangium, **capsule**, usually borne terminally on a slender stalk, **seta**. It is embedded in the gametophyte by a foot, which contains specialised transfer cells for nutrient exchange with the gametophyte. Hornwort and moss sporophytes have some capacity for photosynthesis—for maintenance but not enough for independent growth.

Liverwort sporophytes usually develop a seta that is white and delicate, elongating only when the spore capsule is mature and the spores are ready to be shed (Fig. 37.16a). The brown, spherical capsule either disintegrates or dehisces along one or more, usually longitudinal, lines of weakness, releasing spores and elaters. The haploid spores germinate into new gametophytes. The **elaters** are specialised, elongated, hygroscopic (water-absorbing) cells that have helically arranged wall thickenings. As they dry out, stresses build up because of the differential thickening of the cell walls, causing the elaters to move and flick spores from the capsule.

Hornwort sporophytes (Fig. 37.12) lack a seta. The sporophyte consists of a foot and a long green sporangium. A meristem between the foot and the base of the sporangium is responsible for the growth and elongation of the sporangium. The surface of the sporangium is covered with stomata, making hornworts the simplest living land plants to possess these structures. They shed their spores and elaters through one or more vertical slits that extend down from the apex.

Some mosses, such as *Sphagnum* (peat moss), also lack a seta. However, all mosses lack elaters and advanced mosses possess a capsule that is borne on a stiff, wiry seta that elongates before sporangium maturation. The capsule is generally more elaborate in mosses than in liverworts. It is often crowned by a **calyptra**, which is modified tissue of the old gameto-phytic archegonial neck (Fig. 37.16c). The apical portion of the capsule is a cap-shaped **operculum**. It is shed when the spores are mature to reveal one or two specialised rows of tooth-like structures around the top of the sporangium known as the **peristome** (Fig. 37.17). The peristome shelters the spores when

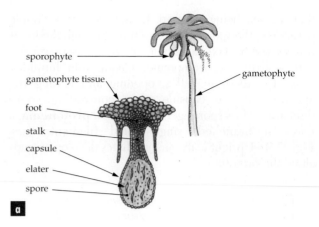

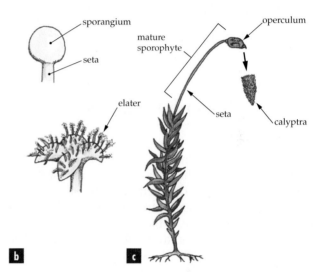

Fig. 37.16 Sporophytes (diploid generation) of bryophytes. **(a)** The liverwort *Marchantia*, consisting of a sporangium, here emerging from an archegonium, elevated on a stalk (archegoniophore). **(b)** The liverwort *Frullania* splits open to release spores. **(c)** Many species of mosses have a stiff seta, with a sporangium (capsule) crowned by a calyptra (part of the old gametophyte archegonial neck), which is shed

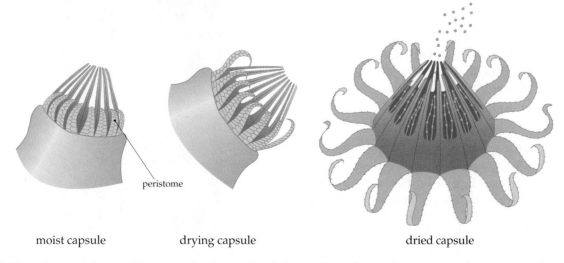

moist capsule   drying capsule   dried capsule

Fig. 37.17 The peristome at the top of the sporangium (capsule) controls the release of spores in most mosses; in some mosses it consists of two rows of interlocking teeth. As the capsule dries out, the teeth fold back to allow spore dispersal by the wind

they are not being released. In mosses with a double peristome, the spores must also filter through the inner row of teeth. The peristome teeth are impregnated with a lignin-like compound. Changes in humidity cause the teeth to reflex, opening and closing the aperture of the capsule, thus controlling spore discharge. Moss spores germinate into a **protonema**, a mass of branched, single-celled filaments (see Fig. 37.9). Upright leafy gametophytes develop as buds along the protonema.

> Bryophyte plants are small, photosynthetic, free-living gametophytes that lack vascular tissue. Sexual reproduction requires water for sperm to swim from an antheridium to an egg in an archegonium. The sporophyte grows on the gametophyte and is dependent on it for nutrition.

Vascular plants

Vascular plants, characterised by the presence of xylem and phloem, include fern allies, ferns and seed plants (Fig. 37.7). Their tracheids and vessels are resistant to decay and have been detected in fossils of early land plants (Box 37.2). In contrast to bryophytes, the sporophyte of vascular plants is the dominant generation of the life cycle, existing as a free-living plant. The gametophyte is usually short-lived and degenerates once the sporophyte is established.

> The dominant generation of living vascular plants is the sporophyte, which has phloem and lignified xylem.

Fern allies

The term '**fern allies**' refers to a loose assemblage of spore-producing vascular plants that are classified in three phlya (Table 37.1): **Psilophyta** (*Psilotum* and *Tmesipteris*), **Lycophyta** (*Lycopodium*, *Selaginella* and *Isoetes*) and **Sphenophyta** (*Equisetum*).

Psilophytes: 'living fossils'

Two closely related genera, *Psilotum* and *Tmesipteris* (sometimes called whisk ferns), are the most primitive living vascular plants (Fig. 37.7). The aerial shoots of the sporophyte branch simply and dichotomously. They lack true leaves but produce small outgrowths of the stem (enations) that bear two or three fused sporangia. They do not produce true roots but are anchored to the substrate by a creeping rhizome covered with rhizoids. *Tmesipteris* is an epiphyte,

often growing on the trunks of tree ferns in temperate Australia (Fig. 37.18a). *Psilotum* is either epiphytic or found in sheltered terrestrial environments. It is common in tropical Australia, often found in trees, hanging from the base of staghorn ferns and epiphytic orchids (Fig. 37.18b). The growth form of both genera, but particularly *Psilotum*, resembles closely the earliest vascular plant fossils, such as *Rhynia* (Box 37.2), and for this reason they are often referred to as living

Fig. 37.18 (a) *Tmesipteris* and **(b)** *Psilotum* are primitive vascular plants found in Australia, often growing as epiphytes on the trunks of tree ferns

BOX 37.2 Early vascular plants

The oldest vascular plant fossils are found in late Silurian sediments around 410 million years old, although much of our knowledge of the structure and diversity of early land plants comes from slightly younger fossils, from the early Devonian (about 400 million years BP). Current research on plant fossils of these ages is revealing that early land floras in different parts of the world were relatively uniform and shared common taxa. Plants were generally quite small, ranging from several centimetres to about a metre in height. The larger specimens may have grown semi-submerged in swampy environments and relied on buoyancy for support. They were all relatively simple in construction, consisting of leafless, dichotomously branched stems with a small central strand of primary vascular tissue. They reproduced by spores.

These extinct plants are classified into three phyla of which the phylum **Rhyniophyta** is the most primitive. Given the critical part these plants play in our understanding of early land plant evolution and the often fragmentary nature of their fossil record, it is not surprising that aspects of their structure are under some dispute.

Rhyniophyta

Rhyniophytes were small plants with creeping rhizomes that produced erect, leafless, dichotomous stems with terminal sporangia. They are the simplest of the early land plants, characterised by the two most widely known examples, *Rhynia* (Fig. a) and *Cooksonia* (Fig. b). Two species of *Rhynia* were described from exquisitely preserved silicified peat beds near the Scottish village of Rhynie. Several features of these plants attest to their existence in a terrestrial habitat. They are covered by a layer of cuticle, they have stomata, their meiotic spores possess a resistant sporopollenin wall and the stems have a darker zone of cells forming a central strand where vascular tissue would be expected. *Cooksonia*, named after the eminent Australian palaeobotanist Isabel Cookson, was a similar but smaller plant.

Close examination of the central strand cells of *Rhynia* and *Cooksonia* has yielded some interesting observations. Firstly, central strand cells of *Cooksonia* and one species of *Rhynia* (*Rhynia major*, now classified as *Aglaophyton*) have smooth, unthickened cell walls. These cannot really be defined as tracheids as we know them in

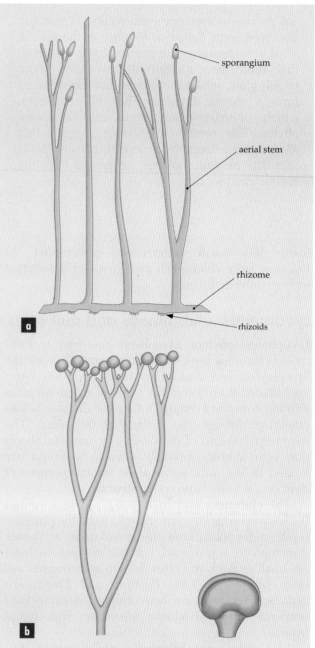

(a) *Rhynia* is one of the simplest of the early, extinct vascular plants. *Rhynia* fossils are of early Devonian age (about 400 million years old). The plant had an aerial branch system about 20 cm high, with a vascular cylinder of tracheids and phloem-like tissue, stomata and cuticle. Spores were produced in sporangia on the ends of aerial branches. There were no roots but a prostrate rhizome system with rhizoids. (b) An early land plant fossil, the rhyniophyte *Cooksonia*

living vascular plants. Central strand cells of the second species of *Rhynia* (*R. gwynne-vaughanii*) possess wall in-growths that form rings or spirals, but these are very different in size and structure to tracheids of modern plants. Consequently, the status of rhyniophytes as true vascular plants is not proven to everyone's satisfaction. There has also been some debate as to whether the *Rhynia* plant was sporophytic or gametophytic. However, their spores have a clearly defined trilete (spore tetrad) mark, which is evidence that they resulted from a meiotic division. They have stomata, which, in living plants, occur only on sporophytes. This combined evidence suggests that *Rhynia* was a sporophyte. Several unusual fossils that may be gametophytes are known from the Rhynie chert locality.

fossils. The small subterranean gametophyte of *Psilotum* lacks chlorophyll and grows in association with a symbiotic fungus.

Lycophytes: clubmosses and quillworts

Lycophytes (phylum Lycophyta) also have a fossil record extending back 400 million years. They are the first plants to have evolved true roots (Fig. 37.7). Roots are initiated deep within the tissue of the stem, often growing downward through it for some distance before emerging through the surface of the plant. The sporophyte consists of dichotomously branched shoots that bear simple, spike-like leaves. Sporangia are located in leaf axils, either along normal portions of stem or condensed into cones, **strobili**.

The **clubmoss** *Lycopodium* (Fig. 37.19a) exhibits the homosporous life cycle, thought to be the primitive condition for plants. Only one type of spore, and hence gametophyte, is produced. *Lycopodium* gametophytes are small thalloid structures that are subterranean and lack chlorophyll. Like *Psilotum* and *Tmesipteris* gametophytes, they are heterotrophic, deriving their nutrients from a symbiotic association with fungal hyphae.

The clubmosses *Selaginella* (spike mosses, Fig. 37.19b) and *Isoetes* (quillworts) have a heterosporous life cycle (Fig. 37.20), with two very different types of spores developing into mega- and microgametophytes. These gameophytes are extremely reduced and are contained entirely within the confines of the spore wall (endosporic). Megaspores undergo a short period of internal cell division to produce the **megagametophyte** (female), which develops archegonia with eggs. The archegonia are exposed when the megaspore wall splits along the trilete (spore tetrad) mark. Microspores divide internally to produce the **microgametophyte** (male). The microgametophyte is essentially a single antheridium. The antheridium produces many biflagellated sperm, which are released when the microspore splits open.

Fig. 37.19 Clubmosses **(a)** *Lycopodium* and **(b)** *Selaginella*

Sphenophytes: horsetails

Horsetails (phylum Sphenophyta) have their origins in the late Devonian (360 million years BP). Today the phylum consists of a single genus, the herbaceous, often weedy *Equisetum* (Fig. 37.21). It mostly occurs in the Northern Hemisphere but horsetails were once diverse and formed an important component of the earth's vegetation. In the Carboniferous (360–290 million years BP), giant horsetails (*Calamites*) were large trees. Together with tree-like lycophytes, these giant horsetails dominated forests in massive coalforming swamps across what is now North America and Europe (Chapter 31).

Equisetum consists of a creeping, underground rhizome that produces upright stems. Small and insignificant leaves are borne in sheath-like whorls at nodes in the stem and the stem surface itself is green and photosynthetic. The stem epidermis is impregnated with crystals of silicate, which give it a coarse, abrasive texture. During the Middle Ages, *Equisetum*

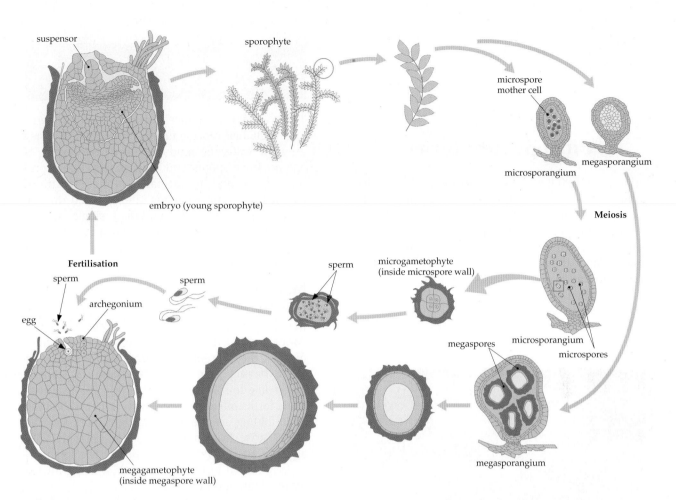

Fig. 37.20 Heterosporous life cycle of *Selaginella* (phylum Lycophyta). The sporophyte (diploid) is the dominant generation. On the sporophyte, two kinds of haploid spores (megaspores and microspores) develop in sporangia. The male gametophyte (microgametophyte) develops within the microspore and produces biflagellate sperm, which swim to the egg cell. The female gametophyte (megagametophyte), which produces archegonia, each with an egg cell, develops within the megaspore. As the megagametophyte grows, the megaspore wall splits and the gametophyte protrudes to the outside. After fertilisation, the young sporophyte develops within the megagametophyte tissue for a time

Fig. 37.21 *Equisetum*, the only living representative of the phylum Sphenophyta (horsetails), some of which were large trees in Carboniferous forests

was extensively collected and used as an abrasive, hence its more antiquated common name, 'scouring rush'.

Sporangia are produced in small strobili (cones) borne terminally on either normal vegetative or specialised reproductive shoots. *Equisetum* spores are unusual. The outer layer of the spore wall unravels to form four sporopollenin threads that remain attached to the spore. These threads are hygroscopically sensitive and their movement assists in spore dispersal. They are termed elaters but they have a different structural origin to the cellular elaters of either liverworts or hornworts. Spore dispersal by movement of structures that are differentially thickened with cellulose and lignin or sporopollenin is an interesting feature that has evolved independently in liverwort and hornwort elaters, moss peristome teeth, *Equisetum* spores and, as we will see in the following, the sporangia of true ferns.

'Fern allies' are a loose assemblage of spore-producing vascular plants. Lycophytes were the first plants to evolve true roots and tree forms of extinct lycophytes and sphenophytes (horsetails) formed Carboniferous forests.

Filicophytes: ferns

With over 12 000 living species, **ferns**, phylum **Filicophyta**, are the most diverse and conspicuous of the spore-producing vascular plants. Although a large proportion of species are tropical, ferns are well represented in cool temperate climates. In Australia and New Zealand, they range in size from small, delicate filmy ferns and floating aquatics (see Box 37.3) to tree ferns exceeding 20 m in height (Fig. 37.22). Epiphytes, climbers and xerophytes add to a remarkable variety of ecological preferences and life strategies.

Fig. 37.22 (a) Australian tree ferns. **(b)** The young rolled-up leaves of ferns are called 'fiddle-heads'

BOX 37.3 Weird, watery and weedy ferns

There are three families of unusual aquatic ferns and each has representatives in Australia. They grow in water or mud and form two types of spores: microspores and megaspores. Spore production occurs in special nut-like structures, sporocarps. Gametophytes are reduced to the development of sperm within a single antheridium within the microspore and a rudimentary megagametophyte with archegonia and egg cells within the megaspore.

Marsilea

Marsilea (family Marsileaceae) includes more than 50 species, many of which occur in tropical Africa and Australia. Australian species are commonly called nardoo. They do not look like other ferns, having leaves with two pairs of leaflets, rather like four-leaf clover (Fig. a). The leaves of some forms often have 'sleep movements', folding or unfolding according to the light intensity. The leaves arise from a long creeping rhizome with adventitious roots at the nodes. The plant is usually rooted in mud in shallow water, with the leaves floating on the surface.

The sporocarps occur near the base of the leaves. They are resistant to desiccation and can survive for up to 20 years when ponds dry out. Common nardoo, *M. drummondii*, is widespread in mainland Australia, especially in drier inland areas. Aboriginal Australians ground up the nutritious sporocarps into a starchy paste to bake as cakes. *M. salvatrix* provided some food for the men on the fateful Burke expedition in central Australia, which is how the species came to be named.

Azolla

Azolla (family Azollaceae) is a cosmopolitan genus, two species of which occur in Australia. *Azolla* is a small, free-floating aquatic fern (Fig. b). It has tiny overlapping leaves, green to red in colour, that cover a branching stem. This gives individual plants a triangular shape. Because cyanobacteria live in a cavity at the base of each leaf and fix atmospheric nitrogen, the fern is grown in rice fields of China and Vietnam as a 'green' fertiliser. *Azolla* multiplies easily by fragmentation and grows in dense populations that can cover extensive areas of still water. It can smother other organisms, such as troublesome mosquito larvae.

Water ferns: **(a)** *Marsilea* (nardoo), **(b)** *Azolla* (red in colour) and *Salvinia molesta* (green)

Salvinia

Salvinia (family Salviniaceae) is also free-floating and is related to *Azolla*. These plants have leaves in whorls of three—two floating and one submerged. The submerged leaf is highly dissected, resembling a root system, but there are no true roots. The upper surface of the floating, boat-shaped leaves are covered with silvery, stiff hairs, which make them waterproof.

Salvinia occurs in tropical regions of the world and *S. molesta* (Fig. b) was introduced into Australia for use in dams, ponds and fish tanks as an ornamental aquatic plant and refuge for fishes. It is sterile and believed to be of hybrid origin. Its effective vegetative reproduction and very fast growth have allowed it to become a major weed species of aquatic ecosystems in many tropical countries.

Ferns are an assemblage of plants that probably consist of several distinct evolutionary lineages, representing stages of evolution between fern allies and seed plants (Fig. 37.7). They have their origins in the late Devonian. The greatest proportion of living species are leptosporangiate ferns. They possess small, delicate sporangia (usually with fewer than 64 spores) that develop from a single cell, compared with the eusporangiate ferns, including fewer species, which possess massive sporangia (with 256 or more spores).

Fern sporophytes

The fern sporophyte, like that of the fern allies, is the dominant, free-living generation (Fig. 37.23). It consists of a stem bearing true leaves and true roots. Stems vary in form but often they are creeping, underground rhizomes, such as in bracken, *Pteridium esculentum*. Only in a few species are stems erect and trunk-like as in tree ferns. Many fern stems have a dense covering, indumentum, of scales or hairs, which are modifications of epidermal cells. Leaves (**fronds**) are prominent parts of most fern sporophytes. The leaf lamina is often deeply dissected or divided into leaflets (**pinnae** and **pinnules**), giving the frond a feathery appearance. A distinctive feature is the unusual manner in which fern fronds unfurl in their final stage of development. The young rolled-up leaves are commonly called 'fiddle-heads' (see Fig. 37.22) because of their resemblance to the end of a violin. Although many ferns contain toxic substances, 'fiddle-heads' of certain species are considered a culinary delicacy in some countries.

Sporangia are usually clustered in **sori**. These are found on the margins or undersurfaces of either normal or specialised reproductive fronds. In many ferns, they are covered by a protective scale-like membrane or **indusium** (Fig. 37.24). Sporangia of leptosporangiate ferns are small delicate structures consisting of a spore sac, with walls that are only one cell thick, and a short, slender stalk. Embedded in the sporangium wall is an encircling ring of enlarged cells, the annulus, with thickened walls. When the sporangium is ripe, stress develops in the annulus causing the sporangium to rupture in a non-thickened zone, the stomium, explosively releasing spores into the environment (Fig. 37.23). This hygroscopic sensitivity is similar to that in liverwort elaters and moss peristomes.

Fern gametophytes are small, free-living, heart-shaped thalloid structures that bear rhizoids and sex organs on their ventral (lower) surface. Globular antheridia release motile sperm that swim to the protruding necks of embedded archegonia containing the eggs. Like bryophytes and fern allies, ferns rely on free water for sexual reproduction. The young sporophyte develops in place on the gametophyte, which dies and decays once the sporophyte is established. Most ferns have a homosporous life cycle but a group of peculiar water ferns are heterosporous (see Box 37.3).

The sporophyte of homosporous ferns is photosynthetic, free-living and the dominant phase. Like bryophytes, ferns have flagellated sperm and rely on free water for sexual reproduction. Fern antheridia and archegonia are borne on a small, free-living, heart-shaped gametophyte.

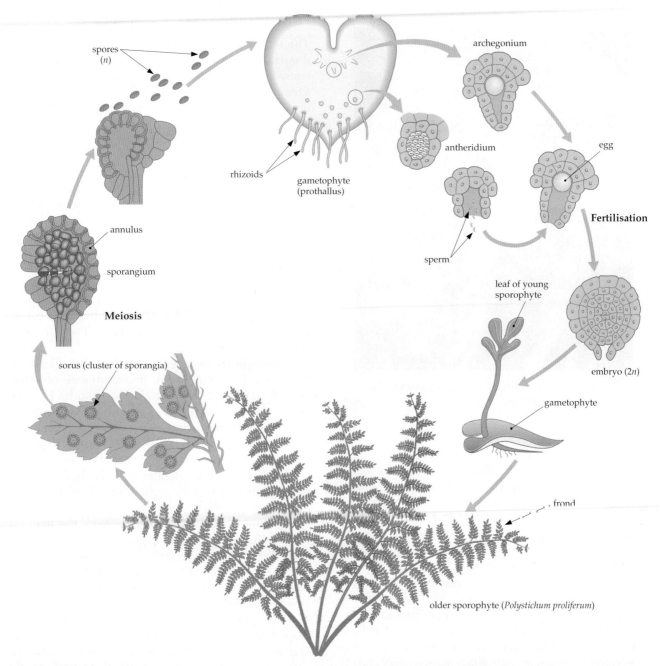

spores
(*n*)

archegonium

rhizoids

antheridium

egg

gametophyte
(prothallus)

Fertilisation

sperm

annulus

leaf of young
sporophyte

sporangium

Meiosis

sorus (cluster of sporangia)

embryo (2*n*)

gametophyte

frond

older sporophyte (*Polystichum proliferum*)

Fig. 37.23 Life cycle of a homosporous fern. Spores are produced by meiosis in sporangia located on the sporophyte (the conspicuous fern plant). The sporangium dehisces by the action of the annulus. Haploid spores germinate and develop into small, delicate, heart-shaped gametophytes (prothallus), which attach to the substrate by rhizoids. Each gametophyte forms antheridia and archegonia on the undersurface. Spirally coiled, multiflagellated sperm swim from an antheridium to the neck of an archegonium to fertilise the egg cell. The zygote divides immediately and a new sporophyte plant begins development, gaining nutrition from the gametophyte for a time

Seed plants

There are five groups of living seed plants. Flowering plants, phylum **Magnoliophyta**, are by far the most numerous and well known of these, with about one-quarter of a million species, although conifers, phylum **Coniferophyta** (with about 600 species), and cycads, phylum **Cycadophyta** (about 150 species), are familiar to most people. The remaining two phyla are the bizarre **Gnetophyta** with three living genera, and **Ginkgophyta** with just a single living species, *Ginkgo biloba*. Cycads, *Ginkgo*, conifers and gnetophytes are commonly called 'gymnosperms', although they are not a single evolutionary lineage.

Seed plants have a number of derived features that are widely considered to represent 'major events' in land plant evolution (Fig. 37.7). These features include:

which produce egg cells. In seed plants, the mega-gametophyte is retained within the megasporangium, which is further surrounded and protected by one or more layers of cells, the **integuments**. Megasporangia of other vascular plants are not so protected. This whole structure (integuments enclosing megasporangium with developing megagametophyte) is the ovule (see Fig. 37.25). After fertilisation and development of the embryo, the ovule matures into a seed (Chapter 14).

Within the megasporangium, four megaspores are produced by meiosis. Three of these degenerate, leaving only one to grow into the female gametophyte. In cycads, *Ginkgo* and conifers, the mature female gametophyte contains as many as several thousand sterile cells and several archegonia; after fertilisation of

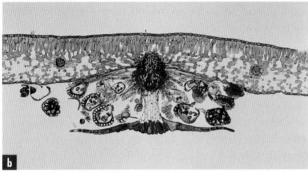

Fig. 37.24 (a) Filmy fern (*Hymenophyllum*) with sporangia clustered in sori partly covered by an indusium. **(b)** Section of a fern sorus with indusium

- enclosure of the female gametophyte and embryo within an **ovule**, which becomes the **seed**
- microspores transported directly to the ovule as pollen (pollination)
- secondary growth (vascular cambium).

Seeds from ovules

Evolution of the seed is one of the key reasons for the success of higher vascular plants. This depended on a heterosporous life cycle. As we saw on page 969, in heterospory there are two different types of spores and the larger megaspores grow into megagametophytes,

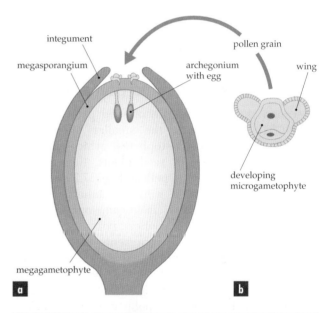

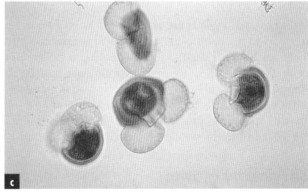

Fig. 37.25 Seed plants are characterised by ovules and pollen. **(a)** An ovule includes the developing (female) megagametophyte (with developing archegonia and egg cells) surrounded by protective integuments. An ovule becomes a seed after fertilisation. **(b)** Pollen grains are the (male) microgametophyte (in which sperm develop). **(c)** In pines, a pollen grain has two bladder-like wings for wind transport to the ovule and to orientate it on the pollination droplet

the egg cell, the embryo utilises the gametophytic remains (nucellus) as a nutritive source. In flowering plants, the megagametophyte is less elaborate, consisting of only eight cells (embryo sac), of which one is the egg cell (Chapter 14).

The seed is significant because it is a protective structure encasing the embryo of the next sporophytic generation until conditions become optimal for its germination and growth. The embryo lies dormant, embedded in nutritive tissue and surrounded by a seed coat or testa derived from the one or two integuments, which thicken and harden. Protected in this way and with a supply of stored food, the dormant embryo can be shed from the parent plant and dispersed by wind or animals.

Seed plants represent the other extreme to mosses and liverworts, in which the gametophyte is photosynthetic and free-living and the sporophyte grows in situ and is nutritionally dependent. In seed plants the opposite is true; the megagametophyte is reduced and retained within the dominant sporophyte.

Pollen transport of male gametes

A second significant feature of seed plants is their pollen. In heterospory, the smaller microspores develop into microgametophytes, which produce sperm (Chapter 14). In seed plants, these microgametophytes develop as pollen (Fig. 37.25b), which can be transported by wind or animals to the immediate vicinity of the female gametophyte before release of sperm. In most conifers, pollen is dispersed by wind. This probably represents the primitive condition, although it is possible that the earliest seed plants had water-dispersed microspores. Cycads are insect pollinated, and flowering plants have further developed a multitude of pollination mechanisms, including insects, birds and small mammals. As a consequence, seed plants are not dependent on free water to provide a medium allowing sperm access to eggs. This has allowed seed plants to develop a far greater ecological tolerance to dry environments and is an important factor allowing their widespread dominance. Although some ferns, fern allies and bryophytes do inhabit quite xerophytic habitats, their opportunities for sexual reproduction are intermittent and limited.

Secondary growth

The production of woody tissue by a vascular cambium is a characteristic feature of seed plants and a few extinct but closely related plants (progymnosperms). Vascular cambium produces large quantities of secondary xylem (wood) to the inside of a stem or root and smaller quantities of secondary phloem (bark) to the outside (see Chapter 17).

Vascular cambium adds girth to stems and roots (the apical meristems generate length), and has a dual function. The tracheids and sieve cells produced by the cambium continually add new capacity for water and food transport within the plant. Older water-conducting elements become non-functional and filled with waste products. These cells, together with any lignified fibres, constitute a rigid support characteristic of tree trunks and branches. These two functions have allowed woody plants to grow to massive sizes (Fig. 37.26).

mportant features of seed plants are: the enclosure of the female gametophyte and embryo within an ovule, which becomes the seed; the microspore transported as pollen (pollination); and secondary growth (vascular cambium).

Cycadophytes: ancient seed plants

Cycads (phylum Cycadophyta) are the most ancient of living seed plants (Fig. 37.7) and have a fossil record extending back early in the Permian (290–251 million years BP). Twenty-three extant species in four genera (*Cycas, Bowenia, Lepidozamia* and *Macrozamia*) occur in Australia.

The most common growth form of cycads consists of a thick, unbranched or sparsely branched trunk and a crown of pinnate leaves (Fig. 37.27), although stems of some species are subterranean. Below ground, cycads have a tap root that produces highly specialised lateral roots, coralloid roots, in which nitrogen-fixing cyanobacteria live.

Superficially, many cycads resemble small palm trees but palms are flowering plants and the two groups are not related. Cycad stems have a vascular cambium that produces a limited amount of secondary growth. However, most of the tissue contributing to the thickness of the trunk is derived from primary thickening by a special meristem that develops at the base of the most recently formed leaves. A meristem somewhat like this

Fig. 37.26 Fossil *Glossopteris* wood (seed plant, 260 million years old) from Antarctica. Secondary growth allowed seed plants to grow to large tree size

Fig. 37.27 (a) Australian cycad, *Macrozamia moorei*, growing in Queensland. **(b)** Female seed cone

is present in palms (see Box 17.1), which retain leaf bases on the trunk, hence the similarity in appearance.

Branching in cycads is dichotomous, that is, the shoot apex divides in two. Axillary buds (buds in the axil of leaves) are completely lacking. In vegetative regions of the stem, these dichotomies are even, resulting in a branch point from which two stems arise. In reproductive regions, dichotomies are uneven, each resulting in a large woody cone and a dormant meristem that will later continue with stem growth. Ovules and microsporangia are borne in these prominent cones on separate female and male plants (Fig. 37.27).

Cycads were long thought to be wind pollinated but it is now clear that many, if not all, cycads are pollinated by beetles, one of the oldest insect groups. Although the boat-shaped pollen grains are transported directly

to the ovules and free water is not required for gamete transfer, cycad sperm are multiflagellated, a condition retained from their spore-producing ancestors. Pollen grains germinate inside the micropyle to produce a microgametophyte of about five cells with a structure known as a 'feeder' pollen tube (haustorium) that penetrates nucellar tissue of the ovule and absorbs nutrients needed for its growth (Fig. 37.28). Two flagellated sperm cells are produced by each microgametophyte and are liberated into the micropylar chamber close to the megagametophyte. The megagametophyte is a mass of starch-filled cells with several archegonia differentiated at the micropylar end of the ovule. Sexual fusion occurs when sperm contact an egg within this aqueous environment.

Like all seed plants, the megagametophyte remains enclosed by the ovule integument and retained on the parent sporophyte. The seed that develops has a fleshy outer coat, coloured red, orange, yellow, brown or bluish, and a thick inner coat. Seeds of *Cycas media*, although toxic, were a staple food for Aboriginal people in Arnhem Land and Cape York Peninsula. They were cracked open and their kernel pounded into flour, which was then leached of toxins and made into dough for baking. Seeds of *Macrozamia* were processed in a similar manner in coastal New South Wales and Queensland.

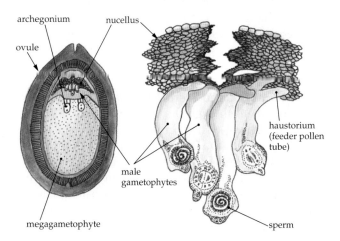

Fig. 37.28 The pollen tube of cycads forms a haustorium, which extracts food reserves from the nucellus for nutrition of the megagametophyte

Ginkgophytes: retain flagellated sperm

Ginkgophytes are represented in the world's modern flora by a single species, *Gingko biloba* (Fig. 37.29), which is endemic to a restricted region of China. It is the sole survivor of a lineage of plants that were prominent in the past. For example, *Ginkgo* fossils are common in mid-Mesozoic (160–100 million years BP) sediments of Europe, North and South America, and Australia.

Fig. 37.29 *Ginkgo biloba*

Ginkgo biloba is a commonly cultivated tree, valued for its shape, delicate foliage and autumn colours.

The growth habit of *Ginkgo* is arborescent (a tree), the result of axillary branching and a well-developed vascular cambium. In this respect it is similar to conifers and woody flowering plants. The leaves are very distinctively wedge-shaped with an apical notch, giving *Ginkgo* its common names—maidenhair tree, because of its resemblance to the fern fronds of the same name, and the Chinese name duck's foot tree. Branches consist of two different kinds of shoots, fast-growing long shoots and slow-growing short shoots. Most leaves are borne on short shoots.

Micro- and megasporangiate structures are also borne on short shoots on separate male and female trees. Loose cones shed large quantities of wind-dispersed pollen. Ovules are borne in pairs at the end of an elongated stalk and produce a pollination droplet that traps pollen grains and draws them through the micropyle into the ovule. Gametophyte growth and fertilisation is very similar to that of cycads. The megagametophyte derives its nutritional requirements from the nucellus of the ovule by way of specialised feeding structures. During this period the megagametophyte matures into a multicellular mass of tissue with several archegonia. When both micro- and megagametophytes are mature, which may be up to six months after pollination, the microgametophyte ruptures to deposit a pair of flagellated sperm into a pool of fluid that covers the archegonia, essentially the same as for cycads. The sperm swim the remaining short distance to the archegonia. *Ginkgo* represents the highest form of vascular plant life to retain motile, multiflagellated sperm. The flagella retain the 9 + 2 structure of microtubules described for algae.

C ycads are the most primitive of the living seed plants. They produce ovules and pollen in cones. Cycads and *Ginkgo biloba* rely on insects or wind for transport of microspores (pollination). They have multiflagellated sperm, which have to swim the final stage of transport to the egg.

Coniferophytes: successful non-flowering seed plants

Conifers (phylum Coniferophyta) are the most diverse and widespread of the non-flowering seed plants. They range from prostrate shrubs, such as the mountain plum pine, *Podocarpus alpina*, of Australian alpine regions (see Chapter 41), to the largest trees in the world, *Sequoia gigantea* of southwestern United States. Their fossil record dates from the Permian (290–251 million years BP) but they were dominant during the Mesozoic era (251–65 million years BP). Even today, conifers are the most important elements in some vegetation types, for example, the extensive pine forest communities of the Northern Hemisphere and *Callitris* woodlands of Australia (see Chapter 41).

Conifer wood is a valuable commercial resource worldwide and in Australia it is the basis for a diverse range of products. Extensive plantations of Monterey pine, *Pinus radiata*, are the basis of the paper industry and provide wood for furniture and building materials. Several native conifers (Box 37.4), such as white cypress pine or Murray pine, *Callitris columellaris*, and Huon pine, *Lagarostrobos franklinii*, are highly valued as specialty timbers. Because conifer wood is composed largely of tracheids, as opposed to flowering plant wood, which has large vessels embedded in a matrix of fibres, it is soft and easily crafted and hence commonly called 'softwood'. Many conifers are extremely slow growing, adding small amounts of new wood in annual increments. This results in the very fine-grained appearance sought after by woodcarvers and decorators.

C onifers are the most diverse and widespread of the living non-flowering seed plants. They are all woody plants, many of which are important softwood timber trees.

Most conifers bear their micro- and megasporangiate structures in **cones** (Fig. 37.30). However, a few conifers have such a highly modified ovule-bearing structure that their cone-like nature can only be seen very early in development. In the Southern Hemisphere, genus *Podocarpus* (family Podocarpaceae) and its relatives bear a single ovule on a fleshy receptacle that represents a swollen cone axis. The receptacle is often brightly coloured (usually red) and attracts birds that eat the fleshy part and disperse the seeds (Fig. 37.31).

Microsporangia (containing pollen) are borne on microsporophylls (fertile leaves) that are aggregated into pollen cones. Pollen grains of conifers are wind-dispersed. Entry of pollen into the ovule is effected in one of several ways. In kauri and Norfolk Island pines (family Araucariaceae) and fir trees (*Abies*,

BOX 37.4 Southern conifers

Conifers such as the family Pinaceae (e.g. *Pinus*) are prominent in the landscape of the Northern Hemisphere, particularly in mountainous regions and cold climates. In contrast, Southern Hemisphere conifers are often described as 'living fossils' or relics of a once more dominant vegetation that has been replaced by flowering plants. About 160 species have a wholly southern distribution. They include species in five families (see table). The families Podocarpaceae and Araucariaceae, for example, were once dominant in ancient Gondwanan forests (see Chapter 41).

Some of the largest forest trees in New Zealand are kauri trees (*Agathis australis*), which grow in the warm parts of the North Island. The podocarp 'rimu' (*Dacrydium cupressinum*) dominates more to the south and mountain cedar (*Libocedrus bidwillii*) forms subalpine timber-line forests. Trees are slow growing and estimated to be up to 1000 years old.

In tropical Queensland, kauri (*Agathis robusta*), hoop pine (*Araucaria cunninghamii*) and bunya pine (*A. bidwillii*) are prized for their timber and many populations have been lost. In 1994, a new genus of conifer related to *Agathis* and *Araucaria* was discovered remarkably less than 200 km from Sydney. It was named *Wollemia nobilis* (wollemi pine) after the Wollemi National Park, where it grows in a few deep, narrow gorges that protect it from bushfires (Fig. a). Podocarps, including *Phyllocladus*, *Podocarpus* and *Lagarostrobos*, are cultivated as ornamentals or for timber. Of the family Taxodiaceae, which was once more diverse, there is only one genus in Australia, *Athrotaxis*, including two species, pencil pine (*A. cupressoides*) and King Billy pine (*A. selaginoides*), which are endemic in western Tasmania.

Cypress pines, of the worldwide family Cupressaceae, are recognised by their small, scale-like leaves. In Australia, *Callitris* is widespread, often on sandy soils in dry areas. All species are killed by fire but regenerate from seed. The wood of *C. columellaris* was used by Aboriginal Australians for canoe paddles and spear throwers, and European settlers used the wood for houses and fences. Two other Australian genera are restricted to the south-east of Western Australia (*Actinostrobus*) and south-west Tasmania (*Diselma*).

Table Southern distribution of conifers

Family	Examples of genera	Southern species: number (%) in the world
Araucariaceae	*Agathis* (kauri)	15 (60)
	Araucaria	19 (100)
	Wollemia (wollemi pine)	1 (100)
Cupressaceae	*Actinostrobus, Callitris*	3 (100)
	(cypress pines)	16 (100)
	Libocedrus (cedar)	5 (100)
Podocarpaceae	*Dacrydium, Lagarostrobos*	16 (52)
	(Huon pine)	2 (100)
	Phyllocladus (celery-top pine)	5 (80)
	Podocarpus (plum pines)	58 (48)
Taxaceae	*Austrotaxus*	1 (100)
Taxodiaceae	*Athrotaxis*	3 (100)

(a) Wollemi pine, *Wollemia nobilis*, in eastern Australia

(b) *Auracaria* in New Caledonia

(c) Podocarpaceae in northern New Zealand

family Pinaceae), pollen lands on the surface of the cone adjacent to the ovule and the pollen tube grows over the surface of, or sometimes through, the cone scales into the micropyle. More commonly, as in the cypress pines (*Callitris* and *Cupressus*, family Cupressaceae), the ovule secretes a sticky drop of fluid that protrudes from the micropyle. Pollen

blowing through the cone is trapped on the surface of the micropylar drop. As the droplet evaporates, it shrinks back through the micropyle, drawing the pollen into the ovule.

Pollen of many conifers has two (sometimes one or three) air bladders or saccae developed from the pollen wall that are thought to aid in both wind dispersal and adhesion to the micropylar droplet. Once in the micropyle of the ovules, pollen grains undergo several cell divisions within the confines of the pollen grain wall to form the microgametophyte. A pollen tube germinates through an aperture in the pollen grain wall, grows through the nucellus (mega-sporangium) and provides a direct route to the archegonia and egg cell. The nucellus acts as a screen for foreign pollen, with the pollen tubes of other species being arrested within this tissue. Such events result in incompatibility between species of pines. In this way, the female cone is able to screen out unwanted potential male partners, selecting only those of its own species.

In conifers, sperm lack flagella and they are transported passively within the tip of a pollen tube. When the tube arrives at the egg, two sperm are released and one fuses with it. The other sperm aborts.

Seed cones of conifers are compound structures of several orders of branching. The structures of the cone that bear ovules, **ovuliferous scales** (Fig. 37.30), although flattened and leaf-like, are thought to be very reduced shoots. Pollen cones, on the other hand, are simpler in structure. In a comparative sense, a seed cone is equivalent to a group of pollen cones.

One unusual feature of conifers (and cycads and *Ginkgo*) is the timing of events involved in reproduction. Ovules are pollinated when they are very small and female gametophyte development is not triggered until after pollination. The megaspore undergoes meiosis and then becomes dormant. The post-meiotic changes to produce the egg within an archegonium are induced by pollen germination. Gametophyte development is slow and in many conifers not complete until the following year. Fertilisation of the egg cell by the sperm cell may occur a year or more after the pollen grains are actually deposited in the micropyle. Development of the embryo to the stage where the seed is ready to be shed can take another year, making the time elapsed from pollination to seed maturity in excess of two years in some cases.

Conifers bear their pollen and seeds in separate cones. Pollen is wind-dispersed and, on germination, produces a pollen tube that delivers non-flagellated sperm cells directly to archegonia.

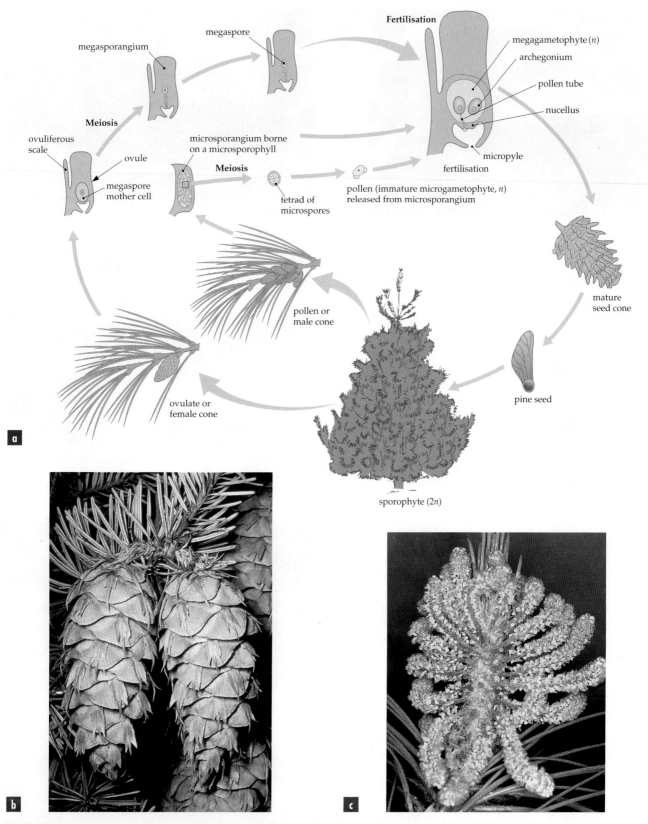

Fig. 37.30 Life cycle of pines. **(a)** Gametophytes are very reduced and enclosed within sporophyte tissue. The ovule, produced in female cones, encloses the megagametophyte. Pollen grains develop in male cones and are transported by wind to the micropyle of the ovule. The tube of the germinating pollen grain (immature microgametophyte) slowly digests its way towards the female gametophyte. After many months the pollen tube reaches the egg cell and the now-mature microgametophyte discharges its two sperm. The fertilised ovule develops into a seed. **(b)** A cluster of mature pollen cones of Monterey pine (*Pinus radiata*). **(c)** Female (seed) cones of Douglas fir (*Pseudotsuga menziesii*)

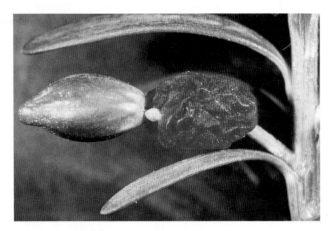

Fig. 37.31 In the conifer *Podocarpus*, an ovule is borne on a fleshy, coloured receptacle, which attracts birds

Gnetophytes: closest relatives of flowering plants

Gnetophytes (phylum Gnetophyta) are a bizarre group of plants consisting of just three extant genera that are markedly different from each other in appearance. Switch bushes, *Ephedra*, are shrubs of arid areas; species of *Gnetum* are lianes or small trees of the tropics; and the single species of *Welwitschia* is a xerophyte of the Namibian desert that was described by one botanist as resembling a giant octopus stranded on the beach (Fig. 37.32). Gnetophytes are of only minor importance in the present-day vegetation of the earth but are botanically very important because they are the closest relatives of the flowering plants (Fig. 37.7), thus providing some clues to the origin and evolution of that group.

Although at first glance the three genera of gnetophytes look very different, their close relationship is evidenced by similar reproductive structures. The micro- and megasporangia are clustered into small flower-like structures that are borne in a cone-like inflorescence (Fig. 37.32c). The droplet produced from the micropyle of the ovule has a high content of sugar and nitrogen and acts as an attractant for insects. In 'flowers' that produce only pollen, a central sterile ovule produces the nectar-like pollen drop necessary to attract pollinators.

The fertilisation process and the structure of xylem cells in gnetophytes have attracted great interest, as both features have marked similarities to flowering plants. In *Ephedra*, there is some evidence for a kind of double fertilisation. Here, one sperm nucleus fuses with the egg nucleus, while the other fuses with the sterile neck canal nucleus of the archegonium. Both fusion events produce an embryo, but only one of these remains viable as the ovule matures into the seed.

Water-conducting cells in the xylem of gnetophytes are open-ended, forming a continuous vessel network throughout the plant, which is far more efficient than

Fig. 37.32 **(a)** *Ephedra* (California), **(b)** *Welwitschia* (Africa) and **(c)** *Gnetum* (tropics) are the only living gnetophytes in the world

the discrete tracheids of other vascular plants. Almost identical xylem vessels are a characteristic of flowering plants.

G netophytes include only three living genera. They are the closest relatives of flowering plants.

Magnoliophytes: flowering plants

The flowering plants, or **angiosperms** (phylum Magnoliophyta), are dominant vascular plants of modern floras. They are major producers in terrestrial ecosystems and provide food, hardwood timber, natural fibres, spices and medicinal drugs that sustain the human population.

Flowering plants are adapted to a great range of habitats, from the tropics to polar regions, including fresh and salt water. The smallest forms are free-floating aquatic duckweeds (Fig. 37.33a), which are highly reduced and not even differentiated into leaf and stem.

The tallest flowering plant is the mountain ash tree, *Eucalyptus regnans*, of south-eastern Australia, which can reach heights of more than 100 m (Fig. 37.33b).

Flowering plants are classified into two major groups (Fig. 37.34): the **monocotyledons** (class Liliopsida) and the **dicotyledons** (class Magnoliopsida). Monocotyledons are generally non-woody plants, with flower parts in threes, parallel leaf veins, scattered vascular bundles in the stem and one cotyledon (seed leaf) in the embryo. They include grasses (e.g. cereals), lilies, grass trees, kangaroo paws (Fig. 37.35a), orchids, palms and duckweeds. Dicotyledons are either woody or herbaceous, with flower parts in fours or fives (Fig. 37.35b), net-like venation, vascular bundles arranged as a cylinder in the stem and two cotyledons. Dicots include magnolias, buttercups,

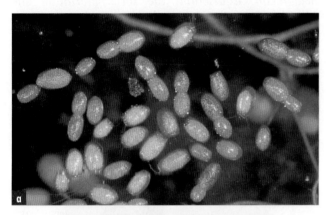

Fig. 37.33 **(a)** The duckweed *Wolffia*, one of the smallest angiosperms. **(b)** The Australian mountain ash, *Eucalyptus regnans*, is the tallest flowering plant in the world

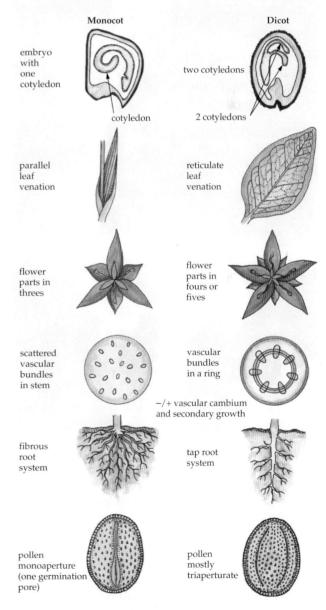

Fig. 37.34 Major characteristics of the two major groups of angiosperms: monocotyledons and dicotyledons

Fig. 37.35 (a) Australian kangaroo paw, *Anigozanthos manglessii*, is a monocot. **(b)** Australian *Thryptomene maissonneuvei* is a dicot

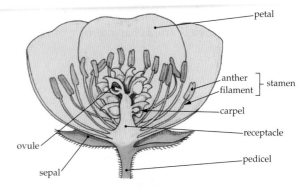

Fig. 37.36 The buttercup flower (family Ranunculaceae) has numerous stamens and free carpels. When mature, each carpel becomes a fruit with seeds developing from ovules

In addition to carpels, angiosperms are character-ised by a number of vegetative features. Phloem sieve tubes lacking nuclei at maturity are associated with nucleated companion cells (Chapter 18). Xylem consists of vessels as well as tracheids. Vessels also occur in *Selaginella*, *Equisetum*, the ferns *Pteridium* and *Marsilea*, as well as gnetophytes, but these seem to have evolved independently. Angiosperm vessels are different in having a range of structural patterns, including ladder-like (scalariform) perforations (Fig. 37.37).

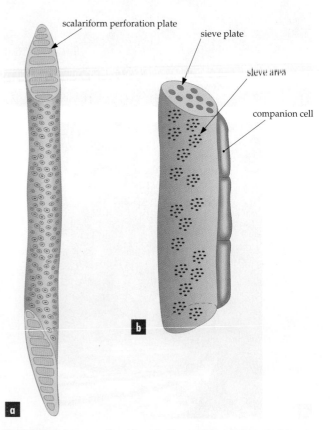

Fig. 37.37 Xylem vessels with scalariform perforations and phloem sieve elements with companion cells are characteristics of angiosperms

daisies, roses, peas, acacias, banksias and the mountain ash eucalypt.

The term angiosperm (literally, a vessel seed) describes one of the most important characteristics of flowering plants—the enclosure of the ovules within a hollow **carpel** (Fig. 37.36). As the ovules develop into seeds, the carpel becomes the **fruit**. This is in contrast to other seed plants, in which seeds are borne 'naked' on the surface of the megasporophylls of the female cone.

Flowers

Chapter 14 describes the process of sexual reproduction in angiosperms. We can now compare angiosperms with other seed plants to discover the homologies of the different floral structures and the evolution of the flower.

You will remember that **flowers** typically consist of four whorls or series of similar parts attached to a central axis (Fig. 37.36). In summary, the outermost whorl is the calyx, consisting of sepals, usually leaf-like in appearance and protective of the inner parts when the flower is in bud. Next is the corolla, consisting of petals, often colourful and adapted to attract animal pollinators. The next whorl consists of the stamens with their anthers (with pollen sacs), collectively the androecium. The innermost whorl consists of the carpels, collectively the gynoecium. Each carpel consists of a stigma, style and basal **ovary** containing ovules. Flowers may have a single carpel or many, either free or fused together. In **hypogynous** flowers (Fig. 37.38a), the ovaries are attached to or above the receptacle and are described as superior. In **epigynous** flowers, the ovaries are buried within the receptacle, below the perianth. These are inferior ovaries (Fig. 37.38b).

Flowers vary in size, colour, and number, shape and arrangement of floral parts. Flowers that have a regular arrangement of parts (radially symmetrical, Fig. 37.38c) are **actinomorphic**, for example, *Eucalyptus* and *Leptospermum* (tea tree). Irregular (asymmetrical) flowers (Fig. 37.38d) are **zygomorphic**, for example, peas and *Grevillea*.

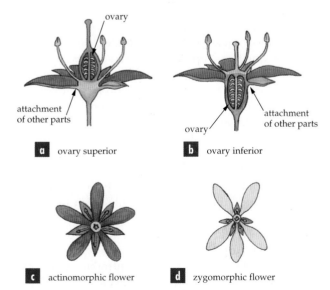

Fig. 37.38 Flowers with **(a)** superior ovary (hypogynous), **(b)** inferior ovary (epigynous), **(c)** regular (actinomorphic) and **(d)** irregular (zygomorphic) arrangement of parts

Fertilisation: a 'double' event

Microspores develop into microgametophytes within the **pollen** sacs of the anther and are shed as pollen grains, as in other seed plants (Fig. 37.39). The microgametophytes of angiosperms have three cells when mature—two sperm, which lack flagella, and one cell that plays a role in the growth of the pollen tube (see Chapter 14) when pollen reaches the female stigma. The pollen tube emerges through one of several thin areas in the pollen grain wall (apertures). Monocots have only one aperture (monosulcate pollen) and most dicots have three (a useful diagnostic character when identifying pollen).

Within the ovule, a functional megaspore (one of the four haploid products of meiosis) develops into the megagametophyte, which is the embryo sac described in Chapter 14. In most flowering plants, the embryo sac consists of only eight nuclei enclosed within seven cells (a result of three mitotic divisions and cell wall formation). At one end of the embryo sac is the egg cell, flanked by two cells; in the middle is one large cell containing two polar nuclei; and at the other end are three cells (Fig. 14.9). The embryo sac is thus a very reduced megagametophyte and, unlike in cycads, *Gingko* and conifers, eggs do not form in archegonia.

Double fertilisation is a unique event in angiosperms. Of the two sperm delivered to the ovule by the growth of the pollen tube, one fuses with the egg to form the zygote and the other fuses with the two polar nuclei to form the triploid endosperm nucleus, which divides by mitosis to produce endosperm tissue (Figs 14.6, 37.39). Endosperm only occurs in angiosperms and supplies the developing embryo with nutrition (Chapter 14).

> In flowering plants, the megagametophyte, which develops in the ovule, is reduced to an embryo sac. The microgametophyte develops in anthers and is shed as pollen. Double fertilisation that leads to endosperm formation is unique to flowering plants.

Origin of the flower

Flowers are determinate shoots, which means that the apical meristem stops growing once the last floral organs are formed: sepals being produced first, followed by petals, stamens and finally carpels. The classic theory of the origin of the flower therefore interprets the various floral parts as modified leaves borne on a shoot that has a limited capacity for elongation. Under this interpretation, stamens and carpels are merely specialised leaf-like appendages (sporophylls) that bear sporangia.

These classic ideas receive some support from the observation that sepals resemble leaves in position,

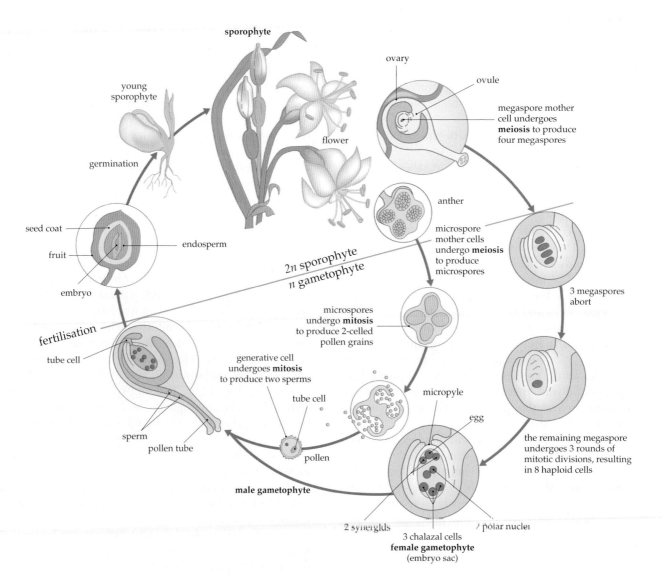

Fig. 37.39 The reproductive cycle of flowering plants (angiosperms) follows the same alternation of generations as conifers

venation, colour, form and early development, and also from recent work in molecular biology, which shows that genes that affect the development of leaves also affect the development of sepals. However, the evolution of other floral structures from leaves is much less clear. Petals, for example, may have arisen in a number of ways. Some petals may be leaves that have been modified directly while others may be modified stamens that have become sterile and function to attract pollinators rather than produce pollen. It has also been suggested that stamens represent a modified, reduced branch system with terminal microsporangia rather than a leaf-like organ. Controversy also surrounds the interpretation of carpels. In many respects they resemble leaves and can be influenced by the same developmental genes that are expressed in sepals, but in some flowering plants the early stages of carpel development are very distinctive and unlike that of leaves or any other floral organ.

Whatever their evolutionary origin, carpels are the most distinctive reproductive structures of angiosperms, and the enclosure of ovules by these structures undoubtedly led to a series of important modifications of the reproductive process. Most obviously, carpels provide additional protection for developing ovules and, after fertilisation, they may also be modified to form a variety of specialised fruit types for animal, wind and water dispersal (see p. 998). A more subtle consequence of the origin of the carpel is that pollen is no longer trapped directly by the ovule, as in most other seed plants, but instead becomes trapped on the stigmatic surface. Pollen tubes must then grow through some of the carpel tissues in order to deliver the male gametes to the embryo sac, thus introducing new possibilities for 'selection' of male gametes by the female plant (see Chapter 14).

Traditionally, flowers such as those of the living members of the magnolia family have been considered the most 'primitive' flower among angiosperms (Fig. 37.40a). Such flowers are bisexual, with numerous, often spirally arranged, perianth parts, stamens and carpels; they are often pollinated by beetles. It has frequently been hypothesised that all other flowers evolved from this type by reduction in number and fusion of parts to form more complex structures, such as corolla tubes. In support of this idea, there is evidence that magnolia-like flowers existed in the mid-Cretaceous (about 95 million years BP) at an early stage in the angiosperm fossil record.

Although the magnolia theory has been widely accepted for almost a century, recent studies have begun to challenge some of its basic tenets. Firstly, there is a great variety of floral types among living flowering plants, which range from very small flowers with only one carpel, one stamen and no petals to the large multiparted magnolia-like forms. Secondly, recent studies of angiosperm relationships using both morphological and molecular data favour the view that the

magnolia group may be an early but perhaps rather specialised offshoot in angiosperm evolution. Molecular analyses recently published in *Nature* indicate that a non-Magnolian shrub from New Caledonia, *Amborella*, may be a living relic of the most ancient lineage of flowering plants; *Austrobaileya*, a climbing vine from the rainforests of north-east Queensland, also has an ancient lineage. Thirdly, the fossil record shows that a great variety of other angiosperm flowers (Fig. 37.40b) were already present by the time magnolia-like forms are first recorded (approximately 100 million years ago). As a result of these studies, recent ideas of floral evolution suggest that the earliest, least specialised flowers may have been small, with relatively few parts and were probably organised on a three-part ground plan as occurs in some primitive monocots. Although the earliest flowers were almost certainly bisexual and insect-pollinated, it is also quite probable that unisexual, perhaps wind-pollinated flowers also developed at a very early stage in angiosperm evolution.

The enclosure of the ovules within a hollow carpel led to important modifications of the reproductive process of angiosperms. Carpels protect ovules and, when mature as fruits, aid seed dispersal. Pollen is trapped directly on the stigma of the carpel.

Fig. 37.40 (a) Flowers of the tulip tree (*Liriodendron tulipifera*, family Magnoliaceae) are considered one of the primitive types of flowers. **(b)** Mid-Cretaceous fossil flowers, however, include small separate male flowers, shown here (*Spanomera mauldinensis*), and female flowers

Pollination

The agents of **pollination** in modern flowering plants are wind, water, insects, birds, bats and other small mammals. Pollen is usually transported between flowers of different individual plants, sometimes separated by considerable distance, promoting out-crossing. Among the flowers that are wind-pollinated are grasses and sedges (Fig. 37.41a). They have inconspicuous flowers that are odourless, greenish in colour, and with petals reduced or absent. They produce abundant pollen, which lands on stigmas by chance. Flowers that attract animals are more effective in ensuring the transfer of pollen. This is of considerable advantage since a one-to-one relationship between a plant and animal species reduces wastage of pollen by ensuring that it is deposited on the 'right' flower.

Animals that act as pollinators search in flowers for a meal of nectar, a sugar-rich liquid secreted by floral nectaries, or pollen. The Australian honey possum is one of the few mammals that specialises in eating flower nectar (Fig. 37.41b). In the process of robbing the plant for food, the pollinator brushes against the sticky stigma, depositing a load of pollen.

Attracting pollinators

Individual plants that produce only a single flower, such as a tulip, are not common. Most plants produce many

Fig. 37.41 **(a)** Grass flowers, such as *Lolium perenne*, are small, clustered in inflorescences, and wind-pollinated. **(b)** The Australian honey possum, *Tarsipes rostratus*, feeding on and pollinating *Dryandra quercifolia*. **(c)** The red colour of the New South Wales waratah, *Telopea speciosissima*, attracts bird pollinators

flowers, usually in discrete clusters or **inflorescences** (Fig. 37.42). A good example is wattle, where each golden head is actually an inflorescence containing a few to many flowers (see Fig. 41.24). Inflorescences provide a mass display of flowers to attract pollinators.

Once in the vicinity of an inflorescence, an insect may be attracted to a flower by its scent. Heavy, sweet-smelling scents are produced by flowers that are pollinated by night-flying moths or bats that do not rely on vision. Some flowers produce foul-smelling scents that mimic rotting meat and attract flies.

Colouration and markings are important attractants in other flowers. Birds do not perceive scents but are often attracted to red flowers, such as the conspicuous flowers of the New South Wales waratahs (Fig. 37.41c). Waratah flowers are long, tubular and slightly curved; their rate of nectar production is relatively high and they are commonly visited by nectar-feeding honey-eaters.

Red is not a distinct colour to insects. Bees, which are especially important as pollinators, are generally attracted to yellow or blue flowers. Petals that look uniformly yellow to our eyes can appear quite different to a bee. In addition to reflecting yellow light, the petals can have markings that reflect UV light, giving a mix-ture of colours called bee's purple. Markings on petals,

Fig. 37.42 Flowers, such as *Grevillea*, are usually clustered as inflorescences, which makes them more conspicuous for pollinators

like lights on an airport runway, help the bee to locate nectaries (Fig. 37.43).

Flower shapes are also varied to ensure pollination. Petals may be fused to form a trumpet shape, with nectaries placed deep in the throat of the flower. This arrangement forces a pollinator to push into the flower to reach its reward. Anthers may be positioned in a way that maximises transfer of pollen to the pollinator. Some insects can also rob flowers without effecting pollination by cutting a hole in the petals and getting at the food source more directly.

Pollination by deceit

There are some orchids whose flowers mimic the shape and colouring of female insects. The mimics are so realistic that male insects will attempt to copulate with the flower, thereby pollinating them. Some of these orchids do not even present a nectar reward for the insect. In the genus *Drakaea*, the hammer orchids of Western Australia, the labellum (central petal) of the flower resembles a wingless insect, with shiny eyes,

hairy thorax and fat body (Fig. 37.44). The flower is held outwards by a hinged arm. When triggered by an insect, the hinged arm moves towards the column of the flower, where the stigma and pollinia (pollen masses) are located. Female thynnine wasps are wingless and spend much of their time underground, emerging to mate. Above ground, they climb to the top of a small shrub or grass stalk and attract a male wasp by releasing a chemical scent (pheromone) and posing in a characteristic manner—mimicked by the orchid! The male wasp swoops on the female, carrying her off to copulate in the air or on a nearby plant. However, if the male attempts to copulate with a hammer orchid, the hinged labellum of the flower flings him against the column, effecting pollination.

Pollination of flowers by animals is an example of how interactions between organisms play a role in the evolution of each group. The topic of species interactions is dealt with further in Chapter 43.

Pollination by animals is an advantage to flowering plants, being less wasteful and more effective than wind pollination. Flower scent, colour, markings, shape and nectaries are important in attracting animals.

Fig. 37.43 (a) *Rhododendron lochae*, a Queensland species, has bright red tubular flowers with no scent or visible markings to guide pollinators. **(b)** However, under ultraviolet light, the outer parts of the petals are reflecting and the inner region of the flower absorbing. The dark central region probably attracts and guides insects and birds

Fig. 37.44 (a) Flying male thynnine wasp, *Megalothynnus klugii*, approaching a wingless female, which lives undergound, emerging only to attract a mate. **(b)** A hammer orchid, *Drakaea glyptodon*, mimics the female wasp, whereby the male is deceived and mates with the orchid flower, effecting pollination

BOX 37.5 Pollination in aquatic flowering plants

In the pond weed *Vallisneria*, male flower buds abscise and rise to the surface. There, they float across the water surface, opening their anthers and eventually colliding with their stigma targets. Female flowers with receptive stigmas are raised to the water surface by remarkable elongation of the flower stalk. A high probability of pollen contacting a stigma is achieved by releasing the male flowers to float across the surface of the water rather than individual pollen grains, which have a much smaller diameter. Surface tension near the female plant causes the pollen mass to be drawn into the flower.

Pollination of the aquatic plants *Ruppia* (family Ruppiaceae) and *Lepilaena* (family Zannichelliaceae) is achieved in a different way. *Ruppia* pollen is released from the anther under water and floats to the surface in a gas bubble (see figure). In *Lepilaena*, the boat-like anthers are released under water and float to the surface, where they then release their pollen. As in *Vallisneria*, the female flowers of these two aquatics have an elongated stalk, so that they also float on the surface.

Angiosperms that have adapted to aquatic habitats have novel mechanisms to ensure pollen reaches the female stigma. In *Ruppia polycarpia*, large pollen grains stick together as rafts, which float on the surface of the water

Fruit and seed dispersal

After successful pollination and fertilisation, the fruit develops as the product of growth in the persistent parts of a flower. Fruits come in all colours and forms, including peanuts and their shells, corn kernels, pea pods, tomatoes, cucumbers, apples, oranges and eucalypt gumnuts (capsules). They are usually classified as simple, aggregate and multiple.

Simple fruits

A simple fruit is derived from the ovary of a single flower with one or more carpels, together with, in some species, adhering sepals. Simple fruits are moist and fleshy, or dry, when mature.

In fleshy fruits, the wall of the ovary increases greatly after fertilisation. The fruit wall, **pericarp**, becomes differentiated into three layers of tissue—the exocarp, mesocarp and endocarp. The **exocarp** is the outermost layer, usually a single layer of epidermal cells; in peaches, it is the skin. The **mesocarp** is the middle layer and varies in thickness; in peaches, it is the thick edible layer. The **endocarp**, the innermost layer, is also variable in thickness and structure; in peaches, it is the stone, whereas the seed is the kernel inside.

Cucumber, grape, orange and tomato are all berries (Fig. 37.45a). These are fleshy fruits in which the seeds, usually many, are enveloped within a fleshy mesocarp and a well-defined outer skin-like exocarp. In an orange, the segments correspond to the carpels. The juicy flesh comprises filaments of cells that arise from the carpel walls.

A **drupe** is a fleshy fruit, such as a plum, containing a single seed enclosed in a hard stony endocarp (Fig. 37.45b). The coconut is a drupe, the shell being the hard endocarp; the mesocarp is fibrous rather than fleshy and is usually removed before coconuts are sold in shops.

A **pome** is a fruit derived from an ovary of an epigynous flower. The fleshy, edible part of the fruit includes non-ovarian tissues. A good example is the apple (Fig. 37.45c). The true fruit arises from an inferior ovary to form only the core, while the bulk of the fruit derives from the receptacle together with the sepal and petal bases of the flower.

Fleshy fruits, particularly those coloured red or orange, may be eaten by birds, who unwittingly transport the seeds to new locations, depositing them by defecation. (Some seeds are covered with material that acts like a laxative!) The fleshy layers of fruits sometimes provide additional nourishment for developing embryos. Aboriginal Australians have always relied on native fruits as 'bush tucker', knowing which ones are edible and which ones are poisonous.

Of the dry simple fruits (Fig. 37.46), some are dehiscent, opening and shedding their seeds at maturity. A **follicle**, which opens on the lower side, is formed from a single carpel, such as in *Grevillea*. Pods or **legumes** open along two sides, as in beans, peas and *Acacia*. The dry fruits of many Australian native plants have thick, woody walls that protect the seeds from the heat of bushfires. In eucalypts, the top of the ovary forms three or four valves, which split open as the fruit, a capsule, matures and dries out. After a fire, a rain of seed is released from fruits held in the canopy of trees, the heat of which causes the valves to open quickly. The

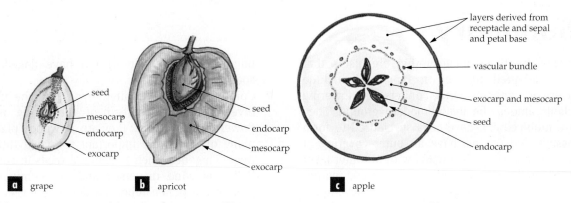

a grape   **b** apricot   **c** apple

Fig. 37.45 Fleshy fruits include **(a)** berries—grape, *Vitis*; **(b)** drupes—apricots and plums, *Prunus*; and **(c)** pomes—apple, *Pyrus malus*. The fruit wall (pericarp) consists of three layers: exocarp, mesocarp and endocarp

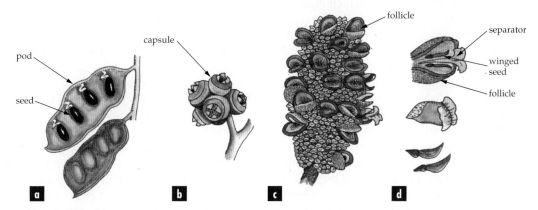

a   **b**   **c**   **d**

Fig. 37.46 Dry fruits include **(a)** pods or legumes, *Acacia*; **(b)** capsules, *Eucalyptus*; and **(c)** follicles, *Banksia*. **(d)** In the *Banksia* follicle, a separator, when wet, pulls the two seeds out of the fruit

woody follicles of banksias are very fire-resistant and open after fire. Each follicle has two seeds attached to a wing-like structure, the separator. When the separator is sufficiently wet by rain following a fire, it expands, pulling the seeds out of the fruit (Fig. 37.46d). Seeds will otherwise remain protected in the fruit until rain falls, providing suitable conditions for germination.

Aggregate and multiple fruits

An aggregate fruit arises from a cluster of separate carpels from one flower. Raspberries and blackberries are familiar examples. The strawberry is an aggregate of numerous seed-like fruits (achenes) from a single flower that develop on a sweet, fleshy receptacle (stalk), which is the edible part (Fig. 37.47).

A multiple fruit, such as pineapple and figs, is a cluster of many carpels produced from several flowers in an inflorescence. In pineapple (Fig. 37.47), the inflorescence axis and various floral organs contribute to its structure.

b

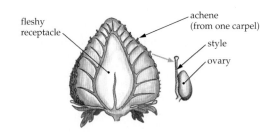

a

Fig. 37.47 **(a)** A strawberry is an aggregate fruit, consisting of numerous small achenes (from individual carpels) on a fleshy receptacle derived from one flower. **(b)** A pineapple is a multiple fruit derived from a number of flowers in an inflorescence

Summary

- Plants are those photosynthic organisms that have adapted to life on land and include liverworts, mosses, ferns, conifers and flowering plants among the major modern groups. They are multicellular, have cells specialised to form tissues and organs, have an outer covering of cuticle, and have sex organs with an outer layer of vegetative cells to protect gametes and, within the female reproductive organ, the developing zygote.

- In plant life cycles, the haploid gametophyte generation forms male antheridia, which produce sperm, and female archegonia, each of which produces a single egg. Male and female gametes fuse to form a zygote, which grows into the diploid sporophyte generation. Meiosis occurs on the sporophyte, resulting in the formation of haploid spores for dispersal. In some ferns, fern allies and seed plants, two types of spores are produced. These plants are heterosporous.

- Bryophytes include liverworts (phylum Hepatophyta), hornworts (Anthocerotophyta) and mosses (Bryophyta). Bryophyte plants have small, photosynthetic, free-living gametophytes, which lack vascular tissue. Sexual reproduction requires water for sperm to swim to an egg in an archegonium. The sporophyte grows on the gametophyte and is dependent on it for nutrition.

- 'Fern allies' refers to a loose assemblage of living, spore-producing, vascular plants that are classified in three phyla: Psilophyta (*Psilotum* and *Tmesipteris*), Lycophyta (*Lycopodium*, *Selaginella* and *Isoetes*) and Sphenophyta (*Equisetum*).

- Homosporous ferns, like fern allies, have a sporophyte that is photosynthetic, free-living and the dominant generation. Like bryophytes, ferns have flagellated sperm and rely on free water for sexual reproduction. Fern antheridia and archegonia are borne on a small, free-living, heart-shaped gametophyte.

- Cycads (phylum Cycadophyta) are the most primitive of the living seed plants. They are palm-like in appearance but produce ovules and pollen in cones. Cycads and *Ginkgo biloba*, the only living representative of the phylum Ginkgophyta, rely on insects or wind for transport of microspores (pollination) but have flagellated sperm.

- Conifers (phylum Coniferophyta) are the most diverse and widespread of the living non-flowering seed plants; they are all woody plants, many of which are important softwood timber trees. They bear their pollen and seeds in cones. Pollen is wind dispersed and on germination produces a pollen tube that delivers non-flagellated sperm cells directly to archegonia. Reproduction may take up to several years.

- Gnetophytes (phylum Gnetophyta) include only three living genera. They are the closest relatives of flowering plants. Their reproductive structures are surrounded by bract-like structures, which make them resemble flowers. Their xylem contains vessels in addition to tracheids.

- Flowering plants or angiosperms (phylum Magnoliophyta) are classified into two major groups: the monocotyledons (class Liliopsida) and dicotyledons (class Magnoliopsida). Monocots are generally non-woody plants, with flower parts in threes, parallel leaf veins, scattered vascular bundles in the stem, and one cotyledon (seed leaf) in the embryo. Dicots are either woody or herbaceous, with flower parts in fours or fives, net-like venation, vascular bundles arranged as a cylinder in the stem and two cotyledons.

- In angiosperms, the male (micro-) and female (mega-) sporangia are located in the flower. The microgametophyte is shed from anthers as pollen. The megagametophyte is the embryo sac, which often has eight nuclei within seven cells.

- Double fertilisation is unique to angiosperms. One sperm fertilises the egg, the other sperm fuses with the two polar nuclei of the embryo sac, resulting in a zygote surrounded by triploid, nutritious endosperm.

- The enclosure of the ovules, containing the megagametophytes, within a hollow carpel led to important modifications of the reproductive process of angiosperms. Carpels provide protection for ovules and, when mature as fruits, they aid seed dispersal. Pollen is trapped directly on the stigma of the carpel.

- Angiosperms have the advantage of attracting animals for pollination and fruit and seed dispersal.

keyterms

actinomorphic (p. 993)	epigynous (p. 993)	Lycophyta (p. 976)	pericarp (p. 998)
alternation of generations (p. 969)	exocarp (p. 998)	Magnoliophyta (p. 982)	peristome (p. 975)
angiosperm (p. 991)	fern (p. 980)	megagametophyte (p. 978)	pinnae and pinnules (p. 981)
antheridium (p. 969)	fern allies (p. 976)	megasporangium (p. 969)	pollen (p. 993)
Anthocerophyta (p. 970)	Filicophyta (p. 980)	megaspore (p. 969)	pollination (p. 995)
archegonium (p. 969)	flower (p. 993)	mesocarp (p. 998)	pome (p. 998)
Bryophyta (p. 970)	follicle (p. 998)	microgametophyte (p. 978)	protonema (p. 976)
calyptra (p. 975)	frond (p. 981)	microsporangium (p. 969)	Psilophyta (p. 976)
capsule (p. 974)	fruit (p. 992)	microspore (p. 969)	rhizoid (p. 968)
carpel (p. 992)	gametophyte (p. 969)	monocotyledon (p. 991)	rhizome (p. 968)
clubmoss (p. 978)	gemmae (p. 970)	moss (p. 970)	Rhyniophyta (p. 977)
Coleochaete (p. 965)	Ginkgophyta (p. 982)	operculum (p. 975)	seed (p. 983)
cone (p. 986)	Gnetophyta (p. 982)	ovary (p. 993)	seta (p. 974)
Coniferophyta (p. 982)	Hepatophyta (p. 970)	ovule (p. 983)	sorus (p. 981)
Cycadophyta (p. 982)	heterospory (p. 969)	ovuliferous scale (p. 988)	Sphenophyta (p. 976)
dicotyledon (p. 991)	homospory (p. 969)	paraphyses (p. 970)	sporophyte (p. 969)
drupe (p. 998)	hornwort (p. 970)		sporopollenin (p. 968)
elater (p. 975)	hypogynous (p. 993)		strobilus (p. 978)
endocarp (p. 998)	indusium (p. 981)		zygomorphic (p. 993)
	inflorescence (p. 996)		
	integument (p. 983)		
	legume (p. 998)		
	liverwort (p. 970)		

Review questions

1. List the evidence that indicates that green algae such as *Coleochaete* (class Charaphyceae) are the closest relatives of land plants.

2. What are the major problems of living on land compared with living in water? What adaptations to living on land are associated with the gametophyte generation and the sporophyte generation in the life cycle of plants?

3. Of the following phyla of living plants, which have vascular tissue: Hepatophyta, Bryophyta, Psilophyta, Filiciphyta, Coniferophyta, Magnoliophyta? Which of these phyla are ferns? Which of these phyla are liverworts?

4. Plant life cycles show an alternation of generations.

    (a) On what stage do sporangia develop?

    (b) Does spore production involve mitosis or meiosis?

    (c) What is the main function of spores?

    (d) What are antheridia and archegonia and on what stage do they develop?

5. What is the age of the oldest land plant fossils? Briefly describe the form of *Rhynia*.

6. What is meant by 'homosporous' and 'heterosporous' plant life cycles? Is a pine tree homosporous or heterosporous?

7. What is a megaspore and a microspore, and what do they develop into?

8. (a) What is a seed and from what structure(s) does it develop?

   (b) Name the major groups of plants that develop seeds and state whether their seeds are borne in cones or fruits.

   (c) In what ways is the seed an important advantage to plants living on land?

9. (a) Name the parts of a flower and state a function for each.

   (b) Name the two major groups of flowering plants and list their major differences.

10. Fruits are classified as simple fruits, including berries, pomes, drupes, follicles, legumes and capsules, aggregate fruits or multiple fruits. Using these terms, classify the following fruits: strawberry, tomato, apple, eucalypt 'gumnut', *Acacia* pod, pineapple, plum and grape.

Extension questions

1. Mosses not only occur in damp situations but some species live in inland Australia or Antarctica. How are such mosses able to live in an arid environment?

2. Compare the transport of sperm to the egg cell in a moss, fern, cycad and conifer.

3. Given that a flower is a determinate shoot, discuss the possible origin of the flower and its various parts.

4. Of the vascular plants, flowering plants dominate modern floras. What features of flowering plants explain their success? How has their evolution been affected by animals?

Suggested further reading

Enright, N. J. and Hill, R. S. (1995). *Ecology of the Southern Conifers.* Melbourne: Melbourne University Press.

Advanced textbook that describes the Southern Hemisphere conifers (their origin, evolution, diversification and ecology), including Australian and New Zealand species.

Jones, D. L. and Clemesha, S. C. (1993). *Australian Ferns and Fern Allies.* Sydney: The Currawong Press.

Provides further information on fern structure and life cycle, with details on Australian plant groups.

Jones, D. L. (1993). *Cycads of the World.* Sydney: Reed Books.

Comprehensive text on cycads of the world, their fossil history, structure, biology, cultivation, conservation and economic importance.

Low, T. (1989). *Bush Tucker—Australia's Wild Food Harvest.* London: Angus & Robertson.

Describes and illustrates many native Australian plants, where they are found and their usefulness as food.

Moore, R., Clark, W. D., Vodopich, D. S. (1998). *Botany.* 2nd edn. Quebec: WCB/McGraw-Hill.

Includes detailed chapters on the biology and economic importance of land plants; international edition.

Morley, B. D. and Toelken, H. R. (1983). *Flowering Plants in Australia.* Adelaide: Rigby.

Provides an overview of both native and introduced flowering plant families as well as cycads and conifers. The book includes descriptions of plants, where they are found and their human uses.

Raven, P. H., Evert, R. F., Eichhorn, S. E. (1999). *Biology of Plants.* 6th edn. New York: W. H. Freeman & Co.

Includes detailed chapters on the biology and evolution of land plants.

CHAPTER
38

Simple animals: sponges to nematodes

All members of kingdom Animalia, sometimes called the kingdom Metazoa, are multicellular. The multicellular organisation of animals, as in plants (Chapter 37), allows division of labour between individual cells or groups of cells. For instance, some animal cells function in locomotion, while others function in nutrition or reproduction. Except in sponges, cells are differentiated into distinct types that aggregate to form tissues, which are in turn organised into structural and functional organs. The cells of animals recognise other cells that belong to the same animal, and thus, if cells are separated in tissue culture, they will reaggregate. This 'self-recognition' depends on the presence of glycoproteins (Chapter 1) on the cell surface.

Animals undergo embryonic development (called ontogeny), where a multicellular individual develops from a single-celled zygote (see Chapter 15). Animals produce gametes in multicellular organs (the gonads) or in aggregations surrounded by non-sexual (i.e. somatic) body cells. Animals are usually diploid, with meiosis occurring only during the production of gametes. Animals are heterotrophs, that is, they feed on other organisms, and are motile during some part of their life cycle.

The animal kingdom includes an amazing variety of species, which are classified into some 30 phyla. Each phylum has characteristic features, which reflect a common body plan. Phyla differ in their pattern of development, body symmetry, body segmentation, body cavity, external covering, internal skeleton, appendages, and in other ways that we will study in Chapters 38–40.

This chapter describes simple animals, including sponges (which are simplest in structure), jellyfish, anemones, corals and comb jellies (which are radially symmetrical), flatworms (which lack a body cavity or coelom) and nematodes (which are roundworms with a pseudocoelom). Chapter 39 describes coelomate animals (those with a mesodermal body cavity), including segmented annelid worms, arthropods characterised by an external skeleton and jointed appendages, and molluscs (which usually have a shell). Chapter 40 describes animals as different as echinoderms, such as starfish, and chordates, which include vertebrates from fishes to humans. Echinoderms and chordates show a similar pattern of embryonic development.

> Animals are multicellular, heterotrophic organisms. They undergo embryonic development from a zygote and have cells, which, with the exception of in sponges, differentiate into tissues, which are specialised for different functions.

Origin and early evolution of animals

Origin of multicellularity

How animals evolved from protistan-like organisms and how phyla are related to one another are not fully understood. Various theories have been proposed for the origin of multicellularity and it may have arisen more than once. One theory is that a multicellular organism arose by partitioning of the cytoplasm of a ciliate protist, producing a flatworm-like organism. Another idea is that animals evolved from a colony of flagellate protists. A third theory suggests that the earliest multicellular animal was a two-layered organism that crept along the sea floor, rather like *Trichoplax*, a marine animal living today that digests organic material trapped between a lower layer of endoderm and the substrate to which it attaches. *Trichoplax* is the least differentiated of all known animals and has the smallest amount of DNA per nucleus of any multicellular animal. To date, there is little hard evidence for any of the theories but analysis of the genetic and molecular make-up of simple organisms may eventually provide greater understanding of their relatedness.

Fossil faunas and what they tell us about the history of animals

Fossil evidence suggests that the first multicellular animals originated in Precambrian seas. Fossils representing the transition from protists to animals are not known. The earliest known fossils that are clearly animal were found first in the Ediacara Hills of South Australia in rocks more than 600 million years old (Chapter 31). Assemblages of these primitive animals are known as the Ediacaran fauna and have subsequently been found on several continents. Most of them resemble the medusa stage of modern jellyfish. Some, reaching a length of 40 cm, have been likened to sea pens, in the Cnidaria (see p. 1007), while others, such as *Dickinsonia*, almost a metre long but only 3 mm thick (Fig. 38.1), resemble annelid worms. The Ediacaran fauna tells us that animals were present and already well diversified in the Precambrian era.

By the Cambrian period a vast diversity of animals had evolved. The most remarkable record of this radiation, especially of arthropods, occurs in shales from the Burgess Pass in the Canadian Rocky Mountains. The Burgess Shales (Chapter 31) were derived from sediments of fine mud, which allowed minute details of animals to be preserved. The fossils of the Burgess Shales suggest that most of the animal phyla evolved within a very short period of time relative

Fig. 38.1 *Dickinsonia costata*, of the Precambrian Ediacaran fauna

There is very little evidence concerning the origin of multicellularity. There is fossil evidence indicating when simple animals appeared. Fossil evidence together with molecular and biochemical data provide insights into the relationship between animal groups.

to the time scale of the evolution of life on earth (Chapter 31). The reason for this is not understood, although there are many theories.

New techniques, particularly the analysis of DNA and RNA sequences, are providing new insights into the relationships between modern animal phyla. Figure 38.2 shows an example of a phylogenetic tree based on a combination of data from both morphology and molecules.

Sponges: phylum Porifera

Sponges, phylum **Porifera**, are aquatic animals (Fig. 38.3) that are similar to other multicellular animals in having three layers of cells. However, in sponges these layers are not organised into tissues or organs. Sponges have no mouth or gut, no circulatory system and no nervous system, and they are the only multicellular animals with this very simple level of organisation. Because of this, the sponges have at various times been classified in their own separate division of the animal kindgom, subkingdom Parazoa, to distinguish them from all other animals, which are classified in the subkingdom **Eumetazoa**.

There are about 5000 species of sponges. Most are marine and colonial. Adults are sessile on a substrate such as rocks but larvae are free-swimming.

In the simplest sponges (Fig. 38.4a), the body is tubular and its wall has three functional layers. The inner layer, lining an internal cavity, consists of flagellated cells,

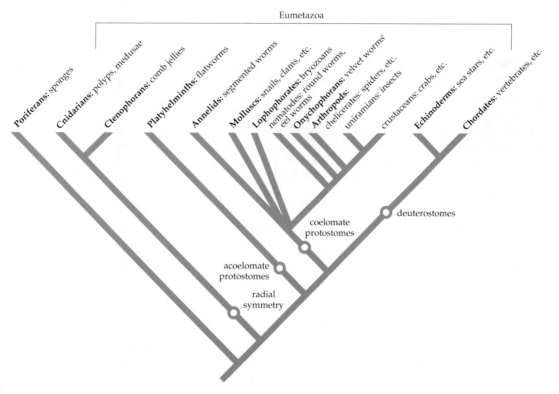

Fig. 38.2 Example of a phylogenetic tree of animal phyla based on both morphological and molecular data. The relationships of some animal groups are not fully known

Fig. 38.3 A cluster of small calcareous sponges of the genus *Callispongia*, at Heron Island

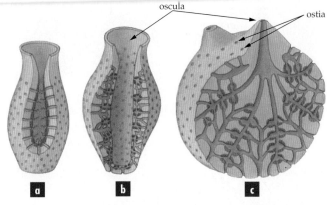

Fig. 38.4 Structure of **(a)** simple, **(b)** intermediate and **(c)** advanced forms of sponges

choanocytes (collar cells), which have a distinctive collar of microvilli surrounding the flagellum (Fig. 38.5a). The outer surface of the sponge consists of a layer of flattened cells, **pinacocytes**, making up the **pinacoderm**, analogous to the epithelium of other animals. Between these two layers is the **mesohyl**, a gelatinous protein matrix containing amoeboid cells, **amoebocytes**, dispersed collagen fibres (a protein common to all animals), and skeletal elements. Some amoeboid cells are large and phagocytic. They are important in digestion but are also capable of giving rise to any other type of cell; that is, they are totipotent.

The skeleton is composed of calcium carbonate or silica **spicules**, or fibres of a coarse collagenous proteinaceous material termed **spongin**, or a combination of siliceous spicules and spongin fibres. Collagen fibres are secreted by amoebocytes as well as by fixed cells. The components of the skeleton, particularly the structure of the spicules, are used in the classification of sponges. Skeletons are sold for bath sponges and other uses!

Sponges have a unique method of filter feeding. Water is drawn in through pores, the **ostia** (sing. ostium), in the external wall of the sponge into a central cavity, the **atrium** or **spongocoel** (Fig. 38.5a). This feeding and respiratory current then leaves through

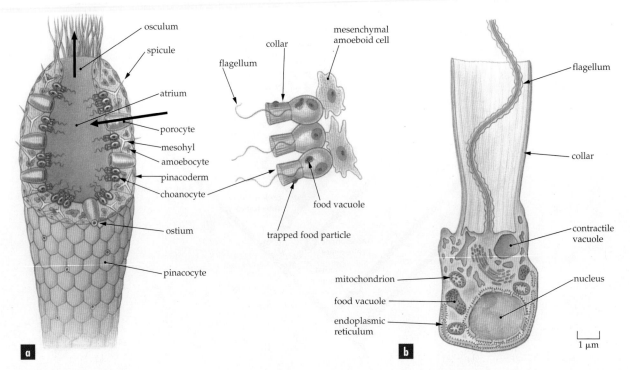

Fig. 38.5 (a) Cutaway view of a simple sponge, showing the arrangement and action of collar cells (choanocytes). **(b)** The ultrastructure of a choanocyte

one or more large openings, the **oscula** (sing. osculum). The current is propelled by the beating of the flagella of the choanocytes lining the passages.

Sponges feed on extremely fine, mostly submicroscopic, particulate matter drawn in by the water current. Larger particles are phagocytosed by pinacocytes lining the inhalant canals but particles of bacterial size and below (less than 1 mm) are trapped on the exterior of the choanocyte collar and ingested in food vacuoles at its base. The structure of a choanocyte, as revealed by electron microscopy, is shown in Figure 38.5b. Both choanocytes and pinacocytes transfer food to amoebocytes, which appear to be the principal sites of digestion.

In advanced sponges (Fig. 38.4c), choanocytes are located in chambers embedded in the mesohyl. Water is drawn through the ostia into inhalant spaces, lined by pinacocytes, and percolates through narrow passages into the choanocyte-lined chambers. From these, exhalant canals allow water to escape through the oscula. This body form increases the efficiency of feeding because a greater volume of water can be pumped but it still moves slowly in the feeding chambers. More efficient feeding allows larger size and greater diversity of shape than is possible in simpler sponges. Intermediate conditions occur (Fig. 38.4b).

> Sponges (phylum Porifera) are simple aquatic filter feeders, with three layers of cells but without tissues, organs, mouth or nervous system.

Radially symmetrical animals

Polyps and medusae: phylum Cnidaria

The simplest of the animals in the remainder of the animal kingdom are found in the phylum **Cnidaria**. The phylum includes jellyfish, sea wasps, hydrozoans, sea anemones and corals. Cnidarians are marine or freshwater animals with radial symmetry (Box 38.1). Some, a few anemones, are radiobilateral; that is, they are primarily radially symmetrical but the development of one or two ciliated grooves at the edges of the mouth means that the animal can only be divided into two equal halves in one or two planes.

The body of a cnidarian is basically a sac with a central **gastrovascular cavity** or **coelenteron**. The single opening to this cavity is fringed with tentacles and functions both as mouth and anus. The gastrovascular cavity is lined by endodermal cells forming a gastrodermis, and is homologous with the alimentary canal of other metazoans. The life cycles of cnidarians include one or two variations on this basic body form—polyp and medusa. A **polyp** is an attached tubular form

BOX 38.1 Radial and bilateral symmetry

Two living metazoan phyla, Cnidaria and Ctenophora, have **radial symmetry**, which means that any longitudinal plane bisects the animal into similar halves (Fig. a). Radial symmetry is characteristic of animals that are attached to a substrate (e.g. polyps) or float freely in water (e.g. medusae or jellyfish). Such animals receive equal stimuli and opportunity to obtain food from all sides. Radial symmetry is seen also in adult echinoderms, starfish for example, but in that case it is secondary because it is acquired as a result of metamorphosis from a bilaterally symmetrical larva (Chapter 40).

In animals with **bilateral symmetry**, only one plane divides the animal into equal halves (Fig. b). Bilateral symmetry is typical of animals that move along a substrate. Locomotory mechanisms are more effective if one part of the animal, the anterior or head end on which sensory receptors can be concentrated, leads the way. Bilaterally symmetrical animals often have a lower ventral surface specialised for movement; the upper dorsal surface may be protective.

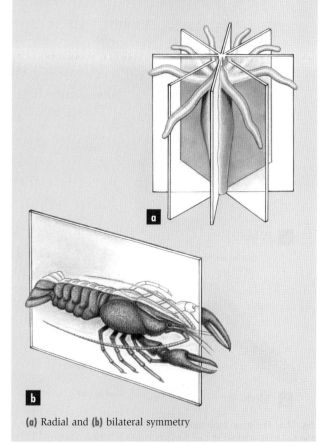

(a) Radial and **(b)** bilateral symmetry

with its mouth directed upwards, for example, a sea anemone; a **medusa** is free-floating (pelagic) and bell-shaped, with its mouth pointing downwards, for example, a jellyfish (Fig. 38.6). The mouth is borne on a projection, termed the **hypostome** in the polyp and the **manubrium** in the medusa.

Cnidarians are characterised by stinging organelles, **nematocysts** (**cnidae**), which function in defence and capture of prey. Each nematocyst is produced in a cell, the **cnidocyte**. There are many types of nematocysts but all basically consist of a capsule and an enclosed, usually hollow, thread that is eversible and often barbed (Fig. 38.7). In penetrant nematocysts, a proteinaceous toxin is injected from the capsule through the thread into the prey or predator.

In Australian waters, people often experience stings from cnidarians. The commonest encounters are those with the blue bottle or Portuguese man o'war, *Physalia utriculus* (Fig. 38.8b) and with stinging jellyfish, which occur all around the Australian coast. In tropical waters and on the Great Barrier Reef, contact with the hydroid polyp, *Aglaophaenia cupressina* (Fig. 38.8a) causes severe pain and stings from the medusae of the sea wasp or box jellyfish, *Chironex fleckeri* (Box 38.2), are dangerous and often fatal.

Cnidarians are described as **diploblastic** because the body wall has two cellular layers, an ectoderm and endoderm. Between them is an intermediate, largely

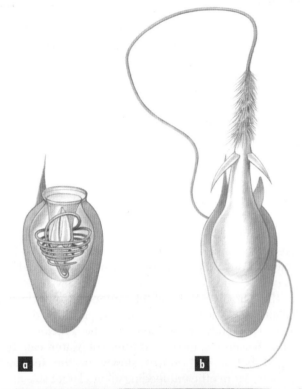
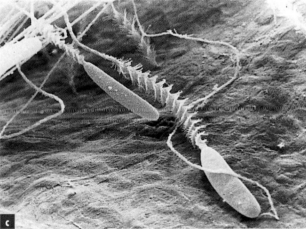

Fig. 38.7 Structure of a penetrant nematocyst: **(a)** undischarged nematocyst; **(b)** nematocyst discharged after contact with the prey. **(c)** Discharged nematocyst photographed under a scanning electron microscope

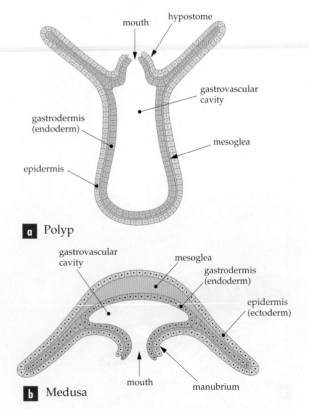

Fig. 38.6 The basic structure of **(a)** a polyp and **(b)** a medusa are the same. A medusa is like an upside-down flattened polyp

gelatinous, layer, the **mesoglea**, which differs from true mesoderm (Chapter 15). The mesoglea forms the bulk of the body mass in medusae.

Most of the organisation of cnidarians is at the level of tissues. In the absence of mesoderm, organ formation is minimal, although tentacles, reproductive structures and sensory structures concerned with balance and light detection are organised as simple organs.

All cnidarians are carnivorous but some members of all three classes, particularly corals, harbour symbiotic intracellular algae (zooxanthellae or **zoochlorellae**, or both), usually in the endoderm. These algae enable

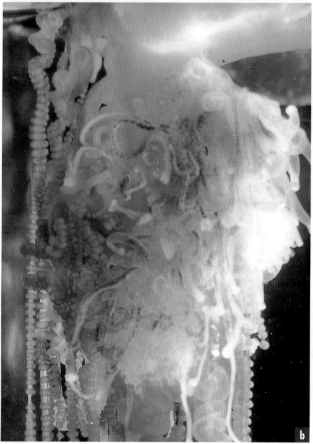

Fig. 38.8 Stinging hydrozoans: **(a)** *Aglaophaenia cupressina*; **(b)** *Physalia utriculus*, known in Australia as a 'blue bottle'

BOX 38.2 Sea wasps in Australian waters

One of the greatest threats to people in coastal waters of tropical Australia is the sea wasp, *Chironex fleckeri* (see Fig. 27.4). Stings from its tentacles cause near-sudden death of, on average, one Australian per year. *Chironex fleckeri* is chiefly restricted to Australia, although it may occur in Indonesian and Malaysian waters to the north. It is an inshore species, not normally occurring in offshore waters, such as those surrounding the Great Barrier Reef.

Populations are sighted in shallow water off the coast of Queensland from late November to May each year, after which they disappear. Young are carried in the bell of adult female medusae and are released in late November when adults move into shallow water.

Like all box jellyfish (class Cubozoa), *C. fleckeri* medusae can be recognised by their cube-shaped bell and tentacles in groups of four. Photosensitive organs are present marginally and stinging cells (nematocysts) occur in bands on the tentacles. A large specimen can have a bell as large as a human head with trailing tentacles reaching a total length of 90 m. The nematocysts discharge a toxin into other animals including fish, which they eat. The toxin elicits sustained contraction of all muscle types. There is variation in the stinging ability of individual sea wasps, but when a human is stung over a large area of the body, death usually results within minutes.

dioxide by the algae assists with calcium carbonate formation, which is why the skeleton of the coral grows faster in the presence of algae and is slowed in corals deprived of algae (e.g. if they are kept in the dark). Prey caught by the coral provides nitrogen and phosphorus to both the coral and the algae.

Cnidarians (polyps and medusae) are characterised by stinging cells, nematocysts (cnidae). They have a body wall consisting of two cellular layers and an intermediate layer, also containing cells, the mesoglea. They have a gastrovascular cavity (coelenteron) with a single opening fringed with tentacles.

Jellyfish: class Scyphozoa

Scyphozoans are exclusively marine, the majority living in coastal waters. The conspicuous phase of their life cycle is the jellyfish or medusa (Fig. 38.9), while the

corals to build reefs, of which the Great Barrier Reef of eastern Australia is the most extensive example in the world. The cnidarian derives nutritional and metabolic advantages from the algae. In return, the animal provides nutrition and a favourable environment. Virtually all reef-building corals possess zooxanthellae and these may account for up to 50% of the nitrogen in the proteins of the coral. A significant part of the carbon fixed in photosynthesis by the algae is passed to the tissues of the coral, mostly as glycerol but also as glucose and the amino acid alanine. Removal of carbon

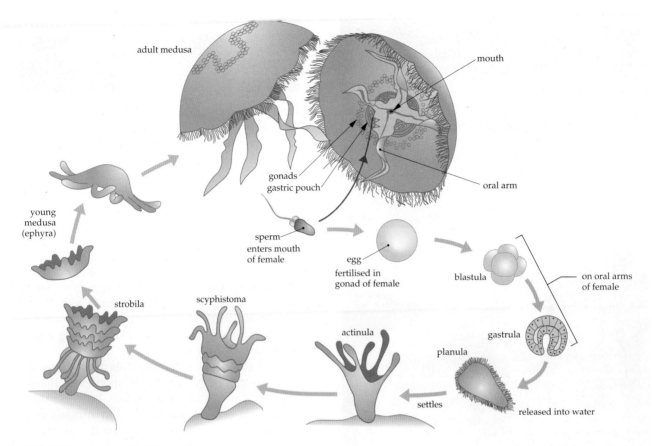

Fig. 38.9 Life cycle of the jellyfish *Aurelia*

polyp is very small or sometimes absent. Scyphozoan medusae include the largest individual cnidarians. Although most are free-swimming, some rest on or attach to the ocean floor. Typically, the medusa consists of a flat or domed bell (Fig. 38.10). The mouth is borne on the manubrium, a trunk-like structure in the centre of the ventral or subumbrellar surface. In *Pelagia*

(Fig. 38.11), a luminescent jellyfish of Australian waters, and *Aurelia*, the margins of the manubrium commonly extend as four **oral arms** (Fig. 38.10). The mouth leads into a central stomach from which radiate four **radial canals**. The radial canals connect with a **ring canal**, which runs around the margin of the bell. The muscular system is ectodermal in origin. The margin bears tentacles armed with nematocysts and is usually scalloped. In the indentations there are often sense organs, the **rhopalia**, which contain balance structures, **statoliths**, and sometimes light-sensitive organs.

Adult medusae, usually of separate sexes, produce eggs and spermatozoa in gonads situated on the septa dividing the stomach or on the walls of the four gastric pouches. When mature, eggs and sperm break into the gastrovascular cavity and pass out the mouth. In some scyphozoans, such as *Aurelia*, eggs lodge and zygotes develop on the oral (mouth) arms. A ciliated **planula larva** is released and later attaches to the substrate to become a small polyp, called an **actinula**. The actinula grows into a **scyphistoma** that produces small medusae called **ephyrae** (sing. ephyra) by horizontal splitting of the body, an asexual reproductive process called **strobilisation** (Fig. 38.9). Attached polyps may persist for years, outliving the individual medusae they produce.

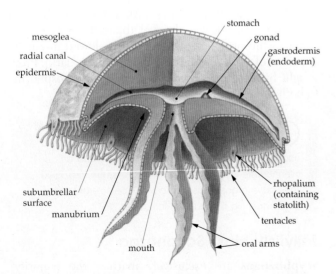

Fig. 38.10 Structure of the jellyfish *Aurelia*. One-quarter of the body has been cut away to show internal features

Fig. 38.11 The medusa of the schyphozoan *Pelagia*

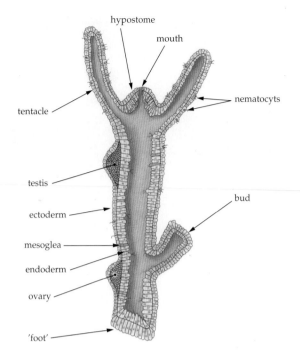

Fig. 38.12 Structure of *Hydra*

Box jellyfish: class Cubozoa

In box jellyfish, class **Cubozoa**, the bell is square in cross-section and the margin bears four extensions from which hang one or more hollow tentacles. The medusae are strong swimmers and their sting is often fatal to humans. They include the sea wasp, *Chironex fleckeri*, of tropical Australian waters (Box 38.2).

> Scyphozoans (jellyfish) are marine animals with the conspicuous life cycle stage being the jellyfish or medusa. Cubozoans (box jellyfish) are also marine, with a cube-shaped bell and tentacles in groups of four. Their stings are extremely painful and often fatal to humans.

Hydras: class Hydrozoa

In the class **Hydrozoa**, the life cycle typically includes a dominant, although often very small, attached polyp stage, which reproduces asexually, and a microscopic pelagic medusa stage, which reproduces sexually. In some species, however, either the polyp or the medusa may be reduced or absent. The freshwater genus *Hydra* is commonly used to illustrate the structure of a hydrozoan polyp (Fig. 38.12).

Hydrozoan polyps are commonly colonial, with an interconnecting gastrovascular cavity. They usually have a chitinous, sometimes calcified, exoskeleton, *Hydra* being one exception.

In contrast with jellyfish (scyphozoan medusae), hydrozoan medusae lack radial partitions and rhopalia (sense organs), and the gametes ripen in the ectoderm. In the life cycle of *Obelia* (Fig. 38.13), the colony of polyps grows by asexual budding. Two types of polyps are produced, feeding **hydranths** and reproductive **gonangia** (sing. gonangium). Free-swimming medusae of both sexes bud off from the gonangia (Fig. 38.14a) and give rise to eggs or sperm, which are released into the water. Fertilisation is external and the zygote divides to form a blastula, which becomes a ciliated planula larva. The planula rapidly attaches to the substrate and forms a new colony by budding.

The medusoid stage may be reduced and permanently attached to the hydroid colony as a gonophore, which produces gametes. In *Tubularia*, the gonophore is situated within the tentacle crown of the hydroid (Fig. 38.14b). In *Hydra* and some other genera, the medusa is totally absent. In a few species, the polyp is reduced or absent and the medusa is conspicuous. The medusae of the luminescent hydrozoan *Aequoria* are washed up on Australian shores as thick, colourless, transparent bells several centimetres in diameter.

Colonial hydrozoans may have members that assume several different forms in the same colony. The blue bottle, *Physalia* (Fig. 38.8b), is an example of a floating colony. The gas-filled float develops first and then buds off both feeding forms with long, fishing, nematocyst-laden tentacles, and elongate female forms

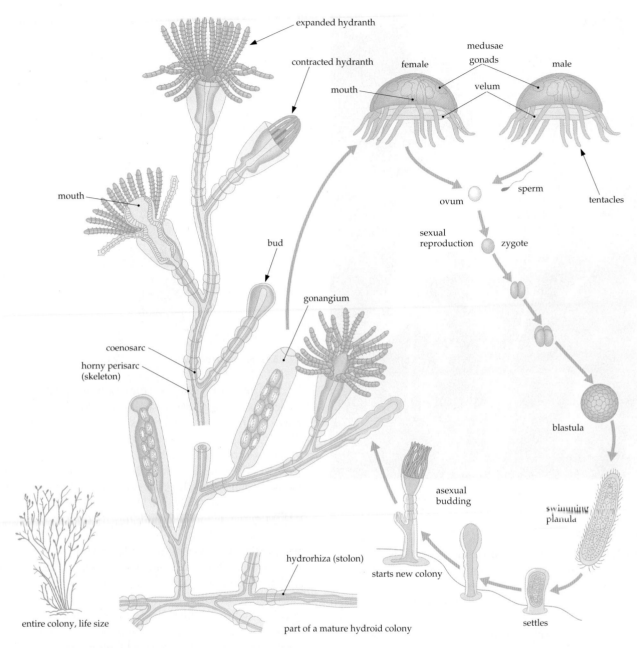

Fig. 38.13 Life cycle of the hydrozoan *Obelia*

(gonophores). Grape-like clusters of male gonophores develop at the base of the female gonophores.

Hydrozoans (hydras) can have both polyps and medusae in their life cycle but usually have a dominant polyp stage. They commonly form colonies with interconnecting coelenterons. The colonies may have polyps of several different types with different functions.

Sea anemones and corals: class Anthozoa

The class **Anthozoa** includes two-thirds of all living cnidarian species: soft corals and their relatives, such as sea fans and sea pens, together with hard corals and sea anemones (Fig. 38.15). Anthozoans consist exclusively of solitary or colonial, mostly sessile (benthic) polyps. There is no medusoid phase and the polyps are generally larger and fleshier than those of hydrozoans. The anthozoan polyp is usually a hollow cylinder topped by an **oral disc** fringed by hollow tentacles (Fig. 38.16b). Towards the centre of the oral disc the body wall reflects downwards to form a tubular pharynx that runs part of the way towards the pedal disc before opening into the gastrovascular cavity (coelenteron), which runs up into the tentacles. The cavity is partially divided by **septa** (mesenteries) disposed radially and symmetrically around the central pharynx. The septa

Fig. 38.14 (a) The hydroid *Pennaria australis*. The medusae are about to be released. **(b)** In *Tubularia*, medusae are permanently attached to gonophores, which release young polyps

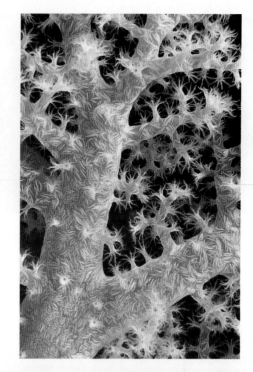

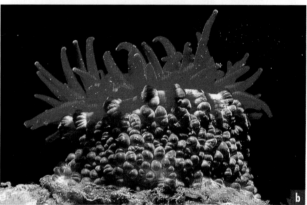

Fig. 38.15 Anthozoans: **(a)** soft coral; **(b)** anemone, *Phlyctenanthus australis*; and **(c)** hard corals, forming part of the Great Barrier Reef

have filaments on their central edges that aid in digestion and longitudinal muscle bands on their surfaces that retract the polyp.

oral disc   tentacle   polyp

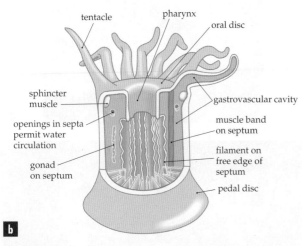

Fig. 38.16 (a) In hard corals, the polyps, here retracted, are embedded in a skeleton of calcium carbonate. **(b)** Structure of a sea anemone. Half the wall of the column has been cut away to show the internal features

The gonads, unlike those of hydrozoans, are endodermal and are located on the radial septa. Sexual reproduction usually involves external fertilisation and dispersal is brought about by pelagic (free-swimming) planula larvae. Some species, however, have internal fertilisation and may brood the young to the polyp stage. Many species also reproduce asexually by budding off new polyps.

Polyps such as anemones are 'naked' or may have a chitinous cuticle but the polyps of hard corals are embedded in a skeleton of calcium carbonate secreted by and external to the ectoderm. Soft corals have spicules in the mesoglea. During the day, coral polyps withdraw into spaces within the skeleton called **corallites** (Fig. 38.16a). Polyps feed by everting and extending their tentacles to catch zooplankton on the nematocysts. The tentacles pass the prey through the mouth opening and into the coelenteron, where digestion occurs. Food passes through connections between individual polyps to provide nutrition for the whole colony.

Coral colonies, such as those that form the Great Barrier Reef, occur in a great variety of colours, shapes

and sizes (Fig. 38.15). Many of the colours come from the pigments of symbiotic zooxanthellae in their tissues. Their shape depends on the way individual polyps build their skeletons and how they bud. In brain corals, polyps divide without forming complete walls so that long lines of polyps have a common wall around them. In staghorn corals, long branches arise from single polyps.

Anthozoans, the soft corals, hard corals and sea anemones, have only polyp forms, no medusae. They reproduce sexually and asexual budding is also common. Hard corals secrete a skeleton of calcium carbonate, forming reefs such as the Great Barrier Reef. They are able to do so because of symbiotic relationships with zooxanthellae.

Comb jellies: phylum Ctenophora

Phylum **Ctenophora** (comb jellies) included pelagic marine animals. There are about 100 species, some of which have transparent, sac-like bodies with a jelly-like mesoglea so that they resemble cnidarian medusae (Fig. 38.17). The fact that one species, *Euchlora rubra*, has its own nematocysts (some obtain them by eating

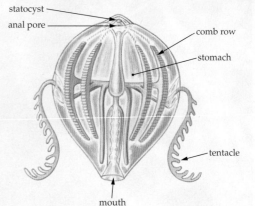

Fig. 38.17 (a) Fragile comb jellies or sea gooseberries move by beating rows of ciliated plates. **(b)** Structure of a globular ctenophore

cnidarians) suggests that ctenophores are related to cnidarians.

Ctenophores are distinguished by eight longitudinal rows of ciliary plates called **comb rows** that are made up of fused cilia. These beat in synchronised waves to propel the animal through the water, mouth forward. Light interference patterns on the comb rows render them iridescent against the background of the almost transparent body (Fig. 38.17). A line of ciliated cells connects each comb row to a complex **apical organ** at the aboral pole, the point on the upper surface that is diametrically opposite the mouth on the lower surface. The most conspicuous structure in the apical organ is a balance organ, the **statocyst**.

Ctenophores have an endodermal digestive system equivalent to the coelenteron of cnidarians. The mouth leads into an elongated tube with greatly folded walls, which secrete enzymes that rapidly break down food. Digested food enters the dilated stomach and is circulated by the action of cilia along narrower longitudinal branches. The tubes form a canal-like system leading to a long branch beneath each comb row. A notable difference from cnidarians is the presence of two posterior openings to the digestive system, the **anal pores**.

Ctenophores prey on other animals usually captured with a pair of branched, retractile, adhesive tentacles. Species with reduced tentacles take food directly into the mouth. Ctenophores are hermaphroditic and, even in species in which the adult body is ovoid, depressed or elongated, the larva is spherical.

> Ctenophores (comb jellies) resemble cnidarian medusae but, except for one species, lack nematocysts. They have eight ciliary comb rows that are used for locomotion.

Protostomes and deuterostomes: two modes of development

Early in animal development the embryonic cavity, the archenteron, develops into the digestive tract (Chapter 15). With the notable exception of sponges, cnidarians and nematodes (p. 1023), most animals can be assigned to one of two groups chiefly on the basis of whether the opening to the archenteron, the blastopore, develops into the mouth or the anus. Distinctions based on events that occur early in embryological development have proved to be useful indicators of the relationships between groups of animals because they are retained so conservatively during evolution. It is argued that major changes are unlikely to occur in the early developmental stages because they will mostly produce outcomes that

compromise the survival of later stages or adult animals. Studying development (ontogeny) is, however, only one form of information that can be brought to bear on questions of animal relationships and it cannot be used in isolation. Some groups, the lophophorate phyla for example (Chapter 39), have characteristics of both protostomes and deuterostomes. The distinction between protostomes and deuterostomes is retained because it is in general agreement with the molecular data, particularly ribosomal RNA sequence analysis (see Fig. 38.2).

Protostomes are animals in which the blastopore becomes the mouth. Cell division of the early embryo has a spiral form (*spiral cleavage*), which means that at the eight-cell stage the four upper cells (blastomeres) of the embryo are rotated 45° relative to the four lower blastomeres (Fig. 38.18). The fates of various parts of the embryo are determined or fixed at a very early stage in cleavage, or even in the egg, so that they show *determinate development*. The mesoderm, for example, derives from a single cell. If a coelom (body cavity) develops, it is schizocoelic, produced by splits in the mesoderm. Major protostome phyla are Platyhelminthes, Annelida, Mollusca and Arthropoda.

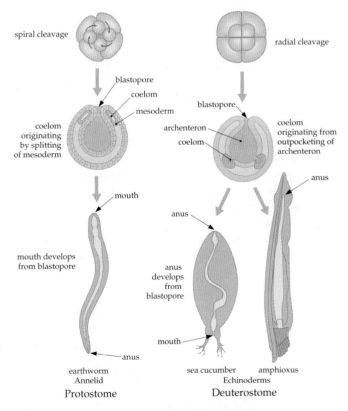

Fig. 38.18 Protostomes and deuterostomes differ in development. In protostomes, the blastopore becomes the mouth, and in deuterostomes, it becomes the site of the anus. If a coelom develops in a protostome, the development is schizocoelic, produced by splitting of the mesoderm. In deuterostomes, such as echinoderms and amphioxus, the coelom arises from outpockets of the embryonic gut (archenteron)

In **deuterostomes**, the anus forms at the site of the blastopore, the mouth forming at a secondary opening. Cell division of the early embryo has a radial form (*radial cleavage*), which means that the upper blastomeres of the embryo at the eight-cell stage are directly above the lower blastomeres (Fig. 38.18). The fate of individual cells is not fixed until a stage in cleavage much later than in protostomes, so that they show *indeterminate development*. The mesoderm and enclosed coelom arise by outpocketing from the embryonic precursor of the gut (except in vertebrates, which are schizocoelic). The two chief deuterostome phyla are Echinodermata and Chordata.

> Protostomes are animals in which the blastopore of the embryonic gut (archenteron) becomes the mouth in the adult. In deuterostomes, the anus forms at the site of the blastopore.

Acoelomate protostomes

Flatworms: phylum Platyhelminthes

Phylum **Platyhelminthes** include dorsoventrally flattened, bilaterally symmetrical animals ('flatworms') termed **acoelomate** because the mesoderm surrounding the endodermal gut contains no large body cavity (no perivisceral cavity). Absence of a coelom in flatworms is generally regarded as a primitive feature.

Flatworms live in marine, freshwater and moist terrestrial habitats. They are abundant as free living scavengers and carnivores but many are parasites. There are about 12 000 species classified in four major classes, including free-living turbellarians, and parasitic monogeneans (skin or gill flukes), trematodes (endoparasitic flukes) and cestodes (tapeworms).

Platyhelminths share with cnidarians a number of characteristics that are probably primitive: no coelom; no anus; no circulatory system; and no respiratory system. Flatworms appear more advanced than cnidarians in having a differentiated excretory system of **protonephridia**, the basic unit of which is the **flame cell** (Fig. 38.19). They also differ from cnidarians in possessing a third 'germ layer', the mesoderm, between ectoderm and endoderm. Mesoderm differs from the intermediate layer (mesoglea) of cnidarians as its cells appear early in embryonic development and are related to development of the endoderm. Animals with a true mesoderm show organ formation on a scale not found in cnidarians and ctenophores.

In primitive flatworms, the nervous system consists of a simple network or plexus with some anterior concentration of neurons. In the more advanced turbellarians it is a ladder-like system, with two or more longitudinal nerve cords and anterior cerebral ganglia. This process, called **cephalisation**, is not evident in cnidarians and is associated with increasing behavioural capability. The simplest eyes are merely pigment spots with associated nerve endings. In most free-living turbellarians and larval trematodes, the eyes are pigment cups enclosing photoreceptors.

Most flatworms are hermaphroditic, that is, they have both sexes in each individual (**monoecious** condition). Separate sexes do occur in some species (**dioecious** condition), for example, in the blood fluke, *Schistosoma*. Cross-fertilisation is usual even in hermaphroditic species but self-fertilisation can occur. Parthenogenesis (reproduction from unfertilised eggs) is also known.

Some turbellarians and tapeworms use hypodermic impregnation, in which the penis is injected through the body wall of the partner. In most cases, however, sperm are passed directly into receptacles in the female reproductive system. Most species release unfertilised eggs (**oviparity**) but development of fertilised eggs (**ovoviviparity**) or live young (**viviparity**) within the female also occur.

There are four classes of Platyhelminthes: turbellarians are the only free-living flatworms; trematodes

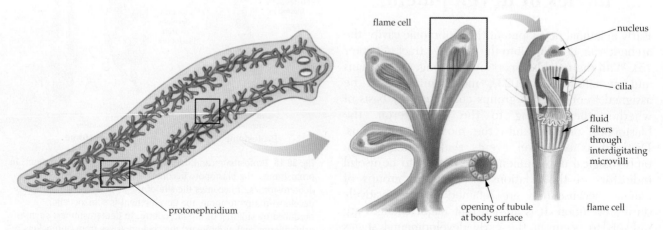

Fig. 38.19 Excretory system of a platyhelminth, showing the structure of a flame cell

and monogeneans are fluke-like parasites; and **cestodes** are parasitic tapeworms.

> Platyhelminths (flatworms) are acoelomate, dorsoventrally flattened, free-living or parasitic worms. Their digestive cavity, when present, has a single opening and lacks an anus. Their excretory system consists of protonephridia, the basic unit of which is the flame cell. There are four classes of Platyhelminthes: turbellarians are free-living flatworms, trematodes, monogeneans and cestodes are all parasites.

Free-living flatworms: class Turbellaria

There is some evidence that the class **Turbellaria** may not be a single evolutionary lineage but the term 'turbellarian' is still a convenient collective name for the free-living flatworms. A commonly used example of a turbellarian is *Dugesia*, a freshwater planarian (Fig. 38.20).

Turbellarians are so thin in the dorsoventral dimension that they have no need for respiratory organs. Even relatively large turbellarians, such as *Bipalium kewense*, a terrestrial form that reaches a length of many centimetres, is able to respire by simple diffusion across the body wall because of its flattened form and large surface area. Both larvae and adults have a ciliated epidermis which, on the ventral surface, is used for locomotion. The epidermis usually has **rhabdites**, rod-like structures secreted by gland cells. These are absent from parasitic classes. Although most turbellarians live in marine or freshwater habitats, some are terrestrial.

In some turbellarians, the alimentary canal is similar to the hydrozoan gastrovascular cavity in that it is sac-like with a simple in-tucking of the external surface forming a simple pharynx. Some turbellarians have a more complex tubular or bulbous pharynx (Fig. 38.20), which can be protruded from the mouth. *Craspedella* (Fig. 38.21) and *Temnocephala* have a bulbous pharynx. They live on the surface of Australian crayfish, feeding on small organisms in the surrounding water. In one group of turbellarians termed the Acoela, once thought to be the most primitive eumetazoans, the alimentary canal is a solid endodermal core (and individual cells phagocytise and digest food particles).

Platyhelminth reproductive systems are more complex than those of the radiate phyla because they develop male and female ducts and copulatory apparatus from the mesoderm. Although all turbellarians have a penis, the more primitive members have no genital ducts; their sperm are hypodermically injected and their fertilised eggs exit through the mouth or body wall.

Unlike the parasitic flatworms, most turbellarians have no larval stages, although a few exceptions have a free-swimming, ciliated larva.

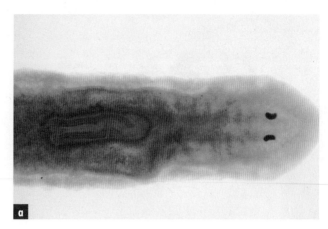

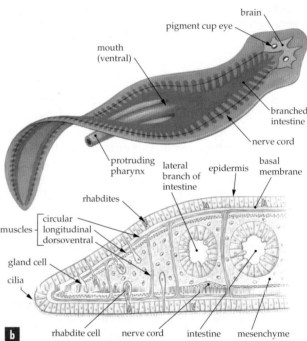

Fig. 38.20 (a) Dorsal view of the flatworm *Dugesia*. The animal is so thin in the dorsoventral dimension that it is possible to see many of the internal structures, including the gut and pharynx and excretory canals. **(b)** Dorsal view of a flatworm, showing digestive and nervous systems, with a cross-section to show the anatomy. Note the lack of a coelom (mesodermal body cavity)

> Turbellarians are free-living, dorsoventrally flattened flatworms. Most have a sac-like gastrovascular cavity and many have a pharynx for capturing prey. They have complex, ducted, mesodermal reproductive systems.

Ectoparasitic flukes: class Monogenea

Monogeneans are ectoparasites and rarely endoparasites, chiefly of aquatic vertebrates. They typically have a posterior holdfast device that attaches the adult parasite to the host. The holdfast bears anchors, suckers or pincers or is fringed with hooks. *Gyrodactylus* attaches to the skin and gills of fishes and can be a serious pest in fish hatcheries. *Polystoma*

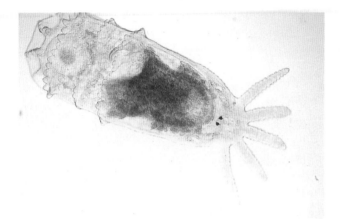

Fig. 38.21 The flatworms *Temnocephala* and *Craspedella spenceri* (shown here) live on the surface of Australian crayfish. They have anterior tentacles used in locomotion and in catching prey, a bulbous pharynx and a sac-like intestine without an anal opening

is unusual in occurring in the urinary (cloacal) bladder of frogs and toads. Reproduction in *Polystoma* is triggered by the reproductive hormones of the frog, with eggs of the parasite being shed in the urine as the frog enters water to breed. This ensures that the larvae of the parasite, which attach to the gills of tadpoles, find a new host. Although a larval stage is typical, monogeneans do not have the series of larval stages seen in trematodes.

Endoparasitic flukes: class Trematoda

Class **Trematoda** includes endoparasitic flukes, often having complicated life cycles involving a number of different hosts. One group of about 7000 species, the digeneans, becomes sexually mature in a vertebrate host, especially fishes. Several larval stages occur in a snail host. A few digeneans are precociously sexually mature in the snail and may dispense with the vertebrate host, for example, the Sydney species, *Parahemiurus bennettae* (Fig. 38.22) in the estuarine snail *Salinator fragilis*. Most flukes of vertebrates are gut parasites but some, for example, *Fasciola hepatica*, the liver fluke of sheep, live in the bile ducts. A few, including *Schistosoma mansoni*, the human blood fluke, occur in blood vessels, and others occur in lungs or other sites.

Adult flukes attach themselves to their vertebrate hosts by suckers. There is usually one oral sucker, around the mouth, and one ventral sucker. Their body has an outer covering, **tegument**, which is highly resistant to the host's enzymes but capable of absorbing some food materials. Their digestive system consists of a mouth, pharynx, very short oesophagus and two intestinal caeca (rarely one caecum), which are usually unbranched but may bear numerous lateral branches, as in *Fasciola hepatica*. Very rarely there may be one or two anal apertures, which evolved independently of the

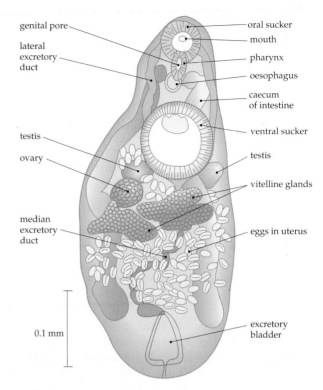

Fig. 38.22 The trematode fluke, *Parahemiurus bennettae*, has dispensed with a vertebrate host and matures in the snail host

anus found in other protostomes. Almost all are hermaphrodites, except for the family Schistosomatidae. In *Schistosoma*, the male has a ventral groove that holds the female.

Liver flukes

The life cycle of *Fasciola hepatica*, the liver fluke of sheep (Fig. 38.23), includes a free-swimming ciliated **miracidium**, which hatches from the fertilised egg. The miracidium penetrates an aquatic snail, losing its ciliated epidermis on entry, and transforms into a sac-like **sporocyst**. Sporocysts give rise to numerous **rediae**, which, with or without the production of secondary rediae, produce free-swimming **cercariae**. A cercaria encysts on vegetation and becomes a **metacercaria**. Grazing sheep become infected with liver flukes by ingesting metacercariae.

In the human liver fluke, *Clonorchis sinensis*, the adult of which lives in the bile ducts of the liver, humans become infected by ingesting metacercariae that have encysted under the scales of fishes.

The blood fluke, *Schistosoma*

Although it is less often lethal, the blood fluke *Schistosoma* rivals the malarial parasite in its impact on human health. Adults of *Schistosoma* species, common in South-East Asia and the tropics, live in blood vessels in the wall of the large intestine (Fig. 38.24) (or of the bladder). Eggs break through the wall of the intestine

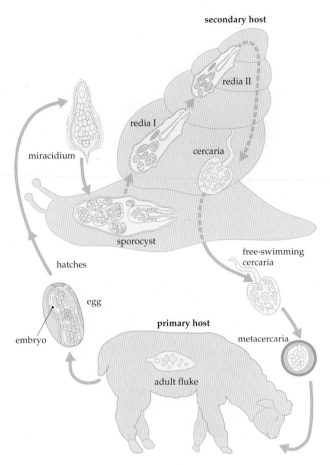

into the lumen and pass out with the host's faeces. If the faeces enter water, the eggs undergo further development. The miracidium hatches and penetrates a freshwater snail. Two generations of sporocysts occur in the snail before release of fork-tailed cercariae. The cercaria loses its tail during penetration of the human skin and becomes a juvenile fluke or **schisto-somulum**. Schistosomula travel via the heart and lungs to the liver. The female is unable to make the journey unaided down the hepatic portal vein to its tributaries on the wall of the intestine and is carried, against the portal blood flow, in the ventral groove of the male.

Schistosoma causes incalculable human suffering. The four major species parasitic in humans infect some 200 million people. Disease symptoms are caused by eggs traversing the wall of the intestine or, in the case of *S. haematobium*, by eggs penetrating the bladder. This causes loss of blood and anaemia. The most severe symptoms arise when eggs are swept along with the hepatic portal flow (Chapter 21) and become trapped there and subsequently in the vascular system in all parts of the body. Swelling of the spleen, liver and abdomen commonly occur and growth of the human host can be severely stunted.

Fig. 38.23 Life history of the liver fluke of sheep, *Fasciola hepatica* (an endoparasite)

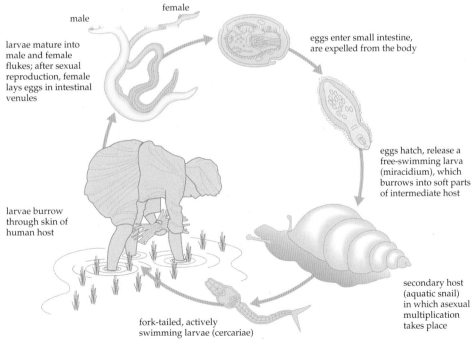

Fig. 38.24 Life cycle of the human blood fluke, *Schistosoma japonicum*, which occurs in South-East Asia. Larvae (cercariae) in water burrow into the skin of the human host. Larvae mature into adults and females lay eggs in the host's intestinal venules. Blood vessels rupture and eggs make their way into the gut, from where they are expelled from the host's anus. The eggs hatch in water and the ciliated, free-swimming larvae (miracidia) burrow into their secondary host, a snail. The parasite reproduces asexually (as sporocysts and cercariae) in the snail, and the life cycle is complete when cercariae escape into the water to reinfect and mature in the human host

Monogeneans are ectoparasitic flukes infecting one host, usually an aquatic vertebrate. Digeneans are endoparasitic flukes (trematodes) that have complicated life cycles with two or more hosts, often a snail and a vertebrate. Digeneans include important parasites of humans.

Tapeworms: class Cestoda

Trematodes, although showing important adaptations to parasitism, do not differ greatly from free-living flatworms, many of which have suckers. Tapeworms (class Cestoda), in contrast, are much more specialised. As adults, most are parasites of the vertebrate gut. There is usually a vertebrate or invertebrate intermediate host.

In general structure, tapeworms retain basic platyhelminth features but they show several striking adaptations. They have a special anterior holdfast, the **scolex**, with suckers and often hooks (Fig. 38.25a). Just behind the scolex, there is a growth zone that continuously buds off segment-like body units, **proglottids** (see Fig. 38.25b), in a process termed strobilisation. The external covering of the body, the tegument, is highly permeable to food materials, including glucose, amino acids, fatty acids and vitamins, in the gut of the vertebrate host and is the sole absorptive surface for food. The body is also covered by a cuticle, which is resistant to enzymes and digestion by the host. Tapeworms have lost the mouth and alimentary canal. Strobilisation produces many proglottids, each with at least one hermaphroditic set of reproductive organs. Mature proglottids can be laden with thousands of eggs.

The pork tapeworm

Taenia solium, the pork tapeworm, produces many proglottids and lacks asexual reproduction. The primary host, in which sexual reproduction occurs, is human and the intermediate host is the pig. The scolex (Fig. 38.25a) attaches to the wall of the human intestine by four suckers and an apical protrusion, the **rostellum**, armed at its base with 32 sickle-shaped hooks. Strobilisation produces a tape, **strobila**, 3 m long (Fig. 38.25c). Younger, more anterior proglottids cross-fertilise with older, more posterior ones. The uterus becomes filled with eggs and expands to occupy most of the proglottid, which is then said to be **gravid**. Gravid proglottids detach from the strobila and pass out in human faeces. If eggs or proglottids in faeces are ingested by pigs, the eggs may hatch in the duodenum and a six-hooked larva, the **hexacanth** or **onchosphere**, is released. This uses its hooks to penetrate the wall of the pig's gut and passes through the vascular system to the voluntary muscles. There it forms a **cysticercus**, a larva consisting of a bladder-like

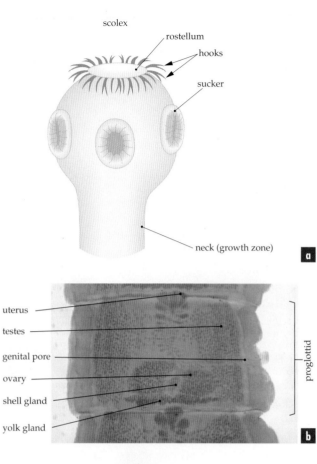

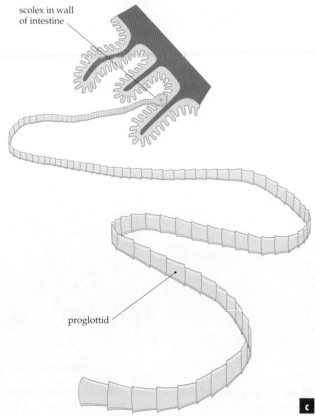

Fig. 38.25 *Taenia* species. **(a)** Scolex; **(b)** mature proglottid; **(c)** scolex and strobila attached to the intestinal wall

structure with an invaginated scolex. The pea-sized cysticerci in the muscle produce what is termed 'measly' pork. If a cysticercus is infected and under-cooked pork is eaten by a human, the scolex evaginates and attaches to the wall of the intestine, the bladder is cast off, and strobilisation occurs, producing the tape. If eggs of *T. solium* are ingested by humans, cysticerci can be produced in the muscles, including those of the heart. An infestation with large numbers of cysticerci may prove fatal to the infected person and thus to the parasite.

Taemiarhynchus saginatus, the beef tapeworm, has a similar life cycle, but the cysticerci occur in cattle. It differs in that eggs do not produce cysticerci in humans and it lacks a rostellum and hooks in the adult worm.

Echinococcus granulosus is unusual in tapeworms in undergoing larval asexual multiplication (see Box 38.3).

BOX 38.3 Hydatid cyst tapeworm

Taenia solium is very rare in Australia but the hydatid cyst tapeworm, *Echinococcus granulosus*, is alarmingly common. Here the primary host is the dog and the intermediate host is the sheep, a kangaroo or human. The scolex of *E. granulosus* closely resembles that of *T. solium* but it is the smallest tapeworm, with only four proglottids, only one of which is gravid at any one time.

E. granulosus compensates for its relatively poor powers of sexual reproduction by producing asexually in the tissues of the intermediate host—it produces a cyst with many invaginated scolices (protoscolices) instead of the single cysticercus of *Taenia*. The wall of the cyst buds off numerous brood capsules, each with protoscolices on its walls (see Fig. b). This hydatid cyst is usually about the size of a golf ball, with hundreds of thousands of protoscolices. One Australian woman was found to have a cyst with a volume of 8 L, which contained an estimated two million protoscolices.

(a) Adult of the hydatid cyst tapeworm, *Echinococcus granulosus*

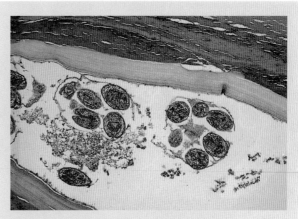

(b) A portion of a hydatid cyst of *Echinococcus granulosus* in the tissues of a sheep, showing a few brood capsules and their protoscolices

Cysts commonly cause pathological symptoms when they press on or grow in other organs and there are about 100 surgical removals a year in Australia. Surgery is difficult, however, not only because rupture of the cyst releases proteins that may cause fatal anaphylactic shock, but because there is also a danger that protoscolices may be released to establish new cysts elsewhere in the body.

Cestodes, tapeworms, are parasites of vertebrates. They have a number of specialised features, including an anterior holdfast, the scolex, and many body segments, proglottids, produced by strobilisation. They include species that infect and harm humans.

Proboscis worms: phylum Nemertinea

Nemertinea or Rhynchocoela contain about 900 species of elongate, sometimes flattened, often highly coloured ribbon or proboscis worms, so-named because of their unique eversible anterior proboscis, which is used in capturing prey. There are many marine representatives (Fig. 38.26a), one freshwater genus, and one terrestrial genus (*Geonemertes*), which occurs in Australian rainforests. A few species are parasitic in marine invertebrates, for example, crabs.

Nemertineans probably share a common ancestry with platyhelminths. Characteristics that they have in common include: protostome type of development; lack of a coelom (acoelomate); a flame cell excretory system (protonephridia); a ciliated epidermis; and a dorsoventrally flattened body with parenchyma between the body wall and the intestine. The presence in some of mucus-producing epidermal rodlets, rhabdites, also links nemertineans and flatworms and the nervous system is also like that of higher turbellarians. The brain has a dorsal and a lateral

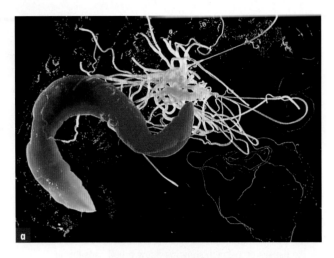

Fig. 38.26 **(a)** Adult of an Australian marine nemertean. **(b)** Diagram of a marine nemertine worm, *Gorgonorhynchus*, with proboscis contracted back into the rhynchocoel and **(c)** with the proboscis everted

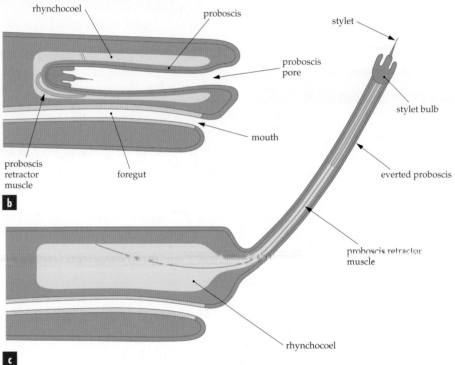

ganglion on each side interconnected to form a ring around the rhynchocoel. A pair of lateral ganglionated nerve cords with, sometimes, additional longitudinal cords gives an arrangement reminiscent of that in some flatworms.

In addition to the possession of the distinctive proboscis, nemertineans are clearly different from flatworms in a number of other ways. They have long, slender bodies (they are sometimes called ribbon worms), which in some forms have multiple layers of circular and longitudinal muscle layers along the whole length. They have a limited closed circulatory system and a one-way digestive system with mouth and anus. The proboscis (Fig. 38.26b, c) is not connected to the gut but is continuous with an ectodermal invagination and is housed in a body cavity, the **rhynchocoel**. This body cavity is located dorsal to the intestine and opens anteriorly to it, and is considered by some to be the homologue of a coelom, or at least a coelom-like cavity. The proboscis is often armed with a spike for penetrating prey, into which toxins may be injected.

Most nemertineans are dioecious but many freshwater and terrestrial species are hermaphroditic. Fertilisation is external or internal. In the more primitive members, there is a ciliated free-swimming larva, which resembles the larva of some turbellarians (p. 1017).

Nemertinea (proboscis worms) are long slender worms with a characteristic, eversible anterior proboscis used in defence and capturing prey. They resemble plathyhelminths in many ways but are clearly different in others.

Pseudocoelomates

Some eight animal phyla are artificially classified together as 'pseudocoelomates' because they possess a body cavity, a pseudocoel, which is not a true coelom (Fig. 38.27). A true coelom is lined on all sides by mesoderm. The inner wall of a pseudocoel is the endoderm of the gut and the outer wall consists of mesoderm, hence the term pseudocoel (false coelom).

Roundworms: phylum Nematoda

The pseudocoelomate group that impacts most obviously upon human activites is the phylum Nematoda. It includes 12 000 known species, many of which are parasites of plants and other animals. **Nematodes** (roundworms and eelworms) are found, often in huge numbers, in virtually every habitat, including soil and water. Almost every species of plant or animal that has been studied has been found to have at least one parasitic species of nematode living in it. Nematode parasites of agricultural crops cause millions of dollars of damage annually.

Nematodes have bilaterally symmetrical, elongate, cylindrical bodies tapering to a point at both ends. They are never segmented and are covered by a thick, semitransparent proteinaceous cuticle secreted by epidermal cells. Most species are similar in form although in parasitic species, the body may be worm-like, pear-shaped, lemon-shaped or sac-like. They lack circular muscle in their body wall but have very high internal pressure against which the longitudinal muscles contract. Because of this most move by vertical undulations, which result in thrashing movements unless they are in contact with a substrate.

Nematodes are herbivorous, carnivorous or saprophagous (sucking up liquid food). Many species are microscopic but *Ascaris suum* (Fig. 38.28), one of the largest roundworms, reaches a length of up to 30 cm. It is a parasite of the pig intestine.

In nematodes, the mouth is anterior and often surrounded by specialised lips. The anterior end of the body bears seta-like, papilla-like, probably tactile, sense organs and chemoreceptors. The mouth leads into the pharynx or stoma (often muscular and glandular), intestine and rectum that opens through a subterminal ventral anus. The excretory system opens through a single, anterior ventromedial pore.

The nervous system consists of a ganglionated ring surrounding the oesophagus. Ventral, dorsal and lateral nerves extend posteriorly from the brain and run within longitudinal cords of epidermis. Longitudinal epidermal cords divide the body wall musculature into two main blocks (Fig. 38.28) and the opposing contraction

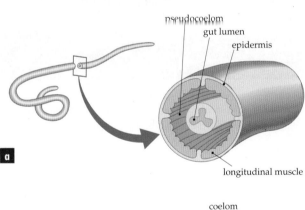

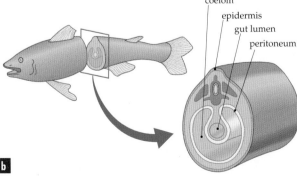

Fig. 38.27 Comparison of the structure of **(a)** a pseudocoelom in a nematode with **(b)** a true coelom in a vertebrate

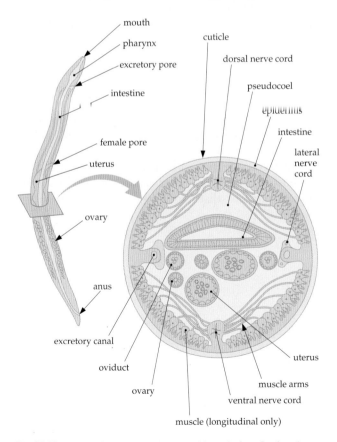

Fig. 38.28 Nematode structure. At up to 30 cm in length, the pig roundworm *Ascaris suum* is one of the largest nematodes. The diagram shows the position of structures in an adult female and a transverse section through the mid region of the body

of these blocks (upper half and lower half) against the high internal pressure accounts for the characteristic locomotory undulations of nematodes. The musculature, which is mesodermal, forms the outer lining of the spacious pseudocoel, the inner wall of which, unlike a coelom, is formed by the endoderm of the gut. Cilia are absent from nematodes, although structures resembling the bases of cilia occur in some sense organs. The excretory system consists either of small gland cells or a single, large H-shaped cell, the arms of which run, as lateral excretory canals, in the longitudinal epidermal cords. The sexes are usually separate.

Nematodes are unusual because they have a fixed number of mitotic divisions during their development and these take place before they hatch. Mitotic divisions do not occur after hatching so they have a set number of cells (**eutelic** condition). Growth occurs by increases in cell size not by increases in cell number. This pattern of growth makes nematodes useful experimental animals for developmental and genetic studies and many important insights in these areas come from studies on an organism known as the vinegar eel, the nematode *Caenorhabditis elegans*, which has around 1000 cells. Vertebrate nematode parasites include the human hookworms, *Ancylostoma duodenale* and *Necator americanus*, which infect more than 50 million people in Asia alone. Hookworms are armed with teeth for gripping the wall of the small intestine, from which blood is sucked continuously. *Enterobius vermicularis*, the pinworm or seatworm, is very common in children. *Ascaris lumbricoides* is a roundworm that also infects humans. This produces enzyme inhibitors that protect the worm from the host's digestive system. Related to *Ascaris* is *Toxocara canis*, a roundworm of the gut of puppies, which can cause serious larval infection of human tissues, especially the nervous system. *Wuchereria bancrofti*, a widespread tropical parasite that blocks the lymphatic drainage system of humans, causes **elephantiasis** or filariasis, grotesque swellings of the body including limbs, genital organs and mammary glands (Fig. 38.29). Another tropical nematode is *Dracunculus medinensis*, a subdermal parasite of humans. *Dracunculus* is credited with being the biblical 'fiery serpent', a testimony to the pain it causes when it comes to the surface of the skin to discharge its larvae into water.

> Nematodes (roundworms and eelworms) are pseudocoelomate animals, many of which are parasites of plants and animals. They are round in cross-section, covered with a cuticle, are never segmented, have high internal pressure and no circular muscle. Nematode parasites cause serious human health problems.

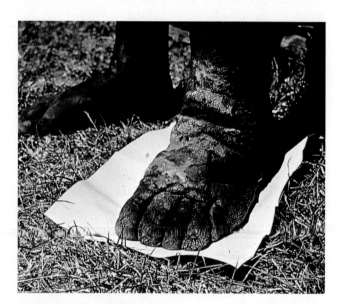

Fig. 38.29 Elephantiasis is caused by a roundworm, *Wuchereria bancrofti*, which blocks lymphatic tissue, causing oedema and often grotesque swelling in the affected regions

Summary

- Animals are multicellular, heterotrophic organisms. They undergo embryonic development from a zygote and have cells and, with the exception of sponges, tissues differentiated for different functions.
- There is very little evidence concerning the origin of multicellularity. There is fossil evidence indicating when simple animals appeared. Fossil evidence together with molecular and biochemical data provide insights into the relationship between animal groups.
- Sponges (phylum Porifera) are simple aquatic filter feeders, with three layers of cells but without tissues, organs, mouth or nervous system.
- Cnidarians (polyps and medusae) are characterised by stinging cells, nematocysts (cnidae). They have a body wall consisting of two cellular layers and an intermediate layer, also containing cells, the mesoglea. They have a gastrovascular cavity (coelenteron).
- Scyphozoans (jellyfish) are marine animals with the conspicuous life cycle stage being the jellyfish or medusa. Cubozoans (box jellyfish), are also marine, with a cube-shaped bell and tentacles in groups of four. Their stings are extremely painful and often fatal to humans.
- Hydrozoans (hydras) can have both polyps and medusae in their life cycle but usually have a dominant polyp stage. They commonly form colonies with interconnecting coelenterons. The colonies may have polyps of several different types with different functions.
- Anthozoans, the soft corals, hard corals and sea anemones, have only polyp forms, no medusae. They reproduce sexually and asexual budding is also common. Hard corals secrete a skeleton of calcium carbonate, forming reefs such as the Great Barrier Reef. They are able to do so because of symbiotic relationships with algae.
- Ctenophores (comb jellies) resemble cnidarian medusae but, except for one species, lack nematocysts. They have eight ciliary comb rows that are used for locomotion.
- Protostomes are animals in which the blastopore of the embryonic gut (archenteron) becomes the mouth in the adult. In deuterostomes, the anus forms at the site of the blastopore.
- Platyhelminths (flatworms) are acoelomate, dorsoventrally flattened, free-living or parasitic worms. Their digestive cavity, when present, has a single opening and lacks an anus. Their excretory system consists of protonephridia, the basic unit of which is the flame cell. There are four classes of Platyhelminthes: turbellarians are free-living flatworms, trematodes, monogeneans and cestodes are all parasites.
- Turbellarians are free-living, dorsoventrally flattened flatworms. Most have a sac-like gastrovascular cavity and many have a pharynx for capturing prey. They have complex, ducted, mesodermal reproductive systems.
- Monogeneans are ectoparasitic flukes infecting one host, usually an aquatic vertebrate. Digeneans are endoparasitic flukes (trematodes) that have complicated life cycles with two or more hosts, often a snail and a vertebrate. Digeneans include important parasites of humans.
- Cestodes, tapeworms, are parasites of vertebrates. They have a number of specialised features, including an anterior holdfast, the scolex, and many body segments, proglottids, produced by strobilisation. They include species that infect and harm humans.
- Nemertineans (proboscis worms) are long slender worms with a characteristic, eversible anterior proboscis used in defence and capturing prey. They resemble plathyhelminths in many ways but are clearly different in others.
- Nematodes (roundworms and eelworms) are pseudocoelomate animals, many of which are parasites of plants and animals. They are round in cross-section, covered with a cuticle, are never segmented, have high internal pressure and no circular muscle. A number of nematode parasites cause serious human health problems.

keyterms

acoelomate (p. 1016)
actinula (p. 1010)
amoebocyte (p. 1006)
anal pore (p. 1015)
Anthozoa (p. 1012)
apical organ
    (p. 1015)
atrium (p. 1006)
bilateral symmetry
    (p. 1007)
cephalisation
    (p. 1016)
cercaria (p. 1018)
cestode (p. 1017)
choanocyte (p. 1006)
cnidae (p. 1008)
Cnidaria (p. 1007)
cnidocyte (p. 1008)
coelenteron (p. 1007)
comb row (p. 1015)
corallites (p. 1014)
Ctenophora (p. 1014)
Cubozoa (p. 1011)
cysticercus (p. 1020)
deuterostome
    (p. 1016)
dioecious (p. 1016)

diploblastic (p. 1008)
elephantiasis
    (p. 1024)
ephyrae (p. 1010)
Eumetazoa (p. 1005)
eutelic (p. 1024)
flame cell (p. 1016)
gastrovascular cavity
    (p. 1007)
gonangium (p. 1011)
gravid (p. 1020)
hexacanth (p. 1020)
hydranth (p. 1011)
Hydrozoa (p. 1011)
hypostome (p. 1008)
manubrium (p. 1008)
medusa (p. 1008)
mesoglea (p. 1008)
mesohyl (p. 1006)
metacercaria
    (p. 1018)
miracidium (p. 1018)
monoecious (p. 1016)
monogenean
    (p. 1017)
nematocyst (p. 1008)
nematode (p. 1023)

nemertinean (p. 1021)
onchosphere
    (p. 1020)
oral arm (p. 1010)
oral disc (p. 1012)
oscula (p. 1007)
ostia (p. 1006)
oviparity (p. 1016)
ovoviviparity
    (p. 1016)
pinacocyte (p. 1006)
pinacoderm (p. 1006)
planula larva
    (p. 1010)
Platyhelminthes
    (p. 1016)
polyp (p. 1007)
Porifera (p. 1005)
proglottid (p. 1020)
protonephridia
    (p. 1016)
protostome (p. 1015)
radial canal (p. 1010)
radial symmetry
    (p. 1007)
rediae (p. 1018)
rhabdite (p. 1017)

rhopalia (p. 1010)
rhynchocoel (p. 1022)
ring canal (p. 1010)
rostellum (p. 1020)
schistosomulum
    (p. 1019)
scolex (p. 1020)
scyphistoma (p. 1010)
scyphozoan (p. 1009)
septa (p. 1012)
spicules (p. 1006)
sponge (p. 1005)
spongin (p. 1006)
spongocoel (p. 1006)
sporocyst (p. 1018)
statocyst (p. 1015)
statolith (p. 1010)
strobila (p. 1020)
strobilisation (p. 1010)
tegument (p. 1018)
Trematoda (p. 1018)
Turbellaria (p. 1017)
viviparity (p. 1016)
zoochlorellae
    (p. 1008)

Review questions

1. List the main features that characterise members of the animal kingdom compared with those from the other eukaryotic kingdoms. What are the advantages of multicellularity for animals?

2. What phylum is characterised by nematocysts? Describe the structure and function of these cells. Name one Australian marine animal that is dangerous to humans because of its nematocysts.

3. What is meant by a symbiotic relationship? How are corals and zooxanthellae symbiotic?

4. Explain the difference between the terms acoelomate, pseudocoelomate and coelomate. Give an example of an animal for each term.

5. What do the terms 'protostome' and 'deuterostome' refer to? Of the following animals, which are protostomes and which are deuterostomes: starfish, nemertine, flatworm, human?

6. Explain the structure and function of a flatworm flame cell.

7. Phylum Platyhelminthes includes a variety of parasitic animals. Which classes in this phylum are parasites? Describe the life cycle of one example that infects humans.

8. In what way are nemertean worms similar to turbellarians? How do they differ?

9. List the characteristics that would enable you to identify an animal as a nematode. In what way do they cause problems for humans?

Extension questions

1. What kind of evidence is there concerning the evolution and relationship of simple animals?
2. In what way are the life functions of a sponge related to water flow through the sponge?
3. Explain what is meant by the symmetry of an animal. Compare radial and bilateral symmetry. Which animals have radial symmetry and which have bilateral symmetry? What is the relationship between body symmetry and lifestyle?
4. A marine biologist surveys the animals on a remote part of the Australian coastline that has not been sampled previously. What features should she look for to identify the following invertebrate phyla: Porifera, Cnidaria, Ctenophora, Platyhelminthes, Nemertinea and Nematoda?

Suggested further reading

Edgar, G. J. (1997). *Australian Marine Life—The Plants and Animals of Temperate Waters.* Melbourne: Reed Books.

An illustrated catalogue of Australian marine animals, particularly from the southern regions. Provides natural history information.

Gould, S. J. (1989). *The Burgess Shale and the Nature of History.* New York: Norton.

Popular book on the fossil history and evolution of animals.

Pechenik, J. A. (2000). *Biology of the Invertebrates.* New York: McGraw Hill.

A very readable text providing more detail on the material covered here and classification and also a number of examples of recent research involving invertebrates in boxes embedded in the text.

Ruppert, E. E. and Barnes, R. D. (1994). *Invertebrate Zoology.* 6th edn. Fort Worth, TX. Saunders College Publishing.

A detailed account of invertebrate zoology with extensive figures and classification.

CHAPTER

39

Annelids, arthropods and molluscs

Annelids (segmented worms), arthropods (spiders, crustaceans, insects and their relatives), and molluscs (snails, clams, octopuses and their relatives) are coelomate protostomes. Each group is highly abundant in total numbers and rich in species diversity. They are structurally more complex than the animals described in Chapter 38.

The phylogeny in Figure 38.2, based on both molecular and morphological data, shows that the three phyla, Annelida, Arthropoda and Mollusca are related to one another and form a monophyletic group (Chapter 30). Annelids and arthropods share, for example, many structural similarities in early development. A glance at an earthworm (Fig. 39.1a) reveals a body made of a series of 'rings' or segments. Typically, each segment contains key structures, such as nephridia (for excretion), ganglia (clusters of nerve cells), muscles and blood vessels. Such a body plan, in which there is linear repetition of functional units, is called **metameric segmentation**. It is also evident in arthropods where it is easily seen in millipedes (Fig. 39.1b). The segmentation of annelids and arthropods is not the same as the repetition of units (strobilisation) seen in tapeworm proglottids. Unlike in tapeworms, where new proglottids are added continuously in adult animals, in annelids and arthropods, new segments are added at the posterior end of the series during development until the fixed number of segments characteristic of the adult is reached.

Repetition of functional units may have allowed primitive segmented animals to become larger because it was associated with development of a vascular system. More complex and diverse body structures evolved as particular segments became specialised for certain tasks (e.g. head end with concentration of sensory structures). Differentiation of segments is most evident in arthropods. The only sign of segmentation in modern molluscs is in **Neopilina**, a limpet-like animal

that was dredged from deep water in the Pacific Ocean off the coast of Costa Rica in 1952. It is the only living example of the Monoplacophora, a class that includes early fossil molluscs. *Neopilina* has a series of repeating structures, including five pairs of gills, six pairs of nephridia and 10 sets of lateral nerve connections, which may indicate segmentation, a feature that would link molluscs to annelids and arthropods.

> The body of annelids and arthropods shows metameric segmentation, in which there is a linear repetition of functional units. Typically, each segment contains key structures such as nephridia (for excretion), ganglia (clusters of nerve cells), muscles and blood vessels. The only suggestion of segmentation in molluscs is in the class Monoplacophora, which includes early fossil molluscs and the extant *Neopilina*.

Segmented worms: phylum Annelida

Annelids are soft-bodied segmented worms of soils and marine and freshwater habitats. They include marine bristle worms, earthworms, leeches and 'jawed' worms that live externally on the bodies of crayfish.

Annelids are protostomes characterised by:

- segmentation
- a spacious, schizocoelic coelom (a coelom that develops by splitting of the mesoderm)
- a closed vascular system
- a through-gut with an anus
- a solid ventral ganglionated nerve cord
- epidermal setae borne in four bundles per segment
- no exoskeleton.

The annelid body has three parts (Fig. 39.2). At the anterior end is the presegmental **prostomium**, which

Fig. 39.1 The external appearance of two metamerically segmented animals. **(a)** An annelid, the megascolecid earthworm; **(b)** an arthropod, a millipede

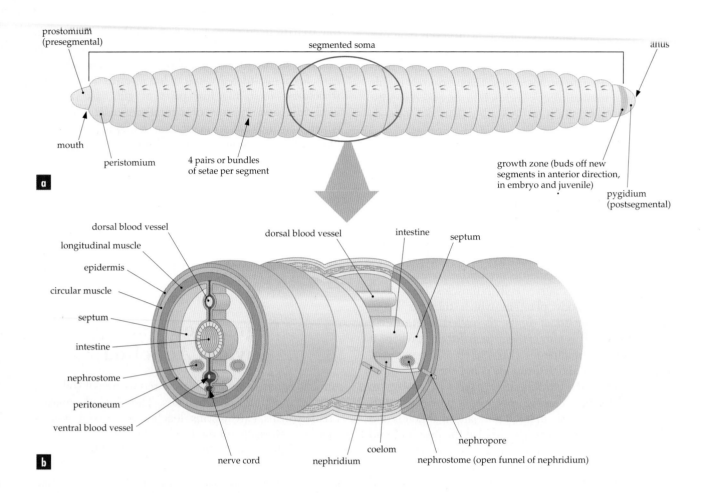

prostomium
(presegmental)

segmented soma

anus

mouth

peristomium

4 pairs or bundles
of setae per segment

growth zone (buds off new
segments in anterior direction,
in embryo and juvenile)

pygidium
(postsegmental)

a

dorsal blood vessel

longitudinal muscle

epidermis

circular muscle

septum

intestine

nephrostome

peritoneum

ventral blood vessel

dorsal blood vessel

intestine

septum

nerve cord

nephridium

coelom

nephropore

nephrostome (open funnel of nephridium)

b

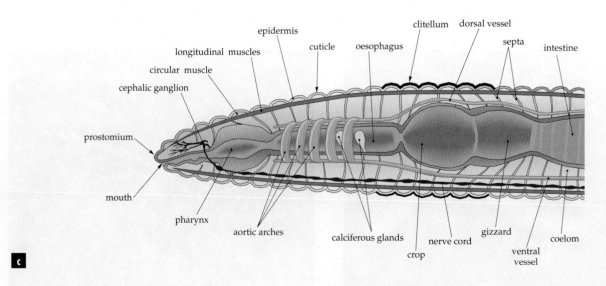

epidermis

clitellum

dorsal vessel

longitudinal muscles

cuticle

oesophagus

septa

intestine

circular muscle

cephalic ganglion

prostomium

mouth

pharynx

aortic arches

calciferous glands

crop

nerve cord

gizzard

ventral
vessel

coelom

c

Fig. 39.2 Structure of an earthworm of the Australian family Megascolecidae. **(a)** General body morphology. **(b)** Diagramatic representation of the three-dimensional structure of individual segments. **(c)** Diagram showing a dorsal view inside the body to reveal the position of major structures

houses the brain. Behind this is the segmented **soma**. At the posterior end is the postsegmental **pygidium**. Immediately anterior to the pygidium there is a **growth zone**, where embryonic growth occurs by addition of new segments at the anterior edge. The first apparent segment is the **peristomium**, surrounding the mouth.

Functional implications of segmentation and a coelom

Locomotion

The longitudinal division of a worm into a series of segments, each with muscles and nerves that control it, and the hydrostatic pressure provided by the coelom, are important in the mechanics of locomotion. It is probable that a coelom evolved in the first annelids as an adaptation for burrowing. Burrowing annelids move through the substrate by successive waves of body contractions that pass along the length of the animal. This process, called **peristaltic locomotion**, depends on the presence of a deformable, fluid-filled body cavity, in this case, the coelom. The development of **transverse septa** (Fig. 39.2), typical of the annelid coelom, is also important because it allows different hydrostatic pressures in various parts of the body. The pressure differences are responsible for the local changes in shape along the worm's body. Local control of body diameter allows fast, effective crawling and burrowing for both escape and foraging.

Respiration and circulation

Because segmental septa divide the coelom, septation may have been accompanied by evolution of a vascular system for circulating body fluid between the compartments as an aid to respiration and nutrition. A vascular system augments simple diffusion of oxygen and carbon dioxide across the body surface and transports food materials, hormones and excretory substances around the body (Chapter 22). Consequently, there is potential for greater size and complexity and increased levels of activity. It has been estimated that, without a circulatory system, the diameter of the annelid body could not exceed 1.5 mm. Some Australian giant earthworms reach 25 mm in diameter.

The respiratory pigment is commonly erythrocruorin (haemoglobin). It is not contained in blood cells and has a molecular weight of approximately 4 million daltons. Alternative pigments, such as chlorocruorin (also iron-containing but green), may occur. In the annelid circulatory system (Fig. 39.2), the general pattern of circulation is anteriorly in the dorsal vessel and posteriorly in the ventral vessel. Oxygenation of blood occurs at or near the epidermis and specialised gas exchange organs have developed in some groups.

Excretion

Excretory products are released into the coelom. **Nephridia**, tubes specialised for transporting the waste products from the coelom to the exterior, are repeated in successive segments (Fig. 39.2). **Protonephridia** end as blind tubes with terminal flagellated cells and occur in the embryos of all classes and in some adult polychaetes. **Metanephridia**, which are open tubes with funnels, occur in many polychaetes, all oligochaetes and leeches.

Digestion

The coelomic space separates the internal organs from the body wall so that gut and body wall movements can be relatively independent. Annelids have a throughgut, one-way flow of food, and specialisation of consecutive regions for successive stages in food digestion.

> Annelids are coelomate protostomes. They are metamerically segmented worms with a digestive system, closed circulatory system, solid ventral ganglionated nervous system and nephridia for excretion. They move by peristaltic locomotion.

Marine bristle worms: class Polychaeta

Most polychaetes are marine but a few are freshwater or terrestrial. They can be divided loosely into either free-moving (errant) or sedentary forms. Many sedentary species construct and live in burrows or tubes (Fig. 39.3). Polychaetes usually have lateral paired appendages, **parapodia**, on each body segment. A parapodium is a lateral fleshy lobe of the body that functions for both respiration and locomotion (Fig. 39.4). The name 'Polychaeta' means many setae

Fig. 39.3 Calcareous tubes of the polychaete, *Galeolaria caespitosa*, mark the mid-tidal level on rocky shores in Southern Australia. The tubes can be closed by a calcareous plug, the operculum, which is borne on one of the prostomial tentacles

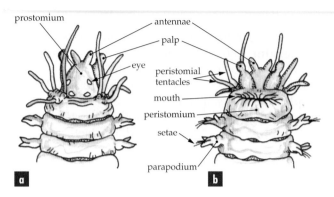

Fig. 39.4 Diagram of the prostomium and anterior segments of the polychaete *Perinereis*: **(a)** dorsal view; **(b)** ventral view

and refers to the chitinous bristles, **setae**, associated with the parapodia. The setae can be elaborate in form and bundles of them tend to spread out in a variety of fan-like patterns. In some tropical polychaetes that inhabit coral reefs, the setae are poisonous and are used for defence. Polychaetes often have sensory structures such as dorsal antennae, eyes and ventral palps on the prostomium (Fig. 39.4).

The structure of the anterior end of the alimentary canal, the pharynx, is modified for different modes of feeding in different polychaetes (Box 39.1). In some, the pharynx can be everted through the mouth to catch prey. In the bloodworm *Marphysa sanguinea*, the pharynx bears cuticular 'jaws', the dorsal maxillae and ventral mandibles (Fig. 39.5), which seize food.

Many polychaetes feed actively by means of a well-developed cylindrical pharynx, which may or may not be armed with cuticular jaws (in some species poisonous). In carnivorous species, the pharynx can be everted rapidly to catch prey. Herbivorous or scavenging polychaetes with jaws use them to tear off pieces of food, such as algae. In *Sabellastarte indica* (Fig. b), a filter feeder, bipinnate prostomial tentacles form a crown, which draws in organic particles to the base for sorting.

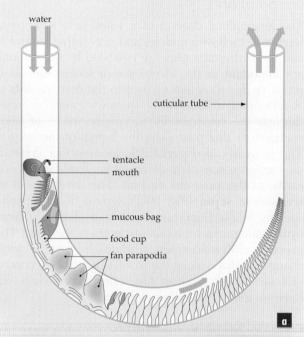

BOX 39.1 Feeding in polychaetes

Polychaete worms show a variety of feeding methods. *Chaetopterus*, for example, is a marine polychaete that lives in a parchment-like tube (Fig. a). Each tube is composed of cuticle elevated above the epidermis of the animal. Except for its ends, the tube is buried in mud. Water is drawn into the tube by the beating of fan-like parapodia of three segments. A mucus net filters out particles from the water and forms a bag, which is gathered by a ciliated cup. Periodically, a food bolus is detached from the bag and placed on a dorsal groove, which takes the bolus forward to the mouth for ingestion.

Other polychaetes feed by means of ciliated, pinnate tentacles that arise from the prostomium and filter particles from water. An example is the sedentary polychaete *Galeolaria caespitosa*, the white calcareous tubes of which form a conspicuous mid-tidal band on rocks and pier pylons in Southern Australia (Fig. 39.3).

(a) Feeding in the polychaete *Chaetopterus*. **(b)** *Sabellastarte indica*, a polychaete of shores from Queensland to Western Australia, feeds by filtering organic particles with its crown of prostomial tentacles, at the base of which they are sorted

Gas exchange occurs across the body surface of polychaetes but many also have gills associated with parapodia or other parts of the body. Polychaetes are mostly dioecious but some, particularly minute worms living buried in sand, are hermaphroditic. They usually

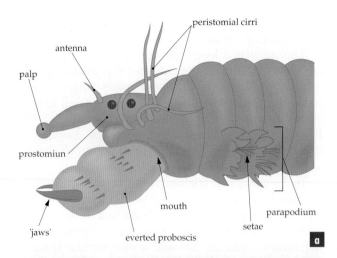

Fig. 39.5 **(a)** Head of a feeding nereid polychaete showing the pharynx everted through the mouth. **(b)** *Marphysa sanguinea*, the common 'bloodworm', favoured by anglers as bait in Moreton Bay, southern Queensland

have a long series of gonads, which are simply swellings containing masses of developing gametes that are shed into the coelom. Gametes escape into the surrounding sea water through special ducts, nephridia, or sometimes through rupture of the body wall. Some polychaetes copulate, either with hypodermic impregnation or insemination by a penis. Many polychaetes have a free-swimming ciliated larva, the **trochophore** (Fig. 39.6), which may be reduced or absent in species with large-yolked eggs.

> Polychaetes are marine worms characterised by numerous bristles, setae, and parapodia, paired appendages that function in locomotion and respiration.

Annelids with a clitellum: class Euclitellata

Earthworms, leeches and their near relatives are annelids with a **clitellum**, a thickening of the epidermis that secretes the **cocoon** in which eggs are

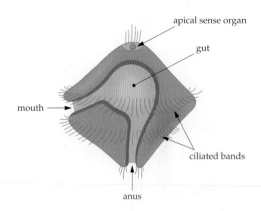

Fig. 39.6 Many polychaetes have a trochophore larva, typical of protostomes

deposited. The clitellum is located around and/or behind the female genital pores. Unlike polychaetes, euclitellates lack parapodia. Setae, although they may be numerous, are never jointed and are absent from advanced leeches. There is no trochophore or other larval stage.

> Euclitellate worms lack parapodia. They are characterised by a clitellum, which secretes a cocoon in which eggs are deposited. Euclitellate worms include oligochaetes, which have few setae on each segment, and leeches, which have none.

Earthworms and relatives: subclass Oligochaeta

The name 'oligochaetes' means 'few setae', and there are usually only four pairs per segment, although some oligochaetes have many more. Oligochaetes have a reduced prostomium and most are burrowing animals. They feed on vegetation or on organic matter in soil, and a few are carnivorous. Although the best known oligochaetes are earthworms and related freshwater worms, increasing numbers of taxa are being found in marine habitats. It has been estimated that there are 20 000 worms per cubic metre in the sublittoral sand of Heron Island on the Great Barrier Reef. There may prove to be more oligochaete species in marine than in other habitats. There are some 25 families of oligochaetes, of which six are native to Australia. Five of these have species restricted to this continent but no family is endemic. The largest Australian oligochaete is *Megascolides australis* from Gippsland, which may exceed 3 m when extended; *Digaster longmani*, from Queensland and New South Wales, is also gigantic (Fig. 39.7).

In contrast to most polychaetes, oligochaetes have gonads in only a few segments, usually from two to four pairs. They are hermaphroditic and, in earthworms, there are usually two pairs of testes, in segments 10 and 11, followed by one pair of ovaries, in segment 13. Reproduction (Fig. 39.8) involves reception of sperm

Fig. 39.7 A giant Australian earthworm, *Digaster longmani*

from a partner, almost always into small sacs, **spermathecae**. Sperm are then ejected from spermathecae into a cocoon secreted on the outside of the body by the clitellum. The cocoon receives eggs from the female genital pore(s) and fertilised eggs develop into juvenile worms in the cocoon after it has been shed from the anterior end of the body. The African night-crawler (family Eudrilidae) has internal fertilisation.

Asexual reproduction is common in aquatic oligochaetes; the parent worm simply divides into two or more worms.

> Oligochaetes (earthworms) are euclitellates typically with few setae on each segment. They are hermaphroditic, with gonads in only a few segments.

Leeches: subclass Hirudinea

Hirudinea are related to oligochaetes and all have a clitellum. Leeches generally have a more specialised diet than polychaetes and oligochaetes. Although a few are carnivores, many are external parasites feeding on host blood. Terrestrial leeches that live in moist Australian forests attach to leaves to await passing prey.

A few species are endoparasitic in the respiratory passages of mammals.

The medicinal leech, *Hirudo*, used from antiquity for 'bleeding' human patients and still used in plastic surgery, is ectoparasitic on amphibians when juvenile and on mammals when mature. Leeches attach to their hosts by posterior and anterior suckers (Fig. 39.9). They use their suckers for locomotion by 'looping', in contrast to the lateral undulations of most polychaetes or the peristaltic creeping of oligochaetes.

In leeches, the coelom is reduced and forms a system of tubes, which in the most advanced leeches replaces the vascular system. Reduction of the coelom is correlated with loss of peristaltic creeping. Correspondingly, septa are lost and a submuscular tissue is developed. Internal segmentation is evident only in the nervous and excretory systems. External segmentation is obscured by the development of secondary rings. The number of segments is fixed at 34 (six in the head region and anterior sucker, 21 body segments, and seven in the posterior sucker). Most leeches have lost setae, probably associated with locomotion by looping.

In leeches, single-celled photoreceptors are clustered as ocelli. Some leeches find a host, for example, a fish, by responding to moving shadows, while those that parasitise endothermic animals are attracted by the host's body heat.

Many leeches have jaws for penetrating the host (Fig. 39.9b). Those that suck blood or tissues from their prey have a crop and intestinal caeca for long-term storage of ingested material. Many blood sucking leeches, such as the medicinal leach *Hirudo*, secrete an anticoagulant protein, **hirudin**, which is a specific inhibitor of the clotting factor thrombin. Hirudin prevents clotting of blood at the wound or, in *Hirudo*, of blood stored in the crop. *Hirudo* also secretes a histamine, which causes dilatation of blood vessels of the host and augments bleeding. There are no proteolytic enzymes in the crop of *Hirudo* and digestion of blood depends upon symbiotic bacteria (*Pseudomonas*), the presence of which appears to prevent putrefaction of the blood.

Leeches differ from almost all oligochaetes in having internal fertilisation. Sperm may be transferred directly by a penis or by hypodermic impregnation. In hypodermic impregnation, a bundle of sperm (a spermatophore) is expelled from one worm and penetrates the body wall of another (Chapter 15).

> Leeches (Hirudinea) have an anterior and posterior sucker, generally lack setae, have reduced segmentation, a reduced but specialised coelom and have lost peristaltic locomotion. Most leeches are specialised as external parasites.

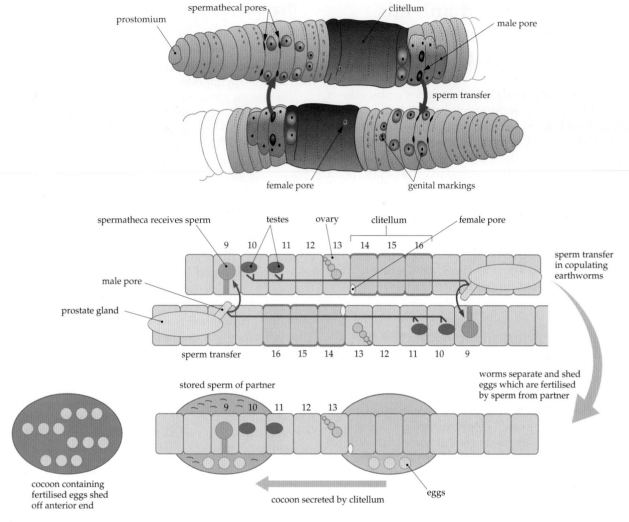

Fig. 39.8 Copulation in earthworms, here exemplified by the Australian genus *Spenceriella*. The worms lie with their anterior regions in contact, facing in opposite directions. The male pores deposit sperm in spermathecae of the partner. The partners separate. Eggs pass out of the female genital pore into a cocoon, which is secreted by the clitellum. The worm wriggles backwards out of the cocoon, which effectively moves forwards and receives stored sperm from the spermathecae. Eggs are fertilised and the cocoon is shed off the anterior end of the body

Fig. 39.9 (a) A leech attached to a human host by its suckers. **(b)** Diagram showing structure of the jaws of a leech

Joint-limbed animals with an exoskeleton: phylum Arthropoda

The phylum Arthropoda, which includes crabs, insects and spiders, is the most species-rich phylum in the animal kingdom. More than one million species have been described. They occur in the sea, in fresh water and on land, and the largest group, insects, have conquered the air. This abundance depends on a combination of features.

Arthropods have:

- a hard, chitinous external skeleton, **exoskeleton**
- metameric segmentation
- jointed limbs ('Arthropoda' meaning 'jointed feet')
- **tagmatisation**—organisation of segments into functional groups, **tagmata** (sing. tagma).

Advantages and disadvantages of an exoskeleton

The exoskeleton protects soft parts of the body, allows development of internal projections to which muscles attach and, in insects, has resulted in moveable wings. Limbs must be jointed because the exoskeleton is rigid, but this rigidity allows greater complexity of form and function of appendages (Fig. 39.10) than is possible in fleshy appendages. For example, the chela (claw) of a crab is able to grasp and crush and is structurally different from a walking limb. However, a rigid exoskeleton restricts growth and has to be shed periodically for an arthropod to grow. This is energetically costly and the animal is potentially vulnerable while it is undergoing a moult, so various behavioural and developmental devices have evolved to compensate.

The arthropod body

Arthropods include three subphyla: **Chelicerata** (spiders, scorpions, ticks, mites and their relatives); **Crustacea** (from water fleas to crabs); and **Uniramia** (chiefly myriapods, onychophorans and insects). Uniramians are so-called because their appendages are unbranched. Appendages of crustaceans and primitive chelicerates are made up of several branches. Uniramians chew with the tips of whole limbs (mandibles). In other arthropods, biting is done by the bases of limbs, termed mandibles in crustaceans and chelicerae in chelicerates.

The coelom, so important for annelid locomotion, is greatly reduced and virtually absent in adult arthropods. There are, however, large spaces in the body that are filled with blood and which form a **haemocoel**.

The arthropod blood vascular system is termed an open system because the blood is not enclosed in vessels throughout the whole of its passage around the body. It is pumped from the heart, an elongate dorsal vessel, along arteries to the periphery. The blood then enters the haemocoel to bathe the tissues directly, returning to the heart by way of a series of sinuses (Fig. 39.11).

The nervous system, like that of annelids, consists of a dorsal anterior brain joined by circumoesophageal connectives to a paired, solid, ventral nerve cord with segmental ganglia.

Evolution of arthropods

There has been a long debate as to whether arthropods are a monophyletic or a polyphyletic group (one or multiple evolutionary lineages). Those supporting a polyphyletic origin propose that Chelicerata, Crustacea and Uniramia have separate origins. They consider that chelicerates, which alone in the arthropods have spiral cleavage, have no close affinities with insects and related forms. Supporters of arthropod monophyly argue that the chitinous exoskeleton and jointed limbs

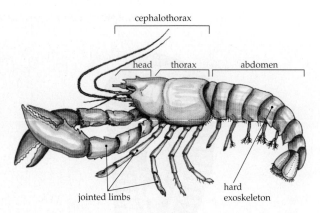

Fig. 39.10 Diagram of a crayfish showing arthropod characteristics: a hard exoskeleton, jointed limbs and body segments grouped into tagmata (e.g. the thorax)

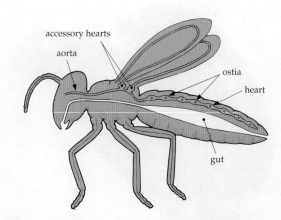

Fig. 39.11 Diagram of an insect showing the open circulatory system (haemocoel)

are unlikely to have evolved more than once. Recent phylogenies based on DNA sequences and developmental evidence do not support the polyphyletic theory and suggest that arthropods are monophyletic (Fig. 38.2). Arthropod fossils have been found in early Cambrian rocks and thus the group is at least this old.

Arthropoda are the most abundant and diverse phylum of animals. Arthropods include subphyla Chelicerata (spiders, scorpions and ticks), Crustacea (from water fleas to crabs) and Uniramia (insects and their relatives). Arthropods have a hard exoskeleton for protection and attachment of muscles, subdivision of the body into groups of segments (tagmata), and jointed limbs, necessary because of the rigid nature of the skeleton. The coelom, as seen in annelids, is reduced and virtually absent. They have an open blood vascular system (the haemocoel).

Spiders, scorpions and their relatives: subphylum Chelicerata

Chelicerates lack antennae and only some primitive forms (horseshoe crabs) have compound eyes, which are typical of crustaceans and uniramians. The body is basically divided into two or three tagmata: an anterior cephalothorax, **prosoma**; an abdomen, **opisthosoma**, with at most 12 metameres; and, in primitive members, a postsegmental part, the **telson**. Typically, the first appendages behind the mouth are a pair of food-handling chelicerae followed by a pair of sensory, prey-catching (and often copulatory) pedipalps and four pairs of legs.

Marine chelicerates

Fossil evidence suggests that chelicerates had a marine origin, beginning in the early Palaeozoic. Now only horseshoe crabs (class Merostomata) and sea spiders (class Pycnogonida) are marine.

Horseshoe crabs, of which there are only five living species (Fig. 39.12), have five or six pairs of **book gills**, modified abdominal appendages, and a long spike-like telson at the posterior end of the body. The prosoma is covered by a large shield, the **carapace**.

The relationship of sea spiders to other chelicerates is uncertain. Their usually narrow body (Fig. 39.13) bears an anterior proboscis and four, five or six pairs of long legs. Organs for gas exchange and excretion are absent, possibly because pycnogonids have a large surface-to-volume ratio, making diffusion possible. The male carries the eggs on a pair of legs located in front of the first walking legs.

Terrestrial chelicerates: the arachnids

The class **Arachnida** includes spiders, scorpions, ticks and mites and their relatives. The first arachnids, including the earliest scorpions, were aquatic, but most modern forms are terrestrial. A waxy epicuticle contributes significantly to the success of arachnids in terrestrial habitats.

The arachnid body

In arachnids, the prosoma is covered by a carapace fused onto the body. The ventral surface of the prosoma has one or more sternal plates or is covered by the **coxae** (basal joints) of the appendages. In scorpions, the long and segmented abdomen is divided into a pre- and postabdomen. In primitive spiders the abdomen may be segmented, but in mites, primary segmentation is lost and the abdomen is fused with the prosoma.

Gas exchange

Large arachnids (scorpions and some spiders, Fig. 39.14) have **book lungs** and small forms (pseudoscorpions, some spiders, and mites) have tracheae (respiratory tubes). The blood contains the copper-containing respiratory pigment **haemocyanin**.

Excretion

Excretion is by *coxal glands* and **Malpighian tubules**. Water conservation is greatly aided by production of guanine as the nitrogenous excretory substance, a material requiring very little water for its excretion.

Fig. 39.12 Horseshoe crabs, *Limulus polyphemus*, coming ashore

Fig. 39.13 A pycnogonid—a sea spider

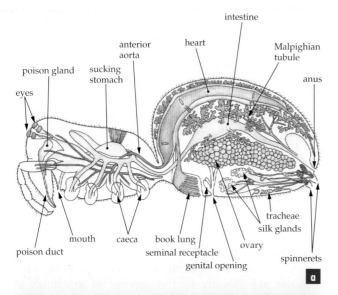

Fig. 39.14 (a) Anatomy of a spider in longitudinal section.
(b) Spiders produce silk, often used to build webs to catch prey

Catching and digesting prey

Most arachnids are carnivorous and poison is often used to immobilise prey. The way in which arachnids deliver poison (see Box 26.4) varies and mechanisms have evolved independently in different groups. In spiders, poison is delivered by the **chelicerae**, in scorpions it is by the terminal abdominal barb, and in pseudoscorpions through fingers of the **pedipalps**. Many spiders use silk, produced in spinning organs, **spinnerets**, at the end of the abdomen, to immobilise and capture prey (Fig. 39.14b). Spiders are essentially fluid feeders. They pour enzymes onto the surface of their prey or inject it inside and then suck out the dissolved tissues. Other arachnids, such as many harvestmen, are scavengers, and many mites are parasites.

Sense organs

Most arachnids have three types of sense organs: sensory hairs, eyes and slit sense organs. Hairs on the body and appendages are sensitive to slight air vibrations and are probably the most important sense organs. Slit sense organs, composed of slits in the cuticle covered by a diaphragm in contact with a basal sensory seta, detect sound vibrations. Many arachnids possess simple eyes but only a few, such as hunting spiders, construct detailed images.

Reproduction

In many arachnids, the male transfers sperm in spermatophores. Spermatophores are picked up by the female, or transferred to her by the male, as in scorpions, pseudoscorpions and some mites (Chapter 15). Often the male shows courtship behaviour to attract a female to the spermatophore. In spiders, sperm are exuded by the pedipalps onto a web and then transferred by the pedipalps to the female genital opening. In harvestmen and many mites, however, sperm transfer is direct by insemination during copulation.

> Chelicerate arthropods have anterior appendages (chelicerae) that function as fangs or pincers and they lack antennae. The body is divided into a cephalothorax, an abdomen and, in primitive forms, a telson.

Prawns, crabs and their relatives: subphylum Crustacea

With approximately 40 000 species, crustaceans (Fig. 39.15) are the only major group of aquatic arthropods and relatively few species, such as wood lice and slaters, are terrestrial. They live at most depths of the sea, in fresh water up to elevations of 3.6 km, in waters from 0°C to 55°C, and in leaf litter of forests. Most are **omnivores**, often **scavengers**; others are filter feeders or **carnivores**.

The crustacean body

In crustaceans, the ancestral segments are organised into three more or less distinctly demarcated tagmata: head, thorax and abdomen. Crustaceans are unique among arthropods in having two pairs of antennae (**antennules** and **antennae**; Fig. 39.16). These are on their first two head segments. Antennae are typically followed by one pair of **mandibles**, with grinding and biting surfaces, and two pairs of accessory feeding appendages, **maxillae**. Trunk specialisations and appendages vary greatly but a carapace covering all or part of the body is common. Crustacean appendages are typically made of two parts (biramous)—**endopodite** and **exopodite** (Fig. 39.16)—a major difference from insects and other uniramians. In filter feeders, closely placed setae on appendages function as filters.

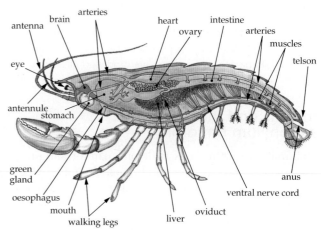

Fig. 39.16 Anatomy and appendages of a crayfish. The two parts of the crustacean appendage can be seen in the trunk appendages here

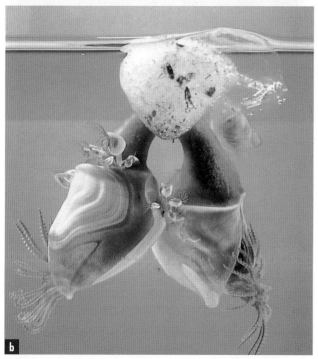

Fig. 39.15 Crustaceans: (a) Soldier crab, *Mictyris longicarpus*; (b) goose barnacle, *Lepas fascicularis*

Circulation and gas exchange

Crustaceans have a heart with perforations (ostia) through which venous blood enters from the pericardial sinus. The blood of most crustaceans is blue owing to the presence of haemocyanin but a few species have red, haemoglobin-containing blood. Many small crustaceans use the body surface for gas exchange but most large ones have evolved thin-walled gills at the surface, often on the limbs.

Excretion

Excretion is by a pair of blind sacs in the head that open onto the bases of the second pair of antennae (**antennal glands**) or the second pair of maxillae (**maxillary glands**).

Nervous system

Typically, the nervous system is a double, solid ventral ganglionated nerve cord (Fig. 39.16). Crustacean eyes are of two main types: a pair of compound eyes and a small median dorsal eye, which consists of three or four closely placed simple eyes, **ocelli**.

Reproduction

Sexes are usually separate but **hermaphroditism** is the rule in sessile barnacles, in some primitive crustaceans and in some parasitic forms. *Parthenogenesis* (Chapter 15) is frequent in water fleas such as *Daphnia* and other members of the class Branchiopoda. Copulation is usual and egg brooding is common. In many crustaceans, sperm lack a flagellum. The earliest hatching stage, when it is not suppressed, is a **larva or nauplius** (Fig. 39.17). The nauplius has only three pairs of appendages, namely antennules, antennae and mandibles. A single eye, the nauplius eye, is at the front of the head. The nauplius is free-living in many lower crustaceans but in higher forms the egg hatches as a more advanced larva. After the nauplius stage in some groups, there are remarkable larvae, such as the free-swimming cypris of barnacles, the phyllosome of spiny lobsters and the zoea and megalopa of crabs.

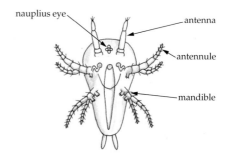

Fig. 39.17 A nauplius larva of a crustacean

Crustaceans are unique among arthropods in having two pairs of antennae, a pair of mandibles and two pairs of maxillae. Their appendages are typically biramous (with two branches). Segments are organised functionally into head, thorax and abdomen.

Insects and their relatives: subphylum Uniramia

Uniramia are arthropods with one pair of antennae. The name refers to their unbranched limbs, which contrast with the biramous appendages of crustaceans. However, some fossil 'uniramians' have branched limbs. Uniramians include millipedes, centipedes and insects. The velvet worms, onychophorans, are also often classified as uniramians but their relationships are still debated (Box 39.2). Onychophorans are slug-like animals with legs. They have many features intermediate between annelids and arthropods. Molecular evidence suggests that they branched off early in arthropod evolution. Uniramians are found worldwide but there are few marine representatives.

Two examples of onychophorans: **(a)** a velvet worm, *Peripatoides*; and **(b)** the recently discovered *Cephalofovea*

BOX 39.2 Onychophorans: arthropods or annelids?

Onychophora are the velvet worms, such as the Australian *Peripatoides* and the remarkable, recently discovered, *Cephalofovea* (see figures). They look rather like slugs with legs and most live in moist habitats, such as tropical rainforests of the world, under logs and litter. Their body is covered by tubercles and has a velvety and iridescent appearance so that they are often called velvet worms.

There has been much debate about the relationships of onychophorans because in some respects they are similar to annelids and in others to arthropods. Like annelids, their body is soft and covered by a flexible, permeable cuticle, not a rigid exoskeleton as in arthropods. Their body plan also is more like that of an annelid, and their legs are not jointed like those of arthropods. However, like arthropods they have mandibles, their body cavity is a haemocoel, and their gas exchange organs are tracheae. Their embryonic development is similar to uniramians, and they have one pair of antennae. Molecular data in which 12 S rRNA sequences were compared for a variety of invertebrates suggest that onychophorans are an early lineage of arthropods. In this case, their similarities to annelids would be interpreted as retention of general primitive characteristics.

Uniramians (millipedes, centipedes and insects) have one pair of antennae and unbranched limbs.

Millipedes and centipedes: classes Diplopoda and Chilopoda

Millipedes and centipedes have a head and a segmented body (Fig. 39.18). Most of the trunk segments bear legs, which is why they are called **myriapods** (many legs). Millipedes appear to have two pairs of legs per segment but each 'segment' is actually a tagma of two fused primordial segments so, like the centipedes, they actually only have one pair of legs per true

Fig. 39.18 Centipedes (class Chilopoda) are uniramian arthropods

segment. Both live in leaf litter and soil but millipedes are mostly herbivores whereas centipedes are carnivores. In millipedes, each segment contains a pair of glands, which secrete an offensive chemical as a defence against predators. Centipedes have poison claws on the first trunk segment.

M illipedes and centipedes have a head and segmented body with limbs on most trunk segments.

Insects: class Insecta

Insects are the largest group of animals. About three-quarters of a million species have been described and insects are among the most numerous inhabitants of terrestrial environments. Their abundance and extensive distribution relates to a number of features but especially their ability to fly and to exploit an aerial environment. Moths, butterflies, bees, wasps and flies are important pollinators of flowering plants and the evolutionary histories of the angiosperms and insects are closely interlinked (Chapter 37).

Body structure

Insects have the uniramian features of unbranched limbs and a single pair of antennae but differ from other classes in having a tagma of three segments that forms a thorax, which is distinct from the head and abdomen (Fig. 39.19).

The thorax bears three pairs of walking legs and, in all but primitive insects, two pairs of wings on the two posterior segments. In flies (order Diptera), the last pair of wings is modified as a pair of mobile knobbed rods, the **halteres**, which assist with balance during flight. The insect abdomen generally lacks appendages apart from a pair of **cerci** (sing. cercus) on the last segment. Short structures, homologous to limbs, are present on the abdomen in bristle tails.

Mouthparts and feeding

The mouthparts consist of three kinds of appendages: a pair of mandibles, a pair of maxillae and a **labium** derived from the fused pair of second maxillae. Each maxilla bears a lateral maxillary palp and the labium has a pair of labial palps. An anterior plate of the exoskeleton of the head, the **labrum**, covers these mouthparts and a tongue-like **hypopharynx** projects behind the mouth.

Primitively, the mouthparts were used for chewing. Modern examples are grasshoppers and many beetles. In many insects, the mouthparts have evolved for specialised feeding. In flies, the labium is a sponge-like pad, used to take up liquid food. In mosquitoes, the mouthparts are elongated as a proboscis. The proboscis consists of a fleshy, grooved labium, overarched by the labrum to enclose the elongate mandibles and maxillae, which form sharp stylets. Stylets penetrate skin for

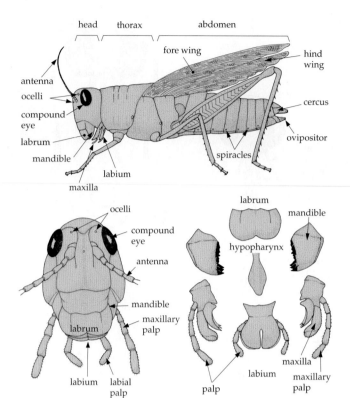

Fig. 39.19 External features of a grasshopper, a typical insect. Details of the head and mouthparts are shown. The head and biting mouthparts of a grasshopper are adapted for cutting and then crushing food between the mandibles. The mandibles of carnivorous insects are sharp and pointed blades

blood sucking in female mosquitoes (Fig. 39.20) or plant vascular tissue for sap feeding in the males. As food is sucked through a channel in the stylet, saliva can also pass down through the slender hypopharynx. It is this exchange of fluid that enables mosquitoes to act as parasite vectors.

Sensory organs

Insects have a pair of lateral **compound eyes**, three dorsal ocelli and a pair of antennae (Fig. 39.19). A variety of sensory organs bearing olfactory receptors and mechanoreceptors, including ones modified for hearing, is distributed over the body, particularly on the antennae and legs. These all involve modification of the cuticle, allowing environmental signals to pass through the exoskeleton to the nervous system.

Gas exchange

Gas exchange (Chapter 20) is by a system of air-filled **tracheae**, the openings of which are *spiracles* or stigmata, located along the sides of the thorax and abdomen (Fig. 39.19). The finest branches of the tracheae, **tracheoles**, are fluid-filled and exchange oxygen and carbon dioxide directly with tissue cells. Tracheoles are enclosed within a tracheal end-cell (Fig. 39.21).

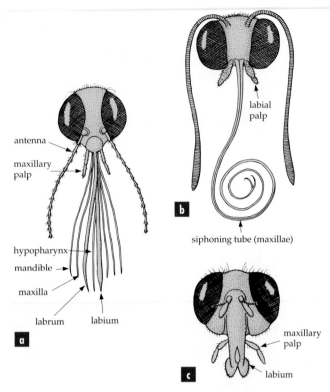

Fig. 39.20 Three types of insect mouthparts. **(a)** The piercing and sucking mouthparts of a mosquito. **(b)** The long coiled sucking proboscis of a butterfly is a highly modified maxilla. The mandibles are absent. At rest the tube is coiled. **(c)** The lobes at the end of the mouthparts in the housefly have grooves on the lower surface that direct liquid towards a central channel. They sponge or lap up surface liquids

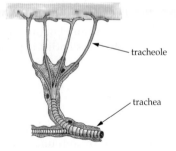

Fig. 39.21 Respiratory system of an insect. Tracheae, with supportive rings in their walls, branch into tracheoles. Tracheoles are closely associated with and supply oxygen to cells of tissues

Excretion

Insects excrete uric acid through Malpighian tubules into the hindgut. Uric acid is non-toxic and because it is insoluble in water, very little is required for its excretion. This mode of excretion and a waxy epicuticle allow insects to conserve water, favouring their survival in terrestrial environments.

Reproduction

Primitive insects transfer sperm indirectly using a **spermatophore** but advanced insects have direct sperm transfer into the female reproductive system.

The fertilised egg undergoes superficial cleavage and hatches as a juvenile stage, which is similar to the adult but lacks wings and is sexually immature. This is termed a **nymph** and undergoes a series of moults, giving a succession of stages, **instars** (Chapter 25).

Insect diversity

The subclass **Apterygota** includes bristletails (orders Protura and Thysanura) and springtails (Collembola), insects that lack wings. More advanced insects with wings constitute the subclass **Pterygota**. Australian examples of some orders in these two subclasses are shown in Figure 39.22.

In the lower pterygote orders, successive instars have increasingly developed wing buds. They have gradual or **incomplete metamorphosis** and are termed **exopterygotes** because the developing wings are externally visible in nymphs. Exopterygote insects include, among others, mayflies (order Ephemeroptera), dragonflies and damselflies (Odonata), grasshoppers and crickets (Orthoptera), cockroaches (Blattodea), mantids (Mantodea) and stick and leaf insects (Phasmatodea), termites (Isoptera), stoneflies (Plecoptera), earwigs (Dermaptera), biting and sucking lice (Anisoptera), bugs (Hemiptera) and lacewings (Neuroptera).

In higher insects, the juvenile or larva is often radically different in appearance and mode of feeding from the adult (contrast a caterpillar with a butterfly,

Fig. 39.22 Examples of some of the insect orders found in Australia: **(a)** order Phasmatodea, violet stick insect, *Didymuria violescens*

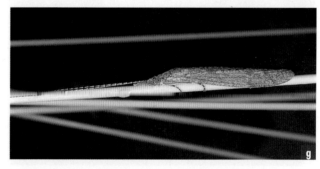

Fig. 39.22 (b) Order Hemiptera, grey-back cane beetle, *Dermolepida albohirtum*; **(c)** order Blattodea, a native cockroach, *Polyzotteria mitchelli*; **(d)** order Orthoptera, northern Australian locust, *Chortoicetes terminifera*; **(e)** order Diptera, hover fly, *Syrphus damastor*; **(f)** order Coleoptera, dung beetle; **(g)** order Trichoptera, caddis fly

Fig. 39.23). A profound metamorphosis to the adult stage, **imago**, occurs in a **pupa**. Such insects have **complete metamorphosis** and, because wing development is concealed in the pupa, are termed endopterygotes. They include, among others, beetles (order Coleoptera), scorpion flies (Mecoptera), caddis flies (Trichoptera), moths and butterflies (Lepidoptera), true two-winged flies (Diptera), ants, bees and wasps (Hymenoptera) and fleas (Siphonaptera).

Insects and humans

Many insects are pests of humans and of their domestic animals and crops, and some are major

Fig. 39.23 Three stages of the wanderer butterfly life cycle: (a) caterpillar; (b) pupa; (c) imago

carriers (**vectors**) of human diseases (Fig. 39.24). The mosquito, *Anopheles*, is the intermediate host for the malarial parasite, *Plasmodium*. A moth fly, *Phlebotomus*, transmits typhus. Fleas, including the genus

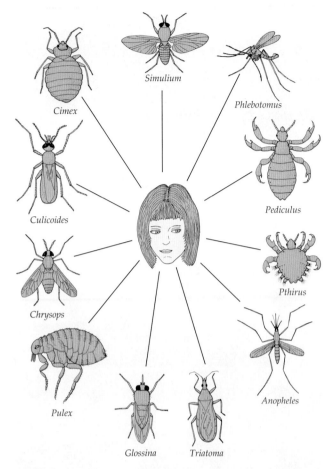

Fig. 39.24 Examples of winged insects that parasitise humans and may carry diseases

Pulex, transmit several diseases, including bubonic plague and tapeworms. *Simulium*, the buffalo gnat, causes blindness by transmitting the nematode *Onchocerca*. The sandfly, *Culicoides*, and the horse fly, *Chrysops*, also carry nematode parasites. The tsetse fly, *Glossina*, transmits the protist *Trypanosoma*, the organism that causes sleeping sickness. The louse *Pediculus* transmits *Rickettsias*, although another louse, *Pthirus*, seems innocent of carrying disease. Insects also provide many benefits for humans. Not only are they vital for pollinating agricultural plants but many insects also prey on other insect pests.

Insects, the largest class of uniramians, have a well-defined head, thorax and abdomen, three pairs of legs attached to the thorax, specialised mouthparts and usually compound eyes as well as ocelli. They respire by tracheae, excrete uric acid through Malpighian tubules, have a waxy cuticle and have evolved flight, all of which make them well adapted for terrestrial environments. Insects impact directly on human activities in both positive and negative ways.

Snails, clams and relatives: phylum Mollusca

After arthropods, molluscs constitute the second largest animal phylum. There are more than 100 000 living species and 35 000 fossil species dating back to the Cambrian. They are aquatic and terrestrial animals, including chitons, gastropods (snails and nudibranchs), bivalves (mussels, oysters and cockles), tusk shells and cephalopods (squids and octopuses).

Molluscan body plan

Morphologically, molluscs are highly diverse and all modern forms appear to have diverged significantly from what is assumed to be the ancestral form. So that we can understand their relationship to each other, their basic structure is often considered in terms of their evolution from a hypothetical primitive molluscan ancestor. It is important to remember that no existing mollusc conforms exactly to this hypothetical plan.

The primitive molluscan ancestor (Fig. 39.25) is envisaged as a dorsoventrally flattened, unsegmented, almost worm-like animal with a head, a ventral foot for gliding locomotion and a visceral mass containing the body organs. Covering the visceral mass was a dorsal **mantle**, a fold of the body wall with a cavity beneath it opening backwards. The mantle secreted a shell, which was simply a chitinous cuticle invested with more or less overlapping scales consisting of a calcareous mineral (aragonite). The mantle cavity contained a pair, possibly several pairs, of respiratory gills and received the excretory, alimentary and reproductive openings. The alimentary canal was straight and possessed a mid-gut with ventrolateral, serial outpockets, which alternated with dorsoventral muscles.

One of the most characteristic features was a broad, tongue-like structure, the **radula** (Fig. 39.26), rows of rasping teeth borne on a cuticular strip present in the floor of the foregut. All modern molluscs, except bivalves, have retained the radula.

The nervous system consisted of paired, dorsal cerebral ganglia, a nerve ring around the oesophagus and two pairs of longitudinal nerve cords innervating the gills and a possibly paired sensory organ associated with the **mantle cavity**. The circulatory system was open and consisted of a dorsoposterior sac, the pericardium, containing a ventricle and a pair of lateral auricles. The animals are thought to have been dioecious with paired gonads discharging into the mantle cavity. Whether the pericardium and the cavity in which the gonads lie are homologous with the coelom of other protostomes is debatable.

Fertilisation was external, cleavage spiral, and development included a pelagic, ciliated trochophore

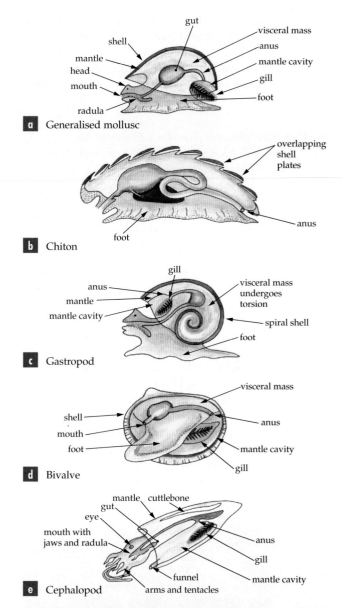

a Generalised mollusc

b Chiton

c Gastropod

d Bivalve

e Cephalopod

Fig. 39.25 (a) Generalised structure of a mollusc with head, visceral mass and flattened foot for gliding locomotion. Modifications of this basic structure in the evolution of **(b)** chitons; **(c)** gastropods; **(d)** bivalves; **(e)** cephalopods

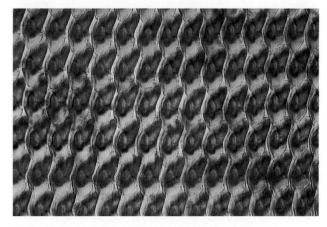

Fig. 39.26 Teeth of the radula of a gastropod

larva. The larva developed into a second stage, the **veliger**, which develops a foot, mantle and shell. The adult animal was marine, living on sea floors and shores, and probably fed by scraping microorganisms off the substrate by means of the protrusible, rasping radula.

P hylum Mollusca is second in size only to the arthropods. Molluscs are a diverse group of aquatic and terrestrial animals, including chitons, snails and nudibranchs, bivalves, tusk shells and squids and octopuses. The primitive molluscan ancestor is envisaged as having had a head, visceral mass, mantle cavity and a ventral foot for gliding locomotion. All modern molluscs, except bivalves, have a rasping tongue, the radula.

Chitons: class Polyplacophora

Polyplacophorans (chitons or coat-of-mail shells) are exclusively marine, bilaterally symmetrical molluscs (Fig. 39.27). They are elongated and dorsoventrally flattened with eight overlapping calcareous shell plates. The shell plates are held in place and sometimes entirely covered by a girdle. The epithelium of the girdle produces a cuticle in which are embedded calcareous or chitinous spicules. The broad foot is surrounded by a groove, the mantle cavity, in which gills (six to 88 pairs) are arranged. The mouth is anterior but there is no distinct head. Sexes are

separate. Although there are usually pelagic larvae, some chitons brood their young in the mantle cavity until the young reach the creeping stage.

C hitons are bilaterally symmetrical, exclusively marine molluscs. They are elongated and dorsoventrally flattened with eight overlapping calcareous shell plates on the dorsal surface.

Torsion and the class Gastropoda

Gastropods (Fig. 39.28) have undergone three major changes relative to the hypothetical primitive mollusc (Fig. 39.25). They show greater development of a head, **cephalisation**. The shell has become spiral. The upper part of the body, the visceral mass, has undergone *torsion*, a twisting through 180° relative to the foot so that the originally posterior mantle cavity, with its two gills and anus, is now anterior. The evolutionary significance of torsion is uncertain. One theory suggests that torsion is specifically a larval adaptation, perhaps allowing protective withdrawal of the head into the mantle cavity. A further view is that it increases the ventilating current over the gills during forward locomotion.

A disadvantage of torsion was that the anus then discharged directly over the head and mouth! Much of the diversity of gastropods relates to different ways of overcoming this sanitation problem. In the most primitive gastropods, slit-limpets and abalone, a slit or a row of holes in the shell (Fig. 39.29a) allows a flow-through ventilating current to sweep wastes out. True limpets (Fig. 39.29b) retain a posteriorly directed ventilating current but lose one or both gills with or without replacement by lateral gills in the mantle

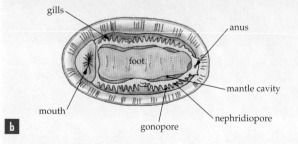

Fig. 39.27 **(a)** Dorsal view of a chiton. **(b)** Diagram of the morphology of a chiton viewed ventrally

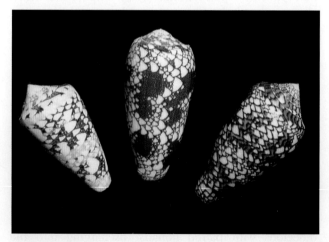

Fig. 39.28 Examples of three shells of the species *Conus*, cone shells. Cone shells are venomous gastropods. The barbed end of a single radula tooth is hurled into prey, which is quickly immobilised by a neurotoxin that enters the wound through a hollow cavity in the tooth. The poison of some South Pacific cone shells is highly toxic to humans

Fig. 39.29 Gastropods. (a) A solution to the problem of sanitation in gastropods is seen in the shell of the abalone *Haliotis*, which has a series of holes for a ventilation current. (b) The limpet *Cellana*. (c) The snail *Nerita*

cavity. Top and turban shells and nerites (Fig. 39.29c) have the respiratory current entering the anterior mantle cavity on the left side of the head, where a single gill is retained, then flowing past the anus to exit on the right side of the mantle cavity.

A further solution to the sanitation problem resulting from torsion has been detorsion. This process normally involves differential growth of body parts during development. It can involve loss of the original gills and movement of the anus to a lateral and ultimately posterior position or, in side-gilled slugs,

movement of both the mantle cavity and anus to the side. In some nudibranchs (Fig. 39.30) or sea slugs, the shell, mantle cavity and original gill have been lost completely; the anus has become posterior where it is surrounded by a circle of gills. Reduction of the shell also occurs, for example, in sea hares and bubble shells.

Land snails and related freshwater forms (subclass Pulmonata, Fig. 39.31) have lost all gills. The mantle margin is fused with the body wall so that the mantle cavity becomes a lung with a single, restricted respiratory opening, hence the name which means lung. The lung cavity is actively ventilated.

Gastropods are marine, freshwater and terrestrial snails and slugs. In gastropods, the visceral mass undergoes torsion in early development so that the mantle cavity, with its two gills and anus, rotates to an anterior position. It is suggested that this is an advantage to the larvae by allowing the head to be withdrawn into the mantle cavity for protection. Subsequent growth into the adult involves modifications to reduce the problems caused by having faecal material voided in the head region.

Filter feeding and the class Bivalvia

Bivalvia, as their name suggests, have two shell plates. These lie either side of the body and enclose it (Fig. 39.32). Most bivalves are sessile or slow moving and they have no distinct head or radula. The foot, like the body, is laterally compressed and in advanced forms is shaped like a hatchet. The general features of the class appear to have arisen in relation to burrowing in soft substrates.

As their descriptive name—**lamellibranch**—suggests, the gills have expanded as sheets or lamellae, extending from the anterior to the posterior end of the long mantle cavity on either side of the foot (Fig. 39.33) and providing a surface area in excess of that needed for gas exchange. This is because most bivalves use cilia on the surface of the gills to filter feed. The ventilating current, which supplies both gases and food, enters posteriorly, passing forwards then upwards through the

Fig. 39.30 An Australian nudibranch, *Chromodoris*

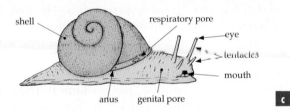

Fig. 39.31 Land snails. **(a)** *Sphaerospira*, a large land snail of Queensland rainforests. **(b)** *Triboniophorus*, a large slug. **(c)** Diagram showing the morphology of the garden snail, *Helix*

Fig. 39.32 A bivalve, showing the exhalent and inhalent siphons and the flattened foot

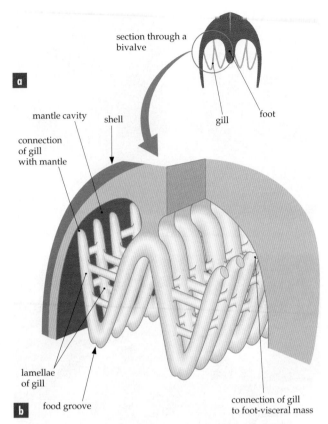

Fig. 39.33 Diagram showing the structure of a bivalve gill. The gills are greatly expanded, forming sheets or lamellae. They function in both gas exchange and filter feeding. **(a)** Position of the gills on either side of the foot—the visceral mass is shown in cross-section. **(b)** Detail showing folding of gill and position of food groove

gills, before exiting posteriorly. Each of the two gills is folded on itself in a V-shape. Cilia on the surface of the gills and mantle cavity also create the ventilating current. Specialised cilia on the gills sweep food particles, filtered from sea water, downwards to a food groove, running along the ventral edge (the bottom edge of the 'V') of the gill. Cilia in the groove convey the particles in an anterior direction to labial palps for sorting and transfer to the mouth.

Evolution of filter feeding led to radiation of lamellibranchs. Although many bivalves live on soft substrates, others live on hard substrates. These anchor one valve of the shell to the substrate by horny (**byssal**) threads, as in the mussel, *Mytilus*, or by fusion to the substrate, as in the oyster, *Crassostrea*. Others bore into rock, coral and even wood. A few bivalves, such as scallops (*Pecten* and *Chlamys*), are unattached and can swim vigorously by clapping the shell valves together when disturbed.

Most bivalves are dioecious. In protobranchs, gametes exit through nephridia but in other bivalves through special gonoducts. Fertilisation is usually external, although it may occur in the mantle cavity. A trochophore and a bivalved veliger larva typically develop. In some freshwater bivalves, such as *Alathyria*, the Australian freshwater mussel, a modified veliger, termed the **glochidium**, is released from the parent and grips onto fishes, which it parasitises until it becomes a cyst. Eventually, an immature mussel falls from the cyst to the bottom and undergoes development to the free-living adult.

> B ivalve molluscs have two bilateral shell plates. They have no distinct head, no radula and a laterally compressed foot for burrowing. Most bivalves are filter feeders, with gills functioning in both gas exchange and food collection.

'Brainy' molluscs: class Cephalopoda

Cephalopoda (Fig. 39.34) include the largest living invertebrate, the giant squid *Architeuthis*, known from South Australian waters. It attains a body length of 16 m, including the tentacles. Cephalopods are bilaterally symmetrical molluscs in which the shell, although primitively external, as in *Nautilus* and fossil **ammonites** (see Chapter 31), is usually internal or even absent. In *Nautilus* (Fig. 39.34a), the most primitive living cephalopod, the animal occupies the largest, last formed chamber of the shell. In the cuttlefish *Sepia*, the shell is shield-shaped and in *Spirula* it is coiled (Fig. 39.34b); it may be reduced to a thin plate, as in the squid *Loligo* (Fig. 39.34a), and occasionally it has been lost, as in *Octopus* (Fig. 39.34c).

Buoyancy and locomotion

In *Sepia* and *Spirula*, as in *Nautilus*, gas-filled chambers of the shell, separated by transverse septa, give buoyancy. Regulation of the amount of gas and of buoyancy is brought about by a cord of body tissue, the **siphuncle**, which runs through all the chambers. Water flows from the mantle cavity through a specialised funnel, the **siphon**, which can produce a jet of water for propulsion. Octopuses, which are benthic and crawl on the sea bottom, use jet propulsion only intermittently, often for escape. In *Nautilus*, which is a slow swimmer, force is generated only by contraction of the funnel, but in most living cephalopods the mantle provides the contractile force that permits very rapid movement. Some cephalopods are even capable of leaping out of the water to glide through the air. It has been suggested that development of rapid swimming, with streamlining of the body by internalisation of the shell, was a selective response to competition from bony fishes, which had replaced **nautiloids** and ammonites as the dominant pelagic animals by the end of the Mesozoic (Chapter 31).

The cephalopod head

The name cephalopod means 'head–foot', because the head is surrounded by a ring of arms and tentacles (usually equipped with suckers or hooks), which represent a modified foot. *Nautilus* has 38 tentacles, without suckers, while most cephalopods have eight arms and two extensible, prehensile tentacles, as in cuttlefish, squids and vampire squids. Octopuses have lost the two tentacles.

In cephalopods the mouth leads to a strong beak, shaped like that of a parrot, with dorsal and ventral mandibles accompanied by a radula. Salivary glands are

Fig. 39.34 Cephalopods include **(a)** **(i)** squid; **(ii)** paper nautilus and **(iii)** octopus

Fig. 39.34 (b) The internal shell of the cephalopod *Spirula* is often washed up on our beaches with attached goose barnacles. **(c)** The common Australian octopus, *Octopus cyaneus*, showing the highly developed eye and respiratory and locomotory siphon. Colour changes occur during courtship and in defensive and offensive situations

usually present and their secretion is sometimes poisonous, as in the blue-ringed octopus, *Hapalochlaena maculosa*, of southern Australian shores.

Nervous system, sense organs and behaviour

Cephalopods have a well-developed brain, which provides advanced capabilities in vision, touch, memory and recognition of shapes, and swimming. While the eyes of *Nautilus* lack lenses, those of squids, cuttlefish and octopuses have a lens and camera-like organisation that in many ways resembles the eyes of vertebrates. Cephalopods have colour vision and are perhaps unique in invertebrates in displaying signals and emotional

states by changes in the colour and pattern of the skin. Pigment granules are located in **chromatophores**, branching cells that expand under the action of tiny muscles attached to their periphery. The muscles are under nervous and hormonal control. Colour changes occur during courtship and defensive responses, merging an animal with its background, or in offensive situations, as when an animal is alarmed. In cephalopods other than *Nautilus* and some deep-water species, an ink-sac opening just inside the anus can release a cloud of alkaloid-containing inky fluid into the water when the animal is alarmed. This acts as a distraction to predators, and is possibly objectionable or even anaesthetic to chemoreceptors or other sense organs.

Gas exchange and circulation

The mantle encloses a ventral cavity, which houses one or two pairs of gills and receives the openings of the alimentary canal, and the excretory and reproductive systems. Cephalopods are the only molluscs with a closed vascular system and are exceptional in the invertebrates in having a true endothelium lining the blood vessels, as in the vertebrates. Blood passes through a muscular heart before entering the gills. Accessory pumping structures, **branchial hearts**, may be present at the bases of the gills. The closed, high-pressure blood system, presence of the respiratory pigment haemocyanin, absence of gill cilia, highly developed eyes, complex nervous system and behaviour, presence of chromatophores and production of 'ink' are features that relate to an active predatory lifestyle, a high metabolic rate and rapid growth rates.

Reproduction

The sexes are separate in cephalopods. There is a single gonad. In the female, the gonad usually has a pair of ducts. In the male, one duct is retained while the other forms a sac that contains sperm packets (spermatophores). Many cephalopods have elaborate and extended courtship behaviour involving chromatophore displays that lead to head-to-head copulation and sperm transfer. Males often have the lower left arm, called the **hectocotylised arm**, specialised for transferring spermatophores into the mantle cavity of the female during copulation. After fertilisation in the female oviduct, yolky eggs are usually deposited on the substrate or shed into the sea. A few cephalopods brood their eggs. Many cephalopods die after producing or fertilising eggs. Development is direct, with no trochophore or veliger larvae.

Cephalopod features, including closed vascular system, complex nervous system and behaviour, chromatophores, ink gland, jet propulsion, mobile arms and highly developed eyes, relate to their active predatory lifestyle, high metabolic rate and rapid growth rates.

Bryozoans and relatives: superphylum Lophophorata

In structure and development, three phyla of animals straddle the borderline between protostomes and deuterostomes. These three phyla are **Phoronida**, **Bryozoa** and **Brachiopoda**, constituting the superphylum **Lophophorata**. The **lophophore**, which gives them their name, is a feeding structure of ciliated tentacles, which contain extensions of the coelom.

Lophophorates can be considered to consist of only three segments. They have three coelomic cavities in longitudinal sequence, the protocoel, mesocoel and metacoel, corresponding with three body divisions, the protosome, mesosome and metasome. The metacoel forms the main body cavity.

Bryozoans, moss animals, form the largest group and are the most common lophophorates, with about 4000 living marine and freshwater species (Fig. 39.35). They are sessile and colonial and have a chitinous covering with or without calcareous layers, with a pore for protrusion of the lophophore. The coelom is spacious and the digestive tract is U-shaped. Bryozoan colonies play an important part in the ecology of most benthic marine communities where they form a thin, film-like lacework on seaweed or on the underside of boulders or look like small, branching plants or corals. They are a major marine fouling organism on boats, sea-water intakes and other structures.

Phoronids are worm-like animals generally regarded as the most primitive lophophorates and there are only about 10, exclusively marine, species. Brachiopods

Fig. 39.35 Bryozoans, the largest and most common group of lophophorate animals. Bryozoans are sessile and colonial, looking like lacework or seaweed

(lamp shells) are today represented by some 280 species in contrast with the 30 000 or so known fossil species of the Palaeozoic and Mesozoic eras.

Molecular data suggest that the lophophorates are more closely related to protostomes than to deuterostomes but it is still likely that their ancestors lay close to the division between those two assemblages.

The superphylum Lophophorata includes three phyla of animals that have few body segments. They are characterised by a lophophore consisting of ciliated feeding tentacles that contain extensions of the coelom. Structurally and developmentally they have features similar to both protostomes and deuterostomes, and their phylogenetic relationships are unclear. Recent molecular evidence suggests that they are closer to the protostomes.

Summary

- The body of annelids and arthropods shows metameric segmentation, in which there is a linear repetition of functional units. Typically, each segment contains key structures such as nephridia (for excretion), ganglia (clusters of nerve cells), muscles and blood vessels. The only suggestion of segmentation in molluscs is in the class Monoplacophora, which includes early fossil molluscs and the extant *Neopilina*.

- Annelids are coelomate protostomes. They are metamerically segmented worms with a digestive system, closed circulatory system, solid ventral ganglionated nervous system and nephridia for excretion. They move by peristaltic locomotion.

- Polychaetes are marine worms characterised by parapodia, paired appendages that function in locomotion and respiration.

- Euclitellate worms lack parapodia. They are characterised by a clitellum, which secretes a cocoon in which eggs are deposited. Euclitellate worms include oligochaetes, which have few setae on each segment, and leeches, which have none. Most leeches are specialised as external parasites and attach to hosts by posterior and anterior suckers. The coelom is reduced and leeches have lost peristaltic locomotion.

- Oligochaetes (earthworms) are euclitellates with typically few setae on each segment. They are hermaphroditic, with gonads in only a few segments.

- Leeches (Hirudinea) have an anterior and posterior sucker, generally lack setae, have reduced segmentation and a reduced but specialised coelom. Most leeches are specialised as external parasites.

- Arthropoda are the most abundant and diverse phylum of animals. Arthropods include subphyla Chelicerata (spiders, scorpions and ticks), Crustacea (from water fleas to crabs) and Uniramia (insects and their relatives). Arthropods have a hard exoskeleton for protection and attachment of muscles, subdivision of the body into groups of segments (tagmata), and jointed limbs, necessary because of the rigid nature of the skeleton. The coelom, as seen in annelids, is reduced and virtually absent. They have an open blood vascular system (the haemocoel).

- Chelicerate arthropods have anterior appendages (chelicerae) that function as fangs or pincers and they lack antennae. The body is divided into a cephalothorax, an abdomen and, in primitive forms, a telson.

- Crustaceans are unique among arthropods in having two pairs of antennae, a pair of mandibles and two pairs of maxillae. Their appendages are typically biramous (with two branches). Segments are organised functionally into head, thorax and abdomen.

- Uniramians (millipedes, centipedes and insects) have one pair of antennae and unbranched limbs.

- Onychophorans are slug-like animals with legs. They have many features intermediate between annelids and arthropods. Molecular evidence suggests that they branched off early in arthropod evolution.

- Millipedes and centipedes have a head and segmented body with limbs on most trunk segments.

- Insects, the largest class of uniramians, have a well-defined head, thorax and abdomen, three pairs of legs attached to the thorax, specialised mouthparts and usually compound eyes as well as ocelli. They respire by tracheae, excrete uric acid through Malpighian tubules, have a waxy cuticle and have evolved flight, all of which make them well adapted for terrestrial environments. Insects impact directly on human activities in both positive and negative ways.

- Phylum Mollusca is second in size only to the arthropods. Molluscs are a diverse group of aquatic and terrestrial animals including chitons, snails and nudibranchs, bivalves, tusk shells and squids and octopuses. The primitive molluscan ancestor is envisaged as having had a head, visceral mass, mantle cavity and a ventral foot for gliding locomotion. All modern molluscs, except bivalves, have a rasping tongue, the radula.

- Chitons are bilaterally symmetrical, exclusively marine molluscs. They are elongated and dorsoventrally flattened with eight overlapping calcareous shell plates on the dorsal surface.

- Gastropods are marine, freshwater and terrestrial snails and slugs. In gastropods, the visceral mass undergoes torsion in early development so that the mantle cavity, with its two gills and anus, rotates to an anterior position. It is suggested that this advantages the larvae by allowing the head to be withdrawn into the mantle cavity for protection. Subsequent growth into the adult involves modifications to reduce the problems caused by having faecal material voided in the head region.

- Bivalve molluscs have two bilateral shell plates. They have no distinct head, no radula, and a laterally compressed foot for burrowing. Most bivalves are filter feeders, with gills functioning in both gas exchange and food collection.
- Cephalopod features, including closed vascular system, complex nervous system and behaviour, chromatophores, ink gland, jet propulsion, mobile arms and highly developed eyes, relate to

their active predatory lifestyle, high metabolic rate and rapid growth rates.

- The superphylum Lophophorata includes three phyla of animals that have few body segments. They are characterised by a lophophore consisting of ciliated feeding tentacles that contain extensions of the coelom. Structurally and developmentally they have features similar to both protostomes and deuterostomes and their phylogenetic relationships are unclear. Recent molecular evidence suggests that they are closer to the protostomes.

keyterms

ammonite (p. 1049)
antenna (p. 1038)
antennal gland (p. 1039)
antennule (p. 1038)
Apterygota (p. 1042)
Arachnida (p. 1037)
book gills (p. 1037)
book lung (p. 1037)
Brachiopoda (p. 1051)
branchial heart (p. 1050)
Bryozoa (p. 1051)
byssal (p. 1048)
carapace (p. 1037)
carnivore (p. 1038)
cephalisation (p. 1046)
cerci (p. 1041)
chelicera (p. 1038)
Chelicerata (p. 1036)
chromatophores (p. 1050)
clitellum (p. 1033)
cocoon (p. 1033)
complete metamorphosis (p. 1043)
compound eye (p. 1041)

coxa (p. 1037)
Crustacea (p. 1036)
endopodite (p. 1038)
exopodite (p. 1038)
exopterygotes (p. 1042)
exoskeleton (p. 1036)
glochidium (p. 1049)
growth zone (p. 1031)
haemocoel (p. 1036)
haemocyanin (p. 1037)
haltere (p. 1041)
hectocotylised arm (p. 1050)
hermaphroditism (p. 1039)
hirudin (p. 1034)
hypopharynx (p. 1041)
imago (p. 1043)
incomplete metamorphosis (p. 1042)
instar (p. 1042)
labium (p. 1041)
labrum (p. 1041)
lamellibranch (p. 1047)
larva (p. 1039)
Lophophorata (p. 1051)

lophophore (p. 1051)
Malpighian tubule (p. 1037)
mandible (p. 1038)
mantle (p. 1045)
mantle cavity (p. 1045)
maxilla (p. 1038)
maxillary gland (p. 1039)
metameric segmentation (p. 1029)
metanephridium (p. 1031)
myriapod (p. 1040)
nauplius (p. 1039)
nautiloid (p. 1049)
Neopilina (p. 1029)
nephridium (p. 1031)
nymph (p. 1042)
ocelli (p. 1039)
omnivore (p. 1038)
opisthosoma (p. 1037)
parapodium (p. 1031)
pedipalp (p. 1038)
peristaltic locomotion (p. 1031)
peristomium (p. 1031)
Phoronida (p. 1051)

prosoma (p. 1037)
prostomium (p. 1029)
protonephridium (p. 1031)
Pterygota (p. 1042)
pupa (p. 1043)
pygidium (p. 1031)
radula (p. 1045)
scavenger (p. 1038)
seta (p. 1032)
siphon (p. 1049)
siphuncle (p. 1049)
soma (p. 1031)
spermatheca (p. 1034)
spermatophore (p. 1042)
spinneret (p. 1038)
tagma (p. 1036)
tagmatisation (p. 1036)
telson (p. 1037)
trachea (p. 1041)
tracheole (p. 1041)
transverse septum (p. 1031)
trochophore (p. 1033)
Uniramia (p. 1036)
vector (p. 1044)
veliger (p. 1046)

Review questions

1. What is meant by metameric segmentation? Refer to an annelid, such as an earthworm, and an arthropod, such as a centipede, to illustrate your explanation.

2. Which animal phyla are coelomate protostomes? Explain what this means.

3. You are collecting in a marine habitat and find an elongated animal with numerous bristles on body segments. Each segment also bears a pair of lateral fleshy lobed appendages. Name the phylum and the class to which this animal probably belongs. Name the type of appendage described and state its function(s).

4. Earthworms move by peristaltic locomotion. Describe what this means and how they do it. Why is internal compartmentation important in the process?

5. What key body features do arthropods have in common?

6. Prepare a table showing the features that distinguish chelicerates, crustaceans and uniramians from one another.

7. Briefly describe the basic mouthparts of an insect and how they are modified in specialist feeders such as a house fly and a female mosquito.

8. List the features of insects that have allowed them to expand into terrestrial environments and indicate why they are significant in this regard.

9. (a) Describe the basic body structure of a mollusc.

   (b) Compare and contrast the method of feeding in a gastropod and a bivalve.

10. In this chapter, cephalopods are described as the 'brainy' molluscs. Explain what is meant by this and highlight the features of cephalopods that relate to their active way of life.

Extension questions

1. Polychaete annelids show enormous diversity in lifestyle. Compare the structure of an active, predatory worm with that of a sedentary, burrowing worm.

2. Discuss the ways in which spiders use silk.

3. Describe some of the adaptations developed by adult gastropods to overcome the disadvantages of anterior voiding of faecal waste (diagrams and illustrations could be useful here).

4. What is meant by direct and indirect development in insects? How do the two differ?

5. The relationship of the lophophorates with other phyla has always been controversial. Why is this?

Suggested further reading

Gee, H. (1995). Lophophorates prove likewise variable. *Nature* 374: 493.

An accessible discussion of the relationship between the older structural and developmental information and the more recent molecular data concerning the phylogenetic position of the lophophorates. Provides references for students who wish to read the original molecular biology papers.

Pechenik, J. A. (2000). *Biology of the Invertebrates.* New York: McGraw-Hill.

A very readable text providing more detail on the material covered here and classification and also a number of examples of recent research involving invertebrates in boxes embedded in the text.

Ruppert, E. E. and Barnes, R. D. (1994). *Invertebrate Zoology.* Philadelphia: Saunders College Publishing.

A detailed account of invertebrate zoology with extensive figures and classification.

Wilmer, P. (1990). *Invertebrate Relationships—Patterns in Animal Evolution.* Cambridge: Cambridge University Press.

A series of high-level, critical evaluations of the evidence for the relationships between the invertebrate phyla.

CHAPTER 40

Echinoderms and chordates

n Chapter 38 we saw that, with the exception of cnidarians and nematodes and some borderline cases in the lophophorates, eumetazoan animals are either protostomes or deuterostomes, depending on their pattern of embryonic development (Chapter 38). Of the coelomate animals, annelids, arthropods and molluscs (Chapter 39) are protostomes. Despite the remarkable differences between an adult sea star and a vertebrate, such as a trout, they are both deuterostomes, in which the anus of the adult develops at the site of the blastopore of the embryonic precursor of the gut (archenteron). Sea stars and their relatives are classified in the phylum **Echinodermata**, while vertebrates and their distant relatives, acorn worms, tunicates and lancelets, are classified in the phylum **Chordata**. The very different body organisations of these coelomate deuterostomes reflect ancient divergence of the two phyla (see Fig. 38.2), probably more than 600 million years ago.

Echinoderms

The phylum Echinodermata includes some 6000 marine species. One of the main sites of radiation of these species is in the Pacific basin and thus Australia has a rich echinoderm fauna. The phylum includes six classes: **Crinoidea**, feather stars and sea lilies; **Asteroidea**, sea stars or star fishes; **Ophiuroidea**, brittle stars and basket stars; **Echinoidea**, sea urchins, sand dollars and heart urchins; **Holothuroidea**, sea cucumbers; and the little known **Concentricycloidea**, which are related to sea stars. Echinoderms are unsegmented coelomate animals. As larvae they are bilaterally symmetrical but develop into radially symmetrical adults, with a basic five-rayed (pentameric) symmetry and no head or brain. Echinoderms are never colonial and are distinguished from all other animals by structural peculiarities of the skeleton and coelom.

The echinoderm skeleton

The skeleton of echinoderms is internal in origin and mesodermal. It is composed of **spicules** or plates, **ossicles**, of protective armour just beneath the skin. Each ossicle represents a single crystal of calcium carbonate (calcite) with small amounts of magnesium carbonate. The crystal, like a spicule, is first formed within a cell and enlarges into an ossicle. Spicules are usually embedded beneath the skin, where they may be single, as in sea cucumbers, articulated, as in sea stars and brittle stars (Fig. 40.1), or sutured together to form a rigid skeleton (test), as in sea urchins and sand dollars. Typically, the body surface has a warty or spiny

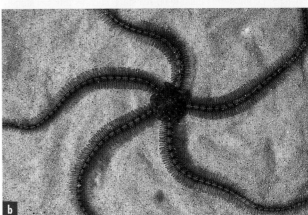

Fig. 40.1 (see caption p. 1058)

Fig. 40.1 Echinoderms include (a) sea stars, such as this sea star, *Coscinasterias calamaria*, from Port Phillip Bay, viewed from below feeding on mussel and showing its tube feet, (b) brittle stars, (c) sea urchins, such as this slate-pencil sea urchin, *Heterocentrotus mammillatus*, and (d) sea cucumbers

appearance because of projecting spines or tubercles, which give the group the name echinoderm, 'spiny skin'.

The water vascular system

Unique to echinoderms is a system of coelomic canals, the **water vascular system** (Fig. 40.2). This includes a circular 'water ring', around the mouth, from which radial canals run to the arms in sea stars and brittle stars or their equivalent areas in other echinoderms. The water ring connects via a single vertical stone canal, so-called because of calcareous deposits in its walls, to a porous disc, the **madreporite**. Water is drawn in through the madreporite by cilia, which form tracts on its surface and line the water vascular system. The radial canals bear rows of **tube feet** (Fig. 40.1). Tube feet function in attachment to the substrate, locomotion, gas exchange and, in predatory forms, such as sea stars, manipulation of prey. Many short, side branches from the radial canals of the water vascular system lead to the tube feet. A tube foot is extended when watery fluid of the water vascular system is forced into it by contraction of a sac, the **ampulla**, at its base (Fig. 40.2). A valve prevents the fluid from being forced back into the lateral canal. The flattened tip of the tube foot forms a sucker. When it comes into contact with the substrate, the centre of the sucker is withdrawn, with the consequent production of a partial vacuum. Adhesion is aided by a sticky secretion.

The tube feet, together with other thin areas of the body, allow excretion of nitrogenous wastes by diffusion. Echinoderms are usually considered to lack excretory organs and the ability to osmoregulate. They are iso-osmotic, conforming to their sea-water environment,

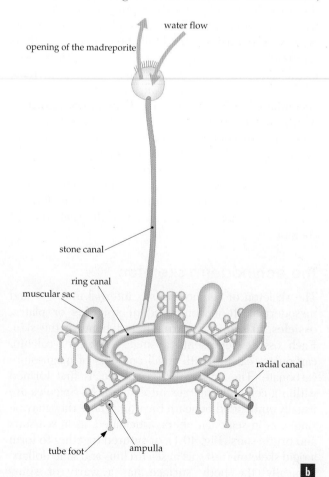

Fig. 40.2 (a) Structure of a sea star, showing **(b)** the water vascular system. The water vascular system operates as a hydraulic system. The ampulla contracts and forces water into the tube foot, which becomes extended. When the flattened, sucker-like end of the tube foot comes into contact with a substrate, the centre of the sucker is withdrawn, creating a vacuum, which results in adhesion. The other parts of the water vascular system (the madreporite, stone canal, ring and radial canals) appear to control the water pressure needed for the ampullae to function

and thus are not found in brackish water. It has recently been suggested, however, that cells termed podocytes, which line part of the water vascular system, act as a type of metanephridium.

Reproduction, development and regeneration

In echinoderms, sexes are usually separate and fertilisation is external after release of gametes directly into the sea. During development the embryo transforms from a single-layered blastula to a two-layered gastrula. In approximately half of all echinoderm species the gastrula becomes a free-swimming **dipleurula** larva (Fig. 40.3). Its winding ciliated bands distinguish it clearly from the trochophore larva of protostomes (Chapter 39). A later larval stage usually develops projecting arms, the arrangement of which distinguishes the various classes. These arms do not survive into the adult and are not homologous with the arms seen in adult sea stars and brittle stars. The larva then undergoes a complete metamorphosis. It settles on its left side, which becomes the oral surface of the adult, the anus being at the other, aboral, pole.

Sea stars and brittle stars have remarkable powers of regeneration and some reproduce asexually by splitting in half and regenerating the missing half. Some holothurians regularly regenerate their entire gut, which they discard as a distraction to predators.

Sea lilies and feather stars: class Crinoidea

Crinoids are the only living echinoderms with the oral surface normally directed upward (Fig. 40.4a), the extended arms being used for suspension feeding on microplankton. They are the only survivors of a Palaeozoic group well represented by fossils (see Fig. 31.12). Sea lilies are sessile, with a five-armed

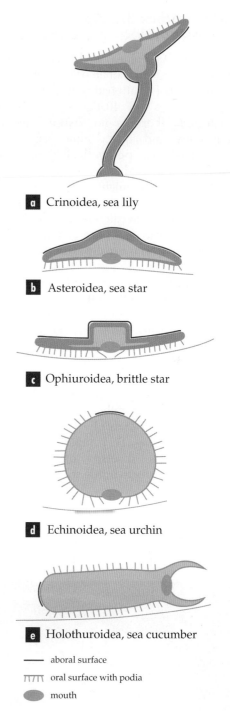

a Crinoidea, sea lily

b Asteroidea, sea star

c Ophiuroidea, brittle star

d Echinoidea, sea urchin

e Holothuroidea, sea cucumber

—— aboral surface

ⴔⴔⴔ oral surface with podia

⬭ mouth

Fig. 40.4 Comparison of the basic body structure of the different classes of echinoderms: **(a)** sea lily; **(b)** sea star; **(c)** brittle star; **(d)** sea urchin; and **(e)** sea cucumber

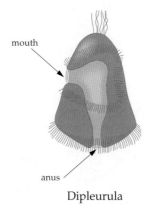

mouth

anus

Dipleurula

Fig. 40.3 An echinoderm larva (a dipleurula) is bilaterally symmetrical, with winding bands of cilia. Compare this larval type to the trochophore of many protostomes (Fig. 39.6)

body attached to the substrate by a stalk. Feather stars, which are free-swimming, detach from their stalk shortly after metamorphosis of the ciliated larva. Crinoids show a great range of often bright colouration, which advertises the fact that they are toxic. They therefore have few predators. They do, however, have numerous parasites and commensals (Chapter 43).

Sea stars and brittle stars: classes Asteroidea and Ophiuroidea

Sea stars (class Asteroidea) and related brittle stars (class Ophiuroidea) are mobile echinoderms that have a body composed of a flattened central disc and radially arranged arms (Figs 40.1, 40.4).

In asteroids there is no distinct demarcation between arms and disc. Some asteroids have pincer-like structures, **pedicellariae**, on their body surface. These are used in protection against small animals or larvae that might settle on the asteroid. *Stylasterias forreri* of the north-east Pacific Ocean uses the pedicellariae to catch small fish. Sea stars are carnivores. Those with short arms swallow prey whole but those with long arms usually evert the stomach and partially digest prey outside the body. *Acanthaster plancei*, the crown-of-thorns sea star (Fig. 40.5), is a major predator of corals, often causing extensive damage to reefs.

Ophiuroids usually have very long arms and a distinct central disc (Fig. 40.4c). They move by horizontal serpentine movements of their arms, which break off very easily, which explains the common names serpent star and brittle star. The movement of the arms is used for locomotion rather than tube feet. They obtain their food in a variety of ways, including deposit and suspension feeding. Their rapid motility, relatively small size and the diversity of food consumed undoubtedly contribute to brittle stars being the largest group of echinoderms.

Sea urchins, sand dollars and heart urchins: class Echinoidea

Echinoids lack arms and the body, with five bands of tube feet, is spherical or flattened along the oral–aboral axis (Fig. 40.4d). Skeletal ossicles form a test or external casing and the animal is typically covered with movable spines (Fig. 40.6). Although sea urchins are radially symmetrical, many soft-bottom dwellers (sand dollars and heart urchins) have varying degrees of secondary bilateral symmetry. Sea urchins feed by rasping algae or other materials from hard substrates using a complex organ of bony appearance, 'Aristotle's lantern', which has five large jaws (Fig. 40.6).

Sea cucumbers: class Holothuroidea

Sea cucumbers, like echinoids, have no arms (Fig. 40.1). Their body has five bands of tube feet and is greatly elongated along the oral–aboral axis (Fig. 40.4e). The resulting worm-like shape means that most sea cucumbers lie with one side of the body in contact with the substrate. As a result, the tube feet in the bands away from the substrate are often reduced. Holothurians are further distinguished from other echinoderms by modification of the tube feet around the mouth as branched tentacles, which are used either in deposit or in suspension feeding. The skeleton consists of spicules (microscopic ossicles) embedded in the body wall.

Phylum Echinodermata includes deuterostome marine animals that have a basic five-rayed symmetry and no head. Their larvae are bilaterally symmetrical but the adults exhibit secondary radial symmetry. A unique feature is their water vascular system connected to tube feet, which function in locomotion, attachment, gas exchange and, in some groups, manipulation of prey.

Fig. 40.5 The crown-of-thorns sea star, *Acanthaster plancei*, is a major predator of corals of the Indian and Pacific Oceans. Population explosions have caused extensive damage to parts of the Great Barrier Reef. An adult sea star everts its stomach over coral and digests the soft-bodied polyps, leaving behind the dead skeleton of the coral. Juvenile sea stars graze on algae. The crown-of-thorns' large size, armour of spines and nocturnal habit enable it to avoid being eaten by most predators

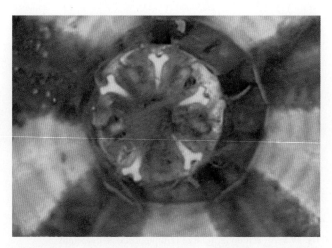

Fig. 40.6 Most sea urchins feed by rasping algae from rocks using a complex organ, Aristotle's lantern, which has five jaws

Chordates

Phylum Chordata includes the subphyla **Hemichordata**, acorn worms; **Urochordata**, tunicates or sea squirts; **Cephalochordata**, lancelets, including *Branchiostoma*, commonly called amphioxus; and **Vertebrata**, fishes, amphibians, reptiles, birds and mammals.

Chordates are so-named because of the presence of a **notochord**, a dorsal, cylindrical rod situated below the nerve cord and above the gut (Chapter 15). The notochord gives the chordate's body longitudinal support and lateral (side-to-side) flexibility. Chordates have **pharyngeal slits**, paired openings in the pharynx, at some stage during development. They are often called gill slits, but the earliest pharyngeal slits were for feeding, acting as exits for a water current entering through the mouth and supplying food. During evolution, pharyngeal slits have become modified for hearing, vocalisation or gas exchange (as true gills) in different chordates. All chordates have a hollow **dorsal nerve cord** that controls body movement.

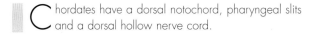
Chordates have a dorsal notochord, pharyngeal slits and a dorsal hollow nerve cord.

Chordates are related to echinoderms

A group of hemichordates termed pterobranchs appears to be the nearest relative of echinoderms. It has been suggested that the water vascular system of echinoderms originated from a lophophore (Chapter 39), which the pterobranchs, alone among the chordates, retain. It is possible, therefore, that chordates and echinoderms are descended from a pterobranch-like ancestor. Recently, a living pterobranch has been discovered, which can be classified in the Palaeozoic group of graptolites. Acorn worms may have lost the lophophore in acquiring a worm-like burrowing body, and tunicates lost the lophophore in favour of filter feeding through gill slits. Larvae of acorn worms are strikingly similar to those of echinoderms (Fig. 40.3), lending further support to an echinoderm–chordate relationship.

Acorn worms and pterobranchs: hemichordates

Acorn worms, the most common hemichordates, live mostly in burrows in shallow marine envirnments. They have three body divisions—a *proboscis*, *collar* and *trunk* (Fig. 40.7). Pharyngeal slits, located in the anterior part of the trunk, filter off excess water from material, often sand and mud, collected by mucus secreted by the proboscis and conveyed by cilia to the mouth. The mouth lies in a groove between the proboscis and collar. The dorsal tubular nerve cord

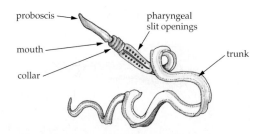

Fig. 40.7 Acorn worms are marine hemichordates, animals possessing gill slits

(neural tube), characteristic of chordates, is situated in the collar. Elsewhere the nervous system resembles that of echinoderms in consisting of a sheet of nerve fibres and cells lying under the epidermis over the entire body. A short structure beneath the dorsal nerve cord at the junction of the proboscis and trunk, and associated with a skeletal plate, is called the **stomochord**. It forms a dorsal diverticulum of the pharynx. The stomochord of hemichordates may not be homologous with a true notochord and, on this basis, the subphylum is often excluded from the Chordata and raised to the rank of an independent, although unquestionably deuterostomatous, phylum.

Blood flows forward in the dorsal blood vessel and backward in the ventral vessel. The dorsal vessel expands into a heart-like sinus, surrounded by a muscular pericardium. The front of the sinus forms a series of glomeruli, which are covered by a region of the proboscis coelom that is specialised for excretion. The entire body surface is ciliated.

Tunicates: urochordates

The relationship between tunicates (subphylum Urochordata) and other chordates is supported by impressive evidence, including the identical mode of development of pharyngeal gill slits. The tadpole-like larva of tunicates (Fig. 40.8a) resembles other chordates, although adult tunicates are very different. Larvae have a hollow dorsal nerve cord and a notochord in the tail, absent in adults.

Most tunicates, totalling about 1300 species, are sessile but others are pelagic. They are marine and worldwide, found in a range of water depths, from intertidal to great depths. They are filter feeders in which the adult body (Fig. 40.8b) is typically covered by a complex **tunic** or supportive and protective 'coat' secreted by ectoderm. The tunic is unique among animals in containing a type of cellulose (tunicin) and, although it lies outside the epidermis, in some animals the tunic includes amoeboid cells and blood vessels.

The pharynx, perforated by numerous pharyngeal slits, typically forms a large basket. Water taken in at the incurrent siphon passes through the pharyngeal slits into a space, the atrium, between the pharynx and

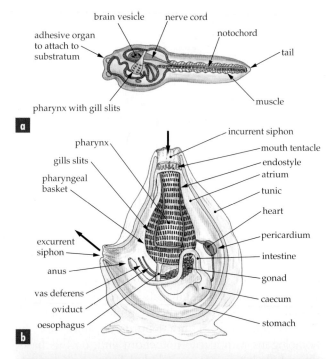

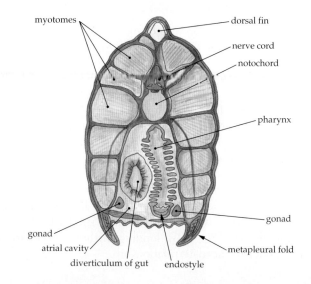

pumps blood through vessels. Most tunicates are hermaphrodites, fertilisation occurring externally or within the atrium.

Lancelets: cephalochordates

Subphylum Cephalochordata includes small marine, fish-like animals (up to 8 cm long) that occur in all tropical oceans (Fig. 40.9). They show typical chordate features: a notochord, pharyngeal slits and hollow dorsal nerve cord. The notochord extends to the tip of the head, where there is a cerebral vesicle but no true brain. Unlike hemichordates and urochordates, lancelets have blocks of muscles, **myotomes**, on each side of the body (Fig. 40.9), as in vertebrate chordates. Although a lancelet swims only sporadically, alternate contraction of the myotomes of the two sides of the body, in a wave from head to tail, propels the animal forward.

The alimentary canal has a liver diverticulum and a pharyngeal endostyle. Numerous pharyngeal slits open from the pharynx into a chamber, the atrium, which opens to the exterior posteriorly at the atriopore and is covered by lateral folds of the body wall, the metapleural folds (Fig. 40.9). A lancelet lies buried in the substrate with buccal cirri exposed and feeds by

Fig. 40.8 (a) The tadpole-like larva of some tunicates resembles chordates and is evidence of their relationship. The larva does not feed and swims for only a short time before undergoing metamorphosis. The tail retracts, parts of the body degrade, the pharynx develops and the pharyngeal slits become functional. **(b)** Structure of an adult tunicate, which is a sessile filter feeder. **(c)** These colonial tunicates, or ascidians, *Polycarpa aurata*, with blue ascidians, *Rhopalea crassa*, are found in coral reefs of Australia and the Indo-Pacific

the body wall before being ejected through an excurrent siphon (Fig. 40.8b). In this way, tunicates can filter an enormous volume of water to extract plankton as food. In addition, a ciliated ventral groove in the pharynx, the endostyle, secretes mucus, which helps to trap plankton. In mobile, non-sessile tunicates, openings for the incurrent and excurrent siphons are at opposite ends of the body and the water current is used for locomotion in addition to feeding and respiration.

The heart, a short U-shaped tube lying in a pericardium, periodically reverses the direction in which it

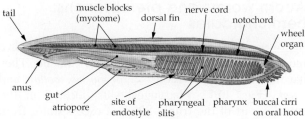

Fig. 40.9 Lancelets, cephalochordates, are small, fish-like marine animals. The structure of *Branchiostoma* is shown here, illustrating true chordate features: notochord, pharyngeal slits, hollow, dorsal nerve cord and muscle blocks (myotomes)

straining minute organisms from the water. The feeding current is maintained by cilia of a 'wheel organ' in the oral hood and by cilia of the gill bars and endostyle. Sensory tentacles are present within the hood. The anus is well anterior to the hind end of the body, leaving a definite postanal tail as is typical of chordate metamerism.

Vertebrates

Vertebrates have a distinct head with a skull (cranium) protecting the brain, eyes and cranial nerves, and are sometimes called the craniate chordates. Most vertebrates, as the name implies, have a backbone composed of **vertebrae**. Vertebrae develop around and replace the notochord during embryonic development (Fig. 40.10). Dorsal projections from the vertebrae form a neural arch that protects the hollow nerve cord. Ventral projections may form a haemal arch, which protects the aorta, the major artery carrying blood posteriorly from the gills and heart. The backbone forms part of the endoskeleton, but most vertebrates have, in addition, an exoskeleton of **dermal bones**, which develop in the skin. Dermal bones include tissues such as dentine and enamel, which often form teeth or tooth-like structures (denticles). Many features unique

to vertebrates develop from an embryonic tissue, the **neural crest**, composed of cells associated with the embryonic nerve cord. Neural crest cells eventually develop into sensory organs and some nerve ganglia, branchial cartilages, dentine of teeth and denticles, and other structures such as the adrenal medulla.

Different groups within the subphylum Vertebrata, including fishes, amphibians, turtles, snakes, lizards, crocodiles, birds and mammals, have characteristic internal organ systems, which are described in Chapters 19 to 22. The following is a comparative overview of the major groups of vertebrates, commencing with the most primitive—jawless fishes.

> All vertebrates have a distinct head with a skull and most have a backbone composed of vertebrae. A key feature is the embryonic neural crest, the cells of which develop into many of the structures characteristic of the subphylum.

Jawless vertebrates

Jawless vertebrates, modern lampreys (Fig. 40.11) and hagfishes, have a notochord, not a vertebral column. They have cartilage but lack bone, although ancient fossil jawless fishes, 4000 million years old, had plates of dermal bone. Modern and fossil forms are classified together in the superclass **Agnatha** solely on the basis of the absence of jaws and the group is an artificial taxon comprising a number of separate evolutionary lineages, lampreys and hagfishes being the only surviving ones.

Fossil agnathans were filter feeders but some lampreys are specialised as parasites on other fishes and hagfishes are scavengers. Lampreys are eel-like in shape and have a mouth like a suction cup. Parasitic forms attach to a host and rasp through the skin with their tongue, sucking blood. Adults of parasitic species are marine but during spring they ascend freshwater streams, where they build nests of stone on the stream bottom in which they spawn. After spawning, adults die. Free-living larvae remain in freshwater streams for some years before they undergo metamorphosis.

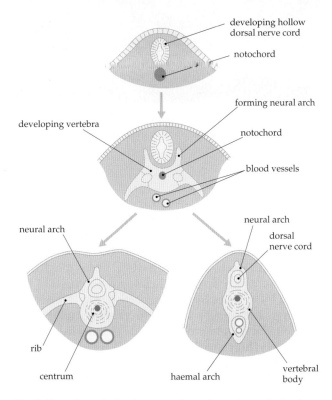

Fig. 40.10 Embryonic development of vertebrae. A vertebral column develops around and replaces the notochord. Paired projections of vertebrae protect the developing dorsal, hollow nerve cord (neural arch; see Chapter 15)

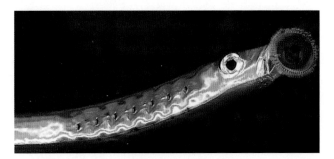

Fig. 40.11 A female pouched lamprey, *Geotria australis*, a jawless vertebrate, with a mouth like a suction cap. Males of the species are unusual in having a pouch in the throat

Jawed vertebrates

Jawed vertebrates (superclass **Gnathostomata**) have jaws derived from anterior branchial arches of the gill system (Fig. 40.12). Evolution of jaws enabled vertebrates to become predators and to attain a wide range of body sizes. Jawed vertebrates include fishes, amphibians, turtles, snakes, lizards, crocodiles, birds and mammals. Within jawed fishes there are two classes—cartilaginous fishes and bony fishes.

> Among vertebrates, agnathans (hagfishes and lampreys) lack jaws. All other vertebrates (gnathostomes) have jaws derived from gill arches.

Cartilaginous fishes

Cartilaginous fishes (class **Chondrichthyes**) include two groups: chimaeras (subclass Holocephali) and sharks, skates and rays (subclass Elasmobranchii). Paired fins, pectoral (anterior) and pelvic (posterior), together with unpaired dorsal, anal and tail fins make sharks agile swimmers. Pectoral fins of skates and rays are extended as wings (Fig. 40.13) and the tail is whip-like, sometimes armed with a poisonous spine. Pectoral and pelvic fins are supported by cartilaginous girdles to which muscles attach. These fin girdles are homologous to the shoulder and hip bones of **tetrapods**.

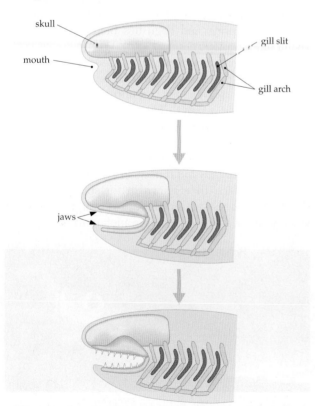

Fig. 40.12 Evolution of jaws by modification of the anterior gill arches of jawless vertebrates

Fig. 40.13 Cartilaginous fishes (class Chondrichthyes) include **(a)** sharks, such as this great white shark, and **(b)** rays, such as this Pacific manta ray

The cartilaginous skeleton is often calcified but it never has true bone. This light skeleton helps chondrichthyans maintain buoyancy despite the lack of a swim bladder or lung. The skin of chondrichthyans is covered with small denticles (p. 1063) and is rough, like sandpaper. Teeth (large denticles) are not fused to the jaws and are successively replaced during life. Nasal openings are usually ventral in position, unlike the double nasal openings on each side of bony fishes. The intestine is short but the surface area for absorption of digestive products is increased by a spiral valve along the length of the intestine. Cartilaginous fishes overcome the problem of conserving water against the dehydrating effects of sea water by osmoregulation: they maintain high blood levels of urea and trimethylamine oxide (see Chapter 22).

All modern forms have internal fertilisation. Pelvic fins of the male are modified as claspers for depositing sperm directly into the female reproductive tract. The embryo is either encapsulated in a leather case or is retained in the oviduct, some sharks giving birth to fully developed young. Both of these methods protect developing young from the osmotic effects of sea water. Chondrichthyans show extreme sensitivity to electric fields, and skates and rays have electric organs that produce either weak or strong discharges.

Bony fishes

All remaining fishes are members of the class **Osteichthyes**, classified into two subgroups: ray-finned fishes, which include most of the world's fishes (subclass Actinopterygii), and fleshy finned or lobe-finned fishes (subclass Sarcopterygii).

The skeleton of bony fishes, as their name Osteichthyes indicates, is made of true bone, at least in part. The skull has sutures; teeth are usually fused to the jaws; fin rays are bony; nasal openings on each side are usually double and lateral in position; and the biting edge of the jaws are usually formed by dermal bones, the maxilla and premaxilla of the upper jaw and the dentary of the lower jaw. The success of bony fishes is attributed in part to the **swim bladder** or lung, a gas-filled sac, which forms as an outgrowth of the pharynx and allows the fish to regulate buoyancy at any water depth and, in some cases, to breathe air.

Internal fertilisation is rare. Even when fertilisation is internal, claspers are absent, although the anal fin is sometimes modified as a gonopodium. Embryos are not encapsulated in a case.

Ray-finned fishes

Ray-finned fishes (subclass **Actinopterygii**) have dermal, ray-like supports within the fins (Fig. 40.14). They include primitive bichirs of African fresh waters, sturgeons of the Northern Hemisphere, paddlefishes of fresh water in North America and China, and the worldwide group, the teleosts.

Primitive ray-finned fishes (bichirs and gars) have unique scales with an outer layer of enamel, termed ganoine, and a lower layer of dentine (normally present in teeth). The structure of the pectoral fin is unique, including an extensive surface for articulation with the shoulder girdle of the endoskeleton. Another feature is the general presence of a single dorsal fin. The internal skeleton does not extend into the fins at the fin base.

Teleosts, with 20 000 species, are by far the most diverse group of the ray-finned fishes (Box 40.1). They include well over half of all vertebrates. Teleosts are distinguished by fusion of vertebrae in the caudal fin and loss of dentine and enamel from the scales.

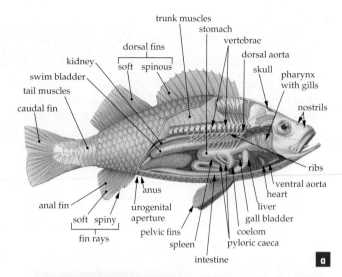

Fig. 40.14 (a) Structure of a ray-finned bony fish (class Osteichthyes). **(b)** An Australian teleost fish, the Murray cod, *Maccullochella peeli*

Coelacanths and lungfishes

Fleshy finned fishes (subclass **Sarcopterygii**), many of which are only known as fossils, include **coelacanths** and **lungfishes**. Coelacanths were known only from Devonian and Cretaceous fossils until a specimen, later named *Latimeria chalumnae* (Fig. 40.15a), was caught off the South African port of East London in 1938. Since then, more specimens have been caught in what appears to be their normal habitat, deep water off the Comoros, islands north of Madagascar. Recently, specimens of an Indonesian population, perhaps a new species of *Latimeria*, have been taken in the local fishery near Manado, north-eastern Sulawesi. *Latimeria* reaches 1.8 m in length and fertilisation of the huge, 9 cm diameter eggs is internal, with birth of well-developed young.

The relationships of *Latimeria* and extinct coelacanths to other vertebrates have been the subject of controversy. Some workers place coelacanths near sharks. However, morphological evidence and new data from DNA and sperm ultrastructure now seem to

BOX 40.1 Teleost diversity

Teleost fishes occur wherever there is permanent water and even as 'annual fishes' in some tropical regions of seasonal drought, where eggs left in drying pools hatch with the advent of the rainy season. Teleosts are classified in about 40 orders, 400 families and 4000 genera.

Primitive teleosts typically lack fin spines and have large cycloid or smooth scales. The pelvic fins, located beneath the dorsal fin in the middle of the body, have more than six rays and the caudal fin has 17 branched rays. There are more than 25 vertebrae, and the upper jaw has two tooth-bearing

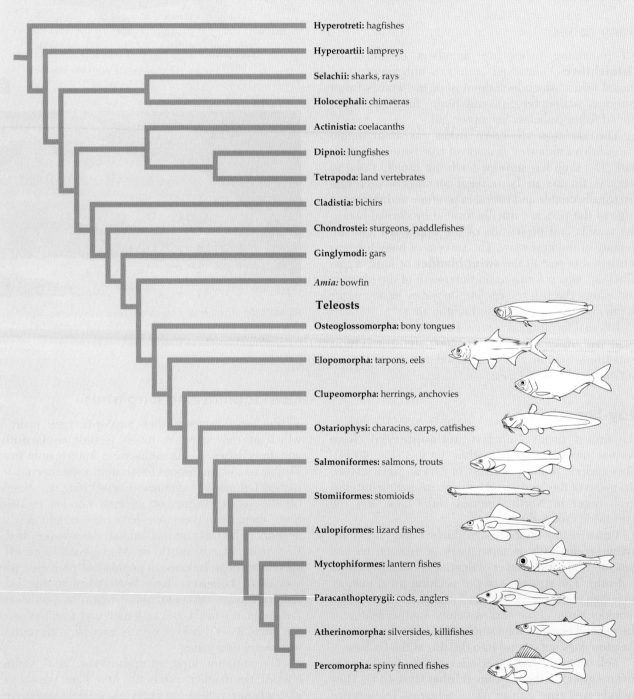

Hyperotreti: hagfishes

Hyperoartii: lampreys

Selachii: sharks, rays

Holocephali: chimaeras

Actinistia: coelacanths

Dipnoi: lungfishes

Tetrapoda: land vertebrates

Cladistia: bichirs

Chondrostei: sturgeons, paddlefishes

Ginglymodi: gars

Amia: bowfin

Teleosts

Osteoglossomorpha: bony tongues

Elopomorpha: tarpons, eels

Clupeomorpha: herrings, anchovies

Ostariophysi: characins, carps, catfishes

Salmoniformes: salmons, trouts

Stomiiformes: stomioids

Aulopiformes: lizard fishes

Myctophiformes: lantern fishes

Paracanthopterygii: cods, anglers

Atherinomorpha: silversides, killifishes

Percomorpha: spiny finned fishes

Phylogeny of fishes

bones (premaxilla and maxilla) on each side, the premaxilla firmly united to the skull.

One large group of primitive teleosts is the Ostariophysi (see figure), so-named because of the bones derived from anterior vertebrae that interconnect the swim bladder and internal ear. Ostariophysans are predominantly freshwater, abundant in all continents except Australia, represented there only by tandan catfishes. They include more than 5000 species of minnows (carps), loaches, electric eels, characins and catfishes. Ostariophysans are best known as the 'tropical fish' of the international aquarium trade. Certain species, such as the goldfish, native to the Far East, have been kept and bred for centuries and are represented by numerous domestic varieties.

Other major groups of primitive teleosts are freshwater bony tongues, including the saratogas of tropical Australia, marine tarpons and true eels; herrings, sprats and anchovies; and salmons, trouts and smelts, including the smelts, *Retropinna*, and native trouts, *Galaxias*, of southern Australia and New Zealand.

Advanced teleosts typically have ctenoid or prickly scales, spines in all fins except the pectorals and caudal; pelvic fins with one spine and five soft rays located anteriorly in the body beneath the pectoral fins; a caudal fin with 15 branched rays; 25 vertebrae; and the upper jaw with a single tooth-bearing bone (premaxilla), freely movable from the skull.

The major group of advanced teleosts is the Percomorpha, including numerous subgroups and more than 10 000 species. The group is named after the genus *Perca*, freshwater perch, of which there are only two closely related species—the yellow perch of eastern North America and the redfin perch of Europe, introduced into southern Australia.

Fig. 40.15 One subclass of bony fishes is the fleshy finned or lobe-finned fishes, including coelocanths. **(a)** *Latimeria chalumnae*, from waters off the Comoros, islands near Madagascar, is the only living example, here shown at a depth of about 200 m off Grande Comore island. **(b)** There are three living genera of lungfishes in the world. One occurs only in Australia, the lungfish, *Neoceratodus forsteri*

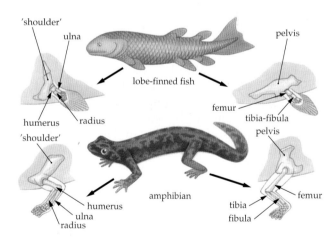

Fig. 40.16 Comparison of the limbs of a fleshy or lobe-finned fish and a tetrapod, showing the homology of structure

unequivocally place *Latimeria* with lungfishes and tetrapods (Fig. 40.16).

There are only three extant genera of lungfishes, one in Africa, one in South America and one, including a single species (*Neoceratodus forsteri*), endemic to Australia (Fig. 40.15b). African lungfishes are freshwater fishes that live in shallow rivers and lakes, which dry up periodically. As the water level recedes, the fish burrows into the mud and makes a protective cocoon of mucus and mud. The single or paired swim bladder, which has become a functional lung, allows the fish to have sufficient gas exchange to survive. The fleshy

lobed fins also allow the fish to 'walk' along the lake bottom.

Of the bony fishes, fleshy finned or lobe-finned fishes, are the closest relatives of four-limbed vertebrates, the tetrapods.

Tetrapods: four-limbed vertebrates

Tetrapods, as their name suggests, have four limbs with separate digits (fingers and toes) instead of the two pairs of fins of fishes. Tetrapods are conventionally divided into the classes **Amphibia**, **Reptilia**, **Aves** (birds) and **Mammalia**. However, the term 'reptile' covers a range of organisms that are not a single evolutionary group. Some reptiles are related to birds and others are related to mammals.

A large number of advanced characteristics is shared between lungfishes and tetrapods, suggesting that they are related. These characteristics include the presence in lungfishes of a pair of internal nostrils (nares), although in lungfishes the nares are not used for breathing as they are in tetrapods. Other shared characteristics include: type of gill arch muscles; ciliation of the larvae; structure of the pelvic girdle; structure of the pelvic and pectoral appendage; dermal bone pattern covering the brain case; and numerous features of the soft anatomy, such as the partially divided conus arteriosus (the anterior continuation of the heart), and division of the atrium of the heart by a septum.

Embryologically, lungfishes resemble amphibians and differ from other fishes in that cleavage is total and gastrulation produces a yolk plug.

Living in water and on land: amphibians

The class Amphibia includes the oldest land vertebrates, although most modern species require water in which to reproduce. Most have an aquatic larval stage. Modern amphibians, the **Lissamphibia** (Fig. 40.17), include salamanders, mud puppies and newts (order Urodela), frogs and toads (order Anura), and worm-like, legless forms of the tropics (order Gymnophiona or caecilians). They are basically quadrupedal, tailed vertebrates that have two occipital condyles, articulating the skull with the neck vertebrae and a single sacral (pelvic) vertebra. The skin is glandular and lacks epidermal structures such as scales, feathers and hair, which are typical of the other vertebrates. It is moist and used for gas exchange (Chapter 20). They are the first vertebrates to have a true tongue. Amphibians have the antidiuretic hormone vasopressin, which decreases the amount of water excreted (Chapter 22). Eggs are laid in fresh water and fertilisation is usually external, although some amphibians have internal fertilisation.

In extant amphibians, the number of bones in the skull is greatly reduced relative to fossil amphibians and there is a number of unique characteristics. Typically,

Fig. 40.17 The oldest tetrapods are amphibians, including **(a)** modern frogs such as *Litoria moorei* and **(b)** the barred tiger salamander

modern amphibians have a maximum of four digits on the forefeet. In some salamanders, hind limbs and pelvic girdles have been lost. Caecilians have lost both sets of limbs and girdles (pectoral and pelvic). Frogs are characterised by elongation and modification of the hind limbs for jumping. Lungs are lost in some salamanders and frogs, and the left lung is reduced in caecilians.

Amphibians are land vertebrates but most require water for reproduction. Their skin is glandular and moist and used, usually with lungs, in gas exchange.

An embryo with a pond of its own: amniotes

All tetrapods other than amphibians are amniotes. In the Amphibia, the whole of the fertilised egg, the zygote, contributes to the structures of the embryo. In amniotes, large regions of the early embryo give rise to a number of extra-embryonic membranes including one, the **amnion**, which encloses the embryo in a fluid-filled sac, providing the embryo with an aquatic environment—a pond of its own (Chapter 15). This is a major adaptation to life on land. Correspondingly, fertilisation is always internal. Many amniotes lay eggs but in others live young are born.

Amniotes have a set of distinctive features in addition to the amnion. The **allantois**, an outgrowth of the embryonic hindgut, is used for excretion during development. The amniote skin is thick and horny, and epidermal scales (or feathers or hair) provide a water-resistant barrier suitable for a terrestrial environment. Unless the animal is live-bearing, eggs have shells and are generally rich in yolk. Cleavage of the fertilised egg is not total, only part of the egg dividing into cells (**meroblastic cleavage**; Chapter 15). Development from egg to juvenile and adult is direct. In the skeleton, **intervertebral discs** are formed and the first two neck vertebrae are differentiated into an anterior **atlas**, supporting the skull, and a posterior **axis**. Usually the atlas is capable of rotation, allowing movement of the head. There are at least two sacral (pelvic) vertebrae, contrasting with only one in Amphibia.

A vexing question of classification and phylogeny of vertebrates is the position of the so-called reptiles within the Amniota. The term reptile signifies to most people 'cold-blooded' amniotes with scales rather than 'warm-blooded' amniotes with feathers or hair but it is improbable that reptiles are a single group. Mammals are commonly regarded as separate from but related to all other amniotes, giving two groups, the Mammalia

and the Sauropsida (Fig. 40.18). Sauropsids include turtles, the New Zealand tuatara, squamates (geckos, lizards and snakes), crocodiles and birds. The name **Sauropsida** is used instead of Reptilia since the birds are within this group (Fig. 40.18). The term 'reptile' should perhaps be restricted to common usage and will here be used in a descriptive sense.

Several other hypotheses of phylogenetic relationships exist and have yet to be disproved. Most workers agree that crocodiles and birds are related, with their extinct precursors, dinosaurs, forming the **Archosauria**. It is less certain the tuatara and squamates constitute a group.

> All tetrapods other than amphibians are amniotes. Extra-embryonic membranes, including the amnion, surround and protect the embryo and are a major adaptation to life on land.

Turtles and their relatives

Turtles (Fig. 40.19) and their relatives (order **Chelonia**) have all the basic amniote features mentioned above. Some of them live in water but others are successful on land. On land they have a sprawling gait, presumed typical of the first four-legged land vertebrates. They have a protective, rigid trunk armour consisting of a dorsal shield, **carapace**, and a ventral shield, **plastron**, covered by horny plates. A feature unique to chelonians is that the shoulder girdle is

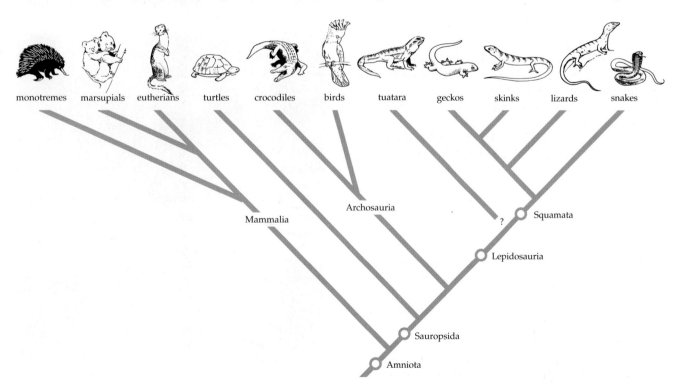

Fig. 40.18 One hypothesis of the relationships of amniotes. Note that 'reptiles' do not form a single evolutionary group (i.e. they are not monophyletic). The name Sauropsida is used for the group that includes turtles, squamates (lizards etc.), crocodiles and birds

Fig. 40.19 Female green turtles remember their terrestrial origin by coming ashore to lay their eggs

snake can open its jaws wide and swallow the prey whole because its jaws can be disarticulated.

The tuatara, *Sphenodon* (Fig. 40.20), is a 'living fossil', the only surviving member of its order, found today on a few offshore islands of New Zealand. It differs from lizards and snakes (squamates) in having

Fig. 40.20 (a) The Australian western blue-tongued lizard and **(b)** the children's python are two Australian examples of squamates

within the rib cage. The skull is **anapsid** (see Fig. 40.21a), lacking the one or two pairs of openings in the temples that accommodate the jaw muscles of other reptiles. The jaws are toothless, being covered by a horny sheath.

The relationship of turtles to other groups of amniotes is a subject of much debate but they are probably at the base of non-mammalian amniotes (Fig. 40.18).

Snakes, lizards and their relatives

Snakes, lizards, goannas and their relatives (**Lepido-sauria**, Figs 40.18, 40.20) have teeth that are firmly fused to the edges of the jaws. A common feature of lizards is the ability to shed the tail at preformed points of fracture in several tail vertebrae, a device that is effective in diverting a predator's attention to the writhing tail, allowing the escape of the tailless individual. In snakes, external limbs have been lost, probably as an adaptation to a subterranean existence. Many snakes are poisonous, manufacturing venom in their salivary glands. After immobilising its prey, a

Fig. 40.20 (c) Distantly related to squamates is the New Zealand tuatara, *Sphenodon*, the only surviving member of its order. It has the diapsid condition of two well-defined temporal openings in the skull

two well-defined temporal openings, the **diapsid** condition (Fig. 40.21b), in its sperm ultrastructure, and in having a rigidly fixed upper jaw with a beak-like process. Its pineal or 'third eye' is better developed than in most other amniotes.

Crocodiles and birds

Crocodiles and birds (Archosauria) share a large number of features. Extinct fossil birds and crocodiles have teeth inserted into deep pits. Teeth, so conspicuous in living crocodiles, are present only in embryos of living birds. A gap (foramen) in the skull over the pineal is closed over. In addition to two temporal openings typical of diapsids there is a further opening in front of the eyes (preorbital opening).

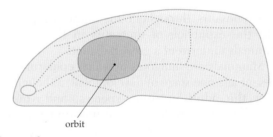

a Anapsid

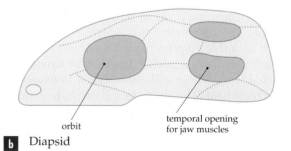

b Diapsid

Fig. 40.21 (a) An anapsid skull is characteristic of chelonians (turtles and their relatives). This is in contrast to the skulls of other sauropsids (lizards, crocodiles and birds), which have **(b)** the diapsid skull, with one or two pairs of openings in the temples to accommodate jaw muscles

Crocodiles (Fig. 40.22), caimans and alligators are large carnivores of the tropics and subtropics. In Australia, there are two species of crocodile, the freshwater crocodile, *Crocodylus johnstoni*, and the larger, more dangerous, saltwater crocodile, *Crocodylus porosus*. They spend most of their time in water and venture on land to catch prey and to lay their eggs. Many peculiar features can be attributed to their semiaquatic mode of life. As examples, the nostrils are united to form one opening at the tip of the snout; the tail is compressed laterally as the main swimming organ; and the pupils are vertically elongated.

Birds (class Aves) are a highly evolved group and are phylogenetically younger than mammals. The oldest bird fossil, *Archaeopteryx lithographica*, was found in limestone dated at 150 million years ago, in the Jurassic (Chapter 31). *Archaeopteryx* had clawed forelimbs (seen on the wings of one modern bird from South America), teeth and a long tail containing vertebrae. Little more than its feathers distinguish it from those dinosaurs (a name that means 'terrible lizards') called theropods.

Most of the peculiar features of birds are connected with the acquisition of flight. Birds, although basically quadrupedal animals, have forelimbs modified as wings and skin covered by **feathers**, which are modified scales (Fig. 40.23). In recent birds there are specialised flight feathers on the wings, and specialised tail feathers are attached to a short, plate-like structure formed by fused caudal vertebrae, the **pygostyle**, unlike the long tail of *Archaeopteryx*. There is a large, bony keeled **sternum** for insertion of the larger, pectoral flight muscles. Bones tend to be hollow, making the skeleton light.

The jaws of modern birds are covered by a horny bill. Like turtles, crocodiles, lizards and snakes, they have only one knob on the back of the skull (the occipital condyle) for articulation with the vertebral column, in contrast with two in amphibians and mammals. All birds are endothermic with a four-chambered heart and right aortic arch, compared with the left retained in mammals. The brain is highly developed and birds have excellent vision, hearing and balance. They lay eggs with a large yolk and a shell and these are incubated by the adult. Many birds are social and have elaborate courtship involving songs, bright plumage and displays (Chapter 28). Colour vision, a basic feature of sauropsids, is correspondingly highly developed.

There are three groups of extant birds (Fig. 40.24). One is the penguins, which have wings modified for swimming. The second group is the Southern Hemisphere **ratites**, mostly flightless birds, including the tinamou, ostrich, rhea, emu and cassowary, all of which lack the keel on the sternum. The third group includes the vast majority of birds, more than 9000 species. They

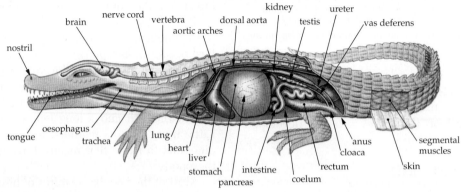

Fig. 40.22 (a) Australian crocodile, *Crocodylus porosus*. Crocodiles and alligators lay eggs on land and their skin has horny scales. They share a number of features with birds, to which they are related. **(b)** Anatomy of a crocodile

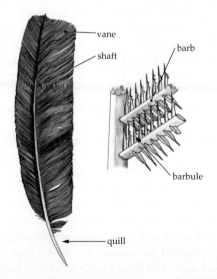

Fig. 40.23 Structure of a feather. The vanes or secondary branches are linked together, allowing the feather to function in flight

have a keeled sternum, like penguins, but unlike both other groups they have contour feathers arranged in a regular fashion.

> Within amniotes, birds are a highly evolved group with unique features associated with flight. They have feathers and a sternum for insertion of flight muscles. Their closest living relatives are crocodiles.

Mammals

Mammals (class Mammalia) appear from the fossil record to have originated from mammal-like reptiles, therapsids. The oldest fossil mammals date from the Triassic period, 190 million years ago. They coexisted with dinosaurs throughout the Mesozoic period but with the extinction of dinosaurs in the Cretaceous, mammals underwent a great radiation.

Mammals are amniotes in which epidermal **hair** is present at some stage of development, although absent from the adult in some species, and young are nourished by secretions of the **mammary glands** of the female. They are endothermic, maintaining a high internal body temperature. Mammals, like crocodiles and birds, have a four-chambered heart and double circulation (Chapter 21). As in birds, the heart is fully divided into two auricles and two ventricles but it is the left aortic arch that empties the left ventricle, the right arch being absent from adults. An efficient circulatory system is able to deliver sufficient oxygen to tissues to maintain body heat. The ability to maintain a high body temperature, together with the insulating effect of hair, probably allowed mammals to survive in a range of habitats.

> Mammals are amniotes that have hair at some stage, and mammary glands for feeding young. Mammals appear to have evolved from mammal-like 'reptiles' called therapsids.

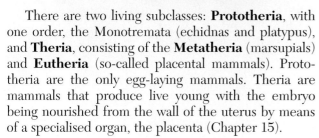

Fig. 40.24 Birds include: **(a)** flightless emus; **(b)** penguins, with wings modified for swimming; **(c)** black swans; and **(d)** songbirds, such as this brown songlark

There are two living subclasses: **Prototheria**, with one order, the Monotremata (echidnas and platypus), and **Theria**, consisting of the **Metatheria** (marsupials) and **Eutheria** (so-called placental mammals). Prototheria are the only egg-laying mammals. Theria are mammals that produce live young with the embryo being nourished from the wall of the uterus by means of a specialised organ, the placenta (Chapter 15).

Monotremes

Monotremes are now represented by the web-footed, duck-billed platypus, *Ornithorhynchus anatinus* (Fig. 40.25), and the echidnas, *Tachyglossus* and *Zaglossus*, which are restricted to Australia, New Guinea and some neighbouring islands. Survival of these monotremes is probably accounted for by their individual peculiarities. Thus, the platypus has an aquatic way of life in an environment where there are few competitors or predators. Echidnas typically have strong protective spines, similar to those of the eutherian hedgehog in the Northern Hemisphere, and

are remarkable in their highly adaptive torpor (Chapter 29). In both echidnas and the platypus, there is a large horny venomous spur on the ankle of the male, evidence that monotremes are a single evolutionary lineage.

Among living mammals, monotremes have the most ancient lineage and retain several basic amniote

Fig. 40.25 The platypus, *Ornithorhynchus anatinus*, is an egg-laying, protatherian mammal (monotreme)

features. These include, among others, the presence of 'reptilian' bones in the pectoral girdles; shell glands in the female reproductive tract; and a cloaca. They have two completely separate uteri and, correlated with this, the penis is bifurcate. There is no scrotum as the testes are retained in the abdomen. Eggs of monotremes are incubated by the parent. Although the young feed on a milky secretion from mammary glands, there are no nipples.

Marsupials

Metatheria, **marsupials** (Fig. 40.26), occur mostly in the Australian region (see Chapter 41) but also in the Americas. The young of marsupials are nourished from the uterine wall via a yolk-sac placenta. However, a simple chorioallantoic placenta typical of eutherians, although lacking villi, occurs in bandicoots. As in monotremes, the two uteri are completely separate, there are two distinct lateral vaginal canals and the penis is bifurcate.

Marsupials are characterised by birth of young at a stage that appears equivalent to an early fetal stage in eutherians. The young then actively find their way, unaided, to the nipples. There is usually a pouch, the **marsupium**, for concealment and carriage of the young (Fig. 40.27). Marsupials are highly adapted to their environment, including carnivorous and herbivorous species (Chapter 41).

Eutherian mammals

Eutheria include a diversity of forms, totalling about 4500 species. Eutherians are viviparous mammals with a chorioallantoic placenta, formed from the chorion and the allantois, both extra-embryonic membranes of the developing young, and a portion of the uterine wall. We will avoid using the common name 'placental mammals' because of the presence of a placenta, albeit of different structure, in marsupials.

Typically in eutherians, and unlike marsupials, finger-like villi project from the embryonic portion of the placenta into the portion formed by the wall of the uterus. The two uteri may be separate to completely fused but there is only a single terminal tube, the vagina. In males, the penis is not bifurcated, a scrotum may be present and, if so, is situated posterior to the penis in all eutherians except rabbits and hares (order Lagomorpha), in which it is anterior, as in marsupials.

Fig. 40.26 Marsupials include: **(a)** the Eastern quolls and **(b)** the bilby

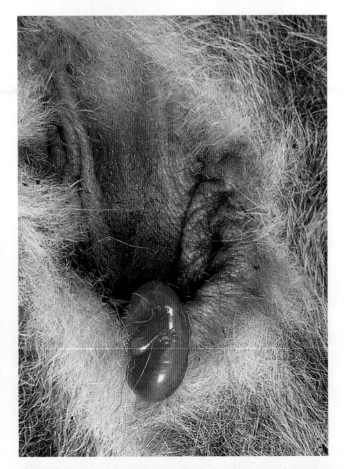

Fig. 40.27 Marsupials are characterised by a pouch. Here the young of a kangaroo is entering a pouch

From fossil evidence it appears that eutherians and marsupials diverged from a common ancestor about 80 to 100 million years ago. The existing diversity of orders of the Eutheria (Fig. 40.28) evolved during the late Cretaceous and early Tertiary periods, about 70 to 45 million years ago from a common stock of small insectivorous mammals. In Australia, marsupials became the predominant mammalian fauna. Eutherians appear to be relatively recent immigrants from the north (Chapter 41) but recently a single eutherian tooth has been found in Cretaceous deposits in Queensland.

Echidnas and platypuses (monotremes) lay eggs and are the only living members of the subclass Prototheria. The subclass Theria, including marsupials and eutherians, produce live young, with the embryo being nourished by a placenta.

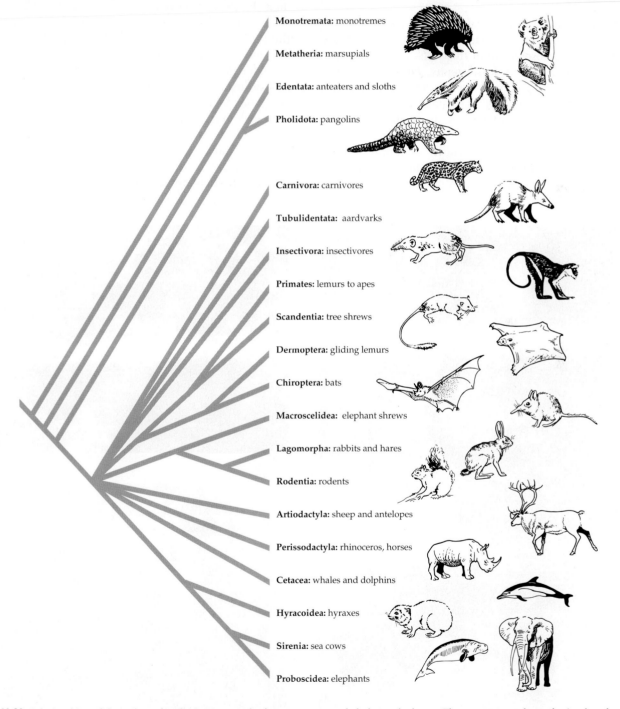

Monotremata: monotremes

Metatheria: marsupials

Edentata: anteaters and sloths

Pholidota: pangolins

Carnivora: carnivores

Tubulidentata: aardvarks

Insectivora: insectivores

Primates: lemurs to apes

Scandentia: tree shrews

Dermoptera: gliding lemurs

Chiroptera: bats

Macroscelidea: elephant shrews

Lagomorpha: rabbits and hares

Rodentia: rodents

Artiodactyla: sheep and antelopes

Perissodactyla: rhinoceros, horses

Cetacea: whales and dolphins

Hyracoidea: hyraxes

Sirenia: sea cows

Proboscidea: elephants

Fig. 40.28 Relationships of the orders of eutherian mammals, from anteaters and sloths to elephants. The monotremes (prototherians) and marsupials (metatherians) are shown at the base of the tree

Evolution of primates

The order **Primates** (Fig. 40.29) was named by Linnaeus, who included in it not only the primates as we define them but also the bats (Chiroptera) and flying lemurs (Dermoptera). In 1781, Thomas Pennant rejected the grouping primates, solely 'because my vanity will not suffer me to rank mankind with Apes, Monkeys, Maucacos and Bats, the companions Linnaeus has alloted [*sic*] us'. The order continued to be recognised, however, although without inclusion of bats.

Early primates were probably small insectivore-like arboreal animals. Whether the living tree shrew, *Tupaia*, is representative of this stock is the subject of continuing debate. A fossil genus, *Purgatorius*, from the Cretaceous–Tertiary boundary of North America, about 65 million years ago, is thought by some to be the oldest known primate. However, its fossils are no more than teeth and jaw fragments and its primate affinities are questionable. The earliest undoubted fossil primates are lemur-like animals from the Eocene of Europe and North America, 55 million years ago. In a recent study, the origin of primates has been estimated to be as early as 90 to 100 million years ago.

Premolar teeth of primates are **bicuspid**, that is, with two cusps, while molars have three to five cusps. The nose is moderate to very short and eyes are well developed, with binocular vision. The brain is usually large, exceptionally so in apes, including humans. Limbs are plantigrade—the whole foot, or hind foot in apes, touches the ground. Digits are prehensile (grasping) and thumbs are apposable to the second digit. Fingers are very mobile and sensitive. Claws are often replaced with nails and there is always a nail on the big toe. A well-developed clavicle (collar bone) is present and allows rotation of the forelimb on the shoulder girdle for swinging from branches, aided by lack of fusion of the long bones of the arm, the radius and ulna. A long tail is a primitive feature of primates such as lemurs. A tail is lost in apes and some others. The uterus is usually bicornuate (two-horned) and the penis is pendent. Births are usually single and there is long parental care and a progressive deferment of maturity, particularly in humans.

Fig. 40.29 Primates include **(a)** this siamang gibbon, hanging from a branch; **(b)** this male gorilla; and **(c)** these black-faced monkeys

Prosimians

Primates are classified into two main groups, the **Strepsirhini** and the **Haplorhini**, named for features of the nostrils (Figs 40.30, 40.31).

Strepsirhini are **prosimians**, meaning precursors of monkeys. A naked nose pad (rhinarium) surrounds both nostrils. The nose pad has a marked median groove and is attached to the gums anteriorly. Each nostril has a crescentic lateral slit, hence the group's name. Examples are lemurs and the aye-aye, from Madagascar, African galagos and South-East Asian lorises and their African relatives, the pottos.

Monkeys, apes and humans

Haplorhini do not have the bilateral slit at the nostrils. They include the anthropoids or human-like animals (**Anthropoidea**). Whether or not tarsiers should also be included in the haplorhines is debatable.

There are two groups of anthropoids. One is the New World monkeys, the **platyrrhines** (meaning flat-nosed), which evolved in South America. They have nostrils far apart and earholes (not a simple tube). Many of these monkeys are arboreal (tree-dwelling) and many have a prehensile tail, a feature never seen in Old World monkeys. Any monkey you see hanging from its tail would be from the New World!

The second group of anthropoids is the **catarrhines** (downward-nosed). They have nostrils close together and the earhole leads to a tube, as in humans. They are the Old World monkeys, apes and humans, some of which are arboreal. All of these forms occur in the Oriental region and only gibbons are absent from the Ethiopian region (Africa, south of the Sahara). Monkeys extend into the Australian geographical region on Sulawesi (Celebes).

The great apes have been traditionally set apart from humans as a separate family, **Pongidae**, but it is apparent from mitochondrial DNA (Chapter 30) that chimpanzees and the gorilla are more closely related to humans than they are to other apes and monkeys. It is clearly inconsistent with the evidence to place humans in a different group from great apes. In a new and somewhat radical taxonomic system, therefore, humans (*Homo*) are grouped with chimpanzees (*Pan*)

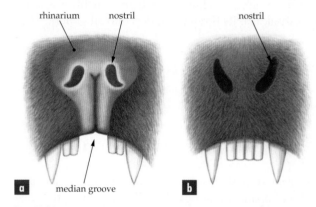

Fig. 40.31 Primates are classified into two groups, Strepsirhini and Haplorhini, on the basis of features of the nostrils. **(a)** In the strepsirhine condition in lemurs and lorises, the nostrils are surrounded by the rhinarium, which has a marked medium groove. **(b)** In haplorhines, hairy maxillary processes have completely obliterated the rhinarium

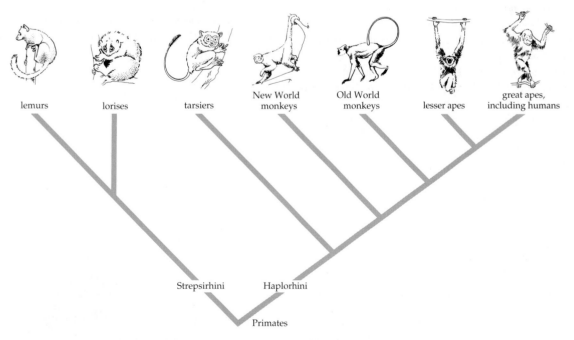

Fig. 40.30 Phylogenetic trees of primates have been constructed from estimates of the genetic distances between the different groups by cross-hybridisation of the total genome of single-copy DNAs and from DNA sequences of β-type globin. The various trees agree closely with one another and with those based on morphological characteristics

and gorilla (*Gorilla*) in the tribe Homini and, with the orang-utan (*Pongo*) and gibbons (*Hylobates*), in the family Hominidae (previously restricted to humans).

From DNA and morphological studies it is deduced that apes descend from a catarrhine ancestor shared with Old World monkeys. The earliest known fossil apes are *Catopithecus* and *Proteopithecus*, described from Egypt in 1990 and of Eocene age (36–55 million years ago). They are believed to have been precursors of early apes that spread from Africa into Eurasia in the Miocene, from about 25 million years ago. From this stock arose the modern apes. Of these, the orang-utans and gibbons remained arboreal, but the ancestor of the gorilla–chimpanzee–human tribe presumably was more terrestrial, like its three living descendant genera.

> Primates have bicuspid teeth, short noses, well-developed eyes and brains, prehensile digits, and usually long parental care of offspring.

Human evolution

We have seen that humans share a common ancestry with chimpanzees and gorillas. The genetic difference between the chimpanzee and humans is, in fact, extremely small and is less than would be expected from morphological differences between the two genera. It appears that in these apes small genetic shifts may give quite large morphological, and behavioural, changes (Chapter 32). To state that humans are the closest living relatives of chimpanzees does not, however, imply that, after divergence of the ancestor of the present-day chimpanzees, the ancestors of modern humans, *Homo sapiens*, may not have given rise to other human-like species that have become extinct. In fact, the fossil evidence suggests that there have been such sidelines in human evolution.

Africa is the place of many fossil deposits, with sites in South and East Africa being most famous. The Rift Valley of East Africa is marked by a series of great lakes, with thousands of metres of accumulated sediments, ideal for trapping fossils. The region includes Olduvai Gorge in Tanzania, explored extensively by Mary and Louis Leakey, and sites in Ethiopia and Kenya.

Australopithecus: upright apes, the first hominids

The oldest known fossils of **hominids**, meaning human-like—walking upright and with hands and teeth similar to ours—are those of ***Australopithecus*** ('southern ape'). The first fossil, discovered in 1924 in a limestone mine site in South Africa, is *A. africanus*. It lived in the African savanna from about three million to

one million years ago. In 1974 a further species, *A. afarensis*, was described from a remarkably complete female skeleton, only about a metre tall, found in the Afar region of Ethiopia. The age of Lucy (Fig. 40.32), as she was nicknamed (after a Beatles song that was being played on a tape at the time of her discovery), has been the subject of controversy but is regarded as about 3.5 million years old. More recently, in 1992, the oldest australopithecine fossil so far found was discovered in the Awash region of Ethiopia, at a site that was once a fairly densely wooded environment. This fossil was named *A. ramidus* and is dated at 4.4 million years old. Based on its teeth, it is the most primitive species and is placed at the base of the hominid phylogenetic tree (Fig. 40.33).

The erect posture of *Australopithecus* has been assumed to be an adaptation for allowing a creature with few defences (it did not even make tools) to detect predators in long savanna grass. But given the evidence that *A. ramidus* was not a savanna dweller, it probably had its origins in using prehensile hands. Functions of hands include grasping food (Fig. 40.34), holding young and possibly nest building, which is seen in the gorilla. Apes, like monkeys, have retained the prehensile abilities of the feet, being able to appose the big toe to the other toes just as the thumb can touch the tips of the other fingers; however, the human foot has lost this ability. The big toe projects forwards and is longer than the other toes, providing thrust in the striding **bipedalism** that distinguishes human locomotion. Even the gorilla and chimpanzee, the most bipedal of the apes, normally progress when on the ground by 'knuckle walking' (Fig. 40.35).

Lucy shows that upright walking developed before expansion of the brain. Her brain had a volume of only 400 mL, although that of *A. africanus* was somewhat

Fig. 40.32 Reconstruction of the fossil 'Lucy', possibly 3.5 million years old

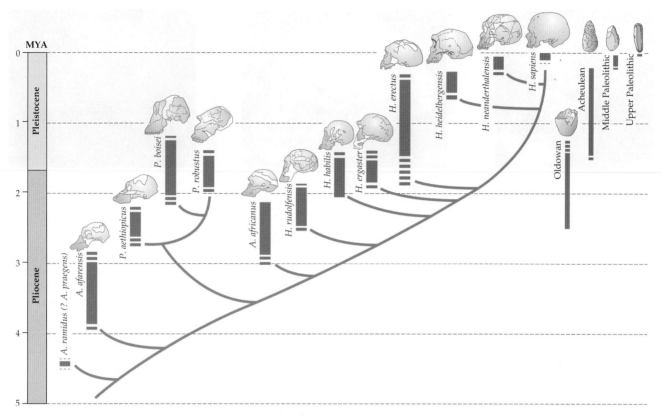

Fig. 40.33 A suggested phylogenetic tree showing the relationships of species of hominids. The stratigraphic range of fossils and the geologic time scale are shown, together with skulls and stone tools (known as Oldowan, Acheulean and Paleolithic tools)

Fig. 40.34 Why did our ancestors take to upright walking? Bipedalism may have allowed our ancestors to grasp objects, as this chimpanzee is doing, and hence carry food away or reach fruit in trees

larger, at 500 mL (Fig. 40.36). They all had a forward-jutting face, poorly developed brow, a more or less pronounced ridge above the eyes, and none had a chin (Fig. 40.36a). Their hands were similar, and their molar teeth relatively flat, like ours. Male australopithecines were much bigger than females, although even males were only about 1.3 m tall.

Paranthropus: robust forms

A group of early hominids that were more robust in form than *Australopithecus* has been placed in a

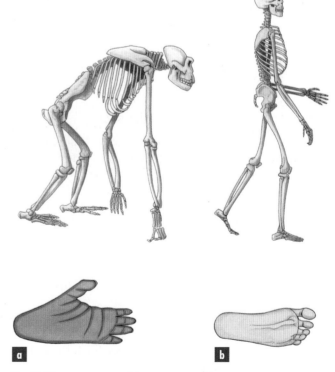

Fig. 40.35 Comparison of **(a)** a gorilla and **(b)** a human, illustrating skeletal changes associated with the evolution of bipedalism. Note the shortening of the pelvis, the S-shape of the vertebral column, and the position of the head in the human. In the human, the big toe is the longest digit and not divergent as in the gorilla

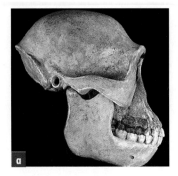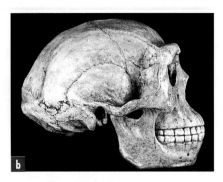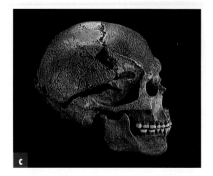

Fig. 40.36 African hominid skulls: **(a)** *Australopithecus africanus*; **(b)** *Homo erectus*; **(c)** *Homo sapiens*. Note the progressive expansion in size of the brain case and gradual reduction in size of the jaw relative to the overall size of the skull

separate genus: **Paranthropus**. (They have been previously classified under *Australopithecus* as well as other generic names.) Three species have been described and dated from 2.8 to 1.6 million years old: *P. robustus*, *P. boisei* and *P. aethiopicus* (Fig. 40.33). These fossils are distinguished by a marked crest on the top of the skull, enormous cheek teeth and a powerful jaw. They had a cranial capacity of 550 mL and are thought to have been vegetarian. Stone tools and pieces of antelope bones, which appear to have been used for digging in the ground for tubers and other plants, occur with these fossils. In the same deposits are fossils of our genus, **Homo**, and it is thought that *Homo* used the stone tools and *Paranthropus* the digging tools. *Paranthropus* appears to be an evolutionary line that became extinct (Fig. 40.33).

> The earliest hominids are *Australopithecus* and *Paranthropus*. Bipedal, they stood upright, but they had a relatively small brain size.

Homo: increase in brain size

A. africanus is considered the closest australopithecine to *Homo*, the genus to which we belong. The two most primitive fossil species of *Homo* are *H. rudolfensis* (the oldest at about 2.5 million years old) and *H. habilis*, first discovered by the Leakeys in Olduvai Gorge. These early forms of *Homo* coexisted with the smaller brained *Australopithecus* and may well have brought about its extinction. *Homo* is larger in size, has a flattened face and no brow ridge, smaller molars and premolars and a significantly larger brain up to 800 mL. Expansion in the brain case was accompanied by gradual reduction in the size of the jaws relative to the overall size of the skull. *H. habilis* was a tool maker, striking flakes off both sides of stones. Such stone tools have been dated at up to 2.6 million years old.

More modern species, with a brain capacity of 800–1300 mL, are *H. ergaster*, and *H. erectus*. *H. ergaster* is known from Africa (Turkana Boy is a near complete skeleton of an 11–12 year old boy found near

Lake Turkana in 1984), while *H. erectus* is the first species found outside Africa, in South-East Asia. Early finds of *H. erectus* were named Java Man (in Indonesia) and Peking Man (in China). Fossils of this species date from 1.5 million years to about 300 000 years ago. The intelligence conferred by the larger brain size of *H. erectus* (Fig. 40.35) equipped it to make more sophisticated tools than those of *H. habilis* and its wide distribution indicates its success.

> *Australopithecus africanus* is the closest relative of the genus *Homo*, the hominid line that leads to our species. The genus *Homo* is characterised by an increase in brain size and tool making. *H. erectus* migrated out of Africa about two million years ago and successfully spread as far as South-East Asia.

'How, when and where did *H. sapiens*, our species, evolve?' is a complex question. Around one million years ago, hominid brain sizes were close to the modern average and fossils from the last 500 000 years are often described as 'archaic' forms of *H. sapiens*. Yet they still looked different from us, with retreating foreheads, brow ridges, large faces and biggish teeth, so they are usually classified as different species (Fig. 40.33). *Homo heidelbergensis* (400 000–200 000 years old) is one distinct form. *Homo neanderthalensis* is another, found originally in the Neander Valley in Germany. **Neanderthals** are first recorded from about 130 000 years ago. They had brains as large as ours and made elaborate tools. The fact that they left evidence of rituals and burials suggests that they were capable of abstract thought. *Homo sapiens* overlapped Neanderthals for many millennia and may have interbred with them. Neanderthals disappeared some 30 000 years ago and fossils since that time are of modern human form.

Perhaps the key to being human is language. It has recently been proposed that, because of the complex involvement of the brain and anatomy of the larynx and face in speech, speech of some kind is at least two to three million years old. It is likely that even *Australopithecus* had a form of speech. However, the evidence of art, symbolism and manipulation of materials that are associated with the anatomically modern human fossils suggest that our species, *H. sapiens*, was the first to possess articulate speech.

Out-of-Africa or a multiregional origin of *Homo sapiens*?

Today there are two main competing theories for the geographic origin of *H. sapiens*. Both theories accept that *H. erectus* populations migrated out of Africa more than one million years ago. However, the *out-of-Africa theory* says that a second wave of migration of anatomically modern humans occurred about 100 000 years ago. Thus, *H. sapiens* evolved in Africa from ancestral stock similar to *H. ergaster* and subsequently replaced all other populations of *Homo*, including Neanderthals.

The competing theory rejects the idea of a second wave of migration and hypothesises that modern humans, *H. sapiens*, evolved semi-independently from *H. erectus*-like stock in a number of different regions of the world—hence the name *multiregional theory* (Fig. 40.37). With some gene flow between adjacent populations, a single species evolved but different races (e.g. caucasoids of Europe, African bushmen and Mbuti pygmies, Arctic Inuit, Asian Mongols, American Indians, South-East Asian Polynesians, New Guineans and Aboriginal Australians) was the result.

> Anatomically more modern fossils of *Homo* date from one million years ago. Two distinct fossil forms are classified as *H. heidelbergensis* and *H. neanderthalensis*. Our species, *H. sapiens*, overlapped with Neanderthals, which disappeared about 30 000 years ago. The out-of-Africa theory says that *H. sapiens* migrated from Africa in the last 100 000 years. The multiregional theory says that *H. sapiens* evolved semi-independently in a number of geographic regions in the world following the earlier migration from Africa of *H. erectus*.

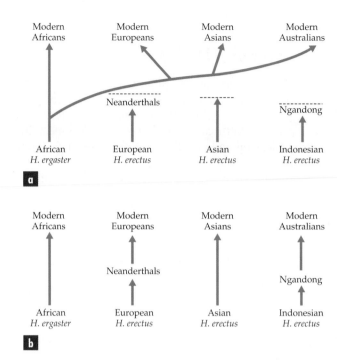

Fig. 40.37 (a) The out-of-Africa theory versus **(b)** the multiregional theory of the geographic origin of modern humans, *Homo sapiens*

method is unreliable beyond 40 000 years of age. Recently, the remains of Mungo Man have been redated using new techniques (uranium series, luminescence and electron spin resonance), which indicate that the fossil is much older at 62 000 years plus or minus 6000 years. This is the earliest known use of pigment for a ceremonial burial role.

These findings have revised debate between those who accept the out-of-Africa theory and those who accept the multiregional theory for the origin of modern humans (see Fig. 40.37). An age of about 60 000 for a fossil human in south-east Australia suggests that *H. sapiens* arrived in northern regions even earlier.

BOX 40.2 Mungo Man

In 1974, sand deposits on the edge of Lake Mungo in the Wallandra Lakes World Heritage area in New South Wales revealed a human skeleton that is known as 'Mungo 3'. It had been covered in red ochre during a burial ceremony, and the hands were clasped in front of the body. Five years earlier, a female skeleton known by local Aboriginal people as 'Mungo Lady' was found in the same area. Carbon dating was originally used to provide an estimated age for the fossils but the

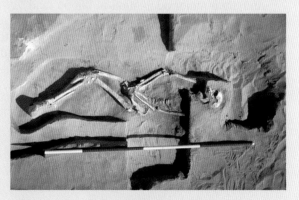

Mungo Man, found at Lake Mungo in New South Wales in 1974

Summary

- Despite their obvious differences in body organisation and their divergence nearly 600 million years ago, echinoderms and chordates are related. Both are coelomate deuterostomes, in which the anus of the adult develops at the site of the blastopore of the embryonic gut.

- Phylum Echinodermata includes marine animals that have a basic five-rayed symmetry with no head. A unique feature is their water vascular system connected to tube feet that function in locomotion, attachment, gas exchange and, in some groups, manipulation of prey.

- Phylum Chordata includes animals characterised by a dorsal notochord, pharyngeal slits and a dorsal hollow nerve cord. It include hemichordates (acorn worms), urochordates (tunicates), cephalochordates (lancelets) and vertebrates, although there is some doubt as to whether hemichordates develop a true notochord.

- A key feature of vertebrates is the embryonic neural crest, the cells of which develop into many of the structures characteristic of the subphylum. All vertebrates have a distinct head with a skull and most have a backbone composed of vertebrae. Among the vertebrates, agnathans (hagfishes and lampreys) lack jaws. All other vertebrates (gnathostomes) have jaws derived from gill arches.

- Of the bony fishes, fleshy finned or lobe-finned fishes are the closest relatives of four-limbed vertebrates, the tetrapods. Amphibians are land vertebrates but most require water for reproduction. Their skin is glandular and moist and used, usually with lungs, in gas exchange.

- All tetrapods other than amphibians are amniotes. Extra-embryonic membranes, including the amnion, surround and protect the embryo and are a major adaptation to life on land.

- The name Sauropsida is used for the group of amniotes that includes turtles, squamates (lizards, etc.), crocodiles and birds. The traditional term 'reptile' is not used because so-called reptiles are not a monophyletic group.

- Within the sauropsids, birds are a highly evolved group with unique features associated with flight. They have feathers (modified scales) and a sternum for insertion of flight muscles. Their closest living relatives are crocodiles.

- Mammals are amniotes that have hair at some stage and mammary glands for feeding young. Mammals appear to have evolved from mammal-like 'reptiles' called therapsids. Echidnas and platypuses (monotremes) lay eggs and are the only living members of the subclass Prototheria. The subclass Theria includes marsupials and eutherian mammals.

- Early primates were probably small insectivore-like animals. Modern forms are shrews, lemurs, monkeys and apes, including humans. Primates are characterised by bicuspid teeth, short nose, well-developed eyes and brain, prehensile digits and usually long parental care of offspring.

- Within the primates, humans are related to the great apes (chimpanzees and gorillas) of Africa.

- The biological past of humans spans the last five million years. The evolutionary lineage that led to modern humans is characterised by bipedalism. The earliest hominid fossils, *Australopithecus*, dating from three to one million years ago, stood upright but had skull features like other apes. *H. habilis*, the oldest fossils known for our genus, had a larger brain size and used stone tools. *H. erectus* had more advanced tools and an even larger brain size. *H. neanderthalensis* (Neanderthals) date from 130 000 years ago. Neanderthals, which disappeared from the fossil record about 30 000 years ago, coexisted with our species, *H. sapiens*. Articulate language characterises being human.

- The out-of-Africa theory says that *H. sapiens* migrated from Africa in the last 100 000 years. The multiregional theory says that *H. sapiens* evolved semi-independently in a number of geographic regions in the world following the earlier migration from Africa of *H. erectus*.

keyterms

Actinopterygii (p. 1065)
Agnatha (p. 1063)
allantois (p. 1069)
amnion (p. 1068)
Amphibia (p. 1068)
ampulla (p. 1058)
anapsid (p. 1070)
Anthropoidea (p. 1077)
Archosauria (p. 1069)
Asteroidea (p. 1057)
atlas (p. 1069)
Australopithecus (p. 1078)
Aves (p. 1068)
axis (p. 1069)
bicuspid (p. 1076)
bipedalism (p. 1078)
carapace (p. 1069)
catarrhine (p. 1077)
Cephalochordata (p. 1061)
Chelonia (p. 1069)
Chondrichthyes (p. 1064)

Chordata (p. 1057)
coelacanth (p. 1065)
Concentricycloidea (p. 1057)
Crinoidea (p. 1057)
dermal bone (p. 1063)
diapsid (p. 1071)
dipleurula (p. 1059)
dorsal nerve cord (p. 1061)
Echinodermata (p. 1057)
Echinoidea (p. 1057)
Eutheria (p. 1073)
feather (p. 1071)
Gnathostomata (p. 1064)
hair (p. 1072)
Haplorhini (p. 1077)
Hemichordata (p. 1061)
Holothuroidea (p. 1057)
hominid (p. 1078)
Homo (p. 1080)

intervertebral disc (p. 1069)
Lepidosaura (p. 1070)
Lissamphibia (p. 1068)
lungfish (p. 1065)
madreporite (p. 1058)
Mammalia (p. 1068)
mammary gland (p. 1072)
marsupial (p. 1074)
marsupium (p. 1074)
meroblastic cleavage (p. 1069)
Metatheria (p. 1073)
monotreme (p. 1073)
myotome (p. 1062)
Neanderthal (p. 1080)
neural crest (p. 1063)
notochord (p. 1061)
Ophiuroidea (p. 1057)
ossicle (p. 1057)
Osteichthyes (p. 1065)
Paranthropus (p. 1080)
pedicellariae (p. 1060)
pharyngeal slit (p. 1061)

plastron (p. 1069)
platyrrhine (p. 1077)
Pongidae (p. 1077)
Primates (p. 1076)
prosimian (p. 1077)
Prototheria (p. 1073)
pygostyle (p. 1071)
ratite (p. 1071)
Reptilia (p. 1068)
Sarcopterygii (p. 1065)
Sauropsida (p. 1069)
spicule (p. 1057)
sternum (p. 1071)
stomochord (p. 1061)
Strepsirhini (p. 1077)
swim bladder (p. 1065)
teleost (p. 1065)
tetrapod (p. 1064)
Theria (p. 1073)
tube feet (p. 1058)
tunic (p. 1061)
Urochordata (p. 1061)
vertebrae (p. 1063)
Vertebrata (p. 1061)
water vascular system (p. 1058)

Review questions

1. Which group of echinoderms has an 'Aristotle's lantern'? What is this structure and what is its function?

2. Briefly describe the structure and function of the water vascular system of a sea star. Which echinoderms do not use tube feet for locomotion?

3. What are the names of the larvae of the Asteroidea (sea stars) and Ophiuroidea (brittle stars) and how do the adults differ in body form?

4. What animals are characterised by pharyngeal gill slits? What functions do gill slits perform?

5. You find a sessile animal living in the intertidal zone of a marine habitat. A colleague suggests that the animal is a tunicate. What features would you look for to determine that it is a tunicate?

6. Which of the following is correct? Tetrapods include:

   A the coelacanth, frogs, turtles, lizards, birds and mammals

   B frogs, turtles, lizards, snakes, birds and mammals

   C turtles, lizards, birds and mammals

   D all amniotes

7. Bony fishes have a swim bladder, believed to have been important in their success. What is the swim bladder or its equivalent? What is its function in a typical teleost fish, a lungfish and a tetrapod?

8. An anapsid skull is typical of turtles and their relatives.

   (a) What does 'anapsid' mean?

   (b) In contrast, what type of skull do lizards, crocodiles and birds have and what does it mean?

9. Describe the features of birds that are associated with flight.
10. What features are unique and characteristic of mammals? Name the three main groups of mammals and explain how you would identify a representative of each.
11. How does the skeleton of australopithecines and other hominids differ from that of the gorilla?
12. Compare and contrast the out-of-Africa and multiregional theories for the origin of modern humans.

Extension questions

1. Echinoderms (phylum Echinodermata) and vertebrates (phylum Chordata) appear to be very different animals but they share a number of embryonic features. Discuss the evidence that indicates that these two phyla are related. What evidence is there for their divergence being about 600 million years ago? (*Note:* see also Chapter 31.)
2. Adult chordates have either a notochord or vertebral column but not both. Explain why not.
3. What is the importance of the embryonic neural crest of vertebrates (see also Chapter 15)?
4. In this book, we did not classify turtles, lizards, snakes and crocodiles together as Reptilia, as in more traditional classifications. With reference to the relationships of the amniotes, explain why the use of Reptilia is problematic.
5. Discuss the differences in brain size, and the implication for behaviour, between *H. habilis*, *H. erectus* and *H. sapiens*. Discuss how we might define what is *H. sapiens* relative to other fossil species of *Homo*.

Suggested further reading

Forey, P. and Janvier, P. (1994). Evolution of the early vertebrates. *American Scientist* November–December.

An account of the evolution and radiation of Devonian fishes.

Kardong, K. V. (1998). *Vertebrates. Comparative Anatomy, Function and Evolution.* International edition. New York: McGraw-Hill.

This book is for those students who want to explore the topic in depth; a useful source book of information.

Tattersall, Ian. (1995). *The Fossil Trail.* New York: Oxford University Press.

This book asks the questions 'what is human?' and 'what do we know about human evolution?'.

Thomas, Herbert. (1994). *The First Humans. The Search for Our Origins.* (English trans. 1995 Paul Bahn). London: Thames and Hudson Ltd.

A scientific and popular account of the evolution of hominids, including details of the discovery of fossils, the scientists involved and various controversies.

CHAPTER

41

Australian biota

hapters 30–40 describe the diversity and evolution of organisms in general. This chapter focuses on the Australian biota and its unique features. The Australian biota is the product of millions of years of evolution influenced by isolation and climatic changes resulting from alteration in the positions of continents and seas. The history of Australia's present-day flora and fauna is directly linked by geologic processes to the history of other southern continents.

The Australian biota: southern connections

It has long been recognised that Australia and other lands in the Southern Hemisphere share many plant and animal taxa. The English botanist Joseph Hooker was one of the first, in 1853, to point out that many Indian plant genera occur in similar monsoonal environments in northern Australia, that Malaysian rainforest genera occur in tropical eastern Australia, and that cool temperate and alpine genera are shared between southern mainland Australia, Tasmania and New Zealand. Southern beech trees, *Nothofagus*, have a modern-day and fossil geographic distribution in

Australia, New Guinea, New Caledonia, New Zealand, Antarctica and South America (Fig. 41.1). Bony-tongue fishes (family Osteoglossidae) occur today in tropical regions of Australia, New Guinea, South-East Asia, South America and Africa (Fig. 41.2). Flightless birds, the **ratites**, include cassowaries (Australia, New Guinea), emus (Australia), extinct moas (New Zealand), kiwis (New Zealand), rheas (South America), extinct elephant birds (Madagascar) and ostriches (Africa), all restricted to the Southern Hemisphere. Other largely Southern Hemisphere distributions of animals include marsupials, parrots, chelid turtles, tree frogs, galaxiid fishes and castniid moths (Fig. 41.3).

In the past, these similarities in biotas were explained by recent long-distance dispersal—over waterways or across ancient land bridges that connected stationary continents. With acceptance of the theories of continental drift, sea-floor spreading and plate tectonics (see Chapter 31), the distributions of *Nothofagus*, bony-tongue fishes, ratites and many other animal and plant groups are now best explained by past connections of the southern continents and their moving apart to their present positions on the earth's surface.

Rainforests of present-day eastern Australia are probably relics of forests similar to those that once covered much of Australia and Antarctica during the

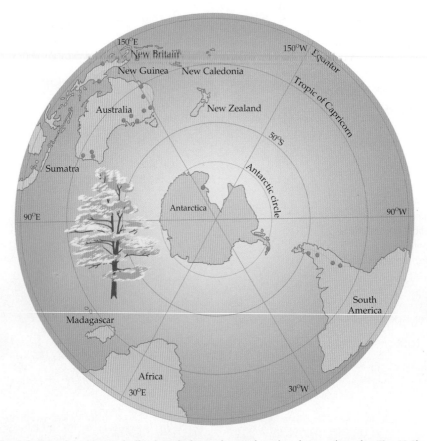

Fig. 41.1 Distribution of fossil (red dots) and living (yellow) *Nothofagus,* the southern beech trees (see also Fig. 42.3)

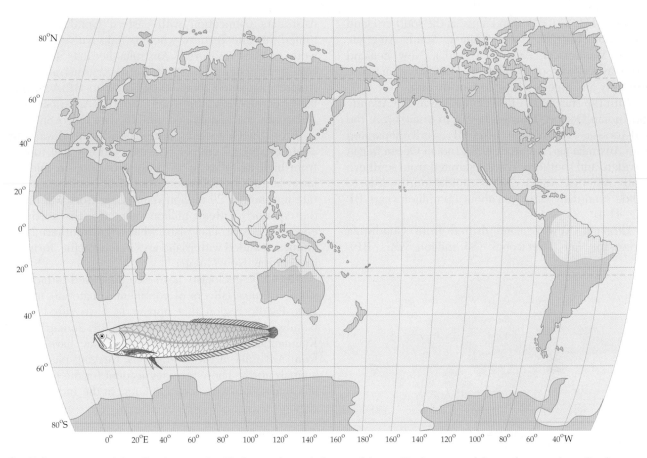

Fig. 41.2 Bony-tongue fishes (family Osteoglossidae) occur in tropical parts of the world—fragments of the southern continent Gondwana

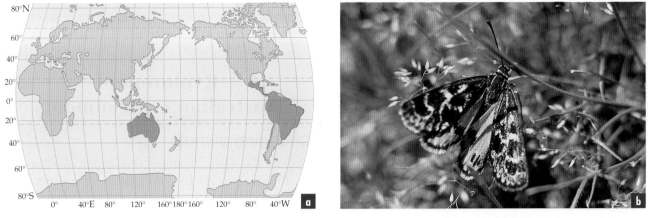

Fig. 41.3 *Synemon* moths are classified in the family Castniidae, which includes about 120 species in Central and South America, about 40 species in Australia and two species in South-East Asia. **(a)** The distribution of *Synemon* and related moths of family Castniidae. This suggests a Gondwanan origin for the group. Many species live in rainforests but Australian forms live in seasonally dry habitats, feeding on sedges and grasses. The greatest diversity of species in Australia is in south-western Western Australia. **(b)** Female of *Synemon plana*, an Australian species whose larvae live in soil and feed on the roots of native wallaby grass (*Danthonia*). With the loss of many areas of native grasslands and the invasion of introduced plants, this moth is now an endangered species

early **Tertiary period**, dating back at least 40 million years ago. Native marsupials, birds, reptiles, frogs, fishes and insects and the vegetation that now characterises the Australian landscape—eucalypts, wattles, banksias, casuarinas, heaths, grass trees and spinifex—have expanded their range in response to increased aridity.

Australia in Gondwana

The last major tectonic–climatic cycle of earth's history started 320 million years ago, with the coalescence of all continents into the supercontinent **Pangaea**—a coalescence that was completed by about 230 million years BP (Fig. 31.4). Within Pangaea, the southern land masses of Australia, New Guinea, South America, Africa, Madagascar, Antarctica, India, New Zealand and maybe parts of South-East Asia were in close proximity and together formed **Gondwana**.

With the start of sea-floor spreading in the Atlantic and Indian Oceans, Pangaea began to break up in the mid-Jurassic (160 million years BP). India separated from Australia and Antarctica, the Indian Ocean began to widen, and southern Africa separated from southern South America. India drifted rapidly northwards to collide eventually with Asia about 50 million years BP (Fig. 41.4).

Although New Zealand drifted from Gondwana during the late Cretaceous (80 million years BP),

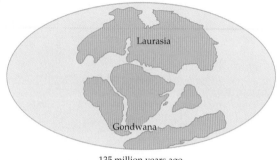

135 million years ago
(early Cretaceous)

65 million years ago
(late Cretaceous)

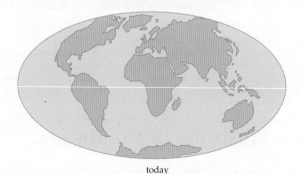

today

Fig. 41.4 With the start of sea-floor spreading, Pangaea began to break up during the Jurassic (160 million years BP). By the Cretaceous, Gondwana was breaking up

Australia's connection with eastern Gondwana continued into the Tertiary, when its links with Antarctica were gradually severed. Australia became an island continent when the strait between Australia and Antarctica opened about 30 million years BP. A narrow land connection between South America and Antarctica was maintained until the late Oligocene.

From the Oligocene, Australia drifted northwards at a rate of about 7 cm per year on a journey that took it from high latitudes to its present position in mid-to-lower latitudes. (If this northward drift continues at this rate for another 20 million years, Cape York will reach the equator!) Fifteen million years ago, the northern edge of the Australian plate, what is now southern New Guinea, collided with island arcs in the Pacific, forming present-day New Guinea.

Recently, geologists have also identified numerous small *terranes* (slivers of continental material) in the Indo-Pacific region that appear to have rifted from Gondwana at various times. Parts of Sumatra and Kalimantan (Borneo), for example, may have rifted from Australia in the Cretaceous, carrying plants and animals north. These terranes account for the presence of Gondwanan-type flora and fauna in the Indo-Pacific region.

Australia was part of the southern supercontinent Gondwana during the Cretaceous and early Tertiary. Australia severed its link with Antarctica and became an island continent by 30 million years ago. Environmental changes and isolation moulded the evolution of the modern Australian biota.

Evolution of Australian environments

Changing climate

During the early Tertiary (Paleocene to Eocene, 65–40 million years BP), the climate of Australia and adjacent Antarctica was humid temperate, and rainforest covered much of the land. Separation of land masses created the Southern Ocean and its current began to circle Antarctica (Fig. 41.5). This current, known as the **circum Antarctic current**, has a profound effect on wind patterns and climate. It does not mix with warmer tropical water, and thus surface water temperature and air temperature are low. With development of the circum Antarctic current, ice began to form on Antarctica, and by the late Miocene (seven million years BP) it had reached its present-day extent. These changes reduced air temperature and rainfall over parts of Australia, heralding the onset of **aridity** and contraction of rainforests dominated by *Notho-fagus* and conifers.

BOX 41.1 Ancient forests

In the mountains and dry valleys of Antarctica, fossils embedded in rocks and coal seams reveal plants and animals of the past. Fossil evidence of ancient Palaeozoic forests dominated by *Glossopteris* seed ferns was collected by Edward Wilson on Captain Robert Scott's last and fatal expedition to the South Pole. These important collections of rocks and fossils were found with their frozen bodies. Preserved leaves and stems of *Glossopteris* are the same as those found in deposits of the same age (250 million years BP) in India, South America, South Africa and Australia.

Glossopteris leaves are tongue-shaped with net-like venation and pronounced mid-rib (see figure). Glossopterids were a diverse group of trees and shrubs of swampy habitats, which favoured the formation of coal, particularly in the eastern part of Australia. Glossopterids had gymnospermous wood and seeds borne in clusters on leaves.

The *Glossopteris* flora dominated the Permian (286–248 million years BP) and replaced earlier forests composed of giant clubmosses and horsetails (see Chapter 37). A variety of insects inhabited the *Glossopteris* forests together with amphibians and reptiles. By the early Triassic (about 230 million years BP), glossopterids disappeared from the fossil record and a new flora developed, characterised by forked-frond seed ferns, *Dicroidium*, early conifers and cycads.

From the Jurassic through to the early Cretaceous (213–100 million years BP), forests were dominated by conifers of the modern-day families Araucariaceae, which includes the kauri pine *Agathis*, and Podocarpaceae, which includes *Podocarpus*. *Ginkgo*, now restricted to China, was also widely distributed during this period. Dinosaurs, too, became established in the Australian part of Gondwana (see Box 41.6).

Following the Jurassic, with its uniformly wet climate and forests of conifers, cycads and ferns, the vegetation changed. By the end of the Cretaceous and the beginning of the Tertiary (65 million years BP), dinosaurs had become extinct and the ancient conifer forests had contracted with the rise to dominance of flowering plants. By this time, the break-up of Gondwana was well underway.

The earliest record of flowering plants in Australia is fossil pollen, of uncertain modern affinity, from Cretaceous sediments. Pollen identifiable as *Nothofagus* and the family Proteaceae is up to 80 million years old. Similar pollen is recorded from New Zealand. As flowering plants diversified, so did the early marsupials, probably feeding on flowers, fruits and insects attracted to flowers.

With the increasing isolation of Australia from Antarctica in the Tertiary, the stage was set for the evolution of a unique Australian flora and fauna.

Glossopteris leaf fossil

Table Summary of changes in the biota from the Permian (Palaeozoic) to the Tertiary (Cenozoic)	
Time	**Biota**
Permian (286–248 my BP)	*Glossopteris* (seed fern) flora, with insects, amphibians, reptiles; coal formation
Triassic (230 my BP)	Forked-frond seed ferns (*Dicroidium*), early conifers and cycads
Jurassic–early Cretaceous (213–100 my BP)	Conifer-dominated forests (e.g. araucarias and podocarps), *Ginkgo*, dinosaurs
End of Cretaceous– flowering Tertiary (65 my BP)	Conifer forests contract, flowering plants dominate, dinosaurs become extinct. Break-up of Gondwana and evolution of unique Australian flora and fauna well underway

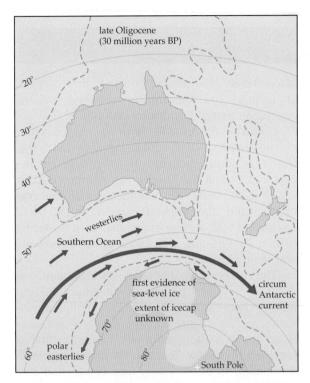

Fig. 41.5 Australia's connection with Antarctica was severed during the late Oligocene. This severance created the Southern Ocean and the circum Antarctic current, which resulted in declining temperatures and ice formation on the Antarctic continent. The small arrows show atmospheric circulation patterns

Fig. 41.6 The ancient, lateritic land surface near Mt Magnet, Western Australia, which developed on Archaean greenstones, has here been partially eroded and is surrounded by younger plains

Changing landforms

Australia is often described as one of the oldest landmasses because of the great age of rocks in regions such as the Pilbara (more than 3.8 billion years old). Landscapes from the Kimberley and Arnhem Land plateaus through the desert regions across to the eastern highlands, largely undisturbed by mountain building and volcanoes, became extensively weathered. Exposed rocks wore down to low hills and ranges. Soils were deeply weathered during warm wet periods, with nutrients, such as phosphorus and nitrogen, removed by millions of years of leaching. By the middle Tertiary, weathered land surfaces included **laterite**, still evident today where ironstone gravel or boulders occur at the surface or subsurface overlying a layer of kaolinitic (bleached white) clay (Fig. 41.6). Where weathering of laterite subsequently occurred, sediments of low fertility were blown over the continent, adding to sandy soils derived from rocks, such as sandstones and coarse granites.

In Miocene times, there were huge inland lakes where deserts now occur. Lake Eyre and Lake Frome basins once contained fresh water and provided environments for crocodiles, flamingoes and plants typical of warm and humid conditions. When these areas became arid, winds eroded and transported

sediments to form extensive desert dune systems, and the aquatic life became extinct. Evaporation of lakes left saline mudflats. Seas, which had invaded low-lying areas across southern Australia, retreated, exposing extensive deposits of limestone formed from shelled invertebrates. The Nullarbor Plain and Murray Basin, previously submerged, were exposed as dry land.

Climatic changes through the Tertiary altered landforms and soil, which in turn influenced the history of Australia's plants and animals.

During the early Tertiary (65–40 million years BP), the climate of Australia and adjacent Antarctica was humid temperate, and rainforest covered much of the land. Separation of land masses created a circum Antarctic current, causing rainfall to decrease over parts of Australia, air temperatures to decrease and conditions to become more arid.

Fire

Fire is an integral part of the present Australian environment (Fig. 41.7). Its history is recorded by charcoal particles preserved in sedimentary deposits. A combination of preserved charcoal and pollen reveals past vegetation–fire relationships.

There is evidence of fire throughout the period of evolution of the Australian biota but its frequency and effects on vegetation have varied. In the early part of the Tertiary, when rainfall was high, occasional fires were caused by volcanic activity and lightning during dry periods. With the dominance of rainforest and rarity of fire-promoting plants, such as eucalypts, fire did not cause large-scale destruction of vegetation. However, with the development of a drier and more variable climate towards the end of the Tertiary period, fires became more frequent and rainforest was replaced by more fire-tolerant and arid-adapted open forests.

Fig. 41.7 Fire is an important part of the present and past Australian environment

Ice ages

About 2.5 million years ago (around the Pliocene–Pleistocene boundary), Australia, for the first time, came under the influence of westerly winds, resulting in wet winters and hot dry summers. This influence also coincided with the beginning of climatic fluctuations that were being experienced worldwide. Throughout Australia during the **Quaternary period**, cool **glacial periods** were associated with aridity and warmer **interglacial periods**, such as that existing today (the Holocene), were associated with higher rainfall. Wetter conditions than today occurred at times of slightly lower temperatures when evaporation was reduced.

Australia did not experience major expansions of ice, covering large areas of continents at high latitudes, as in the Northern Hemisphere, but some glaciation did occur in the Snowy Mountains and Central Highlands of Tasmania. Characteristic of the glacial periods in Australia, dry and windy conditions led to the movement of sand. A lack of continuous vegetation cover facilitated sand movement, and sand dune systems developed well beyond the present arid centre. During certain wet phases, large lakes formed over what are now small salt pans, and river discharge and erosion were much higher.

Variability in climate and associated landscape has prevented the development of continuous sedimentary sequences, which provide palaeobiologists with a full record of environmental changes; however, a good chronology for the Quaternary is preserved in ocean sediments. An excellent example showing the number amplitude and increasing magnitude of climatic cycles through this period is illustrated in Figure 41.8.

The climatic fluctuations of the Quaternary eventually restricted rainforest to patches not much larger than those surviving today. During interglacial periods, forests were dominated by she-oak (*Casuarina*) and

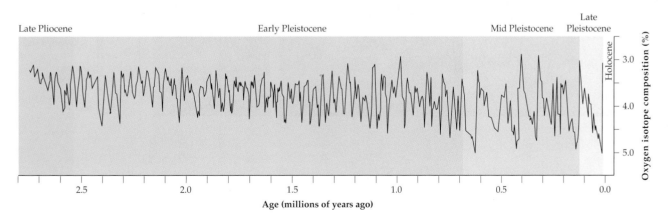

Fig. 41.8 Changes in the oxygen isotope composition of shells preserved in a sediment core from the north Atlantic Ocean, illustrating fluctuating environmental conditions during the last 2.8 million years. High percentage values of oxygen-18 reflect low temperatures and extensive icecap development. Fluctuations increase in magnitude close to the Pliocene–Pleistocene boundary and increase further around the early to mid Pleistocene boundary

during glacial periods, herbaceous plants such as daisies (composites) were conspicuous. This general pattern persisted until the last glacial–interglacial cycle, when it is probable that increased burning due to the activities of Aboriginal people caused a substantial replacement of casuarinas and composites by eucalypts and grasses, as well as reducing rainforest to localised fire shadows. The present vegetation, dominated by eucalypts, achieved its present expansion within the last 100 000 years.

> With the development of a drier climate, fires became more frequent towards the end of the Tertiary. During the last 2.5 million years (the Quaternary), cool, arid glacial periods alternated with warm, wetter interglacial periods.

Arrival of humans

There is a lot of debate over the first humans in Australia, their time of arrival and the impact that they had on the environment. Examination of fossil skulls shows that there were a number of different immigrant groups, most likely derived from both Indonesia and China. Some ocean travel would have been required, although passage to Australia would have been easier from Indonesia when sea levels were much lower during the last glacial period. At this time, New Guinea was connected to Australia. It is generally accepted that the first colonisation occurred at least 40 000 years ago but some archaeologists believe that it may have been much earlier. A major problem in determining the time

of arrival of humans has been that of obtaining absolute dates on archaeological sites beyond 35 000–40 000 years ago—the effective limit of radiocarbon dating. However, recent studies in northern Australia, using an alternative dating method, thermoluminescence, suggest that humans arrived more than 50 000 years ago.

Some indirect evidence for the long residence time of humans is provided by increased levels of charcoal within pollen records, which may be attributed to burning by humans. These increases date from about 38 000 years ago at Lynch's Crater in north-east Queensland and perhaps as early as 128 000 years ago at Lake George near Canberra.

Changes in the pollen record measure the way in which, and degree to which, burning by humans has altered the landscape. Around Lynch's Crater, for example (Fig. 41.9), rainforest, which dominated during wetter interglacial periods, was generally replaced by drier rainforest with some fire-tolerant vegetation during glacial periods. Drier rainforest, evidenced by fossil pollen associated with an increase in charcoal during the last glacial period, was totally replaced by fire-tolerant vegetation. Drier rainforest exists now only as small isolated patches. Some rainforest plants became extinct in this region, including the conifer *Lagarostrobus*.

The extent of environmental change is best illustrated by the number of large animals that became extinct during the period in which people have been in Australia (see also Box 41.6). Most extinctions occurred between 35 000 and 15 000 years ago, at a time of driest

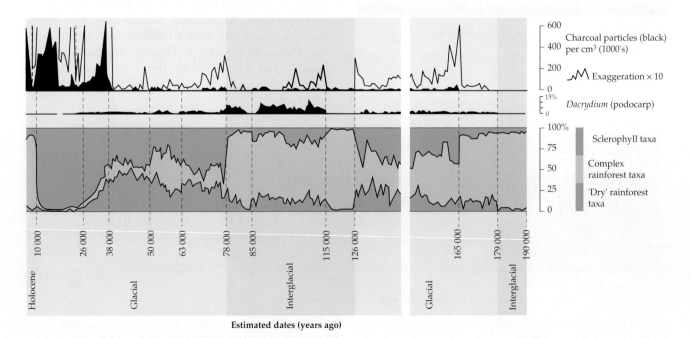

Estimated dates (years ago)

Fig. 41.9 Changes in the abundance of pollen and charcoal from Lynch's Crater in north-east Queensland illustrate vegetation responses to climatic fluctuations in the late Quaternary and changes associated with increased burning

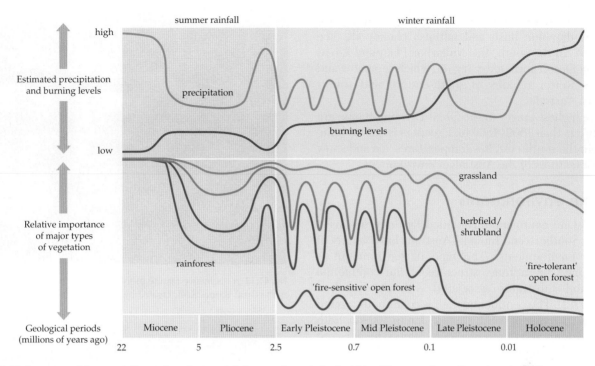

Fig. 41.10 Summary of the vegetation and environmental changes through the last 14 million years in south-eastern Australia

conditions during the last glacial period. Although climate has been considered to be the main cause of these extinctions, there is no evidence to suggest that climatic conditions at this time were more extreme than during the previous glacial phases. However, increased environmental instability resulting from human burning and associated vegetation changes may have sufficiently altered stream flow and lake levels to produce a more drought-prone environment.

A generalised picture summarising the development of the present vegetation and environment in south-eastern Australia is shown in Figure 41.10.

H umans colonised Australia at least 40 000 years ago and caused increased burning.

Modern Australian environments

Terrestrial environments

Terrestrial environments of modern Australia are varied. The continent spans the latitudinal belt 10°–40°S and climate varies with latitude from tropical (in the north), subtropical, warm temperate to cool temperate (in the south). The seasonal distribution and effectiveness of rainfall determine the type of plant and animal communities; the monsoonal climate of the north, with summer maximum rainfall (the 'wet' season), contrasts with the

Mediterranean climate of the south, with winter maximum rainfall (Fig. 41.11).

The Great Dividing Range slices Australia into two unequal parts from east to west. It is a watershed, with moister environments on the eastern side, supporting relict pockets of rainforest. The western side is drier and supports woodlands. The interior of the continent is arid and sparsely vegetated, although, compared with desert regions in other parts of the world, Australia's

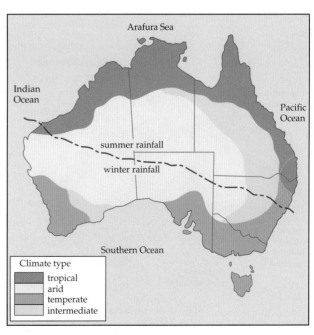

Fig. 41.11 Australian climatic regions and major landforms

arid region is well covered with grasses and shrubs, such as the blue bush and saltbush chenopods (see p. 1098). In the south, the Nullarbor Plain and Great Victoria Desert form a dry region, which separates and isolates wetter regions of south-eastern and south-western Australia.

The highest ranges in the Great Dividing Range are mostly less than 1500–1800 m. Except in isolated alpine areas in the south-east, low temperatures, ice and snow are not features of the Australian environment.

Marine environments

As an island continent surrounded by three oceans—Pacific, Southern and Indian—Australia has a variety of marine environments. The type of environment is determined by latitude, structure of the continental shelf and substrates associated with the sea bed (Fig. 41.12). The continental shelf is continuous around Australia, varying in width from 15 km to 400 km, with its outer limit at a depth of about 150 m. The adjacent continental slope has an incline up to 40° and plunges to abyssal depths greater than 4 km. The Great Barrier Reef, created by living corals and coralline algae, is a distinctive feature of the north-east coast and extends over 2000 km (Fig. 41.13). With many lagoons and cays (small islands), the reef has a range of different environments for marine organisms. Rocky shores in the south and east include sandstones, shales, volcanic rocks and limestones, which provide anchorage for marine benthic algae and animals (see Box 41.2). Mangrove trees trap sediments along sheltered coasts, forming muddy shores for seagrass beds, although sandy shores are more extensive along the Australian coast.

Fig. 41.13 A distinctive feature of the north-east coast of Australia is the Great Barrier Reef, created by living corals

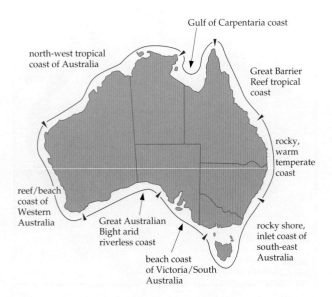

Fig. 41.12 Australian coastal types

BOX 41.2 Marine biodiversity

The impact of Australia's geologic history on the diversity of its biota is usually identified with larger, more obvious, land-based plants and animals. However, the less conspicuous seaweeds and marine animals that form complex intertidal and submarine communities along the region's coasts are equally novel and diverse and constitute a marine biota that is one of the richest in the world.

Marine flora

The red seaweeds (phylum Rhodophyta) from southern Australia include more than 1100 recorded species, with approximately 70% being endemic. Because large areas of the Australian coastline are still unexplored, particularly in the north and north-west of the continent and in Tasmania, the diversity and endemism of Australian red seaweeds will probably prove to be greater than in any other area of similar size in the world.

Red algae are generally less conspicuous in size and overall biomass than other seaweeds. The gelling and emulsifying properties of the 'slimes' (polysaccharides) extracted from their cell walls provide products of considerable commercial value (see Chapter 35). Economically important genera, such as *Gelidium*, *Pterocladia*, *Gracilaria* and *Gigartina*, often attach to rocks in the intertidal zone. Some of these organisms are now studied by biotechnology companies interested in generating genetically improved strains for mass culture of red algal agars and carrageens.

In addition to the Rhodophyta, two other major phyla of benthic seaweeds populate most marine habitats. The largest of all the algae, and most abundant in terms of biomass, are the brown algae (phylum Phaeophyta). Included within this group are the true kelps *Ecklonia* and *Macrocystis* (Fig. a), the bull kelp *Durvillaea*, and the large fucoids, such as *Hormosira*, *Xiphora* and *Cystophora*. All genera are conspicuous on rocky, temperate coastlines in southern Australia, Tasmania and New Zealand. Harvested commercially in a number of locations around the world, including King Island (in Bass Strait) and New Zealand, brown algae provide a range of products, such as gelling agents, fertiliser and food for both livestock and human consumption.

The highly diverse green algae (phylum Chlorophyta) are abundant in most marine habitats but are particularly obvious in tropical regions, including the Great Barrier Reef, where *Chlorodesmis* (turtle grass), *Caulerpa* and the calcified genus *Halimeda* dominate. Although *Caulerpa*, *Codium*, *Ulva* and other genera are used for food in Asian countries and some Pacific islands, green seaweeds are of relatively little economic use and are not commercially exploited in Australia.

In addition to algae, marine flowering plants, which grow submerged on sheltered and soft-bottomed coasts, contribute to primary productivity and marine food webs (Chapter 44). Marine flowering plants are monocotyledons, commonly called seagrasses (Fig. b) because of their ribbon or strap-like leaves arising from nodes on a creeping rhizome. There are 31 species in Australasian waters, including five species of eelgrasses (*Zostera*) and eight species of strapweeds (*Posidonia*).

(b) Australian seagrass beds are important primary producers, such as for this dugong

Marine fauna

Australia has some 3500 species of marine fishes, of which around 2700 are shelf and near-shore species. The highest diversity is found in northern communities, with over half of the species but a low level of endemism. In contrast, the southern temperate fishes are less diverse (around 600 species) but highly endemic (about 85%) or shared with New Zealand (more than 10%). Similar patterns of diversity and biogeography are seen in the molluscs (10% northern endemics; 95% southern endemics) and echinoderms (13% northern endemics; 90% southern endemics). In addition, over 20% of Tasmanian echinoderms are shared with New Zealand. Such trans-Tasman patterns may have come about because many marine species have long, planktonic, larval lives and so could be transported in the general west-to-east surface current flow.

Recent unintentional human transport may have been involved in the distribution of some marine species—via solid ballast and ballast

(a) The bull kelp, *Durvillaea potatorum*

(c) Red morwong, *Cheilodactylus fuscus*, is a commercial fish caught off south-eastern Australia, and has a delicate flavour

water, growth on ships' hulls, or associated with the export of live commercial catches, especially bivalves such as oysters and mussels. We now know that several potentially threatening, introduced marine organisms from the Northern Hemisphere, for example, the Japanese sea star, are colonising southern Australian shores after their accidental introduction, presumably through shipping activities.

Australian flora

The modern Australian terrestrial flora is strikingly different from that of other continents and islands. Numerous species, genera and families are endemic to Australia, while others have their centre of diversity in Australia but are also represented in adjacent areas. The distinctive appearance of most Australian vegetation is due to the dominance of **Eucalyptus** and **Acacia** trees and shrubs over about 70% of the continent. Gymnospermous trees such as conifers are far less common, although there are a number of distinctive endemic Australian taxa, such as *Athrotaxis* (King Billy pine) in Tasmania (see Chapter 37). The she-oak family (Casuarinaceae, Fig. 41.14) is ubiquitous in Australia, with representatives also in tropical regions of Malaysia. The 'grass trees', *Kingia* (Fig. 41.15) and *Xanthorrhoea* (family Xanthorrhoeaceae, related to lilies) are unlike any other monocotyledons; of about 100 species in the family, all but four are endemic to Australia. Endemic hummock grasses, such as *Triodia* and *Plectrachne* (Fig. 41.16), form a characteristic ground cover over vast areas of arid and semiarid Australia.

The major components of the Australian flora have a Gondwanan origin. Today, a relict component is represented by plants confined to moist rainforest

Fig. 41.15 The grass tree *Kingia* from Western Australia has a woody trunk, linear leaves and flowers clustered in globular heads. The related *Xanthorrhoea* was used by Aboriginal Australians as a source of resin, as an adhesive and to make torches to set fire to vegetation to catch game

habitats. However, the major component has diversified since the middle of the Tertiary, predominantly in temperate and more arid environments.

Dominance of sclerophylls

Why do eucalypts, wattles, banksias, casuarinas, grass trees, heaths, native peas, hummock grasses and other familiar 'bush flowers' dominate the modern Australian flora? These plants possess features that allow them to survive low soil nutrients, water stress and fire. All are **sclerophylls** (*sclero*, hard; *phyllus*, leaved), characterised by rigid, often small leaves, short internodes, and often small plant size.

In 1916, E. C. Andrews, and later N. C. W. Beadle in 1954, developed the idea that sclerophylly was primarily an adaptation to infertile soils, even where rainfall was relatively high, because sandy soils dry out and can cause water stress in summer. Sclerophylly is well developed in banksias (family Proteaceae) and the southern heaths (family Epacridaceae), which were present in the early Tertiary when rainfall was high and fires infrequent. Beadle and other workers suggested that sclerophyll plants formed wet heathlands on the

Fig. 41.14 The she-oaks, *Casuarina* and *Allocasuarina*, have leaves reduced to scales or toothed sheaths, needle-like photosynthetic stems and small flowers clustered in cone-like heads

Fig. 41.16 The pungent-leaved hummock grasses *Triodia* and *Plectrachne* are endemic to Australia and have probably undergone fairly recent speciation, occurring in vast areas of the arid and semiarid zones

margins of rainforest where soil fertility was low. Heathlands occur today on the most extremely infertile soils, in both wet and dry climates.

Beadle suggested that low levels of soil phosphorus were particularly important in limiting the growth of rainforest species and favouring sclerophylls. He argued that low phosphorus levels affect physiological processes and lead to reduction in the number of cells formed and length of stem internodes. Such changes result in smaller leaf sizes and smaller plants, which would have been able to survive low and variable rainfall as aridity increased, and also droughts of cool montane environments during the onset of the ice ages.

Sclerophylls of the modern flora are characterised by a number of features in addition to small leaves and short internodes (Fig. 41.17): thick cuticle on the leaf epidermis, which may overarch guard cells; stomates sunken in grooves or pits, or protected by a dense covering of hairs; high proportion of lignified cells such as fibres; and thick-walled hypodermal cells between the epidermis and photosynthetic mesophyll.

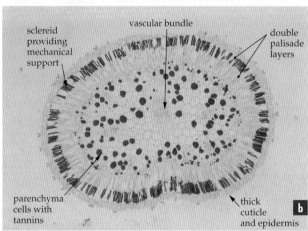

Fig. 41.17 (a) *Hakea sericea* (Proteaceae) has sclerophyll leaves, one of which is **(b)** shown in cross-section. Note the thick cuticle, distribution of photosynthetic tissue and cells with thick walls

With increased frequency of fire towards the end of the Tertiary (p. 1080), sclerophylls became even more widespread. Many of their characteristic features promoted their fire resistance.

Modern species (e.g. eucalypts) survive fire by having thick, insulating bark, at least on the lower half of the main trunk. If the shoot of a tree or shrub is damaged, resprouting after fire may take place from dormant buds on the trunk (Fig. 41.18), or from a mass of dormant buds that form swellings, **lignotubers**, at the base of the stem. Lignotubers are well developed in mallee eucalypts but resprouting ability is absent from species of wet forests such as mountain ash, *Eucalyptus regnans*. Grass trees, *Xanthorrhoea*, are stimulated by fire to flower and produce seed.

Some plants are killed by fire but reproduce from seed that is protected in hard, woody cones (cypress pines) or fruits, such as follicles (hakeas, banksias) and capsules (eucalypts, tea-trees). Seeds of peas and wattles have a hard seed coat, which is cracked by the heat of fire; fast-growing wattles are often among the first woody plants to regenerate after fire.

Many of these characteristics that favour plant survival after fire may have evolved for other reasons. The ability to resprout, for example, may have evolved in response to grazing or drought, while it has been suggested that hard seed coats and woody fruits were a response to low nutrients.

> Sclerophylly was primarily an adaptation to soils of low fertility but features of sclerophyll plants allowed them to survive water stress and fire.

Succulent survivors

The Australian arid flora also includes a number of **succulents**, which, like sclerophylls, are **xerophytes** (plants that live in dry conditions in contrast to **mesophytes**, which live where water is in adequate supply). Succulents survive harsh conditions by having fleshy leaves or stems and a highly mucilaginous cell sap. The outer epidermis is often heavily cutinised and stomates may be sunken but plants are not lignified. Examples of Australian succulents are stoneflowers and pigface (family Aizoaceae), parakelias (family Portulacaceae), and saltbushes (Fig. 41.19) and samphires (family Chenopodiaceae), some of which are C_4 plants (Chapter 5). Succulents occur in dry or salty areas, either inland or on the coast, although they are rarer in Australia than in other parts of the world (e.g. North and South America, where the cactus family occurs).

Fig. 41.18 Many woody Australian sclerophylls, such as this eucalypt, resprout from dormant buds beneath the bark or in lignotubers

Fig. 41.19 *Halosarcia* is a succulent plant that grows near Oodnadatta in South Australia

Characteristics of Australian flowering plants

Myrtaceae: the eucalypt family

Australian eucalypts (*Angophora*, *Corymbia* and *Eucalyptus*), tea-trees (*Leptospermum*), paperbarks (*Melaleuca*), bottlebrushes (*Callistemon*) and lilly pilly (*Acmena*) are five examples of the family **Myrtaceae**. The distribution of the family is primarily in the Southern Hemisphere (Fig. 41.20a), with nearly 50% of genera occurring naturally in Australia.

Myrtaceae have leaves with oil glands, flowers with four or five perianth parts and numerous stamens borne above the inferior ovary (Fig. 41.21). Dry fruited forms are most numerous and widespread in drier parts of Australia. Fleshy fruited forms occur more in wetter tropical habitats, such as rainforests. Many species are used in ornamental horticulture—brush box (*Lophostemon confertus*) is a commonly planted street tree—and cloves (*Syzygium aromaticum* in Indonesia) and allspice (*Pimenta dioica* in South America) provide spices. The most familiar Australian group, the eucalypts, are fast-growing trees planted in many parts of the world for timber, paper pulp, firewood and aromatic oils (Box 41.3).

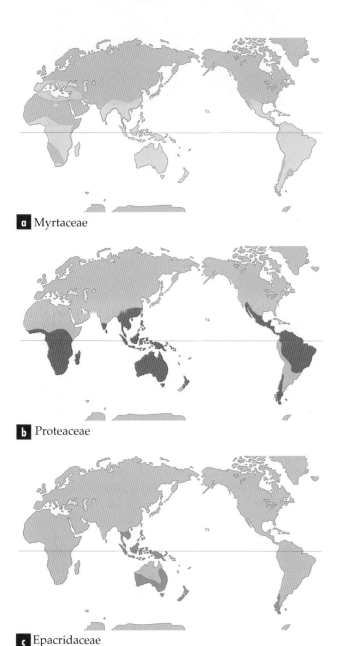

a Myrtaceae

b Proteaceae

c Epacridaceae

Fig. 41.20 Distributions of the families **(a)** Myrtaceae, **(b)** Proteaceae and **(c)** Epacridaceae

Fig. 41.21 *Leptospermum laevigatum* is typical of the family Myrtaceae

Proteaceae: the banksia family

Characteristic of Australian vegetation, the family **Proteaceae**, particularly subfamily Grevilleoideae, includes many taxa endemic to Australia. Well-known Australian genera are *Grevillea*, *Telopea* (waratahs), *Macadamia* and, perhaps the most distinctive of all, *Banksia* (see Fig. 30.14). There are 92 species of *Banksia*, with 61 in south-western Western Australia. Proteaceae are a Gondwanan group, occurring in South Africa, India, South-East Asia and South America, as well as Australia (Fig. 41.20b). The family has a long fossil record in the Southern Hemisphere, including Antarctica. Pollen and macrofossils of *Banksia* have been found, for example, in Victorian and South Australian Oligocene deposits. The greatest number of Australian species occurs in heaths and sclerophyll forests, with primitive genera in rainforests of north-eastern Queensland.

Proteaceae are characterised by flowers with a four-lobed perianth (tepals), four stamens often attached to the tepals, and a one- or two-celled ovary (Fig. 41.22). They can be spectacular when flowering because their flowers cluster in brightly coloured inflorescences and attract bird pollinators (Chapter 37). *Banksia* flowers cluster in dense cylindrical or globular heads.

BOX 41.3 Eucalypts: native Australians

To overseas visitors, eucalypts are as 'Australian' as koalas and kangaroos. Eucalypts dominate Australian landscapes, except rainforests and the arid interior. They form tall forests in high rainfall areas (mountain ash, *Eucalyptus regnans*, in the south-east and jarrah, *E. marginata*, in the south-west), low forests and woodlands in drier regions (dominated by bloodwoods, gums, boxes, peppermints, stringybarks or ironbarks), mallee shrublands in semiarid areas, and subalpine forest and woodlands (snow gums, *E. pauciflora* and *E. coccifera*) at higher altitudes up to 2000 m. A few species occur on islands to the north of Australia.

'Eucalypt' is a common name used for three closely related genera: the large and familiar genus *Eucalyptus*, a small genus *Angophora* (restricted to eastern Australia) and a new genus *Corymbia* (the bloodwood species that are common in northern Australia) named in 1995 in recognition of the close relationship between bloodwoods and angophoras. *Eucalyptus* includes about 600 species and was named in 1788 by Charles Louis L'Héritier de Brutelle, a French botanist working in London at the time. L'Héritier published a description of *E. obliqua* (messmate) based on specimens collected in 1777 from Bruny Island, Tasmania, on Captain Cook's third voyage to Australia. *Angophora* was named later in 1797.

Despite the present-day dominance of eucalypts, little is known of their evolutionary history from the fossil record. Fossil eucalypt

(b) A sapling of *Eucalyptus nitens* (shining gum) showing different forms of leaves, from juvenile to adult

pollen is recorded for the Oligocene and there are younger records of leaves and fruits. There is also evidence of fossil eucalypt pollen from New Zealand, although no extant species occurs there naturally. It is thought, however, that the eucalypts are an ancient Australian group that has increased in dominance relatively recently as the climate developed a marked dry season and fire became more frequent.

Eucalyptus and *Corymbia* flowers lack showy petals and sepals but develop either one or two opercula—caps that enclose the stamens and style of the flower (Fig. a). These opercula are homologous to sepals and petals that have become continuous during floral development. In contrast, the genus *Angophora* has flowers with creamy free petals and green free sepals. Variation in the development and number of opercula, together with variation in ovules, seeds, fruits, leaf hairs and other characteristics, has led botanists to recognise a number of subgroups (usually treated as subgenera).

One feature of eucalypts is their development of different types of foliage as they grow and mature (Fig. b). Young plants have characteristic juvenile foliage, with leaves often held horizontally, sessile and in opposite pairs; saplings develop intermediate leaves and mature plants usually have petiolate leaves hanging in the vertical plane, reducing heating and water loss.

(a) *Eucalyptus pulverulenta* has two opercula: the outer one is the small brown cap being pushed off the flower bud by growth of the inner operculum

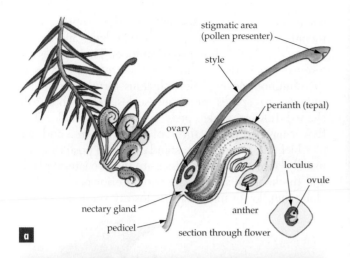

Fig. 41.22 (a) Typical flower of family Proteaceae. **(b)** *Grevillea longistyla*

Fig. 41.23 *Epacris impressa*, typical of the southern heath family Epacridaceae, and floral emblem of Victoria

Epacridaceae: the southern heath family

Epacridaceae are commonly called southern or Australian 'heaths' because of their similarity to Northern Hemisphere heaths and heathers of the related family Ericaceae. Epacridaceae are largely Australian but with species in New Guinea, Indo-China, Malaysia, Indonesia, the Philippines, New Zealand, New Caledonia, Pacific islands and South America (Fig. 41.20c). Epacridaceae are shrubs or small trees with stiff, hard leaves, distinctive leaf venation (palmate or subparallel) and tube-like flowers (Fig. 41.23). Epacrids are common on infertile soils and, like the Myrtaceae, they typically have root mycorrhizae (Chapter 36).

Mimosaceae: the wattle family

Mimosaceae (one of three families of legumes) are best known in Australia for *Acacia*, the **wattles**. There are about 835 Australian species, making *Acacia* the largest genus represented on the continent. *Acacia* is also well represented in Africa and tropical America. It is known in the Australian fossil record only from the early Miocene but it is believed to have had a longer Tertiary history.

Wattles are trees or shrubs with foliage of either compound bipinnate leaves or phyllodes. **Bipinnate** ('twice divided') leaves consist of small leaflets (pinnules) arising from pinnae arranged along a central axis, the rachis. **Phyllodes** are laterally compressed or flattened petioles and rachis (Fig. 41.24a). Usually longitudinal veins (nerves) are visible on their surface and glands may occur along the edges. Phyllodinous acacias (Fig. 41.24b) are almost entirely Australian (and are treated as a separate genus by some taxonomists). They are particularly dominant in the woody vegetation of arid central Australia, for example, mulga, *A. aneura*, although only a small number (118) of *Acacia* species occur there. Phyllodes are a scleromorphic feature and in arid-zone species they are often small in size, sometimes reduced to spines, with stomates sunken in pits.

Acacia flowers are small, clustered in globular heads or spikes. The familiar golden-yellow colour of the flower

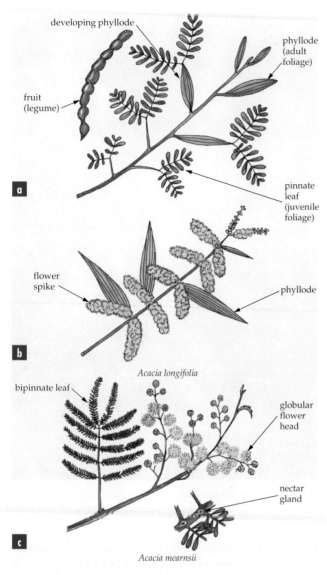

Fig. 41.24 **(a)** Phyllodinous acacias first develop bipinnate, compound leaves and then phyllodes. **(b)** *Acacia longifolia* is a phyllodinous acacia. Note that the flowers are typically clustered in spikes. **(c)** *Acacia mearnsii* is an Australian species that retains bipinnate foliage when mature. Note that its flowers are clustered in globular heads

heads is due to the stamens, which are often numerous. The fruit is a leathery or woody legume (a pod).

Many wattles are grown as ornamentals and some, such as blackwood, *Acacia melanoxylon*, are prized for timber. Seeds and seed pods were roasted and eaten by Aboriginal Australians, who also used the bark of some species as a poison to catch fish. Wattles are ecologically important because they increase the nitrogen content of soil through the activity of symbiotic species of *Rhizobium*, which live in nodules on *Acacia* roots (Box 44.5).

Fabaceae: the pea family

The pea family, **Fabaceae**, is found throughout the world but 1100 species in a large number of genera occur in Australia, especially in the tropics. They include trees, shrubs (e.g. the familiar 'eggs and bacon' plants, Fig. 41.25), herbs and climbers. Their leaves are often compound, commonly with three leaflets, as in the common clover. Sclerophyllous forms have leaves reduced to spines or small scales. Their butterfly shaped flowers make peas easily recognisable. The flower has five petals, three of which are free and two of which are united. The standard is the posterior and usually largest petal, the wings are the two lateral petals and the keel consists of the two anterior petals united along their lower margin to enclose 10 stamens and the

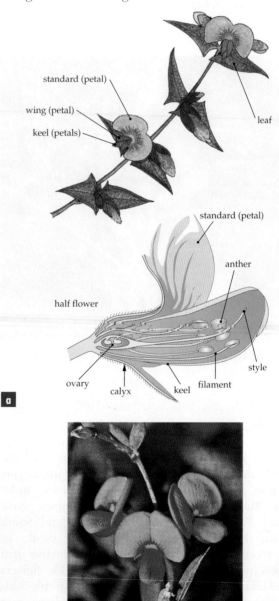

Fig. 41.25 **(a)** Pea flowers have distinctive butterfly shaped flowers. **(b)** *Gastrolobium oxylobioides*, a type of Australian 'eggs and bacon' flower, which is insect pollinated

Fig. 41.25 (c) Sturt's desert pea, *Clianthus formosus*, floral emblem of South Australia, is bird pollinated (see genetic variation in this species in Chapter 32)

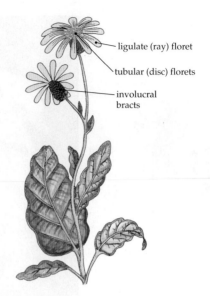

- ligulate (ray) floret
- tubular (disc) florets
- involucral bracts

Fig. 41.26 The familiar daisy 'flower' is a cluster of small flowers (florets), the showy outer ones of which often function to attract insect pollinators

ovary (Fig. 41.25). Pea seeds, like those of acacias, have hard seed coats, which are cracked by the heat of bushfires. Peas also have root nodules and fix nitrogen.

Asteraceae: the daisy family

The daisy family **Asteraceae** (old name Compositae) is one of the largest angiosperm families in the world with more than 25 000 species. Some taxa in the family are confined to the Southern Hemisphere and, within Australia today, native species occur in habitats ranging from alpine herbfields, where silvery carpets of snow daisies of the genus *Celmisia* are conspicuous, to semiarid and arid regions, where there is a large number of species. The abundance of daisies relates to their high reproductive rate and 'plasticity' of form; many species are widely distributed weeds, such as the common ragwort, dandelions, thistles, capeweed and boneseed.

Daisies are easily recognised by their flowers, which are clustered together in an inflorescence called a head (or **capitulum**). Individual flowers are, in fact, small florets, which are of two types. **Ligulate** (ray) **florets** on the outside of the head have five united petals strongly developed to one side, forming the showy rays. These flowers are often unisexual. **Tubular** (disc) **florets** form towards the centre of the daisy head (Fig. 41.26). They usually have both fertile stamens and an ovary. The inflorescence is surrounded by one or more rows of **involucral bracts**, the arrangement, shape and texture of which are important characteristics for identification.

Orchidaceae: the orchid family

Of the monocots represented in Australia, the **Orchidaceae** are a large and diverse family, generally easily recognisable when in flower. The orchid flower has three sepals and three petals, with one of the petals usually modified as the lip or **labellum** (Fig. 41.27). One or two stamens are united with the stigma and style to form the **column**. The striking form of many orchid flowers relates to the variety of ways that they attract insects for pollination, including 'sexual deceit' by mimicking female insects (Chapter 37).

Although the distribution of the family is worldwide, there are 1100 orchid species in Australia. In northern tropical regions there are a large number of epiphytic species growing on the trunks of trees; one example is the Cooktown orchid, *Dendrobium phalaenopsis*, which is the floral emblem of Queensland. Epiphytes and **lithophytes**, which grow among rocks, are xeromorphic, often having stems that are swollen into fleshy pseudobulbs, which appear to function as water-storage organs; some are also C_4 plants (Chapter 5). In the southern more temperate regions of Australia, many native orchids are terrestrials, the majority of which are of Gondwanan origin and related to African taxa. Terrestrials include greenhoods, *Pterostylis* (Fig. 41.27c), sun orchids, *Thelymitra*, and donkey orchids, *Diuris*; some have root–stem tuberoids and survive dry periods by retreating beneath the ground.

Today, the relict Gondwanan component of the Australian flora is represented by plants confined to moist rainforest habitats. The major component has diversified predominantly in temperate and more arid environments and includes, among the flowering plants, *Eucalyptus*, *Acacia*, *Banksia*, heaths, peas, daisies and orchids.

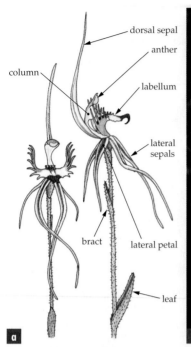

Fig. 41.27 (a) The orchid flower is recognised by one of the petals, the labellum, which is usually very different in shape, size and markings, and the column (fusion of stamens, style and stigma). Orchid flowers usually twist as they develop to present the labellum petal as a landing platform for insects. **(b)** The Cooktown orchid, *Dendrobium phalaenopsis*, the floral emblem of Queensland, and **(c)** the terrestrial greenhood, *Pterostylis revoluta*, are two Australian examples

BOX 41.4 Northern plants

Some plant groups represented in Australia have distribution patterns that suggest origins elsewhere than in Australia. For example, about 120 tropical and subtropical rainforest genera of dicotyledons are each represented in Australia by a single species and in Asia by several to many. There are 250 species of *Cinnamomum* (family Lauraceae) in Asia and New Guinea and only four in Australia. There are 850 species of *Rhododendron* (family Ericaceae), many of which occur in the Himalayas, South-East Asia and Malaysia; 155 species are endemic in New Guinea but only one occurs in Australia. Such patterns are sometimes interpreted as plant migration into Australia from the north, an explanation requiring dispersal of seeds across water barriers. Alternatively, some plants of New Guinea and Australia may have dispersed northwards as the Australian plate made contact with South-East Asia. This is one explanation for the occurrence of a single species of eucalypt, *Eucalyptus deglupta*, on Mindanao in the Philippines.

Some plants are cosmopolitan. For example, there are many species of *Epilobium* (fuchsia family, Onagraceae) in western North America, on arctic and tropical mountains, as well as in Australia. *Epilobium* seeds have a tuft of hairs, making them buoyant in air and easily dispersed; some species have spread as weeds after habitat disturbance this century. Of the cosmopolitan grass family Poaceae, the swamp reed, *Phragmites australis*, is probably the most widespread species of flowering plant in the world. In fact, many widespread species are freshwater aquatics.

Humans are responsible for recent introductions of a variety of plants into Australia. Familiar examples are *Pinus radiata*, *Lantana* and *Opuntia* (prickly pear), deliberately introduced for cultivation, as ornamentals or hedges.

Australian fauna

Like Australian plants, much of the modern terrestrial fauna is markedly different from that found elsewhere in the world. Also, it includes many animal groups that show Gondwanan patterns of relationships (Box 41.5), as well as groups with relationships to Asian and Northern Hemisphere groups.

BOX 41.5 Insects with southern connections

Insects have a long evolutionary history, being known in the fossil record for around 350 million years. The fossil record in Australia stretches from the Upper Carboniferous (300 million years ago) and includes over 400 different species. The modern insect fauna of Australia is represented by over 660 families and around 86 000 species. General patterns of relationships suggest that the primitive members of many groups are descendants of the Gondwanan insect fauna, while the more modern highly evolved members of each group have affinities with Asia. Orders of Australian insects that have clear Gondwanan elements include mayflies (Ephemeroptera, Fig. a); dragonflies and damselflies (Odonata); stoneflies (Plecoptera); termites (Isoptera); bugs (Hemiptera); beetles (Coleoptera); scorpion flies (Mecoptera); flies (Diptera); caddis flies (Trichoptera); moths and butterflies (Lepidoptera); and wasps, bees, ants and sawflies (Hymenoptera).

Many of the ancient groups maintain associations with early plants, which presumably represent coevolved relationships (see Chapter 43) that have existed throughout most of the evolutionary history of herbivore and plant. Examples of such associations in Australia include beetles of the subfamily Paracucujinae, which feed as adults and larvae entirely on the pollen of cycads (Cycadaceae); larvae of the primitive moth family Agathiphagidae, which feed on seeds of *Agathis* (Araucariaceae); cynipoid wasps (subfamily Austrocynipinae), which are associated with seeds of *Araucaria*; and primitive weevils (family Nemonychidae), which feed on pollen of Araucariaceae and Podocarpaceae. Larvae of an as-yet-unnamed genus of Australian leaf miners feed on *Banksia* (Fig. b). Relatives of these insects in Africa feed on proteas, relatives of banksias in the family Proteaceae.

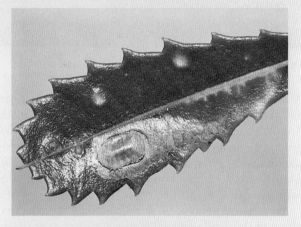

(b) Australian leaf miners are host-specific, feeding only on banksias

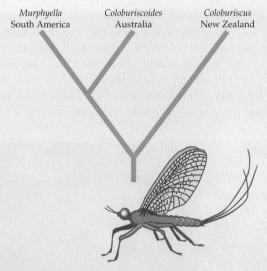

(a) The evolutionary relationships of three genera of mayflies show a geographic pattern that is consistent with our knowledge of the historical sequence of the break-up of Gondwana. The South American genus is more closely related to the Australian genus than it is to the older New Zealand taxon. This pattern is seen in other groups of mayflies as well as other groups of animals and plants

Relationships label in figure:
Murphyella South America — *Coloburiscoides* Australia — *Coloburiscus* New Zealand

Biogeographic patterns of terrestrial vertebrates

For terrestrial animals, major waterways provide a substantial barrier to dispersal so that islands such as Australia and New Zealand often have many unique or endemic species, which commonly include surviving groups that have long since become extinct in the rest of the world. The tuatara (Chapter 40) of New Zealand is an example of island survival of an ancient reptilian group for perhaps as much as 160 million years. During the break-up of Gondwana many such 'islands' of differing ages were created and the biota of these continental fragments often provide important insights into past biotic assemblages. For example, the native vertebrate fauna of New Zealand includes only three species of very primitive frogs, some skinks and geckos as well as the tuatara, no snakes, no terrestrial mammals (except for two species of bat) and a limited bird fauna, although the birds include an ancient group of ratites—the flightless kiwis (Fig. 41.28)—and the recently extinct moas.

Three of the more important island land masses throughout much of the Tertiary were Antarctica,

Fig. 41.28 The flightless kiwi, a ratite endemic to New Zealand

Australia and South America. Decreasing temperatures and development of the southern polar ice sheet made Antarctica an unsuitable place for all but a few animal groups. In South America and Australia, however, highly regionalised faunas developed, essentially in isolation from the rest of the world. For South America, this isolation came to an end with the formation of the Panamanian isthmus some six million years ago. The first effect of this land connection was a mixing of northern and southern faunas and a marked faunal enrichment in both North and South America. Inevitably, the interaction between many ecologically similar species led to large-scale extinctions. Now, many of the animal groups we associate with South America, such as those represented by llamas, jaguars and peccaries, are really recent immigrants from North America.

Australia, since finally rifting from Antarctica around 30 million years ago, has maintained much of its insularity together with much of its unique fauna, although new groups have reached the Australian plate since it has come in contact with South-East Asia.

While much of the fauna shows similar biogeographic patterns of relationships (see, for example, Box 41.5), these patterns are best illustrated by two groups of vertebrates: the amphibians (order Anura) and mammals (class Mammalia), both of which evolved in the early to middle Mesozoic, around the time Gondwana began to break up. In general, these animals have relatively poor powers of dispersal over water so that any global patterns of biogeographic relationships are less likely to be confounded by major dispersal events.

The modern Australian fauna is diverse and includes many unique and endemic groups, some of which evolved around the time of the break-up of Gondwana.

Adaptive radiation of frogs

Modern amphibians are represented by three orders, but only one, the **Anura**, which includes the frogs and toads, is found in Australia. The Australian Anura includes two ancient families, two families derived more recently from Asian stocks and one family recently introduced by Europeans (the introduced cane toad, *Bufo marinus*, now posing one of the most significant threats to tropical Australian ecosystems).

Most of the fauna comprises **myobatrachids** (family Myobatrachidae) and **hylids** (family Hylidae), two families with Gondwanan affinities. Except for one genus, *Heleophryne*, in southern Africa, modern relatives of myobatrachids are restricted to South and Central America. *Indobatrachus*, a fossil frog from India, is also thought to be related to these groups. Within Australia, myobatrachids include a variety of often bizarre species (Fig. 41.29) classified into 22 different genera.

Adaptive radiation, that is, the rapid evolution of a large number of closely related species, within the myobatrachids is manifest in morphological, developmental, behavioural and ecological specialisations. For example, female frogs of the genus *Rheobatrachus* from southern Queensland brood their larvae in the stomach, which functions as a uterus during incubation of the young. When tadpoles complete their larval development, froglets are 'born' through the mouth of the mother (Fig. 15.15). Another example of specialisation of breeding behaviour is shown by *Assa darlingtoni*, a frog of high cool rainforests of northern New South Wales and southern Queensland. The male has specialised lateral 'pockets' to protect developing tadpoles. The male sits among the egg mass that has been laid under the forest leaf litter and, when the eggs hatch, the tadpoles find their way into the male's pouches (Fig. 41.29), where they complete their larval development. These examples are two of the more unusual myobatrachid life-history patterns but a trend

Fig. 41.29 A myobatrachid frog, *Assa darlingtoni*, with tadpoles crawling to the male pouch

towards increasing independence of aquatic environments for larval development is seen in the family.

The family Hylidae, like the myobatrachids, has two centres of diversity—Australia and South America—although the family does have a worldwide distribution. Within Australia, hylids are represented by three genera: *Cyclorana*, *Litoria* and *Nyctimystes*. Except for some specialisations for living in and around torrential streams, life-history patterns throughout the family all follow the primitive pattern of aquatic development from eggs through tadpole stages followed by metamorphosis to a terrestrial frog. Nevertheless, the family has been just as successful as the myobatrachids in utilising most available habitats within Australia.

The highest diversity of frogs is found in mesic and warmer areas of the continent but, given their dependence on the availability of water for survival and breeding, there is a surprising diversity of frogs in the two-thirds of Australia described as arid or semiarid (see Fig. 41.11). In these generally dry environments, where rainfall is unpredictable, frogs avoid extreme conditions by burrowing into the soil (35% of Australian frog fauna burrow) or by seeking shelter under rocks or in cracks in the ground. Many species reduce water loss through the skin. Species of *Cyclorana*, the so-called water-holding frogs, develop a cocoon from shed skin, which, when hardened, becomes relatively impermeable to water and within which the frog survives underground for many months (Fig. 41.30). Most arid-zone species reduce the time spent as eggs and tadpoles, the most vulnerable stages of the life cycle. Whereas some species from mesic environments spend more than a year as tadpoles, the successful species in dry areas commonly complete the aquatic stages of their life in less than one month.

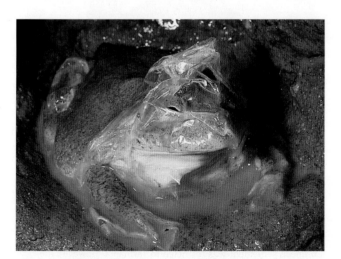

Fig. 41.30 *Cyclorana platycephala*, the water-holding frog, is well adapted to withstanding long periods of drought. It burrows deep into sand and makes a cocoon from its cast-off skin, which it sheds after rain

The remaining two families of amphibians in Australia, the Microhylidae and Ranidae, are of Asian origin, having entered Australia through New Guinea after collision between the Australian and Oriental crustal plates. Throughout the rest of the world, the Ranidae are a successful and diverse family but in Australia they are represented by a single species, *Rana daemeli*, restricted to north-east Queensland and Arnhem Land. Even though microhylids are more diverse, represented by two genera, *Cophixalus* and *Sphenophryne*, and 18 species, their distribution is still restricted to the tropical north of Australia, particularly north-east Queensland.

The largest component of Australia's frog fauna are myobatrachids and hylids, two Gondwanan families that radiated on the continent and adapted to dry environments. Two other families of native amphibians in northern Australia are of Asian origin.

Australian reptiles

Australia had a diverse and ancient reptilian fauna, which included a variety of dinosaurs. Many of those known from early Cretaceous deposits were small- to medium-sized (human-sized) hypsilophodonts, bipedal plant-eaters. One of the largest dinosaurs was *Rhoetosaurus brownei*, a giant plant-eater, that lived in Queensland about 190 million years ago. Some dinosaurs from southern Victoria had large eyes and enlarged optic lobes in the brain, seemingly well adapted to the polar light conditions they experienced in the southern high latitudes (Fig. 41.31a). These reptilian groups all became extinct, leaving no derivative modern descendants. Even so, on some small parts of Gondwana that are part of New Zealand, the tuatara, *Sphenodon punctatus*, now an endangered species, is a living representative of an ancient group of reptiles that was once widespread in Triassic and Jurassic times.

There are many arguments concerning the origins and relationships of the modern reptilian fauna of Australia. The side-necked tortoises (family Chelidae) are a Gondwanan group as are some of the geckos and their relatives (subfamily Diplodactylinae) but most modern forms appear to have been derived from Asian groups and colonised Australia after the Australian plate drifted north.

Even so, Australian dragon lizards (family Agamidae), which traditionally are believed to have been derived from Asia, may have Gondwanan affinities. Recent molecular data suggest that one group of dragons, including the bearded dragon (Fig. 41.31b), may be more closely related to African rather than Asian agamids.

Fig. 41.31 (a) Victorian dinosaur with large eyes adapted to the longer polar night that the animal experienced when Australia was part of Gondwana. **(b)** Australian bearded dragon

Adaptive radiation in mammals

Australia is the only continent where all three subclasses of the Mammalia are represented: the **Prototheria**—egg-laying monotremes; the **Metatheria**—marsupials; and the **Eutheria**—so-called placental mammals. The mammal fauna of modern Australia is extremely diverse and includes five orders of native terrestrial mammals as well as marine mammals (e.g. sea lions, seals, porpoises and whales) and a large variety of introduced species. Nearly all of the introduced mammals, many of which, like the rabbit, fox and house mouse, have become serious pest species, were deliberately or accidentally brought to Australia by Europeans. Even so, the first introduced species, the **dingo**, came with Aboriginal

colonisers and records of its presence in southern Australia date from around 7000 years ago.

Introduced mammals, including humans, have had a profound impact on Australian ecosystems and have led to the extinction of many native species. This issue is discussed further in Chapter 45 but one example is the interaction between the introduced dingo, *Canis lupus dingo*, and its ecological equivalent, the **thylacine** (Tasmanian tiger or marsupial wolf; *Thylacinus cynocephalus*).

T. cynocephalus (Fig. 41.32) was among the largest of the dasyuromorphs and thus the largest mammalian predator in Australia. This dog-like marsupial was the last descendant of a once quite diverse assemblage of dog-like predators (Fig. 41.33). Until the introduction of the dingo, *Thylacinus* was widely distributed across Australia (Fig. 41.32b) but subsequently it became extinct except on Tasmania, an area the dingo did not

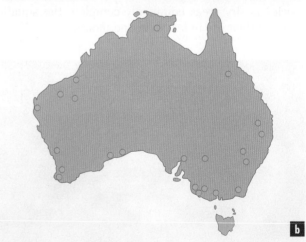

Fig. 41.32 (a) *Thylacinus cynocephalus*, the marsupial wolf. **(b)** Despite numerous reports of 'sightings' of thylacines on the mainland, the known range of *T. cynocephalus* at the time of European settlement was restricted to Tasmania. However, the localities on the map show sites where late Quaternary and Recent fossils, mummified remains or Aboriginal rock art have been recorded. These indicate that the marsupial wolf was once widely distributed on mainland Australia as well as New Guinea

Fig. 41.33 Australian scientists have discovered a wealth of fossils at Riversleigh in north-western Queensland. In these 20 million year old deposits, five types of fossil thylacines have been recognised, ranging from a species the size of a domestic cat to one as large as a medium-sized dog. Once widespread in Australia, this predatory lineage was nevertheless in decline. Deposits 15 million years in age reveal only two fossil species and by eight million years ago only one species, *Thylacinus potens*. By the early Pliocene, this species is replaced in the fossil record by *T. cynocephalus*, which survived until recent times. It was driven to extinction in its final stronghold, Tasmania, by an active program of extermination undertaken because thylacines were seen as serious pests for sheep farmers

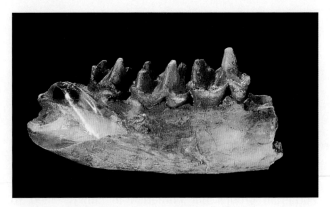

Fig. 41.34 Oldest known Australian mammal fossil: a platypus jaw 120 million years old

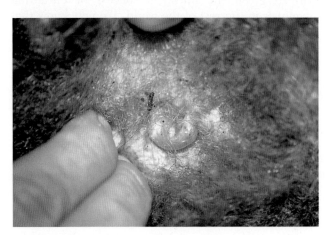

Fig. 41.35 This minute puggle in the pouch of an echidna has just suckled milk, which is visible in its stomach

reach before the island became isolated from the mainland at the close of the last glacial period of the Pleistocene (around 13 000 years ago). It was not until Europeans colonised Australia that *Thylacinus* was finally hunted to extinction. The last known individual died in the Hobart Zoo in 1936.

> The Australian fauna includes species introduced by humans, such as the dingo brought by Aboriginal Australians about 7000 years ago and other species brought in by Europeans in the last 200 years. Introduced species have led to the extinction of many native species.

Prototheria: the platypus and echidnas

The living members of this remarkable group of mammals are endemic to the Australian plate where it has had a long history. One of the oldest known fossils of an Australian mammal is a 120 million year old platypus-like jaw (Fig. 41.34). Fossil prototherians are also known from South America. Prototherians display many primitive features, such as egg laying, secreting milk from glands that do not form nipples (Fig. 41.35), a cloaca and markedly reptilian features in the bones of the limbs and girdles. Nevertheless, the two extant species in Australia, the **platypus**, *Ornithorhynchus anatinus*, and the **echidna**, *Tachyglossus aculeatus*, show some remarkable morphological and ecological specialisations, for example, the recently reported electrosensory system. The platypus, which is aquatic and widespread in most unpolluted streams in mesic

eastern Australia, uses this sixth sense to locate its prey. Specialised sensory pits or pores on the bill of the platypus (see Fig. 7.11) allow it to detect extremely weak electric fields generated by neuromuscular activity of its invertebrate prey. Underwater, with its eyes, ears and nostrils (nares) closed, it has been suggested that the field generated by a tail flick of a freshwater shrimp can be detected by a platypus as far away as 10 cm. More recently, the system has been found in a less well-developed form in the snout of the echidna.

Such specialisations enabled these ancient mammals to survive in the face of increasing competition from other animal groups, particularly other types of mammals. If a diverse monotreme fauna existed in Mesozoic Australia, it was largely displaced by the marsupials.

> Living prototherians, the platypus and echidnas, are endemic to the Australian plate. They have primitive features and are known from fossils as old as 120 million years.

Metatheria: marsupials

Like hylid and myobatrachid frogs, **marsupials** have two centres of diversity, South America and Australia, although the oldest marsupial fossil comes from North America, indicating that the group had a widespread distribution during its early history. Even so, the past distribution of various mammalian groups is unclear, mainly because few fossils of Mesozoic age (200 million years ago) are available for many places, especially Australia and Antarctica. The earliest South American mammal faunas included both metatherian and eutherian groups but eutherian mammals failed to become established in the original mammal fauna of Australia. It is difficult to explain this anomalous pattern given the universal success of eutherian mammals in the rest of the world. We now know that eutherians were part of early Australian mammal faunas, as evidenced by the discovery of a presumed eutherian tooth in early Tertiary deposits (55 million years old) at Tingamarra, Queensland, and the startling recent discovery of the jaws and teeth of a hedgehog-like eutherian in Mesozoic (115 million years old) sediments in West Gippsland, Victoria. While of a similar age to the earliest-known prototherian fossils, this Mesozoic eutherian is twice as old as the most ancient Australian marsupial fossil so it is possible that eutherians became extinct in ancient Australia for reasons unconnected with the successful adaptive radiation of Australian marsupials.

The present marsupial fauna of Australia includes four orders and three other orders exist in South America. Two Australian orders, the Dasyuromorphia and the Peramelemorphia, are called **polyprotodont** marsupials, with more than one pair of incisor teeth in the lower jaw, but in general these animals are characterised by having a large number of sharp, relatively unspecialised teeth (Fig. 41.36). Dasyuromorphia includes carnivorous or insectivorous marsupials,

such as the quolls, dunnarts, planigales, *Antechinus* (Fig. 15.14), Tasmanian devil (Fig. 41.37), marsupial wolf (Fig. 41.32) and numbat. Peramelemorphia includes the omnivorous bandicoots and bilbies. A third order, Notoryctemorphia, contains two species, the bizarre marsupial moles.

Diprotodontia (Fig. 41.38) is the fourth and largest order and includes those marsupials that most people recognise as typical Australian mammals. There are two distinctive suborders: the koala and wombats (Vombatiformes); and a large group (Phalangerida) that contains the possums and gliders (in four superfamilies) and the kangaroos, wallabies, rat kangaroos, bettongs and potoroos (superfamily Macropodoidea). Although these two suborders include a variety of strikingly different types of animals, all of them are related and form a monophyletic group, characterised by several derived characteristics. The most obvious features of **diprotodonts** are a general reduction in the number of teeth, with only one pair of incisors in the lower jaw, together with a tendency to greater specialisation of tooth structure, especially in the development of molars with grinding and crushing surfaces used in browsing and grazing (Fig. 41.39). While the majority of the Diprotodontia are herbivores, fossil material shows that a reversion to a carnivorous lifestyle did occur, although all of these predators, like the carnivorous kangaroo (*Propleopus*), are now extinct. The group also included many extinct species that were part of Australia's megafauna (Box 41.6).

> The present marsupial fauna of Australia includes four orders: the Dasyuromorphia and the Peramelemorphia (polyprotodont marsupials that are carnivores and omnivores, with more than one pair of incisor teeth in the lower jaw); the Diprotodontia (mostly herbivores, with only one pair of incisors in the lower jaw) and the Notoryctemorphia (the marsupial moles).

Fig. 41.36 Skull and teeth of a Tasmanian devil, typical of a polyprotodont marsupial

Fig. 41.37 Tasmanian devil (Dasyuromorphia)

Fig. 41.38 Examples of Diprotodontia: **(a)** the common wombat, **(b)** yellow-bellied glider and **(c)** the red-necked (Bennett's) wallaby

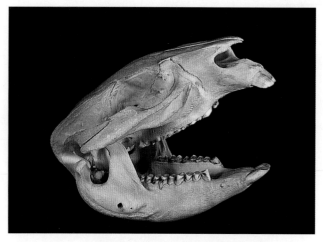

Fig. 41.39 Skull and teeth typical of the Diprotodontia

BOX 41.6 Megafauna

Unlike the rest of the world where mammals are commonly the conspicuous large predators, the largest carnivorous vertebrates in the terrestrial fauna of Australia in the Tertiary were reptiles: a giant lizard, *Megalania*, a large land-dwelling crocodile, *Pallimnarchus*, and a giant snake, *Wonambi* (Fig. a). These predators were part of the Australian **megafauna**, which also included a great many other large and impressive organisms (Fig. b), two of the most spectacular being *Diprotodon optatum*, a lumbering diprotodontid that reached the size of a modern hippopotamus, and the giant mihirung birds (family Dromornithidae), which were flightless emu-like animals that included the largest bird that has ever lived. During the late Pleistocene, the megafauna became extinct; the last *Diprotodon* and mihirung birds died out around 20 000 years ago.

These large-scale extinctions of megafauna over the past one to two million years were a worldwide phenomenon and causal explanations are generally of two kinds. The first implicates human hunting and there is evidence that humans did hunt and kill elements of the megafauna. The second suggests that environmental deterioration accompanying increasing aridity brought about the megafaunal extinctions. Clearly both processes could have been acting in concert. In Australia, humans coexisted with the megafauna for at least 20 000 years so that direct hunting, along with human-induced environmental changes, may have driven some of these large animals to extinction. At the same time, increasing aridity, which

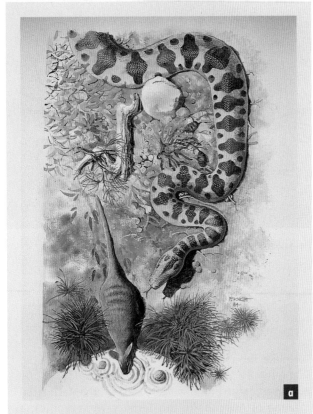

(a) *Wonambi*, a giant python of Australia's extinct megafauna, stalking a young wallaby. **(b)** A large and a small *Palorchestes*, one of Australia's most unusual extinct mammals. It was the size of a large bull and may have inspired the legend of the bunyip

reached its most extreme about 17 000 years ago, led to marked environmental change. For example, Lake Callabonna, to the north-east of the Flinders Ranges in South Australia, is now one of a series of dry salt pans. In the recent past, this area supported large populations of *Diprotodon*—in fact, it is one of the best localities for finding fossils of this animal.

Eutheria: bats and rats

While Australia is thought of as being the 'land of marsupials', the native, terrestrial mammal fauna includes a remarkable diversity of eutherian mammals. Even so, this large number of endemic genera and species belong to only two orders: bats (Chiroptera) and rodents (Rodentia).

Bats (Fig. 41.40) have a long history in Australia, being known as fossils from the Eocene (55 million years ago). Because they are flying mammals, water is less of a barrier to dispersal and the early appearance of bats is perhaps not so surprising. The order Chiroptera includes two suborders and both are well represented in the Australian fauna. The familiar fruit and blossom bats and the flying foxes (suborder Megachiroptera, see Chapter 40) are generally large herbivorous bats. The small predatory bats (suborder Microchiroptera) hunt with a sophisticated mechanism of echo location of prey. In Australia, all microchiropterans are insectivorous except: the ghost bat, *Macroderma gigas*, which, although taking some large insects and other invertebrates, preys on a variety of small vertebrates, including other bats; and *Myotis adversus*, which catches insects and small fishes with its hind feet while flying over water.

Even though Australia has a large and diverse fauna of bats, the native rodents (Fig. 41.41) provide an even

Fig. 41.40 A microchiropteran bat, *Pipistrellus tasmaniensis*, with large ears for echo location

Fig. 41.41 A native rodent, *Hydromys chrysogaster*, a water rat of eastern Australia

more remarkable example of adaptive radiation. All native rodents belong to only one family, Muridae. Because there are so many different types of animals among the 50 or more species of native rodents (ranging from mouse- and rat-like forms to large aquatic animals), it was once thought that this group had had a very long history in Australia. It is now clear that this diverse fauna has arisen over a relatively short period and from as few as two episodes of colonisation.

Rodents first make their appearance in the Australian fossil record in the Pliocene, so native rats and mice have evolved over the past few million years, certainly within the last 15 million years. Most of the rodent fauna (subfamily Hydromyinae) derive from this initial colonisation. The remainder (subfamily Murinac) is made up of a small number of native species of the cosmopolitan genus *Rattus* (excluding the introduced *R. rattus*, and the Melanesian rat, *R. exulans*; together with the house mouse *Mus musculus*) that derive from a much later colonisation event that occurred within the last one million years. Both groups of rodents colonised the Australian plate from Asia. Ancestors of the Hydromyinae may have rafted across a decreasing water gap between South-East Asia and Australia or they may have been part of an existing rodent fauna on Asian islands that were incorporated into the north coast of New Guinea during the collision between the crustal plates. The more recent colonisation of *Rattus* undoubtedly involved chance dispersal across a water gap, although the distances involved may have been quite small.

During the last 2.5 million years the world has been subject to times of very cold conditions (see p. 1091). The importance of glacial periods for dispersal of terrestrial animals is that much of the world's water is bound up as ice and the resulting lower sea levels both connect continental islands to the mainland as well as reduce the extent of seaways between neighbouring land masses. During the last glacial period, continental islands such as Tasmania, New Guinea and Kangaroo

Island were part of mainland Australia. Importantly, during glacial periods the Australian plate was separated from South-East Asia in some places by water gaps only a few kilometres wide. It was across such barriers that ancestors of Australia's native rodents dispersed.

In summary, the biogeographic patterns illustrated by the frogs and mammals are repeated in many groups of the Australian fauna. Usually two distinctive components are evident. One is an old faunal component, which has affinities with similar groups in other fragments of Gondwana, and which has undergone long and extensive adaptive radiation while isolated in Australia. The other is a more recent faunal component, which has affinities with the South-East Asian fauna and which entered Australia when the Australian plate drifted nearer to and came into contact with the Asian plate. The history of other groups of animals, such as the song birds (Box 41.7), is less well known but is being actively investigated.

> The terrestrial eutherian fauna of Australia is diverse but includes only bats (order Chiroptera) and rodents (order Rodentia). Rodents radiated over a relatively short time after dispersal from the north.

BOX 41.7 Australian song birds

Australia has a large and diverse bird fauna, including many taxa that have had a long evolutionary history on the continent. Some, such as the emus and cassowaries, have relationships with other Gondwanan ratites (p. 1086), while in other groups, such as the song birds (Fig. a), the evolutionary relationships are not so easily detected. Using modern techniques to study the structure of DNA and various proteins, systematic zoologists are able to determine the remarkable pattern of adaptive radiation and relationships within and between various groups. Even though early European settlers gave familiar names to many native song birds (e.g. robins, wrens and warblers), the similarity with 'Old World' species is only superficial, the similar phenotypes being the result of convergent evolution through adaptation to exploit similar ecological conditions. A simplified phylogeny of song birds (Fig. b) shows the long history of many of these groups in Australia. The information is derived from comparing the differences in protein structure between the groups together with a knowledge of the rates of change in bird DNA.

(a) The lyrebird, an endemic Australian song bird

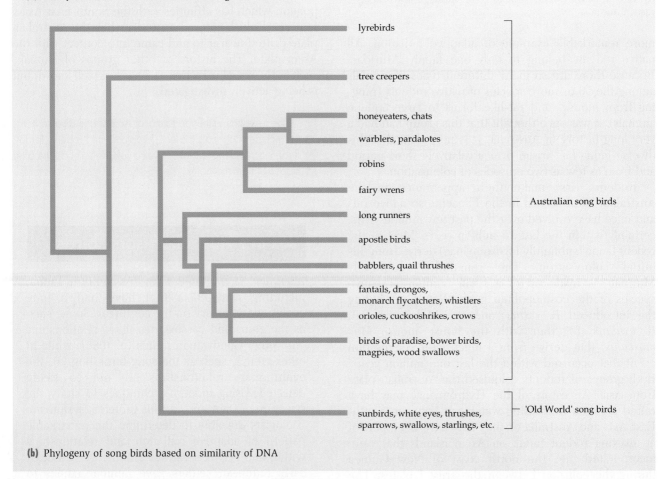

(b) Phylogeny of song birds based on similarity of DNA

Summary

- Australia was part of the southern continent Gondwana during the Cretaceous and early Tertiary. The climate was humid and rainforest covered much of the land.
- Australia severed its links with Antarctica and became an island continent about 30 million years ago. Separation of land masses created a circum Antarctic current, causing rainfall and temperatures to decrease and climate to become more arid. Climate changes altered landforms and soils, and fires became more frequent towards the end of the Tertiary. Together with isolation, these changes influenced the evolution of Australian plants and animals.
- During the last 2.5 million years (the Quaternary), cool, arid glacial periods alternated with warm, wetter interglacial periods. Humans, who colonised Australia at least 40 000 years ago (and possibly much earlier), caused increased burning. These environmental changes correlate with pollen assemblages, indicating that rainforest contracted and fire-tolerant vegetation became dominant. Environmental changes since the arrival of humans also correlate with the extinction of large animals, which occurred between 35 000 and 15 000 years ago.
- Today, the modern Australian biota consists of a number of components, one of which is a relict Gondwanan component represented by plants and animals confined to moist rainforest habitats. The major component has evolved from Gondwanan stock and diversified since the middle of the Tertiary, predominantly in temperate and more arid environments.
- Native plants include xeromorphs, such as the sclerophylls *Eucalyptus* (family Myrtaceae), *Acacia* (Mimosaceae), *Banksia* (Proteaceae), heaths (Epacridaceae), peas (Fabaceae), she-oaks (Casuarinaceae), grass trees (Xanthorrhoeaceae) and hummock grasses (e.g. *Triodia*). Xeromorphs also include epiphytic orchids and succulents such as saltbush.
- Sclerophylls possess features that allow them to survive low soil nutrients, water stress and fire. Sclerophylls are characterised by small leaves, short internodes, thick cuticle, sunken stomates, high proportion of lignified cells and often a dense covering of hairs.
- Like the flora, the modern Australian fauna includes many endemic groups. Mammals, for example, evolved around the time of the break-up of Gondwana and Australia's mammal fauna is distinctive and diverse. There are monotremes (subclass Prototheria), marsupials (Metatheria) and bats and rats (Eutheria). Monotremes and marsupials are ancient Australian mammals but the native bats and rats are relatively recent arrivals, mostly reaching this continent within the last 15 million years after the Australian crustal plate collided with that of Asia.
- Changes in vegetation and range expansion of arid and semiarid flora led to a dominance of grazing rather than browsing herbivores, in particular, wallabies and kangaroos, which, in terms of number of species, numbers of individuals, and extent of distribution, are the most successful group of marsupials.
- Both the flora and fauna include species introduced by humans, such as the dingo brought by Aboriginal Australians about 7000 years ago and other species brought by Europeans in the last 200 years. Introduced species have had an impact on the Australian environment, leading to extinction of many native species.

key terms

Acacia (p. 1096)	*Eucalyptus* (p. 1096)	ligulate floret (p. 1103)	polyprotodont
adaptive radiation	Eutheria (p. 1108)	lithophyte (p. 1103)	(p. 1110)
(p. 1106)	Fabaceae (p. 1102)	marsupial (p. 1110)	Proteaceae (p. 1099)
Anura (p. 1106)	fire (p. 1090)	megafauna (p. 1111)	Prototheria (p. 1108)
aridity (p. 1088)	glacial period	mesophyte (p. 1098)	Quaternary period
Asteraceae (p. 1103)	(p. 1091)	Metatheria (p. 1108)	(p. 1091)
bipinnate (p. 1101)	Gondwana (p. 1088)	Mimosaceae (p. 1101)	ratite (p. 1086)
capitulum (p. 1103)	hylids (p. 1106)	myobatrachids	sclerophyll (p. 1096)
circum Antarctic	interglacial period	(p. 1106)	succulent (p. 1098)
current (p. 1088)	(p. 1091)	Myrtaceae (p. 1099)	Tertiary period
column (p. 1103)	involucral bract	*Nothofagus* (p. 1086)	(p. 1087)
dingo (p. 1108)	(p. 1103)	Orchidaceae (p. 1103)	thylacine (p. 1108)
diprotodont (p. 1110)	labellum (p. 1103)	Pangaea (p. 1087)	tubular floret (p. 1103)
echidna (p. 1109)	laterite (p. 1090)	phyllode (p. 1101)	wattle (p. 1101)
Epacridaceae (p. 1101)	lignotuber (p. 1098)	platypus (p. 1109)	xerophyte (p. 1098)

Review questions

1. Match up the terms on the left with the appropriate term in the list on the right.

   Cretaceous   160 million years ago

   Tertiary   135–65 million years ago

   Gondwana   65 million years ago

   Pangaea began to break up   30 million years ago

   Australia was an island continent   Southern supercontinent

                                           Northern supercontinent

                                           Included all continents

2. On what land masses are living representatives and fossils of *Nothofagus* found? Name two other plant or animal groups that have a Southern Hemisphere distribution. What is the accepted geologic explanation for these biogeographic patterns?

3. Indicate whether each of the following is true or false. The native vertebrate fauna of New Zealand includes:

   (a) no snakes

   (b) an ancient reptilian animal, the tuatara

   (c) many species of frogs

   (d) no terrestrial mammals, except bats

   (e) flightless, ratite birds, kiwis and cassowaries.

4. Define the following terms:

   (a) circum Antarctic current

   (b) laterite

   (c) glacial period

   (d) interglacial period.

5. Briefly explain why the climate of Australia changed from humid temperate to more arid during the Tertiary period. How did this affect the dominant type of vegetation?

6. List a characteristic feature that helps you to identify each of the following angiosperm families:

   (a) Myrtaceae

   (b) Proteaceae

   (c) Fabaceae

   (d) Asteraceae.

   In which of these families are classified:

   (e) native peas?

   (f) banksias?

   (g) eucalypts?

   (h) grevilleas?

7. (a) What kind of evidence have scientists used to interpret the history of fire in Australia?

   (b) Examine Figure 41.10. How has fire frequency changed over the last 22 million years and what affect has it had on the vegetation?

8. Wattles, *Acacia*, are ecologically important because they are able to fix _____ in the soil by having symbiotic species of _____ associated with their roots. Most Australian species of *Acacia* have an unusual type of foliage called a _____. Their fruit is a woody or leathery _____.

9. Briefly discuss the various adaptations that reduce water loss in Australian frogs.

10. Give an example of an animal in each of the following families of diprotodont marsupials:

    (a) Vombatidae

    (b) Phalangeridae

    (c) Tarsipedidae

    (d) Macropodidae.

    What features characterise species of the Diprotodonta?

11. In contrast to other continents, Australia's largest mammalian predators were dog-sized thylacines. Within the recently extinct megafauna of Australia, which animals filled the role of large predators, ecologically similar to tigers, bears and lions?

12. What impact did the arrival of humans at least 40 000 years ago have on Australian environments and biota?

Extension questions

1. The marine biota associated with the continental shelf of Australia is as unique and diverse as the terrestrial plants and animals. Discuss the reason(s) for this high level of endemism among the marine organisms.

2. Living prototherians are only found in Australia and have been here for at least 120 million years. The group displays many features that reflect the mammal's reptilian heritage. Describe four of these primitive characteristics found in prototherians.

3. Australia is known as the 'land of the marsupials', yet many of our native mammals are eutherian mammals. Describe in general terms the composition of the native eutherian fauna and discuss its evolutionary and biogeographical relationships.

4. There is increased attention being focused by governments and the public on Australia's biodiversity. What is meant by 'biodiversity' (see Chapter 30)? Why should people be concerned with documenting and understanding the history and diversity of the world's biotas?

Suggested further reading

Dyne, G. R. and Walton, D. W., eds. (1987). *Fauna of Australia.* Vol. 1A: General Articles. Canberra: Australian Government Publishing Service.

This volume, part of the Fauna of Australia *series supported by the Australian Biological Resources Study (ABRS), describes the evolution of Australian environments, including marine, freshwater and terrestrial. It describes the biology and adaptations of animals, including arthropods, corals, molluscs, echinoderms, fishes and land vertebrates.*

ABRS. *Flora of Australia.* Vol. 1. 1st edn. (1981). 2nd edn. (1999). Canberra. ABRS/CSIRO Publishing.

Both the first and second editions of this volume summarise the evolutionary history of the Australian flora. Keys, including electronic computer-interactive keys, are available with this volume to identify flowering plants.

Hill, R. S., ed. (1994). *History of the Australian Vegetation—Cretaceous to Recent.* Cambridge: Cambridge University Press.

An advanced text summarising the break-up of Gondwana and the fossil and climatic history of Australia. The book deals largely with plants but also mammals.

Robinson, M. (1998). *A Field Guide to Frogs of Australia.* Sydney: New Holland.

This is one example of many field guides to animals, which include keys to identify species, descriptions, coloured photographs and notes on frog biology. The book also covers the topic of frog conservation.

Strahan, R., ed. (1995). *The Mammals of Australia.* Sydney: Reed Books.

This book covers all the Australian kangaroos, wallabies and possums and other mammals in one book, including distribution maps, habitats and information on behaviour, reproduction, growth and diet.

White, Mary. (1988). *Australia's Fossil Plants.* Sydney: Reed Books.

Introductory book to geologic time, fossil formation, collecting fossils and the history of Australian plants since the Devonian. See also other superbly illustrated books by Mary White: The Greening of Gondwana *(1986), Reed Books, Sydney, and* After the Greening: The Browning of Australia *(1994), Kangaroo Press, Sydney.*

Part 7

7

Ecology

CHAPTER

42

Population ecology

Until now this book has been concerned primarily with biological processes, such as the functioning of cells, organs and individual organisms, and the evolution and diversity of species. This section deals with the **ecology** of organisms—the ways in which organisms interact with one another and their environment. We can study the ecology of organisms at three levels of organisation: at the level of **populations** of single species; at the level of **communities**, which consist of groups of species living together; and at the level of **ecosystems**, which are communities together with their non-living surroundings. This chapter examines the first of these.

The primary concern of population ecologists is to understand the dynamics of the distribution and abundance of organisms. This understanding is essential to applied biology, including conservation biology, pest control and the management of natural populations that are harvested to provide resources, such as food, for humans. Population ecology provides the essential building blocks of community and ecosystem ecology. It is important to remember that humans are a species of animal and many of the concepts and phenomena described in this section apply equally well to us.

Population ecologists try to understand the dynamics of the distribution and abundance of organisms, knowledge that is essential to the management of natural populations.

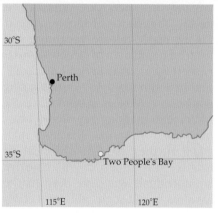

Fig. 42.1 The noisy scrub-bird, *Atrichornis clamosus*, was known from six areas in Western Australia before 1889 but was not sighted again until 1961, when a small population was discovered at Two People's Bay. Since then, the population has expanded and new populations are being formed by translocating individuals into other suitable localities, thus avoiding the possibility that a single catastrophe, such as fire, extinguishes the species

What is a population?

A population is a number of organisms of the same species that inhabit a defined geographic area, for example, noisy scrub-birds, *Atrichornis clamosus*, at Two People's Bay, Western Australia (Fig. 42.1), or mountain plum pines, *Podocarpus lawrencia*, on the Errinundra Plateau, Victoria. Members of species usually occur in a number of local populations. This subdivision often depends on geographic features that tend to dissect an otherwise continuous population into fragments. The definition of a population is particularly important with regard to evolution, since populations, and not individuals, evolve (see Chapter 32). In this section, however, we will concentrate on non-evolutionary processes influencing populations, although remember that a population's evolution and its dynamics are inextricably linked.

Populations have many properties. Some of the most important properties are: number of individuals in the population; area occupied by the population; age structure of the population; and sex ratio of the population. For species with complex life cycles, understanding the ecology of the entire species is more complicated. For example, plants usually exist in a particular locality, not only as growing individuals but also as dormant seeds; barnacles occur as both sessile adults adhering to a rocky shore and as larvae in the ocean. The distribution and abundance of organisms at each phase of the life cycle may be affected by different processes. For example, growing plants of golden wattle, *Acacia pycnantha*, experience herbivory by macropods and are killed by fire, while the seeds of this species are eaten by ants and birds but often require fire to germinate.

Two questions immediately spring to mind: 'Why is a species only present in particular places?' and 'What determines the number of individuals (size) of a population in one particular place?'. In other words, 'What determines **distribution** and **abundance** of populations?'. Ecologists studying, for example, the effects of pollutants on a river would monitor the distribution and abundance of a number of species (aquatic invertebrates such as dragonflies or aquatic plants) and compare these with a similar but unpolluted river.

BOX 42.1 Measuring the size of animal populations using the mark–release–recapture method

It is usually not possible to count all the individuals in a population, especially when organisms are mobile. An alternative method is to estimate the total population from a sample. Stationary organisms, such as plants, can be counted in a small sample area—a quadrat—marked out on the ground. Soil organisms can be sampled in a core of soil, and aquatic organisms in a volume of water. One approach used to estimate the abundance of mobile animal populations is the **mark–release–recapture method**.

Consider a population of sleepy lizards, *Tiliqua rugosa* (see Fig. a), that lives in a grazed mallee woodland in South Australia. During a single day, two population ecologists systematically search a square kilometre marking any sleepy lizards they find with pink dye. Assume that they marked M_1 individuals on the first day. The next day they mark lizards in the same area with blue dye. Some of these individuals will have been marked with pink dye already and some will not. Let M_2 be the number of individuals marked with blue dye that were already marked with pink dye and let N represent the total number of individuals marked with blue dye on the second day. (The number of individuals marked on the second visit that were not marked on the first visit is $N - M_2$.) The proportion of individuals marked on the second day that were already marked should equal the proportion of individuals in the population that are marked:

$$\frac{M_1}{\text{Total population size}} = \frac{M_2}{N}$$

Hence, an estimate of the number of lizards in the mallee woodland, subject to statistical error, is:

$$\text{Total population size} = N \left(\frac{M_1}{M_2} \right)$$

For example, if 52 animals were marked on the first day and 64 on the second (of which 16 had already been marked), then the estimate of the total population size is 208. This estimate relies on many assumptions that can be difficult to satisfy:

1. The population is assumed to be closed. If individuals die or move into and out of the area between the first and second marking days, the estimate of population size will be inaccurate.

2. All individuals are equally likely to be marked. In practice, some individuals may be more likely to be found than others.

3. Marked individuals do not lose their mark (see Fig. b). Despite these problems this method can be a useful means of assessing the population size of a species as it changes from year to year and place to place (Fig. c).

(a) The sleepy lizard, *Tiliqua rugosa*, in threat posture

(b) Ecologists have used natural marks on whales, such as this humpback whale, and dolphins to track animals off the coast of Queensland when estimating population sizes

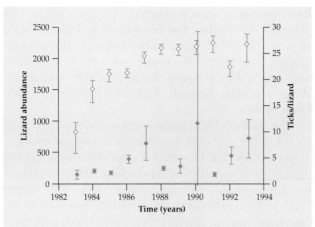

(c) Scientists have been studying a sleepy lizard *Tiliqua rugosa* (open diamonds) using the mark–release–recapture method since a drought in 1982. Using simple counts, they have also monitored the number of ticks on each lizard—a measure of tick abundance (solid diamonds) can be gained by multiplying the number of lizards by the average number of ticks per lizard. Error bars are 95% confidence limits

Distribution and abundance of populations

The environment of an organism can be divided into two parts: the **biotic** (other organisms) and the **abiotic** (the physical inorganic components). Both abiotic and biotic factors influence the distribution and abundance of populations.

Plants spend their entire life in a single locality. For a plant population to persist in a particular locality, it must be able to survive and to reproduce successfully. Successful reproduction of flowering plants may depend on the availability of a specific animal pollinator, a biotic factor. However, the survival of many plants appears to be closely correlated with abiotic factors, particularly light, temperature, rainfall and soil. The way in which these abiotic factors limit the distribution of a species can often be related to physiological processes that directly affect a plant.

To assimilate carbon for growth and reproduction, plant stomata must be open to allow the entry of carbon dioxide. This inevitably entails a loss of water and if a plant loses too much water it may wilt and die. The distribution of many species is limited by the availability of water, which, in turn, is determined by a number of environmental parameters, including temperature, rainfall, soil type, humidity and altitude. In this case, available water can be regarded as a direct environmental factor that is controlled by a number of indirect environmental factors.

By recording the presence or absence of eucalypt species in over 5000 sites in south-eastern New South Wales, and knowing the mean annual temperature at

each of these sites, M. Austin from the CSIRO was able to plot the probability of occurrence of different species as a function of temperature (Fig. 42.2). Plots of species' occurrence against an environmental gradient, such as temperature, typically show a peak in the probability of occurrence. Such curves define the realised distribution of a species in environmental (not

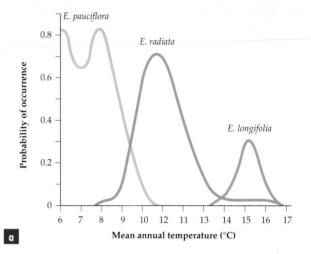

Fig. 42.2 (a) Plots of the probability of occurrence of different eucalypt species in south-eastern New South Wales and Victoria as a function of temperature. Some species show typical bell-shaped curves, such as *Eucalyptus longifolia*, while the response of other species is skewed, such as *E. radiata*. By mapping environmental space onto a geographic map, it is possible to use curves like these to predict the probability of occurrence of a species at any locality in a region. **(b)** *E. pauciflora*, snow gum, is tolerant of low temperatures and snow. It forms subalpine woodlands at high altitudes in south-eastern Australia

geographic) space. The limits to the distribution of the eucalypt species in the environmental space is often determined by a plant's ability to tolerate extreme physiological stress (e.g. very cold nights), although it may also be determined by biotic factors such as herbivory, pathogens or competition with other plant species. More generally, distribution is determined by a complex combination of biotic and abiotic factors such that it is rare to be able to attribute a single cause to the absence of a species from a particular environment.

Potential and realised distributions

The **potential distribution** of a species is the geographic range over which individuals could theoretically survive and reproduce. The **realised distribution** of a species is the range in which it does live and reproduce. The two are rarely the same.

One way of determining the potential distribution of a species is to conduct transplant experiments. By introducing a population of a plant or animal into a number of localities and monitoring the fate of those populations, we can gain some insight into the potential distribution of that species. This process is not as simple as it may first appear, for although it may be easy to monitor the survival of individuals in new localities, the long-term persistence of a population relies on successful reproduction and survival over several generations. Conducting such experiments, especially with long-lived organisms, may be impractical.

A knowledge of the response of a species to environmental gradients can be used to make predictions about the potential distribution and abundance of species.

One indirect method of estimating the potential distribution of a species is to determine the abiotic and biotic properties of the places in which it currently lives and extrapolate these to other areas to determine its potential range. Although this procedure has numerous pitfalls, it may be a useful predictive tool for determining the effect of global warming and climatic change (Chapter 45) on the distribution of species. This is particularly relevant to pests that damage crops. For those pest species with distributions and population growth rates limited by low temperatures, an idea of a species' **fundamental niche** (the total range of environmental conditions under which the members of a species could live and reproduce) helps us to identify those areas in cooler, southern regions of Australia that the pest will invade as climate changes. For example, since its introduction to Queensland at the beginning of the twentieth century, the cane toad,

Bufo marinus, has increased its distribution. Its potential distribution has been modelled on the basis of temperature tolerance of the tadpoles. It continues to spread west to the Northern Territory but appears to have reached its temperature limit in the south. With climatic change, the southern limit is likely to shift (see Chapter 45).

The realised distribution is often a surprisingly small subset of the potential distribution. There are a variety of reasons for this. One is that a species may not occur in an environmentally suitable locality simply because it has not been able to colonise that place. This simple concept often explains the limited distribution of sessile organisms, such as myrtle beech trees, which have poor dispersal abilities (Fig. 42.3). Historical factors, such as past climates, and catastrophic events, such as fires, play a key role in restricting species to subsets of their potential range.

It seems reasonable to expect that all the individuals of a species would attempt to live in places that maximise their chances of survival and reproduction, that is, their 'evolutionary fitness'. This idea is the basis of **habitat selection theory**. Very mobile species, such as migratory animals, can choose between different localities from year to year and season to season. Their absence from a particular place in any one year may reflect the unsuitability of that habitat in that year. Also, in years when a species is abundant, the most favourable habitat may fill up with the territories of individuals of the same species so some are forced to occupy less favourable habitats. To determine the realised distribution of species, we need to know the way populations shift in space and change in size over time.

Biotic factors, such as competition between species and predation, also affect realised distributions, but we will examine these in the next chapter.

E xplaining the realised distribution of a species requires a knowledge of the way in which populations fluctuate in space and time and not just the environmental conditions necessary for reproduction and survival.

To predict and manage populations we need mathematical models of population dynamics built around the abiotic and biotic processes that we have discussed so far in this chapter. In the next section we introduce the fundamentals of population models. Unfortunately, lack of long-term spatial data on the distribution and abundance of species hampers our ability to build realistic models of population dynamics in both space and time. In the first instance, it is easiest to ignore the spatial aspects of populations (distribution) and concentrate on the change of abundance of populations through time.

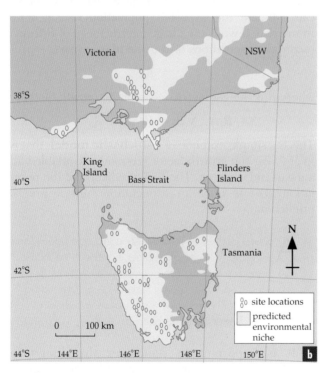

Fig. 42.3 (a) The myrtle beech tree, *Nothofagus cunninghamii*, occurs in cool temperate rainforests in south-eastern Australia and Tasmania. **(b)** Its distribution is of considerable interest to biologists because of its Gondwanan origin (Chapter 32). By means of data on the current distribution of myrtle beech, its niche, based on rainfall, temperature and altitude, has been determined. In general, the species occurred in all locations in which the environment was suitable except for areas in the north-eastern highlands of Victoria and southern New South Wales. It is suggested that the species is slowly migrating north-east along a corridor of suitable habitat

Density-independent population dynamics

The size of a population can only change (increase or decrease) as a consequence of four processes: **birth**, **death**, **immigration** and **emigration**. Adults reproducing within the population and the immigration of individuals into a population from other populations increase the size of a population. The death of individuals in a population and the emigration of individuals out of a population decrease population size. Using these principles, it is possible to formulate an equation that forms the basis for studies of population dynamics:

$$N_{t+1} = N_t + B_t - D_t + I_t - E_t$$

where N_{t+1} is the number of individuals in the population at time $t + 1$, N_t is the number of individuals at time t, B_t is the number of births between time t and time $t + 1$, D_t is the number of deaths between times t and $t + 1$, I_t is the number of immigrants that arrive between times t and $t + 1$, and E_t is the number of individuals leaving the population between times t and $t + 1$. Usually, population size is recorded each year, so t is measured in units of years.

When the per capita birth and death rates of a population are independent of the size of that population, we say that the population dynamics are **density independent**. Conversely, when the per capita birth and death rates depend on population size or density, then the population dynamics of the species are **density dependent**.

Population ecologists distinguish between two types of population: *closed* and *open*. There is no immigration into or emigration out of closed populations, so that I_t and E_t are zero for all time. A closed population with density-independent birth and death rates will experience an exponential increase or decrease in size or remain constant.

Exponential population growth in discrete time

Imagine a closed population living in a place where the environment is constant. Each year each pair of individuals produces a constant number of offspring, say four, and then dies. In other words, the population size changes over a discrete period of time. If the initial number of individuals in the population is N_0, then after one year the number of individuals in the population will have doubled:

$$N_1 = 2N_0$$

After two years the number will have doubled again and, using this concept inductively, we can see that the number of individuals in year t will be:

$$N_1 = 2^t N_0$$

This is known as **exponential growth**. Under these circumstances, the size of the population increases rapidly. An initial population of only 10 pairs will result in more than 10 million pairs after 20 years. Clearly there is potential for enormous growth in any population. Before we consider some of the factors that limit this potential for increase, let us examine the mathematics of exponential growth in more detail.

Modelling density-independent growth

Although populations are often made up of two sexes, it is convenient to consider only the females in the population so, in the following model, the population size is equivalent to the number of females in the population. The number of females in a closed population at time $t + 1$, N_{t+1}, is related to the number at the previous time, N_t, according to the following equation:

$$N_{t+1} = N_t b_t + N_t(1 - d_t)$$

where b_t is the average number of female offspring that one female gives birth to each year and that survive to be counted the following year, and d_t is the proportion of adult females that die each year. Rewriting this equation we find that:

$$N_{t+1} = N_t(1 + b_t - d_t)$$
$$= R_t N_t$$

where R_t is the growth factor of the population in year t. If R_t is constant, then, as in our example before (when the population doubled each year, i.e. $R = 2$):

$$N_t = R^t N_0$$

If R is greater than one, then the population will increase in size indefinitely. If R is less than one, then the population size will approach zero asymptotically and the species will become extinct (Fig. 42.4).

Environmental variability and exponential growth

In a single population, birth and death rates will vary from year to year, so the growth factor will be different every year. If we know the value of R for each year, then we can predict the population size at time t from the initial population size, N_0, by applying the above equation repeatedly:

$$N_t = R_{t-1} R_{t-2} R_{t-3} \dots R_0 N_0$$

where R_{t-1} is the growth rate in the year before year t and so forth.

Because we cannot predict changes in the environment that affect birth and death rates, we are unlikely

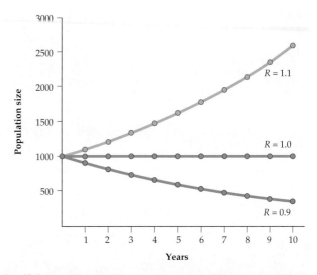

Fig. 42.4 The dynamics of a population with different growth factors: 0.9, 1.0 and 1.1. If the growth rate is less than one, the population declines towards zero. If the growth rate is greater than one, the population will increase geometrically and reach an enormous size surprisingly quickly. If the growth rate is exactly one, the population size remains constant

to know the growth factors for each year. However, we can estimate the expected population size at some future date if we know the geometric mean of R from a record of observations of R made in past years.

$$N_t = R^t N_0$$

where R is the geometric mean of the growth factors:

$$R = \sqrt[t]{R_{t-1} R_{t-2} \dots R_0}$$

This is a simple example of a **stochastic model**, that is, in any year the growth factor is a random variable with a mean and variance. The final prediction of the population size is only an expectation, not a precise prediction.

> A closed population with density-independent death and birth rates will ultimately either increase to an infinite size or decline to extinction depending on whether the geometric mean of its growth factor (R) is greater than or less than one. Theoretically, a population will remain constant if R exactly equals one.

It is important to remember that long-term population dynamics are determined by the geometric mean of the growth factors and not the arithmetic mean (i.e. $[R_{t-1} + R_{t-2} + \dots R_0]/t$). This fact is illustrated by a simple example.

Consider a population that experiences good and bad years with equal probability. In a good year the growth rate is $R = 2$, while in a bad year $R = \frac{1}{8}$. The arithmetic average of the growth rate is 1.0625 (> 1) but the geometric average is $\frac{1}{2}$ (< 1). In Table 42.1, the population size is tabulated for a number of years when

Table 42.1 The geometric mean of the rate of population increase (*R*) correctly predicts population size

Year	Growth rate (*R*)	Population size
0	—	480
1	2	960
2	⅛	120
3	2	240
4	2	480
5	⅛	60
6	2	120
7	⅛	15
8	⅛	< 2
Geometric mean	½	

the initial population size is 480. The geometric mean, not the arithmetic average, correctly predicts the decrease in population size.

Exponential population growth in continuous time

So far we have only considered population growth over discrete time intervals where the population size is 'updated' from year to year. The model we developed is especially appropriate for populations in which births occur within a short space of time. Species that live in environments with strong seasonal conditions behave in this way. For example, mammals, birds and plants living in temperate and sub-Antarctic conditions breed almost exclusively in spring (Fig. 42.5). In contrast, some populations, in particular the human population, change continuously as births and deaths occur throughout the year. The dynamics of these populations are more accurately described using differential equations.

If the per capita birth rate, that is, births per unit time per individual, is b, and the per capita death rate is d, then the rate of change in the population size $N(t)$ is $(b - d)N(t)$. Written as a differential equation this is:

$$\frac{dN(t)}{dt} = (b - d)N(t) = rN(t)$$

where the intrinsic rate of increase of the population is r. This equation can be solved using elementary calculus to generate an expression for the population size as a function of time:

$$N(t) = N(0)\exp(rt)$$

Fig. 42.5 Gentoo penguins at Heard Island. Birds, mammals and plants of sub-Antarctic islands breed almost exclusively in spring, that is, births occur in a short space of time

In the continuous time model of population growth, the intrinsic rate of increase determines the fate of the population. If $r > 0$, then the population grows exponentially; if $r < 0$, the population declines towards zero; and in the special case $r = 0$, the population remains constant.

After the introduction or colonisation of a species to a new area, the population may grow exponentially and often shows a rapid expansion in range. One intriguing quality of this expansion in range, especially in the early stages of an invasion, is that the boundary of the species distribution spreads at a constant rate. A simple model has been produced for this phenomenon, showing that the rate of spread of the population is proportional to the growth rate of the population. This phenomenon has been observed for many plant and animal species. Figure 42.6 illustrates the spread of a species of mangrove in New Zealand. A mathematical model was used to predict successfully the spread of the population using a population growth rate of 1.2 based on life-history data.

Density-dependent population dynamics

Given that all populations have a capacity for exponential growth, and the observation that none, including human populations, increases indefinitely in size, the obvious question is, 'What controls population increase?'.

One mechanism that can limit the indefinite increase of a population is a decrease in birth rate (or an increase in death rate) as the population grows. This

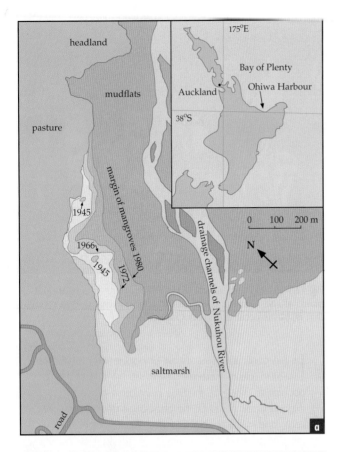

Fig. 42.6 (a) The expansion in the size of a population of *Avicennia marina* mangroves in Ohiwa harbour, New Zealand, from 1945 to 1980. At this locality the species is at the southern limit of its range. **(b)** *Avicennia resinifera* estuary in the Bay of Islands, North Island, New Zealand

is called *negative density-dependent population growth* because the growth rate is negatively correlated with population size.

Negative density dependence is based on the idea of a **limiting resource**. This limiting resource is often food but it could also be nesting sites, space, water or sites in which an organism is safe from predators. When a population is small, resources are unlimited and the population should grow exponentially. As the population becomes large, a deficiency in one or more of the

resources critical to a species' survival causes the death rate to increase or the birth rate to decrease until the growth rate is reduced to an average of one, so that the population is no longer growing. At this point the population should approach an equilibrium state.

There are a number of ways to test experimentally the idea of density-dependent population growth. By adding and removing individuals from a closed population, the effects of changes in density on birth and death rates can be measured directly. A more mechanistic approach involves manipulating the supply of a resource that is believed to be limiting, for example, by providing nest boxes for a black cockatoo population that is limited by the availability of tree hollows. Figure 42.7 illustrates the effect of increasing

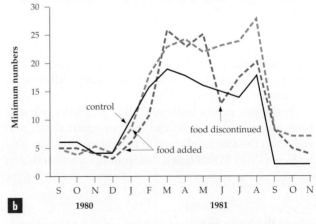

Fig. 42.7 (a) Supplementary food was provided to a marked population of an Australian marsupial *Antechinus stuartii*. **(b)** The population increased in numbers, mean body weight and survival and showed smaller overlaps in home ranges when contrasted to a control population. These responses in *A. stuartii* indicate that the density of the populations studied was limited by food supply

food supply on a population of small marsupial carnivores.

When there is density-dependent population growth, an increase in population size causes a decrease in growth rate. Eventually, an equilibrium may be reached when the number of births and deaths balances. This is the **carrying capacity** of a population in a habitat. The logistic equation is a means of representing the population dynamics of a resource-limited species:

$$\frac{dN}{dt} = rN\left(1 - \frac{N}{K}\right)$$

Now, r is equal to the rate of increase per head of population only when the population size is very low and density-dependent effects have not decreased the rate at which the population grows. The carrying capacity is represented by the parameter K. When the population size reaches K, it ceases to grow (Fig. 42.8).

The logistic growth model is a naive representation of the population dynamics of a species limited by resources. Although it has conceptual value, we would not expect the abundance of wild populations to obey such a simple equation. In Australia, the variability of the environment means that the birth and death rates in the model vary from year to year and there is no such thing as a fixed carrying capacity except under special circumstances (Fig. 42.9). Catastrophic events such as drought or severe predation may keep population sizes well below their carrying capacity most of the time. Under these circumstances, populations will usually exhibit density-independent dynamics.

Whether or not population growth is density dependent has been a subject of controversy for decades.

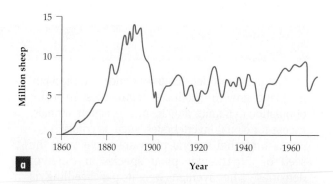

Fig. 42.9 The population dynamics of sheep in the Western Divisions of New South Wales between 1860 and 1972. The initial increase in population size was approximately exponential. Numbers appeared to initially overshoot the carrying capacity of about eight million sheep. The large fluctuations in population size from year to year may reflect environmental changes but are equally likely to be a consequence of changes in the economy

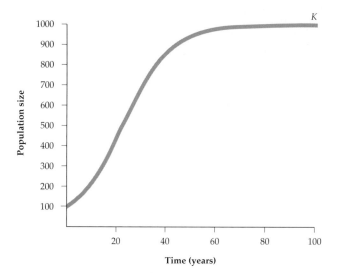

Fig. 42.8 Solving the logistic equation yields an explicit expression for the dynamics of a population in a constant environment. This is an example of logistic growth when the initial population size is 100

A. J. Nicholson, an Australian biologist, used populations of blowflies to test experimentally a simple model of density dependent population growth. The importance of density dependence in controlling the dynamics of species was challenged strongly by H. G. Andrewartha and L. C. Birch in their classic text *The Distribution and Abundance of Animals*. They believed fluctuations in environmental conditions to be the primary influence on population size, that competition for resources was overemphasised and that simple mathematical models of population dynamics were of little value. Other scientists studying plant populations have emphasised the importance of density-dependent population growth (see Box 42.2).

Recent controversy about the importance of density-dependent growth can often be traced to differences in the kinds of organisms that population ecologists study. Vertebrates, especially those high in the food chain, such as birds, often appear to maintain populations that only fluctuate slightly about a fairly constant carrying capacity. Many invertebrates, such as locusts, however, display enormous fluctuations in abundance, completely disappearing from parts of their

BOX 42.2 Self-thinning and the law of constant yield

A combination of density dependence and plasticity in plant growth produces regularities in plant population dynamics that are rarely seen in animals, except for clonal invertebrates, such as corals.

Assume that several plots are sown with the seed of a particular plant species at different densities. The average weight per seedling is measured for each plot a number of times after sowing. For each plot, the logarithm of the average weight of a plant is graphed as a function of the density of plants on the plot (see figure).

At the beginning the seedlings are very small and do not compete with each other for resources. Consequently, the per seedling weight will initially increase at a rate that is independent of the initial sowing density. As time passes, the seedlings in the denser plots begin to compete and the weight per seedling decreases, whereas the seedlings in the sparsely sown plots grow unabated.

If we harvest the denser plots after a fixed time, we find that the yield—plant number times average plant biomass—is constant. This is the rule of constant yield and it can be expressed mathematically:

$$Y = WN$$

where Y is the total yield and is constant for all plots, W is the per individual weight and N is the density of plants. The equation can be rewritten to show that plant weight is inversely correlated with plant density.

The law of constant yield can be used to determine the density at which crop plants are sown. A low sowing density will yield a small number of large plants, while a high sowing density will yield a large number of small plants. Depending on the relationship between plant size and value

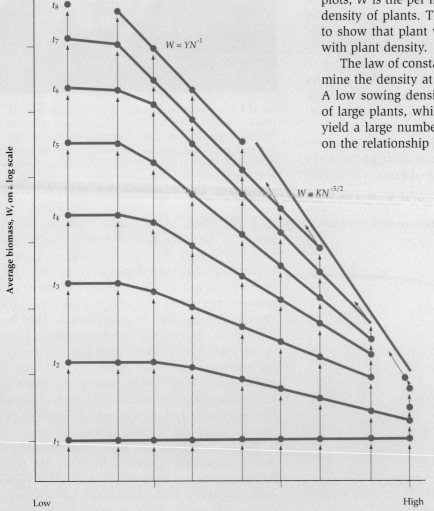

A graph depicting the rule of constant yield and the $-\frac{3}{2}$ self-thinning rule. Following the arrows indicates the changes in average plant biomass and plant density as time passes. The solid lines connect plots of different sowing density that were sown at the same time

to the farmer, and the cost of seeds and sowing, the rule of constant yield helps us to choose the optimal sowing density.

As time passes, individuals on the denser plots begin to die, in addition to growing more slowly. For a particular plot, the average weight per plant becomes related to the density of survivors according to the following law:

$$W = k/N^{-3/2}$$

where k is a constant. This is referred to as the $-\frac{3}{2}$ **self-thinning rule** because plant weight is inversely proportional to the density of plants raised to the power $-\frac{3}{2}$.

Fig. 42.10 These barnacles, packed onto the surface of rocks in the intertidal zone of Australia and New Zealand, are a graphic illustration of competition for space—a limiting resource

range for several years and then reappearing in plague proportions. **Density-vague population dynamics** is a new term that has been used to describe the dynamics of species whose population dynamics usually appear to be independent of population size. In these species, resource limitation may be rare, however, that does not mean it is unimportant.

A finite resource, such as food, may constrain the exponential growth of natural populations; paradoxically, environmental fluctuations in space and time often mean that resources are rarely limiting for many species.

Space-limited populations

For some species the resource that limits population growth is not food but space. This is particularly true for sessile organisms, such as plants and adult barnacles. Many of the inhabitants of the rocky intertidal zone live on particles of food that they filter from water. For these species food may not be limiting, however, space for attachment can be in short supply. Two individuals cannot occupy the same place and the amount of available space puts an upper limit on the number of individuals that can inhabit a particular area. For example, if the average size of a barnacle is 1 cm^2, then a square metre can, at most, contain 10 000 individuals (Fig. 42.10).

The same may be true for plants. Although their canopies often overlap, competition for space may also be critical. The apparent competition for space is caused by true competition for resources like light, nutrients and water that can only be obtained by occupying space.

Two features characterise space-limited sessile populations: they live in habitats that experience disturbance and they have widely dispersing propagules.

For populations limited by the availability of space, disturbance is invariably necessary for recruitment of new individuals. This disturbance may be the death of an individual tree, creating a gap in the canopy, a starfish eating a patch of mussels, or a fire that destroys thousands of hectares of forest. After a disturbance, space is liberated and new individuals can find a place to grow. In marine systems, new individuals often come from far away and the population in a small area is considered open.

Age- and size-structured population dynamics

So far we have assumed that all the individuals in a population are the same age and size with equal probabilities of dying and giving birth. Clearly this is never true. Most detailed population studies have shown that the age and/or size of an individual has a significant effect on its **fecundity** (probability of giving birth) and probability of survival. For example, in human populations, the probability of a female giving birth before the age of 10 or after the age of 50 is negligible; for most plant species the fecundity of an individual (often measured as the number of seeds set) is positively correlated with its size. Figure 42.11 gives an example of how the age structure of a population affects population growth.

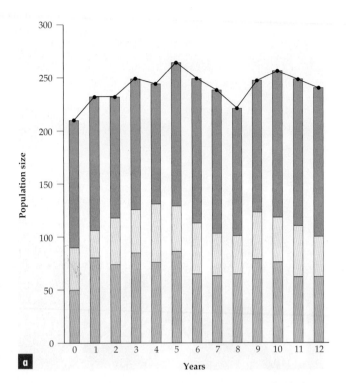

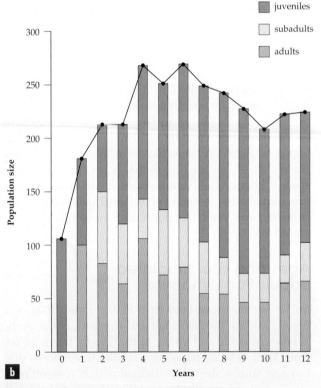

Fig. 42.11 An example of age-structured population dynamics. The graphs show the age structure of a population with three age classes: juveniles, subadults and adults. In both graphs, the birth and death rates are identical. In **(a)**, the initial population has a balanced age structure. The total population size and the age structure (proportion of individuals in each age class) remain relatively constant. In **(b)**, the initial population is 100 adults. Although the parameters of the population are identical, this imbalanced age structure generates population cycles in both the age structure and the total number of individuals

A ge and size affect the fecundity and survival of individuals of most species. Treating all members of a population as identical is only a crude representation of reality.

A useful way of describing the effect of age on survival and reproduction is to construct a **life table** (Table 42.2). A life table consists of age-specific probabilities of survival and expected reproduction (i.e. the average number of young successfully reared by a female of that age). The age-specific survival probabilities can be used to determine a survivorship probability, that is, the probability of surviving to a particular age. The expected number of births per female during an entire lifetime can be estimated by combining the age-specific fecundities and survivals.

Table 42.2 Life table for the Himalayan thar, *Hemitragus jemlahicus* (Fig. 42.12), in New Zealand (females only)

Age class	Fecundity[a]	Probability of dying	Survivorship
0	0.000	0.533	1.00
1	0.005	0.006	0.467
2	0.135	0.028	0.461
3	0.440	0.046	0.433
4	0.420	0.056	0.387
5	0.465	0.062	0.331
6	0.425	0.060	0.269
7	0.460	0.054	0.209
8	0.485	0.046	0.155
9	0.500	0.036	0.109
10	0.500	0.026	0.073
11	0.470	0.018	0.047
>12	< 0.470		

(a) The fecundity of a thar of a particular age is the average number of live births per female per year.

Studying size and age structure can often tell us about the processes affecting a population; however, caution is required. In Figure 42.13, which shows the height frequency of a semi-arid shrub, *Eremophila forrestii*, it is tempting to conclude that there was a large recruitment event that yielded a high frequency of 50–100 cm individuals several years ago. Taking a snapshot of the size and/or age structure of a

Fig. 42.12 Himalayan thar, *Hemitragus jemlahicus* (see Table 42.2)

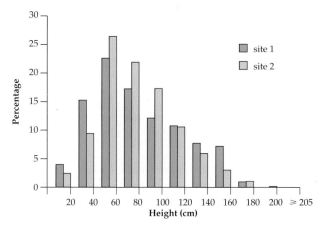

Fig. 42.13 Size frequency distribution of mature individuals of *Eremophila forrestii* in two different sites. The peak frequency around 60 cm might lead us to believe that several years ago many plants were recruited in a pulse and have now grown to this size. However, detailed study by Watson, Westoby and Holm showed that the distribution of heights is more a complex interaction between recruitment events, grazing, size-dependent mortality and growth. Plants grew at different rates and some even shrank

population can give us clues to the dynamics of populations but long-term studies are the key to a deeper understanding of population dynamics. There has been a lack of long-term ecological study in Australia.

Bioeconomics: managing exploited populations

Bioeconomics, or biological resource economics, is the study of the interaction of economic and biological systems. Bioeconomic theory addresses the problem of harvesting a natural resource, often a population (e.g. trout), for maximum economic and social benefit (e.g.

BOX 42.3 Predicting Australia's human population dynamics

One animal population about which we know a lot is the human population. Predicting the population dynamics of human populations is so important to our social, economic and environmental well-being that there is an entire discipline devoted to its study—*human demography*.

Predicting population trends in humans is notoriously difficult and riddled with pitfalls. Most of these pitfalls arise because of rapid and unexpected changes in human behaviour. Who would have predicted the enormous decline in fecundity in China caused by the political decision, in 1968, to encourage families to have only one child?

Age structure plays a central role in human population dynamics. Not only does the age structure of populations change but the age-specific fecundity and survival rates also change over time. Predicting the proportion of the population in different age classes is of economic interest; for example, the proportion of the population of working age or pensionable age is important to the economy.

In Australia, immigration has a large impact on our population dynamics. It is difficult to predict

Predicting the population dynamics of human populations is the discipline of human demography

changes in immigration policy or the number of people wanting to migrate to Australia. However, these predictions can, in turn, be essential to the political processes that determine our immigration policy.

In view of the hazards of predicting trends in human population dynamics, projections by demographers often describe different scenarios based on different assumptions about possible birth rates, immigration rates and other parameters thought to vary unpredictably. For example, using 1987 data, the Australian Bureau of Statistics made the following predictions about Australia's population in the year 2000: the population size will be between 19.2 million and 19.8 million; the proportion of people of working ages (15–64 years years) will be about 63.5%; and the number of people aged 65 is expected to increase greatly. The trend towards an ageing population will become more evident as the twentieth century progresses.

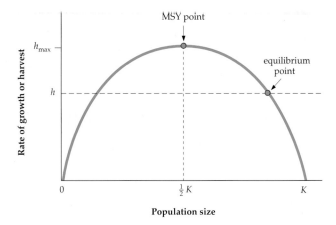

Fig. 42.15 Assume that a biological population grows logistically in the absence of harvesting and that the harvesting rate is h. The rate at which the population is growing can be plotted as a function of the current population size. When the population size is zero, or at the carrying capacity (K), the population ceases to grow. When the population is half its normal carrying capacity ($\frac{1}{2}K$), then the growth rate reaches a maximum. If we harvest the population at this rate, while maintaining the population at half its maximum, then we achieve maximum sustainable yield, MSY. If the population is harvested at a greater rate than the MSY rate, then the population will become extinct. If the population is harvested at less than the MSY rate, then the population size will reach an equilibrium at greater than half its carrying capacity if and only if its initial size is large enough

recreational fishing) (Fig. 42.14). One of the major applications of population ecology is in bioeconomics.

One of the paradigms of bioeconomic theory is **maximum sustainable yield** (**MSY**)—the largest harvesting rate that, in a world without fluctuations and uncertainty, can be maintained indefinitely. Harvesting populations at MSY maximises the long-term supply of resources, although environmental fluctuations and uncertainty about the biology of a species often mean that it is more prudent to harvest at rates below the MSY rate (Fig. 42.15).

Unfortunately, the value of a resource is discounted through time, so maximising long-term harvesting rates does not always make sense economically. If maximising profit in the short term is the basis for a natural resource industry, then it may be more profitable to over-exploit a resource and drive the species in question close to extinction. This has occurred with many species of whale, starting with the largest species, the blue whale, and ending with the over-exploitation of populations of even small whales, such as the minke whale.

Despite over-exploitation, no species of great whale has yet become extinct. As each species was exploited, its catchability, and hence profitability, was reduced to a level such that it was more profitable to switch to a smaller species. This phenomenon has allowed many highly mobile organisms to avoid extinction. For species that are concentrated in small areas (seabirds), or for sessile species, such as trees, reduced population size does not necessarily decrease the profitability of harvesting the species. This situation is illustrated by the continual harvest of old-growth forest around the world despite an enormous reduction in the area of forest. In marine systems, less mobile species such as abalone and scallops are often locally exterminated in the absence of controls. Government intervention is often necessary to ensure that natural resources, particularly regarding fisheries and forests, are harvested prudently.

A bioeconomic strategy that maximises short-term profits invariably leads to the over-exploitation of natural resources.

Fig. 42.14 Commercial fishing of school and gummy sharks in Bass Strait is an example of harvesting a natural resource

BOX 42.4 A conservation strategy for Leadbeater's possum

The conservation of Leadbeater's possum, *Gymnobelideus leadbeateri*, is one of the most contentious forestry issues in Australia. This possum inhabits the tall mountain ash (*Eucalyptus regnans*) forests north-east of Melbourne and favours areas with many hollows in large old or dead trees. Fire and logging interact to reduce the availability of hollows, hence the concern for the survival of the species.

With the fear that existing management strategies were inadequate, the Federal Government and the Government of Victoria supported Dr David Lindenmayer and Professor Hugh Possingham to carry out an in-depth population viability analysis of the species. The purpose of the analysis was to assess the relative merits of different conservation strategies for the species, such as creating new reserves or increasing the time between forest harvesting events.

Successful population viability analysis requires assembling detailed data on the species concerned and the way forest management options will affect the species. Information on the life-history attributes of the species, habitat attributes and dynamics (e.g. how does habitat quality develop after fire and/or logging), catastrophes and forest management options were assembled and input into a population simulation model. The model simulated the dynamics of the species over the next 150 years and recorded the likelihood of extinction of the species for a given conservation strategy.

Lindenmayer and Possingham found that the best conservation option involved setting aside

about 10 reserves of 50–100 ha in each forest block. Each reserve would be grown through to old growth forest and be sufficient to sustain a population resistant to demographic and environmental uncertainty in its own right. The reason so many reserves are needed is to spread the risk from catastrophic wildfires, which are known to affect the region. Interestingly, increasing the time between harvesting a forest stand provided limited benefit to this species compared to permanent reserves. Their recommendations are being enacted by the Government of Victoria.

Because extinction occurs over long time frames and is an uncertain event, mathematical models are essential for projecting in to the future and finding the best ways to manage our threatened species.

Leadbeater's possum, *Gymnobelideus leadbeateri*, a threatened arboreal marsupial restricted to Victoria

Viable population sizes for conservation

One aim of conservation biology is to maintain biodiversity by minimising the possibility that species become extinct. A fundamental question relevant to the maintenance of biodiversity is, 'What is the population size above which a species needs to be maintained to ensure its long-term survival?'. This population size is termed the **minimum viable population size (MVP)** of a species. The precise definition of what is 'viable', together with the characteristics of the ecology and behaviour of a species, have an enormous impact on a species' MVP.

What is an acceptable probability of extinction? There is no population size that can guarantee the persistence of a species indefinitely. All species will eventually become extinct! Consequently, the definition of a minimum viable population size must be accompanied by an acceptable probability of extinction over a fixed time interval. For example, an acceptable extinction probability over a period of 1000 years may be 2%, while an acceptable extinction probability over a period of only 100 years may be 1%. These figures will depend on the value society places on the species concerned and the cost of management practices that increase the chance the species will persist. In general, the probability that a species becomes extinct is a function of its distribution and abundance.

There is no single minimum viable population size that applies to all species under all circumstances.

Before discussing causes of extinction and the contribution they make to the MVP of a species, we need to discuss the related concepts of chance, uncertainty and stochasticity. It is tautological to say that a species becomes extinct when the last individual of that species dies. However, it is worth noting that this extinction event is, by its very nature, a chance event. Precisely when an individual dies is largely a matter of chance. Chance or uncertainty are characteristics of the various processes that may cause extinction. Scientists call processes that include elements of chance and uncertainty stochastic processes.

Four processes of chance (*stochasticity*) may cause, alone or in conjunction, the extinction of a species: genetic stochasticity, demographic stochasticity, environmental stochasticity and catastrophes. These processes interact in complex ways with attributes of the population, such as its age structure and distribution in space, to determine the likelihood that a species becomes extinct over a given period of time. For example, it is estimated that there is an 86% chance that the Sumatran rhinoceos, *Dicerorhinus sumatrensis*, will become extinct within the next 30 years (Fig. 42.16).

Fig. 42.16 The Sumatran rhinoceros, *Dicerorhinus sumatrensis*, is threatened with extinction

The degree of genetic heterozygosity of a population is known to affect the average fecundity and survivorship of natural populations. Small populations lose genetic variability rapidly as a consequence of genetic drift (Chapter 32).

Demographic stochasticity is a process that describes the random nature of births and deaths. The birth and death of individuals are instantaneous events so population sizes do not change smoothly. For example, given two individuals, each with a probability ½ of dying in one year, and each with a probability ½ of giving birth in one year (before death), the probability of the number of individuals next year will be: 4 individuals, $\frac{1}{16}$; 3 individuals, $\frac{4}{16}$; 2 individuals, $\frac{6}{16}$; 1 individual, $\frac{4}{16}$; and 0 individuals, $\frac{1}{16}$. On average, this population should remain constant; however, because of demographic stochasticity, there is a significant chance that the population may become extinct after only one year; indeed, this population is very unlikely to persist for more than 10 years.

The variable nature of the environment and the possibility of catastrophes (e.g. cyclones, epidemics, fire) impose further levels of stochasticity on the dynamics of populations. In the eucalypt forests of Australia, fires may wipe out entire populations of arboreal marsupials (Fig. 42.17). Conversely, maintaining an appropriate fire frequency, and thereby a series of vegetation successions (Chapter 43), is essential to the survival of species, such as the heath-dwelling ground parrot, *Pezoporus wallicus*.

Fig. 42.17 Fire is one factor that causes variability in an environment and imposes a level of stochasticity on the dynamics of populations

Summary

- Population ecology is devoted to a mechanistic understanding of the dynamics of the distribution and abundance of species.
- Experiments identify biotic and abiotic factors that affect the four processes of population change: birth, death, immigration and emigration. Mathematical theory is used to formalise these experimental results and make predictions about population change.
- Theoretically, a closed population with density-independent death and birth rates ultimately either will increase to an infinite size or will decline to extinction depending on whether the geometric mean of its growth rate is greater or less than one.
- In the continuous time model of population growth, the intrinsic rate of increase (r) determines the fate of the population. If $r > 0$,

the population grows exponentially; if $r < 0$, the population declines towards zero; and if $r = 0$, the population remains constant.

- A finite resource, such as food, may constrain the exponential growth of natural populations; however, environmental fluctuations in space and time often mean that resources are rarely limiting for many species.
- Age and size affect the fecundity and survival of individuals of most species. Thus, the age structure of a population affects population growth.
- Mathematical models are used to manage natural resources, control pest species and reduce the probability that species become extinct. There is no single minimum viable population size that applies to all species under all circumstances.

keyterms

abiotic (p. 1123)
abundance (p. 1121)
biotic (p. 1123)
birth (p. 1125)
carrying capacity (p. 1129)
community (p. 1121)
death (p. 1125)
demographic stochasticity (p. 1136)
density-dependent population dynamics (p. 1125)

density-independent population dynamics (p. 1125)
density-vague population dynamics (p. 1131)
distribution (p. 1121)
ecology (p. 1121)
ecosystem (p. 1121)
emigration (p. 1125)
exponential growth (p. 1126)
fecundity (p. 1131)

fundamental niche (p. 1124)
habitat selection theory (p. 1124)
immigration (p. 1125)
life table (p. 1132)
limiting resource (p. 1128)
mark–release–recapture method (p. 1122)
maximum sustainable yield (MSY) (p. 1134)

minimum viable population size (MVP) (p. 1135)
population (p. 1121)
potential distribution (p. 1124)
realised distribution (p. 1124)
self-thinning rule (p. 1131)
stochastic model (p. 1126)

Review questions

1. Name four abiotic factors and four biotic factors that could limit the distribution of a species.

2. What is a closed population and where might you find one?

3. Choose the correct answer. Density-dependent population growth in a closed population occurs when:

   A the rate of population growth depends on the size of its individuals

   B the per-capita birth and death rates are functions of the population size

   C the rate of population growth depends on the density of resources

4. (a) 'Populations that fluctuate are more likely to become extinct than populations that do not.' Is this true or false?

   (b) 'No species can be harvested to extinction because the last individuals are too hard to find.' Provide counter-examples to this statement.

5. How do we know that most populations do experience some density dependence in their population growth?

6. Self-thinning occurs when:

   A some of the plants in a population sacrifice themselves for the good of the species

   B weeds compete with other plants

   C some individuals in a stand die from competition from members of their own species

7. Name two groups of organisms where birth and death rates are affected by the sizes of the individuals in a population.

8. A minimum viable population is one that:

   A has an acceptably low chance of extinction in the long term

   B is increasing in size

   C is capable of performing its role in the ecosystem

   D is small enough to not get in the way of development

Extension questions

1. The Conservation Commission of the Northern Territory of Australia is interested in determining the number of saltwater crocodiles, *Crocodylus porosus*, in a 10 km stretch of the Adelaide River. On the first day they catch and mark 40 crocodiles with a blue tag. On the second day they catch and mark 50 crocodiles with a red tag, 20 of which were already marked with a blue tag. What would be your estimate of the size of the population? During a third visit to the area they catch 40 more crocodiles: 10 have both tags, 10 have no tags, 10 have only a blue tag and 10 have only a red tag. Use this result to generate two more estimates of population size. Discuss the problems associated with this method of estimating the population size.

2. An ecologist transplants 60 snow gums, *Eucalyptus pauciflora*, into three sites in the Australian Alps, 20 plants above the current treeline, 20 at the treeline and 20 below the treeline. After 10 years of careful monitoring the ecologist finds that the snow gums survive equally well at all three sites but that the individuals above the treeline have not grown as fast as those at or below the treeline. Discuss the implications of these results for the environmental distribution of *E. pauciflora*.

3. Soil moisture, temperature and nutrients are environmental gradients that are said to affect plant distribution and abundance directly. Altitude, rainfall, and soil type are said to affect plant distribution and abundance indirectly. Discuss.

4. In a study of frog competition, a zoologist manipulates the numbers of frogs in a number of isolated farm dams. Fifteen initially empty, almost identical dams are stocked with tadpoles at five different densities (three dams for each density) in the early spring. The number of adult frogs in each dam is determined by sampling on three successive years. The mean adult abundance for each treatment is tabulated opposite.

    Discuss these results in the context of intraspecific competition and a variable environment. What other experiments

Initial stocking density	1992	1993	1994
10	8	0	0
20	30	30	80
40	60	40	100
80	60	80	100
160	20	30	60

could be conducted to explore the possibility that resources are limiting?

5. A butterfly population in the mallee is believed to respond directly to summer rainfall. When the summer rainfall is below 100 mm, each adult female produces, on average, 0.1 adult females; when between 100 mm and 200 mm the annual fertility is 1; when above 200 mm, annual fertility is 5. If the probability of summer rainfall being one of these three rainfall classes is equal, will the population increase or decrease in the long term? Do you think that this is a permanent or ephemeral population? If you are familiar with programming, write a computer program to simulate the dynamics of the species and estimate the probability that the population is extinct after 10 years given an initial population size of 100.

6. Consider a biennial herb that sets seed twice in its life. In its first year the average plant produces 100 seeds and in its second year the average plant sets 200 seeds (and then dies), but there is a 25% chance that the herb dies between the first and second seed set. If the chance of a seed germinating and reaching the age at which it first sets seed is 0.7%, on average, how many mature plants does each mature plant produce?

7. A butterfly species lives in two regions. In one region, the species exists as only one population that inhabits a large area capable of supporting 1000 individuals. In the other region the species can inhabit three areas each capable of supporting 100 individuals. Discuss the effects of

catastrophes and migration between subpopulations on the viability of the species in the two regions.

8. Years of study suggest that a closed fish population displays dynamics similar to that predicted by the logistic equation

$$\frac{dN(t)}{dt} = rN\left(1 - \frac{N}{K}\right)$$

with $r = 1$ and $K = 20\,000$. Predict the change in population size when the initial population size is 1000 fish and the harvesting rate is 1000 fish per annum, 500 fish per annum and no fish per annum.

9. After a cyclone, an ecologist monitors the recovery of trees in a tropical rainforest. The number of trees of different size classes is counted in a 1 ha plot and the results tabulated opposite. Discuss the data in the context of self-thinning, the law of constant yield and plant growth.

Size (m³)	Year			
	1950	1960	1970	1980
0.0–0.5	1000	200	100	50
0.5–1.0	10	180	75	40
1.0–2.0	20	5	70	30
2.0–4.0	10	5	2	40
4.0–8.0	6	1	0	0

10. Consider a common species of barnacle that dominates part of the rocky intertidal zone. A starfish predates on the barnacle, selectively choosing patches of about 10 cm² and consuming all individuals in the patch. Discuss the consequences of this predation on patterns in the distribution of adult barnacles at different spatial scales.

Suggested further reading

Begon, M., Harper, J. L., Townsend, C. R. (1990). *Ecology*. 2nd edn. London: Blackwell Scientific Publications.

An excellent undergraduate text for ecology students.

Burgman, M. A. & Lindenmayer, D. B. (1999). *Conservation Biology for the Australian Environment*. Sydney: Surrey Beatty & Sons.

This text discusses principles and methods of conservation biology and provides Australian examples and further references.

Hilborn, R. and Mangel, M. (1997). *The Ecological Detective: Confronting Models with Data*. Princeton, NJ: Princeton University Press.

A useful text for the modeller.

TREE—*Trends in Ecology and Evolution*.

This journal is a useful source of articles on population ecology and conservation biology.

CHAPTER

43

Living in communities

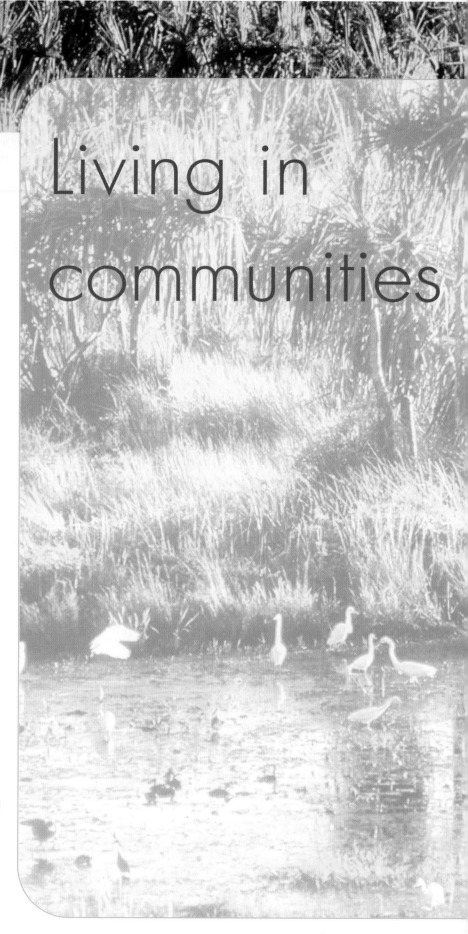

n Chapter 42, we examined the ecology of populations. Populations, however, usually occur together with populations of other species and the neighbours of any one individual may be individuals of other species as well as those of its own. Individuals in these different populations interact in a variety of ways: animals eat and fight, plants shade one another and animals pollinate plants. These interactions include symbiosis, predation and competition, all of which affect the distribution and abundance of species. In the long term, such interactions may lead to evolutionary change. For example, in rainforests in northern Queensland, there is synchronisation of flowering and insect activity: peaks in the flowering of tree species coincide with peaks in the number of insects in the same area.

A group of populations living and interacting together is called a **community** (Fig. 43.1). Some ecologists think of the species in a community as having evolved together, occurring almost as a super-organism with species interacting strongly with each other. Under this view, we would expect to find the same combinations of species occurring consistently whenever we visit the same kind of habitat. Other ecologists believe that species just happen to occur together in assemblages, which are simply collections of organisms living at the same location—they may or may not interact strongly with each other, and under this view we might not always find the same combinations of species.

We start this chapter by thinking about how ecological communities can be described and then about the kinds of ecological interactions that produce the patterns we see. Finally, we will examine larger scale patterns and see what happens when we move from community to community.

Community structure

Species diversity

A community may be described by recording the populations that are present at a particular time—measuring its **diversity**. Measurements of diversity range from counts of the number of species in a sample, **species richness**, to measures that incorporate the number of species and the sizes of individual populations (Box 43.1). Most of these measures allow the community to be summarised with a simple index or a graph.

Species richness (S) is the simplest measure and is a fairly crude summary of diversity. It is restricted usually to a count of the number of species within a particular taxonomic group—the number of canopy tree species in a forest, the number of ant species in a grassland and so on. These measurements are most useful as a broad indication of a community: they allow us to recognise communities that are particularly diverse (Fig. 43.2), such as tropical forests and coral reefs, and also those that have few species, such as the land mass of Antarctica. However, this kind of measurement is not a true community measurement because it does not encompass all the interacting species, just those of a particular ecological group. Species richness is also a poor measure of the diversity within a community because it takes no account of the abundance of individual species. When we look at many ecological communities, we find that they are made up of a few very common species and many rare ones. Consequently, how many species we find often depends on how hard we look. If we take a few samples from a community, we will probably see the common species, and only when we take many samples will we find the rarer species. This makes it hard to compare different communities.

Fig. 43.1 An ecological community is a group of organisms living and interacting with one another, such as in this wetland community at Kakadu National Park

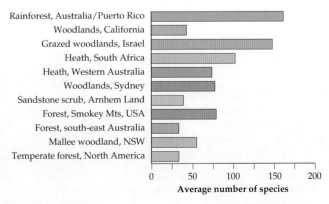

Fig. 43.2 A conventional way of measuring the richness of botanical communities is to count the number of vascular plant species in a 1000 m² (0.1 ha) area. Various people have sampled 1000 m² plots in a range of vegetation types throughout the world, and the ranges of species richness vary greatly as shown here. The data provide a guide to patterns of species richness in Australia and throughout the world

BOX 43.1 Measuring diversity

α-diversity

Diversity within communities, α-diversity, has two components: the number of species and their relative abundances. There are many ways that α-diversity can be expressed quantitatively. One way is Simpson's Index,

$$D = 1 - \sum_{i=1}^{S}\left(\frac{n_i}{N}\right)^2 \qquad [43.1]$$

where n_i is the number of individuals in species i, N is the total number of individuals in the community and S is the number of species present. This index is equivalent to calculating the probability of picking at random two individuals that are different species. It ranges from 0 in communities made up of single species to almost 1 in communities made up of many species, each present in equal numbers.

The data in Table 1 represent the relative abundances of eight small mammal species in three vegetation types in an area within Kakadu National Park. Sandstone forest and mature open forest have three species each and the riparian woodland contains five species. Total abundance, on the other hand, is highest in mature open forest and lowest in riparian woodland.

Even though both sandstone forest and mature open forest have the same number of species, and total abundance is greater in mature open forest, species diversity is greater in sandstone forest. In other words, there is a greater chance of picking up two individuals that are different in the sandstone forest. Species diversity is higher in sandstone forest because the relative abundances of the three species are more even than in mature open forest, in which most individuals are northern brushtail possums. For the same reason, the species diversity values in sandstone forest and riparian woodland are much the same, even though riparian woodland supports more species.

β-diversity

β-diversity is the change in community composition we meet as we move through the landscape. If the turnover of species between sites is high, communities will be relatively different.

There are numerous indices for measuring the similarity between communities in different locations. One common coefficient of similarity is Jaccard's coefficient (S_j). It is employed when the

Table 1 Abundances (numbers) of small mammals in three vegetation types at Jabiluka in Kakadu National Park[a]

Mammal species	Sandstone forest	Mature open forest	Riparian woodland
Antechinus bilarni (sandstone antechinus)	6.3		
Dasyurus hallucatus (northern quoll)	8.6	16.7	1.2
Zyzomys woodwardi (large rock rat)	5.5		
Sminthopsis virginiae (red-cheeked dunnart)		1.1	0.4
Trichosurus arnhemensis (northern brushtail possum)		33.3	
Pseudomys nanus (western chestnut mouse)			3.7
Pseudomys delicatulus (delicate mouse)			0.6
Melomys burtoni (grassland melomys)			6.5
Species richness	3	3	5
Total abundance	20.4	51.1	12.4
Simpson's α-diversity	0.65	0.47	0.62

(a) Values are based on trapping success. Data are used to calculate α-diversity.

only information available about the communities is lists of species. The coefficient is used to compare pairs of sites. Lists can be arranged in a table in which the rows (or columns) are sites and the columns (or rows) are species (Table 2).

Jaccard's coefficient is calculated for pairs of sites from the equation:

$$S_j = \frac{a}{t} \qquad [43.2]$$

where a is the number of species shared by the two sites and t is the total number of species, including those that are unique to either of the sites.

The coefficient may be calculated to represent the degree of similarity between any pair of sites such as those in Table 2. The coefficient between sites 1 and 2 is ½, while between sites 2 and 3 it is ¾. This suggests that the community at site 2 is more similar to the community at site 3 than it is to the community at site 1 because sites 2 and 3 have more species in common.

Table 2 A site by species table for grass, sedge and orchid species growing in three different communities on low granite rocks in the semiarid mallee zone of southern Western Australia[a]

Species	Sites		
	1	2	3
Neurachne alopecuroides (grass)	1		
Gahnia drummondii (sedge)		1	1
Lepidosperma drummondii (sedge)	1		
Lomandra mucronata (needle lomandra)	1		
Stypandra imbricata (lily)	1		1
Borya nitida (resurrection plant)		1	1
Diurus longifolia (orchid)	1	1	1

(a) The 1's indicate the presence of a species in a site.

In such a small data set as that in Table 2, the coefficient is hardly necessary. One can see by inspection that the communities at sites 2 and 3 are relatively similar. The index becomes useful in larger studies where there are many rows and columns of information. The results can then be used to make decisions about which areas to include in a conservation park to ensure that the park includes representative samples of the variety of communities.

More useful measures of diversity use information about species abundance. A native grassland with many similarly abundant species would be considered to be more diverse than a monoculture such as a wheat field, which has relatively few species, with most individuals in the community belonging to a single species and only a few weeds to add variety. Even if there were many species of weeds, they would not usually be very abundant relative to the crop species so they do not add much to the diversity of the community. Such species diversity of a community within a local area is called **α-diversity**.

Another aspect of diversity concerns the variety of species that we are likely to meet when we move from one place to another, either within one type of community or when we move from one community to another. If a wheat crop extends throughout a region, the environment on the large scale is not very diverse. In contrast, as one moves from place to place in a tropical rainforest, one will find many new species at each place. The diversity among different communities in different habitats in spatially separate areas is called **β-diversity**, and ecologists have suggested quite a few different kinds of measurements (Box 43.1).

Diversity within communities can be described by species richness, which is the number of species, or by more complex measures of diversity that take into account not only the number of species but their relative abundances. Diversity of a community within a local area is called α-diversity, while β-diversity measures the diversity between communities separated in space.

Structural diversity

So far we have viewed diversity from the perspective of the species composition of a community. Another aspect of community diversity is **structural diversity**, which focuses on variations in the size and shape of plants irrespective of the species. In a pine plantation, the structural diversity may be low, not because all of the individual trees are of the same species but because they are very often of the same age.

One of the most important characteristics of community structure is projective foliage cover of the plants. This is the percentage of the ground surface above which there is foliage. If we know the percentage cover and at what height the foliage is found, it tells us a lot about the biomass and productivity of the community (see Chapter 44).

Tropical rainforests are tall and have very high values for projective foliage cover (Fig. 43.3a). More than 90% of the surface of a site may be covered by foliage that is more than 30 m above the ground. Temperate eucalypt forests may be as tall or taller than their tropical counterparts but projective foliage cover of the tallest stratum never reaches 90%. Values around 30% are more common (Fig. 43.3b). Arid zone shrublands dominated by saltbush often have a projective foliage cover of less than 10%, including all height classes.

Information on the size and shape of plants, their life forms and growth habits is also used in the classification of communities. In 1970, the Australian ecologist R. L. Specht designed a table for Australian vegetation. Specht's table (Table 43.1) is based on the projective foliage cover and growth form of plants in the tallest stratum (Fig. 43.4) and has been used in slightly varying forms as a basis for vegetation maps of all parts of Australia.

Fig. 43.3 Overhead cover of **(a)** an Australian rainforest compared with **(b)** an open eucalypt forest

Table 43.1 Specht's structural classification of Australian vegetation

Growth form of the tallest stratum	Foliage cover of the tallest stratum			
	> 70%	30–70%	10–30%	< 10%
Tall trees (> 30 m)	Tall closed forest	Tall open forest	Tall woodland	
Medium trees (10–30 m)	Closed forest	Open forest	Woodland	Open woodland
Low trees (< 10 m)	Low closed forest	Low open forest	Low woodland	Low open woodland
Tall shrubs (> 2 m)	Closed scrub	Open scrub	Tall shrubland	Tall open shrubland
Low shrubs (< 2 m)	Closed heath	Open heath	Low shrubland	Low open shrubland
Hummock grasses			Hummock grassland	
Tufted/tussock grasses	Closed tussock grassland	Tussock grassland	Open tussock grassland	Sparse open grassland
Graminoids	Closed sedgeland	Sedgeland	Open sedgeland	
Other herbaceous spp.	Dense sown pasture	Sown pasture	Open herbfield	Sparse open herbfield

Structure in plant communities is recorded usually as percentage projective foliage cover within height classes.

Interactions within communities

When we understand the patterns within a community, our next task is to explain those patterns, and to do that we need to look at how the different populations within the community interact with each other and with their physical environment. The simplest interactions within communities are those involving individuals of only two species, and some of the most important are grouped under symbiosis.

Symbiosis

Symbiosis is a term used for interactions in which two organisms (symbionts) live together in a close relationship that is beneficial to at least one of them. A lichen (Chapter 36) is a symbiotic relationship between an alga and a fungus. A mycorrhiza (Chapter 36) is an association between a fungus and the root of a higher plant.

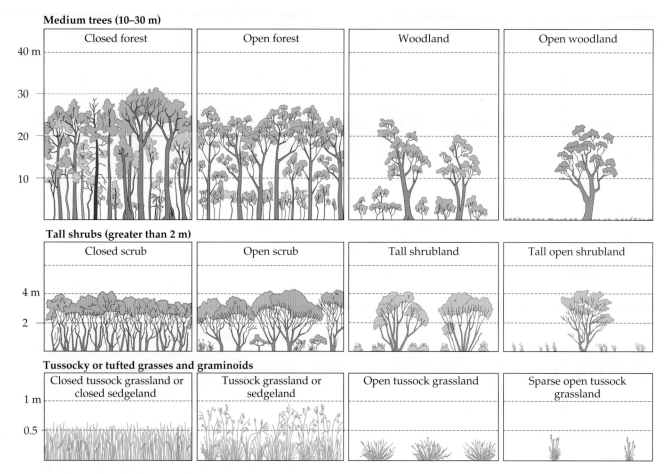

Fig. 43.4 Examples of plant communities classified on the basis of structure—height of the dominant plant and percentage cover: closed forest, open forest, woodland and open woodland are dominated by trees (plants with a single stem and greater than 10 m in height); closed scrub, open scrub, tall shrubland and tall open shrubland are dominated by shrubs (woody plants with multiple stems and greater than 2 m in height); sedgelands and grasslands are dominated by grasses and other herbaceous plants, many of which form tussocks

Where the characteristics of interacting species evolve in concert, resulting in a one-to-one relation, the relationship is regarded as an example of **coevolution**. An orchid flower that mimics the shape of a female wasp, tricking males into copulating with it, thus resulting in pollination (Fig. 37.44), is one such case.

Symbiotic relationships are the rule rather than the exception in most ecological communities, especially in complex communities, such as tropical rainforests and coral reefs. Symbiotic interactions are of three kinds: one species benefits and the other is unaffected—**commensalism**; both species in the relationship benefit from the association—**mutualism**; or one species benefits and the other is harmed—**parasitism**. In some associations, one organism is physiologically dependent on another; in others, organisms can just as easily live apart. Symbiosis usually involves providing protection, food, cleaning or transportation. The nature of the relationships between species in symbiotic relationships may change over time as either partner matures or as the composition of the community in which they live changes.

Symbiosis is where two organisms (symbionts) live together in a close relationship that benefits at least one of them. It includes mutualism (where both species benefit), commensalism (one benefits, one is unaffected) and parasitism (one benefits and one is harmed).

Commensalism

In moist forests, most tree trunks are covered with mosses, small ferns and orchids. These plants are **epiphytes**: they benefit from living on the trunk of a host tree, which provides a substrate, a catchment for rainwater in which mineral and organic nutrients are dissolved, and a position closer to the light. Epiphytes do not appear to affect the host tree adversely. In marine communities, epiphytic algae attach to crabs and molluscs, thus having somewhere to live and gaining free transportation. In some cases, there may be a benefit to the host through camouflage but the host does not depend on the algae.

Interactions between species can change over time. In tropical forests of South-East Asia and Australia, the strangler fig commences its life as an epiphyte. A seed delivered in the droppings of a bird germinates on the host tree and the young fig commences to grow (Fig. 43.5a). In the early stages of the association, the fig benefits and the host is not affected. However, when the fig grows, it extends its roots down to the soil and envelops the host, eventually crushing it (Fig. 43.5b). The relationship changes from commensalism to one of aggressive competition for space (see p. 1155).

Mutualism

The symbiotic relationships of lichens and mycorrhizae (Chapter 36) are examples of mutualism since both partners in each association benefit. Mutualistic relationships are particularly common in coral reef communities.

All reef-building corals (Chapter 38) have abundant symbiotic algae within their tissues—up to 30 000 algal cells per cubic millimetre of coral tissue (Fig. 43.6). The algae are dinoflagellates (Chapter 35), called zooxanthellae, which have yellow-brown pigments that give the coral its colour. Within the cells of coral polyps, zooxanthellae live, reproduce, photosynthesise and utilise the waste products of the animal host (carbon dioxide, phosphorus and nitrogen). In turn, the coral uses oxygen and food produced by the algae during photosynthesis to grow, reproduce and form its hard

Fig. 43.6 Reef-building corals have a mutualistic relationship with algal cells, zooxanthellae, living in their tissues. A close-up view of a branching coral, *Acropora sarmentosa*, branch shows dark colouration of tentacles due to populations of zooxanthellae

Fig. 43.5 (a) This strangling fig in a Queensland rainforest starts off life as an epiphyte (commensalism), germinating from a seed that has landed on the host tree. **(b)** An older fig has enveloped its host tree and is now an aggressive competitor

skeleton, which is the basis of the reef. Without this mutualistic relationship, the formation of coral reefs such as the Great Barrier Reef would be impossible. Corals regulate the number of zooxanthellae they contain, picking up new ones as required and 'spitting out' the excess. Since zooxanthellae need light for photosynthesis, reef-building corals grow only in clear waters less than 100 m deep; disturbance that makes water murky threatens the life of reefs.

Among other coral reef animals, sea anemones have a number of mutualistic relationships. The most well-known example is the association between the sea anemone and the anemonefish (Fig. 43.7). The anemonefish is one fish that is neither stung nor eaten by the anemone. The anemonefish acclimatises itself to the sting of the anemone by repeatedly brushing itself against the tentacles until its own mucous coating inhibits the anemone's sting. The fish is thus protected from predators and feeds on scraps of the anemone's food. The anemone benefits in turn when the anemonefish cleans its host and lures other animals into the anemone's tentacles.

Some species can have more than one symbiotic association. The 'ant plant' *Myrmecodia* is an epiphyte that also has a mutualistic relationship with a species of ant. The plant has a swollen base or tuber, in which there are specialised chambers. The ants form large colonies within these chambers and they carry their excreta and the corpses of their prey to parts of the chambers (a cemetery) where the plant is able to absorb the waste nutrients. This relationship is called **myrmecotrophy** (Box 43.2).

Fig. 43.7 The anemonefish is unharmed by the tentacles of its symbiotic partner, the anemone

BOX 43.2 Ants feeding plants

The flowering plant genera *Myrmecodia* and *Hydnophytum* (family Rubiaceae) from South-East Asia and north-eastern Australia have spectacular swollen, tuber-like organs that are inhabited by ants of the genus *Iridomyrmex*. The ants live in special chambers called **domatia** (see Fig. b). They pack adjacent tunnels with their waste material. Radioactive labelling of prey or honey water given to foraging ants from these colonies shows that nutrient ions are absorbed from the colony waste, translocated and incorporated into the plant's tissue. Bacteria and fungi in the chambers probably facilitate decomposition of the wastes. The mineral nutrients of the plant are thus augmented and the ant colony functions as a second root system.

In South American bromeliads (plants of the pineapple family) that also have myrmecotrophic associations, absorption of such waste nutrients appears to be in areas of plant tissue that lack cuticle or where there are specially absorptive trichomes (hairs, etc.). In other plants where the walls of domatia are thick and apparently impenetrable to nutrients, these ants tend homopteran insects that also live in the domatia; the ants chew through to the vascular tissue of the

(a) The epiphytic 'ant plant', *Myrmecodia tuberosa*, photographed in a forest in Borneo, has specialised chambers in its swollen pseudobulbs, in which ants live in a mutualistic relationship

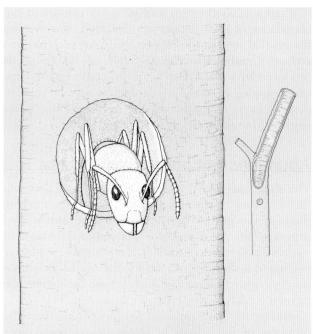

(b) A domatium formed from the hollow stem of *Endospermum myrmecophilum*. The entrance hole has been gnawed through the twig by the ant *Camponotus quadriceps*

host plant so that the homopterans have access to sap. The ants use the insects rather like cattle, cleaning them, feeding them and collecting the honey dew that the insects secrete.

There are about 350 known plant species (in more than 70 families), some of which occur in Australia, that have domatia, that is, structures that accommodate ants. These species occur in tropical and subtropical environments. Many structures function as domatia. Those that are lived in more or less permanently include tubers, swollen thorns (e.g. found in *Acacia*) and hollow stems and twigs (see Fig. b). Leaf pouches and swollen petioles are less permanent (ephemeral) homes.

Benefits of ant–plant associations

Analysis of the costs and benefits of myrmecotrophy is not simple. There is a cost to the plant in producing domatia or having its sap tapped for homopterans to feed. There is the benefit to the plant of increased nutrients from the ants' wastes, especially for a plant that is an epiphyte living in a nutrient-poor environment, such as on the trunk of a tree. Many of the ant species implicated in myrmecotrophy also perform other services for the plant: they defend and protect the plant from herbivores and seed eaters. In turn, some host plants offer food rewards such as nectar secreted in extrafloral nectaries and other food bodies.

We assume that these relationships are mutualistic and that they are the result of coevolution—the plants are certainly structured to accommodate ants. However, we are less sure that the ants' traits are special adaptations because the traits also occur in ant species not involved in myrmecotrophy.

Parasitism

Parasites live in a close obligatory association with one or more host species for most of their lives. They gain sustenance from their host, who is harmed but not often killed. Parasites are usually smaller than their host. They may live on the surface of their host (**ectoparasites**, e.g. lampreys that attach to other fishes, Fig. 40.11) or internally (**endoparasites**, e.g. tapeworms, stem-rotting fungi). **Parasitoids** are insects that are free-living in the adult stage but are parasites as larvae (Fig. 43.8). Adults lay their eggs in a host, usually the larvae of another insect species. Larvae of the parasitoid develop in the host doing no harm for a while but they eventually consume and kill the host before pupation.

We have looked at the life cycles and biology of some parasites in Chapters 35 and 38 but how do parasites affect the abundance of their hosts? Parasitism is a special type of predation and the relationship between host and parasite abundance is similar in some ways to that of predators and their prey.

Fig. 43.8 A female ichneumonid wasp, *Diadromus collaris* (Gravenhorst), ovipositing into the pupa of the cabbage moth, *Plutella xylostella* (Linnaeus), its natural enemy. The wasp was introduced into Australia as a biological control agent

Predation

Predation in a broad sense is the feeding by one organism on another. Typically, **predators** are animals that capture and eat various other animals. **Herbivores** are animals that consume a range of plants or parts of plants. Parasitism is a special case of predation. There are exceptions to these general patterns: some plants consume animals (Fig. 43.9), while the dynamics of feeding by mobile animals, such as sea stars and molluscs, on sedentary, colonial marine animals may be much more like herbivory than animal predation (Fig. 43.10).

Fig. 43.9 The pitcher plant traps and digests insects in highly modified leaves

Fig. 43.10 A predatory snail feeding on a soft coral. The snail will eat some polyps of the coral, and move on, allowing the coral to regenerate. This interaction is more like a herbivore feeding on plants than the action of a true predator

Predators, herbivores and parasites can affect the abundance of their animal or plant food. Herbivorous insects that feed on flowers or seeds, for example, can reduce significantly the number of viable seeds produced by a plant. Larvae of the moth *Agathiphaga queenslandensis*, which is a highly specific feeder, destroy large quantities of the seed of kauri pines (*Agathis*) in rainforests of northern Australia.

Abundance of predators and prey

The abundance of a predator and its prey can fluctuate through time, with the predator numbers tracking those of the prey. When there are large numbers of prey available, the predator population increases in size. As prey are consumed, their numbers decline, leading to a shortage of food for the predators, whose numbers also decline (Fig. 43.11). Such oscillations are termed predator–prey cycles.

Biologists have known for centuries that the numbers of individuals in some natural populations fluctuate in a predictable, regular fashion, and that such cycles can occur in both predators and their prey. In the early part of the twentieth century, ecologists began to develop mathematical models to predict these changes. Numerous models predict fluctuations, ranging from simple damped oscillations to extraordinarily complex, even chaotic, dynamics. Predator–prey cycles are not, however, common. Even when they exist, the fluctuations that are due to interactions between predators and their prey are confounded by fluctuations due to random variations in the environment.

One reason that cycles are not universal is that predators and their prey do not normally exist in simple two-species' communities isolated from other organisms, nor do they exist in a homogeneous environment. Another factor that complicates the relationship between predator and prey is the quality of food. Despite being abundant, plant food may be of such low quality that herbivores barely have enough energy for maintenance, let alone growth and reproduction (see also Chapter 19). Herbivores are affected particularly by the nitrogen content of their food (Fig. 43.12). Peaks in insect populations, including outbreaks of pest species such as psyllids in eucalypt forests, have been related to peaks in the quality of available food measured as soluble nitrogen in phloem sap (see Box 44.4).

Fruit eaters and seed dispersal

Some herbivores, called **frugivores**, specialise as fruit eaters. They include mammals and birds. In a northern Queensland rainforest, a CSIRO scientist noted that about 84% of 774 plant species examined produce fleshy fruits. The abundance of fruit in a rainforest

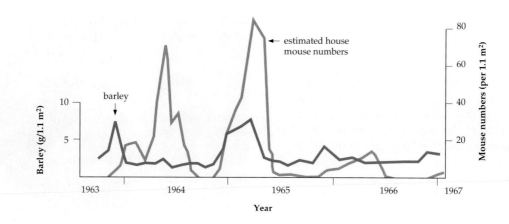

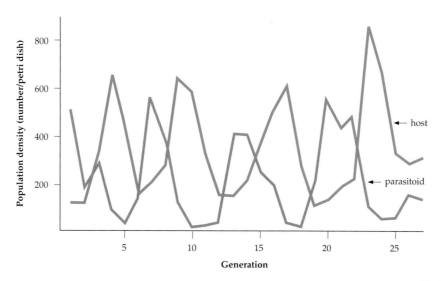

Fig. 43.11 (a) The relationship of predator and prey abundance: the house mouse, *Mus musculus*, and barley, *Hordeum vulgare*, in South Australia 1963-7. **(b)** The number of parasitoid wasps, *Heterospilus prosopidis*, oscillates in relation to its host, the bean weevil, *Callosobruchus chinensis*

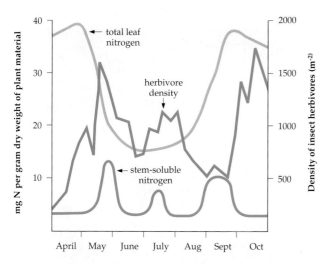

Fig. 43.12 Seasonal changes in the number of insects on the grass *Holcus mollis* in relation to changes in food quality, measured as nitrogen in leaves and stems

BOX 43.3 Predation and biological control

Parasites can be useful for biological control of pest species. Control is best where a predator that is monophagous (feeds on a single prey type) virtually exterminates its prey and then becomes scarce itself but does not become extinct. The survival in small numbers of both species makes it possible for a small population of predator to control the prey indefinitely. Biological control has not been successful in cases where the predator has a broad dietary spectrum.

The most famous example of successful biological control is the use of the moth *Cactoblastis*, which was introduced to control vast outbreaks of prickly pear cactus, *Opuntia*, in Queensland and

New South Wales. The herbivorous larvae of the moth fed on the cactus and reduced the *Opuntia* population to a tiny fraction of its original size. Other Australian examples are water hyacinth and *Salvinia* (Chapter 37), which are introduced aquatic weeds that grow unchecked by local herbivores. One of the organisms used to control water hyacinth, *Eichhornia crassipes*, is the South American weevil *Neochetina eichhorniae*, the larvae of which tunnel into the plant's tissues, chiefly petioles. The Brazilian weevil *Cyrtobagous singularis* is being used as part of a program for biological control of *Salvinia* in Australia.

Attempts at biological control can, of course, also be disastrous. The introduction of the cane toad, *Bufo marinus*, to Australia to control pests of sugar cane led to a population explosion of the toad, which is now a major pest species (see Chapter 45).

means that a diversity of frugivores can be supported, including many birds (Table 43.2). They digest only the soft wall of the fruit and not the seeds and thus ensure the dispersal of seeds, which usually pass unharmed through the animal's digestive tract. Fruit pigeons and doves (Fig. 43.13) and cassowaries have thin-walled gizzards with hardened lumps that massage the flesh

Table 43.2 Major bird frugivores in Australian tropical rainforests[a]

| | Number of plants in their diet | |
Frugivorous birds	Species	Families
Southern cassowary	115	46
Superb fruitdove	54	27
Rose-crowned fruitdove	16	11
Wompoo fruitdove	61	30
Torresian imperial pigeon	37	20
Topknot pigeon	20	10
White-headed pigeon	14	9
Brown cuckoo-dove	32	23

(a) Note the large number of species and families of fruits recorded in their diets.

Fig. 43.13 The rose-crowned fruitdove, *Ptilinopus regina*, is a frugivore and important to a number of plant species for dispersal of seeds

away from the seed but do not grind up the seed, as happens in the gizzards of other birds.

> Frugivores, herbivorous animals that specifically eat fruit, are important for seed dispersal.

Plant defences

Predation of plants has resulted in the evolution of defence mechanisms. Plants have a variety of defence mechanisms to avoid being eaten. Physical defences include tough leaves, spines (Fig. 43.14) and hairs, which generally decrease a plant's susceptibility to insect attack. Tough leaves are relatively difficult to chew and hairs make it difficult for insects to reach the more nutritious leaf tissue. Spines (Fig. 43.14) may damage insect larvae. Most plants are unpalatable or toxic to some extent. A survey of two African rainforests showed that 90% of plants contained either tannins or toxic alkaloids (Fig. 43.15). Tannins are not toxic by themselves but they combine with proteins in leaves in

Fig. 43.14 Spines on plants, such as *Solanum cardleyi* in the Northern Territory, help to deter herbivores

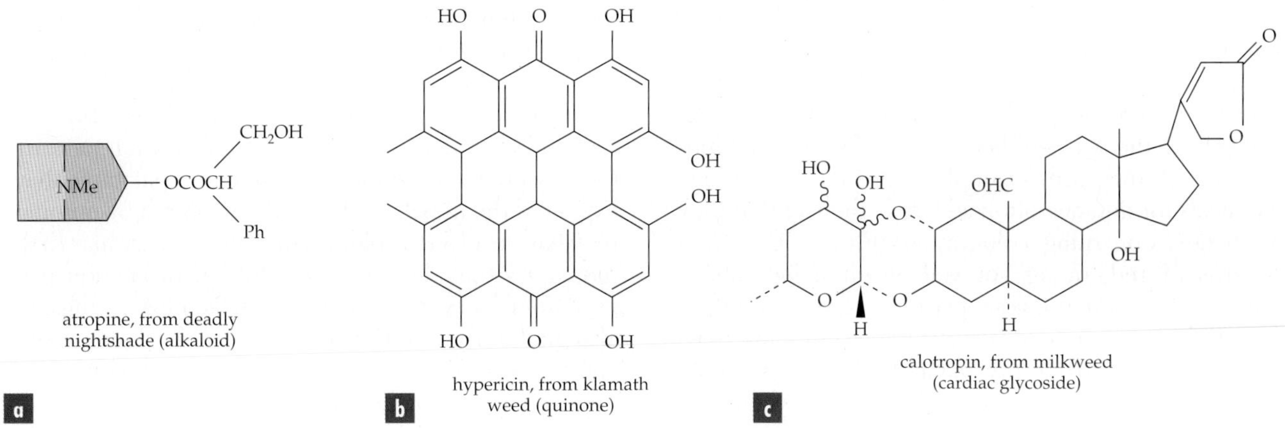

atropine, from deadly
nightshade (alkaloid)

a

hypericin, from klamath
weed (quinone)

b

calotropin, from milkweed
(cardiac glycoside)

c

Fig. 43.15 Plant secondary compounds, that is, compounds that do not play a role in primary metabolism, are important chemical defences against herbivores. There is a variety of compounds, many of which are useful medicinally. **(a)** The potato family (Solanaceae) is rich in alkaloids, one of which is atropine, the principal toxin of deadly nightshade, *Atropa belladonna*. **(b)** Hypericin is a quinone found in the klamath weed, *Hypericum perforatum*. **(c)** Calotropin is the major cardiac glycoside found in milkweeds (*Asclepias*), which monarch butterflies feed on

a way that makes the protein indigestible to caterpillars. The volatile oils of many young eucalypt seedlings makes them less palatable to herbivores (see Chapter 6). Chemical defences can be formed by a plant or by its microbial symbionts. Chemical defence is also found in marine organisms. Many marine red, brown and green algae produce terpenoids and other compounds that deter herbivorous fishes (as well as inhibiting bacterial growth).

Herbivores, however, can cope with plant defences in two ways. They may be generalised feeders, eating a small amount of a variety of plants, or they may be specialised feeders, with a mechanism for avoiding the specific plant defences. For example, as tannin-filled vacuoles are near the leaf surface, leaf-mining insect larvae can avoid tannin by burrowing into leaves and consuming the inner tissue. Most herbivores possess an array of detoxifying enzymes that enable generalised feeders to eat small quantities of each toxic species. Colobine monkeys, for example, have microflora in the forestomach that can detoxify some alkaloids. The western grey kangaroo (*Macropus fuliginosus*) can tolerate poisons present in the leaves of many members of the pea family (Fabaceae). One of the compounds, fluoroacetate, is similar to the highly toxic rabbit poison known as '10/80'. The natural levels of these poisons can be lethal to introduced animals of similar body size, such as sheep and cattle. Similarly, brown algae in eastern Australia have very high levels of phlorotannins, much higher than comparable algae in temperate areas of North America. Local herbivorous snails do not appear to be deterred by the algal chemicals.

> Plants are protected from herbivores by physical or chemical defences, although many animals have evolved ways to cope with plant toxins.

Animal defences

Animal defences against predators include speed and agility, large size, camouflage, toxic chemicals, physical barriers such as thick shells, physical defences such as horns and antlers, and behavioural adaptations such as living in social groups (Chapter 28). Modification of shape, colour and form for camouflage can be extreme in insects. Species of *Phyllium* (family Phasmatidae, 'stick insects') mimic leaves or twigs; the insects have large, leaf-like expansions of the legs and abdomen and flattened green wings. A catydid insect mimics the shape of an *Acacia phyllode* (Fig. 43.16). Black and white larvae of some papilionid butterflies resemble bird droppings. Disruptive black and white colour

Fig. 43.16 A small, unnamed catydid insect from Central Australia on mulga, *Acacia aneura*. As an adult, the catydid mimics the phyllodes of the wattle on which it spends the daylight hours. After dark, the insect becomes active, feeding on the phyllodes and perhaps the flowers of the host. This species is, as yet, unnamed even to the genus level (it is known in the Australian National Insect Collection as Genus 20, Species 1). It is not known to occur on any species other than *A. aneura*

patterns afford camouflage in the dappled light beneath a forest canopy.

Many animals release compounds that kill, damage or at least ward off predators; the secretions of some caterpillars, the stings of bees and the bite of the blue-ringed octopus are examples. Animals that use chemicals for defence often advertise the fact that they are toxic by warning colouration (Fig. 43.17). Bold patterns of red, orange or yellow, often with black stripes, are common colour patterns and it is thought that predators learn to avoid warning colouration. Redback spiders, *Latrodectus basseltii*, have a warning mark on their abdomen, which they expose when they hang upside down from their webs.

The monarch butterfly, *Danaus plexippus* (Fig. 43.18), is unpalatable because, as a larva, it accumulates toxic compounds, cardiac glycosides, from its diet of plants in the milkweed family (Asclepidaceae). It is a specialist feeder that is able to sequester the toxin and store it. The toxin, which passes through

Fig. 43.17 Warning colours are often red, orange or yellow, striped with black. The Australian frog *Pseudophryne corroboree* produces toxic alkaloids in the skin and is distinctively coloured gold with black stripes

Fig. 43.18 Insects have not only become resistant to the chemical defences of plants but some, such as the monarch butterfly, *Danaus plexippus*, use plant food as a source of chemicals for their own defence against predators. Monarch butterflies advertise their unpalatability by warning colouration

to the adult butterfly stage (and even the eggs), will cause a predatory bird to vomit and regurgitate its prey. The bird quickly learns to leave other individuals alone, especially because monarch butterflies, both as caterpillars and adults, have bold colouration patterns. Other species of butterflies that are palatable have exploited the unpalatability of the monarch by mimicry: they have evolved a colouration pattern so similar to the monarch's that they too are afforded protection from predators. Bees, wasps and ants, which have stings, are also models for mimics. One family of flies mimics bees and wasps. Some spiders mimic ants by holding their first pair of walking legs in the air to look like antennae. When harmless species mimic toxic models it is called **Batesian mimicry** (see also Chapter 28).

Animal predators have adapted to the defence mechanisms of their prey. Speed, agility, size, camouflage and adaptations to penetrate physical defences and tolerate chemicals evolve in concert with prey species, resulting in an evolutionary race between predators and prey. For example, the hunting strategies of lions are an adaptation to the sentinal systems and herding behaviour of large herbivores. Sophisticated interactions among members of a pride are necessary to implement tactics successfully.

Animal defences against predators include speed, agility, size, camouflage, toxic chemicals, physical barriers and behavioural adaptations. Toxic animals often advertise their unpalatability by warning colouration. Other species mimic unpalatable forms so that they too are protected from predators. Predators evolve characteristics in response to the evolution of defences in their prey.

Keystone predators

Some predators affect not only their particular prey species but also the overall structure of the community. On some rocky shores along the north-west coast of North America, the conspicuous predatory sea star *Pisaster ochraceus* lives in the intertidal zone and preys on a range of other animals—mussels, barnacles and snails. Their predation on mussels is the most important interaction because the mussels can out-compete most other animals on the shore. Left alone, mussels can exclude most other species. However, by feeding on mussels, the sea star creates clear areas on the shore, which can then be colonised by other species. The sea star was termed a keystone species because its presence or absence has a major effect on the structure and diversity of the rocky shore community (Fig. 43.19).

The various forms of predation are examples in which organisms interact directly with one another within their environment. Indirect interactions, particularly competition for limited resources, also determine the distribution and abundance of species.

Fig. 43.19 Keystone predators, such as (a) the sea star *Pisaster ochraceus*, consume the dominant competitors, such as (b) these mussels, and raise diversity by allowing weaker competitors, such as (c) stalked barnacles, to survive

Competition

All animals and plants have a number of requirements without which they could no longer survive or reproduce. It is easy to think of examples, such as light energy for plants, or the abundance of flowers and fruit for fruit bats. If these factors are in limited supply, individuals within a species will compete for them. We

measure the relative success with which individuals cope with such limitations in terms of the number of offspring they produce compared with their competitors. Competition between members of a species is **intraspecific competition**.

Different species living in the same geographic space also compete for limited resources. This is **interspecific competition**. It happens when, for example, different species of lizards compete for a limited supply of insects or when different species of rainforest trees germinating in a newly created gap in the canopy compete for sunlight.

Organisms may use up all of a resource or reduce it to such low levels that individuals can no longer get enough—plants may use up crucial nutrients, ants might remove all seeds from an area. This competition is called *exploitative competition*. Alternatively, individuals may, by getting access to resources, prevent others from getting to them—territorial birds may gain control of nesting areas, tall trees may shade their neighbours, and sessile animals, such as sponges, may grow over the top of neighbours (Fig. 43.20). When the organisms do not consume the resource but pre-empt it, it is termed *interference competition*. Interference competition is usually easier to detect than exploitation.

Under a given set of conditions, some species are, on average, better competitors than others. In an experiment in the 1950s, Charles Birch observed the population sizes of two species of grain beetles living in wheat. In isolation, populations of both species could persist indefinitely. However, when the species were sharing the same environment, one species was always driven to very low numbers or became extinct (Fig. 43.21). Individuals of the less successful species were out-competed for food by individuals of the species that eventually replaced it.

Such graphic experiments have the ability to colour the thinking of biologists for decades. The phenomenon demonstrated in such experiments became known as the **competitive exclusion principle**: if two species have the same mode of life and use the same kinds of resources, they will be unable to coexist.

Of course, these experiments were based on populations of very simple organisms living in a very simple environment, namely an experimental tank filled with wheat. The real world is much more complex and we should not expect that, for any two species that compete for a resource, one will always replace the other. We only need to look around to observe that similar species coexist. Many species of leaf-eating insect exist on any rainforest tree. Numerous native mammal species roam the Australian semiarid plains, competing for limited supplies of water and edible plants. Simple laboratory experiments and simple ecological models of competition (Box 43.4) suggest what will happen unless the competitive interaction is modified by interactions

Fig. 43.20 (a) Coral on the Great Barrier Reef. Competition for light means that one species can shade its neighbour and interfere with its access to light. **(b)** Two species in the subtidal zone of a south-east Australian seashore involved in aggressive competition for space. The red animal is a species of bryozoan, *Mucropetraliella ellerii*. The white colonial ascidian (a primitive chordate), *Distaplia viridis*, is enveloping the bryozoan and will eventually overgrow and kill it

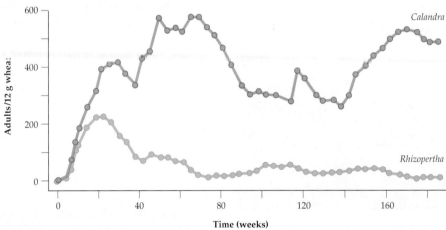

Fig. 43.21 Population abundances of two species of grain beetle, *Calandra* and *Rhizopertha*, living together. Birch found he could reverse the winner and loser of the competition by adjusting the temperature of the environment

BOX 43.4 Predicting the effect of competition

It is possible to describe quantitatively the interactions between members of two species. In Chapter 42, the equation that represents the

growth of a single population in an environment that will support a given number of individuals is:

$$\frac{dN}{dt} = rN\left(\frac{K - N}{K}\right) \qquad [43.3]$$

where N is the population size within a con-fined region, K is the carrying capacity of the environment, and r is the maximum rate at which the population can increase when resources are not limiting. If we use this equation to represent a population, we are saying that one or more of the resources that the individuals in the population require are limited, so that no more than K individuals can ever be supported.

We can represent the competitive interactions between members of two species in an equation analogous to 43.3. We let α represent the amount of a resource taken up by a member of species 2, relative to the amount taken up by a member of species 1. Then the growth of the population of species 1 is given by:

$$\frac{dN_1}{dt} = rN_1 \left(\frac{K - N_1 - \alpha N_2}{K} \right) \qquad [43.4]$$

This equation says that the growth rate of species 1 is determined by the number of individuals of species 1 (N_1), and by the number of individuals of species 2 (N_2), multiplied by a term, α, that represents the strength of the competitive effect of species 2 on species 1. If the effect is small, α will be small and species 2 will not reduce the growth rate of species 1 very much unless it is present in very large numbers.

Usually, α is less than 1, which means that adding another member of species 1 is felt more by species 1 than adding another member of species 2. Most species use the environment in different ways. Even if two species compete for a single limited resource, such as a set of food items, there will be other resources for which they do not compete. For example, there will be food palatable to one species but not the other.

When α is greater than 1, either interference or exploitation competition is implied. In a sense, species 2 is aggressively interfering with the growth of the population of species 1. For example, the ecologist J. Connell, researching intertidal communities in Scotland, showed that when two species of barnacle compete for limited space on rocks, the faster growing species (*Balanus balanoides*) dramatically reduces numbers of the other (*Chthamalus stellatus*) from the lower reaches of the intertidal zone by smothering, undercutting or crushing young individuals before they become established. *B. balanoides* is sensitive to desiccation and is unable to colonise the upper zone where *C. stellatus* persists.

Interspecific competition is competition between members of different species for limited resources. Competitive exclusion occurs when one species out-competes another for a limited resource, resulting in its local extinction. Usually, interspecific competition is less intense than intraspecific competition.

Ecological niche

In 1927, the ecologist C. Elton described the niche of a species as its functional role in the community, especially with regard to trophic interactions. In the late 1950s, G. E. Hutchinson extended the concept and defined the *fundamental niche* as that region of the environment within which a species could persist indefinitely. It is defined by all the biotic and abiotic factors that impinge on the survival and reproduction of the species. Hutchinson saw the environment as composed of a number of independent dimensions, each one representing a factor that could limit the distribution and abundance of a species (see Chapter 42). The range of tolerance of a species to each of these factors determines a multidimensional niche (Fig. 43.22). In practice, a species will not occupy its full potential range across these factors, and the **realised niche** describes the actual range of factors.

Some species can make use of a broader range of resources than others. Two species of small marsupials, the yellow-footed antechinus, *Antechinus flavipes*, and the brown antechinus, *A. stuartii*, coexist in open forest north of Canberra. Both species are nocturnal, partly arboreal and have similar behaviour. As both species eat insects, we can view the available insect population as a resource axis, ranging from very small insects at one end to very large ones at the other (Fig. 43.23). One way to measure the use the species make of the insect population is to count how many of each type of insect

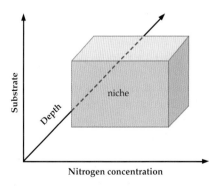

Fig. 43.22 An idealised three-dimensional niche space for the brown seaweed *Macrocystis pyrifera*. Each axis represents a range of environmental conditions for three factors that limit the growth of the species. The edges of the cube represent the limits of each factor that the species can tolerate and the cube represents the fundamental niche for the species

with additional species or the physical environment and make us look for the reasons why some species are not excluded. We explore concepts about coexistence below.

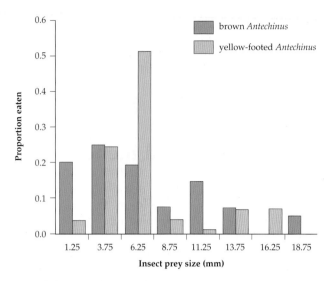

Fig. 43.23 The proportion of insects of different sizes eaten by two species of *Antechinus*. The values for each category of insect prey sizes are the class marks for the classes, each with a range of 2.5 mm

they eat. By comparing their diets, we may see the ways in which these two species use the insect resource axis in slightly different ways. One species might only eat large insects, the other only eat small ones, or they might specialise on different families of insects.

The range and abundance of insects of different sizes eaten by a species will represent its niche breadth on this axis. Figure 43.23 shows that the yellow-footed antechinus concentrates on eating insects mostly in the size class range 5–7.4 mm (class mark 6.25), whereas the brown antechinus takes insects more evenly from the entire range of insect sizes. As a result, the brown antechinus has a greater realised niche breadth than the other species.

Niche overlap and character displacement

In the late 1950s, Robert MacArthur and G. E. Hutchinson noted that species competing for a limited resource appear sometimes to exploit the resource axis in different ways. This is termed resource partitioning. In 1972, MacArthur raised the question of how much overlap in the use of resources species can tolerate to coexist in the same habitat. A series of studies followed in which the amount of niche overlap between species was measured. Many of these studies took a single axis on which it was considered two or more species were competing. The amount of overlap in the use of the resource by the different species was measured, just as in the example of antechinus above.

These studies showed that when two species compete for a limiting resource, they sometimes adapt to minimise competition. The adaptations are sometimes reflected in the morphology of competing

populations, a phenomenon that became known as **character displacement**. The phenomenon has been inferred from populations of animals in which the morphology of a species is more different when it is in competition with a second species than in places where no competition exists. The differences in morphology between different populations of some species are directly attributable to competition for a limiting resource, usually food.

For example, when two species of mud snails (*Hydrobia*) found in Limfjorden, Denmark, live apart (allopatry), their sizes are the same. In places where they coexist (sympatry), there is a marked difference in their sizes (Fig. 43.24). The size differences are related to the size of food eaten by the two species and appear to be the result of selection for resource partitioning leading to character displacement. The differences are not simply due to changes in the availability of food but are the result of the evolution of morphological differences.

> Niche breadth is the range of a resource that a species uses. Niche overlap and competition for a resource sometimes lead to selection for resource partitioning, reflected in character displacement in coexisting populations.

Spatial variability can also make it easier for species to share the same habitat. Changes in the environment may change species' relative competitive ability. Variation in time may be just as important as variation in space. The environment changes from one year to the next and temporal fluctuations provide some species with a competitive advantage in some years, allowing them to persist if they are at a disadvantage in other years, a kind of temporal refuge. Environmental variation can ensure the persistence of several competing species in the same habitat even when, in a constant environment, a competitively superior species would have forced the other species to extinction.

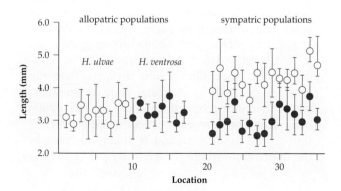

Fig. 43.24 Character displacement in mud snails, *Hydrobia ulvae* and *H. ventrosa*. In allopatric populations there are no important differences in size, but when the species are sympatric, there are significant differences

V ariability in the environment in space and time allows species that naturally compete to exist in the same habitat.

Communities are not constant

Succession

Communities change with time. Individuals are born and others die. Species become locally extinct and others colonise. Even within a single year, the composition and structure of any community will change because the behavioural and physiological characteristics of individuals enable them to adapt to the changing environment.

When communities are monitored over long periods of time, say decades, the relative abundances of different species change. The process of replacement of one species by another within an area of habitat may be gradual. If one plant species is a better competitor for light than another, but only as a seedling, it will have to wait for the adults of its competitor to die before the competitive interaction takes place. The process of replacement is part of the process called succession.

Succession is a theory about how communities change, for example, how species invade and colonise empty patches of habitat. Species interact by eating each other, providing protection and resources for each other, competing with each other for limited resources and so on. The assemblage of species found at any given place is the result of which species were able to colonise the habitat, the tolerance of these species for local conditions and the interactions that have taken place between these species. Primary succession takes place in a habitat that was previously uninhabited and secondary succession takes place in a habitat that has been modified by other species.

The process of replacement of one community by another sometimes proceeds through a series of assemblages or **seres**, each with an identifiable character. If one clears a patch of forest and then leaves it alone, species will colonise from surrounding vegetation (Fig. 43.25). Among the species to arrive first are those that disperse quickly in large numbers and over relatively large distances. Often, individuals of these species are relatively small, produce large numbers of offspring and have relatively short life spans. They are adapted to an opportunistic lifestyle. Examples of these so-called 'disturbance opportunists' include many members of the flowering plant family Asteraceae (daisies and everlastings).

Eventually, the community (the 'climax' community) will be dominated by individuals of species that are longer lived and slower growing. These individuals are often larger and are able to tolerate the rigours of

Fig. 43.25 Succession: species colonise a cleared patch in the Victorian highlands. In the foreground, forest was cleared for timber harvesting. In the background, the standing stems were killed by a wildfire in 1983

limited resources better than members of the opportunistic species but are not as well adapted to dispersal or rapid population increase. Early colonists may be lost entirely if they are unable to tolerate the competition for limited resources. This concept of succession is known as the tolerance model. Examples of tolerant species include the eucalypts, which eventually dominate many Australian landscapes after disturbance.

One can make a qualitative distinction between organisms that employ different strategies to compete with other species. Species have been called 'r' and 'K' strategists in reference to the two parameters of the logistic equation (equation 43.3). r strategists are the opportunists, those species that disperse rapidly, have large numbers of offspring (and hence high population growth rates, r) and short life cycles, for example, many plant weeds and insects that periodically reach plague numbers. **K strategists** are better long-term competitors, slower growing but more likely eventually to encounter intraspecific competition when they approach the carrying capacity of the environment (hence the K). The stages in the development of communities through time are a result of the different strategies adopted by different species. The species making up the so-called climax community are simply the best long-term competitors.

S uccession is the process of replacement of one community by another. Competition between species with different life-history strategies may lead to a climax community dominated by the best long-term competitors.

Many communities are disturbed

In looking at competition and predation, and symbiosis, it is tempting to think of ecological communities that are 'balanced', consisting of finely tuned biological interactions that result in many species being able to coexist. This kind of view is often propagated during public discussions of ecology but it is far from accurate. Most ecologists working in the field will be all too aware that the community on which they work is subject to major environmental fluctuations—temperate woodlands and forests are affected by fire, droughts affect many terrestrial habitats, while storms and major climatic perturbations such as cyclones and El Niño events affect coastal areas, and floods and droughts have major influences on rivers and streams. It is questionable whether many communities ever reach a 'climax' state because of such **disturbances** and changes in environmental conditions.

Many of these disturbances play important ecological roles but they are difficult to study. They act on such large scales of time and space that they cannot be duplicated experimentally. Instead, much of our best information comes from many years of careful observation. In terrestrial environments, the best understood disturbances in the Australian landscape are those resulting from fire.

Fire

Many of the plant and animal species that inhabit the Australian landscape are adapted to periodic fires (see Chapter 41). For example, several species of *Banksia* are killed by fire and their seeds will only germinate in the ash beds left after a fire. Similarly, the seeds of numerous *Acacia* and other species, stored in the soil, depend on the heat and smoke generated by fire to germinate. Fire may enhance species richness and diversity through its interactions with community processes. Vegetative regrowth, flowering and seedling survival after fire in south-western Australian woodlands (Fig. 43.26) are facilitated by the post-fire environment, which includes changes such as improved

Fig. 43.26 This mallee vegetation north of Esperance in Western Australia is regenerating after a wildfire. Repeated fires at short intervals in this region may change open woodland to low open shrubland

soil nutrient and water availability and reduced predation pressure from insect herbivores.

Communities in the subtropical savanna of northern Australia are structured by fire. The long fire season, from April to December, provides a range of burning conditions, which in turn provide a resource that is used in different ways by numerous species of birds (Fig. 43.27) and many other animals. The behaviour and feeding preferences of different species determine when they make intensive use of the vegetation. Brown and black falcons and whistling kites, for example, hunt grasshoppers, lizards and small mammals that escape before the fire front. The hot ash phase after fire is exploited by a second group of birds, including butcherbirds and kingfishers, who forage on vulnerable vertebrates and invertebrates. Corellas, rosellas and galahs, which are granivorous species, do not exploit the habitat heavily until regeneration commences. In general, the abundance and quality of food is enhanced for a large number of bird species for the few months after a fire.

Fire usually burns in a mosaic pattern unless environmental conditions result in very severe fires. The mosaic of fires provides a variety of habitats in space while the dynamic processes after fire, such as succession, provide a variety of habitats through time.

> Fire causes variability in habitats in both time and space and may promote species richness and diversity.

Disturbance can promote diversity

Disturbance can be an important process that prevents competitive exclusion and, by doing so, maintains diversity. If a community has one or a few consistently strong competitors, these species may eventually displace all others. If disturbances happen at shorter intervals than it takes for succession to reach a climax, competitive exclusion may be prevented. In two long series of observations, Joe Connell has described the effects of cyclones and other major storms on two diverse Australian habitats—tropical rainforests of north Queensland and coral reefs of the southern Great Barrier Reef. In the forest, storms create light gaps as trees are blown down, often taking neighbours with them. The seedlings of many canopy trees are on the forest floor beneath these gaps (Fig. 43.28) and when the light levels increase, the seedlings begin growing quickly. The race to the canopy is over when one or two reach the top of the light gap and then begin to shade their neighbours.

On shallow areas of the Great Barrier Reef, Connell found that cyclones pass by every few years and, depending on the exact path they follow, can cause considerable damage, tearing up and breaking corals

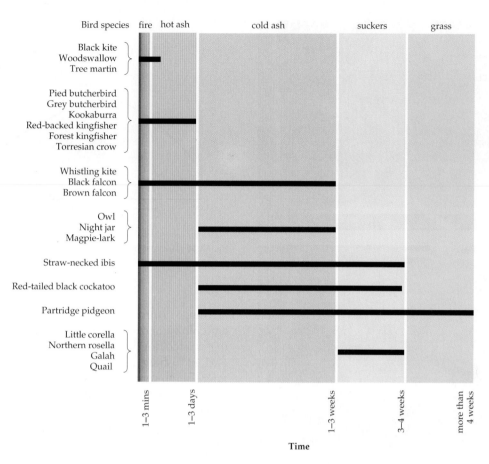

Fig. 43.27 Generalised diagram of the use of fire and the vegetation stages immediately following it by eight groups of birds in Kakadu National Park. The heavy horizontal bars represent periods of intensive use

Fig. 43.28 Tree seedlings on the forest floor at Davies Creek in north Queensland. The seedlings were tagged soon after they germinated, so their ages are known. Some of them have waited 20 years for a light gap to form into which they can grow. Closing the light gap again may take many years

(Fig. 43.29). The loss of corals frees up space for new colonists. The effect of these sporadic, unpredictable events over the whole reef is that they create a mosaic of patches of habitat, some disturbed recently by cyclones and others with assemblages of species that reflect competitive interactions undisturbed for long periods.

Connell suggested that overall diversity may be a function of the frequency and intensity of disturbances—when disturbances are too frequent, only a few species, the 'weedy' species, are likely to survive, and when disturbances are rare, competition can occur, resulting in reductions in diversity because communities become dominated by a few long-lived species that exclude many others. When disturbances are at an intermediate frequency, diversity will be maximised (Fig. 43.30) and the community will consist of a mixture of early and late successional species or a mixture of weedy and highly competitive ones. The disturbance regime will vary between communities and the resulting mix of species with different ecological abilities will reflect the attributes of disturbance processes (their frequency, reliability, extent and intensity). For example, on coral reefs studied by Connell, there does not appear to be a single dominant or climax species but a group of highly competitive species that would form a dominant community in the absence of disturbance. This relationship between disturbance and diversity was suggested independently by a few ecologists in the 1970s and has become known as the **intermediate disturbance hypothesis**.

Fig. 43.29 Corals in shallow water at Heron Island, Great Barrier Reef. Cyclones damage them and allow new colonies to arrive: **(a)** before cyclone disturbance; **(b)** following cyclone disturbance; and **(c)** coral recovery 11 years after the cyclone

Natural disturbances are an intrinsic part of most communities and result in a mosaic of habitat patches in different successional stages. They can act to maintain diversity by freeing resources.

... but humans can act as novel disturbances

In most habitats, human activities can alter communities by changing the structure of the habitat or by

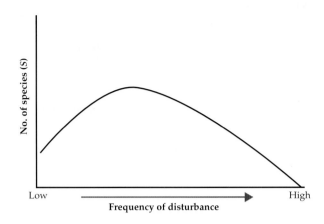

Fig. 43.30 The frequency of disturbance can affect the diversity of a community—the intermediate disturbance hypothesis

creating disturbances: humans clear vegetation for agriculture and urban development, use control burning as a tool for bushfire control, dredge the sea floor as part of fishing activities and so on. These activities might act in ways similar to natural disturbances and, as long as they are not too intense, may not cause too many changes. Often, though, our activities may be novel disturbances, acting on scales of time and space that are quite unlike natural processes.

For example, controlled burning is a part of the management of many natural areas (see Chapter 45), and, in the humid tropical savannas of northern Australia, much of the burning happens during the early dry season, mainly to reduce fuel loads and thereby reduce the severity of fires to protect human life and property. The burns tend to be regular today but prior to human colonisation most fires probably were started by lightning strikes. When Aboriginal Australians arrived, fire was used as a tool for hunting and for land management but the times of burning may have been quite different from the timing of natural burns, with much more frequent burning during the first half of the dry season. The arrival of Europeans produced another change in fire regimes.

In the intertidal zone of rocky shores of temperate Australia, the dominant seaweed, *Hormosira banksii*, occurs in large, dense mats, which provide important habitat for other algae and invertebrates that require moisture and shade (Fig. 43.31). Natural disturbances occur in the form of extreme low tides in summer that cause severe desiccation. Trampling can also act as a disturbance but the two kinds of disturbance act in different ways—a year in which natural disturbances are severe may be a year in which trampling is minor and vice versa. Trampling is relatively uniform over rock platforms separated by hundreds of metres, while natural disturbances usually affect these environments more patchily.

Fig. 43.31 Regenerating Neptune's necklace, *Hormosira banksii*. This brown alga forms extensive mats in the rocky intertidal zones of southern and eastern Australia. They are damaged by extreme midday low tides in summer but also by pedestrian traffic. Here, recovery has begun from a basal area

Human activities can represent novel disturbances that act in different ways from natural disturbances and create new selective pressures and competitive interactions.

Communities are the result of many different interactions

So far, we have described the wide range of interactions that can occur within ecological communities. Community structure is rarely the result of just one or two of these processes—competition fluctuates, disturbances come and go, predator numbers rise and fall. Community structure can be determined by processes involving established organisms, such as competition and predation ('top-down'), by processes affecting nutrient supply to the lowest tropic level ('bottom-up') or by the immigration of new individuals or species into existing communities. Understanding the way all of these processes interact is difficult and time consuming and there are few studies detailed enough to give a comprehensive picture. One of the best known communities is found in the intertidal zone of rocky shores in eastern Australia, especially the area around Sydney studied by Tony Underwood and his colleagues. Rocky shores have provided us with many ecological insights, in part because plants and animals compete for resources, but also because space and light are relatively easy to measure and, most importantly, manipulate experimentally.

On an exposed shore near Sydney, the community is a mosaic. Patches of barnacles alternate with areas of dense algal cover and other areas dominated by herbivorous snails (Fig. 43.32). The physical environment plays an important role. The position taken by an organism on the shore influences its exposure to the air and the physical stresses of wave action but a variety of

Fig. 43.32 Important components of the community on rocky shores near Sydney are patches of **(a)** barnacles, **(b)** herbivores and **(c)** predatory snails, and seaweeds such as *Hormosira* (Fig. 43.31)

biological interactions play a major part. Barnacle clumps are established when larvae settle from the plankton after drifting for a few weeks. Often, there are not enough larvae for the barnacles to cover the space completely. Even when the barnacle larvae attach themselves, many are killed by desiccation and by the bulldozing effects of limpets. As the barnacles grow, they are consumed by the predatory whelk *Morula*. These predators stay in crevices at low tide to avoid drying out, then forage when water covers the platforms. Close to crevices, predation is intense and few barnacles may survive, but further from crevices, whelks forage for shorter times and barnacles survive better.

Elsewhere on the shore, dense mats of algae cover the rocks. The upper limit of these algal patches is ultimately determined by desiccation but the important interactions are between the herbivorous snails, especially the limpets, and the algae. The limpets scrape the surface of the rocks and feed on microscopic algae, including the newly arrived spores of larger seaweeds. In the open areas, large numbers of snails compete with each other for this algal food. With such high levels of herbivory, larger algae cannot establish. In the algal mats, the herbivorous snails do not find much food and they cannot eat larger algae. The algal areas and the bare patches with herbivores can be viewed as alternative communities, part of a dynamic mosaic determined by chance and species interactions. If the number of snails falls, they can no longer consume all the algal spores, and some plants grow to a size where they can no longer be eaten. When this happens, the area becomes an algal patch. Conversely, if a disturbance removes established algae, the limpets and other herbivores can move in and then prevent the algae from re-establishing.

> The communities that we see are not the result of a single process but the combination of a range of biological interactions, modified by the physical environment and random disturbances.

Biomes: communities on a global scale

Regardless of the particular species in a community, the vegetation structure found in different parts of the world is similar in similar environments. On a global scale, communities with the same structure are called biomes (Box 43.5). For example, forests inhabit cool temperate regions of most parts of the world. Regions at high latitudes, close to the poles, and at high altitudes, support tundra, herbfields and low shrubs. Biomes are a result of evolutionary convergence (Chapter 32).

Gradients and ecotones

Gradients represent continuous, usually gradual, changes in environmental variables that occur with geographic distance. Dive in the ocean and you will notice that available light is a simple function of depth. Drive from the coast inland to the top of a mountain range and there will be a more or less gradual change in temperature. Such gradients are the factors that determine the distribution and abundance of species

that change in a more or less continuous fashion from place to place in the landscape.

Apparent abrupt changes in the environment from one place to another are a matter of scale. In fact, it is possible to argue that all changes in environmental conditions are gradual—it is only a matter of how closely you look. Near the treeline on alpine slopes, trees gradually become more and more reduced in height, vigour and density with increasing altitude.

Because individuals respond to environmental conditions in their immediate vicinity, and because these conditions change gradually in space, the interfaces between different habitats and different communities are usually blurred. The boundaries between two different communities are termed **ecotones**. Often, ecotones are species rich because they support individuals from both community types. Some species that persist there are at the edge of their niche, either because of competition or because they barely tolerate the combination of physical conditions. For example, mangrove trees and shrubs are distributed differentially along salinity gradients such that banded zonation patterns form (Fig. 43.33). Despite differences in distribution, most mangrove species grow best under low salinity but differ in the range of salinities that they can tolerate. In general, the greater the salt tolerance of a species, the lower its growth rate under optimal conditions. It appears that the attributes that enable species to tolerate highly saline conditions involve trade-offs that reduce growth and competitive ability under low salinity. Thus, species that dominate in optimal conditions are limited by their susceptibility to highly saline conditions. Other species that can tolerate salt are limited to highly saline environments in which conditions for growth are suboptimal but in which there is limited competition from less tolerant, faster growing species. In this way, interspecific differences in salt tolerance may influence the structure of mangrove forests along natural salinity gradients.

Fig. 43.33 (a) Mangroves in Papua New Guinea

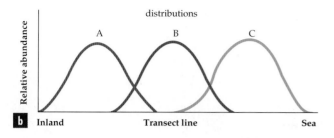

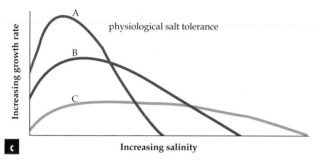

Fig. 43.33 (b) Distribution of three mangrove species, A, B and C, along a line transect from the sea to inland. The species closest to the sea, C, is the most tolerant of saline conditions but is the slowest growing under **(c)** a range of salinities

> Gradients are gradual changes in environmental parameters that occur with geographic distance.

BOX 43.5 World biomes represented in Australia

Biomes are ecological communities delineated by climate. They occur widely across the earth and each has a characteristic structure. Some important biomes that occur in Australia (see Fig. a) are described here. Biomes are broad categories and each includes several structurally different classes of vegetation. Specht's vegetation classification is shown in Table 43.1 and Fig. 43.4.

Rainforest includes tall closed forest and closed forest and occurs where rainfall is high and reliable throughout the year and soils have high concentrations of phosphorus. In Australia, rainforest is restricted to less than 1% of the surface area of the continent; it was much more widespread during the late Cretaceous and Tertiary periods (see Chapter 41). There are a number of types of rainforest, which are described as tropical (north Queensland, see Figs b and c), subtropical (from Mackay, Queensland, to New South Wales), warm temperate (from New South Wales to east Gippsland, Victoria) and cool temperate (Tasmania and Victoria). There is a general reduction in floristic diversity and structural complexity from north to south.

Open forest and woodlands, dominated in Australia by evergreen trees, predominantly *Eucalyptus* (see Fig. d), occur where dry periods are infrequent and rainfall is reliable, ranging from 650 mm in the south to 1500 + mm in the north. There are many types of open forest, with species being endemic to different regions, such as Cape York, Queensland, the south-west region of Western Australia or Tasmania. Tree heights range from 30 m or more to 10 m and overhead canopy cover varies from 70% to 30%. In drier climates, open forests are replaced by woodlands where trees are more sparse.

Alpine zones are those areas in the world between the climatic limit of tree growth (the treeline) and the zone of permanent snow and ice cover. The mean annual temperature of the warmest month is 10°C. In Australia, the alpine biome is restricted to the south-eastern highlands, such as Mt Kosciuszko, where the treeline occurs at about 1830 m altitude, and the 'snow country' of Victoria and Tasmania, a combined area of 0.15% of Australia. The biome includes a number of plant communities such as herbfields (see Fig. e), heaths, fens and bogs. Many of the plants have their renewal buds close to the ground, protected from the cold, or they are annuals with a short life cycle. Snowgum woodlands (*E. pauciflora* or *E. coccifera*) grow at lower altitudes in the subalpine zone.

Savanna is a kind of low open woodland with hummock grasses (see Fig. f). In northern Australia it occurs in a tropical, semihumid climate with seasonal (summer) rainfall. Trees are scattered through the region, including *Brachychiton* (bottle tree), *Callitris* (native pine) and *Eucalyptus* (such as *E. alba*, white gum), and tall grasses form an understorey. Some tree species are deciduous in the dry season. Savanna in the east of Australia experiences a semiarid, subtropical to temperate climate, with some winter rainfall. Different species of eucalypts and acacias, such as brigalow, *Acacia harpophylla*, occur here. In treeless areas within the savanna biome, grasses dominate to form semiarid hummock grassland (*Triodia* and/or *Plectrachne*) or tussock grassland (*Astrebla*, Mitchell grass). The *Triodia* grasslands are sometimes called spinifex.

Mallee is a tall shrubland dominated by mallee-form (multistemmed) eucalypts that are usually less than 10 m tall. It predominates in

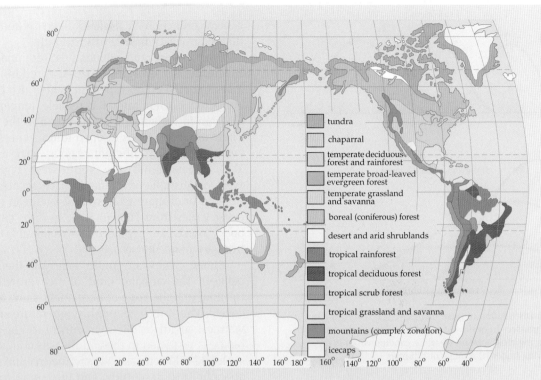

(a) Some of the major world biomes

Legend:
- tundra
- chaparral
- temperate deciduous forest and rainforest
- temperate broad-leaved evergreen forest
- temperate grassland and savanna
- boreal (coniferous) forest
- desert and arid shrublands
- tropical rainforest
- tropical deciduous forest
- tropical scrub forest
- tropical grassland and savanna
- mountains (complex zonation)
- icecaps

(b) The structure of a tropical rainforest is complex with a number of vertical layers, including emergent trees reaching heights of more than 30 m. Little light reaches the forest floor, and many plants are epiphytes or lianes (vines), growing on the trunks of trees. Vegetation profile showing the structure of a tropical rainforest in Brunei

(c) Tall trees of tropical rainforest often buttress at the base

(d) Open forest

(e) Alpine herbfield

(f) Savanna

chenopods; see Chapter 41). The type of community within the biome depends on factors such as rainfall and soil. Typically, heath communities are found on nutrient-deficient sands, whereas mallee communities are found on sandy clays with higher nutrient status, and chenopod shrublands (shrub steppe) dominate on dry and saline or calcareous soils such as the Nullarbor Plain.

Desert occurs in arid areas where rainfall is low (less than 350 mm per year) and where temperatures can be extremely high. Australian deserts are in the interior of the continent and consist of sandhills, sand plains, salt lakes and gibber plains (stony deserts). Deserts are sparsely vegetated, with grasses such as *Triodia* dominating. Parts are treeless but tree species form a tall shrubland or open woodland bordering the desert called mulga. Mulga is dominated by *Acacia*, mainly *A. aneura* (also called mulga), as well as species of *Allocasuarina* and *Eucalyptus*.

extensive areas of the southern part of Australia where the climate is semiarid or Mediterranean. The mallee communities grade into heath and shrublands in which sparse low mallees sometimes persist. Mallee also intergrades with saltbush and bluebush low shrublands (succulent

Summary

- Communities are natural assemblages of species that interact with one another.
- Community structure is often summarised by measures of diversity. Species richness is the number of species within a community, and more complex measures of diversity account for the number of species and their relative abundances within sites, within communities at different sites and between different communities. α-diversity is the diversity of species abundances within communities and β-diversity is species diversity between communities. These measures can be quantified to identify areas of high conservation status.
- Structure in plant communities is recorded usually as percentage projective foliage cover within height classes. The data are used to compare structural diversity between communities and to classify and map vegetation.
- Species interactions within communities include symbiosis, predation and competition.
- Coevolution is where interactions between organisms lead to reciprocal evolutionary change in the characteristics of species.
- Symbiosis is where two organisms live together in a close relationship and have coevolved. It includes mutualism (where both species benefit), commensalism (one benefits, one is unaffected) and parasitism (one benefits and one is harmed).
- The abundance of a predator or herbivore and its prey can show oscillations: predator–prey cycles. However, these are not universal because species interact in numerous ways in communities.
- Frugivores, animals that consume fruit, are important for the dispersal of seed. Frugivores in Australian tropical rainforests are mainly birds and mammals.
- Plants and algae are protected from herbivores by physical defences or secondary chemical defences, although animals, especially insects, have evolved ways to cope with plant toxins. Animal defences against predators include speed, agility, size, camouflage, toxic chemicals, physical barriers and behavioural adaptations. Toxic animals often advertise their unpalatability by warning colouration. Other species mimic unpalatable forms so that they too are avoided by predators.
- Interspecific competition is competition between members of different species for limited resources. Competitive exclusion occurs when one species out-competes another for a limited resource, resulting in its local extinction. Usually, interspecific competition is less intense than intraspecific competition.
- The fundamental niche of a species is that region of the environment within which a species can survive and reproduce. Niche breadth is the range of a resource a species uses. Niche overlap is the length of a resource axis over which the use by different species overlaps. Competition for a resource sometimes leads to selection for resource partitioning, reflected in character displacement in sympatric populations.
- Variability in the environment in space and time promotes coexistence of species with similar resource requirements.
- The process of replacement of one community by another, succession, sometimes proceeds through a series of assemblages. Competition between species with different life-history strategies may lead to a community (the climax community) dominated by the best long-term competitors.
- Natural communities are often subject to disturbance, which may reduce the frequency of competition and can help maintain diversity.
- Fire may provide variability in habitat in both time and space, promoting species richness and diversity. In Australia, fires may be the primary determinant of community structure and composition in terrestrial habitats.
- Many human activities act as disturbances and can often act on different scales of time and space than natural disturbance. When this occurs, humans can change the patterns of natural selection within communities.
- Patterns of community structure are rarely the result of simple processes but reflect the combined effects of the physical environment, biological interactions and disturbances.
- Gradients are gradual changes in environmental parameters that occur with geographic distance. Because these changes are gradual and because individuals respond to the conditions in their immediate vicinity, zones of transition (ecotones) between species or between communities are often blurred.

keyterms

α-diversity (p. 1144)
β-diversity (p. 1144)
Batesian mimicry (p. 1154)
character displacement (p. 1158)
coevolution (p. 1146)
commensalism (p. 1146)
community (p. 1142)
competition (intraspecific and interspecific) (p. 1155)

competitive exclusion principle (p. 1155)
disturbance (p. 1160)
diversity (p. 1142)
domatia (p. 1148)
ecotone (p. 1164)
ectoparasite (p. 1149)
endoparasite (p. 1149)
epiphyte (p. 1146)
frugivore (p. 1150)
herbivore (p. 1149)

intermediate disturbance hypothesis (p. 1161)
K strategist (p. 1159)
mutualism (p. 1146)
myrmecotrophy (p. 1148)
parasite (p. 1149)
parasitism (p. 1146)
parasitoid (p. 1149)
predator (p. 1149)
realised niche (p. 1157)

r strategist (p. 1159)
sere (p. 1159)
species richness (p. 1142)
structural diversity (p. 1144)
succession (p. 1159)
symbiosis (p. 1145)

Review questions

1. Which of the following is the best definition of 'community'?

   A an assemblage of interacting individuals

   B an assemblage of populations of different species

   C the sum total of all species in an area together with the physical environment

   D a particular structural type of vegetation

2. If a plant community is dominated by shrubs (> 2 m tall) and has an overhead cover of 30%, it would be classified in Specht's system as:

   A open heath

   B low shrubland

   C tall shrubland

   D low open woodland

3. (a) What is meant by 'symbiosis'?

   (b) Distinguish between commensalism and mutualism.

   (c) Distinguish between parasitism and predation.

4. What is the significance of 'warning colouration'? Give an example.

5. Why are mimics that are themselves non-toxic usually present in low numbers relative to the distasteful species that they resemble?

6. How could species diversity in a community be determined by the activities of predators?

7. Define the term 'succession' and give an example.

8. List four factors that can be called disturbances in communities. What is the intermediate disturbance hypothesis?

9. How might you expect measures of species richness and diversity to change as you move along an environmental gradient and across an ecotone between forests dominated by different species?

10. Explain the processes (biotic and abiotic) that create the patterns in community composition on temperate Australian seashores.

Extension questions

1. In the tropics, leaf-cutter ants remove pieces of leaves and take them to underground nests. There they chew and innoculate them with fungal spores. The fungi grow on the chewed leaves and are cultivated in this way. The fungi, not the leaves directly, are the main food of the ants. What sort of interaction, between the fungus and ant, do you think this is an example of? Explain your reasoning. How would you test your hypothesis?

2. Two species of similar-sized insectivorous lizards coexist in rocky outcrops scattered in dry sclerophyll forest in a region of eastern Australia. You hypothesise that the two species are competing for food. Discuss observations that could be made and manipulative experiments that could be carried out to test your hypothesis. Discuss the relative merits of different approaches.

3. Describe the multidimensional niche space of a species in isolation from other species (in allopatry) compared with its volume in the same niche space when in competition with another species for food. How will character displacement affect the shape of the niche space?

4. Suggest experiments that would test the hypothesis that a physical difference between two closely related species has been caused by selection resulting from interspecific competition for a limited resource.

5. What role does fire play in determining species richness and diversity in the Australian biota? Given circumstances that require fire control to protect human life and property, how would you go about establishing the most appropriate fire control regimen to manage the flora and fauna?

6. How do the ideas underlying the theory of succession relate to the concepts of environmental gradients and ecotones?

Suggested further reading

Australian Surveying and Land Information Group. (1990). *Atlas of Australian Resources. Vegetation* Vol 6. Canberra: Department of Administrative Services.

This atlas provides an overview of Australian vegetation types and the system used for classification.

Connell, J. H., Hughes, T. P., Wallace, C. C. (1997). 30 years of watching corals on Heron Island in Ecological Monographs.

A summary of disturbance dynamics on the Great Barrier Reef.

Connell, J. H. and Slatyer, R. O. (1977). Mechanisms of succession in natural communities and their role in community stability and organisation. *American Naturalist* 111: 1119–44.

This article describes the processes of succession and disturbance in communities.

Hutchinson, G. E. (1958). Concluding remarks. *Cold Spring Harbor Symposium on Quantitative Biology* 22: 415–27.

This scientific paper gives the original description of the multidimensional concept of niche.

Specht, R. L. and Specht, A. (1999). *Australian Plant Communities.* Melbourne: Oxford University Press.

This book provides an up-to-date overview of the physiological processes that shape Australian plant communities.

Underwood, A. J. and Chapman, M. G. (1995). Rocky shores. In Underwood, A. J. & Chapman, M. G. eds. *Coastal Marine Ecology of Temperate Australia.* Sydney: University of NSW Press.

This text describes the ecological processes on rocky shores.

CHAPTER 44

Ecosystems

ndividual animals or plants aggregate into populations, populations assemble into communities, and assemblages of communities interacting with their physical environments constitute ecosystems. **Ecosystems** are the basic functional units of ecology. They are living, dynamic systems encompassing organisms, their *biotic* (interactions with other organisms) and *abiotic* (physical) environments, and exchanges within and between each of these. Ecosystems are difficult to define more precisely than this because their boundaries are seldom fixed or precise. An ecosystem is neither static nor closed; instead, an ecosystem is a changing, self-modifying ecological system.

> Ecosystems are dynamic, self-modifying ecological systems consisting of interacting biotic and abiotic components.

Ecological pyramids

At a most fundamental level, ecological interactions are exchanges and flows of energy and matter, and these interactions are governed by the ways in which the component organisms obtain their food. As a result, the structure and function of ecosystems are often described in terms of their feeding or *trophic relationships*. **Autotrophs** are organisms that synthesise complex organic compounds from inorganic precursors using external energy sources. Green plants and algae are the largest group by far, manufacturing all their required organic molecules from simple inorganic molecules, using sunlight as the energy source for photosynthesis (Chapter 5). Chemosynthetic bacteria also manufacture their own organic molecules (Chapter 5) but their contribution to the volume of energy flowing through ecosystems is far outweighed by that of plants and algae. Ultimately, autotrophs are the producers of organic energy for all other organisms.

All other organisms are **heterotrophs** or **consumers** because they cannot synthesise organic matter but can only reorganise it (Chapter 19). They must consume other organisms in order to gain the organic molecules they need for life. Heterotrophs include primary consumers (mostly herbivores), which feed directly on producers, secondary consumers (mostly carnivores and parasites), which feed on herbivores, and tertiary and higher order consumers (also carnivores and parasites), which feed on other consumers. Among heterotrophs there are also organisms that feed on dead organisms and organic wastes from several **trophic levels**. These are **degraders**, which include **scavengers** (animals that eat dead organisms) and **detritivores** (animals that eat organic litter or detritus), and **decomposers** (fungi and bacteria that

cause chemical decay of organic matter from all trophic levels) (Fig. 44.1).

Of course, this view is simplified. **Omnivores** (e.g. many ants, humans, many fishes, which feed on both plants and other animals) overlap more than one trophic level, as do degraders such as detritivores. Some organisms change trophic levels during their life cycles. Some insects, for example, are carnivorous as larvae but herbivorous as adults, and many species of fishes move up the trophic scale as they grow. The simple trophic relationships of Figure 44.1 serve to demonstrate some useful general characteristics of ecosystems.

> Trophic interactions are probably the most important relationships determining ecosystem structure and functioning. All organisms are either autotrophs (producers) or heterotrophs (consumers, degraders and decomposers).

Pyramids of numbers

Why are there few large carnivores? Why are there billions of plankton and millions of krill but only hundreds of whales in Antarctic waters? Why can a patch of eucalypt woodland support thousands of leaf-eating insects but only hundreds of insect-eating birds and only one or two bird-eating foxes or cats? Or, as the ecologist Paul Colinvaux has put it, why are big fierce animals rare? An early British ecologist, Charles Elton, was one of the first to consider these questions in an ecosystem context. In his book *Animal Ecology* (first published in 1927), Elton suggested that a trophic chain, like the one in Figure 44.1, represents a 'pyramid of numbers', with large numbers of small herbivorous animals near the base of the pyramid being consumed by successively smaller numbers of increasingly large carnivores toward its apex. He thought that the relationships between organism size and the mechanics of eating and being eaten were the key to this **ecological pyramid** structure. For example, insectivorous birds have to be big enough to kill and eat insects, and foxes and cats have to be big enough to kill and eat birds.

It seems sensible that trophic interactions should result in some sort of pyramidal relationship that can be graphed. After all, each successive trophic level depends on the one below, and you might expect some sort of reduction to occur up successive rungs of a trophic ladder. However, numbers may not be the best way to represent the food value of any particular trophic level. In northern Australia, termites and cattle both eat grass (Box 44.3) but the same number of grass plants can support many more termites than cattle. Pyramids of numbers are particularly prone to such inversion when they include plants and herbivores because their feeding interactions do not necessarily

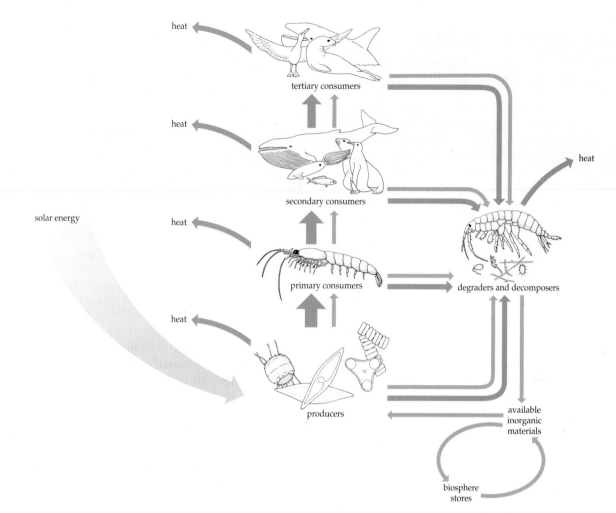

Fig. 44.1 A simplified diagram of the flow of energy (red) and materials (blue) through an ecosystem. Note that although there is a net loss of energy from the system, materials may be recycled

comply with Elton's generalisation about sizes. Herbivores do not have to kill plants to eat them, so plants provide different-sized mouthfuls for different sizes of animals, and pyramids based on plants can support many different sizes and numbers of herbivores. In Figure 44.2, for example, large numbers of pasture plants in a grassy agricultural field support relatively smaller numbers of consumers, but small numbers of trees in an oak forest support relatively larger numbers of consumers.

Pyramids of biomass

The amount of food in any trophic level depends in part on its **biomass**, that is, the total amount of living material present at any one time (see Box 44.1). Large numbers of pasture plants and small numbers of oak trees may represent very similar amounts of food. Diagrams depicting biomass of successive trophic levels in an ecosystem are more frequently pyramidal in shape than diagrams of numbers of organisms (Fig. 44.3).

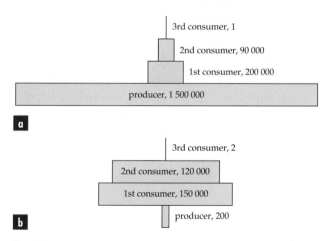

Fig. 44.2 Ecological pyramids of numbers (above-ground individuals per 0.1 ha). **(a)** A grassy agricultural field in Michigan, USA. **(b)** An oak forest near Oxford, UK

Both numbers and biomass are instantaneous measures of ecosystem properties and are liable to change as organisms are born, grow and die. In

BOX 44.1 Biomass, energy and productivity

Biomass ('living mass') is usually expressed as a weight (e.g. grams). Each unit of biomass can also be considered as a parcel of energy, usually measured as joules (J) or calories. The energy represented by a unit of biomass at any particular time may be stored, used in chemical reactions, converted to movement or heat, or consumed by other organisms. The energy content of different types of biomass varies but not by very much. For example, plants contain about 19 kJ of energy per gram of dry weight, algae contain about 22 kJ, insects contain about 23 kJ and vertebrates contain about 24 kJ per gram of dry weight. Thus, biomass can also be expressed as energy equivalents and vice versa.

Of particular importance to ecosystem function is **net primary productivity**. This is the portion of total (or gross) primary productivity that remains after the respiratory losses of primary producers have been accounted for; it is this portion that is available for harvest by the ecosystem's consumers or decomposers. Since energy can be expressed in biomass equivalents, net productivity of an individual plant is usually measured as the biomass (usually dry weight) that it synthesises and accumulates in tissues per unit time. Net primary productivity of an ecosystem is most commonly measured as dry weight of biomass synthesised per unit area of the earth's surface per unit time and is expressed as grams per square metre per year ($g/m^2/year$).

Some part of the net production of a plant or community may be eaten, shed or may die and this must be taken into account when measuring productivity. Hence, the growth of a tree through a year may appear as a 10 kg increase in its weight but the net production by the tree includes this growth plus net production expended in leaves, fruits, flowers, branches and roots that were lost during the year because of herbivory, abscission or senescence.

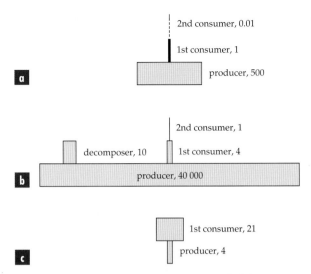

Fig. 44.3 Ecological pyramids of biomass (g/m^2). **(a)** A grassy agricultural field in Georgia, USA. **(b)** A tropical rainforest in Panama. **(c)** Plankton in the English Channel, North Sea

instantaneous biomass indicates. The biomass of the community is dynamic rather than static and food value depends on the rate of flow of energy from one trophic level to the next.

Pyramids of energy

Productivity is the rate at which energy flows between trophic levels, for example, primary productivity indicates the rate at which the main primary producers of an ecosystem trap energy from sunlight. The energy flowing into one trophic level sets an upper limit to the food available to the next. Not all the energy and material taken in by one trophic group is passed on to the next because not all organisms at one trophic level are consumed by the next; there are also losses associated with cellular respiration and elimination of wastes. Thus, there is a net loss of energy and materials as the trophic scale is ascended. Energy flow indicates the food value of trophic levels more accurately than either numbers or biomass and, for most ecosystems, a pyramid of energy flow conforms most closely to Elton's classical ascending pyramid (Fig. 44.4).

> Graphs of the energy flowing into successive trophic levels represent ascending ecological pyramids, but pyramids based on numbers or biomass may sometimes be inverted.

Figure 44.3, fields and forests represent large amounts of relatively long-lived biomass, so their biomass is a reasonably reliable indication of their food value to consumers. In contrast, although the biomass of plankton in the English Channel at any one time is quite small, individual plankton organisms have short life spans and are constantly being replaced. The food value of the plankton community is much larger than its

Productivity of different ecosystems

What is productivity? How much does it vary among ecosystems and what does such variation tell us about

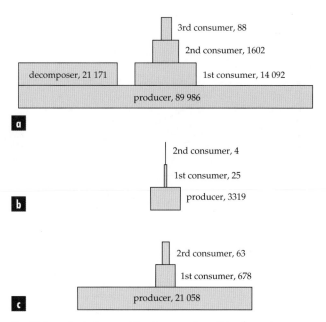

Fig. 44.4 Ecological pyramids of energy flow (kJ/m²/year). **(a)** A freshwater spring in Florida, USA. **(b)** A desert grassland in New Mexico, USA. **(c)** A prairie in Oklahoma, USA

functioning ecosystems? First, productivity is the rate at which the main primary producers of an ecosystem synthesise energy equivalents (indicated by biomass) from sunlight (see Box 44.1). If the productivity of an ecosystem changes little over time, it suggests that either the ecosystem's environment is unchanging or that its organisms are compensating for changes that are occurring. If productivity changes dramatically, it could mean that an important environmental change is

occurring or that there has been an important change in the interactions of organisms within the system.

The amount of energy passing through ecosystems—their *net primary productivity*—is determined by the activities of plants and algae, which lie at the base of most energy flow/productivity pyramids. Since productivity is limited by the availability of all resources, material as well as energy, the average net primary productivity of different terrestrial ecosystems varies in ways that we might have guessed (Fig. 44.5): tropical forests are generally more productive than grasslands, which are generally more productive than deserts. In other words, communities of plants growing in sunny, warm, wet, fertile environments are more productive than those growing in environments in which the availability of sunlight, heat, water or nutrients is sometimes limiting.

What about marine ecosystems? Rather surprisingly, most marine ecosystems are relatively unproductive. This is largely because ocean nutrients (particularly phosphorus and nitrogen) accumulate at great depths where they are unavailable to oceanic algae (mostly phytoplankton), which are restricted to growing in the top 100 m or so of ocean that sunlight can penetrate. The major oceanic fishing grounds are upwelling zones such as the North Sea, the Newfoundland Banks and the Antarctic Ocean. These are more productive because they are regions in which deep sea currents (and their nutrients) are forced upward and thereby continuously replenish the nutrient supply for phytoplankton in the surface layers. Coastal waters are also relatively productive because

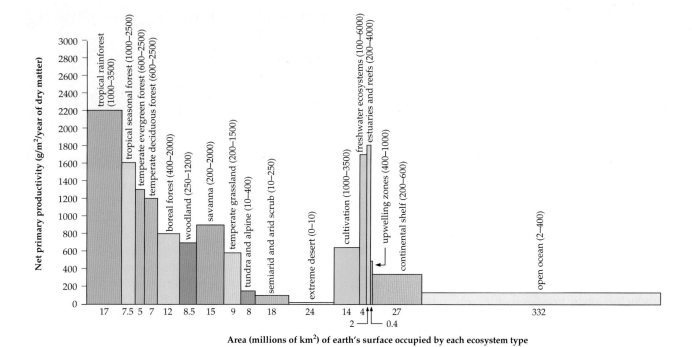

Fig. 44.5 Net primary productivity of world ecosystems

nutrients are continuously provided by freshwater rivers and continuously circulated by coastal currents. However, the total primary productivity of oceans (including the upwelling and coastal zones) is still less than half that of terrestrial ecosystems, even though oceans cover 70% of the world's surface.

For the systems shown in Figure 44.5, productivity and biomass rankings are very similar. However, as we saw when considering ecological pyramids, biomass and productivity are not always directly related. Productivity relates to the rate of turnover of biomass, not just the total amount present at any one time, so biomass can be misleading as an indicator of productivity. The duration and season of measurement of productivity are not accounted for. Daily rates of productivity in tropical rainforests are not very different from daily rates during the growing season in most other forests. Annual rates, however, are much greater in rainforests (Fig. 44.5) because the growing season in the humid tropics lasts for most of the year instead of the six months (or less) that is more usual in other regions.

Productivity through time

As well as seasonal influences on primary productivity, there are trends associated with the age of an ecosystem: for the same amount of available resources, young or regenerating ecosystems are often more productive than older ones because they usually consist of a greater proportion of young, actively growing tissue. However, older ecosystems usually contain more biomass. For example, despite the apparent destruction caused by fire, many plant species in Australia are able to regenerate rapidly from underground root stocks even if most of their aerial parts are killed. After one fire, scientists in South Australia followed the dynamics of a mallee woodland during 12 years of post-fire recovery (Fig. 44.6). As the woodland regenerated, the above-ground productivity declined from about 180 g/m²/year before the fire to about 50 g/m²/year after the fire, but there was nearly a ten-fold increase in the above-ground biomass over the same period.

If resources are limited, regenerating ecosystems are not always more productive than mature ecosystems. Soils under tropical rainforests are often nutrient-poor because of the intensity of tropical weathering processes. Clearing of mature rainforest may lead to sufficient loss of nutrients to cause a shift in species composition in the regenerating ecosystem and may result in a change from rainforest to less productive open grassland (Fig. 44.7). In the mulga (*Acacia aneura*) rangelands of semiarid Australia (Fig. 44.8), prolonged drought can lead to the death of most of the perennial grasses that are utilised as dry season food by grazing animals, such as sheep, cattle and

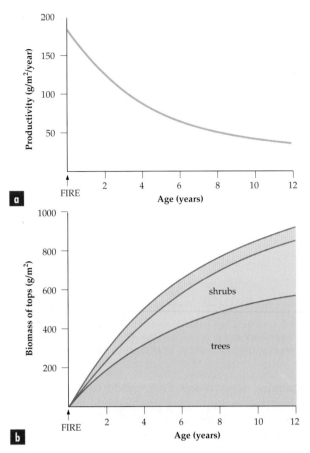

Fig. 44.6 (a) Productivity and **(b)** biomass of a mallee woodland recovering from fire

Fig. 44.7 Grasslands in the lower Jimi Valley of Papua New Guinea. Repeated prehistoric clearing and burning of rainforest has led to the depletion of soil nutrients and replacement of the rainforest by grassland

kangaroos. Under conservative grazing management, these perennial grasses will regenerate when drought-breaking rains fall. However, if ecosystem resources are depleted by heavy grazing, regenerating pastures may come to be dominated by ephemeral herbs rather than perennial grasses, leading to marked reductions in long-term ecosystem productivity.

Fig. 44.8 In the mulga (*Acacia aneura*) rangelands of semiarid Australia, prolonged drought and heavy grazing by sheep, cattle or kangaroos leads to perennial grasses being replaced by herbs, with the ecosystem becoming less productive

The amount of energy flowing through any ecosystem is determined by its net primary productivity. The productivity of forests and freshwater swamps is generally higher than that of grasslands and streams, which exceed that of oceans and deserts.

Food webs

Primary productivity tells us something about the energy base of ecosystems, but to what extent is ecosystem structure affected by energy flow? Primary productivity is limited at least as much by the availability of material resources, such as nutrients and water, as it is by energy. And, while energy reduction up the trophic scale helps explain ecological pyramids, how much more does it tell us about the detail of trophic relationships within ecosystems? A useful way to analyse this question is to consider the flow of energy up the trophic scale as a simple sequence of organisms eating and being eaten; this can be diagrammatically represented as a **food chain** (e.g. Fig. 44.9) or, more realistically (since food chains are not isolated in ecosystems), as **food webs**, which depict patterns of interlocking food chains (e.g. Fig. 44.10).

Although green plants and algae are the ultimate base of most food chains, it is sometimes convenient to separate those food chains that are directly dependent on green plants (**grazing food chains**; e.g. Fig. 44.9a) from those whose primary food base is the mixed debris (or detritus) that is shed from organisms and their remains (**detritus food chains**; e.g. Fig. 44.9b). The relative importance of these food chains varies among ecosystems. In grasslands such as mulga rangelands, grazing food chains are prominent, commencing with pastures and shrubs and proceeding through herbivores such as red and grey kangaroos, sheep, cattle and goats, to consumers such as people, dingoes, foxes and cats. In forested ecosystems, detritus food chains are often more prominent, with much of the primary production proceeding via leaf litter through detritus

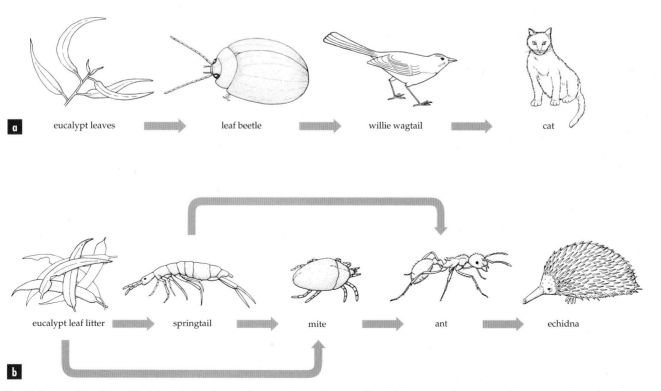

| a | eucalypt leaves | leaf beetle | willie wagtail | cat |

| b | eucalypt leaf litter | springtail | mite | ant | echidna |

Fig. 44.9 Examples of simplified food chains that might occur in eucalypt woodland: **(a)** a grazing food chain; **(b)** a detritus food chain

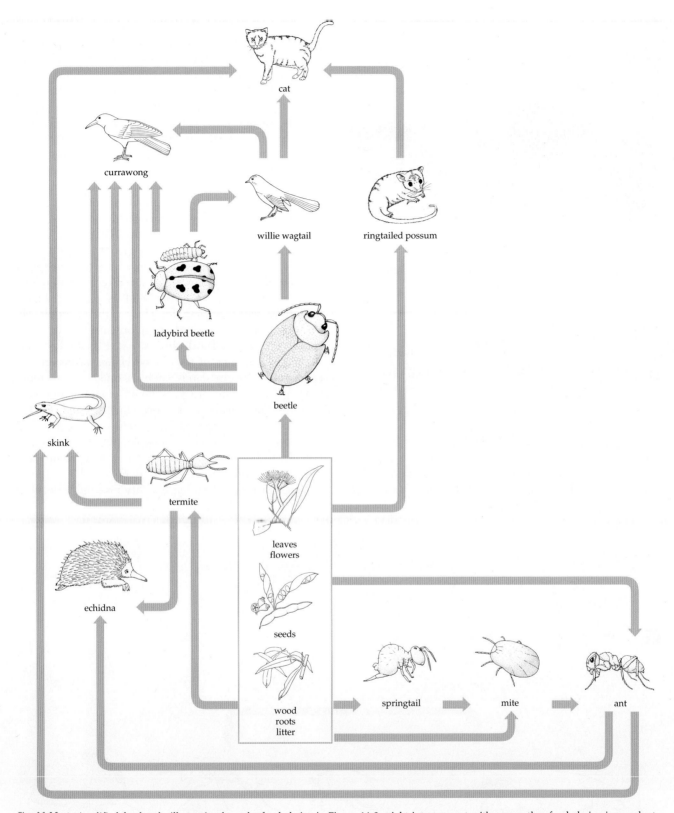

Fig. 44.10 A simplified food web, illustrating how the food chains in Figure 44.9 might interconnect with some other food chains in eucalypt woodland

food chains based on soil arthropods, such as springtails and mites (Fig. 44.9b). However, the distinction is not always clear. For example, in the rangelands, termites are also important primary consumers, so much so that their biomass may exceed that of domestic livestock (Box 44.3), but they eat both living and dead plant material. In this case, the distinction between grazing and detritus food chains is not very meaningful.

Despite their apparent complexity, food webs from different ecosystems show basic patterns:

- Food chains are short, with typically three or four trophic levels.
- Omnivores, defined here as organisms that feed on more than one trophic level, are usually scarce. Typically, food webs have one omnivore per top predator. For example, in Figure 44.10, currawongs are omnivores but echidnas are not.
- Omnivores usually feed on species in adjacent trophic levels. For example, in Figure 44.10, currawongs feed on willy wagtails and beetles but not on eucalypt leaves.
- Insect- and detritivore-dominated food webs are often exceptions to these patterns. For example, in Figure 44.10, the detritus food chain that proceeds from litter through springtails to cats has six levels, and two omnivores—ants and mites.
- Trophic interactions take place only within a particular habitat, but small numbers of sub-groupings (compartments) may exist. For example, in a forest, consumers living in trees may infrequently interact with those on the forest floor. Because skinks live on the ground they are unlikely to feed on leaf-eating beetles, but within their ground habitat they eat any potential prey that they encounter, regardless of whether it depends on a grazing or detritus food chain. Nevertheless, if the pupae of the leaf-eating beetles shelter in litter, then they may be eaten by skinks, bringing about interaction between compartments.
- Food webs are not too complex; because they have a limited number of trophic levels, there is a limited number of relationships in which any one species may be involved.
- There are a number of commonsense patterns in food webs: predators do not exist in the absence of prey; a predator cannot feed on all the species it encounters; and loops (e.g. a species of bird able to eat and be eaten by a species of insect) do not generally occur.

Some of these patterns have been known for a long time; by 1927 Elton had recognised that the length of food chains was usually limited to four or five links. More recently, mathematical ecologists have identified other apparent regularities.

The work that has identified the apparent patterns shown above is still exploratory and there is still much debate about the ecological significance of the patterns. Food webs are obviously restricted by the characteristics of organisms. Some specialist adaptations (e.g. an efficient snout for eating termites and ants; Fig. 44.11a) are likely to preclude an animal from some other activities (e.g. catching and killing large animals; Fig. 44.11b). The commonsense patterns are largely a result of such logistics. The comparative simplicity of

Fig. 44.11 (a) This echidna has a specialised long snout to enable it to catch and eat ants and termites. However, this precludes it from eating large prey compared with **(b)** the dingo, seen here catching an eastern grey kangaroo

most webs may also be a consequence of the biology of organisms. As species are added to an ecosystem, they will tend to have different adaptations or inhabit different local habitats. Consequently, the number of species a predator may exploit will be limited and will not continue to increase linearly with the number of species in the food web.

> Food webs are diagrams depicting trophic interactions between individual organisms in an ecosystem.

Why food chains are short

Not all patterns in ecosystems are likely to be due solely to biological restrictions. The reduction of energy up the trophic scale is often quoted as an explanation for the limited number of trophic levels that are observed in nature (Box 44.2).

BOX 44.2 Potted food webs

What limits the number of links in a food chain? Two alternative explanations are often suggested. The energy limitation hypothesis proposes that the amount of energy entering the bottom of a chain limits the number of consumers it can support. The dynamic constraints hypothesis, in contrast, proposes a top-down limitation; it suggests that species at the top of a chain may not be able to establish after ecosystem disturbances, such as famines, if there are too many links below them.

An Australian ecologist, Roger Kitching, has tested these hypotheses. He reasoned that experimental studies on food chain length should be on relatively simple, common and clearly bounded food webs. Water-filled tree holes and the communities of insect larvae and other organisms that inhabit them fit this description. In subtropical rainforests, water-filled tree holes typically contain three trophic levels: forest detritus (mostly leaf litter) that falls into them and provides the energy available to higher trophic levels; saprophages (insect larvae and microscopic crustaceans and mites) that feed on the detritus; and predators (other insect larvae and crustaceans, and also tadpoles) that feed on the saprophages.

Because the amount of detritus in tree holes determines the energy available at the base of the food chains, we should, by measuring the lengths of food chains in tree holes containing varying amounts of detritus, be able to determine whether food chain length is limited by energy. Unfortunately, it is difficult to perform controlled experiments on natural tree holes since their age, size, shape and volume of water vary enormously. Luckily, however, animal communities that are very similar to the natural ones will establish in water-filled plastic containers and the energy base of these 'potted webs' can easily be manipulated by varying the amount of detritus added to them, while keeping other factors constant.

Experiments in Lamington National Park

Roger Kitching first experimented with potted webs in Lamington National Park in south-eastern Queensland in conjunction with Stuart Pimm. They added leaf litter to some pots to about the natural level, to other pots to about twice the natural level and to a third group of pots to about half the natural level. The water-filled pots were then placed near natural water-filled tree holes in the rainforest and sampled frequently for six months to determine the structure of the animal communities that developed in them.

Two different groups of litter-feeding saprophages, mosquito larva and tiny crustaceans, established quickly in all the pots in numbers and sizes that were independent of the litter (and therefore energy) supplied to each pot. Two different groups of predators, tadpoles of the frog *Lechriodus fletcheri* and a crustacean, also established in all the pots but much more slowly (their numbers were still increasing after six months). The energy supplied to the pots influenced slightly the numbers of these predators; the time the pots had been in the forest was a far more important determinant of the numbers of predators. Pimm and Kitching reasoned that the food chains that established in their experimental pots were probably more limited by the time it took them to establish than by the energy available to them.

Experiments in Dorrigo National Park

What would happen if there had only been a very small supply of litter to start with? Bert Jenkins

Experimental potted webs in rainforest in Dorrigo National Park

tested this by putting out more experimental potted webs in rainforest in Dorrigo National Park in north-eastern New South Wales. He also used three levels of input of leaf litter, one of which was about the same as the natural level in Dorrigo tree holes. But where the lowest level of litter used in the first experiment was only about half the natural average, the second experiment was designed to use levels of 10 times less and 100 times less than natural. The second experiment ran for a year rather than six months to make sure that complex food webs had enough time to establish.

At the end of this second experiment both the numbers of species in the potted webs and the length of their food chains were less in the low-litter pots; in other words, when the energy supply is really low, energy can limit food web structure. However, since the energy supply in this experiment was probably unnaturally low, it is also unlikely that energy is a limiting factor at the base of tree hole food webs in the real world.

Given that only a small proportion of the energy entering a trophic level is transferable to the level above, then, at some stage, lack of available energy must set an upper limit to the number of possible trophic levels that can occur. Given also that the energy base of different ecosystems (determined by their net primary productivity) is enormously variable (Fig. 44.5), then we should expect enormous variation in the lengths of food chains. However, the observed number of trophic levels is uniformly small, regardless of the ecosystem. Something other than energy must limit the length of food chains.

As yet, there is no clear consensus among ecologists about what causes food chains to be short, or about the best explanation for other common patterns in food webs. Most of the patterns are consistent with at least two different sets of mathematical models. One set of models emphasises ecosystem stability: these models assume that the populations of organisms in a food web should return to stable equilibria after disturbance and that populations able to return rapidly to equilibrium should be more likely to persist in variable environments. These models predict that systems with too many linkages (Fig. 44.12a), too many omnivores (Fig. 44.12b) or too many interactions between distant trophic levels (Fig. 44.12c) are less likely to persist over time and through disturbances than are simpler systems.

Another set of mathematical models emphasises more static patterns, particularly in size of organisms: they assume a cascading hierarchy in feeding links; and

that a given species can feed only on those species below it and can itself be fed upon only by those species above it in the hierarchy. Both sets of models constitute reasonable working hypotheses but more data on natural and experimentally manipulated food webs are needed before a unified theory explaining food web patterns can be expected to emerge.

> Food chains are usually short and many food webs share other patterns that may be related to logistic constraints, body size and/or ecosystem stability.

Biogeochemical cycles

Energy flows through the biosphere. It is continuously radiated by the sun and, despite the temporary divergence of a small fraction of solar energy into ecosystems, most is eventually dissipated and lost as heat (Fig. 44.1). The inorganic materials necessary for life, however, are not continuously produced. Instead, the elements from which they are synthesised are withdrawn from world atmospheric and geologic stores to which they are eventually returned. As we have already learnt, the productivity of ecosystems is limited at least as much by the supply of materials as of energy. Thus, the efficiency of cycling of materials (**biogeochemical cycles**) is a major determinant of ecosystem structure and function.

For materials actively cycled through the atmosphere (e.g. water, carbon, nitrogen and sulfur) it is often useful to think of **global cycles** operating on a world scale because atmospheric circulation patterns result in effective transfers between ecosystems. Disturbances of these cycles can have drastic effects far removed from the source of the disturbance.

Other minerals, such as phosphorus, potassium, calcium and magnesium, generally undergo far less mixing; their movements are often considered as **local cycles** because they operate mostly at the local, rather than the global, scale. However, because minerals are transferred between local ecosystems, the availability and efficiency of recycling of minerals is highly variable within any one ecosystem. Accumulation of minerals in the biotic part of an ecosystem often occurs during regeneration (e.g. after a forest fire) or succession (Chapter 43), when productivity is high and biomass is accumulating. A small depletion of minerals (negative balance between inputs and outputs) may result from normal weathering processes in a mature ecosystem. A large deficit in minerals indicates a disruption of the usual ecosystem processes. Energy flows through the biosphere but materials are recycled within it. Materials cycled through the atmosphere circulate in global cycles. Other materials cycle mostly in local ecosystems.

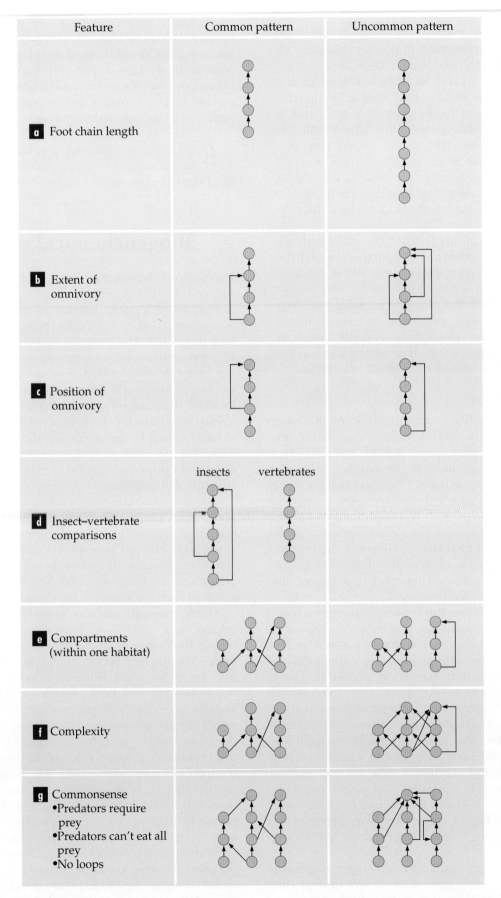

Fig. 44.12 Models of common and uncommon patterns in food webs

Water

We saw at the beginning of this book that water is an essential and major component of all life. The bulk (97%) of water on earth resides in oceans, where it is directly accessible to marine life. Water used by terrestrial life (less than 0.5% of the total) circulates between the oceans and atmosphere through processes of evaporation and precipitation. Within the atmosphere it circulates by convection, resulting in some transfer to land surfaces as precipitation. Soil moisture and ground water constitute two substantial pools (Table 44.1). Water is cycled through these pools in three ways: by deep drainage from the soil moisture into the ground water; by direct evaporation from the soil surface into the atmosphere; and by transpiration through plants into the atmosphere. Eventually, most of the water precipitated over land is transferred back to the oceans through processes of run-off and stream flow (Fig. 44.13; Table 44.1).

Within this larger **water cycle**, subcycles also operate. Some water is returned from the land surface directly to the atmosphere in the processes of

Table 44.1 Active and storage pools within the global water cycle

Pools	Pool capacity (10^{12} moles)
Active	
Atmosphere	720 000
Soil moisture	3 700 000
Stream channels	69 000
Freshwater lakes	6 900 000
Saline lakes	5 800 000
Oceans	74 000 000 000
Total active pools	74 017 189 000
Storage	
Icecaps and glaciers	1 600 000 000
Ground water	460 000 000
Total storage pools	2 060 000 000

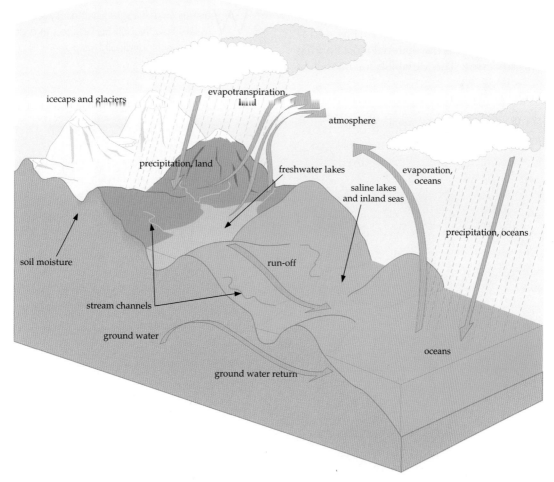

Fig. 44.13 The global water cycle

transpiration and respiration. Some resides in non-oceanic pools as either soil moisture or freshwater bodies; it is from these stores that terrestrial organisms derive the water that they require. Most of the world's water circulates between active pools that are theoretically available to life, but a small fraction (less than 3%) is in relatively accessible storage pools, such as icecaps, glaciers and deep ground water.

Water is not equally available in all ecosystems. At a local ecosystem level, water behaves more like energy than most other materials in that it effectively flows through systems rather than being recycled within them. The amount of water recycled from plants and animals back to soil and local water bodies is negligible and relatively little evaporated or transpired water is recycled locally. Most of it is lost from the local ecosystem to the atmosphere to be precipitated in a distant ecosystem. Thus, for local ecosystems, water is essentially a non-cyclable, periodically exhaustible resource, replenished only by new input.

> Water cycles globally, with the ocean as an enormous active pool. Within local ecosystems, its movement is effectively a one-way flow that is replenished only by new input.

Australia is particularly dry by world standards. Two-thirds of mainland Australia is desert (Fig. 44.14) and in many regions water is the material most limiting for ecosystem processes. The variability of rainfall in Australia (except for a narrow zone along the southern coast) is also particularly high by world standards. This has major implications for the ways in which many Australian ecosystems function. The flow of energy into primary producers occurs via photosynthesis, with water loss an inevitable consequence. Thus, primary productivity in water-controlled ecosystems is closely coupled with water availability. This means that desert primary producers effectively operate as converters of water flow to energy flow.

The seasonal rate of energy flow can usually be regarded as fairly constant for most ecosystems. However, because desert rainfall is infrequent and unpredictable, the driving input into desert ecosystems, and therefore their productivity, is pulsed. Thus, rather than considering desert water cycles as systems of pools and flows, it may be more appropriate to consider them as systems of 'pulses and reserves' (Fig. 44.15). Consumers in these ecosystems must either adopt a pulse-and-reserve pattern themselves (e.g. insects such as grasshoppers), utilise the reserves of other organisms (e.g. seed or wood eaters) or adopt extremely flexible feeding habits, using whatever pulse or reserve is available (Box 44.3).

> Water moving through desert ecosystems does not usually flow through active pools in a steady state but instead moves episodically, as a series of pulses and reserves.

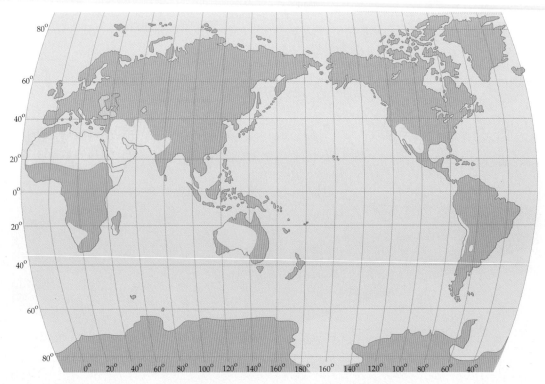

Fig. 44.14 Desert areas (yellow) of the world, showing the high proportion of Australia that is desert

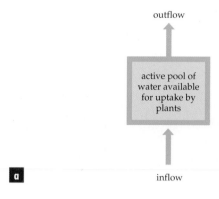

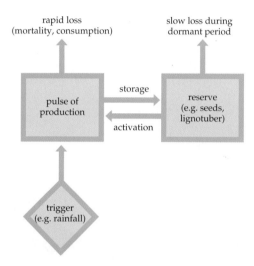

Fig. 44.15 Different ways of representing water flows through a desert ecosystem. **(a)** Level-and-pool model. The level of a resource is assumed to be controlled by uniform rates of inflow and outflow. Thus, there is always an active pool of resource available for uptake and use. **(b)** Pulse-and-reserve model. A rainfall 'trigger' sets off a pulse of production. Much of this pulse is rapidly lost to consumption and mortality but some is diverted back into storage reserves such as seeds or lignotubers. The reserve compartment diminishes only slowly during the no-growth period and from it the next pulse is activated

Trees—top water users

Throughout the world, including Australia, the water cycle has been altered by the clearing of trees and woody vegetation for agricultural uses. Before clearing, deep-rooted perennial plants were major users of the soil moisture that accumulated after rain. The roots of woody plants penetrate several metres into the soil. Normally the water taken up by such roots moves into the above-ground part of the plant and is transpired through the leaves into the atmosphere during photosynthesis. Woody plants will grow under natural conditions to densities at which virtually all the soil moisture is utilised and subsequently transpired. Only tiny proportions of the soil moisture pool drain so deep as to enter the ground water pool. It is not surprising, then, that after clearing more moisture drains deep into the soil and goes into the ground water. The water

infiltrates beyond the root zone of the annual grasses constituting pasture or crops, and 'recharges' the ground water so that it begins to rise through the soil profile towards the surface.

Figure 44.16 shows a dramatic case from a catchment in south-western Australia where ground water has risen more than 20 m through the soil to the surface over a period of 20 years since clearing. The ground water discharges as surface seeps and as drainage into streams. This rising water also can bring salt from the subsoil to the soil surface, leading to salinity problems and degradation of land (see Chapter 45).

Carbon

Most of the world's carbon, unlike its water, exists in relatively inaccessible storage pools; the amount circulating between active pools is only about one five hundredth of this store (Fig. 44.17; Table 44.2). The vast bulk occurs as carbonate in rocks (e.g. chalk, limestone, marble) and as fossil fuels (e.g. coal, oil, natural gas) formed from carbon-rich sediments, with the amount in rocks far outweighing that in fossil fuels (Table 44.2). Although the amount of carbon in the atmosphere is tiny (the concentration of carbon dioxide is only about 0.03% of total atmospheric gases), the atmospheric pool is the most active. In the **carbon cycle**, carbon dioxide is withdrawn from the atmosphere during photosynthesis (at a rate determined by the primary productivity of ecosystems) and returned by cellular respiration, mostly by decomposers. Because it is readily soluble, atmospheric carbon dioxide also dissolves in the oceans, where the largest active pool resides and where it is available in solution to marine plants. Oceanic and atmospheric pools are generally in equilibrium and rates of vaporisation tend to balance rates of solution.

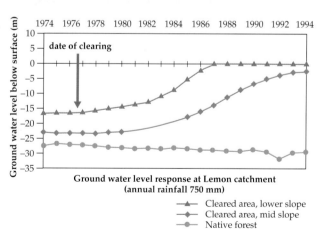

Fig. 44.16 Hydrographs showing rising ground water levels in the lower slope and mid slope of cleared land in the Lemon catchment, Western Australia. The catchment was cleared in 1976. Note that water levels have fallen in forested areas in response to low rainfalls during the period of monitoring

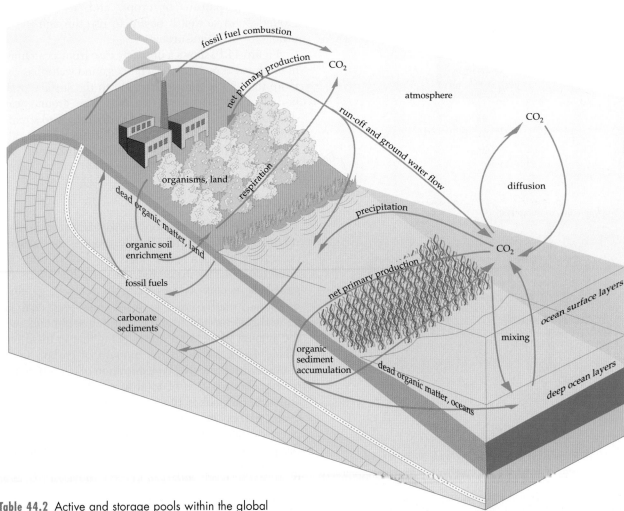

Fig. 44.17 The global carbon cycle

Table 44.2 Active and storage pools within the global carbon cycle

Pools	Pool capacity (10^{12} moles)
Active	
As CO_2 or HCO_3^- in solution	
Atmosphere	58 000
Ocean surface	43 000
Ocean depths	2 900 000
As organic carbon	
Land organisms	38 000
Decaying land organic matter	58 000
Marine organisms	830
Decaying marine organic matter	250 000
Total active pools	3 347 830
Storage	
Carbonate sediments	1 700 000 000
Fossil fuels	830 000
Total storage pools	1 700 830 000

Exchanges do occur between active and storage carbon pools but, until recently, only at rates measurable on geologic time scales. Much of the carbonate in freshwater and oceanic sediments eventually becomes lithified into carbonate rocks and fossil fuels. Carbon returns from storage very slowly during the weathering of carbonate rocks but much more rapidly during the combustion of fossil fuels. Since the Industrial Revolution, fossil fuels have been mined and burned at unprecedented rates, with the result that carbon dioxide is currently being returned to the atmosphere faster than it can be cycled, resulting in a net increase in the concentration of atmospheric carbon dioxide. Some of the implications of this for world climates are discussed in Chapter 45.

Carbon cycles globally through a small but active atmospheric pool from which it is withdrawn by photosynthesising organisms. Most of the world's carbon resides in storage pools in rocks and sediments.

BOX 44.3 Termites—outstanding nutrient cyclers

In moist soils, decomposer organisms such as soil bacteria, fungi and earthworms play a critical role in nutrient recycling. But what happens when soils are too dry and too hot to support active populations of soil organisms? Termites become important! Although termites are commonly called 'white ants' they are not ants at all. They belong to the insect order Isoptera and are more closely related to cockroaches than to ants. Termites are social insects and live in nests (see figure). Termites are able to eat cellulose—a resource not accessible to most other organisms—because their guts contain a variety of symbiotic microorganisms that can break down this compound. Grass, herbs, plant litter and wood are eaten in differing proportions by different species of termites.

Although much of inland Australia is both infertile and dry, plant biomass is often surprisingly high. This is because perennial grasses are able to establish during infrequent but heavy rains. For much of the time this biomass is unavailable to most herbivores because it has insufficient water and nutrients for their requirements. It is, however, an abundant food resource for termites, which also have the advantage in arid regions of the insulating properties of their nests, their co-operative social structure and their ability to store food harvested during pulses of plant production for use in leaner times.

In Australia, the largest numbers and majority of species of termites are found in subtropical and tropical sclerophyll forests, woodlands and savannas. In these habitats their numbers may be extraordinarily high. For example, in lightly timbered pastoral country in Queensland, where cattle are grazed at about one beast to 25 ha (about 24 kg of cattle/ha), CSIRO scientists calculated that the biomass of termites may be as high as 40–120 kg/ha. These scientists estimated that decomposition of plant litter by termites may be responsible for recycling up to 20% of the ecosystem's carbon. Organic matter that termites bring to their mounds is relatively rich in most nutrients and, as a result, termite mounds contain much more nitrogen, phosphorus and other valuable nutrients than do the surrounding soils. Although it may take as long as 30 years, this nutrient-rich mound material is eventually redistributed to surface soils, at rates equivalent to adding 3000 kg of fertiliser per hectare every year.

Termite mounds in subtropical woodland, northern Queensland

Nitrogen

Although nitrogen is the most abundant gas in the atmosphere (78%), the functioning of many ecosystems is limited by the availability of the element. This is because most plants cannot absorb atmospheric nitrogen but obtain their nitrogen as ammonium or nitrate ions from symbiotic soil microorganisms or from soil solution (Chapter 18). Since most animals obtain their nitrogen requirements by eating plants (or animals that have eaten plants), most of the nitrogen requirements for the biotic components of ecosystems therefore depend on active pools of nitrogen cycling in soil (Table 44.3).

In the **nitrogen cycle**, the concentration and composition of soil nitrogen pools are determined mainly by the activities of soil microorganisms (Box 44.4; Fig. 44.18). Some atmospheric nitrogen is also fixed during thunderstorms when nitrogen is converted to oxides by lightning and washed from the

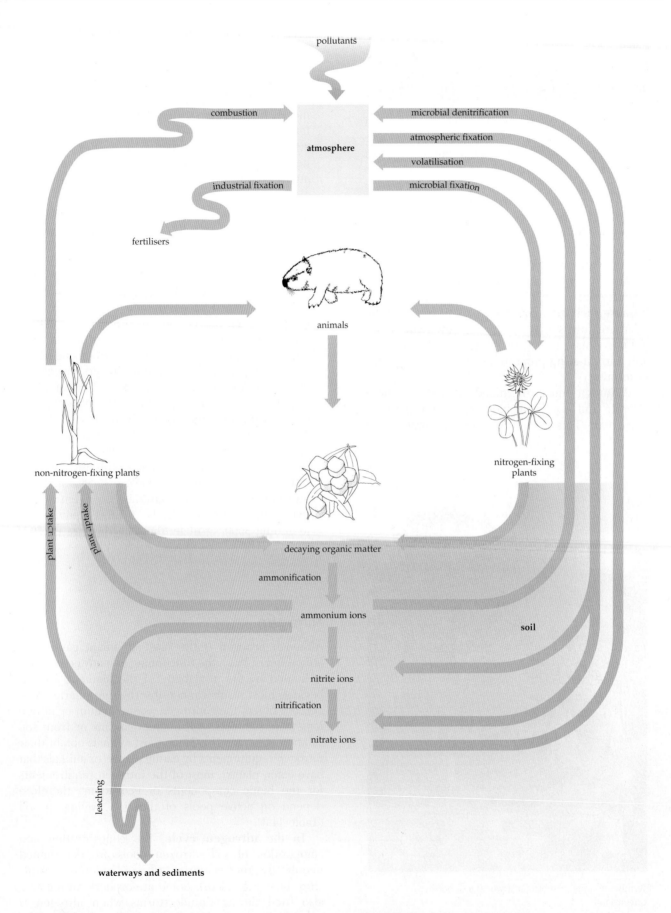

Fig. 44.18 The global/local cycle of nitrogen in a terrestrial ecosystem. Wavy lines indicate distant sources and sinks

Table 44.3 Active and storage pools within the global nitrogen cycle

Pools	Pool capacity (10^{12} moles)
Active	
As organic nitrogen	
Land organisms	870
Marine organisms	69
Decaying organic matter	120 000
As inorganic nitrogen	
Soils	10 000
Marine sediments	7 100
Total active pools	138 039
Storage	
As N_2	
Atmosphere	270 000 000
Oceans	1 400 000
As rock-forming minerals	
Sedimentary rocks	290 000 000
Crustal rocks	1 000 000 000
Total storage pools	1 561 400 000

atmosphere by rain. An increasing amount is fixed industrially, in high-temperature reactions that convert atmospheric nitrogen to chemical fertilisers. A variable amount is lost from local ecosystems by volatilisation of organic wastes. This amount can become major during bushfires and from stockpiles of animal excreta accumulated in intensive livestock operations. Some volatilised and ashed compounds may be deposited later, often in distant ecosystems, as atmospheric fall-out, but much is lost to the atmospheric store. Some industrial processes, particularly automobile use, also contribute volatile nitrogen compounds (particularly oxides) to the atmosphere, which may be precipitated elsewhere as fall-out.

Because the total soil nitrogen pool is so highly dependent on soil microorganisms, nitrogen is most likely to be limiting in environments in which microbial activity is inhibited (e.g. in cold or dry regions) or in soils with low organic contents (e.g. sandy soils). Any ecosystem disturbance large enough to modify soil microclimates or destroy organic matter will have drastic effects on the availability of nitrogen in the ecosystem. For example, cutting and burning of tropical forests completely changes the living microbial community that fixes and conserves soil nitrogen. Ammonium, which would be taken up rapidly by the trees in an unlogged forest, is converted to nitrate; crop

species use only a small fraction of this nitrate and the rest is lost through leaching or denitrification.

S oil microorganisms are the main agents of fixation of atmospheric nitrogen, mineralisation of nitrogen in organic matter and denitrification.

BOX 44.4 Microorganisms drive the nitrogen cycle

Microorganisms causing **nitrogen fixation**, the conversion of unavailable atmospheric nitrogen to forms available for plant uptake, include:

- symbiotic nodule-forming bacteria (mostly *Rhizobium* spp.) on roots of agricultural legumes (e.g. clover, lucerne, beans) and non-agricultural legumes (e.g. *Acacia*, *Cassia*)
- symbiotic nodule-forming bacteria (mostly *Frankia* spp.) on roots of non-leguminous plants (many genera, including she-oaks, *Casuarina* (figure), and alders, *Alnus*)
- symbiotic cyanobacteria, including those in lichens, liverworts, some ferns and root associations on cycads (*Macrozamia*) and a few other genera
- free-living bacteria and cyanobacteria (many genera; low fixation rates but total contribution may be substantial).

Microorganisms causing **nitrogen mineralisation**, the conversion of organic nitrogen compounds in humus to simple ions in soil solution, include:

The roots of *Casuarina* trees are associated with nodule-forming, nitrogen-fixing bacteria

- ammonifying bacteria and fungi, which release ammonium ions by breaking down decaying organic matter in soil
- nitrifying bacteria, chemautotrophs that convert soil ammonium to nitrite (*Nitrosomonas*) and nitrite to nitrate (*Nitrobacter*).

Microorganisms causing **denitrification**, the conversion of nitrate ions to gaseous nitrogen compounds released to the atmosphere, include:

- denitrifying bacteria, anaerobic organisms that are most active in soils with poor aeration.

BOX 44.5 Too much of a good thing: nitrogen and tree dieback

Tree dieback occurs whenever trees are subjected to long or repeated periods of sublethal stress. Sometimes the cause of stress is well understood, although nonetheless damaging and difficult to reverse. For example, jarrah dieback, which affects *Eucalyptus marginata* forests in Western Australia, is primarily caused by the destructive introduced soil-borne pathogen, *Phytophthora cinnamomi* (Chapter 35).

Dieback can be the result of a number of interacting factors. This is the case for **rural dieback**, which is the premature and relatively rapid decline and death of native trees (usually eucalypts) in rural Australia. Rural dieback is connected with changes that agriculture and pastoralism have wrought but the specific factors involved vary in importance from region to region. In the New England and Southern Tablelands of New South Wales, leaf-feeding insects are an important part of the dieback equation (Fig. a) but are not the whole story since the same species of insects also feed on trees growing in natural woodlands where rural dieback does not occur.

Insect damage is probably just one link in a long chain of events that are still not well understood. Nitrogen cycling may also be a crucial factor. Grazing is intensively managed on the Tablelands. Since the 1950s, farmers have been steadily improving the fertility of their pastures by planting nitrogen-fixing legumes and spreading fertilisers, such as superphosphate. This has enabled pastures to carry more and better sheep and has increased wool production.

Sheep are not the only herbivores that benefit from pasture improvement. White curl grubs, which are the larvae of beetles belonging to the scarab family, live in soil and feed on grass roots and other organic matter. Therefore, improving the quantity and quality of pasture available to sheep may also unintentionally benefit these unseen underground herbivores. In some improved pastures, the weight of scarab larvae grazing on pasture roots may actually equal the weight of sheep grazing the tops. This is bad news for the farmer because feeding by scarabs damages and sometimes kills pasture. It is even worse for eucalypt trees growing among or adjacent to pasture. If the weather is favourable during December, swarms of large golden 'Christmas beetles' emerge from pupae of scarab larvae and fly to the nearest eucalypt trees, where they feed voraciously on foliage, sometimes defoliating whole trees. By late summer all the adult Christmas beetles mate, lay their eggs in the soil under the trees and die, giving the trees a chance to produce new foliage to carry them through the winter.

(a) Rural dieback of eucalypts in the New England Tablelands of New South Wales

The trees' problems are not restricted to Christmas beetles. Increased soil fertility, as well as improving the protein content of pasture plants, also enhances the protein content of the foliage of trees growing among the pasture. If sheep congregate under trees, nutrient enhancement of foliage may become very marked as concentrated nutrients are applied to the soil in the form of urine and dung. This means that all the insect herbivores of trees (beetles, caterpillars, leaf hoppers, stick insects and many others) will benefit from feeding on fertilised foliage in the same way that sheep and scarab larvae benefit from fertilised pasture. This is probably why trees under which sheep graze may carry as much as ten times the weight of insects as trees in more natural woodlands, and why much nitrogen creates unexpected problems for eucalypts in rural Australia (Fig. b).

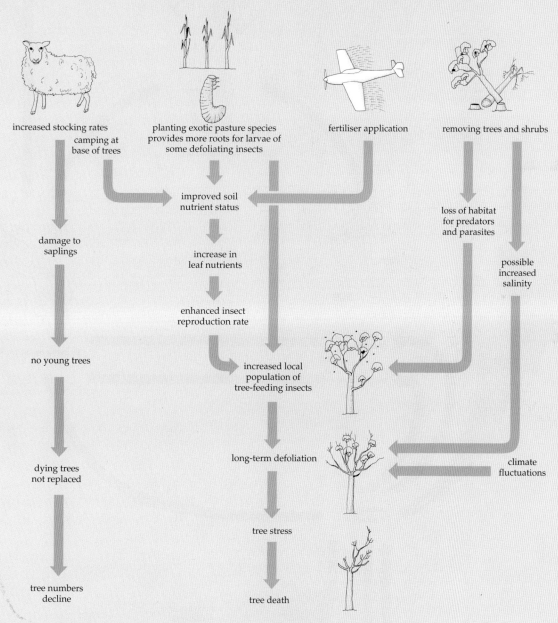

(b) Rural dieback is the result of a complex array of interacting factors

Phosphorus

Phosphorus is essential for life because it is used by all organisms in the form of ATP, but its supply often constrains ecosystem functioning because it is not a common element in the earth's crust. The pathways and storage of elements that are cycled less actively, such as phosphorus, operate through essentially similar processes to those seen for carbon and nitrogen, although the size of storage compartments and rates of flow vary between elements and between ecosystems (Fig. 44.19).

In the **phosphorus cycle**, the rate of phosphorus input from the atmosphere is very low compared with most other nutrient elements, so the supply in soil is precious. Thus, it is important that phosphorus is far less mobile in soils than most other nutrients. Soluble

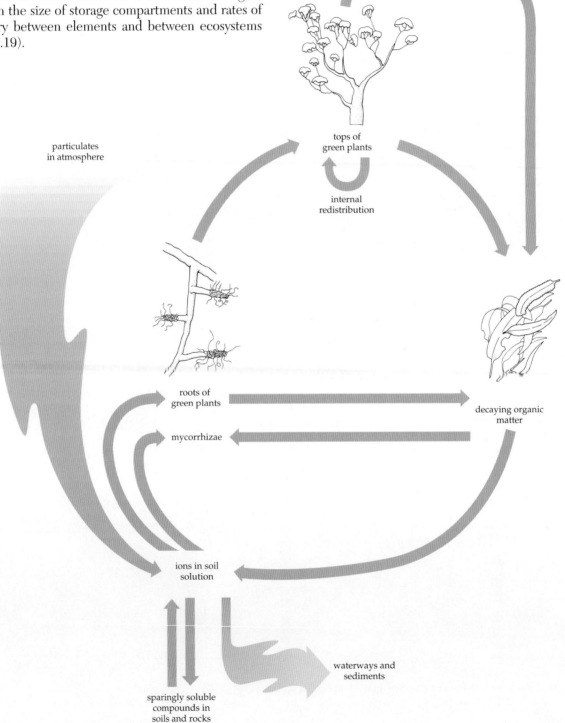

Fig. 44.19 General diagram of the local cycle of a material, such as phosphorus, that is not actively cycled in a terrestrial ecosystem. Wavy lines indicate distant sources and sinks

phosphate ions are readily available to plants. Sparingly soluble phosphates (believed to be ions bound to the surfaces of iron, aluminium and calcium salts) are released slowly into the soil solution and phosphates incorporated into clay minerals or concretions of iron or aluminium are immobilised and largely unavailable for plant growth. During the weathering of soil minerals, an increasingly large proportion of phosphorus becomes immobilised. Phosphorus immobilisation is most extreme in soils of the tropics that experience intense weathering. Because most Australian soils have been subjected to weathering for a long time (Chapter 41), their content of available phosphorus is extremely low by comparison with soils of other continents. This is why most non-native crops in Australia require the addition of large quantities of superphosphate fertiliser to achieve adequate yields. Native plants, however, are adapted to make the most of the naturally available phosphorus levels in Australian ecosystems (Chapter 18).

Associations between plant roots and mycorrhizal fungi are very common in plants on infertile soils. In addition to enhancing the rate of absorption of nutrients, mycorrhizae probably also play an important role in nutrient cycling. The fungal hyphae readily penetrate litter and compete with other decomposers for the utilisation of phosphorus and other nutrients, thereby short-cutting the labile pool of soluble soil nutrients (Fig. 44.19). Efficient extraction from soil and litter, a low absolute requirement for phosphorus, and efficient use and recycling of phosphorus within biomass, are adaptations that allow plants of low phosphorus ecosystems to achieve remarkably high rates of productivity. For example, on sand islands near Brisbane, highly productive eucalypt forests grow on some of the most phosphorus-deficient soils in the world. Most of the phosphorus requirements for primary production are stored and recycled within the living biomass.

Because the productivity of low phosphorus ecosystems usually depends on biomass stores of phosphorus, these ecosystems are easily disrupted. When tropical rainforests growing on extremely infertile soils are cleared, agriculture is possible for the first few years because of mobilisation of phosphorus stores in organic matter and ash from the original forest. Eventually, however, surface organic matter decomposes, soil nutrients are leached away, phosphorus becomes immobilised, crop productivity fails and the rainforest is replaced by far less productive open grasslands (Fig. 44.7).

The main inputs of phosphorus into local ecosystems are from sparingly soluble soil storage pools. In low phosphorus systems, vegetation constitutes the largest active pool of phosphorus and is efficient at internal recycling.

Summary

- Ecosystems are dynamic, self-modifying ecological systems consisting of interacting biotic and abiotic components.
- Trophic interactions are probably the most important relationships determining ecosystem structure and functioning. All organisms can be considered as either autotrophs (producers) or heterotrophs (consumers, degraders and decomposers). Graphs of the energy flowing into successive trophic levels represent ascending ecological pyramids but pyramids based on numbers or biomass may sometimes be inverted.
- The amount of energy flowing through any ecosystem is determined by its net primary productivity, which is also limited by availability of materials. The productivity of forests and freshwater swamps is generally higher than that of grasslands and streams, which exceed that of oceans and deserts, but there are many exceptions to these trends. Net productivity also depends on the duration and season of measurement, and on the age of the ecosystem.
- Food webs are diagrams depicting trophic interactions between individual organisms in an ecosystem. Although food webs appear complex, their component food chains are usually short. Many food webs share other patterns that may be related to logistic constraints, body size and/or ecosystem stability.
- Water cycles globally, with the ocean as an enormous active pool. Within local ecosystems its movement is effectively a one-way flow that is replenished only by new input. Much of Australia is desert, characterised by low and highly variable rainfall. Water moving through desert ecosystems does not usually flow through active pools in a steady state but moves episodically, as a series of pulses and reserves.
- Carbon cycles globally through a small but active atmospheric pool from which it is withdrawn by photosynthesising plants. Most of the world's carbon resides in storage pools in rocks and sediments; post Industrial Revolution combustion of the fossil fuel fraction of these pools has resulted in a global increase in the atmospheric pool of carbon.
- The nitrogen cycle is dependent on local soil and microbial pools. Soil microorganisms are the main agents of fixation of atmospheric nitrogen, mineralisation of nitrogen in organic matter and denitrification, by which soil nitrogen is returned to the atmosphere. External inputs of nitrogen in plant-available forms can cause major ecosystem disturbance.
- The main inputs of phosphorus into local ecosystems are from sparingly soluble soil storage pools. In low phosphorus systems, such as most tropical rainforests and most Australian ecosystems, vegetation constitutes the largest active pool of phosphorus and is very efficient in its uptake, use and internal recycling.

keyterms

autotroph (p. 1172)
biogeochemical cycle (p. 1181)
biomass (p. 1173)
carbon cycle (p. 1185)
consumer (p. 1172)
decomposer (p. 1172)
degrader (p. 1172)
denitrification (p. 1190)

detritivore (p. 1172)
detritus food chain (p. 1177)
ecological pyramid (of numbers, biomass or energy) (p. 1172)
ecosystem (p. 1172)
food chain (p. 1177)
food web (p. 1177)
global cycle (p. 1181)

grazing food chain (p. 1177)
heterotroph (p. 1172)
local cycle (p. 1181)
net primary productivity (p. 1174)
nitrogen cycle (p. 1187)
nitrogen fixation (p. 1189)

nitrogen mineralisation (p. 1189)
omnivore (p. 1172)
phosphorus cycle (p. 1192)
productivity (p. 1174)
rural dieback (p. 1190)
scavenger (p. 1172)
trophic level (p. 1172)
water cycle (p. 1183)

Review questions

1. Explain the difference between the terms 'population', 'community' and 'ecosystem' (see also Chapters 42 and 43).

2. Explain the difference between the ecological meaning of the terms 'biomass', 'energy' and 'productivity'.

3. An aquatic biomass pyramid with phytoplankton at the base is an inverted pyramid because the phytoplankton (producers):

   A are small organisms

   B only occur in small populations

   C do not photosynthesise sufficiently to produce much energy

   D have short life spans and are constantly being replaced

4. Food chains typically have three or four trophic levels.

   (a) What is a trophic level?

   (b) Why are food chains short?

5. Energy and nutrients move differently through ecosystems.

   (a) What are these differences?

   (b) What is the consequences of these differences?

6. The largest storage pool of water within the global water cycle is:

   A icecaps and glaciers

   B the atmosphere

   C oceans

   D ground water and soil moisture

7. In what way(s) have industrialised societies affected the global carbon cycle?

8. Name the processes in the nitrogen cycle that are controlled primarily by soil microorganisms.

Extension questions

1. Which ecosystem would you expect to have the highest annual productivity: a coastal rainforest north of Cairns in Queensland or a monsoonal rainforest in Kakadu National Park in the Northern Territory? Why? What ecosystem properties would you need to know to test your prediction?

2. Construct a food web showing possible feeding interactions between cattle, grass, trees, wood-feeding termites, harvester termites, scavenger termites, legless lizards, kookaburras and humans. What are the least, most and average number of links in the food chains you have drawn?

3. Design an experiment to test whether the amount of leaf litter that accumulates in water-filled tree holes depends on their depth. If your experiment showed that very shallow tree holes accumulated little leaf litter, what consequences might this have for the development of food webs in very shallow tree holes?

4. What are some advantages harvester termites might have over kangaroos as primary consumers of semiarid rangelands? What are some possible disadvantages? Which represents more energy: 100 g of termites or 100 g of kangaroo?

5. Under what circumstances might you expect there to be limited soil nitrogen available for plant uptake?

6. Which fertiliser is most commonly used by Australian farmers? Why do Australian farmers use relatively more of this fertiliser than farmers in Europe and North America? What sorts of problems could result from excessive reliance on fertilisers?

7. Why does rainforest in many tropical regions fail to regenerate after widespread clearing?

Suggested further reading

Attiwill, P. M. and Adams, M., eds. (1996). *Nutrition of Eucalypts*. Melbourne: CSIRO Australia.

This book has chapters on soil phosphorus levels, biomass production and nutrient cycling in Australian forests. The use of fertilisers in forests in a range of countries, particularly in the Southern Hemisphere, is also reviewed.

Attiwill, P. M. and Leeper, G. W. (1987). *Forest Soils and Nutrient Cycles*. Melbourne: Melbourne University Press.

A useful reference text with emphasis on Australian ecosystems.

CSIRO. (1996). *Australia: State of the Environment*. Melbourne: CSIRO Publishing.

This book is a useful source of data and examples, and covers the atmosphere, land ecosystems, inland waters, estuaries and the sea. It documents and discusses the state of the environment and ecosystems in Australia and the issue of ecological sustainability.

Mooney, H. A., Cushman, J. H., Medina, E., Sala, O. E., Schulze, E. (1996). *Functional Roles of Biodiversity: A Global Perspective*. Chichester, UK: John Wiley & Sons.

This book considers the newer issue of food webs and nutrient cycling in relation to biodiversity.

CHAPTER

45

Human impacts

Understanding the principles of ecology that we reviewed in Chapters 42–44 is vital to the future of life on earth. The human population continues to grow rapidly and the rate of habitat removal and modification, with the consequential loss of species, is greater than in any other period of human history. In this chapter, we have chosen the important issues of loss of biodiversity, impact of introduced species, and land and water degradation to highlight the impact of the human species on ecosystems, and to consider positive solutions through the application of our knowledge of biology.

Humans in Australia

As we discussed in Chapter 41, the history of the Australian continent has dictated, to a large extent, the composition of the Australian biota. Isolated as an island continent after its separation from Gondwana, Australia evolved unique animals and plants. As the climate became more arid during the Tertiary period, the typically Australian element of the biota evolved under the new environmental conditions. However, once the continent was close enough to the South-East Asian plate it became possible for some organisms, for example, bats and rodents, to be exchanged between the island continent and the island archipelago to the north.

With proximity to islands and other land masses, it was also possible for people to enter northern Australia. The first humans arrived in Australia possibly as early as 100 000 years ago and their impact on the Australian environment, especially through the use of fire, continues to be widely debated (see Chapter 41). Europeans arrived more recently. Portuguese sailors drew maps in the 1500s that bear a striking likeness to the east coast of Australia, suggesting that they were the first Europeans to sight Australia. William Dampier, a Dutchman, is credited with the first landing in Western Australia during the late 1600s.

Sydney Cove was the site of the first permanent European settlement (Fig. 45.1) with the arrival of five small sailing ships carrying about 1500 people from Great Britain in 1788. The small fleet called at Cape Town, South Africa, for provisions and water. There the settlers obtained plant cuttings and seeds to plant in their new home. The fleet became a veritable Noah's Ark as they took on board sheep, goats, cattle, horses, pigs and poultry.

As well as alien animals and plants, the European colonisers brought with them European attitudes. Australia was seen as a wild and inhospitable place to be conquered and 'civilised'. In 1855, the explorer William Howitt said of Victoria:

Fig. 45.1 The first European settlement at Sydney Cove

the choked-up valleys, dense with scrub and rank grass and weeds, and the equally rank vegetation of swamps, cannot tend to health. All these evils, the axe and the plough, and the fire of settlers, will gradually and eventually remove; and when it is done here, I do not believe that there will be a more healthy country on the globe.

The impact of Europeans on the Australian environment was rapid and extensive. Open grasslands and grassy woodlands were regarded as productive country and quickly turned over to pasture for domestic stock. Heavily wooded country was extensively cleared for agriculture with little regard for the impact on the soil or native wildlife. Cities were built with little planning for effluent disposal and raw sewage was discharged directly into streams.

In Australia today, ecosystems most under threat from human activities are those in agricultural areas, although coastal and urban development is increasingly having an impact. Extinction rates and declines in abundance and range of native flora and fauna have been highest in regions where settlement first occurred (Box 45.1), attributable to inappropriate land and water use, habitat loss and fragmentation, over-exploitation of resources, and the spread of exotic herbivores, predators, weeds and disease.

BOX 45.1 The use of Australian forests

The management of Australia's native forests has attracted public attention and, at times, political controversy during much of the twentieth century. In 1998 the Commonwealth government changed the definition of forests to make it consistent with the definition used internationally. The change means that areas traditionally called 'woodland' in Australia are now called 'open forest'. Open forests generally occur in the drier inland and wet–dry north and mostly are not a viable economic resource for wood extraction and timber production. Rather, intensive forestry activities such as woodchipping are based on the wet eucalypt forests, including those found in south-eastern Australia and parts of south-west Western Australia. Below we discuss these wetter forests.

At the time of European settlement, Australia was covered with about 80 million ha of forest, about 10% of the surface area of the continent. Since then, about half of these forests, approximately 40 million ha, have been cleared, largely for agriculture.

European impacts on vegetation and fauna followed similar patterns across the geographic range of these forests, with initial settlement being followed by almost a century of largely uncontrolled clearing. The first export commodity of the new colony of New South Wales was red cedar, *Toona australis* (see Fig. a), discovered on the banks of the Hawkesbury River in 1790. In Victoria, the discovery of gold near Clunes in 1851 led to a rapid population increase. Forests were heavily exploited to provide timber and fuel for mines and towns.

The Conservator of Forests in Victoria reviewed the condition of the forests in 1896 and emphasised the 'forest vandalism' practised by miners, timber splitters and sleeper hewers who produced more waste than timber. Forest Commissions were established in most Australian states to halt the overuse and degradation of forests, and formal and relatively effective control of forest exploitation was in place by the early 1900s. The first timber reserves in New South Wales were set aside in 1871 to protect forest resources from uncontrolled exploitation. By 1920, most of the forest estate in New South Wales was established. In Tasmania, there was no effective legislative control over forest use until 1920.

The current situation

Of the 40 million ha of forest remaining in Australia, 26 million ha are publicly owned native forest. Of this, around 10 million ha (about 40%) is in national parks and flora and fauna reserves. Forest ecosystems make an important contribution to Australia's biodiversity, especially rainforests and old growth eucalypt forests. This has led to an ongoing debate about the continued logging of

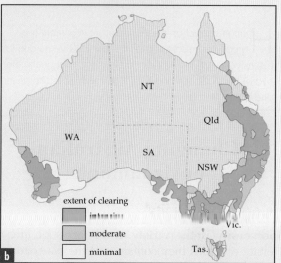

(a) Red cedar, *Toona australis*, grows in the rainforests of New South Wales and Queensland. Because it provides a fine furniture timber, it has been heavily harvested and is now a rare species in its native habitat. (b) This map shows where in Australia forest and woodland have been cleared to varying extents in the 200 years since European settlement; the remaining area is either naturally unforested (central Australia) or has not been affected by clearing (e.g. in south-west Tasmania)

native forests, especially when it has been demonstrated that plantation forests could provide most of the nation's domestic timber needs.

More than 7 million ha of native forest is managed for commercial forestry in Australia. In 1996–97, for example, about $97 million of round and sawn wood products, $516 million of wood-chips (6 million tonnes per year), $370 million of paper and paper products and $64 million of other forest products were exported.

In 1992, Australia's National Forest Policy set out broad conservation and industry goals for the management of Australia's forests agreed between Commonwealth, State and Territory governments. To implement this national policy, governments have opted for an approach where Regional Forest Agreements (RFAs) are negotiated between the Commonwealth and State governments about the long-term management and use of forests in a particular region. RFAs aim to provide a blueprint for the future management of native forests and the basis for an internationally competitive and ecologically sustainable forest products industry. Once an RFA is signed, there is no limit to the amount of woodchips that can be exported from the area covered by the agreement.

Fig. 45.2 The pig-footed bandicoot, *Chaeropus ecuadatus*, is now an extinct Australian mammal

Human impact on the Australian environment has greatly accelerated over the last 200 years due to extensive forest and woodland clearing, changes to the hydrological regime and the introduction of exotic animals and plants.

Changing Australian environments

Losing biodiversity

Biodiversity is the variety of all forms of life, the diversity of genes they contain and the ecosystems of which they are components. Genetic diversity within species is what allows populations to adapt to changes in climate and other local environmental conditions. Species diversity provides essential ecosystem services, functioning in soil formation, nutrient cycling, pollination of crops and maintenance of water cycles.

We learnt in Chapter 30 that species are becoming globally extinct at a rate of 1000 to 10 000 times the natural rate before human intervention and that it is likely that 20% of all species will become extinct within the next 30 years. Of the estimated 25 000 vascular plant species in Australia, 85% are endemic and over 1000 species are considered endangered or vulnerable. Of Australia's birds, mammals and reptiles, about 8% of taxa (145 species and subspecies) are considered endangered or vulnerable. Altogether 19 mammals (Fig. 45.2) and 21 birds are known to have become extinct, although the paradise parrot is the only *species* of bird that is thought to have been lost (Fig. 45.3). Aquatic organisms are also at risk. A high proportion of temperate marine organisms are endemic (see Chapter 41) and seagrass beds have declined significantly with the arrival of Europeans.

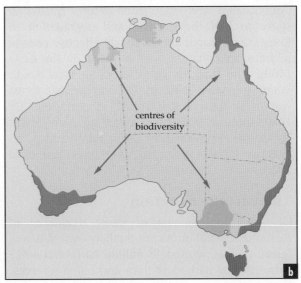

Fig. 45.3 (a) The paradise parrot, *Psephotus pulcherrimus*, is, to date, the only species of Australian bird thought to have become extinct. **(b)** Australia has a number of geographic areas (as shown here) that are recognised centres of species diversity on a world scale. Accelerated rates of extinction are expected for birds

Many species have undergone dramatic reductions in range and abundance. For example, eight mammal species that were once present on the mainland persist only on small offshore islands. Most of these mammals were from the arid zone where they were decimated by a combination of altered fire regimes and the introduction of feral animals that preyed on native fauna and competed with them for food and shelter, particularly during times of drought. Numerous bird species have also undergone major declines in their distribution and there is no reason to believe that any equilibrium has been reached between species conservation and broad-scale land uses such as agriculture and pastoralism. In fact, accelerated rates of extinction are expected, particularly among birds. It is predicted that if patterns of land use do not rapidly change, up to 50% of all terrestrial bird species in Australia will become extinct in the next 100 years.

Why conserve biodiversity?

Does it matter whether we lose species? There are a number of reasons why the answer is 'yes'. The first reason is an *ecological* one. Organisms play roles in ecological systems that are essential for ecosystem survival. Our failure to discern the role of a species within an ecosystem does not mean that the species is unimportant. Ecosystem processes, including species interactions (Chapters 43 and 44), are complex and loss of a key species may have a substantial impact. For example, in Chapter 43 we discussed the role of frugivores, such as cassowaries (Fig. 45.4), which eat the large fleshy fruit of some rainforest trees in northern Queensland. The seeds of these trees are still viable after passing through a bird's digestive tract. Because of this, cassowaries play an important role in the dispersal of these seeds and therefore the ecology

Fig. 45.4 It is essential to conserve organisms that play a pivotal role in ecosystems. The cassowary, *Casuarius casuarius*, is a keystone species (one that if removed would cause important changes in the structure and function of the ecosystem) in Queensland rainforests

of the trees. Cassowaries are thought to be the only species that are capable of eating many of these fruits, but the birds are threatened by rainforest clearing and feral animals. The loss of cassowaries from rainforests could have a substantial effect on the ability of many tree species to disperse and regenerate.

The second reason is a *practical* one: we use organisms for our own benefit (as bioresources) for food, fibre, medicines, beverages and so on. We need a diversity of species to maintain our quality of life. Since so much biodiversity remains undiscovered there must be equivalent quantities of bioresources still to be discovered. The potential of these undiscovered bioresources is being recognised and efforts are underway to identify uses for them. For example, it has been discovered that ants possess specialised glands for producing antibiotics to reduce disease in their colonies and that marine molluscs contain an antibiotic as effective as penicillin. These antibiotics hold great potential for therapeutic use.

The third reason for conserving biodiversity is one of *aesthetics*. We enjoy the beauty of plants and animals, their complexity, their appearance of wildness. There is even uncertainty about whether humans can actually survive psychologically, let alone physically, in a world with reduced biodiversity.

The fourth reason is a matter of *philosophy*. Should we allow a species to become extinct if we can save it? Humans are, after all, just another species. From an ethical point of view, other species have as much right to exist as we do. Some people believe in the maintenance of biodiversity on religious grounds.

The fifth reason is *custodial*. Countries conserve their natural and cultural heritage, which is passed down ('inherited') as a birthright.

On a world scale, about 1.4 million species have been described so far out of an estimated total of 10 million species. We know little about the genetic diversity of these species, their relationships or their biology, and there will never be sufficient time to describe all species. Consequently, in addition to efforts to describe individual species, classifications of communities and ecosystems are being developed based on the *functional* attributes of organisms. By dividing up the animals according to their feeding habits it is possible to work out their ecological relationships without naming all the organisms. For example, rainforest ecologists often use this approach because of the complexity of rainforest ecosystems. Keystone, indicator and vulnerable species can be identified and action taken to conserve them.

Human survival and quality of life depend on conserving biodiversity in order to maintain ecosystem processes, provide food and resources, and maintain the beauty of the natural environment.

BOX 45.2 Clearing tropical forests

The diversity of species in tropical rainforests is unparalleled in any other ecosystem. They cover only 7% of the earth's land surface but they contain at least 50% of all species. Ecologists have recorded 700 tree species in ten 1 ha plots in a Borneo rainforest—equal to the entire diversity of trees in North America. Forty-three species of ant belonging to 26 genera were collected from a single tree in a Peruvian rainforest—about equal to the entire ant fauna of the British Isles. And in one study of just 19 trees in Panama, 80% of the 1200 beetle species discovered were previously unknown to science.

Tropical forests of the world (see Fig. a) are also shrinking faster than any other ecological biome on earth. For example, almost all of the decline in the area of the world's forest and other wooded land from 1980 to 1990 occurred in tropical countries. The conversion of forest to other uses such as cropland and shifting cultivation in these regions was around 100 million hectares—an area about the size of South Australia and more than the entire area of forest found by Europeans when they arrived in Australia.

It is not surprising then that the single greatest cause of species extinction in the next half-century at the global level is thought to be tropical deforestation. Scientists concur that roughly 5–10% of closed tropical forest species will become extinct *per decade* at current rates of tropical forest loss and disturbance. With more than 50% of species occurring in closed tropical forests and a total of roughly 10 million species on earth, this amounts to the phenomenal extinction rate of more than 100 species per day.

Disturbance in tropical rainforests

Disturbance is a natural process in rainforests, as it is in most forests. The most common natural disturbances in Amazonian rainforests are tree falls and forest fires. Gaps created by disturbance provide increased light, water and nutrients, and the forest rapidly recovers through a process of succession, the progressive replacement of one group of species by another (Chapter 43). Rainforests are dynamic ecosystems made up of a mosaic of vegetation patches at various stages of succession.

Rainforest ecosystems are also characterised by rapid nutrient cycling (Chapter 44). Organic material that falls to the forest floor is rapidly

(a) Tropical rainforests are being lost at a rate faster than any other world biome. This forest in Borneo is home to the orang-utan

decomposed by fungi, invertebrates and bacteria in the litter and top layer of soil. Nutrients, which are released by the decomposition process, are mopped up by a network of plant roots that form a mat at the top of the soil. In this way, nutrients are returned to the living biomass without entering the mineral soil and are not subjected to leaching by abundant rainfall. On average, the above-ground biomass in rainforest contains approximately one-half of the total nutrients in the system. This contrasts with most other terrestrial vegetation types where the soil is the major nutrient store, and has important consequences in relation to human disturbance, which is typically more intense and more prolonged than natural disturbance.

The traditional agriculture practised by the indigenous peoples of the Amazonian rainforests is **slash-and-burn agriculture** (Fig. b). This involves clearing a small area of forest (usually about 1 ha) and burning the fallen timber. This releases abundant nutrients from the ashed plant material, enough to support an agricultural crop for two to three years. However, once this supply is exhausted, the soils are too poor to support further agriculture and the farmers move on to another patch of forest.

The re-establishment of rainforest trees on slash-and-burn sites is dependent on dispersal of seed into the clearing from surrounding forest. Some 'pioneer' species, that is, species that colonise disturbed sites, would naturally germinate

(b) Disturbance by slash-and-burn clearing for small-scale agriculture in an Amazonian rainforest

from the soil 'seed bank' where they may have been dormant for many years. However, the constant weeding of farm plots severely depletes the soil seed bank. The seeds of many rainforest trees stay viable only for a short period, as little as six weeks, so they do not accumulate in the soil. These species must be introduced into clearings from surrounding vegetation. However, the clearings produced for slash-and-burn agriculture are large compared with natural gaps produced by tree falls. Many species are dispersed by animals that do not frequent large clearings, further slowing the process of rainforest regeneration. It may take 200 years or more for rainforest to re-establish on an abandoned farm plot in the Amazon Basin.

Much Amazonian rainforest is cleared on a larger scale for cattle pasture. As with slash-and-burn agriculture, pastures on cleared rainforest sites are only productive for a short period of time. The ground is compacted by cattle and seed dispersal must occur over large distances. Natural re-establishment of rainforest on these sites is extremely slow, if it happens at all.

Land clearance and habitat fragmentation

As a result of broad-scale land clearance and land use changes that have occurred since European settlement, much of the remaining native vegetation in southern and eastern Australia exists in small and isolated patches that are surrounded by agricultural, urban or other land use systems. In some areas, less than 5% of the original vegetation remains.

Because of these patterns of land use, **habitat fragmentation** is a major problem (Fig. 45.5). Large organisms are generally more vulnerable to fragmentation and many habitat remnants are too small to support viable populations of resident organisms (see Chapter 42). Thus, the probability of organisms becoming extinct is usually greatest in small patches, although the effects of fragmentation depend on the ecology and dispersal ability of organisms, the management history of the site and the variability of the environment.

The size of the vegetation remnant is important in terms of habitat, especially for vertebrate species. In the wheat belt region of Western Australia, which has been cleared extensively, the impacts of habitat change and loss have been shown to be greatest on resident birds dependent on native vegetation, with 72% of dependent species showing decline in numbers since the early 1900s. Even large remnants of habitat (> 80 ha) are often unable to retain the full compliment of native birds. For plants, smaller remnants can still provide important habitat and rare species have been found in quite degraded remnants in Tasmania.

Fig. 45.5 Habitat fragmentation is a major threat to biodiversity. **(a)** Narrow corridors of native vegetation, shown in the wheat belt of Western Australia, are the only remaining habitat for the survival of native species, such as **(b)** *Banksia cuneata*

The effective size of fragments can be increased by linking up adjacent fragments with **habitat corridors** of similar vegetation. Corridors have a number of potentially beneficial roles including: helping the movement of animals between patches, providing habitat for resident populations of animals and plants; and promoting the exchange of genes between subpopulations. However, corridors also have potential disadvantages, such as helping the spread of weeds, pest animals, diseases, deleterious genes and fires. Therefore, the advantages and disadvantages of establishing networks of wildlife corridors have to be carefully considered before human and financial resources are allocated to them.

Fragment shape is also important. The edge of a remnant fragment in an agricultural landscape experiences different environmental conditions from the interior of the fragment. Edges are typically windier and receive more light and inputs of nutrients from surrounding pastures and crops. Certain species of plant and animal are found more commonly in edge habitats compared with interiors of remnant fragments. A long thin remnant patch has a greater edge-to-area ratio than a circular patch. Corridors that consist entirely of edge habitat are thus unsuitable for the movement of interior species.

> The removal or fragmentation of habitat threatens the biodiversity in an area. The size and shape of remnant patches and whether they are linked by corridors are important for the survival of viable population sizes.

Introducing new species

Some of the greatest impacts that Europeans have had on the Australian environment have been caused by the introduction of new species (Fig. 45.6). Many species have been unintentionally introduced. In Chapter 35 we mentioned the cinnamon 'fungus' the oomycete *Phytophthora cinnamomi*, which was introduced into Australia in the nineteenth century, probably accompanying plant material brought in for agriculture. *P. cinnamomi* causes dieback disease (see Box 35.3), which affects a broad range of ecosystems, including heathlands, shrublands, woodlands and forest. It has been suggested that 20% of the 9000 angiosperms of south-west Western Australia are at risk from dieback disease, including more than 80% of all Proteaceae, which are particularly susceptible. At least 10% of the remaining jarrah forest in south-west Western Australia is infected with dieback disease. The extent of the threat of *P. cinnamomi* to the nation's biodiversity is recognised in the 1992 National Strategy for the Conservation of Biodiversity where it is the only pathogenic taxon specifically cited. The *Commonwealth Endangered Species Act* of 1992 also lists the

Fig. 45.6 Native heathland on French Island, Victoria, invaded by introduced cluster pine, *Pinus pinaster*

disease caused by *Phytopthora* as one of five 'key threatening processes'.

Great efforts are now being taken to prevent the introduction of foreign diseases into Australia. For example, for an animal to be legally introduced into Australia it must be kept in quarantine, out of contact with other animals and people, until it is confirmed that it is not carrying any foreign disease. Some animals must be kept in quarantine for years before they can be declared disease-free! The Australian Quarantine and Inspection Service has also formally adopted a new system, called the Weed Risk Assessment system, for assessing all new plant imports, whether seeds, nursery stock or tissue culture.

Impact of introduced animals

The introduction of many species has been quite deliberate. All our important crop plants and farm animals have been introduced from other countries. In 1861 the Victorian Acclimatisation Society was formed with the principal objective: 'the introduction, acclimatisation and domestication of all innoxious animals, birds, fishes, insects and vegetables whether useful or ornamental'. The key word in this statement is 'innoxious', meaning 'not harmful', but this is a matter of perception. These well-intentioned people, who wanted to make the Australian environment more like 'home' in Britain, were responsible for introducing two of Australia's most serious **pest** animals: European rabbits, *Oryctolagus cuniculus* (Fig. 45.7), and European foxes, *Vulpes vulpes*.

Feral animals are either domestic animals that have gone wild or those that were introduced for control of pests or for recreation. They may damage vegetation and soils, foul water or compete with native

Fig. 45.7 (a) One of Australia's most serious pest species is the introduced European rabbit, *Oryctolagus cuniculus*. **(b)** Native *Callitris* pines in the mallee ecosystem of Australia are killed by fire. Normally they would grow again from seed but the rabbit finds young pine seedlings palatable, preventing regeneration

mussel was described as the most dangerous thing ever to threaten Australia's marine environment, a creature with 'cataclysmic power' over both biodiversity and coastal industries. Given the black-striped mussel's tolerance of temperature and salinity, it could potentially infest a range of sea ports. A major program was put in place to try and eradicate the mussel, focusing on the application of large quantities of chlorine and copper sulfate. While no viable mussels are thought to have survived this treatment, there is a strong chance that it could reappear and a national task force has been formed to combat the pest.

Not all animals successfully establish and become invasive when introduced into a foreign environment. The success of establishment depends on the conditions of the ecosystem and the characteristics of the invading organism. Ecosystems susceptible to invasion are often disturbed and free of competitors and predators of the invading organism. Common characteristics of successful invaders are tolerance of a wide range of environmental conditions, high mobility, high fecundity (either producing many progeny in each reproductive cycle or reproducing frequently) and a broad diet.

Integrated pest management

Integrated pest management involves co-ordinated use of various control techniques and integrating control with other activities. Rabbits are a good example. The major focus of rabbit control has been **biological control** through the release of the myxomatosis virus and more recently the rabbit calicivirus (RCD), which is highly infectious to rabbits and can cause mortalities of about 95% among adults. The initial reduction in rabbit populations may be brought about by an effective outbreak of these viruses or by poisoning or drought. To keep populations low, however, follow-up techniques such as poisoning, warren ripping and fumigation are also recommended. Rabbit control may also be integrated with the control of other pest species, such as foxes and feral cats, and even plant weeds.

The cane toad: a control agent turned pest

Biological control continues to be an important part of pest management and there have been success stories such as the control of prickly pear infestations by the moth *Cactoblastis cactorum*. One of the critical tests for new organisms is whether they will affect other non-target organisms. Stringent tests are enforced so that biological control agents are effective and do not become pests themselves, as happened with the cane toad in the 1930s.

The introduction of the cane toad in 1935 into Queensland from South America to control two insect

animals for habitat and food. Apart from rabbits and foxes, feral animals in Australia include cats, pigs, goats, donkeys, camels, water buffalo, mosquito fish, *Asterias amurensis* (northern Pacific sea star), honey bees and cane toads. In total, about 20 species of mammals, 25 species of birds, several amphibians and about 30 species of freshwater fishes make up the feral vertebrate populations of Australia.

Invasive animals are not confined to vertebrate species. About 500 introduced invertebrate species are thought to be harming the environment, including the honey bee, which may compete with other fauna for nesting sites and other resources. New species continue to appear, such as the recently discovered *Congeria sallei*, a very invasive freshwater mussel from Central America, which was found in Darwin harbour. The

pests of sugar cane was a failure. While the insects were eventually controlled using insecticides and suitable management practices, the cane toad population expanded. The cane toad thrives on a range of available prey, including native invertebrates and small vertebrates, out-competing native frogs and snakes for these foods. In addition, the toad has toxic skin secretions that are known to kill vertebrate predators such as goannas, quolls, skinks and snakes, which otherwise may have controlled its population growth. The toads have been spreading, mainly to the north, since their introduction (Fig. 45.8). In December 1998 they were noticed 500 km south of Darwin and there is widespread concern about their potential impact on the fauna of Kakadu National Park.

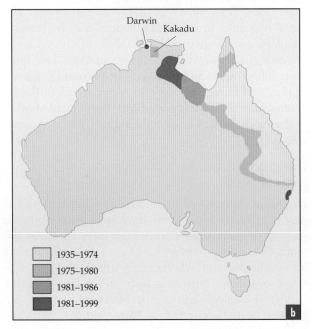

Fig. 45.8 (a) The cane toad is a serious pest in Australia.
(b) Distribution map of the spread of the cane toad following its introduction in 1935. It is moving towards Kakadu National Park

Sterile ferals

The impact of the biological control on the target organism is also of concern, and some people consider the creation of 'sterile ferals' a more humane approach than the release, for example, of calicivirus among rabbits. To address this, the Australian Cooperative Research Centre (CRC) for Vertebrate Biocontrol is developing ways to control rabbit, fox and introduced house mouse populations through fertility control using immunocontraceptive vaccines. The vaccines are delivered by bait or through the agency of a virus or other contagious agent that spreads naturally through the target pest population. Such technology must be treated with caution due to the possibility that the contagious agent might cross to non-target species.

> Ecosystems susceptible to invasion by introduced organisms are often disturbed and free of competitors and predators of the invading species. Biological control is one tool to control pest species.

Impact of weeds

A **weed** is a plant that is growing where it is not wanted (see Box 45.3). Thus, what constitutes a weed is largely a matter of perception. A weed to one person may be dinner to another! Dandelion, *Taraxacum officionale*, is a common urban weed in Australia, the leaves and tubers of which are a popular food in Europe. Many weeds are of agricultural importance since they may compete with crops for light, water and nutrients. Others interfere with wool production by lodging in the fleece of sheep and decreasing its value. Devil's claw, *Proboscidea louisianica*, is a weed originally from the Americas. It is common in eastern Australian grazing country. It produces a large, hooked seed capsule (Fig. 45.9), which can lodge in the knee joint or around the nostrils of sheep, greatly distressing and debilitating them.

Like pest animals, many weed species were introduced into Australia on purpose. The eminent botanist Baron Ferdinand von Mueller went to great lengths to ensure that the European blackberry, *Rubus fruticosus*, was spread throughout Victoria so that a hungry traveller could always obtain a meal of berries. Blackberry is now the most severe weed of streamsides in Victoria, choking native vegetation and providing a haven for rabbits and foxes.

Weeds are often opportunistic organisms. They usually produce many seeds that germinate under a range of conditions, are quick growing, flower early, are self-compatible, can reproduce vegetatively and are good competitors. Weeds usually invade environments that have been disturbed as a result of natural events, such as fire or wind, or human activities, such as forest clearing, fertiliser application, grazing or mining.

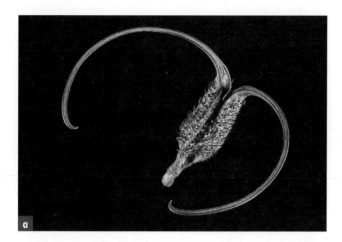

Fig. 45.9 (a) The hooked fruit of the weed devil's claw, *Proboscidea louisianica*, causes damage to sheep. **(b)** Bitou bush, *Chrysanthemoides monilifera* subsp. *rotundata*, is a weed that invades coastal vegetation. **(c)** In Victoria, the European blackbird, *Turdus merula*, is responsible for the spread of sweet pittosporum, *Pittosporum undulatum*, a native species turned weed

BOX 45.3 *Mimosa pigra*: pest plant of the tropics

The giant sensitive plant, *Mimosa pigra* (see figure), gets its name from its ability to quickly close up its leaves when touched. It is native to central America but has spread to many parts of the world including Africa, South-East Asia and northern Australia. It is a plant of the tropics, thriving in hot humid conditions. This plant has been growing in the Darwin region since the late nineteenth century, becoming a troublesome but very minor weed in the early part of the twentieth century around the Darwin Botanic Gardens. However, since the 1950s its population has 'exploded' in the Northern Territory and it now stretches in a 450 km arc from the Moyle River in the west to Arnhem Land in the east, with the total area infested in the order of 80 000 ha. It can form dense impenetrable thickets that exclude native plants and animals, transforming sedgeland and grasslands on floodplains into monospecific tall shrublands. *Mimosa* is now considered such a problem that it has been declared one of Australia's 'weeds of national significance'.

River floodplains in the Northern Territory were heavily grazed by another pest, water buffalo, introduced from Indonesia in the early 1800s. *M. pigra* is thorny and unpalatable to water buffalo, so it thrived while many of the palatable native species suffered. *M. pigra* also has other traits that give it an edge over native species. These include floating seeds, which are widely dispersed by floodwaters, drought tolerance, and a rapid growth rate of up to 1 cm per day under good conditions.

The giant sensitive plant, *Mimosa pigra*, is a pest plant of tropical Australia

M. pigra is in the family Mimosaceae and is related to wattles. It produces large quantities of hard seeds that remain viable for long periods. Ecologists have estimated that soil under thickets contains up to 12 000 *M. pigra* seeds per square metre. They also discovered that the longevity of seed is determined by the depth to which they are buried in the soil. At 1 cm depth, 50% of seeds are viable after nine weeks, whereas at 10 cm depth, 50% of seeds are still viable after 99 weeks. This means that deeply buried seeds may be viable for many years, even after the removal of adult plants.

Why should we worry about an invasive, introduced plant such as *M. pigra*? The dense thickets of this plant stop much light from reaching the ground. This causes the diversity of plant species under the thicket to decline. Many reptiles also find *M. pigra* thickets inhospitable, probably because of the reduced radiation from the sun, essential for the maintenance of their body temperature. Birds such as ducks, magpie geese, brolga and jabiru, whose open sedge and grassland habitat is under threat from *M. pigra*, are also affected. However, some species, in particular small mammals, benefit at least in the short term from *M. pigra* since the thickets provide cover from predatory animals, such as hawks.

Several insects have been introduced for biological control of *M. pigra*, with various degrees of success. Two stem-boring moths were released in 1989 and one of them, *Nuerostrata gunniella*, is now found in almost every *M. pigra* plant in the Northern Territory. It has reduced the growth of *M. pigra*, and possibly seed output, and has also enhanced infection rates by the widespread pathogenic tropical fungus *Botryodiplodia theobromae*. Weevils that can inflict heavy damage on flower buds were released in the early 1990s and two fungal pathogens are currently being assessed as biological control agents. An integrated weed management strategy, combined with control of the feral water buffalo and increased public awareness, has slowed the rate of establishment of this highly invasive weed.

Weeds can sometimes provide food or shelter for native animals when their natural habitat has been cleared. Mostly though, weeds pose a serious threat to the conservation of natural plant communities. Bitou bush, *Chrysanthemoides monilifera* subsp. *rotundata* (Fig. 45.9b), is a serious weed of coastal New South Wales. Introduced from South Africa to help stabilise

sand dunes after mining, it now forms impenetrable thickets on foreshores and chokes native coastal plant species such as *Acacia sophorae*. In South Africa, bitou bush is not considered to be a pest since it has natural insect and fungal predators that keep it in check. Consequently, several biological controls have been introduced into Australia. Of these, the bitou tip moth has been the most successful control agent and is having a significant impact on flowering and seed production of bitou bush. A combination of fire and herbicide also is an effective management tool to help combat infestations.

Many popular ornamental plants can be serious pests if they are allowed to spread into native bush. Some of the worst weeds are plants with fleshy fruit, such as sweet pittosporum, *Pittosporum undulatum*. Sweet pittosporum is a native plant from eastern Victoria and New South Wales but also a popular garden plant, which has become a pest species outside its natural range. The seeds of these plants are dispersed over large distances by birds, which eat the fruit and defecate the seeds into native eucalypt forest. The introduced European blackbird, *Turdus merula* (Fig. 45.9c), is one of the main dispersers of berry-fruited weeds in bushland adjacent to urban areas.

> Weeds are often opportunistic plant species. They threaten the conservation value of native communities by competing with native plant species as well as degrading animal habitats.

Land and water degradation

In addition to the effects of introducing alien species, human activities have caused degradation of the physical environment. In recent years, for example, the world has witnessed ecologically disastrous oil spills (Fig. 45.10), such as when the *Exxon Valdez* fouled the Alaskan coast (1989) and during the Gulf War when

Fig. 45.10 This fish was a victim of a heavy oil spill in the Yarra River, Melbourne, in October 1980

millions of barrels of oil were spilled into the Arabian Gulf (1991), representing the biggest oil spill on record. In comparison, Australia has only suffered minor spills, such as in Sydney Harbour in 1999; but the major spills remind us of the vulnerability of our local coastal ecosystems.

Fresh water—the key to life

Water is a critical resource in Australia, which is the world's driest, permanently inhabited continent. Agricultural and urban development has dramatically altered the flow of many rivers and streams, which can have significant effects on biodiversity. Threatening processes in freshwater systems include pollution, salinity, changes in water temperature and low oxygen levels, conditions often exacerbated by the reduced flow rates associated with river regulation and water harvesting. Changes in river flow may also affect the terrestrial biota surrounding a watercourse, such as the death of a large number of trees recorded along the Murray River, which was associated with the modification of water levels for flood control. The widespread introduction of watering points in the pastoral areas of semiarid and arid Australia has also had an impact on native flora and fauna. The watering points have made much more of the countryside accessible to hard-hoofed grazing animals, which selectively graze certain plants, and have resulted in increased numbers of native herbivores.

Holding on to soil

Erosion, a loss of soil by the action of either wind or water, is a natural process that shapes the surface of the earth. However, the clearing of land for agriculture has, in many cases, greatly increased the rate of erosion, stripping away nutrient-rich topsoil and silting up river systems.

Soil in tree-dominated ecosystems is stabilised by a dense mat of plant roots. The soil surface is also covered by a layer of litter (dead leaves and twigs), which protects the soil surface from the erosive action of wind and water. In a typical rainfall, water is quickly absorbed into the top layers of soil. This is because soil is naturally porous due to the adhesion of soil particles into larger crumbs, producing a matrix with many air spaces into which water can flow. Earthworms and other soil invertebrates assist this process. They burrow through the soil, breaking down organic material and creating a network of tiny channels.

Substantial amounts of soil erosion would usually only occur in tree-dominated ecosystems following intense fire. Fire removes stabilising vegetation cover and the heat of fire can cause the soil surface to become water-repellent. Heavy rainfall after fire can therefore wash away large amounts of soil and associated nutrients.

Agricultural soils, in contrast to soils in forest systems, have little protection from the action of wind and rain, particularly during dry periods when pasture dies back. Grazing by hard-hoofed stock, such as sheep and cattle, compacts the ground and reduces soil porosity. This slows the rate of rainwater absorption, leading to increased run-off and water erosion. The reduced vegetation cover results in greatly increased wind velocities at the surface of the soil, leading to loss of soil through wind erosion. It has been estimated that more than half of Australian agricultural land requires treatment to repair erosion damage (Fig. 45.11).

Wind and water erosion can be greatly reduced by sensible land management. Agricultural methods to improve soil conservation are becoming more widespread as the seriousness of soil loss is realised. Ceasing to cultivate erodible soils can reduce erosion significantly; this is called the 'no-till' system of agriculture. In cropping systems, leaving the stubble (the stalks of the grain) behind after harvest rather than ploughing or burning, as is traditional, is one of the best ways to reduce soil erosion. The standing stubble reduces wind velocities and the organic material helps to cement soil particles together, improving soil structure.

Soil acidification

Soil acidification (induced soil acidity) is a problem that has only emerged since the mid 1970s but it is one of the most serious of all **land degradation** problems in Australia. Soil acidification highlights the danger of solving one problem and, sooner or later, creating another. The yields of most plants begin to decline when soil pH is 5.0 or less. Such high acidity levels generally result from repeated applications of ammonia-based fertilisers without liming and the growing of pastures based on legumes such as subterranean clover. Australia-wide, some 90 million hectares of land are potentially affected: 55 million hectares are moderately to slightly acidic, 35 million hectares highly acidic and 1.5 million hectares have dangerously high

Fig. 45.11 Land degradation caused by wind erosion is a serious problem in Australian agricultural ecosystems

acid levels. Soil acidity can be reversed by better management of soil nutrients and the application of lime, but just to treat the worst 1.5 million hectares would require 2.3 million tonnes of lime and cost some $120 million.

Soil salinity

Salts can be found naturally in most Australian landscapes. Salts have accumulated in subsoils over thousands of years, left from previous cycles of drainage and desiccation or blown in on winds from the sea. Salt levels are highest in dry regions and lower in wetter areas where rain washes the salts out of the soil. Under normal circumstances, salt in the soil or ground water is harmless since it remains below the root zone of the vegetation. However, changes to the water cycle (Chapter 44) may result in the movement of stored salts towards the soil surface or into surface water bodies.

Trees transpire large amounts of water through open stomata when photosynthesising (Chapter 18). The combined process of transpiration from plants and evaporation of water directly from the soil surface is known as **evapotranspiration**. The level to which water rises in the soil (Fig. 45.12) is largely determined by the balance between rainfall and evapotranspiration. When land is cleared of deep-rooted native vegetation, the rate of evapotranspiration greatly decreases and more rainwater reaches the watertable (see Fig. 44.16).

Salinity problems in the surface water and soil therefore occur when more water enters the ground water system (through a process called recharge) than is discharged under natural conditions. The recharge of water to ground water systems has increased in Australia in the last 200 years. In some dryland areas—especially those with winter rainfall patterns—recharge may have increased up to 100 times since clearing and the introduction of crops or pastures. This causes the watertable to rise (Fig. 45.12) and once it is within about 2 m of the surface, salty water is carried to the surface by capillary action. At the surface water evaporates, leaving salt behind (called discharge). This results in **dryland salinity**.

An important feature of dryland salinity is that the recharge and discharge zones may be many kilometres apart. Thus, farmers in one area may be suffering salinisation from over-clearing elsewhere. Salinity must therefore be managed co-operatively on a catchment basis rather than at the level of the individual farm.

Irrigation salinity occurs in much the same way as dryland salinity but the rise in the watertable is largely due to the increased water applied to the soil surface for irrigation.

The first sign of increasing soil salinity is usually death of the most deeply rooted plants. Salt-sensitive species, such as clovers and rye-grasses, are replaced by species of increasing salt tolerance, such as buckshorn plantain, *Plantago coronopus*, and sea barley-grass, *Hordeum leporinum*, until eventually no plants can tolerate the salt and a bare 'salt scald' results. At this stage, a white crust of salt crystals is usually visible on

water, taken up from soil, is transpired by plants

trees removed; transpiration reduced and more water filters through to ground water

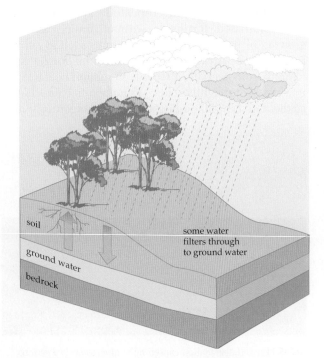
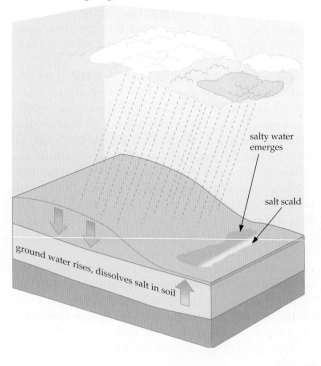

Fig. 45.12 The salinisation process

the soil surface during dry periods. During wet periods the salt scald may become a lake.

The increased amount of salt in the ground water eventually finds its way into streams. Streams become increasingly saline downstream as more and more salt water flows in. The Murray River is a good example of this. It drains more than 1 million km^2 of Queensland, New South Wales, Victoria and South Australia (Fig. 45.13). All the salt that is washed out of the soils in the Murray River catchment passes out the river mouth at Lake Alexandrina—an average of 5.5 tonnes per minute.

The only way to prevent dryland salinity is to stop clearing native vegetation. In situations where salinisation has already occurred, trees can be planted to reduce the amount of water entering the aquifer. In some regions it has been estimated that up to 50% of the catchment needs to be replanted. The challenge therefore is to revegetate farms using deep-rooted plants, such as eucalypts, in plantations or in widely spaced plantings so that they minimise the impact on the normal operation of the farm.

Dryland salinity as a result of clearing is accelerating in temperate Australia. The situation is presently worst in south-western Australia, where 10% of farming land is affected (Fig. 45.14). Unless salinity is mitigated by revegetation or by some engineering means, up to 30% of the land will be affected. The situation in south-eastern Australia has developed more slowly but is just as serious, with as much as 9 million hectares of land at risk from dryland salinity in New South Wales, Victoria and South Australia.

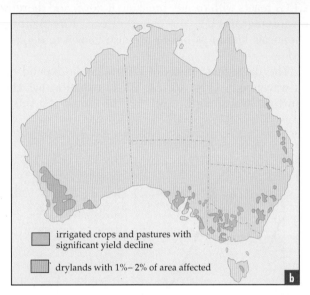

irrigated crops and pastures with significant yield decline

drylands with 1%–2% of area affected

Fig. 45.14 (a) Dryland salinity in mallee of wheat belt of Western Australia. **(b)** Salt-affected areas in Australia

More than half the agricultural land in Australia is affected by land degradation, with salinity, soil acidification and soil erosion being major problems. These problems can be managed by large-scale revegetation and better management of ground water and soil nutrients.

Atmospheric pollution

The influence of humans now reaches the most remote parts of the earth. Even the Arctic icecap is sometimes covered with a fine layer of black soot—air pollution blown from the industrial cities of Europe. Such global effects of pollution are best exemplified by atmospheric changes, such as the ozone 'hole' and global warming.

The ozone layer

The diffuse layer of ozone gas (O_3) in the stratosphere, between 13 and 20 km above the earth, plays an essential role for life on earth, shielding it from destructive ultraviolet (UV) radiation. Ozone gas is constantly created by the effect of sunlight on oxygen in the upper atmosphere. It is also constantly destroyed by the impact of various atmospheric constituents, both natural and artificial. However, in recent years the rate of ozone destruction has exceeded its formation and the

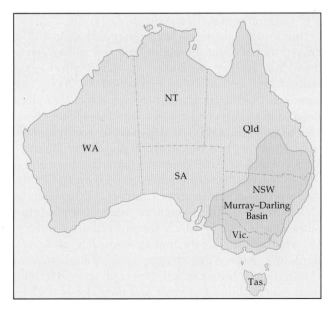

Fig. 45.13 The Murray–Darling Basin drains a large area of eastern Australia and the rivers accumulate salt that is washed out of the soils in the catchment

ozone layer has been thinning, allowing more UV radiation to reach the earth.

Destruction of stratospheric ozone is caused by gases containing chlorine and bromine, whose sources are mainly human-made halocarbon gases. One of the main sources of ozone-depleting substances is **chlorofluorocarbons** (CFCs), which were used as aerosol propellants (freon-11) and in refrigerators, air-conditioners and freezers (freon-12). The chlorine in CFCs reacts with ozone, breaking it down and thereby destroying its UV screening effect. Fire-fighting chemicals known as halons are also known to deplete ozone.

The damaging effects of UV radiation depend on its wavelength (Fig. 45.15). UV-C radiation has the shortest wavelength and is the most harmful.

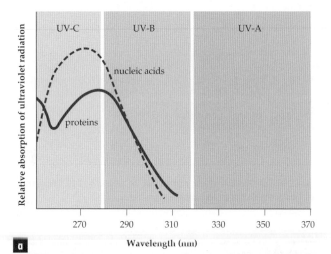

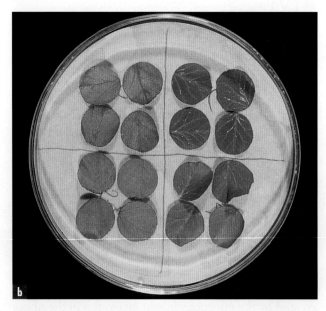

Fig. 45.15 (a) UV-B and UV-C bands of ultraviolet radiation are absorbed by proteins and nucleic acids and therefore are the most damaging to organisms. **(b)** The bronze-coloured leaves on the left have been exposed to UV-B radiation for 20 days relative to control leaves on the right

Fortunately, almost no UV-C radiation reaches the earth's surface since most of it is screened out by the ozone layer. UV-B radiation is the next most damaging form and it also is largely screened out. It is primarily UV-A radiation that causes sunburn and substantial amounts of UV-A radiation reach the earth's surface, even on cloudy days. Both UV-A and UV-B radiation are likely to increase substantially with a thinning of the ozone layer. For each 1% decrease in ozone, UV-B radiation at the earth's surface increases by approximately 2%.

All plants and animals that are exposed to UV radiation are threatened. For example, phytoplankton play an important role in many climatic processes. They scatter and absorb light entering the water, which warms the ocean surface. They release volatile organic compounds, which help clouds form. They fix atmospheric carbon dioxide, including large amounts released by industrial processes each year. Phytoplankton therefore play an important role in mopping up elevated levels of carbon dioxide, yet their productivity is reduced by increased UV radiation.

The thinning of the ozone layer is also a particular problem for Australians. Each spring (September and October) a so-called **ozone 'hole'** develops over Antarctica when the amount of ozone is depleted. This was first reliably measured in the early 1980s and the hole has been getting larger over time. Within two to three months the ozone hole usually breaks down and air with normal levels of ozone sweeps in. As this air sweeps in, ozone-depleted air from the hole drifts up over southern Australia, at times increasing UV-B radiation reaching the earth's surface by 20% or more. Australia has the highest rate of skin cancer of any country in the world, so any increase in cancer-causing UV radiation is of serious concern.

Since ozone depletion is a global problem, in 1987 a large group of countries met in Montreal and agreed to have no further increases in CFC production immediately and to halve production by 2000. This agreement is the Montreal Protocol, which has since been amended several times when new global observations of significant ozone depletion demonstrated the need to strengthen the treaty. Without the Montreal Protocol and its Amendments, continuing human use of CFCs and other compounds would have tripled the stratospheric abundances of chlorine and bromine by about the year 2050. In contrast, under current international agreements, which are now reducing and aim to eventually eliminate human emissions of ozone-depleting gases, the stratospheric abundances of chlorine and bromine are expected to reach their maximum within a few years and then slowly decline. All other things being equal, the ozone layer is expected to return to normal by the middle of the twenty-first century.

Thinning of the ozone layer is allowing more UV radiation to reach the earth's surface, causing problems such as increased occurrence of skin cancer.

The greenhouse effect and global warming

The **greenhouse effect** (Fig. 45.16) is the natural warming of the earth by heat trapped due to the presence of certain heat-absorbing atmospheric gases. On reaching the earth's surface, sunlight, largely short wavelength radiation, is converted into longer wavelength heat energy. Much of this heat is reradiated from the earth but is absorbed by atmospheric gases, the greenhouse gases (CO_2, N_2O, CH_4) and water vapour warming the lower atmosphere. The heat trapped by atmospheric gases is essential for providing suitable conditions for life on earth. Without greenhouse gases to trap heat the temperature at the surface of the earth would average 30°C less than at present (average 15–16°C).

Since industrialisation in the mid 1800s, with the burning of fossil fuels (coal, oil and natural gas), concentrations of several atmospheric gases have been steadily increasing: CO_2 has increased by about 25% (Fig. 45.17); N_2O by 19% and CH_4 by 100%. Ozone in the lower atmosphere and CFCs also act as greenhouse gases. If emissions continue to grow at current rates, it is almost certain that atmospheric levels of carbon dioxide will double from pre-industrial levels during the twenty-first century. If no steps are taken to slow

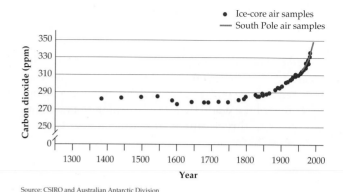

Source: CSIRO and Australian Antarctic Division

Fig. 45.17 Atmospheric carbon dioxide levels since 1300 determined from Antarctic ice cores

greenhouse gas emissions, it is quite possible that levels will triple by the year 2100. Increased gas emissions are enhancing the greenhouse effect and the prospect of global climate change.

Recognising the problem of potential global climate change, the World Meteorological Organization (WMO) and the United Nations Environment Program (UNEP) established the **Intergovernmental Panel on Climate Change** (IPCC) in 1988. In 1998 the IPCC published an internationally co-ordinated assessment of the regional impacts of climate change. The IPCC found that the global mean surface temperature has increased by 0.3–0.6°C since the late nineteenth century. In addition, the 1990s were found to be the warmest decade of the millenium and 1998 the warmest year so far. The Panel concluded 'the balance of evidence suggests a discernible human influence on global climate'. Computer modelling of the impact of increased greenhouse gases is extremely difficult since many factors need to be considered and little is known about how global climate functions. However, based on the best available knowledge, climate models project that the mean annual global surface temperature will increase by 1–3.5°C by 2100, that the global mean sea level will rise by 15–95 cm and that changes in the spatial and temporal patterns of precipitation will occur if the greenhouse effect increases.

How the hydrological cycle will respond to climate change is particularly important. Most models predict more clouds, which have the capacity to enhance the greenhouse effect by trapping more heat or, alternatively, to reduce it by reflecting solar radiation back into space, thereby cooling the earth. The response of the oceans will also be of great importance since ocean currents redistribute energy across the globe and play an important role in determining rainfall patterns.

Increasing sea temperatures, together with increased levels of solar radiation, have been linked to the bleaching of corals. **Coral bleaching** has been

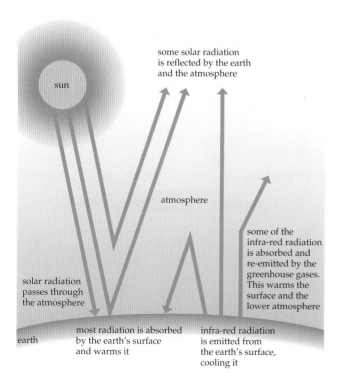

Fig. 45.16 Mechanism of the greenhouse effect

reported in over 30 countries throughout the Caribbean, Indian and Pacific Oceans and up to 88% of near-shore corals in the Great Barrier Reef are affected. Scientists predict that the Reef will be affected by yearly severe bleaching events by the year 2030, and that coral reefs could be eliminated from most areas of the world by 2100. Greenhouse pollutants also pose a direct threat to coral reefs, causing crumbling in a process similar to osteoporosis in humans.

Of great concern is that a warmer climate will lead to higher sea levels due to heating and expansion of seas, and partial melting of icecaps and glaciers. The impact of a rise in sea level would be greatest in developing countries since they have the most extensive regions of highly productive, densely populated, low-lying river deltas. Many large cities, including Bangkok, Calcutta, Shanghai and Hanoi, are on low-lying river banks and are under threat of inundation by rising seas unless major engineering works, such as dykes, are constructed to hold back the rising waters. Engineers in some coastal cities at least are taking the threat of rising sea levels seriously and are making allowances in plans for new coastal developments.

Given the wide range of climates experienced across Australia and the large natural variability of Australia and its climate, it is hard to predict what the impacts of climate change will be and where they will occur. It is possible that there will be positive and negative impacts. These impacts may affect: the distribution of plants and animals; the frequency of storms and floods; and the spread of weeds, pests and diseases, which may influence agriculture and human health.

Like the hole in the ozone layer, the greenhouse effect is a global problem and requires a response at the international level. The **Kyoto Protocol** was adopted in December 1997 and is set to enter into force some time after the year 2000. The legally binding Protocol commits developed countries to reduce their collective emissions of greenhouse gases by at least 5% compared to 1990 levels by the period 2008–2012. The potential to make the greatest reductions in greenhouse emissions is through changes to fossil fuel use and to land use practices, such as agriculture, land clearance and forestry. It is estimated that 30% or more of Australia's anthropogenic greenhouse gas emissions are derived from these sources.

The greatest gains in CO_2 reduction are immediately available through improved energy efficiency. More efficient lighting, economical motor vehicles and energy-efficient industry could greatly reduce the amount of CO_2 released each year. Fossil fuels presently provide 78% of energy requirements. The amount of CO_2 produced for each unit of energy varies according to the fuel. Natural gas is the cleanest, while oil produces 45% more CO_2 per unit of energy and coal 75% more. Thus, changing to 'cleaner' fossil fuels may provide some benefits.

Clean or non-carbon based forms of energy, such as solar thermal (trapping the sun's heat; Fig. 45.18), photovoltaic (direct conversion of sunlight to electricity), wind, geothermal and tidal energy hold the greatest potential for long-term reductions in carbon emissions. The Australian Federal government recently established a 2% renewable electricity target by 2010, providing $45 million in competitive grants to commercial enterprises to develop tidal, solar and wind power projects. At this rate it will be some time before these alternative energies substantially replace fossil fuels.

The greenhouse effect is the warming of the earth's lower atmosphere due to heat-absorbing 'greenhouse' gases, such as CO_2, CH_4 and N_2O. The most effective ways to reduce emission of greenhouse gases are by reduced fossil fuel use, improved energy conservation and a reduction in land clearing.

Fig. 45.18 Use of clean (non-carbon) forms of energy such as solar power has the potential to reduce carbon dioxide emissions and the greenhouse effect

Restoring the balance?

At the international level there has been a series of legally binding treaties that deal with climate change, pollution of the oceans, expanding deserts, damage to the ozone layer and the rapid extinction of species and loss of biodiversity. Countries that have ratified these treaties have a range of responsibilities and obligations.

CITES

The flourishing worldwide trade in exotic wildlife threatens biodiversity (Fig. 45.19). Wildlife trade is controlled through the **Convention on International Trade in Endangered Species of Wild Fauna and Flora** (CITES), to which 150 countries, including Australia and New Zealand, are signatories. Over 20 000 species are listed by CITES, which bans trade in highly endangered species or regulates and monitors trade in others that might become endangered.

Highly protected species include big cats, elephants, rhinoceros and apes; more widely traded species include ginseng, aloe, cacti, carnivorous plants, orchids, corals, crocodiles, birdwing butterflies, sturgeon (including their caviar), and vicunas (including their wool). The rarer the species the more 'collectable' it is to some people. For example, extremely rare giant pitcher plants from Borneo can sell for $1300 each on the black market. Rare Peruvian butterflies may fetch $4000 each. In Japan, South-East Asian clouded leopard skin coats sell for as much as $165 000. Overall, the trade in listed species is in excess of US$12 billion per annum so it is critical to have controls that limit the impact of the trade.

Fig. 45.19 Worldwide trade in exotic wildlife includes endangered species such as leopards and tigers

Legislative and policy framework

Effective legislative and policy frameworks are needed within countries in order to conserve ecosystems and their component organisms. At the level of the Commonwealth in Australia, the **Environment Protection and Biodiversity Conservation Act 1999** provides for the protection of environments of national significance and promotes ecologically sustainable development and the conservation of biodiversity. It also focuses on a co-operative approach to environmental protection and management, recognition and promotion of indigenous knowledge and implementation of Australia's international environmental responsibilities.

Land management is primarily the responsibility of the states and each Australian state or territory has legislative power to protect native species. For example, the Victorian **Native Flora and Fauna Guarantee Act 1987** aims to protect all species of wattles, banksias, grevilleas, orchids and many other species of plant and animal (Fig. 45.20). Conservation is best achieved through protection of whole habitats in Nature Reserves and National Parks, with complementary off-reserve conservation. Australia's first National Park, for example, was established near

Fig. 45.20 The eastern pygmy possum is a rare species of marsupial protected in Victoria under the *Native Flora and Fauna Guarantee Act 1987*

Sydney in 1879 and formal control of forest exploitation was in place by the 1900s (see Box 45.1). Legislation relating to the management of private land varies from state to state but generally relates to five major areas— land clearing and land degradation, heritage agreements, nature conservation, planning controls and Crown Land leases. For example, most Australian state governments have introduced planning regulations that require landowners to obtain a permit before clearing native vegetation.

Managing natural communities for conservation

Management of natural communities to preserve biodiversity and *prevent* degradation is the highest conservation priority. This can be achieved through reserve systems, such as National Parks, World Heritage Areas and Flora and Fauna Reserves, but it is important that such reserves include a comprehensive and representative coverage of ecosystem types. However, the reserve system will only ever cover a relatively small proportion of the landscape and many of the most endangered communities, such as temperate grasslands, occur outside protected areas. Consequently, increasing attention is being paid to conservation of natural communities across a range of land tenures, not just in reserves. Private land covers two-thirds of Australia and in many areas a range of incentives, including assistance with fencing and rate and tax rebates, are being offered to support the management and conservation of natural areas.

Management of reserves requires knowledge of the ecology of natural ecosystems—understanding what factors regulate the abundance and distribution of species (Chapter 42), how different species interact and depend on one another (Chapter 43) and how organisms interact with their physical surroundings and how ecosystems function (Chapter 44). Apart from managing weeds, feral animals and degradation of the physical environment as discussed earlier in this chapter, fire is an important aspect of conservation management (Box 45.4).

Park and Kakadu. The Anangu people are traditional owners of Uluṟu (Ayers Rock) and Kata Tjuṯa (The Olgas), which are part of Uluṟu-Kata Tjuṯa National Park (Figure). Before the arrival of Europeans, the Anangu practised 'fire stick' farming to manage the environment for hunting, to encourage 'bush tucker' vegetation and to provide green feed for animals (see Chapter 41). The Anangu burnt the country in a patchwork fashion, providing a diversity of habitat patches, which is thought to impede the passage of large fires.

When Europeans arrived with cattle, sheep and fences, traditional fire management was abandoned and irregular and extensive wildfires occurred. Extensive areas of vegetation of the same age resulted, limiting the available habitat for animals. These changes, along with the impact of feral animals, are thought to have driven some mammal species to extinction.

In 1985, title to Uluṟu National Park was returned to the Anangu. They leased the park back to the Australian National Parks and Wildlife Service, who set out to manage it, together with the Anangu people, using traditional ideas of stewardship. An integral part of this was reintroducing patch-burning to minimise fuel for wildfires and to maintain a diversity of plant successions and therefore animal habitat. The fire management plan for Uluṟu-Kata Tjuṯa was reviewed in 2000 to examine the effectiveness of this approach.

Management of fire is an important aspect of conservation management, whether it be excluding fire from tropical rainforest or encouraging regular patchwork fires to maintain diversity in coastal heath. Other impacts must also be managed: feral animals and weeds, and disturbance by off-road vehicles, which damage vegetation and cause soil erosion.

BOX 45.4 Aboriginal knowledge and fire management

The contribution of traditional ecological knowledge to the management of natural resources is being increasingly recognised worldwide, and Australia is seen to be a world leader in the joint management of both Uluṟu-Kata Tjuṯa National

Uluṟu-Kata Tjuṯa National Park

Managing land for conservation uses knowledge of the ecology of natural communities, including the role of fire, and the effects of feral animals, weeds and human impacts.

Restoration ecology: resetting the clock

Where natural ecosystems have been highly modified, **restoration** is the process used to produce a healthy, natural, self-regulating system that is a close approximation of the predisturbance condition and that is integrated into the surrounding landscape. It can be likened to resetting the biological clock so that natural processes, such as succession (Chapter 43) and evolution, continue as they did before human interference. However, because each ecosystem is the result of a unique sequence of climatic and biological events, total restoration is unlikely even under the most favourable conditions.

Restoration may be required following a diversity of land uses such as agriculture and mining (Fig. 45.21), so not unexpectedly a diversity of restoration approaches are used. Local people in a region also should be involved in the planning and setting of objectives for restoration. In Western Australia, for example, the wheat belt has been highly modified and the entire landscape requires restoration. Information on the habitat requirements and movement patterns of species of conservation or functional significance is used to design the scale of patches and linkages in the landscape. The incorporation of appropriate disturbance regimes, such as fire, is necessary to ensure different successional stages across the landscape. Cost-effective methods of replanting, such as direct seeding, are needed and local species should be used in

Fig. 45.21 Ecosystem restoration. An aerial view of Fraser Island, where the mined area was seeded with the dominant tree and shrub species three years prior to this photo to re-establish the plant communities

the first instance. Sometimes, however, local moisture, temperature and nutrient regimes are so modified that the original species are unable to cope with the altered environment. If introduced species need to be used, every attempt should be made to ensure that they are non-invasive. The desired end result is that all original vegetation types and strata are represented in the restored landscape.

Ecologically sustainable development

In this chapter we have considered many human impacts on the environment, most of them negative. In the future, we have the opportunity to greatly increase the positive impacts that we have on our environment through the application of biological knowledge. A sustainable society is one in which all human activity takes place within the limitations set by the environment. This includes its capacity to assimilate waste, provide food and supply other resources such as minerals and energy.

Ecologically sustainable development (ESD) means using, conserving and enhancing the communities' resources so that ecological processes, on which life depends, are maintained, and the total quality of life can be increased. Historically, a rising standard of living has been linked to increasing energy and resource usage, environmental degradation and pollution. During the 1980s and 1990s this has been reversed, with a trend towards decreasing the intensity of resource and energy usage. Today, we can do more with much less than we could a decade ago because technology has allowed industries to become more efficient.

In 1992, Australian governments adopted a National Strategy for Ecologically Sustainable Development (NSESD), which is concerned with the idea that future generations should enjoy an environment that is, at least, as healthy, diverse and productive as the one we presently experience. This is called **intergenerational equity**. The main objectives of the NSESD are to:

- enhance individual and community well-being and welfare by following a path of economic development that safeguards the welfare of future generations
- provide for equity within and between generations
- protect biological diversity and maintain essential ecological processes and life-support systems.

To achieve ecologically sustainable development, a number of other objectives should be met:

1. *Community participation.* Strong community involvement is an essential prerequisite for a smooth transition to an ecologically sustainable society.
2. *Constant natural capital.* Natural capital, such as biological diversity, healthy environments and

freshwater supplies, cannot be substituted by human-made capital and should be maintained from one generation to the next.

3. *Anticipatory and precautionary policy approach*. Policies should err on the side of caution. The onus of proof should be placed on developers to demonstrate that their activities are ecologically sustainable.

4. *Limits on natural resource use*. The limits of the environment to absorb wastes and provide resources must be recognised and accommodated.

5. *Pricing environmental values and natural resources*. Wherever possible, prices for natural resources should be set to recover the full social and environmental costs. Many environmental values cannot be expressed in monetary terms, hence pricing policies must form a part of a broader framework for decision making.

6. *Global and regional perspectives*. These are required so that environmental problems are not just moved from country to country.

7. *Efficiency*. Efficient use of resources should be a major objective in economic policy.

One of the most important changes that must occur is decoupling environmental degradation from economic development. Increased recycling of materials can make a substantial contribution towards this since it decreases the amount of resources and energy required to produce goods. 'Think global, act local' was a catch cry of the 1970s and early 1980s. It essentially meant 'think about the global implications of what you do; lots of small changes in people's behaviour can add up to change on a global scale'. If one person uses recycled lecture pads, this will not save many trees, but if every student does the same, this might mean a whole forest is still standing. It is also important for governments and others to consider global and regional perspectives so that environmental problems are not just moved from country to country.

Changes in lifestyle will also be important. A greatly improved public transport system, together with better urban planning, could substantially decrease pollution from transport. Likewise, a shift from non-renewable, highly polluting sources of energy, such as oil, coal and gas, to renewable sources of energy, such as wind, water and solar radiation, can greatly reduce the environmental impact of economic growth. Developing a population policy nationally and globally is also important. Ultimately, a global society developed along ESD principles is the only solution.

Summary

- The impact of humans on the Australian environment has greatly accelerated over the last 200 years since European settlement. Extensive land clearing has resulted in loss and fragmentation of habitat, the single major cause of loss of biodiversity.
- Human survival depends both physically and psychologically on the maintenance of bio-diversity.
- Introduced animals, plants and microorganisms, which have become pests or weeds, have also been detrimental to the environment and affected the conservation value of native biological communities.
- More than half the agricultural land in Australia is affected by land degradation, with salinity, soil erosion and soil acidification being major problems. Dryland salinity can be managed by using vegetation to increase evapotranspiration and lower watertables, and soil acidification by better management of soil nutrients and the application of lime.
- Erosion can be reduced by minimising cultivation in susceptible areas and by maintaining vegetation cover.

- Human impacts are being felt on a global scale. Although the greenhouse effect is a natural process, increased warming of the earth's lower atmosphere is predicted because of the emission of the greenhouse gases CO_2, CH_4 and N_2O. Thinning of the ozone layer, which normally absorbs harmful UV radiation, is of concern in Australia.
- Land management uses biological knowledge of how natural ecosystems function, the role of fire, and the effects of feral animals, weeds and human impacts. Restoration ecology aims to restore degraded lands and habitats to a state that resembles the predisturbance condition and applies ecological principles such as succession.
- A sustainable society is one in which all human activity takes place within the limitations set by the environment. Ecologically sustainable development means using, conserving and enhancing resources so that ecological processes, on which life depends, are maintained.

key terms

biodiversity (p. 1200)
biological control (p. 1205)
chlorofluorocarbon (CFC) (p. 1212)
Convention on International Trade in Endangered Species of Wild Fauna and Flora (CITES) (p. 1215)
coral bleaching (p. 1213)
dryland salinity (p. 1210)

ecologically sustainable development (ESD) (p. 1217)
Environment Protection and Biodiversity Conservation Act 1999 (p. 1215)
erosion (p. 1209)
evapotranspiration (p. 1210)
feral animal (p. 1204)
greenhouse effect (p. 1213)

habitat corridor (p. 1204)
habitat fragmentation (p. 1203)
integrated pest management (p. 1205)
intergenerational equity (p. 1217)
Intergovernmental Panel on Climate Change (IPCC) (p. 1213)
irrigation salinity (p. 1210)

Kyoto Protocol (p. 1214)
land degradation (p. 1209)
Native Flora and Fauna Guarantee Act 1987 (p. 1215)
ozone 'hole' (p. 1212)
pest (p. 1204)
restoration (p. 1217)
slash-and-burn agriculture (p. 1202)
soil acidification (p. 1209)
weed (p. 1206)

Review questions

1. Biologists and conservationists are concerned about the 'biodiversity crisis'. What is biodiversity and why is its conservation considered important?

2. What is meant by CITES? Name three species that are listed under CITES.

3. What are feral animals? Name three feral animals found in your region and research the impact of one of these species.

4. Why was the cane toad introduced into Australia? What factors led to its population 'explosion' and range expansion?

5. What is a weed? What are some of the characteristic features of weedy plants? Explain why a plant that is not a problem in its native country can become a serious weed when introduced into another country.

6. Why are agricultural soils more susceptible to erosion than tree-dominated soils?

7. What is soil acidification and what causes it?

8. What is dryland salinity and what causes it? How can the planting of native vegetation help to reduce salting?

9. Briefly explain the 'greenhouse effect', naming the major causes. List the positive and negative results of the greenhouse effect.

Extension questions

1. The little pygmy possum is found in mallee scrubs and heaths of north-eastern Victoria and south-eastern South Australia. The species lives within certain environmental limits: 33% of its current range is within conservation areas. However, ecologists concerned with the effects of global warming have predicted that, with a rise in temperature of 1°C, the animal's range would shift southwards into areas outside nature reserves. If the potential range of a species can change because of temperature increase, we need to take this into account when designing conservation areas. What sort of design features should we consider to ensure saving species affected by climate change? What, for example, are habitat corridors and what function do they serve for nature conservation?

2. Integrated pest control involves the use of various techniques to control a pest species rather than a single method. Explain what is meant by integrated pest control. Discuss one example, such as a plant weed or rabbits, to illustrate this approach.

3. Explain what is meant by 'fire stick' farming. Why has it been reintroduced at Uluru-Kata Tjuta National Park?

4. Discuss what is meant by ecologically sustainable development.

5. How can you contribute to maintaining biodiversity?

Suggested further reading

Burgman, M. A. and Lindenmayer, D. B. (1998). *Conservation Biology for the Australian Environment.* Sydney: Surrey Beatty & Sons.

This text discusses principles and methods of conservation biology and provides Australian examples and further references.

Groves, R. H., Shepherd, R. C. H., Richardson, R. G., eds. (1995). *The Biology of Australian Weeds.* Melbourne: R. G. & F. J. Richardson.

Use as a source for information on particular weeds.

Hobbs, R. and Yates, C., eds. (2000). *Temperate Eucalypt Woodlands in Australia.* Sydney: Surrey Beatty & Sons.

This text discusses the ecology of woodlands in southern parts of Australia and conservation issues.

Industry Commission. (1998). *A Full Repairing Lease: Inquiry into Ecologically Sustainable Land Management.* Report no. 60. Canberra: Australian Government Publishing Service.

Further information is available in this report on the issues of sustainable management of land in Australia.

Kitching, R. L., ed. (1986). *The Ecology of Exotic Animals and Plants.* Brisbane: Wiley.

Use this book for further information on particular pest species, including a chapter by S. Easteal and R. B. Floyd on the cane toad.

Sindell, B., ed. (2000). *Integrated Weed Management Textbook.* Melbourne: R. G. & F. J. Richardson.

Up-to-date book on issues of weed management practices.

State of the Environment Advisory Council. (1996). *Australia: State of the Environment Report.* Melbourne: CSIRO.

This report summarises environmental data on Australian ecosystems, both terrestrial and aquatic, and discusses human impacts.

Watson, R. T., Zinyowera, M. C., Moss, R. H., Dokken, D. J. (1998). *The Regional Impacts of Climate Change: An Assessment of Vulnerability.* A special report of IPCC Working Group II. Cambridge: Cambridge University Press.

An up-to-date discussion of human-induced climate change and its likely impacts.

The following are examples of websites related to human impacts and natural resource management at national and international levels. Most of the sites have links to many other websites, so the challenge is to use them wisely!

- *Agriculture, Fisheries and Forestry—Australia*: http://www.affa.gov.au/

- *Bureau of Rural Sciences*: http://www.brs.gov.au/

- *CITES*: http://www.wcmc.org.uk/CITES/english/index.html

- *Environment Australia on-line*: http://www.erin.gov.au/erin.html

- *Food and Agriculture Organization*: http://www.fao.org/

- *Society for Ecological Restoration*: http://ser.org/

- *The Ozone Secretariat (including the Montreal Protocol)*: http://www.unep.org/unep/secretar/ozone/home.htm

- *The Australian Greenhouse Office*: http://www.greenhouse.gov.au/

- *The National Land and Water Resources Audit*: http://www.nlwra.gov.au/

- *United Nations Convention on Biological Diversity*: http://www.biodiv.org/

- *United Nations Framework Convention on Climate Change*: http://www.unfccc.de/

Classification of cellular organisms

The following classification is of living, cellular organisms and excludes many fossil groups as well as viruses. Some groups of organisms, particularly within the protists, have not yet been formally named by biologists and are here listed as informal groups. Overall, the order of the taxa reflects our current knowledge of the hierarchy of life.

Prokaryotes

'Super Kingdom' Archaea

Methanogens, thermophiles and halophiles

'Super Kingdom' Bacteria

Spirochaetes, Gram-negative, Gram-positive, purple and sulfur bacteria, rickettsias, chlamydias, mycoplasmas and cyanobacteria

(Alternative classification schemes place all bacteria in the Kingdom Monera.)

Eukaryotes

'Super Kingdom' Eukarya

Kingdom Protista

Photosynthetic protists

Phylum Glaucacystophyta: flagellates with a photosynthetic endosymbiont
Phylum Rhodophyta: red algae
Phylum Chlorophyta: green algae (e.g. *Chlamydomonas*, *Spirogyra*, *Ulva*, desmids and *Chara*)

Heterokonts

Phylum Chrysophyta: golden flagellates
Phylum Prymnesiophyta: haptophytes (e.g. coccolithophorids)
Phylum Bacillariophyta: diatoms
Phylum Phaeophyta: brown algae (e.g. *Durvillea*, *Hormosira*)
Phylum Oomycota: water moulds, downy mildews, *Phytophthora*

Euglenoids and kinetoplasts

Phylum Euglenophyta: flagellates (e.g. *Euglena*); kinetoplasts: flagellate parasites (currently phylum Zoomastigina, class Kinetoplastida) including trypanosomes and *Leishmania*

Pelobiontids

Amoebae lacking mitochondria (currently phylum Zoomastigina, order Pelobiontida)

Diplomonads

Parasitic flagellates: currently phylum Zoomastigina, class Diplomonadida (e.g. *Giardia*)

Alveolates

Phylum Dinophyta: dinoflagellates (e.g. zooxanthellae of corals)
Phylum Apicomplexa: animal parasites (also called sporozoans) including gregarines, coccidians, haematozoa (e.g. *Plasmodium*, which causes malaria)
Phylum Ciliophora: ciliates (e.g. *Paramecium*)

Protists with second-hand chloroplasts

Phylum Cryptophyta: cryptomonads (flagellates)
Phylum Chlorarachnida: *Chlorarachnion*, an ameoboid plasmodium

Slime moulds

Cellular slime moulds: Phylum Acrasea, Phylum Dictyostelida, protostelids
Acellular slime moulds: Phylum Myxomycota

Sponge-like protists

Choanoflagellates

Protists of unknown affinity

Phylum Actinopoda: actinopods (e.g. radiolarians)
    Subphylum Sarcodina: amoebae including rhizopods and foraminiferans
Phylum Zoomastigina
    Class Opalinata: opalinids—large, commensal flagellates
    Class Proteromonadida: proteromonads—commensals
    Class Parabasalia: parabasilids—commensals and parasites

Kingdom Fungi

Phylum Chytridiomycota: chytrids
Phylum Zygomycota: moulds (e.g. *Rhizopus*) and other microfungi
Phylum Ascomycota: sac fungi—cup fungi, yeast, morels, truffles, *Penicillium* and *Neurospora*
Phylum Basidiomycota: club fungi—rusts, smuts, mushrooms, puffballs and bracket fungi
Phylum Deuteromycota: anamorphs or Fungi Imperfecta
Phylum Mycophycota: lichens

Kingdom Plantae

Non-vascular land plants

Phylum Hepatophyta: liverworts
Phylum Anthocerophyta: hornworts
Phylum Bryophyta: mosses

Vascular land plants

Phylum Psilophyta: *Psilotum* and *Tmesipteris*
Phylum Lycophyta: clubmosses and quillworts
Phylum Sphenophyta: horsetails
Phylum Filicophyta: ferns

Seed plants

Phylum Cycadophyta: cycads
Phylum Ginkgophyta: *Ginkgo*
Phylum Coniferophyta: conifers
Phylum Gnetophyta: gnetophytes
Phylum Magnoliophyta: flowering plants (angiosperms)
    Class Liliopsida: monocotyledons
    Class Magnoliopsida: dicotyledons

Kingdom Animalia: the Metazoa

Phylum Porifera: sponges

Radiate phyla

Phylum Cnidaria
    Class Scyphozoa: jellyfish
    Class Cubozoa: sea wasps
    Class Hydrozoa: hydras and siphonophores
    Class Anthozoa: sea anemones and corals
Phylum Ctenophora: comb jellies

Acoelomate protostomes

Phylum Platyhelminthes: flatworms
    Class Turbellaria: free-living flatworms
    Class Monogenea: ectoparasitic flukes
    Class Trematoda: endoparasitic flukes
    Class Cestoda: tapeworms
Phylum Nemertinea: proboscis worms

Pseudocoelomates

Phylum Nematoda

Coelomate protostomes

Phylum Annelida
    Class Polychaeta: marine bristle worms
    Class Euclitellata: clitellum-bearing worms
        Subclass Oligochaeta: earthworms and aquatic relatives
        Subclass Hirudinea: leeches
        Subclass Branchiobdellida: 'jawed' worms
Phylum Arthropoda
    Subphylum Chelicerata
        Class Merostomata: horseshoe crabs
        Class Arachnida: spiders, scorpions, ticks and mites
        Class Pycnogonida: sea spiders
    Subphylum Crustacea
        Class Branchiopoda: water fleas
        Class Copepoda: water fleas
        Class Ostracoda: mussel or seed shrimps
        Class Cirripedia: barnacles
        Class Branchiura: fish lice
        Class Pentastomida: tongue worms
        Class Peracarida: slaters and woodlice

        Class Malacostraca: shrimps, crabs and so on
    Subphylum Uniramia
        Class Diplopoda: millipedes
        Class Chilopoda: centipedes
        Class Insecta: insects
            Subclass Apterygota: wingless insects
            Subclass Pterygota: winged insects
        Class Onychophora: velvet worms (*Peripatus*, *Peripatoides*)
Phylum Mollusca
    Class Aplacophora: primitive, worm-like molluscs
    Class Monoplacophora: *Neopilina*
    Class Polyplacophora: chitons
    Class Gastropoda: snails, limpets and so on
    Class Bivalvia: mussels, oysters, scallops and so on
    Class Scaphopoda: tusk shells
    Class Cephalopoda: *Nautilus*, cuttlefish, squids, octopuses

Lophophorates

Phylum Phoronida: primitive lophophorates
Phylum Brachiopoda: lamp shells
Phylum Bryozoa: bryozoans or moss animals

Deuterostomes

Phylum Echinodermata
    Class Crinoidea: feather stars
    Class Asteroidea: sea stars
    Class Ophiuroidea: brittle stars and basket stars
    Class Echinoidea: sea urchins, sand dollars and heart urchins
    Class Holothuroidea: sea cucumbers
Phylum Chordata
    Subphylum Hemichordata: acorn worms
    Subphylum Urochordata: tunicates or sea squirts
    Subphylum Cephalochordata: lancelets
    Subphylum Vertebrata
        Superclass Agnatha: jawless fishes (lampreys and hagfishes)
        Superclass Gnathostomata: jawed vertebrates
        Class Chondrichthyes: cartilaginous fishes (chimaeras, sharks, skates and rays)
        Class Osteichthyes: bony fishes
            Subclass Actinopterygii: ray-finned fishes including teleosts
            Subclass Sarcopterygii: fleshy-finned fishes (coelocanths and lungfishes)

Tetrapods

Class Amphibia: salamanders, mud puppies, newts, frogs, toads and caecilians
'Sauropsida': including traditional classes Reptilia (turtles, snakes, lizards, crocodiles) and Aves (birds)
Class Mammalia
    Subclass Prototheria: platypus and echidnas
    Subclass Theria: Metatheria (marsupials) and Eutheria (so-called placental mammals, including the order Primates)

Glossary

abiotic Pertaining to physical and inorganic components.

ABO blood group antigen Human group of carbohydrate antigens that most people (except those of blood type AB) have serum antibodies against.

abscisic acid A growth-inhibitory hormone controlling responses of plants to stress (such as drought), frost tolerance and seed dormancy; synthesised in the carotenoid pigment pathway.

abscission zone A zone of tissue at the base of a petiole that regulates abscission or shedding of the leaf.

absorption spectrum Pattern of absorption of photons at different wavelengths of light.

acclimation An adaptive change in an organism in response to a change in a single feature of the environment, for example, acclimation to temperature.

acclimatisation An adaptive change in an organism in response to broadly defined changes in weather conditions.

acetyl CoA A 2-carbon compound that is the substrate for the citric acid cycle; produced in mitochondria during the final stage of glycolysis when O_2 is available, and as a product of β-oxidation.

acid Substance that is a proton donor, releasing hydrogen ions (H^+) into solution; possessing a pH in solution below 7.

acoelomate Containing no coelom (body cavity within mesoderm).

acrosome Secretory vesicle in the head of the sperm containing hydrolytic enzyme, which are released during fertilisation.

actin A globular protein; main structural component of microfilaments.

actin filaments Thin myofilaments composed of two actin molecules (in turn consisting of a chain of globular actin molecules), tropomyosin and troponin molecules.

actinomorphic Describes a flower with parts arranged in a regular way (radially symmetric).

action potential Electrical event that is conducted the full length of an axon without loss of amplitude because it regenerates itself at successive points; triggered by depolarisation that reaches the threshold potential for the membrane; involves a rapid non-linear opening of voltage-dependent sodium channels, followed by opening of voltage-dependent potassium channels.

action spectrum Absorption spectrum of light that activates photosynthesis.

activation Increase in metabolism of an egg following penetration by a sperm cell.

activation energy Energy required to initiate a reaction; is more than the minimal level to break existing bonds at the moment molecules collide.

activators Regulatory proteins of eukaryotes, which bind to promotor elements upstream of the coding region in order for it to be optimally transcribed.

active response Amplification of a local depolarisation of a neuron that dies away with distance from the point of initiation; triggered by depolarisation that reaches the threshold potential for the membrane.

active site Specialised region of an enzyme into which substrate molecules fit; pocket or groove formed by folding of the polypeptide chains of the enzyme (quaternary structure).

active transport Carrier-mediated process requiring energy derived from hydrolysis of ATP; can move substances against a concentration gradient.

activin A protein believed to be responsible for the induction of mesoderm in the amphibian *Xenopus*.

adaptation Characteristic of an organism that fits it for a particular environment (*see also* adaptive evolution).

adapter molecule Transfer-RNA; delivers amino acids to nucleotides during protein synthesis.

adaptive evolution The aquisition of inherited anatomical, physiological and behavioural traits that benefit an individual organism by enabling it to survive and reproduce better than other individuals in a particular environment.

adaptive radiation Rapid evolution and divergence of members of one lineage into different niches.

adduction Movement (e.g. of limb) towards the mid-line of the body.

adenosine triphosphate (ATP) An adenine-containing compound that releases free energy when its terminal phosphate bond is hydrolysed; this energy is used to drive energy-requiring reactions in cells.

adherant junction Junction that forms cross-links between contractile bundles of actin filaments in the cortical cytoplasm of neighbouring cells; important in developmental processes.

adhesion belt A band of microfilaments often found in the apical cytoplasm of epithelial cells and associated with adherant junctions.

adrenal cortex Endocrine gland that secretes steroid hormones—mineralocorticoids (such as aldosterone) and glucocorticoids (such as corticosterone).

adrenal gland A gland located anterior to the kidneys; composed of the adrenal medulla (neurosecretory cells) surrounded by the adrenal cortex (non-neural endocrine cells).

adrenal medulla Neurosecretory gland that secretes catecholamines—adrenaline, noradrenaline and dopamine.

adventitious root Root that arises from deep within the stem of certain plants, such as at the nodes of grasses and palms.

aerenchyma Type of parenchyma found in aquatic plants; spongy in appearance with large gas-filled intercellular spaces.

aerobic metabolism Metabolic processes that require molecular oxygen.

aerobic organism Organism that relies on oxidation of organic substrates for energy.

afferent Leading to.

akinete A spore of a cyanobacterium that is an enlarged cell filled with food reserves and which can remain dormant.

aleurone layer The outer layer(s) of the endosperm of cereal grains that produces enzymes required for endosperm breakdown.

alga (pl. algae) Any photosynthetic, aquatic protist, including unicellular and multicellular forms, such as seaweeds.

allantois In amniotes; outgrowth of the embryonic hindgut used for excretion during development.

allele One of two or more forms of a gene located in the same position on homologous chromosomes.

allele frequency The proportion of a particular allele in a population.

allergen Agent that provokes an over-reaction of the immune system.

allergy Immune response that is excessive; may cause tissue damage or be potentially life-threatening.

allograft Graft tissue from other individuals of the same species.

allopatric speciation Geographic separation of populations leading to divergent evolution and formation of new species.

allopatry Living in different areas.

allopolyploid A polyploid organism with sets of chromosomes originating from different species.

allosteric interaction Occurs when the binding of a compound to a protein induces a shape change in the protein at a site distant to the binding site.

alternation of generations The alternation of haploid and diploid stages in the life cycle of eukaryotes.

altruism Behaviour whereby individuals reduce their own reproductive success to enhance the reproductive success of others (e.g. co-operating to help others raise their young).

alveolus (pl. alveoli) A cup-shaped chamber of extremely small diameter located after repeated branching of the airways in mammals; forms the gas-exchange surface of the lungs.

amino acid An organic molecule with an amino group ($-NH_2$, except for proline, which has an imino group, $-NH-$), an acidic carboxyl group ($-COOH$), a hydrogen atom and a unique side chain (R-group) all bonded to a central carbon atom; structural unit of proteins.

aminoacylation Attachment of an amino acid to its specific tRNA molecule catalysed by specific synthetase enzymes.

aminoacyl-tRNA Transfer RNA charged with the amino acid corresponding to its anticodon.

ammonotelic Pattern of excretion of nitrogen wastes in the form of ammonia.

amnion Exta-embryonic membrane enclosing an amniote embryo in a fluid-filled sac.

amniote All tetrapods, other than amphibians; having an amnion.

amoeboid Able to move by means of pseudopodia (temporary cytoplasmic extensions of a cell).

amphipathic molecule Molecule in which there is a difference in water solubility between one end and the other, such as a phospholipid, which has a phosphate head (hydrophilic) and a fatty acid tail (hydrophobic).

amyloplast A kind of plastid containing large starch granules and very few, if any, membranes within the stroma.

anabolism Metabolic reactions involving the building or synthesis of molecules.

anaemia Low concentration of haemoglobin in the blood.

anaerobic Not requiring oxygen.

anaerobic respiration Cellular respiration occurring in the absence of oxygen; glycolysis, fermentation.

analogous Structures that have a similar function as a result of convergent evolution are called analogous. Compare homologous.

anaphase A phase of mitosis in which the two kinetochores of the centromeres separate and sister chromatids move apart, forming two groups.

anapsid Describing a vertebrate skull that lacks one or two pairs of openings in the temples.

anastomose Hyphal fusion in fungi allowing migration of nuclei from one hyphal cell to another.

anatropous Type of ovule in flowering plants in which the micropyle is located at the base, adjacent to the funicle (stalk).

anchoring junction Junction linking neighbouring cells and providing mechanical support; includes desmosomes, hemidesmosomes and adherant junctions.

androgen Steroid hormone including testosterone and dihydrotestosterone; secreted by the Leydig cells of the testes.

angiosperm Flowering plant.

angle of attack The orientation (angle) of an aerofoil or hydrofoil in relation to the direction of travel. A disc travelling through air edge-on has an angle of attack of 0° and travelling flat surface first of 90°.

animal Multicellular heterotrophic eukaryote; member of kingdom Animalia.

animal pole The pole of the egg containing relatively low concentrations of yolk.

anion A negatively charged ion (e.g. Cl^-).

anisogamy Having gametes that differ in appearance, the male being small and motile, the female large and non-motile.

annual rings Rings of xylem in trees and shrubs that represent secondary growth from one season to the next; can be used to determine the age of trees and climatic changes.

annulus Ring of enlarged cells with lignified wall found in sporangia of leptosporangiate ferns.

anomorphic Refers to the asexual form of a fungus, which may have been given a taxonomic name in addition to the sexual stage, often discovered later.

anoxygenic Photosynthetic bacteria that do not generate oxygen during photosynthesis.

antagonist muscle A muscle that can oppose the action of another.

antenna (pl. antennae) Paired appendages, located on the head of an arthropod and bearing sensory receptors.

antennal gland The paired metanephridial excretory organs of crustaceans, which have excretory openings at the base of the antennae.

anterior determinant A developmental factor that specifies both the head and thoracic segments of *Drosophila*; released from the anterior pole of the egg.

anther In a flower, part of the stamen that houses developing male reproductive cells.

antheridium Male gametangium (sex organ) producing sperm (or male haploid nuclei); antheridia are unicellular in algae and fungi and multicellular in plants (bryophytes and lower vascular plants).

antibiotic Naturally occurring inhibitor of bacterial protein synthesis and thus bacterial growth.

antibody Protein molecule produced by B cells in response to antigen and which reacts specifically with that antigen.

anticodon Trinucleotide sequence occurring at the end of a loop in transfer RNA molecules that is complementary to a specific mRNA codon.

antidiuretic hormone Hormone that increases the permeability of the renal collecting duct to water, and thus osmoconcentrates urine.

antigens Any molecule that can be recognised by one of the specific molecules (antibodies or T-cell receptors) of the immune system.

antipodal cells Nutritive cells of the embryo sac of flowering plants; lie at the end opposite the micropyle.

aorta Main artery that carries oxygenated blood from the heart to the body in higher vertebrates.

aortic arches Main arteries in fishes that stem from the ventral aorta, flow past the gills and join the dorsal aorta. Some aortic arches persist in vertebrates without gills, while others have been lost.

apical cell Single meristematic cell in the shoot and root apices of mosses and ferns.

apical complex A structure of endoparasitic protists (sporozoans) involved in penetration of host cells.

apical dominance Occurs in plants where the apical bud inhibits the growth of axillary buds further down the stem, resulting in a single dominant shoot; thought to be maintained by the auxin hormone, indole-3-acetic acid.

apical meristem Specialised growth region at the tip of shoots and roots; cells divide continually to produce the primary tissues and organs of the plant.

apomixis Type of seed formation in plants where the embryo is derived only from cells in the female ovule rather than from the fusion of male and female gametes.

apomorphic An advanced character, one that has evolved more recently, is called an apomorphic character.

apoplastic pathway Pathway of least resistance of water from soil into the plant; through cell walls and intercellular spaces.

appressoria Swelling at the tip of a hypha of a parasitic fungus that penetrates the cell wall of a host plant.

apterous Without wings; especially in insects.

Archaeobacteria 'Very old bacteria'; one of two major groups of bacteria that includes methanogens, halophiles and thermoacidophiles (*see also* Eubacteria).

Archaezoan Primitive form of eukaryote (protist) with a nucleus but lacking typical eukaryote organelles such as mitochondria.

archegonium Multicellular, female gametangium (sex organ) producing egg cells; in all plants except flowering plants.

archenteron Embryonic cavity in animals that becomes the gut.

Archimedes' principle A body immersed in a fluid is subject to an upward force equal in magnitude to the weight of fluid it displaces.

Aristotle's lantern In sea urchins; complex, bony organ with five teeth and used for feeding.

arterial Of an artery; usually referring to blood flowing in arteries.

arterialised Referring to blood that has been oxygenated in a gas-exchange organ. Thus, arterialised blood flows in the pulmonary vein.

arteriole Small muscular arteries leading to the capillaries.

artery Large blood vessel carrying blood from the heart.

arthrology The study of joints, from the Greek *arthron*, joint, and *logos*, study.

ascocarp Fruiting body formed from vegetative hyphae of an ascomycete fungus; encloses asci with ascospores.

ascogonium Female gametangium enclosing female gametes or female haploid nuclei (e.g. in ascomycete fungi).

ascospore Haploid, sexual spore produced in an ascus of an ascomycete fungus.

ascus (pl. asci) Sac-like cell that produces ascospores (ascomycete fungi).

asexual reproduction Reproduction in which offspring are clones of the parent organism.

aspiration Generation of negative pressure that results in air being sucked in.

assortative mating Mating between organisms with similar phenotypes and genotypes.

aster An organelle associated with nuclear division; comprises bundles of microtubules produced from a central centrosome.

atom The smallest part of an element that can exist and retain the properties of that element; comprising a central nucleus made of protons (positively charged) and neutrons (neutral charge) surrounded by one or more orbiting electrons (negatively charged).

atomic number Number of protons in the nucleus of an atom; characteristic for each element.

atrioventricular bundle (Bundle of His) A group of rapidly conducting cells that leads from the atrioventricular node to the Purkinje fibres; together, they produce a co-ordinated ventricular contraction.

atrioventricular node A patch of modified muscle cells lying between the right atrium and ventricle in higher vertebrates; slows the conduction of excitation between atria and ventricles.

atrium (pl. atria) Heart chamber that receives blood from veins or the sinus venosus and delivers it to the ventricle.

Australian region Biogeographic region including Australian mainland and islands on the continental shelf, such as Tasmania and New Guinea.

auto- From the same or a genetically identical individual; as in autoimmune, autoantigen, autograft.

autocrine hormones Hormones that interact with receptors on the surface of the cell that releases them.

autoimmune disease Disease resulting from development of an immune response to an individual's own antigens; usually chronic diseases, such as diabetes and arthritis.

autonomic nervous system Nervous system in animals that innervates the visceral organs of the body so that their functions are not consciously controlled.

autoregulation Control of blood flow to a tissue caused by direct effects of metabolites on smooth muscle of arterioles and precapillary sphincters.

autosomes Chromosomes that exist in pairs in diploid organisms.

autotroph Organism able to synthesise its own food by photosynthesis or chemosynthesis.

auxin A type of plant hormone controlling stem elongation; synthesised in the growing shoot and root tips of plants; synthetic auxins used as selective herbicides.

axon A long neuronal process, which carries the output of the neuron to the next cell.

axon hillock The first part of an axon, the membrane properties of which allow action potentials to be generated.

axoneme A precise array of microtubules covered by the plasma membrane; structural basis of cilia or flagella.

axopod Long, slender radial projection of cytoplasm bounded by plasma membrane; characteristic of unicellular, actinopod protists.

B cell (B lymphocyte) Lymphocyte that makes antibodies; produced in the bone marrow, spleen or gut lymphoid tissue.

β-oxidation Process involving the release of chemical energy stored as C—C bonds in lipids; results in long-chain fatty acids being degraded by two carbon atoms at a time to form acetyl-CoA.

backcross In plant breeding, a cross between F_1 (heterozygous) individuals and either of their pure-breeding parents.

bacteriophage Virus that infects and multiplies in bacteria; commonly called phage.

bacterium (pl. bacteria) The smallest cellular life form on earth (*see* prokaryote).

balanced polymorphism Genetic polymorphism that is stable and can be maintained in balance in terms of Hardy–Weinberg principle; occurs if heterozygotes for particular alleles are fitter than either homozygote.

baroreceptor (pressoreceptor) Nerve endings in the walls of blood vessels that sense blood pressure.

base Substance that can accept hydrogen ions (H^+) causing a decrease in their number in solution; possessing a pH above 7.

basement lamina Type of extracellular matrix that underlies epithelial cell layers.

basic helix-loop-helix proteins A family of transcriptional regulatory proteins that share particular structural domains, which confer sequence-specific DNA binding.

basidiocarp Fungal fruiting body such as a toadstool or mushroom (basidiomycete fungi).

basidiospore Spore of a basidiomycete fungus borne externally on a basidium.

basidium A club-shaped fungal cell that bears basidiospores on its surface (basidiomycete fungi).

Batesian mimicry Resemblance of one animal (the mimic) to another to the benefit of the mimic; named after the naturalist H. W. Bates.

bilateral symmetry Symmetry of an organism where only one plane divides the organism into two similar halves.

binary fission Process of cell division involving cleavage to create two equal-sized cells each containing one copy of the genetic information and approximately half the cytoplasm.

binomial system The system devised by Linnaeus whereby the name of each kind of organism (each species) consists of two words: genus name and specific epithet.

biodiversity Number, relative abundance and genetic diversity of organisms on earth (*see also* species diversity).

bioeconomics The study of the interaction of economic and biological systems.

biogeochemical cycle The movement of material through an ecosystem, from atmospheric and geologic stores through food webs and back again.

biological control Control of a pest species by biological means, for example, introducing herbivore to consume and control a plant pest.

biomass 'Living mass'; amount, usually expressed as weight, of organisms in a particular area at a particular time.

biome On a global scale, ecological communities with the same structure and delineated by climate (e.g. grasslands of the world).

biomechanics The field of study involving the application of engineering principles to biology.

biota Fauna and flora of a given habitat or region.

biotic Pertaining to organisms.

biramous Referring to appendages consisting of two parts (e.g. in crustaceans).

blastocoel First cavity of the embryo; appears during cleavage.

blastocyst Mammalian blastula.

blastodisc Layer of cells of the avian embryo, forming a disc on the uncleaved yolk mass.

blastomeres Cells of an embryo during cleavage.

blastopore Depression on the surface of the gastrula marking the site at which inward cell movement occurs.

blastula Embryo at the end of cleavage.

Bohr effect A decrease in the oxygen affinity of a respiratory pigment when the partial pressure of CO_2 and/or proton (H^+) concentration increases.

book gill In some chelicerates; abdominal appendage, modified as a gill that has many leaf-like folds (lamellae); for gas exchange.

book lung Gas-exchange organ of spiders and scorpions; similar in structure to a book gill but internal on the ventral side of the abdomen.

bordered pits Pores with an overarching lip; connecting water-conducting tracheids of vascular plants.

Boreal region From or belonging to the north; biogeographical region of the world extending from the Polar Sea southwards and including North American and Eurasian regions (also called Holarctic).

boundary layer A barrier to diffusion that is established in the layer of fluid next to a surface across which a diffusion gradient exists.

bradycardia Decrease in heart rate.

buccal chamber Mouth cavity.

buccal force pump Mechanism that forces gas in the buccal chamber into the lungs under positive pressure.

budding A form of asexual reproduction involving the development of a new individual from outgrowths of the body wall of the parent.

buffer A substance that minimises changes in the pH of a solution by taking up or releasing H^+ ions when extraneous acids or bases are added to the solution.

bundle sheath Layer of cells surrounding veins of leaves of some plants; provides a link between photosynthetic mesophyll cells and vascular tissue.

bursicon (tanning hormone) Brain neurosecretory hormone that produces hardening and darkening of the adult cuticle in insects.

C phase Phase of the cell cycle; cytokinesis (division of the whole cell).

C_3 photosynthesis The process of carbon fixation in most plants in which the 3-carbon compound phosphoglyceric acid is the first stable product.

C_4 photosynthesis The process of carbon fixation in which 4-carbon compounds (e.g. malate) are the first stable product; found in tropical and subtropical grasses and cereals.

callus cells Cells produced at a cut surface of a plant; wound tissue; widely used in in-vitro culture.

calyptra Tissue of the old gametophyte neck that persists on the top of a moss spore capsule.

cambium A secondary (sheet) meristem in vascular plants; vascular cambium increases the girth of stems and roots.

canine teeth Conical teeth used for stabbing and gripping.

capillary Smallest blood vessel where exchange of substances between blood and extravascular fluid occurs.

capillary action Combined effect of cohesion and adhesion; means by which water rises up a fine capillary tube.

capsule A dry simple fruit that opens by valves on the top; fruit typical of eucalypts.

carapace Dorsal, protective covering over the thorax or anterior trunk segments of crustaceans; dorsal, protective shield in turtles.

carbohydrates Most abundant organic compounds in nature, composed of carbon, hydrogen and oxygen; basic unit a sugar molecule.

carbon fixation The capture of atmospheric carbon dioxide and its conversion into carbohydrates; occurs in the stroma of chloroplasts in eukaryotes.

carbonic anhydrase An enzyme that speeds up the hydration of CO_2 by a factor of 109.

cardiac centre A centre in the brain that controls the rate and strength of the heart beat.

cardiac muscle Branching network of individual striated muscle cells linked by many communicating junctions; found in the heart, hence its name.

cardiac output Rate of blood flow (mL/min) from one ventricle of the heart.

carnassial teeth Large molar teeth for shearing flesh and/or bone; found in carnivores.

carnivore Animal that catches other animals for food.

carotenoid Orange, yellow, red or brown fat soluble pigments involved as accessory pigments in photosynthesis; also found in flowers and fruits; carotenes and xanthophylls.

carotid body A peripheral chemoreceptor found in mammals and birds near the bifurcation of the common carotid; responds to changes in the partial pressure of oxygen.

carpel Female reproductive organ of a flowering plant; encloses ovules; ripens to become a fruit.

carrier proteins Membrane proteins involved in the transport of molecules across cellular membranes.

Casparian strip Strip of suberin thickening on the radial walls of the endodermis or exodermis of plant roots; regulates uptake of water and solutes.

caste A set of individuals within a colony of ants that are morphologically distinct and behaviourally specialised; includes queens, workers, soldiers and alates.

catabolism Metabolic reactions involving the breakdown of molecules.

catalase Enzyme involved in breakdown of hydrogen peroxide; present in large amounts in microbodies.

catalysis Process by which the activation energy of a reaction is lowered; affects only the rate of the reaction.

cation A positively charged ion (e.g. Na^+).

cavitation Breaking of the water column in xylem vessels under water stress.

cDNA library cDNA clones spliced into vectors and transformed into eukaryotic host cells.

cell cycle The continuous cycle of growth and division in cells.

cell fate The developmental destiny of a cell.

cell lineage Series of cell divisons that precedes the formation of a particular cell.

cell plate Region of new cell wall that forms during cytokinesis in eukaryotic, walled cells.

cell wall A strong and rigid extracellular matrix outside the plasma membrane of plant, fungal and bacterial cells; in plant cells, the wall is characteristically composed of cellulose.

cell-adhesion molecules Molecules involved in the adhesion between cells, or between a cell, and the extracellular matrix.

cellular immunity Active destruction of foreign cells or of the body's own virally-infected cells by T cells.

cellular respiration Oxidation of fuel molecules, that is, removal of electrons, coupled to synthesis of ATP.

cellulases A group of enzymes able to hydrolyse cellulose molecules; produced by some micro-organisms and a few invertebrates.

cellulose A structural polysaccharide present in cell walls of plants and some protists; composed of long chain of glucose molecules.

Cenozoic The youngest era in the geologic time scale, from 65 million years to present day.

centimorgan (cM) A measure of the extent of linkage between two genes, expressed as the percentage of recombination between the two loci.

central nervous system The brain and spinal cord of vertebrates.

centriole An organelle near the nucleus of animal and protist cells; associated with the spindle during mitosis and meiosis.

centromere A constricted region of a chromosome at which the two chromatids are held together.

centrosome Organelle containing two centrioles at right angles at which microtubules are assembled; the microtubular organising centre.

cephalisation *See* encephalisation.

cephalothorax United head and thorax in crustaceans and spiders.

cercaria (pl. cercariae) The fluke larva that develops from a redia.

CFCs Chlorofluorocarbons; used as aerosol propellants and in refrigerators, air conditioners and freezers; chlorine in CFCs reacts with and breaks down atmospheric ozone.

checkpoint controls Mechanisms that prevent progression into the next cell cycle phase if they sense non-completion of essential cell cycle processes (e.g. DNA replication) or problems with processes such as spindle formation.

chela Claw of a chelicerate (e.g. crab).

chemical digestion Breakdown of food molecules by hydrolytic enzymes into smaller molecules.

chemoautotroph An organism that uses reduced inorganic substrates as sources of energy and reduces carbon dioxide to organic carbon, using water or hydrogen gas as a reductant; certain forms of bacteria.

chemoheterotroph An organism that uses organic substances as a source of both carbon and energy (e.g. animals, certain forms of bacteria).

chemoreceptor A type of receptor that binds to a particular signal molecule, a ligand.

chemotroph An organism that gets its energy chemically, ultimately by oxidation–reduction reactions.

chiasma (pl. chiasmata) Attachment point between chromosomes where crossing over occurs.

chlorenchyma Photosynthetic parenchyma cell.

chloride shift Maintenance of electrical balance within an erythrocyte by the exchange of Cl^- for HCO_3 across the membrane.

chlorophyll A light-absorbing, green pigment involved in photosynthesis.

chloroplast An organelle (plastid) containing membrane-bound light-absorbing pigments; functions in photosynthesis.

choanocyte Collar cell; flagellated cell lining the internal cavity of a sponge.

chorioallantois A vascular extraembryonic membrane consisting of the fusion of the chorion and allantois; underlies the eggshell and functions for respiration and osmotic and ionic regulation in avian and reptilian embryos.

chorion Outermost embryonic membrane in amniotes.

chromatid A single chromosomal strand.

chromatin The DNA–protein complex that makes up eukaryotic chromosomes.

chromatophore Cell that contains pigment granules and expands and contracts under muscular control, allowing body colour change.

chromophore Light-absorbing region of a protein photoreceptor; absorbs light of a particular waelength.

chromoplast Plastid containing carotenoid pigments; responsible for red, orange or yellow colours of some plant organs.

chromosomal inversion Chromosomal mutation where a segment within a chromosome is turned around end to end.

chromosomal translocation Chromosomal mutation where segments are exchanged between non-homologous chromosomes.

chromosome Structure containing a single DNA molecule; in prokaryotic cells and in the nucleus, mitochondria and chloroplasts of eukaryotic cells; nuclear chromosomes are visible during cell division.

cilia Short, thin extensions of cytoplasm that undergo vigorous bending movements from their base, thus providing non-muscular locomotion as a result of a powerstroke.

citric acid cycle Also known as the Krebs cycle; cyclic series of reactions involving oxidation of fuel molecules; occurs in mitochondria in eukaryotes.

cladode A photosynthetic stem in plants with leaves reduced or absent.

cladogram A branching diagram or phylogenetic (evolutionary) tree that shows the relationships of organisms and their descent from a common ancestor.

clamp connection Feature of basidiomycete fungal hyphae that ensures that the two nuclei of the dikaryon remain together following mitosis.

classification A hierarchy of groups and subgroups of organisms reflecting their phylogenetic relationships. Each group (taxon) is given a name and rank: kingdom, phylum, class, order, family, genus, species (*see also* taxonomy).

cleavage Series of mitotic cell divisions that takes place in the egg after fertilisation and that results in a progressive decrease in cell size.

cleistogamy In flowering plants, a process that ensures self-fertilisation; anthers open and self-pollination occurs within unopened flowers.

climacteric Period of increased respiration in fruits that includes a set of changes resulting in fruit ripening.

clitellum Thickened region of the epidermis of a cuclitellate worm (e.g. earthworm) that secretes the cocoon in which eggs are deposited.

clonal selection Process during an immune response in which lymphocytes that encounter their specific antigen are stimulated to proliferate, thus increasing the number of cells reacting to that antigen.

clot (thrombus) An aggregation of blood platelets, fibrin and trapped blood cells, usually at the site of a wound where it prevents blood loss.

cnidocyte Cell in a cnidarian that contains a nematocyst.

co-current flow Flow of two media in the same direction.

coding sequences DNA sequences that encode amino acids incorporated into polypeptides during protein synthesis.

codominance The full expression of two alternative alleles in a heterozygote.

codominant Alleles whose phenotypes are equally recognisable in the heterozygote.

codon A set of three nucleotides in RNA that determines the amino acid incorporated into a growing polypeptide.

coelenteron In cnidarians; gastrovascular cavity lined with endoderm.

coelom Body cavity of an animal, lined on all sides by mesoderm.

coelomoduct A tubular excretory organ that has a ciliated, funnel-like opening in the coelomic cavity to draw coelomic fluid into the tubule; develops from the interior of an animal towards the outside, unlike nephridia.

coenocytic A term used to describe a cell or non-septate hypha containing numerous nuclei.

coenzyme A type of cofactor, required by an enzyme to function as a catalyst; non-protein, complex organic molecule, often with a vitamin as a building unit.

coevolution The evolution of two species in relation to one another, such as flowers and their animal pollinators, parasites and their hosts.

cofactor An additional chemical component, such as a metal ion or organic molecule, required by certain enzymes in order to function.

cohesion The attraction between similar polar molecules, such as hydrogen bonding between water molecules.

colchicine Drug derived from the bulbs of the autumn crocus (*Colchicum*); causes disassembly of spindle microtubules, preventing completion of cell division.

collagen A structural protein of the extracellular matrix that is the most abundant protein of mammals; collagens associate into a strong sheet-like meshwork in basement lamina and form fibrils in interstitial matrices.

collecting duct The terminal portion of the vertebrate nephron, which conveys fluid from the distal convoluted tubule into the renal pelvis.

collenchyma Living plant cells strengthened with primary thickening either at the corners or on the tangential walls; have a support function.

colloid osmotic pressure (oncotic pressure) Osmotic pressure due to large proteins, chiefly albumin, in the blood; involved in the balance between filtration and reabsorption of fluid in tissues.

colony A group of cells derived from a single initial cell. Normally used to describe bacterial or unicellular fungal (e.g. yeast) clusters of cells derived from one cell growing on a nutrient agar plate.

commensalism Symbiotic interaction between two species where one benefits and the other is unaffected; usually one organism living with another for shelter or support (e.g. epiphyte).

communicating junctions (gap junctions) Junctions specialised for chemical and electrical communication between cells.

community In an ecological sense, an assemblage of populations of different species, interacting with one another, living in a particular area (e.g. pond or forest).

companion cell A cell type of phloem; transfer cell involved in the loading of sucrose into the sieve cells.

competency Refers to a host cell that has been treated so that it will take up DNA in the environment (i.e. the cell is competent to be transformed).

competition Individuals of either one species (intraspecific competition) or different species (interspecific competition) striving for the same resource, which is in limited supply.

competitive exclusion When one species outcompetes another for a limited resource, resulting in its local extinction.

complement system A series of about 20 serum proteins that activate sequentially in a cascade of reactions. Triggering of the complement cascade leads to activation of non-specific defensive cells, facilitation of phagocytosis and lysis of cells.

complementary DNA (cDNA) A DNA copy of an RNA transcript.

compliance Relationship between change in volume and change in pressure.

compound A molecule composed of more than one type of atom.

compound leaves Leaves divided into leaflets, each with its own stalk.

condensation reactions Reactions involving removal of water molecules in the assembly of complex molecules from simpler ones.

conformation (of a protein) Three-dimensional shape of a protein molecule.

conidiophore Specialised fungal hypha that forms a stalk and bears spores (conidia).

conidiospore (conidium) Asexual fungal spore formed on a conidiophore; for dispersal and spread of the fungus.

conjugated protein A protein molecule consisting of amino acids and other organic or inorganic components, such as glyco-, lipo-, nucleo- and phosphoproteins.

conjugation Process by which organisms, such as bacteria, make direct contact and transfer DNA via plasmids from one cell to another, leading to genetic variation.

connective tissue Tissue that provides structural, metabolic and defensive support for other tissues; for example, blood, bone and cartilage; extracellular matrix usually more abundant than cells.

constitutive expression Constant expression of a gene.

constitutive secretion Secretion that continues throughout the life of a cell.

consumer An organism that derives its energy by consuming other organisms.

contractile vacuole An organelle of cells that excretes fluid by a pulsating action, first filling the vacuole with fluid then ejecting the fluid from the cell.

convection Mass movement of a fluid; transfer of heat or substances along with a moving fluid, usually air or water.

convection requirement Amount of medium passed over a gas-exchange surface in order to extract a given amount of oxygen.

convergent evolution Evolution whereby organisms from different, distantly related lineages come to resemble one another.

co-operative breeding Occurs when adults in a group do not themselves reproduce, but instead help raise the young of the breeding individuals.

coprophagy (caecotrophy) Reingestion of special faecal pellets (the contents of the caecum, which are very rich in microbes) released at night, allowing utilisation of this high protein source.

copulation Mating between sexes associated with internal fertilisation.

coralloid roots Coral-like upward growth of roots of certain plants (alders, cycads, she-oaks) following root hair infection by the bacterium *Frankia* or cyanobacteria.

cork cambium Meristem responsible for producing bark at the surface of stems and roots of plants.

coronary artery Artery that supplies blood to heart muscle.

corpora allata Non-neural endocrine gland in insects, which secretes juvenile hormone.

corpus In the shoot apex of flowering plants, the inner layers of the apical dome of cells that contribute to stem formation.

corpus luteum Ovulated follicle, which secretes the steroid hormone progesterone.

corpuscle A blood cell.

cortex (adj. cortical) Outer region of structure.

cotyledon Leaf of a plant embryo functioning in food storage and digestion; first leaf to emerge following seed germination.

countercurrent exchange Transfer of heat or a substance from a fluid flowing in one direction to fluid flowing in the opposite direction in an adjacent vessel.

countercurrent flow Flow of two media in opposite directions.

courtship behaviour Interactions between members of the opposite sex that take place before mating.

covalent bond Bond formed between atoms due to sharing of electrons in their outermost orbitals.

crassulacean acid metabolism (CAM) A variation of the C_4 pathway of photosynthesis, in which C_4 and Calvin–Benson cycle reactions occur in the same cells but at different times. CAM plants fix CO_2 at night and convert it to carbohydrate during the day.

cretinism Impairment of growth and development, particularly of the nervous system, as a result of lack of thyroid hormones in children.

cristae Folds of the inner membrane of mitochondria.

cross-current flow Flow of one medium perpendicular to the flow of another.

crossing over Exchange of DNA between chromatids of homologous chromosomes; involves cutting and rejoining strands of DNA.

crozier Curled branch of a dikaryotic, heterokaryotic hypha where mitosis occurs and acsi form in ascomycete fungi.

crustose A flat, 'crusty' growth form (e.g. of a lichen).

cryptic Deceptive defence mechanism by which an animal is well camouflaged and blends into the background substrate, thus reducing the risk of predation.

cuticle Outer water-resistant layer secreted by epidermis.

cyanobacterium Photosynthetic eubacterium that has chlorophyll *a* and produces oxygen as a by-product of photosynthesis.

cyclosis Cytoplasmic streaming. Circulation of cytoplasm inside cells.

cyclosporine Immunosuppressant used in medicine and derived from a fungus (*Tolypocladium inflatum*).

cysticercus Tapeworm larva consisting of a bladder-like structure and inverted scolex.

cytochalasins Anti-actin agents derived from certain fungi, which act by specifically disrupting actin microfilament-based cytoskeletal systems.

cytokinesis Division of the cytoplasm of a cell following mitosis or meiosis.

cytokinins Plant hormones promoting cell division.

cytoplasm The cytosol and organelles of eukaryotic cells, excluding the nucleus.

cytoplasmic determinants Substances present in the egg cytoplasm that are believed to determine embryonic cell fate.

cytoskeleton Network of microtubules and microfilaments in eukaryotic cells; involved in functions such as maintenance and change in cell shape, movement of organelles within the cytoplasm and cell movement.

cytosol An aqueous solution of molecules with a gel-like consistency within the cytoplasm of eukaryotic cells.

cytotoxic cells (T_C cells, killer T cells) T cells that, when stimulated by antigen and lymphokines produced by T_H cells, directly lyse or kill target cells recognised by T_C cells on the basis of their particular antigen.

Darwinian fitness Estimates fitness (adaptive value) of individual genotypes of a species in terms of relative reproductive success.

day-neutral plant Plant not affected by day length for flower initiation.

dead space Volume of air in the conducting airways that takes no part in gas exchange.

decomposer Organism, such as some fungi and bacteria, that consume and break down organic matter for energy, releasing inorganic nutrients.

deletion A mutation that removes one or more nucleotides from the DNA.

dendrites Branching processes of neurons that are generally short and receive information from other cells.

dendritic cell Cell of the immune system that has long branching processes and is able to break down foreign molecules and present them to lymphocytes.

denitrification The conversion of nitrate to nitrite and nitrite to molecular nitrogen; carried out by certain types of bacteria in ecosystems.

density-dependent In ecology, when the per capita birth and death rates of a population depend on the size of the population.

density-independent In ecology, when the per capita birth and death rates of a population are independent of the size of the population; also termed density-vague.

deoxyribonucleic acid (DNA) A nucleic acid that is the hereditary material of an organism, stored as a coded sequence of nitrogenous bases; comprising two complementary double helical strands of nucleotides made up of a pentose sugar, phosphate group and nitrogenous base.

depolarisation Decreased voltage difference across a membrane; brings membrane potential closer to threshold potential and therefore is excitatory.

dermal bone Bone that develops in the skin without going through a cartilaginous phase.

desmosomes Provides structural support by cross-linking between cytoskeletons of neighbouring cells.

determination Process by which the developmental fate of a cell is fixed.

detritivore Animal that feeds on detritus, that is, dead organic matter.

detritus Debris and dead remains of organisms in an ecosystem.

deuterostome Animal in which, during development, the anus forms at the site of the blastopore and the mouth forms as a secondary opening.

development Series of events leading to the formation of an adult organism from a zygote.

diabetes mellitus Insulin-dependent diabetes caused by the destruction of beta cells of pancreatic islets, apparently due to an autoimmune response.

diapause A period of dormancy during development in which growth and differentiation virtually cease.

diaphragm Mammalian structure composed of muscle and tendons; separates the thoracic cavity from the abdominal cavity; a major source of inspiratory force.

diapsid Describes a vertebrate skull with two, well-defined temporal openings.

diastole Phase of the cardiac cycle involving muscle relaxation and filling of a heart chamber.

dichotomous branching Dividing into two equal branches.

dicotyledon One of two major types of flowering plant that typically has two embryonic leaves in the seed.

dideoxynucleoside triphosphates (ddNTPs) Nucleoside triphosphates lacking both the 2′ and 3′ OH groups so that they can be incorporated into a growing DNA chain, but prevent addition of any further addition of nucleotides, permitting rapid DNA sequence determination.

differentiation Process leading to changes in structure and function of cells during development often leading to specialisation; in animal cells such changes are often irreversible.

diffusion Net passive movement of molecules from a region where they are in high concentration to one where they are in low concentration; due to random thermal motion of molecules; passive movement of molecules along their electrochemical gradient.

dihybrid cross A cross involving organisms that are heterozygous at two different loci.

dikaryon Fungal cell containing two haploid nuclei, one from each parent; usually formed after sexual fusion of parent hyphae.

dilate To increase in diameter, as in blood vessels or pupil of the eye.

dioecious Organism in which sperm and eggs are produced by separate individuals.

diploblastic In an animal, having two cell layers.

diploid An organism that carries two sets of chromosomes.

direct development Development in which an animal is born with the general form of the adult.

disaccharide Two monosaccharide molecules joined by a glycosidic bond.

disulfide bond Covalent bond that can cross-link polypeptide chains; formed between the sulfur atoms of two cysteine residues.

divergent evolution Evolution that leads to descendants becoming different in form from their common ancestor.

diving response A co-ordinated set of cardiovascular and metabolic changes that occurs strongly in diving, air-breathing vertebrates.

DNA hybridisation The process whereby two DNA strands from different sources form a double-stranded DNA molecule through complementary base pairing. DNA strands require complementary sequences to be able to hybridise with each other and form double-stranded DNA.

DNA polymerases Enzymes that catalyse the replication (template-dependent synthesis) and repair of DNA.

dolipore septum Septum with complex pore separating adjacent cells in hyphae of basidiomycete fungi.

dominance hierarchy Physical domination of one individual over another; usually established by aggressive behaviour and once established the relationship remains stable without subsequent high levels of aggression.

dominant A homozygous phenotype, such as yellow colour in seeds, that appears in a heterozygous (Yy) organism.

dominant oncogene A mutated version of a normal cellular gene that is overexpressed, misexpressed or produces an altered product to cause tumour formation.

donor DNA DNA that is to be cleaved into segments, ligated into a vector and transformed to produce cloned DNA segments.

dormant State of a mature seed that does not result in germination as a result of rehydration but is broken more by periods of cold or exposure to appropriate levels of red light.

dorsal lip of the blastopore Region of the blastopore which, in amphibian embryos, initiates gastrulation and induces formation of a dorsal axis.

drag Backward component of force acting on a moving body produced by a fluid resistance.

drupe Fleshy fruit, such as a plum, containing a single seed enclosed in a hard stony layer (endocarp).

ductus arteriosus An embryonic blood vessel that connects the pulmonary artery with the aorta so that much of the blood from the placenta bypasses the non-functional lungs of the fetus.

ecdysial gland Non-neural endocrine gland in the head of some insects; secretes ecdysone.

ecdysone Insect hormone, secreted by the ecdysial glands, which stimulates moulting, growth and differentiation of adult tissues.

eclosion Hormonally induced change from pupal state to emergence of the adult in holometabolous organisms.

ecological gradient Continuous or gradual change in environmental variables that occurs over geographic distance.

ecological pyramid Diagram showing the change in energy, biomass or numbers of organisms at successive trophic levels in an ecosystem.

ecological sustainable development (ESD) Using, conserving and enhancing resources so that ecological processes in ecosystems are maintained.

ecosystem An ecological community together with the physical environment with which its members interact.

ecotone Boundary between two different ecological communities.

ecotoxicology Field of biology that deals with the effects of pollutants on organisms and ecosystems.

ectoderm Outermost germ layer of animal embryos, giving rise to the outer body covering and associated structures.

ectotherm Animal whose body temperature is more or less determined by the temperature of its surrounding environment.

Ediacaran fauna Fossil traces of a collection of soft-bodied animals dated at 640–680 million years BP; best evidence that animals had evolved in the Precambrian; found on all continents.

efferent Leading away from.

egg Female gamete.

elastin Structural protein of the extracellular matrix that is unusual because it remains in an unfolded, random coil configuration.

electrocardiogram (ECG) Electrical activity from the heart that can be measured with electrodes on the body surface. The sequence of cardiac events can be interpreted from changes in electrical potential between the electrodes.

electrochemical gradient Combined concentration and electrical gradient along which ions diffuse across a membrane, provided the appropriate ion-selective channel is present and open.

electron transport system A group of membrane-bound enzymes and cofactors, which operate sequentially in a highly organised manner.

element A substance made up of only one type of atom with the same atomic number.

elephantiasis Grotesque swelling of lymphatic tissue caused by the tropical nematode parasite *Wuchereria bancrofti*.

elimination Loss of undigested and unabsorbed food from the digestive tract (not to be confused with excretion).

embryo A developing organism.

embryo sac Female gametophyte of flowering plants typically containing seven cells and eight haploid nuclei.

embryogenesis Series of events between fertilisation and hatching or birth.

embryonic induction The determination of embryonic cell fate as a result of influences from a neighbouring cell or tissue.

encephalisation Over the course of evolution, the progressive aggregation of nerve cells in groups at the anterior end of animals.

endarch xylem Pattern of primary xylem development in which new xylem is added to the outside of the protoxylem in stems of plants.

endemic (species) Describes a species that is unique to a specific geographic region; assumed to have evolved there.

endergonic Reaction in which the change in free energy is positive; energy is needed for the reaction to proceed.

endocrine glands In animals, glands of internal secretion; usually secrete into the circulatory system.

endocytosis Process of invagination of the plasma membrane to form a vesicle containing extracellular material that is transported into the cell.

endoderm In animals, innermost germ layer; lines the archenteron and gives rise to the epithelial mucosa of the gut and associated structures, such as glands, and the lining of the lungs.

endodermis In plants, layer of cells immediately outside the pericycle of a root; regulates uptake of water and solutes into the central vascular cylinder by means of the Casparian strip.

endoplasmic reticulum (ER) Network of membranous sacs (cisternae) extending throughout the cytoplasm of a eukaryotic cell; usually flat and sheet-like but can be linked by tubular cisternae.

endosperm Triploid nutritive tissue in seeds of angiosperms.

endostyle Ciliated ventral groove in the pharynx of tunicates that secretes mucus to trap planktonic food.

endosymbiosis *See* symbiosis.

endothecium Secondarily thickened layer of cells that lies within the epidermis of an anther of a flower, and following loss of water through the filament, causes the anther to open, releasing pollen.

endotherm Animal with a relatively constant body temperature that is usually higher than the temperature of the surrounding environment; maintained by regulation of internal heat production.

energy Capacity to do work; exists in a number of forms, including chemical, heat, sound, electricity and light.

enhancer A sequence in a eukaryotic gene that binds transcription factors to increase gene expression.

enteric nervous system Division of the autonomic nervous system that controls the functions of visceral organs.

entropy Measure of disorder (randomness) in a system; energy becomes lost as heat in every energy conversion, resulting in increased entropy.

environmental niche Total range of conditions under which members of a species live and reproduce.

enzyme Biological catalyst, usually a protein, which increases the rate of a reaction.

epiboly Overgrowth of one cell layer by another layer during gastrulation.

epicotyl Growing meristem or shoot of a germinating seed that lies above the cotyledons.

epidermis Outer cellular layer of a multicellular organism.

epigenetic regulation Inherited states of activity of a gene, independent of the genotype.

epigynous Describes a flower with an inferior ovary, that is buried within the receptacle below the perianth.

epiphyte Plant that grows on another plant for support, but is not parasitic.

epistasis The masking of the phenotype of one gene by the phenotype of a different gene.

epithelium Tissue that forms a continuous layer covering internal or external surfaces of most multicellular organisms.

epitoky Morphological changes in some invertebrates that lead to sexual maturity, from the atoke (non-reproductive) to the epitoke (reproductive) form.

epitope Portion of an antigenic molecule that is recognised by a receptor. A large protein may have hundreds of different epitopes.

equilibrium (of a chemical reaction) When a chemical reaction is in this state, there is no net change in either the concentration of reactants or products.

erythrocyte Mature, anucleate red blood cell; contains haemoglobin.

erythropoiesis Sequence of synthesis of erythrocytes, mainly in bone marrow.

erythropoietin Glycoprotein hormone produced by the kidney in response to low oxygenation of the blood; stimulates erythropoiesis.

essential amino acid Amino acid that cannot be manufactured by an animal and so must come from a dietary source.

ethology The study of animal behaviour.

ethylene (C_2H_2) Gas of low molecular mass, influencing a wide range of processes in plant development, including promoting fruit ripening, shoot growth and flower senescence.

etioplast Plastid that develops in darkness which, on exposure to light, develops into a chloroplast.

Eubacteria 'Good or true bacteria'; one of two major groups of bacteria, with features typical of most bacteria (*see also* Archaeobacteria).

euchromatin Lightly staining regions in an interphase nucleus; consisting of dispersed strands of chromatin that are sites of active gene transcription.

eukaryote Protists, fungi, animals and plants; cellular organism with membrane-bound organelles such as a nucleus, mitochondria and chloroplasts.

eukaryotic cell Cell with a nucleus and other membrane-bound organelles (compare prokaryotic cell).

euryhaline Able to tolerate a broad range of salinities.

eusocial Social group in which individuals co-operate in raising young; essentially sterile workers care for the young of reproductively active individuals (reproductive division of labour).

eusporangiate Describes a fern with large sporangia, containing numerous spores.

evolution Process of change and divergence in populations and taxa.

evolutionarily stable strategy Strategy (or set of behaviours) that, if adopted by most members of a population of interacting individuals, cannot be bettered (in terms of reproductive success) by another.

exarch xylem Pattern of xylem development in roots in which the xylem forms from the outside, filling the centre of the root.

excretion Loss of ions, solutes, metabolic waste products or water from body fluids (not to be confused with elimination).

exergonic Term used to describe a reaction when the change in free energy is negative; energy is released in the reaction.

exine Outer patterned layer of pollen grains.

exocrine glands Glands of external secretion.

exocytosis Fusion of a vesicle with the plasma membrane, expelling its contents from the cell.

exodermis Layer of suberised cells at the junction of epidermis and cortex of certain roots; regulates uptake of water, solutes and ions into the cortex.

exogenous mutagen Environmental agent causing mutation (e.g. radiation and chemicals).

exons Segments of RNA that remain in the mRNA following splicing.

exoskeleton External hard body covering of some animals.

exponential growth Population growth in which the size of the population regularly doubles; population size thus increases rapidly (geometrically).

expressed sequence tag A known sequence of an expressed gene, determined by sequence analysis of cDNA clones.

expressivity The degree to which an allele is expressed phenotypically in an individual.

extracellular environment The physical and chemical environment that exists outside cells; important aspects of the extracellular environment are ion concentrations, organic solute concentrations and total osmotic concentration.

extracellular matrix Forms the extracellular environment of animal cells; a fluid matrix containing an extensive network of proteins and polysaccharides linked together by covalent and non-covalent bonds that fills the spaces between cells.

F_1 progeny The progeny of a cross between two pure-breeding individuals.

F_2 progeny The progeny of a cross between F_1 progeny.

facilitated diffusion Passive movement of molecules across a membrane in which transport by membrane-spanning carrier proteins enables faster movement of molecules than by the diffusion gradient alone.

fatty acid Hydrocarbon chain with a carboxyl group at one end; component of many lipids.

fecundity Probability of giving birth.

fermentation Anaerobic production of alcohol, lactic acid or similar molecules from carbohydrates by the glycolytic pathway.

fertilisation Specific interaction between an egg and sperm leading to formation of a zygote.

fibre (in diet) Cellulose and pectins of plant cell walls, which are not easily digested and which form bulk in the diet.

fibronectin An adhesive protein of the extracellular matrix occurring in interstitial matrices; has a high relative molecular mass (about 460 kD) and two polypeptide chains.

filopodia Long, thin projections seen in many migrating cells and in the growing tips of axons of nerve cells.

filtration Loss of fluid through holes in capillary walls (or membranes) due to hydrostatic pressure. Water and small dissolved substances move through but large proteins and blood cells remain.

First Law of Thermodynamics *Energy can be neither created nor destroyed*; energy can be transformed from one form to another but the total energy of the universe remains constant.

first polar body One of the products of the first meiotic division of the oocyte, a small cell that eventually degenerates.

fitness Biological success as measured by an individual organism's contribution of offspring to the next generation.

flagella Long, thin extensions of cytoplasm that result in non-muscular locomotion when a wave of bending travels from the tip of a long flagellum to its base, or base to tip, forcing water in the opposite direction.

flame cell Cell of excretory organs, protonephridia, found in flatworms and annelids.

flower Sexual reproductive structure of angiosperms; comprises four whorls or layers: sepals, petals, stamens (male organs) and carpels (female organs).

fluid mosaic Describes cell membranes; fluidity referring to the lateral movment of lipid molecules, and mosaic referring to the irregular arrangement of proteins.

foliose Leaf-like, describing some types of lichen.

follicle A dry simple fruit from one carpel and which opens on the lower side (e.g. banksias and grevilleas).

follicle cells Somatic cells surrounding the maturing oocyte and serving a protective and nutritive function.

food chain A sequence of organisms from producer to consumers along which energy flows in an ecosystem; usually with three or four trophic levels.

food web A number of interacting food chains in an ecosystem.

foramen ovale A valved opening between the right and left atria of embryonic mammals that allows blood from the placenta to bypass the non-functional lungs of the fetus.

foregut fermentation Digestion of cellulose by symbiotic micro-organisms located anterior to the true stomach.

fossil Preserved remains of an organism or traces of it, such as footprints (trace fossil), or chemical compounds produced by it (chemical fossil).

frameshift mutation A mutation that removes or adds a number of nucleotides not equal to multiples of three (i.e. mutations that disrupt the normal sequence of codons).

free energy (G) Usable energy in a chemical system.

frond Fern leaf.

frugivore Fruit-eating animal.

fruit Mature ovary of a flowering plant; contains seeds; may be dry or fleshy; simple (from one carpel), aggregate (from a cluster of separate carpels on one flower) or multiple (from a cluster of many carpels from different flowers).

fruiting body Specialised, spore-producing structure of a fungus.

frustule Valve or silica dish, two of which make up the cell wall of a diatom.

fruticose Shrub-like, describing some types of lichen.

fucoxanthin An accessory photosynthetic pigment found in the chloroplasts of chrysophytes (golden flagellates).

fuel molecules Molecules such as carbohydrates and fats with energy-rich chemical bonds that are broken down to give energy.

fulcrum Pivot point of a lever.

fundamental niche That region of the environment within which a species can persist indefinitely; defined by all the abiotic and biotic factors that impinge on the survival and reproduction of the species.

fungus (pl. fungi) Eukaryote with cell walls, lacks chlorophyll and absorbs its food (e.g. moulds, yeasts, mushrooms etc.).

fusion Combination of egg and sperm pronuclei to form the zygote nucleus during fertilisation.

G1 phase Gap 1, usually first and longest phase of synthesis and growth in the cell cycle.

G2 phase Gap 2 phase, the main period of synthesis of cellular molecules that occurs following S phase in the cell cycle.

gait A characteristic pattern of locomotion.

gametes Mature male and female germ cells that fuse to form the zygote.

gametogenesis Formation of the gametes.

gametophyte Haploid stage of a plant life cycle that produces gametes.

ganglion (pl. ganglia) Organised group of neurons.

gap genes Genes expressed in and required for formation of a group of segments.

gastrula Embryonic stage following cleavage; characterised by reorganisation of cells to form the three germ layers and development of bilateral symmetry.

gastrulation Bulk movement of cells leading to the formation of the gastrula.

gene family A set of genes that exhibit sequence similarity and characteristically encode proteins that have related biochemical functions.

gene flow Exchange of genes between individuals, and thus also populations and species.

gene pool All the genes (alleles) in a population or species.

generative cell Male reproductive cell of a pollen grain, formed by asymmetric cell division of the microspore and lies entirely within the vegetative cell; the progenitor of the sperm cells.

genes Discrete hereditary factors that determine traits.

genetic code Relationship between codons of DNA and RNA and specific amino acids.

genetic drift Random change in allele frequencies in small populations of organisms.

genetic engineering Techniques making use of recombinant DNA to produce transgenic (transformed) organisms for applications in pharmaceutical, agricultural, horticultural and veterinary industries.

genetic polymorphism The presence in a population of more than one allelic form of a gene, or two or more genotypes at higher frequencies than would be expected on the basis of mutation.

genome The sum of the genetic material of a cell or organelle.

genomic equivalence Cells of the same organism having the same genes, although they may be differentially expressed in different tissues.

genomic library A collection of cloned genomic DNA fragments. Genomic libraries normally consist of a large number of random genomic fragments cloned into a vector and propagated as plaques or colonies, each plaque or colony carrying one of these random genomic fragments.

genotype The genetic make-up of an organism.

germ cells The line of cells that gives rise to gametes.

germ layers The three basic tissue layers formed during gastrulation—endoderm, mesoderm and ectoderm.

gibberellin A type of plant hormone promoting stem elongation and seed germination; composed of small molecules each containing 19 or 20 carbon atoms; synthesised in the shoot and germinating seeds.

gill Outgrowth of the body surface used in gas exchange.

gill chamber Chamber that houses gills.

glial cells Supporting cells of the nervous system; provide insulation, and mechanical and nutritional support for neurons, and guide their development and repair.

glomerular filtrate Fluid filtered from glomerular capillaries into the renal capsule; primary filtrate.

glomerulus A spherical tuft of capillaries associated with the vertebrate nephron; filtration of fluid from the glomerular capillaries forms the primary filtrate.

glucagon Hormone released by alpha pancreatic islet cells; causes increased blood glucose levels due to breakdown of glycogen and synthesis of glucose from amino acids.

gluconeogenesis Synthesis of glucose from non-carbohydrate sources such as amino acids.

glycocalyx Carbohydrate chains on the outer surface of the plasma.

glycogen A polysaccharide that serves as the principal storage form of carbohydrate in animals.

glycolipid Lipid with a short chain of sugar residues; in membranes, occurs on the non-cytosolic side.

glycolysis Anaerobic catabolism of glucose to pyruvic acid, producing two molecules of ATP.

glycoprotein Chain of sugar molecules attached to protein; occurs on the non-cytosolic side of plasma membranes.

glycosaminoglycans Large polysaccharide molecules

composed of repeating dissaccharide units, usually linked to a protein core; are a major component of the extracellular matrix and are responsible for gel hydration.

glycosidic bond Bond linking two monosaccharides in which the first carbon atom of one sugar molecule reacts with a hydroxyl group of another sugar molecule, with loss of a water molecule.

goitre Enlargement of the thyroid gland caused by lack of iodine in the diet.

Golgi apparatus Stacks of four to 10 disc-shaped cisternae functioning in storage and modification of secretory products.

gonad Testis or ovary.

gonadotrophic hormone Hormone that stimulates gonads to produce gametes.

gonangium (pl. gonangia) A reproductive hydranth polyp.

Gondwana Past supercontinent uniting all southern land masses.

G-proteins (guanosine triphosphate-binding regulatory proteins) Intermediate molecules in many cellular signalling pathways, which can couple to receptors altering the activity of an ion channel or intracellular enzyme.

grana Stacks of thylakoids that form part of the internal membrane system of chloroplasts.

granulocyte A leucocyte produced in bone marrow that migrates to sites of infection where it engulfs and kills foreign organisms.

gravid Filled with eggs.

greenhouse effect Natural warming of the earth by heat trapped due to the presence of certain heat-absorbing gases in the atmosphere.

guanine ($C_5HO_5N_5$) A nitrogen waste product excreted by spiders; formed by nucleic acid metabolism, and from ammonia.

guanotelic Pattern of excretion of nitrogen wastes in the form of guanine.

guard cells Pair of kidney-shaped cells regulating stomata.

habitat selection theory The idea that all individuals of a species attempt to live in places that maximise their chances of survival and reproduction (their evolutionary fitness).

haematocrit Volume fraction of whole blood occupied by the blood cells.

haemocoel Large spaces in the body that are filled with blood.

Haldane effect Decrease in the affinity of respiratory pigments for CO_2 with increased oxygenation of the pigment.

halophyte Plant adapted to saline environments.

haploid A cell possessing only one set of chromosomes (n), as in egg or sperm.

Hardy–Weinberg principle In an infinitely large, interbreeding population, in which there is random mating but no migration, mutation or selection, frequencies of genes and genotypes will be constant in each generation.

hartig net Mycorrhizal fungus mycelium that grows between root cortical cells facilitating nutrient transfer.

heartwood Mature secondary xylem, in which the rays have degenerated, and the vessels and tracheids are filled with secondary organic compounds that make the wood hard and durable.

helper cells (T_H cells) Regulatory T cells that produce and secrete lymphokines.

hemidesmosome Junction formed by cross-linking between the cytoskeleton of a cell and the extracellular matrix; provides structural support by anchoring cells to the matrix.

hemimetabolous development Development in which newly hatched young resemble adults in all except size and sexual maturity.

hemitropous Type of ovule in flowering plants in which the micropyle is located at the side.

hemizygous A gene present in only one copy in a diploid organism.

herbivore Animal that consumes algae or plants as food.

hermaphrodite Type of animal in which both male and female reproductive organs occur within the same individual (also called monoecious). Type of plant in which the flowers contain both male and female organs.

heterochromatin Densely staining regions in an interphase nucleus; consists of aggregated strands of chromatin that are inactive in gene transcription.

heterokaryon Multinucleate vegetative cell of a fungus where the nuclei are genetically different.

heterokont Protist characterised by one smooth flagellum directed posteriorly and one hairy flagellum directed anteriorly.

heterospory Having two types of spore; heterosporous plants develop separate male and female gametophytes.

heterotroph Organism that consumes other organisms as food; unable to synthesise organic molecules from inorganic compounds.

heterozygote An individual carrying different alleles of a gene.

heterozygote advantage Occurs when the Darwinian fitness of heterozygotes for particular alleles is greater than either homozygote.

heterozygous Organisms in which the alleles are different, for example, Yy.

hindgut fermentation Digestion of cellulose by symbiotic micro-organisms located posterior to the true stomach.

holometabolous development Development in which the young have the form of worm-like larvae bearing little resemblance to adults. Larvae enter a quiescent stage during which they undergo complete structural reorganisation (metamorphosis) into adult form.

homeobox Characteristic DNA sequence of about 180 nucleotides that occurs in homeotic genes.

homeodomain proteins Transcriptional regulatory proteins that are frequently involved in developmental regulation and contain a conserved protein domain, termed the homeodomain, that confers sequence-specific DNA binding.

homeostasis Maintenance of a relatively constant internal environment.

homeotic (*Hox*) genes Genes that are expressed in a restricted region of the anterior–posterior axis and determine the developmental fates of segments along the anterior–posterior axis.

homokaryon Multinucleate vegetative cell in fungi where the nuclei are genetically all the same.

homologous chromosomes (homologues) A pair of chromosomes in a diploid individual. One homologue is inherited from each parent.

homology Similarity indicating common ancestry. Homologous structures in different organisms have the same basic plan but not necessarily the same function.

homozygote An individual carrying two copies of the same allele of a gene.

homozygous Organisms in which the alleles are the same, for example, YY or yy.

hormone Chemical messenger secreted by cells of an organism in response to specific stimuli. Hormones modify the activity of cells as a result of interaction with specific receptors.

hot spot Immobile point at the surface of the earth's mantle where a column of hot, upwelling asthenosphere rises, which may form islands.

humoral immunity Immunity mediated by soluble antibody molecules secreted by B cells in the serum or other body fluids.

hybrid Offspring of two different varieties or species.

hybrid vigour The basis of hybrid seed production in crop plants, occurring when two or more pure lines (homozygous) are crossed, and the resulting progeny show increased yield and vigour (heterozygous).

hydathode Pore-like structure on the tips or margins of leaves of rainforest plants that permit water to be extruded when required by high root pressure.

hydranth An individual in a colony of hydroid polyps.

hydrofoil Structure that generates a lifting force when moving through a fluid.

hydrogen bond Relatively weak bond formed between hydrogen and another polar atom such as oxygen.

hydrolysis Reaction involving the addition of water in the breakdown of a complex molecule to simpler ones such as a protein to amino acids.

hydrophilic Substances such as polar molecules that dissolve readily in water because they can readily form hydrogen bonds with water molecules.

hydrophobic Substances such as non-polar molecules that are insoluble in water because the hydrogen bonds between water molecules tend to exclude non-polar molecules.

hydrostatic pressure Pressure exerted by a liquid, such as blood. A misnomer because the pressure can be exerted by liquids other than water, and the liquids can be moving.

hydroxyapitite Main mineral of bone and teeth; complex form of calcium phosphate.

hyperosmotic solution Solution with a higher osmotic concentration than another.

hyperpolarisation Increased voltage difference across a membrane; moves membrane potential further from the threshold potential and therefore is inhibitory.

hypersaline Salt solutions that have a higher salt concentration than sea water.

hypha (pl. hyphae) Microscopic tube of cytoplasm bounded by a tough, waterproof cell wall; form fungal mycelia.

hypocotyl The part of the axis of a germinating seed below the point of attachment of the cotyledons.

hypogynous Describes a flower with a superior ovary, that is, attached to or above the receptacle.

hypo-osmotic solution Solution with a lower osmotic concentration than another.

hypostome Projection on which the mouth of a polyp is borne.

hypothalamus In vertebrates, region in the midbrain surrounding the third ventricle; receives information regarding the well-being of an animal and provides central neural and hormonal control of many functions.

hysteresis The failure of a system to follow identical paths of response upon application of and withdrawal of a forcing agent.

ideal free distribution The distribution of animals between two resource sites; individuals are *free* to choose between the sites, and the distribution is *ideal* because each individual goes to the place that provides the highest returns.

imago Sexually developed adult stage of an insect life cycle.

immunodeficiency Absence of T cells as a result of absence of the thymus in which T cells mature, may be due to certain genetic abnormalities or removal of the thymus early in development.

immunogen Antigen that can stimulate an immune response.

immunoglobulin (Ig) Another term for antibody.

immunological memory Retention of stimulated B cells, called memory cells; allows a rapid immunological response to a subsequent interaction with the same antigen (secondary immune response).

imprinting (behavioural) Occurs when a newborn animal recognises the first moving object it sees (usually its mother) and follows it for the next few weeks; occurs usually during a limited period, the sensitive period.

imprinting (genetic) A 'marking' of a gene during gametogenesis that alters the activity of the gene in the offspring.

incisors Teeth used to grasp, hold and cut food.

incomplete dominance The situation in which a heterozygote exhibits a phenotype that is intermediate between the homozygous phenotypes of the two alleles.

indirect development Development that involves a larval stage followed by metamorphosis into the adult stage.

indole-3-acetic acid (IAA) Plant hormone; main auxin occurring naturally in plants.

induced gene A gene whose expression is induced by the presence of an inducer molecule.

indusium Protective leaf-like structure covering sori of some ferns.

inflorescence A cluster of flowers.

initiation complex Complex formed when a ribosome binds to mRNA and the first specified aminoacyl-tRNA; initiates protein synthesis.

inner cell mass A group of cells at one end of the mammalian blastocyst, part of which gives rise to the embryo proper.

inorganic compound Substance that does not contain carbon atoms; usually associated with non-living sources.

insertion A mutation that inserts one or more nucleotides into the DNA.

inspiration Bulk movement of air from outside the body into the lungs.

instars Juvenile stages between moults during development of arthropods.

insulin Hormone released by alpha cells of the islets of Langerhans. Binds to membrane receptors and increases the membrane permeability to glucose and amino acids. Leads to increased storage of glucose and increased fat production.

integument Layer of cells surrounding megasporangium in seed plants (*see also* ovule).

intercellular Between cells.

interferon Group of proteins secreted by some virus-infected cells that assist uninfected cells to resist infection by that virus.

intermediate filaments Filaments of the cytoskeleton of eukaryotic cells that are intermediate in size between microtubules and microfilaments, 8–10 nm in diameter, and provide mechanical support for the cell.

interneuron Neuron that transmits information from one neuron to another.

internode Portion of stem between successive nodes (site of leaf attachment).

interstitial fluid Liquids in the spaces between cells.

interstitial matrix Type of extracellular matrix prominent in connective tissues.

intine The inner wall layer of microspores and pollen grains of flowering plants that is pectocellulosic in nature, and forms the pollen tube.

intracellular Inside cells.

intracellular environment　Physical and chemical environment that exists within cells; important aspects of the intracellular environment are ion concentrations, organic solute concentrations and total osmotic concentration.

introns　Segments of RNA removed from the primary transcript during splicing.

invagination　Local buckling of an epithelial cell layer resulting in a depression opening to the outside.

involution　A process whereby groups of cells roll under their neighbours and move inwards, for example, during gastrulation of frog embryos.

ion　An atom that loses or gains electrons, becoming positively or negatively charged.

ion channels　Fastest enzymes known; permit passive transport through cellular membranes; highly selective for particular ions; opened by a change in voltage across a membrane, or by binding with specific signal molecules.

ionic bond　Bond formed when ions of opposite charge are attracted to each other.

ionisation　The process in which a substance dissociates in solution to form ions.

ionoconform　To have the same ionic concentrations in the body fluids as occur in the external medium.

ionoregulate　To maintain body fluid ionic concentrations different from those of the external medium.

ischaemia　Severe reduction in blood flow to a tissue.

islets of Langerhans　Endocrine cells of the pancreas that secrete glucagon (alpha cells), insulin (beta cells) and somatostatin (gamma cells).

isogamy　Situation where gametes are similar in appearance but differ in mating type.

isotope　One or more kinds of an element whose atoms have a similar number of protons and electrons but differ in number of neutrons.

juvenile hormone　Insect hormone secreted by the corpora allata; stimulates development of nymphal structures.

juxtamedullary nephron　Nephron with its glomerulus located near the junction of the renal cortex and medulla; has a long loop of Henle.

kidney　Excretory organ of vertebrates; the nephron is the functional excretory unit of the kidney.

kinetic energy　Energy of movement, as in running water.

kinetochore　The two protein discs of a centromere, into which microtubules are inserted.

kinetoplast　Large mass of DNA, composed of thousands of catenated DNA mini-circles (linked as in a chain) present in the mitochondrion of certain flagellate parasites (e.g. trypanosomes).

labellum　Petal modified as the lip of an orchid flower.

lactic acid　In animals; three-carbon molecule derived from pyruvic acid as a product of anaerobic respiration, especially in muscle.

Lamarckism　Theory of evolution by Jean-Baptiste Lamarck that traits acquired during the lifetime of an organism are inherited by subsequent generations.

lamella (pl. lamellae)　Thin layer or plate-like structure.

lamellar bodies　Intracellular storage form of surfactant.

lamellipodia　Veil-like extensions of the plasma membrane of migrating cells or the growing tip of nerve cells.

laminin　An adhesive protein of the extracellular matrix, occurring in basement lamina; has a very high relative molecular mass (about 850 kD) and three polypeptide chains arranged in the form of a cross.

larva　Juvenile stage in the life cycle of many animals.

laterite　Weathered land surface with ironstone gravel occurring at the surface or subsurface overlying a layer of bleached, white clay.

Laurasia　Past supercontinent uniting all northern land masses.

learning　Any change in an individual's behaviour that is due to its experience.

lenticel　Special site for gas exchange in the periderm (outer layer of bark) of woody plants; raised area of cells with extensive intercellular spaces.

leptosporangiate　Fern with small, delicate sporangia.

leucocyte　A class of nucleated white blood cells that protect the body against invasion and collect cellular debris.

lichen　Mutualistic relationship between a fungus and a green alga or cyanobacterium.

life cycle　The sequence of stages in the growth and development of organisms from zygote to reproduction; in sexually reproducing organisms, these show an alternation between diploid and haploid stages.

lift　The component of force acting at right angles to the direction of motion of a hydrofoil.

ligament　A type of connective tissue linking two bones in a joint.

ligand　Signal molecule which is capable of interacting with a receptor.

lignin　Main component of secondary walls and wood of plants; composed of phenylpropanoid units, which provide a rigid matrix for cellulose fibres.

limiting resource　Environmental requirement of an organism that is in limited supply (e.g. food, nest site etc.).

linkage　The transmission of alleles of different genes located on the same chromosome at a frequency greater than that expected for independent assortment.

lipid　Biological compound that functions in membranes, energy storage and transport, and insulation; insoluble in water as a result of the non-polar (hydrophobic) nature of their numerous C—H bonds; composed principally of carbon, hydrogen and oxygen together with phosphorus and nitrogen.

lipid bilayer　Double layer of lipid molecules that forms the basic structure of cell membranes.

lithophyte　Plant that grows among rocks.

locus　The position of a gene on a chromosome; a locus may be occupied by any one of the alleles of a gene.

loop of Henle　Part of the mammalian nephron (and some avian nephrons) that lies between the proximal and distal convoluted tubules; enables the osmoconcentration of urine.

lophophore　Feeding structure of ciliated tentacles containing extensions of the coelom in lophophorates.

lorica　External vase-shaped shell of some chrysophyte protists.

lung　Invaginated gas-exchange surface connected to air outside the body via narrow tubes.

lymph　A transparent fluid formed by filtration of liquid from capillaries into the interstitial space; collected by primary lymphatic vessels and returned to the blood; contains white blood cells that attack invading organisms; transports proteins and fats into the blood.

lymphatic system　A system of interconnected vessels, nodes and spaces through which lymph is transported throughout the body.

lymphocyte (lymphoid cell)　Mononuclear cells that are the predominant cells in immune organs; responsible for the immune response; two principal classes are T cells and B cells.

lymphokines　Molecules secreted by helper T cells that control the development and function of other T and B cells, as well as of accessory cells such as macrophages.

lysosome Cellular organelle; vesicle bounded by membrane; contains hydrolytic enzymes involved in the breakdown and recycling of many types of molecules.

M phase Phase of nuclear division (mitosis) in the cell cycle.

macronutrients Nutrients that are required in large amounts.

macrophage Phagocytic white blood cell.

madreporite Porous disc, opening of the water vascular system of echinoderms.

magnetic reversal Reversal of the earth's magnetic field, from normal (present day) to reversed polarity (north becomes south and south becomes north).

major histocompatibility complex (MHC) Presents (shows) antigen to T cells, without which T cells will not respond to antigen; fundamental to identification of 'self' and in graft rejection.

male germ unit In flowering plants; single transmitting unit for transfer of male gamete to female gamete; consisting of generative or sperm cells and tube nucleus.

Malpighian tubule Blind-ended excretory tubule of arthropods; urine is formed by active K^+ secretion into the tubule and passive solute and water influx. The urine is emptied into the hindgut.

mandibles First pair of insect appendages for feeding; modified for grinding and chewing (jaw-like) or piercing and sucking (stylets).

mantle (mycorrhizal) Thick sheath of a mycorrhizal fungal mycelium surrounding a root, replacing epidermis and root hairs.

mantle (of molluscs) Dorsal fold of the body wall with a cavity beneath it; secretes the shell.

manubrium Projection in a medusa on which the mouth is borne.

mass number Combined number of protons and neutrons in a nucleus.

mate-guarding Male behaviour that ensures the female he mates with does not mate with another male.

maternal inheritance Situation in which the genotype of the female parent alone determines a particular phenotype of the individual.

maternal-effect genes Genes that are transcribed from the genome of the mother and whose products are deposited in the egg but act after formation of the zygote (i.e. after fertilisation).

mating system Defined according to the number of partners each sex may have during its lifetime or during the mating season.

maxilla (pl. maxillae) Second pair of appendages; mouthparts in insects.

maximum sustainable yield (MSY) Harvesting a population at a rate that allows the population size to be maintained at half its maximum.

mechanoreceptor A type of receptor that detects stimuli such as mechanical pressure or stretch.

medusa Free-floating (pelagic), bell-shaped form of a cnidarian, with its mouth pointing downwards; jellyfish.

megasporangium Female sporangium of plants in which megaspores develop.

megaspore Haploid spore of ferns, fern allies and seed plants that germinates into a female gametophyte, which bears egg cells.

meiosis A type of nuclear division that takes place in germ cells and that results in halving of the number of chromosomes;

each daughter cell (gamete) receives one set of chromosomes.

memnospore Sexual spore of a fungus; allows survival during harsh conditions.

memory cell Long-lasting B cell formed after binding of antigen to the specific receptor of a B cell in the presence of helper T cells. Is the basis of immunological memory.

mesenchyme A class of mesodermal cells, characterised by a tendency to migrate as individual cells.

mesoderm One of the three germ layers, lying between the ectoderm and the endoderm; gives rise to many of the internal organs and internal epithelia.

mesoglea Intermediate gelatinous layer, between ectoderm and endoderm, in a cnidarian (e.g. jellyfish).

mesohyl Middle layer of a sponge consisting of a gelatinous protein matrix containing amoeboid cells, collagen fibres and skeletal elements.

mesophile Organism that grows best between 10° and 30°C.

mesophyte Plant that lives where water is in adequate supply (compare xerophyte).

Mesozoic Geologic era from 245 to 65 million years BP.

messenger RNA (mRNA) An RNA molecule with a sequence complementary to a DNA coding sequence; carries the information specifying the amino acid sequence of a given polypeptide.

metabolic rate The rate of metabolism of an organism; usually related to level of activity. Basal metabolic rate refers to an inactive endotherm in a thermoneutral environment, standard metabolic rate refers to an inactive ectotherm at a particular temperature.

metabolism All the chemical processes occurring within the cells of a living organism.

metameric segmentation A body plan in animals in which there is linear repetition of functional units, which are added at the posterior end (e.g. in annelids).

metanephridium (pl. metanephridia) Tubular excretory organ with a ciliated, funnel-like opening in the coelomic cavity that draws coelomic fluid into the tubule.

metaphase A phase of mitosis, in which chromosomes become arranged equatorially on the mitotic spindle.

metaxylem Xylem tissue that forms outside the protoxylem in stems of plants; has larger and thicker-walled cells than protoxylem, with reticulate secondary thickening.

metazoan An animal.

micelle Spherical structure formed when phospholipids are added to water; forms because fatty acid tails of phospholipids are hydrophobic.

microbody Organelle in eukaryotic cells involved in removal of compounds generated within a cell; spherical in shape and surrounded by a single membrane; often contain crystalline inclusions.

microcirculation Circulation and exchange in arterioles, capillaries and venules.

microfilaments Fine fibres composed of filamentous chains of actin molecules; part of the cytoskeleton.

micronutrients (trace elements) Nutrients that are required in small amounts.

microsatellites Segments of DNA that consist of short repeated sequences. Such segments are often polymorphic and can be used in DNA fingerprinting.

microsporangium Male sporangium of plants in which microspores develop.

microspore Haploid spore of ferns, fern allies and seed plants that germinates into a male gametophyte, which produces sperm.

microsporocyte Diploid male spore-forming cell of plants that undergoes meiosis to form microspores and pollen grains.

microsporophyll Fertile leaf that bears microsporangia; aggregated into male (pollen) cones.

microtubules Major tubular scaffolding components of the cytoskeleton; form hollow cylinders 25 nm in diameter; associated with plasma membranes and forming spindle fibres during eukaryotic cell division.

midbody During cytokinesis in animal cells, a contractile ring of actin between the two daughter cells causes cleavage and the fibres running between the poles aggregate into a tight rod, the midbody.

middle lamella A thin layer between the primary wall of a plant cell; rich in pectins.

mineral elements Inorganic elements required by organisms.

minimum viable population size (MVP) Population size above which a species needs to be maintained to ensure its long-term survival.

miracidium Free-swimming ciliated stage of parasitic flukes.

missense mutations A mutation that changes the nucleotide sequence of a codon so that it encodes a different amino acid.

mitochondrion (pl. mitochondria) DNA-containing organelle of eukaryotic cells; surrounded by a highly permeable double membrane; contain circular DNA molecules, RNA and small ribosomes; site of cellular respiration.

mitosis Nuclear division in eukaryotes in which the daughter nuclei are identical to the parent.

mitotic spindle An elaborate cytoskeletal structure that causes chromosomes to move towards the equator at metaphase of mitosis, and the chromatids to separate and move towards the poles at anaphase.

molar concentration Number of moles of solute disolved in a litre of solution.

molar tooth Cheek teeth involved in mechanical processing of foods, such as cutting, grinding and chewing.

molecule Stable association of two or more atoms due to sharing of electrons in their outer orbitals.

momentum The mass of a body multiplied by its velocity.

moneran Member of the prokaryote kingdom Monera (*see* bacterium).

monocistronic transcript Primary transcript from a typical eukaryotic messenger RNA transcript; contains one functional reading frame and codes for one polypeptide.

monocyte A motile leucocyte capable of engulfing old erythrocytes and infectious organisms.

monoecious Type of animal in which both male and female productive organs occur within the same individual (also called hermaphrodite). Type of plant in which male and female organs are present in different flowers on the same plant.

monohybrid cross A cross involving organisms that are heterozygous at a single locus.

monokaryon Fungal cell containing one nucleus.

monophyletic Referring to a group or taxon of organisms that includes all of the lineages descended from a common ancestor (i.e. an entire branch on a phylogenetic tree).

monosaccharide Sweet-tasting simple sugar that cannot be broken down into smaller sugar molecules; most common are 5-carbon pentoses and 6-carbon hexoses.

monosome Functional ribosome composed of a large and a small subunit.

morphogen A molecule that induces pattern formation during development.

morphogenesis Generation of new shape during development.

morphology The study of body form, including embryology, anatomy and the study of both fossil and living organisms.

mosaic development Development in which the fate of many blastomeres is predetermined in the absence of interactions with surrounding cells.

mutation A change in the nucleotide sequence of a DNA molecule resulting normally from errors in DNA synthesis or from chemical or radiation induced damage to DNA.

mutualism Symbiotic interaction between two species where both benefit from the association (e.g. lichen).

mycelium The body of a fungus, generally growing as filamentous hyphae.

mycorrhiza (pl. mycorrhizae) A mutualistic association between certain types of fungi and the roots of plants; enhances nutrient uptake by the plant; includes arbuscular, orchid, epacrid and ectomycorrhizae.

myelin sheath Insulation layer around some axons in vertebrates; formed by wrapping the axon in many layers of glial cell membrane.

myocardium The three-layered muscle composing the heart wall.

myofibril Rod-like bundle of myofilaments found in muscle cells.

myogenic Initiation of the heart beat within cardiac muscle itself, for example, in vertebrate hearts.

myoglobin Oxygen-binding pigment in muscle; has a greater affinity for oxygen than does haemoglobin.

myosin Protein that commonly interacts with actin filaments to generate cytoplasmic movements or changes in cell shape; organised into thick filaments in muscle cells.

myotomes Blocks of muscle on each side of the body.

myrmecotrophy Mutualistic relationship where ants live within special chambers formed in certain plants.

myxoedema Condition associated with accumulation of water as a result of lack of thyroid hormones in adult humans.

natural selection Term proposed by Darwin for 'survival of the fittest': some members of a population with characteristics that enable them to compete successfully in a particular environment are likely to survive and leave more offspring than are members with less favourable characteristics.

nauplius First, free-swimming, planktonic larva of most crustaceans.

negative feedback control Control system where the response produced to a particular stimulus reduces the size of the original disturbance; leads to homeostasis.

negative regulation Regulation of a gene such that the gene is repressed by the presence of a particular molecule.

nematocyst Stinging organelle of cnidarians (e.g. jellyfish), which functions in defence and capture of prey; nematocysts are also called cnidae.

neotropical region Biogeographical region of the world including South America and lower Central America.

nephridium (pl. nephridia) Tubular excretory organ of invertebrates that develops from the body surface into the coelomic cavity; it is either a protonephridium or a meta-nephridium, depending on the structure of the coelomic end.

nephron Tubular excretory unit of the vertebrate kidney; derived from a coelomoduct but lacks a ciliated, funnel-like opening.

net primary productivity Portion of total (gross) primary productivity that remains after the respiratory losses of primary producers are accounted for.

neural crest In vertebrates, a group of cells formed as the neural tube detaches from the overlying ectoderm. These cells migrate individually throughout the embryo and ultimately differentiate into many different cell types. Neural crest tissues are important in the evolution of many vertebrate features.

neural plate Region of dorsal ectoderm in vertebrate embryos that forms the neural tube.

neural tube Structure in vertebrate embryos from which the spinal cord and brain form.

neurogenic Initiation of the heart beat by nerves leading to the heart muscle, for example, in many invertebrate hearts.

neurohaemal organ Aggregation of neurosecretory cells into a discrete, highly vascularised organ.

neuron   Cell specialised for receiving, conducting and transmitting information to other cells; basic unit of the nervous system.

neurosecretion Secretion of hormones by nerve cells.

neurosecretory cells Specialised nerve cells that secrete hormones.

neurotoxins   Chemicals that cause nervous systems to malfunction, produced as offensive or defensive weapons in both plants and animals.

neurotransmitter Water-soluble signal molecule released from nerve endings at a synapse with an effector cell; acts on receptors located on other nerve cells, muscle cells or glands.

neurula   Embryonic stage in vertebrates during which the neural tube forms.

neurulation The process of neural tube formation.

neutral evolution Evolutionary change having no apparent effect on the survival or reproduction of an organism; result of changes fixed in small populations by genetic drift.

Newton's Laws of Motion **(i)** A body remains at rest or moves at constant velocity unless acted on by external forces. **(ii)** An unbalanced force F acting on a body of mass m gives the body an acceleration a in the direction of the force: $F = m \cdot a$. **(iii)** For every action (force) there will be an equal and opposite reaction (force).

niche *See* fundamental niche.

niche breadth For any abiotic or biotic factor, the range of tolerance of the species.

nitrogen fixation Conversion of gaseous, atmospheric nitrogen by certain bacteria to ammonia, nitrites and nitrates.

node Site on stem at which leaves are attached.

nodes of Ranvier   Small bare regions of axon between Schwann cells that form myelin sheaths; allow saltatory conduction of action potentials from node to node.

non-coding DNA DNA sequences that do not encode amino acids incorporated into polypeptides during protein synthesis.

nonsense mutation A mutation that changes the nucleotide sequence of a codon so that it becomes a stop codon.

nuclear envelope Double membrane surrounding the nucleus in eukaryotic cells.

nuclear pore Channel in the nuclear envelope that allows movement of certain molecules between nucleus and cytoplasm.

nucleolus A spherical fibrillar and granular structure within the nucleus of eukaryotic cells; composed chiefly of rRNA in the process of being transcribed from multiple copies of rRNA genes.

nucleosome   Particle about 10 nm in diameter comprising nucleosome core particle (histone proteins) and associated DNA; found in large numbers in chromatin.

nucleosome core particle A complex of eight histone proteins that form a core around which the double helix of DNA is coiled.

nucleotide   Five-carbon sugar, a phosphate group and a nitrogenous base; nucleotides are linked together by phosphodiester bonds between the sugar and phosphate groups to form nucleic acids.

nucleus The principal membrane-bound compartment of the eukaryotic cell; control centre of the cell; contains chromosomal DNA.

nurse cells Found in many invertebrates; a class of cells that provides nutrients and other materials to a maturing oocyte. In insects, nurse cells are cytoplasmically connected to the oocyte.

nutrient Particular substance required by an organism that must be obtained from its environment; includes organic compounds (carbohydrates, amino acids and fats), vitamins and minerals.

occluding junction (tight junction)   Junction where the plasma membranes of adjacent cells are fused tightly, preventing passage through the extracellular space.

oceanic ridge Site where lava upwells from part of the earth's mantle.

ocellus (pl. ocelli) Simple eye of adult arthropods.

oedema   Build-up of fluid in tissues when filtration exceeds reabsorption and lymph flow. Usually associated with vascular disease.

oestrogens Steroid hormones including oestradiol, oestrone and oestriol; produced by ovarian follicles.

omnivore An animal that feeds on a variety of organisms, for example, plants and animals.

onchosphere Six-hooked larva that hatches from the egg of a tapeworm.

oncogene A tumour-causing form of a gene.

oocyte Female germ cell undergoing meiosis within the ovary.

oogenesis Process by which eggs form from primordial germ cells.

oogonia   Diploid female germ cells undergoing mitosis; in animals within the ovary; in protists within oogonia (female reproductive structure).

open reading frame A nucleotide sequence that can encode a polypeptide (i.e. that has a string of codons in frame that encode amino acids).

open-flow ventilation   Unidirectional flow of respiratory medium past a gas-exchange surface.

operculum Cap-shaped covering, for example, apical portion of a moss spore capsule or covering formed from fused sepals or petals of a eucalypt flower bud.

operons In bacteria, transcription units that cover more than one polypeptide coding region.

opisthosoma In arthropods, abdomen, posterior region of the body behind the prosoma.

organ Combination of tissues forming a functional unit.

organelle Specialised part of a cell, such as nucleus or ribosome.

organic compound   Molecule that contains one or more carbon atoms; largely produced by living organisms.

organogenesis Process of organ formation.

origin Start sequence of DNA replication; there is a single origin in the circular DNA molecules of prokaryotes, but multiple origins in eukaryote chromosomes.

osculum (pl. oscula) Opening in the wall of a sponge through which water leaves.

osmoconcentration   Physiological process whereby water is resorbed from the urine by excretory tubules making the urine more osmotically concentrated than body fluids.

osmoconform To have the same osmotic concentration of body fluids and the external medium.

osmoregulate To maintain an osmotic concentration of the body fluids different from that of the external medium.

osmosis The movement of water from a region of low osmotic concentration (high water concentration, high water potential) to one of high osmotic concentration (low water concentration, low water potential), as a result of the random thermal motion of water molecules through a selectively permeable membrane.

osmotic adjustment A plant response to drought, involving an increase in amount of vacuolar solutes; leads to decreased osmotic potential and reduced cell water potential without adversely affecting cell turgor, allowing growth and photosynthesis to continue in drier conditions.

osmotic concentration The total concentration of all solutes dissolved in solution, expressed as moles of solutes per litre of solution, or osmols per litre.

ossicle Crystal of calcium carbonate; first formed within a cell and enlarging to a plate beneath the skin; forming the skeleton of echinoderms.

osteoporosis Depletion of bone calcium in women as a result of decreasing oestrogen levels after menopause.

ostium (pl. ostia) Pore in the wall of a sponge, through which water enters during filter-feeding.

ovulation The release of a mature egg cell from the ovary.

ovule Megagametophyte retained within the megasporangium, which is further surrounded and protected by one or more layers of cells, the integuments; following fertilisation the whole structure (the ovule) develops into a seed.

ovuliferous scale Leaf-like structure (thought to be a reduced shoot) that bears ovules; aggregated into female cones.

oxidation Reactions involving the breakdown of molecules by removal of electrons, especially those in which oxygen combines with, or hydrogen is removed from, a compound.

oxygen affinity The ease with which oxygen binds to a respiratory pigment.

oxygen equilibrium curve Relationship between the percentage of oxygen saturation of a respiratory pigment and the partial pressure of oxygen.

oxytocin Posterior pituitary hormone that influences reproductive functions.

ozone Triatomic form of oxygen, O_3; highly reactive molecule that increases in concentration in the atmosphere as a result of air pollution.

P_{50} P at which a respiratory pigment is 50% saturated with oxygen; is a measure of the oxygen affinity of the pigment.

pair-rule genes Genes expressed in a two-segment periodicity and required for formation of segments.

palaeotropical region Biogeographical region of the world including Africa (Ethiopian region) and India and South-East Asia (Oriental region).

Palaeozoic 'Ancient life'; geologic era 570 to 245 million years BP.

Pangaea The supercontinent uniting all northern (Laurasia) and southern (Gondwana) landmasses that formed, and then fragmented, during the Mesozoic.

parabronchi The smallest airways in the gas-exchange region of the bird lung in which flow is unidirectional; give rise to the air capillaries.

paracrine hormones Animal hormones that usually act over very short distances, travelling by diffusion through extracellular fluid.

parapatric speciation Divergence in populations, initially in geographic isolation and subsequently when the populations again come in contact; leads to formation of new species.

paraphysis (pl. paraphyses) Sterile hair (e.g. associated with reproductive structures in mosses).

parapodium (pl. parapodia) Lateral, paired appendage on the body segments of polychaetes; functions in gas exchange and locomotion.

parasite An organism that lives and feeds on or in another organism, the host, which is usually larger than the parasite.

parasitism Symbiotic interaction between two species where one benefits and the other is harmed.

parasitoid Insect that is free-living as an adult but parasitic as a larva.

parathyroid gland Endocrine gland closely associated with the thyroid gland; secretes parathormone, which is involved in increasing blood calcium levels.

parenchyma Large living cells that form the ground tissue of plants, comprising large, thin-walled cells, large central vacuole, and a peripheral nucleus and cytoplasm.

parthenogenesis A mode of reproduction, found in some sexually reproducing organisms, in which the male plays no role.

partial pressure of a gas Total pressure in a mixture of gases is the sum of the partial pressures of each gas.

partitioning sequence DNA sequence in cells of prokaryotes that attaches to the plasma membrane in the region where the cell is to divide; similar function to the centromere of eukaryotic cells.

pathogen An organism capable of causing disease.

pattern formation The process of generating the pattern of different tissues in the developing embryo.

pedicellaria (pl. pedicellariae) Pincer-like structure on the body surface of some echinoderms.

pedipalps Second pair of appendages of the cephalothorax in arachnids; modified for various functions; corresponding to insect mandibles.

pelagic realm Marine biogeographic region including the surface water of open oceans and planktonic organisms.

penetrance The proportion of individuals of a particular genotype that show a phenotypic effect.

penis Intromittent organ of males used for transferring sperm during internal fertilisation in many animals.

pepsinogen Inactive precursor (zymogen) of the protease, pepsin; released into the stomach.

peptide bond Bond formed when the acidic carboxyl group (–COOH) of an amino acid attaches to the amino group (–NH$_2$) of another, with the release of a molecule of water.

perfusion Convective movement of an internal fluid.

pericardial Of the fluid-filled space and membranes surrounding the heart.

pericarp Fruit wall, comprised of exocarp, mesocarp and endocarp.

pericycle Layer of cells in a root that surrounds the vascular cylinder; site of initiation of lateral roots.

periderm A protective outer tissue that replaces the epidermis in secondary stems and roots of woody plants; corky tissue.

peridinin A xanthophyll; an accessory photosynthetic pigment found in dinoflagellates.

peripheral nervous system Peripheral part of the autonomic nervous system in vertebrates, which comprises ganglia and connecting nerves, and is classified into three subsystems: sympathetic, parasympathetic and enteric divisions.

peripheral resistance Resistance to blood flow due mainly to friction of the blood in small diameter arterioles.

peristaltic locomotion Movement of an animal by alternate constriction and widening of the body; depends on a fluid-filled body cavity, as in annelids.

peristome Specialised rows of teeth-like structures around the top of a moss sporangium; shelters spores.

peristomium First segment of an annelid.

permeability coefficient An estimate of the ease by which a molecule can pass through a cellular membrane; dependent especially on the lipid solubility of the molecule.

peroxisome Type of microbody in eukaryotic cells that contains numerous enzymes; involved in the production and degradation of peroxides and oxidition of amino acids and uric acid.

peudosclerotium (pl. pseudosclerotia) Sclerotium composed of both fungal mycelium and host tissue (*see* sclerotium).

pH The concentration of hydrogen ions (H^+) in solution; measured on a logarithmic scale ranging from 0 to 14.

phage *See* bacteriophage.

phagocyte White blood cell able to engulf micro-organisms or damaged tissues.

phagocytosis Method of ingestion of food by endocytosis utilised by unicellular organisms, such as sponges, whereby a food particle is engulfed in a membrane-bound food vacuole.

pharyngeal slits Paired openings appearing in the pharynx of chordates at some stage of development.

phenocopy Phenotype that resembles a mutant phenotype, even though it is genetically wild type; due to severe environmental effects.

phenotype The set of detectable properties or traits of an organism.

pheromone Hormone released into the external environment for chemical communication between individuals.

phloem Transport tissue of vascular plants comprising several cell types, including sucrose-transporting sieve cells, companion cells and sclerenchyma fibres.

phosphorylation Addition of a phosphate group to a compound; often results in the formation of a high energy bond (e.g. ATP from ADP).

photoautotroph An organism that gets its energy from sunlight and uses carbon dioxide as a carbon source (e.g. photosynthetic plants, algae and cyanobacteria).

photoheterotroph An organism that uses sunlight for energy but organic compounds (rather than carbon dioxide as in plants) as ready-made building blocks for growth and development (e.g. purple and green bacteria).

photon Elementary particle of electromagnetic radiation (light).

photoperiodism Plant response (e.g. flowering) to the length of light and dark periods in a 24-hour cycle.

photoreceptor A type of receptor that detects light by absorbing it at a particular wavelength.

photorespiration A light-activated type of respiration found in the chloroplasts of plants in which Rubisco uses O_2 as a substrate to oxygenate RuBP and produce CO_2 as a product of oxygenation.

photosynthesis The process by which solar energy is harvested and used to convert CO_2 and water into carbohydrates; involves absorption of energy from sunlight by means of pigments, reactivation of pigments, and carbon fixation to produce sucrose in 'dark reactions'.

phragmoplast Structure containing the remnants of spindle fibres orientated at right angles to the new cross-wall forming during cell division in all plant cells and some related green algae (charophytes).

phycoplast System of microtubules orientated in the plane of cell division following the collapse of the spindle during cytokinesis; feature of many green algae.

phyllode Laterally compressed petiole and rachis; foliage that replaces true leaves in most Australian acacias.

phyllotaxy Geometric pattern of leaf arrangement on a stem of plants (e.g. spiral).

phylogenetic tree *see* cladogram.

phylogeny Evolutionary relationships of organisms, usually depicted as a branching tree diagram (phylogenetic tree).

physical digestion Breakdown of food into small particles, such as by grinding or chewing.

phytochrome A plant pigment that absorbs light; exists in two interconvertible forms, P_r (inactive form) and P_{fr} (active form); involved in the timing of a number of processes (e.g. flowering, dormancy and seed germination).

pigment Molecule that appears coloured because it absorbs photons with particular energy levels and reflects others.

pinacoderm Outer surface of a sponge, consisting of a layer of flattened cells (pinacocytes).

pineal gland In vertebrates, an outgrowth in the midline of the roof of the third ventricle, which is used to measure photoperiod. In fishes, amphibians and some reptiles may contain photoreceptor cells (the 'third eye'). In mammals and birds, is a neurosecretory organ releasing melatonin at night.

pinna (pl. pinnae) Leaflet of a compound leaf.

pinocytosis Mode of capillary exchange, whereby large particles and lipid-insoluble materials are exchanged slowly via numerous tiny vesicles.

pistil In a flower, a female reproductive organ, comprising a stigma, style and ovary.

pith Parenchyma cells that lie centrally within the vascular tissue of stems and some roots.

pituitary gland (hypophysis) Gland at the base of, and largely controlled by, the hypothalamus; composed of the median eminence, the anterior pituitary gland (adenohypophysis—an endocrine gland) and the posterior pituitary gland (neurohypophysis—neurosecretory gland).

placenta An organ that enables exchange between an embryo and the maternal circulation; present in marsupial and placental mammals.

plant Multicellular, photosynthetic eukaryote (kingdom Plantae) that lives on land.

plantigrade Whole foot touches the ground, as in primates.

planula A ciliated type of animal larva.

plaque A small circular clearing in a lawn of bacteria caused by lysis of the bacteria. Under normal plating conditions, each plaque arises from a single bacteriophage, which infects a cell and initiates rounds of replication, cell lysis and further infection of adjacent bacteria.

plasma The transparent, slightly yellowish fluid component of blood without the cellular components; obtained by centrifugation.

plasma cell Non-dividing, antibody-secreting B cell formed after binding of antigen to the specific receptor of a B cell in the presence of helper T cells; produces the humoral response.

plasma membrane The boundary of living cells separating a cell from its environment; formed from a phospholipid bilayer.

plasmid Small unattached circular DNA molecule occurring in bacterial cells, often carrying genes conferring antibiotic resistance.

plasmodesma (pl. plasmodesmata) Special channel through

primary cell walls, connecting plasma membranes and cytosols of adjacent plant cells.

plasmodium A multinucleate mass of protoplasm surrounded by a membrane; stage in the life cycle of acellular slime moulds (myxomycetes).

plasmolysis Shrinkage of cytoplasm due to loss of water by osmosis, drawing the plasma membrane away from the wall.

plastid Organelle of plant and algal cells that is surrounded by a double membrane; generally larger in size than mitochondria and contain circular DNA molecules, RNA and small ribosomes; function in processes such as photosynthesis, starch storage and geotrophism.

plastron In turtles and tortoises; a ventral, protective shield covered by horny plates.

plate tectonics Modern geological theory that recognises that the earth's crust and part of the upper mantle (together the lithosphere) are divided into a number of plates that move relative to one another.

platelet Anucleate disc-shaped cell fragments derived from megakaryocytes in bone marrow; involved in several aspects of blood clotting.

pleiotropy The effect of one gene on multiple traits.

plesiomorphic A primitive or ancestral character is called a plesiomorphic character.

pleural cavity Coelomic space surrounding the lungs in mammals; separated from the rest of the visceral coelom by the diaphragm.

pluripotent Stem cells that produce a whole range of cell types, for example, blood-forming stem cells.

pneumatophore Upright aerial root of mangroves; exposed at low tide and function in gas exchange.

pod A fruit (legume) that opens along two sides, as in beans and peas.

poikilohydric Capable of withstanding desiccation, as in resurrection plants.

point mutation A mutation that affects one or a small number of nucleotides.

polar nuclei The two nuclei present in the central cell of the embryo sac of flowering plants.

polarisation Imbalance of electrical charge across the plasma membrane due to uneven distribution of ions, with the inside of the cell usually negative with respect to the outside.

pole cells Cells that form at the posterior end of the *Drosophila* embryo and give rise to the germ line cells.

pollen A collective term for pollen grains; microgametophytes of seed plants, which develop sperm and can be transported by wind or animals.

pollination The process in which pollen of flowering plants is transferred by animals, air or water currents to the stigma for fertilisation.

pollution The addition of materials to air, soil or water that adversely affects the environment.

polyadenylation The process in which a series of A residues are added to the 3′ end of an RNA transcript.

polyarch xylem Pattern of xylem development in roots of monocotyledons among flowering plants, in which the xylem does not fill the centre, but is divided into many ridge-like projections (archs).

polycistronic Refers to a transcription unit that contains more than one reading frame.

polygamy The acquisition of more than one mate during a lifetime or mating season; polygyny refers to more than one female per male; polyandry is the reverse.

polygene A set of genes that together control a quantitative character such as height or mass of an organism.

polygenic A phenotype influenced by multiple genes.

polymerase chain reaction (PCR) Selective and repeated replication of segments of DNA *in vitro*.

polymorphism The presence in a population of different forms.

polyp An attached tubular form of a cnidarian with its mouth upwards.

polypeptide A molecule composed of a chain of amino acid residues.

polypeptide chain A molecule of many amino acids joined together by peptide bonds; a protein is a large polypeptide chain.

polyphyletic Referring to a group or taxon of organisms that has had multiple origins (e.g. viruses).

polyploid An organism with more than two sets of chromosomes.

polysaccharide A carbohydrate composed of many monosaccharides joined in long linear or branched chains (polymers).

polysome A number of ribosomes attached to the same mRNA strand, each at different stages of protein synthesis; may be free in the cytosol or attached to endoplasmic reticulum.

pome A fleshy fruit from an ovary of an epigynous flower.

population A group of organisms of the same species living in a defined geographic area.

portal system System of blood vessels carrying blood between two capillary networks.

positive regulation Regulation of a gene such that the gene is induced by the presence of a particular molecule.

potential distribution Of a species; the range over which individuals could theoretically survive and reproduce.

potential energy Stored energy, such as chemical energy stored in the bonds of atoms and molecules.

Precambrian The oldest era in the geologic time scale; before 570 million years ago.

precapillary sphincter Band of smooth muscle at the entrance of a true capillary; opens or closes the vessel in response to local influences.

predation One kind of animal (a predator) eating another; an animal (herbivore) eating a plant.

predator An organism that catches and kills another organism for food.

predator–prey cycle Oscillations in population size where predator numbers follow those of the prey.

preferential channel Main capillary route for blood flow through a tissue when metabolic demands are minimal.

prehensile Able to grasp, as with a tail or fingers.

pressure potential Energy level of water as a result of hydrostatic pressure or suction.

primary cell wall First wall of a plant cell, composed of cellulose, pectins and non-cellulosic polysaccharides.

primary response Initial immune response to an antigen, usually results in immunological memory, which causes a later immune response to the same antigen to be larger and more rapid.

primary transcript The RNA molecule produced by transcription prior to processing.

primer Short sequence of bases that pairs with a complementary sequence in a strand of DNA and provides a free hydroxyl end for DNA polymerase to commence synthesis of a nucleotide chain.

primordial germ cell Embryonic germ cell prior to its proliferation (mitotic division) in the ovary.

Principle of Independent Assortment Alternative forms of a gene controlling one trait assort into gametes independently of alternative forms of another gene controlling a different trait.

Principle of Segregation Alternative forms of an inherited trait, controlled by alternative forms of a gene (alleles), segregate into gametes after meiosis with equal probability.

prion Infectious agent that is virus-like but appears to lack nucleic acid and consists only of protein.

probe A term given to DNA that is labelled with radioactivity or other markers and hybridised to other DNA sequences to detect complementary DNA sequences.

producer An organism that converts the energy of sunlight (by photosynthesis) or inorganic chemicals (chemosynthesis) to chemical energy in the form of carbohydrates and other molecules.

productivity Rate at which biomass accumulates; primary productivity refers to productivity of producer organisms (e.g. plants).

progesterone Hormone, released by the corpus luteum, which prepares the reproductive tract for embryo implantation or stimulates secretion of coatings for eggs.

proglottid Segment-like body unit of a tapeworm; new segments are added at the anterior end.

programmed cell death (apoptosis) The process of cell suicide that involves a characteristic series of events leading to death of the cell.

prokaryote Bacteria; small cells that lack membrane organelles such as a nucleus or mitochondria or chloroplasts; compare eukaryote.

prokaryotic cell Cell with a simple structure, lacking a nucleus and other internal membrane-bound organelles; bacterial cell.

prometaphase A phase of mitosis in which the nuclear envelope breaks down, allowing the mitotic spindle to interact with and move chromosomes.

promoter Specific sequence of DNA that binds to RNA polymerase, promoting initiation of transcription of the coding region by the enzyme.

pronucleus The male or female nucleus in the zygote, prior to fusion.

prophase The initial phase of mitosis in which dispersed chromatin in the cell condenses into chromosomes composed of paired chromatids, and the mitotic spindle is formed.

proplastid Precursor organelle of all types of plastids.

prosoma In arthropods, anterior part of the body, the head or cephalothorax.

prostomium Anterior, presegmental part of an annelid (segmented worm), which houses the brain.

protandry In animals, situation in which an individual starts life as a male changing to a female at some later stage. In plants, condition where a flower first opens in the male phase (anthers dehisce) and later becomes female (stigmas receptive).

protein *see* polypeptide chain.

proteoglycans O-linked glycoproteins containing many sugar chains; produced in the Golgi apparatus from proteins containing serine, threonine or hydroxyproline residues; mature to form surface slimes and mucus.

proteoid root Cluster of hairy rootlets in some Proteaceae and Fabaceae that forms a dense mat at the soil surface to enhance nutrient uptake in nutrient-poor soils.

prothoracic gland Non-neural endocrine gland in the thorax of some insects that secretes ecdysone.

protist A eukaryote including unicellular, multicellular, photosynthetic and non-photosynthetic organisms. Member of the kingdom Protista, which is a diverse array of organisms, not a single evolutionary group.

protogyny In animals, situation in which an individual starts life as a female, changing to a male at some later stage. In plants, condition where a flower firsts opens in the female phase (stigma receptive) and later becomes male (anthers dehisce).

proton A positively charged particle of an atomic nucleus; a hydrogen ion.

protonephridium (pl. protonephridia) Tubular excretory organ of animals that has a flame cell to filter coelomic fluid into the tubule for excretion.

proto-oncogene A normal cellular gene that when mutated leads to tumour formation.

protoplasm The cytoplasm and nucleus of eukaryotic cells.

protostome Animal in which the blastopore becomes the mouth; primary mouth.

protoxylem The first formed and earliest maturing xylem tissue that forms towards the centre of the stem in plants.

pseudocoel Body cavity of an animal, such as a nematode, that is not lined on all sides by mesoderm.

pseudoparenchyma Mass of branched hyphal mycelium of macrofungi resembling tissue.

pseudoplasmodium Mass of amoebae of a cellular slime mould that aggregate to form a single mobile colony ('slug').

pseudopodia Transient extensions of a cell surface ('false feet' of amoebae).

psychophile Organism, such as a fungus, that can live in cold conditions.

pulmonary Referring to the circulation of blood in the lungs.

pulmonary artery Main artery that carries deoxygenated blood from the heart to the lungs in higher vertebrates.

pulvinus A motility organ at the base of the leaf petiole in certain plants, such as legumes, that controls the position of the leaf.

pupa Developmental stage in some insect life cycles between the larva and adult; non-feeding, immobile and sometimes encapsulated or in a cocoon.

pure breeding Strains in which individuals and their progeny have the same phenotype.

purine Type of nitrogenous organic base (nucleotide), adenine and guanine, which pair with pyrimidine bases in DNA.

pygidium Posterior, postsegmental part of an annelid.

pyrimidine Single ring molecule; forms three of the bases of nucleic acids—cytosine, thymine and uracil.

pyruvate ($C_3H_3O_3$) End product of glycolysis.

Q_{10} Measure of the sensitivity of biochemical processes to temperature, measured as the increase in the rate of a physiological process or reaction for a 10°C rise in temperature.

qualitative character Inherited character that is distinctively different from others and can be used to group organisms into distinct phenotypic classes.

quantitative character Character of an organism, such as height or weight, that shows continuous variation and is controlled by both environment and polygenes.

quantitative trait locus A locus that influences a polygenic phenotype.

radial symmetry Symmetry of an organism such that any plane passing through the central axis bisects the organism into equal halves (e.g. jellyfish).

radicle Root axis of a germinating seed.

radula Tongue-like structure, with rows of rasping teeth; present in the floor of the foregut of molluscs, except bivalves.

ram ventilation Ventilation of gills that results from the forward movement of the body.

raphe Longitudinal slit in the valve of a pennate diatom.

ray initial A type of meristematic cell produced by the vascular cambium differentiating into wood rays.

reabsorption Physiological process in animals whereby solutes are actively transported from the urine or the gut contents back into body fluids.

reaction centre Specialised chlorophyll complex functioning as a photosynthetic unit capable of channelling energised electrons to an acceptor molecule; located on the thylakoid membranes of chloroplasts.

reading frame Pattern by which nucleotide sequences of DNA are decoded as a sequence of triplets (codons) from the 5′ end to the 3′ end of messenger RNA.

realised distribution Of a species; the range over which individuals live and reproduce.

receptor A molecule, usually a protein, in the plasma membrane or within a cell, that undergoes a change as a result of a specific interaction with a signal, leading to a particular response.

recessive A phenotype, such as green colour in seeds, which is apparent in homozygous (yy) individuals, but absent in the heterozygous (Yy) individuals.

recombinant DNA Technology involving cutting DNA into fragments using restriction enzymes, isolating the fragments, identifying genes and multiplying them in bacteria, where the gene products can be expressed.

recombinants Progeny that exhibit different allele combinations than their parents due to recombination.

recombination The mixing of alleles of different genes on homologous chromosomes (homologues), caused by crossing-over between homologues during meiosis.

redia (pl. rediae) Larval stage in the life cycle of parasitic flukes that develops within sporocysts and that forms cercariae.

reduction Reactions involving the breakdown of molecules by addition of electrons, especially those in which oxygen is lost and hydrogen accepted.

regeneration A form of asexual reproduction involving the production and differentiation of new tissues of an organism.

regulated secretion Secretion that only occurs in response to a specific signal.

regulative development Development in which the fate of many blastomeres becomes determined as a result of interactions with surrounding cells.

repetitive DNA DNA that contains the same nucleotide sequence multiple times in the genome.

replication Process of DNA duplication; involves unwinding of the double strand of DNA and synthesis of new complementary strands using the parent strands as templates. Replication is semiconservative, meaning that each of the parental strands pairs with a newly synthesised strand.

replicon Unit region of replication, between origins, in chromosomes of eukaryotes.

repressed gene A gene whose expression is repressed by the presence of an inducer molecule.

repressor protein A protein that binds to the operator sequence and prevents transcription initiation.

reproductive effort The investment of parents in sexual reproduction, which is dependent on their own survival and future reproductive success.

reproductive isolating mechanism Geographical, physical, physiological or behavioural barrier preventing the interbreeding of individuals from different species or populations.

reproductive success The number of surviving offspring produced by an individual.

resource partitioning Where species exploit a limited resource in different ways, reducing direct competition and allowing species coexistence.

respiratory pigments Metalloproteins (hence the colour) that bind oxygen reversibly to the metal; primarily responsible for increasing the amount of oxygen carried in blood.

resting membrane potential Difference in electrical potential across a plasma membrane at electrochemical equilibrium.

restriction (in immune system) Inability of T cell receptors to recognise antigen unless presented by MHC molecules.

restriction endonuclease (restriction enzyme) An enzyme that recognises specific sequences within a double-stranded DNA molecule and cleaves the DNA.

restriction enzyme Site-specific nuclease enzyme (endo-deoxyribonuclease) that recognises short sequences of unmethylated DNA and cleaves the double strands either at specific sites (the recognition sequence) or nearby.

restriction fragment length polymorphisms (RFLPs) Restriction fragments that differ in length in different individuals (i.e. they are polymorphic) because of differences in DNA sequence.

restriction mapping Technique for mapping (or physically locating) sites at which restriction enzymes cut cloned DNA.

rete Network of interdigitated arteries and veins that functions for efficient countercurrent exchange.

reverse transcriptase An RNA-dependent DNA polymerase (i.e. an enzyme that synthesises DNA from an RNA template).

rhabdite Rod-like structure secreted by epidermal gland cells in free-living flatworms.

Rhesus antigen A human blood group antigen that may cause haemolytic reactions, especially during pregnancy. Blood containing this antigen is called Rh positive, while that lacking it is Rh negative.

rhizoid Short root-like structure that anchors bryophytes and some fungi to their substrate.

rhizome Underground stem.

rhizosphere Soil zone immediately surrounding the root hairs in plant roots where interactions occur between plant, soil and micro-organisms.

rhodopsin An important photoreceptor of animals, found in rod cells of the retina of the vertebrate eye, associated with the chromophore retinal, which undergoes a chemical rearrangement after absorbing light.

rhombomeres Segmented structures of the hindbrain neural tube.

rhopalia Sense organ of a jellyfish that contains a statolith or sometimes an eyespot.

rhyncocoel Coelomic body cavity that houses the proboscis of a proboscis worm (nemertine).

ribonucleic acid (RNA) Single-stranded nucleic acid characterised by a ribose sugar in each nucleotide and the bases adenine, cytosine, thyamine and uracil; *see also* mRNA, rRNA and tRNA; involved in transcribing and translating coded information of DNA during the production of proteins.

ribose A five-carbon sugar. Sugar component of RNA.

ribosomal RNA (rRNA) RNA that forms the major part of ribosomes.

ribosome Cytoplasmic organelle where protein synthesis occurs; formed from two rRNA subunits in association with an mRNA molecule.

ribozymes RNA molecules that act as enzymes.

ribulose bisphosphate (RuBP) The 5-carbon sugar that binds to CO_2 in the first step of carbon fixation in photosynthesis.

ribulose bisphosphate carboxylase-oxygenase (Rubisco) Enzyme that catalyses the first step of carbon fixation (*see* RuBP); constitutes 50% of protein in chloroplasts.

RNA polymerase Enzyme that catalyses transcription of a DNA template to form mRNA.

root cap Parenchyma tissue covering a root tip; secretes a mucigel to aid the root penetration through the soil; usually contain numerous starch granules that may act as sensors of gravity.

root cluster (proteoid roots) Group of rootlets that form dense mats at the soil surface in family Proteaceae and certain legumes.

Root effect In gas exchange in mammals, the oxygen affinity of haemoglobin is reduced by carbon dioxide binding reversibly with the pigment.

root nodule Outgrowth of roots in a wide range of plants, especially legumes, in which symbiotic nitrogen-fixing bacteria (*Rhizobium* or *Frankia*) occur.

root pressure Positive pressure generated by roots that is responsible for the exudation of sap from tapped or cut stems in spring.

rough endoplasmic reticulum (RER) Endoplasmic reticulum with attached ribosomes; involved in synthesis of proteins usually destined for export from a cell.

ruminant Foregut fermenter that ruminates, that is, regurgitates food from the first part of the stomach and rechews it to reduce further the particle size.

S phase Phase of the cell cycle in which DNA replication occurs.

salt gland Gland that can secrete a salt solution that is more concentrated than body fluids (e.g. reptiles, birds and mangroves).

saltatory conduction Conduction of action potentials along myelinated axons involving action potentials skipping from node to node.

saltatory evolution Evolutionary change that occurs suddenly followed by long periods of time in which there is little change.

saprophyte An organism that lives on dead organic matter.

sapwood The outer region of wood (secondary xylem) in a tree trunk containing living rays; of lesser strength than heartwood.

sarcolemma Electrically excitable membrane that surrounds muscle cells.

sarcomere Region of myofibril between adjacent Z-lines.

satellite DNA DNA that contains large numbers of small nucleotide sequences repeated in tandem.

satellite virus A type of virus only able to replicate in cells infected with a specific helper virus.

scavenger An animal feeding on carrion (dead organisms).

sclereid Plant cell with thick, lignified wall; stone cells of fruit etc.

sclerenchyma Plant tissue that has a support role; cells with thickened lignified secondary walls that impart rigidity as well as strength; includes sclereids, branched or more-or-less even-shaped stone cells that form the hard tissue of fruits and seed coats and fibres; elongate cells.

sclerophyll Plant characterised by rigid, often small leaves, and short internodes; able to survive low soil nutrients, water stress and fire.

sclerotium (pl. sclerotia) Hard, resistant resting body, composed of masses of tightly compacted mycelium, formed by some soil-inhabiting fungi.

scolex Anterior attachment organ, with suckers and often hooks, of a tapeworm.

scutellum The single cotyledon of monocotyledonous flowering plants; forms an interface tissue between the embryo and endosperm.

Second Law of Thermodynamics *The entropy of the universe is increasing*; thus an input of energy is needed to maintain the ordered state of the universe.

second messengers Activation of first messengers, G-protein-linked receptors by extracellular signals leads to changes in the concentration of one or more small intracellular signalling molecules, second messengers, which then produce a cellular response; examples of second messengers are calcium ions and cyclic AMP.

secondary response Larger, more rapid immune response to a particular antigen as a result of immunological memory due to an earlier primary immune response.

secondary walls During development, plants cells strengthen their walls and make them more rigid by differentiating secondary walls, whose main component is lignin.

seed Structure from which a new plant develops; produced from a fertilised ovule, containing an embryo, a food source (cotyledons or endosperm) and usually a hard outer seed coat (testa).

segment-polarity genes Genes expressed in a portion of each segment and required for formation of proper segmental structures.

selective permeability The ability of membranes to allow passage of some molecules and not others.

self-incompatibility Genetically controlled process preventing self-fertilisation; in flowering plants recognition of self-pollen leads to a rejection response usually in the style, while compatible pollen from another individual of the same species is accepted.

semiconservative replication *See* replication.

septum (pl. septa) Wall that divides, or partially divides a structure or cavity; cross-wall in fungal hypha; septum dividing the ventricle of the heart.

sere Orderly sequence of ecological communities that replace one another over time.

serum The yellowish fluid isolated from clotted blood; plasma without some of the constituents bound to the clot.

sessile Fixed in one position, immobile.

seta (pl. setae) Chitinous bristle.

sex chromosomes Chromosomes that differ in morphology, are present in different numbers in males and females and are involved in sex determination.

sex-linked genes Genes that are located on sex chromosomes.

sexual reproduction Formation of offspring by the fusion of haploid gametes from two different organisms.

sexual selection The differential ability of individuals to acquire mates, involves contests between males or choice by females; leads to the selection of morphological or behavioural traits relating to attracting mates.

short-day plant Plant that will only flower if the daily period of light is shorter than some critical length—the period of continuous darkness to which the plant is exposed; plants that usually flower in spring or autumn

sieve cells Phloem cells that transport sucrose; long and tubular in shape, bounded by a plasma membrane and containing mitochondria and ribosomes, with perforated sieve areas on their end walls.

signal A stimulus, chemical (e.g. food or hormones) or physical (e.g. light or heat), that can be detected by cells leading to a particular response.

signal sequence A sequence within a protein that is recognised by a receptor within the cell so that the protein is targeted to a particular site within the cell.

signal transduction cascade The process of sequential activation of a series of proteins following binding of a ligand to a receptor, ultimately triggering changes to cellular physiology and gene expression.

silencer A sequence in a eukaryotic gene that binds transcription factors to repress expression.

simple leaf Leaf with a single lamina.

sinoatrial node The heart's 'pacemaker'. A small group of non-contractible muscle cells in the right atrium of higher vertebrates that initiates the cardiac cycle; evolutionarily derived from the sinus venouses of fishes.

sinus A fluid- or gas-filled space within the body.

sinus gland In crustacea, highly vascular reservoir for neurosecretions of the x-organ (eye stalk hormones).

sinus venosus The first chamber of the fish heart that collects blood from the major veins and leads to the atrium.

siphon Specialised funnel in cephalopods that can produce a jet of water for propulsion.

siphuncle Structure that regulates the amount of gas and buoyancy in some cephalopods.

skeletal muscle Cylindrical fibres of multinucleate striated muscle, with nuclei located peripherally. Sometimes referred to as voluntary muscle, reflecting its voluntary control.

sliding filament model Model of mechanism of contraction of skeletal muscle, which proposes that the relative motion between actin and myosin myofilaments within muscle cells produces shortening of individual cells, resulting in muscle contraction.

smooth endoplasmic reticulum (SER) Endoplasmic reticulum that lacks attached ribosomes.

smooth muscle Spindle shaped contractile cells with a central nucleus and less regular arrangement of myofilaments than striated muscle cells; lines the walls of internal organs and arteries and veins and is under involuntary control.

solutes Ions or organic molecules dissolved in a liquid.

soma Middle, segmented part of an annelid.

soma cell The cell body of a neuron.

somatic cells All cells in the body of a multicellular organism other than germ cells.

somites Blocks of mesoderm that form adjacent to the notochord in the vertebrate embryo.

soredium (pl. soredia) Structure of a lichen analagous to a spore; consisting of an algal cell embedded in fungal hyphae.

sorocarp Fruiting body produced during the life cycle of slime moulds.

sorus (pl. sori) Cluster of sporangia on the margins or undersurface of a fern frond.

specialisation Occurs during differentiation of cells when certain functions develop at the expense of others.

speciation Formation of new species (see allopatric, sympatric and parapatric speciation).

species As a taxonomic category, the species is the lowest rank in a taxonomic classification (see also binomial system). There is no single definition that is agreed upon by evolutionary biologists. The biological species concept is a group of actually or potentially interbreeding natural populations that are reproductively isolated from other such groups.

species diversity Variability (species richness and abundance) of biota in an area; *alpha diversity*, within a local area; *beta diversity*, diversity among different communities in different areas.

species richness Number of species within a community.

specific heat The amount of heat required to raise the temperature of a substance by 1°C.

sperm The male gamete.

spermatheca (pl. spermathecae) A small sac containing sperm.

spermatocyte Male germ cell in the process of meiosis within the testis.

spermatogenesis Process by which sperm form from primordial germ cells.

spermatogonium Diploid male germ cell in the process of mitosis in the testis.

spermatophore A structure enclosing many sperm (*see also* spermatheca).

spicule Skeletal component in sponges, composed of calcium carbonate or silica.

spindle pole body Dense region of microtubules attached to the nuclear membrane at each spindle pole during mitosis in fungi; analogous to the centriole of other eukaryotes.

spindle The structure formed by microtubules between opposite poles of a cell during mitosis or meiosis and which guides the movement of chromosomes.

spinneret Spinning organ that produces silk in spiders.

spiracle Small external opening of the air-filled gas-exchange system (tracheae) of spiders and insects.

splicing The process whereby regions of RNA from the primary transcript (introns) are removed and flanking exons are joined.

spongin Course collagenous proteinaceous material forming skeletal fibres in sponges.

spongocoel Internal cavity of a sponge; also called atrium.

sporangiospore Haploid asexual spore that develops in a sporangium.

sporangium A sac-like cell or multicellular struture in which asexual spores form.

spore A cell capable of producing a new individual; often a dormant resistant structure or functioning in dispersal.

sporocyst Sac-like structure in the life cycle of parasitic flukes that produces rediae.

sporophyte Diploid stage of a plant life cycle that produces spores.

sporopollenin A polymer, tougher than lignin but with similar properties, composed chiefly of carotenoids; makes spores and pollen grains of plants resistant to biodegradation.

stamen In a flower, the male reproductive organ, comprising a bi-lobed anther on an elongated filament.

starch Insoluble polymer of glucose; composed of amylose (long chains of glucose units) and amylopectin (short branched chains of glucose); chief storage polysaccharide of green algae and plants; formed in chloroplasts and amyloplasts.

Starling principle In capillary exchange, the net fluid movement between the capillary and the interstitial fluid is determined by the balance between the hydrostatic pressure and the colloid osmotic pressure across the capillary wall.

statolith Small calcareous body found in sensory organs for balance.

stele The central vascular cylinder of stems and roots.

stem Main part of the aerial shoot of plants, usually bears leaves, lateral branches and reproductive organs.

stem cells Relatively undifferentiated cells that can divide continuously throughout the lifetime of an animal; daughter cells can either remain as stem cells or differentiate fully.

stenohaline Tolerant of a narrow salinity range only.

stigma The terminal cells of the pistil of a flower that receive and recognise pollen grains during interactions that may lead to fertilisation.

stolon A horizontally growing stem or runner.

stoma (pl. stomata) Specialised pore in epidermis of leaves and stems that allows uptake of CO_2 from the atmosphere for photosynthesis; pore through which transpiration occurs.

stomium Zone of thin-walled cells where sporangia or anthers rupture to release spores or pollen grains.

stomochord Short structure beneath the dorsal nerve cord at the junction of the proboscis and trunk of a hemichordate, forming a dorsal diverticulum of the pharynx.

storage parenchyma Tissue of parenchyma cells containing storage reserves such as starch granules, lipid droplets or protein storage organelles.

striated muscle Muscle with a highly organised array of actin and myosin filaments giving the appearance of cross-striations when viewed under the light microscope; includes skeletal and cardiac muscle.

strobilisation Process of adding new proglottids in the growth of a tapeworm; occurs at the anterior end.

strobilus (pl. strobili) In plants, a cone, which is a collection of sporangia; body of a tapeworm.

stroma (fungal) Mat of fungal hyphae bearing spores.

stroma (of chloroplast) Matrix enclosed within inner membrane.

stromatolite Concentrically layered rock, the layers being formed by successive growth of thin mats of cyanobacteria; fossil and present-day.

style The pathway for pollen tubes between stigma and ovary in the pistil of flowering plants; may comprise solid transmitting tissue or a canal.

subduction Descent of sea floor back into the earth's mantle (at deep-sea trenches).

substitutions A mutation in which a nucleotide in a particular position is changed to a different nucleotide.

succession Process of replacement over time of one ecological community by another.

succulent Xerophytic plant with fleshy leaves or stems and highly mucilaginous cell sap.

surface tension Surface tension at an air–liquid interface is the result of intramolecular attractive forces in the liquid, providing the potential energy that draws molecules from the surface and therefore shrinks the interfacial area.

surfactant A substance predominantly composed of phospholipid that forms a film on air–liquid interfaces and decreases surface tension.

suspensor A filament of cells below an embryo of flowering plants that connects it to the ovule.

swim bladder A gas-filled sac that forms as an outgrowth of the pharynx, allowing fishes to regulate buoyancy; homologous with lungs of land vertebrates.

symbiosis Interactions in which two organisms (symbionts) live together in a close relationship that is beneficial to at least one of them.

sympathetic nervous system Division of the autonomic nervous system that innervates the enteric nervous system and controls vascular changes in organs.

sympatric speciation Populations specialising on different resources diverge and form new species without geographic isolation.

sympatry Living in the same area.

symplastic pathway Pathway in plants for uptake of water, solutes and ions from cell to cell via the cytosol; only pathway for crossing the endodermis of roots.

synapse Small area of close contact between an axon terminal and a post-synaptic cell across which information is transmitted, usually by chemical neurotransmitters. May be excitatory or inhibitory, and also electrical.

synapsis Pairing of homologous chromosomes during prophase of meiosis I.

synaptonemal complex The molecular scaffold on which crossing over occurs between paired chromosomes at prophase of meiosis.

synergids In flowering plants, pair of cells adjacent to the egg at the micropylar end of an embryo sac, one of which acts to receive the pollen tube.

synonymous codons Different codons that encode incorporation of the same amino acid.

synovial fluid Slippery fluid found inside joints, produced by the synovial membrane; lubricates joint surfaces and nourishes articular cartilage.

systemic Referring to the circulation of blood to the body organs except the lungs.

systems In animals, several organs may combine to form systems, which carry out entire processes such as digestion or gas exchange.

systole The phase of the cardiac cycle involving muscle contraction and ejection of blood from a heart chamber.

T cell (T lymphocyte) Lymphocyte that matures in the thymus and recognises antigen by means of the T-cell receptor; functions independently to kill micro-organisms and controls B-cell responses.

tachycardia Increase in heart rate.

tagmatisation The organisation of segments into groups with differing structure and function.

tapetum The inner layer of the anther wall of a flower, comprising cells dedicated to nutrition of the developing microspores.

target cells Cells that respond specifically to a particular hormone by means of specialised receptor molecules located on the surface or inside the cell.

taxon Any formal name (rank) in the classification of living organisms (e.g. phylum, class, order, family, genus, species).

taxonomy Methods and principles of classification of organisms.

T-cell receptor (TCR) The glycoprotein molecule on the surface of T cells that recognises an antigen.

tegument Outer, resistant body coat of parasitic animals such as flukes and tapeworms.

telomere DNA sequences occurring at the ends of chromosomes of eukaryotes.

telomorphic Refers to the sexual form of a fungus.

telophase The final phase of mitosis, in which new nuclear envelopes form, surrounding each group of chromosomes.

telson Posterior tagma (tail end) of arthropods, which bears the anus.

terminus The sequence of DNA that terminates replication.

territory An area defended and occupied exclusively by one (or sometimes more) individuals.

testa The hard outer coat of a seed, formed from the integuments of an ovule.

testcross A cross between an F$_1$ progeny individual and a homozygous recessive individual.

tetrapod Vertebrate with four limbs with separate fingers and toes.

thallus Body of an alga, fungus or plant that lacks special tissue systems or organs.

thermophile Organism that grows best in hot conditions, between 30° and 50°C.

thermoreceptor A type of receptor that detects heat or cold.

threshold potential Potential difference across a membrane at which certain voltage-dependent channels (usually sodium, occasionally calcium) increase their permeability.

thrombin An enzyme that converts soluble fibrinogen into insoluble filaments of fibrin during the formation of a blood clot.

thylakoid Flattened disc-like sac that forms part of the internal membrane system of chloroplasts; site of location of photosynthetic pigments.

thyroid hormones Thyroxine and triiodothyonine; lipid soluble hormones that bind to intracellular receptors where they regulate the production of enzymes important in normal growth and maturation of cells. They stimulate metabolic rate of cells, and growth and metamorphosis in some immature vertebrates.

thyrotoxicosis A condition of elevated metabolic rate caused by excessive thyroid hormone activity.

tidal ventilation Movement of air into and out of an invaginated gas-exchange surface through the same opening; results in some rebreathing.

tissues Groups of similar, differentiated cells and associated extracellular matrix, which together carry out a particular function.

tolerance Failure of the immune systems to respond to a potentially immunogenic antigen. Occurs early in development, when lymphocytes become tolerant to self-antigens.

torpor Physiological state in animals in which metabolism decreases and body temperature is maintained at a lower level than normal.

totipotent The ability of plant cells to regenerate a new plantlet from a single cell.

trace element Micronutrient required by plants in small amounts, such as iron, zinc and copper, to carry out various metabolic functions.

trachea (pl. tracheae) Tiny spirally ringed tube that carries air internally to cells in insects and some other arthropods; in mammals, initial portion of the airway before it branches dichotomously into the bronchi leading to the lungs.

tracheid Type of conducting cell in the xylem of all vascular plants; *see also* vessel element.

tracheole Finest tube arising from an arthropod trachea, where gas exchange occurs directly with body tissues.

trait A characteristic, or phenotype.

transcription The synthesis of RNA from a DNA template.

transcription factors Proteins that interact with the regulatory sequences of genes to control their transcription.

transcription unit A region of DNA from which an RNA molecule is transcribed.

transfer cell Type of parenchyma cell of plants characterised by primary wall ingrowths, resulting in massive increase in surface area of plasma membrane, allowing rapid transfer of molecules to adjacent cells, especially the vascular system.

transfer RNA (tRNA) RNA that carries specific amino acids to ribosomes to add to growing polypeptide chains according to the base sequence in mRNA.

transformation The process of introducing DNA into a cell so that the DNA is stably maintained within that cell.

transition A base substitution (mutation) in DNA in which a pyrimidine or purine is replaced by another pyrimidine or purine respectively.

translation The synthesis of a protein from an mRNA template.

translocation Transport of assimilates (sugar) in the phloem of vascular plants, from the site of production in leaves (source) to other parts of the plant (sink).

transpiration Loss of water from a plant by evaporation through stomata in leaves; requiring energy from incoming solar radiation to vaporise water.

transport proteins Enzymes that catalyse the movement of specific solutes across cellular membranes.

transposable elements Segments of DNA that have the capacity to transpose to new sites within the genome

transversion A base substitution (mutation) in DNA in which a pyrimidine is replaced by a purine or vice versa.

triglyceride A fat molecule, having a backbone of glycerol with three carbon atoms, to each of which is attached a fatty acid chain.

trophic level Position of an organism in a food chain (e.g. primary producer or first-order consumer).

trophoblast Outer layer of epithelial cells in the mammalian blastocyst, gives rise to the chorion and forms the barrier between fetal and maternal tissues.

true capillary Capillary vessels across which exchange occurs; open in response to high metabolic demands.

trypsinogen Inactive precursor (zymogen) of the protease, trypsin, released into the duodenum.

tube foot Short, tubular, external projection of the body wall of echinoderms containing an extension of the radial canals of the water vascular system; functioning in gas exchange, attachment, locomotion and catching prey.

tubulin Protein that forms the major cytoskeletal scaffolding elements, microtubules, composed of equal amounts of two forms: α- and β-tubulin.

tumour suppressor gene A normal cellular gene that, when made non-functional by mutation, results in tumour formation.

tunic Supportive and protective 'coat' of tunicates secreted by ectoderm.

tunica In the shoot apex of flowering plants, the outer one to three layers of the apical dome of cells that contributes to leaf and flower formation.

turgor pressure Hydrostatic pressure within a cell that has a cell wall.

ultimobranchial bodies Endocrine glands that secrete calcitonin, which is involved in decreasing blood calcium levels.

unipotent Stem cells that produce only one type of cell, for example, skin stem cells.

urea (CON$_2$H$_4$) A common nitrogen waste product of terrestrial animals that is formed from ammonia by the urea cycle.

urediniospore Asexual spore of a rust fungus.

ureotelic Pattern of excretion of nitrogen wastes in the form of urea.

uric acid (C$_5$H$_4$O$_3$N$_4$) Nitrogen waste product of insects, reptiles and birds; formed by purine metabolism and from ammonia by uricogenesis.

uricotelic Pattern of excretion of nitrogen wastes in the form of uric acid.

urine Fluid formed and modified by the excretory system to excrete ions, organic solutes or water.

vacuole Large membrane-bound organelle, prominent in plant cells.

van der Waals force A weak attractive force between two polar atoms or molecules; arises due to the non-uniform distribution of electric charge at any instant on an atom.

vaporisation The change from a liquid to a gaseous state, such as water to steam.

variable numbers of tandem repeats (VNTRs) Segments of DNA that differ between different individuals because they have different numbers of direct repeats at a particular locus.

vascular bed The blood vessels in a region of the body, usually an organ, that are supplied from a common artery. Blood flow rates to individual vascular beds are usually independent of one another.

vascular bundle Prominent structural feature of primary growth of shoots, roots and leaves, comprising the transport system; consists of xylem and phloem.

vascular cambium Meristem responsible for producing wood (secondary xylem and secondary phloem).

vascular tone Tension in the walls of blood vessels which partly controls blood pressure and vascular resistance. Venous tone is important in controlling the rate of blood flow into the heart and arteriolar tone is largely responsible for determining the pattern of blood flow to vascular beds.

vasomotor centre Centre in the brain that controls the pressure and distribution of blood by affecting the contraction of smooth muscle in arterioles.

vasopressin Posterior pituitary hormone that influences blood pressure and water balance.

vector An agent, such as an insect, able to transfer a pathogen from one organism to another.

vector DNA DNA that carries an origin of replication and a selectable marker gene into which segments of donor DNA are ligated to create recombinant DNA that can be propagated in a host cell.

vegetal pole The pole of the egg containing high concentrations of yolk.

vegetative cell The largest cell of a pollen grain that contains the generative cell, and in some cases the sperm cells, and produces the pollen tube at pollen germination.

vein In animals, large blood vessel channelling blood towards the heart. In vascular plants, network of vascular bundles in the leaf lamina.

velamen Thick, multilayered epidermis of aerial roots of epiphytic orchids; covers all but the absorptive tip of the root.

veliger Second stage in the life cycle of molluscs; develops from the larva and forms a foot, mantle and shell.

venous Of veins. Usually referring to blood flowing in veins, or deoxygenated blood flowing in arteries or veins.

ventilation Movement of the external medium for purposes of gas exchange.

ventricle A heart chamber with strong muscular walls that develops most of the force necessary to pump blood through the circulatory system.

venule Small blood vessels that collect blood from the capillaries.

vernalisation Induction of flowering in certain plants by exposure to low temperatures.

vertebra (pl. vertebrae) A segment of backbone.

vessel element Tubular conducting cell in the xylem of flowering plants; together with tracheids, conducts water and dissolved mineral nutrients upwards.

vestigial organ Reduced and simpler structure than a corresponding part in another organism but with no apparent function; indicates relatedness of organisms.

virion A complete mature virus particle, which is metabolically inert and is the transmission (infective) phase.

viroid Infectious agent that is virus-like but lacks a protein coat.

virus Subcellular genetic parasite that reproduces only in the cells of a susceptible host and may cause disease.

viscosity Physical property of a fluid that determines its tendency to flow.

vitamin Organic compound required by animals in small amounts for normal growth and maintenance.

vitelline membrane A tough, clear, elastic envelope found in many eggs and embryos, lying adjacent to the cell membrane.

vitellogenesis Incorporation of lipids, proteins and carbohydrates into developing oocytes.

vitellogenic hormones Hormones that stimulate vitellogenesis in egg-laying animals.

viviparity A mode of sexual reproduction in animals in which offspring develop inside the maternal body and are released as live young or eggs; in plants, seeds germinate while still attached to the parent plant (e.g. in mangroves).

water potential Total energy level of water, which is the sum of the osmotic potential and pressure potential of water.

water vascular system System of coelomic canals, lined with ciliated cells and filled with fluid; includes a circular water canal and radial canals leading to tube feet; unique to echinoderms (*see also* tube foot).

wax Esters of fatty acids with long-chain monohydric alcohols, water-repellent coating on plant leaves, animal fur etc.

weed Plant growing where it is not wanted; often an introduced species.

whole blood Blood containing both plasma and blood cells, as drawn directly from a blood vessel.

wild-type allele The common allele (genotype) that occurs in nature compared with mutant type.

wild-type The phenotype found in most individuals in a population.

xenospore Asexual fungal spore for dispersal and spread of the fungus (*see* conidiospore).

xerophyte Plant tolerant of dry conditions.

x-organ In crustacea, neurosecretory cells located in the largest eyestalk ganglion.

xylem A major tissue of vascular plants comprising several cell types, including water-conducting vessels and tracheids, sclerenchyma fibres for support and parenchyma.

xylem sap The liquid contained as a continuous column in the tracheids and vessels of xylem; a dilute solution of inorganic ions and some organic nitrogen compounds.

y-organ In crustacea, non-neural endocrine glands located in the head, usually at the base of the first antennae, which secretes a moult-stimulating hormone.

zona pellucida The vitelline membrane of the mammalian egg and embryo.

zooxanthella (pl. zooxanthellae) Dinoflagellate endosymbiont found in the tissues of corals, sea anemones and molluscs.

zygomorphic Describes a flower with parts arranged asymmetrically.

zygospore Dormant spore with thick cell wall enclosing a zygote.

zygote Diploid cell resulting from the fusion of the male and female gametes.

zygotic genes Genes that are transcribed and act in the developing zygote.

zymogen An inactive precursor of a digestive enzyme released into the gut by cells to prevent damage to the secretory cells. Zymogens are activated by conditions in the gut.

Acknowledgments

Every care has been taken to trace and acknowledge copyright. The publisher apologises for any accidental infringement where copyright has proved untraceable and would be pleased to come to a suitable arrangement with the rightful owner in each case.

Part 1 Cell biology and energetics

Introduction

I.1a: Great Barrier Reef Marine Park Authority
I.1b: Dr Gert Hansen, University of Copenhagen
I.1c: Dr Rick Wetherbee
I.2: Professor Pauline Ladiges
I.3: Jeffrey Wright, Queensland Museum
Box I.2a, b, d: Lochman Transparencies

Chapter 1

1.1: The Biosphere Press
1.10, 1.38a: C. A. Smith & E. J. Wood, 'Molecular and Cell Biochemistry', *Biological Molecules*, 1991, Chapman & Hall, London, p. 67
1.11a: C. A. Henley, Auscape
1.11b: Colin Monteath, Auscape
1.11c: Claude Steelman/Survival, Anglia-Oxford Scientific Films
1.15a: Queensland Sugar
1.16a: M. N. Ausumees et al, *Microbiology*, 1999; 145:1253–62, Fig. 1a
1.16b: Kerrie Ruth, Auscape
1.16c: Bill Bachman, ANT Photo Library
1.16d: Robin Smith, Photolibrary.com
1.17b: Reproduced with permission of Academic Press
1.17c: T. Itoh and R. M. Brown Jnr, *Planta*, 1984; 160: 372–81 (Springer-Verlag)
1.19: *Brock Biology of Microorganisms*, 8th edn, 1997, Madigan/Martinko
1.20a: Dr Patricia J. Schultz/Peter Arnold, Auscape
1.20b: Volker Steger/Peter Arnold, Auscape
1.23a: Courtesy of Yuuji Tsukii, Hosei University, Protist Information Server, http://protist.i.hosei. ac.jp/Protist_menuE.html
1.23b: Bruce Fuhrer
1.23c, 1.39c: Leanne Peters
1.23d: Michael Whitehead, Auscape
1.25: Anne & Jacques Six, Auscape
1.26a: Lipid Research Inc.
1.26b: N. De Deepesh, *Plant Cell Vacuoles: an Introduction*, 2000, CSIRO, p. 120; reproduced by permission of CSIRO Australia and N. De Deepesh
1.35a–c: J. S. Richardson, *Advances in Protein Chemistry*, 1981: 34: 167–339 (Harcourt Brace)

1.36a, b: M. Hrmova, *Journal of Biological Chemistry*, 270: 14556–63, American Society for Biochemistry & Molecular Biology Inc.
1.38b: D. E. Fox, *New Scientist*, 1999; 2183 (24 April): 40
1.38c, d, 1.43c: B. Alberts, J. Bray, J. Lewis, M. Raff, K. Roberts, et al, *Molecular Biology of the Cell*, 3rd edn, 1994, reproduced by permission of Taylor & Francis, Inc., Routledge, Inc., http://www.routledge-ny.com
1.39a: Wayne Lawler, Auscape
1.39b: C. Gautier-Pho.n.e/Auscape
1.39d: Ferrero/Labat, Auscape
1.39e: David Fleetham, OSF/Auscape
1.40a: Joseph A. Buckwalter, MD
1.40b: Lubert Stryer, *Biochemistry*, 4th edn, © 1975, 1981, 1988 and 1995; used with permission of W. H. Freeman and Company
Box 1.1a: Jean-Paul Ferrero, Auscape
Box 1.1b: Andrew Drinnan
Box 1.1c: Kathie Atkinson, Auscape

Chapter 2

2.2a: N. N. Birks, Auscape
2.2b: J. M. Labat, Auscape
2.5a–b: Eldra Pearl Solomon, Claude A. Villee, P. William Davis, *Biology*, international edn, 1985, Saunders College Publishing, reproduced by permission of the publisher
2.11: Jean-Paul Ferrero, Auscape
2.12: Mary K. Campbell, *Biochemistry*, 1991, Saunders College Publishing, reproduced by permission of the publisher

Chapter 3

3.1a, 3.6c, 3.9a, 3.13b: Professor Adrienne Hardham
3.1b: P. A. Janssens
3.1c: L. Marotte
3.5a: The Electron Microscope Unit, University of Sydney
3.9b, 3.12c, d: M. Fitzgerald
3.10a: Photolibrary.com
3.12a, b, 3.15a, 3.16b: B. Gunning & M. Steer, *Plant Cell Biology: an Ultrastructural Approach*, 1975, Jones & Bartlett Publishers
3.15b: D. Barkla
3.18a: Nancy Kedersha, Photolibrary.com
3.18d: R. E. Williamson & G. O. Wasteneys, *Fortschritte der Zoologie*, 1987; 34: 17 (Fig. 2A)
3.19a: Francis Leroy, Photolibrary.com
3.19b: Dr Gerald Schatten, Photolibrary.com
3.20a: Dr Peter Dawson, Photolibrary.com
3.21a: K. A. Pickerd, EM Unit, CSIRO Division of Entomology

3.21b: M. Abbey, Photolibrary.com
3.22b: Chris Bjornberg, Photolibrary.com

Chapter 4

4.9: S. Tyerman
4.10a, b: G. R. Roberts, Documentary Photographs
4.19b: R. Marginson

Chapter 5

5.19: Queensland Sugar
5.23b: K. Winter & C. B. Osmond

Chapter 6

6.1a, b: Ian and Vicky Lovegreen, Lovegreen Photographics Pty Ltd
6.2: J. Chamley-Campbell
6.3a: Dr Robert Trelstad, Lab of Medicine, UMDNJ-RW Johnson Medical School
6.5a: Professor Andrew Staehelin and Dr Barbara Hull
6.6b, 6.8b, c: Professor Andrew Staehelin, Dept of Biology-MCD, University of Colorado
6.8d: Reproduced from *The Journal of Cell Biology*, 1979; 81: 67–82, by copyright permission of The Rockefeller University Press
6.10a–c, 6.11a–c, 6.12a, b: Histology Collection, Department of Anatomy and Cell Biology, University of Melbourne
6.12c: Photolibrary.com, Biophoto Association
6.13a: H. M. Young
6.14a: Professor Adrienne Hardham
6.14b: M. C. McCann, B. Wells, K. Roberts, *Journal of Cell Science*, 1990; 96, The Company of Biologists Ltd
6.16a, b: Robyn Overall
6.17, 6.22d: Ripon Microslides, USA
6.18b: J. Carpenter
6.20a: John Considine, Faculty of Agriculture, University of Western Australia
6.20b: Professor Pauline Ladiges

Chapter 7

7.1a, b: Ron Oldfield & Keith Williams, Macquarie University
7.9: P. Roach, NPIAW
7.10: Bill & Peter Boyle, Auscape
7.11: C. Marcroft

Chapter 8

8.6a–j, 8.11a–d: Professor Jeremy Pickett-Heaps, University of Melbourne

Part 2 Genetics and molecular biology

Chapter 9

9.1a: Archiv Für Kunst und Geschichte, Berlin
9.1b: David Smyth, Monash University
9.5: Australian National Botanic Gardens
9.9b: Vision Training Products Corporation
9.14: Jean-Paul Ferrero, Auscape
9.15a, b: Stirling Macoboy

9.15c: Nigel Dennis, ANT Photo Library
9.17, 9.18a–c: J. Alvarez
9.19: M. Martin & J. Kinnear

Chapter 10

10.1: image was captured on a CytoVision (Applied Imaging Corporation) and is supplied courtesy of E. Baker, Women's and Children's Hospital, Adelaide
10.2: Dr Thomas Ried, National Human Genome Research Institute, National Institute of Health, Bethesda, MD, USA
10.3: George V. Kelvin/Scientific American
10.5a–d: Helena Curtis et al., *Biology*, 5th edn, 1989, Worth Publishers Inc., Chapter 14 (Fig. 14-3), p. 283
10.6: Tobin & Morel, *Asking about Cells*, 1997, Saunders College Publishing, Harcourt Brace
10.9b: P. Sloof, A. Maagdellin & A. Boswinkel, *Journal of Molecular Biology*, 1983; 163: 278, Fig. 6
10.15a: R. G. Wake, 'Circularity of the *Bacillus subtilis* Chromosome and Further Studies on it's Bidirectional Replication', *Journal of Molecular Biology*, 1973; 77: 569–75, Academic Press (London) Ltd
10.17: Professor David Hogness, Stanford University
10.19: Donald E. Riley & Harold Weintraub
Box 10.2: D. T. Suzuki, A. J. F. Griffiths, J. H. Miller, R. C. Lewontin, *An Introduction to Genetic Analysis*, 4th edn, 1989, W. H. Freeman, New York, p. 269
Box 10.4: Robert T. Schimk, 'Gene Amplification and Drug Resistance', *Scientific American*, 1980; 243: 51

Chapter 11

11.18: Bruce Alberts et al., *Molecular Biology of the Cell*, 3rd edn, Garland Publishing: Chapter 9 (Fig. 9.29): p. 420
11.19: Bruce Alberts et al., *Molecular Biology of the Cell*, 3rd edn, Garland Publishing: Chapter 9 (Fig. 9.25): p. 418

Chapter 12

12.5: David Searls, University of Pennsylvania Computational Biology and Informatics Laboratory, 1994
Box 12.1: Reprinted with permission from White et al., *Science*, 1999; 286: 1571–77; copyright American Association of the Advancement of Science
Box 12.3: K. P. White

Chapter 13

13.5a, b: Huiling Xu, University of Melbourne
Box 13.1: Reprinted Courtesy of *The Boston Globe*
Box 13.5: The White House Collection, copyright White House Historical Association
Box 13.6b: Reprinted from *Nature*, 1991; 351: 119, with permission from R. Lovell-Badge

Part 3 Reproduction and development

Chapter 14

14.2b: Jeremy Pickett-Heaps
14.2c: N. Yanagishima in G. P. Chapman et al., *Eukaryote*

Cell Recognition: Concepts and Model Systems, 1999,
Cambridge University Press, Cambridge, p. 52,
Fig. 4.3

14.3a: E. Evans

14.4a, b: Dr David Beardsell, Institute for Horticultural
Development, Department of Natural Resources and
Environment, & Tony Slater

14.6a: S. Blackmore

14.11a, b: I. Lambiris & A. Drinnan

14.13: J. Kenrick

14.25: Dr Arancha Arbeloa & Maria Herrero, Diputacion
General de Aragon, Zaragoza, Spain

14.26a: Steve Read

14.28a, b, 14.29: Professor S. Y. Zee, Department of Botany,
University of Hong Kong

14.31: Otto Ridge, ANT Photo Library

14.32a: G. R. Roberts, Documentary Photographs

14.35b, 14.36a, b: E. G. Williams

14.37a(i), (ii), (iii): Huiling Xu, University of Melbourne

Chapter 15

15.1, 15.7: A. Flowers & L. Newman

15.2: John Cook, © Oxford Scientific Films

15.3: NHPA, ANT Photo Library

15.5a: Marilyn Renfree, University of Melbourne

15.5b: J. Anthony, *Proceedings of The Royal Society of
London* 1980; 349–67: 208, plate 2

15.6, 15.8a, 15.11: D. Parer & E. Parer-Cook, Auscape

15.8b: Kathie Atkinson

15.9: Jan Aldenhoven, Auscape

15.10a: J. M. Labat, Auscape

15.10b: Otto Ridge, ANT Photo Library

15.12, 15.16: Hans & Judy Beste, NHI

15.13: John Cascalosi, Auscape

15.14, 15.26b: C. Andrew Henley, Auscape

15.15: M. J. Tyler, ANT Photo Library

15.17: Dr Russell Shiel, Murray–Darling Freshwater
Research Centre

15.18: Professor Yuh Nung Jan, University of California, San
Francisco

15.21b: Professor Barrie Jamieson, Department of Zoology
and Entomology, University of Queensland

15.26a, 15.27b: G. I. Bernard, Oxford Scientific Films

15.27a: D. Rentz, Australian National Insect Collection,
CSIRO Division of Entomology

15.38: Reprinted with permission from Stephen J. Smith,
'Neuronal Cytomechanics: the Actin-based Motility of
Growth Cones', *Science*, 1988; 242: 708–15 (Fig. 14.2a),
copyright American Association for the Advancement of
Science

15.41a: Alberts et al., *Molecular Biology of the Cell*, 1989,
Garland Press, New York, p. 915, Fig. 16.44b

15.41b: Professor Lewis Wolpert, University College
London

15.44: J. Whittaker, reproduced by permission

Box 15.2b: Geoff Shaw

Box 15.3a: Jamie Oliver/ GBRMPA

Box 15.3 b, c, d: Dr Peter Harrison

Box 15.3e: R. Babcock

Box 15.5a(i)–(iii), b, c(i)–(ii), d(i)–(ii): D. Newgreen

Chapter 16

16.2: David Smyth, Monash University

16.3: P. A. Photos Limited, London

16.4: Reprinted by permission from Jonathan M. Passner,
Hyung Don Ryoo, Leyi Shen, Richard S. Mann, Aneel
K. Aggarwal, 'Structure of DNA-bound Ultrabithorax
Extradenticle Homeodomain Complex', *Nature*, 1999;
397: 714–19: © Macmillan Magazines Ltd

16.5: S. F. Gilbert, *Developmental Biology*, 5th edn, Sinauer
Associates Inc., Fig. 9.9

16.6: Katrina Bell & Dr Andrew Sinclair, Department of
Pediatrics, University of Melbourne and Murdoch
Children's Research Institute; reprinted from K. Bell,
P. Western, A. Sinclair, *Mechanims of Development*,
vol. 94, 'SOX8 Expression During Embryogenesis',
2000, Elsevier Science, pp. 257–60

16.8: The Company of Biologists, UK, & Dr Mary C.
Mullins, University of Pennsylvania

16.9, 16.19a–b: S. F. Gilbert, *Developmental Biology*, 5th
edn, Sinauer Associates Inc., Fig. 3.33

16.13b: Professor Christiane Nüsslein-Volhard, Max Planck
Institut für Entwicklungsbiologie

Box 16.2a: The Company of Biologists, UK

Box 16.2b: Diagram courtesy of Eric P. Spana & Chris Q.
Doe, *Neuron* 1996; 17: 21–6

Part 4 Regulation of the internal environment

Chapter 17

17.1, 17.7b, c, 17.9, 17.25, Box 17.1b: Andrew Drinnan

17.2: L. A. J. Thomson, *ECOS*, 1991; 67: 21

17.3a, b, 17.4a, b, 17.6d, 17.29: P. Taylor

17.8: G. R. Roberts, Documentary Photographs

17.17a: A. Ashford

17.22, 17.23a: P. Grierson

17.23b: R. Gross

17.27a, b: S. A. James

17.31a. b: David Albrecht

Chapter 18

18.8, 18.18a–b: I. Ridge (ed), *Plant Physiology*, 1991,
Cambridge University Press

18.12: J. S. Pate

18.13: Ken Hodgkinson, CSIRO

18.14a, c: J. A. Milburn, *Water Flow In Plants*, Longman,
London, p. 101, Fig. 5.9

18.21: Bill Bachman, ANT Photo Library

18.22a–b: L. D. Incoll

18.24: J. S. Pate

Box 18.1: Andrew Drinnan

Box 18.3: Stirling Macoboy

Chapter 19

19.1a: Peter Atkinson, ANT Photo Library

19.1b: C. & S. Pollitt, ANT Photo Library

19.1c: V. Tunnicliffe

19.2: B. Hills

19.4: D. A. Denton, *The Hunger for Salt*, Springer-Verlag,
1982, Fig. 3.18

19.9b: Histology Collection, Department of Anatomy and Cell Biology, the University of Melbourne
19.9c: Megan Klemmin
19.16: K. F. Liem, *Journal of Morphology*, © 1978; reprinted by permission of Wiley-Liss, Inc., a subsidiary of John Wiley & Sons, Inc.
19.17: Australian Picture Library
19.20, 19.21, 19.22: G. Sanson
19.23a: Stephen Burnell
19.23b: B. Willis
19.28a, b: R. Ortner

Chapter 20

20.1: Peter Greenaway
20.5a: Kelvin Aitken, ANT Photo Library
20.15a: Dr Jill Hallam
20.18a, b: H. R. Duncker
Box 20.2a: J. P. Mortola, P. B. Frappell, P. A. Woolley & T. Phillips, La Trobe University
Box 20.4c(i), (ii): CSIRO Entomology and Melbourne University Press

Chapter 21

Box 21.2c: Australian Picture Library

Chapter 22

22.24: P. Withers

Chapter 23

23.9a–b: Dr P. J. Holt, WAICH

Part 5 Responsiveness and co-ordination

Chapter 24

24.6: P. F. Wareing & I. D. J. Phillips (eds), *The Control of Growth and Differentiation in Plants*, 3rd edn, 1981, Pergamon Press, Sydney, p. 144
24.7: Professor A. Sievers & Dr K. Schroter, 'Versuch einer Kausalanalyse der geotropischen Reaktionskette im Chara-Rhizoid', *Planta*, 1971; 96: 339–53
24.9: Brian & Hemming, *Plant Physiology*, 1955; 8: 669–81
24.17: P. Taylor
24.18c: Dr T. L. Setter, International Rice Research Institute, The Philippines
24.24: Stirling Macoboy

Chapter 25

25.12a: Jean Joss
25.13: C. Donnell Turner & J. T. Bagnara, *General Endocrinology*, 6th edn, © 1976 by Saunders College Publisher, reproduced by permission of the publisher
25.15a, b: Photolibrary.com
25.20b: Professor B. Tuch, Pancreas Transplant Unit, Prince of Wales Hospital

Chapter 26

26.3a: Dr John Heath, University of Newcastle
26.3b: D. Parer & E. Parer-Cook, Auscape

26.4: Dr Sueli Pompolo, Prince Henry's Institute of Medical Research and Professor John Furness, University of Melbourne
Box 26.4a: Darran Leal
Box 26.4b: J. C. Wombey, Auscape
Box 26.4c: R. Hill, Botanic Gardens of Adelaide
Box 26.6a: Bruce G. Thomson, ANT Photo Library
Box 26.6b: Cyril Webster, ANT Photo Library

Chapter 27

27.1a: Mike Tinsley, Auscape
27.3: Alby Ziebell, Auscape
27.4: Ron & Valerie Taylor, ANT Photo Library
27.5a: CSIRO Division of Fisheries
27.6a–b: Alexander R. McNeill, *Dynamics of Dinosaurs and Other Extinct Giants*, 1989, Columbia University Press, New York, Fig. 9.5b, p. 130, reprinted by permission of the publisher
27.11a: Jean-Paul Ferrero, Auscape
27.14b: J. O'Neil, ANT Photo Library

Chapter 28

28.1, 28.5a, 28.5c, 28.18: Kathie Atkinson
28.2: Sybille Kalas & Klaus Kalas
28.4: R. J. Allingham, ANT Photo Library
28.5b: Pictor/Austral
28.6, 28.7, 28.8: Densey Clyne, Mantis Wildlife Films
28.9: N. N. Birks, Auscape
28.11: Jean-Paul Ferrero, Auscape
28.12: D. Parer & E. Parer-Cook, Auscape
28.13: Glen Threlfo, Auscape
28.15: Rudie Kuiter, ANT Photo Library
28.16: Robert Hernandez, Photolibrary.com
28.17: Australian Antarctic Division, photograph by Kevin Sheridan © Commonwealth of Australia, 1992
28.19: J. R. Krebs & N. B. Davis, *Behavioural Ecology*, 3rd edn, 1978, Blackwell Scientific, Melbourne, Fig. 11.1, p. 339; reprinted by permission of Blackwell Science, Inc.
28.20a–f: W. M. Wheeler, *Ants: their Structure, Development and Behaviour*, 1910, Columbia University Press, New York
28.21: Mike Gillam, Auscape
Box 28.1a: Photolibrary.com
Box 28.1b(i): Jan Taylor, ANT Photo Library
Box 28.1b(ii): Silvestris/ANT Photo Library

Chapter 29

29.1: Kerrie Ruth, Auscape
29.2: Kathie Atkinson
29.3: J. Burt, ANT Photo Library
Box 29.1a, b: Roger Seymour, Environmental Biology, University of Adelaide

Part 6 Evolution and biodiversity

Chapter 30

30.2: E. Haeckel, *The Evolution of Man*, vol. 1, 1896, D. Appleton & Co., New York
30.3: Jason Edwards, Bio-Images

30.5a: Dennis Stevenson, New York Botanic Garden
30.5b, c: Professor Pauline Ladiges
30.11a, Box 30.4a(i), (ii), b: Tim Low, Australian Natural History, Australian Museum
30.14a–c: K. Thiele, School of Botany, University of Melbourne
Box 30.1: Australian National Insect Collection, CSIRO Division of Entomology
Box 30.2: Courtesy of the American Philosophical Society
Box 30.3a: Kelvin Aitken, ANT Photo Library
Box 30.3b: Tony Howard, ANT Photo Library

Chapter 31

31.1, Box 31.1: W. Harland et al, *Geologic Timescale*, 1989, Cambridge University Press, Cambridge, p. 164
31.5, 31.14: Andrew Drinnan
31.6, 31.16, 31.19, Box 31.4b: Reproduced with permission from P. V. Rich, G. F. van Tets & F. Knight, *Kadimakara: Extinct Vertebrates of Australia*, 1985, Pioneer Design Studio Pty Ltd, Melbourne
31.7: Dr P. Crane, Field Museum of Natural History, Chicago
31.8, 31.11, 31.12, 31.15: Field Museum of Natural History, Chicago
31.9: Artist: Peter Sawyer, Smithsonian Institution
31.10: N. Pledge, © South Australian Museum
31.13: Kristine Brimmell, © Western Australian Museum
31.18: Natural History Museum, London
Box 31.2b: Linda S. Moore
Box 31.3a: After K. Campbell, *Australian Journal Earth Sciences*, 1990; 37: 247–65
Box 31.3b: Dr D. Briggs, Department of Geology, Queens Rd, Bristol, England

Chapter 32

32.1: Photolibrary.com
32.2: Dr Greg Kirby, Flinders University
32.7: Breck P. Kent
Box 32.3a: Mark A. Clements, Australian National Botanic Gardens
Box 32.3b: D. Whitford, ANT Photo Library
Box 32.3c: Martin Harvey, ANT Photo Library
32.8a: K. W. Dixon, Kings Park & Botanic Gardens
32.8b: M. Hutchins & B. Sleeper, 'Life without Legs: The Pygopodid Lizards', *Australian Natural History*, 1988; 22 (11): 526
32.10a: G. D. Anderson, ANT Photo Library
32.10b: L. H. Schmitt, *Evolution*, 1978; 32: 1–14 (Fig. 2)
Box 32.4a, b: Dr Greg Kirby
32.13a: B. G. Thomson, ANT Photo Library
32.13b: From *Australian Journal of Zoology*, 1990; 37 (2, 3, 4); reproduced by permission of CSIRO Australia
32.14: R. H. Groves, CSIRO Division of Plant Industry
32.16: Craig Moritz, Department of Zoology, University of Queensland
32.17: N.H.P.A/ANT Photo Library

Chapter 33

33.2a–f: Peter Stewart
33.4a: Dr Tony Brain, Photolibrary.com

33.6: Dr Christina Cheers, University of Melbourne
33.9b: Professor Pauline Ladiges
33.10: Yakult Australia Pty Ltd
Box 33.1b, c: Microbiology Department, University of Melbourne

Chapter 34

34.1a, b, 34.3a–e, Box 34.3: Adrian Gibbs
34.2: David Guest

Chapter 35

35.3, 35.32a–b: David Patterson
35.4a–c, 35.5a, b, 35.7, 35.8, 35.11a–c, 35.13a, b, 35.14b, 35.17, 35.19a–c, 35.20b, 35.23, 35.25a, b, 35.26b, c, 35.29, 35.31, Box 35.1: Geoff McFadden
35.6a–d: Tim Entwistle, State Herbarium, Melbourne
35.9, 35.12, 35.28a–b, 35.30, Box 35.4: Gustaaf Hallegraeff
35.10a–c: Dr Peter Beech, Deakin University
35.14a, 35.20a: Biophoto Associates, Bristol
35.15: Rosey Van Driel
35.16: Jeanetta Brewer
35.18a, b: Karen Kabnick
35.27: Harvey Marchant, Australian Antarctic Division
Box 35.2: Paddy Ryan, ANT Photo Library
Box 35.3a: Michael Borowitzka
Box 35.3b: Dr Willi Jahnen-Dechent, University of Aachen, Germany
Box 35.5a, b: D. Reifsynder

Chapter 36

36.2a, 36.10, 36.11, 36.16, 36.21b: D. Guest, University of Melbourne
36.2b–d, 36.13a, b, 36.15, 36.18a–b: Gilbert Bompeix, Laboratoire Biochimie et Pathologie Végétale, Université Pierre et Marie Curie, Paris
36.4: J. H. Burnett, *Fundamentals of Mycology*, 1976, Cambridge University Press
36.5a–d, 36.7a–b: H. Swart
36.6: C. O'Brien
36.8a–b, 36.9, 36.12: R. Guggenheim, REM-Labor, Universitaet Basel-Geowissenschaffen, Basel
36.14: J. W. Deacon, *Modern Mycology*, 3rd edn, 1979, Blackwell Science
36.20a, b: D. Orlovich
36.20c: R. Gross, University of Melbourne
36.21a: P. Keane, La Trobe University
36.24: Norman Nicholls
36.25: B. Fuhrer
36.26: Professor Ross St C. Barnetson, University of Sydney
36.27: G. L. Barrum, *The Nematode Destroying Fungi, Topics in Mycobiology 1*, 1977, Canadian Biological Publications, Guelph
Box 36.2: P. Keane

Chapter 37

37.1, 37.19b, 37.22a, b, 37.24a, 37.27a, 37.32b, c, 37.35a, Box 37.1, Box 37.3a: Professor Pauline Ladiges
37.3: Professor Jeremy Pickett-Heaps, University of Melbourne
37.4, 37.6, 37.8, 37.10a, b, 37.11a–c, 37.13a, b, 37.14,

37.15a, b, 37.18a, b, 37.19a, 37.21, 37.25a–c, 37.26, 37.29, 37.30a–c, 37.31, 37.40a, b, 37.41a, 37.42, 37.43a, b, 37.45, Box 37.4b, c: Andrew Drinnan, University of Melbourne

37.24b, 37.27b, 37.28: M. Regan

37.32a: D. Stevenson

37.33a: C. Humphries

37.33b: D. Ashton

37.35b: K. Thiele

37.41b: Reg Morrison, Auscape

37.41c: Jean-Paul Ferrero, Auscape

37.44b: B. & B. Wells

Box 37.3b: G. Sainty

Box 37.4a: Jamie Plaza, Wildlight Photo Agency Pty Ltd

Box 37.5: Dr Bob Congdon, James Cook University

Chapter 38

38.1, 38.10: N. Pledge, © South Australian Museum

38.3: A. Flowers & L. Newman

38.8a, b, 38.11, 38.14a, b, 38.15b, 38.21, 38.25b: Barrie Jamieson

38.15a: Becca Saunders, Auscape

38.16a, 38.26a: Kathie Atkinson

38.29: Professor Ross St C. Barnetson, University of Sydney

Chapter 39

39.1a: Kathie Atkinson, Auscape

39.1b: C. A. Henley, Auscape

39.3, Box 39.1b, 39.26, 39.27a, 39.28, 39.29b, c, 39.31b, 39.32: Barrie Jamieson

39.5b, 39.12, 39.15a, b, Box 39.2b, 39.23a–c, 39.34b, c: Kathie Atkinson

39.9a: Martin Dohrn, Photolibrary.com

39.13: Kev Deacon, Ocean Earth Images

39.14b: Dr Robert Raven, Queensland Museum

Box 39.2a: R. W. Taylor, Australian National Insect Collection

39.18, 39.22e, g: Otto Ridge, ANT Photo Library

39.22a: Jean-Paul Ferrero, Auscape

39.22b: Klaus Uhlenhut, ANT Photo Library

39.22c, d: John McCammon, Auscape

39.29a, 39.35: A. Flowers & L. Newman

39.30, 39.31a: R. Willan

39.34a(i): Jan Aldenhoven, Auscape

39.34a(ii), (iii): D. Parer & E. Parer-Cook, Auscape

Chapter 40

40.1a: Mark Wellard, ANT Photo Library

40.1b: Kathie Atkinson, Auscape

40.1c: Pete Atkinson, ANT Photo Library

40.5: P. Moran

40.6: © Jon G. Houseman; this image is part of the BIODIDAC image bank

40.8c: Ron & Valerie Taylor, ANT Photo Library

40.11: Babs & Burt Wells

40.13a: Kev Deacon, Auscape

40.13b: Francois Gohier, Auscape

40.14b: Gunther Schmida

40.15a: © Schauer/Fricke

40.15b, 40.20b, 40.26a, 40.27: Jean-Paul Ferrero, Auscape

40.17b: S. L. & J. T. Collins, Photolibrary.com

40.19: Kathie Atkinson

40.20a: D. Rosen

40.22a: Darran Leal

40.24b: D. Cantrill

40.24c: Professor Pauline Ladiges

40.24d: Roger Brown, Auscape

40.25: Australian Tourist Commission

40.26b: D. Parer & E. Parer-Cook, Auscape

40.29a: I. R. Van Nostrand

40.29b: Brett Allatt, Photolibrary.com

40.29c: David Messent, Photolibrary.com

40.32, 40.36b, c: Department of Library Services, American Museum of Natural History

40.33: Ian Tattersall, Department of Anthropology, American Museum of Natural History

40.34: George Holton, Photolibrary.com

40.36a: Julius Kirschner, Courtesy Department of Library Services, The American Museum of Natural History

40.37a, b: Based on Herbert Thomas, *The First Humans: the Search for our Origins*, trans. 1995 Paul Bahn, Editions Gallimard, Paris

Box 40.2: Professor Jim Bowler, University of Melbourne

Chapter 41

41.3b: CSIRO, Division of Entomology

41.6: Dr Charles Butt, Division of Exploration Geoscience, CSIRO

41.13: A. Flowers & L. Newman

41.14: Stirling Macoboy

41.16, 41.19: K. Thiele

41.17a: Murray Fagg, © Australian National Botanic Gardens

41.17b: School of Botany, University of Melbourne

41.18: Professor Pauline Ladiges

41.22b, 41.25b, c, Box 41.3a: Andrew Drinnan

41.23: J. Bruhl

41.27b, c: Dr Mark Clements, CSIRO Centre for Plant Biodiversity Research, Australian National Botanic Gardens

41.28: G. R. Roberts, Documentary Photography

41.29: H. Ehmann/NPIAW

41.30, Box 41.2b: D. Parer & E. Parer-Cook, Auscape

41.31a: Peter Schouten

41.32a: Vincent Serventy, Photolibrary.com

41.33: Professor Michael Archer, University of New South Wales

41.34: John Fields/ Nature Focus, Australian Museum

41.35: Dr Peggy Rismiller & Mike McKelvey

41.36: D. Paul

41.37, 41.38b: Jean-Paul Ferrero, Auscape

41.38a, Box 41.7a: Hans & Judy Beste, NHI

41.38c: John Cancalosi, Auscape

41.39: D. Paul

41.40: L. Lumsdon

41.41: Esther Beaton, Australian Picture Library

Box 41.2c: Ian Shaw, Nature Focus

Box 41.5b: Dr Ebbe Nielsen, CSIRO Department of Entomology

Box 41.6a, b: Reproduced with permission from P. V. Rich, G. F. van Terts & F. Knight, *Kadimakara: Extinct*

Vertebrates of Australia, 1985, Pioneer Design Studio, Melbourne

Part 7 Ecology

Chapter 42

42.1a: Babs & Bert Wells, NPIAW
42.5: D. Cantrill
42.6a: B. R. Burns & J. Ogden, *Australian Journal of Ecology*, 1985; 10: Fig. 1
42.6b: G. R. Roberts, Documentary Photographs
42.7a: C. A. Henley, Auscape
42.7b: G. R. Dickman, *Australian Journal of Ecology*, 1986; 11: Fig. 2
42.9b, Box 42.3: Australian Tourist Commission
42.10: Paddy Ryan, ANT Photo Library
42.12: Sylvian Cordier, JACANA, Auscape
42.13: I. W. Watson, M. Westoby, A. McR. Holm, 'Demography of Two Shrub Species from an Arid Grazed Ecosystem in Western Australia 1983–93', *Journal of Ecology*, 1997; 85: 815–32 (Fig. 7)
42.14: Dave Watts, ANT Photo Library
42.16: Jean-Paul Ferrero, Auscape
Box 42.1a: C. A. Henley, Auscape
Box 42.1b: Ross Isaacs, Ocean Planet Images
Box 42.1c: C. M. Bull, 'Population of Ecology of the Sleepy Lizard, *Tiliqua rugosa*, at Mt Mary, South Australia', *Australian Journal of Ecology*, 20: 393–402
Box 42.4: Fredy Mercay, ANT Photo Library

Chapter 43

43.1: David Hancock
43.3a, b: CSIRO, Division of Plant Industry
43.6: Dr Peter Harrison, Southern Cross University
43.7: Michele Hall, Oxford Scientific Films
43.8: Australian National Insect Collection, CSIRO Division of Entomology
43.9: David Albrecht
43.10, 43.19a–c, 43.20a, b, 43.28, 43.29a, b, 43.32a–d: Dr M. Keough, University of Melbourne
43.11a: A. E. Newsome, A population study of house-mice permanently inhabiting a reed-bed in South Australia, *Journal of Animal Ecology*, 1969; 38: 361–77
43.11b: Ecological Society of America, Washington DC
43.12: S. McNeil & T. R. E. Southwood, 'The Role of Nitrogen in the Development of Insect/plant Relationships', in J. B. Harborne (ed), *Biochemical Aspects of Plant and Animal Coevolution*, Academic Press, London, pp. 77–98
43.13: Glen Threlfo, Auscape
43.14: K. Thiele
43.16: D. Renz, CSIRO Division of Entomology
43.17: Chuck Miller

43.18: G. R. Roberts, Documentary Photographs
43.21: Charles J. Krebs, *Ecology: the Experimental Analysis of Distribution and Abundance*, Addison Wesley Longman, Inc., Fig. 13.9, p. 246; reprinted by permission of publisher
43.25: C. Weston
43.31: M. J. Keough & G. P. Quinn, 'Effects of Periodic Disturbances from Trampling on Rocky Intertidal Algal Beds', *Ecological Applications*, 1998; 8: 141–61 (Ecological Society of America)
Box 43.2a: David Albrecht
Box 43.2b: A. Beattie & C. Turnbull

Chapter 44

44.3a–c: CSIRO Division of Wildlife & Ecology
44.11a: Kathie Atkinson, Oxford Scientific Films
44.11b: Jean-Paul Ferrero, Auscape
Box 44.2: Bert Jenkins
Box 44.4: G. Chapman, CSIRO Division of Wildlife & Ecology
Box 44.4: Surrey W. L. Jacobs; courtesy of Royal Botanic Gardens, Sydney

Chapter 45

45.1: W. Bradley, *Sydney Cove, Port Jackson*, 1788, watercolour, Mitchell Library, State Library of New South Wales
45.2: Karen Wynn-Moylan, NPIAW
45.3a: A. D. & M. C. Traunson, NPIAW
45.3b: International Council for Bird Preservation, *Putting Biodiversity on the Map: Priority Areas for Global Conservation*, BirdLife International, Cambridge, p. 35
45.4: Dr Andy Gillison
45.5a, b: Byron Lamont, Western Australian Herbarium
45.7a, 45.9c: Jean-Michel Labat, Auscape
45.8b: Jann Williams, Department of Botany, University of Melbourne
45.9b: Ted Hutchison, ANT Photo Library
45.10: Ralph & Daphne Keller, ANT Photo Library
45.11: D. H. Ashton
45.14a: Bill Bachman, ANT Photo Library
45.15b: Dr Jan Anderson, CSIRO Division of Plant Industry
45.17: *Australia's Environment Issues and Facts* 1992, AGPS, Canberra, © Commonwealth of Australia, reproduced by permission
45.18, Box 45.4: Australian Tourist Commission
45.19: Belinda Wright, Oxford Scientific Films
45.2: Kathie Atkinson, Oxford Scientific Films
45.21: CRA Corporate Services
Box 45.1a: State Forests of NSW
Box 45.2a: David Albrecht
Box 45.2b: Pavel German, ANT Photo Library
Box 45.3: D. Curl, CSIRO Division of Plant Industry

Index